Theory of Finite Simple Groups II

Commentary on the Classification Problems

This second volume provides a coherent explanation for the existence of the known 26 sporadic simple groups originally arising out of many unrelated contexts. Chapter 1 presents a new algorithm constructing centralizers of 2-central involutions of finite simple groups from indecomposable subgroups of the general linear groups $\mathrm{GL}_n(2)$ using the representation theoretic and algorithmic methods developed in the first volume. It is shown that 25 sporadic simple groups can be constructed by this algorithm. The smallest Mathieu group M_{11} can be omitted for theoretical reasons. The algorithm is not restricted to sporadic simple groups as is shown explicitly in Chapters 2 and 10.

Here the author describes the constructions of Conway's groups Co_3, Co_2, Co_1, Fischer's groups Fi_{22}, Fi_{23}, Fi_{24}', Janko's group J_4, McLaughlin's group McL, Rudvalis' group Ru, Lyons' group Ly, Suzuki's group Suz, and O'Nan's group ON. Their uniqueness is proved whenever possible. The computational existence proofs are documented in the accompanying DVD. Chapter 16 outlines such theoretically possible constructions for the baby monster and the monster from well determined indecomposable subgroups of $\mathrm{GL}_9(2)$ and $\mathrm{GL}_{10}(2)$, respectively. The other sporadic groups were constructed in volume I.

The present mathematical literature does not contain an accessible proof of the announced classification theorem asserting that there are exactly 26 sporadic simple groups. On the other hand the literature has not paid much attention to R. Brauer's warning (published in 1979) that there may be infinitely many sporadic groups. Therefore the author describes Brauer's ideas on a general classification scheme in Chapter 1 and states several related open problems in Chapter 16. Some require new experiments with the author's algorithm.

GERHARD MICHLER is an Emeritus Professor of the Institute of Experimental Mathematics at the University of Duisburg-Essen and Adjunct Professor at Cornell University.

NEW MATHEMATICAL MONOGRAPHS

All the titles listed below can be obtained from good booksellers or from Cambridge University Press. For a complete series listing visit
http://www.cambridge.org/uk/series/sSeries.asp?code=NMM

1 M. Cabanes and M. Enguehard *Representation Theory of Finite Reductive Groups*
2 J. B. Garnett and D. E. Marshall *Harmonic Measure*
3 P. Cohn *Free Ideal Rings and Localization in General Rings*
4 E. Bombieri and W. Gubler *Heights in Diophantine Geometry*
5 Y. J. Ionin and M. S. Shrikhande *Combinatorics of Symmetric Designs*
6 S. Berhanu, P. D. Cordaro and J. Hounie *An Introduction to Involutive Structures*
7 A. Shlapentokh *Hilbert's Tenth Problem*
8 G. Michler *Theory of Finite Simple Groups I*
9 A. Baker and G. Wüstholz *Logarithmic Forms and Diophantine Geometry*
10 P. Kronheimer and T. Mrowka *Monopoles and Three-Manifolds*
11 B. Bekka, P. de la Harpe and A. Valette *Kazhdan's Property (T)*
12 J. Neisendorfer *Algebraic Methods in Unstable Homotopy Theory*
13 M. Grandis *Directed Algebraic Topology*

Theory of Finite Simple Groups II

Commentary on the Classification Problems

GERHARD MICHLER

Institute of Experimental Mathematics
University of Duisburg-Essen

and

Cornell University

CAMBRIDGE UNIVERSITY PRESS
Cambridge, New York, Melbourne, Madrid, Cape Town, Singapore,
São Paulo, Delhi

Cambridge University Press
The Edinburgh Building, Cambridge CB2 8RU, UK

Published in the United States of America by
Cambridge University Press, New York

www.cambridge.org
Information on this title: www.cambridge.org/9780521764919

First published 2010

Printed in the United Kingdom at the University Press, Cambridge

A catalog record for this publication is available from the British Library

ISBN 978-0-521-76491-9 Hardback

In Memory

of

Ruth I. Michler
1967 (Ithaca, NY)–2000 (Boston, MA)

Contents

Acknowledgements

The computational work described in this book was performed at the Institute of Experimental Mathematics of Essen University between 1989 and 2003 and at Cornell University since 2004. The recent research with H. Kim on the large Conway and Fischer groups described in this book has been supported by the grant NFS/SCREMS DMS-0532/06.

The author owes special thanks to his coauthors of various joint articles on some sporadic simple groups dealt with in this book: H. Kim (Yale University), M. Kratzer (Bayerische Staatsbibliothek, Munich), W. Lempken (Essen), A. Previtali (Como), R. Staszewski (Essen), K. Waki (Hirasaki), L. Wang (Peking University) and M. Weller (Essen).

The author also thanks his former secretary, B. Hasel at Essen University, for typing some parts of the manuscript. The complex tables in some of the chapters were printed automatically by using a special program developed by M. Weller, P. Young and H. Kim. The layout of the entire manuscript and the construction of the table of contents, the references and index were carried out by B. Chan and Professor K. Dennis of Cornell University. The accompanying DVD was produced with the help of H. Kim and M. Weller. The author is greatly indebted to all of them for their technical support.

The author thanks the Department of Mathematics of Cornell University for its great hospitality during his stay as a visiting and adjunct professor since 2003. Finally, he would like to thank Roger Astley at Cambridge University Press for his constant assistance and the copy-editor for many suggestions that improved the presentation of the book.

The entire work carried out in Germany and in the United States benefitted greatly from the strong support and understanding of my wife, Waltraud.

Introduction

This is the second and last volume of the author's introduction to the representation theoretic and algorithmic theory of *abstract* finite simple groups. In particular, it yields the theoretical and algorithmic background for uniform existence and uniform uniqueness proofs of the (known) sporadic simple groups. A finite simple group G is called *sporadic* if it is not isomorphic to an alternating group A_n or a finite simple group of Lie type described in R. W. Carter's book [13]. The theoretical results and algorithms presented in the first volume [92] and here hold in general. They are not restricted to sporadic groups at all as will be shown again in this volume.

Many results on abstract finite simple groups are inspired by the celebrated Brauer–Fowler Theorem. It asserts that there are only finitely many simple groups G which possess an involution $z \neq 1$ such that its centralizer $C_G(z)$ is isomorphic to a given group H of even order. But it does not give any hint of how to find such a group H without knowing at least one of the simple groups G. In particular, most of the known sporadic simple groups G were not discovered by a construction from a given centralizer H. In his survey article [37], p. 71, D. Gorenstein wrote in 1979: "Much of the excitement generated by the developments in simple group theory over the last 20 years can be directly attributed to the discovery of over 20 sporadic new simple groups ... The existence of these strange objects ... revealed the richness of the subject and lent an air of mystery to the nature of simple groups." After a brief survey about the "*presently known*" 26 sporadic simple groups he remarks: "There you have the 26 beautiful enigmatic sporadic groups with Janko's fourth group and the Fischer monster still waiting to be born. Arising out of so many unrelated contexts, is it yet possible that there is a single, coherent explanation for their existence? If so, it will require some new vision, seemingly beyond the capabilities of the present generation, to discover it."

It is one purpose of this second volume to answer Gorenstein's question using the representation theoretic and algorithmic methods developed in the first volume [92]. This solution is achieved by means of the author's Algorithm 1.3.8 and its iterated version 1.3.15 described in Chapter 1. They construct centralizers H of 2-central involutions z of certain finite simple groups G from indecomposable subgroups T of the general linear groups $\mathrm{GL}_n(2)$. By Definition 1.3.5 such a subgroup T is called indecomposable if the natural vector space $V = F^n$ is an indecomposable FT-module over the prime field $F = GF(2)$ of characteristic 2. Irreducible subgroups T of $\mathrm{GL}_n(2)$ are defined similarly. In general H is not uniquely determined by T. Several steps of Algorithm 7.4.8 of [92] have been incorporated into Algorithm 1.3.8. Thus it is possible to build the simple target groups G from a previously constructed centralizer H provided all its conditions can be satisfied in the course of the construction. This algorithm constructs (in theory) all simple groups G whose Sylow 2-subgroups S have a non-cyclic elementary abelian normal subgroup A of order 2^n such that $N_G(A)/C_G(A) \cong T \leq \mathrm{GL}_n(2)$ and the centralizer $C_G(A)$ is either A or an iterated FT-module extension. In view of Theorem 4.8.5 of [92] and Remark 1.3.14, these requirements are not serious restrictions. It is shown in this book that these conditions are satisfied by all known sporadic groups. By Kondo's work [77] they hold for all alternating groups A_{4k+r} with $k \geq 2$ and $0 \leq r \leq 3$. They are also true for several simple groups of Lie type whose Sylow 2-subgroups are neither metabelian, dihedral nor semi-dihedral.

Algorithm 1.3.8 provides a new link between modular representation theory and the study of finite simple groups. Since the Sylow 2-subgroups S of the relevant indecomposable groups T of $\mathrm{GL}_n(2)$ are not cyclic, their group algebras FT have infinitely many non-isomorphic finite-dimensional indecomposable FT-modules by a classical theorem of D. G. Higman; see Theorem 1.2.3. This means that there are infinitely many dimensions m such that $\mathrm{GL}_m(2)$ has an indecomposable subgroup T_m which is an epimorphic image of T. In theory Algorithm 1.3.8 can be applied to any of these groups T_m again. Therefore the first section of Chapter 1 contains a summary of two survey articles by R. Brauer [10] and D. Gorenstein [37] addressing the general classification problem of the finite simple groups. They were published in 1979, two years before D. Gorenstein's announcement of the classification theorem; see Theorem 1.1.2 and its Corollary 1.1.3. Gorenstein's approach to the classification problem assumes implicitly that an inductive argument can be produced which shows that there will be exactly 26 sporadic simple groups. On the other hand, R. Brauer addresses all aspects of the classification problem. In particular, he remarks: "It is not even impossible that no classification exists."

The present mathematical literature does not contain an accessible proof of the announced classification theorem; see Theorem 1.1.2. Furthermore, D. G. Higman's Theorem 1.2.3, Algorithm 1.3.8 and all its successful applications in nature provide strong support for R. Brauer's remark. In particular, we show in [92] and in this book that 23 of the known sporadic simple groups can be constructed by means of Algorithms 1.3.8 and 1.3.15 in a uniform way. The smallest sporadic group is not considered here because the Sylow 2-subgroups of Mathieu group M_{11} are semi-dihedral; see Theorem 1.6.7 of [92]. In theory, given enough computational power and efficient implementations of the relevant basic algorithms of MAGMA, an application of Algorithm 1.3.15 to some well determined indecomposable subgroups of $\mathrm{GL}_9(2)$ and $\mathrm{GL}_{10}(2)$ will also construct the baby monster and the monster, respectively; see Chapter 16.

This volume contains 16 chapters, each of which has an introduction summarizing its contents and mentioning the relevant literature. The given proofs build again on the intimate relations between general group theory, ordinary character theory, modular representation theory and algorithmic algebra described in [92].

The last three sections of Chapter 1 contain several new algorithms and theoretical results that improve the applicability of Algorithms 1.3.8 and 1.3.15. Remarks 1.3.11 and 1.3.12 present a survey about the sporadic simple groups dealt with in [92] that can be constructed from an indecomposable subgroup T of some $\mathrm{GL}_n(2)$. They are the four Mathieu groups $\mathsf{M}_{12}, \mathsf{M}_{22}$, M_{23} and M_{24}, Janko's group J_1, Held's group He, Harada's group Ha and Thompson's group Th. All these simple groups G have a Sylow 2-subgroup S containing a maximal elementary abelian normal subgroup A which is self centralizing in G. By Remark 1.3.14, the two Janko groups J_2 and J_3 and the Higman–Sims group HS dealt with in Chapters 8 and 10 of [92] can be constructed from an indecomposable subgroup T of $\mathrm{GL}_2(2)$ by means of Algorithm 1.3.15 using iterated FT-module extensions. In this book we show that the iterated version of Algorithm 1.3.8 can be applied to construct the Tits group, $F_2(4)'$, and the sporadic simple groups Ru of Rudvalis, Ly of Lyons, Suz of M. Suzuki and ON of O'Nan.

Since the Mathieu group M_{11} has a semi-dihedral subgroup it can be neglected by Theorem 1.6.7 of [92] due to J. Alperin, R. Brauer and

D. Gorenstein; see [2]. All other known sporadic simple groups are dealt with in [92] and in this book.

In the course of their constructions, several simple groups of Lie type, such as $\mathsf{Sp}_6(2)$, $\mathrm{U}_4(2)$, $\mathrm{U}_4(3)$, $\mathrm{U}_6(2)$ and $\mathrm{O}_8(2)$ are constructed by means of Algorithm 1.3.8 from the given data of certain constructed subgroups of the particular sporadic groups. Each time, these classical groups could be realized in some of their indecomposable or irreducible representations over their fields of definition; see Carter's book [13].

In Chapter 2, Algorithm 1.3.8 is applied to the irreducible subgroup $T = \mathrm{GL}_3(2)$ of $\mathrm{GL}_3(2)$. Thus one obtains Dickson's simple group $G = \mathsf{G}_2(3)$. We also present in Chapter 2, L. Wang's and the author's proof for Z. Janko's characterization [68] of this simple group by its centralizer $H = C_G(z)$ of a 2-central involution z. Let q be an odd prime power such that $q \equiv 3$ or 5 mod (8). By P. Fong's article [31], there are infinitely many simple Dickson groups $\mathsf{G}_2(q)$, each of which, in theory, can be constructed by means of Algorithm 1.3.8 from the irreducible subgroup $T = \mathrm{GL}_3(2)$ of $\mathrm{GL}_3(2)$; see Corollary 2.1.3. So it cannot be proved that Algorithm 1.3.8 constructs finitely many simple groups from a given indecomposable subgroup T of some $\mathrm{GL}_n(2)$.

Chapter 3 deals with Conway's sporadic simple group Co_3. The presented existence and uniqueness proofs for Co_3 are due to the author. J. Conway discovered this simple group in 1969. He proved its existence in his famous paper on the automorphism group of the Leech lattice [16]. The first uniqueness proof for Co_3 was given by D. Fendel in [28]. It depends on a deep theorem of W. Feit, classifying certain integral sublattices of the Leech lattice; see [27]. We construct in Proposition 3.1.2 a finitely presented group H with center of order 2 by an application of Algorithm 1.3.8 to the irreducible subgroup $T = \mathrm{GL}_4(2)$ of $\mathrm{GL}_4(2)$. Then we build from H a simple subgroup $\mathfrak{G}$ of $\mathrm{GL}_{23}(23)$ such that $H \cong C_{\mathfrak{G}}(\mathfrak{z})$ for a 2-central involution $\mathfrak{z}$ of $\mathfrak{G}$. This matrix group is an important ingredient of the given self-contained uniqueness proof of Co_3; see Theorem 3.3.13. In particular, we do not quote Feit's theorem.

Chapters 4–9 describe H. Kim's and the author's recent applications of Algorithm 1.3.8 to the 10-dimensional and 11-dimensional 2-modular irreducible representations of the Mathieu groups M_{22}, M_{23} and M_{24} over $GF(2)$, respectively. The resulting simple groups are isomorphic to the sporadic Conway groups Co_2, Co_1, the sporadic Fischer groups Fi_{22}, Fi_{23}, Fi_{24}' and the large Janko group J_4 discovered in [16], [29] and [69], respectively.

Whereas the three Conway groups were originally constructed in [16] as subgroups or factor groups of the group $Aut(\Lambda)$ of isometries of the

24-dimensional Leech lattice Λ, Fischer realized his groups as automorphism groups of special graphs defined by means of certain involutions of his simple groups; see [29]. Janko's original article [69] on J_4 employs purely algebraic methods. They have been generalized by the author to obtain uniform existence proofs for all sporadic simple groups. In fact, the combinatorial methods used in Conway's survey article on the Leech lattice groups [17] and the geometric methods described in the books by Aschbacher [3] and Ivanov [64], dealing with the Fischer groups and the large Janko group J_4, respectively, are distinct and unrelated.

Chapters 4 and 5 present H. Kim's and the author's simultaneous existence proofs for the simple sporadic groups Co_2 of Conway and Fi_{22} of Fischer published in [72]. The sporadic simple group Co_2 is constructed as a simple subgroup $\mathfrak{G}_3$ of $\mathrm{GL}_{23}(13)$ by an application of Algorithm 1.3.8 to a well determined irreducible subgroup $T \cong Aut(\mathsf{M}_{22})$ of $\mathrm{GL}_{10}(2)$. In Chapter 5 the simple group Fi_{22} is constructed as a simple subgroup $\mathfrak{G}_2$ of $\mathrm{GL}_{78}(13)$ by applying Algorithm 1.3.8 to another well determined irreducible subgroup $T \cong \mathsf{M}_{22}$ of $\mathrm{GL}_{10}(2)$. Section 5.3 contains a sketch of a uniqueness proof for Fi_{22}. In Section 5.4 we give some examples of irreducible subgroups T of $\mathrm{GL}_{10}(2)$ for which no simple group G exists having a Sylow 2-subgroup S containing a non-cyclic maximal elementary abelian normal subgroup A such that $C_G(A) = A$ and $N_G(A)/A \cong T$.

The constructions of the large sporadic groups Fi_{23}, Co_1, J_4 and Fi_{24}' are technically more demanding because (in 2007) our applications of MAGMA were not able to calculate the cohomological data required for performing Steps 2 and 5 of Algorithm 1.3.8. In order to obtain the presentations of the centralizers H of the 2-central involutions z of the simple target groups G we were forced to construct some local subgroups of H in advance.

Chapter 6 describes Kim's application [73] of Algorithm 1.3.8 to the non-split extension E of the simple Mathieu group M_{23} by a well determined irreducible module V_2 of dimension 11 over $GF(2)$. Proposition 6.2.1 gives a presentation of the centralizer $D = C_E(z)$ of a 2-central involution z of E by means of Step 4 of Algorithm 1.3.8. It also describes the construction of a presentation of another group H_2 having a central subgroup $Z_2 = \langle z \rangle$ such that H_2/Z_2 is isomorphic to the centralizer H_1 of a 2-central involution z of Fi_{22} constructed in Chapter 5. Using the character tables of D and H_2 and Algorithm 7.4.8 of [92], Kim constructed a matrix subgroup $\mathfrak{H}$ in $\mathrm{GL}_{352}(17)$ with center $\mathfrak{Z}$ of order 2 and showed in [73] that $\mathfrak{H}/\mathfrak{Z} \cong \mathsf{Fi}_{22}$. The details are given in Proposition 6.2.2. Proposition 6.2.3 constructs a fairly nice presentation and a faithful permutation representation of $\mathfrak{H}$ of degree 28160. Since all conditions of Step 5 of Algorithm 1.3.8 are verified in Corollary 6.2.4, its Steps 6–11 are applied in the proof of Theorem 6.3.1, realizing Fischer's simple group Fi_{23} as a simple subgroup $\mathfrak{G}$ of $\mathrm{GL}_{782}(17)$. This existence theorem is due to Kim [73]. The constructions by means of Algorithm 1.3.8 of the large Conway's group Co_1, Janko's group J_4 and Fischer's group Fi_{24}' from the two 11-dimensional irreducible 2-modular representations of the Mathieu group M_{24} are summarized in the following diagram:

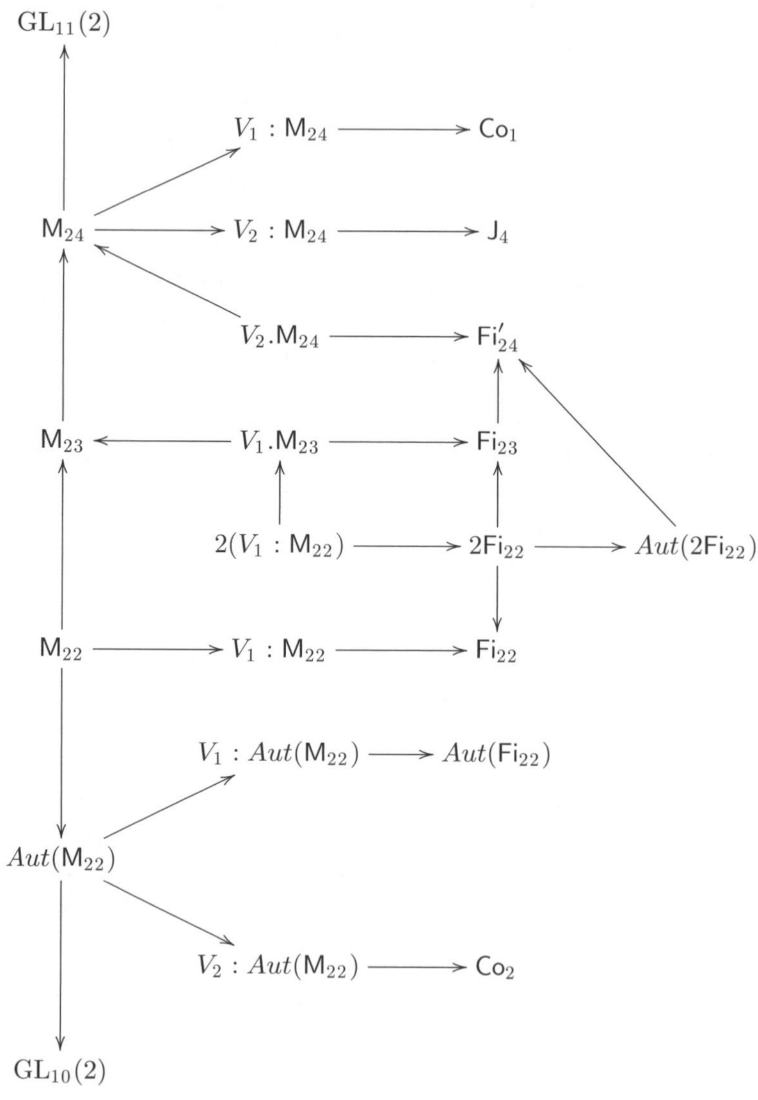

The smallest non-trivial irreducible 2-modular representations of M_{24} denoted by V_1 and V_2 are of dimension 11 over $GF(2)$. Chapter 7 describes H. Kim's and the author's application of Algorithm 1.3.8 to the split extension E_1 of M_{24} by V_1. Using similar methods as the ones described in Chapter 6 we constructed a finitely presented group H having a center $\langle z \rangle$ of order 2 in Proposition 7.2.2 satisfying all conditions of step 5 of Algorithm 1.3.8. It is then used to realize Co_1 as a matrix subgroup $\mathfrak{G}$ of $\mathrm{GL}_{276}(13)$; see Theorem 7.3.2. In particular, $\mathfrak{G}$ has a 2-central involution $\mathfrak{z}$ such that $C_{\mathfrak{G}}(\mathfrak{z}) \cong H$. In Section 7.3 we also prove Soicher's Theorem 7.3.3 of [120], giving a nice presentation of Conway's original group Co_1. Using it and the constructed faithful permutation representation of H of degree 61440 it is shown in Corollary 7.3.4 that H is also isomorphic to a centralizer of a 2-central involution of Co_1. Section 7.4 discusses the uniqueness problem of $\mathfrak{G}$ and Co_1.

Applying Algorithm 1.3.8 to the split extension E_2 of M_{24} by V_2, one obtains Janko's sporadic group J_4 as a simple subgroup $\mathfrak{G}$ of $\mathrm{GL}_{1333}(43)$; see Theorem 8.3.2. Because of lack of space, only the construction of $\mathfrak{G}$ from a constructed presentation of the centralizer H of a 2-central involution z of the simple target group G is described in detail in Chapter 8. This construction and several results of Janko's original article [69] are applied in Section 8.3 to give a uniqueness proof for J_4; see Theorem 8.3.4. In Section 8.4 we present a survey about earlier constructions of J_4 as subgroups of $\mathrm{GL}_{112}(2)$ and $\mathrm{GL}_{1333}(11)$ due to Benson, Conway, Norton, Parker and Thackray [8] (see also [102]) and W. Lempken [84], respectively. We also mention Weller's construction of a faithful permutation representation of Lempken's group G of degree 173067389 in [135]. It has been used by Weller and the author in [98] to construct a 112-dimensional representation J of G in $\mathrm{GL}_{112}(2)$ and to show that J and the original group constructed by Benson, Conway, Norton, Parker and Thackray are conjugate in $\mathrm{GL}_{112}(2)$. All generating matrices of $\mathfrak{G}$, G and J and the corresponding permutations are stored on the accompanying DVD.

Fischer's simple group Fi_{24}' is constructed in Chapter 9 as a simple subgroup $\mathfrak{G}$ of $\mathrm{GL}_{8671}(13)$ following H. Kim's and the author's recent article [74]; see Proposition 9.3.2. Let E_3 be the uniquely determined non-split extension of M_{24} by V_2. For technical reasons described in the introduction of Chapter 9 it is constructed by means of an application of Algorithm 1.3.8 to a non-2-central involution v of E_3 with centralizer $U = C_{E_3}(v)$ of order $2^{19} \cdot 3^2 \cdot 5 \cdot 7 \cdot 11$. In Section 9.4 we construct an isomorphism between the matrix group $\mathfrak{G} = \langle \mathfrak{q}, \mathfrak{y}, \mathfrak{t}, \mathfrak{w} \rangle$ of Proposition 9.3.2 and the commutator subgroup of the finitely presented group G of

Hall and Soicher; see [110], p. 111. Hence $\mathfrak{G}$ is isomorphic to Fischer's simple group Fi_{24}'. Thus it follows that $\mathfrak{G}$ has a faithful permutation representation of degree 306936 with stabilizer $G_1 \cong \mathsf{Fi}_{23}$ constructed in Chapter 6. It is used to calculate the character table of $\mathfrak{G}$ by means of MAGMA. In Section 9.5 we determine a presentation of the centralizer H of a 2-central involution z of $\mathfrak{G}$. It is used in Section 9.6 to prove D. Parrott's Theorem [107] asserting that all simple groups G of Fi_{24}'-type have the same order $G = 2^{21} \cdot 3^{16} \cdot 5^2 \cdot 7^3 \cdot 11 \cdot 13 \cdot 17 \cdot 23 \cdot 29$.

By Remark 1.3.11 the Mathieu groups M_{22}, M_{23} and M_{24} can be constructed from irreducible subgroups of $\mathrm{GL}_4(2)$ and $\mathrm{GL}_6(2)$. Therefore the two large Conway groups Co_2 and Co_1, all three Fischer groups and the large Janko group J_4 can be constructed from these small linear groups by repeated applications of Algorithm 1.3.8.

In Chapter 10 we apply Algorithm 1.3.8 to an *indecomposable* subgroup of $\mathrm{GL}_5(2)$ and obtain a self-contained existence proof of Tits' simple group $G = F_2(4)'$ of Lie type; see Theorem 10.3.1. It is used in the uniqueness proof of Theorem 10.6.2 due to L. Wang and the author; see [97]. This example also shows how difficult it can be to find the indecomposable subgroups T of $\mathrm{GL}_n(2)$ which lead to a simple group. The constructed simple subgroup $\mathfrak{G}$ of $\mathrm{GL}_{26}(73)$ is shown to be isomorphic to the original Tits group by means of Parrott's presentation [105] of a centralizer of a 2-central involution.

The existence and uniqueness of J. McLaughlin's simple group McL [89] is proved in Chapter 11 by deriving a presentation of this group from I. Schur's presentation of the covering group $2A_8$ of the alternating group A_8, which is assumed to be the given centralizer of a 2-central involution z of the simple target group G; see Theorem 11.3.1 due to M. Kratzer, W. Lempken, K. Waki and the author [80]. These results show that McL could also have been constructed by application of Algorithm 1.3.8 to the irreducible subgroup A_7 of $\mathrm{GL}_4(2)$.

The simple groups G studied in the remaining chapters of the book can be constructed by applications of Algorithm 1.3.15 to indecomposable subgroups of some $\mathrm{GL}_n(2)$. In Chapter 12 we apply it to an irreducible subgroup $T \cong \mathrm{GL}_3(2)$ of $\mathrm{GL}_8(2)$ and construct an iterated extension E of T by its irreducible FT-modules W and V of dimensions 8 and 3 over $F = GF(2)$, respectively. In Sections 12.2 and 12.3 we show that the application of all the steps of Algorithm 1.3.15 gives an existence proof of the sporadic simple group Ru discovered by A. Rudvalis [115]; see Theorem 12.2.1. It is used in Section 12.4, where we prove the uniqueness of Ru in Theorem 12.4.1 due to D. Parrott [106].

Chapter 13 provides a self-contained existence proof for the simple sporadic group Ly originally discovered R. Lyons [88]. It is constructed as a simple subgroup $\mathfrak{G}$ of $\mathrm{GL}_{2480}(31)$ from Schur's presentation of the given centralizer $H \cong 2A_{11}$ of an involution z of the simple target group G, which is assumed to be the covering group of the alternating group A_{11}; see Theorem 13.5.2. In Section 13.6, M. Weller's and the author's uniqueness proof for the Lyons group is given; see Theorem 13.5.4.

Chapter 14 contains an existence proof for the sporadic simple Suz of M. Suzuki [124]. The sporadic Suzuki group can be constructed by means of an application of Algorithm 1.3.15 to an irreducible subgroup $L \cong 3A_6$ of $\mathrm{GL}_6(2)$. Because of lack of space we present only our construction of Ru inside $\mathrm{GL}_{143}(13)$ from a presentation of the given centralizer of a 2-central involution. We also give a uniqueness proof. It is self-contained, except for a quotation of Phan's uniqueness theorem of $\mathrm{U}_4(3)$ of [109], which can also be proved completely by means of the methods and results of [92] and this book.

The last sporadic simple group dealt with in this book is O'Nan's simple group ON [104]. Its Sylow 2-subgroup S was originally constructed by J. Alperin [1] in terms of generators and relations. O'Nan derived the structure of the centralizer $H = C_G(z)$ of a 2-central involution z from Alperin's presentation. He proved important results on the structure of the maximal subgroups and determined the character table of his group. The first published existence and uniqueness proof for ON is due to L. Soicher [121]. It depends heavily on O'Nan's work. The existence and uniqueness proof given in Chapter 14 is due to A. Previtali and the author; see [94]. It starts from a given presentation of the group H. It derives from that a uniquely determined presentation of a simple group $\mathfrak{G} \cong$ ON and a faithful permutation representation of degree 2624832 with stabilizer $\mathfrak{J}$ isomorphic to Janko's sporadic group J_1 dealt with in [92].

Most of the sporadic simple groups G treated in this book have been constructed as matrix subgroups of $\mathrm{GL}_n(p)$, where n is the degree of an ordinary irreducible character χ of G which is of p-defect zero. The reason for that choice is Brauer's characterization of characters proved in [92]. It allows us to construct irreducible ordinary characters from the normalizers of cyclic subgroups of prime order. This is an important ingredient of all our uniqueness proofs. It does not have any analog for p-modular representations. By Theorem 3.12.4 of [92], due to R. Brauer, the reduction of an ordinary character of p-defect zero yields a uniquely determined modular representation of G in $\mathrm{GL}_n(p^k)$, where the finite field $GF(p^k)$ is a splitting field for G.

The final chapter contains a number of open problems related to R. Brauer's remarks on the aforementioned classification project. Section 16.2 summarizes a selection of uniqueness problems of some finite simple groups that we either left open in the previous chapters or have not been dealt with in the mathematical literature. In Section 16.3 we describe Thompson's suggestion [127] for finding a possibly new 27th sporadic simple group. Its Sylow 2-subgroup would have order 2^{25}. Therefore an application of Algorithm 1.3.15 to the iterated extension K of $T = \mathrm{GL}_5(2)$ described in R. Griess' Theorem 16.3.2 would be technically very demanding. But it would prove or disprove Thompson's suggestion ignored in [38] and the recent literature. In Section 16.4 we illustrate Brauer's question for a general classification scheme of the finite simple groups by means of a special classification problem. It asks for an abstract classification of all finite simple groups G which can be constructed by an application of Algorithm 1.3.8 to all the indecomposable subgroups of the infinite sequence of general linear groups $\mathrm{GL}_n(2)$. For a fixed integer n we also describe an exhaustive search method for finding the corresponding simple groups. In view of D. G. Higman's Theorem 1.2.3, one cannot bound the dimension n.

The practicality of Algorithm 1.3.8 has been greatly improved by recent implementations of Holt's Algorithm [53] into MAGMA for constructing presentations of the FG-module extensions E of a finitely presented transitive permutation group G. Furthermore, the isomorphism test described in Chapter 6 of [92] has been improved by Cannon and Holt [12]. Recent releases of MAGMA also contain an implementation of a powerful algorithm due to Cannon, Cox and Holt [11], constructing the conjugacy classes of subgroups of a finite permutation group. Many of the group constructions described in this book would not have been possible without these tools. Furthermore, MAGMA [9] is very well documented. Therefore the author quotes its commands that have been used in the specific calculations of a proof or in the constructions of the simple groups dealt with in this book. Of course, that will help only those readers who have a computer at hand. On the other hand, young students, like the author's collaborator H. Kim, can get started in this area of research without great difficulty provided they have a good knowledge of undergraduate mathematics.

The Appendix contains a detailed table of contents of the accompanying DVD. It consists of two parts. On DVD.1 we stored all the large tables of the systems of representatives of the conjugacy classes and the corresponding character tables of the local subgroups of several simple groups dealt with in this book. They are quoted in the text. On DVD.2 we present for each simple group studied in Chapters 3–14 (except for Chapters 10 and 11) a folder containing the four generating matrices mentioned in the existence proof of the particular sporadic group. The folder of Chapter 8 also contains the generating matrices and permutations of some earlier existence proofs for J_4.

Concerning terminology, notation and a list of symbols, the reader is referred to Volume I [92].

1
Simple groups and indecomposable subgroups of $\mathrm{GL}_n(2)$

There are only finitely many non-isomorphic simple groups G having a 2-central involution z such that their centralizers $C_G(z)$ are isomorphic to a given group H of even order by Theorem 2.5.12 of [92] due to Brauer and Fowler. But their theorem does not provide any hint for how to find such a group H without knowing one of the simple groups G. In Section 1.3 an algorithm is described which solves this problem for certain simple groups; see Algorithm 1.3.8. In view of Theorem 4.8.5 of [92] we consider in this book only *abstract* finite simple groups G having a Sylow 2-subgroup S containing a maximal non-cyclic elementary abelian normal subgroup A of S and a 2-central involution z such that $G = \langle H, E \rangle$, where $H = C_G(z)$ and $E = N_G(A)$. Hence A is a vector space over the prime field $F = GF(2)$ of characteristic 2 of dimension $n > 1$. Let $C = C_G(A)$. Then $T = E/C$ is a subgroup of $\mathrm{GL}_n(2)$ and $V = F^n$ is an FT-module. For all known simple groups G dealt with in [92] and in this volume, $V = F^n$ is an indecomposable FT-module for the corresponding subgroups T.

Algorithm 1.3.8, due to the author [93], describes a method for constructing a group H with a center of even order from an indecomposable subgroup T of a general linear group $\mathrm{GL}_n(2)$. Applying then Algorithm 7.4.8 of [92] to H one obtains a simple group G with a 2-central involution z such that $H = C_G(z)$ provided all its steps can be performed. The practicability of Algorithm 1.3.8 is demonstrated in the following chapters of this book.

If $n \geq 3$ then the Sylow 2-subgroups of the indecomposable subgroups T of $\mathrm{GL}_n(2)$ are not necessarily cyclic. If so, then the group algebra FT is of infinite representation type by Theorem 1.2.3, due to D. G. Higman, proved in Section 1.2. That means that there are *infinitely* many general linear groups $\mathrm{GL}_m(2)$ containing indecomposable subgroups T_m which are epimorphic images of T. As can be seen from J. A. Green's theory of vertices and sources of indecomposable modules dealt with in Chapter 3 of [92] and Chapter 4 of [100] there are no known theoretical results for calculating these dimensions m.

Section 1.1 contains a summary of two survey articles by R. Brauer [10] and D. Gorenstein [37], presenting a pessimistic and an optimistic view about a mathematically sound determination of the correct number of finite simple sporadic groups, respectively. Both authors state clearly that this number was not known in 1979. However, two years later, D. Gorenstein announced that there would be exactly 26 sporadic simple groups [38], pp. 134–5. R. Brauer's previous warning that their number may be *infinite* seems to be neglected in the mathematical literature, although it still does not contain an accessible proof for the classification theorem stated in [39], p. 2. There is also no theorem asserting that a successful application of Algorithm 1.3.8 to some indecomposable subgroup of some $\mathrm{GL}_n(2)$ must always construct an alternating group or a simple group of Lie type except for 25 exceptional cases. The 26th sporadic group is the Mathieu group M_{11} not considered here because its Sylow 2-subgroups are semi-dihedral.

The final three sections of this chapter contain some algorithmic results which improve the performance of the application of Algorithm 1.3.8. Section 1.4 contains an efficient algorithm calculating a small set of generators u_j of a subgroup U of a group G which are short words in the given generators g_i of G. Section 1.5 contains an efficient algorithm constructing projective irreducible modules of fairly large finite groups. They are applied in several subsequent chapters and turn out to be powerful tools for constructing finite simple groups as subgroups of general linear groups $\mathrm{GL}_n(p)$ over suitable finite fields of odd characteristic $p > 0$.

Section 1.6 contains a character formula for calculating the coefficients of Thompson's group order formula of Theorem 4.2.1 of [92] from the character tables of the centralizers of the non-G-conjugate involutions of G; see Theorem 1.6.4.

1.1 Two alternative views on the classification problem

The first issue of the New Series of the *Bulletin of the American Mathematical Society* published in 1979 is an important document for the future study of the history of algebra and its applications in other disciplines in mathematics. It is dedicated to the memory of the late

Harvard professor Richard Brauer. It contains two detailed survey articles addressing the problem of determining all finite simple groups, an optimistic one by D. Gorenstein [37] and a pessimistic one by R. Brauer [10]. In fact, Brauer's article is an extended version of a talk he presented at a special session of a regional conference of the American Society on October 30, 1976, shortly before his death on April 17, 1977. Both articles state clearly that the classification problem was not completed at that time.

D. Gorenstein dedicates his article to the memory of "Richard Brauer, for his pioneering studies of finite simple groups." He mentions in particular his intensive and successful cooperation with R. Brauer and J. Alperin on the classification of the simple groups with semi-dihedral and wreathed Sylow 2-subgroups. So it is clear that, at that time, Brauer and Gorenstein had a complete and similar overview about the ongoing classification projects.

Gorenstein's article [37] (p. 52) aims "to present a detailed summary of the proof of the classification of the simple groups, to the extent that this has been accomplished at the present time." After mentioning on p. 50 that "the determination of all simple groups is nearly complete," he remarks on p. 51 that "the ultimate theorem which will assert the classification of simple groups is unlike any other in the history of mathematics; for the complete proof, when it is attained, will run to well over 5000 journal pages." He views "this classification as an entire field of mathematics rather than an attempt to prove a single theorem."

Already, Gorenstein's survey article [37] contains the central definition of the anticipated and used induction argument for the determination of all finite simple groups.

Definition 1.1.1 A finite group whose composition factors are known simple groups is said to be a *K-group.*

Here K-group stands for group of known type, at a given time. So in 1979 D. Gorenstein's article [37] could not state the classification theorem, because at that time there were no existence and uniqueness proofs known for Janko's and Fischer's largest sporadic groups J_4 and the Monster M, respectively. But he gives detailed outlines about the ideas and methods of the classification project which finally culminated in 1981 in the announcement of the following theorem stated as the second form of the main classification theorem in [39], p. 3.

Theorem 1.1.2 *If G is a finite simple group each of whose proper subgroup is a K-group, then G is a K-group.*

The simple K-groups are listed in [38], pp. 134–5.

Corollary 1.1.3 *There are exactly* 26 *sporadic simple groups.*

Since 24 of these groups are dealt with in detail in [92] and in this book, Gorenstein's list is not restated here.

The introduction of Gorenstein's first book [38] begins: "In February 1981, the classification of the finite simple groups was completed.... Involving the combined efforts of several hundred mathematicians from around the world over a period of 30 years, the full proof covered something between 5000 and 10000 journal pages, spread over 300 to 500 individual papers." But he also added on p. 6: "As one can well imagine, the logical interconnections between the several hundred papers making up the classification proof are often subtle, and it is no easy task to present a completely detailed flow diagram." In [39] Gorenstein began a survey of a coherent proof of this theorem. But only the first volume of that book was published before his death in 1992.

Since the finiteness of the number of non-isomorphic simple sporadic groups depends on the correctness of the inductive proof of Theorem 1.1.2, the author finds it important to quote Gorenstein's article [37] once more. On p. 52 he writes: "This is an appropriate moment to add a cautionary word about the meaning of 'proof' in the present context; for it seems beyond human capacity to present a closely reasoned several hundred page argument with absolute accuracy." Later on he adds: "If the arguments are often ad hoc to begin with, how can one guarantee that the 'sieve' has not let slip a configuration which leads to yet another (sporadic) simple group? Unfortunately, there are no guarantees – one must live with this reality." He ends: "...it clearly indicates the strong need for continual reexamination of the existing 'proofs'. This will be especially true on that day when the final classification of simple groups is announced."

Furthermore, Gorenstein could not anticipate that several mathematicians involved in the original classification project have still not published their manuscripts or notes in which they announced deep results. They have been quoted in several important articles by other authors. R. Lyons and R. Solomon write in [40], p. 45: "The most serious problem concerns the sporadic groups, whose development at the time of the completion of the classification theorem was far from satisfactory. The existence and uniqueness of the sporadic groups and the development of their properties form a very elaborate chapter of simple group theory, spread across a large number of journal articles. Moreover, some results are unpublished."

Whereas Gorenstein's approach to the determination of all finite simple sporadic groups assumes from the beginning that one can show that their number is finite, R. Brauer's article [10] addresses all aspects of the classification problem. On p. 31 Brauer states: "So far, it has not been proved that there are only finitely many non isomorphic sporadic groups." This was certainly true in 1976. But worse, even the present literature does not contain a general theorem independent of Theorem 1.1.2 proving this number to be finite. Brauer describes the importance of this general finiteness problem in detail. Certainly, its solution would help to understand why an inductive argument will work.

In the introduction of [10] he also questions the optimism of some group theorists believing in 1977 that the determination of the finite simple groups would be finished within a few years.

Remark 1.1.4 The following remarks might best summarize Brauer's view on that:

(a) "I may add that there may be some doubt about the exact meaning of a *classification* of the finite simple groups." See [10], p. 21.
(b) "The crux of the matter then is the question: *Are there finitely many or infinitely many sporadic groups?*". See [10], p. 22.
(c) "On the other hand, it is quite possible that we have an infinite sequence G_n with orders $|G_n|$ strictly increasing and that infinitely often, entirely new types of groups occur in our-sequence ... Then, indeed, it may become necessary to state more clearly what we mean by a classification of the simple groups. It is not even impossible that no classification exists." See [10], p. 23.

In the main part of his article [10], Brauer describes his requirements for a classification concept. He also considers the theoretically important case that there might be infinitely many non-isomorphic sporadic groups.

So the main problems of the classification of the finite simple groups are as follows:

(a) Are there exactly 26 sporadic simple groups (up to isomorphism)?
(b) Is the number of isomorphism types of sporadic simple groups finite, but not 26?
(c) Is the number of isomorphism types of sporadic simple groups infinite?

By the Brauer–Fowler Theorem 2.5.12 of [92] the following question arises in this context: *How does one construct a centralizer* $H = C_G(z)$ *of a 2-central involution of a simple group* G *from a given pair* $(T \leq \mathrm{GL}_n(2), V = F^n)$*?*

A special answer to this question is provided by the author's Algorithms 1.3.8 and its iterated version 1.3.15 stated in Section 1.3.

1.2 Simple groups are of infinite representation type, $p = 2$

The author's Algorithm 1.3.8 described in Section 1.3 relates the study of finite simple groups to one of the indecomposable modules of group algebras FG over finite fields F of characteristic $p > 0$. Therefore we prove in this section a classical theorem due to D. G. Higman [50] asserting that FG has only finitely many non-isomorphic indecomposable modules if and only if the Sylow p-subgroups of G are cyclic. In particular, if $p = 2$ then Corollary 1.4.18 of [92] implies that the group algebra FG of a finite simple group G is of infinite representation type in the sense of the following definition.

Definition 1.2.1 The finite-dimensional algebra A over the field K is of *finite representation type* if there are only finitely many non-isomorphic indecomposable A-modules. Otherwise, A is said to be of *infinite representation type.*

The proof of the following subsidiary result is due to Assem *et al.* [6], pp. 176–8. In [50] Higman proves a similar result for odd-dimensional modules.

Lemma 1.2.2 *Let* F *be a field of characteristic* $p > 0$. *Let* P *be an elementary abelian* p*-group of order* p^2. *Then for each natural number* $d > 0$ *the group algebra* FP *has a* $2d$*-dimensional indecomposable* FP*-module. In particular,* FP *is of infinite representation type.*

Proof Let $P = \langle x, y \rangle$, $X = \langle x \rangle$ and $Y = \langle y \rangle$. Then $FX \cong FY \cong F[T]/(T^p)$, where $F[T]$ denotes the polynomial ring in the indeterminate

T, and (T^p) denotes the ideal of $F[T]$ generated by T^p. This follows at once from the fact that FX is a local uniserial algebra in the sense of Definition 3.17.1 of [92] with Jacobson radical $J(FX) = (1-x)FX$ having index of nilpotency p. The map $\phi : 1-x \to T$ induces an isomorphism between FX and $F[T]/(T^p)$. Let $T_1 = 1-x$ and $T_2 = 1-y$. Then $FP \cong F[T_1, T_2]/(T_1^p, T_2^p)$, where $B = F[T_1, T_2]$ denotes the polynomial ring in the two commuting variables T_1 and T_2, and $W = (T_1^p, T_2^p)$ is its ideal generated by T_1^p and T_2^p. Clearly, $W \leq V = (T_1, T_2)^2$. Hence there is an epimorphism $\tau : B \to A = B/V$. Thus each A-module is a B-module annihilated by V.

For each integer $d > 0$ let $R_d = F[T]/(T^p)$ and let W_d be the direct sum of two copies of the $F[T]$-module R_d. So $W_d = \{(n_1, n_2) \mid n_i \in R_d, i = 1, 2\}$. On W_d define an A-module action by $(n_1, n_2)T_1 = (0, n_1 T)$ and $(n_1, n_2)T_2 = (0, n_1)$. This action is well defined because T_1^2 T_2^2 and $T_1 T_2$ annihilate W_d. Hence W_d is an A-module of dimension $2d$. Each endomorphism $\alpha \in End_{F[T]}(W_d)$ is represented by a 2×2-matrix given by

$$M_\alpha = \begin{pmatrix} \alpha_{1,1} & \alpha_{1,2} \\ \alpha_{2,1} & \alpha_{2,2} \end{pmatrix},$$

where all $\alpha_{i,j} \in End_{F[T]}(R_d)$. Using the defined A-module structure on W_d it is easy to verify that $\alpha \in End_A(W_d)$ if and only if

$$M_\alpha = \begin{pmatrix} \alpha_{1,1} & 0 \\ \alpha_{2,1} & \alpha_{1,1} \end{pmatrix}.$$

Hence $\alpha^2 = \alpha$ if and only if $M_\alpha^2 = M_\alpha$, which is equivalent to $\alpha_{1,1}^2 = \alpha_{1,1} \in End_{F[T]}(R_d) \cong F[T]/(T^p)$. Since $F[T]/(T^p)$ is a local ring, its identity element is the unique non-zero idempotent. Hence W_d is an indecomposable A-module. Thus it is also an indecomposable FP-module. □

Theorem 1.2.3 (D. G. Higman) *Let F be a field of characteristic $p > 0$. The group algebra FG of the finite group G is of finite representation type if and only if the Sylow p-subgroups of G are cyclic.*

Proof Let P be a Sylow p-subgroup of G and $N = N_G(P)$. Suppose that P is not cyclic. Let $\Phi(P)$ be the Frattini subgroup of P. Then $P/\Phi(P)$ is elementary abelian and non-cyclic. Thus it has an epimorphic image E which is elementary abelian of order p^2. Hence E, and therefore P, have infinitely many non-isomorphic indecomposable FP-modules R_i with strictly increasing dimensions $dim_F R_i$ for $i = 1, 2, \ldots$

by Lemma 1.2.2. Since each indecomposable FG-module is P-projective by Lemma 3.10.3(b) of [92] it follows from the Krull–Schmidt Theorem and Theorem 3.5.11 of [92] that each induced FG-module R_i^G contains an indecomposable direct summand M_i such that R_i is isomorphic to a direct summand of $M_{i|P}$. Thus G is of infinite representation type.

Suppose that P is cyclic. The number of projective indecomposable FG-modules equals the number of p-regular conjugacy classes of G. Theorem 3.10.5 of [92] due to J. A. Green states that each non-projective indecomposable FG-module M has a vertex V and a source S which is uniquely determined up to $N_G(V)$-conjugation. As FP is uniserial FV is also uniserial. Thus it has exactly $|V|$ indecomposable FV-modules. Hence $dim_F(S) \leq |V| \leq |P|$ implies that $dim_F(M) \leq |V||G : V|$. So G is of finite representation type.

Further important results on finite groups of infinite representation type are due to Crawley-Boevey [15] and Erdmann [25]. □

1.3 The algorithm

The main purpose of this section is to describe an algorithm which constructs the centralizer $H = C_G(z)$ of a 2-central involution z of some finite simple groups G from a given indecomposable subgroup U of some general linear group $\mathrm{GL}_n(2)$, provided all its conditions can be satisfied in the course of the construction.

In this book only finite simple groups G are considered whose Sylow 2-subgroups S have a non-cyclic elementary abelian characteristic subgroup A. In view of the classification of the simple groups with dihedral [36] and semi-dihedral [2] Sylow 2-subgroups and the Bender-Suzuki Theorem [7], [126], this is not a serious restriction by Theorem 4.8.5 of [92]. It asserts that a fixed Sylow 2-subgroup S of any other simple group G contains a maximal non-cyclic characteristic elementary abelian subgroup A such that

$$(*) \quad G = \langle N_G(A), C_G(x) \mid x \in I(N) \rangle,$$

where $I(N)$ denotes a set of representatives of the conjugacy classes of involutions of $N = N_G(A)$. This result provides the theoretical background for Algorithm 7.4.6 of [92] constructing simple groups G having a 2-central involution z with a given centralizer $H = C_G(z)$. It constructs the subgroup $G_1 = \langle H, N_G(A) \rangle$ of G from the given centralizer H.

Often G_1 itself turns out to be a simple group. However, G_1 may be a proper subgroup of a larger simple subgroup of G; see Theorem 8.6.6 of [92].

In fact, the following theorem of M. Suzuki [123] provides an infinite family of finite simple groups G having a unique conjugacy class of involutions z such that $S = C_G(z)$ is a Sylow 2-subgroup whose non-cyclic center $Z(S) = A$ is the unique maximal characteristic subgroup of S and there is another Sylow 2-subgroup S_1 of G such that $G_1 = \langle N_G(A), S_1 \rangle = N_G(B) < G$, where $B > A$ is a maximal elementary abelian normal subgroup of S. In particular, G_1 is not simple.

Theorem 1.3.1 (M. Suzuki) *Let G be a finite simple group such that its Sylow 2-subgroups S have non-cyclic centers $Z(S)$ and are not disjoint from their conjugates in G. If the centralizer $C_G(x)$ of each involution $x \in G$ has a normal Sylow 2-subgroup then $G \cong \mathrm{PSL}_3(2^n)$, $n \geq 2$.*

Proof The assertion follows from M. Suzuki's Theorem 8 [123]. □

In many examples dealt with in [92] and in this book a maximal characteristic subgroup A is a maximal elementary abelian normal subgroup of S. It need not be the only maximal elementary abelian normal subgroup of S; see Lemma 11.1.2 of [92]. Furthermore, a maximal characteristic elementary abelian subgroup A of a Sylow 2-subgroup S of a finite simple group G is not necessarily a maximal elementary abelian normal subgroup of S, as the above example $\mathrm{PSL}_3(4)$ shows. Other examples are provided by the simple groups studied in the following two chapters. Therefore we prove an analog of (*) also for this situation.

Lemma 1.3.2 *Let G be a finite simple group having no strongly embedded subgroups. Suppose that $A \neq 1$ is a maximal non-cyclic elementary abelian characteristic subgroup of the Sylow 2-subgroup S of the finite simple group G such that A is not a maximal elementary abelian normal subgroup of S. Then A is the intersection of finitely many maximal elementary abelian normal subgroups B_i of S which form an $Aut(S)$-orbit, $i = 1, 2, .., k$. Let $E = N_G(B_1)$ be of maximal order of the normalizers of the B_i. If $N_G(S) \leq E$ then $G = \langle N_G(B_1), C_G(x) \mid x \in I(E) \rangle$, where $I(E)$ denotes a set of representatives of the conjugacy classes of involutions of $E = N_G(B_1)$.*

Proof Let B be a maximal elementary abelian normal subgroup of S containing A. Let D be the intersection of the orbit $B^{Aut(S)}$, where $Aut(S)$ denotes the automorphism group of S. Then D is a characteristic elementary abelian subgroup of S containing the maximal characteristic elementary abelian normal subgroup A of S. Hence $D = A$.

Let $U = \langle E, C_G(x) \mid x \in I(E) \rangle$. Let a be any involution of U. Then there is a Sylow 2-subgroup T of U such that $a \in T$. By Sylow's Theorem $T^u = S \leq E$ for some $u \in U$. Hence there is an $e \in E$ such that $a^{ue} = x \in I(E)$. Thus $C_G(a) = [C_G(x)]^{(ue)^{-1}} \leq U$. As $N_G(S) \leq E$, by hypothesis Theorem 4.8.4 of [92] implies that $U = G$, because G does not have any strongly embedded subgroup. □

Remark 1.3.3 If a maximal non-cyclic elementary abelian normal subgroup of S is also a maximal elementary abelian normal subgroup of S then the condition $N_G(S) \leq N_G(A)$ is automatically satisfied. Otherwise it has to be checked. In the applications given in the following two chapters, S is self-normalizing in G. So the conditions of Lemma 1.3.2 hold again.

In view of the above remarks and Lemma 1.3.2, we introduce the following notation, which is kept throughout this section.

Notation 1.3.4 Let A be a maximal non-cyclic elementary abelian *normal* subgroup of S of a finite simple group G without strongly embedded subgroups such that $N_G(A)$ is maximal among the normalizers of maximal elementary abelian normal subgroups of S and $N_G(S) \leq N_G(A)$. Let $|A| = 2^n$, $C = C_G(A)$, $E = N_G(A)$, $T = E/C$ and let $F = GF(2)$ be the field with two elements. Then T is isomorphic to a subgroup of $\mathrm{GL}_n(2)$. Furthermore, the FT-module structure of A is uniquely determined by the matrices of the generators t_i of T with respect to a given basis B of the vector space A over F.

The FT-module structure of $A = F^n$ varies.

Definition 1.3.5 A subgroup T of $\mathrm{GL}_n(2)$ is called *indecomposable* if the natural F-vector space V of dimension n is an indecomposable FT-module. If V is an irreducible FT-module then T is called *irreducible* subgroup

There are many finite simple groups G having a unique non-cyclic elementary abelian normal subgroup A which is an irreducible FT-module, e.g. the alternating groups $G = A_{4k}, k \geq 3$ by Proposition 8.6.8 of [92] and Kondo [77]. The following example of an irreducible subgroup of $\mathrm{GL}_3(2)$ showed up in Chapter 9 of [92] dealing with Janko's sporadic simple group J_1.

Example 1.3.6 $\mathrm{GL}_3(2)$ has an irreducible subgroup T of order 21. It is generated by the following two matrices:

$$e = \begin{pmatrix} 0&1&0 \\ 0&0&1 \\ 1&1&0 \end{pmatrix}, \quad d = \begin{pmatrix} 1&0&0 \\ 0&0&1 \\ 0&1&1 \end{pmatrix}.$$

There are finite simple groups G having a Sylow 2-subgroup S such that S has a maximal non-cyclic elementary abelian normal subgroup A which is an indecomposable but *not* an irreducible FT-module. One example is the Tits simple group $G = {}^2F_4(2)'$ dealt with in Chapter 10.

Example 1.3.7 Lemma 10.1.1 provides three generating matrices in $\mathrm{GL}_5(2)$ generating an indecomposable subgroup T of order $2^6 \cdot 3$. From the generating matrices of T it can immediately be seen that $V = F^5$ is an indecomposable FT-module with three composition factors of dimensions 2, 1 and 2.

The main purpose of this section is to describe an algorithm which constructs from a given indecomposable subgroup T of $\mathrm{GL}_n(2)$ simple groups G of the form $G = \langle H, E \rangle$ having a Sylow 2-subgroup S with a maximal elementary abelian normal subgroup A of order 2^n such that $T \cong E/C_G(A)$ of $\mathrm{GL}_n(2)$, where $E = N_G(A)$ and $H = C_G(z)$ for an involution in the center $Z(S)$ of S. It is now stated for the important special case that $C_G(A) = A$.

Algorithm 1.3.8 (Michler) *Let* $F = GF(2)$. *Let* T *be an indecomposable subgroup of* $\mathrm{GL}_n(2)$ *acting on* $V = F^n$ *by matrix multiplication.*

Step 1 *Calculate a faithful permutation representation* PT *of* T *and a finite presentation* $T = \langle t_i | 1 \leq i \leq r \rangle$ *with set* $\mathcal{R}(T)$ *of defining relations.*

Step 2 *Compute all* FT*-module extension groups* E *of* T *by* V *by means of Holt's Algorithm [52]. Determine a complete set* $\mathfrak{S}$ *of non-isomorphic extension groups* E *by means of the Cannon–Holt Algorithm [12].*

Step 3 *Let $E \in \mathfrak{S}$. From the given presentation of E determine a faithful permutation representation PE of E. Using it and Kratzer's Algorithm 5.3.18 of [92] calculate a complete system of representatives of all the conjugacy classes of E.*

Step 4 *Let $z \neq 1$ be a 2-central involution of E such that $D = C_E(z) < E$. Check that $C_E(V) = V$ and that V is a maximal elementary abelian normal subgroup of any fixed Sylow 2-subgroup S of E.*

If it is not maximal, the algorithm terminates.

Step 5 *Construct a group $H > D$ with the following properties:*

(a) *z belongs to the center $Z(H)$ of H;*
(b) *the index $|H : D|$ is odd;*
(c) *the normalizer $N_H(V) = D = C_E(z)$.*

If no such H exists the algorithm terminates.

Otherwise, apply for each constructed group H the following steps of Algorithm 7.4.6 of [92]. By Step 5(c) it may be assumed from now on that $D = H \cap E$.

Step 6 *Compute the character tables of the groups D, H and E, and the fusion patterns of the conjugacy classes of D in H and in E, respectively. Using Kratzer's Algorithm 7.3.10 of [92] calculate the finite set of compatible pairs*

$$\Sigma = \{(\nu, \omega) \in char_{\mathbb{C}}(H) \times char_{\mathbb{C}}(E) \mid \nu_{|D} = \omega_{|D} \in fchar_{\mathbb{C}}(D)\}.$$

Step 7 *Let $(\nu, \omega) \in \Sigma$ be a compatible pair of faithful characters of H and E of minimal degree $\nu(1) = n$ not yet dealt with. Let F be a finite splitting field of characteristic $p > 0$ for the irreducible constituents of the characters of the compatible pair (ν, ω) such that all their irreducible constituents belong to p-blocks of defect zero. Then do the following steps:*

(a) *Construct the faithful semi-simple FH-module $\mathfrak{V}$ corresponding to ν and the faithful FE-module $\mathfrak{W}$ corresponding to ω.*
(b) *Identify H and E with their isomorphic images in $GL_n(F)$ under their representations afforded by $\mathfrak{V}$ and $\mathfrak{W}$, respectively. Determine by means of Theorem 7.2.2 of [92] a double coset*

decomposition

$$C_{GL_n(F)}(D) = \bigcup_{i=1}^{s} C_{GL_n(F)}(H) T_i C_{GL_n(F)}(E)$$

of the centralizer $C_{GL_n(F)}(D)$ *of* D *in* $GL_n(F)$. *For each double coset representative* T_i, *let*

$$G_i = \langle H, T_i^{-1} E T_i \rangle, 1 \leq i \leq s.$$

Then $C_{G_i}(z) \geq H$ *for all* i.

(c) *For each* $i \in \{1, 2, \ldots, s\}$ *compute the orders of some suitable elements of* G_i *and use this information to check whether a Sylow* 2*-subgroup of* G_i *may be isomorphic to* S.
If the Sylow 2*-subgroups of all constructed groups* G_i *are not isomorphic to* S, *then the algorithm ends.*

Otherwise, let $G = \langle H, T^{-1} E T \rangle$ *be any of the groups* G_i *having fulfilled the Sylow* 2*-group test of Step 7(c). Then the canonical* n*-dimensional vector space* $M = F^n$ *is an irreducible* FG*-module with restriction* $M_{|H} \cong \mathfrak{V}$.

Step 8 *Since* $M_{|H}$ *has a proper non-zero* FH*-submodule* U, *one can construct a permutation representation* $\pi : G \longrightarrow S_m$ *into the symmetric group on* m *letters with stabilizer* $\tilde{H} \geq H$ *and a strong base and generating set for* $\pi(G)$ *by Theorem 6.2.1 and Remark 5.2.12 of [92], respectively.*

Step 9 *Check by means of Proposition 5.2.14 of [92] that* π *has stabilizer* $\tilde{H} = H$. *If so, then* $\pi : G \longrightarrow S_m$ *is faithful, and* $|G| = |H| \cdot m$.

In particular, G *has a* 2*-central involution* t *with* $C_G(t) \cong H$ *and one can calculate a presentation of* G *by means of Theorem 5.2.18 of [92].*

Step 10 *Using the faithful permutation representation* $\pi : G \longrightarrow S_m$ *of* G *and Kratzer's Algorithm 5.3.18 of [92] compute a complete set of representatives of all the conjugacy classes of* G.

Step 11 *Compute a character table of* G *by means of MAGMA or the algorithms described in [92]. If no conjugacy class of* G *is in the kernel of some irreducible character then* G *is a simple group. Otherwise, the concrete character table of* G *provides matrix generators of a proper normal subgroup of* G.

Remark 1.3.9 Keep the notation of Algorithm 1.3.8. Sometimes it will not be possible to construct the permutation representation $(1_H)^G$ for technical reasons. Suppose one is able to build a permutation representation $(1_U)^G$ of another subgroup U of G having a smaller index $|G:U|$ than H. Then one can calculate the exact order of $H = C_G(z)$ of the constructed matrix group G by means of Proposition 2.6.6 of [92] and the number of fixed points of the action of z on $(1_U)^G$. The construction of Thompson's sporadic group Th in Chapter 12 of [92] provides such an example.

The following example shows how to construct the centralizer H of a 2-central involution of Janko's smallest sporadic group J_1 from the irreducible subgroup T of $\mathrm{GL}_3(2)$ defined in Example 1.3.6. The group H is used in Chapter 9 of [92] to construct Janko's smallest sporadic group J_1 by means of Steps 5–11 of Algorithm 1.3.8.

Example 1.3.10 Let E be the split extension $E = V : T$ of the irreducible subgroup T of $\mathrm{GL}_3(2)$ defined in Example 1.3.6 by its natural vector space V of dimension 3 over $F = GF(2)$. As T is isomorphic to the Frobenius group of order 21, the extension E has an involution $z \in V$ such that $D = C_E(z) \cong A_4$. Since the alternating group A_4 has index 5 in A_5, Step 5 of Algorithm 1.3.8 can be immediately performed after constructing $D = C_E(z)$. Thus we have the following diagram:

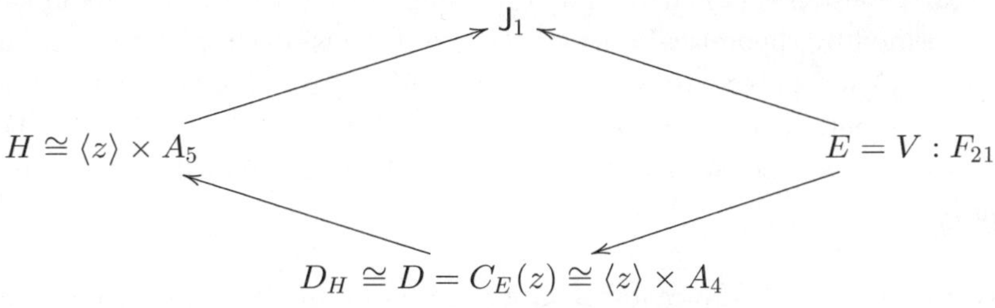

J_1 can be constructed now as in Chapter 9 of [92].

Whenever a simple group G has been constructed from a given centralizer $H = C_G(z)$ of a 2-central involution by means of Algorithm 7.4.8 of [92] and a Sylow 2-subgroup S of G has a *self-centralizing* maximal elementary abelian normal subgroup A of order 2^n then it can also be constructed from a given indecomposable subgroup $T = N_G(A)/A \in \mathrm{GL}_n(2)$ by means of Algorithm 1.3.8. The following remarks present a survey describing the sporadic groups dealt with in [92] which meet these requirements. Of course, the Mathieu group M_{11} can be neglected because its Sylow 2-subgroups are semi-dihedral.

Remark 1.3.11 By Theorem 8.1.9 of [92] the Mathieu group $G = \mathsf{M}_{12}$ has a Sylow 2-subgroup S having a self-centralizing elementary abelian normal subgroup A such that $T_{12} = N_G(A)/A \cong S_4$ is an indecomposable subgroup of $\mathrm{GL}_3(2)$.

The Mathieu groups M_{22}, M_{23} and M_{24} have Sylow 2-subgroups S_{22}, S_{23} and S_{24} containing self-centralizing elementary abelian normal subgroups A_{22}, A_{23} and A_{24} of dimensions 4, 4 and 6 over $F = GF(2)$ such that their normalizer quotient groups $T_{22} \cong S_5$, $T_{23} \cong (3 \times A_5) : 2$ and $T_{24} \cong 3S_6$ are irreducible subgroups of $\mathrm{GL}_4(2)$, $\mathrm{GL}_4(2)$ and $\mathrm{GL}_6(2)$, respectively. All these statements follow from Proposition 8.2.4 of [92]. Hence these four Mathieu groups can be constructed by means of Algorithm 1.3.8 from the four subgroups T_i stated before.

Remark 1.3.12 Janko's group J_1, Held's group He, Harada's group Ha and Thompson's group Th can be constructed by means of Algorithm 1.3.8 from well defined irreducible subgroups of $\mathrm{GL}_3(2)$, $\mathrm{GL}_6(2)$, $\mathrm{GL}_5(2)$ and $\mathrm{GL}_5(2)$ by Theorems 9.3.2, 8.3.2, 11.2.2 and 12.3.1 of [92], respectively.

Definition 1.3.13 Let T be an indecomposable subgroup of a general linear group $\mathrm{GL}_n(2)$ such that some step of Algorithm 1.3.8 cannot be performed by theoretical reasons. Hence there is no simple group G having a Sylow 2-subgroup S containing a self-centralizing non-cyclic maximal elementary abelian normal subgroup A such that $N_G(A)/C_G(A) \cong T$. In such a case, we say that *the application of Algorithm 1.3.8 to T fails.*

Remark 1.3.14 If a Sylow 2-subgroup S of a finite simple group G has a unique maximal non-cyclic elementary abelian normal subgroup A of order 4 then A cannot be self-centralizing because S is not a dihedral group by Theorem 4.8.3 of [92]. The sporadic groups J_2, J_3 of Janko have such a Sylow 2-subgroup. Let $G \in \{\mathsf{J}_2, \mathsf{J}_3\}$, $C = C_G(A)$ and $E = N_G(A)$. Then Propositions 8.5.1 and 8.5.2 of [92] assert that the center $Z(C) = A$, $V = C/A$ is elementary abelian of order 16. Hence E/A is an FT-module extension of an indecomposable subgroup $T = E/C \in \mathrm{GL}_4(2)$ by V. Therefore E is an *iterated* FT-module extension of E/A by A.

In all these cases Steps 2 and 4 of Algorithm 1.3.8 have to be modified by now allowing E to be an iterated FT-module extension. All other steps remain essentially unchanged.

Algorithm 1.3.15 *Let T be an indecomposable subgroup of $\mathrm{GL}_m(2)$ acting faithfully on $V = F^m$ by matrix multiplication where $F = GF(2)$.*

Step 1 *Construct the split extension E_1 of T by V and a faithful permutation representation PE_1 of E_1. Determine then a presentation of E_1 in terms of generators and a set $\mathcal{R}(E_1)$ of defining relations.*

Step 2 *Let n be a positive integer for which there exists an indecomposable FE_1-module A of dimension n on which the normal subgroup V of E_1 acts trivially, and the following constructions can be performed.*

Using PE_1 and $\mathcal{R}(E_1)$ construct all FE_1-module extensions E of E_1 by A by means of Holt's Algorithm [54] and MAGMA. Determine a complete set $\mathfrak{T}_n$ of non-isomorphic iterated extension groups E by means of the Cannon–Holt Algorithm [12].

For each $E \in \mathfrak{T}_n$ determine a faithful permutation representation PE and check the following conditions:

(a) *A is an elementary abelian normal subgroup in any Sylow 2-subgroup S of E and A contains a non-cyclic characteristic subgroup of S;*
(b) *$C = C_E(A)$ is a maximal special or homocyclic normal subgroup of E with center $Z(C) = A$ or $\Omega[Z(C)] = \{c \in C | c^2 = 1\} = A$, respectively;*
(c) *$C/A \cong V$ and E/C is isomorphic to an indecomposable subgroup of $\mathrm{GL}_n(2)$.*

If no such integer n exists the Algorithm terminates. Otherwise, let $\mathfrak{E}_n$ be the set of all extensions $E \in \mathfrak{T}_n$ satisfying these conditions.

Step 3 *Let $E \in \mathfrak{E}_n$. Let PE be its faithful permutation representation. Using it and Kratzer's Algorithm 5.3.18 of [92] calculate a complete system of representatives of all the conjugacy classes of E.*

Check that E has a 2-central involution $z \neq 1$ such that $D = C_E(z) < E$.

Let $\mathfrak{I}_n$ be the set of all extension groups E_n for which such an involution exists. If $\mathfrak{I}_n$ is empty the algorithm terminates.

Otherwise, apply Steps 5–11 of Algorithm 1.3.8. If they can all be successfully performed, then it constructs a finite simple group G having a Sylow 2-subgroup S with an elementary abelian normal subgroup A such that $C_G(A)/A \cong V$, $N_G(A)/C_G(A) \cong T$ and $G = \langle C_G(z), N_G(A)\rangle$.

Remark 1.3.16 The sporadic group HS of Higman–Sims can be constructed by Algorithm 1.3.15 by Lemma 10.1.2 and Proposition 10.2.2 of [92]. In this example choose $T = \mathrm{GL}_3(2)$, $m = n = 3$. Then V and A are isomorphic irreducible FT-modules. Furthermore, E is the split extension of $C = C_E(A)$ by T, C is a direct product of three copies of a cyclic group of order 4, and $\Omega(C) = A$.

Remark 1.3.17 Keep the notation of Algorithm 1.3.15. If the Sylow 2-subgroup of the chosen indecomposable subgroup T of $\mathrm{GL}_m(2)$ is not cyclic then by Theorem 1.2.3 there are infinitely many integers n for which there is an indecomposable subgroup T_1 of $\mathrm{GL}_n(2)$ which is an epimorphic image of T. In particular, it is not possible to determine computationally all integers n for which all conditions of Step 2 of Algorithm 1.3.15 can be verified.

1.4 Documentation of experimental results

In order to do specific calculations in a finite group G with a given set $\{g_k | 1 \le k \le n\}$ of generators, we always construct a faithful permutation representation PG of G with a stabilizer $U = \langle u_j \mid 1 \le j \le m\rangle$ and an isomorphism $\tau : G \to PG$. The generators u_j of the stabilizer U are stated as short words in the generators g_i of G. In general, MAGMA uses random algorithms when it is applied to PG. Its results are often described by finitely many permutations $\pi \in PG$. In order to document them we have to be able to write the elements $w = \tau^{-1}(\pi) \in G$ as short words in the given generators g_k of G. For small groups G the MAGMA command `InverseWordMap(PG)` works well. But for the large groups studied in this book its output would often be so large that it cannot be printed. Therefore we now describe an algorithm which is due to the author's former collaborator R. Staszewski (Essen) and his student H. Kim (Cornell University), whose senior thesis [73] contains the improved version presented here. It has been used efficiently in all large-scale calculations described in [72], [74] and the following chapters. In particular, Kim's implementation can be loaded into any MAGMA session.

Let G be a finite group with faithful permutation representation $\tau : G \to PG$. Let $\{g_1, g_2, \ldots, g_n\}$ be a fixed set of generators g_k of G ordered by their indices. Let $\mathcal{T}(G)$ of G be the complete word tree of G; see Definition 5.3.10 of [92]. The length $\ell(w)$ of a word $w \in \mathcal{T}(G)$ is the length of the path in $\mathcal{T}(G)$ of the initial point g_k of w to w. Let $W_0 = \{1_G\}$, the identity element of G. For each integer $i \geq 1$ let $W_i = \{w \in \mathcal{T}(G) | \ell(w) = i\}$. In particular, $W_1 = \{g_k | 1 \leq k \leq n\}$. The isomorphism $\tau : G \to PG$ induces an evaluation map $\phi : \mathcal{T}(G) \to PG$ defined by $\phi(g_k) = \tau(g_k)$ for $1 \leq k \leq n$ and its multiplicative extension. For each $i \geq 1$ define

$$PV_i = \{\phi(w)^r | w \in W_i, 1 \leq r \leq order[\phi(w)] - 1\}.$$

Algorithm 1.4.1 (Kim–Staszewski) *Let G be a finite group with faithful permutation representation $\tau : G \to PG$. Let $\{g_1, g_2, \ldots, g_n\}$ be a fixed set of generators g_k of G ordered by their indices. Let $\mathcal{T}(G)$ be the complete word tree of G. Let PU be a subgroup of PG. The following algorithms construct a set of generators u_j of the subgroup $U = \tau^{-1}(PU)$ in G such that all its generators u_j are short words in the given generators g_k of G. Let $i \geq 1$ be an integer such that $PU \neq PU_i = \langle PU \cap PV_i \rangle$.*

Step 1 *Let $i \geq 1$ be an integer such that $PU \neq PU_i = \langle PU \cap PV_i \rangle$. Using the faithful permutation representation PG and MAGMA determine a minimal generating set PR_i of PU_i.*

Step 2 *Calculate the set W_{i+1} of all words $w \in \mathcal{T}(G)$ of length $\ell(w) = i + 1$, PV_{i+1} and PU_{i+1}. Using membership tests with MAGMA check whether $PU = PU_{i+1}$. If so, the algorithm ends after having determined a minimal generating set PR_{i+1} of PU.*

Otherwise, go to Step 1.

Since PU is finite, the algorithm terminates and

$$PU = \langle \phi(u_j) | 1 \leq j \leq u \rangle,$$

where the u_j are powers of short words in the given generators g_k of G.

In particular, $U = \langle u_j | 1 \leq j \leq u \rangle$.

Remark 1.4.2 H. Kim's implementation of Algorithm 1.4.1 is efficient because it avoids substantial memory problems by storing the words $w \in W_i$ in the word tree $\mathcal{T}(G)$ of G as sequences of indices of the generators g_k in the order they occur in the words w and not as the actual permutations $\tau(w)$ in PG, e.g. the word $w = g_1^3 \cdot g_2^2 \cdot g_1$ is stored as the sequence $[1, 1, 1, 2, 2, 1]$. Furthermore, doing the membership test $\phi(w)^r \in PU_i$ for all powers of $\phi(w)$ of a new word w in W_i at the same time speeds up the generation process of PU. In fact, the algorithm often finds generators of PU which are powers of $\phi(w)$ of a word $w \in W_i$ although $\phi(w)$ does not belong to PU.

Algorithm 1.4.3 (Kim–Staszewski) *Applying Algorithm 1.4.1 to cyclic subgroups generated by a permutation it can sometimes find a short word of an element g of a finite permutation group $G = \langle g_k \mid 1 \le k \le n \rangle$ in the given generators g_k of G. If it fails then often it is helpful to calculate generators u_j of suitable subgroups U of G containing g and then try to get a word for g in the new generators u_j of U by means of Algorithm 1.4.1. This application of Algorithm 1.4.1 is called* `LookupWord(G,g)` *in Kim's senior thesis [74]. It improves an earlier version due to R. Staszewski.*

Remark 1.4.4 Algorithm 1.4.1 is of great help when it is necessary to document the stabilizer U of a constructed permutation representation PG of a finite group G. The following example will be used in Section 1.5.

Applying the MAGMA command `LowIndexSubgroups(PG,n)` MAGMA returns all conjugacy classes of subgroups PT of PG of index $|PG : PT| \le n$. For each of these subgroups T of G Algorithm 1.4.1 finds a finite set of generators u_j as short words in the given generators g_k of G. Thus all new calculations with the permutation representation $(1_T)^G$ can be documented.

Remark 1.4.5 (P. Young) In the course of the calculations described in this book we have to do a number of demanding coset enumerations. Since MAGMA's command `CosetAction(G,U)` is very restricted, Paul Young and H. Kim developed a command `MyCosetAction(G, U: maxsize: = n)`, which is a batch of standard MAGMA functions for coset enumerations. It allows us to increase the size of the coset table to a large extent. The size is regulated by the parameter n.

Step 2 of Algorithm 1.3.8 requires the construction of extensions of a given group T by an elementary abelian group V. Often this extension splits. In order to be able to quote all the relations in the given generators of T and V in a fixed set $\mathcal{R}(E)$ of defining relations of E in a uniform way, we restate now the following definitions and well known subsidiary results from H. Kim's and the author's article [72].

Definition 1.4.6 Let F be a prime field of characteristic $p > 0$. Let G be a subgroup of $\mathrm{GL}_n(F)$. Let $V = F^n$ be the canonical n-dimensional vector space. Then V is a right FG-module of the group algebra FG of G over F with respect to the multiplication $(v \cdot g)$ defined by the product of the row vector $v \in V$ times the matrix $g \in G$. For each matrix $g \in G$ let $g^* = [g^{-1}]^T$ be the transpose of the inverse matrix g^{-1} of g. Then by Definition 3.4.4 of [92] V becomes a right FG-module under the multiplication $[v, g] := [g^* \cdot (v^T)]^T$, where $\cdot$ denotes the product of the matrix g^* times column vector v^T. This right FG-module V^* is called the *dual* FG-module of V.

The *semidirect product* $V \rtimes G$ of V and G consists of all pairs $(v, g), v \in V, g \in G$. Its multiplication $*$ is defined by

$$(v_1, g_1) * (v_2, g_2) : = (v_1 + [v_2, g_1], g_1 g_2) \quad \text{for } v_i \in V \text{ and } g_i \in G,$$
$$v * g = (v, 1) * (0, g) = (v + [0, 1], 1 * g) = (v, g) \quad \text{for}$$
$$v \in V,\ g \in G.$$

Lemma 1.4.7 *Let $G = \langle x_i | 1 \leq i \leq r \rangle$ be a finitely generated subgroup of $\mathrm{GL}_n(F)$ with the set of defining relations $\mathcal{R}(G)$ in the given generators x_i. Let $V = F^n$ be the canonical n-dimensional vector space with standard basis $\{e_j | 1 \leq j \leq n\}$.*

Let $\mathcal{R}_1(V \rtimes G)$ be the set of all relations:

$$e_j^p = 1 \quad \text{and} \quad e_k * e_j = e_j * e_k \quad \text{for all} \quad 1 \leq j, k \leq n.$$

Let $\mathcal{R}_2(V \rtimes G)$ be the set of all relations:

$$(*) \quad x_i * e_j * x_i^{-1} = {e_1}^{a(i)_{1,j}} * \cdots * {e_n}^{a(i)_{n,j}} \quad \text{for all} \quad 1 \leq i \leq r, 1 \leq j \leq n,$$

where $(a(i)_{j1}, a(i)_{j2}, \ldots, a(i)_{jn}) = [(x_i)^* \cdot ((e_j)^T)]^T$. Then

$$\mathcal{R}(V \rtimes G) = \mathcal{R}_1(V \rtimes G) \cup \mathcal{R}_2(V \rtimes G) \cup \mathcal{R}(G).$$

Proof Clearly $\{x_i | 1 \leq i \leq r\}, \{e_j | 1 \leq j \leq n\}$ is a set of generators of the semidirect product $V \rtimes G$ of V and G. By Definition 1.4.6 the multiplication $*$ agrees in $G = \{(0, g) | g \in G\}$ with the given multiplication of G. Hence $\mathcal{R}(G)$ is a defining set of relations of the complement $G = \langle x_i | 1 \leq i \leq r \rangle$ of V in $\mathcal{R}(V \rtimes G)$. As V is elementary abelian of order $|F|^n$, $\mathcal{R}_1(V \rtimes G)$ yields a presentation of the normal subgroup V of

$\mathcal{R}(V \rtimes G)$. The conjugate action of G on V is described by the relations (*) of $\mathcal{R}_2(V \rtimes G)$. By Definition 1.4.6 the proof is complete. □

Definition 1.4.8 Keep the notation of Lemma 1.4.7. The set $\mathcal{R}_2(V \rtimes G)$ of relations of the semidirect product $(V \rtimes G)$ of the FG-module V and the finitely presented group G is called the set of *essential relations* of $(V \rtimes G)$. The set $\mathcal{R}_1(V \rtimes G)$ of relations of the semidirect product $(V \rtimes G)$ is called the set of *non-essential relations* of $(V \rtimes G)$.

In particular, it suffices to state only the set $\mathcal{R}_2(V \rtimes G)$ of essential relations of a semidirect product $(V \rtimes G)$ because sets $\mathcal{R}(G)$ and $\mathcal{R}_1(V \rtimes G)$ are identical in all semidirect products $(V \rtimes G)$.

1.5 Constructing projective irreducible modular representations

Step 7 of Algorithm 1.3.8 requires the construction of projective irreducible FG-modules of a finite group G over a finite field of characteristic $p > 0$ from its given character table. In this section we describe an algorithm for such a construction using Theorem 5.6.5 of [92] due to M. Weller and the author. It provides a formula for calculating all the character values $\chi_s(g), g \in G$ of the irreducible constituents χ_s of the permutation character $(1_U)^G$ G with stabilizer U. Here we apply the method of proof of that theorem to an irreducible constituent χ of p-defect zero of a permutation character $(1_U)^G$ of G and get the irreducible matrix representation of G corresponding to χ.

Based on Weller's stand-alone implementation of Algorithm 5.7.1 of [92], A. Previtali developed in 2004 an implementation of that algorithm which can be loaded into a MAGMA session. In [94] it has been used to find all the character values of the simple O'Nan group ON; see Chapter 14. In his senior thesis [74] H. Kim has modified Weller's and Previtali's programs to get a program for calculating irreducible p-modular matrix representations over suitable finite fields F of characteristic $p > 0$ corresponding to the irreducible constituents χ_s of p-defect zero of a permutation character $(1_U)^G$ of a finite group G with stabilizer U. It works well provided all values $\chi_s(g)$, $g \in G$, are rational integers. In that case F can be chosen to be the prime field $GF(p)$.

Kim's program uses Previtali's program calculating double coset decompositions of G with respect to its subgroup U and the corresponding adjacency matrices. It applies several steps of Algorithm 5.7.1 of [92].

In view of all its important applications in this book it is now stated in detail.

Algorithm 1.5.1 (Kim–Michler–Previtali–Weller) *Let G be a finite group with faithful permutation representation $\tau : G \to PG$. Let $\{g_1, g_2, \ldots, g_n\}$ be a fixed set of generators g_k of G ordered by their indices. Let $p > 0$ be a prime and let χ be an irreducible character of G of p-defect zero. Let F be a splitting field of characteristic p for the p-modular irreducible representation V afforded by χ. The following algorithm constructs the FG-module V up to isomorphism provided all its steps can be performed.*

Step 1 *Apply the MAGMA command* `LowIndexSubgroups(PG,n)` *to PG for a suitable integer $n > 1$ and, for each constructed subgroup PU_m of PG having index $m \leq n$, calculate the fusion of its classes in PG and the character table of PU_m and the inner product $(\chi, (1_{PU_m})^G)$ until MAGMA returns a subgroup PU_m such that this inner product is non-zero.*

If no such index can be found, the algorithm fails.

Otherwise let n_0 be the smallest index m such that $(\chi, (1_{PU_m})^G) \neq 0$. Let $PU = PU_{n_0}$.

Step 2 *Using Algorithm 1.4.1 determine a set $\{u_j \mid 1 \leq j \leq m\}$ of short generators of U in terms of the generators g_k of G. Determine a right transversal $T = \{g_a \mid 1 \leq a \leq r\}$ of U in G.*

Step 3 *Compute a complete set of double coset representatives x_k, $1 \leq k \leq d$, of $G = \bigcup_{k=1}^{d} U x_k U$.*

Step 4 *Compute the d intersection matrices $D_k = (p_{ij}^k)$ of all double cosets $U x_k U$ of G.*

Step 5 *Determine a basis $A = \{Z_1, Z_2, \ldots, Z_t\}$ of the solution space $\mathcal{Z}(\mathcal{B})$ of the system of linear equations*

$$\text{(U)} \qquad \sum_{k=1}^{d} (p_{ij}^k - p_{kj}^i) q_k = 0$$

described in Theorem 5.5.6 of [92]. Thus each basis vector $Z_j = \sum_{k=1}^{d} q_{jk} D_k$ for uniquely determined $q_{jk} \in \mathbb{Q}$.

Step 6 *Compute the structure constants z_{uv}^j of the center $\mathcal{Z}(\mathcal{B})$ of the intersection algebra $\mathcal{B}$ with respect to the basis A by solving the system* *of linear equations:*

$$Z_u Z_j = \sum_{v=1}^{t} z_{uv}^j Z_v.$$

To each basis element associate its matrix

$$U_j = (z_{uv}^j) \in Mat_t(\mathbb{Q}).$$

Step 7 *Compute a basis* $f_1, f_2, \ldots, f_t$ *of* $\mathcal{Z}(\mathcal{B})$ *consisting of common eigenvectors* f_i *of the* t *pairwise commuting matrices* U_j.

Then there are t *uniquely determined* $c_{is} \in \mathbb{Q}$ *such that* $f_i = \sum_{s=1}^{t} c_{is} Z_s$ *for* $1 \leq i \leq t$.

For each eigenvector f_i *compute a sequence* $h_{i1}, h_{i2}, \ldots, h_{it}$ *of eigenvalues of* $U_1, U_2, \ldots, U_t$, *respectively, such that* $U_j f_i = f_i h_{ij}$. *Then*

$$\left\{ e_i = \frac{1}{q_i} f_i \mid 1 \leq i \leq t \right\}$$

is the complete set of all primitive idempotents e_i *of* $\mathcal{Z}(\mathcal{B})$, *where* $q_i = \sum_{s=1}^{t} h_{is} c_{is}$.

In particular, there are uniquely determined $a_{ik} = \frac{1}{q_i} \sum_{s=1}^{t} c_{is} q_{ik} \in \mathbb{C}$ *such that*

$$e_i = \sum_{k=1}^{d} a_{ik} D_k \qquad \text{for } i = 1, 2, \ldots, t.$$

Step 8 *For each* $i \in \{1, 2, \ldots, t\}$ *compute* $m_i = \sqrt{\dim_{\mathbb{C}}(e_i \mathcal{B})}$, *which is the multiplicity of* χ_i *in* $(1_U)^G$.

Step 9 *For the given* $g \in G$ *compute the integers*

$$m_{gk} = \mid \{g_p \in T \mid g_p g g_p^{-1} \in H x_k H\} \mid \qquad \text{for } 1 \leq k \leq d.$$

Step 10 *For* $i = 1, 2, \ldots, t$ *the character values of* χ_i *are computed by the following formula:*

$$\chi_i(g) = \frac{1}{m_i} \sum_{k=1}^{d} m_{gk} \bar{a}_{ik}.$$

Step 11 *There is a unique index* i *in* $i = 1, 2, \ldots, t$ *such that* $(\chi, \chi_i) \neq 0$ *and so* $\chi = \chi_i$. *For* $k = 1, 2, \ldots, d$ *calculate the first row* $R_{1,k}$ *of the* $r \times r$ *adjacency matrix* A_k *corresponding to the intersection matrix* D_k *by means of Definition 5.1.7 of [92]. Then*

$$V = \sum_{k=1}^{d} a_{ik} R_{1,k} \qquad \text{for } i = 1, 2, \ldots, t$$

is the first row of the central primitive idempotent E_i in the endomorphism ring $End_{FG}(W)$ corresponding to the irreducible constituent $\chi = \chi_i$ of $W = (1_u)^G \otimes F$. Hence $V = E_i(W)$ is the irreducible FG-module corresponding to the p-modular character χ of defect zero.

1.6 Thompson's group order formula revisited

Thompson's group order formula stated in Theorem 4.2.1 of [92] provides an efficient method to calculate the order $|G|$ of a finite group G having at least two distinct conjugacy classes of involutions $\{u_i^G \mid 1 \le i \le t\}$, $t > 1$. In fact, $|G|$ can be computed from the character tables of the centralizers $C_G(u_i)$ and the fusion patterns of the conjugacy classes of their involutions in G. In all applications described in [92] these centralizers $C_G(u_i)$ were so small that the constants of the group order formula could be calculated from their permutation presentations directly without using the character tables.

The calculations of the particular constants of Thompson's group order formula of the large sporadic groups existing in the literature are not uniform and lengthy. Therefore we prove a general character formula for these constants in this section.

Lemma 1.6.1 *Let G be any finite group and let $Irr_{\mathbb{C}}(G)$ be the set of all its distinct irreducible characters. Let x, y, z be any three elements of G. Then the number $p(x, y, z)$ of pairs $\{x_1, y_1) \in (x^G \times y^G)|x_1y_1 = z\}$ is given by the following character formula:*

$$p(x, y, z) = \frac{|G|}{|C_G(x)| \cdot |C_G(y)|} \cdot \sum_{\psi \in Irr_{\mathbb{C}}(G)} \psi(x)\psi(y)\psi(z^{-1})\psi(1)^{-1}.$$

Proof The assertion follows immediately from Theorem 2.4.13 of [92]. □

Definition 1.6.2 Let G be a finite group with at least two distinct conjugacy classes u^G, v^G of involutions. An element w of G is called *strongly (u, v)-real* if there is a pair $(x, y) \in (u^G \times v^G)$ such that $xy = w$.

Remark 1.6.3 With the hypothesis of Definition 1.6.2, Lemma 1.6.1 implies that $w \in G$ is strongly (u, v)-real if and only if

$$p(u, v, w) = \frac{|G|}{|C_G(u)| \cdot |C_G(v)|} \cdot \sum_{\psi \in Irr_{\mathbb{C}}(G)} \psi(u)\psi(v)\psi(w^{-1})\psi(1)^{-1} \neq 0.$$

In particular, all strongly (u, v)-real elements of G are real by Lemma 2.2.12(a) of [92]. Hence all their character values are real numbers.

In particular, the strongly (u, v)-real conjugacy classes can be determined from the character table of the group G.

Theorem 1.6.4 *Let G be a finite group with at least two distinct conjugacy classes u^G, v^G of involutions. Let $u_1 = u$, $u_2 = v$, $u_3, \ldots, u_t$ be representatives of the different conjugacy classes of involutions of G with respective centralizers $U_i = C_G(u_i)$. Suppose that the fusion of all involutions of all centralizers U_i in G has been determined and that the character tables of all centralizers and the power maps of their conjugacy classes are known.*

For each fixed $i \in \{1, 2, \ldots, t\}$ let $\{a_{i,j} | 1 \leq j \leq r_i\}$ and $\{b_{i,k} | 1 \leq k \leq s_i\}$ be sets of representatives of all conjugacy classes of U_i fusing in G with u and v, respectively. In particular,

$$u^G \cap U_i = a_{i,1}^{U_i} \cup a_{i,2}^{U_i} \cdots \cup a_{i,r_i}^{U_i}$$

and

$$v^G \cap U_i = b_{i,1}^{U_i} \cup b_{i,2}^{U_i} \cdots \cup b_{i,s_i}^{U_i}.$$

Let $\{w_n \mid 1 \leq n \leq m\}$ be a set of representatives of all real u_i-special conjugacy classes of U_i. For each index n let

$$d_i(j, k, w_n) = \frac{|a_{i,j}^{U_i}| \cdot |b_{i,k}^{U_i}| \cdot |w_n^{U_i}|}{|U_i|} \cdot \sum_{\psi \in Irr_{\mathbb{C}}(U_i)} \frac{\psi(a_{i,j})\psi(b_{i,k})\psi(w_n)}{\psi(1)}.$$

Then the following assertions hold.

(a) *A conjugacy class w^{U_i} of U_i is strongly $(a_{i,j}, b_{i,k})$-real if and only if*

$$p_i(j, k, w) = \frac{|a_{i,j}^{U_i}| \cdot |b_{i,k}^{U_i}|}{|U_i|} \cdot \sum_{\psi \in Irr_{\mathbb{C}}(U_i)} \frac{\psi(a_{i,j})\psi(b_{i,k})\psi(w^{-1})}{\psi(1)} \neq 0.$$

(b) *The number $r(u, v, u_i) = |\{(x, y) \in u^G \times v^G | (xy)^{n(x,y)} = u_i\}|$ is calculated by the character formula*

$$r(u, v, u_i) = \sum_{j=1}^{r_i} \sum_{k=1}^{s_i} \sum_{n=1}^{m(j,k)} d_i(j, k, w_n).$$

(c) $|G| = \sum_{i=1}^{t} r(u, v, u_i) \frac{|C_G(u)||C_G(v)|}{|C_G(u_i)|}$.

Proof Assertion (a) holds by Lemma 1.6.1.

(b) Let $\mathfrak{Q} = \{(x, y) \in [u^G \cap C_G(u_i)] \times [v^G \cap C_G(u_i)] | (xy)^{n(x,y)} = u_i\}$. Theorem 4.2.1(a) of [92] asserts that $r(u, v, u_i) = |\mathfrak{Q}|$. Let $(x, y) \in \mathfrak{Q}$. Then $w = xy$ is u_i-special and real by Remark 1.6.3. Both properties are inherited by all its conjugates in U_i. Furthermore, there is exactly one pair (j, k) of indices such that $(x, y) \in a_{i,j}^{U_i} \times b_{i,k}^{U_i}$. Thus w is strongly $(a_{i,j}, b_{i,k})$-real. Therefore (a) implies that $w^{U_i} = w_n^{U_i}$ for a unique index $n \in \{1 \le n \le m\}$. Thus $d_i(j, k, w) = d_i(j, k, w_n) > 0$.

Conversely, if $(x, y) \in \{(x, y) \in (u^G \times v^G)\}$ has a product $q = xy$ such that $d_i(x, y, w_n) = |w_n^{U_i}| \cdot p_i(x, y, w_n) \neq 0$ for some u_i-special class $w_n^{U_i}$ then q is u_i-special by Lemma 1.6.1. Hence $(xy)^{n(x,y)} = u_i$ for some integer $n(x, y) > 0$ depending on the pair (x, y). Thus $(x, y) \in \mathfrak{Q}$. Therefore

$$r(u, v, u_i) = |\mathfrak{Q}| = \sum_{j=1}^{r_i} \sum_{k=1}^{s_i} \sum_{n=1}^{m} d_i(j, k, w_n).$$

Assertion (c) follows from (b) and Theorem 4.2.1(b) of [92]. □

2

Dickson group $\mathsf{G}_2(3)$ and related simple groups

It is well known that for $3 \leq n \leq 5$ each simple linear group $T_n = \mathrm{GL}_n(3)$ operating on the vector space $V_n = F^n$ over the prime field $F = GF(2)$ has a non-split extension E_n of T_n by V_n which is uniquely determined up to isomorphism; see [41]. Furthermore, $C_{E_n}(V_n) = V_n$. Hence Algorithm 1.3.8 can be applied to any of these extensions. Theorem 12.3.1 of [92] implies that for $n = 5$ such an application yields Thompson's sporadic simple group Th of order $2^{15} \cdot 3^{10} \cdot 5^3 \cdot 7^2 \cdot 13 \cdot 19 \cdot 31$. In Chapter 3 we show that for $n = 4$ an application of Algorithm 1.3.8 returns Conway's sporadic group Co_3. Both simple groups are uniquely determined up to isomorphism by their centralizers of 2-central involutions.

In this chapter the case $n = 3$ is dealt with. Let E be the uniquely determined non-split extension of $\mathrm{GL}_3(2)$ by $V = V_3$. It follows from P. Fong's main theorem of his article [31] and Theorem 1.6.2 of [92], due to R. Brauer, M. Suzuki and G. E. Wall, that there are infinitely many simple groups $\mathsf{G}_2(q)$ having a 2-central involution z such that $H_q = C_{\mathsf{G}_2(q)}(z)$ has a Sylow 2-subgroup S_q that is isomorphic to a Sylow 2-subgroup S of E because there are infinitely odd prime powers $q = p^k$ such that $q \equiv 3, 5 \bmod (8)$. Furthermore, each S_q has an elementary abelian normal subgroup A_q such that $D_q = N_{\mathsf{G}_2(q)}(A_q) \cong C_E(z)$, where z is assumed to be the generator of the center $Z(S)$ of S. In particular, all indices $|\mathsf{G}_2(q) : H_q|$ are odd. Hence all conditions of Step 5 of Algorithm 1.3.8 are satisfied for infinitely many groups H_q of even order. Thus it cannot be proved that Algorithm 1.3.8 constructs in general only finitely many simple groups.

The infinite series of simple groups $\mathsf{G}_2(q)$ was discovered by L. E. Dickson in 1905 [24]. They are defined over all finite fields order $q = p^k$, where $p > 0$ is any prime. For odd primes p and $q > 3$, Fong characterized any $\mathsf{G}_2(q)$ by the structure of its centralizer of an involution [31]. Fong's uniqueness proof uses J. Tits' theory of (B, N)-pairs [131]. In [68] Z. Janko completed Fong's study for the remaining case $p = q = 3$ by verifying the hypothesis of J. G. Thompson's deep Theorem 8.1 of [128]. Janko's theorem was quoted in the uniqueness proof for Thompson's sporadic group Th in Chapter 12 of [92] without proof. Therefore a self-contained proof for it is given in this chapter. In his recent article [5] M. Aschbacher gives another characterization of $\mathsf{G}_2(3)$ which is independent of [128] and contains Janko's theorem as a corollary.

As $q = p = 3$ Algorithm 1.3.8 can be applied to the non-split extension group E to get a group $H = H_3$ of even order. Weller, Previtali and the author constructed in [134] a simple matrix subgroup $\mathfrak{W} \le \mathrm{GL}_{14}(5)$ having a 2-central involution $\mathfrak{z}$ such that $C_{\mathfrak{W}}(\mathfrak{z}) \cong H_3$. The main result of this chapter is Theorem 2.4.5. It asserts that any simple group G of $\mathsf{G}_2(3)$-type is isomorphic to the simple matrix group $\mathfrak{W}$. The generating matrices are stated as well. This result is due to L. Wang and the author. Its proof builds on Janko's ideas [68]. But Janko's reduction of the uniqueness proof to Thompson's Theorem 8.1 of [128] is replaced here by constructing presentations in terms of generators and relations of all normalizers $N_G(\langle x \rangle)$ of cyclic subgroups $\langle x \rangle$ of prime order of G. Thus their character tables can be calculated. Then Brauer's characterization of characters given in Corollary 2.8.10 and the author's Theorems 4.3.7 and 7.5.1, all from [92], are applied to finish the proof.

2.1 Involution centralizers of Dickson's groups $\mathsf{G}_2(q)$, q odd

An explicit generic presentation for all the Dickson simple groups $\mathsf{G}_2(q)$ is stated in R. Ree's article [111], p. 443. If q is an odd prime power then a Sylow 2-subgroup S_q has a center of order 2. Its centralizer is described in the following definition due to P. Fong [31].

Definition 2.1.1 Let $q = p^k$ be any odd power of a prime $p > 2$. A finite simple group G is called to be of $\mathsf{G}_2(q)$-*type*, if it has a 2-central involution z such that $C_G(z)$ is isomorphic to a finite group H_q with the following properties:

(a) H_q has two subgroups L_1, L_2, such that $L_i \cong \mathsf{SL}_2(q)$, $[L_1, L_2] = 1$, $L_1 \cap L_2 = \langle z \rangle$, and

(b) there is an involution t such that $H = \langle L_1 L_2, t \rangle$ and $L_i^t = L_i$ for $i = 1, 2$.

Using the Steinberg relations for $\mathsf{SL}_2(q)$ it is easy to deduce from Definition 2.1.1 a set $\mathcal{R}(H_q)$ of defining relations of the centralizer H_q. We state without proof the following result due to Fong [31].

Theorem 2.1.2 (P. Fong) *Let $q = p^k > 3$ be any odd prime power. Any finite simple group G of $\mathsf{G}_2(q)$-type is isomorphic to Dickson's simple group $\mathsf{G}_2(q)$.*

Let E be the uniquely determined non-split extension of $\mathrm{GL}_3(2)$ by $V = V_3$. The presentation of E stated in the following result is taken from Lemma 12.4.2(c) of [92]. Using this presentation of E will allow us to replace the quotation of Janko's paper [68] in the proof of Lemma 12.4.2(e) by the main result of this chapter.

Corollary 2.1.3 *Let q be an odd prime power such that $q \equiv 3$ or $5 \bmod (8)$. Let E be a non-split extension of $\mathrm{GL}_3(2)$ by its natural 3-dimensional vector space over $F = GF(2)$. Then the following statements hold.*

(a) *There is an element s_0 in E such that $E = \langle D, s_0\rangle = \langle k_i, d_4, s_0 \mid 1 \leq i \leq 5\rangle$ has the following set $\mathcal{R}(E)$ of defining relations:*

$$\begin{aligned}
&k_1^4 = k_2^4 = k_3^4 = k_4^4 = k_5^2 = d_4^3 = s_0^7 = 1,\\
&k_2^{-1}k_1^2k_2^{-1} = k_3^{-1}k_1^2k_3^{-1} = k_4^{-1}k_1^2k_4^{-1} = k_1^{-1}k_2^{-1}k_1k_2^{-1} = 1,\\
&[k_1, k_3] = [k_2, k_3] = [k_1, k_4] = [k_2, k_4] = 1,\\
&k_3^{-1}k_4^{-1}k_3k_4^{-1} = k_1^{-1}k_5k_2^{-1}k_5 = (k_3^{-1}k_5)^2 = 1,\\
&k_2^{-1}d_4^{-1}k_1d_4 = k_4^{-1}d_4^{-1}k_3d_4 = (k_5d_4^{-1})^2 = 1,\\
&d_4^{-1}k_2^{-1}k_1^{-1}d_4k_1 = k_1^{-1}s_0k_1^2s_0^{-1}k_4^{-1} = 1,\\
&k_5k_4^{-1}k_3^{-1}k_5k_4 = (k_2s_0^{-1}k_5)^2 = s_0k_3^{-1}d_4^{-1}k_2^{-1}s_0k_5 = 1,\\
&k_1^{-1}s_0d_4^{-1}k_4^{-1}s_0d_4^{-1} = s_0^2d_4k_3s_0k_3^{-1} = 1.
\end{aligned}$$

(b) *E has a faithful permutation representation of degree 192 with stabilizer $\langle s_0\rangle$.*

(c) *$z = k_1^2$ generates the center $Z(S)$ of order 2 of the Sylow 2-subgroup $S = \langle k_i \mid 1 \leq i \leq 5\rangle$ of E.*

(d) *$A = \langle z, k_1k_4, k_1k_2k_3^{-1}\rangle$ is a self-centralizing elementary abelian normal subgroup of E of order 8 such that $E/A \cong \mathrm{GL}_3(2)$. It is also a maximal elementary abelian normal subgroup of S.*

(e) *The Sylow 2-subgroups of E are isomorphic to the Sylow 2-subgroups of the Mathieu group* M_{12}.

(f) *The Sylow 2-subgroups S_q of all Dickson groups $\mathsf{G}_2(q)$ are isomorphic to a Sylow 2-subgroup S of E.*

(g) *Let H_q be the centralizer of a 2-central involution z_q of $\mathsf{G}_2(q)$ contained in the Sylow 2-subgroup S_q of H_q. Then there is a maximal elementary abelian normal subgroup A_q of order 8 in S_q and a 2-central involution z in E such that $D_q = N_{H_q}(A_q) \cong C_E(z)$.*

(h) *A system of representatives e_i of the 11 conjugacy classes of E and the corresponding centralizer orders $|C_E(e_i)|$ are given in Table 12.8.5 of [92].*

(i) *The character table of E is stated in Table 12.9.5 of [92].*

Proof (a) This presentation of E is taken from Lemma 12.4.2(c) of [92].

Using MAGMA the assertions (b), (c) and (d) can easily be derived from the presentation given in (a); see [92], p. 575.

(e) This statement follows from Theorem 8.1.9(b) of [92].

(f) Theorem 1.6.2 of [92] implies that the Sylow 2-subgroups V of $T = \mathsf{SL}_2(q)/\langle z_q \rangle$ are elementary abelian of order 4 and $N_T(V)$ is isomorphic to an alternating group A_4. Hence a Sylow 2-subgroup S_q of H_q has order 2^6 by Definition 2.1.1. The Sylow 2-subgroups Q_1 and Q_2 of L_1 and L_2 are quaternion groups of order 8, which are both normalized by the involution t. In particular, $S_q = \langle Q_1 Q_2, t \rangle$ is a Sylow 2-subgroup of H_q. From (a) and (c) it now follows easily that $S_q \cong S$.

(g) By (a) $D = C_E(z) = \langle S, d_4 \rangle$. Since $N_T(V) \cong A_4$ each subgroup L_i of H_q has an element d_i of order 3 such that $d = d_1 d_2^{-1}$ centralizes the central involution z_q. By (f) H_q and E have isomorphic Sylow 2-subgroups. Let S_q be a Sylow 2-subgroup of H_q and let A_q be its elementary abelian normal subgroup corresponding to the normal subgroup A of S. Then z_q belongs to A_q and d to $N_{H_q}(A)$. Since A_q is self-centralizing in H_q the assertion follows. □

In Lemma 12.4.1(d) of [92] the reader also finds an explicit embedding of the centralizer $H = C_{\mathsf{G}_2(3)}(z)$ up to isomorphism into any simple group of Th-type such that the amalgam $H \leftarrow D = C_E(z) \rightarrow E$ satisfies all conditions of Step 5 of Algorithm 1.3.8 as is shown in Lemma 12.4.2 of [92]. Keeping the notation of Corollary 2.1.3 it is shown there that there are two elements d_1 and d_2 of order 3 in $H = \langle L_1 L_2, k_5 \rangle$ such that $L_1 = \langle k_1, k_2, d_1 \rangle$, $L_2 = \langle k_1, k_2, d_2 \rangle$ and $d_4 = d_1 d_2^{-1}$. In particular, the group H has the following presentation in the given generators.

Lemma 2.1.4 *Let* $H = \langle k_i, d_j | 1 \le i \le 5, 1 \le j \le 2\rangle$ *have the following defining relations:*

$$\begin{aligned}
&k_1^4 = k_2^4 = k_3^4 = k_4^4 = k_5^2 = d_1^3 = d_2^3 = 1,\\
&k_1^2 = k_2^2 = k_3^2 = k_4^2,\\
&k_2^{k_1} = k_2^3, \quad k_4^{k_3} = k_4^3, \quad k_1^{k_5} = k_2^3, \quad k_2^{k_5} = k_1^3, \quad k_3^{k_5} = k_3^3, \quad k_4^{k_5} = k_3 k_4,\\
&k_1^{d_1} = k_2, \quad k_2^{d_1} = k_2^3 k_1, \quad k_3^{d_2} = k_3^3 k_4, \quad k_4^{d_2} = k_3,\\
&[k_1, k_3] = [k_2, k_3] = [k_1, k_4] = [k_2, k_4] = [d_1, d_2] = 1,\\
&[k_1, d_2] = [k_2, d_2] = [k_3, d_1] = [k_4, d_1] = 1,\\
&d_1^{k_5} = d_1^2, \quad d_2^{k_5} = d_2^2.
\end{aligned}$$

Then the following statements hold.

(a) $z = k_1^2 = k_2^2 = k_3^2 = k_4^2$ *generates the center* $Z(H)$ *of* H *of order* 2.

(b) *The subgroups* $L_1 = \langle k_1, k_2, d_1\rangle$ *and* $L_2 = \langle k_3, k_4, d_2\rangle$ *are both isomorphic to* $\mathsf{SL}_2(3)$, *and they are both normal in* H.

(c) *The commutator subgroup* H' *of* H *is the central product* $L_1 * L_2$ *with amalgamated central subgroup* $\langle z\rangle$, *and* $H = \langle H', k_5\rangle$.

(d) H *has a faithful permutation representation of degree* 2^6 *with stabilizer* $\langle d_1, d_2\rangle$.

(e) $S = \langle k_i \mid 1 \le i \le 5\rangle$ *is a Sylow* 2*-subgroup of* H.

(f) S *has two maximal elementary abelian normal subgroups* $A = \langle z, k_1k_4, k_1k_2k_3^{-1}\rangle$ *and* $B = \langle z, k_2k_4, k_1k_2k_3\rangle$, *and* $|A| = |B| = 8$.

(g) $V = \langle z, k_1k_2k_3^{-1}\rangle = A \cap B$ *is the unique normal Klein 4-group of* S, $C = C_H(V)$ *has order* 2^5 *and center* $Z(C) = V$.

(h) C *has four elementary abelian normal subgroups of order* 8. *They are* A, B, $A_1 = \langle z, k_3k_5, k_1k_5k_1\rangle$ *and* $A_2 = \langle z, k_5, k_1k_3k_5k_1\rangle$. *The involutions* $t = k_1k_2k_4k_3$ *and* $w = k_1k_3$ *generate both a complement of* C *in* S *and these two complements are not conjugate in* S *but in* H. *Furthermore,* $A_1^t = A_1^w = A_2$.

(i) $D = N_H(A) = \langle S, d_4\rangle$, *where* $d_4 = d_1d_2^{-1}$, *and* $D_1 = N_H(B) = \langle S, d_3\rangle$, *where* $d_3 = d_1d_2$.

(j) *The automorphism groups* $Aut(D)$ *and* $Aut(D_1)$ *are isomorphic and have order* $2^7 \cdot 3$.

(k) *The map* δ *defined by*

$$k_1 \to k_4^3, k_2 \to k_3k_4, k_3 \to k_1k_2^3, k_4 \to k_1^3, k_5 \to zk_5, d_4 \to d_4$$

is an automorphism of $D = N_H(A)$ of order 2. It can be extended to an automorphism α of order 2 of H satisfying also $\alpha(d_1) = d_2^2$ and $\alpha(d_2) = d_1^2$.

(l) *A system of representatives h_i of the 20 conjugacy classes of H and the corresponding centralizers orders $|C_H(h_i)|$ are stated in Table 12.8.4 of [92].*

(m) *The character table of H is stated in Table 12.9.4 of [92].*

Proof The statements (a), (b), (c) and (d) are immediate consequences of the definition of H and its defining relations.

(d) Since $|H| = 2^6 \cdot 3^2$ the subgroup $T = \langle d_1, d_2 \rangle$ is a Sylow 3-subgroup of H. Thus $(1_T)^H$ is a faithful permutation representation of H of degree 2^6.

(f) and (g) Using the faithful permutation representation PH of H constructed in (d) and the MAGMA command

```
subs :=Subgroups(S : Al := Normal, IsElementaryAbelian
  := true),
```

one can verify that S has exactly two elementary abelian normal subgroups A and B of order 8 and a unique normal Klein 4-group V. Furthermore, $V = A \cap B$. The given generators of these normal subgroups have been found by means of Algorithm 1.4.1. Both A and B are maximal elementary abelian normal subgroups of S.

Using MAGMA again, one can check that $C = C_H(V)$ has order 2^5 and center $Z(C) = V$.

(h) Running the subgroup command mentioned above for C it follows that C has exactly four elementary abelian normal subgroups A_i of order 8, two of which are A and B, and the two others, denoted by A_1 and A_2, are not normal in S. Again V is a unique normal Klein 4-group in C. The generators of A_1 and A_2 have been determined by means of Algorithm 1.4.1.

Applying the MAGMA command `HasComplement(S,C)` the complement $K_1 = \langle t \rangle$ of C in S has been found. The second complement $K_2 = \langle w \rangle$ of C in S was obtained by means of the MAGMA command

```
exists(w){x : x in H - CV|Order(x)=2, IsConjugate(S,t,x)
  :=false}.
```

The given words for t and w were obtained by another application of Algorithm 1.4.1. Now it is easy to check the equations $A_1^t = A_1^w = A_2$.

(i) This statement can be verified by means of MAGMA.

(j) Using the the faithful permutation representation $p : H \to PH$ and the MAGMA commands `AD:=AutomorphismGroup(D)`, `h_AD_PAD`, `PAD:=PermutationRepresentation(AD)`, one obtains a faithful permutation representation PAD of degree 24 of the automorphism group AD of D. Similarly one constructs a faithful permutation representation PAD_1 of degree 24 of D_1. MAGMA also asserts that both permutation groups have order $2^7 \cdot 3$. Using then the command `IsIsomorphic(PAd, PAD1)` it follows that $D \cong D_1$.

(k) For each $x \in PAD$ let ax = `x@@h_AD_PAD` in AD. Using the following commands in MAGMA:

```
PX:={x: x in PAD|Order(x)=2 and pd4@ax = pd4 and k1@ax = pk4^3},

PY:={x: x in PX|pk2@ax = pk3pk4,pk3@ax = pk1pk2^3,pk5@ax = pzpk5},
```

it follows that $|PY| = 1$. Hence the automorphism $\delta \in PY$ of D is well defined. The same methods show that the automorphism group AH of H has order $2^8 \cdot 3^2$ and that there is exactly one automorphism α of H whose restriction to D has the same values as δ. Furthermore, it satisfies the additional equations $\alpha(d_1) = d_2^2$ and $\alpha(d_2) = d_1^2$.

The statements (l) and (m) do not require a proof. □

Definition 2.1.5 A finite simple group G is called to be of $G_2(3)$-*type* if it has a 2-central involution z such that $C_G(z)$ is isomorphic to the finitely presented group H stated in Lemma 2.1.4.

In the remainder of this chapter it is shown that any simple group of $G_2(3)$-type is isomorphic to the simple subgroup $\mathfrak{W}$ of $GL_{14}(5)$ stated in the proof of Lemma 12.4.2 of [92].

2.2 Fusion and conjugacy classes of even order

Throughout this section G denotes a finite simple group of $G_2(3)$-*type*. It is shown here that G has a unique conjugacy class of involutions.

Proposition 2.2.1 *Let G be any simple group having a 2-central involution z such that $C_G(z) \cong H$. Then the following statements hold.*

(a) *z is a representative of the unique conjugacy class of involutions of G.*

(b) *The normalizer $N_G(V)$ of the unique normal Klein four subgroup $V = \langle z, k_1k_2k_3^{-1}\rangle$ of the Sylow 2-subgroup $S = \langle k_i \mid 1 \le i \le 5\rangle$ contains an element $g \in N_G(V) - N_H(V)$ of order 3 and $N_G(V)/C_G(V) \cong S_3$.*

(c) *S contains a maximal elementary abelian normal subgroup W such that $g \in N_G(W)$.*

(d) *W can be chosen to be the maximal elementary abelian normal subgroup $A = \langle z, k_1k_4, k_1k_2k_3^{-1}\rangle$, and $N_G(A) \cong E$, where E is the finitely presented group of Corollary 2.1.3(a).*

(e) *$N_H(A) \cong D = C_E(z)$.*

(f) *The amalgam $H \leftarrow D \rightarrow E$ has Goldschmidt index 1.*

(i) *The 3-elements d_1, d_2, $d_3 = d_1d_2$ and $d_4 = d_1d_2^{-1}$ belong to four distinct conjugacy classes in any simple group G of $\mathsf{G}_2(3)$-type.*

Furthermore, a Sylow 2-subgroup of $C_G(d_i)$ is a quaternion group of order 8 for $i = 1, 2$ and cyclic of order 2 for $i = 3, 4$.

Proof (a) Let S be the Sylow 2-subgroup of H given in Lemma 2.1.4(e). The involution k_5 of S does not belong to H' by Lemma 2.1.4(c) and $|C_H(k_5)| = 8$ by Table 12.8.4 of [92]. Furthermore, $z \in H'$, $v = k_1k_2k_3^{-1} \in H'$ and $k_5 \in H - H'$ are representatives of all conjugacy classes of involutions of H by Table 12.8.4 of [92] because $|C_H(v)| = 2^5$ by Lemma 2.1.4(g).

Suppose that z and v are not conjugate in G. Then Glauberman's Z^*-Theorem implies that $z^q = k_5$ for some $q \in G$. Hence $C_H(k_5) = H \cap H^q$. Now $\frac{|H\cap H^q|}{|H'\cap H'^q|}$ divides 4 because $|H : H'| = 2$ by Lemma 2.1.4(c). As $|C_H(k_5)| = 8$ it follows that $|H' \cap H'^q|$ is even. Hence there is an involution $b \in H' \cap H'^q$ and $y = k_5b$ is an involution in $H - H'$. Thus y and k_5 are H-conjugate. Furthermore, $y \in H'^q$, because $k_5 = z^q$ and b belong to H'^q. Therefore $y^h = v^q$ for some $h \in H^q$. In particular, z and v are G-conjugate. This contradiction proves that there is an x in G such that $z^x = v$, and $x \notin H$.

(b) By Lemma 2.1.4(g) $V = \langle z, v\rangle$ is the unique normal Klein 4-group in S and V is the center of $C = C_S(V) = C_H(V) = C_G(V)$. Since v and z fuse in G the subgroup $C = C_S(v)$ of order 2^5 is properly contained in a Sylow 2-subgroup of $C_G(v)$. Hence $N_G(V)/C \cong \mathrm{GL}_2(2)$ by Lemma 1.4.4 of [92]. In particular, there is an element g of order 3 in $N_G(V) - N_H(V)$.

(c) Lemma 2.1.4(f) states S has exactly two elementary abelian normal subgroups of order 8. They are denoted by A and B. By Lemma 2.1.4(h) C has exactly four elementary abelian normal subgroups A, B, A_1 and A_2 and two conjugacy classes of cyclic complements of order 2 represented by $K_1 = \langle t \rangle$ and $K_2 = \langle w \rangle$. Furthermore, $A_1^t = A_1^w = A_2$. Hence $g \in N_G(V) - N_H(V)$ fixes one of the two normal subgroups A or B, and $N_G(V)/C = S_3$ can be represented as the full permutation group of the other three elementary abelian normal subgroups of C. In particular, $W \in \{A, B\}$. Thus (c) holds.

(d) $D = N_H(A) \cong D_1 = N_H(B)$ by Lemma 2.1.4(i), Corollary 2.1.3 (this volume) and Lemma 8.1.8(e) of [92]. Thus $D/A \cong S_4 \cong D_1/B$ are both maximal subgroups of $\mathrm{GL}_3(2)$. Therefore $N_G(A)$ and $N_G(B)$ are both isomorphic to the unique non-split extension E of $\mathrm{GL}_3(2)$ by $V_3 = F^3$. Hence we may choose $W = A$. Thus $N_G(A) \cong E$.

(e) Both subgroups $D = N_H(A)$ and $D' = C_E(z)$ are non-split extensions of S_4 by A. Hence $D \cong D'$ by Lemma 8.1.8(e) of [92].

(f) Faithful permutation representations of H and E are given in Lemma 2.1.4(d) and Corollary 2.1.3(b), respectively. Using them, MAGMA and Kratzer's Algorithm 7.1.10 of [92] it has been checked that the Goldschmidt index of the amalgam $H \leftarrow D \rightarrow E$ is 1.

(h) By (a) any simple group G of $\mathsf{G}_2(3)$-*type* has a unique conjugacy class z^G of involutions. By Table 12.8.4 of [92] the centralizers of d_3 and d_4 in H have cyclic Sylow 2-subgroups of order 2. Lemma 2.1.4(b) implies that a Sylow 2-subgroup of $C_H(d_i)$ is a quaternion group of order 8 for $1 \le i \le 2$. Since $\langle z \rangle$ is a characteristic subgroup in any of these four Sylow 2-subgroups it is easy to see that $C_G(d_i)$ and $C_H(d_i)$ have a common Sylow 2-subgroup. Hence none of the two elements d_1 and d_2 is conjugate to d_3 or d_4 in G. The Sylow 2-subgroups Q_1 and Q_2 of $C_H(d_1)$ and $C_H(d_2)$, respectively, are not conjugate in H by Table 12.8.4 of [92] because each Q_i contains an element v_i of order 4 powering into z such that v_1 and v_2 are not conjugate in H. Therefore, they are not conjugate in G. This shows that d_1 and d_2 belong to distinct G-classes.

Suppose that $d_3^u = d_4$ for some $u \in G$. Then

$$z^u d_4 = z^u d_3^u = (zd_3)^u = (d_3 z)^u = d_4 z^u.$$

Hence $z^u \in C_G(d_4)$. By Sylow's theorem there is an $x \in C_G(d_4)$ such that $z^{ux} = z$. Thus $ux \in H$, and

$$(d_3 z)^{ux} = d_3^{ux} z^{ux} = d_4^x z = d_4 z.$$

But $d_3 z$ and $d_4 z$ are not conjugate in H by Table 12.8.4 of [92]. This contradiction completes the proof. □

Corollary 2.2.2 *Any finite simple group G of $\mathsf{G}_2(3)$-type has 11 conjugacy classes of elements of even order. They are represented by the conjugacy classes 2_a, 4_a, 4_b, 6_a, 6_b, 6_c, 6_d, 8_a, 8_b, 12_a, 12_b of $H = C_G(z)$ given in Table 12.8.4 of [92].*

Proof By Proposition 10.2.2(a) each simple group G of $\mathsf{G}_2(3)$-*type* has a unique conjugacy class z^G of involutions. Looking in the character Table 12.9.4 of [92] at the representatives of the conjugacy classes of elements of even order in $H = C_G(z)$ powering onto z, we see that the 11 classes of H stated in the assertion are all the z-special classes of H. By Theorem 2.2.11 of [92] they represent all even classes of G. □

2.3 The 3-singular conjugacy classes

In this section it is shown that each finite simple group G of $\mathsf{G}_2(3)$-*type* has five conjugacy class of elements 3_i of order 3 and three conjugacy classes of elements of order 9. The structure of the five non-conjugate normalizers $N_i = N_G(3_i)$ of the cyclic subgroups of order 3 of G, their conjugacy classes and their character tables for $1 \leq i \leq 5$ are determined as well. Furthermore, the fusion of the classes of each group N_i in G is given.

The first three lemmas are due to Z. Janko [68]. All other results were obtained by L. Wang and the author.

Lemma 2.3.1 *Let G be any finite simple group of $\mathsf{G}_2(3)$-type with a 2-central involution z such that $C_G(z) = H$. Then the order of a Sylow 3-subgroup of G is at least 27.*

Proof By Proposition 10.2.2 G has only one conjugacy class z^G of involutions. Suppose that a Sylow 3-subgroup of G has order 9. Then $P = \langle d_1, d_2 \rangle$ is a Sylow 3-subgroup of G.

By Proposition 10.2.2(i) G has four conjugacy classes of elements of order 3 represented by d_1, d_2, $d_3 = d_1 d_2$ and $d_4 = d_1 d_2^{-1}$. Let d and e be any of the elements d_1, d_2 and d_3, d_4, respectively. Then by Proposition 10.2.2(i) a Sylow 2-subgroup S_1 of $J = C_H(d) \cong \langle d \rangle \times \mathsf{SL}_2(3)$ is a quaternion group of order 8 with center $\langle z \rangle$, and $S_2 = \langle z \rangle$ is a Sylow 2-subgroup of $K = C_h(e)$. Let $N = C_G(d)$ and $M = C_G(e)$. Let X and Y be the largest normal subgroups of odd order of N and M, respectively. By Lemma 2.1.4 d and e are both inverted by the involution k_5. Hence the Klein 4-group $V = \langle z, k_5 \rangle$ acts on X and Y by conjugation. Applying

Theorem 4.2.2 of [92] and Table 12.8.4 of [92] it follows that $X = \langle d \rangle$ and $Y = \langle P \rangle$. Since S_1 is a quaternion group of order 8 the Brauer–Suzuki Theorem 4.6.5 of [92] implies now that $N = C_G(d) = XC_N(z) = C_N(z) \leq H$. As S_2 is a cyclic group of order 2 Corollary 1.4.18 of [92] implies that $M = C_G(e) = XS_2 = C_X(z) \leq H$.

Let $W = \{h \in H \mid h^k = z \quad \text{for some natural number} \quad k\}$. Let V be the union of W and the set of all elements v of order 3 in H. Then V is a set of special classes of H in G by Definition 2.7.1 of [92]. Looking at Table 12.9.4 of [92] we see that $\psi = \psi_1 + \psi_{18} - \psi_{20}$ is a class function of H vanishing on $H - V$. Since V is a special set of conjugacy classes Theorem 2.7.5 of [92] implies that $[\psi^G, \psi^G] = [\psi, \psi] = 3$ and that the trivial character 1_G is a constituent of ψ^G. Hence there are two irreducible characters χ_i of G and signs ϵ_i, $i = 1, 2$, such that $\psi^G = 1_G + \epsilon_1 \cdot \chi_1 + \epsilon_2 \cdot \chi_2$. Since $\psi^G(1) = 0$ we may assume that $\epsilon_1 = 1$ and $\epsilon_2 = -1$.

Let $z_1 = z, z_2, z_3$ be representatives of the three conjugacy classes of involutions of H. For each irreducible character ψ of H let

$$\psi(\hat{H}) = \sum_{i=1}^{3} |H : C_H(z_i)| \psi(z_i).$$

Then Suzuki's group order formula stated in Theorem 4.1.4 of [92] asserts that

$$|G| \left(1 + \frac{\chi_1(z)^2}{\chi_1(1)} - \frac{\chi_2(z)^2}{\chi_2(1)}\right) = |H| \left(\psi_1(\hat{H})^2 + \frac{\psi_{18}(\hat{H})^2}{\psi_{18}(1)} - \frac{\psi_{20}(\hat{H})^2}{\psi_{20}(1)}\right).$$

Inserting the values given in Table 12.9.4 of [92] and letting $\chi_1(z) = a$ and $\chi_1(1) = x$ we obtain the following equation:

$$|G| \left(1 + \frac{a^2}{x} - \frac{(a+1)^2}{x+1}\right) = |H| \left(91^2 + \frac{8^2}{8} - \frac{99^2}{9}\right).$$

Therefore

$$(*) \quad |G| \cdot \frac{(x-a)^2}{x(x+1)} = 2^{11} \cdot 3^2 \cdot 5^2.$$

By Theorem 2.4.12 of [92] the character degrees x and $x + 1$ divide the group order $|G|$. Therefore $x(x+1)$ is a divisor of $|G|$ because x and $x+1$ are coprime. In particular, $q = |G|/x(x+1)$ is a divisor of $2^{11} \cdot 3^2 \cdot 5^2$ such that its factors of 2- and 3-powers are divisors of 2^6 and 3^2, respectively. Now (*) implies that

$$(**) \quad q \in \{2^j, 2^j \cdot 3^2, 2^j \cdot 5^2, 2^j \cdot 3^2 \cdot 5^2\} \quad \text{for} \quad j \in \{1, 3, 5\}.$$

Let $\pi = \{2,3\}$. Then π is the smallest set of primes containing 2 such that a centralizer of a π-element of G is a π-subgroup of G. By Lemma 4.3.2 of [92] an element r of a subgroup U of G is strongly real if its extended centralizer $C^*_G(r) \leq U$ and the following invariant $d(r)$ is non-zero:

$$d(r) = \left|\{(x,y) \in z^G \times z^G \mid xy = r\}\right| = \frac{1}{|U|} \cdot \sum_{\psi \in Irr_{\mathbb{C}}(U)} \psi(\hat{U})^2 \psi(r^{-1}) \psi(1)^{-1}.$$

This formula has been applied to all representatives of the conjugacy classes y^G of H by means of Tables 12.8.4 and 12.9.4 of [92]. It follows that G has 15 strongly real conjugacy classes r_i^G, $1 \leq i \leq 15$, of π-elements $r_i \neq 1$. Their invariants $d(r_i)$ and centralizer orders $|C_G(r_i)|$ are given in the following table:

Class	$\lvert C_G(r_i)\rvert$	$d(r_i)$	Class	$\lvert C_G(r_i)\rvert$	$d(r_i)$
2_A	$2^6 \cdot 3^2$	90	6_B	$2^3 \cdot 3^2$	36
3_A	$2^3 \cdot 3^2$	36	6_C	$2 \cdot 3^2$	18
3_B	$2^3 \cdot 3^2$	36	6_D	$2 \cdot 3^2$	18
3_C	$2 \cdot 3^2$	18	8_A	2^3	8
3_D	$2 \cdot 3^2$	18	8_B	2^3	8
4_A	$2^5 \cdot 3$	36	12_A	$2^2 \cdot 3$	12
4_B	$2^5 \cdot 3$	36	12_B	$2^2 \cdot 3$	12
6_A	$2^3 \cdot 3^2$	36			

Let t be the number of strongly real π'-classes of G, where π' is the set of all prime divisors q of $|G|$ which do not belong to π. Then Theorem 4.3.7 of [92] asserts that

$$|G| = c|H| + t|H|^2, \text{where} \quad c = 1 + \sum_{i=1}^{15} |H : C_H(r_i)| d(r_i) = 6283.$$

Furthermore, $t < \frac{3}{2}|H|$ by Theorem 4.3.9 of [92]. Proposition 10.2.2 states that 7 is a divisor of $|G|$. Hence there are 124 possible group orders $g_k = |H| r_k$, where

$$r_k = c + (5 + 7 \cdot k) \cdot |H| = 6283 + (5 + 7 \cdot k)576 = 9163 + 4032 \cdot k$$

for $k = 0, 1, ..., 123$. Now $(**)$ implies that

$$(***) \quad g_k = |H| \cdot r_k = 2^6 \cdot 3^2 \cdot (9163 + 4032 \cdot k) = 2^u \cdot 3^v \cdot 5^w \cdot x \cdot (x+1),$$

where $u \in \{1,3,5\}$ and $v, w \in \{0,2\}$.

Suppose that $w \neq 0$. Then 5^2 divides r_k by $(***)$. Using MAGMA it has been checked that this is true only for $k \in \{12, 37, 62, 87, 112\}$ and that $r_{12} = 7 \cdot 8221$, $r_{37} = 7 \cdot 22621$, $r_{62} = 7 \cdot 37021$, $r_{87} = 7 \cdot 51421$

and $r_{112} = 7^2 \cdot 9403$ are factorizations into primes. Furthermore, it has been verified that for these numbers r_k the system $(***)$ has no integer solutions x for all $u \in \{1,3,5\}$ and $v \in \{0,2\}$. Therefore $w = 0$. Thus we have to find the integral solutions of the equations

$$(****) \quad 2^6 \cdot 3^2 \cdot (9163 + 4032 \cdot k) = 2^u \cdot 3^v \cdot x \cdot (x+1),$$

where $u \in \{1,3,5\}$ and $v \in \{0,2\}$.

Suppose that $v \neq 0$. Then $(****)$ implies that

$$2x + 1 = \sqrt{([4 \cdot 2^{6-u} \cdot (9163 + 4032 \cdot k)] + 1)}.$$

Using MAGMA again it has been checked that for $u = 1$ and $u = 3$ this equation has no integral solution x for all k. However, for $u = 5$ it has two solutions, $x = 301$ and $x = 685$, for $k = 9$ and $k = 56$, respectively. Hence $g_9 = 2^6 \cdot 3^2 \cdot 7 \cdot 43 \cdot 151$ and $g_{56} = 2^6 \cdot 3^2 \cdot 5 \cdot 7^3 \cdot 137$.

Since the centralizers of π-elements are π-subgroups the centralizers of cyclic subgroups of orders 151 and 137 have orders prime to 2 and 3. Let y be a generator of a cyclic Sylow 151-subgroup Y of a simple group G with order $|G| = g_9$. By Lemma 4.3.6(a) of [92] $|C_G(Y)| \in \{151, 151 \cdot 7, 151 \cdot 43, 151 \cdot 7 \cdot 43\}$. Also $C_G(Y) \neq N_G(Y)$ by Burnside's theorem; see Theorem 1.4.17 of [92]. Furthermore, $T = N_G(Y)/C_G(Y)$ contains a unique involution t and belongs to the center $Z(T)$ of T by Lemma 4.3.6(e) of [92]. Hence $|T| \in \{2,6\}$. Applying MAGMA it follows that $|G/N_G(Y)| \equiv 1 \bmod (151)$ does not hold in any of all possible cases. Now Sylow's Theorem asserts that g_9 is not a group order.

The case $g_{56} = 2^6 \cdot 3^2 \cdot 5 \cdot 7^3 \cdot 137$ is dismissed similarly. This time $|T| \in \{2,4,8\}$ because 17 does not divide $|H|$. Hence $v = 0$.

Since for all $u \in \{1,3,5\}$ and all $0 \leq k \leq 123$ the square roots of the numbers $2^{2+u} \cdot 9 \cdot (9163 + 4032 \cdot k) + 1$ are not integral, the system $(*)$ has no integral solutions x. This completes the proof. □

Lemma 2.3.2 *Keep the notations of Lemma 2.1.4. Let G be any finite simple group of $\mathsf{G}_2(3)$-type having a 2-central involution z such that $C_G(z) \cong H$. Let $P = \langle d_1, d_2 \rangle$, $C = C_G(P)$, $K = O(C)$ and $V = \langle z, k_5 \rangle$. For $i \in \{1,2\}$ let $C_i = C_G(d_i)$, $N_i = N_G(d_i)$ and $K_i = O(C_i)$. Then the following statements hold.*

(a) *$H_1 = C_H(d_1) = \langle d_1 \rangle \times R_1$, where $R_1 = \langle d_2, k_3, k_4 \rangle \cong \mathsf{SL}_2(3)$, and $Q_1 = \langle k_3, k_4 \rangle$ is a quaternion Sylow 2-subgroup of H_1 of order $|Q_1| = 8$.*

(b) *$Y_1 = \langle k_3, k_4, k_5\rangle$ is a semi-dihedral Sylow 2-subgroup of $N_H(d_1) = \langle H_1, k_5\rangle$ of order $|S_1| = 16$, and $N_H(d_1)$ is a split extension of $T_1 = \langle R_1, k_5\rangle \cong \mathrm{GL}_2(3)$ by $\langle d_1\rangle$.*

(c) *V and $\langle z\rangle$ are Sylow 2-subgroups of $N_G(P)$ and $C = C_G(P)$, respectively. $N_G(P)$ is a split extension of V by K.*

(d) *$N_1 = N_G(d_1)$ is an extension of T_1 by $K_1 = O(C_1) = O(N_1)$.*

(e) *K is an elementary abelian 3-group of order 3^3 or 3^4.*

(f) *$|K_1|$ is a 3-group of exponent 3 and its order is 3^3 or 3^5.*

Proof Assertions (a) and (b) follow at once from Lemma 2.1.4.

(c) By Proposition 10.2.2 $\langle z\rangle$ is a Sylow 2-subgroup of $C = C_G(P)$. Let K be the largest normal subgroup of odd order in C. Then Corollary 1.4.18 of [92] asserts that C is a semidirect product of K and $\langle z\rangle$. K is normal in $N_G(P)$ because it is a characteristic subgroup of C. Furthermore, $N_G(P)/K$ is isomorphic to a subgroup of $Aut(P) \cong \mathrm{GL}_2(3)$ of order 48. If 3 were a divisor of $|N_G(P)/K|$ then there would be an element $g \in N_G(P)$ such that $d_1^g = d_1d_2 = d_3$. But d_1 and d_3 are not conjugate in G by Proposition 10.2.2. Hence $N_G(P)/K$ is a 2-group. Thus K has a complement W in $N_G(P)$ by the Schur–Zassenhaus theorem. As C/K is isomorphic to a subgroup of $\mathsf{SL}_2(3)$, which has a quaternion Sylow 2-subgroup, the Brauer–Suzuki Theorem 4.6.5 of [92] implies that the center $Z(W)$ contains z. Therefore each $n \in N_W(P)$ centralizes z. Thus $N_G(P) = KW = KN_H(P)$. By Lemma 2.1.4 $V = \langle z, k_5\rangle$ is a Sylow 2-subgroup of $N_H(P)$ and $N_H(P) = PV$. Thus $V = W$ is a Sylow 2-subgroup of $N_G(P) = KV$, and $K \cap V = 1$.

(d) By (b) $|N_1 : C_1| = 2$. Hence $O(N_1) = K_1$ because K_1 is a characteristic subgroup of C_1. Thus $N_1/K_1 \cong T_1$.

(e) and (f) In particular, the Klein 4-group $V = \langle z, k_5\rangle$ operates on each of the subgroups $K_1 = O(N_1)$ and K by conjugation. Let X be any of the groups K or K_1. Then Theorems 1.5.8 and 4.2.2 of [92] imply that

$$(*) \quad X = \langle C_X(z), C_X(k_5), C_X(zk_5)\rangle,$$

$$(**) \quad |X| = |C_X(z)||C_X(k_5)||C_X(zk_5)|.$$

By Lemma 2.1.4 $P = \langle d_1, d_2\rangle$ is an elementary abelian Sylow 3-subgroup of order 9 in H. Proposition 10.2.2(a) states that all involutions of G are conjugate in G. Therefore $C_K(z) = P$, $C_{K_1}(z) = \langle d_2\rangle$ and $|C_{K_1}(k_5)| = |C_{K_1}(zk_5)| \le 3^2$. Hence $C_X(z)$ is normal in X and z inverts $X_1 = X/C_X(z)$. Now Burnside's Lemma 1.5.9 of [92] and both

$(*)$ and $(**)$ imply that X_1 is an elementary abelian 3-group of order at most 3^4.

In the case that $X = K$ it follows from $(*)$ that K is elementary abelian, because $K \leq C_G(P)$. Since KK_1 is a Sylow 3-subgroup of C_1 and $|K/K \cap K_1| = 3$ by (d) it follows that $|K| = 3 \cdot |K_1 \cap K|$. Now Lemma 2.3.1 asserts that 3 divides $|N_G(P) : P|$. Therefore $|K/P| \geq 3$ because $N_G(P) = KV$ by (c). Hence $|K| \geq 3^3$. Furthermore, Lemma 2.3.1 asserts that 3 divides $|N_G(P) : P|$. Therefore $|K/P| \geq 3$ because $N_G(P) = KV$ by (c). In particular, $|K| \geq 3^3$. Hence $(**)$ implies that we may assume the existence of some element $q \in C_K(k_5)$ having order 3. Clearly, $q \in K \cap K_1$, but $q \notin \langle d_1 \rangle$.

By Lemma 2.1.4 $k_5^{k_1 k_2} = zk_5$ and $k_5^{k_3} = zk_5$. Thus

$$(***) \quad |K_i| = |C_{K_i}(z)||C_{K_i}(k_5)|^2.$$

Therefore $W_1 = K_1/\langle d_1 \rangle$ is a non-trivial elementary abelian 3-group inverted by z. Hence each composition factor of W_i is a faithful irreducible FR_i-module, where $F = GF(3)$. Now the character table of $R_1 \cong \mathsf{SL}_2(3)$ stated in [58] implies that each composition factor of W_1 is isomorphic to the unique faithful simple FR_1-module M of dimension 2. Thus $|K_1|$ equals either 3^3 or 3^5 by $(***)$, and the multiplicity of M in W_1 is either 1 or 2.

Since d_2 generates a Sylow 3-subgroup of R_1 its Jordan normal form w.r.t. a basis of M shows that d_2 has exactly one fixed point in each copy of M occurring in the FR_1-module W_1. Hence $|K \cap K_1|$ is either 3^2 or 3^3 depending on the multiplicity of M in W_1. Thus $|K| = 3 \cdot |K \cap K_1|$ is either 3^3 or 3^4.

It remains to show that K_1 has exponent 3. Let $D_1 = \langle d_1 \rangle$. Suppose that y is an element of K_1 of order 9. As z inverts W_1 we know that $(D_1 y)^z = D_1 y$. Hence Lemma 1.5.2 of [92] implies the existence of an element $u \in C_{K_1}(z) = D_1$ such that $D_1 y = D_1 u = D_1$, which is impossible. This contradiction completes the proof. □

Lemma 2.3.3 *Keep the notations of Lemma 2.1.4. Let G be any finite simple group of $\mathsf{G}_2(3)$-type having a 2-central involution z such that $C_G(z) \cong H$. Let $P = \langle d_1, d_2 \rangle$, $C = C_G(P)$ and $K = O(C)$. For $i \in \{1, 2\}$ let $C_i = C_G(d_i)$, $N_i = N_G(d_i)$, $K_i = O(C_i)$ and $Z_i = Z(K_i)$. Then the following statements hold:*

(a) $|K| = 3^4$,
(b) $|K_1| = |K_2| = 3^5$,
(c) $|Z_1| = |Z_2| = 3^3$.

Proof (a) We first show that $|Z_i| \geq 3^3$ for $i = 1, 2$. Suppose that $Z_i = \langle d_i \rangle = Z(C_i)$. Let $\{i, j\} = \{1, 2\}$. By Lemma 2.3.2 $S_i = \langle K_i, d_j \rangle$ is a Sylow 3-subgroup of N_i and $|N_i : C_i| = 2$. Hence $Z_i = Z(S_i)$ and $N_G(S_i) \leq N_G(d_i) = N_i$ because $Z(S_i)$ is characteristic in S_i. Thus S_i is a Sylow 3-subgroup of G. Now Sylow's Theorem implies that d_1 is a G-conjugate of d_2. This is a contradiction to Proposition 10.2.2(h). Since Z_i is a characteristic subgroup of C_i it is normal in N_i. Hence $V = \langle z, k_5 \rangle$ operates on Z_i. Therefore $|Z_i/\langle d_i \rangle| \geq 3^2$ by the argument of the proof of Lemma 2.3.2(f). Hence $|Z_i| \geq 3^3$ for $i = 1, 2$.

If (a) were false then $|K| = 3^3$ by Lemma 2.3.2(e). Thus $|K| = 3 \cdot |K \cap K_i|$ and $|K \cap K_i| \geq 3^2$ for $i = 1, 2$ by the proof of Lemma 2.3.2(e). Hence $|K_1| = |K_2| = 3^3$, and $Z_i = K_i$ for $i = 1, 2$. By (**) of the proof of the Lemma 2.3.2 there is an element $y \in C_K(k_5)$ of order 3 such that $K \cap K_i = \langle d_j, y \rangle$, where $\{i, j\} = \{1, 2\}$. Let $y_1 = y^{k_3}$ and $y_2 = y^{k_1}$. Then $y_1 \in C_{K_1}(zk_5)$ and $y_2 \in C_{K_2}(zk_5)$ because $k_5^{k_1 k_2} = zk_5$ and $k_5^{k_3} = zk_5$ by Lemma 2.1.4. In particular, $M_i = \langle y, y_i \rangle \leq Z_i$ is a faithful irreducible FR_i-module of dimension 2 over $F = GF(3)$, and $K_i = \langle d_i \rangle \times M_i$. Furthermore, we may assume that $y_i^{d_j} = yy_i$ because $\langle d_j \rangle$ is a cyclic Sylow 3-subgroup of $R_i \cong \mathsf{SL}_2(3)$. Thus $L = \langle d_1, d_2, y, y_1, y_2 \rangle$ is a subgroup of order 3^5 in $C_G(y)$, and $L_i = \langle d_1, d_2, y, y_i \rangle$ is a Sylow 3-subgroup of N_i of order 3^4 for $i = 1, 2$.

Let $A = \langle P, y \rangle$. Now A is a self-centralizing elementary abelian subgroup of G of order 27 because $C_G(P) = K \times \langle z \rangle$ and $y^z = y^2$. From the above relations involving the five generators of L it follows that:

$$d_3^{y_2 y_1^2} = d_1^{y_2 y_1^2} d_2^{y_2 y_1^2} = y^2 d_1^{y_1} (y^2 d_2)^{y_1} = y^6 d_1 d_2 = d_3.$$

Similarly, we get $d_4^{y_1^2 y_2^2} = d_4$. Hence $y_2 y_1^2 \in N_G(A) \cap C_G(d_3)$ and $y_1^2 y_2^2 \in N_G(A) \cap C_G(d_4)$, but neither y_1 nor y_2 belong to $C_G(d_k)$ for $k \in \{3, 4\}$. Therefore each Sylow 3-subgroup L_k of $C_C(d_k)$ has at least 3^4 elements. Now $N_G(A)/A$ is isomorphic to a subgroup of $\mathrm{GL}_3(3)$ and the matrices M_{y_1} and M_{y_2} of y_1 and y_2 w.r.t. to the basis $\mathbb{B} = \{d_1, d_2, y\}$ of A generate an abelian subgroup of order 9 of the Sylow 3-subgroup U of $\mathrm{GL}_3(3)$ consisting of all upper triangular unipotent matrices. If $|L_k| > 3^4$ then S_k/A would contain an element q with matrix M_q such that $U = \langle M_{y_1}, M_{y_2}, M_q \rangle$. Therefore there is an element $n \in N_G(A)$ such that $d_1^n = d_k$ for some $k \in \{3, 4\}$, which contradicts Proposition 10.2.2(h). Hence a Sylow 3-subgroup of any $C_G(d_i)$, $1 \leq i \leq 4$, has order 3^4.

By Proposition 10.2.2(a) $C_G(zk_5)$ and $H = C_G(z)$ are conjugate in G. Therefore y is G-conjugate to exactly one of the representatives d_j,

$1 \leq j \leq 4$, by Proposition 10.2.2(h). As L has order greater than 3^4 this is impossible.

(b) This assertion is an immediate consequence of (a) and Lemma 2.3.2.

(c) We know that $|Z_i| \geq 3^3$ for $i = 1, 2$. If (c) were wrong then (b) implies that $Z_i = K_i$ for at least one index i because K_i/Z_i is a faithful 2-dimensional FR_i-module. Since $V = \langle z, k_5 \rangle$ operates on K_i, equation (***) of the proof of Lemma 2.3.2 states that

$$|K_i| = |C_{K_i}(z)||C_{K_i}(k_5)|^2 \quad \text{and} \quad C_{K_i}(z) = \langle d_i \rangle.$$

Hence $J_i = C_{K_i}(z)$ is a Sylow 3-subgroup of $C_G(k_5)$ because $C_G(zk_5)$ and $H = C_G(z)$ are conjugate in G. In particular, P and J_i are conjugate in G. Therefore $C_G(P)$ and $C_G(J_i) \geq K_i$ are conjugate. Hence by (b) a Sylow 3-subgroup of $C_G(P)$ has order at least 3^5, which contradicts assertions (c) and (e) of Lemma 2.3.2. This contradiction completes the proof. □

Lemma 2.3.4 *Keep the notations of Lemma 2.1.4. Let* $x_1 = k_3$, $x_2 = k_5$, $x_3 = d_2$, $x_4 = k_4$ *and* $x_9 = d_1$. *Then* $N_H(d_1) = Y = \langle x_1, x_2, x_3, x_4, x_9 \rangle$ *and up to isomorphism there are exactly three extensions* X_i *of* Y *by* $K_i = O(X_i)$ *such that for all three indices* $1 \leq i \leq 3$ *the following statements hold.*

(a) K_i *has order* 3^5 *and exponent* 3.
(b) *The center* $Z_i = Z(K_i)$ *of* K_i *has order* 3^3.
(c) $Z(X_i') \cap Z_i = \langle d \rangle$ *is cyclic of order* 3, $C_{X_i}(d) = X_i'$ *and* $d^{k_5} = d^2$.

Moreover, the groups X_i *have the following presentations.*

(d) $X_1 = \langle\, x_i | 1 \leq i \leq 9 \rangle$ *has the following set* $\mathcal{R}(X_1)$ *of defining relations:*

$$x_1^4 = x_2^2 = x_3^3 = x_4^4 = x_5^3 = x_6^3 = x_7^3 = x_8^3 = x_9^3 = 1,$$
$$x_1^2 = x_4^2, \quad x_1^{x_2} = x_1^3, \quad x_3^{x_2} = x_3^2, \quad x_4^{x_1} = x_4^3, \quad x_4^{x_2} = x_4 x_1^3,$$
$$x_4^{x_3} = x_1, \quad x_5^{x_2} = x_5^2, \quad x_7^{x_1} = x_8^2, \quad x_7^{x_2} = x_8^2,$$
$$x_7^{x_3} = x_7^2 x_8, \quad x_8^{x_1} = x_7, \quad x_8^{x_3} = x_7^2, \quad x_9^{x_2} = x_9^2,$$
$$(x_3, x_5) = (x_5, x_7) = (x_6, x_7) = (x_6, x_8) = (x_7, x_8) = 1,$$
$$(x_1, x_9) = (x_3, x_9) = (x_5, x_9) = (x_6, x_9) = (x_7, x_9) = (x_8, x_9) = 1,$$
$$x_1 x_5^{-1} x_1^{-1} x_6^{-1} x_9^2 = 1, \quad x_1 x_6^{-1} x_1^{-1} x_5 x_9^2 = 1,$$
$$x_4 x_5^{-1} x_4^{-1} x_5^{-1} x_6^{-1} = 1.$$

(e) $X_2 = \langle\, x_i | 1 \leq i \leq 9\rangle$ *has the following set* $\mathcal{R}(X_2)$ *of defining relations:*

$$\begin{aligned}
&x_1^4 = x_2^2 = x_3^3 = x_4^4 = x_5^3 = x_6^3 = x_7^3 = x_8^3 = x_9^3 = 1,\\
&x_1^2 = x_4^2, \quad x_1^{x_2} = x_1^3, \quad x_3^{x_2} = x_3^2, \quad x_4^{x_1} = x_4^3, \quad x_4^{x_2} = x_4 x_1^3,\\
&x_4^{x_3} = x_1, \quad x_5^{x_2} = x_5^2, \quad x_7^{x_1} = x_8^2, \quad x_7^{x_2} = x_8,\\
&x_7^{x_3} = x_7^2 x_8, \quad x_8^{x_1} = x_7, \quad x_8^{x_3} = x_7^2, \quad x_9^{x_2} = x_9^2, \quad (x_2, x_6) = 1,\\
&(x_3, x_5) = (x_5, x_7) = (x_6, x_7) = (x_5, x_8) = (x_6, x_8) = 1,\\
&(x_7, x_8) = (x_1, x_9) = (x_3, x_9) = (x_5, x_9) = (x_6, x_9) = 1,\\
&(x_7, x_9) = (x_8, x_9) = 1,\\
&x_1 x_5^{-1} x_1^{-1} x_6^{-1} x_9 = 1, \quad x_6 x_5^{-1} x_3^{-1} x_6^{-1} x_3 = 1,\\
&x_8 x_7^{-1} x_3^{-1} x_7^{-1} x_3 = 1.
\end{aligned}$$

(f) $X_3 = \langle\, x_i | 1 \leq i \leq 9\rangle$ *has the following set* $\mathcal{R}(X_3)$ *of defining relations:*

$$\begin{aligned}
&x_1^4 = x_2^2 = x_3^3 = x_4^4 = x_5^3 = x_6^3 = x_7^3 = x_8^3 = x_9^3 = 1,\\
&x_1^2 = x_4^2, \quad x_1^{x_2} = x_1^3, \quad x_3^{x_2} = x_3^2, \quad x_4^{x_1} = x_4^3, \quad x_4^{x_2} = x_4 x_1^3,\\
&x_4^{x_3} = x_1, \quad x_6^{x_2} = x_9^2 x_6, \quad x_7^{x_1} = x_8^2, \quad x_7^{x_2} = x_8,\\
&x_7^{x_3} = x_7^2 x_8, \quad x_8^{x_1} = x_7, \quad x_8^{x_3} = x_7^2, \quad x_9^{x_2} = x_9^2,\\
&(x_5, x_7) = (x_6, x_7) = (x_5, x_8) = (x_6, x_8) = (x_7, x_8) = 1,\\
&(x_1, x_9) = (x_3, x_9) = (x_5, x_9) = (x_6, x_9) = (x_7, x_9) = (x_8, x_9) = 1,\\
&x_1 x_6^{-1} x_1^{-1} x_5 x_7 = 1, \quad x_3 x_6^{-1} x_3^{-1} x_6 x_5 = 1.
\end{aligned}$$

Furthermore, the following statements hold.

(g) *The respective Sylow* 3*-subgroups* S_1 *and* S_2 *of* X_1 *and* X_2 *are isomorphic, but* $N_{X_1}(S_1)$ *is not isomorphic to* $N_{X_2}(S_2)$.

(h) S_1 *is not isomorphic to the Sylow* 3*-subgroup* S_3 *of* X_3.

(i) *The system of representatives and the corresponding centralizer orders of the* 31 *conjugacy classes of each of the three groups* X_i *are stated in Tables 2.5.2, 2.5.3 and 2.5.4.*

Proof Since $Y/\langle d_1\rangle \cong \mathrm{GL}_2(3)$ by Lemma 2.3.2 the group Y has up to isomorphism exactly two non-isomorphic faithful simple FY-modules W_1 and W_2 of dimension 2, where $F = GF(3)$. In particular, the unique

2-central involution $z = k_1^2 = k_4^2$ does not belong to the kernels of any W_i. The cohomological dimension of $H^2(Y, W_i)$ is zero for $i = 1, 2$. Furthermore, the two split extensions $W_1 : Y$ and $W_2 : Y$ are isomorphic. Let $Y_1 = W_1 : Y$. Since the normal subgroup W_1 acts trivially on any simple FY_1-module the groups Y_1 and Y have the same simple modules over F up to isomorphism. Hence W_1 and W_2 are the only faithful simple FY_1-modules of dimension 2, and z does not act trivially on each of them. By means of MAGMA one checks that the second cohomology groups $H^2(Y_1, W_i)$ have both dimension 1 over F. For each 2-cocycle $\sigma_i \in H^2(Y_1, W_i)$ there is an extension Y_{σ_i} which is a finitely presented group $Y_{i,j}$, where $1 \leq i \leq 2$ and $1 \leq j \leq 2$. Using suitable faithful permutation representations of each of these six groups and MAGMA's isomorphism testing program it has been checked that exactly four of the extensions are non-isomorphic. Denote them by $Q_i, 1 \leq i \leq 4$. Again there is a one-to-one correspondence between the simple FQ_i-modules and the simple FY-modules. Hence each group Q_i has exactly two 1-dimensional FQ_i-modules, the trivial $V_{i,1}$ and another $V_{i,2}$ on which k_5 acts by multiplication with -1, because $Q_i/Q_i' \cong Y/Y' \cong \langle k_5 \rangle$. Now using the faithful permutation representations of the four groups Q_i and MAGMA one calculates the dimensions of the second cohomologyy groups $H^2(Q_i, V_{i,k})$. The results are stated in the following table, where $e(i, k)$ denotes the number of extensions of Q_i by $V_{i,k}$:

i	k	$dim_F H^2(Q_i, V_{i,k})$	$e(i, k)$	i	k	$dim_F H^2(Q_i, V_{i,k})$	$e(i, k)$
1	1	0	1	1	2	4	81
2	1	1	3	2	2	3	27
3	1	0	1	3	2	2	9
4	1	0	1	4	2	2	9

Using MAGMA again one calculates for each of these 132 extension groups X_s a finite presentation and a faithful permutation representation. Thus MAGMA is able to determine $K_s = O(X_s)$ and its center $Z_s = Z(K_s)$. It turns out that 48 of the groups X_s satisfy the condition $|Z_s| = 3^3$. Applying then the isomorphism test of MAGMA it follows that only 12 of the 48 groups are pairwise non-isomorphic. Checking now the exponent of the subgroup $K_s = O(X_s)$ of the remaining groups X_s it turns out that only the three groups X_i whose presentations are given in the assertion of the lemma have exponent 3. By construction $|K_i| = 3^5$. From the presentations of the three groups X_i follows at once that they also satisfy condition (c) of the statement.

The remaining statements of the assertion have been verified by means of faithful permutation representations of the three groups X_i and MAGMA for $i = 1, 2, 3$. This completes the proof. □

Lemma 2.3.5 *Keep the notations of Lemmas 2.1.4 and 2.3.4. Let G be any simple group G of $\mathsf{G}_2(3)$-type. Then a Sylow 3-subgroup S of G has order 3^6, its center $Z(S)$ is elementary abelian of order 3^2 and $N_G(S) \cong S : W$, where W is a Klein 4-group.*

Proof Let S_1 be a Sylow 3-subgroup of $N_1 = N_G(d_1)$. Then $|S_1| = 3^6$ by Lemmas 2.3.2, 2.3.3 and 2.3.4. Furthermore, $K_1 = O(N_1)$ and its center $Z_i = Z(K_i)$ have orders 3^5 and 3^3, respectively. Since N_1 is isomorphic to one of the three groups X_i, $1 \leq i \leq 3$, described by generators and relations in Lemma 2.3.4 we know that the Klein 4-group $V = \langle z = x_1^2 = k_3^2, k_5 \rangle$ is a subgroup of $Y = N_G(S_1)$ acting on S_1 by conjugation. As G has a unique conjugacy class of involutions, Lemma 2.1.4(e) and Theorem 4.2.2 of [92] imply that the order of $X = O(Y)$ is given by

$$(*) \quad |X| = |C_X(z)||C_X(k_5)||C_X(zk_5)|.$$

Hence $S_1 = O(Y)$, and S_1 is a Sylow 3-subgroup of G.

Using faithful permutation representations and MAGMA it has been checked for all three groups X_i of Lemma 2.3.4 that the center $Z(S_1)$ of S_1 has order $|Z(S_1)| = 9$. Another application of Theorem 4.2.2 of [92] and $(*)$ applied to $X = Z(S_1)$ yields now that in each of the three cases there is an involution e in V such that $|C_{Z(S_1)}(e)| = 1$. In particular, the three involutions of V are not conjugate in Y.

Suppose that V is not a Sylow 2-subgroup of Y. Then there is a subgroup U of Y containing $V = \langle z, k_5 \rangle$ such that $|U : V| = 2$. Hence $t = k_5 z$ is in the center of U. Since $Z(S_1)$ has order 9 and $C_Y[Z(S_1)] = S_1$ it follows that Y/S_1 is isomorphic to a subgroup of a semi-dihedral Sylow 2-subgroup of $\mathrm{GL}_2(3)$. Hence U is a dihedral group of order 8 by Lemma 1.3.9 of [92]. In particular, U has a cyclic normal subgroup generated by an element u of order 4. By $(*)$ $|C_X(zk_5)| = 9$, where $X = O(Y)$. Hence $|C_Y(t)| \geq 8 \cdot 9$. Since t is G-conjugate to z it follows that $H = C_G(z)$ has a subgroup Q of order 72 with a dihedral Sylow 2-subgroup. It has been checked with MAGMA that this is wrong. This contradiction proves that V is a Sylow 2-subgroup of Y. □

Lemma 2.3.6 *Keep the notations of Lemma 2.1.4. Let G be any finite simple group of $\mathsf{G}_2(3)$-type having a 2-central involution z such that $C_G(z) \cong H$. Let $P = \langle d_1, d_2 \rangle$, $C = C_G(P)$ and $K = O(C)$. Then the following assertions hold.*

(a) *The restriction of the automorphism α of H stated in Lemma 2.1.4(k) is an isomorphism $\gamma : N_H(d_1) \to N_H(d_2)$ which extends to an isomorphism $\beta : N_H(d_1) \to N_H(d_2)$.*

(b) *$N_G(d_i) \cong X_3$ for $i = 1, 2$, where X_3 is the finitely presented group of Lemma 2.3.4.*

(c) *K is not normal in any Sylow 3-subgroup of $N_i = N_G(d_i)$ for $i = 1, 2$.*

Proof (a) Lemma 2.3.2 states that $N_H(d_1) = \langle d_1, d_2, k_3, k_4 \rangle$. Thus $N_H(d_2) = \langle d_1, d_2, k_3, k_4 \rangle$. The restriction γ of the automorphism α of H defined in Lemma 2.1.4(k) is an isomorphism $\gamma : N_H(d_1) \to N_H(d_2)$ as is easily checked. Now Lemma 2.3.4 implies that γ extends to an isomorphism $\beta : N_G(d_1) \to N_G(d_2)$.

(b) By Lemma 2.3.4 $N_1 = N_G(d_1)$ is isomorphic to exactly one of the three groups X_i, $1 \le i \le 3$. If G were a counter example to the statement (b), then N_1 would be isomorphic to X_1 or X_2. Suppose that $N_1 \cong X_1$. It has been verified by means of MAGMA and a faithful permutation representation of X_1 that $K = O[C_{X_1}(P)]$ is a normal elementary abelian subgroup of the Sylow 3-subgroup $S_1 = \langle x_9, x_3, x_5, x_6, x_7, x_8 \rangle$ of X_1. By Lemma 2.3.4 $V = \langle z, k_5 \rangle$ is a Klein 4-group of N_1. Now Lemma 2.3.5 implies that $N_G(S_1) = N_{N_1}(S_1) = S_1 : V$. In particular, V operates on S_1 by conjugation. So it does on K by Lemma 2.3.2. Hence Theorem 4.2.2 of [92] implies that

$$(*) \quad |X| = |C_X(z)||C_X(k_5)||C_X(zk_5)|,$$

where X is any of the groups K or S_1.

By (a) there is an isomorphism $\gamma : N_H(d_1) \to N_H(d_2)$ which extends to an isomorphism $\beta : N_G(d_1) \to N_G(d_2)$. Now $(*)$ and Lemma 2.1.4(j) imply that $\beta(K) = K$, and that K is normal in the Sylow 3-subgroup $S_2 = \beta(S_1)$ of $N_G(d_2)$. In particular, both Sylow subgroups of G are conjugate in $U = N_G(K)$ by Lemma 2.3.5 and Sylow's theorem. Thus $S_2 = S_1^q$ for some $q \in U$. By Tables 2.5.2 and 2.5.3 the third powers of the elements x_3x_6 and $(x_3x_6)^q$ of order 9 of the respective Sylow subgroups S_1 and S_2 are equal to d_1 and d_2. Let $n_1 = x_3x_6 \in N_G(d_1)$ be of order 9. Then $n_2 = n_1^q \in S_1^q = S_2$. Hence $d_1^q = (n_1^3)^q = (n_1^q)^3 = d_2$. In particular, d_1 and d_2 are conjugate in G, which contradicts Proposition 10.2.2(h).

The same argument also shows that $N_G(d_1)$ is not isomorphic to X_2 because X_1 and X_2 have isomorphic Sylow 3-subgroups by Lemma 2.3.4(g). Thus $N_G(d_1) \cong X_3 \cong N_G(d)$ by (a).

(c) Using a faithful permutation representation of $N_G(d_1) = X_3$ with stabilizer $\langle x_1, x_2 \rangle$ and MAGMA it has been checked that the Sylow 3-subgroup $S_3 = \langle x_9, x_3, x_5, x_6, x_7, x_8 \rangle$ of X_3 has a unique normal elementary abelian subgroup A of order 81 and that $N_{X_3}(K)$ has order $3^5 \cdot 2^2$. Hence K is not normal in any Sylow 3-subgroup of X_3.

□

Lemma 2.3.7 *Let G be any simple group G of $\mathsf{G}_2(3)$-type. Let $P = \langle d_1, d_2 \rangle$ be the Sylow 3-subgroup of H chosen in Lemma 2.1.4. Let $d_3 = d_1 d_2, d_4 = d_1^2 d_2$. Then d_3 and d_4 are not conjugate in G, but $N_G(d_3) = N_G(d_4) = N_G(P)$. Furthermore, the character table of $N_G(P)$ is stated in Table 2.6.2.*

Proof The elements d_3 and d_4 are not conjugate in G by Proposition 10.2.2. But $C_G(d_3)$ and $C_G(d_4)$ both have a cyclic Sylow 2-subgroup of order 2 generated by $z = k_1^2$. Furthermore, Lemma 2.1.4 asserts that both d_j are inverted by k_5. Hence the Klein 4-group $V = \langle z, k_5 \rangle$ operates on $K_j = O(C_G(d_j))$. Therefore Theorem 4.2.2 of [92] implies that

$$(*) \quad |K_j| = |C_{K_j}(z)||C_{K_j}(k_5)||C_{K_j}(zk_5)|.$$

By Lemma 2.1.4 $P = \langle d_1, d_2 \rangle$ is an elementary abelian Sylow 3-subgroup of order 9 in H. Proposition 10.2.2(a) states that all involutions of G are conjugate in G. Therefore $C_{K_j}(z) = P$ and $|K_j| \le 3^6$. Clearly $C_G(P) \le C_G(d_j)$ for $3 \le j \le 4$ because $P = \langle d_3, d_4 \rangle$. By Lemmas 2.3.2 and 2.3.3 $N_G(P) = K : V$, where $K = O[C_G(P)]$ is an elementary abelian group of order 3^4. As $\langle z \rangle$ is a Sylow 2-subgroup of $C_G(d_j)$ we also know that K_j is a normal 2-complement of $C_G(d_j)$. Thus K is a subgroup of K_j, and so $|K_j| \ge 3^4$ by Lemma 2.3.3. Clearly $d_j \in Z(K_j)$. Let Q be a Sylow 2-subgroup of G containing K_j. By Lemma 2.3.5 and Sylow's Theorem there is a $g \in G$ such that Q^g is a Sylow 3-subgroup of $N_1 = N_G(d_1)$. Hence $(K_j)^g \le C_{N_1}((d_j)^g)$. Since $N_1 \cong X_i$, where X_i is one of the groups constructed in Lemma 2.3.4 and $d_j \in X_i$ for $i = 1, 2, 3$, we can determine in which conjugacy classes of X_i both d_j belong. Using a faithful permutation representation of each group X_i

and MAGMA it has been checked that $|C_{X_i}(d_j)| = 2 \cdot 3^4$ in any of the three cases $1 \leq i \leq 3$. Therefore $|K_j| = 3^4$. Hence $K_j = K$ and $N_G(d_3) = N_G(P) = K : V = N_G(d_4)$. This completes the proof. □

Lemma 2.3.8 *Keep the notations of Lemmas 2.1.4 and 2.3.4. Let* $N_G(d_1) = N_G(x_9) = X_3$ *and let* $S = \langle x_3, x_5, x_6, x_7, x_8, x_9 \rangle$ *be a Sylow 3-subgroup of the simple group* G *of Dickson type with centralizer* $H = C_G(z)$ *of the 2-central involution* z. *Then the following statements hold.*

(a) *S has 43 normal subgroups N_i which are indexed such that $|N_0| = 1$ and $|N_{42}| = 3^6$.*
(b) *S has four normal subgroups N_i of order 3, $1 \leq i \leq 4$.*
(c) *S has seven elementary abelian normal subgroups N_i of order 9, $5 \leq i \leq 11$. Furthermore, N_5 is the center $Z(S) = \langle x_9, u \rangle$ of S, where $u = x_7x_8$.*
(d) *Let $q_1 = x_5x_7x_9$, $q = x_5^2x_8^2x_9^2$, $u = x_7x_8$ and $u_1 = q^{x_1^3}$. Then the subgroups $P = \langle x_3, x_9 \rangle$, $P_1 = \langle u, u_1 \rangle$ and $P_2 = \langle q, q_1 \rangle$ of S with order 9 are centralized in X_3 by the involutions z, x_2 and zx_2, respectively.*
(e) *For each of these 11 normal subgroups $W = N_j$, $1 \leq j \leq 11$, exists a non-cyclic elementary abelian subgroup $U = U(W)$ such that $W \cap U = 1$ and U is centralized by some involution u in G.*
(f) *S has four elementary abelian normal subgroup N_i of order 27, $12 \leq i \leq 15$.*
(g) *S has a unique elementary abelian normal subgroup $A = N_{16} = \langle x_9, u, x_7, x_5 \rangle$ of order 81, and $C_S(A) = A$.*
(h) *S has 13 normal subgroups N_i of order 3^4, $16 \leq i \leq 28$.*
(i) *S has 13 normal subgroups N_i of order 3^5, $29 \leq i \leq 41$.*
(j) *$C_S(N_i) = A$ for $13 \leq i \leq 14$, $C_S(N_{12}) = \langle x_9, x_5, x_6, x_7, x_8 \rangle = N_{40}$ and $C_S(N_{15}) = \langle A, x_3 \rangle$.*
(k) *S has 26 non-abelian normal proper subgroups N_i, $17 \leq i \leq 42$. Their centers $Z(N_i)$ equal the center $Z(S)$ of S for $17 \leq i \leq 38$ and $i = 42$. Furthermore, the centers $Z(N_{39}) = N_{15}$ and $Z_{40} = N_{12}$ are elementary abelian of order 27.*

Proof All these statements have been obtained computationally from the presentation of S given in Lemma 2.3.4 using a faithful permutation representation of $X_3 = \langle S, x_1, x_2, x_4 \rangle$ and MAGMA. The the normal subgroups N_i, $1 \leq i \leq 42$, of S have been obtained by means of the

MAGMA command `NormalSubgroups(S)`. In view of the importance of statements (e), (f), (g), (j) and (k) more details are provided.

(e) Let $N_1 = \langle x_9 u\rangle$, $N_2 = \langle x_9\rangle$, $N_3 = \langle u\rangle$, $N_4 = \langle u x_9^2\rangle$, $N_5 = \langle x_9, u\rangle = Z(S)$, $N_6 = \langle x_7, u\rangle$, $N_7 = \langle x_9 x_7, u\rangle$, $N_8 = \langle x_9^2 x_7, u\rangle$, $N_9 = \langle x_9, x_5 x_8\rangle$, $N_{10} = \langle x_9, x_5 x_7 x_8^2\rangle$, $N_{11} = \langle x_9, x_8 x_5^2 x_7^2\rangle$. Let P, P_1 and P_2 be the three subgroups of order 9 in X_3 defined in (d). Then P_2 does not meet the subgroups N_i for $i \in \{2,3,4,5,6,8,9,12\}$, P does not meet the subgroups N_j for $j \in \{1,2,4,7,8,9\}$ and P_1 does not meet the subgroups N_k for $k \in \{2,3,5,10,11\}$. Hence (e) holds.

(g) Let $A = \langle q_1, q, u, x_9\rangle$. Then using the MAGMA command

```
subs:=Subgroups(S : Al:=Normal, IsElementaryAbelian:=true)
```

it has been checked that A is the unique maximal elementary abelian normal subgroup of S, and that $C_S(A) = A$.

(f) and (j) $N_{12} = \langle x_9, u, q q_1\rangle$, $N_{13} = \langle x_9, u, x_5 x_8 x_7^2\rangle$, $N_{14} = \langle x_9, u, x_5^2 x_8^2\rangle$ and $N_{15} = \langle x_9, u, x_5 x_7 x_8\rangle$ are the four elementary abelian normal subgroups N_i of order 27 of S. It also has been checked that $C_S(N_{12}) = \langle x_9, x_5, x_6, x_7, x_8\rangle = N_{40}$, $C_S(N_i) = A$ for $13 \le i \le 14$ and $C_S(N_{15}) = \langle A, x_3\rangle$.

(k) Using MAGMA it is easy to check the assertions about the centers $Z(N_i)$ of the 26 normal subgroups N_i with indices $17 \le i \le 42$. This completes the proof. □

Proposition 2.3.9 *Keep the notations of Lemmas 2.1.4 and 2.3.4. Let G be any finite simple group of $\mathsf{G}_2(3)$-type having a 2-central involution z such that $C_G(z) \cong H$. Let $P = \langle d_1, d_2\rangle$. Let $N_G(d_1) = X_3 = \langle x_1, x_2, x_3, x_4, x_5, x_6, x_7, x_8, x_9\rangle$, where $d_1 = x_9, k_5 = x_2, k_3 = x_1, x_1^2 = z, k_4 = x_4$. Then $S = \langle x_3, x_5, x_6, x_7, x_8, x_9\rangle$ is a Sylow 3-subgroup of G and the following statements hold.*

(a) *H has four conjugacy classes of elements of order 3 represented by $d_1 = x_9$, $d_2 = x_3$, $d_3 = d_1 d_2 = x_9 x_3$, $d_4 = d_1^2 d_2 = x_9^2 x_3$, which are also not conjugate in G.*
(b) *$K = O[C_G(P)] = \langle x_9, x_3, u, q\rangle$, where $u = x_7 x_8$ and $q = x_5^2 x_8^2 x_9^2$.*
(c) *$N_G(d_1) = \langle x_9, q_1, u, q, x_3, x_2, x_6, x_1\rangle$, where $q_1 = x_5 x_7 x_9 \in X_3$. Furthermore, $N_G(d_1)$ has 13 conjugacy classes of elements of order 3 and three conjugacy classes of elements of order 9. Their representatives and centralizer orders are given in Table 2.5.4.*
(d) *$d_5 = x_9 u$ is not G-conjugate to any element of order 3 in H. Furthermore, $C_G(d_5) = S$ and $N_G(d_5) = N_{X_3}(d_5)$.*

(e) *G has five conjugacy classes of elements of order* 3 *represented by* $d_1 = x_9$, $d_2 = x_3$, $d_3 = d_1d_2 = x_9x_3$, $d_4 = d_1^2d_2 = x_9^2x_3$ *and* $d_5 = x_9u$.

(f) *G has three conjugacy classes of elements of order* 9 *represented by* $n_1 = x_3x_6$, $n_2 = q_1x_3x_6$ *and* $n_3 = n_2^2$.

(g) *Let* $\pi = \{2, 3\}$. *Then* π *is the smallest set of primes containing* 2 *such that a centralizer of a* π*-element of G is a* π*-subgroup of G.*

Proof Lemma 2.3.6 states that $N_G(d_1) \cong X_3$. In order to simplify notation we assume this isomorphism to be an identification. By Lemma 2.3.5 $S = \langle x_3, x_5, x_6, x_7, x_8, x_9 \rangle$ is a Sylow 3-subgroup of G.

(a) holds by Proposition 10.2.2(i).

(b) Lemma 2.3.3 states that $|K| = 3^4$. Its generators have been found by means of a faithful permutation representation of X_3 and MAGMA.

(c) The given generators of $N_G(d_1) = X_3$ and the representatives of its conjugacy classes stated in Table 2.5.4 have been found computationally. The following fusion arguments use the notation of that table.

(d) By Table 2.5.4 $d_1 = x_9$, u and $d_5 = x_9u$ represent all 3-central conjugacy classes of elements of order 3 in $N_G(d_1) = X_3$. Hence G has at most three 3-central conjugacy 3-classes. Suppose that $d_5 = x_9u \in Z(S)$ is G-conjugate to some element of order 3 in H. Then Lemma 2.3.7 and Table 2.5.4 imply that d_5 is G-conjugate to d_1 or to $d_2 = x_3$. Lemma 2.3.6(a) states that there is an isomorphism $\beta : N_G(d_1) \to N_G(d_2)$ such that $\beta(x_9) = x_3^2$ and $\beta(x_3) = x_9^2$. Furthermore, $x_9^{x_2} = x_9^2$ and $x_3^{x_2} = x_3^2$ by Lemma 2.3.4. Hence we may assume that d_5 is G-conjugate to x_9. By Table 2.5.4 X_3 has three conjugacy classes of elements n_i of order 9 which all power onto d_5. Suppose that $d_1 = d_5^g$ for some $g \in G$. Then $n_1^3 = d_5$ implies that $(n_1^g)^3 = (n_1^3)^g = d_5^g = x_9$ and $n_1^g \in C_G(x_9) \le X_3$. But by Table 2.5.4 there is no element of order 9 in X_3 which powers onto x_9. Hence d_1, d_2 and d_5 are pairwise non-conjugate in G.

As G has a unique conjugacy class of involutions it follows that $L = C_G(d_5)$ has odd order. Let $t = zx_2$. By Lemma 2.3.4 $u^{x_2} = u$ and $u^z = u^2$. Hence $d_5^t = d_5^2$. Thus $Y = N_G(d_5) = \langle L, t \rangle$ and $|Y/L| = 2$.

We claim that $L = S = \langle x_3, x_5, x_6, x_7, x_8, x_9 \rangle$. As L has odd order it is a solvable group by the famous Theorem of Feit and Thompson. Let $W \neq 1$ be a minimal normal subgroup of L. Then $|W| = r^m$ for some odd prime r and some positive integer m. If r were different from 3 then the non-cyclic elementary abelian subgroup $P = \langle x_3, x_9 \rangle$ would operate on W by conjugation. Now Theorem 1.5.8 of [92] implies that $W = \langle C_W(a) | 1 \neq a \in P \rangle$. All non-trivial elements $a \in P$ are contained in $H = C_G(z)$ by Lemma 2.3.8(d). Hence $C_G(a)$ is a $\{2,3\}$-group by Tables 2.5.4 and 2.6.2. Thus $W = 1$.

In particular, a minimal elementary abelian normal subgroup W of L is a 3-group. Clearly, $W = \langle d_5 \rangle$ is a minimal elementary abelian normal subgroup of L. Let $W_1 \neq 1$ be a minimal elementary abelian normal subgroup of $L_1 = L/W$. By the proof of Lemma 2.3.8(d) $P \cap \langle d_5 \rangle = 1$. Hence the non-cyclic elementary abelian group $P_1 = PW/W \cong P$ operates on W_1 by conjugation. The previous argument shows again that W_1 is a 3-group. Let W_2 be a normal subgroup of L containing W such that $W_1 = W_2/W$. Then $W_2 \leq S$ and $|W_2| \geq 9$.

If $|W_2| = 9$ then $W_2 = Z(S) = \langle x_9, d_5 \rangle$ by Lemma 2.3.8(c). Hence $C_2 = C_L(W_2)$ is a normal subgroup of L which contains the Sylow 3-subgroup S of L. Furthermore, $S \leq C_2 \leq C_G(x_9) \cap C_G(d_5) = C_{X_3}(d_5) = S$ by Table 2.5.4. Hence S is normal in L. Since L does not have any proper normal subgroups of $3'$-order we have $C_L(S) = Z(S)$. Hence $L/Z(S)$ is isomorphic to a subgroup of the automorphism group AS of S. Applying the MAGMA command `AS:=AutomorphismGroup(S)` it follows that $|AS| = 2^3 \cdot 3^{10}$. Hence $L = S$ in this case.

Therefore we may assume that L contains a proper normal 3-subgroup W with order $|W| > 9$. So Lemma 2.3.8 states that W is one of 31 normal subgroups N_i of S, $12 \leq i \leq 42$. In particular, it contains d_5 and x_9 by Lemma 2.3.8. Thus $C_G(W) \leq C_G(x_9) \cap C_G(d_5) = C_{X_3}(d_5) = S$ by Table 2.5.4. Therefore $C_G(W) = C_L(W) = C_S(W)$ is also a normal subgroup of S and $L/C_L(W) \leq Aut(W)$.

By Lemma 2.3.8(g) N_{16} is the unique elementary abelian normal subgroup $A = \langle x_9, u, q, q_1 \rangle$ of order 81 and $C_S(A) = A$. Lemma 2.3.8(f) asserts that S has four normal subgroups N_i of order 27 with centralizers $C_S(N_i) = A$ for $i = 13, 14$, $C_S(N_{12}) = N_{40} = \langle x_9, x_5, x_6, x_7, x_8 \rangle$ and $C_S(N_{15}) = N_{39} = \langle A, x_3 \rangle$. Since $L = C_G(d_5)$ and $d_5 \in N_i$ it follows that $L/C_L(N_i) \leq \mathrm{GL}_2(3)$ for $12 \leq i \leq 15$. Hence $L = S$ in these four cases also.

If $W = A$ then the same argument shows that $L/A \leq \mathrm{GL}_3(3)$. Now L/A has a cyclic Sylow 3-subgroup of order 3 and $|\mathrm{GL}_3(3)| = 2^4 \cdot 3^3 \cdot 13$. Applying the MAGMA command

```
subs:=Subgroups(GL(3,3): OrderMultipleOf := 3*13)
```

one can verify that there is exactly one conjugacy class of subgroups B in $\mathrm{GL}_3(3)$ with Sylow 3-subgroups of order 3 meeting this order condition. MAGMA also states that $|B| = 2 \cdot 3 \cdot 13$ and that B operates irreducibly on the complement $A_1 = \langle q, q_1, x_9 \rangle$ of $\langle d_5 \rangle$ in A. Hence q and q_1 are conjugate in L and therefore in G. By Lemma 2.3.8(d) the subgroups

$P = \langle x_3, x_9 \rangle$ and $P_1 = \langle q, p_1 \rangle$ are G-conjugate because G has a unique conjugacy class of involutions. Thus x_3 and x_9 are G-conjugate, which contradicts Proposition 10.2.2(i). Thus $L = S$ holds also in this case.

In the remaining cases we know that $|W| > 27$ and $Z(W) < W$. Lemma 2.3.8(k) yields that $Z(W) = Z(S) = \langle d_5, x_9 \rangle$ or $W \in \{N_{39}, N_{40}\}$ and $Z(W) \in \{Z(N_{39}) = N_{15}, Z(N_{40}) = N_{12}\}$. Since $Z(W)$ is a characteristic subgroup of W the group L acts also on its normal subgroup $Z(W)$ by conjugation. If $Z(W) = Z(S)$ then $C_S(Z(S)) = S$ and so $L = S$ because $|Aut(S)| = 2^3 \cdot 3^{10}$. In the final cases, where $W \in \{N_{39}, N_{40}\}$, it follows from Lemma 2.3.8(k) that $Z(W)$ is elementary abelian of order 27 and that $C_S(Z(W)) = W$. As L acts on $Z(W)$ these two cases have been dealt with before. Thus (d) holds.

(e) By Table 2.5.4 the elements x_9, u and $d_5 = x_9 u$ represent the three conjugacy classes of $N_G(d_1)$ of highest defect. From (d) it follows that d_5 is not G-conjugate to any of the elements d_i, where $1 \le i \le 4$. By Lemma 2.3.6(a) there is an isomorphism β between $N_G(x_9)$ and $N_G(x_3)$. The center $Z(S)$ of the Sylow 3-subgroup S is generated by x_9 and u. By (b) $K = C_G(P) = \langle x_9, x_3, u, q \rangle$ and q is centralized by zx_2. Hence the Brauer–Wielandt Theorem 4.2.2 of [92] implies that the center of the Sylow 3-subgroup $S_1 = \beta(S)$ of $N_G(x_3)$ is generated by x_3 and q. By Proposition 10.2.2(i) $d_1 = x_9$ and $d_2 = x_3$ are not conjugate in G. By (d) we know that $C_G(x_9 u) = S$. Thus $C_G(x_3 q) = S_1$. Furthermore, Sylow's Theorem asserts that $S_1 = S^g$ and $Z(S)^g = Z(S_1)$ for some $g \in G$. Therefore $d_5 \sim x_3 q$, $d_1 \sim q$ and $u \sim d_2$, where $\sim$ denotes the fusion of the elements in G.

By Lemma 2.3.8(d) the subgroups $P = \langle x_9, x_3 \rangle$, $P_1 = \langle u, u_1 \rangle$ and $P_2 = \langle q, q_1 \rangle$ are conjugate in G, where $u_1 = q^{x_1^3}$. Therefore (b) implies that the above argument can be applied to P_1 in the same way as to P. Hence $uq \sim x_9 u = d_5$. Using a faithful permutation representation of X_3 and MAGMA it has been checked that $q_1 x_3 \sim q x_3$, $d_4 = x_9^2 x_3 \sim x_9^2 u x_3$ and $q_1 q \sim u$ in $X_3 = N_G(d_1)$. Hence $q_1 x_3 \sim d_5$. Using the isomorphism β and Sylow's Theorem again it follows that $x_9 q_1 x_3 \sim d_4$ and $q_1 q x_3 \sim d_3$ in G.

The centralizers $C_{X_3}(q_1)$ and $C_{X_3}(q_1^2 q)$ have order $2 \cdot 3^4$. Hence q_1 and $q_1^2 q$ are G-conjugate to d_3 and d_4, respectively. As x_3 is G-conjugate to d_2 it follows that all 13 classes of elements of order 3 fuse to the five classes of G stated in the assertion.

(f) By Table 2.5.4 the group $N_G(d_1)$ has three conjugacy classes of elements of order 9 represented by the three elements n_i given in the statement. It has been checked computationally that $n_1^3 = (x_3 x_6)^3 = x_9 u^2 = n_2^3 = n_3^3$ and that this element of order 3 is X_3-conjugate to d_5.

Hence (e) and (d) and Theorem 2.2.11 of [92] imply that G has exactly three conjugacy classes of elements of order 9.

(g) This assertion is an immediate consequence of Lemmas 2.3.6, 2.3.7, 2.3.2 and statement (d). This completes the proof. □

2.4 Janko's characterization of $G_2(3)$

In this section we show that each finite simple group G of $G_2(3)$-*type* has the same order as $G_2(3)$ and that it has a uniquely determined ordinary irreducible character of degree 14. Hence the author's uniqueness criterion stated in Theorem 2.1 of [91] can be applied to show that each finite simple group G of $G_2(3)$-*type* is isomorphic to the simple group $\mathfrak{W}$ constructed in [92], pp. 577–9. Janko's main theorem of [68] follows from this result by Lemma 2.1.4 and Definition 2.1.1.

Proposition 2.4.1 *Each finite simple group G of $G_2(3)$-type has order* $|G| = 2^6 \cdot 3^6 \cdot 7 \cdot 13$.

Proof By Proposition 10.2.2 G has only one conjugacy class z^G of involutions. Let $\pi = \{2,3\}$. Then π is the smallest set of primes containing 2 such that a centralizer of a π-element of G is a π-subgroup of G by Proposition 2.3.9. Furthermore, it asserts that a Sylow 3-subgroup of G has order 3^6 and that G has five conjugacy classes of elements of order 3 and three conjugacy classes of elements of order 9. The character tables of the normalizers of the elements d_1, d_2 and d_3, d_4 of order 3 are stated in Tables 2.6.1 and 2.6.2, respectively. By Proposition 2.3.9 $N_G(d_5) \le N_G(d_1)$. Using these character tables, Table 12.9.4 of [92] and Lemma 4.3.2 of [92] it has been verified that G has 17 strongly real conjugacy classes r_i^G of π-elements. Their invariants $d(r_i)$ and centralizer orders $|C_G(r_i)|$ are given in the following table:

Class	$\lvert C_G(r_i)\rvert$	$d(r_i)$	Class	$\lvert C_G(r_i)\rvert$	$d(r_i)$
2_A	$2^6 \cdot 3^2$	90	6_B	$2^3 \cdot 3^2$	36
3_A	$2^3 \cdot 3^6$	324	6_C	$2 \cdot 3^2$	18
3_B	$2^3 \cdot 3^6$	324	6_D	$2 \cdot 3^2$	18
3_C	$2 \cdot 3^4$	54	8_A	2^3	8
3_D	$2 \cdot 3^4$	54	8_B	2^3	8
3_E	3^6	81	9_C	3^3	27
4_A	$2^5 \cdot 3$	36	12_A	$2^2 \cdot 3$	12
4_B	$2^5 \cdot 3$	36	12_B	$2^2 \cdot 3$	12
6_A	$2^3 \cdot 3^2$	36			

Let π' be the set of all prime divisors q of $|G|$ which do not belong to π. Let $g_{\pi'} = |G|/g_\pi$, where $g_\pi = 2^6 \cdot 3^6$ by Lemmas 2.1.4 and 2.3.5.

Let $c(x_i)$ be the orders of the centralizer $C_G(x_i)$ of the eight conjugacy classes x_i^G of orders 3 and 9. Then the above table implies immediately that

$$F = \sum\nolimits_{i=1}^{8} \frac{g_\pi}{c(x_i)} = 5840.$$

Let t be the number of strongly real π'-classes of G. Then Theorem 4.3.7 of [92] asserts that

$$|G| = c|H| + t|H|^2, \text{ where } \quad c = 1 + \sum_{i=1}^{17} \frac{|H|}{|C_G(r_i)|}|d(r_i).$$

Inserting the values of the invariants $d(r_i)$ given in the above table and the orders of the centralizers of the 17 strongly real π-classes of G it follows that $c = 5643$.

By Proposition 2.3.9 and Lemma 2.1.4 $\Delta = C_G(d_1)$ has a maximal order $2^4 \cdot 3^6$ among the centralizers of all π-elements of G. Thus $t < |\Delta|$ by Theorem 4.3.9 of [92]. Proposition 10.2.2 states that 7 is a divisor of $|G|$. Since $c|H| + 3|H|^2$ is divisible by 7 it follows that there are 186 possible group orders g_k which are divisible by 7 and $g_k = c|H| + (3 + 63k)|H|^2$ for $k = 0, 1, ..., 185$. Hence $g_k = |H| \cdot 3^4 \cdot (91 + 448 \cdot k)$. In particular, $g_\pi = 81 \cdot |H|$ is a divisor of all integers g_k. Let $r_k = 91 + 448 \cdot k$ for all k.

By Theorem 4.5.2 of [92] the factor $g_{\pi'}$ of $|G|$ has to satisfy the Frobenius congruence

$$(*) \quad 1 + g_{\pi'} \left[\sum_{i=1}^{8} \frac{g_\pi}{c(x_i)} \right] \equiv 0 \bmod (3^6),$$

where $c(x_i)$ are the centralizer orders of the eight conjugacy classes x_i^G of orders 3 and 9. By the table stated above it follows that

$$F = \sum\nolimits_{i=1}^{8} \frac{g_\pi}{c(x_i)} = 5840.$$

Applying now the MAGMA command

```
IsIntegral((r_k * F +1)/3^6)
```

for $0 \leq k \leq 185$, it follows that $k = 0$ is the only number k for which this congruence has a solution. Hence $|G| = c|H| + 3|H|^2 = 2^6 \cdot 3^6 \cdot 7 \cdot 13$. This completes the proof. □

Corollary 2.4.2 *Let G be any finite simple group of $\mathsf{G}_2(3)$-type with a 2-central involution z such that $C_G(z) = H$. Then the following assertions hold.*

(a) *G has 23 conjugacy classes.*

(b) *G has three strongly real conjugacy π'-classes: one of elements of order 7 and two of elements of order 13.*

(c) *The Sylow 7-normalizer is a Frobenius group of order 42.*

(d) *The Sylow 13-normalizer is a Frobenius group of order 78.*

Proof All statements are immediate consequences of Lemma 2.1.4, Corollary 14.2.5, Proposition 10.2.2, Proposition 2.3.9, Proposition 10.4.3, Sylow's Theorem and the class equation of G. □

Proposition 2.4.3 *Let G be a finite simple group of $\mathsf{G}_2(3)$-type having a 2-central involution z with centralizer $H = C_G(z)$. The character tables of the local subgroups $C_G(2_A) = H$, $N_G(3_A) = N_G(d_1) \cong N_G(d_2) = N_G(3_B)$, $N_G(7_A)$ and $N_G(13_A)$ are given in Table 12.9.4 of [92] and Tables 2.6.1, 2.6.3 and 2.6.4, respectively. Keeping their notations for the conjugacy classes of these local subgroups of G the following statements hold.*

(a) *$N_G(3_C) = N_G(d_5)$ is a subgroup of $N_G(d_1) = N_G(3_A)$ and $N_G(d_2)$.*

(b) *There is an isomorphism $\beta : N_G(d_1) \to N_G(d_2)$ by Lemma 2.3.6(a), but d_1 and d_2 are not conjugate in G. Using β the fusion of the classes of $N_G(d_2) = N_G(3_B)$ can be retrieved from the fusion of the classes of $N_G(d_1) = N_G(3_A)$ in G.*

(c) *$N_G(3_D) = N_G(d_3) = N_G(d_4) = N_G(3_E)$ is a subgroup of $N_G(3_C) = N_G(3_5)$, but d_3 and d_4 are not conjugate in G.*

(d) *G has 23 conjugacy classes x^G with centralizer orders $|C_G(x)|$ given in the table below. The fusion of the conjugacy classes of the four maximal local subgroups $C_G(2_A)$, $N_G(3_A)$, $N_G(7_A)$ and $N_G(13_A)$ of G in the conjugacy classes of G is described by the following table:*

x	$\lvert C_G(x)\rvert$	$C_G(2_A)$	$N_G(3_A)$	$N_G(7_A)$	$N_G(13_A)$
1_A	$2^6 \cdot 3^6 \cdot 7 \cdot 13$	a	a	a	a
2_A	$2^6 \cdot 3^6$	a, b	a	a	a
3_A	$2^3 \cdot 3^6$	a	a, d		
3_B	$2^3 \cdot 3^6$	b	b, j		
3_C	3^6		c, e, k		
3_D	$2 \cdot 3^4$	c	f, i, m		
3_E	$2 \cdot 3^4$	d	g, h, l	a, b	a, b
4_A	$2^5 \cdot 3$	a, c	a		
4_B	$2^5 \cdot 3$	b, d			
6_A	$2^3 \cdot 3^2$	a	a, d		
6_B	$2^3 \cdot 3^2$	b	b, f		
6_C	$2 \cdot 3^2$	c	e, g		
6_D	$2 \cdot 3^2$	d	c, h	a, b	a, b
7_A	7			a	
8_A	2^3	a	a		
8_B	2^3	b	b		
9_A	3^3		a		
9_B	3^3		b		
9_C	3^3		c		
12_A	$2^2 \cdot 3$	a	a		
12_B	$2^2 \cdot 3$	b			
13_A	13				a
13_B	13				

The orders of the representatives of the conjugacy classses of the four local subgroups of G are omitted, because they can be read off from the first column of the respective table of conjugacy class representatives.

Proof The fusion of the $H = C_G(2_A)$-classes in G follows from Corollary 14.2.5.

By Table 2.5.4 $N_G(d_1) = N_G(3_A)$ has 13 conjugacy classes of elements of order 3 and three conjugacy classes of order 9. Furthermore, $N_G(d_1) \cong N_G(d_2)$ by Lemma 2.3.6 and both groups contain a Sylow 3-subgroup of G by Lemma 2.3.5. Therefore the fusion of the first ten classes of elements of order 3 of $N_G(d_1)$ into the classes of G can easily be established by means of Table 2.5.4 and the proof of Proposition 2.3.9. Using the faithful permutation representation of $X_3 = N_G(d_1)$ and

MAGMA it has been checked that three representatives 3_k, 3_l and 3_m of X_3 have the same elementary abelian centralizer C of order 3^4 in X_3, and that they are X_3-conjugate to x_3uq, x_9x_3uq and $x_9^2x_3uq$, respectively. Furthermore, C is X_3-conjugate to $K = \langle x_3, x_9, u, q\rangle$. Clearly, K is contained in the intersection of $N_G(d_1) = X_3$ and $N_G(d_2)$. By Lemma 2.3.6(a) there is an isomorphism $\beta : X_3 \to N_G(d_2)$ satisfying $\beta(x_3) = x_9^2$ and $\beta(x_9) = x_3^2$. Therefore it induces an automorphism of K by Lemma 2.3.6. In particular, $\beta(u) = q^2$ and $\beta(q) = u^2$. With these changes the classes of $N_G(x_3) = N_G(d_2) = X_4$ can easily be classified as in Table 2.5.4.

Now the roles of x_9 and x_3 are interchanged. Since x_9uq is X_3-conjugate to uq the element x_3uq of X_4 is X_4-conjugate to uq, which is G-conjugate to 3_C, represented by x_9u. It has been checked computationally that the centralizers of all the elements $x_9^ix_3^juq$ and $x_9^ix_3^ju^2q^2$ in X_3, $1 \le i, j \le 2$, have order 3^4. Hence the centralizers of $\beta(x_9x_3uq) = x_9^2x_3^2u^2q^2$ and $\beta(x_9^2x_3uq) = x_3x_9^2u^2q^2$ in X_4 have order 3^4. Therefore both classes 3_l and 3_m of X_3 fuse to 3_D or 3_E.

All 9-classes power into class 3_c of X_3 meeting the center of a Sylow 3-subgroup of G. Hence they represent the three distinct 9-classes of G by Table 2.5.4.

By Proposition 10.2.2 we know that the 7-elements of $E = N_G(A)$ are normalized by conjugates of $d_4 = d_1d_2^2 \in (3_E)^G$. Corollary 2.4.2 states that the Frobenius group $N_G(7_A)$ has order 42. Thus the unique cyclic subgroup of order 6 in $N_G(7_A)$ is generated by a G-conjugate of 6_D represented by the element zd_4 of H. The two representatives 6_a and 6_b of $N_G(7)$ are inverses of each other. As the class 6_D of G is real both 6_a and 6_b fuse with 6_D. Now the power map information of Table 2.6.3 determines the fusion of the 3-classes of $N_G(7)$ in G.

The character table of Frobenius normalizer $N_G(13_A)$ is stated in Table 2.6.4. The fusion of the two classes 6_a and 6_b of $N_G(13_1)$ given in the table has been checked in $\mathfrak{W}$. □

Proposition 2.4.4 *Let G be a finite simple group of $\mathsf{G}_2(3)$-type. Then the minimal degree of a faithful irreducible complex character χ of G is 14, and G has exactly one character $\chi \in Irr_{\mathbb{C}}(G)$ of this degree.*

Proof Let $H = C_G(z)$ be the centralizer of a 2-central involution of G. Let S be the Sylow 2-subgroup of H and let A be the uniquely determined maximal elementary abelian normal subgroup of S given in

Lemma 2.1.4. Let $E = N_G(A)$. By Proposition 10.2.2 any simple group G of $\mathsf{G}_2(3)$-*type* has a unique conjugacy class of involutions. Therefore Corollary 4.8.10 of [92] asserts that $G = \langle H, E\rangle$. Proposition 10.2.2(g) states that the amalgam $H \leftarrow D \rightarrow E$ has Goldschmidt index 1, where $D = N_H(A)$. Therefore the free product $P = H *_D E$ of H and E with amalgamated subgroup D is uniquely determined up to isomorphism. By the proof of Lemma 12.4.2(e) of [92] the smallest degree of a compatible pair is 14, and there is exactly one such pair $(\chi, \tau) \in mfchar_{\mathbb{C}}(H) \times mfchar_{\mathbb{C}}(E)$ of complex conjugate characters. As G is a simple epimorphic image of P it follows that G has at most one faithful character $\chi \in Irr_{\mathbb{C}}(G)$ with minimal degree 14.

Using Brauer's characterization of characters we now prove that any finite simple group G of $\mathsf{G}_2(3)$-*type* has a complex irreducible characters $\chi \in Irr_{\mathbb{C}}(G)$ of degree 14. In order to do so we define a class function $\chi : G \rightarrow \mathbb{C}$ by the following table:

i	1_A	2_A	3_A	3_B	3_C	3_D	3_E	4_A	4_B	6_A	6_B	6_C
$\chi(g_i)$	14	-2	5	5	-4	2	-1	2	2	1	1	-2

i	6_D	7_A	8_A	8_B	9_A	9_B	9_C	12_A	12_B	13_A	13_B
$\chi(g_i)$	1	0	0	0	2	-1	-1	-1	-1	1	1

Let p be any prime divisor of $|G|$. If Y is an over-group of a p-elementary subgroup X, and the restriction $\chi_{|Y}$ of X to Y is a generalized character, then $\chi_{|X}$ is also a generalized character of X.

By Corollary 2.4.2 the Sylow 7-normalizer N_{7A} and the Sylow 13-normalizers N_{13A} and N_{13B} of G are Frobenius groups of orders 42 and 78, respectively: N_{7A} has one faithful irreducible character χ_1 of degree 6 and six linear characters; N_{13A} has two faithful irreducible characters χ_1, χ_2 of degree 6 and six linear characters. By Proposition 2.3.9 and Lemmas 2.3.6 and 2.3.7 we may assume that the normalizer of any cyclic subgroup of order 3 is contained in $N_G(d_1) \cong X_3 \cong N_G(d_2)$.

Therefore by Corollary 2.8.10 of [92] it remains to show that $X_{|Y}$ is a generalized character of Y, whenever Y belongs to the following set of subgroups $\mathfrak{Y} = \{H, N_G(d_1) = X_3, N_{7A}, N_{13A}\}$ of G.

In Proposition 2.4.3 the fusion of the conjugacy classes of the groups $Y \in \mathfrak{Y}$ into the classes of G has been determined. Therefore the inner products $(\chi, \chi)_Y$ can be calculated for each $\chi \in Irr_{\mathbb{C}}(Y)$ and for each

$Y \in \mathfrak{Y}$ by means of the character tables of the local subgroups Y. It turns out that

$$\begin{aligned}
\chi_{|H} &= \chi_8 + \chi_{10} + \chi_{18}, \text{ where } \chi_8, \chi_{10}, \chi_{18} \in Irr_{\mathbb{C}}(H),\\
\chi_{|N_G(d_1)} &= \chi_9 + \chi_{14}, \text{ where } \chi_3, \chi_{13} \in Irr_{\mathbb{C}}(N_1),\\
\chi_{|N_{7A}} &= \chi_4 + \chi_6 + 2\chi_7, \text{ where } \chi_4, \chi_6, \chi_7 \in Irr_{\mathbb{C}}(N_7A),\\
\chi_{|N_{13A}} &= 2\chi_2 + \chi_7 + \chi_8, \text{ where } \chi_2, \chi 7, \chi 8 \in Irr_{\mathbb{C}}(N_{13A}),
\end{aligned}$$

and that the inner product $(\chi, \chi)_G = 1$. Therefore χ is an irreducible character of G with $\chi(1) = 14$ by Corollary 2.8.10 of [92]. This completes the proof. □

Theorem 2.4.5 (Michler–Wang) *Let H be the finite group of even order defined in Lemma 2.1.4. Then each finite simple group G of $\mathsf{G}_2(3)$-type is isomorphic to the simple group $\mathfrak{W} = \langle \mathfrak{k}_1, \mathfrak{d}_1, \mathfrak{s}_0 \rangle \leq \mathrm{GL}_{14}(5)$ of order $|\mathfrak{W}| = 2^6 \cdot 3^6 \cdot 7 \cdot 13$, where*

$$\mathfrak{k}_1 = \begin{pmatrix}
1&0&0&0&0&0&0&0&0&0&0&0&0&0\\
0&1&0&0&0&0&0&0&0&0&0&0&0&0\\
0&0&1&0&0&0&0&0&0&0&0&0&0&0\\
0&0&0&4&2&2&0&0&0&0&0&0&0&0\\
0&0&0&0&3&4&0&0&0&0&0&0&0&0\\
0&0&0&0&3&2&0&0&0&0&0&0&0&0\\
0&0&0&0&0&0&1&4&4&3&1&3&0&2\\
0&0&0&0&0&0&2&2&4&4&2&1&2&3\\
0&0&0&0&0&0&3&1&3&1&1&0&3&3\\
0&0&0&0&0&0&1&1&1&4&4&0&1&2\\
0&0&0&0&0&0&4&2&4&3&2&2&2&3\\
0&0&0&0&0&0&4&4&3&3&1&4&1&3\\
0&0&0&0&0&0&0&4&2&0&2&4&1&1\\
0&0&0&0&0&0&2&1&1&4&0&2&1&3
\end{pmatrix},$$

$$\mathfrak{d}_1 = \begin{pmatrix}
1&0&0&0&0&0&0&0&0&0&0&0&0&0\\
0&1&0&0&0&0&0&0&0&0&0&0&0&0\\
0&0&1&0&0&0&0&0&0&0&0&0&0&0\\
0&0&0&2&2&2&0&0&0&0&0&0&0&0\\
0&0&0&3&2&1&0&0&0&0&0&0&0&0\\
0&0&0&3&1&1&0&0&0&0&0&0&0&0\\
0&0&0&0&0&0&4&3&0&4&2&4&4&0\\
0&0&0&0&0&0&2&1&2&2&4&1&2&4\\
0&0&0&0&0&0&3&4&4&0&3&4&0&0\\
0&0&0&0&0&0&2&3&4&4&0&1&3&3\\
0&0&0&0&0&0&0&2&4&0&1&0&0&4\\
0&0&0&0&0&0&4&0&1&1&3&0&3&3\\
0&0&0&0&0&0&1&3&1&0&2&2&0&3\\
0&0&0&0&0&0&2&2&4&0&2&1&0&3
\end{pmatrix},$$

$$\mathfrak{s}_0 = \begin{pmatrix}
3&2&3&3&4&1&4&4&2&2&2&1&4&4\\
2&3&2&2&1&4&0&0&0&0&0&0&0&0\\
4&1&4&4&2&3&2&0&1&2&0&0&2&2\\
4&1&4&2&1&4&0&0&1&0&2&1&2&1\\
3&2&3&4&2&3&2&1&1&1&0&0&0&2\\
1&4&1&3&4&1&3&4&4&4&0&0&0&3\\
0&4&0&1&4&0&2&2&4&4&2&0&1&0\\
2&2&0&1&4&0&2&4&2&0&2&2&0&1\\
3&4&0&2&4&1&4&4&1&0&2&3&1&0\\
1&3&0&1&2&3&4&1&0&2&1&0&3&2\\
1&3&0&4&0&4&1&4&3&0&2&3&1&3\\
3&3&0&2&3&0&3&2&1&4&2&3&4&2\\
3&2&0&1&1&2&4&0&3&1&4&4&2&1\\
3&0&0&3&1&4&3&1&0&3&0&0&1&4
\end{pmatrix}.$$

Furthermore, $\mathfrak{W}$ has 23 conjugacy classes. A system of their representatives $\mathfrak{w}_i$ and centralizer orders $|C_{\mathfrak{W}}(w_i)|$ is given in Table 2.5.1.

The character table of $\mathfrak{W}$ coincides with that of $\mathsf{G}_2(3)$ in the Atlas [19], p. 60.

Proof By the proof of Lemma 12.4.2 of [92] the simple subgroup

$$\mathfrak{W} = \langle \mathfrak{k}_i, \mathfrak{d}_j, \mathfrak{s}_0 | 1 \le i \le 5, 1 \le j \le 2 \rangle \quad \text{of} \quad \mathrm{GL}_{14}(5)$$

is of $\mathsf{G}_2(3)$-type. It has a faithful permutation representation PW of degree 7371 with stabilizer $C_{\mathfrak{W}}(\mathfrak{z}) = \langle \mathfrak{k}_i, \mathfrak{d}_j | 1 \le i \le 5, 1 \le j \le 2 \rangle$, where $\mathfrak{z} = \mathfrak{k}_1^2$ by [92], pp. 577–9. Using PW and MAGMA it has been verified that it can be generated by the three matrices $\mathfrak{k}_1$, $\mathfrak{d}_1$ and $\mathfrak{s}_0$ stated in the assertion.

Let G be any finite simple group of $\mathsf{G}_2(3)$-*type*. By Proposition 10.2.2 G has exactly one conjugacy class z^G of involutions. We may assume that $C_G(z) = H$. Proposition 10.2.2 asserts that the Sylow 2-subgroup S of H contains a maximal elementary abelian normal subgroup A of order 2^3 such that $D = N_H(A)$ is a proper subgroup of $E = N_G(A)$ such that $E/A \cong \mathrm{GL}_3(2)$ and E does not split over A. Furthermore, the amalgam $H \leftarrow D \rightarrow E$ has Goldschmidt index 1. Hence $G = \langle H, E \rangle$ by Corollary 4.8.10 of [92]. By Proposition 2.4.4 and Maschke's Theorem G has a unique 5-modular irreducible representation over GF(5) of minimal degree 14 because 5 does not divide $|G| = \mathfrak{W}$ by Proposition 10.4.3. Hence $G \cong \mathfrak{W}$ by Theorem 7.5.1 of [92]. This completes the proof. □

Corollary 2.4.6 (Z. Janko) *Each simple group G of $\mathsf{G}_2(3)$-type is isomorphic to $\mathsf{G}_2(3)$.*

Proof The result follows immediately from Lemma 2.1.4 and Theorem 2.4.5. □

2.5 Representatives of conjugacy classes

2.5.1 *Conjugacy classes of* $\mathfrak{W} = \langle k, d, t \rangle$

Class	*Representative*	*Class*	*Centralizer*
2	$(k)^2$	7371	$2^6 \cdot 3^2$
3_1	d	728	$2^3 \cdot 3^6$
3_2	$(ds)^4$	728	$2^3 \cdot 3^6$
3_3	$(dsds^2)^3$	5824	3^6
3_4	$(ks)^2$	26208	$2 \cdot 3^4$
3_5	$(ksdsd)^2$	26208	$2 \cdot 3^4$
4_1	k	44226	$2^5 \cdot 3$
4_2	$(kd^2s)^2$	44226	$2^5 \cdot 3$
6_1	$(ds)^2$	58968	$2^3 \cdot 3^2$
6_2	k^2d	58968	$2^3 \cdot 3^2$
6_3	ks	235872	$2 \cdot 3^2$
6_4	$ksdsd$	235872	$2 \cdot 3^2$
7	s	606528	7
8_1	kd^2s	530712	2^3
8_2	ks^3	530712	2^3
9_1	$dsds^2$	157248	3^3
9_2	$(dsds^2)^2$	157248	3^3
9_3	$dsdsds^2$	157248	3^3
12_1	ds	353808	$2^2 \cdot 3$
12_2	ks^2d	353808	$2^2 \cdot 3$
13_1	kds	326592	13
13_2	$(kds)^2$	326592	13

2.5.2 *Conjugacy classes of* $X_1 = \langle x_i | 1 \le i \le 9 \rangle$

Class	*Representative*	*Class*	*Centralizer*	2P	3P
1	1	1	$2^4 \cdot 3^6$	1	1
2_1	$(x_1)^2$	81	$2^4 \cdot 3^2$	1	2_1
2_2	x_2	324	$2^2 \cdot 3^2$	1	2_2
3_1	x_9	2	$2^3 \cdot 3^6$	3_1	1
3_2	x_7	8	$2 \cdot 3^6$	3_2	1
3_3	x_7x_9	16	3^6	3_3	1
3_4	x_5	24	$2 \cdot 3^5$	3_4	1
3_5	$(x_2x_6x_7)^2$	24	$2 \cdot 3^5$	3_5	1
3_6	$(x_2x_6x_8)^2$	24	$2 \cdot 3^5$	3_6	1
3_7	x_3	72	$2 \cdot 3^4$	3_7	1
3_8	x_3x_5	72	$2 \cdot 3^4$	3_8	1
3_9	$x_3^2x_9$	72	$2 \cdot 3^4$	3_9	1
3_{10}	x_3x_7	144	3^4	3_{10}	1
3_{11}	x_5x_7	144	3^4	3_{11}	1
3_{12}	$x_3x_5x_7$	144	3^4	3_{12}	1
3_{13}	$x_3^2x_7x_9$	144	3^4	3_{13}	1
4	x_1	486	$2^3 \cdot 3$	2_1	4
6_1	$(x_1x_5)^2$	162	$2^3 \cdot 3^2$	3_1	2_1
6_2	x_2x_6	648	$2 \cdot 3^2$	3_4	2_2
6_3	x_2x_7	648	$2 \cdot 3^2$	3_2	2_2
6_4	$x_1^2x_3$	648	$2 \cdot 3^2$	3_7	2_1
6_5	$x_2x_6x_7$	648	$2 \cdot 3^2$	3_5	2_2
6_6	$x_2x_6x_8$	648	$2 \cdot 3^2$	3_6	2_2
6_7	$x_1^2x_3x_6$	648	$2 \cdot 3^2$	3_8	2_1
6_8	$x_1x_3x_1x_5$	648	$2 \cdot 3^2$	3_9	2_1
8_1	x_2x_4	1458	2^3	4	8_1
8_2	$(x_2x_4)^5$	1458	2^3	4	8_2
9_1	x_3x_6	432	3^3	9_1	3_1
9_2	$x_3x_6x_7$	432	3^3	9_2	3_1
9_3	$x_3x_6x_8$	432	3^3	9_3	3_1
12	x_1x_5	972	$2^2 \cdot 3$	6_1	4

2.5.3 *Conjugacy classes of* $X_2 = \langle x_i | 1 \leq i \leq 9 \rangle$

Class	*Representative*	*Class*	*Centralizer*	2P	3P
1	1	1	$2^4 \cdot 3^6$	1	1
2_1	$(x_1)^2$	81	$2^4 \cdot 3^2$	1	2_1
2_2	x_2	324	$2^2 \cdot 3^2$	1	2_2
3_1	x_9	2	$2^3 \cdot 3^6$	3_1	1
3_2	x_7	8	$2 \cdot 3^6$	3_2	1
3_3	x_7x_9	16	3^6	3_3	1
3_4	x_5	24	$2 \cdot 3^5$	3_4	1
3_5	$x_5x_6x_7$	48	3^5	3_5	1
3_6	x_3	72	$2 \cdot 3^4$	3_6	1
3_7	x_3x_9	72	$2 \cdot 3^4$	3_7	1
3_8	x_5x_7	72	$2 \cdot 3^4$	3_8	1
3_9	x_5x_8	72	$2 \cdot 3^4$	3_9	1
3_{10}	$x_3^2x_9$	72	$2 \cdot 3^4$	3_{10}	1
3_{11}	x_3x_7	144	3^4	3_{11}	1
3_{12}	$x_3x_7x_9$	144	3^4	3_{12}	1
3_{13}	$x_3^2x_7x_9$	144	3^4	3_{13}	1
4	x_1	486	$2^3 \cdot 3$	2_1	4
6_1	$(x_1x_5)^2$	162	$2^3 \cdot 3^2$	3_1	2_1
6_2	x_2x_6	648	$2 \cdot 3^2$	3_4	2_2
6_3	x_2x_7	648	$2 \cdot 3^2$	3_2	2_2
6_4	$x_1^2x_3$	648	$2 \cdot 3^2$	3_6	2_1
6_5	$x_2x_6x_7$	648	$2 \cdot 3^2$	3_9	2_2
6_6	$x_1^2x_3x_5$	648	$2 \cdot 3^2$	3_7	2_1
6_7	$x_1x_3x_1x_5$	648	$2 \cdot 3^2$	3_{10}	2_1
6_8	$x_1x_5x_2x_8$	648	$2 \cdot 3^2$	3_8	2_2
8_1	x_2x_4	1458	2^3	4	8_1
8_2	$(x_2x_4)^5$	1458	2^3	4	8_2
9_1	x_3x_6	432	3^3	9_1	3_1
9_2	$x_3x_6x_7$	432	3^3	9_3	3_1
9_3	$(x_3x_6x_7)^2$	432	3^3	9_2	3_1
12	x_1x_5	972	$2^2 \cdot 3$	6_1	4

2.5.4 *Conjugacy classes of $N_G(d_1) = X_3 = \langle x_9, q_1, u, q, x_3, x_2, x_6, x_1\rangle$*

Class	*Representative*	|*Class*|	|*Centralizer*|	2P	3P
1	1	1	$2^4 \cdot 3^6$	1	1
2_a	$(x_1)^2$	81	$2^4 \cdot 3^2$	1	2_1
2_b	x_2	324	$2^2 \cdot 3^2$	1	2_2
3_a	x_9	2	$2^3 \cdot 3^6$	3_1	1
3_b	u	8	$2 \cdot 3^6$	3_2	1
3_c	$x_9 u$	16	3^6	3_3	1
3_d	q	24	$2 \cdot 3^5$	3_4	1
3_e	uq	48	3^5	3_5	1
3_f	q_1	72	$2 \cdot 3^4$	3_6	1
3_g	x_3	72	$2 \cdot 3^4$	3_7	1
3_h	$x_9 x_3$	72	$2 \cdot 3^4$	3_8	1
3_i	$x_9^2 x_3$	72	$2 \cdot 3^4$	3_9	1
3_j	$q_1^2 q$	72	$2 \cdot 3^4$	3_{10}	1
3_k	$q_1 x_3$	144	3^4	3_{11}	1
3_l	$x_9 q_1 x_3$	144	3^4	3_{12}	1
3_m	$q_1 q x_3$	144	3^4	3_{13}	1
4	x_1	486	$2^3 \cdot 3$	2_1	4
6_a	$(x_9 x_1)^2$	162	$2^3 \cdot 3^2$	3_1	2_1
6_b	$u x_2$	648	$2 \cdot 3^2$	3_2	2_2
6_c	$x_2 x_6$	648	$2 \cdot 3^2$	3_6	2_2
6_d	$u x_2 x_6$	648	$2 \cdot 3^2$	3_4	2_2
6_e	$q x_2 x_1$	648	$2 \cdot 3^2$	3_{10}	2_2
6_f	$x_3^2 x_1$	648	$2 \cdot 3^2$	3_7	2_1
6_g	$x_9 x_3^2 x_1$	648	$2 \cdot 3^2$	3_9	2_1
6_h	$x_9 x_3 x_1^2$	648	$2 \cdot 3^2$	3_8	2_1
8_a	$x_3 x_2 x_1$	1458	2^3	4	8_1
8_b	$(x_3 x_2 x_1)^5$	1458	2^3	4	8_2
9_a	$x_3 x_6$	432	3^3	9_1	3_3
9_b	$q_1 x_3 x_6$	432	3^3	9_3	3_3
9_c	$(q_1 x_3 x_6)^2$	432	3^3	9_2	3_3
12	$x_9 x_1$	972	$2^2 \cdot 3$	6_1	4

where $u = x_7 x_8$, $q = x_5^2 x_8^2 x_9^2$ *and* $q_1 = x_5 x_7 x_9$.

2.6 Character tables of local subgroups of $\mathsf{G}_2(3)$

2.6.1 *Character table of $N_G(3_A) \cong N_G(3_B)$*

	1a	2a	2b	3a	3b	3c	3d	3e	3f	3g	3h	3i	3j	3k	3l	3m	4a	6a	6b	6c	6d	6e	6f	6g	6h	8a	8b	9a	9b	9c	12a
2	4	4	2	3	1	.	1	.	1	1	1	1	1	.	.	.	3	3	1	1	1	1	1	1	1	3	3	.	.	.	2
3	6	2	2	6	6	6	5	5	4	4	4	4	4	4	4	4	1	2	2	2	2	2	2	2	2	.	.	3	3	3	1
2P	1a	1a	1a	3a	3b	3c	3d	3e	3f	3g	3h	3i	3j	3k	3l	3m	2a	3a	3h	3i	3j	3d	3f	3b	3g	4a	4a	9c	9b	9a	6a
3P	1a	2a	2b	1a	1a	1a	1a	1a	1a	1a	1a	1a	1a	1a	1a	1a	4a	2a	2a	2a	2a	2b	2b	2b	2b	8a	8b	3c	3c	3c	4a
5P	1a	2a	2b	3a	3b	3c	3d	3e	3f	3g	3h	3i	3j	3k	3l	3m	4a	6a	6b	6c	6d	6e	6f	6g	6h	8b	8a	9c	9b	9a	12a
7P	1a	2a	2b	3a	3b	3c	3d	3e	3f	3g	3h	3i	3j	3k	3l	3m	4a	6a	6b	6c	6d	6e	6f	6g	6h	8b	8a	9a	9b	9c	12a
11P	1a	2a	2b	3a	3b	3c	3d	3e	3f	3g	3h	3i	3j	3k	3l	3m	4a	6a	6b	6c	6d	6e	6f	6g	6h	8a	8b	9c	9b	9a	12a
X.1	1	1	1	1	1	1	1	1	1	1	1	1	1	1	1	1	1	1	1	1	1	1	1	1	1	1	1	1	1	1	1
X.2	1	1	−1	1	1	1	1	1	1	1	1	1	1	1	1	1	1	1	1	1	1	−1	−1	−1	−1	−1	−1	1	1	1	1
X.3	2	2	.	2	2	2	2	2	2	2	−1	−1	−1	−1	−1	−1	2	2	−1	−1	−1	.	.	.	.	.	.	−1	−1	−1	2
X.4	2	−2	.	2	2	2	2	2	2	2	−1	−1	−1	−1	−1	−1	.	−2	1	1	1	.	.	.	.	A	−A	−1	−1	−1	.
X.5	2	−2	.	2	2	2	2	2	2	2	−1	−1	−1	−1	−1	−1	.	−2	1	1	1	.	.	.	.	−A	A	−1	−1	−1	.
X.6	3	3	1	3	3	3	3	3	3	3	.	.	.	.	.	.	−1	3	.	.	.	1	1	1	1	−1	−1	.	.	.	−1
X.7	3	3	−1	3	3	3	3	3	3	3	.	.	.	.	.	.	−1	3	.	.	.	−1	−1	−1	−1	1	1	.	.	.	−1
X.8	4	−4	.	4	4	4	4	4	4	4	1	1	1	1	1	1	.	−4	−1	−1	−1	.	.	.	.	.	.	1	1	1	.
X.9	6	−2	.	−3	6	−3	.	.	.	.	−3	.	3	−3	.	3	2	1	1	−2	1	.	.	.	.	.	.	.	.	.	−1
X.10	6	−2	.	−3	6	−3	.	.	.	.	.	3	−3	.	3	−3	2	1	−2	1	1	.	.	.	.	.	.	.	.	.	−1
X.11	6	−2	.	−3	6	−3	.	.	.	.	3	−3	.	3	−3	.	2	1	1	1	−2	.	.	.	.	.	.	.	.	.	−1
X.12	8	.	−2	8	8	8	−1	−1	−1	−1	2	2	2	2	2	2	.	.	.	.	.	1	1	−2	1	.	.	−1	−1	−1	.
X.13	8	.	2	8	8	8	−1	−1	−1	−1	2	2	2	2	2	2	.	.	.	.	.	−1	−1	2	−1	.	.	−1	−1	−1	.
X.14	8	.	−2	8	−1	−1	5	−4	2	−1	2	2	2	−1	−1	−1	.	.	.	.	.	1	−2	1	1	.	.	−1	2	−1	.
X.15	8	.	2	8	−1	−1	5	−4	2	−1	2	2	2	−1	−1	−1	.	.	.	.	.	−1	2	−1	−1	.	.	−1	2	−1	.
X.16	12	4	.	−6	12	−6	.	.	.	.	.	−3	3	.	−3	3	.	−2	−2	1	1	.	.	.	.	.	.	.	.	.	.
X.17	12	4	.	−6	12	−6	.	.	.	.	3	.	−3	3	.	−3	.	−2	1	−2	1	.	.	.	.	.	.	.	.	.	.
X.18	12	4	.	−6	12	−6	.	.	.	.	−3	3	.	−3	3	.	.	−2	1	1	−2	.	.	.	.	.	.	.	.	.	.
X.19	16	.	.	16	16	16	−2	−2	−2	−2	−2	−2	−2	−2	−2	−2	.	.	.	.	.	.	.	.	.	.	.	1	1	1	.
X.20	16	.	.	16	−2	−2	10	−8	4	−2	−2	−2	−2	1	1	1	.	.	.	.	.	.	.	.	.	.	.	1	−2	1	.
X.21	16	.	.	16	−2	−2	−8	1	4	−2	4	4	4	−2	−2	−2	.	.	.	.	.	.	.	.	.	.	.	1	−2	1	.
X.22	16	.	.	16	−2	−2	−8	1	4	−2	−2	−2	−2	1	1	1	.	.	.	.	.	.	.	.	.	.	.	B	1	/B	.
X.23	16	.	.	16	−2	−2	−8	1	4	−2	−2	−2	−2	1	1	1	.	.	.	.	.	.	.	.	.	.	.	/B	1	B	.
X.24	18	−6	.	−9	18	−9	.	.	.	.	.	.	.	.	.	.	−2	3	.	.	.	.	.	.	.	.	.	.	.	.	1
X.25	24	.	−2	24	−3	−3	6	6	−3	−3	.	.	.	.	.	.	.	.	.	.	.	−2	1	1	1	.	.	.	.	.	.
X.26	24	.	2	24	−3	−3	6	6	−3	−3	.	.	.	.	.	.	.	.	.	.	.	2	−1	−1	−1	.	.	.	.	.	.
X.27	24	.	−2	24	−3	−3	−3	−3	−3	6	.	.	.	.	.	.	.	.	.	.	.	1	1	1	−2	.	.	.	.	.	.
X.28	24	.	2	24	−3	−3	−3	−3	−3	6	.	.	.	.	.	.	.	.	.	.	.	−1	−1	−1	2	.	.	.	.	.	.
X.29	48	.	.	−24	−6	3	.	.	.	.	6	−6	.	−3	3	.	.	.	.	.	.	.	.	.	.	.	.	.	.	.	.
X.30	48	.	.	−24	−6	3	.	.	.	.	−6	.	6	3	.	−3	.	.	.	.	.	.	.	.	.	.	.	.	.	.	.
X.31	48	.	.	−24	−6	3	.	.	.	.	.	6	−6	.	−3	3	.	.	.	.	.	.	.	.	.	.	.	.	.	.	.

where $A = i\sqrt{2}$, $B = -1/2(1 + 3i\sqrt{3})$.

2.6.2 *Character table of $N_G(3_D) \cong N_G(3_E)$*

	1a	2a	2b	2c	3a	3b	3c	3d	3e	3f	3g	3h	3i	3j	3k	3l	3m	3n	3o	3p	3q	3r	3s	3t	3u	3v	3w	6a	6b	6c	6d	6e	6f
2	2	2	2	2	1	1	1	1	1	1	.	.	.	.	.	.	.	.	.	.	.	.	.	.	.	.	.	1	1	1	1	1	1
3	4	2	1	1	4	4	4	4	4	4	4	4	4	4	4	4	4	4	4	4	4	4	4	4	4	4	4	2	2	2	2	1	1
2P	1a	1a	1a	1a	3a	3b	3c	3d	3e	3f	3j	3h	3i	3g	3k	3l	3n	3m	3o	3p	3q	3r	3t	3s	3u	3w	3v	3f	3a	3b	3c	3e	3d
3P	1a	2a	2b	2c	1a	1a	1a	1a	1a	1a	1a	1a	1a	1a	1a	1a	1a	1a	1a	1a	1a	1a	1a	1a	1a	1a	1a	2a	2a	2a	2a	2c	2b
5P	1a	2a	2b	2c	3a	3b	3c	3d	3e	3f	3j	3h	3i	3g	3k	3l	3n	3m	3o	3p	3q	3r	3t	3s	3u	3w	3v	6a	6b	6c	6d	6e	6f
X.1	1	1	1	1	1	1	1	1	1	1	1	1	1	1	1	1	1	1	1	1	1	1	1	1	1	1	1	1	1	1	1	1	1
X.2	1	−1	−1	1	1	1	1	1	1	1	1	1	1	1	1	1	1	1	1	1	1	1	1	1	1	1	1	−1	−1	−1	−1	1	−1
X.3	1	−1	1	−1	1	1	1	1	1	1	1	1	1	1	1	1	1	1	1	1	1	1	1	1	1	1	1	−1	−1	−1	−1	−1	1
X.4	1	1	−1	−1	1	1	1	1	1	1	1	1	1	1	1	1	1	1	1	1	1	1	1	1	1	1	1	1	1	1	1	−1	−1
X.5	2	2	.	.	−1	−1	−1	2	2	2	−1	−1	−1	−1	2	−1	2	2	−1	−1	−1	2	−1	−1	2	−1	−1	2	−1	−1	−1	.	.
X.6	2	−2	.	.	−1	−1	−1	2	2	2	−1	−1	−1	−1	2	−1	2	2	−1	−1	−1	2	−1	−1	2	−1	−1	−2	1	1	1	.	.
X.7	2	2	.	.	−1	−1	2	2	2	−1	2	−1	−1	2	2	−1	−1	−1	−1	2	2	−1	−1	−1	−1	−1	−1	−1	−1	−1	2	.	.
X.8	2	−2	.	.	−1	−1	2	2	2	−1	2	−1	−1	2	2	−1	−1	−1	−1	2	2	−1	−1	−1	−1	−1	−1	1	1	1	−2	.	.
X.9	2	.	.	2	2	2	2	2	−1	2	−1	2	−1	−1	−1	2	−1	−1	−1	2	−1	−1	−1	−1	2	−1	−1	.	.	.	.	−1	.
X.10	2	.	.	−2	2	2	2	2	−1	2	−1	2	−1	−1	−1	2	−1	−1	−1	2	−1	−1	−1	−1	2	−1	−1	.	.	.	.	1	.
X.11	2	.	2	.	2	2	2	−1	2	2	−1	−1	2	−1	−1	−1	−1	−1	2	−1	2	2	−1	−1	−1	−1	−1	.	.	.	.	.	−1
X.12	2	.	−2	.	2	2	2	−1	2	2	−1	−1	2	−1	−1	−1	−1	−1	2	−1	2	2	−1	−1	−1	−1	−1	.	.	.	.	.	1
X.13	2	2	.	.	−1	2	−1	2	2	−1	−1	2	2	−1	2	−1	−1	−1	−1	−1	−1	−1	2	2	−1	−1	−1	−1	−1	2	−1	.	.
X.14	2	−2	.	.	−1	2	−1	2	2	−1	−1	2	2	−1	2	−1	−1	−1	−1	−1	−1	−1	2	2	−1	−1	−1	1	1	−2	1	.	.
X.15	2	2	.	.	2	−1	−1	2	2	−1	−1	−1	−1	−1	2	2	−1	−1	2	−1	−1	−1	−1	−1	−1	2	2	−1	2	−1	−1	.	.
X.16	2	−2	.	.	2	−1	−1	2	2	−1	−1	−1	−1	−1	2	2	−1	−1	2	−1	−1	−1	−1	−1	−1	2	2	1	−2	1	1	.	.
X.17	4	.	.	.	−2	−2	4	−2	−2	−2	1	1	1	1	1	1	A	/A	1	−2	−2	1	/A	A	1	A	/A	.	.	.	.	.	.
X.18	4	.	.	.	−2	−2	4	−2	−2	−2	1	1	1	1	1	1	/A	A	1	−2	−2	1	A	/A	1	/A	A	.	.	.	.	.	.
X.19	4	.	.	.	−2	−2	4	4	−2	−2	−2	−2	1	−2	−2	−2	1	1	1	4	−2	1	1	1	−2	1	1	.	.	.	.	.	.
X.20	4	.	.	.	4	−2	−2	4	−2	−2	1	−2	1	1	−2	4	1	1	−2	−2	1	1	1	1	−2	−2	−2	.	.	.	.	.	.
X.21	4	.	.	.	−2	4	−2	4	−2	−2	1	4	−2	1	−2	−2	1	1	1	−2	1	1	−2	−2	−2	1	1	.	.	.	.	.	.
X.22	4	.	.	.	−2	−2	4	−2	4	−2	−2	1	−2	−2	−2	1	1	1	−2	−2	4	−2	1	1	1	1	1	.	.	.	.	.	.
X.23	4	.	.	.	4	−2	−2	−2	4	−2	1	1	−2	1	−2	−2	1	1	4	1	−2	−2	1	1	1	−2	−2	.	.	.	.	.	.
X.24	4	.	.	.	−2	4	−2	−2	4	−2	1	−2	4	1	−2	1	1	1	−2	1	−2	−2	−2	−2	1	1	1	.	.	.	.	.	.
X.25	4	.	.	.	−2	−2	−2	4	−2	4	1	−2	1	1	−2	−2	−2	−2	1	−2	1	−2	1	1	4	1	1	.	.	.	.	.	.
X.26	4	.	.	.	4	4	4	−2	−2	4	1	−2	−2	1	1	−2	1	1	−2	−2	−2	−2	1	1	−2	1	1	.	.	.	.	.	.
X.27	4	.	.	.	−2	−2	−2	−2	−2	4	A	1	1	/A	1	1	1	1	1	1	1	−2	A	/A	−2	A	/A	.	.	.	.	.	.
X.28	4	.	.	.	−2	−2	−2	−2	−2	4	/A	1	1	A	1	1	1	1	1	1	1	−2	/A	A	−2	/A	A	.	.	.	.	.	.
X.29	4	.	.	.	−2	−2	−2	−2	4	4	1	1	−2	1	−2	1	−2	−2	−2	1	−2	4	1	1	−2	1	1	.	.	.	.	.	.
X.30	4	.	.	.	−2	4	−2	−2	−2	−2	A	−2	−2	/A	1	1	A	/A	1	1	1	1	1	1	1	/A	A	.	.	.	.	.	.
X.31	4	.	.	.	−2	4	−2	−2	−2	−2	/A	−2	−2	A	1	1	/A	A	1	1	1	1	1	1	1	A	/A	.	.	.	.	.	.
X.32	4	.	.	.	4	−2	−2	−2	−2	−2	A	1	1	/A	1	−2	/A	A	−2	1	1	1	/A	A	1	1	1	.	.	.	.	.	.
X.33	4	.	.	.	4	−2	−2	−2	−2	−2	/A	1	1	A	1	−2	A	/A	−2	1	1	1	A	/A	1	1	1	.	.	.	.	.	.

where $A = -1/2(1 - 3i\sqrt{3})$.

2.6.3 *Character table of $N_G(7_A)$*

2	1	1	1	1	1	1	.
3	1	1	1	1	1	1	.
7	1	.	.	.	.	.	1
	1a	2a	3a	3b	6a	6b	7a
2P	1a	1a	3b	3a	3a	3b	7a
3P	1a	2a	1a	1a	2a	2a	7a
7P	1a	2a	3a	3b	6a	6b	1a
X.1	1	1	1	1	1	1	1
X.2	1	−1	1	1	−1	−1	1
X.3	1	−1	A	/A	−/A	−A	1
X.4	1	1	A	/A	/A	A	1
X.5	1	−1	/A	A	−A	−/A	1
X.6	1	1	/A	A	A	/A	1
X.7	6	.	.	.	.	.	−1

where $A = -(1/2)(1 + i\sqrt{3})$.

2.6.4 *Character table of $N_G(13_A) \cong N_G(13_B)$*

2	1	1	1	1	1	1	.	.
13	1	.	.	.	.	.	1	1
3	1	1	1	1	1	1	.	.
	1a	2a	3a	3b	6a	6b	13a	13b
2P	1a	1a	3b	3a	3a	3b	13b	13a
13P	1a	2a	3a	3b	6a	6b	1a	1a
3P	1a	2a	1a	1a	2a	2a	13a	13b
X.1	1	1	1	1	1	1	1	1
X.2	1	−1	1	1	−1	−1	1	1
X.3	1	−1	/A	A	−/A	−A	1	1
X.4	1	1	/A	A	/A	A	1	1
X.5	1	−1	A	/A	−A	−/A	1	1
X.6	1	1	A	/A	A	/A	1	1
X.7	6	.	.	.	.	.	B	C
X.8	6	.	.	.	.	.	C	B

where $A = -(1/2)(1+i\sqrt{3})$, $B = -(1/2)(1+\sqrt{13})$ *and* $C = -(1/2)(1-\sqrt{13})$.

3

Conway's simple group Co_3

In 1969 Conway [16] discovered three sporadic simple groups which he defined in terms of the automorphism group $A = Aut(\Lambda)$ of the 24-dimensional Leech lattice Λ; see also [17]. His results rest on ingenious combinatorial arguments. They do not generalize. Therefore Algorithm 1.3.8 is applied here and in Chapters 4 and 7 to present self-contained algebraic existence proofs for Conway's sporadic simple groups Co_3, Co_2 and Co_1, respectively. Furthermore, in this chapter we give a new proof for D. Fendel's uniqueness theorem of Co_3 [28] which depends on Feit's classification theorem [26] of the finite groups having a 23-dimensional irreducible integral lattice. According to Gorenstein [37], p. 121, Feit's article "is one of the deepest papers ever written in the representation theory of finite groups." Our uniqueness proof is independent of that paper.

A convenient definition of the Leech lattice is given in [13], p. 308. Let $\mathbb{Z}^n$ be a free abelian group of rank n. A positive definite quadratic form of degree n over $\mathbb{Z}$ is a map $q : \mathbb{Z}^n \to \mathbb{Z}$ defined by

$$q(x_1, \dots, x_n) = \sum_{i,j} a_{i,j} x_i x_j,$$

where $a_{i,j} = a_{j,i} \in \mathbb{Z}$ such that $q(x) \geq 0$ for all $x \in \mathbb{Z}^n$ and $q(x) = 0$ if and only if $x = 0 \in \mathbb{Z}^n$. The abelian group $\mathbb{Z}^n$, together with the map $q : \mathbb{Z}^n \to \mathbb{Z}$, is called a *lattice*.

The positive definite quadratic form q is called *unimodular* if $det(a_{i,j}) = 1$. It is *even* if $q(x)$ is an even integer for each $x \in \mathbb{Z}^n$. By [103] q is an even, unimodular, positive definite quadratic form if n is divisible by 8. In his thesis [101] H. V. Niemeyer showed that there is exactly one even, unimodular, positive definite quadratic form q_1 of degree $n = 24$ such that $q(x) \neq 2$ for all $x \in \mathbb{Z}^n$. According to Carter [13], p. 308, the corresponding lattice $\Lambda := (\mathbb{Z}^n, q_1)$ was discovered independently by Leech in [82]. Therefore the pair $\Lambda := (\mathbb{Z}^n, q_1)$ is called a *Leech lattice.*

In [16] Conway showed that the group $A = Aut(\Lambda)$ of all isometries of the Leech lattice Λ has a center $Z(A)$ of order 2 and that $\mathsf{Co}_1 = A/Z(A)$ is a finite simple group. Let $\Gamma = \{x \in \Lambda | q_1(x) = 4\}$ and $\Delta = \{x \in \Lambda | q_1(x) = 6\}$. Then A acts transitively on Γ and Δ. Fixing points $x \in \Gamma$ and $y \in \Delta$ he proved in [16] that the stabilizers A_x and A_y of x and y in A are finite simple groups denoted here by Co_2 and Co_3, respectively. The three simple groups Co_1, Co_2 and Co_3 are the sporadic *Conway* groups.

In Sections 3.1 and 3.2 we present the author's existence proof for Conway's group Co_3 published in [93]. We apply Algorithm 1.3.8 to $\mathrm{GL}_4(2)$ and construct a finitely presented group H with a central involution z such that $H \cong 2\mathsf{Sp}_6(2)$; see Proposition 3.1.2. Applying Algorithm 7.4.8 of [92] to H in Section 3.2 we obtain a simple subgroup $\mathfrak{G}$ of $\mathrm{GL}_{23}(23)$ having a 2-central involution $\mathfrak{z}$ such that $C_{\mathfrak{G}}(\mathfrak{z}) \cong H$; see Theorem 3.2.2. In [93] we verified that the character table of $\mathfrak{G}$ is equivalent to that of Co_3 given in [19], p. 154.

In Section 3.4 it is shown that any finite simple group G having a 2-central involution z such that $C_G(z) \cong H$ is isomorphic to $\mathfrak{G}$; see Theorem 3.3.13. In particular, $\mathsf{Co}_3 \cong \mathfrak{G}$ by Corollary 3.7 of [28] asserting that the centralizer $C_{\mathsf{Co}3}(z)$ of a 2-central involution z of Co_3 is isomorphic to the 2-fold cover $2\mathsf{Sp}_6(2)$ of the symplectic group $\mathsf{Sp}_6(2)$.

In [28] Fendel shows that each finite simple group G having a centralizer of a 2-central involution isomorphic to $C_{\mathsf{Co}3}(z)$ has a rational, absolutely irreducible representation of degree 23. Then he applies Feit's Theorem of [26] to get $G \cong \mathsf{Co}_3$. The author's uniqueness proof given in this chapter does not provide an isomorphism of G to the original Conway group Co_3. Since Co_3 is the automorphism group of an irreducible lattice Γ of degree 23 by [16], it is generated by finitely many integral matrices $M_k \in \mathrm{GL}_{23}(\mathbb{Z})$. Reducing their entries $a^k_{i,j}$ modulo the prime 23, the resulting matrices $m_k \in \mathrm{GL}_{23}(23)$ generate a simple group $\mathfrak{G}_1$ which is still isomorphic to Co_3. This trivial observation helps us to avoid a quotation of Feit's deep theorem in our uniqueness proof which shows that $G \cong \mathfrak{G}_1$.

Sections 3.5 and 3.6 contain the systems of representatives of the conjugacy classes and the character tables the local subgroups dealt with in this chapter. They are essential tools for the given existence and uniqueness proofs of Co_3. The four generating matrices of $\mathfrak{G}$ are documented on the accompanying DVD.

3.1 Construction of the involution centralizer

It is well known that the general linear groups $\Gamma = \mathrm{GL}_n(2)$ can be generated by all elementary matrices $e_{i,j}$. Furthermore, if $n \geq 3$ a special case of a classical theorem of R. Steinberg yields that Γ has the following set $\mathcal{R}(\Gamma)$ of defining relations:

$$\begin{aligned} &e_{i,j}^2 = 1 \quad \text{for all} \quad 1 \leq i, j \leq n, \quad i \neq j \\ &[e_{i,j}, e_{j,l}] = e_{i,l} \quad \text{for} \quad i \neq l, \\ &[e_{i,j}, e_{k,l}] = 1 \quad \text{for} \quad j \neq k \quad \text{and} \quad i \neq l; \end{aligned}$$

see [99], p. 40. In [23] Dennis and Stein reduced Steinberg's sets of defining relations of the classical groups substantially. In the case $n = 4$ dealt with here their presentation is used in statement (a) of the following result.

Lemma 3.1.1 *Let $T = \mathrm{GL}_4(2)$ and $F = GF(2)$. Let X be the set of the ten elementary matrices $\{e_{1,2}, e_{1,3}, e_{2,1}, e_{2,3}, e_{2,4}, e_{3,1}, e_{3,2}, e_{3,4}, e_{4,2}, e_{4,3}\}$. Then the following assertions hold.*

(a) *$T = \langle X \rangle$ has the following set $\mathcal{R}(T)$ of defining relations:*

$$\begin{aligned} &e_{1,2}^2 = e_{1,3}^2 = e_{2,1}^2 = e_{2,3}^2 = e_{2,4}^2 = e_{3,1}^2 = e_{3,2}^2 = e_{3,4}^2 = e_{4,2}^2 = 1, \\ &e_{4,3}^2 = (e_{1,2}, e_{1,3}) = (e_{1,2}, e_{3,2}) = (e_{1,2}, e_{3,4}) = (e_{1,2}, e_{4,2}) = 1, \\ &(e_{1,2}, e_{4,3}) = (e_{1,3}, e_{1,2}) = (e_{1,3}, e_{2,3}) = (e_{1,3}, e_{2,4}) = 1, \\ &(e_{1,3}, e_{4,2}) = (e_{1,3}, e_{4,3}) = (e_{2,1}, e_{2,3}) = (e_{2,1}, e_{2,4}) = 1, \\ &(e_{2,1}, e_{3,1}) = (e_{2,1}, e_{3,4}) = (e_{2,1}, e_{4,3}) = (e_{2,3}, e_{1,3}) = 1, \\ &(e_{2,3}, e_{2,1}) = (e_{2,3}, e_{2,4}) = (e_{2,3}, e_{4,3}) = (e_{2,4}, e_{1,3}) = 1, \\ &(e_{2,4}, e_{2,1}) = (e_{2,4}, e_{2,3}) = (e_{2,4}, e_{3,1}) = (e_{2,4}, e_{3,4}) = 1, \\ &(e_{3,1}, e_{2,1}) = (e_{3,1}, e_{2,4}) = (e_{3,1}, e_{3,2}) = (e_{3,1}, e_{3,4}) = 1, \\ &(e_{3,1}, e_{4,2}) = (e_{3,2}, e_{1,2}) = (e_{3,2}, e_{3,1}) = (e_{3,2}, e_{3,4}) = 1, \\ &(e_{3,2}, e_{4,2}) = (e_{3,4}, e_{1,2}) = (e_{3,4}, e_{2,1}) = (e_{3,4}, e_{2,4}) = 1, \\ &(e_{3,4}, e_{3,1}) = (e_{3,4}, e_{3,2}) = (e_{4,2}, e_{1,2}) = (e_{4,2}, e_{1,3}) = 1, \\ &(e_{4,2}, e_{3,1}) = (e_{4,2}, e_{3,2}) = (e_{4,2}, e_{4,3}) = (e_{4,3}, e_{1,2}) = 1, \\ &(e_{4,3}, e_{1,3}) = (e_{4,3}, e_{2,1}) = (e_{4,3}, e_{2,3}) = (e_{4,3}, e_{4,2}) = 1, \\ &(e_{1,2}, e_{2,3}) = e_{1,3}, \quad (e_{1,3}, e_{3,2}) = e_{1,2}, \quad (e_{2,1}, e_{1,3}) = e_{2,3}, \\ &(e_{2,3}, e_{3,1}) = e_{2,1}, \quad (e_{2,3}, e_{3,4}) = e_{2,4}, \quad (e_{2,4}, e_{4,3}) = e_{2,3}, \end{aligned}$$

$$(e_{3,1}, e_{1,2}) = e_{3,2}, \quad (e_{3,2}, e_{2,1}) = e_{3,1}, \quad (e_{3,2}, e_{2,4}) = e_{3,4},$$
$$(e_{3,4}, e_{4,2}) = e_{3,2}, \quad (e_{4,2}, e_{2,3}) = e_{4,3}, \quad (e_{4,3}, e_{3,2}) = e_{4,2},$$
$$(e_{1,2}e_{2,1}^{-1}e_{1,2})^4 = 1.$$

(b) *T has a faithful permutation representation PT of degree 15 with stabilizer $U = \langle e_{1,2}, e_{1,3}, e_{2,1}, e_{2,3}, e_{3,1}, e_{3,2}, e_{4,3}\rangle$.*

(c) *The simple group T has two non-isomorphic simple modules V_1 and V_2 of dimension 4 over $F = GF(2)$. They are dual to each other, and w.l.o.g we may assume that A_1 is isomorphic to the 4-dimensional F-vector space F^4, and that T acts on V_1 by matrix multiplication.*

(d) *The second irreducible representation V_2 of T is described by the transpose inverse matrices of the generating matrices $e_{i,j}$ of T defining V_1.*

(e) $dim_F[H^2(T, V_1)] = 1 = dim_F[H^2(T, V_2)]$.

(f) *The two non-split extensions E_1 and E_2 of $T = \mathrm{GL}_4(2)$ by V_1 and V_2 are isomorphic groups.*

(g) *Let E be E_1 and let the last four generators of E be generators of V_1. Then the first ten generators e_i of E are chosen so that they map onto the given generators $\{e_{1,2}, e_{1,3}, e_{2,1}, e_{2,3}, e_{2,4}, e_{3,1}, e_{3,2}, e_{3,4}, e_{4,2}, e_{4,3}\}$ of T in this order.*

The non-split extension $E = \langle e_i \mid 1 \le i \le 14\rangle$ of T by V_1 has the following set $\mathcal{R}(E)$ of defining relations:

$$e_1^2 = e_2^2 = e_5^2 = e_{11}^2 = e_{12}^2 = e_{13}^2 = e_{14}^2 = 1,$$
$$(e_r, e_s) = 1 \quad \textit{for} \quad 11 \le s < r \le 14,$$
$$e_1e_{11}e_1e_{11}e_{12} = e_2e_{11}e_2e_{11}e_{13} = e_3^{-1}e_{12}e_3e_{11}e_{12} = 1,$$
$$(e_1, e_{12}) = (e_1, e_{13}) = (e_1, e_{14}) = (e_2, e_{12}) = (e_2, e_{13}) = 1,$$
$$(e_2, e_{14}) = (e_3, e_{11}) = (e_3, e_{13}) = (e_3, e_{14}) = (e_4, e_{11}) = 1,$$
$$(e_4, e_{13}) = (e_4, e_{14}) = (e_5, e_{11}) = (e_5, e_{13}) = (e_5, e_{14}) = 1,$$
$$(e_6, e_{11}) = (e_6, e_{12}) = (e_6, e_{14}) = (e_7, e_{11}) = (e_7, e_{12}) = 1,$$
$$(e_7, e_{14}) = (e_8, e_{11}) = (e_8, e_{12}) = (e_8, e_{14}) = 1,$$
$$(e_9, e_{11}) = (e_9, e_{12}) = (e_9, e_{13}) = (e_{10}, e_{11}) = (e_{10}, e_{12}) = 1,$$
$$(e_{10}, e_{13}) = (e_1, e_7) = (e_1, e_9) = (e_2, e_4) = (e_2, e_{10}) = (e_3, e_4) = 1,$$

$$(e_3, e_{10}) = (e_4, e_2) = (e_4, e_3) = (e_4, e_{10}) = (e_5, e_8) = (e_6, e_7) = 1,$$
$$(e_6, e_8) = (e_7, e_1) = (e_7, e_6) = (e_7, e_8) = (e_7, e_9) = (e_8, e_5) = 1,$$
$$(e_8, e_6) = (e_8, e_7) = (e_9, e_1) = (e_9, e_7) = (e_9, e_{10}) = 1,$$
$$(e_{10}, e_2) = (e_{10}, e_3) = (e_{10}, e_4) = (e_{10}, e_9) = 1,$$
$$e_4^{-1}e_{12}e_4e_{12}e_{13} = e_5e_{12}e_5e_{12}e_{14} = e_6^{-1}e_{13}e_6e_{11}e_{13} = 1,$$
$$e_7^{-1}e_{13}e_7e_{12}e_{13} = e_8^{-1}e_{13}e_8e_{13}e_{14} = e_9^{-1}e_{14}e_9e_{12}e_{14} = 1,$$
$$e_{10}^{-1}e_{14}e_{10}e_{13}e_{14} = e_1e_2e_1e_2e_{12}e_{13} = 1,$$
$$e_1e_4^{-1}e_1e_4e_2e_{11}e_{12}e_{13}e_{14} = e_1e_8^{-1}e_1e_8e_{12}e_{14} = 1,$$
$$e_1e_{10}^{-1}e_1e_{10}e_{12} = e_2e_1e_2e_1e_{12}e_{13} = e_2e_5e_2e_5e_{13} = 1,$$
$$e_2e_7^{-1}e_2e_7e_1e_{11}e_{13}e_{14} = e_2e_9^{-1}e_2e_9e_{13} = e_3^{-1}e_2e_3e_2e_4^{-1}e_{11} = 1,$$
$$e_3^2e_{11} = e_4^2e_{13} = e_6^2e_{11} = e_7^2e_{12} = e_8^2e_{14} = e_9^2e_{12} = e_{10}^2e_{13} = 1,$$
$$e_3^{-1}e_5e_3e_5e_{11} = e_3^{-1}e_6^{-1}e_3e_6e_{11} = e_3^{-1}e_8^{-1}e_3e_8e_{14} = 1,$$
$$e_4^{-1}e_5e_4e_5e_{13} = e_4^{-1}e_6^{-1}e_4e_6e_3^{-1}e_{11}e_{14} = 1,$$
$$e_4^{-1}e_8^{-1}e_4e_8e_5e_{12}e_{13} = e_5e_2e_5e_2e_{13} = e_5e_3^{-1}e_5e_3e_{11} = 1,$$
$$e_5e_4^{-1}e_5e_4e_{13} = e_5e_6^{-1}e_5e_6e_{11}e_{14} = e_5e_{10}^{-1}e_5e_{10}e_4^{-1}e_{14} = 1,$$
$$e_6^{-1}e_1e_6e_1e_7^{-1}e_{11}e_{14} = e_6^{-1}e_3^{-1}e_6e_3e_{11} = e_6^{-1}e_5e_6e_5e_{11}e_{14} = 1,$$
$$e_6^{-1}e_9^{-1}e_6e_9e_{11} = e_7^{-1}e_3^{-1}e_7e_3e_6^{-1}e_{14} = 1,$$
$$e_7^{-1}e_5e_7e_5e_8^{-1}e_{12}e_{14} = e_8^{-1}e_1e_8e_1e_{12}e_{14} = e_8^{-1}e_3^{-1}e_8e_3e_{14} = 1,$$
$$e_8^{-1}e_9^{-1}e_8e_9e_7^{-1}e_{11}e_{12}e_{14} = e_9^{-1}e_2e_9e_2e_{13} = 1,$$
$$e_9^{-1}e_4^{-1}e_9e_4e_{10}^{-1}e_{12} = e_9^{-1}e_6^{-1}e_9e_6e_{11} = e_{10}^{-1}e_1e_{10}e_1e_{12} = 1,$$
$$e_{10}^{-1}e_7^{-1}e_{10}e_7e_9^{-1}e_{12}e_{13} = (e_1e_3^{-1}e_1)^4 = 1.$$

Proof By the introductory remarks we know that (a) is a presentation of the group T.

(b) As $T = \mathrm{GL}_4(2)$ is a matrix group we use the parabolic subgroup U of the statement as stabilizer in order to get a transitive permutation representation PT of T of degree $2^4 - 1$. The MAGMA command

```
h,PT:=CosetAction(T,sub<T|U>)
```

provides also the induced isomorphism h between T and PT.

(c) and (d) These assertions are obvious.

(e), (f) and (g) Let X denote the set of ten generating elementary matrices $e_{i,j}$ of T. Now the MAGMA command `FEalg:=MatrixAlgebra<F, 4|X>` constructs the algebra of all 4×4-matrices over F. MAGMA uses the isomorphism h to define on the natural 4-dimensional vector space

$V_1 = F^4$ a PT-module structure by means of its command `CQ:=GModule (PT,FEalg)`. Now Holt's algorithm [54] implemented in MAGMA can be applied. Thus the command `CohomologicalDimension(PT,CQ,2)` calculates the dimension of the second cohomology group $H^2(T, V_1)$. It is 1. Using the presentation of T given in (a), Holt's algorithm also delivers for each of the two 2-cocycles a presentation of the extension groups E_i, $i = 0, 1$, by means of the MAGMA command `P:=ExtensionProcess(PT,CQ, T)`. The presentation given in statement (g) is the non-split extension `E_1:=Extension(P,[1])` of T by V_1. The same commands applied to the dual module V_2 yield a similar presentation for the non-split extension E_2 of T by V_2. The isomorphism between the two extensions has been verified by means of MAGMA using permutation representations of E_1 and E_2 and the algorithms of Cannon and Holt described in [12]. □

Proposition 3.1.2 *Keep the notation of Lemma 3.1.1. Let* $E = \langle e_i \mid 1 \le i \le 14\rangle$ *be the non-split extension of* $T = \mathrm{GL}_4(2)$ *by its simple module* V_1 *of dimension* 4 *over* $F = GF(2)$. *Then the following statements hold.*

(a) *E has a faithful permutation representation of degree* 168 *with stabilizer* $\langle e_2 e_9 e_6^{-1} e_8 e_3, e_9 e_8^{-1} e_4 e_{10}\rangle$.

(b) $z = e_3^2$ *is a* 2*-central involution of* E *with centralizer* $D = C_E(z)$ *of order* $2^{10} \cdot 3 \cdot 7$. *Furthermore,* D *has a unique normal subgroup* Q *of order* 2^7. *It is extra-special with center* $Z(Q) = \langle z\rangle$ *and* Q *has a complement* $C \cong \mathrm{PSL}_3(2)$ *in* D.

(c) *Let* $x_1 = e_6^{e_{10}}$, $x_2 = e_5^{e_9}$, $c_1 = e_{12}e_3e_8e_6e_{14}e_9$ *and* $c_2 = e_8e_{10}e_{13}e_7e_3 e_5e_4$. *Then* $S = \langle e_3, e_7, e_8, x_1, x_2\rangle$ *is a Sylow* 2*-subgroup of* E *with elementary abelian normal subgroup* $A = \langle e_{11}, e_{12}, e_{13}, e_{14}\rangle$. *Furthermore,* $D = \langle S, e_4\rangle$ *and* $C = \langle c_1, c_2\rangle$.

(d) *The normal subgroup* Q *of* D *is generated by*

$$f_1 = e_8^2, \quad f_2 = x_1, \quad f_3 = e_7^2, \quad f_4 = e_3,$$
$$f_5 = x_2 e_3 x_2 \quad \text{and} \quad f_6 := e_7 e_3 e_7.$$

Let $V = Q/Z(Q)$ *and* $v_i = f_i Z(Q) \in V$, $1 \le i \le 6$. *Then* $\mathcal{B} = \{v_i | 1 \le i \le 6\}$ *is a basis of the elementary abelian group* V. *The conjugate actions of* c_1 *and* c_2 *on* Q *induce endomorphisms* *on V described by the following matrices with respect to the basis $\mathcal{B}$:*

$$Mc_1 = \begin{pmatrix} 1&1&1&0&0&1\\ 0&1&1&0&1&0\\ 0&0&1&0&0&0\\ 1&1&1&1&0&0\\ 0&1&0&0&1&1\\ 0&0&1&0&1&1 \end{pmatrix} \quad \text{and} \quad Mc_2 = \begin{pmatrix} 1&1&1&0&1&1\\ 0&0&1&0&1&1\\ 0&1&1&1&0&0\\ 0&1&0&0&0&0\\ 0&1&1&0&1&0\\ 1&0&1&0&0&1 \end{pmatrix}.$$

(e) *$E_1 = D/Z(Q) = \langle c_1, c_2, v_i | 1 \le i \le 6\rangle$ has the following set $\mathcal{R}(E_1)$ of defining relations:*

$$\begin{aligned}
&c_1^4 = c_2^4 = 1, \quad (c_1^{-1}c_2^{-1})^4 = 1,\\
&v_i^2 = 1 \quad \text{for} \quad 1 \le i \le 6, \quad [v_i, v_j] = 1 \quad \text{for} \quad 1 \le i, j \le 6,\\
&c_2c_1^{-1}c_2^{-1}c_1^{-2}c_2^{-1}c_1^{-1}c_2c_1^{-1} = c_1c_2^{-1}c_1c_2^{-1}c_1^{-1}c_2^2c_1^{-1}c_2^{-1} = 1,\\
&c_1^{-1}v_1c_1(v_1v_3v_4v_5)^{-1} = c_1^{-1}v_2c_1(v_1v_2v_3v_4v_6)^{-1} = 1,\\
&c_1^{-1}v_3c_1(v_3)^{-1} = c_1^{-1}v_4c_1(v_3v_4)^{-1} = c_1^{-1}v_5c_1(v_1v_5v_6)^{-1} = 1,\\
&c_1^{-1}v_6c_1(v_4v_5v_6)^{-1} = c_2^{-1}v_1c_2(v_1v_6)^{-1} = c_2^{-1}v_4c_2(v_3)^{-1} = 1,\\
&c_2^{-1}v_3c_2(v_1v_2v_3v_5v_6)^{-1} = c_2^{-1}v_2c_2(v_1v_3v_4v_5)^{-1} = 1,\\
&c_2^{-1}v_5c_2(v_1v_2v_5)^{-1} = c_2^{-1}v_6c_2(v_1v_2v_6)^{-1} = 1.
\end{aligned}$$

(f) *The element v_4 is a 2-central involution of E_1, and the Fitting subgroup V of $D_1 = C_{E_1}(v_4)$ has a complement $C_1 = \langle t_1, t_2, c_1^2\rangle \cong S_4$, where both $t_1 = c_1^2c_2c_1c_2c_1^{-1}c_2^{-1}c_1^{-1}c_2$ and $t_2 = c_1^{-1}c_2c_1c_2$ have order 3. The conjugate actions of t_1 and t_2 on V induce endomorphisms given by the following matrices with respect to the basis $\mathcal{B}_1 = \{v_4, v_1, v_2, v_3, v_5, v_6\}$:*

$$Mt_1 = \begin{pmatrix} 1&0&0&0&0&0\\ 0&1&1&1&0&0\\ 0&0&0&1&1&1\\ 0&1&0&1&1&1\\ 0&0&1&1&0&1\\ 0&1&1&1&1&1 \end{pmatrix} \quad \text{and} \quad Mt_2 = \begin{pmatrix} 1&1&0&1&1&1\\ 0&1&0&1&1&0\\ 0&0&0&1&1&1\\ 0&1&0&1&1&1\\ 0&0&1&1&0&1\\ 0&1&1&1&1&1 \end{pmatrix}.$$

(g) *Let $MC_1 = \langle Mt_1, Mt_2, (Mc_1)^2\rangle$. Then $C_{\mathrm{GL}_6(2)}(MC_1)$ is generated by the involution*

$$Ma = \begin{pmatrix} 1&0&1&0&1&0\\ 0&1&0&0&0&0\\ 0&0&1&0&0&0\\ 0&0&0&1&0&0\\ 0&0&0&0&1&0\\ 0&0&0&0&0&1 \end{pmatrix}.$$

(h) *The matrix*

$$Mw = \begin{pmatrix} 1&1&1&0&1&1\\ 1&0&1&0&1&0\\ 0&1&1&0&1&0\\ 0&0&1&0&1&0\\ 0&0&0&1&1&0\\ 0&1&0&0&1&1 \end{pmatrix}$$

of order 8 satisfies the following conditions:

(1) *the subgroup* $MP = \langle MC_1, Mc_2, Ma^{Mw} \rangle$ *of* $\mathrm{GL}_6(2)$ *has order* 1451520*;*

(2) $(Mt_1)^{Mw}, (Mt_2)^{Mw}$ *and* $(Mc_1^2)^{Mw}$ *belong to* MP*.*

(i) $Mt = Ma^{Mw}$ *is a* 2*-central involution of* MP*, and the Fitting subgroup* MF_1 *of* $MH_1 = C_{MP}(Mt)$ *is elementary abelian of order* 2^5*. It has a complement* MC_2 *isomorphic to the symmetric group* S_6*.*

(j) *A Sylow* 2*-subgroup* MT *of* MP *has a unique elementary abelian normal subgroup* MA *of order* 2^6 *and* $N_{MP}(MA) \cong E_1$*.*

(k) *The matrices* $Mp_1 = (Mt_1)^{Mw}$ *and* $Mp_2 = Mc_2Mt(Mc_2)^3 MtMc_2$*, of respective orders* 3 *and* 6*, generate* MH_1*.*

(l) $MP = \langle MH_1, Mc_2 \rangle$ *is a simple group isomorphic to the symplectic group* $\mathsf{Sp}_6(2)$*.*

(m) *The simple group* MP *is isomorphic to the finitely presented group* $P = \langle p_1, p_2, c_2 \rangle$ *having the following set* $\mathcal{R}(P)$ *of defining relations:*

$$
\begin{aligned}
&p_1^3 = p_2^6 = c_2^4 = 1,\\
&(p_2c_2p_2)^2 = (c_2^{-1}p_2)^3 = (p_2^{-1}, c_2^{-1})^2 = 1,\\
&p_1c_2^{-1}p_2p_1^{-1}p_2^{-1}p_1^{-1}p_2^{-1}c_2p_1p_2 = p_1^{-1}p_2c_2p_2^{-2}p_1p_2^2c_2^{-1}p_2^{-1},\\
&(p_1p_2^{-1})^6 = (p_1^{-1}p_2^{-2})^4 = p_1^{-1}c_2p_2^{-1}p_1p_2^{-2}p_1c_2p_2^{-1}p_1^{-1}p_2^2,\\
&(p_1^{-1}p_2^{-2}p_1p_2^{-1}p_1^{-1}p_2^{-1})^2 = 1,\\
&c_2p_2^{-2}p_1^{-1}c_2^{-1}p_1^{-1}p_2^{-1}p_1^{-1}c_2p_1c_2^2p_2p_1 = 1.
\end{aligned}
$$

(n) *The 2-fold cover* $H = \langle h_1, h_2, h_3, h_4 \rangle$ *of* P *has the following set* $\mathcal{R}(H)$ *of defining relations:*

$$
\begin{aligned}
&h_1^3 = h_4^2 = 1, \quad h_2^6 = h_3^4 = h_4,\\
&(h_1, h_4) = (h_2, h_4) = (h_3, h_4) = 1,\\
&h_2h_3h_2^2h_3h_2h_4^{-1} = (h_3^{-1}h_2)^3 = h_2h_3h_2^{-1}h_3^{-1}h_2h_3h_2^{-1}h_3^{-1}h_4^{-1} = 1,\\
&h_1h_3^{-1}h_2h_1^{-1}h_2^{-1}h_1^{-1}h_2^{-1}h_3h_1h_2 = h_1^{-1}h_2h_3h_2^{-2}h_1h_2^2h_3^{-1}h_2^{-1} = 1,\\
&(h_1h_2^{-1})^6 = h_1^{-1}h_2^{-2}h_1^{-1}h_2^{-2}h_1^{-1}h_2^{-2}h_1^{-1}h_2^{-2}h_4^{-1} = 1,\\
&h_1^{-1}h_3h_2^{-1}h_1h_2^{-2}h_1h_3h_2^{-1}h_1^{-1}h_2^2 = (h_1^{-1}h_2^{-2}h_1h_2^{-1}h_1^{-1}h_2^{-1})^2 = 1,\\
&h_3h_2^{-2}h_1^{-1}h_3^{-1}h_1^{-1}h_2^{-1}h_1^{-1}h_3h_1h_3^2h_2h_1h_4^{-1} = 1.
\end{aligned}
$$

(o) H *has a faithful permutation representation of degree* 1920 *with stabilizer* $\langle (h_4^2h_1^{-1})^2, (h_2h_1^{-1}k_4^{-1}h_3^{-1})^8 \rangle$*.*

(p) *The Sylow* 2*-subgroups of* H *and* E *are isomorphic.*

Proof (a) The faithful permutation representation PE of degree 168 of $E = \langle e_i \mid 1 \le i \le 14 \rangle$ has been found by means of the presentation of E given in Lemma 3.1.1 and MAGMA.

(b) Using the faithful permutation representation PE of E and MAGMA it has been checked that $z = e_3^2$ is a 2-central involution of E and $D = C_E(z)$ has order 2^{10}. Furthermore, an application of the MAGMA command

```
NormalSubgroups(PD)
```

yields that D has a unique normal subgroup Q of order 2^7. It is extra-special with its center $Z(Q) = \langle z \rangle$, and $V = Q/Z(Q)$ is elementary abelian. Applying then the MAGMA command `C:=HasComplemnet(D,Q)` it follows that Q has a complement C. An obvious isomorphism test with MAGMA shows that $C \cong \mathrm{PSL}_3(2)$.

(c) The five generators of the Sylow 2-subgroup $S = \langle e_3, e_7, e_8, x_1, x_2 \rangle$ have been constructed by means of MAGMA. As $A = \langle e_{11}, e_{12}, e_{13}, e_{14} \rangle$ is an elementary abelian normal subgroup of E contained in D it is also normal in S. Using now the MAGMA command

```
subs:=Subgroups(S : Al:=Normal, IsElementaryAbelian:=true)
```

we see that A is the maximal elementary abelian normal subgroup of S. Therefore condition (2) of Algorithm 1.3.8 is satisfied by E.

The generators of $D = \langle S, e_4 \rangle$ and $C = \langle c_1, c_2 \rangle$ given in the statement have been determined by means of PE and MAGMA.

(d) Similarly it has been verified that $Q = \langle f_i \mid 1 \le i \le 6 \rangle$. Since Q is extra-special of order 2^7 the set $\mathcal{B} = \{v_i = f_i Z(Q) \mid 1 \le i \le 6\}$ is a basis of the elementary abelian group $V = Q/Z(Q)$. Let $F = GF(2)$. Since Q is normal in D the conjugate actions of the generators c_i of C induce endomorphisms of the F-vector space V described with respect to $\mathcal{B}$ by the matrices Mc_i, $i = 1, 2$, as has been established by means of MAGMA.

(e) Using the faithful permutation representation PE of E, and the MAGMA command `FPGroup(C)` to the permutation subgroup of PE generated by the permutations of the 2 generators of $C = \langle c_1, c_2 \rangle$, we get the following set $\mathcal{R}C$ of defining relations of C:

$$
\begin{aligned}
& c_1^4 = c_2^4 = 1, \\
& (c_1^{-1} c_2^{-1})^4 = c_2 c_1^{-1} c_2^{-1} c_1^{-2} c_2^{-1} c_1^{-1} c_2 c_1^{-1} = 1, \\
& c_1 c_2^{-1} c_1 c_2^{-1} c_1^{-1} c_2^2 c_1^{-1} c_2^{-1} = 1.
\end{aligned}
$$

Since $Q = \langle f_i \mid 1 \le i \le 6\rangle$ is extra-special with center $Z(Q) = \langle z\rangle$ and D is the semidirect product of Q and C, the actions of the generators c_j of C on the f_i can be explicitly determined by means of the matrices Mc_i and a MAGMA calculation in the permutation representation PE of E. It follows that $D = \langle c_1, c_2, f_1, f_2, \ldots, f_6\rangle$ has a set $\mathcal{R}D$ of defining relations consisting of $\mathcal{R}C$ and the following relations:

$$\begin{aligned}
&f_3^2 = f_4^2 = f_6^2 = z^2 = 1, \quad f_1^2 = f_2^2 = f_5^2 = z,\\
&c_1^{-1}f_1c_1(f_1f_3f_4f_5)^{-1} = c_1^{-1}f_2c_1(f_1f_2f_3f_4f_6)^{-1} = z,\\
&c_1^{-1}f_3c_1(f_3)^{-1} = c_1^{-1}f_5c_1(f_1f_5f_6)^{-1} = c_1^{-1}f_6c_1(f_4f_5f_6)^{-1} = z,\\
&c_1^{-1}f_4c_1(f_3f_4)^{-1} = c_2^{-1}f_1c_2(f_1f_6)^{-1} = 1,\\
&c_2^{-1}f_2c_2(f_1f_3f_4f_5)^{-1} = c_2^{-1}f_3c_2(f_1f_2f_3f_5f_6)^{-1} = 1,\\
&c_2^{-1}f_4c_2(f_3)^{-1} = c_2^{-1}f_5c_2(f_1f_2f_5)^{-1} = c_2^{-1}f_6c_2(f_1f_2f_6)^{-1} = z,\\
&(f_1, f_2) = (f_1, f_3) = (f_1, f_5) = (f_2, f_4) = (f_2, f_6) = z,\\
&(f_1, f_4) = (f_1, f_6) = (f_2, f_3) = (f_2, f_5) = (f_3, f_4) = (f_3, f_5) = 1,\\
&(f_3, f_6) = (f_4, f_5) = z, \quad \text{and} \quad (f_4, f_6) = (f_5, f_6) = 1.
\end{aligned}$$

Thus $E_1 = D/Z(Q) = V : C$ has the presentation given in the statement. Furthermore, it has a faithful permutation representation PE_1 of degree 64 with stabilizer$C$.

(f) We now apply Algorithm 1.3.8 to the extension E_1 of the F-module V by $C \cong PSL_3(2)$. Using MAGMA it has been checked that V is a uniserial indecomposable FC-module with two non-isomorphic composition factors having both dimension 3 over F. By Proposition 8.6.5 of [92] we know that E_1 is isomorphic to $N_G(A)$, where A is a maximal elementary abelian normal subgroup of a Sylow 2-subgroup of the simple symplectic group $G = \mathsf{Sp}_6(2)$. Therefore we use the conjugate action of C on V to get an embedding of C into $\mathrm{GL}_6(2)$. Using the permutation representation PE_1 of E_1 and MAGMA we see that $C_{E_1}(v_4)$ is a semidirect product of V and that the subgroup $C_1 = \langle t_1, t_2, c_1^2\rangle \cong S_4$, where $t_1 = c_1^2c_2c_1c_2c_1^{-1}c_2^{-1}c_1^{-1}c_2$ and $t_2 = c_1^{-1}c_2c_1c_2$ both have order 3. Furthermore, with respect to $\mathcal{B}_1$ the conjugate actions of t_1 and t_2 on V are described by the matrices Mt_1 and Mt_2 in $\mathrm{GL}_6(2)$, respectively.

(g) Let $MC_1 = \langle Mt_1, Mt_2, (Mc_1)^2\rangle$. Another MAGMA calculation now shows that $C_{\mathrm{GL}_6(2)}(MC_1)$ is generated by the matrix Ma of order 2 given in the statement.

(h) In order to be able to do the following calculation, finding the matrix Mw of the statement, we use the faithful permutation representation $PG62$ of $\mathrm{GL}_6(2)$ with degree 63. Its stabilizer MU is generated by the five elementary matrices $M_{2,1}$, $M_{3,1}$, $M_{4,1}$, $M_{5,1}$, $M_{6,1}$ of $\mathrm{GL}_6(2)$ and the following two matrices:

$$Mg_1 = \begin{pmatrix} 1&0&0&0&0&0\\ 0&1&1&0&0&0\\ 0&0&1&0&0&0\\ 0&0&0&1&0&0\\ 0&0&0&0&1&0\\ 0&0&0&0&0&1 \end{pmatrix} \quad \text{and} \quad Mg_2 = \begin{pmatrix} 1&0&0&0&0&0\\ 0&0&0&0&0&1\\ 0&1&0&0&0&0\\ 0&0&1&0&0&0\\ 0&0&0&1&0&0\\ 0&0&0&0&1&0 \end{pmatrix}.$$

Let $pt_1, pt_2, pc_1, pc_2, pa$ be the respective images of the matrices Mt_1, Mt_2, Mc_1, Mc_2, Ma under the isomorphism $h : \mathrm{GL}_6(2) \to PG62$. Let $PC_1 = \langle pt_1, pt_2, pc_1^2 \rangle$. For each $x \in PG62$ let $W(x) = \langle PC_1, pc_2, pa^x \rangle$, $U(x) = \langle pt_1^x, pt_2^x, (pc_1^2)^x \rangle$ and $D(x) = U(x) \cap W(x)$. Using then the MAGMA command

```
exists(w){x: x in PG62|Order(W(x)) eq 1451520 and D(x) eq  U(x)}
```

we obtain a permutation pw of order 8 in $PG62$ such that $h(Mw) = pw$, where Mw is the matrix of the statement. It satisfies both conditions (1) and (2) of the statement by the definition of w.

(i) All assertions of this statement have been verified by means of the faithful permutation representation $PG62$ and MAGMA.

(j) Let $MP = \langle MC_1, Mc_2, Ma^{Mw} \rangle$. Let $PP = h(MP)$ be the image of the matrix group MP in $PG62$, and let $pt = h(Mt)$, where $Mt = Ma^{Mw}$. Let PT be any Sylow 2-subgroup of $PH_1 = C_{PP}(pt)$. Using the MAGMA command

```
subs:=Subgroups(PT : Al:=Normal, IsElementaryAbelian:=true)
```

we see that PT has a unique maximal elementary abelian normal subgroup PA of order 2^6. Furthermore, an isomorphism test shows that $N_{PP}(PA) \cong E_1$. Since h is an isomorphism the statement holds.

(k) Again this assertion has been verified by means of $PG62$ and MAGMA. The elements $Mp_1 = (Mt_1)^{Mw}$ and $Mp_2 = Mc_2 Mt (Mc_2)^3$ $Mt Mc_2$ generate MH_1 and have orders 3 and 6, respectively.

(l) Another application of MAGMA yields that $PP = \langle PH_1, pc_2 \rangle$ is a simple group isomorphic to $\mathsf{Sp}_6(2)$; see also Proposition 8.6.5 of [92].

(m) Since $MP = \langle Mp_1, Mp_2, Mc_2 \rangle \cong PP$ under the isomorphism h we can apply the MAGMA command `FPGroup(PP)` to the permutation group $PP = \langle pp_1, pp_2, pc_2 \rangle$ of degree 63. This way we obtain the presentation of MP given in the statement.

(n) By (m) the finitely presented group $P = \langle p_1, p_2, c_2 \rangle$ satisfies all conditions of Holt's Algorithm 7.4.5 of [92] implemented in MAGMA [52] and the trivial matrix representation F of PP over F. It follows by means of MAGMA that the second cohomological dimension $dim_F[H^2(H_1, F)] = 1$. Using then the extension process of MAGMA we found the presentation of the unique 2-fold cover H of P with center $Z(H) = \langle h_4 \rangle$ given in (n).

(o) This permutation representation PH of H has been established by means of MAGMA and a stand alone program due to P. Young. Using the Todd–Coxeter Algorithm implemented in MAGMA, statement (o) can easily be checked.

(p) Using now the faithful permutation representations PE and PH of the groups E and H defined in (a) and (o) and MAGMA it can be verified that H and E have isomorphic Sylow 2-subgroups. This completes the proof. □

In order to relate the results of this chapter to Conway's original sporadic simple group Co_3 of his paper [16], we now state Fendel's Corollary 3.7 of [28] without proof.

Remark 3.1.3 (Fendel) Conway's original sporadic simple group Co_3 has a 2-central involution z such that $C_{\mathsf{Co}_3}(z)/\langle z \rangle \cong \mathsf{Sp}_6(2)$.

In view of Proposition 3.1.2 and Remark 3.1.3 we introduce the following definition which is essential for our existence and uniqueness proof of Co_3 given in the following sections.

Definition 3.1.4 A finite simple group G is called to be of Co_3*-type* if it has a 2-central involution z such that $C_G(z)$ is isomorphic to the finitely presented group H stated in Proposition 3.1.2.

3.2 Construction of a simple group of Co_3-type

In this section the main condition of Algorithm 1.3.8 is verified for the amalgam $H \leftarrow D \rightarrow E$ constructed in Sections 3.2 and 3.3. Therefore its Steps 6 to 11 can be applied to give here a self-contained existence proof for a simple group of Co_3-type.

Lemma 3.2.1 *Let $H = H(\mathsf{Co}_3) = \langle h_i | 1 \le i \le 4 \rangle$ be the finitely presented group constructed in Proposition 3.1.2. Let $E = \langle e_i \mid 1 \le i \le 14 \rangle$ be the non-split extension of $T = \mathrm{GL}_4(2)$ by its simple module V_1 of dimension 4 over $F = GF(2)$ constructed in Lemma 3.1.1. Then the following statements hold.*

(a) $H = \langle x, y, h\rangle$, *where* $x = h_3$ *and* $y = h_9h_{14}h_{15}$ *and* $h = h_1$ *have respective orders* 8, 8 *and* 2.

(b) $S = \langle r_i \mid 1 \le i \le 4\rangle$ *is a Sylow* 2*-subgroup of* H *of order* 2^{10}, *where* $r_1 = x^2$, $r_2 = y$, $r_3 = xyx^2y^3x$ *and* $r_4 = xyxy^2(xy)^3x$ *have respective orders* 4, 8, 2 *and* 8.

(c) S *has two maximal elementary abelian normal subgroups* $A = \langle z, v, c, a\rangle$ *and* $B = \langle z, v, c, b\rangle$ *of order* 2^4 *with normalizers* $N_H(A)$ *and* $N_H(B)$ *of respective orders* $2^{10}\cdot 3\cdot 7$ *and* $2^{10}\cdot 3$, *where* $z = x^2$, $v = (x^2y)^4$, $c = r_1r_2^3(r_3r_2)^2r_1r_2^3$, $a = r_1r_2r_1r_4r_2r_4$ *and* $b = r_2^2r_1r_2r_3r_2$.

(d) $V = \langle z, v\rangle$ *is the unique normal Klein* 4*-subgroup of* S *and it is equal to the center* $Z(C_H(V))$ *of the centralizer of* V *in* H.

(e) $C = A \cap B = \langle z, v, c\rangle$ *is the unique maximal elementary abelian characteristic subgroup of* S.

(f) $D = N_H(A) = \langle x, y\rangle$.

(g) $D \cong D_1 = C_E(z)$, *where* $E = \langle e_i \mid 1 \le i \le 14\rangle$ *is the non-split extension of* $T = \mathrm{GL}_4(2)$ *by its simple module* V_1 *of dimension* 4 *over* $F = GF(2)$ *constructed in Lemma 3.1.1 and* $z = e_3^2 \in E$.

(h) *The Goldschmidt index of the amalgam* $H \leftarrow D \rightarrow E$ *is* 1.

(i) *A system of representatives* r_i *of the* 43 *conjugacy classes of* H *and the corresponding centralizer orders* $|C_H(r_i)|$ *are given in Table 3.4.1.*

(j) *A system of representatives* d_i *of the* 30 *conjugacy classes of* D *and the corresponding centralizer orders* $|C_D(d_i)|$ *are given in Table 3.4.3.*

(k) *Let* $\sigma : N_H(A) \rightarrow D_1 = C_E(z)$ *be the isomorphism given in (g). Then there is an element* $e = e_1 \in E$ *of order* 2 *such that* $E = \langle x_1 = \sigma(x), y_1 = \sigma(y), e\rangle$.

A system of representatives e_i *of the* 25 *conjugacy classes of* E *and the corresponding centralizer orders* $|C_E(e_i)|$ *are given in Table 3.4.2.*

(l) *The character tables of* H, D *and* E *are stated in Tables 3.5.3, 3.5.2 and 3.5.1, respectively.*

Proof By Proposition 3.1.2(o) the group H has faithful permutation representations PH of degree 1920. It has been used to perform the following calculations with MAGMA. In particular, these tools suffice to check all the statements on H.

(a) The given words for the generators x, y and h of H have been found using PH, MAGMA and H. Kim's program `GetShortGens(PH, PH)`.

(b) another calculation with MAGMA shows that the order of the subgroup $D = \langle x, y \rangle$ is $2^{10} \cdot 3 \cdot 7$. Thus it contains a Sylow 2-subgroup S. The one given in the statement was obtained by means of MAGMA. Its generators were calculated by means of H. Kim's program `GetShortGens(PH, PS)`, where PS is the image of S in PH.

(c) The elementary abelian normal subgroups of S were found by means of the MAGMA command

```
subs:=Subgroups(PS: Al:=Normal, IsElementaryAbelian:=true).
```

It follows that S has exactly four non-cyclic elementary abelian normal subgroups. They are A and B of order 16, C of order 8 and V of order 4. Their generators have been calculated by means of Algorithm 1.4.1 adapted now to the new subgroups of S.

The assertions (d), (e) and (f) can now easily checked using PH and MAGMA.

(g) By Proposition 3.1.2, $z = e_3^2$ is a 2-central involution of E with centralizer $D_1 = C_E(z)$ of order $2^{10} \cdot 3 \cdot 7$. Furthermore, E has a faithful permutation representation PE of degree 168. The isomorphism $\sigma : PD \to D_1$ has been found by means of an isomorphism test in MAGMA using the faithful permutation representations PE of E and PH of H.

(h) The Goldschmidt index has been calculated by means of Kratzer's Algorithm 7.1.10 of [92].

(i) and (j) The systems of representatives of the conjugacy classes of H and D have been computed by means of PH, MAGMA and Kratzer's Algorithm 5.3.18 of [92].

(k) Let $x_1 = \sigma(x)$ and $y_1 = \sigma(y)$ in $C_E(z_1)$, where σ is the isomorphism mentioned in (g). An application of MAGMA and the permutation representation PE of E shows that $E = \langle x_1, y_1, e \rangle$ for some element e of E of order 2. The system of representatives of the conjugacy classes of E has been computed using PE, MAGMA and Kratzer's Algorithm 5.3.18 of [92].

The character tables of (l) have been calculated by using PH, PE and MAGMA. □

Theorem 3.2.2 (Michler) *Keep the notation of Lemma 3.2.1 and Proposition 3.1.2. Using the notation of the three character Tables 3.5.3, 3.5.2 and 3.5.1 of the groups H, D and E, respectively, the following statements hold.*

(a) *There is exactly one compatible pair* $(\chi, \tau) \in mf\,char_{\mathbb{C}}(H) \times mf\,char_{\mathbb{C}}(E)$ *of degree* 23 *of the groups* $H = \langle D, h\rangle$ *and* $E = \langle D, e\rangle$:

$$(\chi, \tau) = (\chi_{\mathbf{3}} + \chi_4, \tau_1 + \tau_2 + \tau_{\mathbf{4}})$$

with common restriction

$$\tau_{|D} = \chi_{|D} = \psi_1 + \psi_6 + \psi_8 + \psi_{\mathbf{10}},$$

where irreducible characters with bold face indices denote faithful irreducible characters.

(b) *Let* $\mathfrak{V}$ *and* $\mathfrak{W}$ *be the up to isomorphism uniquely determined faithful semi-simple multiplicity-free 23-dimensional modules of* H *and* E *over* $F = GF(23)$ *corresponding to the compatible pair* χ, τ, *respectively.*

Let $\kappa_{\mathfrak{V}} : H \to \mathrm{GL}_{23}(23)$ *and* $\kappa_{\mathfrak{W}} : E \to \mathrm{GL}_{23}(23)$ *be the representations of* H *and* E *afforded by the modules* $\mathfrak{V}$ *and* $\mathfrak{W}$, *respectively.*

Let $\mathfrak{h} = \kappa_{\mathfrak{V}}(h)$, $\mathfrak{x} = \kappa_{\mathfrak{V}}(x)$, $\mathfrak{y} = \kappa_{\mathfrak{V}}(y)$ *in* $\kappa_{\mathfrak{V}}(H) \le \mathrm{GL}_{23}(23)$. *Then the following assertions hold.*

(1) $\mathfrak{V}_{|D} \cong \mathfrak{W}_{|D}$, *and there is a transformation matrix* $\mathcal{T} \in \mathrm{GL}_{23}(23)$ *such that*

$$\mathfrak{x} = \mathcal{T}^{-1}\kappa_{\mathfrak{W}}(x_1)\mathcal{T}, \mathfrak{y} = \mathcal{T}^{-1}\kappa_{\mathfrak{W}}(y_1)\mathcal{T}.$$

Let $\mathfrak{e} = \mathcal{T}^{-1}\kappa_{\mathfrak{W}}(e_1)\mathcal{T} \in \mathrm{GL}_{23}(23)$.

(2) $\mathfrak{G} = \langle \mathfrak{h}, \mathfrak{x}, \mathfrak{y}, \mathfrak{e}\rangle$ *has a faithful permutation representation of degree* 170775 *with stabilizer* $\mathfrak{E} = \langle \mathfrak{x}, \mathfrak{y}, \mathfrak{e}\rangle$.

(3) *The four generating matrices of* $\mathfrak{G}$ *are stored in MAGMA format on the accompanying DVD.*

(4) $\mathfrak{G} = \langle \mathfrak{h}, \mathfrak{x}, \mathfrak{y}, \mathfrak{e}\rangle$, *and* $\mathfrak{G}$ *has* 42 *conjugacy classes* $\mathfrak{g}_i^{\mathfrak{G}}$ *with representatives* $\mathfrak{g}_i$ *and centralizer orders* $|C_{\mathfrak{G}}(\mathfrak{g}_i)|$ *as given in Table* `DVD.1.1.1` *on the accompanying DVD.*

(5) *The character table of* $\mathfrak{G}$ *given in Table* `DVD.1.1.2` *coincides with that of* Co_3 *in the Atlas [19], p. 154.*

(c) $\mathfrak{G}$ *is a finite simple group with 2-central involution* $\kappa_{\mathfrak{V}}(z) = (\mathfrak{x})^4$ *such that*

$$C_{\mathfrak{G}}(\kappa_{\mathfrak{V}}(z)) = \kappa_{\mathfrak{V}}(H), \text{ and } |\mathfrak{G}| = 2^{10} \cdot 3^7 \cdot 5^3 \cdot 7 \cdot 11 \cdot 23.$$

(d) *The simple group* $\mathfrak{G}$ *is isomorphic to Conway's simple group* Co_3.

Proof Throughout we keep the notation of Lemma 3.2.1.

(a) The character tables of the groups $H = \langle x, y, h\rangle$, $D = \langle x, y\rangle$, $D_1 = \langle x_1, y_1\rangle$ and E are stated in Section 3.5. In the following we use their notations. Using MAGMA and these character tables and the fusion of the classes of D in H and D_1 in E the compatible pair stated in assertion (a) has been calculated by means of MAGMA and Kratzer's Algorithm 7.3.10 of [92].

(b) In order to construct the semi-simple faithful representation $\mathfrak{V}$ corresponding to the character $\chi = \chi_3 + \chi_4$ we use the faithful permutation representation PH of H of degree 1920 constructed in Proposition 3.1.2. Since both irreducible characters χ_3, χ_4 of H are not contained in the permutation character of PH the MAGMA command `LowIndexSubgroups(PH, 1000)` was applied. MAGMA provided generating permutations of 34 subgroups U_i of H having index $|H : U_i| \leq 1000$. Furthermore it calculated all their character tables and the fusion between the classes of each U_i and H. Thus it was easy to calculate the inner products $((1_{U_i})^H, \chi_k)$ for all $\{1 \leq i \leq 34\}$ and $k = 3, 4$. It follows that χ_3 and χ_4 are constituents of the permutation characters PU_i of the subgroups U_{10} and U_3 of indices 240 and 72 for $i = 10$ and $i = 3$, respectively. As $H = \langle x, y, h\rangle$ MAGMA provided the permutation matrices $P_i(x)$, $P_i(y)$ and $P_i(h)$ for $i = 10, 3$. Applying the Meataxe algorithm of MAGMA over the field $GF(23)$ one obtains from PU_{10} the 8×8-matrices $M(x)$, $M(y)$ and $M(h)$ corresponding to the faithful irreducible character χ_3. Similarly, PU_3 yields the three 15×15-matrices $N(x), N(y), N(h) \in \mathrm{GL}_{23}(23)$ corresponding to χ_4. Hence $\mathfrak{H} = \langle \mathfrak{x}, \mathfrak{y}, \mathfrak{h}\rangle$, where the generators of $\mathfrak{H}$ are the blocked diagonal matrices $diag(N(x), M(x))$, $diag(N(y), M(y))$, $diag(N(h), M(h))$, respectively. These three matrices describe the semi-simple H-module $\mathfrak{V}$ over $F = GF(23)$.

The faithful semi-simple representation $\mathfrak{W}$ corresponding to the character $\tau = \tau_1 + \tau_2 + \tau_4$ of $E = \langle x_1, y_1, e_1\rangle$ has been obtained similarly from the permutation module PE of degree 168, because all three characters τ_1, τ_2 and τ_4 are contained in the permutation character of PE. Hence we were able to calculate for the three generators x_1, y_1 and e_1 of E their blocked diagonal matrices $\mathfrak{W}(x_1)$, $\mathfrak{W}(y_1)$ and $\mathfrak{W}(e_1)$. They describe the semi-simple FE-module $\mathfrak{W}$.

The matrix groups $\mathfrak{D} = \langle \mathfrak{x}, \mathfrak{y}\rangle$ and $\mathfrak{D}_1 = \langle \mathfrak{W}(x_1), \mathfrak{W}(y_1)\rangle$ are isomorphic, and so are the semi-simple representations $\mathfrak{V}_{|\mathfrak{D}}$ and $\mathfrak{W}_{|\mathfrak{D}_1}$ by the definition of the compatible pair $(\chi, \tau) \in mf char_{\mathbb{C}}(H) \times mf char_{\mathbb{C}}(E)$

and Maschke's Theorem, which is applicable because the orders of H and E are not divisible by 23.

Let $Y = \mathrm{GL}(23, 23) = \mathrm{GL}_{23}(23)$. Applying then Parker's isomorphism test of Proposition 6.1.6 of [92] by means of the MAGMA command

```
IsIsomorphic(GModule(sub<Y|V(x),V(y)>),GModule(sub<Y|W(x1),W(y1)>))
```

we obtain the transformation matrix

$$\mathcal{T} = \left(\begin{array}{ccccccccccccccccccccccc}
\cdot&\cdot&\cdot&\cdot&\cdot&\cdot&17&\cdot&\cdot&\cdot&\cdot&\cdot&\cdot&\cdot&\cdot&\cdot&\cdot&\cdot&\cdot&\cdot&\cdot&\cdot&\cdot\\
14&9&14&14&9&14&\cdot&9&14&\cdot&14&\cdot&\cdot&9&\cdot&\cdot&\cdot&\cdot&\cdot&\cdot&\cdot&\cdot&\cdot\\
\cdot&9&\cdot&14&\cdot&\cdot&\cdot&9&\cdot&\cdot&14&14&\cdot&\cdot&9&\cdot&\cdot&\cdot&\cdot&\cdot&\cdot&\cdot&\cdot\\
14&9&14&\cdot&9&\cdot&\cdot&9&14&14&\cdot&14&9&\cdot&\cdot&\cdot&\cdot&\cdot&\cdot&\cdot&\cdot&\cdot&\cdot\\
\cdot&9&\cdot&\cdot&\cdot&14&\cdot&9&\cdot&14&\cdot&\cdot&9&9&9&\cdot&\cdot&\cdot&\cdot&\cdot&\cdot&\cdot&\cdot\\
14&\cdot&14&\cdot&9&14&\cdot&\cdot&14&\cdot&\cdot&14&\cdot&9&9&\cdot&\cdot&\cdot&\cdot&\cdot&\cdot&\cdot&\cdot\\
14&\cdot&14&14&9&\cdot&\cdot&\cdot&14&14&14&\cdot&9&\cdot&9&\cdot&\cdot&\cdot&\cdot&\cdot&\cdot&\cdot&\cdot\\
\cdot&\cdot&\cdot&14&\cdot&14&5&\cdot&\cdot&14&14&14&9&9&\cdot&\cdot&\cdot&\cdot&\cdot&\cdot&\cdot&\cdot&\cdot\\
14&20&3&20&20&9&\cdot&20&3&9&20&20&3&3&9&\cdot&\cdot&\cdot&14&14&\cdot&\cdot&\cdot\\
20&20&20&3&3&20&\cdot&20&9&3&3&20&20&3&3&\cdot&\cdot&\cdot&\cdot&\cdot&\cdot&14&9\\
3&14&14&3&20&20&\cdot&3&3&20&14&3&3&3&3&9&\cdot&14&14&9&9&\cdot&\cdot\\
3&14&14&9&20&14&\cdot&3&3&3&20&20&9&20&3&\cdot&\cdot&14&\cdot&9&\cdot&\cdot&9\\
20&14&20&3&14&14&\cdot&3&20&20&3&9&3&20&20&\cdot&\cdot&\cdot&\cdot&14&\cdot&\cdot&9\\
20&3&20&14&14&3&\cdot&14&20&20&3&20&3&20&9&\cdot&\cdot&\cdot&\cdot&9&\cdot&\cdot&9\\
20&20&20&20&14&14&\cdot&20&20&20&9&3&3&20&3&\cdot&14&14&14&\cdot&\cdot&9&14\\
3&20&14&3&20&14&\cdot&20&3&14&3&3&20&20&20&\cdot&9&14&14&\cdot&9&14&\cdot\\
3&3&14&20&20&3&\cdot&14&3&14&20&20&20&20&3&14&14&\cdot&\cdot&14&14&\cdot&\cdot\\
3&20&14&3&20&3&\cdot&20&3&3&14&14&9&20&9&14&\cdot&\cdot&14&\cdot&\cdot&9&14\\
20&20&20&3&14&20&\cdot&20&20&3&3&20&9&14&3&9&9&14&14&9&18&\cdot&\cdot\\
3&20&14&20&20&20&\cdot&20&3&20&20&20&3&3&9&\cdot&\cdot&\cdot&\cdot&\cdot&9&9&\cdot\\
20&3&20&20&3&9&\cdot&14&9&3&20&14&20&3&9&14&\cdot&18&9&14&14&\cdot&\cdot\\
3&3&14&3&20&20&\cdot&14&3&20&3&14&3&14&3&\cdot&9&\cdot&14&14&\cdot&\cdot&9\\
14&3&3&14&20&9&\cdot&14&3&20&3&3&14&3&3&\cdot&14&\cdot&\cdot&\cdot&14&9&14
\end{array}\right).$$

satisfying $\mathfrak{V}(x) = (\mathfrak{W}(x_1))^{\mathcal{T}}$ and $\mathfrak{V}(y) = (\mathfrak{W}(y_1))^{\mathcal{T}}$.

Let $\mathfrak{e} = (\mathfrak{W}(e_1))^{\mathcal{T}}$. Then $\mathfrak{E} = \langle \mathfrak{x}, \mathfrak{y}, \mathfrak{e} \rangle \cong E$.

Let $\mathfrak{G} = \langle \mathfrak{h}, \mathfrak{x}, \mathfrak{y}, \mathfrak{e} \rangle$. Using the MAGMA command `CosetAction(G, E)` we obtained a faithful permutation representation PG of $\mathfrak{G}$ of degree 170775 with stabilizer $\mathfrak{E}$. In particular $\mathfrak{G} = 2^{10} \cdot 3^7 \cdot 5^3 \cdot 7 \cdot 11 \cdot 23$.

Using the faithful permutation PG of degree 170775 and Kratzer's Algorithm 5.3.18 of [92] the representatives of all the conjugacy classes of $\mathfrak{G}$ were calculated; see Table `DVD.1.1.1`.

Furthermore, the character table of $\mathfrak{G}$ has been calculated by means of the above permutation representation and MAGMA. It is stated in Table `DVD.1.1.2`. It coincides with that of Co_3 in [19], p. 135.

(c) Let $\mathfrak{z} = (\mathfrak{x})^4$. Then $C_{\mathfrak{G}}(\mathfrak{z})$ contains $\mathfrak{H} = \langle \mathfrak{h}, \mathfrak{x}, \mathfrak{y} \rangle$ which is isomorphic to H.

The character table of $\mathfrak{G}$ implies that $\mathfrak{G}$ is a simple group and that $C_{\mathfrak{G}}(\mathfrak{z}) \cong \mathfrak{H}$.

(d) This assertion follows from (c), Remark 3.1.3 and Fendel's Theorem 3.3.13. This completes the proof. $\square$

3.3 Uniqueness proof

In this section we prove that each finite simple group G of Co_3-type is isomorphic to the simple matrix group $\mathfrak{G}$ of Theorem 3.2.2. Thus we obtain a new self-contained proof for Fendel's Theorem [28].

Throughout this section G denotes any finite simple group G of Co_3-*type*. Thus it has a 2-central involution z such that $C_G(z) = H$ constructed in Proposition 3.1.2. All results stated in Lemma 3.2.1 are now used to show first that any simple group G of Co_3-*type* has two conjugacy classes of involutions and the same order as the matrix group $\mathfrak{G}$ constructed in Theorem 3.2.2.

Lemma 3.3.1 *Let G be a finite simple group of Co_3-type having a 2-central involution z with centralizer $H = C_G(z) = \langle x, y, h\rangle$ described in Lemma 3.2.1. Let $z = x^4$ and $u = (x^2hy)^3$. Then the following statements hold.*

(a) *$H = \langle x, y, h\rangle$ has three conjugacy classes of involutions represented by $z = x^4$, $v = (x^2y)^4$ and $u = (x^2hy)^3$.*
(b) *The centralizer $T = C_H(u)$ of the involution u is a direct product $\langle u\rangle \times X$, where $|X| = 2^6 \cdot 3$ and any Sylow 2-subgroup Y of X has center $Z(Y) = \langle z\rangle = Z(X)$.*
(c) *z and u are not conjugate in G and represent all conjugacy classes of involutions of G.*
(d) *$X \cong C_{\mathsf{M}_{12}}(f)$ for some 2-central involution f of the Mathieu group M_{12}.*

Proof (a) This assertion follows immediately from Lemma 3.2.1 and Table 3.4.1.

(b) $T = C_H(u)$ has order $2^7 \cdot 3$ by Table 3.4.1. Using the faithful permutation representation of H of degree 1920 constructed in Proposition 3.1.2(o) and the MAGMA command `X:=HasComplement(T,sub<T|u>)` one sees that $T = C_H(u) = \langle u\rangle \times X$ and $|X| = 2^6 \cdot 3$. Furthermore, any Sylow 2-subgroup Y of X has a center $Z(Y) = \langle z\rangle$.

(c) Table 3.4.1 states that H has three conjugacy classes of involutions represented by z, $v = (x^2y)^4$ and u. Since G is assumed to be simple, Theorem 4.7.3 of [92] implies that G has at most two conjugacy classes of involutions.

Suppose that u and z are conjugate in G. Then there is a Sylow 2-subgroup S of $U = C_G(u)$ such that $Y < S$. In particular, Y is a proper normal subgroup of index 2 in some subgroup W of S. Thus

$Z(Y) = \langle z \rangle \lhd W$. Hence (b) implies that $W \leq C_G(z) \cap C_G(u) = C_H(u) = T$, a contradiction.

(d) By (b) z generates the center $Z(X)$ of the direct factor X of $T = C_H(u)$. Proposition 8.1.6 of [92] states a presentation of the $C_{\mathsf{M}_{12}}(f)$ of a 2-central involution f of the Mathieu group M_{12} determined in Theorem 8.1.9(c) of [92]. Using it and the faithful permutation representation of H an isomorphism test with MAGMA verifies that $X \cong C_{\mathsf{M}_{12}}(f)$. □

Lemma 3.3.2 *Let G be any finite simple group of Co_3-type having a 2-central involution z with centralizer $H = C_G(z) = \langle x, y, h \rangle$ stated in Lemma 3.2.1. Then the following statements hold.*

(a) *$V = \langle z, v \rangle$ is the unique normal Klein 4-group in the Sylow 2-subgroup $S = \langle r_1, r_2, r_3, r_4 \rangle$ of H, where $z = x^4$, $v = (x^2y)^4$ and where $r_1 = x^2$, $r_2 = y$, $r_3 = xyx^2y^3x$ and $r_4 = xyxy^2(xy)^3x$ have respective orders 4, 8, 2 and 8. Furthermore, V is the center of $C_H(V)$.*

(b) *The normalizer $N = N_G(V)$ is larger than $N_H(V)$, $C_G(V) = C_H(V)$ and $N/C_G(V) \cong \mathrm{GL}_2(2)$.*

(c) *$N = N_G(V)$ is isomorphic to $N = \langle n_i \mid 1 \leq i \leq 13 \rangle$ satisfying the following set $\mathcal{R}(N)$ of defining relations:*

$$n_1^4 = n_2^4 = n_3^2 = n_4^4 = n_5^2 = n_6^2 = n_7^3 = n_8^3 = n_9^2 = 1,$$
$$n_{10}^3 = n_{11}^2 = n_{12}^2 = n_{13}^2 = 1,$$
$$(n_{12}, n_{13}) = (n_1, n_{12}) = (n_1, n_{13}) = (n_2, n_{12}) = (n_2, n_{13}) = 1,$$
$$(n_3, n_{12}) = (n_3, n_{13}) = (n_4, n_{12}) = (n_4, n_{13}) = (n_5, n_{12}) = 1,$$
$$(n_5, n_{13}) = (n_6, n_{12}) = (n_6, n_{13}) = (n_7, n_8) = (n_7, n_{12}) = 1,$$
$$(n_7, n_{13}) = (n_8, n_{12}) = (n_8, n_{13}) = (n_9, n_{12}) = (n_9, n_{13}) = 1,$$
$$n_{10}^{-1} n_{12} n_{10} n_{13} = n_{10}^{-1} n_{13} n_{10} n_{12} n_{13} = n_{11}^{-1} n_{12} n_{11} n_{13} = 1,$$
$$n_{11}^{-1} n_{13} n_{11} n_{12} = n_1^2 n_{12} = n_2^2 n_{12} = n_4^2 n_{13} = n_1 n_2 n_1 n_2 n_{12} = 1,$$
$$n_1 n_3 n_1 n_3 n_{12} n_{13} = n_2 n_3 n_2 n_3 n_{13} = (n_1 n_4)^2 = (n_2 n_4)^2 = 1,$$
$$(n_3 n_4)^2 = n_1 n_5 n_1 n_5 n_{13} = (n_2 n_5)^2 = (n_3 n_5)^2 = 1,$$
$$n_4 n_5 n_4 n_5 n_{12} n_{13} = n_1 n_6 n_1 n_6 n_{13} = n_2 n_6 n_2 n_6 n_{12} n_{13} = 1,$$
$$(n_3 n_6)^2 = (n_4 n_6)^2 = n_5 n_6 n_5 n_6 n_{12} n_{13} = 1,$$
$$(n_6 n_9)^2 = (n_7^{-1} n_9)^2 = (n_8^{-1} n_9)^2 = n_6 n_{10}^{-1} n_6 n_{10} n_{12} n_{13} = 1,$$
$$n_7^{-1} n_{10}^{-1} n_8 n_{10} = n_9 n_{10}^{-1} n_9 n_{10} = n_6 n_{11} n_6 n_{11} n_{12} n_{13} = 1,$$
$$n_7^{-1} n_{11} n_7 n_{11} = (n_9 n_{11})^2 = (n_{10}^{-1} n_{11})^2 = n_{10} n_3 n_1 n_{10}^{-1} n_4 n_{12} = 1,$$

$$n_8^{-1}n_5n_1n_8n_4n_{12} = n_7n_6n_1n_7^{-1}n_4 = n_{10}n_6n_1n_{10}^{-1}n_2 = 1,$$
$$n_{11}n_6n_1n_{11}n_1n_{12}n_{13} = n_7^{-1}n_3n_2n_7n_4n_{13} = 1,$$
$$n_8^{-1}n_3n_2n_1n_8n_5n_{12} = n_{10}n_3n_2n_1n_{10}^{-1}n_5n_{13} = 1,$$
$$n_7^{-1}n_5n_2n_7n_3n_{13} = n_{10}^{-1}n_8^{-1}n_7^{-1}n_{10}n_8^{-1} = 1.$$

(d) *S is a Sylow 2-subgroup of G and $A = \langle z, v, c, a\rangle$ is a maximal elementary abelian normal subgroup such that $D = N_H(A)$ has order $2^{10} \cdot 3^2 \cdot 7$, $C_G(A) = C_H(A) = A$, and there is an element d of order 3 in $N = N_G(V)$ such that $N_G(A) = \langle N_H(A), d\rangle$ and $N_G(A)/A \cong \mathrm{GL}_4(2)$.*

(e) *$N_G(A) \cong E$, where E is the finitely presented group of Lemma 3.1.1.*

Proof (a) This assertion holds by Lemma 3.2.1.

(b) By Lemma 3.3.1 the involutions z and v of H are conjugate in G. Since $C_G(z) = H$ has center $Z(H) = \langle z\rangle$ and $V = \langle z, v\rangle$ it follows that $C_G(V) = C_H(V) = C_H(v)$ has order $2^9 \cdot 3^2$ by Table 3.4.1. Hence $Q = C_S(v)$ has index 2 in the Sylow 2-subgroup S of H given in statement (a). Furthermore, $V = Z(Q)$ by (a). As v is conjugate to z we also know that Q is properly contained in a Sylow 2-subgroup of $C_G(v)$. Thus Lemma 1.4.4 of [92] implies that $N = N_G(V) > N_H(V)$ and that $N/C_G(V) \cong \mathrm{GL}_2(2)$.

(c) In order to construct the normalizer $N = N_G(V)$ of the Klein 4-group V we first construct N/V as a subgroup of the automorphism group of $C = C_G(V)$. As $V = \langle z, v\rangle$ is the center of C we know from Table 3.4.1 that $|C/V| = 2^7 \cdot 3^2$. Let pC be the image of C in the faithful permutation representation PH of H given in Proposition 3.1.2(o). Using the MAGMA commands `AC:=AutomorphismGroup(pC)` and

```
h, pAC:=PermutationRepresentation(AC)
```

one obtains a permutation group pAC of order $2^{10} \cdot 3^3$ of degree 288. MAGMA provides 15 generators g_i of AC and therefore 15 generators $pg_i = g_i@h$ of $pAC = \langle pg_i \mid 1 \le i \le 15\rangle$. It has a normal subgroup pI of order $2^7 \cdot 3^2$ corresponding to the inner automorphisms induced by the elements of C. In the permutation group $R = pAC$ applications of MAGMA work well. Thus we see that the Fitting subgroup pFI of pI is elementary abelian of order 2^6. After choosing six generators pf_j of pFI the MAGMA command `pKT:=HasComplement(pR,pFI)` provides a complement pKT of pFI in pR of order $2^4 \cdot 3^3$. Using MAGMA again it follows that pKT has a normal extra-special Sylow 3-subgroup pS_3 of order 27 and exponent 3. Also pFI has complement pKI in pI of order 18 with elementary abelian normal Sylow 3-subgroup pIS_3 of order 9. Choosing its generators pd_1 and pd_2 another application of MAGMA yields a complement $\langle pt \rangle$ of order 2 in pKI such that $pd_k^{pt} = pd_k^2$ for $k = 1, 2$. Hence $C/V \cong pI = \langle pf_j, pd_1, pd_2, pt \mid 1 \le j \le 6 \rangle$.

Furthermore, $pIS_3 = \langle pd_1, pd_2 \rangle$ is a normal subgroup of pS_3 of index 3. Using MAGMA again one gets an element $pp_3 \in pS_3$ not contained in pIS_3 such that $pd_1^{pd_3} = pd_2^2 pd_1^2$ and $pd_2^{pd_3} = pd_1$. Since $[N_G(V)/V]/[C/V] \cong \mathrm{GL}_2(2) \cong Sym(3)$ by (b) we searched for an element pr of order 2 in pKT inverting pd_3. MAGMA's search was successful. Furthermore, it confirmed that the group

$$pM = \langle pf_1, pf_2, pf_3, pf_4, pf_5, pf_6, pd_1, pd_2, pt, pd_3, pr \rangle$$

has order $2^8 \cdot 3^3$. Applying then the MAGMA command `FPGroup(pM)` we obtained the finitely presented group $M = \langle m_i | 1 \le i \le 11 \rangle$ with the following set of defining relations:

$$\begin{aligned}
&m_1^2 = m_2^2 = m_3^2 = m_4^2 = m_5^2 = m_6^2 = m_7^3 = m_8^3 = m_9^2 = m_{10}^3 = m_{11}^2 = 1,\\
&(m_1m_2)^2 = (m_1m_3)^2 = (m_2m_3)^2 = (m_1m_4)^2 = (m_2m_4)^2 = 1,\\
&(m_3m_4)^2 = (m_1m_5)^2 = (m_2m_5)^2 = (m_3m_5)^2 = (m_4m_5)^2 = 1,\\
&(m_1m_6)^2 = (m_2m_6)^2 = (m_3m_6)^2 = (m_4m_6)^2 = (m_5m_6)^2 = 1,\\
&(m_6m_9)^2 = (m_7^{-1}m_9)^2 = (m_8^{-1}m_9)^2 = m_6m_{10}^{-1}m_6m_{10} = 1,\\
&m_7^{-1}m_{10}^{-1}m_8m_{10} = m_9m_{10}^{-1}m_9m_{10} = m_7^{-1}m_{11}m_7m_{11} = 1,\\
&(m_6m_{11})^2 = (m_9m_{11})^2 = (m_{10}^{-1}m_{11})^2 = 1, \quad (m_7, m_8) = 1,\\
&m_{10}m_3m_1m_{10}^{-1}m_4 = m_8^{-1}m_5m_1m_8m_4 = m_7m_6m_1m_7^{-1}m_4 = 1,\\
&m_{10}m_6m_1m_{10}^{-1}m_2 = m_{11}m_6m_1m_{11}m_1 = m_7^{-1}m_3m_2m_7m_4 = 1,\\
&m_8^{-1}m_3m_2m_1m_8m_5 = m_{10}m_3m_2m_1m_{10}^{-1}m_5 = 1,\\
&m_7^{-1}m_5m_2m_7m_3 = m_{10}^{-1}m_8^{-1}m_7^{-1}m_{10}m_8^{-1} = 1.
\end{aligned}$$

Since $V = Z(C)$ we now have to extend the group M by V such that the extension group N has a Sylow 2-subgroup which is isomorphic to S of H. Fortunately, MAGMA provides the images $aut_{d_3}(z) = v$, $aut_r(z) = v$ and $aut_{d_3}(v) = zv$, $aut_r(v) = z$ of the two generators z and v of V, respectively, where $aut_{d_3} = pd_3@@h$ and $aut_r = pr@@h$. Hence m_{10} and m_{11} are represented in $\mathrm{GL}_2(2)$ with respect to their actions on

V by the matrices

$$M_{10} = \begin{pmatrix} 0 & 1 \\ 1 & 1 \end{pmatrix} \quad \text{and} \quad M_{11} = \begin{pmatrix} 0 & 1 \\ 1 & 0 \end{pmatrix}.$$

In order to apply Holt's Algorithm 7.4.5 of [92] implemented in MAGMA the following MAGMA commands are introduced. Let 1 be the 2×2 identity matrix over $F = GF(2)$. Let

```
FEalg:=MatrixAlgebra<F,2|1,1,1,1,1,1,1,1,1,M_{10},M_{11}>,
CM :=GModule(pM, FEalg).
```

Applying the MAGMA command

```
CohomologicalDimension(pM, CM, 2)
```

we obtained the second cohomological dimension $dim_F[H^2(M,V)] = 1$. Since V does not have a complement in H, only the non-split extension N_1 can be isomorphic to $N_G(V)$. Using the commands

`P :=ExtensionProcess(pM, CM, M)` and `N_a := Extension(P, [a])`,

where $a \in F$, MAGMA returned the presentation of $N_1 = \langle n_k \mid 1 \leq k \leq 13\rangle$ of the unique non-split extension N_1 of M by V. It is stated in the assertion. It implies that $U = \langle n_7, n_8, n_{10}\rangle$ is an extra-special 3-group of order 27 and exponent 3. Another application of MAGMA shows that U is a stabilizer of a faithful permutation representation PN_1 of N_1 and that $|N_1| = 2^{10} \cdot 3^3$. Using PM_1, PH and MAGMA it has been verified that the Sylow 2-subgroups S of H and those of N_1 are isomorphic.

Let $V_1 = \langle n_{12}, n_{13}\rangle$. Then another isomorphism test with MAGMA shows that $N_2 = \langle n_k | 1 \leq k \leq 13 \quad \text{and} \quad k \neq 10\rangle = N_{N_1}(V_1)$ is isomorphic to the normalizer $N_H(V)$ which has order $2^{10} \cdot 3^2$. Hence $N_G(V) \cong N_1$ by Theorem 1.4.15 of [92].

(d) By Lemma 3.2.1(c) $A = \langle z, v, c, a\rangle$ is a maximal elementary abelian normal subgroup of the Sylow 2-subgroup S of H such that $N_H(A)$ has order $2^{10} \cdot 3 \cdot 7$. Clearly S is a Sylow 2-subgroup of G because S has center $Z(S) = \langle z\rangle$ by Table 3.4.1 and z is a 2-central involution of G. Using the faithful permutation representation PH of H and MAGMA it has been checked that $|N_{N_H(V)}(A)| = 2^{10} \cdot 3$.

Using the faithful permutation representation PN_1 of $N_1 = \langle n_k \mid 1 \leq k \leq 13\rangle \cong N_G(V)$ and MAGMA one can check that $S_1 = \langle n_2, n_4, n_6, n_9, n_{11}\rangle$ is a Sylow 2-subgroup of N_1. It has a maximal elementary abelian normal subgroup $A_1 = \langle n_4^2, n_2^2, (n_9 n_4 n_2)^2, n_9 n_4 n_9 n_2\rangle$ of order 16 such that $N_{N_1}(A_1)$ has order $2^{10} \cdot 3^2$. Furthermore, the normalizer $N_{N_2}(A_1)$

of A_1 in $N_2 = \langle n_k | 1 \le k \le 13 \text{ and } k \ne 10\rangle \cong N_H(V)$ has order $2^{10} \cdot 3$. Hence there is an element d of order 3 in $N_G(A)$ not contained in $N_H(A)$.

Another application of the permutation representation PH of H and MAGMA yields that $C_H(A) = A$. Hence $C_G(A) = A$ because $z \in A$. Thus $N_G(A)/A$ is isomorphic to a subgroup U of $\mathrm{GL}_4(2)$ of order $|U| \ge 2^6 \cdot 3^2 \cdot 7$. Thus $U = \mathrm{GL}_4(2)$.

(e) Let $T = N_H(A)$. Then A is also a self-centralizing normal subgroup of T. Using PH and the MAGMA command `HasComplement(T,A)` it follows that T is a non-split extension of T/A by A. As $|N_G(A) : T| = 3$ Theorem 1.4.15 of [92] and (d) imply that $N_G(A)$ is a non-split extension of $\mathrm{GL}_4(2)$ by A. Hence Lemma 3.1.1 asserts that $N_G(A) \cong E$, where E is the finitely presented group with set $\mathcal{R}(E)$ of defining relations given there. This completes the proof. □

Proposition 3.3.3 *Let G be a finite simple group of Co_3-type with a 2-central involution z such that $C_G(z) = H = \langle x, y, h\rangle$ described in Lemma 3.2.1. Then the following statements hold.*

(a) *G has two conjugacy classes of involutions represented by $z = x^4$ and $u = (x^2hy)^3$.*
(b) *G has 20 z-special conjugacy classes represented by the classes of z, 4_1, 4_2, 6_1, 6_2, 6_3, 8_1, 8_2, 8_3, 10_1, 12_1, 12_2, 12_3, 14_1, 18_1, 20_1, 20_2, 24_1, 24_2 and 30_1 in H given in Table 3.4.1.*
(c) *$U = C_G(u) = \langle u\rangle \times Q$, where $Q \cong \mathsf{M}_{12}$.*
(d) *U has a faithful permutation representation PU of degree 14. It has 30 conjugacy classes. Its character table is stated in Table 3.5.4.*
(e) *U has five conjugacy classes of involutions represented by u, z, uz, s and us, where $C_U(s) = \langle u\rangle \times \langle s\rangle \times S_5$.*
(f) *$(uz)^G = s^G = (us)^G = u^G$ and $u^G \ne z^G$.*
(g) *G has six u-special conjugacy classes of even order denoted by u, 6_4, 6_5, 10_2, 22_1 and 22_2. They correspond to the conjugacy classes of odd order in $U/\langle u\rangle \cong \mathsf{M}_{12}$ classified in Theorem 8.1.9(c) of [92].*
(h) *$|G| = 2^{10} \cdot 3^7 \cdot 5^3 \cdot 7 \cdot 11 \cdot 23$.*

Proof (a) This assertion is a restatement of Lemma 3.3.1(c).

(b) This statement follows at once from the power map information given in Table 3.4.1.

(c) Lemma 3.3.1 states that the centralizer $T = C_H(u)$ of the involution u in $H = C_G(z)$ is a direct product $\langle u \rangle \times X$ such that the subgroup X has a center $Z(X) = \langle z \rangle$. Furthermore, X is isomorphic to the centralizer $C_{\mathsf{M}_{12}(f)}$ of a 2-central involution f of the Mathieu group M_{12}. In order to show that $C_G(u) \cong \mathsf{M}_{12}$ we need the following notation and results.

Let PH be the faithful permutation representation of H constructed in Proposition 3.1.2(o). Let $S = \langle r_1, r_2, r_3, r_4 \rangle$ be the Sylow 2-subgroup of H whose generators r_i are defined in Lemma 3.2.1. Let $V = \langle z, v \rangle$ be the unique normal Klein 4-subgroup of S. Then $W = N_H(V)$ has order $2^{10} \cdot 3^2$ by Table 3.4.1. By Lemma 3.3.2(c) $N_G(V)$ is isomorphic to the finitely presented group $N = \langle n_k \mid 1 \leq k \leq 13 \rangle$. It has a faithful permutation representation PN of degree 2^{10} with stabilizer $\langle n_7, n_8, n_{10} \rangle$. Furthermore, W is isomorphic to the subgroup $N_2 = \langle n_k | 1 \leq k \leq 13 \rangle$ and $k \neq 10 \rangle$ of N.

Using the faithful permutation representation PH of H, Kim's program `GetShortGens(PH,X)` and MAGMA, we observed that the subgroup X of $T = C_H(u)$ can be generated by the three elements $x_1 = (xhx^2hxh^2xh^2)^3$, $x_2 = (hxhx^3hxh^2xh)^3$ and $x_3 = (x^2h^2xhxhxhx^2h^2x)^2$ of respective orders 2, 2 and 3. Applying the MAGMA command `IsConjugate(PH,u,b)` and PH it has been verified that $u^c = b = r_2^2 r_1 r_2 r_3 r_2$ for some $c \in H$. By Lemma 3.2.1(c) we know that $b \in C_G(V)$. Another application of MAGMA shows that $|C_W(b)| = 2^7$.

Using PH, PN and MAGMA we found an isomorphism $\sigma : W \longrightarrow N_2$ such that the centralizer $M = C_N(m)$ of $m = \sigma(b)$ is a direct product of $\langle m \rangle$ and a group K of order $2^6 \cdot 3$ which is generated by the elements $q_1 = (n_1 n_{10} n_7)^3$, $q_2 = n_7 n_9 n_{11}$ and $q_3 = n_7 (n_{10})^2$ of N having orders 2, 2 and 3, respectively. The group K has been constructed by application of the MAGMA command

```
K := HasComplement(M,sub<M| m>).
```

Its generators q_i were found by using MAGMA and Kim's aforementioned program. The Fitting subgroup F of K has order 2^5, and its Frattini subgroup $L = \Phi(F)$ is a Klein 4-group. Another application of Kim's program implies that $F = \langle f_1, f_2, f_3 \rangle$ and $L = \langle g_1, g_2 \rangle$, where $f_1 = n_5$, $f_2 = n_1 n_3$ and $f_3 = n_1 n_4$ have respective orders 2, 4 and 2, and where $g_1 = f_2^2$, $g_2 = (f_1 f_3)^2$ are involutions.

Applying MAGMA again we find a complement J of F in K having order 6. It is generated by $j_1 = n_7 n_9 n_{11}$ and $j_2 = n_7 n_{10}^2$ of respective orders 2 and 3. Now we see that $j_2 \notin N_2$ because $n_{10} \notin N_2$ and $n_7 \in N_2$.

MAGMA confirmed the following relations: $g_1^{j_2} = g_2$ and $g_2^{j_2} = g_1 g_2$. Clearly, $K = \langle f_1, f_2, f_3, j_1, j_2 \rangle$ and $S_2 = \langle f_1, f_2, f_3, j_1, j_2 \rangle$ is a Sylow 2-subgroup of K of order 2^6. As $M = C_N(m) = \langle m \rangle \times K$ and $j_2 \notin C_{N_2}(m)$ it follows that $C_G(b)$ contains an element s of order 3 which is not contained in $C_H(b)$. Thus $r = s^{c^{-1}} \in C_G(u)$ but $r \notin T = C_H(u)$.

As $T = C_H(u)$ is a proper subgroup of $C_G(u)$ Theorem 1.4.15 of [92] implies that $C_G(u)$ has a normal subgroup Q containing X such that $C_G(u) = \langle u \rangle \times Q$ and $|Q : X|$ is odd. Hence $C_Q(z) \geq X$ because z generates the center of X. On the other hand

$$C_Q(z) \leq C_G(u) \cap C_G(z) = C_H(u) = \langle u \rangle \times X.$$

Hence $C_Q(z) = X$. By Lemma 3.3.1(d) $X \cong C_{\mathsf{M}_{12}}(f)$ for some 2-central involution f of the Mathieu group M_{12}.

As $\sigma : W \to N_2$ is an isomorphism and $C_W(b)$ is conjugate in H to a Sylow 2-subgroup of X, it follows that X is isomorphic to the complement K of $\langle m = \sigma(b) \rangle$ in $M = C_N(m)$. It has been checked computationally that the element j_2 of K which is not in N_2 operates on the elementary abelian Frattini subgroup $L = \langle g_1, g_2 \rangle$ of order 4 of K as follows: $g_1^{j_2} = g_2$ and $g_2^{j_2} = g_1 g_2$. Hence $z^r \neq z$ and $z^r \in Y$ for some element r in Q of order 3, where $Y = C_W(b)^{c^{-1}}$ is a Sylow 2-subgroup of Q. Now Theorem 8.1.9 of [92], due to W. J. Wong [138], asserts that Q has two conjugacy classes of involutions and either $Q \cong \mathsf{M}_{12}$ or Q is a non-split extension of $\mathrm{PSL}_3(2)$ by its natural vector space F^3 over $F = GF(2)$. In the second case $|Q| = 2^6 \cdot 3 \cdot 7$, which is impossible because r and x_3 generate a 3-subgroup of order at least 9. Hence $Q \cong \mathsf{M}_{12}$.

(d) By Lemma 8.1.2 of [92] $\mathsf{M}_{12} = \langle A, B, C, D, E, F \rangle$, where the six generators are well defined permutations of the alternating group A_{12}. Let $r = ACA$ and $s = DEF$. A MAGMA calculation shows that the elements $r = ACA$ and $s = DEF$ have respective orders 3 and 4 and that they generate M_{12}. Let $u = (13, 14)$. Then U is a subgroup of the symmetric group S_{14} and $(u, s) = (u, r) = 1$. Using this faithful permutation representation PU of U the character table of U has been calculated by MAGMA.

(e) This assertion follows immediately from (c) and Theorem 8.1.9(c) of [92].

(f) Since the center of $X \cong C_{\mathsf{M}_{12}}(f)$ is generated by z the involutions f and z are identified. Hence they have the same conjugacy classes in G. As $uf = uz$ and $C_H(uz) = C_H(u)$ the involutions uz and u fuse in H by Table 3.4.1. By (b) and Theorem 8.1.9(c) of [92] the subgroup Q has exactly one other class s^Q of involutions and $C_Q(s) = \langle s \rangle \times Q_1$, where

$Q_1 \cong Sym(5)$. If s were conjugate to z in G then $H = C_G(z)$ would have subgroup R of order 240 such that $\langle z \rangle$ would have a complement $Q_2 \cong Q_1$. Using MAGMA and the faithful permutation representation PH it has been checked that H has exactly 11 conjugacy classes of subgroups R of order 240, but $\langle z \rangle$ does not split off in any of them. Hence s is conjugate to u in G. The same argument also shows that the remaining class representative us of $C_G(u)$ is G-conjugate to u.

(g) This assertion follows immediately from (c) and Table 3.5.4.

(h) By Proposition 3.1.2 and Lemma 3.3.1 each simple group G of Co_3-type has exactly two conjugacy classes of involutions z^G and u^G and the orders of their centralizers $H = C_G(z)$ and $U = C_G(u)$ are well determined.

Let $r(z,u,z) = |\{(x,y) \in (z^G \cap H) \times (u^G \cap H) | z \in \langle xy \rangle\}|$ and $r(z,u,u) = \left|\{(x,y) \in (z^G \cap U) \times (u^G \cap U) | u \in \langle xy \rangle\}\right|$. Table 3.4.1 and Lemma 3.3.2 imply that $z^G \cap H = \{z\} \cup v^H$, where v represents the class 2_2 of H. Furthermore, $u^G \cap H = u^H$ by (a). Let $\{t_j \mid 1 \leq j \leq 18\}$ be representatives of the 18 real z-special conjugacy classes of H. Then Theorem 1.6.4 asserts that

$$r(z,u,z) = \sum_{j=1}^{18} d(v,u,t_j),$$

where

$$d(v,u,t_j) = \frac{|H|^2}{|C_H(v)| \cdot |C_H(u)| \cdot |C_H(t_j)|} \cdot \sum_{\psi \in Irr_{\mathbb{C}}(H)} \psi(v)\psi(u)\psi(t_j)\psi(1)^{-1}.$$

Inserting the values of the character Table 3.5.3 of H into these formulas yields the following:

$$\begin{aligned}
&d(v,u,t_1) = 0, \quad d(v,u,t_2) = 15120, \quad d(v,u,t_3) = 45360,\\
&d(v,u,t_4) = d(v,u,t_5) = d(v,u,t_6) = 0,\\
&d(v,u,t_7) = 181440, \quad d(v,u,t_8) = 181440, \quad d(v,u,t_9) = 362880,\\
&d(v,u,t_{10}) = d(v,u,t_{11}) = d(v,u,t_{12}) = 0,\\
&d(v,u,t_{13}) = 483840, \quad d(v,u,t_{14}) = d(v,u,t_{15}) = 0,\\
&d(v,u,t_{16}) = d(v,u,t_{17}) = d(v,u,t_{18}) = 0.
\end{aligned}$$

Hence $r(z,u,z) = 1270080$.

By (f) we know that the group U has five conjugacy classes of involutions, four of which fuse with u and the remaining one with z. Let $u_1 = u = 2a$, $u_2 = 2b$, $u_3 = 2c$, $u_4 = 2e$ and $z = 2d$ in the notation of the character Table 3.5.4 of U. Let $\{s_k \mid 1 \leq j \leq 4\}$ be representatives of the four real u-special conjugacy classes of U. Then $u^G \cap U = \{u_1\} \cup u_2^U \cup u_3^U \cup u_4^U$ and $z^G \cap U = z^U$. Hence Theorem 1.6.4 asserts that

$$r(z,u,u) = \sum_{j=1}^{4} d(u_2,z,s_k) + d(u_3,z,s_k) + d(u_4,z,s_k),$$

where

$$d(u_n,z,s_k) = \frac{|U|^2}{|C_U(u_n)| \cdot |C_U(z)| \cdot |C_U(s_k)|} \cdot \sum_{\psi \in Irr_{\mathbb{C}}(U)} \psi(u_n)\psi(z)\psi(s_k)\psi(1)^{-1}$$

for all $1 \le n,k \le 4$. Inserting the values of the character Table 3.5.4 of U into these formulas yields the following:

$$\begin{aligned}
d(u_2,z,t_1) &= d(u_3,z,t_1) = 0, \quad d(u_4,z,t_1) = 495,\\
d(u_2,z,t_2) &= d(u_3,z,t_2) = 0, \quad d(u_4,z,t_2) = 31680,\\
d(u_2,z,t_3) &= d(u_3,z,t_3) = 0, \quad d(u_4,z,t_3) = 7920,\\
d(u_2,z,t_4) &= d(u_3,z,t_4) = 0, \quad d(u_4,z,t_4) = 47520.
\end{aligned}$$

Hence $r(z,u,u) = 87615$. Now Theorem 4.2.1 of [92] and (c) imply that

$$|G| = r(z,u,z) \cdot |C_G(u)| + r(z,u,u) \cdot |C_G(z)| = 2^{10} \cdot 3^7 \cdot 5^3 \cdot 7 \cdot 11 \cdot 23.$$

□

Lemma 3.3.4 *Let G be a finite simple group of* Co3*-type with a 2-central involution z such that $C_G(z) = H = \langle x,y,h\rangle$ described in Lemma 3.2.1. Then the following statements hold.*

(a) *$r_1 = (yh)^4 \in H$ generates the center of a Sylow 3-subgroup S of G and it is a representative of the unique 3-central conjugay class of G of order 3.*

(b) *$N_1 = N_G(\langle r_1\rangle)$ is isomorphic to the finitely presented group $N_1 = \langle k_1,k_2,k_3,k_4,k_5,k_6\rangle$ with the following set $\mathcal{R}(N_1)$ of defining*

relations:

$$
\begin{aligned}
& k_1^6 = k_2^4 = k_3^6 = k_4^4 = k_5^3 = k_6^3 = 1, \\
& (k_1, k_6^{-1}) = (k_2, k_6^{-1}) = (k_3, k_6^{-1}) = (k_5, k_6^{-1}) = 1, \\
& k_4^{-1}k_6k_4k_6 = 1, \quad (k_1, k_2) = 1, \quad k_4^{-1}k_2^2k_4^{-1} = 1, \\
& k_1^{-1}k_5^{-1}k_1k_5^{-1}k_6^{-1} = k_1^{-3}k_2^2 = k_3^{-1}k_1^3k_3^{-2} = 1, \\
& k_3k_1k_3^{-1}k_1^{-1}k_4k_1k_4^{-1} = k_3k_4^{-1}k_3^{-1}k_1^{-1}k_3k_1k_4 = 1, \\
& k_3^{-1}k_5^{-1}k_3^{-1}k_2^2k_5^{-1}k_3^{-1}k_5 = k_2^{-1}k_4^{-1}k_1^{-1}k_3^{-1}k_2k_3k_1k_4 = 1, \\
& k_2^{-1}k_3^{-1}k_2^{-1}k_3^{-1}k_2k_3^{-1}k_2^{-1}k_3^{-1} = k_2^{-1}k_5^{-1}k_2^{-1}k_5^{-1}k_2k_5k_2k_5^{-1}k_6^{-1} = 1, \\
& k_4^{-1}k_5^{-1}k_4^{-1}k_5^{-1}k_4k_5k_4k_5^{-1}k_6^{-1} = 1, \\
& k_3^{-1}k_5^{-1}k_3k_2^{-1}k_3^{-1}k_2k_5^{-1}k_2k_3k_2^{-1}k_6^{-1} = 1, \\
& k_1k_3^{-1}k_5^{-1}k_4^{-1}k_3^{-1}k_1k_5^{-1}k_3^{-1}k_5^{-1}k_4^{-1}k_6 = 1, \\
& k_3^{-1}k_4^{-1}k_5^{-1}k_4^{-1}k_3^{-1}k_2k_5^{-1}k_2^{-1}k_3^{-1}k_5k_6 = 1.
\end{aligned}
$$

(c) *$N_1 = \langle k_1, k_2, k_3, k_4, k_5, k_6 \rangle$ has a faithful permutation representation PN_1 with stabilizer $\langle k_1, k_2, k_3, k_4 \rangle$ of degree 3^5.*

(d) *A system of representatives n_i of the* 54 *conjugacy classes of N_1 and the corresponding centralizer orders $|C_{N_1}(n_i)|$ are given in Table 3.4.4.*

(e) *The character table of N_1 is stated in Table 3.5.5.*

(f) *G has two r_1-special conjugacy classes of order* 9 *represented by the classes 9_a and 9_b of Table 3.5.5.*

(g) *N_1 has four conjugacy classes of involutions denoted by 2_1, 2_2, 2_3 and 2_4 in Table 3.4.4. The elements of the classes 2_1, 2_2 and 2_3, 2_4 of N_1 are G-conjugate to z and u, respectively.*

Proof (b) Let PH be the faithful permutation representation of H constructed in Proposition 3.1.2(o). Table 3.4.1 states that $r_1 = (yh)^4$ is an element of H with order 3 whose centralizer $C_H(r_1)$ in H has order $2^5 \cdot 3^3 \cdot 5$. Let $N = N_H(\langle r_1 \rangle)$. Let $k_1 = (xhxh^2xh)^5$, $k_2 = (hxhxhx^2)^5$, $k_3 = (xhxhx^3h)^4$ and $k_4 = (h^2x^3h^2xh^2xh^2xhxhx)^2$. They have respective orders 6, 4, 6 and 4. Using PH and MAGMA the reader can easily verify that $K = \langle k_1, k_2, k_3, k_4 \rangle$ is a complement of $\langle r_1 \rangle$ in N.

Let X be the largest normal subgroup of $N_1 = N_G(r_1)$ having odd order. As the center of K is generated by the central involution $z = x^4$, Theorem 4.7.3 of [92] due to Glauberman implies that $N_1 = XK$. It has been checked computationally that $v = (k_1k_2k_3)^3$ is an involution such that $|C_H(v)| = 2^9 \cdot 3^2$. By Table 3.4.1 v and vz are conjugate in H.

The Klein 4-subgroup $V = \langle z, v\rangle$ of K operates on X by conjugation. Since $C_X(z) = \langle r_1\rangle$ Theorem 4.2.2 of [92] implies that $|X|$ divides 3^5. Now z inverts $X_1 = X/\langle r1\rangle$. Thus X_1 is an elementary abelian group by Lemma 1.5.9 of [92]. In particular, it is a faithful module of the group algebra FK, where $F = GF(3)$. In order to find the suitable faithful irreducible modules describing X_1 we now construct a faithful permutation representation of K.

Using MAGMA and the faithful permutation representation PH of H the reader can check that the elements $d_1 = k_1^2$ and $d_2 = (k_2k_3k_2k_3^2)^2$ generate an elementary abelian Sylow 3-subgroup S of K. Let PK be the permutation representation of K with stabilizer S. It is faithful and has degree 320. The application of the MAGMA command `K:=FPGroup(PK)` returned a set $\mathcal{R}(K)$ of defining relations of K which consists precisely of all the relations of $\mathcal{R}(N_1)$ stated in assertion (b) not involving the additional generators k_5 and k_6 of N_1. In the following K is identified with this finitely presented group, and PK is its faithful permutation representation of degree 320.

Let qk_i be the image of k_i in PK for $i = 1, 2, 3, 4$. Using the MAGMA command

```
M320ki := PermutationMatrix(FiniteField(3), qki)
```

for $1 \le i \le 4$ and then the Meataxe algorithm implemented in MAGMA one can check that the FK-module $PS \otimes F$ contains 16 non-isomorphic irreducible FK-modules of dimensions. They have dimensions 1, 4, 6, 9 and 12, and their degrees occur with respective multiplicities 2, 6, 2, 4 and 2; K has exactly 18 non-isomorphic simple FK-modules by its character table, four of which are linear. Hence $PS \otimes F$ contains all the non-linear irreducible 3-modular representations of K. Another application of MAGMA yields that $PS \otimes F$ contains exactly two non-isomorphic faithful 4-dimensional modules M_1 and M_2 of K over F and that they are dual to each other. Since z inverts X_1 this module is one of the faithful 4-dimensional FK-modules M_1 or M_2. In particular, $|X| = 3^5$.

Let MK_1 be the subgroup of $\mathrm{GL}_4(3)$ generated by the four matrices $M_1(k_i)$ of the generators k_i of K belonging to M_1:

$$Mk_1 = \begin{pmatrix} 2 & 0 & 2 & 0 \\ 0 & 2 & 2 & 1 \\ 0 & 0 & 2 & 0 \\ 0 & 0 & 0 & 2 \end{pmatrix}, \quad Mk_2 = \begin{pmatrix} 0 & 1 & 2 & 1 \\ 2 & 0 & 2 & 0 \\ 0 & 0 & 1 & 2 \\ 0 & 0 & 2 & 2 \end{pmatrix},$$

$$Mk_3 = \begin{pmatrix} 1 & 1 & 1 & 2 \\ 0 & 2 & 0 & 0 \\ 2 & 2 & 0 & 2 \\ 0 & 1 & 0 & 2 \end{pmatrix} \quad \text{and} \quad Mk_4 = \begin{pmatrix} 0 & 2 & 1 & 1 \\ 0 & 1 & 2 & 0 \\ 0 & 2 & 2 & 0 \\ 2 & 2 & 0 & 2 \end{pmatrix}.$$

Suppose that X_1 is isomorphic to M_1. Since M_1 is an irreducible FK-module, the following MAGMA commands can be applied:

```
W1:=Getvecs(MK_1),  PE1:=Semidir(MK_1,W1).
```

Thus MAGMA constructs the semi-direct product of the FK-module M_1 and the finitely presented group K. It returns this extension $E1$ of order $2^6 \cdot 3^6 \cdot 5$ as a permutation group $PE1$ acting on a set of cardinality 81. Similarly it constructs the split extension $E2$ of M_2 by K. Applying now the MAGMA command `IsIsomorphic(PE1,PE2)` we see that $E1 \cong E2$. In particular, the extension $E1$ of K by X_1 is uniquely determined up to isomorphism. The application of the MAGMA command `FPGroup(PE1)` yields a presentation of $E1 = \langle k_1, k_2, k_3, k_4, k_5 \rangle$ such that the first four generators k_i correspond to the generators of K, and k_5 is a generator of the irreducible FK-module M_1. Furthermore, the set $\mathcal{R}(E1)$ of defining relations of $E1$ is exactly the one obtained from the set $\mathcal{R}(N_1)$ in assertion (b) by removing all relations involving k_6.

In order to construct a presentation of the group $N_1 = N_G(r_1)$ we now extend $E1$ by a linear module described by the action of k_4 on r_1. Thus we assign to k_4 the matrix $M4 = [2] \in \mathrm{GL}_1(3)$ and the identity matrix Id in $\mathrm{GL}_1(3)$ to all other generators of $E1$. This module is described by the MAGMA command

```
FEalg := MatrixAlgebra<FiniteField(3),1|Id,Id,Id,M4,Id>.
```

Using the presentation of $E1$ and the permutation representation $PE1$ of $E1$ MAGMA constructs the cohomology module `CM := GModule(PE1, FEalg)` and calculates the dimension of the second cohomology group of $E1$ with coefficients in the constructed 1-dimensional module by means of the MAGMA command `CohomologicalDimension(PE1, CM, 2)`. It follows that this dimension is 1. Applying the MAGMA commands

```
P := ExtensionProcess(PE1, CM, E1),E11:=Extension(P, [1])
```

one obtains the presentation of $N_1 = E11 \cong N_G(r_1)$ given in statement (b). It is the unique non-split extension of $E1$ by the cyclic group $\langle r_1 \rangle$.

(c) This assertion has been verified by means of the presentation of N_1 stated in (b) and MAGMA.

(d) The system of representatives of the conjugacy classes of N_1 has been computed by means of PN_1, MAGMA and Kratzer's Algorithm 5.3.18 of [92].

(a) By Proposition 3.3.3(g) any Sylow 3-subgroup of G has order 3^7. Using MAGMA and the faithful permutation representation PN_1 it has been checked that k_6 generates the center of a Sylow 3-subgroup of N_1 of order 3^7. Now (a) follows at once from Table 3.4.4.

(e) The character table of N_1 has been calculated by means of $PN1$ and MAGMA.

(f) This assertion holds by Table 3.5.5.

(g) Table 3.4.4 states that the classes 2_1, 2_2, 2_3 and 2_4 are represented by the words k_1^3, $v = (k_1k_2k_3)^3$, $t = (k_1k_2k_3k_2k_4)^3$ and $w = k_2k_3^2k_4k_3$, respectively. Thus they belong to the subgroup K of $H = C_G(z)$. Using their words in the generators x, y and h of H, the faithful permutation representation PH of H and MAGMA, the author has verified that $k_1^2 = z$ and that v is H-conjugate to the class representative $(x^2y)^4$ of H given in Table 3.5.3. It fuses with z in G by Lemma 3.3.2. Proposition 3.3.3 states that $u = (x^2hy)^3$ of H represents the second conjugacy class of involutions of G. Using PH and MAGMA again it has been checked that u fuses with t and w in H. This completes the proof. □

Lemma 3.3.5 *Let G be a finite simple group of Co_3-type with a 2-central involution z such that $C_G(z) = H = \langle x, y, h\rangle$ described in Lemma 3.2.1. Let $N_1 = \langle k_i \mid 1 \le k_i \le 6\rangle$ be the finitely presented group constructed in Lemma 3.3.4. Let $r_2 = (k_1k_2k_3)^2$ and $Y = N_{N_1}(\langle r_2\rangle)$. Then the following statements hold.*

(a) *The Fitting subgroup Q of Y is an elementary abelian normal subgroup of order 3^5 in Y. It is generated by $f_1 = (k_1k_2k_3)^2$, $f_2 = (k_2k_5)^4$, $f_3 = (k_1k_3^2k_2)^2$, $f_4 = (k_1k_2k_5k_3k_5)^2$ and $f_5 = (k_1k_3k_5k_3k_2)^2$. It has a complement T_1 of order 8 in Y generated by $e_1 = (k_2k_3k_1^2)^3$, $e_2 = (k_5k_2k_5k_4k_1)^5$ and $e_3 = (k_1^2k_2k_3k_2k_4)^3$ of orders 2, 4 and 2, respectively.*

(b) *The normalizer $N = N_{N_1}(Q)$ of Q in N_1 contains a Sylow 3-subgroup S of G such that Q is the unique maximal elementary abelian normal subgroup of S. It has a complement T_2 of order 2^5 which is generated by T_1 and the element $w = (k_1k_5k_3k_5^2k_4k_1k_2k_5)^3$ of order 8.*

(c) *T_2 has a normal Sylow 3-subgroup S_1 of order 9 generated by $f = k_3k_1^2k_5k_1k_3^2k_5$ and f^w. The center Z_2 of T_2 is generated by the involution $b = e_1e_2^2e_3$. It has a complement T_3 in T_2 which is generated by f, w and $e = e_1e_2$. The subgroup $\langle w, e\rangle$ is a semi-dihedral Sylow 2-subgroup of T_3.*

(d) *k_5 and r_2 are not conjugate in N_1, but in G. The Fitting subgroup Q_5 of $Y_5 = N_{N_1}(\langle k_5\rangle)$ has order 3^6. It has a unique* *maximal elementary abelian normal subgroup* R_5 *of order* 3^5 *such that* $N_{N_1}(R_5)$ *contains a Sylow 3-subgroup of* G.

(e) $L = N_G(Q)$ *is a semidirect product of* $K = \langle b\rangle \times T$ *by* Q, *where* $T \cong \mathsf{M}_{11}$. *Furthermore,* $N_2 = N_G(\langle r_2\rangle)$ *is a subgroup of* L *of order* $2^4 \cdot 3^6 \cdot 5$.

(f) $N_2 = N_G(\langle r_2\rangle)$ *is isomorphic to the finitely presented group* $N_2 = \langle n_i \mid 1 \le i \le 5\rangle$ *with the following set* $\mathcal{R}(N_2)$ *of defining relations:*

$$n_1^2 = n_2^3 = n_3^3 = n_4^2 = n_5^2 = 1,$$
$$(n_2^{-1}n_1)^2 = n_1n_3^{-1}n_1n_3 = (n_1n_4)^2 = n_2^{-1}n_5n_2n_5 = 1,$$
$$(n_1n_5)^2 = n_3^{-1}n_5n_3n_5 = n_2^{-1}n_3n_2^{-1}n_3^{-1}n_2n_3n_2n_3^{-1} = 1,$$
$$n_2^{-1}n_3n_4n_3^{-1}n_2n_3n_4n_3^{-1} = n_2^{-1}n_4n_2^{-1}n_4n_2n_4n_2n_4 = 1,$$
$$(n_3^{-1}n_4n_5n_4)^2 = (n_4n_5)^4 = (n_4n_3^{-1})^5 = 1,$$
$$n_4n_3^{-1}n_2^{-1}n_3^{-1}n_2^{-1}n_3^{-1}n_2^{-1}n_4n_2^{-1}n_3n_2^{-1}n_3^{-1} = 1.$$

(g) $N_2 = \langle n_i \mid 1 \le i \le 6\rangle$ *has order* $N_1| = 2^6 \cdot 3^7 \cdot 5$ *and a faithful permutation representation* PN_2 *with stabilizer* $\langle n_1, n_2, n_3, n_5\rangle$ *of degree* 180.

(i) *A system of representatives* s_i *of the* 48 *conjugacy classes of* N_2 *and the corresponding centralizer orders* $|C_{N_2}(s_i)|$ *are given in Table 3.4.5.*

(j) *The character table of* N_2 *is stated in Table 3.5.6.*

(k) *The elements* $r_1 = k_6$, $r_2 = (k_1k_2k_3)^2$ *and* $r_3 = (k_1k_2k_3k_2k_4k_5)^2$ *of the normalizer* N_1 *represent all conjugacy classes of elements of order* 3 *of* G. *Furthermore,* r_3 *is not* G*-conjugate to any element of* H.

Proof By Lemma 3.3.4 and Proposition 3.1.2 the groups N_1 and H have faithful permutation representations PN_1 and PH of degrees 3^5 and 1920, respectively. The given generators of the various subgroups U_2 of the finitely generated subgroup U_1 with explicitly given generators have been obtained by means of Kim's program `GetShortGens(U_1,U_2)`. All these tools will now be used frequently. We also know from Proposition 3.3.3 that G has exactly two conjugacy classes of involutions represented by z and u.

(a) The given structure of $Y = QT_2$ can easily be verified by means of PN_1, the tools mentioned above and MAGMA.

(b) Now one constructs the normalizer $N = N_{N_1}(Q)$ of the Fitting subgroup Q of Y and fixes a Sylow 3-subgroup S of N. It has order 3^7

by MAGMA. Thus it is a Sylow 3-subgroup of G by Proposition 3.3.3. Applying the MAGMA command

```
subs:=Subgroups(S : Al:=Normal, IsElementaryAbelian := true)
```

one sees that Q has 35 such subgroups and that Q is the unique one of order 3^5. The given generators of the complement T_2 of Q in N were found by means of MAGMA.

(c) This statement is easily checked computationally.

(d) The elements k_5 and r_2 are not N_1-conjugate by Table 3.4.4. Using PN_1 and MAGMA one sees that $Q_5 = \langle k_1^2, (k_2k_3k_2k_3^2)^2, (k_1k_2k_3^2k_1k_3k_5)^2\rangle$ is a normal Sylow 3-subgroup of $C_5 = C_{N_1}(k_5)$ of order 3^6. Furthermore, $K_5 = \langle b_1, b_2\rangle$ is an abelian complement of order 4, where $b_1 = (k_3^2k_1k_5k_2)^9$ and $b_2 := (k_1k_2k_3k_1k_5k_2k_3k_4)^3$. Since $|C_{N_1}(b_1)| = 2^5 \cdot 3^4$ and $|C_{N_1}(b_2)| = 2^5 \cdot 3^3$ Table 3.4.4 and Proposition 3.3.3(g) imply that b_1 and b_2 are G-conjugate to z and u, respectively. Since b_1 and b_2 commute it follows from Table 3.4.1 that k_5 is G-conjugate to a generator j of the cyclic Sylow 3-subgroup of order 3 of $C_H(u)$. Using the faithful permutation representation PH of H it has been checked that j and h are H-conjugate, where h is defined in Lemma 3.2.1. By the proof of Lemma 3.3.4 we know that r_2 is also contained in H because k_1, k_2, k_3 and k_4 are uniquely determined words in x, y and h. Using PH and MAGMA again one checks that $|C_H(r_2)| = 2^3 \cdot 3^3$. Hence r_2 is H-conjugate to h by Table 3.4.1. In particular, k_5 and r_2 are G-conjugate.

Applying then the subgroup command again for the Sylow 3-subgroup Q_5 of C_5 one confirms that also Q_5 has a unique elementary abelian normal subgroup Q_5 of order 3^5. Thus (d) holds.

(e) Furthermore, $N_2 = N_G(r_2)$ has order at least $2^4 \cdot 3^6$ and it is a subgroup of $L = N_G(Q)$. Since $N_2 > Y$ also $L > N$. By (c) $N = Q : T_2$ and $T_2 = \langle b\rangle \times T_3$, and $R = \langle w, e\rangle$ is a semi-dihedral Sylow 2-subgroup of T_3 of order 16. Hence Theorem 1.4.15 of [92] implies that Q has a complement K in L such that $K = \langle b\rangle \times T$ and $T > T_3$. The application of the MAGMA command `IsConjugate(PN1,b,(k_2k_3^2k_4k3)^3)` has a positive answer. Therefore Lemma 3.3.4(g) and Table 3.4.1 imply that b is G-conjugate to u. Thus Proposition 3.3.3 asserts that T is isomorphic to a proper subgroup of M_{12} whose Sylow 2-subgroup is semi-dihedral of order 16 by (c). Hence $T \cong \mathsf{M}_{11}$ because T_3 is isomorphic to a maximal subgroup of M_{11} by (c).

(f) In order to construct N_2 as a finitely presented group we first construct T as a matrix group. By (b) and (c) we know that $N = N_{N_1}(Q) = Q : T_2 = Q : (\langle b\rangle \times T_3)$ and that $T_3 = \langle f, f^w, w, e\rangle$. Hence

$T_2 = \langle b, f, w, e\rangle$. Since they normalize the elementary abelian group $Q = \langle f_i \mid 1 \le i \le 5\rangle$ of order 3^5 they can be represented over $F = GF(3)$ by the following matrices (w.r.t to the basis of Q):

$$Mf = \begin{pmatrix} 0&0&0&2&0\\ 0&1&2&2&2\\ 1&0&0&1&2\\ 1&0&0&2&0\\ 2&0&1&2&2 \end{pmatrix}, \quad Mw = \begin{pmatrix} 0&0&1&2&2\\ 0&2&0&0&1\\ 2&0&0&0&0\\ 0&0&0&1&2\\ 0&0&0&2&0 \end{pmatrix} \quad \text{and} \quad Me = \begin{pmatrix} 2&0&0&1&0\\ 0&2&0&1&0\\ 0&0&1&2&2\\ 0&0&0&1&0\\ 0&0&0&1&2 \end{pmatrix}$$

and the diagonal matrix Mb with all diagonal entries equal to 2. Since T has order twice the order of M_{11} we now search for a subgroup A of $\mathrm{GL}_5(3)$ containing $MT = \langle Mb, Mf, Mw, Me\rangle$ having order $|A| = 15840$. For that we construct a faithful permutation representation $PG53$ of $\mathrm{GL}_5(3)$ of degree 242 with stabilizer equal to a maximal parabolic subgroup. Let PT be the image of MT in $PG53$. Using the subgroup command

```
subs:=Subgroups(PG53:OrderEqual := 15840)
```

MAGMA produces two conjugacy classes of such subgroups A_1 and A_2. It turns out that only the second conjugacy class contains a subgroup T_5 of order 288 which is conjugate to PT. MAGMA also provides the conjugating permutation g_2. If `h_GL(5,3)_PG53` denotes the isomorphism between the matrix group and its permutation representation $PG53$, then the MAGMA command `Mg2:=g2@@h_GL(5,3)_PG53` yields the conjugating matrix in $\mathrm{GL}_5(3)$. It is

$$Mg_2 = \begin{pmatrix} 2&2&0&1&1\\ 1&0&2&0&0\\ 0&0&1&1&1\\ 2&2&1&0&0\\ 1&2&0&1&2 \end{pmatrix}.$$

Furthermore, in $PG53$ we also have the subgroup $A_2^{g_2}$ containing PT. Choosing a Sylow 5-subgroup and its generator c it follows that $A_2^{g_2} = \langle PT, c\rangle$ and

$$Mc = \texttt{c@@h\_GL(5,3)\_PG53} = \begin{pmatrix} 1&0&2&0&0\\ 1&2&1&1&0\\ 2&0&2&0&1\\ 0&1&1&1&2\\ 2&0&1&2&0 \end{pmatrix}.$$

Let $MW = \langle MT, Mc\rangle$. Using the MAGMA command `L=FPGroup(ML)` one obtains the finitely presented group $W = \langle w_1, w_2, w_3, w_4, w_5\rangle$ with the following set $\mathcal{R}(W)$ of defining relations:

$$\begin{aligned}
&w_2^2 = w_4^3 = w_5^5 = 1,\\
&w_2^{-1}w_1w_2w_1 = w_1w_3^{-1}w_1w_3 = (w_2^{-1}w_4)^2 = 1,\\
&(w_1w_4)^2 = 1, \quad w_1w_5^{-1}w_1w_5 = w_4w_3^3w_4w_3^{-1} = 1,\\
&w_3^{-1}w_2^{-1}w_3^{-1}w_2^{-1}w_3^2w_2 = w_2w_3^-2w_2^{-1}w_3w_2^{-1}w_3 = 1,\\
&w_5^-2w_3^{-1}w_2^{-1}w_5^{-1}w_3w_2^{-1} = (w_2^{-1}w_5w_4w_5^{-1})^2 = 1,\\
&w_3^{-3}w_1w_2^{-1}w_3^{-1}w_5w_4w_5^{-1} = w_5^{-1}w_2^{-1}w_5^{-1}w_1w_2^{-1}w_3^{-1}w_5^{-1}w_3^{-1}w_4 = 1,\\
&w_5w_3w_2^{-1}w_3^{-1}w_2^{-1}w_5w_2w_5^{-1}w_2 = w_3^2w_5w_3w_2^{-1}w_5^{-1}w_3^2w_5 = 1.
\end{aligned}$$

Since $L = N_G(Q)$ is the semidirect product of W by the elementary abelian group Q of order 3^5 we apply Lemma 1.4.7 to get a presentation for L. It yields that $L = \langle w_i, v_j \mid 1 \le i, j \le 5\rangle$ has a set $\mathcal{R}(L)$ of defining relations consisting of $\mathcal{R}(w)$, $\mathcal{R}_1(Q \rtimes W)$ and the following set $\mathcal{R}_2(Q \rtimes W)$ of essential relations:

$$\begin{aligned}
&w_1v_1w_1^{-1}v_1 = w_1v_2w_1^{-1}v_2 = w_1v_3w_1^{-1}v_3 = w_1v_4w_1^{-1}v_4 = 1,\\
&w_1v_5w_1^{-1}v_5 = w_2v_1w_2^{-1}v_3^2v_4^2v_5 = w_2v_2w_2^{-1}v_2^2 = 1,\\
&w_2v_3w_2^{-1}v_2v_5^2 = w_2v_4w_2^{-1}v_1v_2v_3^2v_4v_5 = w_2v_5w_2^{-1}v_2v_3v_5 = 1,\\
&w_3v_1w_3^{-1}v_3 = w_3v_2w_3^{-1}v_2 = w_3v_3w_3^{-1}v_1^2 = w_3v_4w_3^{-1}v_1v_4^2v_5 = 1,\\
&w_3v_5w_3^{-1}v_1v_2^2v_4 = w_4v_1w_4^{-1}v_1 = w_4v_2w_4^{-1}v_2 = w_4v_3w_4^{-1}v_3^2 = 1,\\
&w_4v_4w_4^{-1}v_1^2v_2^2v_3v_4^2v_5^2 = w_4v_5w_4^{-1}v_3v_5 = 1,\\
&w_5v_1w_5^{-1}v_1^2v_2^2v_3v_5 = w_5v_2w_5^{-1}v_2v_4^2 = 1,\\
&w_5v_3w_5^{-1}v_1v_2^2v_3v_4^2v_5^2 = w_5v_4w_5^{-1}v_2^2v_4^2v_5 = w_5v_5w_5^{-1}v_3^2v_4 = 1.
\end{aligned}$$

Clearly the group L has a faithful permutation representation PL of degree 3^5 with stabilizer $W = \langle w_i \mid 1 \le i \le 5\rangle$. Using it the reader can verify that $|L| = 2^5 \cdot 3^7 \cdot 5 \cdot 11$ and that $L = \langle W, v_1\rangle$.

By (a) $r_2 = f_1$ and by the construction of L the vector v_1 corresponds to f_1. Now (e) implies that $N_G(r_2) = N_L(r_2) \cong N_L(v_1) = N_2$. Using the faithful permutation representation PL and Kim's program `GetShortGens(L,N2)` MAGMA confirms that N_2 is generated by $n_1 = w_1$, $n_2 = (w_2v_1)^3$, $n_3 = (w_5w_2)^2$, $n_4 = (w_5w_2w_5)^3$ and $n_5 = (w_2^2w_5^2)^2$. Furthermore, its order $|N_2| = 2^4 \cdot 3^6 \cdot 5$. The set $\mathcal{R}(N_2)$ stated in assertion (f) has been obtained by means of the MAGMA command `N2:=FPGroup(PN2)`, where PN2 denotes the restriction of PL to N_2.

(g) This assertion can be easily verified by means of MAGMA.

(h) The system of representatives of the conjugacy classes of N_2 has been computed by means of PN_2, MAGMA and Kratzer's Algorithm 5.3.18 of [92].

(i) The character table of N_2 has been calculated by means of PN_2 and MAGMA.

(j) By Lemma 3.3.4 $N_1 = \langle k_i | 1 \le i \le 6 \rangle$ contains a Sylow 3-subgroup B of G of order 3^7 with center $Z(B) = \langle k_6 \rangle$ of order 3. Hence N_1 contains representatives of all conjugacy classes of elements of order 3 in G. Table 3.4.4 asserts that N_1 has seven conjugacy classes of elements of order 3 represented by k_6, k_5, $s_3 = k_1^2$, $r_2 = (k_1k_2k_3)^2$, $s_5 = (k_1k_5)^2$, $s_6 = (k_1k_2k_3k_6)^2$ and $r_3 = (k_1k_2k_3k_2k_4k_5)^2$.

By the proof of Lemma 3.3.4 we know that $K = \langle k_1, k_2, k_3, k_4 \rangle$ may be assumed to be a subgroup of $H = \langle x, y, h \rangle$. In fact, $k_1 = (xhxh^2xh)^5$, $k_2 = (hxhxhx^2)^5$, $k_3 = (xhxhx^3h)^4$ and $k_4 = (h^2x^3h^2xh^2xh^2xhxhx)^2$.

By the proof of Lemma 3.3.4 we know that $r_1 = (yh)^4 \in H$ is G-conjugate to k_6 and that $N_G(r_1) \cong N_1$. Hence r_1 and k_6 are G-conjugate. So are r_2 and k_5 by (d). Furthermore, r_1 and r_2 are not G-conjugate by (e) and Lemma 3.3.4(c).

Using PN_1 and MAGMA it follows that $C_5 = C_{N_1}(s_5)$ and $C_6 = C_{N_1}(s_6)$ have normal Sylow 3-subgroups with respective complements K_5 and K_6. Both of them are elementary abelian of order 4. Furthermore, there are involutions x_i and y_i in K_i, $i = 5, 6$, such that $K_i = \langle x_i, y_i \rangle$ and $|C_{N_1}(x_5)| = 2^5 \cdot 3^2$, $|C_{N_1}(y_5)| = |C_{N_1}(y_6)| = 2^6 \cdot 3^3 \cdot 5$ and $|C_{N_1}(x_6)| = 2^5 \cdot 3^2$. By Table 3.4.4 and Lemma 3.3.4(g) the involutions x_5 and x_6 belong to the classes 2_4 and 2_3 of N_1, respectively. Hence they fuse with u in G by Lemma 3.3.4(g). Both involutions y_5 and y_6 belong to 2_1 and fuse to z by Lemma 3.3.4(g). Hence both s_5 and s_6 are G-conjugate to $h \in H$. Thus they are G-conjugate to r_2.

Table 3.4.1 states that $H\langle x, y, h \rangle$ has three conjugacy classes of elements of order 3 represented by $r_1 = (yh)^4$, $b = (xh)^3$ and h. Using the faithful permutation representation PH of H and MAGMA it has been verified that the representative r_2 of N_1 is H-conjugate to h and that $s_3 = k_1^2 \in H \cap K$ is H-conjugate to r_1.

Using PH and MAGMA again it follows that the center Z of $C_H(b)$ is generated by b and z and that $|C_H(b)| = |C_H(c)| = 2^4 \cdot 3^4$, where $c = zb$. As c powers to z Table 3.4.1 asserts that c is a representative of the z-special conjugacy class 6_2. Hence $|C_G(c)| = 2^4 \cdot 3^4$. By Table 3.4.4 the representative $d = (k_1k_2k_3k_5)^3$ of the conjugacy class 6_2 of N_1 has centralizer order $|C_{N_1}(d)| = 2^4 \cdot 3^4$ and powers to r_1. Now Proposition

3.3.3(b), Lemma 3.3.4(g) and Tables 3.4.1 and 3.5.4 imply that c and d are conjugate in G. Hence b is G-conjugate to r_1. Therefore a representative r of any additional conjugacy class of elements of order 3 cannot be centralized by an involution of z^G.

By Table 3.4.4 the centralizer $C_{N_1}(r_3)$ has order $2 \cdot 3^4$. Using the faithful permutation representation PN_1 it has been checked that its Sylow 2-subgroup is generated by an involution w with $|C_{N1}(w)| = 2^5 \cdot 3^3$. Hence it fuses with u by Table 3.4.4 and Lemma 3.3.4. Therefore r_3 is not G-conjugate to r_1 or r_2. This completes the proof. □

Lemma 3.3.6 *Let G be a finite simple group of Co_3-type with a 2-central involution z such that $C_G(z) = H = \langle x, y, h \rangle$ as described in Lemma 3.2.1. Let $N_1 = N_G(r_1) = \langle k_i | 1 \le i \le 6 \rangle$ be the finitely presented group constructed in Lemma 3.3.4. Let $r_3 = (k_1k_2k_3k_2k_4k_5)^2$. Using the representatives of its conjugacy classes given in Table 3.4.4 the following statements hold.*

(a) *$N = N_{N_1}(\langle r_3 \rangle) \cong X \times Y$, where $X \cong S_3$ and Y is a split extension of a cyclic group of order 2 by its extra-special Sylow 3-subgroup T of order 27 and exponent 9.*

(b) *The normalizer $N_3 = N_G(\langle r_3 \rangle)$ of the subgroup $\langle r_3 \rangle$ of order 3 is isomorphic to the finitely presented group $N_3 = \langle q, r, s, t \rangle$ with the following set $\mathcal{R}(N_3)$ of defining relations:*

$$\begin{aligned}
&q^3 = r^3 = s^2 = t^3 = 1, \quad r^s = r^2,\\
&(q^{-1}t^{-1})^6 = 1, (qt^{-1})^6 = 1,\\
&q^{-1}t^{-1}q^{-1}tq^{-1}t^{-1}q^{-1}t^{-1}q^{-1}tqt^{-1}q^{-1}tq^{-1}t = 1,\\
&q^{-1}t^{-1}qtq^{-1}t^{-1}q^{-1}t^{-1}qtq^{-1}tq^{-1}t^{-1}qt = 1,\\
&(q, r) = (q, s) = (t, r) = (t, s) = 1.
\end{aligned}$$

(b) *$N_3 \cong S_3 \times Aut(\mathrm{PSL}_2(8))$, and it has a faithful permutation representation PN_3 of degree 112 with stabilizer $\langle r, tqtqt^2qt^2, (tqt^2qt^2qtqt)^2 \rangle$ of order 3^4.*

(c) *A system of representatives n_i of the 33 conjugacy classes of N_3 and the corresponding centralizers orders $|C_{N_3}(n_i)|$ are given in Table 3.4.6.*

(d) *The character table of N_3 is stated in Table 3.5.7.*

(e) *The elements $r_1 = k_6$, $r_2 = (k_1k_2k_3)^2$ and $r_3 = (k_1k_2k_3k_2k_4k_5)^2$ of the normalizer N_1, represent all conjugacy classes of elements of order 3 of G.*

Proof (a) Let $N = N_{N_1}(\langle r_3 \rangle)$ and $C = C_{N_1}(r_3)$. Using MAGMA and the faithful permutation representation PN_1 of N_1 stated in Lemma 3.3.4(c) one sees that N has order $2^2 \cdot 3^4 = 2|C|$. The elements $b_1 = k_1 k_3^2 k_5 k_3 k_1^2$, $b_2 = (k_2 k_3 k_1 k_2 k_3 k_2 k_5)^2$ and $b_3 = k_1^2 k_3^2 k_1 k_3 k_5 k_2^2$ of C generate a Sylow 3-subgroup S of N. It has a complement in C generated by the involution $b_4 = (k_1 k_4 k_2 k_5 k_3 k_2 k_5)^3$. By means of the MAGMA command

```
exists(w){x:x in N|r3^x eq r3^2 and
    Order(sub<N1|C,x>) eq 324}
```

we found the element $w = (k_1 k_2 k_5 k_2 k_3 k_2 k_3 k_1 k_3^2)^{11}$ of order 18 in N. Let $b_6 = w^9$. It has been checked computationally that $r_3^{b_6} = r_3^2$ and that $T = \langle b_1, b_2 \rangle$ is an extra-special Sylow 3-subgroup of Y of order 27 and exponent 9. Hence $X = \langle r_3, b_6 \rangle \cong S_3$, the symmetric group of order 6. Applying now the MAGMA command `HasComplement(N,X)` one finds a complement Y of X in N. It is generated by b_1, b_2 and the involution $y = (k_2 k_4 k_2 k_5 k_1 k_3 k_5 k_1)^3$. Furthermore, $N = X \times Y$.

(b) Let $N_3 = N_G(\langle r_3 \rangle)$ and $M = C_G(r_3)$. Since N contains a Sylow 3-subgroup of N_3 we have $(|M : Y|, 3) = 1$. Therefore Theorem 1.4.15 of [92] implies that $M = \langle r_3 \rangle \times W$ for some subgroup W of G containing Y. Thus $N_3 = X \times W$ because the symmetric group S_3 is complete.

The orders of $C_{N_1}(b_6)$ and $C_{N_1}(y)$ are $2^5 \cdot 3^4$ and $2^5 \cdot 3^4$, respectively. Hence the involutions b_6 and y belong to the conjugacy classes 2_2 and 2_3 of N_1 listed in Table 3.4.4. Thus Lemma 3.3.4(g) implies that b_6 and y are G-conjugate to z and u, respectively. In particular, r_3 is G-conjugate to an element d in $U = C_G(u)$ of order 3. By Lemma 3.3.5(g) r_3 is not G-conjugate to any 3-element in H. Therefore Proposition 3.3.3 and Theorem 8.1.9(c) of [92] imply that $C_U(d) \cong \langle d \rangle \times (\langle s \rangle \times A_4)$, where s is a non-2-central involution of M_{12}. Hence $C_U(d)$ has order $2^3 \cdot 3^2$, and all its involutions are G-conjugate to u. Thus $|W|$ is a multiple of $2^3 \cdot 3^3$. Since W commutes with b_6 we may assume that W is a subgroup of H. Using its faithful permutation representation PH and the MAGMA command

```
subs:=Subgroups(PH : OrderMultipleOf :=2^3*3^3)
```

we see that H has 60 conjugacy classes of subgroups having such an order. However, among them there are only two conjugacy classes of subgroups having an extra-special Sylow 3-subgroup of order 27 and exponent 9. Their orders are $2^3 \cdot 3^3 \cdot 7$ and $2^4 \cdot 3^3 \cdot 7$. Furthermore, getting the composition factors of these two groups one sees that the first is isomorphic to $W_1 = Aut(\mathrm{PSL}(2,8))$ and the second is the direct product of W_1 with $\langle z \rangle$. Hence $W \cong W_1$ because b_6 belongs to X and $X \cap W = 1$.

Using the permutation representation PH of $H = \langle x, y, h \rangle$ and Kim's program `GetShortGens(H,W_1)` MAGMA returns two generators $q = (x^2hx^3hx^2)^3$ and $t = (xh^2xhx^3h)^2$ of W_1. Another application of the MAGMA command `W = FPGroup(W_1)` provides the following set of defining relations $\mathcal{R}(W)$ of the finitely presented group $W = \langle q, t \rangle$:

$$
\begin{gathered}
q^3 = 1, \quad t^3 = 1, \quad (q^{-1}t^{-1})^6 = 1, \quad (qt^{-1})^6 = 1, \\
q^{-1}t^{-1}q^{-1}tq^{-1}t^{-1}q^{-1}t^{-1}q^{-1}tqt^{-1}q^{-1}tq^{-1}t = 1, \\
q^{-1}t^{-1}qtq^{-1}t^{-1}q^{-1}t^{-1}qtq^{-1}tq^{-1}t^{-1}qt = 1.
\end{gathered}
$$

Since $N_3 = X \times W$ and $X \cong S_3$ the remaining relations of $\mathcal{R}(N_3)$ hold trivially for $r = r_3$ and $s = b_6$.

(b) $N_3 \cong S_3 \times \mathrm{PSL}(2,8)$ holds by the proof of (a). Since N_3 has a small order we first calculated the regular faithful permutation representation pN_3 by means of MAGMA from the given presentation of N_3. Using pN_3 we determined a Sylow 3-subgroup S of N_3 inside the normalizer $N_{N_3}(\langle q, r \rangle)$. By means of the short generator command MAGMA we found the generators r, $x_2 = tqtqt^2qt^2$ and $x_3 = (tqt^2qt^2qtqt)^2$ of S. They have respective orders 3, 9 and 3. Clearly, S is the stabilizer of a faithful permutation representation PN_3 of N_3 of degree 112.

(c) The system of representatives of the conjugacy classes of N_3 has been computed by means of the faithful permutation representation PN_3 given in (b), MAGMA and Kratzer's Algorithm 5.3.18 of [92].

(d) The character table of N_3 has been calculated by means of PN_3 and MAGMA. □

Lemma 3.3.7 *Let G be a finite simple group of* Co_3*-type with a 2-central involution z such that $C_G(z) = H = \langle x, y, h \rangle$ as described in Lemma 3.2.1. Then the following statements hold.*

(a) *$f_1 = (y^2h)^2 \in H$ generates the center of an extra-special Sylow 5-subgroup of G of order 5^3 and exponent 5. Furthermore, it is a representative of the unique 5-central conjugay class of G of order 5.*

(b) *$N_5 = N_G(\langle f_1 \rangle)$ is isomorphic to the finitely presented group $N_5 = \langle q_i \mid 1 \le i \le 5 \rangle$ with the following set $\mathcal{R}(N_2)$ of defining relations:*

$$q_1^4 = q_2^6 = q_3^8 = q_4^5 = q_5^5 = 1,$$
$$(q_1, q_5) = (q_2, q_5) = (q_4, q_5) = (q_2, q_3) = 1,$$
$$q_3^{-1} q_5 q_3 q_5^2 = q_1^{-1} q_2^{-1} q_1 q_2^{-1} = q_1^{-1} q_3^{-1} q_1^{-1} q_3 = 1,$$
$$q_2^{-1} q_1^{-2} q_2^{-2} = q_3^{-2} q_1^2 q_3^{-2} = q_1^{-1} q_4^{-1} q_1^2 q_4^{-1} q_1^{-1} q_5^{-1} = 1,$$
$$q_1 q_4^{-1} q_2 q_1^{-1} q_4^{-2} q_2 q_5 = q_1^{-1} q_4^{-1} q_3^{-1} q_1^{-1} q_4^{-1} q_3 q_4 = 1.$$

(c) *$N_5 = \langle t_i \mid 1 \le i \le 5 \rangle$ has a faithful permutation representation PN_5 with stabilizer $\langle q_1, q_2, q_3 \rangle$ of degree 5^3.*

(d) *A system of representatives n_i of the 26 conjugacy classes of N_5 and the corresponding centralizers orders $|C_{N_5}(n_i)|$ are given in Table 3.4.7.*

(e) *The character table of N_5 is stated in Table 3.5.8.*

(f) *The elements $f_1 = q_5$ and $f_2 = q_4$ of the normalizer N_5 represent all conjugacy classes of elements of order 5 in G. Furthermore, f_2 is not G-conjugate to any element of H.*

Proof (b) Let PH be the faithful permutation representation of $H = \langle x, y, h \rangle$ constructed in Proposition 3.1.2(o). Table 3.4.1 states that $f = (y^2h)^4$ is an element of H with order 5. Using PH and MAGMA one can easily see that its normalizer $N = N_H(f)$ in H has order $2^4 \cdot 3 \cdot 5$ and that it is generated by the three elements $y_1 = (hxhx^2hxhxh^2)^2$, $y_2 = (hxhxh^2xhxhx)^5$ and $y_3 = xh^2xhxhxhx^2h^2xhx^3$. Applying now the MAGMA command `HasComplement(N,sub<H|f>)` one obtains a complement K of the subgroup $\langle f \rangle$ in N. Using MAGMA and the command `GetShortGens(N, K)` it follows that $K = \langle e_1, e_2, e_3 \rangle$, where $e_1 = y_2$, $e_2 = y_3^4$ and $e_3 = y_3 y_1 y_3^2$. They have respective orders 4, 6 and 8, and $f = (y_1 y_3^2)^6$. Furthermore, $f^{e_3} = f^3$.

The restriction of PH to K yields a faithful permutation representation PK of K. Using it and the MAGMA command `K = FPGroup(PK)` one obtains the following set $\mathcal{R}(K)$ of defining relations of the group $K = \langle e_1, e_2, e_3 \rangle$ of order 48:

$$e_1^4 = e_2^4 = e_3^8 = 1, \quad (e_2, e_3) = 1, \quad e_1^{-1} e_2^{-1} e_1 e_2^{-1} = 1,$$
$$e_1^{-1} e_3^{-1} e_1^{-1} e_3 = e_2^{-1} e_1^{-2} e_2^{-2} = e_3^{-2} e_1^2 e_3^{-2} = 1.$$

Let X be the largest normal subgroup of $N_5 = N_G(f)$ having odd order. Let $F = GF(5)$. Then by the argument given in Lemma 3.3.4

it follows that $X_1 = X/\langle f \rangle$ is a faithful irreducible FK-module of dimension 2 over F. From the regular permutation representation PK of degree 48 of K one constructs all irreducible representations of K over F using the Meataxe algorithm and MAGMA. There are 16 irreducible representations. Four of them are faithful. The following three matrices Me_i of the three generators e_i of K describe a faithful irreducible representation A of degree 2 over F:

$$Me_1 = \begin{pmatrix} 3 & 0 \\ 2 & 2 \end{pmatrix}, \quad Me_2 = \begin{pmatrix} 0 & 4 \\ 1 & 1 \end{pmatrix} \quad \text{and} \quad Me_3 = \begin{pmatrix} 1 & 2 \\ 3 & 4 \end{pmatrix}.$$

Let MK be the subgroup of $\mathrm{GL}_2(5)$ generated by these three matrices. Since it is irreducible the following two MAGMA commands `Q := Getvecs(MK)` and `Semidir(MK,Q)` construct the semidirect product of K by A as a finitely presented group $K_1 = \langle e_1, e_2, e_3, e_4 \rangle$ of order $2^4 \cdot 3 \cdot 5^2$ with set $\mathcal{R}(K_1)$ of defining relations consisting of $\mathcal{R}(K)$ and the following relations:

$$e_4^5 = 1, \quad e_1^{-1}e_4^{-1}e_1^2e_4^{-1}e_1^{-1} = 1,$$
$$e_1e_4^{-1}e_2e_1^{-1}e_4^{-2}e_2 = e_1^{-1}e_4^{-1}e_3^{-1}e_1^{-1}e_4^{-1}e_3e_4 = 1.$$

K has two additional faithful irreducible representations of degree 2. Applying the same methods as above one obtains two other extension groups K_2 and K_3. Using isomorphism tests with MAGMA the author checked that $K_1 \cong K_2 \cong K_3$.

In order to get the presentation of the group $N_5 = N_G(f)$ we now extend K_1 by a linear module described by the action of e_3 on f. As $f^{e_3} = f^3$ we assign to e_3 the matrix $M3 = [3] \in \mathrm{GL}_1(5)$ and the identity matrix Id in $\mathrm{GL}_1(5)$ to all other generators of K_1. This module is implemented into MAGMA by the command

```
FEalg := MatrixAlgebra<FiniteField(5),1|Id,Id,M3,Id>.
```

Using the presentation of K_1 and its permutation representation $PK1$ MAGMA constructs the cohomology module CM by means of the command `CM := GModule(PK1, FEalg)`. The dimension of the second cohomology group of K_1 with coefficients in CM is 1, as can easily be verified by an application of the MAGMA command `CohomologicalDimension (PE1, CM, 2)`. Applying the MAGMA commands

`P := ExtensionProcess(PE1, CM, K1)` and

`K11 := Extension(P, [1])`

one obtains the presentation of $N_5 = K11 \cong N_G(f)$ given in statement (b), where the letter e is replaced with q. N_5 is the unique (up to isomorphism) non-split extension of K_1 by the cyclic group $\langle f \rangle$.

(c) This assertion is trivial because N_5 is a split extension of K by a group of order 125.

(d) The system of representatives of the conjugacy classes of N_5 has been computed by means of PN_5, MAGMA and Kratzer's Algorithm 5.3.18 of [92].

(a) This follows at once from Table 3.4.7 and Proposition 3.3.3 because each Sylow 5-subgroup of N_5 is extra-special of order 125 and exponent 5.

(e) The character table of N_5 has been calculated by means of PN_5 and MAGMA.

(f) Proposition 3.3.3 asserts that a Sylow 5-subgroup of G has order 125 and that G has two conjugacy classes of involutions represented by z and u. Thus (a) implies that N_5 contains a Sylow 5-subgroup S of G. It has exponent 5 by Table 3.4.7. This table also states that N_5 has two conjugacy classes of elements of order 5, one of which is 5-central. Since q_6 is G-conjugate to an element f of order 5 in H it is centralized by an element of z^G. The representative q_4 of the second class 5_2 of N_5 has a centralizer $C_{N_5}(q_4)$ with a cyclic Sylow 2-subgroup of order 2 by Table 3.4.7. Let x be its generator. Then it has been checked computationally that x belongs to the class 2_1 of involutions of N_5. That is not G-conjugate to z by Table 3.4.7 because the Sylow 5-subgroups of H are cyclic and all involutions in a Sylow 2-subgroup of $C_H(f)$ are G-conjugate to z. Hence $x \in u^G$. Thus G has two conjugacy classes of elements of order 5. This completes the proof. □

Lemma 3.3.8 *Let G be a finite simple group of Co_3-type with a 2-central involution z such that $C_G(z) = H = \langle x, y, h \rangle$ as described in Lemma 3.2.1. Let $N_5 = \langle q_i | 1 \le i \le 5 \rangle$ be the finitely presented group constructed in Lemma 3.3.7. Using the representatives of its conjugacy classes given in Table 3.4.7 the following statements hold.*

(a) *$f_2 = q_4 \in N_5$ is a representative of a non-5-central conjugay class of G of elements of order 5.*

(b) *$N_G(\langle f_2 \rangle)$ is uniquely determined up to isomorphism. It is isomorphic to the finitely presented group $N_6 = \langle x_1, x_2, x_3, x_4 \rangle$ with the* *following set $\mathcal{R}(N_6)$ of defining relations:*

$$x_1^3 = x_2^3 = x_3^5 = x_4^4 = 1, \quad x_3^{x_4} = x_3^3,$$
$$(x_1, x_3) = (x_1, x_4) = (x_2, x_3) = (x_2, x_4) = 1,$$
$$(x_1 x_2^{-1} x_1^{-1} x_2^{-1})^2 = (x_1^{-1} x_2 x_1^{-1} x_2^{-1})^2 = 1.$$

(c) *A system of representatives n_i of the 20 conjugacy classes of N_6 and the corresponding centralizer orders $|C_{N_6}(n_i)|$ are given in Table 3.4.8.*

(d) *The character table of N_6 is stated in Table 3.5.9.*

Proof (a) By Lemma 3.3.7(a) and (f) $f_2 = q_4$ is not conjugate to $f_1 = q_5$ in G. Furthermore, G has exctly these two conjugacy classes of 5-elements.The Sylow 2-subgroups of $C_{N_5}(\langle f_2 \rangle)$ are cyclic of order 2 by Table 3.4.7.

(b) Using the faithful permutation representation PN_5 of N_5 defined in Lemma 3.3.7(c) and MAGMA the reader can verify again that $M = N_{N_5}(\langle f_2 \rangle)$ is generated by $m_1 = (q_1^2 q_4)^2$, $m_2 = (q_3 q_1 q_4)^2$, $m_3 = q_2 q_1 q_3^2$ and $|M| = 2^3 \cdot 5^2$. These generators of M have respective orders 5, 4 and 2. Furthermore, the following relations hold: $f_2 = (m_2^2 m_3)^6$, $f_2^{m_2^2} = f_2^3$. Thus $X = \langle f_2, m_2^2 \rangle$ of M is a Frobenius group of order 20. The subgroup $Y = \langle u_1, m_1 \rangle$ is dihedral of order 10 because $u_1 = (m_2^2 m_3)^5$ has order 2 and $m_1^{u_1} = m_1^4$. Another application of MAGMA yields that $M = X \times Y$. It also asserts that the involutions u_1 and $z_1 = q_1 q_3^2$ are conjugate in N_5. By Tables 3.4.7 and 3.4.1 z_1 is G-conjugate to the 2-central involution z of G. Now Table 3.4.1 implies that f_2 is G-conjugate to the 5-element $j = (y^2 h)^4$ of $H = \langle x, y, h \rangle$ whose centralizer $C_H(j)$ has order $2^2 \cdot 3 \cdot 5$ and a non-cyclic abelian Sylow 2-subgroup of order 4. Therefore $C_G(f_2)/\langle f_2 \rangle > Y$ has a strongly self-centralizing Sylow 2-subgroup. Hence $C_G(f_2) = \langle f_2 \rangle \times \mathrm{PSL}_2(5)$ by Theorem 1.6.2 of [92] due to Brauer, Suzuki and Wall. $\mathrm{PSL}_2(5) = \langle x_1, x_2 \rangle$ has a set of defining relations consisting of all the relations of $\mathcal{R}(N_6)$ given in the statement except the ones involving the generators x_3 and x_4 of N_6. Letting $x_3 = f_2$ and $x_4 = m_2^2$ the given presentation of $N_6 = N_G(f_2)$ holds.

(c) The system of representatives of the conjugacy classes of N_5 has been computed by means of the regular permutation representation PN_6 of N_6, MAGMA and Kratzer's Algorithm 5.3.18 of [92].

(d) The character table of N_5 has been calculated by means of PN_6 and MAGMA. This completes the proof. □

Lemma 3.3.9 (Fendel) *Let G be any finite simple group of Co_3-type with a 2-central involution z such that $C_G(z) = H\langle x, y, h\rangle$. Then the following assertions hold.*

(a) *The Sylow 7-normalizer N_7 is a direct product of the symmetric group S_3 and the Frobenius group of order 42.*
(b) *The Sylow 11-normalizer N_{11} is a direct product of the cyclic group of order 2 and the Frobenius group of order 22.*
(c) *The Sylow 23-normalizer N_{23} is a Frobenius group of order $23{\cdot}11$.*

Proof (a) By Proposition 3.3.3 each group G of Co_3-type has order $|G| = 2^{10} \cdot 3^7 \cdot 5^3 \cdot 7 \cdot 11 \cdot 23$. Let T be a Sylow 7-subgroup of G. By Lemma 3.3.6 and Sylow's Theorem we may assume that $T \le N_3 = \langle p, q, r, s, t\rangle$ and that $T = \langle p\rangle$. Hence $C_{N_3}(S_7) = S_3 \times T$ and $N_{N_3}(T) = S_3 \times F_{21}$, where $S_3 = \langle r, s\rangle$ and $F_{21} = \langle p, t\rangle$. Since 7 divides $|H|$ Table 3.4.1 and Sylow's Theorem assert that T is conjugate to $T_1 = \langle (xy)^2\rangle$. Using the faithful permutation representation PH of H given in Proposition 3.1.2(o) and MAGMA it has been checked that $N_H(T_1)$ is a direct product of a cyclic group of order 2 and a Frobenius group F_{42} of order 42. Hence $N_G(T) = S_3 \times F_{42}$ by Table 3.4.6.

(b) Proposition 3.3.3 states that G has two conjugacy classes of involutions represented by z and u and that $C_G(u) = U \cong \langle u\rangle \times \mathsf{M}_{12}$. In particular, we may assume that a Sylow 11-subgroup L of G is contained in U. Hence $N_U(L)$ is a direct product of $\langle u\rangle$ and a Frobenius group F_{55} of order 55. By the previous results of this section we know that 11 does not divide the orders of $C_G(z)$ and the ones of all the m-Sylow normalizers for the primes $m \in \{3, 5, 7\}$. Therefore $N_G(L) = N_U(L)$.

(c) Now the previous results and Sylow's Theorem imply that a Sylow 23-subgroup W of G is cyclic and self-centralizing. Furthermore, $N_G(W)$ has order $23 \cdot 11$. □

Lemma 3.3.10 *Let G be any finite simple group of Co_3-type with a 2-central involution z such that $C_G(z) = H$. Then G has 42 conjugacy classes x^G. They can be ordered such that their centralizers $C_G(x)$ have the same orders as the centralizers $C_{\mathfrak{G}}(\mathfrak{x})$ of the corresponding conjugacy classes $\mathfrak{x}^{\mathfrak{G}}$ of the simple group $\mathfrak{G}$ constructed in Theorem 3.2.2. They are stated in Table* `DVD.1.1.1` *on the accompanying DVD.*

Proof (a) By Proposition 3.3.3 each group G of Co_3-type has order $|G| = 2^{10} \cdot 3^7 \cdot 5^3 \cdot 7 \cdot 11 \cdot 23$.

By Proposition 3.3.3 G has 26 conjugacy classes of elements of even order. G has one conjugacy class of elements of order 7, two conjugacy classes of elements of order 11 and two conjugacy classes of elements of order 23 by Lemma 3.3.9.

From the power map information of the classification of the conjugacy classes of the groups N_1, N_2, N_3, N_5 and N_6 given in Section 3.4 and the order of G it follows that G has eight 3-singular classes denoted by 3_A, 3_B, 3_C, 9_A, 9_B, 15_A, 15_B, 21_A and two classes of elements of order 5.

Furthermore, Lemma 3.3.9 and the tables stated in Section 3.4 imply that the centralizers of these 41 non-trivial conjugacy classes have the same centralizer orders as the corresponding conjugacy classes of the constructed group $\mathfrak{G}$ of Theorem 3.2.2; see Table `DVD.1.1.1`. Since G and $\mathfrak{G}$ have the same group order by Proposition 3.3.3 and Theorem 3.2.2 the proof is complete. □

Lemma 3.3.11 (Fendel) *Let G be a finite simple group of Co_3-type with a 2-central involution z such that $C_G(z) = H = \langle x, y, h\rangle$ as described in Lemma 3.2.1. Then H is a maximal subgroup of G.*

Proof Lemma 3.3.11 must be true; otherwise there is a proper subgroup X of G containing $H = C_G(z)$ properly such that X is of minimal order with respect to these two properties. Hence $C_X(z) = H$. Therefore Proposition 3.3.3 implies that X has a proper normal subgroup $Y \neq 1$. Hence $Y \cap H < H$.

By Proposition 3.1.2 the center $Z(H) = \langle z\rangle$ of order 2 is the only non-trivial proper normal subgroup of H because $H/Z(H) \cong \mathsf{Sp}_6(2)$. Therefore either $Y \cap H = 1$ or $Y \cap H = Z(H)$. In the first case Lemma 1.5.9 of [92] and Proposition 3.3.3 assert that Y is an abelian group of order dividing $q = 3^3 \cdot 5^2 \cdot 11 \cdot 23$. The Klein 4-group $V = \langle z, v\rangle$ defined in Lemma 3.3.2 operates on Y. Now Theorem 4.2.2 of [92] and Table 3.4.1 yield that $|Y||C_Y(V)|^2 = |C_Y(z)||C_Y(v)|^2$ because v and vz are conjugate in H. Hence $C_Y(z) = 1$ as H does not have any normal subgroup of odd order. Furthermore, $|C_Y)| \leq 3^2$ and $|C_Y(v)| \leq 3^4 \cdot 5 \cdot 7$ by Table 3.4.1. Thus $|Y|$ divides $3^2 \cdot 5^2$ as $|Y|$ divides q. However, $H \cong 2\mathsf{Sp}_6(2)$ does not have any irreducible 3- or 5-modular representations of degree 2 by [71], its pp. 111–112.

In the second case Corollary 1.4.18 of [92] implies that the cyclic Sylow 2-subgroup of Y has a normal complement Q. As a characteristic subgroup of Y it is normal in X. Hence $Q = 1$ by the previous argument. Furthermore, $Y = Z(H)$ and $X = C_X(z) = H$, a contradiction. This completes the proof. □

Proposition 3.3.12 *Let G be a finite simple group of Co_3-type. Then the minimal degree of a faithful irreducible complex character χ of G is 23 and G has exactly one complex character $\chi \in Irr_{\mathbb{C}}(G)$ of this degree.*

Proof Let $H = C_G(z)$ be the centralizer of a 2-central involution of G. Let S be the Sylow 2-subgroup of H defined in Proposition 10.1.5, and let A be the maximal elementary abelian normal subgroup of S defined in Lemma 3.3.2(d). Then $D = N_H(A)$ is a proper subgroup of $E = N_G(A)$ by Lemma 3.3.2(e). Hence $G = \langle H, E \rangle$ by Lemma 3.3.11. Lemma 3.2.1(h) states that the amalgam $H \leftarrow D \rightarrow E$ has Goldschmidt index 1. Therefore the free product $P = H *_D E$ of H and E with amalgamated subgroup D is uniquely determined up to isomorphism. By Theorem 3.2.2 there is exactly one compatible pair $(\chi, \tau) \in mf\,char_{\mathbb{C}}(H) \times mf\,char_{\mathbb{C}}(E)$ of degree 23. As G is a simple epimorphic image of P it follows that G has at most one character $\chi \in Irr_{\mathbb{C}}(G)$ of minimal degree 23.

The simple group of Co_3-type $\mathfrak{G}$ constructed in Theorem 3.2.2 has the same number of conjugacy classes as G; see Lemma 3.3.10. By Table DVD.1.1.2 the group $\mathfrak{G}$ has an irreducible character of degree 23. Therefore we define a class function $\chi : G \to \mathbb{C}$ by

i	1_A	2_A	2_B	3_A	3_B	3_C	4_A	4_B	5_A	5_B	6_A	6_B
$\chi(g_i)$	23	7	−1	−4	5	−1	−5	3	−2	3	4	−2

i	6_C	6_D	6_E	7_A	8_A	8_B	8_C	9_A	9_B	10_A	10_B
$\chi(g_i)$	1	−1	−1	2	1	−3	1	−1	2	2	−1

i	11_A	11_B	12_A	12_B	12_C	14_A	15_A	15_B	18_A	20_A	20_B
$\chi(g_i)$	1	1	−2	0	1	0	1	0	1	0	0

i	21_A	22_A	22_B	23_A	23_B	24_A	24_B	30_A
$\chi(g_i)$	−1	−1	−1	0	0	0	−2	−1

Let p be any prime divisor of $|G|$. If Y is an over-group of a p-elementary subgroup X, and the restriction $\chi_{|Y}$ of X to Y is a generalized character, then also $\chi_{|X}$ is a generalized character of X. By Proposition 3.3.3 G has a second conjugacy class u^G of involutions and $U = C_G(u)$ is uniquely determined up to isomorphism. Let $N_1 = N_G(3_A)$, $N_2 = N_G(3_B)$, $N_3 = N_G(3_C)$ and $N_5 = N_G(5_A)$, $N_6 = N_G(5_B)$ be the normalizers of the cyclic subgroups of order 3 and 5, respectively. They are uniquely determined up to conjugation in G by Lemmas 3.3.4, 3.3.5, 3.3.6 and 3.3.7, 3.3.8, respectively.

By Lemma 3.3.10 the Sylow 23-normalizer N_{23} of G is a Frobenius group of order $23 \cdot 6$. As $\chi(x) = 0$ and $\chi(1) = 23$ its restriction $\chi_{|N_{23}}$ is a projective character of N_{23} over $GF(23)$. The Sylow 11-normalizer $N_G(11)$ of G is conjugate to a subgroup of U by Lemma 3.3.10 and Proposition 3.3.3. The 7-elementary subgroups of G are contained in N_3. Therefore by Corollary 2.8.10 and Theorem 2.8.9 of [92] it remains to show that $\chi_{|Y}$ is a generalized character of Y whenever Y belongs to the following set of subgroups:

$$\mathfrak{Y} = \{H, U, N_1, N_2, N_3, N_5, N_6\}$$

of G.

By means of the results of this section the fusion of the conjugacy classes of the groups $Y \in \mathfrak{Y}$ in G can be determined easily. Therefore the inner products $(\lambda_i, \chi)_Y$ can be calculated for each $\lambda_i \in Irr_{\mathbb{C}}(Y)$ and each $Y \in \mathfrak{Y}$ using the character tables given in Section 3.6. It turns out that

$$\begin{aligned}
\chi_{|H} &= \chi_3 + \chi_4, \text{ where } \quad \chi_3, \chi_4 \in Irr_{\mathbb{C}}(H),\\
\chi_{|U} &= \chi_2 + \chi_3 + \chi_4, \text{ where } \quad \chi_2, \chi_3, \chi_4 \in Irr_{\mathbb{C}}(U),\\
\chi_{|N_1} &= \chi_6 + \chi_{27}, \text{ where } \quad \chi_6, \chi_{27} \in Irr_{\mathbb{C}}(N_1),\\
\chi_{|N_2} &= \chi_3 + \chi_{15} + \chi_{17}, \text{ where } \quad \chi_3, \chi_{15}, \chi_{17} \in Irr_{\mathbb{C}}(N_2),\\
\chi_{|N_3} &= \chi_{11} + \chi_{25}, \text{ where } \quad \chi_{11}, \chi_{25} \in Irr_{\mathbb{C}}(N_3),\\
\chi_{|N_5} &= \chi_4 + \chi_9 + \chi_{19}, \text{ where } \quad \chi_4, \chi_9, \chi_{19} \in Irr_{\mathbb{C}}(N_5),\\
\chi_{|N_6} &= \chi_1 + \chi_5 + \chi_7 + \chi_{24}, \text{ where } \quad \chi_1, \chi_5, \chi_7, \chi_{24} \in Irr_{\mathbb{C}}(N_6).
\end{aligned}$$

Since the inner product $(\chi, \chi)_G = 1$, χ is an irreducible character of G with $\chi(1) = 23$ by Corollary 2.8.10 of [92] or Corollary 8.12 of [58]. This completes the proof. □

Theorem 3.3.13 (Fendel–Michler) *Let H be the finite group of even order defined in Proposition 3.1.2. Then each finite simple group G of* Co_3*-type is isomorphic to the finite simple group $\mathfrak{G} \le \mathrm{GL}_{23}(23)$ of order $|\mathfrak{G}| = 2^{10} \cdot 3^7 \cdot 5^3 \cdot 11 \cdot 23$ constructed in Theorem 3.2.2.*

Proof By Theorem 3.2.2 there exists a finite simple group $\mathfrak{G}$ of order $|\mathfrak{G}| = 2^{11} \cdot 3^3 \cdot 5^2 \cdot 13$ which is of Co_3-type.

Let G be any finite simple group of Co_3-type. By Proposition 3.3.3 G has exactly one 2-central conjugacy class z^G of involutions and the same order as $\mathfrak{G}$. Lemma 3.3.2 asserts that the Sylow 2-subgroup S of H contains a uniquely determined maximal elementary abelian normal subgroup A such that $D = N_H(A)$ is a proper subgroup of $E = N_G(A)$. Furthermore, E is uniquely determined by H up to isomorphism. Lemma 3.3.11 implies that $G = \langle H, E\rangle$. The amalgam $H \leftarrow D \rightarrow E$ has Goldschmidt index 1 by Lemma 3.2.1. Proposition 10.6.1 states that G has a unique irreducible complex character of degree 23. It is of defect 0 for the prime 23. Therefore each finite group G of Co_3-type has a unique irreducible 23-modular representation of degree 23 by Theorem 3.12.2 of [92]. Hence $G \cong \mathfrak{G}$ by Theorem 7.5.1 of [92]. This completes the proof.

□

3.4 Representatives of conjugacy classes

3.4.1 *Conjugacy classes of $H = \langle x, y, h\rangle$*

Class	*Representative*	*Class*	*Centralizer*	2P	3P	5P	7P
1	1	1	$2^{10}\cdot 3^4\cdot 5\cdot 7$	1	1	1	1
2_1	$(x)^4$	1	$2^{10}\cdot 3^4\cdot 5\cdot 7$	1	2_1	2_1	2_1
2_2	$(x^2y)^4$	630	$2^9\cdot 3^2$	1	2_2	2_2	2_2
2_3	$(x^2hy)^3$	7560	$2^7\cdot 3$	1	2_3	2_3	2_3
3_1	$(yh)^4$	672	$2^5\cdot 3^3\cdot 5$	3_1	1	3_1	3_1
3_2	$(xh)^3$	2240	$2^4\cdot 3^4$	3_2	1	3_2	3_2
3_3	h	13440	$2^3\cdot 3^3$	3_3	1	3_3	3_3
4_1	$(xhy)^3$	126	$2^9\cdot 3^2\cdot 5$	2_1	4_1	4_1	4_1
4_2	$(x)^2$	1890	$2^9\cdot 3$	2_1	4_2	4_2	4_2
4_3	$(xyh)^3$	3780	$2^8\cdot 3$	2_2	4_3	4_3	4_3
4_4	$(xy^2hy)^3$	3780	$2^8\cdot 3$	2_2	4_4	4_4	4_4
4_5	$(x^2y)^2$	22680	2^7	2_2	4_5	4_5	4_5
5	$(y^2h)^4$	48384	$2^2\cdot 3\cdot 5$	5	5	1	5
6_1	$(yh)^2$	672	$2^5\cdot 3^3\cdot 5$	3_1	2_1	6_1	6_1
6_2	$(xh^2)^3$	2240	$2^4\cdot 3^4$	3_2	2_1	6_2	6_2
6_3	$(xhy)^2$	13440	$2^3\cdot 3^3$	3_3	2_1	6_3	6_3
6_4	$(xyh)^2$	20160	$2^4\cdot 3^2$	3_2	2_2	6_4	6_4
6_5	x^3hy	20160	$2^4\cdot 3^2$	3_2	2_2	6_5	6_5
6_6	$xhxh^2y$	20160	$2^4\cdot 3^2$	3_1	2_2	6_6	6_6
6_7	xy^2	80640	$2^2\cdot 3^2$	3_3	2_2	6_7	6_7
6_8	x^2hy	241920	$2^2\cdot 3$	3_3	2_3	6_8	6_8
7	$(xy)^2$	207360	$2\cdot 7$	7	7	7	1
8_1	$(xy^3h)^3$	15120	$2^6\cdot 3$	4_2	8_1	8_1	8_1
8_2	$(x^2yhyxh)^3$	15120	$2^6\cdot 3$	4_2	8_2	8_2	8_2
8_3	x	90720	2^5	4_2	8_3	8_3	8_3
8_4	x^2y	90720	2^5	4_5	8_4	8_4	8_4
8_5	xhy^2	90720	2^5	4_5	8_5	8_5	8_5
8_6	x^2yhy	181440	2^4	4_3	8_6	8_6	8_6
9	xh	161280	$2\cdot 3^2$	9	3_2	9	9
10	$(y^2h)^2$	48384	$2^2\cdot 3\cdot 5$	5	10	2_1	10
12_1	xy^2xyhxh	20160	$2^4\cdot 3^2$	6_1	4_1	12_1	12_1
12_2	yh	60480	$2^4\cdot 3$	6_1	4_2	12_2	12_2
12_3	xhy	80640	$2^2\cdot 3^2$	6_3	4_1	12_3	12_3
12_4	xyh	120960	$2^3\cdot 3$	6_4	4_3	12_4	12_4
12_5	xy^2hy	120960	$2^3\cdot 3$	6_4	4_4	12_5	12_5
14	xy	207360	$2\cdot 7$	7	14	14	2_1
15	$(x^3yh)^2$	96768	$2\cdot 3\cdot 5$	15	5	3_1	15
18	xh^2	161280	$2\cdot 3^2$	9	6_2	18	18
20_1	y^2h	145152	$2^2\cdot 5$	10	20_1	4_1	20_1
20_2	$(y^2h)^{11}$	145152	$2^2\cdot 5$	10	20_2	4_1	20_2
24_1	xy^3h	120960	$2^3\cdot 3$	12_2	8_1	24_1	24_1
24_2	x^2yhyxh	120960	$2^3\cdot 3$	12_2	8_2	24_2	24_2
30	x^3yh	96768	$2\cdot 3\cdot 5$	15	10	6_1	30

3.4.2 *Conjugacy classes of* $E = \langle x, y, e \rangle$

Class	*Representative*	|*Class*|	|*Centralizer*|	2P	3P	5P	7P
1	1	1	$2^{10} \cdot 3^2 \cdot 5 \cdot 7$	1	1	1	1
2_1	$(x)^4$	15	$2^{10} \cdot 3 \cdot 7$	1	2_1	2_1	2_1
2_2	e	840	$2^7 \cdot 3$	1	2_2	2_2	2_2
2_3	$(x^3ey)^3$	840	$2^7 \cdot 3$	1	2_3	2_3	2_3
3_1	$(x^3ey)^2$	1792	$2^2 \cdot 3^2 \cdot 5$	3_1	1	3_1	3_1
3_2	$(ye)^2$	4480	$2^3 \cdot 3^2$	3_2	1	3_2	3_2
4_1	$(x^2yxy)^3$	210	$2^9 \cdot 3$	2_1	4_1	4_1	4_1
4_2	$(x)^2$	630	2^9	2_1	4_2	4_2	4_2
4_3	$(y)^2$	2520	2^7	2_1	4_3	4_3	4_3
4_4	xy^2ey	20160	2^4	2_3	4_4	4_4	4_4
5	xey	21504	$3 \cdot 5$	5	5	1	5
6_1	xy^2	13440	$2^3 \cdot 3$	3_2	2_1	6_1	6_1
6_2	ye	26880	$2^2 \cdot 3$	3_2	2_2	6_2	6_2
6_3	x^3ey	26880	$2^2 \cdot 3$	3_1	2_3	6_3	6_3
7_1	$(xy)^2$	23040	$2 \cdot 7$	7_1	7_2	7_2	1
7_2	$(xy)^6$	23040	$2 \cdot 7$	7_2	7_1	7_1	1
8_1	x^2y	5040	2^6	4_2	8_1	8_1	8_1
8_2	x^2ey	5040	2^6	4_2	8_2	8_2	8_2
8_3	x	10080	2^5	4_2	8_3	8_3	8_3
8_4	y	20160	2^4	4_3	8_4	8_4	8_4
12	x^2yxy	26880	$2^2 \cdot 3$	6_1	4_1	12	12
14_1	xy	23040	$2 \cdot 7$	7_1	14_2	14_2	2_1
14_2	$(xy)^3$	23040	$2 \cdot 7$	7_2	14_1	14_1	2_1
15_1	$xyxey$	21504	$3 \cdot 5$	15_1	5	3_1	15_2
15_2	$(xyxey)^7$	21504	$3 \cdot 5$	15_2	5	3_1	15_1

3.4.3 *Conjugacy classes of* $D = \langle x, y \rangle$

Class	*Representative*	|*Class*|	|*Centralizer*|	2P	3P	7P
1	1	1	$2^{10} \cdot 3 \cdot 7$	1	1	1
2_1	$(x)^4$	1	$2^{10} \cdot 3 \cdot 7$	1	2_1	2_1
2_2	$(x^2y)^4$	14	$2^9 \cdot 3$	1	2_2	2_2
2_3	$x^2yx^2y^3$	56	$2^7 \cdot 3$	1	2_3	2_3
2_4	$x^2y^2xy^2$	168	2^7	1	2_4	2_4
2_5	$x^2yx^2yxy^2$	336	2^6	1	2_5	2_5
3	$(xy^2)^2$	896	$2^3 \cdot 3$	3	1	3
4_1	$(x^2yxy)^3$	14	$2^9 \cdot 3$	2_1	4_1	4_1
4_2	$(x)^2$	42	2^9	2_1	4_2	4_2
4_3	x^2y^3xy	84	2^8	2_2	4_3	4_3
4_4	$(x^2yxyxy^2)^2$	84	2^8	2_2	4_4	4_4
4_5	$(y)^2$	168	2^7	2_1	4_5	4_5
4_6	$(x^2y)^2$	168	2^7	2_2	4_6	4_6
4_7	$x^2yxyx^2y^2$	336	2^6	2_2	4_7	4_7
4_8	$xyxy^3xy^2$	1344	2^4	2_4	4_8	4_8
6_1	$(x^2yxy)^2$	896	$2^3 \cdot 3$	3	2_1	6_1
6_2	xy^2	1792	$2^2 \cdot 3$	3	2_2	6_2
6_3	x^2yxy^2xy	1792	$2^2 \cdot 3$	3	2_3	6_3
7_1	$(xy)^2$	1536	$2 \cdot 7$	7_1	7_2	1
7_2	$(xy)^6$	1536	$2 \cdot 7$	7_2	7_1	1
8_1	x^2yxy^3	336	2^6	4_2	8_1	8_1
8_2	$x^2y^3xy^2xy$	336	2^6	4_2	8_2	8_2
8_3	x	672	2^5	4_2	8_3	8_3
8_4	x^2y	672	2^5	4_6	8_4	8_4
8_5	x^6y	672	2^5	4_6	8_5	8_5
8_6	y	1344	2^4	4_5	8_6	8_6
8_7	x^2yxyxy^2	1344	2^4	4_4	8_7	8_7
12	x^2yxy	1792	$2^2 \cdot 3$	6_1	4_1	12
14_1	xy	1536	$2 \cdot 7$	7_1	14_2	2_1
14_2	$(xy)^3$	1536	$2 \cdot 7$	7_2	14_1	2_1

3.4.4 *Conjugacy classes of* $N_1 = N_G(r_1) = \langle k_i | 1 \leq i \leq 6 \rangle$

Class	*Representative*	*Centralizer*	2P	3P	5P
1	1	$2^6 \cdot 3^7 \cdot 5$	1	1	1
2_1	$(k_1)^3$	$2^6 \cdot 3^3 \cdot 5$	1	2_1	2_1
2_2	$(k_1 k_2 k_3)^3$	$2^5 \cdot 3^4$	1	2_2	2_2
2_3	$(k_1 k_2 k_3 k_2 k_4)^3$	$2^5 \cdot 3^3$	1	2_3	2_3
2_4	$k_2 k_3^2 k_4 k_3$	$2^5 \cdot 3^2$	1	2_4	2_4
3_1	k_6	$2^5 \cdot 3^7 \cdot 5$	3_1	1	3_1
3_2	k_5	$2^2 \cdot 3^6$	3_2	1	3_2
3_3	$(k_1)^2$	$2^3 \cdot 3^5$	3_3	1	3_3
3_4	$(k_1 k_2 k_3)^2$	$2^3 \cdot 3^5$	3_4	1	3_4
3_5	$(k_1 k_5)^2$	$2^2 \cdot 3^5$	3_5	1	3_5
3_6	$(k_1 k_2 k_3 k_6)^2$	$2^2 \cdot 3^5$	3_6	1	3_6
3_7	$(k_1 k_2 k_3 k_2 k_4 k_5)^2$	$2 \cdot 3^4$	3_7	1	3_7
4_1	$(k_2 k_4)^3$	$2^5 \cdot 3^2 \cdot 5$	2_1	4_1	4_1
4_2	k_2	$2^5 \cdot 3^2$	2_1	4_2	4_2
4_3	k_4	$2^5 \cdot 3$	2_1	4_3	4_3
4_4	$(k_1 k_4)^2$	$2^5 \cdot 3$	2_1	4_4	4_4
5	$(k_1 k_3)^2$	$2^2 \cdot 3 \cdot 5$	5	5	1
6_1	$(k_2 k_5)^2$	$2^5 \cdot 3^3 \cdot 5$	3_1	2_1	6_1
6_2	$(k_1 k_2 k_3 k_5)^3$	$2^4 \cdot 3^4$	3_1	2_2	6_2
6_3	k_1	$2^3 \cdot 3^3$	3_3	2_1	6_3
6_4	$(k_1 k_2 k_3 k_4)^2$	$2^3 \cdot 3^3$	3_4	2_1	6_4
6_5	$k_1 k_5$	$2^2 \cdot 3^3$	3_5	2_1	6_5
6_6	$k_1 k_2 k_3$	$2^2 \cdot 3^3$	3_4	2_2	6_6
6_7	$k_1 k_2 k_3 k_6$	$2^2 \cdot 3^3$	3_6	2_2	6_7
6_8	$k_1 k_2 k_3 k_5 k_3$	$2^2 \cdot 3^3$	3_6	2_2	6_8
6_9	$k_1^2 k_2 k_4 k_3 k_4 k_5$	$2^2 \cdot 3^3$	3_2	2_2	6_9
6_{10}	$k_1 k_2 k_3 k_2 k_3 k_4 k_5$	$2^2 \cdot 3^3$	3_2	2_3	6_{10}
6_{11}	$k_1^2 k_2 k_3 k_2 k_3^2 k_5$	$2^2 \cdot 3^3$	3_6	2_1	6_{11}
6_{12}	$k_1 k_2 k_3 k_2 k_4$	$2^2 \cdot 3^2$	3_4	2_3	6_{12}
6_{13}	$k_2 k_3^2 k_4 k_5 k_3$	$2^2 \cdot 3^2$	3_2	2_4	6_{13}
6_{14}	$k_1 k_2 k_3 k_2 k_4 k_5$	$2 \cdot 3^2$	3_7	2_3	6_{14}
8_1	$k_2 k_3$	$2^4 \cdot 3$	4_4	8_1	8_1
8_2	$k_1 k_2 k_3 k_2 k_3$	$2^4 \cdot 3$	4_4	8_2	8_2
8_3	$k_1 k_3 k_2 k_4$	2^5	4_4	8_3	8_4
8_4	$(k_1 k_3 k_2 k_4)^5$	2^5	4_4	8_4	8_3
8_5	$k_1 k_4$	2^4	4_4	8_5	8_5
9_1	$(k_1 k_2 k_3 k_5)^2$	$2 \cdot 3^4$	9_1	3_1	9_1
9_2	$k_3^2 k_5$	3^4	9_2	3_1	9_2
10	$k_1 k_3$	$2^2 \cdot 3 \cdot 5$	5	10	2_1
12_1	$k_2 k_5$	$2^4 \cdot 3^2$	6_1	4_2	12_1
12_2	$(k_2 k_3 k_5)^2$	$2^4 \cdot 3$	6_1	4_4	12_2
12_3	$k_1 k_2$	$2^2 \cdot 3^2$	6_3	4_2	12_3
12_4	$k_2 k_4$	$2^2 \cdot 3^2$	6_3	4_1	12_4
12_5	$k_1 k_2 k_5$	$2^2 \cdot 3^2$	6_5	4_2	12_5
12_6	$(k_1 k_2 k_5)^7$	$2^2 \cdot 3^2$	6_5	4_2	12_6
12_7	$k_1 k_2 k_3 k_4$	$2^2 \cdot 3^2$	6_4	4_1	12_7
12_8	$k_1 k_3 k_4$	$2^2 \cdot 3$	6_3	4_3	12_8
15	$(k_1 k_3 k_6)^2$	$2 \cdot 3 \cdot 5$	15	5	3_1
18	$k_1 k_2 k_3 k_5$	$2 \cdot 3^2$	9_1	6_2	18
20_1	$k_1 k_2 k_4$	$2^2 \cdot 5$	10	20_1	4_1
20_2	$(k_1 k_2 k_4)^{11}$	$2^2 \cdot 5$	10	20_2	4_1
24_1	$k_2 k_3 k_5$	$2^3 \cdot 3$	12_2	8_1	24_1
24_2	$k_1 k_2 k_3 k_2 k_3 k_5$	$2^3 \cdot 3$	12_2	8_2	24_2
30	$k_1 k_3 k_6$	$2 \cdot 3 \cdot 5$	15	10	6_1

3.4.5 *Conjugacy classes of* $N_2 = N_G(r_2) = \langle n_i | 1 \leq i \leq 5\rangle$

Class	*Representative*	|*Centralizer*|	2P	3P	5P
1	1	$2^4 \cdot 3^6 \cdot 5$	1	1	1
2_1	n_5	$2^3 \cdot 3^4$	1	2_1	2_1
2_2	n_4	$2^4 \cdot 3^3$	1	2_2	2_2
2_3	n_1	$2^4 \cdot 3 \cdot 5$	1	2_3	2_3
2_4	n_1n_5	$2^3 \cdot 3^3$	1	2_4	2_4
2_5	n_1n_4	$2^4 \cdot 3^2$	1	2_5	2_5
3_1	$(n_2n_3n_4)^5$	$2^3 \cdot 3^6 \cdot 5$	3_1	1	3_1
3_2	$(n_2n_3)^3$	$2^2 \cdot 3^6$	3_2	1	3_2
3_3	n_2	$2^3 \cdot 3^5$	3_3	1	3_3
3_4	$(n_1n_2n_4n_5)^4$	$2^3 \cdot 3^5$	3_4	1	3_4
3_5	$(n_2n_4)^2$	$2 \cdot 3^6$	3_5	1	3_5
3_6	$(n_1n_2n_4)^2$	$2^2 \cdot 3^5$	3_6	1	3_6
3_7	$(n_1n_2n_3n_4n_3n_4n_5)^4$	$2^2 \cdot 3^5$	3_7	1	3_7
3_8	n_3	$2^2 \cdot 3^3$	3_8	1	3_8
3_9	$n_2n_4n_3n_4$	$2 \cdot 3^3$	3_9	1	3_9
4_1	$n_1n_4n_5$	$2^3 \cdot 3^2$	2_2	4_1	4_1
4_2	n_4n_5	$2^3 \cdot 3$	2_2	4_2	4_2
5	n_3n_4	$2 \cdot 3 \cdot 5$	5	5	1
6_1	$(n_2n_3n_5)^3$	$2^2 \cdot 3^4$	3_2	2_1	6_1
6_2	$(n_2n_4n_5)^2$	$2^3 \cdot 3^3$	3_3	2_2	6_2
6_3	$(n_1n_2n_4n_5)^2$	$2^3 \cdot 3^3$	3_4	2_2	6_3
6_4	$(n_1n_2n_4n_3n_5n_4n_3)^2$	$2^3 \cdot 3^3$	3_1	2_2	6_4
6_5	n_2n_5	$2^2 \cdot 3^3$	3_3	2_1	6_5
6_6	$n_2n_3n_4n_3^2$	$2^2 \cdot 3^3$	3_3	2_2	6_6
6_7	$n_2n_4n_2n_4n_5$	$2^2 \cdot 3^3$	3_4	2_1	6_7
6_8	$n_2n_4n_5n_4n_5$	$2^2 \cdot 3^3$	3_2	2_2	6_8
6_9	$n_1n_2n_4n_3n_4n_5$	$2^2 \cdot 3^3$	3_1	2_4	6_9
6_{10}	$n_2^2n_4n_2n_4n_5$	$2^2 \cdot 3^3$	3_2	2_1	6_{10}
6_{11}	$(n_1n_2n_3n_4n_3n_4n_5)^2$	$2^2 \cdot 3^3$	3_7	2_2	6_{11}
6_{12}	$n_2^2n_3n_2n_3^2n_5$	$2^2 \cdot 3^3$	3_6	2_1	6_{12}
6_{13}	n_2n_4	$2 \cdot 3^3$	3_5	2_2	6_{13}
6_{14}	n_3n_5	$2^2 \cdot 3^2$	3_8	2_1	6_{14}
6_{15}	$n_1n_2n_4$	$2^2 \cdot 3^2$	3_6	2_5	6_{15}
6_{16}	$n_1n_3n_5$	$2^2 \cdot 3^2$	3_8	2_4	6_{16}
6_{17}	$n_1n_2n_4n_5n_4$	$2^2 \cdot 3^2$	3_6	2_4	6_{17}
6_{18}	$n_1n_2n_3n_2n_3^2n_4$	$2^2 \cdot 3^2$	3_4	2_5	6_{18}
6_{19}	$n_1n_2n_4n_3n_5$	$2 \cdot 3^2$	3_9	2_4	6_{19}
6_{20}	n_1n_3	$2^2 \cdot 3$	3_8	2_3	6_{20}
9_1	n_2n_3	$2 \cdot 3^3$	9_1	3_2	9_1
9_2	$n_2n_3n_4n_2n_4$	3^3	9_2	3_2	9_2
10	$n_1n_3n_4$	$2 \cdot 5$	5	10	2_3
12_1	$n_1n_2n_4n_5$	$2^2 \cdot 3^2$	6_3	4_1	12_1
12_2	$n_1n_2n_3n_4n_3n_4n_5$	$2^2 \cdot 3^2$	6_{11}	4_1	12_2
12_3	$(n_1n_2n_3n_4n_3n_4n_5)^7$	$2^2 \cdot 3^2$	6_{11}	4_1	12_3
12_4	$n_1n_2n_4n_3n_5n_4n_3$	$2^2 \cdot 3^2$	6_4	4_1	12_4
12_5	$n_2n_4n_5$	$2^2 \cdot 3$	6_2	4_2	12_5
15	$n_2n_3n_4$	$3 \cdot 5$	15	5	3_1
18	$n_2n_3n_5$	$2 \cdot 3^2$	9_1	6_1	18

3.4.6 *Conjugacy classes of $N_3 = N_G(r_3) = \langle q, r, s, t \rangle$*

Class	*Representative*	*Centralizer*	2P	3P	7P
1	1	$2^4 \cdot 3^4 \cdot 7$	1	1	1
2_1	s	$2^4 \cdot 3^3 \cdot 7$	1	2_1	2_1
2_2	$(pt)^3$	$2^4 \cdot 3^2$	1	2_2	2_2
2_3	$(pts)^3$	$2^4 \cdot 3$	1	2_3	2_3
3_1	r	$2^3 \cdot 3^4 \cdot 7$	3_1	1	3_1
3_2	p	$2 \cdot 3^4$	3_2	1	3_2
3_3	t	$2^2 \cdot 3^3$	3_4	1	3_3
3_4	$(t)^2$	$2^2 \cdot 3^3$	3_3	1	3_4
3_5	pr	3^4	3_5	1	3_5
3_6	tr	$2 \cdot 3^3$	3_7	1	3_6
3_7	$(tr)^2$	$2 \cdot 3^3$	3_6	1	3_7
6_1	$ptptptr$	$2^3 \cdot 3^2$	3_1	2_2	6_1
6_2	ps	$2 \cdot 3^3$	3_2	2_1	6_2
6_3	pt	$2^2 \cdot 3^2$	3_4	2_2	6_3
6_4	$(pt)^5$	$2^2 \cdot 3^2$	3_3	2_2	6_4
6_5	ts	$2^2 \cdot 3^2$	3_4	2_1	6_5
6_6	$(ts)^5$	$2^2 \cdot 3^2$	3_3	2_1	6_6
6_7	ptr	$2 \cdot 3^2$	3_7	2_2	6_7
6_8	$(ptr)^5$	$2 \cdot 3^2$	3_6	2_2	6_8
6_9	pts	$2^2 \cdot 3$	3_4	2_3	6_9
6_{10}	$(pts)^5$	$2^2 \cdot 3$	3_3	2_3	6_{10}
7	$ptpt^2$	$2 \cdot 3 \cdot 7$	7	7	1
9_1	p^2tpt	$2 \cdot 3^3$	9_2	3_2	9_1
9_2	$(p^2tpt)^2$	$2 \cdot 3^3$	9_1	3_2	9_2
9_3	p^2tpt^2	$2 \cdot 3^3$	9_3	3_2	9_3
9_4	p^2tptr	3^3	9_5	3_2	9_4
9_5	$(p^2tptr)^2$	3^3	9_4	3_2	9_5
9_6	p^2tpt^2r	3^3	9_6	3_2	9_6
14	$ptpt^2s$	$2 \cdot 7$	7	14	2_1
18_1	p^2tpts	$2 \cdot 3^2$	9_2	6_2	18_1
18_2	$(p^2tpts)^5$	$2 \cdot 3^2$	9_1	6_2	18_2
18_3	p^2tpt^2s	$2 \cdot 3^2$	9_3	6_2	18_3
21	$ptpt^2r$	$3 \cdot 7$	21	7	3_1

3.4.7 *Conjugacy classes of* $N_5 = N_G(f_1) = \langle q_i \mid 1 \le i \le 5 \rangle$

Class	*Representative*	*Centralizer*	2P	3P	5P
1	1	$2^4 \cdot 3 \cdot 5^3$	1	1	1
2_1	$(q_1)^2$	$2^4 \cdot 3 \cdot 5$	1	2_1	2_1
2_2	$q_1 q_3^2$	$2^3 \cdot 5$	1	2_2	2_2
3	$(q_2)^2$	$2^3 \cdot 3 \cdot 5$	3	1	3
4_1	$(q_3)^2$	$2^4 \cdot 3$	2_1	4_2	4_1
4_2	$(q_3)^6$	$2^4 \cdot 3$	2_1	4_1	4_2
4_3	q_1	$2^3 \cdot 5$	2_1	4_3	4_3
5_1	q_5	$2^2 \cdot 3 \cdot 5^3$	5_1	5_1	1
5_2	q_4	$2 \cdot 5^2$	5_2	5_2	1
6	q_2	$2^3 \cdot 3 \cdot 5$	3	2_1	6
8_1	q_3	$2^3 \cdot 3$	4_1	8_2	8_1
8_2	$(q_3)^3$	$2^3 \cdot 3$	4_2	8_1	8_2
8_3	$q_1 q_3$	2^3	4_1	8_4	8_3
8_4	$(q_1 q_3)^3$	2^3	4_2	8_3	8_4
10_1	$(q_1 q_5)^2$	$2^2 \cdot 3 \cdot 5$	5_1	10_1	2_1
10_2	$q_1 q_3^2 q_4$	$2 \cdot 5$	5_2	10_2	2_2
12_1	$(q_2 q_3)^2$	$2^3 \cdot 3$	6	4_2	12_1
12_2	$(q_2 q_3)^{14}$	$2^3 \cdot 3$	6	4_1	12_2
15	$(q_2 q_5)^2$	$2 \cdot 3 \cdot 5$	15	5_1	3
20_1	$q_1 q_5$	$2^2 \cdot 5$	10_1	20_1	4_3
20_2	$(q_1 q_5)^{11}$	$2^2 \cdot 5$	10_1	20_2	4_3
24_1	$q_2 q_3$	$2^3 \cdot 3$	12_1	8_2	24_1
24_2	$(q_2 q_3)^7$	$2^3 \cdot 3$	12_2	8_1	24_2
24_3	$(q_2 q_3)^{13}$	$2^3 \cdot 3$	12_1	8_2	24_3
24_4	$(q_2 q_3)^{19}$	$2^3 \cdot 3$	12_2	8_1	24_4
30	$q_2 q_5$	$2 \cdot 3 \cdot 5$	15	10_1	6

3.4.8 *Conjugacy classes of* $N_6 = N_G(f_2) = \langle x_i \mid 1 \le i \le 4 \rangle$

Class	*Representative*	*Centralizer*	2P	3P	5P
1	1	$2^4 \cdot 3 \cdot 5^2$	1	1	1
2_1	$(x_4)^2$	$2^4 \cdot 3 \cdot 5$	1	2_1	2_1
2_2	$x_1^2 x_2 x_1 x_2$	$2^4 \cdot 5$	1	2_2	2_2
2_3	$x_1^2 x_2 x_1 x_2 x_4^2$	2^4	1	2_3	2_3
3	x_1	$2^2 \cdot 3 \cdot 5$	3	1	3
4_1	x_4	$2^4 \cdot 3 \cdot 5$	2_1	4_2	4_1
4_2	$(x_1 x_4)^3$	$2^4 \cdot 3 \cdot 5$	2_1	4_1	4_2
4_3	$x_1^2 x_2 x_1 x_2 x_4$	2^4	2_1	4_4	4_3
4_4	$x_1^2 x_2 x_1 x_2 x_4^3$	2^4	2_1	4_3	4_4
5_1	x_3	$2^2 \cdot 3 \cdot 5^2$	5_1	5_1	1
5_2	$x_1 x_2$	$2^2 \cdot 5^2$	5_3	5_3	1
5_3	$x_1^2 x_2$	$2^2 \cdot 5^2$	5_2	5_2	1
5_4	$x_1 x_2 x_3$	5^2	5_5	5_5	1
5_5	$x_1^2 x_2 x_3$	5^2	5_4	5_4	1
6	$(x_1 x_4)^2$	$2^2 \cdot 3$	3	2_1	6
10_1	$(x_1 x_2 x_4)^2$	$2^2 \cdot 5$	5_2	10_2	2_1
10_2	$(x_1^2 x_2 x_4)^2$	$2^2 \cdot 5$	5_3	10_1	2_1
10_3	$x_1^2 x_2 x_1 x_2 x_3$	$2^2 \cdot 5$	5_1	10_3	2_2
12_1	$x_1 x_4$	$2^2 \cdot 3$	6	4_2	12_1
12_2	$x_1 x_4^3$	$2^2 \cdot 3$	6	4_1	12_2
15	$x_1 x_3$	$3 \cdot 5$	15	5_1	3
20_1	$x_1 x_2 x_4$	$2^2 \cdot 5$	10_1	20_4	4_1
20_2	$x_1^2 x_2 x_4$	$2^2 \cdot 5$	10_2	20_3	4_1
20_3	$x_1 x_2 x_4^3$	$2^2 \cdot 5$	10_1	20_2	4_2
20_4	$x_1^2 x_2 x_4^3$	$2^2 \cdot 5$	10_2	20_1	4_2

3.5 Character tables of local subgroups

3.5.1 *Character table of* $E = \langle x, y, e \rangle$

	1a	2a	2b	2c	3a	3b	4a	4b	4c	4d	5a	6a	6b	6c	7a	7b	8a	8b	8c	8d	12a	14a	14b	15a	15b
2	10	10	7	7	2	3	9	9	7	4	.	3	2	2	1	1	6	6	5	4	2	1	1	.	.
3	2	1	1	1	2	2	1	.	.	.	1	1	1	1	.	.	.	.	.	.	1	.	.	1	1
5	1	.	.	.	1	.	.	.	.	.	1	.	.	.	.	.	.	.	.	.	.	.	.	1	1
7	1	1	.	.	.	.	.	.	.	.	.	.	.	.	1	1	.	.	.	.	.	1	1	.	.
2P	1a	1a	1a	1a	3a	3b	2a	2a	2a	2c	5a	3b	3b	3a	7a	7b	4b	4b	4b	4c	6a	7a	7b	15a	15b
3P	1a	2a	2b	2c	1a	1a	4a	4b	4c	4d	5a	2a	2b	2c	7b	7a	8a	8b	8c	8d	4a	14b	14a	5a	5a
5P	1a	2a	2b	2c	3a	3b	4a	4b	4c	4d	1a	6a	6b	6c	7b	7a	8a	8b	8c	8d	12a	14b	14a	3a	3a
7P	1a	2a	2b	2c	3a	3b	4a	4b	4c	4d	5a	6a	6b	6c	1a	1a	8a	8b	8c	8d	12a	2a	2a	15b	15a
X.1	1	1	1	1	1	1	1	1	1	1	1	1	1	1	1	1	1	1	1	1	1	1	1	1	1
X.2	7	7	−1	3	4	1	−1	−1	3	1	2	1	−1	.	.	.	−1	−1	−1	1	−1	.	.	−1	−1
X.3	14	14	6	2	−1	2	6	6	2	.	−1	2	.	−1	.	.	2	2	2	.	.	.	.	−1	−1
X.4	15	−1	−1	3	.	3	−5	3	−1	1	.	−1	−1	.	1	1	1	−3	1	−1	1	−1	−1	.	.
X.5	20	20	4	4	5	−1	4	4	4	.	.	−1	1	1	−1	−1	.	.	.	.	1	−1	−1	.	.
X.6	21	21	−3	1	6	.	−3	−3	1	−1	1	.	.	−2	.	.	1	1	1	−1	.	.	.	1	1
X.7	21	21	−3	1	−3	.	−3	−3	1	−1	1	.	.	1	.	.	1	1	1	−1	.	.	.	A	$\bar{A}$
X.8	21	21	−3	1	−3	.	−3	−3	1	−1	1	.	.	1	.	.	1	1	1	−1	.	.	.	$\bar{A}$	A
X.9	28	28	−4	4	1	1	−4	−4	4	.	−2	1	−1	1	.	.	.	.	.	.	−1	.	.	1	1
X.10	35	35	3	−5	5	2	3	3	−5	−1	.	2	.	1	.	.	−1	−1	−1	−1	.	.	.	.	.
X.11	45	45	−3	−3	.	.	−3	−3	−3	1	.	.	.	.	B	$\bar{B}$	1	1	1	1	.	$\bar{B}$	B	.	.
X.12	45	−3	−3	−3	.	.	9	1	1	1	.	.	.	.	$\bar{B}$	B	3	−1	−1	−1	.	$-B$	$-\bar{B}$	.	.
X.13	45	45	−3	−3	.	.	−3	−3	−3	1	.	.	.	.	$\bar{B}$	B	1	1	1	1	.	B	$\bar{B}$	.	.
X.14	45	−3	−3	−3	.	.	9	1	1	1	.	.	.	.	B	$\bar{B}$	3	−1	−1	−1	.	$-\bar{B}$	$-B$	.	.
X.15	56	56	8	.	−4	−1	8	8	.	.	1	−1	−1	.	.	.	.	.	.	.	−1	.	.	1	1
X.16	64	64	.	.	4	−2	.	.	.	.	−1	−2	.	.	1	1	.	.	.	.	.	1	1	−1	−1
X.17	70	70	−2	2	−5	1	−2	−2	2	.	.	1	1	−1	.	.	−2	−2	−2	.	1	.	.	.	.
X.18	90	−6	−6	6	.	.	−6	10	−2	.	.	.	.	.	−1	−1	−2	−2	2	.	.	1	1	.	.
X.19	105	−7	−7	−3	.	3	13	5	1	−1	.	−1	−1	.	.	.	−1	3	−1	1	1	.	.	.	.
X.20	105	−7	1	9	.	3	5	−3	−3	1	.	−1	1	.	.	.	−1	3	−1	−1	−1	.	.	.	.
X.21	105	−7	1	−3	.	3	−19	5	1	−1	.	−1	1	.	.	.	3	−1	−1	1	−1	.	.	.	.
X.22	120	−8	−8	.	.	−3	8	8	.	.	.	1	1	.	1	1	.	.	.	.	−1	−1	−1	.	.
X.23	210	−14	2	6	.	−3	−14	2	−2	.	.	1	−1	.	.	.	2	2	−2	.	1	.	.	.	.
X.24	315	−21	3	3	.	.	15	−9	−1	−1	.	.	.	.	.	.	1	−3	1	1	.	.	.	.	.
X.25	315	−21	3	−9	.	.	−9	−1	3	1	.	.	.	.	.	.	−3	1	1	−1	.	.	.	.	.

where $A = -\frac{1}{2}(1 + i\sqrt{15}), B = -\frac{1}{2}(1 + i\sqrt{7})$.

3.5.2 *Character table of* $D = \langle x, y \rangle$

	1a	2a	2b	2c	2d	2e	3a	4a	4b	4c	4d	4e	4f	4g	4h	6a	6b	6c	7a	7b	8a	8b	8c	8d	8e	8f	8g	12a	14a	14b
2	10	10	9	7	7	6	3	9	9	8	8	7	7	6	4	3	2	2	1	1	6	6	5	5	5	4	4	2	1	1
3	1	1	1	1	.	.	1	1	.	.	.	.	.	.	.	1	1	1	.	.	.	.	.	.	.	.	.	1	.	.
7	1	1	.	.	.	.	.	.	.	.	.	.	.	.	.	.	.	.	1	1	.	.	.	.	.	.	.	.	1	1
2P	1a	1a	1a	1a	1a	1a	3a	2a	2a	2b	2b	2a	2b	2b	2d	3a	3a	3a	7a	7b	4b	4b	4b	4f	4f	4e	4d	6a	7a	7b
3P	1a	2a	2b	2c	2d	2e	1a	4a	4b	4c	4d	4e	4f	4g	4h	2a	2b	2c	7b	7a	8a	8b	8c	8d	8e	8f	8g	4a	14b	14a
7P	1a	2a	2b	2c	2d	2e	3a	4a	4b	4c	4d	4e	4f	4g	4h	6a	6b	6c	1a	1a	8a	8b	8c	8d	8e	8f	8g	12a	2a	2a
X.1	1	1	1	1	1	1	1	1	1	1	1	1	1	1	1	1	1	1	1	1	1	1	1	1	1	1	1	1	1	1
X.2	3	3	3	3	−1	−1	.	3	3	−1	−1	−1	−1	−1	1	.	.	.	A	$\bar{A}$	−1	−1	1	−1	1	1	1	.	A	$\bar{A}$
X.3	3	3	3	3	−1	−1	.	3	3	−1	−1	−1	−1	−1	1	.	.	.	$\bar{A}$	A	−1	−1	1	−1	1	1	1	.	$\bar{A}$	A
X.4	6	6	6	6	2	2	.	6	6	2	2	2	2	2	.	.	.	.	−1	−1	2	2	.	2	.	.	.	.	−1	−1
X.5	7	7	−1	−1	3	−1	1	−5	3	−1	−1	3	−1	−1	1	1	−1	−1	.	.	1	−3	−1	1	−1	1	−1	1	.	.
X.6	7	7	−1	−1	−1	−1	1	−5	3	3	3	−1	3	−1	−1	1	−1	−1	.	.	−3	1	1	1	1	−1	1	1	.	.
X.7	7	7	7	−1	−1	3	1	−1	−1	3	3	3	−1	−1	−1	1	−1	1	.	.	−1	−1	1	−1	1	1	−1	−1	.	.
X.8	7	7	7	−1	3	−1	1	−1	−1	−1	−1	−1	3	3	1	1	−1	1	.	.	−1	−1	−1	−1	−1	−1	1	−1	.	.
X.9	7	7	7	7	−1	−1	1	7	7	−1	−1	−1	−1	−1	−1	1	1	1	.	.	−1	−1	−1	−1	−1	−1	−1	1	.	.
X.10	8	−8	.	.	.	.	2	.	.	−4	4	.	.	.	.	−2	.	.	1	1	.	.	2	.	−2	.	.	.	−1	−1
X.11	8	8	8	8	.	.	−1	8	8	.	.	.	.	.	.	−1	−1	−1	1	1	.	.	.	.	.	.	.	−1	1	1
X.12	14	14	14	−2	2	2	−1	−2	−2	2	2	2	2	2	.	−1	1	−1	.	.	−2	−2	.	−2	.	.	.	1	.	.
X.13	14	14	−2	−2	2	−2	−1	−10	6	2	2	2	2	−2	.	−1	1	1	.	.	−2	−2	.	2	.	.	.	−1	.	.
X.14	21	21	−3	−3	1	1	.	9	1	1	1	−3	5	−3	−1	.	.	.	.	.	3	−1	−1	−1	−1	1	1	.	.	.
X.15	21	21	−3	−3	−3	−3	.	9	1	5	5	1	1	1	1	.	.	.	.	.	3	−1	1	−1	1	−1	−1	.	.	.
X.16	21	21	−3	−3	−3	1	.	−15	9	1	1	−3	1	1	1	.	.	.	.	.	−1	3	−1	−1	−1	1	−1	.	.	.
X.17	21	21	−3	−3	1	−3	.	9	1	1	1	5	−3	1	−1	.	.	.	.	.	−1	3	−1	−1	−1	1	1	.	.	.
X.18	21	21	−3	−3	1	1	.	−15	9	−3	−3	1	−3	1	−1	.	.	.	.	.	3	−1	1	−1	1	−1	1	.	.	.
X.19	21	21	21	−3	−3	1	.	−3	−3	1	1	1	−3	−3	1	.	.	.	.	.	1	1	−1	1	−1	−1	1	.	.	.
X.20	21	21	−3	−3	5	1	.	9	1	−3	−3	1	1	−3	1	.	.	.	.	.	−1	3	1	−1	1	−1	−1	.	.	.
X.21	21	21	21	−3	1	−3	.	−3	−3	−3	−3	−3	1	1	−1	.	.	.	.	.	1	1	1	1	1	1	−1	.	.	.
X.22	24	−24	.	.	.	.	.	.	.	4	−4	.	.	.	.	.	.	.	$\bar{A}$	A	.	.	2	.	−2	.	.	.	$-\bar{A}$	$-A$
X.23	24	−24	.	.	.	.	.	.	.	4	−4	.	.	.	.	.	.	.	A	$\bar{A}$	.	.	2	.	−2	.	.	.	$-A$	$-\bar{A}$
X.24	28	28	−4	4	4	.	1	−4	−4	4	4	−4	−4	.	.	1	1	−1	.	.	.	.	.	.	.	.	.	−1	.	.
X.25	28	28	−4	4	−4	.	1	−4	−4	−4	−4	4	4	.	.	1	1	−1	.	.	.	.	.	.	.	.	.	−1	.	.
X.26	42	42	−6	−6	−2	2	.	18	2	−2	−2	−2	−2	2	.	.	.	.	.	.	−2	−2	.	2	.	.	.	.	.	.
X.27	48	−48	.	.	.	.	.	.	.	−8	8	.	.	.	.	.	.	.	−1	−1	.	.	.	.	.	.	.	.	1	1
X.28	56	56	−8	8	.	.	−1	−8	−8	.	.	.	.	.	.	−1	−1	1	.	.	.	.	.	.	.	.	.	1	.	.
X.29	56	−56	.	.	.	.	2	.	.	4	−4	.	.	.	.	−2	.	.	.	.	.	.	−2	.	2	.	.	.	.	.
X.30	64	−64	.	.	.	.	−2	.	.	.	.	.	.	.	.	2	.	.	1	1	.	.	.	.	.	.	.	.	−1	−1

where $A = -\frac{1}{2}(1 + i\sqrt{7})$.

3.5.3 *Character table of* $H = \langle x, y, h \rangle$

	1a	2a	2b	2c	3a	3b	3c	4a	4b	4c	4d	4e	5a	6a	6b	6c	6d	6e	6f	6g	6h	7a	8a	8b	8c	8d	8e
2	10	10	9	7	5	4	3	9	9	8	8	7	2	5	4	3	4	4	4	2	2	1	6	6	5	5	5
3	4	4	2	1	3	4	3	2	1	1	1	.	1	3	4	3	2	2	2	2	1	.	1	1	.	.	.
5	1	1	.	.	1	.	.	1	.	.	.	.	1	1	.	.	.	.	.	.	.	.	.	.	.	.	.
7	1	1	.	.	.	.	.	.	.	.	.	.	.	.	.	.	.	.	.	.	.	1	.	.	.	.	.
2P	1a	1a	1a	1a	3a	3b	3c	2a	2a	2b	2b	2b	5a	3a	3b	3c	3b	3b	3a	3c	3c	7a	4b	4b	4b	4e	4e
3P	1a	2a	2b	2c	1a	1a	1a	4a	4b	4c	4d	4e	5a	2a	2a	2a	2b	2b	2b	2b	2c	7a	8a	8b	8c	8d	8e
5P	1a	2a	2b	2c	3a	3b	3c	4a	4b	4c	4d	4e	1a	6a	6b	6c	6d	6e	6f	6g	6h	7a	8a	8b	8c	8d	8e
7P	1a	2a	2b	2c	3a	3b	3c	4a	4b	4c	4d	4e	5a	6a	6b	6c	6d	6e	6f	6g	6h	1a	8a	8b	8c	8d	8e
X.1	1	1	1	1	1	1	1	1	1	1	1	1	1	1	1	1	1	1	1	1	1	1	1	1	1	1	1
X.2	7	7	−1	−1	4	−2	1	−5	3	3	3	−1	2	4	−2	1	2	2	2	−1	−1	.	−3	1	1	1	1
X.3	8	−8	.	.	−4	−1	2	.	.	−4	4	.	−2	4	1	−2	3	.	−3	.	.	1	.	.	−2	.	2
X.4	15	15	7	−1	.	−3	3	−5	3	−1	−1	3	.	.	−3	−3	1	−2	1	1	−1	1	1	−3	−1	1	−1
X.5	21	21	5	−3	6	3	.	−11	5	1	1	1	1	6	3	.	−1	2	−1	2	.	.	−3	−3	−1	1	−1
X.6	21	21	−3	−3	6	3	.	9	1	5	5	1	1	6	3	.	3	.	3	.	.	.	3	−1	1	−1	1
X.7	27	27	3	3	9	.	.	15	7	3	3	−1	2	9	.	.	.	3	.	.	.	−1	5	1	−1	1	−1
X.8	35	35	11	3	5	−1	2	15	7	−1	−1	3	.	5	−1	2	−1	−1	−1	2	.	.	1	5	1	1	1
X.9	35	35	3	3	5	−1	2	−5	−5	7	7	−1	.	5	−1	2	3	−3	3	.	.	.	−1	−1	1	−1	1
X.10	48	−48	.	.	−12	3	.	.	.	−8	8	.	−2	12	−3	.	3	.	−3	.	.	−1	.	.	.	.	.
X.11	56	56	−8	.	11	2	2	−24	8	.	.	.	1	11	2	2	−2	1	−2	−2	.	.	−4	4	.	.	.
X.12	64	−64	.	.	4	−8	−2	.	.	.	.	.	−1	−4	8	2	.	.	.	.	.	1	.	.	.	.	.
X.13	64	−64	.	.	4	−8	−2	.	.	.	.	.	−1	−4	8	2	.	.	.	.	.	1	.	.	.	.	.
X.14	70	70	−10	−2	−5	7	1	−10	6	2	2	2	.	−5	7	1	−1	−1	−1	−1	1	.	2	2	.	−2	.
X.15	84	84	20	4	−6	3	3	4	4	4	4	4	−1	−6	3	3	−1	2	−1	−1	1	.	.	.	.	.	.
X.16	105	105	−7	1	.	6	3	25	9	−3	−3	−3	.	.	6	3	2	−4	2	−1	1	.	−3	−3	−1	1	−1
X.17	105	105	17	−7	.	6	3	5	−3	−3	−3	1	.	.	6	3	2	2	2	−1	−1	.	−1	3	−1	−1	−1
X.18	105	105	1	1	15	−3	−3	−35	5	5	5	1	.	15	−3	−3	1	1	1	1	1	.	−5	−1	−1	−1	−1
X.19	112	−112	.	.	4	4	4	.	.	−8	8	.	2	−4	−4	−4	.	.	.	.	.	.	.	.	.	.	.
X.20	112	−112	.	.	−8	−5	−2	.	.	−8	8	.	2	8	5	2	3	.	−3	.	.	.	.	.	.	.	.
X.21	120	120	−8	.	15	−6	.	40	8	.	.	.	.	15	−6	.	−2	1	−2	−2	.	1	4	−4	.	.	.
X.22	120	−120	.	.	.	3	6	.	.	4	−4	.	.	.	−3	−6	3	.	−3	.	.	1	.	.	2	.	−2
X.23	168	168	8	8	6	6	−3	40	8	.	.	.	−2	6	6	−3	2	2	2	−1	−1	.	.	.	.	.	.
X.24	168	−168	.	.	−24	−3	.	.	.	−4	4	.	−2	24	3	.	−3	.	3	.	.	.	.	.	2	.	−2
X.25	189	189	21	−3	9	.	.	−39	1	−3	−3	1	−1	9	.	.	.	3	.	.	.	.	−1	−5	1	−1	1
X.26	189	189	−3	−3	9	.	.	21	−11	9	9	1	−1	9	.	.	.	−3	.	.	.	.	1	1	−1	1	−1
X.27	189	189	−3	−3	9	.	.	−51	13	−3	−3	−3	−1	9	.	.	.	−3	.	.	.	.	1	1	1	1	1
X.28	210	210	2	−6	15	3	.	50	2	−2	−2	−2	.	15	3	.	−1	−1	−1	2	.	.	2	2	.	−2	.
X.29	210	210	−14	2	−15	−6	3	10	10	6	6	−2	.	−15	−6	3	−2	1	−2	1	−1	.	−2	−2	.	−2	.
X.30	216	216	24	.	−9	.	.	−24	8	.	.	.	1	−9	.	.	.	−3	.	.	.	−1	4	−4	.	.	.
X.31	280	280	−8	8	10	10	1	−40	−8	.	.	.	.	10	10	1	−2	−2	−2	1	−1	.	.	.	.	.	.
X.32	280	280	24	.	−5	−8	−2	40	8	.	.	.	.	−5	−8	−2	.	−3	.	.	.	.	−4	4	.	.	.
X.33	280	−280	.	.	−20	1	4	.	.	4	−4	.	.	20	−1	−4	−3	.	3	.	.	.	.	.	−2	.	2
X.34	315	315	−21	3	.	−9	.	−45	3	−5	−5	3	.	.	−9	.	3	.	3	.	.	.	3	3	−1	−1	−1
X.35	336	336	16	.	6	−6	.	−16	−16	.	.	.	1	6	−6	.	−2	−2	−2	−2	.	.	.	.	.	.	.
X.36	378	378	−6	−6	−9	.	.	−30	2	6	6	−2	−2	−9	.	.	.	3	.	.	.	.	2	2	.	2	.
X.37	405	405	−27	−3	.	.	.	45	−3	−3	−3	5	.	.	.	.	.	.	.	.	.	−1	−3	−3	1	1	1
X.38	420	420	4	4	.	−3	3	20	−12	−4	−4	−4	.	.	−3	3	1	4	1	1	1	.	.	.	.	.	.
X.39	448	−448	.	.	16	16	−2	.	.	.	.	.	−2	−16	−16	2	.	.	.	.	.	.	.	.	.	.	.
X.40	512	−512	.	.	−16	8	−4	.	.	.	.	.	2	16	−8	4	.	.	.	.	.	1	.	.	.	.	.
X.41	512	512	.	.	−16	8	−4	.	.	.	.	.	2	−16	8	−4	.	.	.	.	.	1	.	.	.	.	.
X.42	560	−560	.	.	20	−7	2	.	.	−8	8	.	.	−20	7	−2	−3	.	3	.	.	.	.	.	.	.	.
X.43	720	−720	.	.	.	−9	.	.	.	8	−8	.	.	.	9	.	3	.	−3	.	.	−1	.	.	.	.	.

Character table of H (continued)

	8f	9a	10a	12a	12b	12c	12d	12e	14a	15a	18a	20a	20b	24a	24b	30a
2	4	1	2	4	4	2	3	3	1	1	1	2	2	3	3	1
3	.	2	1	2	1	2	1	1	.	1	2	.	.	1	1	1
5	.	.	1	.	.	.	.	.	.	1	.	1	1	.	.	1
7	.	.	.	.	.	.	.	.	1	.	.	.	.	.	.	.
2P	4c	9a	5a	6a	6a	6c	6d	6d	7a	15a	9a	10a	10a	12b	12b	15a
3P	8f	3b	10a	4a	4b	4a	4c	4d	14a	5a	6b	20a	20b	8a	8b	10a
5P	8f	9a	2a	12a	12b	12c	12d	12e	14a	3a	18a	4a	4a	24a	24b	6a
7P	8f	9a	10a	12a	12b	12c	12d	12e	2a	15a	18a	20a	20b	24a	24b	30a
X.1	1	1	1	1	1	1	1	1	1	1	1	1	1	1	1	1
X.2	−1	1	2	−2	.	1	.	.	.	−1	1	.	.	−2	.	−1
X.3	.	−1	2	.	.	.	1	−1	−1	1	1	.	.	.	.	−1
X.4	1	.	.	−2	.	1	−1	−1	1	.	.	.	.	.	−2	.
X.5	−1	.	1	−2	2	−2	1	1	.	1	.	−1	−1	.	.	1
X.6	−1	.	1	.	−2	.	−1	−1	.	1	.	−1	−1	2	.	1
X.7	1	.	2	3	1	.	.	.	−1	−1	.	.	.	1	−1	−1
X.8	−1	−1	.	3	1	.	−1	−1	.	.	−1	.	.	−1	1	.
X.9	1	−1	.	1	1	−2	1	1	.	.	−1	.	.	−1	−1	.
X.10	.	.	2	.	.	.	−1	1	1	−2	.	.	.	.	.	2
X.11	.	−1	1	−3	−1	.	.	.	.	1	−1	1	1	1	−1	1
X.12	.	1	1	.	.	.	.	.	−1	−1	−1	A	$\bar{A}$	.	.	1
X.13	.	1	1	.	.	.	.	.	−1	−1	−1	$\bar{A}$	A	.	.	1
X.14	.	1	.	−1	3	−1	−1	−1	.	.	1	.	.	−1	−1	.
X.15	.	.	−1	−2	−2	1	1	1	.	−1	.	−1	−1	.	.	−1
X.16	−1	.	.	4	.	1	.	.	.	.	.	.	.	.	.	.
X.17	1	.	.	2	.	−1	.	.	.	.	.	.	.	.	2	.
X.18	1	.	.	1	−1	1	−1	−1	.	.	.	.	.	−1	1	.
X.19	.	1	−2	.	.	.	2	−2	.	−1	−1	.	.	.	.	1
X.20	.	1	−2	.	.	.	−1	1	.	2	−1	.	.	.	.	−2
X.21	.	.	.	1	−1	−2	.	.	1	.	.	.	.	−1	1	.
X.22	.	.	.	.	.	.	−1	1	−1	.	.	.	.	.	.	.
X.23	.	.	−2	−2	2	1	.	.	.	1	.	.	.	.	.	1
X.24	.	.	2	.	.	.	1	−1	.	1	.	.	.	.	.	−1
X.25	−1	.	−1	3	1	.	.	.	.	−1	.	1	1	1	−1	−1
X.26	−1	.	−1	−3	1	.	.	.	.	−1	.	1	1	1	1	−1
X.27	1	.	−1	−3	1	.	.	.	.	−1	.	−1	−1	1	1	−1
X.28	.	.	.	−1	−1	2	1	1	.	.	.	.	.	−1	−1	.
X.29	.	.	.	1	1	1	.	.	.	.	.	.	.	1	1	.
X.30	.	.	1	−3	−1	.	.	.	−1	1	.	1	1	−1	1	1
X.31	.	1	.	2	−2	−1	.	.	.	.	1	.	.	.	.	.
X.32	.	1	.	1	−1	−2	.	.	.	.	1	.	.	1	−1	.
X.33	.	1	.	.	.	.	−1	1	.	.	−1	.	.	.	.	.
X.34	−1	.	.	.	.	.	1	1	.	.	.	.	.	.	.	.
X.35	.	.	1	2	2	2	.	.	.	1	.	−1	−1	.	.	1
X.36	.	.	−2	3	−1	.	.	.	.	1	.	.	.	−1	−1	1
X.37	1	.	.	.	.	.	.	.	−1	.	.	.	.	.	.	.
X.38	.	.	.	−4	.	−1	−1	−1	.	.	.	.	.	.	.	.
X.39	.	1	2	.	.	.	.	.	.	1	−1	.	.	.	.	−1
X.40	.	−1	−2	.	.	.	.	.	−1	−1	1	.	.	.	.	1
X.41	.	−1	2	.	.	.	.	.	1	−1	−1	.	.	.	.	−1
X.42	.	−1	.	.	.	.	−1	1	.	.	1	.	.	.	.	.
X.43	.	.	.	.	.	.	1	−1	1	.	.	.	.	.	.	.

where $A = i\sqrt{5}$.

3.5.4 *Character table of* $U = C_G(u) \cong \langle u \rangle \times Q$

	1a	2a	2b	2c	2d	2e	3a	3b	4a	4b	4c	4d	5a	6a	6b	6c	6d	6e	6f	8a	8b	8c
2	7	7	5	5	7	7	2	3	6	6	6	6	2	2	3	3	3	2	2	4	4	4
3	3	3	1	1	1	1	3	2	.	.	.	.	.	3	2	1	1	1	1	.	.	.
5	1	1	1	1	.	.	.	.	.	.	.	.	1	.	.	.	.	.	.	.	.	.
11	1	1	.	.	.	.	.	.	.	.	.	.	.	.	.	.	.	.	.	.	.	.
2P	1a	1a	1a	1a	1a	1a	3a	3b	2d	2d	2d	2d	5a	3a	3b	3b	3b	3a	3a	4b	4b	4a
3P	1a	2a	2b	2c	2d	2e	1a	1a	4a	4b	4c	4d	5a	2a	2a	2b	2c	2d	2e	8a	8b	8c
5P	1a	2a	2b	2c	2d	2e	3a	3b	4a	4b	4c	4d	1a	6a	6b	6c	6d	6e	6f	8a	8b	8c
11P	1a	2a	2b	2c	2d	2e	3a	3b	4a	4b	4c	4d	5a	6a	6b	6c	6d	6e	6f	8a	8b	8c
X.1	1	1	1	1	1	1	1	1	1	1	1	1	1	1	1	1	1	1	1	1	1	1
X.2	1	−1	1	−1	1	−1	1	1	1	1	−1	−1	1	−1	−1	1	−1	1	−1	1	−1	1
X.3	11	−11	−1	1	3	−3	2	−1	−1	3	1	−3	1	−2	1	−1	1	.	.	1	−1	−1
X.4	11	11	−1	−1	3	3	2	−1	3	−1	3	−1	1	2	−1	−1	−1	.	.	−1	−1	1
X.5	11	−11	−1	1	3	−3	2	−1	3	−1	−3	1	1	−2	1	−1	1	.	.	−1	1	1
X.6	11	11	−1	−1	3	3	2	−1	−1	3	−1	3	1	2	−1	−1	−1	.	.	1	1	−1
X.7	16	16	4	4	.	.	−2	1	.	.	.	.	1	−2	1	1	1	.	.	.	.	.
X.8	16	−16	4	−4	.	.	−2	1	.	.	.	.	1	2	−1	1	−1	.	.	.	.	.
X.9	16	−16	4	−4	.	.	−2	1	.	.	.	.	1	2	−1	1	−1	.	.	.	.	.
X.10	16	16	4	4	.	.	−2	1	.	.	.	.	1	−2	1	1	1	.	.	.	.	.
X.11	45	45	5	5	−3	−3	.	3	1	1	1	1	.	.	3	−1	−1	.	.	−1	−1	−1
X.12	45	−45	5	−5	−3	3	.	3	1	1	−1	−1	.	.	−3	−1	1	.	.	−1	1	−1
X.13	54	54	6	6	6	6	.	.	2	2	2	2	−1	.	.	.	.	.	.	.	.	.
X.14	54	−54	6	−6	6	−6	.	.	2	2	−2	−2	−1	.	.	.	.	.	.	.	.	.
X.15	55	55	−5	−5	−1	−1	1	1	3	−1	3	−1	.	1	1	1	1	−1	−1	1	1	−1
X.16	55	−55	−5	5	−1	1	1	1	3	−1	−3	1	.	−1	−1	1	−1	−1	1	1	−1	−1
X.17	55	55	−5	−5	7	7	1	1	−1	−1	−1	−1	.	1	1	1	1	1	1	−1	−1	−1
X.18	55	55	−5	−5	−1	−1	1	1	−1	3	−1	3	.	1	1	1	1	−1	−1	−1	−1	1
X.19	55	−55	−5	5	7	−7	1	1	−1	−1	1	1	.	−1	−1	1	−1	1	−1	−1	1	−1
X.20	55	−55	−5	5	−1	1	1	1	−1	3	1	−3	.	−1	−1	1	−1	−1	1	−1	1	1
X.21	66	66	6	6	2	2	3	.	−2	−2	−2	−2	1	3	.	.	.	−1	−1	.	.	.
X.22	66	−66	6	−6	2	−2	3	.	−2	−2	2	2	1	−3	.	.	.	−1	1	.	.	.
X.23	99	−99	−1	1	3	−3	.	3	−1	−1	1	1	−1	.	−3	−1	1	.	.	1	−1	1
X.24	99	99	−1	−1	3	3	.	3	−1	−1	−1	−1	−1	.	3	−1	−1	.	.	1	1	1
X.25	120	−120	.	.	−8	8	3	.	.	.	.	.	.	−3	.	.	.	1	−1	.	.	.
X.26	120	120	.	.	−8	−8	3	.	.	.	.	.	.	3	.	.	.	1	1	.	.	.
X.27	144	−144	4	−4	.	.	.	−3	.	.	.	.	−1	.	3	1	−1	.	.	.	.	.
X.28	144	144	4	4	.	.	.	−3	.	.	.	.	−1	.	−3	1	1	.	.	.	.	.
X.29	176	−176	−4	4	.	.	−4	−1	.	.	.	.	1	4	1	−1	1	.	.	.	.	.
X.30	176	176	−4	−4	.	.	−4	−1	.	.	.	.	1	−4	−1	−1	−1	.	.	.	.	.

Character table of $U = C_G(u) \cong \langle u \rangle \times Q$ *(continued)*

	8d	10a	10b	10c	11a	11b	22a	22b
2	4	2	2	2	1	1	1	1
3	.	.	.	.	.	.	.	.
5	.	1	1	1	.	.	.	.
11	.	.	.	.	1	1	1	1
2P	4a	5a	5a	5a	11b	11a	11b	11a
3P	8d	10a	10b	10c	11a	11b	22a	22b
5P	8d	2b	2a	2c	11a	11b	22a	22b
11P	8d	10a	10b	10c	1a	1a	2a	2a
X.1	1	1	1	1	1	1	1	1
X.2	−1	1	−1	−1	1	1	−1	−1
X.3	1	−1	−1	1	.	.	.	.
X.4	1	−1	1	−1	.	.	.	.
X.5	−1	−1	−1	1	.	.	.	.
X.6	−1	−1	1	−1	.	.	.	.
X.7	.	−1	1	−1	A	$\bar{A}$	A	$\bar{A}$
X.8	.	−1	−1	1	$\bar{A}$	A	$-\bar{A}$	$-A$
X.9	.	−1	−1	1	A	$\bar{A}$	$-A$	$-\bar{A}$
X.10	.	−1	1	−1	$\bar{A}$	A	$\bar{A}$	A
X.11	−1	.	.	.	1	1	1	1
X.12	1	.	.	.	1	1	−1	−1
X.13	.	1	−1	1	−1	−1	−1	−1
X.14	.	1	1	−1	−1	−1	1	1
X.15	−1	.	.	.	.	.	.	.
X.16	1	.	.	.	.	.	.	.
X.17	−1	.	.	.	.	.	.	.
X.18	1	.	.	.	.	.	.	.
X.19	1	.	.	.	.	.	.	.
X.20	−1	.	.	.	.	.	.	.
X.21	.	1	1	1	.	.	.	.
X.22	.	1	−1	−1	.	.	.	.
X.23	−1	−1	1	1	.	.	.	.
X.24	1	−1	−1	−1	.	.	.	.
X.25	.	.	.	.	−1	−1	1	1
X.26	.	.	.	.	−1	−1	−1	−1
X.27	.	−1	1	1	1	1	−1	−1
X.28	.	−1	−1	−1	1	1	1	1
X.29	.	1	−1	−1	.	.	.	.
X.30	.	1	1	1	.	.	.	.

where $A = \frac{1}{2}(-1 + i\sqrt{11})$.

3.5.5 *Character table of* $N_1 = N_G(r_1) = \langle k_i | 1 \leq i \leq 6 \rangle$

	1a	2a	2b	2c	2d	3a	3b	3c	3d	3e	3f	3g	4a	4b	4c	4d	5a	6a	6b	6c	6d	6e
2	6	6	5	5	5	5	2	3	3	2	2	1	5	5	5	5	2	5	4	3	3	2
3	7	3	4	3	2	7	6	5	5	5	5	4	2	2	1	1	1	3	4	3	3	3
5	1	1	.	.	.	1	.	.	.	.	.	.	1	.	.	.	1	1	.	.	.	.
2P	1a	1a	1a	1a	1a	3a	3b	3c	3d	3e	3f	3g	2a	2a	2a	2a	5a	3a	3a	3c	3d	3e
3P	1a	2a	2b	2c	2d	1a	1a	1a	1a	1a	1a	1a	4a	4b	4c	4d	5a	2a	2b	2a	2a	2a
5P	1a	2a	2b	2c	2d	3a	3b	3c	3d	3e	3f	3g	4a	4b	4c	4d	1a	6a	6b	6c	6d	6e
X.1	1	1	1	1	1	1	1	1	1	1	1	1	1	1	1	1	1	1	1	1	1	1
X.2	1	1	1	−1	−1	1	1	1	1	1	1	1	−1	1	−1	1	1	1	1	1	1	1
X.3	1	1	−1	−1	1	1	1	1	1	1	1	1	1	−1	−1	1	1	1	−1	1	1	1
X.4	1	1	−1	1	−1	1	1	1	1	1	1	1	−1	−1	1	1	1	1	−1	1	1	1
X.5	5	5	−1	1	−1	5	5	2	−1	2	−1	−1	−5	3	−3	1	.	5	−1	2	−1	2
X.6	5	5	1	−1	−1	5	5	2	−1	2	−1	−1	−5	−3	3	1	.	5	1	2	−1	2
X.7	5	5	1	1	1	5	5	2	−1	2	−1	−1	5	−3	−3	1	.	5	1	2	−1	2
X.8	5	5	−1	−1	1	5	5	2	−1	2	−1	−1	5	3	3	1	.	5	−1	2	−1	2
X.9	5	5	3	3	1	5	5	−1	2	−1	2	2	5	−1	−1	1	.	5	3	−1	2	−1
X.10	5	5	3	−3	−1	5	5	−1	2	−1	2	2	−5	−1	1	1	.	5	3	−1	2	−1
X.11	5	5	−3	3	−1	5	5	−1	2	−1	2	2	−5	1	−1	1	.	5	−3	−1	2	−1
X.12	5	5	−3	−3	1	5	5	−1	2	−1	2	2	5	1	1	1	.	5	−3	−1	2	−1
X.13	8	−8	.	.	.	8	8	−4	2	−4	2	2	.	.	.	.	−2	−8	.	4	−2	4
X.14	8	−8	.	.	.	8	8	2	−4	2	−4	−4	.	.	.	.	−2	−8	.	−2	4	−2
X.15	9	9	3	−3	−1	9	9	.	.	.	.	.	−9	3	−3	1	−1	9	3	.	.	.
X.16	9	9	−3	−3	1	9	9	.	.	.	.	.	9	−3	−3	1	−1	9	−3	.	.	.
X.17	9	9	3	3	1	9	9	.	.	.	.	.	9	3	3	1	−1	9	3	.	.	.
X.18	9	9	−3	3	−1	9	9	.	.	.	.	.	−9	−3	3	1	−1	9	−3	.	.	.
X.19	10	10	−2	2	2	10	10	1	1	1	1	1	−10	2	−2	−2	.	10	−2	1	1	1
X.20	10	10	2	2	−2	10	10	1	1	1	1	1	10	−2	−2	−2	.	10	2	1	1	1
X.21	10	10	−2	−2	−2	10	10	1	1	1	1	1	10	2	2	−2	.	10	−2	1	1	1
X.22	10	10	2	−2	2	10	10	1	1	1	1	1	−10	−2	2	−2	.	10	2	1	1	1
X.23	16	16	.	.	.	16	16	−2	−2	−2	−2	−2	−16	.	.	.	1	16	.	−2	−2	−2
X.24	16	16	.	.	.	16	16	−2	−2	−2	−2	−2	16	.	.	.	1	16	.	−2	−2	−2
X.25	16	−16	.	.	.	16	16	−2	−2	−2	−2	−2	.	.	.	.	1	−16	.	2	2	2
X.26	16	−16	.	.	.	16	16	−2	−2	−2	−2	−2	.	.	.	.	1	−16	.	2	2	2
X.27	18	2	6	.	.	−9	.	−6	6	3	−3	.	.	−2	.	2	−2	−1	−3	2	2	−1
X.28	18	2	−6	.	.	−9	.	−6	6	3	−3	.	.	2	.	2	−2	−1	3	2	2	−1
X.29	20	−20	.	.	.	20	20	2	2	2	2	2	.	.	.	.	.	−20	.	−2	−2	−2
X.30	20	−20	.	.	.	20	20	2	2	2	2	2	.	.	.	.	.	−20	.	−2	−2	−2
X.31	72	−8	.	.	.	−36	.	12	6	−6	−3	.	.	.	.	.	2	4	.	4	−2	−2
X.32	72	−8	.	.	.	−36	.	12	6	−6	−3	.	.	.	.	.	2	4	.	4	−2	−2
X.33	72	−8	.	.	.	−36	.	−6	−12	3	6	.	.	.	.	.	2	4	.	−2	4	1
X.34	72	−8	.	.	.	−36	.	−6	−12	3	6	.	.	.	.	.	2	4	.	−2	4	1
X.35	80	.	−8	−8	8	80	−1	8	8	8	8	−1	.	.	.	.	.	.	−8	.	.	.
X.36	80	.	8	8	8	80	−1	8	8	8	8	−1	.	.	.	.	.	.	8	.	.	.
X.37	80	.	8	−8	−8	80	−1	8	8	8	8	−1	.	.	.	.	.	.	8	.	.	.
X.38	80	.	−8	8	−8	80	−1	8	8	8	8	−1	.	.	.	.	.	.	−8	.	.	.
X.39	90	10	6	.	.	−45	.	−12	−6	6	3	.	.	6	.	2	.	−5	−3	4	−2	−2
X.40	90	10	−18	.	.	−45	.	6	12	−3	−6	.	.	−2	.	2	.	−5	9	−2	4	1
X.41	90	10	−6	.	.	−45	.	−12	−6	6	3	.	.	−6	.	2	.	−5	3	4	−2	−2
X.42	90	10	18	.	.	−45	.	6	12	−3	−6	.	.	2	.	2	.	−5	−9	−2	4	1
X.43	160	.	.	16	.	160	−2	−8	4	−8	4	−5	.	.	.	.	.	.	.	.	.	.
X.44	160	.	16	.	.	160	−2	−8	4	−8	4	4	.	.	.	.	.	.	16	.	.	.
X.45	160	.	−16	.	.	160	−2	−8	4	−8	4	4	.	.	.	.	.	.	−16	.	.	.
X.46	160	.	.	−16	.	160	−2	−8	4	−8	4	−5	.	.	.	.	.	.	.	.	.	.
X.47	162	18	18	.	.	−81	.	.	.	.	.	.	.	−6	.	2	2	−9	−9	.	.	.
X.48	162	18	−18	.	.	−81	.	.	.	.	.	.	.	6	.	2	2	−9	9	.	.	.
X.49	180	20	−12	.	.	−90	.	−6	6	3	−3	.	.	−4	.	−4	.	−10	6	2	2	−1
X.50	180	20	12	.	.	−90	.	−6	6	3	−3	.	.	4	.	−4	.	−10	−6	2	2	−1
X.51	288	32	.	.	.	−144	.	12	−12	−6	6	.	.	.	.	.	−2	−16	.	−4	−4	2
X.52	288	−32	.	.	.	−144	.	12	−12	−6	6	.	.	.	.	.	−2	16	.	4	4	−2
X.53	320	.	.	.	.	320	−4	8	−16	8	−16	2	.	.	.	.	.	.	.	.	.	.
X.54	360	−40	.	.	.	−180	.	−12	12	6	−6	.	.	.	.	.	.	20	.	−4	−4	2

Character table of N_1 (continued)

	6f	6g	6h	6i	6j	6k	6l	6m	6n	8a	8b	8c	8d	8e	9a	9b	10a	12a	12b	12c	12d	12e	12f	12g
2	2	2	2	2	2	2	2	2	1	4	4	5	5	4	1	.	2	4	4	2	2	2	2	2
3	3	3	3	3	3	3	2	2	2	1	1	.	.	.	4	4	1	2	1	2	2	2	2	2
5	.	.	.	.	.	.	.	.	.	.	.	.	.	.	.	.	1	.	.	.	.	.	.	.
2P	3d	3f	3f	3b	3b	3f	3d	3b	3g	4d	4d	4d	4d	4d	9a	9b	5a	6a	6a	6c	6c	6e	6e	6d
3P	2b	2b	2b	2b	2c	2a	2c	2d	2c	8a	8b	8c	8d	8e	3a	3a	10a	4b	4d	4b	4a	4b	4b	4a
5P	6f	6g	6h	6i	6j	6k	6l	6m	6n	8a	8b	8d	8c	8e	9a	9b	2a	12a	12b	12c	12d	12e	12f	12g
X.1	1	1	1	1	1	1	1	1	1	1	1	1	1	1	1	1	1	1	1	1	1	1	1	1
X.2	1	1	1	1	−1	1	−1	−1	−1	1	1	−1	−1	−1	1	1	1	1	1	1	−1	1	1	−1
X.3	−1	−1	−1	−1	−1	1	−1	1	−1	−1	1	1	1	−1	1	1	1	−1	1	−1	1	−1	−1	1
X.4	−1	−1	−1	−1	1	1	1	−1	1	−1	1	−1	−1	1	1	1	1	−1	1	−1	−1	−1	−1	−1
X.5	−1	−1	−1	−1	1	−1	1	−1	1	1	−1	1	1	−1	−1	2	.	3	1	.	−2	.	.	1
X.6	1	1	1	1	−1	−1	−1	−1	−1	−1	−1	1	1	1	−1	2	.	−3	1	.	−2	.	.	1
X.7	1	1	1	1	1	−1	1	1	1	−1	−1	−1	−1	−1	−1	2	.	−3	1	.	2	.	.	−1
X.8	−1	−1	−1	−1	−1	−1	−1	1	−1	1	−1	−1	−1	1	−1	2	.	3	1	.	2	.	.	−1
X.9	.	.	.	3	3	2	.	1	.	1	−1	−1	−1	1	2	−1	.	−1	1	−1	−1	−1	−1	2
X.10	.	.	.	3	−3	2	.	−1	.	1	−1	1	1	−1	2	−1	.	−1	1	−1	1	−1	−1	−2
X.11	.	.	.	−3	3	2	.	−1	.	−1	−1	1	1	1	2	−1	.	1	1	1	1	1	1	−2
X.12	.	.	.	−3	−3	2	.	1	.	−1	−1	−1	−1	−1	2	−1	.	1	1	1	−1	1	1	2
X.13	.	.	.	.	.	−2	.	.	.	.	.	.	.	.	2	−4	2	.	.	.	.	.	.	.
X.14	.	.	.	.	.	4	.	.	.	.	.	.	.	.	−4	2	2	.	.	.	.	.	.	.
X.15	.	.	.	3	−3	.	.	−1	.	−1	1	−1	−1	1	.	.	−1	3	1	.	.	.	.	.
X.16	.	.	.	−3	−3	.	.	1	.	1	1	1	1	1	.	.	−1	−3	1	.	.	.	.	.
X.17	.	.	.	3	3	.	.	1	.	−1	1	1	1	−1	.	.	−1	3	1	.	.	.	.	.
X.18	.	.	.	−3	3	.	.	−1	.	1	1	−1	−1	−1	.	.	−1	−3	1	.	.	.	.	.
X.19	1	1	1	−2	2	1	−1	2	−1	.	.	.	.	.	1	1	.	2	−2	−1	−1	−1	−1	−1
X.20	−1	−1	−1	2	2	1	−1	−2	−1	.	.	.	.	.	1	1	.	−2	−2	1	1	1	1	1
X.21	1	1	1	−2	−2	1	1	−2	1	.	.	.	.	.	1	1	.	2	−2	−1	1	−1	−1	1
X.22	−1	−1	−1	2	−2	1	1	2	1	.	.	.	.	.	1	1	.	−2	−2	1	−1	1	1	−1
X.23	.	.	.	.	.	−2	.	.	.	.	.	.	.	.	−2	−2	1	.	.	.	2	.	.	2
X.24	.	.	.	.	.	−2	.	.	.	.	.	.	.	.	−2	−2	1	.	.	.	−2	.	.	−2
X.25	.	.	.	.	.	2	.	.	.	.	.	.	.	.	−2	−2	−1	.	.	.	.	.	.	.
X.26	.	.	.	.	.	2	.	.	.	.	.	.	.	.	−2	−2	−1	.	.	.	.	.	.	.
X.27	.	−3	3	.	.	−1	.	.	.	2	−2	.	.	.	.	.	2	1	−1	−2	.	1	1	.
X.28	.	3	−3	.	.	−1	.	.	.	−2	−2	.	.	.	.	.	2	−1	−1	2	.	−1	−1	.
X.29	.	.	.	.	.	−2	.	.	.	.	.	B	$\bar{B}$	.	2	2	.	.	.	.	.	.	.	.
X.30	.	.	.	.	.	−2	.	.	.	.	.	$\bar{B}$	B	.	2	2	.	.	.	.	.	.	.	.
X.31	6	−3	−3	.	.	1	.	.	.	.	.	.	.	.	.	.	2	.	.	.	.	.	.	.
X.32	−6	3	3	.	.	1	.	.	.	.	.	.	.	.	.	.	2	.	.	.	.	.	.	.
X.33	.	.	.	.	.	−2	.	.	.	.	.	.	.	.	.	.	2	.	.	.	.	C	$\bar{C}$	.
X.34	.	.	.	.	.	−2	.	.	.	.	.	.	.	.	.	.	2	.	.	.	.	$\bar{C}$	C	.
X.35	−2	−2	−2	1	1	.	−2	−1	1	.	.	.	.	.	−1	−1	.	.	.	.	.	.	.	.
X.36	2	2	2	−1	−1	.	2	−1	−1	.	.	.	.	.	−1	−1	.	.	.	.	.	.	.	.
X.37	2	2	2	−1	1	.	−2	1	1	.	.	.	.	.	−1	−1	.	.	.	.	.	.	.	.
X.38	−2	−2	−2	1	−1	.	2	1	−1	.	.	.	.	.	−1	−1	.	.	.	.	.	.	.	.
X.39	.	−3	3	.	.	1	.	.	.	−2	2	.	.	.	.	.	.	−3	−1	.	.	.	.	.
X.40	.	.	.	.	.	−2	.	.	.	−2	2	.	.	.	.	.	.	1	−1	−2	.	1	1	.
X.41	.	3	−3	.	.	1	.	.	.	2	2	.	.	.	.	.	.	3	−1	.	.	.	.	.
X.42	.	.	.	.	.	−2	.	.	.	2	2	.	.	.	.	.	.	−1	−1	2	.	−1	−1	.
X.43	.	.	.	.	−2	.	−2	.	1	.	.	.	.	.	4	1	.	.	.	.	.	.	.	.
X.44	−2	−2	−2	−2	.	.	.	.	.	.	.	.	.	.	−5	1	.	.	.	.	.	.	.	.
X.45	2	2	2	2	.	.	.	.	.	.	.	.	.	.	−5	1	.	.	.	.	.	.	.	.
X.46	.	.	.	.	2	.	2	.	−1	.	.	.	.	.	4	1	.	.	.	.	.	.	.	.
X.47	.	.	.	.	.	.	.	.	.	−2	−2	.	.	.	.	.	−2	3	−1	.	.	.	.	.
X.48	.	.	.	.	.	.	.	.	.	2	−2	.	.	.	.	.	−2	−3	−1	.	.	.	.	.
X.49	.	−3	3	.	.	−1	.	.	.	.	.	.	.	.	.	.	.	2	2	2	.	−1	−1	.
X.50	.	3	−3	.	.	−1	.	.	.	.	.	.	.	.	.	.	.	−2	2	−2	.	1	1	.
X.51	.	.	.	.	.	2	.	.	.	.	.	.	.	.	.	.	2	.	.	.	.	.	.	.
X.52	.	.	.	.	.	−2	.	.	.	.	.	.	.	.	.	.	−2	.	.	.	.	.	.	.
X.53	.	.	.	.	.	.	.	.	.	.	.	.	.	.	2	−1	.	.	.	.	.	.	.	.
X.54	.	.	.	.	.	2	.	.	.	.	.	.	.	.	.	.	.	.	.	.	.	.	.	.

Character table of N_1 *(continued)*

2	2	1	1	2	2	3	3	1
3	1	1	2	.	.	1	1	1
5	.	1	.	1	1	.	.	1
	12h	15a	18a	20a	20b	24a	24b	30a
2P	6c	15a	9a	10a	10a	12b	12b	15a
3P	4c	5a	6b	20a	20b	8a	8b	10a
5P	12h	3a	18a	4a	4a	24a	24b	6a
X.1	1	1	1	1	1	1	1	1
X.2	−1	1	1	−1	−1	1	1	1
X.3	−1	1	−1	1	1	−1	1	1
X.4	1	1	−1	−1	−1	−1	1	1
X.5	.	.	−1	.	.	1	−1	.
X.6	.	.	1	.	.	−1	−1	.
X.7	.	.	1	.	.	−1	−1	.
X.8	.	.	−1	.	.	1	−1	.
X.9	−1	.	.	.	.	1	−1	.
X.10	1	.	.	.	.	1	−1	.
X.11	−1	.	.	.	.	−1	−1	.
X.12	1	.	.	.	.	−1	−1	.
X.13	.	−2	.	.	.	.	.	2
X.14	.	−2	.	.	.	.	.	2
X.15	.	−1	.	1	1	−1	1	−1
X.16	.	−1	.	−1	−1	1	1	−1
X.17	.	−1	.	−1	−1	−1	1	−1
X.18	.	−1	.	1	1	1	1	−1
X.19	1	.	1	.	.	.	.	.
X.20	1	.	−1	.	.	.	.	.
X.21	−1	.	1	.	.	.	.	.
X.22	−1	.	−1	.	.	.	.	.
X.23	.	1	.	−1	−1	.	.	1
X.24	.	1	.	1	1	.	.	1
X.25	.	1	.	A	$\bar{A}$	.	.	−1
X.26	.	1	.	$\bar{A}$	A	.	.	−1
X.27	.	1	.	.	.	−1	1	−1
X.28	.	1	.	.	.	1	1	−1
X.29	.	.	.	.	.	.	.	.
X.30	.	.	.	.	.	.	.	.
X.31	.	−1	.	.	.	.	.	−1
X.32	.	−1	.	.	.	.	.	−1
X.33	.	−1	.	.	.	.	.	−1
X.34	.	−1	.	.	.	.	.	−1
X.35	.	.	1	.	.	.	.	.
X.36	.	.	−1	.	.	.	.	.
X.37	.	.	−1	.	.	.	.	.
X.38	.	.	1	.	.	.	.	.
X.39	.	.	.	.	.	1	−1	.
X.40	.	.	.	.	.	1	−1	.
X.41	.	.	.	.	.	−1	−1	.
X.42	.	.	.	.	.	−1	−1	.
X.43	.	.	.	.	.	.	.	.
X.44	.	.	1	.	.	.	.	.
X.45	.	.	−1	.	.	.	.	.
X.46	.	.	.	.	.	.	.	.
X.47	.	−1	.	.	.	1	1	1
X.48	.	−1	.	.	.	−1	1	1
X.49	.	.	.	.	.	.	.	.
X.50	.	.	.	.	.	.	.	.
X.51	.	1	.	.	.	.	.	−1
X.52	.	1	.	.	.	.	.	1
X.53	.	.	.	.	.	.	.	.
X.54	.	.	.	.	.	.	.	.

where $A = -2\zeta(20)^{17} - 2\zeta(20)^{13} - \zeta(20)^5$, $B = -2\zeta(8)^3 - 2\zeta(8)$, $C = -3\zeta(4)$.

3.5.6 *Character table of* $N_2 = N_G(r_2) = \langle n_i | 1 \leq i \leq 5 \rangle$

	1a	2a	2b	2c	2d	2e	3a	3b	3c	3d	3e	3f	3g
2	4	3	4	4	3	4	3	2	3	3	1	2	2
3	6	4	3	1	3	2	6	6	5	5	6	5	5
5	1	.	.	1	.	.	1	.	.	.	.	.	.
2P	1a	1a	1a	1a	1a	1a	3a	3b	3c	3d	3e	3f	3g
3P	1a	2a	2b	2c	2d	2e	1a	1a	1a	1a	1a	1a	1a
5P	1a	2a	2b	2c	2d	2e	3a	3b	3c	3d	3e	3f	3g
X.1	1	1	1	1	1	1	1	1	1	1	1	1	1
X.2	1	−1	1	−1	1	−1	1	1	1	1	1	1	1
X.3	1	1	1	−1	−1	−1	1	1	1	1	1	1	1
X.4	1	−1	1	1	−1	1	1	1	1	1	1	1	1
X.5	4	−2	.	4	−2	.	4	4	4	4	4	4	4
X.6	4	2	.	4	2	.	4	4	4	4	4	4	4
X.7	4	−2	.	−4	2	.	4	4	4	4	4	4	4
X.8	4	2	.	−4	−2	.	4	4	4	4	4	4	4
X.9	5	−1	1	5	−1	1	5	5	5	5	5	5	5
X.10	5	1	1	−5	−1	−1	5	5	5	5	5	5	5
X.11	5	1	1	5	1	1	5	5	5	5	5	5	5
X.12	5	−1	1	−5	1	−1	5	5	5	5	5	5	5
X.13	6	.	−2	−6	.	2	6	6	6	6	6	6	6
X.14	6	.	−2	6	.	−2	6	6	6	6	6	6	6
X.15	10	6	2	.	.	.	10	1	−5	−2	1	4	−2
X.16	10	−6	2	.	.	.	10	1	−5	−2	1	4	−2
X.17	12	.	4	.	.	.	−6	−6	.	6	3	.	−3
X.18	12	.	4	.	.	.	−6	−6	.	6	3	.	−3
X.19	12	.	−4	.	.	.	−6	−6	.	6	3	.	−3
X.20	12	.	−4	.	.	.	−6	−6	.	6	3	.	−3
X.21	20	−6	.	.	−2	4	20	2	8	−4	2	−1	−4
X.22	20	−6	.	.	2	−4	20	2	8	−4	2	−1	−4
X.23	20	−8	4	.	.	.	20	−7	2	2	−7	2	2
X.24	20	−4	−4	.	.	.	20	−7	2	2	−7	2	2
X.25	20	6	.	.	2	4	20	2	8	−4	2	−1	−4
X.26	20	6	.	.	−2	−4	20	2	8	−4	2	−1	−4
X.27	20	4	−4	.	.	.	20	−7	2	2	−7	2	2
X.28	20	8	4	.	.	.	20	−7	2	2	−7	2	2
X.29	20	.	4	.	.	.	20	2	−10	−4	2	8	−4
X.30	30	−6	−2	.	.	.	30	3	−15	−6	3	12	−6
X.31	30	6	2	.	.	−4	30	3	−6	3	3	−6	3
X.32	30	.	−2	.	6	.	−15	12	.	6	−6	.	−3
X.33	30	−6	2	.	.	4	30	3	−6	3	3	−6	3
X.34	30	.	6	.	6	.	−15	12	.	6	−6	.	−3
X.35	30	−6	2	.	.	−4	30	3	−6	3	3	−6	3
X.36	30	6	2	.	.	4	30	3	−6	3	3	−6	3
X.37	30	.	−2	.	−6	.	−15	12	.	6	−6	.	−3
X.38	30	.	6	.	−6	.	−15	12	.	6	−6	.	−3
X.39	30	6	−2	.	.	.	30	3	−15	−6	3	12	−6
X.40	40	−4	.	.	.	.	40	−14	4	4	−14	4	4
X.41	40	.	.	.	−4	.	40	4	16	−8	4	−2	−8
X.42	40	4	.	.	.	.	40	−14	4	4	−14	4	4
X.43	40	.	.	.	4	.	40	4	16	−8	4	−2	−8
X.44	48	.	.	.	.	.	−24	−24	.	24	12	.	−12
X.45	60	.	−4	.	.	.	60	6	−12	6	6	−12	6
X.46	60	.	−4	.	.	.	−30	24	.	12	−12	.	−6
X.47	120	.	8	.	.	.	−60	−6	.	−12	3	.	6
X.48	120	.	−8	.	.	.	−60	−6	.	−12	3	.	6

	3h	3i	4a	4b	5a	6a	6b	6c	6d	6e	6f	6g	6h	6i	6j	6k
2	2	1	3	3	1	2	3	3	3	2	2	2	2	2	2	2
3	3	3	2	1	1	4	3	3	3	3	3	3	3	3	3	3
5	.	.	.	.	1	.	.	.	.	.	.	.	.	.	.	.
2P	3h	3i	2b	2b	5a	3b	3c	3d	3a	3c	3c	3d	3b	3a	3b	3g
3P	1a	1a	4a	4b	5a	2a	2b	2b	2b	2a	2b	2a	2b	2d	2a	2b
5P	3h	3i	4a	4b	1a	6a	6b	6c	6d	6e	6f	6g	6h	6i	6j	6k
X.1	1	1	1	1	1	1	1	1	1	1	1	1	1	1	1	1
X.2	1	1	1	−1	1	−1	1	1	1	−1	1	−1	1	1	−1	1
X.3	1	1	−1	1	1	1	1	1	1	1	1	1	1	−1	1	1
X.4	1	1	−1	−1	1	−1	1	1	1	−1	1	−1	1	−1	−1	1
X.5	1	1	.	.	−1	−2	.	.	.	−2	.	−2	.	−2	−2	.
X.6	1	1	.	.	−1	2	.	.	.	2	.	2	.	2	2	.
X.7	1	1	.	.	−1	−2	.	.	.	−2	.	−2	.	2	−2	.
X.8	1	1	.	.	−1	2	.	.	.	2	.	2	.	−2	2	.
X.9	−1	−1	1	1	.	−1	1	1	1	−1	1	−1	1	−1	−1	1
X.10	−1	−1	1	−1	.	1	1	1	1	1	1	1	1	−1	1	1
X.11	−1	−1	−1	−1	.	1	1	1	1	1	1	1	1	1	1	1
X.12	−1	−1	−1	1	.	−1	1	1	1	−1	1	−1	1	1	−1	1
X.13	.	.	.	.	1	.	−2	−2	−2	.	−2	.	−2	.	.	−2
X.14	.	.	.	.	1	.	−2	−2	−2	.	−2	.	−2	.	.	−2
X.15	4	−2	.	2	.	−3	−1	2	2	−3	−1	.	−1	.	3	2
X.16	4	−2	.	−2	.	3	−1	2	2	3	−1	.	−1	.	−3	2
X.17	.	.	−4	.	2	.	4	−2	−2	.	−2	.	−2	.	.	1
X.18	.	.	4	.	2	.	4	−2	−2	.	−2	.	−2	.	.	1
X.19	.	.	.	.	2	.	−4	2	2	.	2	.	2	.	.	−1
X.20	.	.	.	.	2	.	−4	2	2	.	2	.	2	.	.	−1
X.21	2	−1	.	.	.	−6	.	.	.	.	.	.	.	−2	.	.
X.22	2	−1	.	.	.	−6	.	.	.	.	.	.	.	2	.	.
X.23	2	2	.	.	.	1	−2	−2	4	−2	−2	4	1	.	1	−2
X.24	2	2	.	.	.	5	2	2	−4	−4	2	2	−1	.	−1	2
X.25	2	−1	.	.	.	6	.	.	.	.	.	.	.	2	.	.
X.26	2	−1	.	.	.	6	.	.	.	.	.	.	.	−2	.	.
X.27	2	2	.	.	.	−5	2	2	−4	4	2	−2	−1	.	1	2
X.28	2	2	.	.	.	−1	−2	−2	4	2	−2	−4	1	.	−1	−2
X.29	−4	2	.	.	.	.	−2	4	4	.	−2	.	−2	.	.	4
X.30	.	.	.	2	.	3	1	−2	−2	3	1	.	1	.	−3	−2
X.31	.	.	−2	.	.	−3	2	−1	2	.	2	3	−1	.	−3	−1
X.32	.	.	−2	.	.	.	4	−2	1	.	−2	.	4	−3	.	1
X.33	.	.	−2	.	.	3	2	−1	2	.	2	−3	−1	.	3	−1
X.34	.	.	2	.	.	.	.	6	−3	.	.	.	.	−3	.	−3
X.35	.	.	2	.	.	3	2	−1	2	.	2	−3	−1	.	3	−1
X.36	.	.	2	.	.	−3	2	−1	2	.	2	3	−1	.	−3	−1
X.37	.	.	2	.	.	.	4	−2	1	.	−2	.	4	3	.	1
X.38	.	.	−2	.	.	.	.	6	−3	.	.	.	.	3	.	−3
X.39	.	.	.	−2	.	−3	1	−2	−2	−3	1	.	1	.	3	−2
X.40	−2	−2	.	.	.	−4	.	.	.	2	.	2	.	.	2	.
X.41	−2	1	.	.	.	.	.	.	.	.	.	.	.	−4	.	.
X.42	−2	−2	.	.	.	4	.	.	.	−2	.	−2	.	.	−2	.
X.43	−2	1	.	.	.	.	.	.	.	.	.	.	.	4	.	.
X.44	.	.	.	.	−2	.	.	.	.	.	.	.	.	.	.	.
X.45	.	.	.	.	.	.	−4	2	−4	.	−4	.	2	.	.	2
X.46	.	.	.	.	.	.	−4	−4	2	.	2	.	−4	.	.	2
X.47	.	.	.	.	.	.	−4	−4	−4	.	2	.	2	.	.	2
X.48	.	.	.	.	.	.	4	4	4	.	−2	.	−2	.	.	−2

Character table of $N_G(r_2)$ *(continued)*

	6l	6m	6n	6o	6p	6q	6r	6s	6t	9a	9b	10a	12a	12b	12c	12d	12e	15a	18a
2	2	1	2	2	2	2	2	1	2	1	.	1	2	2	2	2	2	.	1
3	3	3	2	2	2	2	2	2	1	3	3	.	2	2	2	2	1	1	2
5	.	.	.	.	.	.	.	.	.	.	.	1	.	.	.	.	.	1	.
2P	3f	3e	3h	3f	3h	3f	3d	3i	3h	9a	9b	5a	6c	6k	6k	6d	6b	15a	9a
3P	2a	2b	2a	2e	2d	2d	2e	2d	2c	3b	3b	10a	4a	4a	4a	4a	4b	5a	6a
5P	6l	6m	6n	6o	6p	6q	6r	6s	6t	9a	9b	2c	12a	12b	12c	12d	12e	3a	18a
X.1	1	1	1	1	1	1	1	1	1	1	1	1	1	1	1	1	1	1	1
X.2	−1	1	−1	−1	1	1	−1	1	−1	1	1	−1	1	1	1	1	−1	1	−1
X.3	1	1	1	−1	−1	−1	−1	−1	−1	1	1	−1	−1	−1	−1	−1	1	1	1
X.4	−1	1	−1	1	−1	−1	1	−1	1	1	1	1	−1	−1	−1	−1	−1	1	−1
X.5	−2	.	1	.	1	−2	.	1	1	1	1	−1	.	.	.	.	.	−1	1
X.6	2	.	−1	.	−1	2	.	−1	1	1	1	−1	.	.	.	.	.	−1	−1
X.7	−2	.	1	.	−1	2	.	−1	−1	1	1	1	.	.	.	.	.	−1	1
X.8	2	.	−1	.	1	−2	.	1	−1	1	1	1	.	.	.	.	.	−1	−1
X.9	−1	1	−1	1	−1	−1	1	−1	−1	−1	−1	.	1	1	1	1	1	.	−1
X.10	1	1	1	−1	−1	−1	−1	−1	1	−1	−1	.	1	1	1	1	−1	.	1
X.11	1	1	1	1	1	1	1	1	−1	−1	−1	.	−1	−1	−1	−1	−1	.	1
X.12	−1	1	−1	−1	1	1	−1	1	1	−1	−1	.	−1	−1	−1	−1	1	.	−1
X.13	.	−2	.	2	.	.	2	.	.	.	.	−1	.	.	.	.	.	1	.
X.14	.	−2	.	−2	.	.	−2	.	.	.	.	1	.	.	.	.	.	1	.
X.15	.	−1	.	.	.	.	.	.	.	−2	1	.	.	.	.	.	−1	.	.
X.16	.	−1	.	.	.	.	.	.	.	−2	1	.	.	.	.	.	1	.	.
X.17	.	1	.	.	.	.	.	.	.	.	.	.	2	−1	−1	2	.	−1	.
X.18	.	1	.	.	.	.	.	.	.	.	.	.	−2	1	1	−2	.	−1	.
X.19	.	−1	.	.	.	.	.	.	.	.	.	.	.	A	$\bar{A}$	.	.	−1	.
X.20	.	−1	.	.	.	.	.	.	.	.	.	.	.	$\bar{A}$	A	.	.	−1	.
X.21	3	.	.	1	−2	1	−2	1	.	2	−1	.	.	.	.	.	.	.	.
X.22	3	.	.	−1	2	−1	2	−1	.	2	−1	.	.	.	.	.	.	.	.
X.23	−2	1	−2	.	.	.	.	.	.	−1	−1	.	.	.	.	.	.	.	1
X.24	2	−1	2	.	.	.	.	.	.	−1	−1	.	.	.	.	.	.	.	−1
X.25	−3	.	.	1	2	−1	−2	−1	.	2	−1	.	.	.	.	.	.	.	.
X.26	−3	.	.	−1	−2	1	2	1	.	2	−1	.	.	.	.	.	.	.	.
X.27	−2	−1	−2	.	.	.	.	.	.	−1	−1	.	.	.	.	.	.	.	1
X.28	2	1	2	.	.	.	.	.	.	−1	−1	.	.	.	.	.	.	.	−1
X.29	.	−2	.	.	.	.	.	.	.	2	−1	.	.	.	.	.	.	.	.
X.30	.	1	.	.	.	.	.	.	.	.	.	.	.	.	.	.	−1	.	.
X.31	.	−1	.	2	.	.	−1	.	.	.	.	.	1	1	1	−2	.	.	.
X.32	.	−2	.	.	.	.	.	.	.	.	.	.	−2	1	1	1	.	.	.
X.33	.	−1	.	−2	.	.	1	.	.	.	.	.	1	1	1	−2	.	.	.
X.34	.	.	.	.	.	.	.	.	.	.	.	.	2	−1	−1	−1	.	.	.
X.35	.	−1	.	2	.	.	−1	.	.	.	.	.	−1	−1	−1	2	.	.	.
X.36	.	−1	.	−2	.	.	1	.	.	.	.	.	−1	−1	−1	2	.	.	.
X.37	.	−2	.	.	.	.	.	.	.	.	.	.	2	−1	−1	−1	.	.	.
X.38	.	.	.	.	.	.	.	.	.	.	.	.	−2	1	1	1	.	.	.
X.39	.	1	.	.	.	.	.	.	.	.	.	.	.	.	.	.	1	.	.
X.40	−4	.	2	.	.	.	.	.	.	1	1	.	.	.	.	.	.	.	−1
X.41	.	.	.	.	2	2	.	−1	.	−2	1	.	.	.	.	.	.	.	.
X.42	4	.	−2	.	.	.	.	.	.	1	1	.	.	.	.	.	.	.	1
X.43	.	.	.	.	−2	−2	.	1	.	−2	1	.	.	.	.	.	.	.	.
X.44	.	.	.	.	.	.	.	.	.	.	.	.	.	.	.	.	.	1	.
X.45	.	2	.	.	.	.	.	.	.	.	.	.	.	.	.	.	.	.	.
X.46	.	2	.	.	.	.	.	.	.	.	.	.	.	.	.	.	.	.	.
X.47	.	−1	.	.	.	.	.	.	.	.	.	.	.	.	.	.	.	.	.
X.48	.	1	.	.	.	.	.	.	.	.	.	.	.	.	.	.	.	.	.

where $A = 3\zeta(4)$.

3.5.7 *Character table of* $N_3 = N_G(r_3) = \langle p, q, r, s, t \rangle$

2	4	4	4	4	3	1	2	2	.	1	1	3	1	2	2	2	2	1	1	2	2	1	1	1	1
3	4	3	2	1	4	4	3	3	4	3	3	2	3	2	2	2	2	2	2	1	1	1	3	3	3
7	1	1	.	.	1	.	.	.	.	.	.	.	.	.	.	.	.	.	.	.	.	1	.	.	.
	1a	2a	2b	2c	3a	3b	3c	3d	3e	3f	3g	6a	6b	6c	6d	6e	6f	6g	6h	6i	6j	7a	9a	9b	9c
2P	1a	1a	1a	1a	3a	3b	3d	3c	3e	3g	3f	3a	3b	3d	3c	3d	3c	3g	3f	3d	3c	7a	9b	9a	9c
3P	1a	2a	2b	2c	1a	1a	1a	1a	1a	1a	1a	2b	2a	2b	2b	2a	2a	2b	2b	2c	2c	7a	3b	3b	3b
7P	1a	2a	2b	2c	3a	3b	3c	3d	3e	3f	3g	6a	6b	6c	6d	6e	6f	6g	6h	6i	6j	1a	9a	9b	9c
X.1	1	1	1	1	1	1	1	1	1	1	1	1	1	1	1	1	1	1	1	1	1	1	1	1	1
X.2	1	−1	1	−1	1	1	1	1	1	1	1	1	−1	1	1	−1	−1	1	1	−1	−1	1	1	1	1
X.3	1	−1	1	−1	1	1	A	$\bar{A}$	1	A	$\bar{A}$	1	−1	A	$\bar{A}$	$-A$	$-\bar{A}$	A	$\bar{A}$	$-A$	$-\bar{A}$	1	$\bar{A}$	A	1
X.4	1	−1	1	−1	1	1	$\bar{A}$	A	1	$\bar{A}$	A	1	−1	$\bar{A}$	A	$-\bar{A}$	$-A$	$\bar{A}$	A	$-\bar{A}$	$-A$	1	A	$\bar{A}$	1
X.5	1	1	1	1	1	1	$\bar{A}$	A	1	$\bar{A}$	A	1	1	$\bar{A}$	A	$\bar{A}$	A	$\bar{A}$	A	$\bar{A}$	A	1	A	$\bar{A}$	1
X.6	1	1	1	1	1	1	A	$\bar{A}$	1	A	$\bar{A}$	1	1	A	$\bar{A}$	A	$\bar{A}$	A	$\bar{A}$	A	$\bar{A}$	1	$\bar{A}$	A	1
X.7	2	.	2	.	−1	2	2	2	−1	−1	−1	−1	.	2	2	.	.	−1	−1	.	.	2	2	2	2
X.8	2	.	2	.	−1	2	B	$\bar{B}$	−1	$-A$	$-\bar{A}$	−1	.	B	$\bar{B}$	.	.	$-A$	$-\bar{A}$	.	.	2	$\bar{B}$	B	2
X.9	2	.	2	.	−1	2	$\bar{B}$	B	−1	$-\bar{A}$	$-A$	−1	.	$\bar{B}$	B	.	.	$-\bar{A}$	$-A$	.	.	2	B	$\bar{B}$	2
X.10	7	−7	−1	1	7	−2	1	1	−2	1	1	−1	2	−1	−1	−1	−1	−1	−1	1	1	.	1	1	1
X.11	7	7	−1	−1	7	−2	1	1	−2	1	1	−1	−2	−1	−1	1	1	−1	−1	−1	−1	.	1	1	1
X.12	7	−7	−1	1	7	−2	$\bar{A}$	A	−2	$\bar{A}$	A	−1	2	$-\bar{A}$	$-A$	$-\bar{A}$	$-A$	$-\bar{A}$	$-A$	$\bar{A}$	A	.	A	$\bar{A}$	1
X.13	7	7	−1	−1	7	−2	A	$\bar{A}$	−2	A	$\bar{A}$	−1	−2	$-A$	$-\bar{A}$	A	$\bar{A}$	$-A$	$-\bar{A}$	$-A$	$-\bar{A}$	.	$\bar{A}$	A	1
X.14	7	−7	−1	1	7	−2	A	$\bar{A}$	−2	A	$\bar{A}$	−1	2	$-A$	$-\bar{A}$	$-A$	$-\bar{A}$	$-A$	$-\bar{A}$	A	$\bar{A}$	.	$\bar{A}$	A	1
X.15	7	7	−1	−1	7	−2	$\bar{A}$	A	−2	$\bar{A}$	A	−1	−2	$-\bar{A}$	$-A$	$\bar{A}$	A	$-\bar{A}$	$-A$	$-\bar{A}$	$-A$	.	A	$\bar{A}$	1
X.16	8	−8	.	.	8	−1	2	2	−1	2	2	.	1	.	.	−2	−2	.	.	.	.	1	−1	−1	−1
X.17	8	8	.	.	8	−1	2	2	−1	2	2	.	−1	.	.	2	2	.	.	.	.	1	−1	−1	−1
X.18	8	8	.	.	8	−1	$\bar{B}$	B	−1	$\bar{B}$	B	.	−1	.	.	$\bar{B}$	B	.	.	.	.	1	$-A$	$-\bar{A}$	−1
X.19	8	8	.	.	8	−1	B	$\bar{B}$	−1	B	$\bar{B}$	.	−1	.	.	B	$\bar{B}$	.	.	.	.	1	$-\bar{A}$	$-A$	−1
X.20	8	−8	.	.	8	−1	$\bar{B}$	B	−1	$\bar{B}$	B	.	1	.	.	$-\bar{B}$	$-B$	.	.	.	.	1	$-A$	$-\bar{A}$	−1
X.21	8	−8	.	.	8	−1	B	$\bar{B}$	−1	B	$\bar{B}$	.	1	.	.	$-B$	$-\bar{B}$	.	.	.	.	1	$-\bar{A}$	$-A$	−1
X.22	14	.	−2	.	−7	−4	2	2	2	−1	−1	1	.	−2	−2	.	.	1	1	.	.	.	2	2	2
X.23	14	.	−2	.	−7	−4	$\bar{B}$	B	2	$-\bar{A}$	$-A$	1	.	$-\bar{B}$	$-B$	.	.	$\bar{A}$	A	.	.	.	B	$\bar{B}$	2
X.24	14	.	−2	.	−7	−4	B	$\bar{B}$	2	$-A$	$-\bar{A}$	1	.	$-B$	$-\bar{B}$	.	.	A	$\bar{A}$	.	.	.	$\bar{B}$	B	2
X.25	16	.	.	.	−8	−2	4	4	1	−2	−2	.	.	.	.	.	.	.	.	.	.	2	−2	−2	−2
X.26	16	.	.	.	−8	−2	C	$\bar{C}$	1	$-B$	$-\bar{B}$	.	.	.	.	.	.	.	.	.	.	2	$-\bar{B}$	$-B$	−2
X.27	16	.	.	.	−8	−2	$\bar{C}$	C	1	$-\bar{B}$	$-B$	.	.	.	.	.	.	.	.	.	.	2	$-B$	$-\bar{B}$	−2
X.28	21	−21	−3	3	21	3	.	.	3	.	.	−3	−3	.	.	.	.	.	.	.	.	.	.	.	.
X.29	21	21	−3	−3	21	3	.	.	3	.	.	−3	3	.	.	.	.	.	.	.	.	.	.	.	.
X.30	27	−27	3	−3	27	.	.	.	.	.	.	3	.	.	.	.	.	.	.	.	.	−1	.	.	.
X.31	27	27	3	3	27	.	.	.	.	.	.	3	.	.	.	.	.	.	.	.	.	−1	.	.	.
X.32	42	.	−6	.	−21	6	.	.	−3	.	.	3	.	.	.	.	.	.	.	.	.	.	.	.	.
X.33	54	.	6	.	−27	.	.	.	.	.	.	−3	.	.	.	.	.	.	.	.	.	−2	.	.	.

Character table of $N_G(r_3)$ *(continued)*

2	.	.	.	1	1	1	1	.
3	3	3	3	.	2	2	2	1
7	.	.	.	1	.	.	.	1
	9d	9e	9f	14a	18a	18b	18c	21a
2P	9e	9d	9f	7a	9b	9a	9c	21a
3P	3b	3b	3b	14a	6b	6b	6b	7a
7P	9d	9e	9f	2a	18a	18b	18c	3a
X.1	1	1	1	1	1	1	1	1
X.2	1	1	1	−1	−1	−1	−1	1
X.3	$\bar{A}$	A	1	−1	$-\bar{A}$	$-A$	−1	1
X.4	A	$\bar{A}$	1	−1	$-A$	$-\bar{A}$	−1	1
X.5	A	$\bar{A}$	1	1	A	$\bar{A}$	1	1
X.6	$\bar{A}$	A	1	1	$\bar{A}$	A	1	1
X.7	−1	−1	−1	.	.	.	.	−1
X.8	$-\bar{A}$	$-A$	−1	.	.	.	.	−1
X.9	$-A$	$-\bar{A}$	−1	.	.	.	.	−1
X.10	1	1	1	.	−1	−1	−1	.
X.11	1	1	1	.	1	1	1	.
X.12	A	$\bar{A}$	1	.	$-A$	$-\bar{A}$	−1	.
X.13	$\bar{A}$	A	1	.	$\bar{A}$	A	1	.
X.14	$\bar{A}$	A	1	.	$-\bar{A}$	$-A$	−1	.
X.15	A	$\bar{A}$	1	.	A	$\bar{A}$	1	.
X.16	−1	−1	−1	−1	1	1	1	1
X.17	−1	−1	−1	1	−1	−1	−1	1
X.18	$-A$	$-\bar{A}$	−1	1	$-A$	$-\bar{A}$	−1	1
X.19	$-\bar{A}$	$-A$	−1	1	$-\bar{A}$	$-A$	−1	1
X.20	$-A$	$-\bar{A}$	−1	−1	A	$\bar{A}$	1	1
X.21	$-\bar{A}$	$-A$	−1	−1	$\bar{A}$	A	1	1
X.22	−1	−1	−1	.	.	.	.	.
X.23	$-A$	$-\bar{A}$	−1	.	.	.	.	.
X.24	$-\bar{A}$	$-A$	−1	.	.	.	.	.
X.25	1	1	1	.	.	.	.	−1
X.26	$\bar{A}$	A	1	.	.	.	.	−1
X.27	A	$\bar{A}$	1	.	.	.	.	−1
X.28	.	.	.	.	.	.	.	.
X.29	.	.	.	.	.	.	.	.
X.30	.	.	.	1	.	.	.	−1
X.31	.	.	.	−1	.	.	.	−1
X.32	.	.	.	.	.	.	.	.
X.33	.	.	.	.	.	.	.	1

where $A = \zeta(3)$, $B = 2A$, $C = 4A$.

3.5.8 *Character table of* $N_5 = N_G(f_1) = \langle q_i | 1 \leq i \leq 5 \rangle$

	1a	2a	2b	3a	4a	4b	4c	5a	5b	6a	8a	8b	8c	8d	10a	10b	12a	12b	15a	20a	20b	24a	24b	24c	24d	30a
2	4	4	3	3	4	4	3	2	1	3	3	3	3	3	2	1	3	3	1	2	2	3	3	3	3	1
3	1	1	.	1	1	1	.	1	.	1	1	1	.	.	1	.	1	1	1	.	.	1	1	1	1	1
5	3	1	1	1	.	.	1	3	2	1	.	.	.	.	1	1	.	.	1	1	1	.	.	.	.	1
2P	1a	1a	1a	3a	2a	2a	2a	5a	5b	3a	4a	4b	4a	4b	5a	5b	6a	6a	15a	10a	10a	12a	12b	12a	12b	15a
3P	1a	2a	2b	1a	4b	4a	4c	5a	5b	2a	8b	8a	8d	8c	10a	10b	4b	4a	5a	20a	20b	8b	8a	8b	8a	10a
5P	1a	2a	2b	3a	4a	4b	4c	1a	1a	6a	8a	8b	8c	8d	2a	2b	12a	12b	3a	4c	4c	24a	24b	24c	24d	6a
X.1	1	1	1	1	1	1	1	1	1	1	1	1	1	1	1	1	1	1	1	1	1	1	1	1	1	1
X.2	1	1	1	1	1	1	1	1	1	1	−1	−1	−1	−1	1	1	1	1	1	1	1	−1	−1	−1	−1	1
X.3	1	1	−1	1	1	1	−1	1	1	1	1	1	−1	−1	1	−1	1	1	1	−1	−1	1	1	1	1	1
X.4	1	1	−1	1	1	1	−1	1	1	1	−1	−1	1	1	1	−1	1	1	1	−1	−1	−1	−1	−1	−1	1
X.5	1	1	1	1	−1	−1	−1	1	1	1	A	$\bar{A}$	$\bar{A}$	A	1	1	−1	−1	1	−1	−1	A	$\bar{A}$	A	$\bar{A}$	1
X.6	1	1	1	1	−1	−1	−1	1	1	1	$\bar{A}$	A	A	$\bar{A}$	1	1	−1	−1	1	−1	−1	$\bar{A}$	A	$\bar{A}$	A	1
X.7	1	1	−1	1	−1	−1	1	1	1	1	A	$\bar{A}$	A	$\bar{A}$	1	−1	−1	−1	1	1	1	A	$\bar{A}$	A	$\bar{A}$	1
X.8	1	1	−1	1	−1	−1	1	1	1	1	$\bar{A}$	A	$\bar{A}$	A	1	−1	−1	−1	1	1	1	$\bar{A}$	A	$\bar{A}$	A	1
X.9	2	2	.	−1	2	2	.	2	2	−1	2	2	.	.	2	.	−1	−1	−1	.	.	−1	−1	−1	−1	−1
X.10	2	2	.	−1	2	2	.	2	2	−1	−2	−2	.	.	2	.	−1	−1	−1	.	.	1	1	1	1	−1
X.11	2	−2	.	2	B	$\bar{B}$	.	2	2	−2	.	.	.	.	−2	.	B	$\bar{B}$	2	.	.	.	.	.	.	−2
X.12	2	2	.	−1	−2	−2	.	2	2	−1	B	$\bar{B}$	.	.	2	.	1	1	−1	.	.	A	$\bar{A}$	A	$\bar{A}$	−1
X.13	2	2	.	−1	−2	−2	.	2	2	−1	$\bar{B}$	B	.	.	2	.	1	1	−1	.	.	$\bar{A}$	A	$\bar{A}$	A	−1
X.14	2	−2	.	2	$\bar{B}$	B	.	2	2	−2	.	.	.	.	−2	.	$\bar{B}$	B	2	.	.	.	.	.	.	−2
X.15	2	−2	.	−1	$\bar{B}$	B	.	2	2	1	.	.	.	.	−2	.	$\bar{A}$	A	−1	.	.	C	$-\bar{C}$	$-C$	$\bar{C}$	1
X.16	2	−2	.	−1	$\bar{B}$	B	.	2	2	1	.	.	.	.	−2	.	$\bar{A}$	A	−1	.	.	$-C$	$\bar{C}$	C	$-\bar{C}$	1
X.17	2	−2	.	−1	B	$\bar{B}$	.	2	2	1	.	.	.	.	−2	.	A	$\bar{A}$	−1	.	.	$\bar{C}$	$-C$	$-\bar{C}$	C	1
X.18	2	−2	.	−1	B	$\bar{B}$	.	2	2	1	.	.	.	.	−2	.	A	$\bar{A}$	−1	.	.	$-\bar{C}$	C	$\bar{C}$	$-C$	1
X.19	20	4	.	−4	.	.	−4	−5	.	4	.	.	.	.	−1	.	.	.	1	1	1	.	.	.	.	−1
X.20	20	4	.	−4	.	.	4	−5	.	4	.	.	.	.	−1	.	.	.	1	−1	−1	.	.	.	.	−1
X.21	20	−4	.	−4	.	.	.	−5	.	−4	.	.	.	.	1	.	.	.	1	D	$\bar{D}$	.	.	.	.	1
X.22	20	−4	.	−4	.	.	.	−5	.	−4	.	.	.	.	1	.	.	.	1	$\bar{D}$	D	.	.	.	.	1
X.23	24	.	4	.	.	.	.	24	−1	.	.	.	.	.	.	−1	.	.	.	.	.	.	.	.	.	.
X.24	24	.	−4	.	.	.	.	24	−1	.	.	.	.	.	.	1	.	.	.	.	.	.	.	.	.	.
X.25	40	−8	.	4	.	.	.	−10	.	4	.	.	.	.	2	.	.	.	−1	.	.	.	.	.	.	−1
X.26	40	8	.	4	.	.	.	−10	.	−4	.	.	.	.	−2	.	.	.	−1	.	.	.	.	.	.	1

where $A = \zeta(4)$, $B = -2\zeta(4)$, $C = -2\zeta(24)^{11} - \zeta(24)^3$, $D = 2\zeta(20)^{17} + 2\zeta(20)^{13} + \zeta(20)^5$.

3.5.9 *Character table of* $N_6 = N_G(f_2) = \langle x_i | 1 \leq i \leq 4 \rangle$

	1a	2a	2b	2c	3a	4a	4b	4c	4d	5a	5b	5c	5d	5e	6a	10a	10b	10c	12a	12b	15a	20a	20b	20c	20d
2	4	4	4	4	2	4	4	4	4	2	2	2	.	.	2	2	2	2	2	2	.	2	2	2	2
3	1	1	.	.	1	1	1	.	.	1	.	.	.	.	1	.	.	.	1	1	1	.	.	.	.
5	2	1	1	.	1	1	1	.	.	2	2	2	2	2	.	1	1	1	.	.	1	1	1	1	1
2P	1a	1a	1a	1a	3a	2a	2a	2a	2a	5a	5c	5b	5e	5d	3a	5b	5c	5a	6a	6a	15a	10a	10b	10a	10b
3P	1a	2a	2b	2c	1a	4b	4a	4d	4c	5a	5c	5b	5e	5d	2a	10b	10a	10c	4b	4a	5a	20d	20c	20b	20a
5P	1a	2a	2b	2c	3a	4a	4b	4c	4d	1a	1a	1a	1a	1a	6a	2a	2a	2b	12a	12b	3a	4a	4a	4b	4b
X.1	1	1	1	1	1	1	1	1	1	1	1	1	1	1	1	1	1	1	1	1	1	1	1	1	1
X.2	1	1	1	1	1	−1	−1	−1	−1	1	1	1	1	1	1	1	1	1	−1	−1	1	−1	−1	−1	−1
X.3	1	−1	1	−1	1	A	$\bar{A}$	A	$\bar{A}$	1	1	1	1	1	−1	−1	−1	1	A	$\bar{A}$	1	A	A	$\bar{A}$	$\bar{A}$
X.4	1	−1	1	−1	1	$\bar{A}$	A	$\bar{A}$	A	1	1	1	1	1	−1	−1	−1	1	$\bar{A}$	A	1	$\bar{A}$	$\bar{A}$	A	A
X.5	3	3	−1	−1	.	−3	−3	1	1	3	B	C	B	C	.	C	B	−1	.	.	.	$-B$	$-C$	$-B$	$-C$
X.6	3	3	−1	−1	.	3	3	−1	−1	3	C	B	C	B	.	B	C	−1	.	.	.	C	B	C	B
X.7	3	3	−1	−1	.	−3	−3	1	1	3	C	B	C	B	.	B	C	−1	.	.	.	$-C$	$-B$	$-C$	$-B$
X.8	3	3	−1	−1	.	3	3	−1	−1	3	B	C	B	C	.	C	B	−1	.	.	.	B	C	B	C
X.9	3	−3	−1	1	.	D	$\bar{D}$	$\bar{A}$	A	3	C	B	C	B	.	$-B$	$-C$	−1	.	.	.	E	F	$\bar{E}$	$\bar{F}$
X.10	3	−3	−1	1	.	$\bar{D}$	D	A	$\bar{A}$	3	C	B	C	B	.	$-B$	$-C$	−1	.	.	.	$\bar{E}$	$\bar{F}$	E	F
X.11	3	−3	−1	1	.	D	$\bar{D}$	$\bar{A}$	A	3	B	C	B	C	.	$-C$	$-B$	−1	.	.	.	F	E	$\bar{F}$	$\bar{E}$
X.12	3	−3	−1	1	.	$\bar{D}$	D	A	$\bar{A}$	3	B	C	B	C	.	$-C$	$-B$	−1	.	.	.	$\bar{F}$	$\bar{E}$	F	E
X.13	4	.	4	.	4	.	.	.	.	−1	4	4	−1	−1	.	.	.	−1	.	.	−1	.	.	.	.
X.14	4	4	.	.	1	−4	−4	.	.	4	−1	−1	−1	−1	1	−1	−1	.	−1	−1	1	1	1	1	1
X.15	4	4	.	.	1	4	4	.	.	4	−1	−1	−1	−1	1	−1	−1	.	1	1	1	−1	−1	−1	−1
X.16	4	−4	.	.	1	G	$\bar{G}$	.	.	4	−1	−1	−1	−1	−1	1	1	.	$\bar{A}$	A	1	A	A	$\bar{A}$	$\bar{A}$
X.17	4	−4	.	.	1	$\bar{G}$	G	.	.	4	−1	−1	−1	−1	−1	1	1	.	A	$\bar{A}$	1	$\bar{A}$	$\bar{A}$	A	A
X.18	5	5	1	1	−1	−5	−5	−1	−1	5	.	.	.	.	−1	.	.	1	1	1	−1	.	.	.	.
X.19	5	5	1	1	−1	5	5	1	1	5	.	.	.	.	−1	.	.	1	−1	−1	−1	.	.	.	.
X.20	5	−5	1	−1	−1	H	$\bar{H}$	A	$\bar{A}$	5	.	.	.	.	1	.	.	1	$\bar{A}$	A	−1	.	.	.	.
X.21	5	−5	1	−1	−1	$\bar{H}$	H	$\bar{A}$	A	5	.	.	.	.	1	.	.	1	A	$\bar{A}$	−1	.	.	.	.
X.22	12	.	−4	.	.	.	.	.	.	−3	I	J	$-B$	$-C$	.	.	.	1	.	.	.	.	.	.	.
X.23	12	.	−4	.	.	.	.	.	.	−3	J	I	$-C$	$-B$	.	.	.	1	.	.	.	.	.	.	.
X.24	16	.	.	.	4	.	.	.	.	−4	−4	−4	1	1	.	.	.	.	.	.	−1	.	.	.	.
X.25	20	.	4	.	−4	.	.	.	.	−5	.	.	.	.	.	.	.	−1	.	.	1	.	.	.	.

where $A = \zeta(4)$, $B = -\zeta(5)^3 - \zeta(5)^2$, $C = \zeta(5)^3 + \zeta(5)^2 + 1$, $D = 3A$, $E = \zeta(20)^{17} + \zeta(20)^{13} + \zeta(20)^5$, $F = -\zeta(20)^{17} - \zeta(20)^{13}$, $G = -4A$, $H = 5A$, $I = 4B$, $J = 4C$.

4
Conway's simple group Co_2

It is the purpose of this chapter to provide an existence proof for the sporadic Conway group Co_2 by means of Algorithm 1.3.8. It is due to H. Kim and the author; see [72].

Section 4.1 summarizes some known facts about M_{22}, its automorphism group A_{22} and their irreducible representations over $F = GF(2)$. These subsidiary results will be used in this chapter and in Chapter 5. By G. James' article [65] each of the groups M_{22} and A_{22} has two simple modules V_1, V_2 and V_3, V_4 of dimension 10, respectively. Lemmas 4.1.1 and 4.1.2 assert there are four split extensions $E_i = V_i \rtimes \mathsf{M}_{22}$, $i = 1, 2$, $E_i = V_i \rtimes A_{22}$, $i = 3, 4$ and one non-split extension E_5 of V_4 by A_{22}. Each of these five groups is described by a presentation in terms of generators and relations. In Lemma 4.1.3 the reader finds for each of these groups a faithful permutation representation.

In [72] H. Kim and the author applied Algorithm 1.3.8 to the extension groups E_3 and E_2 and obtained simple matrix subgroups $\mathfrak{G}_3$ in $\mathrm{GL}_{23}(13)$ and $\mathfrak{G}_2$ in $\mathrm{GL}_{78}(13)$ isomorphic to Conway's simple group Co_2 of [16] and to Fischer's simple group Fi_{22} of [29], respectively. This chapter deals with the second Conway group Co_2. The small Fischer group is constructed in the same way in Chapter 5.

The first case $E = E_3$ is dealt with in Section 4.2 of this chapter. Proposition 4.2.1 describes a construction of a finitely presented group H_3 with center $Z(H_3) = \langle z \rangle$ and Sylow 2-subgroup S_3 such that $H_3/Z(H_3)$ is a split extension of $\mathsf{Sp}_6(2)$ by an elementary abelian group of order 2^8. Furthermore, $D_3 = N_{H_3}(A_3) \cong D = C_{E_3}(z)$, where A_3 is the unique maximal elementary abelian normal subgroup of S_3 of order 2^{10} and z_3 is a representative of the unique conjugacy class of 2-central involutions of E_3. Hence all conditions of Step 5 of Algorithm 1.3.8 are satisfied.

In Section 4.3 the presentation of H_3 is taken as the input of Step 6 and the subsequent steps of Algorithm 1.3.8. It returns a finite simple subgroup $\mathfrak{G}_3$ of $\mathrm{GL}_{23}(13)$ of order $2^{18} \cdot 3^6 \cdot 5^3 \cdot 7 \cdot 11 \cdot 23$ having a 2-central involution $\mathfrak{z}$ such that $C_{\mathfrak{G}_3}(\mathfrak{z}) \cong H_3$; see Theorem 4.3.2. By [72] the character table of $\mathfrak{G}_3$ is equivalent to the one of Conway's second sporadic group Co_2. For applications in Chapter 7 a finite presentation of $\mathfrak{G}_3$ is determined in Corollary 4.3.3, which has the same set of defining relations as the one of Co_2 due to C. Praeger and L. Soicher; see [110].

Section 4.4 deals with the uniqueness problem. Since the Goldschmidt index of the amalgam $E \leftarrow D \rightarrow H_3$ is 2 by Lemma 4.3.1 the author's uniqueness criterion Theorem 7.5.1 of [92] cannot be applied. So a short survey is given here about the relevant literature.

Sections 4.5 and 4.6 contain the systems of representatives of the conjugacy classes and the character tables of the local subgroups dealt with in this chapter. They are essential tools for the given existence proof of Co_2. The four generating matrices of $\mathfrak{G}_3$ are documented on the accompanying DVD.

4.1 Extensions of the Mathieu group M_{22} and $Aut(\mathsf{M}_{22})$

The Mathieu group M_{22} is defined in Definition 8.2.1 of [92] by means of generators and relations. This beautiful presentation is due to J. A. Todd. The irreducible 2-modular representations of the Mathieu group M_{22} were determined by G. James [65]. Here only the two non-isomorphic simple modules V_i, $i = 1, 2$, of dimension 10 over $F = GF(2)$ will be considered. Todd's permutation representations of the Mathieu groups are stated in Lemma 8.2.2 of [92]. Therefore all conditions of Holt's Algorithm [52] implemented in MAGMA are satisfied. Using it, Kim and the author showed in [72] that, for each simple module V_i, there is exactly one extension E_i of M_{22} by V_i, the split extension. The presentations of E_1 and E_2 are given in Lemma 4.1.1.

The automorphism group $A_{22} = Aut(\mathsf{M}_{22})$ has two 10-dimensional irreducible representations V_i, $i = 3, 4$. Lemma 4.1.2 states that A_{22} has one non-split extension E_5 by V_3 besides the two split extensions by V_3 and V_4. All extension groups are described here by generators and relations. Lemma 4.1.3 also documents faithful permutation representations for all five extension groups E_i. All these results will be used in this chapter and in Chapter 5.

In the following $F = GF(2)$ and $p = 2$.

Lemma 4.1.1 *Let* $\mathsf{M}_{22} = \langle a, b, c, d, t, g, h, i \rangle$ *be the finitely presented group of Definition 8.2.1 of [92]. Then the following statements hold.*

(a) *A faithful permutation representation of degree* 24 *of* M_{22} *is stated in Lemma 8.2.2 of [92].*

(b) *The first irreducible representation* V_1 *of* M_{22} *is described by the following matrices:*

$$Ma_1 = \begin{pmatrix} 1&0&1&0&0&0&0&0&0&0 \\ 0&0&0&1&0&0&0&0&0&0 \\ 0&0&1&0&0&0&0&0&0&0 \\ 0&1&0&0&0&0&0&0&0&0 \\ 0&0&1&0&0&0&0&1&0&0 \\ 0&0&0&0&0&0&0&0&1&0 \\ 0&0&1&0&0&0&1&0&0&0 \\ 0&0&1&0&1&0&0&0&0&0 \\ 0&0&0&0&0&1&0&0&0&0 \\ 0&0&1&0&0&0&0&0&0&1 \end{pmatrix}, \quad Mb_1 = \begin{pmatrix} 1&0&1&0&1&0&0&1&0&0 \\ 0&0&0&0&1&1&0&0&0&0 \\ 0&0&1&0&0&0&0&0&0&0 \\ 0&0&1&0&0&0&0&1&1&0 \\ 0&0&0&0&0&0&0&1&0&0 \\ 0&1&0&0&0&0&0&1&0&0 \\ 0&0&1&0&0&0&1&0&0&0 \\ 0&0&0&0&1&0&0&0&0&0 \\ 0&0&1&1&1&0&0&0&0&0 \\ 0&0&1&0&1&0&0&1&0&1 \end{pmatrix},$$

$$Mc_1 = \begin{pmatrix} 1&0&0&0&0&1&0&0&1&0 \\ 0&0&1&0&1&1&0&0&0&0 \\ 0&0&1&0&0&0&0&0&0&0 \\ 0&0&0&0&0&0&0&1&1&0 \\ 0&1&0&0&0&0&0&0&1&0 \\ 0&0&1&0&0&0&0&0&1&0 \\ 0&0&1&0&0&0&1&0&0&0 \\ 0&0&1&1&0&1&0&0&0&0 \\ 0&0&1&0&0&1&0&0&0&0 \\ 0&0&1&0&0&1&0&0&1&1 \end{pmatrix}, \quad Md_1 = \begin{pmatrix} 1&0&0&1&0&1&0&1&0&0 \\ 0&0&1&0&1&0&0&0&1&0 \\ 0&1&0&1&1&1&0&1&1&0 \\ 0&0&0&0&0&1&0&1&0&0 \\ 0&0&0&1&0&0&0&0&1&0 \\ 0&0&0&0&1&0&0&1&1&0 \\ 0&0&0&1&1&1&1&0&0&0 \\ 0&0&0&1&1&0&0&1&1&0 \\ 0&0&0&0&1&1&0&1&0&0 \\ 0&0&0&1&1&1&0&0&0&1 \end{pmatrix},$$

$$Mt_1 = \begin{pmatrix} 1&0&1&1&0&0&0&0&1&0 \\ 0&1&1&0&1&1&0&0&1&0 \\ 0&0&1&1&1&1&0&0&0&0 \\ 0&0&1&1&1&0&0&0&0&0 \\ 0&0&1&0&1&1&0&0&0&0 \\ 0&0&1&1&0&1&0&0&0&0 \\ 0&0&1&0&0&0&1&0&0&0 \\ 0&0&0&1&0&1&0&0&1&0 \\ 0&0&0&1&1&0&0&1&1&0 \\ 0&0&0&0&1&0&0&0&1&1 \end{pmatrix}, \quad Mg_1 = \begin{pmatrix} 1&0&1&0&0&0&0&0&0&0 \\ 0&0&1&0&0&0&1&1&1&0 \\ 0&0&1&0&0&0&0&0&0&0 \\ 0&0&1&1&0&0&0&1&1&0 \\ 0&0&1&0&1&0&0&1&0&0 \\ 0&0&1&0&0&1&0&0&1&0 \\ 0&1&1&0&0&0&0&1&1&0 \\ 0&0&0&0&0&0&0&1&0&0 \\ 0&0&0&0&0&0&0&0&1&0 \\ 0&0&0&0&0&0&0&1&0&1 \end{pmatrix},$$

$$Mh_1 = \begin{pmatrix} 1&0&0&0&0&0&0&0&1&0 \\ 0&1&0&0&0&1&0&0&0&0 \\ 0&0&1&0&0&1&0&0&1&0 \\ 0&0&0&1&0&0&0&0&1&0 \\ 0&0&0&0&0&1&0&1&0&0 \\ 0&0&0&0&0&1&0&0&0&0 \\ 0&0&0&0&0&0&0&0&1&1 \\ 0&0&0&0&1&1&0&0&0&0 \\ 0&0&0&0&0&0&0&0&1&0 \\ 0&0&0&0&0&0&1&0&1&0 \end{pmatrix} \quad \text{and} \quad Mi_1 = \begin{pmatrix} 0&0&0&0&1&1&0&0&0&1 \\ 0&1&1&1&1&0&0&0&0&0 \\ 0&0&1&1&1&1&0&0&0&0 \\ 0&0&1&0&1&0&0&0&0&0 \\ 0&0&1&0&1&1&0&0&0&0 \\ 0&0&0&1&1&0&0&0&0&0 \\ 0&0&0&1&1&1&1&0&0&0 \\ 0&0&0&1&0&1&0&0&1&0 \\ 0&0&1&1&0&0&0&1&0&0 \\ 1&0&1&1&0&1&0&0&0&0 \end{pmatrix}.$$

(c) *The second irreducible representation* V_2 *of* M_{22} *is described by the transpose inverse matrices of the generating matrices of* M_{22} *defining* V_1*:*

$$\begin{aligned} Ma_2 &= [Ma_1^{-1}]^T, Mb_2 = [Mb_1^{-1}]^T, Mc_2 = [Mc_1^{-1}]^T, \\ Md_2 &= [Md_1^{-1}]^T, Mt_2 = [Mt_1^{-1}]^T, Mg_2 = [Mg_1^{-1}]^T, \\ Mh_2 &= [Mh_1^{-1}]^T \text{ and } Mi_2 = [Mi_1^{-1}]^T. \end{aligned}$$

(d) $dim_F[H^2(\mathsf{M}_{22}, V_i)] = 0$ *for* $i = 1, 2$.

(e) $E_1 = V_1 \rtimes \mathsf{M}_{22} = \langle a,b,c,d,t,g,h,i,v_1,v_2,v_3,v_4,v_5,v_6,v_8,v_8,v_9, v_{10}\rangle$ *has a set* $\mathcal{R}(E_1)$ *of defining relations consisting of* $\mathcal{R}(\mathsf{M}_{22})$, $\mathcal{R}_1(V_1 \rtimes \mathsf{M}_{22})$ *and the following set* $\mathcal{R}_2(V_1 \rtimes \mathsf{M}_{22})$ *of essential relations:*

$$
\begin{aligned}
& av_1a^{-1}v_1v_3 = av_2a^{-1}v_4 = av_3a^{-1}v_3 = av_4a^{-1}v_2 = 1,\\
& av_5a^{-1}v_3v_8 = av_6a^{-1}v_9 = av_7a^{-1}v_3v_7 = av_8a^{-1}v_3v_5 = 1,\\
& av_9a^{-1}v_6 = av_{10}a^{-1}v_3v_{10} = bv_1b^{-1}v_1v_3v_5v_8 = 1,\\
& bv_2b^{-1}v_5v_6 = bv_3b^{-1}v_3 = bv_4b^{-1}v_3v_8v_9 = bv_5b^{-1}v_8 = 1,\\
& bv_6b^{-1}v_2v_8 = bv_7b^{-1}v_3v_7 = bv_8b^{-1}v_5 = bv_9b^{-1}v_3v_4v_5 = 1,\\
& bv_{10}b^{-1}v_3v_5v_8v_{10} = cv_1c^{-1}v_1v_6v_9 = cv_2c^{-1}v_3v_5v_6 = 1,\\
& cv_3c^{-1}v_3 = cv_4c^{-1}v_8v_9 = cv_5c^{-1}v_2v_9 = cv_6c^{-1}v_3v_9 = 1,\\
& cv_7c^{-1}v_3v_7 = cv_8c^{-1}v_3v_4v_6 = cv_9c^{-1}v_3v_6 = 1\\
& cv_{10}c^{-1}v_3v_6v_9v_{10} = dv_1d^{-1}v_1v_4v_6v_8 = dv_2d^{-1}v_3v_5v_9 = 1,\\
& dv_3d^{-1}v_2v_4v_5v_6v_8v_9 = dv_4d^{-1}v_6v_8 = dv_5d^{-1}v_4v_9 = 1,\\
& dv_6d^{-1}v_5v_8v_9 = dv_7d^{-1}v_4v_5v_6v_7 = dv_8d^{-1}v_4v_5v_8v_9 = 1,\\
& dv_9d^{-1}v_5v_6v_8 = dv_{10}d^{-1}v_4v_5v_6v_{10} = tv_1t^{-1}v_1v_3v_5v_6v_8 = 1,\\
& tv_2t^{-1}v_2v_4v_8 = tv_3t^{-1}v_4v_5v_6 = tv_4t^{-1}v_3v_5 = tv_5t^{-1}v_3v_6 = 1,\\
& tv_6t^{-1}v_3v_4 = tv_7t^{-1}v_4v_5v_6v_7 = tv_8t^{-1}v_4v_6v_8v_9 = 1,\\
& tv_9t^{-1}v_4v_5v_8 = tv_{10}t^{-1}v_3v_4v_5v_6v_8v_{10} = gv_1g^{-1}v_1v_3 = 1,\\
& gv_2g^{-1}v_3v_7v_8v_9 = gv_3g^{-1}v_3 = gv_4g^{-1}v_3v_4v_8v_9 = 1,\\
& gv_5g^{-1}v_3v_5v_8 = gv_6g^{-1}v_3v_6v_9 = gv_7g^{-1}v_2v_3v_8v_9 = 1,\\
& gv_8g^{-1}v_8 = gv_9g^{-1}v_9 = gv_{10}g^{-1}v_8v_{10} = 1,\\
& hv_1h^{-1}v_1v_9 = hv_2h^{-1}v_2v_6 = hv_3h^{-1}v_3v_6v_9 = hv_4h^{-1}v_4v_9 = 1,\\
& hv_5h^{-1}v_6v_8 = hv_6h^{-1}v_6 = hv_7h^{-1}v_9v_{10} = hv_8h^{-1}v_5v_6 = 1,\\
& hv_9h^{-1}v_9 = hv_{10}h^{-1}v_7v_9 = iv_1i^{-1}v_5v_6v_{10} = 1,\\
& iv_2i^{-1}v_2v_3v_4v_5 = iv_3i^{-1}v_3v_4v_5v_6 = iv_4i^{-1}v_3v_5 = 1,\\
& iv_5i^{-1}v_3v_5v_6 = iv_6i^{-1}v_4v_5 = iv_7i^{-1}v_4v_5v_6v_7 = 1,\\
& iv_8i^{-1}v_4v_6v_9 = iv_9i^{-1}v_3v_4v_8 = iv_{10}i^{-1}v_1v_3v_4v_6 = 1.
\end{aligned}
$$

(f) $E_2 = V_2 \rtimes \mathsf{M}_{22} = \langle a,b,c,d,t,g,h,i,v_1,v_2,v_3,v_4,v_5,v_6,v_8,v_8,v_9, v_{10}\rangle$ *has a set* $\mathcal{R}(E_2)$ *of defining relations consisting of* $\mathcal{R}(\mathsf{M}_{22})$, $\mathcal{R}_1(V_2 \rtimes \mathsf{M}_{22})$ *and the following set* $\mathcal{R}_2(V_2 \rtimes \mathsf{M}_{22})$ *of essential*

relations:

$$av_1a^{-1}v_1 = av_2a^{-1}v_4 = av_3a^{-1}v_1v_3v_5v_7v_8v_{10} = av_4a^{-1}v_2 = 1,$$
$$av_5a^{-1}v_8 = av_6a^{-1}v_9 = av_7a^{-1}v_7 = av_8a^{-1}v_5 = av_9a^{-1}v_6 = 1,$$
$$av_{10}a^{-1}v_{10} = bv_1b^{-1}v_1 = bv_2b^{-1}v_6 = bv_3b^{-1}v_1v_3v_4v_7v_9v_{10} = 1,$$
$$bv_4b^{-1}v_9 = bv_5b^{-1}v_1v_2v_8v_9v_{10} = bv_6b^{-1}v_2 = bv_7b^{-1}v_7 = 1,$$
$$bv_8b^{-1}v_1v_4v_5v_6v_{10} = bv_9b^{-1}v_4 = bv_{10}b^{-1}v_{10} = 1,$$
$$cv_1c^{-1}v_1 = cv_2c^{-1}v_5 = cv_3c^{-1}v_2v_3v_6v_7v_8v_9v_{10} = cv_4c^{-1}v_8 = 1,$$
$$cv_5c^{-1}v_2 = cv_6c^{-1}v_1v_2v_8v_9v_{10} = cv_7c^{-1}v_7 = cv_8c^{-1}v_4 = 1,$$
$$cv_9c^{-1}v_1v_4v_5v_6v_{10} = cv_{10}c^{-1}v_{10} = dv_1d^{-1}v_1 = dv_2d^{-1}v_3 = 1,$$
$$dv_3d^{-1}v_2 = dv_4d^{-1}v_1v_3v_5v_7v_8v_{10} = dv_5d^{-1}v_2v_3v_6v_7v_8v_9v_{10} = 1,$$
$$dv_6d^{-1}v_1v_3v_4v_7v_9v_{10} = dv_7d^{-1}v_7 = dv_8d^{-1}v_1v_3v_4v_6v_8v_9 = 1,$$
$$dv_9d^{-1}v_2v_3v_5v_6v_8 = dv_{10}d^{-1}v_{10} = tv_1t^{-1}v_1 = tv_2t^{-1}v_2 = 1,$$
$$tv_3t^{-1}v_1v_2v_3v_4v_5v_6v_7 = tv_4t^{-1}v_1v_3v_4v_6v_8v_9 = 1,$$
$$tv_5t^{-1}v_2v_3v_4v_5v_9v_{10} = tv_6t^{-1}v_2v_3v_5v_6v_8 = tv_7t^{-1}v_7 = 1,$$
$$tv_8t^{-1}v_9 = tv_9t^{-1}v_1v_2v_8v_9v_{10} = tv_{10}t^{-1}v_{10} = gv_1g^{-1}v_1 = 1,$$
$$gv_2g^{-1}v_7 = gv_3g^{-1}v_1v_2v_3v_4v_5v_6v_7 = gv_4g^{-1}v_4 = gv_5g^{-1}v_5 = 1,$$
$$gv_6g^{-1}v_6 = gv_7g^{-1}v_2 = gv_8g^{-1}v_2v_4v_5v_7v_8v_{10} = 1,$$
$$gv_9g^{-1}v_2v_4v_6v_7v_9 = gv_{10}g^{-1}v_{10} = hv_1h^{-1}v_1 = hv_2h^{-1}v_2 = 1,$$
$$hv_3h^{-1}v_3 = hv_4h^{-1}v_4 = hv_5h^{-1}v_8 = hv_6h^{-1}v_2v_3v_5v_6v_8 = 1,$$
$$hv_7h^{-1}v_{10} = hv_8h^{-1}v_5 = hv_9h^{-1}v_1v_3v_4v_7v_9v_{10} = 1,$$
$$hv_{10}h^{-1}v_7 = iv_1i^{-1}v_{10} = iv_2i^{-1}v_2 = 1,$$
$$iv_3i^{-1}v_2v_3v_4v_5v_9v_{10} = iv_4i^{-1}v_2v_3v_6v_7v_8v_9v_{10} = 1,$$
$$iv_5i^{-1}v_1v_2v_3v_4v_5v_6v_7 = iv_6i^{-1}v_1v_3v_5v_7v_8v_{10} = 1,$$
$$iv_7i^{-1}v_7 = iv_8i^{-1}v_9 = iv_9i^{-1}v_8 = iv_{10}i^{-1}v_1 = 1.$$

Proof The two irreducible $F\mathsf{M}_{22}$-modules V_i, $i = 1, 2$, occur as composition factors with multiplicity 1 in the permutation module $(1_{\mathsf{M}_{21}})^{\mathsf{M}_{22}}$ of degree 22, where $\mathsf{M}_{21} = \langle a, b, c, d, t, g, h, i\rangle$. They are dual to each other. Using the faithful permutation representation of M_{22} stated in (a) and the Meataxe algorithm implemented in MAGMA one obtains the generating matrices of M_{22} stated in (b) defining V_1.

(c) Their dual matrices define V_2.

(d) The cohomological dimensions $d_i = dim_F[H^2(\mathsf{M}_{22}, V_i)]$, $i = 1, 2$, have been calculated by means of MAGMA using Holt's Algorithm 7.4.5

of [92]. Its hypothesis is given by the presentation of M_{22} in Definition 8.2.1 of [92] and all the data stated in (a), (b) and (c). It follows that $d_1 = 0 = d_2$.

(e) Since $d_1 = 0$ there exists only the split extension E_1 of M_{22} by V_1. The presentation of E_1 has been obtained automatically by means of Lemma 1.4.7 and MAGMA.

(f) Since $d_2 = 0$ the split extension E_2 of M_{22} by V_2 is well defined. Its presentation has been obtained by Lemma 1.4.7. □

Lemma 4.1.2 *Let $\mathsf{M}_{22} = \langle a, b, c, d, t, g, h, i \rangle$ be the finitely presented group of Definition 8.2.1 of [92]. Let $A_{22} = Aut(\mathsf{M}_{22})$. Then the following statements hold:*

(a) *A_{22} has a faithful permutation representation of degree 44 and is isomorphic to the subgroup $\langle Pp, Pq \rangle$ of the symmetric group S_{44} generated by the following permutations:*

$$\begin{aligned} Pp = {} & (1,2,7,4)(3,21,44,25)(5,24,22,26)(6,8,39,36)(9,35) \\ & \cdot(10,37,12,13)(11,20,14,17)(18,19)(23,42,41,32) \\ & \cdot(27,29)(28,33)(30,31,43,34), \\ Pq = {} & (1,3,16,44,13,5)(2,15)(4,28,36,22,17,29) \\ & \cdot(6,24,12,43,39,38)(7,30)(8,33,11,25,19,41) \\ & \cdot(9,31,18,21,10,42)(14,27,20,23,35,32)(26,37)(34,40). \end{aligned}$$

(b) *$A_{22} = \langle p, q \rangle$ has the following set $\mathcal{R}(A_{22})$ of defining relations:*

$$\begin{aligned} & p^4 = q^6 = 1, \quad q^{-1}p^{-2}q^{-1}p^{-1}q^{-1}p^{-2}q^{-1}p^{-1}q^{-1}p^2q^{-1}p^{-1} = 1, \\ & pq^{-3}p^{-1}q^{-1}p^{-2}q^{-1}pqp^{-1}qpq^{-1} = (p^{-1}q^{-2}p^{-1}qpq^{-1}p^{-1}q^{-1})^2 = 1, \\ & q^{-1}p^{-2}q^{-1}pq^{-1}p^{-2}q^{-1}pq^{-1}p^2q^{-1}p = (q^{-1}p^{-1}q^{-1})^5 = 1, \\ & qpq^2pqp^{-1}q^{-1}p^{-2}q^{-2}p^2q^{-1}p^{-1}q^{-2}p^{-1} = 1. \end{aligned}$$

(c) *$A_{22} = \langle p, q \rangle$ has (up to isomorphism) two irreducible modules V_3 and V_4 of degree 10 over $F = GF(2)$. V_3 is described by the following matrices:*

$$Mp = \begin{pmatrix} 1&0&0&0&0&0&0&0&0&0 \\ 0&0&0&0&0&0&1&0&0&1 \\ 1&1&0&0&1&0&1&0&0&1 \\ 0&1&1&1&0&0&1&0&0&0 \\ 1&1&0&0&0&0&1&0&0&0 \\ 0&1&0&0&0&0&0&0&1&1 \\ 1&0&0&1&0&1&1&0&0&0 \\ 1&0&0&1&0&0&0&0&0&1 \\ 1&1&0&1&0&0&0&0&0&1 \\ 1&1&0&1&0&0&1&1&0&1 \end{pmatrix} \quad \text{and} \quad Mq = \begin{pmatrix} 0&0&1&0&1&0&0&1&0&0 \\ 0&0&1&0&0&0&0&0&0&0 \\ 0&0&0&0&1&0&0&0&0&0 \\ 1&0&1&0&0&1&0&0&0&0 \\ 0&0&0&0&1&0&0&0&1&0 \\ 0&0&1&0&0&0&1&0&0&0 \\ 0&0&0&0&1&1&0&0&0&0 \\ 0&0&0&1&1&1&0&0&0&0 \\ 0&0&0&0&1&1&0&0&0&1 \\ 0&1&1&0&1&1&0&0&0&0 \end{pmatrix}$$

(d) *V_4 is described by the matrices $Mp_1 = Mp^* = [Mp^{-1}]^T$, $Mq_1 = Mq^* = [Mq^{-1}]^T$ in* $\mathrm{GL}_{10}(2)$.

(e) *$dim_F[H^2(A_{22}, V_3)] = 1$ and $dim_F[H^2(A_{22}, V_4)] = 0$.*

(f) *$E_3 = V_3 \rtimes A_{22} = \langle p, q, v_1, v_2, v_3, v_4, v_5, v_6, v_8, v_8, v_9, v_{10}\rangle$ has a set $\mathcal{R}(E_3)$ of defining relations consisting of $\mathcal{R}(A_{22})$, $\mathcal{R}_1(V_3 \rtimes A_{22})$ and the following set $\mathcal{R}_2(V_3 \rtimes A_{22})$ of essential relations:*

$$pv_1p^{-1}v_1 = pv_2p^{-1}v_8v_9 = pv_3p^{-1}v_1v_2v_4v_9 = pv_4p^{-1}v_2v_5v_9 = 1,$$
$$pv_5p^{-1}v_1v_2v_3v_8v_9 = pv_6p^{-1}v_2v_7v_8 = pv_7p^{-1}v_1v_5v_8v_9 = 1,$$
$$pv_8p^{-1}v_1v_5v_8v_{10} = pv_9p^{-1}v_1v_2v_5v_6 = pv_{10}p^{-1}v_1v_2v_5v_8v_9 = 1,$$
$$qv_1q^{-1}v_2v_3v_4v_7 = qv_2q^{-1}v_2v_7v_{10} = qv_3q^{-1}v_2 = 1,$$
$$qv_4q^{-1}v_7v_8 = qv_5q^{-1}v_3 = qv_6q^{-1}v_3v_7 = qv_7q^{-1}v_2v_6 = 1,$$
$$qv_8q^{-1}v_1v_2v_3 = qv_9q^{-1}v_3v_5 = qv_{10}q^{-1}v_7v_9 = 1.$$

(g) *$E_4 = V_4 \rtimes A_{22} = \langle p_1, q_1, v_1, v_2, v_3, v_4, v_5, v_6, v_8, v_8, v_9, v_{10}\rangle$ has a set $\mathcal{R}(E_4)$ of defining relations consisting of $\mathcal{R}(A_{22})$, $\mathcal{R}_1(V_4 \rtimes A_{22})$ and the following set $\mathcal{R}_2(V_4 \rtimes A_{22})$ of essential relations:*

$$p_1v_1p_1^{-1}v_1v_3v_5v_7v_8v_9v_{10} = p_1v_2p_1^{-1}v_3v_4v_5v_6v_9v_{10} = 1,$$
$$p_1v_3p_1^{-1}v_4 = p_1v_4p_1^{-1}v_4v_7v_8v_9v_{10} = 1,$$
$$p_1v_5p_1^{-1}v_3 = p_1v_6p_1^{-1}v_7 = p_1v_7p_1^{-1}v_2v_3v_4v_5v_7v_{10} = 1,$$
$$p_1v_8p_1^{-1}v_{10} = p_1v_9p_1^{-1}v_6 = p_1v_{10}p_1^{-1}v_2v_3v_6v_8v_9v_{10} = 1,$$
$$q_1v_1q_1^{-1}v_4 = q_1v_2q_1^{-1}v_{10} = q_1v_3q_1^{-1}v_1v_2v_4v_6v_{10} = 1,$$
$$q_1v_4q_1^{-1}v_8 = q_1v_5q_1^{-1}v_1v_3v_5v_7v_8v_9v_{10} = q_1v_6q_1^{-1}v_4v_7v_8v_9v_{10} = 1,$$
$$q_1v_7q_1^{-1}v_6 = q_1v_8q_1^{-1}v_1 = q_1v_9q_1^{-1}v_5 = q_1v_{10}q_1^{-1}v_9 = 1.$$

(h) *The non-split extension $E_5 = \langle p_2, q_2, v_1, v_2, v_3, v_4, v_5, v_6, v_7, v_8, v_9, v_{10}\rangle$ of A_{22} by V_3 has a set $\mathcal{R}(E_5)$ of defining relations consisting of $\mathcal{R}_1(V_3 \rtimes A_{22})$ and the following relations:*

$$p_2^8 = q_2^{12} = 1,$$
$$(p_2, v_1^{-1}) = p_2^{-1}v_2p_2v_7^{-1}v_{10}^{-1} = p_2^{-1}v_3p_2v_1^{-1}v_2^{-1}v_5^{-1}v_7^{-1}v_{10}^{-1} = 1,$$
$$p_2^{-1}v_4p_2v_2^{-1}v_3^{-1}v_4^{-1}v_7^{-1} = p_2^{-1}v_5p_2v_1^{-1}v_2^{-1}v_7^{-1} = 1,$$
$$p_2^{-1}v_6p_2v_2^{-1}v_9^{-1}v_{10}^{-1} = p_2^{-1}v_7p_2v_1^{-1}v_4^{-1}v_6^{-1}v_7^{-1} = 1,$$
$$p_2^{-1}v_8p_2v_1^{-1}v_4^{-1}v_{10}^{-1} = p_2^{-1}v_9p_2v_1^{-1}v_2^{-1}v_4^{-1}v_{10}^{-1} = 1,$$

$$
\begin{aligned}
&p_2^{-1}v_{10}p_2v_1^{-1}v_2^{-1}v_4^{-1}v_7^{-1}v_8^{-1}v_{10}^{-1} = q_2^{-1}v_1q_2v_3^{-1}v_5^{-1}v_8^{-1} = 1,\\
&q_2^{-1}v_2q_2v_3^{-1} = q_2^{-1}v_3q_2v_5^{-1} = q_2^{-1}v_4q_2v_1^{-1}v_3^{-1}v_6^{-1} = 1,\\
&q_2^{-1}v_5q_2v_5^{-1}v_9^{-1} = q_2^{-1}v_6q_2v_3^{-1}v_7^{-1} = q_2^{-1}v_7q_2v_5^{-1}v_6^{-1} = 1,\\
&q_2^{-1}v_8q_2v_4^{-1}v_5^{-1}v_6^{-1} = q_2^{-1}v_9q_2v_5^{-1}v_6^{-1}v_{10}^{-1} = 1,\\
&q_2^{-1}v_{10}q_2v_2^{-1}v_3^{-1}v_5^{-1}v_6^{-1} = q_2^6v_1^{-1}v_4^{-1}v_8^{-1} = 1,\\
&p_2^4v_3^{-1}v_4^{-1}v_5^{-1}v_6^{-1}v_7^{-1}v_8^{-1}v_9^{-1}v_{10}^{-1} = 1,\\
&q_2^{-1}p_2^{-2}q_2^{-1}p_2^{-1}q_2^{-1}p_2^{-2}q_2^{-1}p_2^{-1}q_2^{-1}p_2^2q_2^{-1}p_2^{-1}v_1^{-1}v_2^{-1}v_4^{-1} = 1,\\
&p_2q_2^{-3}p_2^{-1}q_2^{-1}p_2^{-2}q_2^{-1}p_2q_2p_2^{-1}q_2p_2q_2^{-1}v_1^{-1}v_2^{-1}v_5^{-1}v_6^{-1}v_8^{-1}v_{10}^{-1} = 1,\\
&q_2^{-1}p_2^{-2}q_2^{-1}p_2q_2^{-1}p_2^{-2}q_2^{-1}p_2q_2^{-1}p_2^2q_2^{-1}p_2v_5^{-1}v_6^{-1}v_7^{-1}v_9^{-1} = 1,\\
&q_2^{-1}p_2^{-1}q_2^{-2}p_2^{-1}q_2^{-2}p_2^{-1}q_2^{-2}p_2^{-1}q_2^{-2}p_2^{-1}\\
&\quad\cdot q_2^{-1}v_1^{-1}v_4^{-1}v_5^{-1}v_6^{-1}v_7^{-1}v_8^{-1}v_9^{-1} = 1,\\
&p_2^{-1}q_2^{-2}p_2^{-1}q_2p_2q_2^{-1}p_2^{-1}q_2^{-1}p_2^{-1}q_2^{-2}p_2^{-1}q_2p_2q_2^{-1}p_2^{-1}q_2^{-1}\\
&\quad\cdot v_1^{-1}v_2^{-1}v_3^{-1}v_4^{-1}v_5^{-1}v_6^{-1}v_8^{-1}v_9^{-1} = 1,\\
&q_2p_2q_2^2p_2q_2p_2^{-1}q_2^{-1}p_2^{-2}q_2^{-2}p_2^2q_2^{-1}p_2^{-1}q_2^{-2}p_2^{-1}v_1^{-1}v_4^{-1}v_5^{-1}v_9^{-1} = 1.
\end{aligned}
$$

Proof (a) Using Todd's permutation representation PM_{22} of M_{22} stated in Lemma 8.2.2 of [92] and the MAGMA command `AutomorphismGroup (M_22)` the automorphism group $A_{22} = \mathrm{Aut}(\mathsf{M}_{22})$ of M_{22} was calculated. MAGMA returned 11 generators and a faithful permutation representation PA_{22} of degree 44 by means of the command `PermutationGroup (A_{22})`. Using this permutation representation one can show that PA_{22} is generated by the two permutations $Pp, Pq \in S_{44}$ given in the statement. Also it has been checked computationally that the derived subgroup A_{22}' is a simple group which is isomorphic to M_{22} and that $|A_{22} : A_{22}'| = 2$.

(b) The given presentation of $A_{22} = \langle p, q \rangle$ has been obtained by means of the faithful permutation representation of degree 44 and the MAGMA command `FPGroup(A_22)`.

(c) Decomposing the permutation module PA_{22} of degree 44 into irreducible composition factors by means of the Meataxe algorithm the first irreducible representation V_3 of A_{22} over $F = GF(2)$ was calculated in [72].

(d) The second irreducible representation V_4 of A_{22} is the dual of V_3, which has been checked to be non-isomorphic to V_3.

(e) The cohomological dimensions $d_i = dim_F[H^2(A_{22}, V_i)]$, $i = 3, 4$, have been calculated by means of MAGMA using Holt's Algorithm 7.4.5 of [92], the presentation of A_{22} given in (b), the faithful permutation representation PA_{22} constructed in (a) and the two irreducible representations of A_{22} given in (c) and (d). It follows that $d_3 = 1$ and $d_4 = 0$.

(f) and (g) The two presentations of the split extensions E_3 and E_4 are obvious.

(h) The presentation of the non-split extension E_5 has been obtained by means of Holt's Algorithm 7.4.5 of [92] implemented in MAGMA [52]. This completes the proof. □

Lemma 4.1.3 *Keep the notation of Lemmas 4.1.1 and 4.1.2. Then the following statements hold.*

(a) *Each of the four split extensions E_i, $i \in \{1,2,3,4\}$, has a faithful permutation representation of degree 1024. Its stabilizer is the complement of V_i in E_i, which generates E_i together with $v_1 \in V_i$ for all four groups.*

(b) *The non-split extension*

$$E_5 = \langle p_2, q_2, v_1, v_2, v_3, v_4, v_5, v_6, v_7, v_8, v_9, v_{10}\rangle = \langle p_2, q_2\rangle$$

has a faithful permutation representation of degree 88 with stabilizer $U = \langle q_2^2, (p_2 q_2^2)^2\rangle$.

(c) *The split extension E_1 of M_{22} by V_1 has 47 conjugacy classes; $z_1 = (tv_1)^3$ represents the unique conjugacy class of 2-central involutions of E_1 and $|C_{E_1}(z_1)| = 2^{17} \cdot 3^2 \cdot 5$.*

(d) *The split extension E_2 of M_{22} by V_2 has 43 conjugacy classes; $z_2 = (iv_1)^2$ represents the unique conjugacy class of 2-central involutions of E_2 and $|C_{E_2}(z_2)| = 2^{17} \cdot 3 \cdot 5$.*

(e) *The split extension E_3 of A_{22} by V_3 has 79 conjugacy classes; $z_3 = (pq^2v_1)^5$ represents the unique conjugacy class of 2-central involutions of E_3 and $|C_{E_3}(z_3)| = 2^{18} \cdot 3^2 \cdot 5$.*

(f) *The split extension E_4 of A_{22} by V_4 has 77 conjugacy classes; $z_4 = (p_1^2 q_1 v_1)^{10}$ represents the unique conjugacy class of 2-central involutions of E_4 and $|C_{E_4}(z_4)| = 2^{18} \cdot 3 \cdot 5$.*

(g) *The non-split extension E_5 of A_{22} by V_3 has 79 conjugacy classes; $z_5 = p_2^4$ represents the unique conjugacy class of 2-central involutions of E_5 and $|C_{E_5}(z_5)| = 2^{18} \cdot 3^2 \cdot 5$.*

Proof (a) The presentations of the four split extensions E_i are given in Lemmas 4.1.1 and 4.1.2. Taking their complements M_{22} or A_{22} of V_i

as stabilizers MAGMA provides the faithful permutation representation PE_i of E_i of degree 1024 for $i = 1, 2, 3, 4$. The final assertions have been proved in [72] by means of these permutation representations and MAGMA.

(b) The presentation of the non-split extension E_5 is given in Lemma 4.1.1(h). Let $U = \langle q_2^2, (p_2 q_2^2)^2 \rangle$. The faithful permutation representation PE_5 of E_5 with degree 88 and stabilizer $U = \langle q_2^2, (p_2 q_2^2)^2 \rangle$ was found by means of MAGMA. It has been used to show that $E_5 = \langle p_2, q_2 \rangle$.

All remaining assertions were proved as follows. For each extension group E_i, $i \in \{1, 2, 3, 4, 5\}$, we used its faithful permutation representation PE_i, MAGMA and Kratzer's Algorithm 5.3.18 of [92] to determine a system of representatives of all conjugacy classes of E_i in terms of the given generators of the presentation of E_i. In view of the limited space they are not stated except for the given words of the 2-central involutions z_i of E_i, which in each case are uniquely determined up to conjugation in E_i. This completes the proof. □

4.2 Construction of the 2-central involution centralizer

In this section Algorithm 1.3.8 is applied to the extension group E_3 constructed in Lemma 4.1.2 in order to construct a finitely presented group H which is shown later on to be isomorphic to the centralizer of a 2-central involution of the simple group Co_2.

Proposition 4.2.1 *Keep the notation of Lemmas 4.1.1 and 4.1.2. Let $E_3 = \langle p, q, v_1 \rangle$ be the split extension of $A_{22} = Aut(\mathsf{M}_{22})$ by its simple module V_3 of dimension 10 over $F = GF(2)$. Then the following statements hold.*

(a) *$z = (pq^2v_1)^5$ is a 2-central involution of $E_3 = \langle p, q, v_1 \rangle$ with centralizer $D = C_{E_3}(z) = \langle n_i, f_j \mid 1 \le i \le 12, 1 \le j \le 5 \rangle$ where*

$$n_1 = (p^2qp^2q^2v_1)^6, \quad n_2 = (p^2qpv_1qpv_1qv_1)^6, \quad n_3 = (v_1qp^3qpv_1qp)^4,$$
$$n_4 = (pqpq^4pqv_1q)^7, \quad n_5 = (pqpv_1qp)^3, \quad n_6 = (pqp^2q^2v_1)^6,$$
$$n_7 = (pq^2pv_1qp)^6, \quad n_8 = (qpqp^2qv_1)^6, \quad n_9 = (p^3q^3p^2q^2)^4,$$
$$n_{10} = (pqpq^2p^2qpq)^6, \quad n_{11} = (pq^2p^3q^3p)^4, \quad n_{12} = (qp^2q^4pqp)^3,$$
$$f_1 = (p^3q^2v_1qpv_1q)^5, \quad f_2 = (q^2pqv_1qpqpq)^7, \quad f_3 = (pqp^2q^3v_1qpq)^6,$$
$$f_4 = (p^3q^2pqpv_1q)^7, \quad f_5 = (p^2q^3v_1qv_1q^2)^2.$$

(b) *D has a unique extra-special normal subgroup Q of order* 512. *It is generated by the following involutions:*

$$q_1 = n_8 n_{10} n_{11} n_8 n_{12}, \quad q_2 = n_5 n_6 n_7 n_{12}, \quad q_3 = n_5 n_7 n_9 n_6,$$
$$q_4 = n_6 n_8 n_9 n_{10}, \quad q_5 = n_5, \quad q_6 = n_5 n_6 n_7, \quad q_7 = n_5 n_6 n_7 n_8,$$
$$q_8 = n_5 n_7 n_8.$$

Furthermore, Q has the following set $\mathcal{R}(Q)$ of defining relations:

$$q_1^2 = q_2^2 = q_3^2 = q_4^2 = q_5^2 = q_6^2 = q_7^2 = q_8^2 = 1,$$
$$(q_1, q_2) = (q_1, q_3) = (q_1, q_4) = (q_1, q_6) = (q_1, q_8) = 1,$$
$$(q_1, q_5) = (q_1, q_7) = z, \quad (q_2, q_8) = z,$$
$$(q_2, q_3) = (q_2, q_4) = (q_2, q_5) = (q_2, q_6) = (q_2, q_7) = 1,$$
$$(q_3, q_4) = (q_3, q_6) = (q_3, q_7) = 1, \quad (q_3, q_5) = (q_3, q_8) = z,$$
$$(q_4, q_5) = (q_4, q_6) = (q_4, q_7) = (q_4, q_8) = z,$$
$$(q_5, q_6) = (q_5, q_7) = (q_5, q_8) = (q_6, q_7) = (q_6, q_8) = (q_7, q_8) = 1.$$

(c) *Q has a complement W in D of order $|W| = 2^9 \cdot 3^2 \cdot 5$. The Fitting subgroup $B = \langle f_j \mid 1 \le j \le 5\rangle$ of W is elementary abelian of order 2^5 and has a complement $L = \langle s_k \mid 1 \le k \le 3\rangle \cong S_6$ in W, where*

$$s_1 = n_2 n_3 n_2 n_1 n_3 n_4, \quad s_2 = n_1 n_2 n_3 n_1 n_2 n_3, \quad s_3 = (n_3 n_1 n_4)^2.$$

(d) *$D = \langle q_i, f_j, s_k \mid 1 \le i \le 8, 1 \le j \le 5, 1 \le k \le 3\rangle$ has center $Z(D) = \langle z = (q_4 q_8)^2\rangle$ of order 2 and the following set $\mathcal{R}(D)$ of defining relations consisting of $\mathcal{R}(Q)$ and the relations:*

$$s_1^4 = s_2^5 = s_3^3 = 1,$$
$$(s_2 s_1)^2 = s_1 s_3 s_2^{-1} s_1^{-1} s_3^{-1} s_2 = (s_2 s_3)^3 = s_2^2 s_3 s_1 s_2^{-1} s_1 s_3 = 1,$$
$$s_2^{-1} s_1 s_3^{-1} s_2^{-1} s_1 s_3^{-1} s_2 s_3 = 1,$$
$$f_1^2 = f_2^2 = f_3^2 = f_4^2 = f_5^2 = 1,$$
$$f_1^{s_1}(f_1)^{-1} = f_1^{s_2}(f_1)^{-1} = f_1^{s_3}(f_1)^{-1} = 1,$$
$$f_2^{s_1}(f_1 f_4 f_5)^{-1} = f_2^{s_2}(f_1 f_2 f_3 f_5)^{-1} = f_2^{s_3}(f_3 f_5)^{-1} = 1,$$
$$f_3^{s_1}(f_3 f_5)^{-1} = f_3^{s_2}(f_2 f_3)^{-1} = f_3^{s_3}(f_2 f_5)^{-1} = 1,$$
$$f_4^{s_1}(f_1 f_3 f_4 f_5)^{-1} = f_4^{s_2}(f_1 f_3 f_4)^{-1} = f_4^{s_3}(f_1 f_3 f_4)^{-1} = 1,$$
$$f_5^{s_1}(f_2 f_3 f_4)^{-1} = f_5^{s_2}(f_2 f_3 f_4)^{-1} = f_5^{s_3}(f_1 f_2)^{-1} = 1,$$

$$
\begin{aligned}
&(f_1,f_2)=(f_1,f_3)=(f_1,f_4)=(f_1,f_5)=(f_2,f_3)=(f_2,f_4)=1,\\
&(f_2,f_5)=(f_3,f_4)=(f_3,f_5)=(f_4,f_5)=1,\\
&q_1^{s_1}(q_3q_4q_6q_7q_8)^{-1}=(q_4q_8)^2, \quad q_1^{s_2}(q_1q_2q_3q_8)^{-1}=1,\\
&q_1^{s_3}(q_1q_3q_5q_7q_8)^{-1}=q_1^{f_1}(q_1q_5q_7)^{-1}=q_1^{f_2}(q_1)^{-1}=1,\\
&q_1^{f_3}(q_1q_5q_6q_7)^{-1}=q_1^{f_4}(q_1)^{-1}=q_1^{f_5}(q_1q_6q_8)^{-1}=1,\\
&q_2^{s_1}(q_1q_4q_5)^{-1}=(q_4q_8)^2, \quad q_2^{f_1}(q_2q_6)^{-1}=(q_4q_8)^2,\\
&q_2^{s_2}(q_2q_4q_5q_6)^{-1}=q_2^{s_3}(q_1q_2q_4q_5)^{-1}=1,\\
&q_2^{f_2}(q_2)^{-1}=q_2^{f_3}(q_2q_5q_6q_7)^{-1}=1,\\
&q_2^{f_4}(q_2q_5q_7)^{-1}=q_2^{f_5}(q_2q_6q_7)^{-1}=(q_4q_8)^2,\\
&7q_3^{s_1}(q_3q_5q_7q_8)^{-1}=(q_4q_8)^2, \quad q_3^{f_1}(q_3q_7)^{-1}=(q_4q_8)^2,\\
&q_3^{s_2}(q_2q_3q_6q_8)^{-1}=(q_4q_8)^2, \quad q_3^{s_3}(q_1q_5q_6q_7q_8)^{-1}=1,\\
&q_3^{f_2}(q_3q_6)^{-1}=q_3^{f_3}(q_3q_5q_8)^{-1}=1,\\
&q_3^{f_4}(q_3q_5q_6q_7q_8)^{-1}=q_3^{f_5}(q_3q_7)^{-1}=(q_4q_8)^2,\\
&q_4^{s_1}(q_2q_4q_6q_7)^{-1}=q_4^{s_2}(q_1q_3q_5q_7q_8)^{-1}=1,\\
&q_4^{s_3}(q_1q_2q_3q_5q_8)^{-1}=(q_4q_8)^2, \quad q_4^{f_1}(q_4q_6q_8)^{-1}=1,\\
&q_4^{f_2}(q_4q_5q_7)^{-1}=q_4^{f_5}(q_4q_5q_7)^{-1}=(q_4q_8)^2,\\
&q_4^{f_3}(q_4q_5q_8)^{-1}=q_4^{f_4}(q_4)^{-1}=1,\\
&q_5^{s_1}(q_6q_8)^{-1}=q_5^{s_2}(q_5q_7)^{-1}=1,\\
&q_5^{s_3}(q_7)^{-1}=(q_4q_8)^2, \quad q_5^{f_1}(q_5)^{-1}=1,\\
&q_5^{f_2}(q_5)^{-1}=q_5^{f_3}(q_5)^{-1}=q_5^{f_4}(q_5)^{-1}=q_5^{f_5}(q_5)^{-1}=1,\\
&q_6^{s_1}(q_5q_6q_7q_8)^{-1}=q_6^{s_2}(q_8)^{-1}=q_6^{s_3}(q_5q_7q_8)^{-1}=(q_4q_8)^2,\\
&q_6^{f_1}(q_6)^{-1}=q_7^{s_1}(q_7)^{-1}=1,\\
&q_6^{f_2}(q_6)^{-1}=q_6^{f_3}(q_6)^{-1}=q_6^{f_4}(q_6)^{-1}=q_6^{f_5}(q_6)^{-1}=1,\\
&q_7^{s_2}(q_6q_7)^{-1}=q_7^{s_3}(q_5q_7)^{-1}=(q_4q_8)^2,\\
&q_7^{f_1}(q_7)^{-1}=q_7^{f_2}(q_7)^{-1}=q_7^{f_3}(q_7)^{-1}=1\\
&q_7^{f_4}(q_7)^{-1}=q_7^{f_5}(q_7)^{-1}=q_8^{s_3}(q_6q_7q_8)^{-1}=1,\\
&q_8^{s_1}(q_5q_6q_7)^{-1}=q_8^{s_2}(q_5q_8)^{-1}=(q_4q_8)^2,\\
&q_8^{f_1}(q_8)^{-1}=q_8^{f_2}(q_8)^{-1}=q_8^{f_3}(q_8)^{-1}=1,\\
&q_8^{f_4}(q_8)^{-1}=q_8^{f_5}(q_8)^{-1}=1,\\
&((q_4q_8)^2,s_k)=1 \quad \textit{for} \quad k=1,2,3,\\
&((q_4q_8)^2,f_j)=1 \quad \textit{for} \quad 1\le j\le 5,\\
&((q_4q_8)^2,q_i)=1 \quad \textit{for} \quad 1\le i\le 8.
\end{aligned}
$$

(e) *$D_1 = D/Z(D)$ has a faithful permutation representation of degree* 256 *with stabilizer W. Furthermore, $V = Q/Z(D)$ is the unique elementary abelian normal subgroup $V = Q/Z(D)$ of order* 256 *in D_1, and D_1 is a semidirect product of W by V.*

(f) *There is a basis $\mathcal{B} = \{v_i = q_i Z(D) | 1 \le i \le 8\}$ of V such that the three generators s_k of L are represented by the following matrices with respect to $\mathcal{B}$:*

$$M s_1 = \begin{pmatrix} 0&1&0&0&0&0&0&0\\ 0&0&0&1&0&0&0&0\\ 1&0&1&0&0&0&0&0\\ 1&1&0&1&0&0&0&0\\ 0&1&1&0&0&1&0&1\\ 1&0&0&1&1&1&0&1\\ 1&0&1&1&0&1&1&1\\ 1&0&1&0&1&1&0&0 \end{pmatrix},$$

$$M s_2 = \begin{pmatrix} 1&0&0&1&0&0&0&0\\ 1&1&1&0&0&0&0&0\\ 1&0&1&1&0&0&0&0\\ 0&1&0&0&0&0&0&0\\ 0&1&0&1&1&0&0&1\\ 0&1&1&0&0&0&1&0\\ 0&0&0&1&1&0&1&0\\ 1&0&1&1&0&1&0&1 \end{pmatrix} \quad \text{and} \quad M s_3 = \begin{pmatrix} 1&1&1&1&0&0&0&0\\ 0&1&0&1&0&0&0&0\\ 1&0&0&1&0&0&0&0\\ 0&1&0&0&0&0&0&0\\ 1&1&1&1&0&1&1&0\\ 0&0&1&0&0&0&0&1\\ 1&0&1&0&1&1&1&1\\ 1&0&1&1&0&1&0&1 \end{pmatrix}.$$

(g) *$W = \langle u, r, s_1, s_2, s_3 \rangle$, where $u = f_1$ is the central involution of W and $r = f_2 f_4$.*

(h) *The subgroup $S = \langle f_j, a_s, w \mid 1 \le j \le 5, 1 \le s \le 6 \rangle$ is a Sylow 2-subgroup of W which has a unique maximal elementary abelian normal 2-subgroup $A = \langle a_1, a_2, a_3, a_4, a_5 = u, a_6 \rangle$ of order 64, where*

$$w = (s_1 s_2^2 s_3 s_2)^3, \quad a_1 = s_2^2 f_3 s_1 s_3 s_2, \quad a_2 = (f_5 s_2 s_1^3)^3,$$
$$a_3 = (s_1 f_3)^2, \quad a_4 = (s_1)^2, \quad a_6 = (f_2 s_1)^4.$$

(i) *Let MA be the subgroup of $\mathrm{GL}_8(2)$ generated by the matrices Ma_1, Ma_2, Ma_3, Ma_4, Mu, Ma_6 of the generators of A with respect to $\mathcal{B}$. Then $NA = N_{\mathrm{GL}_8(2)}(MA)$ has a cyclic Sylow 7-subgroup generated by the matrix*

$$M s = \begin{pmatrix} 1&1&0&0&0&0&0&0\\ 1&1&0&0&1&0&0&1\\ 1&1&0&0&0&1&0&0\\ 0&1&0&1&0&0&0&0\\ 0&0&1&1&0&0&0&0\\ 0&1&0&0&1&1&0&0\\ 0&1&0&1&0&1&1&1\\ 1&1&1&0&0&0&0&0 \end{pmatrix}.$$

(j) *Let $MW = \langle Mr, Mu, Ms_1, Ms_2, Ms_3 \rangle$. Let MK be the subgroup of $\mathrm{GL}_8(2)$ generated by MW and the matrix Ms. Then MK is* *a simple group of order* $|MK| = 2^9 \cdot 3^4 \cdot 5 \cdot 7$ *which is isomorphic to* $\mathsf{Sp}_6(2)$.

(k) MK *is an irreducible subgroup of* $\mathrm{GL}_8(2)$ *generated by the matrices* $m_1 = Ms$, $m_2 = Mr$, $m_3 = Ms_1$, $m_4 = Ms_2$ *and* $m_5 = Ms_3$ *of respective orders* 7, 2, 4, 5 *and* 3. *With respect to this set of generators* MK *has the following set* $\mathcal{R}(K)$ *of defining relations:*

$$
\begin{aligned}
& m_1^7 = m_2^2 = m_3^4 = m_4^5 = m_5^3 = 1, \\
& m_2 m_4^{-1} m_3^{-1} m_2 m_3 m_4 = (m_3^{-1} m_4^{-1})^2 = 1, \\
& m_3 m_5 m_4^{-1} m_3^{-1} m_5^{-1} m_4 = (m_5^{-1} m_4^{-1})^3 = 1, \\
& (m_1^{-1} m_2)^4 = (m_2 m_3^{-1})^4 = 1, \quad (m_2 m_3 m_2 m_3^{-1})^2 = 1, \\
& m_5^{-1} m_3^{-1} m_4 m_3^{-1} m_5^{-1} m_4^{-2} = m_2 m_3 m_5^{-1} m_3^{-1} m_2 m_3 m_5 m_3^{-1} = 1, \\
& m_4 m_1^{-1} m_4 m_3 m_1 m_4^{-1} m_3 m_4 = 1, \\
& m_4^{-1} m_3 m_5^{-1} m_4^{-1} m_3 m_5^{-1} m_4 m_5 = 1, \\
& (m_5, m_1^{-1}, m_5) = 1, \quad (m_1^{-2} m_2)^3 = 1, \\
& m_5 m_3 m_1 m_2 m_1^{-1} m_5^{-1} m_2 m_5^{-1} m_4^{-1} = 1, \\
& (m_3^{-1} m_1^{-1} m_2)^3 = 1, \quad (m_3 m_1^{-1} m_2)^3 = 1, \\
& m_1 m_3^{-1} m_2 m_3 m_1^{-1} m_4^{-2} m_3 m_5 = 1, \\
& m_1^{-1} m_3 m_2 m_1^3 m_3 m_1^2 m_3^{-1} = m_2 m_1 m_3 m_1^3 m_2 m_1 m_2 m_3^{-1} = 1.
\end{aligned}
$$

(l) *Let* K *be the finitely presented group constructed in (k). Then* V *is an irreducible* 8*-dimensional representation of* K *over* $F = GF(2)$. *Furthermore,* K *has a faithful permutation representation* PK *of degree* 63.

(m) *Let* $H_1 = \langle m_i, v_j | 1 \le i \le 5, 1 \le j \le 8 \rangle$ *be the split extension of* K *by* V. *Then* H_1 *has a set* $\mathcal{R}(H_1)$ *of defining relations consisting of* $\mathcal{R}(K)$, $\mathcal{R}_1(V \rtimes K)$ *and the following set* $\mathcal{R}_2(V \rtimes K)$ *of essential relations:*

$$
\begin{aligned}
& m_1 v_1 m_1^{-1} v_7 v_8 = m_1 v_2 m_1^{-1} v_1 = m_1 v_3 m_1^{-1} v_2 v_8 = 1, \\
& m_1 v_4 m_1^{-1} v_3 = m_1 v_5 m_1^{-1} v_4 v_8 = m_1 v_6 m_1^{-1} v_5 v_8 = 1, \\
& m_1 v_7 m_1^{-1} v_6 = m_1 v_8 m_1^{-1} v_8 = m_2 v_1 m_2^{-1} v_4 v_5 v_6 = 1, \\
& m_2 v_2 m_2^{-1} v_1 v_2 v_3 v_4 v_7 v_8 = m_2 v_3 m_2^{-1} v_5 v_6 v_7 v_8 = 1, \\
& m_2 v_4 m_2^{-1} v_3 v_4 v_5 v_6 v_7 v_8 = m_2 v_5 m_2^{-1} v_1 v_4 v_6 = 1, \\
& m_2 v_6 m_2^{-1} v_3 v_5 v_7 v_8 = m_2 v_7 m_2^{-1} v_1 v_4 v_5 v_6 v_7 = 1, \\
& m_2 v_8 m_2^{-1} v_8 = m_3 v_1 m_3^{-1} v_1 v_2 v_7 v_8 = m_3 v_2 m_3^{-1} v_1 v_3 v_4 v_5 v_7 = 1,
\end{aligned}
$$

$$m_3v_3m_3^{-1}v_1v_2v_3 = m_3v_4m_3^{-1}v_1v_4v_5v_6v_7 = 1,$$
$$m_3v_5m_3^{-1}v_2v_3v_4 = m_3v_6m_3^{-1}v_1v_3v_4 = m_3v_7m_3^{-1}v_1v_3v_5 = 1,$$
$$m_3v_8m_3^{-1}v_8 = m_4v_1m_4^{-1}v_1v_2v_5v_7 = m_4v_2m_4^{-1}v_1v_3v_7v_8 = 1,$$
$$m_4v_3m_4^{-1}v_4v_6v_7v_8 = m_4v_4m_4^{-1}v_1v_3v_4v_5v_6v_7v_8 = 1,$$
$$m_4v_5m_4^{-1}v_1v_2v_3v_6 = m_4v_6m_4^{-1}v_4v_5v_8 = m_4v_7m_4^{-1}v_1v_2v_6v_8 = 1,$$
$$m_4v_8m_4^{-1}v_3v_4v_5v_7 = m_5v_1m_5^{-1}v_1v_4v_8 = m_5v_2m_5^{-1}v_1v_3v_8 = 1,$$
$$m_5v_3m_5^{-1}v_2v_4v_5v_6v_7v_8 = m_5v_4m_5^{-1}v_1v_3v_5v_6v_7v_8 = 1,$$
$$m_5v_5m_5^{-1}v_1v_6 = m_5v_6m_5^{-1}v_4v_5v_6v_8 = 1,$$
$$m_5v_7m_5^{-1}v_1v_2v_6v_7v_8 = m_5v_8m_5^{-1}v_3v_5v_6v_7v_8 = 1.$$

(n) *$dim_F[H^2(H_1, F)] = 2$ and there exists a unique central extension H of H_1 whose Sylow 2-subgroups are isomorphic to the ones of $D = C_{E_3}(z)$. Also, H is a split extension of K by its Fitting subgroup Q, and H is isomorphic to the finitely presented group $H = \langle h_i | 1 \le i \le 14\rangle$ with the following set $\mathcal{R}(H)$ of defining relations:*

$$h_1^7 = h_2^2 = h_3^4 = h_4^5 = h_5^3 = h_6^2 = h_7^2 = h_{10}^2 = h_{13}^2 = h_{14}^2 = 1,$$
$$h_8^2 = h_9^2 = h_{11}^2 = h_{12}^2 = h_{14}, \quad (h_1, h_{14}) = (h_2, h_{14}) = 1,$$
$$(h_3, h_{14}) = (h_4, h_{14}) = (h_5, h_{14}) = (h_6, h_{14}) = (h_7, h_{14}) = 1,$$
$$(h_8, h_{14}) = (h_9, h_{14}) = (h_{10}, h_{14}) = (h_{11}, h_{14}) = (h_{12}, h_{14}) = 1,$$
$$(h_{13}, h_{14}) = 1, \quad h_2h_4^{-1}h_3^{-1}h_2h_3h_4 = (h_3^{-1}h_4^{-1})^2 = 1,$$
$$h_3h_5h_4^{-1}h_3^{-1}h_5^{-1}h_4 = (h_5^{-1}h_4^{-1})^3 = (h_1^{-1}h_2)^4 = (h_2h_3^{-1})^4 = 1,$$
$$(h_2h_3h_2h_3^{-1})^2 = h_5^{-1}h_3^{-1}h_4h_3^{-1}h_5^{-1}h_4^{-2} = 1,$$
$$h_2h_3h_5^{-1}h_3^{-1}h_2h_3h_5h_3^{-1} = 1, \quad (h_5, h_1^{-1}, h_5) = 1,$$
$$h_4h_1^{-1}h_4h_3h_1h_4^{-1}h_3h_4 = h_4^{-1}h_3h_5^{-1}h_4^{-1}h_3h_5^{-1}h_4h_5 = 1,$$
$$(h_1^{-2}h_2)^3 = h_5h_3h_1h_2h_1^{-1}h_5^{-1}h_2h_5^{-1}h_4^{-1} = 1,$$
$$(h_3^{-1}h_1^{-1}h_2)^3 = (h_3h_1^{-1}h_2)^3 = h_1h_3^{-1}h_2h_3h_1^{-1}h_4^{-2}h_3h_5 = 1,$$
$$h_1^{-1}h_3h_2h_1^3h_3h_1^2h_3^{-1} = h_2h_1h_3h_1^3h_2h_1h_2h_3^{-1} = 1,$$
$$h_1h_6h_1^{-1}h_{12}h_{13} = h_1h_7h_1^{-1}h_6 = h_1h_8h_1^{-1}h_7h_{13} = 1,$$
$$h_1h_9h_1^{-1}h_8h_{14} = h_1h_{10}h_1^{-1}h_9h_{13} = h_1h_{11}h_1^{-1}h_{10}h_{13} = 1,$$
$$h_1h_{12}h_1^{-1}h_{11}h_{14} = h_1h_{13}h_1^{-1}h_{13} = h_2h_6h_2^{-1}h_9h_{10}h_{11} = 1,$$
$$h_2h_7h_2^{-1}h_6h_7h_8h_9h_{12}h_{13}h_{14} = h_2h_8h_2^{-1}h_{10}h_{11}h_{12}h_{13} = 1,$$
$$h_2h_9h_2^{-1}h_8h_9h_{10}h_{11}h_{12}h_{13} = h_2h_{10}h_2^{-1}h_6h_9h_{11} = 1,$$

$$h_2h_{11}h_2^{-1}h_8h_{10}h_{12}h_{13} = h_2h_{12}h_2^{-1}h_6h_9h_{10}h_{11}h_{12}h_{14} = 1,$$
$$h_2h_{13}h_2^{-1}h_{13} = h_3h_6h_3^{-1}h_6h_7h_{12}h_{13}h_{14} = 1,$$
$$h_3h_7h_3^{-1}h_6h_8h_9h_{10}h_{12}h_{14} = h_3h_8h_3^{-1}h_6h_7h_8h_{14} = 1,$$
$$h_3h_9h_3^{-1}h_6h_9h_{10}h_{11}h_{12} = h_3h_{10}h_3^{-1}h_7h_8h_9 = 1,$$
$$h_3h_{11}h_3^{-1}h_6h_8h_9 = h_3h_{12}h_3^{-1}h_6h_8h_{10}h_{14} = 1,$$
$$h_3h_{13}h_3^{-1}h_{13} = h_4h_6h_4^{-1}h_6h_7h_{10}h_{12} = 1,$$
$$h_4h_7h_4^{-1}h_6h_8h_{12}h_{13} = h_4h_8h_4^{-1}h_9h_{11}h_{12}h_{13}h_{14} = 1,$$
$$h_4h_9h_4^{-1}h_6h_8h_9h_{10}h_{11}h_{12}h_{13} = h_4h_{10}h_4^{-1}h_6h_7h_8h_{11}h_{14} = 1,$$
$$h_4h_{11}h_4^{-1}h_9h_{10}h_{13}h_{14} = h_4h_{12}h_4^{-1}h_6h_7h_{11}h_{13} = 1,$$
$$h_4h_{13}h_4^{-1}h_8h_9h_{10}h_{12}h_{14} = h_5h_6h_5^{-1}h_6h_9h_{13}h_{14} = 1,$$
$$h_5h_7h_5^{-1}h_6h_8h_{13} = h_5h_8h_5^{-1}h_7h_9h_{10}h_{11}h_{12}h_{13}h_{14} = 1,$$
$$h_5h_9h_5^{-1}h_6h_8h_{10}h_{11}h_{12}h_{13} = h_5h_{10}h_5^{-1}h_6h_{11} = 1,$$
$$h_5h_{11}h_5^{-1}h_9h_{10}h_{11}h_{13}h_{14} = h_5h_{12}h_5^{-1}h_6h_7h_{11}h_{12}h_{13}h_{14} = 1,$$
$$h_5h_{13}h_5^{-1}h_8h_{10}h_{11}h_{12}h_{13} = h_6^{-1}h_7^{-1}h_6h_7h_{14} = 1,$$
$$h_6^{-1}h_8^{-1}h_6h_8h_{14} = h_6^{-1}h_9^{-1}h_6h_9h_{14} = (h_6, h_{10}) = 1,$$
$$h_6^{-1}h_{11}^{-1}h_6h_{11}h_{14} = (h_6, h_{12}) = h_6^{-1}h_{13}^{-1}h_6h_{13}h_{14} = 1,$$
$$h_7^{-1}h_9^{-1}h_7h_9h_{14} = (h_7, h_{10}) = h_7^{-1}h_{11}^{-1}h_7h_{11}h_{14} = 1,$$
$$h_7^{-1}h_{12}^{-1}h_7h_{12}h_{14} = h_7^{-1}h_{13}^{-1}h_7h_{13}h_{14} = 1, \quad (h_7, h_8) = 1,$$
$$(h_8, h_{11}) = (h_8, h_{12}) = (h_9, h_{10}) = (h_9, h_{11}) = (h_9, h_{12}) = 1,$$
$$h_8^{-1}h_9^{-1}h_8h_9h_{14} = h_8^{-1}h_{10}^{-1}h_8h_{10}h_{14} = h_8^{-1}h_{13}^{-1}h_8h_{13}h_{14} = 1,$$
$$h_9^{-1}h_{13}^{-1}h_9h_{13}h_{14} = (h_{10}, h_{11}) = h_{10}^{-1}h_{12}^{-1}h_{10}h_{12}h_{14} = 1,$$
$$h_{10}^{-1}h_{13}^{-1}h_{10}h_{13}h_{14} = h_{11}^{-1}h_{12}^{-1}h_{11}h_{12}h_{14} = 1,$$
$$h_{11}^{-1}h_{13}^{-1}h_{11}h_{13}h_{14} = h_{12}^{-1}h_{13}^{-1}h_{12}h_{13}h_{14} = 1.$$

Proof (a) By Lemma 4.1.3 the split extension $E_3 = V_3 \rtimes A_{22} = \langle p, q, v_1 \rangle$ has a unique conjugacy class of involutions of highest defect. It is represented by $z = (pq^2v_1)^5$ and its centralizer $D = C_{E_3}(z)$ has order $2^{18} \cdot 3^2 \cdot 5$.

Using the faithful permutation representation of E_3 with stabilizer A_{22} and MAGMA it has been checked that D has a unique normal subgroup Q of order 512 with center $Z(Q) = \langle z \rangle$. Furthermore, Q is extra-special and it has a complement W of order $2^9 \cdot 3^2 \cdot 5$. Another application of MAGMA yields that the Fitting subgroup B of W is elementary abelian of order 2^5, and that B has a complement $L \cong S_6$ in W. Using then Algorithm 1.4.1 we found the generators of the subgroups

$B = \langle f_j \mid 1 \le j \le 5 \rangle$, $L = \langle n_l \mid 1 \le l \le 4 \rangle$ and $Q = \langle n_l \mid 5 \le l \le 12 \rangle$ of D. Hence (a) holds.

(b) Since Q is extra-special of order 2^9 and its center $Z(Q) = Z(D) = \langle z \rangle$ we determined, by means of MAGMA, eight involutions q_i generating Q. Using Algorithm 1.4.3 the words in the generators n_l of Q of the eight involutions were calculated. Since their residue classes generate the elementary abelian normal subgroup $V = Q/Z(Q)$ of $D_1 = D/Z(Q)$, one obtains the set $\mathcal{R}(Q)$ of defining relations of $Q = \langle q_i \mid 1 \le i \le 8 \rangle$ given in the statement. In particular, $z = (q_4, q_8) = (q_4 q_8)^2$.

(c) Another application of MAGMA and the permutation representation PE_3 yields that the permutation group PL of the complement $L = \langle n_k \mid 1 \le k \le 4 \rangle$ can be generated by the three elements s_1, s_2 and s_3. For these generators s_i MAGMA provided the following set $\mathcal{R}(L)$ of defining relations:

$$s_1^4 = s_2^5 = s_3^3 = 1, \quad (s_2 s_1)^2 = (s_2 s_3)^3 = 1,$$
$$s_1 s_3 s_2^{-1} s_1^{-1} s_3^{-1} s_2 = s_2^2 s_3 s_1 s_2^{-1} s_1 s_3 = s_2^{-1} s_1 s_3^{-1} s_2^{-1} s_1 s_3^{-1} s_2 s_3 = 1.$$

Using MAGMA again it has been checked that $L \cong S_6$. Thus (c) holds.

(d) Since the elementary abelian subgroup B is normal in the semidirect product W of L by B the presentation of $W = \langle f_j, s_k \rangle$ can easily be calculated by means of $\mathcal{R}(L)$ and the conjugation action of the $s_k \in L$ on B. Hence W has a defining set $\mathcal{R}(W)$ of relations consisting of $\mathcal{R}(L)$ and the following relations:

$$f_1^2 = f_2^2 = f_3^2 = f_4^2 = f_5^2 = 1,$$
$$f_1^{s_1}(f_1)^{-1} = f_1^{s_2}(f_1)^{-1} = f_1^{s_3}(f_1)^{-1} = 1,$$
$$f_2^{s_1}(f_1 f_4 f_5)^{-1} = f_2^{s_2}(f_1 f_2 f_3 f_5)^{-1} = f_2^{s_3}(f_3 f_5)^{-1} = 1,$$
$$f_3^{s_1}(f_3 f_5)^{-1} = f_3^{s_2}(f_2 f_3)^{-1} = f_3^{s_3}(f_2 f_5)^{-1} = 1,$$
$$f_4^{s_1}(f_1 f_3 f_4 f_5)^{-1} = f_4^{s_2}(f_1 f_3 f_4)^{-1} = f_4^{s_3}(f_1 f_3 f_4)^{-1} = 1,$$
$$f_5^{s_1}(f_2 f_3 f_4)^{-1} = f_5^{s_2}(f_2 f_3 f_4)^{-1} = f_5^{s_3}(f_1 f_2)^{-1} = 1,$$
$$(f_1, f_2) = (f_1, f_3) = (f_1, f_4) = (f_1, f_5) = (f_2, f_3) = (f_2, f_4) = (f_2, f_5) = 1,$$
$$(f_3, f_4) = (f_3, f_5) = (f_4, f_5) = 1.$$

As Q is normal in the semidirect product D of W by Q the conjugate action of the f_j and s_k on Q provides the following set of essential relations:

$$q_1^{s_1}(q_3 q_4 q_6 q_7 q_8)^{-1} = z, \quad q_1^{s_2}(q_1 q_2 q_3 q_8)^{-1} = q_1^{s_3}(q_1 q_3 q_5 q_7 q_8)^{-1} = 1,$$

$$q_1^{f_1}(q_1q_5q_7)^{-1} = q_1^{f_2}(q_1)^{-1} = q_1^{f_3}(q_1q_5q_6q_7)^{-1} = q_1^{f_4}(q_1)^{-1} = 1,$$
$$q_1^{f_5}(q_1q_6q_8)^{-1} = q_2^{s_1}(q_1q_4q_5)^{-1} = (q_4q_8)^2,$$
$$q_2^{s_2}(q_2q_4q_5q_6)^{-1} = q_2^{s_3}(q_1q_2q_4q_5)^{-1} = 1,$$
$$q_2^{f_1}(q_2q_6)^{-1} = (q_4q_8)^2, \quad q_2^{f_2}(q_2)^{-1} = q_2^{f_3}(q_2q_5q_6q_7)^{-1} = 1,$$
$$q_2^{f_4}(q_2q_5q_7)^{-1} = q_2^{f_5}(q_2q_6q_7)^{-1} = z, \quad q_3^{s_1}(q_3q_5q_7q_8)^{-1} = z,$$
$$q_3^{s_2}(q_2q_3q_6q_8)^{-1} = (q_4q_8)^2, \quad q_3^{s_3}(q_1q_5q_6q_7q_8)^{-1} = 1,$$
$$q_3^{f_1}(q_3q_7)^{-1} = z, \quad q_3^{f_2}(q_3q_6)^{-1} = q_3^{f_3}(q_3q_5q_8)^{-1} = 1,$$
$$q_3^{f_4}(q_3q_5q_6q_7q_8)^{-1} = q_3^{f_5}(q_3q_7)^{-1} = (q_4q_8)^2,$$
$$q_4^{s_1}(q_2q_4q_6q_7)^{-1} = q_4^{s_2}(q_1q_3q_5q_7q_8)^{-1} = 1, \quad q_4^{s_3}(q_1q_2q_3q_5q_8)^{-1} = z,$$
$$q_4^{f_1}(q_4q_6q_8)^{-1} = 1, \quad q_4^{f_2}(q_4q_5q_7)^{-1} = q_4^{f_5}(q_4q_5q_7)^{-1} = z,$$
$$q_4^{f_3}(q_4q_5q_8)^{-1} = q_4^{f_4}(q_4)^{-1} = 1,$$
$$q_5^{s_1}(q_6q_8)^{-1} = q_5^{s_2}(q_5q_7)^{-1} = 1, \quad q_5^{s_3}(q_7)^{-1} = (q_4q_8)^2,$$
$$q_5^{f_1}(q_5)^{-1} = q_5^{f_2}(q_5)^{-1} = q_5^{f_3}(q_5)^{-1} = q_5^{f_4}(q_5)^{-1} = q_5^{f_5}(q_5)^{-1} = 1,$$
$$q_6^{s_1}(q_5q_6q_7q_8)^{-1} = q_6^{s_2}(q_8)^{-1} = q_6^{s_3}(q_5q_7q_8)^{-1} = z,$$
$$q_6^{f_1}(q_6)^{-1} = 1, \quad q_6^{f_2}(q_6)^{-1} = q_6^{f_3}(q_6)^{-1} = q_6^{f_4}(q_6)^{-1} = q_6^{f_5}(q_6)^{-1} = 1,$$
$$q_7^{s_1}(q_7)^{-1} = 1, \quad q_7^{s_2}(q_6q_7)^{-1} = q_7^{s_3}(q_5q_7)^{-1} = (q_4q_8)^2,$$
$$q_7^{f_1}(q_7)^{-1} = q_7^{f_2}(q_7)^{-1} = q_7^{f_3}(q_7)^{-1} = q_7^{f_4}(q_7)^{-1} = q_7^{f_5}(q_7)^{-1} = 1,$$
$$q_8^{s_1}(q_5q_6q_7)^{-1} = q_8^{s_2}(q_5q_8)^{-1} = (q_4q_8)^2, \quad q_8^{s_3}(q_6q_7q_8)^{-1} = 1,$$
$$q_8^{f_1}(q_8)^{-1} = q_8^{f_2}(q_8)^{-1} = q_8^{f_3}(q_8)^{-1} = q_8^{f_4}(q_8)^{-1} = q_8^{f_5}(q_8)^{-1} = 1.$$

Thus the set $\mathcal{R}(D)$ of defining relations of D consists of $\mathcal{R}(Q)$ given in (b) and the above equations. Using MAGMA again it has been checked in [72] that this is a presentation of D.

(e) All the statements follow immediately from the presentation of D given in (d).

(f) From (c) and (d) it follows that the Fitting subgroup B of W has order 2^5 and a complement $L = \langle s_1, s_2, s_3 \rangle \cong S_6$. Hence W is isomorphic to the centralizer $H(\mathsf{Sp}_6(2))$ of a 2-central involution of $\mathsf{Sp}_6(2)$ by Proposition 8.6.5 of [92]. Thus we may apply Algorithm 7.4.8 of [92] to construct an over group K of W such that $|K : W|$ is odd and K normalizes V. For that we choose the basis $\mathcal{B} = \{v_i = q_i Z(Q) \mid 1 \leq i \leq 8\}$ of V. With respect to $\mathcal{B}$ the conjugate action of the s_k on V is described by the three matrices Ms_k of the statement.

(g) From the presentation of D it follows that $u = f_1$ is an involution of $Z(W)$. Another application of MAGMA yields that $W = \langle u, r, y_1, y_2, y_3 \rangle$

for $r = f_2 f_4$ and that the conjugate actions of r and u on V have the following matrices w.r.t. $\mathcal{B}$:

$$Mu = \begin{pmatrix} 1&0&0&0&0&0&0&0 \\ 0&1&0&0&0&0&0&0 \\ 0&0&1&0&0&0&0&0 \\ 0&0&0&1&0&0&0&0 \\ 1&0&0&0&1&0&0&0 \\ 0&1&0&1&0&1&0&0 \\ 1&0&1&0&0&0&1&0 \\ 0&0&0&1&0&0&0&1 \end{pmatrix} \quad \text{and} \quad Mr = \begin{pmatrix} 1&0&0&0&0&0&0&0 \\ 0&1&0&0&0&0&0&0 \\ 0&0&1&0&0&0&0&0 \\ 0&0&0&1&0&0&0&0 \\ 0&1&1&1&1&0&0&0 \\ 0&0&0&0&0&1&0&0 \\ 0&1&1&1&0&0&1&0 \\ 0&0&1&0&0&0&0&1 \end{pmatrix}.$$

(h) Using the faithful permutation representation PD_1 of D_1 given in (e) and MAGMA we find the generators of the given Sylow 2-subgroup S of W. It has a unique maximal elementary abelian normal subgroup $A = \langle a_1, a_2, a_3, a_4, a_5 = u, a_6 \rangle$ of order 64. Its generators a_i are represented by the following matrices with respect to $\mathcal{B}$:

$$Ma_1 = \begin{pmatrix} 1&0&0&0&0&0&0&0 \\ 0&1&0&0&0&0&0&0 \\ 1&0&1&1&0&0&0&0 \\ 0&0&0&1&0&0&0&0 \\ 0&0&0&0&1&0&0&0 \\ 0&0&0&0&0&1&0&0 \\ 0&0&0&0&1&0&1&1 \\ 0&0&0&0&0&0&0&1 \end{pmatrix}, \quad Ma_2 = \begin{pmatrix} 0&0&0&1&0&0&0&0 \\ 0&1&0&0&0&0&0&0 \\ 0&1&1&0&0&0&0&0 \\ 1&0&0&0&0&0&0&0 \\ 0&0&0&0&0&0&0&1 \\ 0&0&0&0&1&1&0&1 \\ 0&0&0&0&1&1&1&0 \\ 0&0&0&0&1&0&0&0 \end{pmatrix},$$

$$Ma_3 = \begin{pmatrix} 0&0&0&1&0&0&0&0 \\ 1&1&0&1&0&0&0&0 \\ 1&1&1&0&0&0&0&0 \\ 1&0&0&0&0&0&0&0 \\ 0&0&0&0&0&0&0&1 \\ 0&0&0&0&0&1&0&0 \\ 0&0&0&0&0&1&1&0 \\ 0&0&0&0&1&0&0&0 \end{pmatrix}, \quad Ma_4 = \begin{pmatrix} 0&0&0&1&0&0&0&0 \\ 1&1&0&1&0&0&0&0 \\ 1&1&1&0&0&0&0&0 \\ 1&0&0&0&0&0&0&0 \\ 1&0&0&0&0&0&0&1 \\ 1&1&0&0&0&1&0&0 \\ 1&0&1&1&0&1&1&0 \\ 0&0&0&1&1&0&0&0 \end{pmatrix},$$

$$Ma_6 = \begin{pmatrix} 1&0&0&0&0&0&0&0 \\ 0&1&0&0&0&0&0&0 \\ 0&0&1&0&0&0&0&0 \\ 0&0&0&1&0&0&0&0 \\ 1&0&0&1&1&0&0&0 \\ 0&0&0&0&0&1&0&0 \\ 1&1&0&0&0&0&1&0 \\ 1&0&0&1&0&0&0&1 \end{pmatrix}.$$

Observe that the map $\varphi : Px \to Mx$ sending the permutation Px of PD_1 to its matrix Mx with respect to the basis $\mathcal{B}$ yields an anti-epimorphism from PD_1 onto $MW = \langle Mr, Mu, Ms_1, Ms_2, Ms_3 \rangle \leq GL_8(2)$ with kernel PV.

(i) Now $MA = \langle Ma_1, Ma_2, Ma_3, Ma_4, Ma_5 = Mu, Ma_6 \rangle$ is an elementary abelian subgroup of $\Gamma = \mathrm{GL}_8(2)$ of order 2^6. Another application of MAGMA yields that $NA = N_\Gamma(MA)$ has a Sylow 7-subgroup of NA of order 7 generated by the matrix Ms of the statement.

(j) Let $MK = \langle MW, Ms \rangle$. Then an application of MAGMA yields that $MU = NA \cap MK = N_{MK}(MA)$ has order $2^9 \cdot 3 \cdot 7$. Furthermore, $MK = \langle MW, MU \rangle$ is a simple group of order $2^9 \cdot 3^4 \cdot 5 \cdot 7$ and $MK \cong \mathsf{Sp}_6(2)$ by Proposition 8.6.5 of [92].

(k) The matrix group MK has a faithful permutation representation of degree 63 with stabilizer MW. Using it and MAGMA the set $\mathcal{R}(K)$ of the defining relations of $K = \langle m_1, m_2, m_3, m_4, m_5 \rangle$ has been calculated, where $m_1 = Ms$, $m_2 = Mr$, $m_3 = Ms_1$, $m_4 = My_4$ and $m_5 = Ms_3$.

(l) By (k) and (j) K has an irreducible 8-dimensional matrix representation V. The given presentation of the split extension $H_1 = V \rtimes K$ has been calculated by means of MAGMA using the presentation of K given in (k).

(m) Assertion (c) implies that $D_1 = D/Z(D)$ splits over $V = Q/Z(D)$ with complement W. By Algorithm 1.3.8 the centralizer $H = C_G(z)$ of z in the simple (at this time unknown) target group G has to have an odd index $|H : D| = |K : W|$. Therefore Theorem 1.4.15 of [92] implies that H is a central extension of H_1 by a cyclic subgroup $Z(H) = \langle z \rangle$ having a normal subgroup Q containing z such that $V = Q/Z(H)$ has a complement in $H/Z(H)$ isomorphic to K.

By construction, $H_1 = \langle m_1, m_2, m_3, m_4, m_5, v_1, v_2, v_3, v_4, v_5, v_6, v_7, v_8 \rangle$ has a faithful permutation representation PH_1 of degree 256 with stabilizer K. Let FPH_1 be the presentation of H_1 given in (m). Then we can apply Holt's Algorithm implemented in MAGMA [52] to the trivial matrix representation of H_1 over $F = GF(2)$. It yields that the second cohomological dimension $dim_F[H^2(H_1, F)] = 2$. Thus there are three non-split central extensions $E_{0,1}, E_{1,0}, E_{1,1}$ of H_1. As D does not split over $Z(D)$ the target group H can only be isomorphic to a non-split extension.

Let TH_1 := `GModule(PH_1, FEalg)` be the trivial module of the matrix algebra `FEalg` generated by the 13 identity matrices corresponding to the 13 generators of H_1. Using the MAGMA command

```
P_H :=ExtensionProcess(PH_1,TH_1,FPH_1)
```

we construct a presentation for each of the three non-split central extensions

$E_{0,1}$:= `Extension(P_H, [0,1])`, $E_{1,0}$:= `Extension(P_H, [1,0])`,

$E_{1,1}$:= `Extension(P_H, [1,1])`,

corresponding to the three linearly independent 2-cocycles $[a, b] \in F^2$. Each of these three presentations has 14 generators, where the 14th generator corresponds to the new central element z. The first five generators correspond to m_1, m_2, m_3, m_4, m_5 in the factor group H_1. Taking the corresponding generators in $E_{1,0}$ as the stabilizer of a permutation representation $PE_{1,0}$ of $E_{1,0}$, it follows by another application of MAGMA that $PE_{1,0}$ is a faithful permutation of $E_{1,0}$ of degree 512, and that $E_{1,0}$ has an extra-special normal subgroup Q of order 2^9 with complement $K = \langle m_1, m_2, m_3, m_4, m_5\rangle$. We also have checked that the Sylow 2-subgroups of $E_{1,0}$ and D are isomorphic, and that this is not the case for the two other extensions $E_{0,1}$ and $E_{1,1}$. Therefore only $E_{1,0}$ can be isomorphic to the centralizer $H = C_G(z)$. The presentation of $E_{1,0}$ is given in the statement. This completes the proof. □

Definition 4.2.2 A finite simple group G is called to be of Co_2*-type*, if it has a 2-central involution z such that $C_G(z)$ is isomorphic to the finitely presented group H of Proposition 4.2.1(n).

4.3 Construction of Conway's simple group Co_2

The following Lemma 4.3.1 states that the amalgam $H \leftarrow D \rightarrow E_3$ of the three groups constructed in Sections 4.1 and 4.2 satisfies the conditions of Step 5 of Algorithm 1.3.8. In [72] H. Kim and the author applied Algorithm 7.4.8 of [92] to give a new existence proof for Conway's sporadic group Co_2. It is presented in this section.

Lemma 4.3.1 *Keep the notation of Lemmas 4.1.1 and 4.1.2 and Proposition 4.2.1. Let* $H = \langle h_i | 1 \le i \le 14\rangle$ *be the finitely presented group constructed in Proposition 4.2.1. Then the following statements hold.*

(a) *H has a faithful permutation representation of degree 512 with stabilizer $\langle h_1, h_4\rangle$.*
(b) *Each Sylow 2-subgroup S of H has a unique maximal elementary abelian normal subgroup A of order 2^{10} and $N_H(A) \cong D = C_{E_3}(z)$.*
(c) *There is a Sylow 2-subgroup S such that $D = N_H(A) = \langle x, y\rangle$, where $x = h_2h_{16}h_1$ and $y = h_9h_{14}h_{15}$ have respective orders 12 and 6. Furthermore, $H = \langle x, y, h\rangle$, where $h = h_1$ has order 7.*
(d) *The Goldschmidt index of the amalgam $H \leftarrow D \rightarrow E_3$ is 2.*
(e) *A system of representatives r_i of the 100 conjugacy classes of H and the corresponding centralizers orders $|C_H(r_i)|$ are given in Table 4.5.1.*
(f) *A system of representatives d_i of the 148 conjugacy classes of D and the corresponding centralizers orders $|C_D(d_i)|$ are given in Table 4.5.2.*

(g) *Let* $\sigma : N_H(A) \to D = C_{E_3}(z)$ *be the isomorphism given in (b). Then* $e = p \in E_3$ *has order* 4 *and* $E_3 = \langle \sigma(D), e \rangle$. *A system of representatives* e_i *of the* 79 *conjugacy classes of* E_3 *and the corresponding centralizers orders* $|C_{E_3}(e_i)|$ *are given in Table 4.5.3.*

(h) *The character tables of* H, E_3 *and* D *are given in Tables 4.6.2,* `DVD.1.2.2` *and 4.6.1, respectively.*

The proof is routine and therefore omitted.

Theorem 4.3.2 (Kim–Michler) *Keep the notation of Lemma 4.3.1 and Proposition 4.2.1. Using the notation of the three character Tables 4.6.2,* `DVD.1.2.2` *and 4.6.1, of the groups* H, E_3 *and* D, *respectively, the following statements hold.*

(a) *There is exactly one compatible pair* $(\chi, \tau) \in mf\,char_{\mathbb{C}}(H) \times mf\,char_{\mathbb{C}}(E_3)$ *of degree* 23 *of the groups* $H = \langle D, h \rangle$ *and* $E_3 = \langle D, e \rangle$:

$$(\chi, \tau) = (\chi_2 + \chi_{\mathbf{4}}, \tau_2 + \tau_{\mathbf{6}})$$

with common restriction

$$\tau_{|D} = \chi_{|D} = \psi_2 + \psi_8 + \psi_{\mathbf{26}},$$

where irreducible characters with bold face indices denote faithful irreducible characters.

(b) *Let* $\mathfrak{V}$ *and* $\mathfrak{W}$ *be the up to isomorphism uniquely determined faithful semi-simple multiplicity-free 23-dimensional modules of* H *and* E_3 *over* $F = \mathrm{GF}(13)$ *corresponding to the compatible pair* (χ, τ), *respectively.*

Let $\kappa_{\mathfrak{V}} : H \to \mathrm{GL}_{23}(13)$ *and* $\kappa_{\mathfrak{W}} : E_3 \to \mathrm{GL}_{23}(13)$ *be the representations of* H *and* E_3 *afforded by the modules* $\mathfrak{V}$ *and* $\mathfrak{W}$, *respectively.*

Let $\mathfrak{h} = \kappa_{\mathfrak{V}}(h)$, $\mathfrak{x} = \kappa_{\mathfrak{V}}(x)$, $\mathfrak{y} = \kappa_{\mathfrak{V}}(y)$ *in* $\kappa_{\mathfrak{V}}(H) \le \mathrm{GL}_{23}(13)$. *Then the following assertions hold.*

(1) $\mathfrak{V}_{|D} \cong \mathfrak{W}_{|D}$, *and there is a transformation matrix* $\mathcal{T} \in \mathrm{GL}_{23}(13)$ *such that*

$$\mathfrak{x} = \mathcal{T}^{-1}\kappa_{\mathfrak{W}}(x_1)\mathcal{T}, \quad \mathfrak{y} = \mathcal{T}^{-1}\kappa_{\mathfrak{W}}(y_1)\mathcal{T}.$$

Let $\mathfrak{e} = \mathcal{T}^{-1}\kappa_{\mathfrak{W}}(e)\mathcal{T} \in \mathrm{GL}_{23}(13)$.

(2) *In* $\mathfrak{G}_3 = \langle \mathfrak{h}, \mathfrak{x}, \mathfrak{y}, \mathfrak{e} \rangle$ *the subgroup* $\mathfrak{E}_3 = \langle \mathfrak{x}, \mathfrak{y}, \mathfrak{e} \rangle$ *is the stabilizer of a 1-dimensional subspace* $\mathfrak{U}$ *of* $\mathfrak{V}$ *such that the* $\mathfrak{G}_3$*-orbit* $\mathfrak{U}^{\mathfrak{G}}$ *has degree* 46575.

(3) *The four generating matrices of* $\mathfrak{G}_3$ *are stored on the accompanying DVD.*

(4) $\mathfrak{G}_3 = \langle \mathfrak{h}, \mathfrak{x}, \mathfrak{y}, \mathfrak{e} \rangle$, *and* $\mathfrak{G}_3$ *has* 60 *conjugacy classes* $\mathfrak{g}_i^{\mathfrak{G}_3}$ *with representatives* $\mathfrak{g}_i$ *and centralizer orders* $|C_{\mathfrak{G}_3}(\mathfrak{g}_i)|$ *as given in Table* `DVD.1.2.1` *on the accompanying DVD.*

(5) *The character Table of* $\mathfrak{G}_3$ *coincides with that of* Co_2 *in the Atlas [19], pp. 154–155.*

(c) $\mathfrak{G}_3$ *is a finite simple group with* 2*-central involution* $\kappa_{\mathfrak{V}}(z) = (\mathfrak{x}\mathfrak{y}\mathfrak{h})^{15}$ *such that*

$$C_{\mathfrak{G}_3}(\kappa_{\mathfrak{V}}(z)) = \kappa_{\mathfrak{V}}(H), \text{ and } |\mathfrak{G}_3| = 2^{18} \cdot 3^6 \cdot 5^3 \cdot 7 \cdot 11 \cdot 23.$$

Proof (a) The character tables of the groups H, D and E_3 are stated in Section 4.6. In the following we use their notations. Using MAGMA and the character tables of H, D and E_3, and the fusion of the classes of $D(H)$ in H and $D(E_3)$ in E_3, an application of Kratzer's Algorithm 7.3.10 of [92] yields the compatible pair stated in assertion (a). It also shows that the pair (χ, τ) of (b) is the unique compatible pair of degree 23 with respect to the fusion of the D-classes into the H-classes and into the E_3-classes.

(b) Let $\mathfrak{V}$ be the semi-simple faithful representation of H corresponding to the character $\chi = \chi_2 + \chi_4$. By the character table of H we know that χ_4 is the only irreducible character of H of degree 16. It occurs as an irreducible constituent of the faithful permutation representation $(1_T)^H$ of H with stabilizer $T = \langle h_1, h_4 \rangle$ given in Lemma 4.3.1. This can be checked by calculating its inner product with the permutation character $(1_T)^H$. Constructing the permutation matrices of the three generators h, x and y of H over the field $K = GF(13)$ and applying the Meataxe algorithm implemented in MAGMA, one obtains the matrices $\mathfrak{J}x, \mathfrak{J}y, \mathfrak{J}h \in \mathrm{GL}_{16}(13)$ of the generators of H in the 16-dimensional representation $\mathfrak{V}_1$. Furthermore, it has been checked in [72] that $\mathfrak{V}_1$ restricted to $D(H) = \langle x, y \rangle$ is irreducible.

The irreducible module $\mathfrak{V}_2$ corresponding to the character χ_2 of degree 7 is an irreducible constituent of the fourth exterior power $\bigwedge^4 \mathfrak{V}_1$ of dimension 1820 over $K = GF(13)$. Decomposing it into irreducible composition factors it follows that it has three composition factors $\mathfrak{W}_j$, $1 \le j \le 3$, of dimensions $35, 840$ and 945. Let $\mathfrak{W}_1$ be the one of dimension 35. Its second exterior power $\bigwedge^2 \mathfrak{W}_1$ of dimension 595 splits into four irreducible constituents $\mathfrak{X}_k$, $1 \le k \le 4$, of dimensions $7, 21, 189$ and 378. Furthermore, H has exactly one irreducible character of dimension

7 by Table 4.6.2. Hence $\mathfrak{V}_2 = \mathfrak{X}_1$. Choosing a basis in this KH-module the generators of H are represented by the following matrices:

$$\mathfrak{N}x = \begin{pmatrix} 5 & 8 & 8 & 5 & 9 & 11 & 8 \\ 9 & 6 & 11 & 1 & 6 & 3 & 1 \\ 10 & 3 & 2 & 11 & 10 & 7 & 12 \\ 3 & 8 & 10 & 0 & 10 & 12 & 9 \\ 11 & 11 & 6 & 5 & 3 & 7 & 11 \\ 0 & 6 & 6 & 1 & 0 & 8 & 2 \\ 0 & 3 & 3 & 7 & 0 & 11 & 0 \end{pmatrix},$$

$$\mathfrak{N}y = \begin{pmatrix} 4 & 10 & 4 & 1 & 10 & 8 & 5 \\ 8 & 1 & 7 & 6 & 1 & 8 & 12 \\ 10 & 4 & 5 & 11 & 12 & 4 & 1 \\ 5 & 5 & 0 & 4 & 10 & 11 & 0 \\ 12 & 8 & 0 & 3 & 11 & 12 & 0 \\ 12 & 10 & 0 & 1 & 5 & 0 & 0 \\ 6 & 4 & 0 & 8 & 6 & 6 & 12 \end{pmatrix} \quad \text{and} \quad \mathfrak{N}h = \begin{pmatrix} 12 & 10 & 11 & 8 & 5 & 9 & 3 \\ 4 & 0 & 11 & 1 & 11 & 4 & 5 \\ 0 & 3 & 6 & 5 & 6 & 9 & 2 \\ 7 & 5 & 4 & 10 & 0 & 3 & 6 \\ 1 & 4 & 0 & 11 & 11 & 0 & 0 \\ 1 & 3 & 12 & 2 & 2 & 12 & 2 \\ 7 & 2 & 0 & 12 & 12 & 7 & 1 \end{pmatrix}.$$

Using MAGMA and the Meataxe algorithm again we see that the restriction of $\mathfrak{V}_2$ to $D_H = \langle x, y\rangle$ splits into two irreducible KD_H-modules, $\mathfrak{V}_{21}$ and $\mathfrak{V}_{22}$, of dimensions 1 and 6, having respective bases B_1 and B_2. Hence their union B is a basis of $\mathfrak{V}_2$. Let $\mathcal{T}_1 \in \mathrm{GL}_7(13)$ be the matrix of the base change from the canonical basis to the basis B of $\mathfrak{V}_2$. Then $\mathfrak{L}x = \mathcal{T}_1\mathfrak{N}x(\mathcal{T}_1)^{-1}$, $\mathfrak{L}y = \mathcal{T}_1\mathfrak{N}y(\mathcal{T}_1)^{-1}$ and $\mathfrak{L}h = \mathcal{T}_1\mathfrak{N}h(\mathcal{T}_1)^{-1}$ represent the three generators x, y, h of H w.r.t. the basis B of $\mathfrak{V}_2$. It follows that the blocked diagonal 7×7 matrices $\mathfrak{L}x$ and $\mathfrak{L}y$ are the 7×7 matrices in the upper left corner of the matrices $\mathfrak{x}$ and $\mathfrak{y}$ given in the statement. Furthermore, the matrix $\mathfrak{L}h$ of $h \in H$ is the corresponding 7×7 block of the matrix $\mathfrak{h}$ in the statement.

Let $\mathfrak{x}$, $\mathfrak{y}$ and $\mathfrak{h}$ be the diagonal joins of the matrices $\mathfrak{L}x$ and $\mathfrak{J}x$, $\mathfrak{L}y$ and $\mathfrak{J}y$ and $\mathfrak{L}h$ and $\mathfrak{J}h$ in $\mathrm{GL}_{23}(13)$, respectively. Then these three 23×23 matrices are stated in the assertion. Clearly, they satisfy the equations $\mathfrak{x} = \kappa_{\mathfrak{V}}(x)$, $\mathfrak{y} = \kappa_{\mathfrak{V}}(y)$ and $\mathfrak{h} = \kappa_{\mathfrak{V}}(h)$, where $\kappa_{\mathfrak{V}} : H \to \mathrm{GL}_{23}(13)$ denotes the semi-simple representation of $H = \langle x, y, h\rangle$ afforded
by $\mathfrak{V}$.

Now we construct the faithful representation $\mathfrak{W}$ of E_3 corresponding to the character $\tau = \tau_2 + \tau_6$. By the character table of E_3 we know that τ_2 is the only non-trivial linear character of E. Its corresponding representation $\mathfrak{W}_1 : E_3 \to \mathrm{GL}_1(K)$ is given by $\mathfrak{W}_1(x_1) = \mathfrak{W}_1(y_1) = 12$ and $\mathfrak{W}_1(e_1) = 1$. In order to find the irreducible representation $\mathfrak{W}_2$ of τ_6 we determine the irreducible constituents of the faithful permutation representation $(1_{T_1})^{E_3}$ of E_3 of degree 1024 with stabilizer A_{22} constructed in Lemma 4.1.2. An application of MAGMA shows that it has four irreducible constituents of degrees 1, 22, 231 and 770. Using the Meataxe algorithm implemented in MAGMA the matrices $\mathfrak{J}x_1, \mathfrak{J}y_1, \mathfrak{J}e_1 \in \mathrm{GL}_{22}(13)$

of the generators of E_3 in the 22-dimensional irreducible constituent $\mathfrak{X}$ of $(1_{T_1})^E$ over $K = GF(13)$ and their traces have been determined. Now it follows from the character table of E_3 that τ_6 is the character of the tensor product $\mathfrak{X} \otimes \mathfrak{W}_1$. Another application of the Meataxe algorithm and MAGMA implies that the restriction of $\mathfrak{W}_2$ to $D_E = \langle x_1, y_1 \rangle$ splits into two irreducible KD_E-modules, $\mathfrak{W}_{21}$ and $\mathfrak{W}_{22}$, having respective bases B_1 and B_2 such that the generators x_1 and y_1 of D have the following matrices w.r.t. B_1 and B_2, respectively:

$$\mathfrak{W}_{21}(x_1) = \begin{pmatrix} 12 & 1 & 1 & 0 & 0 & 0 \\ 0 & 1 & 0 & 0 & 12 & 1 \\ 12 & 0 & 0 & 0 & 0 & 0 \\ 0 & 0 & 12 & 0 & 0 & 0 \\ 1 & 0 & 0 & 1 & 0 & 12 \\ 0 & 0 & 0 & 1 & 1 & 12 \end{pmatrix}, \quad \mathfrak{W}_{21}(y_1) = \begin{pmatrix} 12 & 0 & 1 & 12 & 0 & 0 \\ 12 & 0 & 0 & 12 & 0 & 1 \\ 0 & 0 & 1 & 0 & 1 & 12 \\ 0 & 0 & 0 & 1 & 1 & 12 \\ 12 & 1 & 0 & 12 & 12 & 1 \\ 0 & 0 & 0 & 0 & 0 & 12 \end{pmatrix},$$

$$\mathfrak{W}_{22}(x_1) = \begin{pmatrix} 1 & 12 & 0 & 0 & 0 & 0 & 1 & 12 & 0 & 12 & 12 & 1 & 1 & 1 & 12 & 12 \\ 2 & 0 & 1 & 0 & 1 & 12 & 1 & 11 & 12 & 12 & 12 & 0 & 1 & 12 & 1 & 12 \\ 3 & 12 & 2 & 1 & 1 & 12 & 0 & 11 & 11 & 12 & 0 & 12 & 1 & 12 & 2 & 11 \\ 3 & 11 & 2 & 2 & 2 & 12 & 0 & 10 & 12 & 12 & 1 & 11 & 0 & 0 & 2 & 10 \\ 0 & 0 & 0 & 0 & 0 & 12 & 0 & 1 & 12 & 0 & 1 & 0 & 12 & 12 & 1 & 0 \\ 12 & 1 & 0 & 12 & 12 & 0 & 1 & 0 & 1 & 0 & 11 & 1 & 1 & 1 & 12 & 1 \\ 0 & 12 & 0 & 1 & 0 & 0 & 0 & 12 & 0 & 0 & 1 & 12 & 0 & 1 & 0 & 12 \\ 2 & 0 & 1 & 1 & 1 & 12 & 12 & 12 & 11 & 0 & 1 & 11 & 0 & 12 & 2 & 12 \\ 1 & 11 & 0 & 1 & 1 & 1 & 1 & 12 & 0 & 12 & 0 & 1 & 1 & 1 & 12 & 11 \\ 2 & 10 & 1 & 2 & 1 & 0 & 0 & 11 & 12 & 12 & 2 & 12 & 1 & 1 & 0 & 10 \\ 2 & 11 & 1 & 1 & 1 & 0 & 1 & 11 & 12 & 12 & 0 & 0 & 1 & 0 & 0 & 11 \\ 3 & 12 & 2 & 0 & 1 & 12 & 1 & 11 & 12 & 12 & 12 & 0 & 1 & 12 & 1 & 11 \\ 2 & 10 & 1 & 3 & 2 & 0 & 12 & 11 & 11 & 0 & 3 & 11 & 0 & 0 & 1 & 10 \\ 1 & 0 & 0 & 0 & 0 & 12 & 0 & 0 & 12 & 0 & 0 & 0 & 0 & 12 & 1 & 0 \\ 0 & 0 & 0 & 0 & 0 & 0 & 0 & 0 & 0 & 1 & 0 & 0 & 0 & 0 & 0 & 0 \\ 0 & 11 & 0 & 1 & 0 & 0 & 0 & 0 & 0 & 0 & 1 & 0 & 12 & 1 & 0 & 12 \end{pmatrix},$$

$$\mathfrak{W}_{22}(y_1) = \begin{pmatrix} 3 & 10 & 2 & 3 & 2 & 12 & 0 & 10 & 11 & 12 & 2 & 11 & 0 & 0 & 2 & 10 \\ 2 & 11 & 2 & 2 & 1 & 12 & 0 & 11 & 12 & 12 & 2 & 11 & 12 & 0 & 2 & 11 \\ 11 & 1 & 12 & 0 & 12 & 0 & 12 & 1 & 0 & 1 & 1 & 0 & 12 & 0 & 0 & 1 \\ 2 & 11 & 1 & 2 & 1 & 0 & 12 & 11 & 12 & 0 & 2 & 11 & 0 & 0 & 1 & 11 \\ 12 & 0 & 0 & 0 & 0 & 0 & 0 & 0 & 0 & 0 & 0 & 1 & 0 & 0 & 12 & 0 \\ 0 & 2 & 0 & 12 & 0 & 0 & 0 & 0 & 0 & 0 & 12 & 0 & 0 & 12 & 1 & 1 \\ 2 & 0 & 2 & 0 & 1 & 12 & 0 & 11 & 12 & 0 & 0 & 11 & 0 & 12 & 2 & 12 \\ 12 & 0 & 0 & 1 & 0 & 0 & 12 & 0 & 12 & 1 & 2 & 12 & 12 & 0 & 0 & 0 \\ 12 & 11 & 12 & 2 & 0 & 1 & 0 & 0 & 1 & 0 & 1 & 0 & 0 & 2 & 12 & 12 \\ 1 & 0 & 1 & 1 & 1 & 12 & 12 & 12 & 12 & 1 & 1 & 11 & 12 & 12 & 2 & 12 \\ 2 & 11 & 2 & 2 & 1 & 12 & 0 & 11 & 12 & 12 & 1 & 11 & 0 & 0 & 2 & 11 \\ 1 & 12 & 1 & 2 & 1 & 12 & 12 & 12 & 12 & 0 & 2 & 11 & 12 & 12 & 2 & 12 \\ 1 & 11 & 1 & 2 & 1 & 0 & 12 & 12 & 12 & 0 & 2 & 11 & 0 & 0 & 1 & 11 \\ 0 & 0 & 1 & 0 & 0 & 12 & 0 & 0 & 12 & 0 & 1 & 0 & 12 & 12 & 1 & 0 \\ 12 & 1 & 0 & 0 & 0 & 0 & 12 & 1 & 0 & 1 & 1 & 12 & 12 & 12 & 1 & 1 \\ 2 & 11 & 1 & 1 & 1 & 0 & 0 & 11 & 12 & 12 & 0 & 0 & 2 & 0 & 0 & 11 \end{pmatrix}.$$

Let $B = B_1 \cup B_2$. Then B is a basis of the irreducible KE_3-module $\mathfrak{W}_2$, and w.r.t. B the third generator e_1 of E_3 has the following matrix:

$$\mathfrak{W}_2(e_1) = \begin{pmatrix}
7&0&6&0&0&0&7&6&7&7&7&0&6&6&6&0&7&6&0&0&7&6\\
6&7&0&6&6&1&7&0&7&6&0&0&0&6&0&0&6&0&7&0&0&0\\
7&6&0&7&7&12&6&6&0&7&0&0&0&0&0&0&7&0&6&7&0&0\\
6&0&7&0&0&0&12&6&6&1&0&0&6&7&0&7&8&6&12&7&0&0\\
7&0&6&0&0&0&0&0&0&0&0&0&6&0&0&7&7&6&0&0&0&0\\
7&0&6&0&0&0&12&6&6&1&0&0&6&7&0&7&8&6&12&7&0&0\\
6&0&7&0&1&0&11&9&6&11&12&0&7&8&1&7&5&8&0&6&12&2\\
7&6&0&7&7&0&7&12&7&7&1&0&1&6&0&12&7&0&7&0&0&12\\
7&0&7&0&0&0&7&11&0&8&7&1&7&12&7&12&7&7&1&8&5&5\\
7&0&7&0&0&0&8&0&2&7&7&12&6&12&5&0&7&5&0&5&8&6\\
7&6&0&7&7&0&5&0&5&6&12&1&1&7&2&0&5&2&7&2&11&0\\
7&12&6&0&0&0&12&0&12&0&0&1&6&1&0&7&7&7&0&0&12&1\\
6&0&7&0&1&0&1&12&1&1&1&0&6&12&12&6&8&5&0&12&1&12\\
7&0&7&0&0&0&5&7&6&6&6&1&7&7&8&0&6&7&0&1&5&7\\
6&7&1&6&7&0&12&1&6&12&5&0&0&8&7&7&6&1&6&7&6&8\\
6&7&1&6&7&0&5&7&12&6&12&1&0&1&1&0&6&1&7&7&12&1\\
6&0&7&0&1&0&0&0&0&0&0&0&7&0&0&6&7&7&0&0&0&0\\
7&0&7&0&0&0&6&7&6&6&6&0&7&7&7&0&6&8&0&0&6&7\\
6&7&1&6&7&0&7&0&7&7&0&0&0&6&0&0&7&12&6&0&1&0\\
7&6&0&7&7&0&7&6&0&7&1&0&1&12&1&12&6&0&7&7&0&12\\
7&0&7&0&0&0&0&7&6&0&0&0&6&7&0&7&7&6&12&6&1&1\\
6&0&7&0&1&0&1&1&1&0&0&12&6&12&12&7&6&6&0&12&1&0
\end{pmatrix}.$$

By construction the KD-modules $\mathfrak{V}_{|D}$ and $\mathfrak{W}_{|D}$ described by the two pairs $(\mathfrak{V}(x), \mathfrak{V}(y))$ and $(\mathfrak{W}(x_1), \mathfrak{W}(y_1))$ of matrices in $\mathrm{GL}_{23}(13)$ are isomorphic. Let $Y = \mathrm{GL}(23, 13) = \mathrm{GL}_{23}(13)$. Applying then Parker's isomorphism test of Proposition 6.1.6 of [92] by means of the MAGMA command

```
IsIsomorphic(GModule(sub<Y|V(x),V(y)>),GModule(sub<Y|W(x1),W(y1)>))
```

we obtain the transformation matrix given by

$$\mathcal{T} = \begin{pmatrix}
1&0&0&0&0&0&0&0&0&0&0&0&0&0&0&0&0&0&0&0&0&0&0\\
0&1&1&11&1&9&0&0&0&0&0&0&0&0&0&0&0&0&0&0&0&0&0\\
0&1&1&11&0&8&0&0&0&0&0&0&0&0&0&0&0&0&0&0&0&0&0\\
0&4&2&5&3&12&2&0&0&0&0&0&0&0&0&0&0&0&0&0&0&0&0\\
0&3&1&7&2&3&11&0&0&0&0&0&0&0&0&0&0&0&0&0&0&0&0\\
0&0&0&0&1&12&0&0&0&0&0&0&0&0&0&0&0&0&0&0&0&0&0\\
0&1&1&9&10&2&11&0&0&0&0&0&0&0&0&0&0&0&0&0&0&0&0\\
0&0&0&0&0&0&0&0&11&0&0&1&0&0&1&1&0&2&12&0&12&0&12\\
0&0&0&0&0&0&0&0&0&11&0&3&2&0&0&11&2&2&11&12&11&3&12\\
0&0&0&0&0&0&0&12&12&3&11&11&11&1&1&3&12&0&3&1&2&9&1\\
0&0&0&0&0&0&0&1&4&7&4&4&3&12&12&7&3&1&7&10&10&10&11\\
0&0&0&0&0&0&0&0&0&2&11&12&0&0&0&2&11&0&2&1&2&10&1\\
0&0&0&0&0&0&0&0&11&2&12&12&11&1&1&2&0&1&2&1&0&9&0\\
0&0&0&0&0&0&0&11&1&1&12&10&10&1&12&3&11&11&3&1&3&9&1\\
0&0&0&0&0&0&0&0&0&1&12&0&12&1&1&1&12&0&0&1&0&12&1\\
0&0&0&0&0&0&0&1&11&0&1&2&1&0&2&0&1&2&11&12&10&3&12\\
0&0&0&0&0&0&0&12&12&2&12&10&10&0&0&4&11&0&3&1&2&9&1\\
0&0&0&0&0&0&0&12&11&1&12&1&0&0&1&2&0&1&1&0&0&12&0\\
0&0&0&0&0&0&0&0&0&12&1&2&1&12&1&12&1&2&11&12&11&3&12\\
0&0&0&0&0&0&0&0&0&0&0&1&0&0&0&0&0&0&0&12&0&1&1\\
0&0&0&0&0&0&0&12&12&2&10&12&12&0&0&2&11&0&3&1&2&9&1\\
0&0&0&0&0&0&0&0&12&1&0&1&12&0&1&0&0&1&0&0&11&0&0\\
0&0&0&0&0&0&0&0&1&12&1&1&1&12&12&12&0&0&12&12&1&2&12
\end{pmatrix},$$

satisfying $\mathfrak{V}(x) = (\mathfrak{W}(x_1))^{\mathcal{T}}$ and $\mathfrak{V}(y) = (\mathfrak{W}(y_1))^{\mathcal{T}}$.

Let $\mathfrak{e} = (\mathfrak{W}(e_1))^{\mathcal{T}}$. Then $\mathfrak{E}_3 = \langle \mathfrak{x}, \mathfrak{y}, \mathfrak{e} \rangle \cong E$. Let $\mathfrak{G}_3 = \langle \mathfrak{h}, \mathfrak{x}, \mathfrak{y}, \mathfrak{e} \rangle$. Using the MAGMA command `CosetAction(G, E)` we obtain a faithful permutation representation of $\mathfrak{G}_3$ of degree 46575 with stabilizer $\mathfrak{E}_3$. In particular, $|\mathfrak{G}_3| = 2^{18} \cdot 3^6 \cdot 5^3 \cdot 7 \cdot 11 \cdot 23$.

Using the faithful permutation representation of $\mathfrak{G}_3$ of degree 46575 and Kratzer's Algorithm 5.3.18 of [92] we calculated the representatives of all the conjugacy classes of $\mathfrak{G}_3$; see Table `DVD.1.2.1`.

Furthermore, the character table of $\mathfrak{G}_3$ has been calculated by means of the above permutation representation and MAGMA. It coincides with the one of Co_2 in [19], pp. 154–155.

(c) Let $\mathfrak{z} = (\mathfrak{x}\mathfrak{y}\mathfrak{h})^{15}$. Then $C_{\mathfrak{G}_3}(\mathfrak{z})$ contains $\mathfrak{H} = \langle \mathfrak{h}, \mathfrak{x}, \mathfrak{y} \rangle$, which is isomorphic to H. By Table `DVD.1.2.1` $|C_{\mathfrak{G}_3}(\mathfrak{z})| = |H|$. Hence $C_{\mathfrak{G}_3}(\mathfrak{z}) \cong \mathfrak{H}$. The character table of $\mathfrak{G}_3$ implies that $\mathfrak{G}_3$ is a simple group. This completes the proof. □

Praeger and Soicher give in [110] a nice presentation for Conway's simple group Co_2. It is used in [72] to show that $\mathfrak{G}_3$ is isomorphic to Co_2.

Corollary 4.3.3 *Keep the notation of Theorem 4.3.2. The finite simple group $\mathfrak{G}_3$ is isomorphic to the finitely presented group $G = \langle a, b, c, d, e, f, g \rangle$ with the following set $\mathcal{R}(G)$ of defining relations:*

$$
\begin{aligned}
&a^2 = b^2 = c^2 = d^2 = e^2 = f^2 = g^2 = 1,\\
&(ab)^3 = (bc)^5 = (cd)^3 = (df)^3 = (fe)^6 = (ae)^4 = 1,\\
&(ec)^3 = (cf)^4 = (fg)^4 = (gb)^4 = 1,\\
&(ac)^2 = (ad)^2 = (af)^2 = (ag)^2 = (bd)^2 = (be)^2 = 1,\\
&(bf)^2 = (cg)^2 = (dg)^2 = (eg)^2 = 1,\\
&a = (cf)^2, \quad e = (bg)^2, \quad b = (ef)^3,\\
&(aecd)^4 = (baefg)^3 = (cef)^7 = 1.
\end{aligned}
$$

Furthermore, G has a faithful permutation representation PG of degree 46575 with stabilizer $Q = \langle a, b, c, d, e, g, (gfdc)^4 \rangle$.

Proof The presentation of $G = \langle a, b, c, d, e, f, g \rangle$ is taken from [110], p. 106. Its subgroup, $Q = \langle a, b, c, d, e, g, (gfdc)^4 \rangle$, is isomorphic to E_3 by Lemma 4.1.2(f) and [110], p. 106. Taking it as a stabilizer MAGMA constructs a faithful permutation representation PG of the group $G = \langle a, b, c, d, e, f, g \rangle$ of degree 46575. An isomorphism test using MAGMA and the faithful permutation representation of $\mathfrak{G}_3$ given in Theorem 4.3.2(c) shows that $\mathfrak{G}_3 \cong G$. □

Corollary 4.3.3 will be used in Chapter 6 in H. Kim's construction of Fischer's sporadic group Fi_{23}.

4.4 On the uniqueness of Co_2

By Lemma 4.3.1 the Goldschmidt index of the amalgam $H \leftarrow D \rightarrow E_3$ is 2. Hence the author's sufficient uniqueness criterion stated as Theorem 7.5.1 in [92] cannot be applied. Therefore we do not give a uniqueness proof.

However, the main result of F. Smith's article [119] asserts that Co_2 is uniquely determined up to isomorphism by the structure of the centralizer $H = C_{\mathsf{Co}_2}(z)$ of a 2-central involution $z \in \mathsf{Co}_2$. Furthermore, the largest normal subgroup O_2 of 2-power order of H is extra-special and $H = H/O_2 \cong \mathsf{Sp}_6(2)$. In particular, the matrix subgroup $\mathfrak{G}_3$ of Theorem 4.3.2 is isomorphic to Co_2.

Smith's proof is involved. He constructs for a simple group G of Co_2-type a graph $\mathcal{G}$ and its automorphism group $Aut(\mathcal{G})$. But his proof does not provide an isomorphism between G and Conway's group Co_2 of [16].

Using Smith's uniqueness theorem T. Yoshida proved an even stronger uniqueness result in [141].

Theorem 4.4.1 (T. Yoshida) *Let* Co_2 *be the simple sporadic Conway group of order* $2^{18}{\cdot}3^6{\cdot}5^3{\cdot}7{\cdot}11{\cdot}23$. *A finite simple group* G *is isomorphic to Conway's group* Co_2 *if it has a Sylow* 2*-subgroup* S *which is isomorphic to the Sylow* 2*-subgroups of* Co_2.

Proof See [141], p. 405. □

In [141] T. Yoshida states a presentation of a Sylow 2-subgroup T of Co_2 in 18 generators. Using it, and the faithful permutation representation of the simple group $\mathfrak{G}_3$ stated in Theorem 4.3.2, it can be verified that Co_2 and $\mathfrak{G}_2$ have isomorphic Sylow 2-subgroups.

4.5 Representatives of conjugacy classes

4.5.1 *Conjugacy classes of* $H(\mathsf{Co}_2) = \langle x, y, h\rangle$

Class	*Representative*	$\lvert$*Class*$\rvert$	$\lvert$*Centralizer*$\rvert$	2P	3P	5P	7P
1	1	1	$2^{18}\cdot 3^4\cdot 5\cdot 7$	1	1	1	1
2_1	$(xyh)^{15}$	1	$2^{18}\cdot 3^4\cdot 5\cdot 7$	1	2_1	2_1	2_1
2_2	$(yh)^7$	270	$2^{17}\cdot 3\cdot 7$	1	2_2	2_2	2_2
2_3	$(y)^3$	1008	$2^{14}\cdot 3^2\cdot 5$	1	2_3	2_3	2_3
2_4	$(xyhy)^5$	1008	$2^{14}\cdot 3^2\cdot 5$	1	2_4	2_4	2_4
2_5	$(xy)^4$	1260	$2^{16}\cdot 3^2$	1	2_5	2_5	2_5
2_6	$(xyxh^2)^6$	1260	$2^{16}\cdot 3^2$	1	2_6	2_6	2_6
2_7	$(x)^6$	15120	$2^{14}\cdot 3$	1	2_7	2_7	2_7
2_8	$(x^2y^4)^3$	15120	$2^{14}\cdot 3$	1	2_8	2_8	2_8
2_9	$(xy^4)^2$	22680	2^{15}	1	2_9	2_9	2_9
2_{10}	$(xh^2)^3$	120960	$2^{11}\cdot 3$	1	2_{10}	2_{10}	2_{10}
3_1	$(x^2h)^4$	143360	$2^6\cdot 3^4$	3_1	1	3_1	3_1
3_2	$(x)^4$	172032	$2^5\cdot 3^3\cdot 5$	3_2	1	3_2	3_2
3_3	$(xhy)^4$	215040	$2^7\cdot 3^3$	3_3	1	3_3	3_3
4_1	$(yh^2)^7$	240	$2^{14}\cdot 3^3\cdot 7$	2_1	4_1	4_1	4_1
4_2	$(x^2yh)^4$	15120	$2^{14}\cdot 3$	2_1	4_2	4_2	4_2
4_3	$(xh^2y)^3$	30240	$2^{13}\cdot 3$	2_2	4_3	4_3	4_3
4_4	$(xyxh^2)^3$	60480	$2^{12}\cdot 3$	2_6	4_4	4_4	4_4
4_5	$(xh^2y^2)^3$	60480	$2^{12}\cdot 3$	2_2	4_5	4_5	4_5
4_6	$(xhyxh^2)^3$	60480	$2^{12}\cdot 3$	2_6	4_6	4_6	4_6
4_7	$(x^2yxhxy)^3$	60480	$2^{12}\cdot 3$	2_2	4_7	4_7	4_7
4_8	$(x^4y^3)^2$	90720	2^{13}	2_2	4_8	4_8	4_8
4_9	$(x^2y)^2$	120960	$2^{11}\cdot 3$	2_5	4_9	4_9	4_9
4_{10}	$(xhy)^3$	120960	$2^{11}\cdot 3$	2_2	4_{10}	4_{10}	4_{10}
4_{11}	$(xy)^2$	181440	2^{12}	2_5	4_{11}	4_{11}	4_{11}
4_{12}	$x^2yxhyxyh$	181440	2^{12}	2_5	4_{12}	4_{12}	4_{12}
4_{13}	$(yh^3)^3$	241920	$2^{10}\cdot 3$	2_5	4_{13}	4_{13}	4_{13}
4_{14}	$(xh^3y)^2$	362880	2^{11}	2_2	4_{14}	4_{14}	4_{14}
4_{15}	$(x^3yhxy)^2$	362880	2^{11}	2_6	4_{15}	4_{15}	4_{15}
4_{16}	$(x)^3$	483840	$2^9\cdot 3$	2_7	4_{16}	4_{16}	4_{16}
4_{17}	$(xy^2)^3$	483840	$2^9\cdot 3$	2_7	4_{17}	4_{17}	4_{17}
4_{18}	xhy^2	483840	$2^9\cdot 3$	2_7	4_{18}	4_{18}	4_{18}
4_{19}	$(yh^4)^3$	483840	$2^9\cdot 3$	2_7	4_{19}	4_{19}	4_{19}
4_{20}	x^2y^2	725760	2^{10}	2_2	4_{20}	4_{20}	4_{20}
4_{21}	x^2hxy	725760	2^{10}	2_5	4_{21}	4_{21}	4_{21}
4_{22}	xy^4	2903040	2^8	2_9	4_{22}	4_{22}	4_{22}
4_{23}	$xyxh^3y$	2903040	2^8	2_7	4_{23}	4_{23}	4_{23}
4_{24}	xh^5y	2903040	2^8	2_7	4_{24}	4_{24}	4_{24}
5	xh	12386304	$2^2\cdot 3\cdot 5$	5	5	1	5
6_1	$(x^2hyh)^3$	143360	$2^6\cdot 3^4$	3_1	2_1	6_1	6_1
6_2	$(xyh)^5$	172032	$2^5\cdot 3^3\cdot 5$	3_2	2_1	6_2	6_2
6_3	$(x^3y)^2$	215040	$2^7\cdot 3^3$	3_3	2_1	6_3	6_3
6_4	$(x^2h)^2$	1290240	$2^6\cdot 3^2$	3_1	2_5	6_4	6_4
6_5	$(xyxh^2)^2$	1290240	$2^6\cdot 3^2$	3_1	2_6	6_5	6_5
6_6	y	2580480	$2^5\cdot 3^2$	3_2	2_3	6_6	6_6
6_7	x^4y	2580480	$2^5\cdot 3^2$	3_2	2_4	6_7	6_7
6_8	x^3hxy	2580480	$2^5\cdot 3^2$	3_3	2_5	6_8	6_8
6_9	x^2yh^3	2580480	$2^5\cdot 3^2$	3_3	2_4	6_9	6_9
6_{10}	x^2y^4h	2580480	$2^5\cdot 3^2$	3_3	2_6	6_{10}	6_{10}
6_{11}	x^2hyhxh	2580480	$2^5\cdot 3^2$	3_2	2_5	6_{11}	6_{11}
6_{12}	x^2h^2yxy	2580480	$2^5\cdot 3^2$	3_2	2_6	6_{12}	6_{12}
6_{13}	xyh^2yh^2	2580480	$2^5\cdot 3^2$	3_3	2_3	6_{13}	6_{13}
6_{14}	$(xhy)^2$	3870720	$2^6\cdot 3$	3_3	2_2	6_{14}	6_{14}

Conjugacy classes of $H(\mathsf{Co}_2) = \langle x, y, h \rangle$ (continued)

Class	*Representative*	|*Class*|	|*Centralizer*|	2P	3P	5P	7P
6_{15}	$(x)^2$	7741440	$2^5 \cdot 3$	3_2	2_7	6_{15}	6_{15}
6_{16}	x^2y^4	7741440	$2^5 \cdot 3$	3_2	2_8	6_{16}	6_{16}
6_{17}	xh^2	15482880	$2^4 \cdot 3$	3_3	2_{10}	6_{17}	6_{17}
7	h	13271040	$2^3 \cdot 7$	7	7	7	1
8_1	$(xy^3h)^3$	967680	$2^8 \cdot 3$	4_1	8_1	8_1	8_1
8_2	$(x^2yh)^2$	1451520	2^9	4_2	8_2	8_2	8_2
8_3	$(xhyh)^2$	1451520	2^9	4_2	8_3	8_3	8_3
8_4	x^4y^3	2903040	2^8	4_8	8_4	8_4	8_4
8_5	$x^2h^2xy^2$	2903040	2^8	4_8	8_5	8_5	8_5
8_6	x^2y	5806080	2^7	4_9	8_6	8_6	8_6
8_7	x^3yhxy	5806080	2^7	4_{15}	8_7	8_7	8_7
8_8	x^2y^2hyh	5806080	2^7	4_8	8_8	8_8	8_8
8_9	xh^4yh	5806080	2^7	4_9	8_9	8_9	8_9
8_{10}	x^2hxy^3h	5806080	2^7	4_{15}	8_{10}	8_{10}	8_{10}
8_{11}	xy	11612160	2^6	4_{11}	8_{11}	8_{11}	8_{11}
8_{12}	xh^3y	11612160	2^6	4_{14}	8_{12}	8_{12}	8_{12}
8_{13}	xh^2y^2h	11612160	2^6	4_4	8_{13}	8_{13}	8_{13}
9	y^2h	41287680	$2 \cdot 3^2$	9	3_1	9	9
10_1	$(xyh)^3$	12386304	$2^2 \cdot 3 \cdot 5$	5	10_1	2_1	10_1
10_2	$xyhy$	37158912	$2^2 \cdot 5$	5	10_2	2_4	10_2
10_3	y^2h^2	37158912	$2^2 \cdot 5$	5	10_3	2_3	10_3
12_1	$xh^2xh^2y^2$	860160	$2^5 \cdot 3^3$	6_1	4_1	12_1	12_1
12_2	x^3y	2580480	$2^5 \cdot 3^2$	6_3	4_1	12_2	12_2
12_3	$xyxh^2$	7741440	$2^5 \cdot 3$	6_5	4_4	12_3	12_3
12_4	xh^2y^2	7741440	$2^5 \cdot 3$	6_{14}	4_5	12_4	12_4
12_5	xy^4h	7741440	$2^5 \cdot 3$	6_1	4_2	12_5	12_5
12_6	$xhyxh^2$	7741440	$2^5 \cdot 3$	6_5	4_6	12_6	12_6
12_7	x^2yxhxy	7741440	$2^5 \cdot 3$	6_{14}	4_7	12_7	12_7
12_8	x	15482880	$2^4 \cdot 3$	6_{15}	4_{16}	12_8	12_8
12_9	x^2h	15482880	$2^4 \cdot 3$	6_4	4_9	12_9	12_9
12_{10}	xy^2	15482880	$2^4 \cdot 3$	6_{15}	4_{17}	12_{10}	12_{10}
12_{11}	xhy	15482880	$2^4 \cdot 3$	6_{14}	4_{10}	12_{11}	12_{11}
12_{12}	xh^2y	15482880	$2^4 \cdot 3$	6_{14}	4_3	12_{12}	12_{12}
12_{13}	yh^3	15482880	$2^4 \cdot 3$	6_4	4_{13}	12_{14}	12_{13}
12_{14}	$(yh^3)^5$	15482880	$2^4 \cdot 3$	6_4	4_{13}	12_{13}	12_{14}
12_{15}	yh^4	15482880	$2^4 \cdot 3$	6_{15}	4_{19}	12_{15}	12_{15}
12_{16}	x^2y^2hy	15482880	$2^4 \cdot 3$	6_{15}	4_{18}	12_{16}	12_{16}
14_1	$(yh^2)^2$	13271040	$2^3 \cdot 7$	7	14_1	14_1	2_1
14_2	yh	26542080	$2^2 \cdot 7$	7	14_3	14_3	2_2
14_3	$(yh)^3$	26542080	$2^2 \cdot 7$	7	14_2	14_2	2_2
15	$(xyh)^2$	24772608	$2 \cdot 3 \cdot 5$	15	5	3_2	15
16_1	x^2yh	23224320	2^5	8_2	16_1	16_1	16_1
16_2	$xhyh$	23224320	2^5	8_3	16_2	16_2	16_2
18	x^2hyh	41287680	$2 \cdot 3^2$	9	6_1	18	18
24	xy^3h	30965760	$2^3 \cdot 3$	12_2	8_1	24	24
28	yh^2	26542080	$2^2 \cdot 7$	14_1	28	28	4_1
30	xyh	24772608	$2 \cdot 3 \cdot 5$	15	10_1	6_2	30

4.5.2 *Conjugacy classes of* $D(\mathsf{Co}_2) = \langle x, y \rangle$

Class	*Representative*	*Class*	*Centralizer*	2P	3P	5P
1	1	1	$2^{18} \cdot 3^2 \cdot 5$	1	1	1
2_1	$(x^3y)^6$	1	$2^{18} \cdot 3^2 \cdot 5$	1	2_1	2_1
2_2	$(x^5y)^5$	16	$2^{14} \cdot 3^2 \cdot 5$	1	2_2	2_2
2_3	$(x^4yxy^2)^5$	16	$2^{14} \cdot 3^2 \cdot 5$	1	2_3	2_3
2_4	$(x^4y^3)^4$	30	$2^{17} \cdot 3$	1	2_4	2_4
2_5	$(xy)^4$	60	$2^{16} \cdot 3$	1	2_5	2_5
2_6	$(x^4yxy)^4$	60	$2^{16} \cdot 3$	1	2_6	2_6
2_7	$(x)^6$	240	$2^{14} \cdot 3$	1	2_7	2_7
2_8	$(x^2y^2)^2$	240	$2^{14} \cdot 3$	1	2_8	2_8
2_9	$(x^2y^4)^3$	240	$2^{14} \cdot 3$	1	2_9	2_9
2_{10}	$(x^4yxy^3)^3$	240	$2^{14} \cdot 3$	1	2_{10}	2_{10}
2_{11}	$(x^2y^3xy^3xy)^3$	240	$2^{14} \cdot 3$	1	2_{11}	2_{11}
2_{12}	$(xy^4)^2$	360	2^{15}	1	2_{12}	2_{12}
2_{13}	$(y)^3$	480	$2^{13} \cdot 3$	1	2_{13}	2_{13}
2_{14}	$(x^4y)^3$	480	$2^{13} \cdot 3$	1	2_{14}	2_{14}
2_{15}	$(x^2yxyxy)^3$	480	$2^{13} \cdot 3$	1	2_{15}	2_{15}
2_{16}	$(xy^2xy^3)^3$	480	$2^{13} \cdot 3$	1	2_{16}	2_{16}
2_{17}	$(x^2yxyxy^2x^2y^2)^2$	720	2^{14}	1	2_{17}	2_{17}
2_{18}	$x^3yxy^2x^2yxyx^2y^3$	1440	2^{13}	1	2_{18}	2_{18}
2_{19}	$(x^2y^3xy)^3$	1920	$2^{11} \cdot 3$	1	2_{19}	2_{19}
2_{20}	$(x^3y^3xy)^2$	2880	2^{12}	1	2_{20}	2_{20}
2_{21}	$x^2yxyxy^2xy^4$	2880	2^{12}	1	2_{21}	2_{21}
2_{22}	$x^3y^4x^2yx^2yxy^2$	2880	2^{12}	1	2_{22}	2_{22}
2_{23}	$xyxyxyxy^4$	5760	2^{11}	1	2_{23}	2_{23}
2_{24}	$x^3yx^2yxyxyx^2y$	5760	2^{11}	1	2_{24}	2_{24}
3_1	$(x^3y)^4$	10240	$2^7 \cdot 3^2$	3_1	1	3_1
3_2	$(x)^4$	40960	$2^5 \cdot 3^2$	3_2	1	3_2
4_1	$(x^3y)^3$	240	$2^{14} \cdot 3$	2_1	4_1	4_1
4_2	$(x^2yxyxy^2)^3$	480	$2^{13} \cdot 3$	2_4	4_2	4_2
4_3	$(x^2yxy^2)^4$	720	2^{14}	2_1	4_3	4_3
4_4	$(x^4y^3xy)^3$	960	$2^{12} \cdot 3$	2_4	4_4	4_4
4_5	$(x^2yxy^3xy^3)^3$	960	$2^{12} \cdot 3$	2_4	4_5	4_5
4_6	$(x^4yx^2y)^2$	1440	2^{13}	2_4	4_6	4_6
4_7	$(x^2yxy^3)^3$	1920	$2^{11} \cdot 3$	2_4	4_7	4_7
4_8	$(xy)^2$	2880	2^{12}	2_5	4_8	4_8
4_9	$(x^4yxy)^2$	2880	2^{12}	2_6	4_9	4_9
4_{10}	x^7yxy	2880	2^{12}	2_4	4_{10}	4_{10}
4_{11}	$x^3yxy^3x^2y^2$	2880	2^{12}	2_5	4_{11}	4_{11}
4_{12}	x^7yxy^4	2880	2^{12}	2_4	4_{12}	4_{12}
4_{13}	x^6yxyxy^3	2880	2^{12}	2_4	4_{13}	4_{13}
4_{14}	$x^2y^2xyxy^4xy$	2880	2^{12}	2_1	4_{14}	4_{14}
4_{15}	$x^5yx^2yx^2yx^2y$	2880	2^{12}	2_4	4_{15}	4_{15}
4_{16}	$x^3yx^3yx^2y^3xy$	2880	2^{12}	2_4	4_{16}	4_{16}
4_{17}	$x^3yxyx^2yx^2y^2xy^2$	2880	2^{12}	2_6	4_{17}	4_{17}
4_{18}	$x^3y^2x^2yxyx^2yxy^2$	2880	2^{12}	2_4	4_{18}	4_{18}
4_{19}	$(x^2y)^2$	5760	2^{11}	2_5	4_{19}	4_{19}
4_{20}	$(x^4y^3)^2$	5760	2^{11}	2_4	4_{20}	4_{20}
4_{21}	x^2yxy^3xy	5760	2^{11}	2_5	4_{21}	4_{21}
4_{22}	$(x^5yx^3y)^2$	5760	2^{11}	2_6	4_{22}	4_{22}
4_{23}	$x^3yxyx^3y^2$	5760	2^{11}	2_5	4_{23}	4_{23}
4_{24}	$x^6yx^2y^2x^2y$	5760	2^{11}	2_4	4_{24}	4_{24}
4_{25}	$x^2yxyxyxy^2x^2y^2$	5760	2^{11}	2_4	4_{25}	4_{25}
4_{26}	$x^5y^5xy^4xy$	5760	2^{11}	2_4	4_{26}	4_{26}
4_{27}	$(x)^3$	7680	$2^9 \cdot 3$	2_7	4_{27}	4_{27}
4_{28}	$(xy^2)^3$	7680	$2^9 \cdot 3$	2_7	4_{28}	4_{28}
4_{29}	$(x^6y)^3$	7680	$2^9 \cdot 3$	2_7	4_{29}	4_{29}
4_{30}	$xyxy^4$	7680	$2^9 \cdot 3$	2_7	4_{30}	4_{30}

Conjugacy classes of $D(\mathsf{Co}_2) = \langle x, y \rangle$ *(continued)*

Class	*Representative*	*Class*	*Centralizer*	2P	3P	5P
4_{31}	x^4y^4	11520	2^{10}	2_8	4_{31}	4_{31}
4_{32}	$(x^3y^2xy^3)^2$	11520	2^{10}	2_8	4_{32}	4_{32}
4_{33}	$x^2y^2xy^4$	11520	2^{10}	2_5	4_{33}	4_{33}
4_{34}	$x^2y^2x^2y^5$	11520	2^{10}	2_8	4_{34}	4_{34}
4_{35}	x^4yx^4yxy	11520	2^{10}	2_5	4_{35}	4_{35}
4_{36}	x^4yxyx^2yxy	11520	2^{10}	2_5	4_{36}	4_{36}
4_{37}	$x^2yxy^2x^2y^2xy$	11520	2^{10}	2_4	4_{37}	4_{37}
4_{38}	$x^2yxy^2xyxy^3$	11520	2^{10}	2_4	4_{38}	4_{38}
4_{39}	x^4yxyx^3yxy	11520	2^{10}	2_4	4_{39}	4_{39}
4_{40}	$x^2yxyx^2y^3xy^2$	11520	2^{10}	2_6	4_{40}	4_{40}
4_{41}	$x^2yxy^4xyxy^2$	11520	2^{10}	2_4	4_{41}	4_{41}
4_{42}	$x^2y^2x^2y^3xy^3$	11520	2^{10}	2_5	4_{42}	4_{42}
4_{43}	$x^4yx^2yx^3y^2xy$	11520	2^{10}	2_8	4_{43}	4_{43}
4_{44}	$(x^3yxyxy^2)^3$	15360	$2^8 \cdot 3$	2_8	4_{44}	4_{44}
4_{45}	x^2y^2	23040	2^9	2_8	4_{45}	4_{45}
4_{46}	$x^3yxyxyx^3y^2$	23040	2^9	2_8	4_{46}	4_{46}
4_{47}	xy^4	46080	2^8	2_{12}	4_{47}	4_{47}
4_{48}	x^3y^3xy	46080	2^8	2_{20}	4_{48}	4_{48}
4_{49}	x^8y	46080	2^8	2_{20}	4_{49}	4_{49}
4_{50}	x^2y^3xyxy	46080	2^8	2_{20}	4_{50}	4_{50}
4_{51}	x^4yxy^2xy	46080	2^8	2_7	4_{51}	4_{51}
4_{52}	x^7yxy^2	46080	2^8	2_{20}	4_{52}	4_{52}
4_{53}	x^6yxyxy	46080	2^8	2_{20}	4_{53}	4_{53}
4_{54}	x^8yx^2y	46080	2^8	2_7	4_{54}	4_{54}
4_{55}	x^7yxy^3	46080	2^8	2_{20}	4_{55}	4_{55}
4_{56}	x^7y^3xy	46080	2^8	2_{20}	4_{56}	4_{56}
4_{57}	x^6y^2xyxy	46080	2^8	2_{20}	4_{57}	4_{57}
4_{58}	$x^2yxyxy^2x^2y^2$	46080	2^8	2_{17}	4_{58}	4_{58}
4_{59}	$x^3yx^3yx^2y^3$	46080	2^8	2_{17}	4_{59}	4_{59}
4_{60}	$x^5yx^3y^3xy$	46080	2^8	2_8	4_{60}	4_{60}
5	xy^3	589824	$2^2 \cdot 5$	5	5	1
6_1	$(x^3y)^2$	10240	$2^7 \cdot 3^2$	3_1	2_1	6_1
6_2	$x^5y^2xy^2$	40960	$2^5 \cdot 3^2$	3_2	2_3	6_2
6_3	$xyxyxyxy^3$	40960	$2^5 \cdot 3^2$	3_2	2_2	6_3
6_4	$x^6y^2x^2y^2$	40960	$2^5 \cdot 3^2$	3_2	2_1	6_4
6_5	$x^3y^2xy^3xy^2$	40960	$2^5 \cdot 3^2$	3_1	2_2	6_5
6_6	$x^2yx^2y^2xyxy^2$	40960	$2^5 \cdot 3^2$	3_1	2_3	6_6
6_7	$(x^2yxy^3)^2$	61440	$2^6 \cdot 3$	3_1	2_4	6_7
6_8	$(x)^2$	122880	$2^5 \cdot 3$	3_2	2_7	6_8
6_9	x^2y^4	122880	$2^5 \cdot 3$	3_2	2_9	6_9
6_{10}	x^4yxy^3	122880	$2^5 \cdot 3$	3_1	2_{10}	6_{10}
6_{11}	$(x^3yxyxy^2)^2$	122880	$2^5 \cdot 3$	3_1	2_8	6_{11}
6_{12}	x^3yxyxy^3	122880	$2^5 \cdot 3$	3_2	2_6	6_{12}
6_{13}	$x^3y^2xy^4$	122880	$2^5 \cdot 3$	3_2	2_5	6_{13}
6_{14}	$x^2y^3xy^3xy$	122880	$2^5 \cdot 3$	3_1	2_{11}	6_{14}
6_{15}	y	245760	$2^4 \cdot 3$	3_2	2_{13}	6_{15}
6_{16}	x^4y	245760	$2^4 \cdot 3$	3_2	2_{14}	6_{16}
6_{17}	x^2yxyxy	245760	$2^4 \cdot 3$	3_2	2_{15}	6_{17}
6_{18}	x^2y^3xy	245760	$2^4 \cdot 3$	3_1	2_{19}	6_{18}
6_{19}	xy^2xy^3	245760	$2^4 \cdot 3$	3_2	2_{16}	6_{19}

Conjugacy classes of $D(\mathsf{Co}_2) = \langle x, y \rangle$ *(continued)*

Class	*Representative*	|*Class*|	|*Centralizer*|	2P	3P	5P
8_1	$(x^3y^2xyxy)^3$	15360	$2^8 \cdot 3$	4_1	8_1	8_1
8_2	$(x^2yxy^2)^2$	23040	2^9	4_3	8_2	8_2
8_3	$(x^3yxy^2)^2$	23040	2^9	4_3	8_3	8_3
8_4	$x^3yx^3y^2$	46080	2^8	4_6	8_4	8_4
8_5	$x^3yx^2yxyxy^2$	46080	2^8	4_3	8_5	8_5
8_6	$x^2yx^2y^2x^2y^3$	46080	2^8	4_1	8_6	8_6
8_7	$x^5yx^2y^2xy^2$	46080	2^8	4_1	8_7	8_7
8_8	$x^4yx^2yxyx^2y$	46080	2^8	4_6	8_8	8_8
8_9	$x^3yx^3y^3x^2y$	46080	2^8	4_3	8_9	8_9
8_{10}	x^2y	92160	2^7	4_{19}	8_{10}	8_{10}
8_{11}	x^4y^3	92160	2^7	4_{20}	8_{11}	8_{11}
8_{12}	x^4yx^2y	92160	2^7	4_6	8_{12}	8_{12}
8_{13}	x^3yxy^3	92160	2^7	4_{20}	8_{13}	8_{13}
8_{14}	x^2yxyxy^3	92160	2^7	4_{20}	8_{14}	8_{14}
8_{15}	x^5yx^3y	92160	2^7	4_{22}	8_{15}	8_{15}
8_{16}	$x^3yxy^2xy^2$	92160	2^7	4_{22}	8_{16}	8_{16}
8_{17}	x^2yxyxy^4	92160	2^7	4_{20}	8_{17}	8_{17}
8_{18}	$x^5y^2x^3y$	92160	2^7	4_{19}	8_{18}	8_{18}
8_{19}	xy	184320	2^6	4_8	8_{19}	8_{19}
8_{20}	x^4yxy	184320	2^6	4_9	8_{20}	8_{20}
8_{21}	$x^3y^2xy^3$	184320	2^6	4_{32}	8_{21}	8_{21}
8_{22}	$x^4y^2xy^3$	184320	2^6	4_{32}	8_{22}	8_{22}
8_{23}	$x^4y^3xy^2$	184320	2^6	4_{31}	8_{23}	8_{23}
8_{24}	$x^4yx^2yxy^2$	184320	2^6	4_{31}	8_{24}	8_{24}
8_{25}	$x^5yx^4y^2xy$	184320	2^6	4_{24}	8_{25}	8_{25}
10_1	x^5y	589824	$2^2 \cdot 5$	5	10_1	2_2
10_2	x^4yxy^2	589824	$2^2 \cdot 5$	5	10_2	2_3
10_3	x^3yxyxy	589824	$2^2 \cdot 5$	5	10_3	2_1
12_1	x^3y	122880	$2^5 \cdot 3$	6_1	4_1	12_1
12_2	x^4y^3xy	122880	$2^5 \cdot 3$	6_7	4_4	12_2
12_3	$x^2yxy^3xy^3$	122880	$2^5 \cdot 3$	6_7	4_5	12_3
12_4	x	245760	$2^4 \cdot 3$	6_8	4_{27}	12_4
12_5	xy^2	245760	$2^4 \cdot 3$	6_8	4_{28}	12_5
12_6	x^6y	245760	$2^4 \cdot 3$	6_8	4_{29}	12_6
12_7	x^2yxy^3	245760	$2^4 \cdot 3$	6_7	4_7	12_7
12_8	x^2yxyxy^2	245760	$2^4 \cdot 3$	6_7	4_2	12_8
12_9	$x^3y^2xy^5$	245760	$2^4 \cdot 3$	6_8	4_{30}	12_9
12_{10}	x^3yxyxy^2	491520	$2^3 \cdot 3$	6_{11}	4_{44}	12_{10}
16_1	x^2yxy^2	368640	2^5	8_2	16_1	16_1
16_2	x^3yxy^2	368640	2^5	8_3	16_2	16_2
24	x^3y^2xyxy	491520	$2^3 \cdot 3$	12_1	8_1	24

4.5.3 *Conjugacy classes of* $E(\mathsf{Co}_2) = \langle x, y, e \rangle$

Class	*Representative*	|*Class*|	|*Centralizer*|	2P	3P	5P	7P	11P
1	1	1	$2^{18}\cdot 3^2\cdot 5\cdot 7\cdot 11$	1	1	1	1	1
2_1	$(xye)^5$	77	$2^{18}\cdot 3^2\cdot 5$	1	2_1	2_1	2_1	2_1
2_2	$(x)^6$	330	$2^{17}\cdot 3\cdot 7$	1	2_2	2_2	2_2	2_2
2_3	$(xe)^{10}$	616	$2^{15}\cdot 3^2\cdot 5$	1	2_3	2_3	2_3	2_3
2_4	$(x^2ey)^7$	2640	$2^{14}\cdot 3\cdot 7$	1	2_4	2_4	2_4	2_4
2_5	$(xy^2e)^7$	2640	$2^{14}\cdot 3\cdot 7$	1	2_5	2_5	2_5	2_5
2_6	$(e)^2$	18480	$2^{14}\cdot 3$	1	2_6	2_6	2_6	2_6
2_7	$(y)^3$	36960	$2^{13}\cdot 3$	1	2_7	2_7	2_7	2_7
2_8	$(xe^3)^5$	44352	$2^{12}\cdot 5$	1	2_8	2_8	2_8	2_8
2_9	$(x^2ye^3y)^2$	55440	2^{14}	1	2_9	2_9	2_9	2_9
3	$(x)^4$	788480	$2^7\cdot 3^2$	3	1	3	3	3
4_1	$(x^3y)^3$	18480	$2^{14}\cdot 3$	2_1	4_1	4_1	4_1	4_1
4_2	$(y^2e^2)^3$	36960	$2^{13}\cdot 3$	2_2	4_2	4_2	4_2	4_2
4_3	$(xe^2)^4$	55440	2^{14}	2_1	4_3	4_3	4_3	4_3
4_4	$(ye)^3$	73920	$2^{12}\cdot 3$	2_2	4_4	4_4	4_4	4_4
4_5	$(xy^2)^3$	73920	$2^{12}\cdot 3$	2_2	4_5	4_5	4_5	4_5
4_6	$(xy)^2$	110880	2^{13}	2_2	4_6	4_6	4_6	4_6
4_7	$(x)^3$	147840	$2^{11}\cdot 3$	2_2	4_7	4_7	4_7	4_7
4_8	x^2yx^2eyxe	221760	2^{12}	2_1	4_8	4_8	4_8	4_8
4_9	$(x^3yey^2)^2$	443520	2^{11}	2_2	4_9	4_9	4_9	4_9
4_{10}	$xyxe^3xe$	443520	2^{11}	2_2	4_{10}	4_{10}	4_{10}	4_{10}
4_{11}	xy^3exe^2	443520	2^{11}	2_2	4_{11}	4_{11}	4_{11}	4_{11}
4_{12}	$(xe)^5$	709632	$2^8\cdot 5$	2_3	4_{12}	4_{12}	4_{12}	4_{12}
4_{13}	e	887040	2^{10}	2_6	4_{13}	4_{13}	4_{13}	4_{13}
4_{14}	$(xy^2e^2)^2$	887040	2^{10}	2_6	4_{14}	4_{14}	4_{14}	4_{14}
4_{15}	$x^2e^2y^3e$	887040	2^{10}	2_6	4_{15}	4_{15}	4_{15}	4_{15}
4_{16}	x^8y	887040	2^{10}	2_6	4_{16}	4_{16}	4_{16}	4_{16}
4_{17}	$(x^3y^2e)^3$	1182720	$2^8\cdot 3$	2_6	4_{17}	4_{17}	4_{17}	4_{17}
4_{18}	x^2y^2	1774080	2^9	2_6	4_{18}	4_{18}	4_{18}	4_{18}
4_{19}	$x^2y^2xe^3$	1774080	2^9	2_6	4_{19}	4_{19}	4_{19}	4_{19}
4_{20}	x^2ye^3y	3548160	2^8	2_9	4_{20}	4_{20}	4_{20}	4_{20}
4_{21}	x^2exeye	3548160	2^8	2_6	4_{21}	4_{21}	4_{21}	4_{21}
4_{22}	xye^2y^3	3548160	2^8	2_9	4_{22}	4_{22}	4_{22}	4_{22}
5	$(xe)^4$	22708224	$2^3\cdot 5$	5	5	1	5	5
6_1	$(x^3y)^2$	788480	$2^7\cdot 3^2$	3	2_1	6_1	6_1	6_1
6_2	x^2y^4	3153920	$2^5\cdot 3^2$	3	2_3	6_2	6_2	6_2
6_3	$xyxe^2y$	3153920	$2^5\cdot 3^2$	3	2_1	6_3	6_3	6_3
6_4	$(x)^2$	4730880	$2^6\cdot 3$	3	2_2	6_4	6_4	6_4
6_5	x^4y	9461760	$2^5\cdot 3$	3	2_4	6_5	6_5	6_5
6_6	x^2yxe	9461760	$2^5\cdot 3$	3	2_6	6_6	6_6	6_6
6_7	x^3e^3	9461760	$2^5\cdot 3$	3	2_5	6_7	6_7	6_7
6_8	y	18923520	$2^4\cdot 3$	3	2_7	6_8	6_8	6_8
7_1	x^2e	32440320	$2^2\cdot 7$	7_1	7_2	7_2	1	7_1
7_2	$(x^2e)^3$	32440320	$2^2\cdot 7$	7_2	7_1	7_1	1	7_2

Conjugacy classes of $E(\mathsf{Co}_2) = \langle x, y, e\rangle$ *(continued)*

Class	*Representative*	|*Class*|	|*Centralizer*|	2P	3P	5P	7P	11P
8_1	$(x^2exe^2)^3$	1182720	$2^8 \cdot 3$	4_1	8_1	8_1	8_1	8_1
8_2	$(xe^2)^2$	1774080	2^9	4_3	8_2	8_2	8_2	8_2
8_3	x^3ye	1774080	2^9	4_3	8_3	8_3	8_3	8_3
8_4	x^2y	3548160	2^8	4_6	8_4	8_4	8_4	8_4
8_5	x^3e^2	3548160	2^8	4_6	8_5	8_5	8_5	8_5
8_6	x^2yey^2	3548160	2^8	4_1	8_6	8_6	8_6	8_6
8_7	x^2eyxe	3548160	2^8	4_3	8_7	8_7	8_7	8_7
8_8	xy^4e^2	3548160	2^8	4_3	8_8	8_8	8_8	8_8
8_9	$x^3y^2e^2y$	3548160	2^8	4_1	8_9	8_9	8_9	8_9
8_{10}	xy	7096320	2^7	4_6	8_{10}	8_{10}	8_{10}	8_{10}
8_{11}	x^2e^2y	14192640	2^6	4_{13}	8_{11}	8_{11}	8_{11}	8_{11}
8_{12}	xy^2e^2	14192640	2^6	4_{14}	8_{12}	8_{12}	8_{12}	8_{12}
8_{13}	$yeye^2$	14192640	2^6	4_{13}	8_{13}	8_{13}	8_{13}	8_{13}
8_{14}	$x^2e^2y^2$	14192640	2^6	4_{14}	8_{14}	8_{14}	8_{14}	8_{14}
8_{15}	x^3yey^2	14192640	2^6	4_9	8_{15}	8_{15}	8_{15}	8_{15}
10_1	$(xe)^2$	22708224	$2^3 \cdot 5$	5	10_1	2_3	10_1	10_1
10_2	xye	45416448	$2^2 \cdot 5$	5	10_2	2_1	10_2	10_2
10_3	xe^3	45416448	$2^2 \cdot 5$	5	10_3	2_8	10_3	10_3
11	xy^2xe	82575360	11	11	11	11	11	1
12_1	ye	9461760	$2^5 \cdot 3$	6_4	4_4	12_1	12_1	12_1
12_2	xy^2	9461760	$2^5 \cdot 3$	6_4	4_5	12_2	12_2	12_2
12_3	x^3y	9461760	$2^5 \cdot 3$	6_1	4_1	12_3	12_3	12_3
12_4	x	18923520	$2^4 \cdot 3$	6_4	4_7	12_4	12_4	12_4
12_5	y^2e^2	18923520	$2^4 \cdot 3$	6_4	4_2	12_5	12_5	12_5
12_6	x^3y^2e	37847040	$2^3 \cdot 3$	6_6	4_{17}	12_6	12_6	12_6
14_1	x^2ey	32440320	$2^2 \cdot 7$	7_1	14_2	14_2	2_4	14_1
14_2	$(x^2ey)^3$	32440320	$2^2 \cdot 7$	7_2	14_1	14_1	2_4	14_2
14_3	xy^2e	32440320	$2^2 \cdot 7$	7_2	14_4	14_4	2_5	14_3
14_4	$(xy^2e)^3$	32440320	$2^2 \cdot 7$	7_1	14_3	14_3	2_5	14_4
14_5	$xeye$	32440320	$2^2 \cdot 7$	7_2	14_6	14_6	2_2	14_5
14_6	$(xeye)^3$	32440320	$2^2 \cdot 7$	7_1	14_5	14_5	2_2	14_6
16_1	xe^2	28385280	2^5	8_2	16_1	16_1	16_1	16_1
16_2	$xyxey$	28385280	2^5	8_3	16_2	16_2	16_2	16_2
20	xe	45416448	$2^2 \cdot 5$	10_1	20	4_{12}	20	20
24	x^2exe^2	37847040	$2^3 \cdot 3$	12_3	8_1	24	24	24

4.6 Character tables of local subgroups of Co_2

4.6.1 *Character table of* $E_3 = E(\mathsf{Co}_2) = \langle x, y, e \rangle$

2	18	18	17	15	14	14	14	13	12	14	7	14	13	14	12	12	13	11	12	11	11	11
3	2	2	1	2	1	1	1	1	.	.	2	1	1	.	1	1	.	1	.	.	.	.
5	1	1	.	1	.	.	.	.	1	.	.	.	.	.	.	.	.	.	.	.	.	.
7	1	.	1	.	1	1	.	.	.	.	.	.	.	.	.	.	.	.	.	.	.	.
11	1	.	.	.	.	.	.	.	.	.	.	.	.	.	.	.	.	.	.	.	.	.
	1a	2a	2b	2c	2d	2e	2f	2g	2h	2i	3a	4a	4b	4c	4d	4e	4f	4g	4h	4i	4j	4k
2P	1a	1a	1a	1a	1a	1a	1a	1a	1a	1a	3a	2a	2b	2a	2b	2b	2b	2b	2a	2b	2b	2b
3P	1a	2a	2b	2c	2d	2e	2f	2g	2h	2i	1a	4a	4b	4c	4d	4e	4f	4g	4h	4i	4j	4k
5P	1a	2a	2b	2c	2d	2e	2f	2g	2h	2i	3a	4a	4b	4c	4d	4e	4f	4g	4h	4i	4j	4k
7P	1a	2a	2b	2c	2d	2e	2f	2g	2h	2i	3a	4a	4b	4c	4d	4e	4f	4g	4h	4i	4j	4k
11P	1a	2a	2b	2c	2d	2e	2f	2g	2h	2i	3a	4a	4b	4c	4d	4e	4f	4g	4h	4i	4j	4k
X.1	1	1	1	1	1	1	1	1	1	1	1	1	1	1	1	1	1	1	1	1	1	1
X.2	1	1	1	1	−1	−1	1	−1	−1	1	1	1	1	1	−1	−1	1	−1	−1	1	1	−1
X.3	21	21	21	21	−7	−7	5	−7	1	5	3	5	5	5	−7	−7	5	−7	1	5	5	1
X.4	21	21	21	21	7	7	5	7	−1	5	3	5	5	5	7	7	5	7	−1	5	5	−1
X.5	22	−10	6	−2	8	−8	6	.	.	−2	4	6	−6	−2	−4	4	2	.	.	2	−2	.
X.6	22	−10	6	−2	−8	8	6	.	.	−2	4	6	−6	−2	4	−4	2	.	.	2	−2	.
X.7	45	45	45	45	−3	−3	−3	−3	5	−3	.	−3	−3	−3	−3	−3	−3	−3	5	−3	−3	5
X.8	45	45	45	45	−3	−3	−3	−3	5	−3	.	−3	−3	−3	−3	−3	−3	−3	5	−3	−3	5
X.9	45	45	45	45	3	3	−3	3	−5	−3	.	−3	−3	−3	3	3	−3	3	−5	−3	−3	−5
X.10	45	45	45	45	3	3	−3	3	−5	−3	.	−3	−3	−3	3	3	−3	3	−5	−3	−3	−5
X.11	55	55	55	55	13	13	7	13	5	7	1	7	7	7	13	13	7	13	5	7	7	5
X.12	55	55	55	55	−13	−13	7	−13	−5	7	1	7	7	7	−13	−13	7	−13	−5	7	7	−5
X.13	99	99	99	99	15	15	3	15	−1	3	.	3	3	3	15	15	3	15	−1	3	3	−1
X.14	99	99	99	99	−15	−15	3	−15	1	3	.	3	3	3	−15	−15	3	−15	1	3	3	1
X.15	154	154	154	154	14	14	10	14	6	10	1	10	10	10	14	14	10	14	6	10	10	6
X.16	154	154	154	154	−14	−14	10	−14	−6	10	1	10	10	10	−14	−14	10	−14	−6	10	10	−6
X.17	210	210	210	210	−14	−14	2	−14	10	2	3	2	2	2	−14	−14	2	−14	10	2	2	10
X.18	210	210	210	210	14	14	2	14	−10	2	3	2	2	2	14	14	2	14	−10	2	2	−10
X.19	231	39	7	−9	−21	−21	7	11	11	−9	6	23	15	7	−5	−5	−1	3	−5	−1	−1	3
X.20	231	39	7	−9	−35	−35	23	−3	−11	7	6	7	15	−9	−3	−3	−1	5	5	−1	−1	−3
X.21	231	231	231	231	−7	−7	7	−7	9	7	−3	7	7	7	−7	−7	7	−7	9	7	7	9
X.22	231	39	7	−9	35	35	23	3	11	7	6	7	15	−9	3	3	−1	−5	−5	−1	−1	3
X.23	231	39	7	−9	21	21	7	−11	−11	−9	6	23	15	7	5	5	−1	−3	5	−1	−1	−3
X.24	231	231	231	231	7	7	7	7	−9	7	−3	7	7	7	7	7	7	7	−9	7	7	−9
X.25	385	385	385	385	21	21	1	21	5	1	−2	1	1	1	21	21	1	21	5	1	1	5
X.26	385	385	385	385	−21	−21	1	−21	−5	1	−2	1	1	1	−21	−21	1	−21	−5	1	1	−5
X.27	440	−200	120	−40	−48	48	24	.	.	−8	8	24	−24	−8	24	−24	8	.	.	8	−8	.
X.28	440	−200	120	−40	48	−48	24	.	.	−8	8	24	−24	−8	−24	24	8	.	.	8	−8	.
X.29	560	560	560	560	.	.	−16	.	.	−16	2	−16	−16	−16	.	.	−16	.	.	−16	−16	.
X.30	770	−30	−14	10	84	−28	34	−4	20	−6	5	−14	−10	10	.	−8	−2	4	4	−2	2	−4
X.31	770	−350	210	−70	56	−56	18	.	.	−6	−4	18	−18	−6	−28	28	6	.	.	6	−6	.
X.32	770	−30	−14	10	28	−84	34	4	−20	−6	5	−14	−10	10	8	.	−2	−4	−4	−2	2	4
X.33	770	−30	−14	10	−84	28	34	4	−20	−6	5	−14	−10	10	.	8	−2	−4	−4	−2	2	4
X.34	770	−30	−14	10	−28	84	34	−4	20	−6	5	−14	−10	10	−8	.	−2	4	4	−2	2	−4
X.35	770	−350	210	−70	−56	56	18	.	.	−6	−4	18	−18	−6	28	−28	6	.	.	6	−6	.
X.36	924	156	28	−36	−84	−84	28	−20	−4	28	6	−4	12	−4	−4	−4	12	12	−4	−4	−4	−4
X.37	924	156	28	−36	−28	−28	−4	36	4	−4	6	28	12	28	−12	−12	12	4	4	−4	−4	4
X.38	924	156	28	−36	84	84	28	20	4	28	6	−4	12	−4	4	4	12	−12	4	−4	−4	4
X.39	924	156	28	−36	28	28	−4	−36	−4	−4	6	28	12	28	12	12	12	−4	−4	−4	−4	−4
X.40	990	−450	270	−90	−24	24	−18	.	.	6	.	−18	18	6	12	−12	−6	.	.	−6	6	.
X.41	990	−450	270	−90	−24	24	−18	.	.	6	.	−18	18	6	12	−12	−6	.	.	−6	6	.
X.42	990	−450	270	−90	24	−24	−18	.	.	6	.	−18	18	6	−12	12	−6	.	.	−6	6	.
X.43	990	−450	270	−90	24	−24	−18	.	.	6	.	−18	18	6	−12	12	−6	.	.	−6	6	.
X.44	1155	195	35	−45	63	63	35	31	15	19	−6	19	27	3	−1	−1	11	−9	−1	−5	−5	7
X.45	1155	195	35	−45	7	7	19	39	15	3	−6	35	27	19	−9	−9	11	−1	−1	−5	−5	7
X.46	1155	195	35	−45	−7	−7	19	−39	−15	3	−6	35	27	19	9	9	11	1	1	−5	−5	−7
X.47	1155	195	35	−45	−63	−63	35	−31	−15	19	−6	19	27	3	1	1	11	9	1	−5	−5	−7
X.48	1386	234	42	−54	42	42	−6	42	−14	26	.	−6	−6	26	−6	−6	26	−6	18	−6	−6	2
X.49	1386	234	42	−54	−42	−42	−6	−42	14	26	.	−6	−6	26	6	6	26	6	−18	−6	−6	−2
X.50	1408	−640	384	−128	−64	64	.	.	.	.	4	.	.	.	32	−32	.	.	.	.	.	.
X.51	1408	−640	384	−128	64	−64	.	.	.	.	4	.	.	.	−32	32	.	.	.	.	.	.
X.52	1540	−60	−28	20	.	.	−28	.	.	20	10	68	−20	−12	.	.	−4	.	.	−4	4	.
X.53	1540	−700	420	−140	.	.	36	.	.	−12	−8	36	−36	−12	.	.	12	.	.	12	−12	.
X.54	2772	−1260	756	−252	.	.	−12	.	.	4	.	−12	12	4	.	.	−4	.	.	−4	4	.
X.55	3080	−120	−56	40	56	56	40	−8	40	8	−7	40	−40	8	−8	−8	−8	8	8	−8	8	−8
X.56	3080	−120	−56	40	−112	112	40	.	.	8	2	40	−40	8	−8	8	−8	.	.	−8	8	.
X.57	3080	−120	−56	40	−56	−56	40	8	−40	8	−7	40	−40	8	8	8	−8	−8	−8	−8	8	8
X.58	3080	−120	−56	40	112	−112	40	.	.	8	2	40	−40	8	8	−8	−8	.	.	−8	8	.
X.59	3465	585	105	−135	105	105	9	9	25	25	.	−39	−15	−23	9	9	1	−15	−23	1	1	1
X.60	3465	585	105	−135	−105	−105	9	−9	−25	25	.	−39	−15	−23	−9	−9	1	15	23	1	1	−1
X.61	3465	585	105	−135	63	63	−39	−33	−25	−23	.	9	−15	25	15	15	1	−9	23	1	1	−1
X.62	3465	585	105	−135	63	63	9	−33	15	−39	.	57	33	9	15	15	−15	−9	−1	1	1	7
X.63	3465	585	105	−135	−105	−105	57	−9	15	9	.	9	33	−39	−9	−9	−15	15	−1	1	1	7
X.64	3465	585	105	−135	−63	−63	−39	33	25	−23	.	9	−15	25	−15	−15	1	9	−23	1	1	1
X.65	3465	585	105	−135	105	105	57	9	−15	9	.	9	33	−39	9	9	−15	−15	1	1	1	−7
X.66	3465	585	105	−135	−63	−63	9	33	−15	−39	.	57	33	9	−15	−15	−15	9	1	1	1	−7
X.67	4620	−180	−84	60	−112	−112	44	16	.	−4	3	−52	4	28	16	16	−12	−16	.	4	−4	.
X.68	4620	−180	−84	60	112	112	44	−16	.	−4	3	−52	4	28	−16	−16	−12	16	.	4	−4	.
X.69	4620	−180	−84	60	56	56	−52	−8	−40	28	3	44	4	−4	−8	−8	−12	8	−8	4	−4	8
X.70	4620	−180	−84	60	−56	−56	−52	8	40	28	3	44	4	−4	8	8	−12	−8	8	4	−4	−8
X.71	6930	1170	210	−270	−42	−42	−30	−42	30	2	.	−30	−30	2	6	6	2	6	−2	2	2	14
X.72	6930	1170	210	−270	42	42	−30	42	−30	2	.	−30	−30	2	−6	−6	2	−6	2	2	2	−14
X.73	6930	−270	−126	90	84	−252	18	12	20	−22	.	−30	6	−6	24	.	14	−12	4	−2	2	−4
X.74	6930	−270	−126	90	−252	84	18	12	20	−22	.	−30	6	−6	.	24	14	−12	4	−2	2	−4
X.75	6930	−270	−126	90	−84	252	18	−12	−20	−22	.	−30	6	−6	−24	.	14	12	−4	−2	2	4
X.76	6930	−270	−126	90	252	−84	18	−12	−20	−22	.	−30	6	−6	.	−24	14	12	−4	−2	2	4
X.77	9240	−360	−168	120	56	56	−8	−8	40	24	−3	−8	8	24	−8	−8	−24	8	8	8	−8	−8
X.78	9240	−360	−168	120	−56	−56	−8	8	−40	24	−3	−8	8	24	8	8	−24	−8	−8	8	−8	8
X.79	13860	−540	−252	180	.	.	−60	.	.	−12	.	36	12	−44	.	.	28	.	.	−4	4	.

Character table of $E(\mathsf{Co}_2)$ *(continued)*

2	8	10	10	10	10	8	9	9	8	8	8	3	7	5	5	6	5	5	5	4	2	2	8	9	9	8	8	8	8	8
3	.	.	.	.	.	1	.	.	.	.	.	.	2	2	2	1	1	1	1	1	.	.	1	.	.	.	.	.	.	.
5	1	.	.	.	.	.	.	.	.	.	.	1	.	.	.	.	.	.	.	.	.	.	.	.	.	.	.	.	.	.
7	.	.	.	.	.	.	.	.	.	.	.	.	.	.	.	.	.	.	.	.	1	1	.	.	.	.	.	.	.	.
11	.	.	.	.	.	.	.	.	.	.	.	.	.	.	.	.	.	.	.	.	.	.	.	.	.	.	.	.	.	.
	4l	4m	4n	4o	4p	4q	4r	4s	4t	4u	4v	5a	6a	6b	6c	6d	6e	6f	6g	6h	7a	7b	8a	8b	8c	8d	8e	8f	8g	8h
2P	2c	2f	2f	2f	2f	2f	2f	2f	2i	2f	2i	5a	3a	3a	3a	3a	3a	3a	3a	3a	7a	7b	4a	4c	4c	4f	4f	4a	4c	4c
3P	4l	4m	4n	4o	4p	4q	4r	4s	4t	4u	4v	5a	2a	2c	2a	2b	2d	2f	2e	2g	7b	7a	8a	8b	8c	8d	8e	8f	8g	8h
5P	4l	4m	4n	4o	4p	4q	4r	4s	4t	4u	4v	1a	6a	6b	6c	6d	6e	6f	6g	6h	7b	7a	8a	8b	8c	8d	8e	8f	8g	8h
7P	4l	4m	4n	4o	4p	4q	4r	4s	4t	4u	4v	5a	6a	6b	6c	6d	6e	6f	6g	6h	1a	1a	8a	8b	8c	8d	8e	8f	8g	8h
11P	4l	4m	4n	4o	4p	4q	4r	4s	4t	4u	4v	5a	6a	6b	6c	6d	6e	6f	6g	6h	7a	7b	8a	8b	8c	8d	8e	8f	8g	8h
X.1	1	1	1	1	1	1	1	1	1	1	1	1	1	1	1	1	1	1	1	1	1	1	1	1	1	1	1	1	1	1
X.2	−1	1	1	−1	−1	−1	1	−1	1	1	−1	1	1	1	1	1	−1	1	−1	−1	1	1	−1	1	1	−1	−1	−1	1	−1
X.3	1	1	1	−3	−3	1	1	−3	1	1	1	1	3	3	3	3	−1	−1	−1	−1	.	.	1	1	1	−3	−3	−3	1	1
X.4	−1	1	1	3	3	−1	1	3	1	1	−1	1	3	3	3	3	1	−1	1	1	.	.	−1	1	1	3	3	3	1	−1
X.5	.	2	2	−4	4	.	−2	.	−2	2	.	2	−4	−2	2	.	2	.	−2	.	1	1	.	2	−2	−2	2	.	2	.
X.6	.	2	2	4	−4	.	−2	.	−2	2	.	2	−4	−2	2	.	−2	.	2	.	1	1	.	2	−2	2	−2	.	2	.
X.7	5	1	1	1	1	−3	1	1	1	1	−3	.	.	.	.	.	.	.	.	.	A	$\bar{A}$	−3	1	1	1	1	1	1	−3
X.8	5	1	1	1	1	−3	1	1	1	1	−3	.	.	.	.	.	.	.	.	.	$\bar{A}$	A	−3	1	1	1	1	1	1	−3
X.9	−5	1	1	−1	−1	3	1	−1	1	1	3	.	.	.	.	.	.	.	.	.	A	$\bar{A}$	3	1	1	−1	−1	−1	1	3
X.10	−5	1	1	−1	−1	3	1	−1	1	1	3	.	.	.	.	.	.	.	.	.	$\bar{A}$	A	3	1	1	−1	−1	−1	1	3
X.11	5	3	3	1	1	1	3	1	−1	−1	1	.	1	1	1	1	1	1	1	1	−1	−1	1	3	3	1	1	1	−1	1
X.12	−5	3	3	−1	−1	−1	3	−1	−1	−1	−1	.	1	1	1	1	−1	1	−1	−1	−1	−1	−1	3	3	−1	−1	−1	−1	−1
X.13	−1	3	3	−1	−1	3	3	−1	−1	−1	3	−1	.	.	.	.	.	.	.	.	1	1	3	3	3	−1	−1	−1	−1	3
X.14	1	3	3	1	1	−3	3	1	−1	−1	−3	−1	.	.	.	.	.	.	.	.	1	1	−3	3	3	1	1	1	−1	−3
X.15	6	−2	−2	2	2	2	−2	2	2	2	2	−1	1	1	1	1	−1	1	−1	−1	.	.	2	−2	−2	2	2	2	2	2
X.16	−6	−2	−2	−2	−2	−2	−2	−2	2	2	−2	−1	1	1	1	1	1	1	1	1	.	.	−2	−2	−2	−2	−2	−2	2	−2
X.17	10	−2	−2	−2	−2	2	−2	−2	−2	−2	2	.	3	3	3	3	1	−1	1	1	.	.	2	−2	−2	−2	−2	−2	−2	2
X.18	−10	−2	−2	2	2	−2	−2	2	−2	−2	−2	.	3	3	3	3	−1	−1	−1	−1	.	.	−2	−2	−2	2	2	2	−2	−2
X.19	−1	−1	−1	−5	−5	3	−1	3	3	−1	−1	1	6	.	.	−2	.	−2	.	2	.	.	3	3	3	−1	−1	3	3	−1
X.20	1	3	3	−7	−7	−3	3	1	−1	3	1	1	6	.	.	−2	−2	2	−2	.	.	.	−3	−1	−1	1	1	1	−1	1
X.21	9	−1	−1	1	1	1	−1	1	−1	−1	1	1	−3	−3	−3	−3	−1	1	−1	−1	.	.	1	−1	−1	1	1	1	−1	1
X.22	−1	3	3	7	7	3	3	−1	−1	3	−1	1	6	.	.	−2	2	2	2	.	.	.	3	−1	−1	−1	−1	−1	−1	−1
X.23	1	−1	−1	5	5	−3	−1	−3	3	−1	1	1	6	.	.	−2	.	−2	.	−2	.	.	−3	3	3	1	1	−3	3	1
X.24	−9	−1	−1	−1	−1	−1	−1	−1	−1	−1	−1	1	−3	−3	−3	−3	1	1	1	1	.	.	−1	−1	−1	−1	−1	−1	−1	−1
X.25	5	1	1	−3	−3	−3	1	−3	1	1	−3	.	−2	−2	−2	−2	.	−2	.	.	.	.	−3	1	1	−3	−3	−3	1	−3
X.26	−5	1	1	3	3	3	1	3	1	1	3	.	−2	−2	−2	−2	.	−2	.	.	.	.	3	1	1	3	3	3	1	3
X.27	.	.	.	8	−8	.	.	.	.	.	.	.	−8	−4	4	.	.	.	.	.	−1	−1	.	.	.	4	−4	.	.	.
X.28	.	.	.	−8	8	.	.	.	.	.	.	.	−8	−4	4	.	.	.	.	.	−1	−1	.	.	.	−4	4	.	.	.
X.29	.	.	.	.	.	.	.	.	.	.	.	.	2	2	2	2	.	2	.	.	.	.	.	.	.	.	.	.	.	.
X.30	.	10	−6	−4	4	4	−2	.	2	2	.	.	−3	1	−3	1	3	1	−1	−1	.	.	−4	−2	2	2	−2	.	−2	.
X.31	.	6	6	4	−4	.	−6	.	2	−2	.	.	4	2	−2	.	2	.	−2	.	.	.	.	6	−6	2	−2	.	−2	.
X.32	.	−6	10	−4	4	−4	−2	.	2	2	.	.	−3	1	−3	1	1	1	−3	1	.	.	4	−2	2	2	−2	.	−2	.
X.33	.	10	−6	4	−4	−4	−2	.	2	2	.	.	−3	1	−3	1	−3	1	1	1	.	.	4	−2	2	−2	2	.	−2	.
X.34	.	−6	10	4	−4	4	−2	.	2	2	.	.	−3	1	−3	1	−1	1	3	−1	.	.	−4	−2	2	−2	2	.	−2	.
X.35	.	6	6	−4	4	.	−6	.	2	−2	.	.	4	2	−2	.	−2	.	2	.	.	.	.	6	−6	−2	2	.	−2	.
X.36	4	4	4	−4	−4	.	4	−4	.	.	.	−1	6	.	.	−2	.	−2	.	−2	.	.	.	−4	−4	.	.	4	.	.
X.37	−4	−4	−4	−4	−4	.	−4	−4	.	.	.	−1	6	.	.	−2	2	2	2	.	.	.	.	4	4	.	.	4	.	.
X.38	−4	4	4	4	4	.	4	4	.	.	.	−1	6	.	.	−2	.	−2	.	2	.	.	.	−4	−4	.	.	−4	.	.
X.39	4	−4	−4	4	4	.	−4	4	.	.	.	−1	6	.	.	−2	−2	2	−2	.	.	.	.	4	4	.	.	−4	.	.
X.40	.	2	2	−4	4	.	−2	.	−2	2	.	.	.	.	.	.	.	.	.	.	A	$\bar{A}$	.	2	−2	−2	2	.	2	.
X.41	.	2	2	−4	4	.	−2	.	−2	2	.	.	.	.	.	.	.	.	.	.	$\bar{A}$	A	.	2	−2	−2	2	.	2	.
X.42	.	2	2	4	−4	.	−2	.	−2	2	.	.	.	.	.	.	.	.	.	.	A	$\bar{A}$	.	2	−2	2	−2	.	2	.
X.43	.	2	2	4	−4	.	−2	.	−2	2	.	.	.	.	.	.	.	.	.	.	$\bar{A}$	A	.	2	−2	2	−2	.	2	.
X.44	−5	3	3	−1	−1	3	3	7	3	−1	−1	.	−6	.	.	2	.	2	.	−2	.	.	3	−1	−1	−1	−1	−1	3	−1
X.45	−5	−1	−1	3	3	3	−1	−5	−1	3	−1	.	−6	.	.	2	−2	−2	−2	.	.	.	3	3	3	−1	−1	3	−1	−1
X.46	5	−1	−1	−3	−3	−3	−1	5	−1	3	1	.	−6	.	.	2	2	−2	2	.	.	.	−3	3	3	1	1	−3	−1	1
X.47	5	3	3	1	1	−3	3	−7	3	−1	1	.	−6	.	.	2	.	2	.	2	.	.	−3	−1	−1	1	1	1	3	1
X.48	−6	−2	−2	−2	−2	−6	−2	−2	−2	−2	2	1	.	.	.	.	.	.	.	.	.	.	−6	−2	−2	2	2	−2	−2	2
X.49	6	−2	−2	2	2	6	−2	2	−2	−2	−2	1	.	.	.	.	.	.	.	.	.	.	6	−2	−2	−2	−2	2	−2	−2
X.50	.	.	.	.	.	.	.	.	.	.	.	−2	−4	−2	2	.	2	.	−2	.	1	1	.	.	.	.	.	.	.	.
X.51	.	.	.	.	.	.	.	.	.	.	.	−2	−4	−2	2	.	−2	.	2	.	1	1	.	.	.	.	.	.	.	.
X.52	.	−4	−4	.	.	.	4	.	−4	−4	.	.	−6	2	−6	2	.	2	.	.	.	.	.	4	−4	.	.	.	4	.
X.53	.	−4	−4	.	.	.	4	.	−4	4	.	.	8	4	−4	.	.	.	.	.	.	.	.	−4	4	.	.	.	4	.
X.54	.	−4	−4	.	.	.	4	.	4	−4	.	2	.	.	.	.	.	.	.	.	.	.	.	−4	4	.	.	.	−4	.
X.55	.	.	.	.	.	8	.	.	.	.	.	.	9	−5	3	1	−1	1	−1	1	.	.	−8	.	.	.	.	.	.	.
X.56	.	.	.	8	−8	.	.	.	.	.	.	.	−6	4	.	−2	2	−2	−2	.	.	.	.	.	.	−4	4	.	.	.
X.57	.	.	.	.	.	−8	.	.	.	.	.	.	9	−5	3	1	1	1	1	−1	.	.	8	.	.	.	.	.	.	.
X.58	.	.	.	−8	8	.	.	.	.	.	.	.	−6	4	.	−2	−2	−2	2	.	.	.	.	.	.	4	−4	.	.	.
X.59	5	1	1	−7	−7	−3	1	1	1	−3	1	.	.	.	.	.	.	.	.	.	.	.	−3	5	5	1	1	1	1	1
X.60	−5	1	1	7	7	3	1	−1	1	−3	−1	.	.	.	.	.	.	.	.	.	.	.	3	5	5	−1	−1	−1	1	−1
X.61	−5	5	5	−5	−5	3	5	3	−3	1	−1	.	.	.	.	.	.	.	.	.	.	.	3	1	1	−1	−1	3	−3	−1
X.62	−5	1	1	−1	−1	3	1	−9	1	−3	−1	.	.	.	.	.	.	.	.	.	.	.	3	−3	−3	3	3	−1	1	−1
X.63	−5	−3	−3	−5	−5	3	−3	3	−3	1	−1	.	.	.	.	.	.	.	.	.	.	.	3	1	1	3	3	−5	−3	−1
X.64	5	5	5	5	5	−3	5	−3	−3	1	1	.	.	.	.	.	.	.	.	.	.	.	−3	1	1	1	1	−3	−3	1
X.65	5	−3	−3	5	5	−3	−3	−3	−3	1	1	.	.	.	.	.	.	.	.	.	.	.	−3	1	1	−3	−3	5	−3	1
X.66	5	1	1	1	1	−3	1	9	1	−3	1	.	.	.	.	.	.	.	.	.	.	.	−3	−3	−3	−3	−3	1	1	1
X.67	.	−4	−4	.	.	4	4	.	.	.	−4	.	3	−3	−3	3	−1	−1	−1	1	.	.	−4	4	−4	.	.	.	.	4
X.68	.	−4	−4	.	.	−4	4	.	.	.	4	.	3	−3	−3	3	1	−1	1	−1	.	.	4	4	−4	.	.	.	.	−4
X.69	.	4	4	.	.	4	−4	.	.	.	4	.	3	−3	−3	3	−1	−1	−1	1	.	.	−4	−4	4	.	.	.	.	−4
X.70	.	4	4	.	.	−4	−4	.	.	.	−4	.	3	−3	−3	3	1	−1	1	−1	.	.	4	−4	4	.	.	.	.	4
X.71	−10	−2	−2	2	2	−6	−2	2	2	2	2	.	.	.	.	.	.	.	.	.	.	.	−6	−2	−2	−2	−2	2	2	2
X.72	10	−2	−2	−2	−2	6	−2	−2	2	2	−2	.	.	.	.	.	.	.	.	.	.	.	6	−2	−2	2	2	−2	2	−2
X.73	.	−6	10	4	−4	.	−2	.	−2	−2	4	.	.	.	.	.	.	.	.	.	.	.	.	−2	2	−2	2	.	2	−4
X.74	.	10	−6	−4	4	.	−2	.	−2	−2	4	.	.	.	.	.	.	.	.	.	.	.	.	−2	2	2	−2	.	2	−4
X.75	.	−6	10	−4	4	.	−2	.	−2	−2	−4	.	.	.	.	.	.	.	.	.	.	.	.	−2	2	2	−2	.	2	4
X.76	.	10	−6	4	−4	.	−2	.	−2	−2	−4	.	.	.	.	.	.	.	.	.	.	.	.	−2	2	−2	2	.	2	4
X.77	.	.	.	.	.	−8	.	.	.	.	.	.	−3	3	3	−3	−1	1	−1	1	.	.	8	.	.	.	.	.	.	.
X.78	.	.	.	.	.	8	.	.	.	.	.	.	−3	3	3	−3	1	1	1	−1	.	.	−8	.	.	.	.	.	.	.
X.79	.	−4	−4	.	.	.	4	.	4	4	.	.	.	.	.	.	.	.	.	.	.	.	.	4	−4	.	.	.	−4	.

Character table of $E(Co_2)$ (continued)

	8i	8j	8k	8l	8m	8n	8o	10a	10b	10c	11a	12a	12b	12c	12d	12e	12f	14a	14b	14c	14d	14e	14f	16a	16b	20a	24a
2	8	7	6	6	6	6	6	3	2	2	.	5	5	5	4	4	3	2	2	2	2	2	2	5	5	2	3
3	.	.	.	.	.	.	.	.	.	.	.	1	1	1	1	1	1	.	.	.	.	.	.	.	.	.	1
5	.	.	.	.	.	.	.	1	1	1	.	.	.	.	.	.	.	.	.	.	.	.	.	.	.	1	.
7	.	.	.	.	.	.	.	.	.	.	.	.	.	.	.	.	.	1	1	1	1	1	1	.	.	.	.
11	.	.	.	.	.	.	.	.	.	.	1	.	.	.	.	.	.	.	.	.	.	.	.	.	.	.	.
2P	4a	4f	4m	4n	4m	4n	4i	5a	5a	5a	11a	6d	6d	6a	6d	6d	6f	7a	7b	7b	7a	7b	7a	8b	8c	10a	12c
3P	8i	8j	8k	8l	8m	8n	8o	10a	10b	10c	11a	4d	4e	4a	4g	4b	4q	14b	14a	14d	14c	14f	14e	16a	16b	20a	8a
5P	8i	8j	8k	8l	8m	8n	8o	2c	2a	2h	11a	12a	12b	12c	12d	12e	12f	14b	14a	14d	14c	14f	14e	16a	16b	4l	24a
7P	8i	8j	8k	8l	8m	8n	8o	10a	10b	10c	11a	12a	12b	12c	12d	12e	12f	2d	2d	2e	2e	2b	2b	16a	16b	20a	24a
11P	8i	8j	8k	8l	8m	8n	8o	10a	10b	10c	1a	12a	12b	12c	12d	12e	12f	14a	14b	14c	14d	14e	14f	16a	16b	20a	24a
X.1	1	1	1	1	1	1	1	1	1	1	1	1	1	1	1	1	1	1	1	1	1	1	1	1	1	1	1
X.2	1	1	−1	−1	1	1	1	1	1	−1	1	−1	−1	1	−1	1	−1	−1	−1	−1	−1	1	1	−1	1	−1	−1
X.3	1	1	−1	−1	−1	−1	1	1	1	1	−1	−1	−1	−1	−1	−1	1	.	.	.	.	.	.	−1	−1	1	1
X.4	1	1	1	1	−1	−1	1	1	1	−1	−1	1	1	−1	1	−1	−1	.	.	.	.	.	.	1	−1	−1	−1
X.5	−2	.	2	−2	.	.	.	−2	.	.	.	2	−2	.	.	.	.	1	1	−1	−1	−1	−1	.	.	.	.
X.6	−2	.	−2	2	.	.	.	−2	.	.	.	−2	2	.	.	.	.	−1	−1	1	1	−1	−1	.	.	.	.
X.7	1	1	−1	−1	−1	−1	1	.	.	.	1	.	.	.	.	.	.	$-A$	$-\bar{A}$	$-\bar{A}$	$-A$	$\bar{A}$	A	−1	−1	.	.
X.8	1	1	−1	−1	−1	−1	1	.	.	.	1	.	.	.	.	.	.	$-\bar{A}$	$-A$	$-A$	$-\bar{A}$	A	$\bar{A}$	−1	−1	.	.
X.9	1	1	1	1	−1	−1	1	.	.	.	1	.	.	.	.	.	.	A	$\bar{A}$	$\bar{A}$	A	$\bar{A}$	A	1	−1	.	.
X.10	1	1	1	1	−1	−1	1	.	.	.	1	.	.	.	.	.	.	$\bar{A}$	A	A	$\bar{A}$	A	$\bar{A}$	1	−1	.	.
X.11	−1	3	−1	−1	1	1	−1	.	.	.	.	1	1	1	1	1	1	−1	−1	−1	−1	−1	−1	−1	1	.	1
X.12	−1	3	1	1	1	1	−1	.	.	.	.	−1	−1	1	−1	1	−1	1	1	1	1	−1	−1	1	1	.	−1
X.13	−1	3	−1	−1	−1	−1	−1	−1	−1	−1	.	.	.	.	.	.	.	1	1	1	1	1	1	−1	−1	−1	.
X.14	−1	3	1	1	−1	−1	−1	−1	−1	1	.	.	.	.	.	.	.	−1	−1	−1	−1	1	1	1	−1	1	.
X.15	2	−2	.	.	.	.	2	−1	−1	1	.	−1	−1	1	−1	1	−1	.	.	.	.	.	.	.	.	1	−1
X.16	2	−2	.	.	.	.	2	−1	−1	−1	.	1	1	1	1	1	1	.	.	.	.	.	.	.	.	−1	1
X.17	−2	−2	.	.	.	.	−2	.	.	.	1	1	1	−1	1	−1	−1	.	.	.	.	.	.	.	.	.	−1
X.18	−2	−2	.	.	.	.	−2	.	.	.	1	−1	−1	−1	−1	−1	1	.	.	.	.	.	.	.	.	.	1
X.19	−1	−1	−1	−1	−1	−1	−1	1	−1	1	.	−2	−2	2	.	.	.	.	.	.	.	.	.	1	1	−1	.
X.20	3	−1	−1	−1	1	1	−1	1	−1	−1	.	.	.	−2	2	.	.	.	.	.	.	.	.	1	−1	1	.
X.21	−1	−1	1	1	−1	−1	−1	1	1	−1	.	−1	−1	1	−1	1	1	.	.	.	.	.	.	1	−1	−1	1
X.22	3	−1	1	1	1	1	−1	1	−1	1	.	.	.	−2	−2	.	.	.	.	.	.	.	.	−1	−1	−1	.
X.23	−1	−1	1	1	−1	−1	−1	1	−1	−1	.	2	2	2	.	.	.	.	.	.	.	.	.	−1	1	1	.
X.24	−1	−1	−1	−1	−1	−1	−1	1	1	1	.	1	1	1	1	1	−1	.	.	.	.	.	.	−1	−1	1	−1
X.25	1	1	1	1	1	1	1	.	.	.	.	.	.	−2	.	−2	.	.	.	.	.	.	.	1	1	.	.
X.26	1	1	−1	−1	1	1	1	.	.	.	.	.	.	−2	.	−2	.	.	.	.	.	.	.	−1	1	.	.
X.27	.	.	.	.	.	.	.	.	.	.	.	.	.	.	.	.	.	1	1	−1	−1	1	1	.	.	.	.
X.28	.	.	.	.	.	.	.	.	.	.	.	.	.	.	.	.	.	−1	−1	1	1	1	1	.	.	.	.
X.29	.	.	.	.	.	.	.	.	.	.	−1	.	.	2	.	2	.	.	.	.	.	.	.	.	.	.	.
X.30	−2	.	.	.	2	−2	.	.	.	.	.	−3	1	1	1	−1	1	.	.	.	.	.	.	.	.	.	−1
X.31	2	.	−2	2	.	.	.	.	.	.	.	2	−2	.	.	.	.	.	.	.	.	.	.	.	.	.	.
X.32	−2	.	.	.	−2	2	.	.	.	.	.	−1	3	1	−1	−1	−1	.	.	.	.	.	.	.	.	.	1
X.33	−2	.	.	.	2	−2	.	.	.	.	.	3	−1	1	−1	−1	−1	.	.	.	.	.	.	.	.	.	1
X.34	−2	.	.	.	−2	2	.	.	.	.	.	1	−3	1	1	−1	1	.	.	.	.	.	.	.	.	.	−1
X.35	2	.	2	−2	.	.	.	.	.	.	.	−2	2	.	.	.	.	.	.	.	.	.	.	.	.	.	.
X.36	.	.	.	.	.	.	.	−1	1	1	.	2	2	2	.	.	.	.	.	.	.	.	.	.	.	−1	.
X.37	.	.	.	.	.	.	.	−1	1	−1	.	.	.	−2	−2	.	.	.	.	.	.	.	.	.	.	1	.
X.38	.	.	.	.	.	.	.	−1	1	−1	.	−2	−2	2	.	.	.	.	.	.	.	.	.	.	.	1	.
X.39	.	.	.	.	.	.	.	−1	1	1	.	.	.	−2	2	.	.	.	.	.	.	.	.	.	.	−1	.
X.40	−2	.	−2	2	.	.	.	.	.	.	.	.	.	.	.	.	.	$-A$	$-\bar{A}$	$\bar{A}$	A	$-\bar{A}$	$-A$	.	.	.	.
X.41	−2	.	−2	2	.	.	.	.	.	.	.	.	.	.	.	.	.	$-\bar{A}$	$-A$	A	$\bar{A}$	$-A$	$-\bar{A}$	.	.	.	.
X.42	−2	.	2	−2	.	.	.	.	.	.	.	.	.	.	.	.	.	A	$\bar{A}$	$-\bar{A}$	$-A$	$-\bar{A}$	$-A$	.	.	.	.
X.43	−2	.	2	−2	.	.	.	.	.	.	.	.	.	.	.	.	.	$\bar{A}$	A	$-A$	$-\bar{A}$	$-A$	$-\bar{A}$	.	.	.	.
X.44	−1	−1	−1	−1	−1	−1	−1	.	.	.	.	2	2	−2	.	.	.	.	.	.	.	.	.	1	1	.	.
X.45	3	−1	1	1	1	1	−1	.	.	.	.	.	.	2	2	.	.	.	.	.	.	.	.	−1	−1	.	.
X.46	3	−1	−1	−1	1	1	−1	.	.	.	.	.	.	2	−2	.	.	.	.	.	.	.	.	1	−1	.	.
X.47	−1	−1	1	1	−1	−1	−1	.	.	.	.	−2	−2	−2	.	.	.	.	.	.	.	.	.	−1	1	.	.
X.48	−2	2	.	.	.	.	2	1	−1	1	.	.	.	.	.	.	.	.	.	.	.	.	.	.	.	−1	.
X.49	−2	2	.	.	.	.	2	1	−1	−1	.	.	.	.	.	.	.	.	.	.	.	.	.	.	.	1	.
X.50	.	.	.	.	.	.	.	2	.	.	.	2	−2	.	.	.	.	−1	−1	1	1	−1	−1	.	.	.	.
X.51	.	.	.	.	.	.	.	2	.	.	.	−2	2	.	.	.	.	1	1	−1	−1	−1	−1	.	.	.	.
X.52	4	.	.	.	.	.	.	.	.	.	.	.	.	2	.	−2	.	.	.	.	.	.	.	.	.	.	.
X.53	−4	.	.	.	.	.	.	.	.	.	.	.	.	.	.	.	.	.	.	.	.	.	.	.	.	.	.
X.54	4	.	.	.	.	.	.	−2	.	.	.	.	.	.	.	.	.	.	.	.	.	.	.	.	.	.	.
X.55	.	.	.	.	.	.	.	.	.	.	.	1	1	1	−1	−1	−1	.	.	.	.	.	.	.	.	.	1
X.56	.	.	.	.	.	.	.	.	.	.	.	−2	2	−2	.	2	.	.	.	.	.	.	.	.	.	.	.
X.57	.	.	.	.	.	.	.	.	.	.	.	−1	−1	1	1	−1	1	.	.	.	.	.	.	.	.	.	−1
X.58	.	.	.	.	.	.	.	.	.	.	.	2	−2	−2	.	2	.	.	.	.	.	.	.	.	.	.	.
X.59	−3	−3	1	1	1	1	1	.	.	.	.	.	.	.	.	.	.	.	.	.	.	.	.	−1	−1	.	.
X.60	−3	−3	−1	−1	1	1	1	.	.	.	.	.	.	.	.	.	.	.	.	.	.	.	.	1	−1	.	.
X.61	1	−3	1	1	−1	−1	1	.	.	.	.	.	.	.	.	.	.	.	.	.	.	.	.	−1	1	.	.
X.62	−3	1	−1	−1	1	1	1	.	.	.	.	.	.	.	.	.	.	.	.	.	.	.	.	1	−1	.	.
X.63	1	1	1	1	−1	−1	1	.	.	.	.	.	.	.	.	.	.	.	.	.	.	.	.	−1	1	.	.
X.64	1	−3	−1	−1	−1	−1	1	.	.	.	.	.	.	.	.	.	.	.	.	.	.	.	.	1	1	.	.
X.65	1	1	−1	−1	−1	−1	1	.	.	.	.	.	.	.	.	.	.	.	.	.	.	.	.	1	1	.	.
X.66	−3	1	1	1	1	1	1	.	.	.	.	.	.	.	.	.	.	.	.	.	.	.	.	−1	−1	.	.
X.67	.	.	.	.	.	.	.	.	.	.	.	1	1	−1	−1	1	1	.	.	.	.	.	.	.	.	.	−1
X.68	.	.	.	.	.	.	.	.	.	.	.	−1	−1	−1	1	1	−1	.	.	.	.	.	.	.	.	.	1
X.69	.	.	.	.	.	.	.	.	.	.	.	1	1	−1	−1	1	1	.	.	.	.	.	.	.	.	.	−1
X.70	.	.	.	.	.	.	.	.	.	.	.	−1	−1	−1	1	1	−1	.	.	.	.	.	.	.	.	.	1
X.71	2	2	.	.	.	.	−2	.	.	.	.	.	.	.	.	.	.	.	.	.	.	.	.	.	.	.	.
X.72	2	2	.	.	.	.	−2	.	.	.	.	.	.	.	.	.	.	.	.	.	.	.	.	.	.	.	.
X.73	2	.	.	.	2	−2	.	.	.	.	.	.	.	.	.	.	.	.	.	.	.	.	.	.	.	.	.
X.74	2	.	.	.	−2	2	.	.	.	.	.	.	.	.	.	.	.	.	.	.	.	.	.	.	.	.	.
X.75	2	.	.	.	2	−2	.	.	.	.	.	.	.	.	.	.	.	.	.	.	.	.	.	.	.	.	.
X.76	2	.	.	.	−2	2	.	.	.	.	.	.	.	.	.	.	.	.	.	.	.	.	.	.	.	.	.
X.77	.	.	.	.	.	.	.	.	.	.	.	1	1	1	−1	−1	1	.	.	.	.	.	.	.	.	.	−1
X.78	.	.	.	.	.	.	.	.	.	.	.	−1	−1	1	1	−1	−1	.	.	.	.	.	.	.	.	.	1
X.79	−4	.	.	.	.	.	.	.	.	.	.	.	.	.	.	.	.	.	.	.	.	.	.	.	.	.	.

where $A = \frac{1}{2}(-1 + i\sqrt{7})$.

4.6.2 *Character table of* $H(Co_2) = \langle x, y, h \rangle$

	1a	2a	2b	2c	2d	2e	2f	2g	2h	2i	2j	3a	3b	3c	4a	4b	4c	4d	4e	4f	4g
2	18	18	17	14	14	16	16	14	14	15	11	6	5	7	14	14	13	12	12	12	12
3	4	4	1	2	2	2	2	1	1	.	1	4	3	3	3	1	1	1	1	1	1
5	1	1	.	1	1	.	.	.	.	.	.	.	1	.	.	.	.	.	.	.	.
7	1	1	1	.	.	.	.	.	.	.	.	.	.	.	1	.	.	.	.	.	.
2P	1a	1a	1a	1a	1a	1a	1a	1a	1a	1a	1a	3a	3b	3c	2a	2a	2b	2f	2b	2f	2b
3P	1a	2a	2b	2c	2d	2e	2f	2g	2h	2i	2j	1a	1a	1a	4a	4b	4c	4d	4e	4f	4g
5P	1a	2a	2b	2c	2d	2e	2f	2g	2h	2i	2j	3a	3b	3c	4a	4b	4c	4d	4e	4f	4g
7P	1a	2a	2b	2c	2d	2e	2f	2g	2h	2i	2j	3a	3b	3c	4a	4b	4c	4d	4e	4f	4g
X.1	1	1	1	1	1	1	1	1	1	1	1	1	1	1	1	1	1	1	1	1	1
X.2	7	7	7	−5	−5	−1	−1	3	3	−1	−1	−2	4	1	7	−1	−5	3	−1	3	−1
X.3	15	15	15	−5	−5	7	7	3	3	7	−1	−3	.	3	15	7	−5	−1	−1	−1	−1
X.4	16	−16	.	4	−4	8	−8	4	−4	.	.	−2	1	4	.	.	.	4	−4	−4	4
X.5	21	21	21	−11	−11	5	5	5	5	5	−3	3	6	.	21	5	−11	1	−3	1	−3
X.6	21	21	21	9	9	−3	−3	1	1	−3	−3	3	6	.	21	−3	9	5	−3	5	−3
X.7	27	27	27	15	15	3	3	7	7	3	3	.	9	.	27	3	15	3	3	3	3
X.8	35	35	35	−5	−5	3	3	−5	−5	3	3	−1	5	2	35	3	−5	7	3	7	3
X.9	35	35	35	15	15	11	11	7	7	11	3	−1	5	2	35	11	15	−1	3	−1	3
X.10	56	56	56	−24	−24	−8	−8	8	8	−8	.	2	11	2	56	−8	−24	.	.	.	.
X.11	70	70	70	−10	−10	−10	−10	6	6	−10	−2	7	−5	1	70	−10	−10	2	−2	2	−2
X.12	84	84	84	4	4	20	20	4	4	20	4	3	−6	3	84	20	4	4	4	4	4
X.13	105	105	105	25	25	−7	−7	9	9	−7	1	6	.	3	105	−7	25	−3	1	−3	1
X.14	105	105	105	5	5	17	17	−3	−3	17	−7	6	.	3	105	17	5	−3	−7	−3	−7
X.15	105	105	105	−35	−35	1	1	5	5	1	1	−3	15	−3	105	1	−35	5	1	5	1
X.16	112	−112	.	−20	20	−8	8	12	−12	.	.	4	4	4	.	.	.	12	4	−12	−4
X.17	120	120	−8	.	.	24	24	.	.	−8	−8	3	.	6	8	8	.	12	8	12	8
X.18	120	120	120	40	40	−8	−8	8	8	−8	.	−6	15	.	120	−8	40	.	.	.	.
X.19	120	120	−8	.	.	24	24	.	.	−8	8	3	.	6	8	8	.	−4	−8	−4	−8
X.20	135	135	7	15	15	39	39	15	15	7	7	.	.	9	−9	−9	−1	3	7	3	7
X.21	168	168	168	40	40	8	8	8	8	8	8	6	6	−3	168	8	40	.	8	.	8
X.22	189	189	189	21	21	−3	−3	−11	−11	−3	−3	.	9	.	189	−3	21	9	−3	9	−3
X.23	189	189	189	−39	−39	21	21	1	1	21	−3	.	9	.	189	21	−39	−3	−3	−3	−3
X.24	189	189	189	−51	−51	−3	−3	13	13	−3	−3	.	9	.	189	−3	−51	−3	−3	−3	−3
X.25	210	210	210	10	10	−14	−14	10	10	−14	2	−6	−15	3	210	−14	10	6	2	6	2
X.26	210	210	210	50	50	2	2	2	2	2	−6	3	15	.	210	2	50	−2	−6	−2	−6
X.27	216	216	216	−24	−24	24	24	8	8	24	.	.	−9	.	216	24	−24	.	.	.	.
X.28	240	−240	.	−20	20	56	−56	12	−12	.	.	6	.	12	.	.	.	−4	4	4	−4
X.29	280	280	280	40	40	24	24	8	8	24	.	−8	−5	−2	280	24	40	.	.	.	.
X.30	280	280	280	−40	−40	−8	−8	−8	−8	−8	8	10	10	1	280	−8	−40	.	8	.	8
X.31	315	315	315	−45	−45	−21	−21	3	3	−21	3	−9	.	.	315	−21	−45	−5	3	−5	3
X.32	336	−336	.	36	−36	−24	24	4	−4	.	.	−6	6	.	.	.	.	20	12	−20	−12
X.33	336	−336	.	−44	44	40	−40	20	−20	.	.	−6	6	.	.	.	.	4	12	−4	−12
X.34	336	336	336	−16	−16	16	16	−16	−16	16	.	−6	6	.	336	16	−16	.	.	.	.
X.35	378	378	378	−30	−30	−6	−6	2	2	−6	−6	.	−9	.	378	−6	−30	6	−6	6	−6
X.36	405	405	405	45	45	−27	−27	−3	−3	−27	−3	.	.	.	405	−27	45	−3	−3	−3	−3
X.37	405	405	21	45	45	−27	−27	−3	−3	5	−3	.	.	.	−27	21	−3	−3	−3	−3	−3
X.38	405	405	21	45	45	−27	−27	−3	−3	5	−3	.	.	.	−27	21	−3	−3	−3	−3	−3
X.39	420	420	420	20	20	4	4	−12	−12	4	4	−3	.	3	420	4	20	−4	4	−4	4
X.40	432	−432	.	60	−60	24	−24	28	−28	.	.	.	9	.	.	.	.	12	−12	−12	12
X.41	512	512	512	.	.	.	.	.	.	.	.	8	−16	−4	512	.	.	.	.	.	.
X.42	560	−560	.	60	−60	88	−88	28	−28	.	.	2	5	8	.	.	.	−4	−12	4	12
X.43	560	−560	.	−20	20	24	−24	−20	20	.	.	2	5	8	.	.	.	28	−12	−28	12
X.44	720	720	−48	.	.	−48	−48	.	.	16	.	−9	.	.	48	−16	.	−8	.	−8	.
X.45	720	720	−48	.	.	−48	−48	.	.	16	.	−9	.	.	48	−16	.	−8	.	−8	.
X.46	810	810	42	90	90	90	90	42	42	26	18	.	.	.	−54	−6	−6	6	18	6	18
X.47	840	840	−56	.	.	−24	−24	.	.	8	8	−6	.	6	56	−8	.	36	−8	36	−8
X.48	840	840	−56	.	.	−24	−24	.	.	8	−8	−6	.	6	56	−8	.	−12	8	−12	8
X.49	896	−896	.	−96	96	−64	64	32	−32	.	.	−4	11	8	.	.	.	.	.	.	.
X.50	945	945	49	−75	−75	105	105	−3	−3	9	−7	.	.	9	−63	−39	5	−3	−7	−3	−7
X.51	945	945	49	−75	−75	−39	−39	45	45	−7	−7	.	.	9	−63	9	5	9	−7	9	−7
X.52	945	945	49	−15	−15	129	129	33	33	33	−7	.	.	9	−63	−15	1	−3	−7	−3	−7
X.53	945	945	49	−15	−15	−15	−15	−15	−15	17	17	.	.	9	−63	33	1	9	17	9	17
X.54	945	945	49	105	105	−15	−15	9	9	17	1	.	.	9	−63	33	−7	−3	1	−3	1
X.55	1080	1080	56	120	120	24	24	24	24	24	8	.	.	−9	−72	24	−8	.	8	.	8
X.56	1120	−1120	.	−40	40	−80	80	24	−24	.	.	−14	−5	4	.	.	.	8	8	−8	−8
X.57	1344	−1344	.	16	−16	160	−160	16	−16	.	.	−6	−6	12	.	.	.	16	−16	−16	16
X.58	1680	1680	−112	.	.	144	144	.	.	−48	.	−12	.	12	112	48	.	−8	.	−8	.
X.59	1680	−1680	.	100	−100	−56	56	36	−36	.	.	−12	.	12	.	.	.	−12	−4	12	4
X.60	1680	1680	−112	.	.	−48	−48	.	.	16	16	15	.	−6	112	−16	.	24	−16	24	−16
X.61	1680	1680	−112	.	.	−48	−48	.	.	16	−16	15	.	−6	112	−16	.	−8	16	−8	16
X.62	1680	−1680	.	−140	140	8	−8	20	−20	.	.	6	15	−12	.	.	.	20	−4	−20	4
X.63	1680	−1680	.	20	−20	136	−136	−12	12	.	.	−12	.	12	.	.	.	−12	28	12	−28
X.64	1890	1890	98	−30	−30	114	114	18	18	50	10	.	.	−9	−126	18	2	6	10	6	10
X.65	1890	1890	98	−150	−150	66	66	42	42	2	−14	.	.	−9	−126	−30	10	6	−14	6	−14
X.66	1920	−1920	.	160	−160	−64	64	32	−32	.	.	12	15	.	.	.	.	.	.	.	.
X.67	2520	2520	−168	.	.	120	120	.	.	−40	−24	9	.	.	168	40	.	−4	24	−4	24
X.68	2520	2520	−168	.	.	120	120	.	.	−40	24	9	.	.	168	40	.	12	−24	12	−24
X.69	2688	−2688	.	160	−160	64	−64	32	−32	.	.	−12	6	−12	.	.	.	.	−32	.	32
X.70	2835	2835	147	135	135	171	171	15	15	11	3	.	.	.	−189	−69	−9	−9	3	−9	3
X.71	2835	2835	147	135	135	27	27	63	63	−5	3	.	.	.	−189	−21	−9	3	3	3	3
X.72	2835	2835	147	−45	−45	−45	−45	−45	−45	51	3	.	.	.	−189	99	3	3	3	3	3
X.73	2835	2835	147	−225	−225	27	27	−9	−9	−5	3	.	.	.	−189	−21	15	−9	3	−9	3
X.74	2835	2835	147	135	135	27	27	−33	−33	−5	−21	.	.	.	−189	−21	−9	3	−21	3	−21
X.75	2835	2835	147	−225	−225	−117	−117	39	39	−21	3	.	.	.	−189	27	15	3	3	3	3
X.76	2835	2835	147	135	135	−117	−117	15	15	−21	−21	.	.	.	−189	27	−9	15	−21	15	−21
X.77	2835	2835	147	−45	−45	99	99	3	3	67	−21	.	.	.	−189	51	3	−9	−21	−9	−21
X.78	3024	−3024	.	−204	204	−24	24	52	−52	.	.	.	9	.	.	.	.	−12	12	12	−12
X.79	3024	−3024	.	−156	156	168	−168	4	−4	.	.	.	9	.	.	.	.	−12	12	12	−12
X.80	3024	−3024	.	84	−84	−24	24	−44	44	.	.	.	9	.	.	.	.	36	12	−36	−12
X.81	3240	3240	−216	.	.	72	72	.	.	−24	−24	.	.	.	216	24	.	36	24	36	24
X.82	3240	3240	−216	.	.	72	72	.	.	−24	24	.	.	.	216	24	.	−12	−24	−12	−24

Character table of $H(\mathsf{Co}_2)$ *(continued)*

	4h	4i	4j	4k	4l	4m	4n	4o	4p	4q	4r	4s	4t	4u	4v	4w	4x	5a	6a	6b	6c	6d	6e	6f	6g
2	13	11	11	12	12	10	11	11	9	9	9	9	10	10	8	8	8	2	6	5	7	6	6	5	5
3	.	1	1	.	.	1	.	.	1	1	1	1	.	.	.	.	.	1	4	3	3	2	2	2	2
5	.	.	.	.	.	.	.	.	.	.	.	.	.	.	.	.	.	1	.	1	.	.	.	.	.
7	.	.	.	.	.	.	.	.	.	.	.	.	.	.	.	.	.	.	.	.	.	.	.	.	.
2P	2b	2e	2b	2e	2e	2e	2b	2f	2g	2g	2g	2g	2b	2e	2i	2g	2g	5a	3a	3b	3c	3a	3a	3b	3b
3P	4h	4i	4j	4k	4l	4m	4n	4o	4p	4q	4r	4s	4t	4u	4v	4w	4x	5a	2a	2a	2a	2e	2f	2c	2d
5P	4h	4i	4j	4k	4l	4m	4n	4o	4p	4q	4r	4s	4t	4u	4v	4w	4x	1a	6a	6b	6c	6d	6e	6f	6g
7P	4h	4i	4j	4k	4l	4m	4n	4o	4p	4q	4r	4s	4t	4u	4v	4w	4x	5a	6a	6b	6c	6d	6e	6f	6g
X.1	1	1	1	1	1	1	1	1	1	1	1	1	1	1	1	1	1	1	1	1	1	1	1	1	1
X.2	3	3	−1	−1	−1	3	3	−1	1	−3	1	−3	−1	−1	−1	1	1	2	−2	4	1	2	2	−2	−2
X.3	3	−1	7	3	3	−1	3	3	−3	1	−3	1	−1	3	3	1	1	.	−3	.	3	1	1	−2	−2
X.4	.	.	.	4	−4	.	.	.	−2	−2	2	2	.	.	.	2	−2	1	2	−1	−4	2	−2	1	−1
X.5	5	1	5	1	1	1	5	1	−3	−3	−3	−3	−3	1	1	1	1	1	3	6	.	−1	−1	−2	−2
X.6	1	5	−3	1	1	5	1	1	−1	3	−1	3	−3	1	1	−1	−1	1	3	6	.	3	3	.	.
X.7	7	3	3	−1	−1	3	7	−1	1	5	1	5	3	−1	−1	1	1	2	.	9	.	.	.	3	3
X.8	−5	7	3	−1	−1	7	−5	−1	−1	−1	−1	−1	3	−1	−1	−1	−1	.	−1	5	2	3	3	1	1
X.9	7	−1	11	3	3	−1	7	3	5	1	5	1	3	3	3	1	1	.	−1	5	2	−1	−1	3	3
X.10	8	.	−8	.	.	.	8	.	4	−4	4	−4	.	.	.	.	.	1	2	11	2	−2	−2	−3	−3
X.11	6	2	−10	2	2	2	6	2	2	2	2	2	−2	2	2	−2	−2	.	7	−5	1	−1	−1	−1	−1
X.12	4	4	20	4	4	4	4	4	.	.	.	.	4	4	4	.	.	−1	3	−6	3	−1	−1	−2	−2
X.13	9	−3	−7	−3	−3	−3	9	−3	−3	−3	−3	−3	1	−3	−3	1	1	.	6	.	3	2	2	4	4
X.14	−3	−3	17	1	1	−3	−3	1	3	−1	3	−1	−7	1	1	−1	−1	.	6	.	3	2	2	2	2
X.15	5	5	1	1	1	5	5	1	−1	−5	−1	−5	1	1	1	−1	−1	.	−3	15	−3	1	1	1	1
X.16	.	.	.	−4	4	.	.	.	−2	6	2	−6	.	.	.	2	−2	2	−4	−4	−4	4	−4	−2	2
X.17	.	−4	.	4	4	−4	.	4	.	.	.	.	.	−4	.	.	.	.	3	.	6	3	3	.	.
X.18	8	.	−8	.	.	.	8	.	−4	4	−4	4	.	.	.	.	.	.	−6	15	.	−2	−2	1	1
X.19	.	12	.	4	4	−4	.	4	.	.	.	.	.	−4	.	.	.	.	3	.	6	3	3	.	.
X.20	−1	3	−1	11	11	3	−1	−5	3	3	3	3	−1	3	−1	3	3	.	.	.	9	.	.	.	.
X.21	8	.	8	.	.	.	8	.	.	.	.	.	8	.	.	.	.	−2	6	6	−3	2	2	−2	−2
X.22	−11	9	−3	1	1	9	−11	1	1	1	1	1	−3	1	1	1	1	−1	.	9	.	.	.	−3	−3
X.23	1	−3	21	1	1	−3	1	1	−5	−1	−5	−1	−3	1	1	−1	−1	−1	.	9	.	.	.	3	3
X.24	13	−3	−3	−3	−3	−3	13	−3	1	1	1	1	−3	−3	−3	1	1	−1	.	9	.	.	.	−3	−3
X.25	10	6	−14	−2	−2	6	10	−2	−2	−2	−2	−2	2	−2	−2	−2	−2	.	−6	−15	3	−2	−2	1	1
X.26	2	−2	2	−2	−2	−2	2	−2	2	2	2	2	−6	−2	−2	−2	−2	.	3	15	.	−1	−1	−1	−1
X.27	8	.	24	.	.	.	8	.	−4	4	−4	4	.	.	.	.	.	1	.	−9	.	.	.	−3	−3
X.28	.	.	.	12	−12	.	.	.	6	−2	−6	2	.	.	.	2	−2	.	−6	.	−12	2	−2	−2	2
X.29	8	.	24	.	.	.	8	.	4	−4	4	−4	.	.	.	.	.	.	−8	−5	−2	.	.	1	1
X.30	−8	.	−8	.	.	.	−8	.	.	.	.	.	8	.	.	.	.	.	10	10	1	−2	−2	2	2
X.31	3	−5	−21	3	3	−5	3	3	3	3	3	3	3	3	3	−1	−1	.	−9	.	.	3	3	.	.
X.32	.	.	.	4	−4	.	.	.	2	−6	−2	6	.	.	.	−2	2	1	6	−6	.	6	−6	.	.
X.33	.	.	.	4	−4	.	.	.	6	6	−6	−6	.	.	.	2	−2	1	6	−6	.	−2	2	−2	2
X.34	−16	.	16	.	.	.	−16	.	.	.	.	.	.	.	.	.	.	1	−6	6	.	−2	−2	2	2
X.35	2	6	−6	−2	−2	6	2	−2	2	2	2	2	−6	−2	−2	2	2	−2	.	−9	.	.	.	3	3
X.36	−3	−3	−27	5	5	−3	−3	5	−3	−3	−3	−3	−3	5	5	1	1	.	.	.	.	.	.	.	.
X.37	13	−3	−3	5	5	−3	−3	−11	−3	−3	−3	−3	5	−3	1	1	1	.	.	.	.	.	.	.	.
X.38	13	−3	−3	5	5	−3	−3	−11	−3	−3	−3	−3	5	−3	1	1	1	.	.	.	.	.	.	.	.
X.39	−12	−4	4	−4	−4	−4	−12	−4	.	.	.	.	4	−4	−4	.	.	.	−3	.	3	1	1	−4	−4
X.40	.	.	.	−4	4	.	.	.	−2	−10	2	10	.	.	.	2	−2	2	.	−9	.	.	.	3	−3
X.41	.	.	.	.	.	.	.	.	.	.	.	.	.	.	.	.	.	2	8	−16	−4	.	.	.	.
X.42	.	.	.	12	−12	.	.	.	−10	−2	10	2	.	.	.	2	−2	.	−2	−5	−8	−2	2	3	−3
X.43	.	.	.	−4	4	.	.	.	2	2	−2	−2	.	.	.	−2	2	.	−2	−5	−8	6	−6	1	−1
X.44	.	−8	.	8	8	8	.	8	.	.	.	.	.	−8	.	.	.	.	−9	.	.	3	3	.	.
X.45	.	−8	.	8	8	8	.	8	.	.	.	.	.	−8	.	.	.	.	−9	.	.	3	3	.	.
X.46	10	6	−6	6	6	6	−6	6	6	6	6	6	2	6	−2	2	2	.	.	.	.	.	.	.	.
X.47	.	−12	.	−4	−4	−12	.	−4	.	.	.	.	.	4	.	.	.	.	−6	.	6	6	6	.	.
X.48	.	36	.	−4	−4	−12	.	−4	.	.	.	.	.	4	.	.	.	.	−6	.	6	6	6	.	.
X.49	.	.	.	.	.	.	.	.	−8	8	8	−8	.	.	.	.	.	1	4	−11	−8	−4	4	−3	3
X.50	13	−3	1	9	9	−3	−3	−7	−9	3	−9	3	1	1	−3	−1	−1	.	.	.	9	.	.	.	.
X.51	−3	9	1	−11	−11	9	−3	5	3	−9	3	−9	1	−3	1	3	3	.	.	.	9	.	.	.	.
X.52	−15	−3	−7	13	13	−3	1	−3	−3	−3	−3	−3	1	5	1	1	1	.	.	.	9	.	.	.	.
X.53	1	9	−7	−7	−7	9	1	9	−3	−3	−3	−3	−7	1	−3	−3	−3	.	.	.	9	.	.	.	.
X.54	25	−3	−7	−11	−11	−3	−7	5	−3	−3	−3	−3	9	−3	1	−3	−3	.	.	.	9	.	.	.	.
X.55	24	.	−8	.	.	.	−8	.	.	.	.	.	8	.	.	.	.	.	.	.	−9	.	.	.	.
X.56	.	.	.	8	−8	.	.	.	−4	−4	4	4	.	.	.	−4	4	.	14	5	−4	−2	2	−1	1
X.57	.	.	.	16	−16	.	.	.	.	.	.	.	.	.	.	.	.	−1	6	6	−12	−2	2	−2	2
X.58	.	−8	.	8	8	8	.	8	.	.	.	.	.	−8	.	.	.	.	−12	.	12	.	.	.	.
X.59	.	.	.	−12	12	.	.	.	6	6	−6	−6	.	.	.	2	−2	.	12	.	−12	4	−4	4	−4
X.60	.	−8	.	8	8	−8	.	8	.	.	.	.	.	−8	.	.	.	.	15	.	−6	3	3	.	.
X.61	.	24	.	8	8	−8	.	8	.	.	.	.	.	−8	.	.	.	.	15	.	−6	3	3	.	.
X.62	.	.	.	4	−4	.	.	.	2	10	−2	−10	.	.	.	−2	2	.	−6	−15	12	2	−2	1	−1
X.63	.	.	.	4	−4	.	.	.	−6	2	6	−2	.	.	.	−2	2	.	12	.	−12	4	−4	2	−2
X.64	−14	6	−14	6	6	6	2	6	−6	−6	−6	−6	−6	6	−2	−2	−2	.	.	.	−9	.	.	.	.
X.65	10	6	2	−2	−2	6	−6	−2	−6	−6	−6	−6	2	−2	−2	2	2	.	.	.	−9	.	.	.	.
X.66	.	.	.	.	.	.	.	.	8	−8	−8	8	.	.	.	.	.	.	−12	−15	.	−4	4	1	−1
X.67	.	12	.	4	4	−4	.	4	.	.	.	.	.	−4	.	.	.	.	9	.	.	−3	−3	.	.
X.68	.	−4	.	4	4	−4	.	4	.	.	.	.	.	−4	.	.	.	.	9	.	.	−3	−3	.	.
X.69	.	.	.	.	.	.	.	.	.	.	.	.	.	.	.	.	.	−2	12	−6	12	4	−4	−2	2
X.70	−1	−9	3	3	3	−9	−1	−13	9	−3	9	−3	−5	−5	−1	−3	−3	.	.	.	.	.	.	.	.
X.71	−17	3	3	−17	−17	3	−1	−1	−3	9	−3	9	−5	−9	3	1	1	.	.	.	.	.	.	.	.
X.72	3	3	−21	3	3	3	3	−13	3	3	3	3	−5	−5	−1	3	3	.	.	.	.	.	.	.	.
X.73	39	−9	3	−5	−5	−9	−9	11	−3	9	−3	9	−5	3	−1	1	1	.	.	.	.	.	.	.	.
X.74	15	3	3	−1	−1	3	−1	15	9	−3	9	−3	3	7	−5	1	1	.	.	.	.	.	.	.	.
X.75	23	3	3	7	7	3	−9	−9	9	−3	9	−3	−5	−1	3	−3	−3	.	.	.	.	.	.	.	.
X.76	−1	15	3	11	11	15	−1	−5	−3	9	−3	9	3	3	−1	−3	−3	.	.	.	.	.	.	.	.
X.77	−13	−9	−21	−9	−9	−9	3	7	3	3	3	3	3	−1	3	−1	−1	.	.	.	.	.	.	.	.
X.78	.	.	.	−12	12	.	.	.	−2	−2	2	2	.	.	.	2	−2	−1	.	−9	.	.	.	−3	3
X.79	.	.	.	4	−4	.	.	.	10	2	−10	−2	.	.	.	−2	2	−1	.	−9	.	.	.	3	−3
X.80	.	.	.	4	−4	.	.	.	−2	−2	2	2	.	.	.	2	−2	−1	.	−9	.	.	.	−3	3
X.81	.	−12	.	−4	−4	−12	.	−4	.	.	.	.	.	4	.	.	.	.	.	.	.	.	.	.	.
X.82	.	36	.	−4	−4	−12	.	−4	.	.	.	.	.	4	.	.	.	.	.	.	.	.	.	.	.

Character table of $H(\mathsf{Co}_2)$ *(continued)*

2	5	5	5	5	5	5	6	5	5	4	3	8	9	9	8	8	7	7	7	7	7	6	6	6	1	2	2	2	5
3	2	2	2	2	2	2	1	1	1	1	.	1	.	.	.	.	.	.	.	.	.	.	.	.	2	1	.	.	3
5	.	.	.	.	.	.	.	.	.	.	.	.	.	.	.	.	.	.	.	.	.	.	.	.	.	1	1	1	.
7	.	.	.	.	.	.	.	.	.	.	1	.	.	.	.	.	.	.	.	.	.	.	.	.	.	.	.	.	.
	6h	6i	6j	6k	6l	6m	6n	6o	6p	6q	7a	8a	8b	8c	8d	8e	8f	8g	8h	8i	8j	8k	8l	8m	9a	10a	10b	10c	12a
2P	3c	3c	3c	3b	3b	3c	3c	3b	3b	3c	7a	4a	4b	4b	4h	4h	4i	4o	4h	4i	4o	4k	4n	4d	9a	5a	5a	5a	6a
3P	2e	2d	2f	2e	2f	2c	2b	2g	2h	2j	7a	8a	8b	8c	8d	8e	8f	8g	8h	8i	8j	8k	8l	8m	3a	10a	10b	10c	4a
5P	6h	6i	6j	6k	6l	6m	6n	6o	6p	6q	7a	8a	8b	8c	8d	8e	8f	8g	8h	8i	8j	8k	8l	8m	9a	2a	2d	2c	12a
7P	6h	6i	6j	6k	6l	6m	6n	6o	6p	6q	1a	8a	8b	8c	8d	8e	8f	8g	8h	8i	8j	8k	8l	8m	9a	10a	10b	10c	12a
X.1	1	1	1	1	1	1	1	1	1	1	1	1	1	1	1	1	1	1	1	1	1	1	1	1	1	1	1	1	1
X.2	−1	1	−1	2	2	1	1	.	.	−1	.	−1	−1	3	−3	1	−1	1	1	−1	1	1	1	−1	1	2	.	.	−2
X.3	1	1	1	−2	−2	1	3	.	.	−1	1	−1	3	−1	1	−3	1	−1	1	1	−1	−1	1	1	.	.	.	.	−3
X.4	2	2	−2	−1	1	−2	.	1	−1	.	2	.	.	.	.	.	2	−2	.	−2	2	.	.	.	1	−1	1	−1	.
X.5	2	−2	2	2	2	−2	.	2	2	.	.	−3	1	1	−3	−3	−1	−1	1	−1	−1	−1	1	−1	.	1	−1	−1	3
X.6	.	.	.	.	.	.	.	−2	−2	.	.	−3	1	5	3	−1	−1	1	−1	−1	1	1	−1	−1	.	1	−1	−1	3
X.7	.	.	.	3	3	.	.	1	1	.	−1	3	−1	3	5	1	1	−1	1	1	−1	−1	1	1	.	2	.	.	.
X.8	.	−2	.	−3	−3	−2	2	1	1	.	.	3	−1	7	−1	−1	1	1	−1	1	1	1	−1	1	−1	.	.	.	−1
X.9	2	.	2	−1	−1	.	2	1	1	.	.	3	3	−1	1	5	−1	1	1	−1	1	1	1	−1	−1	.	.	.	−1
X.10	−2	.	−2	1	1	.	2	−1	−1	.	.	.	.	.	−4	4	.	.	.	.	.	.	.	.	−1	1	1	1	2
X.11	−1	−1	−1	−1	−1	−1	1	3	3	1	.	−2	2	2	2	2	.	.	−2	.	.	.	−2	.	1	.	.	.	7
X.12	−1	1	−1	2	2	1	3	−2	−2	1	.	4	4	4	.	.	.	.	.	.	.	.	.	.	.	−1	−1	−1	3
X.13	−1	1	−1	−4	−4	1	3	.	.	1	.	1	−3	−3	−3	−3	−1	−1	1	−1	−1	−1	1	−1	.	.	.	.	6
X.14	−1	−1	−1	2	2	−1	3	.	.	−1	.	−7	1	−3	−1	3	1	−1	−1	1	−1	−1	−1	1	.	.	.	.	6
X.15	1	1	1	1	1	1	−3	−1	−1	1	.	1	1	5	−5	−1	1	−1	−1	1	−1	−1	−1	1	.	.	.	.	−3
X.16	−2	2	2	−2	2	−2	.	.	.	.	.	.	.	.	.	.	−2	−2	.	2	2	.	.	.	1	−2	.	.	.
X.17	.	.	.	.	.	.	−2	.	.	−2	1	.	.	.	.	.	2	2	.	2	2	−2	.	−2	.	.	.	.	−1
X.18	−2	−2	−2	1	1	−2	.	−1	−1	.	1	.	.	.	4	−4	.	.	.	.	.	.	.	.	.	.	.	.	−6
X.19	.	.	.	.	.	.	−2	.	.	2	1	.	.	.	.	.	2	−2	.	2	−2	2	.	−2	.	.	.	.	−1
X.20	3	3	3	.	.	3	1	.	.	1	2	−1	−1	−1	−1	−1	1	1	−1	1	1	1	−1	1	.	.	.	.	.
X.21	−1	1	−1	2	2	1	−3	2	2	−1	.	8	.	.	.	.	.	.	.	.	.	.	.	.	.	−2	.	.	6
X.22	.	.	.	−3	−3	.	.	1	1	.	.	−3	1	9	1	1	−1	−1	1	−1	−1	−1	1	−1	.	−1	1	1	.
X.23	.	.	.	3	3	.	.	1	1	.	.	−3	1	−3	−1	−5	−1	1	−1	−1	1	1	−1	−1	.	−1	1	1	.
X.24	.	.	.	−3	−3	.	.	1	1	.	.	−3	−3	−3	1	1	1	1	1	1	1	1	1	1	.	−1	−1	−1	.
X.25	1	1	1	1	1	1	3	1	1	−1	.	2	−2	6	−2	−2	.	.	−2	.	.	.	−2	.	.	.	.	.	−6
X.26	2	2	2	−1	−1	2	.	−1	−1	.	.	−6	−2	−2	2	2	.	.	−2	.	.	.	−2	.	.	.	.	.	3
X.27	.	.	.	−3	−3	.	.	−1	−1	.	−1	.	.	.	4	−4	.	.	.	.	.	.	.	.	.	1	1	1	.
X.28	2	2	−2	2	−2	−2	.	.	.	.	2	.	.	.	.	.	2	2	.	−2	−2	.	.	.	.	.	.	.	.
X.29	.	−2	.	−3	−3	−2	−2	−1	−1	.	.	.	.	.	−4	4	.	.	.	.	.	.	.	.	1	.	.	.	−8
X.30	1	−1	1	−2	−2	−1	1	−2	−2	−1	.	8	.	.	.	.	.	.	.	.	.	.	.	.	1	.	.	.	10
X.31	.	.	.	.	.	.	.	.	.	.	.	3	3	−5	3	3	−1	−1	−1	−1	−1	−1	−1	−1	.	.	.	.	−9
X.32	.	.	.	.	.	.	.	−2	2	.	.	.	.	.	.	.	−2	−2	.	2	2	.	.	.	.	−1	−1	1	.
X.33	4	−4	−4	−2	2	4	.	2	−2	.	.	.	.	.	.	.	−2	2	.	2	−2	.	.	.	.	−1	−1	1	.
X.34	−2	2	−2	−2	−2	2	.	2	2	.	.	.	.	.	.	.	.	.	.	.	.	.	.	.	.	1	−1	−1	−6
X.35	.	.	.	3	3	.	.	−1	−1	.	.	−6	−2	6	2	2	.	.	2	.	.	.	2	.	.	−2	.	.	.
X.36	.	.	.	.	.	.	.	.	.	.	−1	−3	5	−3	−3	−3	1	1	1	1	1	1	1	1	.	.	.	.	.
X.37	.	.	.	.	.	.	.	.	.	.	−1	−3	1	1	1	1	1	1	−3	1	1	1	1	1	.	.	.	.	.
X.38	.	.	.	.	.	.	.	.	.	.	−1	−3	1	1	1	1	1	1	−3	1	1	1	1	1	.	.	.	.	.
X.39	1	−1	1	4	4	−1	3	.	.	1	.	4	−4	−4	.	.	.	.	.	.	.	.	.	.	.	.	.	.	−3
X.40	.	.	.	−3	3	.	.	1	−1	.	−2	.	.	.	.	.	2	2	.	−2	−2	.	.	.	.	−2	.	.	.
X.41	.	.	.	.	.	.	−4	.	.	.	1	.	.	.	.	.	.	.	.	.	.	.	.	.	−1	2	.	.	8
X.42	4	.	−4	1	−1	.	.	1	−1	.	.	.	.	.	.	.	−2	−2	.	2	2	.	.	.	−1	.	.	.	.
X.43	.	−4	.	3	−3	4	.	1	−1	.	.	.	.	.	.	.	2	−2	.	−2	2	.	.	.	−1	.	.	.	.
X.44	.	.	.	.	.	.	.	.	.	.	−1	.	.	.	.	.	.	.	.	.	.	.	.	.	.	.	.	.	3
X.45	.	.	.	.	.	.	.	.	.	.	−1	.	.	.	.	.	.	.	.	.	.	.	.	.	.	.	.	.	3
X.46	.	.	.	.	.	.	.	.	.	.	−2	−6	−2	−2	−2	−2	.	.	2	.	.	.	−2	.	.	.	.	.	.
X.47	.	.	.	.	.	.	−2	.	.	2	.	.	.	.	.	.	−2	2	.	−2	2	−2	.	2	.	.	.	.	2
X.48	.	.	.	.	.	.	−2	.	.	−2	.	.	.	.	.	.	−2	−2	.	−2	−2	2	.	2	.	.	.	.	2
X.49	−4	.	4	−1	1	.	.	−1	1	.	.	.	.	.	.	.	.	.	.	.	.	.	.	.	−1	−1	1	−1	.
X.50	−3	3	−3	.	.	3	1	.	.	−1	.	1	5	1	−1	3	1	−1	3	1	−1	−1	−1	1	.	.	.	.	.
X.51	−3	3	−3	.	.	3	1	.	.	−1	.	1	1	−3	3	−1	−1	1	−1	−1	1	1	−1	−1	.	.	.	.	.
X.52	3	−3	3	.	.	−3	1	.	.	−1	.	1	−7	1	1	1	−1	−1	−3	−1	−1	−1	1	−1	.	.	.	.	.
X.53	3	−3	3	.	.	−3	1	.	.	−1	.	1	5	−3	1	1	1	1	1	1	1	1	1	1	.	.	.	.	.
X.54	3	3	3	.	.	3	1	.	.	1	.	−7	1	1	1	1	−1	−1	1	−1	−1	−1	1	−1	.	.	.	.	.
X.55	−3	−3	−3	.	.	−3	−1	.	.	−1	2	−8	.	.	.	.	.	.	.	.	.	.	.	.	.	.	.	.	.
X.56	−2	−2	2	1	−1	2	.	3	−3	.	.	.	.	.	.	.	.	.	.	.	.	.	.	.	1	.	.	.	.
X.57	−2	2	2	−2	2	−2	.	−2	2	.	.	.	.	.	.	.	.	.	.	.	.	.	.	.	.	1	−1	1	.
X.58	.	.	.	.	.	.	−4	.	.	.	.	.	.	.	.	.	.	.	.	.	.	.	.	.	.	.	.	.	4
X.59	−2	2	2	4	−4	−2	.	.	.	.	.	.	.	.	.	.	−2	2	.	2	−2	.	.	.	.	.	.	.	.
X.60	.	.	.	.	.	.	2	.	.	−2	.	.	.	.	.	.	.	.	.	.	.	.	.	.	.	.	.	.	−5
X.61	.	.	.	.	.	.	2	.	.	2	.	.	.	.	.	.	.	.	.	.	.	.	.	.	.	.	.	.	−5
X.62	2	2	−2	−1	1	−2	.	−1	1	.	.	.	.	.	.	.	2	2	.	−2	−2	.	.	.	.	.	.	.	.
X.63	−2	−2	2	−2	2	2	.	.	.	.	.	.	.	.	.	.	2	2	.	−2	−2	.	.	.	.	.	.	.	.
X.64	−3	3	−3	.	.	3	−1	.	.	1	.	2	−2	−2	2	2	.	.	−2	.	.	.	2	.	.	.	.	.	.
X.65	3	−3	3	.	.	−3	−1	.	.	1	.	2	6	−2	2	2	.	.	2	.	.	.	−2	.	.	.	.	.	.
X.66	−4	−4	4	−1	1	4	.	−1	1	.	2	.	.	.	.	.	.	.	.	.	.	.	.	.	.	.	.	.	.
X.67	.	.	.	.	.	.	.	.	.	.	.	.	.	.	.	.	−2	2	.	−2	2	−2	.	2	.	.	.	.	−3
X.68	.	.	.	.	.	.	.	.	.	.	.	.	.	.	.	.	−2	−2	.	−2	−2	2	.	2	.	.	.	.	−3
X.69	−2	2	2	−2	2	−2	.	2	−2	.	.	.	.	.	.	.	.	.	.	.	.	.	.	.	.	2	.	.	.
X.70	.	.	.	.	.	.	.	.	.	.	.	3	7	3	1	−3	−1	1	1	−1	1	1	1	−1	.	.	.	.	.
X.71	.	.	.	.	.	.	.	.	.	.	.	3	3	−1	−3	1	1	−1	−3	1	−1	−1	1	1	.	.	.	.	.
X.72	.	.	.	.	.	.	.	.	.	.	.	3	7	−1	−1	−1	−1	−1	−1	−1	−1	−1	−1	−1	.	.	.	.	.
X.73	.	.	.	.	.	.	.	.	.	.	.	3	−1	3	−3	1	−1	1	−3	−1	1	1	1	−1	.	.	.	.	.
X.74	.	.	.	.	.	.	.	.	.	.	.	3	3	−1	1	−3	1	−1	−3	1	−1	−1	1	1	.	.	.	.	.
X.75	.	.	.	.	.	.	.	.	.	.	.	3	−5	−1	1	−3	1	−1	1	1	−1	−1	1	1	.	.	.	.	.
X.76	.	.	.	.	.	.	.	.	.	.	.	3	−1	−5	−3	1	−1	1	1	−1	1	1	1	−1	.	.	.	.	.
X.77	.	.	.	.	.	.	.	.	.	.	.	3	−5	3	−1	−1	1	1	3	1	1	1	−1	1	.	.	.	.	.
X.78	.	.	.	3	−3	.	.	1	−1	.	.	.	.	.	.	.	2	−2	.	−2	2	.	.	.	.	1	−1	1	.
X.79	.	.	.	−3	3	.	.	1	−1	.	.	.	.	.	.	.	−2	−2	.	2	2	.	.	.	.	1	1	−1	.
X.80	.	.	.	3	−3	.	.	1	−1	.	.	.	.	.	.	.	−2	2	.	2	−2	.	.	.	.	1	1	−1	.
X.81	.	.	.	.	.	.	.	.	.	.	−1	.	.	.	.	.	2	−2	.	2	−2	2	.	−2	.	.	.	.	.
X.82	.	.	.	.	.	.	.	.	.	.	−1	.	.	.	.	.	2	2	.	2	2	−2	.	−2	.	.	.	.	.

Character table of $H(Co_2)$ (continued)

2	5	5	5	5	5	5	4	4	4	4	4	4	4	4	4	3	2	2	1	5	5	1	3	2	1
3	2	1	1	1	1	1	1	1	1	1	1	1	1	1	1	.	.	.	1	.	.	2	1	.	1
5	.	.	.	.	.	.	.	.	.	.	.	.	.	.	.	.	.	.	1	.	.	.	.	.	1
7	.	.	.	.	.	.	.	.	.	.	.	.	.	.	.	1	1	1	.	.	.	.	.	1	.
	12b	12c	12d	12e	12f	12g	12h	12i	12j	12k	12l	12m	12n	12o	12p	14a	14b	14c	15a	16a	16b	18a	24a	28a	30a
2P	6c	6e	6n	6a	6e	6n	6o	6d	6o	6n	6n	6d	6d	6o	6o	7a	7a	7a	15a	8b	8c	9a	12b	14a	15a
3P	4a	4d	4e	4b	4f	4g	4p	4i	4q	4j	4c	4m	4m	4s	4r	14a	14c	14b	5a	16a	16b	6a	8a	28a	10a
5P	12b	12c	12d	12e	12f	12g	12h	12i	12j	12k	12l	12n	12m	12o	12p	14a	14c	14b	3b	16a	16b	18a	24a	28a	6b
7P	12b	12c	12d	12e	12f	12g	12h	12i	12j	12k	12l	12m	12n	12o	12p	2a	2b	2b	15a	16a	16b	18a	24a	4a	30a
X.1	1	1	1	1	1	1	1	1	1	1	1	1	1	1	1	1	1	1	1	1	1	1	1	1	1
X.2	1	.	−1	2	.	−1	−2	.	.	−1	1	.	.	.	−2	.	.	.	−1	1	−1	1	−1	.	−1
X.3	3	−1	−1	1	−1	−1	.	−1	−2	1	1	−1	−1	−2	.	1	1	1	.	−1	1	.	−1	1	.
X.4	.	−2	2	.	2	−2	1	.	1	.	.	.	.	−1	−1	−2	.	.	1	.	.	−1	.	.	−1
X.5	.	1	.	−1	1	.	.	1	.	2	−2	1	1	.	.	.	.	.	1	−1	−1	.	.	.	1
X.6	.	−1	.	3	−1	.	2	−1	.	.	.	−1	−1	.	2	.	.	.	1	1	−1	.	.	.	1
X.7	.	.	.	.	.	.	1	.	−1	.	.	.	.	−1	1	−1	−1	−1	−1	−1	1	.	.	−1	−1
X.8	2	1	.	3	1	.	−1	1	−1	.	−2	1	1	−1	−1	.	.	.	.	1	1	−1	.	.	.
X.9	2	−1	.	−1	−1	.	−1	−1	1	2	.	−1	−1	1	−1	.	.	.	.	1	−1	−1	.	.	.
X.10	2	.	.	−2	.	.	1	.	−1	−2	.	.	.	−1	1	.	.	.	1	.	.	−1	.	.	1
X.11	1	−1	1	−1	−1	1	−1	−1	−1	−1	−1	−1	−1	−1	−1	.	.	.	.	.	.	1	1	.	.
X.12	3	1	1	−1	1	1	.	1	.	−1	1	1	1	.	.	.	.	.	−1	.	.	.	1	.	−1
X.13	3	.	1	2	.	1	.	.	.	−1	1	.	.	.	.	.	.	.	.	−1	−1	.	1	.	.
X.14	3	.	−1	2	.	−1	.	.	2	−1	−1	.	.	2	.	.	.	.	.	−1	1	.	−1	.	.
X.15	−3	−1	1	1	−1	1	−1	−1	1	1	1	−1	−1	1	−1	.	.	.	.	−1	1	.	1	.	.
X.16	.	.	−2	.	.	2	−2	.	.	.	.	.	.	.	2	.	.	.	−1	.	.	−1	.	.	1
X.17	2	3	2	−1	3	2	.	−1	.	.	.	−1	−1	.	.	1	−1	−1	.	.	.	.	.	1	.
X.18	.	.	.	−2	.	.	−1	.	1	−2	−2	.	.	1	−1	1	1	1	.	.	.	.	.	1	.
X.19	2	−1	−2	−1	−1	−2	.	3	.	.	.	−1	−1	.	.	1	−1	−1	.	.	.	.	.	1	.
X.20	−3	.	1	.	.	1	.	.	.	−1	−1	.	.	.	.	2	.	.	.	−1	−1	.	−1	−2	.
X.21	−3	.	−1	2	.	−1	.	.	.	−1	1	.	.	.	.	.	.	.	1	.	.	.	−1	.	1
X.22	.	.	.	.	.	.	1	.	1	.	.	.	.	1	1	.	.	.	−1	−1	−1	.	.	.	−1
X.23	.	.	.	.	.	.	1	.	−1	.	.	.	.	−1	1	.	.	.	−1	1	−1	.	.	.	−1
X.24	.	.	.	.	.	.	1	.	1	.	.	.	.	1	1	.	.	.	−1	1	1	.	.	.	−1
X.25	3	.	−1	−2	.	−1	1	.	1	1	1	.	.	1	1	.	.	.	.	.	.	.	−1	.	.
X.26	.	1	.	−1	1	.	−1	1	−1	2	2	1	1	−1	−1	.	.	.	.	.	.	.	.	.	.
X.27	.	.	.	.	.	.	−1	.	1	.	.	.	.	1	−1	−1	−1	−1	1	.	.	.	.	−1	1
X.28	.	2	−2	.	−2	2	.	.	−2	.	.	.	.	2	.	−2	.	.	.	.	.	.	.	.	.
X.29	−2	.	.	.	.	.	1	.	−1	.	−2	.	.	−1	1	.	.	.	.	.	.	1	.	.	.
X.30	1	.	−1	−2	.	−1	.	.	.	1	−1	.	.	.	.	.	.	.	.	.	.	1	−1	.	.
X.31	.	1	.	3	1	.	.	1	.	.	.	1	1	.	.	.	.	.	.	−1	−1	.	.	.	.
X.32	.	2	.	.	−2	.	2	.	.	.	.	.	.	.	−2	.	.	.	1	.	.	.	.	.	−1
X.33	.	−2	.	.	2	.	.	.	.	.	.	.	.	.	.	.	.	.	1	.	.	.	.	.	−1
X.34	.	.	.	−2	.	.	.	.	.	−2	2	.	.	.	.	.	.	.	1	.	.	.	.	.	1
X.35	.	.	.	.	.	.	−1	.	−1	.	.	.	.	−1	−1	.	.	.	1	.	.	.	.	.	1
X.36	.	.	.	.	.	.	.	.	.	.	.	.	.	.	.	−1	−1	−1	.	1	1	.	.	−1	.
X.37	.	.	.	.	.	.	.	.	.	.	.	.	.	.	.	−1	A	$\bar{A}$	.	−1	−1	.	.	1	.
X.38	.	.	.	.	.	.	.	.	.	.	.	.	.	.	.	−1	$\bar{A}$	A	.	−1	−1	.	.	1	.
X.39	3	−1	1	1	−1	1	.	−1	.	1	−1	−1	−1	.	.	.	.	.	.	.	.	.	1	.	.
X.40	.	.	.	.	.	.	1	.	−1	.	.	.	.	1	−1	2	.	.	−1	.	.	.	.	.	1
X.41	−4	.	.	.	.	.	.	.	.	.	.	.	.	.	.	1	1	1	−1	.	.	−1	.	1	−1
X.42	.	2	.	.	−2	.	−1	.	1	.	.	.	.	−1	1	.	.	.	.	.	.	1	.	.	.
X.43	.	−2	.	.	2	.	−1	.	−1	.	.	.	.	1	1	.	.	.	.	.	.	1	.	.	.
X.44	.	1	.	−1	1	.	.	1	.	.	.	B	$\bar{B}$	.	.	−1	1	1	.	.	.	.	.	−1	.
X.45	.	1	.	−1	1	.	.	1	.	.	.	$\bar{B}$	B	.	.	−1	1	1	.	.	.	.	.	−1	.
X.46	.	.	.	.	.	.	.	.	.	.	.	.	.	.	.	−2	.	.	.	.	.	.	.	2	.
X.47	2	.	−2	−2	.	−2	.	.	.	.	.	.	.	.	.	.	.	.	.	.	.	.	.	.	.
X.48	2	.	2	−2	.	2	.	.	.	.	.	.	.	.	.	.	.	.	.	.	.	.	.	.	.
X.49	.	.	.	.	.	.	1	.	−1	.	.	.	.	1	−1	.	.	.	1	.	.	1	.	.	−1
X.50	−3	.	−1	.	.	−1	.	.	.	1	−1	.	.	.	.	.	.	.	.	1	−1	.	1	.	.
X.51	−3	.	−1	.	.	−1	.	.	.	1	−1	.	.	.	.	.	.	.	.	−1	1	.	1	.	.
X.52	−3	.	−1	.	.	−1	.	.	.	−1	1	.	.	.	.	.	.	.	.	1	1	.	1	.	.
X.53	−3	.	−1	.	.	−1	.	.	.	−1	1	.	.	.	.	.	.	.	.	−1	−1	.	1	.	.
X.54	−3	.	1	.	.	1	.	.	.	−1	−1	.	.	.	.	.	.	.	.	1	1	.	−1	.	.
X.55	3	.	−1	.	.	−1	.	.	.	1	1	.	.	.	.	2	.	.	.	.	.	.	1	−2	.
X.56	.	2	2	.	−2	−2	−1	.	−1	.	.	.	.	1	1	.	.	.	.	.	.	−1	.	.	.
X.57	.	−2	2	.	2	−2	.	.	.	.	.	.	.	.	.	.	.	.	−1	.	.	.	.	.	1
X.58	4	−2	.	.	−2	.	.	−2	.	.	.	2	2	.	.	.	.	.	.	.	.	.	.	.	.
X.59	.	.	2	.	.	−2	.	.	.	.	.	.	.	.	.	.	.	.	.	.	.	.	.	.	.
X.60	−2	−3	2	−1	−3	2	.	1	.	.	.	1	1	.	.	.	.	.	.	.	.	.	.	.	.
X.61	−2	1	−2	−1	1	−2	.	−3	.	.	.	1	1	.	.	.	.	.	.	.	.	.	.	.	.
X.62	.	2	2	.	−2	−2	−1	.	1	.	.	.	.	−1	1	.	.	.	.	.	.	.	.	.	.
X.63	.	.	−2	.	.	2	.	.	2	.	.	.	.	−2	.	.	.	.	.	.	.	.	.	.	.
X.64	3	.	1	.	.	1	.	.	.	1	−1	.	.	.	.	.	.	.	.	.	.	.	−1	.	.
X.65	3	.	1	.	.	1	.	.	.	−1	1	.	.	.	.	.	.	.	.	.	.	.	−1	.	.
X.66	.	.	.	.	.	.	−1	.	1	.	.	.	.	−1	1	−2	.	.	.	.	.	.	.	.	.
X.67	.	−1	.	1	−1	.	.	3	.	.	.	−1	−1	.	.	.	.	.	.	.	.	.	.	.	.
X.68	.	3	.	1	3	.	.	−1	.	.	.	−1	−1	.	.	.	.	.	.	.	.	.	.	.	.
X.69	.	.	−2	.	.	2	.	.	.	.	.	.	.	.	.	.	.	.	1	.	.	.	.	.	−1
X.70	.	.	.	.	.	.	.	.	.	.	.	.	.	.	.	.	.	.	.	−1	1	.	.	.	.
X.71	.	.	.	.	.	.	.	.	.	.	.	.	.	.	.	.	.	.	.	1	−1	.	.	.	.
X.72	.	.	.	.	.	.	.	.	.	.	.	.	.	.	.	.	.	.	.	1	1	.	.	.	.
X.73	.	.	.	.	.	.	.	.	.	.	.	.	.	.	.	.	.	.	.	−1	1	.	.	.	.
X.74	.	.	.	.	.	.	.	.	.	.	.	.	.	.	.	.	.	.	.	1	−1	.	.	.	.
X.75	.	.	.	.	.	.	.	.	.	.	.	.	.	.	.	.	.	.	.	1	−1	.	.	.	.
X.76	.	.	.	.	.	.	.	.	.	.	.	.	.	.	.	.	.	.	.	−1	1	.	.	.	.
X.77	.	.	.	.	.	.	.	.	.	.	.	.	.	.	.	.	.	.	.	−1	−1	.	.	.	.
X.78	.	.	.	.	.	.	1	.	1	.	.	.	.	−1	−1	.	.	.	−1	.	.	.	.	.	1
X.79	.	.	.	.	.	.	1	.	−1	.	.	.	.	1	−1	.	.	.	−1	.	.	.	.	.	1
X.80	.	.	.	.	.	.	1	.	1	.	.	.	.	−1	−1	.	.	.	−1	.	.	.	.	.	1
X.81	.	.	.	.	.	.	.	.	.	.	.	.	.	.	.	−1	1	1	.	.	.	.	.	−1	.
X.82	.	.	.	.	.	.	.	.	.	.	.	.	.	.	.	−1	1	1	.	.	.	.	.	−1	.

Character table of $H(Co_2)$ *(continued)*

	1a	2a	2b	2c	2d	2e	2f	2g	2h	2i	2j	3a	3b	3c	4a	4b	4c	4d	4e	4f	4g
2	18	18	17	14	14	16	16	14	14	15	11	6	5	7	14	14	13	12	12	12	12
3	4	4	1	2	2	2	2	1	1	.	1	4	3	3	3	1	1	1	1	1	1
5	1	1	.	1	1	.	.	.	.	.	.	.	1	.	.	.	.	.	.	.	.
7	1	1	1	.	.	.	.	.	.	.	.	.	.	.	1	.	.	.	.	.	.
2P	1a	1a	1a	1a	1a	1a	1a	1a	1a	1a	1a	3a	3b	3c	2a	2a	2b	2f	2b	2f	2b
3P	1a	2a	2b	2c	2d	2e	2f	2g	2h	2i	2j	1a	1a	1a	4a	4b	4c	4d	4e	4f	4g
5P	1a	2a	2b	2c	2d	2e	2f	2g	2h	2i	2j	3a	3b	3c	4a	4b	4c	4d	4e	4f	4g
7P	1a	2a	2b	2c	2d	2e	2f	2g	2h	2i	2j	3a	3b	3c	4a	4b	4c	4d	4e	4f	4g
X.83	3360	−3360	.	40	−40	−112	112	40	−40	.	.	12	−15	12	.	.	.	24	−8	−24	8
X.84	3360	−3360	.	200	−200	16	−16	8	−8	.	.	−6	15	.	.	.	.	−8	24	8	−24
X.85	3456	−3456	.	−96	96	192	−192	32	−32	.	.	.	−9	.	.	.	.	.	.	.	.
X.86	3780	3780	196	−60	−60	132	132	−60	−60	4	4	.	.	9	−252	−60	4	12	4	12	4
X.87	3780	3780	196	−60	−60	−156	−156	36	36	−28	4	.	.	9	−252	36	4	−12	4	−12	4
X.88	4480	−4480	.	−160	160	−64	64	−32	32	.	.	−20	10	4	.	.	.	.	−32	.	32
X.89	4480	−4480	.	160	−160	192	−192	32	−32	.	.	16	−5	−8	.	.	.	.	.	.	.
X.90	5040	5040	−336	.	.	48	48	.	.	−16	.	18	.	.	336	16	.	−24	.	−24	.
X.91	5040	−5040	.	−180	180	−168	168	12	−12	.	.	18	.	.	.	.	.	−20	−12	20	12
X.92	5376	−5376	.	−64	64	128	−128	−64	64	.	.	12	6	.	.	.	.	.	.	.	.
X.93	5670	5670	294	270	270	−90	−90	−18	−18	−26	6	.	.	.	−378	6	−18	−6	6	−6	6
X.94	6048	−6048	.	−120	120	−48	48	8	−8	.	.	.	−9	.	.	.	.	24	24	−24	−24
X.95	6480	−6480	.	180	−180	−216	216	−12	12	.	.	.	.	.	.	.	.	−12	12	12	−12
X.96	6720	6720	−448	.	.	−192	−192	.	.	64	.	6	.	12	448	−64	.	.	.	.	.
X.97	6720	−6720	.	80	−80	32	−32	−48	48	.	.	6	.	12	.	.	.	−16	−16	16	16
X.98	7560	7560	392	−120	−120	−24	−24	−24	−24	−24	8	.	.	−9	−504	−24	8	.	8	.	8
X.99	7680	7680	−512	.	.	.	.	.	.	.	.	−24	.	−12	512	.	.	.	.	.	.
X.100	8192	−8192	.	.	.	.	.	.	.	.	.	−16	−16	−16	.	.	.	.	.	.	.

Character table of $H(Co_2)$ *(continued)*

	4h	4i	4j	4k	4l	4m	4n	4o	4p	4q	4r	4s	4t	4u	4v	4w	4x	5a	6a	6b	6c	6d	6e	6f	6g
2	13	11	11	12	12	10	11	11	9	9	9	9	10	10	8	8	8	2	6	5	7	6	6	5	5
3	.	1	1	.	.	1	.	.	1	1	1	1	.	.	.	.	.	1	4	3	3	2	2	2	2
5	.	.	.	.	.	.	.	.	.	.	.	.	.	.	.	.	.	1	.	1	.	.	.	.	.
7	.	.	.	.	.	.	.	.	.	.	.	.	.	.	.	.	.	.	.	.	.	.	.	.	.
2P	2b	2e	2b	2e	2e	2e	2b	2f	2g	2g	2g	2g	2b	2e	2i	2g	2g	5a	3a	3b	3c	3a	3a	3b	3b
3P	4h	4i	4j	4k	4l	4m	4n	4o	4p	4q	4r	4s	4t	4u	4v	4w	4x	5a	2a	2a	2a	2e	2f	2c	2d
5P	4h	4i	4j	4k	4l	4m	4n	4o	4p	4q	4r	4s	4t	4u	4v	4w	4x	1a	6a	6b	6c	6d	6e	6f	6g
7P	4h	4i	4j	4k	4l	4m	4n	4o	4p	4q	4r	4s	4t	4u	4v	4w	4x	5a	6a	6b	6c	6d	6e	6f	6g
X.83	.	.	.	−8	8	.	.	.	4	4	−4	−4	.	.	.	−4	4	.	−12	15	−12	−4	4	1	−1
X.84	.	.	.	−8	8	.	.	.	−4	−4	4	4	.	.	.	−4	4	.	6	−15	.	−2	2	−1	1
X.85	.	.	.	.	.	.	.	.	8	−8	−8	8	.	.	.	.	.	1	.	9	.	.	.	−3	3
X.86	4	12	4	−4	−4	12	4	−4	.	.	.	.	4	−4	4	.	.	.	.	.	9	.	.	.	.
X.87	−28	−12	4	4	4	−12	4	4	.	.	.	.	4	4	−4	.	.	.	.	.	9	.	.	.	.
X.88	.	.	.	.	.	.	.	.	.	.	.	.	.	.	.	.	.	.	20	−10	−4	−4	4	2	−2
X.89	.	.	.	.	.	.	.	.	−8	8	8	−8	.	.	.	.	.	.	−16	5	8	.	.	1	−1
X.90	.	−24	.	−8	−8	24	.	−8	.	.	.	.	.	8	.	.	.	.	18	.	.	6	6	.	.
X.91	.	.	.	12	−12	.	.	.	−6	−6	6	6	.	.	.	−2	2	.	−18	.	.	6	−6	.	.
X.92	.	.	.	.	.	.	.	.	.	.	.	.	.	.	.	.	.	1	−12	−6	.	−4	4	2	−2
X.93	14	−6	6	2	2	−6	−2	2	−6	−6	−6	−6	−10	2	2	2	2	.	.	.	.	.	.	.	.
X.94	.	.	.	−8	8	.	.	.	−4	−4	4	4	.	.	.	4	−4	−2	.	9	.	.	.	3	−3
X.95	.	.	.	20	−20	.	.	.	6	6	−6	−6	.	.	.	2	−2	.	.	.	.	.	.	.	.
X.96	.	.	.	.	.	.	.	.	.	.	.	.	.	.	.	.	.	.	6	.	12	−6	−6	.	.
X.97	.	.	.	−16	16	.	.	.	.	.	.	.	.	.	.	.	.	.	−6	.	−12	2	−2	−4	4
X.98	−24	.	8	.	.	.	8	.	.	.	.	.	8	.	.	.	.	.	.	.	−9	.	.	.	.
X.99	.	.	.	.	.	.	.	.	.	.	.	.	.	.	.	.	.	.	−24	.	−12	.	.	.	.
X.100	.	.	.	.	.	.	.	.	.	.	.	.	.	.	.	.	.	2	16	16	16	.	.	.	.

Character table of $H(\mathsf{Co}_2)$ (continued)

2	5	5	5	5	5	5	6	5	5	4	3	8	9	9	8	8	7	7	7	7	7	6	6	6	1	2	2	2	5
3	2	2	2	2	2	2	1	1	1	1	.	1	.	.	.	.	.	.	.	.	.	.	.	.	2	1	.	.	3
5	.	.	.	.	.	.	.	.	.	.	.	.	.	.	.	.	.	.	.	.	.	.	.	.	.	1	1	1	.
7	.	.	.	.	.	.	.	.	.	.	1	.	.	.	.	.	.	.	.	.	.	.	.	.	.	.	.	.	.
	6h	6i	6j	6k	6l	6m	6n	6o	6p	6q	7a	8a	8b	8c	8d	8e	8f	8g	8h	8i	8j	8k	8l	8m	9a	10a	10b	10c	12a
2P	3c	3c	3c	3b	3b	3c	3c	3b	3b	3c	7a	4a	4b	4b	4h	4h	4i	4o	4h	4i	4o	4k	4n	4d	9a	5a	5a	5a	6a
3P	2e	2d	2f	2e	2f	2c	2b	2g	2h	2j	7a	8a	8b	8c	8d	8e	8f	8g	8h	8i	8j	8k	8l	8m	3a	10a	10b	10c	4a
5P	6h	6i	6j	6k	6l	6m	6n	6o	6p	6q	7a	8a	8b	8c	8d	8e	8f	8g	8h	8i	8j	8k	8l	8m	9a	2a	2d	2c	12a
7P	6h	6i	6j	6k	6l	6m	6n	6o	6p	6q	1a	8a	8b	8c	8d	8e	8f	8g	8h	8i	8j	8k	8l	8m	9a	10a	10b	10c	12a
X.83	2	2	−2	−1	1	−2	.	1	−1	.	.	.	.	.	.	.	.	.	.	.	.	.	.	.	.	.	.	.	.
X.84	4	4	−4	1	−1	−4	.	−1	1	.	.	.	.	.	.	.	.	.	.	.	.	.	.	.	.	.	.	.	.
X.85	.	.	.	3	−3	.	.	−1	1	.	−2	.	.	.	.	.	.	.	.	.	.	.	.	.	.	−1	1	−1	.
X.86	−3	−3	−3	.	.	−3	1	.	.	1	.	−4	−4	−4	.	.	.	.	.	.	.	.	.	.	.	.	.	.	.
X.87	−3	−3	−3	.	.	−3	1	.	.	1	.	−4	4	4	.	.	.	.	.	.	.	.	.	.	.	.	.	.	.
X.88	2	−2	−2	2	−2	2	.	−2	2	.	.	.	.	.	.	.	.	.	.	.	.	.	.	.	1	.	.	.	.
X.89	.	−4	.	3	−3	4	.	−1	1	.	.	.	.	.	.	.	.	.	.	.	.	.	.	.	1	.	.	.	.
X.90	.	.	.	.	.	.	.	.	.	.	.	.	.	.	.	.	.	.	.	.	.	.	.	.	.	.	.	.	−6
X.91	.	.	.	.	.	.	.	.	.	.	.	.	.	.	.	.	−2	2	.	2	−2	.	.	.	.	.	.	.	.
X.92	−4	4	4	2	−2	−4	.	2	−2	.	.	.	.	.	.	.	.	.	.	.	.	.	.	.	.	−1	−1	1	.
X.93	.	.	.	.	.	.	.	.	.	.	.	6	−6	2	2	2	.	.	2	.	.	.	−2	.	.	.	.	.	.
X.94	.	.	.	−3	3	.	.	−1	1	.	.	.	.	.	.	.	.	.	.	.	.	.	.	.	.	2	.	.	.
X.95	.	.	.	.	.	.	.	.	.	.	−2	.	.	.	.	.	2	−2	.	−2	2	.	.	.	.	.	.	.	.
X.96	.	.	.	.	.	.	−4	.	.	.	.	.	.	.	.	.	.	.	.	.	.	.	.	.	.	.	.	.	−2
X.97	2	−2	−2	−4	4	2	.	.	.	.	.	.	.	.	.	.	.	.	.	.	.	.	.	.	.	.	.	.	.
X.98	3	3	3	.	.	3	−1	.	.	−1	.	−8	.	.	.	.	.	.	.	.	.	.	.	.	.	.	.	.	.
X.99	.	.	.	.	.	.	4	.	.	.	1	.	.	.	.	.	.	.	.	.	.	.	.	.	.	.	.	.	8
X.100	.	.	.	.	.	.	.	.	.	.	2	.	.	.	.	.	.	.	.	.	.	.	.	.	−1	−2	.	.	.

Character table of $H(\mathsf{Co}_2)$ (continued)

2	5	5	5	5	5	5	4	4	4	4	4	4	4	4	4	3	2	2	1	5	5	1	3	2	1
3	2	1	1	1	1	1	1	1	1	1	1	1	1	1	1	.	.	.	1	.	.	2	1	.	1
5	.	.	.	.	.	.	.	.	.	.	.	.	.	.	.	.	.	.	1	.	.	.	.	.	1
7	.	.	.	.	.	.	.	.	.	.	.	.	.	.	.	1	1	1	.	.	.	.	.	1	.
	12b	12c	12d	12e	12f	12g	12h	12i	12j	12k	12l	12m	12n	12o	12p	14a	14b	14c	15a	16a	16b	18a	24a	28a	30a
2P	6c	6e	6n	6a	6e	6n	6o	6d	6o	6n	6n	6d	6d	6o	6o	7a	7a	7a	15a	8b	8c	9a	12b	14a	15a
3P	4a	4d	4e	4b	4f	4g	4p	4i	4q	4j	4c	4m	4m	4s	4r	14a	14c	14b	5a	16a	16b	6a	8a	28a	10a
5P	12b	12c	12d	12e	12f	12g	12h	12i	12j	12k	12l	12n	12m	12o	12p	14a	14c	14b	3b	16a	16b	18a	24a	28a	6b
7P	12b	12c	12d	12e	12f	12g	12h	12i	12j	12k	12l	12m	12n	12o	12p	2a	2b	2b	15a	16a	16b	18a	24a	4a	30a
X.83	.	.	−2	.	.	2	1	.	1	.	.	.	.	−1	−1	.	.	.	.	.	.	.	.	.	.
X.84	.	−2	.	.	2	.	−1	.	−1	.	.	.	.	1	1	.	.	.	.	.	.	.	.	.	.
X.85	.	.	.	.	.	.	−1	.	1	.	.	.	.	−1	1	2	.	.	1	.	.	.	.	.	−1
X.86	−3	.	1	.	.	1	.	.	.	1	1	.	.	.	.	.	.	.	.	.	.	.	−1	.	.
X.87	−3	.	1	.	.	1	.	.	.	1	1	.	.	.	.	.	.	.	.	.	.	.	−1	.	.
X.88	.	.	−2	.	.	2	.	.	.	.	.	.	.	.	.	.	.	.	.	.	.	−1	.	.	.
X.89	.	.	.	.	.	.	1	.	−1	.	.	.	.	1	−1	.	.	.	.	.	.	−1	.	.	.
X.90	.	.	.	−2	.	.	.	.	.	.	.	.	.	.	.	.	.	.	.	.	.	.	.	.	.
X.91	.	−2	.	.	2	.	.	.	.	.	.	.	.	.	.	.	.	.	.	.	.	.	.	.	.
X.92	.	.	.	.	.	.	.	.	.	.	.	.	.	.	.	.	.	.	1	.	.	.	.	.	−1
X.93	.	.	.	.	.	.	.	.	.	.	.	.	.	.	.	.	.	.	.	.	.	.	.	.	.
X.94	.	.	.	.	.	.	−1	.	−1	.	.	.	.	1	1	.	.	.	1	.	.	.	.	.	−1
X.95	.	.	.	.	.	.	.	.	.	.	.	.	.	.	.	2	.	.	.	.	.	.	.	.	.
X.96	4	.	.	2	.	.	.	.	.	.	.	.	.	.	.	.	.	.	.	.	.	.	.	.	.
X.97	.	2	2	.	−2	−2	.	.	.	.	.	.	.	.	.	.	.	.	.	.	.	.	.	.	.
X.98	3	.	−1	.	.	−1	.	.	.	−1	−1	.	.	.	.	.	.	.	.	.	.	.	1	.	.
X.99	−4	.	.	.	.	.	.	.	.	.	.	.	.	.	.	1	−1	−1	.	.	.	.	.	1	.
X.100	.	.	.	.	.	.	.	.	.	.	.	.	.	.	.	−2	.	.	−1	.	.	1	.	.	1

where $A = i\sqrt{7}$, $B = -1 - i\sqrt{3}$.

5
Fischer's simple group Fi_{22}

In 1970 B. Fischer found three sporadic simple groups by characterizing the finite groups G which can be generated by a conjugacy class $D = z^G$ of 3-*transpositions*. That means that the product of two involutions of D has order 1, 2 or 3; see [30]. He proved that, besides the symmetric groups S_n, the symplectic groups $\mathsf{Sp}_n(2)$, the projective unitary groups $\mathrm{U}_n(2)$ over the field with four elements and certain orthogonal groups, his two simple groups Fi_{22}, Fi_{23}, and the automorphism group Fi_{24} of his third sporadic group Fi_{24}' comprise all 3-transposition groups [29]. For such a group $G = \langle D \rangle$ Fischer constructed a graph $\mathcal{G}$ and its action on $\mathcal{G}$. As its vertices he took the 3-transpositions x of D. Two vertices $x, y \in \mathcal{G}$ are *connected* if x and y commute in G. In that case they are joined by an *edge* (x, y) in $\mathcal{G}$. Fischer showed in his lecture notes [30] that each of the groups G considered in his theorem has a natural representation as an automorphism group of its graph $\mathcal{G}$. Furthermore, Theorem 17.2.3 of his lecture notes [30] contains the first existence proof for his sporadic group Fi_{22}.

It is the purpose of this chapter to provide an existence proof for the smallest sporadic Fischer group Fi_{22} by means of Algorithm 1.3.8. This proof is due to H. Kim and the author [72]. The methods presented here are purely algebraic and do not require any geometric insight. In particular, they are independent of [30] and [4].

In Section 5.1 Algorithm 1.3.8 is applied to the split extension E_2 of the Mathieu group M_{22} by its simple module V_2 defined in Lemma 4.1.1. Proposition 5.1.1 describes a construction of a finitely presented group H_2 with center $Z(H_2) = \langle z, v \rangle$ of order 4 and exponent 2. It has a Sylow 2-subgroup S_2 containing a unique elementary abelian normal subgroup A_2 of order 2^{10} such that $D_2 = N_{H_2}(A_2) \cong D = C_{E_2}(z)$. Hence all conditions of Step 5 of Algorithm 1.3.8 are satisfied.

In Section 5.2 we take the constructed presentation of H_2 as the input of Algorithm 7.4.8 of [92]. It returns a finite simple subgroup of $\mathfrak{G}_2$ $\mathrm{GL}_{78}(13)$ having order $2^{17} \cdot 3^9 \cdot 5^2 \cdot 7 \cdot 11 \cdot 13$ and a 2-central involution $\mathfrak{z}$ such that $C_{\mathfrak{G}_2}(\mathfrak{z}) \cong H_2$; see Theorem 5.2.1. In [72] H. Kim and the author verified that the character table of $\mathfrak{G}_2$ is equivalent to the one of Fischer's simple sporadic group Fi_{22}. For applications in Chapters 6 and 9 a nice presentation of $\mathfrak{G}_2$ is determined in Corollary 5.2.2, which, by C. Praeger's and L. Soicher's book [110], is also satisfied by Fischer's original simple group Fi_{22}.

Section 5.3 contains a sketch of a uniqueness proof for Fi_{22}. This result is mainly due to D. Parrott [107] and D. Hunt [55]. In particular, it can be shown that any simple group G having a 2-central involution z with centralizer $C_G(z) \cong H_2$ is isomorphic to the simple matrix subgroup $\mathfrak{G}_2$ of $\mathrm{GL}_{78}(13)$.

In Section 5.4 it is shown that Algorithm 1.3.8 constructs the automorphism group $Aut(\mathsf{Fi}_{22})$ of the Fischer group Fi_{22} when it is applied to the extension group E_4 constructed in Lemma 4.1.1; see Corollary 5.4.1. However, it is also mentioned in Section 5.4 that it fails to construct a centralizer H_i of a group when it is applied to the extensions E_i, where $i = 1$ or 5.

The systems of representatives of conjugacy classes of the local subgroups E_2, H_2 and D_2 are stated in Section 5.5. The final section contains the character tables of these groups. The four generating matrices of the matrix group $\mathfrak{G}_2 \cong \mathsf{Fi}_{22}$ and a system of representatives of its conjugacy classes are documented on the accompanying DVD.

5.1 Construction of the 2-central involution centralizers

In this section we apply Algorithm 1.3.8 to the extension group E_2 of Lemma 4.1.2 in order to construct a group H_2 which is isomorphic to the centralizer of a 2-central involution of the simple group Fi_{22}.

Proposition 5.1.1 *Let $E_2 = \langle a, b, c, d, t, g, h, i, v_1 \rangle$ be the split extension of M_{22} by its simple module V_2 of dimension 10 over $F = GF(2)$ defined in Lemma 4.1.1. Then the following statements hold.*

(a) *$z = (iv_1)^2$ is a 2-central involution of E_2 with centralizer $D = C_{E_2}(z)$.*

(b) *D is a finitely presented group $D = \langle y_i | 1 \le i \le 11 \rangle$ having the following set $\mathcal{R}(D)$ of defining relations:*

$$
\begin{aligned}
&y_2^2 = y_3^2 = y_5^2 = y_6^2 = y_7^2 = y_8^2 = y_9^2 = y_{11}^2 = 1,\\
&(y_5y_9)^2 = (y_5y_{11})^2 = (y_9y_{11})^2 = (y_{10}y_3y_5y_4)^3 = 1,\\
&y_{11}y_3y_5y_4y_{10}y_3y_5y_4y_{10}y_5^{-1}y_{10}^{-1}y_4^{-1}y_5^{-1}y_3^{-1}y_{10}^{-1}y_4^{-1}y_5^{-1}y_3^{-1} = 1,\\
&y_{11}y_3y_5y_4y_{10}y_3y_5y_4y_{10}y_9^{-1}y_{10}^{-1}y_4^{-1}y_5^{-1}y_3^{-1}y_{10}^{-1}y_4^{-1}y_5^{-1}y_3^{-1}y_5 = 1,\\
&y_{11}y_{10}^{-1}y_4^{-1}y_5^{-1}y_3^{-1}y_{10}^{-1}y_4^{-1}y_5^{-1}y_3^{-1}y_9^{-1}y_5^{-1}y_3y_5y_4y_{10}y_3y_5y_4y_{10} = 1,\\
&y_{10}y_5^{-1}y_{10}^{-1}y_4^{-1}y_5^{-1}y_3^{-1}y_5y_9y_3y_5y_4 = y_{11}y_3y_5y_4y_{10}y_9^{-1}y_{10}^{-1}y_4^{-1}y_5^{-1}y_3^{-1}y_5y_9 = 1,\\
&y_{11}y_{10}^{-1}y_4^{-1}y_5^{-1}y_3^{-1}y_5^{-1}y_3y_5y_4y_{10} = y_9y_4^{-1}y_5^{-1}y_4 = y_9y_4^{-1}y_9^{-1}y_4y_5 = 1,\\
&y_{11}y_4^{-1}y_{11}^{-1}y_4y_5 = y_{11}y_3y_5y_4y_{10}y_4^{-1}y_3y_5y_4y_{10}y_4^{-1}y_9 = 1,\\
&y_{10}y_4^{-1}y_{10}^{-1}y_4^{-1}y_5^{-1}y_3^{-1}y_4^{-1}y_3y_5y_4y_{10}y_4y_3y_5y_4 = 1,\\
&y_{11}y_3y_5y_4y_{10}y_3y_5y_4y_{10}y_4^{-2}y_{10}^{-1}y_4^{-1}y_5^{-1}y_3^{-1}y_4y_9 = 1,\\
&y_{11}y_{10}y_5^{-1}y_{10}^{-1}y_{11}^{-1}y_3^{-1}y_5y_9y_3y_5y_4y_{10}y_3y_5y_4y_{10}y_4^{-1}y_{10}^{-1}y_4^{-1}y_5^{-1} = 1,\\
&y_{11}y_{10}y_9^{-1}y_{10}^{-1}y_{11}^{-1}y_3^{-1}y_5y_3y_5y_4y_{10}y_4^{-1}y_3 = 1,\\
&y_{11}y_{10}^{-1}y_{11}^{-1}y_3^{-1}y_4y_{10}^{-1}y_4^{-1}y_5^{-1}y_3^{-1}y_{11}^{-1}y_9^{-1}y_5^{-1}y_3y_{11}y_{10} = 1,\\
&y_{11}y_4^{-1}y_5^{-1}y_3^{-1}y_{10}^{-1}y_4^{-1}y_5^{-1}y_3^{-1}y_{10}^{-1}y_{11}^{-1}y_3^{-1}y_5y_9y_{11}y_3y_5y_4y_{10}y_3y_5y_4y_{10}y_3 = 1,\\
&y_{11}y_4^{-1}y_5^{-1}y_3^{-1}y_4y_{10}^{-1}y_4^{-1}y_5^{-1}y_3^{-1}y_{11}^{-1}y_9^{-1}y_{10}^{-1}y_{11}^{-1}y_3^{-1}\\
&\quad\cdot y_5y_3y_5y_4y_{10}y_3y_5y_4y_{10}y_4^{-1}y_3 = 1,\\
&y_{11}y_{10}y_3y_{11}y_{10}y_5y_4^{-1}y_3 = y_{11}y_{10}y_3y_5y_9y_{10}^{-1}y_5y_3y_5y_4y_{10}y_4^{-1}y_{10}^{-1}y_4^{-1}y_5^{-1} = 1,\\
&y_{11}y_3y_5y_4y_{10}y_3y_5y_4y_{10}y_4^{-1}y_3y_5y_{10}^{-1}y_4^{-1}y_5^{-1}y_3^{-1}y_{10}^{-1}y_4^{-1}y_5^{-1}y_3^{-1}y_5^{-1}y_3^{-1}y_5 = 1,\\
&y_{11}y_4^{-1}y_{10}^{-1}y_4^{-1}y_5^{-1}y_3^{-1}y_5^{-1}y_3^{-1}y_{10}^{-1}y_{11}^{-1}y_3^{-1}y_{11}^{-1}y_9^{-1}y_5^{-1}y_3y_5^2 = 1,\\
&y_{11}y_{10}y_5^{-1}y_3^{-1}y_{10}^{-1}y_{11}^{-1}y_3^{-1}y_4y_9^{-1}y_3y_5y_9y_3 = 1,\\
&y_{11}y_5^{-1}y_3^{-1}y_{10}^{-1}y_{11}^{-1}y_5y_4y_{10}y_4y_{11}^{-1}y_3y_5y_9 = (y_5y_3y_5)^2 = 1,\\
&y_{11}y_3y_5y_4y_{10}y_4^{-1}y_3y_5y_4y_{10}y_3y_5y_4y_3y_5y_4^{-1}y_5^{-1}y_3^{-1}y_5 = 1,\\
&y_9y_3y_5y_4^{-1}y_2y_3^{-1}y_2^{-1}y_5 = y_9y_4^{-1}y_2y_4^{-1}y_2^{-1} = y_9y_2y_5^{-1}y_2^{-1} = 1,\\
&y_9y_2^{-1}y_5^{-1}y_2 = y_{10}y_2^{-1}y_5^{-1}y_4^{-1}y_{10}^{-1}y_2 = y_{11}y_2^{-1}y_{11}^{-1}y_9^{-1}y_5^{-1}y_2 = 1,\\
&y_9y_2y_1y_2y_1y_2^{-1}y_1^{-1}y_2^{-1}y_1^{-1}y_5 = y_9y_4^{-1}y_1y_2y_1y_3^{-1}y_1^{-1}y_2^{-1}y_1^{-1}y_5y_3 = 1,\\
&y_9y_4^{-1}y_1y_2y_1y_4^{-1}y_1^{-1}y_2^{-1}y_1^{-1}y_5 = y_9y_1y_2y_1y_5^{-1}y_1^{-1}y_2^{-1}y_1^{-1} = 1,\\
&y_9y_1^{-1}y_2^{-1}y_1^{-1}y_5^{-1}y_1y_2y_1 = y_{10}y_1^{-1}y_2^{-1}y_1^{-1}y_4^{-1}y_9^{-1}y_{10}^{-1}y_9^{-1}y_1y_2y_1 = 1,\\
&y_{11}y_1^{-1}y_2^{-1}y_1^{-1}y_{11}^{-1}y_9^{-1}y_5^{-1}y_1y_2y_1 = y_9y_1y_2y_1^2y_2y_1y_5 = y_9y_4y_1y_2y_1^2y_2^{-1}y_1^{-1} = 1,\\
&y_{11}y_4^{-1}y_1y_3^{-1}y_1^{-1}y_9y_{11}y_3y_9 = (y_4^{-1}, y_1) = y_9y_1y_5^{-1}y_1^{-1} = y_9y_1^{-1}y_9^{-1}y_5^{-1}y_1 = 1,\\
&y_{10}y_1^{-1}y_4y_{10}y_1 = y_{11}y_1^{-1}y_{11}^{-1}y_5^{-1}y_1 = y_9y_1y_4^{-1}y_5^{-1}y_1y_5 = y_4y_1^{-2} = 1,\\
&y_{11}y_2y_1y_6y_3y_{11}y_{10}y_7y_1^{-1}y_2^{-1}y_4^{-1}y_{10}^{-1}y_3^{-1}y_5^{-1}y_7^{-1}y_{10}^{-1}y_{11}^{-1}y_3^{-1}y_6^{-1}y_5y_9y_{11}y_{10}^2 = 1,\\
&y_{11}y_{10}^{-1}y_9y_2y_1y_2y_1y_6y_3y_{11}y_{10}y_7y_1^{-1}y_2^{-1}y_1^{-1}y_2^{-1}y_4y_9^{-1}y_{10}y_4^{-1}\\
&\quad\cdot y_{11}^{-1}y_5^{-1}y_7^{-1}y_{10}^{-1}y_{11}^{-1}y_3^{-1}y_6^{-1} = 1,\\
&y_{11}y_{10}y_7y_4y_{10}^{-1}y_4y_5^{-1}y_7^{-1}y_{10}^{-1}y_{11}^{-1}y_3^{-1}y_6^{-1}y_5y_9y_{11}y_{10}^2y_6y_3 = 1,\\
&y_{11}y_4^{-1}y_{10}^{-1}y_5y_2y_6y_3y_{11}y_{10}y_7y_2^{-1}y_3^{-1}y_{10}y_3^{-1}y_7^{-1}y_{10}^{-1}y_{11}^{-1}y_3^{-1}y_6^{-1}y_9 = 1,\\
&y_{11}y_3y_4y_9y_{10}y_4^{-1}y_6y_3y_{11}y_{10}y_7y_4^{-1}y_{10}^{-1}y_9^{-1}y_3^{-1}y_{11}^{-1}y_9^{-1}y_5^{-1}y_7^{-1}y_{10}^{-1}y_{11}^{-1}y_3^{-1}y_6^{-1} = 1,\\
&y_{11}y_4^{-1}y_{10}y_9y_7^{-1}y_{10}^{-1}y_{11}^{-1}y_3^{-1}y_6^{-1}y_4^{-1}y_{10}^{-1}y_4^{-1}y_{11}^{-1}y_9^{-1}y_6y_3y_{11}y_{10}y_7 = 1,\\
&y_{11}y_7^{-1}y_{10}^{-1}y_{11}^{-1}y_3^{-1}y_6^{-1}y_{11}^{-1}y_9^{-1}y_6y_3y_{11}y_{10}y_7y_9 = 1,\\
&y_{11}y_{10}y_7y_{10}^{-1}y_6y_3y_{11}y_{10}y_7y_5y_9y_{10}^{-1}y_4^{-1}y_{10}y_6y_3 = 1,\\
&y_{11}y_3y_2y_1y_2y_1y_6y_3y_{11}y_{10}y_7y_3y_{10}y_6y_3y_{11}y_{10}y_7y_4y_7^{-1}y_{10}^{-1}y_{11}^{-1}y_3^{-1}y_6^{-1}y_5y_3y_4 = 1,
\end{aligned}
$$

$$y_{11}y_{10}y_7y_4^{-1}y_2^{-1}y_7^{-1}y_{10}^{-1}y_{11}^{-1}y_3^{-1}y_6^{-1}y_5y_9y_{10}y_3y_{10}y_1y_2y_1^2y_6y_3y_{11}y_{10}y_7y_9y_6y_3 = 1,$$
$$y_{11}y_3y_{11}y_{10}y_6y_3y_{11}y_{10}y_7y_5y_6y_3y_{11}y_{10}y_7y_9^{-1}y_4^{-1}y_7^{-1}y_{10}^{-1}y_{11}^{-1}y_3^{-1}y_6^{-1}y_9 = 1,$$
$$y_7y_4^{-1}y_2^{-1}y_6^{-1}y_2^{-1}y_4y_5^{-1} = y_9y_6y_2^2y_8^{-1}y_5 = y_9y_6y_2y_5^{-1}y_2^{-1}y_6^{-1} = 1,$$
$$y_9y_2^{-1}y_6^{-1}y_5^{-1}y_6y_2 = y_{11}y_2^{-1}y_6^{-1}y_{11}^{-1}y_9^{-1}y_5^{-1}y_6y_2 = 1,$$
$$y_{11}y_{10}y_7y_3y_4y_2^{-1}y_6^{-1}y_3^{-1}y_7^{-1}y_{10}^{-1}y_{11}^{-1}y_3^{-1}y_6^{-1}y_1^{-1}y_2^{-1}y_1^{-1}y_2^{-1}y_4^{-1}y_{10}^{-1}y_3^{-1}$$
$$\cdot y_9^{-1}y_6y_2y_5y_3y_4y_{10}^{-1}y_6y_3 = 1,$$
$$y_{11}y_4y_{10}y_2y_1y_2y_1y_6y_3y_{11}y_{10}y_7y_2y_3y_2^{-1}y_6^{-1}y_{10}y_7^{-1}y_{10}^{-1}y_{11}^{-1}y_3^{-1}y_6^{-1}y_1^{-1}y_2^{-1}y_1^{-1}$$
$$\cdot y_9^{-1}y_{10}y_4^{-1}y_6y_2y_5y_9 = 1,$$
$$y_{11}y_{10}y_7y_2^{-1}y_6^{-1}y_9^{-1}y_7^{-1}y_{10}^{-1}y_{11}^{-1}y_3^{-1}y_6^{-1}y_1^{-2}y_2^{-1}y_1^{-1}y_2^{-1}$$
$$\cdot y_9^{-1}y_{10}^{-1}y_{11}^{-1}y_3^{-1}y_9^{-1}y_5^{-1}y_6y_2y_5y_3y_{11}y_{10}y_9y_2y_1y_2y_1^2y_6y_3 = 1,$$
$$y_{11}y_{10}y_7y_4y_9y_6y_2y_1^{-1}y_2^{-1}y_1^{-1}y_2^{-1}y_{10}^{-1}y_5^{-1}y_{10}^{-1}y_2^{-1}y_6^{-1}y_2y_1y_2y_1y_6y_3 = 1,$$
$$y_{11}y_3y_4y_{10}y_4y_2y_1y_2y_1y_6y_3y_{11}y_{10}y_7y_3y_4y_6y_2y_1^{-1}y_3^{-1}y_9^{-1}y_{10}$$
$$\cdot y_9^{-1}y_5^{-1}y_2^{-1}y_6^{-1}y_5y_9 = 1,$$
$$y_{11}y_3y_{11}y_{10}y_6^{-1}y_{10}y_7^{-1}y_{10}^{-1}y_{11}^{-1}y_3^{-1}y_6^{-1}y_2^{-1}y_3^{-1}y_{10}^{-1}y_{11}^{-1}y_9^{-1}y_6y_2 = 1,$$
$$y_{11}y_{10}y_2^{-1}y_6^{-1}y_{10}y_7^{-1}y_{10}^{-1}y_{11}^{-1}y_3^{-1}y_6^{-1}y_4^{-1}y_9^{-1}y_{10}^{-1}y_5^{-1}y_6y_2y_5y_3y_4^{-1} = 1,$$
$$(y_6y_2^2)^2 = y_9y_4^{-1}y_2y_6y_2y_6y_2y_4^{-1}y_2^{-1}y_6^{-1} = 1,$$
$$y_{11}y_3y_5y_2y_1y_2y_1^2y_6y_3y_{11}y_{10}y_7y_2y_6y_2y_4y_{10}^{-1}y_6y_2y_{10}y_2^{-1}y_6^{-1}y_9 = 1,$$
$$y_{11}y_{10}y_7y_2y_{10}^{-1}y_6y_2^2y_{10}^{-1}y_6y_2y_5^{-1}y_3^{-1}y_2^{-1}y_6^{-1}y_5y_9y_4y_{10}y_9y_2y_6y_3 = 1.$$

(c) *D has a faithful permutation representation PD of degree 1024 with stabilizer $U = \langle y_1, y_4, y_{10}\rangle$.*

(d) *D has a unique normal non-abelian subgroup Q of order 1024 with center $Z(Q) = \langle z_1, z_2\rangle$ of order 4, where $z_1 = y_9y_1y_2y_1y_2$, $z_2 = y_9y_{11}y_4y_1$. Furthermore, Q has a complement $W = \langle w_1, w_2, w_3\rangle$ in D of order $|W| = 2^7{\cdot}3{\cdot}5$, where $w_1 = y_{11}y_{10}^2y_4^{-1}y_1y_6y_{10}^{-1}y_8y_{10}y_8y_2$, $w_2 = y_{11}y_{10}^{-1}y_1y_6y_{10}y_4y_8y_{10}y_7$ and $w_3 = y_{11}y_{10}y_3y_4y_1y_6y_{10}y_4y_8y_{10}y_7$.*

(e) *Let $\alpha : D \to D_1 = D/Z(Q)$ be the canonical epimorphism with kernel $ker(\alpha) = Z(Q)$. Let $V = \alpha(Q)$ and let $u_i = \alpha(y_i) \in D_1$ for $i = 1, 2, ..., 11$. Then V is an elementary abelian normal subgroup of order 2^8 of D_1 having a complement $W_1 = \langle k_1, k_2, k_3\rangle \cong W$, where $k_j = \alpha(w_j)$ for $j = 1, 2, 3$. Furthermore, D_1 has a faithful permutation representation of degree 256 with stabilizer W_1, and $V = \langle q_l | 1 \le l \le 8\rangle$, where*

$$q_1 = u_5u_9u_4u_{10}^{-1}u_4^{-1}u_1u_6u_3u_1u_{10},$$
$$q_2 = u_3u_{11}u_4u_1u_7u_3, \quad q_3 = u_9u_7, \quad q_4 = u_{11}u_1u_6u_4,$$
$$q_5 = u_9u_{11}u_3u_5u_1u_7u_{10}u_8u_3u_6,$$
$$q_6 = u_9u_{11}u_3u_{11}u_{10}u_1u_6u_3u_1^{-1}u_3u_7,$$

$$q_7 = u_5 u_{11} u_3 u_9 u_{11} u_4^{-1} u_6 u_3 u_8 u_{10}^{-1} u_7,$$
$$q_8 = u_9 u_3 u_4^{-1} u_1 u_6 u_3 u_8 u_{10}^{-1} u_7 u_1.$$

(f) *The conjugate action of the three generators k_j of W_1 on V w.r.t. the basis $\mathcal{B} = \{q_l | 1 \leq l \leq 8\}$ of V is given by the following matrices:*

$$Mk_1 = \begin{pmatrix} 1&1&0&0&0&0&0&0\\ 0&1&0&0&0&0&0&0\\ 0&0&1&1&0&0&0&0\\ 0&0&0&1&0&0&0&0\\ 0&0&0&0&0&0&1&0\\ 0&1&0&0&1&1&1&1\\ 0&0&0&0&1&0&0&0\\ 0&0&0&0&0&0&0&1 \end{pmatrix}, \quad Mk_2 = \begin{pmatrix} 1&0&1&0&0&0&0&0\\ 1&1&1&1&0&0&0&0\\ 1&1&0&1&0&0&0&0\\ 1&0&0&1&0&0&0&0\\ 0&1&0&0&0&1&0&0\\ 0&0&0&1&0&0&1&1\\ 0&1&0&0&1&0&1&1\\ 1&0&1&0&0&0&1&0 \end{pmatrix},$$

$$Mk_3 = \begin{pmatrix} 1&1&0&0&0&0&0&0\\ 1&0&0&0&0&0&0&0\\ 1&0&1&1&0&0&0&0\\ 1&1&1&0&0&0&0&0\\ 1&0&0&0&0&1&0&0\\ 0&1&0&0&1&1&0&0\\ 0&0&0&0&1&0&1&1\\ 1&0&0&0&0&1&1&0 \end{pmatrix}.$$

(g) *The Fitting subgroup A of W_1 is elementary abelian of order* 16 *and generated by $a_1 = k_1 k_3 k_2 k_1 k_2^3$, $a_2 = k_2 k_3^2 k_2 k_3^2$, $a_3 = k_2 k_1 k_2^3 k_1 k_3$, $a_4 = k_1 k_3 k_2 k_3 k_2 k_1 k_2$. It has a complement $L = \langle k_1, k_2 \rangle$ in W_1 which is isomorphic to the symmetric group S_5.*

(h) *A is a maximal elementary abelian normal subgroup of the Sylow 2-subgroup $S = \langle A, k_1, r \rangle$ of W_1 with center $Z(S) = \langle u \rangle$, where $r = k_2 k_1 k_3 k_2^2 k_1 k_2 k_3$ and $u = k_1 (k_3 k_2)^2 k_1 k_2$. The centralizer $C_{W_1}(u)$ of u has order $2^7 \cdot 3$. It is generated by S and the element $d = (k_1 k_2)^2 (k_2 k_1)^2$ of order* 3.

(i) *Let MW_1 be the subgroup of $\mathrm{GL}_8(2)$ generated by the matrices of the conjugate action of the generators of W_1 on V with respect to the basis $\mathcal{B}$. Let Ma_i, Mu and Md be the corresponding matrix of a_i, u and d, respectively. Let $MX = C_{\mathrm{GL}_8(2)}(Mu) \cap C_{\mathrm{GL}_8(2)}(Md)$. Then MX has an abelian Sylow 3-subgroup MT of order 3^3 and MT contains a matrix*

$$Mx = \begin{pmatrix} 0&1&0&0&0&0&0&0\\ 1&1&0&0&0&0&0&0\\ 1&1&0&1&0&0&0&1\\ 1&0&1&1&0&1&1&1\\ 0&0&0&1&1&1&0&0\\ 0&1&1&0&1&0&0&0\\ 1&1&1&1&1&0&0&0\\ 0&1&1&1&0&0&0&0 \end{pmatrix}$$

of order 3 *such that $MC = \langle C_{MW_1}(Mu), Mx \rangle$ has order $2^7 \cdot 3^2$.*

(j) *Let* $MK = \langle MC, MW_1 \rangle$. *Then the derived subgroup* $MK^{'}$ *of* MK *is a simple group of order* $|MK^{'}| = 2^6 \cdot 3^4 \cdot 5$, *which is isomorphic to the unitary group* $U_4(2)$.

(k) MK *is an irreducible subgroup of* $\mathrm{GL}_8(2)$ *generated by the matrices* $m_1 = Mk_1$, $m_2 = Mk_2$, $m_3 = Mk_3$, $m_4 = Mx$ *of respective orders* 2, 5, 3 *and* 3. *With respect to this set of generators* MK *has the following set* $\mathcal{R}(K)$ *of defining relations:*

$$\begin{aligned}
&m_1^2 = m_2^5 = m_3^3 = m_4^3 = 1, \quad (m_3, m_4) = 1, \quad (m_2^{-1}m_1)^4 = 1,\\
&m_3^{-1}m_2m_1m_2^{-1}m_3m_1 = (m_2^{-1}m_1)^4 = 1,\\
&m_3^{-1}m_2m_3^{-1}m_1m_2m_3m_1m_2^{-1} = (m_2m_3^{-1})^4 = 1,\\
&(m_4, m_2^{-1}, m_4) = 1, \quad m_1m_4^{-1}m_1m_4^{-1}m_1m_4m_1m_4 = 1,\\
&m_2^{-1}m_3^{-1}m_1m_2^{-2}m_3m_4m_1m_4^{-1} = 1,\\
&m_2m_3^{-1}m_2m_4m_2^2m_3^{-1}m_2^{-1}m_3m_4 = 1.
\end{aligned}$$

(l) *Let* K *be the finitely presented group constructed in (k). Then* V *is an irreducible* 8*-dimensional representation of* K *over* $F = GF(2)$ *of second cohomological dimension* $dim_F[H^2(K, V)] = 0$. *Furthermore,* K *has a faithful permutation representation* PK *of degree* 640 *having a stabilizer which is the Sylow* 3*-subgroup* $\langle (m_2m_4)^2, (m_1m_3^2m_4)^2 \rangle$ *of* K.

(m) *Let* $H_1 = \langle m_i, q_j | 1 \le i \le 4, 1 \le j \le 8 \rangle$ *be the split extension of* K *by* V. *Then* H_1 *has a set* $\mathcal{R}(H_1)$ *of defining relations consisting of* $\mathcal{R}(K)$, $\mathcal{R}_1(V \rtimes K)$ *and the following set* $\mathcal{R}_2(V \rtimes K)$ *of essential relations:*

$$\begin{aligned}
&m_1q_1m_1^{-1}q_1q_2 = 1, \quad m_1q_2m_1^{-1}q_2 = 1, \quad m_1q_3m_1^{-1}q_3q_4 = 1,\\
&m_1q_4m_1^{-1}q_4 = 1, \quad m_1q_5m_1^{-1}q_7 = 1, \quad m_1q_6m_1^{-1}q_2q_5q_6q_7q_8 = 1,\\
&m_1q_7m_1^{-1}q_5 = 1, \quad m_1q_8m_1^{-1}q_8 = 1, \quad m_2q_1m_2^{-1}q_1q_2q_3 = 1,\\
&m_2q_2m_2^{-1}q_3q_4 = 1, \quad m_2q_3m_2^{-1}q_2q_3 = 1, \quad m_2q_4m_2^{-1}q_1q_2q_3q_4 = 1,\\
&m_2q_5m_2^{-1}q_1q_2q_6q_7 = 1, \quad m_2q_6m_2^{-1}q_3q_4q_5 = 1, \quad m_2q_7m_2^{-1}q_1q_8 = 1,\\
&m_2q_8m_2^{-1}q_2q_3q_4q_6q_8 = 1, \quad m_3q_1m_3^{-1}q_2 = 1, \quad m_3q_2m_3^{-1}q_1q_2 = 1,\\
&m_3q_3m_3^{-1}q_1q_4 = 1, \quad m_3q_4m_3^{-1}q_1q_2q_3q_4 = 1, \quad m_3q_5m_3^{-1}q_1q_5q_6 = 1,\\
&m_3q_6m_3^{-1}q_2q_5 = 1, \quad m_3q_7m_3^{-1}q_5q_8 = 1, \quad m_3q_8m_3^{-1}q_1q_6q_7q_8 = 1,\\
&m_4q_1m_4^{-1}q_1q_2 = 1, \quad m_4q_2m_4^{-1}q_1 = 1, \quad m_4q_3m_4^{-1}q_2q_6q_7q_8 = 1,\\
&m_4q_4m_4^{-1}q_1q_2q_6q_7 = 1, \quad m_4q_5m_4^{-1}q_1q_2q_7q_8 = 1, \quad m_4q_6m_4^{-1}q_5q_6q_8 = 1,\\
&m_4q_7m_4^{-1}q_1q_2q_3q_4q_5q_7 = 1, \quad m_4q_8m_4^{-1}q_1q_3q_6q_7 = 1.
\end{aligned}$$

(n) *$dim_F[H^2(H_1, F)] = 4$ and there exists a unique central extension H of H_1 of order $|H| = 2^{17} \cdot 3^4 \cdot 5$ whose Sylow 2-subgroups are isomorphic to a Sylow 2-subgroup of D. Also, H is a split extension of K by its Fitting subgroup Q, and H is isomorphic to the finitely presented group $H = \langle h_i | 1 \leq i \leq 14 \rangle$ with the following set $\mathcal{R}(H)$ of defining relations:*

$$
\begin{aligned}
&h_1^2 = h_2^5 = h_3^3 = h_4^3 = h_5^2 = h_6^2 = h_7^2 = h_8^2 = h_9^2 = 1,\\
&h_{11}^2 = h_{12}^2 = h_{13}^2 = h_{14}^2 = 1, \quad h_{10}^2 h_{14}^{-1} = 1,\\
&(h_i, h_{14}) = 1 \quad \textit{for} \quad 1 \leq i \leq 13,\\
&(h_j, h_{13}) = 1 \quad \textit{for} \quad 2 \leq j \leq 12,\\
&(h_3, h_4) = (h_5, h_6) = (h_5, h_7) = (h_5, h_8) = (h_5, h_9) = (h_5, h_{12}) = 1,\\
&(h_6, h_7) = (h_8, h_{10}) = (h_{10}, h_{12}) = (h_{11}, h_{12}) = 1,\\
&h_1^{-1} h_{13} h_1 h_{13}^{-1} h_{14}^{-1} = 1, \quad (h_2 h_3^{-1})^4 = 1, \quad (h_4, h_2^{-1}, h_4) = 1,\\
&h_3^{-1} h_2 h_1 h_2^{-1} h_3 h_1 = (h_2^{-1} h_1)^4 = h_3^{-1} h_2 h_3^{-1} h_1 h_2 h_3 h_1 h_2^{-1} = 1,\\
&h_1 h_4^{-1} h_1 h_4^{-1} h_1 h_4 h_1 h_4 = h_2^{-1} h_3^{-1} h_1 h_2^{-2} h_3 h_4 h_1 h_4^{-1} = 1,\\
&h_2 h_3^{-1} h_2 h_4 h_2^2 h_3^{-1} h_2^{-1} h_3 h_4 = h_1 h_5 h_1^{-1} h_5 h_6 h_{13}^{-1} = 1,\\
&h_1 h_6 h_1^{-1} h_6 h_{14}^{-1} = h_1 h_7 h_1^{-1} h_7 h_8 h_{14}^{-1} = 1,\\
&h_1 h_8 h_1^{-1} h_8 = h_1 h_9 h_1^{-1} h_{11} h_{13}^{-1} h_{14}^{-1} = 1,\\
&h_1 h_{10} h_1^{-1} h_6 h_9 h_{10} h_{11} h_{12} h_{13}^{-1} = h_1 h_{11} h_1^{-1} h_9 h_{13}^{-1} = 1,\\
&h_1 h_{12} h_1^{-1} h_{12} h_{14}^{-1} = h_2 h_5 h_2^{-1} h_5 h_6 h_7 h_{14}^{-1} = 1,\\
&h_2 h_6 h_2^{-1} h_7 h_8 h_{13}^{-1} h_{14}^{-1} = h_2 h_7 h_2^{-1} h_6 h_7 h_{14}^{-1} = 1,\\
&h_2 h_8 h_2^{-1} h_5 h_6 h_7 h_8 = h_2 h_9 h_2^{-1} h_5 h_6 h_{10} h_{11} h_{13}^{-1} h_{14}^{-1} = 1,\\
&h_2 h_{10} h_2^{-1} h_7 h_8 h_9 h_{13}^{-1} = h_2 h_{11} h_2^{-1} h_5 h_{12} h_{14}^{-1} = 1,\\
&h_2 h_{12} h_2^{-1} h_6 h_7 h_8 h_{10} h_{12} = h_3 h_5 h_3^{-1} h_6 h_{14}^{-1} = 1,\\
&h_3 h_6 h_3^{-1} h_5 h_6 h_{13}^{-1} = h_3 h_7 h_3^{-1} h_5 h_8 = h_3 h_8 h_3^{-1} h_5 h_6 h_7 h_8 = 1,\\
&h_3 h_9 h_3^{-1} h_5 h_9 h_{10} h_{14}^{-1} = h_3 h_{10} h_3^{-1} h_6 h_9 = 1,\\
&h_3 h_{11} h_3^{-1} h_9 h_{12} h_{14}^{-1} = h_3 h_{12} h_3^{-1} h_5 h_{10} h_{11} h_{12} h_{13}^{-1} h_{14}^{-1} = 1,\\
&h_4 h_5 h_4^{-1} h_5 h_6 h_{13}^{-1} h_{14}^{-1} = h_4 h_6 h_4^{-1} h_5 h_{14}^{-1} = 1,\\
&h_4 h_7 h_4^{-1} h_6 h_{10} h_{11} h_{12} h_{14}^{-1} = 1,\\
&h_4 h_8 h_4^{-1} h_5 h_6 h_{10} h_{11} h_{13}^{-1} h_{14}^{-1} = 1,\\
&h_4 h_9 h_4^{-1} h_5 h_6 h_{11} h_{12} h_{14}^{-1} = 1,\\
&h_4 h_{10} h_4^{-1} h_9 h_{10} h_{12} h_{13}^{-1} h_{14}^{-1} = 1,
\end{aligned}
$$

$$h_4h_{11}h_4^{-1}h_5h_6h_7h_8h_9h_{11}h_{13}^{-1} = h_4h_{12}h_4^{-1}h_5h_7h_{10}h_{11}h_{14}^{-1} = 1,$$
$$h_5^{-1}h_{10}^{-1}h_5h_{10}h_{14}^{-1} = h_5^{-1}h_{11}^{-1}h_5h_{11}h_{14}^{-1} = 1,$$
$$(h_6, h_8) = h_6^{-1}h_9^{-1}h_6h_9h_{14}^{-1} = h_6^{-1}h_{10}^{-1}h_6h_{10}h_{14}^{-1} = 1,$$
$$h_6^{-1}h_{11}^{-1}h_6h_{11}h_{14}^{-1} = (h_6, h_{12}) = (h_7, h_8) = (h_7, h_9) = 1,$$
$$h_7^{-1}h_{10}^{-1}h_7h_{10}h_{14}^{-1} = h_7^{-1}h_{11}^{-1}h_7h_{11}h_{14}^{-1} = 1,$$
$$h_7^{-1}h_{12}^{-1}h_7h_{12}h_{14}^{-1} = h_8^{-1}h_9^{-1}h_8h_9h_{14}^{-1} = 1,$$
$$h_8^{-1}h_{11}^{-1}h_8h_{11}h_{14}^{-1} = (h_8, h_{12}) = (h_9, h_{10}) = 1,$$
$$(h_9, h_{11}) = (h_9, h_{12}) = h_{10}^{-1}h_{11}^{-1}h_{10}h_{11}h_{14}^{-1} = 1.$$

Proof (a) By Table 5.5.3 the split extension $E_2 = V_2 \rtimes \mathsf{M}_{22}$ has a unique conjugacy class of involutions of highest defect. It is represented by $z = (iv_1)^2$ and its centralizer $D = C_{E_2}(z)$ has order $2^{17} \cdot 3 \cdot 5$.

(b) Using the faithful permutation representation of E_2 with stabilizer M_{22} and the MAGMA command `FPGroupStrong(PD)` the presentation of D given in the statement has been calculated.

(c) Another application of MAGMA yields that the finitely presented group given in (b) has a faithful permutation representation PD of degree 1024 with stabilizer $X = \langle y_1, y_4, y_{10}\rangle$.

(d) Using this permutation representation PD of D and the MAGMA command `NormalSubgroups(PD)` it follows that D has a unique non-abelian normal subgroup Q of order 2^{10} with center $Z(Q) = \langle z_1, z_2\rangle$ of order 4. The words of z_1, z_2 in the generators y_i of D given in the statement have also been obtained computationally. Another application of MAGMA determined the generators of the stated complement $W = \langle w_1, w_2, w_3\rangle$ of Q in D.

(e) Since Q is a characteristic subgroup of D its center $Z(Q)$ is normal in D. Let $\alpha : D \to D_1 = D/Z(Q)$ be the canonical epimorphism with kernel $ker(\alpha) = Z(Q)$. Let $V = \alpha(Q)$ and let $u_i = \alpha(y_i) \in D_1$ for $i = 1, 2, ..., 11$. Then $W_1 = \alpha(W) = \langle k_1, k_2, k_3\rangle$ is a complement of V, where $k_j = \alpha(w_j)$. Furthermore, D_1 has a faithful permutation representation PD_1 of degree 256 with stabilizer W_1. Another application of MAGMA now yields that the V is elementary abelian and generated by the eight elements q_l given in the statement.

(f) This assertion has been verified computationally.

(g) Using the faithful permutation representation PD_1 and MAGMA it is straightforward to see that the Fitting subgroup $A = \langle a_1, a_2, a_3, a_4\rangle$ of W_1 is elementary abelian of order 16, and that $L = \langle k_1, k_2\rangle$ is a complement of A in W_1. Also, the generators a_i of A and the isomorphism $L \cong S_5$ have been obtained by this application of MAGMA.

(h) Another calculation with MAGMA in the permutation group PD_1 shows that $S = \langle A, k_1, r \rangle$ is a Sylow 2-subgroup of W_1 of order 2^7, where $r = k_2 k_1 k_3 k_2^2 k_1 k_2 k_3$. Furthermore, A is the unique maximal elementary abelian normal subgroup of S, and the involution $u = k_1 (k_3 k_2)^2 k_1 k_2$ generates the center $Z(S)$ of S. Its centralizer $C_{W_1}(u) = \langle S, d \rangle$ has order $2^7 \cdot 3$, and $d = (k_1 k_2)^2 (k_2 k_1)^2$ generates a cyclic Sylow 3-subgroup of $C_{W_1}(u)$.

(i) In [72] H. Kim and the author apply Algorithm 1.3.8 to construct a larger centralizer of u by using the anti-isomorphism between PW_1 and its image in $\mathrm{GL}_8(2)$ induced by the conjugate action of W_1 on V. In particular, the matrix d w.r.t the basis $\mathcal{B}$ is $Md = (Mk_1 Mk_2)^2 (Mk_2 Mk_1)^2$ and $Mu = Mk_2 Mk_1 (Mk_2 Mk_3)^2 Mk_1$. Let MW_1 be the subgroup of $\mathrm{GL}_8(2)$ generated by the matrices Mk_1, Mk_2 and Mk_3 in $\mathrm{GL}_8(2)$.

Another application of MAGMA in $\mathrm{GL}_8(2)$ shows that

$$MX = C_{\mathrm{GL}_8(2)}(Mu) \cap C_{\mathrm{GL}_8(2)}(Md)$$

has an abelian Sylow 3-subgroup MT of order 3^3 and $|MX| = 2^{12} \cdot 3^3 \cdot 5$. Furthermore, it follows that the matrix Mx of the statement belongs to the set of all matrices $My \in MX$ of order 3 such that $|\langle C_{MW_1}(Mu), My \rangle| = 2^7 \cdot 3^2$. In particular, Mu is the unique central involution of

$$MC = \langle C_{MW_1}(Mu), Mx \rangle \quad \text{and} \quad |MC| = 2^7 \cdot 3^2.$$

(j) This statement has been verified by means of the previous results and MAGMA.

(k) Let MK be the subgroup of $\mathrm{GL}_8(2)$ generated by the matrices $m_1 = Mk_1$, $m_2 = Mk_2$, $m_3 = Mk_3$, $m_4 = Mx$. Taking a Sylow 3-subgroup of MK as a stabilizer one obtains a faithful permutation representation PK of this matrix group having degree 640. Then the presentation of the finitely generated group $K = \langle m_1, m_2, m_3, m_4 \rangle$ has been calculated by means of the MAGMA command `FPGroup(PK)`.

(l) All assertions of this statement follow easily from (j), (k) and Holt's Algorithm 7.4.5 of [92] implemented in MAGMA.

(m) The presentation of the split extension $H_1 = V \rtimes K$ has been calculated by means of Lemma 1.4.7 using the presentation of K given in (k).

(n) Assertion (e) states that $D_1 = D/Z(D)$ splits over $V = Q/Z(D)$ with complement W_1. By Step 5 of Algorithm 1.3.8 the required centralizer $H = C_G(z)$ of $z = z_1$ in a simple (at this time unknown) target group G has to have an odd index $|H : D| = |K : W|$. Theorem 1.4.15 of [92] implies that such a group H is a central extension of H_1 by a central subgroup $Z(H) = \langle z_1, z_2 \rangle$ having a normal complement Q containing $Z(H)$ such that $V = Q/Z(H)$ has a complement isomorphic to K.

Since $Z(H) = Z(D) = \langle z_1, z_2 \rangle$ is a Klein 4-group the central extension H has to be constructed in two steps.

Clearly, $H_1 = \langle m_1, m_2, m_3, m_4, v_1, v_2, v_3, v_4, v_5, v_6, v_7, v_8 \rangle$ has a faithful permutation PH_1 of degree 256 with stabilizer K. Let FPH_1 be the presentation of H_1 given in (m). Then we apply Holt's Algorithm 7.4.5 of [92] implemented in MAGMA [52] to the trivial matrix representation of H_1 over $F = GF(2)$. It yields that the second cohomological dimension $dim_F[H^2(H_1, F)] = 4$. Thus the first central extension H_2 of H_1 by a central involution $y \in Z(D)$ is one of the 16 central extensions $E_{a,b,c,d}$, $0 \le a, b, c, d \le 1$. As D does not split over $Z(D)$ the group H_2 can only be isomorphic to a non-split extension. Let $CQ :=$ `GModule(PH_1, FEalg)` be the trivial module of the matrix algebra `FEalg` generated by the 12 identity matrices corresponding to the 12 generators of H_1. Using the MAGMA commands

```
P_H :=ExtensionProcess(PH_1,CQ,FPH_1)
```

and

```
E_{a,b,c,d}:=Extension(P,[a,b,c,d]),
```

with $a, b, c, d \in F$, we obtain a presentation for each of the 15 non-split central extensions $E_{a,b,c,d}$, where $(a, b, c, d) \neq (0, 0, 0, 0)$. Another application of Theorem 1.4.15 of [92] implies that H_2 has a normal complement Q_2 containing $Z_2 = \langle y \rangle$ such that $Q_1 = Q_2/Z_2$. In particular, H_2 has a faithful permutation representation of degree 512, and its Sylow 2-subgroups are isomorphic to the ones of D/Z_2. Constructing the corresponding permutation representations of the groups $E_{a,b,c,d}$ it follows that only the three groups $E_{0,1,0,0}$, $E_{1,0,0,0}$ and $E_{1,1,0,0}$ have a faithful permutation representation of degree 512 whose stabilizer is a subgroup isomorphic to L. As $y \in \{z_1, z_2, z_3 = z_1 z_2\}$ another application of MAGMA yields that only $E_{1,0,0,0}$ has a Sylow 2-subgroup which is isomorphic to the ones of exactly one factor group $D/\langle z_k\rangle$, where $1 \leq k \leq 3$. In fact, this $k = 1$.

By construction, the extension group $H_2 = E_{1,0,0,0}$ has a complement and therefore a faithful permutation representation of degree 512. Its central involution is the new 13th generator. Applying Holt's Algorithm 7.4.5 of [92] implemented in MAGMA [52] again to the trivial matrix representation of H_2 over $F = GF(2)$ it follows that the second cohomological dimension $dim_F[H^2(H_1, F)] = 4$. After constructing the corresponding permutation representations of the 15 non-split extension groups $F_{a,b,c,d}$, H. Kim and the author observed in [72] that only the three groups $F_{0,0,1,0}$, $F_{1,0,0,0}$ and $F_{1,0,1,0}$ have a faithful permutation representation of degree 1024 whose stabilizer is a subgroup isomorphic to K. The isomorphism tests with MAGMA yield that only $F_{1,0,1,0}$ has a Sylow 2-subgroup which is isomorphic to a Sylow 2-subgroup of D. Hence $H = C_G(z) := F_{1,0,1,0}$. Its presentation is given in the statement. This completes the proof. □

Lemma 5.1.2 *Keep the notation of Lemma 4.1.1 and Proposition 5.1.1. Let $H = \langle h_i | 1 \leq i \leq 14\rangle$ be the finitely presented group constructed in Proposition 5.1.1. Then the following statements hold.*

(a) *H has a faithful permutation representation of degree 1024 with stabilizer $\langle h_1, h_2, h_3, h_4\rangle$.*

(b) *Each Sylow 2-subgroup S of H has a unique maximal elementary abelian normal subgroup A of order 2^{10} and $N_H(A) \cong D = C_{E_2}(z)$.*

(c) *There is a Sylow 2-subgroup S such that $D = N_H(A) = \langle x, y\rangle$, where*

$$x = [h_2h_4h_{12}h_4(h_2h_4)^4]^2,$$
$$y = ((h_2h_4^2)^2h_{12}h_4^2h_2h_1h_4^2h_2h_4h_1h_4h_2)^2(h_2h_4^2h_2)^3h_4^2h_{12}h_4^2 h_2h_1h_4h_{12}h_4$$

have respective orders 2 and 6. Furthermore, $H = \langle x, y, h\rangle$, where $h = h_4$ has order 3.

(d) *The amalgam $H \leftarrow D \rightarrow E_2$ has Goldschmidt index 1.*

(e) *A system of representatives r_i of the 115 conjugacy classes of H and the corresponding centralizers orders $|C_H(r_i)|$ are given in Table 5.5.1.*

(f) *A system of representatives d_i of the 97 conjugacy classes of D and the corresponding centralizers orders $|C_D(d_i)|$ are given in Table 5.5.2.*

(g) *Let $\sigma : N_H(A) \to D = C_{E_2}$ be the isomorphism given in (b) and $e = t \in E_2$. Then e has order 3 and $E_2 = \langle \sigma(D), e \rangle$. A system of representatives e_i of the 43 conjugacy classes of E_2 and the corresponding centralizers orders $|C_{E_2}(e_i)|$ are given in Table 5.5.3.*

(h) *The character tables of H, E_2 and D are given in Tables 5.6.2 and 5.6.1 of Section 5.6 and in Table* `DVD.1.3.2` *of the accompanying DVD, respectively.*

Proof (a) By Proposition 5.1.1(n) H has a faithful permutation representation PH of degree 1024.

Using it and MAGMA it is straightforward to verify statements (b), (c) and (g). The words for the generators x and y of D in the generators h_i of H have been found by means of Algorithm 1.4.3.

(d) The Goldschmidt index has been calculated by means of Kratzer's Algorithm 7.1.10 of [92].

The systems of representatives of the conjugacy classes of H, D and E_2 have been calculated by means of PH, MAGMA and Kratzer's Algorithm 5.3.18 of [92].

(h) The three character tables have been calculated using PH and MAGMA. □

5.2 Construction of Fischer's simple group Fi_{22}

By Lemma 5.1.2 the amalgam $H \leftarrow D \rightarrow E_2$ constructed in Section 5.1 satisfies the conditions of Step 5 of Algorithm 1.3.8. Therefore H. Kim and the author applied Algorithm 7.4.8 of [92] in [72] to give a new existence proof for Fischer's sporadic group Fi_{22}. It is presented in this section.

Theorem 5.2.1 (Kim–Michler) *Keep the notation of Lemma 5.1.2 and Proposition 5.1.1. Using the notation of the three character Tables 5.6.2,* `DVD.1.3.2` *and 5.6.1 of the groups H, D and E_2, respectively, the following statements hold.*

(a) *There is a unique compatible pair* $(\chi, \tau) \in mf\,char_{\mathbb{C}}(H) \times mf\,char_{\mathbb{C}}(E_2)$ *of degree* 78 *of the groups* $H = \langle D, h\rangle$ *and* $E_2 = \langle D, e\rangle$:

$$(\chi, \tau) = (\chi_4 + \chi_{20} + \chi_{\mathbf{17}}, \tau_1 + \tau_{\mathbf{6}})$$

with common restriction

$$\tau_{|D} = \chi_{|D} = \psi_1 + \psi_8 + \psi_{41} + \psi_{\mathbf{33}},$$

where irreducible characters with bold face indices denote faithful irreducible characters.

(b) *Let* $\mathfrak{V}$ *and* $\mathfrak{W}$ *be the up to isomorphism uniquely determined faithful semi-simple multiplicity-free* 78*-dimensional modules of* H *and* E_2 *over* $F = GF(13)$ *corresponding to the compatible pair* (χ, τ), *respectively.*

Let $\kappa_{\mathfrak{V}} : H \to \mathrm{GL}_{78}(13)$ *and* $\kappa_{\mathfrak{W}} : E_2 \to \mathrm{GL}_{78}(13)$ *be the representations of* H *and* E_2 *afforded by the modules* $\mathfrak{V}$ *and* $\mathfrak{W}$, *respectively.*

Let $\mathfrak{h} = \kappa_{\mathfrak{V}}(h)$, $\mathfrak{x} = \kappa_{\mathfrak{V}}(x)$, $\mathfrak{y} = \kappa_{\mathfrak{V}}(y)$ *in* $\kappa_{\mathfrak{V}}(H) \leq \mathrm{GL}_{78}(13)$. *Then the following assertions hold.*

(1) $\mathfrak{V}_{|D} \cong \mathfrak{W}_{|D}$, *and there is a transformation matrix* $\mathcal{T} \in \mathrm{GL}_{23}(13)$ *such that*

$$\mathfrak{x} = \mathcal{T}^{-1}\kappa_{\mathfrak{W}}(x_1)\mathcal{T}, \qquad \mathfrak{y} = \mathcal{T}^{-1}\kappa_{\mathfrak{W}}(y_1)\mathcal{T}.$$

Let $\mathfrak{e} = \mathcal{T}^{-1}\kappa_{\mathfrak{W}}(e)\mathcal{T} \in \mathrm{GL}_{78}(13)$.

(2) *In* $\mathfrak{G}_2 = \langle \mathfrak{h}, \mathfrak{x}, \mathfrak{y}, \mathfrak{e}\rangle$ *the subgroup* $\mathfrak{E} = \langle \mathfrak{x}, \mathfrak{y}, \mathfrak{e}\rangle$ *is the stabilizer of a* 1*-dimensional subspace* $\mathfrak{U}$ *of* $\mathfrak{V}$ *such that the* $\mathfrak{G}_2$*-orbit* $\mathfrak{U}^{\mathfrak{G}_2}$ *has degree* 142155.

(3) *The four generating matrices of* $\mathfrak{G}_2$ *are stated in the accompanying DVD.*

(4) $\mathfrak{G}_2 = \langle \mathfrak{h}, \mathfrak{x}, \mathfrak{y}, \mathfrak{e}\rangle$, *and* $\mathfrak{G}_2$ *has* 65 *conjugacy classes* $\mathfrak{g}_i^{\mathfrak{G}_2}$ *with representatives* $\mathfrak{g}_i$ *and centralizer orders* $|C_{\mathfrak{G}_2}(\mathfrak{g}_i)|$ *as given in Table* `DVD.1.3.1` *on the accompanying DVD.*

(5) *The character table of* $\mathfrak{G}_2$ *coincides with that of* Fi_{22} *in the Atlas [19], pp. 156–157.*

(c) $\mathfrak{G}_2$ *is a finite simple group with* 2*-central involution* $\kappa_{\mathfrak{V}}(z) = (\mathfrak{x}\mathfrak{y}\mathfrak{x}\mathfrak{h})^{10}$ *such that*

$$C_{\mathfrak{G}_2}(\kappa_{\mathfrak{V}}(z)) = \kappa_{\mathfrak{V}}(H) \text{ and } |\mathfrak{G}_2| = 2^{17} \cdot 3^9 \cdot 5^2 \cdot 7 \cdot 11 \cdot 13.$$

Proof (a) The character tables of the groups H, E_2 and D are stated in Section 5.6 and on the DVD. In the following we use their notations. Using the faithful permutation representations PH and PE_2 of H and E_2 given in Lemmas 5.1.2 and 4.1.1, respectively, MAGMA calculates these three character tables, the fusion of the classes of $D(H)$ in H and $D(E_2)$ in E_2. Then an application of Kratzer's Algorithm 7.3.10 of [92] yields the compatible pair stated in assertion (a). It also follows that the given pair (χ, τ) is the unique compatible pair of degree 78 with respect to the fusion of the D-classes into the H- and into the E_2-classes.

(b) In order to construct the semi-simple faithful representation $\mathfrak{V}$ corresponding to the character $\chi = \chi_4 + \chi_{20} + \chi_{\mathbf{17}}$ one determines in [72] the irreducible constituents of the faithful permutation representation $(1_T)^H$ of H with stabilizer $T = \langle h_1, h_2, h_3, h_4 \rangle$ given in Lemma 5.1.2. Calculating inner products with the irreducible characters of H it follows that the irreducible character $\chi_{\mathbf{17}}$ of H of degree 32 occurs in the permutation character $(1_T)^H$. Constructing the permutation matrices of the three generators h, x and y of H over the field $K = GF(13)$ and applying the Meataxe algorithm implemented in MAGMA one obtains the matrices $\mathcal{X}_2, \mathcal{Y}_2, \mathcal{H}_2 \in \mathrm{GL}_{32}(13)$ of the generators of H in the 32-dimensional representation $\mathfrak{V}_1$. These three matrices are stated in Appendix C of [72]. Furthermore, it has been checked that $\mathfrak{V}_1$ restricted to $D(H) = \langle x, y \rangle$ is irreducible.

The irreducible modules $\mathfrak{V}_2$ and $\mathfrak{V}_3$ corresponding to the characters χ_4 of degree 6 and χ_{20} of degree 40 have been obtained as constituents of the sixth tensor power of $\mathfrak{V}_1$. A character calculation shows that the irreducible characters χ_{20} and χ_4 can be constructed by means of the following character products.

In $\chi_{\mathbf{17}}^2$ occurs the 135-dimensional irreducible character χ_{42}. Its tensor product $\chi_{42} \otimes \chi_{\mathbf{17}}$ contains the 640-dimensional character χ_{80}, and χ_{59} of degree 270 is an irreducible constituent of $\chi_{80} \otimes \chi_{\mathbf{17}}$. The tensor product $\chi_{59} \otimes \chi_{\mathbf{17}}$ has the 192-dimensional constituent χ_{47}. The required character χ_4 of degree 6 is then an irreducible composition factor of $\chi_{47} \otimes \chi_{\mathbf{17}}$.

In $\chi_{\mathbf{17}}^3$ occurs the faithful 480-dimensional irreducible character $\chi_{\mathbf{74}}$. Its tensor product $\chi_{\mathbf{74}} \otimes \chi_{\mathbf{17}}$ contains the 135-dimensional character χ_{41}, and χ_{45} of degree 160 is an irreducible constituent of $\chi_{41} \otimes \chi_{\mathbf{17}}$. Its tensor product $\chi_{45} \otimes \chi_{\mathbf{17}}$ has the required character χ_{20} of degree 40 as an irreducible composition factor.

An application of the Meataxe algorithm implemented in MAGMA produces the six matrices of the three generators of H w.r.t. fixed bases in $\mathfrak{V}_2$ and $\mathfrak{V}_3$. Their 46-dimensional diagonal joins $\mathcal{H}_1$, $\mathcal{X}_1$ and $\mathcal{Y}_1$ are stated in Appendix C of [72].

Let $\mathfrak{W}$ be the faithful representation of E_2 corresponding to the character $\tau = \tau_1 + \tau_{\mathbf{6}}$. Clearly, τ_1 corresponds to the trivial representation $\mathfrak{W}_1$ of E_2. The irreducible representation $\mathfrak{W}_2$ of $\tau_{\mathbf{6}}$ occurs as an irreducible constituent of the faithful permutation representation $(1_{T_1})^{E_2}$ of E_2 of degree 1024 with stabilizer M_{22} constructed in Lemma 4.1.1. Using the Meataxe algorithm implemented in MAGMA the 77-dimensional irreducible constituent $\mathfrak{W}_2$ of $(1_{T_1})^E$ over $K = GF(13)$ has been determined. Its restriction $(\mathfrak{W}_2)_{|D}$ to $D = \langle x_1, y_1 \rangle$ decomposes into three irreducible constituents of dimensions 5, 40 and 32. Fixing a basis B_1 of the 1-dimensional restriction $(\mathfrak{W}_1)_{|D}$ to D and bases B_2, B_2 and B_4 in the three irreducible constituents of $(\mathfrak{W}_2)_{|D}$ one obtains the two matrices $\mathfrak{W}(x_1)$ and $\mathfrak{W}(y_1)$ of the generators x_1 and y_1 of $D = D_{E_2}$ in $\mathrm{GL}_{78}(13)$. Let $B = B_1 \cup B_2 \cup B_3 \cup B_4$. Then B is a basis of the semisimple KE_2-module $\mathfrak{W}$. Let $\mathfrak{W}(e_1)$ be the matrix of the third generator e_1 of E_2 w.r.t. B.

The two KD-modules $\mathfrak{V}_{|D}$ and $\mathfrak{W}_{|D}$ described by the pairs $(\mathfrak{V}(x), \mathfrak{V}(y))$ and $(\mathfrak{W}(x_1), \mathfrak{W}(y_1))$ of matrices in $\mathrm{GL}_{78}(13)$ are isomorphic by construction. Let $Y = \mathrm{GL}(78, 13) = \mathrm{GL}_{78}(13)$. Applying then Parker's isomorphism test of Proposition 6.1.6 of [92] by means of the MAGMA command

```
IsIsomorphic(GModule(sub<Y|V(x),V(y)>),
  GModule(sub<Y|W(x1),W(y1)>))
```

one obtains a transformation matrix $\mathcal{T}$ satisfying $\mathfrak{V}(x) = (\mathfrak{W}(x_1))^{\mathcal{T}}$ and $\mathfrak{V}(y) = (\mathfrak{W}(y_1))^{\mathcal{T}}$. In view of its size this transformation matrix is decomposed into four block matrices $\mathcal{A} \in Mat_{38,38}$, $\mathcal{B} \in Mat_{38,40}$, $\mathcal{C} \in Mat_{40,38}$ and $\mathcal{D} \in Mat_{40,40}$ such that

$$\mathcal{T} := \begin{pmatrix} \mathcal{A} & \mathcal{B} \\ \mathcal{C} & \mathcal{D} \end{pmatrix}$$

has the following four block matrices.

Matrix $\mathcal{A}$:

12	.	.	.	.	.	.	.	.	.	.	.	.	.	.	.	.	.	.	.	.	.	.	.	.	.	.	.	.	.	.	.	.	.	.	.	.	.
.	11	2	9	10	1	.	.	.	.	.	.	.	.	.	.	.	.	.	.	.	.	.	.	.	.	.	.	.	.	.	.	.	.	.	.	.	.
.	6	12	3	3	7	.	.	.	.	.	.	.	.	.	.	.	.	.	.	.	.	.	.	.	.	.	.	.	.	.	.	.	.	.	.	.	.
.	11	10	5	12	4	.	.	.	.	.	.	.	.	.	.	.	.	.	.	.	.	.	.	.	.	.	.	.	.	.	.	.	.	.	.	.	.
.	11	7	12	6	11	.	.	.	.	.	.	.	.	.	.	.	.	.	.	.	.	.	.	.	.	.	.	.	.	.	.	.	.	.	.	.	.
.	11	5	9	10	1	.	.	.	.	.	.	.	.	.	.	.	.	.	.	.	.	.	.	.	.	.	.	.	.	.	.	.	.	.	.	.	.
.	.	.	.	.	.	2	9	2	5	2	3	5	6	4	12	7	7	6	2	8	8	10	4	11	9	1	6	9	6	1	12	.	.	10	8	8	1
.	.	.	.	.	.	.	8	7	.	5	7	8	3	3	9	5	11	9	1	.	12	8	1	1	.	.	11	3	7	6	1	3	12	8	12	2	8
.	.	.	.	.	.	.	5	4	11	7	8	1	3	10	1	11	2	2	3	4	8	10	4	8	7	8	9	4	7	3	5	5	11	8	7	10	8
.	.	.	.	.	.	3	3	8	3	9	.	1	9	1	2	9	3	3	5	6	11	3	6	7	10	5	6	10	5	10	2	6	12	3	3	7	3
.	.	.	.	.	.	.	.	.	4	3	9	3	3	9	9	7	6	3	10	6	11	12	8	6	5	2	7	.	10	8	.	7	.	2	11	8	11
.	.	.	.	.	.	.	.	3	10	7	6	4	8	5	3	12	6	5	6	1	9	2	.	12	.	7	2	3	.	1	4	5	.	9	8	7	6
.	.	.	.	.	.	12	11	7	4	10	11	.	12	1	11	1	3	9	6	11	6	2	9	3	12	5	5	8	3	3	1	1	6	12	3	.	10
.	.	.	.	.	.	.	8	7	.	3	10	3	5	7	8	5	4	3	.	2	10	7	9	2	6	5	1	7	5	3	8	12	9	5	12	2	.
.	.	.	.	.	.	12	6	3	10	12	4	11	12	9	12	.	12	12	1	2	9	7	10	12	6	12	1	12	3	1	11	4	7	.	2	5	7
.	.	.	.	.	.	12	6	2	7	7	7	2	9	5	4	3	12	4	9	.	2	1	10	1	12	6	8	11	10	5	5	7	1	4	12	2	4
.	.	.	.	.	.	.	5	5	5	11	7	1	12	.	4	8	5	4	9	7	4	1	5	2	5	9	8	3	1	1	8	3	.	12	.	12	8
.	.	.	.	.	.	.	8	5	11	2	1	11	3	3	5	4	9	3	11	8	8	11	12	9	5	10	6	11	10	.	7	11	.	7	8	9	4
.	.	.	.	.	.	12	3	12	9	8	10	11	11	3	8	.	.	.	6	7	10	5	7	6	11	2	5	9	9	.	1	10	8	3	10	9	10
.	.	.	.	.	.	.	8	7	4	2	4	7	3	8	4	3	2	4	6	4	3	4	7	.	6	5	2	11	10	10	.	.	8	2	3	2	.
.	.	.	.	.	.	1	7	1	12	7	1	1	8	10	.	8	3	.	5	6	6	6	2	1	10	7	6	11	4	8	5	12	11	3	4	1	.
.	.	.	.	.	.	.	5	8	2	4	5	2	5	12	2	5	10	10	9	1	5	1	3	3	12	9	3	9	6	.	2	6	4	9	10	9	9
.	.	.	.	.	.	1	2	5	2	12	11	3	5	7	6	11	2	9	7	.	8	6	10	8	6	7	3	12	6	2	3	.	4	5	5	10	9
.	.	.	.	.	.	2	9	.	7	7	9	5	6	3	5	9	.	3	2	11	.	12	8	12	11	12	5	8	9	8	.	6	12	8	.	2	4
.	.	.	.	.	.	12	6	5	8	4	3	10	3	4	5	2	7	6	1	6	4	5	3	.	11	11	4	7	8	7	9	10	10	3	5	9	5
.	.	.	.	.	.	.	8	9	2	10	12	8	2	.	5	2	11	3	12	12	10	3	12	5	4	6	12	1	5	7	6	10	10	9	.	4	9
.	.	.	.	.	.	1	7	8	5	8	1	9	.	10	2	3	11	1	5	12	3	6	9	12	3	12	11	9	11	6	10	9	1	5	.	4	12
.	.	.	.	.	.	.	.	.	.	6	3	9	4	5	10	7	5	6	11	10	5	11	3	10	4	8	11	10	11	9	2	10	1	9	12	8	5
.	.	.	.	.	.	1	7	11	2	.	8	9	5	10	4	6	11	1	11	5	12	9	4	2	12	9	10	.	4	9	.	6	5	6	8	5	8
.	.	.	.	.	.	.	8	4	3	1	6	1	.	7	10	.	4	3	3	5	8	3	1	.	3	3	12	11	1	8	9	10	5	5	11	7	6
.	.	.	.	.	.	1	12	3	7	11	5	7	8	8	5	9	8	2	4	9	.	2	5	4	4	7	10	9	2	9	5	8	6	6	9	4	3
.	.	.	.	.	.	11	4	5	1	11	12	3	6	5	1	4	8	12	2	3	8	5	8	1	12	5	4	5	9	7	8	6	4	8	11	10	5
.	.	.	.	.	.	12	6	8	1	4	9	6	7	.	3	9	1	10	9	7	11	12	1	7	6	3	7	1	3	12	1	11	5	2	1	4	3
.	.	.	.	.	.	1	12	6	9	6	8	8	2	4	3	7	4	1	3	10	1	12	11	11	5	5	9	.	4	5	8	3	10	7	.	9	.
.	.	.	.	.	.	.	.	.	.	4	2	12	5	8	.	9	8	5	5	11	11	7	7	5	5	10	7	4	5	12	10	1	7	3	2	6	.
.	.	.	.	.	.	.	.	10	3	11	9	8	12	11	8	3	8	2	1	3	10	9	2	1	1	6	2	8	4	1	10	3	2	.	9	6	5
.	.	.	.	.	.	.	.	.	.	4	12	6	5	12	12	4	5	8	4	6	2	5	6	9	2	11	4	2	9	1	5	7	4	1	12	9	12
.	.	.	.	.	.	1	2	.	7	5	11	7	4	12	5	7	7	6	6	12	10	3	6	.	9	2	.	11	10	7	10	6	7	3	7	11	.

Matrix $\mathcal{B}$:

.	.	.	.	.	.	.	.	.	.	.	.	.	.	.	.	.	.	.	.	.	.	.	.	.	.	.	.	.	.	.	.	.	.	.	.	.	.
.	.	.	.	.	.	.	.	.	.	.	.	.	.	.	.	.	.	.	.	.	.	.	.	.	.	.	.	.	.	.	.	.	.	.	.	.	.
.	.	.	.	.	.	.	.	.	.	.	.	.	.	.	.	.	.	.	.	.	.	.	.	.	.	.	.	.	.	.	.	.	.	.	.	.	.
.	.	.	.	.	.	.	.	.	.	.	.	.	.	.	.	.	.	.	.	.	.	.	.	.	.	.	.	.	.	.	.	.	.	.	.	.	.
.	.	.	.	.	.	.	.	.	.	.	.	.	.	.	.	.	.	.	.	.	.	.	.	.	.	.	.	.	.	.	.	.	.	.	.	.	.
.	.	.	.	.	.	.	.	.	.	.	.	.	.	.	.	.	.	.	.	.	.	.	.	.	.	.	.	.	.	.	.	.	.	.	.	.	.
9	12	6	8	2	5	4	10	.	.	.	.	.	.	.	.	.	.	.	.	.	.	.	.	.	.	.	.	.	.	.	.	.	.	.	.	.	.
6	11	4	6	1	8	8	10	.	.	.	.	.	.	.	.	.	.	.	.	.	.	.	.	.	.	.	.	.	.	.	.	.	.	.	.	.	.
12	4	4	6	7	11	3	3	.	.	.	.	.	.	.	.	.	.	.	.	.	.	.	.	.	.	.	.	.	.	.	.	.	.	.	.	.	.
8	3	6	6	9	4	6	2	.	.	.	.	.	.	.	.	.	.	.	.	.	.	.	.	.	.	.	.	.	.	.	.	.	.	.	.	.	.
9	11	3	6	11	5	12	11	.	.	.	.	.	.	.	.	.	.	.	.	.	.	.	.	.	.	.	.	.	.	.	.	.	.	.	.	.	.
3	1	10	12	10	10	12	4	.	.	.	.	.	.	.	.	.	.	.	.	.	.	.	.	.	.	.	.	.	.	.	.	.	.	.	.	.	.
6	4	4	1	3	2	2	11	.	.	.	.	.	.	.	.	.	.	.	.	.	.	.	.	.	.	.	.	.	.	.	.	.	.	.	.	.	.
9	8	8	4	11	7	12	6	.	.	.	.	.	.	.	.	.	.	.	.	.	.	.	.	.	.	.	.	.	.	.	.	.	.	.	.	.	.
.	3	8	1	8	10	6	12	.	.	.	.	.	.	.	.	.	.	.	.	.	.	.	.	.	.	.	.	.	.	.	.	.	.	.	.	.	.
4	8	10	10	11	9	6	1	.	.	.	.	.	.	.	.	.	.	.	.	.	.	.	.	.	.	.	.	.	.	.	.	.	.	.	.	.	.
1	4	7	5	7	9	5	8	.	.	.	.	.	.	.	.	.	.	.	.	.	.	.	.	.	.	.	.	.	.	.	.	.	.	.	.	.	.
11	1	10	1	1	.	2	8	.	.	.	.	.	.	.	.	.	.	.	.	.	.	.	.	.	.	.	.	.	.	.	.	.	.	.	.	.	.
5	12	.	8	9	1	4	10	.	.	.	.	.	.	.	.	.	.	.	.	.	.	.	.	.	.	.	.	.	.	.	.	.	.	.	.	.	.
7	12	4	6	6	12	1	10	.	.	.	.	.	.	.	.	.	.	.	.	.	.	.	.	.	.	.	.	.	.	.	.	.	.	.	.	.	.
3	3	10	3	6	8	3	4	.	.	.	.	.	.	.	.	.	.	.	.	.	.	.	.	.	.	.	.	.	.	.	.	.	.	.	.	.	.
12	1	11	3	2	1	.	2	.	.	.	.	.	.	.	.	.	.	.	.	.	.	.	.	.	.	.	.	.	.	.	.	.	.	.	.	.	.
.	5	2	8	5	8	2	6	.	.	.	.	.	.	.	.	.	.	.	.	.	.	.	.	.	.	.	.	.	.	.	.	.	.	.	.	.	.
9	1	.	10	1	3	3	4	.	.	.	.	.	.	.	.	.	.	.	.	.	.	.	.	.	.	.	.	.	.	.	.	.	.	.	.	.	.
6	7	9	11	1	7	1	6	.	.	.	.	.	.	.	.	.	.	.	.	.	.	.	.	.	.	.	.	.	.	.	.	.	.	.	.	.	.
8	4	2	1	3	7	3	9	.	.	.	.	.	.	.	.	.	.	.	.	.	.	.	.	.	.	.	.	.	.	.	.	.	.	.	.	.	.
7	.	6	9	11	7	11	.	.	.	.	.	.	.	.	.	.	.	.	.	.	.	.	.	.	.	.	.	.	.	.	.	.	.	.	.	.	.
1	8	8	6	3	1	7	.	.	.	.	.	.	.	.	.	.	.	.	.	.	.	.	.	.	.	.	.	.	.	.	.	.	.	.	.	.	.
1	9	3	8	7	3	5	6	.	.	.	.	.	.	.	.	.	.	.	.	.	.	.	.	.	.	.	.	.	.	.	.	.	.	.	.	.	.
2	3	10	11	7	5	2	9	.	.	.	.	.	.	.	.	.	.	.	.	.	.	.	.	.	.	.	.	.	.	.	.	.	.	.	.	.	.
.	10	4	1	1	4	7	6	.	.	.	.	.	.	.	.	.	.	.	.	.	.	.	.	.	.	.	.	.	.	.	.	.	.	.	.	.	.
5	1	1	8	5	2	6	3	.	.	.	.	.	.	.	.	.	.	.	.	.	.	.	.	.	.	.	.	.	.	.	.	.	.	.	.	.	.
3	10	11	3	9	9	5	5	.	.	.	.	.	.	.	.	.	.	.	.	.	.	.	.	.	.	.	.	.	.	.	.	.	.	.	.	.	.
2	6	8	7	4	11	4	7	.	.	.	.	.	.	.	.	.	.	.	.	.	.	.	.	.	.	.	.	.	.	.	.	.	.	.	.	.	.
10	5	9	7	7	.	8	6	.	.	.	.	.	.	.	.	.	.	.	.	.	.	.	.	.	.	.	.	.	.	.	.	.	.	.	.	.	.
5	2	11	2	12	12	8	1	.	.	.	.	.	.	.	.	.	.	.	.	.	.	.	.	.	.	.	.	.	.	.	.	.	.	.	.	.	.
8	10	.	8	2	7	4	6	.	.	.	.	.	.	.	.	.	.	.	.	.	.	.	.	.	.	.	.	.	.	.	.	.	.	.	.	.	.
6	.	3	11	3	12	11	10	.	.	.	.	.	.	.	.	.	.	.	.	.	.	.	.	.	.	.	.	.	.	.	.	.	.	.	.	.	.

Matrix $\mathcal{C}$:

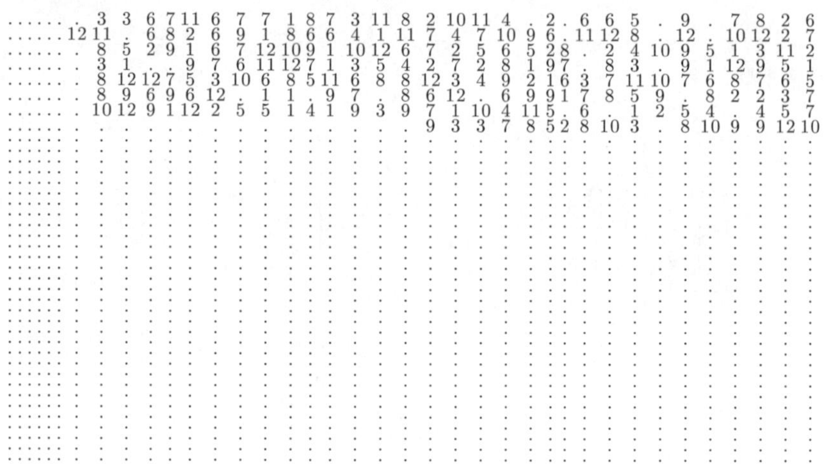

.	.	.	.	.	.	.	3	3	6	7	11	6	7	7	1	8	7	3	11	8	2	10	11	4	.	2	.	6	6	5	.	9	.	7	8	2	6
.	.	.	.	.	.	12	11	.	6	8	2	6	9	1	8	6	6	4	1	11	7	4	7	10	9	6	.	11	12	8	.	12	.	10	12	2	7
.	.	.	.	.	.	.	8	5	2	9	1	6	7	12	10	9	1	10	12	6	7	2	5	6	5	2	8	.	2	4	10	9	5	1	3	11	2
.	.	.	.	.	.	.	3	1	.	.	9	7	6	11	12	7	1	3	5	4	2	7	2	8	1	9	7	.	8	3	.	9	1	12	9	5	1
.	.	.	.	.	.	.	8	12	12	7	5	3	10	6	8	5	11	6	8	8	12	3	4	9	2	1	6	3	7	11	10	7	6	8	7	6	5
.	.	.	.	.	.	.	8	9	6	9	6	12	.	1	1	.	9	7	.	8	6	12	.	6	9	9	1	7	8	5	9	.	8	2	2	3	7
.	.	.	.	.	.	.	10	12	9	1	12	2	5	5	1	4	1	9	3	9	7	1	10	4	11	5	.	6	.	1	2	5	4	.	4	5	7
.	.	.	.	.	.	.	.	.	.	.	.	.	.	.	.	.	.	.	.	.	9	3	3	7	8	5	2	8	10	3	.	8	10	9	9	12	10
.	.	.	.	.	.	.	.	.	.	.	.	.	.	.	.	.	.	.	.	.	.	.	.	.	.	.	.	.	.	.	.	.	.	.	.	.	.
.	.	.	.	.	.	.	.	.	.	.	.	.	.	.	.	.	.	.	.	.	.	.	.	.	.	.	.	.	.	.	.	.	.	.	.	.	.
.	.	.	.	.	.	.	.	.	.	.	.	.	.	.	.	.	.	.	.	.	.	.	.	.	.	.	.	.	.	.	.	.	.	.	.	.	.
.	.	.	.	.	.	.	.	.	.	.	.	.	.	.	.	.	.	.	.	.	.	.	.	.	.	.	.	.	.	.	.	.	.	.	.	.	.
.	.	.	.	.	.	.	.	.	.	.	.	.	.	.	.	.	.	.	.	.	.	.	.	.	.	.	.	.	.	.	.	.	.	.	.	.	.
.	.	.	.	.	.	.	.	.	.	.	.	.	.	.	.	.	.	.	.	.	.	.	.	.	.	.	.	.	.	.	.	.	.	.	.	.	.
.	.	.	.	.	.	.	.	.	.	.	.	.	.	.	.	.	.	.	.	.	.	.	.	.	.	.	.	.	.	.	.	.	.	.	.	.	.
.	.	.	.	.	.	.	.	.	.	.	.	.	.	.	.	.	.	.	.	.	.	.	.	.	.	.	.	.	.	.	.	.	.	.	.	.	.
.	.	.	.	.	.	.	.	.	.	.	.	.	.	.	.	.	.	.	.	.	.	.	.	.	.	.	.	.	.	.	.	.	.	.	.	.	.
.	.	.	.	.	.	.	.	.	.	.	.	.	.	.	.	.	.	.	.	.	.	.	.	.	.	.	.	.	.	.	.	.	.	.	.	.	.
.	.	.	.	.	.	.	.	.	.	.	.	.	.	.	.	.	.	.	.	.	.	.	.	.	.	.	.	.	.	.	.	.	.	.	.	.	.
.	.	.	.	.	.	.	.	.	.	.	.	.	.	.	.	.	.	.	.	.	.	.	.	.	.	.	.	.	.	.	.	.	.	.	.	.	.
.	.	.	.	.	.	.	.	.	.	.	.	.	.	.	.	.	.	.	.	.	.	.	.	.	.	.	.	.	.	.	.	.	.	.	.	.	.
.	.	.	.	.	.	.	.	.	.	.	.	.	.	.	.	.	.	.	.	.	.	.	.	.	.	.	.	.	.	.	.	.	.	.	.	.	.
.	.	.	.	.	.	.	.	.	.	.	.	.	.	.	.	.	.	.	.	.	.	.	.	.	.	.	.	.	.	.	.	.	.	.	.	.	.
.	.	.	.	.	.	.	.	.	.	.	.	.	.	.	.	.	.	.	.	.	.	.	.	.	.	.	.	.	.	.	.	.	.	.	.	.	.
.	.	.	.	.	.	.	.	.	.	.	.	.	.	.	.	.	.	.	.	.	.	.	.	.	.	.	.	.	.	.	.	.	.	.	.	.	.
.	.	.	.	.	.	.	.	.	.	.	.	.	.	.	.	.	.	.	.	.	.	.	.	.	.	.	.	.	.	.	.	.	.	.	.	.	.
.	.	.	.	.	.	.	.	.	.	.	.	.	.	.	.	.	.	.	.	.	.	.	.	.	.	.	.	.	.	.	.	.	.	.	.	.	.
.	.	.	.	.	.	.	.	.	.	.	.	.	.	.	.	.	.	.	.	.	.	.	.	.	.	.	.	.	.	.	.	.	.	.	.	.	.
.	.	.	.	.	.	.	.	.	.	.	.	.	.	.	.	.	.	.	.	.	.	.	.	.	.	.	.	.	.	.	.	.	.	.	.	.	.
.	.	.	.	.	.	.	.	.	.	.	.	.	.	.	.	.	.	.	.	.	.	.	.	.	.	.	.	.	.	.	.	.	.	.	.	.	.
.	.	.	.	.	.	.	.	.	.	.	.	.	.	.	.	.	.	.	.	.	.	.	.	.	.	.	.	.	.	.	.	.	.	.	.	.	.
.	.	.	.	.	.	.	.	.	.	.	.	.	.	.	.	.	.	.	.	.	.	.	.	.	.	.	.	.	.	.	.	.	.	.	.	.	.
.	.	.	.	.	.	.	.	.	.	.	.	.	.	.	.	.	.	.	.	.	.	.	.	.	.	.	.	.	.	.	.	.	.	.	.	.	.
.	.	.	.	.	.	.	.	.	.	.	.	.	.	.	.	.	.	.	.	.	.	.	.	.	.	.	.	.	.	.	.	.	.	.	.	.	.
.	.	.	.	.	.	.	.	.	.	.	.	.	.	.	.	.	.	.	.	.	.	.	.	.	.	.	.	.	.	.	.	.	.	.	.	.	.
.	.	.	.	.	.	.	.	.	.	.	.	.	.	.	.	.	.	.	.	.	.	.	.	.	.	.	.	.	.	.	.	.	.	.	.	.	.
.	.	.	.	.	.	.	.	.	.	.	.	.	.	.	.	.	.	.	.	.	.	.	.	.	.	.	.	.	.	.	.	.	.	.	.	.	.
.	.	.	.	.	.	.	.	.	.	.	.	.	.	.	.	.	.	.	.	.	.	.	.	.	.	.	.	.	.	.	.	.	.	.	.	.	.
.	.	.	.	.	.	.	.	.	.	.	.	.	.	.	.	.	.	.	.	.	.	.	.	.	.	.	.	.	.	.	.	.	.	.	.	.	.
.	.	.	.	.	.	.	.	.	.	.	.	.	.	.	.	.	.	.	.	.	.	.	.	.	.	.	.	.	.	.	.	.	.	.	.	.	.

Matrix $\mathcal{D}$:

10	4	9	3	4	.	3	11	.	.	.	.	.	.	.	.	.	.	.	.	.	.	.	.	.	.	.	.	.	.	.	.	.	.	.	.	.	.	.	.
1	1	5	7	12	3	5	12	.	.	.	.	.	.	.	.	.	.	.	.	.	.	.	.	.	.	.	.	.	.	.	.	.	.	.	.	.	.	.	.
7	1	9	4	6	7	7	1	.	.	.	.	.	.	.	.	.	.	.	.	.	.	.	.	.	.	.	.	.	.	.	.	.	.	.	.	.	.	.	.
12	1	12	.	9	3	2	12	.	.	.	.	.	.	.	.	.	.	.	.	.	.	.	.	.	.	.	.	.	.	.	.	.	.	.	.	.	.	.	.
.	8	3	4	2	9	11	4	.	.	.	.	.	.	.	.	.	.	.	.	.	.	.	.	.	.	.	.	.	.	.	.	.	.	.	.	.	.	.	.
6	1	8	7	2	3	2	11	.	.	.	.	.	.	.	.	.	.	.	.	.	.	.	.	.	.	.	.	.	.	.	.	.	.	.	.	.	.	.	.
2	8	1	3	6	2	7	9	.	.	.	.	.	.	.	.	.	.	.	.	.	.	.	.	.	.	.	.	.	.	.	.	.	.	.	.	.	.	.	.
5	5	6	10	2	3	7	6	.	.	.	.	.	.	.	.	.	.	.	.	.	.	.	.	.	.	.	.	.	.	.	.	.	.	.	.	.	.	.	.
.	.	.	.	.	.	.	.	12	9	7	1	10	5	5	2	9	1	8	4	10	5	6	9	5	7	2	3	5	12	2	8	1	2	10	3	7	8	3	4
.	.	.	.	.	.	.	.	10	7	1	4	12	2	4	2	2	1	1	12	3	9	2	9	4	1	9	4	1	7	4	1	1	7	2	9	9	12	4	1
.	.	.	.	.	.	.	.	5	3	9	.	.	5	8	5	5	3	.	8	8	8	4	9	9	4	1	7	7	5	1	8	1	3	5	1	.	9	.	8
.	.	.	.	.	.	.	.	10	8	5	8	.	5	.	.	2	.	.	5	8	.	5	5	3	8	.	5	8	5	5	.	.	.	10	.	8	5	10	8
.	.	.	.	.	.	.	.	1	8	10	10	2	1	6	7	5	7	2	9	3	7	3	11	11	10	10	9	1	10	11	5	3	4	1	8	.	5	11	12
.	.	.	.	.	.	.	.	2	.	7	1	6	3	8	10	.	11	2	11	7	7	9	2	9	.	.	4	2	11	10	1	12	4	9	3	7	3	10	5
.	.	.	.	.	.	.	.	9	2	10	.	11	.	.	5	11	10	.	10	7	8	12	.	3	12	9	9	12	7	.	4	6	3	1	7	1	8	6	9
.	.	.	.	.	.	.	.	.	.	8	8	8	.	.	.	.	.	8	.	3	8	11	1	10	5	8	8	1	2	11	.	8	8	5	3	6	5	5	5
.	.	.	.	.	.	.	.	8	10	.	10	3	5	8	3	2	1	5	11	3	.	3	1	11	.	3	11	.	3	3	7	2	8	1	5	9	2	.	5
.	.	.	.	.	.	.	.	4	9	9	9	9	9	1	9	9	9	4	9	9	9	1	1	4	4	9	9	6	12	1	9	9	9	4	9	1	4	9	9
.	.	.	.	.	.	.	.	11	1	1	12	9	12	2	3	5	5	1	2	9	8	4	5	7	8	1	10	11	4	.	9	12	.	9	8	11	11	5	9
.	.	.	.	.	.	.	.	10	7	10	2	4	3	8	.	1	5	2	11	4	10	7	3	11	3	3	1	9	10	11	3	3	10	7	5	8	4	.	12
.	.	.	.	.	.	.	.	1	3	10	9	11	6	3	8	12	10	1	10	2	8	6	3	6	1	4	7	6	7	4	4	11	10	8	4	10	9	8	10
.	.	.	.	.	.	.	.	2	3	5	12	4	4	7	3	10	10	2	6	8	6	2	1	12	12	9	8	5	6	2	1	7	4	4	11	12	5	.	4
.	.	.	.	.	.	.	.	2	12	5	10	11	6	11	5	8	10	2	10	10	8	5	6	5	.	12	7	2	1	8	3	7	9	3	12	.	4	3	9
.	.	.	.	.	.	.	.	2	4	4	10	12	12	2	8	6	5	7	9	8	8	5	.	7	6	4	8	10	6	3	4	5	8	12	3	1	12	11	5
.	.	.	.	.	.	.	.	6	9	8	3	2	7	2	3	5	8	11	11	3	5	8	7	3	8	9	6	3	2	5	2	6	12	5	9	8	9	10	12
.	.	.	.	.	.	.	.	11	4	4	10	5	8	.	7	11	6	12	8	1	10	6	5	9	6	9	7	3	8	4	1	3	7	2	9	11	9	.	1
.	.	.	.	.	.	.	.	6	1	9	8	6	1	11	5	7	5	6	12	.	5	8	8	11	2	4	.	8	10	5	1	8	.	9	5	7	6	10	8
.	.	.	.	.	.	.	.	7	3	3	1	4	3	8	11	9	.	5	2	4	8	7	12	7	8	5	11	12	9	12	12	12	1	6	1	7	9	11	12
.	.	.	.	.	.	.	.	.	8	5	8	.	5	.	8	.	3	.	.	5	.	8	.	10	8	10	.	8	5	.	10	.	.	.	5	.	.	5	.
.	.	.	.	.	.	.	.	8	7	.	5	5	8	.	5	5	7	5	3	8	3	5	3	8	5	11	1	10	.	3	5	.	10	8	8	3	12	3	5
.	.	.	.	.	.	.	.	9	12	1	8	7	6	3	8	2	8	1	7	7	4	.	6	12	.	8	7	9	3	7	3	6	9	1	11	9	4	4	9
.	.	.	.	.	.	.	.	6	5	3	5	5	7	5	10	2	5	12	9	7	10	11	6	4	8	9	5	10	11	10	5	9	11	5	6	10	10	.	5
.	.	.	.	.	.	.	.	5	.	5	8	5	.	.	2	.	11	.	5	.	5	.	.	10	8	.	5	3	5	.	3	.	5	7	.	.	8	5	.
.	.	.	.	.	.	.	.	10	10	5	.	5	.	.	2	5	11	5	5	.	.	.	.	.	8	.	8	.	5	.	3	.	5	10	5	5	5	5	.
.	.	.	.	.	.	.	.	10	.	5	8	.	.	.	5	5	11	.	5	.	5	8	.	5	.	.	8	8	.	.	.	.	5	5	10	5	5	10	5
.	.	.	.	.	.	.	.	5	8	.	8	.	.	.	.	8	3	.	10	.	5	.	5	5	.	.	5	8	.	5	8	8	.	10	3	.	8	.	.
.	.	.	.	.	.	.	.	4	4	4	4	12	9	1	9	12	7	4	9	9	9	4	9	1	9	9	9	12	4	4	4	4	9	9	9	9	4	4	4
.	.	.	.	.	.	.	.	.	6	.	.	.	.	.	.	.	.	3	.	.	10	10	4	6	10	.	4	.	3	7	6	.	3	.	.	7	9	.	3
.	.	.	.	.	.	.	.	3	7	.	10	.	.	.	3	.	7	10	3	.	3	.	3	.	.	.	9	7	.	3	7	.	.	6	.	.	4	3	10
.	.	.	.	.	.	.	.	3	10	3	10	.	.	.	.	.	4	.	3	3	3	.	9	10	.	3	3	.	.	6	10	.	.	3	3	9	10	3	.

Let $\mathfrak{e} = (\mathfrak{W}(e_1))^{\mathcal{T}}$. Then $\mathfrak{E}_2 = \langle \mathfrak{x}, \mathfrak{y}, \mathfrak{e} \rangle \cong E_2$. Let $\mathfrak{G}_2 = \langle \mathfrak{h}, \mathfrak{x}, \mathfrak{y}, \mathfrak{e} \rangle$. The four generating matrices of $\mathfrak{G}_2$ are stated on the accompanying DVD.

Using the MAGMA command `CosetAction(G, E)` one obtains a faithful permutation representation of $\mathfrak{G}_2$ of degree 142155 with stabilizer $\mathfrak{E}_2$. In particular, $\mathfrak{G}_2 = 2^{17} \cdot 3^9 \cdot 5^2 \cdot 7 \cdot 11 \cdot 13$.

Using the faithful permutation representation of $\mathfrak{G}_2$ of degree 142155 and Kratzer's Algorithm 5.3.18 of [92] H. Kim and the author calculated representatives of the 65 conjugacy classes of $\mathfrak{G}_2$ in [72]. They are stated in Table `DVD.1.3.1`.

Furthermore, the character table of $\mathfrak{G}_2$ has been calculated by means of the above permutation representation and MAGMA. It coincides with the one of Fi_{22} in [19], pp. 156–157.

(d) Let $\mathfrak{z} = (\mathfrak{x}\mathfrak{y}\mathfrak{x}\mathfrak{h})^{10}$. Then $C_{\mathfrak{G}_2}(\mathfrak{z})$ contains $\mathfrak{H} = \langle \mathfrak{h}, \mathfrak{x}, \mathfrak{y} \rangle$, which is isomorphic to H. Now $|C_{\mathfrak{G}_2}(\mathfrak{z})| = |H|$ by Table `DVD.1.3.1`. Hence $C_{\mathfrak{G}_2}(\mathfrak{z}) \cong \mathfrak{H}$. The character table of $\mathfrak{G}_2$ implies that $\mathfrak{G}_2$ is a simple group. This completes the proof. □

In [110] one finds a nice presentation for Fischer's simple group Fi_{22}, which now is verified for $\mathfrak{G}_2$ too.

Corollary 5.2.2 *Keep the notation of Theorem 5.2.1. The finite simple group $\mathfrak{G}_2$ is isomorphic to the finitely presented group $G = \langle a, b, c, d, e, f, g, h, i \rangle$ with the following set $\mathcal{R}(G)$ of defining relations:*

$$\begin{aligned}
&a^2 = b^2 = c^2 = d^2 = e^2 = f^2 = g^2 = h^2 = i^2 = 1,\\
&(ab)^3 = (bc)^3 = (cd)^3 = (de)^3 = (ef)^3 = (fg)^3 = (dh)^3 = (hi)^3 = 1,\\
&(ac)^2 = (ad)^2 = (ae)^2 = (af)^2 = (ag)^2 = (ah)^2 = (ai)^2 = 1,\\
&(bd)^2 = (be)^2 = (bf)^2 = (bg)^2 = (bh)^2 = (bi)^2 = 1,\\
&(ce)^2 = (cf)^2 = (cg)^2 = (ch)^2 = (ci)^2 = (df)^2 = (dg)^2 = (di)^2 = 1,\\
&(eg)^2 = (eh)^2 = (ei)^2 = (fh)^2 = (fi)^2 = (gh)^2 = (gi)^2 = 1,\\
&(dcbdefdhi)^{10} = (abcdefh)^9 = (bcdefgh)^9 = 1.
\end{aligned}$$

Proof The presentation stated in the assertion is due to Praeger and Soicher [110], p. 110. Taking its subgroup

$$Q = \langle a, c, e, g, h, bacb, dced, fegf, dchd, dehd, (cdehi)^4 \rangle$$

as a stabilizer, Kim and the author obtained in [72] a faithful permutation representation of degree 142155 for their group $G = \langle a, b, c, d, e, f, g \rangle$. Using the faithful permutation representation of the simple group $\mathfrak{G}_2$ given in Theorem 5.2.1(c) of degree 142155 an isomorphism test in MAGMA showed then that $\mathfrak{G}_2 \cong G$. □

5.3 Sketch of a uniqueness proof

In view of the size of the book it is not possible to give here all the details of a uniqueness proof for Fischer's group Fi_{22}. It is similar to the one of Theorem 3.3.13. Therefore we mention here the relevant literature, state the uniqueness theorem and split the proof into a series of (non-trivial) exercises. It is independent of [30] and [4].

Definition 5.3.1 A finite simple group G is said to be of Fi_{22}-*type* if it has a 2-central involution z such that $C_G(z)$ is isomorphic to the finitely presented group H stated in Proposition 5.1.1.

Fischer constructed his simple group $P = \mathsf{Fi}_{22}$ as an automorphism group of a certain graph in Theorem 17.2.3 of [30]. It has a conjugacy class d^P of 3-transpositions, i.e., the order of the product xy of two elements $x, y \in D$ is 1, 2 or 3. Using the faithful permutation representation of degree 3510 of the simple group $\mathfrak{G}_2 = \langle \mathfrak{x}, \mathfrak{y}, \mathfrak{e} \rangle$ constructed in Theorem 5.2.1 one can check that $\mathfrak{D} = u^{\mathfrak{G}_2}$ is a conjugacy class of 3-transpositions generating $\mathfrak{G}_2$, where $\mathfrak{u} = (\mathfrak{x}\mathfrak{y}\mathfrak{e})^7$. However, $\mathfrak{U} = C_{\mathfrak{G}_2}(u)$ has order $2^{16} \cdot 3^6 \cdot 5 \cdot 7 \cdot 11$ by Table `DVD.1.3.1`. Hence $\mathfrak{u}$ is *not* a 2-central involution of $\mathfrak{G}_2$. So its centralizer is not the one of Definition 5.3.1.

Hunt showed in [55] the following uniqueness result.

Theorem 5.3.2 (D. C. Hunt) *Let $P = \mathsf{Fi}_{22}$ be the simple group constructed by Fischer in [30]. Let d be a fixed element in the conjugacy class $D = d^P$ of 3-transpositions of P. Let $U = C_P(d)$. Let G be any finite simple group containing an involution u such that $C = C_G(u) \cong U$. Then $G \cong P$.*

Hunt's proof uses this explicit isomorphism to show that $D_1 = u^G$ is a conjugacy class of 3-transpositions. This is not obvious at all. He determines a Sylow 2-subgroup S of order 2^{17}. Then he shows that G has three conjugacy classes of involutions u^G, n^G and the 2-central class z^G. He also determines the centralizers $H = C_G(z)$, $N = C_G(n)$. Using Thompson's group order formula he then shows that $|G| = 2^{17} \cdot 3^9 \cdot 5^2 \cdot 7 \cdot 11 \cdot 13$. This enables him to prove that G has three double cosets with respect to U. Therefore G acts as a rank-3 permutation group on the conjugacy class D_1. Now Fischer's main theorem of [29] asserts that $G \cong P$. Hunt also calculated the character table of $P = \mathsf{Fi}_{22}$ in [57].

Using Hunt's Theorem 6.4.2D. Parrott proved in [107] the following result stated only for simple groups.

Theorem 5.3.3 (D. Parrott) *Let G be a finite simple group, z an involution in G and $H = C_G(z)$. Suppose that $J = O_2(H)$ is the direct product of a group of order 2 with an extra-special group of order 2^9 such that $J' = \langle z \rangle$. If $C_H(J) \leq J$ and $H/J \cong Aut(\mathrm{U}_4(2)$ then $G \cong \mathsf{Fi}_{22}$.*

Proof See [107], Sect. 4. □

Remark 5.3.4 (a) Proposition 5.1.1 and Theorem 5.2.1 imply that the simple matrix subgroup $\mathfrak{G}_2$ has a centralizer $\mathfrak{H} = C_{\mathfrak{G}_2}(\mathfrak{z})$ of a 2-central involution $\mathfrak{z}$ which satisfies all conditions of Parrott's Theorem 5.3.3. Hence it would imply that $\mathfrak{G}_2 \cong \mathsf{Fi}_{22}$ if there were only one group H satisfying all the hypothesis of Theorem 5.3.3.

(b) By the last step of the proof of Proposition 5.1.1 the reader knows that that there are at least two non-isomorphic groups H satisfying all the conditions of Parrott's Theorem 5.3.3, but they still have non-isomorphic Sylow 2-subgroups.

(c) So Parrott's hypothesis is replaced in the following result by assuming that G be a simple group of Fi_{22}-type.

Theorem 5.3.5 *A finite simple group G of Fi_{22}-type is isomorphic to the simple subgroup $\mathfrak{G}_2$ of $\mathrm{GL}_{78}(13)$ constructed in Theorem 5.2.1.*

Proof (Sketch of proof) Let z be a 2-central involution of G. Then $C_G(z) = H = \langle x, y, h \rangle$ defined in Lemma 5.1.2. In particular, H has a Sylow 2-subgroup S having a unique maximal elementary abelian normal subgroup A of order 2^{10} such that $D = N_H(A) = \langle x, y \rangle$ has odd index in H. Furthermore, H has a faithful permutation representation of degree 1024. Its conjugacy classes are classified in Table 5.5.1. Hence we may assume that $z = (xyxh)^{10} \in H$. Since H has 13 conjugacy classes of involutions by Table 5.5.1 one first finds short generators of A by means of Algorithm 1.4.1. By a fusion argument of [107] and Theorem 4.7.3 of [92] it can be shown that z is G-conjugate to an element a of A. Hence $N_G(A) > D$ by Corollary 4.8.8 of [92].

As D/A is a maximal subgroup of M_{22} one can show that $N_G(A) \cong E_2$, where E_2 is the finitely presented group of Lemma 4.1.1. So we may assume that $E_2 = \langle x, y, e \rangle$ for some element e of E_2 of order 3. By Table 5.5.3 E_2 has five conjugacy classes of involutions. Let $u = (xye)^7$ and $w = (ye^2)$ in E. Using the fusion of D in H and of D in E one can show that z, u and w represent all conjugacy classes of involutions in G.

In order to construct $U = C_G(u)$ and $W = C_G(w)$ one first builds $C_{E_2}(u)$ and $C_{E_2}(w)$ and then applies all the steps of Algorithm 1.3.8 to get U and W as finitely presented groups. For each of the groups U and W one calculates a faithful permutation representation, a system of representatives of its conjugacy classes and its character table. Isomorphism tests with MAGMA show that $U \cong C_{\mathfrak{G}_2}(\mathfrak{u})$ and $W \cong C_{\mathfrak{G}_2}(\mathfrak{w})$, where $\mathfrak{u} = (\mathfrak{x}\mathfrak{y}\mathfrak{e})^7$ and $\mathfrak{w} = (\mathfrak{x}\mathfrak{h})^6$ are the representatives of the conjugacy classes of involutions of $\mathfrak{G}_2$ given in Table `DVD.1.3.1`. In particular, $|U| = 2^{16} \cdot 3^6 \cdot 5 \cdot 7 \cdot 11$ and $|W| = 2^{16} \cdot 3^3$.

Now determine the fusion of the conjugacy classes of involutions of H, U and W in G. Then Theorem 1.6.4 can be applied to determine the order of G. One has to show also that G has four conjugacy classes 3_i of elements of order 3, one class 5 of order 5, one class of order 7, two classes 11_k of order 11 and two classes 13_ℓ of order 13. For each of these primes p one has to construct a presentation of the normalizers $N_G(\langle p_s \rangle)$. Since these groups are small it is easy to get faithful permutation representations for them. Thus one obtains their systems of conjugacy classes and their character tables. Finally one has to determine the fusion of their classes in G.

As in Lemma 3.3.11 one shows that H is a maximal subgroup of G. Hence $G = \langle H, E_2 \rangle$. Therefore one can prove, as in Proposition 10.6.1, that G has an irreducible complex character χ of degree 78 having the same values at the conjugacy classes as the character χ_2 of Fi_{22} of the Atlas [19], p. 156. It is here where one needs the character tables of all the local subgroups constructed before.

Now the proof of Theorem 3.3.13 can be adapted easily to show that $G \cong \mathfrak{G}_2$. □

5.4 The remaining cases E_1, E_4 and E_5

In this section we summarize our results on the remaining cases. Using Theorem 5.2.1 we prove that the application Algorithm 1.3.8 to the extension group E_4 constructs the automorphism group $Aut(\mathsf{Fi}_{22})$. In the cases of E_1 and E_5 the algorithm terminates without success.

Corollary 5.4.1 *Keep the notation of Lemma 4.1.2. Let $E_4 = \langle p_1, q_1, v_1 \rangle$ be the split extension of $A_{22} = Aut(\mathsf{M}_{22})$ by its simple module V_4 of dimension 10 over $F = GF(2)$. Let $\mathfrak{A}_2$ be the automorphism group of the simple group $\mathfrak{G}_2$, and let $\mathfrak{H} = C_{\mathfrak{G}_2}(\mathfrak{z})$ be the centralizer of a 2-central involution $\mathfrak{z}$ of $\mathfrak{G}_2$. Then there is a unique maximal elementary abelian normal subgroup $\mathfrak{B}$ of a Sylow 2-subgroup $\mathfrak{S}$ of $\mathfrak{H}$ such that $N_{\mathfrak{G}_2}(\mathfrak{B}) \cong E_4$.*

Proof In order to simplify the notation in the proof we replace Gothic letters by Roman letters and $\mathfrak{G}_2$, $\mathfrak{A}_2$ by G, A, respectively. Then $G = \langle x, y, h, e\rangle$ by Theorem 5.2.1 and $t = (xye)^7$ is an involution of G with a centralizer $C_G(t)$ of index $|G : C_G(t)| = 3510$ by Table `DVD.1.3.1`. Since G is simple it has a faithful permutation representation PG of degree 3510. Using it MAGMA has been able to calculate the automorphism group A of G and its order $|A| = 2^{18} \cdot 3^9 \cdot 5^2 \cdot 7 \cdot 11 \cdot 13$. Using the command `PermutationRepresentation(AutG)` MAGMA produces a faithful permutation representation PA of A of degree 3510. Let PU be the derived subgroup of PA. Then MAGMA asserts that $PU \cong PG$ and that $|PA : PU| = 2$. Let PS be a Sylow 2-subgroup of PU and let pz be an involution in the center of PS. Let $PH = C_{PU}(pz)$. An isomorphism test shows that $PH \cong C_G(z)$, where z is a 2-central involution of the simple group G. Let $NH = N_{PA}(PH)$. Then $|NH : PH| = 2$. Furthermore, PH has a complement $PC = \langle pu\rangle$ of order 2 in NH, as has been checked computationally. Now Sylow's Theorem asserts that pu can be chosen so that $PS = PS^{pu}$. Another application of MAGMA shows that PS has a unique elementary abelian normal subgroup PB of order 2^{10}. Since PB is a characteristic subgroup of PS it is normalized by pu. Furthermore, $PA = \langle PU, pu\rangle$. MAGMA also asserts that $PD = N_{PA}(PB)$ has order $2^{18} \cdot 3 \cdot 5$. Another isomorphism test verifies that $PD \cong C_{E_4}(y)$, where y is the 2-central involution $y = (p_1^2 q_1 v)^{10}$ of E_4 given in Lemma 4.1.3. Therefore Theorem 5.2.1 implies that an application of that Algorithm 1.3.8 to the extension group E_4 returns the automorphism group A of the simple group G. □

Remark 5.4.2 Let $E_1 = V_1 \rtimes \mathsf{M}_{22} = \langle a, b, c, d, t, g, h, i, v_1\rangle$ be the split extension of M_{22} by its simple module V_1 of dimension 10 over $F = GF(2)$ defined in Lemma 4.1.1. By Lemma 4.1.3 E_1 has a unique class of 2-central involutions represented by $z = (tv_1)^3$. Its centralizer $D = C_{E_1}(z)$ has a uniquely determined non-abelian normal subgroup Q of order 512 such that $V = Q/Z(Q)$ is elementary abelian of order 2^8, where $Z(Q) = \langle z\rangle$ denotes the center of Q. Furthermore, Q has a complement W in D. The Fitting subgroup B of W has order 16 and its complement L in W is isomorphic to the alternating group A_6 and not to S_5 as in Proposition 5.1.1. Since the center of W has order 2 we applied Algorithm 7.4.8 of [92] to construct a subgroup K of $\mathrm{GL}_8(2)$ such that $|K : W|$ is odd. However, this application was not successful.

Remark 5.4.3 Let $E_5 = \langle p_2, q_2 \rangle$ be the non-split extension of A_{22} by its simple module V_3 of dimension 10 over $F = GF(2)$ defined in Lemma 4.1.2. By Lemma 4.1.3 E_5 has a unique class of 2-central involutions represented by $z = p_2^4$. Its centralizer $D = C_{E_5}(z)$ has a uniquely determined non-abelian normal subgroup Q of order 512 such that $V = Q/Z(Q)$ is elementary abelian of order 2^8, where $Z(Q) = \langle z \rangle$ denotes the center of Q. This time Q does not have a complement in D. The Fitting subgroup B of $W = D/Q$ has order 32, and its factor group $L = W/B$ is isomorphic to the symmetric group S_6. Since the center of W has order 2 we applied Algorithm 7.4.8 of [92] to construct a subgroup K of $\mathrm{GL}_8(2)$ such that $|K : W|$ is odd. However, that application was not successful.

5.5 Representatives of conjugacy classes

5.5.1 *Conjugacy classes of* $H(\mathsf{Fi}_{22}) = \langle x, y, h \rangle$

Class	*Representative*	|*Class*|	|*Centralizer*|	2P	3P	5P
1	1	1	$2^{17} \cdot 3^4 \cdot 5$	1	1	1
2_1	$(xyxh)^{10}$	1	$2^{17} \cdot 3^4 \cdot 5$	1	2_1	2_1
2_2	$(xhy^2)^9$	2	$2^{16} \cdot 3^4 \cdot 5$	1	2_2	2_2
2_3	$(xh)^6$	180	$2^{15} \cdot 3^2$	1	2_3	2_3
2_4	$(xyhyxh^2)^3$	180	$2^{15} \cdot 3^2$	1	2_4	2_4
2_5	$(yh)^6$	270	$2^{16} \cdot 3$	1	2_5	2_5
2_6	$(xy^2xy^2xh)^4$	270	$2^{16} \cdot 3$	1	2_6	2_6
2_7	$(xyxyxh)^3$	360	$2^{14} \cdot 3^2$	1	2_7	2_7
2_8	$(y)^3$	1152	$2^{10} \cdot 3^2 \cdot 5$	1	2_8	2_8
2_9	x	4320	$2^{12} \cdot 3$	1	2_9	2_9
2_{10}	$(xhyh)^6$	4320	$2^{12} \cdot 3$	1	2_{10}	2_{10}
2_{11}	$xyhxyhxyhyh$	6480	2^{13}	1	2_{11}	2_{11}
2_{12}	$(xyxyxy^2h^2)^3$	8640	$2^{11} \cdot 3$	1	2_{12}	2_{12}
2_{13}	$(xhy^2hy)^3$	17280	$2^{10} \cdot 3$	1	2_{13}	2_{13}
3_1	h	5120	$2^7 \cdot 3^4$	3_1	1	3_1
3_2	$(yh)^4$	7680	$2^8 \cdot 3^3$	3_2	1	3_2
3_3	$(y)^2$	61440	$2^5 \cdot 3^3$	3_3	1	3_3
4_1	$(xy^3h)^6$	480	$2^{12} \cdot 3^3$	2_1	4_1	4_1
4_2	$(xyxh)^5$	1152	$2^{10} \cdot 3^2 \cdot 5$	2_1	4_2	4_2
4_3	$(xyh^2xh)^4$	4320	$2^{12} \cdot 3$	2_1	4_3	4_3
4_4	$(yh)^3$	17280	$2^{10} \cdot 3$	2_5	4_4	4_4
4_5	$(xyxhy^2)^3$	17280	$2^{10} \cdot 3$	2_1	4_5	4_5
4_6	$(xy^2xh^2)^3$	17280	$2^{10} \cdot 3$	2_5	4_6	4_6
4_7	$(xy^2h^2y)^3$	17280	$2^{10} \cdot 3$	2_5	4_7	4_7
4_8	$(xhyhy^2)^3$	17280	$2^{10} \cdot 3$	2_5	4_8	4_8
4_9	y^2hyh^2	17280	$2^{10} \cdot 3$	2_5	4_9	4_9
4_{10}	$(xy^2xh^2y^2)^3$	17280	$2^{10} \cdot 3$	2_5	4_{10}	4_{10}
4_{11}	$(xhy)^2$	25920	2^{11}	2_5	4_{11}	4_{11}
4_{12}	$(xy^2hyh^2)^2$	25920	2^{11}	2_5	4_{12}	4_{12}
4_{13}	$(xh)^3$	34560	$2^9 \cdot 3$	2_3	4_{13}	4_{13}
4_{14}	$(xyxy^2)^2$	34560	$2^9 \cdot 3$	2_3	4_{14}	4_{14}
4_{15}	$(xy^3xhy)^3$	34560	$2^9 \cdot 3$	2_7	4_{15}	4_{15}

Conjugacy classes of $H(\mathsf{Fi}_{22}) = \langle x, y, h \rangle$ (continued)

Class	*Representative*	*|Class|*	*|Centralizer|*	2P	3P	5P
4_{16}	$(xyhxh^2y)^3$	34560	$2^9 \cdot 3$	2_7	4_{16}	4_{16}
4_{17}	$(xy)^3$	69120	$2^8 \cdot 3$	2_9	4_{17}	4_{17}
4_{18}	$(xhyh)^3$	69120	$2^8 \cdot 3$	2_{10}	4_{18}	4_{18}
4_{19}	xhy^2hyh	103680	2^9	2_5	4_{19}	4_{19}
4_{20}	$xhyhyhy$	103680	2^9	2_5	4_{20}	4_{20}
4_{21}	$(xy^2xy^2xh)^2$	103680	2^9	2_6	4_{21}	4_{21}
4_{22}	$xy^2hxhyhyh$	103680	2^9	2_5	4_{22}	4_{22}
4_{23}	xy^3	207360	2^8	2_9	4_{23}	4_{23}
4_{24}	$xyhyh$	207360	2^8	2_9	4_{24}	4_{24}
4_{25}	$xyxhxh^2$	207360	2^8	2_{10}	4_{25}	4_{25}
4_{26}	$xyhxyhxh$	207360	2^8	2_9	4_{26}	4_{26}
4_{27}	xy^2xyhyh	414720	2^7	2_9	4_{27}	4_{27}
5	xy^2	1327104	$2^3 \cdot 5$	5	5	1
6_1	$(xy^2h)^3$	5120	$2^7 \cdot 3^4$	3_1	2_1	6_1
6_2	$(xhy^2)^3$	5120	$2^7 \cdot 3^4$	3_1	2_2	6_3
6_3	$(xhy^2)^{15}$	5120	$2^7 \cdot 3^4$	3_1	2_2	6_2
6_4	$(xy^3h)^4$	7680	$2^8 \cdot 3^3$	3_2	2_1	6_4
6_5	$xyxhyhyhy$	15360	$2^7 \cdot 3^3$	3_2	2_2	6_5
6_6	$(xh)^2$	46080	$2^7 \cdot 3^2$	3_1	2_3	6_6
6_7	$(xy^3xhy)^2$	46080	$2^7 \cdot 3^2$	3_1	2_7	6_8
6_8	$(xy^3xhy)^{10}$	46080	$2^7 \cdot 3^2$	3_1	2_7	6_7
6_9	xy^2h^2yxyh	46080	$2^7 \cdot 3^2$	3_1	2_4	6_9
6_{10}	$(xy^2xhyh)^2$	61440	$2^5 \cdot 3^3$	3_3	2_1	6_{10}
6_{11}	$xyxh^2xhy$	92160	$2^6 \cdot 3^2$	3_2	2_3	6_{11}
6_{12}	$xyhxyhyhyh$	92160	$2^6 \cdot 3^2$	3_2	2_4	6_{12}
6_{13}	xyh^2xh^2yh	122880	$2^4 \cdot 3^3$	3_3	2_2	6_{13}
6_{14}	$(yh)^2$	138240	$2^7 \cdot 3$	3_2	2_5	6_{14}
6_{15}	$xyxyhxy^2h$	138240	$2^7 \cdot 3$	3_2	2_6	6_{15}
6_{16}	y^2hyh	184320	$2^5 \cdot 3^2$	3_2	2_8	6_{16}
6_{17}	$xyxyxh$	184320	$2^5 \cdot 3^2$	3_2	2_7	6_{17}
6_{18}	$xyhxhy$	368640	$2^4 \cdot 3^2$	3_3	2_7	6_{19}
6_{19}	$(xyhxhy)^5$	368640	$2^4 \cdot 3^2$	3_3	2_7	6_{18}
6_{20}	$xyhyxh^2$	368640	$2^4 \cdot 3^2$	3_3	2_4	6_{20}
6_{21}	$xyxy^2h^2y$	368640	$2^4 \cdot 3^2$	3_3	2_3	6_{21}
6_{22}	$(xy)^2$	552960	$2^5 \cdot 3$	3_3	2_9	6_{22}
6_{23}	$(xhyh)^2$	552960	$2^5 \cdot 3$	3_3	2_{10}	6_{23}
6_{24}	xhy^2hy	552960	$2^5 \cdot 3$	3_2	2_{13}	6_{24}
6_{25}	y	737280	$2^3 \cdot 3^2$	3_3	2_8	6_{25}
6_{26}	$xyxyxy^2h^2$	1105920	$2^4 \cdot 3$	3_3	2_{12}	6_{26}
8_1	$(xy^3h)^3$	138240	$2^7 \cdot 3$	4_1	8_1	8_1
8_2	$(xyxy^2h)^3$	138240	$2^7 \cdot 3$	4_1	8_2	8_2
8_3	xhy	414720	2^7	4_{11}	8_3	8_3
8_4	$xhyhy$	414720	2^7	4_{11}	8_4	8_4
8_5	$xyxy^3$	414720	2^7	4_{11}	8_5	8_5
8_6	$(xyh^2xh)^2$	414720	2^7	4_3	8_6	8_6
8_7	xy^2xy^3	414720	2^7	4_{11}	8_7	8_7
8_8	xy^2hyh^2	829440	2^6	4_{12}	8_8	8_8
8_9	xyh	1658880	2^5	4_{13}	8_9	8_9
8_{10}	$xyxy^2$	1658880	2^5	4_{14}	8_{10}	8_{10}
8_{11}	xy^2xy^2xh	1658880	2^5	4_{21}	8_{11}	8_{11}

Conjugacy classes of $H(\mathsf{Fi}_{22}) = \langle x, y, h \rangle$ *(continued)*

Class	*Representative*	*|Class|*	*|Centralizer|*	2P	3P	5P
9	$(xy^2h)^2$	1474560	$2^2 \cdot 3^2$	9	3_1	9
10_1	$(xyxh)^2$	1327104	$2^3 \cdot 5$	5	10_1	2_1
10_2	xyh^2	2654208	$2^2 \cdot 5$	5	10_2	2_8
10_3	xhy^2h^2	2654208	$2^2 \cdot 5$	5	10_3	2_2
12_1	y^2hyhyh^2	61440	$2^5 \cdot 3^3$	6_1	4_1	12_1
12_2	$(xy^3h)^2$	92160	$2^6 \cdot 3^2$	6_4	4_1	12_2
12_3	$xyxhy^2hyh$	92160	$2^6 \cdot 3^2$	6_4	4_1	12_3
12_4	$xyxhyhxhy$	184320	$2^5 \cdot 3^2$	6_4	4_2	12_4
12_5	yh	552960	$2^5 \cdot 3$	6_{14}	4_4	12_5
12_6	$xyxhy^2$	552960	$2^5 \cdot 3$	6_4	4_5	12_6
12_7	xy^2xh^2	552960	$2^5 \cdot 3$	6_{14}	4_6	12_7
12_8	xy^2h^2y	552960	$2^5 \cdot 3$	6_{14}	4_7	12_8
12_9	$xhyhy^2$	552960	$2^5 \cdot 3$	6_{14}	4_8	12_9
12_{10}	xy^3xhy	552960	$2^5 \cdot 3$	6_7	4_{15}	12_{11}
12_{11}	$(xy^3xhy)^5$	552960	$2^5 \cdot 3$	6_8	4_{15}	12_{10}
12_{12}	xy^2h^2xh	552960	$2^5 \cdot 3$	6_1	4_3	12_{12}
12_{13}	$xyhxh^2y$	552960	$2^5 \cdot 3$	6_7	4_{16}	12_{14}
12_{14}	$(xyhxh^2y)^5$	552960	$2^5 \cdot 3$	6_8	4_{16}	12_{13}
12_{15}	$xy^2xh^2y^2$	552960	$2^5 \cdot 3$	6_{14}	4_{10}	12_{15}
12_{16}	$xyhxhxh^2$	552960	$2^5 \cdot 3$	6_{14}	4_9	12_{16}
12_{17}	xy^2xhyh	737280	$2^3 \cdot 3^2$	6_{10}	4_2	12_{17}
12_{18}	xh	1105920	$2^4 \cdot 3$	6_6	4_{13}	12_{18}
12_{19}	xy^2xh	1105920	$2^4 \cdot 3$	6_6	4_{14}	12_{19}
12_{20}	xy	2211840	$2^3 \cdot 3$	6_{22}	4_{17}	12_{20}
12_{21}	$xhyh$	2211840	$2^3 \cdot 3$	6_{23}	4_{18}	12_{21}
16_1	xyh^2xh	1658880	2^5	8_6	16_1	16_2
16_2	$(xyh^2xh)^5$	1658880	2^5	8_6	16_2	16_1
18_1	xy^2h	1474560	$2^2 \cdot 3^2$	9	6_1	18_1
18_2	xhy^2	1474560	$2^2 \cdot 3^2$	9	6_2	18_3
18_3	$(xhy^2)^5$	1474560	$2^2 \cdot 3^2$	9	6_3	18_2
20	$xyxh$	2654208	$2^2 \cdot 5$	10_1	20	4_2
24_1	xy^3h	1105920	$2^4 \cdot 3$	12_2	8_1	24_1
24_2	$xyxy^2h$	1105920	$2^4 \cdot 3$	12_2	8_2	24_2

5.5.2 *Conjugacy classes of* $D(\mathsf{Fi}_{22}) = \langle x, y \rangle$

Class	*Representative*	*|Class|*	*|Centralizer|*	2P	3P	5P
1	1	1	$2^{17} \cdot 3 \cdot 5$	1	1	1
2_1	$(xyxyxy^2xy^3)^8$	1	$2^{17} \cdot 3 \cdot 5$	1	2_1	2_1
2_2	$(xyxyxyxy^2xy^5)^3$	2	$2^{16} \cdot 3 \cdot 5$	1	2_2	2_2
2_3	$(xyxy^2)^4$	20	$2^{15} \cdot 3$	1	2_3	2_3
2_4	$(xyxyxyxy^3)^3$	20	$2^{15} \cdot 3$	1	2_4	2_4
2_5	$(xyxy^3)^4$	30	2^{16}	1	2_5	2_5
2_6	$(xyxy^3xy^4)^4$	30	2^{16}	1	2_6	2_6
2_7	$(xyxyxyxyxy^2)^2$	40	$2^{14} \cdot 3$	1	2_7	2_7
2_8	$(xy)^6$	160	$2^{12} \cdot 3$	1	2_8	2_8
2_9	$(xyxyxyxy^2xy^4)^6$	160	$2^{12} \cdot 3$	1	2_9	2_9
2_{10}	$(xyxy^2xy^4)^4$	240	2^{13}	1	2_{10}	2_{10}
2_{11}	$(xyxyxyxy^3xy^2xy^4)^2$	240	2^{13}	1	2_{11}	2_{11}
2_{12}	$xyxyxyxy^2xy^5xy^2xyxyxyxy^3$	240	2^{13}	1	2_{12}	2_{12}

Conjugacy classes of $D(\mathsf{Fi}_{22}) = \langle x, y\rangle$ *(continued)*

Class	*Representative*	|*Class*|	|*Centralizer*|	2P	3P	5P
2_{13}	$(xyxyxyxy^2xyxy^4xy^2)^3$	320	$2^{11}\cdot 3$	1	2_{13}	2_{13}
2_{14}	$xyxyxy^2xy^4xy^2$	480	2^{12}	1	2_{14}	2_{14}
2_{15}	$(y)^3$	640	$2^{10}\cdot 3$	1	2_{15}	2_{15}
2_{16}	x	960	2^{11}	1	2_{16}	2_{16}
2_{17}	$(xyxyxyxyxy^5)^2$	960	2^{11}	1	2_{17}	2_{17}
2_{18}	$xyxy^2xy^4xy^2xy^4$	1920	2^{10}	1	2_{18}	2_{18}
2_{19}	$xyxyxyxyxy^2xyxy^2xyxy^2$	1920	2^{10}	1	2_{19}	2_{19}
3	$(y)^2$	20480	$2^5\cdot 3$	3	1	3
4_1	$(xyxyxy^2xy^3)^4$	480	2^{12}	2_1	4_1	4_1
4_2	$(xyxyxy^3xy^4)^2$	480	2^{12}	2_1	4_2	4_2
4_3	$(xyxy^2xy^3xy^2xy^3)^3$	640	$2^{10}\cdot 3$	2_1	4_3	4_3
4_4	$(xyxy^3)^2$	960	2^{11}	2_5	4_4	4_4
4_5	$(xyxyxy^2xyxy^4)^2$	960	2^{11}	2_5	4_5	4_5
4_6	$xyxy^3xy^4xy^3$	1920	2^{10}	2_5	4_6	4_6
4_7	$(xyxyxyxyxy^2xyxy^4)^2$	1920	2^{10}	2_5	4_7	4_7
4_8	$xyxy^3xyxy^3xy^2xy^3$	1920	2^{10}	2_5	4_8	4_8
4_9	$xyxy^3xy^2xy^3xy^2xy^3$	1920	2^{10}	2_5	4_9	4_9
4_{10}	$xyxyxy^3xy^2xy^2xyxy^4$	1920	2^{10}	2_5	4_{10}	4_{10}
4_{11}	$xyxyxyxy^4xyxy^2xy^5$	1920	2^{10}	2_5	4_{11}	4_{11}
4_{12}	$xyxyxyxyxyxy^3xyxyxy^5$	1920	2^{10}	2_1	4_{12}	4_{12}
4_{13}	$xyxyxyxy^2xy^2xy^3xyxy^2xy^3$	1920	2^{10}	2_5	4_{13}	4_{13}
4_{14}	$xyxyxyxyxy^2xyxyxyxy^2xyxy^3$	1920	2^{10}	2_5	4_{14}	4_{14}
4_{15}	$(xy)^3$	2560	$2^8\cdot 3$	2_8	4_{15}	4_{15}
4_{16}	$(xyxyxyxy^2xy^4)^3$	2560	$2^8\cdot 3$	2_9	4_{16}	4_{16}
4_{17}	$(xyxy^2)^2$	3840	2^9	2_3	4_{17}	4_{17}
4_{18}	$(xyxy^4)^2$	3840	2^9	2_3	4_{18}	4_{18}
4_{19}	$(xyxy^2xy^4)^2$	3840	2^9	2_{10}	4_{19}	4_{19}
4_{20}	$xyxyxyxyxy^2$	3840	2^9	2_7	4_{20}	4_{20}
4_{21}	$(xyxy^3xy^4)^2$	3840	2^9	2_6	4_{21}	4_{21}
4_{22}	$(xyxyxyxy^3xy^5)^2$	3840	2^9	2_{10}	4_{22}	4_{22}
4_{23}	$xyxyxy^2xy^2xy^3xyxy^3$	3840	2^9	2_5	4_{23}	4_{23}
4_{24}	$xyxyxyxy^2xyxy^5xy^3$	3840	2^9	2_5	4_{24}	4_{24}
4_{25}	$xyxyxyxy^2xy^2xyxy^2xy^3xy^2$	3840	2^9	2_5	4_{25}	4_{25}
4_{26}	$xyxyxyxy^2xyxy^2xy^3xy^2xy^3$	3840	2^9	2_7	4_{26}	4_{26}
4_{27}	$xyxyxy^2xyxy^2$	7680	2^8	2_{10}	4_{27}	4_{27}
4_{28}	$xyxyxyxyxyxy^4$	7680	2^8	2_8	4_{28}	4_{28}
4_{29}	$xyxyxyxy^5xy^5$	7680	2^8	2_{10}	4_{29}	4_{29}
4_{30}	$xyxyxy^2xyxy^4xy^3$	7680	2^8	2_6	4_{30}	4_{30}
4_{31}	$xyxy^2xy^3xyxy^3xy^2$	7680	2^8	2_8	4_{31}	4_{31}
4_{32}	$xyxy^2xyxy^2xy^4xy^3$	7680	2^8	2_{10}	4_{32}	4_{32}
4_{33}	$xyxyxy^2xy^2xyxy^2xyxy^3$	7680	2^8	2_{10}	4_{33}	4_{33}
4_{34}	$xyxyxyxyxy^2xyxy^5xy^2$	7680	2^8	2_{10}	4_{34}	4_{34}
4_{35}	$xyxyxyxy^2xy^5xy^2xy^3$	7680	2^8	2_9	4_{35}	4_{35}
4_{36}	$xyxyxyxy^5xy^2xyxy^2xy^5$	7680	2^8	2_8	4_{36}	4_{36}
4_{37}	$xyxyxyxy^3xy^2xy^4$	15360	2^7	2_{11}	4_{37}	4_{37}
4_{38}	$xyxyxyxyxy^2xy^2xyxy^3$	15360	2^7	2_8	4_{38}	4_{38}
4_{39}	$xyxyxyxy^2xyxy^3xy^2xy^3$	15360	2^7	2_{10}	4_{39}	4_{39}
4_{40}	xy^3	30720	2^6	2_{16}	4_{40}	4_{40}
4_{41}	$xyxyxyxyxy^5$	30720	2^6	2_{17}	4_{41}	4_{41}
5	xy^2	98304	$2^2\cdot 5$	5	5	1
6_1	$(xy)^2$	20480	$2^5\cdot 3$	3	2_8	6_1
6_2	$(xyxyxyxy^2xy^4)^2$	20480	$2^5\cdot 3$	3	2_9	6_2
6_3	$xyxyxyxy^2xy^3xy^2$	20480	$2^5\cdot 3$	3	2_1	6_3
6_4	$xyxyxyxy^3$	40960	$2^4\cdot 3$	3	2_4	6_4
6_5	$xyxy^2xy^5$	40960	$2^4\cdot 3$	3	2_7	6_6
6_6	$(xyxy^2xy^5)^5$	40960	$2^4\cdot 3$	3	2_7	6_5

Conjugacy classes of $D(\mathsf{Fi}_{22}) = \langle x, y \rangle$ *(continued)*

Class	*Representative*	$\lvert$*Class*$\rvert$	$\lvert$*Centralizer*$\rvert$	2P	3P	5P
6_7	$xyxyxyxy^2xy^5$	40960	$2^4 \cdot 3$	3	2_2	6_7
6_8	$xyxy^2xyxy^2xy^4$	40960	$2^4 \cdot 3$	3	2_3	6_8
6_9	$xyxyxyxy^2xyxy^4xy^2$	40960	$2^4 \cdot 3$	3	2_{13}	6_9
6_{10}	y	81920	$2^3 \cdot 3$	3	2_{15}	6_{10}
8_1	$xyxy^3$	15360	2^7	4_4	8_1	8_1
8_2	xy^2xy^3	15360	2^7	4_4	8_2	8_2
8_3	$(xyxyxy^2xy^3)^2$	15360	2^7	4_1	8_3	8_3
8_4	$xyxyxy^3xy^4$	15360	2^7	4_2	8_4	8_4
8_5	$xyxyxyxyxy^2xy^4$	15360	2^7	4_4	8_5	8_5
8_6	$xyxyxyxy^2xy^2xy^4$	15360	2^7	4_4	8_6	8_6
8_7	$xyxyxyxy^2xy^3xy^2xy^3$	15360	2^7	4_2	8_7	8_7
8_8	$xyxyxy^2xyxy^3$	30720	2^6	4_1	8_8	8_8
8_9	$xyxyxy^2xyxy^4$	30720	2^6	4_5	8_9	8_9
8_{10}	$xyxyxyxyxy^2xyxy^4$	30720	2^6	4_7	8_{11}	8_{10}
8_{11}	$(xyxyxyxyxy^2xyxy^4)^3$	30720	2^6	4_7	8_{10}	8_{11}
8_{12}	$xyxy^2$	61440	2^5	4_{17}	8_{12}	8_{12}
8_{13}	$xyxy^4$	61440	2^5	4_{18}	8_{13}	8_{13}
8_{14}	$xyxy^2xy^4$	61440	2^5	4_{19}	8_{14}	8_{14}
8_{15}	$xyxy^3xy^4$	61440	2^5	4_{21}	8_{15}	8_{15}
8_{16}	$xyxyxyxy^3xy^5$	61440	2^5	4_{22}	8_{16}	8_{16}
10_1	$xyxyxy^3xy^5$	98304	$2^2 \cdot 5$	5	10_1	2_1
10_2	$xyxy^2xy^3xyxy^3$	98304	$2^2 \cdot 5$	5	10_3	2_2
10_3	$(xyxy^2xy^3xyxy^3)^3$	98304	$2^2 \cdot 5$	5	10_2	2_2
12_1	xy	81920	$2^3 \cdot 3$	6_1	4_{15}	12_1
12_2	$xyxyxyxy^2xy^4$	81920	$2^3 \cdot 3$	6_2	4_{16}	12_2
12_3	$xyxy^2xy^3xy^2xy^3$	81920	$2^3 \cdot 3$	6_3	4_3	12_3
16_1	$xyxyxy^2xy^3$	61440	2^5	8_3	16_1	16_2
16_2	$(xyxyxy^2xy^3)^5$	61440	2^5	8_3	16_2	16_1

5.5.3 *Conjugacy classes of* $E(\mathsf{Fi}_{22}) = \langle x, y, e \rangle$

Class	*Representative*	|*Class*|	|*Centralizer*|	2P	3P	5P	7P	11P
1	1	1	$2^{17} \cdot 3^2 \cdot 5 \cdot 7 \cdot 11$	1	1	1	1	1
2_1	$(xye)^7$	22	$2^{16} \cdot 3^2 \cdot 5 \cdot 7$	1	2_1	2_1	2_1	2_1
2_2	$(xy)^6$	231	$2^{17} \cdot 3 \cdot 5$	1	2_2	2_2	2_2	2_2
2_3	$(ye^2)^6$	770	$2^{16} \cdot 3^2$	1	2_3	2_3	2_3	2_3
2_4	x	18480	$2^{13} \cdot 3$	1	2_4	2_4	2_4	2_4
2_5	$(xeyxe^2)^2$	55440	2^{13}	1	2_5	2_5	2_5	2_5
3	$(y)^2$	788480	$2^6 \cdot 3^2$	3	1	3	3	3
4_1	$(xyey)^4$	110880	2^{12}	2_2	4_1	4_1	4_1	4_1
4_2	$(xyxy^3)^2$	110880	2^{12}	2_2	4_2	4_2	4_2	4_2
4_3	$(xy)^3$	147840	$2^{10} \cdot 3$	2_2	4_3	4_3	4_3	4_3
4_4	$(ye^2)^3$	295680	$2^9 \cdot 3$	2_3	4_4	4_4	4_4	4_4
4_5	$xeyey^2e^2$	443520	2^{10}	2_2	4_5	4_5	4_5	4_5
4_6	$(xy^2e^2)^2$	887040	2^9	2_4	4_6	4_6	4_6	4_6
4_7	$(xy^2xe^2)^2$	887040	2^9	2_4	4_7	4_7	4_7	4_7
4_8	$xeye^2$	1774080	2^8	2_4	4_8	4_8	4_8	4_8
4_9	xy^3	3548160	2^7	2_4	4_9	4_9	4_9	4_9
4_{10}	$xeyxe^2$	3548160	2^7	2_5	4_{10}	4_{10}	4_{10}	4_{10}
5	xy^2	22708224	$2^2 \cdot 5$	5	5	1	5	5
6_1	$(ye^2)^2$	788480	$2^6 \cdot 3^2$	3	2_3	6_1	6_1	6_1
6_2	$xyxye$	3153920	$2^4 \cdot 3^2$	3	2_1	6_2	6_2	6_2
6_3	$xyeyey$	3153920	$2^4 \cdot 3^2$	3	2_3	6_3	6_3	6_3
6_4	$(xy)^2$	4730880	$2^5 \cdot 3$	3	2_2	6_4	6_4	6_4
6_5	y	9461760	$2^4 \cdot 3$	3	2_4	6_5	6_5	6_5
7_1	$(xye)^2$	32440320	$2 \cdot 7$	7_1	7_2	7_2	1	7_1
7_2	$(xye)^6$	32440320	$2 \cdot 7$	7_2	7_1	7_1	1	7_2
8_1	$(xyey)^2$	3548160	2^7	4_1	8_1	8_1	8_1	8_1
8_2	$xyxy^3$	3548160	2^7	4_2	8_2	8_2	8_2	8_2
8_3	$xyxy^4e$	3548160	2^7	4_2	8_3	8_3	8_3	8_3
8_4	xe^2ye^2	7096320	2^6	4_1	8_4	8_4	8_4	8_4
8_5	$xyxy^2$	14192640	2^5	4_4	8_5	8_5	8_5	8_5
8_6	xy^2e^2	14192640	2^5	4_6	8_6	8_6	8_6	8_6
8_7	xy^2xe^2	14192640	2^5	4_7	8_7	8_7	8_7	8_7
10_1	y^2e	22708224	$2^2 \cdot 5$	5	10_2	2_1	10_2	10_1
10_2	$(y^2e)^3$	22708224	$2^2 \cdot 5$	5	10_1	2_1	10_1	10_2
10_3	xy^2e	22708224	$2^2 \cdot 5$	5	10_3	2_2	10_3	10_3
11_1	xe	41287680	11	11_2	11_1	11_1	11_2	1
11_2	$(xe)^2$	41287680	11	11_1	11_2	11_2	11_1	1
12_1	ye^2	9461760	$2^4 \cdot 3$	6_1	4_4	12_1	12_1	12_1
12_2	xy	18923520	$2^3 \cdot 3$	6_4	4_3	12_2	12_2	12_2
14_1	xye	32440320	$2 \cdot 7$	7_1	14_2	14_2	2_1	14_1
14_2	$(xye)^3$	32440320	$2 \cdot 7$	7_2	14_1	14_1	2_1	14_2
16_1	$xyey$	14192640	2^5	8_1	16_1	16_2	16_2	16_1
16_2	$(xyey)^5$	14192640	2^5	8_1	16_2	16_1	16_1	16_2

5.6 Character tables of local subgroups of Fi_{22}

5.6.1 *Character table of* $E_2 = E(\mathsf{Fi}_{22}) = \langle x, y, e \rangle$

2	17	16	17	16	13	13	6	12	12	10	9	10	9	9	8	7	7	2	6	4	4	5	4	1	1	7
3	2	2	1	2	1	.	2	.	.	1	1	.	.	.	.	.	.	.	2	2	2	1	1	.	.	.
5	1	1	1	.	.	.	.	.	.	.	.	.	.	.	.	.	.	1	.	.	.	.	.	.	.	.
7	1	1	.	.	.	.	.	.	.	.	.	.	.	.	.	.	.	.	.	.	.	.	.	1	1	.
11	1	.	.	.	.	.	.	.	.	.	.	.	.	.	.	.	.	.	.	.	.	.	.	.	.	.
	1a	2a	2b	2c	2d	2e	3a	4a	4b	4c	4d	4e	4f	4g	4h	4i	4j	5a	6a	6b	6c	6d	6e	7a	7b	8a
2P	1a	1a	1a	1a	1a	1a	3a	2b	2b	2b	2c	2b	2d	2d	2d	2d	2e	5a	3a	3a	3a	3a	3a	7a	7b	4a
3P	1a	2a	2b	2c	2d	2e	1a	4a	4b	4c	4d	4e	4f	4g	4h	4i	4j	5a	2c	2a	2c	2b	2d	7b	7a	8a
5P	1a	2a	2b	2c	2d	2e	3a	4a	4b	4c	4d	4e	4f	4g	4h	4i	4j	1a	6a	6b	6c	6d	6e	7b	7a	8a
7P	1a	2a	2b	2c	2d	2e	3a	4a	4b	4c	4d	4e	4f	4g	4h	4i	4j	5a	6a	6b	6c	6d	6e	1a	1a	8a
11P	1a	2a	2b	2c	2d	2e	3a	4a	4b	4c	4d	4e	4f	4g	4h	4i	4j	5a	6a	6b	6c	6d	6e	7a	7b	8a
X.1	1	1	1	1	1	1	1	1	1	1	1	1	1	1	1	1	1	1	1	1	1	1	1	1	1	1
X.2	21	21	21	21	5	5	3	5	5	5	5	5	1	1	1	1	1	1	3	3	3	3	−1	.	.	1
X.3	45	45	45	45	−3	−3	.	−3	−3	−3	−3	−3	1	1	1	1	1	.	.	.	.	.	.	A	$\bar{A}$	1
X.4	45	45	45	45	−3	−3	.	−3	−3	−3	−3	−3	1	1	1	1	1	.	.	.	.	.	.	$\bar{A}$	A	1
X.5	55	55	55	55	7	7	1	7	7	7	7	7	3	3	3	−1	−1	.	1	1	1	1	1	−1	−1	3
X.6	77	−35	13	−3	13	−3	5	−3	5	−7	1	1	−3	5	1	1	1	2	−3	1	−3	1	1	.	.	1
X.7	99	99	99	99	3	3	.	3	3	3	3	3	3	3	3	−1	−1	−1	.	.	.	.	.	1	1	3
X.8	154	154	154	154	10	10	1	10	10	10	10	10	−2	−2	−2	2	2	−1	1	1	1	1	1	.	.	−2
X.9	210	210	210	210	2	2	3	2	2	2	2	2	−2	−2	−2	−2	−2	.	3	3	3	3	−1	.	.	−2
X.10	231	231	231	231	7	7	−3	7	7	7	7	7	−1	−1	−1	−1	−1	1	−3	−3	−3	−3	1	.	.	−1
X.11	280	280	280	280	−8	−8	1	−8	−8	−8	−8	−8	.	.	.	.	.	.	1	1	1	1	1	.	.	.
X.12	280	280	280	280	−8	−8	1	−8	−8	−8	−8	−8	.	.	.	.	.	.	1	1	1	1	1	.	.	.
X.13	330	90	10	−6	26	10	6	−6	2	6	−2	−2	−2	6	2	2	2	.	6	.	.	−2	2	1	1	−2
X.14	385	385	385	385	1	1	−2	1	1	1	1	1	1	1	1	1	1	.	−2	−2	−2	−2	−2	.	.	1
X.15	385	−175	65	−15	17	1	−2	1	9	−11	5	−3	5	−3	1	1	1	.	6	−4	.	2	2	.	.	1
X.16	385	−175	65	−15	17	1	7	1	9	−11	5	−3	5	−3	1	1	1	.	−9	5	−3	−1	−1	.	.	1
X.17	616	−56	−24	8	24	−8	4	8	−8	.	.	.	4	4	−4	4	−4	1	−4	−2	2	.	.	.	.	.
X.18	616	−56	−24	8	24	−8	4	8	−8	.	.	.	4	4	−4	−4	4	1	−4	−2	2	.	.	.	.	.
X.19	616	−280	104	−24	8	8	−5	8	8	−8	8	−8	.	.	.	.	.	1	3	−1	3	−1	−1	.	.	.
X.20	616	−280	104	−24	8	8	−5	8	8	−8	8	−8	.	.	.	.	.	1	3	−1	3	−1	−1	.	.	.
X.21	693	−315	117	−27	21	5	.	5	13	−15	9	−7	−3	5	1	1	1	−2	.	.	.	.	.	.	.	1
X.22	770	−350	130	−30	−14	18	5	18	2	2	10	−14	−2	−2	−2	−2	−2	.	−3	1	−3	1	1	.	.	−2
X.23	990	270	30	−18	−18	−2	.	14	6	−6	−6	2	−6	2	−2	2	2	.	.	.	.	.	.	$\bar{A}$	A	2
X.24	990	270	30	−18	−18	−2	.	14	6	−6	−6	2	−6	2	−2	2	2	.	.	.	.	.	.	A	$\bar{A}$	2
X.25	1155	−525	195	−45	35	−13	3	−13	11	−17	−1	7	−5	3	−1	−1	−1	.	3	−3	−3	3	−1	.	.	−1
X.26	1155	−525	195	−45	−13	3	3	3	−5	7	−1	−1	7	−1	3	−1	−1	.	3	−3	−3	3	−1	.	.	3
X.27	1980	540	60	−36	60	28	.	−4	12	12	−12	−4	4	4	4	.	.	.	.	.	.	.	.	−1	−1	−4
X.28	2310	−1050	390	−90	22	−10	−3	−10	6	−10	−2	6	2	2	2	−2	−2	.	−3	3	3	−3	1	.	.	2
X.29	2310	630	70	−42	−10	6	6	22	14	−6	−14	2	2	−6	−2	−2	−2	.	6	.	.	−2	2	.	.	2
X.30	2310	630	70	−42	22	38	6	−10	−18	−6	2	2	2	−6	−2	2	2	.	6	.	.	−2	−2	.	.	2
X.31	2310	630	70	−42	22	−26	6	−10	14	18	2	−6	−6	2	−2	−2	−2	.	6	.	.	−2	−2	.	.	2
X.32	2640	720	80	−48	16	16	−6	16	16	.	−16	.	.	.	.	.	.	.	−6	.	.	2	−2	1	1	.
X.33	3465	−1575	585	−135	−39	9	.	9	−15	21	−3	−3	−3	5	1	1	1	.	.	.	.	.	.	.	.	1
X.34	3465	−1575	585	−135	9	−7	.	−7	1	−3	−3	5	1	−7	−3	1	1	.	.	.	.	.	.	.	.	−3
X.35	4620	1260	140	−84	44	12	−6	−20	−4	12	4	−4	−4	−4	−4	.	.	.	−6	.	.	2	2	.	.	4
X.36	5544	−504	−216	72	24	−8	.	8	−8	.	.	.	4	4	−4	4	−4	−1	.	.	.	.	.	.	.	.
X.37	5544	−504	−216	72	24	−8	.	8	−8	.	.	.	4	4	−4	−4	4	−1	.	.	.	.	.	.	.	.
X.38	6160	−560	−240	80	48	−16	4	16	−16	.	.	.	−8	−8	8	.	.	.	−4	−2	2	.	.	.	.	.
X.39	6160	−560	−240	80	−48	16	4	−16	16	.	.	.	.	.	.	.	.	.	−4	−2	2	.	.	.	.	.
X.40	6160	−560	−240	80	−48	16	4	−16	16	.	.	.	.	.	.	.	.	.	−4	−2	2	.	.	.	.	.
X.41	6930	1890	210	−126	−30	18	.	2	−22	−18	6	6	−2	6	2	−2	−2	.	.	.	.	.	.	.	.	−2
X.42	6930	1890	210	−126	−30	−46	.	2	10	6	6	−2	6	−2	2	2	2	.	.	.	.	.	.	.	.	−2
X.43	9856	−896	−384	128	.	.	−8	.	.	.	.	.	.	.	.	.	.	1	8	4	−4	.	.	.	.	.

Character table of $E(\mathsf{Fi}_{22})$ *(continued)*

2	7	7	6	5	5	5	2	2	2	.	.	4	3	1	1	5	5
3	.	.	.	.	.	.	.	.	.	.	.	1	1	.	.	.	.
5	.	.	.	.	.	.	1	1	1	.	.	.	.	.	.	.	.
7	.	.	.	.	.	.	.	.	.	.	.	.	.	1	1	.	.
11	.	.	.	.	.	.	.	.	.	1	1	.	.	.	.	.	.
	8b	8c	8d	8e	8f	8g	10a	10b	10c	11a	11b	12a	12b	14a	14b	16a	16b
2P	4b	4b	4a	4d	4f	4g	5a	5a	5a	11b	11a	6a	6d	7a	7b	8a	8a
3P	8b	8c	8d	8e	8f	8g	10b	10a	10c	11a	11b	4d	4c	14b	14a	16a	16b
5P	8b	8c	8d	8e	8f	8g	2a	2a	2b	11a	11b	12a	12b	14b	14a	16b	16a
7P	8b	8c	8d	8e	8f	8g	10b	10a	10c	11b	11a	12a	12b	2a	2a	16b	16a
11P	8b	8c	8d	8e	8f	8g	10a	10b	10c	1a	1a	12a	12b	14a	14b	16a	16b
X.1	1	1	1	1	1	1	1	1	1	1	1	1	1	1	1	1	1
X.2	1	1	1	1	−1	−1	1	1	1	−1	−1	−1	−1	.	.	−1	−1
X.3	1	1	1	1	−1	−1	.	.	.	1	1	.	.	A	$\bar{A}$	−1	−1
X.4	1	1	1	1	−1	−1	.	.	.	1	1	.	.	$\bar{A}$	A	−1	−1
X.5	3	3	−1	−1	1	1	.	.	.	.	.	1	1	−1	−1	1	1
X.6	1	−3	1	−1	1	1	.	.	−2	.	.	1	−1	.	.	−1	−1
X.7	3	3	−1	−1	−1	−1	−1	−1	−1	.	.	.	.	1	1	−1	−1
X.8	−2	−2	2	2	.	.	−1	−1	−1	.	.	1	1	.	.	.	.
X.9	−2	−2	−2	−2	.	.	.	.	.	1	1	−1	−1	.	.	.	.
X.10	−1	−1	−1	−1	−1	−1	1	1	1	.	.	1	1	.	.	−1	−1
X.11	.	.	.	.	.	.	.	.	.	B	$\bar{B}$	1	1	.	.	.	.
X.12	.	.	.	.	.	.	.	.	.	$\bar{B}$	B	1	1	.	.	.	.
X.13	−2	2	−2	.	.	.	.	.	.	.	.	−2	.	−1	−1	.	.
X.14	1	1	1	1	1	1	.	.	.	.	.	−2	−2	.	.	1	1
X.15	−3	1	1	−1	−1	−1	.	.	.	.	.	2	−2	.	.	1	1
X.16	−3	1	1	−1	−1	−1	.	.	.	.	.	−1	1	.	.	1	1
X.17	.	.	.	.	−2	2	−1	−1	1	.	.	.	.	.	.	.	.
X.18	.	.	.	.	2	−2	−1	−1	1	.	.	.	.	.	.	.	.
X.19	.	.	.	.	.	.	C	$-C$	−1	.	.	−1	1	.	.	.	.
X.20	.	.	.	.	.	.	$-C$	C	−1	.	.	−1	1	.	.	.	.
X.21	1	−3	1	−1	1	1	.	.	2	.	.	.	.	.	.	−1	−1
X.22	2	2	−2	2	.	.	.	.	.	.	.	1	−1	.	.	.	.
X.23	−2	2	−2	.	.	.	.	.	.	.	.	.	.	$-\bar{A}$	$-A$	.	.
X.24	−2	2	−2	.	.	.	.	.	.	.	.	.	.	$-A$	$-\bar{A}$	.	.
X.25	3	−1	−1	1	−1	−1	.	.	.	.	.	−1	1	.	.	1	1
X.26	−5	−1	−1	1	1	1	.	.	.	.	.	−1	1	.	.	−1	−1
X.27	.	.	.	.	.	.	.	.	.	.	.	.	.	1	1	.	.
X.28	−2	−2	−2	2	.	.	.	.	.	.	.	1	−1	.	.	.	.
X.29	2	−2	2	.	.	.	.	.	.	.	.	−2	.	.	.	.	.
X.30	2	−2	−2	.	.	.	.	.	.	.	.	2	.	.	.	.	.
X.31	−2	2	2	.	.	.	.	.	.	.	.	2	.	.	.	.	.
X.32	.	.	.	.	.	.	.	.	.	.	.	2	.	−1	−1	.	.
X.33	1	−3	1	−1	−1	−1	.	.	.	.	.	.	.	.	.	1	1
X.34	1	5	1	−1	1	1	.	.	.	.	.	.	.	.	.	−1	−1
X.35	.	.	.	.	.	.	.	.	.	.	.	−2	.	.	.	.	.
X.36	.	.	.	.	2	−2	1	1	−1	.	.	.	.	.	.	.	.
X.37	.	.	.	.	−2	2	1	1	−1	.	.	.	.	.	.	.	.
X.38	.	.	.	.	.	.	.	.	.	.	.	.	.	.	.	.	.
X.39	.	.	.	.	.	.	.	.	.	.	.	.	.	.	.	D	$\bar{D}$
X.40	.	.	.	.	.	.	.	.	.	.	.	.	.	.	.	$\bar{D}$	D
X.41	−2	2	2	.	.	.	.	.	.	.	.	.	.	.	.	.	.
X.42	2	−2	−2	.	.	.	.	.	.	.	.	.	.	.	.	.	.
X.43	.	.	.	.	.	.	−1	−1	1	.	.	.	.	.	.	.	.

where $A = \frac{1}{2}(-1+i\sqrt{7})$, $B = \frac{1}{2}(-1+i\sqrt{11})$, $C = -i\sqrt{5}$, $D = -2(\zeta(8)_8^3 + \zeta(8)_8)$.

5.6.2 *Character table of* $H(\mathsf{Fi}_{22}) = \langle x, y, h \rangle$

	1a	2a	2b	2c	2d	2e	2f	2g	2h	2i	2j	2k	2l	2m	3a	3b	3c	4_1	4_2	4_3	4_4	4_5
2	17	17	16	15	15	16	16	14	10	12	12	13	11	10	7	8	5	12	10	12	10	10
3	4	4	4	2	2	1	1	2	2	1	1	.	1	1	4	3	3	3	2	1	1	1
5	1	1	1	.	.	.	.	.	1	.	.	.	.	.	.	.	.	.	1	.	.	.
2P	1a	1a	1a	1a	1a	1a	1a	1a	1a	1a	1a	1a	1a	1a	3a	3b	3c	2a	2a	2a	2e	2a
3P	1a	2a	2b	2c	2d	2e	2f	2g	2h	2i	2j	2k	2l	2m	1a	1a	1a	4_1	4_2	4_3	4_4	4_5
5P	1a	2a	2b	2c	2d	2e	2f	2g	2h	2i	2j	2k	2l	2m	3a	3b	3c	4_1	4_2	4_3	4_4	4_5
X.1	1	1	1	1	1	1	1	1	1	1	1	1	1	1	1	1	1	1	1	1	1	1
X.2	1	1	1	1	1	1	1	1	−1	1	1	1	1	−1	1	1	1	1	−1	1	−1	−1
X.3	6	6	6	−2	−2	6	6	−2	−4	2	2	−2	2	.	−3	.	3	6	−4	−2	−4	.
X.4	6	6	6	−2	−2	6	6	−2	4	2	2	−2	2	.	−3	.	3	6	4	−2	4	.
X.5	10	10	10	−6	−6	10	10	−6	.	2	2	−6	2	.	1	4	−2	10	.	−6	.	.
X.6	15	15	15	−1	−1	15	15	−1	5	−1	−1	−1	−1	−3	6	.	3	15	5	−1	5	−3
X.7	15	15	15	−1	−1	15	15	−1	−5	−1	−1	−1	−1	3	6	.	3	15	−5	−1	−5	3
X.8	15	15	15	7	7	15	15	7	−5	3	3	7	3	−1	−3	3	.	15	−5	7	−5	−1
X.9	15	15	15	7	7	15	15	7	5	3	3	7	3	1	−3	3	.	15	5	7	5	1
X.10	20	20	20	4	4	20	20	4	.	−4	−4	4	−4	.	−7	2	2	20	.	4	.	.
X.11	20	20	20	4	4	20	20	4	−10	4	4	4	4	−2	2	−1	5	20	−10	4	−10	−2
X.12	20	20	20	4	4	20	20	4	10	4	4	4	4	2	2	−1	5	20	10	4	10	2
X.13	24	24	24	8	8	24	24	8	4	.	.	8	.	4	6	3	.	24	4	8	4	4
X.14	24	24	24	8	8	24	24	8	−4	.	.	8	.	−4	6	3	.	24	−4	8	−4	−4
X.15	30	30	30	−10	−10	30	30	−10	10	2	2	−10	2	−2	3	3	3	30	10	−10	10	−2
X.16	30	30	30	−10	−10	30	30	−10	−10	2	2	−10	2	2	3	3	3	30	−10	−10	−10	2
X.17	32	−32	.	16	−16	.	.	.	.	8	−8	.	.	.	−4	8	2	.	.	.	.	.
X.18	40	40	−40	16	16	8	−8	−16	−10	4	4	.	−4	−6	4	7	1	.	10	.	2	6
X.19	40	40	−40	16	16	8	−8	−16	10	4	4	.	−4	6	4	7	1	.	−10	.	−2	−6
X.20	40	40	−40	−16	−16	8	−8	16	10	4	4	.	−4	−2	4	7	1	.	−10	.	−2	2
X.21	40	40	−40	−16	−16	8	−8	16	−10	4	4	.	−4	2	4	7	1	.	10	.	2	−2
X.22	60	60	60	−4	−4	60	60	−4	−10	4	4	−4	4	−2	6	−3	−3	60	−10	−4	−10	−2
X.23	60	60	60	−4	−4	60	60	−4	10	4	4	−4	4	2	6	−3	−3	60	10	−4	10	2
X.24	60	60	60	12	12	60	60	12	.	4	4	12	4	.	−3	.	−6	60	.	12	.	.
X.25	64	64	64	.	.	64	64	.	−16	.	.	.	.	.	−8	−2	4	64	−16	.	−16	.
X.26	64	64	64	.	.	64	64	.	16	.	.	.	.	.	−8	−2	4	64	16	.	16	.
X.27	80	80	−80	.	.	16	−16	.	20	8	8	.	−8	4	8	−4	2	.	−20	.	−4	−4
X.28	80	80	80	−16	−16	80	80	−16	.	.	.	−16	.	.	−10	2	−4	80	.	−16	.	.
X.29	80	80	−80	.	.	16	−16	.	−20	8	8	.	−8	−4	8	−4	2	.	20	.	4	4
X.30	81	81	81	9	9	81	81	9	9	−3	−3	9	−3	−3	.	.	.	81	9	9	9	−3
X.31	81	81	81	9	9	81	81	9	−9	−3	−3	9	−3	3	.	.	.	81	−9	9	−9	3
X.32	90	90	90	−6	−6	90	90	−6	.	−6	−6	−6	−6	.	9	.	.	90	.	−6	.	.
X.33	120	120	−120	16	16	24	−24	−16	10	−4	−4	.	4	−10	12	3	3	.	−10	.	−2	10
X.34	120	120	−120	16	16	24	−24	−16	−10	−4	−4	.	4	10	12	3	3	.	10	.	2	−10
X.35	120	120	120	24	24	−8	−8	24	.	.	.	−8	.	8	3	6	.	8	.	8	.	8
X.36	120	120	120	24	24	−8	−8	24	.	.	.	−8	.	8	3	6	.	8	.	8	.	8
X.37	120	120	−120	−16	−16	24	−24	16	−10	−4	−4	.	4	2	12	3	3	.	10	.	2	−2
X.38	120	120	−120	−16	−16	24	−24	16	10	−4	−4	.	4	−2	12	3	3	.	−10	.	−2	2
X.39	120	120	120	24	24	−8	−8	24	.	.	.	−8	.	−8	3	6	.	8	.	8	.	−8
X.40	120	120	120	24	24	−8	−8	24	.	.	.	−8	.	−8	3	6	.	8	.	8	.	−8
X.41	135	135	135	−33	−33	7	7	−33	−15	3	3	−1	3	5	.	9	.	−9	−15	15	1	5
X.42	135	135	135	39	39	7	7	39	−15	15	15	7	15	−7	.	9	.	−9	−15	−9	1	−7
X.43	135	135	135	39	39	7	7	39	15	15	15	7	15	7	.	9	.	−9	15	−9	−1	7
X.44	135	135	135	−33	−33	7	7	−33	15	3	3	−1	3	−5	.	9	.	−9	15	15	−1	−5
X.45	160	−160	.	−48	48	.	.	.	.	8	−8	.	.	.	−2	16	−2	.	.	.	.	.
X.46	160	−160	.	−48	48	.	.	.	.	8	−8	.	.	.	−2	16	−2	.	.	.	.	.
X.47	192	−192	.	−32	32	.	.	.	.	16	−16	.	.	.	12	.	6	.	.	.	.	.
X.48	216	216	−216	−48	−48	−8	8	48	−6	12	12	.	−12	−2	.	9	.	.	6	.	−2	2
X.49	216	216	−216	48	48	−8	8	−48	6	12	12	.	−12	10	.	9	.	.	−6	.	2	−10
X.50	216	216	−216	48	48	−8	8	−48	−6	12	12	.	−12	−10	.	9	.	.	6	.	−2	10
X.51	216	216	−216	−48	−48	−8	8	48	6	12	12	.	−12	2	.	9	.	.	−6	.	2	−2
X.52	240	240	−240	−32	−32	48	−48	32	40	8	8	.	−8	.	−12	.	3	.	−40	.	−8	.
X.53	240	240	−240	32	32	48	−48	−32	40	8	8	.	−8	.	−12	.	3	.	−40	.	−8	.
X.54	240	240	240	48	48	−16	−16	48	.	.	.	−16	.	.	6	12	.	16	.	16	.	.
X.55	240	240	−240	−32	−32	48	−48	32	−40	8	8	.	−8	.	−12	.	3	.	40	.	8	.
X.56	240	240	−240	32	32	48	−48	−32	−40	8	8	.	−8	.	−12	.	3	.	40	.	8	.
X.57	240	240	240	−48	−48	−16	−16	−48	.	.	.	16	.	.	6	12	.	16	.	−16	.	.
X.58	240	240	240	−48	−48	−16	−16	−48	.	.	.	16	.	.	6	12	.	16	.	−16	.	.
X.59	270	270	270	6	6	14	14	6	30	18	18	6	18	2	.	−9	.	−18	30	6	−2	2
X.60	270	270	270	6	6	14	14	6	−30	18	18	6	18	−2	.	−9	.	−18	−30	6	2	−2
X.61	320	320	−320	−64	−64	64	−64	64	.	.	.	.	.	.	−4	14	−4	.	.	.	.	.
X.62	320	320	−320	64	64	64	−64	−64	.	.	.	.	.	.	−4	14	−4	.	.	.	.	.
X.63	320	−320	.	32	−32	.	.	.	.	−16	16	.	.	.	14	8	2	.	.	.	.	.
X.64	320	−320	.	32	−32	.	.	.	.	−16	16	.	.	.	14	8	2	.	.	.	.	.
X.65	405	405	405	45	45	21	21	45	45	9	9	13	9	9	.	.	.	−27	45	−3	−3	9
X.66	405	405	405	−27	−27	21	21	−27	−45	−3	−3	5	−3	3	.	.	.	−27	−45	21	3	3
X.67	405	405	405	−27	−27	21	21	−27	45	−3	−3	5	−3	−3	.	.	.	−27	45	21	−3	−3
X.68	405	405	405	45	45	21	21	45	−45	9	9	13	9	−9	.	.	.	−27	−45	−3	3	−9
X.69	480	480	−480	32	32	96	−96	−32	40	.	.	.	.	8	12	−3	.	.	−40	.	−8	−8
X.70	480	480	−480	.	.	96	−96	.	.	−16	−16	.	16	.	−24	.	6	.	.	.	.	.
X.71	480	480	−480	−32	−32	96	−96	32	−40	.	.	.	.	8	12	−3	.	.	40	.	8	−8
X.72	480	480	−480	−32	−32	96	−96	32	40	.	.	.	.	−8	12	−3	.	.	−40	.	−8	8
X.73	480	480	−480	32	32	96	−96	−32	−40	.	.	.	.	−8	12	−3	.	.	40	.	8	8
X.74	480	−480	.	−16	16	.	.	.	.	−8	8	.	.	.	−24	.	6	.	.	.	.	.
X.75	480	−480	.	112	−112	.	.	.	.	24	−24	.	.	.	12	24	.	.	.	.	.	.
X.76	540	540	540	60	60	28	28	60	30	−12	−12	−4	−12	−2	.	9	.	−36	30	−36	−2	−2
X.77	540	540	540	60	60	28	28	60	−30	−12	−12	−4	−12	2	.	9	.	−36	−30	−36	2	2
X.78	540	540	540	−84	−84	28	28	−84	30	12	12	−20	12	−2	.	9	.	−36	30	12	−2	−2
X.79	540	540	540	−84	−84	28	28	−84	−30	12	12	−20	12	2	.	9	.	−36	−30	12	2	2
X.80	640	−640	.	64	−64	.	.	.	.	32	−32	.	.	.	−8	−8	10	.	.	.	.	.
X.81	640	640	−640	.	.	128	−128	.	.	.	.	.	.	.	−8	−8	−8	.	.	.	.	.
X.82	720	720	720	−48	−48	−48	−48	−48	.	.	.	16	.	.	−9	.	.	48	.	−16	.	.
X.83	720	720	720	−48	−48	−48	−48	−48	.	.	.	16	.	.	−9	.	.	48	.	−16	.	.
X.84	720	720	720	−48	−48	−48	−48	−48	.	.	.	16	.	.	−9	.	.	48	.	−16	.	.

Character table of $H(\mathsf{Fi}_{22})$ *(continued)*

	4_6	4_7	4_8	4_9	4_{10}	4_{11}	4_{12}	4_{13}	4_{14}	4_{15}	4_{16}	4_{17}	4_{18}	4_{19}	4_{20}	4_{21}	4_{22}	4_{23}	4_{24}	4_{25}	4_{26}	4_1	$5a$	$6a$	$6b$
2	10	10	10	10	10	11	11	9	9	9	9	8	8	9	9	9	9	8	8	8	8	7	3	7	7
3	1	1	1	1	1	.	.	1	1	1	1	1	1	.	.	.	.	.	.	.	.	.	.	4	4
5	.	.	.	.	.	.	.	.	.	.	.	.	.	.	.	.	.	.	.	.	.	.	1	.	.
2P	2e	2e	2e	2e	2e	2e	2e	2c	2c	2g	2g	2i	2j	2e	2e	2f	2e	2i	2i	2j	2i	2i	5a	3a	3a
3P	4_6	4_7	4_8	4_9	4_{10}	4_{11}	4_{12}	4_{13}	4_{14}	4_{15}	4_{16}	4_{17}	4_{18}	4_{19}	4_{20}	4_{21}	4_{22}	4_{23}	4_{24}	4_{25}	4_{26}	4_1	5a	2a	2b
5P	4_6	4_7	4_8	4_9	4_{10}	4_{11}	4_{12}	4_{13}	4_{14}	4_{15}	4_{16}	4_{17}	4_{18}	4_{19}	4_{20}	4_{21}	4_{22}	4_{23}	4_{24}	4_{25}	4_{26}	4_1	1a	6a	6c
X.1	1	1	1	1	1	1	1	1	1	1	1	1	1	1	1	1	1	1	1	1	1	1	1	1	1
X.2	1	−1	−1	−1	1	1	1	1	1	1	1	−1	−1	−1	−1	1	1	−1	1	−1	1	1	1	1	1
X.3	−2	.	.	−4	−2	2	2	2	2	2	2	2	2	.	.	2	2	−2	.	−2	.	.	1	−3	−3
X.4	−2	.	.	4	−2	2	2	2	2	2	2	−2	−2	.	.	2	2	2	.	2	.	.	1	−3	−3
X.5	−6	.	.	.	−6	2	2	2	2	2	2	.	.	.	.	2	2	.	−2	.	−2	−2	.	1	1
X.6	−1	−3	−3	5	−1	−1	−1	3	3	3	3	1	1	−3	−3	−1	−1	1	−1	1	−1	−1	.	6	6
X.7	−1	3	3	−5	−1	−1	−1	3	3	3	3	−1	−1	3	3	−1	−1	−1	−1	−1	−1	−1	.	6	6
X.8	7	−1	−1	−5	7	3	3	−1	−1	−1	−1	−3	−3	−1	−1	3	3	1	1	1	1	1	.	−3	−3
X.9	7	1	1	5	7	3	3	−1	−1	−1	−1	3	3	1	1	3	3	−1	1	−1	1	1	.	−3	−3
X.10	4	.	.	.	4	−4	−4	4	4	4	4	.	.	.	.	−4	−4	.	.	.	.	.	.	−7	−7
X.11	4	−2	−2	−10	4	4	4	.	.	.	.	−2	−2	−2	−2	4	4	−2	.	−2	.	.	.	2	2
X.12	4	2	2	10	4	4	4	.	.	.	.	2	2	2	2	4	4	2	.	2	.	.	.	2	2
X.13	8	4	4	4	8	.	.	.	.	.	.	.	.	4	4	.	.	.	.	.	.	.	−1	6	6
X.14	8	−4	−4	−4	8	.	.	.	.	.	.	.	.	−4	−4	.	.	.	.	.	.	.	−1	6	6
X.15	−10	−2	−2	10	−10	2	2	−2	−2	−2	−2	−4	−4	−2	−2	2	2	.	.	.	.	.	.	3	3
X.16	−10	2	2	−10	−10	2	2	−2	−2	−2	−2	4	4	2	2	2	2	.	.	.	.	.	.	3	3
X.17	.	.	.	.	.	.	.	.	.	4	−4	.	.	.	.	.	.	.	4	.	−4	.	2	4	.
X.18	4	−2	2	−2	−4	4	−4	.	.	.	.	−4	4	−2	2	.	.	.	2	.	2	−2	.	4	−4
X.19	4	2	−2	2	−4	4	−4	.	.	.	.	4	−4	2	−2	.	.	.	2	.	2	−2	.	4	−4
X.20	−4	−6	6	2	4	4	−4	.	.	.	.	−4	4	2	−2	.	.	.	−2	.	−2	2	.	4	−4
X.21	−4	6	−6	−2	4	4	−4	.	.	.	.	4	−4	−2	2	.	.	.	−2	.	−2	2	.	4	−4
X.22	−4	−2	−2	−10	−4	4	4	.	.	.	.	2	2	−2	−2	4	4	2	.	2	.	.	.	6	6
X.23	−4	2	2	10	−4	4	4	.	.	.	.	−2	−2	2	2	4	4	−2	.	−2	.	.	.	6	6
X.24	12	.	.	.	12	4	4	4	4	4	4	.	.	.	.	4	4	.	.	.	.	.	.	−3	−3
X.25	.	.	.	−16	.	.	.	.	.	.	.	.	.	.	.	.	.	.	.	.	.	.	−1	−8	−8
X.26	.	.	.	16	.	.	.	.	.	.	.	.	.	.	.	.	.	.	.	.	.	.	−1	−8	−8
X.27	.	−4	4	4	.	8	−8	.	.	.	.	.	.	4	−4	.	.	.	.	.	.	.	.	8	−8
X.28	−16	.	.	.	−16	.	.	.	.	.	.	.	.	.	.	.	.	.	.	.	.	.	.	−10	−10
X.29	.	4	−4	−4	.	8	−8	.	.	.	.	.	.	−4	4	.	.	.	.	.	.	.	.	8	−8
X.30	9	−3	−3	9	9	−3	−3	−3	−3	−3	−3	3	3	−3	−3	−3	−3	−1	−1	−1	−1	−1	1	.	.
X.31	9	3	3	−9	9	−3	−3	−3	−3	−3	−3	−3	−3	3	3	−3	−3	1	−1	1	−1	−1	1	.	.
X.32	−6	.	.	.	−6	−6	−6	2	2	2	2	.	.	.	.	−6	−6	.	2	.	2	2	.	9	9
X.33	4	2	−2	2	−4	−4	4	.	.	.	.	4	−4	−6	6	.	.	.	−2	.	−2	2	.	12	−12
X.34	4	−2	2	−2	−4	−4	4	.	.	.	.	−4	4	6	−6	.	.	.	−2	.	−2	2	.	12	−12
X.35	.	−8	−8	.	.	.	.	−4	12	−4	−4	.	.	.	.	.	.	.	.	.	.	.	.	3	3
X.36	.	−8	−8	.	.	.	.	12	−4	−4	−4	.	.	.	.	.	.	.	.	.	.	.	.	3	3
X.37	−4	−10	10	−2	4	−4	4	.	.	.	.	4	−4	6	−6	.	.	.	2	.	2	−2	.	12	−12
X.38	−4	10	−10	2	4	−4	4	.	.	.	.	−4	4	−6	6	.	.	.	2	.	2	−2	.	12	−12
X.39	.	8	8	.	.	.	.	−4	12	−4	−4	.	.	.	.	.	.	.	.	.	.	.	.	3	3
X.40	.	8	8	.	.	.	.	12	−4	−4	−4	.	.	.	.	.	.	.	.	.	.	.	.	3	3
X.41	−1	5	5	1	−1	3	3	3	3	3	3	3	3	−3	−3	3	−5	−1	−3	−1	−3	−3	.	.	.
X.42	−1	−7	−7	1	−1	−1	−1	3	3	3	3	−3	−3	1	1	−1	−1	−3	3	−3	3	3	.	.	.
X.43	−1	7	7	−1	−1	−1	−1	3	3	3	3	3	3	−1	−1	−1	−1	3	3	3	3	3	.	.	.
X.44	−1	−5	−5	−1	−1	3	3	3	3	3	3	−3	−3	3	3	3	−5	1	−3	1	−3	−3	.	.	.
X.45	.	.	.	.	.	.	.	.	.	4	−4	.	.	.	.	.	.	.	−4	.	4	.	.	2	A
X.46	.	.	.	.	.	.	.	.	.	4	−4	.	.	.	.	.	.	.	−4	.	4	.	.	2	$\bar{A}$
X.47	.	.	.	.	.	.	.	.	.	8	−8	.	.	.	.	.	.	.	.	.	.	.	2	−12	.
X.48	4	10	−10	2	−4	−4	4	.	.	.	.	.	.	2	−2	.	.	4	−2	−4	−2	2	1	.	.
X.49	−4	−2	2	−2	4	−4	4	.	.	.	.	.	.	−2	2	.	.	4	2	−4	2	−2	1	.	.
X.50	−4	2	−2	2	4	−4	4	.	.	.	.	.	.	2	−2	.	.	−4	2	4	2	−2	1	.	.
X.51	4	−10	10	−2	−4	−4	4	.	.	.	.	.	.	−2	2	.	.	−4	−2	4	−2	2	1	.	.
X.52	−8	.	.	8	8	8	−8	.	.	.	.	−8	8	.	.	.	.	.	.	.	.	.	.	−12	12
X.53	8	.	.	8	−8	8	−8	.	.	.	.	8	−8	.	.	.	.	.	.	.	.	.	.	−12	12
X.54	.	.	.	.	.	.	.	−8	−8	8	8	.	.	.	.	.	.	.	.	.	.	.	.	6	6
X.55	−8	.	.	−8	8	8	−8	.	.	.	.	8	−8	.	.	.	.	.	.	.	.	.	.	−12	12
X.56	8	.	.	−8	−8	8	−8	.	.	.	.	−8	8	.	.	.	.	.	.	.	.	.	.	−12	12
X.57	.	.	.	.	.	.	.	.	.	.	.	.	.	.	.	.	.	.	.	.	.	.	.	6	6
X.58	.	.	.	.	.	.	.	.	.	.	.	.	.	.	.	.	.	.	.	.	.	.	.	6	6
X.59	−2	2	2	−2	−2	2	2	6	6	6	6	.	.	2	2	2	−6	4	.	4	.	.	.	.	.
X.60	−2	−2	−2	2	−2	2	2	6	6	6	6	.	.	−2	−2	2	−6	−4	.	−4	.	.	.	.	.
X.61	−16	.	.	.	16	.	.	.	.	.	.	.	.	.	.	.	.	.	.	.	.	.	.	−4	4
X.62	16	.	.	.	−16	.	.	.	.	.	.	.	.	.	.	.	.	.	.	.	.	.	.	−4	4
X.63	.	.	.	.	.	.	.	.	.	8	−8	.	.	.	.	.	.	.	.	.	.	.	.	−14	$\bar{A}$
X.64	.	.	.	.	.	.	.	.	.	8	−8	.	.	.	.	.	.	.	.	.	.	.	.	−14	A
X.65	−3	9	9	−3	−3	9	9	−3	−3	−3	−3	3	3	1	1	−7	1	−1	−1	−1	−1	−1	.	.	.
X.66	−3	3	3	3	−3	13	13	−3	−3	−3	−3	3	3	−5	−5	−3	−3	3	1	3	1	1	.	.	.
X.67	−3	−3	−3	−3	−3	13	13	−3	−3	−3	−3	−3	−3	5	5	−3	−3	−3	1	−3	1	1	.	.	.
X.68	−3	−9	−9	3	−3	9	9	−3	−3	−3	−3	−3	−3	−1	−1	−7	1	1	−1	1	−1	−1	.	.	.
X.69	8	8	−8	8	−8	.	.	.	.	.	.	.	.	.	.	.	.	.	.	.	.	.	.	12	−12
X.70	.	.	.	.	.	−16	16	.	.	.	.	.	.	.	.	.	.	.	.	.	.	.	.	−24	24
X.71	−8	8	−8	−8	8	.	.	.	.	.	.	.	.	.	.	.	.	.	.	.	.	.	.	12	−12
X.72	−8	−8	8	8	8	.	.	.	.	.	.	.	.	.	.	.	.	.	.	.	.	.	.	12	−12
X.73	8	−8	8	−8	−8	.	.	.	.	.	.	.	.	.	.	.	.	.	.	.	.	.	.	12	−12
X.74	.	.	.	.	.	.	.	.	.	12	−12	.	.	.	.	.	.	.	−4	.	4	.	.	24	.
X.75	.	.	.	.	.	.	.	.	.	−4	4	.	.	.	.	.	.	.	4	.	−4	.	.	−12	.
X.76	4	−2	−2	−2	4	4	4	.	.	.	.	6	6	−2	−2	4	−4	−2	.	−2	.	.	.	.	.
X.77	4	2	2	2	4	4	4	.	.	.	.	−6	−6	2	2	4	−4	2	.	2	.	.	.	.	.
X.78	4	−2	−2	−2	4	−4	−4	.	.	.	.	−6	−6	−2	−2	−4	4	2	.	2	.	.	.	.	.
X.79	4	2	2	2	4	−4	−4	.	.	.	.	6	6	2	2	−4	4	−2	.	−2	.	.	.	.	.
X.80	.	.	.	.	.	.	.	.	.	.	.	.	.	.	.	.	.	.	.	.	.	.	.	8	.
X.81	.	.	.	.	.	.	.	.	.	.	.	.	.	.	.	.	.	.	.	.	.	.	.	−8	8
X.82	.	.	.	.	.	.	.	24	−8	−8	−8	.	.	.	.	.	.	.	.	.	.	.	.	−9	−9
X.83	.	.	.	.	.	.	.	−8	24	−8	−8	.	.	.	.	.	.	.	.	.	.	.	.	−9	−9
X.84	.	.	.	.	.	.	.	−8	−8	8	8	.	.	.	.	.	.	.	.	.	.	.	.	−9	−9

Character table of $H(\mathsf{Fi}_{22})$ *(continued)*

	6c	6d	6e	6f	6g	6h	6i	6j	6k	6l	6m	6n	6o	6p	6q	6r	6s	6t	6u	6v	6w	6x	6y	6z	8a	8b	8c	8d	8e
2	7	8	7	7	7	7	7	5	6	6	4	7	7	5	5	4	4	4	4	5	5	5	3	4	7	7	7	7	7
3	4	3	3	2	2	2	2	3	2	2	3	1	1	2	2	2	2	2	2	1	1	1	2	1	1	1	.	.	.
5	.	.	.	.	.	.	.	.	.	.	.	.	.	.	.	.	.	.	.	.	.	.	.	.	.	.	.	.	.
2P	3a	3b	3b	3a	3a	3a	3a	3c	3b	3b	3c	3b	3b	3b	3b	3c	3c	3c	3c	3c	3c	3b	3c	3c	4_{1}	4_{1}	4_{11}	4_{11}	4_{11}
3P	2b	2a	2b	2c	2g	2g	2d	2a	2c	2d	2b	2e	2f	2h	2g	2g	2g	2d	2c	2i	2j	2m	2h	2l	8a	8b	8c	8d	8e
5P	6b	6d	6e	6f	6h	6g	6i	6j	6k	6l	6m	6n	6o	6p	6q	6s	6r	6t	6u	6v	6w	6x	6y	6z	8a	8b	8c	8d	8e
X.1	1	1	1	1	1	1	1	1	1	1	1	1	1	1	1	1	1	1	1	1	1	1	1	1	1	1	1	1	1
X.2	1	1	1	1	1	1	1	1	1	1	1	1	1	−1	1	1	1	1	1	1	1	−1	−1	1	−1	−1	−1	1	1
X.3	−3	.	.	1	1	1	1	3	−2	−2	3	.	.	2	−2	1	1	1	1	−1	−1	.	−1	−1	.	.	2	.	.
X.4	−3	.	.	1	1	1	1	3	−2	−2	3	.	.	−2	−2	1	1	1	1	−1	−1	.	1	−1	.	.	−2	.	.
X.5	1	4	4	−3	−3	−3	−3	−2	.	.	−2	4	4	.	.	.	.	.	.	2	2	.	.	2	.	.	.	−2	−2
X.6	6	.	.	2	2	2	2	3	2	2	3	.	.	2	2	−1	−1	−1	−1	−1	−1	.	−1	−1	−3	−3	1	−1	−1
X.7	6	.	.	2	2	2	2	3	2	2	3	.	.	−2	2	−1	−1	−1	−1	−1	−1	.	1	−1	3	3	−1	−1	−1
X.8	−3	3	3	1	1	1	1	.	1	1	.	3	3	1	1	−2	−2	−2	−2	.	.	−1	−2	.	−1	−1	−3	1	1
X.9	−3	3	3	1	1	1	1	.	1	1	.	3	3	−1	1	−2	−2	−2	−2	.	.	1	2	.	1	1	3	1	1
X.10	−7	2	2	1	1	1	1	2	−2	−2	2	2	2	.	−2	−2	−2	−2	−2	2	2	.	.	2	.	.	.	.	.
X.11	2	−1	−1	−2	−2	−2	−2	5	1	1	5	−1	−1	−1	1	1	1	1	1	1	1	1	−1	1	−2	−2	−2	.	.
X.12	2	−1	−1	−2	−2	−2	−2	5	1	1	5	−1	−1	1	1	1	1	1	1	1	1	−1	1	1	2	2	2	.	.
X.13	6	3	3	2	2	2	2	.	−1	−1	.	3	3	1	−1	2	2	2	2	.	.	1	−2	.	4	4	.	.	.
X.14	6	3	3	2	2	2	2	.	−1	−1	.	3	3	−1	−1	2	2	2	2	.	.	−1	2	.	−4	−4	.	.	.
X.15	3	3	3	−1	−1	−1	−1	3	−1	−1	3	3	3	1	−1	−1	−1	−1	−1	−1	−1	1	1	−1	−2	−2	−4	.	.
X.16	3	3	3	−1	−1	−1	−1	3	−1	−1	3	3	3	−1	−1	−1	−1	−1	−1	−1	−1	−1	−1	−1	2	2	4	.	.
X.17	.	−8	.	4	.	.	−4	−2	4	−4	.	.	.	.	.	.	.	2	−2	2	−2	.	.	.	.	.	.	.	.
X.18	−4	7	−7	4	−4	−4	4	1	1	1	−1	−1	1	−1	−1	−1	−1	1	1	1	1	−3	−1	−1	−2	2	.	2	−2
X.19	−4	7	−7	4	−4	−4	4	1	1	1	−1	−1	1	1	−1	−1	−1	1	1	1	1	3	1	−1	2	−2	.	2	−2
X.20	−4	7	−7	−4	4	4	−4	1	−1	−1	−1	−1	1	1	1	1	1	−1	−1	1	1	1	1	−1	−2	2	.	−2	2
X.21	−4	7	−7	−4	4	4	−4	1	−1	−1	−1	−1	1	−1	1	1	1	−1	−1	1	1	−1	−1	−1	2	−2	.	−2	2
X.22	6	−3	−3	2	2	2	2	−3	−1	−1	−3	−3	−3	−1	−1	−1	−1	−1	−1	1	1	1	−1	1	−2	−2	2	.	.
X.23	6	−3	−3	2	2	2	2	−3	−1	−1	−3	−3	−3	1	−1	−1	−1	−1	−1	1	1	−1	1	1	2	2	−2	.	.
X.24	−3	.	.	−3	−3	−3	−3	−6	.	.	−6	.	.	.	.	.	.	.	.	−2	−2	.	.	−2	.	.	.	.	.
X.25	−8	−2	−2	.	.	.	.	4	.	.	4	−2	−2	2	.	.	.	.	.	.	.	.	2	.	.	.	.	.	.
X.26	−8	−2	−2	.	.	.	.	4	.	.	4	−2	−2	−2	.	.	.	.	.	.	.	.	−2	.	.	.	.	.	.
X.27	−8	−4	4	.	.	.	.	2	.	.	−2	4	−4	2	.	.	.	.	.	2	2	−2	2	−2	.	.	.	.	.
X.28	−10	2	2	2	2	2	2	−4	2	2	−4	2	2	.	2	2	2	2	2	.	.	.	.	.	.	.	.	.	.
X.29	−8	−4	4	.	.	.	.	2	.	.	−2	4	−4	−2	.	.	.	.	.	2	2	2	−2	−2	.	.	.	.	.
X.30	.	.	.	.	.	.	.	.	.	.	.	.	.	.	.	.	.	.	.	.	.	.	.	.	−3	−3	3	−1	−1
X.31	.	.	.	.	.	.	.	.	.	.	.	.	.	.	.	.	.	.	.	.	.	.	.	.	3	3	−3	−1	−1
X.32	9	.	.	−3	−3	−3	−3	.	.	.	.	.	.	.	.	.	.	.	.	.	.	.	.	.	.	.	.	2	2
X.33	−12	3	−3	4	−4	−4	4	3	1	1	−3	3	−3	1	−1	−1	−1	1	1	−1	−1	−1	1	1	−2	2	.	−2	2
X.34	−12	3	−3	4	−4	−4	4	3	1	1	−3	3	−3	−1	−1	−1	−1	1	1	−1	−1	1	−1	1	2	−2	.	−2	2
X.35	3	6	6	3	3	3	3	.	.	.	.	−2	−2	.	.	.	.	.	.	.	.	2	.	.	.	.	.	.	.
X.36	3	6	6	3	3	3	3	.	.	.	.	−2	−2	.	.	.	.	.	.	.	.	2	.	.	.	.	.	.	.
X.37	−12	3	−3	−4	4	4	−4	3	−1	−1	−3	3	−3	−1	1	1	1	−1	−1	−1	−1	−1	−1	1	−2	2	.	2	−2
X.38	−12	3	−3	−4	4	4	−4	3	−1	−1	−3	3	−3	1	1	1	1	−1	−1	−1	−1	1	1	1	2	−2	.	2	−2
X.39	3	6	6	3	3	3	3	.	.	.	.	−2	−2	.	.	.	.	.	.	.	.	−2	.	.	.	.	.	.	.
X.40	3	6	6	3	3	3	3	.	.	.	.	−2	−2	.	.	.	.	.	.	.	.	−2	.	.	.	.	.	.	.
X.41	.	9	9	.	.	.	.	.	3	3	.	1	1	−3	3	.	.	.	.	.	.	−1	.	.	1	1	−1	1	1
X.42	.	9	9	.	.	.	.	.	3	3	.	1	1	−3	3	.	.	.	.	.	.	−1	.	.	1	1	1	−1	−1
X.43	.	9	9	.	.	.	.	.	3	3	.	1	1	3	3	.	.	.	.	.	.	1	.	.	−1	−1	−1	−1	−1
X.44	.	9	9	.	.	.	.	.	3	3	.	1	1	3	3	.	.	.	.	.	.	1	.	.	−1	−1	1	1	1
X.45	$\bar{A}$	−16	.	−6	B	$\bar{B}$	6	2	.	.	.	.	.	.	.	$\bar{B}$	B	.	.	2	−2	.	.	.	.	.	.	.	.
X.46	A	−16	.	−6	$\bar{B}$	B	6	2	.	.	.	.	.	.	.	B	$\bar{B}$	.	.	2	−2	.	.	.	.	.	.	.	.
X.47	.	.	.	4	.	.	−4	−6	−8	8	.	.	.	.	.	.	.	2	−2	−2	2	.	.	.	.	.	.	.	.
X.48	.	9	−9	.	.	.	.	.	−3	−3	.	1	−1	−3	3	.	.	.	.	.	.	1	.	.	−2	2	.	2	−2
X.49	.	9	−9	.	.	.	.	.	3	3	.	1	−1	3	−3	.	.	.	.	.	.	1	.	.	−2	2	.	−2	2
X.50	.	9	−9	.	.	.	.	.	3	3	.	1	−1	−3	−3	.	.	.	.	.	.	−1	.	.	2	−2	.	−2	2
X.51	.	9	−9	.	.	.	.	.	−3	−3	.	1	−1	3	3	.	.	.	.	.	.	−1	.	.	2	−2	.	2	−2
X.52	12	.	.	4	−4	−4	4	3	−2	−2	−3	.	.	−2	2	−1	−1	1	1	−1	−1	.	1	1	.	.	.	.	.
X.53	12	.	.	−4	4	4	−4	3	2	2	−3	.	.	−2	−2	1	1	−1	−1	−1	−1	.	1	1	.	.	.	.	.
X.54	6	12	12	6	6	6	6	.	.	.	.	−4	−4	.	.	.	.	.	.	.	.	.	.	.	.	.	.	.	.
X.55	12	.	.	4	−4	−4	4	3	−2	−2	−3	.	.	2	2	−1	−1	1	1	−1	−1	.	−1	1	.	.	.	.	.
X.56	12	.	.	−4	4	4	−4	3	2	2	−3	.	.	2	−2	1	1	−1	−1	−1	−1	.	−1	1	.	.	.	.	.
X.57	6	12	12	−6	−6	−6	−6	.	.	.	.	−4	−4	.	.	.	.	.	.	.	.	.	.	.	.	.	.	.	.
X.58	6	12	12	−6	−6	−6	−6	.	.	.	.	−4	−4	.	.	.	.	.	.	.	.	.	.	.	.	.	.	.	.
X.59	.	−9	−9	.	.	.	.	.	−3	−3	.	−1	−1	−3	−3	.	.	.	.	.	.	−1	.	.	−2	−2	.	.	.
X.60	.	−9	−9	.	.	.	.	.	−3	−3	.	−1	−1	3	−3	.	.	.	.	.	.	1	.	.	2	2	.	.	.
X.61	4	14	−14	−4	4	4	−4	−4	2	2	4	−2	2	.	−2	−2	−2	2	2	.	.	.	.	.	.	.	.	.	.
X.62	4	14	−14	4	−4	−4	4	−4	−2	−2	4	−2	2	.	2	2	2	−2	−2	.	.	.	.	.	.	.	.	.	.
X.63	A	−8	.	2	$\bar{A}$	A	−2	−2	−4	4	.	.	.	.	.	.	.	−2	2	2	−2	.	.	.	.	.	.	.	.
X.64	$\bar{A}$	−8	.	2	A	$\bar{A}$	−2	−2	−4	4	.	.	.	.	.	.	.	−2	2	2	−2	.	.	.	.	.	.	.	.
X.65	.	.	.	.	.	.	.	.	.	.	.	.	.	.	.	.	.	.	.	.	.	.	.	.	−3	−3	−1	3	3
X.66	.	.	.	.	.	.	.	.	.	.	.	.	.	.	.	.	.	.	.	.	.	.	.	.	3	3	−1	−3	−3
X.67	.	.	.	.	.	.	.	.	.	.	.	.	.	.	.	.	.	.	.	.	.	.	.	.	−3	−3	1	−3	−3
X.68	.	.	.	.	.	.	.	.	.	.	.	.	.	.	.	.	.	.	.	.	.	.	.	.	3	3	1	3	3
X.69	−12	−3	3	−4	4	4	−4	.	−1	−1	.	−3	3	1	1	−2	−2	2	2	.	.	−1	−2	.	4	−4	.	.	.
X.70	24	.	.	.	.	.	.	6	.	.	−6	.	.	.	.	.	.	.	.	2	2	.	.	−2	.	.	.	.	.
X.71	−12	−3	3	4	−4	−4	4	.	1	1	.	−3	3	−1	−1	2	2	−2	−2	.	.	−1	2	.	4	−4	.	.	.
X.72	−12	−3	3	4	−4	−4	4	.	1	1	.	−3	3	1	−1	2	2	−2	−2	.	.	1	−2	.	−4	4	.	.	.
X.73	−12	−3	3	−4	4	4	−4	.	−1	−1	.	−3	3	−1	1	−2	−2	2	2	.	.	1	2	.	−4	4	.	.	.
X.74	.	.	.	8	.	.	−8	−6	8	−8	.	.	.	.	.	.	.	−2	2	−2	2	.	.	.	.	.	.	.	.
X.75	.	−24	.	4	.	.	−4	.	4	−4	.	.	.	.	.	.	.	−4	4	.	.	.	.	.	.	.	.	.	.
X.76	.	9	9	.	.	.	.	.	−3	−3	.	1	1	−3	−3	.	.	.	.	.	.	1	.	.	2	2	−2	.	.
X.77	.	9	9	.	.	.	.	.	−3	−3	.	1	1	3	−3	.	.	.	.	.	.	−1	.	.	−2	−2	2	.	.
X.78	.	9	9	.	.	.	.	.	−3	−3	.	1	1	−3	−3	.	.	.	.	.	.	1	.	.	2	2	2	.	.
X.79	.	9	9	.	.	.	.	.	−3	−3	.	1	1	3	−3	.	.	.	.	.	.	−1	.	.	−2	−2	−2	.	.
X.80	.	8	.	−8	.	.	8	−10	4	−4	.	.	.	.	.	.	.	2	−2	2	−2	.	.	.	.	.	.	.	.
X.81	8	−8	8	.	.	.	.	−8	.	.	8	8	−8	.	.	.	.	.	.	.	.	.	.	.	.	.	.	.	.
X.82	−9	.	.	3	3	3	3	.	.	.	.	.	.	.	.	.	.	.	.	.	.	.	.	.	.	.	.	.	.
X.83	−9	.	.	3	3	3	3	.	.	.	.	.	.	.	.	.	.	.	.	.	.	.	.	.	.	.	.	.	.
X.84	−9	.	.	3	3	3	3	.	.	.	.	.	.	.	.	.	.	.	.	.	.	.	.	.	.	.	.	.	.

Character table of $H(\mathsf{Fi}_{22})$ (continued)

	8f	8g	8h	8i	8j	8k	9a	10a	10b	10c	12a	12b	12c	12d	12e	12f	12g	12h	12i	12j	12k	12l	12m	12n	12o	12p
2	7	7	6	5	5	5	2	3	2	2	5	6	6	5	5	5	5	5	5	5	5	5	5	5	5	5
3	.	.	.	.	.	.	2	.	.	.	3	2	2	2	1	1	1	1	1	1	1	1	1	1	1	1
5	.	.	.	.	.	.	.	1	1	1	.	.	.	.	.	.	.	.	.	.	.	.	.	.	.	.
2P	4_3	4_{11}	4_{12}	4_{13}	4_{14}	4_{21}	9a	5a	5a	5a	6a	6d	6d	6d	6n	6d	6n	6n	6n	6g	6h	6a	6g	6h	6n	6n
3P	8f	8g	8h	8i	8j	8k	3a	10a	10b	10c	4_1	4_1	4_1	4_2	4_4	4_5	4_6	4_7	4_8	4_{15}	4_{15}	4_3	4_{16}	4_{16}	4_{10}	4_9
5P	8f	8g	8h	8i	8j	8k	9a	2a	2h	2b	12a	12b	12c	12d	12e	12f	12g	12h	12i	12k	12j	12l	12n	12m	12o	12p
X.1	1	1	1	1	1	1	1	1	1	1	1	1	1	1	1	1	1	1	1	1	1	1	1	1	1	1
X.2	1	−1	−1	−1	−1	1	1	1	−1	1	1	1	1	−1	−1	−1	1	−1	−1	1	1	1	1	1	1	−1
X.3	2	−2	−2	.	.	.	.	1	1	1	−3	.	.	2	2	.	−2	.	.	−1	−1	1	−1	−1	−2	2
X.4	2	2	2	.	.	.	.	1	−1	1	−3	.	.	−2	−2	.	−2	.	.	−1	−1	1	−1	−1	−2	−2
X.5	2	.	.	.	.	−2	1	.	.	.	1	4	4	.	.	.	.	.	.	−1	−1	−3	−1	−1	.	.
X.6	3	1	1	−1	−1	−1	.	.	.	.	6	.	.	2	2	.	2	.	.	.	.	2	.	.	2	2
X.7	3	−1	−1	1	1	−1	.	.	.	.	6	.	.	−2	−2	.	2	.	.	.	.	2	.	.	2	−2
X.8	−1	1	1	1	1	1	.	.	.	.	−3	3	3	1	1	−1	1	−1	−1	−1	−1	1	−1	−1	1	1
X.9	−1	−1	−1	−1	−1	1	.	.	.	.	−3	3	3	−1	−1	1	1	1	1	−1	−1	1	−1	−1	1	−1
X.10	4	.	.	.	.	.	−1	.	.	.	−7	2	2	.	.	.	−2	.	.	1	1	1	1	1	−2	.
X.11	.	−2	−2	.	.	.	−1	.	.	.	2	−1	−1	−1	−1	1	1	1	1	.	.	−2	.	.	1	−1
X.12	.	2	2	.	.	.	−1	.	.	.	2	−1	−1	1	1	−1	1	−1	−1	.	.	−2	.	.	1	1
X.13	.	.	.	.	.	.	.	−1	−1	−1	6	3	3	1	1	1	−1	1	1	.	.	2	.	.	−1	1
X.14	.	.	.	.	.	.	.	−1	1	−1	6	3	3	−1	−1	−1	−1	−1	−1	.	.	2	.	.	−1	−1
X.15	−2	.	.	.	.	.	.	.	.	.	3	3	3	1	1	1	−1	1	1	1	1	−1	1	1	−1	1
X.16	−2	.	.	.	.	.	.	.	.	.	3	3	3	−1	−1	−1	−1	−1	−1	1	1	−1	1	1	−1	−1
X.17	.	.	.	.	.	.	2	−2	.	.	.	.	.	.	.	.	.	.	.	−2	−2	.	2	2	.	.
X.18	.	.	.	.	.	.	1	.	.	.	.	3	−3	1	−1	3	1	1	−1	.	.	.	.	.	−1	1
X.19	.	.	.	.	.	.	1	.	.	.	.	3	−3	−1	1	−3	1	−1	1	.	.	.	.	.	−1	−1
X.20	.	.	.	.	.	.	1	.	.	.	.	3	−3	−1	1	−1	−1	−3	3	.	.	.	.	.	1	−1
X.21	.	.	.	.	.	.	1	.	.	.	.	3	−3	1	−1	1	−1	3	−3	.	.	.	.	.	1	1
X.22	.	2	2	.	.	.	.	.	.	.	6	−3	−3	−1	−1	1	−1	1	1	.	.	2	.	.	−1	−1
X.23	.	−2	−2	.	.	.	.	.	.	.	6	−3	−3	1	1	−1	−1	−1	−1	.	.	2	.	.	−1	1
X.24	4	.	.	.	.	.	.	.	.	.	−3	.	.	.	.	.	.	.	.	1	1	−3	1	1	.	.
X.25	.	.	.	.	.	.	1	−1	−1	−1	−8	−2	−2	2	2	.	.	.	.	.	.	.	.	.	.	2
X.26	.	.	.	.	.	.	1	−1	1	−1	−8	−2	−2	−2	−2	.	.	.	.	.	.	.	.	.	.	−2
X.27	.	.	.	.	.	.	−1	.	.	.	.	.	.	−2	2	2	.	2	−2	.	.	.	.	.	.	−2
X.28	.	.	.	.	.	.	−1	.	.	.	−10	2	2	.	.	.	2	.	.	.	.	2	.	.	2	.
X.29	.	.	.	.	.	.	−1	.	.	.	.	.	.	2	−2	−2	.	−2	2	.	.	.	.	.	.	2
X.30	−3	−1	−1	1	1	−1	.	1	−1	1	.	.	.	.	.	.	.	.	.	.	.	.	.	.	.	.
X.31	−3	1	1	−1	−1	−1	.	1	1	1	.	.	.	.	.	.	.	.	.	.	.	.	.	.	.	.
X.32	2	.	.	.	.	2	.	.	.	.	9	.	.	.	.	.	.	.	.	−1	−1	−3	−1	−1	.	.
X.33	.	.	.	.	.	.	.	.	.	.	.	3	−3	−1	1	1	1	−1	1	.	.	.	.	.	−1	−1
X.34	.	.	.	.	.	.	.	.	.	.	.	3	−3	1	−1	−1	1	1	−1	.	.	.	.	.	−1	1
X.35	.	.	.	−2	2	.	.	.	.	.	−1	2	2	.	.	2	.	−2	−2	−1	−1	−1	−1	−1	.	.
X.36	.	.	.	2	−2	.	.	.	.	.	−1	2	2	.	.	2	.	−2	−2	−1	−1	−1	−1	−1	.	.
X.37	.	.	.	.	.	.	.	.	.	.	.	3	−3	1	−1	1	−1	−1	1	.	.	.	.	.	1	1
X.38	.	.	.	.	.	.	.	.	.	.	.	3	−3	−1	1	−1	−1	1	−1	.	.	.	.	.	1	−1
X.39	.	.	.	2	−2	.	.	.	.	.	−1	2	2	.	.	−2	.	2	2	−1	−1	−1	−1	−1	.	.
X.40	.	.	.	−2	2	.	.	.	.	.	−1	2	2	.	.	−2	.	2	2	−1	−1	−1	−1	−1	.	.
X.41	−1	3	−1	1	1	1	.	.	.	.	.	−3	−3	−3	1	−1	−1	−1	−1	.	.	.	.	.	−1	1
X.42	−1	1	1	−1	−1	−1	.	.	.	.	.	−3	−3	−3	1	−1	−1	−1	−1	.	.	.	.	.	−1	1
X.43	−1	−1	−1	1	1	−1	.	.	.	.	.	−3	−3	3	−1	1	−1	1	1	.	.	.	.	.	−1	−1
X.44	−1	−3	1	−1	−1	1	.	.	.	.	.	−3	−3	3	−1	1	−1	1	1	.	.	.	.	.	−1	−1
X.45	.	.	.	.	.	.	1	.	.	.	.	.	.	.	.	.	.	.	.	C	$\bar{C}$	.	$-C$	$-\bar{C}$	.	.
X.46	.	.	.	.	.	.	1	.	.	.	.	.	.	.	.	.	.	.	.	$\bar{C}$	C	.	$-\bar{C}$	$-C$	.	.
X.47	.	.	.	.	.	.	.	−2	.	.	.	.	.	.	.	.	.	.	.	2	2	.	−2	−2	.	.
X.48	.	.	.	.	.	.	.	1	−1	−1	.	−3	3	3	1	−1	1	1	−1	.	.	.	.	.	−1	−1
X.49	.	.	.	.	.	.	.	1	1	−1	.	−3	3	−3	−1	−1	−1	1	−1	.	.	.	.	.	1	1
X.50	.	.	.	.	.	.	.	1	−1	−1	.	−3	3	3	1	1	−1	−1	1	.	.	.	.	.	1	−1
X.51	.	.	.	.	.	.	.	1	1	−1	.	−3	3	−3	−1	1	1	−1	1	.	.	.	.	.	−1	1
X.52	.	.	.	.	.	.	.	.	.	.	.	.	.	2	−2	.	−2	.	.	.	.	.	.	.	2	2
X.53	.	.	.	.	.	.	.	.	.	.	.	.	.	2	−2	.	2	.	.	.	.	.	.	.	−2	2
X.54	.	.	.	.	.	.	.	.	.	.	−2	4	4	.	.	.	.	.	.	2	2	−2	2	2	.	.
X.55	.	.	.	.	.	.	.	.	.	.	.	.	.	−2	2	.	−2	.	.	.	.	.	.	.	2	−2
X.56	.	.	.	.	.	.	.	.	.	.	.	.	.	−2	2	.	2	.	.	.	.	.	.	.	−2	−2
X.57	.	.	.	.	.	.	.	.	.	.	−2	4	4	.	.	.	.	.	.	.	.	2	.	.	.	.
X.58	.	.	.	.	.	.	.	.	.	.	−2	4	4	.	.	.	.	.	.	.	.	2	.	.	.	.
X.59	−2	−4	.	.	.	.	.	.	.	.	.	3	3	−3	1	−1	1	−1	−1	.	.	.	.	.	1	1
X.60	−2	4	.	.	.	.	.	.	.	.	.	3	3	3	−1	1	1	1	1	.	.	.	.	.	1	−1
X.61	.	.	.	.	.	.	−1	.	.	.	.	6	−6	.	.	.	2	.	.	.	.	.	.	.	−2	.
X.62	.	.	.	.	.	.	−1	.	.	.	.	6	−6	.	.	.	−2	.	.	.	.	.	.	.	2	.
X.63	.	.	.	.	.	.	−1	.	.	.	.	.	.	.	.	.	.	.	.	$-\bar{C}$	$-C$	.	$\bar{C}$	C	.	.
X.64	.	.	.	.	.	.	−1	.	.	.	.	.	.	.	.	.	.	.	.	$-C$	$-\bar{C}$	.	C	$\bar{C}$	.	.
X.65	1	3	−1	−1	−1	−1	.	.	.	.	.	.	.	.	.	.	.	.	.	.	.	.	.	.	.	.
X.66	1	−1	−1	−1	−1	1	.	.	.	.	.	.	.	.	.	.	.	.	.	.	.	.	.	.	.	.
X.67	1	1	1	1	1	1	.	.	.	.	.	.	.	.	.	.	.	.	.	.	.	.	.	.	.	.
X.68	1	−3	1	1	1	−1	.	.	.	.	.	.	.	.	.	.	.	.	.	.	.	.	.	.	.	.
X.69	.	.	.	.	.	.	.	.	.	.	.	−3	3	−1	1	1	−1	−1	1	.	.	.	.	.	1	−1
X.70	.	.	.	.	.	.	.	.	.	.	.	.	.	.	.	.	.	.	.	.	.	.	.	.	.	.
X.71	.	.	.	.	.	.	.	.	.	.	.	−3	3	1	−1	1	1	−1	1	.	.	.	.	.	−1	1
X.72	.	.	.	.	.	.	.	.	.	.	.	−3	3	−1	1	−1	1	1	−1	.	.	.	.	.	−1	−1
X.73	.	.	.	.	.	.	.	.	.	.	.	−3	3	1	−1	−1	−1	1	−1	.	.	.	.	.	1	1
X.74	.	.	.	.	.	.	.	.	.	.	.	.	.	.	.	.	.	.	.	.	.	.	.	.	.	.
X.75	.	.	.	.	.	.	.	.	.	.	.	.	.	.	.	.	.	.	.	2	2	.	−2	−2	.	.
X.76	.	−2	2	.	.	.	.	.	.	.	.	−3	−3	−3	1	1	1	1	1	.	.	.	.	.	1	1
X.77	.	2	−2	.	.	.	.	.	.	.	.	−3	−3	3	−1	−1	1	−1	−1	.	.	.	.	.	1	−1
X.78	.	2	−2	.	.	.	.	.	.	.	.	−3	−3	−3	1	1	1	1	1	.	.	.	.	.	1	1
X.79	.	−2	2	.	.	.	.	.	.	.	.	−3	−3	3	−1	−1	1	−1	−1	.	.	.	.	.	1	−1
X.80	.	.	.	.	.	.	−2	.	.	.	.	.	.	.	.	.	.	.	.	.	.	.	.	.	.	.
X.81	.	.	.	.	.	.	1	.	.	.	.	.	.	.	.	.	.	.	.	.	.	.	.	.	.	.
X.82	.	.	.	.	.	.	.	.	.	.	3	.	.	.	.	.	.	.	.	1	1	−1	1	1	.	.
X.83	.	.	.	.	.	.	.	.	.	.	3	.	.	.	.	.	.	.	.	1	1	−1	1	1	.	.
X.84	.	.	.	.	.	.	.	.	.	.	3	.	.	.	.	.	.	.	.	F	$\bar{F}$	−1	F	$\bar{F}$	.	.

Character table of $H(\mathsf{Fi}_{22})$ *(continued)*

	12q	12r	12s	12t	12u	16a	16b	18a	18b	18c	20a	24a	24b
2	3	4	4	3	3	5	5	2	2	2	2	4	4
3	2	1	1	1	1	.	.	2	2	2	.	1	1
5	.	.	.	.	.	.	.	.	.	.	1	.	.
2P	6j	6f	6f	6v	6w	8f	8f	9a	9a	9a	10a	12b	12b
3P	4_2	4_{13}	4_{14}	4_{17}	4_{18}	16a	16b	6a	6b	6c	20a	8a	8b
5P	12q	12r	12s	12t	12u	16b	16a	18a	18c	18b	4_2	24a	24b
X.1	1	1	1	1	1	1	1	1	1	1	1	1	1
X.2	−1	1	1	−1	−1	−1	−1	1	1	1	−1	−1	−1
X.3	−1	−1	−1	−1	−1	.	.	.	.	.	1	.	.
X.4	1	−1	−1	1	1	.	.	.	.	.	−1	.	.
X.5	.	−1	−1	.	.	.	.	1	1	1	.	.	.
X.6	−1	.	.	1	1	−1	−1	.	.	.	.	.	.
X.7	1	.	.	−1	−1	1	1	.	.	.	.	.	.
X.8	−2	−1	−1	.	.	1	1	.	.	.	.	−1	−1
X.9	2	−1	−1	.	.	−1	−1	.	.	.	.	1	1
X.10	.	1	1	.	.	.	.	−1	−1	−1	.	.	.
X.11	−1	.	.	1	1	.	.	−1	−1	−1	.	1	1
X.12	1	.	.	−1	−1	.	.	−1	−1	−1	.	−1	−1
X.13	−2	.	.	.	.	.	.	.	.	.	−1	1	1
X.14	2	.	.	.	.	.	.	.	.	.	1	−1	−1
X.15	1	1	1	−1	−1	.	.	.	.	.	.	1	1
X.16	−1	1	1	1	1	.	.	.	.	.	.	−1	−1
X.17	.	.	.	.	.	.	.	−2	.	.	.	.	.
X.18	1	.	.	−1	1	.	.	1	−1	−1	.	1	−1
X.19	−1	.	.	1	−1	.	.	1	−1	−1	.	−1	1
X.20	−1	.	.	−1	1	.	.	1	−1	−1	.	1	−1
X.21	1	.	.	1	−1	.	.	1	−1	−1	.	−1	1
X.22	−1	.	.	−1	−1	.	.	.	.	.	.	1	1
X.23	1	.	.	1	1	.	.	.	.	.	.	−1	−1
X.24	.	1	1	.	.	.	.	.	.	.	.	.	.
X.25	2	.	.	.	.	.	.	1	1	1	−1	.	.
X.26	−2	.	.	.	.	.	.	1	1	1	1	.	.
X.27	−2	.	.	.	.	.	.	−1	1	1	.	.	.
X.28	.	.	.	.	.	.	.	−1	−1	−1	.	.	.
X.29	2	.	.	.	.	.	.	−1	1	1	.	.	.
X.30	.	.	.	.	.	1	1	.	.	.	−1	.	.
X.31	.	.	.	.	.	−1	−1	.	.	.	1	.	.
X.32	.	−1	−1	.	.	.	.	.	.	.	.	.	.
X.33	−1	.	.	1	−1	.	.	.	.	.	.	1	−1
X.34	1	.	.	−1	1	.	.	.	.	.	.	−1	1
X.35	.	−1	3	.	.	.	.	.	.	.	.	.	.
X.36	.	3	−1	.	.	.	.	.	.	.	.	.	.
X.37	1	.	.	1	−1	.	.	.	.	.	.	1	−1
X.38	−1	.	.	−1	1	.	.	.	.	.	.	−1	1
X.39	.	−1	3	.	.	.	.	.	.	.	.	.	.
X.40	.	3	−1	.	.	.	.	.	.	.	.	.	.
X.41	.	.	.	.	.	−1	−1	.	.	.	.	1	1
X.42	.	.	.	.	.	1	1	.	.	.	.	1	1
X.43	.	.	.	.	.	−1	−1	.	.	.	.	−1	−1
X.44	.	.	.	.	.	1	1	.	.	.	.	−1	−1
X.45	.	.	.	.	.	.	.	−1	D	$\bar{D}$	.	.	.
X.46	.	.	.	.	.	.	.	−1	$\bar{D}$	D	.	.	.
X.47	.	.	.	.	.	.	.	.	.	.	.	.	.
X.48	.	.	.	.	.	.	.	.	.	.	1	1	−1
X.49	.	.	.	.	.	.	.	.	.	.	−1	1	−1
X.50	.	.	.	.	.	.	.	.	.	.	1	−1	1
X.51	.	.	.	.	.	.	.	.	.	.	−1	−1	1
X.52	−1	.	.	1	−1	.	.	.	.	.	.	.	.
X.53	−1	.	.	−1	1	.	.	.	.	.	.	.	.
X.54	.	−2	−2	.	.	.	.	.	.	.	.	.	.
X.55	1	.	.	−1	1	.	.	.	.	.	.	.	.
X.56	1	.	.	1	−1	.	.	.	.	.	.	.	.
X.57	.	.	.	.	.	E	$\bar{E}$	.	.	.	.	.	.
X.58	.	.	.	.	.	$\bar{E}$	E	.	.	.	.	.	.
X.59	.	.	.	.	.	.	.	.	.	.	.	1	1
X.60	.	.	.	.	.	.	.	.	.	.	.	−1	−1
X.61	.	.	.	.	.	.	.	−1	1	1	.	.	.
X.62	.	.	.	.	.	.	.	−1	1	1	.	.	.
X.63	.	.	.	.	.	.	.	1	$\bar{D}$	D	.	.	.
X.64	.	.	.	.	.	.	.	1	D	$\bar{D}$	.	.	.
X.65	.	.	.	.	.	1	1	.	.	.	.	.	.
X.66	.	.	.	.	.	1	1	.	.	.	.	.	.
X.67	.	.	.	.	.	−1	−1	.	.	.	.	.	.
X.68	.	.	.	.	.	−1	−1	.	.	.	.	.	.
X.69	2	.	.	.	.	.	.	.	.	.	.	1	−1
X.70	.	.	.	.	.	.	.	.	.	.	.	.	.
X.71	−2	.	.	.	.	.	.	.	.	.	.	1	−1
X.72	2	.	.	.	.	.	.	.	.	.	.	−1	1
X.73	−2	.	.	.	.	.	.	.	.	.	.	−1	1
X.74	.	.	.	.	.	.	.	.	.	.	.	.	.
X.75	.	.	.	.	.	.	.	.	.	.	.	.	.
X.76	.	.	.	.	.	.	.	.	.	.	.	−1	−1
X.77	.	.	.	.	.	.	.	.	.	.	.	1	1
X.78	.	.	.	.	.	.	.	.	.	.	.	−1	−1
X.79	.	.	.	.	.	.	.	.	.	.	.	1	1
X.80	.	.	.	.	.	.	.	2	.	.	.	.	.
X.81	.	.	.	.	.	.	.	1	−1	−1	.	.	.
X.82	.	−3	1	.	.	.	.	.	.	.	.	.	.
X.83	.	1	−3	.	.	.	.	.	.	.	.	.	.
X.84	.	1	1	.	.	.	.	.	.	.	.	.	.

Character table of $H(\mathsf{Fi}_{22})$ *(continued)*

	$1a$	$2a$	$2b$	$2c$	$2d$	$2c$	$2f$	$2g$	$2h$	$2i$	$2j$	$2k$	$2l$	$2m$	$3a$	$3b$	$3c$	4_1	4_2	4_3	4_4	4_5
2	17	17	16	15	15	16	16	14	10	12	12	13	11	10	7	8	5	12	10	12	10	10
3	4	4	4	2	2	1	1	2	2	1	1	.	1	1	4	3	3	3	2	1	1	1
5	1	1	1	.	.	.	.	.	1	.	.	.	.	.	.	.	.	.	1	.	.	.
$2P$	$1a$	$1a$	$1a$	$1a$	$1a$	$1a$	$1a$	$1a$	$1a$	$1a$	$1a$	$1a$	$1a$	$1a$	$3a$	$3b$	$3c$	$2a$	$2a$	$2a$	$2e$	$2a$
$3P$	$1a$	$2a$	$2b$	$2c$	$2d$	$2e$	$2f$	$2g$	$2h$	$2i$	$2j$	$2k$	$2l$	$2m$	$1a$	$1a$	$1a$	4_1	4_2	4_3	4_4	4_5
$5P$	$1a$	$2a$	$2b$	$2c$	$2d$	$2e$	$2f$	$2g$	$2h$	$2i$	$2j$	$2k$	$2l$	$2m$	$3a$	$3b$	$3c$	4_1	4_2	4_3	4_4	4_5
$X.85$	720	720	720	−48	−48	−48	−48	−48	.	.	.	16	.	.	−9	.	.	48	.	−16	.	.
$X.86$	768	−768	.	128	−128	.	.	.	.	.	.	.	.	.	−24	24	.	.	.	.	.	.
$X.87$	810	810	810	90	90	42	42	90	.	18	18	26	18	.	.	.	.	−54	.	−6	.	.
$X.88$	810	810	810	−54	−54	42	42	−54	.	−6	−6	10	−6	.	.	.	.	−54	.	42	.	.
$X.89$	810	810	810	18	18	42	42	18	.	−18	−18	18	−18	12	.	.	.	−54	.	18	.	12
$X.90$	810	810	810	18	18	42	42	18	.	−18	−18	18	−18	−12	.	.	.	−54	.	18	.	−12
$X.91$	864	864	−864	96	96	−32	32	−96	24	.	.	.	.	8	.	9	.	.	−24	.	8	−8
$X.92$	864	864	−864	−96	−96	−32	32	96	−24	.	.	.	.	8	.	9	.	.	24	.	−8	−8
$X.93$	864	864	−864	−96	−96	−32	32	96	24	.	.	.	.	−8	.	9	.	.	−24	.	8	8
$X.94$	864	864	−864	96	96	−32	32	−96	−24	.	.	.	.	−8	.	9	.	.	24	.	−8	8
$X.95$	960	960	960	.	.	−64	−64	.	.	.	.	.	.	16	24	−6	.	64	.	.	.	16
$X.96$	960	960	960	.	.	−64	−64	.	.	.	.	.	.	−16	24	−6	.	64	.	.	.	−16
$X.97$	960	−960	.	−160	160	.	.	.	.	16	−16	.	.	.	−12	24	6	.	.	.	.	.
$X.98$	960	−960	.	96	−96	.	.	.	.	16	−16	.	.	.	6	.	−6	.	.	.	.	.
$X.99$	960	−960	.	96	−96	.	.	.	.	16	−16	.	.	.	6	.	−6	.	.	.	.	.
$X.100$	1080	1080	−1080	48	48	−40	40	−48	30	12	12	.	−12	10	.	−9	.	.	−30	.	10	−10
$X.101$	1080	1080	−1080	−48	−48	−40	40	48	−30	12	12	.	−12	−2	.	−9	.	.	30	.	−10	2
$X.102$	1080	1080	1080	−24	−24	56	56	−24	60	.	.	−24	.	−4	.	−9	.	−72	60	−24	−4	−4
$X.103$	1080	1080	1080	−24	−24	56	56	−24	−60	.	.	−24	.	4	.	−9	.	−72	−60	−24	4	4
$X.104$	1080	1080	−1080	−48	−48	−40	40	48	30	12	12	.	−12	2	.	−9	.	.	−30	.	10	−2
$X.105$	1080	1080	−1080	48	48	−40	40	−48	−30	12	12	.	−12	−10	.	−9	.	.	30	.	−10	10
$X.106$	1280	−1280	.	−128	128	.	.	.	.	.	.	.	.	.	20	8	−4	.	.	.	.	.
$X.107$	1280	−1280	.	−128	128	.	.	.	.	.	.	.	.	.	20	8	−4	.	.	.	.	.
$X.108$	1296	1296	−1296	.	.	−48	48	.	36	−24	−24	.	24	−12	.	.	.	.	−36	.	12	12
$X.109$	1296	1296	−1296	.	.	−48	48	.	−36	−24	−24	.	24	12	.	.	.	.	36	.	−12	−12
$X.110$	1440	1440	1440	96	96	−96	−96	96	.	.	.	−32	.	.	−18	.	.	96	.	32	.	.
$X.111$	1440	−1440	.	−48	48	.	.	.	.	−24	24	.	.	.	−18	.	.	.	.	.	.	.
$X.112$	1440	−1440	.	−48	48	.	.	.	.	−24	24	.	.	.	−18	.	.	.	.	.	.	.
$X.113$	1920	−1920	.	−64	64	.	.	.	.	32	−32	.	.	.	−24	−24	−6	.	.	.	.	.
$X.114$	2048	−2048	.	.	.	.	.	.	.	.	.	.	.	.	32	−16	8	.	.	.	.	.
$X.115$	2592	−2592	.	144	−144	.	.	.	.	−24	24	.	.	.	.	.	.	.	.	.	.	.

Character table of $H(\mathsf{Fi}_{22})$ *(continued)*

	4_6	4_7	4_8	4_9	4_{10}	4_{11}	4_{12}	4_{13}	4_{14}	4_{15}	4_{16}	4_{17}	4_{18}	4_{19}	4_{20}	4_{21}	4_{22}	4_{23}	4_{24}	4_{25}	4_{26}	4_1	$5a$	$6a$	$6b$
2	10	10	10	10	10	11	11	9	9	9	9	8	8	9	9	9	9	8	8	8	8	7	3	7	7
3	1	1	1	1	1	.	.	1	1	1	1	1	1	.	.	.	.	.	.	.	.	.	.	4	4
5	.	.	.	.	.	.	.	.	.	.	.	.	.	.	.	.	.	.	.	.	.	.	1	.	.
$2P$	$2e$	$2e$	$2e$	$2e$	$2e$	$2e$	$2e$	$2c$	$2c$	$2g$	$2g$	$2i$	$2j$	$2e$	$2e$	$2f$	$2e$	$2i$	$2i$	$2j$	$2i$	$2i$	$5a$	$3a$	$3a$
$3P$	4_6	4_7	4_8	4_9	4_{10}	4_{11}	4_{12}	4_{13}	4_{14}	4_{15}	4_{16}	4_{17}	4_{18}	4_{19}	4_{20}	4_{21}	4_{22}	4_{23}	4_{24}	4_{25}	4_{26}	4_1	$5a$	$2a$	$2b$
$5P$	4_6	4_7	4_8	4_9	4_{10}	4_{11}	4_{12}	4_{13}	4_{14}	4_{15}	4_{16}	4_{17}	4_{18}	4_{19}	4_{20}	4_{21}	4_{22}	4_{23}	4_{24}	4_{25}	4_{26}	4_1	$1a$	$6a$	$6c$
$X.85$	.	.	.	.	.	.	.	−8	−8	8	8	.	.	.	.	.	.	.	.	.	.	.	.	−9	−9
$X.86$	.	.	.	.	.	.	.	.	.	.	.	.	.	.	.	.	.	.	.	.	.	.	−2	24	.
$X.87$	−6	.	.	.	−6	−14	−14	−6	−6	−6	−6	.	.	.	.	2	2	.	−2	.	−2	−2	.	.	.
$X.88$	−6	.	.	.	−6	−6	−6	−6	−6	−6	−6	.	.	.	.	10	−6	.	2	.	2	2	.	.	.
$X.89$	−6	12	12	.	−6	−2	−2	6	6	6	6	−6	−6	−4	−4	−2	6	−2	.	−2	.	.	.	.	.
$X.90$	−6	−12	−12	.	−6	−2	−2	6	6	6	6	6	6	4	4	−2	6	2	.	2	.	.	.	.	.
$X.91$	−8	8	−8	−8	8	.	.	.	.	.	.	.	.	.	.	.	.	.	.	.	.	.	−1	.	.
$X.92$	8	8	−8	8	−8	.	.	.	.	.	.	.	.	.	.	.	.	.	.	.	.	.	−1	.	.
$X.93$	8	−8	8	−8	−8	.	.	.	.	.	.	.	.	.	.	.	.	.	.	.	.	.	−1	.	.
$X.94$	−8	−8	8	8	8	.	.	.	.	.	.	.	.	.	.	.	.	.	.	.	.	.	−1	.	.
$X.95$	.	−16	−16	.	.	.	.	.	.	.	.	.	.	.	.	.	.	.	.	.	.	.	.	24	24
$X.96$	.	16	16	.	.	.	.	.	.	.	.	.	.	.	.	.	.	.	.	.	.	.	.	24	24
$X.97$	.	.	.	.	.	.	.	.	.	−8	8	.	.	.	.	.	.	.	.	.	.	.	.	12	.
$X.98$	.	.	.	.	.	.	.	.	.	8	−8	.	.	.	.	.	.	.	.	.	.	.	.	−6	G
$X.99$	.	.	.	.	.	.	.	.	.	8	−8	.	.	.	.	.	.	.	.	.	.	.	.	−6	$\bar{G}$
$X.100$	−4	−2	2	−10	4	−4	4	.	.	.	.	.	.	−2	2	.	.	−4	−2	4	−2	2	.	.	.
$X.101$	4	10	−10	10	−4	−4	4	.	.	.	.	.	.	2	−2	.	.	−4	2	4	2	−2	.	.	.
$X.102$	8	−4	−4	−4	8	.	.	.	.	.	.	.	.	−4	−4	.	.	.	.	.	.	.	.	.	.
$X.103$	8	4	4	4	8	.	.	.	.	.	.	.	.	4	4	.	.	.	.	.	.	.	.	.	.
$X.104$	4	−10	10	−10	−4	−4	4	.	.	.	.	.	.	−2	2	.	.	4	2	−4	2	−2	.	.	.
$X.105$	−4	2	−2	10	4	−4	4	.	.	.	.	.	.	2	−2	.	.	4	−2	−4	−2	2	.	.	.
$X.106$	.	.	.	.	.	.	.	.	.	.	.	.	.	.	.	.	.	.	.	.	.	.	.	−20	H
$X.107$	.	.	.	.	.	.	.	.	.	.	.	.	.	.	.	.	.	.	.	.	.	.	.	−20	$\bar{H}$
$X.108$	.	12	−12	−12	.	8	−8	.	.	.	.	.	.	4	−4	.	.	.	.	.	.	.	1	.	.
$X.109$	.	−12	12	12	.	8	−8	.	.	.	.	.	.	−4	4	.	.	.	.	.	.	.	1	.	.
$X.110$	.	.	.	.	.	.	.	.	.	.	.	.	.	.	.	.	.	.	.	.	.	.	.	−18	−18
$X.111$	.	.	.	.	.	.	.	.	.	4	−4	.	.	.	.	.	.	.	4	.	−4	.	.	18	G
$X.112$	.	.	.	.	.	.	.	.	.	4	−4	.	.	.	.	.	.	.	4	.	−4	.	.	18	$\bar{G}$
$X.113$	.	.	.	.	.	.	.	.	.	.	.	.	.	.	.	.	.	.	.	.	.	.	.	24	.
$X.114$	.	.	.	.	.	.	.	.	.	.	.	.	.	.	.	.	.	.	.	.	.	.	−2	−32	.
$X.115$	.	.	.	.	.	.	.	.	.	−12	12	.	.	.	.	.	.	.	−4	.	4	.	2	.	.

Character table of $H(\mathsf{Fi}_{22})$ *(continued)*

	6c	6d	6e	6f	6g	6h	6i	6j	6k	6l	6m	6n	6o	6p	6q	6r	6s	6t	6u	6v	6w	6x	6y	6z	8a	8b	8c	8d	8e
2	7	8	7	7	7	7	7	5	6	6	4	7	7	5	5	4	4	4	4	5	5	5	3	4	7	7	7	7	7
3	4	3	3	2	2	2	2	3	2	2	3	1	1	2	2	2	2	2	2	1	1	1	2	1	1	1	.	.	.
5	.	.	.	.	.	.	.	.	.	.	.	.	.	.	.	.	.	.	.	.	.	.	.	.	.	.	.	.	.
2P	3a	3b	3b	3a	3a	3a	3a	3c	3b	3b	3c	3b	3b	3b	3b	3c	3c	3c	3c	3c	3c	3b	3c	3c	4_1	4_1	4_{11}	4_{11}	4_{11}
3P	2b	2a	2b	2c	2g	2g	2d	2a	2c	2d	2b	2e	2f	2h	2g	2g	2g	2d	2c	2i	2j	2m	2h	2l	8a	8b	8c	8d	8e
5P	6b	6d	6e	6f	6h	6g	6i	6j	6k	6l	6m	6n	6o	6p	6q	6s	6r	6t	6u	6v	6w	6x	6y	6z	8a	8b	8c	8d	8e
X.85	−9	.	.	3	3	3	3	.	.	.	.	.	.	.	.	.	.	.	.	.	.	.	.	.	.	.	.	.	.
X.86	.	−24	.	8	.	.	−8	.	−4	4	.	.	.	.	.	.	.	4	−4	.	.	.	.	.	.	.	.	.	.
X.87	.	.	.	.	.	.	.	.	.	.	.	.	.	.	.	.	.	.	.	.	.	.	.	.	.	.	.	−2	−2
X.88	.	.	.	.	.	.	.	.	.	.	.	.	.	.	.	.	.	.	.	.	.	.	.	.	.	.	.	2	2
X.89	.	.	.	.	.	.	.	.	.	.	.	.	.	.	.	.	.	.	.	.	.	.	.	.	.	.	2	.	.
X.90	.	.	.	.	.	.	.	.	.	.	.	.	.	.	.	.	.	.	.	.	.	.	.	.	.	.	−2	.	.
X.91	.	9	−9	.	.	.	.	.	−3	−3	.	1	−1	3	3	.	.	.	.	.	.	−1	.	.	−4	4	.	.	.
X.92	.	9	−9	.	.	.	.	.	3	3	.	1	−1	−3	−3	.	.	.	.	.	.	−1	.	.	−4	4	.	.	.
X.93	.	9	−9	.	.	.	.	.	3	3	.	1	−1	3	−3	.	.	.	.	.	.	1	.	.	4	−4	.	.	.
X.94	.	9	−9	.	.	.	.	.	−3	−3	.	1	−1	−3	3	.	.	.	.	.	.	1	.	.	4	−4	.	.	.
X.95	24	−6	−6	.	.	.	.	.	.	.	.	2	2	.	.	.	.	.	.	.	.	−2	.	.	.	.	.	.	.
X.96	24	−6	−6	.	.	.	.	.	.	.	.	2	2	.	.	.	.	.	.	.	.	2	.	.	.	.	.	.	.
X.97	.	−24	.	−4	.	.	4	−6	−4	4	.	.	.	.	.	.	.	−2	2	−2	2	.	.	.	.	.	.	.	.
X.98	$\bar{G}$	.	.	−6	$\bar{B}$	B	6	6	.	.	.	.	.	.	.	B	$\bar{B}$	.	.	−2	2	.	.	.	.	.	.	.	.
X.99	G	.	.	−6	B	$\bar{B}$	6	6	.	.	.	.	.	.	.	$\bar{B}$	B	.	.	−2	2	.	.	.	.	.	.	.	.
X.100	.	−9	9	.	.	.	.	.	3	3	.	−1	1	−3	−3	.	.	.	.	.	.	1	.	.	−2	2	.	2	−2
X.101	.	−9	9	.	.	.	.	.	−3	−3	.	−1	1	3	3	.	.	.	.	.	.	1	.	.	−2	2	.	−2	2
X.102	.	−9	−9	.	.	.	.	.	3	3	.	−1	−1	3	3	.	.	.	.	.	.	−1	.	.	4	4	.	.	.
X.103	.	−9	−9	.	.	.	.	.	3	3	.	−1	−1	−3	3	.	.	.	.	.	.	1	.	.	−4	−4	.	.	.
X.104	.	−9	9	.	.	.	.	.	−3	−3	.	−1	1	−3	3	.	.	.	.	.	.	−1	.	.	2	−2	.	−2	2
X.105	.	−9	9	.	.	.	.	.	3	3	.	−1	1	3	−3	.	.	.	.	.	.	−1	.	.	2	−2	.	2	−2
X.106	$\bar{H}$	−8	.	4	I	$\bar{I}$	−4	4	4	−4	.	.	.	.	.	B	$\bar{B}$	2	−2	.	.	.	.	.	.	.	.	.	.
X.107	H	−8	.	4	$\bar{I}$	I	−4	4	4	−4	.	.	.	.	.	$\bar{B}$	B	2	−2	.	.	.	.	.	.	.	.	.	.
X.108	.	.	.	.	.	.	.	.	.	.	.	.	.	.	.	.	.	.	.	.	.	.	.	.	.	.	.	.	.
X.109	.	.	.	.	.	.	.	.	.	.	.	.	.	.	.	.	.	.	.	.	.	.	.	.	.	.	.	.	.
X.110	−18	.	.	−6	−6	−6	−6	.	.	.	.	.	.	.	.	.	.	.	.	.	.	.	.	.	.	.	.	.	.
X.111	$\bar{G}$	.	.	−6	$\bar{A}$	A	6	.	.	.	.	.	.	.	.	.	.	.	.	.	.	.	.	.	.	.	.	.	.
X.112	G	.	.	−6	A	$\bar{A}$	6	.	.	.	.	.	.	.	.	.	.	.	.	.	.	.	.	.	.	.	.	.	.
X.113	.	24	.	8	.	.	−8	6	−4	4	.	.	.	.	.	.	.	−2	2	2	−2	.	.	.	.	.	.	.	.
X.114	.	16	.	.	.	.	.	−8	.	.	.	.	.	.	.	.	.	.	.	.	.	.	.	.	.	.	.	.	.
X.115	.	.	.	.	.	.	.	.	.	.	.	.	.	.	.	.	.	.	.	.	.	.	.	.	.	.	.	.	.

Character table of $H(\mathsf{Fi}_{22})$ *(continued)*

	8f	8g	8h	8i	8j	8k	9a	10a	10b	10c	12a	12b	12c	12d	12e	12f	12g	12h	12i	12j	12k	12l	12m	12n	12o	12p
2	7	7	6	5	5	5	2	3	2	2	5	6	6	5	5	5	5	5	5	5	5	5	5	5	5	5
3	.	.	.	.	.	.	2	.	.	.	3	2	2	2	1	1	1	1	1	1	1	1	1	1	1	1
5	.	.	.	.	.	.	.	1	1	1	.	.	.	.	.	.	.	.	.	.	.	.	.	.	.	.
2P	4_3	4_{11}	4_{12}	4_{13}	4_{14}	4_{21}	9a	5a	5a	5a	6a	6d	6d	6d	6n	6d	6n	6n	6n	6g	6h	6a	6g	6h	6n	6n
3P	8f	8g	8h	8i	8j	8k	3a	10a	10b	10c	4_1	4_1	4_1	4_2	4_4	4_5	4_6	4_7	4_8	4_{15}	4_{15}	4_3	4_{16}	4_{16}	4_{10}	4_9
5P	8f	8g	8h	8i	8j	8k	9a	2a	2h	2b	12a	12b	12c	12d	12e	12f	12g	12h	12i	12k	12j	12l	12n	12m	12o	12p
X.85	.	.	.	.	.	.	.	.	.	.	3	.	.	.	.	.	.	.	.	$\bar{F}$	F	−1	$\bar{F}$	F	.	.
X.86	.	.	.	.	.	.	.	2	.	.	.	.	.	.	.	.	.	.	.	.	.	.	.	.	.	.
X.87	2	.	.	.	.	2	.	.	.	.	.	.	.	.	.	.	.	.	.	.	.	.	.	.	.	.
X.88	2	.	.	.	.	−2	.	.	.	.	.	.	.	.	.	.	.	.	.	.	.	.	.	.	.	.
X.89	−2	−2	2	.	.	.	.	.	.	.	.	.	.	.	.	.	.	.	.	.	.	.	.	.	.	.
X.90	−2	2	−2	.	.	.	.	.	.	.	.	.	.	.	.	.	.	.	.	.	.	.	.	.	.	.
X.91	.	.	.	.	.	.	.	−1	−1	1	.	−3	3	−3	−1	1	1	−1	1	.	.	.	.	.	−1	1
X.92	.	.	.	.	.	.	.	−1	1	1	.	−3	3	3	1	1	−1	−1	1	.	.	.	.	.	1	−1
X.93	.	.	.	.	.	.	.	−1	−1	1	.	−3	3	−3	−1	−1	−1	1	−1	.	.	.	.	.	1	1
X.94	.	.	.	.	.	.	.	−1	1	1	.	−3	3	3	1	−1	1	1	−1	.	.	.	.	.	−1	−1
X.95	.	.	.	.	.	.	.	.	.	.	−8	−2	−2	.	.	−2	.	2	2	.	.	.	.	.	.	.
X.96	.	.	.	.	.	.	.	.	.	.	−8	−2	−2	.	.	2	.	−2	−2	.	.	.	.	.	.	.
X.97	.	.	.	.	.	.	.	.	.	.	.	.	.	.	.	.	.	.	.	−2	−2	.	2	2	.	.
X.98	.	.	.	.	.	.	.	.	.	.	.	.	.	.	.	.	.	.	.	$-\bar{C}$	$-C$	.	$\bar{C}$	C	.	.
X.99	.	.	.	.	.	.	.	.	.	.	.	.	.	.	.	.	.	.	.	$-C$	$-\bar{C}$	.	C	$\bar{C}$	.	.
X.100	.	.	.	.	.	.	.	.	.	.	.	3	−3	3	1	−1	−1	1	−1	.	.	.	.	.	1	−1
X.101	.	.	.	.	.	.	.	.	.	.	.	3	−3	−3	−1	−1	1	1	−1	.	.	.	.	.	−1	1
X.102	.	.	.	.	.	.	.	.	.	.	.	3	3	3	−1	−1	−1	−1	−1	.	.	.	.	.	−1	−1
X.103	.	.	.	.	.	.	.	.	.	.	.	3	3	−3	1	1	−1	1	1	.	.	.	.	.	−1	1
X.104	.	.	.	.	.	.	.	.	.	.	.	3	−3	3	1	1	1	−1	1	.	.	.	.	.	−1	−1
X.105	.	.	.	.	.	.	.	.	.	.	.	3	−3	−3	−1	1	−1	−1	1	.	.	.	.	.	1	1
X.106	.	.	.	.	.	.	−1	.	.	.	.	.	.	.	.	.	.	.	.	.	.	.	.	.	.	.
X.107	.	.	.	.	.	.	−1	.	.	.	.	.	.	.	.	.	.	.	.	.	.	.	.	.	.	.
X.108	.	.	.	.	.	.	.	1	1	−1	.	.	.	.	.	.	.	.	.	.	.	.	.	.	.	.
X.109	.	.	.	.	.	.	.	1	−1	−1	.	.	.	.	.	.	.	.	.	.	.	.	.	.	.	.
X.110	.	.	.	.	.	.	.	.	.	.	6	.	.	.	.	.	.	.	.	.	.	2	.	.	.	.
X.111	.	.	.	.	.	.	.	.	.	.	.	.	.	.	.	.	.	.	.	$\bar{C}$	C	.	$-\bar{C}$	$-C$	.	.
X.112	.	.	.	.	.	.	.	.	.	.	.	.	.	.	.	.	.	.	.	C	$\bar{C}$	.	$-C$	$-\bar{C}$	.	.
X.113	.	.	.	.	.	.	.	.	.	.	.	.	.	.	.	.	.	.	.	.	.	.	.	.	.	.
X.114	.	.	.	.	.	.	2	2	.	.	.	.	.	.	.	.	.	.	.	.	.	.	.	.	.	.
X.115	.	.	.	.	.	.	.	−2	.	.	.	.	.	.	.	.	.	.	.	.	.	.	.	.	.	.

Character table of $H(Fi_{22})$ (continued)

2	3	4	4	3	3	5	5	2	2	2	2	4	4
3	2	1	1	1	1	.	.	2	2	2	.	1	1
5	.	.	.	.	.	.	.	.	.	.	1	.	.
	12q	12r	12s	12t	12u	16a	16b	18a	18b	18c	20a	24a	24b
2P	6j	6f	6f	6v	6w	8f	8f	9a	9a	9a	10a	12b	12b
3P	4_2	4_{13}	4_{14}	4_{17}	4_{18}	16a	16b	6a	6b	6c	20a	8a	8b
5P	12q	12r	12s	12t	12u	16b	16a	18a	18c	18b	4_2	24a	24b
X.85	.	1	1	.	.	.	.	.	.	.	.	.	.
X.86	.	.	.	.	.	.	.	.	.	.	.	.	.
X.87	.	.	.	.	.	.	.	.	.	.	.	.	.
X.88	.	.	.	.	.	.	.	.	.	.	.	.	.
X.89	.	.	.	.	.	.	.	.	.	.	.	.	.
X.90	.	.	.	.	.	.	.	.	.	.	.	.	.
X.91	.	.	.	.	.	.	.	.	.	.	1	−1	1
X.92	.	.	.	.	.	.	.	.	.	.	−1	−1	1
X.93	.	.	.	.	.	.	.	.	.	.	1	1	−1
X.94	.	.	.	.	.	.	.	.	.	.	−1	1	−1
X.95	.	.	.	.	.	.	.	.	.	.	.	.	.
X.96	.	.	.	.	.	.	.	.	.	.	.	.	.
X.97	.	.	.	.	.	.	.	.	.	.	.	.	.
X.98	.	.	.	.	.	.	.	.	.	.	.	.	.
X.99	.	.	.	.	.	.	.	.	.	.	.	.	.
X.100	.	.	.	.	.	.	.	.	.	.	.	1	−1
X.101	.	.	.	.	.	.	.	.	.	.	.	1	−1
X.102	.	.	.	.	.	.	.	.	.	.	.	1	1
X.103	.	.	.	.	.	.	.	.	.	.	.	−1	−1
X.104	.	.	.	.	.	.	.	.	.	.	.	−1	1
X.105	.	.	.	.	.	.	.	.	.	.	.	−1	1
X.106	.	.	.	.	.	.	.	1	D	$\bar{D}$	.	.	.
X.107	.	.	.	.	.	.	.	1	$\bar{D}$	D	.	.	.
X.108	.	.	.	.	.	.	.	.	.	.	−1	.	.
X.109	.	.	.	.	.	.	.	.	.	.	1	.	.
X.110	.	.	.	.	.	.	.	.	.	.	.	.	.
X.111	.	.	.	.	.	.	.	.	.	.	.	.	.
X.112	.	.	.	.	.	.	.	.	.	.	.	.	.
X.113	.	.	.	.	.	.	.	.	.	.	.	.	.
X.114	.	.	.	.	.	.	.	−2	.	.	.	.	.
X.115	.	.	.	.	.	.	.	.	.	.	.	.	.

where $A = -6i\sqrt{3}$, $B = 2i\sqrt{3}$, $C = 1+i\sqrt{3}$, $D = -i\sqrt{3}$, $E = -2(\zeta(8)^3 + \zeta(8))$, $F = -1 + 2i\sqrt{3}$, $G = -18i\sqrt{3}$, $H = 12i\sqrt{3}$, $I = 4i\sqrt{3}$.

6

Fischer's simple group Fi_{23}

This chapter presents Kim's recent construction [73] of Fischer's sporadic simple group Fi_{23} by means of Algorithm 1.3.8. The first existence proof is due to Fischer; see Theorem 18.2.7 of [30]. The Mathieu group M_{23} has two 11-dimensional irreducible 2-modular representations. They are denoted by V_1 and V_2. Kim showed in [73] that there is a uniquely determined non-split extension E of M_{23} by V_1 and that $dim_F[H^2(\mathsf{M}_{23}, V_2)] = 0$. These (known) results are presented in Section 6.1.

In Section 6.2 we describe Kim's construction of a group H with center $Z(H) = \langle z \rangle$ of order 2 such that $H/Z(H) \cong \mathsf{Fi}_{22}$. Unfortunately, MAGMA did not construct the central extensions of Fi_{22} automatically. By Algorithm 1.3.8 a Sylow 2-subgroup S of the target group H is known because S is a Sylow 2-subgroup of $D = C_E(z)$. It is used in Proposition 6.2.1 to construct another amalgam $H_2 \leftarrow D_2 \rightarrow D$ whose free product $H_2 *_{D_2} D$ with amalgamated subgroup D_2 has a 352-dimensional irreducible representation over $GF(17)$. This matrix group $\mathfrak{H}$ is proved to be isomorphic to the 2-fold cover H of Fi_{22}; see Proposition 6.2.3. It follows that all conditions of Step 5 of Algorithm 1.3.8 are satisfied; see Corollary 6.2.4.

The amalgam $E \leftarrow D \rightarrow H$ has a unique compatible pair of degree 782 over $GF(17)$ which is not multiplicity-free at the D-level. Using Theorem 7.2.2 of [92] Kim showed that the free product $H *_D E$ with amalgamated subgroup D has exactly one irreducible representation of degree 782 over the prime field $GF(17)$ satisfying the Sylow 2-subgroup test of Step 5(c) of Algorithm 7.4.8 of [92]; see Theorem 6.3.1. Its image $\mathfrak{G}$ in $\mathrm{GL}_{782}(17)$ is proved to be a simple group of order $2^{18} \cdot 3^{13} \cdot 5^2 \cdot 7 \cdot 11 \cdot 13 \cdot 17 \cdot 23$. It has a 2-central involution $\mathfrak{z}$ such that $C_{\mathfrak{G}}(\mathfrak{z}) \cong H$; see Theorem 6.3.1. The character table of $\mathfrak{G}$ agrees with the one of Fi_{23} in the Atlas [19].

Since $\mathfrak{G}$ has three conjugacy classes of 2-central involutions the uniqueness of Fischer's simple group Fi_{23} cannot be proved by means of the author's uniqueness criterion Theorem 7.5.1 of [92]. In [56] D. Hunt characterizes Fi_{23} by means of a centralizer of an involution which is isomorphic to the group H. It depends on Fischer's classification of all simple groups having a conjugacy class of 3-transpositions; see [29]. Hunt's result is stated in Section 6.4.

Section 6.5 contains the tables of representatives of the conjugacy classes of $\mathfrak{G}$ and of the local subgroups of H, D and E. The character tables of these local subgroups are stated in Section 6.6. The corresponding tables of the other local subgroups are stored on the accompanying DVD.

6.1 Extensions of the Mathieu group M_{23}

The Mathieu group M_{23} is defined in Definition 8.2.1 of [92] by means of generators and relations. This beautiful presentation is due to J. A. Todd. The irreducible 2-modular representations of the Mathieu group M_{23} were determined by G. James [65]. Here only the two non-isomorphic simple modules V_i, $i = 1, 2$, of dimension 11 over $F = GF(2)$ will be considered. Todd's permutation representations of the Mathieu groups are stated in Lemma 8.2.2 of [92]. Therefore all conditions of Holt's Algorithm [52] implemented in MAGMA are satisfied. It is applied here. Thus we show that for the simple module V_1 there are exactly two extensions of M_{23} by V_1, the split extension E_1 and the non-split extension E and that M_{23} has only the split extension E_2 by V_2. In [73] it is shown that the applications of Algorithm 1.3.8 to E_1 and E_2 fail. Therefore only the constructed presentation of E is given in Lemma 6.1.1.

Lemma 6.1.1 *Let* $\mathsf{M}_{23} = \langle a, b, c, d, t, g, h, i, j \rangle$ *be the finitely presented group with the set of defining relations* $\mathcal{R}(\mathsf{M}_{23})$ *given in Definition 8.2.1 of [92]. Then the following statements hold.*

(a) *A faithful permutation representation of degree* 23 *of* M_{23} *is stated in Lemma 8.2.2 of [92].*

(b) *The first irreducible representation* V_1 *of* M_{23} *is described by the following matrices:*

$$a_1 = \begin{pmatrix} 0&1&0&0&1&1&1&1&1&0&0 \\ 0&1&1&0&0&0&1&1&1&1&1 \\ 0&0&0&0&0&0&1&0&0&0&0 \\ 0&0&1&1&0&1&1&1&0&1&0 \\ 1&1&0&0&0&1&1&0&0&1&1 \\ 0&0&0&0&0&1&0&0&0&0&0 \\ 0&0&1&0&0&0&0&0&0&0&0 \\ 0&0&0&0&0&0&0&1&0&0&0 \\ 0&0&0&0&0&0&0&0&1&0&0 \\ 0&0&0&0&0&0&0&0&0&1&0 \\ 0&0&0&0&0&0&0&0&0&0&1 \end{pmatrix}, \quad b_1 = \begin{pmatrix} 0&1&0&0&0&0&0&0&0&0&0 \\ 1&0&0&0&0&0&0&0&0&0&0 \\ 0&0&0&0&1&0&0&0&0&0&0 \\ 1&1&0&1&0&1&0&1&1&1&0 \\ 0&0&1&0&0&0&0&0&0&0&0 \\ 0&0&0&0&0&1&0&0&0&0&0 \\ 1&1&0&0&0&1&1&0&0&1&1 \\ 0&0&0&0&0&0&0&1&0&0&0 \\ 0&0&0&0&0&0&0&0&1&0&0 \\ 0&0&0&0&0&0&0&0&0&1&0 \\ 0&0&0&0&0&0&0&0&0&0&1 \end{pmatrix},$$

$$c_1 = \begin{pmatrix} 0&0&0&0&1&0&0&0&0&0&0\\ 0&0&1&0&0&0&0&0&0&0&0\\ 0&1&0&0&0&0&0&0&0&0&0\\ 1&0&0&1&1&1&0&1&0&0&1\\ 1&0&0&0&0&0&0&0&0&0&0\\ 0&0&0&0&0&1&0&0&0&0&0\\ 0&1&1&0&0&0&1&1&1&1&1\\ 0&0&0&0&0&0&0&1&0&0&0\\ 0&0&0&0&0&0&0&0&1&0&0\\ 0&0&0&0&0&0&0&0&0&1&0\\ 0&0&0&0&0&0&0&0&0&0&1 \end{pmatrix}, \quad d_1 = \begin{pmatrix} 0&1&1&1&1&0&1&0&0&0&1\\ 1&0&1&1&1&0&1&1&1&0&0\\ 0&0&1&1&0&1&1&1&0&1&0\\ 0&0&0&0&0&0&1&0&0&0&0\\ 0&0&0&1&1&1&1&0&1&1&1\\ 0&0&0&0&0&1&0&0&0&0&0\\ 0&0&0&1&0&0&0&0&0&0&0\\ 0&0&0&0&0&0&0&1&0&0&0\\ 0&0&0&0&0&0&0&0&1&0&0\\ 0&0&0&0&0&0&0&0&0&1&0\\ 0&0&0&0&0&0&0&0&0&0&1 \end{pmatrix},$$

$$t_1 = \begin{pmatrix} 0&1&0&0&0&0&0&0&1&0&0\\ 1&0&0&1&1&1&0&1&1&0&1\\ 0&1&0&0&1&1&1&1&0&0&0\\ 1&0&1&1&1&0&1&1&0&0&0\\ 0&1&1&1&1&0&1&0&1&0&1\\ 0&0&0&0&0&0&0&0&1&0&0\\ 1&1&0&1&0&1&0&1&0&1&0\\ 0&0&0&0&0&0&0&1&1&0&0\\ 0&0&0&0&0&1&0&0&1&0&0\\ 0&0&0&0&0&0&0&0&1&1&0\\ 0&0&0&0&0&0&0&0&1&0&1 \end{pmatrix}, \quad g_1 = \begin{pmatrix} 0&1&0&1&0&0&0&0&0&0&0\\ 1&0&0&1&0&0&0&0&0&0&0\\ 0&1&0&1&1&1&1&1&1&0&0\\ 0&0&0&1&0&0&0&0&0&0&0\\ 1&1&0&1&0&1&1&0&0&1&1\\ 1&0&1&0&1&0&1&1&1&0&0\\ 0&1&1&1&0&0&1&1&1&1&1\\ 1&0&1&0&0&0&0&0&1&1&1\\ 0&0&0&0&1&1&1&0&1&1&1\\ 0&0&0&1&0&0&0&0&0&1&0\\ 0&0&0&1&0&0&0&0&0&0&1 \end{pmatrix},$$

$$h_1 = \begin{pmatrix} 1&0&0&0&0&1&0&0&0&0&0\\ 1&0&0&1&1&0&0&1&0&0&1\\ 0&0&1&1&0&0&1&1&0&1&0\\ 0&0&0&0&0&1&1&0&0&0&0\\ 1&1&0&0&0&0&1&0&0&1&1\\ 0&0&0&0&0&1&0&0&0&0&0\\ 0&0&0&1&0&1&0&0&0&0&0\\ 0&0&0&0&0&1&0&0&0&1&0\\ 0&0&0&0&0&1&0&0&1&0&0\\ 0&0&0&0&0&1&0&1&0&0&0\\ 0&0&0&0&0&1&0&0&0&0&1 \end{pmatrix}, \quad i_1 = \begin{pmatrix} 1&0&0&1&1&0&0&1&0&0&1\\ 0&1&0&0&0&1&0&0&0&0&0\\ 1&1&0&1&0&0&0&1&1&1&0\\ 0&0&0&1&0&1&0&0&0&0&0\\ 0&0&0&0&1&1&0&0&0&0&0\\ 0&0&0&0&0&1&0&0&0&0&0\\ 0&1&0&0&1&0&1&1&1&0&0\\ 0&0&0&0&0&1&0&1&0&0&0\\ 0&0&0&0&0&1&0&0&1&0&0\\ 1&1&1&0&1&0&0&0&1&0&1\\ 0&0&0&0&0&1&0&0&0&0&1 \end{pmatrix},$$

$$j_1 = \begin{pmatrix} 0&1&0&0&0&0&0&0&0&0&0\\ 1&0&0&0&0&0&0&0&0&0&0\\ 0&0&0&0&1&0&0&0&0&0&0\\ 0&0&0&1&0&0&0&0&0&0&0\\ 0&0&1&0&0&0&0&0&0&0&0\\ 0&0&0&0&0&0&0&0&1&0&0\\ 0&1&1&0&0&0&1&1&1&1&1\\ 0&0&0&0&0&0&0&1&0&0&0\\ 0&0&0&0&0&1&0&0&0&0&0\\ 0&0&0&0&0&0&0&0&0&1&0\\ 1&1&1&0&1&1&0&0&1&0&1 \end{pmatrix}.$$

(c) *The second irreducible representation V_2 of M_{23} is described by the transpose inverse matrices of the generating matrices of M_{23} defining V_1:*

$$a_2 = [a_1^{-1}]^T, b_2 = [b_1^{-1}]^T, c_2 = [c_1^{-1}]^T, d_2 = [d_1^{-1}]^T, t_2 = [t_1^{-1}]^T,$$
$$g_2 = [g_1^{-1}]^T, h_2 = [h_1^{-1}]^T, i_2 = [i_1^{-1}]^T \textit{ and } i_2 = [j_1^{-1}]^T.$$

(d) $dim_F[H^2(\mathsf{M}_{23}, V_1)] = 1$ *and* $dim_F[H^2(\mathsf{M}_{23}, V_2)] = 0$.

(e) *The presentations of the split extensions E_1 and E_2 of M_{23} by V_1 and V_2, respectively, can be constructed immediately from the matrices of (b) and (c) by means of Lemma 1.4.7.*

(f) *The unique non-split extension E of M_{23} by V_1 has the presentation*

$$E = \langle a, b, c, d, t, g, h, i, j, v_i \mid 1 \le i \le 11 \rangle$$

with the set $\mathcal{R}(E)$ of defining relations consisting of the following *set of relations:*

$$a^2 = b^2 = c^2 = d^2 = t^3 = g^4 = h^4 = j^2 = 1,$$
$$(b,a) = (c,a) = (d,a) = (h,a) = (c,b) = (d,b) = (d,c) = 1,$$
$$(j,b) = (j,c) = (j,g) = 1,$$
$$v_r^2 = 1 \quad \textit{for} \quad 1 \le r \le 11,$$
$$(v_r, v_s) = 1 \quad \textit{for} \quad 1 \le r,s \le 11,$$
$$a^{-1}v_1av_2^{-1}v_5^{-1}v_6^{-1}v_7^{-1}v_8^{-1}v_9^{-1} = 1,$$
$$a^{-1}v_2av_2^{-1}v_3^{-1}v_7^{-1}v_8^{-1}v_9^{-1}v_{10}^{-1}v_{11}^{-1} = 1,$$
$$a^{-1}v_3av_7^{-1} = a^{-1}v_7av_3^{-1} = 1,$$
$$a^{-1}v_4av_3^{-1}v_4^{-1}v_6^{-1}v_7^{-1}v_8^{-1}v_{10}^{-1} = 1,$$
$$a^{-1}v_5av_1^{-1}v_2^{-1}v_6^{-1}v_7^{-1}v_{10}^{-1}v_{11}^{-1} = 1,$$
$$(a,v_6) = (a,v_8) = (a,v_9) = (a,v_{10}) = (a,v_{11}) = 1,$$
$$b^{-1}v_1bv_2^{-1} = b^{-1}v_2bv_1^{-1} = b^{-1}v_3bv_5^{-1} = b^{-1}v_5bv_3^{-1} = 1,$$
$$b^{-1}v_4bv_1^{-1}v_2^{-1}v_4^{-1}v_6^{-1}v_8^{-1}v_9^{-1}v_{10}^{-1} = 1,$$
$$(b,v_6^{-1}) = (b,v_8^{-1}) = (b,v_9^{-1}) = (b,v_{10}^{-1}) = (b,v_{11}^{-1}) = 1,$$
$$b^{-1}v_7bv_1^{-1}v_2^{-1}v_6^{-1}v_7^{-1}v_{10}^{-1}v_{11}^{-1} = 1,$$
$$c^{-1}v_1cv_5^{-1} = c^{-1}v_2cv_3^{-1} = c^{-1}v_3cv_2^{-1} = c^{-1}v_5cv_1^{-1} = 1,$$
$$c^{-1}v_4cv_1^{-1}v_4^{-1}v_5^{-1}v_6^{-1}v_8^{-1}v_{11}^{-1} = 1,$$
$$(c,v_6^{-1}) = (c,v_8^{-1}) = (c,v_9^{-1}) = (c,v_{10}^{-1}) = (c,v_{11}^{-1}) = 1,$$
$$c^{-1}v_7cv_2^{-1}v_3^{-1}v_7^{-1}v_8^{-1}v_9^{-1}v_{10}^{-1}v_{11}^{-1} = 1,$$
$$d^{-1}v_1dv_2^{-1}v_3^{-1}v_4^{-1}v_5^{-1}v_7^{-1}v_{11}^{-1} = 1,$$
$$d^{-1}v_2dv_1^{-1}v_3^{-1}v_4^{-1}v_5^{-1}v_7^{-1}v_8^{-1}v_9^{-1} = 1,$$
$$d^{-1}v_3dv_3^{-1}v_4^{-1}v_6^{-1}v_7^{-1}v_8^{-1}v_{10}^{-1} = 1,$$
$$d^{-1}v_4dv_7^{-1} = d^{-1}v_7dv_4^{-1} = 1,$$
$$d^{-1}v_5dv_4^{-1}v_5^{-1}v_6^{-1}v_7^{-1}v_9^{-1}v_{10}^{-1}v_{11}^{-1} = 1,$$
$$(d,v_6^{-1}) = (d,v_8^{-1}) = (d,v_9^{-1}) = (d,v_{10}^{-1}) = (d,v_{11}^{-1}) = 1,$$
$$t^{-1}v_1tv_2^{-1}v_9^{-1} = t^{-1}v_6tv_9^{-1} = 1,$$
$$t^{-1}v_2tv_1^{-1}v_4^{-1}v_5^{-1}v_6^{-1}v_8^{-1}v_9^{-1}v_{11}^{-1} = 1,$$
$$t^{-1}v_3tv_2^{-1}v_5^{-1}v_6^{-1}v_7^{-1}v_8^{-1} = 1,$$
$$t^{-1}v_4tv_1^{-1}v_3^{-1}v_4^{-1}v_5^{-1}v_7^{-1}v_8^{-1} = 1,$$
$$t^{-1}v_5tv_2^{-1}v_3^{-1}v_4^{-1}v_5^{-1}v_7^{-1}v_9^{-1}v_{11}^{-1} = 1,$$
$$t^{-1}v_7tv_1^{-1}v_2^{-1}v_4^{-1}v_6^{-1}v_8^{-1}v_{10}^{-1} = 1,$$

$$
\begin{aligned}
&t^{-1}v_8tv_8^{-1}v_9^{-1} = t^{-1}v_9tv_6^{-1}v_9^{-1} = 1,\\
&t^{-1}v_{10}tv_9^{-1}v_{10}^{-1} = t^{-1}v_{11}tv_9^{-1}v_{11}^{-1} = 1,\\
&g^{-1}v_1gv_2^{-1}v_4^{-1} = g^{-1}v_2gv_1^{-1}v_4^{-1} = 1,\\
&g^{-1}v_3gv_2^{-1}v_4^{-1}v_5^{-1}v_6^{-1}v_7^{-1}v_8^{-1}v_9^{-1} = 1, \quad (g, v_4) = 1,\\
&g^{-1}v_5gv_1^{-1}v_2^{-1}v_4^{-1}v_6^{-1}v_7^{-1}v_{10}^{-1}v_{11}^{-1} = 1,\\
&g^{-1}v_6gv_1^{-1}v_3^{-1}v_5^{-1}v_7^{-1}v_8^{-1}v_9^{-1} = 1,\\
&g^{-1}v_7gv_2^{-1}v_3^{-1}v_4^{-1}v_7^{-1}v_8^{-1}v_9^{-1}v_{10}^{-1}v_{11}^{-1} = 1,\\
&g^{-1}v_8gv_1^{-1}v_3^{-1}v_9^{-1}v_{10}^{-1}v_{11}^{-1} = 1,\\
&g^{-1}v_9gv_5^{-1}v_6^{-1}v_7^{-1}v_9^{-1}v_{10}^{-1}v_{11}^{-1} = 1,\\
&g^{-1}v_{10}gv_4^{-1}v_{10}^{-1} = g^{-1}v_{11}gv_4^{-1}v_{11}^{-1} = 1,\\
&h^{-1}v_1hv_1^{-1}v_6^{-1} = h^{-1}v_4hv_6^{-1}v_7^{-1} = 1, \quad (h, v_6) = 1,\\
&h^{-1}v_2hv_1^{-1}v_4^{-1}v_5^{-1}v_8^{-1}v_{11}^{-1} = 1,\\
&h^{-1}v_3hv_3^{-1}v_4^{-1}v_7^{-1}v_8^{-1}v_{10}^{-1} = 1,\\
&h^{-1}v_5hv_1^{-1}v_2^{-1}v_7^{-1}v_{10}^{-1}v_{11}^{-1} = 1,\\
&h^{-1}v_7hv_4^{-1}v_6^{-1} = h^{-1}v_8hv_6^{-1}v_{10}^{-1} = 1,\\
&h^{-1}v_9hv_6^{-1}v_9^{-1} = h^{-1}v_{10}hv_6^{-1}v_8^{-1} = h^{-1}v_{11}hv_6^{-1}v_{11}^{-1} = 1,\\
&i^{-1}v_1iv_1^{-1}v_4^{-1}v_5^{-1}v_8^{-1}v_{11}^{-1} = 1,\\
&i^{-1}v_2iv_2^{-1}v_6^{-1} = i^{-1}v_4iv_4^{-1}v_6^{-1} = i^{-1}v_5iv_5^{-1}v_6^{-1} = 1,\\
&i^{-1}v_3iv_1^{-1}v_2^{-1}v_4^{-1}v_8^{-1}v_9^{-1}v_{10}^{-1} = 1, \quad (i, v_6) = 1,\\
&i^{-1}v_7iv_2^{-1}v_5^{-1}v_7^{-1}v_8^{-1}v_9^{-1} = i^{-1}v_{11}iv_6^{-1}v_{11}^{-1} = 1,\\
&i^{-1}v_8iv_6^{-1}v_8^{-1} = i^{-1}v_9iv_6^{-1}v_9^{-1} = 1,\\
&i^{-1}v_{10}iv_1^{-1}v_2^{-1}v_3^{-1}v_5^{-1}v_9^{-1}v_{11}^{-1} = 1,\\
&j^{-1}v_1jv_2^{-1} = j^{-1}v_2jv_1^{-1} = j^{-1}v_3jv_5^{-1} = (j, v_4) = (j, v_8) = 1,\\
&(j, v_{10}) = j^{-1}v_5jv_3^{-1} = j^{-1}v_6jv_9^{-1} = j^{-1}v_9jv_6^{-1} = 1,\\
&j^{-1}v_7jv_2^{-1}v_3^{-1}v_7^{-1}v_8^{-1}v_9^{-1}v_{10}^{-1}v_{11}^{-1} = 1,\\
&j^{-1}v_{11}jv_1^{-1}v_2^{-1}v_3^{-1}v_5^{-1}v_6^{-1}v_9^{-1}v_{11}^{-1} = 1,\\
&t^{-1}atd^{-1}c^{-1}v_6^{-1}v_9^{-1} = t^{-1}btd^{-1}a^{-1}v_6^{-1}v_9^{-1} = 1,\\
&t^{-1}ctd^{-1}b^{-1}v_6^{-1}v_9^{-1} = 1,\\
&t^{-1}dtc^{-1}b^{-1}a^{-1}v_1^{-1}v_2^{-1}v_3^{-1}v_5^{-1}v_9^{-1} = 1,\\
&g^2v_1^{-1}v_3^{-1}v_8^{-1}v_9^{-1}v_{10}^{-1}v_{11}^{-1} = 1,\\
&(ga)^3v_{10}^{-1}v_{11}^{-1} = (gb)^3v_1^{-1}v_2^{-1}v_3^{-1}v_5^{-1}v_6^{-1}v_9^{-1} = 1,\\
&(gc)^3 = 1, \quad (ij)^3 = 1
\end{aligned}
$$

$$(gt)^2 v_1^{-1} v_3^{-1} v_8^{-1} v_{10}^{-1} v_{11}^{-1} = 1,$$
$$h^2 v_6^{-1} = 1, \quad h^{-1} bhd^{-1} b^{-1} a^{-1} v_1^{-1} v_2^{-1} v_3^{-1} v_5^{-1} = 1,$$
$$h^{-1} chc^{-1} a^{-1} v_1^{-1} v_2^{-1} v_3^{-1} v_5^{-1} v_6^{-1} v_9^{-1} = 1,$$
$$h^{-1} dhd^{-1} v_6^{-1} = h^{-1} thtv_9^{-1} = 1,$$
$$(gh)^3 v_3^{-1} v_7^{-1} v_9^{-1} v_{10}^{-1} v_{11}^{-1} = 1,$$
$$i^{-1} aid^{-1} c^{-1} v_1^{-1} v_2^{-1} v_3^{-1} v_5^{-1} v_6^{-1} v_9^{-1} v_{10}^{-1} v_{11}^{-1} = 1,$$
$$i^{-1} bid^{-1} a^{-1} v_1^{-1} v_2^{-1} v_3^{-1} v_5^{-1} v_6^{-1} v_9^{-1} v_{10}^{-1} v_{11}^{-1} = 1,$$
$$i^{-1} cid^{-1} c^{-1} b^{-1} a^{-1} v_1^{-1} v_2^{-1} v_3^{-1} v_5^{-1} v_6^{-1} v_9^{-1} v_{10}^{-1} v_{11}^{-1} = 1,$$
$$i^{-1} did^{-1} c^{-1} b^{-1} = 1, \quad i^{-1} tit = 1,$$
$$i^{-1} gig^{-1} t^{-1} v_5^{-1} v_6^{-1} v_7^{-1} v_9^{-1} v_{10}^{-1} v_{11}^{-1} = 1,$$
$$(hi)^3 v_1^{-1} v_2^{-1} v_3^{-1} v_5^{-1} v_6^{-1} v_8^{-1} v_{10}^{-1} = 1,$$
$$j^{-1} ajc^{-1} b^{-1} a^{-1} = 1,$$
$$j^{-1} djd^{-1} c^{-1} v_1^{-1} v_2^{-1} v_3^{-1} v_5^{-1} v_6^{-1} v_9^{-1} = 1,$$
$$j^{-1} tjt = 1, \quad j^{-1} hjh^{-1} t^{-1} = 1.$$

Proof The two irreducible $F\mathsf{M}_{23}$-modules V_i, $i = 1, 2$, occur as composition factors with multiplicity 1 in the permutation module $(1_{\mathsf{M}_{22}})^{\mathsf{M}_{23}}$ of degree 23, where $\mathsf{M}_{22} = \langle a, b, c, d, t, g, h, i \rangle$. They are dual to each other. Using the faithful permutation representation of M_{23} stated in (a) and the Meataxe algorithm implemented in MAGMA one obtains the generating matrices of M_{23} defining V_1. They are stated in (b).

(c) Their dual matrices define V_2.

(d) The cohomological dimensions $d_i = dim_F[H^2(\mathsf{M}_{23}, V_i)]$, $i = 1, 2$, have been calculated by means of MAGMA using Holt's Algorithm 7.4.5 of [92]. Its hypothesis is satisfied by the presentation of M_{23} stated in Definition 8.2.1 of [92] and all the data of (a), (b) and (c). It follows that $d_1 = 1$ and $d_2 = 0$.

(e) This statement is clear by Lemma 1.4.7.

(g) The presentation of the non-split extension E has been obtained by means of the command `ExtensionProcess` of Holt's Algorithm 7.4.5 of [92] implemented in MAGMA [52]. This completes the proof.

□

The first statement of the following subsidiary lemma is mainly due to Paul Young.

Lemma 6.1.2 *With the notation of Lemma 6.1.1 the following statements hold.*

(a) *The non-split extension* E *of* M_{23} *by its simple module* V_1 *has a faithful permutation representation of degree* 1012 *with stabilizer* $U = \langle s_1 = (i^{-1}j^{-1}bg)^2, s_2 = (i^{-1}bth^{-1})^4\rangle$.

(b) $E = \langle a, b, c, d, t, g, h, i, j\rangle$.

(c) E *has three conjugacy classes of* 2*-central involutions. They are represented by* $z = (dg)^5$, $z_1 = h^2$ *and* $z_2 = g^2$. *Their centralizers have orders* $|C_E(z)| = 2^{18} \cdot 3^2 \cdot 5 \cdot 7 \cdot 11$, $|C_E(z_1)| = 2^{18} \cdot 3^2 \cdot 5 \cdot 7$ *and* $|C_E(z_2)| = 2^{18} \cdot 3^2 \cdot 5$.

(d) $D = C_E(z) = \langle x, y\rangle$ *and* $E = \langle D, e\rangle$, *where* $x = a$, $y = bghi$ *and* $e = j$ *have respective orders* 2, 14 *and* 2.

(e) $E = \langle x, y, e\rangle$ *has* 56 *conjugacy classes. A system of representatives is given in Table 6.5.1.*

Proof (a) The generators s_1 and s_2 of the stabilizer U have been found by means of a stand alone program due to P. Young. Using the MAGMA command `CosetAction(E,U)` one obtains a faithful permutation representation PE of E with stabilizer U and degree 1012.

(b), (c) and (d) These statements are easily checked by means of the permutation representation PE and MAGMA.

(e) In [73] H. Kim used the faithful permutation representation PE of E, MAGMA and Kratzer's Algorithm 5.3.18 of [92] to calculate a system of representatives of all conjugacy classes of D and E in terms of their generators given in (d). The results are stated in Table 6.5.1. □

6.2 Construction of a 2-central involution centralizer

Let z denote the central involution of the extension E of M_{23} defined in Lemma 6.1.1 and let $D = C_E(z)$. In the following subsidiary result it is shown that $D_1 = D/\langle z\rangle$ is isomorphic to the extension E_2 constructed in Lemma 4.1.1. By the results of Chapter 5 the reader knows that Fischer's simple group $G_1 \cong \mathsf{Fi}_{22}$ is constructed from the group D_1 by Algorithm 1.3.8. Unfortunately, (in 2007) an application of the implementation of Holt's Algorithm 7.4.5 of [92] in MAGMA failed to construct a 2-fold cover H from the nice presentation of G_1 given in Corollary 5.2.2. Therefore Kim constructed in [73] all the central extensions H_2 of the centralizer $H_1 = C_{G_1}(t)$ of a 2-central involution t of G_1 by a cyclic group of order 2. He showed that only one of them has a Sylow 2-subgroup which is isomorphic to the ones of D. Thus H_2 is uniquely determined up to isomorphism.

The target group H with center $Z(H) = \langle z \rangle$ is then constructed by means of Algorithm 7.4.8 of [92] as a matrix subgroup $\mathfrak{H}$ of $\mathrm{GL}_{352}(17)$. It enabled Kim to calculate a presentation of H. It follows from Corollary 5.2.2 that $H/\langle z \rangle \cong \mathsf{Fi}_{22}$ and that H has a faithful permutation representation representation of degree 28160.

Proposition 6.2.1 *Keep the notations of Lemmas 6.1.1 and 6.1.2. Let $E = \langle a, b, c, d, t, g, h, i, j \rangle$ be the non-split extension of M_{23} by its simple module V_1 of dimension 11 over $F = GF(2)$ and $D = C_E(z)$. Then the following statements hold.*

(a) $z = (dg)^5$ *is a 2-central involution of E with centralizer $D = C_E(z)$ of order* $2^{18} \cdot 3^2 \cdot 5 \cdot 7 \cdot 11$.

(b) $D = \langle x, y \rangle$, $z = (xy^2)^7$ *and* $E = \langle D, e \rangle$, *where* $x = a$, $y = bghi$ *and* $e = j$ *have respective orders* 2, 14 *and* 2.

(c) V_1 *is the unique normal subgroup of order* 2^{11} *of* D. *Also,* V_1 *has a basis* $\mathcal{B} = \{z, v_i \mid 1 \le i \le 10\}$ *of* V_1, *where*

$$v_1 = y^7, \quad v_2 = (xyx)^7, \quad v_3 = (xyxy^2)^8, \quad v_4 = (xy^2xy)^8,$$
$$v_5 = (yxy^2x)^8, \quad v_6 = (xyxy^3)^6, \quad v_7 = (xy^3xy)^6,$$
$$v_8 = (xy^5)^7, \quad v_9 = (yxyxy^2)^6, \quad v_{10} = (y^2xy^3)^7.$$

(d) $V = V_1/\langle z \rangle$ *has a complement* W *in* $D_1 = D/\langle z \rangle$ *such that* $W \cong \mathsf{M}_{22}$.

In particular, $D_1 \cong E(\mathsf{Fi}_{22})$ *defined in Lemma 4.1.1.*

(e) *Let* $z = (xy^2)^7$ *and let* v_i *be the ten basis elements of* $\mathcal{B}$ *given in (c). Then* $D = \langle x, y \rangle$ *has the following set* $\mathcal{R}(D)$ *of defining relations:*

$$z^2 = 1, (x, z) = (y, z) = 1,$$
$$v_i^2 = 1, \quad \text{and} \quad (v_j, v_k) = 1 \quad \text{for all} \quad 1 \le i, j, k \le 10,$$
$$x^2 = y^{-7}v_1 = (xy^{-1})^7 = (y^{-1}xyx)^5 = (y^{-2}x)^7 z = 1,$$
$$(y^{-2}xy^2xy^{-1}x)^3 v_2 = (xy^{-2}xy)^5 v_1 v_2 v_3 v_4 v_7 v_8 v_{10} = 1,$$
$$(yxyxy^2xyxy^{-1}xy^2)^2 v_1 v_3 v_5 v_6 v_9 z = 1,$$
$$xy^{-1}xyxy^2xy^{-1}xy^{-1}xy^2xyxy^{-1}xy^2xy^2xyv_1v_3v_6v_8z,$$
$$yxy^3xyxy^{-2}xy^{-3}xyxy^3xy^{-1}xy^{-3}xv_4v_8v_{10} = 1,$$

$$xv_1x^{-1}v_2 = xv_2x^{-1}v_1 = xv_3x^{-1}v_5 = xv_4x^{-1}v_3v_4v_5 = 1,$$
$$xv_5x^{-1}v_3 = xv_6x^{-1}v_1v_2v_6 = xv_7x^{-1}v_1v_2v_7 = 1,$$
$$xv_8x^{-1}v_4v_6v_9 = xv_9x^{-1}v_1v_2v_3v_4v_5v_6v_8 = 1,$$
$$xv_{10}x^{-1}v_1v_3v_8v_9v_{10}z = yv_1y^{-1}v_1 = yv_2y^{-1}v_3v_4v_5v_{10} = 1,$$
$$yv_3y^{-1}v_1v_3v_4z = yv_4y^{-1}v_5 = yv_5y^{-1}v_2v_6v_8v_9(xy^2)^7 = 1,$$
$$yv_6y^{-1}v_9 = yv_7y^{-1}v_1v_2v_6 = yv_8y^{-1}v_1v_2v_3v_4z = 1,$$
$$yv_9y^{-1}v_1v_6v_7v_9z = yv_{10}y^{-1}v_2v_4v_5v_7z = 1.$$

(f) *Let $H_1 = \langle h_i | 1 \le i \le 14\rangle \cong H(\mathsf{Fi}_{22})$ be the finitely presented group constructed in Proposition 5.1.1(n). Then H_1 has a central extension $H_2 = \langle k_j \mid 1 \le j \le 15\rangle$ of order $2^{18} \cdot 3^4 \cdot 5$ having the following set of $\mathcal{R}(H_2)$ of defining relations:*

$$k_1^2 = k_2^5 = k_3^3 = k_4^3 = k_5^2 = k_6^2 = k_7^2 = 1,$$
$$k_8^2 = k_9^2 = k_{10}^4 = k_{11}^2 = k_{12}^2 = k_{13}^2 = k_{14}^2 = k_{15}^2 = 1,$$
$$(k_i, k_{15}) = 1 \quad \textit{for} \quad 1 \le i \le 14,$$
$$(k_i, k_{14}) = 1 \quad \textit{for} \quad 1 \le i \le 13,$$
$$(k_i, k_{13}) = 1 \quad \textit{for} \quad 2 \le i \le 12,$$
$$k_1^{-1}k_{13}k_1k_{13}^{-1}k_{14}^{-1} = 1,$$
$$(k_3, k_4) = 1, \quad k_3^{-1}k_2k_1k_2^{-1}k_3k_1 = 1,$$
$$(k_2^{-1}k_1)^4 = 1, \quad k_3^{-1}k_2k_3^{-1}k_1k_2k_3k_1k_2^{-1} = 1,$$
$$(k_2k_3^{-1})^4 = 1, \quad (k_4, k_2^{-1}, k_4) = 1,$$
$$k_1k_4^{-1}k_1k_4^{-1}k_1k_4k_1k_4 = 1, \quad k_2^{-1}k_3^{-1}k_1k_2^{-2}k_3k_4k_1k_4^{-1} = 1,$$
$$k_2k_3^{-1}k_2k_4k_2^2k_3^{-1}k_2^{-1}k_3k_4 = 1, \quad k_1k_5k_1^{-1}k_5k_6k_{13}^{-1} = 1,$$
$$k_1k_6k_1^{-1}k_6k_{14}^{-1} = k_1k_7k_1^{-1}k_7k_8k_{14}^{-1} = k_1k_8k_1^{-1}k_8 = 1,$$
$$k_1k_9k_1^{-1}k_{11}k_{13}^{-1}k_{14}^{-1}k_{15}^{-1} = 1,$$
$$k_1k_{10}k_1^{-1}k_6k_9k_{10}k_{11}k_{12}k_{13}^{-1}k_{15}^{-1} = 1,$$
$$k_1k_{11}k_1^{-1}k_9k_{13}^{-1}k_{15}^{-1} = k_1k_{12}k_1^{-1}k_{12}k_{14}^{-1} = 1,$$
$$k_2k_5k_2^{-1}k_5k_6k_7k_{14}^{-1}k_{15}^{-1} = k_2k_6k_2^{-1}k_7k_8k_{13}^{-1}k_{14}^{-1} = 1,$$
$$k_2k_7k_2^{-1}k_6k_7k_{14}^{-1}k_{15}^{-1} = k_2k_8k_2^{-1}k_5k_6k_7k_8 = 1,$$
$$k_2k_9k_2^{-1}k_5k_6k_{10}k_{11}k_{13}^{-1}k_{14}^{-1}k_{15}^{-1} = 1,$$
$$k_2k_{10}k_2^{-1}k_7k_8k_9k_{13}^{-1} = k_2k_{11}k_2^{-1}k_5k_{12}k_{14}^{-1}k_{15}^{-1} = 1,$$
$$k_2k_{12}k_2^{-1}k_6k_7k_8k_{10}k_{12} = k_3k_5k_3^{-1}k_6k_{14}^{-1} = 1,$$

$$k_3k_6k_3^{-1}k_5k_6k_{13}^{-1} = k_3k_7k_3^{-1}k_5k_8k_{15}^{-1} = 1,$$
$$k_3k_8k_3^{-1}k_5k_6k_7k_8 = k_3k_9k_3^{-1}k_5k_9k_{10}k_{14}^{-1} = 1,$$
$$k_3k_{10}k_3^{-1}k_6k_9k_{15}^{-1} = k_3k_{11}k_3^{-1}k_9k_{12}k_{14}^{-1}k_{15}^{-1} = 1,$$
$$k_3k_{12}k_3^{-1}k_5k_{10}k_{11}k_{12}k_{13}^{-1}k_{14}^{-1} = 1,$$
$$k_4k_5k_4^{-1}k_5k_6k_{13}^{-1}k_{14}^{-1} = 1,$$
$$k_4k_6k_4^{-1}k_5k_{14}^{-1} = k_4k_7k_4^{-1}k_6k_{10}k_{11}k_{12}k_{14}^{-1}k_{15}^{-1} = 1,$$
$$k_4k_8k_4^{-1}k_5k_6k_{10}k_{11}k_{13}^{-1}k_{14}^{-1}k_{15}^{-1} = 1,$$
$$k_4k_9k_4^{-1}k_5k_6k_{11}k_{12}k_{14}^{-1} = 1,$$
$$k_4k_{10}k_4^{-1}k_9k_{10}k_{12}k_{13}^{-1}k_{14}^{-1}k_{15}^{-1} = 1,$$
$$k_4k_{11}k_4^{-1}k_5k_6k_7k_8k_9k_{11}k_{13}^{-1}k_{15}^{-1} = 1,$$
$$k_4k_{12}k_4^{-1}k_5k_7k_{10}k_{11}k_{14}^{-1} = k_{10}^2k_{14}^{-1}k_{15}^{-1} = 1,$$
$$(k_5, k_6) = (k_5, k_7) = (k_5, k_8) = (k_5, k_9) = 1,$$
$$k_5^{-1}k_{10}^{-1}k_5k_{10}k_{14}^{-1}k_{15}^{-1} = k_5^{-1}k_{11}^{-1}k_5k_{11}k_{14}^{-1}k_{15}^{-1} = 1,$$
$$(k_5, k_{12}) = (k_6, k_7) = (k_6, k_8) = 1,$$
$$k_6^{-1}k_9^{-1}k_6k_9k_{14}^{-1}k_{15}^{-1} = k_6^{-1}k_{10}^{-1}k_6k_{10}k_{14}^{-1}k_{15}^{-1} = 1,$$
$$k_6^{-1}k_{11}^{-1}k_6k_{11}k_{14}^{-1}k_{15}^{-1} = 1,$$
$$(k_6, k_{12}) = (k_7, k_8) = (k_7, k_9) = (k_8, k_{10}) = (k_8, k_{12}) = 1,$$
$$k_7^{-1}k_{10}^{-1}k_7k_{10}k_{14}^{-1}k_{15}^{-1} = k_7^{-1}k_{11}^{-1}k_7k_{11}k_{14}^{-1}k_{15}^{-1} = 1,$$
$$k_7^{-1}k_{12}^{-1}k_7k_{12}k_{14}^{-1}k_{15}^{-1} = k_8^{-1}k_9^{-1}k_8k_9k_{14}^{-1}k_{15}^{-1} = 1,$$
$$k_8^{-1}k_{11}^{-1}k_8k_{11}k_{14}^{-1}k_{15}^{-1} = 1,$$
$$(k_9, k_{10}) = (k_9, k_{11}) = (k_9, k_{12}) = (k_{10}, k_{12}) = (k_{11}, k_{12}) = 1,$$
$$k_{10}^{-1}k_{11}^{-1}k_{10}k_{11}k_{14}^{-1}k_{15}^{-1} = 1.$$

(g) *H_2 has a faithful permutation representation of degree* 2048 *with stabilizer $U_2 = \langle k_1, k_2, k_3, k_4\rangle$.*

(h) *H_2 has a Sylow* 2*-subgroup S_2 which is generated by the four involutions $m_1 = k_1$, $m_2 = (k_2k_1k_2k_4k_5)^6$, $m_3 = (k_2^3k_1k_2)^2$ and $m_4 = (k_4^2k_2k_4k_2k_4)^6$.*

S_2 contains a unique maximal elementary abelian normal subgroup A_2 of order 2048 *such that $D_2 = N_{H_2}(A_2) = \langle p_2, q_2\rangle$, where $p_2 = k_1k_2$ and $q_2 = k_1(k_1k_5k_4k_2)^5$. They have respective orders* 4 *and* 6*.*

(i) *$H_2 = \langle D_2, h\rangle = \langle p_2, q_2, h\rangle$, where $h = k_4$ has order* 3*.*

(j) *The involution $t = (xyxy^3)^6$ of D has a centralizer $T = C_D(t)$ of order $2^{18} \cdot 3 \cdot 5$ such that there is an isomorphism $\sigma : D_2 \to T$*

and $\sigma(p_2) = p_1 = (s_1 s_2 s_4 s_7 s_8)^3$, $\sigma(q_2) = q_1 = (r_1 r_2 r_3 r_4 r_2 r_3 r_4)^5$, *where the elements* s_i *and* r_j *of* $D\langle x, y\rangle$ *are given by the following equations:*

$$\begin{aligned}
&s_1 = (xy^2)^7, \quad s_2 = (xyxy^3)^6, \quad s_3 = (y^2xy^3)^7, \quad s_4 = (y^3xy^2)^7,\\
&s_5 = xyxyxyxy^2xyxy, \quad s_6 = (xy^2xy^5xyxyxy)^4,\\
&s_7 = (y^3xy^2xy^5xy^3xy)^2, \quad s_8 = (y^5xy^2xyxyxyxy^2x)^4,\\
&r_1 = (xyx)^7,\\
&r_2 = ((xyx)^4(xyxy^3)^4(xyx)^2(xyxy^3)^4xyx)^2,\\
&r_3 = (xyx)^2(xyxy^3)^2(xyx)^3(xyxy^3)^4(xyx)^3,\\
&r_4 = ((xyx)^4(xyxy^3)^4(xyx)^2(xyxy^3)^2(xyx)^3(xyxy^3)^2(xyx)^3)^2.
\end{aligned}$$

(k) *A system of representatives* w_i *of the* 189 *conjugacy classes of* $H_2 = \langle p_2, q_2, h\rangle$ *and the corresponding centralizer orders* $|C_{H_2}(r_i)|$ *are given in Table* `DVD.1.4.1` *stored on the accompanying DVD.*

(l) *A system of representatives* d_i *of the* 151 *conjugacy classes of* $D_2 = \langle p_2, q_2\rangle$ *and the corresponding centralizer orders* $|C_{D_2}(d_i)|$ *are given in Table* `DVD.1.4.2`.

(m) *A system of representatives* e_i *of the* 69 *conjugacy classes of* $D = \langle p_1, q_1, x, y\rangle = \langle x, y\rangle$ *and the corresponding centralizer orders* $|C_D(e_i)|$ *are given in Table 6.5.3.*

(n) *The character tables of* H_2, D_2 *and* D *are given in Tables* `DVD.1.4.4`, `DVD.1.4.3` *and 6.6.2, respectively.*

Proof The assertions (a) and (b) are restatements of assertions (c) and (d) of Lemma 6.1.2, respectively.

(c) Using the faithful permutation representation PE of E given in statement (a) of Lemma 6.1.2 and the MAGMA command `NormalSubgroups(D)` Kim verified in [73] that V_1 is the unique normal subgroup of order 2^{11} of $D = C_E(z)$. It has a basis $\mathcal{B} = \{z, b_i \mid 1 \le i \le 10\}$ of V_1, where the v_i are stated in the assertion. These basis elements of V_1 have been found by Kim using the program `GetShortGens(D,V_1)`.

(d) As $z \in V_1$ it follows that $V = V_1/\langle z\rangle$ is the unique normal subgroup of $D_1 = D/\langle z\rangle$ of order 2^{10}. Clearly, it is elementary abelian.

Applying the MAGMA command `CompositionFactors(D)` one sees that $D_1/V \cong D/V_1 \cong \mathsf{M}_{22}$. Thus Lemma 4.1.1(d) asserts that D_1 splits over V.

Let Mx and My be the matrices of the generators x and y of D w.r.t. the basis $\mathcal{B}$ of V_1. Then

$$Mx = \begin{pmatrix} 1&0&0&0&0&0&0&0&0&0&0 \\ 0&0&1&0&0&0&0&0&0&0&0 \\ 0&1&0&0&0&0&0&0&0&0&0 \\ 0&0&0&0&0&1&0&0&0&0&0 \\ 0&0&0&1&1&1&0&0&0&0&0 \\ 0&0&0&1&0&0&0&0&0&0&0 \\ 0&1&1&0&0&0&1&0&0&0&0 \\ 0&1&1&0&0&0&0&1&0&0&0 \\ 0&0&0&0&1&0&1&0&0&1&0 \\ 0&1&1&1&1&1&1&0&1&0&0 \\ 1&1&0&1&0&0&0&0&1&1&1 \end{pmatrix}, \quad My = \begin{pmatrix} 1&0&0&0&0&0&0&0&0&0&0 \\ 0&1&0&0&0&0&0&0&0&0&0 \\ 0&0&0&1&0&0&0&0&1&0&0 \\ 1&1&0&1&1&0&1&1&0&1&1 \\ 0&0&0&0&1&0&1&1&0&1&1 \\ 0&0&0&0&1&0&0&0&0&0&0 \\ 0&1&0&1&0&0&0&1&1&0&0 \\ 1&0&0&1&0&0&1&1&1&1&0 \\ 1&1&0&0&0&1&1&1&0&0&0 \\ 0&0&0&0&0&0&1&0&0&0&0 \\ 1&1&1&1&1&0&0&0&0&0&0 \end{pmatrix},$$

Both matrices are blocked lower triangular matrices with upper left diagonal blocks equal to 1 and lower diagonal 10×10 blocks Mx_1 and My_1 in $\mathrm{GL}_{10}(2)$. Hence $D_1/V \cong W$, where $W = \langle Mx_1, My_1 \rangle \leq \mathrm{GL}_{10}(2)$. Applying the MAGMA command `FPGroup(sub<GL(10,2)|Mx1,My1>)` to W one obtains the following set $\mathcal{R}(W)$ of defining relations of $W = \langle x_1, y_1 \rangle$:

$$\begin{aligned}
&x_1^2 = 1, \quad y_1^7 = 1, \\
&(x_1 y_1^{-1})^7 = 1, \quad (y_1^{-1} x_1 y_1 x_1)^5 = 1, \quad (y_1^{-2} x_1)^7 = 1, \\
&(y_1^{-2} x_1 y_1^2 x_1 y_1^{-1} x_1) = 1, \quad (x_1 y_1^{-2} x_1 y_1)^5 = 1, \\
&(y_1 x_1 y_1 x_1 y_1^2 x_1 y_1 x_1 y_1^{-1} x_1 y_1^2)^2 = 1, \\
&x_1 y_1^{-1} x_1 y_1 x_1 y_1^2 x_1 y_1^{-1} x_1 y_1^{-1} x_1 y_1^2 x_1 y_1 x_1 y_1^{-1} x_1 y_1^2 x_1 y_1^2 x_1 y_1 = 1, \\
&y_1 x_1 y_1^3 x_1 y_1 x_1 y_1^{-2} x_1 y_1^{-3} x_1 y_1 x_1 y_1^3 x_1 y_1^{-1} x_1 y_1^{-3} x_1 = 1.
\end{aligned}$$

Let PD_1 be its faithful permutation representation of degree 1024 with stabilizer W. Let $E_2 = E(\mathsf{Fi}_{22})$ be the finitely presented group of Lemma 4.1.1(f) with faithful permutation representation PE_2 defined in Lemma 4.1.3(a). By means of the MAGMA command `IsIsomorphic (PE_2,PD_1)` Kim verified [73] that D_1 is isomorphic to E_2.

(e) By Lemma 1.4.7 the semidirect product $D_1 = \langle x_1, y_1, v_i \mid 1 \leq i \leq 10 \rangle$ of W by $V = \langle v_i \mid 1 \leq i \leq 10 \rangle$ has a set of defining relations $\mathcal{R}(D_1)$ consisting of $\mathcal{R}(W)$ and the following set of relations:

$$\begin{aligned}
&v_i^2 = 1, (v_j, v_k) = 1 \quad \text{for all} \quad 1 \leq i, j, k \leq 10, \\
&x_1 v_1 x_1^{-1} v_2 = x_1 v_2 x_1^{-1} v_1 = x_1 v_3 x_1^{-1} v_5 = 1, \\
&x_1 v_4 x_1^{-1} v_3 v_4 v_5 = x_1 v_5 x_1^{-1} v_3 = 1,
\end{aligned}$$

$$x_1v_6x_1^{-1}v_1v_2v_6, x_1v_7x_1^{-1}v_1v_2v_7 = 1,$$
$$x_1v_8x_1^{-1}v_4v_6v_9, x_1v_9x_1^{-1}v_1v_2v_3v_4v_5v_6v_8 = 1,$$
$$x_1v_{10}x_1^{-1}v_1v_3v_8v_9v_{10} = y_1v_1y_1^{-1}v_1 = 1,$$
$$y_1v_2y_1^{-1}v_3v_4v_5v_{10} = y_1v_3y_1^{-1}v_1v_3v_4 = 1,$$
$$y_1v_4y_1^{-1}v_5 = y_1v_5y_1^{-1}v_2v_6v_8v_9 = 1,$$
$$y_1v_6y_1^{-1}v_9 = y_1v_7y_1^{-1}v_1v_2v_6 = 1,$$
$$y_1v_8y_1^{-1}v_1v_2v_3v_4 = y_1v_9y_1^{-1}v_1v_6v_7v_9 = y_1v_{10}y_1^{-1}v_2v_4v_5v_7 = 1.$$

By (c) each v_i is a word in the generators x and y of D. Evaluating the relations of $\mathcal{R}(D_1)$ in D one obtains the defining set of relations $\mathcal{R}(D)$.

(f) By Table `DVD.1.3.1` the simple group Fi_{22} has a unique conjugacy class of 2-central involutions u. Theorem 5.2.1 asserts that $H_1 = C_{\mathsf{Fi}_{22}}(u)$ is isomorphic to the finitely presented group $H_1 = \langle h_i \mid 1 \leq i \leq 14\rangle$ constructed in Proposition 5.1.1(n). Furthermore, H_1 has a faithful permutation representation PH_1 of degree 1024 with stabilizer $\langle h_1, h_2, h_3, h_4\rangle$ by Lemma 5.1.2.

Let $F = GF(2)$, and let M be the trivial FH_1-module. Using the faithful permutation representation PH_1 and the MAGMA command

```
CohomologicalDimension(PH_1, M, 2)
```

Kim showed in [73] that this dimension is 3. For each of the eight 2-cocycles (a, b, c) with $a, b, c \in F$ MAGMA constructs a finitely presented group $E_{(a,b,c)}$ by means of

```
P = ExtensionProcess(PH_1,M,H_1)
```

and

```
E_{a,b,c} = Extension(P, [a,b,c]).
```

Since D is a non-split extension of D_1 by $\langle z\rangle$ the split extension $E_{(0,0,0)}$ can be neglected. In [73] Kim determined for the non-split extension $E_{(1,1,0)}$ a faithful permutation representation of degree 1012. Using it and MAGMA he verified that $E_{(1,1,0)}$ is a non-split extension of order $2^{18} \cdot 3^4 \cdot 5$ of H_1 having a Sylow 2-subgroup which is isomorphic to any of the Sylow 2-subgroups of D. The presentation of $H_2 = E_{(1,1,0)}$ is given in the statement.

(g) Kim verified in [73] this assertion by using the MAGMA command `CosetAction(H_2,U_2)` for the subgroup $U_2 = \langle k_1, k_2, k_3, k_4\rangle$ of the finitely presented group H_2.

(h) Let PH_2 be the faithful permutation representation of H_2 of (g). Using MAGMA and his stand alone program `GetShortGenerators` Kim showed in [73] that the four involutions $m_1 = k_1$, $m_2 = (k_2k_1k_2k_4k_5)^6$, $m_3 = (k_2^3k_1k_2)^2$ and $m_4 = (k_4^2k_2k_4k_2k_4)^6$ generate a Sylow 2-subgroup S_2 of H_2. Applying the MAGMA command

```
Subgroups(S_2 : Al:=Normal, IsElementaryAbelian:=true)
```

one verifies that S_2 has a unique maximal elementary abelian normal subgroup A_2 of order 2^{11}. Using the program `GetShortGenerators` again Kim found the following 11 generators of A_2:

$$m_4, \quad (m_2m_4)^2, \quad (m_3m_4)^2, \quad (m_1m_3m_4)^4, \quad (m_2m_3m_4)^4,$$
$$(m_1m_2m_3m_4)^8, \quad (m_1m_3m_1m_4)^2, \quad (m_2m_3m_2m_4)^2,$$
$$(m_2m_3m_4m_2)^2, \quad (m_1m_2m_3m_1m_4)^4, \quad (m_1m_2m_3m_2m_4)^4.$$

Another calculation with MAGMA asserts that $D_2 = N_{H_2}(A_2)$ is generated by the elements $p_2 = k_1k_2$, $q_2 = k_1(k_1k_5k_4k_2)^5$ of H_2 having respective orders 4 and 6. Furthermore, $|D_2| = 2^{18} \cdot 3 \cdot 5$.

(i) Using MAGMA Kim verified in [73] that $H_2 = \langle T, h\rangle = \langle p_2, q_2, h\rangle$, and that $h = k_4$ has order 3.

(j) Looking now at Table 6.5.3 we see that the centralizer $T = C_D(t)$ of the involution $t = (xyxy^3)^6$ of D has order $2^{18}\cdot3\cdot5$. Using the two faithful permutation representations PH_2 and PD one finds an isomorphism $\sigma : D_2 \to T$ by means of the MAGMA command `IsIsomorphic(D_2,T)`. The image $D_1 = \sigma(D_2)$ of D_2 in D is generated by $p_1 = \sigma(p_2)$ and $q_1 = \sigma(q_2)$, where p_1 and q_1 are words in x and y.

In order to find these two words Kim calculated in [73] short generators for the subgroups $P_1 = N_D(\langle p_1\rangle)$ and $Q_1 = N_D(\langle q_1^3\rangle)$ of respective orders 2^8 and $2^{17} \cdot 3^2 \cdot 5 \cdot 7$.

Using MAGMA and the program `GetShortGens(sub<D|x,y>,P_1)` Kim showed that $P_1 = \langle s_i | 1 \leq i \leq 8\rangle$ for the eight generators s_i of P_1 given in statement (j).

An application of the program `LookupWord(P_1, p_1)` yields the equation $p_1 = (s_1s_2s_4s_7s_8)^3$.

Using MAGMA and the program `GetShortGens(sub<D|x,y>,Q_1)` Kim observed in [73] that $u_1 = y^7$, $u_2 = xyx$, $u_3 = (xyxy^2)^8$, $u_4 = (yxyxy)^2$ and $u_5 = (xyxy^3)^2$ generate Q_1. Applying the same method to $Q_2 = N_{Q_1}(\langle q_1\rangle)$ he obtained the following generators of Q_2:

$$r_1 = (u_2)^7, \quad r_2 = (u_2^4u_5^2u_2^2u_5^2u_2)^2,$$
$$r_3 = u_2^2u_5u_2^3u_5^2u_2^3, \quad r_4 = (u_2^4u_5^2u_2^2u_5u_2^3u_5u_2^3)^2.$$

Inserting the words for u_2 and u_5 into these four terms Kim obtained the words r_j in x and y given in statement (j).

Then he found the word $q_1 = (r_1r_2r_3r_4r_2r_3r_4)^5$ by another application of MAGMA and the program

```
LookupWord(sub<Q_2|[ r_1,r_2,r_3,r_4]>, q_1).
```

(k), (l) and (m) The systems of representatives of the conjugacy classes of H_2, D_2 and D have been calculated by means of the permutation representations PH_2 and PE, MAGMA and Kratzer's Algorithm 5.3.18 of [92].

The character tables of (n) have been obtained in [73] by means of PH_2, PD and MAGMA. □

Proposition 6.2.2 *Keep the notation of Proposition 6.2.1. Using the notation of the character Tables* DVD.1.4.4, DVD.1.4.3 *and 6.6.2 of the groups $H_2 = \langle p_2, q_2, h\rangle$, $D_2 = \langle p_2, q_2\rangle \cong T = \langle p_1, q_1\rangle$ and $D = \langle T, x, y\rangle$, respectively, the following statements hold.*

(a) *There is a compatible pair $(\chi, \tau) \in mf\,char_{\mathbb{C}}(H_2) \times mf\,char_{\mathbb{C}}(D)$ of degree 352 of the groups H_2 and D,*

$$(\chi, \tau) = (\chi_{\mathbf{40}} + \chi_{\mathbf{41}} + \chi_{57}, \tau_{\mathbf{9}} + \tau_{\mathbf{10}}),$$

with common restriction given by

$$\tau_{|D} = \chi_{|D} = \psi_{\mathbf{26}} + \psi_{\mathbf{27}} + \psi_{59} + \psi_{63} + \psi_{70} + \psi_{73} \in mf\,char_{\mathbb{C}}(T),$$

where irreducible characters with bold face indices denote faithful irreducible characters.

(b) *Let $\mathfrak{V}$ and $\mathfrak{W}$ be the up to isomorphism uniquely determined faithful semi-simple multiplicity-free 352-dimensional modules of H_2 and D over $F = \mathrm{GF}(17)$ corresponding to the compatible pair (χ, τ), respectively.*

Let $\kappa_{\mathfrak{V}} : H_2 \to \mathrm{GL}_{352}(17)$ and $\kappa_{\mathfrak{W}} : D \to \mathrm{GL}_{352}(17)$ be the representations of H_2 and D afforded by the modules $\mathfrak{V}$ and $\mathfrak{W}$, respectively.

Let $\mathfrak{h} = \kappa_{\mathfrak{V}}(h)$, $\mathfrak{p}_2 = \kappa_{\mathfrak{V}}(p_2)$, $\mathfrak{q}_2 = \kappa_{\mathfrak{V}}(q_2)$ in $\kappa_{\mathfrak{V}}(H_2) \le \mathrm{GL}_{352}(17)$. Then the following assertions hold.

(1) *There is a transformation matrix $\mathcal{T} \in \mathrm{GL}_{352}(17)$ such that*

$$\mathfrak{p}_2 = \mathcal{T}^{-1}\kappa_{\mathfrak{W}}(p_1)\mathcal{T}, \quad \mathfrak{q}_2 = \mathcal{T}^{-1}\kappa_{\mathfrak{W}}(q_1)\mathcal{T},$$

$$\mathfrak{x} = \mathcal{T}^{-1}\kappa_{\mathfrak{W}}(x)\mathcal{T} \quad \text{and} \quad \mathfrak{y} = \mathcal{T}^{-1}\kappa_{\mathfrak{W}}(y)\mathcal{T}.$$

(2) $\mathfrak{H} = \langle \mathfrak{h}, \mathfrak{x}, \mathfrak{y}\rangle$.

Proof (a) The character tables of the groups H_2, $D_2 \cong T$ and D are stated in the accompanying DVD and Section 6.6, respectively. In the following we use their notations. Using MAGMA and the faithful permutation representations PH_2 and PD of H_2 and D constructed in Proposition 6.2.1(e) and Lemma 6.1.2(a) H. Kim determined in [73] the fusion of the classes of D_2 in H_2 and T in D, respectively. By means of the character tables of H_2, D_2 and D and Kratzer's Algorithm 7.3.10 of [92] he found two compatible pairs of degree 352. Only one of them provides an irreducible representation of degree 352 of the free product $H_2 *_{D_2} D$ with amalgamated subgroup D_2 satisfying the Sylow 2-subgroup test of Algorithm 7.4.8, Step 5(c), of [92]. It is stated in the assertion.

(b) By Proposition 6.2.1 there are well determined elements p_2, q_2 of D_2, h of H_2 and an isomorphism $\sigma : D_2 \to T$ such that $D_2 = \langle p_2, q_2 \rangle$, $H_2 = \langle D_2, h \rangle$ and $T = \langle p_1, q_1 \rangle$, where $p_1 = \sigma(p_2)$ and $q_1 = \sigma(q_2)$ are well determined words in x and y; see the proof of Proposition 6.2.1(j). Furthermore, $D = \langle p_1, q_1, x, y \rangle$.

The semi-simple faithful representation $\mathfrak{V}$ of $H_2 = \langle p_2, q_2, h \rangle$ corresponding to the character $\chi = \chi_{\mathbf{41}} + \chi_{\mathbf{41}} + \chi_{57}$ decomposes into the three non-isomorphic irreducible constituents $\mathfrak{V}(40)$, $\mathfrak{V}(41)$ and $\mathfrak{V}(57)$ corresponding to the irreducible characters $\chi_{\mathbf{40}}$, $\tau_{\mathbf{41}}$ and χ_{57}, respectively. In order to get these three representations Kim used in [73] the faithful permutation representation PD of $D = \langle p_1, q_1, x, y \rangle$ and the MAGMA command `LowIndexSubgroups(PD, k)` searching for conjugacy classes of subgroups Y of index $|D : Y| = k$. He found three subgroups Y_i, $i \in \{40, 41, 57\}$, of respective indices $k_{40} = 512$, $k_{41} = 512$ and $k_{57} = 480$, such that χ_i is an irreducible constituent of the permutation character $1_{Y_i}^H$.

Kim constructed for each $i \in \{40, 41, 57\}$ the permutation matrices $P(i, h)$, $P(i, p_2)$ and $P(i, q_2)$ of the three generators h, p_2 and q_2 of H_2 over the field $F = \mathrm{GF}(17)$, respectively. Then he applied the Meataxe algorithm implemented in MAGMA. Thus he obtained the three triples of diagonal block matrices $\mathcal{P}_2(i), \mathcal{Q}_2(i), \mathcal{H}(i) \in \mathrm{GL}_{n_i}(17)$ of the generators of $H_2 = \langle p_2, q_2, h \rangle$ describing the three n_i-dimensional irreducible constituents $\mathfrak{V}(i)$ corresponding to χ_i of the semi-simple representation $\mathfrak{V}$ of H_2, where $(n_{40}, n_{41}, n_{57}) = (96, 96, 160)$ by Table `DVD.1.4.4`.

The semi-simple faithful representation $\mathfrak{W}$ corresponding to the character $\tau = \tau_{\mathbf{9}} + \tau_{\mathbf{10}}$ decomposes into the two non-isomorphic irreducible constituents $\mathfrak{W}(9)$ and $\mathfrak{W}(10)$ corresponding to the irreducible characters $\tau_{\mathbf{9}}$ and $\tau_{\mathbf{10}}$, respectively. In order to get these two representations

Kim used in [73] the faithful permutation representation PD of $D = \langle p_1, q_1, x, y \rangle$ and the MAGMA command `LowIndexSubgroups(PD, k)` searching for conjugacy classes of subgroups X of index $|D : X| = k$. Thus he observed that the permutation characters $(1_{X_9})^D$ and $(1_{X_{10}})^D$ of the subgroups $X_9 = \langle (q_1 p_1^3)^3, (x q_1 p_1 q_1 x)^3, (x q_1 x p_1 q_1)^2 \rangle$ and $X_{10} = \langle (p_1 x q_1)^4, x q_1 x, (p_1 q_1 x p_1)^2 \rangle$ of index $k = 352$ contain τ_9 and τ_{10} as irreducible constituents, respectively. Applying then the Meataxe program of MAGMA he obtained two sets of four matrices in $\mathrm{GL}_{176}(17)$ of the generators of $D = \langle p_1, q_1, x, y \rangle$ describing the irreducible representations $\mathfrak{W}_9$ and $\mathfrak{W}_{10}$ over F.

Choose fixed bases $\mathcal{H}$ and $\mathcal{D}$ in the restrictions of $\mathfrak{V}$ to D_2 and $\mathfrak{W}$ to T, respectively, such that they are unions of bases of their six direct summands. Calculate then the matrices $\mathfrak{h} = \kappa_{\mathfrak{V}}(h)$, $\mathfrak{p}_2 = \kappa_{\mathfrak{V}}(p_2)$ and $\kappa_{\mathfrak{W}}(p_1)$, $\kappa_{\mathfrak{W}}(q_1)$, $\kappa_{\mathfrak{W}}(x)$, $\kappa_{\mathfrak{W}}(y)$ with respect to the bases $\mathcal{H}$ and $\mathcal{D}$, respectively.

By Proposition 6.2.1 there is a fixed isomorphism σ between the two groups D_2 and T satisfying $\sigma(p_2) = p_1$ and $\sigma(q_2) = q_1$. Therefore the two FT-modules $\mathfrak{V}_{|T}$ and $\mathfrak{W}_{|T}$ described by the pairs $(\mathfrak{p}_2, \mathfrak{q}_2)$ and $(\kappa_{\mathfrak{W}}(p_1), \kappa_{\mathfrak{W}}(q_1))$ of matrices in $\mathrm{GL}_{352}(17)$ are isomorphic. Let $Y = \mathrm{GL}_{352}(17)$. Applying now Parker's isomorphism test of Proposition 6.1.6 of [92] by means of the MAGMA command

```
IsIsomorphic(GModule(sub<Y|V(p2),V(q2)>),
  GModule(sub<Y|W(p1), W(q1)>))
```

one obtains the transformation matrix $\mathcal{T}_1$ satisfying $\mathfrak{p}_2 = \kappa_{\mathfrak{W}}(p_1)^{\mathcal{T}_1}$ and $\mathfrak{q}_2 = \kappa_{\mathfrak{W}}(q_1)^{\mathcal{T}_1}$.

By assertion (a) and Corollary 7.2.4 of [92] this transformation matrix $\mathcal{T}_1$ has to be multiplied by a diagonal matrix $\mathcal{D}$ of $\mathrm{GL}_{352}(17)$. In order to calculate its entries one has to deduce the composition factors of the restrictions $\chi_{i|D_2}$ and $\tau_{j|T}$ to D_2 and T, respectively. From the fusion and the three character Tables `DVD.1.4.4`, `DVD.1.4.3` and 6.6.2 it follows that:

$$\begin{aligned}
&\chi_{40}|_{D_2} = \psi_{27} + \psi_{63}, \quad \chi_{41}|_{D_2} = \psi_{26} + \psi_{59},\\
&\chi_{57}|_{D_2} == \psi_{70} + \psi_{73}, \quad \text{and}\\
&\tau_9|_T = \psi_{26} + \psi_{63} + \psi_{73}, \quad \tau_{10}|_T = \psi_{27} + \psi_{59} + \psi_{70}.
\end{aligned}$$

By Schur's Lemma and the degrees of these characters the following linear system in the variables $a, b, c, d, e, f \in F$ holds:

$$96 = 16b + 80d, \quad 96 = 16a + 80c, \quad 160 = 80e + 80f \quad \text{and}$$
$$176 = 16a + 80d + 80f, \quad 176 = 16b + 80c + 80e.$$

It has the following solution: $c = 8 - 7a$, $d = 8 - 7b$, $e = 7a - 7b + 1$ and $f = -7a + 7b + 1$, where a and b run through all non-zero elements of F. Hence the diagonal matrix $\mathcal{D}$ has the form

$$\mathcal{D}_{(a,b)} = diag(a^{16}, b^{16}, [8-7a]^{80}, [8-7b]^{80}, [7a-7b+1]^{80}, [-7a+7b+1]^{80})$$

for suitable elements $a, b \in F$. Let

$$\mathfrak{x}_{(b,d)} = \mathcal{D}_{(b,d)}^{-1} \mathcal{T}_1^{-1} \kappa_{\mathfrak{W}}(x) \mathcal{T}_1 \mathcal{D}_{(b,d)},$$

$$\mathfrak{y}_{(b,d)} = \mathcal{D}_{(b,d)}^{-1} \mathcal{T}_1^{-1} \kappa_{\mathfrak{W}}(y) \mathcal{T}_1 \mathcal{D}_{(b,d)}.$$

Then the field elements a, b have to be chosen so that all entries of the diagonal matrix $\mathcal{D}_{a,b}$ are non-zero and that the matrix group

$$\mathfrak{H}_{a,b} = \langle \mathfrak{p}_2, \mathfrak{q}_2, \mathfrak{h}, \mathfrak{x}_{a,b}, \mathfrak{y}_{a,b} \rangle$$

satisfies the Sylow 2-subgroup test of Algorithm 7.4.8, Step 5(c), of [92]. Running through all possible pairs $(a, b) \in F^2$ it follows that this test is only successful for the pair $(a, b) = (15, 9)$.

Let $\mathfrak{x} = \mathfrak{x}_{(15,9)}$, $\mathfrak{y} = \mathfrak{y}_{(15,9)}$ and $\mathfrak{H} = \langle \mathfrak{p}_2, \mathfrak{q}_2, \mathfrak{h}, \mathfrak{x}, \mathfrak{y} \rangle$. By Proposition 6.2.1(i) p_2 and q_2 are words in x and y. Hence $\mathfrak{H} = \langle \mathfrak{x}, \mathfrak{y}, \mathfrak{h} \rangle$. □

Proposition 6.2.3 (H. Kim) *In $\mathfrak{H} = \langle \mathfrak{x}, \mathfrak{y}, \mathfrak{h} \rangle \le \mathrm{GL}_{352}(17)$ constructed in Proposition 6.2.2 let $Mx = \mathfrak{x}$, $My = \mathfrak{y}$ and $Mh = \mathfrak{h}$. Furthermore, define the following:*

$$ma = (MxMyMx)^7, \quad mb = [(MhMy)^2 MhMy^3 Mh^2 My^3 MhMy]^7,$$
$$mc = (My^2 MxMyMxMy^3)^5,$$
$$md = (MhMyMh^2 MyMhMyMhMyMhMy^2 Mh^2)^{15},$$
$$me = (MyMxMy^5 Mx)^5,$$
$$mf = (MyMhMyMh^2 MyMh^2 My^2 MhMy^4 Mh^2)^5,$$
$$mg = (MxMy^2 MxMy^3 Mx)^7, \quad mh = (My^5 MxMyMx)^5,$$
$$mi = (Mh^2 My^2 MhMyMh^2)^7 \quad \text{and}$$
$$mj = (mc \cdot md \cdot me \cdot mh)^3, \quad mk = (mc \cdot md \cdot me \cdot mf \cdot mg)^2,$$
$$ml = (ma \cdot mb \cdot mc \cdot md \cdot me \cdot mh)^5,$$

$$mo = (ml \cdot mb \cdot mk \cdot mj \cdot mb \cdot mi)^6,$$
$$mp = (mj \cdot mk \cdot mj \cdot ml \cdot mb \cdot mi \cdot mj \cdot mi)^5,$$
$$mr = mf \cdot me \cdot mg \cdot mf, \quad mt = (mc \cdot md \cdot me \cdot mh \cdot mi)^4,$$
$$mu = mb \cdot ma \cdot mc \cdot mb, \quad mw = md \cdot me \cdot mh \cdot md.$$

Then the following statements hold.

(a) *The finitely presented group* $H = \langle a, b, c, d, e, f, g, h, i, z\rangle$ *satisfying the following set* $\mathcal{R}(H)$ *of defining relations:*

$$\begin{aligned}
&a^2 = b^2 = c^2 = d^2 = e^2 = f^2 = g^2 = h^2 = i^2 = z^2 = 1,\\
&(z,a) = (z,b) = (z,c) = (z,d) = (z,e) = 1,\\
&(z,f) = (z,g) = (z,h) = (z,i) = 1,\\
&(ab)^3 = (cd)^3 = (de)^3 = 1, \quad (bc)^3 = (ef)^3 = (fg)^3 = z,\\
&(ac)^2 = (ad)^2 = (ae)^2 = (af)^2 = (ag)^2 = (ah)^2 = (ai)^2 = 1,\\
&(bd)^2 = (be)^2 = (bf)^2 = (bg)^2 = (bh)^2 = (bi)^2 = 1,\\
&(ce)^2 = (cf)^2 = (cg)^2 = (ch)^2 = (ci)^2 = 1,\\
&(df)^2 = (dg)^2 = (eg)^2 = (eh)^2 = (ei)^2 = (dh)^3 = (hi)^3 = 1,\\
&(di)^2 = (fh)^2 = (fi)^2 = (gh)^2 = (gi)^2 = 1,\\
&(dcbdefdhi)^{10} = (abcdefh)^9 = (bcdefgh)^9 = 1,
\end{aligned}$$

has a faithful permutation representation of degree 28160 *with stabilizer* $U = \langle bz, c, d, e, fz, g, h, i\rangle$.

(b) $\mathfrak{H} = \langle Mx, My, Mh\rangle$ *is isomorphic to the finitely presented group* $H = \langle a, b, c, d, e, f, g, h, i, z\rangle$. *Furthermore,*

$$Mx = (R_1R_2R_3)^3,$$
$$My = (S_1S_2S_4S_1S_4S_2S_4)^5(T_1T_3^2T_1T_3T_1T_2T_1T_3)^{20},$$
$$Mh = [(mp \cdot mo \cdot mj \cdot mo \cdot mk)^4 \cdot (mj \cdot mk \cdot mo \cdot mk^2 \cdot mp)^4]^4,$$
where $R_1 = (mw \cdot mu \cdot mt \cdot mu \cdot mr \cdot mt \cdot mr \cdot mt)^4,$
$$R_2 = (mw \cdot mt \cdot mw \cdot mu \cdot mw \cdot mr \cdot mt \cdot mr)^4,$$
$$R_3 = (mr \cdot mw \cdot mt \cdot mu \cdot mr \cdot mw \cdot mt \cdot mr)^2,$$
$$S_1 = mu \cdot mw \cdot mu, \quad S_2 = (mu \cdot mw \cdot mr)^2,$$
$$S_3 = (mu \cdot mw \cdot mt \cdot mu)^2, \quad S_4 = (mu \cdot mt \cdot mw \cdot mr)^4,$$
$$T_1 = S_2S_4S_2^2, \quad T_2 = (S_4S_2S_1S_3S_4)^2, \quad T_3 = (S_4S_2S_4S_2S_1S_3)^2.$$

(c) $\mathfrak{H}$ *has a center* $Z(H) = \langle (Mx \cdot My^2)^7 \rangle$ *of order* 2 *and* $\mathfrak{H}/Z(\mathfrak{H}) \cong \mathsf{Fi}_{22}$.

(d) $|\mathfrak{H}| = 2^{18} \cdot 3^9 \cdot 5^2 \cdot 7 \cdot 11 \cdot 13$.

Proof (a) and (b) Corollary 5.2.2 contains a presentation for Fischer's simple group Fi_{22}. Let $G = \langle a, b, c, d, e, f, g, h, i \rangle$ be the finitely presented group with the set of defining relations $\mathcal{R}(G)$ given there. Its subgroup $Q = \langle a, c, d, e, f, g, h, i, (abcdeh)^5 \rangle$ has index 3510 by [110], p. 110. This has also been checked by means of the MAGMA command `CosetAction(G,Q)`. Since G is simple the corresponding permutation representation PG of G is faithful.

Let $D = \langle x, y \rangle$ be the group defined in Proposition 6.2.1(b). Let $z = (xy^2)^7$ and $D_1 = D/\langle z \rangle$. Denote $x' = x\langle z \rangle$ and $y' = y\langle z \rangle$ in D_1. By the proof of Corollary 5.2.2 and Proposition 6.2.1(d) we know that $D_1 \cong 2^{10} : \mathsf{M}_{22} \cong M$, where the subgroup $M = \langle a, c, e, g, h, u, v, w, r, s, t \rangle$ of G is generated by the given generators a, c, e, g, h and the words $u = bacb$, $v = dced$, $w = dehd$, $r = fegf$, $s = dchd$ and $t = (cdehi)^4$. In particular, M has order $2^{17} \cdot 3^2 \cdot 5 \cdot 7 \cdot 11$.

Let PD_1 be the faithful permutation representation of D_1 of degree 1024 given in the proof of Proposition 6.2.1(d). Using it, PG and the MAGMA command `IsIsomorphic(D_1,M)` Kim found in [73] an isomorphism $\alpha : D_1 \to M$ such that

$$x_0 = \alpha(x') = (r_1 r_2 r_3)^3$$

and

$$y_0 = \alpha(y') = (s_1 s_2 s_4 s_1 s_4 s_2 s_4)^5 (t_1 t_3^2 t_1 t_3 t_1 t_2 t_1 t_3)^{20},$$

where

$$\begin{aligned} r_1 &= (wuturtrt)^4, \quad r_2 = (wtwuwrtr)^4, \quad r_3 = (rwturwtr)^2, \\ s_1 &= uwu, \quad s_2 = (uwr)^2, \quad s_3 = (uwtu)^2, \quad s_4 = (utwr)^4, \\ t_1 &= s_2 s_4 s_2^2, \quad t_2 = (s_4 s_2 s_1 s_3 s_4)^2, \quad t_3 = (s_4 s_2 s_4 s_2 s_1 s_3)^2. \end{aligned}$$

In $D_1 = \langle x', y' \rangle$ let $t' = (x'y'x'y'^3)^6$ and $T' = C_{D_1}(t')$. By Proposition 6.2.1(j) T' is generated by two well determined words p_1' and q_1' in x' and y'. Furthermore, T' has order $2^{17} \cdot 3 \cdot 5$. Hence t' is a 2-central involution of D_1. Therefore $t_0 = \alpha(t')$ is a 2-central involution of G. As $G \cong \mathsf{Fi}_{22}$ Table `DVD.1.3.1` implies that $H' = C_G(t')$ has order $2^{17} \cdot 3^4 \cdot 5$. Using the program `GetShortGens(G,H')` Kim showed in [73] that $H' = \langle j, k, o, p \rangle$, where $j = (cdeh)^3$, $k = (cdefg)^2$, $l = (abcdeh)^5$, $o = (lbkjbi)^6$, $p = (jkjlbiji)^5$, $v_1 = (pojok)^4$ and $v_2 = (jkok^2p)^4$.

The finitely presented group $H_1 = \langle h_i \mid 1 \le i \le 14\rangle$ given in Proposition 5.1.1(n) has a faithful permutation representation PH_1 of degree 1024 by Lemma 5.1.2. Proposition 6.2.1(f), (h) and (i) state that H_1 is generated by $p_2' = h_1h_2$, $q_2' = h_1(h_1h_5h_4h_2)^5$ and $h' = h_4$. Using the permutation representations PG and PH_1 and the MAGMA command `IsIsomorphic(H_1,H')` Kim has found an isomorphism $\beta : H_1 \to H'$ such that $\beta(p_2') = \alpha(p_1')$, $\beta(q_2') = \alpha(q_1')$ and $h_0 = \beta(h') = (v_1v_2)^4$. By another application of PG and MAGMA he verified that $G = \langle H', \alpha(D')\rangle = \langle \alpha(p_1), \alpha(q_1), h_0, x_0, y_0\rangle = \langle x_0, y_0, h_0\rangle$. Furthermore, $H' \cap \alpha(D_1) = \alpha(T')$. In particular, the finitely presented group G is an epimorphic image of the matrix group $\mathfrak{H}$.

In order to find a nice presentation for $H = \langle x, y, h\rangle$, Kim determined short words for the original generators of G in terms of its new generators x_0, y_0 and h_0 by means of the faithful permutation representation PG, MAGMA and Algorithm 1.4.3. Thus he showed that

$$\begin{aligned}
&a = (x_0y_0x_0)^7, \quad b = (h_0y_0h_0y_0h_0y_0^3h_0^2y_0^3h_0y_0)^7, \quad c = (y_0^2x_0y_0x_0y_0^3)^5,\\
&d = (h_0y_0h_0^2y_0h_0y_0h_0y_0h_0y_0^2h_0^2)^{15}, \quad e = (y_0x_0y_0^5x_0)^5,\\
&f = (y_0h_0y_0h_0^2y_0h_0^2y_0^2h_0y_0^4h_0^2)^5, \quad g = (x_0y_0^2x_0y_0^3x_0)^7,\\
&h = (y_0^5x_0y_0x_0)^5, \quad i = (h_0^2y_0^2h_0y_0h_0^2)^7.
\end{aligned}$$

The matrices ma, mb, ..., mi of $\mathfrak{H} = \langle Mx, My, Mh\rangle \le \mathrm{GL}_{352}(17)$ given in the statement are obtained from these equations by replacing x_0, y_0 and h_0 by the matrices Mx, My and Mh, respectively.

Let $mz = (MxMy^2)^7$ in H. Kim checked in [73] that mz commutes with the three generators Mx, My, Mh of $\mathfrak{H}$. He also verified that the matrices ma, mb, ..., mi and mz satisfy the relations of statement (a) by evaluating them in $\mathrm{GL}_{782}(17)$. In particular, $(mbmc)^3 = (memf)^3 = (mfmg)^3 = mz$, and all other relations have value 1 in $\mathrm{GL}_{782}(17)$.

Clearly, $\mathfrak{H} = \langle Mx, My, Mh\rangle \ge \langle ma, mb, mc, md, me, mf, mg, mh, mi\rangle$. In [73] Kim also verified in the matrix group H that $Mx = (mr_1mr_2\, mr_3)^3$, $My = (ms_1ms_2ms_4ms_1ms_4ms_2ms_4)^5(mt_1mt_3^2mt_1mt_3mt_1mt_2\, mt_1mt_3)^{20}$ and $Mh = (mv_1mv_2)^4$, where the matrices ms_i, mt_j and mv_k are constructed from the words s_i, t_k and v_k given above by inserting the matrices ma, mb, ..., mi in place of the original generators $a, b, \dots, i$, respectively. Hence $\mathfrak{H} = \langle ma, mb, mc, md, me, mf, mg, mh, mi\rangle$ and $\mathfrak{H} \cong H$.

By [110], [19] and Corollary 5.2.2 it is known that the subgroup $U_1 = \langle b, c, d, e, f, g, h, i\rangle$ of $H_1 \cong \mathsf{Fi}_{22}$ has index 14080. Let $\epsilon_i \in \{1, 2\}$ for $1 \le i \le 8$. Using the MAGMA command `CosetAction(H,` sub<H|U_{\delta}>) Kim calculated for all possible 2^8 subgroups

$$U_\delta = \langle bz^{\epsilon_1}, cz^{\epsilon_2}, dz^{\epsilon_3}, ez^{\epsilon_4}, fz^{\epsilon_5}, gz^{\epsilon_6}, hz^{\epsilon_7}, iz^{\epsilon_8} \rangle$$

their indices in the finitely presented group $H = \langle a, b, c, d, e, f, g, h, i, z \rangle$ defined in (a). Thus he found that $[1, 2, 2, 2, 1, 2, 2, 2]$ is the only 8-tuple δ of the ϵ_i for which this index is 28160. In all other cases δ' the involution z belongs to the subgroup $U_{\delta'}$. Hence $U = \langle bz, c, d, e, fz, g, h, i \rangle$ is the stabilizer of a faithful permutation representation PH of H with degree 28160 and $|H| = 2^{18} \cdot 3^9 \cdot 5^2 \cdot 7 \cdot 11 \cdot 13$.

(c) and (d) By the proof of (a) we know that $mz = (MxMy^2)^7$ belongs to the center $Z(H)$ of H. Setting $z = 1$ in the presentation of H given in (a) it follows from (a) and Corollary 5.2.2 that $H/\langle mz \rangle \cong \mathsf{Fi}_{22}$. Theorem 5.2.1(d) states that Fi_{22} is a simple group. Hence $Z(H) = \langle mz \rangle$. This completes the proof. □

In the following the generators x and y of $D = C_E(z)$ constructed in Lemma 6.1.2 will be denoted by x_1 and y_1, respectively. Thus $z = (xy^2)^7$ has to be replaced with $z_1 = (x_1 {y_1}^2)^7$ and $D = C_E(z_1)$.

Corollary 6.2.4 *Let $D = \langle x_1, y_1 \rangle = C_E(z_1)$ be the centralizer of the 2-central involution z_1 of the extension group $E = \langle x_1, y_1, e \rangle$ constructed in Lemma 6.1.2. Let $H = \langle a, b, c, d, e, f, g, h, i, z \rangle$ be the finitely presented group defined in statement (a) of Proposition 6.2.3. Then there are three words x, y and h in the original generators $a, b, \ldots, i$ of H such that the following statements hold.*

(a) *$H = \langle x, y, h \rangle$, and $z = (xy^2)^7$ generates the center $Z(H)$ of H.*
(b) *$D_H = \langle x, y \rangle$ has order $2^{18} \cdot 3^2 \cdot 5 \cdot 7 \cdot 11$, and $H = \langle D_H, h \rangle$.*
(c) *H has an irreducible representation $\kappa : H \to \mathrm{GL}_{352}(17)$ such that $\kappa(x) = \mathfrak{x}$, $\kappa y = \mathfrak{y}$ and $\kappa(h) = \mathfrak{h}$ generate the subgroup $\mathfrak{H}$ of $\mathrm{GL}_{352}(17)$ constructed in Proposition 6.2.2.*
(d) *There is an isomorphism $\delta : D_H \to D$ such that $D = \langle x_1 = \delta(x), y_1 = \delta(y) \rangle$.*
(e) *H has a Sylow 2-subgroup S having a maximal elementary abelian normal subgroup A of order 2^{11} such that $N_H(A) = \langle x, y \rangle = D_H$.*
(f) *A system of representatives h_i of the 114 conjugacy classes of $H = \langle x, y, h \rangle$ and the corresponding centralizer orders $|C_H(h_i)|$ are given in Table 6.5.2.*
(g) *The character table of H is stated in Table 6.6.3.*

Proof (a) By Proposition 6.2.3 and its proof there are words R_i, S_j and V_k in the original generators $a, b, \ldots, i$ of H such that $x = (R_1R_2R_3)^3$, $y = (S_1S_2S_4S_1S_4S_2S_4)^5(S_5S_7^2S_5S_7S_5S_6S_5S_7)^{20}$ and $h = (V_1V_2)^4$. Furthermore, it has been shown there that $H = \langle x, y, h\rangle$ and that its center $Z(H)$ is generated by $z = (xy^2)^7$.

(b) Using the faithful permutation representation PH of degree 28160 stated in Proposition 6.2.3(a) it has been checked in [73] that $D_H = \langle x, y\rangle$ has order $2^{18} \cdot 3^2 \cdot 5 \cdot 7 \cdot 11$ and $H = \langle D_H, h\rangle$.

(c) There it is also shown that the map $\kappa : H \to \mathfrak{H}$ defined by $\kappa(x) = Mx = \mathfrak{x}$, $\kappa y = My = \mathfrak{y}$, $\kappa(h) = Mh = \mathfrak{h}$ defines an irreducible representation κ of H of dimension 352 over $F = GF(17)$ with image $\kappa(H) = \mathfrak{H}$ constructed in Proposition 6.2.2.

(d) Proposition 6.2.1(c) states that $D = C_E(z_1)$ has a unique normal subgroup V_1 of order 2^{12}. It is the unique elementary abelian normal subgroup of any Sylow 2-subgroup S_1 of D. By Lemma 6.1.2 D is generated by elements x_1 and y_1 of respective orders 2 and 14, and $z_1 = (x_1y_1^2)^7$. Furthermore, D has a faithful permutation representation PD of degree 1012. Using the faithful permutation representations PH and PD it has been verified in [73] that the map $\delta : D_H \to D$ defined by $\delta(x) = x_1$ and $\delta(y) = y_1$ is an isomorphism.

(e) In particular, D_H and D have isomorphic Sylow 2-subgroups. Therefore $A = \delta^{-1}(V_1)$ is the unique maximal elementary abelian normal subgroup of any Sylow 2-subgroup S of D_H containing it. Furthermore, A is normal in D_H. By another application of PH and MAGMA the reader can check that $N_H(A)$ has the same order as D_H. Hence $N_H(A) = D_H \cong D$.

(f) The system of representatives of the conjugacy classes of H has been calculated by Kim in [73] by means of the permutation representation PH given in Proposition 6.2.3, MAGMA and Kratzer's Algorithm 5.3.18 of [92].

(g) The character table of H has been computed in [73] with PH and MAGMA. □

Definition 6.2.5 A finite simple group G is said to be of Fi_{23}*-type* if it has a 2-central involution z such that $C_G(z)$ is isomorphic to the finitely presented group H stated in Proposition 6.2.1.

6.3 Construction of Fischer's simple group Fi_{23}

The amalgam $H \leftarrow D \rightarrow E$ constructed in Corollary 6.2.4 satisfies the main condition of Algorithm 1.3.8. Therefore Algorithm 7.4.8 of [92] can be applied to give an existence proof for Fischer's sporadic group Fi_{23}. It is due to H. Kim; see [73].

Theorem 6.3.1 (H. Kim) *Keep the notations of Propositions 6.2.1 and 6.2.2. Let $H \leftarrow D \rightarrow E$ be the amalgam constructed in Corollary 6.2.4 with isomorphism $\delta : D_H = \langle x, y\rangle \rightarrow D = C_E(z_1) = \langle x_1, y_1\rangle$, where $\delta(x) = x_1$, $\delta(y) = y_1$ and $z_1 = (x_1y_1^2)^7$. Let F^* be the multiplicative group of the prime field $F = GF(17)$. Let $Y = \mathrm{GL}_{782}(17)$.*

Using the notations of the three character Tables 6.6.3, 6.6.2 and 6.6.1 of the groups H, D and E, respectively, the following statements hold.

(a) *There is a compatible pair $(\chi, \tau) \in mf\,char_{\mathbb{C}}(H) \times mf\,char_{\mathbb{C}}(E)$ of degree 782 of the groups $H = \langle x, y, h\rangle$ and $E = \langle D, e\rangle$ such that*

$$(\chi, \tau) = (\chi_1 + \chi_{\mathbf{3}} + \chi_4, \tau_1 + \tau_2 + \tau_{\mathbf{10}} + \tau_{\mathbf{11}})$$

and

$$\tau_{|D} = \chi_{|D} = 2 \times \psi_1 + \psi_2 + \psi_6 + \psi_{\mathbf{9}} + \psi_{\mathbf{10}} + \psi_{15} \in char_{\mathbb{C}}(D),$$

where irreducible characters with bold face indices denote faithful irreducible characters and the characters ψ_1, ψ_6, ψ_9, ψ_{10} and ψ_{15} have respective degrees 1, 21, 77, 176, 176 and 330.

(b) *Let $\mathfrak{V}$ and $\mathfrak{W}$ be the up to isomorphism uniquely determined faithful semi-simple multiplicity-free 782-dimensional modules of H and E over F corresponding to the compatible pair (χ, τ), respectively. Let $\kappa_{\mathfrak{V}} : H \rightarrow \mathrm{GL}_{782}(17)$ and $\kappa_{\mathfrak{W}} : E \rightarrow \mathrm{GL}_{782}(17)$ be the representations of H and E afforded by the modules $\mathfrak{V}$ and $\mathfrak{W}$, respectively.*

Let $\mathfrak{h} = \kappa_{\mathfrak{V}}(h)$, $\mathfrak{x} = \kappa_{\mathfrak{V}}(x)$, $\mathfrak{y} = \kappa_{\mathfrak{V}}(y)$ in $\kappa_{\mathfrak{V}}(H) \le \mathrm{GL}_{782}(17)$. Then $\mathfrak{V}_{|D} \cong \mathfrak{W}_{|D}$, and there is a transformation matrix $\mathcal{T} \in \mathrm{GL}_{782}(17)$ such that

$$\mathfrak{x} = \mathcal{T}^{-1}\kappa_{\mathfrak{W}}(\delta(x))\mathcal{T}, \qquad \mathfrak{y} = \mathcal{T}^{-1}\kappa_{\mathfrak{W}}(\delta(y))\mathcal{T}.$$

(c) *Let $\mathfrak{D} = \langle \mathfrak{x}, \mathfrak{y}\rangle$ and $\mathfrak{H} = \langle \mathfrak{x}, \mathfrak{y}, \mathfrak{h}\rangle$. Let $\mathcal{D} = C_Y(\mathfrak{D})$ and $\mathcal{H} = C_Y(\mathfrak{H})$. Let $\mathfrak{e}_1 = \mathcal{T}^{-1}\kappa_{\mathfrak{W}}(e)\mathcal{T}$. Let $\mathfrak{E} = \langle \mathfrak{D}, \mathfrak{e}_1\rangle$ and $\mathcal{E} = C_Y(\mathfrak{E})$. Then the following statements hold.*

   (1) *There is an isomorphism*

$$\alpha : \mathcal{D} \rightarrow \mathcal{D}_1 = \mathrm{GL}_2(17) \times F^{*5} \le \mathrm{GL}_7(17).$$

(2) $\mathcal{H}_1 = \alpha(\mathcal{H})$ *is generated by the three diagonal matrices*

$$a_1 = diag(3,1,1,1,1,1,1), a_2 = diag(1,1,1,1,3,3,1),$$
$$a_3 = diag(1,3,3,3,1,1,3).$$

(3) $\mathcal{E}_1 = \alpha(\mathcal{E})$ *is generated by the four diagonal matrices*

$$b_1 = a_1, b_2 = diag(1,3,3,1,1,1,1),$$
$$b_3 = diag(1,1,1,3,3,1,1),$$
$$b_4 = diag(1,1,1,1,1,3,3).$$

(4) $\mathcal{D}$ *has* 321×16 $\mathcal{H}$-$\mathcal{E}$ *double cosets.*

(5) *The free product* $H *_D E$ *of* H *and* E *with amalgamated subgroup* D *has exactly one irreducible* 782*-dimensional representation over* F *whose Sylow* 2*-subgroups have the same exponent as the ones of* H*. It corresponds to the* $\mathcal{H}$-$\mathcal{E}$ *double coset representative*

$$\mathcal{F} = diag(u, 1^{21}, 1^{77}, 1^{176}, 1^{176}, 16^{330}) \in \mathrm{GL}_{782}(17),$$

where

$$u = \begin{pmatrix} 1 & 1 \\ 9 & 14 \end{pmatrix}.$$

Let $\mathfrak{e} = \mathcal{F}^{-1}\mathfrak{e}_1\mathcal{F}$ *and* $\mathfrak{G} = \langle \mathfrak{x}, \mathfrak{y}, \mathfrak{h}, \mathfrak{e} \rangle$*. The four generating matrices of* $\mathfrak{G}$ *are stored on the accompanying DVD.*

(d) $\mathfrak{G}$ *has a faithful permutation representation* $P\mathfrak{G}$ *of degree* 31671 *with stabilizer* $\mathfrak{H} = \langle \mathfrak{x}, \mathfrak{y}, \mathfrak{h} \rangle$.

(e) $\mathfrak{G}$ *is a finite simple group of order* $2^{18} \cdot 3^{13} \cdot 5^2 \cdot 7 \cdot 11 \cdot 13 \cdot 17 \cdot 23$ *with centralizer* $C_{\mathfrak{G}}(\mathfrak{z}) = \mathfrak{H} \cong H$ *of the involution* $\mathfrak{z} = (\mathfrak{x}\mathfrak{y}^2)^7$.

(f) $\mathfrak{G}$ *has* 98 *conjugacy classes* $\mathfrak{g}_i^{\mathfrak{G}}$ *with representatives* $\mathfrak{g}_i$ *and centralizer orders* $|C_{\mathfrak{G}}(\mathfrak{g}_i)|$ *as given in Table 6.5.4.*

(g) *The character table of* $\mathfrak{G}$ *coincides with that of* Fi_{23} *in the Atlas [19], pp. 178–179.*

Proof (a) Let $H \leftarrow D \rightarrow E$ be the amalgam constructed in Corollary 6.2.4. The character tables of the groups H, D and E are stated in Section 6.6. In the following we use their notations. Using the character tables of H, D, and E, and the fusion of the classes of D_H in H and D in E, Kim found the compatible pair (χ, τ) stated in (a) by an application of Kratzer's Algorithm 7.3.10 of [92] and MAGMA; see [73].

(b) Let $\mathfrak{V}$ be the semi-simple faithful representation over F corresponding to the character $\chi = \chi_1 + \chi_3 + \chi_4$ of H. It splits into three direct summands $\mathfrak{V}_k$ corresponding to the irreducible constituents χ_k of χ; $\mathfrak{V}_1$ is the trivial representation of H. By Proposition 6.2.2 the three generating matrices $\mathfrak{x}_3$, $\mathfrak{y}_3$ and $\mathfrak{h}_3$ in $\mathrm{GL}_{352}(17)$ describing the representation $\mathfrak{V}_3$ of H are known. Thus it remains to construct $\mathfrak{V}_4$.

In [73] Kim calculated the inner product of χ_4 with the permutation character of the faithful permutation representation $(1_T)^H$ of H of degree 28160 determined in Proposition 6.2.2(d). Since it is 1 he calculated the permutation matrices of the three generators h, x and y of H over the field $F = GF(17)$. Applying then the Meataxe algorithm implemented in MAGMA Kim obtained the matrices $\mathfrak{x}_4$, $\mathfrak{y}_4$ and $\mathfrak{h}_4$ in $\mathrm{GL}_{429}(17)$ describing the irreducible representation $\mathfrak{V}_4$ of H.

Let $\mathfrak{W}$ be the semi-simple faithful representation over F corresponding to the character $\tau = \tau_1 + \tau_2 + \tau_{10} + \tau_{11}$ of E. It splits into four direct summands $\mathfrak{W}_j$ corresponding to the irreducible constituents τ_j of τ. Clearly $\mathfrak{W}_1$ is the trivial representation of E. By Lemma 6.1.2 the group E has a faithful permutation representation PE of degree 1012. Calculating inner products of its permutation character with the non-trivial irreducible characters τ_j Kim observed in [73] that each of them is an irreducible constituent of PE. He constructed the permutation matrices of the three generators $x_1 = \delta(x)$, $y_1 = \delta(y)$ and e_1 of E over the field F. Applying then the Meataxe algorithm implemented in MAGMA he obtained the matrices $\mathfrak{x}_{1,j}$, $\mathfrak{y}_{1,j}$ and $\mathfrak{e}_{1,j}$ in $\mathrm{GL}_{m_j}(17)$ describing the irreducible representations $\mathfrak{W}_j$ of E, where $m_2 = 22$, $m_{10} = 253$ and $m_{11} = 506$ by Table 6.6.1. Now it is easy to construct the three generating matrices $\mathfrak{x}_1$, $\mathfrak{y}_1$ and $\mathfrak{e}_1$ describing the 782-dimensional semi-simple representation $\mathfrak{W}$ of E as blocked diagonal matrices.

The two FD-modules $\mathfrak{V}_{|D}$ and $\mathfrak{W}_{|D}$ described by the pairs $(\mathfrak{V}(x), \mathfrak{V}(y))$ and $(\mathfrak{W}(x_1), \mathfrak{W}(y_1))$ of matrices in $\mathrm{GL}_{782}(17)$ are isomorphic by construction. Let $Y = \mathrm{GL}_{782}(17)$. Applying then Parker's isomorphism test of Proposition 6.1.6 of [92] by means of the MAGMA command

```
IsIsomorphic(GModule(sub<Y|V(x),V(y)>),GModule(sub<Y|W(x1),
  W(y1)>))
```

one obtains the transformation matrix $\mathcal{T} \in \mathrm{GL}_{782}(17)$ satisfying $\mathfrak{V}(x) = (\mathfrak{W}(x_1))^{\mathcal{T}}$ and $\mathfrak{V}(y) = (\mathfrak{W}(y_1))^{\mathcal{T}}$; see [73]. Let $\mathfrak{e}_1 == \mathcal{T}^{-1}\kappa_{\mathfrak{W}}(e_1)\mathcal{T}$. Then $\mathfrak{E} = \langle \mathfrak{D}, \mathfrak{e}_1 \rangle$ and $\mathfrak{H} \cap \mathfrak{E} = \mathfrak{D} = \langle \mathfrak{x}, \mathfrak{y} \rangle$.

(c) Let $Y = \mathrm{GL}_{782}(17)$, $\mathcal{D} = C_Y(\mathfrak{D})$, $\mathcal{H} = C_Y(\mathfrak{H})$ and $\mathcal{E} = C_Y(\mathfrak{E})$. By (a) the restrictions of the compatible characters (χ, τ) are not multiplicity free. Therefore Theorem 7.2.2 of [92] asserts that one has to determine the $\mathcal{H}$-$\mathcal{E}$ double cosets of $\mathcal{D}$ in order to find all the suitable representations of degree 782 of the free product $H *_D E$ with amalgamated subgroup D. Since α is an isomorphism it suffices to determine the $\mathcal{H}_1$-$\mathcal{E}_1$ double cosets of $\mathcal{D}_1$.

For each integer k let v^k denote the diagonal matrix of $\mathrm{GL}_k(17)$ having all diagonal entries equal to $v \in F^*$.

By (a), Table 6.6.2 and the Theorem of Artin–Wedderburn we know that each element $\mathcal{V}$ of $\mathcal{D}$ can be represented as a blocked diagonal matrix given by

$$\mathcal{V} = diag(a, b^{21}, c^{77}, d^{176}, e^{176}, f^{330}) \in \mathrm{GL}_{782}(17),$$

where $a \in \mathrm{GL}_2(17)$ and all $b, c, d, e, f \in F^*$ are uniquely determined by $\mathcal{V}$.

The map $\alpha : \mathcal{D} \to \mathcal{D}_1 = \mathrm{GL}_2(17) \times F^{*5}$ defined by

$$\alpha(\mathcal{V}) = diag(a, b, c, d, e, f) \in \mathrm{GL}_7(17)$$

is an isomorphism. Thus assertion (1) of (c) holds.

Let $\mathcal{H}_1 = \alpha(\mathcal{H})$, $\mathcal{E}_1 = \alpha(\mathcal{E})$. From the fusion of the conjugacy classes of $D_H = \langle x, y\rangle$ in $H = \langle D_H, h\rangle$ and the two character Tables 6.6.2 and 6.6.3 it follows that:

$$\chi_1|_{D_H} = \psi_1, \quad \chi_{\mathbf{3}}|_{D_H} = \psi_{\mathbf{9}} + \psi_{\mathbf{10}}$$

and

$$\chi_4|_{D_H} = \psi_1 + \psi_2 + \psi_6 + \psi_{15}.$$

Similarly one verifies the following restrictions of the irreducible constituents of $\tau \in mf\mathrm{char}_{\mathbb{C}}(E)$ to the subgroup D of E:

$$\tau_1|_D = \psi_1, \quad \tau_2|_D = \psi_1 + \psi_2,$$
$$\tau_{\mathbf{10}}|_D = \psi_6 + \psi_{\mathbf{9}}, \quad \tau_{\mathbf{11}}|_D = \psi_{\mathbf{10}} + \psi_{15}.$$

Since $3 \in F$ generates the multiplicative group F^* of F it follows from these restrictions that $\mathcal{H}_1 = \langle a_1, a_2, a_3\rangle$ and $\mathcal{E}_1 = \langle b_1, b_2, b_3, b_4\rangle$, where the generators a_i and b_j denote the diagonal matrices of $\mathrm{GL}_7(17)$ given in assertions (2) and (3) of statement (c), respectively.

Since α is an isomorphism it suffices to determine the $\mathcal{H}_1$-$\mathcal{E}_1$ double cosets of $\mathcal{D}_1$. Let $\mathcal{A}$ and $\mathcal{B}$ be the direct factors $\mathrm{GL}_2(17)$ and F^{*5} of $\mathcal{D}_1$, respectively. Let $\mathcal{Z}$ be the center of $\mathcal{A}$ and $\mathcal{C} = \mathcal{Z} \times \mathcal{B}$. Then $\mathcal{C}$ is a normal subgroup of $\mathcal{D}_1$ and $\mathcal{H}_1\mathcal{C}/\mathcal{C} = \mathcal{E}_1\mathcal{C}/\mathcal{C} = \{m_a\mathcal{Z} \in \mathcal{A}/\mathcal{Z} | a \in F^*\}$, where

$$m_a = \begin{pmatrix} a & 0 \\ 0 & 1 \end{pmatrix}.$$

Let $\mathcal{A}_1 = \{m_a | a \in F^*\}$. Then the following 321 matrices of $A = \mathrm{GL}_2(17)$ yield a complete list of all $\mathcal{A}_1$-$\mathcal{Z}$ double coset representatives in $\mathcal{A}$:

$$d = \begin{pmatrix} 0 & 1 \\ 1 & 0 \end{pmatrix}, \quad u_c = \begin{pmatrix} 1 & 1 \\ c & 0 \end{pmatrix}, \quad v_c = \begin{pmatrix} 1 & 0 \\ 1 & c \end{pmatrix}, \quad w_c = \begin{pmatrix} 1 & 1 \\ 0 & c \end{pmatrix}, \quad x_c = \begin{pmatrix} 0 & 1 \\ 1 & c \end{pmatrix},$$
$$y_c = \begin{pmatrix} 1 & 0 \\ 0 & c \end{pmatrix}, \quad z_{e,f} = \begin{pmatrix} 1 & 1 \\ e & f \end{pmatrix},$$

where c, e and f run through $GF(17)^*$, but $e \neq f$.

Let $\mathcal{H}_2$ and $\mathcal{E}_2$ be the projections of $\mathcal{H}_1$ and $\mathcal{E}_2$ into $\mathcal{B}$, respectively. Let $\mathcal{X} = \mathcal{H}_1\mathcal{E}_1$. Then

$$|\mathcal{D}_1 : \mathcal{X}\mathcal{A}| = |\mathcal{D}_1/\mathcal{A} : \mathcal{X}\mathcal{A}/\mathcal{A}| = |\mathcal{B} : \mathcal{H}_2\mathcal{E}_2| = 16$$

and the cosets of $\mathcal{H}_2\mathcal{E}_2$ in $\mathcal{B}$ can be represented by the 16 tuples

$$\{(1,1,1,1,f) | f \in F^*\}.$$

Since $\mathcal{B}$ is abelian $\mathcal{H}_2$-$\mathcal{E}_2$ double cosets in $\mathcal{B}$ are cosets of the subgroup $\mathcal{H}_2\mathcal{E}_2$ of $\mathcal{B}$.

In particular, there are exactly 321×16 $\mathcal{H}_1$-$\mathcal{E}_1$ double cosets in $\mathcal{D}_1$. They are represented by the 82176 matrices

$$R(m,f) = \begin{pmatrix} m_{1,1} & m_{1,2} & 0 & 0 & 0 & 0 & 0 \\ m_{2,1} & m_{2,2} & 0 & 0 & 0 & 0 & 0 \\ 0 & 0 & 1 & 0 & 0 & 0 & 0 \\ 0 & 0 & 0 & 1 & 0 & 0 & 0 \\ 0 & 0 & 0 & 0 & 1 & 0 & 0 \\ 0 & 0 & 0 & 0 & 0 & 1 & 0 \\ 0 & 0 & 0 & 0 & 0 & 0 & f \end{pmatrix}$$

of $\mathrm{GL}_7(17)$, where f runs through all elements of F^* and

$$m = \begin{pmatrix} m_{1,1} & m_{1,2} \\ m_{2,1} & m_{2,2} \end{pmatrix}$$

runs through the 321 matrices of $\mathcal{A} = \mathrm{GL}_2(17)$ described above.

Let $\mathfrak{e}_{m,f} = [\alpha^{-1}(R(m,f))]^{-1}\mathfrak{e}_1[\alpha^{-1}(R(m,f))]$ and

$$\mathfrak{G}_{m,f} = \langle \mathfrak{x}, \mathfrak{y}, \mathfrak{h}, \mathfrak{e}_{m,f} \rangle.$$

In [73] Kim verified that the Sylow 2-subgroup test of Step 5(c) of Algorithm 7.4.8 of [92] was satisfied only by the subgroup $\mathfrak{G}_{m_1,16}$ of $\mathrm{GL}_{782}(17)$ with constant $f = 16$ and matrix

$$m_1 = \begin{pmatrix} 1 & 1 \\ 9 & 14 \end{pmatrix}.$$

In particular, the block diagonal matrix $\mathcal{F}$ of assertion (6) equals the matrix $\alpha^{-1}(R(m_1, 16))$, and the matrix group $\mathfrak{G} = \mathfrak{G}_{m_1,16}$ has the same 2-adic exponent as $\mathfrak{H}$.

(d) Using a stand alone implementation of the algorithm described in the proof of Theorem 6.2.1 of [92] and MAGMA, Kim constructed a faithful permutation representation $P\mathfrak{G}$ of $\mathfrak{G}$ of degree 31671 with stabilizer $\mathfrak{H}$. In particular, $|\mathfrak{G}| = 2^{18} \cdot 3^{13} \cdot 5^2 \cdot 7 \cdot 11 \cdot 13 \cdot 17 \cdot 23$.

(e) Let $\mathfrak{z} = (\mathfrak{x}\mathfrak{y}^2)^7$. Then $C_{\mathfrak{G}}(\mathfrak{z})$ contains $\mathfrak{H} = \langle \mathfrak{h}, \mathfrak{x}, \mathfrak{y} \rangle$, which is isomorphic to H. Using the faithful permutation representation $P\mathfrak{G}$ of degree 31671, Kim verified in [73] that $|C_{\mathfrak{G}}(\mathfrak{z})| = |H|$. Hence $C_{\mathfrak{G}_2}(\mathfrak{z}) \cong \mathfrak{H}$.

(f) Using the faithful permutation representation $P\mathfrak{G}$ of degree 31671 and Kratzer's Algorithm 5.3.18 of [92] Kim calculated the representatives of all the conjugacy classes of $\mathfrak{G}$; see Table 6.5.4.

(g) Furthermore, Kim computed the character table of $\mathfrak{G}$ in [73] by means of the above permutation representation $P\mathfrak{G}$ and MAGMA. It coincides with the one of Fi_{23} in [19], pp. 178–179. The character table of $\mathfrak{G}$ implies that $\mathfrak{G}$ is a simple group. This completes the proof. □

6.4 On the uniqueness of Fi_{23}

By Table 6.5.4 the constructed matrix group of Fi_{23}-type has three conjugacy classes of 2-central involutions. Therefore the author's uniqueness criterion stated in Theorem 7.5.1 in [92] cannot be applied. In fact, the proof of Theorem 6.3.1 shows that there are three non-isomorphic free products $H_i *_{D_i} E$, where H_i, $1 \le i \le 3$, are the three non-isomorphic centralizers of the three whose orders are given in Table 6.5.4 and where the groups E and D_i are defined in Lemma 6.1.2. One has to apply Algorithm 7.4.8 of [92]. If all its conditions can be satisfied then one would get three groups $\mathfrak{G}_i$, $1 \le i \le 3$, where $\mathfrak{G}_1 = \mathfrak{G}$. Using their faithful permutation representations one still has to show that all three groups are isomorphic.

However, the literature contains special uniqueness results due to Hunt [56]. For its statements we need the following definition.

Definition 6.4.1 A finite simple group G is said to be of *special* Fi_{23}-*type*, if it has a 2-central involution z such that $C_G(z)$ is isomorphic to the centralizer of a 3-transposition z of Fischer's sporadic simple group Fi_{23} constructed in [30].

For a definition of a conjugacy class of 3-transpositions see Section 5.3.

Theorem 6.4.2 (D. Hunt) *Let $P = \mathsf{Fi}_{23}$ be the simple group constructed by B. Fischer in [30]. Let d be a fixed element in the conjugacy class $D = d^P$ of 3-transpositions of P. Let $H = C_P(d) \cong 2\mathsf{Fi}_{22}$ be its centralizer in the Fischer group P. Let G be any finite simple group containing an involution z such that $C = C_G(z) \cong H$. Then $G \cong P$.*

Proof See Hunt's main theorem in [56]. □

Corollary 6.4.3 *Let H be the finitely presented group of Proposition 6.2.3(a). Let $\mathfrak{G} = \langle \mathfrak{e}, \mathfrak{x}, \mathfrak{y}, \mathfrak{h} \rangle$ be the simple subgroup of* $\mathrm{GL}_{782}(17)$ *of order* $2^{18} \cdot 3^{13} \cdot 5^2 \cdot 7 \cdot 11 \cdot 13 \cdot 17 \cdot 23$ *with centralizer $C_{\mathfrak{G}}(\mathfrak{z}) = \langle \mathfrak{x}, \mathfrak{y}, \mathfrak{h} \rangle \cong H$ of the 2-central involution $\mathfrak{z} = (\mathfrak{x}\mathfrak{y}^2)^7$ constructed in Kim's Theorem 6.3.1. Then $\mathfrak{G} \cong \mathsf{Fi}_{23}$.*

Proof By Proposition 6.2.3(c) H is isomorphic to the covering group $2\mathsf{Fi}_{23}$. Hence $\mathfrak{G}$ *congP* by Theorems 6.3.1 and 6.4.2. □

Remark 6.4.4 By the character table of Fi_{23} this simple group G has three distinct conjugacy classes of involutions z_i, $i = 1, 2, 3$, such that $H_1 = C_G(z_1)$, $H_2 = C_G(z_2)$ and $H_3 = C_G(z_3)$ have respective orders $2^{18} \cdot 3^9 \cdot 5^2 \cdot 7 \cdot 11 \cdot 13$, $2^{18} \cdot 3^6 \cdot 5 \cdot 7 \cdot 11$ and $2^{18} \cdot 3^5 \cdot 5$. Neither Kim, nor Hunt or Parrott, constructed a presentation or permutation representation for the groups H_2 or H_3 in their papers; see [72], [56] and [107].

(a) Parrott applied Hunt's Theorem 6.4.2 and showed in Theorem C of [107] that a simple group G having a 2-central involution w such that $C_G(z) \cong H_3$ is isomorphic to Fi_{23}.

(b) Neither Hunt nor Parrott considered the remaining uniqueness problem corresponding to the centralizer H_2.

6.5 Representatives of conjugacy classes

6.5.1 *Conjugacy classes of* $E(\mathsf{Fi}_{23}) = \langle x, y, e \rangle$

Class	*Representative*	*Centralizer*	2P	3P	5P	7P	11P	23P
1	1	$2^{18} \cdot 3^2 \cdot 5 \cdot 7 \cdot 11 \cdot 23$	1	1	1	1	1	1
2_1	$(y)^7$	$2^{18} \cdot 3^2 \cdot 5 \cdot 7 \cdot 11$	1	2_1	2_1	2_1	2_1	2_1
2_2	$(y^3e)^6$	$2^{18} \cdot 3^2 \cdot 5 \cdot 7$	1	2_2	2_2	2_2	2_2	2_2
2_3	$(xye)^8$	$2^{18} \cdot 3^2 \cdot 5$	1	2_3	2_3	2_3	2_3	2_3
2_4	e	$2^{14} \cdot 3 \cdot 7$	1	2_4	2_4	2_4	2_4	2_4
2_5	x	$2^{14} \cdot 3$	1	2_5	2_5	2_5	2_5	2_5
3	$(y^3e)^4$	$2^7 \cdot 3^2 \cdot 5$	3	1	3	3	3	3
4_1	$(xy^2xey^2)^7$	$2^{11} \cdot 3 \cdot 7$	2_2	4_1	4_1	4_1	4_1	4_1
4_2	$(xye)^4$	$2^{12} \cdot 3$	2_3	4_2	4_2	4_2	4_2	4_2
4_3	$(y^2e)^2$	$2^{12} \cdot 3$	2_3	4_3	4_3	4_3	4_3	4_3
4_4	$(y^3e)^3$	$2^{11} \cdot 3$	2_2	4_4	4_4	4_4	4_4	4_4
4_5	$(xey)^2$	2^9	2_5	4_5	4_5	4_5	4_5	4_5
4_6	$(xyey^3)^2$	2^9	2_5	4_6	4_6	4_6	4_6	4_6
4_7	xe	2^8	2_4	4_7	4_7	4_7	4_7	4_7
5	$(xyxye)^2$	$2^3 \cdot 3 \cdot 5$	5	5	1	5	5	5
6_1	$(xyxey^2)^5$	$2^7 \cdot 3^2 \cdot 5$	3	2_3	6_1	6_1	6_1	6_1
6_2	$(xyexe)^2$	$2^7 \cdot 3^2$	3	2_3	6_2	6_2	6_2	6_2
6_3	$xyey^3e$	$2^7 \cdot 3^2$	3	2_1	6_3	6_3	6_3	6_3
6_4	$(y^3e)^2$	$2^6 \cdot 3^2$	3	2_2	6_4	6_4	6_4	6_4
6_5	$xyxexey^2$	$2^6 \cdot 3^2$	3	2_3	6_5	6_5	6_5	6_5
6_6	y^4e	$2^5 \cdot 3$	3	2_4	6_6	6_6	6_6	6_6
6_7	$xyxeyxe$	$2^5 \cdot 3$	3	2_5	6_7	6_7	6_7	6_7
7_1	$(y)^2$	$2^3 \cdot 7$	7_1	7_2	7_2	1	7_1	7_1
7_2	$(ye)^2$	$2^3 \cdot 7$	7_2	7_1	7_1	1	7_2	7_2
8_1	$(xye)^2$	2^7	4_2	8_1	8_1	8_1	8_1	8_1
8_2	y^2e	2^7	4_3	8_2	8_2	8_2	8_2	8_2
8_3	xy^2ey^2e	2^7	4_3	8_3	8_3	8_3	8_3	8_3
8_4	xey	2^5	4_5	8_4	8_4	8_4	8_4	8_4
8_5	$xyey^3$	2^5	4_6	8_5	8_5	8_5	8_5	8_5
10_1	$(xyxey^2)^3$	$2^3 \cdot 3 \cdot 5$	5	10_1	2_3	10_1	10_1	10_1
10_2	$xyxye$	$2^3 \cdot 5$	5	10_2	2_2	10_2	10_2	10_2
10_3	$xyxyey$	$2^3 \cdot 5$	5	10_3	2_1	10_3	10_3	10_3
11_1	$(xy^3)^2$	$2 \cdot 11$	11_2	11_1	11_1	11_2	1	11_1
11_2	$(xy^3)^4$	$2 \cdot 11$	11_1	11_2	11_2	11_1	1	11_2
12_1	$xyexe$	$2^5 \cdot 3$	6_2	4_2	12_1	12_1	12_1	12_1
12_2	$xyxyxey^2$	$2^5 \cdot 3$	6_2	4_3	12_2	12_2	12_2	12_2
12_3	y^3e	$2^4 \cdot 3$	6_4	4_4	12_3	12_3	12_3	12_3
12_4	y^3ey^2e	$2^4 \cdot 3$	6_4	4_1	12_4	12_4	12_4	12_4
14_1	$xyey^2$	$2^3 \cdot 7$	7_2	14_2	14_2	2_2	14_1	14_1
14_2	xy^5	$2^3 \cdot 7$	7_1	14_1	14_1	2_2	14_2	14_2
14_3	y	$2^2 \cdot 7$	7_1	14_4	14_4	2_1	14_3	14_3
14_4	ye	$2^2 \cdot 7$	7_2	14_3	14_3	2_1	14_4	14_4
14_5	$xyxe$	$2^2 \cdot 7$	7_2	14_6	14_6	2_4	14_5	14_5
14_6	xy^3xe	$2^2 \cdot 7$	7_1	14_5	14_5	2_4	14_6	14_6
15_1	xy^3e	$2 \cdot 3 \cdot 5$	15_1	5	3	15_2	15_2	15_1
15_2	$(xy^3ey)^2$	$2 \cdot 3 \cdot 5$	15_2	5	3	15_1	15_1	15_2
16_1	xye	2^5	8_1	16_1	16_2	16_2	16_1	16_2
16_2	xey^3	2^5	8_1	16_2	16_1	16_1	16_2	16_1
22_1	xy^3	$2 \cdot 11$	11_1	22_1	22_1	22_2	2_1	22_1
22_2	xey^2	$2 \cdot 11$	11_2	22_2	22_2	22_1	2_1	22_2
23_1	$xeyey$	23	23_1	23_1	23_2	23_2	23_2	1
23_2	xy^2eyxe	23	23_2	23_2	23_1	23_1	23_1	1
28_1	xy^2xey^2	$2^2 \cdot 7$	14_1	28_2	28_2	4_1	28_1	28_1
28_2	xey^3ey	$2^2 \cdot 7$	14_2	28_1	28_1	4_1	28_2	28_2
30_1	$xyxey^2$	$2 \cdot 3 \cdot 5$	15_1	10_1	6_1	30_2	30_2	30_1
30_2	xy^3ey	$2 \cdot 3 \cdot 5$	15_2	10_1	6_1	30_1	30_1	30_2

6.5.2 *Conjugacy classes of* $H(\mathsf{Fi}_{23}) = \langle x, y, h\rangle$

Class	*Representative*	*Centralizer*	2P	3P	5P	7P	11P	13P
1	1	$2^{18}\cdot 3^9\cdot 5^2\cdot 7\cdot 11\cdot 13$	1	1	1	1	1	1
2_1	$(xy^2)^7$	$2^{18}\cdot 3^9\cdot 5^2\cdot 7\cdot 11\cdot 13$	1	2_1	2_1	2_1	2_1	2_1
2_2	$(y)^7$	$2^{17}\cdot 3^6\cdot 5\cdot 7\cdot 11$	1	2_2	2_2	2_2	2_2	2_2
2_3	$(xyhy)^{11}$	$2^{17}\cdot 3^6\cdot 5\cdot 7\cdot 11$	1	2_3	2_3	2_3	2_3	2_3
2_4	x	$2^{18}\cdot 3^4\cdot 5$	1	2_4	2_4	2_4	2_4	2_4
2_5	$(yh)^{10}$	$2^{18}\cdot 3^4\cdot 5$	1	2_5	2_5	2_5	2_5	2_5
2_6	$(xh)^6$	$2^{16}\cdot 3^3$	1	2_6	2_6	2_6	2_6	2_6
3_1	$(xyxh)^{10}$	$2^9\cdot 3^7\cdot 5\cdot 7$	3_1	1	3_1	3_1	3_1	3_1
3_2	h	$2^8\cdot 3^9$	3_2	1	3_2	3_2	3_2	3_2
3_3	$(xyxy^3)^4$	$2^7\cdot 3^7$	3_3	1	3_3	3_3	3_3	3_3
3_4	$(xyhyh)^6$	$2^4\cdot 3^7$	3_4	1	3_4	3_4	3_4	3_4
4_1	$(xy^2xh)^3$	$2^{12}\cdot 3^3$	2_4	4_1	4_1	4_1	4_1	4_1
4_2	$(yh)^5$	$2^{11}\cdot 3^2\cdot 5$	2_5	4_2	4_2	4_2	4_2	4_2
4_3	$(xhy)^5$	$2^{11}\cdot 3^2\cdot 5$	2_5	4_3	4_3	4_3	4_3	4_3
4_4	$(y^2h)^4$	$2^{12}\cdot 3$	2_4	4_4	4_4	4_4	4_4	4_4
4_5	$(xh)^3$	$2^{10}\cdot 3^2$	2_6	4_5	4_5	4_5	4_5	4_5
4_6	$(xyh)^2$	$2^{10}\cdot 3^2$	2_6	4_6	4_6	4_6	4_6	4_6
4_7	$(xh^2y^3)^3$	$2^{10}\cdot 3$	2_5	4_7	4_7	4_7	4_7	4_7
5	$(yh)^4$	$2^4\cdot 3\cdot 5^2$	5	5	1	5	5	5
6_1	$(y^3h^2)^5$	$2^9\cdot 3^7\cdot 5\cdot 7$	3_1	2_1	6_1	6_1	6_1	6_1
6_2	$(xy^2h^2yhy)^3$	$2^8\cdot 3^9$	3_2	2_1	6_2	6_2	6_2	6_2
6_3	$(xyxh)^5$	$2^8\cdot 3^5\cdot 5$	3_1	2_2	6_3	6_3	6_3	6_3
6_4	$(xhy^3)^5$	$2^8\cdot 3^5\cdot 5$	3_1	2_3	6_4	6_4	6_4	6_4
6_5	$xyh^2yh^2xh^2xh$	$2^7\cdot 3^7$	3_3	2_1	6_5	6_5	6_5	6_5
6_6	$(xyxyh)^3$	$2^8\cdot 3^6$	3_2	2_3	6_6	6_6	6_6	6_6
6_7	$(xyxhy)^3$	$2^8\cdot 3^6$	3_2	2_2	6_7	6_7	6_7	6_7
6_8	$(xyhyh)^3$	$2^4\cdot 3^7$	3_4	2_1	6_8	6_8	6_8	6_8
6_9	$(xy^2xh)^2$	$2^8\cdot 3^4$	3_2	2_4	6_9	6_9	6_9	6_9
6_{10}	$(xy^2xyh)^3$	$2^8\cdot 3^4$	3_2	2_5	6_{10}	6_{10}	6_{10}	6_{10}
6_{11}	$(yhyh^2)^4$	$2^9\cdot 3^3$	3_1	2_4	6_{11}	6_{11}	6_{11}	6_{11}
6_{12}	$(xh^2y^3)^2$	$2^9\cdot 3^3$	3_1	2_5	6_{12}	6_{12}	6_{12}	6_{12}
6_{13}	y^7h	$2^5\cdot 3^5$	3_3	2_2	6_{13}	6_{13}	6_{13}	6_{13}
6_{14}	$xyxyxhyxhy^2$	$2^5\cdot 3^5$	3_3	2_3	6_{14}	6_{14}	6_{14}	6_{14}
6_{15}	$(xh)^2$	$2^8\cdot 3^3$	3_2	2_6	6_{15}	6_{15}	6_{15}	6_{15}
6_{16}	$(xhxhy^2h)^2$	$2^8\cdot 3^3$	3_2	2_6	6_{16}	6_{16}	6_{16}	6_{16}
6_{17}	$xyhxhxh$	$2^7\cdot 3^3$	3_3	2_6	6_{17}	6_{17}	6_{17}	6_{17}
6_{18}	$(xy^2xhyhy)^2$	$2^7\cdot 3^3$	3_3	2_6	6_{18}	6_{18}	6_{18}	6_{18}
6_{19}	$xyxy^2xyxyhy^2$	$2^7\cdot 3^3$	3_1	2_6	6_{19}	6_{19}	6_{19}	6_{19}
6_{20}	$(xyxy^3)^2$	$2^6\cdot 3^3$	3_3	2_5	6_{20}	6_{20}	6_{20}	6_{20}
6_{21}	xh^2yh^2y	$2^6\cdot 3^3$	3_3	2_4	6_{21}	6_{21}	6_{21}	6_{21}
6_{22}	$xyxhy^2xyh$	$2^5\cdot 3^3$	3_3	2_6	6_{22}	6_{22}	6_{22}	6_{22}
6_{23}	$xyxh^2yhyh$	$2^5\cdot 3^3$	3_3	2_6	6_{23}	6_{23}	6_{23}	6_{23}
6_{24}	$(xyxyhyh)^2$	$2^4\cdot 3^3$	3_4	2_6	6_{24}	6_{24}	6_{24}	6_{24}
6_{25}	xyh^2y^2xh	$2^4\cdot 3^3$	3_4	2_6	6_{25}	6_{25}	6_{25}	6_{25}
7	$(y)^2$	$2^2\cdot 3\cdot 7$	7	7	7	1	7	7
8_1	$(yhyh^2)^3$	$2^7\cdot 3$	4_1	8_1	8_1	8_1	8_1	8_1
8_2	$(xyhy^2h)^3$	$2^7\cdot 3$	4_1	8_2	8_2	8_2	8_2	8_2
8_3	$(y^2h)^2$	2^7	4_4	8_3	8_3	8_3	8_3	8_3
8_4	xyh	2^6	4_6	8_4	8_4	8_4	8_4	8_4
8_5	xyh^2yh	2^6	4_6	8_5	8_5	8_5	8_5	8_5
9_1	$(xyxyh)^2$	$2^3\cdot 3^4$	9_1	3_2	9_1	9_1	9_1	9_1
9_2	$(xyhy^3)^2$	$2^2\cdot 3^4$	9_2	3_2	9_2	9_2	9_2	9_2
9_3	$(xyhyh)^2$	$2\cdot 3^3$	9_3	3_4	9_3	9_3	9_3	9_3
10_1	$(y^3h^2)^3$	$2^4\cdot 3\cdot 5^2$	5	10_1	2_1	10_1	10_1	10_1
10_2	yh^2	$2^3\cdot 3\cdot 5$	5	10_2	2_2	10_2	10_2	10_2
10_3	$(xhy^3)^3$	$2^3\cdot 3\cdot 5$	5	10_3	2_3	10_3	10_3	10_3
10_4	$(yh)^2$	$2^4\cdot 5$	5	10_4	2_5	10_4	10_4	10_4

Conjugacy classes of $H(\mathsf{Fi}_{23}) = \langle x, y, h\rangle$ *(continued)*

Class	*Representative*	*Centralizer*	2P	3P	5P	7P	11P	13P
10_5	xh^2y	$2^4\cdot 5$	5	10_5	2_4	10_5	10_5	10_5
11_1	$(xy^3)^2$	$2^2\cdot 11$	11_2	11_1	11_1	11_2	1	11_2
11_2	$(xy^3)^4$	$2^2\cdot 11$	11_1	11_2	11_2	11_1	1	11_1
12_1	$(yhyh^2)^2$	$2^7\cdot 3^2$	6_{11}	4_1	12_1	12_1	12_1	12_1
12_2	$(xyhy^2h)^2$	$2^7\cdot 3^2$	6_{11}	4_1	12_2	12_2	12_2	12_2
12_3	xy^2xh	$2^5\cdot 3^3$	6_9	4_1	12_3	12_3	12_3	12_3
12_4	xy^2xy^3h	$2^6\cdot 3^2$	6_{12}	4_3	12_4	12_4	12_4	12_4
12_5	$xyxy^3xh^2$	$2^6\cdot 3^2$	6_{12}	4_2	12_5	12_5	12_5	12_5
12_6	$xy^2xhxhyhy$	$2^6\cdot 3^2$	6_{11}	4_1	12_6	12_6	12_6	12_6
12_7	xh	$2^5\cdot 3^2$	6_{15}	4_5	12_7	12_7	12_7	12_7
12_8	$xyxh^2y$	$2^5\cdot 3^2$	6_{15}	4_6	12_8	12_8	12_8	12_8
12_9	$xhxhy^2h$	$2^5\cdot 3^2$	6_{16}	4_6	12_9	12_9	12_9	12_9
12_{10}	xy^2xhyhy	$2^5\cdot 3^2$	6_{18}	4_5	12_{10}	12_{10}	12_{10}	12_{10}
12_{11}	$xhxh^2y^4$	$2^5\cdot 3^2$	6_{18}	4_6	12_{11}	12_{11}	12_{11}	12_{11}
12_{12}	xhy^4hyh	$2^5\cdot 3^2$	6_{16}	4_5	12_{12}	12_{12}	12_{12}	12_{12}
12_{13}	$xyxy^3$	$2^4\cdot 3^2$	6_{20}	4_2	12_{13}	12_{13}	12_{13}	12_{13}
12_{14}	$xyxhy^3h$	$2^4\cdot 3^2$	6_{20}	4_3	12_{14}	12_{14}	12_{14}	12_{14}
12_{15}	xh^2y^3	$2^5\cdot 3$	6_{12}	4_7	12_{15}	12_{15}	12_{15}	12_{15}
12_{16}	xy^3hy^2h	$2^5\cdot 3$	6_9	4_4	12_{16}	12_{16}	12_{16}	12_{16}
12_{17}	$xyxyhyh$	$2^3\cdot 3^2$	6_{24}	4_6	12_{17}	12_{17}	12_{17}	12_{17}
12_{18}	xy^3h^2xh	$2^3\cdot 3^2$	6_{24}	4_5	12_{18}	12_{18}	12_{18}	12_{18}
13_1	$(xy^3h)^2$	$2\cdot 13$	13_2	13_1	13_2	13_2	13_2	1
13_2	$(xy^3h)^4$	$2\cdot 13$	13_1	13_2	13_1	13_1	13_1	1
14_1	xy^2	$2^2\cdot 3\cdot 7$	7	14_1	14_1	2_1	14_1	14_1
14_2	y	$2^2\cdot 7$	7	14_2	14_2	2_2	14_2	14_2
14_3	xh^2y^2	$2^2\cdot 7$	7	14_3	14_3	2_3	14_3	14_3
15	$(xyxh)^2$	$2^2\cdot 3\cdot 5$	15	5	3_1	15	15	15
16_1	y^2h	2^5	8_3	16_1	16_2	16_2	16_1	16_2
16_2	$xyxyhy$	2^5	8_3	16_2	16_1	16_1	16_2	16_1
18_1	xhy^2hy^2h	$2^3\cdot 3^4$	9_1	6_2	18_1	18_1	18_1	18_1
18_2	xy^2h^2yhy	$2^2\cdot 3^4$	9_2	6_2	18_2	18_2	18_2	18_2
18_3	$xyxyh$	$2^3\cdot 3^3$	9_1	6_6	18_5	18_3	18_5	18_3
18_4	$xyxhy$	$2^3\cdot 3^3$	9_1	6_7	18_6	18_4	18_6	18_4
18_5	xy^2xhy	$2^3\cdot 3^3$	9_1	6_6	18_3	18_5	18_3	18_5
18_6	xhy^3h^2	$2^3\cdot 3^3$	9_1	6_7	18_4	18_6	18_4	18_6
18_7	$xyhy^3$	$2^2\cdot 3^3$	9_2	6_6	18_7	18_7	18_7	18_7
18_8	$xyxh^2xh$	$2^2\cdot 3^3$	9_2	6_7	18_8	18_8	18_8	18_8
18_9	xy^2xyh	$2^3\cdot 3^2$	9_1	6_{10}	18_9	18_9	18_9	18_9
18_{10}	xh^2yhy	$2^3\cdot 3^2$	9_1	6_9	18_{10}	18_{10}	18_{10}	18_{10}
18_{11}	$xyhyh$	$2\cdot 3^3$	9_3	6_8	18_{11}	18_{11}	18_{11}	18_{11}
20_1	yh	$2^3\cdot 5$	10_4	20_1	4_2	20_1	20_1	20_1
20_2	xhy	$2^3\cdot 5$	10_4	20_2	4_3	20_2	20_2	20_2
21	$xhyhy$	$2\cdot 3\cdot 7$	21	7	21	3_1	21	21
22_1	xy^3	$2^2\cdot 11$	11_1	22_1	22_1	22_3	2_1	22_3
22_2	$xyhy$	$2^2\cdot 11$	11_1	22_2	22_2	22_5	2_3	22_5
22_3	$xyxh^2$	$2^2\cdot 11$	11_2	22_3	22_3	22_1	2_1	22_1
22_4	$xyxhy^2$	$2^2\cdot 11$	11_2	22_4	22_4	22_6	2_2	22_6
22_5	$xyhxh^2$	$2^2\cdot 11$	11_2	22_5	22_5	22_2	2_3	22_2
22_6	xy^2xy^2h	$2^2\cdot 11$	11_1	22_6	22_6	22_4	2_2	22_4
24_1	$yhyh^2$	$2^4\cdot 3$	12_1	8_1	24_1	24_1	24_1	24_1
24_2	$xyhy^2h$	$2^4\cdot 3$	12_2	8_2	24_2	24_2	24_2	24_2
26_1	xy^3h	$2\cdot 13$	13_1	26_1	26_2	26_2	26_2	2_1
26_2	xhy^2h	$2\cdot 13$	13_2	26_2	26_1	26_1	26_1	2_1
30_1	$xyxh$	$2^2\cdot 3\cdot 5$	15	10_2	6_3	30_1	30_1	30_1
30_2	xhy^3	$2^2\cdot 3\cdot 5$	15	10_3	6_4	30_2	30_2	30_2
30_3	y^3h^2	$2^2\cdot 3\cdot 5$	15	10_1	6_1	30_3	30_3	30_3
42	xhy^2hy	$2\cdot 3\cdot 7$	21	14_1	42	6_1	42	42

6.5.3 *Conjugacy classes of $D(\mathsf{Fi}_{23}) = \langle x, y \rangle$*

Class	*Representative*	*Centralizer*	2P	3P	5P	7P	11P
1	1	$2^{18} \cdot 3^2 \cdot 5 \cdot 7 \cdot 11$	1	1	1	1	1
2_1	$(xy^2)^7$	$2^{18} \cdot 3^2 \cdot 5 \cdot 7 \cdot 11$	1	2_1	2_1	2_1	2_1
2_2	$(y)^7$	$2^{17} \cdot 3^2 \cdot 5 \cdot 7$	1	2_2	2_2	2_2	2_2
2_3	$(xy^5)^7$	$2^{17} \cdot 3^2 \cdot 5 \cdot 7$	1	2_3	2_3	2_3	2_3
2_4	$(xyxy^2)^8$	$2^{18} \cdot 3 \cdot 5$	1	2_4	2_4	2_4	2_4
2_5	$(xyxy^3)^6$	$2^{18} \cdot 3 \cdot 5$	1	2_5	2_5	2_5	2_5
2_6	$(xyxy^2xy^2)^4$	$2^{16} \cdot 3^2$	1	2_6	2_6	2_6	2_6
2_7	x	$2^{14} \cdot 3$	1	2_7	2_7	2_7	2_7
2_8	$(xy^3xy^4)^3$	$2^{14} \cdot 3$	1	2_8	2_8	2_8	2_8
2_9	$(xy^2xy^3xy^2xy^5)^2$	2^{13}	1	2_9	2_9	2_9	2_9
3	$(xyxy^3)^4$	$2^7 \cdot 3^2$	3	1	3	3	3
4_1	$(xyxy^3)^3$	$2^{11} \cdot 3$	2_5	4_1	4_1	4_1	4_1
4_2	$(xyxy^3xyxy^4)^3$	$2^{11} \cdot 3$	2_5	4_2	4_2	4_2	4_2
4_3	$(xyxy^2)^4$	2^{12}	2_4	4_3	4_3	4_3	4_3
4_4	$(xyxyxy^2xy^6)^2$	2^{12}	2_4	4_4	4_4	4_4	4_4
4_5	$(xyxy^2xy^2)^2$	$2^{10} \cdot 3$	2_6	4_5	4_5	4_5	4_5
4_6	$(xyxy^2xyxy^6)^3$	$2^{10} \cdot 3$	2_6	4_6	4_6	4_6	4_6
4_7	$xy^2xy^7xy^5$	2^{10}	2_5	4_7	4_7	4_7	4_7
4_8	$(xyxy^3xy^2)^2$	2^9	2_7	4_8	4_8	4_8	4_8
4_9	$(xyxy^2xy^4)^2$	2^9	2_7	4_9	4_9	4_9	4_9
4_{10}	$xy^2xy^3xy^5$	2^8	2_8	4_{10}	4_{10}	4_{10}	4_{10}
4_{11}	$xy^2xy^3xy^2xy^5$	2^8	2_9	4_{11}	4_{11}	4_{11}	4_{11}
4_{12}	$xyxyxy^4xy^4xy^2$	2^8	2_9	4_{12}	4_{12}	4_{12}	4_{12}
4_{13}	$xyxyxyxy^2xy^2$	2^7	2_8	4_{13}	4_{13}	4_{13}	4_{13}
5	$(xyxy^5)^2$	$2^3 \cdot 5$	5	5	1	5	5
6_1	$(xyxy^2xyxy^6)^2$	$2^7 \cdot 3^2$	3	2_6	6_1	6_1	6_1
6_2	$xy^2xy^2xy^6xy^3$	$2^7 \cdot 3^2$	3	2_6	6_2	6_2	6_2
6_3	$xyxy^2xy^3xy^2xy^5$	$2^7 \cdot 3^2$	3	2_1	6_3	6_3	6_3
6_4	$xyxyxy^2xyxyxy^3$	$2^5 \cdot 3^2$	3	2_6	6_4	6_4	6_4
6_5	$xyxy^2xy^6xy^4$	$2^5 \cdot 3^2$	3	2_6	6_5	6_5	6_5
6_6	$xyxy^3xyxy^4xy^3$	$2^5 \cdot 3^2$	3	2_3	6_6	6_6	6_6
6_7	$xyxyxy^4xy^3xyxy^2$	$2^5 \cdot 3^2$	3	2_2	6_7	6_7	6_7
6_8	$(xyxy^3)^2$	$2^6 \cdot 3$	3	2_5	6_8	6_8	6_8
6_9	$xyxy^5xy^2$	$2^6 \cdot 3$	3	2_4	6_9	6_9	6_9
6_{10}	xy^3xy^4	$2^5 \cdot 3$	3	2_8	6_{10}	6_{10}	6_{10}
6_{11}	xy^4xy^6	$2^5 \cdot 3$	3	2_7	6_{11}	6_{11}	6_{11}
7_1	$(y)^2$	$2^2 \cdot 7$	7_1	7_2	7_2	1	7_1
7_2	$(xy^2)^2$	$2^2 \cdot 7$	7_2	7_1	7_1	1	7_2
8_1	$(xyxy^2)^2$	2^7	4_3	8_1	8_1	8_1	8_1
8_2	$xyxyxy^2xy^6$	2^7	4_4	8_2	8_2	8_2	8_2
8_3	$xyxy^6xy^3xy^4$	2^7	4_4	8_3	8_3	8_3	8_3
8_4	$xyxy^2xy^2$	2^6	4_5	8_4	8_4	8_4	8_4
8_5	$xy^2xy^3xy^3$	2^6	4_3	8_5	8_5	8_5	8_5
8_6	$xy^2xy^2xy^2xy^3$	2^6	4_5	8_6	8_6	8_6	8_6
8_7	$xyxy^3xy^2$	2^5	4_8	8_7	8_7	8_7	8_7
8_8	$xyxy^2xy^4$	2^5	4_9	8_8	8_8	8_8	8_8
10_1	$xyxy^5$	$2^3 \cdot 5$	5	10_2	2_3	10_2	10_1
10_2	$xyxy^6$	$2^3 \cdot 5$	5	10_1	2_3	10_1	10_2
10_3	$xyxy^3xy^3$	$2^3 \cdot 5$	5	10_3	2_5	10_3	10_3
10_4	$xyxy^2xyxy^3$	$2^3 \cdot 5$	5	10_6	2_2	10_6	10_4
10_5	$xy^2xy^3xy^4$	$2^3 \cdot 5$	5	10_5	2_4	10_5	10_5
10_6	$xy^2xy^4xy^3$	$2^3 \cdot 5$	5	10_4	2_2	10_4	10_6
10_7	$xyxyxy^4xy^3xy^2$	$2^3 \cdot 5$	5	10_7	2_1	10_7	10_7

Conjugacy classes of $D(\mathsf{Fi}_{23}) = \langle x, y \rangle$ *(continued)*

Class	*Representative*	*Centralizer*	2P	3P	5P	7P	11P
11_1	$(xy^3)^2$	$2 \cdot 11$	11_2	11_1	11_1	11_2	1
11_2	$(xy^3)^4$	$2 \cdot 11$	11_1	11_2	11_2	11_1	1
12_1	$xyxy^2xyxy^6$	$2^5 \cdot 3$	6_1	4_6	12_1	12_1	12_1
12_2	$xyxyxy^2xy^7$	$2^5 \cdot 3$	6_1	4_5	12_2	12_2	12_2
12_3	$xyxy^3$	$2^4 \cdot 3$	6_8	4_1	12_3	12_3	12_3
12_4	$xyxy^3xyxy^4$	$2^4 \cdot 3$	6_8	4_2	12_4	12_4	12_4
14_1	y	$2^2 \cdot 7$	7_1	14_6	14_6	2_2	14_1
14_2	xy^2	$2^2 \cdot 7$	7_2	14_5	14_5	2_1	14_2
14_3	xy^5	$2^2 \cdot 7$	7_1	14_4	14_4	2_3	14_3
14_4	xy^2xy^3	$2^2 \cdot 7$	7_2	14_3	14_3	2_3	14_4
14_5	xy^2xy^4	$2^2 \cdot 7$	7_1	14_2	14_2	2_1	14_5
14_6	xy^3xy^5	$2^2 \cdot 7$	7_2	14_1	14_1	2_2	14_6
16_1	$xyxy^2$	2^5	8_1	16_1	16_2	16_2	16_1
16_2	$xyxyxy^2$	2^5	8_1	16_2	16_1	16_1	16_2
22_1	xy^3	$2 \cdot 11$	11_1	22_1	22_1	22_2	2_1
22_2	$xyxyxy^2xy^2$	$2 \cdot 11$	11_2	22_2	22_2	22_1	2_1

6.5.4 *Conjugacy classes of* $\mathsf{Fi}_{23} = \langle x, h, y, e\rangle$

Class	*Representative*	*Centralizer*	2P	3P	5P	7P	11P	13P	17P	23P
1	1	$2^{18}\cdot 3^{13}\cdot 5^2\cdot 7\cdot 11\cdot 13\cdot 17\cdot 23$	1	1	1	1	1	1	1	1
2_1	$(y)^7$	$2^{18}\cdot 3^9\cdot 5^2\cdot 7\cdot 11\cdot 13$	1	2_1	2_1	2_1	2_1	2_1	2_1	2_1
2_2	e	$2^{18}\cdot 3^6\cdot 5\cdot 7\cdot 11$	1	2_2	2_2	2_2	2_2	2_2	2_2	2_2
2_3	x	$2^{18}\cdot 3^5\cdot 5$	1	2_3	2_3	2_3	2_3	2_3	2_3	2_3
3_1	$(he)^4$	$2^9\cdot 3^{10}\cdot 5\cdot 7\cdot 13$	3_1	1	3_1	3_1	3_1	3_1	3_1	3_1
3_2	h	$2^{10}\cdot 3^{13}$	3_2	1	3_2	3_2	3_2	3_2	3_2	3_2
3_3	$(xyeh)^4$	$2^7\cdot 3^{10}\cdot 5$	3_3	1	3_3	3_3	3_3	3_3	3_3	3_3
3_4	$(xyhyh)^6$	$2^4\cdot 3^{10}$	3_4	1	3_4	3_4	3_4	3_4	3_4	3_4
4_1	$(xhy)^5$	$2^{11}\cdot 3^4\cdot 5\cdot 7$	2_2	4_1	4_1	4_1	4_1	4_1	4_1	4_1
4_2	$(xyh)^2$	$2^{12}\cdot 3^4$	2_3	4_2	4_2	4_2	4_2	4_2	4_2	4_2
4_3	xe	$2^{11}\cdot 3^2\cdot 5$	2_2	4_3	4_3	4_3	4_3	4_3	4_3	4_3
4_4	$(xh)^3$	$2^{12}\cdot 3^2$	2_3	4_4	4_4	4_4	4_4	4_4	4_4	4_4
5	$(yh)^4$	$2^4\cdot 3^2\cdot 5^2\cdot 7$	5	5	1	5	5	5	5	5
6_1	$(xyxh)^5$	$2^9\cdot 3^7\cdot 5\cdot 7$	3_1	2_1	6_1	6_1	6_1	6_1	6_1	6_1
6_2	$(xyxhy)^3$	$2^8\cdot 3^9$	3_2	2_1	6_2	6_2	6_2	6_2	6_2	6_2
6_3	$(he)^2$	$2^9\cdot 3^5\cdot 5$	3_1	2_2	6_3	6_3	6_3	6_3	6_3	6_3
6_4	$(yehe)^3$	$2^9\cdot 3^6$	3_2	2_2	6_4	6_4	6_4	6_4	6_4	6_4
6_5	$xheh^2e$	$2^7\cdot 3^7$	3_3	2_1	6_5	6_5	6_5	6_5	6_5	6_5
6_6	$(xh)^2$	$2^{10}\cdot 3^5$	3_2	2_3	6_6	6_6	6_6	6_6	6_6	6_6
6_7	$(xehy)^5$	$2^7\cdot 3^5\cdot 5$	3_3	2_3	6_7	6_7	6_7	6_7	6_7	6_7
6_8	$(xhye)^3$	$2^8\cdot 3^5$	3_2	2_3	6_8	6_8	6_8	6_8	6_8	6_8
6_9	$(xhe)^2$	$2^9\cdot 3^4$	3_1	2_3	6_9	6_9	6_9	6_9	6_9	6_9
6_{10}	$(xyhyh)^3$	$2^4\cdot 3^7$	3_4	2_1	6_{10}	6_{10}	6_{10}	6_{10}	6_{10}	6_{10}
6_{11}	$(xyeh)^2$	$2^6\cdot 3^5$	3_3	2_2	6_{11}	6_{11}	6_{11}	6_{11}	6_{11}	6_{11}
6_{12}	$(xyexe)^2$	$2^7\cdot 3^4$	3_3	2_3	6_{12}	6_{12}	6_{12}	6_{12}	6_{12}	6_{12}
6_{13}	xh^2ehe	$2^6\cdot 3^4$	3_3	2_3	6_{13}	6_{13}	6_{13}	6_{13}	6_{13}	6_{13}
6_{14}	$xyhyehe$	$2^4\cdot 3^5$	3_4	2_3	6_{14}	6_{14}	6_{14}	6_{14}	6_{14}	6_{14}
6_{15}	$(xh^2ey)^2$	$2^4\cdot 3^4$	3_4	2_3	6_{15}	6_{15}	6_{15}	6_{15}	6_{15}	6_{15}
7	$(y)^2$	$2^3\cdot 3\cdot 5\cdot 7$	7	7	7	1	7	7	7	7
8_1	xyh	$2^7\cdot 3$	4_2	8_1	8_1	8_1	8_1	8_1	8_1	8_1
8_2	$(xye)^2$	$2^7\cdot 3$	4_4	8_2	8_2	8_2	8_2	8_2	8_2	8_2
8_3	xey	$2^7\cdot 3$	4_2	8_3	8_3	8_3	8_3	8_3	8_3	8_3
9_1	$(xyh^2e)^3$	$2^3\cdot 3^7$	9_1	3_2	9_1	9_1	9_1	9_1	9_1	9_1
9_2	$(xhye)^2$	$2^3\cdot 3^6$	9_2	3_2	9_2	9_2	9_2	9_2	9_2	9_2
9_3	$(xyhey)^4$	$2^3\cdot 3^6$	9_3	3_2	9_3	9_3	9_3	9_3	9_3	9_3
9_4	$(yh^2eh^2)^2$	$2\cdot 3^6$	9_4	3_2	9_4	9_4	9_4	9_4	9_4	9_4
9_5	$(xyhyh)^2$	$2\cdot 3^4$	9_5	3_4	9_5	9_5	9_5	9_5	9_5	9_5
10_1	yh^2	$2^4\cdot 3\cdot 5^2$	5	10_1	2_1	10_1	10_1	10_1	10_1	10_1
10_2	$(yh)^2$	$2^4\cdot 3\cdot 5$	5	10_2	2_2	10_2	10_2	10_2	10_2	10_2
10_3	xh^2y	$2^4\cdot 3\cdot 5$	5	10_3	2_3	10_3	10_3	10_3	10_3	10_3
11	$(xy^3)^2$	$2^2\cdot 11$	11	11	11	11	1	11	11	11
12_1	$(xeyh^2)^5$	$2^6\cdot 3^3\cdot 5$	6_3	4_1	12_1	12_1	12_1	12_1	12_1	12_1
12_2	$(xheh)^2$	$2^7\cdot 3^3$	6_9	4_2	12_2	12_2	12_2	12_2	12_2	12_2
12_3	$xyxh^2y$	$2^7\cdot 3^3$	6_6	4_2	12_3	12_3	12_3	12_3	12_3	12_3
12_4	$xehe$	$2^5\cdot 3^4$	6_8	4_2	12_4	12_4	12_4	12_4	12_4	12_4
12_5	$(xyhey)^3$	$2^5\cdot 3^4$	6_4	4_1	12_5	12_5	12_5	12_5	12_5	12_5
12_6	xh	$2^7\cdot 3^2$	6_6	4_4	12_6	12_6	12_6	12_6	12_6	12_6
12_7	xhe	$2^7\cdot 3^2$	6_9	4_2	12_7	12_7	12_7	12_7	12_7	12_7
12_8	$xyxhxhe$	$2^5\cdot 3^3$	6_{12}	4_2	12_8	12_8	12_8	12_8	12_8	12_8
12_9	he	$2^6\cdot 3^2$	6_3	4_3	12_9	12_9	12_9	12_9	12_9	12_9
12_{10}	$xyeh$	$2^4\cdot 3^3$	6_{11}	4_1	12_{10}	12_{10}	12_{10}	12_{10}	12_{10}	12_{10}
12_{11}	$xyexe$	$2^5\cdot 3^2$	6_{12}	4_4	12_{11}	12_{11}	12_{11}	12_{11}	12_{11}	12_{11}
12_{12}	$xhexeh$	$2^5\cdot 3^2$	6_8	4_4	12_{12}	12_{12}	12_{12}	12_{12}	12_{12}	12_{12}

Conjugacy classes of $\mathsf{Fi}_{23} = \langle x, y, h, e \rangle$ *(continued)*

Class	*Representative*	*Centralizer*	2P	3P	5P	7P	11P	13P	17P	23P
12_{13}	xh^2ey	$2^3 \cdot 3^3$	6_{15}	4_2	12_{13}	12_{13}	12_{13}	12_{13}	12_{13}	12_{13}
12_{14}	y^3e	$2^4 \cdot 3^2$	6_{11}	4_3	12_{14}	12_{14}	12_{14}	12_{14}	12_{14}	12_{14}
12_{15}	$xeyhey$	$2^3 \cdot 3^2$	6_{15}	4_4	12_{15}	12_{15}	12_{15}	12_{15}	12_{15}	12_{15}
13_1	yeh	$2 \cdot 3 \cdot 13$	13_2	13_1	13_2	13_2	13_2	1	13_1	13_1
13_2	$(xy^2eh)^2$	$2 \cdot 3 \cdot 13$	13_1	13_2	13_1	13_1	13_1	1	13_2	13_2
14_1	y	$2^2 \cdot 3 \cdot 7$	7	14_1	14_1	2_1	14_1	14_1	14_1	14_1
14_2	$xyxe$	$2^3 \cdot 7$	7	14_2	14_2	2_2	14_2	14_2	14_2	14_2
15_1	$(xyxh)^2$	$2^3 \cdot 3^2 \cdot 5$	15_1	5	3_1	15_1	15_1	15_1	15_1	15_1
15_2	$(xehy)^2$	$2 \cdot 3^2 \cdot 5$	15_2	5	3_3	15_2	15_2	15_2	15_2	15_2
16_1	xye	2^5	8_2	16_1	16_2	16_2	16_1	16_2	16_1	16_2
16_2	$yheh$	2^5	8_2	16_2	16_1	16_1	16_2	16_1	16_2	16_1
17	$xhey$	17	17	17	17	17	17	17	1	17
18_1	$xyxhy$	$2^3 \cdot 3^4$	9_2	6_2	18_1	18_1	18_1	18_1	18_1	18_1
18_2	$xeyxeh$	$2^2 \cdot 3^4$	9_3	6_2	18_2	18_2	18_2	18_2	18_2	18_2
18_3	$xhye$	$2^3 \cdot 3^3$	9_2	6_8	18_3	18_3	18_3	18_3	18_3	18_3
18_4	$yehe$	$2^3 \cdot 3^3$	9_2	6_4	18_4	18_4	18_4	18_4	18_4	18_4
18_5	$(xyhey)^2$	$2^3 \cdot 3^3$	9_3	6_4	18_5	18_5	18_5	18_5	18_5	18_5
18_6	$xhehy$	$2^3 \cdot 3^3$	9_1	6_8	18_6	18_6	18_6	18_6	18_6	18_6
18_7	$xyhyh$	$2 \cdot 3^3$	9_5	6_{10}	18_7	18_7	18_7	18_7	18_7	18_7
18_8	yh^2eh^2	$2 \cdot 3^3$	9_4	6_8	18_8	18_8	18_8	18_8	18_8	18_8
20_1	xhy	$2^3 \cdot 3 \cdot 5$	10_2	20_1	4_1	20_1	20_1	20_1	20_1	20_1
20_2	yh	$2^3 \cdot 5$	10_2	20_2	4_3	20_2	20_2	20_2	20_2	20_2
21	yhe	$2 \cdot 3 \cdot 7$	21	7	21	3_1	21	21	21	21
22_1	xy^3	$2^2 \cdot 11$	11	22_1	22_1	22_3	2_1	22_3	22_3	22_1
22_2	$xyhy$	$2^2 \cdot 11$	11	22_2	22_2	22_2	2_2	22_2	22_2	22_2
22_3	xey^2	$2^2 \cdot 11$	11	22_3	22_3	22_1	2_1	22_1	22_1	22_3
23_1	$xyeh^2$	23	23_1	23_1	23_2	23_2	23_2	23_1	23_2	1
23_2	yh^2ye	23	23_2	23_2	23_1	23_1	23_1	23_2	23_1	1
24_1	$xheh$	$2^4 \cdot 3$	12_2	8_3	24_1	24_1	24_1	24_1	24_1	24_1
24_2	$xyhyxe$	$2^4 \cdot 3$	12_6	8_2	24_2	24_2	24_2	24_2	24_2	24_2
24_3	$xyhy^2h$	$2^4 \cdot 3$	12_7	8_1	24_3	24_3	24_3	24_3	24_3	24_3
26_1	xy^3h	$2 \cdot 13$	13_1	26_1	26_2	26_2	26_2	2_1	26_1	26_1
26_2	xy^2eh	$2 \cdot 13$	13_2	26_2	26_1	26_1	26_1	2_1	26_2	26_2
27	xyh^2e	3^3	27	9_1	27	27	27	27	27	27
28	$xhyhxe$	$2^2 \cdot 7$	14_2	28	28	4_1	28	28	28	28
30_1	xhy^3	$2^3 \cdot 3 \cdot 5$	15_1	10_2	6_3	30_1	30_1	30_1	30_1	30_1
30_2	$xyxh$	$2^2 \cdot 3 \cdot 5$	15_1	10_1	6_1	30_2	30_2	30_2	30_2	30_2
30_3	$xehy$	$2 \cdot 3 \cdot 5$	15_2	10_3	6_7	30_3	30_3	30_3	30_3	30_3
35	y^2he	$5 \cdot 7$	35	35	7	5	35	35	35	35
36_1	$xyhey$	$2^2 \cdot 3^2$	18_5	12_5	36_1	36_1	36_1	36_1	36_1	36_1
36_2	xy^2eyh	$2^2 \cdot 3^2$	18_6	12_4	36_2	36_2	36_2	36_2	36_2	36_2
39_1	$xhyhe$	$3 \cdot 13$	39_2	13_2	39_2	39_2	39_2	3_1	39_1	39_1
39_2	xh^2xey	$3 \cdot 13$	39_1	13_1	39_1	39_1	39_1	3_1	39_2	39_2
42	$yeyeh$	$2 \cdot 3 \cdot 7$	21	14_1	42	6_1	42	42	42	42
60	$xeyh^2$	$2^2 \cdot 3 \cdot 5$	30_1	20_1	12_1	60	60	60	60	60

6.6 Character tables of local subgroups of Fi_{23}

6.6.1 *Character table of* $E = E(\mathsf{Fi}_{23}) = \langle x, y, e \rangle$

2	18	18	18	18	14	14	7	11	12	12	11	9	9	8	3	7	7	7	6	6	5	5	3	3
3	2	2	2	2	1	1	2	1	1	1	1	.	.	.	1	2	2	2	2	2	1	1	.	.
5	1	1	1	1	.	.	1	.	.	.	.	.	.	.	1	1	.	.	.	.	.	.	.	.
7	1	1	1	.	1	.	.	1	.	.	.	.	.	.	.	.	.	.	.	.	.	.	1	1
11	1	1	.	.	.	.	.	.	.	.	.	.	.	.	.	.	.	.	.	.	.	.	.	.
23	1	.	.	.	.	.	.	.	.	.	.	.	.	.	.	.	.	.	.	.	.	.	.	.
	1a	2a	2b	2c	2d	2e	3a	4a	4b	4c	4d	4e	4f	4g	5a	6a	6b	6c	6d	6e	6f	6g	7a	7b
2P	1a	1a	1a	1a	1a	1a	3a	2b	2c	2c	2b	2e	2e	2d	5a	3a	3a	3a	3a	3a	3a	3a	7a	7b
3P	1a	2a	2b	2c	2d	2e	1a	4a	4b	4c	4d	4e	4f	4g	5a	2c	2c	2a	2b	2c	2d	2e	7b	7a
5P	1a	2a	2b	2c	2d	2e	3a	4a	4b	4c	4d	4e	4f	4g	1a	6a	6b	6c	6d	6e	6f	6g	7b	7a
7P	1a	2a	2b	2c	2d	2e	3a	4a	4b	4c	4d	4e	4f	4g	5a	6a	6b	6c	6d	6e	6f	6g	1a	1a
11P	1a	2a	2b	2c	2d	2e	3a	4a	4b	4c	4d	4e	4f	4g	5a	6a	6b	6c	6d	6e	6f	6g	7a	7b
23P	1a	2a	2b	2c	2d	2e	3a	4a	4b	4c	4d	4e	4f	4g	5a	6a	6b	6c	6d	6e	6f	6g	7a	7b
X.1	1	1	1	1	1	1	1	1	1	1	1	1	1	1	1	1	1	1	1	1	1	1	1	1
X.2	22	22	22	22	6	6	4	6	6	6	6	2	2	2	2	4	4	4	4	4	.	.	1	1
X.3	45	45	45	45	−3	−3	.	−3	−3	−3	−3	1	1	1	.	.	.	.	.	.	.	.	A	$\bar{A}$
X.4	45	45	45	45	−3	−3	.	−3	−3	−3	−3	1	1	1	.	.	.	.	.	.	.	.	$\bar{A}$	A
X.5	230	230	230	230	22	22	5	22	22	22	22	2	2	2	.	5	5	5	5	5	1	1	−1	−1
X.6	231	231	231	231	7	7	6	7	7	7	7	−1	−1	−1	1	6	6	6	6	6	−2	−2	.	.
X.7	231	231	231	231	7	7	−3	7	7	7	7	−1	−1	−1	1	−3	−3	−3	−3	−3	1	1	.	.
X.8	231	231	231	231	7	7	−3	7	7	7	7	−1	−1	−1	1	−3	−3	−3	−3	−3	1	1	.	.
X.9	253	253	253	253	13	13	1	13	13	13	13	1	1	1	−2	1	1	1	1	1	1	1	1	1
X.10	253	−99	29	−3	29	−3	10	−15	−3	5	1	5	−3	1	3	.	−6	.	2	.	2	.	1	1
X.11	506	154	26	−6	42	10	11	14	−6	2	−2	6	−2	2	1	9	3	1	−1	−3	3	1	2	2
X.12	770	770	770	770	−14	−14	5	−14	−14	−14	−14	−2	−2	−2	.	5	5	5	5	5	1	1	.	.
X.13	770	770	770	770	−14	−14	5	−14	−14	−14	−14	−2	−2	−2	.	5	5	5	5	5	1	1	.	.
X.14	896	896	896	896	.	.	−4	.	.	.	.	.	.	.	1	−4	−4	−4	−4	−4	.	.	.	.
X.15	896	896	896	896	.	.	−4	.	.	.	.	.	.	.	1	−4	−4	−4	−4	−4	.	.	.	.
X.16	990	990	990	990	−18	−18	.	−18	−18	−18	−18	2	2	2	.	.	.	.	.	.	.	.	$\bar{A}$	A
X.17	990	990	990	990	−18	−18	.	−18	−18	−18	−18	2	2	2	.	.	.	.	.	.	.	.	A	$\bar{A}$
X.18	1035	1035	1035	1035	27	27	.	27	27	27	27	−1	−1	−1	.	.	.	.	.	.	.	.	−1	−1
X.19	1288	−56	−56	8	56	−8	10	.	8	−8	.	4	4	−4	3	−10	2	−2	−2	2	2	−2	.	.
X.20	1518	−594	174	−18	62	−2	15	−34	−2	14	−2	2	2	2	3	−15	−9	9	3	−3	−1	1	−1	−1
X.21	2024	2024	2024	2024	8	8	−1	8	8	8	8	.	.	.	−1	−1	−1	−1	−1	−1	−1	−1	1	1
X.22	2530	−990	290	−30	−46	18	10	18	18	2	−14	−2	−2	−2	.	.	−6	.	2	.	2	.	A	$\bar{A}$
X.23	2530	−990	290	−30	−46	18	10	18	18	2	−14	−2	−2	−2	.	.	−6	.	2	.	2	.	$\bar{A}$	A
X.24	3542	−1386	406	−42	70	6	5	−42	6	22	−10	2	2	2	−3	15	−3	−9	1	3	1	3	.	.
X.25	3542	1078	182	−42	70	−26	14	42	−10	14	−6	2	−6	−2	2	6	6	−2	2	−6	−2	−2	.	.
X.26	3542	−1386	406	−42	70	6	5	−42	6	22	−10	2	2	2	−3	−15	−3	9	1	−3	1	−3	.	.
X.27	3795	−1485	435	−45	83	−13	15	−41	−13	11	7	3	−5	−1	.	15	−9	−9	3	3	−1	−1	1	1
X.28	3795	−1485	435	−45	−13	19	15	−1	19	11	−17	−5	3	−1	.	−15	−9	9	3	−3	−1	1	1	1
X.29	5313	−2079	609	−63	49	17	−15	−35	17	25	−19	−3	5	1	3	15	9	−9	−3	3	1	−1	.	.
X.30	7084	2156	364	−84	140	76	19	28	−20	−4	−4	4	4	4	−1	21	3	5	−5	−3	−1	1	.	.
X.31	8855	−3465	1015	−105	7	39	−10	−21	39	31	−37	3	−5	−1	.	.	6	.	−2	.	−2	.	.	.
X.32	10120	3080	520	−120	168	40	−5	56	−24	8	−8	.	.	.	.	−15	3	−7	7	−3	3	1	−2	−2
X.33	10626	3234	546	−126	−14	−46	6	14	2	10	−2	−2	6	2	1	−6	6	−6	6	−6	−2	2	.	.
X.34	10626	3234	546	−126	−14	−46	−3	14	2	10	−2	−2	6	2	1	3	−3	3	−3	3	1	−1	.	.
X.35	10626	3234	546	−126	−14	−46	−3	14	2	10	−2	−2	6	2	1	3	−3	3	−3	3	1	−1	.	.
X.36	11385	−4455	1305	−135	−87	9	.	45	9	−15	−3	5	−3	1	.	.	.	.	.	.	.	.	A	$\bar{A}$
X.37	11385	−4455	1305	−135	−87	9	.	45	9	−15	−3	5	−3	1	.	.	.	.	.	.	.	.	$\bar{A}$	A
X.38	12880	−560	−560	80	112	−16	10	.	16	−16	.	8	8	−8	.	−10	2	−2	−2	2	−2	2	.	.
X.39	12880	−560	−560	80	−112	16	10	.	−16	16	.	.	.	.	.	−10	2	−2	−2	2	2	−2	.	.
X.40	12880	−560	−560	80	−112	16	10	.	−16	16	.	.	.	.	.	−10	2	−2	−2	2	2	−2	.	.
X.41	14168	−616	−616	88	168	−24	20	.	24	−24	.	−4	−4	4	3	−20	4	−4	−4	4	.	.	.	.
X.42	14168	4312	728	−168	56	−72	11	56	−8	24	−8	.	.	.	−2	9	3	1	−1	−3	−1	−3	.	.
X.43	17710	5390	910	−210	−98	62	25	−70	14	−26	10	−6	2	−2	.	15	9	−1	1	−9	1	−1	.	.
X.44	20608	−896	−896	128	.	.	−20	.	.	.	.	.	.	.	3	20	−4	4	4	−4	.	.	.	.
X.45	20608	−896	−896	128	.	.	−20	.	.	.	.	.	.	.	3	20	−4	4	4	−4	.	.	.	.
X.46	22770	−8910	2610	−270	162	−30	.	−78	−30	18	18	−2	−2	−2	.	.	.	.	.	.	.	.	−1	−1
X.47	22770	6930	1170	−270	−126	−30	.	−42	18	−6	6	6	−2	2	.	.	.	.	.	.	.	.	F	$\bar{F}$
X.48	22770	6930	1170	−270	−126	−30	.	−42	18	−6	6	6	−2	2	.	.	.	.	.	.	.	.	$\bar{F}$	F
X.49	26565	−10395	3045	−315	−91	5	15	49	5	−19	1	−3	5	1	.	15	−9	−9	3	3	−1	−1	.	.
X.50	28336	8624	1456	−336	112	112	−14	.	−16	−16	.	.	.	.	1	−6	−6	2	−2	6	−2	−2	.	.
X.51	30360	−11880	3480	−360	−8	−8	−15	8	−8	−8	8	.	.	.	.	−15	9	9	−3	−3	1	1	1	1
X.52	32384	9856	1664	−384	.	.	−16	.	.	.	.	.	.	.	−1	−24	.	−8	8	.	.	.	2	2
X.53	35420	10780	1820	−420	28	−36	5	28	−4	12	−4	−4	−4	−4	.	15	−3	7	−7	3	1	3	.	.
X.54	56672	−2464	−2464	352	224	−32	−10	.	32	−32	.	.	.	.	−3	10	−2	2	2	−2	2	−2	.	.
X.55	57960	−2520	−2520	360	−168	24	.	.	−24	24	.	4	4	−4	.	.	.	.	.	.	.	.	.	.
X.56	70840	−3080	−3080	440	−56	8	10	.	−8	8	.	−4	−4	4	.	−10	2	−2	−2	2	−2	2	.	.

Character table of $E(\mathsf{Fi}_{23})$ *(continued)*

2	7	7	7	5	5	3	3	3	1	1	5	5	4	4	3	3	2	2	2	2	1	1	5	5
3	.	.	.	.	.	1	.	.	.	.	1	1	1	1	.	.	.	.	.	.	1	1	.	.
5	.	.	.	.	.	1	1	1	.	.	.	.	.	.	.	.	.	.	.	.	1	1	.	.
7	.	.	.	.	.	.	.	.	.	.	.	.	.	.	1	1	1	1	1	1	.	.	.	.
11	.	.	.	.	.	.	.	.	1	1	.	.	.	.	.	.	.	.	.	.	.	.	.	.
23	.	.	.	.	.	.	.	.	.	.	.	.	.	.	.	.	.	.	.	.	.	.	.	.
	8a	8b	8c	8d	8e	10a	10b	10c	11a	11b	12a	12b	12c	12d	14a	14b	14c	14d	14e	14f	15a	15b	16a	16b
2P	4b	4c	4c	4e	4f	5a	5a	5a	11b	11a	6b	6b	6d	6d	7b	7a	7a	7b	7b	7a	15a	15b	8a	8a
3P	8a	8b	8c	8d	8e	10a	10b	10c	11a	11b	4b	4c	4d	4a	14b	14a	14d	14c	14f	14e	5a	5a	16a	16b
5P	8a	8b	8c	8d	8e	2c	2b	2a	11a	11b	12a	12b	12c	12d	14b	14a	14d	14c	14f	14e	3a	3a	16b	16a
7P	8a	8b	8c	8d	8e	10a	10b	10c	11b	11a	12a	12b	12c	12d	2b	2b	2a	2a	2d	2d	15b	15a	16b	16a
11P	8a	8b	8c	8d	8e	10a	10b	10c	1a	1a	12a	12b	12c	12d	14a	14b	14c	14d	14e	14f	15b	15a	16a	16b
23P	8a	8b	8c	8d	8e	10a	10b	10c	11a	11b	12a	12b	12c	12d	14a	14b	14c	14d	14e	14f	15a	15b	16b	16a
X.1	1	1	1	1	1	1	1	1	1	1	1	1	1	1	1	1	1	1	1	1	1	1	1	1
X.2	2	2	2	.	.	2	2	2	.	.	.	.	.	.	1	1	1	1	−1	−1	−1	−1	.	.
X.3	1	1	1	−1	−1	.	.	.	1	1	.	.	.	.	$\bar{A}$	A	A	$\bar{A}$	$-\bar{A}$	$-A$	.	.	−1	−1
X.4	1	1	1	−1	−1	.	.	.	1	1	.	.	.	.	A	$\bar{A}$	$\bar{A}$	A	$-A$	$-\bar{A}$	.	.	−1	−1
X.5	2	2	2	.	.	.	.	.	−1	−1	1	1	1	1	−1	−1	−1	−1	1	1	.	.	.	.
X.6	−1	−1	−1	−1	−1	1	1	1	.	.	−2	−2	−2	−2	.	.	.	.	.	.	1	1	−1	−1
X.7	−1	−1	−1	−1	−1	1	1	1	.	.	1	1	1	1	.	.	.	.	.	.	B	$\bar{B}$	−1	−1
X.8	−1	−1	−1	−1	−1	1	1	1	.	.	1	1	1	1	.	.	.	.	.	.	$\bar{B}$	B	−1	−1
X.9	1	1	1	−1	−1	−2	−2	−2	.	.	1	1	1	1	1	1	1	1	−1	−1	1	1	−1	−1
X.10	1	−3	1	1	1	−3	−1	1	.	.	.	2	−2	.	1	1	−1	−1	1	1	.	.	−1	−1
X.11	−2	2	−2	.	.	−1	1	−1	.	.	−3	−1	1	−1	−2	−2	.	.	.	.	1	1	.	.
X.12	−2	−2	−2	.	.	.	.	.	.	.	1	1	1	1	.	.	.	.	.	.	.	.	.	.
X.13	−2	−2	−2	.	.	.	.	.	.	.	1	1	1	1	.	.	.	.	.	.	.	.	.	.
X.14	.	.	.	.	.	1	1	1	D	$\bar{D}$	.	.	.	.	.	.	.	.	.	.	1	1	.	.
X.15	.	.	.	.	.	1	1	1	$\bar{D}$	D	.	.	.	.	.	.	.	.	.	.	1	1	.	.
X.16	2	2	2	.	.	.	.	.	.	.	.	.	.	.	A	$\bar{A}$	$\bar{A}$	A	A	$\bar{A}$	.	.	.	.
X.17	2	2	2	.	.	.	.	.	.	.	.	.	.	.	$\bar{A}$	A	A	$\bar{A}$	$\bar{A}$	A	.	.	.	.
X.18	−1	−1	−1	1	1	.	.	.	1	1	.	.	.	.	−1	−1	−1	−1	−1	−1	.	.	1	1
X.19	.	.	.	−2	2	3	−1	−1	1	1	2	−2	.	.	.	.	.	.	.	.	.	.	.	.
X.20	2	−2	−2	.	.	−3	−1	1	.	.	1	−1	1	−1	−1	−1	1	1	−1	−1	.	.	.	.
X.21	.	.	.	.	.	−1	−1	−1	.	.	−1	−1	−1	−1	1	1	1	1	1	1	−1	−1	.	.
X.22	−2	2	2	.	.	.	.	.	.	.	.	2	−2	.	$\bar{A}$	A	$-A$	$-\bar{A}$	$\bar{A}$	A	.	.	.	.
X.23	−2	2	2	.	.	.	.	.	.	.	.	2	−2	.	A	$\bar{A}$	$-\bar{A}$	$-A$	A	$\bar{A}$	.	.	.	.
X.24	2	−2	−2	.	.	3	1	−1	.	.	3	1	−1	−3	.	.	.	.	.	.	.	.	.	.
X.25	2	2	−2	.	.	−2	2	−2	.	.	2	2	.	.	.	.	.	.	.	.	−1	−1	.	.
X.26	2	−2	−2	.	.	3	1	−1	.	.	−3	1	−1	3	.	.	.	.	.	.	.	.	.	.
X.27	−1	−1	3	−1	−1	.	.	.	.	.	−1	−1	1	1	1	1	−1	−1	−1	−1	.	.	1	1
X.28	−1	3	−1	−1	−1	.	.	.	.	.	1	−1	1	−1	1	1	−1	−1	1	1	.	.	1	1
X.29	1	1	−3	−1	−1	−3	−1	1	.	.	−1	1	−1	1	.	.	.	.	.	.	.	.	1	1
X.30	−4	.	.	.	.	1	−1	1	.	.	1	−1	−1	1	.	.	.	.	.	.	−1	−1	.	.
X.31	−1	−1	3	1	1	.	.	.	.	.	.	−2	2	.	.	.	.	.	.	.	.	.	−1	−1
X.32	.	.	.	.	.	.	.	.	.	.	−3	−1	1	−1	2	2	.	.	.	.	.	.	.	.
X.33	−2	−2	2	.	.	−1	1	−1	.	.	2	−2	−2	2	.	.	.	.	.	.	1	1	.	.
X.34	−2	−2	2	.	.	−1	1	−1	.	.	−1	1	1	−1	.	.	.	.	.	.	B	$\bar{B}$	.	.
X.35	−2	−2	2	.	.	−1	1	−1	.	.	−1	1	1	−1	.	.	.	.	.	.	$\bar{B}$	B	.	.
X.36	1	−3	1	−1	−1	.	.	.	.	.	.	.	.	.	$\bar{A}$	A	$-A$	$-\bar{A}$	$-\bar{A}$	$-A$	.	.	1	1
X.37	1	−3	1	−1	−1	.	.	.	.	.	.	.	.	.	A	$\bar{A}$	$-\bar{A}$	$-A$	$-A$	$-\bar{A}$	.	.	1	1
X.38	.	.	.	.	.	.	.	.	−1	−1	−2	2	.	.	.	.	.	.	.	.	.	.	.	.
X.39	.	.	.	.	.	.	.	.	−1	−1	2	−2	.	.	.	.	.	.	.	.	.	.	E	$\bar{E}$
X.40	.	.	.	.	.	.	.	.	−1	−1	2	−2	.	.	.	.	.	.	.	.	.	.	$\bar{E}$	E
X.41	.	.	.	2	−2	3	−1	−1	.	.	.	.	.	.	.	.	.	.	.	.	.	.	.	.
X.42	.	.	.	.	.	2	−2	2	.	.	1	3	1	−1	.	.	.	.	.	.	1	1	.	.
X.43	2	−2	2	.	.	.	.	.	.	.	−1	1	1	−1	.	.	.	.	.	.	.	.	.	.
X.44	.	.	.	.	.	3	−1	−1	D	$\bar{D}$	.	.	.	.	.	.	.	.	.	.	.	.	.	.
X.45	.	.	.	.	.	3	−1	−1	$\bar{D}$	D	.	.	.	.	.	.	.	.	.	.	.	.	.	.
X.46	−2	2	2	.	.	.	.	.	.	.	.	.	.	.	−1	−1	1	1	1	1	.	.	.	.
X.47	−2	2	−2	.	.	.	.	.	.	.	.	.	.	.	$-\bar{F}$	$-F$	.	.	.	.	.	.	.	.
X.48	−2	2	−2	.	.	.	.	.	.	.	.	.	.	.	$-F$	$-\bar{F}$	.	.	.	.	.	.	.	.
X.49	1	1	−3	1	1	.	.	.	.	.	−1	−1	1	1	.	.	.	.	.	.	.	.	−1	−1
X.50	.	.	.	.	.	−1	1	−1	.	.	2	2	.	.	.	.	.	.	.	.	1	1	.	.
X.51	.	.	.	.	.	.	.	.	.	.	1	1	−1	−1	1	1	−1	−1	−1	−1	.	.	.	.
X.52	.	.	.	.	.	1	−1	1	.	.	.	.	.	.	−2	−2	.	.	.	.	−1	−1	.	.
X.53	4	.	.	.	.	.	.	.	.	.	−1	−3	−1	1	.	.	.	.	.	.	.	.	.	.
X.54	.	.	.	.	.	−3	1	1	.	.	2	−2	.	.	.	.	.	.	.	.	.	.	.	.
X.55	.	.	.	2	−2	.	.	.	1	1	.	.	.	.	.	.	.	.	.	.	.	.	.	.
X.56	.	.	.	−2	2	.	.	.	.	.	−2	2	.	.	.	.	.	.	.	.	.	.	.	.

Character table of $E(\mathsf{Fi}_{23})$ *(continued)*

2	1	1	.	.	2	2	1	1
3	.	.	.	.	.	.	1	1
5	.	.	.	.	.	.	1	1
7	.	.	.	.	1	1	.	.
11	1	1	.	.	.	.	.	.
23	.	.	1	1	.	.	.	.
	22a	22b	23a	23b	28a	28b	30a	30b
2P	11a	11b	23a	23b	14a	14b	15a	15b
3P	22a	22b	23a	23b	28b	28a	10a	10a
5P	22a	22b	23b	23a	28b	28a	6a	6a
7P	22b	22a	23b	23a	4a	4a	30b	30a
11P	2a	2a	23b	23a	28a	28b	30b	30a
23P	22a	22b	1a	1a	28a	28b	30a	30b
X.1	1	1	1	1	1	1	1	1
X.2	.	.	−1	−1	−1	−1	−1	−1
X.3	1	1	−1	−1	$-\bar{A}$	$-A$	.	.
X.4	1	1	−1	−1	$-A$	$-\bar{A}$	.	.
X.5	−1	−1	.	.	1	1	.	.
X.6	.	.	1	1	.	.	1	1
X.7	.	.	1	1	.	.	B	$\bar{B}$
X.8	.	.	1	1	.	.	$\bar{B}$	B
X.9	.	.	.	.	−1	−1	1	1
X.10	.	.	.	.	−1	−1	.	.
X.11	.	.	.	.	.	.	−1	−1
X.12	.	.	C	$\bar{C}$	.	.	.	.
X.13	.	.	$\bar{C}$	C	.	.	.	.
X.14	$\bar{D}$	D	−1	−1	.	.	1	1
X.15	D	$\bar{D}$	−1	−1	.	.	1	1
X.16	.	.	1	1	A	$\bar{A}$	.	.
X.17	.	.	1	1	$\bar{A}$	A	.	.
X.18	1	1	.	.	−1	−1	.	.
X.19	−1	−1	.	.	.	.	.	.
X.20	.	.	.	.	1	1	.	.
X.21	.	.	.	.	1	1	−1	−1
X.22	.	.	.	.	$-\bar{A}$	$-A$	.	.
X.23	.	.	.	.	$-A$	$-\bar{A}$	.	.
X.24	.	.	.	.	.	.	.	.
X.25	.	.	.	.	.	.	1	1
X.26	.	.	.	.	.	.	.	.
X.27	.	.	.	.	1	1	.	.
X.28	.	.	.	.	−1	−1	.	.
X.29	.	.	.	.	.	.	.	.
X.30	.	.	.	.	.	.	1	1
X.31	.	.	.	.	.	.	.	.
X.32	.	.	.	.	.	.	.	.
X.33	.	.	.	.	.	.	−1	−1
X.34	.	.	.	.	.	.	$-B$	$-\bar{B}$
X.35	.	.	.	.	.	.	$-\bar{B}$	$-B$
X.36	.	.	.	.	$\bar{A}$	A	.	.
X.37	.	.	.	.	A	$\bar{A}$	.	.
X.38	1	1	.	.	.	.	.	.
X.39	1	1	.	.	.	.	.	.
X.40	1	1	.	.	.	.	.	.
X.41	.	.	.	.	.	.	.	.
X.42	.	.	.	.	.	.	−1	−1
X.43	.	.	.	.	.	.	.	.
X.44	$-\bar{D}$	$-D$	.	.	.	.	.	.
X.45	$-D$	$-\bar{D}$	.	.	.	.	.	.
X.46	.	.	.	.	−1	−1	.	.
X.47	.	.	.	.	.	.	.	.
X.48	.	.	.	.	.	.	.	.
X.49	.	.	.	.	.	.	.	.
X.50	.	.	.	.	.	.	−1	−1
X.51	.	.	.	.	1	1	.	.
X.52	.	.	.	.	.	.	1	1
X.53	.	.	.	.	.	.	.	.
X.54	.	.	.	.	.	.	.	.
X.55	−1	−1	.	.	.	.	.	.
X.56	.	.	.	.	.	.	.	.

where $A = -\zeta(7)^4 - \zeta(7)^2 - \zeta(7) - 1$, $B = 2\zeta(15)^{14} + 2\zeta(15)^{11} + \zeta(15)^5 + \zeta(15)^9 + \zeta(15)^6$, $C = \zeta(23)^{18} + \zeta(23)^{16} + \zeta(23)^{13} + \zeta(23)^{12} + \zeta(23)^9 + \zeta(23)^8 + \zeta(23)^6 + \zeta(23)^4 + \zeta(23)^3 + \zeta(23)^2 + \zeta(23)$, $D = \frac{1}{2}(-1 + i\sqrt{11})$, $E = -2\zeta(8)^3 - 2\zeta(8)$, $F = 2A$.

6.6.2 *Character table of* $D(\mathsf{Fi}_{23}) = \langle x, y \rangle$

2	18	18	17	17	18	18	16	14	14	13	7	11	11	12	12	10	10	10	9	9	8	8
3	2	2	2	2	1	1	2	1	1	.	2	1	1	.	.	1	1	.	.	.	.	.
5	1	1	1	1	1	1	.	.	.	.	.	.	.	.	.	.	.	.	.	.	.	.
7	1	1	1	1	.	.	.	.	.	.	.	.	.	.	.	.	.	.	.	.	.	.
11	1	1	.	.	.	.	.	.	.	.	.	.	.	.	.	.	.	.	.	.	.	.
	1a	2a	2b	2c	2d	2e	2f	2g	2h	2i	3a	4a	4b	4c	4d	4e	4f	4g	4h	4i	4j	4k
2P	1a	1a	1a	1a	1a	1a	1a	1a	1a	1a	3a	2e	2e	2d	2d	2f	2f	2e	2g	2g	2h	2i
3P	1a	2a	2b	2c	2d	2e	2f	2g	2h	2i	1a	4a	4b	4c	4d	4e	4f	4g	4h	4i	4j	4k
5P	1a	2a	2b	2c	2d	2e	2f	2g	2h	2i	3a	4a	4b	4c	4d	4e	4f	4g	4h	4i	4j	4k
7P	1a	2a	2b	2c	2d	2e	2f	2g	2h	2i	3a	4a	4b	4c	4d	4e	4f	4g	4h	4i	4j	4k
11P	1a	2a	2b	2c	2d	2e	2f	2g	2h	2i	3a	4a	4b	4c	4d	4e	4f	4g	4h	4i	4j	4k
X.1	1	1	1	1	1	1	1	1	1	1	1	1	1	1	1	1	1	1	1	1	1	1
X.2	21	21	21	21	21	21	21	5	5	5	3	5	5	5	5	5	5	5	1	1	1	1
X.3	45	45	45	45	45	45	45	-3	-3	-3	.	-3	-3	-3	-3	-3	-3	-3	1	1	1	1
X.4	45	45	45	45	45	45	45	-3	-3	-3	.	-3	-3	-3	-3	-3	-3	-3	1	1	1	1
X.5	55	55	55	55	55	55	55	7	7	7	1	7	7	7	7	7	7	7	3	3	3	-1
X.6	77	77	-35	-35	13	13	-3	13	13	-3	5	-7	-7	-3	5	1	1	1	-3	5	1	1
X.7	99	99	99	99	99	99	99	3	3	3	.	3	3	3	3	3	3	3	3	3	3	-1
X.8	154	154	154	154	154	154	154	10	10	10	1	10	10	10	10	10	10	10	-2	-2	-2	2
X.9	176	-176	-64	64	-16	16	.	-16	16	.	5	8	-8	.	.	4	-4	.	.	.	.	4
X.10	176	-176	64	-64	-16	16	.	-16	16	.	5	-8	8	.	.	4	-4	.	.	.	.	4
X.11	210	210	210	210	210	210	210	2	2	2	3	2	2	2	2	2	2	2	-2	-2	-2	-2
X.12	231	231	231	231	231	231	231	7	7	7	-3	7	7	7	7	7	7	7	-1	-1	-1	-1
X.13	280	280	280	280	280	280	280	-8	-8	-8	1	-8	-8	-8	-8	-8	-8	-8	.	.	.	.
X.14	280	280	280	280	280	280	280	-8	-8	-8	1	-8	-8	-8	-8	-8	-8	-8	.	.	.	.
X.15	330	330	90	90	10	10	-6	26	26	10	6	6	6	-6	2	-2	-2	-2	-2	6	2	2
X.16	385	385	-175	-175	65	65	-15	17	17	1	-2	-11	-11	1	9	5	5	-3	5	-3	1	1
X.17	385	385	385	385	385	385	385	1	1	1	-2	1	1	1	1	1	1	1	1	1	1	1
X.18	385	385	-175	-175	65	65	-15	17	17	1	7	-11	-11	1	9	5	5	-3	5	-3	1	1
X.19	616	616	-56	-56	-24	-24	8	24	24	-8	4	.	.	8	-8	.	.	.	4	4	-4	4
X.20	616	616	-56	-56	-24	-24	8	24	24	-8	4	.	.	8	-8	.	.	.	4	4	-4	-4
X.21	616	616	-280	-280	104	104	-24	8	8	8	-5	-8	-8	8	8	8	8	-8	.	.	.	.
X.22	616	616	-280	-280	104	104	-24	8	8	8	-5	-8	-8	8	8	8	8	-8	.	.	.	.
X.23	672	-672	.	.	32	-32	.	-32	32	.	6	.	.	.	.	-8	8	.	.	.	.	.
X.24	693	693	-315	-315	117	117	-27	21	21	5	.	-15	-15	5	13	9	9	-7	-3	5	1	1
X.25	770	770	-350	-350	130	130	-30	-14	-14	18	5	2	2	18	2	10	10	-14	-2	-2	-2	-2
X.26	990	990	270	270	30	30	-18	-18	-18	-2	.	-6	-6	14	6	-6	-6	2	-6	2	-2	2
X.27	990	990	270	270	30	30	-18	-18	-18	-2	.	-6	-6	14	6	-6	-6	2	-6	2	-2	2
X.28	1056	-1056	-384	384	-96	96	.	-32	32	.	3	16	-16	.	.	8	-8	.	.	.	.	.
X.29	1056	-1056	384	-384	-96	96	.	-32	32	.	3	-16	16	.	.	8	-8	.	.	.	.	.
X.30	1155	1155	-525	-525	195	195	-45	35	35	-13	3	-17	-17	-13	11	-1	-1	7	-5	3	-1	-1
X.31	1155	1155	-525	-525	195	195	-45	-13	-13	3	3	7	7	3	-5	-1	-1	-1	7	-1	3	-1
X.32	1760	-1760	640	-640	-160	160	.	32	-32	.	5	16	-16	.	.	-8	8	.	.	.	.	.
X.33	1760	-1760	-640	640	-160	160	.	32	-32	.	5	-16	16	.	.	-8	8	.	.	.	.	.
X.34	1760	-1760	640	-640	-160	160	.	32	-32	.	5	16	-16	.	.	-8	8	.	.	.	.	.
X.35	1760	-1760	-640	640	-160	160	.	32	-32	.	5	-16	16	.	.	-8	8	.	.	.	.	.
X.36	1980	1980	540	540	60	60	-36	60	60	28	.	12	12	-4	12	-12	-12	-4	4	4	4	.
X.37	2310	2310	630	630	70	70	-42	22	22	-26	6	18	18	-10	14	2	2	-6	-6	2	-2	-2
X.38	2310	2310	-1050	-1050	390	390	-90	22	22	-10	-3	-10	-10	-10	6	-2	-2	6	2	2	2	-2
X.39	2310	2310	630	630	70	70	-42	22	22	38	6	-6	-6	-10	-18	2	2	2	2	-6	-2	2
X.40	2310	2310	630	630	70	70	-42	-10	-10	6	6	-6	-6	22	14	-14	-14	2	2	-6	-2	-2
X.41	2464	-2464	-896	896	-224	224	.	-32	32	.	7	16	-16	.	.	8	-8	.	.	.	.	.
X.42	2464	-2464	896	-896	-224	224	.	-32	32	.	-2	-16	16	.	.	8	-8	.	.	.	.	.
X.43	2464	-2464	-896	896	-224	224	.	-32	32	.	-2	16	-16	.	.	8	-8	.	.	.	.	.
X.44	2464	-2464	896	-896	-224	224	.	-32	32	.	7	-16	16	.	.	8	-8	.	.	.	.	.
X.45	2640	-2640	960	-960	-240	240	.	16	-16	.	3	8	-8	.	.	-4	4	.	.	.	.	-4
X.46	2640	2640	720	720	80	80	-48	16	16	16	-6	.	.	16	16	-16	-16	.	.	.	.	.
X.47	2640	-2640	-960	960	-240	240	.	16	-16	.	3	-8	8	.	.	-4	4	.	.	.	.	-4
X.48	3360	-3360	.	.	160	-160	.	-32	32	.	-6	.	.	.	.	-8	8	.	.	.	.	.
X.49	3360	-3360	.	.	160	-160	.	-32	32	.	-6	.	.	.	.	-8	8	.	.	.	.	.
X.50	3465	3465	-1575	-1575	585	585	-135	9	9	-7	.	-3	-3	-7	1	-3	-3	5	1	-7	-3	1
X.51	3465	3465	-1575	-1575	585	585	-135	-39	-39	9	.	21	21	9	-15	-3	-3	-3	-3	5	1	1
X.52	3696	-3696	-1344	1344	-336	336	.	-16	16	.	-3	8	-8	.	.	4	-4	.	.	.	.	-4
X.53	3696	-3696	1344	-1344	-336	336	.	-16	16	.	-3	-8	8	.	.	4	-4	.	.	.	.	-4
X.54	4620	4620	1260	1260	140	140	-84	44	44	12	-6	12	12	-20	-4	4	4	-4	-4	-4	-4	.
X.55	5544	5544	-504	-504	-216	-216	72	24	24	-8	.	.	.	8	-8	.	.	.	4	4	-4	-4
X.56	5544	5544	-504	-504	-216	-216	72	24	24	-8	.	.	.	8	-8	.	.	.	4	4	-4	4
X.57	6160	-6160	-2240	2240	-560	560	.	16	-16	.	-5	-8	8	.	.	-4	4	.	.	.	.	4
X.58	6160	6160	-560	-560	-240	-240	80	48	48	-16	4	.	.	16	-16	.	.	.	-8	-8	8	.
X.59	6160	-6160	2240	-2240	-560	560	.	16	-16	.	-5	8	-8	.	.	-4	4	.	.	.	.	4
X.60	6160	6160	-560	-560	-240	-240	80	-48	-48	16	4	.	.	-16	16	.	.	.	.	.	.	.
X.61	6160	6160	-560	-560	-240	-240	80	-48	-48	16	4	.	.	-16	16	.	.	.	.	.	.	.
X.62	6720	-6720	.	.	320	-320	.	-64	64	.	6	.	.	.	.	-16	16	.	.	.	.	.
X.63	6720	-6720	.	.	320	-320	.	64	-64	.	6	.	.	.	.	16	-16	.	.	.	.	.
X.64	6930	6930	1890	1890	210	210	-126	-30	-30	-46	.	6	6	2	10	6	6	-2	6	-2	2	2
X.65	6930	6930	1890	1890	210	210	-126	-30	-30	18	.	-18	-18	2	-22	6	6	6	-2	6	2	-2
X.66	7392	-7392	.	.	352	-352	.	32	-32	.	-6	.	.	.	.	8	-8	.	.	.	.	.
X.67	8064	-8064	.	.	384	-384	.	.	.	.	.	.	.	.	.	.	.	.	.	.	.	.
X.68	8064	-8064	.	.	384	-384	.	.	.	.	.	.	.	.	.	.	.	.	.	.	.	.
X.69	9856	9856	-896	-896	-384	-384	128	.	.	.	-8	.	.	.	.	.	.	.	.	.	.	.

Character table of $D(\mathsf{Fi}_{23})$ *(continued)*

2	8	7	3	7	7	7	5	5	5	5	6	6	5	5	2	2	7	7	7	6	6	6	5	5	3	3	3	3	3	3
3	.	.	.	2	2	2	2	2	2	2	1	1	1	1	.	.	.	.	.	.	.	.	.	.	.	.	.	.	.	.
5	.	.	1	.	.	.	.	.	.	.	.	.	.	.	.	.	.	.	.	.	.	.	.	.	1	1	1	1	1	1
7	.	.	.	.	.	.	.	.	.	.	.	.	.	.	1	1	.	.	.	.	.	.	.	.	.	.	.	.	.	.
11	.	.	.	.	.	.	.	.	.	.	.	.	.	.	.	.	.	.	.	.	.	.	.	.	.	.	.	.	.	.
	4l	4m	5a	6a	6b	6c	6d	6e	6f	6g	6h	6i	6j	6k	7a	7b	8a	8b	8c	8d	8e	8f	8g	8h	10a	10b	10c	10d	10e	10f
2P	2i	2h	5a	3a	3a	3a	3a	3a	3a	3a	3a	3a	3a	3a	7a	7b	4c	4d	4d	4e	4c	4e	4h	4i	5a	5a	5a	5a	5a	5a
3P	4l	4m	5a	2f	2f	2a	2f	2f	2c	2b	2e	2d	2h	2g	7b	7a	8a	8b	8c	8d	8e	8f	8g	8h	10b	10a	10c	10f	10e	10d
5P	4l	4m	1a	6a	6b	6c	6d	6e	6f	6g	6h	6i	6j	6k	7b	7a	8a	8b	8c	8d	8e	8f	8g	8h	2c	2c	2e	2b	2d	2b
7P	4l	4m	5a	6a	6b	6c	6d	6e	6f	6g	6h	6i	6j	6k	1a	1a	8a	8b	8c	8d	8e	8f	8g	8h	10b	10a	10c	10f	10e	10d
11P	4l	4m	5a	6a	6b	6c	6d	6e	6f	6g	6h	6i	6j	6k	7a	7b	8a	8b	8c	8d	8e	8f	8g	8h	10a	10b	10c	10d	10e	10f
X.1	1	1	1	1	1	1	1	1	1	1	1	1	1	1	1	1	1	1	1	1	1	1	1	1	1	1	1	1	1	1
X.2	1	1	1	3	3	3	3	3	3	3	3	3	−1	−1	.	.	1	1	1	1	1	1	−1	−1	1	1	1	1	1	1
X.3	1	1	.	.	.	.	.	.	.	.	.	.	.	.	A	Ā	1	1	1	1	1	1	−1	−1	.	.	.	.	.	.
X.4	1	1	.	.	.	.	.	.	.	.	.	.	.	.	Ā	A	1	1	1	1	1	1	−1	−1	.	.	.	.	.	.
X.5	−1	−1	.	1	1	1	1	1	1	1	1	1	1	1	−1	−1	3	3	3	−1	−1	−1	1	1	.	.	.	.	.	.
X.6	1	1	2	−3	−3	5	−3	−3	1	1	1	1	1	1	.	.	1	−3	1	−1	1	−1	1	1	.	.	−2	.	−2	.
X.7	−1	−1	−1	.	.	.	.	.	.	.	.	.	.	.	1	1	3	3	3	−1	−1	−1	−1	−1	−1	−1	−1	−1	−1	−1
X.8	2	2	−1	1	1	1	1	1	1	1	1	1	1	1	.	.	−2	−2	−2	2	2	2	.	.	−1	−1	−1	−1	−1	−1
X.9	−4	.	1	−3	3	−5	3	−3	1	−1	1	−1	1	−1	1	1	.	.	.	−2	.	2	.	.	−1	−1	1	1	−1	1
X.10	−4	.	1	−3	3	−5	−3	3	−1	1	1	−1	1	−1	1	1	.	.	.	2	.	−2	.	.	1	1	1	−1	−1	−1
X.11	−2	−2	.	3	3	3	3	3	3	3	3	3	−1	−1	.	.	−2	−2	−2	−2	−2	−2	.	.	.	.	.	.	.	.
X.12	−1	−1	1	−3	−3	−3	−3	−3	−3	−3	−3	−3	1	1	.	.	−1	−1	−1	−1	−1	−1	−1	−1	1	1	1	1	1	1
X.13	.	.	.	1	1	1	1	1	1	1	1	1	1	1	.	.	.	.	.	.	.	.	.	.	.	.	.	.	.	.
X.14	.	.	.	1	1	1	1	1	1	1	1	1	1	1	.	.	.	.	.	.	.	.	.	.	.	.	.	.	.	.
X.15	2	2	.	6	6	6	.	.	.	.	−2	−2	2	2	1	1	−2	2	−2	.	−2	.	.	.	.	.	.	.	.	.
X.16	1	1	.	6	6	−2	.	.	−4	−4	2	2	2	2	.	.	1	1	−3	−1	1	−1	−1	−1	.	.	.	.	.	.
X.17	1	1	.	−2	−2	−2	−2	−2	−2	−2	−2	−2	−2	−2	.	.	1	1	1	1	1	1	1	1	.	.	.	.	.	.
X.18	1	1	.	−9	−9	7	−3	−3	5	5	−1	−1	−1	−1	.	.	1	1	−3	−1	1	−1	−1	−1	.	.	.	.	.	.
X.19	4	−4	1	−4	−4	4	2	2	−2	−2	.	.	.	.	.	.	.	.	.	.	.	.	2	−2	−1	−1	1	−1	1	−1
X.20	−4	4	1	−4	−4	4	2	2	−2	−2	.	.	.	.	.	.	.	.	.	.	.	.	−2	2	−1	−1	1	−1	1	−1
X.21	.	.	1	3	3	−5	3	3	−1	−1	−1	−1	−1	−1	.	.	.	.	.	.	.	.	.	.	C	−C	−1	−C	−1	C
X.22	.	.	1	3	3	−5	3	3	−1	−1	−1	−1	−1	−1	.	.	.	.	.	.	.	.	.	.	−C	C	−1	C	−1	−C
X.23	.	.	2	6	−6	−6	.	.	.	.	−2	2	2	−2	.	.	.	.	.	.	.	.	.	.	.	.	−2	.	2	.
X.24	1	1	−2	.	.	.	.	.	.	.	.	.	.	.	.	.	1	−3	1	−1	1	−1	1	1	.	.	2	.	2	.
X.25	−2	−2	.	−3	−3	5	−3	−3	1	1	1	1	1	1	.	.	−2	2	2	2	−2	2	.	.	.	.	.	.	.	.
X.26	2	2	.	.	.	.	.	.	.	.	.	.	.	.	A	Ā	2	2	−2	.	−2	.	.	.	.	.	.	.	.	.
X.27	2	2	.	.	.	.	.	.	.	.	.	.	.	.	Ā	A	2	2	−2	.	−2	.	.	.	.	.	.	.	.	.
X.28	.	.	1	3	−3	−3	3	−3	−3	3	3	−3	−1	1	−1	−1	.	.	.	.	.	.	.	.	−1	−1	1	1	−1	1
X.29	.	.	1	3	−3	−3	−3	3	3	−3	3	−3	−1	1	−1	−1	.	.	.	.	.	.	.	.	1	1	1	−1	−1	−1
X.30	−1	−1	.	3	3	3	−3	−3	−3	−3	3	3	−1	−1	.	.	−1	−1	3	1	−1	1	−1	−1	.	.	.	.	.	.
X.31	−1	−1	.	3	3	3	−3	−3	−3	−3	3	3	−1	−1	.	.	3	−1	−5	1	−1	1	1	1	.	.	.	.	.	.
X.32	.	.	.	−3	3	−5	−3	3	−1	1	1	−1	1	−1	A	Ā	.	.	.	.	.	.	.	.	.	.	.	.	.	.
X.33	.	.	.	−3	3	−5	3	−3	1	−1	1	−1	1	−1	Ā	A	.	.	.	.	.	.	.	.	.	.	.	.	.	.
X.34	.	.	.	−3	3	−5	−3	3	−1	1	1	−1	1	−1	Ā	A	.	.	.	.	.	.	.	.	.	.	.	.	.	.
X.35	.	.	.	−3	3	−5	3	−3	1	−1	1	−1	1	−1	A	Ā	.	.	.	.	.	.	.	.	.	.	.	.	.	.
X.36	.	.	.	.	.	.	.	.	.	.	.	.	.	.	−1	−1	−4	.	.	.	.	.	.	.	.	.	.	.	.	.
X.37	−2	−2	.	6	6	6	.	.	.	.	−2	−2	−2	−2	.	.	2	2	−2	.	2	.	.	.	.	.	.	.	.	.
X.38	−2	−2	.	−3	−3	−3	3	3	3	3	−3	−3	1	1	.	.	2	−2	−2	2	−2	2	.	.	.	.	.	.	.	.
X.39	2	2	.	6	6	6	.	.	.	.	−2	−2	−2	−2	.	.	2	−2	2	.	−2	.	.	.	.	.	.	.	.	.
X.40	−2	−2	.	6	6	6	.	.	.	.	−2	−2	2	2	.	.	2	−2	2	.	2	.	.	.	.	.	.	.	.	.
X.41	.	.	−1	−9	9	−7	3	−3	5	−5	−1	1	−1	1	.	.	.	.	.	.	.	.	.	.	1	1	−1	−1	1	−1
X.42	.	.	−1	6	−6	2	.	.	4	−4	2	−2	2	−2	.	.	.	.	.	.	.	.	.	.	−1	−1	−1	1	1	1
X.43	.	.	−1	6	−6	2	.	.	−4	4	2	−2	2	−2	.	.	.	.	.	.	.	.	.	.	1	1	−1	−1	1	−1
X.44	.	.	−1	−9	9	−7	−3	3	−5	5	−1	1	−1	1	.	.	.	.	.	.	.	.	.	.	−1	−1	−1	1	1	1
X.45	4	.	.	3	−3	−3	−3	3	3	−3	3	−3	−1	1	1	1	.	.	.	−2	.	2	.	.	.	.	.	.	.	.
X.46	.	.	.	−6	−6	−6	.	.	.	.	2	2	−2	−2	1	1	.	.	.	.	.	.	.	.	.	.	.	.	.	.
X.47	4	.	.	3	−3	−3	3	−3	−3	3	3	−3	−1	1	1	1	.	.	.	2	.	−2	.	.	.	.	.	.	.	.
X.48	.	.	.	−6	6	6	.	.	.	.	2	−2	2	−2	.	.	.	.	.	.	.	.	.	.	.	.	.	.	.	.
X.49	.	.	.	−6	6	6	.	.	.	.	2	−2	2	−2	.	.	.	.	.	.	.	.	.	.	.	.	.	.	.	.
X.50	1	1	.	.	.	.	.	.	.	.	.	.	.	.	.	.	−3	5	1	−1	1	−1	1	1	.	.	.	.	.	.
X.51	1	1	.	.	.	.	.	.	.	.	.	.	.	.	.	.	1	−3	1	−1	1	−1	−1	−1	.	.	.	.	.	.
X.52	4	.	1	−3	3	3	−3	3	3	−3	−3	3	1	−1	.	.	.	.	.	2	.	−2	.	.	−1	−1	1	1	−1	1
X.53	4	.	1	−3	3	3	3	−3	−3	3	−3	3	1	−1	.	.	.	.	.	−2	.	2	.	.	1	1	1	−1	−1	−1
X.54	.	.	.	−6	−6	−6	.	.	.	.	2	2	2	2	.	.	4	.	.	.	.	.	.	.	.	.	.	.	.	.
X.55	−4	4	−1	.	.	.	.	.	.	.	.	.	.	.	.	.	.	.	.	.	.	.	2	−2	1	1	−1	1	−1	1
X.56	4	−4	−1	.	.	.	.	.	.	.	.	.	.	.	.	.	.	.	.	.	.	.	−2	2	1	1	−1	1	−1	1
X.57	−4	.	.	3	−3	5	−3	3	−1	1	−1	1	−1	1	.	.	.	.	.	−2	.	2	.	.	.	.	.	.	.	.
X.58	.	.	.	−4	−4	4	2	2	−2	−2	.	.	.	.	.	.	.	.	.	.	.	.	.	.	.	.	.	.	.	.
X.59	−4	.	.	3	−3	5	3	−3	1	−1	−1	1	−1	1	.	.	.	.	.	2	.	−2	.	.	.	.	.	.	.	.
X.60	.	.	.	−4	−4	4	2	2	−2	−2	.	.	.	.	.	.	.	.	.	.	.	.	.	.	.	.	.	.	.	.
X.61	.	.	.	−4	−4	4	2	2	−2	−2	.	.	.	.	.	.	.	.	.	.	.	.	.	.	.	.	.	.	.	.
X.62	.	.	.	6	−6	−6	.	.	.	.	−2	2	−2	2	.	.	.	.	.	.	.	.	.	.	.	.	.	.	.	.
X.63	.	.	.	6	−6	−6	.	.	.	.	−2	2	2	−2	.	.	.	.	.	.	.	.	.	.	.	.	.	.	.	.
X.64	2	2	.	.	.	.	.	.	.	.	.	.	.	.	.	.	−2	−2	2	.	−2	.	.	.	.	.	.	.	.	.
X.65	−2	−2	.	.	.	.	.	.	.	.	.	.	.	.	.	.	−2	2	−2	.	2	.	.	.	.	.	.	.	.	.
X.66	.	.	2	−6	6	6	.	.	.	.	2	−2	−2	2	.	.	.	.	.	.	.	.	.	.	.	.	−2	.	2	.
X.67	.	.	−1	.	.	.	.	.	.	.	.	.	.	.	.	.	.	.	.	.	.	.	.	.	C	−C	1	C	−1	−C
X.68	.	.	−1	.	.	.	.	.	.	.	.	.	.	.	.	.	.	.	.	.	.	.	.	.	−C	C	1	−C	−1	C
X.69	.	.	1	8	8	−8	−4	−4	4	4	.	.	.	.	.	.	.	.	.	.	.	.	.	.	−1	−1	1	−1	1	−1

Character table of $D(\mathsf{Fi}_{23})$ *(continued)*

2	3	1	1	5	5	4	4	2	2	2	2	2	2	5	5	1	1
3	.	.	.	1	1	1	1	.	.	.	.	.	.	.	.	.	.
5	1	.	.	.	.	.	.	.	.	.	.	.	.	.	.	.	.
7	.	.	.	.	.	.	.	1	1	1	1	1	1	.	.	.	.
11	.	1	1	.	.	.	.	.	.	.	.	.	.	.	.	1	1
	10g	11a	11b	12a	12b	12c	12d	14a	14b	14c	14d	14e	14f	16a	16b	22a	22b
2P	5a	11b	11a	6a	6a	6h	6h	7a	7b	7a	7b	7a	7b	8a	8a	11a	11b
3P	10g	11a	11b	4f	4e	4a	4b	14f	14e	14d	14c	14b	14a	16a	16b	22a	22b
5P	2a	11a	11b	12a	12b	12c	12d	14f	14e	14d	14c	14b	14a	16b	16a	22a	22b
7P	10g	11b	11a	12a	12b	12c	12d	2b	2a	2c	2c	2a	2b	16b	16a	22b	22a
11P	10g	1a	1a	12a	12b	12c	12d	14a	14b	14c	14d	14e	14f	16a	16b	2a	2a
X.1	1	1	1	1	1	1	1	1	1	1	1	1	1	1	1	1	1
X.2	1	−1	−1	−1	−1	−1	−1	.	.	.	.	.	.	−1	−1	−1	−1
X.3	.	1	1	.	.	.	.	A	$\bar{A}$	A	$\bar{A}$	A	$\bar{A}$	−1	−1	1	1
X.4	.	1	1	.	.	.	.	$\bar{A}$	A	$\bar{A}$	A	$\bar{A}$	A	−1	−1	1	1
X.5	.	.	.	1	1	1	1	−1	−1	−1	−1	−1	−1	1	1	.	.
X.6	2	.	.	1	1	−1	−1	.	.	.	.	.	.	−1	−1	.	.
X.7	−1	.	.	.	.	.	.	1	1	1	1	1	1	−1	−1	.	.
X.8	−1	.	.	1	1	1	1	.	.	.	.	.	.	.	.	.	.
X.9	−1	.	.	−1	1	−1	1	−1	−1	1	1	−1	−1	.	.	.	.
X.10	−1	.	.	−1	1	1	−1	1	−1	−1	−1	−1	1	.	.	.	.
X.11	.	1	1	−1	−1	−1	−1	.	.	.	.	.	.	.	.	1	1
X.12	1	.	.	1	1	1	1	.	.	.	.	.	.	−1	−1	.	.
X.13	.	B	$\bar{B}$	1	1	1	1	.	.	.	.	.	.	.	.	$\bar{B}$	B
X.14	.	$\bar{B}$	B	1	1	1	1	.	.	.	.	.	.	.	.	B	$\bar{B}$
X.15	.	.	.	−2	−2	.	.	−1	1	−1	−1	1	−1	.	.	.	.
X.16	.	.	.	2	2	−2	−2	.	.	.	.	.	.	1	1	.	.
X.17	.	.	.	−2	−2	−2	−2	.	.	.	.	.	.	1	1	.	.
X.18	.	.	.	−1	−1	1	1	.	.	.	.	.	.	1	1	.	.
X.19	1	.	.	.	.	.	.	.	.	.	.	.	.	.	.	.	.
X.20	1	.	.	.	.	.	.	.	.	.	.	.	.	.	.	.	.
X.21	1	.	.	−1	−1	1	1	.	.	.	.	.	.	.	.	.	.
X.22	1	.	.	−1	−1	1	1	.	.	.	.	.	.	.	.	.	.
X.23	−2	1	1	2	−2	.	.	.	.	.	.	.	.	.	.	−1	−1
X.24	−2	.	.	.	.	.	.	.	.	.	.	.	.	−1	−1	.	.
X.25	.	.	.	1	1	−1	−1	.	.	.	.	.	.	.	.	.	.
X.26	.	.	.	.	.	.	.	$-A$	$\bar{A}$	$-A$	$-\bar{A}$	A	$-\bar{A}$	.	.	.	.
X.27	.	.	.	.	.	.	.	$-\bar{A}$	A	$-\bar{A}$	$-A$	$\bar{A}$	$-A$	.	.	.	.
X.28	−1	.	.	1	−1	1	−1	1	1	−1	−1	1	1	.	.	.	.
X.29	−1	.	.	1	−1	−1	1	−1	1	1	1	1	−1	.	.	.	.
X.30	.	.	.	−1	−1	1	1	.	.	.	.	.	.	1	1	.	.
X.31	.	.	.	−1	−1	1	1	.	.	.	.	.	.	−1	−1	.	.
X.32	.	.	.	−1	1	1	−1	A	$-\bar{A}$	$-A$	$-\bar{A}$	$-A$	$\bar{A}$	.	.	.	.
X.33	.	.	.	−1	1	−1	1	$-\bar{A}$	$-A$	$\bar{A}$	A	$-\bar{A}$	$-A$	.	.	.	.
X.34	.	.	.	−1	1	1	−1	$\bar{A}$	$-A$	$-\bar{A}$	$-A$	$-\bar{A}$	A	.	.	.	.
X.35	.	.	.	−1	1	−1	1	$-A$	$-\bar{A}$	A	$\bar{A}$	$-A$	$-\bar{A}$	.	.	.	.
X.36	.	.	.	.	.	.	.	1	−1	1	1	−1	1	.	.	.	.
X.37	.	.	.	2	2	.	.	.	.	.	.	.	.	.	.	.	.
X.38	.	.	.	1	1	−1	−1	.	.	.	.	.	.	.	.	.	.
X.39	.	.	.	2	2	.	.	.	.	.	.	.	.	.	.	.	.
X.40	.	.	.	−2	−2	.	.	.	.	.	.	.	.	.	.	.	.
X.41	1	.	.	1	−1	1	−1	.	.	.	.	.	.	.	.	.	.
X.42	1	.	.	−2	2	2	−2	.	.	.	.	.	.	.	.	.	.
X.43	1	.	.	−2	2	−2	2	.	.	.	.	.	.	.	.	.	.
X.44	1	.	.	1	−1	−1	1	.	.	.	.	.	.	.	.	.	.
X.45	.	.	.	1	−1	−1	1	1	−1	−1	−1	−1	1	.	.	.	.
X.46	.	.	.	2	2	.	.	−1	1	−1	−1	1	−1	.	.	.	.
X.47	.	.	.	1	−1	1	−1	−1	−1	1	1	−1	−1	.	.	.	.
X.48	.	B	$\bar{B}$	2	−2	.	.	.	.	.	.	.	.	.	.	$-\bar{B}$	$-B$
X.49	.	$\bar{B}$	B	2	−2	.	.	.	.	.	.	.	.	.	.	$-B$	$-\bar{B}$
X.50	.	.	.	.	.	.	.	.	.	.	.	.	.	−1	−1	.	.
X.51	.	.	.	.	.	.	.	.	.	.	.	.	.	1	1	.	.
X.52	−1	.	.	−1	1	−1	1	.	.	.	.	.	.	.	.	.	.
X.53	−1	.	.	−1	1	1	−1	.	.	.	.	.	.	.	.	.	.
X.54	.	.	.	−2	−2	.	.	.	.	.	.	.	.	.	.	.	.
X.55	−1	.	.	.	.	.	.	.	.	.	.	.	.	.	.	.	.
X.56	−1	.	.	.	.	.	.	.	.	.	.	.	.	.	.	.	.
X.57	.	.	.	1	−1	1	−1	.	.	.	.	.	.	.	.	.	.
X.58	.	.	.	.	.	.	.	.	.	.	.	.	.	.	.	.	.
X.59	.	.	.	1	−1	−1	1	.	.	.	.	.	.	.	.	.	.
X.60	.	.	.	.	.	.	.	.	.	.	.	.	.	D	$\bar{D}$	.	.
X.61	.	.	.	.	.	.	.	.	.	.	.	.	.	$\bar{D}$	D	.	.
X.62	.	−1	−1	−2	2	.	.	.	.	.	.	.	.	.	.	1	1
X.63	.	−1	−1	2	−2	.	.	.	.	.	.	.	.	.	.	1	1
X.64	.	.	.	.	.	.	.	.	.	.	.	.	.	.	.	.	.
X.65	.	.	.	.	.	.	.	.	.	.	.	.	.	.	.	.	.
X.66	−2	.	.	−2	2	.	.	.	.	.	.	.	.	.	.	.	.
X.67	1	1	1	.	.	.	.	.	.	.	.	.	.	.	.	−1	−1
X.68	1	1	1	.	.	.	.	.	.	.	.	.	.	.	.	−1	−1
X.69	1	.	.	.	.	.	.	.	.	.	.	.	.	.	.	.	.

where $A = -\zeta(7)^4 - \zeta(7)^2 - \zeta(7) - 1$, $B = -\zeta(11)^9 - \zeta(11)^5 - \zeta(11)^4 - \zeta(11)^3 - \zeta(11) - 1$, $C = 2\zeta(5)^3 + 2\zeta(5)^2 + 1$, $D = -2\zeta(8)^3 - 2\zeta(8)$.

6.6.3 *Character table of* $H(\mathsf{Fi}_{23}) = \langle x, y, h \rangle$

2	18	18	17	17	18	18	16	9	8	7	4	12	11	11	12	10	10	10
3	9	9	6	6	4	4	3	7	9	7	7	3	2	2	1	2	2	1
5	2	2	1	1	1	1	.	1	.	.	.	.	1	1	.	.	.	.
7	1	1	1	1	.	.	.	1	.	.	.	.	.	.	.	.	.	.
11	1	1	1	1	.	.	.	.	.	.	.	.	.	.	.	.	.	.
13	1	1	.	.	.	.	.	.	.	.	.	.	.	.	.	.	.	.
	1a	2a	2b	2c	2d	2e	2f	3a	3b	3c	3d	4a	4b	4c	4d	4e	4f	4g
2P	1a	1a	1a	1a	1a	1a	1a	3a	3b	3c	3d	2d	2e	2e	2d	2f	2f	2e
3P	1a	2a	2b	2c	2d	2e	2f	1a	1a	1a	1a	4a	4b	4c	4d	4e	4f	4g
5P	1a	2a	2b	2c	2d	2e	2f	3a	3b	3c	3d	4a	4b	4c	4d	4e	4f	4g
7P	1a	2a	2b	2c	2d	2e	2f	3a	3b	3c	3d	4a	4b	4c	4d	4e	4f	4g
11P	1a	2a	2b	2c	2d	2e	2f	3a	3b	3c	3d	4a	4b	4c	4d	4e	4f	4g
13P	1a	2a	2b	2c	2d	2e	2f	3a	3b	3c	3d	4a	4b	4c	4d	4e	4f	4g
X.1	1	1	1	1	1	1	1	1	1	1	1	1	1	1	1	1	1	1
X.2	78	78	−34	−34	14	14	−2	15	−3	6	−3	6	−6	−6	−2	2	2	2
X.3	352	−352	.	.	−32	32	.	−8	28	10	1	.	.	.	.	−8	8	.
X.4	429	429	77	77	45	45	13	6	24	15	−3	13	5	5	−3	5	5	5
X.5	1001	1001	−231	−231	41	41	−7	56	29	2	2	1	−11	−11	9	5	5	−3
X.6	1430	1430	−154	−154	86	86	6	−1	−28	26	−1	14	−6	−6	6	10	10	2
X.7	2080	−2080	−384	384	−96	96	.	64	−26	19	1	.	16	−16	.	−8	8	.
X.8	2080	−2080	384	−384	−96	96	.	64	−26	19	1	.	−16	16	.	−8	8	.
X.9	3003	3003	539	539	59	59	−37	105	6	15	6	11	11	11	−5	3	3	−5
X.10	3080	3080	616	616	136	136	40	119	2	20	2	24	24	24	8	.	.	8
X.11	5824	−5824	896	−896	−64	64	.	154	−8	−8	−8	.	−16	16	.	.	.	.
X.12	5824	−5824	−896	896	−64	64	.	154	−8	−8	−8	.	16	−16	.	.	.	.
X.13	5824	−5824	−896	896	−64	64	.	154	−8	−8	−8	.	16	−16	.	.	.	.
X.14	5824	−5824	896	−896	−64	64	.	154	−8	−8	−8	.	−16	16	.	.	.	.
X.15	10725	10725	−715	−715	165	165	−43	15	114	24	6	29	−15	−15	−11	9	9	−7
X.16	13650	13650	1330	1330	210	210	114	105	123	33	15	−14	10	10	2	10	10	10
X.17	13728	−13728	−1408	1408	−224	224	.	120	120	39	12	.	16	−16	.	−8	8	.
X.18	13728	−13728	1408	−1408	−224	224	.	120	120	39	12	.	−16	16	.	−8	8	.
X.19	27456	−27456	.	.	−448	448	.	−120	−84	60	−3	.	.	.	.	−16	16	.
X.20	30030	30030	1694	1694	526	526	62	−21	−102	60	6	38	26	26	−2	18	18	2
X.21	32032	32032	−2464	−2464	544	544	−32	91	−44	64	10	64	−24	−24	.	.	.	8
X.22	43680	43680	−4256	−4256	416	416	−32	399	−60	48	−6	.	−24	−24	.	.	.	8
X.23	45045	45045	4389	4389	309	309	133	441	90	−18	9	29	41	41	5	1	1	1
X.24	48048	−48048	4928	−4928	−272	272	.	546	96	−3	15	.	−40	40	.	−4	4	.
X.25	48048	−48048	−4928	4928	−272	272	.	546	96	−3	15	.	40	−40	.	−4	4	.
X.26	48048	48048	−1232	−1232	432	432	48	−84	258	6	−12	−16	.	.	−16	16	16	.
X.27	50050	50050	770	770	130	130	−126	−35	235	19	−8	−14	10	10	18	18	18	−6
X.28	50050	50050	−5390	−5390	450	450	−46	595	73	10	19	10	−50	−50	−14	−2	−2	6
X.29	75075	75075	1155	1155	835	835	131	−210	150	51	−12	83	−5	−5	19	3	3	11
X.30	75075	75075	7315	7315	515	515	51	735	−93	−3	−12	−5	55	55	−13	7	7	−1
X.31	75075	75075	−5005	−5005	−125	−125	83	420	−12	42	15	−5	15	15	3	−1	−1	−9
X.32	81081	81081	3465	3465	633	633	−87	.	162	81	.	33	5	5	−7	−3	−3	−3
X.33	105600	−105600	.	.	640	−640	.	120	−24	48	−24	.	.	.	.	.	.	.
X.34	105600	−105600	.	.	640	−640	.	120	−24	48	−24	.	.	.	.	.	.	.
X.35	114400	114400	−8800	−8800	480	480	32	685	28	−8	−26	64	−40	−40	.	.	.	−8
X.36	123200	−123200	.	.	−960	960	.	−280	404	62	−28	.	.	.	.	−16	16	.
X.37	133056	−133056	.	.	192	−192	.	.	−108	−54	54	.	.	.	.	−16	16	.
X.38	138600	138600	−9240	−9240	360	360	−24	630	−153	−45	9	8	−40	−40	8	8	8	−8
X.39	138600	138600	−9240	−9240	360	360	−24	630	−153	−45	9	8	−40	−40	8	8	8	−8
X.40	146432	−146432	11264	−11264	−1024	1024	.	776	−16	56	−16	.	−64	64	.	.	.	.
X.41	146432	−146432	−11264	11264	−1024	1024	.	776	−16	56	−16	.	64	−64	.	.	.	.
X.42	150150	150150	8470	8470	70	70	54	525	57	48	−24	−34	−30	−30	22	−6	−6	10
X.43	205920	205920	1056	1056	864	864	160	−279	−144	72	18	64	24	24	.	.	.	−8
X.44	228800	−228800	.	.	−320	320	.	−160	380	−52	29	.	.	.	.	−16	16	.
X.45	235872	−235872	−8064	8064	−1056	1056	.	.	−324	81	.	.	16	−16	.	−24	24	.
X.46	235872	−235872	8064	−8064	−1056	1056	.	.	−324	81	.	.	−16	16	.	−24	24	.
X.47	289575	289575	12375	12375	615	615	183	405	162	−81	.	15	35	35	23	3	3	−5
X.48	300300	300300	−7700	−7700	1420	1420	−84	−210	−291	114	−21	−4	−20	−20	−20	20	20	−4
X.49	320320	320320	14784	14784	1344	1344	192	406	−116	64	−8	.	16	16	.	.	.	16
X.50	320320	−320320	9856	−9856	576	−576	.	406	−116	64	−8	.	16	−16	.	.	.	.
X.51	320320	−320320	−9856	9856	576	−576	.	406	−116	64	−8	.	−16	16	.	.	.	.
X.52	360855	360855	18711	18711	1431	1431	279	729	.	.	.	−9	39	39	−9	−9	−9	7
X.53	370656	370656	−15840	−15840	1248	1248	−96	405	324	.	.	.	−40	−40	.	.	.	−8
X.54	400400	−400400	−24640	24640	−560	560	.	1610	−64	−37	17	.	40	−40	.	4	−4	.
X.55	400400	400400	6160	6160	1040	1040	16	−280	−307	−10	17	−48	.	.	16	16	16	.
X.56	400400	−400400	24640	−24640	−560	560	.	1610	−64	−37	17	.	−40	40	.	4	−4	.
X.57	400400	400400	6160	6160	1040	1040	16	−280	−307	−10	17	−48	.	.	16	16	16	.
X.58	400400	400400	−18480	−18480	−240	−240	80	980	98	−28	−10	16	.	.	−16	.	.	.
X.59	400400	400400	−18480	−18480	−240	−240	80	980	98	−28	−10	16	.	.	−16	.	.	.
X.60	436800	−436800	17920	−17920	320	−320	.	840	210	−42	−33	.	.	.	.	−16	16	.
X.61	436800	−436800	−17920	17920	320	−320	.	840	210	−42	−33	.	.	.	.	−16	16	.
X.62	450450	450450	25410	25410	210	210	−350	1575	171	−18	9	−6	30	30	2	−2	−2	−10
X.63	450450	450450	6930	6930	1170	1170	−110	−315	−72	9	9	130	10	10	−30	−14	−14	−6
X.64	480480	−480480	−9856	9856	−1696	1696	.	−336	−174	69	15	.	−16	16	.	8	−8	.
X.65	480480	−480480	9856	−9856	−1696	1696	.	−336	−174	69	15	.	16	−16	.	8	−8	.
X.66	576576	576576	14784	14784	576	576	−320	126	180	72	18	.	−16	−16	.	.	.	−16
X.67	577368	577368	−21384	−21384	216	216	−72	729	.	.	.	72	−24	−24	24	.	.	−8
X.68	579150	579150	−6930	−6930	1230	1230	174	−405	324	81	.	−18	−10	−10	30	6	6	−10
X.69	582400	582400	−8960	−8960	−1280	−1280	256	280	172	64	10	.	.	.	.	.	.	.
X.70	582400	582400	−8960	−8960	−1280	−1280	256	280	172	64	10	.	.	.	.	.	.	.
X.71	600600	600600	21560	21560	920	920	−136	525	−96	12	−42	72	40	40	24	.	.	−8
X.72	600600	600600	9240	9240	−1000	−1000	280	210	−15	57	−15	−8	−40	−40	−8	−8	−8	−8
X.73	600600	600600	9240	9240	−1000	−1000	280	210	−15	57	−15	−8	−40	−40	−8	−8	−8	−8
X.74	675675	675675	10395	10395	−165	−165	411	.	135	−54	54	75	−45	−45	11	−5	−5	3
X.75	686400	−686400	28160	−28160	−960	960	.	960	6	78	−21	.	.	.	.	16	−16	.
X.76	686400	−686400	−28160	28160	−960	960	.	960	6	78	−21	.	.	.	.	16	−16	.
X.77	720720	720720	−33264	−33264	1104	1104	−112	1386	−18	36	−18	−80	−16	−16	16	.	.	16
X.78	800800	800800	−12320	−12320	800	800	−416	−245	520	16	34	.	40	40	.	.	.	8
X.79	852930	852930	−29646	−29646	1026	1026	18	729	.	.	.	−54	6	6	18	−18	−18	−2
X.80	915200	−915200	28160	−28160	−1280	1280	.	440	548	8	8	.	.	.	.	.	.	.
X.81	915200	−915200	−28160	28160	−1280	1280	.	440	548	8	8	.	.	.	.	.	.	.
X.82	938223	938223	−2673	−2673	1647	1647	207	−729	.	.	.	63	15	15	15	−9	−9	−1
X.83	972972	972972	−24948	−24948	1836	1836	−180	.	−243	.	.	60	−36	−36	−20	−12	−12	12
X.84	982800	982800	15120	15120	−240	−240	−240	.	−135	54	27	−48	.	.	16	−16	−16	.
X.85	982800	982800	15120	15120	−240	−240	−240	.	−135	54	27	−48	.	.	16	−16	−16	.
X.86	1029600	−1029600	.	.	−1440	1440	.	−720	252	90	9	.	.	.	.	−8	8	.
X.87	1201200	−1201200	24640	−24640	−1680	1680	.	−210	132	−39	24	.	−40	40	.	12	−12	.

Character table of $H(\mathsf{Fi}_{23})$ *(continued)*

	$5a$	6_1	6_2	6_3	6_4	6_5	6_6	6_7	6_8	6_9	6_{10}	6_{11}	6_{12}	6_{13}	6_{14}	6_{15}	6_{16}	6_{17}	6_{18}	6_{19}	6_{20}	6_{21}
2	4	9	8	8	8	7	8	8	4	8	8	9	9	5	5	8	8	7	7	7	6	6
3	1	7	9	5	5	7	6	6	7	4	4	3	3	5	5	3	3	3	3	3	3	3
5	2	1	.	1	1	.	.	.	.	.	.	.	.	.	.	.	.	.	.	.	.	.
7	.	1	.	.	.	.	.	.	.	.	.	.	.	.	.	.	.	.	.	.	.	.
11	.	.	.	.	.	.	.	.	.	.	.	.	.	.	.	.	.	.	.	.	.	.
13	.	.	.	.	.	.	.	.	.	.	.	.	.	.	.	.	.	.	.	.	.	.
	$5a$	6_1	6_2	6_3	6_4	6_5	6_6	6_7	6_8	6_9	6_{10}	6_{11}	6_{12}	6_{13}	6_{14}	6_{15}	6_{16}	6_{17}	6_{18}	6_{19}	6_{20}	6_{21}
$2P$	$5a$	$3a$	$3b$	$3a$	$3a$	$3c$	$3b$	$3b$	$3d$	$3b$	$3b$	$3a$	$3a$	$3c$	$3c$	$3b$	$3b$	$3c$	$3c$	$3a$	$3c$	$3c$
$3P$	$5a$	$2a$	$2a$	$2b$	$2c$	$2a$	$2c$	$2b$	$2a$	$2d$	$2e$	$2d$	$2e$	$2b$	$2c$	$2f$	$2f$	$2f$	$2f$	$2f$	$2e$	$2d$
$5P$	$1a$	6_1	6_2	6_3	6_4	6_5	6_6	6_7	6_8	6_9	6_{10}	6_{11}	6_{12}	6_{13}	6_{14}	6_{15}	6_{16}	6_{17}	6_{18}	6_{19}	6_{20}	6_{21}
$7P$	$5a$	6_1	6_2	6_3	6_4	6_5	6_6	6_7	6_8	6_9	6_{10}	6_{11}	6_{12}	6_{13}	6_{14}	6_{15}	6_{16}	6_{17}	6_{18}	6_{19}	6_{20}	6_{21}
$11P$	$5a$	6_1	6_2	6_3	6_4	6_5	6_6	6_7	6_8	6_9	6_{10}	6_{11}	6_{12}	6_{13}	6_{14}	6_{15}	6_{16}	6_{17}	6_{18}	6_{19}	6_{20}	6_{21}
$13P$	$5a$	6_1	6_2	6_3	6_4	6_5	6_6	6_7	6_8	6_9	6_{10}	6_{11}	6_{12}	6_{13}	6_{14}	6_{15}	6_{16}	6_{17}	6_{18}	6_{19}	6_{20}	6_{21}
$X.1$	1	1	1	1	1	1	1	1	1	1	1	1	1	1	1	1	1	1	1	1	1	1
$X.2$	3	15	−3	−7	−7	6	−7	−7	−3	5	5	−1	−1	2	2	1	1	−2	−2	1	2	2
$X.3$	2	8	−28	.	.	−10	.	.	−1	4	−4	−8	8	.	.	.	.	6	−6	.	2	−2
$X.4$	4	6	24	14	14	15	−4	−4	−3	.	.	6	6	5	5	4	4	7	7	−2	3	3
$X.5$	1	56	29	−6	−6	2	−15	−15	2	5	5	8	8	−6	−6	−7	−7	2	2	2	2	2
$X.6$	5	−1	−28	−19	−19	26	8	8	−1	−4	−4	−1	−1	8	8	.	.	−6	−6	−3	2	2
$X.7$	5	−64	26	−24	24	−19	6	−6	−1	−6	6	.	.	3	−3	6	−6	−3	3	.	3	−3
$X.8$	5	−64	26	24	−24	−19	−6	6	−1	−6	6	.	.	−3	3	−6	6	−3	3	.	3	−3
$X.9$	3	105	6	17	17	15	26	26	6	14	14	−7	−7	−1	−1	2	2	−1	−1	−7	−1	−1
$X.10$	5	119	2	31	31	20	22	22	2	10	10	7	7	4	4	−2	−2	4	4	7	4	4
$X.11$	−1	−154	8	−4	4	8	−32	32	8	8	−8	2	−2	−4	4	.	.	.	.	.	4	−4
$X.12$	−1	−154	8	4	−4	8	32	−32	8	8	−8	2	−2	4	−4	.	.	.	.	.	4	−4
$X.13$	−1	−154	8	4	−4	8	32	−32	8	8	−8	2	−2	4	−4	.	.	.	.	.	4	−4
$X.14$	−1	−154	8	−4	4	8	−32	32	8	8	−8	2	−2	−4	4	.	.	.	.	.	4	−4
$X.15$	.	15	114	5	5	24	14	14	6	−6	−6	15	15	−4	−4	−10	−10	8	8	5	.	.
$X.16$	.	105	123	25	25	33	7	7	15	3	3	9	9	7	7	15	15	9	9	9	−3	−3
$X.17$	3	−120	−120	−40	40	−39	4	−4	−12	−8	8	−8	8	−13	13	−12	12	9	−9	.	−1	1
$X.18$	3	−120	−120	40	−40	−39	−4	4	−12	−8	8	−8	8	13	−13	12	−12	9	−9	.	−1	1
$X.19$	6	120	84	.	.	−60	.	.	3	20	−20	8	−8	.	.	.	.	−12	12	.	4	−4
$X.20$	5	−21	−102	29	29	60	−34	−34	6	−14	−14	−5	−5	2	2	−10	−10	−4	−4	5	4	4
$X.21$	7	91	−44	−79	−79	64	20	20	10	4	4	−5	−5	2	2	4	4	−8	−8	1	4	4
$X.22$	5	399	−60	−71	−71	48	−44	−44	−6	20	20	−1	−1	10	10	4	4	−8	−8	1	−4	−4
$X.23$	−5	441	90	−21	−21	−18	42	42	9	−6	−6	9	9	6	6	10	10	−2	−2	19	6	6
$X.24$	−2	−546	−96	−4	4	3	−68	68	−15	16	−16	10	−10	5	−5	12	−12	3	−3	.	5	−5
$X.25$	−2	−546	−96	4	−4	3	68	−68	−15	16	−16	10	−10	−5	5	−12	12	3	−3	.	5	−5
$X.26$	−2	−84	258	−44	−44	6	10	10	−12	18	18	12	12	−8	−8	−6	−6	6	6	−12	6	6
$X.27$	.	−35	235	5	5	19	−13	−13	−8	−5	−5	13	13	5	5	3	3	3	3	−3	−5	−5
$X.28$	.	595	73	−35	−35	10	−71	−71	19	9	9	3	3	−8	−8	−7	−7	2	2	5	6	6
$X.29$	.	−210	150	30	30	51	−6	−6	−12	−2	−2	−2	−2	3	3	2	2	11	11	−10	7	7
$X.30$	.	735	−93	25	25	−3	79	79	−12	11	11	−1	−1	−11	−11	−9	−9	−3	−3	9	5	5
$X.31$	.	420	−12	−10	−10	42	−64	−64	15	28	28	4	4	−10	−10	8	8	2	2	−10	−2	−2
$X.32$	6	.	162	90	90	81	−18	−18	.	−6	−6	.	.	9	9	6	6	9	9	−6	−3	−3
$X.33$	.	−120	24	.	.	−48	.	.	24	−8	8	−8	8	.	.	.	.	.	.	.	8	−8
$X.34$	.	−120	24	.	.	−48	.	.	24	−8	8	−8	8	.	.	.	.	.	.	.	8	−8
$X.35$	.	685	28	−25	−25	−8	−52	−52	−26	12	12	−3	−3	2	2	−4	−4	8	8	−1	.	.
$X.36$	.	280	−404	.	.	−62	.	.	28	12	−12	−24	24	.	.	.	.	18	−18	.	6	−6
$X.37$	6	.	108	.	.	54	.	.	−54	12	−12	.	.	.	.	.	.	6	−6	.	−6	6
$X.38$	.	630	−153	30	30	−45	−33	−33	9	−9	−9	6	6	3	3	15	15	3	3	6	3	3
$X.39$	.	630	−153	30	30	−45	−33	−33	9	−9	−9	6	6	3	3	15	15	3	3	6	3	3
$X.40$	7	−776	16	104	−104	−56	−32	32	16	−16	16	8	−8	−4	4	.	.	.	.	.	4	−4
$X.41$	7	−776	16	−104	104	−56	32	−32	16	−16	16	8	−8	4	−4	.	.	.	.	.	4	−4
$X.42$	.	525	57	55	55	48	73	73	−24	25	25	13	13	−8	−8	9	9	.	.	−9	−8	−8
$X.43$	−5	−279	−144	−69	−69	72	−24	−24	18	.	.	9	9	12	12	−8	−8	−8	−8	−5	.	.
$X.44$	.	160	−380	.	.	52	.	.	−29	4	−4	−32	32	.	.	.	.	−12	12	.	−4	4
$X.45$	−3	.	324	−72	72	−81	−36	36	.	−12	12	.	.	9	−9	12	−12	−9	9	.	−3	3
$X.46$	−3	.	324	72	−72	−81	36	−36	.	−12	12	.	.	−9	9	−12	12	−9	9	.	−3	3
$X.47$	.	405	162	−45	−45	−81	−18	−18	.	−6	−6	21	21	9	9	6	6	−9	−9	3	3	3
$X.48$	.	−210	−291	−50	−50	114	49	49	−21	−11	−11	−2	−2	4	4	9	9	−6	−6	6	−2	−2
$X.49$	−5	406	−116	114	114	64	−12	−12	−8	12	12	6	6	6	6	−12	−12	.	.	−6	.	.
$X.50$	−5	−406	116	−44	44	−64	−28	28	8	36	−36	−18	18	10	−10	−12	12	.	.	.	.	.
$X.51$	−5	−406	116	44	−44	−64	28	−28	8	36	−36	−18	18	−10	10	12	−12	.	.	.	.	.
$X.52$	5	729	.	81	81	.	.	.	.	.	.	9	9	.	.	.	.	.	.	9	.	.
$X.53$	6	405	324	−45	−45	.	36	36	.	−12	−12	21	21	−18	−18	−12	−12	.	.	3	.	.
$X.54$	.	−1610	64	20	−20	37	124	−124	−17	16	−16	−14	14	−7	7	12	−12	−3	3	.	−1	1
$X.55$	.	−280	−307	40	40	−10	−23	−23	17	5	5	8	8	−14	−14	1	1	−2	−2	−8	2	2
$X.56$	.	−1610	64	−20	20	37	−124	124	−17	16	−16	−14	14	7	−7	−12	12	−3	3	.	−1	1
$X.57$	.	−280	−307	40	40	−10	−23	−23	17	5	5	8	8	−14	−14	1	1	−2	−2	−8	2	2
$X.58$	.	980	98	60	60	−28	−66	−66	−10	−6	−6	−12	−12	6	6	−10	−10	−4	−4	−4	.	.
$X.59$	.	980	98	60	60	−28	−66	−66	−10	−6	−6	−12	−12	6	6	−10	−10	−4	−4	−4	.	.
$X.60$	.	−840	−210	−80	80	42	−46	46	33	14	−14	8	−8	−8	8	18	−18	−6	6	.	−2	2
$X.61$	.	−840	−210	80	−80	42	46	−46	33	14	−14	8	−8	8	−8	−18	18	−6	6	.	−2	2
$X.62$	.	1575	171	−15	−15	−18	111	111	9	3	3	−9	−9	12	12	7	7	−2	−2	−23	6	6
$X.63$	.	−315	−72	45	45	9	−36	−36	9	.	.	21	21	−9	−9	4	4	1	1	−11	−3	−3
$X.64$	5	336	174	−136	136	−69	−26	26	−15	14	−14	−16	16	−1	1	6	−6	3	−3	.	1	−1
$X.65$	5	336	174	136	−136	−69	26	−26	−15	14	−14	−16	16	1	−1	−6	6	3	−3	.	1	−1
$X.66$	1	126	180	114	114	72	−12	−12	18	36	36	−18	−18	6	6	4	4	−8	−8	10	.	.
$X.67$	−7	729	.	81	81	.	.	.	.	.	.	9	9	.	.	.	.	.	.	9	.	.
$X.68$	.	−405	324	−45	−45	81	36	36	.	−12	−12	−21	−21	9	9	−12	−12	9	9	3	−3	−3
$X.69$	.	280	172	40	40	64	4	4	10	−20	−20	−8	−8	−14	−14	4	4	−8	−8	−8	4	4
$X.70$	.	280	172	40	40	64	4	4	10	−20	−20	−8	−8	−14	−14	4	4	−8	−8	−8	4	4
$X.71$	.	525	−96	5	5	12	−40	−40	−42	−16	−16	−19	−19	−22	−22	8	8	−4	−4	5	−4	−4
$X.72$	.	210	−15	−30	−30	57	33	33	−15	17	17	2	2	−3	−3	1	1	1	1	10	5	5
$X.73$	.	210	−15	−30	−30	57	33	33	−15	17	17	2	2	−3	−3	1	1	1	1	10	5	5
$X.74$	.	.	135	.	.	−54	27	27	54	15	15	.	.	.	.	3	3	−6	−6	.	−6	−6
$X.75$	.	−960	−6	80	−80	−78	−26	26	21	−6	6	.	.	−10	10	6	−6	−6	6	.	−6	6
$X.76$	.	−960	−6	−80	80	−78	26	−26	21	−6	6	.	.	10	−10	−6	6	−6	6	.	−6	6
$X.77$	−5	1386	−18	−54	−54	36	−54	−54	−18	6	6	−6	−6	.	.	2	2	−4	−4	2	.	.
$X.78$	.	−245	520	−35	−35	16	−8	−8	34	8	8	11	11	−8	−8	−8	−8	−8	−8	13	−4	−4
$X.79$	5	729	.	−81	−81	.	.	.	.	.	.	9	9	.	.	.	.	.	.	−9	.	.
$X.80$	.	−440	−548	80	−80	−8	28	−28	−8	−20	20	−8	8	26	−26	12	−12	.	.	.	−4	4
$X.81$	.	−440	−548	−80	80	−8	−28	28	−8	−20	20	−8	8	−26	26	−12	12	.	.	.	−4	4
$X.82$	−2	−729	.	−81	−81	.	.	.	.	.	.	−9	−9	.	.	.	.	.	.	−9	.	.
$X.83$	−3	.	−243	.	.	.	81	81	.	−27	−27	.	.	.	.	9	9	.	.	.	.	.
$X.84$	.	.	−135	.	.	54	−27	−27	27	−15	−15	.	.	.	.	−3	−3	6	6	.	6	6
$X.85$	.	.	−135	.	.	54	−27	−27	27	−15	−15	.	.	.	.	−3	−3	6	6	.	6	6
$X.86$	.	720	−252	.	.	−90	.	.	−9	36	−36	48	−48	.	.	.	.	6	−6	.	−6	6
$X.87$	.	210	−132	−20	20	39	92	−92	−24	12	−12	6	−6	−11	11	12	−12	−9	9	.	9	−9

Character table of $H(\mathsf{Fi}_{23})$ *(continued)*

2	5	5	4	4	2	7	7	7	6	6	3	2	1	4	3	3	4	4	2	2	7	7	5	6	6	6	5	5
3	3	3	3	3	1	1	1	.	.	.	4	4	3	1	1	1	.	.	.	.	2	2	3	2	2	2	2	2
5	.	.	.	.	.	.	.	.	.	.	.	.	.	2	1	1	1	1	.	.	.	.	.	.	.	.	.	.
7	.	.	.	.	1	.	.	.	.	.	.	.	.	.	.	.	.	.	.	.	.	.	.	.	.	.	.	.
11	.	.	.	.	.	.	.	.	.	.	.	.	.	.	.	.	.	.	1	1	.	.	.	.	.	.	.	.
13	.	.	.	.	.	.	.	.	.	.	.	.	.	.	.	.	.	.	.	.	.	.	.	.	.	.	.	.
	6_{22}	6_{23}	6_{24}	6_{25}	$7a$	$8a$	$8b$	$8c$	$8d$	$8e$	$9a$	$9b$	$9c$	$10a$	$10b$	$10c$	$10d$	$10e$	$11a$	$11b$	$12a$	$12b$	$12c$	$12d$	$12e$	$12f$	$12g$	$12h$
$2P$	$3c$	$3c$	$3d$	$3d$	$7a$	$4a$	$4a$	$4d$	$4f$	$4f$	$9a$	$9b$	$9c$	$5a$	$5a$	$5a$	$5a$	$5a$	$11b$	$11a$	6_{11}	6_{11}	6_9	6_{12}	6_{12}	6_{11}	6_{15}	6_{15}
$3P$	$2f$	$2f$	$2f$	$2f$	$7a$	$8a$	$8b$	$8c$	$8d$	$8e$	$3b$	$3b$	$3d$	$10a$	$10b$	$10c$	$10d$	$10e$	$11a$	$11b$	$4a$	$4a$	$4a$	$4c$	$4b$	$4a$	$4e$	$4f$
$5P$	6_{22}	6_{23}	6_{24}	6_{25}	$7a$	$8a$	$8b$	$8c$	$8d$	$8e$	$9a$	$9b$	$9c$	$2a$	$2b$	$2c$	$2e$	$2d$	$11a$	$11b$	$12a$	$12b$	$12c$	$12d$	$12e$	$12f$	$12g$	$12h$
$7P$	6_{22}	6_{23}	6_{24}	6_{25}	$1a$	$8a$	$8b$	$8c$	$8d$	$8e$	$9a$	$9b$	$9c$	$10a$	$10b$	$10c$	$10d$	$10e$	$11b$	$11a$	$12a$	$12b$	$12c$	$12d$	$12e$	$12f$	$12g$	$12h$
$11P$	6_{22}	6_{23}	6_{24}	6_{25}	$7a$	$8a$	$8b$	$8c$	$8d$	$8e$	$9a$	$9b$	$9c$	$10a$	$10b$	$10c$	$10d$	$10e$	$1a$	$1a$	$12a$	$12b$	$12c$	$12d$	$12e$	$12f$	$12g$	$12h$
$13P$	6_{22}	6_{23}	6_{24}	6_{25}	$7a$	$8a$	$8b$	$8c$	$8d$	$8e$	$9a$	$9b$	$9c$	$10a$	$10b$	$10c$	$10d$	$10e$	$11b$	$11a$	$12a$	$12b$	$12c$	$12d$	$12e$	$12f$	$12g$	$12h$
X.1	1	1	1	1	1	1	1	1	1	1	1	1	1	1	1	1	1	1	1	1	1	1	1	1	1	1	1	1
X.2	−2	−2	1	1	1	2	−2	2	.	.	3	.	.	3	1	1	−1	−1	1	1	3	3	−3	−3	−3	−3	−1	−1
X.3	.	.	−3	3	2	.	.	.	.	.	−2	4	1	−2	.	.	2	−2	.	.	.	.	.	.	.	.	−2	2
X.4	1	1	1	1	2	1	1	1	1	1	.	3	.	4	2	2	.	.	.	.	−2	−2	4	2	2	−2	2	2
X.5	2	2	2	2	.	−3	1	1	−1	−1	2	2	−1	1	−1	−1	1	1	.	.	4	4	1	−2	−2	−2	−1	−1
X.6	.	.	3	3	2	−2	2	2	.	.	−1	−1	2	5	1	1	1	1	.	.	−1	−1	−4	−3	−3	5	−2	−2
X.7	−3	3	−3	3	1	.	.	.	.	.	4	−2	1	−5	1	−1	1	−1	1	1	.	.	.	−4	4	.	−2	2
X.8	3	−3	−3	3	1	.	.	.	.	.	4	−2	1	−5	−1	1	1	−1	1	1	.	.	.	4	−4	.	4	−4
X.9	−1	−1	2	2	.	−1	−1	3	−1	−1	3	.	.	3	−1	−1	−1	−1	.	.	5	5	2	5	5	5	.	.
X.10	4	4	−2	−2	.	4	4	.	.	.	5	−1	−1	5	1	1	1	1	.	.	3	3	6	3	3	3	.	.
X.11	.	.	.	.	.	.	.	.	.	.	1	1	1	1	1	−1	−1	1	A	$\bar{A}$	6	−6	.	−2	2	.	.	.
X.12	.	.	.	.	.	.	.	.	.	.	1	1	1	1	−1	1	−1	1	$\bar{A}$	A	6	−6	.	2	−2	.	.	.
X.13	.	.	.	.	.	.	.	.	.	.	1	1	1	1	−1	1	−1	1	A	$\bar{A}$	6	−6	.	2	−2	.	.	.
X.14	.	.	.	.	.	.	.	.	.	.	1	1	1	1	1	−1	−1	1	$\bar{A}$	A	6	−6	.	−2	2	.	.	.
X.15	−4	−4	2	2	1	1	−3	1	−1	−1	.	3	.	.	.	.	.	.	.	.	−1	−1	2	−3	−3	5	.	.
X.16	3	3	3	3	.	−2	−2	−2	2	2	.	3	.	.	.	.	.	.	−1	−1	1	1	−5	1	1	1	1	1
X.17	−3	3	.	.	1	.	.	.	.	.	3	3	.	−3	−3	3	−1	1	.	.	.	.	.	−4	4	.	−2	2
X.18	3	−3	.	.	1	.	.	.	.	.	3	3	.	−3	3	−3	−1	1	.	.	.	.	.	4	−4	.	−2	2
X.19	.	.	−3	3	2	.	.	.	.	.	−6	.	.	−6	.	.	−2	2	.	.	.	.	.	.	.	.	2	−2
X.20	2	2	2	2	.	2	−2	2	.	.	−3	−3	.	5	−1	−1	1	1	.	.	−1	−1	2	5	5	−7	.	.
X.21	−2	−2	−2	−2	.	4	−4	.	.	.	1	1	−2	7	1	1	−1	−1	.	.	−5	−5	−8	−3	−3	1	.	.
X.22	−2	−2	−2	−2	.	−4	4	.	.	.	3	3	.	5	−1	−1	1	1	−1	−1	3	3	.	−3	−3	−3	.	.
X.23	−2	−2	1	1	.	1	5	1	−1	−1	.	.	.	−5	−1	−1	−1	−1	.	.	5	5	2	−1	−1	−1	−2	−2
X.24	3	−3	3	−3	.	.	.	.	2	−2	3	.	.	2	−2	2	2	−2	.	.	6	−6	.	−2	2	.	2	−2
X.25	−3	3	3	−3	.	.	.	.	−2	2	3	.	.	2	2	−2	2	−2	.	.	6	−6	.	2	−2	.	2	−2
X.26	.	.	.	.	.	.	.	.	.	.	.	3	.	−2	−2	−2	2	2	.	.	−4	−4	2	.	.	−4	−2	−2
X.27	−3	−3	.	.	.	2	2	−2	−2	−2	1	1	1	.	.	.	.	.	.	.	1	1	−5	1	1	1	3	3
X.28	−4	−4	−1	−1	.	2	−2	−2	.	.	4	−2	1	.	.	.	.	.	.	.	7	7	1	1	1	1	1	1
X.29	−1	−1	−4	−4	.	3	3	−1	−1	−1	−6	.	.	.	.	.	.	.	.	.	2	2	2	−2	−2	2	.	.
X.30	−3	−3	.	.	.	−5	−1	−1	1	1	3	.	.	.	.	.	.	.	.	.	7	7	−5	1	1	1	1	1
X.31	2	2	−1	−1	.	−1	3	3	1	1	−3	−3	.	.	.	.	.	.	.	.	4	4	4	−6	−6	−2	2	2
X.32	−3	−3	.	.	.	−7	−3	1	−1	−1	.	.	.	6	.	.	−2	−2	.	.	.	.	6	2	2	−6	.	.
X.33	.	.	.	.	−2	.	.	.	.	.	−3	.	.	.	.	.	.	.	.	.	.	.	.	.	.	.	.	.
X.34	.	.	.	.	−2	.	.	.	.	.	−3	.	.	.	.	.	.	.	.	.	.	.	.	.	.	.	.	.
X.35	2	2	2	2	−1	4	−4	.	.	.	4	−2	1	.	.	.	.	.	.	.	1	1	−8	−1	−1	−5	.	.
X.36	.	.	.	.	.	.	.	.	.	.	−4	2	−1	.	.	.	.	.	.	.	.	.	.	.	.	.	2	−2
X.37	.	.	6	−6	.	.	.	.	.	.	.	.	.	−6	.	.	−2	2	.	.	.	.	.	.	.	.	2	−2
X.38	3	3	−3	−3	.	.	.	.	.	.	.	.	.	.	.	.	.	.	.	.	2	2	−1	2	2	2	−1	−1
X.39	3	3	−3	−3	.	.	.	.	.	.	.	.	.	.	.	.	.	.	.	.	2	2	−1	2	2	2	−1	−1
X.40	.	.	.	.	−1	.	.	.	.	.	5	−1	−1	−7	−1	1	−1	1	.	.	.	.	.	4	−4	.	.	.
X.41	.	.	.	.	−1	.	.	.	.	.	5	−1	−1	−7	1	−1	−1	1	.	.	.	.	.	−4	4	.	.	.
X.42	.	.	.	.	.	2	−2	2	.	.	−3	.	.	.	.	.	.	.	.	.	5	5	−7	3	3	−1	−3	−3
X.43	4	4	−2	−2	1	−4	4	.	.	.	.	.	.	−5	1	1	−1	−1	.	.	1	1	−8	3	3	−5	.	.
X.44	.	.	−3	3	−2	.	.	.	.	.	2	2	−1	.	.	.	.	.	.	.	.	.	.	.	.	.	2	−2
X.45	3	−3	.	.	.	.	.	.	.	.	.	.	.	3	1	−1	1	−1	−1	−1	.	.	.	−4	4	.	.	.
X.46	−3	3	.	.	.	.	.	.	.	.	.	.	.	3	−1	1	1	−1	−1	−1	.	.	.	4	−4	.	.	.
X.47	−3	−3	.	.	−1	−1	−5	3	1	1	.	.	.	.	.	.	.	.	.	.	−3	−3	6	−1	−1	3	.	.
X.48	.	.	3	3	.	.	.	−4	.	.	−3	−3	.	.	.	.	.	.	.	.	2	2	5	−2	−2	2	−1	−1
X.49	6	6	.	.	.	.	.	.	.	.	4	−2	1	−5	−1	−1	−1	−1	.	.	−6	−6	.	−2	−2	6	.	.
X.50	6	−6	.	.	.	.	.	.	.	.	4	−2	1	5	1	−1	−1	1	.	.	−6	6	.	2	−2	.	.	.
X.51	−6	6	.	.	.	.	.	.	.	.	4	−2	1	5	−1	1	−1	1	.	.	−6	6	.	−2	2	.	.	.
X.52	.	.	.	.	−2	−1	−1	−1	−1	−1	.	.	.	5	1	1	1	1	.	.	−3	−3	.	−3	−3	−3	.	.
X.53	6	6	.	.	−1	−4	4	.	.	.	.	.	.	6	.	.	−2	−2	.	.	−3	−3	.	−1	−1	3	.	.
X.54	3	−3	−3	3	.	.	.	.	2	−2	−1	−1	−1	.	.	.	.	.	.	.	6	−6	.	2	−2	.	−2	2
X.55	−2	−2	1	1	.	.	.	.	.	.	−1	−1	−1	.	.	.	.	.	.	.	.	.	−3	.	.	.	−5	−5
X.56	−3	3	−3	3	.	.	.	.	−2	2	−1	−1	−1	.	.	.	.	.	.	.	6	−6	.	−2	2	.	−2	2
X.57	−2	−2	1	1	.	.	.	.	.	.	−1	−1	−1	.	.	.	.	.	.	.	.	.	−3	.	.	.	7	7
X.58	2	2	2	2	.	.	.	.	.	.	−4	2	−1	.	.	.	.	.	.	.	4	4	−2	.	.	4	.	.
X.59	2	2	2	2	.	.	.	.	.	.	−4	2	−1	.	.	.	.	.	.	.	4	4	−2	.	.	4	.	.
X.60	.	.	3	−3	.	.	.	.	.	.	−3	.	.	.	.	.	.	.	1	1	.	.	.	.	.	.	2	−2
X.61	.	.	3	−3	.	.	.	.	.	.	−3	.	.	.	.	.	.	.	1	1	.	.	.	.	.	.	−4	4
X.62	4	4	1	1	.	2	−2	−2	.	.	.	.	.	.	.	.	.	.	.	.	3	3	3	−3	−3	−3	1	1
X.63	−5	−5	1	1	.	2	2	−2	2	2	.	.	.	.	.	.	.	.	.	.	1	1	4	1	1	1	−2	−2
X.64	−3	3	3	−3	.	.	.	.	.	.	−3	−3	.	−5	−1	1	1	−1	.	.	.	.	.	4	−4	.	−4	4
X.65	3	−3	3	−3	.	.	.	.	.	.	−3	−3	.	−5	1	−1	1	−1	.	.	.	.	.	−4	4	.	2	−2
X.66	−2	−2	−2	−2	.	.	.	.	.	.	.	.	.	1	−1	−1	1	1	.	.	−6	−6	.	2	2	6	.	.
X.67	.	.	.	.	1	−4	−4	.	.	.	.	.	.	−7	1	1	1	1	.	.	−3	−3	.	−3	−3	−3	.	.
X.68	−3	−3	.	.	−2	2	2	2	2	2	.	.	.	.	.	.	.	.	.	.	3	3	.	−1	−1	3	.	.
X.69	−2	−2	−2	−2	.	.	.	.	.	.	1	1	1	.	.	.	.	.	$\bar{A}$	A	.	.	.	.	.	.	.	.
X.70	−2	−2	−2	−2	.	.	.	.	.	.	1	1	1	.	.	.	.	.	A	$\bar{A}$	.	.	.	.	.	.	.	.
X.71	2	2	2	2	.	4	4	.	.	.	3	3	.	.	.	.	.	.	.	.	−3	−3	.	1	1	−3	.	.
X.72	1	1	1	1	.	.	.	.	.	.	−3	.	.	.	.	.	.	.	.	.	−2	−2	1	2	2	−2	1	1
X.73	1	1	1	1	.	.	.	.	.	.	−3	.	.	.	.	.	.	.	.	.	−2	−2	1	2	2	−2	1	1
X.74	.	.	6	6	.	3	3	−1	−1	−1	.	.	.	.	.	.	.	.	.	.	.	.	3	.	.	.	1	1
X.75	6	−6	3	−3	1	.	.	.	.	.	.	3	.	.	.	.	.	.	.	.	.	.	.	.	.	.	−2	2
X.76	−6	6	3	−3	1	.	.	.	.	.	.	3	.	.	.	.	.	.	.	.	.	.	.	.	.	.	4	−4
X.77	−4	−4	2	2	.	.	.	.	.	.	.	.	.	−5	1	1	−1	−1	.	.	−2	−2	10	2	2	−2	.	.
X.78	4	4	−2	−2	.	4	−4	.	.	.	−5	1	1	.	.	.	.	.	.	.	3	3	.	1	1	−3	.	.
X.79	.	.	.	.	1	−2	2	−2	.	.	.	.	.	5	−1	−1	1	1	1	1	−3	−3	.	3	3	3	.	.
X.80	−6	6	.	.	−1	.	.	.	.	.	−1	−1	2	.	.	.	.	.	.	.	.	.	.	.	.	.	.	.
X.81	6	−6	.	.	−1	.	.	.	.	.	−1	−1	2	.	.	.	.	.	.	.	.	.	.	.	.	.	.	.
X.82	.	.	.	.	−1	−5	−5	−1	−1	−1	.	.	.	−2	2	2	2	2	.	.	3	3	.	3	3	3	.	.
X.83	.	.	.	.	.	.	.	4	.	.	.	.	.	−3	−3	−3	1	1	.	.	.	.	−3	.	.	.	3	3
X.84	.	.	3	3	.	.	.	.	.	.	.	.	.	.	.	.	.	.	$\bar{A}$	A	.	.	−3	.	.	.	−1	−1
X.85	.	.	3	3	.	.	.	.	.	.	.	.	.	.	.	.	.	.	A	$\bar{A}$	.	.	−3	.	.	.	−1	−1
X.86	.	.	−3	3	−2	.	.	.	.	.	.	.	.	.	.	.	.	.	.	.	.	.	.	.	.	.	−2	2
X.87	3	−3	.	.	.	.	.	.	−2	2	.	3	.	.	.	.	.	.	.	.	−6	6	.	−2	2	.	.	.

Character table of $H(\mathsf{Fi}_{23})$ *(continued)*

	12i	12j	12k	12l	12m	12n	12o	12p	12q	12r	13a	13b	14a	14b	14c	15a	16a	16b	18a	18b	18c	18d	18e	18f	18g	18h	18i
2	5	5	5	5	4	4	5	5	3	3	1	1	2	2	2	2	5	5	3	2	3	3	3	3	2	2	3
3	2	2	2	2	2	2	1	1	2	2	.	.	1	.	.	1	.	.	4	4	3	3	3	3	3	3	2
5	.	.	.	.	.	.	.	.	.	.	.	.	.	.	.	1	.	.	.	.	.	.	.	.	.	.	.
7	.	.	.	.	.	.	.	.	.	.	.	.	1	1	1	.	.	.	.	.	.	.	.	.	.	.	.
11	.	.	.	.	.	.	.	.	.	.	.	.	.	.	.	.	.	.	.	.	.	.	.	.	.	.	.
13	.	.	.	.	.	.	.	.	.	.	1	1	.	.	.	.	.	.	.	.	.	.	.	.	.	.	.
	12i	12j	12k	12l	12m	12n	12o	12p	12q	12r	13a	13b	14a	14b	14c	15a	16a	16b	18a	18b	18c	18d	18e	18f	18g	18h	18i
2P	6_{16}	6_{18}	6_{18}	6_{16}	6_{20}	6_{20}	6_{12}	6_{9}	6_{24}	6_{24}	13b	13a	7a	7a	7a	15a	8c	8c	9a	9b	9a	9a	9a	9a	9b	9b	9a
3P	4f	4e	4f	4e	4b	4c	4g	4d	4f	4e	13a	13b	14a	14b	14c	5a	16a	16b	6_{2}	6_{2}	6_{6}	6_{7}	6_{6}	6_{7}	6_{6}	6_{7}	6_{10}
5P	12i	12j	12k	12l	12m	12n	12o	12p	12q	12r	13b	13a	14a	14b	14c	3a	16b	16a	18a	18b	18e	18f	18c	18d	18g	18h	18i
7P	12i	12j	12k	12l	12m	12n	12o	12p	12q	12r	13b	13a	2a	2b	2c	15a	16b	16a	18a	18b	18c	18d	18e	18f	18g	18h	18i
11P	12i	12j	12k	12l	12m	12n	12o	12p	12q	12r	13b	13a	14a	14b	14c	15a	16a	16b	18a	18b	18e	18f	18c	18d	18g	18h	18i
13P	12i	12j	12k	12l	12m	12n	12o	12p	12q	12r	1a	1a	14a	14b	14c	15a	16b	16a	18a	18b	18c	18d	18e	18f	18g	18h	18i
X.1	1	1	1	1	1	1	1	1	1	1	1	1	1	1	1	1	1	1	1	1	1	1	1	1	1	1	1
X.2	−1	2	2	−1	.	.	−1	1	−1	−1	.	.	1	1	1	.	.	.	3	.	−1	−1	−1	−1	2	2	−1
X.3	2	−2	2	−2	.	.	.	.	−1	1	1	1	−2	.	.	2	.	.	2	−4	.	.	.	.	.	.	2
X.4	2	−1	−1	2	−1	−1	2	.	−1	−1	.	.	2	.	.	1	−1	−1	.	3	2	2	2	2	−1	−1	.
X.5	−1	2	2	−1	−2	−2	.	−3	2	2	.	.	.	.	.	1	1	1	2	2	.	.	.	.	.	.	2
X.6	−2	−2	−2	−2	.	.	−1	.	1	1	.	.	2	.	.	−1	.	.	−1	−1	−1	−1	−1	−1	−1	−1	−1
X.7	−4	1	−1	4	1	−1	.	.	−1	1	.	.	−1	1	−1	−1	.	.	−4	2	.	.	.	.	.	.	.
X.8	2	1	−1	−2	−1	1	.	.	−1	1	.	.	−1	−1	1	−1	.	.	−4	2	.	.	.	.	.	.	.
X.9	.	3	3	.	−1	−1	1	−2	.	.	.	.	.	.	.	.	−1	−1	3	.	−1	−1	−1	−1	2	2	−1
X.10	.	.	.	.	.	.	−1	2	.	.	−1	−1	.	.	.	−1	.	.	5	−1	1	1	1	1	1	1	1
X.11	.	.	.	.	2	−2	.	.	.	.	.	.	.	.	.	−1	.	.	−1	−1	1	−1	1	−1	1	−1	1
X.12	.	.	.	.	−2	2	.	.	.	.	.	.	.	.	.	−1	.	.	−1	−1	−1	1	−1	1	−1	1	1
X.13	.	.	.	.	−2	2	.	.	.	.	.	.	.	.	.	−1	.	.	−1	−1	−1	1	−1	1	−1	1	1
X.14	.	.	.	.	2	−2	.	.	.	.	.	.	.	.	.	−1	.	.	−1	−1	1	−1	1	−1	1	−1	1
X.15	.	.	.	.	.	.	−1	−2	.	.	.	.	1	−1	−1	.	−1	−1	.	3	2	2	2	2	−1	−1	.
X.16	1	1	1	1	1	1	1	−1	1	1	.	.	.	.	.	.	.	.	.	3	−2	−2	−2	−2	1	1	.
X.17	2	1	−1	−2	1	−1	.	.	2	−2	.	.	−1	−1	1	.	.	.	−3	−3	1	−1	1	−1	1	−1	−1
X.18	2	1	−1	−2	−1	1	.	.	2	−2	.	.	−1	1	−1	.	.	.	−3	−3	−1	1	−1	1	−1	1	−1
X.19	−2	−4	4	2	.	.	.	.	1	−1	.	.	−2	.	.	.	.	.	6	.	.	.	.	.	.	.	−2
X.20	.	.	.	.	2	2	−1	−2	.	.	.	.	.	.	.	−1	.	.	−3	−3	−1	−1	−1	−1	−1	−1	1
X.21	.	.	.	.	.	.	−1	.	.	.	.	.	.	.	.	1	.	.	1	1	−1	−1	−1	−1	−1	−1	1
X.22	.	.	.	.	.	.	−1	.	.	.	.	.	.	.	.	−1	.	.	3	3	1	1	1	1	1	1	−1
X.23	−2	−2	−2	−2	2	2	1	2	1	1	.	.	.	.	.	1	−1	−1	.	.	.	.	.	.	.	.	.
X.24	−2	−1	1	2	−1	1	.	.	1	−1	.	.	.	.	.	1	.	.	−3	.	1	−1	1	−1	−2	2	−1
X.25	−2	−1	1	2	1	−1	.	.	1	−1	.	.	.	.	.	1	.	.	−3	.	−1	1	−1	1	2	−2	−1
X.26	−2	−2	−2	−2	.	.	.	2	−2	−2	.	.	.	.	.	1	.	.	.	3	−2	−2	−2	−2	1	1	.
X.27	3	3	3	3	1	1	−3	3	.	.	.	.	.	.	.	.	.	.	1	1	−1	−1	−1	−1	−1	−1	1
X.28	1	−2	−2	1	−2	−2	3	1	1	1	.	.	.	.	.	.	.	.	4	−2	−2	−2	−2	−2	−2	−2	.
X.29	.	3	3	.	1	1	2	−2	.	.	.	.	.	.	.	.	1	1	−6	.	.	.	.	.	.	.	−2
X.30	1	1	1	1	1	1	−1	−1	−2	−2	.	.	.	.	.	.	1	1	3	.	1	1	1	1	−2	−2	−1
X.31	2	2	2	2	.	.	.	.	−1	−1	.	.	.	.	.	.	−1	−1	−3	−3	−1	−1	−1	−1	−1	−1	1
X.32	.	−3	−3	.	−1	−1	.	2	.	.	.	.	.	.	.	.	1	1	.	.	.	.	.	.	.	.	.
X.33	.	.	.	.	.	.	.	.	.	.	1	1	2	.	.	.	.	.	3	.	B	$\bar{B}$	$\bar{B}$	B	.	.	−1
X.34	.	.	.	.	.	.	.	.	.	.	1	1	2	.	.	.	.	.	3	.	$\bar{B}$	B	B	$\bar{B}$	.	.	−1
X.35	.	.	.	.	2	2	1	.	.	.	.	.	−1	−1	−1	.	.	.	4	−2	2	2	2	2	2	2	.
X.36	−2	2	−2	2	.	.	.	.	−2	2	−1	−1	.	.	.	.	.	.	4	−2	.	.	.	.	.	.	.
X.37	−2	2	−2	2	.	.	.	.	−2	2	1	1	.	.	.	.	.	.	.	.	.	.	.	.	.	.	.
X.38	−1	−1	−1	−1	−1	−1	−2	−1	−1	−1	C	D	.	.	.	.	.	.	.	.	.	.	.	.	.	.	.
X.39	−1	−1	−1	−1	−1	−1	−2	−1	−1	−1	D	C	.	.	.	.	.	.	.	.	.	.	.	.	.	.	.
X.40	.	.	.	.	2	−2	.	.	.	.	.	.	1	1	−1	1	.	.	−5	1	1	−1	1	−1	1	−1	1
X.41	.	.	.	.	−2	2	.	.	.	.	.	.	1	−1	1	1	.	.	−5	1	−1	1	−1	1	−1	1	1
X.42	−3	.	.	−3	.	.	1	1	.	.	.	.	.	.	.	.	.	.	−3	.	1	1	1	1	−2	−2	1
X.43	.	.	.	.	.	.	1	.	.	.	.	.	1	−1	−1	1	.	.	.	.	.	.	.	.	.	.	.
X.44	−2	−4	4	2	.	.	.	.	1	−1	.	.	2	.	.	.	.	.	−2	−2	.	.	.	.	.	.	2
X.45	.	3	−3	.	1	−1	.	.	.	.	.	.	.	.	.	.	.	.	.	.	.	.	.	.	.	.	.
X.46	.	3	−3	.	−1	1	.	.	.	.	.	.	.	.	.	.	.	.	.	.	.	.	.	.	.	.	.
X.47	.	3	3	.	−1	−1	1	2	.	.	.	.	−1	−1	−1	.	1	1	.	.	.	.	.	.	.	.	.
X.48	−1	2	2	−1	−2	−2	2	1	−1	−1	.	.	.	.	.	.	.	.	−3	−3	1	1	1	1	1	1	1
X.49	.	.	.	.	−2	−2	−2	.	.	.	.	.	.	.	.	1	.	.	4	−2	.	.	.	.	.	.	.
X.50	.	.	.	.	−2	2	.	.	.	.	.	.	.	.	.	1	.	.	−4	2	2	−2	2	−2	2	−2	.
X.51	.	.	.	.	2	−2	.	.	.	.	.	.	.	.	.	1	.	.	−4	2	−2	2	−2	2	−2	2	.
X.52	.	.	.	.	.	.	1	.	.	.	1	1	−2	.	.	−1	−1	−1	.	.	.	.	.	.	.	.	.
X.53	.	.	.	.	2	2	1	.	.	.	.	.	−1	1	1	.	.	.	.	.	.	.	.	.	.	.	.
X.54	2	1	−1	−2	1	−1	.	.	−1	1	.	.	.	.	.	.	.	.	1	1	1	−1	1	−1	1	−1	−1
X.55	7	−2	−2	7	.	.	.	1	1	1	.	.	.	.	.	.	.	.	−1	−1	1	1	1	1	1	1	−1
X.56	2	1	−1	−2	−1	1	.	.	−1	1	.	.	.	.	.	.	.	.	1	1	−1	1	−1	1	−1	1	−1
X.57	−5	−2	−2	−5	.	.	.	1	1	1	.	.	.	.	.	.	.	.	−1	−1	1	1	1	1	1	1	−1
X.58	.	.	.	.	.	.	.	2	.	.	.	.	.	.	.	.	E	$\bar{E}$	−4	2	.	.	.	.	.	.	.
X.59	.	.	.	.	.	.	.	2	.	.	.	.	.	.	.	.	$\bar{E}$	E	−4	2	.	.	.	.	.	.	.
X.60	4	2	−2	−4	.	.	.	.	1	−1	.	.	.	.	.	.	.	.	3	.	−1	1	−1	1	2	−2	1
X.61	−2	2	−2	2	.	.	.	.	1	−1	.	.	.	.	.	.	.	.	3	.	1	−1	1	−1	−2	2	1
X.62	1	−2	−2	1	.	.	−1	−1	1	1	.	.	.	.	.	.	.	.	.	.	.	.	.	.	.	.	.
X.63	−2	1	1	−2	1	1	−3	.	1	1	.	.	.	.	.	.	.	.	.	.	.	.	.	.	.	.	.
X.64	−2	−1	1	2	−1	1	.	.	1	−1	.	.	.	.	.	−1	.	.	3	3	1	−1	1	−1	1	−1	1
X.65	4	−1	1	−4	1	−1	.	.	1	−1	.	.	.	.	.	−1	.	.	3	3	−1	1	−1	1	−1	1	1
X.66	.	.	.	.	2	2	2	.	.	.	.	.	.	.	.	1	.	.	.	.	.	.	.	.	.	.	.
X.67	.	.	.	.	.	.	1	.	.	.	−1	−1	1	1	1	−1	.	.	.	.	.	.	.	.	.	.	.
X.68	.	−3	−3	.	−1	−1	−1	.	.	.	.	.	−2	.	.	.	.	.	.	.	.	.	.	.	.	.	.
X.69	.	.	.	.	.	.	.	.	.	.	.	.	.	.	.	.	.	.	1	1	1	1	1	1	1	1	1
X.70	.	.	.	.	.	.	.	.	.	.	.	.	.	.	.	.	.	.	1	1	1	1	1	1	1	1	1
X.71	.	.	.	.	−2	−2	1	.	.	.	.	.	.	.	.	.	.	.	3	3	−1	−1	−1	−1	−1	−1	−1
X.72	1	1	1	1	−1	−1	−2	1	1	1	.	.	.	.	.	.	.	.	−3	.	B	B	$\bar{B}$	$\bar{B}$	.	.	−1
X.73	1	1	1	1	−1	−1	−2	1	1	1	.	.	.	.	.	.	.	.	−3	.	$\bar{B}$	$\bar{B}$	B	B	.	.	−1
X.74	1	−2	−2	1	.	.	.	−1	−2	−2	.	.	.	.	.	.	1	1	.	.	.	.	.	.	.	.	.
X.75	−4	−2	2	4	.	.	.	.	−1	1	.	.	−1	−1	1	.	.	.	.	−3	−2	2	−2	2	1	−1	.
X.76	2	−2	2	−2	.	.	.	.	−1	1	.	.	−1	1	−1	.	.	.	.	−3	2	−2	2	−2	−1	1	.
X.77	.	.	.	.	2	2	−2	−2	.	.	.	.	.	.	.	1	.	.	.	.	.	.	.	.	.	.	.
X.78	.	.	.	.	−2	−2	−1	.	.	.	.	.	.	.	.	.	.	.	−5	1	1	1	1	1	1	1	−1
X.79	.	.	.	.	.	.	1	.	.	.	.	.	1	−1	−1	−1	.	.	.	.	.	.	.	.	.	.	.
X.80	.	.	.	.	.	.	.	.	.	.	.	.	1	−1	1	.	.	.	1	1	1	−1	1	−1	1	−1	−1
X.81	.	.	.	.	.	.	.	.	.	.	.	.	1	1	−1	.	.	.	1	1	−1	1	−1	1	−1	1	−1
X.82	.	.	.	.	.	.	−1	.	.	.	.	.	−1	1	1	1	−1	−1	.	.	.	.	.	.	.	.	.
X.83	3	.	.	3	.	.	.	1	.	.	.	.	.	.	.	.	.	.	.	.	.	.	.	.	.	.	.
X.84	−1	2	2	−1	.	.	.	1	−1	−1	.	.	.	.	.	.	.	.	.	.	.	.	.	.	.	.	.
X.85	−1	2	2	−1	.	.	.	1	−1	−1	.	.	.	.	.	.	.	.	.	.	.	.	.	.	.	.	.
X.86	2	−2	2	−2	.	.	.	.	−1	1	.	.	2	.	.	.	.	.	.	.	.	.	.	.	.	.	.
X.87	.	3	−3	.	−1	1	.	.	.	.	.	.	.	.	.	.	.	.	.	−3	2	−2	2	−2	−1	1	.

Character table of $H(\mathsf{Fi}_{23})$ *(continued)*

	18j	18k	20a	20b	21a	22a	22b	22c	22d	22e	22f	24a	24b	26a	26b	30a	30b	30c	42a
2	3	1	3	3	1	2	2	2	2	2	2	4	4	1	1	2	2	2	1
3	2	3	.	.	1	.	.	.	.	.	.	1	1	.	.	1	1	1	1
5	.	.	1	1	.	.	.	.	.	.	.	.	.	.	.	1	1	1	.
7	.	.	.	.	1	.	.	.	.	.	.	.	.	.	.	.	.	.	1
11	.	.	.	.	.	1	1	1	1	1	1	.	.	.	.	.	.	.	.
13	.	.	.	.	.	.	.	.	.	.	.	.	.	1	1	.	.	.	.
2P	9a	9c	10d	10d	21a	11a	11a	11b	11b	11b	11a	12a	12b	13a	13b	15a	15a	15a	21a
3P	6_9	6_8	20a	20b	7a	22a	22b	22c	22d	22e	22f	8a	8b	26a	26b	10b	10c	10a	14a
5P	18j	18k	4b	4c	21a	22a	22b	22c	22d	22e	22f	24a	24b	26b	26a	6_3	6_4	6_1	42a
7P	18j	18k	20a	20b	3a	22c	22e	22a	22f	22b	22d	24a	24b	26b	26a	30a	30b	30c	6_1
11P	18j	18k	20a	20b	21a	2a	2c	2a	2b	2c	2b	24a	24b	26b	26a	30a	30b	30c	42a
13P	18j	18k	20a	20b	21a	22c	22e	22a	22f	22b	22d	24a	24b	2a	2a	30a	30b	30c	42a
X.1	1	1	1	1	1	1	1	1	1	1	1	1	1	1	1	1	1	1	1
X.2	−1	.	−1	−1	1	1	−1	1	−1	−1	−1	−1	1	.	.	−2	−2	.	1
X.3	−2	−1	.	.	−1	.	.	.	.	.	.	.	.	−1	−1	.	.	−2	1
X.4	.	.	.	.	−1	.	.	.	.	.	.	−2	−2	.	.	−1	−1	1	−1
X.5	2	−1	−1	−1	.	.	.	.	.	.	.	.	−2	.	.	−1	−1	1	.
X.6	−1	2	−1	−1	−1	.	.	.	.	.	.	1	−1	.	.	1	1	−1	−1
X.7	.	−1	1	−1	1	−1	−1	−1	1	−1	1	.	.	.	.	1	−1	1	−1
X.8	.	−1	−1	1	1	−1	1	−1	−1	1	−1	.	.	.	.	−1	1	1	−1
X.9	−1	.	1	1	.	.	.	.	.	.	.	−1	−1	.	.	2	2	.	.
X.10	1	−1	−1	−1	.	.	.	.	.	.	.	1	1	−1	−1	1	1	−1	.
X.11	−1	−1	−1	1	.	$-\bar{A}$	$-\bar{A}$	$-A$	A	$-A$	$\bar{A}$	.	.	.	.	1	−1	1	.
X.12	−1	−1	1	−1	.	$-A$	A	$-\bar{A}$	$-\bar{A}$	$\bar{A}$	$-A$	.	.	.	.	−1	1	1	.
X.13	−1	−1	1	−1	.	$-\bar{A}$	$\bar{A}$	$-A$	$-A$	A	$-\bar{A}$	.	.	.	.	−1	1	1	.
X.14	−1	−1	−1	1	.	$-A$	$-A$	$-\bar{A}$	$\bar{A}$	$-\bar{A}$	A	.	.	.	.	1	−1	1	.
X.15	.	.	.	.	1	.	.	.	.	.	.	1	3	.	.	.	.	.	1
X.16	.	.	.	.	.	−1	−1	−1	−1	−1	−1	1	1	.	.	.	.	.	.
X.17	1	.	1	−1	1	.	.	.	.	.	.	.	.	.	.	.	.	.	−1
X.18	1	.	−1	1	1	.	.	.	.	.	.	.	.	.	.	.	.	.	−1
X.19	2	.	.	.	−1	.	.	.	.	.	.	.	.	.	.	.	.	.	1
X.20	1	.	1	1	.	.	.	.	.	.	.	−1	1	.	.	−1	−1	−1	.
X.21	1	−2	1	1	.	.	.	.	.	.	.	1	−1	.	.	1	1	1	.
X.22	−1	.	1	1	.	−1	1	−1	1	1	1	−1	1	.	.	−1	−1	−1	.
X.23	.	.	1	1	.	.	.	.	.	.	.	1	−1	.	.	−1	−1	1	.
X.24	1	.	.	.	.	.	.	.	.	.	.	.	.	.	.	1	−1	−1	.
X.25	1	.	.	.	.	.	.	.	.	.	.	.	.	.	.	−1	1	−1	.
X.26	.	.	.	.	.	.	.	.	.	.	.	.	.	.	.	1	1	1	.
X.27	1	1	.	.	.	.	.	.	.	.	.	−1	−1	.	.	.	.	.	.
X.28	.	1	.	.	.	.	.	.	.	.	.	−1	1	.	.	.	.	.	.
X.29	−2	.	.	.	.	.	.	.	.	.	.	.	.	.	.	.	.	.	.
X.30	−1	.	.	.	.	.	.	.	.	.	.	1	−1	.	.	.	.	.	.
X.31	1	.	.	.	.	.	.	.	.	.	.	2	.	.	.	.	.	.	.
X.32	.	.	.	.	.	.	.	.	.	.	.	2	.	.	.	.	.	.	.
X.33	1	.	.	.	1	.	.	.	.	.	.	.	.	−1	−1	.	.	.	−1
X.34	1	.	.	.	1	.	.	.	.	.	.	.	.	−1	−1	.	.	.	−1
X.35	.	1	.	.	−1	.	.	.	.	.	.	1	−1	.	.	.	.	.	−1
X.36	.	1	.	.	.	.	.	.	.	.	.	.	.	1	1	.	.	.	.
X.37	.	.	.	.	.	.	.	.	.	.	.	.	.	−1	−1	.	.	.	.
X.38	.	.	.	.	.	.	.	.	.	.	.	.	.	D	C	.	.	.	.
X.39	.	.	.	.	.	.	.	.	.	.	.	.	.	C	D	.	.	.	.
X.40	−1	1	1	−1	−1	.	.	.	.	.	.	.	.	.	.	−1	1	−1	1
X.41	−1	1	−1	1	−1	.	.	.	.	.	.	.	.	.	.	1	−1	−1	1
X.42	1	.	.	.	.	.	.	.	.	.	.	−1	1	.	.	.	.	.	.
X.43	.	.	−1	−1	1	.	.	.	.	.	.	−1	1	.	.	1	1	1	1
X.44	−2	1	.	.	1	.	.	.	.	.	.	.	.	.	.	.	.	.	−1
X.45	.	.	1	−1	.	1	1	1	−1	1	−1	.	.	.	.	−2	2	.	.
X.46	.	.	−1	1	.	1	−1	1	1	−1	1	.	.	.	.	2	−2	.	.
X.47	.	.	.	.	−1	.	.	.	.	.	.	−1	1	.	.	.	.	.	−1
X.48	1	.	.	.	.	.	.	.	.	.	.	.	.	.	.	.	.	.	.
X.49	.	1	1	1	.	.	.	.	.	.	.	.	.	.	.	−1	−1	1	.
X.50	.	−1	1	−1	.	.	.	.	.	.	.	.	.	.	.	1	−1	−1	.
X.51	.	−1	−1	1	.	.	.	.	.	.	.	.	.	.	.	−1	1	−1	.
X.52	.	.	−1	−1	1	.	.	.	.	.	.	−1	−1	1	1	1	1	−1	1
X.53	.	.	.	.	−1	.	.	.	.	.	.	−1	1	.	.	.	.	.	−1
X.54	1	1	.	.	.	.	.	.	.	.	.	.	.	.	.	.	.	.	.
X.55	−1	−1	.	.	.	.	.	.	.	.	.	.	.	.	.	.	.	.	.
X.56	1	1	.	.	.	.	.	.	.	.	.	.	.	.	.	.	.	.	.
X.57	−1	−1	.	.	.	.	.	.	.	.	.	.	.	.	.	.	.	.	.
X.58	.	−1	.	.	.	.	.	.	.	.	.	.	.	.	.	.	.	.	.
X.59	.	−1	.	.	.	.	.	.	.	.	.	.	.	.	.	.	.	.	.
X.60	−1	.	.	.	.	−1	−1	−1	1	−1	1	.	.	.	.	.	.	.	.
X.61	−1	.	.	.	.	−1	1	−1	−1	1	−1	.	.	.	.	.	.	.	.
X.62	.	.	.	.	.	.	.	.	.	.	.	−1	1	.	.	.	.	.	.
X.63	.	.	.	.	.	.	.	.	.	.	.	−1	−1	.	.	.	.	.	.
X.64	−1	.	−1	1	.	.	.	.	.	.	.	.	.	.	.	−1	1	1	.
X.65	−1	.	1	−1	.	.	.	.	.	.	.	.	.	.	.	1	−1	1	.
X.66	.	.	−1	−1	.	.	.	.	.	.	.	.	.	.	.	−1	−1	1	.
X.67	.	.	1	1	1	.	.	.	.	.	.	−1	−1	−1	−1	1	1	−1	1
X.68	.	.	.	.	1	.	.	.	.	.	.	−1	−1	.	.	.	.	.	1
X.69	1	1	.	.	.	A	A	$\bar{A}$	$\bar{A}$	$\bar{A}$	A	.	.	.	.	.	.	.	.
X.70	1	1	.	.	.	$\bar{A}$	$\bar{A}$	A	A	A	$\bar{A}$	.	.	.	.	.	.	.	.
X.71	−1	.	.	.	.	.	.	.	.	.	.	1	1	.	.	.	.	.	.
X.72	−1	.	.	.	.	.	.	.	.	.	.	.	.	.	.	.	.	.	.
X.73	−1	.	.	.	.	.	.	.	.	.	.	.	.	.	.	.	.	.	.
X.74	.	.	.	.	.	.	.	.	.	.	.	.	.	.	.	.	.	.	.
X.75	.	.	.	.	1	.	.	.	.	.	.	.	.	.	.	.	.	.	−1
X.76	.	.	.	.	1	.	.	.	.	.	.	.	.	.	.	.	.	.	−1
X.77	.	.	−1	−1	.	.	.	.	.	.	.	.	.	.	.	1	1	1	.
X.78	−1	1	.	.	.	.	.	.	.	.	.	1	−1	.	.	.	.	.	.
X.79	.	.	1	1	1	1	−1	1	−1	−1	−1	1	−1	.	.	−1	−1	−1	1
X.80	1	−2	.	.	−1	.	.	.	.	.	.	.	.	.	.	.	.	.	1
X.81	1	−2	.	.	−1	.	.	.	.	.	.	.	.	.	.	.	.	.	1
X.82	.	.	.	.	−1	.	.	.	.	.	.	1	1	.	.	−1	−1	1	−1
X.83	.	.	−1	−1	.	.	.	.	.	.	.	.	.	.	.	.	.	.	.
X.84	.	.	.	.	.	A	$-A$	$\bar{A}$	$-\bar{A}$	$-\bar{A}$	$-A$	.	.	.	.	.	.	.	.
X.85	.	.	.	.	.	$\bar{A}$	$-\bar{A}$	A	$-A$	$-A$	$-\bar{A}$	.	.	.	.	.	.	.	.
X.86	.	.	.	.	1	.	.	.	.	.	.	.	.	.	.	.	.	.	−1
X.87	.	.	.	.	.	.	.	.	.	.	.	.	.	.	.	.	.	.	.

Character table of $H(\mathsf{Fi}_{23})$ *(continued)*

2	18	18	17	17	18	18	16	9	8	7	4	12	11	11	12	10	10	10
3	9	9	6	6	4	4	3	7	9	7	7	3	2	2	1	2	2	1
5	2	2	1	1	1	1	.	1	.	.	.	.	1	1	.	.	.	.
7	1	1	1	1	.	.	.	1	.	.	.	.	.	.	.	.	.	.
11	1	1	1	1	.	.	.	.	.	.	.	.	.	.	.	.	.	.
13	1	1	.	.	.	.	.	.	.	.	.	.	.	.	.	.	.	.
	1a	2a	2b	2c	2d	2e	2f	3a	3b	3c	3d	4a	4b	4c	4d	4e	4f	4g
2P	1a	1a	1a	1a	1a	1a	1a	3a	3b	3c	3d	2d	2e	2e	2d	2f	2f	2e
3P	1a	2a	2b	2c	2d	2e	2f	1a	1a	1a	1a	4a	4b	4c	4d	4e	4f	4g
5P	1a	2a	2b	2c	2d	2e	2f	3a	3b	3c	3d	4a	4b	4c	4d	4e	4f	4g
7P	1a	2a	2b	2c	2d	2e	2f	3a	3b	3c	3d	4a	4b	4c	4d	4e	4f	4g
11P	1a	2a	2b	2c	2d	2e	2f	3a	3b	3c	3d	4a	4b	4c	4d	4e	4f	4g
13P	1a	2a	2b	2c	2d	2e	2f	3a	3b	3c	3d	4a	4b	4c	4d	4e	4f	4g
X.88	1201200	1201200	−30800	−30800	560	560	176	420	−30	−48	51	−16	.	.	−16	16	16	.
X.89	1201200	−1201200	−24640	24640	−1680	1680	.	−210	132	−39	24	.	40	−40	.	12	−12	.
X.90	1297296	−1297296	44352	−44352	−1200	1200	.	1134	−324	−81	.	.	−40	40	.	−12	12	.
X.91	1297296	−1297296	−44352	44352	−1200	1200	.	1134	−324	−81	.	.	40	−40	.	−12	12	.
X.92	1360800	1360800	30240	30240	1440	1440	288	.	486	.	.	−96	.	.	−32	.	.	.
X.93	1372800	1372800	49280	49280	640	640	128	1560	−312	−24	12	.	.	.	.	.	.	.
X.94	1441792	1441792	.	.	.	.	.	−512	640	64	−8	.	.	.	.	.	.	.
X.95	1441792	−1441792	.	.	.	.	.	−512	640	64	−8	.	.	.	.	.	.	.
X.96	1663200	−1663200	.	.	2400	−2400	.	.	108	54	27	.	.	.	.	−8	8	.
X.97	1663200	−1663200	.	.	2400	−2400	.	.	108	54	27	.	.	.	.	−8	8	.
X.98	1791153	1791153	−5103	−5103	−2511	−2511	81	.	.	.	.	81	9	9	33	9	9	9
X.99	1876446	1876446	37422	37422	−1890	−1890	−306	729	.	.	.	54	−30	−30	−18	18	18	10
X.100	2027025	2027025	−24255	−24255	−1455	−1455	33	.	−324	81	.	9	85	85	−15	−3	−3	−3
X.101	2050048	2050048	.	.	2048	2048	.	−1232	−224	−80	−8	.	.	.	.	.	.	.
X.102	2196480	−2196480	11264	−11264	1024	−1024	.	−456	−240	−24	−24	.	64	−64	.	.	.	.
X.103	2196480	−2196480	−11264	11264	1024	−1024	.	−456	−240	−24	−24	.	−64	64	.	.	.	.
X.104	2316600	2316600	−43560	−43560	−840	−840	24	405	−162	.	.	−24	40	40	−8	.	.	−8
X.105	2358720	−2358720	8064	−8064	−1344	1344	.	−1134	−324	.	.	.	−16	16	.	.	.	.
X.106	2358720	−2358720	−8064	8064	−1344	1344	.	−1134	−324	.	.	.	16	−16	.	.	.	.
X.107	2402400	2402400	12320	12320	−160	−160	160	−735	−384	48	−6	−64	−40	−40	.	.	.	−8
X.108	2402400	−2402400	49280	−49280	1760	−1760	.	840	−60	66	21	.	80	−80	.	8	−8	.
X.109	2402400	−2402400	−49280	49280	1760	−1760	.	840	−60	66	21	.	−80	80	.	8	−8	.
X.110	2555904	2555904	−32768	−32768	.	.	.	−384	192	−96	−24	.	.	.	.	.	.	.
X.111	2555904	−2555904	32768	−32768	.	.	.	−384	192	−96	−24	.	.	.	.	.	.	.
X.112	2555904	−2555904	−32768	32768	.	.	.	−384	192	−96	−24	.	.	.	.	.	.	.
X.113	2555904	2555904	32768	32768	.	.	.	−384	192	−96	−24	.	.	.	.	.	.	.
X.114	2729376	2729376	7776	7776	−864	−864	−288	−729	.	.	.	.	−24	−24	.	.	.	8

Character table of $H(\mathsf{Fi}_{23})$ *(continued)*

2	4	9	8	8	8	7	8	8	4	8	8	9	9	5	5	8	8	7	7	7	6	6	5
3	1	7	9	5	5	7	6	6	7	4	4	3	3	5	5	3	3	3	3	3	3	3	3
5	2	1	.	1	1	.	.	.	.	.	.	.	.	.	.	.	.	.	.	.	.	.	.
7	.	1	.	.	.	.	.	.	.	.	.	.	.	.	.	.	.	.	.	.	.	.	.
11	.	.	.	.	.	.	.	.	.	.	.	.	.	.	.	.	.	.	.	.	.	.	.
13	.	.	.	.	.	.	.	.	.	.	.	.	.	.	.	.	.	.	.	.	.	.	.
	5a	6_{1}	6_{2}	6_{3}	6_{4}	6_{5}	6_{6}	6_{7}	6_{8}	6_{9}	6_{10}	6_{11}	6_{12}	6_{13}	6_{14}	6_{15}	6_{16}	6_{17}	6_{18}	6_{19}	6_{20}	6_{21}	6_{22}
2P	5a	3a	3b	3a	3a	3c	3b	3b	3d	3b	3b	3a	3a	3c	3c	3b	3b	3c	3c	3a	3c	3c	3c
3P	5a	2a	2a	2b	2c	2a	2c	2b	2a	2d	2e	2d	2e	2b	2c	2f	2f	2f	2f	2f	2e	2d	2f
5P	1a	6_{1}	6_{2}	6_{3}	6_{4}	6_{5}	6_{6}	6_{7}	6_{8}	6_{9}	6_{10}	6_{11}	6_{12}	6_{13}	6_{14}	6_{15}	6_{16}	6_{17}	6_{18}	6_{19}	6_{20}	6_{21}	6_{22}
7P	5a	6_{1}	6_{2}	6_{3}	6_{4}	6_{5}	6_{6}	6_{7}	6_{8}	6_{9}	6_{10}	6_{11}	6_{12}	6_{13}	6_{14}	6_{15}	6_{16}	6_{17}	6_{18}	6_{19}	6_{20}	6_{21}	6_{22}
11P	5a	6_{1}	6_{2}	6_{3}	6_{4}	6_{5}	6_{6}	6_{7}	6_{8}	6_{9}	6_{10}	6_{11}	6_{12}	6_{13}	6_{14}	6_{15}	6_{16}	6_{17}	6_{18}	6_{19}	6_{20}	6_{21}	6_{22}
13P	5a	6_{1}	6_{2}	6_{3}	6_{4}	6_{5}	6_{6}	6_{7}	6_{8}	6_{9}	6_{10}	6_{11}	6_{12}	6_{13}	6_{14}	6_{15}	6_{16}	6_{17}	6_{18}	6_{19}	6_{20}	6_{21}	6_{22}
X.88	.	420	−30	−20	−20	−48	34	34	51	2	2	−28	−28	−2	−2	2	2	8	8	−4	−4	−4	2
X.89	.	210	−132	20	−20	39	−92	92	−24	12	−12	6	−6	11	−11	−12	12	−9	9	.	9	−9	−3
X.90	−4	−1134	324	−36	36	81	36	−36	.	−12	12	6	−6	−9	9	−12	12	9	−9	.	3	−3	−3
X.91	−4	−1134	324	36	−36	81	−36	36	.	−12	12	6	−6	9	−9	12	−12	9	−9	.	3	−3	3
X.92	.	.	486	.	.	.	−54	−54	.	−18	−18	.	.	.	.	18	18	.	.	.	.	.	.
X.93	.	1560	−312	−40	−40	−24	32	32	12	−8	−8	−8	−8	−4	−4	−16	−16	8	8	8	−8	−8	−4
X.94	−8	−512	640	.	.	64	.	.	−8	.	.	.	.	.	.	.	.	.	.	.	.	.	.
X.95	−8	512	−640	.	.	−64	.	.	8	.	.	.	.	.	.	.	.	.	.	.	.	.	.
X.96	.	.	−108	.	.	−54	.	.	−27	−12	12	.	.	.	.	.	.	−6	6	.	6	−6	.
X.97	.	.	−108	.	.	−54	.	.	−27	−12	12	.	.	.	.	.	.	−6	6	.	6	−6	.
X.98	3	.	.	.	.	.	.	.	.	.	.	.	.	.	.	.	.	.	.	.	.	.	.
X.99	−4	729	.	−81	−81	.	.	.	.	.	.	9	9	.	.	.	.	.	.	−9	.	.	.
X.100	.	.	−324	−90	−90	81	−36	−36	.	12	12	.	.	−9	−9	12	12	9	9	6	−3	−3	3
X.101	−2	−1232	−224	.	.	−80	.	.	−8	32	32	−16	−16	.	.	.	.	.	.	.	8	8	.
X.102	5	456	240	104	−104	24	−32	32	24	16	−16	−8	8	−4	4	.	.	.	.	.	−4	4	.
X.103	5	456	240	−104	104	24	32	−32	24	16	−16	−8	8	4	−4	.	.	.	.	.	−4	4	.
X.104	.	405	−162	45	45	.	18	18	.	6	6	21	21	18	18	−6	−6	.	.	−3	.	.	−6
X.105	−5	1134	324	−36	36	.	36	−36	.	−12	12	−6	6	18	−18	−12	12	.	.	.	.	.	6
X.106	−5	1134	324	36	−36	.	−36	36	.	−12	12	−6	6	−18	18	12	−12	.	.	.	.	.	−6
X.107	.	−735	−384	35	35	48	8	8	−6	−16	−16	17	17	8	8	−8	−8	−8	−8	−5	−4	−4	4
X.108	.	−840	60	−40	40	−66	−32	32	−21	−4	4	8	−8	−4	4	.	.	6	−6	.	−2	2	.
X.109	.	−840	60	40	−40	−66	32	−32	−21	−4	4	8	−8	4	−4	.	.	6	−6	.	−2	2	.
X.110	4	−384	192	64	64	−96	64	64	−24	.	.	.	.	−8	−8	.	.	.	.	.	.	.	.
X.111	4	384	−192	−64	64	96	64	−64	24	.	.	.	.	8	−8	.	.	.	.	.	.	.	.
X.112	4	384	−192	64	−64	96	−64	64	24	.	.	.	.	−8	8	.	.	.	.	.	.	.	.
X.113	4	−384	192	−64	−64	−96	−64	−64	−24	.	.	.	.	8	8	.	.	.	.	.	.	.	.
X.114	1	−729	.	81	81	.	.	.	.	.	.	−9	−9	.	.	.	.	.	.	9	.	.	.

Character table of $H(\mathsf{Fi}_{23})$ *(continued)*

	6_{23}	6_{24}	6_{25}	$7a$	$8a$	$8b$	$8c$	$8d$	$8e$	$9a$	$9b$	$9c$	$10a$	$10b$	$10c$	$10d$	$10e$	$11a$	$11b$	$12a$	$12b$	$12c$	$12d$	$12e$	$12f$	$12g$	$12h$	$12i$
2	5	4	4	2	7	7	7	6	6	3	2	1	4	3	3	4	4	2	2	7	7	5	6	6	6	5	5	5
3	3	3	3	1	1	1	.	.	.	4	4	3	1	1	1	.	.	.	.	2	2	3	2	2	2	2	2	2
5	.	.	.	.	.	.	.	.	.	.	.	.	2	1	1	1	1	.	.	.	.	.	.	.	.	.	.	.
7	.	.	.	1	.	.	.	.	.	.	.	.	.	.	.	.	.	.	.	.	.	.	.	.	.	.	.	.
11	.	.	.	.	.	.	.	.	.	.	.	.	.	.	.	.	.	1	1	.	.	.	.	.	.	.	.	.
13	.	.	.	.	.	.	.	.	.	.	.	.	.	.	.	.	.	.	.	.	.	.	.	.	.	.	.	.
2P	$3c$	$3d$	$3d$	$7a$	$4a$	$4a$	$4d$	$4f$	$4f$	$9a$	$9b$	$9c$	$5a$	$5a$	$5a$	$5a$	$5a$	$11b$	$11a$	6_{11}	6_{11}	6_9	6_{12}	6_{12}	6_{11}	6_{15}	6_{15}	6_{16}
3P	$2f$	$2f$	$2f$	$7a$	$8a$	$8b$	$8c$	$8d$	$8e$	$3b$	$3b$	$3d$	$10a$	$10b$	$10c$	$10d$	$10e$	$11a$	$11b$	$4a$	$4a$	$4a$	$4c$	$4b$	$4a$	$4e$	$4f$	$4f$
5P	6_{23}	6_{24}	6_{25}	$7a$	$8a$	$8b$	$8c$	$8d$	$8e$	$9a$	$9b$	$9c$	$2a$	$2b$	$2c$	$2e$	$2d$	$11a$	$11b$	$12a$	$12b$	$12c$	$12d$	$12e$	$12f$	$12g$	$12h$	$12i$
7P	6_{23}	6_{24}	6_{25}	$1a$	$8a$	$8b$	$8c$	$8d$	$8e$	$9a$	$9b$	$9c$	$10a$	$10b$	$10c$	$10d$	$10e$	$11b$	$11a$	$12a$	$12b$	$12c$	$12d$	$12e$	$12f$	$12g$	$12h$	$12i$
11P	6_{23}	6_{24}	6_{25}	$7a$	$8a$	$8b$	$8c$	$8d$	$8e$	$9a$	$9b$	$9c$	$10a$	$10b$	$10c$	$10d$	$10e$	$1a$	$1a$	$12a$	$12b$	$12c$	$12d$	$12e$	$12f$	$12g$	$12h$	$12i$
13P	6_{23}	6_{24}	6_{25}	$7a$	$8a$	$8b$	$8c$	$8d$	$8e$	$9a$	$9b$	$9c$	$10a$	$10b$	$10c$	$10d$	$10e$	$11b$	$11a$	$12a$	$12b$	$12c$	$12d$	$12e$	$12f$	$12g$	$12h$	$12i$
$X.88$	2	−1	−1	.	.	.	.	.	.	3	.	.	.	.	.	.	.	.	.	−4	−4	2	.	.	−4	−2	−2	−2
$X.89$	3	.	.	.	.	.	.	2	−2	.	3	.	.	.	.	.	.	.	.	−6	6	.	2	−2	.	.	.	.
$X.90$	3	.	.	.	.	.	.	2	−2	.	.	.	4	2	−2	.	.	.	.	−6	6	.	−2	2	.	.	.	.
$X.91$	−3	.	.	.	.	.	.	−2	2	.	.	.	4	−2	2	.	.	.	.	−6	6	.	2	−2	.	.	.	.
$X.92$	.	.	.	.	.	.	.	.	.	.	.	.	.	.	.	.	.	1	1	.	.	−6	.	.	.	.	.	.
$X.93$	−4	−4	−4	2	.	.	.	.	.	−3	.	.	.	.	.	.	.	.	.	.	.	.	.	.	.	.	.	.
$X.94$	.	.	.	2	.	.	.	.	.	4	−2	−2	−8	.	.	.	.	.	.	.	.	.	.	.	.	.	.	.
$X.95$	.	.	.	2	.	.	.	.	.	4	−2	−2	8	.	.	.	.	.	.	.	.	.	.	.	.	.	.	.
$X.96$	.	3	−3	.	.	.	.	.	.	.	.	.	.	.	.	.	.	.	.	.	.	.	.	.	.	−2	2	2
$X.97$	.	3	−3	.	.	.	.	.	.	.	.	.	.	.	.	.	.	.	.	.	.	.	.	.	.	−2	2	2
$X.98$	.	.	.	.	−3	−3	−3	1	1	.	.	.	3	−3	−3	−1	−1	1	1	.	.	.	.	.	.	.	.	.
$X.99$	.	.	.	−2	−2	2	2	.	.	.	.	.	−4	2	2	.	.	.	.	−3	−3	.	3	3	3	.	.	.
$X.100$	3	.	.	.	1	−3	1	−1	−1	.	.	.	.	.	.	.	.	.	.	.	.	.	−2	−2	6	.	.	.
$X.101$	.	.	.	.	.	.	.	.	.	−2	4	1	−2	.	.	−2	−2	.	.	.	.	.	.	.	.	.	.	.
$X.102$	.	.	.	−1	.	.	.	.	.	3	3	.	−5	−1	1	1	−1	.	.	.	.	.	−4	4	.	.	.	.
$X.103$	.	.	.	−1	.	.	.	.	.	3	3	.	−5	1	−1	1	−1	.	.	.	.	.	4	−4	.	.	.	.
$X.104$	−6	.	.	−1	4	4	.	.	.	.	.	.	.	.	.	.	.	.	.	−3	−3	−6	1	1	−3	.	.	.
$X.105$	−6	.	.	.	.	.	.	.	.	.	.	.	5	−1	1	−1	1	1	1	6	−6	.	−2	2	.	.	.	.
$X.106$	6	.	.	.	.	.	.	.	.	.	.	.	5	1	−1	−1	1	1	1	6	−6	.	2	−2	.	.	.	.
$X.107$	4	−2	−2	.	4	−4	.	.	.	3	3	.	.	.	.	.	.	.	.	5	5	8	−1	−1	−1	.	.	.
$X.108$	.	−3	3	.	.	.	.	.	.	−3	.	.	.	.	.	.	.	.	.	.	.	.	4	−4	.	2	−2	−2
$X.109$	.	−3	3	.	.	.	.	.	.	−3	.	.	.	.	.	.	.	.	.	.	.	.	−4	4	.	2	−2	−2
$X.110$	.	.	.	1	.	.	.	.	.	.	−3	.	4	2	2	.	.	−1	−1	.	.	.	.	.	.	.	.	.
$X.111$	.	.	.	1	.	.	.	.	.	.	−3	.	−4	−2	2	.	.	−1	−1	.	.	.	.	.	.	.	.	.
$X.112$	.	.	.	1	.	.	.	.	.	.	−3	.	−4	2	−2	.	.	−1	−1	.	.	.	.	.	.	.	.	.
$X.113$	.	.	.	1	.	.	.	.	.	.	−3	.	4	−2	−2	.	.	−1	−1	.	.	.	.	.	.	.	.	.
$X.114$	.	.	.	−1	−4	4	.	.	.	.	.	.	1	1	1	1	1	1	1	3	3	.	−3	−3	−3	.	.	.

Character table of $H(\mathsf{Fi}_{23})$ *(continued)*

	$12j$	$12k$	$12l$	$12m$	$12n$	$12o$	$12p$	$12q$	$12r$	$13a$	$13b$	$14a$	$14b$	$14c$	$15a$	$16a$	$16b$	$18a$	$18b$	$18c$	$18d$	$18e$	$18f$	$18g$	$18h$	$18i$
2	5	5	5	4	4	5	5	3	3	1	1	2	2	2	2	5	5	3	2	3	3	3	3	2	2	3
3	2	2	2	2	2	1	1	2	2	.	.	1	.	.	1	.	.	4	4	3	3	3	3	3	3	2
5	.	.	.	.	.	.	.	.	.	.	.	.	.	.	1	.	.	.	.	.	.	.	.	.	.	.
7	.	.	.	.	.	.	.	.	.	.	.	1	1	1	.	.	.	.	.	.	.	.	.	.	.	.
11	.	.	.	.	.	.	.	.	.	.	.	.	.	.	.	.	.	.	.	.	.	.	.	.	.	.
13	.	.	.	.	.	.	.	.	.	1	1	.	.	.	.	.	.	.	.	.	.	.	.	.	.	.
2P	6_{18}	6_{18}	6_{16}	6_{20}	6_{20}	6_{12}	6_9	6_{24}	6_{24}	$13b$	$13a$	$7a$	$7a$	$7a$	$15a$	$8c$	$8c$	$9a$	$9b$	$9a$	$9a$	$9a$	$9a$	$9b$	$9b$	$9a$
3P	$4e$	$4f$	$4e$	$4b$	$4c$	$4g$	$4d$	$4f$	$4e$	$13a$	$13b$	$14a$	$14b$	$14c$	$5a$	$16a$	$16b$	6_2	6_2	6_6	6_7	6_6	6_7	6_6	6_7	6_{10}
5P	$12j$	$12k$	$12l$	$12m$	$12n$	$12o$	$12p$	$12q$	$12r$	$13b$	$13a$	$14a$	$14b$	$14c$	$3a$	$16b$	$16a$	$18a$	$18b$	$18e$	$18f$	$18c$	$18d$	$18g$	$18h$	$18i$
7P	$12j$	$12k$	$12l$	$12m$	$12n$	$12o$	$12p$	$12q$	$12r$	$13b$	$13a$	$2a$	$2b$	$2c$	$15a$	$16b$	$16a$	$18a$	$18b$	$18c$	$18d$	$18e$	$18f$	$18g$	$18h$	$18i$
11P	$12j$	$12k$	$12l$	$12m$	$12n$	$12o$	$12p$	$12q$	$12r$	$13b$	$13a$	$14a$	$14b$	$14c$	$15a$	$16a$	$16b$	$18a$	$18b$	$18e$	$18f$	$18c$	$18d$	$18g$	$18h$	$18i$
13P	$12j$	$12k$	$12l$	$12m$	$12n$	$12o$	$12p$	$12q$	$12r$	$1a$	$1a$	$14a$	$14b$	$14c$	$15a$	$16b$	$16a$	$18a$	$18b$	$18c$	$18d$	$18e$	$18f$	$18g$	$18h$	$18i$
$X.88$	4	4	−2	.	.	.	2	1	1	.	.	.	.	.	.	.	.	3	.	1	1	1	1	−2	−2	−1
$X.89$	3	−3	.	1	−1	.	.	.	.	.	.	.	.	.	.	.	.	.	−3	−2	2	−2	2	1	−1	.
$X.90$	−3	3	.	−1	1	.	.	.	.	.	.	.	.	.	−1	.	.	.	.	.	.	.	.	.	.	.
$X.91$	−3	3	.	1	−1	.	.	.	.	.	.	.	.	.	−1	.	.	.	.	.	.	.	.	.	.	.
$X.92$	.	.	.	.	.	.	−2	.	.	−1	−1	.	.	.	.	.	.	.	.	.	.	.	.	.	.	.
$X.93$	.	.	.	.	.	.	.	.	.	.	.	2	.	.	.	.	.	−3	.	−1	−1	−1	−1	2	2	1
$X.94$	.	.	.	.	.	.	.	.	.	1	1	2	.	.	−2	.	.	4	−2	.	.	.	.	.	.	.
$X.95$	.	.	.	.	.	.	.	.	.	1	1	−2	.	.	−2	.	.	−4	2	.	.	.	.	.	.	.
$X.96$	−2	2	−2	.	.	.	.	−1	1	$-C$	$-D$	.	.	.	.	.	.	.	.	.	.	.	.	.	.	.
$X.97$	−2	2	−2	.	.	.	.	−1	1	$-D$	$-C$	.	.	.	.	.	.	.	.	.	.	.	.	.	.	.
$X.98$	.	.	.	.	.	.	.	.	.	.	.	.	.	.	.	−1	−1	.	.	.	.	.	.	.	.	.
$X.99$	.	.	.	.	.	1	.	.	.	.	.	−2	.	.	−1	.	.	.	.	.	.	.	.	.	.	.
$X.100$	−3	−3	.	1	1	.	.	.	.	.	.	.	.	.	.	1	1	.	.	.	.	.	.	.	.	.
$X.101$	.	.	.	.	.	.	.	.	.	.	.	.	.	.	−2	.	.	−2	4	.	.	.	.	.	.	2
$X.102$	.	.	.	−2	2	.	.	.	.	.	.	1	1	−1	−1	.	.	−3	−3	1	−1	1	−1	1	−1	−1
$X.103$	.	.	.	2	−2	.	.	.	.	.	.	1	−1	1	−1	.	.	−3	−3	−1	1	−1	1	−1	1	−1
$X.104$	.	.	.	−2	−2	1	−2	.	.	.	.	−1	1	1	.	.	.	.	.	.	.	.	.	.	.	.
$X.105$	.	.	.	2	−2	.	.	.	.	.	.	.	.	.	1	.	.	.	.	.	.	.	.	.	.	.
$X.106$	.	.	.	−2	2	.	.	.	.	.	.	.	.	.	1	.	.	.	.	.	.	.	.	.	.	.
$X.107$	.	.	.	2	2	1	.	.	.	.	.	.	.	.	.	.	.	3	3	−1	−1	−1	−1	−1	−1	−1
$X.108$	2	−2	2	2	−2	.	.	1	−1	.	.	.	.	.	.	.	.	3	.	1	−1	1	−1	−2	2	1
$X.109$	2	−2	2	−2	2	.	.	1	−1	.	.	.	.	.	.	.	.	3	.	−1	1	−1	1	2	−2	1
$X.110$	.	.	.	.	.	.	.	.	.	.	.	1	−1	−1	1	.	.	.	−3	−2	−2	−2	−2	1	1	.
$X.111$	.	.	.	.	.	.	.	.	.	.	.	−1	1	−1	1	.	.	.	3	−2	2	−2	2	1	−1	.
$X.112$	.	.	.	.	.	.	.	.	.	.	.	−1	−1	1	1	.	.	.	3	2	−2	2	−2	−1	1	.
$X.113$	.	.	.	.	.	.	.	.	.	.	.	1	1	1	1	.	.	.	−3	2	2	2	2	−1	−1	.
$X.114$	.	.	.	.	.	−1	.	.	.	.	.	−1	−1	−1	1	.	.	.	.	.	.	.	.	.	.	.

Character table of $H(\mathsf{Fi}_{23})$ *(continued)*

2	3	1	3	3	1	2	2	2	2	2	2	4	4	1	1	2	2	2	1
3	2	3	.	.	1	.	.	.	.	.	.	1	1	.	.	1	1	1	1
5	.	.	1	1	.	.	.	.	.	.	.	.	.	.	.	1	1	1	.
7	.	.	.	.	1	.	.	.	.	.	.	.	.	.	.	.	.	.	1
11	.	.	.	.	.	1	1	1	1	1	1	.	.	.	.	.	.	.	.
13	.	.	.	.	.	.	.	.	.	.	.	.	.	1	1	.	.	.	.
	$18j$	$18k$	$20a$	$20b$	$21a$	$22a$	$22b$	$22c$	$22d$	$22e$	$22f$	$24a$	$24b$	$26a$	$26b$	$30a$	$30b$	$30c$	$42a$
$2P$	$9a$	$9c$	$10d$	$10d$	$21a$	$11a$	$11a$	$11b$	$11b$	$11b$	$11a$	$12a$	$12b$	$13a$	$13b$	$15a$	$15a$	$15a$	$21a$
$3P$	6_9	6_8	$20a$	$20b$	$7a$	$22a$	$22b$	$22c$	$22d$	$22e$	$22f$	$8a$	$8b$	$26a$	$26b$	$10b$	$10c$	$10a$	$14a$
$5P$	$18j$	$18k$	$4b$	$4c$	$21a$	$22a$	$22b$	$22c$	$22d$	$22e$	$22f$	$24a$	$24b$	$26b$	$26a$	6_3	6_4	6_1	$42a$
$7P$	$18j$	$18k$	$20a$	$20b$	$3a$	$22c$	$22e$	$22a$	$22f$	$22b$	$22d$	$24a$	$24b$	$26b$	$26a$	$30a$	$30b$	$30c$	6_1
$11P$	$18j$	$18k$	$20a$	$20b$	$21a$	$2a$	$2c$	$2a$	$2b$	$2c$	$2b$	$24a$	$24b$	$26b$	$26a$	$30a$	$30b$	$30c$	$42a$
$13P$	$18j$	$18k$	$20a$	$20b$	$21a$	$22c$	$22e$	$22a$	$22f$	$22b$	$22d$	$24a$	$24b$	$2a$	$2a$	$30a$	$30b$	$30c$	$42a$
$X.88$	−1	.	.	.	.	.	.	.	.	.	.	.	.	.	.	.	.	.	.
$X.89$	.	.	.	.	.	.	.	.	.	.	.	.	.	.	.	.	.	.	.
$X.90$	.	.	.	.	.	.	.	.	.	.	.	.	.	.	.	−1	1	1	.
$X.91$	.	.	.	.	.	.	.	.	.	.	.	.	.	.	.	1	−1	1	.
$X.92$	.	.	.	.	.	1	1	1	1	1	1	.	.	−1	−1	.	.	.	.
$X.93$	1	.	.	.	−1	.	.	.	.	.	.	.	.	.	.	.	.	.	−1
$X.94$	.	−2	.	.	−1	.	.	.	.	.	.	.	.	1	1	.	.	−2	−1
$X.95$	.	2	.	.	−1	.	.	.	.	.	.	.	.	−1	−1	.	.	2	1
$X.96$	.	.	.	.	.	.	.	.	.	.	.	.	.	D	C	.	.	.	.
$X.97$	.	.	.	.	.	.	.	.	.	.	.	.	.	C	D	.	.	.	.
$X.98$	.	.	−1	−1	.	1	1	1	1	1	1	.	.	.	.	.	.	.	.
$X.99$	.	.	.	.	1	.	.	.	.	.	.	1	−1	.	.	−1	−1	−1	1
$X.100$	.	.	.	.	.	.	.	.	.	.	.	−2	.	.	.	.	.	.	.
$X.101$	2	1	.	.	.	.	.	.	.	.	.	.	.	.	.	.	.	−2	.
$X.102$	1	.	−1	1	−1	.	.	.	.	.	.	.	.	.	.	−1	1	1	1
$X.103$	1	.	1	−1	−1	.	.	.	.	.	.	.	.	.	.	1	−1	1	1
$X.104$	.	.	.	.	−1	.	.	.	.	.	.	1	1	.	.	.	.	.	−1
$X.105$	.	.	−1	1	.	−1	−1	−1	1	−1	1	.	.	.	.	−1	1	−1	.
$X.106$	.	.	1	−1	.	−1	1	−1	−1	1	−1	.	.	.	.	1	−1	−1	.
$X.107$	−1	.	.	.	.	.	.	.	.	.	.	1	−1	.	.	.	.	.	.
$X.108$	−1	.	.	.	.	.	.	.	.	.	.	.	.	.	.	.	.	.	.
$X.109$	−1	.	.	.	.	.	.	.	.	.	.	.	.	.	.	.	.	.	.
$X.110$	.	.	.	.	1	−1	1	−1	1	1	1	.	.	.	.	−1	−1	1	1
$X.111$	.	.	.	.	1	1	1	1	−1	1	−1	.	.	.	.	1	−1	−1	−1
$X.112$	.	.	.	.	1	1	−1	1	1	−1	1	.	.	.	.	−1	1	−1	−1
$X.113$	.	.	.	.	1	−1	−1	−1	−1	−1	−1	.	.	.	.	1	1	1	1
$X.114$	.	.	1	1	−1	1	−1	1	−1	−1	−1	−1	1	.	.	1	1	1	−1

where $A = \zeta(3432)_{11}^{9} + \zeta(3432)_{11}^{5} + \zeta(3432)_{11}^{4} + \zeta(3432)_{11}^{3} + \zeta(3432)_{11}$, $B = -6\zeta(3432)_3 - 3$, $C = \zeta(3432)_{13}^{11} + \zeta(3432)_{13}^{8} + \zeta(3432)_{13}^{7} + \zeta(3432)_{13}^{6} + \zeta(3432)_{13}^{5} + \zeta(3432)_{13}^{2} + 1$, $D = -\zeta(3432)_{13}^{11} - \zeta(3432)_{13}^{8} - \zeta(3432)_{13}^{7} - \zeta(3432)_{13}^{6} - \zeta(3432)_{13}^{5} - \zeta(3432)_{13}^{2}$, $E = -2\zeta(3432)_{8}^{3} - 2\zeta(3432)_{8}$.

7

Conway's simple group Co_1

In [16] J. Conway described his simple group Co_1 as a quotient group of the group $Aut(\Lambda)$ of the isometries of the 24-dimensional Leech lattice Λ by the unique normal subgroup of order 2 in $Aut(\Lambda)$. In [17], p. 243, he states a long list of subgroups of $Aut(\Lambda)$ and Co_1 which he determined by means of well defined sublattices of the Leech lattice. Thus he also discovered the two sporadic groups Co_2 and Co_3 already discussed in Chapters 3 and 4. Conway and his collaborators also determined the character tables of these three sporadic groups; see [19]. It also contains important information about the structure of the centralizers $H(\mathsf{Co}_i)$ of 2-central involutions z_i of Co_i for $i = 1, 2, 3$. For further details the reader is referred to M. Aschbacher's book [4] and A. Reifart's article [113]. Using Conway's original results, L. Soicher established beautiful presentations for both groups $Aut(\Lambda)$ and Co_1; see [110], pp. 121–122, and [120]. For one presentation of Co_1 Soicher also gives generators of the stabilizer of a faithful permutation representation of Co_1 of degree 1545600.

However, all these methods do not generalize to arbitrary finite simple groups. Therefore in this chapter H. Kim's and the author's recent existence proof for the simple Conway group Co_1 is presented; see [75]. This time Algorithm 1.3.8 is applied to a well determined conjugacy class of irreducible subgroups T of $\mathrm{GL}_{11}(2)$ which are isomorphic to the Mathieu group M_{24}.

In Section 7.1 we collect some known facts about the simple Mathieu group M_{24} defined in [92], p. 388. Let $F = GF(2)$. By [65] M_{24} has two non-isomorphic simple modules V_1, V_2 of dimension 11 over F. We show that the second cohomological dimensions $dim_F[H^2(\mathsf{M}_{24}, V_1)]$ and $dim_F[H^2(\mathsf{M}_{24}, V_2)]$ are 0 and 1, respectively. Therefore there are two split extensions $E_i = V_i \rtimes \mathsf{M}_{24}$, $i = 1, 2$, and one non-split extension E_3 of V_2 by M_{24}. For each of the three groups E_i we state a presentation in terms of generators and relations and a faithful permutation representation, see Lemma 7.1.1. In all three cases it has been checked that the elementary abelian normal subgroup V_i is a maximal elementary abelian subgroup of a Sylow 2-subgroup S_i of E_i. Hence all conditions of Step 4 of Algorithm 1.3.8 are satisfied for all three extension groups.

In this chapter we apply Algorithm 1.3.8 to the extension group E_1 to provide an existence proof for Conway's largest sporadic group Co_1. The other extension groups E_2 and E_3 will be used in Chapters 8 and 9 to construct the sporadic simple groups J_4 of Janko and Fi_{24}' of Fischer, respectively.

The group $E := E_1$ has a unique conjugacy class z^E of 2-central involutions. In Section 7.2 we describe H. Kim's and the author's application [75] of Algorithm 1.3.8 to the subgroup $D = C_E(z)$. In order to construct a finitely presented group H with center $Z(H) = \langle z \rangle$ satisfying all conditions of its Step 5 we employ the same methods as in the proof of Proposition 4.2.1 of Chapter 4. However, this time (in 2007) the implementation of Holt's Algorithm [54] in MAGMA was not able to calculate the second cohomological dimensions for purely technical reasons.

In Proposition 7.2.1 this problem is split into two parts. We construct a subgroup U of order $2^{21} \cdot 3^3$ and of odd index in D with center $Z(U) = \langle z \rangle$. This subgroup U is used to construct an over-group H_1 with center $Z(H_1) = \langle z \rangle$ and odd index $|H_1 : U|$ such that H is an epimorphic image of the free product $H_1 *_U D$ with amalgamated subgroup U. Proposition 7.2.2 realizes the group H with center $Z(H) = \langle z \rangle$ as an irreducible subgroup of $\mathrm{GL}_{128}(23)$. Furthermore, it provides a presentation and a faithful permutation representation of H of degree 276480 with a documented stabilizer. Thus we were able to verify all conditions of Step 5 of Algorithm 1.3.8 and calculate a character table and a system of representatives of the 154 conjugacy classes of H.

Section 7.3 describes H. Kim's and the author's application [75] of Algorithm 7.4.8 of [92] to the finitely presented group H. It returns a finite simple group $\mathfrak{G}$ inside $\mathrm{GL}_{276}(23)$ of order $2^{21} \cdot 3^9 \cdot 5^4 \cdot 7^2 \cdot 11 \cdot 13 \cdot 23$ having a 2-central involution $\mathfrak{z}$ such that $C_{\mathfrak{G}}(\mathfrak{z}) \cong H$; see Theorem 7.3.2. We also construct a faithful permutation representation $P\mathfrak{G}$ of $\mathfrak{G}$ of degree 98280 and provide four short words in the four generating matrices of $\mathfrak{G}$ which generate the stabilizer $\mathfrak{C}$ of $P\mathfrak{G}$. The character table of $\mathfrak{G}$ has been calculated with $P\mathfrak{G}$ and MAGMA. It is equivalent to that of Conway's sporadic group Co_1 in the Atlas [19]. In Corollary 7.3.4 we show that $\mathfrak{G} \cong \mathsf{Co}_1$.

In Section 7.4 we discuss whether any finite simple group G having a 2-central involution z such that $C_G(z) \cong H$ is isomorphic to $\mathfrak{G}$.

7.1 Extensions of the Mathieu group M_{24}

The Mathieu group M_{24} is defined in Definition 8.2.1 of [92] by means of generators and relations. This beautiful presentation is due to J. A. Todd. The irreducible 2-modular representations of the Mathieu group M_{24} were determined by G. James [65]. Here only the two non-isomorphic simple modules V_i, $i = 1, 2$, of dimension 11 over $F = GF(2)$ will be considered. Todd's permutation representations of the Mathieu groups are stated in Lemma 8.2.2 of [92]. Therefore all conditions of Holt's Algorithm [52] implemented in MAGMA are satisfied. It is applied here. Thus we show that for the simple module V_1 there is exactly one extension E_1 of M_{24} by V_1, the split extension. The presentation of E_1 is given in Lemma 7.1.1. However, there are two non-isomorphic extensions of M_{24} by V_2, the split extension E_2 and the non-split extension E_3. The presentations of all three extensions E_i are given in the following Lemma.

In the following $F = GF(2)$ and $p = 2$.

Lemma 7.1.1 *Let* $\mathsf{M}_{24} = \langle a, b, c, d, t, g, h, i, j, k \rangle$ *be the finitely presented group of Definition 8.2.1 of [92]. Then the following statements hold.*

(a) *A faithful permutation representation of degree* 24 *of* M_{24} *is stated in Lemma 8.2.2 of [92].*

(b) *The first irreducible representation* V_1 *of* M_{24} *is described by the following matrices:*

$$a_1 = \begin{pmatrix}
1&0&0&0&0&0&0&0&0&0&0\\
0&1&0&0&0&0&0&0&1&0&0\\
0&0&1&0&0&0&0&0&1&0&0\\
0&0&0&1&0&0&0&0&1&0&0\\
0&0&0&0&0&1&0&0&0&0&0\\
0&0&0&0&1&0&0&0&0&0&0\\
0&0&0&0&0&0&0&0&0&1&0\\
0&0&0&0&0&0&0&0&1&0&1\\
0&0&0&0&0&0&0&0&1&0&0\\
0&0&0&0&0&0&1&0&0&0&0\\
0&0&0&0&0&0&0&1&1&0&0
\end{pmatrix}, \quad b_1 = \begin{pmatrix}
1&0&0&0&0&0&0&1&1&0&1\\
0&1&0&0&0&0&0&1&1&0&1\\
0&0&1&0&0&0&0&1&1&0&1\\
0&0&0&1&0&0&0&0&1&0&0\\
0&0&0&0&0&0&1&1&0&0&0\\
0&0&0&0&0&0&0&0&1&1&1\\
0&0&0&0&1&0&0&0&0&0&1\\
0&0&0&0&0&0&0&0&0&0&1\\
0&0&0&0&0&0&0&0&1&0&0\\
0&0&0&0&0&1&0&1&1&0&0\\
0&0&0&0&0&0&0&1&0&0&0
\end{pmatrix},$$

$$c_1 = \begin{pmatrix} 1&0&0&0&0&0&1&0&0&1&0\\ 0&1&0&0&0&0&1&0&0&1&0\\ 0&0&1&0&0&0&1&0&1&1&0\\ 0&0&0&1&0&0&0&0&1&0&0\\ 0&0&0&0&0&0&1&1&1&0&0\\ 0&0&0&0&0&0&0&0&0&1&1\\ 0&0&0&0&0&0&0&0&1&1&0\\ 0&0&0&0&1&0&0&0&0&1&0\\ 0&0&0&0&0&0&0&0&1&0&0\\ 0&0&0&0&0&0&1&0&1&0&0\\ 0&0&0&0&0&1&1&0&1&0&0 \end{pmatrix}, \quad d_1 = \begin{pmatrix} 1&0&0&0&0&0&1&0&0&1&0\\ 0&1&0&0&0&1&1&0&0&0&1\\ 0&0&1&0&0&1&1&1&0&0&0\\ 0&0&0&1&0&1&1&1&0&0&0\\ 0&0&0&0&0&0&0&1&1&1&0\\ 0&0&0&0&0&0&1&0&0&0&1\\ 0&0&0&0&0&0&0&1&0&1&1\\ 0&0&0&0&0&1&0&0&0&1&0\\ 0&0&0&0&1&1&1&1&0&1&1\\ 0&0&0&0&0&0&1&1&0&0&1\\ 0&0&0&0&0&1&0&1&0&1&1 \end{pmatrix},$$

$$t_1 = \begin{pmatrix} 1&0&0&0&0&0&1&0&0&1&0\\ 0&1&0&0&0&1&0&0&1&1&0\\ 0&0&1&0&0&0&0&1&0&1&0\\ 0&0&0&1&0&0&0&0&1&0&0\\ 0&0&0&0&1&0&1&1&1&1&0\\ 0&0&0&0&0&1&0&1&1&0&0\\ 0&0&0&0&0&1&1&0&1&0&0\\ 0&0&0&0&0&0&1&1&1&0&0\\ 0&0&0&0&0&1&1&1&1&0&0\\ 0&0&0&0&0&1&0&1&0&1&1\\ 0&0&0&0&0&1&1&0&0&1&0 \end{pmatrix}, \quad g_1 = \begin{pmatrix} 1&0&0&0&0&0&0&0&0&1&1\\ 0&1&0&0&0&0&0&0&1&0&0\\ 0&0&1&0&0&0&0&0&0&0&1\\ 0&0&0&0&1&0&0&0&1&1&1\\ 0&0&0&1&0&0&0&0&1&1&1\\ 0&0&0&0&0&1&0&0&1&1&1\\ 0&0&0&0&0&0&1&0&1&1&0\\ 0&0&0&0&0&0&0&1&1&0&1\\ 0&0&0&0&0&0&0&0&1&0&0\\ 0&0&0&0&0&0&0&0&0&1&0\\ 0&0&0&0&0&0&0&0&0&0&1 \end{pmatrix},$$

$$h_1 = \begin{pmatrix} 1&0&0&0&0&0&1&0&0&1&0\\ 0&1&0&0&0&0&0&0&0&1&0\\ 0&0&0&1&0&0&0&0&0&1&0\\ 0&0&1&0&0&0&0&0&0&1&0\\ 0&0&0&0&1&0&1&0&0&0&0\\ 0&0&0&0&0&1&0&0&0&1&0\\ 0&0&0&0&0&0&1&0&0&0&0\\ 0&0&0&0&0&0&1&0&0&0&1\\ 0&0&0&0&0&0&1&0&1&1&0\\ 0&0&0&0&0&0&0&0&0&1&0\\ 0&0&0&0&0&0&1&1&0&0&0 \end{pmatrix}, \quad i_1 = \begin{pmatrix} 1&0&0&0&0&0&0&0&0&0&0\\ 0&0&1&0&0&0&1&1&0&0&0\\ 0&1&0&0&0&1&1&0&1&0&0\\ 0&0&0&1&0&1&1&1&0&0&0\\ 0&0&0&0&1&1&0&1&1&0&0\\ 0&0&0&0&0&0&0&1&1&0&0\\ 0&0&0&0&0&1&0&1&0&0&0\\ 0&0&0&0&0&0&1&1&1&0&0\\ 0&0&0&0&0&1&1&1&1&0&0\\ 0&0&0&0&0&1&0&0&1&0&1\\ 0&0&0&0&0&1&1&0&0&1&0 \end{pmatrix},$$

$$j_1 = \begin{pmatrix} 0&1&0&0&0&1&0&0&0&0&0\\ 1&0&0&0&0&1&0&0&0&0&0\\ 0&0&1&0&0&1&0&0&1&0&0\\ 0&0&0&1&0&0&0&0&1&0&0\\ 0&0&0&0&1&0&0&0&1&0&0\\ 0&0&0&0&0&1&0&0&0&0&0\\ 0&0&0&0&0&1&1&0&1&0&0\\ 0&0&0&0&0&1&0&1&0&0&0\\ 0&0&0&0&0&0&0&0&1&0&0\\ 0&0&0&0&0&0&0&0&1&0&1\\ 0&0&0&0&0&0&0&0&1&1&0 \end{pmatrix} \quad \textit{and} \quad k_1 = \begin{pmatrix} 1&1&1&1&1&1&1&0&1&1&1\\ 0&1&0&0&0&1&0&1&0&0&0\\ 0&0&1&0&0&1&1&1&0&0&0\\ 0&0&0&1&0&1&1&1&0&0&0\\ 0&0&0&0&1&0&1&0&0&0&0\\ 0&0&0&0&0&0&0&1&0&0&0\\ 0&0&0&0&0&0&1&0&0&0&0\\ 0&0&0&0&0&1&0&0&0&0&0\\ 0&0&0&0&0&1&1&1&1&0&0\\ 0&0&0&0&0&0&1&1&0&0&1\\ 0&0&0&0&0&1&1&0&0&1&0 \end{pmatrix}.$$

(c) *The second irreducible representation* V_2 *of* M_{24} *is described by the transpose inverse matrices of the generating matrices of* M_{24} *defining* V_1*:*

$$a_2 = [a_1^{-1}]^T, b_2 = [b_1^{-1}]^T, c_2 = [c_1^{-1}]^T, d_2 = [d_1^{-1}]^T, t_2 = [t_1^{-1}]^T,$$
$$g_2 = [g_1^{-1}]^T, h_2 = [h_1^{-1}]^T, i_2 = [i_1^{-1}]^T, j_2 = [j_1^{-1}]^T, and k_2 = [k_1^{-1}]^T.$$

(d) $dim_F[H^2(\mathsf{M}_{24}, V_1)] = 0$ *and* $dim_F[H^2(\mathsf{M}_{24}, V_2)] = 1$.

(e) *The split extension*

$$E_1 = \langle a, b, c, d, t, g, h, i, j, k, v_1, v_2, v_3, v_4, v_5, v_6, v_8, v_8, v_9, v_{10}, v_{11} \rangle$$

of M_{24} *by* V_1 *has a set* $\mathcal{R}(E_1)$ *of defining relations consisting of* $\mathcal{R}(\mathsf{M}_{24})$, $\mathcal{R}_1(V_1 \rtimes \mathsf{M}_{24})$ *and the following set* $\mathcal{R}_2(V_1 \rtimes \mathsf{M}_{24})$ *of essential relations:*

$$av_1a^{-1}v_1 = av_2a^{-1}v_2v_9 = av_3a^{-1}v_3v_9 = av_4a^{-1}v_4v_9 = 1,$$
$$av_5a^{-1}v_6 = av_6a^{-1}v_5 = av_7a^{-1}v_{10} = av_8a^{-1}v_9v_{11} = 1,$$
$$av_9a^{-1}v_9 = av_{10}a^{-1}v_7 = av_{11}a^{-1}v_8v_9 = bv_1b^{-1}v_1v_8v_9v_{11} = 1,$$
$$bv_2b^{-1}v_2v_8v_9v_{11} = bv_3b^{-1}v_3v_8v_9v_{11} = bv_4b^{-1}v_4v_9 = 1,$$
$$bv_5b^{-1}v_7v_8 = bv_6b^{-1}v_9v_{10}v_{11} = bv_7b^{-1}v_5v_{11} = 1,$$
$$bv_8b^{-1}v_{11} = bv_9b^{-1}v_9 = bv_{10}b^{-1}v_6v_8v_9 = bv_{11}b^{-1}v_8 = 1,$$
$$cv_1c^{-1}v_1v_7v_{10} = cv_2c^{-1}v_2v_7v_{10} = cv_3c^{-1}v_3v_7v_9v_{10} = 1,$$
$$cv_4c^{-1}v_4v_9 = cv_5c^{-1}v_7v_8v_9 = cv_6c^{-1}v_{10}v_{11} = cv_7c^{-1}v_9v_{10} = 1,$$
$$cv_8c^{-1}v_5v_{10} = cv_9c^{-1}v_9 = cv_{10}c^{-1}v_7v_9 = cv_{11}c^{-1}v_6v_7v_9 = 1,$$
$$dv_1d^{-1}v_1v_7v_{10} = dv_2d^{-1}v_2v_6v_7v_{11} = dv_3d^{-1}v_3v_6v_7v_8 = 1,$$
$$dv_4d^{-1}v_4v_6v_7v_8 = dv_5d^{-1}v_8v_9v_{10} = dv_6d^{-1}v_7v_{11} = 1,$$
$$dv_7d^{-1}v_8v_{10}v_{11} = dv_8d^{-1}v_6v_{10} = 1,$$
$$dv_9d^{-1}v_5v_6v_7v_8v_{10}v_{11} = dv_{10}d^{-1}v_7v_8v_{11} = 1,$$
$$dv_{11}d^{-1}v_6v_8v_{10}v_{11} = tv_1t^{-1}v_1v_8v_9v_{11} = tv_2t^{-1}v_2v_7v_8v_9v_{11} = 1,$$
$$tv_3t^{-1}v_3v_6v_7v_8v_9v_{11} = tv_4t^{-1}v_4v_6v_7v_8 = tv_5t^{-1}v_5v_6v_{11} = 1,$$
$$tv_6t^{-1}v_8v_9 = tv_7t^{-1}v_6v_9 = tv_8t^{-1}v_7v_9 = tv_9t^{-1}v_6v_7v_8 = 1,$$
$$tv_{10}t^{-1}v_6v_8v_{11} = tv_{11}t^{-1}v_6v_7v_{10}v_{11} = gv_1g^{-1}v_1v_{10}v_{11} = 1,$$
$$gv_2g^{-1}v_2v_9 = gv_3g^{-1}v_3v_{11} = gv_4g^{-1}v_5v_9v_{10}v_{11} = 1,$$
$$gv_5g^{-1}v_4v_9v_{10}v_{11} = gv_6g^{-1}v_6v_9v_{10}v_{11} = gv_7g^{-1}v_7v_9v_{10} = 1,$$
$$gv_8g^{-1}v_8v_9v_{11} = gv_9g^{-1}v_9 = gv_{10}g^{-1}v_{10} = gv_{11}g^{-1}v_{11} = 1,$$
$$hv_1h^{-1}v_1v_7v_{10} = hv_2h^{-1}v_2v_{10} = hv_3h^{-1}v_4v_{10} = hv_4h^{-1}v_3v_{10} = 1,$$
$$hv_5h^{-1}v_5v_7 = hv_6h^{-1}v_6v_{10} = hv_7h^{-1}v_7 = hv_8h^{-1}v_7v_{11} = 1,$$
$$hv_9h^{-1}v_7v_9v_{10} = hv_{10}h^{-1}v_{10} = hv_{11}h^{-1}v_7v_8 = 1,$$
$$iv_1i^{-1}v_1 = iv_2i^{-1}v_3v_7v_8 = iv_3i^{-1}v_2v_6v_7v_9 = iv_4i^{-1}v_4v_6v_7v_8 = 1,$$
$$iv_5i^{-1}v_5v_6v_8v_9 = iv_6i^{-1}v_8v_9 = iv_7i^{-1}v_6v_8 = iv_8i^{-1}v_7v_8v_9 = 1,$$

$$iv_9i^{-1}v_6v_7v_8v_9 = iv_{10}i^{-1}v_6v_9v_{11} = iv_{11}i^{-1}v_6v_7v_{10} = 1,$$
$$jv_1j^{-1}v_2v_6 = jv_2j^{-1}v_1v_6 = jv_3j^{-1}v_3v_6v_9 = jv_4j^{-1}v_4v_9 = 1,$$
$$jv_5j^{-1}v_5v_9 = jv_6j^{-1}v_6 = jv_7j^{-1}v_6v_7v_9 = jv_8j^{-1}v_6v_8 = 1,$$
$$jv_9j^{-1}v_9 = jv_{10}j^{-1}v_9v_{11} = jv_{11}j^{-1}v_9v_{10} = 1,$$
$$kv_1k^{-1}v_1v_2v_3v_4v_5v_6v_7v_9v_{10}v_{11} = kv_2k^{-1}v_2v_6v_8 = 1,$$
$$kv_3k^{-1}v_3v_6v_7v_8 = kv_4k^{-1}v_4v_6v_7v_8 = kv_5k^{-1}v_5v_7 = 1,$$
$$kv_6k^{-1}v_8 = kv_7k^{-1}v_7 = kv_8k^{-1}v_6 = 1,$$
$$kv_9k^{-1}v_6v_7v_8v_9 = kv_{10}k^{-1}v_7v_8v_{11} = kv_{11}k^{-1}v_6v_7v_{10} = 1.$$

(f) *The split extension*

$$E_2 = \langle a, b, c, d, t, g, h, i, j, k, v_1, v_2, v_3, v_4, v_5, v_6, v_8, v_8, v_9, v_{10}, v_{11}\rangle$$

of M_{24} *by* V_2 *has a set* $\mathcal{R}(E_2)$ *of defining relations consisting of* $\mathcal{R}(\mathsf{M}_{24})$, $\mathcal{R}_1(V_2 \rtimes \mathsf{M}_{24})$ *and the following set* $\mathcal{R}_2(V_2 \rtimes \mathsf{M}_{24})$ *of essential relations:*

$$av_1a^{-1}v_1v_2v_5v_6v_7v_8v_{10} = av_2a^{-1}v_{10} = av_3a^{-1}v_4 = av_4a^{-1}v_3 = 1,$$
$$av_5a^{-1}v_6 = av_6a^{-1}v_5 = av_7a^{-1}v_8 = av_8a^{-1}v_7 = av_9a^{-1}v_9 = 1,$$
$$av_{10}a^{-1}v_2 = av_{11}a^{-1}v_{11} = bv_1b^{-1}v_6 = bv_2b^{-1}v_8 = 1,$$
$$bv_3b^{-1}v_1v_2v_4v_6v_{10}v_{11} = bv_4b^{-1}v_1v_3v_6v_7v_8v_{11} = 1,$$
$$bv_5b^{-1}v_1v_2v_5v_6v_7v_8v_{10} = bv_6b^{-1}v_1 = bv_7b^{-1}v_{10} = bv_8b^{-1}v_2 = 1,$$
$$bv_9b^{-1}v_9 = bv_{10}b^{-1}v_7 = bv_{11}b^{-1}v_{11} = 1,$$
$$cv_1c^{-1}v_{10} = cv_2c^{-1}v_1v_2v_5v_6v_7v_8v_{10} = cv_3c^{-1}v_2v_4v_5v_8v_9v_{10} = 1,$$
$$cv_4c^{-1}v_2v_3v_6v_7v_9v_{10} = cv_5c^{-1}v_8 = cv_6c^{-1}v_7 = cv_7c^{-1}v_6 = 1,$$
$$cv_8c^{-1}v_5 = cv_9c^{-1}v_9 = cv_{10}c^{-1}v_1 = cv_{11}c^{-1}v_{11} = 1,$$
$$dv_1d^{-1}v_3v_5v_7v_8v_9v_{10}v_{11} = dv_2d^{-1}v_1v_3v_6v_7v_8v_{11} = dv_3d^{-1}v_7 = 1,$$
$$dv_4d^{-1}v_8 = dv_5d^{-1}v_2v_3v_6v_7v_9v_{10} = dv_6d^{-1}v_2v_4v_5v_8v_9v_{10} = 1,$$
$$dv_7d^{-1}v_3 = dv_8d^{-1}v_4 = dv_9d^{-1}v_9 = dv_{10}d^{-1}v_1v_2v_4v_6v_{10}v_{11} = 1,$$
$$dv_{11}d^{-1}v_{11} = tv_1t^{-1}v_2v_3v_6v_7v_{10} = tv_2t^{-1}v_3v_5v_7v_8v_{10}v_{11} = 1,$$
$$tv_3t^{-1}v_1v_2v_5v_6v_7v_8v_9v_{10} = tv_4t^{-1}v_1v_2v_4v_6v_9v_{10}v_{11} = 1,$$
$$tv_5t^{-1}v_1v_3v_6v_7v_8v_9v_{11} = tv_6t^{-1}v_1v_9 = tv_7t^{-1}v_3v_9 = 1,$$
$$tv_8t^{-1}v_5v_9 = tv_9t^{-1}v_9v_{11} = tv_{10}t^{-1}v_2v_9 = tv_{11}t^{-1}v_9 = 1,$$
$$gv_1g^{-1}v_4v_6 = gv_2g^{-1}v_4v_{10} = gv_3g^{-1}v_3v_4 = gv_4g^{-1}v_4 = 1,$$

$$gv_5g^{-1}v_4v_8 = gv_6g^{-1}v_1v_4 = gv_7g^{-1}v_1v_2v_4v_5v_6v_7v_8v_{10} = 1,$$
$$gv_8g^{-1}v_4v_5 = gv_9g^{-1}v_2v_5v_8v_9v_{10} = gv_{10}g^{-1}v_2v_4 = 1,$$
$$gv_{11}g^{-1}v_1v_2v_6v_{10}v_{11} = hv_1h^{-1}v_1v_9 = hv_2h^{-1}v_9v_{10} = 1,$$
$$hv_3h^{-1}v_7v_9 = hv_4h^{-1}v_8v_9 = hv_5h^{-1}v_2v_4v_5v_8v_{10} = 1,$$
$$hv_6h^{-1}v_2v_3v_6v_7v_{10} = hv_7h^{-1}v_3v_9 = hv_8h^{-1}v_4v_9 = hv_9h^{-1}v_9 = 1,$$
$$hv_{10}h^{-1}v_2v_9 = hv_{11}h^{-1}v_9v_{11} = iv_1i^{-1}v_2v_3v_6v_7v_{10} = 1,$$
$$iv_2i^{-1}v_3v_5v_7v_8v_{10}v_{11} = iv_3i^{-1}v_5v_9 = iv_4i^{-1}v_4v_9 = 1,$$
$$iv_5i^{-1}v_3v_9 = iv_6i^{-1}v_6v_9 = iv_7i^{-1}v_1v_3v_6v_7v_8v_9v_{11} = 1,$$
$$iv_8i^{-1}v_1v_2v_5v_6v_7v_8v_9v_{10} = iv_9i^{-1}v_9 = iv_{10}i^{-1}v_9v_{10} = 1,$$
$$iv_{11}i^{-1}v_9v_{11} = jv_1j^{-1}v_6 = jv_2j^{-1}v_1v_2v_5v_6v_7v_8v_{10} = 1,$$
$$jv_3j^{-1}v_3v_5v_7v_8v_9v_{10}v_{11} = jv_4j^{-1}v_4 = jv_5j^{-1}v_8 = 1,$$
$$jv_6j^{-1}v_1 = jv_7j^{-1}v_{10} = jv_8j^{-1}v_5 = jv_9j^{-1}v_{11} = 1,$$
$$jv_{10}j^{-1}v_7 = jv_{11}j^{-1}v_9 = kv_1k^{-1}v_3v_5v_7v_8v_{10}v_{11} = 1,$$
$$kv_2k^{-1}v_2v_3v_6v_7v_{10} = kv_3k^{-1}v_7v_9 = kv_4k^{-1}v_4v_9 = 1,$$
$$kv_5k^{-1}v_1v_3v_6v_7v_8v_9v_{11} = kv_6k^{-1}v_9v_{10} = kv_7k^{-1}v_3v_9 = 1,$$
$$kv_8k^{-1}v_8v_9 = kv_9k^{-1}v_9 = kv_{10}k^{-1}v_6v_9 = kv_{11}k^{-1}v_9v_{11} = 1.$$

(g) *The non-split extension*

$$E_3 = \langle a, b, c, d, t, g, h, i, j, k, v_1, v_2, v_3, v_4, v_5, v_6, v_8, v_8, v_9, v_{10}, v_{11} \rangle$$

of M_{24} *by* V_2 *has a set* $\mathcal{R}(E_3)$ *of defining relations consisting of* $\mathcal{R}_1(V_2 \rtimes \mathsf{M}_{24})$ *and the following relations:*

$$a^{-1}v_1av_1^{-1}v_2^{-1}v_5^{-1}v_6^{-1}v_7^{-1}v_8^{-1}v_{10}^{-1} = a^{-1}v_2av_{10}^{-1} = 1,$$
$$a^{-1}v_3av_4^{-1} = a^{-1}v_4av_3^{-1} = a^{-1}v_5av_6^{-1} = a^{-1}v_6av_5^{-1} = 1,$$
$$a^{-1}v_7av_8^{-1} = a^{-1}v_8av_7^{-1} = a^{-1}v_9av_9^{-1} = a^{-1}v_{10}av_2^{-1} = 1,$$
$$a^{-1}v_{11}av_{11}^{-1} = b^{-1}v_1bv_6^{-1} = b^{-1}v_2bv_8^{-1} = 1,$$
$$b^{-1}v_3bv_1^{-1}v_2^{-1}v_4^{-1}v_6^{-1}v_{10}^{-1}v_{11}^{-1} = 1,$$
$$b^{-1}v_4bv_1^{-1}v_3^{-1}v_6^{-1}v_7^{-1}v_8^{-1}v_{11}^{-1} = 1,$$
$$b^{-1}v_5bv_1^{-1}v_2^{-1}v_5^{-1}v_6^{-1}v_7^{-1}v_8^{-1}v_{10}^{-1} = 1,$$
$$b^{-1}v_6bv_1^{-1} = b^{-1}v_7bv_{10}^{-1} = b^{-1}v_8bv_2^{-1} = b^{-1}v_9bv_9^{-1} = 1,$$
$$b^{-1}v_{10}bv_7^{-1} = b^{-1}v_{11}bv_{11}^{-1} = c^{-1}v_1cv_{10}^{-1} = 1,$$
$$c^{-1}v_2cv_1^{-1}v_2^{-1}v_5^{-1}v_6^{-1}v_7^{-1}v_8^{-1}v_{10}^{-1} = 1,$$

$$c^{-1}v_3cv_2^{-1}v_4^{-1}v_5^{-1}v_8^{-1}v_9^{-1}v_{10}^{-1}=1,$$
$$c^{-1}v_4cv_2^{-1}v_3^{-1}v_6^{-1}v_7^{-1}v_9^{-1}v_{10}^{-1}=1,$$
$$c^{-1}v_5cv_8^{-1}=c^{-1}v_6cv_7^{-1}=c^{-1}v_7cv_6^{-1}=1,$$
$$c^{-1}v_8cv_5^{-1}=c^{-1}v_9cv_9^{-1}=c^{-1}v_{10}cv_1^{-1}=c^{-1}v_{11}cv_{11}^{-1}=1,$$
$$d^{-1}v_1dv_3^{-1}v_5^{-1}v_7^{-1}v_8^{-1}v_9^{-1}v_{10}^{-1}v_{11}^{-1}=1,$$
$$d^{-1}v_2dv_1^{-1}v_3^{-1}v_6^{-1}v_7^{-1}v_8^{-1}v_{11}^{-1}=1,$$
$$d^{-1}v_3dv_7^{-1}=d^{-1}v_4dv_8^{-1}=d^{-1}v_5dv_2^{-1}v_3^{-1}v_6^{-1}v_7^{-1}v_9^{-1}v_{10}^{-1}=1,$$
$$d^{-1}v_6dv_2^{-1}v_4^{-1}v_5^{-1}v_8^{-1}v_9^{-1}v_{10}^{-1}=d^{-1}v_7dv_3^{-1}=d^{-1}v_8dv_4^{-1}=1,$$
$$d^{-1}v_9dv_9^{-1}=d^{-1}v_{10}dv_1^{-1}v_2^{-1}v_4^{-1}v_6^{-1}v_{10}^{-1}v_{11}^{-1}=1,$$
$$d^{-1}v_{11}dv_{11}^{-1}=t^{-1}v_1tv_6^{-1}v_{11}^{-1}=t^{-1}v_2tv_{10}^{-1}v_{11}^{-1}=1,$$
$$t^{-1}v_3tv_7^{-1}v_{11}^{-1}=t^{-1}v_4tv_2^{-1}v_4^{-1}v_5^{-1}v_8^{-1}v_9^{-1}v_{10}^{-1}v_{11}^{-1}=1,$$
$$t^{-1}v_5tv_8^{-1}v_{11}^{-1}=t^{-1}v_6tv_2^{-1}v_3^{-1}v_6^{-1}v_7^{-1}v_9^{-1}v_{10}^{-1}v_{11}^{-1}=1,$$
$$t^{-1}v_7tv_1^{-1}v_2^{-1}v_5^{-1}v_6^{-1}v_7^{-1}v_8^{-1}v_{10}^{-1}v_{11}^{-1}=1,$$
$$t^{-1}v_8tv_1^{-1}v_3^{-1}v_6^{-1}v_7^{-1}v_8^{-1}=t^{-1}v_9tv_{11}^{-1}=1,$$
$$t^{-1}v_{10}tv_3^{-1}v_5^{-1}v_7^{-1}v_8^{-1}v_9^{-1}v_{10}^{-1}=t^{-1}v_{11}tv_9^{-1}v_{11}^{-1}=1,$$
$$g^{-1}v_1gv_4^{-1}v_6^{-1}=g^{-1}v_2gv_4^{-1}v_{10}^{-1}=g^{-1}v_3gv_3^{-1}v_4^{-1}=1,$$
$$g^{-1}v_4gv_4^{-1}=g^{-1}v_5gv_4^{-1}v_8^{-1}=g^{-1}v_6gv_1^{-1}v_4^{-1}=1,$$
$$g^{-1}v_7gv_1^{-1}v_2^{-1}v_4^{-1}v_5^{-1}v_6^{-1}v_7^{-1}v_8^{-1}v_{10}^{-1}=1,$$
$$g^{-1}v_8gv_4^{-1}v_5^{-1}=g^{-1}v_9gv_2^{-1}v_5^{-1}v_8^{-1}v_9^{-1}v_{10}^{-1}=1,$$
$$g^{-1}v_{10}gv_2^{-1}v_4^{-1}=g^{-1}v_{11}gv_1^{-1}v_2^{-1}v_6^{-1}v_{10}^{-1}v_{11}^{-1}=1,$$
$$h^{-1}v_1hv_1^{-1}v_9^{-1}=h^{-1}v_2hv_9^{-1}v_{10}^{-1}=h^{-1}v_3hv_7^{-1}v_9^{-1}=1,$$
$$h^{-1}v_4hv_8^{-1}v_9^{-1}=h^{-1}v_5hv_2^{-1}v_4^{-1}v_5^{-1}v_8^{-1}v_{10}^{-1}=1,$$
$$h^{-1}v_6hv_2^{-1}v_3^{-1}v_6^{-1}v_7^{-1}v_{10}^{-1}=h^{-1}v_7hv_3^{-1}v_9^{-1}=1,$$
$$h^{-1}v_8hv_4^{-1}v_9^{-1}=h^{-1}v_9hv_9^{-1}=h^{-1}v_{10}hv_2^{-1}v_9^{-1}=1,$$
$$h^{-1}v_{11}hv_9^{-1}v_{11}^{-1}=i^{-1}v_1iv_2^{-1}v_3^{-1}v_6^{-1}v_7^{-1}v_{10}^{-1}=1,$$
$$i^{-1}v_2iv_3^{-1}v_5^{-1}v_7^{-1}v_8^{-1}v_{10}^{-1}v_{11}^{-1}=1,$$
$$i^{-1}v_3iv_5^{-1}v_9^{-1}=i^{-1}v_4iv_4^{-1}v_9^{-1}=i^{-1}v_5iv_3^{-1}v_9^{-1}=1,$$
$$i^{-1}v_6iv_6^{-1}v_9^{-1}=i^{-1}v_7iv_1^{-1}v_3^{-1}v_6^{-1}v_7^{-1}v_8^{-1}v_9^{-1}v_{11}^{-1}=1,$$
$$i^{-1}v_8iv_1^{-1}v_2^{-1}v_5^{-1}v_6^{-1}v_7^{-1}v_8^{-1}v_9^{-1}v_{10}^{-1}=1,$$
$$i^{-1}v_9iv_9^{-1}=i^{-1}v_{10}iv_9^{-1}v_{10}^{-1}=i^{-1}v_{11}iv_9^{-1}v_{11}^{-1}=j^{-1}v_1jv_6^{-1}=1,$$
$$j^{-1}v_2jv_1^{-1}v_2^{-1}v_5^{-1}v_6^{-1}v_7^{-1}v_8^{-1}v_{10}^{-1}=1,$$
$$j^{-1}v_3jv_3^{-1}v_5^{-1}v_7^{-1}v_8^{-1}v_9^{-1}v_{10}^{-1}v_{11}^{-1}=j^{-1}v_4jv_4^{-1}=j^{-1}v_5jv_8^{-1}=1,$$

$$j^{-1}v_6jv_1^{-1} = j^{-1}v_7jv_{10}^{-1} = j^{-1}v_8jv_5^{-1} = j^{-1}v_9jv_{11}^{-1} = 1,$$
$$j^{-1}v_{10}jv_7^{-1} = j^{-1}v_{11}jv_9^{-1} = k^{-1}v_2kv_2^{-1}v_3^{-1}v_6^{-1}v_7^{-1}v_{10}^{-1} = 1,$$
$$k^{-1}v_1kv_3^{-1}v_5^{-1}v_7^{-1}v_8^{-1}v_{10}^{-1}v_{11}^{-1} = 1,$$
$$k^{-1}v_3kv_7^{-1}v_9^{-1} = k^{-1}v_4kv_4^{-1}v_9^{-1} = 1,$$
$$k^{-1}v_5kv_1^{-1}v_3^{-1}v_6^{-1}v_7^{-1}v_8^{-1}v_9^{-1}v_{11}^{-1} = k^{-1}v_6kv_9^{-1}v_{10}^{-1} = 1,$$
$$k^{-1}v_7kv_3^{-1}v_9^{-1} = k^{-1}v_8kv_8^{-1}v_9^{-1} = k^{-1}v_9kv_9^{-1} = 1,$$
$$k^{-1}v_{10}kv_6^{-1}v_9^{-1} = k^{-1}v_{11}kv_9^{-1}v_{11}^{-1} = a^2v_5^{-1}v_6^{-1}v_7^{-1}v_8^{-1} = 1,$$
$$b^2v_1^{-1}v_3^{-1}v_4^{-1}v_6^{-1}v_7^{-1}v_8^{-1}v_{11}^{-1} = c^2v_1^{-1}v_5^{-1}v_6^{-1}v_7^{-1}v_8^{-1}v_{10}^{-1} = 1,$$
$$d^2v_1^{-1}v_2^{-1}v_3^{-1}v_6^{-1}v_7^{-1}v_8^{-1}v_{11}^{-1} = b^{-1}aba^{-1}v_1^{-1}v_3^{-1}v_4^{-1}$$
$$\cdot v_6^{-1}v_7^{-1}v_{10}^{-1} = c^{-1}aca^{-1}v_2^{-1}v_{10}^{-1} = d^{-1}ada^{-1}v_5^{-1}v_6^{-1}v_9^{-1} = 1,$$
$$c^{-1}bcb^{-1}v_1^{-1}v_3^{-1}v_4^{-1}v_8^{-1}v_9^{-1} = d^{-1}bdb^{-1}v_2^{-1}v_4^{-1} = 1,$$
$$d^{-1}cdc^{-1}v_2^{-1}v_3^{-1}v_5^{-1}v_{10}^{-1}v_{11}^{-1} = t^3v_1^{-1}v_3^{-1}v_8^{-1} = 1,$$
$$t^{-1}atd^{-1}c^{-1}v_2^{-1}v_3^{-1}v_4^{-1}v_5^{-1}v_7^{-1}v_8^{-1}v_9^{-1}v_{11}^{-1} = 1,$$
$$t^{-1}btd^{-1}a^{-1}v_2^{-1}v_3^{-1}v_5^{-1}v_6^{-1}v_9^{-1}v_{11}^{-1} = 1,$$
$$t^{-1}ctd^{-1}b^{-1}v_1^{-1}v_2^{-1}v_4^{-1}v_5^{-1}v_8^{-1}v_9^{-1}v_{11}^{-1} = 1,$$
$$t^{-1}dtc^{-1}b^{-1}a^{-1}v_2^{-1}v_3^{-1}v_5^{-1}v_7^{-1}v_8^{-1}v_9^{-1}v_{11}^{-1} = 1,$$
$$g^2v_5^{-1}v_8^{-1} = gagagav_2^{-1}v_3^{-1}v_4^{-1}v_7^{-1}v_8^{-1}v_{11}^{-1} = 1,$$
$$gbgbgbv_1^{-1}v_3^{-1}v_4^{-1}v_8^{-1}v_9^{-1}v_{10}^{-1} = gcgcgcv_1^{-1}v_2^{-1}v_5^{-1}v_6^{-1}v_9^{-1} = 1,$$
$$gtgtv_4^{-1}v_{11}^{-1} = h^2v_4^{-1}v_8^{-1}v_9^{-1} = h^{-1}aha^{-1}v_2^{-1}v_3^{-1}v_4^{-1}v_9^{-1}v_{10}^{-1} = 1,$$
$$h^{-1}bhd^{-1}b^{-1}a^{-1}v_4^{-1}v_5^{-1}v_7^{-1}v_8^{-1}v_{10}^{-1} = 1,$$
$$h^{-1}chc^{-1}a^{-1}v_1^{-1}v_3^{-1}v_4^{-1}v_5^{-1}v_7^{-1}v_{10}^{-1}v_{11}^{-1} = 1,$$
$$h^{-1}dhd^{-1}v_2^{-1}v_4^{-1}v_8^{-1}v_9^{-1}v_{10}^{-1} = 1,$$
$$h^{-1}thtv_1^{-1}v_2^{-1}v_3^{-1}v_4^{-1}v_5^{-1}v_6^{-1}v_7^{-1}v_{10}^{-1}v_{11}^{-1} = 1,$$
$$ghghghv_1^{-1}v_3^{-1}v_4^{-1}v_6^{-1}v_8^{-1}v_9^{-1} = i^2v_1^{-1}v_5^{-1}v_6^{-1}v_8^{-1}v_9^{-1}v_{11}^{-1} = 1,$$
$$i^{-1}aid^{-1}c^{-1}v_3^{-1}v_5^{-1}v_7^{-1}v_9^{-1}v_{10}^{-1} = 1,$$
$$i^{-1}bid^{-1}a^{-1}v_2^{-1}v_3^{-1}v_4^{-1}v_6^{-1}v_9^{-1}v_{11}^{-1} = 1,$$
$$i^{-1}cid^{-1}c^{-1}b^{-1}a^{-1}v_3^{-1}v_5^{-1}v_7^{-1}v_9^{-1}v_{10}^{-1}v_{11}^{-1} = 1,$$
$$i^{-1}did^{-1}c^{-1}b^{-1}v_1^{-1}v_3^{-1}v_4^{-1}v_9^{-1}v_{10}^{-1}v_{11}^{-1} = 1,$$
$$i^{-1}titv_1^{-1}v_5^{-1}v_6^{-1}v_8^{-1} = i^{-1}gig^{-1}t^{-1}v_1^{-1}v_2^{-1}v_3^{-1}v_8^{-1}v_{10}^{-1}v_{11}^{-1} = 1,$$
$$hihihiv_1^{-1}v_2^{-1}v_4^{-1}v_5^{-1}v_6^{-1}v_7^{-1}v_{10}^{-1}v_{11}^{-1} = j^2v_9^{-1}v_{11}^{-1} = 1,$$
$$j^{-1}ajc^{-1}b^{-1}a^{-1}v_1^{-1}v_3^{-1}v_5^{-1}v_6^{-1}v_7^{-1}v_8^{-1}v_9^{-1} = 1,$$
$$j^{-1}bjb^{-1}v_9^{-1}v_{11}^{-1} = j^{-1}cjc^{-1}v_1^{-1}v_5^{-1}v_6^{-1}v_7^{-1}v_8^{-1}v_{10}^{-1} = 1,$$

$$j^{-1}djd^{-1}c^{-1}v_2^{-1}v_5^{-1}v_8^{-1} = j^{-1}tjtv_3^{-1}v_5^{-1}v_8^{-1}v_{10}^{-1} = 1,$$
$$j^{-1}gjg^{-1}v_2^{-1}v_4^{-1}v_5^{-1}v_7^{-1}v_8^{-1}v_9^{-1}v_{11}^{-1} = 1,$$
$$j^{-1}hjh^{-1}t^{-1}v_3^{-1}v_4^{-1}v_5^{-1}v_6^{-1}v_8^{-1}v_{11}^{-1} = 1,$$
$$ijijijv_1^{-1}v_3^{-1}v_4^{-1}v_5^{-1}v_7^{-1}v_8^{-1} = k^2v_1^{-1}v_5^{-1}v_6^{-1}v_8^{-1}v_9^{-1}v_{11}^{-1} = 1,$$
$$k^{-1}akd^{-1}a^{-1}v_4^{-1}v_6^{-1}v_8^{-1}v_9^{-1} = k^{-1}bkd^{-1}c^{-1}v_1^{-1}v_2^{-1}v_7^{-1}v_9^{-1} = 1,$$
$$k^{-1}ckd^{-1}b^{-1}v_2^{-1}v_3^{-1}v_5^{-1}v_8^{-1}v_9^{-1}v_{10}^{-1}v_{11}^{-1} = 1,$$
$$k^{-1}dkd^{-1}v_2^{-1}v_3^{-1}v_5^{-1}v_7^{-1} = k^{-1}tktv_2^{-1}v_5^{-1}v_6^{-1}v_8^{-1}v_9^{-1}v_{11}^{-1} = 1,$$
$$k^{-1}gkg^{-1}t^{-1}v_2^{-1}v_3^{-1}v_4^{-1}v_5^{-1}v_6^{-1}v_9^{-1}v_{10}^{-1} = 1,$$
$$k^{-1}hkh^{-1}v_2^{-1}v_4^{-1}v_8^{-1}v_{10}^{-1} = 1,$$
$$k^{-1}iki^{-1}v_2^{-1}v_7^{-1}v_8^{-1}v_{10}^{-1}v_{11}^{-1} = jkjkjkv_2^{-1}v_5^{-1}v_8^{-1}v_9^{-1}v_{10}^{-1} = 1.$$

(h) *The split extensions E_1 and E_2 both have a faithful permutation representation of degree* 2048 *with stabilizer $T = \mathrm{M}_{24}$. The elements $z_1 = v_1$ and $z_2 = (gv_1)^2$ represent the unique conjugacy classes of* 2*-central involutions of E_1 and E_2, respectively.*

(i) *E_3 has a faithful permutation representation PE_3 of degree* 1518 *with stabilizer $T = \langle g, h, i, (dg)^5, (dhjk)^3, (ijkj)^2, (dhjidg)^3 \rangle$.*

(j) *$z_3 = a^2$ is a* 2*-central involution of E_3 with centralizer $D = C_{E_3}(z_3) = \langle x, y \rangle$ of order $2^{21} \cdot 3^3 \cdot 5$, where $x = a(agik)^3$ and $y = d(cgihj)^4$, and their orders are* 4 *and* 6*, respectively. Furthermore, $E_3 = \langle x, y, e \rangle$, where $e = g$ has order* 4*.*

(k) *$E_3 = \langle x, y, e \rangle$ has* 72 *conjugacy classes and $z = (xy^3)^{12}$. A system of their representatives and the character table of E_3 are given in Tables 9.7.2 and 9.8.3, respectively.*

(l) *$D = C_{E_3}(z) = \langle x, y \rangle$ has* 160 *conjugacy classes. A system of their representatives and the character table of D are given in Tables* `DVD.1.7.1` *and* `DVD.1.7.4` *of the accompanying DVD, respectively.*

(m) *V_1 is a self-centralizing maximal elementary abelian subgroup of order 2^{11} of each Sylow* 2*-subgroup S of E_1.*

(n) *V_2 is a self-centralizing maximal elementary abelian subgroup of order 2^{11} of each Sylow* 2*-subgroup S of any of the extension groups E_2 or E_3.*

Proof The two irreducible $F\mathrm{M}_{24}$-modules V_i, $i = 1, 2$, occur as composition factors with multiplicity 1 in the permutation module $(1_{\mathrm{M}_{23}})^{\mathrm{M}_{24}}$

and can easily be constructed using the faithful permutation representation of M_{24} stated in (a) and the Meataxe algorithm implemented in MAGMA. The corresponding matrices of the generators of M_{24} with respect to the first irreducible representation of M_{24} are stated in (b). The second irreducible representation V_2 of M_{24} is dual to V_1 and so it is defined by the equations given in (c).

(d) The cohomological dimensions $d_i = dim_F[H^2(\mathsf{M}_{24}, V_i)]$, $i = 1, 2$, have been calculated by means of MAGMA using Holt's Algorithm 7.4.5 of [92], the presentation of M_{24} of Definition 8.2.1 of [92] and all the data stated in (a), (b) and (c). It follows that $d_1 = 0$ and $d_2 = 1$.

(e) Since $d_1 = 0$ there exists only the split extension E_1 of M_{24} by V_1. The presentation of E_1 has been obtained automatically by Lemma 1.4.7 and MAGMA.

(f) Since $d_2 = 1$ there is a split extension E_2 of M_{24} by V_2 and a non-split extension E_3 of M_{24} by V_2. Again, the presentation of E_2 has been obtained by Lemma 1.4.7 and MAGMA.

(g) The presentation of E_3 has been established by means of Step 3 of Holt's Algorithm 7.4.5 of [92] and MAGMA.

(h) The first assertion is an immediate consequence of (e) and (f). Using Kratzer's Algorithm 5.3.18 of [92], the faithful permutation representations PE_i and MAGMA it has been checked that the groups E_1 and E_2 have 80 and 72 conjugacy classes, respectively, and that $z_1 = v_1$ and $z_2 = gv_1$ represent their respective unique conjugacy classes of 2-central involutions.

(i) Using a stand alone program due Paul Young, Kim and the author found in [74] a faithful permutation representation pE_3 of E_3 of degree 24288 with stabilizer $\langle (i^{-1}j^{-1}(bg)^2, (i^{-1}bth^{-1})^4 \rangle$. Applying the MAGMA command `DegreeReduction(pE))` we obtained the faithful permutation representation PE of degree 1518. The given generators of its stabilizer T were obtained by means of Kim's program `GetShortGens (PE,BasicStabilizer(PE,2))` described in Algorithm 1.4.1.

(j), (k) and (l) Using Kratzer's Algorithm 5.3.18 of [92], the faithful permutation representation PE_3 and MAGMA it can verified that E_3 has 72 conjugacy classes and that $z_3 = a^2$ represents its unique conjugacy classes of 2-central involutions. Another application of Algorithm 1.4.1 establishes that $D = C_{E_3}(z_3)$ is generated by the words x and y given in statement (j).

Using PE_3 and MAGMA again it is easily checked that the original generator g of order 4 of E_3 given in (g) satisifies $E_3 = \langle D, g \rangle$. We let $e = g$.

The representatives of the conjugacy classes of E_3 and $D = \langle x, y \rangle$ have been calculated using Kratzer's Algorithm 5.3.18 of [92] and the faithful permutation representation PE_3. The character table of E_3 was automatically calculated by MAGMA using PE_3.

(m) and (n) These two statements have been checked by means of the faithful permutation representations PE_1, PE_2 and PE_3 and the MAGMA command

```
Subgroups(PS_i: IsElementaryAbelian :=true,
  OrderEqual :=2048),
```

where PS_i is a chosen Sylow 2-subgroup of PE_i, $i = 1, 2, 3$. □

7.2 Construction of the 2-central involution centralizer of Co_1

In this section we apply Algorithm 1.3.8 to the extension group E_1 of Lemma 7.1.1 and construct a finitely presented group H, which will be shown in Section 7.3 to be isomorphic to the centralizers of a 2-central involution of the simple Conway group Co_1.

Let $E = E_1$ and $D = C_E(z)$, where z is a fixed representative of the unique class of 2-central involutions of E. It is shown below that D has exactly one extra-special normal subgroup Q of order 2^9 with center $Z(Q) = \langle z \rangle$. Let $W = D/Q$. Let t be a 2-central involution of W and $D_2 = C_W(t)$. In [75] Kim and the author applied Algorithm 7.4.8 of [92] to D_2 and constructed a simple subgroup K of $\mathrm{GL}_8(2)$ such that $|K : W| = 135$ and $N_K(B) \cong D_2$, where B is a self-centralizing maximal elementary abelian normal subgroup of D_2 of order 2^6. As in Chapter 4 we tried to build all extensions H_1 of K by V in terms of generators and relations. But (in 2007) MAGMA was not able to calculate the second cohomological dimension due to technical problems. Therefore we split this problem now.

In Proposition 7.2.1 we construct a subgroup U of order $2^{21} \cdot 3^3$ in D such that $U/Q \cong D_2$ and $Z(U) = Z(Q)$. It is used to build a finitely presented group $H_1 > U$ with $Z(H_1) = Z(Q)$ of odd index $|H_1 : U|$. The target group H is an epimorphic image of the free product $H_1 *_U D$ with amalgamated subgroup U; H is realized as an irreducible subgroup of $\mathrm{GL}_{128}(23)$. It has a faithful permutation representation of H of degree 276480 which is used to calculate the character table of H and a system of representatives of its conjugacy classes.

Proposition 7.2.1 *Keep the notation of Lemma 7.1.1. Let*

$$E = E_1 = \langle a, b, c, d, t, g, h, i, j, k, v_i | 1 \le i \le 11 \rangle$$

be the split extension of M_{24} *by its simple module* V_1 *of dimension* 11 *over* $F = GF(2)$*. Then the following statements hold.*

(a) $E = \langle x, y, e \rangle$, *where* $x = (cgjhi)^7 bt$, $y = (bv)^2 jkj$ *and* $e = b$ *have orders* 7, 4 *and* 2, *respectively.*

(b) $z = v_1 = (xy^3)^{14}$ *is a* 2*-central involution of* E *with centralizer* $D = C_E(z) = \langle x, y \rangle$ *of order* $2^{21} \cdot 3^2 \cdot 5 \cdot 7$.

(c) D *has a unique normal subgroup* Q *of order* 2^9. *It is extra-special and generated by the following eight elements:*

$$q_1 = y^2, \quad q_2 = (xy)^7, \quad q_3 = (yx)^7, \quad q_4 = (xy^2)^7, \quad q_5 = (yxy)^7,$$
$$q_6 = (x^5yx)^7, \quad q_7 = (x^4yx^2)^7, \quad q_8 = (x^4yxy)^6.$$

Furthermore, the elements q_i *satisfy the following set* $\mathcal{R}(Q)$ *of relations:*

$$q_1^2 = q_2^2 = q_3^2 = q_4^2 = q_5^2 = q_6^4 = q_7^4 = q_8^2 = 1,$$
$$(q_1, q_2) = (q_1, q_3) = z, \quad (q_1, q_4) = (q_1, q_5) = (q_1, q_6) = 1,$$
$$(q_1, q_7) = (q_1, q_8) = z, \quad (q_2, q_3) = z, \quad (q_2, q_4) = (q_2, q_5) = 1,$$
$$(q_2, q_6) = z, \quad (q_2, q_7) = 1, \quad (q_2, q_8) = z,$$
$$(q_3, q_4) = (q_3, q_5) = (q_3, q_6) = 1, \quad (q_3, q_7) = (q_3, q_8) = z,$$
$$(q_4, q_5) = (q_4, q_6) = (q_4, q_7) = 1, \quad (q_4, q_8) = z,$$
$$(q_5, q_6) = (q_5, q_7) = z, \quad (q_5, q_8) = 1,$$
$$(q_6, q_7) = (q_6, q_8) = z, \quad (q_7, q_8) = z.$$

(d) *Let* $\alpha : D \to D_1 = D/Z(Q)$ *be the canonical epimorphism with kernel* $ker(\alpha) = Z(Q) = \langle z \rangle$. *Let* $V = \alpha(Q) = \langle v_i = \alpha(q_i) \in D_1 \mid 1 \le i \le 8 \rangle$. *Then* V *is an elementary abelian normal subgroup of order* 2^8 *of* $D_1 = \langle \alpha(x), \alpha(y) \rangle$.

With respect to $\mathcal{B} = \{v_i | 1 \le i \le 8\}$ *the conjugate actions of* $\alpha(x)$ *and* $\alpha(y)$ *on* V *have the following matrices:*

$$Mx = \begin{pmatrix} 1&0&0&1&1&0&0&0 \\ 0&0&1&0&0&0&0&0 \\ 1&1&0&1&1&0&1&0 \\ 0&0&0&1&0&0&0&0 \\ 0&1&0&0&0&1&1&0 \\ 0&0&0&0&0&0&1&0 \\ 1&1&1&1&0&1&0&0 \\ 1&0&0&1&0&1&1&1 \end{pmatrix} \quad and \quad My = \begin{pmatrix} 1&0&0&0&0&0&0&0 \\ 0&0&1&0&0&0&0&0 \\ 0&1&0&0&0&0&0&0 \\ 0&0&0&0&1&0&0&0 \\ 0&0&0&1&0&0&0&0 \\ 0&0&1&0&1&0&0&1 \\ 1&1&0&1&0&1&1&1 \\ 0&1&0&1&0&1&0&0 \end{pmatrix}$$

in $\mathrm{GL}_8(2)$.

(e) *The map* $\varphi : D \to \mathrm{GL}_8(2)$ *defined by* $\varphi(x) = Mx$ *and* $\varphi(y) = My$ *is a group epimorphism onto* $MD = \langle Mx, My\rangle$ *with kernel* Q. *The group* MD *has an elementary abelian Fitting subgroup* MY *of order* 2^6 *generated by* $\{\varphi(m_j) \mid 1 \le j \le 6\}$, *where*

$$m_1 = (x^4(yx)^2y)^5, \quad m_2 = (x^3(yx)^3)^5, \quad m_3 = (x^2(yx)^2yx^2)^5,$$
$$m_4 = (xyx^4yxy)^5, \quad m_5 = (yx^4yxyx)^5, \quad m_6 = (x^3yx^3yxy)^5.$$

Furthermore, $MC = \langle\varphi(x), \varphi(c)\rangle$ *and* $C = \langle x, c\rangle$ *are complements of* MY *in* MD *and of* $M = \langle Q, m_j \mid 1 \le j \le 6\rangle$ *in* D, *respectively, which both are isomorphic to the alternating group* A_8. *In particular,* D *has a faithful permutation representation* PD *of degree* 2^{15} *with stabilizer* C, *where* $c = (x^2yx^2yxyxy^2xy)^3$ *has order* 2.

(f) D_1 *has a faithful permutation representation* PD_1 *of degree* 2^{14} *with stabilizer* $\alpha(C) = \langle\alpha(x), \alpha(c)\rangle$. *Furthermore,* D_1 *is a non-split extension of* D/Q *by* V *and* $C_{D_1}(V) = V$.

(g) D *is a finitely presented group* $D = \langle Q, x, y\rangle$ *having a set* $\mathcal{R}(D)$ *of defining relations consisting of* $\mathcal{R}(Q)$ *and the following relations:*

$$x^7 = y^4 = 1, \quad z = (xy^3)^{14}, \quad z^2 = 1, \quad (x, z) = (y, z) = 1,$$
$$(yx^{-1})^7 = q_1q_2q_4, \quad (xyx^{-1}yx)^4 = q_3q_6q_7,$$
$$(xyx^{-1}yx^{-1}y)^4 = q_4q_6q_7q_8,$$
$$(xyx^{-1}y)^6 = q_1, \quad (xyx^{-1}yx^{-1}yx)^4 = q_5z,$$
$$x^{-1}yx^{-2}yx^2yxyx^{-1}yx^3yx^{-1}yxyx^2yx^{-2}yx^{-1}y = q_2q_7z,$$
$$x^{-2}yx^{-1}yxyx^{-2}yx^{-1}yx^{-3}yx^2yx^2yxyx^{-1}yx^{-1}yx^{-1} = q_2q_4q_5q_8,$$
$$yxyx^{-1}yx^{-2}yx^2yx^{-1}yx^{-2}yxyx^{-2}yx^2yx^{-2}yx^2 = q_5q_6q_8,$$
$$x^{-1}q_1x = q_1q_4q_5q_6^2, \quad x^{-1}q_2x = q_3, \quad x^{-1}q_3x = q_1q_2q_4q_5q_7,$$
$$x^{-1}q_4x = q_4, \quad x^{-1}q_5x = q_2q_6q_7, \quad x^{-1}q_6x = q_7,$$
$$x^{-1}q_7x = q_1q_3q_2q_4q_6, \quad x^{-1}q_8x = q_1q_4q_6q_8q_7,$$
$$q_1^y = q_1, \quad y^{-1}q_2yq_3 = (q_1q_2)^2, \quad y^{-1}q_3yq_2 = 1,$$
$$y^{-1}q_4yq_5 = 1, \quad y^{-1}q_5yq_4 = 1, \quad y^{-1}q_6yq_3q_5q_8 = 1,$$
$$y^{-1}q_7yq_1q_2q_4q_6q_7q_8 = 1, \quad y^{-1}q_8yq_2q_4q_6 = (q_1q_2)^2.$$

(h) MY *contains a 2-central involution* $Mt = \varphi[(x^3yx^3yx^2)^3]$ *of* MD *with centralizer* $C_{MD}(Mt)$ *of order* $2^{12}\cdot 3^2$. *Furthermore,* MY *has a complement* MC_1 *in* $C_{MD}(Mt)$ *having an elementary abelian Fitting subgroup* $MW_1 = \langle\varphi(f_i) \mid 1 \le i \le 4\rangle$ *of order* 2^4 *which again has a complement* $MC_2 = \langle \varphi(r_k), \varphi(t_k) \mid 1 \le k \le 2 \rangle$ *of order* $2^2 \cdot 3^2$, *where*

$$\begin{aligned}
f_1 &= (yx^3yxyx^2yxyxy^2)^2, \quad f_2 = (x^2yxyx^2yxyxyx^5y)^3, \\
f_3 &= (xyx^2yx^2yxyx^5yx^2)^2, \quad f_4 = (x^2yx^4yx^3yx^3y^2xy)^3, \\
r_1 &= (yx^5yx^5yx^2yx^2)^2, \quad r_2 = (x^3yxyx^2yx^4yxyx^2y)^3, \\
t_1 &= (xyx^3yx^2y^2x^2yxyx^3yx)^5(x^5yx^3yxyx^2yx^3yxy)^3 \\
&\quad \cdot (xyx^3yx^2y^2x^2yxyx^3yx)^{10}(x^5yx^3yxyx^2yx^3yxy)^3, \\
t_2 &= ((xyx^3yx^2y^2x^2yxyx^3yx)^5(x^5yx^3yxyx^2yx^3yxy)^3)^3,
\end{aligned}$$

and where r_1 *and* t_1 *have order* 3 *and all other elements are involutions. Furthermore,* $r_1^{r_2} = r_1^2$ *and* $t_1^{t_2} = t_1^2$.

(i) *The centralizer* $MX = C_{\mathrm{GL}_8(2)}(Mt)$ *of* Mt *in* $\mathrm{GL}_8(2)$ *contains the matrix*

$$Mh = \begin{pmatrix}
1&1&0&1&1&1&0&1\\
1&0&0&1&1&1&0&0\\
0&0&1&0&0&0&0&0\\
1&1&0&1&1&1&0&0\\
1&0&0&1&1&0&0&1\\
1&1&0&0&1&0&0&0\\
0&1&0&0&1&1&1&1\\
0&1&0&1&0&1&1&1
\end{pmatrix}$$

of order 3 *such that the subgroup* $MO = \langle MD, Mh \rangle$ *of* $\mathrm{GL}_8(2)$ *has order* $2^{12} \cdot 3^5 \cdot 5^2 \cdot 7$ *and the subgroup*

$$\begin{aligned}
MH_3 &= C_{MO}(Mt) = \langle C_{MD}(Mt), Mh \rangle \\
&= \langle Mm_j, Mf_i, Mr_k, Mt_k, Mh \rangle
\end{aligned}$$

of $\mathrm{GL}_8(2)$ *has order* $2^{12} \cdot 3^3$, *where* $1 \le j \le 6, 1 \le i \le 4, 1 \le k \le 2$. *Furthermore,* MO *is isomorphic to the simple group* $\mathrm{O}_8^+(2)$.

(j) *The Fitting subgroup* MY_3 *of* MH_3 *is extra-special of order* 2^9 *with center* $Z(Y_3) = Mt$, *and*

$$MY_3 = \langle Mm_3, Mm_4, Mm_6, Mf_1, Mf_2, Mf_3, Mf_4 \rangle,$$

where $Mf_5 = Mm_3$, $Mf_6 = Mm_4$, $Mf_7 = Mm_6$ *and* $Mf_8 = (Mm_5 Mf_2)^2$. *Furthermore,* MY_3 *is isomorphic to the finitely presented group* $Y_3 = \langle f_i \mid 1 \le i \le 8 \rangle$ *with the following set* $\mathcal{R}(Y_3)$ *of defining relations:*

$$\begin{aligned}
&f_1^2 = f_2^2 = f_3^2 = f_4^2 = f_5^2 = f_6^2 = f_7^2 = f_8^2 = 1, \\
&(f_1, f_2) = (f_1, f_3) = (f_2, f_3) = (f_1, f_4) = (f_2, f_4) = (f_3, f_4) = 1, \\
&(f_1, f_5) = (f_2, f_5) = (f_3, f_5) = (f_1, f_6) = (f_5, f_6) = (f_5, f_7) = 1,
\end{aligned}$$

$$(f_6,f_7)=(f_2,f_8)=(f_4,f_8)=(f_5,f_8)=(f_6,f_8)=(f_7,f_8)=1,$$
$$(f_1f_7f_2)^2=(f_2f_6f_3)^2=(f_1f_7f_3)^2=(f_1f_8f_3)^2=1,$$
$$(f_2f_6f_4)^2=(f_1f_7f_4)^2=1,$$
$$f_4f_6f_5f_4f_5f_6=f_4f_7f_5f_4f_5f_7=f_3f_8f_6f_3f_6f_8=1.$$

$MT=\langle Mr_1, Mt_1, Ma_1\rangle$ is an elementary abelian Sylow 3-subgroup of MH_3 of order 27, where $Ma_1 = (Mt_1)^2Mh(Mt_1)^2 Mh^2(Mt_1)^2Mh$.

The matrix $Ma_2=Mf_2Mm_2Mm_3Mm_4Mm_6Mf_2Mm_1Mm_2 Mm_6$ is an involution commuting with the subgroup $\langle Mr_1, Mr_2, Mt_1, Mt_2\rangle$ and satisfying $Ma_1^{Ma_2}=Ma_1^2$.

$MK_3=\langle MT, Mr_2, Mt_2, Ma_1, Ma_2\rangle$ is a complement of MY_3 in MH_3.

MH_3 is isomorphic to the finitely presented group

$$H_3=\langle a_1,a_2,r_1,r_2,t_1,t_2,f_i \mid 1\le i\le 8\rangle$$

with the set $\mathcal{R}(H_3)$ of defining relations consisting of $\mathcal{R}(Y_3)$ and the following relations:

$$a_1^3=a_2^2=r_1^3=r_2^2=t_1^3=t_2^2=1,$$
$$(a_i,r_j)=(a_i,t_k)=(r_j,t_k)=1 \quad \textit{for all} \quad 1\le i,j,k\le 2,$$
$$a_1^{a_2}=a_1^2,\quad r_1^{r_2}=r_1^2,\quad t_1^{t_2}=t_1^2,$$
$$f_1^{r_1}=f_2f_3f_4,\quad f_1^{r_2}=f_2f_3f_4,\quad f_1^{t_1}=f_1f_4,\quad f_1^{t_2}=f_1f_4,$$
$$f_1^{a_1}=f_5f_6,\quad f_1^{a_2}=f_1f_5f_6,\quad f_2^{r_1}=f_4,\quad f_2^{r_2}=f_2,$$
$$f_2^{t_1}=f_1f_2f_3f_4,\quad f_2^{t_2}=f_2,\quad f_2^{a_1}=f_8,\quad f_2^{a_2}=f_2f_8,$$
$$f_3^{r_1}=f_1f_3f_4,\quad f_3^{r_2}=f_1f_4,\quad f_3^{t_1}=f_2f_4,\quad f_3^{t_2}=f_2f_3f_4,$$
$$f_3^{a_1}=f_4f_6f_4f_8,\quad f_3^{a_2}=f_6f_3f_8,\quad f_4^{r_1}=f_2f_4,\quad f_4^{r_2}=f_2f_4,$$
$$f_4^{t_1}=f_1,\quad f_4^{t_2}=f_4,\quad f_4^{a_1}=f_6f_7f_8,\quad f_4^{a_2}=f_4f_6f_7f_8,$$
$$f_5^{r_1}=f_4f_5f_4f_7,\quad f_5^{r_2}=f_5f_8,\quad f_5^{t_1}=f_7,\quad f_5^{t_2}=f_5f_8,$$
$$f_5^{a_1}=f_1f_2f_5f_8f_3f_8,\quad f_5^{a_2}=f_5,\quad f_6^{r_1}=f_5f_8,$$
$$f_6^{r_2}=f_4f_5f_4f_7,\quad f_6^{t_1}=f_5f_8,\quad f_6^{t_2}=f_7,\quad f_6^{a_1}=f_2f_6f_3,$$
$$f_6^{a_2}=f_6,\quad f_7^{r_1}=f_5,\quad f_7^{r_2}=f_4f_5f_4f_6f_8,\quad f_7^{t_1}=f_4f_5f_4f_7,$$
$$f_7^{t_2}=f_6,\quad f_7^{a_1}=f_3f_7f_4,\quad f_7^{a_2}=f_7,\quad f_8^{r_1}=f_6f_7f_8,$$
$$f_8^{r_2}=f_8,\quad f_8^{t_1}=f_4f_5f_4f_6f_7f_8,\quad f_8^{t_2}=f_8,\quad f_8^{a_1}=f_2f_8,$$
$$f_8^{a_2}=f_8.$$

(k) *The subgroup*

$$U = \langle a_2, r_1, r_2, t_1, t_2, f_j, q_i \mid 1 \leq i, j \leq 8 \rangle$$

of D has center $Z(U) = \langle z = q_6^2 \rangle$ and satisfies the set $\mathcal{R}(U)$ of defining relations consisting of $\mathcal{R}(Q)$ and the following relations:

$a_2^2 = r_1^3 = r_2^2 = t_1^3 = t_2^2 = 1,$
$f_j^2 = 1, \quad \textit{for} \quad 1 \leq j \leq 8,$
$r_1^{r_2} = r_1^2, \quad t_1^{t_2} = t_1^2,$
$(a_2, r_1) = (a_2, r_2) = (a_2, t_1) = (a_2, t_2) = 1,$
$(r_1, t_1) = (r_1, t_2) = (r_2, t_1) = (r_2, t_2) = 1,$
$f_1^{r_1}(f_2f_3f_4)^{-1} = f_1^{r_2}(f_2f_3f_4)^{-1} = f_1^{t_1}(f_1f_4)^{-1} = f_1^{t_2}(f_1f_4)^{-1} = 1,$
$f_1^{a_2}(f_1f_5f_6)^{-1} = q_5,$
$f_2^{r_1}(f_4)^{-1} = f_2^{r_2}(f_2)^{-1} = f_2^{t_1}(f_1f_2f_3f_4)^{-1} = f_2^{t_2}(f_2)^{-1} = 1,$
$f_2^{a_2}(f_2f_8)^{-1} = f_3^{r_1}(f_1f_3f_4)^{-1} = f_3^{r_2}(f_1f_4)^{-1} = f_3^{t_1}(f_2f_4)^{-1} = 1,$
$f_3^{t_2}(f_2f_3f_4)^{-1} = 1, \quad f_3^{a_2}(f_6f_3f_8)^{-1} = q_4,$
$f_4^{r_1}(f_2f_4)^{-1} = f_4^{r_2}(f_2f_4)^{-1} = f_4^{t_1}(f_1)^{-1} = f_4^{t_2}(f_4)^{-1} = 1,$
$f_4^{a_2}(f_4f_6f_7f_8)^{-1} = q_1q_2q_4q_6q_7, \quad f_5^{r_1}(f_4f_5f_4f_7)^{-1} = q_2q_4q_7q_6,$
$f_5^{r_2}(f_5f_8)^{-1} = q_2q_4q_6q_7, \quad f_5^{t_1}(f_7)^{-1} = q_8q_1q_8,$
$f_5^{t_2}(f_5f_8)^{-1} = q_2q_4q_5q_6q_7, \quad f_5^{a_2}(f_5)^{-1} = 1,$
$f_6^{r_1}(f_5f_8)^{-1} = q_2q_5q_6q_7, \quad f_6^{r_2}(f_4f_5f_4f_7)^{-1} = q_2q_4q_5q_7q_6,$
$f_6^{t_1}(f_5f_8)^{-1} = q_2q_4q_6q_7, \quad f_6^{t_2}(f_7)^{-1} = q_5q_8q_1q_8,$
$f_6^{a_2}(f_6)^{-1} = 1, \quad f_7^{r_1}(f_5)^{-1} = q_8q_1q_8,$
$f_7^{r_2}(f_4f_5f_4f_6f_8)^{-1} = q_5, \quad f_7^{t_1}(f_4f_5f_4f_7)^{-1} = q_1,$
$f_7^{t_2}(f_6)^{-1} = q_1q_2q_4q_7q_6, \quad f_7^{a_2}(f_7)^{-1} = 1,$
$f_8^{r_1}(f_6f_7f_8)^{-1} = q_1q_5, \quad f_8^{r_2}(f_8)^{-1} = 1,$
$f_8^{t_1}(f_4f_5f_4f_6f_7f_8)^{-1} = q_1q_4, \quad f_8^{t_2}(f_8)^{-1} = 1, \quad f_8^{a_2}(f_8)^{-1} = 1,$
$(f_1, f_2) = (f_1, f_3) = (f_2, f_3) = (f_1, f_4) = (f_2, f_4) = (f_3, f_4) = 1,$
$(f_2, f_5) = (f_3, f_5) = (f_1, f_6) = (f_5, f_6) = (f_5, f_7) = (f_6, f_7) = 1,$
$(f_2, f_8) = 1, \quad (f_1f_5)^2 = q_1q_4q_5, \quad (f_4f_8)^2 = q_2q_4q_5q_7q_6,$
$(f_5, f_8) = (f_6, f_8) = (f_7, f_8) = 1,$
$(f_1f_7f_2)^2 = q_1q_4q_5, \quad (f_2f_6f_3)^2 = q_2q_4q_5q_6q_7,$
$(f_1f_7f_3)^2 = q_1q_2q_6q_7, \quad (f_1f_8f_3)^2 = q_2q_4q_5q_6q_7,$
$(f_2f_6f_4)^2 = q_2q_4q_5q_6q_7, \quad (f_1f_7f_4)^2 = q_1q_2q_6q_7,$

$$f_4f_6f_5f_4f_5f_6 = q_2q_4q_5q_7q_6, \quad f_4f_7f_5f_4f_5f_7 = q_2q_4q_5q_6q_7,$$
$$f_3f_8f_6f_3f_6f_8 = 1,$$
$$q_1^{a_2} = q_1, \quad q_1^{r_1} = q_1q_5, \quad q_1^{r_2} = q_1q_5,$$
$$q_1^{t_1} = q_1q_2q_4q_5q_6q_7, \quad q_1^{t_2} = q_1q_2q_4q_5q_6q_7, \quad q_1^{f_1} = q_1,$$
$$q_1^{f_2} = q_1q_2q_4q_5q_6q_7, \quad q_1^{f_3} = q_2q_6q_7, \quad q_1^{f_4} = q_1,$$
$$q_1^{f_5} = q_1^{f_6} = q_1^{f_7} = q_1^{f_8} = q_1,$$
$$q_2^{a_2} = q_1q_6q_7, \quad q_2^{r_1} = q_2q_5, \quad q_2^{r_2} = q_2q_5,$$
$$q_2^{t_1} = q_2q_3q_4q_5q_6q_7, \quad q_2^{t_2} = q_3q_7q_6, \quad q_2^{f_1} = q_4q_5q_8q_6,$$
$$q_2^{f_2} = q_4q_6q_7q_8, \quad q_2^{f_3} = q_1q_7q_6, \quad q_2^{f_4} = q_4q_5q_8q_6,$$
$$q_2^{f_5} = q_1q_4q_6q_7, \quad q_2^{f_6} = q_1q_5q_6q_7, \quad q_2^{f_7} = q_1q_4q_5q_7q_6,$$
$$q_2^{f_8} = q_5q_8q_2q_8, \quad q_3^{a_2} = q_2q_3q_4q_5q_6q_7, \quad q_3^{r_1} = q_3q_5,$$
$$q_3^{r_2} = q_3q_5, \quad q_3^{t_1} = q_1q_3q_4q_5q_6q_7, \quad q_3^{t_2} = q_3,$$
$$q_3^{f_1} = q_2q_3q_4q_5q_6q_8, \quad q_3^{f_2} = q_2q_3q_4q_5q_7q_6,$$
$$q_3^{f_3} = q_1q_3q_4q_7, \quad q_3^{f_4} = q_3, \quad q_3^{f_5} = q_2q_3q_4q_6q_7,$$
$$q_3^{f_6} = q_2q_3q_5q_6q_7, \quad q_3^{f_7} = q_2q_3q_5q_7q_6, \quad q_3^{f_8} = q_3,$$
$$q_4^{a_2} = q_4, \quad q_4^{r_1} = q_5, \quad q_4^{r_2} = q_4q_5, \quad q_4^{t_1} = q_4,$$
$$q_4^{t_2} = q_4, \quad q_4^{f_1} = q_1q_5, \quad q_4^{f_2} = q_2q_5q_6q_7, \quad q_4^{f_3} = q_4,$$
$$q_4^{f_4} = q_2q_5q_6q_7, \quad q_4^{f_5} = q_4^{f_6} = q_4^{f_7} = q_4^{f_8} = q_4,$$
$$q_5^{a_2} = q_5, \quad q_5^{r_1} = q_4q_5, \quad q_5^{r_2} = q_5^{t_1} = q_5^{t_2} = q_5,$$
$$q_5^{f_1} = q_1q_4, \quad q_5^{f_2} = q_5, \quad q_5^{f_3} = q_1q_2q_5q_6q_7,$$
$$q_5^{f_4} = q_2q_4q_6q_7, \quad q_5^{f_5} = q_5^{f_6} = q_5^{f_7} = q_5^{f_8} = q_5,$$
$$q_6^{a_2} = q_1q_2q_4q_7, \quad q_6^{r_1} = q_2q_5q_8, \quad q_6^{r_2} = q_5q_6q_8,$$
$$q_6^{t_1} = q_3q_4q_5q_7, \quad q_6^{t_2} = q_2q_7q_3, \quad q_6^{f_1} = q_1q_2q_8,$$
$$q_6^{f_2} = q_5q_6q_8, \quad q_6^{f_3} = q_1q_2q_7, \quad q_6^{f_4} = q_6q_8q_7, \quad q_6^{f_5} = q_4q_6,$$
$$q_6^{f_6} = q_1q_4q_6, \quad q_6^{f_7} = q_1q_6^3, \quad q_6^{f_8} = q_2q_4q_7,$$
$$q_7^{a_2} = q_4q_7^3, \quad q_7^{r_1} = q_2q_5q_6q_8q_7, \quad q_7^{r_2} = q_5q_7q_8,$$
$$q_7^{t_1} = q_1q_6q_2, \quad q_7^{t_2} = q_7, \quad q_7^{f_1} = q_1q_4q_5q_7, \quad q_7^{f_2} = q_7,$$
$$q_7^{f_3} = q_1q_6q_2, \quad q_7^{f_4} = q_2q_4q_6q_5, \quad q_7^{f_5} = q_1q_2q_6,$$
$$q_7^{f_6} = q_2q_4q_6q_5, \quad q_7^{f_7} = q_2q_4q_6q_5, \quad q_7^{f_8} = q_2q_4q_6q_5,$$
$$q_8^{a_2} = q_5q_7^2q_8, \quad q_8^{r_1} = q_2q_6q_8, \quad q_8^{r_2} = q_8^{t_1} = q_8^{t_2} = q_8,$$
$$q_8^{f_1} = q_1q_4q_5q_8, \quad q_8^{f_2} = q_8, \quad q_8^{f_3} = q_1q_2q_6q_7q_8,$$
$$q_8^{f_4} = q_2q_4q_5q_6q_7q_8, \quad q_8^{f_5} = q_2q_4q_5q_6q_8q_7,$$
$$q_8^{f_6} = q_1q_2q_6q_8q_7, \quad q_8^{f_7} = q_1q_4q_5q_8, \quad q_8^{f_8} = q_8.$$

U has center $Z(U) = \langle q_6^2 \rangle$ and a faithful permutation representation PU of degree 2048 with stabilizer Y generated by r_2r_1, $(t_2q_6)^2$, $(q_6r_1f_4)^4$, $(q_6f_8r_1)^4$, $(r_1^2q_6f_8)^3$, $(r_1t_1t_2f_4)^3$ and $(r_1t_1q_6f_8)^2$.

(1) *The finitely presented group H_3 has a non-split extension*

$$H_2 = \langle a_1, a_2, r_1, r_2, t_1, t_2, f_j, v_i \mid 1 \le i, j \le 8 \rangle$$

by the elementary abelian group $V = \langle v_i \mid 1 \le i \le 8 \rangle$ with the following set $\mathcal{R}(H_2)$ of defining relations:

$$\begin{aligned}
&a_1^3 = a_2^2 = r_1^3 = r_2^2 = t_1^3 = t_2^2 = 1,\\
&a_1^{a_2} = a_1^2, \quad r_1^{r_2} = r_1^2, \quad t_1^{t_2} = t_1^2,\\
&(a_k, r_1) = (a_k, r_2) = (a_k, t_1) = (a_k, t_2) = 1, \quad \textit{for} \quad 1 \le k \le 2,\\
&(r_1, t_1) = (r_1, t_2) = (r_2, t_1) = (r_2, t_2) = 1,\\
&v_i^2 = 1 \quad \textit{for} \quad 1 \le i \le 8,\\
&f_j^2 = 1 \quad \textit{for} \quad 1 \le j \le 8,\\
&(v_i, v_j) = 1 \quad \textit{for} \quad 1 \le i, j \le 8,\\
&a_1^{-1}v_1a_1v_1^{-1}v_2^{-1}v_4^{-1}v_5^{-1}v_6^{-1}v_8^{-1} = a_1^{-1}v_2a_1v_4^{-1}v_5^{-1}v_6^{-1}v_8^{-1} = 1,\\
&a_1^{-1}v_3a_1v_2^{-1}v_3^{-1}v_4^{-1}v_5^{-1}v_6^{-1}v_8^{-1} = a_1^{-1}v_4a_1v_2^{-1}v_6^{-1} = 1,\\
&a_1^{-1}v_5a_1v_8^{-1} = a_1^{-1}v_6a_1v_2^{-1}v_5^{-1}v_8^{-1} = a_1^{-1}v_8a_1v_5^{-1}v_8^{-1} = 1,\\
&a_1^{-1}v_7a_1v_2^{-1}v_5^{-1}v_6^{-1}v_7^{-1}v_8^{-1} = 1,\\
&(a_2, v_1^{-1}) = (a_2, v_2^{-1}) = (a_2, v_3^{-1}) = (a_2, v_4^{-1}) = (a_2, v_5^{-1}) = 1,\\
&a_2^{-1}v_6a_2v_4^{-1}v_6^{-1} = a_2^{-1}v_7a_2v_4^{-1}v_7^{-1} = 1,\\
&a_2^{-1}v_8a_2v_5^{-1}v_8^{-1} = r_1^{-1}v_1r_1v_1^{-1}v_5^{-1} = r_1^{-1}v_2r_1v_2^{-1}v_5^{-1} = 1,\\
&r_1^{-1}v_3r_1v_3^{-1}v_5^{-1} = r_1^{-1}v_4r_1v_5^{-1} = r_1^{-1}v_5r_1v_4^{-1}v_5^{-1} = 1,\\
&r_1^{-1}v_6r_1v_2^{-1}v_5^{-1}v_8^{-1} = r_1^{-1}v_7r_1v_2^{-1}v_5^{-1}v_6^{-1}v_7^{-1}v_8^{-1} = 1,\\
&r_1^{-1}v_8r_1v_2^{-1}v_6^{-1}v_8^{-1} = r_2^{-1}v_1r_2v_1^{-1}v_5^{-1} = 1,\\
&r_2^{-1}v_2r_2v_2^{-1}v_5^{-1} = r_2^{-1}v_3r_2v_3^{-1}v_5^{-1} = 1,\\
&r_2^{-1}v_4r_2v_4^{-1}v_5^{-1} = 1, \quad (r_2, v_5^{-1}) = 1,\\
&r_2^{-1}v_6r_2v_5^{-1}v_6^{-1}v_8^{-1} = 1,\\
&r_2^{-1}v_7r_2v_5^{-1}v_7^{-1}v_8^{-1} = 1, \quad (r_2, v_8^{-1}) = 1,\\
&t_1^{-1}v_1t_1v_1^{-1}v_2^{-1}v_4^{-1}v_5^{-1}v_6^{-1}v_7^{-1} = 1,\\
&t_1^{-1}v_2t_1v_2^{-1}v_3^{-1}v_4^{-1}v_5^{-1}v_6^{-1}v_7^{-1} = 1,\\
&t_1^{-1}v_3t_1v_1^{-1}v_3^{-1}v_4^{-1}v_5^{-1}v_6^{-1}v_7^{-1} = 1,
\end{aligned}$$

$$(t_1, v_4^{-1}) = (t_1, v_5^{-1}) = t_1^{-1}v_6t_1v_3^{-1}v_4^{-1}v_5^{-1}v_7^{-1} = 1,$$
$$t_1^{-1}v_7t_1v_1^{-1}v_2^{-1}v_6^{-1} = 1, \quad (t_1, v_8^{-1}) = 1,$$
$$t_2^{-1}v_1t_2v_1^{-1}v_2^{-1}v_4^{-1}v_5^{-1}v_6^{-1}v_7^{-1} = 1,$$
$$t_2^{-1}v_2t_2v_3^{-1}v_6^{-1}v_7^{-1} = 1, \quad (t_2, v_3^{-1}) = (t_2, v_4^{-1}) = (t_2, v_5^{-1}) = 1,$$
$$t_2^{-1}v_6t_2v_2^{-1}v_3^{-1}v_7^{-1} = 1, \quad (t_2, v_7^{-1}) = (t_2, v_8^{-1}) = (f_1, v_1^{-1}) = 1,$$
$$f_1^{-1}v_2f_1v_4^{-1}v_5^{-1}v_6^{-1}v_8^{-1} = f_1^{-1}v_3f_1v_2^{-1}v_3^{-1}v_4^{-1}v_5^{-1}v_6^{-1}v_8^{-1} = 1,$$
$$f_1^{-1}v_4f_1v_1^{-1}v_5^{-1} = f_1^{-1}v_5f_1v_1^{-1}v_4^{-1} = f_1^{-1}v_6f_1v_1^{-1}v_2^{-1}v_8^{-1} = 1,$$
$$f_1^{-1}v_7f_1v_1^{-1}v_4^{-1}v_5^{-1}v_7^{-1} = f_1^{-1}v_8f_1v_1^{-1}v_4^{-1}v_5^{-1}v_8^{-1} = 1,$$
$$f_2^{-1}v_1f_2v_1^{-1}v_2^{-1}v_4^{-1}v_5^{-1}v_6^{-1}v_7^{-1} = 1,$$
$$f_2^{-1}v_2f_2v_4^{-1}v_6^{-1}v_7^{-1}v_8^{-1} = f_2^{-1}v_3f_2v_2^{-1}v_3^{-1}v_4^{-1}v_5^{-1}v_6^{-1}v_7^{-1} = 1,$$
$$f_2^{-1}v_4f_2v_2^{-1}v_5^{-1}v_6^{-1}v_7^{-1} = f_2^{-1}v_6f_2v_5^{-1}v_6^{-1}v_8^{-1} = 1,$$
$$(f_2, v_5) = (f_2, v_7) = (f_2, v_8) = (f_3, v_4) = (f_4, v_1) = (f_4, v_3) = 1,$$
$$(f_5, v_1) = (f_5, v_4) = (f_5, v_5) = (f_6, v_1) = (f_6, v_4) = (f_6, v_5) = 1,$$
$$(f_7, v_1) = (f_8, v_1) = (f_7, v_4) = (f_7, v_5) = (f_8, v_3) = (f_8, v_4) = 1,$$
$$(f_8, v_5) = (f_8, v_8) = 1, \quad f_3^{-1}v_1f_3v_2^{-1}v_6^{-1}v_7^{-1} = 1,$$
$$f_3^{-1}v_2f_3v_1^{-1}v_6^{-1}v_7^{-1} = f_3^{-1}v_3f_3v_1^{-1}v_3^{-1}v_4^{-1}v_7^{-1} = 1,$$
$$f_3^{-1}v_5f_3v_1^{-1}v_2^{-1}v_5^{-1}v_6^{-1}v_7^{-1} = f_3^{-1}v_6f_3v_1^{-1}v_2^{-1}v_7^{-1} = 1,$$
$$f_3^{-1}v_7f_3v_1^{-1}v_2^{-1}v_6^{-1} = f_3^{-1}v_8f_3v_1^{-1}v_2^{-1}v_6^{-1}v_7^{-1}v_8^{-1} = 1,$$
$$f_4^{-1}v_2f_4v_4^{-1}v_5^{-1}v_6^{-1}v_8^{-1} = f_4^{-1}v_4f_4v_2^{-1}v_5^{-1}v_6^{-1}v_7^{-1} = 1,$$
$$f_4^{-1}v_5f_4v_2^{-1}v_4^{-1}v_6^{-1}v_7^{-1} = f_4^{-1}v_6f_4v_6^{-1}v_7^{-1}v_8^{-1} = 1,$$
$$f_4^{-1}v_7f_4v_2^{-1}v_4^{-1}v_5^{-1}v_6^{-1} = f_4^{-1}v_8f_4v_2^{-1}v_4^{-1}v_5^{-1}v_6^{-1}v_7^{-1}v_8^{-1} = 1,$$
$$f_5^{-1}v_2f_5v_1^{-1}v_4^{-1}v_6^{-1}v_7^{-1} = f_5^{-1}v_3f_5v_2^{-1}v_3^{-1}v_4^{-1}v_6^{-1}v_7^{-1} = 1,$$
$$f_5^{-1}v_6f_5v_4^{-1}v_6^{-1} = f_5^{-1}v_7f_5v_1^{-1}v_2^{-1}v_6^{-1} = 1,$$
$$f_5^{-1}v_8f_5v_2^{-1}v_4^{-1}v_5^{-1}v_6^{-1}v_7^{-1}v_8^{-1} = 1,$$
$$f_6^{-1}v_2f_6v_1^{-1}v_5^{-1}v_6^{-1}v_7^{-1} = f_6^{-1}v_3f_6v_2^{-1}v_3^{-1}v_5^{-1}v_6^{-1}v_7^{-1} = 1,$$
$$f_6^{-1}v_6f_6v_1^{-1}v_4^{-1}v_6^{-1} = f_6^{-1}v_7f_6v_2^{-1}v_4^{-1}v_5^{-1}v_6^{-1} = 1,$$
$$f_6^{-1}v_8f_6v_1^{-1}v_2^{-1}v_6^{-1}v_7^{-1}v_8^{-1} = f_7^{-1}v_2f_7v_1^{-1}v_4^{-1}v_5^{-1}v_6^{-1}v_7^{-1} = 1,$$
$$f_7^{-1}v_3f_7v_2^{-1}v_3^{-1}v_5^{-1}v_6^{-1}v_7^{-1} = 1,$$
$$f_7^{-1}v_6f_7v_1^{-1}v_6^{-1} = f_7^{-1}v_7f_7v_2^{-1}v_4^{-1}v_5^{-1}v_6^{-1} = 1,$$
$$f_7^{-1}v_8f_7v_1^{-1}v_4^{-1}v_5^{-1}v_8^{-1} = f_8^{-1}v_2f_8v_2^{-1}v_5^{-1} = 1,$$
$$f_8^{-1}v_6f_8v_2^{-1}v_4^{-1}v_7^{-1} = f_8^{-1}v_7f_8v_2^{-1}v_4^{-1}v_5^{-1}v_6^{-1} = 1,$$
$$r_1^{-1}f_1r_1f_4^{-1}f_3^{-1}f_2^{-1}v_2^{-1}v_4^{-1}v_5^{-1}v_6^{-1}v_7^{-1} = 1,$$

$$r_2^{-1}f_1r_2f_4^{-1}f_3^{-1}f_2^{-1}v_2^{-1}v_4^{-1}v_5^{-1}v_6^{-1}v_7^{-1} = 1,$$
$$t_1^{-1}f_1t_1f_4^{-1}f_1^{-1} = t_2^{-1}f_1t_2f_4^{-1}f_1^{-1} = 1,$$
$$a_1^{-1}f_1a_1f_6^{-1}f_5^{-1}v_1^{-1}v_4^{-1}v_5^{-1} = 1,$$
$$a_2^{-1}f_1a_2f_6^{-1}f_5^{-1}f_1^{-1}v_1^{-1}v_4^{-1}v_5^{-1} = 1,$$
$$r_1^{-1}f_2r_1f_4^{-1} = 1, \quad (r_2, f_2^{-1}) = (t_2, f_2^{-1}) = 1,$$
$$t_1^{-1}f_2t_1f_4^{-1}f_3^{-1}f_2^{-1}f_1^{-1}v_1^{-1}v_2^{-1}v_6^{-1}v_7^{-1} = 1,$$
$$a_1^{-1}f_2a_1f_8^{-1} = a_2^{-1}f_2a_2f_8^{-1}f_2^{-1} = 1,$$
$$r_1^{-1}f_3r_1f_4^{-1}f_3^{-1}f_1^{-1}v_1^{-1}v_2^{-1}v_6^{-1}v_7^{-1} = 1,$$
$$r_2^{-1}f_3r_2f_4^{-1}f_1^{-1} = t_1^{-1}f_3t_1f_4^{-1}f_2^{-1}v_2^{-1}v_4^{-1}v_5^{-1}v_6^{-1}v_7^{-1} = 1,$$
$$t_2^{-1}f_3t_2f_4^{-1}f_3^{-1}f_2^{-1}v_2^{-1}v_4^{-1}v_5^{-1}v_6^{-1}v_7^{-1} = 1,$$
$$a_1^{-1}f_3a_1f_8^{-1}f_4^{-1}f_6^{-1}f_4^{-1}v_2^{-1}v_4^{-1}v_5^{-1}v_6^{-1}v_7^{-1} = 1,$$
$$a_2^{-1}f_3a_2f_8^{-1}f_3^{-1}f_6^{-1} = r_1^{-1}f_4r_1f_4^{-1}f_2^{-1}v_2^{-1}v_4^{-1}v_5^{-1}v_6^{-1}v_7^{-1} = 1,$$
$$r_2^{-1}f_4r_2f_4^{-1}f_2^{-1}v_2^{-1}v_4^{-1}v_5^{-1}v_6^{-1}v_7^{-1} = 1,$$
$$t_1^{-1}f_4t_1f_1^{-1} = 1, \quad (t_2, f_4^{-1}) = 1,$$
$$a_1^{-1}f_4a_1f_8^{-1}f_7^{-1}f_6^{-1} = a_2^{-1}f_4a_2f_8^{-1}f_7^{-1}f_6^{-1}f_4^{-1} = 1,$$
$$r_1^{-1}f_5r_1f_7^{-1}f_4^{-1}f_5^{-1}f_4^{-1}v_1^{-1}v_2^{-1}v_6^{-1}v_7^{-1} = 1,$$
$$r_2^{-1}f_5r_2f_8^{-1}f_5^{-1}v_2^{-1}v_4^{-1}v_5^{-1}v_6^{-1}v_7^{-1} = t_1^{-1}f_5t_1f_7^{-1} = 1,$$
$$t_2^{-1}f_5t_2f_8^{-1}f_5^{-1}v_2^{-1}v_4^{-1}v_5^{-1}v_6^{-1}v_7^{-1} = 1,$$
$$a_1^{-1}f_5a_1f_8^{-1}f_3^{-1}f_8^{-1}f_5^{-1}f_2^{-1}f_1^{-1}v_2^{-1}v_4^{-1}v_5^{-1}v_6^{-1}v_7^{-1} = 1,$$
$$(a_2, f_5^{-1}) = (a_2, f_6^{-1}) = (a_2, f_7^{-1}) = 1,$$
$$r_1^{-1}f_6r_1f_8^{-1}f_5^{-1}v_2^{-1}v_4^{-1}v_5^{-1}v_6^{-1}v_7^{-1} = 1,$$
$$r_2^{-1}f_6r_2f_7^{-1}f_4^{-1}f_5^{-1}f_4^{-1}v_1^{-1}v_2^{-1}v_6^{-1}v_7^{-1} = 1,$$
$$t_1^{-1}f_6t_1f_8^{-1}f_5^{-1}v_2^{-1}v_4^{-1}v_5^{-1}v_6^{-1}v_7^{-1} = 1,$$
$$t_2^{-1}f_6t_2f_7^{-1} = a_1^{-1}f_6a_1f_3^{-1}f_6^{-1}f_2^{-1} = r_1^{-1}f_7r_1f_5^{-1} = 1,$$
$$r_2^{-1}f_7r_2f_8^{-1}f_6^{-1}f_4^{-1}f_5^{-1}f_4^{-1}v_1^{-1}v_4^{-1}v_5^{-1} = 1,$$
$$t_1^{-1}f_7t_1f_7^{-1}f_4^{-1}f_5^{-1}f_4^{-1}v_1^{-1}v_2^{-1}v_6^{-1}v_7^{-1} = t_2^{-1}f_7t_2f_6^{-1} = 1,$$
$$a_1^{-1}f_7a_1f_4^{-1}f_7^{-1}f_3^{-1} = r_1^{-1}f_8r_1f_8^{-1}f_7^{-1}f_6^{-1} = 1,$$
$$(r_2, f_8^{-1}) = (t_2, f_8^{-1}) = (a_2, f_8^{-1}) = 1,$$
$$t_1^{-1}f_8t_1f_8^{-1}f_7^{-1}f_6^{-1}f_4^{-1}f_5^{-1}f_4^{-1} = a_1^{-1}f_8a_1f_8^{-1}f_2^{-1} = 1,$$
$$(f_1f_2)^2 = (f_1f_3)^2 = (f_2f_3)^2 = (f_1f_4)^2 = (f_2f_4)^2 = (f_3f_4)^2 = 1,$$
$$f_1f_5f_1f_5v_1^{-1}v_4^{-1}v_5^{-1} = f_2f_5f_2f_5v_2^{-1}v_4^{-1}v_5^{-1}v_6^{-1}v_7^{-1} = 1,$$
$$f_3f_5f_3f_5v_1^{-1}v_2^{-1}v_6^{-1}v_7^{-1} = f_1f_6f_1f_6v_1^{-1}v_4^{-1}v_5^{-1} = 1,$$

$$(f_5f_6)^2 = (f_5f_7)^2 = (f_6f_7)^2 = (f_2f_8)^2 = 1,$$
$$f_4f_8f_4f_8v_2^{-1}v_4^{-1}v_5^{-1}v_6^{-1}v_7^{-1} = 1,$$
$$(f_5f_8)^2 = (f_6f_8)^2 = (f_7f_8)^2 = 1,$$
$$f_1f_7f_2f_1f_7f_2v_1^{-1}v_4^{-1}v_5^{-1} = 1,$$
$$(f_2f_6f_3)^2 = f_1f_7f_3f_1f_7f_3v_1^{-1}v_2^{-1}v_6^{-1}v_7^{-1} = 1,$$
$$f_1f_8f_3f_1f_8f_3v_2^{-1}v_4^{-1}v_5^{-1}v_6^{-1}v_7^{-1} = 1,$$
$$f_2f_6f_4f_2f_6f_4v_2^{-1}v_4^{-1}v_5^{-1}v_6^{-1}v_7^{-1} = 1,$$
$$f_1f_7f_4f_1f_7f_4v_1^{-1}v_2^{-1}v_6^{-1}v_7^{-1} = f_4f_6f_5f_4f_5f_6 = 1,$$
$$f_4f_7f_5f_4f_5f_7v_2^{-1}v_4^{-1}v_5^{-1}v_6^{-1}v_7^{-1} = f_3f_8f_6f_3f_6f_8 = 1.$$

(m) *H_2 has an extension H_1 by $Z(Q) = \langle z \rangle$ which is isomorphic to the finitely presented group*

$$H_1 = \langle a_1, a_2, r_1, r_2, t_1, t_2, f_j, q_i \mid 1 \leq i, j \leq 8 \rangle$$

with center $Z(H_1) = \langle z = q_6^2 \rangle$ having the set $\mathcal{R}(H_1)$ of defining relations consisting of $\mathcal{R}(U)$ and the following relations:

$$a_1^3 = 1, \quad a_1^{a_2} = a_1^2, \quad (a_1, r_1) = (a_1, r_2) = (a_1, t_1) = (a_1, t_2) = 1,$$
$$a_1^{-1}q_1a_1(q_1q_2q_4q_5q_6q_8)^{-1} = a_1^{-1}q_2a_1(q_4q_5q_6q_8)^{-1} = 1,$$
$$a_1^{-1}q_3a_1(q_2q_3q_4q_5q_6q_8)^{-1}q_6^2 = a_1^{-1}q_4a_1(q_2q_6)^{-1} = 1,$$
$$a_1^{-1}q_5a_1(q_8)^{-1} = a_1^{-1}q_6a_1(q_2q_5q_8)^{-1} = 1,$$
$$a_1^{-1}q_7a_1(q_2q_5q_6q_7q_8)^{-1}q_6^2 = a_1^{-1}q_8a_1(q_5q_8)^{-1}q_6^2 = 1,$$
$$a_1^{-1}f_1a_1f_6^{-1}f_5^{-1}(q_5)^{-1}q_6^2 = 1,$$
$$a_1^{-1}f_2a_1f_8^{-1}(q_2q_4q_5q_6q_7)^{-1}q_6^2 = 1,$$
$$a_1^{-1}f_3a_1f_8^{-1}f_4^{-1}f_6^{-1}f_4^{-1}(q_1q_2q_4q_6q_7)^{-1}q_6^2 = 1,$$
$$a_1^{-1}f_4a_1f_8^{-1}f_7^{-1}f_6^{-1}(q_1q_2q_4q_6q_7)^{-1}q_6^2 = 1,$$
$$a_1^{-1}f_5a_1f_8^{-1}f_3^{-1}f_8^{-1}f_5^{-1}f_2^{-1}f_1^{-1}(q_2q_4q_5q_6q_8)^{-1}q_6^2 = 1,$$
$$a_1^{-1}f_6a_1f_3^{-1}f_6^{-1}f_2^{-1}(q_1q_4q_7)^{-1}q_6^2 = 1,$$
$$a_1^{-1}f_7a_1f_4^{-1}f_7^{-1}f_3^{-1}(q_1q_4q_5)^{-1} = 1,$$
$$a_1^{-1}f_8a_1f_8^{-1}f_2^{-1}(q_2q_4q_5q_6q_7)^{-1}q_6^2 = 1.$$

In particular, $H_1 \leftarrow U \rightarrow D$ is an amalgam of H_1 and D with common subgroup U.

(n) *H_1 and U have faithful permutation representations PH_1 and PU of degrees 6144 and 2048, respectively, with common stabilizer Y generated by r_2r_1, $(t_2q_6)^2$, $(q_6r_1f_4)^4$, $(q_6f_8r_1)^4$, $(r_1^2q_6f_8)^3$, $(r_1t_1t_2f_4)^3$ and $(r_1t_1q_6f_8)^2$.*

Proof (a) By Lemma 7.1.1(h) and (i) the split extension $E = E_1$ of M_{24} has a faithful permutation representation PE of degree 2048 and a unique conjugacy class of involutions of highest defect. It is represented by $z = v_1$. Its centralizer $D = C_E(z)$ has order $2^{21} \cdot 3^2 \cdot 5 \cdot 7$. Using the faithful permutation representation PE and MAGMA it has been checked that the three generators x, y and e of E given in the statement satisfy $D = C_G(z) = \langle x, y\rangle$ and $E = \langle D, e\rangle$.

(b) Using Kratzer's Algorithm 5.3.18 of [92], the faithful permutation representation PE and MAGMA it has been checked that E has 80 conjugacy classes and that $z = a^2 = (xy^3)^{14}$.

(c) The restriction of PE to $D = \langle x, y\rangle$ is a faithful permutation representation of D. Using it and the MAGMA command `NormalSubgroups (PD)` it follows that D has a unique non-abelian normal subgroup Q of order 2^9 with center $Z(Q) = \langle z\rangle$ of order 2. Furthermore, Q is extra-special. The short words in x and y for its eight generators q_i, $1 \le i \le 8$, have been found by means of the program `GetShortGensHKim(PD, PQ)`. It turns out that all eight generators q_i satisfy the set $\mathcal{R}(Q)$ of defining relations given in (c).

(d) Since Q is a characteristic subgroup of D its center $Z(Q)$ is normal in D. Let $\alpha : D \to D_1 = D/Z(Q)$ be the canonical epimorphism with kernel $ker(\alpha) = Z(Q)$. Let $V = \alpha(Q)$ and let $v_i = \alpha(q_i) \in D_1$ for $i = 1, 2, ..., 8$. In particular, $\mathcal{B} = \{v_i | 1 \le i \le 8\}$ is a basis of the vector space V over $F = GF(2)$. Calculating now $\alpha(q_i^x) \in V$ and $\alpha(q_i^y) \in V$ as short words in the elements of the basis $\mathcal{B}$ one obtains the two matrices Mx and My of x and y in $\mathrm{GL}_8(2)$ stated in the assertion.

Clearly, $D_1/V \cong D/Q$. Applying the MAGMA command `HasComplement(D_1,V)` in the permutation representation of D_1 it follows that D_1 is a non-split extension of D/Q by V.

(e) Let $MD = \langle Mx, My\rangle$ be the subgroup of $\mathrm{GL}_8(2)$ generated by these two matrices. Then the map $\varphi : D \to MD$ defined by $x \to Mx$ and $y \to My$ is an epimorphism from D onto MD with kernel Q by (c). It is well known that $\mathrm{GL}_8(2)$ has a faithful permutation representation $PL8$ of degree 255 whose stabilizer is a parabolic subgroup of order $2^7 | \mathrm{GL}_7(2)|$. Using it and MAGMA one checks that the Fitting subgroup MY of MD has order 2^6, and that $MC = \langle \varphi(x), \varphi(c), \rangle$ is a complement of MY in MD, where $c = (x^2yx^2yxyxy^2xy)^3$ has order 2. Again the six generators $\varphi(m_j)$ of MY stated in assertion (e) have been found by means of the program `GetShortGensHKim(PE, PD)`. Another application of MAGMA yields that $MC \cong A_8$. Using the faithful permutation representation PE and the MAGMA command `NormalSubgroups(PD)` again one sees that D has a unique normal subgroup M of order 2^{15}. It is generated by Q and the six elements m_j given in the statement. Furthermore, an application of MAGMA command `HasComplement(PD,M)` yields that $C = \langle x, c\rangle$ is a complement of M in D. Hence $MY = \varphi(M)$ and $MC = \varphi(C) \cong C \cong A_8$. In particular, D has a faithful permutation representation of degree 2^{15} with stabilizer C.

(f) As $ker(\alpha) = Z(Q) \le Q = ker(\varphi)$ it follows that $\alpha(C)$ is also a complement of $\alpha(M) = \langle V, \alpha(m_j) \mid 1 \le j \le 6\rangle$ in D_1. In particular, D_1 has a faithful permutation representation of degree 2^{14}. Using it one verifies that $C_{D_1}(V) = V$.

Clearly, $D_1/V \cong D/Q$. Applying the MAGMA command `HasComplement(D_1,V)` in the permutation representation PD_1 it follows that D_1 is a non-split extension of D/Q by V.

(g) Since $|MD| = 2^8 \cdot 3^2 \cdot 5 \cdot 7$ is fairly small, MAGMA is able to provide the following set $\mathcal{R}(MD)$ of defining relations of $MD = \langle Mx, My\rangle$ by means of its command `FPMD := FPGroup(sub<GL(12,2)|Mx,My>)`:

$$\begin{aligned}
&x_1^7 = y_1^4 = 1,\\
&(y_1x_1^{-1})^7 = 1, \quad (x_1y_1x_1^{-1}y_1x_1)^4 = 1, \quad (x_1y_1x_1^{-1}y_1x_1^{-1}y_1)^4 = 1,\\
&(x_1y_1x_1^{-1}y_1)^6 = 1, \quad (x_1y_1x_1^{-1}y_1x_1^{-1}y_1x_1)^4 = 1,\\
&x_1^{-1}y_1x_1^{-2}y_1x_1^2y_1x_1y_1x_1^{-1}y_1x_1^3y_1x_1^{-1}y_1x_1y_1x_1^2y_1x_1^{-2}y_1x_1^{-1}y_1 = 1,\\
&x_1^{-2}y_1x_1^{-1}y_1x_1y_1x_1^{-2}y_1x_1^{-1}y_1x_1^{-3}y_1x_1^2y_1x_1^2y_1x_1y_1x_1^{-1}y_1x_1^{-1}y_1x_1^{-1} = 1,\\
&y_1x_1y_1x_1^{-1}y_1x_1^{-2}y_1x_1^2y_1x_1^{-1}y_1x_1^{-2}y_1x_1y_1x_1^{-2}y_1x_1^2y_1x_1^{-2}y_1x_1^2 = 1,
\end{aligned}$$

where $x_1 = \varphi(x)$ and $y_1 = \varphi(x)$. As $ker(\varphi) = Q$ all equations of $\mathcal{R}(MD)$ can be lifted to D by replacing x_1 and y_1 by x and y, respectively, and calculating the right hand side of these equations in x and y as words in the generators of q_i of Q. Thus we obtained the printed nine relations of statement (g).

Since Q is normal in D, all $q_i^x, q_i^y \in Q$. Hence an application of the program `LookupWord(PD, q_i^d)` with $d \in \{x, y\}$ yields all the relations of (g) involving q_i^x and q_i^y for $1 \le i \le 8$. Now the set $\mathcal{R}(Q)$ stated in (c) completes the set $\mathcal{R}(D)$ of defining relations of D.

(h) All statements of this assertion have been obtained computationally using the faithful permutation representations $PL8$ of $\mathrm{GL}_8(2)$ and PE, MAGMA and the program `GetShortGensHKim(PE, PD)`.

(i) It is well known that $\mathrm{GL}_8(2)$ has a faithful permutation representation $PL8$ of degree $2^8 - 1 = 255$ with stabilizer equal to a maximal parabolic subgroup. Let ψ be the group isomorphism between $\mathrm{GL}_8(2)$

and its image $PL8$ in the symmetric group of degree 255. Then ψ induces an embedding of MD into $PL8$ and $PX = C_{PL8}(\psi(Mt))$ is the image of $MX = C = \mathrm{GL}_8(2)(Mt)$ in $PL8$. Using then the faithful permutation representation $PL8$ and the MAGMA command

```
exists(h){x:x in PX|Order(sub<PL8|MD, x>) eq 174182400}
```

we obtain the matrix $Mh \in \mathrm{GL}_8(2)$ of order 3 given in the statement.

It also has been checked that $MO = \langle MD, Mh\rangle$ has order $2^{12} \cdot 3^5 \cdot 5^2 \cdot 7$ and that $C_{MO}(Mt) = \langle C_{MD}(Mt), Mh\rangle$ has order $2^{12} \cdot 3^3$.

From (e) and (h) it follows immediately that

$$C_{MO}(Mt) = \langle Mm_j, Mf_i, Mr_1, Mr_2, Mt_1, Mt_2, Mh \mid 1 \le i \le 4, 1 \le j \le 6\rangle.$$

The isomorphism $MO \cong \mathrm{O}_8^+(2)$ has been found by an application of the MAGMA command `CompositionFactors(MO)`.

(j) Another application of MAGMA yields that the Fitting subgroup MY_3 of $MH_3 = C_{MO}(Mt)$ is extra-special of order 2^9 and that it is generated by the matrices given in the statement. The set $\mathcal{R}(Y_3)$ of defining relations has been calculated by means of MAGMA. In particular $(Mf_4 Mf_7)^2$ is the center of MY_3.

Let $MR = \langle r_1, r_2, t_1, t_2\rangle$ and $T_1 = \langle r_1, t_1\rangle$. The third generator Ma_1 of the Sylow 3-subgroup MT of MH_3 has been found by means of the command

```
exists(a){x:x in C_{MH_3}(MR)|Order(sub<MR | T_1, x >)
  eq 3^3}.
```

A similar existence command provided the matrix Ma_2 in $C_{MD}(MR)$ satisfying $Ma_1^{Ma_2} = Ma_1^2$. The word of the element $a_2 \in D$ satisfying $\varphi(a_2) = Ma_2$ has been obtained by an application of `LookupWord(PD, a_2)` command. Using the faithful permutation representation $PL8$ it has been checked that $MK_3 = \langle MR, Ma_1, Ma_2\rangle$ is a complement of MY_3 in MH_3. Using the structure of the subgroups MK_3 and its action on the normal subgroup MY_3 of MH_3 it is easy to verify all the relations of the finitely presented group H_3 in the matrix group MH_3. Another application of MAGMA yields that both groups MH_3 and H_3 are isomorphic.

(k) By the previous statements we know that the generators of

$$U = \langle a_2, r_1, r_2, t_1, t_2, f_j, q_i \mid 1 \le i, j \le 8\rangle$$

are words in the generators x and y of D. Therefore U is a subgroup of D containing the extra-special normal subgroup Q of order 2^9. Let $P = \langle Q, f_j \mid 1 \le j \le 8\rangle$. Then by (j) $P/Q \cong MY_3$ is extra-special of order 2^9 as well. Using the faithful permutation representation PE of D it has been verified that $K = \langle a_2, r_1, r_2, t_1, t_2\rangle$ is a complement of P in U. Furthermore, it has been used together with MAGMA to determine the set $\mathcal{R}(U)$ of defining relations of U given in the statement. The stabilizer Y of the faithful permutation representation PU of degree 2048 given in the statement has been obtained by application of the MAGMA command `BasicStabilizer(U,2)` to the finitely generated group U with the 21 generators given above.

(l) The subgroup $MH_3 \cong H_3$ of $\mathrm{GL}_8(2)$ has a faithful permutation representation inside $PL8$. In view of its presentation given in (j) it satisfies all hypotheses of Holt's Algorithm 7.4.5 of [92] implemented in MAGMA. Its application in MAGMA yields that the second cohomological dimension $dim_F(H^2(H_3, V)) = 1$, where $V = Q/Z(Q) = \langle v_i \mid 1 \le i \le 8\rangle$. Since $D/Q \cong MD$ by (e), and D_1 is a non-split extension of D/Q by V by (f), we constructed the unique non-split extension H_2 of H_3 by V. The presentation of the non-split extension H_2 has also been calculated by means of Holt's Algorithm using the matrices of the generators of H_3 and the presentation of H_3 given in (j).

Furthermore, $U/Z(Q)$ is isomorphic to the subgroup

$$U_2 = \langle a_2, r_1, r_2, t_1, t_2, f_j, v_i \mid 1 \le i \le 8, 1 \le j \le 6\rangle$$

of H_2. In particular, U_2 has a presentation with set $\mathcal{R}(U_2)$ consisting of all relations of $\mathcal{R}(H_2)$ not involving a_1. All these relations can easily be lifted to U by replacing the generators v_i by q_i, for $1 \le i \le 12$, and using the program `GetShortGensHKim(PE, PD)` to check whether the lifted word on the left hand side of a relation of U_2 equals either 1 or z in $Z(Q)$. Hence $\mathcal{R}(U)$ consists of all these lifted relations and $\mathcal{R}(Q)$.

(m) Since MAGMA was not able to calculate the second cohomology group $H^2(H_2, 1_{FH_2})$ with coefficients in the trivial FH_2-module 1_{FH_2}, we constructed the central extension H_1 of H_2 by the center $Z(Q) = \langle z = q_6^2\rangle$ as follows.

Let H_1 be any central extension of the finitely presented group H_2 by $Z(Q)$. Then the finitely presented group U constructed in (k) may be considered to be a subgroup of H_1 containing the common normal subgroup $Q = \langle q_i \mid 1 \le i \le 8\rangle$. Hence $\mathcal{R}(U)$ is a subset of the set $\mathcal{R}(H_1)$ of defining relations of H_1. Furthermore, by (k), (l), (j) and Theorem 1.4.5 of [92], $K_3 = \langle r_1, r_2, t_1, t_2, a_1, a_2\rangle$ is isomorphic to a complement of the normal subgroup $P = \langle Q, f_j \mid 1 \leq j \leq 8 \rangle$ of H_1 of order $|P| = 2^{18}$. Thus the set $\mathcal{R}(K_3)$ of defining relations of K_3 is part of $\mathcal{R}(H_1)$. Therefore it remains to lift those 16 relations of H_2 given in (1) that involve a_1 and are different from the set of defining relations $\mathcal{R}(K_3)$.

For that purpose a faithful permutation representation PU of U of degree 2048 has been constructed by means of the faithful permutation representation of D described in (e). Its stabilizer Y is generated by the seven elements of U stated in assertion (k).

In particular, Y is a subgroup of H_1. Therefore SU is also a stabilizer of a faithful permutation representation PH_1 of degree 6144 of H_1 because $|H_1 : U| = 3$. Any possible central extension H_1 of H_2 by $Z(Q)$ has a set $\mathcal{R}(H_1)$ of defining relations consisting of $\mathcal{R}(U)$, $\mathcal{R}(K_3)$ and one of the following 2^{16} sets $\mathcal{R}(i)$ of relations:

$$
\begin{aligned}
&a_1^{-1}q_1a_1(q_1q_2q_4q_5q_6q_8)^{-1}[z[i][1]],\\
&a_1^{-1}q_2a_1(q_4q_5q_6q_8)^{-1}[z[i][2]],\\
&a_1^{-1}q_3a_1(q_2q_3q_4q_5q_6q_8)^{-1}[z[i][3]],\\
&a_1^{-1}q_4a_1(q_2q_6)^{-1}[z[i][4]],\\
&a_1^{-1}q_5a_1(q_8)^{-1}[z[i][5]],\\
&a_1^{-1}q_6a_1(q_2q_5q_8)^{-1}[z[i][6]],\\
&a_1^{-1}q_7a_1(q_2q_5q_6q_7q_8)^{-1}[z[i][7]],\\
&a_1^{-1}q_8a_1(q_5q_8)^{-1}[z[i][8]],\\
&a_1^{-1}f_1a_1f_6^{-1}f_5^{-1}(q_5)^{-1}[z[i][9]],\\
&a_1^{-1}f_2a_1f_8^{-1}(q_2q_4q_5q_6q_7)^{-1}[z[i][10]],\\
&a_1^{-1}f_3a_1f_8^{-1}f_4^{-1}f_6^{-1}f_4^{-1}(q_1q_2q_4q_6q_7)^{-1}[z[i][11]],\\
&a_1^{-1}f_4a_1f_8^{-1}f_7^{-1}f_6^{-1}(q_1q_2q_4q_6q_7)^{-1}[z[i][12]],\\
&a_1^{-1}f_5a_1f_8^{-1}f_3^{-1}f_8^{-1}f_5^{-1}f_2^{-1}f_1^{-1}(q_2q_4q_5q_6q_8)^{-1}[z[i][13]],\\
&a_1^{-1}f_6a_1f_3^{-1}f_6^{-1}f_2^{-1}(q_1q_4q_7)^{-1}[z[i][14]],\\
&a_1^{-1}f_7a_1f_4^{-1}f_7^{-1}f_3^{-1}(q_1q_4q_5)^{-1}[z[i][15]],\\
&a_1^{-1}f_8a_1f_8^{-1}f_2^{-1}(q_2q_4q_5q_6q_7)^{-1}[z[i][16]],
\end{aligned}
$$

where $z[i]$ runs through all possible sequences $z[i]$ whose entries $[z[i][j]]$, $1 \leq j \leq 16$, are either 1 or z in $Z(Q) = \langle q_6^2 \rangle$. If $\mathcal{R}(H_1(i)) = \mathcal{R}(U) \cup R(S_3) \cup \mathcal{R}(i)$ describes a set of defining relations of a central extension $H_1(i)$ of H_2 by $Z(Q)$, then Y is a stabilizer of a faithful permutation representation $PH_1(i)$ of degree 6144.

Using MAGMA we checked the degree of $PH_1(i)$ for each of the 2^{16} presentations $H_1(i)$. It turned out that there is exactly one index i where the sequence $z[i]$ provides a central extension $H_1(i)$ of H_2 by $Z(Q)$ having a faithful permutation representation of degree 6144 with stabilizer Y. Using MAGMA it has been checked that the Sylow 2-subgroups of this $H_1(i)$ and D are isomorphic. Thus H_1 is uniquely determined. Its presentation is given in the statement.

(n) The generators of the common stabilizer Y in H_1 and U have been constructed in the proof of (k). This completes the proof. □

Proposition 7.2.2 *Keep the notations of Lemma 7.1.1 and Proposition 7.2.1. Let* $E = \langle x, y, e\rangle$ *be the split extension of* M_{24} *by its simple module* V_1 *of dimension* 11 *over* $F = GF(2)$ *and let* $D = \langle x, y\rangle$. *Let* U *be the subgroup of* D *constructed in Proposition 7.2.1(k). Let* $H_1 \leftarrow U \rightarrow D$ *be the amalgam constructed in Proposition 7.2.1, where* $H_1 = \langle U, a_1\rangle$. *Let* $\sigma : U \rightarrow D$ *denote the corresponding monomorphism of* U *into* D. *Then the following statements hold.*

(a) $U = \langle y, j, k\rangle$ *and* $H_1 = \langle y, j, k, h\rangle$, *where* $y = a_2q_7r_1t_1t_2q_7r_2q_7$, $j = r_2f_4$, $k = t_2q_6$, $h = a_1$, $j_1 = y\,j$, $k_1 = y\,k$ *and* $h_1 = yh$.

(b) *The Goldschmidt index of the amalgam* $H_1 \leftarrow U \rightarrow D$ *is* 1.

(c) *The group* H_1 *has* 376 *conjugacy classes classified in Table* `DVD.1.5.2`. *Its character Table* `DVD.1.5.7` *is stored on the accompanying DVD.*

(d) *The group* U *has* 382 *conjugacy classes classified in Table* `DVD.1.5.3`. *Its character Table* `DVD.1.5.8` *is stored on the accompanying DVD.*

(e) *The* 155 *conjugacy classes of* D *and the* 80 *conjugacy classes of* E *are classified in Tables* `DVD.1.5.1` *and 7.5.2, respectively.*

(f) *The character table of* D *is given in Table* `DVD.1.5.6`.

(g) *There is exactly one compatible pair* $(\chi, \tau) \in mf\,char_{\mathbb{C}}(H_1) \times mf\,char_{\mathbb{C}}(D)$ *of degree* 128 *of the groups* $H_1 = \langle U, h\rangle$ *and* $D = \langle U, x\rangle$: $(\chi_{\mathbf{239}}, \tau_{\mathbf{39}})$, *with common restriction* $\chi_{|U} = \tau_{|U} = \psi_{\mathbf{270}}$, *where irreducible characters with bold face indices denote faithful irreducible characters.*

(h) *Let* $\mathfrak{V}$ *and* $\mathfrak{W}$ *be the up to isomorphism uniquely determined faithful irreducible* 128*-dimensional modules of* H_1 *and* D *over* $F = GF(23)$ *corresponding to the compatible pair* χ *and* τ, *respectively.*

Let $\kappa_{\mathfrak{V}} : H_1 \to \mathrm{GL}_{128}(23)$ *and* $\kappa_{\mathfrak{W}} : D \to \mathrm{GL}_{128}(23)$ *be the representations of* H_1 *and* D *afforded by the modules* $\mathfrak{V}$ *and* $\mathfrak{W}$, *respectively.*

Let $\mathfrak{h} = \kappa_{\mathfrak{V}}(h)$, $\mathfrak{j} = \kappa_{\mathfrak{V}}(j)$, $\mathfrak{k} = \kappa_{\mathfrak{V}}(k)$ *and* $\mathfrak{y} = \kappa_{\mathfrak{B}}(y)$ *in* $\kappa_{\mathfrak{V}}(H_1) \leq \mathrm{GL}_{128}(23)$. *Then the following assertions hold.*

(1) $\mathfrak{V}_{|\mathfrak{U}} \cong \mathfrak{W}_{|\mathfrak{U}}$, *and there is a transformation matrix* $\mathcal{T} \in \mathrm{GL}_{128}(23)$ *such that*

$$\mathfrak{j} = \mathcal{T}^{-1}\kappa_{\mathfrak{W}}(\sigma(j))\mathcal{T},\ \mathfrak{k} = \mathcal{T}^{-1}\kappa_{\mathfrak{W}}(\sigma(k))\mathcal{T}$$
$$\text{and}\quad \mathfrak{y} = \mathcal{T}^{-1}\kappa_{\mathfrak{W}}(\sigma(y))\mathcal{T}.$$

Let $\mathfrak{x} = \mathcal{T}^{-1}\kappa_{\mathfrak{W}}(x)\mathcal{T}$ *in* $\mathrm{GL}_{128}(23)$.

(2) *Let* $\mathfrak{H} = \langle \mathfrak{h}, \mathfrak{y}, \mathfrak{x} \rangle$. *Its generating matrices can be recalculated from the stored generating matrices of the simple group* $\mathfrak{G}$ *constructed in Theorem 7.3.2.*

(i) $\mathfrak{H}$ *is isomorphic to the finitely presented group* $H = \langle h_i | 1 \leq i \leq 8 \rangle$ *with the following set* $\mathcal{R}(H)$ *of defining relations:*

$$
\begin{aligned}
&h_1^4 = h_2^4 = h_3^4 = h_4^3 = h_6^7 = h_7^6 = h_8^6 = 1,\\
&(h_7^{-1}h_6^{-1})^3 = 1,\\
&h_8^{-2}h_6^{-1}h_7^{-1}h_6h_7^{-1}h_8^{-1}h_7 = 1,\\
&h_7h_8h_6^{-1}h_7^{-1}h_6h_7^2h_8^{-1} = 1,\\
&(h_6h_7^{-1}h_8h_7^{-1})^2 = 1, \quad (h_6h_8^{-1})^4 = 1,\\
&(h_7^{-1}h_8^{-1}h_6h_8^{-1})^2 = 1, \quad (h_7h_8^{-1})^4 = 1,\\
&h_7h_8h_6^{-3}h_8^{-2}h_7^{-1}h_6 = h_6h_8h_7h_6^3h_7h_8h_7h_8^{-1} = 1,\\
&h_1h_6h_7h_6h_7^4h_6h_7h_8^3h_1 = h_6^{-1}h_1^{-1}h_6^2h_7^2h_8h_7h_6^3h_1 = 1,\\
&h_7^{-1}h_1^{-1}h_6^3h_7h_6h_7^2h_6h_7h_8^2h_6^2h_1 = 1,\\
&h_8^{-1}h_1^{-1}h_6^4h_7h_6^5h_7h_6h_8h_7^2h_1 = 1,\\
&h_3h_5h_6h_7h_6h_7^2h_6h_7^3h_8h_1h_5^{-1}h_3 = 1,\\
&h_5h_2^{-2}h_6h_7^2h_6h_7^2h_6^2h_8h_7^2h_2h_1h_6^{-1} = 1,\\
&h_1h_4^{-1}h_2^{-1}h_6h_7h_6^2h_7h_6h_7h_6^2h_8h_1h_2h_1^{-1}h_4^{-1} = 1,\\
&h_2h_5h_2^{-1}h_6^3h_7h_6^2h_7h_6h_7^4h_8h_6h_7h_2^{-1}h_7^{-1} = 1,\\
&h_2^{-1}h_5^{-1}h_2h_6h_7h_6h_7^3h_8^2h_6h_8h_2h_1^{-1}h_6^{-1} = 1,\\
&h_3^{-1}h_8^{-1}h_2h_7h_6h_7^2h_6^2h_7h_8h_6h_1h_2h_5^2 = 1,\\
&h_2^{-1}h_1h_3^{-1}h_6^5h_7^4h_6h_7h_8h_6h_7h_1h_2h_7h_6 = 1,
\end{aligned}
$$

$$h_5h_2^{-1}h_3^{-1}h_7h_6h_7^3h_6^2h_7^2h_8h_6^2h_2h_5h_6^{-1} = 1,$$
$$h_1h_3^{-2}h_6^2h_7^2h_6^4h_7h_8h_1h_3^2h_1 = 1, \quad h_2^{-1}h_3^{-2}h_2h_3^2 = 1,$$
$$h_4^{-1}h_3^{-2}h_6^4h_7h_6h_7h_6^2h_7h_6^4h_8h_4^{-1}h_1 = 1,$$
$$h_6^{-1}h_3^{-2}h_6^2h_7^4h_6h_7h_6h_7h_3^2 = 1,$$
$$h_8h_3^{-2}h_6^4h_8h_6h_8h_3^2h_1 = 1,$$
$$h_1h_4^{-1}h_3^{-1}h_6^5h_7h_6h_7^3h_6^2h_7^2h_8h_1h_3h_4h_1 = 1,$$
$$h_3^{-1}h_4h_3^{-1}h_6^6h_7^3h_6h_1h_4h_1^{-1} = 1,$$
$$h_3h_5^{-1}h_3^{-1}h_6^4h_7^3h_6^3h_3h_5h_3 = 1,$$
$$h_3h_6^{-1}h_3^{-1}h_6^2h_7^3h_6^5h_3h_6h_3 = 1,$$
$$h_1h_6h_3^{-1}h_6^2h_7^2h_6^2h_8h_6h_8h_6h_1h_2h_1^{-1}h_2 = 1,$$
$$h_3h_6h_3^{-1}h_6^5h_7h_6^3h_7^2h_8h_1h_3h_6^{-1}h_3 = 1,$$
$$h_7^{-2}h_3^{-1}h_7^3h_3^{-1}h_7^{-1} = 1,$$
$$h_4h_1h_3h_6^5h_7h_6^4h_7h_6h_7h_6h_8h_1h_3h_4h_1 = 1,$$
$$h_3^{-1}h_4^{-1}h_3^2h_4h_3^{-1} = 1,$$
$$h_1h_6^{-1}h_3h_6^6h_7^4h_6h_7^2h_6h_7h_8h_6h_7h_2^{-1}h_8^{-1}h_5^{-1} = 1,$$
$$h_2h_7h_3h_7^2h_6h_7^2h_6^4h_7h_8h_6h_7h_5^{-1}h_8 = 1,$$
$$h_4^{-1}h_1^{-1}h_4^{-1}h_6^5h_7h_8h_6h_8h_3^2h_7 = 1,$$
$$h_1h_3^{-1}h_4^{-1}h_6^5h_7h_6h_7^3h_6^2h_7^2h_8h_1h_4h_3h_1 = 1,$$
$$h_2^{-1}h_3^{-1}h_4^{-1}h_6h_7h_6^2h_7h_6h_7h_6^2h_8h_1h_4h_2h_3 = 1,$$
$$h_4h_3^{-1}h_4^{-1}h_6^5h_7^3h_6^2h_2^{-1}h_3h_2 = 1,$$
$$h_3h_5^{-1}h_4^{-1}h_7^3h_4h_5h_3 = 1,$$
$$h_2h_1h_4h_6h_7h_6^3h_7^3h_6h_7h_8^2h_3^{-1}h_8^{-1}h_4 = 1,$$
$$h_6h_1^{-1}h_5^{-1}h_6^4h_7^3h_6h_7h_6h_2^{-1}h_3h_5 = 1,$$
$$h_5^{-1}h_2^{-1}h_5^{-1}h_7h_6h_7^3h_6^2h_7h_6h_1h_3^{-1}h_5h_7 = 1,$$
$$h_6h_2^{-1}h_5^{-1}h_6h_7h_6h_7h_6^2h_7h_6^2h_7^2h_1h_2h_8^{-1}h_2 = 1,$$
$$h_8^{-1}h_3h_5^{-1}h_7^4h_6^2h_7^2h_8^2h_6h_3h_7h_5 = 1,$$
$$h_1h_4^{-1}h_5^{-1}h_7^3h_5h_1^{-1}h_4^{-1} = 1,$$
$$h_1h_5^{-2}h_6^4h_7^3h_6h_8h_6h_8h_1h_2^{-1}h_1h_5^{-1} = 1,$$
$$h_2^{-1}h_5^{-2}h_6^6h_7h_8h_7^3h_2h_5^{-1}h_6 = 1,$$
$$h_5h_6^{-1}h_5^{-1}h_7h_6h_7^2h_6h_8^2h_6h_3h_8^{-1}h_1^{-1} = 1,$$
$$h_7h_6^{-1}h_5^{-1}h_7h_6^2h_7^2h_6h_8h_7h_6h_1h_3h_6h_8 = 1,$$
$$h_8h_6^{-1}h_5^{-1}h_6^2h_7^4h_6h_7^2h_6h_7^2h_8^2h_3^{-1}h_1h_7^{-1} = 1,$$

$$h_8^{-1}h_7h_5^{-1}h_6^2h_7^3h_6^4h_7h_6h_8h_7h_8h_1h_3h_8^{-1}h_2^{-1}=1,$$
$$h_3^{-1}h_8h_5^{-1}h_6h_7h_6h_7^4h_6^4h_7h_8h_2h_5^{-1}h_3^{-1}=1,$$
$$h_1h_2h_5h_6^2h_7^4h_6h_7h_6h_7^2h_6^2h_8h_1h_5^{-1}h_1h_2=1,$$
$$h_1^{-1}h_3^{-1}h_5h_6^2h_7^2h_6^2h_7^2h_8h_7h_3h_5h_8^{-1}=1,$$
$$h_2h_4h_5h_6h_7^2h_6h_7^2h_6h_7^2h_5^{-1}h_4^{-1}h_2=1,$$
$$h_3h_5h_4^{-1}h_2^{-1}h_6^5h_7h_6^3h_7^2h_8h_1h_2h_4h_5^{-1}h_3=1,$$
$$h_7^{-2}h_4^{-1}h_2^{-1}h_6h_7^4h_6h_7h_6h_7h_2h_4h_7^{-1}=1,$$
$$h_3h_7h_4^{-1}h_2^{-1}h_6h_7h_6h_7^4h_6h_7h_2h_4h_7^{-1}h_3=1,$$
$$h_2h_6h_4h_2^{-1}h_6^3h_7h_6^5h_7^2h_6h_2h_4^{-1}h_6^{-1}h_2=1,$$
$$h_7h_4^{-1}h_7^{-1}h_2^{-1}h_7h_6^3h_7h_6^3h_7h_8h_1h_2h_4^{-1}h_7^{-1}h_4=1,$$
$$h_2^{-1}h_4h_3^{-1}h_2h_6h_7h_6^2h_7h_6h_7^3h_8^2h_1h_2h_5^{-1}h_1^{-1}h_4^{-1}=1,$$
$$h_3h_8h_4h_2h_6h_7h_6h_7^4h_6h_7h_8^3h_2^{-1}h_4^{-1}h_8^{-1}h_3=1,$$
$$h_1h_6^{-1}h_4^{-1}h_3^{-1}h_6^5h_7h_6h_7^3h_6^2h_7^2h_8h_1h_3h_4h_6h_1=1,$$
$$h_1h_6h_4^{-1}h_3^{-1}h_6^5h_7h_6h_7^3h_6^2h_7^2h_8h_1h_3h_4h_6^{-1}h_1=1,$$
$$h_1h_7^{-1}h_4^{-1}h_3^{-1}h_6^5h_7h_6h_7^3h_6^2h_7^2h_8h_1h_3h_4h_7h_1=1,$$
$$h_1h_7h_4^{-1}h_3^{-1}h_6^5h_7h_6h_7^3h_6^2h_7^2h_8h_1h_3h_4h_7^{-1}h_1=1,$$
$$h_3^{-1}h_7h_4^{-1}h_3^{-1}h_6^2h_7^2h_6^2h_7h_6h_8h_1h_2^2h_4h_7^{-1}=1,$$
$$h_1h_8^{-1}h_4^{-1}h_3^{-1}h_6^5h_7h_6h_7^3h_6^2h_7^2h_8h_1h_3h_4h_8h_1=1,$$
$$h_1h_8h_4^{-1}h_3^{-1}h_6^5h_7h_6h_7^3h_6^2h_7^2h_8h_1h_3h_4h_8^{-1}h_1=1,$$
$$h_6h_5^{-1}h_1^{-1}h_4^{-1}h_6h_7h_6^5h_7^2h_6h_8^2h_6h_1h_2h_4^{-1}h_5^{-1}h_1^{-1}=1,$$
$$h_3^{-1}h_8^{-1}h_1h_4^{-1}h_6^4h_7h_6h_7h_6^2h_7h_6^4h_8h_4^{-1}h_8h_3^{-1}=1,$$
$$h_3^{-1}h_7^{-1}h_3^{-1}h_4^{-1}h_6h_7h_6h_7^4h_6h_7h_3^{-1}h_4h_7h_3^{-1}=1,$$
$$h_4^{-1}h_7^{-1}h_3h_4^{-1}h_6^2h_7^2h_6^5h_7h_1h_2h_6^{-1}h_4h_7=1,$$
$$h_1^{-2}h_5^{-1}h_4^{-1}h_6^5h_7h_6h_7^3h_6^2h_7^2h_8h_1h_4h_1^2h_5=1,$$
$$h_1h_2^{-1}h_5h_4^{-1}h_6^2h_7h_6^5h_7h_6^4h_7h_4h_5^{-1}h_2h_1=1,$$
$$h_1h_4h_5h_4^{-1}h_6^4h_7h_6h_7h_6^2h_7h_6^4h_8h_1h_4h_5^{-1}h_1^{-1}h_4=1,$$
$$h_1h_4^{-1}h_6^{-1}h_4^{-1}h_6^4h_7h_6h_7h_6^2h_7h_6^4h_8h_1h_4h_6h_4h_1=1,$$
$$h_1h_4h_6^{-1}h_4^{-1}h_6^5h_7h_6^4h_7h_6h_7h_6h_8h_1h_4h_6h_4^{-1}h_1=1,$$
$$h_6h_4h_6^{-1}h_4^{-1}h_6h_7^4h_6h_7h_8^2h_6h_2^{-1}h_4^{-1}h_2^{-1}h_4^{-1}=1,$$
$$h_6^{-1}h_2^{-1}h_6h_4^{-1}h_7^4h_6^2h_7h_6h_8^2h_6h_8h_1h_2h_4^{-1}h_3h_1^{-1}=1,$$
$$h_6^{-1}h_3^{-1}h_6h_4^{-1}h_6^4h_7^4h_8h_6h_7h_3^{-1}h_6h_4^{-1}h_1^{-1}=1,$$
$$h_3h_5^{-1}h_6h_4^{-1}h_7^3h_4h_6^{-1}h_5h_3=1,$$

$$
\begin{aligned}
&h_1h_3^{-1}h_7^{-1}h_4^{-1}h_6^2h_7h_6^5h_7h_6^4h_7h_4h_7h_3h_1 = 1,\\
&h_6h_1h_7h_4^{-1}h_6h_7h_6h_7^3h_6h_8h_7^2h_8h_4h_5h_2h_3 = 1,\\
&h_2^{-2}h_7h_4^{-1}h_6^2h_7h_6h_7h_6^2h_7h_6^2h_4h_7^{-1}h_3 = 1,\\
&h_3^{-1}h_2^{-1}h_7h_4^{-1}h_6^3h_7h_6^3h_7h_6^4h_7h_8h_1h_4h_7^{-1}h_2^{-1} = 1,\\
&h_1h_2^{-1}h_8h_4^{-1}h_6^2h_7h_6^5h_7h_6^4h_7h_4h_8^{-1}h_2h_1 = 1,\\
&h_7^{-1}h_2^{-1}h_1^{-1}h_4h_6^4h_8h_7^3h_2h_4h_7h_3^{-1} = 1,\\
&h_7^{-1}h_5^{-1}h_1^{-1}h_4h_6^6h_7h_6h_7^4h_6h_7h_6^2h_1h_4h_5h_7^{-2} = 1,\\
&h_4h_6h_1h_4h_6^6h_7h_6h_7^3h_6h_7h_6h_7h_8h_6h_4h_7^{-1}h_8^{-1}h_3 = 1,\\
&h_3^{-1}h_8h_2^{-1}h_4h_7h_6^4h_7h_6^4h_8h_6h_2h_4^{-1}h_3h_8^{-1} = 1,\\
&h_6^{-1}h_5h_3^{-1}h_4h_7^2h_6^5h_7h_8h_7h_6h_3^{-1}h_7^{-1}h_4^{-1}h_5 = 1,\\
&h_6h_5h_3h_4h_6h_7h_6^4h_7^2h_6h_7^2h_8h_6h_2^{-1}h_4^{-1}h_5^{-1}h_8 = 1,\\
&h_3h_1^{-1}h_5^{-1}h_4h_6^5h_7h_6h_7h_6^4h_7h_6^2h_8h_1h_4^{-1}h_5h_1h_3 = 1,\\
&h_5h_1h_5^{-1}h_4h_6h_7h_6^2h_7h_6^3h_7^2h_8h_7h_8h_1h_3h_7h_3h_4^{-1} = 1,\\
&h_1h_2h_5^{-1}h_4h_6^5h_7^3h_6^2h_4^{-1}h_5h_2^{-1}h_1 = 1,\\
&h_2^{-1}h_3^{-1}h_5^{-1}h_4h_6^3h_7h_6^3h_7h_6^3h_7h_4^{-1}h_5h_2h_3^{-1} = 1,\\
&h_5^{-1}h_2h_6^{-1}h_4h_6^3h_7^3h_6h_8h_7^3h_4h_2h_3h_8 = 1,\\
&h_2^{-1}h_3^{-1}h_6^{-1}h_4h_6^3h_7h_6^3h_7h_6^3h_7h_4^{-1}h_6h_2h_3^{-1} = 1,\\
&h_2h_3^{-1}h_6^{-1}h_4h_6h_7^4h_6h_7h_6h_7h_4^{-1}h_6h_2^{-1}h_3^{-1} = 1,\\
&h_5h_3^{-1}h_6^{-1}h_4h_7h_6h_7h_6^4h_7h_6^3h_1h_4^{-1}h_5h_3^{-1}h_6^{-1} = 1,\\
&h_2h_3^{-1}h_6h_4h_6^2h_7^3h_6^5h_4^{-1}h_6^{-1}h_2^{-1}h_3 = 1,\\
&h_6^{-1}h_3^{-1}h_6h_4h_6^5h_7h_6h_7^2h_6^2h_7h_6^2h_8h_7h_3h_2h_4^{-1}h_2^{-1} = 1,\\
&h_6h_1h_8^{-1}h_4h_6h_7^2h_6h_7h_6h_8h_6^2h_1h_4^{-1}h_5^{-1}h_8h_2^{-1} = 1,\\
&h_1h_5h_8^{-1}h_4h_6^5h_7^3h_6^2h_4^{-1}h_8h_5^{-1}h_1 = 1,\\
&h_1h_5h_4^{-1}h_5^{-1}h_6h_7h_6h_7^4h_6h_7h_8^3h_5h_4h_5^{-1}h_1 = 1,\\
&h_3h_8h_4^{-1}h_5^{-1}h_6^5h_7h_6h_7^4h_6h_7h_6^3h_5h_4h_8^{-1}h_3 = 1,\\
&h_2^{-1}h_5^{-1}h_4^{-1}h_1^{-1}h_2^{-1}h_6^4h_7^2h_6h_7^2h_6h_7h_8h_6h_4h_6^{-1}h_3^{-1}h_8 = 1,\\
&h_3h_6h_4^{-1}h_1^{-1}h_2^{-1}h_6^5h_7h_6h_7h_6^2h_7h_6^3h_8h_4^{-1}h_5h_2h_6 = 1.
\end{aligned}
$$

(j) *H has a faithful permutation representation of degree* 276480 *with stabilizer* $\langle h_5, (h_2^3h_5^2h_2h_5)^6, (h_2^2h_5^2h_2^2h_5h_2)^4, (h_2^2h_5h_2^3h_5^3)^6\rangle$.

(k) *Each Sylow 2-subgroup S of H has a unique maximal elementary abelian normal subgroup V of order 2^{11} and $N_H(V) \cong D = C_E(z)$.*

Proof (a) The two generators j and k of U have been obtained by means of MAGMA and the faithful permutation presentation PU of degree 2048 constructed in Proposition 7.2.1(k).

(b) The Goldschmidt index has been calculated by means of Kratzer's Algorithm 7.1.10 of [92], the faithful permutation representations PH_1, PU, PD and MAGMA.

(c), (d) and (e) The systems of representatives of the conjugacy classes of H_1, U and D have been calculated by means of PH_1, PU and PD, MAGMA and Kratzer's Algorithm 5.3.18 of [92].

(f) The three character tables have been obtained by means of MAGMA using the faithful permutation representations of H_1, U and D. The character tables of H_1, D and U are stated in the accompanying DVD.

(g) Using MAGMA and the faithful permutation representations again we have determined the fusion of the classes of U in H and $U(D) = \sigma(U)$ in D. A MAGMA calculation employing Kratzer's Algorithm 7.3.10 of [92] shows then that the amalgam $H_1 \leftarrow U \rightarrow D$ has a unique compatible pair (χ, τ) of degree 128. It is stated in the assertion.

(h) In order to construct the faithful irreducible representation $\mathfrak{V}$ corresponding to the character $\chi = \chi_{\mathbf{239}}$ we determine the irreducible constituents of the faithful permutation representation $(1_Y)^{H_1}$ of H_1 of degree 6144 with stabilizer Y given in Proposition 7.2.1(m). Calculating inner products with the irreducible characters of H_1 it follows $(1_Y)^{H_1}$ has 30 irreducible constituents including the character $\chi_{\mathbf{239}}$ of degree 128. Since the permutation module $(1_Y)^{H_1}$ is only of degree 6144 the Meataxe algorithm can be applied to obtain the corresponding irreducible module $\mathfrak{V}$ because 23 does not divide the order of H_1 and the character $\chi_{\mathbf{239}}$ has rational values.

In order to find the irreducible module $\mathfrak{W}$ corresponding to the irreducible character $\tau = \tau_{39}$ of D with degree 128 we determined the seven irreducible constituents of the permutation module $(1_T)^D$ of D of degree 2048 with stabilizer $T =$ `BasicStabilizer(PD,2)`. It follows that τ_{39} belongs to them. Applying the Meataxe algorithm to the permutation module one obtains the irreducible representation $\mathfrak{W}$ belonging to the constituent $\tau_{\mathbf{39}}$ of the compatible pair (χ, τ).

By construction the KU-modules $\mathfrak{V}_{|U}$ and $\mathfrak{W}_{|U}$ described by the two triples

$$(\mathfrak{V}(j), \mathfrak{V}(k), \mathfrak{V}(y)) \quad \text{and} \quad (\mathfrak{W}(\sigma(j), \mathfrak{W}(\sigma(k)), \mathfrak{W}(\sigma(y)))$$

of matrices in $\mathrm{GL}_{128}(23)$ are isomorphic. Let $Y = \mathrm{GL}(128, 23)$ and let V and W be the KU- and $K(\sigma(U))$-modules described by the first and the second triple of matrices of Y. Applying Parker's isomorphism test of Proposition 6.1.6 of [92] by means of the MAGMA command

```
IsIsomorphic(GModule(sub<Y|W>),GModule(sub<Y|V>)),
```

one obtains the transformation matrix $\mathcal{T}$ in $\mathrm{GL}_{128}(23)$ satisfying

$$\mathfrak{V}(y) = (\mathfrak{W}(\sigma(y)))^{\mathcal{T}}, \mathfrak{V}(j) = (\mathfrak{W}(\sigma(j)))^{\mathcal{T}}, \mathfrak{V}(k) = (\mathfrak{W}(\sigma(k)))^{\mathcal{T}}.$$

Using the faithful permutation representation PD of $D = \langle x, y\rangle$ with stabilizer T it has been checked that x does not belong to $\sigma(U)$ and that $D = \langle\sigma(U), x\rangle = \langle\sigma(y), x\rangle$. Let $\mathfrak{x} = (\mathfrak{W}(x))^{\mathcal{T}}$ and $\mathfrak{y} = (\mathfrak{V}(y))$. Then $\mathfrak{D} = \langle\mathfrak{x}, \mathfrak{y}\rangle \cong D$. Let $\mathfrak{H} = \langle\mathfrak{h}, \mathfrak{y}, \mathfrak{x}\rangle$.

(i) Denote the matrix group $\mathfrak{H}$ by MH. Applications of the MAGMA commands `RandomSchreier(MH)` and `BasicStabilizerChain(MH)` provided a base with three base points and a chain of three basic stabilizers $B(k)$ of orders 89181388800, 43008 and 21504. Using the command `FPGroupStrong(MH)` we obtained the presentation of a finitely presented group $H \cong MH$ with the eight generators h_i and their relations given in the statement.

(j) Using the calculated strong base and generating set of MH, MAGMA and the program `GetShortGensHKim(PE, PD)` we verified that the basic stabilizer B_2 is generated by h_6, h_7, h_8 and $(h_2^3 h_5^2 h_2 h_5)^6$. Applying then the Todd–Coxeter Algorithm implemented in MAGMA with respect to the subgroup B_2 of H we obtained a faithful permutation representation P_1H of degree 4147200. It has been used to obtain the given four generators of the documented stabilizer of the faithful permutation representation PH of H having degree 276480.

(k) By Proposition 7.2.1(m) H has a subgroup U of odd index which is isomorphic to the subgroup $\sigma(U)$ of D with index $|D : \sigma(U)| = 35$. Hence it contains a Sylow 2-subgroup S_1 of H which is isomorphic to the ones of $D = C_E(z)$. As V_1 is the unique maximal elementary abelian normal subgroup of the Sylow 2-subgroup $\sigma(S_1)$ in D by Lemma 7.1.1, S_1 also has a unique elementary abelian normal subgroup V of order 2^{11}. Using the faithful permutation representation PH of H constructed in (j) it has been checked that $N_H(V) \cong D$. This completes the proof. □

Definition 7.2.3 A finite simple group G is said to be of Co_1*-type* if it has a 2-central involution z such that $C_G(z)$ is isomorphic to the finitely presented group H of Proposition 7.2.2.

7.3 Construction of Conway's simple group Co_1

By Proposition 7.2.2 the amalgam $H \leftarrow D \rightarrow E$ constructed in Section 7.2 satisfies the conditions of Step 5 of Algorithm 1.3.8. Therefore Kim and the author applied this algorithm to give in [75] a new existence proof for Conway's sporadic group Co_1. It is realized as a matrix group $\mathfrak{G}$ inside $\mathrm{GL}_{276}(23)$. For that construction, we constructed a faithful permutation representation of the finitely presented group H stated in Proposition 7.2.2 of degree 61440. This is achieved by constructing a monomorphism from the centralizer $H(\mathsf{Co}_2)$ of a 2-central involution of the second Conway group Co_2 constructed in Proposition 4.2.1 into H.

Proposition 7.3.1 *Keep the notation of Lemma 7.1.1 and Propositions 7.2.2 and 7.2.1. Let* $H = \langle h_i \mid 1 \le i \le 8\rangle = \langle x, y, h\rangle$ *be the finitely presented group of Proposition 7.2.2 with center* $z = (x^5yx)^{14}$*, where* $x = h_5$*,* $y = h_1$ *and* $h = h_4$*. Let*

$$q_1 = y^2, \quad q_2 = (xy)^7, \quad q_3 = (yx)^7, \quad q_4 = (xy^2)^7, \quad q_5 = (yxy)^7,$$
$$q_6 = (x^5yx)^7, \quad q_7 = (x^4yx^2)^7, \quad q_8 = (x^4yxy)^6.$$

Then the following statements hold.

(a) $Q = \langle q_i \mid 1 \le i \le 8\rangle$ *is an extra-special normal subgroup of* H *with center* $Z(Q) = \langle q_6^2\rangle$.

(b) *The subgroup* K *generated by*

$$(xh^2x^2y)^6, \quad (h^2xyhx)^6, \quad (hxyhyxy)^6,$$
$$(xyhx^2yx^2)^4, \quad (xyhxh^2x^2)^6, \quad (xyh^2x^3h)^4$$

is a simple group of order $2^9 \cdot 3^4 \cdot 5 \cdot 7$. *It is isomorphic to* $\mathsf{Sp}_6(2)$ *and* $Q \cap K = 1$.

(c) H *has a faithful permutation representation* PH *of degree* 61440 *with stabilizer* K.

(d) $C = QK$ *has center* $Z(C) = Z(Q)$, *and* C *is isomorphic to the centralizer* $H(\mathsf{Co}_2)$ *of a 2-central involution* t *of the simple group* Co_2 *constructed in Proposition 4.2.1.*

(e) H *has* 154 *conjugacy classes* $\mathfrak{h}_i^{\mathfrak{H}}$ *with representatives* $\mathfrak{h}_i$ *and centralizer orders* $|C_{\mathfrak{H}}(\mathfrak{h}_i)|$ *as given in Table 7.5.2.*

(f) *The character table of* H *is stated in Table 7.6.2.*

(g) *The Goldschmidt index of the amalgam* $H \leftarrow D \rightarrow E$ *is* 1.

Proof (a) This assertion is a restatement of Proposition 7.2.1.

(b) Using the faithful permutation representation of Proposition 7.2.2(j) and MAGMA we observed that H has exactly three conjugacy classes of subgroups U of order $2^{18} \cdot 3^4 \cdot 5 \cdot 7$ containing Q. Applying the MAGMA command `HasComplement(U,Q)` we saw that Q has a complement K only in one class of subgroups U. The generators of K have been found by means of the program `GetShortGensHKim(PE, PD)`. Another application of MAGMA yields that K is simple and isomorphic to $\mathsf{Sp}_6(2)$. In particular, $K \cap Q = 1$.

(c) Furthermore, H has a faithful permutation representation PH of degree $|H : K| = 61440$ with stabilizer K.

(d) Clearly $C = QK$ is a subgroup of H because Q is normal in H. Using PH and MAGMA it is easy to see that $Z(C) = Z(Q)$. The remaining statements follow from an isomorphism test between $C = HK$ and $H(\mathsf{Co}_2)$ using MAGMA and the faithful permutation representations PH of H and the one of $H(\mathsf{Co}_2)$ given in Lemma 3.4(a) of [72].

(e) The system of representatives of the conjugacy classes of H has been calculated by means of PH, MAGMA and Kratzer's Algorithm 5.3.18 of [92].

(f) The character table of H has been computed by means of MAGMA using the faithful permutation representation PH of H.

(g) The Goldschmidt index of the amalgam $H \leftarrow D \rightarrow E$ has been calculated by means of Kratzer's Algorithm 7.1.10 of [92], the faithful permutation representations PH, PD, PE and MAGMA. □

Theorem 7.3.2 (Kim–Michler) *Keep the notation of Lemma 7.1.1 and Propositions 7.2.2 and 7.3.1. Let $K = GF(23)$. Let $\sigma : D \to E$ denote the corresponding monomorphism of U into D. Using the notation of the three character Tables 7.6.2,* `DVD.5.1.6` *and 7.6.1 of the groups H, D and E, respectively, the following statements hold.*

(a) *There is exactly one compatible pair $(\chi, \tau) \in mf\,char_{\mathbb{C}}(H) \times mf\,char_{\mathbb{C}}(E)$ of degree 276 of the groups $H = \langle D, h\rangle$ and $E = \langle D, e\rangle$:*

$$\chi_2 + \chi_{10} + \chi_{\mathbf{11}} = \tau_{10}$$

with common restriction

$$\tau_{|D} = \chi_{|D} = \psi_9 + \psi_{36} + \psi_{\mathbf{39}},$$

where irreducible characters with bold face indices denote faithful irreducible characters.

(b) *Let $\mathfrak{V}$ and $\mathfrak{W}$ be the up to isomorphism uniquely determined faithful semi-simple multiplicity-free 276-dimensional modules of H and E over $F = GF(23)$ corresponding to the compatible pair χ, τ, respectively.*

Let $\kappa_{\mathfrak{V}} : H \to \mathrm{GL}_{276}(23)$ and $\kappa_{\mathfrak{W}} : E \to \mathrm{GL}_{276}(13)$ be the representations of H and E afforded by the modules $\mathfrak{V}$ and $\mathfrak{W}$, respectively.

Let $\mathfrak{h} = \kappa_{\mathfrak{V}}(h)$, $\mathfrak{x} = \kappa_{\mathfrak{V}}(x)$, $\mathfrak{y} = \kappa_{\mathfrak{V}}(y)$ in $\kappa_{\mathfrak{V}}(H) \leq \mathrm{GL}_{276}(23)$. Then the following assertions hold.

(1) *$\mathfrak{V}_{|D} \cong \mathfrak{W}_{|D}$, and there is a matrix $\mathcal{T} \in \mathrm{GL}_{276}(23)$ such that*

$$\mathfrak{x} = \mathcal{T}^{-1}\kappa_{\mathfrak{W}}(\sigma(x))\mathcal{T}, \qquad \mathfrak{y} = \mathcal{T}^{-1}\kappa_{\mathfrak{W}}(\sigma(y))\mathcal{T}.$$

Let $\mathfrak{e} = \mathcal{T}^{-1}\kappa_{\mathfrak{W}}(e)\mathcal{T} \in \mathrm{GL}_{276}(23)$.
Let $\mathfrak{G} = \langle \mathfrak{h}, \mathfrak{x}, \mathfrak{y}, \mathfrak{e} \rangle$.

(2) *The four generating matrices of $\mathfrak{G}$ are stored on the accompanying DVD.*

(3) *In $\mathfrak{G}$ the subgroup $\mathfrak{C} = \langle \mathfrak{p}, \mathfrak{y}, \mathfrak{r}, \mathfrak{e} \rangle$ is the stabilizer of a 23-dimensional $\mathfrak{C}$-invariant subspace $\mathfrak{A}$ of $\mathfrak{V} = K^{276}$ such that the $\mathfrak{G}$-orbit $\mathfrak{A}^{\mathfrak{G}}$ has degree 98280, where $\mathfrak{p} = ((\mathfrak{x}\mathfrak{y})^2\mathfrak{x})^3$ and $\mathfrak{r} = (\mathfrak{x}\mathfrak{y}\mathfrak{h})^5$. Furthermore, $\mathfrak{C}$ is isomorphic to the sporadic Conway group Co_2. The 2-central involution centralizer of $\mathfrak{C}$ is the subgroup $H(\mathfrak{C}) = \langle \mathfrak{p}, \mathfrak{y}, \mathfrak{r} \rangle$ of $\mathfrak{C}$.*

(4) *The character table of $\mathfrak{G}$ coincides with that of Co_1 in the Atlas [19], pp. 184–185.*

(c) *$\mathfrak{G}$ is a simple group with 2-central involution $\kappa_{\mathfrak{V}}(z) = (\mathfrak{x}^5\mathfrak{y}\mathfrak{x})^{14}$ such that*

$$C_{\mathfrak{G}}(\kappa_{\mathfrak{V}}(z)) = \kappa_{\mathfrak{V}}(H) \text{ and } |\mathfrak{G}| = 2^{21} \cdot 3^9 \cdot 5^4 \cdot 7^2 \cdot 11 \cdot 13 \cdot 23.$$

Proof (a) Using MAGMA and the faithful permutation representations of PE and PH determined in Lemma 7.1.1 and Proposition 7.3.1, respectively, we determined the fusion of the classes of D in H and in E. A MAGMA calculation employing Kratzer's Algorithm 7.3.10 of [92] shows then that the amalgam $H \leftarrow D \rightarrow E$ has a unique compatible pair (χ, τ) of degree 276. It is stated in the assertion.

Using MAGMA and the faithful permutation representations again we have determined the fusion of the classes of D in H and $\sigma(D)$ in E. A MAGMA calculation employing Kratzer's Algorithm 7.3.10 of [92] shows then that the amalgam $H_1 \leftarrow U \rightarrow D$ has a unique compatible pair (χ, τ) of degree 128. It is stated in the assertion.

(b) The faithful irreducible constituent $\mathfrak{V}_1 = \mathfrak{V}(\chi_{11})$ of dimension 128 of the faithful irreducible representation $\mathfrak{V}$ corresponding to the semi-simple character χ of the compatible pair has already been constructed in Proposition 7.2.2(g). Its second tensor power contains each of the irreducible representations $\mathfrak{V}_2$ and $\mathfrak{V}_3$ corresponding to χ_2 and χ_{10}, respectively, with multiplicity 1. Since the second tensor power of $\mathfrak{V}_1$ has degree 16384 over $K = GF(23)$ the Meataxe algorithm can be applied to obtain the two irreducible representations $\mathfrak{V}_2$ and $\mathfrak{V}_3$.

In order to construct the faithful irreducible representation $\mathfrak{W}$ corresponding to the character $\tau = \tau_{10}$ of degree 276 we employed the MAGMA command `LowIndexSubgroups(PE, 1000)` using the faithful permutation presentation PE of E of degree 2048. MAGMA found a subgroup U of index 552 such that τ_{10} is a constituent of the permutation character $(1_U)^E$. Using then Kim's implementation of Algorithm 1.5.1 and MAGMA we constructed the irreducible representation W over K corresponding to τ_{10}.

Let σ denote the group isomorphism between the subgroup $D = \langle x, y\rangle$ of H and subgroup $D = \langle \sigma(x), \sigma(y)\rangle$ of E. Then the KD-modules $\mathfrak{V}_{|D}$ and $\mathfrak{W}_{|D}$ described by the two pairs

$$(\mathfrak{V}(x), \mathfrak{V}(y)) \quad \text{and} \quad (\mathfrak{W}(\sigma(x), \mathfrak{W}(\sigma(y))) \in \mathrm{GL}_{276}(23)$$

are isomorphic. Let $Y = \mathrm{GL}_{256}(23)$ and let V and W be the KD- and $K(\sigma(D))$-modules described by the first and the second pair of matrices of Y. Applying Parker's isomorphism test of Proposition 6.1.6 of [92] by means of the MAGMA command

```
IsIsomorphic(GModule(sub<Y|V>),GModule(sub<Y|W>)),
```

one obtains the transformation matrix $\mathcal{T}$ in $\mathrm{GL}_{276}(23)$ satisfying

$$\mathfrak{V}(y) = (\mathfrak{W}(\sigma(y)))^{\mathcal{T}} \quad \text{and} \quad \mathfrak{V}(x) = (\mathfrak{W}(\sigma(x)))^{\mathcal{T}}.$$

Since $E = \langle \sigma(x), \sigma(y), e\rangle$ we let $\mathfrak{e} = \mathcal{T}^{-1}\mathfrak{W}(e)\mathcal{T}$ in $\mathrm{GL}_{276}(23)$. Furthermore, $\mathfrak{G} = \langle \mathfrak{x}, \mathfrak{y}, \mathfrak{h}, \mathfrak{e}\rangle$.

In order to construct a small faithful permutation representation of $\mathfrak{G}$ we now embed the sporadic group Co_2 into $\mathfrak{G}$. By Proposition 7.3.1 $z = (x^5yx)^{14}$ generates the center $Z(H)$ of $H = \langle x, y, h\rangle$ and there is a subgroup $H(C) = QK$ of H whose generators are given in statements (a) and (b) of Proposition 7.3.1 such that $H(C)$ is isomorphic to the 2-central involution centralizer $H(\mathsf{Co}_2)$ constructed in [72]. Let $\mathfrak{p} = ((\mathfrak{x}\mathfrak{y})^2\mathfrak{x})^3$ and $\mathfrak{r} = (\mathfrak{x}\mathfrak{y}\mathfrak{h})^5$. Using the faithful permutation presentation PH of degree 61440 of H it has been checked that $H(C) = QK = \langle p, y, r\rangle$, where $p = ((xy)^2x)^3$ and $r = (xyh)^5$. Another application of MAGMA in the faithful permutation representation PE of E of degree 2048 yields that $E_1 = \langle y, p, e\rangle$ is a subgroup of E which is isomorphic to the split extension $E(CO_2)$ of the automorphism group A_{22} by its 10-dimensional irreducible $GF(2)$-module V_3 described in Lemma 2.6(f) of [72]. By Theorem 4.2 of [72] the simple sporadic group Co_2 is an epimorphic image of the free product $H(C) *_{D_1} E_1$ of $H(C)$ and E_1 with amalgamated subgroup $D_1 = \langle y, p\rangle$. Therefore we let $\mathfrak{C} = \langle \mathfrak{p}, \mathfrak{y}, \mathfrak{r}, \mathfrak{e}\rangle$. It is a subgroup of $\mathfrak{G}$, because its four generators are words in the generating matrices $\mathfrak{x}$, $\mathfrak{y}$, $\mathfrak{h}$ and $\mathfrak{e}$ of $\mathfrak{G}$.

Using the Meataxe algorithm it follows that $\mathfrak{V}|_{\mathfrak{C}}$ splits into two irreducible $K\mathfrak{C}$-modules $\mathfrak{A}$ and $\mathfrak{B}$ of dimensions 23 and 256, respectively. Applying then Weller's Algorithm described in Theorem 6.2.1 of [92] and MAGMA we obtained a faithful permutation representation $P\mathfrak{G}$ of $\mathfrak{G}$ of degree 98280. Using then Proposition 5.2.14 of [92] and MAGMA it follows that $\mathfrak{C}$ is the stabilizer of $P\mathfrak{G}$. Furthermore, $P\mathfrak{G}$ and MAGMA have been used to obtain the order $|\mathfrak{C}| = 2^{18} \cdot 3^6 \cdot 5^3 \cdot 7 \cdot 11 \cdot 23$. Hence $|\mathfrak{G}| = |\mathfrak{C}| \cdot 98280$.

The system of representatives of the conjugacy classes of $\mathfrak{G}$ has been calculated by means of $PH\mathfrak{G}$, MAGMA and Kratzer's Algorithm 5.3.18 of [92].

The character table of $\mathfrak{G}$ has been computed by means of MAGMA using the faithful permutation representation $P\mathfrak{G}$ of $\mathfrak{G}$.

(c) The matrix group $\mathfrak{G}$ is simple by its character table. Since $Z(H) = \langle z\rangle$, $H = \langle x, y, h\rangle$ and $z = (x^5yx)^{14}$ we have $C_{\mathfrak{G}}(\mathfrak{z}) \geq \langle \mathfrak{x}, \mathfrak{y}, \mathfrak{h}\rangle$. Counting the number of fixed points of the involution $\mathfrak{z}$ on the permutation representation $P\mathfrak{G}$, application of Proposition 2.6.6 of [92] yields that $C_{\mathfrak{G}}(\mathfrak{z}) = \langle \mathfrak{x}, \mathfrak{y}, \mathfrak{h}\rangle \cong H$. □

In order to show that the centralizer $\mathfrak{H} = C_{\mathfrak{G}}(\mathfrak{z})$ of the simple group $\mathfrak{G}$ constructed in Theorem 7.3.2 is isomorphic to the centralizer of a 2-central involution of the original Conway group Co_1 we use the following result of L. Soicher's article [120]. He verified the presentation of Co_1 given below. It is stated as a conjecture in the first edition of the Atlas [19], p. 183.

Theorem 7.3.3 (L. Soicher) *Conway's simple group* Co_1 *is isomorphic to the finitely presented group* $G = \langle a, b, c, d, e, f, g, h, i\rangle$ *with the following set* $\mathcal{R}(G)$ *of defining relations:*

$$
\begin{aligned}
&a^2 = b^2 = c^2 = d^2 = e^2 = f^2 = g^2 = h^2 = i^2 = 1,\\
&(ab)^3 = (bc)^3 = (cd)^8 = (de)^3 = (ef)^3 = (fg)^3 = (gh)^3 = (hi)^3 = 1,\\
&(ac)^2 = (ad)^2 = (ae)^2 = (af)^2 = (ag)^2 = (ah)^2 = (ai)^2 = 1,\\
&(bd)^2 = (be)^2 = (bf)^2 = (bg)^2 = (bh)^2 = (bi)^2 = 1,\\
&(ce)^2 = (cf)^2 = (cg)^2 = (ch)^2 = (ci)^2 = (df)^2 = (dg)^2 = 1,
\end{aligned}
$$

$$(dh)^2 = (di)^2 = (eg)^2 = (eh)^2 = (ei)^2 = (fh)^2 = (fi)^2 = (gi)^2 = 1,$$
$$a = (cd)^4, \quad (bcde)^8 = 1, \quad ((bcdcdefgh)^{13}i)^3 = 1.$$

Furthermore, G has a faithful permutation representation PG of degree 1545600 with stabilizer $Q = \langle a, b, c, d, e, f, g, h\rangle$.

Proof Using the program `MyCosetAction(G,Q: maxsize:=100000000)` Kim and the author established a faithful permutation representation PG of degree 1545600 with stabilizer Q. Thus we observed that G has the same order as the simple group $\mathfrak{G}$ constructed in Theorem 7.3.2. The isomorphism between Co_1 and G follows from Soicher's article [120]. □

Corollary 7.3.4 *Keep the notations of Theorems 7.3.2 and 7.3.3. Let $x_1 = c$, $x_3 = (cde)^3$, $x_5 = (cdefg)^5$, $x_7 = (cdefghi)^7$ and $h_1 = (x_3b)^4$, $h_2 = (x_1x_7b)^4$, $h_3 = (x_1x_3x_5b)^4$, $h_4 = (x_7bx_5b)^5$. Then the following hold.*

(a) *$a = (h_1^2h_2h_1h_4h_2h_4)^4$ is a 2-central involution of the finitely presented group G such that*

$$C_G(a) = \langle h_i \mid 1 \le i \le 4\rangle \cong \mathfrak{H} = C_{\mathfrak{G}}(\mathfrak{z}),$$

where $\mathfrak{z}$ is the 2-central involution of the simple group $\mathfrak{G} = \langle \mathfrak{x}, \mathfrak{y}, \mathfrak{h}, \mathfrak{e}\rangle$ constructed in Theorem 7.3.2.

(b)

$$\mathfrak{G} \cong G.$$

Proof (a) Let PG be the faithful permutation representation of Soicher's group G defined in Theorem 7.3.3. Using it and MAGMA Kim and the author observed in [75] that $H = C_G(a)$ is a centralizer of the 2-central involution a of G. Applying the program `GetShortGens(G,H)` we found the given generators of H.

Let $P\mathfrak{H}$ be the faithful permutation representation of $\mathfrak{H} = \langle \mathfrak{x}, \mathfrak{y}, \mathfrak{h}\rangle$ of degree 61440 constructed in Proposition 7.3.1. Using the respective permutation representations PG and $P\mathfrak{H}$ of H and $H_1 := \mathfrak{H}$ and the MAGMA command `IsIsomorphic(H_1,H)` we verified the assertion.

(b) By Theorems 7.3.2 and 7.3.3, Proposition 7.3.1 and assertion (a), the groups $\mathfrak{G}$ and G satisfy all conditions of Theorem 7.5.1 of [92]. Hence (b) holds. □

7.4 On the uniqueness of Co_1

The Leech lattice Λ of degree 24 is unique up to isomorphism by Niemeyer's article [101]. J. Conway gave another proof for this fact in [18]. Therefore the group $Aut(\Lambda)$ of all isometries of Λ is uniquely determined, and so is Conway's simple group $\mathsf{Co}_1 = Aut(\Lambda)/Z$, where Z is the center of $Aut(\Lambda)$ of order 2. The literature does not contain an analog for Co_1 of D. Fendel's uniqueness theorem [28] mentioned in Chapter 3, although Feit's article [27] proves the following result.

Theorem 7.4.1 (W. Feit) *Let G be a finite group of order $|G|$ divisible by 23. Suppose that G has a faithful rational representation of degree 24. Then one of the following holds:*

(a) *G has a subgroup of index 23, 24, or 25;*
(b) *the derived subgroup $G' \leq Aut(\Lambda)$.*

This deep result suggests that any simple group G of Co_1-type might have a unique 2-fold cover $2G \cong Aut(\Lambda)$.

In fact, this is a harder problem than satisfying all conditions of the author's uniqueness criterion; see Theorem 7.5.1 of [92]. A successful application of that result would yield the following statement: each finite simple group G of Co_1-type is isomorphic to the simple group $\mathfrak{G} \leq \mathrm{GL}_{276}(23)$ of order $|\mathfrak{G}| = 2^{21} \cdot 3^9 \cdot 5^4 \cdot 7^2 \cdot 11 \cdot 13 \cdot 23$ constructed in Theorem 7.3.2.

This assertion may be true; see also [112]. But the author does not have a complete proof, although Chapters 2 and 3 of [92] provide all the necessary theoretical results to find the correct answers to the questions below.

Remark 7.4.2 Suppose that G is a finite simple group with a 2-central involution z such that $C_G(z) \cong H$, where H is the finitely presented group of Proposition 7.2.2. Therefore all results about $H = \langle x, y, h \rangle$ proved in Proposition 7.3.1 and Table 7.5.2 hold. Furthermore, we know a faithful permutation representation PH of degree 61440 and the character Table 7.6.2 of H. Thus we can calculate a system of generators s_i of a fixed Sylow 2-subgroup S of H. Proposition 7.2.2(k) states that S has a unique maximal elementary abelian normal subgroup A of order 2^{11}.

Remark 7.4.3 Keep the notations of Theorem 7.3.2, Proposition 7.3.1 and Remark 7.4.2. In $\mathfrak{G} = \langle \mathfrak{h}, \mathfrak{x}, \mathfrak{y}, \mathfrak{e} \rangle$ let $\mathfrak{a} = \mathfrak{x}\mathfrak{e}\mathfrak{h}\mathfrak{x}$, $\mathfrak{b} = (\mathfrak{x}\mathfrak{e}\mathfrak{y}^3\mathfrak{x}\mathfrak{e}\mathfrak{h}^2)^{11}$, $\mathfrak{a}_1 = \mathfrak{y}\mathfrak{e}\mathfrak{h}$ and $\mathfrak{b}_1 = (\mathfrak{x}\mathfrak{h}^2\mathfrak{e}\mathfrak{x}\mathfrak{e}\mathfrak{y}\mathfrak{h}\mathfrak{e}\mathfrak{h})^5$.

Are the following statements true?

(a) G has exactly three conjugacy classes of involutions represented by the involutions $z = (x^5yx)^{14}$, $z_2 = (x^2yhxy)^5$ and $z_3 = (xyhxh^2)^6$ of H.
(b) $B = C_G(z_2) \cong C_{\mathfrak{G}}((\mathfrak{x}\mathfrak{e}\mathfrak{h}\mathfrak{e})^4)$ and $|B| = 2^{15} \cdot 3^3 \cdot 5^2 \cdot 7 \cdot 13$.
(c) $C = C_G(z_3) \cong C_{\mathfrak{G}}((\mathfrak{y}\mathfrak{e}\mathfrak{h})^5)$ and $|C| = 2^{18} \cdot 3^3 \cdot 5 \cdot 11$.
(d) The fusion pattern of the conjugacy classes of B in G agrees with the one of $\mathfrak{B} = C_{\mathfrak{G}}((\mathfrak{x}\mathfrak{e}\mathfrak{h}\mathfrak{e})^4) = \langle \mathfrak{a}, \mathfrak{b} \rangle$ in $\mathfrak{G}$.
(e) The fusion pattern of the conjugacy classes of C in G agrees with the one of $\mathfrak{C} = C_{\mathfrak{G}}((\mathfrak{x}\mathfrak{e}\mathfrak{h})^5) = \langle \mathfrak{a}_1, \mathfrak{b}_1 \rangle$ in $\mathfrak{G}$.

If the statements of Remark 7.4.3 can be verified then all simple groups G of Co_1-type have the same order $|\mathfrak{G}|$ by the following corollary of Theorem 7.3.2. Its proof, due to Kim, uses Theorem 1.6.4.

Corollary 7.4.4 (H. Kim) *Keep the notation of Remark 7.4.3. Let $E = \langle x, y, e \rangle$ be the group defined in Proposition 7.2.1(a). Then the following assertions hold.*

(a) *A system of representatives of the* 112 *conjugacy classes of $\mathfrak{B} = \langle \mathfrak{a}, \mathfrak{b} \rangle$ and its character table are given in Tables* `DVD.1.5.4` *and* `DVD.5.1.9` *of the accompanying DVD, respectively.*
(b) *A system of representatives of the* 96 *conjugacy classes of $\mathfrak{C} = \langle \mathfrak{a}_1, \mathfrak{b}_1 \rangle$ and its character table are given in Tables* `DVD.1.5.5` *and* `DVD.5.1.10` *of the accompanying DVD, respectively.*
(c)
$$|\mathfrak{G}| = 1438456320 \cdot |B| + 12583935 \cdot |H| + \left(304128 \cdot \frac{|H||B|}{|C|}\right)$$
$$= 2^{21} \cdot 3^9 \cdot 5^4 \cdot 7^2 \cdot 11 \cdot 13 \cdot 23.$$

Remark 7.4.5 Keep the notations of Remarks 7.4.2 and 7.4.3. Then Remark 7.4.2 and Algorithm 7.4.8 of [92] imply that $E(G) = N_G(A) \cong E = \langle x, y, e \rangle$ for any finite simple group G of Co_1-type. Let $D = H \cap E(G)$. Furthermore, Proposition 7.3.1(g) asserts that the amalgam $H \leftarrow D \rightarrow E(G)$ has Goldschmidt index 1. So by Theorem 7.5.1 of [92] it requires to show that any simple group G of Co_1-type has 101 conjugacy classes which can be ordered such that the class representatives x_i of G with index i have a centralizer $C_G(x_i)$ which is isomorphic to the centralizer $C_{\mathfrak{G}}(\mathfrak{x}_i)$ of the corresponding class representative $\mathfrak{x}_i$ of $\mathfrak{G}$. Using this information we finally have to show that each simple group G of Co_1-type has an irreducible character χ of degree 276.

7.5 Representatives of conjugacy classes

7.5.1 *Conjugacy classes of* $E(\mathsf{Co}_1) = \langle x, y, e\rangle$

Class	*Representative*	*Centralizer*	2P	3P	5P	7P	11P	23P
1	1	$2^{21}\cdot 3^3\cdot 5\cdot 7\cdot 11\cdot 23$	1	1	1	1	1	1
2_1	$(y)^2$	$2^{21}\cdot 3^2\cdot 5\cdot 7$	1	2_1	2_1	2_1	2_1	2_1
2_2	$(xey^2)^{11}$	$2^{18}\cdot 3^3\cdot 5\cdot 11$	1	2_2	2_2	2_2	2_2	2_2
2_3	e	$2^{17}\cdot 3\cdot 7$	1	2_3	2_3	2_3	2_3	2_3
2_4	$(x^2ey)^3$	$2^{15}\cdot 3\cdot 5$	1	2_4	2_4	2_4	2_4	2_4
2_5	$(xexey)^3$	$2^{15}\cdot 3\cdot 5$	1	2_5	2_5	2_5	2_5	2_5
2_6	$(xyxeyxye)^2$	$2^{17}\cdot 3$	1	2_6	2_6	2_6	2_6	2_6
3_1	$(x^2y)^{10}$	$2^8\cdot 3^3\cdot 5$	3_1	1	3_1	3_1	3_1	3_1
3_2	$(x^2ey)^2$	$2^6\cdot 3^2\cdot 7$	3_2	1	3_2	3_2	3_2	3_2
4_1	y^2e	$2^{17}\cdot 3\cdot 7$	2_1	4_1	4_1	4_1	4_1	4_1
4_2	$(x^3yxe)^4$	$2^{17}\cdot 3$	2_1	4_2	4_2	4_2	4_2	4_2
4_3	y	$2^{14}\cdot 3$	2_1	4_3	4_3	4_3	4_3	4_3
4_4	$(xyey^2)^3$	$2^{14}\cdot 3$	2_1	4_4	4_4	4_4	4_4	4_4
4_5	$(x^2yxeye)^2$	2^{15}	2_1	4_5	4_5	4_5	4_5	4_5
4_6	$(x^2yx^2yeye)^2$	2^{15}	2_1	4_6	4_6	4_6	4_6	4_6
4_7	$(x^2y^2ey)^3$	$2^{10}\cdot 3\cdot 5$	2_2	4_7	4_7	4_7	4_7	4_7
4_8	$(x^4yxy)^3$	$2^{11}\cdot 3$	2_3	4_8	4_8	4_8	4_8	4_8
4_9	$(xyeyxey)^3$	$2^{11}\cdot 3$	2_3	4_9	4_9	4_9	4_9	4_9
4_{10}	ye	2^{11}	2_3	4_{10}	4_{10}	4_{10}	4_{10}	4_{10}
4_{11}	$(xyexexe)^2$	2^{11}	2_3	4_{11}	4_{11}	4_{11}	4_{11}	4_{11}
4_{12}	$xyxeyxye$	2^{10}	2_6	4_{12}	4_{12}	4_{12}	4_{12}	4_{12}
4_{13}	$x^2exyeyxye$	2^{10}	2_6	4_{13}	4_{13}	4_{13}	4_{13}	4_{13}
4_{14}	$(xyxe)^3$	$2^8\cdot 3$	2_4	4_{14}	4_{14}	4_{14}	4_{14}	4_{14}
4_{15}	$(x^2yxye)^3$	$2^8\cdot 3$	2_5	4_{15}	4_{15}	4_{15}	4_{15}	4_{15}
5	$(x^2y)^6$	$2^5\cdot 3\cdot 5$	5	5	1	5	5	5
6_1	$(x^2y)^5$	$2^7\cdot 3^2\cdot 5$	3_1	2_1	6_1	6_1	6_1	6_1
6_2	$xyexyey^2$	$2^7\cdot 3^3$	3_1	2_2	6_2	6_2	6_2	6_2
6_3	$(xye)^2$	$2^8\cdot 3^2$	3_1	2_1	6_3	6_3	6_3	6_3
6_4	$(x^2y^2ey)^2$	$2^6\cdot 3^2$	3_2	2_2	6_4	6_4	6_4	6_4
6_5	xy^3e	$2^6\cdot 3$	3_1	2_3	6_5	6_5	6_5	6_5
6_6	x^6exy	$2^6\cdot 3$	3_1	2_6	6_6	6_6	6_6	6_6
6_7	x^2ey	$2^5\cdot 3$	3_2	2_4	6_7	6_7	6_7	6_7
6_8	$xexey$	$2^5\cdot 3$	3_2	2_5	6_8	6_8	6_8	6_8
7_1	x	$2^3\cdot 3\cdot 7$	7_1	7_2	7_2	1	7_1	7_1
7_2	$(x)^3$	$2^3\cdot 3\cdot 7$	7_2	7_1	7_1	1	7_2	7_2
8_1	$(x^2exyxe)^3$	$2^{11}\cdot 3$	4_1	8_1	8_1	8_1	8_1	8_1
8_2	$(x^2eyxyxy)^3$	$2^{11}\cdot 3$	4_1	8_2	8_2	8_2	8_2	8_2
8_3	$(x^3yxe)^2$	2^{11}	4_2	8_3	8_3	8_3	8_3	8_3
8_4	$(xyexey)^2$	2^{11}	4_2	8_4	8_4	8_4	8_4	8_4
8_5	x^5ex^2y	2^{10}	4_1	8_5	8_5	8_5	8_5	8_5
8_6	x^5yex^3e	2^{10}	4_2	8_6	8_6	8_6	8_6	8_6
8_7	x^2yxeye	2^8	4_5	8_7	8_7	8_7	8_7	8_7
8_8	$xyexyey$	2^8	4_3	8_8	8_8	8_8	8_8	8_8
8_9	x^2yx^2yeye	2^8	4_6	8_9	8_9	8_9	8_9	8_9
8_{10}	x^3exe	2^6	4_{10}	8_{10}	8_{10}	8_{10}	8_{10}	8_{10}
8_{11}	$xyexexe$	2^6	4_{11}	8_{11}	8_{11}	8_{11}	8_{11}	8_{11}
10_1	$(x^3yexe)^2$	$2^5\cdot 5$	5	10_1	2_2	10_1	10_1	10_1
10_2	$(x^2y)^3$	$2^3\cdot 3\cdot 5$	5	10_2	2_1	10_2	10_2	10_2
10_3	$xyxyexe$	$2^4\cdot 5$	5	10_3	2_5	10_3	10_3	10_3
10_4	x^3exyxe	$2^4\cdot 5$	5	10_4	2_4	10_4	10_4	10_4
11	xe	$2\cdot 11$	11	11	11	11	1	11
12_1	$(x^2exyxe)^2$	$2^6\cdot 3$	6_3	4_1	12_1	12_1	12_1	12_1
12_2	xy^3ey^2	$2^6\cdot 3$	6_3	4_2	12_2	12_2	12_2	12_2

Conjugacy classes of $E(\mathsf{Co}_1) = \langle x, y, e \rangle$ *(continued)*

Class	*Representative*	*Centralizer*	2P	3P	5P	7P	11P	23P
12_3	xye	$2^5 \cdot 3$	6_3	4_3	12_3	12_3	12_3	12_3
12_4	$xyey^2$	$2^5 \cdot 3$	6_3	4_4	12_4	12_4	12_4	12_4
12_5	x^2y^2ey	$2^4 \cdot 3$	6_4	4_7	12_5	12_5	12_5	12_5
12_6	x^4yxy	$2^4 \cdot 3$	6_5	4_8	12_6	12_6	12_6	12_6
12_7	$xyeyxey$	$2^4 \cdot 3$	6_5	4_9	12_7	12_7	12_7	12_7
12_8	$xyxe$	$2^3 \cdot 3$	6_7	4_{14}	12_8	12_8	12_8	12_8
12_9	x^2yxye	$2^3 \cdot 3$	6_8	4_{15}	12_9	12_9	12_9	12_9
14_1	xy^2	$2^3 \cdot 7$	7_1	14_2	14_2	2_1	14_1	14_1
14_2	$(xy^2)^3$	$2^3 \cdot 7$	7_2	14_1	14_1	2_1	14_2	14_2
14_3	xy	$2^2 \cdot 7$	7_1	14_4	14_4	2_3	14_3	14_3
14_4	$(xy)^3$	$2^2 \cdot 7$	7_2	14_3	14_3	2_3	14_4	14_4
15_1	$(x^2y)^2$	$2 \cdot 3 \cdot 5$	15_1	5	3_1	15_2	15_2	15_1
15_2	$(x^2y)^{14}$	$2 \cdot 3 \cdot 5$	15_2	5	3_1	15_1	15_1	15_2
16_1	x^3yxe	2^6	8_3	16_1	16_1	16_1	16_1	16_1
16_2	$xyexey$	2^6	8_4	16_2	16_2	16_2	16_2	16_2
20	x^3yexe	$2^3 \cdot 5$	10_1	20	4_7	20	20	20
21_1	x^3ey	$3 \cdot 7$	21_1	7_1	21_2	3_2	21_1	21_1
21_2	$(x^3ey)^5$	$3 \cdot 7$	21_2	7_2	21_1	3_2	21_2	21_2
22	xey^2	$2 \cdot 11$	11	22	22	22	2_2	22
23_1	x^2eye	23	23_1	23_1	23_2	23_2	23_2	1
23_2	$(x^2eye)^5$	23	23_2	23_2	23_1	23_1	23_1	1
24_1	x^2exyxe	$2^4 \cdot 3$	12_1	8_1	24_1	24_1	24_1	24_1
24_2	$x^2eyxyxy$	$2^4 \cdot 3$	12_1	8_2	24_2	24_2	24_2	24_2
28_1	xy^3	$2^2 \cdot 7$	14_1	28_2	28_2	4_1	28_1	28_1
28_2	$(xy^3)^3$	$2^2 \cdot 7$	14_2	28_1	28_1	4_1	28_2	28_2
30_1	x^2y	$2 \cdot 3 \cdot 5$	15_1	10_2	6_1	30_2	30_2	30_1
30_2	$(x^2y)^7$	$2 \cdot 3 \cdot 5$	15_2	10_2	6_1	30_1	30_1	30_2

7.5.2 *Conjugacy classes of* $H(\mathsf{Co}_1) = \langle x, y, h \rangle$

Class	*Representative*	$\lvert$*Class*$\rvert$	$\lvert$*Centralizer*$\rvert$	2P	3P	5P	7P
1	1	1	$2^{21} \cdot 3^5 \cdot 5^2 \cdot 7$	1	1	1	1
2_1	$(yh)^4$	1	$2^{21} \cdot 3^5 \cdot 5^2 \cdot 7$	1	2_1	2_1	2_1
2_2	$(y)^2$	270	$2^{20} \cdot 3^2 \cdot 5 \cdot 7$	1	2_2	2_2	2_2
2_3	$(xyhxh^2)^6$	12600	$2^{18} \cdot 3^3$	1	2_3	2_3	2_3
2_4	$(xyh)^5$	60480	$2^{15} \cdot 3^2 \cdot 5$	1	2_4	2_4	2_4
2_5	$(xh^2y)^5$	60480	$2^{15} \cdot 3^2 \cdot 5$	1	2_5	2_5	2_5
2_6	$(x^2h^2xh^2xh)^3$	113400	$2^{18} \cdot 3$	1	2_6	2_6	2_6
2_7	$(x^2yhxy)^5$	181440	$2^{15} \cdot 3 \cdot 5$	1	2_7	2_7	2_7
2_8	$(xyhyxh)^3$	1814400	$2^{14} \cdot 3$	1	2_8	2_8	2_8
3_1	$(x^2hy)^{10}$	8960	$2^{13} \cdot 3^5 \cdot 5$	3_1	1	3_1	3_1
3_2	$(x^2y)^{10}$	573440	$2^7 \cdot 3^5 \cdot 5$	3_2	1	3_2	3_2
3_3	$(x^2h^2)^5$	573440	$2^7 \cdot 3^5 \cdot 5$	3_3	1	3_3	3_3
3_4	h	4300800	$2^8 \cdot 3^4$	3_4	1	3_4	3_4
3_5	$(xhy)^3$	5734400	$2^6 \cdot 3^5$	3_5	1	3_5	3_5
4_1	$(yh)^2$	240	$2^{17} \cdot 3^4 \cdot 5 \cdot 7$	2_1	4_1	4_1	4_1
4_2	$(x^4h^2y)^3$	30240	$2^{16} \cdot 3^2 \cdot 5$	2_2	4_2	4_2	4_2
4_3	$(x^4hxh)^4$	75600	$2^{17} \cdot 3^2$	2_1	4_3	4_3	4_3
4_4	$(xh^2)^5$	120960	$2^{14} \cdot 3^2 \cdot 5$	2_1	4_4	4_4	4_4
4_5	$(x^3yh)^3$	302400	$2^{15} \cdot 3^2$	2_1	4_5	4_5	4_5
4_6	$(x^2yx^2hy)^6$	453600	$2^{16} \cdot 3$	2_2	4_6	4_6	4_6
4_7	$(xyhxh^2)^3$	604800	$2^{14} \cdot 3^2$	2_3	4_7	4_7	4_7
4_8	$(x^4yxy)^3$	604800	$2^{14} \cdot 3^2$	2_2	4_8	4_8	4_8
4_9	$(x^2hx^2hy)^3$	604800	$2^{14} \cdot 3^2$	2_3	4_9	4_9	4_9
4_{10}	y	1814400	$2^{14} \cdot 3$	2_2	4_{10}	4_{10}	4_{10}
4_{11}	$(xh)^3$	1814400	$2^{14} \cdot 3$	2_2	4_{11}	4_{11}	4_{11}
4_{12}	$(x^3hx^2y)^3$	1814400	$2^{14} \cdot 3$	2_2	4_{12}	4_{12}	4_{12}
4_{13}	$(x^3yxh^2)^2$	3628800	$2^{13} \cdot 3$	2_3	4_{13}	4_{13}	4_{13}
4_{14}	$(x^2yx^2h)^2$	10886400	2^{13}	2_2	4_{14}	4_{14}	4_{14}
4_{15}	$(xyxh^2xh)^2$	10886400	2^{13}	2_3	4_{15}	4_{15}	4_{15}
4_{16}	$(x^3hyx^2h)^2$	10886400	2^{13}	2_3	4_{16}	4_{16}	4_{16}
4_{17}	x^5hxy^2xh	21772800	2^{12}	2_3	4_{17}	4_{17}	4_{17}
4_{18}	$(xhxhy)^3$	58060800	$2^9 \cdot 3$	2_4	4_{18}	4_{18}	4_{18}
4_{19}	$(x^3yhxh^2)^3$	58060800	$2^9 \cdot 3$	2_4	4_{19}	4_{19}	4_{19}
4_{20}	$x^5hx^2hy^2$	87091200	2^{10}	2_6	4_{20}	4_{20}	4_{20}
4_{21}	$(x^4hxyh)^3$	116121600	$2^8 \cdot 3$	2_7	4_{21}	4_{21}	4_{21}
4_{22}	x^4hx^2y	348364800	2^8	2_8	4_{22}	4_{22}	4_{22}
5_1	$(x^2hy)^6$	9289728	$2^7 \cdot 3 \cdot 5^2$	5_1	5_1	1	5_1
5_2	$(x^2y)^6$	148635648	$2^3 \cdot 3 \cdot 5^2$	5_2	5_2	1	5_2
5_3	$(xh^2)^4$	148635648	$2^3 \cdot 3 \cdot 5^2$	5_3	5_3	1	5_3
6_1	$(x^2yxhy)^{10}$	8960	$2^{13} \cdot 3^5 \cdot 5$	3_1	2_1	6_1	6_1
6_2	$(x^2hy)^5$	483840	$2^{12} \cdot 3^2 \cdot 5$	3_1	2_2	6_2	6_2
6_3	$(x^2y)^5$	573440	$2^7 \cdot 3^5 \cdot 5$	3_2	2_1	6_3	6_3
6_4	$(x^3hy)^5$	573440	$2^7 \cdot 3^5 \cdot 5$	3_3	2_1	6_4	6_4
6_5	$(xyhxhxhyh^2)^2$	3225600	$2^{10} \cdot 3^3$	3_1	2_3	6_5	6_5
6_6	$(x^3h)^4$	4300800	$2^8 \cdot 3^4$	3_4	2_1	6_6	6_6
6_7	$(xyh^2)^3$	5734400	$2^6 \cdot 3^5$	3_5	2_1	6_7	6_7
6_8	$(x^2yxyxh^2)^2$	25804800	$2^7 \cdot 3^3$	3_2	2_3	6_8	6_8
6_9	$x^2yxhyxhxhy$	25804800	$2^7 \cdot 3^3$	3_4	2_3	6_9	6_9
6_{10}	$x^5yxhyxhy$	25804800	$2^7 \cdot 3^3$	3_2	2_3	6_{10}	6_{10}
6_{11}	$(xyhxh^2)^2$	51609600	$2^6 \cdot 3^3$	3_5	2_3	6_{11}	6_{11}
6_{12}	$(x^2hxyxh)^2$	51609600	$2^6 \cdot 3^3$	3_5	2_3	6_{12}	6_{12}
6_{13}	$(x^4yhx^3y)^2$	51609600	$2^6 \cdot 3^3$	3_3	2_3	6_{13}	6_{13}
6_{14}	$(x^4hxyh)^2$	58060800	$2^9 \cdot 3$	3_1	2_7	6_{14}	6_{14}
6_{15}	$(xh)^2$	77414400	$2^7 \cdot 3^2$	3_4	2_2	6_{15}	6_{15}
6_{16}	x^2yhyxh^2	103219200	$2^5 \cdot 3^3$	3_4	2_3	6_{16}	6_{16}

Conjugacy classes of $H(\mathsf{Co}_1) = \langle x, y, h \rangle$ *(continued)*

Class	*Representative*	|*Class*|	|*Centralizer*|	2P	3P	5P	7P
6_{17}	x^3yxh^2xy	103219200	$2^5 \cdot 3^3$	3_4	2_3	6_{17}	6_{17}
6_{18}	$(xhxhy)^2$	154828800	$2^6 \cdot 3^2$	3_2	2_4	6_{18}	6_{18}
6_{19}	xy^2hy^2	154828800	$2^6 \cdot 3^2$	3_4	2_4	6_{19}	6_{19}
6_{20}	$xyhyh^2$	154828800	$2^6 \cdot 3^2$	3_4	2_5	6_{20}	6_{20}
6_{21}	$xyxhy^2xh$	154828800	$2^6 \cdot 3^2$	3_2	2_5	6_{21}	6_{21}
6_{22}	$x^2h^2xh^2xh$	232243200	$2^7 \cdot 3$	3_4	2_6	6_{22}	6_{22}
6_{23}	$xyhyxh$	928972800	$2^5 \cdot 3$	3_4	2_8	6_{23}	6_{23}
7	x	1592524800	$2^3 \cdot 7$	7	7	7	1
8_1	$(x^3h)^3$	967680	$2^{11} \cdot 3^2 \cdot 5$	4_1	8_1	8_1	8_1
8_2	yh	14515200	$2^{11} \cdot 3$	4_1	8_2	8_2	8_2
8_3	$(x^4hxh)^2$	14515200	$2^{11} \cdot 3$	4_3	8_3	8_3	8_3
8_4	$(x^2yx^2hy)^3$	29030400	$2^{10} \cdot 3$	4_6	8_4	8_4	8_4
8_5	$(x^2yx^2yh^2)^3$	29030400	$2^{10} \cdot 3$	4_2	8_5	8_5	8_5
8_6	$(xyxhxhy)^2$	43545600	2^{11}	4_3	8_6	8_6	8_6
8_7	x^2yxh^2xh	87091200	2^{10}	4_2	8_7	8_7	8_7
8_8	x^3yhyxh^2	87091200	2^{10}	4_6	8_8	8_8	8_8
8_9	$(xyxhy)^3$	116121600	$2^8 \cdot 3$	4_4	8_9	8_9	8_9
8_{10}	$(x^2yx^2h^2)^3$	116121600	$2^8 \cdot 3$	4_5	8_{10}	8_{10}	8_{10}
8_{11}	$xyxh^2xh$	348364800	2^8	4_{15}	8_{11}	8_{11}	8_{11}
8_{12}	x^3hxhxy	348364800	2^8	4_{15}	8_{12}	8_{12}	8_{12}
8_{13}	x^2yxy^2xh	348364800	2^8	4_{10}	8_{13}	8_{13}	8_{13}
8_{14}	$xyxhxyxhy$	348364800	2^8	4_{11}	8_{14}	8_{14}	8_{14}
8_{15}	x^5yxyxy	348364800	2^8	4_{12}	8_{15}	8_{15}	8_{15}
8_{16}	x^2yx^2h	696729600	2^7	4_{14}	8_{16}	8_{16}	8_{16}
8_{17}	x^3yxh^2	696729600	2^7	4_{13}	8_{17}	8_{17}	8_{17}
8_{18}	$xyhxhxh$	696729600	2^7	4_7	8_{18}	8_{18}	8_{18}
8_{19}	x^3hyx^2h	696729600	2^7	4_{16}	8_{19}	8_{19}	8_{19}
9_1	$(x^2hxyh)^4$	412876800	$2^3 \cdot 3^3$	9_1	3_5	9_1	9_1
9_2	xhy	1651507200	$2 \cdot 3^3$	9_2	3_5	9_2	9_2
9_3	$xyxyh$	1651507200	$2 \cdot 3^3$	9_3	3_5	9_3	9_3
10_1	$(x^2yxhy)^6$	9289728	$2^7 \cdot 3 \cdot 5^2$	5_1	10_1	2_1	10_1
10_2	$(x^2hy)^3$	92897280	$2^6 \cdot 3 \cdot 5$	5_1	10_2	2_2	10_2
10_3	$(x^2y)^3$	148635648	$2^3 \cdot 3 \cdot 5^2$	5_2	10_3	2_1	10_3
10_4	$(xh^2)^2$	148635648	$2^3 \cdot 3 \cdot 5^2$	5_3	10_4	2_1	10_4
10_5	x^2yhxy	1114767360	$2^4 \cdot 5$	5_1	10_5	2_7	10_5
10_6	xyh	2229534720	$2^3 \cdot 5$	5_2	10_6	2_4	10_6
10_7	xh^2y	2229534720	$2^3 \cdot 5$	5_2	10_7	2_5	10_7
12_1	$(x^2yxhy)^5$	645120	$2^{10} \cdot 3^3 \cdot 5$	6_1	4_1	12_1	12_1
12_2	$(xyhxyhxh)^2$	9676800	$2^{10} \cdot 3^2$	6_1	4_3	12_2	12_2
12_3	$(x^2yx^2h^2)^2$	19353600	$2^9 \cdot 3^2$	6_1	4_5	12_3	12_3
12_4	$(x^2yx^2yh^2)^2$	19353600	$2^9 \cdot 3^2$	6_2	4_2	12_4	12_4
12_5	$(x^2hxyh)^3$	34406400	$2^5 \cdot 3^4$	6_7	4_1	12_5	12_5
12_6	$x^4hxhyxh^2$	38707200	$2^8 \cdot 3^2$	6_2	4_8	12_6	12_6
12_7	$(x^3h)^2$	51609600	$2^6 \cdot 3^3$	6_6	4_1	12_7	12_7
12_8	$(x^2yx^2hy)^2$	58060800	$2^9 \cdot 3$	6_2	4_6	12_8	12_8
12_9	$xyhxhxhyh^2$	77414400	$2^7 \cdot 3^2$	6_5	4_7	12_9	12_9
12_{10}	xy^3hxhxh^2y	77414400	$2^7 \cdot 3^2$	6_5	4_9	12_{10}	12_{10}
12_{11}	$x^4h^2x^2h$	154828800	$2^6 \cdot 3^2$	6_6	4_3	12_{11}	12_{11}
12_{12}	x^3yh	309657600	$2^5 \cdot 3^2$	6_6	4_5	12_{12}	12_{12}
12_{13}	$(xyxhy)^2$	309657600	$2^5 \cdot 3^2$	6_4	4_4	12_{13}	12_{13}
12_{14}	$xyhxh^2$	309657600	$2^5 \cdot 3^2$	6_{11}	4_7	12_{14}	12_{14}
12_{15}	x^4yxy	309657600	$2^5 \cdot 3^2$	6_{15}	4_8	12_{15}	12_{15}
12_{16}	x^4h^2y	309657600	$2^5 \cdot 3^2$	6_{15}	4_2	12_{16}	12_{16}
12_{17}	x^3yx^2y	309657600	$2^5 \cdot 3^2$	6_6	4_4	12_{17}	12_{17}
12_{18}	x^2hx^2hy	309657600	$2^5 \cdot 3^2$	6_{11}	4_9	12_{18}	12_{18}
12_{19}	$x^2hx^2h^2$	309657600	$2^5 \cdot 3^2$	6_{15}	4_8	12_{19}	12_{19}

Conjugacy classes of $H(\mathsf{Co}_1) = \langle x, y, h\rangle$ *(continued)*

Class	*Representative*	|*Class*|	|*Centralizer*|	2P	3P	5P	7P
12_{20}	x^2yxyxh^2	309657600	$2^5 \cdot 3^2$	6_8	4_7	12_{20}	12_{20}
12_{21}	x^3h^2xyxy	309657600	$2^5 \cdot 3^2$	6_8	4_9	12_{21}	12_{21}
12_{22}	$x^2yx^2hx^2h$	309657600	$2^5 \cdot 3^2$	6_7	4_3	12_{22}	12_{22}
12_{23}	x^4yhx^3y	309657600	$2^5 \cdot 3^2$	6_{13}	4_7	12_{23}	12_{23}
12_{24}	x^4hyx^3y	309657600	$2^5 \cdot 3^2$	6_{13}	4_9	12_{24}	12_{24}
12_{25}	xh	928972800	$2^5 \cdot 3$	6_{15}	4_{11}	12_{25}	12_{25}
12_{26}	$xyxh$	928972800	$2^5 \cdot 3$	6_{15}	4_{10}	12_{26}	12_{26}
12_{27}	x^3hx^2y	928972800	$2^5 \cdot 3$	6_{15}	4_{12}	12_{27}	12_{27}
12_{28}	x^4hxyh	928972800	$2^5 \cdot 3$	6_{14}	4_{21}	12_{28}	12_{28}
12_{29}	$xhxhy$	1857945600	$2^4 \cdot 3$	6_{18}	4_{18}	12_{29}	12_{29}
12_{30}	x^2hxyxh	1857945600	$2^4 \cdot 3$	6_{12}	4_{13}	12_{30}	12_{30}
12_{31}	x^3yhxh^2	1857945600	$2^4 \cdot 3$	6_{18}	4_{19}	12_{31}	12_{31}
14_1	$(xy^3)^2$	1592524800	$2^3 \cdot 7$	7	14_1	14_1	2_1
14_2	xy	3185049600	$2^2 \cdot 7$	7	14_3	14_3	2_2
14_3	xy^2	3185049600	$2^2 \cdot 7$	7	14_2	14_2	2_2
15_1	$(x^2hy)^2$	743178240	$2^3 \cdot 3 \cdot 5$	15_1	5_1	3_1	15_1
15_2	$(x^2y)^2$	2972712960	$2 \cdot 3 \cdot 5$	15_2	5_2	3_2	15_2
15_3	x^2h^2	2972712960	$2 \cdot 3 \cdot 5$	15_3	5_3	3_3	15_3
16_1	x^4hxh	1393459200	2^6	8_3	16_1	16_1	16_1
16_2	$xyxhxhy$	1393459200	2^6	8_6	16_2	16_2	16_2
18_1	$(x^2hxyh)^2$	412876800	$2^3 \cdot 3^3$	9_1	6_7	18_1	18_1
18_2	xyh^2	1651507200	$2 \cdot 3^3$	9_2	6_7	18_2	18_2
18_3	x^2hyxh	1651507200	$2 \cdot 3^3$	9_3	6_7	18_3	18_3
20_1	$(x^2yxhy)^3$	185794560	$2^5 \cdot 3 \cdot 5$	10_1	20_1	4_1	20_1
20_2	$x^2yhxhyh$	1114767360	$2^4 \cdot 5$	10_2	20_2	4_2	20_2
20_3	xh^2	4459069440	$2^2 \cdot 5$	10_4	20_3	4_4	20_3
24_1	$x^5hx^2h^2y$	154828800	$2^6 \cdot 3^2$	12_1	8_1	24_1	24_1
24_2	$xyhxyhxh$	464486400	$2^6 \cdot 3$	12_2	8_3	24_2	24_2
24_3	x^3h	619315200	$2^4 \cdot 3^2$	12_7	8_1	24_3	24_3
24_4	x^2yx^2hy	928972800	$2^5 \cdot 3$	12_8	8_4	24_4	24_4
24_5	$x^2yx^2h^2$	928972800	$2^5 \cdot 3$	12_3	8_{10}	24_5	24_5
24_6	$x^2yx^2yh^2$	928972800	$2^5 \cdot 3$	12_4	8_5	24_6	24_6
24_7	$xhxh^2$	1857945600	$2^4 \cdot 3$	12_7	8_2	24_7	24_7
24_8	$xyxhy$	3715891200	$2^3 \cdot 3$	12_{13}	8_9	24_8	24_8
28	xy^3	3185049600	$2^2 \cdot 7$	14_1	28	28	4_1
30_1	$(x^2yxhy)^2$	743178240	$2^3 \cdot 3 \cdot 5$	15_1	10_1	6_1	30_1
30_2	x^2hy	1486356480	$2^2 \cdot 3 \cdot 5$	15_1	10_2	6_2	30_3
30_3	x^2h^2xy	1486356480	$2^2 \cdot 3 \cdot 5$	15_1	10_2	6_2	30_2
30_4	x^2y	2972712960	$2 \cdot 3 \cdot 5$	15_2	10_3	6_3	30_4
30_5	x^3hy	2972712960	$2 \cdot 3 \cdot 5$	15_3	10_4	6_4	30_5
36	x^2hxyh	2477260800	$2^2 \cdot 3^2$	18_1	12_5	36	36
40	x^3yhxh	2229534720	$2^3 \cdot 5$	20_1	40	8_1	40
60	x^2yxhy	1486356480	$2^2 \cdot 3 \cdot 5$	30_1	20_1	12_1	60

7.6 Character tables of local subgroups of Co_1

7.6.1 *Character table of* $E(\mathsf{Co}_1) = \langle x, y, e \rangle$

2	21	21	18	17	15	15	17	8	6	17	17	14	14	15	15	10	11	11
3	3	2	3	1	1	1	1	3	2	1	1	1	1	.	.	1	1	1
5	1	1	1	.	1	1	.	1	.	.	.	.	.	.	.	1	.	.
7	1	1	.	1	.	.	.	.	1	1	.	.	.	.	.	.	.	.
11	1	.	1	.	.	.	.	.	.	.	.	.	.	.	.	.	.	.
23	1	.	.	.	.	.	.	.	.	.	.	.	.	.	.	.	.	.
	1a	2a	2b	2c	2d	2e	2f	3a	3b	4a	4b	4c	4d	4e	4f	4g	4h	4i
2P	1a	1a	1a	1a	1a	1a	1a	3a	3b	2a	2a	2a	2a	2a	2a	2b	2c	2c
3P	1a	2a	2b	2c	2d	2e	2f	1a	1a	4a	4b	4c	4d	4e	4f	4g	4h	4i
5P	1a	2a	2b	2c	2d	2e	2f	3a	3b	4a	4b	4c	4d	4e	4f	4g	4h	4i
7P	1a	2a	2b	2c	2d	2e	2f	3a	3b	4a	4b	4c	4d	4e	4f	4g	4h	4i
11P	1a	2a	2b	2c	2d	2e	2f	3a	3b	4a	4b	4c	4d	4e	4f	4g	4h	4i
23P	1a	2a	2b	2c	2d	2e	2f	3a	3b	4a	4b	4c	4d	4e	4f	4g	4h	4i
X.1	1	1	1	1	1	1	1	1	1	1	1	1	1	1	1	1	1	1
X.2	23	23	23	7	−1	−1	7	5	−1	7	7	7	7	−1	−1	−1	−1	−1
X.3	45	45	45	−3	5	5	−3	.	3	−3	−3	−3	−3	5	5	5	−3	−3
X.4	45	45	45	−3	5	5	−3	.	3	−3	−3	−3	−3	5	5	5	−3	−3
X.5	231	231	231	7	−9	−9	7	−3	.	7	7	7	7	−9	−9	−9	−1	−1
X.6	231	231	231	7	−9	−9	7	−3	.	7	7	7	7	−9	−9	−9	−1	−1
X.7	252	252	252	28	12	12	28	9	.	28	28	28	28	12	12	12	4	4
X.8	253	253	253	13	−11	−11	13	10	1	13	13	13	13	−11	−11	−11	−3	−3
X.9	276	20	−12	20	−12	12	−12	15	.	36	4	4	−4	4	−4	.	−4	4
X.10	276	20	−12	36	12	−12	4	15	.	20	−12	4	−4	−4	4	.	4	−4
X.11	483	483	483	35	3	3	35	6	.	35	35	35	35	3	3	3	3	3
X.12	770	770	770	−14	10	10	−14	5	−7	−14	−14	−14	−14	10	10	10	2	2
X.13	770	770	770	−14	10	10	−14	5	−7	−14	−14	−14	−14	10	10	10	2	2
X.14	990	990	990	−18	−10	−10	−18	.	3	−18	−18	−18	−18	−10	−10	−10	6	6
X.15	990	990	990	−18	−10	−10	−18	.	3	−18	−18	−18	−18	−10	−10	−10	6	6
X.16	1035	1035	1035	27	35	35	27	.	6	27	27	27	27	35	35	35	3	3
X.17	1035	1035	1035	−21	−5	−5	−21	.	−3	−21	−21	−21	−21	−5	−5	−5	3	3
X.18	1035	1035	1035	−21	−5	−5	−21	.	−3	−21	−21	−21	−21	−5	−5	−5	3	3
X.19	1265	1265	1265	49	−15	−15	49	5	8	49	49	49	49	−15	−15	−15	−7	−7
X.20	1771	1771	1771	−21	11	11	−21	16	7	−21	−21	−21	−21	11	11	11	3	3
X.21	1771	−21	11	−21	11	51	11	16	7	91	−5	−5	3	−5	3	−1	3	−5
X.22	1771	−21	11	91	51	11	−5	16	7	−21	11	−5	3	3	−5	−1	11	3
X.23	2024	2024	2024	8	24	24	8	−1	8	8	8	8	8	24	24	24	8	8
X.24	2277	2277	2277	21	−19	−19	21	.	6	21	21	21	21	−19	−19	−19	−3	−3
X.25	3312	3312	3312	48	16	16	48	.	−6	48	48	48	48	16	16	16	.	.
X.26	3520	3520	3520	64	.	.	64	10	−8	64	64	64	64	.	.	.	.	.
X.27	5313	5313	5313	49	9	9	49	−15	.	49	49	49	49	9	9	9	1	1
X.28	5544	5544	5544	−56	24	24	−56	9	.	−56	−56	−56	−56	24	24	24	−8	−8
X.29	5796	5796	5796	−28	36	36	−28	−9	.	−28	−28	−28	−28	36	36	36	−4	−4
X.30	5796	420	−252	84	12	−12	−76	45	.	196	36	20	−20	−4	4	.	4	−4
X.31	5796	420	−252	196	−12	12	36	45	.	84	−76	20	−20	4	−4	.	−4	4
X.32	8855	−105	55	−105	55	15	55	35	−7	231	7	−9	−1	7	−1	−5	−1	7
X.33	8855	−105	55	7	95	−25	39	−10	14	119	23	−9	−1	15	−9	−5	7	15
X.34	8855	−105	55	119	−25	95	23	−10	14	7	39	−9	−1	−9	15	−5	−9	−1
X.35	8855	−105	55	231	15	55	7	35	−7	−105	55	−9	−1	−1	7	−5	−1	7
X.36	10395	10395	10395	−21	−45	−45	−21	.	.	−21	−21	−21	−21	−45	−45	−45	3	3
X.37	10626	−126	66	−14	−54	−54	50	−3	.	−14	50	2	−14	10	10	−6	2	−14
X.38	10626	−126	66	−14	−54	−54	50	−3	.	−14	50	2	−14	10	10	−6	2	−14
X.39	12420	900	−540	−60	−60	60	36	.	.	−108	−12	−12	12	20	−20	.	12	−12
X.40	12420	900	−540	−60	−60	60	36	.	.	−108	−12	−12	12	20	−20	.	12	−12
X.41	12420	900	−540	−108	60	−60	−12	.	.	−60	36	−12	12	−20	20	.	−12	12
X.42	12420	900	−540	−108	60	−60	−12	.	.	−60	36	−12	12	−20	20	.	−12	12
X.43	15180	1100	−660	300	60	−60	76	15	.	92	−132	28	−28	−20	20	.	4	−4
X.44	15180	1100	−660	92	−60	60	−132	15	.	300	76	28	−28	20	−20	.	−4	4
X.45	15939	−189	99	259	99	−21	35	9	.	−77	83	−13	−5	19	−5	−9	19	11
X.46	15939	−189	99	−77	−21	99	83	9	.	259	35	−13	−5	−5	19	−9	−5	−13
X.47	17710	−210	110	126	−90	−10	62	25	7	−98	94	−2	−18	6	22	−10	−10	−10
X.48	17710	−210	110	−98	−10	−90	94	25	7	126	62	−2	−18	22	6	−10	−2	−2
X.49	21252	−252	132	196	132	132	68	−6	.	196	68	−28	4	4	4	−12	12	12
X.50	27324	1980	−1188	204	−12	12	108	.	.	−36	−132	12	−12	4	−4	.	12	−12
X.51	27324	1980	−1188	−36	12	−12	−132	.	.	204	108	12	−12	−4	4	.	−12	12
X.52	28336	−336	176	112	16	16	112	−14	−14	112	112	−16	−16	16	16	−16	.	.
X.53	31878	−378	198	−42	−66	−66	22	45	.	406	−42	−26	22	−2	−2	6	−2	−2
X.54	31878	−378	198	182	78	78	118	−9	.	182	118	−26	−10	14	14	−18	14	−2
X.55	31878	−378	198	406	−66	−66	−42	45	.	−42	22	−26	22	−2	−2	6	−10	−10
X.56	42504	3080	−1848	392	72	−72	72	15	.	168	−152	40	−40	−24	24	.	8	−8
X.57	42504	3080	−1848	168	−72	72	−152	15	.	392	72	40	−40	24	−24	.	−8	8
X.58	53130	−630	330	154	−30	−30	218	−15	.	154	218	−22	−38	34	34	−30	−14	2
X.59	57960	4200	−2520	168	−120	120	104	45	.	−56	−120	8	−8	40	−40	.	−8	8
X.60	57960	4200	−2520	−56	120	−120	−120	45	.	168	104	8	−8	−40	40	.	8	−8
X.61	63756	4620	−2772	140	108	−108	−84	−45	.	252	28	28	−28	−36	36	.	4	−4
X.62	63756	4620	−2772	252	−108	108	28	−45	.	140	−84	28	−28	36	−36	.	−4	4
X.63	79695	−945	495	399	175	−105	−81	.	21	63	−33	−33	39	31	−25	−5	15	−9
X.64	79695	−945	495	63	55	255	−33	.	21	−273	15	15	−9	−25	15	−5	−9	15
X.65	79695	−945	495	63	−105	175	−33	.	21	399	−81	−33	39	−25	31	−5	15	−9
X.66	79695	−945	495	−273	255	55	15	.	21	63	−33	15	−9	15	−25	−5	−33	−9
X.67	106260	7700	−4620	−140	−60	60	−172	−30	.	196	164	4	−4	20	−20	.	12	−12
X.68	106260	7700	−4620	196	60	−60	164	−30	.	−140	−172	4	−4	−20	20	.	−12	12
X.69	127512	−1512	792	504	−24	−24	−8	45	.	−392	120	−8	−8	−24	−24	24	−8	−8
X.70	127512	−1512	792	−392	−24	−24	120	45	.	504	−8	−8	−8	−24	−24	24	8	8
X.71	154560	11200	−6720	−448	.	.	64	30	.	−448	64	−64	64	.	.	.	.	.
X.72	159390	−1890	990	462	230	150	−114	.	−21	−210	−18	−18	30	6	−10	−10	6	6
X.73	159390	−1890	990	−210	150	230	−18	.	−21	462	−114	−18	30	−10	6	−10	−18	−18
X.74	159390	−1890	990	14	−90	−90	78	−45	.	462	14	−34	14	−26	−26	30	−2	−2
X.75	159390	−1890	990	462	−90	−90	14	−45	.	14	78	−34	14	−26	−26	30	−10	−10
X.76	191268	−2268	1188	−252	84	84	132	.	.	−252	132	36	−60	−44	−44	36	12	12
X.77	239085	−2835	1485	−483	−75	45	−3	.	.	−147	−51	45	−27	5	29	−15	21	−3
X.78	239085	−2835	1485	−147	45	−315	−51	.	.	189	−99	−3	21	61	−11	−15	−3	21
X.79	239085	−2835	1485	189	−315	45	−99	.	.	−147	−51	−3	21	−11	61	−15	−3	21
X.80	239085	−2835	1485	−147	45	−75	−51	.	.	−483	−3	45	−27	29	5	−15	−3	−27

Character table of $E(\mathsf{Co}_1)$ (continued)

2	11	11	10	10	8	8	5	7	7	8	6	6	6	5	5	3	3	11	11	11	11	10	10	8	8	8
3	.	.	.	.	1	1	1	2	3	2	2	1	1	1	1	1	1	1	1	.	.	.	.	.	.	.
5	.	.	.	.	.	.	1	1	.	.	.	.	.	.	.	.	.	.	.	.	.	.	.	.	.	.
7	.	.	.	.	.	.	.	.	.	.	.	.	.	.	.	1	1	.	.	.	.	.	.	.	.	.
11	.	.	.	.	.	.	.	.	.	.	.	.	.	.	.	.	.	.	.	.	.	.	.	.	.	.
23	.	.	.	.	.	.	.	.	.	.	.	.	.	.	.	.	.	.	.	.	.	.	.	.	.	.
	4j	4k	4l	4m	4n	4o	5a	6a	6b	6c	6d	6e	6f	6g	6h	7a	7b	8a	8b	8c	8d	8e	8f	8g	8h	8i
2P	2c	2c	2f	2f	2d	2e	5a	3a	3a	3a	3b	3a	3a	3b	3b	7a	7b	4a	4a	4b	4b	4a	4b	4e	4c	4f
3P	4j	4k	4l	4m	4n	4o	5a	2a	2b	2a	2b	2c	2f	2d	2e	7b	7a	8a	8b	8c	8d	8e	8f	8g	8h	8i
5P	4j	4k	4l	4m	4n	4o	1a	6a	6b	6c	6d	6e	6f	6g	6h	7b	7a	8a	8b	8c	8d	8e	8f	8g	8h	8i
7P	4j	4k	4l	4m	4n	4o	5a	6a	6b	6c	6d	6e	6f	6g	6h	1a	1a	8a	8b	8c	8d	8e	8f	8g	8h	8i
11P	4j	4k	4l	4m	4n	4o	5a	6a	6b	6c	6d	6e	6f	6g	6h	7a	7b	8a	8b	8c	8d	8e	8f	8g	8h	8i
23P	4j	4k	4l	4m	4n	4o	5a	6a	6b	6c	6d	6e	6f	6g	6h	7a	7b	8a	8b	8c	8d	8e	8f	8g	8h	8i
X.1	1	1	1	1	1	1	1	1	1	1	1	1	1	1	1	1	1	1	1	1	1	1	1	1	1	1
X.2	3	3	3	−1	−1	−1	3	5	5	5	−1	1	1	−1	−1	2	2	−1	−1	3	3	3	−1	−1	3	−1
X.3	1	1	1	−3	1	1	.	.	.	.	3	.	.	−1	−1	A	$\bar{A}$	−3	−3	1	1	1	−3	1	1	1
X.4	1	1	1	−3	1	1	.	.	.	.	3	.	.	−1	−1	$\bar{A}$	A	−3	−3	1	1	1	−3	1	1	1
X.5	−1	−1	−1	−1	3	3	1	−3	−3	−3	.	1	1	.	.	.	.	−1	−1	−1	−1	−1	−1	3	−1	3
X.6	−1	−1	−1	−1	3	3	1	−3	−3	−3	.	1	1	.	.	.	.	−1	−1	−1	−1	−1	−1	3	−1	3
X.7	4	4	4	4	.	.	2	9	9	9	.	1	1	.	.	.	.	4	4	4	4	4	4	.	4	.
X.8	1	1	1	−3	1	1	3	10	10	10	1	−2	−2	1	1	1	1	−3	−3	1	1	1	−3	1	1	1
X.9	4	−4	.	.	.	.	6	5	−3	−1	.	−1	−3	.	.	3	3	−4	4	8	.	−4	.	.	.	.
X.10	8	.	−4	.	.	.	6	5	−3	−1	.	3	1	.	.	3	3	4	−4	4	−4	.	.	.	.	.
X.11	3	3	3	3	3	3	−2	6	6	6	.	2	2	.	.	.	.	3	3	3	3	3	3	3	3	3
X.12	−2	−2	−2	2	−2	−2	.	5	5	5	−7	1	1	1	1	.	.	2	2	−2	−2	−2	2	−2	−2	−2
X.13	−2	−2	−2	2	−2	−2	.	5	5	5	−7	1	1	1	1	.	.	2	2	−2	−2	−2	2	−2	−2	−2
X.14	2	2	2	6	−2	−2	.	.	.	.	3	.	.	−1	−1	$\bar{A}$	A	6	6	2	2	2	6	−2	2	−2
X.15	2	2	2	6	−2	−2	.	.	.	.	3	.	.	−1	−1	A	$\bar{A}$	6	6	2	2	2	6	−2	2	−2
X.16	−1	−1	−1	3	3	3	.	.	.	.	6	.	.	2	2	−1	−1	3	3	−1	−1	−1	3	3	−1	3
X.17	3	3	3	3	−1	−1	.	.	.	.	−3	.	.	1	1	D	$\bar{D}$	3	3	3	3	3	3	−1	3	−1
X.18	3	3	3	3	−1	−1	.	.	.	.	−3	.	.	1	1	$\bar{D}$	D	3	3	3	3	3	3	−1	3	−1
X.19	1	1	1	−7	−3	−3	.	5	5	5	8	1	1	.	.	−2	−2	−7	−7	1	1	1	−7	−3	1	−3
X.20	−5	−5	−5	3	−1	−1	1	16	16	16	7	.	.	−1	−1	.	.	3	3	−5	−5	−5	3	−1	−5	−1
X.21	−5	3	−1	−1	−1	7	1	−6	2	.	−1	.	2	−1	3	.	.	3	11	7	−1	3	−1	−1	−1	−1
X.22	7	−1	3	−1	7	−1	1	−6	2	.	−1	4	−2	3	−1	.	.	−5	3	−5	3	−1	−1	−1	−1	−1
X.23	.	.	.	8	.	.	−1	−1	−1	−1	8	−1	−1	.	.	1	1	8	8	.	.	.	8	.	.	.
X.24	1	1	1	−3	−3	−3	−3	.	.	.	6	.	.	2	2	2	2	−3	−3	1	1	1	−3	−3	1	−3
X.25	.	.	.	.	.	.	−3	.	.	.	−6	.	.	−2	−2	1	1	.	.	.	.	.	.	.	.	.
X.26	.	.	.	.	.	.	.	10	10	10	−8	−2	−2	.	.	−1	−1	.	.	.	.	.	.	.	.	.
X.27	−3	−3	−3	1	−3	−3	3	−15	−15	−15	.	1	1	.	.	.	.	1	1	−3	−3	−3	1	−3	−3	−3
X.28	.	.	.	−8	.	.	−1	9	9	9	.	1	1	.	.	.	.	−8	−8	.	.	.	−8	.	.	.
X.29	4	4	4	−4	.	.	1	−9	−9	−9	.	−1	−1	.	.	.	.	−4	−4	4	4	4	−4	.	4	.
X.30	.	−8	4	.	.	.	6	15	−9	−3	.	−3	−1	.	.	.	.	4	−4	12	4	−8	.	.	.	.
X.31	12	4	−8	.	.	.	6	15	−9	−3	.	1	3	.	.	.	.	−4	4	.	−8	4	.	.	.	.
X.32	−9	−1	−5	−5	−5	−1	.	−15	1	3	1	3	1	1	−3	.	.	7	−1	11	3	7	3	3	−1	−1
X.33	3	−5	−1	−5	3	−9	.	.	−8	6	−2	−2	.	2	2	.	.	−1	−9	−1	7	3	3	3	−1	−1
X.34	−1	7	3	3	−9	3	.	.	−8	6	−2	2	−4	2	2	.	.	15	7	3	−5	−1	−5	−1	−1	3
X.35	11	3	7	3	−1	−5	.	−15	1	3	1	3	1	−3	1	.	.	7	−1	−9	−1	−5	−5	−1	−1	3
X.36	−1	−1	−1	3	3	3	.	.	.	.	.	.	.	.	.	.	.	3	3	−1	−1	−1	3	3	−1	3
X.37	−2	−2	−2	2	6	6	1	3	3	−3	.	1	−1	.	.	.	.	−14	2	−2	−2	−2	2	−2	2	−2
X.38	−2	−2	−2	2	6	6	1	3	3	−3	.	1	−1	.	.	.	.	−14	2	−2	−2	−2	2	−2	2	−2
X.39	4	−4	.	.	.	.	.	.	.	.	.	.	.	.	.	E	$\bar{E}$	12	−12	8	.	−4	.	.	.	.
X.40	4	−4	.	.	.	.	.	.	.	.	.	.	.	.	.	$\bar{E}$	E	12	−12	8	.	−4	.	.	.	.
X.41	8	.	−4	.	.	.	.	.	.	.	.	.	.	.	.	E	$\bar{E}$	−12	12	4	−4	.	.	.	.	.
X.42	8	.	−4	.	.	.	.	.	.	.	.	.	.	.	.	$\bar{E}$	E	−12	12	4	−4	.	.	.	.	.
X.43	4	12	−8	.	.	.	.	5	−3	−1	.	3	1	.	.	−3	−3	4	−4	.	8	−4	.	.	.	.
X.44	.	8	−4	.	.	.	.	5	−3	−1	.	−1	−3	.	.	−3	−3	−4	4	4	12	−8	.	.	.	.
X.45	3	11	7	−1	3	3	−1	−9	−9	9	.	1	−1	.	.	.	.	−13	−5	−1	−9	−5	−1	3	−1	−5
X.46	−1	−9	−5	−1	3	3	−1	−9	−9	9	.	1	−1	.	.	.	.	11	19	3	11	7	−1	−5	−1	3
X.47	2	2	2	6	−2	2	.	−15	−7	9	−1	−3	5	3	−1	.	.	−2	−2	−6	−6	−6	−2	−2	2	2
X.48	−6	−6	−6	−2	2	−2	.	−15	−7	9	−1	1	1	−1	3	.	.	−10	−10	2	2	2	6	2	2	−2
X.49	4	4	4	−4	.	.	2	6	6	−6	.	−2	2	.	.	.	.	12	12	4	4	4	−4	.	−4	.
X.50	.	8	−4	.	.	.	−6	.	.	.	.	.	.	.	.	3	3	12	−12	4	12	−8	.	.	.	.
X.51	4	12	−8	.	.	.	−6	.	.	.	.	.	.	.	.	3	3	−12	12	.	8	−4	.	.	.	.
X.52	.	.	.	.	.	.	1	−6	−22	18	2	−2	−2	−2	−2	.	.	.	.	.	.	.	.	.	.	.
X.53	−6	10	2	6	6	−6	3	−15	9	−3	.	−3	1	.	.	.	.	−10	−10	10	−6	2	−2	−2	−2	2
X.54	2	2	2	−2	6	6	−2	9	9	−9	.	−1	1	.	.	.	.	−2	14	2	2	2	−2	−2	−2	−2
X.55	10	−6	2	−2	−6	6	3	−15	9	−3	.	1	−3	.	.	.	.	−2	−2	−6	10	2	6	2	−2	−2
X.56	8	−8	.	.	.	.	−6	5	−3	−1	.	−1	−3	.	.	.	.	8	−8	.	−16	8	.	.	.	.
X.57	.	−16	8	.	.	.	−6	5	−3	−1	.	3	1	.	.	.	.	−8	8	8	−8	.	.	.	.	.
X.58	−2	−2	−2	2	−6	−6	.	15	15	−15	.	1	−1	.	.	.	.	2	−14	−2	−2	−2	2	2	2	2
X.59	−8	8	.	.	.	.	.	15	−9	−3	.	−3	−1	.	.	.	.	−8	8	−16	.	8	.	.	.	.
X.60	−16	.	8	.	.	.	.	15	−9	−3	.	1	3	.	.	.	.	8	−8	−8	8	.	.	.	.	.
X.61	−4	4	.	.	.	.	6	−15	9	3	.	−1	−3	.	.	.	.	4	−4	−8	.	4	.	.	.	.
X.62	−8	.	4	.	.	.	6	−15	9	3	.	3	1	.	.	.	.	−4	4	−4	4	.	.	.	.	.
X.63	−1	−9	−5	3	−1	−5	.	.	.	.	−3	.	.	1	−3	.	.	−9	15	−5	3	−1	−5	−1	3	3
X.64	−5	3	−1	3	−1	7	.	.	.	.	−3	.	.	1	−3	.	.	−9	−33	7	−1	3	3	−1	−1	−1
X.65	−5	3	−1	−5	−5	−1	.	.	.	.	−3	.	.	−3	1	.	.	−9	15	−1	−9	−5	3	3	3	−1
X.66	7	−1	3	3	7	−1	.	.	.	.	−3	.	.	−3	1	.	.	15	−9	−5	3	−1	3	−1	−1	−1
X.67	12	4	−8	.	.	.	.	−10	6	2	.	−2	2	.	.	.	.	12	−12	.	−8	4	.	.	.	.
X.68	.	−8	4	.	.	.	.	−10	6	2	.	−2	2	.	.	.	.	−12	12	12	4	−8	.	.	.	.
X.69	.	.	.	−8	.	.	−3	−15	9	−3	.	−3	1	.	.	.	.	8	8	.	.	.	8	.	.	.
X.70	.	.	.	8	.	.	−3	−15	9	−3	.	1	−3	.	.	.	.	−8	−8	.	.	.	−8	.	.	.
X.71	.	.	.	.	.	.	.	10	−6	−2	.	2	−2	.	.	.	.	.	.	.	.	.	.	.	.	.
X.72	−6	−6	−6	6	−2	2	.	.	.	.	3	.	.	−1	3	.	.	−18	−18	2	2	2	−2	−2	2	2
X.73	2	2	2	−2	2	−2	.	.	.	.	3	.	.	3	−1	.	.	6	6	−6	−6	−6	6	2	2	−2
X.74	10	−6	2	6	−6	6	.	15	−9	3	.	−1	3	.	.	.	.	−10	−10	−6	10	2	−2	2	−2	−2
X.75	−6	10	2	−2	6	−6	.	15	−9	3	.	3	−1	.	.	.	.	−2	−2	10	−6	2	6	−2	−2	2
X.76	−4	−4	−4	−4	.	.	3	.	.	.	.	.	.	.	.	.	.	12	12	−4	−4	−4	−4	.	4	.
X.77	9	1	5	−7	−3	−3	.	.	.	.	.	.	.	.	.	.	.	−27	−3	−3	5	1	9	−3	−3	5
X.78	−3	5	1	−7	−3	9	.	.	.	.	.	.	.	.	.	.	.	21	−3	1	−7	−3	1	−3	1	1
X.79	1	−7	−3	1	9	−3	.	.	.	.	.	.	.	.	.	.	.	21	−3	−3	5	1	−7	1	1	−3
X.80	−3	5	1	9	−3	−3	.	.	.	.	.	.	.	.	.	.	.	−3	21	9	1	5	−7	5	−3	−3

Character table of $E(\mathsf{Co}_1)$ *(continued)*

	8j	8k	10a	10b	10c	10d	11a	12a	12b	12c	12d	12e	12f	12g	12h	12i	14a	14b	14c	14d	15a	15b	16a	16b	20a
2	6	6	5	3	4	4	1	6	6	5	5	4	4	4	3	3	3	3	2	2	1	1	6	6	3
3	.	.	.	1	.	.	.	1	1	1	1	1	1	1	1	1	.	.	.	.	1	1	.	.	.
5	.	.	1	1	1	1	.	.	.	.	.	.	.	.	.	.	.	.	.	.	1	1	.	.	1
7	.	.	.	.	.	.	.	.	.	.	.	.	.	.	.	.	1	1	1	1	.	.	.	.	.
11	.	.	.	.	.	.	1	.	.	.	.	.	.	.	.	.	.	.	.	.	.	.	.	.	.
23	.	.	.	.	.	.	.	.	.	.	.	.	.	.	.	.	.	.	.	.	.	.	.	.	.
2P	4j	4k	5a	5a	5a	5a	11a	6c	6c	6c	6c	6d	6e	6e	6g	6h	7a	7b	7a	7b	15a	15b	8c	8d	10a
3P	8j	8k	10a	10b	10c	10d	11a	4a	4b	4c	4d	4g	4h	4i	4n	4o	14b	14a	14d	14c	5a	5a	16a	16b	20a
5P	8j	8k	2b	2a	2e	2d	11a	12a	12b	12c	12d	12e	12f	12g	12h	12i	14b	14a	14d	14c	3a	3a	16a	16b	4g
7P	8j	8k	10a	10b	10c	10d	11a	12a	12b	12c	12d	12e	12f	12g	12h	12i	2a	2a	2c	2c	15b	15a	16a	16b	20a
11P	8j	8k	10a	10b	10c	10d	1a	12a	12b	12c	12d	12e	12f	12g	12h	12i	14a	14b	14c	14d	15b	15a	16a	16b	20a
23P	8j	8k	10a	10b	10c	10d	11a	12a	12b	12c	12d	12e	12f	12g	12h	12i	14a	14b	14c	14d	15a	15b	16a	16b	20a
X.1	1	1	1	1	1	1	1	1	1	1	1	1	1	1	1	1	1	1	1	1	1	1	1	1	1
X.2	1	1	3	3	−1	−1	1	1	1	1	1	−1	−1	−1	−1	−1	2	2	.	.	.	.	1	1	−1
X.3	−1	−1	.	.	.	.	1	.	.	.	.	−1	.	.	1	1	A	$\bar{A}$	$-A$	$-\bar{A}$	.	.	−1	−1	.
X.4	−1	−1	.	.	.	.	1	.	.	.	.	−1	.	.	1	1	$\bar{A}$	A	$-\bar{A}$	$-A$	.	.	−1	−1	.
X.5	−1	−1	1	1	1	1	.	1	1	1	1	.	−1	−1	.	.	.	.	.	.	B	$\bar{B}$	−1	−1	1
X.6	−1	−1	1	1	1	1	.	1	1	1	1	.	−1	−1	.	.	.	.	.	.	$\bar{B}$	B	−1	−1	1
X.7	.	.	2	2	2	2	−1	1	1	1	1	.	1	1	.	.	.	.	.	.	−1	−1	.	.	2
X.8	−1	−1	3	3	−1	−1	.	−2	−2	−2	−2	1	.	.	1	1	1	1	−1	−1	.	.	−1	−1	−1
X.9	.	.	−2	.	2	−2	1	3	1	1	−1	.	−1	1	.	.	−1	−1	−1	−1	.	.	−2	2	.
X.10	2	−2	−2	.	−2	2	1	−1	−3	1	−1	.	1	−1	.	.	−1	−1	1	1	.	.	.	.	.
X.11	−1	−1	−2	−2	−2	−2	−1	2	2	2	2	.	.	.	.	.	.	.	.	.	1	1	−1	−1	−2
X.12	.	.	.	.	.	.	.	1	1	1	1	1	−1	−1	1	1	.	.	.	.	.	.	.	.	.
X.13	.	.	.	.	.	.	.	1	1	1	1	1	−1	−1	1	1	.	.	.	.	.	.	.	.	.
X.14	.	.	.	.	.	.	.	.	.	.	.	−1	.	.	1	1	$\bar{A}$	A	$\bar{A}$	A	.	.	.	.	.
X.15	.	.	.	.	.	.	.	.	.	.	.	−1	.	.	1	1	A	$\bar{A}$	A	$\bar{A}$	.	.	.	.	.
X.16	1	1	.	.	.	.	1	.	.	.	.	2	.	.	.	.	−1	−1	−1	−1	.	.	1	1	.
X.17	−1	−1	.	.	.	.	1	.	.	.	.	1	.	.	−1	−1	D	$\bar{D}$	.	.	.	.	−1	−1	.
X.18	−1	−1	.	.	.	.	1	.	.	.	.	1	.	.	−1	−1	$\bar{D}$	D	.	.	.	.	−1	−1	.
X.19	1	1	.	.	.	.	.	1	1	1	1	.	−1	−1	.	.	−2	−2	.	.	.	.	1	1	.
X.20	−1	−1	1	1	1	1	.	.	.	.	.	−1	.	.	−1	−1	.	.	.	.	1	1	−1	−1	1
X.21	−1	−1	1	−1	1	1	.	4	−2	−2	.	−1	.	−2	−1	1	.	.	.	.	1	1	1	1	−1
X.22	1	1	1	−1	1	1	.	.	2	−2	.	−1	2	.	1	−1	.	.	.	.	1	1	−1	−1	−1
X.23	.	.	−1	−1	−1	−1	.	−1	−1	−1	−1	.	−1	−1	.	.	1	1	1	1	−1	−1	.	.	−1
X.24	−1	−1	−3	−3	1	1	.	.	.	.	.	2	.	.	.	.	2	2	.	.	.	.	−1	−1	1
X.25	.	.	−3	−3	1	1	1	.	.	.	.	−2	.	.	.	.	1	1	−1	−1	.	.	.	.	1
X.26	.	.	.	.	.	.	.	−2	−2	−2	−2	.	.	.	.	.	−1	−1	1	1	.	.	.	.	.
X.27	−1	−1	3	3	−1	−1	.	1	1	1	1	.	1	1	.	.	.	.	.	.	.	.	−1	−1	−1
X.28	.	.	−1	−1	−1	−1	.	1	1	1	1	.	1	1	.	.	.	.	.	.	−1	−1	.	.	−1
X.29	.	.	1	1	1	1	−1	−1	−1	−1	−1	.	−1	−1	.	.	.	.	.	.	1	1	.	.	1
X.30	−2	2	−2	.	−2	2	−1	1	3	−1	1	.	1	−1	.	.	.	.	.	.	.	.	.	.	.
X.31	.	.	−2	.	2	−2	−1	−3	−1	−1	1	.	−1	1	.	.	.	.	.	.	.	.	2	−2	.
X.32	−1	−1	.	.	.	.	.	3	1	−3	−1	1	−1	1	1	−1	.	.	.	.	.	.	1	1	.
X.33	1	1	.	.	.	.	.	2	−4	.	2	−2	−2	.	.	.	.	.	.	.	.	.	−1	−1	.
X.34	−1	−1	.	.	.	.	.	−2	.	.	2	−2	.	2	.	.	.	.	.	.	.	.	1	1	.
X.35	1	1	.	.	.	.	.	3	1	−3	−1	1	−1	1	−1	1	.	.	.	.	.	.	−1	−1	.
X.36	1	1	.	.	.	.	.	.	.	.	.	.	.	.	.	.	.	.	.	.	.	.	1	1	.
X.37	.	.	1	−1	1	1	.	1	−1	−1	1	.	−1	1	.	.	.	.	.	.	B	$\bar{B}$	.	.	−1
X.38	.	.	1	−1	1	1	.	1	−1	−1	1	.	−1	1	.	.	.	.	.	.	$\bar{B}$	B	.	.	−1
X.39	.	.	.	.	.	.	1	.	.	.	.	.	.	.	.	.	$-\bar{A}$	$-A$	$\bar{A}$	A	.	.	2	−2	.
X.40	.	.	.	.	.	.	1	.	.	.	.	.	.	.	.	.	$-A$	$-\bar{A}$	A	$\bar{A}$	.	.	2	−2	.
X.41	−2	2	.	.	.	.	1	.	.	.	.	.	.	.	.	.	$-\bar{A}$	$-A$	$-\bar{A}$	$-A$	.	.	.	.	.
X.42	−2	2	.	.	.	.	1	.	.	.	.	.	.	.	.	.	$-A$	$-\bar{A}$	$-A$	$-\bar{A}$	.	.	.	.	.
X.43	.	.	.	.	.	.	.	−1	−3	1	−1	.	1	−1	.	.	1	1	−1	−1	.	.	−2	2	.
X.44	2	−2	.	.	.	.	.	3	1	1	−1	.	−1	1	.	.	1	1	1	1	.	.	.	.	.
X.45	−1	−1	−1	1	−1	−1	.	1	−1	−1	1	.	1	−1	.	.	.	.	.	.	−1	−1	1	1	1
X.46	1	1	−1	1	−1	−1	.	1	−1	−1	1	.	1	−1	.	.	.	.	.	.	−1	−1	−1	−1	1
X.47	.	.	.	.	.	.	.	1	1	1	−3	−1	−1	−1	1	−1	.	.	.	.	.	.	.	.	.
X.48	.	.	.	.	.	.	.	−3	5	1	−3	−1	1	1	−1	1	.	.	.	.	.	.	.	.	.
X.49	.	.	2	−2	2	2	.	−2	2	2	−2	.	.	.	.	.	.	.	.	.	−1	−1	.	.	−2
X.50	−2	2	2	.	2	−2	.	.	.	.	.	.	.	.	.	.	−1	−1	1	1	.	.	.	.	.
X.51	.	.	2	.	−2	2	.	.	.	.	.	.	.	.	.	.	−1	−1	−1	−1	.	.	2	−2	.
X.52	.	.	1	−1	1	1	.	−2	−2	2	2	2	.	.	.	.	.	.	.	.	1	1	.	.	−1
X.53	.	.	3	−3	−1	−1	.	1	−3	1	1	.	1	1	.	.	.	.	.	.	.	.	.	.	1
X.54	.	.	−2	2	−2	−2	.	−1	1	1	−1	.	−1	1	.	.	.	.	.	.	1	1	.	.	2
X.55	.	.	3	−3	−1	−1	.	−3	1	1	1	.	−1	−1	.	.	.	.	.	.	.	.	.	.	1
X.56	.	.	2	.	−2	2	.	3	1	1	−1	.	−1	1	.	.	.	.	.	.	.	.	.	.	.
X.57	.	.	2	.	2	−2	.	−1	−3	1	−1	.	1	−1	.	.	.	.	.	.	.	.	.	.	.
X.58	.	.	.	.	.	.	.	1	−1	−1	1	.	1	−1	.	.	.	.	.	.	.	.	.	.	.
X.59	.	.	.	.	.	.	1	1	3	−1	1	.	1	−1	.	.	.	.	.	.	.	.	.	.	.
X.60	.	.	.	.	.	.	1	−3	−1	−1	1	.	−1	1	.	.	.	.	.	.	.	.	.	.	.
X.61	.	.	−2	.	2	−2	.	3	1	1	−1	.	1	−1	.	.	.	.	.	.	.	.	2	−2	.
X.62	−2	2	−2	.	−2	2	.	−1	−3	1	−1	.	−1	1	.	.	.	.	.	.	.	.	.	.	.
X.63	−1	−1	.	.	.	.	.	.	.	.	.	1	.	.	−1	1	.	.	.	.	.	.	1	1	.
X.64	1	1	.	.	.	.	.	.	.	.	.	1	.	.	−1	1	.	.	.	.	.	.	−1	−1	.
X.65	1	1	.	.	.	.	.	.	.	.	.	1	.	.	1	−1	.	.	.	.	.	.	−1	−1	.
X.66	−1	−1	.	.	.	.	.	.	.	.	.	1	.	.	1	−1	.	.	.	.	.	.	1	1	.
X.67	.	.	.	.	.	.	.	−2	2	−2	2	.	.	.	.	.	.	.	.	.	.	.	−2	2	.
X.68	2	−2	.	.	.	.	.	−2	2	−2	2	.	.	.	.	.	.	.	.	.	.	.	.	.	.
X.69	.	.	−3	3	1	1	.	1	−3	1	1	.	1	1	.	.	.	.	.	.	.	.	.	.	−1
X.70	.	.	−3	3	1	1	.	−3	1	1	1	.	−1	−1	.	.	.	.	.	.	.	.	.	.	−1
X.71	.	.	.	.	.	.	−1	2	−2	2	−2	.	.	.	.	.	.	.	.	.	.	.	.	.	.
X.72	.	.	.	.	.	.	.	.	.	.	.	−1	.	.	1	−1	.	.	.	.	.	.	.	.	.
X.73	.	.	.	.	.	.	.	.	.	.	.	−1	.	.	−1	1	.	.	.	.	.	.	.	.	.
X.74	.	.	.	.	.	.	.	3	−1	−1	−1	.	1	1	.	.	.	.	.	.	.	.	.	.	.
X.75	.	.	.	.	.	.	.	−1	3	−1	−1	.	−1	−1	.	.	.	.	.	.	.	.	.	.	.
X.76	.	.	3	−3	−1	−1	.	.	.	.	.	.	.	.	.	.	.	.	.	.	.	.	.	.	1
X.77	−1	−1	.	.	.	.	.	.	.	.	.	.	.	.	.	.	.	.	.	.	.	.	1	1	.
X.78	1	1	.	.	.	.	.	.	.	.	.	.	.	.	.	.	.	.	.	.	.	.	−1	−1	.
X.79	−1	−1	.	.	.	.	.	.	.	.	.	.	.	.	.	.	.	.	.	.	.	.	1	1	.
X.80	1	1	.	.	.	.	.	.	.	.	.	.	.	.	.	.	.	.	.	.	.	.	−1	−1	.

Character table of $E(\mathsf{Co}_1)$ *(continued)*

2	.	.	1	.	.	4	4	2	2	1	1
3	1	1	.	.	.	1	1	.	.	1	1
5	.	.	.	.	.	.	.	.	.	1	1
7	1	1	.	.	.	.	.	1	1	.	.
11	.	.	1	.	.	.	.	.	.	.	.
23	.	.	.	1	1	.	.	.	.	.	.
	21a	21b	22a	23a	23b	24a	24b	28a	28b	30a	30b
2P	21a	21b	11a	23a	23b	12a	12a	14a	14b	15a	15b
3P	7a	7b	22a	23a	23b	8a	8b	28b	28a	10b	10b
5P	21b	21a	22a	23b	23a	24a	24b	28b	28a	6a	6a
7P	3b	3b	22a	23b	23a	24a	24b	4a	4a	30b	30a
11P	21a	21b	2b	23b	23a	24a	24b	28a	28b	30b	30a
23P	21a	21b	22a	1a	1a	24a	24b	28a	28b	30a	30b
X.1	1	1	1	1	1	1	1	1	1	1	1
X.2	−1	−1	1	.	.	−1	−1	.	.	.	.
X.3	$\bar{A}$	A	1	−1	−1	.	.	$-A$	$-\bar{A}$	.	.
X.4	A	$\bar{A}$	1	−1	−1	.	.	$-\bar{A}$	$-A$	.	.
X.5	.	.	.	1	1	−1	−1	.	.	B	$\bar{B}$
X.6	.	.	.	1	1	−1	−1	.	.	$\bar{B}$	B
X.7	.	.	−1	−1	−1	1	1	.	.	−1	−1
X.8	1	1	.	.	.	.	.	−1	−1	.	.
X.9	.	.	−1	.	.	−1	1	1	1	.	.
X.10	.	.	−1	.	.	1	−1	−1	−1	.	.
X.11	.	.	−1	.	.	.	.	.	.	1	1
X.12	.	.	.	C	$\bar{C}$	−1	−1	.	.	.	.
X.13	.	.	.	$\bar{C}$	C	−1	−1	.	.	.	.
X.14	A	$\bar{A}$	.	1	1	.	.	$\bar{A}$	A	.	.
X.15	$\bar{A}$	A	.	1	1	.	.	A	$\bar{A}$	.	.
X.16	−1	−1	1	.	.	.	.	−1	−1	.	.
X.17	$-A$	$-\bar{A}$	1	.	.	.	.	.	.	.	.
X.18	$-\bar{A}$	$-A$	1	.	.	.	.	.	.	.	.
X.19	1	1	.	.	.	−1	−1	.	.	.	.
X.20	.	.	.	.	.	.	.	.	.	1	1
X.21	.	.	.	.	.	.	2	.	.	−1	−1
X.22	.	.	.	.	.	−2	.	.	.	−1	−1
X.23	1	1	.	.	.	−1	−1	1	1	−1	−1
X.24	−1	−1	.	.	.	.	.	.	.	.	.
X.25	1	1	1	.	.	.	.	−1	−1	.	.
X.26	−1	−1	.	1	1	.	.	1	1	.	.
X.27	.	.	.	.	.	1	1	.	.	.	.
X.28	.	.	.	1	1	1	1	.	.	−1	−1
X.29	.	.	−1	.	.	−1	−1	.	.	1	1
X.30	.	.	1	.	.	1	−1	.	.	.	.
X.31	.	.	1	.	.	−1	1	.	.	.	.
X.32	.	.	.	.	.	1	−1	.	.	.	.
X.33	.	.	.	.	.	2	.	.	.	.	.
X.34	.	.	.	.	.	.	−2	.	.	.	.
X.35	.	.	.	.	.	1	−1	.	.	.	.
X.36	.	.	.	−1	−1	.	.	.	.	.	.
X.37	.	.	.	.	.	1	−1	.	.	$-B$	$-\bar{B}$
X.38	.	.	.	.	.	1	−1	.	.	$-\bar{B}$	$-B$
X.39	.	.	−1	.	.	.	.	$-\bar{A}$	$-A$	.	.
X.40	.	.	−1	.	.	.	.	$-A$	$-\bar{A}$	.	.
X.41	.	.	−1	.	.	.	.	$\bar{A}$	A	.	.
X.42	.	.	−1	.	.	.	.	A	$\bar{A}$	.	.
X.43	.	.	.	.	.	1	−1	1	1	.	.
X.44	.	.	.	.	.	−1	1	−1	−1	.	.
X.45	.	.	.	.	.	−1	1	.	.	1	1
X.46	.	.	.	.	.	−1	1	.	.	1	1
X.47	.	.	.	.	.	1	1	.	.	.	.
X.48	.	.	.	.	.	−1	−1	.	.	.	.
X.49	.	.	.	.	.	.	.	.	.	1	1
X.50	.	.	.	.	.	.	.	−1	−1	.	.
X.51	.	.	.	.	.	.	.	1	1	.	.
X.52	.	.	.	.	.	.	.	.	.	−1	−1
X.53	.	.	.	.	.	−1	−1	.	.	.	.
X.54	.	.	.	.	.	1	−1	.	.	−1	−1
X.55	.	.	.	.	.	1	1	.	.	.	.
X.56	.	.	.	.	.	−1	1	.	.	.	.
X.57	.	.	.	.	.	1	−1	.	.	.	.
X.58	.	.	.	.	.	−1	1	.	.	.	.
X.59	.	.	−1	.	.	1	−1	.	.	.	.
X.60	.	.	−1	.	.	−1	1	.	.	.	.
X.61	.	.	.	.	.	1	−1	.	.	.	.
X.62	.	.	.	.	.	−1	1	.	.	.	.
X.63	.	.	.	.	.	.	.	.	.	.	.
X.64	.	.	.	.	.	.	.	.	.	.	.
X.65	.	.	.	.	.	.	.	.	.	.	.
X.66	.	.	.	.	.	.	.	.	.	.	.
X.67	.	.	.	.	.	.	.	.	.	.	.
X.68	.	.	.	.	.	.	.	.	.	.	.
X.69	.	.	.	.	.	−1	−1	.	.	.	.
X.70	.	.	.	.	.	1	1	.	.	.	.
X.71	.	.	1	.	.	.	.	.	.	.	.
X.72	.	.	.	.	.	.	.	.	.	.	.
X.73	.	.	.	.	.	.	.	.	.	.	.
X.74	.	.	.	.	.	−1	−1	.	.	.	.
X.75	.	.	.	.	.	1	1	.	.	.	.
X.76	.	.	.	.	.	.	.	.	.	.	.
X.77	.	.	.	.	.	.	.	.	.	.	.
X.78	.	.	.	.	.	.	.	.	.	.	.
X.79	.	.	.	.	.	.	.	.	.	.	.
X.80	.	.	.	.	.	.	.	.	.	.	.

where $A = \zeta(7)^4 + \zeta(7)^2 + \zeta(7)$, $B = 2\zeta(15)_3\zeta(15)_5^3 + 2\zeta(15)_3\zeta(15)_5^2 + \zeta(15)_3 + \zeta(15)_5^3 + \zeta(15)_5^2$, $C = \zeta(23)^{18} + \zeta(23)^{16} + \zeta(23)^{13} + \zeta(23)^{12} + \zeta(23)^9 + \zeta(23)^8 + \zeta(23)^6 + \zeta(23)^4 + \zeta(23)^3 + \zeta(23)^2 + \zeta(23)$, $D = -2(A+1)$, $E = -3(A+1)$.

7.6.2 *Character table of* $H(\mathsf{Co}_1) = \langle x, y, h \rangle$

	1a	2a	2b	2c	2d	2e	2f	2g	2h	3a	3b	3c	3d	3e	4a	4b	4c	4d	4e	4f
2	21	21	20	18	15	15	18	15	14	13	7	7	8	6	17	16	17	14	15	16
3	5	5	2	3	2	2	1	1	1	5	5	5	4	5	4	2	2	2	2	1
5	2	2	1	.	1	1	.	1	.	1	1	1	.	.	1	1	.	1	.	.
7	1	1	1	.	.	.	.	.	.	.	.	.	.	.	1	.	.	.	.	.
2P	1a	1a	1a	1a	1a	1a	1a	1a	1a	3a	3b	3c	3d	3e	2a	2b	2a	2a	2a	2b
3P	1a	2a	2b	2c	2d	2e	2f	2g	2h	1a	1a	1a	1a	1a	4a	4b	4c	4d	4e	4f
5P	1a	2a	2b	2c	2d	2e	2f	2g	2h	3a	3b	3c	3d	3e	4a	4b	4c	4d	4e	4f
7P	1a	2a	2b	2c	2d	2e	2f	2g	2h	3a	3b	3c	3d	3e	4a	4b	4c	4d	4e	4f
X.1	1	1	1	1	1	1	1	1	1	1	1	1	1	1	1	1	1	1	1	1
X.2	28	28	28	−4	4	4	−4	4	−4	10	10	10	1	1	28	4	−4	4	4	4
X.3	35	35	35	3	−5	−5	3	11	3	14	5	5	2	−1	35	11	3	−5	11	11
X.4	35	35	35	3	−5	−5	3	−5	3	5	5	14	2	−1	35	−5	3	11	−5	−5
X.5	35	35	35	3	11	11	3	−5	3	5	14	5	2	−1	35	−5	3	−5	−5	−5
X.6	50	50	50	18	10	10	18	10	2	5	5	5	5	−4	50	10	18	10	10	10
X.7	84	84	84	20	4	4	20	4	4	−6	−6	21	3	3	84	4	20	20	4	4
X.8	84	84	84	20	20	20	20	4	4	−6	21	−6	3	3	84	4	20	4	4	4
X.9	84	84	84	20	4	4	20	20	4	21	−6	−6	3	3	84	20	20	4	20	20
X.10	120	120	−8	24	.	.	−8	8	−8	36	.	.	6	3	8	32	8	.	−8	.
X.11	128	−128	.	.	16	−16	.	.	.	32	5	−4	8	2	.	.	.	.	.	.
X.12	135	135	7	39	15	15	7	−9	7	27	.	.	9	.	−9	31	−9	15	7	−1
X.13	175	175	175	−17	15	15	−17	15	−1	−5	−5	−5	4	13	175	15	−17	15	15	15
X.14	210	210	210	−14	10	10	−14	10	2	−15	−15	39	3	−6	210	10	−14	26	10	10
X.15	210	210	210	−14	10	10	−14	26	2	39	−15	−15	3	−6	210	26	−14	10	26	26
X.16	210	210	210	−14	26	26	−14	10	2	−15	39	−15	3	−6	210	10	−14	10	10	10
X.17	300	300	300	12	20	20	12	20	12	30	30	30	−6	3	300	20	12	20	20	20
X.18	350	350	350	−2	−10	−10	−2	−10	−2	35	35	35	−1	−1	350	−10	−2	−10	−10	−10
X.19	525	525	525	45	5	5	45	5	−19	30	30	30	3	12	525	5	45	5	5	5
X.20	567	567	567	−9	−9	−9	−9	−9	−9	.	.	81	.	.	567	−9	−9	39	−9	−9
X.21	567	567	567	−9	−9	−9	−9	39	−9	81	.	.	.	.	567	39	−9	−9	39	39
X.22	567	567	567	−9	39	39	−9	−9	−9	.	81	.	.	.	567	−9	−9	−9	−9	−9
X.23	700	700	700	−4	−20	−20	−4	−20	12	10	10	55	4	7	700	−20	−4	60	−20	−20
X.24	700	700	700	92	20	20	92	20	−4	−20	−20	−20	7	−2	700	20	92	20	20	20
X.25	700	700	700	−4	−20	−20	−4	60	12	55	10	10	4	7	700	60	−4	−20	60	60
X.26	700	700	700	−4	60	60	−4	−20	12	10	55	10	4	7	700	−20	−4	−20	−20	−20
X.27	840	840	840	8	−40	−40	8	−40	8	30	30	−24	3	3	840	−40	8	24	−40	−40
X.28	840	840	840	8	−40	−40	8	24	8	−24	30	30	3	3	840	24	8	−40	24	24
X.29	840	840	840	8	24	24	8	−40	8	30	−24	30	3	3	840	−40	8	−40	−40	−40
X.30	840	840	−56	−24	.	.	8	40	8	144	.	.	6	−6	56	80	−8	.	−8	−16
X.31	896	−896	.	.	−16	16	.	.	.	128	11	−16	8	−4	.	.	.	.	.	.
X.32	945	945	49	−15	−15	−15	17	−23	17	108	.	.	9	.	−63	97	33	−15	25	1
X.33	972	972	972	108	36	36	108	36	12	.	.	.	.	.	972	36	108	36	36	36
X.34	1050	1050	1050	58	−30	−30	58	−30	−6	15	15	15	6	−3	1050	−30	58	50	−30	−30
X.35	1050	1050	1050	58	−30	−30	58	50	−6	15	15	15	6	−3	1050	50	58	−30	50	50
X.36	1050	1050	1050	58	50	50	58	−30	−6	15	15	15	6	−3	1050	−30	58	−30	−30	−30
X.37	1344	1344	1344	64	64	64	64	.	.	−24	84	−24	−6	−6	1344	.	64	.	.	.
X.38	1344	1344	1344	64	.	.	64	64	.	84	−24	−24	−6	−6	1344	64	64	.	64	64
X.39	1344	1344	1344	64	.	.	64	.	.	−24	−24	84	−6	−6	1344	.	64	64	.	.
X.40	1400	1400	1400	−72	40	40	−72	40	−8	50	50	50	5	−4	1400	40	−72	40	40	40
X.41	1575	1575	1575	−57	15	15	−57	15	−9	90	−45	−45	.	9	1575	15	−57	15	15	15
X.42	1575	1575	1575	−57	15	15	−57	15	−9	−45	90	−45	.	9	1575	15	−57	15	15	15
X.43	1575	1575	1575	−57	15	15	−57	15	−9	−45	−45	90	.	9	1575	15	−57	15	15	15
X.44	1792	−1792	.	.	96	−96	.	.	.	−32	31	4	16	−8	.	.	.	.	.	.
X.45	1800	1800	−120	168	.	.	−56	40	8	.	.	.	18	−9	120	80	56	.	−8	−16
X.46	1890	1890	98	258	90	90	66	−6	18	−27	.	.	18	.	−126	74	−30	90	26	10
X.47	2100	2100	2100	52	20	20	52	20	4	−60	−60	75	3	−6	2100	20	52	−60	20	20
X.48	2100	2100	2100	52	−60	−60	52	20	4	−60	75	−60	3	−6	2100	20	52	20	20	20
X.49	2100	2100	2100	52	20	20	52	−60	4	75	−60	−60	3	−6	2100	−60	52	20	−60	−60
X.50	2240	2240	2240	−64	.	.	−64	64	.	−4	−40	−40	2	−10	2240	64	−64	.	64	64
X.51	2240	2240	2240	−64	64	64	−64	.	.	−40	−4	−40	2	−10	2240	.	−64	.	.	.
X.52	2240	2240	2240	−64	.	.	−64	.	.	−40	−40	−4	2	−10	2240	.	−64	64	.	.
X.53	2268	2268	2268	−36	−36	−36	−36	−36	12	.	.	81	.	.	2268	−36	−36	12	−36	−36
X.54	2268	2268	2268	−36	12	12	−36	−36	12	.	81	.	.	.	2268	−36	−36	−36	−36	−36
X.55	2268	2268	2268	−36	−36	−36	−36	12	12	81	.	.	.	.	2268	12	−36	−36	12	12
X.56	2520	2520	−168	120	.	.	−40	72	24	216	.	.	.	9	168	128	40	.	−8	−32
X.57	2520	2520	−168	−72	.	.	24	−8	24	216	.	.	.	9	168	48	−24	.	−24	16
X.58	2560	−2560	.	.	64	−64	.	.	.	160	34	−20	−8	4	.	.	.	.	.	.
X.59	2700	2700	140	204	60	60	76	−20	28	135	.	.	−9	.	−180	140	12	60	44	12
X.60	2835	2835	147	−45	−45	−45	51	11	3	162	.	.	.	.	−189	51	99	−45	27	19
X.61	2835	2835	2835	−45	−45	−45	−45	−45	3	.	.	−81	.	.	2835	−45	−45	51	−45	−45
X.62	2835	2835	2835	−45	51	51	−45	−45	3	.	−81	.	.	.	2835	−45	−45	−45	−45	−45
X.63	2835	2835	2835	−45	−45	−45	−45	51	3	−81	.	.	.	.	2835	51	−45	−45	51	51
X.64	2835	2835	147	−45	−45	−45	51	11	3	−81	.	.	.	.	−189	51	99	−45	27	19
X.65	2835	2835	147	−45	−45	−45	51	11	3	−81	.	.	.	.	−189	51	99	−45	27	19
X.66	3200	3200	3200	128	.	.	128	.	.	−40	−40	−40	−4	14	3200	.	128	.	.	.
X.67	3240	3240	−216	72	.	.	−24	40	−24	324	.	.	.	.	216	240	24	.	−72	16
X.68	3584	−3584	.	.	−64	64	.	.	.	32	14	−40	8	2	.	.	.	.	.	.
X.69	3584	−3584	.	.	−64	64	.	.	.	320	14	−4	8	2	.	.	.	.	.	.
X.70	3780	3780	196	132	−60	−60	4	−76	4	270	.	.	9	.	−252	244	−60	−60	52	−12
X.71	3780	3780	196	−156	60	60	−28	−36	−28	270	.	.	9	.	−252	124	36	60	28	−4
X.72	3780	3780	196	−60	−60	−60	68	−12	20	27	.	.	9	.	−252	148	132	−60	52	20
X.73	4096	4096	4096	.	.	.	.	.	.	64	64	64	−8	−8	4096	.	.	.	.	.
X.74	4200	4200	−280	264	.	.	−88	40	−24	180	.	.	12	−3	280	240	88	.	−72	16
X.75	4200	4200	4200	−24	40	40	−24	40	8	−30	−30	−30	−3	15	4200	40	−24	40	40	40
X.76	4200	4200	−280	72	.	.	−24	−40	−24	180	.	.	12	−3	280	−160	24	.	40	.
X.77	4725	4725	245	213	45	45	117	85	−27	135	.	.	18	.	−315	−115	69	45	5	45
X.78	4725	4725	245	405	45	45	53	−75	−11	135	.	.	18	.	−315	205	−123	45	37	−19
X.79	4725	4725	245	117	−75	−75	21	45	21	135	.	.	18	.	−315	−155	−27	165	−35	5
X.80	4725	4725	245	117	165	165	21	45	21	135	.	.	18	.	−315	−155	−27	−75	−35	5
X.81	4725	4725	245	−171	45	45	−11	−75	5	135	.	.	18	.	−315	205	69	45	37	−19
X.82	6075	6075	6075	27	−45	−45	27	−45	−21	.	.	.	.	.	6075	−45	27	−45	−45	−45
X.83	6075	6075	315	27	−45	−45	123	75	−21	.	.	.	.	.	−405	−45	171	−45	27	51
X.84	6075	6075	315	27	−45	−45	123	75	−21	.	.	.	.	.	−405	−45	171	−45	27	51
X.85	6400	−6400	.	.	160	−160	.	.	.	160	25	−20	40	−8	.	.	.	.	.	.
X.86	6720	6720	−448	−192	.	.	64	128	.	396	.	.	12	6	448	192	−64	.	.	−64
X.87	7168	−7168	.	.	128	−128	.	.	.	−128	16	16	−8	−32	.	.	.	.	.	.
X.88	7560	7560	392	456	120	120	200	56	8	−108	.	.	−9	.	−504	56	72	120	56	56

Character table of $H(\mathrm{Co}_1)$ (continued)

2	14	14	14	14	14	14	13	13	13	13	12	9	9	10	8	8	7	3	3	13	12	7	7
3	2	2	2	1	1	1	1	.	.	.	.	1	1	.	1	.	1	1	1	5	2	5	5
5	.	.	.	.	.	.	.	.	.	.	.	.	.	.	.	.	2	2	2	1	1	1	1
7	.	.	.	.	.	.	.	.	.	.	.	.	.	.	.	.	.	.	.	.	.	.	.
	4g	4h	4i	4j	4k	4l	4m	4n	4o	4p	4q	4r	4s	4t	4u	4v	5a	5b	5c	6a	6b	6c	6d
2P	2c	2b	2c	2b	2b	2b	2c	2b	2c	2c	2c	2d	2d	2f	2g	2h	5a	5b	5c	3a	3a	3b	3c
3P	4g	4h	4i	4j	4k	4l	4m	4n	4o	4p	4q	4r	4s	4t	4u	4v	5a	5b	5c	2a	2b	2a	2a
5P	4g	4h	4i	4j	4k	4l	4m	4n	4o	4p	4q	4r	4s	4t	4u	4v	1a	1a	1a	6a	6b	6c	6d
7P	4g	4h	4i	4j	4k	4l	4m	4n	4o	4p	4q	4r	4s	4t	4u	4v	5a	5b	5c	6a	6b	6c	6d
X.1	1	1	1	1	1	1	1	1	1	1	1	1	1	1	1	1	1	1	1	1	1	1	1
X.2	8	−4	8	−4	4	4	8	−4	.	.	.	.	.	.	.	.	3	3	3	10	10	10	10
X.3	7	3	7	3	−5	−5	7	3	−1	−1	−1	−1	−1	−1	3	−1	5	.	.	14	14	5	5
X.4	7	3	7	3	−5	11	7	3	−1	−1	−1	−1	−1	−1	−1	−1	.	.	5	5	5	5	14
X.5	7	3	7	3	11	−5	7	3	−1	−1	−1	3	3	−1	−1	−1	.	5	.	5	5	14	5
X.6	−2	18	−2	2	10	10	−2	2	6	6	6	2	2	6	2	2	.	.	.	5	5	5	5
X.7	4	20	4	4	4	20	4	4	4	4	4	.	.	4	.	.	−1	−1	4	−6	−6	−6	21
X.8	4	20	4	4	20	4	4	4	4	4	4	4	4	4	.	.	−1	4	−1	−6	−6	21	−6
X.9	4	20	4	4	4	4	4	4	4	4	4	.	.	4	4	.	4	−1	−1	21	21	−6	−6
X.10	12	.	12	8	.	.	−4	.	4	4	−4	.	.	.	.	.	10	.	.	36	4	.	.
X.11	16	.	−16	.	.	.	.	.	.	.	.	4	−4	.	.	.	8	3	−2	−32	.	−5	4
X.12	3	−1	3	7	−1	−1	3	−1	−5	11	3	3	3	−1	−1	−1	5	.	.	27	−5	.	.
X.13	−1	−17	−1	−1	15	15	−1	−1	−1	−1	−1	−1	−1	−1	−1	3	.	.	.	−5	−5	−5	−5
X.14	6	−14	6	2	10	26	6	2	−2	−2	−2	−2	−2	−2	−2	2	.	.	5	−15	−15	−15	39
X.15	6	−14	6	2	10	10	6	2	−2	−2	−2	−2	−2	−2	2	2	5	.	.	39	39	−15	−15
X.16	6	−14	6	2	26	10	6	2	−2	−2	−2	2	2	−2	−2	2	.	5	.	−15	−15	39	−15
X.17	8	12	8	12	20	20	8	12	.	.	.	.	.	.	.	.	.	.	.	30	30	30	30
X.18	26	−2	26	−2	−10	−10	26	−2	2	2	2	−2	−2	2	−2	2	.	.	.	35	35	35	35
X.19	−7	45	−7	−19	5	5	−7	−19	1	1	1	−3	−3	1	−3	1	.	.	.	30	30	30	30
X.20	15	−9	15	−9	−9	39	15	−9	−1	−1	−1	3	3	−1	3	−1	−3	−3	7	.	.	.	81
X.21	15	−9	15	−9	−9	−9	15	−9	−1	−1	−1	3	3	−1	−1	−1	7	−3	−3	81	81	.	.
X.22	15	−9	15	−9	39	−9	15	−9	−1	−1	−1	−1	−1	−1	3	−1	−3	7	−3	.	.	81	.
X.23	−4	−4	−4	12	−20	60	−4	12	−4	−4	−4	.	.	−4	.	.	.	.	.	10	10	10	55
X.24	.	92	.	−4	20	20	.	−4	8	8	8	.	.	8	.	.	.	.	.	−20	−20	−20	−20
X.25	−4	−4	−4	12	−20	−20	−4	12	−4	−4	−4	.	.	−4	4	.	.	.	.	55	55	10	10
X.26	−4	−4	−4	12	60	−20	−4	12	−4	−4	−4	4	4	−4	.	.	.	.	.	10	10	55	10
X.27	16	8	16	8	−40	24	16	8	.	.	.	.	.	.	.	.	.	.	−5	30	30	30	−24
X.28	16	8	16	8	−40	−40	16	8	.	.	.	.	.	.	.	.	−5	.	.	−24	−24	30	30
X.29	16	8	16	8	24	−40	16	8	.	.	.	.	.	.	.	.	.	−5	.	30	30	−24	30
X.30	36	.	36	−8	.	.	−12	.	−4	−4	4	.	.	.	4	.	20	.	.	144	16	.	.
X.31	48	.	−48	.	.	.	.	.	.	.	.	−4	4	.	.	.	16	1	−4	−128	.	−11	16
X.32	9	−7	9	17	1	1	9	−7	9	−7	1	−3	−3	−3	−3	−3	10	.	.	108	−20	.	.
X.33	.	108	.	12	36	36	.	12	8	8	8	.	.	8	.	.	−3	−3	−3	.	.	.	.
X.34	−10	58	−10	−6	−30	50	−10	−6	−2	−2	−2	−2	−2	−2	−2	−2	.	.	.	15	15	15	15
X.35	−10	58	−10	−6	−30	−30	−10	−6	−2	−2	−2	−2	−2	−2	2	−2	.	.	.	15	15	15	15
X.36	−10	58	−10	−6	50	−30	−10	−6	−2	−2	−2	2	2	−2	−2	−2	.	.	.	15	15	15	15
X.37	.	64	.	.	64	.	.	.	.	.	.	.	.	.	.	.	4	−1	4	−24	−24	84	−24
X.38	.	64	.	.	.	.	.	.	.	.	.	.	.	.	.	.	−1	4	4	84	84	−24	−24
X.39	.	64	.	.	.	64	.	.	.	.	.	.	.	.	.	.	4	4	−1	−24	−24	−24	84
X.40	−16	−72	−16	−8	40	40	−16	−8	.	.	.	.	.	.	.	.	.	.	.	50	50	50	50
X.41	11	−57	11	−9	15	15	11	−9	3	3	3	−1	−1	3	−1	−1	.	.	.	90	90	−45	−45
X.42	11	−57	11	−9	15	15	11	−9	3	3	3	−1	−1	3	−1	−1	.	.	.	−45	−45	90	−45
X.43	11	−57	11	−9	15	15	11	−9	3	3	3	−1	−1	3	−1	−1	.	.	.	−45	−45	−45	90
X.44	32	.	−32	.	.	.	.	.	.	.	.	8	−8	.	.	.	−8	7	2	32	.	−31	−4
X.45	−12	.	−12	−8	.	.	4	.	12	12	−12	.	.	.	−4	.	.	.	.	.	.	.	.
X.46	6	−14	6	18	−6	−6	6	2	−2	30	14	6	6	−2	−2	2	−5	.	.	−27	5	.	.
X.47	12	52	12	4	20	−60	12	4	−4	−4	−4	.	.	−4	.	.	.	.	.	−60	−60	−60	75
X.48	12	52	12	4	−60	20	12	4	−4	−4	−4	−4	−4	−4	.	.	.	.	.	−60	−60	75	−60
X.49	12	52	12	4	20	20	12	4	−4	−4	−4	.	.	−4	−4	.	.	.	.	75	75	−60	−60
X.50	.	−64	.	.	.	.	.	.	.	.	.	.	.	.	.	.	−5	.	.	−4	−4	−40	−40
X.51	.	−64	.	.	64	.	.	.	.	.	.	.	.	.	.	.	.	−5	.	−40	−40	−4	−40
X.52	.	−64	.	.	.	64	.	.	.	.	.	.	.	.	.	.	.	.	−5	−40	−40	−40	−4
X.53	−12	−36	−12	12	−36	12	−12	12	4	4	4	.	.	4	.	.	3	3	−2	.	.	.	81
X.54	−12	−36	−12	12	12	−36	−12	12	4	4	4	−4	−4	4	.	.	3	−2	3	.	.	81	.
X.55	−12	−36	−12	12	−36	−36	−12	12	4	4	4	.	.	4	−4	.	−2	3	3	81	81	.	.
X.56	12	.	12	−24	.	.	−4	.	4	4	−4	.	.	.	.	.	10	.	.	216	24	.	.
X.57	60	.	60	−24	.	.	−20	.	4	4	−4	.	.	.	−4	.	10	.	.	216	24	.	.
X.58	64	.	−64	.	.	.	.	.	.	.	.	.	.	.	.	.	.	5	.	−160	.	−34	20
X.59	12	−20	12	28	−4	−4	12	−4	12	12	12	.	.	−4	−4	.	.	.	.	135	−25	.	.
X.60	3	−21	3	3	3	3	3	−5	−13	3	−5	3	3	−1	−1	3	5	.	.	162	−30	.	.
X.61	3	−45	3	3	−45	51	3	3	3	3	3	3	3	3	3	−1	.	.	5	.	.	.	−81
X.62	3	−45	3	3	51	−45	3	3	3	3	3	−5	−5	3	3	−1	.	5	.	.	.	−81	.
X.63	3	−45	3	3	−45	−45	3	3	3	3	3	3	3	3	−5	−1	5	.	.	−81	−81	.	.
X.64	3	−21	3	3	3	3	3	−5	−13	3	−5	3	3	−1	−1	3	5	.	.	−81	15	.	.
X.65	3	−21	3	3	3	3	3	−5	−13	3	−5	3	3	−1	−1	3	5	.	.	−81	15	.	.
X.66	.	128	.	.	.	.	.	.	.	.	.	.	.	.	.	.	.	.	.	−40	−40	−40	−40
X.67	36	.	36	24	.	.	−12	.	−4	−4	4	.	.	.	−4	.	20	.	.	324	36	.	.
X.68	64	.	−64	.	.	.	.	.	.	.	.	.	.	.	.	.	−16	−1	−6	−32	.	−14	40
X.69	64	.	−64	.	.	.	.	.	.	.	.	.	.	.	.	.	24	−1	4	−320	.	−14	4
X.70	12	4	12	4	4	4	12	4	−4	−4	−4	.	.	4	.	4	15	.	.	270	−50	.	.
X.71	24	4	24	−28	−4	−4	24	4	.	.	.	.	.	.	.	.	15	.	.	270	−50	.	.
X.72	12	−28	12	20	4	4	12	−12	−4	−4	−4	.	.	−4	−4	.	−10	.	.	27	−5	.	.
X.73	.	.	.	.	.	.	.	.	.	.	.	.	.	.	.	.	−4	−4	−4	64	64	64	64
X.74	−12	.	−12	24	.	.	4	.	12	12	−12	.	.	.	4	.	.	.	.	180	20	.	.
X.75	−8	−24	−8	8	40	40	−8	8	−8	−8	−8	.	.	−8	.	.	.	.	.	−30	−30	−30	−30
X.76	84	.	84	24	.	.	−28	.	−4	−4	4	.	.	.	.	.	.	.	.	180	20	.	.
X.77	−15	−35	−15	−27	−3	−3	−15	13	9	−7	1	−3	−3	5	5	1	.	.	.	135	−25	.	.
X.78	−15	−3	−15	−11	−3	−3	−15	−3	−23	25	1	−3	−3	−3	1	−3	.	.	.	135	−25	.	.
X.79	21	−3	21	21	5	−11	21	−3	5	−11	−3	−3	−3	1	1	1	.	.	.	135	−25	.	.
X.80	21	−3	21	21	−11	5	21	−3	5	−11	−3	9	9	1	1	1	.	.	.	135	−25	.	.
X.81	−15	−3	−15	5	−3	−3	−15	13	9	−7	1	−3	−3	−3	1	1	.	.	.	135	−25	.	.
X.82	−9	27	−9	−21	−45	−45	−9	−21	−1	−1	−1	3	3	−1	3	3	.	.	.	.	.	.	.
X.83	−9	−45	−9	−21	3	3	−9	3	−17	−1	−9	3	3	3	3	−1	.	.	.	.	.	.	.
X.84	−9	−45	−9	−21	3	3	−9	3	−17	−1	−9	3	3	3	3	−1	.	.	.	.	.	.	.
X.85	−32	.	32	.	.	.	.	.	.	.	.	8	−8	.	.	.	.	.	.	−160	.	−25	20
X.86	.	.	.	.	.	.	.	.	.	.	.	.	.	.	8	.	10	.	.	396	44	.	.
X.87	.	.	.	.	.	.	.	.	.	.	.	.	.	.	.	.	8	−2	−2	128	.	−16	−16
X.88	.	−56	.	8	−8	−8	.	8	16	16	16	.	.	.	.	.	5	.	.	−108	20	.	.

Character table of $H(\mathsf{Co}_1)$ *(continued)*

2	10	8	6	7	7	7	6	6	6	9	7	5	5	6	6	6	6	7	5	3	11	11	11	10	10
3	3	4	5	3	3	3	3	3	3	1	2	3	3	2	2	2	2	1	1	.	2	1	1	1	1
5	.	.	.	.	.	.	.	.	.	.	.	.	.	.	.	.	.	.	.	.	1	.	.	.	.
7	.	.	.	.	.	.	.	.	.	.	.	.	.	.	.	.	.	.	.	1	.	.	.	.	.
	6e	6f	6g	6h	6i	6j	6k	6l	6m	6n	6o	6p	6q	6r	6s	6t	6u	6v	6w	7a	8a	8b	8c	8d	8e
2P	3a	3d	3e	3b	3d	3b	3e	3e	3c	3a	3d	3d	3d	3b	3d	3d	3b	3d	3d	7a	4a	4a	4c	4f	4b
3P	2c	2a	2a	2c	2c	2c	2c	2c	2c	2g	2b	2c	2c	2d	2d	2e	2e	2f	2h	7a	8a	8b	8c	8d	8e
5P	6e	6f	6g	6h	6i	6j	6k	6l	6m	6n	6o	6p	6q	6r	6s	6t	6u	6v	6w	7a	8a	8b	8c	8d	8e
7P	6e	6f	6g	6h	6i	6j	6k	6l	6m	6n	6o	6p	6q	6r	6s	6t	6u	6v	6w	1a	8a	8b	8c	8d	8e
X.1	1	1	1	1	1	1	1	1	1	1	1	1	1	1	1	1	1	1	1	1	1	1	1	1	1
X.2	2	1	1	2	−1	2	5	5	2	−2	1	−1	−1	−2	1	1	−2	−1	−1	.	4	−4	8	.	.
X.3	6	2	−1	−3	.	−3	3	3	−3	2	2	.	.	1	−2	−2	1	.	.	.	11	3	7	3	3
X.4	−3	2	−1	−3	.	−3	3	3	6	1	2	.	.	1	−2	−2	1	.	.	.	−5	3	7	−1	−1
X.5	−3	2	−1	6	.	6	3	3	−3	1	2	.	.	2	2	2	2	.	.	.	−5	3	7	−1	−1
X.6	−3	5	−4	−3	3	−3	.	.	−3	1	5	3	3	1	1	1	1	3	−1	1	10	2	−2	2	2
X.7	2	3	3	2	−1	2	−1	−1	5	−2	3	5	−1	−2	1	1	−2	−1	1	.	4	4	4	.	.
X.8	2	3	3	5	−1	5	−1	−1	2	−2	3	−1	5	5	−1	−1	5	−1	1	.	4	4	4	.	.
X.9	5	3	3	2	5	2	−1	−1	2	5	3	−1	−1	−2	1	1	−2	5	1	.	20	4	4	4	4
X.10	12	6	3	.	6	.	3	3	.	−4	−2	.	.	.	.	.	.	−2	−2	1	.	.	.	8	4
X.11	.	−8	−2	−3	.	3	−6	6	.	.	.	.	.	1	4	−4	−1	.	.	2	.	.	.	.	.
X.12	3	9	.	.	9	.	.	.	.	3	1	3	3	.	3	3	.	1	1	2	−1	−1	−1	7	−5
X.13	−5	4	13	−5	−2	−5	1	1	−5	3	4	−2	−2	3	.	.	3	−2	2	.	15	−1	−1	−1	−1
X.14	1	3	−6	1	1	1	−2	−2	7	1	3	−5	1	1	1	1	1	1	−1	.	10	2	6	−2	−2
X.15	7	3	−6	1	−5	1	−2	−2	1	−1	3	1	1	1	1	1	1	−5	−1	.	26	2	6	2	2
X.16	1	3	−6	7	1	7	−2	−2	1	1	3	1	−5	−1	−1	−1	−1	1	−1	.	10	2	6	−2	−2
X.17	6	−6	3	6	.	6	3	3	6	2	−6	.	.	2	2	2	2	.	.	−1	20	12	8	.	.
X.18	−5	−1	−1	−5	1	−5	7	7	−5	−1	−1	1	1	−1	−1	−1	−1	1	1	.	−10	−2	26	−2	−2
X.19	6	3	12	6	3	6	.	.	6	2	3	3	3	2	−1	−1	2	3	−1	.	5	−19	−7	−3	−3
X.20	.	.	.	.	.	.	.	.	9	.	.	.	.	.	.	.	.	.	.	.	−9	−9	15	3	3
X.21	9	.	.	.	.	.	.	.	.	−3	.	.	.	.	.	.	.	.	.	.	39	−9	15	−1	−1
X.22	.	.	.	9	.	9	.	.	.	.	.	.	.	−3	.	.	−3	.	.	.	−9	−9	15	3	3
X.23	2	4	7	2	2	2	−1	−1	−1	−2	4	−4	2	−2	−2	−2	−2	2	.	.	−20	12	−4	.	.
X.24	−4	7	−2	−4	−1	−4	2	2	−4	−4	7	−1	−1	−4	−1	−1	−4	−1	−1	.	20	−4	.	.	.
X.25	−1	4	7	2	−4	2	−1	−1	2	3	4	2	2	−2	−2	−2	−2	−4	.	.	60	12	−4	4	4
X.26	2	4	7	−1	2	−1	−1	−1	2	−2	4	2	−4	3	.	.	3	2	.	.	−20	12	−4	.	.
X.27	−10	3	3	−10	−1	−10	−1	−1	8	2	3	−1	−1	2	−1	−1	2	−1	−1	.	−40	8	16	.	.
X.28	8	3	3	−10	−1	−10	−1	−1	−10	.	3	−1	−1	2	−1	−1	2	−1	−1	.	24	8	16	.	.
X.29	−10	3	3	8	−1	8	−1	−1	−10	2	3	−1	−1	.	3	3	.	−1	−1	.	−40	8	16	.	.
X.30	24	6	−6	.	−6	.	6	6	.	4	−2	.	.	.	.	.	.	2	2	.	.	.	.	4	8
X.31	.	−8	4	3	.	−3	−12	12	.	.	.	.	.	−1	−4	4	1	.	.	.	.	.	.	.	.
X.32	12	9	.	.	−9	.	.	.	.	4	1	3	3	.	−3	−3	.	−1	−1	.	−7	1	−3	9	−3
X.33	.	.	.	.	.	.	.	.	.	.	.	.	.	.	.	.	.	.	.	−1	36	12	.	.	.
X.34	7	6	−3	7	−2	7	1	1	−17	3	6	4	−2	3	.	.	3	−2	.	.	−30	−6	−10	−2	−2
X.35	−17	6	−3	7	4	7	1	1	7	−1	6	−2	−2	3	.	.	3	4	.	.	50	−6	−10	2	2
X.36	7	6	−3	−17	−2	−17	1	1	7	3	6	−2	4	−1	2	2	−1	−2	.	.	−30	−6	−10	−2	−2
X.37	−8	−6	−6	4	−2	4	−2	−2	−8	.	−6	−2	4	4	−2	−2	4	−2	.	.	.	.	.	.	.
X.38	4	−6	−6	−8	4	−8	−2	−2	−8	4	−6	−2	−2	.	.	.	.	4	.	.	64	.	.	.	.
X.39	−8	−6	−6	−8	−2	−8	−2	−2	4	.	−6	4	−2	.	.	.	.	−2	.	.	.	.	.	.	.
X.40	−6	5	−4	−6	−3	−6	.	.	−6	−2	5	−3	−3	−2	1	1	−2	−3	1	.	40	−8	−16	.	.
X.41	−6	.	9	3	.	3	−3	−3	3	−6	.	.	.	3	.	.	3	.	.	.	15	−9	11	−1	−1
X.42	3	.	9	−6	.	−6	−3	−3	3	3	.	.	.	−6	.	.	−6	.	.	.	15	−9	11	−1	−1
X.43	3	.	9	3	.	3	−3	−3	−6	3	.	.	.	3	.	.	3	.	.	.	15	−9	11	−1	−1
X.44	.	−16	8	−9	.	9	.	.	.	.	.	.	.	3	.	.	−3	.	.	.	.	.	.	.	.
X.45	−24	18	−9	.	6	.	3	3	.	4	−6	.	.	.	.	.	.	−2	2	1	.	.	.	4	8
X.46	−3	18	.	.	.	.	.	.	.	−3	2	6	6	.	.	.	.	.	.	.	−14	−6	−2	2	2
X.47	4	3	−6	4	1	4	−2	−2	−5	−4	3	1	1	−4	−1	−1	−4	1	1	.	20	4	12	.	.
X.48	4	3	−6	−5	1	−5	−2	−2	4	−4	3	1	1	3	3	3	3	1	1	.	20	4	12	.	.
X.49	−5	3	−6	4	1	4	−2	−2	4	3	3	1	1	−4	−1	−1	−4	1	1	.	−60	4	12	−4	−4
X.50	−4	2	−10	8	−4	8	2	2	8	4	2	2	2	.	.	.	.	−4	.	.	64	.	.	.	.
X.51	8	2	−10	−4	2	−4	2	2	8	.	2	2	−4	4	−2	−2	4	2	.	.	.	.	.	.	.
X.52	8	2	−10	8	2	8	2	2	−4	.	2	−4	2	.	.	.	.	2	.	.	.	.	.	.	.
X.53	.	.	.	.	.	.	.	.	9	.	.	.	.	.	.	.	.	.	.	.	−36	12	−12	.	.
X.54	.	.	.	9	.	9	.	.	.	.	.	.	.	−3	.	.	−3	.	.	.	−36	12	−12	.	.
X.55	9	.	.	.	.	.	.	.	.	−3	.	.	.	.	.	.	.	.	.	.	12	12	−12	−4	−4
X.56	24	.	9	.	12	.	−3	−3	.	.	.	.	.	.	.	.	.	−4	.	.	.	.	.	.	12
X.57	.	.	9	.	.	.	9	9	.	4	.	.	.	.	.	.	.	.	.	.	.	.	.	−4	−8
X.58	.	8	−4	−6	.	6	−12	12	.	.	.	.	.	−2	4	−4	2	.	.	−2	.	.	.	.	.
X.59	15	−9	.	.	9	.	.	.	.	7	−1	−3	−3	.	3	3	.	1	1	−2	−20	−4	−4	4	4
X.60	18	.	.	.	.	.	.	.	.	2	.	.	.	.	.	.	.	.	.	.	−21	3	−1	−5	7
X.61	.	.	.	.	.	.	.	.	−9	.	.	.	.	.	.	.	.	.	.	.	−45	3	3	3	3
X.62	.	.	.	−9	.	−9	.	.	.	.	.	.	.	3	.	.	3	.	.	.	−45	3	3	3	3
X.63	−9	.	.	.	.	.	.	.	.	3	.	.	.	.	.	.	.	.	.	.	51	3	3	−5	−5
X.64	−9	.	.	.	.	.	.	.	.	−1	.	.	.	.	.	.	.	.	.	.	−21	3	−1	−5	7
X.65	−9	.	.	.	.	.	.	.	.	−1	.	.	.	.	.	.	.	.	.	.	−21	3	−1	−5	7
X.66	8	−4	14	8	−4	8	2	2	8	.	−4	−4	−4	.	.	.	.	−4	.	1	.	.	.	.	.
X.67	36	.	.	.	.	.	.	.	.	−8	.	.	.	.	.	.	.	.	.	−1	.	.	.	12	.
X.68	.	−8	−2	6	.	−6	−6	6	.	.	.	.	.	2	−4	4	−2	.	.	.	.	.	.	.	.
X.69	.	−8	−2	6	.	−6	−6	6	.	.	.	.	.	2	−4	4	−2	.	.	.	.	.	.	.	.
X.70	6	9	.	.	9	.	.	.	.	2	1	−3	−3	.	−3	−3	.	1	1	.	−4	−4	−4	12	−12
X.71	6	9	.	.	−9	.	.	.	.	−6	1	−3	−3	.	3	3	.	−1	−1	.	−4	4	−8	.	.
X.72	3	9	.	.	−9	.	.	.	.	3	1	3	3	.	−3	−3	.	−1	−1	.	−28	4	−4	4	4
X.73	.	−8	−8	.	.	.	.	.	.	.	−8	.	.	.	.	.	.	.	.	1	.	.	.	.	.
X.74	−12	12	−3	.	12	.	−3	−3	.	−8	−4	.	.	.	.	.	.	−4	.	.	.	.	.	12	.
X.75	−6	−3	15	−6	3	−6	3	3	−6	−2	−3	3	3	−2	1	1	−2	3	−1	.	40	8	−8	.	.
X.76	−36	12	−3	.	.	.	9	9	.	−4	−4	.	.	.	.	.	.	.	.	.	.	.	.	−8	−4
X.77	15	18	.	.	.	.	.	.	.	7	2	6	6	.	.	.	.	.	.	.	−35	−3	5	−11	1
X.78	−9	18	.	.	18	.	.	.	.	3	2	.	.	.	.	.	.	2	−2	.	5	5	5	5	−7
X.79	−9	18	.	.	.	.	.	.	.	3	2	.	.	.	−6	−6	.	.	.	.	5	−3	−7	−7	5
X.80	−9	18	.	.	.	.	.	.	.	3	2	.	.	.	6	6	.	.	.	.	5	−3	−7	−7	5
X.81	−9	18	.	.	−18	.	.	.	.	3	2	.	.	.	.	.	.	−2	2	.	5	−11	5	5	−7
X.82	.	.	.	.	.	.	.	.	.	.	.	.	.	.	.	.	.	.	.	−1	−45	−21	−9	3	3
X.83	.	.	.	.	.	.	.	.	.	.	.	.	.	.	.	.	.	.	.	−1	−45	3	3	3	−9
X.84	.	.	.	.	.	.	.	.	.	.	.	.	.	.	.	.	.	.	.	−1	−45	3	3	3	−9
X.85	.	−40	8	9	.	−9	.	.	.	.	.	.	.	1	4	−4	−1	.	.	2	.	.	.	.	.
X.86	12	12	6	.	−12	.	−6	−6	.	8	−4	.	.	.	.	.	.	4	.	.	.	.	.	.	.
X.87	.	8	32	.	.	.	.	.	.	.	.	.	.	8	−4	4	−8	.	.	.	.	.	.	.	.
X.88	−12	−9	.	.	−9	.	.	.	.	−4	−1	−3	−3	.	−3	−3	.	−1	−1	.	−56	−8	.	.	.

Character table of $H(\mathsf{Co}_1)$ *(continued)*

	8f	8g	8h	8i	8j	8k	8l	8m	8n	8o	8p	8q	8r	8s	9a	9b	9c	10a	10b	10c	10d	10e	10f	10g	12_1	12_2
2	11	10	10	8	8	8	8	8	8	8	7	7	7	7	3	1	1	7	6	3	3	4	3	3	10	10
3	.	.	.	1	1	.	.	.	.	.	.	.	.	.	3	3	3	1	1	1	1	.	.	.	3	2
5	.	.	.	.	.	.	.	.	.	.	.	.	.	.	.	.	.	2	1	2	2	1	1	1	1	.
7	.	.	.	.	.	.	.	.	.	.	.	.	.	.	.	.	.	.	.	.	.	.	.	.	.	.
2P	4c	4b	4f	4d	4e	4o	4o	4j	4k	4l	4n	4m	4g	4p	9a	9b	9c	5a	5a	5b	5c	5a	5b	5b	6a	6a
3P	8f	8g	8h	8i	8j	8k	8l	8m	8n	8o	8p	8q	8r	8s	3e	3e	3e	10a	10b	10c	10d	10e	10f	10g	4a	4c
5P	8f	8g	8h	8i	8j	8k	8l	8m	8n	8o	8p	8q	8r	8s	9a	9b	9c	2a	2b	2a	2a	2g	2d	2e	12_1	12_2
7P	8f	8g	8h	8i	8j	8k	8l	8m	8n	8o	8p	8q	8r	8s	9a	9b	9c	10a	10b	10c	10d	10e	10f	10g	12_1	12_2
X.1	1	1	1	1	1	1	1	1	1	1	1	1	1	1	1	1	1	1	1	1	1	1	1	1	1	1
X.2	.	.	.	.	.	2	2	.	.	.	.	−2	−2	2	1	1	1	3	3	3	3	−1	−1	−1	10	2
X.3	−1	3	3	−1	3	1	1	−1	−1	−1	−1	1	1	1	2	−1	−1	5	5	.	.	1	.	.	14	6
X.4	−1	−1	−1	3	−1	1	1	−1	−1	3	−1	1	1	1	−1	2	−1	.	.	.	5	.	.	.	5	−3
X.5	−1	−1	−1	−1	−1	1	1	−1	3	−1	−1	1	1	1	−1	−1	2	.	.	5	.	.	1	1	5	−3
X.6	6	2	2	2	2	.	.	2	2	2	2	.	.	.	−1	−1	−1	.	.	.	.	.	.	.	5	−3
X.7	4	.	.	4	.	.	.	.	.	4	.	.	.	.	.	.	.	−1	−1	−1	4	−1	−1	−1	−6	2
X.8	4	.	.	.	.	.	.	.	4	.	.	.	.	.	.	.	.	−1	−1	4	−1	−1	.	.	−6	2
X.9	4	4	4	.	4	.	.	.	.	.	.	.	.	.	.	.	.	4	4	−1	−1	.	−1	−1	21	5
X.10	.	−4	.	.	.	2	2	.	.	.	.	2	−2	−2	3	.	.	10	2	.	.	−2	.	.	−4	−4
X.11	.	.	.	.	.	4	−4	.	.	.	.	.	.	.	2	−1	2	−8	.	−3	2	.	1	−1	.	.
X.12	−1	3	−1	3	−1	1	1	3	−1	−1	−1	1	1	1	.	.	.	5	−3	.	.	1	.	.	3	3
X.13	−1	−1	−1	−1	−1	−1	−1	3	−1	−1	3	−1	−1	−1	1	1	1	.	.	.	.	.	.	.	−5	−5
X.14	−2	−2	−2	2	−2	.	.	2	−2	2	2	.	.	.	.	.	.	.	.	.	5	.	.	.	−15	1
X.15	−2	2	2	−2	2	.	.	2	−2	−2	2	.	.	.	.	.	.	5	5	.	.	1	.	.	39	7
X.16	−2	−2	−2	−2	−2	.	.	2	2	−2	2	.	.	.	.	.	.	.	.	5	.	.	1	1	−15	1
X.17	.	.	.	.	.	−2	−2	.	.	.	.	2	2	−2	.	.	.	.	.	.	.	.	.	.	30	6
X.18	2	−2	−2	−2	−2	.	.	2	−2	−2	2	.	.	.	−1	−1	−1	.	.	.	.	.	.	.	35	−5
X.19	1	−3	−3	−3	−3	−1	−1	1	−3	−3	1	−1	−1	−1	.	.	.	.	.	.	.	.	.	.	30	6
X.20	−1	3	3	−1	3	−1	−1	−1	3	−1	−1	−1	−1	−1	.	.	.	−3	−3	−3	7	1	1	1	.	.
X.21	−1	−1	−1	3	−1	−1	−1	−1	3	3	−1	−1	−1	−1	.	.	.	7	7	−3	−3	−1	1	1	81	9
X.22	−1	3	3	3	3	−1	−1	−1	−1	3	−1	−1	−1	−1	.	.	.	−3	−3	7	−3	1	−1	−1	.	.
X.23	−4	.	.	4	.	.	.	.	.	4	.	.	.	.	1	−2	1	.	.	.	.	.	.	.	10	2
X.24	8	.	.	.	.	−2	−2	.	.	.	.	2	2	−2	1	1	1	.	.	.	.	.	.	.	−20	−4
X.25	−4	4	4	.	4	.	.	.	.	.	.	.	.	.	−2	1	1	.	.	.	.	.	.	.	55	−1
X.26	−4	.	.	.	.	.	.	.	4	.	.	.	.	.	1	1	−2	.	.	.	.	.	.	.	10	2
X.27	.	.	.	.	.	.	.	.	.	.	.	.	.	.	.	.	.	.	.	.	−5	.	.	.	30	−10
X.28	.	.	.	.	.	.	.	.	.	.	.	.	.	.	.	.	.	−5	−5	.	.	−1	.	.	−24	8
X.29	.	.	.	.	.	.	.	.	.	.	.	.	.	.	.	.	.	.	.	−5	.	.	−1	−1	30	−10
X.30	.	.	−4	.	−4	2	2	.	.	.	.	−2	2	−2	3	.	.	20	4	.	.	.	.	.	−16	−8
X.31	.	.	.	.	.	4	−4	.	.	.	.	.	.	.	2	−1	−1	−16	.	−1	4	.	−1	1	.	.
X.32	5	5	1	−3	−3	1	1	1	1	1	1	1	1	1	.	.	.	10	−6	.	.	2	.	.	12	12
X.33	8	.	.	.	.	2	2	.	.	.	.	−2	−2	2	.	.	.	−3	−3	−3	−3	1	1	1	.	.
X.34	−2	−2	−2	2	−2	.	.	−2	−2	2	−2	.	.	.	.	.	.	.	.	.	.	.	.	.	15	7
X.35	−2	2	2	−2	2	.	.	−2	−2	−2	−2	.	.	.	.	.	.	.	.	.	.	.	.	.	15	−17
X.36	−2	−2	−2	−2	−2	.	.	−2	2	−2	−2	.	.	.	.	.	.	.	.	.	.	.	.	.	15	7
X.37	.	.	.	.	.	.	.	.	.	.	.	.	.	.	.	.	.	4	4	−1	4	.	−1	−1	−24	−8
X.38	.	.	.	.	.	.	.	.	.	.	.	.	.	.	.	.	.	−1	−1	4	4	−1	.	.	84	4
X.39	.	.	.	.	.	.	.	.	.	.	.	.	.	.	.	.	.	4	4	4	−1	.	.	.	−24	−8
X.40	.	.	.	.	.	.	.	.	.	.	.	.	.	.	−1	−1	−1	.	.	.	.	.	.	.	50	−6
X.41	3	−1	−1	−1	−1	1	1	−1	−1	−1	−1	1	1	1	.	.	.	.	.	.	.	.	.	.	90	−6
X.42	3	−1	−1	−1	−1	1	1	−1	−1	−1	−1	1	1	1	.	.	.	.	.	.	.	.	.	.	−45	3
X.43	3	−1	−1	−1	−1	1	1	−1	−1	−1	−1	1	1	1	.	.	.	.	.	.	.	.	.	.	−45	3
X.44	.	.	.	.	.	.	.	.	.	.	.	.	.	.	−2	1	1	8	.	−7	−2	.	1	−1	.	.
X.45	.	.	−4	.	4	−2	−2	.	.	.	.	2	−2	2	.	.	.	.	.	.	.	.	.	.	.	8
X.46	−10	2	2	6	−2	.	.	2	−2	−2	−2	.	.	.	.	.	.	−5	3	.	.	−1	.	.	−3	−3
X.47	−4	.	.	−4	.	.	.	.	.	−4	.	.	.	.	.	.	.	.	.	.	.	.	.	.	−60	4
X.48	−4	.	.	.	.	.	.	.	−4	.	.	.	.	.	.	.	.	.	.	.	.	.	.	.	−60	4
X.49	−4	−4	−4	.	−4	.	.	.	.	.	.	.	.	.	.	.	.	.	.	.	.	.	.	.	75	−5
X.50	.	.	.	.	.	.	.	.	.	.	.	.	.	.	2	−1	−1	−5	−5	.	.	−1	.	.	−4	−4
X.51	.	.	.	.	.	.	.	.	.	.	.	.	.	.	−1	−1	2	.	.	−5	.	.	−1	−1	−40	8
X.52	.	.	.	.	.	.	.	.	.	.	.	.	.	.	−1	2	−1	.	.	.	−5	.	.	.	−40	8
X.53	4	.	.	−4	.	.	.	.	.	−4	.	.	.	.	.	.	.	3	3	3	−2	−1	−1	−1	.	.
X.54	4	.	.	.	.	.	.	.	−4	.	.	.	.	.	.	.	.	3	3	−2	3	−1	2	2	.	.
X.55	4	−4	−4	.	−4	.	.	.	.	.	.	.	.	.	.	.	.	−2	−2	3	3	2	−1	−1	81	9
X.56	.	4	−8	.	.	−2	−2	.	.	.	.	−2	2	2	.	.	.	10	2	.	.	2	.	.	−24	−8
X.57	.	.	4	.	4	2	2	.	.	.	.	−2	2	−2	.	.	.	10	2	.	.	2	.	.	−24	.
X.58	.	.	.	.	.	.	.	.	.	.	.	.	.	.	−2	1	1	.	.	−5	.	.	−1	1	.	.
X.59	−4	4	4	.	−4	.	.	.	.	.	.	.	.	.	.	.	.	.	.	.	.	.	.	.	15	15
X.60	7	−1	3	3	−1	−1	−1	−1	−1	−1	−1	−1	−1	−1	.	.	.	5	−3	.	.	1	.	.	18	18
X.61	3	3	3	−5	3	−1	−1	−1	3	−5	−1	−1	−1	−1	.	.	.	.	.	.	5	.	.	.	.	.
X.62	3	3	3	3	3	−1	−1	−1	−5	3	−1	−1	−1	−1	.	.	.	.	.	5	.	.	1	1	.	.
X.63	3	−5	−5	3	−5	−1	−1	−1	3	3	−1	−1	−1	−1	.	.	.	5	5	.	.	1	.	.	−81	−9
X.64	7	−1	3	3	−1	−1	−1	−1	−1	−1	−1	−1	−1	−1	.	.	.	5	−3	.	.	1	.	.	−9	−9
X.65	7	−1	3	3	−1	−1	−1	−1	−1	−1	−1	−1	−1	−1	.	.	.	5	−3	.	.	1	.	.	−9	−9
X.66	.	.	.	.	.	.	.	.	.	.	.	.	.	.	−1	−1	−1	.	.	.	.	.	.	.	−40	8
X.67	.	−8	4	.	4	−2	−2	.	.	.	.	2	−2	2	.	.	.	20	4	.	.	.	.	.	−36	−12
X.68	.	.	.	.	.	.	.	.	.	.	.	.	.	.	−4	−1	−1	16	.	1	6	.	1	−1	.	.
X.69	.	.	.	.	.	.	.	.	.	.	.	.	.	.	2	2	−1	−24	.	1	−4	.	1	−1	.	.
X.70	−4	4	−4	.	.	.	.	−4	.	.	.	.	.	.	.	.	.	15	−9	.	.	−1	.	.	30	6
X.71	.	.	.	.	.	2	2	.	.	.	.	−2	−2	2	.	.	.	15	−9	.	.	−1	.	.	30	6
X.72	12	4	4	.	−4	.	.	.	.	.	.	.	.	.	.	.	.	−10	6	.	.	−2	.	.	3	3
X.73	.	.	.	.	.	.	.	.	.	.	.	.	.	.	1	1	1	−4	−4	−4	−4	.	.	.	64	.
X.74	.	−8	4	.	−4	2	2	.	.	.	.	−2	2	−2	−3	.	.	.	.	.	.	.	.	.	−20	4
X.75	−8	.	.	.	.	.	.	.	.	.	.	.	.	.	.	.	.	.	.	.	.	.	.	.	−30	−6
X.76	.	4	.	.	.	2	2	.	.	.	.	2	−2	−2	−3	.	.	.	.	.	.	.	.	.	−20	12
X.77	−11	−7	−3	−3	5	−1	−1	−3	1	1	1	−1	−1	−1	.	.	.	.	.	.	.	.	.	.	15	15
X.78	5	1	−3	−3	1	−1	−1	1	1	1	1	−1	−1	−1	.	.	.	.	.	.	.	.	.	.	15	−9
X.79	1	−3	1	9	1	1	1	−3	1	−3	1	1	1	1	.	.	.	.	.	.	.	.	.	.	15	−9
X.80	1	−3	1	−3	1	1	1	−3	−3	1	1	1	1	1	.	.	.	.	.	.	.	.	.	.	15	−9
X.81	5	1	−3	−3	1	−1	−1	5	1	1	−3	−1	−1	−1	.	.	.	.	.	.	.	.	.	.	15	−9
X.82	−1	3	3	3	3	1	1	3	3	3	3	1	1	1	.	.	.	.	.	.	.	.	.	.	.	.
X.83	3	−1	−5	3	3	1	1	3	−1	−1	−1	1	1	1	.	.	.	.	.	.	.	.	.	.	.	.
X.84	3	−1	−5	3	3	1	1	3	−1	−1	−1	1	1	1	.	.	.	.	.	.	.	.	.	.	.	.
X.85	.	.	.	.	.	.	.	.	.	.	.	.	.	.	−2	1	−2	.	.	.	.	.	.	.	.	.
X.86	.	.	.	.	−8	.	.	.	.	.	.	.	.	.	−3	.	.	10	2	.	.	−2	.	.	−44	−4
X.87	.	.	.	.	.	.	.	.	.	.	.	.	.	.	−2	1	1	−8	.	2	2	.	−2	2	.	.
X.88	−16	.	.	.	.	.	.	.	.	.	.	.	.	.	.	.	.	5	−3	.	.	1	.	.	−12	−12

Character table of $H(Co_1)$ *(continued)*

	12_3	12_4	12_5	12_6	12_7	12_8	12_9	12_{10}	12_{11}	12_{12}	12_{13}	12_{14}	12_{15}	12_{16}	12_{17}	12_{18}	12_{19}	12_{20}	12_{21}	12_{22}	12_{23}
2	9	9	5	8	6	9	7	7	6	5	5	5	5	5	5	5	5	5	5	5	5
3	2	2	4	2	3	1	2	2	2	2	2	2	2	2	2	2	2	2	2	2	2
5	.	.	.	.	.	.	.	.	.	.	.	.	.	.	.	.	.	.	.	.	.
7	.	.	.	.	.	.	.	.	.	.	.	.	.	.	.	.	.	.	.	.	.
2P	6a	6b	6g	6b	6f	6b	6e	6e	6f	6f	6d	6k	6o	6o	6f	6k	6o	6h	6h	6g	6m
3P	4e	4b	4a	4h	4a	4f	4g	4i	4c	4e	4d	4g	4h	4b	4d	4i	4h	4g	4i	4c	4g
5P	12_3	12_4	12_5	12_6	12_7	12_8	12_9	12_{10}	12_{11}	12_{12}	12_{13}	12_{14}	12_{15}	12_{16}	12_{17}	12_{18}	12_{19}	12_{20}	12_{21}	12_{22}	12_{23}
7P	12_3	12_4	12_5	12_6	12_7	12_8	12_9	12_{10}	12_{11}	12_{12}	12_{13}	12_{14}	12_{15}	12_{16}	12_{17}	12_{18}	12_{19}	12_{20}	12_{21}	12_{22}	12_{23}
X.1	1	1	1	1	1	1	1	1	1	1	1	1	1	1	1	1	1	1	1	1	1
X.2	−2	−2	1	2	1	−2	2	2	−1	1	−2	−1	−1	1	1	−1	−1	2	2	5	2
X.3	2	2	−1	6	2	2	−2	−2	.	2	1	1	.	2	−2	1	.	1	1	3	1
X.4	1	1	−1	−3	2	1	1	1	.	−2	2	1	.	−2	2	1	.	1	1	3	−2
X.5	1	1	−1	−3	2	1	1	1	.	−2	1	1	.	−2	−2	1	.	−2	−2	3	1
X.6	1	1	−4	−3	5	1	1	1	3	1	1	−2	3	1	1	−2	3	1	1	.	1
X.7	−2	−2	3	2	3	−2	−2	−2	−1	1	5	1	−1	1	−1	1	5	−2	−2	−1	1
X.8	−2	−2	3	2	3	−2	−2	−2	−1	1	−2	1	5	1	1	1	−1	1	1	−1	−2
X.9	5	5	3	5	3	5	1	1	5	−1	−2	1	−1	−1	1	1	−1	−2	−2	−1	−2
X.10	4	8	−1	.	2	.	.	.	2	−2	.	3	.	2	.	3	.	.	.	−1	.
X.11	.	.	.	.	.	.	4	−4	.	.	.	−2	.	.	.	2	.	1	−1	.	−2
X.12	−5	7	.	−1	−3	−1	3	3	−3	1	.	.	−1	1	3	.	−1	.	.	.	.
X.13	3	3	13	−5	4	3	−1	−1	−2	.	3	−1	−2	.	.	−1	−2	−1	−1	1	−1
X.14	1	1	−6	1	3	1	−3	−3	1	1	−1	.	1	1	−1	.	−5	−3	−3	−2	3
X.15	−1	−1	−6	7	3	−1	3	3	−5	−1	1	.	1	−1	1	.	1	−3	−3	−2	−3
X.16	1	1	−6	1	3	1	−3	−3	1	1	1	.	−5	1	1	.	1	3	3	−2	−3
X.17	2	2	3	6	−6	2	2	2	.	2	2	−1	.	2	2	−1	.	2	2	3	2
X.18	−1	−1	−1	−5	−1	−1	−1	−1	1	−1	−1	−1	1	−1	−1	−1	1	−1	−1	7	−1
X.19	2	2	12	6	3	2	2	2	3	−1	2	2	3	−1	−1	2	3	2	2	.	2
X.20	.	.	.	.	.	.	.	.	.	.	−3	.	.	.	.	.	.	.	.	.	−3
X.21	−3	−3	.	9	.	−3	−3	−3	.	.	.	.	.	.	.	.	.	.	.	.	.
X.22	.	.	.	.	.	.	.	.	.	.	.	.	.	.	.	.	.	−3	−3	.	.
X.23	−2	−2	7	2	4	−2	2	2	2	−2	3	−1	2	−2	.	−1	−4	2	2	−1	−1
X.24	−4	−4	−2	−4	7	−4	.	.	−1	−1	−4	.	−1	−1	−1	.	−1	.	.	2	.
X.25	3	3	7	−1	4	3	−1	−1	−4	.	−2	−1	2	.	−2	−1	2	2	2	−1	2
X.26	−2	−2	7	2	4	−2	2	2	2	−2	−2	−1	−4	−2	−2	−1	2	−1	−1	−1	2
X.27	2	2	3	−10	3	2	−2	−2	−1	−1	.	1	−1	−1	3	1	−1	−2	−2	−1	4
X.28	.	.	3	8	3	.	4	4	−1	3	2	1	−1	3	−1	1	−1	−2	−2	−1	−2
X.29	2	2	3	−10	3	2	−2	−2	−1	−1	2	1	−1	−1	−1	1	−1	4	4	−1	−2
X.30	4	−4	2	.	2	−4	.	.	−2	−2	.	.	.	2	.	.	.	.	.	−2	.
X.31	.	.	.	.	.	.	.	.	.	.	.	.	.	.	.	.	.	3	−3	.	.
X.32	4	4	.	−4	−3	4	.	.	3	1	.	.	−1	1	−3	.	−1	.	.	.	.
X.33	.	.	.	.	.	.	.	.	.	.	.	.	.	.	.	.	.	.	.	.	.
X.34	3	3	−3	7	6	3	−1	−1	−2	.	−1	−1	−2	.	2	−1	4	−1	−1	1	−1
X.35	−1	−1	−3	−17	6	−1	−1	−1	4	2	3	−1	−2	2	.	−1	−2	−1	−1	1	−1
X.36	3	3	−3	7	6	3	−1	−1	−2	.	3	−1	4	.	.	−1	−2	−1	−1	1	−1
X.37	.	.	−6	−8	−6	.	.	.	−2	.	.	.	4	.	.	.	−2	.	.	−2	.
X.38	4	4	−6	4	−6	4	.	.	4	−2	.	.	−2	−2	.	.	−2	.	.	−2	.
X.39	.	.	−6	−8	−6	.	.	.	−2	.	4	.	−2	.	−2	.	4	.	.	−2	.
X.40	−2	−2	−4	−6	5	−2	2	2	−3	1	−2	2	−3	1	1	2	−3	2	2	.	2
X.41	−6	−6	9	−6	.	−6	2	2	.	.	3	−1	.	.	.	−1	.	−1	−1	−3	−1
X.42	3	3	9	3	.	3	−1	−1	.	.	3	−1	.	.	.	−1	.	2	2	−3	−1
X.43	3	3	9	3	.	3	−1	−1	.	.	−6	−1	.	.	.	−1	.	−1	−1	−3	2
X.44	.	.	.	.	.	.	−4	4	.	.	.	−4	.	.	.	4	.	−1	1	.	2
X.45	4	−4	3	.	6	−4	.	.	2	−2	.	−3	.	2	.	−3	.	.	.	−1	.
X.46	5	−7	.	1	−6	1	−3	−3	.	2	.	.	−2	2	.	.	−2	.	.	.	.
X.47	−4	−4	−6	4	3	−4	.	.	1	−1	3	.	1	−1	3	.	1	.	.	−2	3
X.48	−4	−4	−6	4	3	−4	.	.	1	−1	−4	.	1	−1	−1	.	1	3	3	−2	.
X.49	3	3	−6	−5	3	3	3	3	1	3	−4	.	1	3	−1	.	1	.	.	−2	.
X.50	4	4	−10	−4	2	4	.	.	−4	−2	.	.	2	−2	.	.	2	.	.	2	.
X.51	.	.	−10	8	2	.	.	.	2	.	.	.	−4	.	.	.	2	.	.	2	.
X.52	.	.	−10	8	2	.	.	.	2	.	4	.	2	.	−2	.	−4	.	.	2	.
X.53	.	.	.	.	.	.	.	.	.	.	−3	.	.	.	.	.	.	.	.	.	−3
X.54	.	.	.	.	.	.	.	.	.	.	.	.	.	.	.	.	.	−3	−3	.	.
X.55	−3	−3	.	9	.	−3	−3	−3	.	.	.	.	.	.	.	.	.	.	.	.	.
X.56	16	8	−3	.	.	−8	.	.	4	4	.	3	.	−4	.	3	.	.	.	1	.
X.57	−12	−12	−3	.	.	4	.	.	.	.	.	−3	.	.	.	−3	.	.	.	−3	.
X.58	.	.	.	.	.	.	4	−4	.	.	.	4	.	.	.	−4	.	−2	2	.	−2
X.59	−1	11	.	−5	3	3	3	3	−3	−1	.	.	1	−1	3	.	1	.	.	.	.
X.60	18	−6	.	−6	.	10	−6	−6	.	.	.	.	.	.	.	.	.	.	.	.	.
X.61	.	.	.	.	.	.	.	.	.	.	3	.	.	.	.	.	.	.	.	.	3
X.62	.	.	.	.	.	.	.	.	.	.	.	.	.	.	.	.	.	3	3	.	.
X.63	3	3	.	−9	.	3	3	3	.	.	.	.	.	.	.	.	.	.	.	.	.
X.64	−9	3	.	3	.	−5	3	3	.	.	.	.	.	.	.	.	.	.	.	.	.
X.65	−9	3	.	3	.	−5	3	3	.	.	.	.	.	.	.	.	.	.	.	.	.
X.66	.	.	14	8	−4	.	.	.	−4	.	.	.	−4	.	.	.	−4	.	.	2	.
X.67	.	12	.	.	.	4	.	.	.	.	.	.	.	.	.	.	.	.	.	.	.
X.68	.	.	.	.	.	.	4	−4	.	.	.	−2	.	.	.	2	.	−2	2	.	4
X.69	.	.	.	.	.	.	−8	8	.	.	.	−2	.	.	.	2	.	−2	2	.	−2
X.70	−14	10	.	−2	−3	−6	−6	−6	−3	1	.	.	1	1	−3	.	1	.	.	.	.
X.71	10	−14	.	−2	−3	2	6	6	3	1	.	.	1	1	3	.	1	.	.	.	.
X.72	−5	7	.	−1	−3	−1	3	3	3	1	.	.	−1	1	−3	.	−1	.	.	.	.
X.73	.	.	−8	.	−8	.	.	.	.	.	.	.	.	.	.	.	.	.	.	.	.
X.74	.	12	1	.	4	4	.	.	4	.	.	−3	.	.	.	−3	.	.	.	1	.
X.75	−2	−2	15	−6	−3	−2	−2	−2	3	1	−2	1	3	1	1	1	3	−2	−2	3	−2
X.76	4	8	1	.	4	.	.	.	.	4	.	3	.	−4	.	3	.	.	.	−3	.
X.77	−1	11	.	−5	−6	3	3	3	.	2	.	.	−2	2	.	.	−2	.	.	.	.
X.78	−5	7	.	3	−6	−1	3	3	−6	−2	.	.	.	−2	.	.	.	.	.	.	.
X.79	−5	7	.	3	−6	−1	3	3	.	−2	.	.	.	−2	6	.	.	.	.	.	.
X.80	−5	7	.	3	−6	−1	3	3	.	−2	.	.	.	−2	−6	.	.	.	.	.	.
X.81	−5	7	.	3	−6	−1	3	3	6	−2	.	.	.	−2	.	.	.	.	.	.	.
X.82	.	.	.	.	.	.	.	.	.	.	.	.	.	.	.	.	.	.	.	.	.
X.83	.	.	.	.	.	.	.	.	.	.	.	.	.	.	.	.	.	.	.	.	.
X.84	.	.	.	.	.	.	.	.	.	.	.	.	.	.	.	.	.	.	.	.	.
X.85	.	.	.	.	.	.	4	−4	.	.	.	4	.	.	.	−4	.	1	−1	.	−2
X.86	.	−12	−2	.	4	−4	.	.	−4	.	.	.	.	.	.	.	.	.	.	2	.
X.87	.	.	.	.	.	.	.	.	.	.	.	.	.	.	.	.	.	.	.	.	.
X.88	−4	−4	.	4	3	−4	.	.	3	−1	.	.	1	−1	−3	.	1	.	.	.	.

Character table of $H(\mathsf{Co}_1)$ (continued)

2	5	5	5	5	5	4	4	4	3	2	2	3	1	1	6	6	3	1	1	5	4	2	6
3	2	1	1	1	1	1	1	1	.	.	.	1	1	1	.	.	3	3	3	1	.	.	2
5	.	.	.	.	.	.	.	.	.	.	.	1	1	1	.	.	.	.	.	1	1	1	.
7	.	.	.	.	.	.	.	.	1	1	1	.	.	.	.	.	.	.	.	.	.	.	.
	12_{24}	12_{25}	12_{26}	12_{27}	12_{28}	12_{29}	12_{30}	12_{31}	$14a$	$14b$	$14c$	$15a$	$15b$	$15c$	$16a$	$16b$	$18a$	$18b$	$18c$	$20a$	$20b$	$20c$	$24a$
$2P$	$6m$	$6o$	$6o$	$6o$	$6n$	$6r$	$6l$	$6r$	$7a$	$7a$	$7a$	$15a$	$15b$	$15c$	$8c$	$8f$	$9a$	$9b$	$9c$	$10a$	$10b$	$10d$	12_1
$3P$	$4i$	$4k$	$4j$	$4l$	$4u$	$4r$	$4m$	$4s$	$14a$	$14c$	$14b$	$5a$	$5b$	$5c$	$16a$	$16b$	$6g$	$6g$	$6g$	$20a$	$20b$	$20c$	$8a$
$5P$	12_{24}	12_{25}	12_{26}	12_{27}	12_{28}	12_{29}	12_{30}	12_{31}	$14a$	$14c$	$14b$	$3a$	$3b$	$3c$	$16a$	$16b$	$18a$	$18b$	$18c$	$4a$	$4b$	$4d$	$24a$
$7P$	12_{24}	12_{25}	12_{26}	12_{27}	12_{28}	12_{29}	12_{30}	12_{31}	$2a$	$2b$	$2b$	$15a$	$15b$	$15c$	$16a$	$16b$	$18a$	$18b$	$18c$	$20a$	$20b$	$20c$	$24a$
$X.1$	1	1	1	1	1	1	1	1	1	1	1	1	1	1	1	1	1	1	1	1	1	1	1
$X.2$	2	1	−1	1	.	.	−1	.	.	.	.	.	.	.	−2	2	1	1	1	3	−1	−1	−2
$X.3$	1	−2	.	−2	.	−1	1	−1	.	.	.	−1	.	.	1	1	2	−1	−1	5	1	.	2
$X.4$	−2	−2	.	2	−1	−1	1	−1	.	.	.	.	.	−1	1	1	−1	2	−1	.	.	1	1
$X.5$	1	2	.	−2	−1	.	1	.	.	.	.	.	−1	.	1	1	−1	−1	2	.	.	.	1
$X.6$	1	1	−1	1	−1	−1	−2	−1	1	1	1	.	.	.	.	.	−1	−1	−1	.	.	.	1
$X.7$	1	1	1	−1	.	.	1	.	.	.	.	−1	−1	1	.	.	.	.	.	−1	−1	.	−2
$X.8$	−2	−1	1	1	.	1	1	1	.	.	.	−1	1	−1	.	.	.	.	.	−1	−1	−1	−2
$X.9$	−2	1	1	1	1	.	1	.	.	.	.	1	−1	−1	.	.	.	.	.	4	.	−1	5
$X.10$	.	.	2	.	.	.	−1	.	1	−1	−1	1	.	.	.	.	3	.	.	−2	2	.	.
$X.11$	2	.	.	.	.	1	.	−1	−2	.	.	2	.	1	.	.	−2	1	−2	.	.	.	.
$X.12$	.	−1	1	−1	−1	.	.	.	2	.	.	2	.	.	−1	−1	.	.	.	1	1	.	−1
$X.13$	−1	.	2	.	−1	−1	−1	−1	.	.	.	.	.	.	−1	−1	1	1	1	.	.	.	3
$X.14$	3	1	−1	−1	1	1	.	1	.	.	.	.	.	−1	.	.	.	.	.	.	.	1	1
$X.15$	−3	1	−1	1	−1	1	.	1	.	.	.	−1	.	.	.	.	.	.	.	5	1	.	−1
$X.16$	−3	−1	−1	1	1	−1	.	−1	.	.	.	.	−1	.	.	.	.	.	.	.	.	.	1
$X.17$	2	2	.	2	.	.	−1	.	−1	−1	−1	.	.	.	2	−2	.	.	.	.	.	.	2
$X.18$	−1	−1	1	−1	1	1	−1	1	.	.	.	.	.	.	.	.	−1	−1	−1	.	.	.	−1
$X.19$	2	−1	−1	−1	.	.	2	.	.	.	.	.	.	.	−1	−1	.	.	.	.	.	.	2
$X.20$	−3	.	.	.	.	.	.	.	.	.	.	.	.	1	−1	−1	.	.	.	−3	1	−1	.
$X.21$	.	.	.	.	−1	.	.	.	.	.	.	1	.	.	−1	−1	.	.	.	7	−1	1	−3
$X.22$	.	.	.	.	.	−1	.	−1	.	.	.	.	1	.	−1	−1	.	.	.	−3	1	1	.
$X.23$	−1	−2	.	.	.	.	−1	.	.	.	.	.	.	.	.	.	1	−2	1	.	.	.	−2
$X.24$	.	−1	−1	−1	.	.	.	.	.	.	.	.	.	.	2	−2	1	1	1	.	.	.	−4
$X.25$	2	−2	.	−2	1	.	−1	.	.	.	.	.	.	.	.	.	−2	1	1	.	.	.	3
$X.26$	2	.	.	−2	.	1	−1	1	.	.	.	.	.	.	.	.	1	1	−2	.	.	.	−2
$X.27$	4	−1	−1	3	.	.	1	.	.	.	.	.	.	1	.	.	.	.	.	.	.	−1	2
$X.28$	−2	−1	−1	−1	.	.	1	.	.	.	.	1	.	.	.	.	.	.	.	−5	−1	.	.
$X.29$	−2	3	−1	−1	.	.	1	.	.	.	.	.	1	.	.	.	.	.	.	.	.	.	2
$X.30$	.	.	−2	.	−2	.	.	.	.	.	.	−1	.	.	.	.	3	.	.	−4	.	.	.
$X.31$	.	.	.	.	.	−1	.	1	.	.	.	−2	1	−1	.	.	−2	1	1	.	.	.	.
$X.32$	.	1	−1	1	.	.	.	.	.	.	.	−2	.	.	−1	−1	.	.	.	2	2	.	−4
$X.33$	.	.	.	.	.	.	.	.	−1	−1	−1	.	.	.	−2	2	.	.	.	−3	1	1	.
$X.34$	−1	.	.	2	1	1	−1	1	.	.	.	.	.	.	.	.	.	.	.	.	.	.	3
$X.35$	−1	.	.	.	−1	1	−1	1	.	.	.	.	.	.	.	.	.	.	.	.	.	.	−1
$X.36$	−1	2	.	.	1	−1	−1	−1	.	.	.	.	.	.	.	.	.	.	.	.	.	.	3
$X.37$	.	−2	.	.	.	.	.	.	.	.	.	1	−1	1	.	.	.	.	.	4	.	.	.
$X.38$	.	.	.	.	.	.	.	.	.	.	.	−1	1	1	.	.	.	.	.	−1	−1	.	4
$X.39$	.	.	.	−2	.	.	.	.	.	.	.	1	1	−1	.	.	.	.	.	4	.	−1	.
$X.40$	2	1	1	1	.	.	2	.	.	.	.	.	.	.	.	.	−1	−1	−1	.	.	.	−2
$X.41$	−1	.	.	.	2	−1	−1	−1	.	.	.	.	.	.	1	1	.	.	.	.	.	.	−6
$X.42$	−1	.	.	.	−1	2	−1	2	.	.	.	.	.	.	1	1	.	.	.	.	.	.	3
$X.43$	2	.	.	.	−1	−1	−1	−1	.	.	.	.	.	.	1	1	.	.	.	.	.	.	3
$X.44$	−2	.	.	.	.	−1	.	1	.	.	.	−2	1	−1	.	.	2	−1	−1	.	.	.	.
$X.45$	.	.	−2	.	2	.	1	.	1	−1	−1	.	.	.	.	.	.	.	.	.	.	.	.
$X.46$	.	.	.	.	1	.	.	.	.	.	.	−2	.	.	.	.	.	.	.	−1	−1	.	1
$X.47$	3	−1	1	3	.	.	.	.	.	.	.	.	.	.	.	.	.	.	.	.	.	.	−4
$X.48$	.	3	1	−1	.	−1	.	−1	.	.	.	.	.	.	.	.	.	.	.	.	.	.	−4
$X.49$	.	−1	1	−1	−1	.	.	.	.	.	.	.	.	.	.	.	.	.	.	.	.	.	3
$X.50$	.	.	.	.	.	.	.	.	.	.	.	1	.	.	.	.	2	−1	−1	−5	−1	.	4
$X.51$	.	−2	.	.	.	.	.	.	.	.	.	.	1	.	.	.	−1	−1	2	.	.	.	.
$X.52$	.	.	.	−2	.	.	.	.	.	.	.	.	.	1	.	.	−1	2	−1	.	.	−1	.
$X.53$	−3	.	.	.	.	.	.	.	.	.	.	.	.	1	.	.	.	.	.	3	−1	2	.
$X.54$	.	.	.	.	.	−1	.	−1	.	.	.	.	1	.	.	.	.	.	.	3	−1	−1	.
$X.55$	.	.	.	.	−1	.	.	.	.	.	.	1	.	.	.	.	.	.	.	−2	2	−1	−3
$X.56$	.	.	.	.	.	.	−1	.	.	.	.	1	.	.	.	.	.	.	.	−2	−2	.	.
$X.57$	.	.	.	.	2	.	1	.	.	.	.	1	.	.	.	.	.	.	.	−2	−2	.	.
$X.58$	2	.	.	.	.	.	.	.	2	.	.	.	−1	.	.	.	2	−1	−1	.	.	.	.
$X.59$	.	−1	1	−1	−1	.	.	.	−2	.	.	.	.	.	.	.	.	.	.	.	.	.	−5
$X.60$	.	.	.	.	2	.	.	.	.	.	.	2	.	.	1	1	.	.	.	1	1	.	−6
$X.61$	3	.	.	.	.	.	.	.	.	.	.	.	.	−1	−1	−1	.	.	.	.	.	1	.
$X.62$	.	.	.	.	.	1	.	1	.	.	.	.	−1	.	−1	−1	.	.	.	.	.	.	.
$X.63$	.	.	.	.	1	.	.	.	.	.	.	−1	.	.	−1	−1	.	.	.	5	1	.	3
$X.64$	.	.	.	.	−1	.	.	.	.	.	.	−1	.	.	1	1	.	.	.	1	1	.	3
$X.65$	.	.	.	.	−1	.	.	.	.	.	.	−1	.	.	1	1	.	.	.	1	1	.	3
$X.66$	.	.	.	.	.	.	.	.	1	1	1	.	.	.	.	.	−1	−1	−1	.	.	.	.
$X.67$	.	.	.	.	2	.	.	.	−1	1	1	−1	.	.	.	.	.	.	.	−4	.	.	.
$X.68$	−4	.	.	.	.	.	.	.	.	.	.	2	−1	.	.	.	4	1	1	.	.	.	.
$X.69$	2	.	.	.	.	.	.	.	.	.	.	.	−1	1	.	.	−2	−2	1	.	.	.	.
$X.70$	.	1	1	1	.	.	.	.	.	.	.	.	.	.	.	.	.	.	.	3	−1	.	2
$X.71$	.	−1	−1	−1	.	.	.	.	.	.	.	.	.	.	2	−2	.	.	.	3	−1	.	2
$X.72$	.	1	−1	1	−1	.	.	.	.	.	.	2	.	.	.	.	.	.	.	−2	−2	.	−1
$X.73$	.	.	.	.	.	.	.	.	1	1	1	−1	−1	−1	.	.	1	1	1	−4	.	.	.
$X.74$	.	.	.	.	−2	.	1	.	.	.	.	.	.	.	.	.	−3	.	.	.	.	.	.
$X.75$	−2	1	−1	1	.	.	1	.	.	.	.	.	.	.	.	.	.	.	.	.	.	.	−2
$X.76$	.	.	.	.	.	.	−1	.	.	.	.	.	.	.	.	.	−3	.	.	.	.	.	.
$X.77$	.	.	.	.	−1	.	.	.	.	.	.	.	.	.	1	1	.	.	.	.	.	.	−5
$X.78$	.	.	−2	.	1	.	.	.	.	.	.	.	.	.	1	1	.	.	.	.	.	.	−1
$X.79$	.	2	.	−2	1	.	.	.	.	.	.	.	.	.	−1	−1	.	.	.	.	.	.	−1
$X.80$	.	−2	.	2	1	.	.	.	.	.	.	.	.	.	−1	−1	.	.	.	.	.	.	−1
$X.81$	.	.	2	.	1	.	.	.	.	.	.	.	.	.	1	1	.	.	.	.	.	.	−1
$X.82$	.	.	.	.	.	.	.	.	−1	−1	−1	.	.	.	1	1	.	.	.	.	.	.	.
$X.83$	.	.	.	.	.	.	.	.	−1	B	$\bar{B}$	.	.	.	−1	−1	.	.	.	.	.	.	.
$X.84$	.	.	.	.	.	.	.	.	−1	$\bar{B}$	B	.	.	.	−1	−1	.	.	.	.	.	.	.
$X.85$	2	.	.	.	.	−1	.	1	−2	.	.	.	.	.	.	.	2	−1	2	.	.	.	.
$X.86$	.	.	.	.	2	.	.	.	.	.	.	1	.	.	.	.	−3	.	.	−2	2	.	.
$X.87$	.	.	.	.	.	.	.	.	.	.	.	2	1	1	.	.	2	−1	−1	.	.	.	.
$X.88$	.	1	−1	1	.	.	.	.	.	.	.	2	.	.	.	.	.	.	.	1	1	.	4

Character table of $H(Co_1)$ *(continued)*

	24b	24c	24d	24e	24f	24g	24h	28a	30a	30b	30c	30d	30e	36a	40a	60a
2	6	4	5	5	5	4	3	2	3	2	2	1	1	2	3	2
3	1	2	1	1	1	1	1	.	1	1	1	1	1	2	.	1
5	.	.	.	.	.	.	.	.	1	1	1	1	1	.	1	1
7	.	.	.	.	.	.	.	1	.	.	.	.	.	.	.	.
2P	12_2	12_7	12_8	12_3	12_4	12_7	12_{13}	14a	15a	15a	15a	15b	15c	18a	20a	30a
3P	8c	8a	8d	8j	8e	8b	8i	28a	10a	10b	10b	10c	10d	12_5	40a	20a
5P	24b	24c	24d	24e	24f	24g	24h	28a	6a	6b	6b	6c	6d	36a	8a	12_1
7P	24b	24c	24d	24e	24f	24g	24h	4a	30a	30c	30b	30d	30e	36a	40a	60a
X.1	1	1	1	1	1	1	1	1	1	1	1	1	1	1	1	1
X.2	2	1	.	.	.	-1	.	.	.	.	.	.	.	1	-1	.
X.3	-2	2	.	.	.	.	-1	.	-1	-1	-1	.	.	2	1	-1
X.4	1	-2	-1	-1	-1	.	.	.	.	.	.	.	-1	-1	.	.
X.5	1	-2	-1	-1	-1	.	-1	.	.	.	.	-1	.	-1	.	.
X.6	1	1	-1	-1	-1	-1	-1	1	.	.	.	.	.	-1	.	.
X.7	-2	1	.	.	.	1	1	.	-1	-1	-1	-1	1	.	-1	-1
X.8	-2	1	.	.	.	1	.	.	-1	-1	-1	1	-1	.	-1	-1
X.9	1	-1	1	1	1	1	.	.	1	1	1	-1	-1	.	.	1
X.10	.	.	2	.	-2	.	.	1	1	-1	-1	.	.	-1	.	1
X.11	.	.	.	.	.	.	.	.	-2	.	.	.	-1	.	.	.
X.12	-1	-1	1	-1	1	-1	.	-2	2	.	.	.	.	.	-1	-2
X.13	-1	.	-1	-1	-1	2	-1	.	.	.	.	.	.	1	.	.
X.14	-3	1	1	1	1	-1	-1	.	.	.	.	.	-1	.	.	.
X.15	3	-1	-1	-1	-1	-1	1	.	-1	-1	-1	.	.	.	1	-1
X.16	-3	1	1	1	1	-1	1	.	.	.	.	-1	.	.	.	.
X.17	2	2	.	.	.	.	.	-1	.	.	.	.	.	.	.	.
X.18	-1	-1	1	1	1	1	1	.	.	.	.	.	.	-1	.	.
X.19	2	-1	.	.	.	-1	.	.	.	.	.	.	.	.	.	.
X.20	.	.	.	.	.	.	-1	.	.	.	.	.	1	.	1	.
X.21	-3	.	-1	-1	-1	.	.	.	1	1	1	.	.	.	-1	1
X.22	.	.	.	.	.	.	.	.	.	.	.	1	.	.	1	.
X.23	2	-2	.	.	.	.	1	.	.	.	.	.	.	1	.	.
X.24	.	-1	.	.	.	-1	.	.	.	.	.	.	.	1	.	.
X.25	-1	.	1	1	1	.	.	.	.	.	.	.	.	-2	.	.
X.26	2	-2	.	.	.	.	.	.	.	.	.	.	.	1	.	.
X.27	-2	-1	.	.	.	-1	.	.	.	.	.	.	1	.	.	.
X.28	4	3	.	.	.	-1	.	.	1	1	1	.	.	.	-1	1
X.29	-2	-1	.	.	.	-1	.	.	.	.	.	1	.	.	.	.
X.30	.	.	-2	2	2	.	.	.	-1	1	1	.	.	-1	.	-1
X.31	.	.	.	.	.	.	.	.	2	.	.	-1	1	.	.	.
X.32	.	-1	.	.	.	1	.	.	-2	.	.	.	.	.	-2	2
X.33	.	.	.	.	.	.	.	-1	.	.	.	.	.	.	1	.
X.34	-1	.	1	1	1	.	-1	.	.	.	.	.	.	.	.	.
X.35	-1	2	-1	-1	-1	.	1	.	.	.	.	.	.	.	.	.
X.36	-1	.	1	1	1	.	1	.	.	.	.	.	.	.	.	.
X.37	.	.	.	.	.	.	.	.	1	1	1	-1	1	.	.	1
X.38	.	-2	.	.	.	.	.	.	-1	-1	-1	1	1	.	-1	-1
X.39	.	.	.	.	.	.	.	.	1	1	1	1	-1	.	.	1
X.40	2	1	.	.	.	1	.	.	.	.	.	.	.	-1	.	.
X.41	2	.	2	2	2	.	-1	.	.	.	.	.	.	.	.	.
X.42	-1	.	-1	-1	-1	.	-1	.	.	.	.	.	.	.	.	.
X.43	-1	.	-1	-1	-1	.	2	.	.	.	.	.	.	.	.	.
X.44	.	.	.	.	.	.	.	.	2	.	.	-1	1	.	.	.
X.45	.	.	-2	-2	2	.	.	1	.	.	.	.	.	.	.	.
X.46	1	-2	-1	1	-1	.	.	.	-2	.	.	.	.	.	1	2
X.47	.	-1	.	.	.	1	-1	.	.	.	.	.	.	.	.	.
X.48	.	-1	.	.	.	1	.	.	.	.	.	.	.	.	.	.
X.49	3	3	-1	-1	-1	1	.	.	.	.	.	.	.	.	.	.
X.50	.	-2	.	.	.	.	.	.	1	1	1	.	.	2	-1	1
X.51	.	.	.	.	.	.	.	.	.	.	.	1	.	-1	.	.
X.52	.	.	.	.	.	.	.	.	.	.	.	.	1	-1	.	.
X.53	.	.	.	.	.	.	-1	.	.	.	.	.	1	.	-1	.
X.54	.	.	.	.	.	.	.	.	.	.	.	1	.	.	-1	.
X.55	-3	.	-1	-1	-1	.	.	.	1	1	1	.	.	.	2	1
X.56	.	.	.	.	.	.	.	.	1	-1	-1	.	.	.	.	1
X.57	.	.	2	-2	-2	.	.	.	1	-1	-1	.	.	.	.	1
X.58	.	.	.	.	.	.	.	.	.	.	.	1	.	.	.	.
X.59	-1	1	1	-1	1	-1	.	2	.	.	.	.	.	.	.	.
X.60	2	.	-2	2	-2	.	.	.	2	.	.	.	.	.	-1	-2
X.61	.	.	.	.	.	.	1	.	.	.	.	.	-1	.	.	.
X.62	.	.	.	.	.	.	.	.	.	.	.	-1	.	.	.	.
X.63	3	.	1	1	1	.	.	.	-1	-1	-1	.	.	.	1	-1
X.64	-1	.	1	-1	1	.	.	.	-1	A	$\bar{A}$	.	.	.	-1	1
X.65	-1	.	1	-1	1	.	.	.	-1	$\bar{A}$	A	.	.	.	-1	1
X.66	.	.	.	.	.	.	.	1	.	.	.	.	.	-1	.	.
X.67	.	.	.	-2	.	.	.	-1	-1	1	1	.	.	.	.	-1
X.68	.	.	.	.	.	.	.	.	-2	.	.	1	.	.	.	.
X.69	.	.	.	.	.	.	.	.	.	.	.	1	-1	.	.	.
X.70	2	-1	.	.	.	-1	.	.	.	.	.	.	.	.	1	.
X.71	-2	-1	.	.	.	1	.	.	.	.	.	.	.	.	1	.
X.72	-1	-1	1	-1	1	1	.	.	2	.	.	.	.	.	2	-2
X.73	.	.	.	.	.	.	.	1	-1	-1	-1	-1	-1	1	.	-1
X.74	.	.	.	2	.	.	.	.	.	.	.	.	.	1	.	.
X.75	-2	1	.	.	.	-1	.	.	.	.	.	.	.	.	.	.
X.76	.	.	-2	.	2	.	.	.	.	.	.	.	.	1	.	.
X.77	-1	-2	1	-1	1	.	.	.	.	.	.	.	.	.	.	.
X.78	-1	2	-1	1	-1	2	.	.	.	.	.	.	.	.	.	.
X.79	-1	2	-1	1	-1	.	.	.	.	.	.	.	.	.	.	.
X.80	-1	2	-1	1	-1	.	.	.	.	.	.	.	.	.	.	.
X.81	-1	2	-1	1	-1	-2	.	.	.	.	.	.	.	.	.	.
X.82	.	.	.	.	.	.	.	-1	.	.	.	.	.	.	.	.
X.83	.	.	.	.	.	.	.	1	.	.	.	.	.	.	.	.
X.84	.	.	.	.	.	.	.	1	.	.	.	.	.	.	.	.
X.85	.	.	.	.	.	.	.	.	.	.	.	.	.	.	.	.
X.86	.	.	.	-2	.	.	.	.	1	-1	-1	.	.	1	.	1
X.87	.	.	.	.	.	.	.	.	-2	.	.	-1	-1	.	.	.
X.88	.	1	.	.	.	1	.	.	2	.	.	.	.	.	-1	-2

Character table of $H(\mathsf{Co}_1)$ *(continued)*

2	21	21	20	18	15	15	18	15	14	13	7	7	8	6	17	16	17
3	5	5	2	3	2	2	1	1	1	5	5	5	4	5	4	2	2
5	2	2	1	.	1	1	.	1	.	1	1	1	.	.	1	1	.
7	1	1	1	.	.	.	.	.	.	.	.	.	.	.	1	.	.
	1a	2a	2b	2c	2d	2e	2f	2g	2h	3a	3b	3c	3d	3e	4a	4b	4c
2P	1a	1a	1a	1a	1a	1a	1a	1a	1a	3a	3b	3c	3d	3e	2a	2b	2a
3P	1a	2a	2b	2c	2d	2e	2f	2g	2h	1a	1a	1a	1a	1a	4a	4b	4c
5P	1a	2a	2b	2c	2d	2e	2f	2g	2h	3a	3b	3c	3d	3e	4a	4b	4c
7P	1a	2a	2b	2c	2d	2e	2f	2g	2h	3a	3b	3c	3d	3e	4a	4b	4c
X.89	8400	8400	−560	−240	.	.	80	80	16	−180	.	.	6	21	560	160	−80
X.90	8640	8640	448	192	.	.	192	64	.	108	.	.	−18	.	−576	64	192
X.91	8960	−8960	.	.	224	−224	.	.	.	−160	74	20	8	14	.	.	.
X.92	9450	9450	490	−54	210	210	10	−30	26	−135	.	.	9	.	−630	50	42
X.93	9450	9450	490	138	−30	−30	202	50	10	−135	.	.	9	.	−630	130	234
X.94	9450	9450	490	−54	−30	−30	10	−30	26	−135	.	.	9	.	−630	50	42
X.95	9450	9450	490	522	210	210	74	−30	10	−135	.	.	9	.	−630	50	−150
X.96	9450	9450	490	522	−30	−30	74	−30	10	−135	.	.	9	.	−630	50	−150
X.97	10080	10080	−672	480	.	.	−160	32	−32	−216	.	.	18	9	672	128	160
X.98	10752	−10752	.	.	64	−64	.	.	.	−192	−30	−84	24	6	.	.	.
X.99	10752	−10752	.	.	64	−64	.	.	.	672	−30	24	24	6	.	.	.
X.100	12600	12600	−840	408	.	.	−136	−40	56	.	.	.	18	18	840	−80	136
X.101	12600	12600	−840	24	.	.	−8	120	−8	540	.	.	−18	−9	840	80	8
X.102	12600	12600	−840	−168	.	.	56	40	−8	.	.	.	18	18	840	320	−56
X.103	13440	−13440	.	.	16	−16	.	.	.	480	21	−60	−24	−6	.	.	.
X.104	14400	14400	−960	−192	.	.	64	.	.	540	.	.	.	−18	960	320	−64
X.105	16128	−16128	.	.	96	−96	.	.	.	−288	−45	−72	.	−18	.	.	.
X.106	16128	−16128	.	.	96	−96	.	.	.	−288	90	36	.	−18	.	.	.
X.107	16128	−16128	.	.	96	−96	.	.	.	576	−45	36	.	−18	.	.	.
X.108	18900	18900	980	372	−180	−180	−12	−20	−12	−270	.	.	18	.	−1260	140	−204
X.109	18900	18900	980	−204	60	60	−76	−100	20	540	.	.	−9	.	−1260	380	−12
X.110	18900	18900	980	468	180	180	84	−60	36	135	.	.	−36	.	−1260	100	−108
X.111	18900	18900	980	84	−60	−60	−44	20	−44	540	.	.	−9	.	−1260	20	−108
X.112	18900	18900	980	−492	180	180	−108	−60	−12	−270	.	.	18	.	−1260	260	84
X.113	20160	20160	−1344	192	.	.	−64	.	−64	216	.	.	−18	18	1344	320	64
X.114	20736	−20736	.	.	−96	96	.	.	.	.	81	.	.	.	.	.	.
X.115	22400	−22400	.	.	240	−240	.	.	.	320	95	−40	−16	8	.	.	.
X.116	22400	−22400	.	.	240	−240	.	.	.	−160	−25	20	32	26	.	.	.
X.117	22680	22680	−1512	−72	.	.	24	232	24	324	.	.	.	.	1512	288	−24
X.118	22680	22680	−1512	504	.	.	−168	88	24	324	.	.	.	.	1512	−48	168
X.119	22680	22680	−1512	−72	.	.	24	−152	24	324	.	.	.	.	1512	−288	−24
X.120	25200	25200	−1680	48	.	.	−16	−80	48	540	.	.	.	9	1680	160	16
X.121	25200	25200	−1680	−336	.	.	112	80	−16	−540	.	.	18	−18	1680	320	−112
X.122	25920	25920	−1728	576	.	.	−192	128	.	−324	.	.	.	.	1728	192	192
X.123	28350	28350	1470	−162	−90	−90	30	−170	−18	405	.	.	.	.	−1890	390	126
X.124	28350	28350	1470	−162	−90	−90	30	70	30	405	.	.	.	.	−1890	−330	126
X.125	33600	33600	−2240	−192	.	.	64	.	−64	360	.	.	6	30	2240	−320	−64
X.126	33600	33600	−2240	576	.	.	−192	.	.	−180	.	.	−12	−24	2240	320	192
X.127	34020	34020	1764	−540	180	180	−156	36	−60	.	.	.	.	.	−2268	36	36
X.128	34020	34020	1764	324	−180	−180	−60	−84	36	.	.	.	.	.	−2268	396	−252
X.129	37800	37800	−2520	−504	.	.	168	120	−24	.	.	.	.	−27	2520	.	−168
X.130	37800	37800	1960	168	−120	−120	−88	200	8	270	.	.	9	.	−2520	−440	−216
X.131	37800	37800	1960	−408	120	120	−152	120	40	270	.	.	9	.	−2520	−200	−24
X.132	38400	−38400	.	.	−320	320	.	.	.	−480	60	60	24	6	.	.	.
X.133	40320	40320	−2688	384	.	.	−128	−128	.	216	.	.	.	−18	2688	−512	128
X.134	42525	42525	2205	−675	45	45	−3	−75	−3	.	.	.	.	.	−2835	45	333
X.135	42525	42525	2205	189	−315	−315	93	45	−3	.	.	.	.	.	−2835	−315	45
X.136	42525	42525	2205	1053	45	45	189	−75	−51	.	.	.	.	.	−2835	45	−243
X.137	42525	42525	2205	189	405	405	93	45	−3	.	.	.	.	.	−2835	−315	45
X.138	44800	−44800	.	.	−160	160	.	.	.	−320	−5	−140	−8	−2	.	.	.
X.139	44800	−44800	.	.	−160	160	.	.	.	640	55	−80	40	16	.	.	.
X.140	44800	−44800	.	.	−160	160	.	.	.	1120	−5	40	−8	−2	.	.	.
X.141	45360	45360	−3024	−144	.	.	48	80	48	−324	.	.	.	.	3024	.	−48
X.142	48600	48600	−3240	−648	.	.	216	−120	24	.	.	.	.	.	3240	.	−216
X.143	50400	50400	−3360	96	.	.	−32	−160	−32	.	.	.	18	−9	3360	−320	32
X.144	51840	−51840	.	.	336	−336	.	.	.	.	81	.	.	.	.	.	.
X.145	53760	−53760	.	.	64	−64	.	.	.	480	−6	−60	−24	−24	.	.	.
X.146	56700	56700	2940	−324	−180	−180	60	−100	12	−405	.	.	.	.	−3780	60	252
X.147	57344	−57344	.	.	.	.	.	.	.	512	−64	80	−16	32	.	.	.
X.148	57344	−57344	.	.	.	.	.	.	.	−640	−64	−64	−16	32	.	.	.
X.149	60480	60480	3136	−192	.	.	−192	64	.	−540	.	.	−18	.	−4032	64	−192
X.150	61440	61440	−4096	.	.	.	.	.	.	−576	.	.	−24	24	4096	.	.
X.151	65536	−65536	.	.	.	.	.	.	.	−512	64	64	−32	16	.	.	.
X.152	67200	−67200	.	.	80	−80	.	.	.	−480	−75	60	24	−30	.	.	.
X.153	72576	−72576	.	.	240	−240	.	.	.	.	−81	.	.	.	.	.	.
X.154	89600	−89600	.	.	−320	320	.	.	.	−160	20	20	8	−22	.	.	.

Character table of $H(\mathsf{Co}_1)$ *(continued)*

2	14	15	16	14	14	14	14	14	14	13	13	13	13	12	9	9	10	8	8	7	3	3
3	2	2	1	2	2	2	1	1	1	1	.	.	.	.	1	1	.	1	.	1	1	1
5	1	.	.	.	.	.	.	.	.	.	.	.	.	.	.	.	.	.	.	2	2	2
7	.	.	.	.	.	.	.	.	.	.	.	.	.	.	.	.	.	.	.	.	.	.
	4d	4e	4f	4g	4h	4i	4j	4k	4l	4m	4n	4o	4p	4q	4r	4s	4t	4u	4v	5a	5b	5c
2P	2a	2a	2b	2c	2b	2c	2b	2b	2b	2c	2b	2c	2c	2c	2d	2d	2f	2g	2h	5a	5b	5c
3P	4d	4e	4f	4g	4h	4i	4j	4k	4l	4m	4n	4o	4p	4q	4r	4s	4t	4u	4v	5a	5b	5c
5P	4d	4e	4f	4g	4h	4i	4j	4k	4l	4m	4n	4o	4p	4q	4r	4s	4t	4u	4v	1a	1a	1a
7P	4d	4e	4f	4g	4h	4i	4j	4k	4l	4m	4n	4o	4p	4q	4r	4s	4t	4u	4v	5a	5b	5c
X.89	.	−16	−32	24	.	24	−16	.	.	−8	.	8	8	−8	.	.	.	.	.	.	.	.
X.90	.	64	64	.	−64	.	.	.	.	.	.	.	.	.	.	.	.	.	.	−5	.	.
X.91	.	.	.	32	.	−32	.	.	.	.	.	.	.	.	8	−8	.	.	.	.	5	.
X.92	−30	2	−14	6	−6	6	26	−14	2	6	10	14	−18	−2	6	6	−2	2	2	.	.	.
X.93	−30	82	66	6	−70	6	10	2	2	6	−6	14	−18	−2	−6	−6	−2	−2	−2	.	.	.
X.94	210	2	−14	6	−6	6	26	2	−14	6	10	14	−18	−2	−6	−6	−2	2	2	.	.	.
X.95	−30	2	−14	6	−6	6	10	−14	2	6	−6	−18	14	−2	6	6	−2	2	−2	.	.	.
X.96	210	2	−14	6	−6	6	10	2	−14	6	−6	−18	14	−2	−6	−6	−2	2	−2	.	.	.
X.97	.	−32	.	48	.	48	32	.	.	−16	.	16	16	−16	.	.	.	.	.	−10	.	.
X.98	.	.	.	64	.	−64	.	.	.	.	.	.	.	.	.	.	.	.	.	−8	−3	−8
X.99	.	.	.	64	.	−64	.	.	.	.	.	.	.	.	.	.	.	.	.	32	−3	2
X.100	.	8	16	−36	.	−36	−56	.	.	12	.	4	4	−4	.	.	.	4	.	.	.	.
X.101	.	40	−80	60	.	60	8	.	.	−20	.	4	4	−4	.	.	.	4	.	.	.	.
X.102	.	−104	32	−36	.	−36	8	.	.	12	.	−12	−12	12	.	.	.	.	.	.	.	.
X.103	.	.	.	80	.	−80	.	.	.	.	.	.	.	.	4	−4	.	.	.	.	−5	.
X.104	.	−128	64	.	.	.	.	.	.	.	.	.	.	.	.	.	.	−8	.	.	.	.
X.105	.	.	.	32	.	−32	.	.	.	.	.	.	.	.	−8	8	.	.	.	8	3	−2
X.106	.	.	.	32	.	−32	.	.	.	.	.	.	.	.	−8	8	.	.	.	8	3	−2
X.107	.	.	.	32	.	−32	.	.	.	.	.	.	.	.	−8	8	.	.	.	8	3	−2
X.108	−180	44	12	24	20	24	−12	12	12	24	20	.	.	.	.	.	.	.	.	.	.	.
X.109	60	92	−4	.	20	.	20	−4	−4	.	−12	−8	−8	−8	.	.	8	.	.	.	.	.
X.110	180	4	−28	12	−12	12	36	−12	−12	12	4	−4	−4	−4	.	.	−4	4	.	.	.	.
X.111	−60	20	20	36	20	36	−44	4	4	36	20	4	4	4	.	.	−4	.	−4	.	.	.
X.112	180	68	4	12	20	12	−12	−12	−12	12	−12	−4	−4	−4	.	.	4	.	−4	.	.	.
X.113	.	−128	64	.	.	.	64	.	.	.	.	.	.	.	.	.	.	.	.	−20	.	.
X.114	.	.	.	96	.	−96	.	.	.	.	.	.	.	.	−8	8	.	.	.	−24	1	6
X.115	.	.	.	−16	.	16	.	.	.	.	.	.	.	.	−4	4	.	.	.	.	.	.
X.116	.	.	.	−16	.	16	.	.	.	.	.	.	.	.	−4	4	.	.	.	.	.	.
X.117	.	24	−128	−36	.	−36	−24	.	.	12	.	−12	−12	12	.	.	.	.	.	−10	.	.
X.118	.	72	−80	−36	.	−36	−24	.	.	12	.	4	4	−4	.	.	.	−4	.	−10	.	.
X.119	.	24	64	108	.	108	−24	.	.	−36	.	4	4	−4	.	.	.	.	.	−10	.	.
X.120	.	−112	96	−24	.	−24	−48	.	.	8	.	−8	−8	8	.	.	.	.	.	.	.	.
X.121	.	−80	.	72	.	72	16	.	.	−24	.	−8	−8	8	.	.	.	.	.	.	.	.
X.122	.	.	−64	.	.	.	.	.	.	.	.	.	.	.	.	.	.	−8	.	10	.	.
X.123	−90	54	−58	−30	−18	−30	−18	6	6	−30	−2	26	−6	10	6	6	−6	2	−2	.	.	.
X.124	−90	−90	−10	42	−18	42	30	6	6	42	−18	−14	18	2	−6	−6	2	2	2	.	.	.
X.125	.	128	−64	.	.	.	64	.	.	.	.	.	.	.	.	.	.	.	.	.	.	.
X.126	.	−128	64	.	.	.	.	.	.	.	.	.	.	.	.	.	.	8	.	.	.	.
X.127	180	36	36	36	36	36	−60	−12	−12	36	4	4	4	4	.	.	−4	.	4	−15	.	.
X.128	−180	108	12	.	36	.	36	12	12	.	4	−8	−8	−8	.	.	8	.	.	−15	.	.
X.129	.	72	−96	−60	.	−60	24	.	.	20	.	12	12	−12	.	.	.	.	.	.	.	.
X.130	−120	−56	72	−24	40	−24	8	8	8	−24	8	8	8	8	.	.	−8	.	.	.	.	.
X.131	120	−8	56	−48	40	−48	40	−8	−8	−48	−24	.	.	.	.	.	.	.	.	.	.	.
X.132	.	.	.	64	.	−64	.	.	.	.	.	.	.	.	.	.	.	.	.	.	.	.
X.133	.	128	.	.	.	.	.	.	.	.	.	.	.	.	.	.	.	.	.	10	.	.
X.134	45	−27	−51	−27	−27	−27	−3	−3	−3	−27	37	−27	21	−3	−3	−3	1	−3	1	.	.	.
X.135	405	−99	−27	9	−27	9	−3	21	−27	9	5	1	−15	−7	9	9	5	−3	−3	.	.	.
X.136	45	−27	−51	−27	−27	−27	−51	−3	−3	−27	−11	5	−11	−3	−3	−3	1	−3	5	.	.	.
X.137	−315	−99	−27	9	−27	9	−3	−27	21	9	5	1	−15	−7	−3	−3	5	−3	−3	.	.	.
X.138	.	.	.	32	.	−32	.	.	.	.	.	.	.	.	8	−8	.	.	.	.	.	.
X.139	.	.	.	−96	.	96	.	.	.	.	.	.	.	.	−8	8	.	.	.	.	.	.
X.140	.	.	.	32	.	−32	.	.	.	.	.	.	.	.	8	−8	.	.	.	.	.	.
X.141	.	48	−64	72	.	72	−48	.	.	−24	.	−8	−8	8	.	.	.	.	.	−20	.	.
X.142	.	−72	96	−36	.	−36	−24	.	.	12	.	20	20	−20	.	.	.	.	.	.	.	.
X.143	.	32	64	−48	.	−48	32	.	.	16	.	−16	−16	16	.	.	.	.	.	.	.	.
X.144	.	.	.	−48	.	48	.	.	.	.	.	.	.	.	4	−4	.	.	.	.	−5	.
X.145	.	.	.	−64	.	64	.	.	.	.	.	.	.	.	.	.	.	.	.	.	5	.
X.146	−180	−36	−68	12	−36	12	12	12	12	12	−20	12	12	12	.	.	−4	4	.	.	.	.
X.147	.	.	.	.	.	.	.	.	.	.	.	.	.	.	.	.	.	.	.	−16	4	−6
X.148	.	.	.	.	.	.	.	.	.	.	.	.	.	.	.	.	.	.	.	24	4	4
X.149	.	64	64	.	64	.	.	.	.	.	.	.	.	.	.	.	.	.	.	15	.	.
X.150	.	.	.	.	.	.	.	.	.	.	.	.	.	.	.	.	.	.	.	20	.	.
X.151	.	.	.	.	.	.	.	.	.	.	.	.	.	.	.	.	.	.	.	16	−4	−4
X.152	.	.	.	16	.	−16	.	.	.	.	.	.	.	.	4	−4	.	.	.	.	.	.
X.153	.	.	.	48	.	−48	.	.	.	.	.	.	.	.	−4	4	.	.	.	−24	−4	6
X.154	.	.	.	−64	.	64	.	.	.	.	.	.	.	.	.	.	.	.	.	.	.	.

Character table of $H(Co_1)$ *(continued)*

	6a	6b	6c	6d	6e	6f	6g	6h	6i	6j	6k	6l	6m	6n	6o	6p	6q	6r	6s	6t	6u	6v	6w	7a
2	13	12	7	7	10	8	6	7	7	7	6	6	6	9	7	5	5	6	6	6	6	7	5	3
3	5	2	5	5	3	4	5	3	3	3	3	3	3	1	2	3	3	2	2	2	2	1	1	.
5	1	1	1	1	.	.	.	.	.	.	.	.	.	.	.	.	.	.	.	.	.	.	.	.
7	.	.	.	.	.	.	.	.	.	.	.	.	.	.	.	.	.	.	.	.	.	.	.	1
2P	3a	3a	3b	3c	3a	3d	3e	3b	3d	3b	3e	3e	3c	3a	3d	3d	3d	3b	3d	3d	3b	3d	3d	7a
3P	2a	2b	2a	2a	2c	2a	2a	2c	2c	2c	2c	2c	2c	2g	2b	2c	2c	2d	2d	2e	2e	2f	2h	7a
5P	6a	6b	6c	6d	6e	6f	6g	6h	6i	6j	6k	6l	6m	6n	6o	6p	6q	6r	6s	6t	6u	6v	6w	7a
7P	6a	6b	6c	6d	6e	6f	6g	6h	6i	6j	6k	6l	6m	6n	6o	6p	6q	6r	6s	6t	6u	6v	6w	1a
X.89	−180	−20	.	.	−12	6	21	.	−6	.	−3	−3	.	−4	−2	.	.	.	.	.	.	2	−2	.
X.90	108	−20	.	.	12	−18	.	.	.	.	.	.	.	4	−2	−6	−6	.	.	.	.	.	.	2
X.91	160	.	−74	−20	.	−8	−14	−6	.	6	6	−6	.	.	.	.	.	2	−4	4	−2	.	.	.
X.92	−135	25	.	.	9	9	.	.	9	.	.	.	.	−3	1	9	−9	.	−3	−3	.	1	−1	.
X.93	−135	25	.	.	−15	9	.	.	9	.	.	.	.	−7	1	3	3	.	3	3	.	1	1	.
X.94	−135	25	.	.	9	9	.	.	9	.	.	.	.	−3	1	−9	9	.	3	3	.	1	−1	.
X.95	−135	25	.	.	9	9	.	.	−9	.	.	.	.	−3	1	−9	9	.	−3	−3	.	−1	1	.
X.96	−135	25	.	.	9	9	.	.	−9	.	.	.	.	−3	1	9	−9	.	3	3	.	−1	1	.
X.97	−216	−24	.	.	24	18	9	.	−6	.	−3	−3	.	8	−6	.	.	.	.	.	.	2	−2	.
X.98	192	.	30	84	.	−24	−6	−6	.	6	6	−6	.	.	.	.	.	−2	4	−4	2	.	.	.
X.99	−672	.	30	−24	.	−24	−6	−6	.	6	6	−6	.	.	.	.	.	−2	4	−4	2	.	.	.
X.100	.	.	.	.	24	18	18	.	−6	.	6	6	.	−4	−6	.	.	.	.	.	.	2	2	.
X.101	540	60	.	.	12	−18	−9	.	6	.	3	3	.	.	6	.	.	.	.	.	.	−2	−2	.
X.102	.	.	.	.	−48	18	18	.	−6	.	6	6	.	−8	−6	.	.	.	.	.	.	2	−2	.
X.103	−480	.	−21	60	.	24	6	−3	.	3	−6	6	.	.	.	.	.	1	4	−4	−1	.	.	.
X.104	540	60	.	.	12	.	−18	.	−12	.	−6	−6	.	.	.	.	.	.	.	.	.	4	.	1
X.105	288	.	45	72	.	.	18	3	.	−3	6	−6	.	.	.	.	.	3	.	.	−3	.	.	.
X.106	288	.	−90	−36	.	.	18	−6	.	6	6	−6	.	.	.	.	.	−6	.	.	6	.	.	.
X.107	−576	.	45	−36	.	.	18	3	.	−3	6	−6	.	.	.	.	.	3	.	.	−3	.	.	.
X.108	−270	50	.	.	−6	18	.	.	.	.	.	.	.	−2	2	−6	−6	.	.	.	.	.	.	.
X.109	540	−100	.	.	12	−9	.	.	−9	.	.	.	.	−4	−1	3	3	.	3	3	.	−1	−1	.
X.110	135	−25	.	.	−9	−36	.	.	.	.	.	.	.	3	−4	.	.	.	.	.	.	.	.	.
X.111	540	−100	.	.	12	−9	.	.	9	.	.	.	.	−4	−1	3	3	.	−3	−3	.	1	1	.
X.112	−270	50	.	.	−6	18	.	.	.	.	.	.	.	6	2	−6	−6	.	.	.	.	.	.	.
X.113	216	24	.	.	24	−18	18	.	−6	.	6	6	.	.	6	.	.	.	.	.	.	2	2	.
X.114	.	.	−81	.	.	.	.	9	.	−9	.	.	.	.	.	.	.	−3	.	.	3	.	.	2
X.115	−320	.	−95	40	.	16	−8	−9	.	9	.	.	.	.	.	.	.	3	.	.	−3	.	.	.
X.116	160	.	25	−20	.	−32	−26	15	.	−15	−6	6	.	.	.	.	.	3	.	.	−3	.	.	.
X.117	324	36	.	.	−36	.	.	.	.	.	.	.	.	4	.	.	.	.	.	.	.	.	.	.
X.118	324	36	.	.	36	.	.	.	.	.	.	.	.	−8	.	.	.	.	.	.	.	.	.	.
X.119	324	36	.	.	−36	.	.	.	.	.	.	.	.	4	.	.	.	.	.	.	.	.	.	.
X.120	540	60	.	.	−12	.	9	.	12	.	−3	−3	.	4	.	.	.	.	.	.	.	−4	.	.
X.121	−540	−60	.	.	12	18	−18	.	6	.	−6	−6	.	−4	−6	.	.	.	.	.	.	−2	2	.
X.122	−324	−36	.	.	−36	.	.	.	.	.	.	.	.	8	.	.	.	.	.	.	.	.	.	−1
X.123	405	−75	.	.	−27	.	.	.	.	.	.	.	.	1	.	.	.	.	.	.	.	.	.	.
X.124	405	−75	.	.	−27	.	.	.	.	.	.	.	.	1	.	.	.	.	.	.	.	.	.	.
X.125	360	40	.	.	−24	6	30	.	6	.	−6	−6	.	.	−2	.	.	.	.	.	.	−2	2	.
X.126	−180	−20	.	.	−36	−12	−24	.	.	.	.	.	.	.	4	.	.	.	.	.	.	.	.	.
X.127	.	.	.	.	.	.	.	.	.	.	.	.	.	.	.	.	.	.	.	.	.	.	.	.
X.128	.	.	.	.	.	.	.	.	.	.	.	.	.	.	.	.	.	.	.	.	.	.	.	.
X.129	.	.	.	.	.	.	−27	.	.	.	9	9	.	.	.	.	.	.	.	.	.	.	.	.
X.130	270	−50	.	.	6	9	.	.	−9	.	.	.	.	2	1	−3	−3	.	3	3	.	−1	−1	.
X.131	270	−50	.	.	6	9	.	.	9	.	.	.	.	−6	1	−3	−3	.	−3	−3	.	1	1	.
X.132	480	.	−60	−60	.	−24	−6	12	.	−12	6	−6	.	.	.	.	.	4	4	−4	−4	.	.	−2
X.133	216	24	.	.	−24	.	−18	.	−12	.	−6	−6	.	−8	.	.	.	.	.	.	.	4	.	.
X.134	.	.	.	.	.	.	.	.	.	.	.	.	.	.	.	.	.	.	.	.	.	.	.	.
X.135	.	.	.	.	.	.	.	.	.	.	.	.	.	.	.	.	.	.	.	.	.	.	.	.
X.136	.	.	.	.	.	.	.	.	.	.	.	.	.	.	.	.	.	.	.	.	.	.	.	.
X.137	.	.	.	.	.	.	.	.	.	.	.	.	.	.	.	.	.	.	.	.	.	.	.	.
X.138	320	.	5	140	.	8	2	3	.	−3	6	−6	.	.	.	.	.	−1	−4	4	1	.	.	.
X.139	−640	.	−55	80	.	−40	−16	−9	.	9	.	.	.	.	.	.	.	−1	−4	4	1	.	.	.
X.140	−1120	.	5	−40	.	8	2	3	.	−3	6	−6	.	.	.	.	.	−1	−4	4	1	.	.	.
X.141	−324	−36	.	.	36	.	.	.	.	.	.	.	.	−4	.	.	.	.	.	.	.	.	.	.
X.142	.	.	.	.	.	.	.	.	.	.	.	.	.	.	.	.	.	.	.	.	.	.	.	−1
X.143	.	.	.	.	48	18	−9	.	6	.	3	3	.	8	−6	.	.	.	.	.	.	−2	−2	.
X.144	.	.	−81	.	.	.	.	9	.	−9	.	.	.	.	.	.	.	−3	.	.	3	.	.	−2
X.145	−480	.	6	60	.	24	24	18	.	−18	.	.	.	.	.	.	.	−2	4	−4	2	.	.	.
X.146	−405	75	.	.	27	.	.	.	.	.	.	.	.	−1	.	.	.	.	.	.	.	.	.	.
X.147	−512	.	64	−80	.	16	−32	.	.	.	.	.	.	.	.	.	.	.	.	.	.	.	.	.
X.148	640	.	64	64	.	16	−32	.	.	.	.	.	.	.	.	.	.	.	.	.	.	.	.	.
X.149	−540	100	.	.	−12	−18	.	.	.	.	.	.	.	4	−2	6	6	.	.	.	.	.	.	.
X.150	−576	−64	.	.	.	−24	24	.	.	.	.	.	.	.	8	.	.	.	.	.	.	.	.	1
X.151	512	.	−64	−64	.	32	−16	.	.	.	.	.	.	.	.	.	.	.	.	.	.	.	.	2
X.152	480	.	75	−60	.	−24	30	−3	.	3	−6	6	.	.	.	.	.	−7	−4	4	7	.	.	.
X.153	.	.	81	.	.	.	.	−9	.	9	.	.	.	.	.	.	.	3	.	.	−3	.	.	.
X.154	160	.	−20	−20	.	−8	22	−12	.	12	−6	6	.	.	.	.	.	4	4	−4	−4	.	.	.

Character table of $H(\mathsf{Co}_1)$ *(continued)*

	8a	8b	8c	8d	8e	8f	8g	8h	8i	8j	8k	8l	8m	8n	8o	8p	8q	8r	8s	9a	9b	9c	10a	10b	10c	10d
2	11	11	11	10	10	11	10	10	8	8	8	8	8	8	8	7	7	7	7	3	1	1	7	6	3	3
3	2	1	1	1	1	.	.	.	1	1	.	.	.	.	.	.	.	.	.	3	3	3	1	1	1	1
5	1	.	.	.	.	.	.	.	.	.	.	.	.	.	.	.	.	.	.	.	.	.	2	1	2	2
7	.	.	.	.	.	.	.	.	.	.	.	.	.	.	.	.	.	.	.	.	.	.	.	.	.	.
2P	4a	4a	4c	4f	4b	4c	4b	4f	4d	4e	4o	4o	4j	4k	4l	4n	4m	4g	4p	9a	9b	9c	5a	5a	5b	5c
3P	8a	8b	8c	8d	8e	8f	8g	8h	8i	8j	8k	8l	8m	8n	8o	8p	8q	8r	8s	3e	3e	3e	10a	10b	10c	10d
5P	8a	8b	8c	8d	8e	8f	8g	8h	8i	8j	8k	8l	8m	8n	8o	8p	8q	8r	8s	9a	9b	9c	2a	2b	2a	2a
7P	8a	8b	8c	8d	8e	8f	8g	8h	8i	8j	8k	8l	8m	8n	8o	8p	8q	8r	8s	9a	9b	9c	10a	10b	10c	10d
X.89	.	.	.	−8	−16	.	.	8	.	.	.	.	.	.	.	.	.	.	.	3	.	.	.	.	.	.
X.90	−64	.	.	.	.	.	.	.	.	.	.	.	.	.	.	.	.	.	.	.	.	.	−5	3	.	.
X.91	.	.	.	.	.	.	.	.	.	.	.	.	.	.	.	.	.	.	.	2	−1	−1	.	.	−5	.
X.92	10	−14	−2	−2	−2	6	−2	−2	−6	2	.	.	2	−2	2	−2	.	.	.	.	.	.	.	.	.	.
X.93	−70	2	−2	2	2	6	2	2	−6	−2	.	.	−2	2	2	2	.	.	.	.	.	.	.	.	.	.
X.94	10	−14	−2	−2	−2	6	−2	−2	6	2	.	.	2	2	−2	−2	.	.	.	.	.	.	.	.	.	.
X.95	10	2	−2	−2	−2	6	−2	−2	−6	2	.	.	−2	−2	2	2	.	.	.	.	.	.	.	.	.	.
X.96	10	2	−2	−2	−2	6	−2	−2	6	2	.	.	−2	2	−2	2	.	.	.	.	.	.	.	.	.	.
X.97	.	.	.	.	.	.	.	.	.	.	.	.	.	.	.	.	.	.	.	.	.	.	−10	−2	.	.
X.98	.	.	.	.	.	.	.	.	.	.	.	.	.	.	.	.	.	.	.	.	.	.	8	.	3	8
X.99	.	.	.	.	.	.	.	.	.	.	.	.	.	.	.	.	.	.	.	.	.	.	−32	.	3	−2
X.100	.	.	.	−4	−8	.	.	4	.	−4	−2	−2	.	.	.	.	2	−2	2	.	.	.	.	.	.	.
X.101	.	.	.	−12	.	.	8	−4	.	−4	−2	−2	.	.	.	.	2	−2	2	.	.	.	.	.	.	.
X.102	.	.	.	.	12	.	4	−8	.	.	−2	−2	.	.	.	.	−2	2	2	.	.	.	.	.	.	.
X.103	.	.	.	.	.	.	.	.	.	.	−4	4	.	.	.	.	.	.	.	.	.	.	.	.	5	.
X.104	.	.	.	.	.	.	.	.	.	8	.	.	.	.	.	.	.	.	.	.	.	.	.	.	.	.
X.105	.	.	.	.	.	.	.	.	.	.	.	.	.	.	.	.	.	.	.	.	.	.	−8	.	−3	2
X.106	.	.	.	.	.	.	.	.	.	.	.	.	.	.	.	.	.	.	.	.	.	.	−8	.	−3	2
X.107	.	.	.	.	.	.	.	.	.	.	.	.	.	.	.	.	.	.	.	.	.	.	−8	.	−3	2
X.108	−20	−12	−8	.	.	.	.	.	.	.	−2	−2	.	.	.	.	2	2	−2	.	.	.	.	.	.	.
X.109	−20	4	.	.	.	−8	.	.	.	.	−2	−2	.	.	.	.	2	2	−2	.	.	.	.	.	.	.
X.110	20	−12	−4	−4	−4	12	−4	−4	.	4	.	.	.	.	.	.	.	.	.	.	.	.	.	.	.	.
X.111	−20	−4	−12	−12	12	4	−4	4	.	.	.	.	4	.	.	.	.	.	.	.	.	.	.	.	.	.
X.112	−20	12	−4	−12	12	−4	−4	4	.	.	.	.	4	.	.	.	.	.	.	.	.	.	.	.	.	.
X.113	.	.	.	.	.	.	.	.	.	.	.	.	.	.	.	.	.	.	.	.	.	.	−20	−4	.	.
X.114	.	.	.	.	.	.	.	.	.	.	.	.	.	.	.	.	.	.	.	.	.	.	24	.	−1	−6
X.115	.	.	.	.	.	.	.	.	.	.	−4	4	.	.	.	.	.	.	.	2	−1	−1	.	.	.	.
X.116	.	.	.	.	.	.	.	.	.	.	−4	4	.	.	.	.	.	.	.	2	−1	2	.	.	.	.
X.117	.	.	.	8	4	.	−4	.	.	.	2	2	.	.	.	.	2	−2	−2	.	.	.	−10	−2	.	.
X.118	.	.	.	−12	.	.	8	−4	.	4	2	2	.	.	.	.	−2	2	−2	.	.	.	−10	−2	.	.
X.119	.	.	.	8	4	.	−4	.	.	.	−2	−2	.	.	.	.	−2	2	2	.	.	.	−10	−2	.	.
X.120	.	.	.	−8	−16	.	.	8	.	.	.	.	.	.	.	.	.	.	.	.	.	.	.	.	.	.
X.121	.	.	.	−16	−8	.	8	.	.	.	.	.	.	.	.	.	.	.	.	.	.	.	.	.	.	.
X.122	.	.	.	.	.	.	.	.	.	8	.	.	.	.	.	.	.	.	.	.	.	.	10	2	.	.
X.123	30	6	10	−2	−2	2	−2	−2	6	2	.	.	−2	−2	−2	2	.	.	.	.	.	.	.	.	.	.
X.124	30	6	−14	−2	−2	−6	−2	−2	−6	2	.	.	2	2	2	−2	.	.	.	.	.	.	.	.	.	.
X.125	.	.	.	.	.	.	.	.	.	.	.	.	.	.	.	.	.	.	.	3	.	.	.	.	.	.
X.126	.	.	.	.	.	.	.	.	.	−8	.	.	.	.	.	.	.	.	.	3	.	.	.	.	.	.
X.127	−36	12	−12	12	−12	4	4	−4	.	.	.	.	−4	.	.	.	.	.	.	.	.	.	−15	9	.	.
X.128	−36	−12	.	.	.	−8	.	.	.	.	2	2	.	.	.	.	−2	−2	2	.	.	.	−15	9	.	.
X.129	.	.	.	.	−12	.	−4	8	.	.	−2	−2	.	.	.	.	−2	2	2	.	.	.	.	.	.	.
X.130	−40	−8	8	.	.	8	.	.	.	.	.	.	.	.	.	.	.	.	.	.	.	.	.	.	.	.
X.131	−40	8	16	.	.	.	.	.	.	.	.	.	.	.	.	.	.	.	.	.	.	.	.	.	.	.
X.132	.	.	.	.	.	.	.	.	.	.	.	.	.	.	.	.	.	.	.	.	.	.	.	.	.	.
X.133	.	.	.	.	.	.	.	.	.	.	.	.	.	.	.	.	.	.	.	.	.	.	10	2	.	.
X.134	45	−27	9	−3	9	1	1	5	−3	−3	1	1	−3	1	1	1	1	1	1	.	.	.	.	.	.	.
X.135	45	−3	−3	9	−3	−3	5	1	−3	−3	−1	−1	1	−3	1	1	−1	−1	−1	.	.	.	.	.	.	.
X.136	45	21	9	−3	9	1	1	5	−3	−3	1	1	1	1	1	−3	1	1	1	.	.	.	.	.	.	.
X.137	45	−3	−3	9	−3	−3	5	1	9	−3	−1	−1	1	1	−3	1	−1	−1	−1	.	.	.	.	.	.	.
X.138	.	.	.	.	.	.	.	.	.	.	.	.	.	.	.	.	.	.	.	4	1	1	.	.	.	.
X.139	.	.	.	.	.	.	.	.	.	.	.	.	.	.	.	.	.	.	.	−2	1	1	.	.	.	.
X.140	.	.	.	.	.	.	.	.	.	.	.	.	.	.	.	.	.	.	.	−2	−2	1	.	.	.	.
X.141	.	.	.	16	8	.	−8	.	.	.	.	.	.	.	.	.	.	.	.	.	.	.	−20	−4	.	.
X.142	.	.	.	.	12	.	4	−8	.	.	2	2	.	.	.	.	2	−2	−2	.	.	.	.	.	.	.
X.143	.	.	.	.	.	.	.	.	.	.	.	.	.	.	.	.	.	.	.	.	.	.	.	.	.	.
X.144	.	.	.	.	.	.	.	.	.	.	4	−4	.	.	.	.	.	.	.	.	.	.	.	.	5	.
X.145	.	.	.	.	.	.	.	.	.	.	.	.	.	.	.	.	.	.	.	.	.	.	.	.	−5	.
X.146	60	12	−4	−4	−4	−4	−4	−4	.	4	.	.	.	.	.	.	.	.	.	.	.	.	.	.	.	.
X.147	.	.	.	.	.	.	.	.	.	.	.	.	.	.	.	.	.	.	.	2	2	−1	16	.	−4	6
X.148	.	.	.	.	.	.	.	.	.	.	.	.	.	.	.	.	.	.	.	−4	−1	−1	−24	.	−4	−4
X.149	−64	.	.	.	.	.	.	.	.	.	.	.	.	.	.	.	.	.	.	.	.	.	15	−9	.	.
X.150	.	.	.	.	.	.	.	.	.	.	.	.	.	.	.	.	.	.	.	−3	.	.	20	4	.	.
X.151	.	.	.	.	.	.	.	.	.	.	.	.	.	.	.	.	.	.	.	−2	1	1	−16	.	4	4
X.152	.	.	.	.	.	.	.	.	.	.	−4	4	.	.	.	.	.	.	.	.	.	.	.	.	.	.
X.153	.	.	.	.	.	.	.	.	.	.	4	−4	.	.	.	.	.	.	.	.	.	.	24	.	4	−6
X.154	.	.	.	.	.	.	.	.	.	.	.	.	.	.	.	.	.	.	.	2	−1	−1	.	.	.	.

Character table of $H(\mathrm{Co}_1)$ *(continued)*

2	4	3	3	10	10	9	9	5	8	6	9	7	7	6	5	5	5	5	5	5	5
3	.	.	.	3	2	2	2	4	2	3	1	2	2	2	2	2	2	2	2	2	2
5	1	1	1	1	.	.	.	.	.	.	.	.	.	.	.	.	.	.	.	.	.
7	.	.	.	.	.	.	.	.	.	.	.	.	.	.	.	.	.	.	.	.	.
	10e	10f	10g	12_1	12_2	12_3	12_4	12_5	12_6	12_7	12_8	12_9	12_{10}	12_{11}	12_{12}	12_{13}	12_{14}	12_{15}	12_{16}	12_{17}	12_{18}
2P	5a	5b	5b	6a	6a	6a	6b	6g	6b	6f	6b	6e	6e	6f	6f	6d	6k	6o	6o	6f	6k
3P	10e	10f	10g	4a	4c	4e	4b	4a	4h	4a	4f	4g	4i	4c	4e	4d	4g	4h	4b	4d	4i
5P	2g	2d	2e	12_1	12_2	12_3	12_4	12_5	12_6	12_7	12_8	12_9	12_{10}	12_{11}	12_{12}	12_{13}	12_{14}	12_{15}	12_{16}	12_{17}	12_{18}
7P	10e	10f	10g	12_1	12_2	12_3	12_4	12_5	12_6	12_7	12_8	12_9	12_{10}	12_{11}	12_{12}	12_{13}	12_{14}	12_{15}	12_{16}	12_{17}	12_{18}
X.89	.	.	.	20	4	20	16	−7	.	2	−8	.	.	−2	2	.	−3	.	−2	.	−3
X.90	−1	.	.	12	12	4	4	.	−4	6	4	.	.	.	−2	.	.	2	−2	.	.
X.91	.	−1	1	.	.	.	.	.	.	.	.	−4	4	.	.	.	2	.	.	.	−2
X.92	.	.	.	−15	9	5	−7	.	−3	−3	1	−3	−3	−3	−1	.	.	3	−1	3	.
X.93	.	.	.	−15	−15	1	−11	.	5	−3	−3	−3	−3	−3	1	.	.	−1	1	3	.
X.94	.	.	.	−15	9	5	−7	.	−3	−3	1	−3	−3	−3	−1	.	.	−3	−1	−3	.
X.95	.	.	.	−15	9	5	−7	.	−3	−3	1	−3	−3	3	−1	.	.	−3	−1	3	.
X.96	.	.	.	−15	9	5	−7	.	−3	−3	1	−3	−3	3	−1	.	.	3	−1	−3	.
X.97	2	.	.	24	−8	−8	−16	−3	.	6	.	.	.	−2	−2	.	3	.	2	.	3
X.98	.	−1	1	.	.	.	.	.	.	.	.	−8	8	.	.	.	−2	.	.	.	2
X.99	.	−1	1	.	.	.	.	.	.	.	.	4	−4	.	.	.	−2	.	.	.	2
X.100	.	.	.	.	−8	−4	4	−6	.	6	4	.	.	−2	2	.	.	.	−2	.	.
X.101	.	.	.	−60	−4	−8	−4	3	.	−6	4	.	.	2	−2	.	−3	.	2	.	−3
X.102	.	.	.	.	16	−8	8	−6	.	6	8	.	.	−2	−2	.	.	.	2	.	.
X.103	.	1	−1	.	.	.	.	.	.	.	.	−4	4	.	.	.	2	.	.	.	−2
X.104	.	.	.	−60	−4	−8	−4	6	.	.	4	.	.	−4	4	.	.	.	−4	.	.
X.105	.	1	−1	.	.	.	.	.	.	.	.	−4	4	.	.	.	2	.	.	.	−2
X.106	.	1	−1	.	.	.	.	.	.	.	.	−4	4	.	.	.	2	.	.	.	−2
X.107	.	1	−1	.	.	.	.	.	.	.	.	8	−8	.	.	.	2	.	.	.	−2
X.108	.	.	.	−30	−6	14	−10	.	2	−6	6	6	6	.	2	.	.	2	2	.	.
X.109	.	.	.	60	12	−4	−4	.	−4	3	−4	.	.	3	−1	.	.	−1	−1	3	.
X.110	.	.	.	15	−9	−5	7	.	3	12	−1	3	3	.	4	.	.	.	4	.	.
X.111	.	.	.	60	12	−4	−4	.	−4	3	−4	.	.	−3	−1	.	.	−1	−1	−3	.
X.112	.	.	.	−30	−6	−10	14	.	2	−6	−2	−6	−6	.	2	.	.	2	2	.	.
X.113	.	.	.	−24	−8	16	8	−6	.	−6	−8	.	.	−2	−2	.	.	.	2	.	.
X.114	.	−1	1	.	.	.	.	.	.	.	.	.	.	.	.	.	.	.	.	.	.
X.115	.	.	.	.	.	.	.	.	.	.	.	8	−8	.	.	.	−4	.	.	.	4
X.116	.	.	.	.	.	.	.	.	.	.	.	−4	4	.	.	.	2	.	.	.	−2
X.117	2	.	.	−36	12	12	.	.	.	.	−8	.	.	.	.	.	.	.	.	.	.
X.118	−2	.	.	−36	−12	.	12	.	.	.	4	.	.	.	.	.	.	.	.	.	.
X.119	−2	.	.	−36	12	12	.	.	.	.	−8	.	.	.	.	.	.	.	.	.	.
X.120	.	.	.	−60	4	−4	−8	−3	.	.	.	.	.	4	−4	.	3	.	4	.	3
X.121	.	.	.	60	−4	4	8	6	.	6	.	.	.	2	−2	.	.	.	2	.	.
X.122	−2	.	.	36	12	.	−12	.	.	.	−4	.	.	.	.	.	.	.	.	.	.
X.123	.	.	.	45	−27	9	−3	.	9	.	5	−3	−3	.	.	.	.	.	.	.	.
X.124	.	.	.	45	−27	9	−3	.	9	.	5	−3	−3	.	.	.	.	.	.	.	.
X.125	.	.	.	−40	8	−16	−8	−10	.	2	8	.	.	2	2	.	.	.	−2	.	.
X.126	.	.	.	20	12	−8	−4	8	.	−4	4	.	.	.	4	.	.	.	−4	.	.
X.127	1	.	.	.	.	.	.	.	.	.	.	.	.	.	.	.	.	.	.	.	.
X.128	1	.	.	.	.	.	.	.	.	.	.	.	.	.	.	.	.	.	.	.	.
X.129	.	.	.	.	.	.	.	9	.	.	.	.	.	.	.	.	3	.	.	.	3
X.130	.	.	.	30	6	−14	10	.	−2	−3	−6	−6	−6	3	1	.	.	1	1	3	.
X.131	.	.	.	30	6	10	−14	.	−2	−3	2	6	6	−3	1	.	.	1	1	−3	.
X.132	.	.	.	.	.	.	.	.	.	.	.	4	−4	.	.	.	−2	.	.	.	2
X.133	2	.	.	−24	8	8	16	6	.	.	.	.	.	−4	−4	.	.	.	4	.	.
X.134	.	.	.	.	.	.	.	.	.	.	.	.	.	.	.	.	.	.	.	.	.
X.135	.	.	.	.	.	.	.	.	.	.	.	.	.	.	.	.	.	.	.	.	.
X.136	.	.	.	.	.	.	.	.	.	.	.	.	.	.	.	.	.	.	.	.	.
X.137	.	.	.	.	.	.	.	.	.	.	.	.	.	.	.	.	.	.	.	.	.
X.138	.	.	.	.	.	.	.	.	.	.	.	8	−8	.	.	.	2	.	.	.	−2
X.139	.	.	.	.	.	.	.	.	.	.	.	.	.	.	.	.	.	.	.	.	.
X.140	.	.	.	.	.	.	.	.	.	.	.	−4	4	.	.	.	2	.	.	.	−2
X.141	.	.	.	36	−12	−12	.	.	.	.	8	.	.	.	.	.	.	.	.	.	.
X.142	.	.	.	.	.	.	.	.	.	.	.	.	.	.	.	.	.	.	.	.	.
X.143	.	.	.	.	−16	8	−8	3	.	6	−8	.	.	2	2	.	−3	.	−2	.	−3
X.144	.	1	−1	.	.	.	.	.	.	.	.	.	.	.	.	.	.	.	.	.	.
X.145	.	−1	1	.	.	.	.	.	.	.	.	−4	4	.	.	.	−4	.	.	.	4
X.146	.	.	.	−45	27	−9	3	.	−9	.	−5	3	3	.	.	.	.	.	.	.	.
X.147	.	.	.	.	.	.	.	.	.	.	.	.	.	.	.	.	.	.	.	.	.
X.148	.	.	.	.	.	.	.	.	.	.	.	.	.	.	.	.	.	.	.	.	.
X.149	−1	.	.	−60	−12	4	4	.	4	6	4	.	.	.	−2	.	.	−2	−2	.	.
X.150	.	.	.	64	.	.	.	−8	.	−8	.	.	.	.	.	.	.	.	.	.	.
X.151	.	.	.	.	.	.	.	.	.	.	.	.	.	.	.	.	.	.	.	.	.
X.152	.	.	.	.	.	.	.	.	.	.	.	4	−4	.	.	.	−2	.	.	.	2
X.153	.	.	.	.	.	.	.	.	.	.	.	.	.	.	.	.	.	.	.	.	.
X.154	.	.	.	.	.	.	.	.	.	.	.	−4	4	.	.	.	2	.	.	.	−2

Character table of $H(Co_1)$ *(continued)*

2	5	5	5	5	5	5	5	5	5	5	4	4	4	3	2	2	3	1	1	6	6
3	2	2	2	2	2	2	1	1	1	1	1	1	1	.	.	.	1	1	1	.	.
5	.	.	.	.	.	.	.	.	.	.	.	.	.	.	.	.	1	1	1	.	.
7	.	.	.	.	.	.	.	.	.	.	.	.	.	1	1	1	.	.	.	.	.
	12_{19}	12_{20}	12_{21}	12_{22}	12_{23}	12_{24}	12_{25}	12_{26}	12_{27}	12_{28}	12_{29}	12_{30}	12_{31}	$14a$	$14b$	$14c$	$15a$	$15b$	$15c$	$16a$	$16b$
2P	$6o$	$6h$	$6h$	$6g$	$6m$	$6m$	$6o$	$6o$	$6o$	$6n$	$6r$	$6l$	$6r$	$7a$	$7a$	$7a$	$15a$	$15b$	$15c$	$8c$	$8f$
3P	$4h$	$4g$	$4i$	$4c$	$4g$	$4i$	$4k$	$4j$	$4l$	$4u$	$4r$	$4m$	$4s$	$14a$	$14c$	$14b$	$5a$	$5b$	$5c$	$16a$	$16b$
5P	12_{19}	12_{20}	12_{21}	12_{22}	12_{23}	12_{24}	12_{25}	12_{26}	12_{27}	12_{28}	12_{29}	12_{30}	12_{31}	$14a$	$14c$	$14b$	$3a$	$3b$	$3c$	$16a$	$16b$
7P	12_{19}	12_{20}	12_{21}	12_{22}	12_{23}	12_{24}	12_{25}	12_{26}	12_{27}	12_{28}	12_{29}	12_{30}	12_{31}	$2a$	$2b$	$2b$	$15a$	$15b$	$15c$	$16a$	$16b$
X.89	.	.	.	1	.	.	.	2	.	.	.	1	.	.	.	.	.	.	.	.	.
X.90	2	.	.	.	.	.	.	.	.	.	.	.	.	2	.	.	−2	.	.	.	.
X.91	.	2	−2	.	2	−2	.	.	.	.	2	.	−2	.	.	.	.	−1	.	.	.
X.92	−3	.	.	.	.	.	1	−1	−1	−1	.	.	.	.	.	.	.	.	.	.	.
X.93	−1	.	.	.	.	.	−1	1	−1	1	.	.	.	.	.	.	.	.	.	.	.
X.94	3	.	.	.	.	.	−1	−1	1	−1	.	.	.	.	.	.	.	.	.	.	.
X.95	3	.	.	.	.	.	1	1	−1	−1	.	.	.	.	.	.	.	.	.	.	.
X.96	−3	.	.	.	.	.	−1	1	1	−1	.	.	.	.	.	.	.	.	.	.	.
X.97	.	.	.	1	.	.	.	2	.	.	.	−1	.	.	.	.	−1	.	.	.	.
X.98	.	−2	2	.	−2	2	.	.	.	.	.	.	.	.	.	.	−2	.	1	.	.
X.99	.	−2	2	.	4	−4	.	.	.	.	.	.	.	.	.	.	2	.	−1	.	.
X.100	.	.	.	−2	.	.	.	−2	.	−2	.	.	.	.	.	.	.	.	.	.	.
X.101	.	.	.	−1	.	.	.	2	.	−2	.	1	.	.	.	.	.	.	.	.	.
X.102	.	.	.	−2	.	.	.	2	.	.	.	.	.	.	.	.	.	.	.	.	.
X.103	.	5	−5	.	2	−2	.	.	.	.	1	.	−1	.	.	.	.	1	.	.	.
X.104	.	.	.	2	.	.	.	.	.	−2	.	.	.	1	−1	−1	.	.	.	.	.
X.105	.	−1	1	.	−4	4	.	.	.	.	1	.	−1	.	.	.	2	.	−2	.	.
X.106	.	2	−2	.	2	−2	.	.	.	.	−2	.	2	.	.	.	2	.	1	.	.
X.107	.	−1	1	.	2	−2	.	.	.	.	1	.	−1	.	.	.	−4	.	1	.	.
X.108	2	.	.	.	.	.	.	.	.	.	.	.	.	.	.	.	.	.	.	−2	2
X.109	−1	.	.	.	.	.	−1	−1	−1	.	.	.	.	.	.	.	.	.	.	−2	2
X.110	.	.	.	.	.	.	.	.	.	1	.	.	.	.	.	.	.	.	.	.	.
X.111	−1	.	.	.	.	.	1	1	1	.	.	.	.	.	.	.	.	.	.	.	.
X.112	2	.	.	.	.	.	.	.	.	.	.	.	.	.	.	.	.	.	.	.	.
X.113	.	.	.	−2	.	.	.	−2	.	.	.	.	.	.	.	.	1	.	.	.	.
X.114	.	−3	3	.	.	.	.	.	.	.	1	.	−1	−2	.	.	.	1	.	.	.
X.115	.	−1	1	.	−4	4	.	.	.	.	−1	.	1	.	.	.	.	.	.	.	.
X.116	.	−1	1	.	2	−2	.	.	.	.	−1	.	1	.	.	.	.	.	.	.	.
X.117	.	.	.	.	.	.	.	.	.	.	.	.	.	.	.	.	−1	.	.	.	.
X.118	.	.	.	.	.	.	.	.	.	2	.	.	.	.	.	.	−1	.	.	.	.
X.119	.	.	.	.	.	.	.	.	.	.	.	.	.	.	.	.	−1	.	.	.	.
X.120	.	.	.	1	.	.	.	.	.	.	.	−1	.	.	.	.	.	.	.	.	.
X.121	.	.	.	2	.	.	.	−2	.	.	.	.	.	.	.	.	.	.	.	.	.
X.122	.	.	.	.	.	.	.	.	.	−2	.	.	.	−1	1	1	1	.	.	.	.
X.123	.	.	.	.	.	.	.	.	.	−1	.	.	.	.	.	.	.	.	.	.	.
X.124	.	.	.	.	.	.	.	.	.	−1	.	.	.	.	.	.	.	.	.	.	.
X.125	.	.	.	2	.	.	.	−2	.	.	.	.	.	.	.	.	.	.	.	.	.
X.126	.	.	.	.	.	.	.	.	.	2	.	.	.	.	.	.	.	.	.	.	.
X.127	.	.	.	.	.	.	.	.	.	.	.	.	.	.	.	.	.	.	.	.	.
X.128	.	.	.	.	.	.	.	.	.	.	.	.	.	.	.	.	.	.	.	2	−2
X.129	.	.	.	−3	.	.	.	.	.	.	.	−1	.	.	.	.	.	.	.	.	.
X.130	1	.	.	.	.	.	−1	−1	−1	.	.	.	.	.	.	.	.	.	.	.	.
X.131	1	.	.	.	.	.	1	1	1	.	.	.	.	.	.	.	.	.	.	.	.
X.132	.	4	−4	.	−2	2	.	.	.	.	.	.	.	2	.	.	.	.	.	.	.
X.133	.	.	.	2	.	.	.	.	.	.	.	.	.	.	.	.	1	.	.	.	.
X.134	.	.	.	.	.	.	.	.	.	.	.	.	.	.	.	.	.	.	.	−1	−1
X.135	.	.	.	.	.	.	.	.	.	.	.	.	.	.	.	.	.	.	.	1	1
X.136	.	.	.	.	.	.	.	.	.	.	.	.	.	.	.	.	.	.	.	−1	−1
X.137	.	.	.	.	.	.	.	.	.	.	.	.	.	.	.	.	.	.	.	1	1
X.138	.	−1	1	.	2	−2	.	.	.	.	−1	.	1	.	.	.	.	.	.	.	.
X.139	.	3	−3	.	.	.	.	.	.	.	1	.	−1	.	.	.	.	.	.	.	.
X.140	.	−1	1	.	−4	4	.	.	.	.	−1	.	1	.	.	.	.	.	.	.	.
X.141	.	.	.	.	.	.	.	.	.	.	.	.	.	.	.	.	1	.	.	.	.
X.142	.	.	.	.	.	.	.	.	.	.	.	.	.	−1	1	1	.	.	.	.	.
X.143	.	.	.	−1	.	.	.	2	.	.	.	1	.	.	.	.	.	.	.	.	.
X.144	.	−3	3	.	.	.	.	.	.	.	1	.	−1	2	.	.	.	1	.	.	.
X.145	.	2	−2	.	2	−2	.	.	.	.	.	.	.	.	.	.	.	−1	.	.	.
X.146	.	.	.	.	.	.	.	.	.	1	.	.	.	.	.	.	.	.	.	.	.
X.147	.	.	.	.	.	.	.	.	.	.	.	.	.	.	.	.	2	1	.	.	.
X.148	.	.	.	.	.	.	.	.	.	.	.	.	.	.	.	.	.	1	1	.	.
X.149	−2	.	.	.	.	.	.	.	.	.	.	.	.	.	.	.	.	.	.	.	.
X.150	.	.	.	.	.	.	.	.	.	.	.	.	.	1	−1	−1	−1	.	.	.	.
X.151	.	.	.	.	.	.	.	.	.	.	.	.	.	−2	.	.	−2	−1	−1	.	.
X.152	.	1	−1	.	−2	2	.	.	.	.	1	.	−1	.	.	.	.	.	.	.	.
X.153	.	3	−3	.	.	.	.	.	.	.	−1	.	1	.	.	.	.	−1	.	.	.
X.154	.	−4	4	.	2	−2	.	.	.	.	.	.	.	.	.	.	.	.	.	.	.

Character table of $H(\mathsf{Co}_1)$ (continued)

	18a	18b	18c	20a	20b	20c	24a	24b	24c	24d	24e	24f	24g	24h	28a	30a	30b	30c	30d	30e	36a	40a	60a
2	3	1	1	5	4	2	6	6	4	5	5	5	4	3	2	3	2	2	1	1	2	3	2
3	3	3	3	1	.	.	2	1	2	1	1	1	1	1	.	1	1	1	1	1	2	.	1
5	.	.	.	1	1	1	.	.	.	.	.	.	.	.	.	1	1	1	1	1	.	1	1
7	.	.	.	.	.	.	.	.	.	.	.	.	.	.	1	.	.	.	.	.	.	.	.
2P	9a	9b	9c	10a	10b	10d	12_1	12_2	12_7	12_8	12_3	12_4	12_7	12_{13}	14a	15a	15a	15a	15b	15c	18a	20a	30a
3P	6g	6g	6g	20a	20b	20c	8a	8c	8a	8d	8j	8e	8b	8i	28a	10a	10b	10b	10c	10d	12_5	40a	20a
5P	18a	18b	18c	4a	4b	4d	24a	24b	24c	24d	24e	24f	24g	24h	28a	6a	6b	6b	6c	6d	36a	8a	12_1
7P	18a	18b	18c	20a	20b	20c	24a	24b	24c	24d	24e	24f	24g	24h	4a	30a	30c	30b	30d	30e	36a	40a	60a
X.89	3	.	.	.	.	.	.	.	.	−2	.	2	.	.	.	.	.	.	.	.	−1	.	.
X.90	.	.	.	−1	−1	.	−4	.	2	.	.	.	.	.	−2	−2	.	.	.	.	.	1	2
X.91	−2	1	1	.	.	.	.	.	.	.	.	.	.	.	.	.	.	.	1	.	.	.	.
X.92	.	.	.	.	.	.	1	1	1	1	−1	1	1	.	.	.	.	.	.	.	.	.	.
X.93	.	.	.	.	.	.	5	1	−1	−1	1	−1	−1	.	.	.	.	.	.	.	.	.	.
X.94	.	.	.	.	.	.	1	1	1	1	−1	1	1	.	.	.	.	.	.	.	.	.	.
X.95	.	.	.	.	.	.	1	1	1	1	−1	1	−1	.	.	.	.	.	.	.	.	.	.
X.96	.	.	.	.	.	.	1	1	1	1	−1	1	−1	.	.	.	.	.	.	.	.	.	.
X.97	.	.	.	2	−2	.	.	.	.	.	.	.	.	.	.	−1	1	1	.	.	.	.	−1
X.98	.	.	.	.	.	.	.	.	.	.	.	.	.	.	.	2	.	.	.	−1	.	.	.
X.99	.	.	.	.	.	.	.	.	.	.	.	.	.	.	.	−2	.	.	.	1	.	.	.
X.100	.	.	.	.	.	.	.	.	.	2	2	−2	.	.	.	.	.	.	.	.	.	.	.
X.101	.	.	.	.	.	.	.	.	.	.	2	.	.	.	.	.	.	.	.	.	.	.	.
X.102	.	.	.	.	.	.	.	.	.	.	.	.	.	.	.	.	.	.	.	.	.	.	.
X.103	.	.	.	.	.	.	.	.	.	.	.	.	.	.	.	.	.	.	−1	.	.	.	.
X.104	.	.	.	.	.	.	.	.	.	.	2	.	.	.	1	.	.	.	.	.	.	.	.
X.105	.	.	.	.	.	.	.	.	.	.	.	.	.	.	.	−2	.	.	.	2	.	.	.
X.106	.	.	.	.	.	.	.	.	.	.	.	.	.	.	.	−2	.	.	.	−1	.	.	.
X.107	.	.	.	.	.	.	.	.	.	.	.	.	.	.	.	4	.	.	.	−1	.	.	.
X.108	.	.	.	.	.	.	−2	−2	−2	.	.	.	.	.	.	.	.	.	.	.	.	.	.
X.109	.	.	.	.	.	.	4	.	1	.	.	.	1	.	.	.	.	.	.	.	.	.	.
X.110	.	.	.	.	.	.	−1	−1	−4	−1	1	−1	.	.	.	.	.	.	.	.	.	.	.
X.111	.	.	.	.	.	.	4	.	1	.	.	.	−1	.	.	.	.	.	.	.	.	.	.
X.112	.	.	.	.	.	.	−2	2	−2	.	.	.	.	.	.	.	.	.	.	.	.	.	.
X.113	.	.	.	4	.	.	.	.	.	.	.	.	.	.	.	1	−1	−1	.	.	.	.	1
X.114	.	.	.	.	.	.	.	.	.	.	.	.	.	.	.	.	.	.	−1	.	.	.	.
X.115	−2	1	1	.	.	.	.	.	.	.	.	.	.	.	.	.	.	.	.	.	.	.	.
X.116	−2	1	−2	.	.	.	.	.	.	.	.	.	.	.	.	.	.	.	.	.	.	.	.
X.117	.	.	.	2	−2	.	.	.	.	2	.	−2	.	.	.	−1	1	1	.	.	.	.	−1
X.118	.	.	.	2	2	.	.	.	.	.	−2	.	.	.	.	−1	1	1	.	.	.	.	−1
X.119	.	.	.	2	2	.	.	.	.	2	.	−2	.	.	.	−1	1	1	.	.	.	.	−1
X.120	.	.	.	.	.	.	.	.	.	−2	.	2	.	.	.	.	.	.	.	.	.	.	.
X.121	.	.	.	.	.	.	.	.	.	2	.	−2	.	.	.	.	.	.	.	.	.	.	.
X.122	.	.	.	−2	2	.	.	.	.	.	2	.	.	.	−1	1	−1	−1	.	.	.	.	1
X.123	.	.	.	.	.	.	−3	1	.	1	−1	1	.	.	.	.	.	.	.	.	.	.	.
X.124	.	.	.	.	.	.	−3	1	.	1	−1	1	.	.	.	.	.	.	.	.	.	.	.
X.125	3	.	.	.	.	.	.	.	.	.	.	.	.	.	.	.	.	.	.	.	−1	.	.
X.126	3	.	.	.	.	.	.	.	.	.	−2	.	.	.	.	.	.	.	.	.	−1	.	.
X.127	.	.	.	−3	1	.	.	.	.	.	.	.	.	.	.	.	.	.	.	.	.	−1	.
X.128	.	.	.	−3	1	.	.	.	.	.	.	.	.	.	.	.	.	.	.	.	.	−1	.
X.129	.	.	.	.	.	.	.	.	.	.	.	.	.	.	.	.	.	.	.	.	.	.	.
X.130	.	.	.	.	.	.	2	2	−1	.	.	.	1	.	.	.	.	.	.	.	.	.	.
X.131	.	.	.	.	.	.	2	−2	−1	.	.	.	−1	.	.	.	.	.	.	.	.	.	.
X.132	.	.	.	.	.	.	.	.	.	.	.	.	.	.	.	.	.	.	.	.	.	.	.
X.133	.	.	.	−2	−2	.	.	.	.	.	.	.	.	.	.	1	−1	−1	.	.	.	.	1
X.134	.	.	.	.	.	.	.	.	.	.	.	.	.	.	.	.	.	.	.	.	.	.	.
X.135	.	.	.	.	.	.	.	.	.	.	.	.	.	.	.	.	.	.	.	.	.	.	.
X.136	.	.	.	.	.	.	.	.	.	.	.	.	.	.	.	.	.	.	.	.	.	.	.
X.137	.	.	.	.	.	.	.	.	.	.	.	.	.	.	.	.	.	.	.	.	.	.	.
X.138	−4	−1	−1	.	.	.	.	.	.	.	.	.	.	.	.	.	.	.	.	.	.	.	.
X.139	2	−1	−1	.	.	.	.	.	.	.	.	.	.	.	.	.	.	.	.	.	.	.	.
X.140	2	2	−1	.	.	.	.	.	.	.	.	.	.	.	.	.	.	.	.	.	.	.	.
X.141	.	.	.	4	.	.	.	.	.	−2	.	2	.	.	.	1	−1	−1	.	.	.	.	1
X.142	.	.	.	.	.	.	.	.	.	.	.	.	.	.	−1	.	.	.	.	.	.	.	.
X.143	.	.	.	.	.	.	.	.	.	.	.	.	.	.	.	.	.	.	.	.	.	.	.
X.144	.	.	.	.	.	.	.	.	.	.	.	.	.	.	.	.	.	.	−1	.	.	.	.
X.145	.	.	.	.	.	.	.	.	.	.	.	.	.	.	.	.	.	.	1	.	.	.	.
X.146	.	.	.	.	.	.	3	−1	.	−1	1	−1	.	.	.	.	.	.	.	.	.	.	.
X.147	−2	−2	1	.	.	.	.	.	.	.	.	.	.	.	.	−2	.	.	−1	.	.	.	.
X.148	4	1	1	.	.	.	.	.	.	.	.	.	.	.	.	.	.	.	−1	−1	.	.	.
X.149	.	.	.	3	−1	.	−4	.	2	.	.	.	.	.	.	.	.	.	.	.	.	1	.
X.150	−3	.	.	−4	.	.	.	.	.	.	.	.	.	.	1	−1	1	1	.	.	1	.	−1
X.151	2	−1	−1	.	.	.	.	.	.	.	.	.	.	.	.	2	.	.	1	1	.	.	.
X.152	.	.	.	.	.	.	.	.	.	.	.	.	.	.	.	.	.	.	.	.	.	.	.
X.153	.	.	.	.	.	.	.	.	.	.	.	.	.	.	.	.	.	.	1	.	.	.	.
X.154	−2	1	1	.	.	.	.	.	.	.	.	.	.	.	.	.	.	.	.	.	.	.	.

where $A = 4\zeta(15)_3\zeta(15)_5^3 + 4\zeta(15)_3\zeta(15)_5^2 + 2\zeta(15)_3 + 2\zeta(15)_5^3 + 2\zeta(15)_5^2 + 1,\ B = 2\zeta(7)^4 + 2\zeta(7)^2 + 2\zeta(7) + 1.$

8

Janko's group J_4

Z. Janko discovered his largest sporadic group J_4 in 1976. Its main properties and definition are described in his fundamental article [67]. He defined his simple group G by means of the group structure of a given centralizer $H = C_G(z)$ of a 2-central involution z of G. He showed that a Sylow 2-subgroup S of H has a unique maximal elementary abelian normal subgroup A of order 2^{11} such that A has a complement M in $N_G(A)$ which is isomorphic to the simple Mathieu group M_{24}. In fact, $N_G(A) \cong E_2$, the finitely presented group of Lemma 7.1.1(f). The main result of [67] states that the order $|G| = 86775571046077562880$ is uniquely determined by the given structure of H. Janko's article also gives a parametrization of the 62 conjugacy classes of J_4 and the corresponding centralizer orders. Using these results J. Conway, S. Norton, J. G. Thompson and D. Hunt determined the character table of J_4 stated in the Atlas [19], pp. 188–189. The first existence proof was given by Benson, Conway, Norton, Parker and Thackray in 1980. They constructed J_4 as a subgroup of $GL_{112}(2)$; see [8] and [102]. Using geometric methods A. A. Ivanov published a uniqueness proof for Janko's group J_4 in [59]. Ivanov and Meierfrankenfeld gave a computer-free existence and uniqueness proof in [61]. For a coherent account of their methods and the history of the literature on Janko's largest sporadic group J_4 the reader is referred to Ivanov's recent book [64].

In view of Janko's result stating that the normalizer $N_G(A)$ is a split extension of M_{24} by its irreducible $GF(2)$-module A, it is clear that J_4 can also be constructed by an application of Algorithm 1.3.8 to the split extension E_2 of Lemma 7.1.1(f). Since the Sylow 2-subgroups of E_2 have order 2^{21} the construction of the 2-centralizer $H = C_G(z)$ is as difficult as the one for Conway's group Co_1 dealt with in Section 7.2 of Chapter 7. We state in Section 8.1 an improved version of the obtained presentation of the constructed group H. The group structure of H is described in Proposition 8.1.1. In particular, all conditions of Janko's Theorem A of [67] are satisfied. We also give a documented faithful permutation representation of H of degree 42240. It has been used to calculate a system of representatives of the conjugacy classes and the character table of H; see Tables 8.5.1 and 8.6.1. Using these results we give in this chapter an existence and uniqueness proof for Janko's simple group J_4 by means of Algorithm 7.4.8 of [92].

In Section 8.2 we show that each simple group G having a 2-central involution z with $C_G(z) = H$ has a maximal elementary abelian subgroup $A \leq H$ of order 2^{11} such that the amalgam $H \leftarrow D \rightarrow E$ has Goldschmidt index 1, where $D = N_H(A)$ and $E = N_G(A)$. Furthermore, $E \cong E_2$ as constructed in Lemma 7.1.1(f). This result implies that G has two conjugacy classes of involutions represented by z and some fixed involution $u \in E$. We show that $U = C_G(u) = C_E(u)$ and calculate its character table. Thus we can apply Theorem 1.6.4 to give a short proof for Janko's Theorem about the uniqueness and the value of the order of G; see Theorem 8.2.3. It is then easy to show that G has 62 conjugacy classes.

Using Brauer's characterization of characters we show in Proposition 8.3.3 that G has unique pair of complex conjugate irreducible characters χ of degree 1333. By Theorem 3.12.4 of [92] it is of 43-defect zero and defines a uniquely determined irreducible representation $\kappa : G \rightarrow GL_{1333}(43)$ of the simple group G. H. Kim's and the author's construction of $\mathfrak{G} = \kappa(G)$ is presented in Section 8.3; see Theorem 8.3.2. Its four generating matrices are stored on the accompanying DVD. In Theorem 8.3.4 we show that each finite simple group G of J_4-type is isomorphic to the simple subgroup $\mathfrak{G}$ of $GL_{1333}(43)$.

In Section 8.4 we give a survey about earlier constructions of J_4 as subgroups of $GL_{112}(2)$ and $GL_{1333}(11)$. In particular, we mention Lempken's construction of J_4 in [84] as a subgroup $G = \langle x, y\rangle$ of $GL_{1333}(11)$ and Weller's construction of a faithful permutation representation of G of degree 173067389 in [135]. It has been used by Weller and the author in [98] to construct a 112-dimensional representation J of G in $GL_{112}(2)$. Furthermore, it is shown there that J and the original group constructed by Benson, Conway, Norton, Parker and Thackray are conjugate in $GL_{112}(2)$. So all these groups, and the group $\mathfrak{G}$ constructed in Theorem 8.3.2, are isomorphic by Theorem 8.3.4. The generating matrices of G and J and their corresponding permutations are also stored on the accompanying DVD.

8.1 Structure of the given centralizer

The author applied Algorithm 1.3.8 to the split extension E_2 of the Mathieu group M_{24} by its second irreducible representation V_2 constructed in Lemma 7.1.1(f). To simplify the notation we denote E_2 by E throughout this chapter. Lemma 7.1.1(h) states that E has a unique conjugacy class of 2-central involutions represented by $z = z_2 = (gv_1)^2$. Let $D = C_E(z)$. A system of representatives of its conjugacy classes and its character table are stored on the accompanying DVD; see Tables `DVD.1.6.1` and `DVD.1.6.3`, respectively. As in Chapter 7 we obtained a finitely presented group H with a Sylow 2-subgroup S of order 2^{21} having a unique maximal elementary abelian normal subgroup A of order 2^{11} such that $D_H = N_H(A) \cong D$. Unfortunately, the presentation of H again has many relations. Therefore it is not restated here. However, we were able to reduce the number of relations drastically by using a system of relations of a subgroup $U \cong 6Aut(\mathsf{M}_{22})$ of H obtained by the author's former student V. Gebhardt [34]. The combined system is stated in the following result.

Proposition 8.1.1 *Let $H = \langle a, b, c, d, e, f\rangle$ be the finitely presented group having the following set $\mathcal{R}(H)$ of defining relations:*

$$
\begin{aligned}
& a^3 = b^2 = c^6 = d^2 = e^3 = f^3 = 1, \quad (a^2b)^2 = (dbd)^2 = 1, \\
& [a, c] = [a, d] = [a, e] = [e, d] = [e, b] = [b, c] = 1, \quad [a, b] = a, \\
& [b, d]^2 = 1, [dbd, c^3] = 1, ([c^{-1}, d]c)^2 = d, \\
& (dbc)^6 = 1, \quad [dbd, b] = 1, \quad [e, c^3] = dbdc^{-2}dbdc^{-1}, \\
& [ec^{-1}dc^{-1}e, dbdc^{-1}] = dcbdbc^3dcd, \\
& [ecdc^{-1}dce, bc^{-1}dbcdc^{-1}db] = (c^2dbdb)^3a^{-1}, \\
& c^{-1}dc^{-1}dbdcdce^{-1}c^{-1}dc^{-1}dbdcdce^2 = 1, \\
& [f, a^{-1}c^{-2}ec^{-1}decbdc] = c^2bdce^{-1}cdbc^{-1}e^{-1}c^{-1}de^{-2}(c^2dbdb)^3, \\
& [f, ac^{-1}dc^{-1}e^{-1}cdc^{-1}dec^{-1}dc^{-1}e^{-1}dc^{-1}] \\
& \quad = ba^{-1}e^{-1}c^{-1}dc^{-1}e^{-1}c^{-1}dec^2ecadec^2e^{-1}cdecdbdc, \\
& [fdf, ac^{-1}e^{-1}c^{-1}dc^{-1}ecec^{-1}bdca^{-1}] \\
& \quad = bc^2ecadc^{-1}dec^{-1}ece^{-1}c^{-2}e^{-1}dcebc^{-1}ecdc^{-1}e^{-1}(c^2dbdb)^3, \\
& [fdf, ac^{-1}e^{-1}c^{-1}dcdc^{-1}e^{-1}ce^{-1}c^{-1}dce^{-1}bdca^{-1}] \\
& \quad = ae^{-1}c^{-1}de^{-1}c^{-1}e^{-1}adcde^{-1}ce^{-1}(c^2dbdb)^3,
\end{aligned}
$$

$$[fdaf, a^{-1}dc^{-1}ec^{-1}e^{-1}c^2eba^{-1}]$$
$$= c^2ecdc^{-1}dbc^{-1}e^{-1}c^{-1}dec^2dbecdc^{-2}e^{-1},$$
$$[fdaf, a^{-1}bc^2e^{-1}ce^{-1}dc^{-2}] = ecae^{-1}c^3dc^{-1}de,$$
$$c^2ec^{-1}dc^{-1}e^{-1}f^{-1}dc^{-1}efc^{-1}e^{-1}cf = 1,$$
$$fab^{-1}a^{-1}f^{-1}a^{-1}f^{-1}a^{-1}bfa^2 = 1.$$

Then the following assertions hold.

(a) H *has a faithful permutation representation* PH *of degree* $2^8 \cdot 3 \cdot 5 \cdot 11$ *with stabilizer* $\langle a, b, c, e, fecf\rangle$ *of order* $2^{13} \cdot 3^2 \cdot 7$.

(b) $S = \langle s_1, s_2, s_3, s_4, s_5\rangle$ *of order* $|S| = 2^{21}$ *is a Sylow* 2*-subgroup of* H, *where*

$$s_1 = bce^{-1}fbcfc, \quad s_2 = bc^{-1}dcdf^{-1}dc^{-1}dc,$$
$$s_3 = bc^{-1}dcdf^{-1}bdc^{-1}dc, \quad s_4 = bc^{-1}dcdf^{-1}dbc^{-1}dc,$$
$$s_5 = bcefa^2dfc^2bef^{-1}d^{-1}e^{-1}a^{-1}.$$

(c) $H = \langle x, y, h\rangle$, *where* $x = s_1as_3b$, $y = e$ *and* $h = c$ *have orders* 12, 3 *and* 6, *respectively. Furthermore, the involution* $z = (x^2y)^{15} \in S$ *generates the center of* H *and also the center of* S.

(d) S *has a unique normal Klein 4-subgroup* $W = \langle z, t\rangle$, *where* $t = x^6$.

(e) S *has a unique elementary abelian subgroup* A *of order* $|A| = 2^{11}$. *It is generated by* z, $t = (s_1as_3b)^6$ *and the following nine involutions;* $a_1 = s_2^4$, $a_2 = (s_1s_2)^4$, $a_3 = (s_1s_5)^2$, $a_4 = (s_1s_2s_5)^4$, $a_5 = (s_1s_5s_2)^4$, $a_6 = s_1^3s_5$, $a_7 = (s_1^2s_2s_5)^4$, $a_8 = s_1^2s_5s_1$ *and* $a_9 = s_1s_2s_1s_2s_1s_2s_5s_2$.

(f) $D = N_H(A) = \langle x, y\rangle$ *and* $|D| = 2^{21} \cdot 3^3 \cdot 5$.

(g) *The normal subgroup* A *of* D *has a complement* K *which is generated by* $k_1 = (x^3y^2)^2$, $k_2 = (x^3yxyx)^8$ *and* $k_3 = (xy^2xyx^3y)^3$ *of respective orders* $o(k_1) = 2$, $o(k_2) = 3$ *and* $o(k_3) = 2$.

(h) *The Fitting subgroup* B *of* K *is elementary abelian and* B *is generated by the six involutions* $b_1 = (k_1k_2k_1k_2^2)^5$, $b_2 = (k_1k_2^2k_1k_2)^5$, $b_3 = (k_2k_1k_2k_1k_2)^5$, $b_4 = (k_1k_2k_1k_2k_1k_2k_1)^5$, $b_5 = (k_1k_2k_1k_2k_1k_2^2)^3$ *and* $b_6 = (k_1k_2k_1k_2^2k_1k_2)^3$.

B *has a complement* $L \cong 3S_6$ *in* K *which is generated by the elements* $c_1 = k_2$, $c_2 = (k_3k_2k_3k_1)^3$ *and* $c_3 = k_3k_1k_3k_1k_3$ *of respective orders* 3, 4 *and* 2.

(i) $U = \langle a, b, c, d, e\rangle \cong 6Aut(M_{22})$ *and* U *has center* $Z(U) = \langle z\rangle$.

(j) H *has a unique normal subgroup* Q *of order* 2^{13}. *It is extraspecial,* $H = QU$, $Q \cap U = Z(H)$ *and* $C_H(Q) \leq Q$.

(k) *A system of representatives* h_i *of the* 113 *conjugacy classes of* H *and the corresponding centralizers orders* $|C_H(h_i)|$ *are given in Table 8.5.1. The character table of* H *is given in Table 8.6.1.*

(l) *A system of representatives* d_i *of the* 160 *conjugacy classes of* D *and the corresponding centralizer orders* $|C_D(d_i)|$ *are given in Table* `DVD.1.6.1`. *The character table of* D *is given in Table* `DVD.1.6.3`. *Both tables are stored on the accompanying DVD.*

Proof (a) This statement has been verified by P. Young's program

```
MyCosetAction(H, sub<H|a,b,c,e,fecf> : maxsize:=100000000).
```

(b) The Sylow 2-subgroup S of H was found by means of the faithful permutation representation PH of H constructed in (a) and the MAGMA command `SylowSubgroup(PH,2)`. Applying the program `GetShortGens(H, S)` we found the five generators s_i of S with the given orders.

(c) All assertions of this statement can be verified by standard applications of MAGMA.

(d) and (e) The elementary abelian normal subgroups W and A of S have been determined by means of the MAGMA command

```
Subgroups(PS: Al:=Normal, IsElementaryAbelian := true)
```

It follows that S has exactly one normal Klein 4-subgroup W and one elementary abelian normal subgroup A of order 2^{11}. Their given generators have been found by the program `GetShortGens(S, X)` with $X \in \{W, A\}$. For a proof that A is in fact the unique maximal elementary abelian subgroup of S, see the proof of Proposition 7 of Janko's article [69].

(f) Using PH and MAGMA the reader can verify that $D = N_H(A) = \langle x, y\rangle$ and that D has the given order.

(g) The existence of the complement K of A in D was obtained computationally using the MAGMA command `HasComplement(D,A)`. Its generators were found by means of the program `GetShortGens(D, K)`.

(h) The Fitting subgroup B of K was determined by means of the command `FittingSubgroup(K)`. Its generators were calculated using the program `GetShortGenerators(K,B)`. The complement $L \cong 3S_6$ and its generators were obtained by the methods used in (g).

(i) Using PH and MAGMA we obtained the description of the structure of the subgroup $U = \langle a, b, c, d, e\rangle$ of $H = \langle a, b, c, d, e, f, g\rangle$.

(j) Another application of MAGMA yields that H has a unique normal subgroup Q of order 2^{13} and that it is extra-special. Furthermore, MAGMA yields that $H = QU$ and $U \cap Q = Z(H)$.

(k) and (l) The systems of representatives of the conjugacy classes of H and D have been calculated by means of PH, MAGMA and Kratzer's Algorithm 5.3.18 of [92]. The character tables of H and D have been computed by means of MAGMA using the faithful permutation representation PH of H. □

Definition 8.1.2 A finite simple group G is said to be of J_4-*type* if it has a 2-central involution z such that $C_G(z)$ is isomorphic to the finitely presented group H of Proposition 8.1.1.

8.2 Conjugacy classes and group order

In this section we give a parametrization of the conjugacy classes of any finite simple group G of J_4-type. Furthermore, we give another proof for Janko's Theorem determining the order of G.

Proposition 8.2.1 *Let $E = E_2 = \langle a, b, c, d, t, g, h, i, j, k, v_i \mid 1 \le i \le 11\rangle$ be the non-split extension of M_{24} by its simple module V_2 of dimension 11 over $F = GF(2)$ given in Lemma 7.1.1. Let $H = \langle a, b, c, d, e, f, g\rangle = \langle x, y, h\rangle$ be the finite group defined in Proposition 8.1.1. Let S be the Sylow 2-subgroup of H constructed in Proposition 8.1.1(b). Let A be its unique maximal elementary abelian normal subgroup and let $D_H = N_H(A) = \langle x, y\rangle$, as constructed in Proposition 8.1.1(e) and (f). Then the following statements hold.*

- (a) *$z = (gv_1)^2$ is a 2-central involution of E such that $D_E = C_E(z)$ has order $2^{21} \cdot 3^3 \cdot 5$ and there is an isomorphism $\varphi : D_H \to D_E$ such that $D_E = \langle x_1, y_1\rangle$ and $E = \langle x_1, y_1, e\rangle$, where $x_1 = \varphi(x)$, $y_1 = \varphi(y)$ and $e = a \in E$.*
- (b) *$V_2 = \varphi(A)$ is the unique maximal elementary abelian normal subgroup of any Sylow 2-subgroup of E.*
- (c) *The Goldschmidt index of the amalgam $H \leftarrow D \rightarrow E$ is 1.*
- (d) *A system of representatives e_i of the 72 conjugacy classes of $E = \langle x_1, y_1, e\rangle$ and the corresponding centralizer orders $|C_E(e_i)|$ are given in Table 8.5.2.*
- (e) *The character table of E is given in Table 8.6.2.*

Proof (a) By Lemma 7.1.1(h) E has a faithful permutation representation PE of degree 2048. Furthermore, $z = (gv_1)^2$ is a 2-central involution of E. Using PE and MAGMA one sees that $D_E = C_E(z)$ has order $2^{21} \cdot 3^3 \cdot 5$.

Let PH be the faithful permutation representation of H given in Proposition 8.1.1(a). By its assertion (f) we know that $D_H = N_H(A) = \langle x, y \rangle$, where A is the unique maximal elementary abelian subgroup of the fixed Sylow 2-subgroup S of H. Using the faithful permutation representations PH and PE, MAGMA establishes an explicit isomorphism $\varphi : D_H \to D_E$ by means of the command `IsIsomorphic(D_H, D_E)`. Let $x_1 = \varphi(x)$ and $y_1 = \varphi(y)$. Then $D_E = \langle x_1, y_1 \rangle$. Another application of PE and MAGMA shows that $E = \langle x_1, y_1, e \rangle$, where $e = a$, the original generator a of E.

(b) By construction of E the elementary abelian subgroup V_2 of E is normal in E and therefore normal in any Sylow 2-subgroup of E. Using PE and the MAGMA command

```
Subgroups(PS: Al:=Normal, IsElementaryAbelian := true)
```

it follows that V_2 is the unique maximal normal elementary abelian subgroup in any Sylow 2-subgroup of E. Hence $\varphi(A) = V_2$.

(c) The Goldschmidt index has been calculated by means of Kratzer's Algorithm 7.1.10 of [92], the faithful permutation representations PH and PE and MAGMA.

(e) The system of representatives of the conjugacy classes of E has been calculated by means of PE, MAGMA and Kratzer's Algorithm 5.3.18 of [92].

(f) The character table of E has been computed by means of MAGMA using the faithful permutation representation PE of E. □

By Proposition 8.2.1 the amalgam $H \leftarrow D \rightarrow E$ satisfies the main condition of Algorithm 1.3.8. Therefore we apply Algorithm 7.4.8 of [92] in Section 8.3 to give an existence proof for Janko's largest sporadic group J_4; see [69].

Lemma 8.2.2 *Keep the notation of Proposition 8.1.1. Let G be a finite simple group of J_4-type with a 2-central involution z such that $C_G(z) = H = \langle x, y, h \rangle$.*

Let A be the maximal elementary abelian normal subgroup of the Sylow 2-subgroup S of H constructed in Proposition 8.1.1(e). Let

$E = \langle x_1, y_1, e\rangle$ be the extension of M_{24} by its simple $GF(2)$-module V_2 constructed in Proposition 8.2.1(a).

Then the following assertions hold.

(a) $N_G(A) > N_H(A)$.
(b) *$N_G(A)$ is isomorphic to the split extension $E = E_2$ of M_{24} by its 11-dimensional irreducible $GF(2)$-module V_2 constructed in Lemma 7.1.1(f).*
(c) *G has two conjugacy classes of involutions represented by z and $w = (xhy)^6$ of H.*
(d) *Using the notations of Tables 8.6.2 and 8.5.1 the conjugacy classes 2_1, 2_4 and 2_6 of involutions of $E = N_G(A)$ and the conjugacy classes 2_3, 2_5, 2_6, 2_9 and 2_{10} of $H = C_G(z)$ fuse with the conjugacy class u^G of G.*

    The conjugacy classes 2_2, 2_3 and 2_5 of involutions of $E = N_G(A)$ and the conjugacy classes 2_1, 2_2, 2_4, 2_7 and 2_8 of $H = C_G(z)$ fuse with the conjugacy class z^G of G.
(e) *w is G-conjugate to $u = (xy^2e)^{11}$ and $U = C_G(u) = C_E(u) = \langle p, q, r\rangle$, where $E = N_G(A) = \langle x, y, e\rangle$ and $p = [(ye)^5(yx)^3(ye)^{10}(exe)^3]^3$, $q = [(ye)^5(yx)^3(exe)^3(yx)^3(ye)^5]^2$ and $r = (ye)^5$.*
(f) *A system of representatives u_i of the 99 conjugacy classes of $U = \langle p, q, r\rangle$ and the corresponding centralizer orders $|C_U(u_i)|$ are given in Table* `DVD.1.6.2`*. The character table of U is given in Table* `DVD.1.6.4`*. Both tables are stored on the accompanying DVD.*
(g) *$U = C_G(u)$ has 12 conjugacy classes of involutions. Using the notation of Table* `DVD.1.6.2` *its classes 2_3, 2_5, 2_6, 2_7 and 2_{10} fuse with the conjugacy class z^G of G, and the classes 2_1, 2_2, 2_4, 2_8, 2_9, 2_{11} and 2_{12} fuse with the conjugacy class u^G of G.*
(h) *G has 27 z-special conjugacy classes represented by the conjugacy classes 2_1, 4_1, 4_2, 6_1, 6_4, 8_1, 8_2, 8_3, 10_1, 12_1, 12_2, 14_1, 14_2, 16_1, 20_1, 20_2, 22_1, 24_1, 24_4, 30_1, 40_1, 40_2, 42_1, 42_2,44_1, 66_1 and 66_2 of H.*
(i) *G has ten z-special conjugacy classes represented by the conjugacy classes 2_1, 4_2, 6_3, 10_3, 12_9, 14_1, 14_2, 22_1, 28_1 and 28_2 of U.*

Proof (a) In view of Proposition 8.1.1, Janko's fusion arguments of the proof of Proposition 8 of [69] can be adapted.

(b) Let $D = D_H(A) = \langle x, y, \rangle$. By Proposition 8.1.1(e) A has a complement $K = \langle k_1, k_2, k_3\rangle$ in D. Using the 11 generators of the elementary abelian normal subgroup given in Proposition 8.1.1(g) as a basis $\mathcal{B}$ of the $GF(2)$-vector space A we represent the generators k_i of K as matrices $Mk_i \in \mathrm{GL}_{11}(2)$ by means of their conjugate action on A. In particular, the map $\eta : k_i \to Mk_i$, for $i = 1, 2, 3$, is an isomorphism from K onto $\Delta = \langle Mk_1, Mk_2, Mk_3 \rangle \leq \mathrm{GL}_{11}(2)$. By Proposition 8.1.1(h) and Proposition 8.2.4(r) of [92] we know that $\Delta \cong N_{\mathsf{M}_{24}}(B)$, where B is a maximal elementary abelian normal subgroup of a Sylow 2-subgroup of M_{24}. Applying Algorithm 1.3.8 to Δ inside $\mathrm{GL}_{11}(2)$ one constructs the centralizer $H_3 = C_{\mathsf{M}_{24}}(z_3)$ of a 2-central involution z_3 of M_{24} as a matrix subgroup Π of $\mathrm{GL}_{11}(2)$ and checks with MAGMA's isomorphism test and Lemma 8.1.2 of [92] that $\Phi = \langle \Pi, \Delta \rangle \cong \mathsf{M}_{24}$. Using a faithful permutation representation of $GL = \mathrm{GL}_{11}(2)$ of degree 2047 and MAGMA's command `Subgroups(GL: OrderMultipleOf := 2^{10}*3^3*5)` and (a) one shows that $N_G(A)/A \cong \Phi$ is uniquely determined up to isomorphism. Now Theorem 1.4.15 of [92] and Lemma 7.1.1(f) imply the assertion.

(c) and (d) By (b) we know that $N_G(A) \cong E = \langle x_1, y_1, e \rangle$ constructed in Proposition 8.2.1(a). Its assertion (c) states that the amalgam $H \leftarrow D \rightarrow E$ has Goldschmidt index 1. Therefore we can identify $D_H = \langle x, y \rangle$ with $D_E = \langle x_1, y_1 \rangle$ using the isomorphism φ constructed in the proof of Proposition 8.2.1(a). Thus we assume that $D = \langle x, y \rangle = H \cap E$. A system of representatives of the 160 conjugacy classes of D is stated in Table `DVD.1.6.1`. It states that D has 21 conjugacy classes of involutions. They are represented by short words in x and y. Using the faithful permutation representations PH and PE of H and E described in Proposition 8.1.1 and Lemma 7.1.1(h) it is easy to check the membership of d^E and d^H in a conjugacy class of involutions of H and E classified in Tables 8.6.2 and 8.5.1, respectively. Thus one obtains the following fusion pattern. Here the classes are denoted by the numbers of the indices plus 1 of their representatives of the involutions in the three tables. In particular, number 2 represents the first involution in each table. The classes of E are stated before those of D. The last entries denote the classes of D.

$$2A := \langle [3, 4, 6], [2, 4, 5, 7, 9, 10, 12, 15, 17, 19], [2, 3, 5, 8, 9] \rangle,$$
$$2B := \langle [2, 5, 7], [3, 6, 8, 11, 13, 14, 16, 18, 20, 21, 22], [4, 6, 7, 10, 11] \rangle.$$

(e) The first statement follows immediately from Tables 8.6.2 and 8.5.1 and the above fusion pattern. Let $U_1 = C_E(u)$, where $u = (xy^2e)^{11} \in E$. Using PE and MAGMA we find a complement K_1 of A in U_1 such that $K_1 \cong Aut(M_{22})$ and $V = A/\langle u \rangle$ is an elementary abelian normal subgroup of $U_2 = U_1/\langle u \rangle$. Furthermore, U_2 does not split over V. Hence it is isomorphic to the extension E_5 constructed in Lemma 4.1.2(h).

Using the arguments of Remark 5.4.3 we can show that $C_G(u)$ cannot be larger than $C_E(u)$. Using PE and MAGMA one can check that $U = C_G(u) = \langle p, q, r \rangle$ for the given generators p, q and r.

(f) The system of representatives of the conjugacy classes of $U = C_G(u) = \langle p, q, r \rangle$ has been calculated by means of PE, MAGMA and Kratzer's Algorithm 5.3.18 of [92]. The character table of U has been computed by means of MAGMA using the faithful permutation representation PE of E.

(g) This assertion can be verified computationally by determination of the conjugacy of the 21 representatives of the conjugacy classes of involutions of U stated in Table `DVD.1.6.2` with one of the six representatives of the conjugacy classes of E stated in Table 8.6.2 because $U = C_E(u)$ is a subgroup of E.

(h) and (i) Both statements follow from the power map information given in Tables 8.5.1 and `DVD.1.6.2`. □

The following theorem is Proposition 23 of Z. Janko's article [69]. The proof given here is due to Kim and the author. It rests on Theorem 1.6.4.

Theorem 8.2.3 (Z. Janko) *Let G be any finite simple group of J_4-type having a 2-central involution z with centralizer $H = C_G(z) = \langle x, y, h \rangle$ stated in Proposition 8.1.1(c). Then the following statements hold.*

(a) *G has two conjugacy classes of involutions represented by $z = (x^2y)^{15}$ and $u = (xhy)^6$ of H. Furthermore, $U = C_G(u) \cong C_E(u_1)$, where $u_1 = (xy^2e)^{11} \in E = \langle x, y, e \rangle \cong N_G(A)$ is the normalizer of the unique maximal elementary abelian normal subgroup of the Sylow 2-subgroup S defined in Lemma 8.1.1(b).*

(b) $G = 2132400816 \cdot |\mathfrak{U}| + 4388805476055 \cdot |\mathfrak{H}| = 2^{21} \cdot 3^{13} \cdot 5^2 \cdot 7 \cdot 11 \cdot 13 \cdot 17 \cdot 23 \cdot 29$.

Proof (a) This assertion holds by Lemma 8.2.2.

(b) Let

$$r(z, u, z) = \left|\left\{(x, y) \in (z^G \cap H) \times (u^G \cap H) \middle| z \in \langle xy \rangle\right\}\right|$$

and

$$r(z, u, u) = \left|\left\{(x, y) \in (z^G \cap U) \times (u^G \cap U) \middle| u \in \langle xy \rangle\right\}\right|.$$

By Table 8.5.1 H has 27 real z-special conjugacy classes with class numbers 2, 14, 15, 39, 42, 55, 56, 57, 68, 73, 76, 85, 86, 92, 93, 94, 99, 100,

103, 104, 107, 108, 109, 110, 111, 112 and 113. Here their representatives are denoted by $\{t_j \mid 1 \leq j \leq 45\}$. Let z_i and u_k be the representatives of the H-classes of involutions fusing to z and u in G, respectively. They are known by Lemma 8.2.2(d). For each triple (z_i, u_k, t_j) let

$$d(z_i, u_k, t_j) = \frac{|H|^2}{|C_H(z_i)| \cdot |C_H(u_k)| \cdot |C_H(t_j)|} \times \sum_{\psi \in Irr_{\mathbb{C}}(H)} \psi(z_i)\psi(u_k)\psi(t_j)\psi(1)^{-1}.$$

Then Theorem 1.6.4 and (g) imply that

$$r(z, u, z) = \sum_{i=1}^{6} \sum_{k=1}^{3} \sum_{j=1}^{27} d(z_i, u_k, t_j).$$

Using (g) and the values of the character Table 8.6.1 of H these formulas yield $r(z, u, z) = 29236337328$.

By Table `DVD.1.6.2` $U = C_G(u) = C_E(e)$ has ten real u-special conjugacy classes with class numbers $2, 16, 44, 70, 84, 85, 86, 95, 98, 99$ and 105. Denote their representatives by $\{s_n \mid 1 \leq n \leq 10\}$. Let u_i and z_k be representatives of the 12 U-classes of involutions fusing to u and z in G, respectively. They are known by Lemma 8.2.2(g). For each triple (u_i, z_k, s_n) let

$$d(u_i, z_k, s_n) = \frac{|U|^2}{|C_U(u_i)| \cdot |C_U(z_k)| \cdot |C_U(s_n)|} \times \sum_{\psi \in Irr_{\mathbb{C}}(U)} \psi(u_i)\psi(z_k)\psi(s_n)\psi(1)^{-1}.$$

Then Theorem 1.6.4 and (h) imply that

$$r(z, u, u) = \sum_{i=1}^{4} \sum_{k=1}^{3} \sum_{n=1}^{10} d(u_i, z_k, s_n).$$

Using (h) and the values of the character Table `DVD.1.6.4` of U these formulas yield $r(z, u, u) = 1544188503$. Now Theorem 4.2.1 of [92] implies that

$$\begin{aligned} |G| &= r(z, u, z) \cdot |C_G(u)| + r(z, u, u) \cdot |C_G(z)| \\ &= 2^{21} \cdot 3^3 \cdot 5 \cdot 7 \cdot 11^3 \cdot 23 \cdot 29 \cdot 31 \cdot 37 \cdot 43. \end{aligned}$$

□

Corollary 8.2.4 (Z. Janko) *Each finite simple group G of J_4-type has self-centralizing Sylow p-subgroups P_p of prime order $|P_p| = p$ for the primes $p \in \{23, 29, 31, 37, 43\}$. Their normalizers $N_p = N_G(P_p)$ are Frobenius groups of respective orders $|N_{23}| = 23 \cdot 22$, $|N_{29}| = 29 \cdot 28$, $|N_{31}| = 31 \cdot 10$, $|N_{37}| = 37 \cdot 12$ and $|N_{43}| = 43 \cdot 14$.*

Proof All assertions are immediate consequences of Theorem 8.2.3, Sylow's Theorem and Burnside's Theorem 1.4.17 of [92]. □

Because of lack of space we quote the following result of Janko's original paper [67] without proof, although it can be derived from Theorem 8.2.3 and Proposition 8.1.1 from scratch.

Lemma 8.2.5 (Z. Janko) *Let G be any finite simple group of J_4-type having a 2-central involution z with centralizer $H = C_G(z) = \langle x, y, h \rangle$ stated in Proposition 8.1.1(c). Then the following statements hold.*

(a) *The Sylow 2-subgroup S of H defined in Proposition 8.1.1(b) contains a special subgroup L of order 2^{15} with an elementary abelian center $Z(L)$ of order 2^3 such that $M = N_G(L)$ is isomorphic to a non-split extension of $M_1 = S_5 \times \mathrm{PSL}_3(2)$ by L.*
(b) *$M = N_G(L)$ contains a normalizer $N_G(p)$ of the cyclic Sylow p-subgroups of order $p \in \{5, 7\}$.*
(c) *G has one conjugacy class of elements of order 5 and two conjugacy classes of elements of order 7.*
(d) *A Sylow 11-subgroup S_{11} of G is extra-special of order 11^3 and exponent 11.*
(e) *$N_{11} = N_G(S_{11})$ is a split extension of the direct product P of a cyclic group of order 5 and $\mathrm{GL}_2(3)$.*
(f) *G has two conjugacy classes of elements of order 11.*
(g) *G has 25 conjugacy classes of elements of odd order. They can be denoted by means of their orders as follows: 1_A, 13_A, 5_A, 7_A, 7_B, 11_A, 11_B, 15_A, 21_A, 21_B, 23_A, 29_A, 31_A, 31_B, 31_C, 33_A, 33_B, 35_A, 35_B, 37_A, 37_B, 37_C, 43_A, 43_B, 43_C.*

8.3 Existence and uniqueness proofs

In this section it is shown that Janko's simple group J_4 is uniquely determined by its centralizer H of a 2-central involution defined in Proposition 8.1.1(c). We also give an existence proof due to Kim and the author.

Lemma 8.3.1 *Let G be a finite simple group of J_4-type with a 2-central involution z such that $C_G(z) = H = \langle x, y, h\rangle$ described in Proposition 8.1.1(c). Then H is a maximal subgroup of G.*

Proof This assertion follows from Proposition 8.2.1 and Theorem 8.2.3 using an extended version of the argument of Lemma 3.3.11. □

Theorem 8.3.2 (Kim–Michler) *Let G be a finite simple group of J_4-type. Keep the notation of Propositions 8.1.1 and 8.2.1. Let $H \leftarrow D \rightarrow E$ be the amalgam constructed in Proposition 8.2.1. Let $\sigma D \rightarrow E$ be the corresponding monomorphism. Using the notation of the three character Tables 8.6.1,* `DVD.1.6.3` *and 8.6.2 of the groups $H = \langle D, h\rangle$, $D = \langle x, y\rangle$ and $E = \langle D, e\rangle$, respectively, the following statements hold.*

(a) *There are four compatible pairs $(\chi, \tau) \in mf\,char_{\mathbb{C}}(H) \times mf\,char_{\mathbb{C}}(E_3)$ of degree 1333 of the groups $H = \langle D, h\rangle$ and $E_3 = \langle D, e\rangle$:*

$$\chi_{\mathbf{32}} + \chi_{38} = \tau_4 + \tau_{\mathbf{20}},$$
$$\chi_{\mathbf{32}} + \chi_{38} = \tau_3 + \tau_{\mathbf{20}},$$
$$\chi_{\mathbf{31}} + \chi_{38} = \tau_4 + \tau_{\mathbf{20}},$$
$$\chi_{\mathbf{31}} + \chi_{38} = \tau_3 + \tau_{\mathbf{20}},$$

with common restriction

$$\tau_{|D} = \chi_{|D} = \psi_{32} + \psi_{71} + \psi_{80} + \psi_{\mathbf{89}},$$

where irreducible characters with bold face indices denote faithful irreducible characters.

(b) *Let $\mathfrak{V}$ and $\mathfrak{W}$ be the up to isomorphism uniquely determined faithful semi-simple multiplicity free 1333-dimensional modules of H and E over $F = GF(43)$ corresponding to the compatible pair*

$$(\chi = \chi_{\mathbf{31}} + \chi_{38}, \quad \tau = \tau_4 + \tau_{\mathbf{20}}),$$

respectively.

Let $\kappa_{\mathfrak{V}} : H \rightarrow \mathrm{GL}_{1333}(43)$ and $\kappa_{\mathfrak{W}} : E \rightarrow \mathrm{GL}_{1333}(43)$ be the representations of H and E afforded by the modules $\mathfrak{V}$ and $\mathfrak{W}$, respectively.

Let $\mathfrak{h} = \kappa_{\mathfrak{V}}(h)$, $\mathfrak{x} = \kappa_{\mathfrak{V}}(x)$ and $\mathfrak{y} = \kappa_{\mathfrak{V}}(y)$ in $\kappa_{\mathfrak{V}}(H) \leq \mathrm{GL}_{1333}(43)$. Then the following assertions hold.

(1) $\mathfrak{V}_{|D} \cong \mathfrak{W}_{|D}$, *and there is a transformation matrix* $\mathcal{T} \in \mathrm{GL}_{1333}(43)$ *such that*

$$\mathfrak{x} = \mathcal{T}^{-1}\kappa_{\mathfrak{W}}(x_1)\mathcal{T}, \qquad \mathfrak{y} = \mathcal{T}^{-1}\kappa_{\mathfrak{W}}(y_1)\mathcal{T}.$$

Let $\mathfrak{e} = \mathcal{T}^{-1}\kappa_{\mathfrak{W}}(e)\mathcal{T} \in \mathrm{GL}_{1333}(43)$. *Let* $\mathfrak{G} = \langle \mathfrak{h}, \mathfrak{x}, \mathfrak{y}, \mathfrak{e} \rangle$.

(2) *The four generating matrices of* $\mathfrak{G}$ *are stored on the accompanying DVD.*

(3) $G \cong \mathfrak{G}$.

(c) $\mathfrak{G}$ *is a finite simple group with the* 2*-central involution* $\mathfrak{z} = (\mathfrak{x}^2\mathfrak{y})^{15}$ *such that*

$$C_{\mathfrak{G}}(\mathfrak{z}) = \langle \mathfrak{h}, \mathfrak{x}, \mathfrak{y} \rangle \text{ and } |\mathfrak{G}| = 2^{21} \cdot 3^3 \cdot 5 \cdot 7 \cdot 11^3 \cdot 23 \cdot 29 \cdot 31 \cdot 37 \cdot 43.$$

Proof (a) Using MAGMA and the faithful permutation representations of PE and PH determined in Lemma 7.1.1 and Proposition 8.1.1, respectively, we determined the fusion of the classes of D in H and in E. A MAGMA calculation employing Kratzer's Algorithm 7.3.10 of [92] shows then that the amalgam $H \leftarrow D \rightarrow E$ has four compatible pairs (χ, τ) of degree 1333. They are stated in the assertion.

(b) In order to construct the faithful semi-simple representation $\mathfrak{W}$ corresponding to the character $\tau = \tau_4 + \tau_{20}$ of degree 1333 we employed the MAGMA command

```
LowIndexSubgroups(PE, n)
```

using the faithful permutation presentation PE of E of degree 2048. Thus we obtained the subgroups

$$U_1 = \langle (x^2exey)^3, (xyexyx)^3, (yx^2y^2x)^3 \rangle \text{ and } U_2 = \langle (xeye)^7, yx^3, (x^4e)^4 \rangle$$

of indices $n_1 = 40320$ and $n_2 = 2048$. Calculating the inner products of their permutation characters with τ_4 and τ_{20} we verified that τ_4 and τ_{20} are constituents of the respective permutation characters $(1_{U_1})^E$ and $(1_{U_2})^E$. We obtained the corresponding irreducible representations $M(4)$ and $M(20)$ of respective dimensions 45 and 1288 over $K = GF(43)$ by application of Algorithm 1.5.1 to the permutation modules. Hence $\mathfrak{W} = M(4) \oplus M(20)$.

In order to construct the faithful semi-simple representation $\mathfrak{V}$ corresponding to the character $\chi = \chi_{31} + \chi_{38}$ of degree 1333 we employed the same methods. This time we found a subgroup $R = \langle (x^2y^2)^5, y^2x^2, (yx^3yx^2y)^3 \rangle$ of index $n = 157696$ such that χ_{31} and χ_{38} are constituents of the permutation character $(1_R)^H$. Another application of

Algorithm 1.5.1 to this permutation module yields the irreducible representations $N(31)$ and $N(38)$ of respective dimensions 640 and 693 over K belonging to these two irreducible constituents of the permutation character. Hence $\mathfrak{V} = N(31) \oplus N(38)$.

By construction the KD-modules $\mathfrak{V}_{|D}$ and $\mathfrak{W}_{|D}$ described by the two pairs

$$(\mathfrak{V}(x), \mathfrak{V}(y)) \quad \text{and} \quad (\mathfrak{W}(\sigma(x), \mathfrak{W}(\sigma(y)))$$

of matrices in $\mathrm{GL}_{1333}(43)$ are isomorphic, where σ denotes the group isomorphism between the subgroup $D = \langle x, y\rangle$ of H and subgroup $D = \langle \sigma(x), \sigma(y)\rangle$ of E. Let $Y = \mathrm{GL}_{1333}(43)$ and let V and W be the KD- and $K(\sigma(D))$-modules described by the first and the second pair of matrices of Y. Applying Parker's isomorphism test of Proposition 6.1.6 of [92] by means of the MAGMA command

```
IsIsomorphic(GModule(sub<Y|V>),GModule(sub<Y|W>)),
```

one gets the transformation matrix $\mathcal{T}_1$ in $\mathrm{GL}_{1333}(43)$ satisfying $\mathfrak{V}(y) = (\mathfrak{W}(\sigma(y)))^{\mathcal{T}_1}$ and $\mathfrak{V}(x) = (\mathfrak{W}(\sigma(x)))^{\mathcal{T}_1}$.

By Corollary 7.2.4 of [92] the transformation matrix $\mathcal{T}$ of the statement is the product matrix $\mathcal{D}\mathcal{T}_1$, where

$$\mathcal{D} = diag(1^{45}, 2^{288}, 26^{300}, 1^{640}) \in \mathrm{GL}_{1333}(43).$$

Since $E = \langle \sigma(x), \sigma(y), e\rangle$ we let $\mathfrak{e} = \mathcal{T}^{-1}\mathfrak{W}(e)\mathcal{T}$ in $\mathrm{GL}_{1333}(43)$. Furthermore, let $\mathfrak{G} = \langle \mathfrak{x}, \mathfrak{y}, \mathfrak{h}, \mathfrak{e}\rangle$.

By Lemma 8.3.1 $G = \langle H, E\rangle$. Hence G is an epimorphic image of P. Now the proof of Theorem 7.5.1 of [92] yields that $G \cong \mathfrak{G}$ because the amalgam $H \leftarrow D \rightarrow E$ has Goldschmidt index 1, by Proposition 8.2.1(c), and G is simple.

(c) This assertion is an immediate consequence of statement (3) of assertion (b), Theorem 8.2.3 and Proposition 8.1.1(c). □

Because of lack of space we were not able to construct presentations of the normalizers $M = N_G(L)$ and N_{11} described in Lemma 8.2.5 due to Z. Janko. Thus we also cannot quote their character tables. Therefore we sketch the proof of the following result.

Proposition 8.3.3 *Each finite simple group G of J_4-type has* 62 *conjugacy classes and a unique pair of complex conjugate characters $\chi : G \rightarrow \mathbb{C}$ of degree $\chi(1) = 1333$.*

Proof Let $H = C_G(z)$ be the centralizer of a 2-central involution of G. In Proposition 8.1.1 we fixed a Sylow 2-subgroup S of H and showed that it has a unique maximal elementary abelian normal subgroup A. By Proposition 8.2.1, $D = N_H(A)$ is a proper subgroup of $E = N_G(A)$ and the amalgam $H \leftarrow D \rightarrow E$ has Goldschmidt index 1. Therefore the free product $P = H *_D E$ of H and E with amalgamated subgroup D is uniquely determined up to isomorphism. By the proof of Theorem 8.3.2(b) there is exactly one pair of conjugate complex compatible pairs $(\chi, \tau) \in mf\text{char}_{\mathbb{C}}(H) \times mf\text{char}_{\mathbb{C}}(E)$ of degree 1333 for which the Sylow 2-subgroup test of Step 5(c) of Algorithm 7.4.8 of [92] can be verified. It is the compatible pair

$$(\chi = \chi_{\mathbf{31}} + \chi_{38}, \quad \tau = \tau_4 + \tau_{\mathbf{20}}).$$

By Lemma 8.3.1, $G = \langle H, E \rangle$. Therefore we try to extend the semisimple character χ of H to a character of G by means of Brauer's characterization of characters. Lemma 8.2.2 states that G has two conjugacy classes of involutions, represented by z and $u = (xy^2e)^{11} \in E$ and $C_G(u) = C_E(u)$, and that G has 27 z-special and ten u-special classes with representatives given in Lemma 8.2.2 by means of the notations of Tables 8.5.1 and `DVD.1.6.2` of the accompanying DVD. Using the character Tables 8.6.1 and `DVD.1.6.4` we define the values of the extended class function $\chi :\rightarrow \mathbb{C}$ on all 37 even classes of G. In view of Theorem 3.12.4 of [92] we let $\chi(p) = 0$ for all p-elements of orders 31 and 43 because $1333 = 31 \cdot 43$. Since the cyclic Sylow p-subgroups for the primes $p \in \{23, 29, 37\}$ are self-centralizing Theorem 3.12.4 of [92] suggests to define $\chi(p)$ to be 1 or -1 according to the congruence of 1333 modulo p. The values of $\chi(g)$ at the 7- and 11-singular classes are determined by means of the character tables of H and E and the normalizers M and N_{11} of Lemma 8.2.5.

Let $b_7 = -\frac{1}{2}(1 + i\sqrt{7})$. In view of all the above information we now define the class function $\chi : g \rightarrow \chi(g)$ for all $g \in G$ by the following:

g	1_A	2_A	2_B	3_A	4_A	4_B	4_C	5_A	6_A	6_B	6_C	7_A	7_B	8_A
$\chi(g)$	1333	53	-11	10	-11	5	-3	3	-10	2	-2	b_7	$\overline{b_7}$	1

g	8_B	8_C	10_A	10_B	11_A	11_B	12_A	12_B	12_C	14_A	14_B	14_C
$\chi(g)$	-1	1	3	-1	2	2	-2	2	0	$-b_7$	$-\overline{b_7}$	b_7

g	14_D	15_A	16_A	20_A	20_B	21_A	21_B	22_A	22_B	23_A	24_A	24_B
$\chi(g)$	$\overline{b_7}$	0	-1	-1	-1	b_7	$\overline{b_7}$	-2	0	-1	0	0

g	28_A	28_B	29_A	30_A	31_A	31_B	31_C	33_A	33_B	35_A	35_B	37_A
$\chi(g)$	$-b_7$	$-\overline{b_7}$	-1	0	0	0	0	-1	-1	b_7	$\overline{b_7}$	1

g	37_B	37_C	40_A	40_B	42_A	42_B	43_A	43_B	43_C	44_A	66_A	66_B
$\chi(g)$	1	1	1	1	$-b_7$	$-\overline{b_7}$	0	0	0	0	1	1

By Theorem 2.8.9 and Corollary 2.8.10, both of [92], it remains to show that $\chi_{|Y}$ is a generalized character of Y, whenever Y belongs to the following set of subgroups:

$$\mathfrak{Y} = \{H, E, M, N_{23}, N_{29}, N_{31}, N_{37}, N_{43}\}$$

of G.

The calculations of the inner products of the restrictions of χ to these local subgroups are routine. Since the inner product $(\chi, \chi)_G = 1$, χ is an irreducible character of G with $\chi(1) = 23$ by Corollary 2.8.10 of [92]. This completes the proof. □

Theorem 8.3.4 *Let H be the finite group of even order defined in Proposition 8.2.1. Then each finite simple group G of J_4-type is isomorphic to the finite simple group $\mathfrak{G} \le \mathrm{GL}_{1333}(43)$ of order $|\mathfrak{G}| = 2^{21} \cdot 3^3 \cdot 5^2 \cdot 7 \cdot 11 \cdot 13 \cdot 23 \cdot 29 \cdot 31 \cdot 37 \cdot 43$ constructed in Theorem 8.3.2.*

Proof Using Lemma 8.3.1, Theorem 8.3.2, Proposition 8.3.3 and Theorem 7.5.1 of [92] the proof of Theorem 3.3.13 can be adapted. □

8.4 Other constructions in $\mathrm{GL}_{1333}(11)$ and $\mathrm{GL}_{112}(2)$

As mentioned in the introduction to this chapter the first construction of Janko's simple group J_4 was given by Benson, Conway, Norton, Parker and Thackray in 1980; see [8] and [102]. They realized it as a subgroup G_1 of $\mathrm{GL}_{112}(2)$.

In Theorem 3.16 and Remark 3.21 of [84], Lempken describes another representation of J_4. Using his characterization of J_4 in [83] he constructed two matrices $x, y \in \mathrm{GL}_{1333}(11)$ of orders $o(x) = 42$ and $o(y) = 10$, respectively, such that $V = GF(11)^{112}$ is an irreducible $GF(11)$-module for $G = \langle x, y \rangle$. He showed that $z = (y^5 x^{28})^2 x^{14} (x^{14} y^5)^2$ is a 2-central involution of G and that $E = \langle x^3, y, z \rangle$ is isomorphic to a split extension of M_{24} by an elementary abelian group A of order 2^{11}. Theorem 8.3 of [21] asserts that G is a simple group with centralizer

$H = C_G(z)$ satisfying all hypothesis of Theorem A of Janko's article [67]. Therefore Lempken's group G provides another existence proof for J_4. His generating matrices x and y are stored on the accompanying DVD.

But why should these two groups G_1 and G be isomorphic? What are their orders? Since the Sylow 2-subgroups and the Sylow 11-subgroups of J_4 are not cyclic these two representations cannot be obtained from the character table of J_4 using only theoretical results because the 2- and 11-modular decomposition numbers are not known.

In order to solve these questions M. Weller constructed a faithful permutation representation PG of Lempken's matrix group $G = \langle x, y\rangle \le \mathrm{GL}_{1333}(11)$ of degree 173067389 with stabilizer $E = \langle x^3, y, z\rangle$ on supercomputers at the Theory Center of Cornell University and Karlsruhe University in 1998 and 1999. Using Proposition 5.2.14 of [92] and special implementations of the algorithm described in Theorem 6.2.1 of [92] on supercomputers, Weller verified that E is the stabilizer of PG. The extensive calculations are described in [135]. The resulting permutations $\pi(x)$ and $\pi(y)$ in the symmetric group $S_{173067389}$ corresponding to the matrices x and y are stored on the accompanying DVD.

Let $\pi(G) = \langle \pi(x), \pi(y)\rangle$. Weller showed in [135] that $\pi(G)$ is a simple group of order $2^{21}\cdot 3^3\cdot 5\cdot 7\cdot 11^3\cdot 23\cdot 29\cdot 31\cdot 37\cdot 43$ such that $C_{\pi(G)}(\pi(z)) \cong H$, the given centralizer of any simple group of J_4-type; see also [98].

Weller's permutation representation PG of Lempken's group G has also been used to calculate the character table of G by means of Algorithm 5.7.1 of [92] and the LLL-Algorithm of [85]. It coincides with the character table of J_4 given in the Atlas [19], pp. 188–189.

In [98] Weller and the author used the large permutation representation PG of G to determine the structure of the endomorphism ring $End_{FG}(P)$ of its permutation module P over $F = GF(2)$. In Theorem 2.1 of that article we state the intersection matrices of the seven double cosets Eg_iE in G. Therefore the integral endomorphism ring $End_{\mathbb{Z}G}(P)$ of PG is well defined by Theorem 5.4.3 of [92]. The reduction of its generating matrices modulo 2 induces an epimorphism of $End_{\mathbb{Z}G}(P)$ onto $End_{FG}(P)$. In Theorem 3.1 of [98] we construct an explicit endomorphism $\varphi \in End_{F\pi(G)}(P)$ such that $W = \varphi(P)$ is a 112-dimensional FG-module over F. It induces a group monomorphism $\kappa : \pi(G) \to \mathrm{GL}_{112}(2)$. Let $X = \kappa(\pi(x))$, $Y = \kappa(\pi(y)) \in \mathrm{GL}_{112}(2)$ and $J = \langle X, Y\rangle$. Then one has the following commutative diagram of isomorphic simple subgroups:

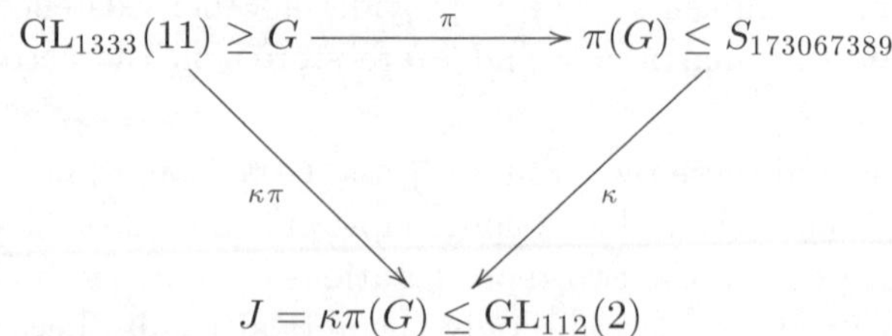

The matrices X and Y in $GL_{112}(2)$ are stored on the accompanying DVD as well. For computations with elements in the Janko group J_4 they are very useful. In [98] Weller and the author have shown that J is conjugate in $GL_{112}(2)$ to the original matrix group $G_1 \leq GL_{112}(2)$ of Benson, Conway, Norton, Parker and Thackray. Furthermore, we give a presentation of J by means of Theorem 5.2.18 of [92].

By Theorem 8.3 of [21], due to Cooperman, Lempken, Weller and the author, and Proposition 8.1.1 and Theorem 8.3.4, Lempken's matrix group G is also isomorphic to the group $\mathfrak{G} \leq GL_{1333}(43)$ of J_4-type constructed in Theorem 8.3.2.

8.5 Representatives of conjugacy classes

8.5.1 *Conjugacy classes of* $H = C_G(z) = \langle x, y, h \rangle$

Class	*Representative*	$\lvert$*Class*$\rvert$	$\lvert$*Centralizer*$\rvert$	2P	3P	5P	7P	11P
1	1	1	$2^{21} \cdot 3^3 \cdot 5 \cdot 7 \cdot 11$	1	1	1	1	1
2_1	$(x^2y)^{15}$	1	$2^{21} \cdot 3^3 \cdot 5 \cdot 7 \cdot 11$	1	2_1	2_1	2_1	2_1
2_2	$(x)^6$	1386	$2^{20} \cdot 3 \cdot 5$	1	2_2	2_2	2_2	2_2
2_3	$(xhy)^6$	2772	$2^{19} \cdot 3 \cdot 5$	1	2_3	2_3	2_3	2_3
2_4	$(h)^3$	18480	$2^{17} \cdot 3^2$	1	2_4	2_4	2_4	2_4
2_5	$(y^2h)^2$	18480	$2^{17} \cdot 3^2$	1	2_5	2_5	2_5	2_5
2_6	$(xh^2)^3$	63360	$2^{14} \cdot 3 \cdot 7$	1	2_6	2_6	2_6	2_6
2_7	$(xhy^2)^7$	63360	$2^{14} \cdot 3 \cdot 7$	1	2_7	2_7	2_7	2_7
2_8	$(xy)^6$	110880	$2^{16} \cdot 3$	1	2_8	2_8	2_8	2_8
2_9	$(x^2h^2)^2$	221760	$2^{15} \cdot 3$	1	2_9	2_9	2_9	2_9
2_{10}	$(xh^2y^2)^5$	532224	$2^{13} \cdot 5$	1	2_{10}	2_{10}	2_{10}	2_{10}
3_1	$(x^2y)^{10}$	8192	$2^8 \cdot 3^3 \cdot 5 \cdot 7 \cdot 11$	3_1	1	3_1	3_1	3_1
3_2	$(x)^4$	9461760	$2^8 \cdot 3^2$	3_2	1	3_2	3_2	3_2
4_1	$(xhyh)^{10}$	4032	$2^{15} \cdot 3 \cdot 5 \cdot 11$	2_1	4_1	4_1	4_1	4_1
4_2	$(xy^2h)^4$	221760	$2^{15} \cdot 3$	2_1	4_2	4_2	4_2	4_2
4_3	$(xh)^5$	532224	$2^{13} \cdot 5$	2_2	4_3	4_3	4_3	4_3
4_4	$(x)^3$	887040	$2^{13} \cdot 3$	2_2	4_4	4_4	4_4	4_4
4_5	$(xh^4)^2$	1330560	2^{14}	2_2	4_5	4_5	4_5	4_5
4_6	$(xyhxhy)^2$	1330560	2^{14}	2_2	4_6	4_6	4_6	4_6
4_7	$(x^2hxh)^2$	2661120	2^{13}	2_2	4_7	4_7	4_7	4_7
4_8	$(xyh)^2$	3548160	$2^{11} \cdot 3$	2_4	4_8	4_8	4_8	4_8
4_9	$(xhy)^3$	3548160	$2^{11} \cdot 3$	2_3	4_9	4_9	4_9	4_9

Conjugacy classes of $H = \langle x, y, h \rangle$ (continued)

Class	*Representative*	|*Class*|	|*Centralizer*|	2P	3P	5P	7P	11P
4_{10}	yh^2	3548160	$2^{11} \cdot 3$	2_4	4_{10}	4_{10}	4_{10}	4_{10}
4_{11}	$(x^3h)^3$	3548160	$2^{11} \cdot 3$	2_2	4_{11}	4_{11}	4_{11}	4_{11}
4_{12}	$(xy^2xh)^3$	3548160	$2^{11} \cdot 3$	2_3	4_{12}	4_{12}	4_{12}	4_{12}
4_{13}	$xyhxhyxhy^2$	5322240	2^{12}	2_2	4_{13}	4_{13}	4_{13}	4_{13}
4_{14}	$(xy)^3$	7096320	$2^{10} \cdot 3$	2_8	4_{14}	4_{14}	4_{14}	4_{14}
4_{15}	$(x^2yhxy^2)^3$	7096320	$2^{10} \cdot 3$	2_8	4_{15}	4_{15}	4_{15}	4_{15}
4_{16}	x^2hxyh^2yh	10644480	2^{11}	2_3	4_{16}	4_{16}	4_{16}	4_{16}
4_{17}	y^2h	14192640	$2^9 \cdot 3$	2_5	4_{17}	4_{17}	4_{17}	4_{17}
4_{18}	$(xh^2y)^3$	14192640	$2^9 \cdot 3$	2_5	4_{18}	4_{18}	4_{18}	4_{18}
4_{19}	x^3y	21288960	2^{10}	2_8	4_{19}	4_{19}	4_{19}	4_{19}
4_{20}	$(xyxy^2)^2$	21288960	2^{10}	2_8	4_{20}	4_{20}	4_{20}	4_{20}
4_{21}	x^3hx^2h	21288960	2^{10}	2_4	4_{21}	4_{21}	4_{21}	4_{21}
4_{22}	x^2h^2	42577920	2^9	2_9	4_{22}	4_{22}	4_{22}	4_{22}
4_{23}	x^3y^2	42577920	2^9	2_9	4_{23}	4_{23}	4_{23}	4_{23}
4_{24}	$xyxy^2h$	42577920	2^9	2_8	4_{24}	4_{24}	4_{24}	4_{24}
5	$(xh)^4$	22708224	$2^6 \cdot 3 \cdot 5$	5	5	1	5	5
6_1	$(x^2y)^5$	8192	$2^8 \cdot 3^3 \cdot 5 \cdot 7 \cdot 11$	3_1	2_1	6_1	6_1	6_1
6_2	h	9461760	$2^8 \cdot 3^2$	3_2	2_4	6_2	6_2	6_2
6_3	$(xh^2y)^2$	9461760	$2^8 \cdot 3^2$	3_2	2_5	6_3	6_3	6_3
6_4	$(xh^3)^4$	9461760	$2^8 \cdot 3^2$	3_2	2_1	6_4	6_4	6_4
6_5	$(x^4h)^4$	9461760	$2^8 \cdot 3^2$	3_1	2_4	6_5	6_5	6_5
6_6	$(x^2y^2h)^2$	9461760	$2^8 \cdot 3^2$	3_1	2_5	6_6	6_6	6_6
6_7	$(xy)^2$	56770560	$2^7 \cdot 3$	3_2	2_8	6_7	6_7	6_7
6_8	$(xhy)^2$	56770560	$2^7 \cdot 3$	3_2	2_3	6_8	6_8	6_8
6_9	x^2h	75694080	$2^5 \cdot 3^2$	3_2	2_4	6_9	6_9	6_9
6_{10}	$xyxh$	75694080	$2^5 \cdot 3^2$	3_2	2_5	6_{10}	6_{10}	6_{10}
6_{11}	$(x)^2$	113541120	$2^6 \cdot 3$	3_2	2_2	6_{11}	6_{11}	6_{11}
6_{12}	x^2yxh^2xy	113541120	$2^6 \cdot 3$	3_2	2_9	6_{12}	6_{12}	6_{12}
6_{13}	xh^2	227082240	$2^5 \cdot 3$	3_2	2_6	6_{13}	6_{13}	6_{13}
6_{14}	x^3yh	227082240	$2^5 \cdot 3$	3_2	2_7	6_{14}	6_{14}	6_{14}
7_1	yh	259522560	$2^2 \cdot 3 \cdot 7$	7_1	7_2	7_2	1	7_1
7_2	x^2yh	259522560	$2^2 \cdot 3 \cdot 7$	7_2	7_1	7_1	1	7_2
8_1	$(xhyh)^5$	17031168	$2^8 \cdot 5$	4_1	8_1	8_1	8_1	8_1
8_2	$(xh^3)^3$	28385280	$2^8 \cdot 3$	4_2	8_2	8_2	8_2	8_2
8_3	$(xy^2h)^2$	42577920	2^9	4_2	8_3	8_3	8_3	8_3
8_4	xh^4	85155840	2^8	4_5	8_4	8_4	8_4	8_4
8_5	x^2yxy^2	85155840	2^8	4_5	8_5	8_5	8_5	8_5
8_6	$x^2h^2y^2$	85155840	2^8	4_5	8_6	8_6	8_6	8_6
8_7	$xyhxhy$	85155840	2^8	4_6	8_7	8_7	8_7	8_7
8_8	x^3hyhy	85155840	2^8	4_6	8_8	8_8	8_8	8_8
8_9	$(x^4h)^3$	113541120	$2^6 \cdot 3$	4_{10}	8_9	8_9	8_9	8_9
8_{10}	x^2hxh	170311680	2^7	4_7	8_{10}	8_{10}	8_{10}	8_{10}
8_{11}	xyh	340623360	2^6	4_8	8_{11}	8_{11}	8_{11}	8_{11}
8_{12}	$xyxy^2$	340623360	2^6	4_{20}	8_{12}	8_{12}	8_{12}	8_{12}
8_{13}	xy^2h^2	340623360	2^6	4_{20}	8_{13}	8_{13}	8_{13}	8_{13}
10_1	$(x^2y)^3$	22708224	$2^6 \cdot 3 \cdot 5$	5	10_1	2_1	10_1	10_1
10_2	$(xh)^2$	136249344	$2^5 \cdot 5$	5	10_2	2_2	10_2	10_2
10_3	x^3hxh	272498688	$2^4 \cdot 5$	5	10_3	2_3	10_3	10_3
10_4	xh^2y^2	544997376	$2^3 \cdot 5$	5	10_4	2_{10}	10_4	10_4
11	$(x^2yhy)^3$	82575360	$2^3 \cdot 3 \cdot 11$	11	11	11	11	1

Conjugacy classes of $H = \langle x, y, h \rangle$ *(continued)*

Class	*Representative*	|*Class*|	|*Centralizer*|	2P	3P	5P	7P	11P
12_1	$(xh^3)^2$	113541120	$2^6 \cdot 3$	6_4	4_2	12_1	12_1	12_1
12_2	$(x^4h)^2$	113541120	$2^6 \cdot 3$	6_5	4_{10}	12_2	12_2	12_2
12_3	x^4h^2	113541120	$2^6 \cdot 3$	6_5	4_8	12_3	12_3	12_3
12_4	$x^2yh^2y^2$	113541120	$2^6 \cdot 3$	6_4	4_1	12_4	12_4	12_4
12_5	xy	227082240	$2^5 \cdot 3$	6_7	4_{14}	12_5	12_5	12_5
12_6	x^2yhxy^2	227082240	$2^5 \cdot 3$	6_7	4_{15}	12_6	12_6	12_6
12_7	x	454164480	$2^4 \cdot 3$	6_{11}	4_4	12_7	12_7	12_7
12_8	xhy	454164480	$2^4 \cdot 3$	6_8	4_9	12_8	12_8	12_8
12_9	x^3h	454164480	$2^4 \cdot 3$	6_{11}	4_{11}	12_9	12_9	12_9
12_{10}	xh^2y	454164480	$2^4 \cdot 3$	6_3	4_{18}	12_{10}	12_{10}	12_{10}
12_{11}	x^2y^2h	454164480	$2^4 \cdot 3$	6_6	4_{17}	12_{11}	12_{11}	12_{11}
12_{12}	xy^2xh	454164480	$2^4 \cdot 3$	6_8	4_{12}	12_{12}	12_{12}	12_{12}
14_1	$(x^2h^2y)^3$	259522560	$2^2 \cdot 3 \cdot 7$	7_2	14_2	14_2	2_1	14_1
14_2	$(x^2h^2y)^9$	259522560	$2^2 \cdot 3 \cdot 7$	7_1	14_1	14_1	2_1	14_2
14_3	xhy^2	778567680	$2^2 \cdot 7$	7_1	14_4	14_4	2_7	14_3
14_4	x^3hy	778567680	$2^2 \cdot 7$	7_2	14_3	14_3	2_7	14_4
14_5	x^2hxy	778567680	$2^2 \cdot 7$	7_2	14_6	14_6	2_6	14_5
14_6	x^3h^2y	778567680	$2^2 \cdot 7$	7_1	14_5	14_5	2_6	14_6
15	$(x^2y)^2$	726663168	$2 \cdot 3 \cdot 5$	15	5	3_1	15	15
16	xy^2h	681246720	2^5	8_3	16	16	16	16
20_1	$(xhyh)^2$	136249344	$2^5 \cdot 5$	10_1	20_2	4_1	20_2	20_1
20_2	x^2hyh	136249344	$2^5 \cdot 5$	10_1	20_1	4_1	20_1	20_2
20_3	xh	272498688	$2^4 \cdot 5$	10_2	20_4	4_3	20_4	20_3
20_4	xhy^2h	272498688	$2^4 \cdot 5$	10_2	20_3	4_3	20_3	20_4
21_1	$(x^2h^2y)^2$	519045120	$2 \cdot 3 \cdot 7$	21_1	7_2	21_2	3_1	21_1
21_2	$(xhxhy)^2$	519045120	$2 \cdot 3 \cdot 7$	21_2	7_1	21_1	3_1	21_2
22	$(x^3hxy)^2$	82575360	$2^3 \cdot 3 \cdot 11$	11	22	22	22	2_1
24_1	xh^3	454164480	$2^4 \cdot 3$	12_1	8_2	24_4	24_4	24_1
24_2	x^4h	454164480	$2^4 \cdot 3$	12_2	8_9	24_3	24_3	24_2
24_3	x^2h^3	454164480	$2^4 \cdot 3$	12_2	8_9	24_2	24_2	24_3
24_4	xh^2yh	454164480	$2^4 \cdot 3$	12_1	8_2	24_1	24_1	24_4
30	x^2y	726663168	$2 \cdot 3 \cdot 5$	15	10_1	6_1	30	30
33_1	x^2yhy	330301440	$2 \cdot 3 \cdot 11$	33_1	11	33_2	33_2	3_1
33_2	$xhxh^3$	330301440	$2 \cdot 3 \cdot 11$	33_2	11	33_1	33_1	3_1
40_1	$xhyh$	544997376	$2^3 \cdot 5$	20_1	40_2	8_1	40_2	40_1
40_2	x^2yhxh	544997376	$2^3 \cdot 5$	20_2	40_1	8_1	40_1	40_2
42_1	x^2h^2y	519045120	$2 \cdot 3 \cdot 7$	21_1	14_1	42_2	6_1	42_1
42_2	$xhxhy$	519045120	$2 \cdot 3 \cdot 7$	21_2	14_2	42_1	6_1	42_2
44	x^3hxy	495452160	$2^2 \cdot 11$	22	44	44	44	4_1
66_1	xyh^2xh	330301440	$2 \cdot 3 \cdot 11$	33_1	22	66_2	66_2	6_1
66_2	x^2hxyxh	330301440	$2 \cdot 3 \cdot 11$	33_2	22	66_1	66_1	6_1

8.5.2 *Conjugacy classes of* $E = N_G(A) = \langle x, y, e \rangle$

Class	*Representative*	*|Class|*	*|Centralizer|*	2P	3P	5P	7P	11P	23P
1	1	1	$2^{21} \cdot 3^3 \cdot 5 \cdot 7 \cdot 11 \cdot 23$	1	1	1	1	1	1
2_1	$(xy^2e)^{11}$	276	$2^{19} \cdot 3^2 \cdot 5 \cdot 7 \cdot 11$	1	2_1	2_1	2_1	2_1	2_1
2_2	$(x)^6$	1771	$2^{21} \cdot 3^3 \cdot 5$	1	2_2	2_2	2_2	2_2	2_2
2_3	e	182160	$2^{17} \cdot 3 \cdot 7$	1	2_3	2_3	2_3	2_3	2_3
2_4	$(xey)^6$	1020096	$2^{15} \cdot 3 \cdot 5$	1	2_4	2_4	2_4	2_4	2_4
2_5	$(x^3y)^2$	1020096	$2^{15} \cdot 3 \cdot 5$	1	2_5	2_5	2_5	2_5	2_5
2_6	$(y^2eye)^2$	1275120	$2^{17} \cdot 3$	1	2_6	2_6	2_6	2_6	2_6
3_1	$(xy)^4$	14508032	$2^8 \cdot 3^3 \cdot 5$	3_1	1	3_1	3_1	3_1	3_1
3_2	$(x)^4$	124354560	$2^6 \cdot 3^2 \cdot 7$	3_2	1	3_2	3_2	3_2	3_2
4_1	$(x)^3$	2040192	$2^{14} \cdot 3 \cdot 5$	2_2	4_1	4_1	4_1	4_1	4_1
4_2	$(xy)^3$	5100480	$2^{15} \cdot 3$	2_2	4_2	4_2	4_2	4_2	4_2
4_3	$(x^3yxy)^6$	5100480	$2^{15} \cdot 3$	2_2	4_3	4_3	4_3	4_3	4_3
4_4	$(x^3ey)^3$	11658240	$2^{11} \cdot 3 \cdot 7$	2_1	4_4	4_4	4_4	4_4	4_4
4_5	$(xyxy^2)^2$	61205760	2^{13}	2_2	4_5	4_5	4_5	4_5	4_5
4_6	$(xy^2)^3$	81607680	$2^{11} \cdot 3$	2_3	4_6	4_6	4_6	4_6	4_6
4_7	$(x^5y)^3$	81607680	$2^{11} \cdot 3$	2_3	4_7	4_7	4_7	4_7	4_7
4_8	$(x^2e)^2$	244823040	2^{11}	2_3	4_8	4_8	4_8	4_8	4_8
4_9	$(x^3e)^2$	244823040	2^{11}	2_3	4_9	4_9	4_9	4_9	4_9
4_{10}	y^2eye	489646080	2^{10}	2_6	4_{10}	4_{10}	4_{10}	4_{10}	4_{10}
4_{11}	$x^2y^2xy^2ey$	489646080	2^{10}	2_6	4_{11}	4_{11}	4_{11}	4_{11}	4_{11}
4_{12}	$(xey)^3$	652861440	$2^8 \cdot 3$	2_4	4_{12}	4_{12}	4_{12}	4_{12}	4_{12}
4_{13}	x^3y	652861440	$2^8 \cdot 3$	2_5	4_{13}	4_{13}	4_{13}	4_{13}	4_{13}
5	$(xe)^4$	1044578304	$2^5 \cdot 3 \cdot 5$	5	5	1	5	5	5
6_1	$(ye)^5$	14508032	$2^8 \cdot 3^3 \cdot 5$	3_1	2_2	6_1	6_1	6_1	6_1
6_2	$(xy)^2$	217620480	$2^8 \cdot 3^2$	3_1	2_2	6_2	6_2	6_2	6_2
6_3	$(x^3ey)^2$	217620480	$2^8 \cdot 3^2$	3_1	2_1	6_3	6_3	6_3	6_3
6_4	$(x)^2$	870481920	$2^6 \cdot 3^2$	3_2	2_2	6_4	6_4	6_4	6_4
6_5	$(xy^2)^2$	2611445760	$2^6 \cdot 3$	3_1	2_3	6_5	6_5	6_5	6_5
6_6	$xyxy^2e$	2611445760	$2^6 \cdot 3$	3_1	2_6	6_6	6_6	6_6	6_6
6_7	$(xey)^2$	5222891520	$2^5 \cdot 3$	3_2	2_4	6_7	6_7	6_7	6_7
6_8	$(x^2yxy)^2$	5222891520	$2^5 \cdot 3$	3_2	2_5	6_8	6_8	6_8	6_8
7_1	$(xye)^2$	2984509440	$2^3 \cdot 3 \cdot 7$	7_1	7_2	7_2	1	7_1	7_1
7_2	$(x^2ye)^2$	2984509440	$2^3 \cdot 3 \cdot 7$	7_2	7_1	7_1	1	7_2	7_2
8_1	$(x^3yxy)^3$	652861440	$2^8 \cdot 3$	4_3	8_1	8_1	8_1	8_1	8_1
8_2	$(x^3exe)^2$	979292160	2^9	4_3	8_2	8_2	8_2	8_2	8_2
8_3	x^4yx^2y	1958584320	2^8	4_2	8_3	8_3	8_3	8_3	8_3
8_4	$xyxy^2$	3917168640	2^7	4_5	8_4	8_4	8_4	8_4	8_4
8_5	x^2e	7834337280	2^6	4_8	8_5	8_5	8_5	8_5	8_5
8_6	x^3e	7834337280	2^6	4_9	8_6	8_6	8_6	8_6	8_6
10_1	$(xe)^2$	1044578304	$2^5 \cdot 3 \cdot 5$	5	10_1	2_2	10_1	10_1	10_1
10_2	xy^2ey	6267469824	$2^4 \cdot 5$	5	10_2	2_4	10_2	10_2	10_2
10_3	$x^2y^2ey^2$	6267469824	$2^4 \cdot 5$	5	10_3	2_5	10_3	10_3	10_3
10_4	x^2yxex^2e	6267469824	$2^4 \cdot 5$	5	10_4	2_1	10_4	10_4	10_4
11	x^2ey	22790799360	$2 \cdot 11$	11	11	11	11	1	11
12_1	xy	2611445760	$2^6 \cdot 3$	6_2	4_2	12_1	12_1	12_1	12_1
12_2	$(x^3yxy)^2$	2611445760	$2^6 \cdot 3$	6_2	4_3	12_2	12_2	12_2	12_2
12_3	x	10445783040	$2^4 \cdot 3$	6_4	4_1	12_3	12_3	12_3	12_3
12_4	xy^2	10445783040	$2^4 \cdot 3$	6_5	4_6	12_4	12_4	12_4	12_4
12_5	x^3ey	10445783040	$2^4 \cdot 3$	6_3	4_4	12_5	12_5	12_5	12_5
12_6	x^5y	10445783040	$2^4 \cdot 3$	6_5	4_7	12_6	12_6	12_6	12_6

Conjugacy classes of $E = \langle x, y, e \rangle$ *(continued)*

Class	*Representative*	|*Class*|	|*Centralizer*|	2P	3P	5P	7P	11P	23P
12_7	xey	20891566080	$2^3 \cdot 3$	6_7	4_{12}	12_7	12_7	12_7	12_7
12_8	x^2yxy	20891566080	$2^3 \cdot 3$	6_8	4_{13}	12_8	12_8	12_8	12_8
14_1	$(x^2yey)^2$	8953528320	$2^3 \cdot 7$	7_1	14_2	14_2	2_1	14_1	14_1
14_2	$(xexey)^2$	8953528320	$2^3 \cdot 7$	7_2	14_1	14_1	2_1	14_2	14_2
14_3	xye	17907056640	$2^2 \cdot 7$	7_1	14_4	14_4	2_3	14_3	14_3
14_4	x^2ye	17907056640	$2^2 \cdot 7$	7_2	14_3	14_3	2_3	14_4	14_4
15_1	$(ye)^2$	16713252864	$2 \cdot 3 \cdot 5$	15_1	5	3_1	15_2	15_2	15_1
15_2	$(y^2e)^2$	16713252864	$2 \cdot 3 \cdot 5$	15_2	5	3_1	15_1	15_1	15_2
16	x^3exe	15668674560	2^5	8_2	16	16	16	16	16
20_1	xe	6267469824	$2^4 \cdot 5$	10_1	20_2	4_1	20_2	20_1	20_2
20_2	x^5e	6267469824	$2^4 \cdot 5$	10_1	20_1	4_1	20_1	20_2	20_1
21_1	$xyey$	23876075520	$3 \cdot 7$	21_1	7_1	21_2	3_2	21_1	21_1
21_2	xey^2	23876075520	$3 \cdot 7$	21_2	7_2	21_1	3_2	21_2	21_2
22	xy^2e	22790799360	$2 \cdot 11$	11	22	22	22	2_1	22
23_1	x^2exy	21799895040	23	23_1	23_1	23_2	23_2	23_2	1
23_2	xey^2e	21799895040	23	23_2	23_2	23_1	23_1	23_1	1
24_1	x^3yxy	10445783040	$2^4 \cdot 3$	12_2	8_1	24_2	24_2	24_1	24_1
24_2	$xyeyxe$	10445783040	$2^4 \cdot 3$	12_2	8_1	24_1	24_1	24_2	24_2
28_1	x^2yey	17907056640	$2^2 \cdot 7$	14_1	28_2	28_2	4_4	28_1	28_1
28_2	$xexey$	17907056640	$2^2 \cdot 7$	14_2	28_1	28_1	4_4	28_2	28_2
30_1	ye	16713252864	$2 \cdot 3 \cdot 5$	15_1	10_1	6_1	30_2	30_2	30_1
30_2	y^2e	16713252864	$2 \cdot 3 \cdot 5$	15_2	10_1	6_1	30_1	30_1	30_2

8.6 Character tables of local subgroups

8.6.1 *Character table of $H(\mathsf{J}_4) = \langle x, y, h \rangle$*

2	21	21	20	19	17	17	14	14	16	15	13	8	8	15	15	13	13	14	14
3	3	3	1	1	2	2	1	1	1	1	.	3	2	1	1	.	1	.	.
5	1	1	1	1	.	.	.	.	.	.	1	1	.	1	.	1	.	.	.
7	1	1	.	.	.	.	1	1	.	.	.	1	.	.	.	.	.	.	.
11	1	1	.	.	.	.	.	.	.	.	.	1	.	1	.	.	.	.	.
	1a	2a	2b	2c	2d	2e	2f	2g	2h	2i	2j	3a	3b	4a	4b	4c	4d	4e	4f
2P	1a	1a	1a	1a	1a	1a	1a	1a	1a	1a	1a	3a	3b	2a	2a	2b	2b	2b	2b
3P	1a	2a	2b	2c	2d	2e	2f	2g	2h	2i	2j	1a	1a	4a	4b	4c	4d	4e	4f
5P	1a	2a	2b	2c	2d	2e	2f	2g	2h	2i	2j	3a	3b	4a	4b	4c	4d	4e	4f
7P	1a	2a	2b	2c	2d	2e	2f	2g	2h	2i	2j	3a	3b	4a	4b	4c	4d	4e	4f
11P	1a	2a	2b	2c	2d	2e	2f	2g	2h	2i	2j	3a	3b	4a	4b	4c	4d	4e	4f
X.1	1	1	1	1	1	1	1	1	1	1	1	1	1	1	1	1	1	1	1
X.2	1	1	1	1	1	1	−1	−1	1	1	−1	1	1	1	1	−1	−1	1	1
X.3	21	21	21	21	5	5	−7	−7	5	5	1	21	3	21	5	1	−7	5	5
X.4	21	21	21	21	5	5	7	7	5	5	−1	21	3	21	5	−1	7	5	5
X.5	42	42	42	42	10	10	.	.	10	10	.	−21	.	42	10	.	.	10	10
X.6	45	45	45	45	−3	−3	−3	−3	−3	−3	5	45	.	45	−3	5	−3	−3	−3
X.7	45	45	45	45	−3	−3	3	3	−3	−3	−5	45	.	45	−3	−5	3	−3	−3
X.8	45	45	45	45	−3	−3	3	3	−3	−3	−5	45	.	45	−3	−5	3	−3	−3
X.9	45	45	45	45	−3	−3	−3	−3	−3	−3	5	45	.	45	−3	5	−3	−3	−3
X.10	55	55	55	55	7	7	−13	−13	7	7	−5	55	1	55	7	−5	−13	7	7
X.11	55	55	55	55	7	7	13	13	7	7	5	55	1	55	7	5	13	7	7
X.12	90	90	90	90	−6	−6	.	.	−6	−6	.	−45	.	90	−6	.	.	−6	−6
X.13	90	90	90	90	−6	−6	.	.	−6	−6	.	−45	.	90	−6	.	.	−6	−6
X.14	99	99	99	99	3	3	15	15	3	3	−1	99	.	99	3	−1	15	3	3
X.15	99	99	99	99	3	3	−15	−15	3	3	1	99	.	99	3	1	−15	3	3
X.16	154	154	154	154	10	10	14	14	10	10	6	154	1	154	10	6	14	10	10
X.17	154	154	154	154	10	10	−14	−14	10	10	−6	154	1	154	10	−6	−14	10	10
X.18	198	198	198	198	6	6	.	.	6	6	.	−99	.	198	6	.	.	6	6
X.19	210	210	210	210	2	2	14	14	2	2	−10	210	3	210	2	−10	14	2	2
X.20	210	210	210	210	2	2	−14	−14	2	2	10	210	3	210	2	10	−14	2	2
X.21	210	210	210	210	18	18	.	.	18	18	.	−105	.	210	18	.	.	18	18
X.22	210	210	210	210	18	18	.	.	18	18	.	−105	.	210	18	.	.	18	18
X.23	231	231	231	231	7	7	7	7	7	7	−9	231	−3	231	7	−9	7	7	7
X.24	231	231	231	231	7	7	−7	−7	7	7	9	231	−3	231	7	9	−7	7	7
X.25	385	385	385	385	1	1	21	21	1	1	5	385	−2	385	1	5	21	1	1
X.26	385	385	385	385	1	1	−21	−21	1	1	−5	385	−2	385	1	−5	−21	1	1
X.27	420	420	420	420	4	4	.	.	4	4	.	−210	.	420	4	.	.	4	4
X.28	462	462	462	462	−18	−18	.	.	−18	−18	.	−231	.	462	−18	.	.	−18	−18
X.29	462	462	462	462	14	14	.	.	14	14	.	−231	.	462	14	.	.	14	14
X.30	560	560	560	560	−16	−16	.	.	−16	−16	.	560	2	560	−16	.	.	−16	−16
X.31	640	−640	.	.	32	−32	−32	32	.	.	.	10	4	.	.	.	.	.	.
X.32	640	−640	.	.	32	−32	−32	32	.	.	.	10	4	.	.	.	.	.	.
X.33	640	−640	.	.	32	−32	32	−32	.	.	.	10	4	.	.	.	.	.	.
X.34	640	−640	.	.	32	−32	32	−32	.	.	.	10	4	.	.	.	.	.	.
X.35	660	660	660	660	−12	−12	.	.	−12	−12	.	−330	.	660	−12	.	.	−12	−12
X.36	693	693	53	−11	69	69	−35	−35	37	5	−11	.	6	−11	−11	−11	−3	−11	5
X.37	693	693	53	−11	69	69	35	35	37	5	11	.	6	−11	−11	11	3	−11	5
X.38	693	693	53	−11	21	21	21	21	53	−11	−11	.	6	−11	5	−11	−11	5	−11
X.39	693	693	53	−11	21	21	−21	−21	53	−11	11	.	6	−11	5	11	11	5	−11
X.40	768	768	768	768	.	.	.	.	.	.	.	−384	.	768	.	.	.	.	.
X.41	1386	1386	−22	42	90	90	−28	−28	−6	26	−20	.	3	−22	−22	−20	4	10	−6
X.42	1386	1386	−22	42	90	90	28	28	−6	26	20	.	3	−22	−22	20	−4	10	−6
X.43	2016	2016	−32	−32	96	96	.	.	−32	−32	32	.	6	32	32	−32	.	.	.
X.44	2016	2016	−32	−32	96	96	.	.	−32	−32	−32	.	6	32	32	32	.	.	.
X.45	2772	2772	212	−44	84	84	−84	−84	20	52	−4	.	6	−44	20	−4	−20	−12	20
X.46	2772	2772	212	−44	−12	−12	28	28	52	20	−4	.	6	−44	52	−4	−36	20	−12
X.47	2772	2772	212	−44	84	84	84	84	20	52	4	.	6	−44	20	4	20	−12	20
X.48	2772	2772	212	−44	−12	−12	−28	−28	52	20	4	.	6	−44	52	4	36	20	−12
X.49	3465	3465	265	−55	57	57	−7	−7	89	25	−15	.	−6	−55	41	−15	−39	9	−7
X.50	3465	3465	265	−55	105	105	63	63	73	41	15	.	−6	−55	25	15	31	−7	9
X.51	3465	3465	265	−55	57	57	7	7	89	25	15	.	−6	−55	41	15	39	9	−7
X.52	3465	3465	265	−55	105	105	−63	−63	73	41	−15	.	−6	−55	25	−15	−31	−7	9
X.53	3584	−3584	.	.	−128	128	.	.	.	.	.	56	8	.	.	.	.	.	.
X.54	3584	−3584	.	.	−128	128	.	.	.	.	.	56	8	.	.	.	.	.	.
X.55	4158	4158	318	−66	−18	−18	−42	−42	−18	78	14	.	.	−66	78	14	−42	14	14
X.56	4158	4158	318	−66	−18	−18	42	42	−18	78	−14	.	.	−66	78	−14	42	14	14
X.57	5544	5544	−88	168	72	72	−56	−56	72	8	−40	.	3	−88	8	−40	8	8	8
X.58	5544	5544	−88	168	72	72	56	56	72	8	40	.	3	−88	8	40	−8	8	8
X.59	6930	6930	−110	210	162	162	28	28	66	34	20	.	−3	−110	−14	20	−4	18	2
X.60	6930	6930	−110	210	66	66	84	84	−30	2	−20	.	6	−110	−46	−20	−12	18	2
X.61	6930	6930	−110	210	66	66	−84	−84	−30	2	20	.	6	−110	−46	20	12	18	2
X.62	6930	6930	−110	210	162	162	−28	−28	66	34	−20	.	−3	−110	−14	−20	4	18	2
X.63	7680	−7680	.	.	−128	128	64	−64	.	.	.	120	12	.	.	.	.	.	.
X.64	7680	−7680	.	.	−128	128	−64	64	.	.	.	120	12	.	.	.	.	.	.
X.65	8316	8316	−132	252	−36	−36	.	.	156	−36	.	.	.	−132	60	.	.	−4	28
X.66	8448	−8448	.	.	−192	192	.	.	.	.	.	−66	.	.	.	.	.	.	.
X.67	8448	−8448	.	.	−192	192	.	.	.	.	.	−66	.	.	.	.	.	.	.
X.68	10395	10395	795	−165	27	27	105	105	−69	27	25	.	.	−165	−21	25	9	−21	27
X.69	10395	10395	795	−165	−117	−117	−63	−63	−21	−21	25	.	.	−165	27	25	33	27	−21
X.70	10395	10395	795	−165	171	171	105	105	75	−21	−15	.	.	−165	−69	−15	9	−37	11
X.71	10395	10395	795	−165	27	27	63	63	123	−69	15	.	.	−165	−21	15	−33	11	−37
X.72	10395	10395	795	−165	−117	−117	63	63	−21	−21	−25	.	.	−165	27	−25	−33	27	−21
X.73	10395	10395	795	−165	27	27	−63	−63	123	−69	−15	.	.	−165	−21	−15	33	11	−37
X.74	10395	10395	795	−165	171	171	−105	−105	75	−21	15	.	.	−165	−69	15	−9	−37	11
X.75	10395	10395	795	−165	27	27	−105	−105	−69	27	−25	.	.	−165	−21	−25	−9	−21	27
X.76	13440	−13440	.	.	160	−160	224	−224	.	.	.	210	12	.	.	.	.	.	.
X.77	13440	−13440	.	.	160	−160	−224	224	.	.	.	210	12	.	.	.	.	.	.
X.78	13860	13860	−220	420	−156	−156	−56	−56	36	−28	40	.	3	−220	68	40	8	−28	4
X.79	13860	13860	−220	420	132	132	112	112	−60	68	.	.	3	−220	−28	.	−16	4	−28
X.80	13860	13860	−220	420	132	132	.	.	−60	4	.	.	−6	−220	−92	.	.	36	4
X.81	13860	13860	−220	420	−156	−156	56	56	36	−28	−40	.	3	−220	68	−40	−8	−28	4
X.82	13860	13860	−220	420	132	132	−112	−112	−60	68	.	.	3	−220	−28	.	16	4	−28
X.83	15360	−15360	.	.	−256	256	.	.	.	.	.	−120	.	.	.	.	.	.	.
X.84	16128	−16128	.	.	192	−192	.	.	.	.	.	252	.	.	.	.	.	.	.
X.85	16128	−16128	.	.	192	−192	.	.	.	.	.	−126	.	.	.	.	.	.	.
X.86	16128	−16128	.	.	192	−192	.	.	.	.	.	−126	.	.	.	.	.	.	.

Character table of $H(J_4)$ *(continued)*

2	13	11	11	11	11	11	12	10	10	11	9	9	10	10	10	9	9	9	6	8	8	8	8	8
3	.	1	1	1	1	1	.	1	1	.	1	1	.	.	.	.	.	.	1	3	2	2	2	2
5	.	.	.	.	.	.	.	.	.	.	.	.	.	.	.	.	.	.	1	1	.	.	.	.
7	.	.	.	.	.	.	.	.	.	.	.	.	.	.	.	.	.	.	.	1	.	.	.	.
11	.	.	.	.	.	.	.	.	.	.	.	.	.	.	.	.	.	.	.	1	.	.	.	.
	4g	4h	4i	4j	4k	4l	4m	4n	4o	4p	4q	4r	4s	4t	4u	4v	4w	4x	5a	6a	6b	6c	6d	6e
2P	2b	2d	2c	2d	2b	2c	2b	2h	2h	2c	2e	2e	2h	2h	2d	2i	2i	2h	5a	3a	3b	3b	3b	3a
3P	4g	4h	4i	4j	4k	4l	4m	4n	4o	4p	4q	4r	4s	4t	4u	4v	4w	4x	5a	2a	2d	2e	2a	2d
5P	4g	4h	4i	4j	4k	4l	4m	4n	4o	4p	4q	4r	4s	4t	4u	4v	4w	4x	1a	6a	6b	6c	6d	6e
7P	4g	4h	4i	4j	4k	4l	4m	4n	4o	4p	4q	4r	4s	4t	4u	4v	4w	4x	5a	6a	6b	6c	6d	6e
11P	4g	4h	4i	4j	4k	4l	4m	4n	4o	4p	4q	4r	4s	4t	4u	4v	4w	4x	5a	6a	6b	6c	6d	6e
X.1	1	1	1	1	1	1	1	1	1	1	1	1	1	1	1	1	1	1	1	1	1	1	1	1
X.2	1	1	−1	1	−1	1	−1	−1	−1	−1	1	−1	−1	1	−1	1	−1	1	1	1	1	1	1	1
X.3	5	1	−7	1	−7	5	1	1	1	1	1	1	−3	1	−3	1	−3	1	1	21	−1	−1	3	5
X.4	5	1	7	1	7	5	−1	−1	−1	−1	1	−1	3	1	3	1	3	1	1	21	−1	−1	3	5
X.5	10	2	.	2	.	10	.	.	.	.	2	.	.	2	.	2	.	2	2	−21	4	4	.	−5
X.6	−3	1	−3	1	−3	−3	5	−3	−3	5	1	−3	1	1	1	1	1	1	.	45	.	.	.	−3
X.7	−3	1	3	1	3	−3	−5	3	3	−5	1	3	−1	1	−1	1	−1	1	.	45	.	.	.	−3
X.8	−3	1	3	1	3	−3	−5	3	3	−5	1	3	−1	1	−1	1	−1	1	.	45	.	.	.	−3
X.9	−3	1	−3	1	−3	−3	5	−3	−3	5	1	−3	1	1	1	1	1	1	.	45	.	.	.	−3
X.10	7	3	−13	3	−13	7	−5	−1	−1	−5	−1	−1	−1	3	−1	3	−1	−1	.	55	1	1	1	7
X.11	7	3	13	3	13	7	5	1	1	5	−1	1	1	3	1	3	1	−1	.	55	1	1	1	7
X.12	−6	2	.	2	.	−6	.	.	.	.	2	.	.	2	.	2	.	2	.	−45	.	.	.	3
X.13	−6	2	.	2	.	−6	.	.	.	.	2	.	.	2	.	2	.	2	.	−45	.	.	.	3
X.14	3	3	15	3	15	3	−1	3	3	−1	−1	3	−1	3	−1	3	−1	−1	−1	99	.	.	.	3
X.15	3	3	−15	3	−15	3	1	−3	−3	1	−1	−3	1	3	1	3	1	−1	−1	99	.	.	.	3
X.16	10	−2	14	−2	14	10	6	2	2	6	2	2	2	−2	2	−2	2	2	−1	154	1	1	1	10
X.17	10	−2	−14	−2	−14	10	−6	−2	−2	−6	2	−2	−2	−2	−2	−2	−2	2	−1	154	1	1	1	10
X.18	6	6	.	6	.	6	.	.	.	.	−2	.	.	6	.	6	.	−2	−2	−99	.	.	.	−3
X.19	2	−2	14	−2	14	2	−10	−2	−2	−10	−2	−2	2	−2	2	−2	2	−2	.	210	−1	−1	3	2
X.20	2	−2	−14	−2	−14	2	10	2	2	10	−2	2	−2	−2	−2	−2	−2	−2	.	210	−1	−1	3	2
X.21	18	2	.	2	.	18	.	.	.	.	2	.	.	2	.	2	.	2	.	−105	.	.	.	−9
X.22	18	2	.	2	.	18	.	.	.	.	2	.	.	2	.	2	.	2	.	−105	.	.	.	−9
X.23	7	−1	7	−1	7	7	−9	−1	−1	−9	−1	−1	−1	−1	−1	−1	−1	−1	1	231	1	1	−3	7
X.24	7	−1	−7	−1	−7	7	9	1	1	9	−1	1	1	−1	1	−1	1	−1	1	231	1	1	−3	7
X.25	1	1	21	1	21	1	5	−3	−3	5	1	−3	−3	1	−3	1	−3	1	.	385	−2	−2	−2	1
X.26	1	1	−21	1	−21	1	−5	3	3	−5	1	3	3	1	3	1	3	1	.	385	−2	−2	−2	1
X.27	4	−4	.	−4	.	4	.	.	.	.	−4	.	.	−4	.	−4	.	−4	.	−210	4	4	.	−2
X.28	−18	6	.	6	.	−18	.	.	.	.	−2	.	.	6	.	6	.	−2	2	−231	.	.	.	9
X.29	14	−2	.	−2	.	14	.	.	.	.	−2	.	.	−2	.	−2	.	−2	2	−231	−4	−4	.	−7
X.30	−16	.	.	.	.	−16	.	.	.	.	.	.	.	.	.	.	.	.	.	560	2	2	2	−16
X.31	.	−8	.	8	.	.	.	−8	8	.	.	.	.	.	.	.	.	.	.	−10	−4	4	−4	2
X.32	.	−8	.	8	.	.	.	−8	8	.	.	.	.	.	.	.	.	.	.	−10	−4	4	−4	2
X.33	.	−8	.	8	.	.	.	8	−8	.	.	.	.	.	.	.	.	.	.	−10	−4	4	−4	2
X.34	.	−8	.	8	.	.	.	8	−8	.	.	.	.	.	.	.	.	.	.	−10	−4	4	−4	2
X.35	−12	−4	.	−4	.	−12	.	.	.	.	4	.	.	−4	.	−4	.	4	.	−330	.	.	.	6
X.36	5	9	5	9	−3	−3	5	−3	−3	−3	9	−3	−7	1	−7	1	1	1	3	.	6	6	6	.
X.37	5	9	−5	9	3	−3	−5	3	3	3	9	3	7	1	7	1	−1	1	3	.	6	6	6	.
X.38	5	−3	−3	−3	5	−3	5	−3	−3	−3	−3	−3	5	5	5	−3	−3	5	3	.	−6	−6	6	.
X.39	5	−3	3	−3	−5	−3	−5	3	3	3	−3	3	−5	5	−5	−3	3	5	3	.	−6	−6	6	.
X.40	.	.	.	.	.	.	.	.	.	.	.	.	.	.	.	.	.	.	−2	−384	.	.	.	.
X.41	−6	6	−4	6	4	2	−4	−4	−4	4	6	−4	8	−2	−8	−2	.	−2	6	.	3	3	3	.
X.42	−6	6	4	6	−4	2	4	4	4	−4	6	4	−8	−2	8	−2	.	−2	6	.	3	3	3	.
X.43	.	.	.	.	.	.	.	−8	−8	.	.	8	.	.	.	.	.	.	6	.	6	6	6	.
X.44	.	.	.	.	.	.	.	8	8	.	.	−8	.	.	.	.	.	.	6	.	6	6	6	.
X.45	4	12	12	12	−4	−12	−4	.	.	−4	.	.	−4	−4	−4	4	−4	.	−3	.	−6	−6	6	.
X.46	4	−12	−4	−12	12	−12	−4	.	.	−4	.	.	4	4	4	−4	4	.	−3	.	6	6	6	.
X.47	4	12	−12	12	4	−12	4	.	.	4	.	.	4	−4	4	4	4	.	−3	.	−6	−6	6	.
X.48	4	−12	4	−12	−12	−12	4	.	.	4	.	.	−4	4	−4	−4	−4	.	−3	.	6	6	6	.
X.49	9	−3	1	−3	9	−15	1	−3	−3	−7	9	−3	−3	5	−3	−3	5	1	.	.	−6	−6	−6	.
X.50	9	9	−9	9	−1	−15	−1	3	3	7	−3	3	−1	1	−1	1	7	5	.	.	6	6	−6	.
X.51	9	−3	−1	−3	−9	−15	−1	3	3	7	9	3	3	5	3	−3	−5	1	.	.	−6	−6	−6	.
X.52	9	9	9	9	1	−15	1	−3	−3	−7	−3	−3	1	1	1	1	−7	5	.	.	6	6	−6	.
X.53	.	.	.	.	.	.	.	.	.	.	.	.	.	.	.	.	.	.	4	−56	−8	8	−8	−8
X.54	.	.	.	.	.	.	.	.	.	.	.	.	.	.	.	.	.	.	4	−56	−8	8	−8	−8
X.55	−2	−6	6	−6	6	−18	−18	6	6	−2	−6	6	2	−6	2	2	2	−6	3	.	.	.	.	.
X.56	−2	−6	−6	−6	−6	−18	18	−6	−6	2	−6	−6	−2	−6	−2	2	−2	−6	3	.	.	.	.	.
X.57	−24	.	−8	.	8	8	−8	−8	−8	8	.	−8	.	.	.	.	.	.	−6	.	3	3	3	.
X.58	−24	.	8	.	−8	8	8	8	8	−8	.	8	.	.	.	.	.	.	−6	.	3	3	3	.
X.59	−30	6	4	6	−4	10	4	4	4	−4	6	4	8	−2	−8	−2	.	−2	.	.	−3	−3	−3	.
X.60	2	6	12	6	−12	−6	−4	.	.	4	−6	.	−8	−2	8	−2	.	2	.	.	−6	−6	6	.
X.61	2	6	−12	6	12	−6	4	.	.	−4	−6	.	8	−2	−8	−2	.	2	.	.	−6	−6	6	.
X.62	−30	6	−4	6	4	10	−4	−4	−4	4	6	−4	−8	−2	8	−2	.	−2	.	.	−3	−3	−3	.
X.63	.	.	.	.	.	.	.	−16	16	.	.	.	.	.	.	.	.	.	.	−120	4	−4	−12	−8
X.64	.	.	.	.	.	.	.	16	−16	.	.	.	.	.	.	.	.	.	.	−120	4	−4	−12	−8
X.65	−36	−12	.	−12	.	12	.	.	.	.	−12	.	.	4	.	4	.	4	6	.	.	.	.	.
X.66	.	−16	.	16	.	.	.	.	.	.	.	.	.	.	.	.	.	.	8	66	.	.	.	6
X.67	.	−16	.	16	.	.	.	.	.	.	.	.	.	.	.	.	.	.	8	66	.	.	.	6
X.68	−5	3	−15	3	9	3	−23	−3	−3	1	−9	−3	−7	11	−7	−5	1	−1	.	.	.	.	.	.
X.69	−5	15	9	15	−15	3	−23	−3	−3	1	3	−3	5	7	5	−1	−3	−5	.	.	.	.	.	.
X.70	11	−9	−15	−9	9	3	1	−3	−3	−7	3	3	5	−1	5	−1	−3	−5	.	.	.	.	.	.
X.71	11	3	−9	3	15	3	−1	3	3	7	−9	3	−1	−5	−1	3	−9	−1	.	.	.	.	.	.
X.72	−5	15	−9	15	15	3	23	3	3	−1	3	3	−5	7	−5	−1	3	−5	.	.	.	.	.	.
X.73	11	3	9	3	−15	3	1	−3	−3	−7	−9	−3	1	−5	1	3	9	−1	.	.	.	.	.	.
X.74	11	−9	15	−9	−9	3	−1	3	3	7	3	3	−5	−1	−5	−1	3	−5	.	.	.	.	.	.
X.75	−5	3	15	3	−9	3	23	3	3	−1	−9	3	7	11	7	−5	−1	−1	.	.	.	.	.	.
X.76	.	−8	.	8	.	.	.	−8	8	.	.	.	.	.	.	.	.	.	.	−210	4	−4	−12	10
X.77	.	−8	.	8	.	.	.	8	−8	.	.	.	.	.	.	.	.	.	.	−210	4	−4	−12	10
X.78	4	12	−8	12	8	4	8	−4	−4	−8	.	−4	.	−4	.	−4	.	.	.	.	3	3	3	.
X.79	4	−12	16	−12	−16	4	.	−4	−4	.	.	−4	.	4	.	4	.	.	.	.	3	3	3	.
X.80	4	12	.	12	.	−12	.	.	.	.	−12	.	.	−4	.	−4	.	4	.	.	6	6	−6	.
X.81	4	12	8	12	−8	4	−8	4	4	8	.	4	.	−4	.	−4	.	.	.	.	3	3	3	.
X.82	4	−12	−16	−12	16	4	.	4	4	.	.	4	.	4	.	4	.	.	.	.	3	3	3	.
X.83	.	.	.	.	.	.	.	.	.	.	.	.	.	.	.	.	.	.	.	120	−16	16	.	8
X.84	.	16	.	−16	.	.	.	.	.	.	.	.	.	.	.	.	.	.	8	−252	.	.	.	12
X.85	.	16	.	−16	.	.	.	.	.	.	.	.	.	.	.	.	.	.	8	126	.	.	.	−6
X.86	.	16	.	−16	.	.	.	.	.	.	.	.	.	.	.	.	.	.	8	126	.	.	.	−6

Character table of $H(\mathsf{J}_4)$ (continued)

	6f	6g	6h	6i	6j	6k	6l	6m	6n	7a	7b	8a	8b	8c	8d	8e	8f	8g	8h	8i	8j	8k	8l	8m	10a	10b	10c
2	8	7	7	5	5	6	6	5	5	2	2	8	8	9	8	8	8	8	8	6	7	6	6	6	6	5	4
3	2	1	1	2	2	1	1	1	1	1	1	.	1	.	.	.	.	.	.	1	.	.	.	.	1	.	.
5	.	.	.	.	.	.	.	.	.	.	.	1	.	.	.	.	.	.	.	.	.	.	.	.	1	1	1
7	.	.	.	.	.	.	.	.	.	1	1	.	.	.	.	.	.	.	.	.	.	.	.	.	.	.	.
11	.	.	.	.	.	.	.	.	.	.	.	.	.	.	.	.	.	.	.	.	.	.	.	.	.	.	.
2P	3a	3b	3b	3b	3b	3b	3b	3b	3b	7a	7b	4a	4b	4b	4e	4e	4e	4f	4f	4j	4g	4h	4t	4t	5a	5a	5a
3P	2e	2h	2c	2d	2e	2b	2i	2f	2g	7b	7a	8a	8b	8c	8d	8e	8f	8g	8h	8i	8j	8k	8l	8m	10a	10b	10c
5P	6f	6g	6h	6i	6j	6k	6l	6m	6n	7b	7a	8a	8b	8c	8d	8e	8f	8g	8h	8i	8j	8k	8l	8m	2a	2b	2c
7P	6f	6g	6h	6i	6j	6k	6l	6m	6n	1a	1a	8a	8b	8c	8d	8e	8f	8g	8h	8i	8j	8k	8l	8m	10a	10b	10c
11P	6f	6g	6h	6i	6j	6k	6l	6m	6n	7a	7b	8a	8b	8c	8d	8e	8f	8g	8h	8i	8j	8k	8l	8m	10a	10b	10c
X.1	1	1	1	1	1	1	1	1	1	1	1	1	1	1	1	1	1	1	1	1	1	1	1	1	1	1	1
X.2	1	1	1	1	1	1	1	−1	−1	1	1	−1	−1	1	−1	−1	1	1	−1	1	−1	−1	1	−1	1	1	1
X.3	5	−1	3	−1	−1	3	−1	−1	−1	.	.	1	1	1	1	−3	1	1	1	−1	−3	−1	−1	−1	1	1	1
X.4	5	−1	3	−1	−1	3	−1	1	1	.	.	−1	−1	1	−1	3	1	1	−1	−1	3	1	−1	1	1	1	1
X.5	−5	4	.	−2	−2	.	4	.	.	.	.	.	.	2	.	.	2	2	.	−2	.	.	−2	.	2	2	2
X.6	−3	.	.	.	.	.	.	.	.	A	$\bar{A}$	5	−3	1	−3	1	1	1	−3	−1	1	−1	−1	−1	.	.	.
X.7	−3	.	.	.	.	.	.	.	.	$\bar{A}$	A	−5	3	1	3	−1	1	1	3	−1	−1	1	−1	1	.	.	.
X.8	−3	.	.	.	.	.	.	.	.	A	$\bar{A}$	−5	3	1	3	−1	1	1	3	−1	−1	1	−1	1	.	.	.
X.9	−3	.	.	.	.	.	.	.	.	$\bar{A}$	A	5	−3	1	−3	1	1	1	−3	−1	1	−1	−1	−1	.	.	.
X.10	7	1	1	1	1	1	1	−1	−1	−1	−1	−5	−1	3	−1	−1	−1	−1	−1	1	−1	1	1	1	.	.	.
X.11	7	1	1	1	1	1	1	1	1	−1	−1	5	1	3	1	1	−1	−1	1	1	1	−1	1	−1	.	.	.
X.12	3	.	.	.	.	.	.	.	.	B	$\bar{B}$	.	.	2	.	.	2	2	.	−2	.	.	−2	.	.	.	.
X.13	3	.	.	.	.	.	.	.	.	$\bar{B}$	B	.	.	2	.	.	2	2	.	−2	.	.	−2	.	.	.	.
X.14	3	.	.	.	.	.	.	.	.	1	1	−1	3	3	3	−1	−1	−1	3	−1	−1	−1	−1	−1	−1	−1	−1
X.15	3	.	.	.	.	.	.	.	.	1	1	1	−3	3	−3	1	−1	−1	−3	−1	1	1	−1	1	−1	−1	−1
X.16	10	1	1	1	1	1	1	−1	−1	.	.	6	2	−2	2	2	2	2	2	.	2	.	.	.	−1	−1	−1
X.17	10	1	1	1	1	1	1	1	1	.	.	−6	−2	−2	−2	−2	2	2	−2	.	−2	.	.	.	−1	−1	−1
X.18	−3	.	.	.	.	.	.	.	.	2	2	.	.	6	.	.	−2	−2	.	−2	.	.	−2	.	−2	−2	−2
X.19	2	−1	3	−1	−1	3	−1	−1	−1	.	.	−10	−2	−2	−2	2	−2	−2	−2	.	2	.	.	.	.	.	.
X.20	2	−1	3	−1	−1	3	−1	1	1	.	.	10	2	−2	2	−2	−2	−2	2	.	−2	.	.	.	.	.	.
X.21	−9	.	.	.	.	.	.	.	.	.	.	.	.	2	.	.	2	2	.	2	.	.	2	.	.	.	.
X.22	−9	.	.	.	.	.	.	.	.	.	.	.	.	2	.	.	2	2	.	2	.	.	2	.	.	.	.
X.23	7	1	−3	1	1	−3	1	1	1	.	.	−9	−1	−1	−1	−1	−1	−1	−1	−1	−1	−1	−1	−1	1	1	1
X.24	7	1	−3	1	1	−3	1	−1	−1	.	.	9	1	−1	1	1	−1	−1	1	−1	1	1	−1	1	1	1	1
X.25	1	−2	−2	−2	−2	−2	−2	.	.	.	.	5	−3	1	−3	−3	1	1	−3	1	−3	1	1	1	.	.	.
X.26	1	−2	−2	−2	−2	−2	−2	.	.	.	.	−5	3	1	3	3	1	1	3	1	3	−1	1	−1	.	.	.
X.27	−2	4	.	−2	−2	.	4	.	.	.	.	.	.	−4	.	.	−4	−4	.	.	.	.	.	.	.	.	.
X.28	9	.	.	.	.	.	.	.	.	.	.	.	.	6	.	.	−2	−2	.	2	.	.	2	.	2	2	2
X.29	−7	−4	.	2	2	.	−4	.	.	.	.	.	.	−2	.	.	−2	−2	.	−2	.	.	−2	.	2	2	2
X.30	−16	2	2	2	2	2	2	.	.	.	.	.	.	.	.	.	.	.	.	.	.	.	.	.	.	.	.
X.31	−2	.	.	2	−2	.	.	−2	2	$\bar{A}$	A	.	.	.	.	.	.	.	.	.	.	.	.	.	.	.	.
X.32	−2	.	.	2	−2	.	.	−2	2	A	$\bar{A}$	.	.	.	.	.	.	.	.	.	.	.	.	.	.	.	.
X.33	−2	.	.	2	−2	.	.	2	−2	A	$\bar{A}$	.	.	.	.	.	.	.	.	.	.	.	.	.	.	.	.
X.34	−2	.	.	2	−2	.	.	2	−2	$\bar{A}$	A	.	.	.	.	.	.	.	.	.	.	.	.	.	.	.	.
X.35	6	.	.	.	.	.	.	.	.	2	2	.	.	−4	.	.	4	4	.	.	.	.	.	.	.	.	.
X.36	.	−2	−2	.	.	2	2	−2	−2	.	.	1	−3	−3	1	1	1	−3	1	3	1	−1	−1	−1	3	3	−1
X.37	.	−2	−2	.	.	2	2	2	2	.	.	−1	3	−3	−1	−1	1	−3	−1	3	−1	1	−1	1	3	3	−1
X.38	.	2	−2	.	.	2	−2	.	.	.	.	1	−3	1	1	−3	−3	1	1	−3	1	1	1	1	3	3	−1
X.39	.	2	−2	.	.	2	−2	.	.	.	.	−1	3	1	−1	3	−3	1	−1	−3	−1	−1	1	−1	3	3	−1
X.40	.	.	.	.	.	.	.	.	.	−2	−2	.	.	.	.	.	.	.	.	.	.	.	.	.	−2	−2	−2
X.41	.	3	3	3	3	−1	−1	−1	−1	.	.	.	4	2	.	.	−2	2	.	.	.	−2	.	2	6	−2	2
X.42	.	3	3	3	3	−1	−1	1	1	.	.	.	−4	2	.	.	−2	2	.	.	.	2	.	−2	6	−2	2
X.43	.	−2	−2	.	.	−2	−2	.	.	.	.	.	.	.	.	.	.	.	.	.	.	.	.	.	6	−2	−2
X.44	.	−2	−2	.	.	−2	−2	.	.	.	.	.	.	.	.	.	.	.	.	.	.	.	.	.	6	−2	−2
X.45	.	2	−2	.	.	2	−2	.	.	.	.	4	.	−4	.	4	.	.	.	.	.	.	.	.	−3	−3	1
X.46	.	−2	−2	.	.	2	2	−2	−2	.	.	4	.	4	.	−4	.	.	.	.	.	.	.	.	−3	−3	1
X.47	.	2	−2	.	.	2	−2	.	.	.	.	−4	.	−4	.	−4	.	.	.	.	.	.	.	.	−3	−3	1
X.48	.	−2	−2	.	.	2	2	2	2	.	.	−4	.	4	.	4	.	.	.	.	.	.	.	.	−3	−3	1
X.49	.	2	2	.	.	−2	−2	2	2	.	.	5	−3	1	1	−3	1	−3	1	3	1	−1	−1	−1	.	.	.
X.50	.	−2	2	.	.	−2	2	.	.	.	.	−5	3	−3	−1	−1	−3	1	−1	−3	−1	−1	1	−1	.	.	.
X.51	.	2	2	.	.	−2	−2	−2	−2	.	.	−5	3	1	−1	3	1	−3	−1	3	−1	1	−1	1	.	.	.
X.52	.	−2	2	.	.	−2	2	.	.	.	.	5	−3	−3	1	1	−3	1	1	−3	1	1	1	1	.	.	.
X.53	8	.	.	4	−4	.	.	.	.	.	.	.	.	.	.	.	.	.	.	.	.	.	.	.	−4	.	.
X.54	8	.	.	4	−4	.	.	.	.	.	.	.	.	.	.	.	.	.	.	.	.	.	.	.	−4	.	.
X.55	.	.	.	.	.	.	.	.	.	.	.	6	6	2	−2	2	2	2	−2	.	−2	.	.	.	3	3	−1
X.56	.	.	.	.	.	.	.	.	.	.	.	−6	−6	2	2	−2	2	2	2	.	2	.	.	.	3	3	−1
X.57	.	3	3	3	3	−1	−1	1	1	.	.	.	8	.	.	.	.	.	.	.	.	.	.	.	−6	2	−2
X.58	.	3	3	3	3	−1	−1	−1	−1	.	.	.	−8	.	.	.	.	.	.	.	.	.	.	.	−6	2	−2
X.59	.	−3	−3	−3	−3	1	1	1	1	.	.	.	−4	2	.	.	−2	2	.	.	.	−2	.	2	.	.	.
X.60	.	−6	6	.	.	−2	2	.	.	.	.	.	.	2	4	.	2	−2	−4	.	.	−2	.	2	.	.	.
X.61	.	−6	6	.	.	−2	2	.	.	.	.	.	.	2	−4	.	2	−2	4	.	.	2	.	−2	.	.	.
X.62	.	−3	−3	−3	−3	1	1	−1	−1	.	.	.	4	2	.	.	−2	2	.	.	.	2	.	−2	.	.	.
X.63	8	.	.	−2	2	.	.	−2	2	1	1	.	.	.	.	.	.	.	.	.	.	.	.	.	.	.	.
X.64	8	.	.	−2	2	.	.	2	−2	1	1	.	.	.	.	.	.	.	.	.	.	.	.	.	.	.	.
X.65	.	.	.	.	.	.	.	.	.	.	.	.	.	−4	.	.	4	−4	.	.	.	.	.	.	6	−2	2
X.66	−6	.	.	.	.	.	.	.	.	B	$\bar{B}$	.	.	.	.	.	.	.	.	.	.	.	.	.	−8	.	.
X.67	−6	.	.	.	.	.	.	.	.	$\bar{B}$	B	.	.	.	.	.	.	.	.	.	.	.	.	.	−8	.	.
X.68	.	.	.	.	.	.	.	.	.	.	.	5	−3	−1	1	1	−1	3	1	3	1	1	−1	1	.	.	.
X.69	.	.	.	.	.	.	.	.	.	.	.	5	−3	−5	1	−3	3	−1	1	−3	1	−1	1	−1	.	.	.
X.70	.	.	.	.	.	.	.	.	.	.	.	5	−3	3	1	5	3	−1	1	−3	−3	−1	1	−1	.	.	.
X.71	.	.	.	.	.	.	.	.	.	.	.	−5	3	−1	−1	−1	−1	3	−1	3	3	−1	−1	−1	.	.	.
X.72	.	.	.	.	.	.	.	.	.	.	.	−5	3	−5	−1	3	3	−1	−1	−3	−1	1	1	1	.	.	.
X.73	.	.	.	.	.	.	.	.	.	.	.	5	−3	−1	1	1	−1	3	1	3	−3	1	−1	1	.	.	.
X.74	.	.	.	.	.	.	.	.	.	.	.	−5	3	3	−1	−5	3	−1	−1	−3	3	1	1	1	.	.	.
X.75	.	.	.	.	.	.	.	.	.	.	.	−5	3	−1	−1	−1	−1	3	−1	3	−1	−1	−1	−1	.	.	.
X.76	−10	.	.	−2	2	.	.	2	−2	.	.	.	.	.	.	.	.	.	.	.	.	.	.	.	.	.	.
X.77	−10	.	.	−2	2	.	.	−2	2	.	.	.	.	.	.	.	.	.	.	.	.	.	.	.	.	.	.
X.78	.	3	3	−3	−3	−1	−1	1	1	.	.	.	4	4	4	.	.	.	−4	.	.	.	.	.	.	.	.
X.79	.	3	3	−3	−3	−1	−1	1	1	.	.	.	4	−4	−4	.	.	.	4	.	.	.	.	.	.	.	.
X.80	.	6	−6	.	.	2	−2	.	.	.	.	.	.	4	.	.	4	−4	.	.	.	.	.	.	.	.	.
X.81	.	3	3	−3	−3	−1	−1	−1	−1	.	.	.	−4	4	−4	.	.	.	4	.	.	.	.	.	.	.	.
X.82	.	3	3	−3	−3	−1	−1	−1	−1	.	.	.	−4	−4	4	.	.	.	−4	.	.	.	.	.	.	.	.
X.83	−8	.	.	−4	4	.	.	.	.	2	2	.	.	.	.	.	.	.	.	.	.	.	.	.	.	.	.
X.84	−12	.	.	.	.	.	.	.	.	.	.	.	.	.	.	.	.	.	.	.	.	.	.	.	−8	.	.
X.85	6	.	.	.	.	.	.	.	.	.	.	.	.	.	.	.	.	.	.	.	.	.	.	.	−8	.	.
X.86	6	.	.	.	.	.	.	.	.	.	.	.	.	.	.	.	.	.	.	.	.	.	.	.	−8	.	.

Character table of $H(J_4)$ *(continued)*

	10d	11a	12a	12b	12c	12d	12e	12f	12g	12h	12i	12j	12k	12l	14a	14b	14c	14d	14e	14f	15a	16a	20a	20b	20c
2	3	3	6	6	6	6	5	5	4	4	4	4	4	4	2	2	2	2	2	2	1	5	5	5	4
3	.	1	1	1	1	1	1	1	1	1	1	1	1	1	1	1	.	.	.	.	1	.	.	.	.
5	1	.	.	.	.	.	.	.	.	.	.	.	.	.	.	.	.	.	.	.	1	.	1	1	1
7	.	.	.	.	.	.	.	.	.	.	.	.	.	.	1	1	1	1	1	1	.	.	.	.	.
11	.	1	.	.	.	.	.	.	.	.	.	.	.	.	.	.	.	.	.	.	.	.	.	.	.
2P	5a	11a	6d	6e	6e	6d	6g	6g	6k	6h	6k	6c	6f	6h	7b	7a	7a	7b	7b	7a	15a	8c	10a	10a	10b
3P	10d	11a	4b	4j	4h	4a	4n	4o	4d	4i	4k	4r	4q	4l	14b	14a	14d	14c	14f	14e	5a	16a	20b	20a	20d
5P	2j	11a	12a	12b	12c	12d	12e	12f	12g	12h	12i	12j	12k	12l	14b	14a	14d	14c	14f	14e	3a	16a	4a	4a	4c
7P	10d	11a	12a	12b	12c	12d	12e	12f	12g	12h	12i	12j	12k	12l	2a	2a	2g	2g	2f	2f	15a	16a	20b	20a	20d
11P	10d	1a	12a	12b	12c	12d	12e	12f	12g	12h	12i	12j	12k	12l	14a	14b	14c	14d	14e	14f	15a	16a	20a	20b	20c
X.1	1	1	1	1	1	1	1	1	1	1	1	1	1	1	1	1	1	1	1	1	1	1	1	1	1
X.2	−1	1	1	1	1	1	−1	−1	−1	−1	−1	−1	1	1	1	1	−1	−1	−1	−1	1	−1	1	1	−1
X.3	1	−1	−1	1	1	3	1	1	−1	−1	−1	1	1	−1	.	.	.	.	.	.	1	−1	1	1	1
X.4	−1	−1	−1	1	1	3	−1	−1	1	1	1	−1	1	−1	.	.	.	.	.	.	1	1	1	1	−1
X.5	.	−2	4	−1	−1	.	.	.	.	.	.	.	−1	−2	.	.	.	.	.	.	−1	.	2	2	.
X.6	.	1	.	1	1	.	.	.	.	.	.	.	1	.	$\bar{A}$	A	$-A$	$-\bar{A}$	$-\bar{A}$	$-A$	.	−1	.	.	.
X.7	.	1	.	1	1	.	.	.	.	.	.	.	1	.	A	$\bar{A}$	$\bar{A}$	A	A	$\bar{A}$	.	1	.	.	.
X.8	.	1	.	1	1	.	.	.	.	.	.	.	1	.	$\bar{A}$	A	A	$\bar{A}$	$\bar{A}$	A	.	1	.	.	.
X.9	.	1	.	1	1	.	.	.	.	.	.	.	1	.	A	$\bar{A}$	$-\bar{A}$	$-A$	$-A$	$-\bar{A}$	.	−1	.	.	.
X.10	.	.	1	3	3	1	−1	−1	−1	−1	−1	−1	−1	1	−1	−1	1	1	1	1	.	1	.	.	.
X.11	.	.	1	3	3	1	1	1	1	1	1	1	−1	1	−1	−1	−1	−1	−1	−1	.	−1	.	.	.
X.12	.	2	.	−1	−1	.	.	.	.	.	.	.	−1	.	$\bar{B}$	B	.	.	.	.	.	.	.	.	.
X.13	.	2	.	−1	−1	.	.	.	.	.	.	.	−1	.	B	$\bar{B}$	.	.	.	.	.	.	.	.	.
X.14	−1	.	.	3	3	.	.	.	.	.	.	.	−1	.	1	1	1	1	1	1	−1	−1	−1	−1	−1
X.15	1	.	.	3	3	.	.	.	.	.	.	.	−1	.	1	1	−1	−1	−1	−1	−1	1	−1	−1	1
X.16	1	.	1	−2	−2	1	−1	−1	−1	−1	−1	−1	2	1	.	.	.	.	.	.	−1	.	−1	−1	1
X.17	−1	.	1	−2	−2	1	1	1	1	1	1	1	2	1	.	.	.	.	.	.	−1	.	−1	−1	−1
X.18	.	.	.	−3	−3	.	.	.	.	.	.	.	1	.	2	2	.	.	.	.	1	.	−2	−2	.
X.19	.	1	−1	−2	−2	3	1	1	−1	−1	−1	1	−2	−1	.	.	.	.	.	.	.	.	.	.	.
X.20	.	1	−1	−2	−2	3	−1	−1	1	1	1	−1	−2	−1	.	.	.	.	.	.	.	.	.	.	.
X.21	.	1	.	−1	−1	.	.	.	.	.	.	.	−1	.	.	.	.	.	.	.	.	.	.	.	.
X.22	.	1	.	−1	−1	.	.	.	.	.	.	.	−1	.	.	.	.	.	.	.	.	.	.	.	.
X.23	1	.	1	−1	−1	−3	−1	−1	1	1	1	−1	−1	1	.	.	.	.	.	.	1	−1	1	1	1
X.24	−1	.	1	−1	−1	−3	1	1	−1	−1	−1	1	−1	1	.	.	.	.	.	.	1	1	1	1	−1
X.25	.	.	−2	1	1	−2	.	.	.	.	.	.	1	−2	.	.	.	.	.	.	.	1	.	.	.
X.26	.	.	−2	1	1	−2	.	.	.	.	.	.	1	−2	.	.	.	.	.	.	.	−1	.	.	.
X.27	.	2	4	2	2	.	.	.	.	.	.	.	2	−2	.	.	.	.	.	.	.	.	.	.	.
X.28	.	.	.	−3	−3	.	.	.	.	.	.	.	1	.	.	.	.	.	.	.	−1	.	2	2	.
X.29	.	.	−4	1	1	.	.	.	.	.	.	.	1	2	.	.	.	.	.	.	−1	.	2	2	.
X.30	.	−1	2	.	.	2	.	.	.	.	.	.	.	2	.	.	.	.	.	.	.	.	.	.	.
X.31	.	2	.	2	−2	.	−2	2	.	.	.	.	.	.	$-A$	$-\bar{A}$	$-\bar{A}$	$-A$	A	$\bar{A}$	.	.	.	.	.
X.32	.	2	.	2	−2	.	−2	2	.	.	.	.	.	.	$-\bar{A}$	$-A$	$-A$	$-\bar{A}$	$\bar{A}$	A	.	.	.	.	.
X.33	.	2	.	2	−2	.	2	−2	.	.	.	.	.	.	$-\bar{A}$	$-A$	A	$\bar{A}$	$-\bar{A}$	$-A$	.	.	.	.	.
X.34	.	2	.	2	−2	.	2	−2	.	.	.	.	.	.	$-A$	$-\bar{A}$	$\bar{A}$	A	$-A$	$-\bar{A}$	.	.	.	.	.
X.35	.	.	.	2	2	.	.	.	.	.	.	.	−2	.	2	2	.	.	.	.	.	.	.	.	.
X.36	−1	.	−2	.	.	−2	.	.	.	2	.	.	.	.	.	.	.	.	.	.	.	1	−1	−1	−1
X.37	1	.	−2	.	.	−2	.	.	.	−2	.	.	.	.	.	.	.	.	.	.	.	−1	−1	−1	1
X.38	−1	.	2	.	.	−2	.	.	−2	.	2	.	.	.	.	.	.	.	.	.	.	−1	−1	−1	−1
X.39	1	.	2	.	.	−2	.	.	2	.	−2	.	.	.	.	.	.	.	.	.	.	1	−1	−1	1
X.40	.	−2	.	.	.	.	.	.	.	.	.	.	.	.	−2	−2	.	.	.	.	1	.	−2	−2	.
X.41	.	.	−1	.	.	−1	−1	−1	1	−1	1	−1	.	−1	.	.	.	.	.	.	.	.	−2	−2	.
X.42	.	.	−1	.	.	−1	1	1	−1	1	−1	1	.	−1	.	.	.	.	.	.	.	.	−2	−2	.
X.43	2	3	2	.	.	2	−2	−2	.	.	.	2	.	.	.	.	.	.	.	.	.	.	2	2	−2
X.44	−2	3	2	.	.	2	2	2	.	.	.	−2	.	.	.	.	.	.	.	.	.	.	2	2	2
X.45	1	.	2	.	.	−2	.	.	−2	.	2	.	.	.	.	.	.	.	.	.	.	.	1	1	1
X.46	1	.	−2	.	.	−2	.	.	.	2	.	.	.	.	.	.	.	.	.	.	.	.	1	1	1
X.47	−1	.	2	.	.	−2	.	.	2	.	−2	.	.	.	.	.	.	.	.	.	.	.	1	1	−1
X.48	−1	.	−2	.	.	−2	.	.	.	−2	.	.	.	.	.	.	.	.	.	.	.	.	1	1	−1
X.49	.	.	2	.	.	2	.	.	.	−2	.	.	.	.	.	.	.	.	.	.	.	1	.	.	.
X.50	.	.	−2	.	.	2	.	.	−2	.	2	.	.	.	.	.	.	.	.	.	.	1	.	.	.
X.51	.	.	2	.	.	2	.	.	.	2	.	.	.	.	.	.	.	.	.	.	.	−1	.	.	.
X.52	.	.	−2	.	.	2	.	.	2	.	−2	.	.	.	.	.	.	.	.	.	.	−1	.	.	.
X.53	.	−2	.	.	.	.	.	.	.	.	.	.	.	.	.	.	.	.	.	.	1	.	.	.	E
X.54	.	−2	.	.	.	.	.	.	.	.	.	.	.	.	.	.	.	.	.	.	1	.	.	.	$-E$
X.55	−1	.	.	.	.	.	.	.	.	.	.	.	.	.	.	.	.	.	.	.	.	.	−1	−1	−1
X.56	1	.	.	.	.	.	.	.	.	.	.	.	.	.	.	.	.	.	.	.	.	.	−1	−1	1
X.57	.	.	−1	.	.	−1	1	1	−1	1	−1	1	.	−1	.	.	.	.	.	.	.	.	2	2	.
X.58	.	.	−1	.	.	−1	−1	−1	1	−1	1	−1	.	−1	.	.	.	.	.	.	.	.	2	2	.
X.59	.	.	1	.	.	1	1	1	−1	1	−1	1	.	1	.	.	.	.	.	.	.	.	.	.	.
X.60	.	.	2	.	.	−2	.	.	.	.	.	.	.	.	.	.	.	.	.	.	.	.	.	.	.
X.61	.	.	2	.	.	−2	.	.	.	.	.	.	.	.	.	.	.	.	.	.	.	.	.	.	.
X.62	.	.	1	.	.	1	−1	−1	1	−1	1	−1	.	1	.	.	.	.	.	.	.	.	.	.	.
X.63	.	2	.	.	.	.	2	−2	.	.	.	.	.	.	−1	−1	−1	−1	1	1	.	.	.	.	.
X.64	.	2	.	.	.	.	−2	2	.	.	.	.	.	.	−1	−1	1	1	−1	−1	.	.	.	.	.
X.65	.	.	.	.	.	.	.	.	.	.	.	.	.	.	.	.	.	.	.	.	.	.	−2	−2	.
X.66	.	.	.	−2	2	.	.	.	.	.	.	.	.	.	$-\bar{B}$	$-B$	.	.	.	.	−1	.	.	.	.
X.67	.	.	.	−2	2	.	.	.	.	.	.	.	.	.	$-B$	$-\bar{B}$	.	.	.	.	−1	.	.	.	.
X.68	.	.	.	.	.	.	.	.	.	.	.	.	.	.	.	.	.	.	.	.	.	−1	.	.	.
X.69	.	.	.	.	.	.	.	.	.	.	.	.	.	.	.	.	.	.	.	.	.	1	.	.	.
X.70	.	.	.	.	.	.	.	.	.	.	.	.	.	.	.	.	.	.	.	.	.	1	.	.	.
X.71	.	.	.	.	.	.	.	.	.	.	.	.	.	.	.	.	.	.	.	.	.	1	.	.	.
X.72	.	.	.	.	.	.	.	.	.	.	.	.	.	.	.	.	.	.	.	.	.	−1	.	.	.
X.73	.	.	.	.	.	.	.	.	.	.	.	.	.	.	.	.	.	.	.	.	.	−1	.	.	.
X.74	.	.	.	.	.	.	.	.	.	.	.	.	.	.	.	.	.	.	.	.	.	−1	.	.	.
X.75	.	.	.	.	.	.	.	.	.	.	.	.	.	.	.	.	.	.	.	.	.	1	.	.	.
X.76	.	−2	.	2	−2	.	−2	2	.	.	.	.	.	.	.	.	.	.	.	.	.	.	.	.	.
X.77	.	−2	.	2	−2	.	2	−2	.	.	.	.	.	.	.	.	.	.	.	.	.	.	.	.	.
X.78	.	.	−1	.	.	−1	−1	−1	−1	1	−1	−1	.	1	.	.	.	.	.	.	.	.	.	.	.
X.79	.	.	−1	.	.	−1	−1	−1	−1	1	−1	−1	.	1	.	.	.	.	.	.	.	.	.	.	.
X.80	.	.	−2	.	.	2	.	.	.	.	.	.	.	.	.	.	.	.	.	.	.	.	.	.	.
X.81	.	.	−1	.	.	−1	1	1	1	−1	1	1	.	1	.	.	.	.	.	.	.	.	.	.	.
X.82	.	.	−1	.	.	−1	1	1	1	−1	1	1	.	1	.	.	.	.	.	.	.	.	.	.	.
X.83	.	4	.	.	.	.	.	.	.	.	.	.	.	.	−2	−2	.	.	.	.	.	.	.	.	.
X.84	.	2	.	−4	4	.	.	.	.	.	.	.	.	.	.	.	.	.	.	.	2	.	.	.	.
X.85	.	2	.	2	−2	.	.	.	.	.	.	.	.	.	.	.	.	.	.	.	−1	.	.	.	.
X.86	.	2	.	2	−2	.	.	.	.	.	.	.	.	.	.	.	.	.	.	.	−1	.	.	.	.

Character table of $H(\mathsf{J}_4)$ (continued)

2	4	1	1	3	4	4	4	4	1	1	1	3	3	1	1	2	1	1
3	.	1	1	1	1	1	1	1	1	1	1	.	.	1	1	.	1	1
5	1	.	.	.	.	.	.	.	1	.	.	1	1	.	.	.	.	.
7	.	1	1	.	.	.	.	.	.	.	.	.	.	1	1	.	.	.
11	.	.	.	1	.	.	.	.	.	1	1	.	.	.	.	1	1	1
	20d	21a	21b	22a	24a	24b	24c	24d	30a	33a	33b	40a	40b	42a	42b	44a	66a	66b
2P	10b	21a	21b	11a	12a	12b	12b	12a	15a	33a	33b	20a	20b	21a	21b	22a	33a	33b
3P	20c	7b	7a	22a	8b	8i	8i	8b	10a	11a	11a	40b	40a	14a	14b	44a	22a	22a
5P	4c	21b	21a	22a	24d	24c	24b	24a	6a	33b	33a	8a	8a	42b	42a	44a	66b	66a
7P	20c	3a	3a	22a	24d	24c	24b	24a	30a	33b	33a	40b	40a	6a	6a	44a	66b	66a
11P	20d	21a	21b	2a	24a	24b	24c	24d	30a	3a	3a	40a	40b	42a	42b	4a	6a	6a
X.1	1	1	1	1	1	1	1	1	1	1	1	1	1	1	1	1	1	1
X.2	−1	1	1	1	−1	1	1	−1	1	1	1	−1	−1	1	1	1	1	1
X.3	1	.	.	−1	1	−1	−1	1	1	−1	−1	1	1	.	.	−1	−1	−1
X.4	−1	.	.	−1	−1	−1	−1	−1	1	−1	−1	−1	−1	.	.	−1	−1	−1
X.5	.	.	.	−2	.	1	1	.	−1	1	1	.	.	.	.	−2	1	1
X.6	.	A	$\bar{A}$	1	.	−1	−1	.	.	1	1	.	.	A	$\bar{A}$	1	1	1
X.7	.	$\bar{A}$	A	1	.	−1	−1	.	.	1	1	.	.	$\bar{A}$	A	1	1	1
X.8	.	A	$\bar{A}$	1	.	−1	−1	.	.	1	1	.	.	A	$\bar{A}$	1	1	1
X.9	.	$\bar{A}$	A	1	.	−1	−1	.	.	1	1	.	.	$\bar{A}$	A	1	1	1
X.10	.	−1	−1	.	−1	1	1	−1	.	.	.	.	.	−1	−1	.	.	.
X.11	.	−1	−1	.	1	1	1	1	.	.	.	.	.	−1	−1	.	.	.
X.12	.	$-A$	$-\bar{A}$	2	.	1	1	.	.	−1	−1	.	.	$-A$	$-\bar{A}$	2	−1	−1
X.13	.	$-\bar{A}$	$-A$	2	.	1	1	.	.	−1	−1	.	.	$-\bar{A}$	$-A$	2	−1	−1
X.14	−1	1	1	.	.	−1	−1	.	−1	.	.	−1	−1	1	1	.	.	.
X.15	1	1	1	.	.	−1	−1	.	−1	.	.	1	1	1	1	.	.	.
X.16	1	.	.	.	−1	.	.	−1	−1	.	.	1	1	.	.	.	.	.
X.17	−1	.	.	.	1	.	.	1	−1	.	.	−1	−1	.	.	.	.	.
X.18	.	−1	−1	.	.	1	1	.	1	.	.	.	.	−1	−1	.	.	.
X.19	.	.	.	1	1	.	.	1	.	1	1	.	.	.	.	1	1	1
X.20	.	.	.	1	−1	.	.	−1	.	1	1	.	.	.	.	1	1	1
X.21	.	.	.	1	.	−1	−1	.	.	C	D	.	.	.	.	1	C	D
X.22	.	.	.	1	.	−1	−1	.	.	D	C	.	.	.	.	1	D	C
X.23	1	.	.	.	−1	−1	−1	−1	1	.	.	1	1	.	.	.	.	.
X.24	−1	.	.	.	1	−1	−1	1	1	.	.	−1	−1	.	.	.	.	.
X.25	.	.	.	.	.	1	1	.	.	.	.	.	.	.	.	.	.	.
X.26	.	.	.	.	.	1	1	.	.	.	.	.	.	.	.	.	.	.
X.27	.	.	.	2	.	.	.	.	.	−1	−1	.	.	.	.	2	−1	−1
X.28	.	.	.	.	.	−1	−1	.	−1	.	.	.	.	.	.	.	.	.
X.29	.	.	.	.	.	1	1	.	−1	.	.	.	.	.	.	.	.	.
X.30	.	.	.	−1	.	.	.	.	.	−1	−1	.	.	.	.	−1	−1	−1
X.31	.	$\bar{A}$	A	−2	.	.	.	.	.	−1	−1	.	.	$-\bar{A}$	$-A$	.	1	1
X.32	.	A	$\bar{A}$	−2	.	.	.	.	.	−1	−1	.	.	$-A$	$-\bar{A}$	.	1	1
X.33	.	A	$\bar{A}$	−2	.	.	.	.	.	−1	−1	.	.	$-A$	$-\bar{A}$	.	1	1
X.34	.	$\bar{A}$	A	−2	.	.	.	.	.	−1	−1	.	.	$-\bar{A}$	$-A$	.	1	1
X.35	.	−1	−1	.	.	.	.	.	.	.	.	.	.	−1	−1	.	.	.
X.36	−1	.	.	.	.	.	.	.	.	.	.	1	1	.	.	.	.	.
X.37	1	.	.	.	.	.	.	.	.	.	.	−1	−1	.	.	.	.	.
X.38	−1	.	.	.	.	.	.	.	.	.	.	1	1	.	.	.	.	.
X.39	1	.	.	.	.	.	.	.	.	.	.	−1	−1	.	.	.	.	.
X.40	.	1	1	−2	.	.	.	.	1	1	1	.	.	1	1	−2	1	1
X.41	.	.	.	.	1	.	.	1	.	.	.	.	.	.	.	.	.	.
X.42	.	.	.	.	−1	.	.	−1	.	.	.	.	.	.	.	.	.	.
X.43	−2	.	.	3	.	.	.	.	.	.	.	.	.	.	.	−1	.	.
X.44	2	.	.	3	.	.	.	.	.	.	.	.	.	.	.	−1	.	.
X.45	1	.	.	.	.	.	.	.	.	.	.	−1	−1	.	.	.	.	.
X.46	1	.	.	.	.	.	.	.	.	.	.	−1	−1	.	.	.	.	.
X.47	−1	.	.	.	.	.	.	.	.	.	.	1	1	.	.	.	.	.
X.48	−1	.	.	.	.	.	.	.	.	.	.	1	1	.	.	.	.	.
X.49	.	.	.	.	.	.	.	.	.	.	.	.	.	.	.	.	.	.
X.50	.	.	.	.	.	.	.	.	.	.	.	.	.	.	.	.	.	.
X.51	.	.	.	.	.	.	.	.	.	.	.	.	.	.	.	.	.	.
X.52	.	.	.	.	.	.	.	.	.	.	.	.	.	.	.	.	.	.
X.53	$-E$	.	.	2	.	.	.	.	−1	1	1	.	.	.	.	.	−1	−1
X.54	E	.	.	2	.	.	.	.	−1	1	1	.	.	.	.	.	−1	−1
X.55	−1	.	.	.	.	.	.	.	.	.	.	1	1	.	.	.	.	.
X.56	1	.	.	.	.	.	.	.	.	.	.	−1	−1	.	.	.	.	.
X.57	.	.	.	.	−1	.	.	−1	.	.	.	.	.	.	.	.	.	.
X.58	.	.	.	.	1	.	.	1	.	.	.	.	.	.	.	.	.	.
X.59	.	.	.	.	−1	.	.	−1	.	.	.	.	.	.	.	.	.	.
X.60	.	.	.	.	.	.	.	.	.	.	.	.	.	.	.	.	.	.
X.61	.	.	.	.	.	.	.	.	.	.	.	.	.	.	.	.	.	.
X.62	.	.	.	.	1	.	.	1	.	.	.	.	.	.	.	.	.	.
X.63	.	1	1	−2	.	.	.	.	.	−1	−1	.	.	−1	−1	.	1	1
X.64	.	1	1	−2	.	.	.	.	.	−1	−1	.	.	−1	−1	.	1	1
X.65	.	.	.	.	.	.	.	.	.	.	.	.	.	.	.	.	.	.
X.66	.	$-A$	$-\bar{A}$	.	.	.	.	.	1	.	.	.	.	A	$\bar{A}$	.	.	.
X.67	.	$-\bar{A}$	$-A$	.	.	.	.	.	1	.	.	.	.	$\bar{A}$	A	.	.	.
X.68	.	.	.	.	.	.	.	.	.	.	.	.	.	.	.	.	.	.
X.69	.	.	.	.	.	.	.	.	.	.	.	.	.	.	.	.	.	.
X.70	.	.	.	.	.	.	.	.	.	.	.	.	.	.	.	.	.	.
X.71	.	.	.	.	.	.	.	.	.	.	.	.	.	.	.	.	.	.
X.72	.	.	.	.	.	.	.	.	.	.	.	.	.	.	.	.	.	.
X.73	.	.	.	.	.	.	.	.	.	.	.	.	.	.	.	.	.	.
X.74	.	.	.	.	.	.	.	.	.	.	.	.	.	.	.	.	.	.
X.75	.	.	.	.	.	.	.	.	.	.	.	.	.	.	.	.	.	.
X.76	.	.	.	2	.	.	.	.	.	1	1	.	.	.	.	.	−1	−1
X.77	.	.	.	2	.	.	.	.	.	1	1	.	.	.	.	.	−1	−1
X.78	.	.	.	.	1	.	.	1	.	.	.	.	.	.	.	.	.	.
X.79	.	.	.	.	1	.	.	1	.	.	.	.	.	.	.	.	.	.
X.80	.	.	.	.	.	.	.	.	.	.	.	.	.	.	.	.	.	.
X.81	.	.	.	.	−1	.	.	−1	.	.	.	.	.	.	.	.	.	.
X.82	.	.	.	.	−1	.	.	−1	.	.	.	.	.	.	.	.	.	.
X.83	.	−1	−1	−4	.	.	.	.	.	1	1	.	.	1	1	.	−1	−1
X.84	.	.	.	−2	.	.	.	.	−2	−1	−1	.	.	.	.	.	1	1
X.85	.	.	.	−2	.	.	.	.	1	$-D$	$-C$	.	.	.	.	.	D	C
X.86	.	.	.	−2	.	.	.	.	1	$-C$	$-D$	.	.	.	.	.	C	D

Character table of $H(\mathsf{J}_4)$ *(continued)*

2	21	21	20	19	17	17	14	14	16	15	13	8	8	15	15	13	13	14
3	3	3	1	1	2	2	1	1	1	1	.	3	2	1	1	.	1	.
5	1	1	1	1	.	.	.	.	.	.	1	1	.	1	.	1	.	.
7	1	1	.	.	.	.	1	1	.	.	.	1	.	.	.	.	.	.
11	1	1	.	.	.	.	.	.	.	.	.	1	.	1	.	.	.	.
	$1a$	$2a$	$2b$	$2c$	$2d$	$2e$	$2f$	$2g$	$2h$	$2i$	$2j$	$3a$	$3b$	$4a$	$4b$	$4c$	$4d$	$4e$
$2P$	$1a$	$1a$	$1a$	$1a$	$1a$	$1a$	$1a$	$1a$	$1a$	$1a$	$1a$	$3a$	$3b$	$2a$	$2a$	$2b$	$2b$	$2b$
$3P$	$1a$	$2a$	$2b$	$2c$	$2d$	$2e$	$2f$	$2g$	$2h$	$2i$	$2j$	$1a$	$1a$	$4a$	$4b$	$4c$	$4d$	$4e$
$5P$	$1a$	$2a$	$2b$	$2c$	$2d$	$2e$	$2f$	$2g$	$2h$	$2i$	$2j$	$3a$	$3b$	$4a$	$4b$	$4c$	$4d$	$4e$
$7P$	$1a$	$2a$	$2b$	$2c$	$2d$	$2e$	$2f$	$2g$	$2h$	$2i$	$2j$	$3a$	$3b$	$4a$	$4b$	$4c$	$4d$	$4e$
$11P$	$1a$	$2a$	$2b$	$2c$	$2d$	$2e$	$2f$	$2g$	$2h$	$2i$	$2j$	$3a$	$3b$	$4a$	$4b$	$4c$	$4d$	$4e$
$X.87$	19712	−19712	.	.	64	−64	.	.	.	.	.	308	8	.	.	.	.	.
$X.88$	20160	20160	−320	−320	192	192	.	.	−64	−64	.	.	−12	320	64	.	.	.
$X.89$	20160	20160	−320	−320	−192	−192	.	.	64	64	.	.	6	320	−64	.	.	.
$X.90$	20160	20160	−320	−320	−192	−192	.	.	64	64	.	.	6	320	−64	.	.	.
$X.91$	20160	20160	−320	−320	192	192	.	.	−64	−64	.	.	6	320	64	.	.	.
$X.92$	20160	20160	−320	−320	192	192	.	.	−64	−64	.	.	6	320	64	.	.	.
$X.93$	20790	20790	−330	630	−90	−90	−84	−84	6	−90	20	.	.	−330	−42	20	12	22
$X.94$	20790	20790	1590	−330	−90	−90	42	42	−90	6	−30	.	.	−330	6	−30	42	6
$X.95$	20790	20790	1590	−330	−90	−90	−42	−42	−90	6	30	.	.	−330	6	30	−42	6
$X.96$	20790	20790	−330	630	−90	−90	84	84	6	−90	−20	.	.	−330	−42	−20	−12	22
$X.97$	21120	−21120	.	.	32	−32	160	−160	.	.	.	330	−12	.	.	.	.	.
$X.98$	21120	−21120	.	.	32	−32	−160	160	.	.	.	330	−12	.	.	.	.	.
$X.99$	22176	22176	−352	−352	−96	−96	.	.	32	32	−32	.	−6	352	−32	32	.	.
$X.100$	22176	22176	−352	−352	−96	−96	.	.	32	32	32	.	−6	352	−32	−32	.	.
$X.101$	24192	24192	−384	−384	.	.	.	.	.	.	64	.	.	384	.	−64	.	.
$X.102$	24192	24192	−384	−384	.	.	.	.	.	.	64	.	.	384	.	−64	.	.
$X.103$	24192	24192	−384	−384	.	.	.	.	.	.	−64	.	.	384	.	64	.	.
$X.104$	24192	24192	−384	−384	.	.	.	.	.	.	−64	.	.	384	.	64	.	.
$X.105$	26880	−26880	.	.	320	−320	.	.	.	.	.	−210	.	.	.	.	.	.
$X.106$	26880	−26880	.	.	−192	192	.	.	.	.	.	−210	.	.	.	.	.	.
$X.107$	26880	−26880	.	.	−192	192	.	.	.	.	.	−210	.	.	.	.	.	.
$X.108$	27720	27720	−440	840	−24	−24	56	56	−24	40	40	.	−3	−440	40	40	−8	−24
$X.109$	27720	27720	−440	840	−24	−24	−56	−56	−24	40	−40	.	−3	−440	40	−40	8	−24
$X.110$	28160	−28160	.	.	−128	128	64	−64	.	.	.	440	−4	.	.	.	.	.
$X.111$	28160	−28160	.	.	−128	128	−64	64	.	.	.	440	−4	.	.	.	.	.
$X.112$	42240	−42240	.	.	64	−64	.	.	.	.	.	−330	.	.	.	.	.	.
$X.113$	49152	−49152	.	.	.	.	.	.	.	.	.	−384	.	.	.	.	.	.

Character table of $H(\mathsf{J}_4)$ *(continued)*

2	14	13	11	11	11	11	11	12	10	10	11	9	9	10	10	10	9	9	9	6	8	8	8	8
3	.	.	1	1	1	1	1	.	1	1	.	1	1	.	.	.	.	.	.	1	3	2	2	2
5	.	.	.	.	.	.	.	.	.	.	.	.	.	.	.	.	.	.	.	1	1	.	.	.
7	.	.	.	.	.	.	.	.	.	.	.	.	.	.	.	.	.	.	.	.	1	.	.	.
11	.	.	.	.	.	.	.	.	.	.	.	.	.	.	.	.	.	.	.	.	1	.	.	.
	$4f$	$4g$	$4h$	$4i$	$4j$	$4k$	$4l$	$4m$	$4n$	$4o$	$4p$	$4q$	$4r$	$4s$	$4t$	$4u$	$4v$	$4w$	$4x$	$5a$	$6a$	$6b$	$6c$	$6d$
$2P$	$2b$	$2b$	$2d$	$2c$	$2d$	$2b$	$2c$	$2b$	$2h$	$2h$	$2c$	$2e$	$2e$	$2h$	$2h$	$2d$	$2i$	$2i$	$2h$	$5a$	$3a$	$3b$	$3b$	$3b$
$3P$	$4f$	$4g$	$4h$	$4i$	$4j$	$4k$	$4l$	$4m$	$4n$	$4o$	$4p$	$4q$	$4r$	$4s$	$4t$	$4u$	$4v$	$4w$	$4x$	$5a$	$2a$	$2d$	$2e$	$2a$
$5P$	$4f$	$4g$	$4h$	$4i$	$4j$	$4k$	$4l$	$4m$	$4n$	$4o$	$4p$	$4q$	$4r$	$4s$	$4t$	$4u$	$4v$	$4w$	$4x$	$1a$	$6a$	$6b$	$6c$	$6d$
$7P$	$4f$	$4g$	$4h$	$4i$	$4j$	$4k$	$4l$	$4m$	$4n$	$4o$	$4p$	$4q$	$4r$	$4s$	$4t$	$4u$	$4v$	$4w$	$4x$	$5a$	$6a$	$6b$	$6c$	$6d$
$11P$	$4f$	$4g$	$4h$	$4i$	$4j$	$4k$	$4l$	$4m$	$4n$	$4o$	$4p$	$4q$	$4r$	$4s$	$4t$	$4u$	$4v$	$4w$	$4x$	$5a$	$6a$	$6b$	$6c$	$6d$
$X.87$	.	.	16	.	−16	.	.	.	.	.	.	.	.	.	.	.	.	.	.	−8	−308	−8	8	−8
$X.88$	.	.	.	.	.	.	.	.	.	.	.	.	.	.	.	.	.	.	.	.	.	12	12	−12
$X.89$	.	.	.	.	.	.	.	.	−16	−16	.	.	16	.	.	.	.	.	.	.	.	6	6	6
$X.90$	.	.	.	.	.	.	.	.	16	16	.	.	−16	.	.	.	.	.	.	.	.	6	6	6
$X.91$	.	.	.	.	.	.	.	.	.	.	.	.	.	.	.	.	.	.	.	.	.	−6	−6	6
$X.92$	.	.	.	.	.	.	.	.	.	.	.	.	.	.	.	.	.	.	.	.	.	−6	−6	6
$X.93$	38	6	−6	−12	−6	12	−18	4	.	.	−4	6	.	−8	2	8	2	.	−2	.	.	.	.	.
$X.94$	6	−10	−6	−6	−6	−6	6	2	6	6	−14	6	6	−2	−6	−2	2	−2	6	.	.	.	.	.
$X.95$	6	−10	−6	6	−6	6	6	−2	−6	−6	14	6	−6	2	−6	2	2	2	6	.	.	.	.	.
$X.96$	38	6	−6	12	−6	−12	−18	−4	.	.	4	6	.	8	2	−8	2	.	−2	.	.	.	.	.
$X.97$	.	.	−8	.	8	.	.	.	8	−8	.	.	.	.	.	.	.	.	.	.	−330	−4	4	12
$X.98$	.	.	−8	.	8	.	.	.	−8	8	.	.	.	.	.	.	.	.	.	.	−330	−4	4	12
$X.99$	.	.	.	.	.	.	.	.	−8	−8	.	.	8	.	.	.	.	.	.	6	.	−6	−6	−6
$X.100$	.	.	.	.	.	.	.	.	8	8	.	.	−8	.	.	.	.	.	.	6	.	−6	−6	−6
$X.101$	.	.	.	.	.	.	.	.	.	.	.	.	.	.	.	.	.	.	.	−3	.	.	.	.
$X.102$	.	.	.	.	.	.	.	.	.	.	.	.	.	.	.	.	.	.	.	−3	.	.	.	.
$X.103$	.	.	.	.	.	.	.	.	.	.	.	.	.	.	.	.	.	.	.	−3	.	.	.	.
$X.104$	.	.	.	.	.	.	.	.	.	.	.	.	.	.	.	.	.	.	.	−3	.	.	.	.
$X.105$	.	.	−16	.	16	.	.	.	.	.	.	.	.	.	.	.	.	.	.	.	210	−16	16	.
$X.106$	.	.	16	.	−16	.	.	.	.	.	.	.	.	.	.	.	.	.	.	.	210	.	.	.
$X.107$	.	.	16	.	−16	.	.	.	.	.	.	.	.	.	.	.	.	.	.	.	210	.	.	.
$X.108$	−24	8	.	8	.	−8	8	8	−8	−8	−8	.	−8	.	.	.	.	.	.	.	.	−3	−3	−3
$X.109$	−24	8	.	−8	.	8	8	−8	8	8	8	.	8	.	.	.	.	.	.	.	.	−3	−3	−3
$X.110$	.	.	.	.	.	.	.	.	16	−16	.	.	.	.	.	.	.	.	.	.	−440	4	−4	4
$X.111$	.	.	.	.	.	.	.	.	−16	16	.	.	.	.	.	.	.	.	.	.	−440	4	−4	4
$X.112$	.	.	−16	.	16	.	.	.	.	.	.	.	.	.	.	.	.	.	.	.	330	16	−16	.
$X.113$	.	.	.	.	.	.	.	.	.	.	.	.	.	.	.	.	.	.	.	−8	384	.	.	.

Character table of $H(J_4)$ (continued)

	6e	6f	6g	6h	6i	6j	6k	6l	6m	6n	7a	7b	8a	8b	8c	8d	8e	8f	8g	8h	8i	8j	8k	8l	8m	10a	10b	10c
2	8	8	7	7	5	5	6	6	5	5	2	2	8	8	9	8	8	8	8	8	6	7	6	6	6	6	5	4
3	2	2	1	1	2	2	1	1	1	1	1	1	.	1	.	.	.	.	.	.	1	.	.	.	.	1	.	.
5	.	.	.	.	.	.	.	.	.	.	.	.	1	.	.	.	.	.	.	.	.	.	.	.	.	1	1	1
7	.	.	.	.	.	.	.	.	.	.	1	1	.	.	.	.	.	.	.	.	.	.	.	.	.	.	.	.
11	.	.	.	.	.	.	.	.	.	.	.	.	.	.	.	.	.	.	.	.	.	.	.	.	.	.	.	.
2P	3a	3a	3b	3b	3b	3b	3b	3b	3b	3b	7a	7b	4a	4b	4b	4e	4e	4e	4f	4f	4j	4g	4h	4t	4t	5a	5a	5a
3P	2d	2e	2h	2c	2d	2e	2b	2i	2f	2g	7b	7a	8a	8b	8c	8d	8e	8f	8g	8h	8i	8j	8k	8l	8m	10a	10b	10c
5P	6e	6f	6g	6h	6i	6j	6k	6l	6m	6n	7b	7a	8a	8b	8c	8d	8e	8f	8g	8h	8i	8j	8k	8l	8m	2a	2b	2c
7P	6e	6f	6g	6h	6i	6j	6k	6l	6m	6n	1a	1a	8a	8b	8c	8d	8e	8f	8g	8h	8i	8j	8k	8l	8m	10a	10b	10c
11P	6e	6f	6g	6h	6i	6j	6k	6l	6m	6n	7a	7b	8a	8b	8c	8d	8e	8f	8g	8h	8i	8j	8k	8l	8m	10a	10b	10c
X.87	4	−4	.	.	4	−4	.	.	.	.	.	.	.	.	.	.	.	.	.	.	.	.	.	.	.	8	.	.
X.88	.	.	−4	4	.	.	4	−4	.	.	.	.	.	.	.	.	.	.	.	.	.	.	.	.	.	.	.	.
X.89	.	.	−2	−2	.	.	−2	−2	.	.	.	.	.	.	.	.	.	.	.	.	.	.	.	.	.	.	.	.
X.90	.	.	−2	−2	.	.	−2	−2	.	.	.	.	.	.	.	.	.	.	.	.	.	.	.	.	.	.	.	.
X.91	.	.	2	−2	.	.	−2	2	.	.	.	.	.	.	.	.	.	.	.	.	.	.	.	.	.	.	.	.
X.92	.	.	2	−2	.	.	−2	2	.	.	.	.	.	.	.	.	.	.	.	.	.	.	.	.	.	.	.	.
X.93	.	.	.	.	.	.	.	.	.	.	.	.	.	.	−2	−4	.	−2	2	4	.	.	−2	.	2	.	.	.
X.94	.	.	.	.	.	.	.	.	.	.	.	.	10	6	2	−2	−2	−2	−2	−2	.	2	.	.	.	.	.	.
X.95	.	.	.	.	.	.	.	.	.	.	.	.	−10	−6	2	2	2	−2	−2	2	.	−2	.	.	.	.	.	.
X.96	.	.	.	.	.	.	.	.	.	.	.	.	.	.	−2	4	.	−2	2	−4	.	.	2	.	−2	.	.	.
X.97	2	−2	.	.	2	−2	.	.	−2	2	1	1	.	.	.	.	.	.	.	.	.	.	.	.	.	.	.	.
X.98	2	−2	.	.	2	−2	.	.	2	−2	1	1	.	.	.	.	.	.	.	.	.	.	.	.	.	.	.	.
X.99	.	.	2	2	.	.	2	2	.	.	.	.	.	.	.	.	.	.	.	.	.	.	.	.	.	6	−2	−2
X.100	.	.	2	2	.	.	2	2	.	.	.	.	.	.	.	.	.	.	.	.	.	.	.	.	.	6	−2	−2
X.101	.	.	.	.	.	.	.	.	.	.	.	.	.	.	.	.	.	.	.	.	.	.	.	.	.	−3	1	1
X.102	.	.	.	.	.	.	.	.	.	.	.	.	.	.	.	.	.	.	.	.	.	.	.	.	.	−3	1	1
X.103	.	.	.	.	.	.	.	.	.	.	.	.	.	.	.	.	.	.	.	.	.	.	.	.	.	−3	1	1
X.104	.	.	.	.	.	.	.	.	.	.	.	.	.	.	.	.	.	.	.	.	.	.	.	.	.	−3	1	1
X.105	−10	10	.	.	−4	4	.	.	.	.	.	.	.	.	.	.	.	.	.	.	.	.	.	.	.	.	.	.
X.106	6	−6	.	.	.	.	.	.	.	.	.	.	.	.	.	.	.	.	.	.	.	.	.	.	.	.	.	.
X.107	6	−6	.	.	.	.	.	.	.	.	.	.	.	.	.	.	.	.	.	.	.	.	.	.	.	.	.	.
X.108	.	.	−3	−3	3	3	1	1	−1	−1	.	.	.	8	.	.	.	.	.	.	.	.	.	.	.	.	.	.
X.109	.	.	−3	−3	3	3	1	1	1	1	.	.	.	−8	.	.	.	.	.	.	.	.	.	.	.	.	.	.
X.110	−8	8	.	.	−2	2	.	.	−2	2	−1	−1	.	.	.	.	.	.	.	.	.	.	.	.	.	.	.	.
X.111	−8	8	.	.	−2	2	.	.	2	−2	−1	−1	.	.	.	.	.	.	.	.	.	.	.	.	.	.	.	.
X.112	−2	2	.	.	4	−4	.	.	.	.	2	2	.	.	.	.	.	.	.	.	.	.	.	.	.	.	.	.
X.113	.	.	.	.	.	.	.	.	.	.	−2	−2	.	.	.	.	.	.	.	.	.	.	.	.	.	8	.	.

Character table of $H(J_4)$ (continued)

	10d	11a	12a	12b	12c	12d	12e	12f	12g	12h	12i	12j	12k	12l	14a	14b	14c	14d	14e	14f	15a	16a	20a	20b	20c
2	3	3	6	6	6	6	5	5	4	4	4	4	4	4	2	2	2	2	2	2	1	5	5	5	4
3	.	1	1	1	1	1	1	1	1	1	1	1	1	1	1	1	.	.	.	.	1	.	.	.	.
5	1	.	.	.	.	.	.	.	.	.	.	.	.	.	.	.	.	.	.	.	1	.	1	1	1
7	.	.	.	.	.	.	.	.	.	.	.	.	.	.	1	1	1	1	1	1	.	.	.	.	.
11	.	1	.	.	.	.	.	.	.	.	.	.	.	.	.	.	.	.	.	.	.	.	.	.	.
2P	5a	11a	6d	6e	6e	6d	6g	6g	6k	6h	6k	6c	6f	6h	7b	7a	7a	7b	7b	7a	15a	8c	10a	10a	10b
3P	10d	11a	4b	4j	4h	4a	4n	4o	4d	4i	4k	4r	4q	4l	14b	14a	14d	14c	14f	14e	5a	16a	20b	20a	20d
5P	2j	11a	12a	12b	12c	12d	12e	12f	12g	12h	12i	12j	12k	12l	14b	14a	14d	14c	14f	14e	3a	16a	4a	4a	4c
7P	10d	11a	12a	12b	12c	12d	12e	12f	12g	12h	12i	12j	12k	12l	2a	2a	2g	2g	2f	2f	15a	16a	20b	20a	20d
11P	10d	1a	12a	12b	12c	12d	12e	12f	12g	12h	12i	12j	12k	12l	14a	14b	14c	14d	14e	14f	15a	16a	20a	20b	20c
X.87	.	.	.	−4	4	.	.	.	.	.	.	.	.	.	.	.	.	.	.	.	−2	.	.	.	.
X.88	.	−3	4	.	.	−4	.	.	.	.	.	.	.	.	.	.	.	.	.	.	.	.	.	.	.
X.89	.	−3	2	.	.	2	2	2	.	.	.	−2	.	.	.	.	.	.	.	.	.	.	.	.	.
X.90	.	−3	2	.	.	2	−2	−2	.	.	.	2	.	.	.	.	.	.	.	.	.	.	.	.	.
X.91	.	−3	−2	.	.	2	.	.	.	.	.	.	.	.	.	.	.	.	.	.	.	.	.	.	.
X.92	.	−3	−2	.	.	2	.	.	.	.	.	.	.	.	.	.	.	.	.	.	.	.	.	.	.
X.93	.	.	.	.	.	.	.	.	.	.	.	.	.	.	.	.	.	.	.	.	.	.	.	.	.
X.94	.	.	.	.	.	.	.	.	.	.	.	.	.	.	.	.	.	.	.	.	.	.	.	.	.
X.95	.	.	.	.	.	.	.	.	.	.	.	.	.	.	.	.	.	.	.	.	.	.	.	.	.
X.96	.	.	.	.	.	.	.	.	.	.	.	.	.	.	.	.	.	.	.	.	.	.	.	.	.
X.97	.	.	.	2	−2	.	2	−2	.	.	.	.	.	.	−1	−1	1	1	−1	−1	.	.	.	.	.
X.98	.	.	.	2	−2	.	−2	2	.	.	.	.	.	.	−1	−1	−1	−1	1	1	.	.	.	.	.
X.99	−2	.	−2	.	.	−2	−2	−2	.	.	.	2	.	.	.	.	.	.	.	.	.	.	2	2	2
X.100	2	.	−2	.	.	−2	2	2	.	.	.	−2	.	.	.	.	.	.	.	.	.	.	2	2	−2
X.101	−1	3	.	.	.	.	.	.	.	.	.	.	.	.	.	.	.	.	.	.	.	.	G	H	1
X.102	−1	3	.	.	.	.	.	.	.	.	.	.	.	.	.	.	.	.	.	.	.	.	H	G	1
X.103	1	3	.	.	.	.	.	.	.	.	.	.	.	.	.	.	.	.	.	.	.	.	H	G	−1
X.104	1	3	.	.	.	.	.	.	.	.	.	.	.	.	.	.	.	.	.	.	.	.	G	H	−1
X.105	.	−4	.	−2	2	.	.	.	.	.	.	.	.	.	.	.	.	.	.	.	.	.	.	.	.
X.106	.	−4	.	2	−2	.	.	.	.	.	.	.	.	.	.	.	.	.	.	.	.	.	.	.	.
X.107	.	−4	.	2	−2	.	.	.	.	.	.	.	.	.	.	.	.	.	.	.	.	.	.	.	.
X.108	.	.	1	.	.	1	1	1	1	−1	1	1	.	−1	.	.	.	.	.	.	.	.	.	.	.
X.109	.	.	1	.	.	1	−1	−1	−1	1	−1	−1	.	−1	.	.	.	.	.	.	.	.	.	.	.
X.110	.	.	.	.	.	.	−2	2	.	.	.	.	.	.	1	1	−1	−1	1	1	.	.	.	.	.
X.111	.	.	.	.	.	.	2	−2	.	.	.	.	.	.	1	1	1	1	−1	−1	.	.	.	.	.
X.112	.	.	.	−2	2	.	.	.	.	.	.	.	.	.	−2	−2	.	.	.	.	.	.	.	.	.
X.113	.	4	.	.	.	.	.	.	.	.	.	.	.	.	2	2	.	.	.	.	1	.	.	.	.

Character table of $H(\mathsf{J}_4)$ *(continued)*

2	4	1	1	3	4	4	4	4	1	1	1	3	3	1	1	2	1	1
3	.	1	1	1	1	1	1	1	1	1	1	.	.	1	1	.	1	1
5	1	.	.	.	.	.	.	.	1	.	.	1	1	.	.	.	.	.
7	.	1	1	.	.	.	.	.	.	.	.	.	.	1	1	.	.	.
11	.	.	.	1	.	.	.	.	.	1	1	.	.	.	.	1	1	1
	20*d*	21*a*	21*b*	22*a*	24*a*	24*b*	24*c*	24*d*	30*a*	33*a*	33*b*	40*a*	40*b*	42*a*	42*b*	44*a*	66*a*	66*b*
2*P*	10*b*	21*a*	21*b*	11*a*	12*a*	12*b*	12*b*	12*a*	15*a*	33*a*	33*b*	20*a*	20*b*	21*a*	21*b*	22*a*	33*a*	33*b*
3*P*	20*c*	7*b*	7*a*	22*a*	8*b*	8*i*	8*i*	8*b*	10*a*	11*a*	11*a*	40*b*	40*a*	14*a*	14*b*	44*a*	22*a*	22*a*
5*P*	4*c*	21*b*	21*a*	22*a*	24*d*	24*c*	24*b*	24*a*	6*a*	33*b*	33*a*	8*a*	8*a*	42*b*	42*a*	44*a*	66*b*	66*a*
7*P*	20*c*	3*a*	3*a*	22*a*	24*d*	24*c*	24*b*	24*a*	30*a*	33*b*	33*a*	40*b*	40*a*	6*a*	6*a*	44*a*	66*b*	66*a*
11*P*	20*d*	21*a*	21*b*	2*a*	24*a*	24*b*	24*c*	24*d*	30*a*	3*a*	3*a*	40*a*	40*b*	42*a*	42*b*	4*a*	6*a*	6*a*
$X.87$	.	.	.	.	.	.	.	.	2	.	.	.	.	.	.	.	.	.
$X.88$	.	.	.	-3	.	.	.	.	.	.	.	.	.	.	.	1	.	.
$X.89$	.	.	.	-3	.	.	.	.	.	.	.	.	.	.	.	1	.	.
$X.90$	.	.	.	-3	.	.	.	.	.	.	.	.	.	.	.	1	.	.
$X.91$	.	.	.	-3	F	.	.	$-F$	.	.	.	.	.	.	.	1	.	.
$X.92$	.	.	.	-3	$-F$	.	.	F	.	.	.	.	.	.	.	1	.	.
$X.93$	.	.	.	.	.	.	.	.	.	.	.	.	.	.	.	.	.	.
$X.94$	.	.	.	.	.	.	.	.	.	.	.	.	.	.	.	.	.	.
$X.95$	.	.	.	.	.	.	.	.	.	.	.	.	.	.	.	.	.	.
$X.96$	.	.	.	.	.	.	.	.	.	.	.	.	.	.	.	.	.	.
$X.97$	.	1	1	.	.	.	.	.	.	.	.	.	.	-1	-1	.	.	.
$X.98$	.	1	1	.	.	.	.	.	.	.	.	.	.	-1	-1	.	.	.
$X.99$	2	.	.	.	.	.	.	.	.	.	.	.	.	.	.	.	.	.
$X.100$	-2	.	.	.	.	.	.	.	.	.	.	.	.	.	.	.	.	.
$X.101$	1	.	.	3	.	.	.	.	.	.	.	I	$-I$	.	.	-1	.	.
$X.102$	1	.	.	3	.	.	.	.	.	.	.	$-I$	I	.	.	-1	.	.
$X.103$	-1	.	.	3	.	.	.	.	.	.	.	I	$-I$	.	.	-1	.	.
$X.104$	-1	.	.	3	.	.	.	.	.	.	.	$-I$	I	.	.	-1	.	.
$X.105$	.	.	.	4	.	.	.	.	.	-1	-1	.	.	.	.	.	1	1
$X.106$	.	.	.	4	.	$-F$	F	.	.	-1	-1	.	.	.	.	.	1	1
$X.107$	.	.	.	4	.	F	$-F$	.	.	-1	-1	.	.	.	.	.	1	1
$X.108$	.	.	.	.	-1	.	.	-1	.	.	.	.	.	.	.	.	.	.
$X.109$	.	.	.	.	1	.	.	1	.	.	.	.	.	.	.	.	.	.
$X.110$	.	-1	-1	.	.	.	.	.	.	.	.	.	.	1	1	.	.	.
$X.111$	.	-1	-1	.	.	.	.	.	.	.	.	.	.	1	1	.	.	.
$X.112$	.	-1	-1	.	.	.	.	.	.	.	.	.	.	1	1	.	.	.
$X.113$	.	1	1	-4	.	.	.	.	-1	1	1	.	.	-1	-1	.	-1	-1

where $A = \zeta(7)^4 + \zeta(7)^2 + \zeta(7)$, $B = 2A$, $C = 2\zeta(33)_3\zeta(33)_{11}^9 + 2\zeta(33)_3\zeta(33)_{11}^5 + 2\zeta(33)_3\zeta(33)_{11}^4 + 2\zeta(33)_3\zeta(33)_{11}^3 + 2\zeta(33)_3\zeta(33)_{11} + \zeta(33)_3 + \zeta(33)_{11}^9 + \zeta(33)_{11}^5 + \zeta(33)_{11}^4 + \zeta(33)_{11}^3 + \zeta(33)_{11}$, $D = -(C+1)$, $E = 4\zeta(5)^3 + 4\zeta(5)^2 + 2$, $F = 4\zeta(12)_4\zeta(12)_3 + 2\zeta(12)_4$, $G = -(E+1)$, $H = E - 1$, $I = 2\zeta(5)^3 + 2\zeta(5)^2 + 1$.

8.6.2 Character table of $E = N_G(A) = \langle x, y, e \rangle$

2	1	1	.	.	.	.	.	.	.	.	.	.	.	.	.	.	.	.	.	.
3	1	.	.	.	.	.	.	.	.	.	.	.	.	.	.	.	.	.	.	.
5	21	19	21	17	15	15	17	8	6	14	15	15	11	13	11	11	11	11	10	10
7	3	2	3	1	1	1	1	3	2	1	1	1	1	.	1	1	.	.	.	.
11	1	1	1	.	1	1	.	1	.	1	.	.	.	.	.	.	.	.	.	.
23	1	1	.	1	.	.	.	.	1	.	.	.	1	.	.	.	.	.	.	.
	1a	2a	2b	2c	2d	2e	2f	3a	3b	4a	4b	4c	4d	4e	4f	4g	4h	4i	4j	4k
2P	1a	1a	1a	1a	1a	1a	1a	3a	3b	2b	2b	2b	2a	2b	2c	2c	2c	2c	2f	2f
3P	1a	2a	2b	2c	2d	2e	2f	1a	1a	4a	4b	4c	4d	4e	4f	4g	4h	4i	4j	4k
5P	1a	2a	2b	2c	2d	2e	2f	3a	3b	4a	4b	4c	4d	4e	4f	4g	4h	4i	4j	4k
7P	1a	2a	2b	2c	2d	2e	2f	3a	3b	4a	4b	4c	4d	4e	4f	4g	4h	4i	4j	4k
11P	1a	2a	2b	2c	2d	2e	2f	3a	3b	4a	4b	4c	4d	4e	4f	4g	4h	4i	4j	4k
23P	1a	2a	2b	2c	2d	2e	2f	3a	3b	4a	4b	4c	4d	4e	4f	4g	4h	4i	4j	4k
X.1	1	1	1	1	1	1	1	1	1	1	1	1	1	1	1	1	1	1	1	1
X.2	23	23	23	7	-1	-1	7	5	-1	-1	7	7	7	-1	-1	-1	3	3	3	-1
X.3	45	45	45	-3	5	5	-3	.	3	5	-3	-3	-3	5	-3	-3	1	1	1	-3
X.4	45	45	45	-3	5	5	-3	.	3	5	-3	-3	-3	5	-3	-3	1	1	1	-3
X.5	231	231	231	7	-9	-9	7	-3	.	-9	7	7	7	-9	-1	-1	-1	-1	-1	-1
X.6	231	231	231	7	-9	-9	7	-3	.	-9	7	7	7	-9	-1	-1	-1	-1	-1	-1
X.7	252	252	252	28	12	12	28	9	.	12	28	28	28	12	4	4	4	4	4	4
X.8	253	253	253	13	-11	-11	13	10	1	-11	13	13	13	-11	-3	-3	1	1	1	-3
X.9	483	483	483	35	3	3	35	6	.	3	35	35	35	3	3	3	3	3	3	3
X.10	759	55	-9	71	15	15	7	21	.	15	7	-9	-1	-1	7	7	11	-5	3	-1
X.11	770	770	770	-14	10	10	-14	5	-7	10	-14	-14	-14	10	2	2	-2	-2	-2	2
X.12	770	770	770	-14	10	10	-14	5	-7	10	-14	-14	-14	10	2	2	-2	-2	-2	2
X.13	990	990	990	-18	-10	-10	-18	.	3	-10	-18	-18	-18	-10	6	6	2	2	2	6
X.14	990	990	990	-18	-10	-10	-18	.	3	-10	-18	-18	-18	-10	6	6	2	2	2	6
X.15	1035	1035	1035	27	35	35	27	.	6	35	27	27	27	35	3	3	-1	-1	-1	3
X.16	1035	1035	1035	-21	-5	-5	-21	.	-3	-5	-21	-21	-21	-5	3	3	3	3	3	3
X.17	1035	1035	1035	-21	-5	-5	-21	.	-3	-5	-21	-21	-21	-5	3	3	3	3	3	3
X.18	1265	1265	1265	49	-15	-15	49	5	8	-15	49	49	49	-15	-7	-7	1	1	1	-7
X.19	1288	-56	8	56	48	-16	-8	10	7	-16	-8	8	.	.	8	-8	4	4	-4	.
X.20	1288	-56	8	56	-16	48	-8	10	7	-16	-8	8	.	.	-8	8	4	4	-4	.
X.21	1771	1771	1771	-21	11	11	-21	16	7	11	-21	-21	-21	11	3	3	-5	-5	-5	3
X.22	2024	2024	2024	8	24	24	8	-1	8	24	8	8	8	24	8	8	.	.	.	8
X.23	2277	2277	2277	21	-19	-19	21	.	6	-19	21	21	21	-19	-3	-3	1	1	1	-3
X.24	3312	3312	3312	48	16	16	48	.	-6	16	48	48	48	16	.	.	.	.	.	.
X.25	3520	3520	3520	64	.	.	64	10	-8	.	64	64	64	.	.	.	.	.	.	.
X.26	5313	385	-63	161	-15	-15	-31	39	.	-15	33	-15	-7	1	-7	-7	9	-7	1	1
X.27	5313	5313	5313	49	9	9	49	-15	.	9	49	49	49	9	1	1	-3	-3	-3	1
X.28	5544	5544	5544	-56	24	24	-56	9	.	24	-56	-56	-56	24	-8	-8	.	.	.	-8
X.29	5796	5796	5796	-28	36	36	-28	-9	.	36	-28	-28	-28	36	-4	-4	4	4	4	-4
X.30	10395	10395	10395	-21	-45	-45	-21	.	.	-45	-21	-21	-21	-45	3	3	-1	-1	-1	3
X.31	10626	770	-126	210	90	90	82	24	.	90	18	-14	-14	-6	18	18	6	6	6	2
X.32	11385	825	-135	265	-15	-15	73	45	.	-15	9	-39	1	1	-7	-7	13	-3	5	1
X.33	15180	1100	-180	300	60	60	44	15	.	60	44	-20	-20	-4	4	4	4	4	4	4
X.34	15939	1155	-189	35	-45	-45	-29	36	.	-45	35	19	-21	3	3	3	-5	11	3	-5
X.35	15939	1155	-189	35	-45	-45	-29	-18	.	-45	35	19	-21	3	3	3	-5	11	3	-5
X.36	15939	1155	-189	35	-45	-45	-29	-18	.	-45	35	19	-21	3	3	3	-5	11	3	-5
X.37	21252	1540	-252	196	-60	-60	-60	21	.	-60	68	4	-28	4	-4	-4	4	4	4	-4
X.38	26565	1925	-315	-203	45	45	117	60	.	45	-11	69	-35	-3	-3	-3	-15	1	-7	5
X.39	28336	-1232	176	336	-32	-32	-48	40	-14	32	-48	48	.	.	.	.	8	8	-8	.
X.40	34155	2475	-405	-165	-45	-45	27	.	.	-45	27	75	-45	3	3	3	7	-9	-1	-5
X.41	34155	2475	-405	-213	75	75	-21	.	.	75	-21	27	3	-5	-21	-21	11	-5	3	3
X.42	34155	2475	-405	-213	75	75	-21	.	.	75	-21	27	3	-5	-21	-21	11	-5	3	3
X.43	34155	2475	-405	-165	-45	-45	27	.	.	-45	27	75	-45	3	3	3	7	-9	-1	-5
X.44	41216	-1792	256	.	128	128	.	-40	14	-128	.	.	.	.	.	.	.	.	.	.
X.45	42504	3080	-504	168	120	120	168	-39	.	120	40	40	-56	-8	8	8	.	.	.	8
X.46	48576	3520	-576	64	.	.	64	-6	.	.	64	64	-64	.	.	.	.	.	.	.
X.47	53130	3850	-630	154	-30	-30	26	-15	.	-30	90	58	-70	2	-14	-14	-2	-2	-2	2
X.48	57960	-2520	360	-168	-80	240	24	.	21	-80	24	-24	.	.	24	-24	4	4	-4	.
X.49	57960	-2520	360	-168	240	-80	24	.	21	-80	24	-24	.	.	-24	24	4	4	-4	.
X.50	68310	4950	-810	582	30	30	198	.	.	30	6	-90	6	-2	6	6	2	2	2	-10
X.51	69552	-3024	432	336	96	96	-48	.	.	-96	-48	48	.	.	.	.	8	8	-8	.
X.52	69552	-3024	432	336	96	96	-48	.	.	-96	-48	48	.	.	.	.	8	8	-8	.
X.53	70840	-3080	440	392	-240	80	-56	10	7	80	-56	56	.	.	-8	8	-4	-4	4	.
X.54	70840	-3080	440	392	80	-240	-56	10	7	80	-56	56	.	.	8	-8	-4	-4	4	.
X.55	79695	5775	-945	399	15	15	-177	45	.	15	79	-65	7	-1	7	7	-1	-17	-9	-1
X.56	79695	5775	-945	-161	135	135	31	45	.	135	-33	15	7	-9	-1	-1	-13	3	-5	-9
X.57	79695	5775	-945	63	-105	-105	255	45	.	-105	-65	-17	7	7	-17	-17	-5	11	3	7
X.58	85008	-3696	528	112	288	-96	-16	30	.	-96	-16	16	.	.	16	-16	-8	-8	8	.
X.59	85008	-3696	528	112	-96	288	-16	30	.	-96	-16	16	.	.	-16	16	-8	-8	8	.
X.60	91080	6600	-1080	104	120	120	104	-45	.	120	-24	-24	8	-8	-8	-8	.	.	.	-8
X.61	127512	-5544	792	168	16	-48	-24	.	21	16	-24	24	.	.	-24	24	-4	-4	4	.
X.62	127512	-5544	792	168	-48	16	-24	.	21	16	-24	24	.	.	24	-24	-4	-4	4	.
X.63	141680	-6160	880	-112	-160	-160	16	20	14	160	16	-16	.	.	.	.	8	8	-8	.
X.64	154560	-6720	960	-448	.	.	64	30	.	.	64	-64	.	.	.	.	.	.	.	.
X.65	154560	-6720	960	-448	.	.	64	30	.	.	64	-64	.	.	.	.	.	.	.	.
X.66	159390	11550	-1890	462	-90	-90	78	-45	.	-90	14	-82	14	6	-10	-10	-6	-6	-6	6
X.67	185472	-8064	1152	.	192	-64	.	.	-21	-64	.	.	.	.	.	.	.	.	.	.
X.68	185472	-8064	1152	.	-64	192	.	.	-21	-64	.	.	.	.	.	.	.	.	.	.
X.69	226688	-9856	1408	.	-192	64	.	-40	-7	64	.	.	.	.	.	.	.	.	.	.
X.70	226688	-9856	1408	.	64	-192	.	-40	-7	64	.	.	.	.	.	.	.	.	.	.
X.71	239085	17325	-2835	-147	45	45	-339	.	.	45	45	-3	21	-3	-3	-3	-3	13	5	5
X.72	239085	17325	-2835	-483	-75	-75	93	.	.	-75	-99	45	21	5	21	21	9	-7	1	-3

Character table of $E = N_G(A) = \langle x, y, e\rangle$ *(continued)*

	4l	4m	5a	6a	6b	6c	6d	6e	6f	6g	6h	7a	7b	8a	8b	8c	8d	8e	8f	10a	10b	10c	10d	11a	12a	12b	12c
2	.	.	.	.	.	.	.	.	.	.	.	.	.	.	.	.	.	.	.	.	.	.	.	1	.	.	.
3	.	.	.	.	.	.	.	.	.	.	.	.	.	.	.	.	.	.	.	.	.	.	.	.	.	.	.
5	8	8	5	8	8	8	6	6	6	5	5	3	3	8	9	8	7	6	6	5	4	4	4	1	6	6	4
7	1	1	1	3	2	2	2	1	1	1	1	1	1	1	.	.	.	.	.	1	.	.	.	.	1	1	1
11	.	.	1	1	.	.	.	.	.	.	.	.	.	.	.	.	.	.	.	1	1	1	1	.	.	.	.
23	.	.	.	.	.	.	.	.	.	.	.	1	1	.	.	.	.	.	.	.	.	.	.	.	.	.	.
	4l	4m	5a	6a	6b	6c	6d	6e	6f	6g	6h	7a	7b	8a	8b	8c	8d	8e	8f	10a	10b	10c	10d	11a	12a	12b	12c
2P	2d	2e	5a	3a	3a	3a	3b	3a	3a	3b	3b	7a	7b	4c	4c	4b	4e	4h	4i	5a	5a	5a	5a	11a	6b	6b	6d
3P	4l	4m	5a	2b	2b	2a	2b	2c	2f	2d	2e	7b	7a	8a	8b	8c	8d	8e	8f	10a	10b	10c	10d	11a	4b	4c	4a
5P	4l	4m	1a	6a	6b	6c	6d	6e	6f	6g	6h	7b	7a	8a	8b	8c	8d	8e	8f	2b	2d	2e	2a	11a	12a	12b	12c
7P	4l	4m	5a	6a	6b	6c	6d	6e	6f	6g	6h	1a	1a	8a	8b	8c	8d	8e	8f	10a	10b	10c	10d	11a	12a	12b	12c
11P	4l	4m	5a	6a	6b	6c	6d	6e	6f	6g	6h	7a	7b	8a	8b	8c	8d	8e	8f	10a	10b	10c	10d	1a	12a	12b	12c
23P	4l	4m	5a	6a	6b	6c	6d	6e	6f	6g	6h	7a	7b	8a	8b	8c	8d	8e	8f	10a	10b	10c	10d	11a	12a	12b	12c
X.1	1	1	1	1	1	1	1	1	1	1	1	1	1	1	1	1	1	1	1	1	1	1	1	1	1	1	1
X.2	−1	−1	3	5	5	5	−1	1	1	−1	−1	2	2	−1	3	3	−1	1	1	3	−1	−1	3	1	1	1	−1
X.3	1	1	.	.	.	.	3	.	.	−1	−1	A	$\bar{A}$	−3	1	1	1	−1	−1	.	.	.	.	1	.	.	−1
X.4	1	1	.	.	.	.	3	.	.	−1	−1	$\bar{A}$	A	−3	1	1	1	−1	−1	.	.	.	.	1	.	.	−1
X.5	3	3	1	−3	−3	−3	.	1	1	.	.	.	.	−1	−1	−1	3	−1	−1	1	1	1	1	.	1	1	.
X.6	3	3	1	−3	−3	−3	.	1	1	.	.	.	.	−1	−1	−1	3	−1	−1	1	1	1	1	.	1	1	.
X.7	.	.	2	9	9	9	.	1	1	.	.	.	.	4	4	4	.	.	.	2	2	2	2	−1	1	1	.
X.8	1	1	3	10	10	10	1	−2	−2	1	1	1	1	−3	1	1	1	−1	−1	3	−1	−1	3	.	−2	−2	1
X.9	3	3	−2	6	6	6	.	2	2	.	.	.	.	3	3	3	3	−1	−1	−2	−2	−2	−2	−1	2	2	.
X.10	3	3	4	9	−3	1	.	5	1	.	.	3	3	−1	−1	−1	−1	1	1	−4	.	.	.	.	1	−3	.
X.11	−2	−2	.	5	5	5	−7	1	1	1	1	.	.	2	−2	−2	−2	.	.	.	.	.	.	.	1	1	1
X.12	−2	−2	.	5	5	5	−7	1	1	1	1	.	.	2	−2	−2	−2	.	.	.	.	.	.	.	1	1	1
X.13	−2	−2	.	.	.	.	3	.	.	−1	−1	A	$\bar{A}$	6	2	2	−2	.	.	.	.	.	.	.	.	.	−1
X.14	−2	−2	.	.	.	.	3	.	.	−1	−1	$\bar{A}$	A	6	2	2	−2	.	.	.	.	.	.	.	.	.	−1
X.15	3	3	.	.	.	.	6	.	.	2	2	−1	−1	3	−1	−1	3	1	1	.	.	.	.	1	.	.	2
X.16	−1	−1	.	.	.	.	−3	.	.	1	1	D	$\bar{D}$	3	3	3	−1	−1	−1	.	.	.	.	1	.	.	1
X.17	−1	−1	.	.	.	.	−3	.	.	1	1	$\bar{D}$	D	3	3	3	−1	−1	−1	.	.	.	.	1	.	.	1
X.18	−3	−3	.	5	5	5	8	1	1	.	.	−2	−2	−7	1	1	−3	1	1	.	.	.	.	.	1	1	.
X.19	4	−4	3	−10	2	−2	−1	2	−2	3	−1	.	.	.	.	.	.	2	−2	3	3	−1	−1	1	−2	2	−1
X.20	−4	4	3	−10	2	−2	−1	2	−2	−1	3	.	.	.	.	.	.	2	−2	3	−1	3	−1	1	−2	2	−1
X.21	−1	−1	1	16	16	16	7	.	.	−1	−1	.	.	3	−5	−5	−1	−1	−1	1	1	1	1	.	.	.	−1
X.22	.	.	−1	−1	−1	−1	8	−1	−1	.	.	1	1	8	.	.	.	.	.	−1	−1	−1	−1	.	−1	−1	.
X.23	−3	−3	−3	.	.	.	6	.	.	2	2	2	2	−3	1	1	−3	−1	−1	−3	1	1	−3	.	.	.	2
X.24	.	.	−3	.	.	.	−6	.	.	−2	−2	1	1	.	.	.	.	.	.	−3	1	1	−3	1	.	.	−2
X.25	.	.	.	10	10	10	−8	−2	−2	.	.	−1	−1	.	.	.	.	.	.	.	.	.	.	.	−2	−2	.
X.26	−3	−3	8	−9	−9	7	.	−1	−1	.	.	.	.	1	5	−3	1	1	1	−8	.	.	.	.	3	3	.
X.27	−3	−3	3	−15	−15	−15	.	1	1	.	.	.	.	1	−3	−3	−3	−1	−1	3	−1	−1	3	.	1	1	.
X.28	.	.	−1	9	9	9	.	1	1	.	.	.	.	−8	.	.	.	.	.	−1	−1	−1	−1	.	1	1	.
X.29	.	.	1	−9	−9	−9	.	−1	−1	.	.	.	.	−4	4	4	.	.	.	1	1	1	1	−1	−1	−1	.
X.30	3	3	.	.	.	.	.	.	.	.	.	.	.	3	−1	−1	3	1	1	.	.	.	.	.	.	.	.
X.31	6	6	−4	36	.	−4	.	.	4	.	.	.	.	−6	−2	−2	−2	.	.	4	.	.	.	.	.	4	.
X.32	−3	−3	.	45	−3	−3	.	1	1	.	.	3	3	1	−7	1	1	−1	−1	.	.	.	.	.	−3	−3	.
X.33	.	.	.	−45	−9	11	.	3	−1	.	.	−3	−3	−4	4	−4	.	.	.	.	.	.	.	.	−1	−5	.
X.34	3	3	4	−36	−12	12	.	−4	4	.	.	.	.	3	−1	−1	−1	−1	−1	−4	.	.	.	.	−4	4	.
X.35	3	3	4	18	6	−6	.	2	−2	.	.	.	.	3	−1	−1	−1	−1	−1	−4	.	.	.	.	2	−2	.
X.36	3	3	4	18	6	−6	.	2	−2	.	.	.	.	3	−1	−1	−1	−1	−1	−4	.	.	.	.	2	−2	.
X.37	.	.	−8	9	−3	1	.	1	−3	.	.	.	.	4	4	−4	.	.	.	8	.	.	.	.	5	1	.
X.38	−3	−3	.	.	−12	8	.	4	.	.	.	.	.	−3	−3	5	1	−1	−1	.	.	.	.	.	4	.	.
X.39	.	.	6	−40	8	−8	2	.	.	−2	−2	.	.	.	.	.	.	.	.	6	−2	−2	−2	.	.	.	2
X.40	3	3	.	.	.	.	.	.	.	.	.	E	$\bar{E}$	3	−5	3	−1	1	1	.	.	.	.	.	.	.	.
X.41	3	3	.	.	.	.	.	.	.	.	.	$\bar{E}$	E	3	−1	−1	−1	−1	−1	.	.	.	.	.	.	.	.
X.42	3	3	.	.	.	.	.	.	.	.	.	E	$\bar{E}$	3	−1	−1	−1	−1	−1	.	.	.	.	.	.	.	.
X.43	3	3	.	.	.	.	.	.	.	.	.	$\bar{E}$	E	3	−5	3	−1	1	1	.	.	.	.	.	.	.	.
X.44	.	.	6	40	−8	8	−2	.	.	2	2	.	.	.	.	.	.	.	.	6	−2	−2	−2	−1	.	.	−2
X.45	.	.	4	9	9	−7	.	−3	−3	.	.	.	.	−8	.	.	.	.	.	−4	.	.	.	.	1	1	.
X.46	.	.	−4	−54	−6	10	.	−2	−2	.	.	3	3	.	.	.	.	.	.	4	.	.	.	.	−2	−2	.
X.47	−6	−6	.	45	9	−11	.	1	5	.	.	.	.	2	6	−2	2	.	.	.	.	.	.	.	−3	1	.
X.48	−4	4	.	.	.	.	−3	.	.	1	−3	.	.	.	.	.	.	−2	2	.	.	.	.	1	.	.	1
X.49	4	−4	.	.	.	.	−3	.	.	−3	1	.	.	.	.	.	.	−2	2	.	.	.	.	1	.	.	1
X.50	−6	−6	.	.	.	.	.	.	.	.	.	−3	−3	6	−6	2	2	.	.	.	.	.	.	.	.	.	.
X.51	.	.	−3	.	.	.	.	.	.	.	.	.	.	.	.	.	.	.	.	−3	1	1	1	−1	.	.	.
X.52	.	.	−3	.	.	.	.	.	.	.	.	.	.	.	.	.	.	.	.	−3	1	1	1	−1	.	.	.
X.53	4	−4	.	−10	2	−2	−1	2	−2	3	−1	.	.	.	.	.	.	−2	2	.	.	.	.	.	−2	2	−1
X.54	−4	4	.	−10	2	−2	−1	2	−2	−1	3	.	.	.	.	.	.	−2	2	.	.	.	.	.	−2	2	−1
X.55	3	3	.	45	−3	−3	.	−3	−3	.	.	.	.	−1	3	3	−1	−1	−1	.	.	.	.	.	1	1	.
X.56	3	3	.	45	−3	−3	.	1	1	.	.	.	.	7	7	−1	−1	1	1	.	.	.	.	.	−3	−3	.
X.57	3	3	.	45	−3	−3	.	−3	−3	.	.	.	.	−1	−1	−1	−1	1	1	.	.	.	.	.	1	1	.
X.58	.	.	3	−30	6	−6	.	−2	2	.	.	.	.	.	.	.	.	.	.	3	3	−1	−1	.	2	−2	.
X.59	.	.	3	−30	6	−6	.	−2	2	.	.	.	.	.	.	.	.	.	.	3	−1	3	−1	.	2	−2	.
X.60	.	.	.	−45	3	3	.	−1	−1	.	.	3	3	8	.	.	.	.	.	.	.	.	.	.	3	3	.
X.61	−4	4	−3	.	.	.	−3	.	.	1	−3	.	.	.	.	.	.	2	−2	−3	1	−3	1	.	.	.	1
X.62	4	−4	−3	.	.	.	−3	.	.	−3	1	.	.	.	.	.	.	2	−2	−3	−3	1	1	.	.	.	1
X.63	.	.	.	−20	4	−4	−2	−4	4	2	2	.	.	.	.	.	.	.	.	.	.	.	.	.	4	−4	−2
X.64	.	.	.	−30	6	−6	.	2	−2	.	.	.	.	.	.	.	.	.	.	.	.	.	.	−1	−2	2	.
X.65	.	.	.	−30	6	−6	.	2	−2	.	.	.	.	.	.	.	.	.	.	.	.	.	.	−1	−2	2	.
X.66	6	6	.	−45	3	3	.	3	3	.	.	.	.	−2	2	2	−2	.	.	.	.	.	.	.	−1	−1	.
X.67	8	−8	−3	.	.	.	3	.	.	3	−1	.	.	.	.	.	.	.	.	−3	−3	1	1	1	.	.	−1
X.68	−8	8	−3	.	.	.	3	.	.	−1	3	.	.	.	.	.	.	.	.	−3	1	−3	1	1	.	.	−1
X.69	8	−8	3	40	−8	8	1	.	.	−3	1	.	.	.	.	.	.	.	.	3	3	−1	−1	.	.	.	1
X.70	−8	8	3	40	−8	8	1	.	.	1	−3	.	.	.	.	.	.	.	.	3	−1	3	−1	.	.	.	1
X.71	−3	−3	.	.	.	.	.	.	.	.	.	.	.	−3	−7	1	1	1	1	.	.	.	.	.	.	.	.
X.72	−3	−3	.	.	.	.	.	.	.	.	.	.	.	−3	5	−3	1	−1	−1	.	.	.	.	.	.	.	.

Character table of $E = N_G(A) = \langle x, y, e \rangle$ (continued)

	12d	12e	12f	12g	12h	14a	14b	14c	14d	15a	15b	16a	20a	20b	21a	21b	22a	23a	23b	24a	24b	28a	28b	30a	30b
2	.	.	.	.	.	.	.	.	.	.	.	.	.	.	.	.	1	.	.	.	.	.	.	.	.
3	.	.	.	.	.	.	.	.	.	.	.	.	.	.	.	.	.	1	1	.	.	.	.	.	.
5	4	4	4	3	3	3	3	2	2	1	1	5	4	4	.	.	1	.	.	4	4	2	2	1	1
7	1	1	1	1	1	.	.	.	.	1	1	.	.	.	1	1	.	.	.	1	1	.	.	1	1
11	.	.	.	.	.	.	.	.	.	1	1	.	1	1	.	.	.	.	.	.	.	.	.	1	1
23	.	.	.	.	.	1	1	1	1	.	.	.	.	.	1	1	.	.	.	.	.	1	1	.	.
2P	6e	6c	6e	6g	6h	7a	7b	7a	7b	15a	15b	8b	10a	10a	21a	21b	11a	23a	23b	12b	12b	14a	14b	15a	15b
3P	4f	4d	4g	4l	4m	14b	14a	14d	14c	5a	5a	16a	20b	20a	7a	7b	22a	23a	23b	8a	8a	28b	28a	10a	10a
5P	12d	12e	12f	12g	12h	14b	14a	14d	14c	3a	3a	16a	4a	4a	21b	21a	22a	23b	23a	24b	24a	28b	28a	6a	6a
7P	12d	12e	12f	12g	12h	2a	2a	2c	2c	15b	15a	16a	20b	20a	3b	3b	22a	23b	23a	24b	24a	4d	4d	30b	30a
11P	12d	12e	12f	12g	12h	14a	14b	14c	14d	15b	15a	16a	20a	20b	21a	21b	2a	23b	23a	24a	24b	28a	28b	30b	30a
23P	12d	12e	12f	12g	12h	14a	14b	14c	14d	15a	15b	16a	20b	20a	21a	21b	22a	1a	1a	24a	24b	28a	28b	30a	30b
X.1	1	1	1	1	1	1	1	1	1	1	1	1	1	1	1	1	1	1	1	1	1	1	1	1	1
X.2	−1	1	−1	−1	−1	2	2	.	.	.	.	1	−1	−1	−1	−1	1	.	.	−1	−1	.	.	.	.
X.3	.	.	.	1	1	A	$\bar{A}$	$-A$	$-\bar{A}$	.	.	−1	.	.	$\bar{A}$	A	1	−1	−1	.	.	$-A$	$-\bar{A}$	.	.
X.4	.	.	.	1	1	$\bar{A}$	A	$-\bar{A}$	$-A$	.	.	−1	.	.	A	$\bar{A}$	1	−1	−1	.	.	$-\bar{A}$	$-A$	.	.
X.5	−1	1	−1	.	.	.	.	.	.	B	$\bar{B}$	−1	1	1	.	.	.	1	1	−1	−1	.	.	B	$\bar{B}$
X.6	−1	1	−1	.	.	.	.	.	.	$\bar{B}$	B	−1	1	1	.	.	.	1	1	−1	−1	.	.	$\bar{B}$	B
X.7	1	1	1	.	.	.	.	.	.	−1	−1	.	2	2	.	.	−1	−1	−1	1	1	.	.	−1	−1
X.8	.	−2	.	1	1	1	1	−1	−1	.	.	−1	−1	−1	1	1	.	.	.	.	.	−1	−1	.	.
X.9	.	2	.	.	.	.	.	.	.	1	1	−1	−2	−2	.	.	−1	.	.	.	.	.	.	1	1
X.10	1	−1	1	.	.	−1	−1	1	1	1	1	−1	.	.	.	.	.	.	.	−1	−1	−1	−1	−1	−1
X.11	−1	1	−1	1	1	.	.	.	.	.	.	.	.	.	.	.	.	C	$\bar{C}$	−1	−1	.	.	.	.
X.12	−1	1	−1	1	1	.	.	.	.	.	.	.	.	.	.	.	.	$\bar{C}$	C	−1	−1	.	.	.	.
X.13	.	.	.	1	1	A	$\bar{A}$	A	$\bar{A}$	.	.	.	.	.	$\bar{A}$	A	.	1	1	.	.	A	$\bar{A}$	.	.
X.14	.	.	.	1	1	$\bar{A}$	A	$\bar{A}$	A	.	.	.	.	.	A	$\bar{A}$	.	1	1	.	.	$\bar{A}$	A	.	.
X.15	.	.	.	.	.	−1	−1	−1	−1	.	.	1	.	.	−1	−1	1	.	.	.	.	−1	−1	.	.
X.16	.	.	.	−1	−1	D	$\bar{D}$	.	.	.	.	−1	.	.	$-A$	$-\bar{A}$	1	.	.	.	.	.	.	.	.
X.17	.	.	.	−1	−1	$\bar{D}$	D	.	.	.	.	−1	.	.	$-\bar{A}$	$-A$	1	.	.	.	.	.	.	.	.
X.18	−1	1	−1	.	.	−2	−2	.	.	.	.	1	.	.	1	1	.	.	.	−1	−1	.	.	.	.
X.19	2	.	−2	1	−1	.	.	.	.	.	.	.	−1	−1	.	.	−1	.	.	.	.	.	.	.	.
X.20	−2	.	2	−1	1	.	.	.	.	.	.	.	−1	−1	.	.	−1	.	.	.	.	.	.	.	.
X.21	.	.	.	−1	−1	.	.	.	.	1	1	−1	1	1	.	.	.	.	.	.	.	.	.	1	1
X.22	−1	−1	−1	.	.	1	1	1	1	−1	−1	.	−1	−1	1	1	.	.	.	−1	−1	1	1	−1	−1
X.23	.	.	.	.	.	2	2	.	.	.	.	−1	1	1	−1	−1	.	.	.	.	.	.	.	.	.
X.24	.	.	.	.	.	1	1	−1	−1	.	.	.	1	1	1	1	1	.	.	.	.	−1	−1	.	.
X.25	.	−2	.	.	.	−1	−1	1	1	.	.	.	.	.	−1	−1	.	1	1	.	.	1	1	.	.
X.26	−1	−1	−1	.	.	.	.	.	.	−1	−1	−1	.	.	.	.	.	.	.	1	1	.	.	1	1
X.27	1	1	1	.	.	.	.	.	.	.	.	−1	−1	−1	.	.	.	.	.	1	1	.	.	.	.
X.28	1	1	1	.	.	.	.	.	.	−1	−1	.	−1	−1	.	.	.	1	1	1	1	.	.	−1	−1
X.29	−1	−1	−1	.	.	.	.	.	.	1	1	.	1	1	.	.	−1	.	.	−1	−1	.	.	1	1
X.30	.	.	.	.	.	.	.	.	.	.	.	1	.	.	.	.	.	−1	−1	.	.	.	.	.	.
X.31	.	−2	.	.	.	.	.	.	.	−1	−1	.	.	.	.	.	.	.	.	.	.	.	.	1	1
X.32	−1	1	−1	.	.	−1	−1	−1	−1	.	.	1	.	.	.	.	.	.	.	1	1	1	1	.	.
X.33	1	1	1	.	.	1	1	−1	−1	.	.	.	.	.	.	.	.	.	.	−1	−1	1	1	.	.
X.34	.	.	.	.	.	.	.	.	.	1	1	1	.	.	.	.	.	.	.	.	.	.	.	−1	−1
X.35	.	.	.	.	.	.	.	.	.	$\bar{B}$	B	1	.	.	.	.	.	.	.	.	.	.	.	$-\bar{B}$	$-B$
X.36	.	.	.	.	.	.	.	.	.	B	$\bar{B}$	1	.	.	.	.	.	.	.	.	.	.	.	$-B$	$-\bar{B}$
X.37	−1	−1	−1	.	.	.	.	.	.	1	1	.	.	.	.	.	.	.	.	1	1	.	.	−1	−1
X.38	.	−2	.	.	.	.	.	.	.	.	.	1	.	.	.	.	.	.	.	.	.	.	.	.	.
X.39	.	.	.	.	.	.	.	.	.	.	.	.	2	2	.	.	.	.	.	.	.	.	.	.	.
X.40	.	.	.	.	.	$-A$	$-\bar{A}$	A	$\bar{A}$	.	.	−1	.	.	.	.	.	.	.	.	.	$-A$	$-\bar{A}$	.	.
X.41	.	.	.	.	.	$-\bar{A}$	$-A$	$-\bar{A}$	$-A$	.	.	1	.	.	.	.	.	.	.	.	.	$\bar{A}$	A	.	.
X.42	.	.	.	.	.	$-A$	$-\bar{A}$	$-A$	$-\bar{A}$	.	.	1	.	.	.	.	.	.	.	.	.	A	$\bar{A}$	.	.
X.43	.	.	.	.	.	$-\bar{A}$	$-A$	$\bar{A}$	A	.	.	−1	.	.	.	.	.	.	.	.	.	$-\bar{A}$	$-A$	.	.
X.44	.	.	.	.	.	.	.	.	.	.	.	.	2	2	.	.	1	.	.	.	.	.	.	.	.
X.45	−1	1	−1	.	.	.	.	.	.	1	1	.	.	.	.	.	.	.	.	1	1	.	.	−1	−1
X.46	.	2	.	.	.	−1	−1	1	1	−1	−1	.	.	.	.	.	.	.	.	.	.	−1	−1	1	1
X.47	1	−1	1	.	.	.	.	.	.	.	.	.	.	.	.	.	.	.	.	−1	−1	.	.	.	.
X.48	.	.	.	−1	1	.	.	.	.	.	.	.	.	.	.	.	−1	.	.	.	.	.	.	.	.
X.49	.	.	.	1	−1	.	.	.	.	.	.	.	.	.	.	.	−1	.	.	.	.	.	.	.	.
X.50	.	.	.	.	.	1	1	1	1	.	.	.	.	.	.	.	.	.	.	.	.	−1	−1	.	.
X.51	.	.	.	.	.	.	.	.	.	.	.	.	F	G	.	.	1	.	.	.	.	.	.	.	.
X.52	.	.	.	.	.	.	.	.	.	.	.	.	G	F	.	.	1	.	.	.	.	.	.	.	.
X.53	−2	.	2	1	−1	.	.	.	.	.	.	.	.	.	.	.	.	.	.	.	.	.	.	.	.
X.54	2	.	−2	−1	1	.	.	.	.	.	.	.	.	.	.	.	.	.	.	.	.	.	.	.	.
X.55	1	1	1	.	.	.	.	.	.	.	.	1	.	.	.	.	.	.	.	−1	−1	.	.	.	.
X.56	−1	1	−1	.	.	.	.	.	.	.	.	−1	.	.	.	.	.	.	.	1	1	.	.	.	.
X.57	1	1	1	.	.	.	.	.	.	.	.	−1	.	.	.	.	.	.	.	−1	−1	.	.	.	.
X.58	−2	.	2	.	.	.	.	.	.	.	.	.	−1	−1	.	.	.	.	.	.	.	.	.	.	.
X.59	2	.	−2	.	.	.	.	.	.	.	.	.	−1	−1	.	.	.	.	.	.	.	.	.	.	.
X.60	1	−1	1	.	.	−1	−1	−1	−1	.	.	.	.	.	.	.	.	.	.	−1	−1	1	1	.	.
X.61	.	.	.	−1	1	.	.	.	.	.	.	.	1	1	.	.	.	.	.	.	.	.	.	.	.
X.62	.	.	.	1	−1	.	.	.	.	.	.	.	1	1	.	.	.	.	.	.	.	.	.	.	.
X.63	.	.	.	.	.	.	.	.	.	.	.	.	.	.	.	.	.	.	.	.	.	.	.	.	.
X.64	.	.	.	.	.	.	.	.	.	.	.	.	.	.	.	.	1	.	.	H	$-H$	.	.	.	.
X.65	.	.	.	.	.	.	.	.	.	.	.	.	.	.	.	.	1	.	.	$-H$	H	.	.	.	.
X.66	−1	−1	−1	.	.	.	.	.	.	.	.	.	.	.	.	.	.	.	.	1	1	.	.	.	.
X.67	.	.	.	−1	1	.	.	.	.	.	.	.	1	1	.	.	−1	.	.	.	.	.	.	.	.
X.68	.	.	.	1	−1	.	.	.	.	.	.	.	1	1	.	.	−1	.	.	.	.	.	.	.	.
X.69	.	.	.	−1	1	.	.	.	.	.	.	.	−1	−1	.	.	.	.	.	.	.	.	.	.	.
X.70	.	.	.	1	−1	.	.	.	.	.	.	.	−1	−1	.	.	.	.	.	.	.	.	.	.	.
X.71	.	.	.	.	.	.	.	.	.	.	.	−1	.	.	.	.	.	.	.	.	.	.	.	.	.
X.72	.	.	.	.	.	.	.	.	.	.	.	1	.	.	.	.	.	.	.	.	.	.	.	.	.

where $A = -\zeta(7)^4 - \zeta(7)^2 - \zeta(7) - 1$, $B = 2\zeta(15)_3\zeta(15)_5^3 + 2\zeta(15)_3\zeta(15)_5^2 + \zeta(15)_3 + \zeta(15)_5^3 + \zeta(15)_5^2$, $C = -\zeta(23)^{18} - \zeta(23)^{16} - \zeta(23)^{13} - \zeta(23)^{12} - \zeta(23)^9 - \zeta(23)^8 - \zeta(23)^6 - \zeta(23)^4 - \zeta(23)^3 - \zeta(23)^2 - \zeta(23) - 1$, $D = 2\zeta(7)^4 + 2\zeta(7)^2 + 2\zeta(7)$, $E = -3\zeta(7)^4 - 3\zeta(7)^2 - 3\zeta(7) - 3$, $F = 4\zeta(5)^3 + 4\zeta(5)^2 + 1$, $G = -4\zeta(5)^3 - 4\zeta(5)^2 - 3$, $H = -4\zeta(12)_4\zeta(12)_3 - 2\zeta(12)_4$.

9
Fischer's simple group Fi_{24}'

This chapter describes H. Kim's and the author's recent construction [74] of Fischer's sporadic simple group Fi_{24}'. Originally we applied Algorithm 1.3.8 to the irreducible subgroup M_{24} of $\mathrm{GL}_{11}(2)$ and its non-split extension E_3 given in Lemma 7.1.1(g). We constructed a finitely presented group H containing a Sylow 2-subgroup S_1 having a maximal elementary abelian normal subgroup A such that $N_H(A) \cong D = C_{E_3}(z)$ and $|H : N_H(A)|$ is odd. In our application of Algorithm 7.4.8 of [92] we saw that the free product $Q = H *_D E$ has 939 080 024 064 non-isomorphic irreducible representations of minimal dimension 8671 over $GF(13)$. In view of our time constraints we decided not to perform all the necessary calculations of Step 5(c) of Algorithm 7.4.8 of [92] in order to find an irreducible representation of Q which has a Sylow 2-subgroup of order 2^{21}. However, the results of this chapter show that this experiment would have been successful and provided a self-contained existence proof of Fi_{24}' by means of Algorithm 1.3.8.

In [74] H. Kim and the author construct Fi_{24}' as an epimorphic image of the free product $P = G_1 *_{H_1} A_1$ of Fischer's simple group $G_1 \cong \mathsf{Fi}_{23}$ and the 2-fold cover A_1 of the automorphism group $Aut(\mathsf{Fi}_{22})$ of Fi_{22} with amalgamated subgroup $H_1 = A_1'$. For this construction described here, Chapters 5 and 6 provide all prerequisites.

In Section 9.1 the 2-fold cover A_1 of $Aut(\mathsf{Fi}_{22})$ is built as a finitely presented group; see Lemma 9.1.1. A suitable permutation representation PA_1 of A_1 has been used to calculate the character table of A_1 stated in Table 9.8.1. Lemma 9.1.2 states that the centralizer H_1 of a 2-central involution u of $G_1 \cong \mathsf{Fi}_{23}$ is isomorphic to the commutator subgroup A_1' of A_1. Furthermore, the amalgam $G_1 \leftarrow H_1 \rightarrow A_1$ has eight compatible pairs (σ_i, λ_i), $1 \le i \le 8$, of semi-simple characters of $G_1 \cong \mathsf{Fi}_{23}$ and A_1 of minimal degree 8671. The irreducible constituents of their restrictions to the subgroup H_1 are described by well determined characters of Table 6.5.2.

Each semi-simple character σ_i is a sum of the two irreducible characters τ_3 and τ_4 of G_1 of 13-defect zero and respective degrees 3588 and 5083. Sections 9.2 and 9.3 describe H. Kim's and the author's construction [74] of the irreducible representations $\mathfrak{V}_3$ and $\mathfrak{V}_4$ of G_1 over $GF(13)$ corresponding to τ_3 and τ_4, respectively. By Theorem 6.3.1(d) G_1 has a faithful permutation representation PG_1 of degree 31671 and four generators x, y, q and w such that $H_1 = \langle x, y, q \rangle$. It is used to find two subgroups mH and mE of G_1 such that $G_1 = \langle mH, mE \rangle$ and $mD = mH \cap mE \cong G_2(3) \times Sym(3)$. Lemma 9.2.1 provides short words u, v, r and s in terms of the four original generators of G_1 such that $mD = \langle u, v \rangle$, $mH = \langle mD, r \rangle$ and $mE = \langle mD, s \rangle$. By means of their character Tables `DVD.1.7.2` and `DVD.1.7.3` of the accompanying DVD it is proved in Lemma 9.2.1 that each of the irreducible constituents of the restrictions of τ_3 and τ_4 to mH and mE is an irreducible constituent of a permutation character of mH and mE, respectively. Therefore Algorithm 1.5.1 can be applied to construct the irreducible constituents of the restrictions of $\mathfrak{V}_3$ and $\mathfrak{V}_4$ to the subgroups mH and mE of G_1.

In order to get the irreducible representations of G_1 in $\mathrm{GL}_{3588}(13)$ and $\mathrm{GL}_{5083}(13)$ the amalgamation method of Step 5(c) of Algorithm 7.4.8 of [92] has to be applied. In [74] this problem is reduced to finding solutions of several large systems of not necessarily linear equations with coefficients in $GF(13)$. This method is described in the proofs of Propositions 9.2.2 and 9.2.3 providing the generating matrices for the original generators q, x, y and w of G_1 in $\mathrm{GL}_{8671}(13)$ as blocked diagonal matrices with diagonal blocks belonging to $\mathrm{GL}_k(13)$ for $k \in \{3588, 5083\}$.

Since $H_1 = \langle y, q \rangle$ is a normal subgroup of index 2 in A_1, Clifford's Theorem 2.6.15 of [92] is applied in Section 9.4 to construct the eight semi-simple representations of A_1 corresponding to the eight semi-simple characters λ_i of A_1. Thus one obtains eight matrices t_i in $\mathrm{GL}_{8671}(13)$ and so eight matrix groups $K_i = \langle q, y, w, t_i \rangle$. Remark 9.3.3 shows that only K_3 has a chance to have a Sylow 2-subgroup isomorphic to the one of the non-split extension E_3 of M_{24} by $V = F^{11}$ constructed in Lemma 7.1.1(g). The matrix subgroup $G = K_3 = \langle q, y, w, t \rangle$ with $t = t_3$ is built in Proposition 9.3.2.

Using the graph $\mathcal{G}_{24}$ of Fischer's 3-transposition group Fi_{24}, J. Hall and L. S. Soicher constructed a presentation P of this non-simple group. In Section 9.4 it is shown that the matrix group G is isomorphic to the commutator subgroup $P' \cong \mathsf{Fi}_{24}'$ of P; see Lemma 9.4.2. It is used to prove that the subgroup $G = \langle q, y, w, t\rangle$ of $\mathrm{GL}_{8671}(13)$ is a simple group of order $2^{21} \cdot 3^{16} \cdot 5^2 \cdot 7^3 \cdot 11 \cdot 13 \cdot 17 \cdot 23 \cdot 29$ having a faithful permutation representation PG of degree 306936 with stabilizer $G_1 = \langle q, y, w\rangle$; see Theorem 9.4.3. The permutation representation PG enabled Kim and the author to calculate the character table of G automatically with MAGMA in [74]. It agrees with the one of Fi_{24}' stated in the Atlas [19].

The main result of Section 9.5 provides Kim's and the author's presentation of the centralizer $H = C_G(z)$ of a 2-central involution z of G in terms of short words in the four generators of the matrix group G of [74]; see Proposition 9.5.1. Furthermore, it gives a faithful permutation representation PH of degree 287048 of H with a documented stabilizer. It has been used to calculate a system of representatives of its 167 conjugacy classes and its character table; see Tables 9.7.3 and 9.8.2, respectively.

A finite simple group G is said to be of Fi_{24}'*-type* if it has a 2-central involution z such that $C_G(z) \cong H$, where H is the finitely presented group of Proposition 9.5.1. In Section 9.6 it is shown that each simple group G of Fi_{24}'-type has order $2^{21} \cdot 3^{13} \cdot 5^2 \cdot 7 \cdot 11 \cdot 13 \cdot 17 \cdot 23 \cdot 29$; see Theorem 9.6.2. This is an important step in a uniqueness proof by means of Theorem 7.5.1 of [92], which we cannot complete here for lack of space.

Sections 9.7 and 9.8 contain the systems of representatives of conjugacy classes and the character tables of the local subgroups of the matrix groups G and H which have been used to construct them.

9.1 The 2-fold cover of the automorphism group $Aut(\mathsf{Fi}_{22})$

By Table 9.7.2 the involution $z_1 = e^2$ of $E = \langle x, y, e\rangle$ has a centralizer $D_1 = C_E(u)$ of order $2^{19} \cdot 3^2 \cdot 5 \cdot 7 \cdot 11$. It is the largest centralizer of any non-2-central involution of E. Applying Algorithm 1.3.8 to this group D_1 one obtains a group A_1 with a central involution z_1 and a Sylow 2-subgroup S_1 of order 2^{19} having a maximal elementary abelian normal subgroup A of order 2^{11} such that $N_{H_1}(A) \cong D_1$. Furthermore, the commutator subgroup $A_1' \cong 2\mathsf{Fi}_{22}$. This 2-fold cover of Fischer's simple group Fi_{22} is isomorphic to the centralizer H_1 of a 2-central involution z_2 of Fischer's simple group Fi_{23} by Proposition 6.2.3. In view of the limited space we derive here only a presentation of the group A_1 from Kim's presentation of H_1 given in Proposition 6.2.3 and the presentation of $Aut(\mathsf{Fi}_{22})$ given in [110], p. 111.

Theorem 6.3.1 states that the simple group $G_1 \cong \mathsf{Fi}_{23}$ has a faithful permutation representation PG_1 of degree 31671. Using it and the character tables of G_1 and A_1 we also show here that the amalgam $A_1 \leftarrow H_1 \rightarrow G_1$ has eight compatible pairs of semi-simple characters of degree 8671.

Lemma 9.1.1 *Let A_1 be the 2-fold cover of the automorphism group $Aut(\mathsf{Fi}_{22})$ of Fischer's simple group Fi_{22} and let $H_1 = A_1'$ be its derived subgroup. Then the following assertions hold.*

(a) *$H_1 = \langle a, b, c, d, e, f, g, h, i, z\rangle$ has the following set $\mathcal{R}(H_1)$ of defining relations:*

$$\begin{aligned}
&a^2 = b^2 = c^2 = d^2 = e^2 = f^2 = g^2 = h^2 = i^2 = z^2 = 1,\\
&(z,a) = (z,b) = (z,c) = (z,d) = (z,e) = (z,f) = (z,g) = 1,\\
&(z,h) = (z,i) = 1, \quad (ab)^3 = 1, \quad (bc)^3 = z, \quad (cd)^3 = (de)^3 = 1,\\
&(ef)^3 = (fg)^3 = z, (ac)^2 = (ad)^2 = (ae)^2 = (af)^2 = (ag)^2 = 1,\\
&(ah)^2 = (ai)^2 = (bd)^2 = (be)^2 = (bf)^2 = (bg)^2 = (bh)^2 = 1,\\
&(bi)^2 = (ce)^2 = (cf)^2 = (cg)^2 = (ch)^2 = (ci)^2 = 1,\\
&(df)^2 = (dg)^2 = (eg)^2 = (eh)^2 = (ei)^2 = 1,\\
&(dh)^3 = (hi)^3 = (di)^2 = (fh)^2 = (fi)^2 = (gh)^2 = (gi)^2 = 1,\\
&(dcbdefdhi)^{10} = (abcdefh)^9 = (bcdefgh)^9 = 1.
\end{aligned}$$

(b) *$A_1 = \langle H_1, t\rangle$ has a set $\mathcal{R}(A_1)$ of defining relations consisting of $\mathcal{R}(H_1)$ and the following relations: $t^2 = 1$, $(z,t) = a^t g = 1$, $b^t f = c^t e = (dt)^2 = (ht)^2 = (it)^2 = z$.*

(c) *$A_1 \cong 2Aut(\mathsf{Fi}_{22})$ has a faithful permutation representation PA_1 of degree 56320 with stabilizer $\langle bz, c, d, e, fz, g, h, i\rangle$.*

(d) *A system of representatives a_i of the 150 conjugacy classes of A_1 and the corresponding centralizer orders $|C_A(a_i)|$ are given in Table 9.7.1.*

(e) *The character table of A_1 is given in Table 9.8.1.*

Proof (a) The presentation of H_1 is a restatement of Proposition 6.2.3(a).

(b) By that result we also know that H_1 has a faithful permutation representation PH_1 of degree 28160 with stabilizer $U = \langle bz, c, d, e, fz, g, h, i\rangle$. Using it and the MAGMA command `AutomorphismGroup(H_1)` we see that $|Aut(H_1)| = |H_1|$. As $H_1/\langle z\rangle \cong \mathsf{Fi}_{22}$ has the same presentation as Fi_{22} given in [110], p. 110, we can quote the presentation of $Aut(\mathsf{Fi}_{22})$ given in [110], p. 111, where z is replaced by 1. Now (b) follows from (a) and 2^6 MAGMA calculations with PH_1 checking whether 1 or z has to be on the right hand side of the six new relations stated in (b). It follows that there is exactly one solution.

(c) This statement has been verified by means of the MAGMA command `CosetAction(A_1,U)`.

(d) The system of representatives of the conjugacy classes of A_1 has been calculated by means of the permutation representations PA_1 of $A_1 = 2Aut(H_1)$, MAGMA and Kratzer's Algorithm 5.3.18 of [92].

(e) The character table of A_1 has been calculated by means of PA_1 and MAGMA. □

Lemma 9.1.2 *Keep the notation of Lemma 9.1.1. Let* $G_1 = \langle x, y, q, w\rangle \cong \mathsf{Fi}_{23}$ *be the simple subgroup of* $\mathrm{GL}_{782}(17)$ *of order* $2^{18} \cdot 3^{13} \cdot 5^2 \cdot 7 \cdot 11 \cdot 13 \cdot 17 \cdot 23$ *with centralizer* $H_1 = \langle x, y, q\rangle = C_{G_1}(z)$ *of the 2-central involution* $z = (xy^2)^7$*; see Kim's Theorem 6.3.1. Let* $A_1 = 2Aut(H_1)$*. Then the following assertions hold.*

(a) $H_1 = \langle a, b, c, d, e, f, g, h, i, z\rangle$, *where*

$$\begin{aligned} a &= (xyx)^7, b = [(qy)^2qy^3q^2y^3qy]^7, c = (y^2xyxy^3)^5,\\ d &= (qyq^2yqyqyqy^2q^2)^{15}, e = (yxy^5x)^5,\\ f &= (yqyq^2yq^2y^2qy^4q^2)^5, g = (xy^2xy^3x)^7, h = (y^5xyx)^5, \text{ and}\\ i &= (q^2y^2qyq^2)^7. \end{aligned}$$

(b) $H_1 = \langle a, b, c, d, e, f, g, h, i, z\rangle$ *satisfies the set* $\mathcal{R}(H_1)$ *of defining relations stated in Lemma 9.1.1(a). Its character table is given in Table 6.6.3.*

(c) *For the character table of* G_1 *see the Atlas [19], pp. 178–179.*

(d) *The amalgam* $A_1 \leftarrow H_1 \rightarrow G_1$ *has Goldschmidt index* 1.

(e) *The amalgam* $A_1 \leftarrow H_1 \rightarrow G_1$ *has eight compatible pairs*

$$(\chi, \tau) \in mf\,char_{\mathbb{C}}(A_1) \times mf\,char_{\mathbb{C}}(G_1)$$

of degree 8671. *All have the same restriction*

$$\delta_2 + \delta_6 + \delta_7 + \delta_8 + \delta_9 \in mf\,char_{\mathbb{C}}(H_1).$$

They are:

(1) $(\chi_3 + \chi_{11} + \chi_{13} + \chi_{\mathbf{17}}, \quad \tau_{\mathbf{3}} + \tau_{\mathbf{4}})$,
(2) $(\chi_3 + \chi_{11} + \chi_{14} + \chi_{\mathbf{17}}, \quad \tau_{\mathbf{3}} + \tau_{\mathbf{4}})$,
(3) $(\chi_3 + \chi_{12} + \chi_{13} + \chi_{\mathbf{17}}, \quad \tau_{\mathbf{3}} + \tau_{\mathbf{4}})$,
(4) $(\chi_3 + \chi_{12} + \chi_{14} + \chi_{\mathbf{17}}, \quad \tau_{\mathbf{3}} + \tau_{\mathbf{4}})$,

(5) $(\chi_4 + \chi_{11} + \chi_{13} + \chi_{17}, \quad \tau_3 + \tau_4)$,
(6) $(\chi_4 + \chi_{11} + \chi_{14} + \chi_{17}, \quad \tau_3 + \tau_4)$,
(7) $(\chi_4 + \chi_{12} + \chi_{13} + \chi_{17}, \quad \tau_3 + \tau_4)$,
(8) $(\chi_4 + \chi_{12} + \chi_{14} + \chi_{17}, \quad \tau_3 + \tau_4)$.

Proof By Theorem 6.3.1 the simple subgroup $G_1 \cong \mathsf{Fi}_{23}$ of $\mathrm{GL}_{782}(17)$ has a faithful permutation representation PG_1 of degree 31671 with stabilizer H_1.

(a) The words of the new generators a, b, etc. of H_1 in terms of the given generators x, y and q of H_1 are quoted from Proposition 6.2.3.

(b) Using PG_1 and MAGMA it has been checked that the new generators a, b etc. of H_1 given in statement (a) satisfy all the relations of $\mathcal{R}(H_1)$ of Lemma 9.1.1(a).

(c) This assertion is a restatement of Theorem 6.3.1.

(d) Using the faithful permutation representation PA_1 of A_1 stated in Lemma 9.1.1(c) MAGMA established that the outer automorphism groups $Out(H_1)$ and $Out(A_1)$ of H_1 and A_1 are both cyclic of order 2. Hence the Goldschmidt index of the amalgam $A_1 \leftarrow H_1 \rightarrow G_1$ is 1 by Step 3 of Algorithm 7.1.10 of [92].

(e) The eight compatible pairs of degree 8671 of the amalgam $A_1 \leftarrow H_1 \rightarrow G_1$ were computed by means of MAGMA, the faithful permutation representations PA_1 and PG_1 and Kratzer's Algorithm 7.3.10 of [92]. □

9.2 A semi-simple representation of Fi_{23} in $\mathrm{GL}_{8671}(13)$

In this section we present Kim's and the author's construction [74] of the two irreducible representations of degrees 3588 and 5083 of $G_1 \cong \mathsf{Fi}_{23}$ over the prime field $GF(13)$ corresponding to the two irreducible characters τ_3 and τ_4 of 13-defect zero of G_1 occurring in the eight compatible pairs constructed in Lemma 9.1.2(e).

Lemma 9.2.1 *Keep the notation of Lemmas 9.1.1 and 9.1.2. Let* $G_1 = \mathsf{Fi}_{23} = \langle x, y, q, w\rangle$ *be the simple subgroup of* $\mathrm{GL}_{782}(17)$ *of order* $2^{18}\cdot 3^{13}\cdot 5^2\cdot 7\cdot 11\cdot 13\cdot 17\cdot 23$ *with faithful permutation representation* PG_1 *of degree* 31671 *and stabilizer* $H_1 = \langle x, y, q\rangle$ *constructed in Theorem 6.3.1. Let* $r = s_1 = (yqy^2q)^7$, $s_2 = (yqyqy)^7$, $s_3 = (yeyeq)^{21}$, $s_4 = (qyqey^2)^7$, $u = s_4^2(s_1s_2s_4)^3$, $s = qyqey^2$ *and* $v = (s_1s_3s_1s_4)^2(s_1s_3s_1s_3s_1s_4)^3s_3s_4s_3s_4^2s_3$. *Then the following assertions hold.*

(a) $mH = \langle s_1, s_2, s_3, s_4 \rangle = \langle r, u, v \rangle \cong \mathrm{O}_8^+(3) : S_3$.

(b) *The character table of* mH *is given in Table* `DVD.1.7.2` *of the attached DVD.*

(c) $mE = \langle u, v, s \rangle \cong O_7(3) \times S_3$.

(d) *The character table of* mE *is Table* `DVD.1.7.3` *of the accompanying DVD.*

(e) $mD = mH \cap mE = \langle u, v \rangle \cong \mathrm{G}_2(3) \times S_3$.

(f) $G_1 = \langle u, v, r, s \rangle$, *and the original generators* q, y, w *and* x *of* G_1 *are equal to the following words in its generators* u, v, r *and* s:

$$q = [(us^2vsr)^9[(svs^3rsr)^{12}(vrsrsvu)^{12}]^2]^{24},$$
$$w = [(w_2w_4w_2w_4w_3w_2w_4)^3(w_1w_2w_3w_1w_4w_3w_1w_4)^7]^3,$$
$$y = (m_1m_2m_1m_2m_3)^7[(m_1m_2)^5(n_3n_1n_2n_3n_1n_3^2n_1)^5]^4,$$
$$x = [(yq^2yqyq^2)^{11}(q^2y^2qyqy)^{11}(qy^2qyqyqyq)^4]^{12},$$

where

$$w_1 = (vuvs^2)^9, \quad w_2 = (v^2uruvr)^{18}, \quad w_3 = (v^2s^2urv)^{11},$$
$$w_4 = (uvrv^4r)^{10}, \quad n_1 = m_1m_2m_3m_2^2m_1, \quad n_2 = m_2m_3m_1m_3m_2^3,$$
$$n_3 = (m_3m_1m_2m_3m_2m_1m_2)^2, \quad m_1 = (t_1t_3t_1t_3t_1^2t_3t_1)^2,$$
$$m_2 = (t_1t_3t_1t_3t_1^2t_2t_3t_2)^2, \quad m_3 = (t_2t_1t_3t_2t_3t_1t_2t_3t_2)^9,$$
$$t_1 = (srsrsrs)^5, \quad t_2 = (s^2rs^4)^{11}, \quad \text{and} \quad t_3 = (rs^7)^3.$$

(g) *The restrictions of the irreducible character* τ_3 *of degree* 3588 *of* G_1 *to* mH *and* mE *are* $\pi_8 + \pi_{16} \in mf\,char_{\mathbb{C}}(mH)$ *and* $\psi_4 + \psi_{10} + \psi_{19} + \psi_{36} + \psi_{61} \in mf\,char_{\mathbb{C}}(mE)$, *respectively.*

(h) *The restrictions of the irreducible character* τ_4 *of degree* 5083 *of* G_1 *to* mH *and* mE *are* $\pi_{12} + \pi_{15} \in mf\,char_{\mathbb{C}}(mH)$ *and* $\psi_{14} + \psi_{22} + \psi_{45} + \psi_{74}) \in mf\,char_{\mathbb{C}}(mE)$, *respectively.*

(i) *The irreducible characters* π_9, π_{11} *and* π_{15} *of* mH *are constituents of the permutation characters* $1_{mH_9}^{mH}$, $1_{mH_{11}}^{mH}$ *and* $1_{mH_{15}}^{mH}$ *of the subgroups*

$$mH_9 = \langle (vru^2)^{13}, (rvur)^2, (uvrv^2)^4 \rangle,$$
$$mH_{11} = \langle (uvurv)^4, (v^5ur)^9, (ururv^2u^2)^{13} \rangle$$

and

$$mH_{15} = (uvuru)^{13}, (uv^2uv)^2, (u^5v)^3 \rangle$$

of mH *with indices* 3240, 72800 *and* 2274480, *respectively.*

(j) *The linear character* π_2 *of* mH *has values* -1 *and* 1 *at* v *and* u, r, *respectively. Furthermore,* $\pi_8 = \pi_2 \otimes \pi_9$, $\pi_{12} = \pi_2 \otimes \pi_{11}$ *and* $\pi_{16} = \pi_2 \otimes \pi_{15}$.

(k) *The irreducible characters* ψ_5 *and* ψ_{10} *of* mE *are constituents of the permutation character* $1_{mE_1}^{mE}$ *of the subgroup* $mE_1 = \langle (s^2vs)^9, (svsv^2s^2)^2 \rangle$ *of index* 2106.

(l) *The irreducible characters* ψ_{14}, ψ_{22}, ψ_{36}, ψ_{45}, ψ_{61} *and* ψ_{74} *of* mE *are constituents of the permutation characters* $1_{mE_{14}}^{mE}$, $1_{mE_{22}}^{mE}$, $1_{mE_{36}}^{mE}$, $1_{mE_{45}}^{mE}$, $1_{mE_{61}}^{mE}$ *and* $1_{mE_{74}}^{mE}$ *of the subgroups*

$$\begin{aligned} mE_{14} &= \langle (s^2us)^6, (su^2s^2)^4, (susu^2)^2, (us^3us)^7 \rangle, \\ mE_{22} &= \langle (us^4u^2)^6, usu^2susus \rangle, \\ mE_{36} &= \langle (s^2vs)^9, (v^2svsv^2)^5, (vsv^2svs)^2 \rangle, \\ mE_{45} &= \langle (s^2v^2)^7, (v^5s)^3, (vsv^2sv^3sv)^6 \rangle, \\ mE_{61} &= \langle (vs^2)^7, (v^4s^2v^2)^{12}, (v^2svsv^4)^{10}, (vs^2v^3s^3)^{30} \rangle \end{aligned}$$

and

$$mE_{74} = \langle (s^2v)^7, (v^2s^2)^{21}, (s^2v^2)^{21}, (svsv^2sv)^3 \rangle$$

of mE *with indices* 702, 2160, 19656, 7280, 85293 *and* 29484, *respectively.*

(m) *The linear character* ψ_2 *of* mE *has values* -1 *and* 1 *at* v *and* u, s, *respectively. Furthermore,* $\psi_4 = \psi_2 \otimes \psi_5$ *and* $\psi_{19} = \psi_2 \otimes \psi_{22}$.

(n) *Both* r *and* $f = (u^3vsv)^9$ *are involutions of* G_1 *such that* $(r, f) = 1$, $rf \notin mH$ *and* $rf \notin mE$.

Proof (a) The subgroup mH of G_1 has been constructed by means of the faithful permutation representation PG_1 of G_1 of degree 31671 and the MAGMA command `LowIndexSubgroups(PG_1, 137632)`. The four generators s_i of mH, $1 \le i \le 4$, were calculated with Kim's program `GetShortGenerators(PG_1, mH)`. Another application of MAGMA determined the composition factors of mH.

(b) The character table of mH was calculated by MAGMA using PG_1.

(c) and (e) By Table 6.5.4 $|C_{G_1}((qw)^4)| = 2^9 \cdot 3^{10} \cdot 5 \cdot 7 \cdot 13$. Let $mX = N_{G_1}(\langle (qw)^4 \rangle)$. Using PG_1 and MAGMA we searched for an element $x \in mX$ of order 3 such that $|N_{mH}(\langle x \rangle)| = 2^7 \cdot 3^7 \cdot 7 \cdot 13$. MAGMA found such an element and stated that $mD = N_{mH}(\langle x \rangle) = \langle u, v \rangle$ and that $mE = N_{G_1}(\langle x \rangle) = \langle mD, s \rangle$, where u, v and s are defined in the statement of this lemma. The composition factors of mD and mE have been determined by means of MAGMA.

(d) The character table of mE was calculated by means of PG_1 and MAGMA.

(f) Using the faithful permutation representation PG_1 of G_1 and MAGMA one verifies that $G_1 = \langle mH, mE\rangle$. Hence $G_1 = \langle u, v, r, s\rangle$ by (a) and (c). The words for q, y, w and x can easily be checked computationally.

(g) G_1 has a unique character τ_3 of degree 3588 by the character table of $G_1 \cong \mathsf{Fi}_{23}$; see [19], p. 178. Its restrictions to mH and mE given in the statements have been determined by means of PG_1, the character tables of the subgroups mH and mE of G_1 and MAGMA.

(h) This assertion is proved in the same way as (g).

(i) Using the MAGMA command `LowIndexSubgroups(mH, k)` we searched for conjugacy classes of subgroups H_k of index $|mH : mH_k| = m_k$ such that π_k is an irreducible constituent of the permutation character $1_{mH_k}^{mH}$ for $k \in \{9, 11, 15\}$. Thus we found three subgroups mH_k of respective indices $m_9 = 3240$, $m_{11} = 72800$ and $m_{15} = 2274480$. Their given generators have been obtained by means of Kim's program `GetShortGenerators(mH, mH_k)`.

(j) π_2 is the unique non-trivial linear character of mH by its character table. The character equations of the statement are easily verified by means of Table `DVD.1.7.2` of the accompanying DVD.

Statements (k) and (l) are proved in the same way as (i).

(m) mE has a unique non-trivial linear character ψ_2, see Table `DVD.1.7.3` of the accompanying DVD. The character equations of the statement are easily verified by means of Table `DVD.1.7.3`.

(n) Using PG_1 and MAGMA we checked that r and f are commuting involutions such that $rf \notin mH$ and $rf \notin mE$. □

Proposition 9.2.2 *Keep the notation of Lemmas 9.1.1, 9.1.2 and 9.2.1. Let PG_1 be the faithful permutation representation of the simple group $G_1 = \langle x, y, q, w\rangle = \langle u, v, r, s\rangle$ of degree 31671 with stabilizer $H_1 = \langle x, y, q\rangle$. Let $mH = \langle u, v, r\rangle$, $mD = \langle u, v\rangle$ and $mE = \langle u, v, s\rangle$. Let F^* be the multiplicative group of the prime field $F = GF(13)$. Let $Y = \mathrm{GL}_{3588}(13)$.*

Let $\mathfrak{V}$ and $\mathfrak{W}$ be the up to isomorphism uniquely determined faithful semi-simple 3588-dimensional modules of mH and mE over F corresponding to the restrictions $\tau_{3|mH}$ and $\tau_{3|mE}$ of the irreducible character τ_3 of G_1, respectively.

Let $\kappa_{\mathfrak{V}} : mH \to \mathrm{GL}_{3588}(13)$ and $\kappa_{\mathfrak{W}} : mE \to \mathrm{GL}_{3588}(13)$ be the representations of mH and mE afforded by the modules $\mathfrak{V}$ and $\mathfrak{W}$, respectively.

Let $\mathfrak{r} = \kappa_{\mathfrak{V}}(r)$, $\mathfrak{u} = \kappa_{\mathfrak{V}}(u)$, $\mathfrak{v} = \kappa_{\mathfrak{V}}(v)$ *in* $\kappa_{\mathfrak{V}}(mH) \le \mathrm{GL}_{3588}(13)$.

Then $\mathfrak{V}_{|mD} \cong \mathfrak{W}_{|mD}$, *and there is a transformation matrix* $\mathcal{T}_1 \in \mathrm{GL}_{3588}(13)$ *such that*

$$\mathfrak{u} = \mathcal{T}_1^{-1}\kappa_{\mathfrak{W}}(u)\mathcal{T}_1, \qquad \mathfrak{v} = \mathcal{T}_1^{-1}\kappa_{\mathfrak{W}}(v)\mathcal{T}_1.$$

Let $\mathfrak{mD} = \langle \mathfrak{u}, \mathfrak{v}\rangle$, $\mathfrak{mH} = \langle \mathfrak{u}, \mathfrak{v}, \mathfrak{r}\rangle$. *Let* $\mathcal{D} = C_Y(\mathfrak{mD})$ *and* $\mathcal{H} = C_Y(\mathfrak{mH})$. *Let* $\mathfrak{s}_1 = \mathcal{T}_1^{-1}\kappa_{\mathfrak{W}}(s)\mathcal{T}_1$. *Let* $\mathfrak{mE} = \langle \mathfrak{mD}, \mathfrak{s}_1\rangle$ *and* $\mathcal{E} = C_Y(\mathfrak{mE})$. *Then the following statements hold.*

(a) *There is an isomorphism*

$$\alpha : \mathcal{D} \to \mathcal{D}_1 = \mathrm{GL}_2(13) \times \mathrm{GL}_2(13) \times F^{*5} \le \mathrm{GL}_9(13).$$

(b) $\mathcal{H}_1 = \alpha(\mathcal{H})$ *is generated by the two blocked diagonal matrices*

$$a_1 = diag(\left(\begin{smallmatrix} 2 & 0 \\ 0 & 1 \end{smallmatrix}\right), \left(\begin{smallmatrix} 2 & 0 \\ 0 & 1 \end{smallmatrix}\right), 2, 1, 2, 1, 1),$$
$$a_2 = diag(\left(\begin{smallmatrix} 1 & 0 \\ 0 & 2 \end{smallmatrix}\right), \left(\begin{smallmatrix} 1 & 0 \\ 0 & 2 \end{smallmatrix}\right), 1, 2, 1, 2, 2),$$

(c) $\mathcal{E}_1 = \alpha(\mathcal{E})$ *is generated by the five blocked diagonal matrices*

$$b_1 = diag(\left(\begin{smallmatrix} 2 & 0 \\ 0 & 1 \end{smallmatrix}\right), \left(\begin{smallmatrix} 1 & 0 \\ 0 & 1 \end{smallmatrix}\right), 1, 1, 1, 1, 1),$$
$$b_2 = diag(\left(\begin{smallmatrix} 1 & 0 \\ 0 & 1 \end{smallmatrix}\right), \left(\begin{smallmatrix} 2 & 0 \\ 0 & 1 \end{smallmatrix}\right), 1, 1, 1, 1, 1),$$
$$b_3 = diag(\left(\begin{smallmatrix} 1 & 0 \\ 0 & 2 \end{smallmatrix}\right), \left(\begin{smallmatrix} 1 & 0 \\ 0 & 1 \end{smallmatrix}\right), 1, 2, 1, 1, 1),$$
$$b_4 = diag(\left(\begin{smallmatrix} 1 & 0 \\ 0 & 1 \end{smallmatrix}\right), \left(\begin{smallmatrix} 1 & 0 \\ 0 & 1 \end{smallmatrix}\right), 2, 1, 1, 2, 1),$$
$$b_5 = diag(\left(\begin{smallmatrix} 1 & 0 \\ 0 & 1 \end{smallmatrix}\right), \left(\begin{smallmatrix} 1 & 0 \\ 0 & 2 \end{smallmatrix}\right), 1, 1, 2, 1, 2).$$

(d) $\mathcal{D}$ *has* $2184^2 \times 12 = 57238272$ $\mathcal{H}$-$\mathcal{E}$ *double cosets.*

(e) *The free product* $mH *_{mD} mE$ *of* mH *and* mE *with amalgamated subgroup* mD *has an irreducible* 3588*-dimensional representation over* F *which induces an irreducible representation of* G_1. *It corresponds to the* $\mathcal{H}$-$\mathcal{E}$ *double coset representative*

$$\mathcal{F} = diag(w, z, 1^{182}, 1^{182}, 1^{364}, 1^{728}, 1^{1664}) \in \mathrm{GL}_{3588}(13),$$

where

$$w = \begin{pmatrix} 1^{78} & 4^{78} \\ 12^{78} & 2^{78} \end{pmatrix}, \qquad z = \begin{pmatrix} 1^{156} & 6^{156} \\ 5^{156} & 10^{156} \end{pmatrix},$$

and a^n *denotes a diagonal* $n \times n$ *matrix with unique diagonal non-zero entry* $a \in GF(13)$.

Let $\mathfrak{s} = \mathcal{F}^{-1}\mathfrak{s}_1\mathcal{F}$ *and* $\mathfrak{G}_1 = \langle \mathfrak{u}, \mathfrak{v}, \mathfrak{r}, \mathfrak{s}\rangle$. *Inserting these four generating matrices of* $\mathfrak{G}_1$ *into the formulas of Lemma 9.2.1(f) one obtains the matrices* $\mathfrak{x}_{3588}$ $\mathfrak{y}_{3588}$, $\mathfrak{q}_{3588}$ *and* $\mathfrak{w}_{3588}$ *of the original generators* x, y, q *and* w *of* $G_1 = \mathsf{Fi}_{23}$ *as words in the generators*

u, v, r and s. The matrices $\mathfrak{x}_{3588}$, $\mathfrak{y}_{3588}$, $\mathfrak{q}_{3588}$ and $\mathfrak{w}_{3588}$ are the respective upper left block matrices of the matrices $\mathfrak{x}$, $\mathfrak{y}$, $\mathfrak{q}$ and $\mathfrak{w}$ of $\mathrm{GL}_{8671}(13)$ *stored on the accompanying DVD.*

Proof Let $\mathfrak{V}$ be the up to isomorphism uniquely determined faithful semi-simple 3588-dimensional module of mH over $F = GF(13)$ corresponding to the restriction $\tau_{\mathbf{3}|mH}$. By Lemma 9.2.1(g) $\tau_{\mathbf{3}|mH} = \pi_8 + \pi_{16}$. Lemma 9.2.1(j) states that $\pi_8 = \pi_2 \otimes \pi_9$.

The irreducible characters π_9 and π_{15} of mH are constituents of the permutation characters $1_{mH_9}^{mH}$ and $1_{mH_{15}}^{mH}$ of the respective subgroups mH_9 and mH_{15} of mH determined in Lemma 9.2.1(i). Using Algorithm 1.5.1 Kim and the author calculated the primitive idempotents of the endomorphism rings of these permutation modules in [74]. Thus we obtained the corresponding irreducible representations $M(\pi_9)$ and $M(\pi_{15})$ of the respective dimensions 780 and 2808 over F. The irreducible FmH-modules $M(\pi_8)$ and $M(\pi_{16})$ are the tensor products of $M(\pi_9)$ and $M(\pi_{15})$ with the linear character π_2 of mH over F. Thus $\mathfrak{V} = M(\pi_8) \oplus M(\pi_{16})$.

Let $\mathfrak{W}$ be the up to isomorphism uniquely determined faithful semi-simple 3588-dimensional module of mE over F corresponding to the restriction $\tau_{\mathbf{3}|mE}$. By Lemma 9.2.1(g) $\tau_{\mathbf{3}|mE} = \psi_4 + \psi_{10} + \psi_{19} + \psi_{36} + \psi_{61}$. Lemma 9.2.1(m) states that $\psi_4 = \psi_2 \otimes \pi_5$ and $\psi_{19} = \psi_2 \otimes \pi_{22}$.

The irreducible characters ψ_5 and ψ_{10} of mE are constituents of the permutation character $1_{mE_1}^{mE}$ by Lemma 9.2.1(k). The irreducible characters ψ_{22}, ψ_{36} and ψ_{61} of mE are constituents of the permutation characters $1_{mE_{22}}^{mE}$, $1_{mE_{36}}^{mE}$ and $1_{mE_{61}}^{mE}$, respectively; see Lemma 9.2.1(l). Using Algorithm 1.5.1. Kim and the author calculated the primitive idempotents of the endomorphism rings of these four permutation modules in [74]. Thus we obtained the corresponding irreducible representations $N(\psi_5)$, $N(\psi_{10})$, $N(\psi_{22})$, $N(\psi_{36})$ and $N(\psi_{61})$ of the respective dimensions 78, 156, 260, 910 and 2184 over F. The irreducible FmE-modules $N(\psi_4)$ and $N(\psi_{19})$ are the tensor products of $N(\psi_5)$ and $N(\psi_{22})$ with the linear character ψ_2 of mE over F. Thus

$$\mathfrak{W} = N(\psi_4) \oplus N(\psi_{10}) \oplus N(\psi_{19}) \oplus N(\psi_{36}) \oplus N(\psi_{61}).$$

Fixing a basis in each irreducible constituent $M\pi_k$ of $\mathfrak{V}$, we get a basis $\mathcal{B}_V$ of $\mathfrak{V}$. It induces a representation $\kappa_{\mathfrak{V}} : mH \to \mathrm{GL}_{3588}(13)$ of mH. Let $\mathfrak{r} = \kappa_{\mathfrak{V}}(r)$, $\mathfrak{u} = \kappa_{\mathfrak{V}}(u)$, $\mathfrak{v} = \kappa_{\mathfrak{V}}(v)$ in $\kappa_{\mathfrak{V}}(mH) \le \mathrm{GL}_{3588}(13)$.

Fixing a basis in each irreducible constituent $N\psi_j$ of $\mathfrak{W}$, we get a basis $\mathcal{B}_W$ of $\mathfrak{W}$. It induces a representation $\kappa_{\mathfrak{W}} : mE \to \mathrm{GL}_{3588}(13)$ of mE. By Lemma 9.2.1(g) $\mathfrak{V}_{|mD} \cong \mathfrak{W}_{|mD}$. Let $Y = \mathrm{GL}_{3588}(13)$. Applying now

Parker's isomorphism test of Proposition 6.1.6 of [92] by means of the MAGMA command

```
IsIsomorphic(GModule(sub<Y|V(u),V(v)>),GModule(sub<Y|W(u),W(v)>))
```

one obtains the transformation matrix $\mathcal{T}_1$ satisfying $\mathfrak{u} = \kappa_{\mathfrak{W}}(u)^{\mathcal{T}_1}$ and $\mathfrak{v} = \kappa_{\mathfrak{W}}(v)^{\mathcal{T}_1}$.

(a) Let $\mathfrak{mD} = \langle \mathfrak{u}, \mathfrak{v} \rangle$ and $\mathfrak{mH} = \langle \mathfrak{u}, \mathfrak{v}, \mathfrak{r} \rangle$. Let $\mathcal{D} = C_Y(\mathfrak{mD})$ and $\mathcal{H} = C_Y(\mathfrak{mH})$. Let δ_{12a}, δ_{12b} and δ_{23a}, δ_{23b} be two distinct copies of the irreducible characters δ_{12} and δ_{23} of mD, respectively. Using PG_1 and MAGMA it can be checked that the irreducible characters π_8 and π_{16} of mH have the following restrictions to $mD = \langle u, v \rangle$:

$$\pi_{8\,|mD} = \delta_{12a} + \delta_{23a} + \delta_{27} + \delta_{39}, \quad \pi_{16\,|mD} = \delta_{12b} + \delta_{23b} + \delta_{29} + \delta_{54} + \delta_{69},$$

where the irreducible characters δ_{12}, δ_{23}, δ_{27}, δ_{29}, δ_{39}, δ_{54} and δ_{69} of $mD \cong G_2(3) \times S_3$ have degrees 78, 156, 182, 182, 364, 728 and 1664, respectively.

Since $\mathfrak{V}_{|mD}$ is a semi-simple FmD-module Theorem 2.1.27 of [92] implies that there is an isomorphism $\alpha : \mathcal{D} \to \mathcal{D}_1 = \mathrm{GL}_2(13) \times \mathrm{GL}_2(13) \times F^{*5}$.

(b) Furthermore, Schur's Lemma asserts that $\mathcal{H}_1 = \alpha(\mathcal{H})$ is generated by the two blocked diagonal matrices given in the statement because 2 is a primitive element of the multiplicative group F^* of $F = GF(13)$.

(c) Let $\mathfrak{s}_1 = \mathcal{T}_1^{-1}\kappa_{\mathfrak{W}}(s)\mathcal{T}_1$. Let $\mathfrak{mE} = \langle \mathfrak{mD}, \mathfrak{s}_1 \rangle$ and $\mathcal{E} = C_Y(\mathfrak{mE})$. Let δ_{12a}, δ_{12b} and δ_{23a}, δ_{23b} be two distinct copies of the irreducible characters δ_{12} and δ_{23} of mD, respectively. Using PG_1 and MAGMA it can be checked that the irreducible characters ψ_4, ψ_{10}, ψ_{19}, ψ_{36} and π_{61} of mE have the following restrictions to $mD = \langle u, v \rangle$:

$$\psi_{4\,|_m D} = \delta_{12a}, \quad \psi_{10\,|_m D} = \delta_{23a}, \quad \psi_{19\,|_m D} = \delta_{12b} + \delta_{29},$$
$$\psi_{36\,|_m D} = \delta_{27} + \delta_{54}, \quad \psi_{61\,|_m D} = \delta_{23b} + \delta_{39} + \delta_{69}.$$

Now Schur's Lemma implies that $\mathcal{E}_1 = \alpha(\mathcal{E})$ is generated by the five blocked diagonal matrices b_j given in the statement.

(d) An $\mathcal{H}$-$\mathcal{E}$ double coset representative is of the form $diag(A, B, 1, 1, 1, 1, v)$ for some $A, B \in Y$ and $v \in F^*$. By multiplying from left and right, we observe that $diag(A, B, 1, 1, 1, 1, v)$ and $diag(A', B', 1, 1, 1, 1, v)$ represent the same double coset if and only if the first columns of A and B are each a scalar multiple of the first columns of A' and B', respectively. So, we have 12 choices for v, and $|\,\mathrm{GL}(2, 13)|/12) = 2184$ choices for A and B. Thus there are $2184^2 \cdot 12 = 57\,238\,272$ $\mathcal{H}$-$\mathcal{E}$ double cosets.

(e) By Theorem 7.2.2 of [92] the irreducible representations of the free product $mH *_{mD} mE$ of the groups mH and mE with amalgamated subgroup mD are described by the $\mathcal{H}$-$\mathcal{E}$ double coset representatives T of $\mathcal{D}$. The elements r and $f = (u^3vsv)^9$ are two commuting involutions of $G_1 \cong \mathsf{Fi}_{23}$ by Lemma 9.2.1(o). Let $\mathfrak{u}$ and $\mathfrak{f}$ be their matrices in $\mathfrak{m}\mathfrak{H}$ and $\mathfrak{m}\mathfrak{E}$, respectively. If $T = diag((\begin{smallmatrix} a & c \\ b & d \end{smallmatrix}), (\begin{smallmatrix} p & t \\ q & u \end{smallmatrix}), 1, 1, 1, 1, v)$ describes a 3588-dimensional representation of G_1 over F then (*) $(\mathfrak{r}, \mathcal{T}^{-1}\mathfrak{f}\mathcal{T}) = 1$ holds, where $\mathcal{T} \in \mathrm{GL}_{3588}(13)$ corresponds to T.

Since $\mathfrak{V}_{|mD} \cong \mathfrak{W}_{|mD}$ is a direct sum of nine irreducible FmD-modules both matrices $\mathfrak{r}$ and $\mathfrak{f}$ consist of 81 blocks $R_{i,j}$ and $F_{i,j}$, $1 \le i, j \le 9$, respectively, such that all diagonal blocks $R_{i,i}$ and $F_{i,i}$ are non-zero. Furthermore a non-diagonal block $R_{i,j}$ of $\mathfrak{r}$ is non-zero if and only if the ith irreducible and the jth irreducible representations of mD appear in the restriction of an irreducible representation of mH to mD. A similar description holds for the blocks of $\mathfrak{f}$. Hence the system of equations in the proofs of (a) and (c) imply that

$$\mathfrak{r} = \begin{pmatrix} R_{1,1} & . & R_{1,3} & . & R_{1,5} & . & R_{1,7} & . & . \\ . & R_{2,2} & . & R_{2,4} & . & R_{2,6} & . & R_{2,8} & R_{2,9} \\ R_{3,1} & . & R_{3,3} & . & R_{3,5} & . & R_{3,7} & . & . \\ . & R_{4,2} & . & R_{4,4} & . & R_{4,6} & . & R_{4,8} & R_{4,9} \\ R_{5,1} & . & R_{5,3} & . & R_{5,5} & . & R_{5,7} & . & . \\ . & R_{6,2} & . & R_{6,4} & . & R_{6,6} & . & R_{6,8} & R_{6,9} \\ R_{7,1} & . & R_{7,3} & . & R_{7,5} & . & R_{7,7} & . & . \\ . & R_{8,2} & . & R_{8,4} & . & R_{8,6} & . & R_{8,8} & R_{8,9} \\ . & R_{9,2} & . & R_{9,4} & . & R_{9,6} & . & R_{9,8} & R_{9,9} \end{pmatrix},$$

$$\mathfrak{f} = \begin{pmatrix} F_{1,1} & . & . & . & . & . & . & . & . \\ . & F_{2,2} & . & . & . & F_{2,6} & . & . & . \\ . & . & F_{3,3} & . & . & . & . & . & . \\ . & . & . & F_{4,4} & . & . & F_{4,7} & . & F_{4,9} \\ . & . & . & . & F_{5,5} & . & . & F_{5,8} & . \\ . & F_{6,2} & . & . & . & F_{6,6} & . & . & . \\ . & . & . & F_{7,4} & . & . & F_{7,7} & . & F_{7,9} \\ . & . & . & . & F_{8,5} & . & . & F_{8,8} & . \\ . & . & . & F_{9,4} & . & . & F_{9,7} & . & F_{9,9} \end{pmatrix}.$$

Let $e = (ad - bc)^{-1}$ and $g = (pu - tq)^{-1}$. Then $e \ne 0 \ne g$. For each integer k let I_k denote the $k \times k$ identity matrix over F. Then

$$\mathcal{T}^{-1} = \begin{pmatrix} ed(I_{78}) & -ec(I_{78}) & . & . & . & . & . & . & . \\ -eb(I_{78}) & ea(I_{78}) & . & . & . & . & . & . & . \\ . & . & gu(I_{156}) & -gt(I_{156}) & . & . & . & . & . \\ . & . & -gq(I_{156}) & gp(I_{156}) & . & . & . & . & . \\ . & . & . & . & I_{182} & . & . & . & . \\ . & . & . & . & . & I_{182} & . & . & . \\ . & . & . & . & . & . & I_{364} & . & . \\ . & . & . & . & . & . & . & I_{728} & . \\ . & . & . & . & . & . & . & . & v^{-1}(I_{1664}) \end{pmatrix}.$$

Hence $\mathfrak{f}' = \mathcal{T}^{-1}\mathfrak{f}\mathcal{T}$ equals the matrix

$$\begin{pmatrix} G_{1,1} & G_{1,2} & . & . & . & -ecF_{2,6} & . & . & . \\ G_{2,1} & G_{2,2} & . & . & . & eaF_{2,6} & . & . & . \\ . & . & G_{3,3} & G_{3,4} & . & . & -gtF_{4,7} & . & -vgtF_{4,9} \\ . & . & G_{4,3} & G_{4,4} & . & . & gpF_{4,7} & . & vgpF_{4,9} \\ . & . & . & . & F_{5,5} & . & . & F_{5,8} & . \\ G_{6,1} & G_{6,2} & . & . & . & F_{6,6} & . & . & . \\ . & . & G_{7,3} & G_{7,4} & . & . & F_{7,7} & . & vF_{7,9} \\ . & . & . & . & F_{8,5} & . & . & F_{8,8} & . \\ . & . & G_{9,3} & G_{9,4} & . & . & v^{-1}F_{9,7} & . & F_{9,9} \end{pmatrix},$$

where

$$\begin{aligned} G_{1,1} &= e(adF_{1,1} - bcF_{2,2}), \quad G_{1,2} = ecd(F_{1,1} - F_{2,2}), \\ G_{2,1} &= -eab(F_{1,1} - F_{2,2}), \quad G_{2,2} = -e(bcF_{1,1} - adF_{2,2}), \\ G_{6,1} &= bF_{6,2}, \quad G_{6,2} = dF_{6,2}, \\ G_{3,3} &= g(puF_{3,3} - qtF_{4,4}), \quad G_{3,4} = gut(F_{3,3} - F_{4,4}), \\ G_{4,3} &= -gpq(F_{3,3} - F_{4,4}), \quad G_{4,4} = -g(tqF_{3,3} - puF_{4,4}), \\ G_{7,3} &= qF_{7,4}, \quad G_{7,4} = uF_{7,4}, \\ G_{9,3} &= qv^{-1}F_{9,4}, \quad G_{9,4} = uv^{-1}F_{9,4}. \end{aligned}$$

Now $(*)$ implies the following equations

$$\begin{aligned} (1,1) &: G_{1,1}R_{1,1} = R_{1,1}G_{1,1}, \\ (1,2) &: G_{1,2}R_{2,2} + (-ec)F_{2,6}R_{6,2} = R_{1,1}G_{1,2}, \\ (6,1) &: G_{6,1}R_{1,1} = R_{6,2}G_{2,1} + R_{6,6}G_{6,1}, \\ (8,1) &: F_{8,5}R_{5,1} = R_{8,2}G_{2,1} + R_{8,6}G_{6,1}. \end{aligned}$$

Inserting the previous equations into these four equations yields

$$\begin{aligned} (1,1) &: e(adF_{1,1} - bcF_{2,2})R_{1,1} = eR_{1,1}(adF_{1,1} - bcF_{2,2}), \\ (1,2) &: ec(d(F_{1,1} - F_{2,2})R_{2,2} - F_{2,6}R_{6,2}) = ecdR_{1,1}(F_{1,1} - F_{2,2}), \\ (6,1) &: bF_{6,2}R_{1,1} = b(-eaR_{6,2}(F_{1,1} - F_{2,2}) + R_{6,6}F_{6,2}), \\ (8,1) &: F_{8,5}R_{5,1} = b(-eaR_{8,2}(F_{1,1} - F_{2,2}) + R_{8,6}F_{6,2}). \end{aligned}$$

Since $e^{-1} = ad - bc \neq 0$, at least one of ad or bc is non-zero. Suppose ad is zero, and bc is non-zero. Then $(1,1)$ yields $F_{1,1}R_{1,1} = R_{1,1}F_{1,1}$. A MAGMA calculation disproves this equation. Hence $ad \neq 0$. If bc is zero and ad is non-zero, then equation $(1,1)$ implies $F_{2,2}R_{1,1} = R_{1,1}F_{2,2}$, which is also wrong by MAGMA. Therefore all a, b, c, d are non-zero. We modify T so that $a = 1$ by multiplying some power of $diag((\begin{smallmatrix} 2 & 0 \\ 0 & 1 \end{smallmatrix}), (\begin{smallmatrix} 1 & 0 \\ 0 & 1 \end{smallmatrix}), 1, 1, 1, 1, 1)$ from the right.

Since ec is non-zero, it can be cancelled on both sides of equation $(1,2)$. Hence

$$(1,2) : d(F_{1,1} - F_{2,2})R_{2,2} - F_{2,6}R_{6,2} = dR_{1,1}(F_{1,1} - F_{2,2}).$$

Using MAGMA it can be verified that this equation holds only for $d = 2$. As $b \neq 0$ and $a = 1$ equation (6,1) implies

$$F_{6,2}R_{1,1} = -eR_{6,2}(F_{1,1} - F_{2,2}) + R_{6,6}F_{6,2}.$$

By MAGMA it has the solution $e = 11$. Now equation (8,1) implies that

$$F_{8,5}R_{5,1} = b(-11R_{8,2}(F_{1,1} - F_{2,2}) + R_{8,6}F_{6,2}).$$

This equation has the solution $b = 4$ by MAGMA. From the equation $ad - bc = e^{-1}$ we now deduce that $c = 12$.

In order to determine the remaining coefficients of the matrix T we use the following matrix equations derived from $(*)$:

$$(9,8) : G_{9,4}R_{4,8} + F_{9,9}R_{9,8} = R_{9,8}F_{8,8},$$
$$(9,9) : G_{9,4}R_{4,9} + F_{9,9}R_{9,9} = vgpR_{9,4}F_{4,9} + R_{9,9}F_{9,9},$$
$$(4,5) : G_{4,3}R_{3,5} + gpF_{4,7}R_{7,5} = R_{4,8}F_{8,5}.$$

Inserting the first set of equations yields

$$(9,8) : uv^{-1}F_{9,4}R_{4,8} + F_{9,9}R_{9,8} = R_{9,8}F_{8,8},$$
$$(9,9) : uv^{-1}F_{9,4}R_{4,9} + F_{9,9}R_{9,9} = vgpR_{9,4}F_{4,9} + R_{9,9}F_{9,9},$$
$$(4,5) : -gpq(F_{3,3} - F_{4,4})R_{3,5} + gpF_{4,7}R_{7,5} = R_{4,8}F_{8,5}.$$

By MAGMA equation (9, 8) has the solution $uv^{-1} = 10$. Now MAGMA asserts that equation (9, 9) has the solution $vgp = 11$. Hence $p \neq 0$. By multiplying some power of $diag(\left(\begin{smallmatrix}2&0\\0&1\end{smallmatrix}\right), \left(\begin{smallmatrix}1&0\\0&1\end{smallmatrix}\right), 1, 1, 1, 1, 1)$ from the right we can modify T so that $p = 1$. Thus $vg = 11$ and the third equation (4, 5) implies that

$$(4,5) : g(-q(F_{3,3} - F_{4,4})R_{3,5} + F_{4,7}R_{7,5}) = R_{4,8}F_{8,5}.$$

This is a non-linear equation in the unknowns q and g and it has a unique solution $q = 6$ and $g = 11$ as has been checked by running with MAGMA through all 13^2 cases. From $6 = g^{-1} = pu - tq = 10 - 6t$ we now deduce that $t = 5$. This completes the determination of the coefficients of the two matrices w and z of statement (e). The remaining assertions are now clear. □

Proposition 9.2.3 *Keep the notation of Lemmas 9.1.1, 9.1.2 and 9.2.1. Let PG_1 be the faithful permutation representation of the simple group*

$G_1 = \langle x, y, q, w\rangle = \langle u, v, r, s\rangle$ of degree 31671 with stabilizer $H_1 = \langle x, y, q\rangle$. Let $mH = \langle u, v, r\rangle$, $mD = \langle u, v, r\rangle$ and $mE = \langle u, v, s\rangle$. Let F^* be the multiplicative group of the prime field $F = GF(13)$. Let $Y = \mathrm{GL}_{5083}(13)$.

Let $\mathfrak{V}$ and $\mathfrak{W}$ be the up to isomorphism uniquely determined faithful semi-simple 5083-dimensional modules of mH and mE over F corresponding to the restrictions $\tau_{4|mH}$ and $\tau_{4|mE}$ of the irreducible character τ_4 of G_1, respectively.

Let $\kappa_{\mathfrak{V}} : mH \to \mathrm{GL}_{5083}(13)$ and $\kappa_{\mathfrak{W}} : mE \to \mathrm{GL}_{5083}(13)$ be the representations of mH and mE afforded by the modules $\mathfrak{V}$ and $\mathfrak{W}$, respectively.

Let $\mathfrak{r} = \kappa_{\mathfrak{V}}(r)$, $\mathfrak{u} = \kappa_{\mathfrak{V}}(u)$, $\mathfrak{v} = \kappa_{\mathfrak{V}}(v)$ in $\kappa_{\mathfrak{V}}(mH) \le \mathrm{GL}_{5083}(13)$.

Then $\mathfrak{V}_{|mD} \cong \mathfrak{W}_{|mD}$, and there is a transformation matrix $\mathcal{T} \in \mathrm{GL}_{5083}(13)$ such that

$$\mathfrak{u} = \mathcal{T}^{-1}\kappa_{\mathfrak{W}}(u)\mathcal{T}, \qquad \mathfrak{v} = \mathcal{T}^{-1}\kappa_{\mathfrak{W}}(v)\mathcal{T}.$$

Let $\mathfrak{mD} = \langle \mathfrak{u}, \mathfrak{v}\rangle$, $\mathfrak{mH} = \langle \mathfrak{u}, \mathfrak{v}, \mathfrak{r}\rangle$. Let $\mathcal{D} = C_Y(\mathfrak{mD})$ and $\mathcal{H} = C_Y(\mathfrak{mH})$. Let $\mathfrak{s}_1 = \mathcal{T}^{-1}\kappa_{\mathfrak{W}}(s)\mathcal{T}$. Let $\mathfrak{mE} = \langle \mathfrak{mD}, \mathfrak{s}_1\rangle$ and $\mathcal{E} = C_Y(\mathfrak{mE})$. Then the following statements hold.

(a) *There is an isomorphism*

$$\alpha : \mathcal{D} \to \mathcal{D}_1 = \mathrm{GL}_2(13) \times F^{*8} \le \mathrm{GL}_{10}(13).$$

(b) *$\mathcal{H}_1 = \alpha(\mathcal{H})$ is generated by the two blocked diagonal matrices*

$$\begin{aligned} a_1 &= diag(\left(\begin{smallmatrix} 2 & 0 \\ 0 & 1 \end{smallmatrix}\right), 1, 2, 1, 2, 2, 1, 2, 1), \\ a_2 &= diag(\left(\begin{smallmatrix} 1 & 0 \\ 0 & 2 \end{smallmatrix}\right), 2, 1, 2, 1, 1, 2, 1, 2). \end{aligned}$$

(c) *$\mathcal{E}_1 = \alpha(\mathcal{E})$ is generated by the four blocked diagonal matrices*

$$\begin{aligned} b_1 &= diag(\left(\begin{smallmatrix} 2 & 0 \\ 0 & 1 \end{smallmatrix}\right), 1, 1, 1, 1, 1, 1, 1, 1), \\ b_2 &= diag(\left(\begin{smallmatrix} 1 & 0 \\ 0 & 2 \end{smallmatrix}\right), 2, 1, 1, 1, 1, 1, 1, 1), \\ b_3 &= diag(\left(\begin{smallmatrix} 1 & 0 \\ 0 & 1 \end{smallmatrix}\right), 1, 2, 1, 2, 2, 2, 1, 1), \\ b_4 &= diag(\left(\begin{smallmatrix} 1 & 0 \\ 0 & 1 \end{smallmatrix}\right), 1, 1, 2, 1, 1, 1, 2, 2). \end{aligned}$$

(d) *$\mathcal{D}$ has $2184 \times 12^4 = 4587424$ $\mathcal{H}$-$\mathcal{E}$ double cosets.*

(e) *The free product $mH *_{mD} mE$ of mH and mE with amalgamated subgroup mD has an irreducible 5083-dimensional representation over F which induces an irreducible representation of G_1. It corresponds to the $\mathcal{H}$-$\mathcal{E}$ double coset representative*

$$\mathcal{F} = diag(w, 1^{78}, 1^{91}, 11^{156}, 6^{273}, 11^{273}, 1^{728}, 1^{1456}, 1^{1664}) \in Y,$$

where

$$w = \begin{pmatrix} 1^{182} & 9^{182} \\ 11^{182} & 9^{182} \end{pmatrix},$$

and a^n *denotes a diagonal* $n \times n$ *matrix with unique diagonal non-zero entry* $a \in GF(13)$.

Let $\mathfrak{s} = \mathcal{F}^{-1}\mathfrak{s}_1\mathcal{F}$ *and* $\mathfrak{G}_1 = \langle \mathfrak{u}, \mathfrak{v}, \mathfrak{r}, \mathfrak{s} \rangle$. *Inserting these four generating matrices of* $\mathfrak{G}_1$ *into the formulas of Lemma 9.2.1(f) one obtains the matrices* $\mathfrak{x}_{5083}$, $\mathfrak{y}_{5083}$, $\mathfrak{q}_{5083}$ *and* $\mathfrak{w}_{5083}$ *of the original generators* x, y, q *and* w *of* $G_1 = \mathsf{Fi}_{23}$ *as words in the generators* u, v, r *and* s. *The matrices* $\mathfrak{x}_{5083}$, $\mathfrak{y}_{5083}$, $\mathfrak{q}_{5083}$ *and* $\mathfrak{w}_{5083}$ *are the respective lower right block matrices of the matrices* $\mathfrak{x}$, $\mathfrak{y}$, $\mathfrak{q}$ *and* $\mathfrak{w}$ *of* $\mathrm{GL}_{8671}(13)$ *stored on the accompanying DVD.*

Proof Let $\mathfrak{V}$ be the up to isomorphism uniquely determined faithful semi-simple 5083-dimensional module of mH over $F = GF(13)$ corresponding to the restriction $\tau_{4|mH}$. By Lemma 9.2.1(g) $\tau_{4|mH} = \pi_{12} + \pi_{15}$.

The irreducible characters π_{11} and π_{15} of mH are constituents of the permutation characters $1_{mH_{11}}^{mH}$ and $1_{mH_{15}}^{mH}$ of the respective subgroups mH_{11} and mH_{15} of mH determined in Lemma 9.2.1(i). Using Algorithm 1.5.1 Kim and the author calculated the primitive idempotents of the endomorphism rings of these permutation modules in [74]. Thus we obtained the corresponding irreducible representations $M(\pi_{11})$ and $M(\pi_{15})$ of the respective dimensions 2275 and 2808 over $F = GF(13)$. The irreducible FmH-module $M(\pi_{12})$ is the tensor product of $M(\pi_{11})$ with the linear character π_2 of mH over F by Lemma 9.2.1(j). Thus $\mathfrak{V} = M(\pi_{12}) \oplus M(\pi_{15})$.

Let $\mathfrak{W}$ be the up to isomorphism uniquely determined faithful semi-simple 5083-dimensional module of mE over F corresponding to the restriction $\tau_{4|mE} = \psi_{14} + \psi_{22} + \psi_{45} + \psi_{74}$; see Lemma 9.2.1(i).

The irreducible characters ψ_{14}, ψ_{22}, ψ_{45} and ψ_{74} of mE are constituents of the permutation characters $1_{mE_{14}}^{mE}$, $1_{mE_{22}}^{mE}$, $1_{mE_{45}}^{mE}$ and $1_{mE_{74}}^{mE}$, respectively, by Lemma 9.2.1(l). Using Algorithm 1.5.1 we calculated the primitive idempotents of the endomorphism rings of these four permutation modules. Thus we obtained the corresponding irreducible representations $N(\psi_{14})$, $N(\psi_{22})$, $N(\psi_{45})$ and $N(\psi_{74})$ of the respective dimensions 182, 260, 1365 and 3276 over F. Hence

$$\mathfrak{W} = N(\psi_{14}) \oplus N(\psi_{22}) \oplus N(\psi_{45}) \oplus N(\psi_{74}).$$

Fixing a basis in each irreducible constituent $M\pi_k$ of $\mathfrak{V}$ we get a basis $\mathcal{B}_V$ of $\mathfrak{V}$. It induces a representation $\kappa_{\mathfrak{V}} : mH \to \mathrm{GL}_{5083}(13)$ of mH. Let $\mathfrak{r} = \kappa_{\mathfrak{V}}(r)$, $\mathfrak{u} = \kappa_{\mathfrak{V}}(u)$ and $\mathfrak{v} = \kappa_{\mathfrak{V}}(v)$ in $\kappa_{\mathfrak{V}}(mH) \le \mathrm{GL}_{5083}(13)$.

Fixing a basis in each irreducible constituent $N\psi_j$ of $\mathfrak{W}$ we get a basis $\mathcal{B}_W$ of $\mathfrak{W}$. It induces a representation $\kappa_{\mathfrak{W}} : mE \to \mathrm{GL}_{5083}(13)$ of mE. By Lemma 9.2.1(h) $\mathfrak{V}_{|mD} \cong \mathfrak{W}_{|mD}$. Let $Y = \mathrm{GL}_{5083}(13)$. Applying now Parker's isomorphism test of Proposition 6.1.6 of [92] by means of the MAGMA command

```
IsIsomorphic(GModule(sub<Y|V(u),V(v)>),GModule(sub<Y|W(u),W(v)>))
```

one obtains the transformation matrix $\mathcal{T}_1$ satisfying $\mathfrak{u} = \kappa_{\mathfrak{W}}(u)^{\mathcal{T}_1}$ and $\mathfrak{v} = \kappa_{\mathfrak{W}}(v)^{\mathcal{T}_1}$.

(a) Let $\mathfrak{m}\mathfrak{D} = \langle \mathfrak{u}, \mathfrak{v} \rangle$ and $\mathfrak{m}\mathfrak{H} = \langle \mathfrak{u}, \mathfrak{v}, \mathfrak{r} \rangle$. Let $\mathcal{D} = C_Y(\mathfrak{m}\mathfrak{D})$ and $\mathcal{H} = C_Y(\mathfrak{m}\mathfrak{H})$. Let δ_{32a} and δ_{32b} be two distinct copies of the irreducible character δ_{32} of mD. Using PG_1 and MAGMA it can be checked that the irreducible characters π_{12} and π_{15} of mH have the following restrictions to $mD = \langle u, v \rangle$:

$$\pi_{12\,|mD} = \delta_{16} + \delta_{32a} + \delta_{34} + \delta_{37} + \delta_{66},$$
$$\pi_{15\,|mD} = \delta_{11} + \delta_{23} + \delta_{32b} + \delta_{53} + \delta_{69},$$

where the irreducible characters δ_{11}, δ_{16}, δ_{23}, δ_{32}, δ_{34}, δ_{37}, δ_{53}, δ_{66} and δ_{69} of $mD \cong G_2(3) \times S_3$ have degrees 78, 91, 156, 182, 273, 273, 728, 1456 and 1664, respectively.

Since $\mathfrak{V}_{|mD}$ is a semi-simple FmD-module Theorem 2.1.27 of [92] implies that there is an isomorphism

$$\alpha : \mathcal{D} \to \mathcal{D}_1 = \mathrm{GL}_2(13) \times F^{*8} \leq \mathrm{GL}_{10}(13).$$

(b) Schur's Lemma asserts that $\mathcal{H}_1 = \alpha(\mathcal{H})$ is generated by the two blocked diagonal matrices a_1 and a_2 given in the statement because 2 is a primitive element in the multiplicative group F^* of $F = GF(13)$.

(c) Let $\mathfrak{s}_1 = \mathcal{T}_1^{-1}\kappa_{\mathfrak{W}}(s)\mathcal{T}_1$. Let $\mathfrak{m}\mathfrak{E} = \langle \mathfrak{m}\mathfrak{D}, \mathfrak{s}_1 \rangle$ and $\mathcal{E} = C_Y(\mathfrak{m}\mathfrak{E})$. Let δ_{32a} and δ_{32b} be two distinct copies of the irreducible character δ_{32} of mD. Using PG_1 and MAGMA it can be checked that the irreducible characters ψ_{14}, ψ_{22}, ψ_{45} and ψ_{74} of mE have the following restrictions to $mD = \langle u, v \rangle$:

$$\psi_{14\,|_m D} = \delta_{32a}, \quad \psi_{22\,|_m D} = \delta_{11} + \delta_{32b},$$
$$\psi_{45\,|_m D} = \delta_{16} + \delta_{34} + \delta_{37} + \delta_{53}, \quad \psi_{74\,|_m D} = \delta_{23} + \delta_{66} + \delta_{69}.$$

Now Schur's Lemma implies that $\mathcal{E}_1 = \alpha(\mathcal{E})$ is generated by the four blocked diagonal matrices b_j given in the statement.

(d) Every $\mathcal{H}$-$\mathcal{E}$ double coset representative is of the form

$$diag(A, 1, v_2, v_3, v_4, v_5, 1, 1, 1)$$

for some $A \in Y$ and $v \in F^*$. By multiplying from left and right, we can find out that $diag(A, 1, v_2, v_3, v_4, v_5, 1, 1, 1)$ and $diag(A', 1, v_2, v_3, v_4, v_5, 1, 1, 1)$ represent the same double coset if and only if the first column of A is a scalar multiple of that of A'. So, we have 12^4 choices for v_i, and $|\operatorname{GL}(2, 13)|/12 = 2184$ choices for A. Thus there are $2184 \cdot 12^4 = 4587424$ $\mathcal{H}$-$\mathcal{E}$ double cosets.

(e) By Theorem 7.2.2 of [92] the irreducible representations of the free product $mH *_{mD} mE$ of the groups mH and mE with amalgamated subgroup mD are described by the $\mathcal{H}$-$\mathcal{E}$ double coset representatives T of $\mathcal{D}$. The elements r and $f = (u^3vsv)^9$ are two commuting involutions of $G_1 \cong \mathsf{Fi}_{23}$ by Lemma 9.2.1(o). Let $\mathfrak{r}$ and $\mathfrak{f}$ be their matrices in $\mathfrak{m}\mathfrak{H}$ and $\mathfrak{m}\mathfrak{E}$, respectively. If $T_1 = diag(\left(\begin{smallmatrix} a & c \\ b & d \end{smallmatrix}\right), 1, v_2, v_3, v_4, v_5, 1, 1, 1)$ describes a 5083-dimensional irreducible representation of G_1 over F then

$$(**)(\mathfrak{r}, \mathcal{T}_1^{-1}\mathfrak{f}\mathcal{T}_1) = 1$$

holds.

Since $\mathfrak{V}_{|mD} \cong \mathfrak{W}_{|mD}$ is a direct sum of ten irreducible FmD-modules, both matrices $\mathfrak{r}$ and $\mathfrak{f}$ consist of 100 blocks $R_{i,j}$ and $F_{i,j}$, $1 \le i, j \le 10$, such that all diagonal blocks $R_{i,i}$ and $F_{i,i}$ are non-trivial. Furthermore, a non-diagonal block $R_{i,j}$ of $\mathfrak{r}$ is non-zero if and only if the ith irreducible and the jth irreducible representations of mD appear in the restriction of an irreducible representation of mH to mD. A similar description holds for the blocks of $\mathfrak{f}$. Hence the system of equations in the proofs of (a) and (c) imply that

$$\mathfrak{r} = \begin{pmatrix}
R_{1,1} & \cdot & \cdot & R_{1,4} & \cdot & R_{1,6} & R_{1,7} & \cdot & R_{1,9} & \cdot \\
\cdot & R_{2,2} & R_{2,3} & \cdot & R_{2,5} & \cdot & \cdot & R_{2,8} & \cdot & R_{2,10} \\
\cdot & R_{3,2} & R_{3,3} & \cdot & R_{3,5} & \cdot & \cdot & R_{3,8} & \cdot & R_{3,10} \\
R_{4,1} & \cdot & \cdot & R_{4,4} & \cdot & R_{4,6} & R_{4,7} & \cdot & R_{4,9} & \cdot \\
\cdot & R_{5,2} & R_{5,3} & \cdot & R_{5,5} & \cdot & \cdot & R_{5,8} & \cdot & R_{5,10} \\
R_{6,1} & \cdot & \cdot & R_{6,4} & \cdot & R_{6,6} & R_{6,7} & \cdot & R_{6,9} & \cdot \\
R_{7,1} & \cdot & \cdot & R_{7,4} & \cdot & R_{7,6} & R_{7,7} & \cdot & R_{7,9} & \cdot \\
\cdot & R_{8,2} & R_{8,3} & \cdot & R_{8,5} & \cdot & \cdot & R_{8,8} & \cdot & R_{8,10} \\
R_{9,1} & \cdot & \cdot & R_{9,4} & \cdot & R_{9,6} & R_{9,7} & \cdot & R_{9,9} & \cdot \\
\cdot & R_{10,2} & R_{10,3} & \cdot & R_{10,5} & \cdot & \cdot & R_{10,8} & \cdot & R_{10,10}
\end{pmatrix},$$

$$\mathfrak{f} = \begin{pmatrix}
F_{1,1} & \cdot & \cdot & \cdot & \cdot & \cdot & \cdot & \cdot & \cdot & \cdot \\
\cdot & F_{2,2} & F_{2,3} & \cdot & \cdot & \cdot & \cdot & \cdot & \cdot & \cdot \\
\cdot & F_{3,2} & F_{3,3} & \cdot & \cdot & \cdot & \cdot & \cdot & \cdot & \cdot \\
\cdot & \cdot & \cdot & F_{4,4} & \cdot & F_{4,6} & F_{4,7} & F_{4,8} & \cdot & \cdot \\
\cdot & \cdot & \cdot & \cdot & F_{5,5} & \cdot & \cdot & \cdot & F_{5,9} & F_{5,10} \\
\cdot & \cdot & \cdot & F_{6,4} & \cdot & F_{6,6} & F_{6,7} & F_{6,8} & \cdot & \cdot \\
\cdot & \cdot & \cdot & F_{7,4} & \cdot & F_{7,6} & F_{7,7} & F_{7,8} & \cdot & \cdot \\
\cdot & \cdot & \cdot & F_{8,4} & \cdot & F_{8,6} & F_{8,7} & F_{8,8} & \cdot & \cdot \\
\cdot & \cdot & \cdot & \cdot & F_{9,5} & \cdot & \cdot & \cdot & F_{9,9} & F_{9,10} \\
\cdot & \cdot & \cdot & \cdot & F_{10,5} & \cdot & \cdot & \cdot & F_{10,9} & F_{10,10}
\end{pmatrix}.$$

Let $e = (ad - bc)^{-1}$ and $g = (pu - tq)^{-1}$. Then $e \neq 0 \neq g$. For each integer k let I_k denote the $k \times k$ identity matrix over F. Then

$$\mathcal{T}^{-1} = \begin{pmatrix} ed(I_{182}) & -ec(I_{182}) & . & . & . & . & . & . & . & . \\ -eb(I_{182}) & ea(I_{182}) & . & . & . & . & . & . & . & . \\ . & . & I_{78} & . & . & . & . & . & . & . \\ . & . & . & v_2^{-1}(I_{91}) & . & . & . & . & . & . \\ . & . & . & . & v_3^{-1}(I_{156}) & . & . & . & . & . \\ . & . & . & . & . & v_4^{-1}(I_{273}) & . & . & . & . \\ . & . & . & . & . & . & v_5^{-1}(I_{273}) & . & . & . \\ . & . & . & . & . & . & . & I_{728} & . & . \\ . & . & . & . & . & . & . & . & I_{1456} & . \\ . & . & . & . & . & . & . & . & . & I_{1664} \end{pmatrix}.$$

Hence $\mathfrak{f}' = \mathcal{T}^{-1}\mathfrak{f}\mathcal{T}$ equals the matrix

$$\begin{pmatrix} G_{1,1} & G_{1,2} & G_{1,3} & . & . & . & . & . & . & . \\ G_{2,1} & G_{2,2} & G_{2,3} & . & . & . & . & . & . & . \\ bF_{3,2} & dF_{3,2} & F_{3,3} & . & . & . & . & . & . & . \\ . & . & . & F_{4,4} & . & G_{4,6} & G_{4,7} & v_2^{-1}F_{4,8} & . & . \\ . & . & . & . & F_{5,5} & . & . & . & v_3^{-1}F_{5,9} & v_3^{-1}F_{5,10} \\ . & . & . & G_{6,4} & . & F_{6,6} & G_{6,7} & v_4^{-1}F_{6,8} & . & . \\ . & . & . & G_{7,4} & . & G_{7,6} & F_{7,7} & v_5^{-1}F_{7,8} & . & . \\ . & . & . & v_2F_{8,4} & . & v_4F_{8,6} & v_5F_{8,7} & F_{8,8} & . & . \\ . & . & . & . & v_3F_{9,5} & . & . & . & F_{9,9} & F_{9,10} \\ . & . & . & . & v_3F_{10,5} & . & . & . & F_{10,9} & F_{10,10} \end{pmatrix},$$

where

$$G_{1,1} = e(adF_{1,1} - bcF_{2,2}), \quad G_{1,2} = ecd(F_{1,1} - F_{2,2}), \quad G_{1,3} = -ecF_{2,3},$$
$$G_{2,1} = -eab(F_{1,1} - F_{2,2}), \quad G_{2,2} = -e(bcF_{1,1} - adF_{2,2}), \quad G_{2,3} = eaF_{2,3},$$
$$G_{4,6} = v_2^{-1}v_4F_{4,6}, \quad G_{4,7} = v_2^{-1}v_5F_{4,7}, \quad G_{6,4} = v_4^{-1}v_2F_{6,4},$$
$$G_{6,7} = v_4^{-1}v_5F_{6,7}, \quad G_{7,4} = v_5^{-1}v_2F_{7,4}, \quad G_{7,6} = v_5^{-1}v_4F_{7,6}.$$

Now $(**)$ implies the following equations:

$$(10,1) : F_{10,9}R_{9,1} = R_{10,2}G_{2,1} + R_{10,3}(bF_{3,2}),$$
$$(9,1) : F_{9,9}R_{9,1} = R_{9,1}G_{1,1},$$
$$(9,2) : v_3F_{9,5}R_{5,2} + F_{9,10}R_{10,2} = R_{9,1}G_{1,2},$$
$$(8,1) : v_2F_{8,4}R_{4,1} + v_4F_{8,6}R_{6,1} + v_5F_{8,7}R_{7,1} = R_{8,2}G_{2,1} + R_{8,3}(bF_{3,2}).$$

Inserting the first set of equations yields

$$\begin{aligned}
(10,1) &: F_{10,9}R_{9,1} = b[-eaR_{10,2}(F_{1,1} - F_{2,2}) + R_{10,3}F_{3,2}],\\
(9,1) &: F_{9,9}R_{9,1} = eR_{9,1}(adF_{1,1} - bcF_{2,2}),\\
(9,2) &: v_3F_{9,5}R_{5,2} + F_{9,10}R_{10,2} = ecdR_{9,1}(F_{1,1} - F_{2,2}).\\
(8,1) &: v_2F_{8,4}R_{4,1} + v_4F_{8,6}R_{6,1} + v_5F_{8,7}R_{7,1}\\
&= b[-eaR_{8,2}(F_{1,1} - F_{2,2}) + R_{8,3}F_{3,2}].
\end{aligned}$$

Equation $(10,1)$ has only one solution $b = 9$, $ea = 1$ in $F = GF(13)$, as has been checked by means of MAGMA. Hence $a \neq 0$. By multiplying some power of $diag(\left(\begin{smallmatrix} 2 & 0 \\ 0 & 1 \end{smallmatrix}\right), 1, 1, 1, 1, 1, 1, 1, 1)$ from the right we can modify the matrix T so that $a = 1$. Thus also $e = 1$. Now equation $(9,1)$ becomes

$$(9,1) : F_{9,9}R_{9,1} = R_{9,1}(dF_{1,1} - 9cF_{2,2}).$$

Another MAGMA calculation running through 13^2 pairs (c, d) of elements of F shows that $(9,1)$ has the unique solution $c = 11$ and $d = 9$.

Inserting the known values for c, d and e into equation $(9,2)$ yields that $v_3 = 11$, as has been checked by means of MAGMA. Finally, inserting the known values for a, b and e into equation $(8,1)$ yields

$$\begin{aligned}
(8,1) &: v_2F_{8,4}R_{4,1} + v_4F_{8,6}R_{6,1} + v_5F_{8,7}R_{7,1}\\
&= 9[-R_{8,2}(F_{1,1} - F_{2,2}) + R_{8,3}F_{3,2}].
\end{aligned}$$

A MAGMA calculation shows that this equation has the unique solution $v_2 = 1$, $v_4 = 6$, $v_5 = 11$. This completes the proof because all remaining statements of (e) are straightforward. □

9.3 Construction of the irreducible subgroup $\mathfrak{G}$ of $\mathrm{GL}_{8671}(13)$

In this section we construct the eight semi-simple representations of the 2-fold cover A_1 of the automorphism group $Aut(\mathsf{Fi}22)$ of Fi_{22} corresponding to the eight compatible pairs of Lemma 9.1.2(e). Since $H_1 = \langle q, y \rangle = C_{G_1}(u) \cong A_1'$ for some involution u of $G_1 = \langle q, y, w \rangle$, and A_1' has index 2 in A_1, it is not difficult to construct the irreducible constituents of these representations of A_1 from the 3588-dimensional irreducible representation of $G_1 \cong \mathsf{Fi}_{23}\langle q, y, w \rangle$ by means of Clifford's Theorem. In fact, we construct eight new matrices t_i of order 2 such that $K_i = \langle G_1, t_i \rangle$ is an irreducible subgroup of $\mathrm{GL}_{8671}(13)$. It turns out that only K_3 corresponding to the compatible pair (3) of Lemma 9.1.2(e) may have a Sylow

2-subgroup which is isomorphic to a Sylow 2-subgroup of the extension group E of Lemma 7.1.1.

Proposition 9.3.1 *Let* $\mathfrak{G}_1 = \langle \mathfrak{y}_{3588}, \mathfrak{q}_{3588}, \mathfrak{w}_{3588} \rangle$ *be the simple subgroup of* $Y = \mathrm{GL}_{3588}(13)$ *of order* $2^{18} \cdot 3^{13} \cdot 5^2 \cdot 7 \cdot 11 \cdot 13 \cdot 17 \cdot 23$ *constructed in Proposition 9.2.2. Let* $y_1 = \mathfrak{y}_{3588}$ *and* $q_1 = \mathfrak{q}_{3588}$*. Let* $\mathfrak{H}_1 = \langle y_1, q_1 \rangle$.

Let $A_1 = 2Aut(\mathsf{Fi}_{22}) = \langle a, b, c, d, e, f, g, h, i, z, t \rangle$ *be the finitely presented group defined in Lemma 9.1.1 and let* $H_1 = \langle a, b, c, d, e, f, g, h, i, z \rangle$*. Then the following assertions hold.*

(a) *There is an isomorphism* $\phi : H_1 \to \mathfrak{H}_1$.

(b) $A_1 = \langle y, q, t \rangle$*, where* $y = \phi^{-1}(y_1)$ *and* $q = \phi^{-1}(q_1)$.

(c) *There is a transformation matrix* $\mathcal{T} \in Y$ *such that*

$$\mathcal{T}^{-1}\mathfrak{y}\mathcal{T} = diag(\mathfrak{y}_{78}, \mathfrak{y}_{1430}, \mathfrak{y}_{2080}),$$
$$\mathcal{T}^{-1}\mathfrak{q}\mathcal{T} = diag(\mathfrak{q}_{78}, \mathfrak{q}_{1430}, \mathfrak{q}_{2080}) \in Y,$$

where $y_k, q_k \in \mathrm{GL}_k(13)$ *for* $k \in \{78, 1430, 2080\}$.

(d) A_1 *has a faithful irreducible representation* $\lambda : A_1 \to \mathrm{GL}_{4160}(13)$ *such that*

$$\lambda(y) = diag(\mathfrak{y}_{2080}, \phi(y^t)_{2080}), \quad \lambda(q) = diag(\mathfrak{h}_{2080}, \phi(q^t)_{2080}),$$
$$\lambda(t) = \begin{pmatrix} 0 & I_{2080} \\ I_{2080} & 0 \end{pmatrix},$$

where I_{2080} *denotes the identity matrix of* $\mathrm{GL}_{2080}(13)$.

(e) *The irreducible characters* χ_3, χ_4, χ_{11} *and* χ_{12} *of Table 9.8.1 of respective degrees* 78, 78, 1430 *and* 1430 *are constituents of the permutation character* $1_U^{A_1}$ *of the subgroup*

$$U = \langle (q_1^2 y_1^3 q_1 y_1^3)^4, (y_1^2 q_1 y_1^3 q_1^2 y_1 q_1)^6, (y_1^4 q_1 y_1 q_1 y_1 q_1 y_1^2)^2 \rangle$$

of A_1 *of index* 2358720.

(f) *The tensor product* $\chi_3 \otimes \chi_3$ *contains a copy of the irreducible character* χ_{13}*. Furthermore,* $\chi_{14} = \chi_{13} \otimes \chi_2$*, where* χ_2 *is the non-trivial linear character of* A_1.

Proof (a) In the simple subgroup $\mathfrak{G}_1$ of $Y = \mathrm{GL}_{3588}(13)$ let

$$x_1 = [(y_1 q_1^2 y_1 q_1 y_1 q_1^2)^{11} (q_1^2 y_1^2 q_1 y_1 q_1 y_1)^{11} (q_1 y_1^2 q_1 y_1 q_1 y_1 q_1 y_1 q_1)^4]^{12},$$
$$a_1 = (x_1 y_1 x_1)^7, \quad b_1 = [(q_1 y_1)^2 q_1 y_1^3 q_1^2 y_1^3 q_1 y_1]^7, \quad c_1 = (y_1^2 x_1 y_1 x_1 y_1^3)^5,$$
$$d_1 = (q_1 y_1 q_1^2 y_1 q_1 y_1 q_1 y_1 q_1 y_1^2 q_1^2)^{15}, \quad e_1 = (y_1 x_1 y_1^5 x_1)^5,$$
$$f_1 = (y_1 q_1 y_1 q_1^2 y_1 q_1^2 y_1^2 q_1 y_1^4 q_1^2)^5, \quad g_1 = (x_1 y_1^2 x_1 y_1^3 x_1)^7,$$
$$h_1 = (y_1^5 x_1 y_1 x_1)^5, \quad i_1 = (q_1^2 y_1^2 q_1 y_1 q_1^2)^7.$$

By Lemma 9.1.2 and Proposition 9.2.2 the map θ sending x, y, q and w of the subgroup $G_1 = \langle x, y, q, w \rangle$ of $\mathrm{GL}_{782}(17)$ to x_1, y_1, q_1 and w_{3588} of the subgroup $\mathfrak{G}_1 = \langle \mathfrak{y}_{3588}, \mathfrak{q}_{3588}, \mathfrak{w}_{3588} \rangle$ of $\mathrm{GL}_{3588}(13)$, respectively, is an isomorphism. Thus Lemma 9.1.2(b) implies that the normal subgroup H_1 of A_1 is isomorphic to $\mathfrak{H}_1 = \langle a_1, b_1, c_1, d_1, e_1, f_1, g_1, h_1, i_1, z_1 \rangle$, where $z_1 = (x_1 y_1^2)^7$. In particular, the map $\phi : H_1 \to \mathfrak{H}_1$ sending a, b, ..., z to a_1, b_1, ..., z_1 is an isomorphism such that $x = \phi^{-1}(x_1)$, $y = \phi^{-1}(y_1)$ and $q = \phi^{-1}(q_1)$.

Statement (b) is an immediate consequence of (a) and Lemma 9.1.1(b).

(c) The natural F-vector space $V = F^{3588}$ is an $F\mathfrak{H}_1$-module because $\mathfrak{H}_1 = \langle y_1, q_1 \rangle$. Applying the Meataxe algorithm it follows that V has three composition factors V_{78}, V_{1430} and V_{2080} of dimensions 78, 1430 and 2080, respectively. Since all three dimensions are divisible by 13, and 13 divides $|\mathfrak{H}_1| = |2\mathsf{Fi}_{22}|$ to the first power only, all three simple composition factors of V are projective $F\mathfrak{H}$-modules by Theorems 3.12.2 and 3.12.4 of [92]. Hence V is isomorphic to their direct sum. Thus (c) holds.

(d) The group A_1 of Lemma 9.1.1 has a unique irreducible character χ_{17} of degree 4160; see Table 9.8.1. Clifford's Theorem 2.6.15 of [92] asserts that its restriction to H_1 is a sum of two inequivalent irreducible characters ν and ν^t of degree 2080. In particular, the induced FA_1-module $V_{2080}^{A_1}$ is the reduction modulo 13 of a lattice which affords the irreducible character χ_{17} of A_1. It corresponds to the irreducible representation $\lambda : A_1 \to \mathrm{GL}_{4160}(13)$ of 13-defect zero defined in statement (d). It is well defined because Lemma 9.1.1(b) implies that $a^t = g^{-1}$, $b^t = f^{-1}$, $c^t = e^{-1}$, $d^t = d^{-1}$, $h^t = h^{-1}$ and $i^t = i^{-1}$. Therefore $\phi(y)_{2080}$ and $\phi(q)_{2080}$ are well defined by (a) and (c).

(e) Using the command `LowIndexSubgroups(A_1, m)` we searched for conjugacy classes of subgroups U of index $|A_1 : U| = m$ such that χ_k is an irreducible constituent of the permutation character $1_U^{A_1}$ for $k \in \{3, 4, 11, 12\}$. Thus we found a subgroup U of index $m = 2358720$ such that its permutation character contains all four irreducible characters χ_k. Its generators were obtained by means of the program `GetShortGenerators(A_1, U)`.

Statement (f) can be verified by means of the character table of A_1. □

Proposition 9.3.2 *Keep the notation of Lemma 9.1.1 and Propositions 9.2.2 and 9.2.3. Let $A_1 \leftarrow H_1 \to G_1$ be the amalgam constructed in*

Lemma 9.1.2, where $G_1 \cong \mathsf{Fi}_{23}$. *Let* $\sigma : H_1 \to A_1$ *denote its monomorphism of* H_1 *into* G_1. *Let* $Y = \mathrm{GL}_{8671}(13)$. *Let* $\sigma(y) = diag(\mathfrak{y}_{3588}, \mathfrak{y}_{5083})$, $\sigma(q) = diag(\mathfrak{q}_{3588}, \mathfrak{q}_{5083})$ *and* $\mathfrak{w}_1 = diag(\mathfrak{w}_{3588}, \mathfrak{w}_{5083})$ *in* Y. *Let* $\mathfrak{H}_1 = \langle \sigma(y), \sigma(q) \rangle$ *and* $\mathfrak{G}_1 = \langle \mathfrak{H}_1, \mathfrak{w}_1 \rangle$. *Keep the notation of Tables 9.8.1 and 6.6.3 and of the character table of* Fi_{23}*; see [19], pp. 178–179.*

Then the following statements hold.

(a) *There is a compatible pair of characters*

$$(\chi, \tau) = (\chi_3 + \chi_{12} + \chi_{13} + \chi_{\mathbf{17}}, \tau_{\mathbf{3}} + \tau_{\mathbf{4}}) \in mf\,char_{\mathbb{C}}(A_1) \times mf\,char_{\mathbb{C}}(G_1)$$

of degree 8671 *of the groups* $A_1 = \langle H_1, t \rangle$ *and* $G_1 = \langle H_1, e_1 \rangle$ *with common restriction*

$$\tau_{|H_1} = \chi_{|H_1} = \delta_2 + \delta_6 + \delta_{\mathbf{7}} + \delta_{\mathbf{8}} + \delta_9 \in mf\,char_{\mathbb{C}}(H_1),$$

where irreducible characters with bold face indices denote faithful irreducible characters.

(b) *Let* $\mathfrak{V}$ *and* $\mathfrak{W}$ *be the up to isomorphism uniquely determined faithful semi-simple multiplicity-free* 8671*-dimensional modules of* A_1 *and* G_1 *over* $F = GF(13)$ *corresponding to the compatible pair* χ, τ*, respectively.*

Let $\kappa_{\mathfrak{V}} : H \to \mathrm{GL}_{8671}(13)$ *and* $\kappa_{\mathfrak{W}} : E \to \mathrm{GL}_{8671}(13)$ *be the representations of* A_1 *and* G_1 *afforded by the modules* $\mathfrak{V}$ *and* $\mathfrak{W}$*, respectively.*

Let $\mathfrak{q} = \kappa_{\mathfrak{V}}(q)$, $\mathfrak{y} = \kappa_{\mathfrak{V}}(y)$ *and* $\mathfrak{t} = \kappa_{\mathfrak{V}}(t)$ *in* $\kappa_{\mathfrak{V}}(A_1) \le \mathrm{GL}_{8671}(13)$. *Then the following assertions hold.*

(1) $\mathfrak{V}_{|\mathfrak{H}_1} \cong \mathfrak{W}_{|\mathfrak{H}_1}$, *and there is a transformation matrix* $\mathcal{T} \in \mathrm{GL}_{8671}(13)$ *such that*

$$\mathfrak{q} = \mathcal{T}^{-1}\kappa_{\mathfrak{W}}(\sigma(q))\mathcal{T}, \qquad \mathfrak{y} = \mathcal{T}^{-1}\kappa_{\mathfrak{W}}(\sigma(y))\mathcal{T}.$$

Let $\mathfrak{w} = \mathcal{T}^{-1}\kappa_{\mathfrak{W}}(w_1)\mathcal{T} \in \mathrm{GL}_{8671}(13)$.

(2) *The subgroup* $\mathfrak{G} = \langle \mathfrak{q}, \mathfrak{y}, \mathfrak{t}, \mathfrak{w} \rangle$ *of* Y *is the uniquely determined irreducible representation of the free product* $P = G_1 *_{H_1} A_1$ *of* G_1 *and* A_1 *with amalgamated subgroup* H_1 *corresponding to the compatible pair (3) of Lemma 9.1.2(e). Its four generating matrices of* $\mathfrak{G}$ *are stored on the accompanying DVD.*

Proof (a) By Lemma 9.1.2(e) the amalgam $A_1 \leftarrow H_1 \rightarrow G_1$ has eight compatible pairs of degree 8671. We constructed the corresponding semi-simple representations of $P = G_1 *_{H_1} A_1$ for each of them. But we give a proof only for that pair (3) of Lemma 9.1.2(e). It belongs to a group of a suitable order.

(b) By Propositions 9.2.2 and 9.2.3 the semi-simple FG_1-module $\mathfrak{W} = \mathfrak{W}_3 \oplus \mathfrak{W}_4$ of dimension 8671 corresponding to the multiplicity free character $\tau_3 + \tau_4$ is described by the three matrices $\mathfrak{y} = \sigma(y)$, $\mathfrak{q} = \sigma(q)$ and $\mathfrak{w}_1$ of Y. The semi-simple FA_1-module $\mathfrak{V}$ of the same dimension corresponding to the multiplicity free character $\chi_3 + \chi_{12} + \chi_{14} + \chi_{17}$ of A_1 is defined by three blocked diagonal matrices $\mathfrak{q} = diag(\mathfrak{q}_3, \mathfrak{q}_{12}, \mathfrak{q}_{13}, \mathfrak{q}_{17})$, $\mathfrak{y} = diag(\mathfrak{y}_3, \mathfrak{y}_{12}, \mathfrak{y}_{13}, \mathfrak{y}_{17})$ and $\mathfrak{t} = diag(\mathfrak{t}_3, \mathfrak{t}_{12}, \mathfrak{t}_{13}, \mathfrak{t}_{17})$, whose entries can be calculated by means of Proposition 9.3.1 as follows.

Assertion (d) of Proposition 9.3.1 states that $\mathfrak{q}_{17} = \lambda(q)$, $\mathfrak{y}_{17} = \lambda(y)$ and $\mathfrak{t}_{17} = \lambda(t)$ in $\mathrm{GL}_{4160}(13)$. By Proposition 9.3.1(e) the irreducible characters χ_3 and χ_{12} are constituents of the permutation character $1_U^{A_1}$ of a well determined subgroup U of A_1 of index 2358720. Let P_U be the corresponding permutation module over $F = GF(13)$. Using Algorithm 1.5.1 we calculated the primitive idempotents E_{χ_3} and $E_{\chi_{12}}$ of the endomorphism ring $End_{FA_1}(P_U)$ of P_U. Since the irreducible characters χ_3 and χ_{12} are of 13-defect zero the FA_1-modules $\mathfrak{V}_3 = E_{\chi_3} P_U$ and $\mathfrak{V}_{12} = E_{\chi_{12}} P_U$ are irreducible by Theorems 3.12.2 and 3.12.4 of [92]. After fixing a basis in each of them the actions of the generators q, y and t of A_1 on P_U induce the matrices $\mathfrak{q}_3$, $\mathfrak{q}_{12}$, $\mathfrak{y}_3$, $\mathfrak{y}_{12}$ and $\mathfrak{t}_3$, $\mathfrak{t}_{12}$, respectively.

The tensor product $\chi_3 \otimes \chi_3$ contains a copy of the irreducible character χ_{13} by Proposition 9.3.1(f). Since $\mathfrak{V}_3$ has dimension 78 the Meataxe algorithm implemented in MAGMA can be applied to the tensor product $\mathfrak{V}_3 \otimes \mathfrak{V}_3$. This application provides the three 3003×3003 matrices $\mathfrak{q}_{13}$, $\mathfrak{y}_{13}$ and $\mathfrak{t}_{13}$ corresponding to the irreducible FA_1-module $\mathfrak{V}_{13}$. Hence $\mathfrak{V} = \mathfrak{V}_3 \oplus \mathfrak{V}_{12} \oplus \mathfrak{V}_{13} \oplus \mathfrak{V}_{17}$.

By (a) the restrictions $\mathfrak{V}_{3|\mathfrak{H}_1}$, $\mathfrak{V}_{12|\mathfrak{H}_1}$ and $\mathfrak{V}_{13|\mathfrak{H}_1}$ to $\mathfrak{H}_1$ are irreducible. By the proof of Proposition 9.3.1(d) we know that

$$\begin{aligned} \mathfrak{V}_{17|\mathfrak{H}_1} &= \mathfrak{V}_{2080} \oplus \mathfrak{V}_{2080} \otimes t, \\ \mathfrak{V}_{3|\mathfrak{H}_1} \oplus \mathfrak{V}_{12|\mathfrak{H}_1} \oplus \mathfrak{V}_{2080} &\cong \mathfrak{W}_{3|\mathfrak{H}_1}, \\ \mathfrak{V}_{13|\mathfrak{H}_1} \oplus \mathfrak{V}_{2080} \otimes t &\cong \mathfrak{W}_{4|\mathfrak{H}_1}. \end{aligned}$$

Let $X_3 = \mathrm{GL}_{3588}(13)$, $X_4 = \mathrm{GL}_{5083}(13)$, $V_3 = \mathfrak{V}_{3|\mathfrak{H}_1} \oplus \mathfrak{V}_{12|\mathfrak{H}_1}$, $V_4 = \mathfrak{V}_{13|\mathfrak{H}_1} \oplus \mathfrak{V}_{2080} \otimes t$, $W_3 = \mathfrak{W}_{3|\mathfrak{H}_1}$ and $W_4 = \mathfrak{W}_{4|\mathfrak{H}_1}$. By the proof of Proposition 9.3.1(d) $V_3 \cong W_3$ and $V_4 \cong W_4$ as $F\mathfrak{H}_1$-modules of respective dimensions 3588 and 5083. Applying now Parker's isomorphism test of Proposition 6.1.6 of [92] by means of the MAGMA command

```
IsIsomorphic(GModule(sub<X_i|V_i(h),V_i(y)>),
  GModule(sub<X_i|W_i(h),W_i(y)>)),
```

$i \in \{3, 4\}$, one obtains the transformation matrices $\mathcal{T}_3 \in X_3$ and $\mathcal{T}_4 \in X_4$ such that $\mathcal{T} = diag(\mathcal{T}_3, \mathcal{T}_4) \in Y$ satisfies $\mathfrak{q} = \kappa_{\mathfrak{W}}(\sigma(q))^{\mathcal{T}}$ and $\mathfrak{y} = \kappa_{\mathfrak{W}}(\sigma(y))^{\mathcal{T}}$.

Let $\mathfrak{w} = \mathcal{T}^{-1}\kappa_{\mathfrak{W}}(w_1)\mathcal{T} \in \mathrm{GL}_{8671}(13)$. Corollary 7.2.4 of [92] asserts that the matrix group $\mathfrak{G} = \langle \mathfrak{q}, \mathfrak{y}, \mathfrak{t}, \mathfrak{w} \rangle$ is uniquely determined by the compatible pair given in (a). □

Remark 9.3.3 Using Proposition 9.3.1 we constructed for each of the eight compatible pairs (k) of Lemma 9.1.2(e) a matrix $\mathfrak{t}_k$ for the new generator t of $A_1 = \langle q, y, w, t \rangle$ using the methods of the proof of Proposition 9.3.2. Thus we obtained eight subgroups $\mathfrak{G}_k = \langle \mathfrak{q}, \mathfrak{y}, \mathfrak{t}_k, \mathfrak{w} \rangle$ of $\mathrm{GL}_{8671}(13)$. In each of them we tried to calculate the orders of the following products of the generators:

Group name	$\mathfrak{w}\mathfrak{t}_k$	$\mathfrak{y}\mathfrak{w}\mathfrak{t}_k$	$\mathfrak{q}\mathfrak{w}\mathfrak{t}_k$	$\mathfrak{y}\mathfrak{w}\mathfrak{t}_k\mathfrak{q}$	$\mathfrak{y}\mathfrak{t}_k\mathfrak{w}\mathfrak{q}\mathfrak{t}_k$
$\mathfrak{G}_1$	12	fail	–	–	–
$\mathfrak{G}_2$	24	fail	–	–	–
$\mathfrak{G}_3$	4	24	24	21	33
$\mathfrak{G}_4$	8	24	fail	–	–
$\mathfrak{G}_5$	8	24	fail	–	–
$\mathfrak{G}_6$	4	24	24	42	66
$\mathfrak{G}_7$	24	fail	–	–	–
$\mathfrak{G}_8$	12	fail	–	–	–

where "fail" means the product has an order which is greater than 100. The group $\mathfrak{G}$ of Proposition 9.3.2 is $\mathfrak{G}_3$. Looking at the orders of many random elements we saw that all such orders were bounded by 60. In particular, $\mathfrak{p} = \mathfrak{y}^2 \cdot \mathfrak{t}_3 \cdot \mathfrak{w} \cdot \mathfrak{q}$ has order 29.

Therefore we prove in the remainder of this chapter that $\mathfrak{G}$ is isomorphic to Fischer's simple group Fi_{24}'. In the other cases, Kim and the author found in [75] words in the generating matrices having orders that do not occur in the known simple group Fi_{24}' by the Atlas [19].

9.4 $\mathfrak{G}$ is isomorphic to Fischer's simple group Fi_{24}'

In this section we construct an isomorphism between the matrix group $\mathfrak{G} = \langle \mathfrak{q}, \mathfrak{y}, \mathfrak{t}, \mathfrak{w} \rangle$ of Proposition 9.3.2 and the commutator subgroup of the finitely presented group G of Hall and Soicher; see [110], p. 111. Hence $\mathfrak{G}$ is isomorphic to Fischer's simple group Fi_{24}'.

Proposition 9.4.1 *Let* $\mathfrak{G} = \langle \mathfrak{q}, \mathfrak{y}, \mathfrak{t}, \mathfrak{w} \rangle$ *be the subgroup of* $\mathrm{GL}_{8671}(13)$ *constructed in Proposition 9.3.2. Let* $\mathfrak{H}_1 = \langle \mathfrak{y}, \mathfrak{q} \rangle$, $\mathfrak{A}_1 = \langle \mathfrak{H}_1, \mathfrak{t} \rangle$ *and* $\mathfrak{G}_1 = \langle \mathfrak{H}_1, \mathfrak{w} \rangle$.

Let $E = \langle a, b, c, d, t, g, h, i, j, k, v_i \mid 1 \le i \le 11 \rangle$ *be the non-split extension of the Mathieu group* M_{24} *by its simple* $GF(2)$*-module* V_2 *constructed in Lemma 7.1.1(g), and let* $E_{23} = \langle a, b, c, d, t, g, h, i, j \rangle$.

Let $\mathfrak{x} = [(\mathfrak{y}\mathfrak{q}^2\mathfrak{y}\mathfrak{q}\mathfrak{y}\mathfrak{q}^2)^{11}(\mathfrak{q}^2\mathfrak{y}^2\mathfrak{q}\mathfrak{y}\mathfrak{q}\mathfrak{y})^{11}(\mathfrak{q}\mathfrak{y}^2\mathfrak{q}\mathfrak{y}\mathfrak{q}\mathfrak{y}\mathfrak{q}\mathfrak{y}\mathfrak{q})^4]^{12}$, $\mathfrak{u}_1 = (\mathfrak{x}\mathfrak{y}\mathfrak{x})^7$, $\mathfrak{u}_2 = (\mathfrak{x}\mathfrak{y}\mathfrak{x}\mathfrak{y}\mathfrak{x}\mathfrak{y}\mathfrak{x})^4$, $\mathfrak{u}_3 = (\mathfrak{x}\mathfrak{y}\mathfrak{x}\mathfrak{y}^2\mathfrak{x}\mathfrak{y}^2\mathfrak{x}\mathfrak{y}\mathfrak{x}\mathfrak{y})^2$, $\mathfrak{u}_4 = (\mathfrak{x}\mathfrak{y}\mathfrak{x}\mathfrak{y}^5\mathfrak{x}\mathfrak{y}^4)^2$ *and* $\mathfrak{u}_5 = [\mathfrak{y}(\mathfrak{s}\mathfrak{w})^2\mathfrak{y}\mathfrak{w}]^7$. *Then the following assertions hold.*

(a) *The subgroup* $\mathfrak{T}_1 = \langle \mathfrak{u}_i \mid 1 \le i \le 4 \rangle$ *of* $\mathfrak{D} = \langle \mathfrak{x}, \mathfrak{y} \rangle$ *is a Sylow 2-subgroup of* $\mathfrak{G}_1 = \langle \mathfrak{q}, \mathfrak{y}, \mathfrak{w} \rangle$ *of order* 2^{18}.

(b) $\mathfrak{T}_1$ *has a unique maximal elementary abelian normal subgroup* $\mathfrak{B}$ *of order* 2^{11}. *It is generated by the 11 involutions*

$$\mathfrak{u}_1, \quad \mathfrak{u}_2^2, \quad (\mathfrak{u}_1\mathfrak{u}_2)^2, \quad (\mathfrak{u}_1\mathfrak{u}_3)^2, \quad (\mathfrak{u}_1\mathfrak{u}_4)^2, \quad (\mathfrak{u}_2\mathfrak{u}_4)^4,$$
$$(\mathfrak{u}_1\mathfrak{u}_2\mathfrak{u}_3)^4, \quad (\mathfrak{u}_1\mathfrak{u}_2\mathfrak{u}_4)^4, \quad (\mathfrak{u}_1\mathfrak{u}_3\mathfrak{u}_4)^4, \quad (\mathfrak{u}_2^2\mathfrak{u}_3)^2, \quad (\mathfrak{u}_1\mathfrak{u}_2\mathfrak{u}_4\mathfrak{u}_2)^2.$$

(c) $\mathfrak{s} = (\mathfrak{y}^5\mathfrak{t})^7$ *is an involution of* $\mathfrak{A}_1$ *such that* $\mathfrak{A}_1 = \langle \mathfrak{H}_1, \mathfrak{s} \rangle$, $\mathfrak{T}_1^{\mathfrak{s}} = \mathfrak{T}_1$ *and* $\mathfrak{B}^{\mathfrak{s}} = \mathfrak{B}$, *and* $\mathfrak{S} = \langle \mathfrak{T}_1, \mathfrak{s} \rangle$ *is a Sylow 2-subgroup of* $\mathfrak{A}_1$ *of order* 2^{19}.

(d) $\mathfrak{N}_1 = N_{\mathfrak{G}_1}(\mathfrak{B}) = \langle \mathfrak{x}, \mathfrak{y}, \mathfrak{w} \rangle$ *is isomorphic to a non-split extension of* M_{23} *by* $V_2|\mathsf{M}_{23}$ *and* $\mathfrak{D}_1 = N_{\mathfrak{A}_1}(\mathfrak{B}) = \langle \mathfrak{x}, \mathfrak{y}, \mathfrak{s} \rangle$

(e) *There is an isomorphism* ρ *between* $\mathfrak{N}_1$ *and the subgroup* E_{23} *of* E *such that* $\rho(\mathfrak{y}) = (y_2^5 y_3 y_2 y_3)^3 (w_3 w_1 w_2 w_1 w_2 w_1 w_3 w_1^2 w_3 w_2 w_3 w_2 w_1 w_3)^{20}$, $\rho(\mathfrak{x}) = (x_1 x_2 x_4 x_5 x_4 x_2 x_5)^3$ *and* $\rho(\mathfrak{w}) = (e_2 e_3 e_2 e_3^2)^7$, *where*

$$x_1 = (ij)^3, \quad x_2 = (gahigai)^2, \quad x_3 = (aghijagh)^4,$$
$$x_4 = (jhighaji)^4, \quad x_5 = (aighjigai)^4,$$
$$y_1 = i, \quad y_2 = ag, \quad y_3 = (ahj)^3,$$
$$w_1 = (y_2 y_3^2)^2, \quad w_2 = (y_1 y_2 y_1 y_2 y_3)^3, \quad w_3 = (y_1 y_2 y_3 y_2^2)^3,$$
$$e_1 = (agijih)^4, \quad e_2 = (ag^3ihj)^7, \quad e_3 = ghghiai.$$

(f) *There is an isomorphism* μ *between* $\mathfrak{D}_1 = N_{\mathfrak{A}_1}(\mathfrak{B})$ *and the centralizer* $C_E(u)$ *of the involution* $u = (\rho(\mathfrak{x})\rho(\mathfrak{y})^2)^7$ *of* E *such that*

$\mu(\mathfrak{x}) = \rho(\mathfrak{x})$, $\mu(\mathfrak{y}) = \rho(\mathfrak{y})$ and $\mu(\mathfrak{s}) = (m_1^4 m_2 m_1 m_2)^2$, where $m_1 = agahj$, $m_2 = (ijhkj)^2$ and $m_3 = (ahjagk)^5$.

(g) *The subgroup $\mathfrak{E} = \langle \mathfrak{x}, \mathfrak{y}, \mathfrak{w}, \mathfrak{s} \rangle$ of $\mathfrak{G}$ has a faithful permutation representation $P\mathfrak{E}$ of degree 1518 with stabilizer $\langle (\mathfrak{y}\mathfrak{s})^7, (\mathfrak{w}\mathfrak{y}\mathfrak{s}\mathfrak{y})^3, (\mathfrak{s}\mathfrak{y}^3)^2, (\mathfrak{y}^2\mathfrak{w}\mathfrak{y}^2)^3 \rangle$.*

(h) *The groups $\mathfrak{E}$ and E are isomorphic.*

(i) *$\mathfrak{z} = (\mathfrak{x}\mathfrak{y}\mathfrak{w})^8$ is a 2-central involution of $\mathfrak{E}$ with centralizer $C_{\mathfrak{E}}(\mathfrak{z})$ of order $2^{21} \cdot 3^3 \cdot 5$ generated by the elements $\mathfrak{r}_1 = (\mathfrak{s}\mathfrak{y}^3)^3$, $\mathfrak{r}_2 = (\mathfrak{y}^2\mathfrak{w}\mathfrak{y}\mathfrak{s})^6$, $\mathfrak{r}_3 = (\mathfrak{s}\mathfrak{y}\mathfrak{w}\mathfrak{y}\mathfrak{s})^2$ and $\mathfrak{r}_4 = (\mathfrak{s}\mathfrak{w}\mathfrak{y}\mathfrak{s}\mathfrak{w})^6$ with respective orders 2, 4, 4 and 2.*

(j) *$C_{\mathfrak{G}_1}(\mathfrak{z})$ has order $2^{18} \cdot 3^5 \cdot 5$. It is generated by $\mathfrak{f}_1 = \mathfrak{x}\mathfrak{y}\mathfrak{w}$, $\mathfrak{f}_2 = (\mathfrak{x}\mathfrak{y}\mathfrak{x}\mathfrak{w})^7$, $\mathfrak{f}_3 = (\mathfrak{x}\mathfrak{w}\mathfrak{y}\mathfrak{w}\mathfrak{y}^2)^7$ and $\mathfrak{v} = (\mathfrak{w}\mathfrak{q}\mathfrak{w}\mathfrak{y}\mathfrak{q}\mathfrak{y})^7$.*

(k) *The subgroup $\langle \mathfrak{v}\mathfrak{f}_2\mathfrak{v}, (\mathfrak{f}_1\mathfrak{f}_2\mathfrak{f}_1\mathfrak{v}\mathfrak{f}_1)^4, (\mathfrak{f}_2\mathfrak{f}_1\mathfrak{v}\mathfrak{f}_1\mathfrak{v})^4 \rangle$ of $C_{\mathfrak{G}_1}(\mathfrak{z})$ has index 512. Furthermore, it does not contain $\mathfrak{z}$.*

(l) *$\mathfrak{B}$ is the Fitting subgroup of $\mathfrak{E}$. It is also the unique maximal elementary abelian normal subgroup of the Sylow 2-subgroup $\mathfrak{S} = \langle \mathfrak{u}_i \mid 1 \le i \le 5 \rangle$ of $\mathfrak{E}$ contained in $C_{\mathfrak{E}}(\mathfrak{z})$.*

Proof In order to simplify the notation of the proof we replace the German letters by Roman letters. In particular, we let $ME = \langle x, y, w, s \rangle$ be the subgroup $\mathfrak{E}$.

(a) Let PA_1 be the faithful permutation representation of A_1 of degree 56320 constructed in Lemma 9.1.1(c). By Lemma 9.1.2 and Lemma 9.2.1(f) we know that $H_1 = \langle y, q \rangle = \langle x, y, q \rangle$. Now Corollary 6.2.4(b) asserts that $D = \langle x, y \rangle$ has odd index in H_1 and therefore in G_1. Thus D contains a Sylow 2-subgroup of G_1. The given Sylow 2-subgroup T_1 of G_1 and its generators t_i have been found by using MAGMA, the permutation representation PA_1 and the program `GetShortGens(H_1,T_1)`.

(b) Applying the MAGMA command

```
Subgroups(T_1: Al:=Normal, IsElementaryAbelian := true)
```

we observed that T_1 has 44 elementary abelian normal subgroups. Exactly one of them is maximal and has order 2^{11}. It is denoted by B. Its given generators have been calculated by means of the program `GetShortGens(T_1, B)`.

(c) Since $|A_1 : H_1| = 2$ a Sylow 2-subgroup of A_1 has order 2^{19}. Applying PA_1 and MAGMA it is easy to verify that $s = (y^5 t)^7$ is an involution of $N_{A_1}(T_1)$ not contained in H_1 and that $S = \langle T_1, s \rangle$ has order 2^{19}. Hence $B^s = B$ holds trivially by (b).

(d) It can easily be verified by another application of PA_1 and MAGMA that $N_{A_1}(B) = \langle x, y, s\rangle$. Similarly one establishes that $N_1 = N_{G_1}(B) = \langle x, y, w\rangle$ by means of the faithful permutation representation PG_1 of degree 31671 with stabilizer H_1 constructed in [73]. Hence N_1 is a non-split extension of M_{23} by Lemma 6.1.2 and Kim's Theorem 6.3.1.

(e) By Lemma 7.1.1(h) $E = \langle a, b, c, d, t, g, h, i, j, k\rangle$ has a faithful permutation representation PE with stabilizer $U_3 = \langle g, h, i, (dg)^5, (dhjk)^3, (ijkj)^2, (dhjidg)^3\rangle$. Lemma 8.2.2 of [92] states that its subgroup $E_{23} = \langle a, b, c, d, t, g, h, i, j\rangle$ has index $|E : E_{23}| = 24$. Applying the command `IsIsomorphic(N_1,E_{23})` MAGMA establishes an isomorphism $\rho : N_1 \to E_{23}$. The words of the images $\rho(x)$, $\rho(y)$ and $\rho(w)$ of the generators x, y and w of N_1 are constructed as follows. Let $C = C_{E_{23}}(\rho(x))$. Using PE and MAGMA one sees that $|C| = 2^{14}$. The five generators $x_1 = (ij)^3$, $x_2 = (gahigai)^2$, $x_3 = (aghijagh)^4$, $x_4 = (jhighaji)^4$, $x_5 = (aighjigai)^4$ of C were obtained computationally by means of the program `GetShortGens(E_{23},C)`. Using the program `LookupWord(C, \rho(x))` MAGMA returned $\rho(x) = (x_1x_2x_4x_5x_4x_2x_5)^3$. The expressions for $\rho(y)$ and $\rho(w)$ are obtained similarly.

(f) By Kim's Theorem 6.3.1 we know that $z_1 = (xy^2)^7$ is a 2-central involution of G_1. Clearly $u = (\rho(x)\rho(y)^2)^7$ is an involution of E_{23}. Applying MAGMA and PE the reader can verify that $C_u = C_{E_{23}}(u)$ has order $2^{19}\cdot 3^2\cdot 5\cdot 7\cdot 11$. Similarly one observes that $D_1 = N_{A_1}(B) = \langle x, y, s\rangle$ has the same order. Using PA_1, PE and the command `IsIsomorphic(D_1, C_u)` MAGMA establishes an isomorphism $\mu : D_1 \to C_u$. Applying the MAGMA command `IsConjugate(C_u,\mu(y),\rho(y))` we found an element $c_1 \in C_u$ such that $\mu(y)^{c_1} = \rho(y)$. Using the command

```
exists(c){q: q in C_{C_u}(\rho(y))| (\mu(x)^{c_1})^q
  = \rho(x)}
```

one gets an element $c_2 \in C_u$ such that $\mu(x)^{c_1c_2} = \rho(x)$. Hence $\mu' : D_1 \to C_u$ defined by $\mu'(d) = \mu(d)^{c_1c_2}, d \in D_1$, is an isomorphism between D_1 and C_u such that $\mu'(x) = \rho(x)$ and $\mu'(y) = \rho(y)$. It has been checked that $C_E(\langle\rho(y), \rho(x)\rangle) = \langle u\rangle$. Furthermore, $\mu'(s)$ has a centralizer $C_{\mathsf{M}_{23}}(\mu'(\mathrm{s}))$ of order 2^{17} which is generated by the three elements m_1, m_2 and m_3 of M_{23} given in the statement. Another application of the `Lookup` command yields the word $\mu'(s) = (m_1^4m_2m_1m_2)^2$. Hence the map $\mu' : D_1 \to C_E(u)$ satisfies all conditions of (f).

(g) Using (e), (f), PE and MAGMA it has been verified that $E = \langle\rho(x), \rho(y), \rho(w), \mu(s)\rangle$. Applying the program `GetShortGens(E,U_3)` w.r.t. these new generators of E MAGMA returns the following result:

$$U_3 = \langle (\rho(y)\mu(s))^7, (\rho(w)\rho(y)\mu(s)\rho(y))^3, (\mu(s)\rho(y)^3)^2,$$
$$(\rho(y)^2\rho(w)\rho(y)^2)^3\rangle.$$

Thus $MU = \langle (ys)^7, (wysy)^3, (sy^3)^2, (y^2wy^2)^3\rangle$ is a subgroup of $ME = \langle x, y, w, s\rangle$, which is isomorphic to U_3. Let V be the 8671-dimensional vector space over $F = GF(13)$. Using the Meataxe Algorithm implemented in MAGMA Kim and the author verified in [74] that the restriction V_{MU} of V to the subgroup MU has a 7-dimensional FMU-submodule W which has a complement of dimension 8664. Applying the algorithm described in Theorem 6.2.1 of [92] we obtained in [74] a faithful permutation representation PME of the matrix group ME of degree 1518 with stabilizer MU.

(h) Using PE, PME and the isomorphism test `IsIsomorphic(PE,PME)` MAGMA established that $ME \cong E$.

(i) By (d) and Table 6.5.1 we know that $z = (xyw)^8$ is an involution of $N_1 = N_{G_1}(B) \cong E_{23}$ with centralizer $C_{N_1}(z)$ of order $2^{18} \cdot 3^2 \cdot 5$. Therefore we calculate $C_E(z)$ by means of PE and MAGMA. It follows that $|C_E(z)| = 2^{21} \cdot 3^3 \cdot 5$. Hence E has a Sylow 2-subgroup S_3 of order 2^{21} with center $Z(S_3) = \langle z\rangle$ by Table 9.7.2. The given generators r_i of $C_z = C_E(z)$ have been determined by means of MAGMA and the program `GetShortGens(E, C_z)`.

(j) Table 6.5.4 implies that $|C_{G_1}(z)| = 2^{18} \cdot 3^5 \cdot 5$ because $z = (xyw)^8 \in G_1 = \langle x, y, w, q\rangle$. Using MAGMA and the faithful permutation representation PG_1 of G_1 we found the involution $v = (wqwyqy)^7$ such that $(z, v) = 1$, and $C_{G_1}(z) = \langle f_1, f_2, f_3, v\rangle$ for the elements $f_i \in G_1$ given in the statement.

(k) All assertions of the statement are easily checked by means of MAGMA and the faithful permutation representation PG_1 of G_1.

(l) By (b) and (c) the elementary abelian subgroup B is normal in ME. Hence it is the Fitting subgroup ME by (h) and Lemma 7.1.1. Using the faithful permutation representation of ME given in (g) the remaining assertions can be verified by means of MAGMA. □

The following presentation of the 3-transposition Fischer group $P = \mathsf{Fi}_{24}$ is taken from [110], p. 124. It is due to J. Hall and L. S. Soicher [45].

Lemma 9.4.2 *Let* $\mathfrak{G} = \langle \mathfrak{y}, \mathfrak{q}, \mathfrak{t}, \mathfrak{w}\rangle$ *be the subgroup of* $\mathrm{GL}_{8671}(13)$ *constructed in Proposition 9.3.2. Let* $\mathfrak{H}_1 = \langle \mathfrak{y}, \mathfrak{q}\rangle$, $\mathfrak{G}_1 = \langle \mathfrak{H}_1, \mathfrak{w}\rangle$ *and* $\mathfrak{A}_1 =$

$\langle \mathfrak{H}_1, \mathfrak{t} \rangle$. *Let* $\mathfrak{s} = (\mathfrak{y}^5\mathfrak{t})^7$ *and* $\mathfrak{x} = [(\mathfrak{y}\mathfrak{q}^2\mathfrak{y}\mathfrak{q}\mathfrak{y}\mathfrak{q}^2)^{11}(\mathfrak{q}^2\mathfrak{y}^2\mathfrak{q}\mathfrak{y}\mathfrak{q}\mathfrak{y})^{11}(\mathfrak{q}\mathfrak{y}^2\mathfrak{q}\mathfrak{y}\mathfrak{q}\mathfrak{y}\mathfrak{q}\mathfrak{y}\mathfrak{q})^4]^{12}$. *Let* $\mathfrak{E} = \langle \mathfrak{x}, \mathfrak{y}, \mathfrak{w}, \mathfrak{s} \rangle$.

Let $P = \langle a, b, c, d, e, f, g, h, i, j, k, l \rangle$ *be the finitely generated group with the following set* $\mathcal{R}(P)$ *of defining relations:*

$$
\begin{gathered}
l^2 = k^2 = a^2 = b^2 = c^2 = d^2 = e^2 = f^2 = g^2 = j^2 = h^2 = i^2 = 1,\\
(lk)^3 = (ka)^3 = (ab)^3 = (bc)^3 = (cd)^3 = (de)^3 = (ef)^3 = (fg)^3 = 1,\\
(gj)^3 = (la)^2 = (lb)^2 = (lc)^2 = (ld)^2 = (le)^2 = (lf)^2 = (lg)^2 = (lj)^2 = 1,\\
(lh)^2 = (li)^2 = (kb)^2 = (kc)^2 = (kd)^2 = (ke)^2 = (kf)^2 = (kg)^2 = 1,\\
(kj)^2 = (kh)^2 = (ki)^2 = (ac)^2 = (ad)^2 = (ae)^2 = (af)^2 = (ag)^2 = 1,\\
(aj)^2 = (ah)^2 = (ai)^2 = (bd)^2 = (be)^2 = (bf)^2 = (bg)^2 = (bj)^2 = 1,\\
(bh)^2 = (bi)^2 = (ce)^2 = (cf)^2 = (cg)^2 = (cj)^2 = (ch)^2 = (ci)^2 = 1,\\
(df)^2 = (dg)^2 = (dj)^2 = (eg)^2 = (ej)^2 = (eh)^2 = (ei)^2 = (fj)^2 = 1,\\
(fh)^2 = (fi)^2 = (gh)^2 = (gi)^2 = (jh)^2 = (ji)^2 = (dh)^3 = (hi)^3 = 1,\\
(di)^2 = 1, \quad (abcdefh)^9 = (dcbakldefgjdhi)^{17} = 1.
\end{gathered}
$$

Then the following statements hold.

(a) *P has a faithful permutation representation PP of degree* 306936 *with stabilizer* $M = \langle a, b, c, d, e, f, g, h, i, j, l \rangle$.

(b) *The commutator subgroup $G = P'$ is a finite simple group of order* $2^{21} \cdot 3^{16} \cdot 5^2 \cdot 7^3 \cdot 11 \cdot 13 \cdot 17 \cdot 23 \cdot 29$.

(c) $G = \langle b_1, c_1, d_1, e_1, f_1, g_1, h_1, i_1, j_1, k_1 \rangle$, *where* $b_1 = ab$, $c_1 = ac$, $d_1 = ad$, $e_1 = ae$, $f_1 = af$, $g_1 = ag$, $h_1 = ah$, $i_1 = ai$, $j_1 = aj$ *and* $k_1 = ak$.

Furthermore, G has a faithful permutation representation PG of degree 306936 *with stabilizer* $M_1 = \langle b_1, c_1, d_1, e_1, f_1, g_1, h_1, i_1, j_1 \rangle$, *and M_1 is a simple group of order* $2^{18} \cdot 3^{13} \cdot 5^2 \cdot 7 \cdot 11 \cdot 13 \cdot 17 \cdot 23$.

(d) *The centralizer $C_1 = C_G(c_1)$ of the involution c_1 of G has order* $2^{19} \cdot 3^9 \cdot 5^2 \cdot 7 \cdot 11 \cdot 13$, *and* $C_1 = \langle e_1, g_1, i_1, j_1, m_1, n_1 \rangle$, *where* $m_1 = (c_1d_1e_1h_1)^3$ *and* $n_1 = (b_1c_1d_1k_1f_1)^3$ *are involutions.*

Furthermore, C_1 has a faithful permutation representation PC_1 of degree 56320 *with stabilizer* $\langle i_1, j_1, (e_1g_1n_1)^2, n_1m_1n_1 \rangle$.

(e) *There is an isomorphism ϕ between $\mathfrak{A}_1 = \langle \mathfrak{y}, \mathfrak{q}, \mathfrak{t} \rangle$ and the finitely presented group* $A_1 = \langle a, b, c, d, e, f, g, h, i, z, t \rangle$ *constructed in*

Lemma 9.1.1(b) such that

$$\begin{aligned} y &= \phi(\mathfrak{y}) = (S_1S_2S_4S_1S_4S_2S_4)^5(T_1T_3^2T_1T_3T_1T_2T_1T_3)^{20}, \\ q &= \phi(\mathfrak{q}) = [(p\cdot o\cdot j\cdot o\cdot k)^4\cdot(j\cdot k\cdot o\cdot k^2p)^4]^4, \\ t &= \phi(\mathfrak{t}), \end{aligned}$$

where

$$\begin{aligned} S_1 &= (b\cdot a\cdot c\cdot b)\cdot(d\cdot e\cdot h\cdot d)\cdot(b\cdot a\cdot c\cdot b), \\ S_2 &= [(b\cdot a\cdot c\cdot b)\cdot(d\cdot e\cdot h\cdot d)\cdot(f\cdot e\cdot g\cdot f)]^2, \\ S_3 &= [(b\cdot a\cdot c\cdot b)\cdot(d\cdot e\cdot h\cdot d)\cdot(c\cdot d\cdot e\cdot h\cdot i)^4\cdot(b\cdot a\cdot c\cdot b)]^2, \\ S_4 &= [(b\cdot a\cdot c\cdot b)\cdot(c\cdot d\cdot e\cdot h\cdot i)^4\cdot(d\cdot e\cdot h\cdot d)\cdot(f\cdot e\cdot g\cdot f)]^4, \\ T_1 &= S_2S_4S_2^2, \quad T_2 = (S_4S_2S_1S_3S_4)^2, \quad T_3 = (S_4S_2S_4S_2S_1S_3)^2, \\ j &= (c\cdot d\cdot e\cdot h)^3, \\ k &= (c\cdot d\cdot e\cdot f\cdot g)^2, \\ l &= (a\cdot b\cdot c\cdot d\cdot e\cdot h)^5, \\ o &= (l\cdot b\cdot k\cdot j\cdot b\cdot i)^6, \\ p &= (j\cdot k\cdot j\cdot l\cdot b\cdot i\cdot j\cdot i)^5. \end{aligned}$$

(f) *There is an isomorphism*

$$\rho : A_1 = \langle a,b,c,d,e,f,g,h,i,z,t\rangle \to C_1 = \langle e_1,g_1,i_1,j_1,m_1,n_1\rangle$$

such that

$$\begin{aligned} \rho(t) &= (m_1n_1)^3, \\ \rho(a) &= [(i_1m_1n_1)^6\cdot(j_1n_1m_1n_1)\cdot(g_1n_1e_1g_1n_1)^2\cdot(i_1m_1n_1)^6 \\ &\qquad \cdot(e_1g_1n_1m_1n_1e_1)^2\cdot(g_1n_1e_1g_1n_1)^2]^7, \\ \rho(b) &= [(i_1n_1g_1j_1m_1)^{12}\cdot(e_1g_1j_1m_1i_1n_1)^4\cdot(i_1n_1g_1j_1m_1)^{12} \\ &\qquad \cdot(e_1g_1j_1m_1i_1n_1)^{12}\cdot(i_1n_1g_1j_1m_1)^{12}\cdot(e_1g_1j_1m_1i_1n_1)^8]^{11}, \\ \rho(c) &= [(j_1n_1m_1n_1)\cdot(e_1n_1g_1n_1e_1)\cdot(e_1g_1i_1j_1m_1n_1)^{18}]^{11}, \\ \rho(d) &= [(j_1m_1)\cdot(n_1g_1n_1)\cdot(e_1i_1j_1n_1g_1m_1)^{18}\cdot(j_1m_1) \\ &\qquad \cdot(e_1i_1j_1n_1g_1m_1)^{18}\cdot(j_1m_1)\cdot(e_1i_1j_1n_1g_1m_1)^{18}]^{11}, \\ \rho(e) &= (bc)^3\cdot c^t, \quad \rho(f) = (bc)^3\cdot b^t, \quad \rho(g) = a^t, \\ \rho(h) &= [e_1\cdot i_1\cdot(m_1i_1n_1m_1n_1)^3\cdot(e_1g_1n_1g_1j_1m_1)^6\cdot i_1 \\ &\qquad \cdot(e_1g_1n_1g_1j_1m_1)^6]^{11}, \\ \rho(i) &= [(m_1i_1m_1)\cdot(e_1g_1j_1n_1m_1)^{12}]^9. \end{aligned}$$

In particular, $\psi = \rho\circ\phi : \mathfrak{A}_1 \to C_1$ *is an isomorphism such that* $y_1 = \psi(\mathfrak{y}) = \rho(y)$, $q_1 = \psi(\mathfrak{q}) = \rho(q)$ *and* $t_1 = \psi(\mathfrak{t}) = \rho(t)$ *generate* C_1.

(g) *In* C_1 *let*

$$\begin{aligned} x_1 &= \psi(\mathfrak{x}), s_1 = \psi(\mathfrak{s}) = (y_1^5 t_1)^7, u_1 = (x_1 y_1 x_1)^7, \\ u_2 &= (x_1 y_1 x_1 y_1 x_1 y_1 x_1)^4, u_3 = (x_1 y_1 x_1 y_1^2 x_1 y_1^2 x_1 y_1 x_1 y_1)^2, \\ u_4 &= (x_1 y_1 x_1 y_1^5 x_1 y_1^4)^2. \end{aligned}$$

Then $S_1 = \langle u_1, u_2, u_3, u_4, s_1 \rangle$ *is a Sylow* 2*-subgroup of* C_1. *Moreover,* S_1 *has a unique maximal elementary abelian normal subgroup* B_1. *It is generated by the* 11 *involutions* u_1, u_2^2, $(u_1 u_2)^2$, $(u_1 u_3)^2$, $(u_1 u_4)^2$, $(u_2 u_4)^4$, $(u_1 u_2 u_3)^4$, $(u_1 u_2 u_4)^4$, $(u_1 u_3 u_4)^4$, $(u_2^2 u_3)^2$ and $(u_1 u_2 u_4 u_2)^2$.

(h) $N_1 = N_G(B_1) = \langle x_1, y_1, s_1, o_1 \rangle$, *where* $o_1 = [d_1 \cdot y_1 \cdot s_1 \cdot (y_1)^2 \cdot s_1)]^{10}$ *and* $|N_1| = 2^{21} \cdot 3^3 \cdot 5 \cdot 7 \cdot 11 \cdot 23$.

(i) *There is an isomorphism* $\mu : \mathfrak{E} \to N_1$ *such that* $\mu(\mathfrak{x}) = \psi(\mathfrak{x}) = x_1$, $\mu(\mathfrak{y}) = \psi(\mathfrak{y}) = y_1$, $\mu(\mathfrak{s}) = \psi(\mathfrak{t}) = s_1$ *and*

$$\mu(w) = w_1 = [(s_1 y_1 x_1 y_1)^2 \cdot (x_1 y_1 o_1 y_1 s_1 y_1^2)^5]^7.$$

(j) $G = \langle C_1, N_1 \rangle = \langle q_1, y_1, w_1, t_1 \rangle$.

(k) *The map* $\kappa : G \to \mathfrak{G}$ *given by* $\kappa(q_1) = \mathfrak{q}$, $\kappa(y_1) = \mathfrak{y}$, $\kappa(w_1) = \mathfrak{w}$ *and* $\kappa(s_1) = \mathfrak{s}$ *is a group isomorphism. In particular,* $\mathfrak{G}$ *is a simple group of order* $2^{21} \cdot 3^{16} \cdot 5^2 \cdot 7^3 \cdot 11 \cdot 13 \cdot 17 \cdot 23 \cdot 29$.

Proof (a) This statement can be verified by running the MAGMA command `CosetAction(P,M)`.

(b) Using (a) and MAGMA it has been checked that $G = P'$ is a simple group of the given order.

(c) Applying the program `GetShortGens(P,G)` one finds the given generators of G. Using the faithful permutation representation PP of (a) and MAGMA it has been checked in [74] that $M_1 = \langle b_1, c_1, d_1, e_1, f_1, g_1, h_1, i_1, j_1 \rangle$ is a simple group of order $2^{18} \cdot 3^{13} \cdot 5^2 \cdot 7 \cdot 11 \cdot 13 \cdot 17 \cdot 23$. It is the stabilizer of the faithful permutation representation PG of G of degree 306936, as has been checked by means of MAGMA and the command `CosetAction(G,M_1)`.

(d) The order of $C_1 = C_G(c_1)$ and its generators have been calculated in [74] by means of the permutation representation PG of G, MAGMA and the program `GetShortGens(G,C_1)`. Applying the MAGMA command `DegreeReduction` we got a faithful permutation representation PC_1 of C_1 having degree 56320. Its stabilizer U_1 has been found by means of the MAGMA command `BasicStabilizer(PC_1,2)`. Its given generators were found by another application of `GetShortGens(C_1, U_1)`.

(e) This statement follows immediately from Lemma 9.1.1(b), Proposition 6.2.3 and Proposition 9.3.2.

(f) By Lemma 9.1.1(c) the finitely presented group A_1 has a faithful permutation representation PA_1 of degree 56320. Let PC_1 be the faithful permutation representation of C_1 constructed in (d). Using the MAGMA command `IsIsomorphic(PA_1,PC_1)` Kim and the author obtained in [74] an isomorphism $\eta : \mathfrak{A}_1 \to C_1$. Applying the command `exists(a){x : x in C_1| \eta(t)^x = (m_1n_1)^3}` we found an element $a \in C_1$ such that $\eta(t)^a = (m_1n_1)^3$. Let α be the inner automorphism of C_1 induced by a. Then the map $\rho = \alpha \circ \eta$ is an isomorphism from A_1 onto C_1 such that $\rho(t) = (m_1n_1)^3$. The centralizers of the images of $\rho(a), \rho(b), \ldots, \rho(i)$ in C_1 all have the same order $2^{17} \cdot 3^6 \cdot 5 \cdot 7 \cdot 11$. Their generators and the words for the images were obtained in [74] computationally using the programs `GetShortGens` and `LookupWord`, respectively, as follows:

(1) $C_{C_1}(\rho(a))$ is generated by $a_1 = (i_1m_1n_1)^6$, $a_2 = j_1n_1m_1n_1$, $a_3 = (g_1n_1e_1g_1n_1)^2$, $a_4 = (e_1g_1n_1m_1n_1e_1)^2$, and $\rho(a) = (a_1a_2a_3a_1a_4a_3)^7$.
(2) $C_{C_1}(\rho(b))$ is generated by $b_1 = (i_1n_1g_1j_1m_1)^{12}$, $b_2 = (e_1g_1j_1m_1i_1n_1)^4$, and $\rho(b) = (b_1b_2b_1b_2^3b_1b_2^2)^{11}$.
(3) $C_{C_1}(\rho(c))$ is generated by $c_1 = j_1n_1m_1n_1$, $c_2 = e_1n_1g_1n_1e_1$, $c_3 = (e_1g_1i_1j_1m_1n_1)^{18}$, and $\rho(c) = (c_1c_2c_3)^{11}$.
(4) $C_{C_1}(\rho(d))$ is generated by $d_1 = j_1m_1$, $d_2 = n_1g_1n_1$, $d_3 = (e_1i_1j_1n_1g_1m_1)^{18}$, and $\rho(d) = (d_1d_2d_3d_1d_3d_1d_3)^{11}$.
(5) $C_{C_1}(\rho(h))$ is generated by $h_1 = e_1$, $h_2 = i_1$, $h_3 = (m_1i_1n_1m_1n_1)^3$, $h_4 = (e_1g_1n_1g_1j_1m_1)^6$, and $\rho(h) = (h_1h_2h_3h_4h_2h_4)^{11}$.
(6) $C_{C_1}(\rho(i))$ is generated by $i_1 = m_1i_1m_1$, $i_2 = j_1n_1m_1n_1$, $i_3 = (e_1g_1j_1n_1m_1)^6$, and $\rho(i) = (i_1i_3^2)^9$.

The given words for the images $\rho(e)$, $\rho(f)$ and $\rho(g)$ can now be calculated from these images and the relations of Lemma 9.1.1(b).

Statement (e) implies that the composition of the isomorphisms $\phi : \mathfrak{A}_1 \to A_1$ and $\rho : A_1 \to C_1$ is an isomorphism $\psi : \mathfrak{A}_1 \to C_1$. Furthermore, the given images $\psi(\mathfrak{q})$, $\psi(\mathfrak{q})$ and $\psi(\mathfrak{q})$ in C_1 of the three generators of $\mathfrak{A}_1$ are well defined.

(g) Since $\mathfrak{x} \in \mathfrak{A}_1$ its image

$$\psi(\mathfrak{x}) = [(y_1q_1^2y_1q_1y_1q_1^2)^{11}(q_1^2y_1^2q_1y_1q_1y_1)^{11}(q_1y_1^2q_1y_1q_1y_1q_1y_1q_1)^4]^{12} = x_1$$

is a well defined element of C_1.

Let $\mathfrak{u}_i$ be the generators of the Sylow 2-subgroup $\mathfrak{T}_1$ of the subgroup $\mathfrak{G}_1 = \langle \mathfrak{q}, \mathfrak{y}, \mathfrak{w} \rangle$ of $\mathfrak{G}$ given in Proposition 9.4.1(a). Hence $\mathfrak{T}_1$ is a subgroup of $\mathfrak{A}_1$ because its generators $\mathfrak{u}_i$ are words in $\mathfrak{x}$ and $\mathfrak{y}$ by Proposition 9.4.1. Therefore their images $u_i = \psi(\mathfrak{u}_i)$, $1 \le i \le 4$, are well defined. So is $s_1 = \psi(\mathfrak{s}) = (y_1^5 t_1)^7$. Hence $S_1 = \psi(\mathfrak{S}) = \langle u_1, u_2, u_3, u_4, s_1 \rangle$ is a Sylow 2-subgroup of C_1 by Proposition 9.4.1(c) and (f).

Let $B_1 = \psi(\mathfrak{B})$, where $\mathfrak{B}$ is the unique maximal elementary abelian normal subgroup of $\mathfrak{S}$ defined in Proposition 9.4.1(b) and (d). Hence B_1 is generated by the 11 involutions given in the statement.

(h) Let $N_1 = N_G(B_1)$. Using the faithful permutation representation PG of G and MAGMA the reader can verify that $|N_1| = 2^{21} \cdot 3^3 \cdot 5 \cdot 7 \cdot 11 \cdot 23$ and that N_1 is generated by x_1, y_1, s_1 and the element $o_1 \in G$ given in the statement.

(i) By Proposition 9.4.1(g) the subgroup $\mathfrak{E}$ of $\mathfrak{G}$ has a faithful permutation representation $P\mathfrak{E}$ of degree 1518 with a well defined stabilizer. Let PN_1 be the reduction of PG to $N_1 = N_G(B_1)$. An application of the MAGMA command `IsIsomorphic(PE,PN_1)` provides an isomorphism $\tau : \mathfrak{E} \to N_1$. As in the proof of (f) we found in [74] an element $a \in N_1$ such that $(\tau(\mathfrak{y}))^a = y_1 = \psi(\mathfrak{y})$. Using MAGMA again we verified that $C_{N_1}(y_1)$ has order 56. Searching through its elements we found an element $b \in C_{N_1}(y_1)$ such that $(\tau(\mathfrak{x}))^{ab} = \psi(\mathfrak{x}) = x_1$ and $(\tau(\mathfrak{s}))^{ab} = \psi(\mathfrak{s}) = s_1$. Let β denote the inner automorphism of N_1 induced by conjugation with ab. Then the map $\mu = \beta \circ \nu$ is an isomorphism from $\mathfrak{E}$ onto $N_1 = N_G(B_1)$ such that $\mu(\mathfrak{x}) = x_1$, $\mu(\mathfrak{y}) = y_1$ and $\mu(\mathfrak{s}) = s_1$. Let $w_1 = \mu(\mathfrak{w})$. By another application of MAGMA and PN_1 we observed that $N_1 = \langle x_1, y_1, s_1, w_1 \rangle$.

The word for w_1 in the generators of G was obtained in [74] as follows. Let $C_2 = C_{N_1}(w_1)$. The generators of N_1 given in (h), MAGMA and the program `GetShortGens(N_1,C_2)` yielded that $C_2 = \langle v_i \mid 1 \le i \le 3 \rangle$, where $v_1 = (s_1 y_1 x_1 y_1)^2$, $v_2 = (y_1 o_1 x_1 s_1 o_1)^4$ and $v_3 = (x_1 y_1 o_1 y_1 s_1 y_1^2)^5$. Applying the command `LookupWord(C_2, w_1)` MAGMA provided the solution $w_1 = (v_1 v_3)^7$.

(j) Using the faithful permutation representation PG of G and MAGMA the reader can verify that $G = \langle C_1, N_1 \rangle$. Hence (f) and (i) imply that $G = \langle q_1, y_1, t_1, x_1, s_1, w_1 \rangle = \langle q_1, y_1, t_1, w_1 \rangle$.

(k) In $\mathfrak{G}$ let $\mathfrak{E}_{23} = \langle \mathfrak{x}, \mathfrak{y}, \mathfrak{w} \rangle$ and $\mathfrak{H}_1 = \langle \mathfrak{q}, \mathfrak{y} \rangle$. Then Proposition 9.3.2 and Theorem 6.3.1 imply that $\mathfrak{G}_1 = \langle \mathfrak{H}_1, \mathfrak{E}_{23} \rangle$ is a simple subgroup of $\mathfrak{G}$ such that $\mathfrak{D}_1 = \mathfrak{E}_{23} \cap \mathfrak{H}_1 = \langle \mathfrak{x}, \mathfrak{y} \rangle$ and $\mathfrak{H}_1 = C_{\mathfrak{G}_1}(\mathfrak{z}_1)$, where $\mathfrak{z}_1 = (\mathfrak{x}\mathfrak{y}^2)^7$ is a 2-central involution of $\mathfrak{G}_1$.

Let $E_{23} = \mu(\mathfrak{E}_{23})$, $H_1 = \psi(\mathfrak{H}_1)$ and $G_1 = \langle E_{23}, H_1 \rangle$ in G. As μ and ψ agree on $\mathfrak{H}_1$ by (j) we have $D_1 = \langle x_1, y_1 \rangle = E_{23} \cap H_1$. Using PG and MAGMA it has been checked that G_1 is a simple group of order $2^{18} \cdot 3^{13} \cdot 5^2 \cdot 7 \cdot 11 \cdot 13 \cdot 17 \cdot 23$. Furthermore, $C_{G_1}(z_1) = H_1$, where $z_1 = (x_1y_1^2)^7 = \psi(\mathfrak{z}_1)$ is a 2-central involution of G_1. Thus Theorem 6.3.1 implies that $G_1 \cong \mathsf{Fi}_{23}$. In particular, the map $\kappa_1 : G_1 \to \mathfrak{G}_1$ defined by $\kappa_1(q_1) = \mathfrak{q}$, $\kappa_1(y_1) = \mathfrak{x}$, and $\kappa_1(w_1) = \mathfrak{w}$ is an isomorphism. By Proposition 9.3.2 and Lemma 9.1.2(e) the amalgam $\mathfrak{A}_1 \leftarrow \mathfrak{H}_1 \to \mathfrak{G}_1$ has Goldschmidt index 1. Therefore (f) and (i) imply that the free products $\mathfrak{A}_1 *_{\mathfrak{H}_1} \mathfrak{G}_1$ and $C_1 *_{H_1} G_1$, with respective amalgamated subgroups $\mathfrak{H}_1$ and H_1, are isomorphic by Corollary 7.1.9 of [92]. Thus (j) and Proposition 9.3.2 imply that the map $\kappa : G \to \mathfrak{G}$ defined by $\kappa(q_1) = \mathfrak{q}$, $\kappa(y_1) = \mathfrak{y}$, $\kappa(w_1) = \mathfrak{w}$ and $\kappa(t_1) = \mathfrak{t}$ is an irreducible 8671-dimensional representation of the group G over $GF(13)$. Hence $\kappa : G \to \mathfrak{G}$ is an isomorphism because κ is surjective and G is simple by (b). □

Theorem 9.4.3 (Kim–Michler) *Let $\mathfrak{G} = \langle \mathfrak{q}, \mathfrak{y}, \mathfrak{t}, \mathfrak{w} \rangle$ be the subgroup of $\mathrm{GL}_{8671}(13)$ constructed in Proposition 9.3.2. Then the following statements hold.*

(a) *$\mathfrak{G} = \langle \mathfrak{q}, \mathfrak{y}, \mathfrak{t}, \mathfrak{w} \rangle$ has a faithful permutation representation of degree 306936 with stabilizer $\mathfrak{G}_1 = \langle \mathfrak{q}, \mathfrak{y}, \mathfrak{w} \rangle$.*

(b) *$\mathfrak{G}$ is a finite simple group of order*

$$|\mathfrak{G}| = 2^{21} \cdot 3^{16} \cdot 5^2 \cdot 7^3 \cdot 11 \cdot 13 \cdot 17 \cdot 23 \cdot 29.$$

(c) *$\mathfrak{G}$ is isomorphic to the commutator subgroup P' of the finitely presented group $P = \langle a, b, c, d, e, f, g, h, i, j, k, l \rangle$ with the set of defining relations $\mathcal{R}(P)$ stated in Lemma 9.4.2.*

(d) *The character table of $\mathfrak{G}$ is equivalent to that of Fi_{24}' [19], pp. 200–202.*

Proof The first three statements hold by Lemma 9.4.2(j) and (k).

(d) Let $G = P'$ be the commutator subgroup of the finitely presented group P. Then $\mathfrak{G} \cong G$ by (c). The character table of G has been calculated by means of MAGMA and the faithful permutation representation PG of G with stabilizer $M = \langle b_1, c_1, d_1, e_1, f_1, g_1, h_1, i_1, j_1 \rangle$ given in Lemma 9.4.2(c). □

9.5 Presentation of 2-central involution centralizer

In this section we determine generators and defining relations of the centralizer $\mathfrak{H} = C_{\mathfrak{G}}(\mathfrak{z})$ of a 2-central involution $\mathfrak{z}$ of the simple subgroup $\mathfrak{G} = \langle \mathfrak{q}, \mathfrak{y}, \mathfrak{w}, \mathfrak{t} \rangle$ of $\mathrm{GL}_{8671}(13)$. Thus we are able to show that $\mathfrak{G}$ satisfies all conditions of Algorithm 1.3.8. In particular, $\mathfrak{G} = \langle \mathfrak{H}, \mathfrak{E} \rangle$ and $\mathfrak{D} = N_{\mathfrak{H}}(\mathfrak{B}) = C_{\mathfrak{E}}(\mathfrak{z})$, where $\mathfrak{B}$ is the unique maximal elementary abelian normal subgroup of a well defined Sylow 2-subgroup $\mathfrak{S}$ of $\mathfrak{G}$ and $\mathfrak{E} = N_{\mathfrak{G}}(\mathfrak{B})$. However, the free product $P = H *_D E$ of the groups $H = \mathfrak{H}$ and $E = \mathfrak{E}$ with amalgamated subgroup $D = \mathfrak{D}$ has too many irreducible representations of degree 8671. Therefore Kim and the author could not find one in [74] satisfying the Sylow 2-subgroup test of Step 5(c) of Algorithm 7.4.8 of [92].

Proposition 9.5.1 *Let* $\mathfrak{G} = \langle \mathfrak{q}, \mathfrak{y}, \mathfrak{t}, \mathfrak{w} \rangle$ *be the subgroup of* $\mathrm{GL}_{8671}(13)$ *defined in Proposition 9.3.2. Let* $\mathfrak{x} = [(\mathfrak{y}\mathfrak{q}^2\mathfrak{y}\mathfrak{q}\mathfrak{y}\mathfrak{q}^2)^{11}(\mathfrak{q}^2\mathfrak{y}^2\mathfrak{q}\mathfrak{y}\mathfrak{q}\mathfrak{y})^{11}(\mathfrak{q}\mathfrak{y}^2\mathfrak{q}\mathfrak{y}\mathfrak{q}\mathfrak{y}\mathfrak{q}\mathfrak{y}\mathfrak{q})^4]^{12}$ *and* $\mathfrak{s} = (\mathfrak{y}\mathfrak{t})^7$. *Let* $\mathfrak{r}_1 = (\mathfrak{s}\mathfrak{y}^3)^3$, $\mathfrak{r}_2 = (\mathfrak{y}^2\mathfrak{w}\mathfrak{y}\mathfrak{s})^6$, $\mathfrak{r}_3 = (\mathfrak{s}\mathfrak{y}\mathfrak{w}\mathfrak{y}\mathfrak{s})^2$, $\mathfrak{r}_4 = (\mathfrak{s}\mathfrak{w}\mathfrak{y}\mathfrak{s}\mathfrak{w})^6$ *and* $\mathfrak{v} = (\mathfrak{w}\mathfrak{q}\mathfrak{w}\mathfrak{y}\mathfrak{q}\mathfrak{y})^7$. *Let* $\mathfrak{H} = \langle \mathfrak{r}_1, \mathfrak{r}_2, \mathfrak{r}_3, \mathfrak{r}_4, \mathfrak{v} \rangle$.
Then the following statements hold.

(a) *The element* $\mathfrak{z} = (\mathfrak{x}\mathfrak{y}\mathfrak{w})^8$ *is a 2-central involution* $\mathfrak{z} = (\mathfrak{x}\mathfrak{y}\mathfrak{w})^8$ *of* $\mathfrak{G}$ *such that* $C_{\mathfrak{G}}(\mathfrak{z}) = \mathfrak{H}$ *has order* $2^{21} \cdot 3^7 \cdot 5 \cdot 7$.

(b) $\mathfrak{H}$ *has a faithful permutation representation* $P\mathfrak{H}$ *of degree* 258048 *with stabilizer* $\mathfrak{U}_1 = \langle \mathfrak{v}\mathfrak{f}_2\mathfrak{v}, (\mathfrak{f}_1\mathfrak{f}_2\mathfrak{f}_1\mathfrak{v}\mathfrak{f}_1)^4, (\mathfrak{f}_2\mathfrak{f}_1\mathfrak{v}\mathfrak{f}_1\mathfrak{v})^4 \rangle$, *where* $\mathfrak{f}_1 = \mathfrak{x}\mathfrak{y}\mathfrak{w}$, $\mathfrak{f}_2 = (\mathfrak{x}\mathfrak{y}\mathfrak{x}\mathfrak{w})^7$ *and* $\mathfrak{f}_3 = (\mathfrak{x}\mathfrak{w}\mathfrak{y}\mathfrak{w}\mathfrak{y}^2)^7$.

(c) $\mathfrak{S} = \langle \mathfrak{s}, \mathfrak{u}_i \mid 1 \le i \le 5 \rangle$ *is a Sylow 2-subgroup of* $\mathfrak{E}$ *contained in* $\mathfrak{H}$ *with center* $Z(\mathfrak{H}) = \langle \mathfrak{z} \rangle$, *where* $\mathfrak{u}_1 = (\mathfrak{x}\mathfrak{y}\mathfrak{x})^7$, $\mathfrak{u}_2 = (\mathfrak{x}\mathfrak{y}\mathfrak{x}\mathfrak{y}\mathfrak{x}\mathfrak{y}\mathfrak{x})^4$, $\mathfrak{u}_3 = (\mathfrak{x}\mathfrak{y}\mathfrak{x}\mathfrak{y}^2\mathfrak{x}\mathfrak{y}^2\mathfrak{x}\mathfrak{y}\mathfrak{x}\mathfrak{y})^2$, $\mathfrak{u}_4 = (\mathfrak{x}\mathfrak{y}\mathfrak{x}\mathfrak{y}^5\mathfrak{x}\mathfrak{y}^4)^2$ *and* $\mathfrak{u}_5 = (\mathfrak{y}\mathfrak{s}\mathfrak{w}\mathfrak{s}\mathfrak{w}\mathfrak{y}\mathfrak{w})^7$.

(d) $\mathfrak{B} = \langle \mathfrak{b}_i \mid 1 \le i \le 11 \rangle$ *is the unique maximal elementary abelian subgroup of* $\mathfrak{S}$, *where* $\mathfrak{b}_1 = \mathfrak{u}_1$, $\mathfrak{b}_2 = \mathfrak{u}_2^2$, $\mathfrak{b}_3 = (\mathfrak{u}_1\mathfrak{u}_2)^2$, $\mathfrak{b}_4 = (\mathfrak{u}_1\mathfrak{u}_3)^2$, $\mathfrak{b}_5 = (\mathfrak{u}_1\mathfrak{u}_4)^2$, $\mathfrak{b}_6 = (\mathfrak{u}_2\mathfrak{u}_4)^4$, $\mathfrak{b}_7 = (\mathfrak{u}_1\mathfrak{u}_2\mathfrak{u}_3)^4$, $\mathfrak{b}_8 = (\mathfrak{u}_1\mathfrak{u}_2\mathfrak{u}_4)^4$, $\mathfrak{b}_9 = (\mathfrak{u}_1\mathfrak{u}_3\mathfrak{u}_4)^4$, $\mathfrak{b}_{10} = (\mathfrak{u}_2^2\mathfrak{u}_3)^2$ *and* $\mathfrak{b}_{11} = (\mathfrak{u}_1\mathfrak{u}_2\mathfrak{u}_4\mathfrak{u}_2)^2$.

(e) $N_{\mathfrak{G}}(\mathfrak{B}) = \langle \mathfrak{x}, \mathfrak{y}, \mathfrak{w}, \mathfrak{s} \rangle = \mathfrak{E}$.

(f) $\mathfrak{D} = \langle \mathfrak{r}_i \mid 1 \le i \le 4 \rangle = N_{\mathfrak{H}}(\mathfrak{B}) = C_{\mathfrak{E}}(\mathfrak{z})$.

(g) *The Fitting subgroup* $\mathfrak{O}$ *of* $\mathfrak{H}$ *is extra-special of order* 2^{13} *and center* $Z(\mathfrak{O}) = \langle \mathfrak{z} \rangle$. *It is generated by the following* 12 *involutions:*

$$\mathfrak{p}_1 = (\mathfrak{r}_2)^2, \quad \mathfrak{p}_2 = (\mathfrak{r}_1\mathfrak{r}_2)^4, \quad \mathfrak{p}_3 = (\mathfrak{r}_3\mathfrak{r}_4)^3, \quad \mathfrak{p}_4 = (\mathfrak{r}_1\mathfrak{r}_2\mathfrak{r}_1)^2,$$

$$\mathfrak{p}_5 = (\mathfrak{r}_1\mathfrak{r}_3\mathfrak{r}_4)^6, \quad \mathfrak{p}_6 = (\mathfrak{r}_2^2\mathfrak{r}_4)^2, \quad \mathfrak{p}_7 = (r_3\mathfrak{r}_4\mathfrak{r}_1)^6, \quad \mathfrak{p}_8 = (\mathfrak{r}_4\mathfrak{r}_1\mathfrak{r}_3)^6,$$

$$\mathfrak{p}_9 = (\mathfrak{r}_1\mathfrak{r}_2\mathfrak{r}_3^2)^4, \quad \mathfrak{p}_{10} = (\mathfrak{r}_1\mathfrak{r}_2\mathfrak{r}_3\mathfrak{r}_4)^4, \quad \mathfrak{p}_{11} = (r_1\mathfrak{r}_2\mathfrak{r}_1\mathfrak{r}_2\mathfrak{r}_4)^5,$$

$$\mathfrak{p}_{12} = (\mathfrak{r}_1\mathfrak{r}_2\mathfrak{r}_3^2\mathfrak{r}_4)^4 \quad \text{and} \quad \mathfrak{z} = (\mathfrak{p}_1\mathfrak{p}_5)^2.$$

(h) $\mathfrak{O}/Z(\mathfrak{O})$ *has a complement* $\mathfrak{K} \cong 3U_4(3) : 2$ *in* $\mathfrak{H}/Z(\mathfrak{O})$.

(i) $\mathfrak{H} = \langle \mathfrak{a}, \mathfrak{b}, \mathfrak{p}_i \mid 1 \le i \le 12 \rangle = \langle \mathfrak{b}, \mathfrak{c}, \mathfrak{d}, \mathfrak{f}, \mathfrak{z}, \mathfrak{p}_i \mid 1 \le i \le 12 \rangle$, *where* $\mathfrak{a} = \mathfrak{r}_1\mathfrak{r}_3^3\mathfrak{v}\mathfrak{r}_3$, $\mathfrak{b} = [\mathfrak{r}_1\mathfrak{r}_3\mathfrak{r}_1\mathfrak{r}_3^2\mathfrak{v}]^6[\mathfrak{r}_1\mathfrak{r}_3\mathfrak{r}_1\mathfrak{r}_3\mathfrak{v}\mathfrak{r}_3\mathfrak{v}]^{12}$, $\mathfrak{c} = (\mathfrak{a}\mathfrak{b})^2$, $\mathfrak{d} = (\mathfrak{b}\mathfrak{c})^7$ *and* $\mathfrak{f} = [(\mathfrak{b}\mathfrak{a}^3)^5 \cdot (\mathfrak{a}^4\mathfrak{b}^2\mathfrak{a})^9 \cdot (\mathfrak{a}^2\mathfrak{b}^3\mathfrak{a}^4\mathfrak{b})^3 \cdot (\mathfrak{b}\mathfrak{a}^3)^5]^3$.

(j) $\mathfrak{H}$ *is isomorphic to the finitely presented group* $H = \langle b, c, d, f, z, p_i \mid 1 \leq i \leq 12\rangle$ *having the following set* $\mathcal{R}(H)$ *of defining relations:*

$$
\begin{aligned}
& b^6 = c^9 = d^3 = f^2 = z^2 = 1, \quad p_i^2 = 1 \quad \textit{for} \quad 1 \leq i \leq 12, \\
& (z,b) = (z,c) = (z,d) = (z,f) = 1, \quad (z,p_i) = 1 \quad \textit{for} \quad 1 \leq i \leq 12, \\
& (b,d) = (c,d) = 1, \quad b^f = b^5, \quad c^f = d(b^3cb^2c^6bcbc), \quad d^f = d^2, \\
& (c^{-1}b^{-1})^7 d = 1, \quad (c^{-1}b)^9 = z, \\
& b^{-1}c^{-1}b^{-3}c^{-1}b^3c^{-1}b^3c^{-1}b^{-2}d = 1, \\
& (bc^{-2}b)^4 d^2 = (bc^{-3}bc^2b)^2 d = bc^3b^{-2}cb^3c^{-1}bc^{-1}bcb^{-1}cbd^2 = 1, \\
& b^{-1}c^{-3}b^{-1}c^{-1}bc^{-1}b^2cb^{-1}cbc^3b^{-1}d = 1, \\
& b^{-1}c^3bc^{-1}bc^{-1}b^2cb^{-1}cb^{-1}c^{-3}b^{-1}d = 1, \\
& c^{-2}b^{-2}c^{-1}bc^{-1}b^{-3}c^{-1}bcb^{-2}cbc^{-1}bd^2 = z, \\
& b^{-2}c^{-1}bc^{-1}b^{-2}c^{-1}b^{-2}cb^{-1}cb^{-1}cb^2c^2d^2 = 1, \\
& cb^{-1}c^4bc^{-1}b^{-2}c^{-1}bcb^{-1}c^2bcbd^2 = 1, \\
& c^{-2}b^{-1}c^2b^{-1}c^{-2}b^{-2}c^2bcb^2c^{-2}b^{-1}d = z, \\
& b^{-1}c^{-3}bc^2bc^{-1}b^{-1}cb^{-1}cb^{-1}c^2b^{-1}c^{-2} = z, \\
& c^2b^{-1}c^{-1}bc^{-1}b^{-1}c^{-1}b^3c^2b^{-1}c^{-1}bc^{-1}b^2c = z, \\
& bc^{-2}bc^{-2}bc^{-1}b^3c^{-1}b^2c^{-1}b^{-1}cb^2c^{-1} = 1, \\
& b^{-1}c^{-2}b^{-1}c^2bc^{-1}b^3cb^{-1}c^{-2}bc^2b^{-2} = 1, \\
& b^{-2}c^2b^{-2}c^{-2}b^3c^{-1}b^{-1}c^{-2}bc^{-1}b^{-1}c^{-2}d = 1, \\
& cbcb^{-1}cb^{-1}c^{-2}b^3c^2b^{-1}cb^{-1}c^2bcd = 1, \\
& (p_1,p_2) = (p_1,p_3) = (p_1,p_4) = 1, \\
& (p_1,p_6) = (p_1,p_7) = (p_1,p_8) = (p_1,p_9) = (p_1,p_{10}) = 1, \\
& (p_1,p_5) = (p_1,p_{11}) = (p_1,p_{12}) = (p_2,p_3) = z, \\
& (p_2,p_4) = (p_2,p_5) = (p_2,p_7) = (p_2,p_9) = (p_2,p_{11}) = 1, \\
& (p_2,p_6) = (p_2,p_8) = (p_2,p_{10}) = (p_2,p_{12}) = (p_3,p_5) = z, \\
& (p_3,p_4) = (p_3,p_6) = (p_3,p_{10}) = (p_3,p_{12}) = 1, \\
& (p_3,p_7) = (p_3,p_8) = (p_3,p_9) = (p_3,p_{11}) = (p_4,p_7) = z, \\
& (p_4,p_5) = (p_4,p_6) = (p_4,p_9) = (p_4,p_{10}) = 1, \\
& (p_4,p_8) = (p_4,p_{11}) = (p_4,p_{12}) = (p_5,p_6) = (p_5,p_{11}) = z, \\
& (p_5,p_7) = (p_5,p_8) = (p_5,p_9) = (p_5,p_{10}) = (p_5,p_{12}) = 1, \\
& (p_6,p_7) = (p_6,p_{10}) = (p_6,p_{11}) = (p_6,p_{12}) = 1, \\
& (p_6,p_8) = (p_6,p_9) = (p_7,p_8) = (p_7,p_9) = (p_7,p_{10}) = z,
\end{aligned}
$$

$$(p_7, p_{11}) = (p_7, p_{12}) = (p_8, p_{10}) = (p_9, p_{12}) = 1,$$
$$(p_8, p_9) = (p_8, p_{11}) = (p_8, p_{12}) = (p_9, p_{10}) = (p_9, p_{11}) = z,$$
$$(p_{10}, p_{11}) = (p_{10}, p_{12}) = (p_{11}, p_{12}) = z,$$
$$p_1^b p_1 p_5 p_6 p_7 p_8 p_{12} = p_2^b p_5 p_8 p_9 p_{12} = 1,$$
$$p_3^b p_2 p_3 p_4 p_5 p_6 p_7 p_8 p_9 p_{10} p_{11} = 1,$$
$$p_4^b p_1 p_2 p_4 p_5 p_8 p_9 p_{10} p_{11} p_{12} = p_5^b p_7 p_8 p_9 p_{12} = z,$$
$$p_6^b p_1 p_3 p_6 p_7 p_8 p_9 = p_{10}^b p_1 p_4 p_5 p_7 p_8 p_9 p_{10} = 1,$$
$$p_7^b p_2 p_5 p_6 p_9 p_{10} p_{11} = p_8^b p_1 p_2 p_5 p_7 p_{10} p_{11} p_{12} = 1,$$
$$p_9^b p_3 p_5 p_7 p_{10} p_{11} p_{12} = p_{11}^b p_2 p_4 p_6 p_7 p_8 p_9 p_{11} = z,$$
$$p_{12}^b p_1 p_2 p_4 p_5 p_9 p_{10} p_{11} = p_1^c p_1 p_2 p_4 p_6 p_7 p_8 = 1,$$
$$p_2^c p_1 p_3 p_4 p_7 p_8 p_9 = p_4^c p_3 p_4 p_5 p_6 p_8 p_{10} p_{11} = p_5^c p_1 p_2 p_5 p_7 p_9 p_{12} = z,$$
$$p_3^c p_1 p_3 p_6 p_7 p_8 p_9 p_{12} = p_6^c p_3 p_6 p_{11} p_{12} = 1,$$
$$p_7^c p_1 p_2 p_3 p_6 p_7 p_8 p_9 p_{10} = p_9^c p_1 p_2 p_3 p_4 p_7 p_{11} = z,$$
$$p_8^c p_2 p_3 p_4 p_5 p_6 p_9 p_{12} = p_{12}^c p_1 p_2 p_8 p_9 = 1,$$
$$p_{10}^c p_2 p_3 p_5 p_7 p_9 p_{10} p_{12} = p_{11}^c p_1 p_4 p_7 p_8 p_{11} p_{12} = z,$$
$$p_1^d p_2 p_6 p_9 p_{10} = p_2^d p_1 p_4 p_5 p_8 p_9 p_{12} = p_3^d p_1 p_2 p_3 p_4 p_6 p_9 = z,$$
$$p_4^d p_1 p_2 p_3 p_4 p_9 p_{10} = p_5^d p_1 p_3 p_5 p_7 p_{10} p_{12} = p_6^d p_2 p_3 p_6 p_9 = z,$$
$$p_7^d p_3 p_4 p_5 p_7 p_8 p_{10} p_{11} = p_8^d p_2 p_3 p_4 p_5 p_6 p_8 p_{11} p_{12} = z,$$
$$p_9^d p_2 p_3 p_5 p_8 p_{12} = p_{12}^d p_1 p_2 p_4 p_7 p_8 p_9 p_{10} p_{11} = z,$$
$$p_{10}^d p_2 p_4 p_9 p_{10} = p_{11}^d p_1 p_8 p_9 p_{10} p_{11} = p_5^f p_1 p_5 p_6 p_8 p_{10} p_{12} = 1,$$
$$p_1^f p_3 p_9 p_{11} p_{12} = p_2^f p_1 p_5 p_6 p_9 p_{12} = p_6^f p_4 p_5 p_7 p_9 = z,$$
$$p_3^f p_1 p_2 p_3 p_7 p_8 p_9 = p_4^f p_1 p_2 p_4 p_6 p_7 p_8 p_9 p_{10} = 1,$$
$$p_9^f p_2 p_3 p_6 p_7 p_8 p_9 p_{10} p_{11} p_{12} = p_{11}^f p_1 p_4 p_6 p_9 p_{10} p_{11} = z,$$
$$p_7^f p_3 p_4 p_5 p_8 p_{10} p_{11} = p_8^f p_3 p_4 p_8 p_{11} p_{12} = 1,$$
$$p_{10}^f p_4 p_5 p_6 p_7 p_9 p_{10} = p_{12}^f p_1 p_4 p_9 p_{12} = 1.$$

(k) *$H = \langle b, c, d, f, z, p_i \mid 1 \le i \le 12 \rangle$ has a faithful permutation representation of degree 258048 with stabilizer $L_1 = \langle b, c^3, f, p_5 p_6 p_7 p_9 p_{10} p_{11} \rangle$.*

(l) *In $H = \langle b, c, d, f, z, p_i \mid 1 \le i \le 12 \rangle$ let $r = [(fp_1)^2 (cfb)^3 (fp_1)^2 (fc^4)]^6$ and $p = [(fp_1)^2 (cfb)^3 (fp_1)^2 (b^2 fc)]^5$. Then $H = \langle b, p, r \rangle$, and it has 167 conjugacy classes whose representatives are given in Table 9.7.3.*

(m) *Table 9.8.2 is the character table of H.*

Proof (a) By Lemma 9.4.2 there is an isomorphism $\sigma : \mathfrak{G} \to G$ such that $\sigma(\mathfrak{q}) = q_1$, $\sigma(\mathfrak{y}) = y_1$, $\sigma(\mathfrak{w}) = w_1$ and $\sigma(\mathfrak{t}) = t_1$. In particular, $G = \langle q_1, y_1, w_1, t_1 \rangle$. Let x_1 be as in Lemma 9.4.2(i). Let $z_1 = \sigma(\mathfrak{z}) = (x_1 y_1 w_1)^8$. Using the faithful permutation representation PG of G given in Lemma 9.4.2(a) and MAGMA one sees that $C_G(z_1)$ has order $2^{21} \cdot 3^7 \cdot 5 \cdot 7$. Hence $\mathfrak{z}$ is a 2-central involution of $\mathfrak{G}$ by Lemma 9.4.2(k).

In G let $r_1 = s_1 y_1^3$, $r_2 = y_1^2 w_1 y_1 s_1$, $r_3 = (s_1 y_1 w_1 s_1)^2$, $r_4 = (s_1 w_1 y_1 s_1 w_1)^6$ and $v = (w_1 q_1 w_1 y_1 q_1 y_1)^7$. Let $H = \langle r_1, r_2, r_3, r_4, v \rangle = \sigma(\mathfrak{H})$. Then Proposition 9.4.1(i), (j) and Lemma 9.4.2(l) imply that $H \le C_G(z_1)$. Another application of PG and MAGMA yields that $C_G(z_1) = H$ and that $Z(H) = \langle z_1 \rangle$.

(b) Proposition 9.4.1(j) implies that $U = C_{G_1}(z_1)$ is generated by $f_1 = x_1 y_1 w_1$, $f_2 = (x_1 y_1 x_1 w_1)^7$ and $f_3 = (x_1 w_1 y_1 w_1 y_1^2)^7$ and that $|U| = 2^{18} \cdot 3^5 \cdot 5$. By Proposition 9.4.1(k) U has a subgroup U_1 of order $2^9 \cdot 3^5 \cdot 5$ which does not contain z. It is generated by the elements of the statement. Hence $H_1 = C_G(z_1)$ has a faithful permutation representation PH_1 of degree 258048 by (a).

(c) By Proposition 9.4.1(l) the subgroup $\mathfrak{S}$ is a Sylow 2-subgroup of $\mathfrak{E} = \langle \mathfrak{x}, \mathfrak{y}, \mathfrak{w}, \mathfrak{s} \rangle$. Lemma 9.4.2(h) states that $\mathfrak{E} \cong E$, where E is the finitely presented group of Lemma 7.1.1(g). Thus $|\mathfrak{S}| = 2^{21}$ by Table 9.8.3. Hence $\mathfrak{S}$ is a Sylow 2-subgroup of $\mathfrak{G}$ by (a). The equality $Z(\mathfrak{H}) = \langle \mathfrak{z} \rangle$ has been checked computationally in PG.

(d) This statement follows now immediately from Proposition 9.4.1(m).

(e) This assertion is true by (d) and Lemma 9.4.2(h), (i) and (k).

(f) By Proposition 9.4.1(i) and Lemma 9.4.2(l) we know that $\mathfrak{D} = C_{\mathfrak{E}}(\mathfrak{z}) = \langle \mathfrak{r}_i \mid 1 \le i \le 4 \rangle$, where $\mathfrak{z} = \kappa(z_1)$. Thus it suffices to check that $N_H(B_1) = \langle r_1, r_2, r_3, r_4 \rangle$. This has been done using the faithful permutation representation PG and MAGMA.

(g) We verified computationally that the Fitting subgroup O of H is extra-special of order 2^{13} and has center $Z(O) = \langle z \rangle$. The 12 involutions p_i generating the subgroup O have been calculated with MAGMA by means of PG and the program `GetShortGens(H,O)`. We also verified that $z = (p_1 p_5)^2$.

(h) Let $\alpha : H \to H/Z(H) = H_1$ be the canonical epimorphism of $H = C_G(z)$ with kernel $Z(H) = \langle z \rangle$. By (c) H has a faithful permutation representation PH with stabilizer $U_1 = \langle v f_2 v, (f_1 f_2 f_1 v f_1)^4, (f_2 f_1 v f_1 v)^4 \rangle$. As z does not belong to U_1 its cosets provide a faithful permutation representation PH of H having degree 258048 by (b). Using MAGMA we checked that the subgroup $\alpha(U_1)$ of H_1 is a stabilizer of a faithful permutation representation PH_1 of H_1 of degree 129024. Applying PH_1 and the MAGMA command `DegreeReduction(H_1)` MAGMA calculated a faithful permutation representation PH_{11} of $H_1 = \langle \alpha(r_i), \alpha(v) \rangle$ of degree 504 with stabilizer $U_{11} = \langle \alpha(r_1 r_3 r_1), [\alpha(r_1 v r_3 v)]^3, [\alpha(r_1 v r_1 r_3 v r_3)]^3 \rangle$.

Clearly, $V = O/Z(H)$ is an elementary abelian normal subgroup of H_1 of order 2^{12}. Using PH_{11} and the command `HasComplement(H_1,V)` MAGMA established a complement K_1 of V in H_1. By means of the command `CompositionFactors(K_1)` we saw that $|K_1 : K_1'| = 2$, $|Z(K_1')| = 3$ and $K'/Z(K') \cong U_4(3)$.

(i) Using PH_{11} a MAGMA calculation confirmed that H_1 is generated by

$$a_1 = \alpha(r_1)(\alpha(r_3)^3 \alpha(v) \alpha(r_3))$$

and

$$b_1 = [\alpha(r_1)\alpha(r_3)\alpha(r_1)\alpha(r_3)^2\alpha(v)]^6 [\alpha(r_1)\alpha(r_3)\alpha(r_1)\alpha(r_3)\alpha(v) \\ \alpha(r_3)\alpha(v)]^{12}.$$

Both generators have order 6. Furthermore, $K_1' = \langle b_1, c_1 \rangle$, where $c_1 = (a_1 b_1)^2$. Using the command `GetShortGens(K_1',Z(K_1'))` we observed that the center $Z(K_1')$ of K_1' is generated by the element $d_1 = (b_1 c_1)^7$ of order 3.

Let $\beta : K_1' \to K_1'/Z(K_1') = K_2$ be the canonical epimorphism of K_1 with kernel $Z(K_1') = \langle d_1 \rangle$. Let $U_{12} = U_{11} \cap K_1'$ and $U_{13} = \langle U_{12}, d_1 \rangle$. Then K_2 has a faithful permutation representation PK_2 of degree 126 with stabilizer $\beta(U_{13})$. Let $a_1 = \beta(b_1)$ and $a_2 = \beta(c_1)$. Then $K_2 = \langle a_1, a_2 \rangle$. Using the command `FPGroup(K_2)` MAGMA calculates the following set $\mathcal{R}(K_2)$ of defining relations of K_2:

$$\begin{aligned}
&a_1^6 = 1, \quad a_2^9 = 1,\\
&(a_2^{-1}a_1^{-1})^7 = 1, \quad (a_1 a_2^{-2} a_1)^4 = 1, \quad (a_2^{-1} a_1)^9 = 1,\\
&a_1^{-1} a_2^{-1} a_1^{-3} a_2^{-1} a_1^3 a_2^{-1} a_1^3 a_2^{-1} a_1^{-2} = 1,\\
&a_1 a_2^3 a_1^{-2} a_2 a_1^3 a_2^{-1} a_1 a_2^{-1} a_1 a_2 a_1^{-1} a_2 a_1 = 1,\\
&a_1^{-1} a_2^{-3} a_1^{-1} a_2^{-1} a_1 a_2^{-1} a_1^2 a_2 a_1^{-1} a_2 a_1 a_2^3 a_1^{-1} = 1,\\
&a_1^{-1} a_2^3 a_1 a_2^{-1} a_1 a_2^{-1} a_1^2 a_2 a_1^{-1} a_2 a_1^{-1} a_2^{-3} a_1^{-1} = (a_1 a_2^{-3} a_1 a_2^2 a_1)^2 = 1,\\
&a_2^{-2} a_1^{-2} a_2^{-1} a_1 a_2^{-1} a_1^{-3} a_2^{-1} a_1 a_2 a_1^{-2} a_2 a_1 a_2^{-1} a_1 = 1,\\
&a_1^{-2} a_2^{-1} a_1 a_2^{-1} a_1^{-2} a_2^{-1} a_1^{-2} a_2 a_1^{-1} a_2 a_1^{-1} a_2 a_1^2 a_2^2 = 1,\\
&a_2 a_1^{-1} a_2^4 a_1 a_2^{-1} a_1^{-2} a_2^{-1} a_1 a_2 a_1^{-1} a_2^2 a_1 a_2 a_1 = 1,\\
&a_2^{-2} a_1^{-1} a_2^2 a_1^{-1} a_2^{-2} a_1^{-2} a_2^2 a_1 a_2 a_1^2 a_2^{-2} a_1^{-1} = 1,\\
&a_1^{-1} a_2^{-3} a_1 a_2^2 a_1 a_2^{-1} a_1^{-1} a_2 a_1^{-1} a_2 a_1^{-1} a_2^2 a_1^{-1} a_2^{-2} = 1,\\
&a_2^2 a_1^{-1} a_2^{-1} a_1 a_2^{-1} a_1^{-1} a_2^{-1} a_1^3 a_2^2 a_1^{-1} a_2^{-1} a_1 a_2^{-1} a_1^2 a_2 = 1,
\end{aligned}$$

$$a_1 a_2^{-2} a_1 a_2^{-2} a_1 a_2^{-1} a_1^3 a_2^{-1} a_1^2 a_2^{-1} a_1^{-1} a_2 a_1^2 a_2^{-1} = 1,$$
$$a_1^{-1} a_2^{-2} a_1^{-1} a_2^2 a_1 a_2^{-1} a_1^3 a_2 a_1^{-1} a_2^{-2} a_1 a_2^2 a_1^{-2} = 1,$$
$$a_1^{-2} a_2^2 a_1^{-2} a_2^{-2} a_1^3 a_2^{-1} a_1^{-1} a_2^{-2} a_1 a_2^{-1} a_1^{-1} a_2^{-2} = 1,$$
$$a_2 a_1 a_2 a_1^{-1} a_2 a_1^{-1} a_2^{-2} a_1^3 a_2^2 a_1^{-1} a_2 a_1^{-1} a_2^2 a_1 a_2 = 1.$$

Since $K_1' = \langle b_1, c\rangle$, $Z(K_1') = \langle d_1\rangle$, $d_1 = (b_1 c_1)^7 \neq 1$, $d_1^3 = 1$ we lift the relations of $\mathcal{R}(K_2)$ to K_1' by evaluating them in the permutation representation of $K_1' = \langle b_1, c_1, d_1\rangle$. Thus we obtain the following set $\mathcal{R}(K_1')$ of defining relations of K_1':

$$b_1^6 = c_1^9 = d_1^3 = 1, \quad (b_1, d_1) = (c_1, d_1) = 1,$$
$$(c_1^{-1} b_1^{-1})^7 d_1 = 1, \quad (c_1^{-1} b_1)^9 = 1,$$
$$b_1^{-1} c_1^{-1} b_1^{-3} c_1^{-1} b_1^3 c_1^{-1} b_1^3 c_1^{-1} b_1^{-2} d_1 = 1,$$
$$(b_1 c_1^{-2} b_1)^4 d_1^2 = (b_1 c_1^{-3} b_1 c_1^2 b_1)^2 d_1 = 1,$$
$$b_1 c_1^3 b_1^{-2} c_1 b_1^3 c_1^{-1} b_1 c_1^{-1} b_1 c_1 b_1^{-1} c_1 b_1 d_1^2 = 1,$$
$$b_1^{-1} c_1^{-3} b_1^{-1} c_1^{-1} b_1 c_1^{-1} b_1^2 c_1 b_1^{-1} c_1 b_1 c_1^3 b_1^{-1} d_1 = 1,$$
$$b_1^{-1} c_1^3 b_1 c_1^{-1} b_1 c_1^{-1} b_1^2 c_1 b_1^{-1} c_1 b_1^{-1} c_1^{-3} b_1^{-1} d_1 = 1,$$
$$c_1^{-2} b_1^{-2} c_1^{-1} b_1 c_1^{-1} b_1^{-3} c_1^{-1} b_1 c_1 b_1^{-2} c_1 b_1 c_1^{-1} b_1 d_1^2 = 1,$$
$$b_1^{-2} c_1^{-1} b_1 c_1^{-1} b_1^{-2} c_1^{-1} b_1^{-2} c_1 b_1^{-1} c_1 b_1^{-1} c_1 b_1^2 c_1^2 d_1^2 = 1,$$
$$c_1 b_1^{-1} c_1^4 b_1 c_1^{-1} b_1^{-2} c_1^{-1} b_1 c_1 b_1^{-1} c_1^2 b_1 c_1 b_1 d_1^2 = 1,$$
$$c_1^{-2} b_1^{-1} c_1^2 b_1^{-1} c_1^{-2} b_1^{-2} c_1^2 b_1 c_1 b_1^2 c_1^{-2} b_1^{-1} d_1 = 1,$$
$$b_1^{-1} c_1^{-3} b_1 c_1^2 b_1 c_1^{-1} b_1^{-1} c_1 b_1^{-1} c_1 b_1^{-1} c_1^2 b_1^{-1} c_1^{-2} = 1,$$
$$c_1^2 b_1^{-1} c_1^{-1} b_1 c_1^{-1} b_1^{-1} c_1^{-1} b_1^3 c_1^2 b_1^{-1} c_1^{-1} b_1 c_1^{-1} b_1^2 c_1 = 1,$$
$$b_1 c_1^{-2} b_1 c_1^{-2} b_1 c_1^{-1} b_1^3 c_1^{-1} b_1^2 c_1^{-1} b_1^{-1} c_1 b_1^2 c_1^{-1} = 1,$$
$$b_1^{-1} c_1^{-2} b_1^{-1} c_1^2 b_1 c_1^{-1} b_1^3 c_1 b_1^{-1} c_1^{-2} b_1 c_1^2 b_1^{-2} = 1,$$
$$b_1^{-2} c_1^2 b_1^{-2} c_1^{-2} b_1^3 c_1^{-1} b_1^{-1} c_1^{-2} b_1 c_1^{-1} b_1^{-1} c_1^{-2} d_1 = 1,$$
$$c_1 b_1 c_1 b_1^{-1} c_1 b_1^{-1} c_1^{-2} b_1^3 c_1^2 b_1^{-1} c_1 b_1^{-1} c_1^2 b_1 c_1 d_1 = 1.$$

Using the faithful permutation representation PH_{11} of H_1 and the MAGMA command `HasComplement(K_1, K_1')` we see that K_1' has a complement $\langle f_1\rangle$ of order 2 such that $b_1^{f_1} \in \{b_1, b_1^{-1}\}$. In order to find a generator f_1 of a suitable complement we searched for an involution f_1 of K_1 so that at least one of the elements $c_1^{f_1}, d_1 \cdot c_1^{f_1}$ or $d_1^2 \cdot c_1^{f_1}$ is in the collection of all words in b_1 and c_1 of length at most 16. This search was successful. The involution $f_1 = [(b_1 a_1^3)^5 \cdot (a_1^4 b_1^2 a_1)^9 \cdot (a_1^2 b_1^3 a_1^4 b_1)^3 \cdot (b_1 a_1^3)^5]^3$ of $K_1 = \langle a_1, b_1\rangle$ satisfies the following set $\mathcal{R}(f_1)$ of relations: $f_1^2 = 1$, $b_1^{f_1} = b_1^5$, $c_1^{f_1} = d_1(b_1^3 c_1 b_1^2 c_1^6 b_1 c_1 b_1 c_1)$, $d_1^{f_1} = d_1^2$. Hence $K_1 = \langle a_1, b_1\rangle =$

$\langle b_1, c_1, d_1, f_1 \rangle$ has a set $\mathcal{R}(K_1)$ of defining relations consisting of $\mathcal{R}(K_1')$ and $\mathcal{R}(f_1)$.

The elementary abelian normal subgroup $V = \alpha(O)$ is generated by the involutions $q_i = \alpha(p_i)$, which are the images of the 12 generating involutions r_i of the Fitting subgroup O of H. Using the faithful permutation representation PH_{11} we calculated the images q_i^x in V for all four generators $x \in \{b_1, c_1, d_1, f_1\}$ of K_1. Thus we obtained the following set of essential relations $\mathcal{R}_2(V \rtimes K_1)$ of the semidirect product $(V \rtimes K_1)$:

$$
\begin{aligned}
&q_1^{b_1} q_1 q_5 q_6 q_7 q_8 q_{12} = q_2^{b_1} q_5 q_8 q_9 q_{12} = q_3^{b_1} q_2 q_3 q_4 q_5 q_6 q_7 q_8 q_9 q_{10} q_{11} = 1,\\
&q_4^{b_1} q_1 q_2 q_4 q_5 q_8 q_9 q_{10} q_{11} q_{12} = q_5^{b_1} q_7 q_8 q_9 q_{12} = q_6^{b_1} q_1 q_3 q_6 q_7 q_8 q_9 = 1,\\
&q_7^{b_1} q_2 q_5 q_6 q_9 q_{10} q_{11} = q_8^{b_1} q_1 q_2 q_5 q_7 q_{10} q_{11} q_{12} = q_9^{b_1} q_3 q_5 q_7 q_{10} q_{11} q_{12} = 1,\\
&q_{10}^{b_1} q_1 q_4 q_5 q_7 q_8 q_9 q_{10} = q_{11}^{b_1} q_2 q_4 q_6 q_7 q_8 q_9 q_{11} = q_{12}^{b_1} q_1 q_2 q_4 q_5 q_9 q_{10} q_{11} = 1,\\
&q_1^{c_1} q_1 q_2 q_4 q_6 q_7 q_8 = q_2^{c_1} q_1 q_3 q_4 q_7 q_8 q_9 = q_3^{c_1} q_1 q_3 q_6 q_7 q_8 q_9 q_{12} = 1,\\
&q_4^{c_1} q_3 q_4 q_5 q_6 q_8 q_{10} q_{11} = q_5^{c_1} q_1 q_2 q_5 q_7 q_9 q_{12} = q_6^{c_1} q_3 q_6 q_{11} q_{12} = 1,\\
&q_7^{c_1} q_1 q_2 q_3 q_6 q_7 q_8 q_9 q_{10} = q_8^{c_1} q_2 q_3 q_4 q_5 q_6 q_9 q_{12} = q_9^{c_1} q_1 q_2 q_3 q_4 q_7 q_{11} = 1,\\
&q_{10}^{c_1} q_2 q_3 q_5 q_7 q_9 q_{10} q_{12} = q_{11}^{c_1} q_1 q_4 q_7 q_8 q_{11} q_{12} = q_{12}^{c_1} q_1 q_2 q_8 q_9 = 1,\\
&q_1^{d_1} q_2 q_6 q_9 q_{10} = q_2^{d_1} q_1 q_4 q_5 q_8 q_9 q_{12} = q_3^{d_1} q_1 q_2 q_3 q_4 q_6 q_9 = 1,\\
&q_4^{d_1} q_1 q_2 q_3 q_4 q_9 q_{10} = q_5^{d_1} q_1 q_3 q_5 q_7 q_{10} q_{12} = q_6^{d_1} q_2 q_3 q_6 q_9 = 1,\\
&q_7^{d_1} q_3 q_4 q_5 q_7 q_8 q_{10} q_{11} = q_8^{d_1} q_2 q_3 q_4 q_5 q_6 q_8 q_{11} q_{12} = q_9^{d_1} q_2 q_3 q_5 q_8 q_{12} = 1,\\
&q_{10}^{d_1} q_2 q_4 q_9 q_{10} = q_{11}^{d_1} q_1 q_8 q_9 q_{10} q_{11} = q_{12}^{d_1} q_1 q_2 q_4 q_7 q_8 q_9 q_{10} q_{11} = 1,\\
&q_1^{f_1} q_3 q_9 q_{11} q_{12} = q_2^{f_1} q_1 q_5 q_6 q_9 q_{12} = q_3^{f_1} q_1 q_2 q_3 q_7 q_8 q_9 = 1,\\
&q_4^{f_1} q_1 q_2 q_4 q_6 q_7 q_8 q_9 q_{10} = q_5^{f_1} q_1 q_5 q_6 q_8 q_{10} q_{12} = q_6^{f_1} q_4 q_5 q_7 q_9 = 1,\\
&q_7^{f_1} q_3 q_4 q_5 q_8 q_{10} q_{11} = q_8^{f_1} q_3 q_4 q_8 q_{11} q_{12} = q_9^{f_1} q_2 q_3 q_6 q_7 q_8 q_9 q_{10} q_{11} q_{12} = 1,\\
&q_{10}^{f_1} q_4 q_5 q_6 q_7 q_9 q_{10} = q_{11}^{f_1} q_1 q_4 q_6 q_9 q_{10} q_{11} = q_{12}^{f_1} q_1 q_4 q_9 q_{12} = 1.
\end{aligned}
$$

Hence the set $\mathcal{R}(H_1)$ of defining relations of the semidirect product $H_1 = (V \rtimes K_1)$ consists of $\mathcal{R}(K_1)$, $\mathcal{R}_2(V \rtimes K_1)$ and the following relations:

$$
\begin{aligned}
q_j^2 &= 1 \quad \text{for all} \quad 1 \le j \le 12,\\
q_k \cdot q_j &= q_j \cdot q_k \quad \text{for all} \quad 1 \le j, k \le 12.
\end{aligned}
$$

In order to get a presentation of H we lift the generators a_1 and b_1 of K_1 to H. Clearly, $a = r_1 r_3^3 v r_3$ and $b = (r_1 r_3 r_1 r_3^2 v)^6 (r_1 r_3 r_1 r_3 v r_3 v)^{12}$ of H map onto a_1 and b_1 of H_1, respectively. Let $c = (ab)^2$, $d = (bc)^7$ and $f = [(ba^3)^5 \cdot (a^4 b^2 a)^9 \cdot (a^2 b^3 a^4 b)^3 \cdot (ba^3)^5]^3$. Then c_1, d_1 and f_1 in H_1 are images of c, d and f in H under α, respectively. Since $z = (p_1 p_5)^2$

generates the center $Z(O)$ of the Frattini subgroup $O = \langle p_i \mid 1 \leq i \leq 12\rangle$ it follows that

$$H = \langle a, b, p_i \mid 1 \leq i \leq 12\rangle = \langle b, c, d, f, p_i \mid 1 \leq i \leq 12\rangle.$$

The set of defining relations $\mathcal{R}(H)$ of $H = \langle b, c, d, f, p_i \mid 1 \leq i \leq 12\rangle$ has been obtained by evaluating the lifted equations of the presentation $\mathcal{R}(H_1)$ of H_1 in the permutation representation PH with stabilizer U_1 defined in the proof of (c). The resulting equations of $\mathcal{R}(H)$ are stated in assertion (k).

The map $\mathfrak{H} \to H$ sending each generator $\mathfrak{x}$ of $\mathfrak{H}$ in (i) to the corresponding generator $x \in H$ in (j) is an isomorphism by (a) and the order of H.

(k) The given stabilizer of the group $H = \langle b, c, d, f, p_i \mid 1 \leq i \leq 12\rangle$ has been found as follows. In the original permutation representation of the finitely presented group G of degree 306936 we checked that the subgroup $L = \langle b, c^3, f\rangle$ has index 1032192 in H and that $z \notin L$. Using then MAGMA and the command `MyCosetAction(H,L: maxsize:=10000000)` we verified that L has the same index in the finitely presented group $H = \langle b, c, d, f, p_i \mid 1 \leq i \leq 12\rangle$. In the corresponding permutation representation pH of this group we searched then for an element $p \in O = \langle p_i\rangle$ such that $z \notin L_1 = \langle L, p\rangle$. MAGMA found an involution $p \in O$ with these properties. Since O is extra-special of order 2^{13} the command `LookupWord(O,p)` worked well. The word of p is stated in the assertion. Using the command `MyCosetAction(H,L_1:maxsize:=10000000)` MAGMA established, in 70 seconds, the index $|H : L_1| = 258048$.

(l) Both elements r and p of H have order 6. Using the faithful permutation representation PH of H with stabilizer L_1 and MAGMA it has been verified that $H = \langle r, p, b\rangle$. Since $H_1 = H/\langle z\rangle$ has a faithful permutation representation of degree 504 we used it and Kratzer's Algorithm 5.3.18 of [92] to calculate a system of representatives of the classes of $H_1 = \langle \alpha(r), \alpha(p), \alpha(b)\rangle$; H_1 has 123 conjugacy classes. Their representatives have been lifted to H. Using PH we have checked the conjugacy of the lifted representatives and the products with the central involution z of H.

(m) The character table of H was calculated automatically by MAGMA using PH. □

Definition 9.5.2 A finite simple group G is said to be of Fi_{24}'*-type* if it has a 2-central involution z such that $C_G(z)$ is isomorphic to the finitely presented group H of Proposition 9.5.1(j).

9.6 On the uniqueness of Fi_{24}'

Since Fischer constructed his 3-transposition group $G = \mathrm{Fi}_{24}$ as the automorphism group of the graph $\mathcal{G}_{24}$ his group is uniquely determined, and so is its commutator subgroup Fi_{24}'. Therefore he did not address in [30] the question whether $G = \mathrm{Fi}_{24}'$ is uniquely determined by the centralizer $H = C_G(z)$ of a 2-central involution z of G. In particular, his lecture notes do not even describe the structure of $C_G(z)$. According to Aschbacher [3], p. 112, David Parrott was the first mathematician who characterized Fi_{24}' by means of the structure of its centralizer $H = C_G(z)$ in [107]. But he did not give a presentation of H. He only assumed that the Fitting subgroup O of H is extra-special of order 2^{13}, $H/O \cong 3U_4(3):2$, and $C_H(O) \leq O$. In order to avoid any ambiguities we use Definition 9.5.2 for the uniqueness results of this section.

The following subsidiary result is mainly due to D. Parrott [107].

Lemma 9.6.1 *Let G be any finite simple group of Fi_{24}'-type having a 2-central involution z with centralizer $H = C_G(z) = \langle r, p, b\rangle$ stated in Proposition 9.5.1(j). Then the following statements hold.*

(a) *H has nine conjugacy classes 2_i of involutions classified in Table 9.7.3. Its representatives $z = (p^4r^2)^6$ and $v = r^6$ fuse in G.*
(b) *Each Sylow 2-subgroup S of H is a Sylow 2-subgroup of G, and S has a unique maximal elementary abelian subgroup A of order 2^{11}. It is self-centralizing and weakly closed in S.*
(c) *$N_G(A) > N_H(A)$ and $N_G(A) \cong E$, where E is the non-split extension of the Mathieu group M_{24} by its 2-modular irreducible $GF(2)$-module V_2 of dimension 11 constructed in Lemma 7.1.1(g).*
(d) *G has two conjugacy classes of involutions represented by the involutions $z = (p^4r^2)^6$ and $u = (pr)^6$ of H.*
(e) *$U = C_G(u) \cong A_1$, where A_1 is the finitely presented group of Lemma 9.1.1(b).*
(f) *H has nine conjugacy classes 2_i of involutions classified in Table 9.7.3. The classes 2_1, 2_3, 2_5, 2_6, 2_8 and 2_9 of H fuse with z in G. Its classes 2_2, 2_4 and 2_7 fuse with u in $\mathfrak{G}$.*
(g) *U has seven conjugacy classes 2_j of involutions classified in Table 9.7.1. The classes 2_1, 2_2, 2_3 and 2_4 of U fuse with u in G. Its classes 2_5, 2_6 and 2_7 fuse with z in G.*

Proof (a) This non-trivial statement follows from Lemma 10 of [107] and Table 9.7.3.

(b) The first assertion is trivial because z is a 2-central involution of G. The second statement holds by Proposition 9.5.1(j) and (d). In particular, the maximal elementary abelian subgroup A of S is weakly closed in S because A and $A^g \leq S$ for some $g \in G$ implies that $A = A^g$.

(c) By (a) and (b) we can apply Corollary 4.8.8(c) of [92]. Hence $N_G(A) > N_H(A) = D$. Since D/A is isomorphic to a maximal subgroup $2^6 : 3S_6$ of M_{24}, and D does not split over A, assertion (c) holds by Lemma 7.1.1 and Theorem 1.4.15 of [92].

(d) and (f) Table 9.7.2 shows that E has five conjugacy classes of involutions and that $z_1 = x^2$ is the representative of the unique 2-central conjugacy class of $E = \langle x, y, e \rangle$. The subgroup $D = N_H(A)$ is isomorphic to $D_1 = C_E(z_1)$ by Proposition 9.5.1(d). It has 18, conjugacy classes 2_k, $1 \leq k \leq 18$, of involutions. Using MAGMA and the faithful permutation representations PH and PE of Propositions 9.5.1(c) and 9.4.1(h), respectively, we calculated the fusion of the classes 2_k of D in H and also of D_1 in E. Thus we obtained a fusion graph $\mathcal{G}(H)$ of the fusion of the H-classes and E-classes of order 2 in the group G. It follows that G has two conjugacy classes of involutions represented by z and u belonging to the classes 2_1 and 2_2 of H in Table 9.7.3, respectively. Furthermore, the results of (f) follow from $\mathcal{G}(H)$.

(e) By Table 9.7.2 the involution $z_1 = e^2$ of $E = \langle x, y, e \rangle \cong N_G(A)$ has a centralizer $E_1 = C_E(z_1)$ of order $2^{19} \cdot 3^2 \cdot 5 \cdot 7 \cdot 11$. It also has been checked computationally that $E_1/\langle z_1 \rangle$ is isomorphic to the split extension E_4 of $Aut(\mathsf{M}_{22})$ by its 10-dimensional simple $GF(2)$-module V_4; see Lemma 4.1.2(g). In particular, E_1' is perfect of index 2 in E_1.

Let $H_1 = C_H(u)$. Using the faithful permutation representation PH of H it has been checked that H_1' has index 4 in H_1 and that $H_1'/F \cong U_4(2)$, where F denotes the Fitting subgroup of H. Furthermore, there is a subgroup H_2 of index 2 in H such that $H_3 = H_2/\langle u \rangle$ is isomorphic to the centralizer of a 2-central involution of Fischer's simple group Fi_{22} described in Proposition 5.1.1. We may assume that $u = z_1 \in A \leq H \cap N_G(A)$. Hence $K = \langle H_2, E_1' \rangle$ is a perfect subgroup of $U = C_G(u)$ such that $C_K(u) = H_2$.

Let T be the least term of a derived series of U. Then $K \leq T$. Let X be a proper normal subgroup of order larger than 2. If $|X|$ is odd then operating with the Klein 4-group $V = \langle u, z \rangle$ on X we see that $X = 1$ by Theorem 4.2.2 of [92] because H_2 has no normal subgroups of odd order. Thus we may assume that $u \in X$ and $X_1 = X/\langle u \rangle$ is a normal subgroup of $T_1 = T/\langle u \rangle$ of even order. Since E_1' contains a Sylow 2-subgroup of T_1 and M_{22} operates irreducibly on $A/\langle u \rangle$ it follows that $|T : X|$ is odd.

Hence $T = X$ by the Feit–Thompson Theorem. Thus T_1 is simple with $H_3 = C_{T_1}(z_3)$ of a 2-central involution z_3 of T_1. Therefore $T_1 \cong \mathsf{Fi}_{22}$ by Theorem 5.3.5. Hence $U \cong A_1 = 2Aut(\mathsf{Fi}_{22})$ by Lemma 9.1.1(b) because T is normal in U.

(g) This assertion is proved similarly to (f) using (e) instead of (d). □

The following theorem is due to Parrott [107], p. 337. The proof given here is due to Kim and the author [75].

Theorem 9.6.2 *Let G be any finite simple group of Fi_{24}'-type having a 2-central involution z with centralizer $H = C_G(z) = \langle r, p, b\rangle$ stated in Proposition 9.5.1(j). Then the following statements hold.*

(a) *G has two conjugacy classes of involutions represented by $z = (p^4r^2)^6$ and $u = (pr)^6$ of H. Furthermore, $U = C_G(u) \cong A_1$, where A_1 is the finitely presented group of Lemma 9.1.1(b).*
(b) $G = 2132400816 \cdot |\mathfrak{U}| + 4388805476055 \cdot |\mathfrak{H}| = 2^{21} \cdot 3^{13} \cdot 5^2 \cdot 7 \cdot 11 \cdot 13 \cdot 17 \cdot 23 \cdot 29.$

Proof (a) This assertion holds by Lemma 9.6.1(d) and (e).

(b) Let

$$r(z,u,z) = \big|\{(x,y) \in (z^G \cap H) \times (u^G \cap H) \big| z \in \langle xy\rangle\}\big|$$

and

$$r(z,u,u) = \big|\{(x,y) \in (z^G \cap U) \times (u^G \cap U) \big| u \in \langle xy\rangle\}\big|.$$

By Table 9.8.2 H has 45 real z-special conjugacy classes with class numbers 2, 18, 19, 39, 40, 41, 42, 43, 46, 58, 77, 78, 79, 94, 100, 101, 102, 103, 104, 107, 108, 109, 113, 127, 134, 136, 137, 138, 139, 141, 142, 147, 153, 154, 155, 156, 158, 159, 160, 162, 163, 164, 165, 166, 167. Here their representatives are denoted by $\{t_j \mid 1 \le j \le 45\}$. Let z_i and u_k be representatives of the H-classes of involutions fusing to z and u in G, respectively. They are known by Lemma 9.6.1(f). For each triple (z_i, u_k, t_j) let

$$d(z_i,u_k,t_j) = \frac{|H|^2}{|C_H(z_i)| \cdot |C_H(u_k)| \cdot |C_H(t_j)|} \times \sum_{\psi \in Irr_{\mathbb{C}}(H)} \psi(z_i)\psi(u_k)\psi(t_j)\psi(1)^{-1}.$$

Then Theorem 1.6.4 and Lemma 9.6.1(f) imply that

$$r(z,u,z) = \sum_{i=1}^{6}\sum_{k=1}^{3}\sum_{j=1}^{45} d(z_i,u_k,t_j).$$

Using Lemma 9.6.1(f) and the values of the character Table 9.8.2 of H these formulas yield $r(z,u,z) = 2132400816$.

By Table 9.8.1, $U = C_G(u) = A_1$ has 22 real u-special conjugacy classes with class numbers 2, 13, 26, 27, 28, 34, 69, 76, 79, 91, 105, 111, 112, 120, 124, 129, 139, 140, 141, 146, 148, 150. Denote their representatives by $\{s_n \mid 1 \leq n \leq 22\}$. Let u_i and z_k be representatives of the U-classes of involutions fusing to u and z in G, respectively. They are known by Lemma 9.6.1(g). For each triple (u_i, z_k, s_n) let

$$d(u_i,z_k,s_n) = \frac{|U|^2}{|C_U(u_i)|\cdot|C_U(z_k)|\cdot|C_U(s_n)|} \times \sum_{\psi\in Irr_{\mathbb{C}}(U)} \psi(u_i)\psi(z_k)\psi(s_n)\psi(1)^{-1}.$$

Then Theorem 1.6.4 andd (h) imply that

$$r(z,u,u) = \sum_{i=1}^{4}\sum_{k=1}^{3}\sum_{n=1}^{22} d(u_i,z_k,s_n).$$

Using Lemma 9.61(g) and the values of the character Table 9.8.1 of U these formulas yield $r(z,u,u) = 4388805476055$. Now Theorem 4.2.1 of [92] due to J. G. Thompson implies that

$$|G| = r(z,u,z)\cdot|C_G(u)| + r(z,u,u)\cdot|C_G(z)| = 2^{21}\cdot 3^{13}\cdot 5^2\cdot 7\cdot 11\cdot 13\cdot 17\cdot 23\cdot 29.$$

□

9.7 Representatives of conjugacy classes

9.7.1 *Conjugacy classes of* $A_1 = 2Aut(\mathsf{Fi}_{22}) = \langle q, y, s \rangle$

Class	*Representative*	*Centralizer*	2P	3P	5P	7P	11P	13P
1	1	$2^{19} \cdot 3^9 \cdot 5^2 \cdot 7 \cdot 11 \cdot 13$	1	1	1	1	1	1
2_1	$(ys)^{14}$	$2^{19} \cdot 3^9 \cdot 5^2 \cdot 7 \cdot 11 \cdot 13$	1	2_1	2_1	2_1	2_1	2_1
2_2	$(y)^7$	$2^{17} \cdot 3^6 \cdot 5 \cdot 7 \cdot 11$	1	2_2	2_2	2_2	2_2	2_2
2_3	s	$2^{14} \cdot 3^6 \cdot 5^2 \cdot 7$	1	2_3	2_3	2_3	2_3	2_3
2_4	$(yq)^{10}$	$2^{19} \cdot 3^4 \cdot 5$	1	2_4	2_4	2_4	2_4	2_4
2_5	$(y^2q)^8$	$2^{19} \cdot 3^4 \cdot 5$	1	2_5	2_5	2_5	2_5	2_5
2_6	$(y^3qs)^9$	$2^{14} \cdot 3^4 \cdot 5$	1	2_6	2_6	2_6	2_6	2_6
2_7	$(ysq)^{12}$	$2^{17} \cdot 3^3$	1	2_7	2_7	2_7	2_7	2_7
3_1	$(qs)^4$	$2^{10} \cdot 3^7 \cdot 5 \cdot 7$	3_1	1	3_1	3_1	3_1	3_1
3_2	q	$2^9 \cdot 3^9$	3_2	1	3_2	3_2	3_2	3_2
3_3	$(ysq)^8$	$2^8 \cdot 3^7$	3_3	1	3_3	3_3	3_3	3_3
3_4	$(y^3qs)^6$	$2^5 \cdot 3^7$	3_4	1	3_4	3_4	3_4	3_4
4_1	$(ys)^7$	$2^{11} \cdot 3^4 \cdot 5 \cdot 7$	2_1	4_1	4_1	4_1	4_1	4_1
4_2	$(yqs)^9$	$2^{14} \cdot 3^4$	2_5	4_2	4_2	4_2	4_2	4_2
4_3	$(y^3sq^2)^5$	$2^{14} \cdot 3 \cdot 5$	2_5	4_3	4_3	4_3	4_3	4_3
4_4	$(yqyq^2)^6$	$2^{13} \cdot 3^3$	2_5	4_4	4_4	4_4	4_4	4_4
4_5	$(yq)^5$	$2^{11} \cdot 3^2 \cdot 5$	2_4	4_5	4_5	4_5	4_5	4_5
4_6	$(y^2q)^4$	$2^{13} \cdot 3$	2_5	4_6	4_6	4_6	4_6	4_6
4_7	$(qs)^3$	$2^{11} \cdot 3^2$	2_4	4_7	4_7	4_7	4_7	4_7
4_8	$(ysq)^6$	$2^{11} \cdot 3^2$	2_7	4_8	4_8	4_8	4_8	4_8
4_9	y^2sys	$2^{11} \cdot 3^2$	2_7	4_9	4_9	4_9	4_9	4_9
4_{10}	$(y^4sq)^3$	$2^{12} \cdot 3$	2_5	4_{10}	4_{10}	4_{10}	4_{10}	4_{10}
4_{11}	$(y^2sy^2sq)^3$	$2^{11} \cdot 3$	2_4	4_{11}	4_{11}	4_{11}	4_{11}	4_{11}
4_{12}	$(y^2qsq)^3$	$2^{10} \cdot 3$	2_7	4_{12}	4_{12}	4_{12}	4_{12}	4_{12}
5	$(yq)^4$	$2^5 \cdot 3 \cdot 5^2$	5	5	1	5	5	5
6_1	$(y^3q^2)^5$	$2^{10} \cdot 3^7 \cdot 5 \cdot 7$	3_1	2_1	6_1	6_1	6_1	6_1
6_2	$(y^4qys)^6$	$2^9 \cdot 3^9$	3_2	2_1	6_2	6_2	6_2	6_2
6_3	$(y^4s)^2$	$2^8 \cdot 3^7$	3_3	2_1	6_3	6_3	6_3	6_3
6_4	$(yqsys)^5$	$2^8 \cdot 3^5 \cdot 5$	3_1	2_2	6_4	6_4	6_4	6_4
6_5	q^2sqsqs	$2^8 \cdot 3^5 \cdot 5$	3_1	2_3	6_5	6_5	6_5	6_5
6_6	$(y^3sq^2s)^3$	$2^8 \cdot 3^6$	3_2	2_2	6_6	6_6	6_6	6_6
6_7	$(y^3q^2s)^7$	$2^8 \cdot 3^4 \cdot 7$	3_1	2_3	6_8	6_7	6_8	6_7
6_8	$(y^2qysq^2)^7$	$2^8 \cdot 3^4 \cdot 7$	3_1	2_3	6_7	6_8	6_7	6_8
6_9	$(y^2sqsq^2)^3$	$2^5 \cdot 3^7$	3_4	2_1	6_9	6_9	6_9	6_9
6_{10}	$(yqs)^6$	$2^9 \cdot 3^4$	3_2	2_5	6_{10}	6_{10}	6_{10}	6_{10}
6_{11}	$(y^2qsy^2s)^3$	$2^9 \cdot 3^4$	3_2	2_4	6_{11}	6_{11}	6_{11}	6_{11}
6_{12}	$(qs)^2$	$2^{10} \cdot 3^3$	3_1	2_4	6_{12}	6_{12}	6_{12}	6_{12}
6_{13}	$(yqyq^2)^4$	$2^{10} \cdot 3^3$	3_1	2_5	6_{13}	6_{13}	6_{13}	6_{13}
6_{14}	$(y^2sq^2)^3$	$2^5 \cdot 3^6$	3_2	2_3	6_{14}	6_{14}	6_{14}	6_{14}
6_{15}	y^3s	$2^5 \cdot 3^5$	3_3	2_3	6_{15}	6_{15}	6_{15}	6_{15}
6_{16}	y^7q	$2^5 \cdot 3^5$	3_3	2_2	6_{16}	6_{16}	6_{16}	6_{16}
6_{17}	$(ysq)^4$	$2^8 \cdot 3^3$	3_3	2_7	6_{17}	6_{17}	6_{17}	6_{17}
6_{18}	y^2qsys	$2^8 \cdot 3^3$	3_3	2_7	6_{18}	6_{18}	6_{18}	6_{18}
6_{19}	$(yqysys)^2$	$2^8 \cdot 3^3$	3_2	2_7	6_{19}	6_{19}	6_{19}	6_{19}
6_{20}	$y^3sqy^2sq^2s$	$2^8 \cdot 3^3$	3_1	2_6	6_{20}	6_{20}	6_{20}	6_{20}
6_{21}	$y^2q^2sqsq^2yq$	$2^8 \cdot 3^3$	3_1	2_7	6_{21}	6_{21}	6_{21}	6_{21}
6_{22}	$(y^4sq)^2$	$2^7 \cdot 3^3$	3_3	2_5	6_{22}	6_{22}	6_{22}	6_{22}
6_{23}	$(y^3q^2ys)^2$	$2^7 \cdot 3^3$	3_3	2_4	6_{23}	6_{23}	6_{23}	6_{23}
6_{24}	$(y^3qs)^3$	$2^5 \cdot 3^4$	3_4	2_6	6_{25}	6_{24}	6_{25}	6_{24}
6_{25}	$(y^3sqyq)^3$	$2^5 \cdot 3^4$	3_4	2_6	6_{24}	6_{25}	6_{24}	6_{25}
6_{26}	y^3q^2syq	$2^5 \cdot 3^4$	3_4	2_3	6_{27}	6_{26}	6_{27}	6_{26}
6_{27}	y^4q^2sqsys	$2^5 \cdot 3^4$	3_4	2_3	6_{26}	6_{27}	6_{26}	6_{27}
6_{28}	$(y^2qsq)^2$	$2^5 \cdot 3^3$	3_4	2_7	6_{28}	6_{28}	6_{28}	6_{28}

Conjugacy classes of $A_1 = 2Aut(\mathsf{Fi}_{22}) = \langle q, y, s \rangle$ *(continued)*

Class	*Representative*	*Centralizer*	2P	3P	5P	7P	11P	13P
6_{29}	y^3qsqs	$2^5 \cdot 3^3$	3_4	2_7	6_{29}	6_{29}	6_{29}	6_{29}
6_{30}	y^4qy^3q	$2^5 \cdot 3^3$	3_3	2_7	6_{30}	6_{30}	6_{30}	6_{30}
6_{31}	$y^3sysqys$	$2^5 \cdot 3^3$	3_3	2_6	6_{31}	6_{31}	6_{31}	6_{31}
7	$(y)^2$	$2^3 \cdot 3 \cdot 7$	7	7	7	1	7	7
8_1	y^2s	$2^8 \cdot 3$	4_6	8_1	8_1	8_1	8_1	8_1
8_2	$(yqyq^2)^3$	$2^8 \cdot 3$	4_4	8_2	8_2	8_2	8_2	8_2
8_3	$(y^4qs)^3$	$2^8 \cdot 3$	4_4	8_3	8_3	8_3	8_3	8_3
8_4	$(ysq^2sq)^3$	$2^8 \cdot 3$	4_4	8_4	8_4	8_4	8_4	8_4
8_5	y^3qy^2s	$2^8 \cdot 3$	4_4	8_5	8_5	8_5	8_5	8_5
8_6	$(y^2q)^2$	2^8	4_6	8_6	8_6	8_6	8_6	8_6
8_7	$(ysq)^3$	$2^6 \cdot 3$	4_8	8_7	8_7	8_7	8_7	8_7
8_8	y^2sqys	2^6	4_9	8_8	8_8	8_8	8_8	8_8
9_1	$(yqs)^4$	$2^4 \cdot 3^4$	9_1	3_2	9_1	9_1	9_1	9_1
9_2	$(y^2sq^2)^2$	$2^3 \cdot 3^4$	9_2	3_2	9_2	9_2	9_2	9_2
9_3	$(y^3qs)^2$	$2^2 \cdot 3^3$	9_3	3_4	9_3	9_3	9_3	9_3
10_1	$(y^3q^2)^3$	$2^5 \cdot 3 \cdot 5^2$	5	10_1	2_1	10_1	10_1	10_1
10_2	$(y^4qysq)^3$	$2^4 \cdot 3 \cdot 5^2$	5	10_2	2_3	10_2	10_2	10_2
10_3	$(yq)^2$	$2^5 \cdot 5$	5	10_3	2_4	10_3	10_3	10_3
10_4	$(y^3sq^2)^2$	$2^5 \cdot 5$	5	10_4	2_5	10_4	10_4	10_4
10_5	yq^2	$2^3 \cdot 3 \cdot 5$	5	10_5	2_2	10_5	10_5	10_5
10_6	y^2q^2s	$2^4 \cdot 5$	5	10_6	2_6	10_6	10_6	10_6
11	$(y^3q)^2$	$2^2 \cdot 11$	11	11	11	11	1	11
12_1	$(yq^2sq)^5$	$2^6 \cdot 3^3 \cdot 5$	6_1	4_1	12_1	12_1	12_1	12_1
12_2	yq^2ysysq^2ys	$2^8 \cdot 3^3$	6_{13}	4_2	12_2	12_2	12_2	12_2
12_3	$(yqs)^3$	$2^5 \cdot 3^4$	6_{10}	4_2	12_3	12_3	12_3	12_3
12_4	$(y^4qys)^3$	$2^5 \cdot 3^4$	6_2	4_1	12_4	12_4	12_4	12_4
12_5	$(yqyq^2)^2$	$2^8 \cdot 3^2$	6_{13}	4_4	12_5	12_5	12_5	12_5
12_6	$(y^4qs)^2$	$2^8 \cdot 3^2$	6_{13}	4_4	12_6	12_6	12_6	12_6
12_7	$y^2qsyqys$	$2^6 \cdot 3^3$	6_{10}	4_4	12_7	12_7	12_7	12_7
12_8	yq^2sqysq^2	$2^7 \cdot 3^2$	6_{13}	4_4	12_8	12_8	12_8	12_8
12_9	$y^2sysqsqyq$	$2^7 \cdot 3^2$	6_{13}	4_2	12_9	12_9	12_9	12_9
12_{10}	y^4qy^3qys	$2^5 \cdot 3^3$	6_{22}	4_2	12_{10}	12_{10}	12_{10}	12_{10}
12_{11}	$y^2qy^2qy^2sq$	$2^8 \cdot 3$	6_{13}	4_3	12_{11}	12_{11}	12_{11}	12_{11}
12_{12}	qs	$2^6 \cdot 3^2$	6_{12}	4_7	12_{12}	12_{12}	12_{12}	12_{12}
12_{13}	$(ysq)^2$	$2^6 \cdot 3^2$	6_{17}	4_8	12_{13}	12_{13}	12_{13}	12_{13}
12_{14}	$y^2sqsqyq$	$2^6 \cdot 3^2$	6_{12}	4_5	12_{14}	12_{14}	12_{14}	12_{14}
12_{15}	y^3sqsyq^2	$2^6 \cdot 3^2$	6_{17}	4_9	12_{15}	12_{15}	12_{15}	12_{15}
12_{16}	y^4s	$2^4 \cdot 3^3$	6_3	4_1	12_{16}	12_{16}	12_{16}	12_{16}
12_{17}	$yqysys$	$2^5 \cdot 3^2$	6_{19}	4_9	12_{17}	12_{17}	12_{17}	12_{17}
12_{18}	y^4sysq	$2^5 \cdot 3^2$	6_{19}	4_8	12_{18}	12_{18}	12_{18}	12_{18}
12_{19}	$y^2qsqsqs$	$2^5 \cdot 3^2$	6_{11}	4_7	12_{19}	12_{19}	12_{19}	12_{19}
12_{20}	$(yqys)^2$	$2^6 \cdot 3$	6_{10}	4_6	12_{20}	12_{20}	12_{20}	12_{20}
12_{21}	y^2sy^2sq	$2^6 \cdot 3$	6_{12}	4_{11}	12_{21}	12_{21}	12_{21}	12_{21}
12_{22}	y^3q^2ys	$2^4 \cdot 3^2$	6_{23}	4_7	12_{22}	12_{22}	12_{22}	12_{22}
12_{23}	y^2sqysq	$2^4 \cdot 3^2$	6_{23}	4_5	12_{23}	12_{23}	12_{23}	12_{23}
12_{24}	$yqysqys$	$2^4 \cdot 3^2$	6_{28}	4_9	12_{24}	12_{24}	12_{24}	12_{24}
12_{25}	$yqysyq^2s$	$2^4 \cdot 3^2$	6_{28}	4_8	12_{25}	12_{25}	12_{25}	12_{25}
12_{26}	y^4sq	$2^5 \cdot 3$	6_{22}	4_{10}	12_{26}	12_{26}	12_{26}	12_{26}
12_{27}	y^2qsq	$2^4 \cdot 3$	6_{28}	4_{12}	12_{28}	12_{27}	12_{28}	12_{27}
12_{28}	y^3qsq	$2^4 \cdot 3$	6_{28}	4_{12}	12_{27}	12_{28}	12_{27}	12_{28}
13	y^4q	$2 \cdot 13$	13	13	13	13	13	1
14_1	$(ys)^2$	$2^3 \cdot 3 \cdot 7$	7	14_1	14_1	2_1	14_1	14_1
14_2	$yqsq$	$2^2 \cdot 3 \cdot 7$	7	14_2	14_2	2_3	14_2	14_2

Conjugacy classes of $A_1 = 2Aut(\mathsf{Fi}_{22}) = \langle q, y, s \rangle$ *(continued)*

Class	*Representative*	|*Centralizer*|	2P	3P	5P	7P	11P	13P
14_3	y	$2^2 \cdot 7$	7	14_3	14_3	2_2	14_3	14_3
15	$(y^3q^2)^2$	$2^3 \cdot 3 \cdot 5$	15	5	3_1	15	15	15
16_1	y^2q	2^5	8_6	16_1	16_1	16_1	16_1	16_1
16_2	y^2qs	2^5	8_6	16_2	16_2	16_2	16_2	16_2
18_1	$y^4syqyqys$	$2^4 \cdot 3^4$	9_1	6_2	18_1	18_1	18_1	18_1
18_2	$(y^4qys)^2$	$2^3 \cdot 3^4$	9_2	6_2	18_2	18_2	18_2	18_2
18_3	y^4qsq	$2^3 \cdot 3^3$	9_1	6_{14}	18_6	18_3	18_6	18_3
18_4	y^3sq^2s	$2^3 \cdot 3^3$	9_1	6_6	18_5	18_4	18_5	18_4
18_5	y^4qsys	$2^3 \cdot 3^3$	9_1	6_6	18_4	18_5	18_4	18_5
18_6	$y^2q^2syq^2$	$2^3 \cdot 3^3$	9_1	6_{14}	18_3	18_6	18_3	18_6
18_7	$(yqs)^2$	$2^4 \cdot 3^2$	9_1	6_{10}	18_7	18_7	18_7	18_7
18_8	y^2qsy^2s	$2^4 \cdot 3^2$	9_1	6_{11}	18_8	18_8	18_8	18_8
18_9	y^2sq^2	$2^2 \cdot 3^3$	9_2	6_{14}	18_9	18_9	18_9	18_9
18_{10}	y^2sqsq^2	$2^2 \cdot 3^3$	9_3	6_9	18_{10}	18_{10}	18_{10}	18_{10}
18_{11}	$ysqysq^2$	$2^2 \cdot 3^3$	9_2	6_6	18_{11}	18_{11}	18_{11}	18_{11}
18_{12}	y^3qs	$2^2 \cdot 3^2$	9_3	6_{24}	18_{13}	18_{12}	18_{13}	18_{12}
18_{13}	y^3sqyq	$2^2 \cdot 3^2$	9_3	6_{25}	18_{12}	18_{13}	18_{12}	18_{13}
20_1	yq^2ys	$2^3 \cdot 3 \cdot 5$	10_1	20_1	4_1	20_1	20_1	20_1
20_2	yq	$2^3 \cdot 5$	10_3	20_2	4_5	20_2	20_2	20_2
20_3	y^3sq^2	$2^3 \cdot 5$	10_4	20_3	4_3	20_3	20_3	20_3
21	$(y^3q^2s)^2$	$2^2 \cdot 3 \cdot 7$	21	7	21	3_1	21	21
22_1	y^3q	$2^2 \cdot 11$	11	22_1	22_1	22_3	2_2	22_3
22_2	$ysqs$	$2^2 \cdot 11$	11	22_2	22_2	22_2	2_1	22_2
22_3	y^2sqs	$2^2 \cdot 11$	11	22_3	22_3	22_1	2_2	22_1
24_1	$yqyq^2$	$2^5 \cdot 3$	12_5	8_2	24_1	24_1	24_1	24_1
24_2	y^4qs	$2^5 \cdot 3$	12_6	8_3	24_2	24_2	24_2	24_2
24_3	ysq^2sq	$2^5 \cdot 3$	12_6	8_4	24_3	24_3	24_3	24_3
24_4	y^2qysyq	$2^5 \cdot 3$	12_5	8_5	24_4	24_4	24_4	24_4
24_5	ysq	$2^4 \cdot 3$	12_{13}	8_7	24_7	24_7	24_5	24_5
24_6	$yqys$	$2^4 \cdot 3$	12_{20}	8_1	24_8	24_8	24_6	24_6
24_7	ysq^2	$2^4 \cdot 3$	12_{13}	8_7	24_5	24_5	24_7	24_7
24_8	y^2qys	$2^4 \cdot 3$	12_{20}	8_1	24_6	24_6	24_8	24_8
26	y^5q	$2 \cdot 13$	13	26	26	26	26	2_1
28	ys	$2^2 \cdot 7$	14_1	28	28	4_1	28	28
30_1	y^3q^2	$2^3 \cdot 3 \cdot 5$	15	10_1	6_1	30_1	30_1	30_1
30_2	$yqsys$	$2^2 \cdot 3 \cdot 5$	15	10_5	6_4	30_2	30_2	30_2
30_3	y^4qysq	$2^2 \cdot 3 \cdot 5$	15	10_2	6_5	30_3	30_3	30_3
36_1	yqs	$2^3 \cdot 3^2$	18_7	12_3	36_1	36_1	36_1	36_1
36_2	$ysysqs$	$2^3 \cdot 3^2$	18_7	12_3	36_2	36_2	36_2	36_2
36_3	y^4qys	$2^2 \cdot 3^2$	18_2	12_4	36_3	36_3	36_3	36_3
42_1	y^3q^2s	$2^2 \cdot 3 \cdot 7$	21	14_2	42_3	6_7	42_3	42_1
42_2	y^3sqs	$2^2 \cdot 3 \cdot 7$	21	14_1	42_2	6_1	42_2	42_2
42_3	y^2qysq^2	$2^2 \cdot 3 \cdot 7$	21	14_2	42_1	6_8	42_1	42_3
60	yq^2sq	$2^2 \cdot 3 \cdot 5$	30_1	20_1	12_1	60	60	60

9.7.2 *Conjugacy classes of* $E(\mathsf{Fi}_{24}) = \langle x, y, e\rangle$

Class	*Representative*	$\lvert$*Class*$\rvert$	$\lvert$*Centralizer*$\rvert$	2P	3P	5P	7P	11P	23P
1	1	1	$2^{21}\cdot 3^3\cdot 5\cdot 7\cdot 11\cdot 23$	1	1	1	1	1	1
2_1	$(e)^2$	276	$2^{19}\cdot 3^2\cdot 5\cdot 7\cdot 11$	1	2_1	2_1	2_1	2_1	2_1
2_2	$(x)^2$	1771	$2^{21}\cdot 3^3\cdot 5$	1	2_2	2_2	2_2	2_2	2_2
2_3	$(y)^3$	182160	$2^{17}\cdot 3\cdot 7$	1	2_3	2_3	2_3	2_3	2_3
2_4	$(ye^2)^3$	1275120	$2^{17}\cdot 3$	1	2_4	2_4	2_4	2_4	2_4
2_5	$(xey)^3$	2040192	$2^{14}\cdot 3\cdot 5$	1	2_5	2_5	2_5	2_5	2_5
3_1	$(y)^2$	14508032	$2^8\cdot 3^3\cdot 5$	3_1	1	3_1	3_1	3_1	3_1
3_2	$(xy)^4$	124354560	$2^6\cdot 3^2\cdot 7$	3_2	1	3_2	3_2	3_2	3_2
4_1	$(xy)^3$	1020096	$2^{15}\cdot 3\cdot 5$	2_2	4_1	4_1	4_1	4_1	4_1
4_2	$(xyey)^3$	1020096	$2^{15}\cdot 3\cdot 5$	2_2	4_2	4_2	4_2	4_2	4_2
4_3	$(xe)^3$	5100480	$2^{15}\cdot 3$	2_2	4_3	4_3	4_3	4_3	4_3
4_4	$(xye)^6$	5100480	$2^{15}\cdot 3$	2_2	4_4	4_4	4_4	4_4	4_4
4_5	e	11658240	$2^{11}\cdot 3\cdot 7$	2_1	4_5	4_5	4_5	4_5	4_5
4_6	x	61205760	2^{13}	2_2	4_6	4_6	4_6	4_6	4_6
4_7	$(x^2yxy^2)^3$	81607680	$2^{11}\cdot 3$	2_3	4_7	4_7	4_7	4_7	4_7
4_8	$(x^2yxy^2e^2)^3$	81607680	$2^{11}\cdot 3$	2_3	4_8	4_8	4_8	4_8	4_8
4_9	$(xyeyey)^2$	244823040	2^{11}	2_4	4_9	4_9	4_9	4_9	4_9
4_{10}	$(x^3exey)^2$	244823040	2^{11}	2_4	4_{10}	4_{10}	4_{10}	4_{10}	4_{10}
4_{11}	x^3y^2xe	489646080	2^{10}	2_3	4_{11}	4_{11}	4_{11}	4_{11}	4_{11}
4_{12}	$xyxyxey$	489646080	2^{10}	2_4	4_{12}	4_{12}	4_{12}	4_{12}	4_{12}
5	$(xy^2)^3$	1044578304	$2^5\cdot 3\cdot 5$	5	5	1	5	5	5
6_1	$(xye^2y)^5$	14508032	$2^8\cdot 3^3\cdot 5$	3_1	2_2	6_1	6_1	6_1	6_1
6_2	$(xe)^2$	217620480	$2^8\cdot 3^2$	3_1	2_2	6_2	6_2	6_2	6_2
6_3	x^2y^2	217620480	$2^8\cdot 3^2$	3_1	2_1	6_3	6_3	6_3	6_3
6_4	$(xy)^2$	870481920	$2^6\cdot 3^2$	3_2	2_2	6_4	6_4	6_4	6_4
6_5	y	2611445760	$2^6\cdot 3$	3_1	2_3	6_5	6_5	6_5	6_5
6_6	ye^2	2611445760	$2^6\cdot 3$	3_1	2_4	6_6	6_6	6_6	6_6
6_7	xey	10445783040	$2^4\cdot 3$	3_2	2_5	6_7	6_7	6_7	6_7
7_1	$(xyxe)^2$	2984509440	$2^3\cdot 3\cdot 7$	7_1	7_2	7_2	1	7_1	7_1
7_2	$(xyexe)^2$	2984509440	$2^3\cdot 3\cdot 7$	7_2	7_1	7_1	1	7_2	7_2
8_1	$(xye)^3$	652861440	$2^8\cdot 3$	4_4	8_1	8_1	8_1	8_1	8_1
8_2	$(xey^2)^3$	652861440	$2^8\cdot 3$	4_1	8_2	8_2	8_2	8_2	8_2
8_3	$(xyxy^2)^3$	652861440	$2^8\cdot 3$	4_2	8_3	8_3	8_3	8_3	8_3
8_4	$(xexey)^2$	979292160	2^9	4_4	8_4	8_4	8_4	8_4	8_4
8_5	xy^2xe	1958584320	2^8	4_3	8_5	8_5	8_5	8_5	8_5
8_6	$xyxy^2xy^2$	3917168640	2^7	4_6	8_6	8_6	8_6	8_6	8_6
8_7	$xyeyey$	7834337280	2^6	4_9	8_7	8_7	8_7	8_7	8_7
8_8	x^3exey	7834337280	2^6	4_{10}	8_8	8_8	8_8	8_8	8_8
10_1	$(xye^2y)^3$	1044578304	$2^5\cdot 3\cdot 5$	5	10_1	2_2	10_1	10_1	10_1
10_2	y^2e	6267469824	$2^4\cdot 5$	5	10_2	2_1	10_2	10_2	10_2
10_3	$xyxeyey$	6267469824	$2^4\cdot 5$	5	10_4	2_5	10_4	10_3	10_4
10_4	$xyxexe^3$	6267469824	$2^4\cdot 5$	5	10_3	2_5	10_3	10_4	10_3
11	$(ye)^2$	22790799360	$2\cdot 11$	11	11	11	11	1	11
12_1	xe	2611445760	$2^6\cdot 3$	6_2	4_3	12_1	12_1	12_1	12_1
12_2	$(xye)^2$	2611445760	$2^6\cdot 3$	6_2	4_4	12_2	12_2	12_2	12_2
12_3	xy	5222891520	$2^5\cdot 3$	6_4	4_1	12_3	12_3	12_3	12_3
12_4	$xyey$	5222891520	$2^5\cdot 3$	6_4	4_2	12_4	12_4	12_4	12_4
12_5	y^3e	10445783040	$2^4\cdot 3$	6_3	4_5	12_5	12_5	12_5	12_5
12_6	x^2yxy^2	10445783040	$2^4\cdot 3$	6_5	4_7	12_6	12_6	12_6	12_6
12_7	$x^2yxy^2e^2$	10445783040	$2^4\cdot 3$	6_5	4_8	12_7	12_7	12_7	12_7
14_1	$(xyey^2)^2$	8953528320	$2^3\cdot 7$	7_2	14_2	14_2	2_1	14_1	14_1
14_2	$(xye^2xe)^2$	8953528320	$2^3\cdot 7$	7_1	14_1	14_1	2_1	14_2	14_2
14_3	$xyxe$	17907056640	$2^2\cdot 7$	7_1	14_4	14_4	2_3	14_3	14_3
14_4	$xyexe$	17907056640	$2^2\cdot 7$	7_2	14_3	14_3	2_3	14_4	14_4
15_1	xy^2	16713252864	$2\cdot 3\cdot 5$	15_1	5	3_1	15_2	15_2	15_1
15_2	xy^4	16713252864	$2\cdot 3\cdot 5$	15_2	5	3_1	15_1	15_1	15_2
16	$xexey$	15668674560	2^5	8_4	16	16	16	16	16
20_1	$xyxexe$	6267469824	$2^4\cdot 5$	10_1	20_1	4_2	20_1	20_1	20_1
20_2	xy^2xy^4	6267469824	$2^4\cdot 5$	10_1	20_2	4_1	20_2	20_2	20_2

Conjugacy classes of $E(\mathsf{Fi}_{24}) = \langle x, y, e \rangle$ *(continued)*

Class	*Representative*	|*Class*|	|*Centralizer*|	2P	3P	5P	7P	11P	23P
21_1	$xeye$	23876075520	$3 \cdot 7$	21_1	7_1	21_2	3_2	21_1	21_1
21_2	xy^2eye	23876075520	$3 \cdot 7$	21_2	7_2	21_1	3_2	21_2	21_2
22	ye	22790799360	$2 \cdot 11$	11	22	22	22	2_1	22
23_1	xy^2e	21799895040	23	23_1	23_1	23_2	23_2	23_2	1
23_2	xey^3	21799895040	23	23_2	23_2	23_1	23_1	23_1	1
24_1	xye	10445783040	$2^4 \cdot 3$	12_2	8_1	24_2	24_2	24_1	24_1
24_2	xy^3	10445783040	$2^4 \cdot 3$	12_2	8_1	24_1	24_1	24_2	24_2
24_3	xey^2	20891566080	$2^3 \cdot 3$	12_3	8_2	24_3	24_3	24_3	24_3
24_4	$xyxy^2$	20891566080	$2^3 \cdot 3$	12_4	8_3	24_4	24_4	24_4	24_4
28_1	$xyey^2$	17907056640	$2^2 \cdot 7$	14_1	28_2	28_2	4_5	28_1	28_1
28_2	xye^2xe	17907056640	$2^2 \cdot 7$	14_2	28_1	28_1	4_5	28_2	28_2
30_1	xye^2y	16713252864	$2 \cdot 3 \cdot 5$	15_1	10_1	6_1	30_2	30_2	30_1
30_2	$x^2y^2xy^2$	16713252864	$2 \cdot 3 \cdot 5$	15_2	10_1	6_1	30_1	30_1	30_2

9.7.3 *Conjugacy classes of* $H(\mathsf{Fi}'_{24}) = \langle r, p, b\rangle$

No.	*Class*	*Representative*	*Centralizer*	2P	3P	5P	7P
1	1	1	$2^{21}\cdot 3^7\cdot 5^1\cdot 7^1$	1	1	1	1
2	2_1	$z=(p^4r^2)^6$	$2^{21}\cdot 3^7\cdot 5^1\cdot 7^1$	1	2_1	2_1	2_1
3	2_2	$(pr)^6$	$2^{19}\cdot 3^4\cdot 5^1$	1	2_2	2_2	2_2
4	2_3	r^6	$2^{20}\cdot 3^2\cdot 5^1$	1	2_3	2_3	2_3
5	2_4	$(p^2br)^9$	$2^{14}\cdot 3^4\cdot 5^1$	1	2_4	2_4	2_4
6	2_5	$z(p^2br)^9$	$2^{14}\cdot 3^4\cdot 5^1$	1	2_5	2_5	2_5
7	2_6	p^6	$2^{17}\cdot 3^3$	1	2_6	2_6	2_6
8	2_7	zp^6	$2^{17}\cdot 3^3$	1	2_7	2_7	2_7
9	2_8	$(rb^3)^4$	$2^{16}\cdot 3^1$	1	2_8	2_8	2_8
10	2_9	$(p^2r)^3$	$2^{13}\cdot 3^2$	1	2_9	2_9	2_9
11	3_1	p^4	$2^8\cdot 3^7\cdot 5^1\cdot 7^1$	3_1	1	3_1	3_1
12	3_2	$z(prpb^2)^4$	$2^{12}\cdot 3^6$	3_2	1	3_2	3_2
13	3_3	$z(pr^2b)^4$	$2^{10}\cdot 3^7$	3_3	1	3_3	3_3
14	3_4	$z(pb^2)^3$	$2^5\cdot 3^7$	3_4	1	3_4	3_4
15	3_5	b^2	$2^8\cdot 3^5$	3_5	1	3_5	3_5
16	3_6	$(prbr)^2$	$2^5\cdot 3^6$	3_6	1	3_6	3_6
17	3_7	r^4	$2^6\cdot 3^4$	3_7	1	3_7	3_7
18	4_1	$(p^3br)^6$	$2^{15}\cdot 3^5\cdot 5^1$	2_1	4_1	4_1	4_1
19	4_2	$(rb)^4$	$2^{15}\cdot 3^2$	2_1	4_2	4_2	4_2
20	4_3	$(prb)^5$	$2^{13}\cdot 3^1\cdot 5^1$	2_3	4_3	4_3	4_3
21	4_4	$(pr)^3$	$2^{11}\cdot 3^2\cdot 5^1$	2_2	4_4	4_4	4_4
22	4_5	r^3	$2^{13}\cdot 3^2$	2_3	4_5	4_5	4_5
23	4_6	$(p^2brb)^2$	$2^{14}\cdot 3^1$	2_3	4_6	4_6	4_6
24	4_7	p^3	$2^{11}\cdot 3^2$	2_6	4_7	4_7	4_7
25	4_8	$(p^4rprb)^3$	$2^{11}\cdot 3^2$	2_2	4_8	4_8	4_8
26	4_9	zp^3	$2^{11}\cdot 3^2$	2_6	4_9	4_9	4_9
27	4_{10}	$(p^2rpr^2)^2$	2^{14}	2_3	4_{10}	4_{10}	4_{10}
28	4_{11}	$(p^2r^2bprb)^3$	$2^{11}\cdot 3^1$	2_2	4_{11}	4_{11}	4_{11}
29	4_{12}	p^4r^3	2^{12}	2_3	4_{12}	4_{12}	4_{12}
30	4_{13}	$z(p^2b^3r)^3$	$2^{10}\cdot 3^1$	2_8	4_{13}	4_{13}	4_{13}
31	4_{14}	$z(p^2rb)^3$	$2^{10}\cdot 3^1$	2_6	4_{14}	4_{14}	4_{14}
32	4_{15}	$(p^2b^3r)^3$	$2^{10}\cdot 3^1$	2_8	4_{15}	4_{15}	4_{15}
33	4_{16}	$(p^2b^3)^3$	$2^9\cdot 3^1$	2_7	4_{16}	4_{16}	4_{16}
34	4_{17}	$(p^2bprb)^3$	$2^9\cdot 3^1$	2_7	4_{17}	4_{17}	4_{17}
35	4_{18}	$(rb^3)^2$	2^{10}	2_8	4_{18}	4_{18}	4_{18}
36	4_{19}	p^2r^3	2^{10}	2_8	4_{19}	4_{19}	4_{19}
37	4_{20}	$prpr^3$	2^9	2_8	4_{20}	4_{20}	4_{20}
38	5	$(p^2b)^6$	$2^6\cdot 3^1\cdot 5^1$	5	5	1	5
39	6_1	$(p^2b)^5$	$2^8\cdot 3^7\cdot 5^1\cdot 7^1$	3_1	2_1	6_1	6_1
40	6_2	$(prpb^2)^4$	$2^{12}\cdot 3^6$	3_2	2_1	6_2	6_2
41	6_3	$(p^2brpr)^3$	$2^{10}\cdot 3^7$	3_3	2_1	6_3	6_3
42	6_4	$(pb^2)^3$	$2^5\cdot 3^7$	3_4	2_1	6_4	6_4
43	6_5	$(p^4rb^2r)^2$	$2^8\cdot 3^5$	3_5	2_1	6_5	6_5
44	6_6	$(prbpbr)^3$	$2^9\cdot 3^4$	3_3	2_2	6_6	6_6
45	6_7	$(p^2r^2bprb)^2$	$2^{10}\cdot 3^3$	3_2	2_2	6_7	6_7
46	6_8	$(pr^2b^2pb^2)^2$	$2^5\cdot 3^6$	3_6	2_1	6_8	6_8
47	6_9	$(p^4bprpb)^2$	$2^{11}\cdot 3^2$	3_2	2_3	6_9	6_9
48	6_{10}	p^2rbpbr^2bpr	$2^8\cdot 3^3$	3_2	2_6	6_{10}	6_{10}
49	6_{11}	$(p^2b^3)^2$	$2^8\cdot 3^3$	3_1	2_7	6_{11}	6_{11}
50	6_{12}	$z(p^3bpbrb)$	$2^8\cdot 3^3$	3_2	2_5	6_{12}	6_{12}
51	6_{13}	p^3bpbrb	$2^8\cdot 3^3$	3_2	2_4	6_{13}	6_{13}
52	6_{14}	p^2	$2^8\cdot 3^3$	3_1	2_6	6_{14}	6_{14}
53	6_{15}	$(p^2b^2pb^2)^2$	$2^8\cdot 3^3$	3_3	2_6	6_{15}	6_{15}
54	6_{16}	$z(p^2b^2pb^2)^2$	$2^8\cdot 3^3$	3_3	2_7	6_{16}	6_{16}
55	6_{17}	b	$2^8\cdot 3^3$	3_5	2_7	6_{17}	6_{17}

Conjugacy classes of $H(\mathsf{Fi}_{24}') = \langle r, p, b\rangle$ *(continued)*

No	*Class*	*Representative*	*Centralizer*	2P	3P	5P	7P
56	6_{18}	zb	$2^8 \cdot 3^3$	3_5	2_6	6_{18}	6_{18}
57	6_{19}	$z(p^2rbpbr^2bpr)$	$2^8 \cdot 3^3$	3_2	2_7	6_{19}	6_{19}
58	6_{20}	$(p^3br)^4$	$2^6 \cdot 3^4$	3_7	2_1	6_{20}	6_{20}
59	6_{21}	$(pr)^2$	$2^7 \cdot 3^3$	3_5	2_2	6_{21}	6_{21}
60	6_{22}	$(p^2br)^3$	$2^5 \cdot 3^4$	3_4	2_4	6_{23}	6_{22}
61	6_{23}	$(p^2br)^{15}$	$2^5 \cdot 3^4$	3_4	2_4	6_{22}	6_{23}
62	6_{24}	$z(p^2br)^{15}$	$2^5 \cdot 3^4$	3_4	2_5	6_{25}	6_{24}
63	6_{25}	$z(p^2br)^3$	$2^5 \cdot 3^4$	3_4	2_5	6_{24}	6_{25}
64	6_{26}	p^3r^2prpb	$2^7 \cdot 3^2$	3_2	2_9	6_{26}	6_{26}
65	6_{27}	p^5r	$2^5 \cdot 3^3$	3_5	2_4	6_{27}	6_{27}
66	6_{28}	$z(p^2rb)^2$	$2^5 \cdot 3^3$	3_4	2_7	6_{28}	6_{28}
67	6_{29}	p^4b	$2^5 \cdot 3^3$	3_5	2_7	6_{29}	6_{29}
68	6_{30}	$z(p^4b)$	$2^5 \cdot 3^3$	3_5	2_6	6_{30}	6_{30}
69	6_{31}	$(p^2r^2bpb)^2$	$2^5 \cdot 3^3$	3_4	2_6	6_{31}	6_{31}
70	6_{32}	$z(p^5r)$	$2^5 \cdot 3^3$	3_5	2_5	6_{32}	6_{32}
71	6_{33}	$prbr$	$2^5 \cdot 3^3$	3_6	2_7	6_{33}	6_{33}
72	6_{34}	$z(prbr)$	$2^5 \cdot 3^3$	3_6	2_6	6_{34}	6_{34}
73	6_{35}	$(p^2b^3r)^2$	$2^7 \cdot 3^1$	3_5	2_8	6_{35}	6_{35}
74	6_{36}	r^2	$2^5 \cdot 3^2$	3_7	2_3	6_{36}	6_{36}
75	6_{37}	p^2r	$2^3 \cdot 3^2$	3_7	2_9	6_{37}	6_{37}
76	7	$(pb)^3$	$2^1 \cdot 3^1 \cdot 7^1$	7	7	7	1
77	8_1	$(p^3br)^3$	$2^8 \cdot 3^2$	4_1	8_1	8_1	8_1
78	8_2	$(pr^2b)^3$	$2^9 \cdot 3^1$	4_2	8_2	8_2	8_2
79	8_3	$(prbpb^2)^3$	$2^8 \cdot 3^1$	4_2	8_3	8_3	8_3
80	8_4	p^2bpbr	2^8	4_{10}	8_4	8_4	8_4
81	8_5	p^2rpr^2	2^8	4_{10}	8_5	8_5	8_5
82	8_6	p^2r^2brbr	2^8	4_6	8_6	8_6	8_6
83	8_7	p^3br^4	2^8	4_{10}	8_7	8_7	8_7
84	8_8	p^2brb	2^8	4_6	8_8	8_8	8_8
85	8_9	$(pr^2)^3$	$2^6 \cdot 3^1$	4_7	8_9	8_9	8_9
86	8_{10}	rb^3	2^6	4_{18}	8_{10}	8_{10}	8_{10}
87	8_{11}	p^2bpb	2^6	4_{18}	8_{11}	8_{11}	8_{11}
88	8_{12}	pr^3	2^6	4_9	8_{12}	8_{12}	8_{12}
89	9_1	$(prbpbr)^2$	$2^5 \cdot 3^4$	9_1	3_3	9_1	9_1
90	9_2	$(p^2b^2rbr)^4$	$2^3 \cdot 3^4$	9_2	3_3	9_2	9_2
91	9_3	$(p^2brpr)^2$	$2^1 \cdot 3^4$	9_3	3_3	9_3	9_3
92	9_4	$(pb^2)^{10}$	$2^2 \cdot 3^3$	9_5	3_4	9_5	9_4
93	9_5	$(pb^2)^2$	$2^2 \cdot 3^3$	9_4	3_4	9_4	9_5
94	10_1	$(p^2b)^3$	$2^6 \cdot 3^1 \cdot 5^1$	5	10_1	2_1	10_1
95	10_2	$(prb)^6$	$2^5 \cdot 5^1$	5	10_3	2_3	10_3
96	10_3	$(prb)^2$	$2^5 \cdot 5^1$	5	10_2	2_3	10_2
97	10_4	$(p^3rb)^2$	$2^5 \cdot 5^1$	5	10_4	2_2	10_4
98	10_5	$pbrb$	$2^4 \cdot 5^1$	5	10_5	2_4	10_5
99	10_6	$z(pbrb)$	$2^4 \cdot 5^1$	5	10_6	2_5	10_6
100	12_1	$(prpb^2)^2$	$2^9 \cdot 3^5$	6_2	4_1	12_1	12_1
101	12_2	$(pbpb^5)^3$	$2^7 \cdot 3^5$	6_3	4_1	12_2	12_2
102	12_3	$(p^2rbrprb)^2$	$2^9 \cdot 3^3$	6_2	4_1	12_3	12_3
103	12_4	p^4rb^2r	$2^6 \cdot 3^4$	6_5	4_1	12_4	12_4
104	12_5	$pr^2b^2pb^2$	$2^4 \cdot 3^5$	6_8	4_1	12_5	12_5
105	12_6	$z(p^4bprpb)$	$2^8 \cdot 3^2$	6_9	4_5	12_6	12_6
106	12_7	p^4bprpb	$2^8 \cdot 3^2$	6_9	4_5	12_7	12_7
107	12_8	$(pr^2b)^2$	$2^7 \cdot 3^2$	6_3	4_2	12_8	12_8
108	12_9	$(p^3br)^2$	$2^5 \cdot 3^3$	6_{20}	4_1	12_9	12_9
109	12_{10}	$(pb^2rb)^2$	$2^5 \cdot 3^3$	6_{20}	4_1	12_{10}	12_{10}
110	12_{11}	$p^2rpr^2b^2$	$2^6 \cdot 3^2$	6_7	4_4	12_{11}	12_{11}
111	12_{12}	p	$2^6 \cdot 3^2$	6_{14}	4_7	12_{12}	12_{12}

Conjugacy classes of $H(\mathsf{Fi}_{24}') = \langle r, p, b\rangle$ *(continued)*

No	*Class*	*Representative*	*Centralizer*	2P	3P	5P	7P
112	12_{13}	p^4rprb	$2^6\cdot 3^2$	6_7	4_8	12_{13}	12_{13}
113	12_{14}	$(prbpb^2)^2$	$2^6\cdot 3^2$	6_5	4_2	12_{14}	12_{14}
114	12_{15}	zp	$2^6\cdot 3^2$	6_{14}	4_9	12_{15}	12_{15}
115	12_{16}	p^2r^2bpr	$2^7\cdot 3^1$	6_9	4_3	12_{16}	12_{16}
116	12_{17}	p^2rp^2br	$2^7\cdot 3^1$	6_9	4_6	12_{17}	12_{17}
117	12_{18}	$z(p^2b^2pb^2)$	$2^5\cdot 3^2$	6_{15}	4_7	12_{18}	12_{18}
118	12_{19}	$p^2rp^2b^2r$	$2^5\cdot 3^2$	6_6	4_8	12_{19}	12_{19}
119	12_{20}	$p^2b^2pb^2$	$2^5\cdot 3^2$	6_{15}	4_9	12_{20}	12_{20}
120	12_{21}	p^2r^2bprb	$2^6\cdot 3^1$	6_7	4_{11}	12_{21}	12_{21}
121	12_{22}	pr	$2^4\cdot 3^2$	6_{21}	4_4	12_{22}	12_{22}
122	12_{23}	$p^2b^2pr^2$	$2^4\cdot 3^2$	6_{21}	4_8	12_{23}	12_{23}
123	12_{24}	r	$2^4\cdot 3^2$	6_{36}	4_5	12_{24}	12_{24}
124	12_{25}	$z(p^2r^2bpb)$	$2^4\cdot 3^2$	6_{31}	4_7	12_{25}	12_{25}
125	12_{26}	p^2r^2bpb	$2^4\cdot 3^2$	6_{31}	4_9	12_{26}	12_{26}
126	12_{27}	zr	$2^4\cdot 3^2$	6_{36}	4_5	12_{27}	12_{27}
127	12_{28}	p^5rbr	$2^4\cdot 3^2$	6_8	4_2	12_{28}	12_{28}
128	12_{29}	p^2b^3r	$2^5\cdot 3^1$	6_{35}	4_{15}	12_{29}	12_{29}
129	12_{30}	$z(p^2b^3r)$	$2^5\cdot 3^1$	6_{35}	4_{13}	12_{30}	12_{30}
130	12_{31}	p^2bprb	$2^4\cdot 3^1$	6_{17}	4_{17}	12_{31}	12_{31}
131	12_{32}	$z(p^2rb)^5$	$2^4\cdot 3^1$	6_{31}	4_{14}	12_{34}	12_{32}
132	12_{33}	p^2b^3	$2^4\cdot 3^1$	6_{11}	4_{16}	12_{33}	12_{33}
133	12_{34}	$z(p^2rb)$	$2^4\cdot 3^1$	6_{31}	4_{14}	12_{32}	12_{34}
134	14	$z(pb)^3$	$2^1\cdot 3^1\cdot 7^1$	7	14	14	2_1
135	15	$(p^2b)^2$	$2^1\cdot 3^1\cdot 5^1$	15	5	3_1	15
136	16	rb	2^5	8_2	16	16	16
137	18_1	$(pbpb^5)^2$	$2^5\cdot 3^4$	9_1	6_3	18_1	18_1
138	18_2	$(p^2b^2rbr)^2$	$2^3\cdot 3^4$	9_2	6_3	18_2	18_2
139	18_3	p^2brpr	$2^1\cdot 3^4$	9_3	6_3	18_3	18_3
140	18_4	$prbpbr$	$2^4\cdot 3^2$	9_1	6_6	18_4	18_4
141	18_5	pb^2	$2^2\cdot 3^3$	9_5	6_4	18_6	18_5
142	18_6	$(pb^2)^5$	$2^2\cdot 3^3$	9_4	6_4	18_5	18_6
143	18_7	p^2br	$2^2\cdot 3^2$	9_5	6_{22}	18_9	18_7
144	18_8	$z(p^2br)^5$	$2^2\cdot 3^2$	9_4	6_{24}	18_{10}	18_8
145	18_9	$(p^2br)^5$	$2^2\cdot 3^2$	9_4	6_{23}	18_7	18_9
146	18_{10}	$z(p^2br)$	$2^2\cdot 3^2$	9_5	6_{25}	18_8	18_{10}
147	20_1	$prbrb$	$2^4\cdot 5^1$	10_1	20_1	4_1	20_1
148	20_2	$(prb)^3$	$2^3\cdot 5^1$	10_2	20_4	4_3	20_4
149	20_3	p^3rb	$2^3\cdot 5^1$	10_4	20_3	4_4	20_3
150	20_4	prb	$2^3\cdot 5^1$	10_3	20_2	4_3	20_2
151	21_1	pb	$2^1\cdot 3^1\cdot 7^1$	21_2	7	21_1	3_1
152	21_2	$(pb)^2$	$2^1\cdot 3^1\cdot 7^1$	21_1	7	21_2	3_1
153	24_1	$p^2rbrprb$	$2^5\cdot 3^2$	12_3	8_1	24_1	24_1
154	24_2	$prpb^2$	$2^5\cdot 3^2$	12_1	8_1	24_2	24_2
155	24_3	pb^2rb	$2^3\cdot 3^2$	12_{10}	8_1	24_3	24_3
156	24_4	p^3br	$2^3\cdot 3^2$	12_9	8_1	24_4	24_4
157	24_5	$(pr^2)^5$	$2^4\cdot 3^1$	12_{12}	8_9	24_9	24_9
158	24_6	$prbpb^2$	$2^4\cdot 3^1$	12_{14}	8_3	24_7	24_7
159	24_7	$(prbpb^2)^5$	$2^4\cdot 3^1$	12_{14}	8_3	24_6	24_6
160	24_8	pr^2b	$2^4\cdot 3^1$	12_8	8_2	24_8	24_8
161	24_9	pr^2	$2^4\cdot 3^1$	12_{12}	8_9	24_5	24_5
162	30	p^2b	$2^1\cdot 3^1\cdot 5^1$	15	10_1	6_1	30
163	36_1	$pbpb^5$	$2^4\cdot 3^3$	18_1	12_2	36_1	36_1
164	36_2	p^3r^2b	$2^4\cdot 3^3$	18_1	12_2	36_2	36_2
165	36_3	p^2b^2rbr	$2^2\cdot 3^3$	18_2	12_2	36_3	36_3
166	42_1	$z(pb)^2$	$2^1\cdot 3^1\cdot 7^1$	21_1	14	42_1	6_1
167	42_2	$z(pb)$	$2^1\cdot 3^1\cdot 7^1$	21_2	14	42_2	6_1

9.8 Character tables of local subgroups

9.8.1 *Character table of* $A_1 = 2Aut(\mathsf{Fi}_{22}) = \langle q, y, s \rangle$

2	19	19	17	14	19	19	14	17	10	9	8	5	11	14	14	13
3	9	9	6	6	4	4	4	3	7	9	7	7	4	4	1	3
5	2	2	1	2	1	1	1	.	1	.	.	.	1	.	1	.
7	1	1	1	1	.	.	.	.	1	.	.	.	1	.	.	.
11	1	1	1	.	.	.	.	.	.	.	.	.	.	.	.	.
13	1	1	.	.	.	.	.	.	.	.	.	.	.	.	.	.
	$1a$	$2a$	$2b$	$2c$	$2d$	$2e$	$2f$	$2g$	$3a$	$3b$	$3c$	$3d$	$4a$	$4b$	$4c$	$4d$
$2P$	$1a$	$1a$	$1a$	$1a$	$1a$	$1a$	$1a$	$1a$	$3a$	$3b$	$3c$	$3d$	$2a$	$2e$	$2e$	$2e$
$3P$	$1a$	$2a$	$2b$	$2c$	$2d$	$2e$	$2f$	$2g$	$1a$	$1a$	$1a$	$1a$	$4a$	$4b$	$4c$	$4d$
$5P$	$1a$	$2a$	$2b$	$2c$	$2d$	$2e$	$2f$	$2g$	$3a$	$3b$	$3c$	$3d$	$4a$	$4b$	$4c$	$4d$
$7P$	$1a$	$2a$	$2b$	$2c$	$2d$	$2e$	$2f$	$2g$	$3a$	$3b$	$3c$	$3d$	$4a$	$4b$	$4c$	$4d$
$11P$	$1a$	$2a$	$2b$	$2c$	$2d$	$2e$	$2f$	$2g$	$3a$	$3b$	$3c$	$3d$	$4a$	$4b$	$4c$	$4d$
$13P$	$1a$	$2a$	$2b$	$2c$	$2d$	$2e$	$2f$	$2g$	$3a$	$3b$	$3c$	$3d$	$4a$	$4b$	$4c$	$4d$
$X.1$	1	1	1	1	1	1	1	1	1	1	1	1	1	1	1	1
$X.2$	1	1	1	−1	1	1	−1	1	1	1	1	1	−1	−1	−1	1
$X.3$	78	78	−34	22	14	14	6	−2	15	−3	6	−3	−6	−10	6	6
$X.4$	78	78	−34	−22	14	14	−6	−2	15	−3	6	−3	6	10	−6	6
$X.5$	352	−352	.	.	32	−32	.	.	−8	28	10	1	.	.	.	.
$X.6$	352	−352	.	.	32	−32	.	.	−8	28	10	1	.	.	.	.
$X.7$	429	429	77	79	45	45	15	13	6	24	15	−3	7	15	15	13
$X.8$	429	429	77	−79	45	45	−15	13	6	24	15	−3	−7	−15	−15	13
$X.9$	1001	1001	−231	49	41	41	−31	−7	56	29	2	2	21	−15	1	1
$X.10$	1001	1001	−231	−49	41	41	31	−7	56	29	2	2	−21	15	−1	1
$X.11$	1430	1430	−154	−170	86	86	6	6	−1	−28	26	−1	−14	22	−26	14
$X.12$	1430	1430	−154	170	86	86	−6	6	−1	−28	26	−1	14	−22	26	14
$X.13$	3003	3003	539	−203	59	59	21	−37	105	6	15	6	21	−43	−11	11
$X.14$	3003	3003	539	203	59	59	−21	−37	105	6	15	6	−21	43	11	11
$X.15$	3080	3080	616	−280	136	136	−56	40	119	2	20	2	−56	−56	−24	24
$X.16$	3080	3080	616	280	136	136	56	40	119	2	20	2	56	56	24	24
$X.17$	4160	−4160	.	.	192	−192	.	.	128	−52	38	2	.	.	.	.
$X.18$	10725	10725	−715	−575	165	165	−15	−43	15	114	24	6	85	33	−15	29
$X.19$	10725	10725	−715	575	165	165	15	−43	15	114	24	6	−85	−33	15	29
$X.20$	11648	−11648	.	.	128	−128	.	.	308	−16	−16	−16	.	.	.	.
$X.21$	11648	−11648	.	.	128	−128	.	.	308	−16	−16	−16	.	.	.	.
$X.22$	13650	13650	1330	−350	210	210	−30	114	105	123	33	15	−70	−30	−30	−14
$X.23$	13650	13650	1330	350	210	210	30	114	105	123	33	15	70	30	30	−14
$X.24$	27456	−27456	.	.	448	−448	.	.	240	240	78	24	.	.	.	.
$X.25$	27456	−27456	.	.	448	−448	.	.	−120	−84	60	−3	.	.	.	.
$X.26$	27456	−27456	.	.	448	−448	.	.	−120	−84	60	−3	.	.	.	.
$X.27$	30030	30030	1694	1330	526	526	66	62	−21	−102	60	6	14	82	66	38
$X.28$	30030	30030	1694	−1330	526	526	−66	62	−21	−102	60	6	−14	−82	−66	38
$X.29$	32032	32032	−2464	1568	544	544	96	−32	91	−44	64	10	−56	−128	64	64
$X.30$	32032	32032	−2464	−1568	544	544	−96	−32	91	−44	64	10	56	128	−64	64
$X.31$	43680	43680	−4256	1120	416	416	−96	−32	399	−60	48	−6	56	−128	64	.
$X.32$	43680	43680	−4256	−1120	416	416	96	−32	399	−60	48	−6	−56	128	−64	.
$X.33$	45045	45045	4389	945	309	309	129	133	441	90	−18	9	189	81	1	29
$X.34$	45045	45045	4389	−945	309	309	−129	133	441	90	−18	9	−189	−81	−1	29
$X.35$	48048	48048	−1232	448	432	432	192	48	−84	258	6	−12	−112	.	.	−16
$X.36$	48048	48048	−1232	−448	432	432	−192	48	−84	258	6	−12	112	.	.	−16
$X.37$	50050	50050	770	−350	130	130	130	−126	−35	235	19	−8	−70	2	−30	−14
$X.38$	50050	50050	−5390	1050	450	450	170	−46	595	73	10	19	−210	−102	10	10
$X.39$	50050	50050	770	350	130	130	−130	−126	−35	235	19	−8	70	−2	30	−14
$X.40$	50050	50050	−5390	−1050	450	450	−170	−46	595	73	10	19	210	102	−10	10
$X.41$	75075	75075	7315	−875	515	515	165	51	735	−93	−3	−12	105	−107	5	−5
$X.42$	75075	75075	1155	2625	835	835	225	131	−210	150	51	−12	105	33	65	83
$X.43$	75075	75075	1155	−2625	835	835	−225	131	−210	150	51	−12	−105	−33	−65	83
$X.44$	75075	75075	−5005	−875	−125	−125	165	83	420	−12	42	15	−175	85	5	−5
$X.45$	75075	75075	7315	875	515	515	−165	51	735	−93	−3	−12	−105	107	−5	−5
$X.46$	75075	75075	−5005	875	−125	−125	−165	83	420	−12	42	15	175	−85	−5	−5
$X.47$	81081	81081	3465	2331	633	633	−117	−87	.	162	81	.	−105	123	75	33
$X.48$	81081	81081	3465	−2331	633	633	117	−87	.	162	81	.	105	−123	−75	33
$X.49$	96096	−96096	.	.	544	−544	.	.	1092	192	−6	30	.	.	.	.
$X.50$	105600	−105600	.	.	−640	640	.	.	120	−24	48	−24	.	.	.	.
$X.51$	105600	−105600	.	.	−640	640	.	.	120	−24	48	−24	.	.	.	.
$X.52$	105600	−105600	.	.	−640	640	.	.	120	−24	48	−24	.	.	.	.
$X.53$	105600	−105600	.	.	−640	640	.	.	120	−24	48	−24	.	.	.	.
$X.54$	114400	114400	−8800	2400	480	480	160	32	685	28	−8	−26	−120	−192	.	64
$X.55$	114400	114400	−8800	−2400	480	480	−160	32	685	28	−8	−26	120	192	.	64
$X.56$	123200	−123200	.	.	960	−960	.	.	−280	404	62	−28	.	.	.	.
$X.57$	123200	−123200	.	.	960	−960	.	.	−280	404	62	−28	.	.	.	.
$X.58$	133056	−133056	.	.	−192	192	.	.	.	−108	−54	54	.	.	.	.
$X.59$	133056	−133056	.	.	−192	192	.	.	.	−108	−54	54	.	.	.	.
$X.60$	150150	150150	8470	350	70	70	270	54	525	57	48	−24	210	−2	−50	−34
$X.61$	150150	150150	8470	−350	70	70	−270	54	525	57	48	−24	−210	2	50	−34
$X.62$	205920	205920	1056	4320	864	864	−96	160	−279	−144	72	18	216	.	64	64
$X.63$	205920	205920	1056	−4320	864	864	96	160	−279	−144	72	18	−216	.	−64	64
$X.64$	228800	−228800	.	.	320	−320	.	.	−160	380	−52	29	.	.	.	.
$X.65$	228800	−228800	.	.	320	−320	.	.	−160	380	−52	29	.	.	.	.
$X.66$	277200	277200	−18480	.	720	720	.	−48	1260	−306	−90	18	.	.	.	16
$X.67$	289575	289575	12375	225	615	615	−495	183	405	162	−81	.	−195	33	−15	15
$X.68$	289575	289575	12375	−225	615	615	495	183	405	162	−81	.	195	−33	15	15
$X.69$	292864	−292864	.	.	2048	−2048	.	.	1552	−32	112	−32	.	.	.	.
$X.70$	300300	300300	−7700	−4900	1420	1420	60	−84	−210	−291	114	−21	140	124	−100	−4
$X.71$	300300	300300	−7700	4900	1420	1420	−60	−84	−210	−291	114	−21	−140	−124	100	−4
$X.72$	320320	320320	14784	−4480	1344	1344	−256	192	406	−116	64	−8	−336	−192	−64	.
$X.73$	320320	320320	14784	4480	1344	1344	256	192	406	−116	64	−8	336	192	64	.
$X.74$	360855	360855	18711	−3645	1431	1431	−189	279	729	.	.	.	−189	−189	−61	−9
$X.75$	360855	360855	18711	3645	1431	1431	189	279	729	.	.	.	189	189	61	−9
$X.76$	370656	370656	−15840	−3744	1248	1248	288	−96	405	324	.	.	−120	192	.	.
$X.77$	370656	370656	−15840	3744	1248	1248	−288	−96	405	324	.	.	120	−192	.	.
$X.78$	450450	450450	25410	−1050	210	210	150	−350	1575	171	−18	9	210	−90	−10	−6
$X.79$	450450	450450	6930	−7350	1170	1170	−150	−110	−315	−72	9	9	210	−150	10	130
$X.80$	450450	450450	6930	7350	1170	1170	150	−110	−315	−72	9	9	−210	150	−10	130
$X.81$	450450	450450	25410	1050	210	210	−150	−350	1575	171	−18	9	−210	90	10	−6
$X.82$	471744	−471744	.	.	2112	−2112	.	.	.	−648	162	.	.	.	.	.
$X.83$	576576	576576	14784	−5376	576	576	384	−320	126	180	72	18	336	−192	−64	.
$X.84$	576576	576576	14784	5376	576	576	−384	−320	126	180	72	18	−336	192	64	.

Character table of $A_1 = 2Aut(\mathsf{Fi}_{22})$ (continued)

	4e	4f	4g	4h	4i	4j	4k	4l	5a	6_1	6_2	6_3	6_4	6_5	6_6	6_7	6_8	6_9	6_{10}	6_{11}	6_{12}
2	11	13	11	11	11	12	11	10	5	10	9	8	8	8	8	8	8	5	9	9	10
3	2	1	2	2	2	1	1	1	1	7	9	7	5	5	6	4	4	7	4	4	3
5	1	.	.	.	.	.	.	.	2	1	.	.	1	1	.	.	.	.	.	.	.
7	.	.	.	.	.	.	.	.	.	1	.	.	.	.	.	1	1	.	.	.	.
11	.	.	.	.	.	.	.	.	.	.	.	.	.	.	.	.	.	.	.	.	.
13	.	.	.	.	.	.	.	.	.	.	.	.	.	.	.	.	.	.	.	.	.
2P	2d	2e	2d	2g	2g	2e	2d	2g	5a	3a	3b	3c	3a	3a	3b	3a	3a	3d	3b	3b	3a
3P	4e	4f	4g	4h	4i	4j	4k	4l	5a	2a	2a	2a	2b	2c	2b	2c	2c	2a	2e	2d	2d
5P	4e	4f	4g	4h	4i	4j	4k	4l	1a	6_1	6_2	6_3	6_4	6_5	6_6	6_8	6_7	6_9	6_{10}	6_{11}	6_{12}
7P	4e	4f	4g	4h	4i	4j	4k	4l	5a	6_1	6_2	6_3	6_4	6_5	6_6	6_7	6_8	6_9	6_{10}	6_{11}	6_{12}
11P	4e	4f	4g	4h	4i	4j	4k	4l	5a	6_1	6_2	6_3	6_4	6_5	6_6	6_8	6_7	6_9	6_{10}	6_{11}	6_{12}
13P	4e	4f	4g	4h	4i	4j	4k	4l	5a	6_1	6_2	6_3	6_4	6_5	6_6	6_7	6_8	6_9	6_{10}	6_{11}	6_{12}
X.1	1	1	1	1	1	1	1	1	1	1	1	1	1	1	1	1	1	1	1	1	1
X.2	1	1	−1	1	1	−1	1	−1	1	1	1	1	1	−1	1	−1	−1	1	1	1	1
X.3	−6	−2	2	2	2	−2	2	2	3	15	−3	6	−7	−5	−7	1	1	−3	5	5	−1
X.4	−6	−2	−2	2	2	2	2	−2	3	15	−3	6	−7	5	−7	−1	−1	−3	5	5	−1
X.5	.	.	.	−8	8	.	.	.	2	8	−28	−10	.	.	.	A	$\bar{A}$	−1	4	−4	8
X.6	.	.	.	−8	8	.	.	.	2	8	−28	−10	.	.	.	$\bar{A}$	A	−1	4	−4	8
X.7	5	−3	7	5	5	−1	5	−1	4	6	24	15	14	16	−4	−8	−8	−3	.	.	6
X.8	5	−3	−7	5	5	1	5	1	4	6	24	15	14	−16	−4	8	8	−3	.	.	6
X.9	−11	9	−3	5	5	9	−3	−3	1	56	29	2	−6	4	−15	−14	−14	2	5	5	8
X.10	−11	9	3	5	5	−9	−3	3	1	56	29	2	−6	−4	−15	14	14	2	5	5	8
X.11	−6	6	−6	10	10	−2	2	2	5	−1	−28	26	−19	19	8	1	1	−1	−4	−4	−1
X.12	−6	6	6	10	10	2	2	−2	5	−1	−28	26	−19	−19	8	−1	−1	−1	−4	−4	−1
X.13	11	−5	5	3	3	5	−5	−3	3	105	6	15	17	−5	26	7	7	6	14	14	−7
X.14	11	−5	−5	3	3	−5	−5	3	3	105	6	15	17	5	26	−7	−7	6	14	14	−7
X.15	24	8	−8	.	.	−8	8	.	5	119	2	20	31	−19	22	−7	−7	2	10	10	7
X.16	24	8	8	.	.	8	8	.	5	119	2	20	31	19	22	7	7	2	10	10	7
X.17	.	.	.	−16	16	.	.	.	10	−128	52	−38	.	.	.	.	.	−2	−12	12	.
X.18	−15	−11	−3	9	9	9	−7	−3	.	15	114	24	5	−35	14	−29	−29	6	−6	−6	15
X.19	−15	−11	3	9	9	−9	−7	3	.	15	114	24	5	35	14	29	29	6	−6	−6	15
X.20	.	.	.	.	.	.	.	.	−2	−308	16	16	.	.	.	.	.	16	16	−16	−4
X.21	.	.	.	.	.	.	.	.	−2	−308	16	16	.	.	.	.	.	16	16	−16	−4
X.22	10	2	−6	10	10	−14	10	2	.	105	123	33	25	−35	7	−35	−35	15	3	3	9
X.23	10	2	6	10	10	14	10	−2	.	105	123	33	25	35	7	35	35	15	3	3	9
X.24	.	.	.	−16	16	.	.	.	6	−240	−240	−78	.	.	.	.	.	−24	−16	16	16
X.25	.	.	.	−16	16	.	.	.	6	120	84	−60	.	.	.	$\bar{A}$	A	3	20	−20	−8
X.26	.	.	.	−16	16	.	.	.	6	120	84	−60	.	.	.	A	$\bar{A}$	3	20	−20	−8
X.27	26	−2	6	18	18	10	2	6	5	−21	−102	60	29	−11	−34	7	7	6	−14	−14	−5
X.28	26	−2	−6	18	18	−10	2	−6	5	−21	−102	60	29	11	−34	−7	−7	6	−14	−14	−5
X.29	−24	.	8	.	.	.	8	.	7	91	−44	64	−79	−61	20	−7	−7	10	4	4	−5
X.30	−24	.	−8	.	.	.	8	.	7	91	−44	64	−79	61	20	7	7	10	4	4	−5
X.31	−24	.	−8	.	.	.	8	.	5	399	−60	48	−71	−59	−44	7	7	−6	20	20	−1
X.32	−24	.	8	.	.	.	8	.	5	399	−60	48	−71	59	−44	−7	−7	−6	20	20	−1
X.33	41	5	21	1	1	9	1	5	−5	441	90	−18	−21	−9	42	21	21	9	−6	−6	9
X.34	41	5	−21	1	1	−9	1	−5	−5	441	90	−18	−21	9	42	−21	−21	9	−6	−6	9
X.35	.	−16	16	16	16	.	.	.	−2	−84	258	6	−44	16	10	−56	−56	−12	18	18	12
X.36	.	−16	−16	16	16	.	.	.	−2	−84	258	6	−44	−16	10	56	56	−12	18	18	12
X.37	10	18	10	18	18	2	−6	−6	.	−35	235	19	5	−35	−13	49	49	−8	−5	−5	13
X.38	−50	−14	6	−2	−2	−14	6	−2	.	595	73	10	−35	15	−71	21	21	19	9	9	3
X.39	10	18	−10	18	18	−2	−6	6	.	−35	235	19	5	35	−13	−49	−49	−8	−5	−5	13
X.40	−50	−14	−6	−2	−2	14	6	2	.	595	73	10	−35	−15	−71	−21	−21	19	9	9	3
X.41	55	−13	1	7	7	13	−1	9	.	735	−93	−3	25	25	79	7	7	−12	11	11	−1
X.42	−5	19	25	3	3	1	11	9	.	−210	150	51	30	60	−6	.	.	−12	−2	−2	−2
X.43	−5	19	−25	3	3	−1	11	−9	.	−210	150	51	30	−60	−6	.	.	−12	−2	−2	−2
X.44	15	3	9	−1	−1	−3	−9	−7	.	420	−12	42	−10	−20	−64	−14	−14	15	28	28	4
X.45	55	−13	−1	7	7	−13	−1	−9	.	735	−93	−3	25	−25	79	−7	−7	−12	11	11	−1
X.46	15	3	−9	−1	−1	3	−9	7	.	420	−12	42	−10	20	−64	14	14	15	28	28	4
X.47	5	−7	−1	−3	−3	−13	−3	−9	6	.	162	81	90	90	−18	.	.	.	−6	−6	.
X.48	5	−7	1	−3	−3	13	−3	9	6	.	162	81	90	−90	−18	.	.	.	−6	−6	.
X.49	.	.	.	−8	8	.	.	.	−4	−1092	−192	6	.	.	.	.	.	−30	32	−32	−20
X.50	.	.	.	.	.	.	.	.	.	−120	24	−48	.	.	.	$\bar{A}$	A	24	−8	8	8
X.51	.	.	.	.	.	.	.	.	.	−120	24	−48	.	.	.	A	$\bar{A}$	24	−8	8	8
X.52	.	.	.	.	.	.	.	.	.	−120	24	−48	.	.	.	A	$\bar{A}$	24	−8	8	8
X.53	.	.	.	.	.	.	.	.	.	−120	24	−48	.	.	.	$\bar{A}$	A	24	−8	8	8
X.54	−40	.	8	.	.	.	−8	.	.	685	28	−8	−25	15	−52	−15	−15	−26	12	12	−3
X.55	−40	.	−8	.	.	.	−8	.	.	685	28	−8	−25	−15	−52	15	15	−26	12	12	−3
X.56	.	.	.	−16	16	.	.	.	.	280	−404	−62	.	.	.	G	$\bar{G}$	28	12	−12	24
X.57	.	.	.	−16	16	.	.	.	.	280	−404	−62	.	.	.	$\bar{G}$	G	28	12	−12	24
X.58	.	.	.	−16	16	.	.	.	6	.	108	54	.	.	.	.	.	−54	12	−12	.
X.59	.	.	.	−16	16	.	.	.	6	.	108	54	.	.	.	.	.	−54	12	−12	.
X.60	−30	22	−6	−6	−6	6	10	−6	.	525	57	48	55	−55	73	35	35	−24	25	25	13
X.61	−30	22	6	−6	−6	−6	10	6	.	525	57	48	55	55	73	−35	−35	−24	25	25	13
X.62	24	.	24	.	.	.	−8	.	−5	−279	−144	72	−69	−99	−24	15	15	18	.	.	9
X.63	24	.	−24	.	.	.	−8	.	−5	−279	−144	72	−69	99	−24	−15	−15	18	.	.	9
X.64	.	.	.	−16	16	.	.	.	.	160	−380	52	.	.	.	J	$\bar{J}$	−29	4	−4	32
X.65	.	.	.	−16	16	.	.	.	.	160	−380	52	.	.	.	$\bar{J}$	J	−29	4	−4	32
X.66	−80	16	.	16	16	.	−16	.	.	1260	−306	−90	60	.	−66	.	.	18	−18	−18	12
X.67	35	23	5	3	3	−23	−5	−3	.	405	162	−81	−45	−45	−18	−27	−27	.	−6	−6	21
X.68	35	23	−5	3	3	23	−5	3	.	405	162	−81	−45	45	−18	27	27	.	−6	−6	21
X.69	.	.	.	.	.	.	.	.	14	−1552	32	−112	.	.	.	.	.	32	−32	32	−16
X.70	−20	−20	−4	20	20	12	−4	4	.	−210	−291	114	−50	50	49	14	14	−21	−11	−11	−2
X.71	−20	−20	4	20	20	−12	−4	−4	.	−210	−291	114	−50	−50	49	−14	−14	−21	−11	−11	−2
X.72	16	.	−16	.	.	.	16	.	−5	406	−116	64	114	−34	−12	14	14	−8	12	12	6
X.73	16	.	16	.	.	.	16	.	−5	406	−116	64	114	34	−12	−14	−14	−8	12	12	6
X.74	39	−9	3	−9	−9	3	7	3	5	729	.	.	81	−81	.	27	27	.	.	.	9
X.75	39	−9	−3	−9	−9	−3	7	−3	5	729	.	.	81	81	.	−27	−27	.	.	.	9
X.76	−40	.	8	.	.	.	−8	.	6	405	324	.	−45	−45	36	−27	−27	.	−12	−12	21
X.77	−40	.	−8	.	.	.	−8	.	6	405	324	.	−45	45	36	27	27	.	−12	−12	21
X.78	30	2	26	−2	−2	−18	−10	−14	.	1575	171	−18	−15	−15	111	−21	−21	9	3	3	−9
X.79	10	−30	2	−14	−14	10	−6	2	.	−315	−72	9	45	−15	−36	21	21	9	.	.	21
X.80	10	−30	−2	−14	−14	−10	−6	−2	.	−315	−72	9	45	15	−36	−21	−21	9	.	.	21
X.81	30	2	−26	−2	−2	18	−10	14	.	1575	171	−18	−15	15	111	21	21	9	3	3	−9
X.82	.	.	.	−48	48	.	.	.	−6	.	648	−162	.	.	.	.	.	.	−24	24	.
X.83	−16	.	16	.	.	.	−16	.	1	126	180	72	114	−66	−12	−42	−42	18	36	36	−18
X.84	−16	.	−16	.	.	.	−16	.	1	126	180	72	114	66	−12	42	42	18	36	36	−18

Character table of $A_1 = 2Aut(Fi_{22})$ (continued)

2	10	5	5	5	8	8	8	8	8	7	7	5	5	5	5	5	5	5	5	3	8	8	8	8	8
3	3	6	5	5	3	3	3	3	3	3	3	4	4	4	4	3	3	3	3	1	1	1	1	1	1
5	.	.	.	.	.	.	.	.	.	.	.	.	.	.	.	.	.	.	.	.	.	.	.	.	.
7	.	.	.	.	.	.	.	.	.	.	.	.	.	.	.	.	.	.	.	1	.	.	.	.	.
11	.	.	.	.	.	.	.	.	.	.	.	.	.	.	.	.	.	.	.	.	.	.	.	.	.
13	.	.	.	.	.	.	.	.	.	.	.	.	.	.	.	.	.	.	.	.	.	.	.	.	.
	6_{13}	6_{14}	6_{15}	6_{16}	6_{17}	6_{18}	6_{19}	6_{20}	6_{21}	6_{22}	6_{23}	6_{24}	6_{25}	6_{26}	6_{27}	6_{28}	6_{29}	6_{30}	6_{31}	$7a$	$8a$	$8b$	$8c$	$8d$	$8e$
2P	$3a$	$3b$	$3c$	$3c$	$3c$	$3c$	$3b$	$3a$	$3a$	$3c$	$3c$	$3d$	$3d$	$3d$	$3d$	$3d$	$3d$	$3c$	$3c$	$7a$	$4f$	$4d$	$4d$	$4d$	$4d$
3P	$2e$	$2c$	$2c$	$2b$	$2g$	$2g$	$2g$	$2f$	$2g$	$2e$	$2d$	$2f$	$2f$	$2c$	$2c$	$2g$	$2g$	$2g$	$2f$	$7a$	$8a$	$8b$	$8c$	$8d$	$8e$
5P	6_{13}	6_{14}	6_{15}	6_{16}	6_{17}	6_{18}	6_{19}	6_{20}	6_{21}	6_{22}	6_{23}	6_{25}	6_{24}	6_{27}	6_{26}	6_{28}	6_{29}	6_{30}	6_{31}	$7a$	$8a$	$8b$	$8c$	$8d$	$8e$
7P	6_{13}	6_{14}	6_{15}	6_{16}	6_{17}	6_{18}	6_{19}	6_{20}	6_{21}	6_{22}	6_{23}	6_{24}	6_{25}	6_{26}	6_{27}	6_{28}	6_{29}	6_{30}	6_{31}	$1a$	$8a$	$8b$	$8c$	$8d$	$8e$
11P	6_{13}	6_{14}	6_{15}	6_{16}	6_{17}	6_{18}	6_{19}	6_{20}	6_{21}	6_{22}	6_{23}	6_{25}	6_{24}	6_{27}	6_{26}	6_{28}	6_{29}	6_{30}	6_{31}	$7a$	$8a$	$8b$	$8c$	$8d$	$8e$
13P	6_{13}	6_{14}	6_{15}	6_{16}	6_{17}	6_{18}	6_{19}	6_{20}	6_{21}	6_{22}	6_{23}	6_{24}	6_{25}	6_{26}	6_{27}	6_{28}	6_{29}	6_{30}	6_{31}	$7a$	$8a$	$8b$	$8c$	$8d$	$8e$
X.1	1	1	1	1	1	1	1	1	1	1	1	1	1	1	1	1	1	1	1	1	1	1	1	1	1
X.2	1	−1	−1	1	1	1	1	−1	1	1	1	−1	−1	−1	−1	1	1	1	−1	1	−1	1	−1	1	−1
X.3	−1	−5	4	2	−2	−2	1	3	1	2	2	−3	−3	1	1	1	1	−2	.	1	2	2	2	−2	−2
X.4	−1	5	−4	2	−2	−2	1	−3	1	2	2	3	3	−1	−1	1	1	−2	.	1	−2	2	−2	−2	2
X.5	−8	.	.	.	−6	6	.	.	.	−2	2	B	$\bar{B}$	C	$\bar{C}$	−3	3	.	.	2	.	.	.	.	.
X.6	−8	.	.	.	−6	6	.	.	.	−2	2	$\bar{B}$	B	$\bar{C}$	C	−3	3	.	.	2	.	.	.	.	.
X.7	6	−2	7	5	7	7	4	.	−2	3	3	−3	−3	1	1	1	1	1	3	2	3	1	3	1	3
X.8	6	2	−7	5	7	7	4	.	−2	3	3	3	3	−1	−1	1	1	1	−3	2	−3	1	−3	1	−3
X.9	8	−5	4	−6	2	2	−7	−4	2	2	2	−4	−4	4	4	2	2	2	−4	.	1	−3	−3	1	1
X.10	8	5	−4	−6	2	2	−7	4	2	2	2	4	4	−4	−4	2	2	2	4	.	−1	−3	3	1	−1
X.11	−1	−8	−8	8	−6	−6	.	3	−3	2	2	−3	−3	1	1	3	3	.	.	2	−6	−2	−2	2	2
X.12	−1	8	8	8	−6	−6	.	−3	−3	2	2	3	3	−1	−1	3	3	.	.	2	6	−2	2	2	−2
X.13	−7	−14	−5	−1	−1	−1	2	3	−7	−1	−1	−6	−6	−2	−2	2	2	−1	3	.	−3	−1	1	−1	1
X.14	−7	14	5	−1	−1	−1	2	−3	−7	−1	−1	6	6	2	2	2	2	−1	−3	.	3	−1	−1	−1	−1
X.15	7	−10	−10	4	4	4	−2	−11	7	4	4	−2	−2	2	2	−2	−2	4	−2	.	.	4	−4	4	−4
X.16	7	10	10	4	4	4	−2	11	7	4	4	2	2	−2	−2	−2	−2	4	2	.	.	4	4	4	4
X.17	.	.	.	.	6	−6	.	.	.	−6	6	.	.	.	.	−6	6	.	.	2	.	.	.	.	.
X.18	15	−8	−8	−4	8	8	−10	−3	5	.	.	−6	−6	−2	−2	2	2	−4	.	1	−3	1	5	−3	1
X.19	15	8	8	−4	8	8	−10	3	5	.	.	6	6	2	2	2	2	−4	.	1	3	1	−5	−3	−1
X.20	4	.	.	.	.	.	.	.	.	−8	8	.	.	.	.	.	.	.	.	.	.	.	.	.	.
X.21	4	.	.	.	.	.	.	.	.	−8	8	.	.	.	.	.	.	.	.	.	.	.	.	.	.
X.22	9	1	1	7	9	9	15	−3	9	−3	−3	−3	−3	1	1	3	3	3	−3	.	−2	−2	−2	−2	−2
X.23	9	−1	−1	7	9	9	15	3	9	−3	−3	3	3	−1	−1	3	3	3	3	.	2	−2	2	−2	2
X.24	−16	.	.	.	−18	18	.	.	.	2	−2	.	.	.	.	.	.	.	.	2	.	.	.	.	.
X.25	8	.	.	.	12	−12	.	.	.	−4	4	F	$\bar{F}$	$\bar{C}$	C	−3	3	.	.	2	.	.	.	.	.
X.26	8	.	.	.	12	−12	.	.	.	−4	4	$\bar{F}$	F	C	$\bar{C}$	−3	3	.	.	2	.	.	.	.	.
X.27	−5	−20	16	2	−4	−4	−10	−3	5	4	4	−6	−6	−2	−2	2	2	2	.	.	6	2	−2	−2	2
X.28	−5	20	−16	2	−4	−4	−10	3	5	4	4	6	6	2	2	2	2	2	.	.	−6	2	2	−2	−2
X.29	−5	2	20	2	−8	−8	4	3	1	4	4	6	6	2	2	−2	−2	−2	.	.	.	4	4	−4	−4
X.30	−5	−2	−20	2	−8	−8	4	−3	1	4	4	−6	−6	−2	−2	−2	−2	−2	.	.	.	4	−4	−4	4
X.31	−1	−14	4	10	−8	−8	4	−3	1	−4	−4	−6	−6	−2	−2	−2	−2	−2	.	.	.	−4	−4	4	4
X.32	−1	14	−4	10	−8	−8	4	3	1	−4	−4	6	6	2	2	−2	−2	−2	.	.	.	−4	4	4	−4
X.33	9	.	.	6	−2	−2	10	15	19	6	6	3	3	3	3	1	1	−2	.	.	−3	1	−3	5	1
X.34	9	.	.	6	−2	−2	10	−15	19	6	6	−3	−3	−3	−3	1	1	−2	.	.	3	1	3	5	−1
X.35	12	16	−2	−8	6	6	−6	.	−12	6	6	−6	−6	−2	−2	.	.	.	6	.	.	.	.	.	.
X.36	12	−16	2	−8	6	6	−6	.	−12	6	6	6	6	2	2	.	.	.	−6	.	.	.	.	.	.
X.37	13	1	1	5	3	3	3	−11	−3	−5	−5	4	4	4	4	.	.	−3	1	.	−2	2	2	2	2
X.38	3	−3	6	−8	3	2	−7	23	5	6	6	−1	−1	3	3	−1	−1	−4	2	.	−2	2	2	−2	−2
X.39	13	−1	−1	5	3	3	3	11	−3	−5	−5	−4	−4	−4	−4	.	.	−3	−1	.	2	2	−2	2	−2
X.40	3	3	−6	−8	3	2	−7	−23	5	6	6	1	1	−3	−3	−1	−1	−4	−2	.	2	2	−2	−2	2
X.41	−1	−11	−11	−11	−3	−3	−9	9	9	5	5	−6	−6	−2	−2	.	.	−3	−3	.	1	−5	5	−1	1
X.42	−2	6	15	3	11	11	2	−12	−10	7	7	.	.	.	.	−4	−4	−1	3	.	−3	3	1	3	1
X.43	−2	−6	−15	3	11	11	2	12	−10	7	7	.	.	.	.	−4	−4	−1	−3	.	3	3	−1	3	−1
X.44	4	16	−2	−10	2	2	8	12	−10	−2	−2	3	3	−5	−5	−1	−1	2	−6	.	−3	−1	1	3	−3
X.45	−1	11	11	−11	−3	−3	−9	−9	9	5	5	6	6	2	2	.	.	−3	3	.	−1	−5	−5	−1	−1
X.46	4	−16	2	−10	2	2	8	−12	−10	−2	−2	−3	−3	5	5	−1	−1	2	6	.	3	−1	−1	3	3
X.47	.	−18	9	9	9	9	6	−6	−6	−3	−3	.	.	.	.	.	.	−3	−3	.	3	−7	3	−3	−1
X.48	.	18	−9	9	9	9	6	6	−6	−3	−3	.	.	.	.	.	.	−3	3	.	−3	−7	−3	−3	1
X.49	20	.	.	.	−6	6	.	.	.	−10	10	.	.	.	.	6	−6	.	.	.	.	.	.	.	.
X.50	−8	.	.	.	.	.	.	.	.	−8	8	.	.	$\bar{A}$	A	.	.	.	.	−2	.	.	.	.	.
X.51	−8	.	.	.	.	.	.	.	.	−8	8	.	.	A	$\bar{A}$	.	.	.	.	−2	.	.	.	.	.
X.52	−8	.	.	.	.	.	.	.	.	−8	8	.	.	A	$\bar{A}$	.	.	.	.	−2	.	.	.	.	.
X.53	−8	.	.	.	.	.	.	.	.	−8	8	.	.	$\bar{A}$	A	.	.	.	.	−2	.	.	.	.	.
X.54	−3	−30	6	2	8	8	−4	7	−1	.	.	−2	−2	−6	−6	2	2	2	−2	−1	.	4	−4	−4	4
X.55	−3	30	−6	2	8	8	−4	−7	−1	.	.	2	2	6	6	2	2	2	2	−1	.	4	4	−4	−4
X.56	−24	.	.	.	−18	18	.	.	.	−6	6	H	$\bar{H}$	D	$\bar{D}$	.	.	.	.	.	.	.	.	.	.
X.57	−24	.	.	.	−18	18	.	.	.	−6	6	$\bar{H}$	H	$\bar{D}$	D	.	.	.	.	.	.	.	.	.	.
X.58	.	.	.	.	−6	6	.	.	.	6	−6	.	.	.	.	6	−6	.	.	.	.	.	.	.	.
X.59	.	.	.	.	−6	6	.	.	.	6	−6	.	.	.	.	6	−6	.	.	.	.	.	.	.	.
X.60	13	−1	8	−8	.	.	9	9	−9	−8	−8	.	.	8	8	.	.	.	.	.	2	2	2	−2	−2
X.61	13	1	−8	−8	.	.	9	−9	−9	−8	−8	.	.	−8	−8	.	.	.	.	.	−2	2	−2	−2	2
X.62	9	.	.	12	−8	−8	−8	−3	−5	.	.	−6	−6	6	6	−2	−2	4	.	1	.	−4	−4	4	4
X.63	9	.	.	12	−8	−8	−8	3	−5	.	.	6	6	−6	−6	−2	−2	4	.	1	.	−4	4	4	−4
X.64	−32	.	.	.	12	−12	.	.	.	4	−4	B	$\bar{B}$	B	$\bar{B}$	−3	3	.	.	−2	.	.	.	.	.
X.65	−32	.	.	.	12	−12	.	.	.	4	−4	$\bar{B}$	B	B	B	−3	3	.	.	−2	.	.	.	.	.
X.66	12	.	.	6	6	6	30	.	12	6	6	.	.	.	.	−6	−6	6	.	.	.	.	.	.	.
X.67	21	−18	9	9	−9	−9	6	3	3	3	3	.	.	.	.	.	.	−3	−3	−1	−3	−1	1	−5	5
X.68	21	18	−9	9	−9	−9	6	−3	3	3	3	.	.	.	.	.	.	−3	3	−1	3	−1	−1	−5	−5
X.69	16	.	.	.	.	.	.	.	.	−8	8	.	.	.	.	.	.	.	.	−2	.	.	.	.	.
X.70	−2	−13	−4	4	−6	−6	9	−6	6	−2	−2	−3	−3	5	5	3	3	.	.	.	−4	.	.	.	.
X.71	−2	13	4	4	−6	−6	9	6	6	−2	−2	3	3	−5	−5	3	3	.	.	.	4	.	.	.	.
X.72	6	2	2	6	.	.	−12	−10	−6	.	.	−4	−4	−4	−4	.	.	6	2	.	.	.	.	.	.
X.73	6	−2	−2	6	.	.	−12	10	−6	.	.	4	4	4	4	.	.	6	−2	.	.	.	.	.	.
X.74	9	.	.	.	.	.	.	−9	9	.	.	.	.	.	.	.	.	.	.	−2	3	−1	−5	−1	−5
X.75	9	.	.	.	.	.	.	9	9	.	.	.	.	.	.	.	.	.	.	−2	−3	−1	5	−1	5
X.76	21	−18	−18	−18	.	.	−12	3	3	.	.	.	.	.	.	.	.	6	6	−1	.	−4	4	4	−4
X.77	21	18	18	−18	.	.	−12	−3	3	.	.	.	.	.	.	.	.	6	−6	−1	.	−4	−4	4	4
X.78	−9	3	−6	12	−2	−2	7	9	−23	6	6	−3	−3	−3	−3	1	1	4	−6	.	2	2	−2	−2	2
X.79	21	−6	−15	−9	1	1	4	9	−11	−3	−3	3	3	3	3	1	1	−5	−3	.	6	2	2	2	2
X.80	21	6	15	−9	1	1	4	−9	−11	−3	−3	−3	−3	−3	−3	1	1	−5	3	.	−6	2	−2	2	−2
X.81	−9	−3	6	12	−2	−2	7	−9	−23	6	6	3	3	3	3	1	1	4	6	.	−2	2	2	−2	−2
X.82	.	.	.	.	18	−18	.	.	.	6	−6	.	.	.	.	.	.	.	.	.	.	.	.	.	.
X.83	−18	−30	6	6	−8	−8	4	6	10	.	.	6	6	−6	−6	−2	−2	−2	6	.	.	.	.	.	.
X.84	−18	30	−6	6	−8	−8	4	−6	10	.	.	−6	−6	6	6	−2	−2	−2	−6	.	.	.	.	.	.

Character table of $A_1 = 2Aut(\mathsf{Fi}_{22})$ (continued)

2	8	6	6	4	3	2	5	4	5	5	3	4	2	6	8	5	5	8	8	6	7	7	5	8
3	.	1	.	4	4	3	1	1	.	.	1	.	.	3	3	4	4	2	2	3	2	2	3	1
5	.	.	.	.	.	.	2	2	1	1	1	1	.	1	.	.	.	.	.	.	.	.	.	.
7	.	.	.	.	.	.	.	.	.	.	.	.	.	.	.	.	.	.	.	.	.	.	.	.
11	.	.	.	.	.	.	.	.	.	.	.	.	1	.	.	.	.	.	.	.	.	.	.	.
13	.	.	.	.	.	.	.	.	.	.	.	.	.	.	.	.	.	.	.	.	.	.	.	.
	8f	8g	8h	9a	9b	9c	10a	10b	10c	10d	10e	10f	11a	12_1	12_2	12_3	12_4	12_5	12_6	12_7	12_8	12_9	12_{10}	12_{11}
2P	4f	4h	4i	9a	9b	9c	5a	5a	5a	5a	5a	5a	11a	6_1	6_{13}	6_{10}	6_2	6_{13}	6_{13}	6_{10}	6_{13}	6_{13}	6_{22}	6_{13}
3P	8f	8g	8h	3b	3b	3d	10a	10b	10c	10d	10e	10f	11a	4a	4b	4b	4a	4d	4d	4d	4d	4b	4b	4c
5P	8f	8g	8h	9a	9b	9c	2a	2c	2d	2e	2b	2f	11a	12_1	12_2	12_3	12_4	12_5	12_6	12_7	12_8	12_9	12_{10}	12_{11}
7P	8f	8g	8h	9a	9b	9c	10a	10b	10c	10d	10e	10f	11a	12_1	12_2	12_3	12_4	12_5	12_6	12_7	12_8	12_9	12_{10}	12_{11}
11P	8f	8g	8h	9a	9b	9c	10a	10b	10c	10d	10e	10f	1a	12_1	12_2	12_3	12_4	12_5	12_6	12_7	12_8	12_9	12_{10}	12_{11}
13P	8f	8g	8h	9a	9b	9c	10a	10b	10c	10d	10e	10f	11a	12_1	12_2	12_3	12_4	12_5	12_6	12_7	12_8	12_9	12_{10}	12_{11}
X.1	1	1	1	1	1	1	1	1	1	1	1	1	1	1	1	1	1	1	1	1	1	1	1	1
X.2	1	−1	1	1	1	1	1	−1	1	1	1	−1	1	−1	−1	−1	−1	1	1	1	1	−1	−1	−1
X.3	2	.	.	3	.	.	3	−3	−1	−1	1	1	1	−3	5	−1	3	3	3	−3	−3	−1	2	−3
X.4	2	.	.	3	.	.	3	3	−1	−1	1	−1	1	3	−5	1	−3	3	3	−3	−3	1	−2	3
X.5	.	.	.	−2	4	1	−2	.	2	−2	.	.	.	.	.	.	.	.	.	.	.	.	.	.
X.6	.	.	.	−2	4	1	−2	.	2	−2	.	.	.	.	.	.	.	.	.	.	.	.	.	.
X.7	1	−1	1	.	3	.	4	4	.	.	2	.	.	4	.	6	−2	−2	−2	4	−2	.	3	.
X.8	1	1	1	.	3	.	4	−4	.	.	2	.	.	−4	.	−6	2	−2	−2	4	−2	.	−3	.
X.9	1	3	−1	2	2	−1	1	−1	1	1	−1	−1	.	6	6	3	3	4	4	1	−2	.	.	−2
X.10	1	−3	−1	2	2	−1	1	1	1	1	−1	1	.	−6	−6	−3	−3	4	4	1	−2	.	.	2
X.11	2	.	.	−1	−1	2	5	5	1	1	1	1	.	1	1	4	4	−1	−1	−4	5	−5	−2	1
X.12	2	.	.	−1	−1	2	5	−5	1	1	1	−1	.	−1	−1	−4	−4	−1	−1	−4	5	5	2	−1
X.13	3	1	−1	3	.	.	3	−3	−1	−1	−1	1	.	3	−13	2	−6	5	5	2	5	−1	−1	−5
X.14	3	−1	−1	3	.	.	3	3	−1	−1	−1	−1	.	−3	13	−2	6	5	5	2	5	1	1	5
X.15	.	.	.	5	−1	−1	5	−5	1	1	1	−1	.	−11	−11	−2	−2	3	3	6	3	1	−2	−3
X.16	.	.	.	5	−1	−1	5	5	1	1	1	1	.	11	11	2	2	3	3	6	3	−1	2	3
X.17	.	.	.	8	−4	2	−10	.	2	−2	.	.	2	.	.	.	.	.	.	.	.	.	.	.
X.18	1	3	−1	.	3	.	.	.	.	.	.	.	.	−5	3	−12	4	−1	−1	2	5	−3	.	3
X.19	1	−3	−1	.	3	.	.	.	.	.	.	.	.	5	−3	12	−4	−1	−1	2	5	3	.	−3
X.20	.	.	.	2	2	2	2	.	−2	2	.	.	−1	.	.	.	.	12	−12	.	.	.	.	.
X.21	.	.	.	2	2	2	2	.	−2	2	.	.	−1	.	.	.	.	12	−12	.	.	.	.	.
X.22	−2	−2	2	.	3	.	.	.	.	.	.	.	−1	5	−3	−3	−7	1	1	−5	1	−3	−3	−3
X.23	−2	2	2	.	3	.	.	.	.	.	.	.	−1	−5	3	3	7	1	1	−5	1	3	3	3
X.24	.	.	.	6	6	.	−6	.	−2	2	.	.	.	.	.	.	.	.	.	.	.	.	.	.
X.25	.	.	.	−6	.	.	−6	.	−2	2	.	.	.	.	.	.	.	.	.	.	.	.	.	.
X.26	.	.	.	−6	.	.	−6	.	−2	2	.	.	.	.	.	.	.	.	.	.	.	.	.	.
X.27	2	.	.	−3	−3	.	5	5	1	1	−1	1	.	−1	−5	−8	−4	−1	−1	2	−7	1	−2	3
X.28	2	.	.	−3	−3	.	5	−5	1	1	−1	−1	.	1	5	8	4	−1	−1	2	−7	−1	2	−3
X.29	.	.	.	1	1	−2	7	−7	−1	−1	1	1	.	−11	1	−2	−2	−5	−5	−8	1	−5	4	1
X.30	.	.	.	1	1	−2	7	7	−1	−1	1	−1	.	11	−1	2	2	−5	−5	−8	1	5	−4	−1
X.31	.	.	.	3	3	.	5	−5	1	1	−1	−1	−1	11	19	−2	2	3	3	.	−3	1	4	−5
X.32	.	.	.	3	3	.	5	5	1	1	−1	1	−1	−11	−19	2	−2	3	3	.	−3	−1	−4	5
X.33	1	3	−1	.	.	.	−5	−5	−1	−1	−1	−1	.	9	9	.	.	5	5	2	−1	3	.	1
X.34	1	−3	−1	.	.	.	−5	5	−1	−1	−1	1	.	−9	−9	.	.	5	5	2	−1	−3	.	−1
X.35	.	.	.	.	3	.	−2	−2	2	2	−2	2	.	−4	.	.	−4	−4	−4	2	−4	.	.	.
X.36	.	.	.	.	3	.	−2	2	2	2	−2	−2	.	4	.	.	4	−4	−4	2	−4	.	.	.
X.37	−2	2	−2	1	1	1	.	.	.	.	.	.	.	5	5	−7	−7	1	1	−5	1	−7	−1	−3
X.38	−2	.	.	4	−2	1	.	.	.	.	.	.	.	−15	9	−3	−3	7	7	1	1	3	−6	1
X.39	−2	−2	−2	1	1	1	.	.	.	.	.	.	.	−5	−5	7	7	1	1	−5	1	7	1	3
X.40	−2	.	.	4	−2	1	.	.	.	.	.	.	.	15	−9	3	3	7	7	1	1	−3	6	−1
X.41	−1	−1	1	3	.	.	.	.	.	.	.	.	.	15	−17	1	−3	7	7	−5	1	1	1	−1
X.42	−1	1	−1	−6	.	.	.	.	.	.	.	.	.	.	−12	6	6	2	2	2	2	.	−3	−4
X.43	−1	−1	−1	−6	.	.	.	.	.	.	.	.	.	.	12	−6	−6	2	2	2	2	.	3	4
X.44	3	3	1	−3	−3	.	.	.	.	.	.	.	.	−10	−14	4	−4	4	4	4	−2	4	4	2
X.45	−1	1	1	3	.	.	.	.	.	.	.	.	.	−15	17	−1	3	7	7	−5	1	−1	−1	1
X.46	3	−3	1	−3	−3	.	.	.	.	.	.	.	.	10	14	−4	4	4	4	4	−2	−4	−4	−2
X.47	1	1	−1	.	.	.	6	6	−2	−2	.	−2	.	.	.	6	−6	.	.	6	−6	−6	3	.
X.48	1	−1	−1	.	.	.	6	−6	−2	−2	.	2	.	.	.	−6	6	.	.	6	−6	6	−3	.
X.49	.	.	.	6	.	.	4	.	4	−4	.	.	.	.	.	.	.	12	−12	.	.	.	.	.
X.50	.	.	.	−3	.	.	.	.	.	.	.	.	.	.	.	.	.	.	.	.	.	.	.	.
X.51	.	.	.	−3	.	.	.	.	.	.	.	.	.	.	.	.	.	.	.	.	.	.	.	.
X.52	.	.	.	−3	.	.	.	.	.	.	.	.	.	.	.	.	.	.	.	.	.	.	.	.
X.53	.	.	.	−3	.	.	.	.	.	.	.	.	.	.	.	.	.	.	.	.	.	.	.	.
X.54	.	.	.	4	−2	1	.	.	.	.	.	.	.	−15	21	6	6	1	1	−8	−5	3	.	−3
X.55	.	.	.	4	−2	1	.	.	.	.	.	.	.	15	−21	−6	−6	1	1	−8	−5	−3	.	3
X.56	.	.	.	−4	2	−1	.	.	.	.	.	.	.	.	.	.	.	.	.	.	.	.	.	.
X.57	.	.	.	−4	2	−1	.	.	.	.	.	.	.	.	.	.	.	.	.	.	.	.	.	.
X.58	.	.	.	.	.	.	−6	.	−2	2	.	.	.	.	.	.	.	.	.	.	.	.	.	.
X.59	.	.	.	.	.	.	−6	.	−2	2	.	.	.	.	.	.	.	.	.	.	.	.	.	.
X.60	2	.	.	−3	.	.	.	.	.	.	.	.	.	15	−5	7	3	5	5	−7	−1	1	−8	−5
X.61	2	.	.	−3	.	.	.	.	.	.	.	.	.	−15	5	−7	−3	5	5	−7	−1	−1	8	5
X.62	.	.	.	.	.	.	−5	−5	−1	−1	1	−1	.	−9	−9	.	.	1	1	−8	−5	−3	.	7
X.63	.	.	.	.	.	.	−5	5	−1	−1	1	1	.	9	9	.	.	1	1	−8	−5	3	.	−7
X.64	.	.	.	2	2	−1	.	.	.	.	.	.	.	.	.	.	.	.	.	.	.	.	.	.
X.65	.	.	.	2	2	−1	.	.	.	.	.	.	.	.	.	.	.	.	.	.	.	.	.	.
X.66	.	.	.	.	.	.	.	.	.	.	.	.	.	.	.	.	.	4	4	−2	4	.	.	.
X.67	3	−1	1	.	.	.	.	.	.	.	.	.	.	−15	−3	6	−6	−3	−3	6	3	3	−3	−3
X.68	3	1	1	.	.	.	.	.	.	.	.	.	.	15	3	−6	6	−3	−3	6	3	−3	3	3
X.69	.	.	.	10	−2	−2	−14	.	−2	2	.	.	.	.	.	.	.	.	.	.	.	.	.	.
X.70	−4	.	.	−3	−3	.	.	.	.	.	.	.	.	−10	10	7	−13	2	2	5	2	−2	4	2
X.71	−4	.	.	−3	−3	.	.	.	.	.	.	.	.	10	−10	−7	13	2	2	5	2	2	−4	−2
X.72	.	.	.	4	−2	1	−5	−5	−1	−1	−1	−1	.	−6	−6	6	6	−6	−6	.	6	−6	.	2
X.73	.	.	.	4	−2	1	−5	5	−1	−1	−1	1	.	6	6	−6	−6	−6	−6	.	6	6	.	−2
X.74	−1	3	−1	.	.	.	5	5	1	1	1	1	.	−9	−9	.	.	−3	−3	.	−3	3	.	−1
X.75	−1	−3	−1	.	.	.	5	−5	1	1	1	−1	.	9	9	.	.	−3	−3	.	−3	−3	.	1
X.76	.	.	.	.	.	.	6	6	−2	−2	.	−2	.	−15	−3	−6	6	−3	−3	.	3	3	.	−3
X.77	.	.	.	.	.	.	6	−6	−2	−2	.	2	.	15	3	6	−6	−3	−3	.	3	−3	.	3
X.78	−2	.	.	.	.	.	.	.	.	.	.	.	.	15	−9	−9	3	3	3	3	−3	−3	.	−1
X.79	−2	−2	2	.	.	.	.	.	.	.	.	.	.	−15	9	−6	−6	1	1	4	1	−3	−3	1
X.80	−2	2	2	.	.	.	.	.	.	.	.	.	.	15	−9	6	6	1	1	4	1	3	3	−1
X.81	−2	.	.	.	.	.	.	.	.	.	.	.	.	−15	9	9	−3	3	3	3	−3	3	.	1
X.82	.	.	.	.	.	.	6	.	2	−2	.	.	−2	.	.	.	.	.	.	.	.	.	.	.
X.83	.	.	.	.	.	.	1	−1	1	1	−1	−1	.	6	−6	6	−6	−6	−6	.	6	−6	.	2
X.84	.	.	.	.	.	.	1	1	1	1	−1	1	.	−6	6	−6	6	−6	−6	.	6	6	.	−2

Character table of $A_1 = 2Aut(\mathsf{Fi}_{22})$ *(continued)*

2	6	6	6	6	4	5	5	5	6	6	4	4	4	4	5	4	4	1	3	2
3	2	2	2	2	3	2	2	2	1	1	2	2	2	2	1	1	1	.	1	1
5	.	.	.	.	.	.	.	.	.	.	.	.	.	.	.	.	.	.	.	.
7	.	.	.	.	.	.	.	.	.	.	.	.	.	.	.	.	.	.	1	1
11	.	.	.	.	.	.	.	.	.	.	.	.	.	.	.	.	.	.	.	.
13	.	.	.	.	.	.	.	.	.	.	.	.	.	.	.	.	.	1	.	.
	12_{12}	12_{13}	12_{14}	12_{15}	12_{16}	12_{17}	12_{18}	12_{19}	12_{20}	12_{21}	12_{22}	12_{23}	12_{24}	12_{25}	12_{26}	12_{27}	12_{28}	$13a$	$14a$	$14b$
2P	6_{12}	6_{17}	6_{12}	6_{17}	6_{3}	6_{19}	6_{19}	6_{11}	6_{10}	6_{12}	6_{23}	6_{23}	6_{28}	6_{28}	6_{22}	6_{28}	6_{28}	$13a$	$7a$	$7a$
3P	$4g$	$4h$	$4e$	$4i$	$4a$	$4i$	$4h$	$4g$	$4f$	$4k$	$4g$	$4e$	$4i$	$4h$	$4j$	$4l$	$4l$	$13a$	$14a$	$14b$
5P	12_{12}	12_{13}	12_{14}	12_{15}	12_{16}	12_{17}	12_{18}	12_{19}	12_{20}	12_{21}	12_{22}	12_{23}	12_{24}	12_{25}	12_{26}	12_{28}	12_{27}	$13a$	$14a$	$14b$
7P	12_{12}	12_{13}	12_{14}	12_{15}	12_{16}	12_{17}	12_{18}	12_{19}	12_{20}	12_{21}	12_{22}	12_{23}	12_{24}	12_{25}	12_{26}	12_{27}	12_{28}	$13a$	$2a$	$2c$
11P	12_{12}	12_{13}	12_{14}	12_{15}	12_{16}	12_{17}	12_{18}	12_{19}	12_{20}	12_{21}	12_{22}	12_{23}	12_{24}	12_{25}	12_{26}	12_{28}	12_{27}	$13a$	$14a$	$14b$
13P	12_{12}	12_{13}	12_{14}	12_{15}	12_{16}	12_{17}	12_{18}	12_{19}	12_{20}	12_{21}	12_{22}	12_{23}	12_{24}	12_{25}	12_{26}	12_{27}	12_{28}	$1a$	$14a$	$14b$
X.1	1	1	1	1	1	1	1	1	1	1	1	1	1	1	1	1	1	1	1	1
X.2	−1	1	1	1	−1	1	1	−1	1	1	−1	1	1	1	−1	−1	−1	1	1	−1
X.3	−1	2	−3	2	.	−1	−1	−1	1	−1	2	.	−1	−1	−2	−1	−1	.	1	1
X.4	1	2	−3	2	.	−1	−1	1	1	−1	−2	.	−1	−1	2	1	1	.	1	−1
X.5	.	−2	.	2	.	2	−2	.	.	.	.	.	−1	1	.	C	$\bar{C}$	1	−2	.
X.6	.	−2	.	2	.	2	−2	.	.	.	.	.	−1	1	.	$\bar{C}$	C	1	−2	.
X.7	4	−1	2	−1	1	2	2	−2	.	2	1	−1	−1	−1	−1	−1	−1	.	2	2
X.8	−4	−1	2	−1	−1	2	2	2	.	2	−1	−1	−1	−1	1	1	1	.	2	−2
X.9	.	2	−2	2	.	−1	−1	3	−3	.	.	−2	2	2	.	.	.	.	.	.
X.10	.	2	−2	2	.	−1	−1	−3	−3	.	.	−2	2	2	.	.	.	.	.	.
X.11	3	−2	−3	−2	−2	−2	−2	.	.	−1	.	.	1	1	−2	−1	−1	.	2	−2
X.12	−3	−2	−3	−2	2	−2	−2	.	.	−1	.	.	1	1	2	1	1	.	2	2
X.13	−1	3	5	3	3	.	.	2	−2	1	−1	−1	.	.	−1	.	.	.	.	.
X.14	1	3	5	3	−3	.	.	−2	−2	1	1	−1	.	.	1	.	.	.	.	.
X.15	1	.	3	.	−2	.	.	−2	2	−1	−2	.	.	.	−2	.	.	−1	.	.
X.16	−1	.	3	.	2	.	.	2	2	−1	2	.	.	.	2	.	.	−1	.	.
X.17	.	2	.	−2	.	−2	2	.	.	.	.	.	−2	2	.	.	.	.	−2	.
X.18	−3	.	−3	.	4	.	.	.	−2	−1	.	.	.	.	.	.	.	.	1	−1
X.19	3	.	−3	.	−4	.	.	.	−2	−1	.	.	.	.	.	.	.	.	1	1
X.20	.	.	.	.	.	.	.	.	.	.	.	.	.	.	.	.	.	.	.	.
X.21	.	.	.	.	.	.	.	.	.	.	.	.	.	.	.	.	.	.	.	.
X.22	−3	1	1	1	−1	1	1	−3	−1	1	3	1	1	1	1	−1	−1	.	.	.
X.23	3	1	1	1	1	1	1	3	−1	1	−3	1	1	1	−1	1	1	.	.	.
X.24	.	2	.	−2	.	4	−4	.	.	.	.	.	4	−4	.	.	.	.	−2	.
X.25	.	−4	.	4	.	−2	2	.	.	.	.	.	1	−1	.	C	$\bar{C}$	.	−2	.
X.26	.	−4	.	4	.	−2	2	.	.	.	.	.	1	−1	.	$\bar{C}$	C	.	−2	.
X.27	−3	.	5	.	2	.	.	.	−2	−1	.	2	.	.	−2	.	.	.	.	.
X.28	3	.	5	.	−2	.	.	.	−2	−1	.	2	.	.	2	.	.	.	.	.
X.29	−1	.	−3	.	−2	.	.	2	.	−1	2	.	.	.	.	.	.	.	.	.
X.30	1	.	−3	.	2	.	.	−2	.	−1	−2	.	.	.	.	.	.	.	.	.
X.31	1	.	−3	.	2	.	.	−2	.	−1	−2	.	.	.	.	.	.	.	.	.
X.32	−1	.	−3	.	−2	.	.	2	.	−1	2	.	.	.	.	.	.	.	.	.
X.33	3	−2	−1	−2	.	−2	−2	.	2	1	.	2	1	1	.	−1	−1	.	.	.
X.34	−3	−2	−1	−2	.	−2	−2	.	2	1	.	2	1	1	.	1	1	.	.	.
X.35	4	−2	.	−2	−4	−2	−2	4	2	.	−2	.	−2	−2	.	.	.	.	.	.
X.36	−4	−2	.	−2	4	−2	−2	−4	2	.	2	.	−2	−2	.	.	.	.	.	.
X.37	1	3	1	3	−1	3	3	1	3	−3	1	1	.	.	−1	.	.	.	.	.
X.38	3	−2	1	−2	.	1	1	−3	1	3	.	−2	1	1	−2	1	1	.	.	.
X.39	−1	3	1	3	1	3	3	−1	3	−3	−1	1	.	.	1	.	.	.	.	.
X.40	−3	−2	1	−2	.	1	1	3	1	3	.	−2	1	1	2	−1	−1	.	.	.
X.41	1	1	1	1	−3	1	1	1	−1	−1	1	1	−2	−2	1	.	.	.	.	.
X.42	4	3	−2	3	3	.	.	−2	−2	2	1	1	.	.	1	.	.	.	.	.
X.43	−4	3	−2	3	−3	.	.	2	−2	2	−1	1	.	.	−1	.	.	.	.	.
X.44	.	2	−6	2	2	2	2	.	.	.	.	.	−1	−1	.	−1	−1	.	.	.
X.45	−1	1	1	1	3	1	1	−1	−1	−1	−1	1	−2	−2	−1	.	.	.	.	.
X.46	.	2	−6	2	−2	2	2	.	.	.	.	.	−1	−1	.	1	1	.	.	.
X.47	2	−3	2	−3	−3	.	.	2	2	.	−1	−1	.	.	−1	.	.	.	.	.
X.48	−2	−3	2	−3	3	.	.	−2	2	.	1	−1	.	.	1	.	.	.	.	.
X.49	.	−2	.	2	.	−4	4	.	.	.	.	.	2	−2	.	.	.	.	.	.
X.50	.	.	.	.	.	.	.	.	.	.	.	.	.	.	.	.	.	1	2	.
X.51	.	.	.	.	.	.	.	.	.	.	.	.	.	.	.	.	.	1	2	.
X.52	.	.	.	.	.	.	.	.	.	.	.	.	.	.	.	.	.	1	2	.
X.53	.	.	.	.	.	.	.	.	.	.	.	.	.	.	.	.	.	1	2	.
X.54	−1	.	−1	.	.	.	.	2	.	1	2	2	.	.	.	.	.	.	−1	−1
X.55	1	.	−1	.	.	.	.	−2	.	1	−2	2	.	.	.	.	.	.	−1	1
X.56	.	2	.	−2	.	−2	2	.	.	.	.	.	−2	2	.	.	.	−1	.	.
X.57	.	2	.	−2	.	−2	2	.	.	.	.	.	−2	2	.	.	.	−1	.	.
X.58	.	2	.	−2	.	−2	2	.	.	.	.	.	−2	2	.	.	.	1	.	.
X.59	.	2	.	−2	.	−2	2	.	.	.	.	.	−2	2	.	.	.	1	.	.
X.60	−3	.	3	.	.	−3	−3	3	1	1	.	.	.	.	.	.	.	.	.	.
X.61	3	.	3	.	.	−3	−3	−3	1	1	.	.	.	.	.	.	.	.	.	.
X.62	−3	.	3	.	.	.	.	.	.	1	.	.	.	.	.	.	.	.	1	1
X.63	3	.	3	.	.	.	.	.	.	1	.	.	.	.	.	.	.	.	1	−1
X.64	.	−4	.	4	.	−2	2	.	.	.	.	.	1	−1	.	$\bar{C}$	C	.	2	.
X.65	.	−4	.	4	.	−2	2	.	.	.	.	.	1	−1	.	C	$\bar{C}$	.	2	.
X.66	.	−2	4	−2	.	−2	−2	.	−2	−4	.	−2	−2	−2	.	.	.	1	.	.
X.67	−1	3	−1	3	3	.	.	2	2	1	−1	−1	.	.	1	.	.	.	−1	1
X.68	1	3	−1	3	−3	.	.	−2	2	1	1	−1	.	.	−1	.	.	.	−1	−1
X.69	.	.	.	.	.	.	.	.	.	.	.	.	.	.	.	.	.	.	2	.
X.70	2	2	−2	2	2	−1	−1	−1	1	2	2	−2	−1	−1	.	1	1	.	.	.
X.71	−2	2	−2	2	−2	−1	−1	1	1	2	−2	−2	−1	−1	.	−1	−1	.	.	.
X.72	2	.	−2	.	.	.	.	2	.	−2	2	−2	.	.	.	.	.	.	.	.
X.73	−2	.	−2	.	.	.	.	−2	.	−2	−2	−2	.	.	.	.	.	.	.	.
X.74	3	.	−3	.	.	.	.	.	.	1	.	.	.	.	.	.	.	1	−2	2
X.75	−3	.	−3	.	.	.	.	.	.	1	.	.	.	.	.	.	.	1	−2	−2
X.76	−1	.	−1	.	.	.	.	2	.	1	2	2	.	.	.	.	.	.	−1	1
X.77	1	.	−1	.	.	.	.	−2	.	1	−2	2	.	.	.	.	.	.	−1	−1
X.78	5	−2	−3	−2	.	1	1	−1	−1	−1	2	.	1	1	.	1	1	.	.	.
X.79	5	1	1	1	3	−2	−2	2	.	−3	−1	1	1	1	1	−1	−1	.	.	.
X.80	−5	1	1	1	−3	−2	−2	−2	.	−3	1	1	1	1	−1	1	1	.	.	.
X.81	−5	−2	−3	−2	.	1	1	1	−1	−1	−2	.	1	1	.	−1	−1	.	.	.
X.82	.	6	.	−6	.	.	.	.	.	.	.	.	.	.	.	.	.	.	.	.
X.83	−2	.	2	.	.	.	.	−2	.	2	−2	2	.	.	.	.	.	.	.	.
X.84	2	.	2	.	.	.	.	2	.	2	2	2	.	.	.	.	.	.	.	.

Character table of $A_1 = 2Aut(\mathsf{Fi}_{22})$ (continued)

	14c	15a	16a	16b	18a	18b	18c	18d	18e	18f	18g	18h	18i	18j	18k	18l	18m	20a	20b	20c	21a	22a	22b	22c
2	2	3	5	5	4	3	3	3	3	3	4	4	2	2	2	2	2	3	3	3	2	2	2	2
3	.	1	.	.	4	4	3	3	3	3	2	2	3	3	3	2	2	1	.	.	1	.	.	.
5	.	1	.	.	.	.	.	.	.	.	.	.	.	.	.	.	.	1	1	1	.	.	.	.
7	1	.	.	.	.	.	.	.	.	.	.	.	.	.	.	.	.	.	.	.	1	.	.	.
11	.	.	.	.	.	.	.	.	.	.	.	.	.	.	.	.	.	.	.	.	.	1	1	1
13	.	.	.	.	.	.	.	.	.	.	.	.	.	.	.	.	.	.	.	.	.	.	.	.
2P	7a	15a	8f	8f	9a	9b	9a	9a	9a	9a	9a	9a	9b	9c	9b	9c	9c	10a	10c	10d	21a	11a	11a	11a
3P	14c	5a	16a	16b	6_2	6_2	6_{14}	6_6	6_6	6_{14}	6_{10}	6_{11}	6_{14}	6_9	6_6	6_{24}	6_{25}	20a	20b	20c	7a	22a	22b	22c
5P	14c	3a	16a	16b	18a	18b	18f	18e	18d	18c	18g	18h	18i	18j	18k	18m	18l	4a	4e	4c	21a	22a	22b	22c
7P	2b	15a	16a	16b	18a	18b	18c	18d	18e	18f	18g	18h	18i	18j	18k	18l	18m	20a	20b	20c	3a	22c	22b	22a
11P	14c	15a	16a	16b	18a	18b	18f	18e	18d	18c	18g	18h	18i	18j	18k	18m	18l	20a	20b	20c	21a	2b	2a	2b
13P	14c	15a	16a	16b	18a	18b	18c	18d	18e	18f	18g	18h	18i	18j	18k	18l	18m	20a	20b	20c	21a	22c	22b	22a
X.1	1	1	1	1	1	1	1	1	1	1	1	1	1	1	1	1	1	1	1	1	1	1	1	1
X.2	1	1	1	−1	1	1	−1	1	1	−1	1	1	−1	1	1	−1	−1	−1	1	−1	1	1	1	1
X.3	1	.	.	.	3	.	1	−1	−1	1	−1	−1	−2	.	2	.	.	−1	−1	1	1	−1	1	−1
X.4	1	.	.	.	3	.	−1	−1	−1	−1	−1	−1	2	.	2	.	.	1	−1	−1	1	−1	1	−1
X.5	.	2	.	.	2	−4	D	.	.	$\bar{D}$	−2	2	.	−1	.	C	$\bar{C}$	.	.	.	−1	.	.	.
X.6	.	2	.	.	2	−4	$\bar{D}$	.	.	D	−2	2	.	−1	.	$\bar{C}$	C	.	.	.	−1	.	.	.
X.7	.	1	−1	1	.	3	−2	2	2	−2	.	.	1	.	−1	.	.	2	.	.	−1	.	.	.
X.8	.	1	−1	−1	.	3	2	2	2	2	.	.	−1	.	−1	.	.	−2	.	.	−1	.	.	.
X.9	.	1	1	1	2	2	−2	.	.	−2	2	2	−2	−1	.	−1	−1	1	−1	1	.	.	.	.
X.10	.	1	1	−1	2	2	2	.	.	2	2	2	2	−1	.	1	1	−1	−1	−1	.	.	.	.
X.11	.	−1	.	.	−1	−1	1	−1	−1	1	−1	−1	1	2	−1	.	.	1	−1	−1	−1	.	.	.
X.12	.	−1	.	.	−1	−1	−1	−1	−1	−1	−1	−1	−1	2	−1	.	.	−1	−1	1	−1	.	.	.
X.13	.	.	−1	1	3	.	1	−1	−1	1	−1	−1	−2	.	2	.	.	1	1	−1	.	.	.	.
X.14	.	.	−1	−1	3	.	−1	−1	−1	−1	−1	−1	2	.	2	.	.	−1	1	1	.	.	.	.
X.15	.	−1	.	.	5	−1	−1	1	1	−1	1	1	−1	−1	1	1	1	−1	−1	1	.	.	.	.
X.16	.	−1	.	.	5	−1	1	1	1	1	1	1	1	−1	1	−1	−1	1	−1	−1	.	.	.	.
X.17	.	−2	.	.	−8	4	.	.	.	.	.	.	.	−2	.	.	.	.	.	.	2	.	−2	.
X.18	−1	.	−1	−1	.	3	−2	2	2	−2	.	.	1	.	−1	.	.	.	.	.	1	.	.	.
X.19	−1	.	−1	1	.	3	2	2	2	2	.	.	−1	.	−1	.	.	.	.	.	1	.	.	.
X.20	.	−2	.	.	−2	−2	.	.	.	.	−2	2	.	−2	.	.	.	.	.	.	.	E	1	$\bar{E}$
X.21	.	−2	.	.	−2	−2	.	.	.	.	−2	2	.	−2	.	.	.	.	.	.	.	$\bar{E}$	1	E
X.22	.	.	.	.	.	3	−2	−2	−2	−2	.	.	1	.	1	.	.	.	.	.	.	−1	−1	−1
X.23	.	.	.	.	.	3	2	−2	−2	2	.	.	−1	.	1	.	.	.	.	.	.	−1	−1	−1
X.24	.	.	.	.	−6	−6	.	.	.	.	2	−2	.	.	.	.	.	.	.	.	2	.	.	.
X.25	.	.	.	.	6	.	$\bar{D}$	.	.	D	2	−2	.	.	.	.	.	.	.	.	−1	.	.	.
X.26	.	.	.	.	6	.	D	.	.	$\bar{D}$	2	−2	.	.	.	.	.	.	.	.	−1	.	.	.
X.27	.	−1	.	.	−3	−3	1	−1	−1	1	1	1	1	.	−1	.	.	−1	1	1	.	.	.	.
X.28	.	−1	.	.	−3	−3	−1	−1	−1	−1	1	1	−1	.	−1	.	.	1	1	−1	.	.	.	.
X.29	.	1	.	.	1	1	−1	−1	−1	−1	1	1	−1	−2	−1	.	.	−1	1	−1	.	.	.	.
X.30	.	1	.	.	1	1	1	−1	−1	1	1	1	1	−2	−1	.	.	1	1	1	.	.	.	.
X.31	.	−1	.	.	3	3	1	1	1	1	−1	−1	1	.	1	.	.	1	1	−1	.	1	−1	1
X.32	.	−1	.	.	3	3	−1	1	1	−1	−1	−1	−1	.	1	.	.	−1	1	1	.	1	−1	1
X.33	.	1	−1	−1	.	.	.	.	.	.	.	.	.	.	.	.	.	−1	1	1	.	.	.	.
X.34	.	1	−1	1	.	.	.	.	.	.	.	.	.	.	.	.	.	1	1	−1	.	.	.	.
X.35	.	1	.	.	.	3	−2	−2	−2	−2	.	.	1	.	1	.	.	−2	.	.	.	.	.	.
X.36	.	1	.	.	.	3	2	−2	−2	2	.	.	−1	.	1	.	.	2	.	.	.	.	.	.
X.37	.	.	.	.	1	1	1	−1	−1	1	1	1	1	1	−1	1	1	.	.	.	.	.	.	.
X.38	.	.	.	.	4	−2	.	−2	−2	.	.	.	.	1	−2	−1	−1	.	.	.	.	.	.	.
X.39	.	.	.	.	1	1	−1	−1	−1	−1	1	1	−1	1	−1	−1	−1	.	.	.	.	.	.	.
X.40	.	.	.	.	4	−2	.	−2	−2	.	.	.	.	1	−2	1	1	.	.	.	.	.	.	.
X.41	.	.	1	−1	3	.	1	1	1	1	−1	−1	−2	.	−2	.	.	.	.	.	.	.	.	.
X.42	.	.	1	−1	−6	.	.	.	.	.	−2	−2	.	.	.	.	.	.	.	.	.	.	.	.
X.43	.	.	1	1	−6	.	.	.	.	.	−2	−2	.	.	.	.	.	.	.	.	.	.	.	.
X.44	.	.	−1	1	−3	−3	1	−1	−1	1	1	1	1	.	−1	.	.	.	.	.	.	.	.	.
X.45	.	.	1	1	3	.	−1	1	1	−1	−1	−1	2	.	−2	.	.	.	.	.	.	.	.	.
X.46	.	.	−1	−1	−3	−3	−1	−1	−1	−1	1	1	−1	.	−1	.	.	.	.	.	.	.	.	.
X.47	.	.	1	−1	.	.	.	.	.	.	.	.	.	.	.	.	.	.	.	.	.	.	.	.
X.48	.	.	1	1	.	.	.	.	.	.	.	.	.	.	.	.	.	.	.	.	.	.	.	.
X.49	.	2	.	.	−6	.	.	.	.	.	2	−2	.	.	.	.	.	.	.	.	.	.	.	.
X.50	.	.	.	.	3	.	$\bar{C}$	$\bar{B}$	B	C	1	−1	.	.	.	.	.	.	.	.	1	.	.	.
X.51	.	.	.	.	3	.	C	B	$\bar{B}$	$\bar{C}$	1	−1	.	.	.	.	.	.	.	.	1	.	.	.
X.52	.	.	.	.	3	.	C	$\bar{B}$	B	$\bar{C}$	1	−1	.	.	.	.	.	.	.	.	1	.	.	.
X.53	.	.	.	.	3	.	$\bar{C}$	B	$\bar{B}$	C	1	−1	.	.	.	.	.	.	.	.	1	.	.	.
X.54	−1	.	.	.	4	−2	.	2	2	.	.	.	.	1	2	1	1	.	.	.	−1	.	.	.
X.55	−1	.	.	.	4	−2	.	2	2	.	.	.	.	1	2	−1	−1	.	.	.	−1	.	.	.
X.56	.	.	.	.	4	−2	D	.	.	$\bar{D}$	.	.	.	1	.	$\bar{C}$	C	.	.	.	.	.	.	.
X.57	.	.	.	.	4	−2	$\bar{D}$	.	.	D	.	.	.	1	.	C	$\bar{C}$	.	.	.	.	.	.	.
X.58	.	.	.	.	.	.	.	.	.	.	.	.	.	.	.	.	.	.	.	.	.	.	.	.
X.59	.	.	.	.	.	.	.	.	.	.	.	.	.	.	.	.	.	.	.	.	.	.	.	.
X.60	.	.	.	.	−3	.	−1	1	1	−1	1	1	2	.	−2	.	.	.	.	.	.	.	.	.
X.61	.	.	.	.	−3	.	1	1	1	1	1	1	−2	.	−2	.	.	.	.	.	.	.	.	.
X.62	−1	1	.	.	.	.	.	.	.	.	.	.	.	.	.	.	.	1	−1	−1	1	.	.	.
X.63	−1	1	.	.	.	.	.	.	.	.	.	.	.	.	.	.	.	−1	−1	1	1	.	.	.
X.64	.	.	.	.	−2	−2	.	.	.	.	−2	2	.	1	.	C	$\bar{C}$	.	.	.	1	.	.	.
X.65	.	.	.	.	−2	−2	.	.	.	.	−2	2	.	1	.	$\bar{C}$	C	.	.	.	1	.	.	.
X.66	.	.	.	.	.	.	.	.	.	.	.	.	.	.	.	.	.	.	.	.	.	.	.	.
X.67	−1	.	1	−1	.	.	.	.	.	.	.	.	.	.	.	.	.	.	.	.	−1	.	.	.
X.68	−1	.	1	1	.	.	.	.	.	.	.	.	.	.	.	.	.	.	.	.	−1	.	.	.
X.69	.	2	.	.	−10	2	.	.	.	.	−2	2	.	2	.	.	.	.	.	.	−2	.	.	.
X.70	.	.	.	.	−3	−3	−1	1	1	−1	1	1	−1	.	1	.	.	.	.	.	.	.	.	.
X.71	.	.	.	.	−3	−3	1	1	1	1	1	1	1	.	1	.	.	.	.	.	.	.	.	.
X.72	.	1	.	.	4	−2	2	.	.	2	.	.	2	1	.	−1	−1	−1	1	1	.	.	.	.
X.73	.	1	.	.	4	−2	−2	.	.	−2	.	.	−2	1	.	1	1	1	1	−1	.	.	.	.
X.74	.	−1	−1	−1	.	.	.	.	.	.	.	.	.	.	.	.	.	1	−1	−1	1	.	.	.
X.75	.	−1	−1	1	.	.	.	.	.	.	.	.	.	.	.	.	.	−1	−1	1	1	.	.	.
X.76	1	.	.	.	.	.	.	.	.	.	.	.	.	.	.	.	.	.	.	.	−1	.	.	.
X.77	1	.	.	.	.	.	.	.	.	.	.	.	.	.	.	.	.	.	.	.	−1	.	.	.
X.78	.	.	.	.	.	.	.	.	.	.	.	.	.	.	.	.	.	.	.	.	.	.	.	.
X.79	.	.	.	.	.	.	.	.	.	.	.	.	.	.	.	.	.	.	.	.	.	.	.	.
X.80	.	.	.	.	.	.	.	.	.	.	.	.	.	.	.	.	.	.	.	.	.	.	.	.
X.81	.	.	.	.	.	.	.	.	.	.	.	.	.	.	.	.	.	.	.	.	.	.	.	.
X.82	.	.	.	.	.	.	.	.	.	.	.	.	.	.	.	.	.	.	.	.	.	.	2	.
X.83	.	1	.	.	.	.	.	.	.	.	.	.	.	.	.	.	.	1	−1	1	.	.	.	.
X.84	.	1	.	.	.	.	.	.	.	.	.	.	.	.	.	.	.	−1	−1	−1	.	.	.	.

Character table of $A_1 = 2Aut(\mathsf{Fi}_{22})$ *(continued)*

2	5	5	5	5	4	4	4	4	1	2	3	2	2	3	3	2	2	2	2	2
3	1	1	1	1	1	1	1	1	.	.	1	1	1	2	2	2	1	1	1	1
5	.	.	.	.	.	.	.	.	.	.	1	1	1	.	.	.	.	.	.	1
7	.	.	.	.	.	.	.	.	.	1	.	.	.	.	.	.	1	1	1	.
11	.	.	.	.	.	.	.	.	.	.	.	.	.	.	.	.	.	.	.	.
13	.	.	.	.	.	.	.	.	1	.	.	.	.	.	.	.	.	.	.	.
	24a	24b	24c	24d	24e	24f	24g	24h	26a	28a	30a	30b	30c	36a	36b	36c	42a	42b	42c	60a
2P	12_5	12_6	12_6	12_5	12_{13}	12_{20}	12_{13}	12_{20}	13a	14a	15a	15a	15a	18g	18g	18b	21a	21a	21a	30a
3P	8b	8c	8d	8e	8g	8a	8g	8a	26a	28a	10a	10e	10b	12_3	12_3	12_4	14b	14a	14b	20a
5P	24a	24b	24c	24d	24g	24h	24e	24f	26a	28a	6_1	6_4	6_5	36a	36b	36c	42c	42b	42a	12_1
7P	24a	24b	24c	24d	24g	24h	24e	24f	26a	4a	30a	30b	30c	36a	36b	36c	6_7	6_1	6_8	60a
11P	24a	24b	24c	24d	24e	24f	24g	24h	26a	28a	30a	30b	30c	36a	36b	36c	42c	42b	42a	60a
13P	24a	24b	24c	24d	24e	24f	24g	24h	2a	28a	30a	30b	30c	36a	36b	36c	42a	42b	42c	60a
X.1	1	1	1	1	1	1	1	1	1	1	1	1	1	1	1	1	1	1	1	1
X.2	1	−1	1	−1	−1	−1	−1	−1	1	−1	1	1	−1	−1	−1	−1	−1	1	−1	−1
X.3	−1	−1	1	1	.	−1	.	−1	.	1	.	−2	.	−1	−1	.	1	1	1	2
X.4	−1	1	1	−1	.	1	.	1	.	−1	.	−2	.	1	1	.	−1	1	−1	−2
X.5	.	.	.	.	.	.	.	.	−1	.	−2	.	.	.	.	.	$\bar{C}$	1	C	.
X.6	.	.	.	.	.	.	.	.	−1	.	−2	.	.	.	.	.	C	1	$\bar{C}$	.
X.7	−2	.	−2	.	−1	.	−1	.	.	.	1	−1	1	.	.	1	−1	−1	−1	−1
X.8	−2	.	−2	.	1	.	1	.	.	.	1	−1	−1	.	.	−1	1	−1	1	1
X.9	.	.	−2	−2	.	1	.	1	.	.	1	−1	−1	.	.	.	.	.	.	1
X.10	.	.	−2	2	.	−1	.	−1	.	.	1	−1	1	.	.	.	.	.	.	−1
X.11	1	1	−1	−1	.	.	.	.	.	.	−1	1	−1	1	1	1	1	−1	1	1
X.12	1	−1	−1	1	.	.	.	.	.	.	−1	1	1	−1	−1	−1	−1	−1	−1	−1
X.13	−1	1	−1	1	1	.	1	.	.	.	.	2	.	−1	−1	.	.	.	.	−2
X.14	−1	−1	−1	−1	−1	.	−1	.	.	.	.	2	.	1	1	.	.	.	.	2
X.15	1	−1	1	−1	.	.	.	.	−1	.	−1	1	1	1	1	1	.	.	.	−1
X.16	1	1	1	1	.	.	.	.	−1	.	−1	1	−1	−1	−1	−1	.	.	.	1
X.17	.	.	.	.	.	.	.	.	.	.	2	.	.	.	.	.	.	−2	.	.
X.18	1	−1	3	1	.	.	.	.	.	1	.	.	.	.	.	1	−1	1	−1	.
X.19	1	1	3	−1	.	.	.	.	.	−1	.	.	.	.	.	−1	1	1	1	.
X.20	.	.	.	.	.	.	.	.	.	.	2	.	.	.	.	.	.	.	.	.
X.21	.	.	.	.	.	.	.	.	.	.	2	.	.	.	.	.	.	.	.	.
X.22	1	1	1	1	1	1	1	1	.	.	.	.	.	.	.	−1	.	.	.	.
X.23	1	−1	1	−1	−1	−1	−1	−1	.	.	.	.	.	.	.	1	.	.	.	.
X.24	.	.	.	.	.	.	.	.	.	.	.	.	.	.	.	.	.	−2	.	.
X.25	.	.	.	.	.	.	.	.	.	.	.	.	.	.	.	.	C	1	$\bar{C}$	.
X.26	.	.	.	.	.	.	.	.	.	.	.	.	.	.	.	.	$\bar{C}$	1	C	.
X.27	−1	1	1	−1	.	.	.	.	.	.	−1	−1	−1	1	1	−1	.	.	.	−1
X.28	−1	−1	1	1	.	.	.	.	.	.	−1	−1	1	−1	−1	1	.	.	.	1
X.29	1	1	−1	−1	.	.	.	.	.	.	1	1	−1	1	1	1	.	.	.	−1
X.30	1	−1	−1	1	.	.	.	.	.	.	1	1	1	−1	−1	−1	.	.	.	1
X.31	−1	1	1	−1	.	.	.	.	.	.	−1	−1	1	1	1	−1	.	.	.	1
X.32	−1	−1	1	1	.	.	.	.	.	.	−1	−1	−1	−1	−1	1	.	.	.	−1
X.33	1	3	−1	1	.	.	.	.	.	.	1	−1	1	.	.	.	.	.	.	−1
X.34	1	−3	−1	−1	.	.	.	.	.	.	1	−1	−1	.	.	.	.	.	.	1
X.35	.	.	.	.	.	.	.	.	.	.	1	1	1	.	.	−1	.	.	.	1
X.36	.	.	.	.	.	.	.	.	.	.	1	1	−1	.	.	1	.	.	.	−1
X.37	−1	−1	−1	−1	−1	1	−1	1	.	.	.	.	.	−1	−1	−1	.	.	.	.
X.38	−1	−1	1	1	.	1	.	1	.	.	.	.	.	.	.	.	.	.	.	.
X.39	−1	1	−1	1	1	−1	1	−1	.	.	.	.	.	1	1	1	.	.	.	.
X.40	−1	1	1	−1	.	−1	.	−1	.	.	.	.	.	.	.	.	.	.	.	.
X.41	1	−1	−1	1	−1	1	−1	1	.	.	.	.	.	1	1	.	.	.	.	.
X.42	.	−2	.	−2	1	.	1	.	.	.	.	.	.	.	.	.	.	.	.	.
X.43	.	2	.	2	−1	.	−1	.	.	.	.	.	.	.	.	.	.	.	.	.
X.44	2	−2	.	.	.	.	.	.	.	.	.	.	.	1	1	−1	.	.	.	.
X.45	1	1	−1	−1	1	−1	1	−1	.	.	.	.	.	−1	−1	.	.	.	.	.
X.46	2	2	.	.	.	.	.	.	.	.	.	.	.	−1	−1	1	.	.	.	.
X.47	2	.	.	2	1	.	1	.	.	.	.	.	.	.	.	.	.	.	.	.
X.48	2	.	.	−2	−1	.	−1	.	.	.	.	.	.	.	.	.	.	.	.	.
X.49	.	.	.	.	.	.	.	.	.	.	−2	.	.	.	.	.	.	.	.	.
X.50	.	.	.	.	.	.	.	.	−1	.	.	.	.	3	−3	.	C	−1	$\bar{C}$	.
X.51	.	.	.	.	.	.	.	.	−1	.	.	.	.	3	−3	.	$\bar{C}$	−1	C	.
X.52	.	.	.	.	.	.	.	.	−1	.	.	.	.	−3	3	.	$\bar{C}$	−1	C	.
X.53	.	.	.	.	.	.	.	.	−1	.	.	.	.	−3	3	.	C	−1	$\bar{C}$	.
X.54	1	−1	−1	1	.	.	.	.	.	−1	.	.	.	.	.	.	−1	−1	−1	.
X.55	1	1	−1	−1	.	.	.	.	.	1	.	.	.	.	.	.	1	−1	1	.
X.56	.	.	.	.	.	.	.	.	1	.	.	.	.	.	.	.	.	.	.	.
X.57	.	.	.	.	.	.	.	.	1	.	.	.	.	.	.	.	.	.	.	.
X.58	.	.	.	.	I	.	$-I$	.	−1	.	.	.	.	.	.	.	.	.	.	.
X.59	.	.	.	.	$-I$	.	I	.	−1	.	.	.	.	.	.	.	.	.	.	.
X.60	−1	−1	1	1	.	−1	.	−1	.	.	.	.	.	1	1	.	.	.	.	.
X.61	−1	1	1	−1	.	1	.	1	.	.	.	.	.	−1	−1	.	.	.	.	.
X.62	−1	−1	1	1	.	.	.	.	.	−1	1	1	1	.	.	.	1	1	1	1
X.63	−1	1	1	−1	.	.	.	.	.	1	1	1	−1	.	.	.	−1	1	−1	−1
X.64	.	.	.	.	.	.	.	.	.	.	.	.	.	.	.	.	C	−1	$\bar{C}$	.
X.65	.	.	.	.	.	.	.	.	.	.	.	.	.	.	.	.	$\bar{C}$	−1	C	.
X.66	.	.	.	.	.	.	.	.	1	.	.	.	.	.	.	.	.	.	.	.
X.67	−1	1	1	−1	−1	.	−1	.	.	1	.	.	.	.	.	.	1	−1	1	.
X.68	−1	−1	1	1	1	.	1	.	.	−1	.	.	.	.	.	.	−1	−1	−1	.
X.69	.	.	.	.	.	.	.	.	.	.	−2	.	.	.	.	.	.	2	.	.
X.70	.	.	.	.	.	−1	.	−1	.	.	.	.	.	1	1	−1	.	.	.	.
X.71	.	.	.	.	.	1	.	1	.	.	.	.	.	−1	−1	1	.	.	.	.
X.72	.	.	.	.	.	.	.	.	.	.	1	−1	1	.	.	.	.	.	.	−1
X.73	.	.	.	.	.	.	.	.	.	.	1	−1	−1	.	.	.	.	.	.	1
X.74	−1	1	−1	1	.	.	.	.	1	.	−1	1	−1	.	.	.	−1	1	−1	1
X.75	−1	−1	−1	−1	.	.	.	.	1	.	−1	1	1	.	.	.	1	1	1	−1
X.76	−1	1	1	−1	.	.	.	.	.	−1	.	.	.	.	.	.	1	−1	1	.
X.77	−1	−1	1	1	.	.	.	.	.	1	.	.	.	.	.	.	−1	−1	−1	.
X.78	−1	1	1	−1	.	−1	.	−1	.	.	.	.	.	.	.	.	.	.	.	.
X.79	−1	−1	−1	−1	1	.	1	.	.	.	.	.	.	.	.	.	.	.	.	.
X.80	−1	1	−1	1	−1	.	−1	.	.	.	.	.	.	.	.	.	.	.	.	.
X.81	−1	−1	1	1	.	1	.	1	.	.	.	.	.	.	.	.	.	.	.	.
X.82	.	.	.	.	.	.	.	.	.	.	.	.	.	.	.	.	.	.	.	.
X.83	.	.	.	.	.	.	.	.	.	.	1	−1	−1	.	.	.	.	.	.	1
X.84	.	.	.	.	.	.	.	.	.	.	1	−1	1	.	.	.	.	.	.	−1

Character table of $A_1 = 2Aut(\mathsf{Fi}_{22})$ (continued)

2	19	19	17	14	19	19	14	17	10	9	8	5	11	14	14
3	9	9	6	6	4	4	4	3	7	9	7	7	4	4	1
5	2	2	1	2	1	1	1	.	1	.	.	.	1	.	1
7	1	1	1	1	.	.	.	.	1	.	.	.	1	.	.
11	1	1	1	.	.	.	.	.	.	.	.	.	.	.	.
13	1	1	.	.	.	.	.	.	.	.	.	.	.	.	.
	1a	2a	2b	2c	2d	2e	2f	2g	3a	3b	3c	3d	4a	4b	4c
2P	1a	1a	1a	1a	1a	1a	1a	1a	3a	3b	3c	3d	2a	2e	2e
3P	1a	2a	2b	2c	2d	2e	2f	2g	1a	1a	1a	1a	4a	4b	4c
5P	1a	2a	2b	2c	2d	2e	2f	2g	3a	3b	3c	3d	4a	4b	4c
7P	1a	2a	2b	2c	2d	2e	2f	2g	3a	3b	3c	3d	4a	4b	4c
11P	1a	2a	2b	2c	2d	2e	2f	2g	3a	3b	3c	3d	4a	4b	4c
13P	1a	2a	2b	2c	2d	2e	2f	2g	3a	3b	3c	3d	4a	4b	4c
X.85	577368	577368	−21384	−5832	216	216	216	−72	729	.	.	.	216	216	56
X.86	577368	577368	−21384	5832	216	216	−216	−72	729	.	.	.	−216	−216	−56
X.87	579150	579150	−6930	−5850	1230	1230	−90	174	−405	324	81	.	−210	102	−90
X.88	579150	579150	−6930	5850	1230	1230	90	174	−405	324	81	.	210	−102	90
X.89	600600	600600	21560	−7000	920	920	−120	−136	525	−96	12	−42	−280	−248	40
X.90	600600	600600	21560	7000	920	920	120	−136	525	−96	12	−42	280	248	−40
X.91	640640	−640640	.	.	−1152	1152	.	.	812	−232	128	−16	.	.	.
X.92	675675	675675	10395	4725	−165	−165	405	411	.	135	−54	54	525	21	−75
X.93	675675	675675	10395	−4725	−165	−165	−405	411	.	135	−54	54	−525	−21	75
X.94	720720	720720	−33264	−1680	1104	1104	−336	−112	1386	−18	36	−18	336	48	−16
X.95	720720	720720	−33264	1680	1104	1104	336	−112	1386	−18	36	−18	−336	−48	16
X.96	800800	−800800	.	.	1120	−1120	.	.	3220	−128	−74	34	.	.	.
X.97	800800	800800	12320	.	2080	2080	.	32	−560	−614	−20	34	.	.	.
X.98	800800	800800	−12320	−5600	800	800	−160	−416	−245	520	16	34	280	64	.
X.99	800800	800800	−12320	5600	800	800	160	−416	−245	520	16	34	−280	−64	.
X.100	800800	800800	−36960	.	−480	−480	.	160	1960	196	−56	−20	.	.	.
X.101	852930	852930	−29646	2430	1026	1026	−594	18	729	.	.	.	594	−162	46
X.102	852930	852930	−29646	−2430	1026	1026	594	18	729	.	.	.	−594	162	−46
X.103	873600	−873600	.	.	−640	640	.	.	1680	420	−84	−66	.	.	.
X.104	938223	938223	−2673	9477	1647	1647	−27	207	−729	.	.	.	−27	−27	5
X.105	938223	938223	−2673	−9477	1647	1647	27	207	−729	.	.	.	27	27	−5
X.106	960960	−960960	.	.	3392	−3392	.	.	−672	−348	138	30	.	.	.
X.107	972972	972972	−24948	9828	1836	1836	324	−180	.	−243	.	.	−252	−252	36
X.108	972972	972972	−24948	−9828	1836	1836	−324	−180	.	−243	.	.	252	252	−36
X.109	1029600	−1029600	.	.	1440	−1440	.	.	−720	252	90	9	.	.	.
X.110	1029600	−1029600	.	.	1440	−1440	.	.	−720	252	90	9	.	.	.
X.111	1164800	1164800	−17920	.	−2560	−2560	.	512	560	344	128	20	.	.	.
X.112	1201200	1201200	−30800	2800	560	560	240	176	420	−30	−48	51	.	−80	−80
X.113	1201200	1201200	18480	.	−2000	−2000	.	560	420	−30	114	−30	.	.	.
X.114	1201200	1201200	−30800	−2800	560	560	−240	176	420	−30	−48	51	.	80	80
X.115	1360800	1360800	30240	.	1440	1440	.	288	.	486	.	.	.	.	.
X.116	1360800	1360800	30240	.	1440	1440	.	288	.	486	.	.	.	.	.
X.117	1372800	1372800	49280	−6400	640	640	.	128	1560	−312	−24	12	.	−256	.
X.118	1372800	−1372800	.	.	1920	−1920	.	.	1920	12	156	−42	.	.	.
X.119	1372800	1372800	49280	6400	640	640	.	128	1560	−312	−24	12	.	256	.
X.120	1441792	1441792	.	−8192	.	.	.	.	−512	640	64	−8	.	.	.
X.121	1441792	1441792	.	8192	.	.	.	.	−512	640	64	−8	.	.	.
X.122	1441792	−1441792	.	.	.	.	.	.	−512	640	64	−8	.	.	.
X.123	1441792	−1441792	.	.	.	.	.	.	−512	640	64	−8	.	.	.
X.124	1791153	1791153	−5103	5103	−2511	−2511	−81	81	.	.	.	.	567	−81	−81
X.125	1791153	1791153	−5103	−5103	−2511	−2511	81	81	.	.	.	.	−567	81	81
X.126	1830400	−1830400	.	.	2560	−2560	.	.	880	1096	16	16	.	.	.
X.127	1876446	1876446	37422	−5346	−1890	−1890	270	−306	729	.	.	.	378	−162	110
X.128	1876446	1876446	37422	5346	−1890	−1890	−270	−306	729	.	.	.	−378	162	−110
X.129	1965600	1965600	30240	.	−480	−480	.	−480	.	−270	108	54	.	.	.
X.130	2027025	2027025	−24255	−7875	−1455	−1455	45	33	.	−324	81	.	105	93	45
X.131	2027025	2027025	−24255	7875	−1455	−1455	−45	33	.	−324	81	.	−105	−93	−45
X.132	2050048	2050048	.	10752	2048	2048	512	.	−1232	−224	−80	−8	.	.	.
X.133	2050048	2050048	.	−10752	2048	2048	−512	.	−1232	−224	−80	−8	.	.	.
X.134	2316600	2316600	−43560	−1800	−840	−840	−360	24	405	−162	.	.	120	24	120
X.135	2316600	2316600	−43560	1800	−840	−840	360	24	405	−162	.	.	−120	−24	−120
X.136	2402400	−2402400	.	.	3360	−3360	.	.	−420	264	−78	48	.	.	.
X.137	2402400	2402400	12320	5600	−160	−160	−480	160	−735	−384	48	−6	−280	64	.
X.138	2402400	2402400	12320	−5600	−160	−160	480	160	−735	−384	48	−6	280	−64	.
X.139	2555904	2555904	32768	4096	.	.	.	.	−384	192	−96	−24	512	.	.
X.140	2555904	2555904	−32768	−4096	.	.	.	.	−384	192	−96	−24	512	.	.
X.141	2555904	2555904	−32768	4096	.	.	.	.	−384	192	−96	−24	−512	.	.
X.142	2555904	2555904	32768	−4096	.	.	.	.	−384	192	−96	−24	−512	.	.
X.143	2594592	−2594592	.	.	2400	−2400	.	.	2268	−648	−162	.	.	.	.
X.144	2729376	2729376	7776	7776	−864	−864	−864	−288	−729	.	.	.	216	.	−64
X.145	2729376	2729376	7776	−7776	−864	−864	864	−288	−729	.	.	.	−216	.	64
X.146	3326400	−3326400	.	.	−4800	4800	.	.	.	216	108	54	.	.	.
X.147	4392960	−4392960	.	.	−2048	2048	.	.	−912	−480	−48	−48	.	.	.
X.148	4717440	−4717440	.	.	2688	−2688	.	.	−2268	−648	.	.	.	.	.
X.149	4804800	−4804800	.	.	−3520	3520	.	.	1680	−120	132	42	.	.	.
X.150	5111808	−5111808	.	.	.	.	.	.	−768	384	−192	−48	.	.	.

Character table of $A_1 = 2Aut(Fi_{22})$ *(continued)*

2	13	11	13	11	11	11	12	11	10	5	10	9	8	8	8	8	8	8	5	9
3	3	2	1	2	2	2	1	1	1	1	7	9	7	5	5	6	4	4	7	4
5	.	1	.	.	.	.	.	.	.	2	1	.	.	1	1	.	.	.	.	.
7	.	.	.	.	.	.	.	.	.	.	1	.	.	.	.	.	1	1	.	.
11	.	.	.	.	.	.	.	.	.	.	.	.	.	.	.	.	.	.	.	.
13	.	.	.	.	.	.	.	.	.	.	.	.	.	.	.	.	.	.	.	.
	4d	4e	4f	4g	4h	4i	4j	4k	4l	5a	6_1	6_2	6_3	6_4	6_5	6_6	6_7	6_8	6_9	6_{10}
2P	2e	2d	2e	2d	2g	2g	2e	2d	2g	5a	3a	3b	3c	3a	3a	3b	3a	3a	3d	3b
3P	4d	4e	4f	4g	4h	4i	4j	4k	4l	5a	2a	2a	2a	2b	2c	2b	2c	2c	2a	2e
5P	4d	4e	4f	4g	4h	4i	4j	4k	4l	1a	6_1	6_2	6_3	6_4	6_5	6_6	6_8	6_7	6_9	6_{10}
7P	4d	4e	4f	4g	4h	4i	4j	4k	4l	5a	6_1	6_2	6_3	6_4	6_5	6_6	6_7	6_8	6_9	6_{10}
11P	4d	4e	4f	4g	4h	4i	4j	4k	4l	5a	6_1	6_2	6_3	6_4	6_5	6_6	6_8	6_7	6_9	6_{10}
13P	4d	4e	4f	4g	4h	4i	4j	4k	4l	5a	6_1	6_2	6_3	6_4	6_5	6_6	6_7	6_8	6_9	6_{10}
X.85	72	−24	24	−24	.	.	−24	−8	.	−7	729	.	.	81	−81	.	27	27	.	.
X.86	72	−24	24	24	.	.	24	−8	.	−7	729	.	.	81	81	.	−27	−27	.	.
X.87	−18	−10	30	14	6	6	−10	−10	6	.	−405	324	81	−45	−45	36	27	27	.	−12
X.88	−18	−10	30	−14	6	6	10	−10	−6	.	−405	324	81	−45	45	36	−27	−27	.	−12
X.89	72	40	24	24	.	.	−8	−8	.	.	525	−96	12	5	65	−40	−7	−7	−42	−16
X.90	72	40	24	−24	.	.	8	−8	.	.	525	−96	12	5	−65	−40	7	7	−42	−16
X.91	.	.	.	.	.	.	.	.	.	−10	−812	232	−128	.	.	.	.	.	16	72
X.92	75	−45	11	−3	−5	−5	−11	3	−3	.	.	135	−54	.	.	27	.	.	54	15
X.93	75	−45	11	3	−5	−5	11	3	3	.	.	135	−54	.	.	27	.	.	54	15
X.94	−80	−16	16	−16	.	.	16	16	.	−5	1386	−18	36	−54	66	−54	−42	−42	−18	6
X.95	−80	−16	16	16	.	.	−16	16	.	−5	1386	−18	36	−54	−66	−54	42	42	−18	6
X.96	.	.	.	.	8	−8	.	.	.	.	−3220	128	74	.	.	.	.	.	−34	32
X.97	−96	.	32	.	32	32	.	.	.	.	−560	−614	−20	80	.	−46	.	.	34	10
X.98	.	40	.	24	.	.	.	8	.	.	−245	520	16	−35	25	−8	7	7	34	8
X.99	.	40	.	−24	.	.	.	8	.	.	−245	520	16	−35	−25	−8	−7	−7	34	8
X.100	32	.	−32	.	.	.	.	.	.	.	1960	196	−56	120	.	−132	.	.	−20	−12
X.101	−54	6	18	−6	−18	−18	6	−2	−6	5	729	.	.	−81	−81	.	−27	−27	.	.
X.102	−54	6	18	6	−18	−18	−6	−2	6	5	729	.	.	−81	81	.	27	27	.	.
X.103	.	.	.	.	−32	32	.	.	.	.	−1680	−420	84	.	.	.	.	.	66	28
X.104	63	15	15	21	−9	−9	21	−1	−3	−2	−729	.	.	−81	−81	.	27	27	.	.
X.105	63	15	15	−21	−9	−9	−21	−1	3	−2	−729	.	.	−81	81	.	−27	−27	.	.
X.106	.	.	.	.	16	−16	.	.	.	10	672	348	−138	.	.	.	.	.	−30	28
X.107	60	−36	−20	−12	−12	−12	−12	12	12	−3	.	−243	.	.	.	81	.	.	.	−27
X.108	60	−36	−20	12	−12	−12	12	12	−12	−3	.	−243	.	.	.	81	.	.	.	−27
X.109	.	.	.	.	−8	8	.	.	.	.	720	−252	−90	.	.	.	$\bar{J}$	J	−9	36
X.110	.	.	.	.	−8	8	.	.	.	.	720	−252	−90	.	.	.	J	$\bar{J}$	−9	36
X.111	.	.	.	.	.	.	.	.	.	.	560	344	128	80	.	8	.	.	20	−40
X.112	−16	.	−16	.	16	16	−16	.	−16	.	420	−30	−48	−20	100	34	28	28	51	2
X.113	−16	−80	−16	.	−16	−16	.	−16	.	.	420	−30	114	−60	.	66	.	.	−30	34
X.114	−16	.	−16	.	16	16	16	.	16	.	420	−30	−48	−20	−100	34	−28	−28	51	2
X.115	−96	.	−32	.	.	.	.	.	.	.	.	486	.	.	.	−54	.	.	.	−18
X.116	−96	.	−32	.	.	.	.	.	.	.	.	486	.	.	.	−54	.	.	.	−18
X.117	.	.	.	.	.	.	.	.	.	.	1560	−312	−24	−40	80	32	8	8	12	−8
X.118	.	.	.	.	32	−32	.	.	.	.	−1920	−12	−156	.	.	.	.	.	42	−12
X.119	.	.	.	.	.	.	.	.	.	.	1560	−312	−24	−40	−80	32	−8	−8	12	−8
X.120	.	.	.	.	.	.	.	.	.	−8	−512	640	64	.	−128	.	64	64	−8	.
X.121	.	.	.	.	.	.	.	.	.	−8	−512	640	64	.	128	.	−64	−64	−8	.
X.122	.	.	.	.	.	.	.	.	.	−8	512	−640	−64	.	.	.	K	$\bar{K}$	8	.
X.123	.	.	.	.	.	.	.	.	.	−8	512	−640	−64	.	.	.	$\bar{K}$	K	8	.
X.124	81	9	33	−9	9	9	−33	9	15	3	.	.	.	.	.	.	.	.	.	.
X.125	81	9	33	9	9	9	33	9	−15	3	.	.	.	.	.	.	.	.	.	.
X.126	.	.	.	.	.	.	.	.	.	.	−880	−1096	−16	.	.	.	.	.	−16	−40
X.127	54	−30	−18	−30	18	18	6	10	−6	−4	729	.	.	−81	81	.	27	27	.	.
X.128	54	−30	−18	30	18	18	−6	10	6	−4	729	.	.	−81	−81	.	−27	−27	.	.
X.129	−96	.	32	.	−32	−32	.	.	.	.	.	−270	108	.	.	−54	.	.	54	−30
X.130	9	85	−15	17	−3	−3	−11	−3	9	.	.	−324	81	−90	90	−36	.	.	.	12
X.131	9	85	−15	−17	−3	−3	11	−3	−9	.	.	−324	81	−90	−90	−36	.	.	.	12
X.132	.	.	.	.	.	.	.	.	.	−2	−1232	−224	−80	.	−48	.	.	.	−8	32
X.133	.	.	.	.	.	.	.	.	.	−2	−1232	−224	−80	.	48	.	.	.	−8	32
X.134	−24	40	−8	8	.	.	−24	−8	.	.	405	−162	.	45	−45	18	27	27	.	6
X.135	−24	40	−8	−8	.	.	24	−8	.	.	405	−162	.	45	45	18	−27	−27	.	6
X.136	.	.	.	.	24	−24	.	.	.	.	420	−264	78	.	.	.	.	.	−48	24
X.137	−64	−40	.	−24	.	.	.	−8	.	.	−735	−384	48	35	−25	8	−7	−7	−6	−16
X.138	−64	−40	.	24	.	.	.	−8	.	.	−735	−384	48	35	25	8	7	7	−6	−16
X.139	.	.	.	.	.	.	.	.	.	4	−384	192	−96	−64	64	−64	64	64	−24	.
X.140	.	.	.	.	.	.	.	.	.	4	−384	192	−96	64	−64	64	−64	−64	−24	.
X.141	.	.	.	.	.	.	.	.	.	4	−384	192	−96	64	64	64	64	64	−24	.
X.142	.	.	.	.	.	.	.	.	.	4	−384	192	−96	−64	−64	−64	−64	−64	−24	.
X.143	.	.	.	.	−24	24	.	.	.	−8	−2268	648	162	.	.	.	.	.	.	−24
X.144	.	−24	.	24	.	.	.	8	.	1	−729	.	.	81	81	.	27	27	.	.
X.145	.	−24	.	−24	.	.	.	8	.	1	−729	.	.	81	−81	.	−27	−27	.	.
X.146	.	.	.	.	−16	16	.	.	.	.	.	−216	−108	.	.	.	.	.	−54	−24
X.147	.	.	.	.	.	.	.	.	.	10	912	480	48	.	.	.	.	.	48	32
X.148	.	.	.	.	.	.	.	.	.	−10	2268	648	.	.	.	.	.	.	.	−24
X.149	.	.	.	.	16	−16	.	.	.	.	−1680	120	−132	.	.	.	.	.	−42	−8
X.150	.	.	.	.	.	.	.	.	.	8	768	−384	192	.	.	.	.	.	48	.

Character table of $A_1 = 2Aut(\mathsf{Fi}_{22})$ (continued)

	6_{11}	6_{12}	6_{13}	6_{14}	6_{15}	6_{16}	6_{17}	6_{18}	6_{19}	6_{20}	6_{21}	6_{22}	6_{23}	6_{24}	6_{25}	6_{26}	6_{27}	6_{28}	6_{29}	6_{30}	6_{31}	$7a$	$8a$
2	9	10	10	5	5	5	8	8	8	8	8	7	7	5	5	5	5	5	5	5	5	3	8
3	4	3	3	6	5	5	3	3	3	3	3	3	3	4	4	4	4	3	3	3	3	1	1
5	.	.	.	.	.	.	.	.	.	.	.	.	.	.	.	.	.	.	.	.	.	.	.
7	.	.	.	.	.	.	.	.	.	.	.	.	.	.	.	.	.	.	.	.	.	1	.
11	.	.	.	.	.	.	.	.	.	.	.	.	.	.	.	.	.	.	.	.	.	.	.
13	.	.	.	.	.	.	.	.	.	.	.	.	.	.	.	.	.	.	.	.	.	.	.
2P	3b	3a	3a	3b	3c	3c	3c	3c	3b	3a	3a	3c	3c	3d	3d	3d	3d	3d	3d	3c	3c	7a	4f
3P	2d	2d	2e	2c	2c	2b	2g	2g	2g	2f	2g	2e	2d	2f	2f	2c	2c	2g	2g	2g	2f	7a	8a
5P	6_{11}	6_{12}	6_{13}	6_{14}	6_{15}	6_{16}	6_{17}	6_{18}	6_{19}	6_{20}	6_{21}	6_{22}	6_{23}	6_{25}	6_{24}	6_{27}	6_{26}	6_{28}	6_{29}	6_{30}	6_{31}	7a	8a
7P	6_{11}	6_{12}	6_{13}	6_{14}	6_{15}	6_{16}	6_{17}	6_{18}	6_{19}	6_{20}	6_{21}	6_{22}	6_{23}	6_{24}	6_{25}	6_{26}	6_{27}	6_{28}	6_{29}	6_{30}	6_{31}	1a	8a
11P	6_{11}	6_{12}	6_{13}	6_{14}	6_{15}	6_{16}	6_{17}	6_{18}	6_{19}	6_{20}	6_{21}	6_{22}	6_{23}	6_{25}	6_{24}	6_{27}	6_{26}	6_{28}	6_{29}	6_{30}	6_{31}	7a	8a
13P	6_{11}	6_{12}	6_{13}	6_{14}	6_{15}	6_{16}	6_{17}	6_{18}	6_{19}	6_{20}	6_{21}	6_{22}	6_{23}	6_{24}	6_{25}	6_{26}	6_{27}	6_{28}	6_{29}	6_{30}	6_{31}	7a	8a
X.85	.	9	9	.	.	.	.	.	.	−9	9	.	.	.	.	.	.	.	.	.	.	1	.
X.86	.	9	9	.	.	.	.	.	.	9	9	.	.	.	.	.	.	.	.	.	.	1	.
X.87	−12	−21	−21	−18	9	9	9	9	−12	3	3	−3	−3	.	.	.	.	.	.	−3	−3	−2	−6
X.88	−12	−21	−21	18	−9	9	9	9	−12	−3	3	−3	−3	.	.	.	.	.	.	−3	3	−2	6
X.89	−16	−19	−19	20	−16	−22	−4	−4	8	−15	5	−4	−4	6	6	2	2	2	2	2	.	.	.
X.90	−16	−19	−19	−20	16	−22	−4	−4	8	15	5	−4	−4	−6	−6	−2	−2	2	2	2	.	.	.
X.91	−72	36	−36	.	.	.	.	.	.	.	.	.	.	.	.	.	.	.	.	.	.	.	.
X.92	15	.	.	27	.	.	−6	−6	3	.	.	−6	−6	.	.	.	.	6	6	.	.	.	1
X.93	15	.	.	−27	.	.	−6	−6	3	.	.	−6	−6	.	.	.	.	6	6	.	.	.	−1
X.94	6	−6	−6	−6	12	.	−4	−4	2	−6	2	.	.	6	6	−6	−6	2	2	−4	.	.	.
X.95	6	−6	−6	6	−12	.	−4	−4	2	6	2	.	.	−6	−6	6	6	2	2	−4	.	.	.
X.96	−32	28	−28	.	.	.	6	−6	.	.	.	2	−2	.	.	.	.	−6	6	.	.	.	.
X.97	10	16	16	.	.	−28	−4	−4	2	.	−16	4	4	.	.	.	.	2	2	−4	.	.	.
X.98	8	11	11	16	−2	−8	−8	−8	−8	−7	13	−4	−4	2	2	−2	−2	−2	−2	4	2	.	.
X.99	8	11	11	−16	2	−8	−8	−8	−8	7	13	−4	−4	−2	−2	2	2	−2	−2	4	−2	.	.
X.100	−12	−24	−24	.	.	12	−8	−8	−20	.	−8	.	.	.	.	.	.	4	4	4	.	.	.
X.101	.	9	9	.	.	.	.	.	.	−9	−9	.	.	.	.	.	.	.	.	.	.	1	−6
X.102	.	9	9	.	.	.	.	.	.	9	−9	.	.	.	.	.	.	.	.	.	.	1	6
X.103	−28	−16	16	.	.	.	12	−12	.	.	.	4	−4	.	.	.	.	6	−6	.	.	.	.
X.104	.	−9	−9	.	.	.	.	.	.	−9	−9	.	.	.	.	.	.	.	.	.	.	−1	−3
X.105	.	−9	−9	.	.	.	.	.	.	9	−9	.	.	.	.	.	.	.	.	.	.	−1	3
X.106	−28	32	−32	.	.	.	−6	6	.	.	.	−2	2	.	.	.	.	6	−6	.	.	.	.
X.107	−27	.	.	27	.	.	.	.	9	.	.	.	.	.	.	.	.	.	.	.	.	.	−4
X.108	−27	.	.	−27	.	.	.	.	9	.	.	.	.	.	.	.	.	.	.	.	.	.	4
X.109	−36	−48	48	.	.	.	−6	6	.	.	.	6	−6	F	$\bar{F}$	$\bar{B}$	B	−3	3	.	.	−2	.
X.110	−36	−48	48	.	.	.	−6	6	.	.	.	6	−6	$\bar{F}$	F	B	$\bar{B}$	−3	3	.	.	−2	.
X.111	−40	−16	−16	.	.	−28	−16	−16	8	.	−16	8	8	.	.	.	.	−4	−4	−4	.	.	.
X.112	2	−28	−28	−8	10	−2	8	8	2	−12	−4	−4	−4	−3	−3	1	1	−1	−1	2	6	.	.
X.113	34	4	4	.	.	−6	2	2	2	.	20	10	10	.	.	.	.	2	2	2	.	.	.
X.114	2	−28	−28	8	−10	−2	8	8	2	12	−4	−4	−4	3	3	−1	−1	−1	−1	2	−6	.	.
X.115	−18	.	.	.	.	.	.	.	18	.	.	.	.	.	.	.	.	.	.	.	.	.	.
X.116	−18	.	.	.	.	.	.	.	18	.	.	.	.	.	.	.	.	.	.	.	.	.	.
X.117	−8	−8	−8	−28	8	−4	8	8	−16	.	8	−8	−8	.	.	8	8	−4	−4	−4	.	2	.
X.118	12	.	.	.	.	.	12	−12	.	.	.	12	−12	.	.	.	.	6	−6	.	.	2	.
X.119	−8	−8	−8	28	−8	−4	8	8	−16	.	8	−8	−8	.	.	−8	−8	−4	−4	−4	.	2	.
X.120	.	.	.	16	16	.	.	.	.	.	.	.	.	.	.	−8	−8	.	.	.	.	2	.
X.121	.	.	.	−16	−16	.	.	.	.	.	.	.	.	.	.	8	8	.	.	.	.	2	.
X.122	.	.	.	.	.	.	.	.	.	.	.	.	.	.	.	A	$\bar{A}$	.	.	.	.	2	.
X.123	.	.	.	.	.	.	.	.	.	.	.	.	.	.	.	$\bar{A}$	A	.	.	.	.	2	.
X.124	.	.	.	.	.	.	.	.	.	.	.	.	.	.	.	.	.	.	.	.	.	.	3
X.125	.	.	.	.	.	.	.	.	.	.	.	.	.	.	.	.	.	.	.	.	.	.	−3
X.126	40	16	−16	.	.	.	.	.	.	.	.	8	−8	.	.	.	.	.	.	.	.	−2	.
X.127	.	9	9	.	.	.	.	.	.	9	−9	.	.	.	.	.	.	.	.	.	.	−2	−6
X.128	.	9	9	.	.	.	.	.	.	−9	−9	.	.	.	.	.	.	.	.	.	.	−2	6
X.129	−30	.	.	.	.	.	12	12	−6	.	.	12	12	.	.	.	.	6	6	.	.	.	.
X.130	12	.	.	−18	9	−9	9	9	12	−6	6	−3	−3	.	.	.	.	.	.	3	−3	.	−3
X.131	12	.	.	18	−9	−9	9	9	12	6	6	−3	−3	.	.	.	.	.	.	3	3	.	3
X.132	32	−16	−16	−48	−12	.	.	.	.	−16	.	8	8	8	8	.	.	.	.	.	−4	.	.
X.133	32	−16	−16	48	12	.	.	.	.	16	.	8	8	−8	−8	.	.	.	.	.	4	.	.
X.134	6	21	21	−18	−18	18	.	.	−6	3	−3	.	.	.	.	.	.	.	.	−6	6	−1	.
X.135	6	21	21	18	18	18	.	.	−6	−3	−3	.	.	.	.	.	.	.	.	−6	−6	−1	.
X.136	−24	−12	12	.	.	.	18	−18	.	.	.	−18	18	.	.	.	.	.	.	.	.	.	.
X.137	−16	17	17	−16	2	8	−8	−8	−8	15	−5	−4	−4	6	6	2	2	−2	−2	4	6	.	.
X.138	−16	17	17	16	−2	8	−8	−8	−8	−15	−5	−4	−4	−6	−6	−2	−2	−2	−2	4	−6	.	.
X.139	.	.	.	−8	−8	8	.	.	.	.	.	.	.	.	.	−8	−8	.	.	.	.	1	.
X.140	.	.	.	8	8	−8	.	.	.	.	.	.	.	.	.	8	8	.	.	.	.	1	.
X.141	.	.	.	−8	−8	−8	.	.	.	.	.	.	.	.	.	−8	−8	.	.	.	.	1	.
X.142	.	.	.	8	8	8	.	.	.	.	.	.	.	.	.	8	8	.	.	.	.	1	.
X.143	24	−12	12	.	.	.	−18	18	.	.	.	−6	6	.	.	.	.	.	.	.	.	.	.
X.144	.	−9	−9	.	.	.	.	.	.	9	9	.	.	.	.	.	.	.	.	.	.	−1	.
X.145	.	−9	−9	.	.	.	.	.	.	−9	9	.	.	.	.	.	.	.	.	.	.	−1	.
X.146	24	.	.	.	.	.	12	−12	.	.	.	−12	12	.	.	.	.	6	−6	.	.	.	.
X.147	−32	16	−16	.	.	.	.	.	.	.	.	8	−8	.	.	.	.	.	.	.	.	−2	.
X.148	24	12	−12	.	.	.	.	.	.	.	.	.	.	.	.	.	.	.	.	.	.	.	.
X.149	8	−16	16	.	.	.	−12	12	.	.	.	4	−4	.	.	.	.	−6	6	.	.	.	.
X.150	.	.	.	.	.	.	.	.	.	.	.	.	.	.	.	.	.	.	.	.	.	2	.

Character table of $A_1 = 2Aut(\mathsf{Fi}_{22})$ *(continued)*

2	8	8	8	8	8	6	6	4	3	2	5	4	5	5	3	4	2	6	8	5	5	8	8	6	7
3	1	1	1	1	.	1	.	4	4	3	1	1	.	.	1	.	.	3	3	4	4	2	2	3	2
5	.	.	.	.	.	.	.	.	.	.	2	2	1	1	1	1	.	1	.	.	.	.	.	.	.
7	.	.	.	.	.	.	.	.	.	.	.	.	.	.	.	.	.	.	.	.	.	.	.	.	.
11	.	.	.	.	.	.	.	.	.	.	.	.	.	.	.	.	1	.	.	.	.	.	.	.	.
13	.	.	.	.	.	.	.	.	.	.	.	.	.	.	.	.	.	.	.	.	.	.	.	.	.
	8b	8c	8d	8e	8f	8g	8h	9a	9b	9c	10a	10b	10c	10d	10e	10f	11a	12_1	12_2	12_3	12_4	12_5	12_6	12_7	12_8
2P	4d	4d	4d	4d	4f	4h	4i	9a	9b	9c	5a	5a	5a	5a	5a	5a	11a	6_1	6_{13}	6_{10}	6_2	6_{13}	6_{13}	6_{10}	6_{13}
3P	8b	8c	8d	8e	8f	8g	8h	3b	3b	3d	10a	10b	10c	10d	10e	10f	11a	4a	4b	4b	4a	4d	4d	4d	4d
5P	8b	8c	8d	8e	8f	8g	8h	9a	9b	9c	2a	2c	2d	2e	2b	2f	11a	12_1	12_2	12_3	12_4	12_5	12_6	12_7	12_8
7P	8b	8c	8d	8e	8f	8g	8h	9a	9b	9c	10a	10b	10c	10d	10e	10f	11a	12_1	12_2	12_3	12_4	12_5	12_6	12_7	12_8
11P	8b	8c	8d	8e	8f	8g	8h	9a	9b	9c	10a	10b	10c	10d	10e	10f	1a	12_1	12_2	12_3	12_4	12_5	12_6	12_7	12_8
13P	8b	8c	8d	8e	8f	8g	8h	9a	9b	9c	10a	10b	10c	10d	10e	10f	11a	12_1	12_2	12_3	12_4	12_5	12_6	12_7	12_8
X.85	−4	4	−4	4	.	.	.	.	.	.	−7	−7	1	1	1	1	.	−9	−9	.	.	−3	−3	.	−3
X.86	−4	−4	−4	−4	.	.	.	.	.	.	−7	7	1	1	1	−1	.	9	9	.	.	−3	−3	.	−3
X.87	2	2	2	2	2	−2	2	.	.	.	.	.	.	.	.	.	.	15	3	−6	6	3	3	.	3
X.88	2	−2	2	−2	2	2	2	.	.	.	.	.	.	.	.	.	.	−15	−3	6	−6	3	3	.	3
X.89	4	4	4	4	.	.	.	3	3	.	.	.	.	.	.	.	.	5	1	4	8	−3	−3	.	−3
X.90	4	−4	4	−4	.	.	.	3	3	.	.	.	.	.	.	.	.	−5	−1	−4	−8	−3	−3	.	−3
X.91	.	.	.	.	.	.	.	8	−4	2	10	.	−2	2	.	.	.	.	.	.	.	−12	12	.	.
X.92	3	−3	3	−3	−1	−3	−1	.	.	.	.	.	.	.	.	.	.	.	.	3	3	.	.	3	.
X.93	3	3	3	3	−1	3	−1	.	.	.	.	.	.	.	.	.	.	.	.	−3	−3	.	.	3	.
X.94	.	.	.	.	.	.	.	.	.	.	−5	−5	−1	−1	1	−1	.	6	−6	−6	−6	−2	−2	10	−2
X.95	.	.	.	.	.	.	.	.	.	.	−5	5	−1	−1	1	1	.	−6	6	6	6	−2	−2	10	−2
X.96	.	.	.	.	.	.	.	−2	−2	−2	.	.	.	.	.	.	.	.	.	.	.	12	−12	.	.
X.97	.	.	.	.	.	.	.	−2	−2	−2	.	.	.	.	.	.	.	.	.	.	.	.	.	−6	.
X.98	4	−4	−4	4	.	.	.	−5	1	1	.	.	.	.	.	.	.	−5	7	−8	−8	3	3	.	−3
X.99	4	4	−4	−4	.	.	.	−5	1	1	.	.	.	.	.	.	.	5	−7	8	8	3	3	.	−3
X.100	.	.	.	.	.	.	.	−8	4	−2	.	.	.	.	.	.	.	.	.	.	.	8	8	−4	8
X.101	−2	2	2	−2	−2	.	.	.	.	.	5	5	1	1	−1	1	1	9	9	.	.	−3	−3	.	3
X.102	−2	−2	2	2	−2	.	.	.	.	.	5	−5	1	1	−1	−1	1	−9	−9	.	.	−3	−3	.	3
X.103	.	.	.	.	.	.	.	−6	.	.	.	.	.	.	.	.	2	.	.	.	.	.	.	.	.
X.104	−5	1	−5	1	−1	−3	−1	.	.	.	−2	2	2	2	2	−2	.	−9	−9	.	.	3	3	.	3
X.105	−5	−1	−5	−1	−1	3	−1	.	.	.	−2	−2	2	2	2	2	.	9	9	.	.	3	3	.	3
X.106	.	.	.	.	.	.	.	−6	−6	.	−10	.	2	−2	.	.	.	.	.	.	.	.	.	.	.
X.107	.	.	.	.	4	.	.	.	.	.	−3	3	1	1	−3	−1	.	.	.	−9	−9	.	.	−3	.
X.108	.	.	.	.	4	.	.	.	.	.	−3	−3	1	1	−3	1	.	.	.	9	9	.	.	−3	.
X.109	.	.	.	.	.	.	.	.	.	.	.	.	.	.	.	.	.	.	.	.	.	.	.	.	.
X.110	.	.	.	.	.	.	.	.	.	.	.	.	.	.	.	.	.	.	.	.	.	.	.	.	.
X.111	.	.	.	.	.	.	.	2	2	2	.	.	.	.	.	.	−1	.	.	.	.	.	.	.	.
X.112	.	.	.	.	.	.	.	3	.	.	.	.	.	.	.	.	.	.	4	−8	.	−4	−4	2	−4
X.113	.	.	.	.	.	.	.	−6	.	.	.	.	.	.	.	.	.	.	.	.	.	−4	−4	2	−4
X.114	.	.	.	.	.	.	.	3	.	.	.	.	.	.	.	.	.	.	−4	8	.	−4	−4	2	−4
X.115	.	.	.	.	.	.	.	.	.	.	.	.	.	.	.	.	1	.	.	.	.	.	.	−6	.
X.116	.	.	.	.	.	.	.	.	.	.	.	.	.	.	.	.	1	.	.	.	.	.	.	−6	.
X.117	.	.	.	.	.	.	.	−3	.	.	.	.	.	.	.	.	.	.	−16	−4	.	.	.	.	.
X.118	.	.	.	.	.	.	.	.	6	.	.	.	.	.	.	.	.	.	.	.	.	.	.	.	.
X.119	.	.	.	.	.	.	.	−3	.	.	.	.	.	.	.	.	.	.	16	4	.	.	.	.	.
X.120	.	.	.	.	.	.	.	4	−2	−2	−8	8	.	.	.	.	.	.	.	.	.	.	.	.	.
X.121	.	.	.	.	.	.	.	4	−2	−2	−8	−8	.	.	.	.	.	.	.	.	.	.	.	.	.
X.122	.	.	.	.	.	.	.	4	−2	−2	8	.	.	.	.	.	.	.	.	.	.	.	.	.	.
X.123	.	.	.	.	.	.	.	4	−2	−2	8	.	.	.	.	.	.	.	.	.	.	.	.	.	.
X.124	−3	3	−3	3	−3	3	1	.	.	.	3	3	−1	−1	−3	−1	1	.	.	.	.	.	.	.	.
X.125	−3	−3	−3	−3	−3	−3	1	.	.	.	3	−3	−1	−1	−3	1	1	.	.	.	.	.	.	.	.
X.126	.	.	.	.	.	.	.	−2	−2	4	.	.	.	.	.	.	.	.	.	.	.	.	.	.	.
X.127	−2	−2	2	2	2	.	.	.	.	.	−4	4	.	.	2	.	.	−9	−9	.	.	−3	−3	.	3
X.128	−2	2	2	−2	2	.	.	.	.	.	−4	−4	.	.	2	.	.	9	9	.	.	−3	−3	.	3
X.129	.	.	.	.	.	.	.	.	.	.	.	.	.	.	.	.	−1	.	.	.	.	.	.	−6	.
X.130	1	−3	−3	−7	1	−1	−1	.	.	.	.	.	.	.	.	.	.	.	.	−6	6	.	.	.	6
X.131	1	3	−3	7	1	1	−1	.	.	.	.	.	.	.	.	.	.	.	.	6	−6	.	.	.	6
X.132	.	.	.	.	.	.	.	−2	4	1	−2	2	−2	−2	.	2	.	.	.	.	.	.	.	.	.
X.133	.	.	.	.	.	.	.	−2	4	1	−2	−2	−2	−2	.	−2	.	.	.	.	.	.	.	.	.
X.134	4	−4	4	−4	.	.	.	.	.	.	.	.	.	.	.	.	.	15	3	6	−6	−3	−3	−6	−3
X.135	4	4	4	4	.	.	.	.	.	.	.	.	.	.	.	.	.	−15	−3	−6	6	−3	−3	−6	−3
X.136	.	.	.	.	.	.	.	.	6	.	.	.	.	.	.	.	.	.	.	.	.	−12	12	.	.
X.137	4	−4	−4	4	.	.	.	3	3	.	.	.	.	.	.	.	.	5	−11	−8	8	5	5	8	−1
X.138	4	4	−4	−4	.	.	.	3	3	.	.	.	.	.	.	.	.	−5	11	8	−8	5	5	8	−1
X.139	.	.	.	.	.	.	.	.	−3	.	4	−4	.	.	−2	.	−1	−16	.	.	8	.	.	.	.
X.140	.	.	.	.	.	.	.	.	−3	.	4	4	.	.	2	.	−1	−16	.	.	8	.	.	.	.
X.141	.	.	.	.	.	.	.	.	−3	.	4	−4	.	.	2	.	−1	16	.	.	−8	.	.	.	.
X.142	.	.	.	.	.	.	.	.	−3	.	4	4	.	.	−2	.	−1	16	.	.	−8	.	.	.	.
X.143	.	.	.	.	.	.	.	.	.	.	8	.	.	.	.	.	.	.	.	.	.	−12	12	.	.
X.144	−4	4	4	−4	.	.	.	.	.	.	1	1	1	1	1	1	1	−9	−9	.	.	3	3	.	−3
X.145	−4	−4	4	4	.	.	.	.	.	.	1	−1	1	1	1	−1	1	9	9	.	.	3	3	.	−3
X.146	.	.	.	.	.	.	.	.	.	.	.	.	.	.	.	.	.	.	.	.	.	.	.	.	.
X.147	.	.	.	.	.	.	.	6	6	.	−10	.	2	−2	.	.	.	.	.	.	.	.	.	.	.
X.148	.	.	.	.	.	.	.	.	.	.	10	.	−2	2	.	.	2	.	.	.	.	12	−12	.	.
X.149	.	.	.	.	.	.	.	−6	.	.	.	.	.	.	.	.	.	.	.	.	.	.	.	.	.
X.150	.	.	.	.	.	.	.	.	−6	.	−8	.	.	.	.	.	−2	.	.	.	.	.	.	.	.

Character table of $A_1 = 2Aut(\mathsf{Fi}_{22})$ (continued)

	12_{9}	12_{10}	12_{11}	12_{12}	12_{13}	12_{14}	12_{15}	12_{16}	12_{17}	12_{18}	12_{19}	12_{20}	12_{21}	12_{22}	12_{23}	12_{24}	12_{25}	12_{26}	12_{27}	12_{28}
2	7	5	8	6	6	6	6	4	5	5	5	6	6	4	4	4	4	5	4	4
3	2	3	1	2	2	2	2	3	2	2	2	1	1	2	2	2	2	1	1	1
5	.	.	.	.	.	.	.	.	.	.	.	.	.	.	.	.	.	.	.	.
7	.	.	.	.	.	.	.	.	.	.	.	.	.	.	.	.	.	.	.	.
11	.	.	.	.	.	.	.	.	.	.	.	.	.	.	.	.	.	.	.	.
13	.	.	.	.	.	.	.	.	.	.	.	.	.	.	.	.	.	.	.	.
2P	6_{13}	6_{22}	6_{13}	6_{12}	6_{17}	6_{12}	6_{17}	6_{3}	6_{19}	6_{19}	6_{11}	6_{10}	6_{12}	6_{23}	6_{23}	6_{28}	6_{28}	6_{22}	6_{28}	6_{28}
3P	$4b$	$4b$	$4c$	$4g$	$4h$	$4e$	$4i$	$4a$	$4i$	$4h$	$4g$	$4f$	$4k$	$4g$	$4e$	$4i$	$4h$	$4j$	$4l$	$4l$
5P	12_{9}	12_{10}	12_{11}	12_{12}	12_{13}	12_{14}	12_{15}	12_{16}	12_{17}	12_{18}	12_{19}	12_{20}	12_{21}	12_{22}	12_{23}	12_{24}	12_{25}	12_{26}	12_{28}	12_{27}
7P	12_{9}	12_{10}	12_{11}	12_{12}	12_{13}	12_{14}	12_{15}	12_{16}	12_{17}	12_{18}	12_{19}	12_{20}	12_{21}	12_{22}	12_{23}	12_{24}	12_{25}	12_{26}	12_{27}	12_{28}
11P	12_{9}	12_{10}	12_{11}	12_{12}	12_{13}	12_{14}	12_{15}	12_{16}	12_{17}	12_{18}	12_{19}	12_{20}	12_{21}	12_{22}	12_{23}	12_{24}	12_{25}	12_{26}	12_{28}	12_{27}
13P	12_{9}	12_{10}	12_{11}	12_{12}	12_{13}	12_{14}	12_{15}	12_{16}	12_{17}	12_{18}	12_{19}	12_{20}	12_{21}	12_{22}	12_{23}	12_{24}	12_{25}	12_{26}	12_{27}	12_{28}
X.85	3	.	−1	3	.	−3	.	.	.	.	.	.	1	.	.	.	.	.	.	.
X.86	−3	.	1	−3	.	−3	.	.	.	.	.	.	1	.	.	.	.	.	.	.
X.87	3	3	3	−1	−3	−1	−3	−3	.	.	2	.	−1	−1	−1	.	.	−1	.	.
X.88	−3	−3	−3	1	−3	−1	−3	3	.	.	−2	.	−1	1	−1	.	.	1	.	.
X.89	1	−2	1	−3	.	1	.	2	.	.	.	.	1	.	−2	.	.	−2	.	.
X.90	−1	2	−1	3	.	1	.	−2	.	.	.	.	1	.	−2	.	.	2	.	.
X.91	.	.	.	.	.	.	.	.	.	.	.	.	.	.	.	.	.	.	.	.
X.92	.	6	.	.	−2	.	−2	6	1	1	3	−1	.	.	.	−2	−2	−2	.	.
X.93	.	−6	.	.	−2	.	−2	−6	1	1	−3	−1	.	.	.	−2	−2	2	.	.
X.94	6	−6	2	2	.	2	.	.	.	.	2	−2	−2	2	2	.	.	−2	.	.
X.95	−6	6	−2	−2	.	2	.	.	.	.	−2	−2	−2	−2	2	.	.	2	.	.
X.96	.	.	.	.	2	.	−2	.	4	−4	.	.	.	.	.	−2	2	.	.	.
X.97	.	.	.	.	−4	.	−4	.	2	2	.	2	.	.	.	2	2	.	.	.
X.98	1	4	−9	−3	.	1	.	−2	.	.	.	.	−1	.	−2	.	.	.	.	.
X.99	−1	−4	9	3	.	1	.	2	.	.	.	.	−1	.	−2	.	.	.	.	.
X.100	.	.	.	.	.	.	.	.	.	.	.	4	.	.	.	.	.	.	.	.
X.101	3	.	1	3	.	3	.	.	.	.	.	.	1	.	.	.	.	.	.	.
X.102	−3	.	−1	−3	.	3	.	.	.	.	.	.	1	.	.	.	.	.	.	.
X.103	.	.	.	.	4	.	−4	.	2	−2	.	.	.	.	.	2	−2	.	.	.
X.104	3	.	−1	3	.	3	.	.	.	.	.	.	−1	.	.	.	.	.	.	.
X.105	−3	.	1	−3	.	3	.	.	.	.	.	.	−1	.	.	.	.	.	.	.
X.106	.	.	.	.	−2	.	2	.	2	−2	.	.	.	.	.	2	−2	.	.	.
X.107	.	.	.	.	.	.	.	.	3	3	3	1	.	.	.	.	.	.	.	.
X.108	.	.	.	.	.	.	.	.	3	3	−3	1	.	.	.	.	.	.	.	.
X.109	.	.	.	.	−2	.	2	.	2	−2	.	.	.	.	.	−1	1	.	$\bar{C}$	C
X.110	.	.	.	.	−2	.	2	.	2	−2	.	.	.	.	.	−1	1	.	C	$\bar{C}$
X.111	.	.	.	.	.	.	.	.	.	.	.	.	.	.	.	.	.	.	.	.
X.112	4	−2	4	.	4	.	4	.	−2	−2	.	2	.	.	.	1	1	2	−1	−1
X.113	.	.	.	.	2	4	2	.	2	2	.	2	−4	.	−2	2	2	.	.	.
X.114	−4	2	−4	.	4	.	4	.	−2	−2	.	2	.	.	.	1	1	−2	1	1
X.115	.	.	.	.	.	.	.	.	.	.	.	−2	.	.	.	.	.	.	.	.
X.116	.	.	.	.	.	.	.	.	.	.	.	−2	.	.	.	.	.	.	.	.
X.117	8	8	.	.	.	.	.	.	.	.	.	.	.	.	.	.	.	.	.	.
X.118	.	.	.	.	−4	.	4	.	−2	2	.	.	.	.	.	−2	2	.	.	.
X.119	−8	−8	.	.	.	.	.	.	.	.	.	.	.	.	.	.	.	.	.	.
X.120	.	.	.	.	.	.	.	.	.	.	.	.	.	.	.	.	.	.	.	.
X.121	.	.	.	.	.	.	.	.	.	.	.	.	.	.	.	.	.	.	.	.
X.122	.	.	.	.	.	.	.	.	.	.	.	.	.	.	.	.	.	.	.	.
X.123	.	.	.	.	.	.	.	.	.	.	.	.	.	.	.	.	.	.	.	.
X.124	.	.	.	.	.	.	.	.	.	.	.	.	.	.	.	.	.	.	.	.
X.125	.	.	.	.	.	.	.	.	.	.	.	.	.	.	.	.	.	.	.	.
X.126	.	.	.	.	.	.	.	.	.	.	.	.	.	.	.	.	.	.	.	.
X.127	−3	.	−1	−3	.	3	.	.	.	.	.	.	1	.	.	.	.	.	.	.
X.128	3	.	1	3	.	3	.	.	.	.	.	.	1	.	.	.	.	.	.	.
X.129	.	.	.	.	4	.	4	.	−2	−2	.	2	.	.	.	−2	−2	.	.	.
X.130	−6	−3	.	2	−3	−2	−3	3	.	.	2	.	.	−1	1	.	.	1	.	.
X.131	6	3	.	−2	−3	−2	−3	−3	.	.	−2	.	.	1	1	.	.	−1	.	.
X.132	.	.	.	.	.	.	.	.	.	.	.	.	.	.	.	.	.	.	.	.
X.133	.	.	.	.	.	.	.	.	.	.	.	.	.	.	.	.	.	.	.	.
X.134	3	.	3	−1	.	1	.	.	.	.	2	−2	1	2	−2	.	.	.	.	.
X.135	−3	.	−3	1	.	1	.	.	.	.	−2	−2	1	−2	−2	.	.	.	.	.
X.136	.	.	.	.	6	.	−6	.	.	.	.	.	.	.	.	.	.	.	.	.
X.137	−5	4	−3	3	.	−1	.	2	.	.	.	.	1	.	2	.	.	.	.	.
X.138	5	−4	3	−3	.	−1	.	−2	.	.	.	.	1	.	2	.	.	.	.	.
X.139	.	.	.	.	.	.	.	−4	.	.	.	.	.	.	.	.	.	.	.	.
X.140	.	.	.	.	.	.	.	−4	.	.	.	.	.	.	.	.	.	.	.	.
X.141	.	.	.	.	.	.	.	4	.	.	.	.	.	.	.	.	.	.	.	.
X.142	.	.	.	.	.	.	.	4	.	.	.	.	.	.	.	.	.	.	.	.
X.143	.	.	.	.	−6	.	6	.	.	.	.	.	.	.	.	.	.	.	.	.
X.144	−3	.	−1	−3	.	−3	.	.	.	.	.	.	−1	.	.	.	.	.	.	.
X.145	3	.	1	3	.	−3	.	.	.	.	.	.	−1	.	.	.	.	.	.	.
X.146	.	.	.	.	−4	.	4	.	4	−4	.	.	.	.	.	−2	2	.	.	.
X.147	.	.	.	.	.	.	.	.	.	.	.	.	.	.	.	.	.	.	.	.
X.148	.	.	.	.	.	.	.	.	.	.	.	.	.	.	.	.	.	.	.	.
X.149	.	.	.	.	4	.	−4	.	−4	4	.	.	.	.	.	2	−2	.	.	.
X.150	.	.	.	.	.	.	.	.	.	.	.	.	.	.	.	.	.	.	.	.

Character table of $A_1 = 2Aut(\mathsf{Fi}_{22})$ *(continued)*

2	1	3	2	2	3	5	5	4	3	3	3	3	3	4	4	2	2	2	2	2	3	3	3	2
3	.	1	1	.	1	.	.	4	4	3	3	3	3	2	2	3	3	3	2	2	1	.	.	1
5	.	.	.	.	1	.	.	.	.	.	.	.	.	.	.	.	.	.	.	.	1	1	1	.
7	.	1	1	1	.	.	.	.	.	.	.	.	.	.	.	.	.	.	.	.	.	.	.	1
11	.	.	.	.	.	.	.	.	.	.	.	.	.	.	.	.	.	.	.	.	.	.	.	.
13	1	.	.	.	.	.	.	.	.	.	.	.	.	.	.	.	.	.	.	.	.	.	.	.
	13a	14a	14b	14c	15a	16a	16b	18a	18b	18c	18d	18e	18f	18g	18h	18i	18j	18k	18l	18m	20a	20b	20c	21a
2P	13a	7a	7a	7a	15a	8f	8f	9a	9b	9a	9a	9a	9a	9a	9a	9b	9c	9b	9c	9c	10a	10c	10d	21a
3P	13a	14a	14b	14c	5a	16a	16b	6_2	6_2	6_{14}	6_6	6_6	6_{14}	6_{10}	6_{11}	6_{14}	6_9	6_6	6_{24}	6_{25}	20a	20b	20c	7a
5P	13a	14a	14b	14c	3a	16a	16b	18a	18b	18f	18e	18d	18c	18g	18h	18i	18j	18k	18m	18l	4a	4e	4c	21a
7P	13a	2a	2c	2b	15a	16a	16b	18a	18b	18c	18d	18e	18f	18g	18h	18i	18j	18k	18l	18m	20a	20b	20c	3a
11P	13a	14a	14b	14c	15a	16a	16b	18a	18b	18f	18e	18d	18c	18g	18h	18i	18j	18k	18m	18l	20a	20b	20c	21a
13P	1a	14a	14b	14c	15a	16a	16b	18a	18b	18c	18d	18e	18f	18g	18h	18i	18j	18k	18l	18m	20a	20b	20c	21a
X.85	−1	1	−1	1	−1	.	.	.	.	.	.	.	.	.	.	.	.	.	.	.	1	1	1	1
X.86	−1	1	1	1	−1	.	.	.	.	.	.	.	.	.	.	.	.	.	.	.	−1	1	−1	1
X.87	.	−2	2	.	.	.	.	.	.	.	.	.	.	.	.	.	.	.	.	.	.	.	.	1
X.88	.	−2	−2	.	.	.	.	.	.	.	.	.	.	.	.	.	.	.	.	.	.	.	.	1
X.89	.	.	.	.	.	.	.	3	3	−1	−1	−1	−1	−1	−1	−1	.	−1	.	.	.	.	.	.
X.90	.	.	.	.	.	.	.	3	3	1	−1	−1	1	−1	−1	1	.	−1	.	.	.	.	.	.
X.91	.	.	.	.	2	.	.	−8	4	.	.	.	.	.	.	.	−2	.	.	.	.	.	.	.
X.92	.	.	.	.	.	1	−1	.	.	.	.	.	.	.	.	.	.	.	.	.	.	.	.	.
X.93	.	.	.	.	.	1	1	.	.	.	.	.	.	.	.	.	.	.	.	.	.	.	.	.
X.94	.	.	.	.	1	.	.	.	.	.	.	.	.	.	.	.	.	.	.	.	1	−1	−1	.
X.95	.	.	.	.	1	.	.	.	.	.	.	.	.	.	.	.	.	.	.	.	−1	−1	1	.
X.96	.	.	.	.	.	.	.	2	2	.	.	.	.	2	−2	.	2	.	.	.	.	.	.	.
X.97	.	.	.	.	.	.	.	−2	−2	.	2	2	.	−2	−2	.	−2	2	.	.	.	.	.	.
X.98	.	.	.	.	.	.	.	−5	1	1	1	1	1	−1	−1	1	1	1	−1	−1	.	.	.	.
X.99	.	.	.	.	.	.	.	−5	1	−1	1	1	−1	−1	−1	−1	1	1	1	1	.	.	.	.
X.100	.	.	.	.	.	.	.	−8	4	.	.	.	.	.	.	.	−2	.	.	.	.	.	.	.
X.101	.	1	1	−1	−1	.	.	.	.	.	.	.	.	.	.	.	.	.	.	.	−1	1	1	1
X.102	.	1	−1	−1	−1	.	.	.	.	.	.	.	.	.	.	.	.	.	.	.	1	1	−1	1
X.103	.	.	.	.	.	.	.	6	.	.	.	.	.	−2	2	.	.	.	.	.	.	.	.	.
X.104	.	−1	−1	1	1	−1	1	.	.	.	.	.	.	.	.	.	.	.	.	.	−2	.	.	−1
X.105	.	−1	1	1	1	−1	−1	.	.	.	.	.	.	.	.	.	.	.	.	.	2	.	.	−1
X.106	.	.	.	.	−2	.	.	6	6	.	.	.	.	−2	2	.	.	.	.	.	.	.	.	.
X.107	.	.	.	.	.	.	.	.	.	.	.	.	.	.	.	.	.	.	.	.	3	−1	1	.
X.108	.	.	.	.	.	.	.	.	.	.	.	.	.	.	.	.	.	.	.	.	−3	−1	−1	.
X.109	.	2	.	.	.	.	.	.	.	.	.	.	.	.	.	.	.	.	.	.	.	.	.	1
X.110	.	2	.	.	.	.	.	.	.	.	.	.	.	.	.	.	.	.	.	.	.	.	.	1
X.111	.	.	.	.	.	.	.	2	2	.	2	2	.	2	2	.	2	2	.	.	.	.	.	.
X.112	.	.	.	.	.	.	.	3	.	1	1	1	1	−1	−1	−2	.	−2	.	.	.	.	.	.
X.113	.	.	.	.	.	.	.	−6	.	.	.	.	.	−2	−2	.	.	.	.	.	.	.	.	.
X.114	.	.	.	.	.	.	.	3	.	−1	1	1	−1	−1	−1	2	.	−2	.	.	.	.	.	.
X.115	−1	.	.	.	.	.	.	.	.	.	.	.	.	.	.	.	.	.	.	.	.	.	.	.
X.116	−1	.	.	.	.	.	.	.	.	.	.	.	.	.	.	.	.	.	.	.	.	.	.	.
X.117	.	2	−2	.	.	.	.	−3	.	−1	−1	−1	−1	1	1	2	.	2	.	.	.	.	.	−1
X.118	.	−2	.	.	.	.	.	.	−6	.	.	.	.	.	.	.	.	.	.	.	.	.	.	2
X.119	.	2	2	.	.	.	.	−3	.	1	−1	−1	1	1	1	−2	.	2	.	.	.	.	.	−1
X.120	1	2	−2	.	−2	.	.	4	−2	−2	.	.	−2	.	.	−2	−2	.	.	.	.	.	.	−1
X.121	1	2	2	.	−2	.	.	4	−2	2	.	.	2	.	.	2	−2	.	.	.	.	.	.	−1
X.122	1	−2	.	.	−2	.	.	−4	2	D	.	.	$\bar{D}$	.	.	.	2	.	.	.	.	.	.	−1
X.123	1	−2	.	.	−2	.	.	−4	2	$\bar{D}$	.	.	D	.	.	.	2	.	.	.	.	.	.	−1
X.124	.	.	.	.	.	−1	1	.	.	.	.	.	.	.	.	.	.	.	.	.	−3	−1	−1	.
X.125	.	.	.	.	.	−1	−1	.	.	.	.	.	.	.	.	.	.	.	.	.	3	−1	1	.
X.126	.	2	.	.	.	.	.	2	2	.	.	.	.	2	−2	.	−4	.	.	.	.	.	.	−2
X.127	.	−2	2	.	−1	.	.	.	.	.	.	.	.	.	.	.	.	.	.	.	−2	.	.	1
X.128	.	−2	−2	.	−1	.	.	.	.	.	.	.	.	.	.	.	.	.	.	.	2	.	.	1
X.129	.	.	.	.	.	.	.	.	.	.	.	.	.	.	.	.	.	.	.	.	.	.	.	.
X.130	.	.	.	.	.	1	1	.	.	.	.	.	.	.	.	.	.	.	.	.	.	.	.	.
X.131	.	.	.	.	.	1	−1	.	.	.	.	.	.	.	.	.	.	.	.	.	.	.	.	.
X.132	.	.	.	.	−2	.	.	−2	4	.	.	.	.	2	2	.	1	.	−1	−1	.	.	.	.
X.133	.	.	.	.	−2	.	.	−2	4	.	.	.	.	2	2	.	1	.	1	1	.	.	.	.
X.134	.	−1	−1	1	.	.	.	.	.	.	.	.	.	.	.	.	.	.	.	.	.	.	.	−1
X.135	.	−1	1	1	.	.	.	.	.	.	.	.	.	.	.	.	.	.	.	.	.	.	.	−1
X.136	.	.	.	.	.	.	.	.	−6	.	.	.	.	.	.	.	.	.	.	.	.	.	.	.
X.137	.	.	.	.	.	.	.	3	3	−1	−1	−1	−1	−1	−1	−1	.	−1	.	.	.	.	.	.
X.138	.	.	.	.	.	.	.	3	3	1	−1	−1	1	−1	−1	1	.	−1	.	.	.	.	.	.
X.139	.	1	1	1	1	.	.	.	−3	−2	2	2	−2	.	.	1	.	−1	.	.	2	.	.	1
X.140	.	1	−1	−1	1	.	.	.	−3	2	−2	−2	2	.	.	−1	.	1	.	.	2	.	.	1
X.141	.	1	1	−1	1	.	.	.	−3	−2	−2	−2	−2	.	.	1	.	1	.	.	−2	.	.	1
X.142	.	1	−1	1	1	.	.	.	−3	2	2	2	2	.	.	−1	.	−1	.	.	−2	.	.	1
X.143	.	.	.	.	−2	.	.	.	.	.	.	.	.	.	.	.	.	.	.	.	.	.	.	.
X.144	.	−1	−1	−1	1	.	.	.	.	.	.	.	.	.	.	.	.	.	.	.	1	1	1	−1
X.145	.	−1	1	−1	1	.	.	.	.	.	.	.	.	.	.	.	.	.	.	.	−1	1	−1	−1
X.146	−1	.	.	.	.	.	.	.	.	.	.	.	.	.	.	.	.	.	.	.	.	.	.	.
X.147	.	2	.	.	−2	.	.	−6	−6	.	.	.	.	2	−2	.	.	.	.	.	.	.	.	−2
X.148	.	.	.	.	2	.	.	.	.	.	.	.	.	.	.	.	.	.	.	.	.	.	.	.
X.149	.	.	.	.	.	.	.	6	.	.	.	.	.	−2	2	.	.	.	.	.	.	.	.	.
X.150	.	−2	.	.	2	.	.	.	6	.	.	.	.	.	.	.	.	.	.	.	.	.	.	2

Character table of $A_1 = 2Aut(\mathsf{Fi}_{22})$ *(continued)*

	22a	22b	22c	24a	24b	24c	24d	24e	24f	24g	24h	26a	28a	30a	30b	30c	36a	36b	36c	42a	42b	42c	60a
2	2	2	2	5	5	5	5	4	4	4	4	1	2	3	2	2	3	3	2	2	2	2	2
3	.	.	.	1	1	1	1	1	1	1	1	.	.	1	1	1	2	2	2	1	1	1	1
5	.	.	.	.	.	.	.	.	.	.	.	.	.	1	1	1	.	.	.	.	.	.	1
7	.	.	.	.	.	.	.	.	.	.	.	.	1	.	.	.	.	.	.	1	1	1	.
11	1	1	1	.	.	.	.	.	.	.	.	.	.	.	.	.	.	.	.	.	.	.	.
13	.	.	.	.	.	.	.	.	.	.	.	1	.	.	.	.	.	.	.	.	.	.	.
2P	11a	11a	11a	12_5	12_6	12_6	12_5	12_{13}	12_{20}	12_{13}	12_{20}	13a	14a	15a	15a	15a	18g	18g	18b	21a	21a	21a	30a
3P	22a	22b	22c	8b	8c	8d	8e	8g	8a	8g	8a	26a	28a	10a	10e	10b	12_3	12_3	12_4	14b	14a	14b	20a
5P	22a	22b	22c	24a	24b	24c	24d	24g	24h	24e	24f	26a	28a	6_1	6_4	6_5	36a	36b	36c	42c	42b	42a	12_1
7P	22c	22b	22a	24a	24b	24c	24d	24g	24h	24e	24f	26a	4a	30a	30b	30c	36a	36b	36c	6_7	6_1	6_8	60a
11P	2b	2a	2b	24a	24b	24c	24d	24e	24f	24g	24h	26a	28a	30a	30b	30c	36a	36b	36c	42c	42b	42a	60a
13P	22c	22b	22a	24a	24b	24c	24d	24e	24f	24g	24h	2a	28a	30a	30b	30c	36a	36b	36c	42a	42b	42c	60a
X.85	.	.	.	−1	1	−1	1	.	.	.	.	−1	−1	−1	1	−1	.	.	.	−1	1	−1	1
X.86	.	.	.	−1	−1	−1	−1	.	.	.	.	−1	1	−1	1	1	.	.	.	1	1	1	−1
X.87	.	.	.	−1	−1	−1	−1	1	.	1	.	.	.	.	.	.	.	.	.	−1	1	−1	.
X.88	.	.	.	−1	1	−1	1	−1	.	−1	.	.	.	.	.	.	.	.	.	1	1	1	.
X.89	.	.	.	1	1	1	1	.	.	.	.	.	.	.	.	.	1	1	−1	.	.	.	.
X.90	.	.	.	1	−1	1	−1	.	.	.	.	.	.	.	.	.	−1	−1	1	.	.	.	.
X.91	.	.	.	.	.	.	.	.	.	.	.	.	.	−2	.	.	.	.	.	.	.	.	.
X.92	.	.	.	.	.	.	.	.	1	.	1	.	.	.	.	.	.	.	.	.	.	.	.
X.93	.	.	.	.	.	.	.	.	−1	.	−1	.	.	.	.	.	.	.	.	.	.	.	.
X.94	.	.	.	.	.	.	.	.	.	.	.	.	.	1	1	1	.	.	.	.	.	.	1
X.95	.	.	.	.	.	.	.	.	.	.	.	.	.	1	1	−1	.	.	.	.	.	.	−1
X.96	.	.	.	.	.	.	.	.	.	.	.	.	.	.	.	.	.	.	.	.	.	.	.
X.97	.	.	.	.	.	.	.	.	.	.	.	.	.	.	.	.	.	.	.	.	.	.	.
X.98	.	.	.	1	−1	−1	1	.	.	.	.	.	.	.	.	.	1	1	1	.	.	.	.
X.99	.	.	.	1	1	−1	−1	.	.	.	.	.	.	.	.	.	−1	−1	−1	.	.	.	.
X.100	.	.	.	.	.	.	.	.	.	.	.	.	.	.	.	.	.	.	.	.	.	.	.
X.101	−1	1	−1	1	−1	−1	1	.	.	.	.	.	−1	−1	−1	−1	.	.	.	1	1	1	−1
X.102	−1	1	−1	1	1	−1	−1	.	.	.	.	.	1	−1	−1	1	.	.	.	−1	1	−1	1
X.103	.	−2	.	.	.	.	.	.	.	.	.	.	.	.	.	.	.	.	.	.	.	.	.
X.104	.	.	.	1	1	1	1	.	.	.	.	.	1	1	−1	−1	.	.	.	−1	−1	−1	1
X.105	.	.	.	1	−1	1	−1	.	.	.	.	.	−1	1	−1	1	.	.	.	1	−1	1	−1
X.106	.	.	.	.	.	.	.	.	.	.	.	.	.	2	.	.	.	.	.	.	.	.	.
X.107	.	.	.	.	.	.	.	.	−1	.	−1	.	.	.	.	.	.	.	.	.	.	.	.
X.108	.	.	.	.	.	.	.	.	1	.	1	.	.	.	.	.	.	.	.	.	.	.	.
X.109	.	.	.	.	.	.	.	.	.	.	.	.	.	.	.	.	.	.	.	$\bar{C}$	−1	C	.
X.110	.	.	.	.	.	.	.	.	.	.	.	.	.	.	.	.	.	.	.	C	−1	$\bar{C}$	.
X.111	−1	−1	−1	.	.	.	.	.	.	.	.	.	.	.	.	.	.	.	.	.	.	.	.
X.112	.	.	.	.	.	.	.	.	.	.	.	.	.	.	.	.	1	1	.	.	.	.	.
X.113	.	.	.	.	.	.	.	.	.	.	.	.	.	.	.	.	.	.	.	.	.	.	.
X.114	.	.	.	.	.	.	.	.	.	.	.	.	.	.	.	.	−1	−1	.	.	.	.	.
X.115	1	1	1	.	.	.	.	.	I	.	$-I$	−1	.	.	.	.	.	.	.	.	.	.	.
X.116	1	1	1	.	.	.	.	.	$-I$	.	I	−1	.	.	.	.	.	.	.	.	.	.	.
X.117	.	.	.	.	.	.	.	.	.	.	.	.	.	.	.	.	−1	−1	.	1	−1	1	.
X.118	.	.	.	.	.	.	.	.	.	.	.	.	.	.	.	.	.	.	.	.	−2	.	.
X.119	.	.	.	.	.	.	.	.	.	.	.	.	.	.	.	.	1	1	.	−1	−1	−1	.
X.120	.	.	.	.	.	.	.	.	.	.	.	1	.	−2	.	2	.	.	.	1	−1	1	.
X.121	.	.	.	.	.	.	.	.	.	.	.	1	.	−2	.	−2	.	.	.	−1	−1	−1	.
X.122	.	.	.	.	.	.	.	.	.	.	.	−1	.	2	.	.	.	.	.	C	1	$\bar{C}$	.
X.123	.	.	.	.	.	.	.	.	.	.	.	−1	.	2	.	.	.	.	.	$\bar{C}$	1	C	.
X.124	1	1	1	.	.	.	.	.	.	.	.	.	.	.	.	.	.	.	.	.	.	.	.
X.125	1	1	1	.	.	.	.	.	.	.	.	.	.	.	.	.	.	.	.	.	.	.	.
X.126	.	.	.	.	.	.	.	.	.	.	.	.	.	.	.	.	.	.	.	.	2	.	.
X.127	.	.	.	1	1	−1	−1	.	.	.	.	.	.	−1	−1	1	.	.	.	−1	1	−1	1
X.128	.	.	.	1	−1	−1	1	.	.	.	.	.	.	−1	−1	−1	.	.	.	1	1	1	−1
X.129	1	−1	1	.	.	.	.	.	.	.	.	.	.	.	.	.	.	.	.	.	.	.	.
X.130	.	.	.	−2	.	.	2	−1	.	−1	.	.	.	.	.	.	.	.	.	.	.	.	.
X.131	.	.	.	−2	.	.	−2	1	.	1	.	.	.	.	.	.	.	.	.	.	.	.	.
X.132	.	.	.	.	.	.	.	.	.	.	.	.	.	−2	.	2	.	.	.	.	.	.	.
X.133	.	.	.	.	.	.	.	.	.	.	.	.	.	−2	.	−2	.	.	.	.	.	.	.
X.134	.	.	.	1	−1	1	−1	.	.	.	.	.	1	.	.	.	.	.	.	−1	−1	−1	.
X.135	.	.	.	1	1	1	1	.	.	.	.	.	−1	.	.	.	.	.	.	1	−1	1	.
X.136	.	.	.	.	.	.	.	.	.	.	.	.	.	.	.	.	.	.	.	.	.	.	.
X.137	.	.	.	1	−1	−1	1	.	.	.	.	.	.	.	.	.	1	1	−1	.	.	.	.
X.138	.	.	.	1	1	−1	−1	.	.	.	.	.	.	.	.	.	−1	−1	1	.	.	.	.
X.139	−1	−1	−1	.	.	.	.	.	.	.	.	.	1	1	1	−1	.	.	−1	1	1	1	−1
X.140	1	−1	1	.	.	.	.	.	.	.	.	.	1	1	−1	1	.	.	−1	−1	1	−1	−1
X.141	1	−1	1	.	.	.	.	.	.	.	.	.	−1	1	−1	−1	.	.	1	1	1	1	1
X.142	−1	−1	−1	.	.	.	.	.	.	.	.	.	−1	1	1	1	.	.	1	−1	1	−1	1
X.143	.	.	.	.	.	.	.	.	.	.	.	.	.	2	.	.	.	.	.	.	.	.	.
X.144	−1	1	−1	−1	1	1	−1	.	.	.	.	.	−1	1	1	1	.	.	.	−1	−1	−1	1
X.145	−1	1	−1	−1	−1	1	1	.	.	.	.	.	1	1	1	−1	.	.	.	1	−1	1	−1
X.146	.	.	.	.	.	.	.	.	.	.	.	1	.	.	.	.	.	.	.	.	.	.	.
X.147	.	.	.	.	.	.	.	.	.	.	.	.	.	2	.	.	.	.	.	.	2	.	.
X.148	.	−2	.	.	.	.	.	.	.	.	.	.	.	−2	.	.	.	.	.	.	.	.	.
X.149	.	.	.	.	.	.	.	.	.	.	.	.	.	.	.	.	.	.	.	.	.	.	.
X.150	.	2	.	.	.	.	.	.	.	.	.	.	.	−2	.	.	.	.	.	.	−2	.	.

where $A = 16\zeta(3) + 8$, $B = 6\zeta(3) + 3$, $C = -2\zeta(3) - 1$, $D = 4\zeta(3) + 2$, $E = -2\zeta(11)^9 - 2\zeta(11)^5 - 2\zeta(11)^4 - 2\zeta(11)^3 - 2\zeta(11) - 1$, $F = 18\zeta(3) + 9$, $G = 112\zeta(3) + 56$, $H = 12\zeta(3) + 6$, $I = 4\zeta(12)_4\zeta(12)_3 + 2\zeta(12)_4$, $J = 96\zeta(3) + 48$, $K = -128\zeta(3) - 64$.

9.8.2 *Character table of* $H(\mathsf{Fi}_{24}') = \langle r, p, b \rangle$

2	21	21	19	20	14	14	17	17	16	13	8	12	10	5	8	5	6	15	15
3	7	7	4	2	4	4	3	3	1	2	7	6	7	7	5	6	4	5	2
5	1	1	1	1	1	1	.	.	.	.	1	.	.	.	.	.	.	1	.
7	1	1	.	.	.	.	.	.	.	.	1	.	.	.	.	.	.	.	.
	1a	2a	2b	2c	2d	2e	2f	2g	2h	2i	3a	3b	3c	3d	3e	3f	3g	4a	4b
2P	1a	1a	1a	1a	1a	1a	1a	1a	1a	1a	3a	3b	3c	3d	3e	3f	3g	2a	2a
3P	1a	2a	2b	2c	2d	2e	2f	2g	2h	2i	1a	1a	1a	1a	1a	1a	1a	4a	4b
5P	1a	2a	2b	2c	2d	2e	2f	2g	2h	2i	3a	3b	3c	3d	3e	3f	3g	4a	4b
7P	1a	2a	2b	2c	2d	2e	2f	2g	2h	2i	3a	3b	3c	3d	3e	3f	3g	4a	4b
X.1	1	1	1	1	1	1	1	1	1	1	1	1	1	1	1	1	1	1	1
X.2	1	1	1	1	−1	−1	1	1	1	−1	1	1	1	1	1	1	1	1	1
X.3	21	21	21	21	9	9	5	5	5	1	21	3	−6	−6	3	3	3	21	5
X.4	21	21	21	21	−9	−9	5	5	5	−1	21	3	−6	−6	3	3	3	21	5
X.5	30	30	30	30	.	.	−2	−2	−2	.	−15	6	−6	12	.	−3	.	30	−2
X.6	35	35	35	35	15	15	3	3	3	−1	35	−1	8	8	8	−1	−1	35	3
X.7	35	35	35	35	−5	−5	3	3	3	−5	35	8	8	8	−1	8	−1	35	3
X.8	35	35	35	35	5	5	3	3	3	5	35	8	8	8	−1	8	−1	35	3
X.9	35	35	35	35	−15	−15	3	3	3	1	35	−1	8	8	8	−1	−1	35	3
X.10	42	42	42	42	.	.	10	10	10	.	−21	12	−3	6	.	−6	.	42	10
X.11	90	90	90	90	−30	−30	10	10	10	−6	90	9	9	9	9	9	.	90	10
X.12	90	90	90	90	30	30	10	10	10	6	90	9	9	9	9	9	.	90	10
X.13	140	140	140	140	−20	−20	12	12	12	−4	140	−4	5	5	−4	−4	5	140	12
X.14	140	140	140	140	20	20	12	12	12	4	140	−4	5	5	−4	−4	5	140	12
X.15	189	189	189	189	−9	−9	−3	−3	−3	−9	189	.	27	27	.	.	.	189	−3
X.16	189	189	189	189	9	9	−3	−3	−3	9	189	.	27	27	.	.	.	189	−3
X.17	210	210	210	210	.	.	18	18	18	.	−105	6	−15	30	.	−3	.	210	18
X.18	210	210	210	210	30	30	2	2	2	−10	210	3	21	21	3	3	3	210	2
X.19	210	210	210	210	.	.	−14	−14	−14	.	−105	6	−15	30	.	−3	.	210	−14
X.20	210	210	210	210	−30	−30	2	2	2	10	210	3	21	21	3	3	3	210	2
X.21	210	210	210	210	.	.	18	18	18	.	−105	24	12	−24	.	−12	.	210	18
X.22	280	280	280	280	−40	−40	−8	−8	−8	8	280	1	10	10	10	1	1	280	−8
X.23	280	280	280	280	−40	−40	−8	−8	−8	8	280	1	10	10	10	1	1	280	−8
X.24	280	280	280	280	40	40	−8	−8	−8	−8	280	1	10	10	10	1	1	280	−8
X.25	280	280	280	280	40	40	−8	−8	−8	−8	280	1	10	10	10	1	1	280	−8
X.26	315	315	315	315	−15	−15	11	11	11	9	315	18	−9	−9	−9	18	.	315	11
X.27	315	315	315	315	75	75	11	11	11	3	315	−9	−9	−9	18	−9	.	315	11
X.28	315	315	315	315	15	15	11	11	11	−9	315	18	−9	−9	−9	18	.	315	11
X.29	315	315	315	315	−75	−75	11	11	11	−3	315	−9	−9	−9	18	−9	.	315	11
X.30	378	378	58	−6	36	36	42	42	10	12	.	54	27	.	9	.	.	−6	−6
X.31	378	378	58	−6	−36	−36	42	42	10	−12	.	54	27	.	9	.	.	−6	−6
X.32	420	420	420	420	60	60	4	4	4	−4	420	6	−39	−39	6	6	−3	420	4
X.33	420	420	420	420	−60	−60	4	4	4	4	420	6	−39	−39	6	6	−3	420	4
X.34	420	420	420	420	.	.	4	4	4	.	−210	30	−3	6	.	−15	.	420	4
X.35	560	560	560	560	.	.	−16	−16	−16	.	560	20	20	20	2	20	2	560	−16
X.36	560	560	560	560	.	.	−16	−16	−16	.	560	2	−34	−34	2	2	2	560	−16
X.37	560	560	560	560	.	.	−16	−16	−16	.	560	2	−34	−34	2	2	2	560	−16
X.38	630	630	630	630	.	.	−10	−10	−10	.	−315	18	36	−72	.	−9	.	630	−10
X.39	672	672	672	672	.	.	32	32	32	.	−336	12	6	−12	.	−6	.	672	32
X.40	720	720	720	720	.	.	16	16	16	.	−360	−18	18	−36	.	9	.	720	16
X.41	720	720	720	720	.	.	16	16	16	.	−360	−18	18	−36	.	9	.	720	16
X.42	729	729	729	729	−81	−81	9	9	9	−9	729	.	.	.	.	.	.	729	9
X.43	729	729	729	729	81	81	9	9	9	9	729	.	.	.	.	.	.	729	9
X.44	768	−768	.	.	.	.	−64	64	.	.	−6	96	−24	−6	.	6	.	.	.
X.45	768	768	768	768	.	.	.	.	.	.	−384	24	−24	48	.	−12	.	768	.
X.46	840	840	840	840	.	.	8	8	8	.	−420	−12	−33	66	.	6	.	840	8
X.47	896	896	896	896	64	64	.	.	.	.	896	−4	32	32	−4	−4	−4	896	.
X.48	896	896	896	896	−64	−64	.	.	.	.	896	−4	32	32	−4	−4	−4	896	.
X.49	1260	1260	1260	1260	.	.	12	12	12	.	−630	−18	−9	18	.	9	.	1260	12
X.50	1280	1280	1280	1280	.	.	.	.	.	.	1280	−16	−16	−16	−16	−16	2	1280	.
X.51	1280	−1280	.	.	.	.	64	−64	.	.	20	32	56	−7	8	−4	8	.	.
X.52	1280	−1280	.	.	.	.	64	−64	.	.	20	32	56	−7	8	−4	8	.	.
X.53	1458	1458	1458	1458	.	.	18	18	18	.	−729	.	.	.	.	.	.	1458	18
X.54	1512	1512	1512	1512	.	.	−24	−24	−24	.	−756	.	−27	54	.	.	.	1512	−24
X.55	1701	1701	−27	37	27	27	117	117	21	19	.	81	.	.	.	.	9	−27	−27
X.56	1701	1701	−27	37	−27	−27	117	117	21	−19	.	81	.	.	.	.	9	−27	−27
X.57	1890	1890	1890	1890	.	.	−30	−30	−30	.	−945	.	27	−54	.	.	.	1890	−30
X.58	2016	2016	−32	−32	.	.	96	96	−32	32	.	120	36	.	6	3	6	32	32
X.59	2016	2016	−32	−32	.	.	96	96	−32	−32	.	120	36	.	6	3	6	32	32
X.60	2268	2268	348	−36	−144	−144	60	60	−4	.	.	.	−81	.	27	.	.	−36	−36
X.61	2268	2268	348	−36	144	144	60	60	−4	.	.	.	−81	.	27	.	.	−36	−36
X.62	3584	−3584	.	.	−192	192	−128	128	.	.	56	32	−16	2	44	−4	8	.	.
X.63	3584	−3584	.	.	128	−128	−128	128	.	.	56	176	−16	2	8	−22	8	.	.
X.64	3584	−3584	.	.	−128	128	−128	128	.	.	56	176	−16	2	8	−22	8	.	.
X.65	3584	−3584	.	.	192	−192	−128	128	.	.	56	32	−16	2	44	−4	8	.	.
X.66	3780	3780	580	−60	.	.	36	36	−28	.	.	216	27	.	−18	.	.	−60	−60
X.67	4480	−4480	.	.	160	−160	−32	32	.	.	70	112	−128	16	28	−14	−8	.	.
X.68	4480	−4480	.	.	−160	160	−32	32	.	.	70	112	−128	16	28	−14	−8	.	.
X.69	4480	−4480	.	.	160	−160	−32	32	.	.	70	−32	88	−11	28	4	−8	.	.
X.70	4480	−4480	.	.	160	−160	−32	32	.	.	70	−32	88	−11	28	4	−8	.	.
X.71	4480	−4480	.	.	−160	160	−32	32	.	.	70	−32	88	−11	28	4	−8	.	.
X.72	4480	−4480	.	.	−160	160	−32	32	.	.	70	−32	88	−11	28	4	−8	.	.
X.73	5670	5670	870	−90	180	180	150	150	54	12	.	162	−81	.	.	.	.	−90	6
X.74	5670	5670	870	−90	180	180	−42	−42	−10	−36	.	.	162	.	27	.	.	−90	6
X.75	5670	5670	870	−90	−180	−180	−42	−42	−10	36	.	.	162	.	27	.	.	−90	6
X.76	5670	5670	870	−90	−180	−180	150	150	54	−12	.	162	−81	.	.	.	.	−90	6
X.77	7560	7560	1160	−120	.	.	−120	−120	8	.	.	108	−189	.	18	.	.	−120	72
X.78	7560	7560	1160	−120	−360	−360	168	168	40	−24	.	−54	54	.	45	.	.	−120	−24
X.79	7560	7560	1160	−120	360	360	168	168	40	24	.	−54	54	.	45	.	.	−120	−24
X.80	7680	−7680	.	.	.	.	−128	128	.	.	120	−96	−96	12	−24	12	12	.	.
X.81	7680	−7680	.	.	.	.	−128	128	.	.	120	−96	−96	12	−24	12	12	.	.
X.82	8064	8064	−128	−128	.	.	.	.	.	64	.	156	144	.	24	12	6	128	.
X.83	8064	8064	−128	−128	.	.	.	.	.	−64	.	156	144	.	24	12	6	128	.
X.84	8505	8505	−135	185	135	135	153	153	57	23	.	162	.	.	.	.	−9	−135	9
X.85	8505	8505	−135	185	−135	−135	153	153	57	−23	.	162	.	.	.	.	−9	−135	9
X.86	8505	8505	−135	185	135	135	153	153	57	23	.	−81	.	.	.	.	18	−135	9
X.87	8505	8505	−135	185	−135	−135	153	153	57	−23	.	−81	.	.	.	.	18	−135	9
X.88	8960	−8960	.	.	.	.	−64	64	.	.	140	224	176	−22	−16	−28	−16	.	.

Character table of $H(\mathsf{Fi}_{24}')$ (continued)

	4c	4d	4e	4f	4g	4h	4i	4j	4k	4l	4m	4n	4o	4p	4q	4r	4s	4t	5a	6_1	6_2	6_3	6_4
2	13	11	13	14	11	11	11	14	11	12	10	10	10	9	9	10	10	9	6	8	12	10	5
3	1	2	2	1	2	2	2	.	1	.	1	1	1	1	1	.	.	.	1	7	6	7	7
5	1	1	.	.	.	.	.	.	.	.	.	.	.	.	.	.	.	.	1	1	.	.	.
7	.	.	.	.	.	.	.	.	.	.	.	.	.	.	.	.	.	.	.	1	.	.	.
2P	2c	2b	2c	2c	2f	2b	2f	2c	2b	2c	2h	2f	2h	2g	2g	2h	2h	2h	5a	3a	3b	3c	3d
3P	4c	4d	4e	4f	4g	4h	4i	4j	4k	4l	4m	4n	4o	4p	4q	4r	4s	4t	5a	2a	2a	2a	2a
5P	4c	4d	4e	4f	4g	4h	4i	4j	4k	4l	4m	4n	4o	4p	4q	4r	4s	4t	1a	6_1	6_2	6_3	6_4
7P	4c	4d	4e	4f	4g	4h	4i	4j	4k	4l	4m	4n	4o	4p	4q	4r	4s	4t	5a	6_1	6_2	6_3	6_4
X.1	1	1	1	1	1	1	1	1	1	1	1	1	1	1	1	1	1	1	1	1	1	1	1
X.2	−1	−1	−1	1	1	1	1	1	−1	−1	−1	−1	−1	1	−1	1	−1	1	1	1	1	1	1
X.3	9	9	1	5	1	5	1	5	1	1	1	−3	1	1	1	1	−3	1	1	21	3	−6	−6
X.4	−9	−9	−1	5	1	5	1	5	−1	−1	−1	3	−1	1	−1	1	3	1	1	21	3	−6	−6
X.5	.	.	.	−2	6	−2	6	−2	.	.	.	.	.	−2	.	6	.	−2	.	−15	6	−6	12
X.6	15	15	−1	3	3	3	3	3	−1	−1	3	−1	3	−1	3	3	−1	−1	.	35	−1	8	8
X.7	−5	−5	−5	3	3	3	3	3	−5	−5	−1	3	−1	−1	−1	3	3	−1	.	35	8	8	8
X.8	5	5	5	3	3	3	3	3	5	5	1	−3	1	−1	1	3	−3	−1	.	35	8	8	8
X.9	−15	−15	1	3	3	3	3	3	1	1	−3	1	−3	−1	−3	3	1	−1	.	35	−1	8	8
X.10	.	.	.	10	2	10	2	10	.	.	.	.	.	2	.	2	.	2	2	−21	12	−3	6
X.11	−30	−30	−6	10	−2	10	−2	10	−6	−6	−2	−2	−2	2	−2	−2	−2	2	.	90	9	9	9
X.12	30	30	6	10	−2	10	−2	10	6	6	2	2	2	2	2	−2	2	2	.	90	9	9	9
X.13	−20	−20	−4	12	4	12	4	12	−4	−4	.	−4	.	.	.	4	−4	.	.	140	−4	5	5
X.14	20	20	4	12	4	12	4	12	4	4	.	4	.	.	.	4	4	.	.	140	−4	5	5
X.15	−9	−9	−9	−3	5	−3	5	−3	−9	−9	3	−1	3	1	3	5	−1	1	−1	189	.	27	27
X.16	9	9	9	−3	5	−3	5	−3	9	9	−3	1	−3	1	−3	5	1	1	−1	189	.	27	27
X.17	.	.	.	18	2	18	2	18	.	.	.	.	.	2	.	2	.	2	.	−105	6	−15	30
X.18	30	30	−10	2	−2	2	−2	2	−10	−10	−2	2	−2	−2	−2	−2	2	−2	.	210	3	21	21
X.19	.	.	.	−14	10	−14	10	−14	.	.	.	.	.	2	.	10	.	2	.	−105	6	−15	30
X.20	−30	−30	10	2	−2	2	−2	2	10	10	2	−2	2	−2	2	−2	−2	−2	.	210	3	21	21
X.21	.	.	.	18	2	18	2	18	.	.	.	.	.	2	.	2	.	2	.	−105	24	12	−24
X.22	−40	−40	8	−8	.	−8	.	−8	8	8	.	.	.	.	.	.	.	.	.	280	1	10	10
X.23	−40	−40	8	−8	.	−8	.	−8	8	8	.	.	.	.	.	.	.	.	.	280	1	10	10
X.24	40	40	−8	−8	.	−8	.	−8	−8	−8	.	.	.	.	.	.	.	.	.	280	1	10	10
X.25	40	40	−8	−8	.	−8	.	−8	−8	−8	.	.	.	.	.	.	.	.	.	280	1	10	10
X.26	−15	−15	9	11	−1	11	−1	11	9	9	1	5	1	−1	1	−1	5	−1	.	315	18	−9	−9
X.27	75	75	3	11	−1	11	−1	11	3	3	3	−1	3	−1	3	−1	−1	−1	.	315	−9	−9	−9
X.28	15	15	−9	11	−1	11	−1	11	−9	−9	−1	−5	−1	−1	−1	−1	−5	−1	.	315	18	−9	−9
X.29	−75	−75	−3	11	−1	11	−1	11	−3	−3	−3	1	−3	−1	−3	−1	1	−1	.	315	−9	−9	−9
X.30	4	−4	12	10	6	2	6	−6	4	−4	4	.	4	6	4	−2	.	−2	3	.	54	27	.
X.31	−4	4	−12	10	6	2	6	−6	−4	4	−4	.	−4	6	−4	−2	.	−2	3	.	54	27	.
X.32	60	60	−4	4	4	4	4	4	−4	−4	.	4	.	.	.	4	4	.	.	420	6	−39	−39
X.33	−60	−60	4	4	4	4	4	4	4	4	.	−4	.	.	.	4	−4	.	.	420	6	−39	−39
X.34	.	.	.	4	−4	4	−4	4	.	.	.	.	.	−4	.	−4	.	−4	.	−210	30	−3	6
X.35	.	.	.	−16	.	−16	.	−16	.	.	.	.	.	.	.	.	.	.	.	560	20	20	20
X.36	.	.	.	−16	.	−16	.	−16	.	.	.	.	.	.	.	.	.	.	.	560	2	−34	−34
X.37	.	.	.	−16	.	−16	.	−16	.	.	.	.	.	.	.	.	.	.	.	560	2	−34	−34
X.38	.	.	.	−10	6	−10	6	−10	.	.	.	.	.	−2	.	6	.	−2	.	−315	18	36	−72
X.39	.	.	.	32	.	32	.	32	.	.	.	.	.	.	.	.	.	.	2	−336	12	6	−12
X.40	.	.	.	16	.	16	.	16	.	.	.	.	.	.	.	.	.	.	.	−360	−18	18	−36
X.41	.	.	.	16	.	16	.	16	.	.	.	.	.	.	.	.	.	.	.	−360	−18	18	−36
X.42	−81	−81	−9	9	−3	9	−3	9	−9	−9	3	3	3	1	3	−3	3	1	−1	729	.	.	.
X.43	81	81	9	9	−3	9	−3	9	9	9	−3	−3	−3	1	−3	−3	−3	1	−1	729	.	.	.
X.44	.	.	.	.	−16	.	16	.	.	.	.	.	.	.	.	.	.	.	8	6	−96	24	6
X.45	.	.	.	.	.	.	.	.	.	.	.	.	.	.	.	.	.	.	−2	−384	24	−24	48
X.46	.	.	.	8	8	8	8	8	.	.	.	.	.	.	.	8	.	.	.	−420	−12	−33	66
X.47	64	64	.	.	.	.	.	.	.	.	.	.	.	.	.	.	.	.	1	896	−4	32	32
X.48	−64	−64	.	.	.	.	.	.	.	.	.	.	.	.	.	.	.	.	1	896	−4	32	32
X.49	.	.	.	12	4	12	4	12	.	.	.	.	.	−4	.	4	.	−4	.	−630	−18	−9	18
X.50	.	.	.	.	.	.	.	.	.	.	.	.	.	.	.	.	.	.	.	1280	−16	−16	−16
X.51	.	.	.	.	−16	.	16	.	.	.	.	.	.	.	.	.	.	.	.	−20	−32	−56	7
X.52	.	.	.	.	−16	.	16	.	.	.	.	.	.	.	.	.	.	.	.	−20	−32	−56	7
X.53	.	.	.	18	−6	18	−6	18	.	.	.	.	.	2	.	−6	.	2	−2	−729	.	.	.
X.54	.	.	.	−24	−8	−24	−8	−24	.	.	.	.	.	.	.	−8	.	.	2	−756	.	−27	54
X.55	−5	3	19	−11	9	−3	9	5	−5	3	3	15	3	9	3	1	−1	1	6	.	81	.	.
X.56	5	−3	−19	−11	9	−3	9	5	5	−3	−3	−15	−3	9	−3	1	1	1	6	.	81	.	.
X.57	.	.	.	−30	2	−30	2	−30	.	.	.	.	.	2	.	2	.	2	.	−945	.	27	−54
X.58	.	.	−32	.	.	.	.	.	.	.	−8	.	−8	.	8	.	.	.	6	.	120	36	.
X.59	.	.	32	.	.	.	.	.	.	.	8	.	8	.	−8	.	.	.	6	.	120	36	.
X.60	−16	16	.	28	12	−4	12	−4	.	.	−4	.	−4	.	−4	−4	.	.	3	.	.	−81	.
X.61	16	−16	.	28	12	−4	12	−4	.	.	4	.	4	.	4	−4	.	.	3	.	.	−81	.
X.62	.	.	.	.	.	.	.	.	.	.	−16	.	16	.	.	.	.	.	4	−56	−32	16	−2
X.63	.	.	.	.	.	.	.	.	.	.	.	.	.	.	.	.	.	.	4	−56	−176	16	−2
X.64	.	.	.	.	.	.	.	.	.	.	.	.	.	.	.	.	.	.	4	−56	−176	16	−2
X.65	.	.	.	.	.	.	.	.	.	.	16	.	−16	.	.	.	.	.	4	−56	−32	16	−2
X.66	.	.	.	36	12	−12	12	4	.	.	.	.	.	−12	.	−4	.	4	.	.	216	27	.
X.67	.	.	.	.	−8	.	8	.	.	.	8	.	−8	.	.	.	.	.	.	−70	−112	128	−16
X.68	.	.	.	.	−8	.	8	.	.	.	−8	.	8	.	.	.	.	.	.	−70	−112	128	−16
X.69	.	.	.	.	−8	.	8	.	.	.	8	.	−8	.	.	.	.	.	.	−70	32	−88	11
X.70	.	.	.	.	−8	.	8	.	.	.	8	.	−8	.	.	.	.	.	.	−70	32	−88	11
X.71	.	.	.	.	−8	.	8	.	.	.	−8	.	8	.	.	.	.	.	.	−70	32	−88	11
X.72	.	.	.	.	−8	.	8	.	.	.	−8	.	8	.	.	.	.	.	.	−70	32	−88	11
X.73	20	−20	12	22	−6	14	−6	−26	4	−4	.	.	.	6	.	2	.	−2	.	.	162	−81	.
X.74	20	−20	−36	−10	18	−2	18	6	−12	12	4	.	4	−6	4	−6	.	2	.	.	.	162	.
X.75	−20	20	36	−10	18	−2	18	6	12	−12	−4	.	−4	−6	−4	−6	.	2	.	.	.	162	.
X.76	−20	20	−12	22	−6	14	−6	−26	−4	4	.	.	.	6	.	2	.	−2	.	.	162	−81	.
X.77	.	.	.	−56	24	8	24	8	.	.	.	.	.	.	.	−8	.	.	.	.	108	−189	.
X.78	−40	40	−24	40	.	8	.	−24	−8	8	−8	.	−8	.	−8	.	.	.	.	.	−54	54	.
X.79	40	−40	24	40	.	8	.	−24	8	−8	8	.	8	.	8	.	.	.	.	.	−54	54	.
X.80	.	.	.	.	.	.	.	.	.	.	.	.	.	.	.	.	.	.	.	−120	96	96	−12
X.81	.	.	.	.	.	.	.	.	.	.	.	.	.	.	.	.	.	.	.	−120	96	96	−12
X.82	.	.	−64	.	.	.	.	.	.	.	.	.	.	.	.	.	.	.	−6	.	156	144	.
X.83	.	.	64	.	.	.	.	.	.	.	.	.	.	.	.	.	.	.	−6	.	156	144	.
X.84	−25	15	23	−7	9	−15	9	9	−1	7	3	−9	3	−3	3	1	7	5	.	.	162	.	.
X.85	25	−15	−23	−7	9	−15	9	9	1	−7	−3	9	−3	−3	−3	1	−7	5	.	.	162	.	.
X.86	−25	15	23	−7	9	−15	9	9	−1	7	3	−9	3	−3	3	1	7	5	.	.	−81	.	.
X.87	25	−15	−23	−7	9	−15	9	9	1	−7	−3	9	−3	−3	−3	1	−7	5	.	.	−81	.	.
X.88	.	.	.	.	−16	.	16	.	.	.	.	.	.	.	.	.	.	.	.	−140	−224	−176	22

Character table of $H(Fi_{24}')$ *(continued)*

2	8	9	10	5	11	8	8	8	8	8	8	8	8	8	8	6	7	5	5	5	5	7	5
3	5	4	3	6	2	3	3	3	3	3	3	3	3	3	3	4	3	4	4	4	4	2	3
5	.	.	.	.	.	.	.	.	.	.	.	.	.	.	.	.	.	.	.	.	.	.	.
7	.	.	.	.	.	.	.	.	.	.	.	.	.	.	.	.	.	.	.	.	.	.	.
	6_5	6_6	6_7	6_8	6_9	6_{10}	6_{11}	6_{12}	6_{13}	6_{14}	6_{15}	6_{16}	6_{17}	6_{18}	6_{19}	6_{20}	6_{21}	6_{22}	6_{23}	6_{24}	6_{25}	6_{26}	6_{27}
2P	3e	3c	3b	3f	3b	3b	3a	3b	3b	3a	3c	3c	3e	3e	3b	3g	3e	3d	3d	3d	3d	3b	3e
3P	2a	2b	2b	2a	2c	2f	2g	2e	2d	2f	2f	2g	2g	2f	2g	2a	2b	2d	2d	2e	2e	2i	2d
5P	6_5	6_6	6_7	6_8	6_9	6_{10}	6_{11}	6_{12}	6_{13}	6_{14}	6_{15}	6_{16}	6_{17}	6_{18}	6_{19}	6_{20}	6_{21}	6_{23}	6_{22}	6_{25}	6_{24}	6_{26}	6_{27}
7P	6_5	6_6	6_7	6_8	6_9	6_{10}	6_{11}	6_{12}	6_{13}	6_{14}	6_{15}	6_{16}	6_{17}	6_{18}	6_{19}	6_{20}	6_{21}	6_{22}	6_{23}	6_{24}	6_{25}	6_{26}	6_{27}
X.1	1	1	1	1	1	1	1	1	1	1	1	1	1	1	1	1	1	1	1	1	1	1	1
X.2	1	1	1	1	1	1	1	−1	−1	1	1	1	1	1	1	1	1	−1	−1	−1	−1	−1	−1
X.3	3	−6	3	3	3	−1	5	−3	−3	5	2	2	−1	−1	−1	3	3	.	.	.	.	1	3
X.4	3	−6	3	3	3	−1	5	3	3	5	2	2	−1	−1	−1	3	3	.	.	.	.	−1	−3
X.5	.	−6	6	−3	6	−2	1	.	.	1	−2	−2	4	4	−2	.	.	.	.	.	.	.	.
X.6	8	8	−1	−1	−1	3	3	3	3	3	.	.	.	.	3	−1	8	6	6	6	6	−1	.
X.7	−1	8	8	8	8	.	3	−2	−2	3	.	.	3	3	.	−1	−1	4	4	4	4	−2	1
X.8	−1	8	8	8	8	.	3	2	2	3	.	.	3	3	.	−1	−1	−4	−4	−4	−4	2	−1
X.9	8	8	−1	−1	−1	3	3	−3	−3	3	.	.	.	.	3	−1	8	−6	−6	−6	−6	1	.
X.10	.	−3	12	−6	12	4	−5	.	.	−5	1	1	4	4	4	.	.	.	.	.	.	.	.
X.11	9	9	9	9	9	1	10	−3	−3	10	1	1	1	1	1	.	9	−3	−3	−3	−3	−3	−3
X.12	9	9	9	9	9	1	10	3	3	10	1	1	1	1	1	.	9	3	3	3	3	3	3
X.13	−4	5	−4	−4	−4	.	12	−2	−2	12	−3	−3	.	.	.	5	−4	7	7	7	7	2	−2
X.14	−4	5	−4	−4	−4	.	12	2	2	12	−3	−3	.	.	.	5	−4	−7	−7	−7	−7	−2	2
X.15	.	27	.	.	.	.	−3	.	.	−3	3	3	.	.	.	.	.	−9	−9	−9	−9	.	.
X.16	.	27	.	.	.	.	−3	.	.	−3	3	3	.	.	.	.	.	9	9	9	9	.	.
X.17	.	−15	6	−3	6	6	−9	.	.	−9	−3	−3	.	.	6	.	.	.	.	.	.	.	.
X.18	3	21	3	3	3	−1	2	3	3	2	5	5	−1	−1	−1	3	3	3	3	3	3	−1	3
X.19	.	−15	6	−3	6	−2	7	.	.	7	1	1	4	4	−2	.	.	.	.	.	.	.	.
X.20	3	21	3	3	3	−1	2	−3	−3	2	5	5	−1	−1	−1	3	3	−3	−3	−3	−3	1	−3
X.21	.	12	24	−12	24	.	−9	.	.	−9	.	.	.	.	.	.	.	.	.	.	.	.	.
X.22	10	10	1	1	1	1	−8	−1	−1	−8	−2	−2	−2	−2	1	1	10	A	$\bar{A}$	$\bar{A}$	A	−1	2
X.23	10	10	1	1	1	1	−8	−1	−1	−8	−2	−2	−2	−2	1	1	10	$\bar{A}$	A	A	$\bar{A}$	−1	2
X.24	10	10	1	1	1	1	−8	1	1	−8	−2	−2	−2	−2	1	1	10	$-\bar{A}$	$-A$	$-A$	$-\bar{A}$	1	−2
X.25	10	10	1	1	1	1	−8	1	1	−8	−2	−2	−2	−2	1	1	10	$-A$	$-\bar{A}$	$-\bar{A}$	$-A$	1	−2
X.26	−9	−9	18	18	18	2	11	.	.	11	−1	−1	−1	−1	2	.	−9	3	3	3	3	.	−3
X.27	18	−9	−9	−9	−9	−1	11	−3	−3	11	−1	−1	2	2	−1	.	18	3	3	3	3	−3	.
X.28	−9	−9	18	18	18	2	11	.	.	11	−1	−1	−1	−1	2	.	−9	−3	−3	−3	−3	.	3
X.29	18	−9	−9	−9	−9	−1	11	3	3	11	−1	−1	2	2	−1	.	18	−3	−3	−3	−3	3	.
X.30	9	−5	−2	.	6	6	.	6	6	.	3	3	9	9	6	.	1	.	.	.	.	6	3
X.31	9	−5	−2	.	6	6	.	−6	−6	.	3	3	9	9	6	.	1	.	.	.	.	−6	−3
X.32	6	−39	6	6	6	−2	4	−6	−6	4	1	1	−2	−2	−2	−3	6	−3	−3	−3	−3	2	.
X.33	6	−39	6	6	6	−2	4	6	6	4	1	1	−2	−2	−2	−3	6	3	3	3	3	−2	.
X.34	.	−3	30	−15	30	−2	−2	.	.	−2	1	1	4	4	−2	.	.	.	.	.	.	.	.
X.35	2	20	20	20	20	−4	−16	.	.	−16	−4	−4	2	2	−4	2	2	.	.	.	.	.	.
X.36	2	−34	2	2	2	2	−16	.	.	−16	2	2	2	2	2	2	2	D	$\bar{D}$	$\bar{D}$	D	.	.
X.37	2	−34	2	2	2	2	−16	.	.	−16	2	2	2	2	2	2	2	$\bar{D}$	D	D	$\bar{D}$	.	.
X.38	.	36	18	−9	18	2	5	.	.	5	−4	−4	−4	−4	2	.	.	.	.	.	.	.	.
X.39	.	6	12	−6	12	−4	−16	.	.	−16	2	2	−4	−4	−4	.	.	.	.	.	.	.	.
X.40	.	18	−18	9	−18	−2	−8	.	.	−8	−2	−2	4	4	−2	.	.	.	.	.	.	.	.
X.41	.	18	−18	9	−18	−2	−8	.	.	−8	−2	−2	4	4	−2	.	.	.	.	.	.	.	.
X.42	.	.	.	.	.	.	9	.	.	9	.	.	.	.	.	.	.	.	.	.	.	.	.
X.43	.	.	.	.	.	.	9	.	.	9	.	.	.	.	.	.	.	.	.	.	.	.	.
X.44	.	.	.	−6	.	8	−2	.	.	2	−4	4	16	−16	−8	.	.	.	.	.	.	.	.
X.45	.	−24	24	−12	24	.	.	.	.	.	.	.	.	.	.	.	.	.	.	.	.	.	.
X.46	.	−33	−12	6	−12	−4	−4	.	.	−4	−1	−1	−4	−4	−4	.	.	.	.	.	.	.	.
X.47	−4	32	−4	−4	−4	.	.	4	4	.	.	.	.	.	.	−4	−4	−8	−8	−8	−8	.	−2
X.48	−4	32	−4	−4	−4	.	.	−4	−4	.	.	.	.	.	.	−4	−4	8	8	8	8	.	2
X.49	.	−9	−18	9	−18	6	−6	.	.	−6	3	3	.	.	6	.	.	.	.	.	.	.	.
X.50	−16	−16	−16	−16	−16	.	.	.	.	.	.	.	.	.	.	2	−16	.	.	.	.	.	.
X.51	−8	.	.	4	.	−8	−4	.	.	4	4	−4	8	−8	8	−8	.	H	$\bar{H}$	H	$\bar{H}$	.	.
X.52	−8	.	.	4	.	−8	−4	.	.	4	4	−4	8	−8	8	−8	.	$\bar{H}$	H	$\bar{H}$	H	.	.
X.53	.	.	.	.	.	.	−9	.	.	−9	.	.	.	.	.	.	.	.	.	.	.	.	.
X.54	.	−27	.	.	.	.	12	.	.	12	−3	−3	.	.	.	.	.	.	.	.	.	.	.
X.55	.	.	9	.	1	9	.	9	9	.	.	.	.	.	9	9	.	.	.	.	.	1	.
X.56	.	.	9	.	1	9	.	−9	−9	.	.	.	.	.	9	9	.	.	.	.	.	−1	.
X.57	.	27	.	.	.	.	15	.	.	15	3	3	.	.	.	.	.	.	.	.	.	.	.
X.58	6	4	−8	3	−8	.	.	.	.	.	12	12	6	6	.	6	−2	.	.	.	.	8	.
X.59	6	4	−8	3	−8	.	.	.	.	.	12	12	6	6	.	6	−2	.	.	.	.	−8	.
X.60	27	15	.	.	.	−12	.	12	12	.	3	3	−9	−9	−12	.	3	.	.	.	.	.	−3
X.61	27	15	.	.	.	−12	.	−12	−12	.	3	3	−9	−9	−12	.	3	.	.	.	.	.	3
X.62	−44	.	.	4	.	−8	8	.	.	−8	−8	8	−4	4	8	−8	.	6	6	−6	−6	.	−6
X.63	−8	.	.	22	.	4	8	4	−4	−8	−8	8	8	−8	−4	−8	.	2	2	−2	−2	.	8
X.64	−8	.	.	22	.	4	8	−4	4	−8	−8	8	8	−8	−4	−8	.	−2	−2	2	2	.	−8
X.65	−44	.	.	4	.	−8	8	.	.	−8	−8	8	−4	4	8	−8	.	−6	−6	6	6	.	6
X.66	−18	−5	−8	.	24	.	.	.	.	.	−9	−9	18	18	.	.	−2	.	.	.	.	.	.
X.67	−28	.	.	14	.	4	2	20	−20	−2	16	−16	−4	4	−4	8	.	−2	−2	2	2	.	−2
X.68	−28	.	.	14	.	4	2	−20	20	−2	16	−16	−4	4	−4	8	.	2	2	−2	−2	.	2
X.69	−28	.	.	−4	.	−8	2	−16	16	−2	4	−4	−4	4	8	8	.	I	$\bar{I}$	$-\bar{I}$	$-I$	.	−2
X.70	−28	.	.	−4	.	−8	2	−16	16	−2	4	−4	−4	4	8	8	.	$\bar{I}$	I	$-I$	$-\bar{I}$	.	−2
X.71	−28	.	.	−4	.	−8	2	16	−16	−2	4	−4	−4	4	8	8	.	$-\bar{I}$	$-I$	I	$\bar{I}$	.	2
X.72	−28	.	.	−4	.	−8	2	16	−16	−2	4	−4	−4	4	8	8	.	$-I$	$-\bar{I}$	$\bar{I}$	I	.	2
X.73	.	15	−6	.	18	6	.	−6	−6	.	3	3	.	.	6	.	.	.	.	.	.	6	6
X.74	27	−30	.	.	.	12	.	12	12	.	6	6	−9	−9	12	.	3	.	.	.	.	.	−3
X.75	27	−30	.	.	.	12	.	−12	−12	.	6	6	−9	−9	12	.	3	.	.	.	.	.	3
X.76	.	15	−6	.	18	6	.	6	6	.	3	3	.	.	6	.	.	.	.	.	.	−6	−6
X.77	18	35	−4	.	12	−12	.	.	.	.	3	3	18	18	−12	.	2	.	.	.	.	.	.
X.78	45	−10	2	.	−6	6	.	−6	−6	.	−6	−6	9	9	6	.	5	.	.	.	.	6	−3
X.79	45	−10	2	.	−6	6	.	6	6	.	−6	−6	9	9	6	.	5	.	.	.	.	−6	3
X.80	24	.	.	−12	.	−8	8	.	.	−8	16	−16	8	−8	8	−12	.	.	.	.	.	.	.
X.81	24	.	.	−12	.	−8	8	.	.	−8	16	−16	8	−8	8	−12	.	.	.	.	.	.	.
X.82	24	16	4	12	−20	.	.	.	.	.	.	.	.	.	.	6	−8	.	.	.	.	4	.
X.83	24	16	4	12	−20	.	.	.	.	.	.	.	.	.	.	6	−8	.	.	.	.	−4	.
X.84	.	.	18	.	2	18	.	18	18	.	.	.	.	.	18	−9	.	.	.	.	.	2	.
X.85	.	.	18	.	2	18	.	−18	−18	.	.	.	.	.	18	−9	.	.	.	.	.	−2	.
X.86	.	.	−9	.	−1	−9	.	−9	−9	.	.	.	.	.	−9	18	.	.	.	.	.	−1	.
X.87	.	.	−9	.	−1	−9	.	9	9	.	.	.	.	.	−9	18	.	.	.	.	.	1	.
X.88	16	.	.	28	.	8	4	.	.	−4	8	−8	16	−16	−8	16	.	.	.	.	.	.	.

Character table of $H(\mathrm{Fi}_{24}')$ (continued)

	6_{28}	6_{29}	6_{30}	6_{31}	6_{32}	6_{33}	6_{34}	6_{35}	6_{36}	6_{37}	$7a$	$8a$	$8b$	$8c$	$8d$	$8e$	$8f$	$8g$	$8h$	$8i$	$8j$	$8k$	$8l$	$9a$	$9b$	$9c$	$9d$	$9e$
2	5	5	5	5	5	5	5	7	5	3	1	8	9	8	8	8	8	8	8	6	6	6	6	5	3	1	2	2
3	3	3	3	3	3	3	3	1	2	2	1	2	1	1	.	.	.	.	.	1	.	.	.	4	4	4	3	3
5	.	.	.	.	.	.	.	.	.	.	.	.	.	.	.	.	.	.	.	.	.	.	.	.	.	.	.	.
7	.	.	.	.	.	.	.	.	.	.	1	.	.	.	.	.	.	.	.	.	.	.	.	.	.	.	.	.
2P	3d	3e	3e	3d	3e	3f	3f	3e	3g	3g	7a	4a	4b	4b	4j	4j	4f	4j	4f	4g	4r	4r	4i	9a	9b	9c	9e	9d
3P	2g	2g	2f	2f	2e	2g	2f	2h	2c	2i	7a	8a	8b	8c	8d	8e	8f	8g	8h	8i	8j	8k	8l	3c	3c	3c	3d	3d
5P	6_{28}	6_{29}	6_{30}	6_{31}	6_{32}	6_{33}	6_{34}	6_{35}	6_{36}	6_{37}	7a	8a	8b	8c	8d	8e	8f	8g	8h	8i	8j	8k	8l	9a	9b	9c	9e	9d
7P	6_{28}	6_{29}	6_{30}	6_{31}	6_{32}	6_{33}	6_{34}	6_{35}	6_{36}	6_{37}	1a	8a	8b	8c	8d	8e	8f	8g	8h	8i	8j	8k	8l	9a	9b	9c	9d	9e
X.1	1	1	1	1	1	1	1	1	1	1	1	1	1	1	1	1	1	1	1	1	1	1	1	1	1	1	1	1
X.2	1	1	1	1	-1	1	1	1	1	-1	1	-1	1	-1	-1	-1	1	1	-1	1	-1	1	-1	1	1	1	1	1
X.3	2	-1	-1	2	3	-1	-1	-1	3	1	.	1	1	1	-3	1	1	1	1	-1	-1	-1	-1	.	.	.	.	.
X.4	2	-1	-1	2	-3	-1	-1	-1	3	-1	.	-1	1	-1	3	-1	1	1	-1	-1	1	-1	1	.	.	.	.	.
X.5	4	-2	-2	4	.	1	1	4	.	.	2	.	6	.	.	.	-2	-2	.	2	.	2	.	.	-3	3	.	.
X.6	.	.	.	.	.	3	3	.	-1	-1	.	-1	3	3	-1	3	-1	-1	3	-1	-1	-1	-1	-1	-1	-1	2	2
X.7	.	3	3	.	1	.	.	3	-1	1	.	-5	3	-1	3	-1	-1	-1	-1	-1	-1	-1	-1	2	2	2	-1	-1
X.8	.	3	3	.	-1	.	.	3	-1	-1	.	5	3	1	-3	1	-1	-1	1	-1	1	-1	1	2	2	2	-1	-1
X.9	.	.	.	.	.	3	3	.	-1	1	.	1	3	-3	1	-3	-1	-1	-3	-1	1	-1	1	-1	-1	-1	2	2
X.10	-2	-2	-2	-2	.	-2	-2	4	.	.	.	.	2	.	.	.	2	2	.	-2	.	-2	.	3	-3	.	.	.
X.11	1	1	1	1	-3	1	1	1	.	.	-1	-6	-2	-2	-2	-2	2	2	-2	.	.	.	.	.	.	.	.	.
X.12	1	1	1	1	3	1	1	1	.	.	-1	6	-2	2	2	2	2	2	2	.	.	.	.	.	.	.	.	.
X.13	-3	.	.	-3	-2	.	.	.	5	-1	.	-4	4	.	-4	.	.	.	.	.	.	.	.	-1	-1	-1	-1	-1
X.14	-3	.	.	-3	2	.	.	.	5	1	.	4	4	.	4	.	.	.	.	.	.	.	.	-1	-1	-1	-1	-1
X.15	3	.	.	3	.	.	.	.	.	.	.	-9	5	3	-1	3	1	1	3	1	-1	1	-1	.	.	.	.	.
X.16	3	.	.	3	.	.	.	.	.	.	.	9	5	-3	1	-3	1	1	-3	1	1	1	1	.	.	.	.	.
X.17	6	.	.	6	.	-3	-3	.	.	.	.	.	2	.	.	.	2	2	.	2	.	2	.	-3	3	.	.	.
X.18	5	-1	-1	5	3	-1	-1	-1	3	-1	.	-10	-2	-2	2	-2	-2	-2	-2	.	.	.	.	.	.	.	.	.
X.19	-2	-2	-2	-2	.	1	1	4	.	.	.	.	10	.	.	.	2	2	.	-2	.	-2	.	-3	3	.	.	.
X.20	5	-1	-1	5	-3	-1	-1	-1	3	1	.	10	-2	2	-2	2	-2	-2	2	.	.	.	.	.	.	.	.	.
X.21	.	.	.	.	.	.	.	.	.	.	.	.	2	.	.	.	2	2	.	2	.	2	.	3	.	-3	.	.
X.22	-2	-2	-2	-2	2	1	1	-2	1	-1	.	8	.	.	.	.	.	.	.	.	.	.	.	1	1	1	B	$\bar{B}$
X.23	-2	-2	-2	-2	2	1	1	-2	1	-1	.	8	.	.	.	.	.	.	.	.	.	.	.	1	1	1	$\bar{B}$	B
X.24	-2	-2	-2	-2	-2	1	1	-2	1	1	.	-8	.	.	.	.	.	.	.	.	.	.	.	1	1	1	$\bar{B}$	B
X.25	-2	-2	-2	-2	-2	1	1	-2	1	1	.	-8	.	.	.	.	.	.	.	.	.	.	.	1	1	1	B	$\bar{B}$
X.26	-1	-1	-1	-1	-3	2	2	-1	.	.	.	9	-1	1	5	1	-1	-1	1	1	-1	1	-1	.	.	.	.	.
X.27	-1	2	2	-1	.	-1	-1	2	.	.	.	3	-1	3	-1	3	-1	-1	3	1	1	1	1	.	.	.	.	.
X.28	-1	-1	-1	-1	3	2	2	-1	.	.	.	-9	-1	-1	-5	-1	-1	-1	-1	1	1	1	1	.	.	.	.	.
X.29	-1	2	2	-1	.	-1	-1	2	.	.	.	-3	-1	-3	1	-3	-1	-1	-3	1	-1	1	-1	.	.	.	.	.
X.30	.	3	3	.	3	.	.	1	.	.	.	.	2	-4	.	.	2	-2	.	.	-2	.	2	9	.	.	.	.
X.31	.	3	3	.	-3	.	.	1	.	.	.	.	2	4	.	.	2	-2	.	.	2	.	-2	9	.	.	.	.
X.32	1	-2	-2	1	.	-2	-2	-2	-3	-1	.	-4	4	.	4	.	.	.	.	.	.	.	.	.	.	.	.	.
X.33	1	-2	-2	1	.	-2	-2	-2	-3	1	.	4	4	.	-4	.	.	.	.	.	.	.	.	.	.	.	.	.
X.34	-2	-2	-2	-2	.	1	1	4	.	.	.	.	-4	.	.	.	-4	-4	.	.	.	.	.	.	3	-3	.	.
X.35	-4	2	2	-4	.	-4	-4	2	2	.	.	.	.	.	.	.	.	.	.	.	.	.	.	-1	-1	-1	2	2
X.36	2	2	2	2	.	2	2	2	2	.	.	.	.	.	.	.	.	.	.	.	.	.	.	-1	-1	-1	-1	-1
X.37	2	2	2	2	.	2	2	2	2	.	.	.	.	.	.	.	.	.	.	.	.	.	.	-1	-1	-1	-1	-1
X.38	8	2	2	8	.	-1	-1	-4	.	.	.	.	6	.	.	.	-2	-2	.	-2	.	-2	.	.	.	.	.	.
X.39	-4	2	2	-4	.	2	2	-4	.	.	.	.	.	.	.	.	.	.	.	.	.	.	.	-3	.	3	.	.
X.40	4	-2	-2	4	.	1	1	4	.	.	-1	.	.	.	.	.	.	.	.	.	.	.	.	.	.	.	.	.
X.41	4	-2	-2	4	.	1	1	4	.	.	-1	.	.	.	.	.	.	.	.	.	.	.	.	.	.	.	.	.
X.42	.	.	.	.	.	.	.	.	.	.	1	-9	-3	3	3	3	1	1	3	-1	1	-1	1	.	.	.	.	.
X.43	.	.	.	.	.	.	.	.	.	.	1	9	-3	-3	-3	-3	1	1	-3	-1	-1	-1	-1	.	.	.	.	.
X.44	-2	4	-4	2	.	-2	2	.	.	.	-2	.	.	.	.	.	.	.	.	.	.	.	.	12	.	-3	.	.
X.45	.	.	.	.	.	.	.	.	.	.	-2	.	.	.	.	.	.	.	.	.	.	.	.	.	-3	3	.	.
X.46	2	2	2	2	.	2	2	-4	.	.	.	.	8	.	.	.	.	.	.	.	.	.	.	3	.	-3	.	.
X.47	.	.	.	.	-2	.	.	.	-4	.	.	.	.	.	.	.	.	.	.	.	.	.	.	-1	-1	-1	-1	-1
X.48	.	.	.	.	2	.	.	.	-4	.	.	.	.	.	.	.	.	.	.	.	.	.	.	-1	-1	-1	-1	-1
X.49	-6	.	.	-6	.	-3	-3	.	.	.	.	.	4	.	.	.	-4	-4	.	.	.	.	.	.	.	.	.	.
X.50	.	.	.	.	.	.	.	.	2	.	-1	.	.	.	.	.	.	.	.	.	.	.	.	2	2	2	2	2
X.51	-1	-4	4	1	.	-4	4	.	.	.	-1	.	.	.	.	.	.	.	.	.	.	.	.	-4	2	-1	-1	-1
X.52	-1	-4	4	1	.	-4	4	.	.	.	-1	.	.	.	.	.	.	.	.	.	.	.	.	-4	2	-1	-1	-1
X.53	.	.	.	.	.	.	.	.	.	.	2	.	-6	.	.	.	2	2	.	-2	.	-2	.	.	.	.	.	.
X.54	6	.	.	6	.	.	.	.	.	.	.	.	-8	.	.	.	.	.	.	.	.	.	.	.	.	.	.	.
X.55	.	.	.	.	.	.	.	.	1	1	.	-1	-3	3	-1	-1	-3	1	-1	3	1	-1	1	.	.	.	.	.
X.56	.	.	.	.	.	.	.	.	1	-1	.	1	-3	-3	1	1	-3	1	1	3	-1	-1	-1	.	.	.	.	.
X.57	-6	.	.	-6	.	.	.	.	.	.	.	.	2	.	.	.	2	2	.	2	.	2	.	.	.	.	.	.
X.58	.	.	.	.	.	3	3	-2	-2	2	.	.	.	.	.	.	.	.	.	.	.	.	.	6	3	.	.	.
X.59	.	.	.	.	.	3	3	-2	-2	-2	.	.	.	.	.	.	.	.	.	.	.	.	.	6	3	.	.	.
X.60	.	3	3	.	-3	.	.	-1	.	.	.	.	4	4	.	-4	.	.	4	.	.	.	.	.	.	.	.	.
X.61	.	3	3	.	3	.	.	-1	.	.	.	.	4	-4	.	4	.	.	-4	.	.	.	.	.	.	.	.	.
X.62	2	2	-2	-2	6	-4	4	.	.	.	.	.	.	.	.	.	.	.	.	.	.	.	.	-4	2	-1	2	2
X.63	2	-4	4	-2	-8	2	-2	.	.	.	.	.	.	.	.	.	.	.	.	.	.	.	.	8	-4	2	-1	-1
X.64	2	-4	4	-2	8	2	-2	.	.	.	.	.	.	.	.	.	.	.	.	.	.	.	.	8	-4	2	-1	-1
X.65	2	2	-2	-2	-6	-4	4	.	.	.	.	.	.	.	.	.	.	.	.	.	.	.	.	-4	2	-1	2	2
X.66	.	.	.	.	.	.	.	2	.	.	.	.	4	.	.	.	-4	4	.	.	.	.	.	9	.	.	.	.
X.67	-4	2	-2	4	2	2	-2	.	.	.	.	.	.	.	.	.	.	.	.	.	.	.	.	4	-2	1	1	1
X.68	-4	2	-2	4	-2	2	-2	.	.	.	.	.	.	.	.	.	.	.	.	.	.	.	.	4	-2	1	1	1
X.69	-1	2	-2	1	2	-4	4	.	.	.	.	.	.	.	.	.	.	.	.	.	.	.	.	4	-2	1	B	$\bar{B}$
X.70	-1	2	-2	1	2	-4	4	.	.	.	.	.	.	.	.	.	.	.	.	.	.	.	.	4	-2	1	$\bar{B}$	B
X.71	-1	2	-2	1	-2	-4	4	.	.	.	.	.	.	.	.	.	.	.	.	.	.	.	.	4	-2	1	$\bar{B}$	B
X.72	-1	2	-2	1	-2	-4	4	.	.	.	.	.	.	.	.	.	.	.	.	.	.	.	.	4	-2	1	B	$\bar{B}$
X.73	.	-6	-6	.	6	.	.	.	.	.	.	.	-2	.	.	-4	2	-2	4	.	2	.	-2	.	.	.	.	.
X.74	.	-3	-3	.	-3	.	.	-1	.	.	.	.	6	-4	.	.	-2	2	.	.	2	.	-2	.	.	.	.	.
X.75	.	-3	-3	.	3	.	.	-1	.	.	.	.	6	4	.	.	-2	2	.	.	-2	.	2	.	.	.	.	.
X.76	.	-6	-6	.	-6	.	.	.	.	.	.	.	-2	.	.	4	2	-2	-4	.	-2	.	2	.	.	.	.	.
X.77	.	-6	-6	.	.	.	.	2	.	.	.	.	8	.	.	.	.	.	.	.	.	.	.	-9	.	.	.	.
X.78	.	3	3	.	-3	.	.	1	.	.	.	.	.	8	.	.	.	.	.	.	.	.	.	-9	.	.	.	.
X.79	.	3	3	.	3	.	.	1	.	.	.	.	.	-8	.	.	.	.	.	.	.	.	.	-9	.	.	.	.
X.80	-4	-4	4	4	.	-4	4	.	.	.	1	.	.	.	.	.	.	.	.	.	.	.	.	.	.	.	.	.
X.81	-4	-4	4	4	.	-4	4	.	.	.	1	.	.	.	.	.	.	.	.	.	.	.	.	.	.	.	.	.
X.82	.	.	.	.	.	.	.	.	-2	-2	.	.	.	.	.	.	.	.	.	.	.	.	.	6	3	.	.	.
X.83	.	.	.	.	.	.	.	.	-2	2	.	.	.	.	.	.	.	.	.	.	.	.	.	6	3	.	.	.
X.84	.	.	.	.	.	.	.	.	-1	-1	.	-5	-3	3	-1	-1	1	-3	-1	-3	-1	1	-1	.	.	.	.	.
X.85	.	.	.	.	.	.	.	.	-1	1	.	5	-3	-3	1	1	1	-3	1	-3	1	1	1	.	.	.	.	.
X.86	.	.	.	.	.	.	.	.	2	2	.	-5	-3	3	-1	-1	1	-3	-1	-3	-1	1	-1	.	.	.	.	.
X.87	.	.	.	.	.	.	.	.	2	-2	.	5	-3	-3	1	1	1	-3	1	-3	1	1	1	.	.	.	.	.
X.88	-2	-8	8	2	.	4	-4	.	.	.	.	.	.	.	.	.	.	.	.	.	.	.	.	-4	2	-1	2	2

Character table of $H(\mathsf{Fi}_{24}')$ *(continued)*

2	6	5	5	5	4	4	9	7	9	6	4	8	8	7	5	5	6	6	6	6	6	7
3	1	.	.	.	.	.	5	5	3	4	5	2	2	2	3	3	2	2	2	2	2	1
5	1	1	1	1	1	1	.	.	.	.	.	.	.	.	.	.	.	.	.	.	.	.
7	.	.	.	.	.	.	.	.	.	.	.	.	.	.	.	.	.	.	.	.	.	.
	10a	10b	10c	10d	10e	10f	12_1	12_2	12_3	12_4	12_5	12_6	12_7	12_8	12_9	12_{10}	12_{11}	12_{12}	12_{13}	12_{14}	12_{15}	12_{16}
2P	5a	5a	5a	5a	5a	5a	6_2	6_3	6_2	6_5	6_8	6_9	6_9	6_3	6_{20}	6_{20}	6_7	6_{14}	6_7	6_5	6_{14}	6_9
3P	10a	10c	10b	10d	10e	10f	4a	4a	4a	4a	4a	4e	4e	4b	4a	4a	4d	4g	4h	4b	4i	4c
5P	2a	2c	2c	2b	2d	2e	12_1	12_2	12_3	12_4	12_5	12_6	12_7	12_8	12_9	12_{10}	12_{11}	12_{12}	12_{13}	12_{14}	12_{15}	12_{16}
7P	10a	10c	10b	10d	10e	10f	12_1	12_2	12_3	12_4	12_5	12_6	12_7	12_8	12_9	12_{10}	12_{11}	12_{12}	12_{13}	12_{14}	12_{15}	12_{16}
X.1	1	1	1	1	1	1	1	1	1	1	1	1	1	1	1	1	1	1	1	1	1	1
X.2	1	1	1	1	−1	−1	1	1	1	1	1	−1	−1	1	1	1	−1	1	1	1	1	−1
X.3	1	1	1	1	−1	−1	3	−6	3	3	3	1	1	2	3	3	−3	1	−1	−1	1	−3
X.4	1	1	1	1	1	1	3	−6	3	3	3	−1	−1	2	3	3	3	1	−1	−1	1	3
X.5	.	.	.	.	.	.	6	−6	6	.	−3	.	.	−2	.	.	.	−3	−2	4	−3	.
X.6	.	.	.	.	.	.	−1	8	−1	8	−1	−1	−1	.	−1	−1	3	3	3	.	3	3
X.7	.	.	.	.	.	.	8	8	8	−1	8	−2	−2	.	−1	−1	−2	3	.	3	3	−2
X.8	.	.	.	.	.	.	8	8	8	−1	8	2	2	.	−1	−1	2	3	.	3	3	2
X.9	.	.	.	.	.	.	−1	8	−1	8	−1	1	1	.	−1	−1	−3	3	3	.	3	−3
X.10	2	2	2	2	.	.	12	−3	12	.	−6	.	.	1	.	.	.	−1	4	4	−1	.
X.11	.	.	.	.	.	.	9	9	9	9	9	−3	−3	1	.	.	−3	−2	1	1	−2	−3
X.12	.	.	.	.	.	.	9	9	9	9	9	3	3	1	.	.	3	−2	1	1	−2	3
X.13	.	.	.	.	.	.	−4	5	−4	−4	−4	2	2	−3	5	5	−2	4	.	.	4	−2
X.14	.	.	.	.	.	.	−4	5	−4	−4	−4	−2	−2	−3	5	5	2	4	.	.	4	2
X.15	−1	−1	−1	−1	1	1	.	27	.	.	.	.	.	3	.	.	.	5	.	.	5	.
X.16	−1	−1	−1	−1	−1	−1	.	27	.	.	.	.	.	3	.	.	.	5	.	.	5	.
X.17	.	.	.	.	.	.	6	−15	6	.	−3	.	.	−3	.	.	.	−1	6	.	−1	.
X.18	.	.	.	.	.	.	3	21	3	3	3	−1	−1	5	3	3	3	−2	−1	−1	−2	3
X.19	.	.	.	.	.	.	6	−15	6	.	−3	.	.	1	.	.	.	−5	−2	4	−5	.
X.20	.	.	.	.	.	.	3	21	3	3	3	1	1	5	3	3	−3	−2	−1	−1	−2	−3
X.21	.	.	.	.	.	.	24	12	24	.	−12	.	.	.	.	.	.	−1	.	.	−1	.
X.22	.	.	.	.	.	.	1	10	1	10	1	−1	−1	−2	1	1	−1	.	1	−2	.	−1
X.23	.	.	.	.	.	.	1	10	1	10	1	−1	−1	−2	1	1	−1	.	1	−2	.	−1
X.24	.	.	.	.	.	.	1	10	1	10	1	1	1	−2	1	1	1	.	1	−2	.	1
X.25	.	.	.	.	.	.	1	10	1	10	1	1	1	−2	1	1	1	.	1	−2	.	1
X.26	.	.	.	.	.	.	18	−9	18	−9	18	.	.	−1	.	.	.	−1	2	−1	−1	.
X.27	.	.	.	.	.	.	−9	−9	−9	18	−9	−3	−3	−1	.	.	−3	−1	−1	2	−1	−3
X.28	.	.	.	.	.	.	18	−9	18	−9	18	.	.	−1	.	.	.	−1	2	−1	−1	.
X.29	.	.	.	.	.	.	−9	−9	−9	18	−9	3	3	−1	.	.	3	−1	−1	2	−1	3
X.30	3	−1	−1	3	1	1	18	3	−6	−3	.	6	6	3	.	.	2	.	2	−3	.	−2
X.31	3	−1	−1	3	−1	−1	18	3	−6	−3	.	−6	−6	3	.	.	−2	.	2	−3	.	2
X.32	.	.	.	.	.	.	6	−39	6	6	6	2	2	1	−3	−3	−6	4	−2	−2	4	−6
X.33	.	.	.	.	.	.	6	−39	6	6	6	−2	−2	1	−3	−3	6	4	−2	−2	4	6
X.34	.	.	.	.	.	.	30	−3	30	.	−15	.	.	1	.	.	.	2	−2	4	2	.
X.35	.	.	.	.	.	.	20	20	20	2	20	.	.	−4	2	2	.	.	−4	2	.	.
X.36	.	.	.	.	.	.	2	−34	2	2	2	.	.	2	2	2	.	.	2	2	.	.
X.37	.	.	.	.	.	.	2	−34	2	2	2	.	.	2	2	2	.	.	2	2	.	.
X.38	.	.	.	.	.	.	18	36	18	.	−9	.	.	−4	.	.	.	−3	2	−4	−3	.
X.39	2	2	2	2	.	.	12	6	12	.	−6	.	.	2	.	.	.	.	−4	−4	.	.
X.40	.	.	.	.	.	.	−18	18	−18	.	9	.	.	−2	.	.	.	.	−2	4	.	.
X.41	.	.	.	.	.	.	−18	18	−18	.	9	.	.	−2	.	.	.	.	−2	4	.	.
X.42	−1	−1	−1	−1	−1	−1	.	.	.	.	.	.	.	.	.	.	.	−3	.	.	−3	.
X.43	−1	−1	−1	−1	1	1	.	.	.	.	.	.	.	.	.	.	.	−3	.	.	−3	.
X.44	−8	.	.	.	.	.	.	.	.	.	.	.	.	.	.	.	.	2	.	.	−2	.
X.45	−2	−2	−2	−2	.	.	24	−24	24	.	−12	.	.	.	.	.	.	.	.	.	.	.
X.46	.	.	.	.	.	.	−12	−33	−12	.	6	.	.	−1	.	.	.	−4	−4	−4	−4	.
X.47	1	1	1	1	−1	−1	−4	32	−4	−4	−4	.	.	.	−4	−4	4	.	.	.	.	4
X.48	1	1	1	1	1	1	−4	32	−4	−4	−4	.	.	.	−4	−4	−4	.	.	.	.	−4
X.49	.	.	.	.	.	.	−18	−9	−18	.	9	.	.	3	.	.	.	−2	6	.	−2	.
X.50	.	.	.	.	.	.	−16	−16	−16	−16	−16	.	.	.	2	2	.	.	.	.	.	.
X.51	.	.	.	.	.	.	.	.	.	.	.	.	.	.	.	.	.	−4	.	.	4	.
X.52	.	.	.	.	.	.	.	.	.	.	.	.	.	.	.	.	.	−4	.	.	4	.
X.53	−2	−2	−2	−2	.	.	.	.	.	.	.	.	.	.	.	.	.	3	.	.	3	.
X.54	2	2	2	2	.	.	.	−27	.	.	.	.	.	−3	.	.	.	4	.	.	4	.
X.55	6	2	2	−2	2	2	−27	.	−3	.	.	1	1	.	−3	−3	−3	.	−3	.	.	1
X.56	6	2	2	−2	−2	−2	−27	.	−3	.	.	−1	−1	.	−3	−3	3	.	−3	.	.	−1
X.57	.	.	.	.	.	.	.	27	.	.	.	.	.	3	.	.	.	−1	.	.	−1	.
X.58	6	−2	−2	−2	.	.	8	−4	8	2	−1	−8	−8	−4	2	2	.	.	.	2	.	.
X.59	6	−2	−2	−2	.	.	8	−4	8	2	−1	8	8	−4	2	2	.	.	.	2	.	.
X.60	3	−1	−1	3	1	1	.	−9	.	−9	.	.	.	3	.	.	4	.	−4	3	.	−4
X.61	3	−1	−1	3	−1	−1	.	−9	.	−9	.	.	.	3	.	.	−4	.	−4	3	.	4
X.62	−4	.	.	.	−2	2	.	.	.	.	.	.	.	.	.	.	.	.	.	.	.	.
X.63	−4	.	.	.	−2	2	.	.	.	.	.	12	−12	.	.	.	.	.	.	.	.	.
X.64	−4	.	.	.	2	−2	.	.	.	.	.	−12	12	.	.	.	.	.	.	.	.	.
X.65	−4	.	.	.	2	−2	.	.	.	.	.	.	.	.	.	.	.	.	.	.	.	.
X.66	.	.	.	.	.	.	72	3	−24	6	.	.	.	−9	.	.	.	.	.	−6	.	.
X.67	.	.	.	.	.	.	.	.	.	.	.	12	−12	.	.	.	.	−2	.	.	2	.
X.68	.	.	.	.	.	.	.	.	.	.	.	−12	12	.	.	.	.	−2	.	.	2	.
X.69	.	.	.	.	.	.	.	.	.	.	.	.	.	.	.	.	.	−2	.	.	2	.
X.70	.	.	.	.	.	.	.	.	.	.	.	.	.	.	.	.	.	−2	.	.	2	.
X.71	.	.	.	.	.	.	.	.	.	.	.	.	.	.	.	.	.	−2	.	.	2	.
X.72	.	.	.	.	.	.	.	.	.	.	.	.	.	.	.	.	.	−2	.	.	2	.
X.73	.	.	.	.	.	.	54	−9	−18	.	.	6	6	3	.	.	−2	.	2	.	.	2
X.74	.	.	.	.	.	.	.	18	.	−9	.	.	.	6	.	.	4	.	4	3	.	−4
X.75	.	.	.	.	.	.	.	18	.	−9	.	.	.	6	.	.	−4	.	4	3	.	4
X.76	.	.	.	.	.	.	54	−9	−18	.	.	−6	−6	3	.	.	2	.	2	.	.	−2
X.77	.	.	.	.	.	.	36	−21	−12	−6	.	.	.	3	.	.	.	.	−4	−6	.	.
X.78	.	.	.	.	.	.	−18	6	6	−15	.	6	6	−6	.	.	−2	.	2	−3	.	2
X.79	.	.	.	.	.	.	−18	6	6	−15	.	−6	−6	−6	.	.	2	.	2	−3	.	−2
X.80	.	.	.	.	.	.	.	.	.	.	.	.	.	.	.	.	.	.	.	.	.	.
X.81	.	.	.	.	.	.	.	.	.	.	.	.	.	.	.	.	.	.	.	.	.	.
X.82	−6	2	2	2	.	.	32	−16	8	8	−4	−4	−4	.	2	2	.	.	.	.	.	.
X.83	−6	2	2	2	.	.	32	−16	8	8	−4	4	4	.	2	2	.	.	.	.	.	.
X.84	.	.	.	.	.	.	−54	.	−6	.	.	2	2	.	3	3	−6	.	−6	.	.	2
X.85	.	.	.	.	.	.	−54	.	−6	.	.	−2	−2	.	3	3	6	.	−6	.	.	−2
X.86	.	.	.	.	.	.	27	.	3	.	.	−1	−1	.	−6	−6	3	.	3	.	.	−1
X.87	.	.	.	.	.	.	27	.	3	.	.	1	1	.	−6	−6	−3	.	3	.	.	1
X.88	.	.	.	.	.	.	.	.	.	.	.	.	.	.	.	.	.	−4	.	.	4	.

Character table of $H(\mathsf{Fi}_{24}')$ (continued)

2	7	5	5	5	6	4	4	4	4	4	4	4	5	5	4	4	4	4	1
3	1	2	2	2	1	2	2	2	2	2	2	2	1	1	1	1	1	1	1
5	.	.	.	.	.	.	.	.	.	.	.	.	.	.	.	.	.	.	.
7	.	.	.	.	.	.	.	.	.	.	.	.	.	.	.	.	.	.	1
	12_{17}	12_{18}	12_{19}	12_{20}	12_{21}	12_{22}	12_{23}	12_{24}	12_{25}	12_{26}	12_{27}	12_{28}	12_{29}	12_{30}	12_{31}	12_{32}	12_{33}	12_{34}	$14a$
$2P$	6_9	6_{15}	6_6	6_{15}	6_7	6_{21}	6_{21}	6_{36}	6_{31}	6_{31}	6_{36}	6_8	6_{35}	6_{35}	6_{17}	6_{31}	6_{11}	6_{31}	$7a$
$3P$	$4f$	$4g$	$4h$	$4i$	$4k$	$4d$	$4h$	$4e$	$4g$	$4i$	$4e$	$4b$	$4o$	$4m$	$4q$	$4n$	$4p$	$4n$	$14a$
$5P$	12_{17}	12_{18}	12_{19}	12_{20}	12_{21}	12_{22}	12_{23}	12_{24}	12_{25}	12_{26}	12_{27}	12_{28}	12_{29}	12_{30}	12_{31}	12_{34}	12_{33}	12_{32}	$14a$
$7P$	12_{17}	12_{18}	12_{19}	12_{20}	12_{21}	12_{22}	12_{23}	12_{24}	12_{25}	12_{26}	12_{27}	12_{28}	12_{29}	12_{30}	12_{31}	12_{32}	12_{33}	12_{34}	$2a$
$X.1$	1	1	1	1	1	1	1	1	1	1	1	1	1	1	1	1	1	1	1
$X.2$	1	1	1	1	−1	−1	1	−1	1	1	−1	1	−1	−1	−1	−1	1	−1	1
$X.3$	−1	−2	2	−2	1	3	−1	1	−2	−2	1	−1	1	1	1	.	1	.	.
$X.4$	−1	−2	2	−2	−1	−3	−1	−1	−2	−2	−1	−1	−1	−1	−1	.	1	.	.
$X.5$	−2	.	−2	.	.	.	−2	.	.	.	.	1	.	.	.	.	1	.	2
$X.6$	3	.	.	.	−1	.	.	−1	.	.	−1	3	.	.	.	2	−1	2	.
$X.7$	.	.	.	.	−2	1	3	1	.	.	1	.	−1	−1	−1	.	−1	.	.
$X.8$	.	.	.	.	2	−1	3	−1	.	.	−1	.	1	1	1	.	−1	.	.
$X.9$	3	.	.	.	1	.	.	1	.	.	1	3	.	.	.	−2	−1	−2	.
$X.10$	4	−1	1	−1	.	.	−2	.	2	2	.	−2	.	.	.	.	−1	.	.
$X.11$	1	1	1	1	−3	−3	1	.	1	1	.	1	1	1	1	1	2	1	−1
$X.12$	1	1	1	1	3	3	1	.	1	1	.	1	−1	−1	−1	−1	2	−1	−1
$X.13$	.	1	−3	1	2	−2	.	−1	1	1	−1	.	.	.	.	−1	.	−1	.
$X.14$	.	1	−3	1	−2	2	.	1	1	1	1	.	.	.	.	1	.	1	.
$X.15$	.	−1	3	−1	.	.	.	.	−1	−1	.	.	.	.	.	−1	1	−1	.
$X.16$	.	−1	3	−1	.	.	.	.	−1	−1	.	.	.	.	.	1	1	1	.
$X.17$	6	−1	−3	−1	.	.	.	.	2	2	.	−3	.	.	.	.	−1	.	.
$X.18$	−1	1	5	1	−1	3	−1	−1	1	1	−1	−1	1	1	1	−1	−2	−1	.
$X.19$	−2	1	1	1	.	.	−2	.	−2	−2	.	1	.	.	.	.	−1	.	.
$X.20$	−1	1	5	1	1	−3	−1	1	1	1	1	−1	−1	−1	−1	1	−2	1	.
$X.21$	.	2	.	2	.	.	.	.	−4	−4	.	.	.	.	.	.	−1	.	.
$X.22$	1	.	−2	.	−1	2	−2	−1	.	.	−1	1	.	.	.	.	.	.	.
$X.23$	1	.	−2	.	−1	2	−2	−1	.	.	−1	1	.	.	.	.	.	.	.
$X.24$	1	.	−2	.	1	−2	−2	1	.	.	1	1	.	.	.	.	.	.	.
$X.25$	1	.	−2	.	1	−2	−2	1	.	.	1	1	.	.	.	.	.	.	.
$X.26$	2	−1	−1	−1	.	−3	−1	.	−1	−1	.	2	1	1	1	−1	−1	−1	.
$X.27$	−1	−1	−1	−1	−3	.	2	.	−1	−1	.	−1	.	.	.	−1	−1	−1	.
$X.28$	2	−1	−1	−1	.	3	−1	.	−1	−1	.	2	−1	−1	−1	1	−1	1	.
$X.29$	−1	−1	−1	−1	3	.	2	.	−1	−1	.	−1	.	.	.	1	−1	1	.
$X.30$	−2	3	−1	3	−2	−1	−1	.	.	.	.	.	1	1	1	.	.	.	.
$X.31$	−2	3	−1	3	2	1	−1	.	.	.	.	.	−1	−1	−1	.	.	.	.
$X.32$	−2	1	1	1	2	.	−2	−1	1	1	−1	−2	.	.	.	1	.	1	.
$X.33$	−2	1	1	1	−2	.	−2	1	1	1	1	−2	.	.	.	−1	.	−1	.
$X.34$	−2	−1	1	−1	.	.	−2	.	2	2	.	1	.	.	.	.	2	.	.
$X.35$	−4	.	−4	.	.	.	2	.	.	.	.	−4	.	.	.	.	.	.	.
$X.36$	2	.	2	.	.	.	2	.	.	.	.	2	.	.	.	.	.	.	.
$X.37$	2	.	2	.	.	.	2	.	.	.	.	2	.	.	.	.	.	.	.
$X.38$	2	.	−4	.	.	.	2	.	.	.	.	−1	.	.	.	.	1	.	.
$X.39$	−4	.	2	.	.	.	2	.	.	.	.	2	.	.	.	.	.	.	.
$X.40$	−2	.	−2	.	.	.	−2	.	.	.	.	1	.	.	.	.	.	.	−1
$X.41$	−2	.	−2	.	.	.	−2	.	.	.	.	1	.	.	.	.	.	.	−1
$X.42$	.	.	.	.	.	.	.	.	.	.	.	.	.	.	.	.	1	.	1
$X.43$	.	.	.	.	.	.	.	.	.	.	.	.	.	.	.	.	1	.	1
$X.44$	.	2	.	−2	.	.	.	.	2	−2	.	.	.	.	.	.	.	.	2
$X.45$	.	.	.	.	.	.	.	.	.	.	.	.	.	.	.	.	.	.	−2
$X.46$	−4	−1	−1	−1	.	.	2	.	2	2	.	2	.	.	.	.	.	.	.
$X.47$	.	.	.	.	.	−2	.	.	.	.	.	.	.	.	.	.	.	.	.
$X.48$	.	.	.	.	.	2	.	.	.	.	.	.	.	.	.	.	.	.	.
$X.49$	6	1	3	1	.	.	.	.	−2	−2	.	−3	.	.	.	.	2	.	.
$X.50$	.	.	.	.	.	.	.	.	.	.	.	.	.	.	.	.	.	.	−1
$X.51$	.	2	.	−2	.	.	.	.	−1	1	.	.	.	.	.	$\bar{E}$	.	E	1
$X.52$	.	2	.	−2	.	.	.	.	−1	1	.	.	.	.	.	E	.	$\bar{E}$	1
$X.53$	.	.	.	.	.	.	.	.	.	.	.	.	.	.	.	.	−1	.	2
$X.54$	.	1	−3	1	.	.	.	.	−2	−2	.	.	.	.	.	.	.	.	.
$X.55$	1	.	.	.	1	.	.	1	.	.	1	.	.	.	.	.	.	.	.
$X.56$	1	.	.	.	−1	.	.	−1	.	.	−1	.	.	.	.	.	.	.	.
$X.57$	.	−1	3	−1	.	.	.	.	2	2	.	.	.	.	.	.	−1	.	.
$X.58$	.	.	.	.	.	.	.	−2	.	.	−2	−1	−2	−2	2	.	.	.	.
$X.59$	.	.	.	.	.	.	.	2	.	.	2	−1	2	2	−2	.	.	.	.
$X.60$	4	−3	−1	−3	.	1	−1	.	.	.	.	.	−1	−1	−1	.	.	.	.
$X.61$	4	−3	−1	−3	.	−1	−1	.	.	.	.	.	1	1	1	.	.	.	.
$X.62$	.	.	.	.	.	.	.	.	.	.	.	.	−2	2	.	.	.	.	.
$X.63$	.	.	.	.	.	.	.	.	.	.	.	.	.	.	.	.	.	.	.
$X.64$	.	.	.	.	.	.	.	.	.	.	.	.	.	.	.	.	.	.	.
$X.65$	.	.	.	.	.	.	.	.	.	.	.	.	2	−2	.	.	.	.	.
$X.66$	.	−3	3	−3	.	.	.	.	.	.	.	.	.	.	.	.	.	.	.
$X.67$	.	4	.	−4	.	.	.	.	−2	2	.	.	−2	2	.	.	.	.	.
$X.68$	.	4	.	−4	.	.	.	.	−2	2	.	.	2	−2	.	.	.	.	.
$X.69$	.	−2	.	2	.	.	.	.	1	−1	.	.	−2	2	.	$\bar{E}$	.	E	.
$X.70$	.	−2	.	2	.	.	.	.	1	−1	.	.	−2	2	.	E	.	$\bar{E}$	.
$X.71$	.	−2	.	2	.	.	.	.	1	−1	.	.	2	−2	.	$\bar{E}$	.	E	.
$X.72$	.	−2	.	2	.	.	.	.	1	−1	.	.	2	−2	.	E	.	$\bar{E}$	.
$X.73$	−2	−3	−1	−3	−2	−2	2	.	.	.	.	.	.	.	.	.	.	.	.
$X.74$	−4	.	−2	.	.	1	1	.	.	.	.	.	1	1	1	.	.	.	.
$X.75$	−4	.	−2	.	.	−1	1	.	.	.	.	.	−1	−1	−1	.	.	.	.
$X.76$	−2	−3	−1	−3	2	2	2	.	.	.	.	.	.	.	.	.	.	.	.
$X.77$	4	3	−1	3	.	.	2	.	.	.	.	.	.	.	.	.	.	.	.
$X.78$	−2	.	2	.	−2	1	−1	.	.	.	.	.	1	1	1	.	.	.	.
$X.79$	−2	.	2	.	2	−1	−1	.	.	.	.	.	−1	−1	−1	.	.	.	.
$X.80$	.	.	.	.	.	.	.	−6	.	.	6	.	.	.	.	.	.	.	−1
$X.81$	.	.	.	.	.	.	.	6	.	.	−6	.	.	.	.	.	.	.	−1
$X.82$	.	.	.	.	.	.	.	2	.	.	2	.	.	.	.	.	.	.	.
$X.83$	.	.	.	.	.	.	.	−2	.	.	−2	.	.	.	.	.	.	.	.
$X.84$	2	.	.	.	2	.	.	−1	.	.	−1	.	.	.	.	.	.	.	.
$X.85$	2	.	.	.	−2	.	.	1	.	.	1	.	.	.	.	.	.	.	.
$X.86$	−1	.	.	.	−1	.	.	2	.	.	2	.	.	.	.	.	.	.	.
$X.87$	−1	.	.	.	1	.	.	−2	.	.	−2	.	.	.	.	.	.	.	.
$X.88$	.	−4	.	4	.	.	.	.	2	−2	.	.	.	.	.	.	.	.	.

Character table of $H(\mathsf{Fi}_{24}')$ *(continued)*

2	1	5	5	3	1	4	2	2	2	2	2	2	4	3	3	3	1	1	5	5	3	3	4	4
3	1	.	4	4	4	2	3	3	2	2	2	2	.	.	.	.	1	1	2	2	2	2	1	1
5	1	.	.	.	.	.	.	.	.	.	.	.	1	1	1	1	.	.	.	.	.	.	.	.
7	.	.	.	.	.	.	.	.	.	.	.	.	.	.	.	.	1	1	.	.	.	.	.	.
	15*a*	16*a*	18*a*	18*b*	18*c*	18*d*	18*e*	18*f*	18*g*	18*h*	18*i*	18*j*	20*a*	20*b*	20*c*	20*d*	21*a*	21*b*	24*a*	24*b*	24*c*	24*d*	24*e*	24*f*
2*P*	15*a*	8*b*	9*a*	9*b*	9*c*	9*a*	9*e*	9*d*	9*e*	9*d*	9*d*	9*e*	10*a*	10*b*	10*d*	10*c*	21*b*	21*a*	12_3	12_1	12_{10}	12_9	12_{12}	12_{14}
3*P*	5*a*	16*a*	6_3	6_3	6_3	6_6	6_4	6_4	6_{22}	6_{24}	6_{23}	6_{25}	20*a*	20*d*	20*c*	20*b*	7*a*	7*a*	8*a*	8*a*	8*a*	8*a*	8*i*	8*c*
5*P*	3*a*	16*a*	18*a*	18*b*	18*c*	18*d*	18*f*	18*e*	18*i*	18*j*	18*g*	18*h*	4*a*	4*c*	4*d*	4*c*	21*a*	21*b*	24*a*	24*b*	24*c*	24*d*	24*i*	24*g*
7*P*	15*a*	16*a*	18*a*	18*b*	18*c*	18*d*	18*e*	18*f*	18*g*	18*h*	18*i*	18*j*	20*a*	20*d*	20*c*	20*b*	3*a*	3*a*	24*a*	24*b*	24*c*	24*d*	24*i*	24*g*
$X.1$	1	1	1	1	1	1	1	1	1	1	1	1	1	1	1	1	1	1	1	1	1	1	1	1
$X.2$	1	−1	1	1	1	1	1	1	−1	−1	−1	−1	1	−1	−1	−1	1	1	−1	−1	−1	−1	1	−1
$X.3$	1	−1	.	.	.	.	.	.	.	.	.	.	1	−1	−1	−1	.	.	1	1	1	1	−1	1
$X.4$	1	1	.	.	.	.	.	.	.	.	.	.	1	1	1	1	.	.	−1	−1	−1	−1	−1	−1
$X.5$	.	.	.	−3	3	.	.	.	.	.	.	.	.	.	.	.	−1	−1	.	.	.	.	−1	.
$X.6$	.	−1	−1	−1	−1	−1	2	2	.	.	.	.	.	.	.	.	.	.	−1	−1	−1	−1	−1	.
$X.7$	.	−1	2	2	2	2	−1	−1	1	1	1	1	.	.	.	.	.	.	−2	−2	1	1	−1	−1
$X.8$	.	1	2	2	2	2	−1	−1	−1	−1	−1	−1	.	.	.	.	.	.	2	2	−1	−1	−1	1
$X.9$	.	1	−1	−1	−1	−1	2	2	.	.	.	.	.	.	.	.	.	.	1	1	1	1	−1	.
$X.10$	−1	.	3	−3	.	3	.	.	.	.	.	.	2	.	.	.	.	.	.	.	.	.	1	.
$X.11$	.	.	.	.	.	.	.	.	.	.	.	.	.	.	.	.	−1	−1	−3	−3	.	.	.	1
$X.12$	.	.	.	.	.	.	.	.	.	.	.	.	.	.	.	.	−1	−1	3	3	.	.	.	−1
$X.13$	.	.	−1	−1	−1	−1	−1	−1	1	1	1	1	.	.	.	.	.	.	2	2	−1	−1	.	.
$X.14$	.	.	−1	−1	−1	−1	−1	−1	−1	−1	−1	−1	.	.	.	.	.	.	−2	−2	1	1	.	.
$X.15$	−1	−1	.	.	.	.	.	.	.	.	.	.	−1	1	1	1	.	.	.	.	.	.	1	.
$X.16$	−1	1	.	.	.	.	.	.	.	.	.	.	−1	−1	−1	−1	.	.	.	.	.	.	1	.
$X.17$	.	.	−3	3	.	−3	.	.	.	.	.	.	.	.	.	.	.	.	.	.	.	.	−1	.
$X.18$	.	.	.	.	.	.	.	.	.	.	.	.	.	.	.	.	.	.	−1	−1	−1	−1	.	1
$X.19$	.	.	−3	3	.	−3	.	.	.	.	.	.	.	.	.	.	.	.	.	.	.	.	1	.
$X.20$	.	.	.	.	.	.	.	.	.	.	.	.	.	.	.	.	.	.	1	1	1	1	.	−1
$X.21$	.	.	3	.	−3	3	.	.	.	.	.	.	.	.	.	.	.	.	.	.	.	.	−1	.
$X.22$	.	.	1	1	1	1	B	$\bar{B}$	C	$\bar{C}$	$\bar{C}$	C	.	.	.	.	.	.	−1	−1	−1	−1	.	.
$X.23$	.	.	1	1	1	1	$\bar{B}$	B	$\bar{C}$	C	C	$\bar{C}$	.	.	.	.	.	.	−1	−1	−1	−1	.	.
$X.24$	.	.	1	1	1	1	$\bar{B}$	B	$-\bar{C}$	$-C$	$-C$	$-\bar{C}$	.	.	.	.	.	.	1	1	1	1	.	.
$X.25$	.	.	1	1	1	1	B	$\bar{B}$	$-C$	$-\bar{C}$	$-\bar{C}$	$-C$	.	.	.	.	.	.	1	1	1	1	.	.
$X.26$	.	−1	.	.	.	.	.	.	.	.	.	.	.	.	.	.	.	.	.	.	.	.	1	1
$X.27$	.	1	.	.	.	.	.	.	.	.	.	.	.	.	.	.	.	.	−3	−3	.	.	1	.
$X.28$	.	1	.	.	.	.	.	.	.	.	.	.	.	.	.	.	.	.	.	.	.	.	1	−1
$X.29$	.	−1	.	.	.	.	.	.	.	.	.	.	.	.	.	.	.	.	3	3	.	.	1	.
$X.30$	.	.	9	.	.	1	.	.	.	.	.	.	−1	−1	1	−1	.	.	.	.	.	.	.	−1
$X.31$	.	.	9	.	.	1	.	.	.	.	.	.	−1	1	−1	1	.	.	.	.	.	.	.	1
$X.32$	.	.	.	.	.	.	.	.	.	.	.	.	.	.	.	.	.	.	2	2	−1	−1	.	.
$X.33$	.	.	.	.	.	.	.	.	.	.	.	.	.	.	.	.	.	.	−2	−2	1	1	.	.
$X.34$	.	.	.	3	−3	.	.	.	.	.	.	.	.	.	.	.	.	.	.	.	.	.	.	.
$X.35$	.	.	−1	−1	−1	−1	2	2	.	.	.	.	.	.	.	.	.	.	.	.	.	.	.	.
$X.36$	.	.	−1	−1	−1	−1	−1	−1	E	$\bar{E}$	$\bar{E}$	E	.	.	.	.	.	.	.	.	.	.	.	.
$X.37$	.	.	−1	−1	−1	−1	−1	−1	$\bar{E}$	E	E	$\bar{E}$	.	.	.	.	.	.	.	.	.	.	.	.
$X.38$	.	.	.	.	.	.	.	.	.	.	.	.	.	.	.	.	.	.	.	.	.	.	1	.
$X.39$	−1	.	−3	.	3	−3	.	.	.	.	.	.	2	.	.	.	.	.	.	.	.	.	.	.
$X.40$	.	.	.	.	.	.	.	.	.	.	.	.	.	.	.	.	F	G	.	.	.	.	.	.
$X.41$	.	.	.	.	.	.	.	.	.	.	.	.	.	.	.	.	G	F	.	.	.	.	.	.
$X.42$	−1	1	.	.	.	.	.	.	.	.	.	.	−1	−1	−1	−1	1	1	.	.	.	.	−1	.
$X.43$	−1	−1	.	.	.	.	.	.	.	.	.	.	−1	1	1	1	1	1	.	.	.	.	−1	.
$X.44$	−1	.	−12	.	3	.	.	.	.	.	.	.	.	.	.	.	1	1	.	.	.	.	.	.
$X.45$	1	.	.	−3	3	.	.	.	.	.	.	.	−2	.	.	.	1	1	.	.	.	.	.	.
$X.46$	.	.	3	.	−3	3	.	.	.	.	.	.	.	.	.	.	.	.	.	.	.	.	.	.
$X.47$	1	.	−1	−1	−1	−1	−1	−1	1	1	1	1	1	−1	−1	−1	.	.	.	.	.	.	.	.
$X.48$	1	.	−1	−1	−1	−1	−1	−1	−1	−1	−1	−1	1	1	1	1	.	.	.	.	.	.	.	.
$X.49$	.	.	.	.	.	.	.	.	.	.	.	.	.	.	.	.	.	.	.	.	.	.	.	.
$X.50$	.	.	2	2	2	2	2	2	.	.	.	.	.	.	.	.	−1	−1	.	.	.	.	.	.
$X.51$	.	.	4	−2	1	.	1	1	E	E	$\bar{E}$	$\bar{E}$	.	.	.	.	−1	−1	.	.	.	.	.	.
$X.52$	.	.	4	−2	1	.	1	1	$\bar{E}$	$\bar{E}$	E	E	.	.	.	.	−1	−1	.	.	.	.	.	.
$X.53$	1	.	.	.	.	.	.	.	.	.	.	.	−2	.	.	.	−1	−1	.	.	.	.	1	.
$X.54$	−1	.	.	.	.	.	.	.	.	.	.	.	2	.	.	.	.	.	.	.	.	.	.	.
$X.55$	.	−1	.	.	.	.	.	.	.	.	.	.	−2	.	−2	.	.	.	−1	−1	−1	−1	.	.
$X.56$	.	1	.	.	.	.	.	.	.	.	.	.	−2	.	2	.	.	.	1	1	1	1	.	.
$X.57$	.	.	.	.	.	.	.	.	.	.	.	.	.	.	.	.	.	.	.	.	.	.	−1	.
$X.58$	.	.	6	3	.	−2	.	.	.	.	.	.	2	.	.	.	.	.	.	.	.	.	.	.
$X.59$	.	.	6	3	.	−2	.	.	.	.	.	.	2	.	.	.	.	.	.	.	.	.	.	.
$X.60$	.	.	.	.	.	.	.	.	.	.	.	.	−1	−1	1	−1	.	.	.	.	.	.	.	1
$X.61$	.	.	.	.	.	.	.	.	.	.	.	.	−1	1	−1	1	.	.	.	.	.	.	.	−1
$X.62$	1	.	4	−2	1	.	−2	−2	.	.	.	.	.	.	.	.	.	.	.	.	.	.	.	.
$X.63$	1	.	−8	4	−2	.	1	1	−1	1	−1	1	.	.	.	.	.	.	.	.	.	.	.	.
$X.64$	1	.	−8	4	−2	.	1	1	1	−1	1	−1	.	.	.	.	.	.	.	.	.	.	.	.
$X.65$	1	.	4	−2	1	.	−2	−2	.	.	.	.	.	.	.	.	.	.	.	.	.	.	.	.
$X.66$	.	.	9	.	.	1	.	.	.	.	.	.	.	.	.	.	.	.	.	.	.	.	.	.
$X.67$	.	.	−4	2	−1	.	−1	−1	1	−1	1	−1	.	.	.	.	.	.	.	.	.	.	.	.
$X.68$	.	.	−4	2	−1	.	−1	−1	−1	1	−1	1	.	.	.	.	.	.	.	.	.	.	.	.
$X.69$	.	.	−4	2	−1	.	$-B$	$-\bar{B}$	$-C$	$\bar{C}$	$-\bar{C}$	C	.	.	.	.	.	.	.	.	.	.	.	.
$X.70$	.	.	−4	2	−1	.	$-\bar{B}$	$-B$	$-\bar{C}$	C	$-C$	$\bar{C}$	.	.	.	.	.	.	.	.	.	.	.	.
$X.71$	.	.	−4	2	−1	.	$-\bar{B}$	$-B$	$\bar{C}$	$-C$	C	$-\bar{C}$	.	.	.	.	.	.	.	.	.	.	.	.
$X.72$	.	.	−4	2	−1	.	$-B$	$-\bar{B}$	C	$-\bar{C}$	$\bar{C}$	$-C$	.	.	.	.	.	.	.	.	.	.	.	.
$X.73$	.	.	.	.	.	.	.	.	.	.	.	.	.	.	.	.	.	.	.	.	.	.	.	.
$X.74$	.	.	.	.	.	.	.	.	.	.	.	.	.	.	.	.	.	.	.	.	.	.	.	−1
$X.75$	.	.	.	.	.	.	.	.	.	.	.	.	.	.	.	.	.	.	.	.	.	.	.	1
$X.76$	.	.	.	.	.	.	.	.	.	.	.	.	.	.	.	.	.	.	.	.	.	.	.	.
$X.77$	.	.	−9	.	.	−1	.	.	.	.	.	.	.	.	.	.	.	.	.	.	.	.	.	.
$X.78$	.	.	−9	.	.	−1	.	.	.	.	.	.	.	.	.	.	.	.	.	.	.	.	.	−1
$X.79$	.	.	−9	.	.	−1	.	.	.	.	.	.	.	.	.	.	.	.	.	.	.	.	.	1
$X.80$	.	.	.	.	.	.	.	.	.	.	.	.	.	.	.	.	1	1	.	.	.	.	.	.
$X.81$	.	.	.	.	.	.	.	.	.	.	.	.	.	.	.	.	1	1	.	.	.	.	.	.
$X.82$	.	.	6	3	.	−2	.	.	.	.	.	.	−2	.	.	.	.	.	−6	6	.	.	.	.
$X.83$	.	.	6	3	.	−2	.	.	.	.	.	.	−2	.	.	.	.	.	6	−6	.	.	.	.
$X.84$	.	1	.	.	.	.	.	.	.	.	.	.	.	.	.	.	.	.	−2	−2	1	1	.	.
$X.85$	.	−1	.	.	.	.	.	.	.	.	.	.	.	.	.	.	.	.	2	2	−1	−1	.	.
$X.86$	.	1	.	.	.	.	.	.	.	.	.	.	.	.	.	.	.	.	1	1	−2	−2	.	.
$X.87$	.	−1	.	.	.	.	.	.	.	.	.	.	.	.	.	.	.	.	−1	−1	2	2	.	.
$X.88$	.	.	4	−2	1	.	−2	−2	.	.	.	.	.	.	.	.	.	.	.	.	.	.	.	.

Character table of $H(\mathsf{Fi}'_{24})$ (continued)

2	4	4	4	1	4	4	2	1	1
3	1	1	1	1	3	3	3	1	1
5	.	.	.	1	.	.	.	.	.
7	.	.	.	.	.	.	.	1	1
	$24g$	$24h$	$24i$	$30a$	$36a$	$36b$	$36c$	$42a$	$42b$
$2P$	12_{14}	12_8	12_{12}	$15a$	$18a$	$18a$	$18b$	$21a$	$21b$
$3P$	$8c$	$8b$	$8i$	$10a$	12_2	12_2	12_2	$14a$	$14a$
$5P$	$24f$	$24h$	$24e$	6_1	$36a$	$36b$	$36c$	$42a$	$42b$
$7P$	$24f$	$24h$	$24e$	$30a$	$36a$	$36b$	$36c$	6_1	6_1
$X.1$	1	1	1	1	1	1	1	1	1
$X.2$	−1	1	1	1	1	1	1	1	1
$X.3$	1	−2	−1	1	.	.	.	.	.
$X.4$	−1	−2	−1	1	.	.	.	.	.
$X.5$	.	.	−1	.	.	.	−3	−1	−1
$X.6$	.	.	−1	.	−1	−1	−1	.	.
$X.7$	−1	.	−1	.	2	2	2	.	.
$X.8$	1	.	−1	.	2	2	2	.	.
$X.9$	.	.	−1	.	−1	−1	−1	.	.
$X.10$	.	−1	1	−1	3	3	−3	.	.
$X.11$	1	1	.	.	.	.	.	−1	−1
$X.12$	−1	1	.	.	.	.	.	−1	−1
$X.13$	.	1	.	.	−1	−1	−1	.	.
$X.14$	.	1	.	.	−1	−1	−1	.	.
$X.15$	.	−1	1	−1	.	.	.	.	.
$X.16$	.	−1	1	−1	.	.	.	.	.
$X.17$	.	−1	−1	.	−3	−3	3	.	.
$X.18$	1	1	.	.	.	.	.	.	.
$X.19$	.	1	1	.	−3	−3	3	.	.
$X.20$	−1	1	.	.	.	.	.	.	.
$X.21$	.	2	−1	.	3	3	.	.	.
$X.22$	.	.	.	.	1	1	1	.	.
$X.23$	.	.	.	.	1	1	1	.	.
$X.24$	.	.	.	.	1	1	1	.	.
$X.25$	.	.	.	.	1	1	1	.	.
$X.26$	1	−1	1	.	.	.	.	.	.
$X.27$	.	−1	1	.	.	.	.	.	.
$X.28$	−1	−1	1	.	.	.	.	.	.
$X.29$	.	−1	1	.	.	.	.	.	.
$X.30$	−1	−1	.	.	−3	−3	.	.	.
$X.31$	1	−1	.	.	−3	−3	.	.	.
$X.32$	.	1	.	.	.	.	.	.	.
$X.33$	.	1	.	.	.	.	.	.	.
$X.34$	.	−1	.	.	.	.	3	.	.
$X.35$	.	.	.	.	−1	−1	−1	.	.
$X.36$	.	.	.	.	−1	−1	−1	.	.
$X.37$	.	.	.	.	−1	−1	−1	.	.
$X.38$	.	.	1	.	.	.	.	.	.
$X.39$	.	.	.	−1	−3	−3	.	.	.
$X.40$	.	.	.	.	.	.	.	G	F
$X.41$	.	.	.	.	.	.	.	F	G
$X.42$	.	.	−1	−1	.	.	.	1	1
$X.43$	.	.	−1	−1	.	.	.	1	1
$X.44$	.	.	.	1	.	.	.	−1	−1
$X.45$	.	.	.	1	.	.	−3	1	1
$X.46$	.	−1	.	.	3	3	.	.	.
$X.47$	.	.	.	1	−1	−1	−1	.	.
$X.48$	.	.	.	1	−1	−1	−1	.	.
$X.49$	.	1	.	.	.	.	.	.	.
$X.50$	.	.	.	.	2	2	2	−1	−1
$X.51$	.	.	.	.	.	.	.	1	1
$X.52$	.	.	.	.	.	.	.	1	1
$X.53$	.	.	1	1	.	.	.	−1	−1
$X.54$	.	1	.	−1	.	.	.	.	.
$X.55$	.	.	.	.	.	.	.	.	.
$X.56$	.	.	.	.	.	.	.	.	.
$X.57$	.	−1	−1	.	.	.	.	.	.
$X.58$	.	.	.	.	2	2	−1	.	.
$X.59$	.	.	.	.	2	2	−1	.	.
$X.60$	1	1	.	.	.	.	.	.	.
$X.61$	−1	1	.	.	.	.	.	.	.
$X.62$	.	.	.	−1	.	.	.	.	.
$X.63$	.	.	.	−1	.	.	.	.	.
$X.64$	.	.	.	−1	.	.	.	.	.
$X.65$	.	.	.	−1	.	.	.	.	.
$X.66$	.	1	.	.	−3	−3	.	.	.
$X.67$	.	.	.	.	.	.	.	.	.
$X.68$	.	.	.	.	.	.	.	.	.
$X.69$	.	.	.	.	.	.	.	.	.
$X.70$	.	.	.	.	.	.	.	.	.
$X.71$	.	.	.	.	.	.	.	.	.
$X.72$	.	.	.	.	.	.	.	.	.
$X.73$	.	1	.	.	.	.	.	.	.
$X.74$	−1	.	.	.	.	.	.	.	.
$X.75$	1	.	.	.	.	.	.	.	.
$X.76$	.	1	.	.	.	.	.	.	.
$X.77$	.	−1	.	.	3	3	.	.	.
$X.78$	−1	.	.	.	3	3	.	.	.
$X.79$	1	.	.	.	3	3	.	.	.
$X.80$	.	.	.	.	.	.	.	−1	−1
$X.81$	.	.	.	.	.	.	.	−1	−1
$X.82$	.	.	.	.	2	2	−1	.	.
$X.83$	.	.	.	.	2	2	−1	.	.
$X.84$	.	.	.	.	.	.	.	.	.
$X.85$	.	.	.	.	.	.	.	.	.
$X.86$	.	.	.	.	.	.	.	.	.
$X.87$	.	.	.	.	.	.	.	.	.
$X.88$	.	.	.	.	.	.	.	.	.

Character table of $H(\mathsf{Fi}_{24}')$ *(continued)*

	1a	2a	2b	2c	2d	2e	2f	2g	2h	2i	3a	3b	3c	3d	3e	3f	3g
2	21	21	19	20	14	14	17	17	16	13	8	12	10	5	8	5	6
3	7	7	4	2	4	4	3	3	1	2	7	6	7	7	5	6	4
5	1	1	1	1	1	1	.	.	.	.	1	.	.	.	.	.	.
7	1	1	.	.	.	.	.	.	.	.	1	.	.	.	.	.	.
2P	1a	1a	1a	1a	1a	1a	1a	1a	1a	1a	3a	3b	3c	3d	3e	3f	3g
3P	1a	2a	2b	2c	2d	2e	2f	2g	2h	2i	1a	1a	1a	1a	1a	1a	1a
5P	1a	2a	2b	2c	2d	2e	2f	2g	2h	2i	3a	3b	3c	3d	3e	3f	3g
7P	1a	2a	2b	2c	2d	2e	2f	2g	2h	2i	3a	3b	3c	3d	3e	3f	3g
X.89	9072	9072	1392	−144	144	144	48	48	48	48	.	162	162	.	.	.	.
X.90	9072	9072	1392	−144	−144	−144	48	48	48	−48	.	162	162	.	.	.	.
X.91	10080	10080	−160	−160	.	.	96	96	−32	−32	.	−48	180	.	30	15	−6
X.92	10080	10080	−160	−160	.	.	96	96	−32	32	.	−48	180	.	30	15	−6
X.93	10752	−10752	.	.	.	.	128	−128	.	.	−84	192	−120	−30	.	12	.
X.94	11340	11340	1740	−180	−360	−360	12	12	−52	24	.	162	81	.	27	.	.
X.95	11340	11340	1740	−180	360	360	12	12	−52	−24	.	162	81	.	27	.	.
X.96	12096	12096	−192	−192	.	.	−192	−192	64	.	.	72	216	.	36	18	.
X.97	13608	13608	−216	296	216	216	72	72	72	8	.	−81	.	.	.	.	−9
X.98	13608	13608	−216	296	−216	−216	72	72	72	−8	.	−81	.	.	.	.	−9
X.99	13608	13608	−216	296	216	216	72	72	72	8	.	−81	.	.	.	.	−9
X.100	13608	13608	−216	296	−216	−216	72	72	72	−8	.	−81	.	.	.	.	−9
X.101	15309	15309	−243	333	−243	−243	189	189	93	−27	.	.	.	.	.	.	.
X.102	15309	15309	−243	333	243	243	189	189	93	27	.	.	.	.	.	.	.
X.103	15360	−15360	.	.	.	.	−256	256	.	.	−120	480	−48	−12	.	30	.
X.104	16128	−16128	.	.	.	.	−320	320	.	.	−126	288	144	36	.	18	.
X.105	17010	17010	−270	370	−270	−270	−126	−126	66	26	.	81	.	.	.	.	9
X.106	17010	17010	−270	370	270	270	−126	−126	66	−26	.	81	.	.	.	.	9
X.107	20160	20160	−320	−320	.	.	192	192	−64	.	.	480	36	.	−12	−15	−12
X.108	20160	20160	−320	−320	.	.	192	192	−64	.	.	−168	36	.	−12	−15	6
X.109	20160	20160	−320	−320	.	.	192	192	−64	.	.	−168	36	.	−12	−15	6
X.110	22680	22680	3480	−360	−360	−360	120	120	−8	−24	.	−162	162	.	−27	.	.
X.111	22680	22680	3480	−360	360	360	120	120	−8	24	.	−162	162	.	−27	.	.
X.112	22680	22680	3480	−360	.	.	216	216	88	.	.	.	−81	.	−54	.	.
X.113	24192	24192	3712	−384	−576	−576	.	.	.	.	.	−108	−216	.	36	.	.
X.114	24192	24192	3712	−384	576	576	.	.	.	.	.	−108	−216	.	36	.	.
X.115	25515	25515	−405	555	−135	−135	−117	−117	−21	57	.	243	.	.	.	.	.
X.116	25515	25515	−405	555	135	135	−117	−117	−21	−57	.	243	.	.	.	.	.
X.117	25515	25515	−405	555	135	135	315	315	27	15	.	243	.	.	.	.	.
X.118	25515	25515	−405	555	−135	−135	315	315	27	−15	.	243	.	.	.	.	.
X.119	26880	−26880	.	.	.	.	320	−320	.	.	420	96	−336	42	24	−12	24
X.120	26880	−26880	.	.	.	.	−192	192	.	.	−210	−96	−192	−48	.	−6	.
X.121	30240	30240	4640	−480	.	.	−96	−96	−96	.	.	108	−270	.	−36	.	.
X.122	30618	30618	4698	−486	324	324	−54	−54	42	−36	.	.	.	.	.	.	.
X.123	30618	30618	4698	−486	−324	−324	−54	−54	42	36	.	.	.	.	.	.	.
X.124	32256	−32256	.	.	768	−768	−128	128	.	.	504	−144	−144	18	72	18	.
X.125	32256	−32256	.	.	−768	768	−128	128	.	.	504	−144	−144	18	72	18	.
X.126	32256	−32256	.	.	192	−192	−128	128	.	.	504	288	−144	18	−36	−36	.
X.127	32256	−32256	.	.	−192	192	−128	128	.	.	504	288	−144	18	−36	−36	.
X.128	34020	34020	5220	−540	.	.	−252	−252	−60	.	.	.	243	.	.	.	.
X.129	34560	−34560	.	.	.	.	192	−192	.	.	540	.	216	−27	.	.	.
X.130	34560	−34560	.	.	.	.	192	−192	.	.	540	.	216	−27	.	.	.
X.131	34560	−34560	.	.	.	.	192	−192	.	.	−270	.	216	54	.	.	.
X.132	34560	−34560	.	.	.	.	192	−192	.	.	−270	.	216	54	.	.	.
X.133	35840	−35840	.	.	−640	640	−256	256	.	.	560	32	272	−34	8	−4	8
X.134	35840	−35840	.	.	640	−640	−256	256	.	.	560	32	272	−34	8	−4	8
X.135	40320	40320	−640	−640	.	.	.	.	.	−64	.	−120	72	.	−24	24	3
X.136	40320	40320	−640	−640	.	.	.	.	.	−64	.	−120	72	.	−24	24	3
X.137	40320	40320	−640	−640	.	.	.	.	.	64	.	−120	72	.	−24	24	3
X.138	40320	−40320	.	.	−480	480	224	−224	.	.	630	144	144	−18	36	−18	.
X.139	40320	40320	−640	−640	.	.	.	.	.	64	.	−120	72	.	−24	24	3
X.140	40320	40320	−640	−640	.	.	−384	−384	128	.	.	240	−252	.	12	6	12
X.141	40320	−40320	.	.	480	−480	224	−224	.	.	630	144	144	−18	36	−18	.
X.142	40320	40320	−640	−640	.	.	.	.	.	64	.	204	72	.	−24	24	−6
X.143	40320	40320	−640	−640	.	.	.	.	.	−64	.	204	72	.	−24	24	−6
X.144	40320	40320	−640	−640	.	.	384	384	−128	.	.	240	−252	.	12	6	12
X.145	43008	−43008	.	.	.	.	512	−512	.	.	−336	192	−48	−12	.	12	.
X.146	49152	−49152	.	.	.	.	.	.	.	.	−384	384	192	48	.	24	.
X.147	51030	51030	−810	1110	−270	−270	198	198	6	42	.	−243	.	.	.	.	.
X.148	51030	51030	−810	1110	270	270	198	198	6	−42	.	−243	.	.	.	.	.
X.149	53760	−53760	.	.	.	.	−384	384	.	.	−420	384	−168	−42	.	24	.
X.150	57344	−57344	.	.	512	−512	.	.	.	.	896	−64	−256	32	−16	8	−16
X.151	57344	−57344	.	.	−512	512	.	.	.	.	896	−64	−256	32	−16	8	−16
X.152	60480	60480	−960	−960	.	.	−192	−192	64	.	.	144	108	.	−36	−45	.
X.153	60480	60480	−960	−960	.	.	192	192	−64	.	.	−72	108	.	18	−18	.
X.154	60480	60480	−960	−960	.	.	192	192	−64	.	.	−72	108	.	18	−18	.
X.155	60480	60480	−960	−960	.	.	−192	−192	64	.	.	−72	108	.	18	−18	.
X.156	60480	60480	−960	−960	.	.	−192	−192	64	.	.	−72	108	.	18	−18	.
X.157	76545	76545	−1215	1665	−405	−405	−351	−351	−63	27	.	.	.	.	.	.	.
X.158	76545	76545	−1215	1665	−405	−405	81	81	−15	−45	.	.	.	.	.	.	.
X.159	76545	76545	−1215	1665	405	405	−351	−351	−63	−27	.	.	.	.	.	.	.
X.160	76545	76545	−1215	1665	405	405	81	81	−15	45	.	.	.	.	.	.	.
X.161	80640	−80640	.	.	.	.	−576	576	.	.	−630	−288	72	18	.	−18	.
X.162	80640	80640	−1280	−1280	.	.	.	.	.	.	.	−168	−504	.	24	12	−12
X.163	80640	−80640	.	.	.	.	−64	64	.	.	−630	−288	72	18	.	−18	.
X.164	80640	−80640	.	.	.	.	−64	64	.	.	−630	−288	72	18	.	−18	.
X.165	81920	−81920	.	.	.	.	.	.	.	.	1280	−256	128	−16	−64	32	8
X.166	107520	−107520	.	.	.	.	256	−256	.	.	−840	−96	−336	−84	.	−6	.
X.167	107520	−107520	.	.	.	.	256	−256	.	.	−840	192	96	24	.	12	.

Character table of $H(\mathsf{Fi}_{24}')$ (continued)

	4a	4b	4c	4d	4e	4f	4g	4h	4i	4j	4k	4l	4m	4n	4o	4p	4q	4r	4s	4t	5a
2	15	15	13	11	13	14	11	11	11	14	11	12	10	10	10	9	9	10	10	9	6
3	5	2	1	2	2	1	2	2	2	.	1	.	1	1	1	1	1	.	.	.	1
5	1	.	1	1	.	.	.	.	.	.	.	.	.	.	.	.	.	.	.	.	1
7	.	.	.	.	.	.	.	.	.	.	.	.	.	.	.	.	.	.	.	.	.
2P	2a	2a	2c	2b	2c	2c	2f	2b	2f	2c	2b	2c	2h	2f	2h	2g	2g	2h	2h	2h	5a
3P	4a	4b	4c	4d	4e	4f	4g	4h	4i	4j	4k	4l	4m	4n	4o	4p	4q	4r	4s	4t	5a
5P	4a	4b	4c	4d	4e	4f	4g	4h	4i	4j	4k	4l	4m	4n	4o	4p	4q	4r	4s	4t	1a
7P	4a	4b	4c	4d	4e	4f	4g	4h	4i	4j	4k	4l	4m	4n	4o	4p	4q	4r	4s	4t	5a
X.89	−144	48	16	−16	48	−16	.	16	.	−16	16	−16	.	.	.	.	.	.	.	.	−3
X.90	−144	48	−16	16	−48	−16	.	16	.	−16	−16	16	.	.	.	.	.	.	.	.	−3
X.91	160	32	.	.	32	.	.	.	.	.	.	.	−8	.	−8	.	8	.	.	.	.
X.92	160	32	.	.	−32	.	.	.	.	.	.	.	8	.	8	.	−8	.	.	.	.
X.93	.	.	.	.	.	.	−32	.	32	.	.	.	.	.	.	.	.	.	.	.	−8
X.94	−180	−84	−40	40	24	44	−12	−20	−12	12	8	−8	4	.	4	.	4	4	.	.	.
X.95	−180	−84	40	−40	−24	44	−12	−20	−12	12	−8	8	−4	.	−4	.	−4	4	.	.	.
X.96	192	−64	.	.	.	.	.	.	.	.	.	.	.	.	.	.	.	.	.	.	6
X.97	−216	72	−40	24	8	8	.	−24	.	8	8	8	.	.	.	.	.	.	.	.	3
X.98	−216	72	40	−24	−8	8	.	−24	.	8	−8	−8	.	.	.	.	.	.	.	.	3
X.99	−216	72	−40	24	8	8	.	−24	.	8	8	8	.	.	.	.	.	.	.	.	3
X.100	−216	72	40	−24	−8	8	.	−24	.	8	−8	−8	.	.	.	.	.	.	.	.	3
X.101	−243	45	45	−27	−27	−3	9	−27	9	13	−3	−11	−3	−15	−3	9	−3	1	1	1	−6
X.102	−243	45	−45	27	27	−3	9	−27	9	13	3	11	3	15	3	9	3	1	−1	1	−6
X.103	.	.	.	.	.	.	.	.	.	.	.	.	.	.	.	.	.	.	.	.	.
X.104	.	.	.	.	.	.	−16	.	16	.	.	.	.	.	.	.	.	.	.	.	8
X.105	−270	162	50	−30	26	34	−18	−30	−18	2	−22	−6	6	6	6	−6	6	−2	6	−6	.
X.106	−270	162	−50	30	−26	34	−18	−30	−18	2	22	6	−6	−6	−6	−6	−6	−2	−6	−6	.
X.107	320	64	.	.	.	.	.	.	.	.	.	.	.	.	.	.	.	.	.	.	.
X.108	320	64	.	.	.	.	.	.	.	.	.	.	.	.	.	.	.	.	.	.	.
X.109	320	64	.	.	.	.	.	.	.	.	.	.	.	.	.	.	.	.	.	.	.
X.110	−360	−72	−40	40	−24	56	.	−8	.	−8	−8	8	8	.	8	.	8	.	.	.	.
X.111	−360	−72	40	−40	24	56	.	−8	.	−8	8	−8	−8	.	−8	.	−8	.	.	.	.
X.112	−360	24	.	.	.	24	24	24	24	−40	.	.	.	.	.	.	.	−8	.	.	.
X.113	−384	.	−64	64	.	.	.	.	.	.	.	.	.	.	.	.	.	.	.	.	−3
X.114	−384	.	64	−64	.	.	.	.	.	.	.	.	.	.	.	.	.	.	.	.	−3
X.115	−405	27	25	−15	57	11	27	3	27	−5	−15	9	−3	−15	−3	−9	−3	3	1	−1	.
X.116	−405	27	−25	15	−57	11	27	3	27	−5	15	−9	3	15	3	−9	3	3	−1	−1	.
X.117	−405	−117	−25	15	15	−37	−9	3	−9	11	−9	−1	3	−21	3	3	3	−1	−5	−5	.
X.118	−405	−117	25	−15	−15	−37	−9	3	−9	11	9	1	−3	21	−3	3	−3	−1	5	−5	.
X.119	.	.	.	.	.	.	−16	.	16	.	.	.	.	.	.	.	.	.	.	.	.
X.120	.	.	.	.	.	.	−48	.	48	.	.	.	.	.	.	.	.	.	.	.	.
X.121	−480	−96	.	.	.	32	.	−32	.	32	.	.	.	.	.	.	.	.	.	.	.
X.122	−486	90	36	−36	−36	−54	−18	18	−18	−6	−12	12	.	.	.	−6	.	6	.	2	3
X.123	−486	90	−36	36	36	−54	−18	18	−18	−6	12	−12	.	.	.	−6	.	6	.	2	3
X.124	.	.	.	.	.	.	.	.	.	.	.	.	.	.	.	.	.	.	.	.	−4
X.125	.	.	.	.	.	.	.	.	.	.	.	.	.	.	.	.	.	.	.	.	−4
X.126	.	.	.	.	.	.	.	.	.	.	.	.	−16	.	16	.	.	.	.	.	−4
X.127	.	.	.	.	.	.	.	.	.	.	.	.	16	.	−16	.	.	.	.	.	−4
X.128	−540	36	.	.	.	−60	12	−12	12	36	.	.	.	.	.	12	.	−4	.	−4	.
X.129	.	.	.	.	.	.	−16	.	16	.	.	.	.	.	.	.	.	.	.	.	.
X.130	.	.	.	.	.	.	−16	.	16	.	.	.	.	.	.	.	.	.	.	.	.
X.131	.	.	.	.	.	.	−16	.	16	.	.	.	.	.	.	.	.	.	.	.	.
X.132	.	.	.	.	.	.	−16	.	16	.	.	.	.	.	.	.	.	.	.	.	.
X.133	.	.	.	.	.	.	.	.	.	.	.	.	.	.	.	.	.	.	.	.	.
X.134	.	.	.	.	.	.	.	.	.	.	.	.	.	.	.	.	.	.	.	.	.
X.135	640	.	.	.	64	.	.	.	.	.	.	.	.	.	.	.	.	.	.	.	.
X.136	640	.	.	.	64	.	.	.	.	.	.	.	.	.	.	.	.	.	.	.	.
X.137	640	.	.	.	−64	.	.	.	.	.	.	.	.	.	.	.	.	.	.	.	.
X.138	.	.	.	.	.	.	24	.	−24	.	.	.	8	.	−8	.	.	.	.	.	.
X.139	640	.	.	.	−64	.	.	.	.	.	.	.	.	.	.	.	.	.	.	.	.
X.140	640	−128	.	.	.	.	.	.	.	.	.	.	.	.	.	.	.	.	.	.	.
X.141	.	.	.	.	.	.	24	.	−24	.	.	.	−8	.	8	.	.	.	.	.	.
X.142	640	.	.	.	−64	.	.	.	.	.	.	.	.	.	.	.	.	.	.	.	.
X.143	640	.	.	.	64	.	.	.	.	.	.	.	.	.	.	.	.	.	.	.	.
X.144	640	128	.	.	.	.	.	.	.	.	.	.	.	.	.	.	.	.	.	.	.
X.145	.	.	.	.	.	.	.	.	.	.	.	.	.	.	.	.	.	.	.	.	8
X.146	.	.	.	.	.	.	.	.	.	.	.	.	.	.	.	.	.	.	.	.	−8
X.147	−810	−90	50	−30	42	−26	18	6	18	6	−6	10	−6	6	−6	−6	−6	2	6	−6	.
X.148	−810	−90	−50	30	−42	−26	18	6	18	6	6	−10	6	−6	6	−6	6	2	−6	−6	.
X.149	.	.	.	.	.	.	32	.	−32	.	.	.	.	.	.	.	.	.	.	.	.
X.150	.	.	.	.	.	.	.	.	.	.	.	.	.	.	.	.	.	.	.	.	4
X.151	.	.	.	.	.	.	.	.	.	.	.	.	.	.	.	.	.	.	.	.	4
X.152	960	−64	.	.	.	.	.	.	.	.	.	.	.	.	.	.	.	.	.	.	.
X.153	960	64	.	.	.	.	.	.	.	.	.	.	16	.	16	.	−16	.	.	.	.
X.154	960	64	.	.	.	.	.	.	.	.	.	.	−16	.	−16	.	16	.	.	.	.
X.155	960	−64	.	.	.	.	.	.	.	.	.	.	.	.	.	.	.	.	.	.	.
X.156	960	−64	.	.	.	.	.	.	.	.	.	.	.	.	.	.	.	.	.	.	.
X.157	−1215	81	75	−45	27	33	9	9	9	−15	3	11	3	3	3	9	3	1	−13	1	.
X.158	−1215	−63	75	−45	−45	−15	−27	9	−27	1	27	3	3	−9	3	−3	3	−3	7	5	.
X.159	−1215	81	−75	45	−27	33	9	9	9	−15	−3	−11	−3	−3	−3	9	−3	1	13	1	.
X.160	−1215	−63	−75	45	45	−15	−27	9	−27	1	−27	−3	−3	9	−3	−3	−3	−3	−7	5	.
X.161	.	.	.	.	.	.	−16	.	16	.	.	.	.	.	.	.	.	.	.	.	.
X.162	1280	.	.	.	.	.	.	.	.	.	.	.	.	.	.	.	.	.	.	.	.
X.163	.	.	.	.	.	.	16	.	−16	.	.	.	.	.	.	.	.	.	.	.	.
X.164	.	.	.	.	.	.	16	.	−16	.	.	.	.	.	.	.	.	.	.	.	.
X.165	.	.	.	.	.	.	.	.	.	.	.	.	.	.	.	.	.	.	.	.	.
X.166	.	.	.	.	.	.	.	.	.	.	.	.	.	.	.	.	.	.	.	.	.
X.167	.	.	.	.	.	.	.	.	.	.	.	.	.	.	.	.	.	.	.	.	.

Character table of $H(\mathsf{Fi}_{24}')$ *(continued)*

	6_1	6_2	6_3	6_4	6_5	6_6	6_7	6_8	6_9	6_{10}	6_{11}	6_{12}	6_{13}	6_{14}	6_{15}	6_{16}	6_{17}	6_{18}	6_{19}	6_{20}	6_{21}
2	8	12	10	5	8	9	10	5	11	8	8	8	8	8	8	8	8	8	8	6	7
3	7	6	7	7	5	4	3	6	2	3	3	3	3	3	3	3	3	3	3	4	3
5	1	.	.	.	.	.	.	.	.	.	.	.	.	.	.	.	.	.	.	.	.
7	1	.	.	.	.	.	.	.	.	.	.	.	.	.	.	.	.	.	.	.	.
2P	3a	3b	3c	3d	3e	3c	3b	3f	3b	3b	3a	3b	3b	3a	3c	3c	3e	3e	3b	3g	3e
3P	2a	2a	2a	2a	2a	2b	2b	2a	2c	2f	2g	2e	2d	2f	2f	2g	2g	2f	2g	2a	2b
5P	6_1	6_2	6_3	6_4	6_5	6_6	6_7	6_8	6_9	6_{10}	6_{11}	6_{12}	6_{13}	6_{14}	6_{15}	6_{16}	6_{17}	6_{18}	6_{19}	6_{20}	6_{21}
7P	6_1	6_2	6_3	6_4	6_5	6_6	6_7	6_8	6_9	6_{10}	6_{11}	6_{12}	6_{13}	6_{14}	6_{15}	6_{16}	6_{17}	6_{18}	6_{19}	6_{20}	6_{21}
X.89	.	162	162	.	.	−30	−6	.	18	−6	.	6	6	.	6	6	.	.	−6	.	.
X.90	.	162	162	.	.	−30	−6	.	18	−6	.	−6	−6	.	6	6	.	.	−6	.	.
X.91	.	−48	180	.	30	20	32	15	−16	.	.	.	.	.	12	12	6	6	.	−6	−10
X.92	.	−48	180	.	30	20	32	15	−16	.	.	.	.	.	12	12	6	6	.	−6	−10
X.93	84	−192	120	30	.	.	.	−12	.	−16	4	.	.	−4	−4	4	16	−16	16	.	.
X.94	.	162	81	.	27	−15	−6	.	18	−6	.	−6	−6	.	−3	−3	−9	−9	−6	.	3
X.95	.	162	81	.	27	−15	−6	.	18	−6	.	6	6	.	−3	−3	−9	−9	−6	.	3
X.96	.	72	216	.	36	24	24	18	−24	.	.	.	.	.	−24	−24	−12	−12	.	.	−12
X.97	.	−81	.	.	.	.	−9	.	−1	−9	.	−9	−9	.	.	.	.	.	−9	−9	.
X.98	.	−81	.	.	.	.	−9	.	−1	−9	.	9	9	.	.	.	.	.	−9	−9	.
X.99	.	−81	.	.	.	.	−9	.	−1	−9	.	−9	−9	.	.	.	.	.	−9	−9	.
X.100	.	−81	.	.	.	.	−9	.	−1	−9	.	9	9	.	.	.	.	.	−9	−9	.
X.101	.	.	.	.	.	.	.	.	.	.	.	.	.	.	.	.	.	.	.	.	.
X.102	.	.	.	.	.	.	.	.	.	.	.	.	.	.	.	.	.	.	.	.	.
X.103	120	−480	48	12	.	.	.	−30	.	8	−8	.	.	8	8	−8	16	−16	−8	.	.
X.104	126	−288	−144	−36	.	.	.	−18	.	−8	−10	.	.	10	−8	8	−16	16	8	.	.
X.105	.	81	.	.	.	.	9	.	1	9	.	−9	−9	.	.	.	.	.	9	9	.
X.106	.	81	.	.	.	.	9	.	1	9	.	9	9	.	.	.	.	.	9	9	.
X.107	.	480	36	.	−12	4	−32	−15	−32	.	.	.	.	.	−12	−12	12	12	.	−12	4
X.108	.	−168	36	.	−12	4	40	−15	−8	.	.	.	.	.	−12	−12	12	12	.	6	4
X.109	.	−168	36	.	−12	4	40	−15	−8	.	.	.	.	.	−12	−12	12	12	.	6	4
X.110	.	−162	162	.	−27	−30	6	.	−18	−6	.	−6	−6	.	6	6	9	9	−6	.	−3
X.111	.	−162	162	.	−27	−30	6	.	−18	−6	.	6	6	.	6	6	9	9	−6	.	−3
X.112	.	.	−81	.	−54	15	.	.	.	.	.	.	.	.	−9	−9	−18	−18	.	.	−6
X.113	.	−108	−216	.	36	40	4	.	−12	.	.	12	12	.	.	.	.	.	.	.	4
X.114	.	−108	−216	.	36	40	4	.	−12	.	.	−12	−12	.	.	.	.	.	.	.	4
X.115	.	243	.	.	.	.	27	.	3	−9	.	9	9	.	.	.	.	.	−9	.	.
X.116	.	243	.	.	.	.	27	.	3	−9	.	−9	−9	.	.	.	.	.	−9	.	.
X.117	.	243	.	.	.	.	27	.	3	−9	.	−9	−9	.	.	.	.	.	−9	.	.
X.118	.	243	.	.	.	.	27	.	3	−9	.	9	9	.	.	.	.	.	−9	.	.
X.119	−420	−96	336	−42	−24	.	.	12	.	8	−20	.	.	20	8	−8	−8	8	−8	−24	.
X.120	210	96	192	48	.	.	.	6	.	24	−6	.	.	6	.	.	.	.	−24	.	.
X.121	.	108	−270	.	−36	50	−4	.	12	12	.	.	.	.	6	6	.	.	12	.	−4
X.122	.	.	.	.	.	.	.	.	.	.	.	.	.	.	.	.	.	.	.	.	.
X.123	.	.	.	.	.	.	.	.	.	.	.	.	.	.	.	.	.	.	.	.	.
X.124	−504	144	144	−18	−72	.	.	−18	.	4	8	12	−12	−8	−8	8	8	−8	−4	.	.
X.125	−504	144	144	−18	−72	.	.	−18	.	4	8	−12	12	−8	−8	8	8	−8	−4	.	.
X.126	−504	−288	144	−18	36	.	.	36	.	−8	8	.	.	−8	−8	8	−4	4	8	.	.
X.127	−504	−288	144	−18	36	.	.	36	.	−8	8	.	.	−8	−8	8	−4	4	8	.	.
X.128	.	.	243	.	.	−45	.	.	.	.	.	.	.	.	−9	−9	.	.	.	.	.
X.129	−540	.	−216	27	.	.	.	.	.	.	−12	.	.	12	−12	12	.	.	.	.	.
X.130	−540	.	−216	27	.	.	.	.	.	.	−12	.	.	12	−12	12	.	.	.	.	.
X.131	270	.	−216	−54	.	.	.	.	.	.	6	.	.	−6	−12	12	.	.	.	.	.
X.132	270	.	−216	−54	.	.	.	.	.	.	6	.	.	−6	−12	12	.	.	.	.	.
X.133	−560	−32	−272	34	−8	.	.	4	.	8	16	16	−16	−16	8	−8	−8	8	−8	−8	.
X.134	−560	−32	−272	34	−8	.	.	4	.	8	16	−16	16	−16	8	−8	−8	8	−8	−8	.
X.135	.	−120	72	.	−24	8	8	24	8	.	.	.	.	.	.	.	.	.	.	3	8
X.136	.	−120	72	.	−24	8	8	24	8	.	.	.	.	.	.	.	.	.	.	3	8
X.137	.	−120	72	.	−24	8	8	24	8	.	.	.	.	.	.	.	.	.	.	3	8
X.138	−630	−144	−144	18	−36	.	.	18	.	−4	−14	12	−12	14	8	−8	4	−4	4	.	.
X.139	.	−120	72	.	−24	8	8	24	8	.	.	.	.	.	.	.	.	.	.	3	8
X.140	.	240	−252	.	12	−28	−16	6	−16	.	.	.	.	.	−12	−12	12	12	.	12	−4
X.141	−630	−144	−144	18	−36	.	.	18	.	−4	−14	−12	12	14	8	−8	4	−4	4	.	.
X.142	.	204	72	.	−24	8	−28	24	−4	.	.	.	.	.	.	.	.	.	.	−6	8
X.143	.	204	72	.	−24	8	−28	24	−4	.	.	.	.	.	.	.	.	.	.	−6	8
X.144	.	240	−252	.	12	−28	−16	6	−16	.	.	.	.	.	12	12	−12	−12	.	12	−4
X.145	336	−192	48	12	.	.	.	−12	.	−16	16	.	.	−16	8	−8	16	−16	16	.	.
X.146	384	−384	−192	−48	.	.	.	−24	.	.	.	.	.	.	.	.	.	.	.	.	.
X.147	.	−243	.	.	.	.	−27	.	−3	9	.	−9	−9	.	.	.	.	.	9	.	.
X.148	.	−243	.	.	.	.	−27	.	−3	9	.	9	9	.	.	.	.	.	9	.	.
X.149	420	−384	168	42	.	.	.	−24	.	.	−12	.	.	12	−12	12	.	.	.	.	.
X.150	−896	64	256	−32	16	.	.	−8	.	.	.	16	−16	.	.	.	.	.	.	16	.
X.151	−896	64	256	−32	16	.	.	−8	.	.	.	−16	16	.	.	.	.	.	.	16	.
X.152	.	144	108	.	−36	12	48	−45	−48	.	.	.	.	.	12	12	−12	−12	.	.	12
X.153	.	−72	108	.	18	12	−24	−18	24	.	.	.	.	.	−12	−12	−6	−6	.	.	−6
X.154	.	−72	108	.	18	12	−24	−18	24	.	.	.	.	.	−12	−12	−6	−6	.	.	−6
X.155	.	−72	108	.	18	12	−24	−18	24	.	.	.	.	.	12	12	6	6	.	.	−6
X.156	.	−72	108	.	18	12	−24	−18	24	.	.	.	.	.	12	12	6	6	.	.	−6
X.157	.	.	.	.	.	.	.	.	.	.	.	.	.	.	.	.	.	.	.	.	.
X.158	.	.	.	.	.	.	.	.	.	.	.	.	.	.	.	.	.	.	.	.	.
X.159	.	.	.	.	.	.	.	.	.	.	.	.	.	.	.	.	.	.	.	.	.
X.160	.	.	.	.	.	.	.	.	.	.	.	.	.	.	.	.	.	.	.	.	.
X.161	630	288	−72	−18	.	.	.	18	.	−24	−18	.	.	18	12	−12	.	.	24	.	.
X.162	.	−168	−504	.	24	−56	40	12	−8	.	.	.	.	.	.	.	.	.	.	−12	−8
X.163	630	288	−72	−18	.	.	.	18	.	8	−2	.	.	2	−4	4	16	−16	−8	.	.
X.164	630	288	−72	−18	.	.	.	18	.	8	−2	.	.	2	−4	4	16	−16	−8	.	.
X.165	−1280	256	−128	16	64	.	.	−32	.	.	.	.	.	.	.	.	.	.	.	−8	.
X.166	840	96	336	84	.	.	.	6	.	−8	8	.	.	−8	−8	8	−16	16	8	.	.
X.167	840	−192	−96	−24	.	.	.	−12	.	16	8	.	.	−8	16	−16	−16	16	−16	.	.

Character table of $H(\mathsf{Fi}_{24}')$ (continued)

2	5	5	5	5	7	5	5	5	5	5	5	5	5	7	5	3	1	8	9	8	8	8	8	8	8	6
3	4	4	4	4	2	3	3	3	3	3	3	3	3	1	2	2	1	2	1	1	.	.	.	.	.	1
5	.	.	.	.	.	.	.	.	.	.	.	.	.	.	.	.	.	.	.	.	.	.	.	.	.	.
7	.	.	.	.	.	.	.	.	.	.	.	.	.	.	.	.	1	.	.	.	.	.	.	.	.	.
	6_{22}	6_{23}	6_{24}	6_{25}	6_{26}	6_{27}	6_{28}	6_{29}	6_{30}	6_{31}	6_{32}	6_{33}	6_{34}	6_{35}	6_{36}	6_{37}	$7a$	$8a$	$8b$	$8c$	$8d$	$8e$	$8f$	$8g$	$8h$	$8i$
2P	$3d$	$3d$	$3d$	$3d$	$3b$	$3e$	$3d$	$3e$	$3e$	$3d$	$3e$	$3f$	$3f$	$3e$	$3g$	$3g$	$7a$	$4a$	$4b$	$4b$	$4j$	$4j$	$4f$	$4j$	$4f$	$4g$
3P	$2d$	$2d$	$2e$	$2e$	$2i$	$2d$	$2g$	$2g$	$2f$	$2f$	$2e$	$2g$	$2f$	$2h$	$2c$	$2i$	$7a$	$8a$	$8b$	$8c$	$8d$	$8e$	$8f$	$8g$	$8h$	$8i$
5P	6_{23}	6_{22}	6_{25}	6_{24}	6_{26}	6_{27}	6_{28}	6_{29}	6_{30}	6_{31}	6_{32}	6_{33}	6_{34}	6_{35}	6_{36}	6_{37}	$7a$	$8a$	$8b$	$8c$	$8d$	$8e$	$8f$	$8g$	$8h$	$8i$
7P	6_{22}	6_{23}	6_{24}	6_{25}	6_{26}	6_{27}	6_{28}	6_{29}	6_{30}	6_{31}	6_{32}	6_{33}	6_{34}	6_{35}	6_{36}	6_{37}	$1a$	$8a$	$8b$	$8c$	$8d$	$8e$	$8f$	$8g$	$8h$	$8i$
X.89	.	.	.	.	6	−6	.	6	6	.	−6	.	.	.	.	.	.	.	.	.	.	.	.	.	.	.
X.90	.	.	.	.	−6	6	.	6	6	.	6	.	.	.	.	.	.	.	.	.	.	.	.	.	.	.
X.91	.	.	.	.	−8	.	.	.	.	.	.	3	3	−2	2	−2	.	.	.	.	.	.	.	.	.	.
X.92	.	.	.	.	8	.	.	.	.	.	.	3	3	−2	2	2	.	.	.	.	.	.	.	.	.	.
X.93	.	.	.	.	.	.	−2	4	−4	2	.	4	−4	.	.	.	.	.	.	.	.	.	.	.	.	.
X.94	.	.	.	.	−6	−3	.	−3	−3	.	−3	.	.	−1	.	.	.	.	−4	−4	.	−4	.	.	4	.
X.95	.	.	.	.	6	3	.	−3	−3	.	3	.	.	−1	.	.	.	.	−4	4	.	4	.	.	−4	.
X.96	.	.	.	.	.	.	.	.	.	.	.	−6	−6	4	.	.	.	.	.	.	.	.	.	.	.	.
X.97	.	.	.	.	−1	.	.	.	.	.	.	.	.	.	−1	−1	.	−8	.	.	.	.	.	.	.	.
X.98	.	.	.	.	1	.	.	.	.	.	.	.	.	.	−1	1	.	8	.	.	.	.	.	.	.	.
X.99	.	.	.	.	−1	.	.	.	.	.	.	.	.	.	−1	−1	.	−8	.	.	.	.	.	.	.	.
X.100	.	.	.	.	1	.	.	.	.	.	.	.	.	.	−1	1	.	8	.	.	.	.	.	.	.	.
X.101	.	.	.	.	.	.	.	.	.	.	.	.	.	.	.	.	.	9	−3	−3	1	1	−3	1	1	3
X.102	.	.	.	.	.	.	.	.	.	.	.	.	.	.	.	.	.	−9	−3	3	−1	−1	−3	1	−1	3
X.103	.	.	.	.	.	.	4	4	−4	−4	.	−2	2	.	.	.	2	.	.	.	.	.	.	.	.	.
X.104	.	.	.	.	.	.	−4	−4	4	4	.	2	−2	.	.	.	.	.	.	.	.	.	.	.	.	.
X.105	.	.	.	.	−1	.	.	.	.	.	.	.	.	.	1	−1	.	10	6	6	−2	−2	2	2	−2	.
X.106	.	.	.	.	1	.	.	.	.	.	.	.	.	.	1	1	.	−10	6	−6	2	2	2	2	2	.
X.107	.	.	.	.	.	.	.	.	.	.	.	−3	−3	−4	4	.	.	.	.	.	.	.	.	.	.	.
X.108	.	.	.	.	.	.	.	.	.	.	.	−3	−3	−4	−2	.	.	.	.	.	.	.	.	.	.	.
X.109	.	.	.	.	.	.	.	.	.	.	.	−3	−3	−4	−2	.	.	.	.	.	.	.	.	.	.	.
X.110	.	.	.	.	6	−3	.	−3	−3	.	−3	.	.	1	.	.	.	.	.	−8	.	.	.	.	.	.
X.111	.	.	.	.	−6	3	.	−3	−3	.	3	.	.	1	.	.	.	.	.	8	.	.	.	.	.	.
X.112	.	.	.	.	.	.	.	.	.	.	.	.	.	−2	.	.	.	.	8	.	.	.	.	.	.	.
X.113	.	.	.	.	.	6	.	.	.	.	6	.	.	.	.	.	.	.	.	.	.	.	.	.	.	.
X.114	.	.	.	.	.	−6	.	.	.	.	−6	.	.	.	.	.	.	.	.	.	.	.	.	.	.	.
X.115	.	.	.	.	−3	.	.	.	.	.	.	.	.	.	.	.	.	−3	−9	−3	1	1	3	−1	1	3
X.116	.	.	.	.	3	.	.	.	.	.	.	.	.	.	.	.	.	3	−9	3	−1	−1	3	−1	−1	3
X.117	.	.	.	.	3	.	.	.	.	.	.	.	.	.	.	.	.	3	3	3	3	−1	−1	3	−1	−3
X.118	.	.	.	.	−3	.	.	.	.	.	.	.	.	.	.	.	.	−3	3	−3	−3	1	−1	3	1	−3
X.119	.	.	.	.	.	.	−2	4	−4	2	.	4	−4	.	.	.	.	.	.	.	.	.	.	.	.	.
X.120	.	.	.	.	.	.	.	.	.	.	.	−6	6	.	.	.	.	.	.	.	.	.	.	.	.	.
X.121	.	.	.	.	.	.	.	6	6	.	.	.	.	.	.	.	.	.	.	.	.	.	.	.	.	.
X.122	.	.	.	.	.	.	.	.	.	.	.	.	.	.	.	.	.	.	−6	.	.	−4	−2	2	4	.
X.123	.	.	.	.	.	.	.	.	.	.	.	.	.	.	.	.	.	.	−6	.	.	4	−2	2	−4	.
X.124	−6	−6	6	6	.	.	2	−4	4	−2	.	2	−2	.	.	.	.	.	.	.	.	.	.	.	.	.
X.125	6	6	−6	−6	.	.	2	−4	4	−2	.	2	−2	.	.	.	.	.	.	.	.	.	.	.	.	.
X.126	−6	−6	6	6	.	6	2	2	−2	−2	−6	−4	4	.	.	.	.	.	.	.	.	.	.	.	.	.
X.127	6	6	−6	−6	.	−6	2	2	−2	−2	6	−4	4	.	.	.	.	.	.	.	.	.	.	.	.	.
X.128	.	.	.	.	.	.	.	.	.	.	.	.	.	.	.	.	.	.	4	.	.	.	4	−4	.	.
X.129	M	$\bar{M}$	M	$\bar{M}$	.	.	3	.	.	−3	.	.	.	.	.	.	1	.	.	.	.	.	.	.	.	.
X.130	$\bar{M}$	M	$\bar{M}$	M	.	.	3	.	.	−3	.	.	.	.	.	.	1	.	.	.	.	.	.	.	.	.
X.131	.	.	.	.	.	.	−6	.	.	6	.	.	.	.	.	.	1	.	.	.	.	.	.	.	.	.
X.132	.	.	.	.	.	.	−6	.	.	6	.	.	.	.	.	.	1	.	.	.	.	.	.	.	.	.
X.133	−10	−10	10	10	.	−4	−2	4	−4	2	4	4	−4	.	.	.	.	.	.	.	.	.	.	.	.	.
X.134	10	10	−10	−10	.	4	−2	4	−4	2	−4	4	−4	.	.	.	.	.	.	.	.	.	.	.	.	.
X.135	.	.	.	.	8	.	.	.	.	.	.	.	.	.	−1	−1	.	.	.	.	.	.	.	.	.	.
X.136	.	.	.	.	8	.	.	.	.	.	.	.	.	.	−1	−1	.	.	.	.	.	.	.	.	.	.
X.137	.	.	.	.	−8	.	.	.	.	.	.	.	.	.	−1	1	.	.	.	.	.	.	.	.	.	.
X.138	6	6	−6	−6	.	6	−2	−2	2	2	−6	−2	2	.	.	.	.	.	.	.	.	.	.	.	.	.
X.139	.	.	.	.	−8	.	.	.	.	.	.	.	.	.	−1	1	.	.	.	.	.	.	.	.	.	.
X.140	.	.	.	.	.	.	.	.	.	.	.	6	6	−4	−4	.	.	.	.	.	.	.	.	.	.	.
X.141	−6	−6	6	6	.	−6	−2	−2	2	2	6	−2	2	.	.	.	.	.	.	.	.	.	.	.	.	.
X.142	.	.	.	.	4	.	.	.	.	.	.	.	.	.	2	−2	.	.	.	.	.	.	.	.	.	.
X.143	.	.	.	.	−4	.	.	.	.	.	.	.	.	.	2	2	.	.	.	.	.	.	.	.	.	.
X.144	.	.	.	.	.	.	.	.	.	.	.	−6	−6	4	−4	.	.	.	.	.	.	.	.	.	.	.
X.145	.	.	.	.	.	.	4	4	−4	−4	.	4	−4	.	.	.	.	.	.	.	.	.	.	.	.	.
X.146	.	.	.	.	.	.	.	.	.	.	.	.	.	.	.	.	−2	.	.	.	.	.	.	.	.	.
X.147	.	.	.	.	3	.	.	.	.	.	.	.	.	.	.	.	.	−6	−6	−6	−2	2	2	2	2	.
X.148	.	.	.	.	−3	.	.	.	.	.	.	.	.	.	.	.	.	6	−6	6	2	−2	2	2	−2	.
X.149	.	.	.	.	.	.	−6	.	.	6	.	.	.	.	.	.	.	.	.	.	.	.	.	.	.	.
X.150	8	8	−8	−8	.	−4	.	.	.	.	4	.	.	.	.	.	.	.	.	.	.	.	.	.	.	.
X.151	−8	−8	8	8	.	4	.	.	.	.	−4	.	.	.	.	.	.	.	.	.	.	.	.	.	.	.
X.152	.	.	.	.	.	.	.	.	.	.	.	3	3	4	.	.	.	.	.	.	.	.	.	.	.	.
X.153	.	.	.	.	.	.	.	.	.	.	.	6	6	2	.	.	.	.	.	.	.	.	.	.	.	.
X.154	.	.	.	.	.	.	.	.	.	.	.	6	6	2	.	.	.	.	.	.	.	.	.	.	.	.
X.155	.	.	.	.	.	.	.	.	.	.	.	−6	−6	−2	.	.	.	.	.	.	.	.	.	.	.	.
X.156	.	.	.	.	.	.	.	.	.	.	.	−6	−6	−2	.	.	.	.	.	.	.	.	.	.	.	.
X.157	.	.	.	.	.	.	.	.	.	.	.	.	.	.	.	.	.	−9	−3	3	3	−1	−3	1	−1	−3
X.158	.	.	.	.	.	.	.	.	.	.	.	.	.	.	.	.	.	−9	9	3	−1	−1	1	−3	−1	3
X.159	.	.	.	.	.	.	.	.	.	.	.	.	.	.	.	.	.	9	−3	−3	−3	1	−3	1	1	−3
X.160	.	.	.	.	.	.	.	.	.	.	.	.	.	.	.	.	.	9	9	−3	1	1	1	−3	1	3
X.161	.	.	.	.	.	.	6	.	.	−6	.	6	−6	.	.	.	.	.	.	.	.	.	.	.	.	.
X.162	.	.	.	.	.	.	.	.	.	.	.	.	.	.	4	.	.	.	.	.	.	.	.	.	.	.
X.163	.	.	.	.	.	.	−2	4	−4	2	.	−2	2	.	.	.	.	.	.	.	.	.	.	.	.	.
X.164	.	.	.	.	.	.	−2	4	−4	2	.	−2	2	.	.	.	.	.	.	.	.	.	.	.	.	.
X.165	.	.	.	.	.	.	.	.	.	.	.	.	.	.	.	.	−1	.	.	.	.	.	.	.	.	.
X.166	.	.	.	.	.	.	−4	−4	4	4	.	2	−2	.	.	.	.	.	.	.	.	.	.	.	.	.
X.167	.	.	.	.	.	.	8	−4	4	−8	.	−4	4	.	.	.	.	.	.	.	.	.	.	.	.	.

Character table of $H(\mathsf{Fi}_{24}')$ *(continued)*

2	6	6	6	5	3	1	2	2	6	5	5	5	4	4	9	7	9	6	4	8	8	7	5	5
3	.	.	.	4	4	4	3	3	1	.	.	.	.	.	5	5	3	4	5	2	2	2	3	3
5	.	.	.	.	.	.	.	.	1	1	1	1	1	1	.	.	.	.	.	.	.	.	.	.
7	.	.	.	.	.	.	.	.	.	.	.	.	.	.	.	.	.	.	.	.	.	.	.	.
	8j	8k	8l	9a	9b	9c	9d	9e	10a	10b	10c	10d	10e	10f	12_1	12_2	12_3	12_4	12_5	12_6	12_7	12_8	12_9	12_{10}
2P	4r	4r	4i	9a	9b	9c	9e	9d	5a	5a	5a	5a	5a	5a	6_2	6_3	6_2	6_5	6_8	6_9	6_9	6_3	6_{20}	6_{20}
3P	8j	8k	8l	3c	3c	3c	3d	3d	10a	10c	10b	10d	10e	10f	4a	4a	4a	4a	4a	4e	4e	4b	4a	4a
5P	8j	8k	8l	9a	9b	9c	9e	9d	2a	2c	2c	2b	2d	2e	12_1	12_2	12_3	12_4	12_5	12_6	12_7	12_8	12_9	12_{10}
7P	8j	8k	8l	9a	9b	9c	9d	9e	10a	10c	10b	10d	10e	10f	12_1	12_2	12_3	12_4	12_5	12_6	12_7	12_8	12_9	12_{10}
X.89	.	.	.	.	.	.	.	.	−3	1	1	−3	−1	−1	54	18	−18	.	.	6	6	6	.	.
X.90	.	.	.	.	.	.	.	.	−3	1	1	−3	1	1	54	18	−18	.	.	−6	−6	6	.	.
X.91	.	.	.	−6	−3	.	.	.	.	.	.	.	.	.	40	−20	−8	10	−5	8	8	−4	−2	−2
X.92	.	.	.	−6	−3	.	.	.	.	.	.	.	.	.	40	−20	−8	10	−5	−8	−8	−4	−2	−2
X.93	.	.	.	.	−6	−3	.	.	8	.	.	.	.	.	.	.	.	.	.	.	.	.	.	.
X.94	.	.	.	.	.	.	.	.	.	.	.	.	.	.	54	9	−18	−9	.	−6	−6	−3	.	.
X.95	.	.	.	.	.	.	.	.	.	.	.	.	.	.	54	9	−18	−9	.	6	6	−3	.	.
X.96	.	.	.	.	.	.	.	.	6	−2	−2	−2	.	.	48	−24	.	12	−6	.	.	8	.	.
X.97	.	.	.	.	.	.	.	.	3	J	K	−1	1	1	27	.	3	.	.	−1	−1	.	3	3
X.98	.	.	.	.	.	.	.	.	3	K	J	−1	−1	−1	27	.	3	.	.	1	1	.	3	3
X.99	.	.	.	.	.	.	.	.	3	K	J	−1	1	1	27	.	3	.	.	−1	−1	.	3	3
X.100	.	.	.	.	.	.	.	.	3	J	K	−1	−1	−1	27	.	3	.	.	1	1	.	3	3
X.101	−1	−1	−1	.	.	.	.	.	−6	−2	−2	2	2	2	.	.	.	.	.	.	.	.	.	.
X.102	1	−1	1	.	.	.	.	.	−6	−2	−2	2	−2	−2	.	.	.	.	.	.	.	.	.	.
X.103	.	.	.	12	6	.	.	.	.	.	.	.	.	.	.	.	.	.	.	.	.	.	.	.
X.104	.	.	.	.	.	.	.	.	−8	.	.	.	.	.	.	.	.	.	.	.	.	.	.	.
X.105	.	.	.	.	.	.	.	.	.	.	.	.	.	.	−27	.	−3	.	.	−1	−1	.	−3	−3
X.106	.	.	.	.	.	.	.	.	.	.	.	.	.	.	−27	.	−3	.	.	1	1	.	−3	−3
X.107	.	.	.	6	−6	.	.	.	.	.	.	.	.	.	32	−4	32	−4	5	.	.	4	−4	−4
X.108	.	.	.	−3	3	.	.	.	.	.	.	.	.	.	32	−4	−16	−4	5	.	.	4	−10	14
X.109	.	.	.	−3	3	.	.	.	.	.	.	.	.	.	32	−4	−16	−4	5	.	.	4	14	−10
X.110	.	.	.	.	.	.	.	.	.	.	.	.	.	.	−54	18	18	9	.	6	6	6	.	.
X.111	.	.	.	.	.	.	.	.	.	.	.	.	.	.	−54	18	18	9	.	−6	−6	6	.	.
X.112	.	.	.	.	.	.	.	.	.	.	.	.	.	.	.	−9	.	18	.	.	.	−9	.	.
X.113	.	.	.	9	.	.	.	.	−3	1	1	−3	−1	−1	−36	−24	12	−12	.	.	.	.	.	.
X.114	.	.	.	9	.	.	.	.	−3	1	1	−3	1	1	−36	−24	12	−12	.	.	.	.	.	.
X.115	1	−1	1	.	.	.	.	.	.	.	.	.	.	.	−81	.	−9	.	.	−3	−3	.	.	.
X.116	−1	−1	−1	.	.	.	.	.	.	.	.	.	.	.	−81	.	−9	.	.	3	3	.	.	.
X.117	1	1	1	.	.	.	.	.	.	.	.	.	.	.	−81	.	−9	.	.	3	3	.	.	.
X.118	−1	1	−1	.	.	.	.	.	.	.	.	.	.	.	−81	.	−9	.	.	−3	−3	.	.	.
X.119	.	.	.	.	.	.	.	.	.	.	.	.	.	.	.	.	.	.	.	.	.	.	.	.
X.120	.	.	.	−12	.	3	.	.	.	.	.	.	.	.	.	.	.	.	.	.	.	.	.	.
X.121	.	.	.	−9	.	.	.	.	.	.	.	.	.	.	36	−30	−12	12	.	.	.	6	.	.
X.122	−2	.	2	.	.	.	.	.	3	−1	−1	3	−1	−1	.	.	.	.	.	.	.	.	.	.
X.123	2	.	−2	.	.	.	.	.	3	−1	−1	3	1	1	.	.	.	.	.	.	.	.	.	.
X.124	.	.	.	.	.	.	.	.	4	.	.	.	−2	2	.	.	.	.	.	−12	12	.	.	.
X.125	.	.	.	.	.	.	.	.	4	.	.	.	2	−2	.	.	.	.	.	12	−12	.	.	.
X.126	.	.	.	.	.	.	.	.	4	.	.	.	2	−2	.	.	.	.	.	.	.	.	.	.
X.127	.	.	.	.	.	.	.	.	4	.	.	.	−2	2	.	.	.	.	.	.	.	.	.	.
X.128	.	.	.	.	.	.	.	.	.	.	.	.	.	.	.	27	.	.	.	.	.	−9	.	.
X.129	.	.	.	.	.	.	.	.	.	.	.	.	.	.	.	.	.	.	.	.	.	.	.	.
X.130	.	.	.	.	.	.	.	.	.	.	.	.	.	.	.	.	.	.	.	.	.	.	.	.
X.131	.	.	.	.	.	.	.	.	.	.	.	.	.	.	.	.	.	.	.	.	.	.	.	.
X.132	.	.	.	.	.	.	.	.	.	.	.	.	.	.	.	.	.	.	.	.	.	.	.	.
X.133	.	.	.	−4	2	−1	−1	−1	.	.	.	.	.	.	.	.	.	.	.	.	.	.	.	.
X.134	.	.	.	−4	2	−1	−1	−1	.	.	.	.	.	.	.	.	.	.	.	.	.	.	.	.
X.135	.	.	.	3	−3	.	.	.	.	.	.	.	.	.	−8	−8	−8	−8	−8	−8	−8	.	7	−5
X.136	.	.	.	3	−3	.	.	.	.	.	.	.	.	.	−8	−8	−8	−8	−8	−8	−8	.	−5	7
X.137	.	.	.	3	−3	.	.	.	.	.	.	.	.	.	−8	−8	−8	−8	−8	8	8	.	7	−5
X.138	.	.	.	.	.	.	.	.	.	.	.	.	.	.	.	.	.	.	.	−12	12	.	.	.
X.139	.	.	.	3	−3	.	.	.	.	.	.	.	.	.	−8	−8	−8	−8	−8	8	8	.	−5	7
X.140	.	.	.	−6	−3	.	.	.	.	.	.	.	.	.	16	28	16	4	−2	.	.	4	4	4
X.141	.	.	.	.	.	.	.	.	.	.	.	.	.	.	.	.	.	.	.	12	−12	.	.	.
X.142	.	.	.	−6	6	.	.	.	.	.	.	.	.	.	−8	−8	16	−8	−8	−4	−4	.	−2	−2
X.143	.	.	.	−6	6	.	.	.	.	.	.	.	.	.	−8	−8	16	−8	−8	4	4	.	−2	−2
X.144	.	.	.	−6	−3	.	.	.	.	.	.	.	.	.	16	28	16	4	−2	.	.	−4	4	4
X.145	.	.	.	−12	.	3	.	.	−8	.	.	.	.	.	.	.	.	.	.	.	.	.	.	.
X.146	.	.	.	.	6	3	.	.	8	.	.	.	.	.	.	.	.	.	.	.	.	.	.	.
X.147	.	.	.	.	.	.	.	.	.	.	.	.	.	.	81	.	9	.	.	3	3	.	.	.
X.148	.	.	.	.	.	.	.	.	.	.	.	.	.	.	81	.	9	.	.	−3	−3	.	.	.
X.149	.	.	.	−12	−6	.	.	.	.	.	.	.	.	.	.	.	.	.	.	.	.	.	.	.
X.150	.	.	.	−4	2	−1	−1	−1	−4	.	.	.	2	−2	.	.	.	.	.	.	.	.	.	.
X.151	.	.	.	−4	2	−1	−1	−1	−4	.	.	.	−2	2	.	.	.	.	.	.	.	.	.	.
X.152	.	.	.	.	.	.	.	.	.	.	.	.	.	.	96	−12	.	−12	15	.	.	−4	.	.
X.153	.	.	.	.	.	.	.	.	.	.	.	.	.	.	−48	−12	.	6	6	.	.	4	.	.
X.154	.	.	.	.	.	.	.	.	.	.	.	.	.	.	−48	−12	.	6	6	.	.	4	.	.
X.155	.	.	.	.	.	.	.	.	.	.	.	.	.	.	−48	−12	.	6	6	.	.	−4	.	.
X.156	.	.	.	.	.	.	.	.	.	.	.	.	.	.	−48	−12	.	6	6	.	.	−4	.	.
X.157	−1	1	−1	.	.	.	.	.	.	.	.	.	.	.	.	.	.	.	.	.	.	.	.	.
X.158	1	−1	1	.	.	.	.	.	.	.	.	.	.	.	.	.	.	.	.	.	.	.	.	.
X.159	1	1	1	.	.	.	.	.	.	.	.	.	.	.	.	.	.	.	.	.	.	.	.	.
X.160	−1	−1	−1	.	.	.	.	.	.	.	.	.	.	.	.	.	.	.	.	.	.	.	.	.
X.161	.	.	.	.	.	.	.	.	.	.	.	.	.	.	.	.	.	.	.	.	.	.	.	.
X.162	.	.	.	6	3	.	.	.	.	.	.	.	.	.	32	56	−16	8	−4	.	.	.	−4	−4
X.163	.	.	.	.	.	.	.	.	.	.	.	.	.	.	.	.	.	.	.	.	.	.	.	.
X.164	.	.	.	.	.	.	.	.	.	.	.	.	.	.	.	.	.	.	.	.	.	.	.	.
X.165	.	.	.	8	−4	2	2	2	.	.	.	.	.	.	.	.	.	.	.	.	.	.	.	.
X.166	.	.	.	12	6	.	.	.	.	.	.	.	.	.	.	.	.	.	.	.	.	.	.	.
X.167	.	.	.	.	−6	−3	.	.	.	.	.	.	.	.	.	.	.	.	.	.	.	.	.	.

Character table of $H(\mathsf{Fi}_{24}')$ (continued)

2	6	6	6	6	6	7	7	5	5	5	6	4	4	4	4	4	4	4	5	5
3	2	2	2	2	2	1	1	2	2	2	1	2	2	2	2	2	2	2	1	1
5	.	.	.	.	.	.	.	.	.	.	.	.	.	.	.	.	.	.	.	.
7	.	.	.	.	.	.	.	.	.	.	.	.	.	.	.	.	.	.	.	.
	12_{11}	12_{12}	12_{13}	12_{14}	12_{15}	12_{16}	12_{17}	12_{18}	12_{19}	12_{20}	12_{21}	12_{22}	12_{23}	12_{24}	12_{25}	12_{26}	12_{27}	12_{28}	12_{29}	12_{30}
2P	6_7	6_{14}	6_7	6_5	6_{14}	6_9	6_9	6_{15}	6_6	6_{15}	6_7	6_{21}	6_{21}	6_{36}	6_{31}	6_{31}	6_{36}	6_8	6_{35}	6_{35}
3P	$4d$	$4g$	$4h$	$4b$	$4i$	$4c$	$4f$	$4g$	$4h$	$4i$	$4k$	$4d$	$4h$	$4e$	$4g$	$4i$	$4e$	$4b$	$4o$	$4m$
5P	12_{11}	12_{12}	12_{13}	12_{14}	12_{15}	12_{16}	12_{17}	12_{18}	12_{19}	12_{20}	12_{21}	12_{22}	12_{23}	12_{24}	12_{25}	12_{26}	12_{27}	12_{28}	12_{29}	12_{30}
7P	12_{11}	12_{12}	12_{13}	12_{14}	12_{15}	12_{16}	12_{17}	12_{18}	12_{19}	12_{20}	12_{21}	12_{22}	12_{23}	12_{24}	12_{25}	12_{26}	12_{27}	12_{28}	12_{29}	12_{30}
X.89	2	.	-2	.	.	-2	2	.	-2	.	-2	2	-2	.	.	.	.	.	.	.
X.90	-2	.	-2	.	.	2	2	.	-2	.	2	-2	-2	.	.	.	.	.	.	.
X.91	.	.	.	2	.	.	.	.	.	.	.	.	.	2	.	.	2	-1	-2	-2
X.92	.	.	.	2	.	.	.	.	.	.	.	.	.	-2	.	.	-2	-1	2	2
X.93	.	4	.	.	-4	.	.	-2	.	2	.	.	.	.	-2	2	.	.	.	.
X.94	-2	.	-2	3	.	2	2	3	1	3	2	1	1	.	.	.	.	.	1	1
X.95	2	.	-2	3	.	-2	2	3	1	3	-2	-1	1	.	.	.	.	.	-1	-1
X.96	.	.	.	-4	.	.	.	.	.	.	.	.	.	.	.	.	.	2	.	.
X.97	3	.	3	.	.	-1	-1	.	.	.	-1	.	.	-1	.	.	-1	.	.	.
X.98	-3	.	3	.	.	1	-1	.	.	.	1	.	.	1	.	.	1	.	.	.
X.99	3	.	3	.	.	-1	-1	.	.	.	-1	.	.	-1	.	.	-1	.	.	.
X.100	-3	.	3	.	.	1	-1	.	.	.	1	.	.	1	.	.	1	.	.	.
X.101	.	.	.	.	.	.	.	.	.	.	.	.	.	.	.	.	.	.	.	.
X.102	.	.	.	.	.	.	.	.	.	.	.	.	.	.	.	.	.	.	.	.
X.103	.	.	.	.	.	.	.	.	.	.	.	.	.	.	.	.	.	.	.	.
X.104	.	2	.	.	-2	.	.	-4	.	4	.	.	.	.	-4	4	.	.	.	.
X.105	3	.	-3	.	.	-1	1	.	.	.	-1	.	.	-1	.	.	-1	.	.	.
X.106	-3	.	-3	.	.	1	1	.	.	.	1	.	.	1	.	.	1	.	.	.
X.107	.	.	.	4	.	.	.	.	.	.	.	.	.	.	.	.	.	1	.	.
X.108	.	.	.	4	.	.	.	.	.	.	.	.	.	.	.	.	.	1	.	.
X.109	.	.	.	4	.	.	.	.	.	.	.	.	.	.	.	.	.	1	.	.
X.110	-2	.	-2	-3	.	2	2	.	-2	.	-2	1	1	.	.	.	.	.	-1	-1
X.111	2	.	-2	-3	.	-2	2	.	-2	.	2	-1	1	.	.	.	.	.	1	1
X.112	.	.	.	6	.	.	.	3	3	3	.	.	.	.	.	.	.	.	.	.
X.113	4	.	.	.	.	-4	.	.	.	.	.	-2	.	.	.	.	.	.	.	.
X.114	-4	.	.	.	.	4	.	.	.	.	.	2	.	.	.	.	.	.	.	.
X.115	-3	.	3	.	.	1	-1	.	.	.	-3	.	.	.	.	.	.	.	.	.
X.116	3	.	3	.	.	-1	-1	.	.	.	3	.	.	.	.	.	.	.	.	.
X.117	3	.	3	.	.	-1	-1	.	.	.	3	.	.	.	.	.	.	.	.	.
X.118	-3	.	3	.	.	1	-1	.	.	.	-3	.	.	.	.	.	.	.	.	.
X.119	.	-4	.	.	4	.	.	-4	.	4	.	.	.	.	2	-2	.	.	.	.
X.120	.	6	.	.	-6	.	.	.	.	.	.	.	.	.	.	.	.	.	.	.
X.121	.	.	4	.	.	.	-4	.	-2	.	.	.	-2	.	.	.	.	.	.	.
X.122	.	.	.	.	.	.	.	.	.	.	.	.	.	.	.	.	.	.	.	.
X.123	.	.	.	.	.	.	.	.	.	.	.	.	.	.	.	.	.	.	.	.
X.124	.	.	.	.	.	.	.	.	.	.	.	.	.	.	.	.	.	.	.	.
X.125	.	.	.	.	.	.	.	.	.	.	.	.	.	.	.	.	.	.	.	.
X.126	.	.	.	.	.	.	.	.	.	.	.	.	.	.	.	.	.	.	-2	2
X.127	.	.	.	.	.	.	.	.	.	.	.	.	.	.	.	.	.	.	2	-2
X.128	.	.	.	.	.	.	.	-3	3	-3	.	.	.	.	.	.	.	.	.	.
X.129	.	-4	.	.	4	.	.	2	.	-2	.	.	.	.	-1	1	.	.	.	.
X.130	.	-4	.	.	4	.	.	2	.	-2	.	.	.	.	-1	1	.	.	.	.
X.131	.	2	.	.	-2	.	.	2	.	-2	.	.	.	.	2	-2	.	.	.	.
X.132	.	2	.	.	-2	.	.	2	.	-2	.	.	.	.	2	-2	.	.	.	.
X.133	.	.	.	.	.	.	.	.	.	.	.	.	.	.	.	.	.	.	.	.
X.134	.	.	.	.	.	.	.	.	.	.	.	.	.	.	.	.	.	.	.	.
X.135	.	.	.	.	.	.	.	.	.	.	.	.	.	1	.	.	1	.	.	.
X.136	.	.	.	.	.	.	.	.	.	.	.	.	.	1	.	.	1	.	.	.
X.137	.	.	.	.	.	.	.	.	.	.	.	.	.	-1	.	.	-1	.	.	.
X.138	.	6	.	.	-6	.	.	.	.	.	.	.	.	.	.	.	.	.	-2	2
X.139	.	.	.	.	.	.	.	.	.	.	.	.	.	-1	.	.	-1	.	.	.
X.140	.	.	.	4	.	.	.	.	.	.	.	.	.	.	.	.	.	-2	.	.
X.141	.	6	.	.	-6	.	.	.	.	.	.	.	.	.	.	.	.	.	2	-2
X.142	.	.	.	.	.	.	.	.	.	.	.	.	.	2	.	.	2	.	.	.
X.143	.	.	.	.	.	.	.	.	.	.	.	.	.	-2	.	.	-2	.	.	.
X.144	.	.	.	-4	.	.	.	.	.	.	.	.	.	.	.	.	.	2	.	.
X.145	.	.	.	.	.	.	.	.	.	.	.	.	.	.	.	.	.	.	.	.
X.146	.	.	.	.	.	.	.	.	.	.	.	.	.	.	.	.	.	.	.	.
X.147	3	.	-3	.	.	-1	1	.	.	.	3	.	.	.	.	.	.	.	.	.
X.148	-3	.	-3	.	.	1	1	.	.	.	-3	.	.	.	.	.	.	.	.	.
X.149	.	-4	.	.	4	.	.	2	.	-2	.	.	.	.	2	-2	.	.	.	.
X.150	.	.	.	.	.	.	.	.	.	.	.	.	.	.	.	.	.	.	.	.
X.151	.	.	.	.	.	.	.	.	.	.	.	.	.	.	.	.	.	.	.	.
X.152	.	.	.	-4	.	.	.	.	.	.	.	.	.	.	.	.	.	-1	.	.
X.153	.	.	.	-2	.	.	.	.	.	.	.	.	.	.	.	.	.	-2	-2	-2
X.154	.	.	.	-2	.	.	.	.	.	.	.	.	.	.	.	.	.	-2	2	2
X.155	.	.	.	2	.	.	.	.	.	.	.	.	.	.	.	.	.	2	.	.
X.156	.	.	.	2	.	.	.	.	.	.	.	.	.	.	.	.	.	2	.	.
X.157	.	.	.	.	.	.	.	.	.	.	.	.	.	.	.	.	.	.	.	.
X.158	.	.	.	.	.	.	.	.	.	.	.	.	.	.	.	.	.	.	.	.
X.159	.	.	.	.	.	.	.	.	.	.	.	.	.	.	.	.	.	.	.	.
X.160	.	.	.	.	.	.	.	.	.	.	.	.	.	.	.	.	.	.	.	.
X.161	.	2	.	.	-2	.	.	2	.	-2	.	.	.	.	2	-2	.	.	.	.
X.162	.	.	.	.	.	.	.	.	.	.	.	.	.	.	.	.	.	.	.	.
X.163	.	-2	.	.	2	.	.	-2	.	2	.	.	.	.	-2	2	.	.	.	.
X.164	.	-2	.	.	2	.	.	-2	.	2	.	.	.	.	-2	2	.	.	.	.
X.165	.	.	.	.	.	.	.	.	.	.	.	.	.	.	.	.	.	.	.	.
X.166	.	.	.	.	.	.	.	.	.	.	.	.	.	.	.	.	.	.	.	.
X.167	.	.	.	.	.	.	.	.	.	.	.	.	.	.	.	.	.	.	.	.

Character table of $H(\mathsf{Fi}_{24}')$ *(continued)*

	12_{31}	12_{32}	12_{33}	12_{34}	$14a$	$15a$	$16a$	$18a$	$18b$	$18c$	$18d$	$18e$	$18f$	$18g$	$18h$	$18i$	$18j$	$20a$	$20b$	$20c$	$20d$	$21a$	$21b$	$24a$
2	4	4	4	4	1	1	5	5	3	1	4	2	2	2	2	2	2	4	3	3	3	1	1	5
3	1	1	1	1	1	1	.	4	4	4	2	3	3	2	2	2	2	.	.	.	.	1	1	2
5	.	.	.	.	.	1	.	.	.	.	.	.	.	.	.	.	.	1	1	1	1	.	.	.
7	.	.	.	.	1	.	.	.	.	.	.	.	.	.	.	.	.	.	.	.	.	1	1	.
2P	6_{17}	6_{31}	6_{11}	6_{31}	$7a$	$15a$	$8b$	$9a$	$9b$	$9c$	$9a$	$9e$	$9d$	$9e$	$9d$	$9d$	$9e$	$10a$	$10b$	$10d$	$10c$	$21b$	$21a$	12_3
3P	$4q$	$4n$	$4p$	$4n$	$14a$	$5a$	$16a$	6_3	6_3	6_3	6_6	6_4	6_4	6_{22}	6_{24}	6_{23}	6_{25}	$20a$	$20d$	$20c$	$20b$	$7a$	$7a$	$8a$
5P	12_{31}	12_{34}	12_{33}	12_{32}	$14a$	$3a$	$16a$	$18a$	$18b$	$18c$	$18d$	$18f$	$18e$	$18i$	$18j$	$18g$	$18h$	$4a$	$4c$	$4d$	$4c$	$21a$	$21b$	$24a$
7P	12_{31}	12_{32}	12_{33}	12_{34}	$2a$	$15a$	$16a$	$18a$	$18b$	$18c$	$18d$	$18e$	$18f$	$18g$	$18h$	$18i$	$18j$	$20a$	$20d$	$20c$	$20b$	$3a$	$3a$	$24a$
$X.89$	.	.	.	.	.	.	.	.	.	.	.	.	.	.	.	.	.	1	1	−1	1	.	.	.
$X.90$	.	.	.	.	.	.	.	.	.	.	.	.	.	.	.	.	.	1	−1	1	−1	.	.	.
$X.91$	2	.	.	.	.	.	.	−6	−3	.	2	.	.	.	.	.	.	.	.	.	.	.	.	.
$X.92$	−2	.	.	.	.	.	.	−6	−3	.	2	.	.	.	.	.	.	.	.	.	.	.	.	.
$X.93$	.	.	.	.	.	1	.	.	6	3	.	.	.	.	.	.	.	.	.	.	.	.	.	.
$X.94$	1	.	.	.	.	.	.	.	.	.	.	.	.	.	.	.	.	.	.	.	.	.	.	.
$X.95$	−1	.	.	.	.	.	.	.	.	.	.	.	.	.	.	.	.	.	.	.	.	.	.	.
$X.96$	.	.	.	.	.	.	.	.	.	.	.	.	.	.	.	.	.	2	.	.	.	.	.	.
$X.97$	.	.	.	.	.	.	.	.	.	.	.	.	.	.	.	.	.	−1	L	−1	$-L$	.	.	1
$X.98$	.	.	.	.	.	.	.	.	.	.	.	.	.	.	.	.	.	−1	L	1	$-L$	.	.	−1
$X.99$	.	.	.	.	.	.	.	.	.	.	.	.	.	.	.	.	.	−1	$-L$	−1	L	.	.	1
$X.100$	.	.	.	.	.	.	.	.	.	.	.	.	.	.	.	.	.	−1	$-L$	1	L	.	.	−1
$X.101$	.	.	.	.	.	.	1	.	.	.	.	.	.	.	.	.	.	2	.	−2	.	.	.	.
$X.102$	.	.	.	.	.	.	−1	.	.	.	.	.	.	.	.	.	.	2	.	2	.	.	.	.
$X.103$	.	.	.	.	−2	.	.	−12	−6	.	.	.	.	.	.	.	.	.	.	.	.	−1	−1	.
$X.104$	.	.	.	.	.	−1	.	.	.	.	.	.	.	.	.	.	.	.	.	.	.	.	.	.
$X.105$	.	.	.	.	.	.	.	.	.	.	.	.	.	.	.	.	.	.	.	.	.	.	.	1
$X.106$	.	.	.	.	.	.	.	.	.	.	.	.	.	.	.	.	.	.	.	.	.	.	.	−1
$X.107$	.	.	.	.	.	.	.	6	−6	.	−2	.	.	.	.	.	.	.	.	.	.	.	.	.
$X.108$	.	.	.	.	.	.	.	−3	3	.	1	.	.	.	.	.	.	.	.	.	.	.	.	.
$X.109$	.	.	.	.	.	.	.	−3	3	.	1	.	.	.	.	.	.	.	.	.	.	.	.	.
$X.110$	−1	.	.	.	.	.	.	.	.	.	.	.	.	.	.	.	.	.	.	.	.	.	.	.
$X.111$	1	.	.	.	.	.	.	.	.	.	.	.	.	.	.	.	.	.	.	.	.	.	.	.
$X.112$	.	.	.	.	.	.	.	.	.	.	.	.	.	.	.	.	.	.	.	.	.	.	.	.
$X.113$	.	.	.	.	.	.	.	9	.	.	1	.	.	.	.	.	.	1	1	−1	1	.	.	.
$X.114$	.	.	.	.	.	.	.	9	.	.	1	.	.	.	.	.	.	1	−1	1	−1	.	.	.
$X.115$	.	.	.	.	.	.	−1	.	.	.	.	.	.	.	.	.	.	.	.	.	.	.	.	3
$X.116$	.	.	.	.	.	.	1	.	.	.	.	.	.	.	.	.	.	.	.	.	.	.	.	−3
$X.117$	.	.	.	.	.	.	−1	.	.	.	.	.	.	.	.	.	.	.	.	.	.	.	.	−3
$X.118$	.	.	.	.	.	.	1	.	.	.	.	.	.	.	.	.	.	.	.	.	.	.	.	3
$X.119$	.	.	.	.	.	.	.	.	.	.	.	.	.	.	.	.	.	.	.	.	.	.	.	.
$X.120$	.	.	.	.	.	.	.	12	.	−3	.	.	.	.	.	.	.	.	.	.	.	.	.	.
$X.121$	.	.	.	.	.	.	.	−9	.	.	−1	.	.	.	.	.	.	.	.	.	.	.	.	.
$X.122$	.	.	.	.	.	.	.	.	.	.	.	.	.	.	.	.	.	−1	1	−1	1	.	.	.
$X.123$	.	.	.	.	.	.	.	.	.	.	.	.	.	.	.	.	.	−1	−1	1	−1	.	.	.
$X.124$	.	.	.	.	.	−1	.	.	.	.	.	.	.	.	.	.	.	.	.	.	.	.	.	.
$X.125$	.	.	.	.	.	−1	.	.	.	.	.	.	.	.	.	.	.	.	.	.	.	.	.	.
$X.126$	.	.	.	.	.	−1	.	.	.	.	.	.	.	.	.	.	.	.	.	.	.	.	.	.
$X.127$	.	.	.	.	.	−1	.	.	.	.	.	.	.	.	.	.	.	.	.	.	.	.	.	.
$X.128$	.	.	.	.	.	.	.	.	.	.	.	.	.	.	.	.	.	.	.	.	.	.	.	.
$X.129$	.	E	.	$\bar{E}$	−1	.	.	.	.	.	.	.	.	.	.	.	.	.	.	.	.	1	1	.
$X.130$	.	$\bar{E}$	.	E	−1	.	.	.	.	.	.	.	.	.	.	.	.	.	.	.	.	1	1	.
$X.131$	.	.	.	.	−1	.	.	.	.	.	.	.	.	.	.	.	.	.	.	.	.	$-F$	$-G$	.
$X.132$	.	.	.	.	−1	.	.	.	.	.	.	.	.	.	.	.	.	.	.	.	.	$-G$	$-F$	.
$X.133$	.	.	.	.	.	.	.	4	−2	1	.	1	1	−1	1	−1	1	.	.	.	.	.	.	.
$X.134$	.	.	.	.	.	.	.	4	−2	1	.	1	1	1	−1	1	−1	.	.	.	.	.	.	.
$X.135$	.	.	.	.	.	.	.	3	−3	.	−1	.	.	.	.	.	.	.	.	.	.	.	.	.
$X.136$	.	.	.	.	.	.	.	3	−3	.	−1	.	.	.	.	.	.	.	.	.	.	.	.	.
$X.137$	.	.	.	.	.	.	.	3	−3	.	−1	.	.	.	.	.	.	.	.	.	.	.	.	.
$X.138$	.	.	.	.	.	.	.	.	.	.	.	.	.	.	.	.	.	.	.	.	.	.	.	.
$X.139$	.	.	.	.	.	.	.	3	−3	.	−1	.	.	.	.	.	.	.	.	.	.	.	.	.
$X.140$	.	.	.	.	.	.	.	−6	−3	.	2	.	.	.	.	.	.	.	.	.	.	.	.	.
$X.141$	.	.	.	.	.	.	.	.	.	.	.	.	.	.	.	.	.	.	.	.	.	.	.	.
$X.142$	.	.	.	.	.	.	.	−6	6	.	2	.	.	.	.	.	.	.	.	.	.	.	.	6
$X.143$	.	.	.	.	.	.	.	−6	6	.	2	.	.	.	.	.	.	.	.	.	.	.	.	−6
$X.144$	.	.	.	.	.	.	.	−6	−3	.	2	.	.	.	.	.	.	.	.	.	.	.	.	.
$X.145$	.	.	.	.	.	−1	.	12	.	−3	.	.	.	.	.	.	.	.	.	.	.	.	.	.
$X.146$	.	.	.	.	2	1	.	.	−6	−3	.	.	.	.	.	.	.	.	.	.	.	1	1	.
$X.147$	.	.	.	.	.	.	.	.	.	.	.	.	.	.	.	.	.	.	.	.	.	.	.	−3
$X.148$	.	.	.	.	.	.	.	.	.	.	.	.	.	.	.	.	.	.	.	.	.	.	.	3
$X.149$	.	.	.	.	.	.	.	12	6	.	.	.	.	.	.	.	.	.	.	.	.	.	.	.
$X.150$	.	.	.	.	.	1	.	4	−2	1	.	1	1	−1	1	−1	1	.	.	.	.	.	.	.
$X.151$	.	.	.	.	.	1	.	4	−2	1	.	1	1	1	−1	1	−1	.	.	.	.	.	.	.
$X.152$	.	.	.	.	.	.	.	.	.	.	.	.	.	.	.	.	.	.	.	.	.	.	.	.
$X.153$	2	.	.	.	.	.	.	.	.	.	.	.	.	.	.	.	.	.	.	.	.	.	.	.
$X.154$	−2	.	.	.	.	.	.	.	.	.	.	.	.	.	.	.	.	.	.	.	.	.	.	.
$X.155$	.	.	.	.	.	.	.	.	.	.	.	.	.	.	.	.	.	.	.	.	.	.	.	.
$X.156$	.	.	.	.	.	.	.	.	.	.	.	.	.	.	.	.	.	.	.	.	.	.	.	.
$X.157$	.	.	.	.	.	.	1	.	.	.	.	.	.	.	.	.	.	.	.	.	.	.	.	.
$X.158$	.	.	.	.	.	.	−1	.	.	.	.	.	.	.	.	.	.	.	.	.	.	.	.	.
$X.159$	.	.	.	.	.	.	−1	.	.	.	.	.	.	.	.	.	.	.	.	.	.	.	.	.
$X.160$	.	.	.	.	.	.	1	.	.	.	.	.	.	.	.	.	.	.	.	.	.	.	.	.
$X.161$	.	.	.	.	.	.	.	.	.	.	.	.	.	.	.	.	.	.	.	.	.	.	.	.
$X.162$	.	.	.	.	.	.	.	6	3	.	−2	.	.	.	.	.	.	.	.	.	.	.	.	.
$X.163$	.	.	.	.	.	.	.	.	.	.	.	.	.	.	.	.	.	.	.	.	.	.	.	.
$X.164$	.	.	.	.	.	.	.	.	.	.	.	.	.	.	.	.	.	.	.	.	.	.	.	.
$X.165$	.	.	.	.	1	.	.	−8	4	−2	.	−2	−2	.	.	.	.	.	.	.	.	−1	−1	.
$X.166$	.	.	.	.	.	.	.	−12	−6	.	.	.	.	.	.	.	.	.	.	.	.	.	.	.
$X.167$	.	.	.	.	.	.	.	.	6	3	.	.	.	.	.	.	.	.	.	.	.	.	.	.

Character table of $H(\mathsf{Fi}'_{24})$ *(continued)*

2	5	3	3	4	4	4	4	4	1	4	4	2	1	1
3	2	2	2	1	1	1	1	1	1	3	3	3	1	1
5	.	.	.	.	.	.	.	.	1	.	.	.	.	.
7	.	.	.	.	.	.	.	.	.	.	.	.	1	1
	24b	24c	24d	24e	24f	24g	24h	24i	30a	36a	36b	36c	42a	42b
2P	12_1	12_{10}	12_9	12_{12}	12_{14}	12_{14}	12_8	12_{12}	15a	18a	18a	18b	21a	21b
3P	8a	8a	8a	8i	8c	8c	8b	8i	10a	12_2	12_2	12_2	14a	14a
5P	24b	24c	24d	24i	24g	24f	24h	24e	6_1	36a	36b	36c	42a	42b
7P	24b	24c	24d	24i	24g	24f	24h	24e	30a	36a	36b	36c	6_1	6_1
X.89	.	.	.	.	.	.	.	.	.	.	.	.	.	.
X.90	.	.	.	.	.	.	.	.	.	.	.	.	.	.
X.91	.	.	.	.	.	.	.	.	.	−2	−2	1	.	.
X.92	.	.	.	.	.	.	.	.	.	−2	−2	1	.	.
X.93	.	.	.	.	.	.	.	.	−1	.	.	.	.	.
X.94	.	.	.	.	−1	−1	−1	.	.	.	.	.	.	.
X.95	.	.	.	.	1	1	−1	.	.	.	.	.	.	.
X.96	.	.	.	.	.	.	.	.	.	.	.	.	.	.
X.97	1	1	1	.	.	.	.	.	.	.	.	.	.	.
X.98	−1	−1	−1	.	.	.	.	.	.	.	.	.	.	.
X.99	1	1	1	.	.	.	.	.	.	.	.	.	.	.
X.100	−1	−1	−1	.	.	.	.	.	.	.	.	.	.	.
X.101	.	.	.	.	.	.	.	.	.	.	.	.	.	.
X.102	.	.	.	.	.	.	.	.	.	.	.	.	.	.
X.103	.	.	.	.	.	.	.	.	.	.	.	.	1	1
X.104	.	.	.	.	.	.	.	.	1	.	.	.	.	.
X.105	1	1	1	.	.	.	.	.	.	.	.	.	.	.
X.106	−1	−1	−1	.	.	.	.	.	.	.	.	.	.	.
X.107	.	.	.	.	.	.	.	.	.	2	2	2	.	.
X.108	.	.	.	.	.	.	.	.	.	−7	5	−1	.	.
X.109	.	.	.	.	.	.	.	.	.	5	−7	−1	.	.
X.110	.	.	.	.	1	1	.	.	.	.	.	.	.	.
X.111	.	.	.	.	−1	−1	.	.	.	.	.	.	.	.
X.112	.	.	.	.	.	.	−1	.	.	.	.	.	.	.
X.113	.	.	.	.	.	.	.	.	.	−3	−3	.	.	.
X.114	.	.	.	.	.	.	.	.	.	−3	−3	.	.	.
X.115	3	.	.	.	.	.	.	.	.	.	.	.	.	.
X.116	−3	.	.	.	.	.	.	.	.	.	.	.	.	.
X.117	−3	.	.	.	.	.	.	.	.	.	.	.	.	.
X.118	3	.	.	.	.	.	.	.	.	.	.	.	.	.
X.119	.	.	.	.	.	.	.	.	.	.	.	.	.	.
X.120	.	.	.	.	.	.	.	.	.	.	.	.	.	.
X.121	.	.	.	.	.	.	.	.	.	3	3	.	.	.
X.122	.	.	.	.	.	.	.	.	.	.	.	.	.	.
X.123	.	.	.	.	.	.	.	.	.	.	.	.	.	.
X.124	.	.	.	.	.	.	.	.	1	.	.	.	.	.
X.125	.	.	.	.	.	.	.	.	1	.	.	.	.	.
X.126	.	.	.	.	.	.	.	.	1	.	.	.	.	.
X.127	.	.	.	.	.	.	.	.	1	.	.	.	.	.
X.128	.	.	.	.	.	.	1	.	.	.	.	.	.	.
X.129	.	.	.	.	.	.	.	.	.	.	.	.	−1	−1
X.130	.	.	.	.	.	.	.	.	.	.	.	.	−1	−1
X.131	.	.	.	.	.	.	.	.	.	.	.	.	G	F
X.132	.	.	.	.	.	.	.	.	.	.	.	.	F	G
X.133	.	.	.	.	.	.	.	.	.	.	.	.	.	.
X.134	.	.	.	.	.	.	.	.	.	.	.	.	.	.
X.135	.	3	−3	.	.	.	.	.	.	−5	7	1	.	.
X.136	.	−3	3	.	.	.	.	.	.	7	−5	1	.	.
X.137	.	−3	3	.	.	.	.	.	.	−5	7	1	.	.
X.138	.	.	.	.	.	.	.	.	.	.	.	.	.	.
X.139	.	3	−3	.	.	.	.	.	.	7	−5	1	.	.
X.140	.	.	.	.	.	.	.	.	.	−2	−2	1	.	.
X.141	.	.	.	.	.	.	.	.	.	.	.	.	.	.
X.142	−6	.	.	.	.	.	.	.	.	−2	−2	−2	.	.
X.143	6	.	.	.	.	.	.	.	.	−2	−2	−2	.	.
X.144	.	.	.	.	.	.	.	.	.	−2	−2	1	.	.
X.145	.	.	.	.	.	.	.	.	1	.	.	.	.	.
X.146	.	.	.	.	.	.	.	.	−1	.	.	.	−1	−1
X.147	−3	.	.	.	.	.	.	.	.	.	.	.	.	.
X.148	3	.	.	.	.	.	.	.	.	.	.	.	.	.
X.149	.	.	.	.	.	.	.	.	.	.	.	.	.	.
X.150	.	.	.	.	.	.	.	.	−1	.	.	.	.	.
X.151	.	.	.	.	.	.	.	.	−1	.	.	.	.	.
X.152	.	.	.	.	.	.	.	.	.	.	.	.	.	.
X.153	.	.	.	.	.	.	.	.	.	.	.	.	.	.
X.154	.	.	.	.	.	.	.	.	.	.	.	.	.	.
X.155	.	.	.	.	N	−N	.	.	.	.	.	.	.	.
X.156	.	.	.	.	−N	N	.	.	.	.	.	.	.	.
X.157	.	.	.	.	.	.	.	.	.	.	.	.	.	.
X.158	.	.	.	.	.	.	.	.	.	.	.	.	.	.
X.159	.	.	.	.	.	.	.	.	.	.	.	.	.	.
X.160	.	.	.	.	.	.	.	.	.	.	.	.	.	.
X.161	.	.	.	.	.	.	.	.	.	.	.	.	.	.
X.162	.	.	.	.	.	.	.	.	.	2	2	−1	.	.
X.163	.	.	.	−N	.	.	.	N	.	.	.	.	.	.
X.164	.	.	.	N	.	.	.	−N	.	.	.	.	.	.
X.165	.	.	.	.	.	.	.	.	.	.	.	.	1	1
X.166	.	.	.	.	.	.	.	.	.	.	.	.	.	.
X.167	.	.	.	.	.	.	.	.	.	.	.	.	.	.

where $A = 12\zeta(3)+2$, $B = -3\zeta(3)-2$, $C = -\zeta(3)$, $D = -12\zeta(3)-6$, $E = -2\zeta(3)-1$, $F = 2\zeta(21)_3\zeta(21)_7^4 + 2\zeta(21)_3\zeta(21)_7^2 + 2\zeta(21)_3\zeta(21)_7 + \zeta(21)_3 + \zeta(21)_7^4 + \zeta(21)_7^2 + \zeta(21)_7 + 1$, $G = -F+1$, $H = 6\zeta(3)+3$, $I = 6\zeta(3)+1$, $J = -4\zeta(5)^3 - 4\zeta(5)^2 - 1$, $K = 4\zeta(5)^3 + 4\zeta(5)^2 + 3$, $L = 2\zeta(5)^3 + 2\zeta(5)^2 + 1$, $M = 18\zeta(3)+9$, $N = -4\zeta(12)_4\zeta(12)_3 - 2\zeta(12)_4$.

9.8.3 *Character table of* $E(\mathsf{Fi}_{24}') = \langle x, y, e \rangle$

2	21	19	21	17	17	14	8	6	15	15	15	15	11	13	11	11	11	11	10	10
3	3	2	3	1	1	1	3	2	1	1	1	1	1	.	1	1	.	.	.	.
5	1	1	1	.	.	1	1	.	1	1	.	.	.	.	.	.	.	.	.	.
7	1	1	.	1	.	.	.	1	.	.	.	.	1	.	.	.	.	.	.	.
11	1	1	.	.	.	.	.	.	.	.	.	.	.	.	.	.	.	.	.	.
23	1	.	.	.	.	.	.	.	.	.	.	.	.	.	.	.	.	.	.	.
	1a	2a	2b	2c	2d	2e	3a	3b	4a	4b	4c	4d	4e	4f	4g	4h	4i	4j	4k	4l
2P	1a	1a	1a	1a	1a	1a	3a	3b	2b	2b	2b	2b	2a	2b	2c	2c	2d	2d	2c	2d
3P	1a	2a	2b	2c	2d	2e	1a	1a	4a	4b	4c	4d	4e	4f	4g	4h	4i	4j	4k	4l
5P	1a	2a	2b	2c	2d	2e	3a	3b	4a	4b	4c	4d	4e	4f	4g	4h	4i	4j	4k	4l
7P	1a	2a	2b	2c	2d	2e	3a	3b	4a	4b	4c	4d	4e	4f	4g	4h	4i	4j	4k	4l
11P	1a	2a	2b	2c	2d	2e	3a	3b	4a	4b	4c	4d	4e	4f	4g	4h	4i	4j	4k	4l
23P	1a	2a	2b	2c	2d	2e	3a	3b	4a	4b	4c	4d	4e	4f	4g	4h	4i	4j	4k	4l
X.1	1	1	1	1	1	1	1	1	1	1	1	1	1	1	1	1	1	1	1	1
X.2	23	23	23	7	7	−1	5	−1	−1	−1	7	7	7	−1	−1	−1	3	3	3	−1
X.3	45	45	45	−3	−3	5	.	3	5	5	−3	−3	−3	5	−3	−3	1	1	1	−3
X.4	45	45	45	−3	−3	5	.	3	5	5	−3	−3	−3	5	−3	−3	1	1	1	−3
X.5	231	231	231	7	7	−9	−3	.	−9	−9	7	7	7	−9	−1	−1	−1	−1	−1	−1
X.6	231	231	231	7	7	−9	−3	.	−9	−9	7	7	7	−9	−1	−1	−1	−1	−1	−1
X.7	252	252	252	28	28	12	9	.	12	12	28	28	28	12	4	4	4	4	4	4
X.8	253	253	253	13	13	−11	10	1	−11	−11	13	13	13	−11	−3	−3	1	1	1	−3
X.9	483	483	483	35	35	3	6	.	3	3	35	35	35	3	3	3	3	3	3	3
X.10	759	55	−9	71	7	15	21	.	15	15	7	−9	−1	−1	7	7	11	−5	3	−1
X.11	770	770	770	−14	−14	10	5	−7	10	10	−14	−14	−14	10	2	2	−2	−2	−2	2
X.12	770	770	770	−14	−14	10	5	−7	10	10	−14	−14	−14	10	2	2	−2	−2	−2	2
X.13	990	990	990	−18	−18	−10	.	3	−10	−10	−18	−18	−18	−10	6	6	2	2	2	6
X.14	990	990	990	−18	−18	−10	.	3	−10	−10	−18	−18	−18	−10	6	6	2	2	2	6
X.15	1035	1035	1035	27	27	35	.	6	35	35	27	27	27	35	3	3	−1	−1	−1	3
X.16	1035	1035	1035	−21	−21	−5	.	−3	−5	−5	−21	−21	−21	−5	3	3	3	3	3	3
X.17	1035	1035	1035	−21	−21	−5	.	−3	−5	−5	−21	−21	−21	−5	3	3	3	3	3	3
X.18	1265	1265	1265	49	49	−15	5	8	−15	−15	49	49	49	−15	−7	−7	1	1	1	−7
X.19	1288	−56	8	56	−8	−16	10	7	48	−16	−8	8	.	.	8	−8	4	4	−4	.
X.20	1288	−56	8	56	−8	−16	10	7	−16	48	−8	8	.	.	−8	8	4	4	−4	.
X.21	1771	1771	1771	−21	−21	11	16	7	11	11	−21	−21	−21	11	3	3	−5	−5	−5	3
X.22	2024	2024	2024	8	8	24	−1	8	24	24	8	8	8	24	8	8	.	.	.	8
X.23	2277	2277	2277	21	21	−19	.	6	−19	−19	21	21	21	−19	−3	−3	1	1	1	−3
X.24	3312	3312	3312	48	48	16	.	−6	16	16	48	48	48	16	.	.	.	.	.	.
X.25	3520	3520	3520	64	64	.	10	−8	.	.	64	64	64	.	.	.	.	.	.	.
X.26	5313	5313	5313	49	49	9	−15	.	9	9	49	49	49	9	1	1	−3	−3	−3	1
X.27	5313	385	−63	161	−31	−15	39	.	−15	−15	33	−15	−7	1	−7	−7	9	−7	1	1
X.28	5544	5544	5544	−56	−56	24	9	.	24	24	−56	−56	−56	24	−8	−8	.	.	.	−8
X.29	5796	5796	5796	−28	−28	36	−9	.	36	36	−28	−28	−28	36	−4	−4	4	4	4	−4
X.30	10395	10395	10395	−21	−21	−45	.	.	−45	−45	−21	−21	−21	−45	3	3	−1	−1	−1	3
X.31	10626	770	−126	210	82	90	24	.	90	90	18	−14	−14	−6	18	18	6	6	6	2
X.32	11385	825	−135	265	73	−15	45	.	−15	−15	9	−39	1	1	−7	−7	13	−3	5	1
X.33	15180	1100	−180	300	44	60	15	.	60	60	44	−20	−20	−4	4	4	4	4	4	4
X.34	15939	1155	−189	35	−29	−45	36	.	−45	−45	35	19	−21	3	3	3	−5	11	3	−5
X.35	15939	1155	−189	35	−29	−45	−18	.	−45	−45	35	19	−21	3	3	3	−5	11	3	−5
X.36	15939	1155	−189	35	−29	−45	−18	.	−45	−45	35	19	−21	3	3	3	−5	11	3	−5
X.37	21252	1540	−252	196	−60	−60	21	.	−60	−60	68	4	−28	4	−4	−4	4	4	4	−4
X.38	26565	1925	−315	−203	117	45	60	.	45	45	−11	69	−35	−3	−3	−3	−15	1	−7	5
X.39	28336	−1232	176	336	−48	32	40	−14	−32	−32	−48	48	.	.	.	.	8	8	−8	.
X.40	34155	2475	−405	−213	−21	75	.	.	75	75	−21	27	3	−5	−21	−21	11	−5	3	3
X.41	34155	2475	−405	−165	27	−45	.	.	−45	−45	27	75	−45	3	3	3	7	−9	−1	−5
X.42	34155	2475	−405	−165	27	−45	.	.	−45	−45	27	75	−45	3	3	3	7	−9	−1	−5
X.43	34155	2475	−405	−213	−21	75	.	.	75	75	−21	27	3	−5	−21	−21	11	−5	3	3
X.44	41216	−1792	256	.	.	−128	−40	14	128	128	.	.	.	.	.	.	.	.	.	.
X.45	42504	3080	−504	168	168	120	−39	.	120	120	40	40	−56	−8	8	8	.	.	.	8
X.46	48576	3520	−576	64	64	.	−6	.	.	.	64	64	−64	.	.	.	.	.	.	.
X.47	53130	3850	−630	154	26	−30	−15	.	−30	−30	90	58	−70	2	−14	−14	−2	−2	−2	2
X.48	57960	−2520	360	−168	24	−80	.	21	−80	240	24	−24	.	.	24	−24	4	4	−4	.
X.49	57960	−2520	360	−168	24	−80	.	21	240	−80	24	−24	.	.	−24	24	4	4	−4	.
X.50	68310	4950	−810	582	198	30	.	.	30	30	6	−90	6	−2	6	6	2	2	2	−10
X.51	69552	−3024	432	336	−48	−96	.	.	96	96	−48	48	.	.	.	.	8	8	−8	.
X.52	69552	−3024	432	336	−48	−96	.	.	96	96	−48	48	.	.	.	.	8	8	−8	.
X.53	70840	−3080	440	392	−56	80	10	7	80	−240	−56	56	.	.	8	−8	−4	−4	4	.
X.54	70840	−3080	440	392	−56	80	10	7	−240	80	−56	56	.	.	−8	8	−4	−4	4	.
X.55	79695	5775	−945	63	255	−105	45	.	−105	−105	−65	−17	7	7	−17	−17	−5	11	3	7
X.56	79695	5775	−945	−161	31	135	45	.	135	135	−33	15	7	−9	−1	−1	−13	3	−5	−9
X.57	79695	5775	−945	399	−177	15	45	.	15	15	79	−65	7	−1	7	7	−1	−17	−9	−1
X.58	85008	−3696	528	112	−16	−96	30	.	−96	288	−16	16	.	.	−16	16	−8	−8	8	.
X.59	85008	−3696	528	112	−16	−96	30	.	288	−96	−16	16	.	.	16	−16	−8	−8	8	.
X.60	91080	6600	−1080	104	104	120	−45	.	120	120	−24	−24	8	−8	−8	−8	.	.	.	−8
X.61	127512	−5544	792	168	−24	16	.	21	−48	16	−24	24	.	.	24	−24	−4	−4	4	.
X.62	127512	−5544	792	168	−24	16	.	21	16	−48	−24	24	.	.	−24	24	−4	−4	4	.
X.63	141680	−6160	880	−112	16	160	20	14	−160	−160	16	−16	.	.	.	.	8	8	−8	.
X.64	154560	−6720	960	−448	64	.	30	.	.	.	64	−64	.	.	.	.	.	.	.	.
X.65	154560	−6720	960	−448	64	.	30	.	.	.	64	−64	.	.	.	.	.	.	.	.
X.66	159390	11550	−1890	462	78	−90	−45	.	−90	−90	14	−82	14	6	−10	−10	−6	−6	−6	6
X.67	185472	−8064	1152	.	.	−64	.	−21	192	−64	.	.	.	.	.	.	.	.	.	.
X.68	185472	−8064	1152	.	.	−64	.	−21	−64	192	.	.	.	.	.	.	.	.	.	.
X.69	226688	−9856	1408	.	.	64	−40	−7	−192	64	.	.	.	.	.	.	.	.	.	.
X.70	226688	−9856	1408	.	.	64	−40	−7	64	−192	.	.	.	.	.	.	.	.	.	.
X.71	239085	17325	−2835	−483	93	−75	.	.	−75	−75	−99	45	21	5	21	21	9	−7	1	−3
X.72	239085	17325	−2835	−147	−339	45	.	.	45	45	45	−3	21	−3	−3	−3	−3	13	5	5

Character table of $E(\mathsf{Fi}_{24}')$ (continued)

	5a	6a	6b	6c	6d	6e	6f	6g	7a	7b	8a	8b	8c	8d	8e	8f	8g	8h	10a	10b	10c	10d	11a	12a	12b	12c	12d
2	5	8	8	8	6	6	6	4	3	3	8	8	8	9	8	7	6	6	5	4	4	4	1	6	6	5	5
3	1	3	2	2	2	1	1	1	1	1	1	1	1	.	.	.	.	.	1	.	.	.	.	1	1	1	1
5	1	1	.	.	.	.	.	.	.	.	.	.	.	.	.	.	.	.	1	1	1	1	.	.	.	.	.
7	.	.	.	.	.	.	.	.	1	1	.	.	.	.	.	.	.	.	.	.	.	.	.	.	.	.	.
11	.	.	.	.	.	.	.	.	.	.	.	.	.	.	.	.	.	.	.	.	.	.	1	.	.	.	.
23	.	.	.	.	.	.	.	.	.	.	.	.	.	.	.	.	.	.	.	.	.	.	.	.	.	.	.
2P	5a	3a	3a	3a	3b	3a	3a	3b	7a	7b	4d	4a	4b	4d	4c	4f	4i	4j	5a	5a	5a	5a	11a	6b	6b	6d	6d
3P	5a	2b	2b	2a	2b	2c	2d	2e	7b	7a	8a	8b	8c	8d	8e	8f	8g	8h	10a	10b	10d	10c	11a	4c	4d	4a	4b
5P	1a	6a	6b	6c	6d	6e	6f	6g	7b	7a	8a	8b	8c	8d	8e	8f	8g	8h	2b	2a	2e	2e	11a	12a	12b	12c	12d
7P	5a	6a	6b	6c	6d	6e	6f	6g	1a	1a	8a	8b	8c	8d	8e	8f	8g	8h	10a	10b	10d	10c	11a	12a	12b	12c	12d
11P	5a	6a	6b	6c	6d	6e	6f	6g	7a	7b	8a	8b	8c	8d	8e	8f	8g	8h	10a	10b	10c	10d	1a	12a	12b	12c	12d
23P	5a	6a	6b	6c	6d	6e	6f	6g	7a	7b	8a	8b	8c	8d	8e	8f	8g	8h	10a	10b	10d	10c	11a	12a	12b	12c	12d
X.1	1	1	1	1	1	1	1	1	1	1	1	1	1	1	1	1	1	1	1	1	1	1	1	1	1	1	1
X.2	3	5	5	5	−1	1	1	−1	2	2	−1	−1	−1	3	3	−1	1	1	3	3	−1	−1	1	1	1	−1	−1
X.3	.	.	.	.	3	.	.	−1	A	$\bar{A}$	−3	1	1	1	1	1	−1	−1	.	.	.	.	1	.	.	−1	−1
X.4	.	.	.	.	3	.	.	−1	$\bar{A}$	A	−3	1	1	1	1	1	−1	−1	.	.	.	.	1	.	.	−1	−1
X.5	1	−3	−3	−3	.	1	1	.	.	.	−1	3	3	−1	−1	3	−1	−1	1	1	1	1	.	1	1	.	.
X.6	1	−3	−3	−3	.	1	1	.	.	.	−1	3	3	−1	−1	3	−1	−1	1	1	1	1	.	1	1	.	.
X.7	2	9	9	9	.	1	1	.	.	.	4	.	.	4	4	.	.	.	2	2	2	2	−1	1	1	.	.
X.8	3	10	10	10	1	−2	−2	1	1	1	−3	1	1	1	1	1	−1	−1	3	3	−1	−1	.	−2	−2	1	1
X.9	−2	6	6	6	.	2	2	.	.	.	3	3	3	3	3	3	−1	−1	−2	−2	−2	−2	−1	2	2	.	.
X.10	4	9	−3	1	.	5	1	.	3	3	−1	3	3	−1	−1	−1	1	1	−4	.	.	.	.	1	−3	.	.
X.11	.	5	5	5	−7	1	1	1	.	.	2	−2	−2	−2	−2	−2	.	.	.	.	.	.	.	1	1	1	1
X.12	.	5	5	5	−7	1	1	1	.	.	2	−2	−2	−2	−2	−2	.	.	.	.	.	.	.	1	1	1	1
X.13	.	.	.	.	3	.	.	−1	A	$\bar{A}$	6	−2	−2	2	2	−2	.	.	.	.	.	.	.	.	.	−1	−1
X.14	.	.	.	.	3	.	.	−1	$\bar{A}$	A	6	−2	−2	2	2	−2	.	.	.	.	.	.	.	.	.	−1	−1
X.15	.	.	.	.	6	.	.	2	−1	−1	3	3	3	−1	−1	3	1	1	.	.	.	.	1	.	.	2	2
X.16	.	.	.	.	−3	.	.	1	D	$\bar{D}$	3	−1	−1	3	3	−1	−1	−1	.	.	.	.	1	.	.	1	1
X.17	.	.	.	.	−3	.	.	1	$\bar{D}$	D	3	−1	−1	3	3	−1	−1	−1	.	.	.	.	1	.	.	1	1
X.18	.	5	5	5	8	1	1	.	−2	−2	−7	−3	−3	1	1	−3	1	1	.	.	.	.	.	1	1	.	.
X.19	3	−10	2	−2	−1	2	−2	−1	.	.	.	−4	4	.	.	.	−2	2	3	−1	−1	−1	1	−2	2	3	−1
X.20	3	−10	2	−2	−1	2	−2	−1	.	.	.	4	−4	.	.	.	−2	2	3	−1	−1	−1	1	−2	2	−1	3
X.21	1	16	16	16	7	.	.	−1	.	.	3	−1	−1	−5	−5	−1	−1	−1	1	1	1	1	.	.	.	−1	−1
X.22	−1	−1	−1	−1	8	−1	−1	.	1	1	8	.	.	.	.	.	.	.	−1	−1	−1	−1	.	−1	−1	.	.
X.23	−3	.	.	.	6	.	.	2	2	2	−3	−3	−3	1	1	−3	−1	−1	−3	−3	1	1	.	.	.	2	2
X.24	−3	.	.	.	−6	.	.	−2	1	1	.	.	.	.	.	.	.	.	−3	−3	1	1	1	.	.	−2	−2
X.25	.	10	10	10	−8	−2	−2	.	−1	−1	.	.	.	.	.	.	.	.	.	.	.	.	.	−2	−2	.	.
X.26	3	−15	−15	−15	.	1	1	.	.	.	1	−3	−3	−3	−3	−3	−1	−1	3	3	−1	−1	.	1	1	.	.
X.27	8	−9	−9	7	.	−1	−1	.	.	.	1	−3	−3	5	−3	1	1	1	−8	.	.	.	.	3	3	.	.
X.28	−1	9	9	9	.	1	1	.	.	.	−8	.	.	.	.	.	.	.	−1	−1	−1	−1	.	1	1	.	.
X.29	1	−9	−9	−9	.	−1	−1	.	.	.	−4	.	.	4	4	.	.	.	1	1	1	1	−1	−1	−1	.	.
X.30	.	.	.	.	.	.	.	.	.	.	3	3	3	−1	−1	3	1	1	.	.	.	.	.	.	.	.	.
X.31	−4	36	.	−4	.	.	4	.	.	.	−6	6	6	−2	−2	−2	.	.	4	.	.	.	.	.	4	.	.
X.32	.	45	−3	−3	.	1	1	.	3	3	1	−3	−3	−7	1	1	−1	−1	.	.	.	.	.	−3	−3	.	.
X.33	.	−45	−9	11	.	3	−1	.	−3	−3	−4	.	.	4	−4	.	.	.	.	.	.	.	.	−1	−5	.	.
X.34	4	−36	−12	12	.	−4	4	.	.	.	3	3	3	−1	−1	−1	−1	−1	−4	.	.	.	.	−4	4	.	.
X.35	4	18	6	−6	.	2	−2	.	.	.	3	3	3	−1	−1	−1	−1	−1	−4	.	.	.	.	2	−2	.	.
X.36	4	18	6	−6	.	2	−2	.	.	.	3	3	3	−1	−1	−1	−1	−1	−4	.	.	.	.	2	−2	.	.
X.37	−8	9	−3	1	.	1	−3	.	.	.	4	.	.	4	−4	.	.	.	8	.	.	.	.	5	1	.	.
X.38	.	.	−12	8	.	4	.	.	.	.	−3	−3	−3	−3	5	1	−1	−1	.	.	.	.	.	4	.	.	.
X.39	6	−40	8	−8	2	.	.	2	.	.	.	.	.	.	.	.	.	.	6	−2	2	2	.	.	.	−2	−2
X.40	.	.	.	.	.	.	.	.	E	$\bar{E}$	3	3	3	−1	−1	−1	−1	−1	.	.	.	.	.	.	.	.	.
X.41	.	.	.	.	.	.	.	.	E	$\bar{E}$	3	3	3	−5	3	−1	1	1	.	.	.	.	.	.	.	.	.
X.42	.	.	.	.	.	.	.	.	$\bar{E}$	E	3	3	3	−5	3	−1	1	1	.	.	.	.	.	.	.	.	.
X.43	.	.	.	.	.	.	.	.	$\bar{E}$	E	3	3	3	−1	−1	−1	−1	−1	.	.	.	.	.	.	.	.	.
X.44	6	40	−8	8	−2	.	.	−2	.	.	.	.	.	.	.	.	.	.	6	−2	2	2	−1	.	.	2	2
X.45	4	9	9	−7	.	−3	−3	.	.	.	−8	.	.	.	.	.	.	.	−4	.	.	.	.	1	1	.	.
X.46	−4	−54	−6	10	.	−2	−2	.	3	3	.	.	.	.	.	.	.	.	4	.	.	.	.	−2	−2	.	.
X.47	.	45	9	−11	.	1	5	.	.	.	2	−6	−6	6	−2	2	.	.	.	.	.	.	.	−3	1	.	.
X.48	.	.	.	.	−3	.	.	1	.	.	.	4	−4	.	.	.	2	−2	.	.	.	.	1	.	.	1	−3
X.49	.	.	.	.	−3	.	.	1	.	.	.	−4	4	.	.	.	2	−2	.	.	.	.	1	.	.	−3	1
X.50	.	.	.	.	.	.	.	.	−3	−3	6	−6	−6	−6	2	2	.	.	.	.	.	.	.	.	.	.	.
X.51	−3	.	.	.	.	.	.	.	.	.	.	.	.	.	.	.	.	.	−3	1	F	G	−1	.	.	.	.
X.52	−3	.	.	.	.	.	.	.	.	.	.	.	.	.	.	.	.	.	−3	1	G	F	−1	.	.	.	.
X.53	.	−10	2	−2	−1	2	−2	−1	.	.	.	4	−4	.	.	.	2	−2	.	.	.	.	.	−2	2	−1	3
X.54	.	−10	2	−2	−1	2	−2	−1	.	.	.	−4	4	.	.	.	2	−2	.	.	.	.	.	−2	2	3	−1
X.55	.	45	−3	−3	.	−3	−3	.	.	.	−1	3	3	−1	−1	−1	1	1	.	.	.	.	.	1	1	.	.
X.56	.	45	−3	−3	.	1	1	.	.	.	7	3	3	7	−1	−1	1	1	.	.	.	.	.	−3	−3	.	.
X.57	.	45	−3	−3	.	−3	−3	.	.	.	−1	3	3	3	3	−1	−1	−1	.	.	.	.	.	1	1	.	.
X.58	3	−30	6	−6	.	−2	2	.	.	.	.	.	.	.	.	.	.	.	3	−1	−1	−1	.	2	−2	.	.
X.59	3	−30	6	−6	.	−2	2	.	.	.	.	.	.	.	.	.	.	.	3	−1	−1	−1	.	2	−2	.	.
X.60	.	−45	3	3	.	−1	−1	.	3	3	8	.	.	.	.	.	.	.	.	.	.	.	.	3	3	.	.
X.61	−3	.	.	.	−3	.	.	1	.	.	.	−4	4	.	.	.	−2	2	−3	1	1	1	.	.	.	−3	1
X.62	−3	.	.	.	−3	.	.	1	.	.	.	4	−4	.	.	.	−2	2	−3	1	1	1	.	.	.	1	−3
X.63	.	−20	4	−4	−2	−4	4	−2	.	.	.	.	.	.	.	.	.	.	.	.	.	.	.	4	−4	2	2
X.64	.	−30	6	−6	.	2	−2	.	.	.	.	.	.	.	.	.	.	.	.	.	.	.	−1	−2	2	.	.
X.65	.	−30	6	−6	.	2	−2	.	.	.	.	.	.	.	.	.	.	.	.	.	.	.	−1	−2	2	.	.
X.66	.	−45	3	3	.	3	3	.	.	.	−2	6	6	2	2	−2	.	.	.	.	.	.	.	−1	−1	.	.
X.67	−3	.	.	.	3	.	.	−1	.	.	.	−8	8	.	.	.	.	.	−3	1	1	1	1	.	.	3	−1
X.68	−3	.	.	.	3	.	.	−1	.	.	.	8	−8	.	.	.	.	.	−3	1	1	1	1	.	.	−1	3
X.69	3	40	−8	8	1	.	.	1	.	.	.	−8	8	.	.	.	.	.	3	−1	−1	−1	.	.	.	−3	1
X.70	3	40	−8	8	1	.	.	1	.	.	.	8	−8	.	.	.	.	.	3	−1	−1	−1	.	.	.	1	−3
X.71	.	.	.	.	.	.	.	.	.	.	−3	−3	−3	5	−3	1	−1	−1	.	.	.	.	.	.	.	.	.
X.72	.	.	.	.	.	.	.	.	.	.	−3	−3	−3	−7	1	1	1	1	.	.	.	.	.	.	.	.	.

Character table of $E(\mathsf{Fi}_{24}')$ *(continued)*

2	4	4	4	3	3	2	2	1	1	5	4	4	.	.	1	.	.	4	4	3	3	2	2	1	1
3	1	1	1	.	.	.	.	1	1	.	.	.	1	1	.	.	.	1	1	1	1	.	.	1	1
5	.	.	.	.	.	.	.	1	1	.	1	1	.	.	.	.	.	.	.	.	.	.	.	1	1
7	.	.	.	1	1	1	1	.	.	.	.	.	1	1	.	.	.	.	.	.	.	1	1	.	.
11	.	.	.	.	.	.	.	.	.	.	.	.	.	.	1	.	.	.	.	.	.	.	.	.	.
23	.	.	.	.	.	.	.	.	.	.	.	.	.	.	.	1	1	.	.	.	.	.	.	.	.
	12e	12f	12g	14a	14b	14c	14d	15a	15b	16a	20a	20b	21a	21b	22a	23a	23b	24a	24b	24c	24d	28a	28b	30a	30b
2P	6c	6e	6e	7b	7a	7a	7b	15a	15b	8d	10a	10a	21a	21b	11a	23a	23b	12b	12b	12c	12d	14a	14b	15a	15b
3P	4e	4g	4h	14b	14a	14d	14c	5a	5a	16a	20a	20b	7a	7b	22a	23a	23b	8a	8a	8b	8c	28b	28a	10a	10a
5P	12e	12f	12g	14b	14a	14d	14c	3a	3a	16a	4b	4a	21b	21a	22a	23b	23a	24b	24a	24c	24d	28b	28a	6a	6a
7P	12e	12f	12g	2a	2a	2c	2c	15b	15a	16a	20a	20b	3b	3b	22a	23b	23a	24b	24a	24c	24d	4e	4e	30b	30a
11P	12e	12f	12g	14a	14b	14c	14d	15b	15a	16a	20a	20b	21a	21b	2a	23b	23a	24a	24b	24c	24d	28a	28b	30b	30a
23P	12e	12f	12g	14a	14b	14c	14d	15a	15b	16a	20a	20b	21a	21b	22a	1a	1a	24a	24b	24c	24d	28a	28b	30a	30b
X.1	1	1	1	1	1	1	1	1	1	1	1	1	1	1	1	1	1	1	1	1	1	1	1	1	1
X.2	1	−1	−1	2	2	.	.	.	.	1	−1	−1	−1	−1	1	.	.	−1	−1	−1	−1	.	.	.	.
X.3	.	.	.	$\bar{A}$	A	$-A$	$-\bar{A}$	.	.	−1	.	.	$\bar{A}$	A	1	−1	−1	.	.	1	1	$-\bar{A}$	$-A$	.	.
X.4	.	.	.	A	$\bar{A}$	$-\bar{A}$	$-A$	.	.	−1	.	.	A	$\bar{A}$	1	−1	−1	.	.	1	1	$-A$	$-\bar{A}$	.	.
X.5	1	−1	−1	.	.	.	.	B	$\bar{B}$	−1	1	1	.	.	.	1	1	−1	−1	.	.	.	.	B	$\bar{B}$
X.6	1	−1	−1	.	.	.	.	$\bar{B}$	B	−1	1	1	.	.	.	1	1	−1	−1	.	.	.	.	$\bar{B}$	B
X.7	1	1	1	.	.	.	.	−1	−1	.	2	2	.	.	−1	−1	−1	1	1	.	.	.	.	−1	−1
X.8	−2	.	.	1	1	−1	−1	.	.	−1	−1	−1	1	1	.	.	.	.	.	1	1	−1	−1	.	.
X.9	2	.	.	.	.	.	.	1	1	−1	−2	−2	.	.	−1	.	.	.	.	.	.	.	.	1	1
X.10	−1	1	1	−1	−1	1	1	1	1	−1	.	.	.	.	.	.	.	−1	−1	.	.	−1	−1	−1	−1
X.11	1	−1	−1	.	.	.	.	.	.	.	.	.	.	.	.	C	$\bar{C}$	−1	−1	1	1	.	.	.	.
X.12	1	−1	−1	.	.	.	.	.	.	.	.	.	.	.	.	$\bar{C}$	C	−1	−1	1	1	.	.	.	.
X.13	.	.	.	$\bar{A}$	A	A	$\bar{A}$	.	.	.	.	.	$\bar{A}$	A	.	1	1	.	.	1	1	$\bar{A}$	A	.	.
X.14	.	.	.	A	$\bar{A}$	$\bar{A}$	A	.	.	.	.	.	A	$\bar{A}$	.	1	1	.	.	1	1	A	$\bar{A}$	.	.
X.15	.	.	.	−1	−1	−1	−1	.	.	1	.	.	−1	−1	1	.	.	.	.	.	.	−1	−1	.	.
X.16	.	.	.	$\bar{D}$	D	.	.	.	.	−1	.	.	$-A$	$-\bar{A}$	1	.	.	.	.	−1	−1	.	.	.	.
X.17	.	.	.	D	$\bar{D}$	.	.	.	.	−1	.	.	$-\bar{A}$	$-A$	1	.	.	.	.	−1	−1	.	.	.	.
X.18	1	−1	−1	−2	−2	.	.	.	.	1	.	.	1	1	.	.	.	−1	−1	.	.	.	.	.	.
X.19	.	2	−2	.	.	.	.	.	.	.	−1	3	.	.	−1	.	.	.	.	−1	1	.	.	.	.
X.20	.	−2	2	.	.	.	.	.	.	.	3	−1	.	.	−1	.	.	.	.	1	−1	.	.	.	.
X.21	.	.	.	.	.	.	.	1	1	−1	1	1	.	.	.	.	.	.	.	−1	−1	.	.	1	1
X.22	−1	−1	−1	1	1	1	1	−1	−1	.	−1	−1	1	1	.	.	.	−1	−1	.	.	1	1	−1	−1
X.23	.	.	.	2	2	.	.	.	.	−1	1	1	−1	−1	.	.	.	.	.	.	.	.	.	.	.
X.24	.	.	.	1	1	−1	−1	.	.	.	1	1	1	1	1	.	.	.	.	.	.	−1	−1	.	.
X.25	−2	.	.	−1	−1	1	1	.	.	.	.	.	−1	−1	.	1	1	.	.	.	.	1	1	.	.
X.26	1	1	1	.	.	.	.	.	.	−1	−1	−1	.	.	.	.	.	1	1	.	.	.	.	.	.
X.27	−1	−1	−1	.	.	.	.	−1	−1	−1	.	.	.	.	.	.	.	1	1	.	.	.	.	1	1
X.28	1	1	1	.	.	.	.	−1	−1	.	−1	−1	.	.	.	1	1	1	1	.	.	.	.	−1	−1
X.29	−1	−1	−1	.	.	.	.	1	1	.	1	1	.	.	−1	.	.	−1	−1	.	.	.	.	1	1
X.30	.	.	.	.	.	.	.	.	.	1	.	.	.	.	.	−1	−1	.	.	.	.	.	.	.	.
X.31	−2	.	.	.	.	.	.	−1	−1	.	.	.	.	.	.	.	.	.	.	.	.	.	.	1	1
X.32	1	−1	−1	−1	−1	−1	−1	.	.	1	.	.	.	.	.	.	.	1	1	.	.	1	1	.	.
X.33	1	1	1	1	1	−1	−1	.	.	.	.	.	.	.	.	.	.	−1	−1	.	.	1	1	.	.
X.34	.	.	.	.	.	.	.	1	1	1	.	.	.	.	.	.	.	.	.	.	.	.	.	−1	−1
X.35	.	.	.	.	.	.	.	$\bar{B}$	B	1	.	.	.	.	.	.	.	.	.	.	.	.	.	$-\bar{B}$	$-B$
X.36	.	.	.	.	.	.	.	B	$\bar{B}$	1	.	.	.	.	.	.	.	.	.	.	.	.	.	$-B$	$-\bar{B}$
X.37	−1	−1	−1	.	.	.	.	1	1	.	.	.	.	.	.	.	.	1	1	.	.	.	.	−1	−1
X.38	−2	.	.	.	.	.	.	.	.	1	.	.	.	.	.	.	.	.	.	.	.	.	.	.	.
X.39	.	.	.	.	.	.	.	.	.	.	−2	−2	.	.	.	.	.	.	.	.	.	.	.	.	.
X.40	.	.	.	$-A$	$-\bar{A}$	$-\bar{A}$	$-A$	.	.	1	.	.	.	.	.	.	.	.	.	.	.	A	$\bar{A}$	.	.
X.41	.	.	.	$-A$	$-\bar{A}$	$\bar{A}$	A	.	.	−1	.	.	.	.	.	.	.	.	.	.	.	$-A$	$-\bar{A}$	.	.
X.42	.	.	.	$-\bar{A}$	$-A$	A	$\bar{A}$	.	.	−1	.	.	.	.	.	.	.	.	.	.	.	$-\bar{A}$	$-A$	.	.
X.43	.	.	.	$-\bar{A}$	$-A$	$-A$	$-\bar{A}$	.	.	1	.	.	.	.	.	.	.	.	.	.	.	$\bar{A}$	A	.	.
X.44	.	.	.	.	.	.	.	.	.	.	−2	−2	.	.	1	.	.	.	.	.	.	.	.	.	.
X.45	1	−1	−1	.	.	.	.	1	1	.	.	.	.	.	.	.	.	1	1	.	.	.	.	−1	−1
X.46	2	.	.	−1	−1	1	1	−1	−1	.	.	.	.	.	.	.	.	.	.	.	.	−1	−1	1	1
X.47	−1	1	1	.	.	.	.	.	.	.	.	.	.	.	.	.	.	−1	−1	.	.	.	.	.	.
X.48	.	.	.	.	.	.	.	.	.	.	.	.	.	.	−1	.	.	.	.	1	−1	.	.	.	.
X.49	.	.	.	.	.	.	.	.	.	.	.	.	.	.	−1	.	.	.	.	−1	1	.	.	.	.
X.50	.	.	.	1	1	1	1	.	.	.	.	.	.	.	.	.	.	.	.	.	.	−1	−1	.	.
X.51	.	.	.	.	.	.	.	.	.	.	1	1	.	.	1	.	.	.	.	.	.	.	.	.	.
X.52	.	.	.	.	.	.	.	.	.	.	1	1	.	.	1	.	.	.	.	.	.	.	.	.	.
X.53	.	2	−2	.	.	.	.	.	.	.	.	.	.	.	.	.	.	.	.	1	−1	.	.	.	.
X.54	.	−2	2	.	.	.	.	.	.	.	.	.	.	.	.	.	.	.	.	−1	1	.	.	.	.
X.55	1	1	1	.	.	.	.	.	.	−1	.	.	.	.	.	.	.	−1	−1	.	.	.	.	.	.
X.56	1	−1	−1	.	.	.	.	.	.	−1	.	.	.	.	.	.	.	1	1	.	.	.	.	.	.
X.57	1	1	1	.	.	.	.	.	.	1	.	.	.	.	.	.	.	−1	−1	.	.	.	.	.	.
X.58	.	2	−2	.	.	.	.	.	.	.	3	−1	.	.	.	.	.	.	.	.	.	.	.	.	.
X.59	.	−2	2	.	.	.	.	.	.	.	−1	3	.	.	.	.	.	.	.	.	.	.	.	.	.
X.60	−1	1	1	−1	−1	−1	−1	.	.	.	.	.	.	.	.	.	.	−1	−1	.	.	1	1	.	.
X.61	.	.	.	.	.	.	.	.	.	.	1	−3	.	.	.	.	.	.	.	−1	1	.	.	.	.
X.62	.	.	.	.	.	.	.	.	.	.	−3	1	.	.	.	.	.	.	.	1	−1	.	.	.	.
X.63	.	.	.	.	.	.	.	.	.	.	.	.	.	.	.	.	.	.	.	.	.	.	.	.	.
X.64	.	.	.	.	.	.	.	.	.	.	.	.	.	.	1	.	.	H	$-H$	.	.	.	.	.	.
X.65	.	.	.	.	.	.	.	.	.	.	.	.	.	.	1	.	.	$-H$	H	.	.	.	.	.	.
X.66	−1	−1	−1	.	.	.	.	.	.	.	.	.	.	.	.	.	.	1	1	.	.	.	.	.	.
X.67	.	.	.	.	.	.	.	.	.	.	1	−3	.	.	−1	.	.	.	.	1	−1	.	.	.	.
X.68	.	.	.	.	.	.	.	.	.	.	−3	1	.	.	−1	.	.	.	.	−1	1	.	.	.	.
X.69	.	.	.	.	.	.	.	.	.	.	−1	3	.	.	.	.	.	.	.	1	−1	.	.	.	.
X.70	.	.	.	.	.	.	.	.	.	.	3	−1	.	.	.	.	.	.	.	−1	1	.	.	.	.
X.71	.	.	.	.	.	.	.	.	.	1	.	.	.	.	.	.	.	.	.	.	.	.	.	.	.
X.72	.	.	.	.	.	.	.	.	.	−1	.	.	.	.	.	.	.	.	.	.	.	.	.	.	.

where $A = \zeta(7)^4 + \zeta(7)^2 + \zeta(7)$, $B = -2\zeta(15)_3\zeta(15)_5^3 - 2\zeta(15)_3\zeta(15)_5^2 - \zeta(15)_3 - \zeta(15)_5^3 - \zeta(15)_5^2 - 1$, $C = -\zeta(23)^{18} - \zeta(23)^{16} - \zeta(23)^{13} - \zeta(23)^{12} - \zeta(23)^9 - \zeta(23)^8 - \zeta(23)^6 - \zeta(23)^4 - \zeta(23)^3 - \zeta(23)^2 - \zeta(23) - 1$, $D = -2\zeta(7)^4 - 2\zeta(7)^2 - 2\zeta(7) - 2$, $E = -3\zeta(7)^4 - 3\zeta(7)^2 - 3\zeta(7) - 3$, $F = -4\zeta(5)^3 - 4\zeta(5)^2 - 3$, $G = 4\zeta(5)^3 + 4\zeta(5)^2 + 1$, $H = -4\zeta(12)_4\zeta(12)_3 - 2\zeta(12)_4$.

10
Tits' group ${}^2F_4(2)'$

In 1964 J. Tits discovered his simple group T as a subgroup of index 2 in the finite group of Lie type ${}^2F_4(2)$ for which he also published a beautiful presentation in [132]. D. Parrott derived new presentations of the Tits group T and of its centralizer $H_1 = C_T(z)$ of a 2-central involution z of T from this presentation in [105]. Furthermore, he gave a proof for the uniqueness of the Tits group in terms of the group structure of H.

In [96] L. Wang and the author published a self-contained existence and uniqueness proof for Tits' group $T = {}^2F_4(2)'$ using Algorithm 7.4.8 and Theorem 7.5.1 of [92]. In particular, it is shown there that a Sylow 2-subgroup S of T has a unique elementary abelian normal subgroup A of order 2^5 which is not normal in $H_1 = C_T(z)$. It is self-centralizing, and $U = N_T(A)/T$ is an indecomposable subgroup of $\mathrm{GL}_5(2)$.

In Section 10.1, Algorithm 1.3.8 is applied to the indecomposable subgroup U of $\mathrm{GL}_5(2)$ described by three generating matrices. Lemma 10.1.1 states that the algorithm returns a finitely presented group U_1 of order $2^{11} \cdot 5$ with a center $Z(U_1)$ of order 2 and a non-split extension E of U by the natural F-vector space V of dimension 5 such that all conditions of Step 5 of Algorithm 1.3.8 are satisfied. In Proposition 10.1.3 it is proved that U_1 is isomorphic to Parrott's centralizer $H_1 = C_T(z)$ of a 2-central involution z of the Tits group T. Furthermore, U_1 is also isomorphic to the 2-central involution centralizer H constructed by Wang and the author in [96]. Proposition 10.1.5 describes the group structure of the involution centralizer H.

A finite simple group G is said to be of *Tits-type* if it has a 2-central involution z such that its centralizer $C_G(z)$ is isomorphic to the finitely presented group H; see Definition 10.1.4. Proposition 10.2.2 of

Section 10.2 states that each simple group G of Tits-type has an elementary abelian normal subgroup A of order 2^5 such that $N_G(A)$ is uniquely determined by H and that $N_G(A) \cong E$.

Section 10.3 contains Wang's and the author's existence proof of a group of Tits-type given in [96] using Algorithm 7.4.8 of [92]. It constructs a simple subgroup $\mathfrak{G}$ of $\mathrm{GL}_{26}(73)$ having a 2-central involution $\mathfrak{z}$ such that $\mathfrak{H} = C_{\mathfrak{G}}(\mathfrak{z}) \cong H$, and $\mathfrak{G}$ has a faithful permutation representation of degree 1755 with stabilizer $\mathfrak{H}$; see Theorem 10.3.1.

In Sections 10.4, 10.5 and 10.6 we show that any simple group G of Tits-type is isomorphic to $\mathfrak{G}$; see Theorem 10.6.2. This is done by showing first that a non-2-central involution $u \in E$ has a centralizer $U = C_G(u) = C_E(u)$. Thus we can apply Thompson's group order formula to get the order of G. Then we determine the structure of the normalizers $N_G(\langle p \rangle)$ of the cyclic groups of odd prime order $p \in \{3, 5, 13\}$ as finitely presented groups and calculate their character tables. This information allows us to apply Brauer's characterization of characters. It implies that any simple group G of Tits-type has a pair of conjugate complex irreducible characters of degree 26. Since 73 does not divide $|G|$ they remain irreducible modulo the prime 73. Now Theorem 7.5.1 of [92] implies that $G \cong \mathfrak{G}$.

Systems of representatives of the conjugacy classes of the local subgroups of a simple group of Tits-type and their character tables are stated in Setions 10.7 and 10.8, respectively.

10.1 Construction of the 2-central involution centralizer

In this section we apply Algorithm 1.3.8 to an indecomposable subgroup U of $\mathrm{GL}_5(2)$ and construct a presentation of a group H with center $Z(H)$ of order 2 which is isomorphic to the centralizer $C_T(z)$ of a 2-central involution z of the simple Tits group $T = {}^2F_4(2)'$.

It is well known that $\mathrm{GL}_5(2)$ has exactly three conjugacy classes of irreducible subgroups. They consist of cyclic groups of order 31, Frobenius groups of order 155 and $\mathrm{GL}_5(2)$; see [47]. However, L. Wang (Peking University) found 464 conjugacy classes of non-irreducible, indecomposable subgroups in $\mathrm{GL}_5(2)$. Among them there are 80 indecomposable subgroups having order $2^6 \cdot p$, where $p \in \{3, 5, 7\}$. Here we apply Algorithm 1.3.8 to the following subgroup U of order $2^6 \cdot 3$.

Lemma 10.1.1 *Let U be the subgroup of* $\mathrm{GL}_5(2)$ *generated by the following matrices:*

$$a = \begin{pmatrix} 1&0&0&0&0\\ 0&1&0&0&0\\ 1&0&1&0&0\\ 1&0&1&1&0\\ 0&0&0&0&1 \end{pmatrix}, \quad b = \begin{pmatrix} 0&1&0&0&0\\ 1&0&0&0&0\\ 0&0&1&0&0\\ 0&0&0&1&0\\ 0&0&0&1&1 \end{pmatrix}, \quad c = \begin{pmatrix} 0&1&0&0&0\\ 1&1&0&0&0\\ 0&1&1&0&0\\ 1&1&0&0&1\\ 0&1&0&1&1 \end{pmatrix}.$$

of orders 2, 4 and 3, respectively. Then the following assertions hold.

(a) *The natural 5-dimensional vector space V over $F = GF(2)$ is an indecomposable FU-module of composition length 3 with irreducible composition factors of dimensions 2, 1 and 2.*

(b) *$U = \langle a, b, c\rangle$ has the following set $\mathcal{R}(U)$ of defining relations:*

$$a^4 = b^2 = c^3 = 1, \quad (bc^{-1})^2 = 1,$$
$$a^{-1}cba^{-1}bc^{-1} = 1, \quad (c^{-1}a^{-1})^3 = 1, \quad baba^{-2}ca^{-1}c^{-1}a = 1.$$

(c) *The cohomological dimension of U with coefficients in the FU-module V is 3, and there are eight non-isomorphic extensions E_i of U by V.*

(d) *V is not a maximal elementary abelian normal subgroup of some Sylow 2-subgroup S in the extensions E_1 and E_3. Each other extension E_j has a unique conjugacy class of 2-central involutions z_j such that $D_j = C_{E_j}(z_j)$ is a Sylow 2-subgroup of E_j.*

(e) *The automorphism groups $Aut(Q)$ of the normal subgroups Q of order 2^9 in D_j are 2-groups for all $4 \le j \le 8$.*

(f) *The extension group $E_2 = \langle e_i \mid 1 \le i \le 8\rangle$ has the following set $\mathcal{R}(E_2)$ of defining relations:*

$$e_1^4 = e_2^2 = e_3^3 = e_4^2 = e_5^2 = e_6^2 = e_7^2 = e_8^2 = 1,$$
$$(e_4, e_5) = (e_4, e_6) = (e_5, e_6) = (e_4, e_7) = (e_5, e_7) = 1,$$
$$(e_6, e_7) = (e_4, e_8) = (e_5, e_8) = (e_6, e_8) = (e_7, e_8) = 1,$$
$$(e_1, e_4) = (e_1, e_5) = (e_1, e_8) = (e_2, e_6) = (e_2, e_7) = 1,$$
$$e_1^{-1}e_6e_1e_4^{-1}e_6^{-1} = 1, \quad e_1^{-1}e_7e_1e_4^{-1}e_6^{-1}e_7^{-1} = 1,$$
$$e_2^{-1}e_4e_2e_5^{-1} = 1, \quad e_2^{-1}e_5e_2e_4^{-1} = 1,$$
$$e_2^{-1}e_8e_2e_7^{-1}e_8^{-1} = 1, \quad e_3^{-1}e_4e_3e_5^{-1} = 1,$$
$$e_3^{-1}e_5e_3e_4^{-1}e_5^{-1} = 1, \quad e_3^{-1}e_6e_3e_5^{-1}e_6^{-1} = 1,$$
$$e_3^{-1}e_7e_3e_4^{-1}e_5^{-1}e_8^{-1} = 1, \quad e_3^{-1}e_8e_3e_5^{-1}e_7^{-1}e_8^{-1} = 1,$$
$$(e_2e_3^{-1})^2 = 1, \quad e_1^{-1}e_3e_2e_1^{-1}e_2e_3^{-1} = 1, \quad (e_3^{-1}e_1^{-1})^3 = 1,$$
$$e_2e_1e_2e_1^{-2}e_3e_1^{-1}e_3^{-1}e_1e_4^{-1}e_5^{-1}e_8^{-1} = 1.$$

Furthermore, E_2 has a faithful permutation representation of degree 128 with stabilizer $\langle e_2, (e_1^2 e_3 e_1^2 e_3 e_1 e_2)^2, (e_1 e_2 e_1^2 e_2 e_3 e_1 e_3)^2 \rangle$.

(g) *$z = (e_1 e_2)^8$ is a 2-central involution of E_2 and $D = C_{E_2}(z) = \langle a, b \rangle$ is a Sylow 2-subgroup of E_2 of order 2^{11}, where $a = e_1$ and $b = e_2$. Furthermore, D has exactly one normal subgroup C of order 2^9 such that $|Aut(C)| = 2^{16} \cdot 5$. Its normal subgroup C is generated by $c_1 = b$, $c_2 = (ab)^4$, $c_3 = (a^2 b)^2$ and $c_4 = a^3 ba$. It has a complement $\langle a \rangle$ in D of order 4.*

(h) *Let $c_5 = a$. Then $D = \langle c_i \mid 1 \le i \le 5 \rangle$ has the following set $\mathcal{R}(D)$ of defining relations:*

$$
\begin{aligned}
&c_1^2 = c_2^4 = c_3^4 = c_4^2 = c_5^4 = 1,\\
&(c_1 c_3^{-1})^2 = 1, \quad c_4 c_5^{-1} c_1 c_5 = 1,\\
&c_2^{-1} c_1 c_2^2 c_1 c_2^{-1} = 1, \quad c_2^{-1} c_3^{-1} c_2^2 c_3 c_2^{-1} = 1,\\
&c_2^{-1} c_4 c_2^2 c_4 c_2^{-1} = 1, \quad c_2^{-1} c_5^{-1} c_2^2 c_5 c_2^{-1} = 1,\\
&c_5 c_3 c_1 c_5^{-1} c_4 = 1, \quad c_2^{-1} c_5 c_1 c_2 c_1 c_5^{-1} = 1,\\
&c_2^{-1} c_3^{-} 2 c_2 c_3^2 = 1, \quad (c_2^{-1} c_4 c_3^{-1})^2 = 1,\\
&c_2^{-1} c_4 c_3 c_2 c_4 c_3 = 1, \quad c_2^{-1} c_1 c_2^{-1} c_1 c_2 c_1 c_2 c_1 = 1,\\
&c_3^{-1} c_1 c_2^{-1} c_1 c_2^{-1} c_4 c_3 c_4 = 1, \quad c_3 c_1 c_4 c_1 c_2^{-1} c_5 c_1 c_5^{-1} = 1.
\end{aligned}
$$

(i) *$Aut(C)$ contains a conjugacy class of subgroups U of order 5120 such that the normal subgroup $I = In(C)$ of inner automorphisms of C has a complement K in U which is isomorphic to the Frobenius group of order 20. Furthermore, $U = \langle u_j \mid 1 \le j \le 6 \rangle$ has the following set $\mathcal{R}(U)$ of defining relations:*

$$
\begin{aligned}
&u_1^2 = u_2^2 = u_3^4 = u_4^2 = u_5^5 = u_6^4 = 1,\\
&(u_2 u_3^{-1})^2 = 1, \quad u_1 u_5^{-1} u_2 u_5 = 1, \quad u_4 u_6^{-1} u_1 u_6 = 1,\\
&u_2 u_6 u_1 u_6^{-1} u_3 = 1, \quad u_1 u_3^{-2} u_1 u_3^2 = 1, \quad (u_1 u_4 u_3^{-1})^2 = 1,\\
&u_5 u_4 u_3^{-1} u_1 u_5^{-1} u_4 = 1, \quad (u_1 u_4 u_3)^2 = 1, \quad u_6^{-1} u_5^{-2} u_6 u_5 = 1,\\
&u_1 u_3^{-1} u_2 u_1 u_5^{-1} u_1 u_5 = 1, \quad u_2 u_4 u_2 u_1 u_5^{-1} u_3 u_5 = 1,\\
&u_3^{-1} u_4 u_2 u_1 u_6 u_3 u_6^{-1} = 1.
\end{aligned}
$$

(j) *U has seven non-isomorphic non-split 2-central extensions U_i, $1 \le i \le 7$. The Sylow 2-subgroups S_i of U_i are pairwise non-isomorphic.*

(k) *D is isomorphic to the Sylow 2-subgroup of the extension $U_1 = \langle f_k \mid 1 \le k \le 7 \rangle$ having the following set $\mathcal{R}(U_1)$ of defining*

relations:

$$f_1^2 = f_2^2 = f_3^4 = f_4^2 = f_5^5 = f_6^4 = f_7^2 = 1,$$
$$(f_1, f_7) = (f_2, f_7) = (f_3, f_7) = (f_4, f_7) = (f_5, f_7) = (f_6, f_7) = 1,$$
$$(f_2 f_3^{-1})^2 = 1, \quad f_1 f_5^{-1} f_2 f_5 = 1, \quad f_4 f_6^{-1} f_1 f_6 = 1,$$
$$f_2 f_6 f_1 f_6^{-1} f_3 = 1, \quad f_1 f_3^{-2} f_1 f_3^2 = 1,$$
$$(f_1 f_4 f_3^{-1})^2 = 1, \quad f_5 f_4 f_3^{-1} f_1 f_5^{-1} f_4 f_7 = 1,$$
$$f_1 f_4 f_3 f_1 f_4 f_3 f_7 = 1, \quad f_6^{-1} f_5^{-2} f_6 f_5 = 1,$$
$$f_1 f_3^{-1} f_2 f_1 f_5^{-1} f_1 f_5 f_7 = 1, \quad f_2 f_4 f_2 f_1 f_5^{-1} f_3 f_5 f_7 = 1,$$
$$f_3^{-1} f_4 f_2 f_1 f_6 f_3 f_6^{-1} = 1.$$

(l) *$S = \langle f_1, f_2, f_3, f_4, f_6, f_7 \rangle$ is a Sylow 2-subgroup of U_1, and $A = \langle f_2, f_3^2, (f_2 f_4)^2, (f_3 f_4)^2, (f_4 f_3)^2 \rangle$ is a maximal elementary abelian normal subgroup of S such that $N_{U_1}(A) = S \cong D = C_{E_2}(z)$.*

(m) *S has exactly two maximal elementary abelian normal subgroups A and $B = \langle f_3^2, (f_1 f_2)^2, (f_1 f_3)^2, (f_1 f_4)^2, (f_2 f_4)^2 \rangle$.*

(n) *$C = \langle f_1, f_2, f_3, f_4 \rangle$ is the Fitting subgroup of U_1. It has order 2^9, and its derived subgroup $C' = B$. In particular, C is not extraspecial.*

Proof (a) Clearly, V is a uniserial FU-module with three composition factors of dimensions 2, 1 and 2, as can be seen from the generating matrices.

(b) The given presentation of $U = \langle a, b, c \rangle$ has been obtained by means of the MAGMA command `FU:=FPGroup(U)`.

(c) Using the command `CosetAction(U,Q)` MAGMA returns a faithful permutation PU of U with stabilizer $Q = \langle c \rangle$. Let V be the natural 5-dimensional vector space over $F = GF(2)$. Hence the hypothesis of Holt's Algorithm 7.4.5 of [92] is satisfied. Using it MAGMA establishes that the second cohomological dimension $dim_F[H^2(U, V)] = 3$. Furthermore, it returns all eight extensions E_i of U by V as finitely presented groups. For each E_i MAGMA calculates a faithful permutation representation PE_i with trivial stabilizer. All E_i are pairwise non-isomorphic. E_1 is the split extension.

(d) For each group E_i the first four steps of Algorithm 1.3.8 can easily be performed. The normal subgroup V is not a maximal elementary abelian normal subgroup in a Sylow 2-subgroup of the extension groups E_1 and E_3. In the remaining six cases, each group E_i has exactly one conjugacy class of involutions z_i which is 2-central, and $D_i = C_{E_i}(z_i)$ is a Sylow 2-subgroup of E_i for $i = 2$ and all $4 \leq i \leq 8$.

(e) Using the MAGMA command `NormalSubgroups(D_i)` one observes that each D_i has exactly three normal subgroups C of order 2^9 with cyclic center $Z(C) = \langle z_i \rangle$. For each of them one calculates $Aut(C)$. It turns out that for $4 \leq i \leq 8$ all these automorphism groups are 2-groups.

(f) Since D_2 has a normal subgroup with an odd outer automorphism we state the presentation of the non-split extension E_2 of U by V. A routine MAGMA calculation verifies that the given stabilizer subgroup of the faithful permutation representation PE_2 of E_2 has index 128.

(g) Using MAGMA we see that $z = (e_1e_2)^8$ is a 2-central involution of E_2 and that $D = C_{E_2}(z)$ is a Sylow 2-subgroup of E_2. Let C be the unique normal subgroup of order 2^9 of D with $|Aut(C)| = 2^{11} \cdot 5$. Another application of MAGMA shows that $C_D(C) = Z(C) = \langle z \rangle$. The given generators of C have been found by means of the program `GetShortGens(D,C)`. Applying the command `HasComplement(D,C)` MAGMA establishes that C has a cyclic complement $\langle a \rangle$ of order 4 in $D = \langle a, b \rangle$.

(h) Let $c_5 = a$. Then $D = \langle c_i \mid 1 \leq i \leq 5 \rangle$. The given presentation of D has been calculated by means of the faithful permutation representation PE_2 and the MAGMA command `FPGroup(D)`.

(i) Using the faithful permutation representation PE_2 of (f) and the commands `A := AutomorphisGroup(C)` and `PA:=PermutationGroup(A)` MAGMA establishes that PA is a permutation group acting on a set of cardinality 80 and that $|A| = 2^{16} \cdot 5$. Furthermore, it gives nine generators of the subgroup I of A and an isomorphism $h : A \to PA$. Let $PI = \langle h(I) \rangle$. Then PI is a normal subgroup of PA of order 2^8. Another application of `GetShortGens(PI,PI)` with MAGMA verifies that PI is generated by four elements of orders 2, 2, 4 and 2, respectively.

Applying the command `Subgroups(PA: OrderEqual:=5120)` MAGMA calculates representatives of the 31 conjugacy classes of such subgroups PU_k of PA. Searching now for the subgroups PU_k with $PI = PU_k \cap PI$ MAGMA establishes that there are exactly seven conjugacy classes of subgroups PU_k of PA satisfying this intersection property. They have indices $k \in \{2, 5, 8, 11, 17, 24, 27\}$. The subgroups PU_2 and PU_8 are isomorphic; so are PU_{11} and PU_{17} and also PU_{24} and PU_{27}. Among the remaining four subgroups PU_2, PU_5, PU_{11} and PU_{24}, only in the group PU_{24} does the normal subgroup PI have a complement PK that is isomorphic to the Frobenius group of order 20. All these facts are easily established with routine MAGMA calculations in the permutation group PA. Another application of the command `FPGroup(PU_{24})` provides the following presentation of the group PU_{24}. It is isomorphic to the finitely presented group $U_{24} = \langle u_1, u_2, u_3, u_4, u_5, u_6 \rangle$ with the following set $\mathcal{R}(U_{24})$ of defining relations:

$$\begin{aligned}
&u_1^2 = u_2^2 = u_3^4 = u_4^2 = u_5^5 = u_6^4 = 1, \\
&(u_2u_3^{-1})^2 = 1, \quad u_1u_5^{-1}u_2u_5 = 1, \quad u_4u_6^{-1}u_1u_6 = 1, \\
&u_2u_6u_1u_6^{-1}u_3 = 1, \quad u_1u_3^{-2}u_1u_3^2 = 1, \quad (u_1u_4u_3^{-1})^2 = 1, \\
&u_5u_4u_3^{-1}u_1u_5^{-1}u_4 = 1, \quad (u_1u_4u_3)^2 = 1, \\
&u_6^{-1}u_5^{-2}u_6u_5 = 1, \quad u_1u_3^{-1}u_2u_1u_5^{-1}u_1u_5 = 1, \\
&u_2u_4u_2u_1u_5^{-1}u_3u_5 = 1, \quad u_3^{-1}u_4u_2u_1u_6u_3u_6^{-1} = 1.
\end{aligned}$$

(j) U_{24} has a faithful permutation representation PU_{24} of degree 1024 with stabilizer $\langle u_5 \rangle$. Let

```
MGL1_2:=GL(1,2),M1:=MGL1_2![1]
```

and

```
FEalg :=MatrixAlgebra<FiniteField(2),1|M1,M1,M1,M1,M1,M1>.
```

Then `CM:=GModule(PU24,FEalg)` denotes the cohomology module. Applying the command

```
CohomologicalDimension(PU_(24}, CM, 2)
```

MAGMA establishes that the second cohomological dimension is 3. Thus there are seven non-split 2-central extensions U_i of U_{24}. The command `P := ExtensionProcess(PU24,CM,U24)` was used to establish a presentation for each of them. We checked computationally that the Sylow 2-subgroups S_i of the extensions U_i are pairwise non-isomorphic.

(k) Using the MAGMA command `IsIsomorphic(S_1,D)` it has been verified that D is isomorphic to the Sylow 2-subgroup $S_1 = \langle f_1, f_2, f_3, f_4, f_6, f_7 \rangle$ of the extension U_1. Therefore only the presentation of U_1 is given.

(l) Clearly, $S = \langle f_1, f_2, f_3, f_4, f_6, f_7 \rangle$ is a Sylow 2-subgroup of U_1. Using PU_1 and the MAGMA command

```
Subgroups(S: Al:=Normal, IsElementaryAbelian := true)
```

one verifies that S has exactly two maximal elementary abelian normal subgroups A and B of order 32. Another application of the short generator program provides the generators of A and B given in the statement.

Furthermore, routine MAGMA calculations verify that $N_{U_1}(A) = S \cong C_{E_2}$.

(m) MAGMA also confirms that $Q = \langle f_1, f_2, f_3, f_4 \rangle$ is the Fitting subgroup of U_1 and that $|Q| = 2^9$.

(n) It has been checked with MAGMA that the derived subgroup Q' of Q equals the normal subgroup B of S. In particular, Q is not extra-special. □

The following subsidiary result is due to Wang and the author [96].

Lemma 10.1.2 *Let $E = \langle e_i \mid 1 \le i \le 8 \rangle$ be the finitely presented group constructed in Lemma 10.1.1(f). Let $s = e_2(e_1^2e_2)^4(e_1^3e_2e_1e_2)^2(e_1^2e_2e_1e_2e_1)^2$, $h = (e_2e_1e_2e_1e_2e_1^3e_2)^2e_1e_2e_1^3e_2e_1^2e_2e_1e_2$ and $r = e_3$. Then the following assertions hold.*

(a) *$E = \langle s, h, r \rangle$ has the following set of defining relations $\mathcal{R}(E)$:*

$$s^2 = h^4 = r^3 = 1,$$
$$hr^{-1}sh^{-1}shr = 1, \quad (sr^{-1})^4 = 1, \quad (h^{-1}r^{-1}h^{-1}rs)^2 = 1,$$
$$h^{-1}rh^{-1}sh^{-2}r^{-1}h^{-1}r^{-1}sh^{-1} = 1.$$

(b) *E has a faithful permutation representation of degree 2048 with stabilizer $\langle r \rangle$.*

(c) *$S = \langle s, h \rangle$ is a Sylow 2-subgroup of E.*

Proof (a) Let $X = \langle s_1, h_1, r_1 \rangle$ be the finitely presented group with defining set of relations $\mathcal{R}(X)$ equal to $\mathcal{R}(E)$ except for the indices of its generators. Then X has a faithful permutation representation PX of degree 2^{11} with stabilizer $\langle r_1 \rangle$. Let PE be the faithful permutation representation of $E = E_2$ given in Lemma 10.1.1(f). Applying the command `IsIsomorphic(PX,PE)` MAGMA establishes an isomorphism $\sigma : X \to E$. Calculating sets of short generators x_i, y_j and r_k of the centralizers $C_E(\sigma(s))$, $C_E(\sigma(h))$ and $C_E(\sigma(r)$ and applying then the `LookupWord` program for the three images in their centralizers in E the author found the presentations of the elements $s = \sigma(s_1)$, $h = \sigma(h_1)$ and $r = \sigma(r_1)$ stated in the hypothesis.

(b) This assertion is trivial by the proof of (a).

(c) This statement has been checked with MAGMA. □

We now show that the constructed group U_1 is isomorphic to the centralizer $C_T(z)$ of a 2-central involution z of the Tits group ${}^2F_4(2)'$. For that proof we restate Parrott's finitely presented group H_1 of [105] and the finitely presented group H constructed by Wang and the author in [96]. Parrott derived the presentation of H_1 from Tits' original presentation of the group of Lie type ${}^2F_4(2)'$; see [132].

Proposition 10.1.3 *Let $U_1 = \langle f_1, f_2, f_3, f_4, f_5, f_6, f_7\rangle$ be the finitely presented group of order $2^{11}\cdot 5$ constructed in Lemma 10.1.1. Then the following statements hold.*

(a) *U_1 is isomorphic to the finitely presented group $H = \langle h_i \mid 1 \le i \le 11\rangle$ of [96] with the following set $\mathcal{R}(H)$ of defining relations:*

$$
\begin{aligned}
&h_1^5 = h_2^4 = h_7^2 = h_8^2 = h_9^2 = h_{10}^2 = h_{11}^2 = 1,\\
&(h_5, h_{10}^{-1}) = (h_5, h_{11}^{-1}) = (h_6, h_7^{-1}) = 1,\\
&(h_7, h_8) = (h_7, h_9) = (h_8, h_9) = (h_7, h_{10}) = 1,\\
&(h_8, h_{10}) = (h_9, h_{10}) = (h_7, h_{11}) = (h_8, h_{11}) = 1,\\
&(h_9, h_{11}) = (h_{10}, h_{11}) = (h_1, h_{11}^{-1}) = (h_2, h_{11}^{-1}) = 1,\\
&(h_3, h_8^{-1}) = (h_3, h_9^{-1}) = (h_3, h_{11}^{-1}) = (h_4, h_8^{-1}) = 1,\\
&(h_4, h_9^{-1}) = (h_4, h_{10}^{-1}) = (h_4, h_{11}^{-1}) = (h_5, h_7^{-1}) = 1,\\
&(h_6, h_{10}^{-1}) = (h_6, h_{11}^{-1}) = (h_5, h_9^{-1}) = 1,\\
&h_3^2 h_{11}^{-1} = h_4^2 h_{11}^{-1} = h_5^2 h_{11}^{-1} = h_6^2 h_{11}^{-1} = 1,\\
&h_2^{-1} h_1^{-2} h_2 h_1 = h_2^{-1} h_5 h_2 h_5 h_{10}^{-1} = h_1 h_3 h_1^{-1} h_4 h_{11}^{-1} = 1,\\
&h_1 h_4 h_1^{-1} h_5 h_{11}^{-1} = h_2^{-1} h_4 h_2 h_6 h_{11}^{-1} = h_2^{-1} h_8 h_2 h_8^{-1} h_9^{-1} = 1,\\
&h_1^{-1} h_9 h_1 h_8^{-1} h_9^{-1} = h_4^{-1} h_7 h_4 h_7^{-1} h_{11}^{-1} = h_5^{-1} h_8 h_5 h_8^{-1} h_{11}^{-1} = 1,\\
&h_6^{-1} h_8 h_6 h_8^{-1} h_{11}^{-1} = h_6^{-1} h_9 h_6 h_9^{-1} h_{11}^{-1} = h_3 h_4 h_3 h_4 h_8^{-1} h_{11}^{-1} = 1,\\
&h_5 h_6 h_5 h_6 h_7^{-1} h_{11}^{-1} = h_1^{-1} h_{10} h_1 h_8^{-1} h_{11}^{-1} = h_2^{-1} h_9 h_2 h_{10}^{-1} h_{11}^{-1} = 1,\\
&h_2^{-1} h_{10} h_2 h_7^{-1} h_{11}^{-1} = h_3^{-1} h_7 h_3 h_7^{-1} h_{11}^{-1} = h_3^{-1} h_{10} h_3 h_{10}^{-1} h_{11}^{-1} = 1,\\
&h_2^{-1} h_3 h_2 h_4 h_8^{-1} h_{11}^{-1} = h_1^{-1} h_8 h_1 h_7^{-1} h_{10}^{-1} h_{11}^{-1} = 1,\\
&h_1^{-1} h_3 h_1 h_6 h_8^{-1} h_9^{-1} h_{10}^{-1} = h_2^{-1} h_7 h_2 h_7^{-1} h_9^{-1} h_{10}^{-1} h_{11}^{-1} = 1,\\
&h_1^{-1} h_7 h_1 h_9^{-1} h_{10}^{-1} h_{11}^{-1} = h_5 h_4 h_3 h_1^{-1} h_3 h_6 h_1 h_7^{-1} h_9^{-1} h_{10}^{-1} h_{11}^{-1} = 1,
\end{aligned}
$$

(b) *U_1 is isomorphic to Parrott's group $H_1 = \langle z, t, v, u, w, a, b, c, d, x, y, r\rangle$ of [105] with the following set $\mathcal{R}(H_1)$ of defining relations:*

$$
\begin{aligned}
&z^2 = t^2 = v^2 = u^2 = w^2 = a^2 = r^2 = 1, \quad x^4 = 1,\\
&(ry)^5 = d^{-1}u^{-1}du = y^{-1}a^{-1}ya = d^{-1}w^{-1}dw = c^{-1}t^{-1}ct = 1,\\
&c^{-1}u^{-1}cu = c^{-1}w^{-1}cw = a^{-1}x^{-1}ax = x^{-1}txt = b^{-1}wbw = 1,
\end{aligned}
$$

$$d^{-1}t^{-1}dt = c^{-1}v^{-1}cv = z, \quad a^{-1}d^{-1}ad = x^{-1}wxw = u,$$
$$a^{-1}b^{-1}ab = x^{-1}vxv = t, \quad a^{-1}c^{-1}ac = vt,$$
$$c^{-1}d^{-1}cd = c^2 = wu, \quad b^{-1}c^{-1}bc = uv, \quad y^{-1}d^{-1}yd = bw,$$
$$b^{-1}x^{-1}bx = a, \quad y^{-1}c^{-1}yc = atz, \quad x^{-1}c^{-1}xc = abuv,$$
$$x^{-1}d^{-1}xd = abcuv, \quad d^{-1}b^{-1}db = b^2 = v, \quad x^{-1}uxu = v,$$
$$r^{-1}tr = wuvz, \quad r^{-1}vr = uv, \quad r^{-1}ur = u, \quad r^{-1}wr = vtz,$$
$$r^{-1}ar = du, \quad r^{-1}dr = au, \quad r^{-1}cr = cdau, \quad rxr = (yr)^2x,$$
$$r^{-1}br = dcbvt, \quad x^2 = yz, \quad a^{-1}waw = b^{-1}ubu = z.$$

In particular, any of the three groups U_1, H and H_1 is isomorphic to the centralizer $C_T(z)$ of a 2-central involution z of the Tits group $T = {}^2F_4(2)'$.

Proof (a) The group H has order $2^{11} \cdot 5$, as can be checked by means of the MAGMA command `CosetAction(H, sub<H|>)`. In particular, H has a faithful permutation representation PH of degree $2^{11} \cdot 5$. By the proof of Lemma 10.1.1 we know that U_1 has a faithful permutation representation PU_1 of degree 2^{11}. Now we apply the MAGMA command `IsIsomorphic(U_1,H)`. MAGMA confirms the assertion.

(b) Both groups H and H_1 have a regular permutation representation PH and PH_1 with stabilizer 1. Using the above isomorphism test program of MAGMA we checked in [96] that $H \cong H_1$. Now (a) completes the proof.

The statement $H_1 \cong C_T(z)$ holds by [105]. □

Definition 10.1.4 A finite simple group G is said to be of *Tits-type* if it has a 2-central involution z such that its centralizer $C_G(z)$ is isomorphic to the finitely presented group H of Proposition 10.1.3.

Proposition 10.1.5 *Let $H = \langle h_1, h_2, h_3, h_4, h_5, h_6, h_7\rangle$ be the finitely presented group of order $2^{11} \cdot 5$ constructed in Lemma 10.1.1(k) with the set of defining relations $\mathcal{R}(H)$ adapted according to the notations of the generators of the groups H and U_1. Then the following statements hold.*

(a) *$H = \langle s, h, x\rangle$, where $s = h_1$, $h = h_6$ and $x = h_5$ have respective orders 2, 4 and 5.*
(b) *$S = \langle s, h\rangle$ is a Sylow 2-subgroup of H with center $Z(S) = \langle z\rangle = Z(H)$, where $z = (sh)^8$.*
(c) *$V = \langle z, t\rangle$ is the unique normal Klein 4-group of S, where $t = (shsh^2)^4$.*

(d) *$A = \langle z, t, a_1, a_2, a_3 \rangle$ is the unique normal elementary abelian subgroup of order 32 in S which is not normal in H, where $a_1 = (h^2sh^2sh^3sh)^2$, $a_2 = (sh^2)^4$ and $a_3 = h^2sh^2shsh^2sh$.*

(e) *$B = \langle z, t, a_1, a_2, b_3 \rangle$ is the unique normal elementary abelian subgroup of order 32 in S which is normal in H, where $b_3 = (sh^3sh)^2$.*

(f) *Both maximal elementary abelian normal subgroups A and B are self-centralizing in H.*

(g) *$D = N_H(A) = S$.*

(h) *$C = C_H(V)$ has order 2^{10} and center $Z(C) = V$.*

(i) *H has a faithful permutation representation PH of degree 2^9 with stabilizer $\langle x, h \rangle$, the Frobenius subgroup.*

(j) *A system of representatives of the 28 conjugacy classes of $H = \langle h, s, x \rangle$ is given in Table 10.7.1.*

(k) *A system of representatives of the 35 conjugacy classes of $D = \langle h, s \rangle$ is given in Table 10.7.3.*

(l) *The character table of H is Table 10.8.1.*

(m) *The character table of D is Table 10.8.2.*

Proof All these statements have been checked with MAGMA and the faithful permutation representation of H given in (i). The various generators of the subgroups of H have been found by the methods used already in the proofs of the previous results. □

10.2 Fusion

In this section the conjugacy classes of elements of even order of a finite simple group G of Tits-type are determined. Furthermore, it is shown that the normalizer $N_G(A)$ of the maximal elementary abelian normal subgroup A of a fixed Sylow 2-subgroup S of H is uniquely determined up to isomorphism in all such simple groups G.

Lemma 10.2.1 *Let G be a finite simple group of Tits-type having a 2-central involution z with centralizer $H = C_G(z) = \langle h, s, x \rangle$ defined in Proposition 10.1.5. Let $u = (h^2s)^4$. Then the following statements hold.*

(a) *u is an involution of the Sylow 2-group S defined in Proposition 10.1.5 with centralizer $C_H(u)$ of order 2^9.*

(b) *$T = C_H(u)$ is a self-normalizing subgroup of $C_G(u)$.*

(c) *T is a Sylow 2-subgroup of $C_G(u)$.*

(d) *z and u are not conjugate in G.*

Proof (a) Using MAGMA and the faithful permutation representation of H of degree 512 described in Proposition 10.1.5(g), it is easy to see that $u \in S$, and $|C_H(u)| = 2^9$.

(b) Suppose that a subgroup R of $U = C_G(u)$ normalizes T. By MAGMA $Z(T) \cap T' = \langle z \rangle$, so $\langle z \rangle \lhd R$. Hence $R \leq C_G(z) \cap C_G(u) = C_H(u) = T$. Thus $R \leq T$.

(c) Follows immediately from (b)

(d) If z and u were conjugate in G, then $|C_G(u)| = |C_G(z)| = 2^{11} \cdot 5$ by Proposition 10.1.5, which is impossible by assertions (a) and (c). □

Proposition 10.2.2 *Let G be a finite simple group of Tits-type with a 2-central involution z such that $C_G(z) = H = \langle h, s, x \rangle$ defined in Proposition 10.1.5. Let $V = \langle z, t \rangle$ and $A = \langle z, t, a_1, a_2, a_3 \rangle$ be the elementary abelian normal subgroups in the fixed Sylow 2-subgroup S of H of orders 4 and 2^5 defined in Proposition 10.1.5(c) and (d), respectively. Then the following statements hold.*

(a) $N_G(V)/C_G(V) \cong \mathrm{GL}_2(2)$.
(b) *There is an element $r \in N_G(V) - N_H(V)$ of order 3 such that $z^r = t$, and $r \in N_G(A)$.*
(c) *$N_G(A)$ is isomorphic to the finitely presented group $E = \langle h, s, r \rangle$ with set $\mathcal{R}(E)$ stated in Lemma 10.1.2*
(d) *$E = N_G(A)$ has a faithful permutation representation of degree 2048 with stabilizer $\langle r \rangle$.*
(e) *The amalgam $H \leftarrow D \rightarrow E$ has Goldschmidt index 1.*
(f) *G has two conjugacy classes z^G and u^G of involutions represented by z and $u = (h^2 s)^4$.*
(g) *H has six conjugacy classes of involutions. Using the indices of all the classes of H given in Table 10.7.1 the involution classes of H with indices 1, 2 and 6 fuse to z, and the ones with indices 3, 4 and 5 fuse to u.*
(h) *Representatives m_i of the 27 conjugacy classes m_i^E of E and the corresponding centralizer orders $|C_E(m_i)|$ are given in Table 10.7.4.*
(i) *Using the indices of all the classes of E given in Table 10.7.4 the involution classes of E with indices 1 and 5 fuse to z, and the ones with indices 2, 3 and 4 fuse to u.*
(j) *The character table of E is Table 10.8.3.*

Proof (a) By Table 10.7.1 the group H has six conjugacy classes of involutions represented by z, $v = (h^2shs)^4 = zt$, $u = (h^2s)^4$, $g = h^2$, s and $q = h^2sx$. Furthermore, the centralizers $Y = C_H(s)$ and $X = C_H(g)$ both have order 2^7. Using the faithful permutation representation of H given in Proposition 10.1.5(i) it has been checked that s and q do not belong to the Frattini subgroup F of the Sylow 2-subgroup of $S = \langle s, h \rangle$ of H, whereas the representatives of the four other conjugacy classes of H belong to F. Hence Thompson's transfer Lemma 1.4.3 of [92] implies that s is G-conjugate to one of the involutions z, v, u or g. If the Sylow 2-subgroups of $C_G(s)$ had order 2^7, then Table 10.7.1 would imply that X and Y would be isomorphic. However, another application of MAGMA proved that to be false. Therefore there is a 2-subgroup Y_1 of $C_G(s)$ not contained in H such that $|Y_1 : Y| = 2$. The characteristic subgroup $K = Y' \cap Z(Y)$ of Y is generated by the involution $s_1 = (h^2shsh)^2(h^2sh^2sh^3sh)^2$ calculated by the short generator program and MAGMA. As K is normal in Y_1 there is a $y \in Y_1 - H$ such that $s_1^y = s_1$. In particular, $t = s_1^w = s_1^{yw} = t^{w^{-1}yw}$. Thus $y^w \in C_G(t) - C_H(t)$. The subgroup $W = C_H(V) = C_G(V)$ of H has center V by Proposition 10.1.5. Thus Lemma 1.4.4 of [92] implies that assertion (a) holds, because $|W| = 2^{10}$ and each Sylow 2-subgroup of G has order 2^{11}.

(b) Lemma 1.4.4 of [92] also states that there is an element r of $N_G(V) - N_H(V)$ such that $z^r = t$. As S splits over W by MAGMA Theorem 1.4.15 of [92] states that $N_G(V)$ is a split extension of $W = C_G(V)$ by $\mathrm{GL}_2(2)$. In particular, the element r can be chosen to be of order 3. Using the MAGMA command

```
Subgroups(W : Al:=Normal, IsElementaryAbelian := true)
```

it can be verified that W has exactly four elementary abelian normal subgroups of order 32, and that $A = \langle z, t, a_1, a_2, a_3 \rangle$ is one of them. Furthermore, the same MAGMA command applied to the commutator subgroup W' yields that A is the unique elementary abelian normal subgroup of W' of order 32. Hence it is a characteristic subgroup of W. Thus A is normal in $N_G(V)$. Hence $r \in N_G(A)$.

(c) Clearly, W is a normal subgroup of S. Applying the MAGMA command `HasComplement(S,W)` one obtains a complement K of order 2. Using the program `GetShortGens(S,X)` for $X \in \{K, W\}$ one gets the generators $c_1 = h$, $c_2 = (sh)^2$ of W and s of K, respectively. Furthermore, the conjugate action of s and h on the elementary abelian normal subgroup A of S w.r.t. to the basis $\{z, t, a_1, a_2, a_3\}$ is described by the following matrices:

$$Ms = \begin{pmatrix} 1&0&0&0&0 \\ 1&1&0&0&0 \\ 0&0&1&0&0 \\ 0&0&0&1&0 \\ 1&0&1&1&1 \end{pmatrix}, \quad Mh = \begin{pmatrix} 1&0&0&0&0 \\ 0&1&0&0&0 \\ 0&0&1&1&0 \\ 1&1&0&1&0 \\ 1&1&0&1&1 \end{pmatrix}.$$

Since A is self-normalizing in H it follows that the subgroup $MS = \langle Ms, Mh \rangle$ of $\mathrm{GL}_5(2)$ has order 2^6 and W/A is isomorphic to the subgroup $MW = \langle Ms, (MsMh)^2 \rangle$ of MS. The automorphism group $Aut(W)$ of W has order $2^{13} \cdot 3$ calculated by the MAGMA command `AutomorphismGroup(W)`. Hence it contains an automorphism τ of order 3. Since A is a characteristic subgroup of W, τ normalizes A. Hence $MN = N_G(MW)$ in $G = \mathrm{GL}_5(2)$ contains an image of τ. Using the MAGMA command

```
exists(Mg){x: x in MN| Order(x) eq 3 and x^Ms eq x^2}
```

the author found the following matrix:

$$Mg = \begin{pmatrix} 0&1&0&0&0 \\ 1&1&0&0&0 \\ 0&1&0&0&1 \\ 0&0&0&1&0 \\ 1&0&1&1&1 \end{pmatrix}.$$

The subgroup $MX = \langle Mg, Ms, MW \rangle$ of G has order $2^6 \cdot 3$. Using the MAGMA command `IsIsomorphic(MX,MU)` the author verified that MX is isomorphic to the subgroup $MU = \langle a, b, c \rangle$ of G defined in Lemma 10.1.1. It and Lemma 10.1.2 imply that $N_G(A) \cong E$ is uniquely determined up to isomorphism.

(d) This is a restatement of Lemma 10.1.2(b).

(e) The amalgam is well defined by Lemma 10.1.2(c) and Proposition 10.1.5. Its Goldschmidt index was calculated by means of Kratzer's Algorithm 7.1.10 of [92] in [96].

(h) The system of representatives of the conjugacy classes of E was computed in [96] by means of PE, MAGMA and Kratzer's Algorithm 5.3.18 of [92].

(f) This assertion follows from Lemma 10.2.1(d) and the fusion results of assertions (g) and (i), which have been verified computationally by means of conjugation tests of the representatives of involutions of the common subgroup $D = \langle s, h \rangle$ given in Table 10.7.3 with the representatives of H in Table 10.7.1 and of E in Table 10.7.4, respectively.

The character table of (j) was calculated in [96] by means PE and MAGMA. This completes the proof. $\square$

10.3 Existence proof of Tits' simple group inside $\mathrm{GL}_{26}(73)$

In this section we present Wang's and the author's existence proof of a simple group $\mathfrak{G}$ of Tits-type of [96].

Theorem 10.3.1 (Michler–Wang) *Let A be the elementary abelian normal subgroup of order 2^5 in the Sylow 2-subgroup S of $H = \langle x, h, s\rangle$ defined in Proposition 10.1.5. Let $D = N_H(A) = \langle h, s\rangle$. Let $E = \langle h, s, r\rangle$ be the finite group with the set $\mathcal{R}(E)$ of defining relations given in Lemma 10.1.2. Then the following statements hold.*

(a) *The smallest degree of a non-trivial pair $(\chi, \tau) \in\in mf\,char_{\mathbb{C}}(H) \times mf\,char_{\mathbb{C}}(E)$ of compatible characters is 26.*

(b) *There are two compatible pairs $(\chi, \tau) \in mf\,char_{\mathbb{C}}(H) \times mf\,char_{\mathbb{C}}(E)$ of degree 26 of the groups $H = \langle x, h, s\rangle$ and $E = \langle h, s, r\rangle$:*

$$(\chi, \tau) = (\chi_{12} + \chi_{\mathbf{22}}, \quad \tau_4 + \tau_{\mathbf{25}}),$$

$$(\chi, \tau) = (\chi_{13} + \chi_{\mathbf{22}}, \quad \tau_5 + \tau_{\mathbf{25}}),$$

with common restrictions

$$\chi_{|D} = \tau_{|D} = \psi_{11} + \psi_{29} + \psi_{\mathbf{33}}$$

to $D = \langle h, s\rangle$, where χ_i, τ_j and ψ_k denote the irreducible characters of the groups H, E and D with indices given in the character Tables 10.8.1, 10.8.2 and 10.8.2, respectively. The indices of the faithful irreducible characters are typed in bold face.

(c) *Let $\mathfrak{V}$ and $\mathfrak{W}$ be the up to isomorphism uniquely determined faithful semi-simple multiplicity free 26-dimensional modules of H and E over the prime field $F = GF(73)$ corresponding to the first compatible pair $\chi = \chi_{14} + \chi_{\mathbf{18}}$ and $\tau = \tau_{14} + \tau_{\mathbf{33}}$, respectively. Then the dual modules $\mathfrak{V}^*$ and $\mathfrak{W}^*$ correspond to the second compatible pair.*

Let $\kappa_{\mathfrak{V}} : H \to \mathrm{GL}_{26}(73)$ and $\kappa_{\mathfrak{W}} : N \to \mathrm{GL}_{26}(73)$ be the representations of H and E afforded by the modules $\mathfrak{V}$ and $\mathfrak{W}$, respectively.

Let $\mathfrak{x} = \kappa_{\mathfrak{V}}(x)$, $\mathfrak{h} = \kappa_{\mathfrak{V}}(h)$ and $\mathfrak{s} = \kappa_{\mathfrak{V}}(s) \in \kappa_{\mathfrak{V}}(H) \le \mathrm{GL}_{26}(73)$. Then the following assertions hold:

(1) *$\mathfrak{V}_{|D} \cong \mathfrak{W}_{|D}$, and there is a transformation matrix $\mathcal{T} \in \mathrm{GL}_{26}(73)$ such that*

$$\mathfrak{h} = \mathcal{T}^{-1}\kappa_{\mathfrak{W}}(h)\mathcal{T} \text{ and } \mathfrak{s} = \mathcal{T}^{-1}\kappa_{\mathfrak{W}}(s)\mathcal{T}.$$

(2) *Let* $\mathfrak{r} = \mathcal{T}^{-1}\kappa_{\mathfrak{W}}(r)\mathcal{T} \in \mathrm{GL}_{26}(73)$. *Then the group*

$$\mathfrak{G} = \langle \mathfrak{x}, \mathfrak{h}, \mathfrak{s}, \mathfrak{r} \rangle \text{ has order } |\mathfrak{G}| = 2^{11} \cdot 3^3 \cdot 5^2 \cdot 13.$$

(3) *In* $\mathfrak{V}$ *and* $\mathfrak{W}$ *one can choose* F*-vector space bases such that:*

$$\mathfrak{x} = \begin{pmatrix}
0&0&0&0&0& 0&0&0&1&0& 0&0&0&0&0& 0&0&0&0&0& 0&0&0&0&0& 0\\
0&0&0&0&0& 0&0&0&0&1& 0&0&0&0&0& 0&0&0&0&0& 0&0&0&0&0& 0\\
1&0&0&0&0& 0&0&0&0&0& 0&0&0&0&0& 0&0&0&0&0& 0&0&0&0&0& 0\\
0&1&0&0&0& 0&0&0&0&0& 0&0&0&0&0& 0&0&0&0&0& 0&0&0&0&0& 0\\
0&0&1&0&0& 0&0&0&0&0& 0&0&0&0&0& 0&0&0&0&0& 0&0&0&0&0& 0\\
0&0&0&1&0& 0&0&0&0&0& 0&0&0&0&0& 0&0&0&0&0& 0&0&0&0&0& 0\\
0&0&0&0&1& 0&0&0&0&0& 0&0&0&0&0& 0&0&0&0&0& 0&0&0&0&0& 0\\
0&0&0&0&0& 1&0&0&0&0& 0&0&0&0&0& 0&0&0&0&0& 0&0&0&0&0& 0\\
0&0&0&0&0& 0&1&0&0&0& 0&0&0&0&0& 0&0&0&0&0& 0&0&0&0&0& 0\\
0&0&0&0&0& 0&0&1&0&0& 0&0&0&0&0& 0&0&0&0&0& 0&0&0&0&0& 0\\
0&0&0&0&0& 0&0&0&0&0& 0&0&0&0&0& 0&1&0&0&0& 0&0&0&0&0& 0\\
0&0&0&0&0& 0&0&0&0&0& 0&0&0&0&0& 0&0&0&0&0& 0&0&0&0&1& 0\\
0&0&0&0&0& 0&0&0&0&0& 72&0&0&0&0& 0&0&0&0&0& 0&0&0&0&0& 0\\
0&0&0&0&0& 0&0&0&0&0& 0&0&0&0&0& 0&0&0&72&0& 0&0&0&0&0& 0\\
0&0&0&0&0& 0&0&0&0&0& 0&0&0&0&0& 0&0&72&0&0& 0&0&0&0&0& 0\\
0&0&0&0&0& 0&0&0&0&0& 0&0&0&0&0& 0&0&0&0&0& 0&0&0&0&0& 72\\
0&0&0&0&0& 0&0&0&0&0& 0&1&0&0&0& 0&0&0&0&0& 0&0&0&0&0& 0\\
0&0&0&0&0& 0&0&0&0&0& 0&0&0&0&0& 0&0&0&0&72& 0&0&0&0&0& 0\\
0&0&0&0&0& 0&0&0&0&0& 0&0&0&0&0& 1&0&0&0&0& 0&0&0&0&0& 0\\
0&0&0&0&0& 0&0&0&0&0& 0&0&0&0&0& 0&0&0&0&0& 0&0&0&72&0& 0\\
0&0&0&0&0& 0&0&0&0&0& 0&0&0&1&0& 0&0&0&0&0& 0&0&0&0&0& 0\\
0&0&0&0&0& 0&0&0&0&0& 0&0&0&0&0& 0&0&0&0&0& 0&1&0&0&0& 0\\
0&0&0&0&0& 0&0&0&0&0& 0&0&0&0&72& 0&0&0&0&0& 0&0&0&0&0& 0\\
0&0&0&0&0& 0&0&0&0&0& 0&0&0&0&0& 0&0&0&0&0& 0&0&1&0&0& 0\\
0&0&0&0&0& 0&0&0&0&0& 0&0&72&0&0& 0&0&0&0&0& 0&0&0&0&0& 0\\
0&0&0&0&0& 0&0&0&0&0& 0&0&0&0&0& 0&0&0&0&0& 1&0&0&0&0& 0
\end{pmatrix},$$

$$\mathfrak{h} = \begin{pmatrix}
0&22&0&0&0& 0&0&0&0&0& 0&0&0&0&0& 0&0&0&0&0& 0&0&0&0&0& 0\\
63&0&0&0&0& 0&0&0&0&0& 0&0&0&0&0& 0&0&0&0&0& 0&0&0&0&0& 0\\
0&0&0&0&0& 0&0&22&0&0& 0&0&0&0&0& 0&0&0&0&0& 0&0&0&0&0& 0\\
0&0&0&0&0& 0&63&0&0&0& 0&0&0&0&0& 0&0&0&0&0& 0&0&0&0&0& 0\\
0&0&0&22&0& 0&0&0&0&0& 0&0&0&0&0& 0&0&0&0&0& 0&0&0&0&0& 0\\
0&0&63&0&0& 0&0&0&0&0& 0&0&0&0&0& 0&0&0&0&0& 0&0&0&0&0& 0\\
0&0&0&0&0& 0&0&0&0&22& 0&0&0&0&0& 0&0&0&0&0& 0&0&0&0&0& 0\\
0&0&0&0&0& 0&0&0&63&0& 0&0&0&0&0& 0&0&0&0&0& 0&0&0&0&0& 0\\
0&0&0&0&0& 22&0&0&0&0& 0&0&0&0&0& 0&0&0&0&0& 0&0&0&0&0& 0\\
0&0&0&0&63& 0&0&0&0&0& 0&0&0&0&0& 0&0&0&0&0& 0&0&0&0&0& 0\\
0&0&0&0&0& 0&0&0&0&0& 0&0&1&0&0& 0&0&0&0&0& 0&0&0&0&0& 0\\
0&0&0&0&0& 0&0&0&0&0& 72&0&0&0&0& 0&0&0&0&0& 0&0&0&0&0& 0\\
0&0&0&0&0& 0&0&0&0&0& 0&0&0&0&0& 0&1&0&0&0& 0&0&0&0&0& 0\\
0&0&0&0&0& 0&0&0&0&0& 0&0&0&0&72& 0&0&0&0&0& 0&0&0&0&0& 0\\
0&0&0&0&0& 0&0&0&0&0& 0&0&0&0&0& 1&0&0&0&0& 0&0&0&0&0& 0\\
0&0&0&0&0& 0&0&0&0&0& 0&0&0&0&0& 0&0&72&0&0& 0&0&0&0&0& 0\\
0&0&0&0&0& 0&0&0&0&0& 0&72&0&0&0& 0&0&0&0&0& 0&0&0&0&0& 0\\
0&0&0&0&0& 0&0&0&0&0& 0&0&0&1&0& 0&0&0&0&0& 0&0&0&0&0& 0\\
0&0&0&0&0& 0&0&0&0&0& 0&0&0&0&0& 0&0&0&0&0& 0&0&0&72&0& 0\\
0&0&0&0&0& 0&0&0&0&0& 0&0&0&0&0& 0&0&0&0&0& 0&0&0&0&0& 72\\
0&0&0&0&0& 0&0&0&0&0& 0&0&0&0&0& 0&0&0&0&72& 0&0&0&0&0& 0\\
0&0&0&0&0& 0&0&0&0&0& 0&0&0&0&0& 0&0&0&0&0& 0&72&0&0&0& 0\\
0&0&0&0&0& 0&0&0&0&0& 0&0&0&0&0& 0&0&0&0&0& 1&0&0&0&0& 0\\
0&0&0&0&0& 0&0&0&0&0& 0&0&0&0&0& 0&0&0&72&0& 0&0&0&0&0& 0\\
0&0&0&0&0& 0&0&0&0&0& 0&0&0&0&0& 0&0&0&0&0& 0&0&0&0&72& 0\\
0&0&0&0&0& 0&0&0&0&0& 0&0&0&0&0& 0&0&0&0&0& 0&0&1&0&0& 0
\end{pmatrix},$$

$$\mathfrak{s} = \begin{pmatrix}
0&1&0&0&0& 0&0&0&0&0& 0&0&0&0&0& 0&0&0&0&0& 0&0&0&0&0& 0\\
1&0&0&0&0& 0&0&0&0&0& 0&0&0&0&0& 0&0&0&0&0& 0&0&0&0&0& 0\\
0&0&0&1&0& 0&0&0&0&0& 0&0&0&0&0& 0&0&0&0&0& 0&0&0&0&0& 0\\
0&0&1&0&0& 0&0&0&0&0& 0&0&0&0&0& 0&0&0&0&0& 0&0&0&0&0& 0\\
0&0&0&0&0& 27&0&0&0&0& 0&0&0&0&0& 0&0&0&0&0& 0&0&0&0&0& 0\\
0&0&0&0&46& 0&0&0&0&0& 0&0&0&0&0& 0&0&0&0&0& 0&0&0&0&0& 0\\
0&0&0&0&0& 0&1&0&0&0& 0&0&0&0&0& 0&0&0&0&0& 0&0&0&0&0& 0\\
0&0&0&0&0& 0&0&1&0&0& 0&0&0&0&0& 0&0&0&0&0& 0&0&0&0&0& 0\\
0&0&0&0&0& 0&0&0&0&27& 0&0&0&0&0& 0&0&0&0&0& 0&0&0&0&0& 0\\
0&0&0&0&0& 0&0&0&46&0& 0&0&0&0&0& 0&0&0&0&0& 0&0&0&0&0& 0\\
0&0&0&0&0& 0&0&0&0&0& 0&0&0&0&0& 0&0&0&0&0& 0&1&0&0&0& 0\\
0&0&0&0&0& 0&0&0&0&0& 0&0&0&0&0& 0&0&0&0&0& 1&0&0&0&0& 0\\
0&0&0&0&0& 0&0&0&0&0& 0&0&0&0&0& 0&0&0&0&72& 0&0&0&0&0& 0\\
0&0&0&0&0& 0&0&0&0&0& 0&0&0&0&0& 0&0&0&1&0& 0&0&0&0&0& 0\\
0&0&0&0&0& 0&0&0&0&0& 0&0&0&0&0& 0&0&0&0&0& 0&0&0&0&0& 1\\
0&0&0&0&0& 0&0&0&0&0& 0&0&0&0&0& 0&0&0&0&0& 0&0&0&0&72& 0\\
0&0&0&0&0& 0&0&0&0&0& 0&0&0&0&0& 0&0&0&0&0& 0&0&0&72&0& 0\\
0&0&0&0&0& 0&0&0&0&0& 0&0&0&0&0& 0&0&0&0&0& 0&0&72&0&0& 0\\
0&0&0&0&0& 0&0&0&0&0& 0&0&0&1&0& 0&0&0&0&0& 0&0&0&0&0& 0\\
0&0&0&0&0& 0&0&0&0&0& 0&0&72&0&0& 0&0&0&0&0& 0&0&0&0&0& 0\\
0&0&0&0&0& 0&0&0&0&0& 0&1&0&0&0& 0&0&0&0&0& 0&0&0&0&0& 0\\
0&0&0&0&0& 0&0&0&0&0& 1&0&0&0&0& 0&0&0&0&0& 0&0&0&0&0& 0\\
0&0&0&0&0& 0&0&0&0&0& 0&0&0&0&0& 0&0&72&0&0& 0&0&0&0&0& 0\\
0&0&0&0&0& 0&0&0&0&0& 0&0&0&0&0& 0&72&0&0&0& 0&0&0&0&0& 0\\
0&0&0&0&0& 0&0&0&0&0& 0&0&0&0&0& 72&0&0&0&0& 0&0&0&0&0& 0\\
0&0&0&0&0& 0&0&0&0&0& 0&0&0&0&1& 0&0&0&0&0& 0&0&0&0&0& 0
\end{pmatrix},$$

$$\mathfrak{r} = \begin{pmatrix}
59&57&0&0&0&0&0&0&0&0&0&0&0&0&0&0&0&0&0&0&0&0&0&0&0&0\\
16&13&0&0&0&0&0&0&0&0&0&0&0&0&0&0&0&0&0&0&0&0&0&0&0&0\\
0&0&0&0&0&0&0&0&0&0&0&0&0&0&0&0&0&0&0&72&46&0&72&0&0&27\\
0&0&0&0&0&0&0&0&0&0&0&0&0&0&0&0&0&0&0&27&1&0&27&0&0&72\\
0&0&0&0&0&0&0&0&0&0&0&0&0&0&0&0&0&0&51&0&0&10&0&22&63&0\\
0&0&0&0&0&0&0&0&0&0&0&0&0&0&0&0&0&0&22&0&0&10&0&51&63&0\\
0&0&0&0&0&0&0&0&0&0&0&0&0&0&0&0&0&0&10&0&0&22&0&10&22&0\\
0&0&0&0&0&0&0&0&0&0&0&0&0&0&0&0&0&0&10&0&0&51&0&10&51&0\\
0&0&0&0&0&0&0&0&0&0&0&0&0&0&0&0&0&0&0&46&72&0&27&0&0&72\\
0&0&0&0&0&0&0&0&0&0&0&0&0&0&0&0&0&0&0&1&27&0&72&0&0&27\\
0&0&0&0&39&34&39&34&0&0&0&0&0&0&0&0&0&0&0&0&0&0&0&0&0&0\\
0&0&55&25&0&0&0&0&18&48&0&0&0&0&0&0&0&0&0&0&0&0&0&0&0&0\\
0&0&48&18&0&0&0&0&48&18&0&0&0&0&0&0&0&0&0&0&0&0&0&0&0&0\\
0&0&0&0&42&42&31&31&0&0&0&0&0&0&0&0&0&0&0&0&0&0&0&0&0&0\\
0&0&55&25&0&0&0&0&55&25&0&0&0&0&0&0&0&0&0&0&0&0&0&0&0&0\\
0&0&0&0&34&39&39&34&0&0&0&0&0&0&0&0&0&0&0&0&0&0&0&0&0&0\\
0&0&0&0&31&31&31&31&0&0&0&0&0&0&0&0&0&0&0&0&0&0&0&0&0&0\\
0&0&25&55&0&0&0&0&48&18&0&0&0&0&0&0&0&0&0&0&0&0&0&0&0&0\\
0&0&0&0&0&0&0&0&0&0&36&0&0&36&0&37&36&0&0&0&0&0&0&0&0&0\\
0&0&0&0&0&0&0&0&0&0&0&36&36&0&36&0&0&36&0&0&0&0&0&0&0&0\\
0&0&0&0&0&0&0&0&0&0&0&37&36&0&36&0&0&37&0&0&0&0&0&0&0&0\\
0&0&0&0&0&0&0&0&0&0&37&0&0&37&0&37&36&0&0&0&0&0&0&0&0&0\\
0&0&0&0&0&0&0&0&0&0&0&36&37&0&36&0&0&37&0&0&0&0&0&0&0&0\\
0&0&0&0&0&0&0&0&0&0&37&0&0&36&0&36&36&0&0&0&0&0&0&0&0&0\\
0&0&0&0&0&0&0&0&0&0&37&0&0&36&0&37&37&0&0&0&0&0&0&0&0&0\\
0&0&0&0&0&0&0&0&0&0&0&37&37&0&36&0&0&36&0&0&0&0&0&0&0&0
\end{pmatrix}.$$

(d) $\mathfrak{G} = \langle \mathfrak{x}, \mathfrak{r} \rangle$ *and* $\mathfrak{G}$ *has* 22 *conjugacy classes* $\mathfrak{g}_i{}^{\mathfrak{G}}$ *with representatives* $\mathfrak{g}_i$ *and centralizer orders* $|C_{\mathfrak{G}}(\mathfrak{g}_i)|$ *as given below:*

Class	*Representative*	$\vert$*Class*$\vert$	$\vert$*Centralizer*$\vert$
1	1	1	$2^{11} \cdot 3^3 \cdot 5^2 \cdot 13$
2_1	$(\mathfrak{x}\mathfrak{s})^5$	1755	$2^{11} \cdot 5$
2_2	$(\mathfrak{h})^2$	11700	$2^9 \cdot 3$
3	$\mathfrak{r}$	166400	$2^2 \cdot 3^3$
4_1	$\mathfrak{h}$	93600	$2^6 \cdot 3$
4_2	$(\mathfrak{h}\mathfrak{s})^4$	140400	2^7
4_3	$(\mathfrak{s}\mathfrak{r})^2$	280800	2^6
5	$\mathfrak{x}$	359424	$2 \cdot 5^2$
6	$(\mathfrak{x}\mathfrak{r})^2$	1497600	$2^2 \cdot 3$
8_1	$(\mathfrak{h}\mathfrak{s})^2$	561600	2^5
8_2	$(\mathfrak{h}\mathfrak{s})^6$	561600	2^5
8_3	$\mathfrak{s}\mathfrak{r}$	1123200	2^4
8_4	$\mathfrak{h}^2\mathfrak{s}$	1123200	2^4
10	$\mathfrak{x}\mathfrak{s}$	1797120	$2 \cdot 5$
12_1	$\mathfrak{x}\mathfrak{r}$	1497600	$2^2 \cdot 3$
12_2	$(\mathfrak{x}\mathfrak{r})^5$	1497600	$2^2 \cdot 3$
13_1	$\mathfrak{x}^2\mathfrak{r}$	1382400	13
13_2	$(\mathfrak{x}^2\mathfrak{r})^2$	1382400	13
16_1	$\mathfrak{h}\mathfrak{s}$	1123200	2^4
16_2	$(\mathfrak{h}\mathfrak{s})^3$	1123200	2^4
16_3	$(\mathfrak{h}\mathfrak{s})^5$	1123200	2^4
16_4	$(\mathfrak{h}\mathfrak{s})^7$	1123200	2^4

(e) $\mathfrak{G}$ *has the same character table as the Tits group given in the Atlas [19], p. 74.*

(f) $\mathfrak{G} = \langle \mathfrak{x}, \mathfrak{h}, \mathfrak{s}, \mathfrak{r} \rangle$ *is a finite simple group with* 2*-central involution* $\mathfrak{z} := (\mathfrak{h}\mathfrak{s})^8$ *such that* $C_{\mathfrak{G}}(\mathfrak{z}) = \langle \mathfrak{h}, \mathfrak{s}, \mathfrak{x} \rangle = \kappa_{\mathfrak{V}}(H)$.

Proof (a) and (b) These assertions follow from Propositions 10.1.3, 10.2.2 and Kratzer's Algorithm 7.3.10 of [92].

(c) Since $V_D \cong W_D$, we apply the MAGMA command `AHom(V, W)` to construct a transformation matrix $\mathfrak{T} \in \mathrm{GL}_{26}(73)$ such that $\mathfrak{h} = \mathfrak{T}^{-1}k_w(h)\mathfrak{T}$ and $\mathfrak{s} = \mathfrak{T}^{-1}k_w(s)\mathfrak{T}$.

Let M be a vector space with dimension 26 over $GF(73)$. The construction of the four matrices $\mathfrak{x}, \mathfrak{h}, \mathfrak{s}, \mathfrak{r} \in \mathrm{GL}_{23}(73)$ generating $\mathfrak{G}$ has been described in the statement. Hence M is a $\mathfrak{G}$-module. Let N be an indecomposable component of $M_{\mathfrak{H}}$ with dimension 10, where $\mathfrak{H} = \langle \mathfrak{h}, \mathfrak{s}, \mathfrak{x} \rangle$. Then the MAGMA command `OrbitAction(G,N)` provides a faithful permutation representation of $\mathfrak{G}$ with degree 1755. Using the MAGMA command `Verify` it follows that the stabilizer of this permutation representation is $\mathfrak{H}$. Hence $|\mathfrak{G}| = |H| \cdot 1755 = 2^{11} \cdot 3^3 \cdot 5^2 \cdot 13$.

(d) The table has been calculated by means of the faithful permutation representation of $\mathfrak{G}$ constructed in (c), MAGMA and Kratzer's Algorithm 5.3.18 of [92].

(e) The character table of $\mathfrak{G}$ has been computed by MAGMA using the the faithful permutation representation of $\mathfrak{G}$ constructed in (c).

(f) The simplicity of $\mathfrak{G}$ can be read off from its character table given in (e). By (c)(2) and Table 10.7.1 $\kappa_{\mathfrak{V}}(z) = (\mathfrak{h}\mathfrak{s})^8$ is a 2-central involution of $\mathfrak{G}$. The equation $C_{\mathfrak{G}}(\mathfrak{z}) = \mathfrak{V}(H) \simeq H$ is an immediate consequence of (d) and Table 10.7.1. Therefore $\mathfrak{G}$ is a group of Tits-type. □

10.4 Group order

In this section we determine the structure of the centralizer $U = C_G(u)$ of a non-2-central involution u in a finite simple group G of Tits-type. It is unique up to group isomorphism. The fusion of its involutions and its character table are calculated. Thus Thompson's group order formula can be applied to show that all simple groups G of Tits-type have the same order as the group $\mathfrak{G}$ of Theorem 10.3.1.

Lemma 10.4.1 *Let G be a finite group with a strongly self-centralizing elementary abelian Sylow 2-subgroup V of order 4 such that $D = N_G(V)$ is a proper subgroup of G, and D is isomorphic to the alternating group A_4. Then $G \cong A_5$.*

Proof Let π be the smallest set of the prime numbers p containing 2 such that $C_G(x)$ is a π-subgroup for all π-elements $x \neq 1$ of G. Then $\pi = \{2\}$ because $C_G(x) = V$ for all involutions $x \in V$. Let t be the number of strongly real conjugacy classes of G of odd order.

Since $D \cong A_4$ has a unique conjugacy class of involutions, and V is an abelian Sylow 2-subgroup of G, Burnside's Lemma 1.4.2 of [92] asserts that G has a unique conjugacy class of involutions z^G. Hence it may be assumed that $z \in D$. Let

$$(*) \quad d(z) = |\{(x, y) \in z^G \times z^G \mid xy = z\}|,$$

and let $c = 1 + d(z)$. Then Theorem 4.3.7 of [92] states that $|G| = c|V| + t|V|^2$. Furthermore, the Harada–Miyamoto Theorem 4.3.9 of [92] states that

$$(**) \quad t < \max\left\{|V| - \frac{c}{|V|},\ \frac{3}{2}|V|\right\} = 6.$$

For each irreducible character ψ of D let $\psi(\hat{D}) = |D : C_D(z)|\psi(z)$. Then Lemma 4.3.2 of [92] asserts that

$$d(z) = \frac{1}{|D|} \sum_{\psi \in Irr_{\mathbb{C}}(D)} \frac{\psi(\hat{D})^2 \psi(z)}{\psi(1)}.$$

Therefore $d(z) = 2$ by the character table of A_4. Now (*) and (**) imply that $c = 3$ and $|G| \in \{12, 28, 44, 60, 76, 92\}$. Since A_4 is a proper subgroup of G it follows that $|G| = 60$. As V is a strongly self-centralizing Sylow 2-subgroup of G no Sylow 5-subgroup of G can be a direct factor of G. Therefore Sylow's theorem applied to the prime $p = 5$ implies that G is simple. Hence $G \cong A_5$. This completes the proof. □

Proposition 10.4.2 *Let G be a finite simple group of Tits-type with a 2-central involution z such that $C_G(z) = H = \langle h, s, x\rangle$. Let $A = \langle z, a, b, c, d\rangle$ be the elementary abelian self-centralizing normal subgroup of order 2^5 in the Sylow 2-subgroup S of H defined in Proposition 10.1.5. Let $D = N_H(A) = \langle h, s\rangle$. By Proposition 10.2.2 there is an element $r \in G$ such that $N_G(A) = \langle h, s, r\rangle$ and the amalgam $H \leftarrow D \rightarrow E$ are uniquely determined by H up to equivalence. Let $u = (h^2 s)^4$, $g = h^2$ and $t = h^2 s h^{-1} s r h^{-1} \in E$. Let $U = C_E(u)$. Then the following statements hold.*

(a) *The involutions z and u represent all conjugacy classes involutions of G.*
(b) *$U = \langle t, g, s\rangle$ and $|U| = 2^9 \cdot 3$.*
(c) *Representatives u_i of the 31 conjugacy classes u_i^U of U and the corresponding centralizer orders $|C_U(u_i)|$ are given in Table 10.7.5.*

(d) *U has 11 conjugacy classes of involutions. Using the indices of all the classes of U given in Table 10.7.5 the involution classes of U with indices 3, 7 and 9 fuse to z, and the ones with indices 1, 2, 4, 5, 6, 8, 10 and 11 fuse to u.*

(e) $C_G(u) = C_E(u)$.

Proof (a) By Proposition 10.2.2 each simple group G of Tits-type has two conjugacy classes of involutions. They are represented by z and u.

(b) Using the faithful permutation representation of E described in Proposition 10.2.2 and MAGMA it has been checked that $U = C_E(u) = \langle t, g, s \rangle$, where $t = h^2sh^{-1}srh^{-1}$.

(c) Furthermore, a system of representatives of the conjugacy classes U has been calculated by means of Kratzer's Algorithm 5.3.19 of [92].

(d) The fusion of the U-classes into the ones of the E-classes has been determined computationally using the faithful permutation representation of E, the representatives of the conjugacy classes of E and U. The results of (d) then follow from Proposition 10.2.2(g) and (h).

(e) Using the faithful permutation representation of E described in Proposition 10.2.2 and MAGMA it has been checked that $T = H \cap U$ is a Sylow 2-subgroup of U and therefore of $C = C_G(u)$ because $|C : U|$ is odd.

Let F be the Fitting subgroup of U. Then another application of MAGMA yields that $|F| = 2^8$ and that $e = h^{-1}sr^{-1}srsh \in U$ is an involution $e \in T$ such that $T = \langle F, e \rangle$, $|C_U(e)| = 2^5$ and e is conjugate in U to s. In particular, e is conjugate in C to s and $|C_T(e) = C_C(e) \cap T| = 2^5$. By Table 10.7.5 all conjugacy classes of involutions y of the maximal subgroup F of T have centralizers $C_U(y) \leq C_C(y)$ of order at least 2^7, because the conjugacy class 2_{11} of U with representative t^2stg is not conjugate in U to any y in F, as has also been checked by MAGMA. Therefore Thompson's Lemma 1.4.3 of [92] implies that C has a maximal subgroup M of index $|C : M| = 2$ and $e \notin M$. Therefore F is a Sylow 2-subgroup of M, and $t \in M$.

Suppose that X is a normal subgroup of M of odd order. Using MAGMA again it has been checked that the center Z_1 of F is generated by $\{u, z, w = (ts)^4\}$. Let $q = uzw$. Then $W = \langle z, q \rangle$ is a Klein 4-subgroup of F which operates on X by conjugation. Another application of MAGMA yields that $|C_U(z)| = |C_U(q)| = 2^9$ and that $z^t = q$, $q^t = qz$. Furthermore, all involutions of W belong to the conjugacy class 2_3 of U; see Table 10.7.5. Now Table 10.7.1, Proposition 10.2.2 and statement (d) imply that $|C_H(q)| = 2^{10}$. Thus $|C_X(W)| = 1$ because $C_X(W) \leq$

$C_G(z) \cap C_G(q) \leq C_H(q)$. Therefore the Brauer–Wielandt Theorem 4.4.2 of [92] asserts that $|X| = |C_X(z)||C_X(q)||C_X(zq)| = |C_X(z)|^3$. Since $C_M(z) \leq C_C(z)$ and $C_C(z) = C_G(z) \cap C_G(u) = C_U(z)$ has order 2^9 it follows that $C_X(z) = X \cap C_M(z) = 1$. Thus $X = 1$. In particular, $O(M) = 1$.

We now claim that the center Z_1 of F is weakly closed with respect to M. Suppose that $Z_1^m \leq F$ for some $m \in M$. Using MAGMA again it has been checked that $|C_U(x)| = 2^7$ for all 72 involutions $x \in F - Z_1$. Therefore the Sylow 2-subgroups of $C_M(x)$ have order 2^6 for all x. For all $y \in Z_1$ a Sylow 2-subgroup of $C_M(y)$ has order greater or equal to 2^8 by Table 10.7.5. Therefore $Z_1^m = Z_1$ because no element of Z_1 can be conjugate in M to any element of $F - Z_1$. Hence Z_1 is weakly closed in F.

As $z \in Z_1$ and $T = C_U(z)$ is a Sylow 2-subgroup of U it follows that $C_M(Z_1) \leq U \cap C_G(z) = C_H(u) = T$. Hence $C_M(Z_1) = F = M \cap T$. Thus $N_M(Z_1)/C_M(Z_1)$ is isomorphic to a subgroup of $\mathrm{PSL}_3(2)$, and $N_M(F)/Z_1$ is isomorphic to a subgroup of $Aut(F)$. Applying the MAGMA command `AutomorphismGroup(F)` to the permutation subgroup F of E we have checked that $Aut(F)$ is a group of order 2^8. In particular, $N_M(Z_1) = C_M(Z_1) = F$ and $N_M(F) = F$ because F is a Sylow 2-subgroup of M. Since Z_1 is weakly closed in F with respect to M, Gruen's Theorem 7.1.8 of [81], p. 168, states that M has a normal subgroup K such that $M/K \cong F/F'$. Hence $t \in K$. Another application of MAGMA now asserts that the commutator subgroup F' of F equals Z_1. In particular Z_1 is a Sylow 2-subgroup of K, and $W = \langle z, q \rangle$ is isomorphic to a Sylow 2-subgroup W_1 of $K_1 = K/\langle u \rangle$. Thus K_1 is a group having a Klein 4-group W_1 as a Sylow 2-subgroup such that $C_{K_1}(w) = W_1$ for all $w \in W_1$ because $C_K(v) = K \cap C_M(v) = K \cap F = Z_1$ for all $v \in Z_1$. Furthermore, $N_1 = N_{K_1}(V_1) = \langle W_1, t \rangle \cong A_4$.

Suppose that $K_1 \neq N_1$. Then Lemma 10.4.1 implies that $K_1 \cong A_5$. In particular, K_1 has a unique conjugacy class of elements of order 3 represented by t. Hence $t^{-1} = t^x$ for some involution x of K_1. Since K_1 has a unique conjugacy class of involutions x is conjugate to z in K_1 and therefore in G. Using MAGMA again it has been checked that $|\{b \in U | t^b = t^2, order(b) = 2\}| = 12$ and that $|C_U(b)| = 32$ for each of these 12 involutions. Hence each b belongs either to the conjugacy class 2_{10} or 2_{11} of U by Table 10.7.5. Now statement (d) implies that each involution x of M inverting t is conjugate to u in G. But z is not conjugate to u in G by (a). This contradiction shows that $K_1 = N_1$.

This completes the proof because

$$|C| = 2|M| = 4|K_1||F : Z_1| = 2^9 \cdot 3 = |U|.$$

□

Proposition 10.4.3 *Let G be a finite simple group of Tits-type with a 2-central involution z such that $C_G(z) = H = \langle h, s, x\rangle$. Then the following statements hold.*

(a) $|G| = 2^{11} \cdot 3^3 \cdot 5^2 \cdot 13$.
(b) $G = \langle H, E\rangle = \langle H, U\rangle$.

Proof (a) By Proposition 10.4.2(a) each simple group G of Tits-type has two conjugacy classes of involutions z^G and u^G such that

$$|H| = |C_G(z)| = 2^{11} \cdot 5 \text{ and } |U| = |C_G(u)| = 2^6 \cdot 3.$$

Using the faithful permutation representations of H and of E described in Propositions 10.1.5 and 10.2.2, respectively, and the fusion patterns of the conjugacy classes of involutions of H and U in G given in Propositions 10.2.2 another application of MAGMA yields that

$$r(z,u,z) = \left|\{(x,y) \in (z^G \cap H) \times (u^G \cap H) \big| z \in \langle xy\rangle\}\right| = 5840,$$
$$r(z,u,u) = \left|\{(x,y) \in (z^G \cap U) \times (u^G \cap U) \big| u \in \langle xy\rangle\}\right| = 879.$$

Now Thompson's order formula implies that

$$\begin{aligned}|G| &= r(z,u,z) \cdot |C_G(u)| + r(z,u,u) \cdot |C_G(z)| \\ &= 2^{11} \cdot 3^3 \cdot 5^2 \cdot 13.\end{aligned}$$

(b) This assertion follows immediately from Proposition 10.4.2(a) and Corollary 4.8.10 of [92]. □

10.5 The 3-, 5- and 13-singular conjugacy classes

In this section we show that each finite simple group G of Tits-type has one conjugacy class of elements of each order 3 and 5. It has two conjugacy classes of elements of order 13. We also describe the structure of the normalizers $N_p = N_G(p)$, their conjugacy classes and their character tables for $p \in \{3, 5, 13\}$. Furthermore, the fusion of the classes of each group N_p in G is determined.

Lemma 10.5.1 *Let G be a finite simple group of Tits-type with a 2-central involution z such that $H = C_G(z) = \langle h, s, x\rangle$, where the generators h, s and x are defined in Proposition 10.1.5. By Proposition 10.2.2 there is an element $r \in G$ of order 3 such that $N_G(A) = \langle h, s, r\rangle$, where A is the maximal elementary abelian normal subgroup of the Sylow 2-subgroup S defined in Proposition 10.1.5. Let $u = (h^2 s)^4$, $g = h^2$ and $t = h^2 sh^{-1} srh^{-1} \in E$. By Proposition 10.4.2 $U = C_G(u) = \langle t, g, s\rangle$ is the centralizer of the non-2-central involution u of G. Let $n = hr^{-1}hr^{-1}hrh^{-1}s, w = (ts)^4$ and $q = uzw$. Then the following statements hold.*

(a) *$K_3 := \langle q, t, z, n\rangle = N_U(t)$ and has the following set of defining relations:*

$$t^3 = n^2 = q^2 = z^2 = u^2 = 1,$$
$$z^t = q, z^n = z, q^n = qz, t^n = t^2, q^t = qz, [q, z] = 1.$$

(b) *In G there are two elements q_1, q_2 of order 3 in N_3 and an element $v = (r^{-1}h)^3$ of order 4 in $C_U(t)$ such that $v^2 = u$, $v^n = v^{-1}$ and $N_3 = \langle n, v, t, q_1, q_2\rangle$ has the following set of defining relations:*

$$v^4 = n^2 = t^3 = q_1^3 = q_2^3 = 1,$$
$$[v, t] = 1, v^n = v^{-1}, t^n = t^{-1}, [q_1, t] = 1, [q_2, t] = 1, q_1^n = q_1,$$
$$q_2^n = q_2^2, q_1^v = q_2, q_2^v = q_1^2, [t, q_1] = [t, q_2] = 1, [q_1, q_2] = t.$$

(c) *$N_3 = N_G(t)$ has a normal Sylow 3-subgroup S_3 which is extra-special of order 27 and exponent 3 and $C_3 = \langle n, q, z, u\rangle$ is a complement of S_3 in N_3.*

(d) *A system of representatives n_i and the corresponding centralizer orders $|C_{N_3}(n_i)|$ of the 13 conjugacy classes $n_i^{N_3}$ of N_3 is given in Table 10.7.6.*

(e) *t is a representative of the unique conjugacy class of elements of order 3 in G.*

(f) *The character table of N_3 is stated in Table 10.8.5.*

Proof (a) By Proposition 10.2.2 $E = N_G(A)$ has a faithful permutation representation of degree 2048. Since the element t of order 3 belongs to $U = CE(u) \leq E$ MAGMA has been applied to this permutation representation to compute the given systems of generators and defining relations of $N_U(\langle t\rangle) = K_3$.

(b) By Proposition 10.4.3 each group G of Tits-type has order $|G| = 2^{11} \cdot 3^3 \cdot 5^2 \cdot 13$. Now (a) and Sylow's Theorem imply that a Sylow 3-subgroup S_3 of G containing t has order 3^3 and that $|N_3 = N_G(S_3)| = 2^3 \cdot 3^3$. The element v of order 4 in $C_U(t)$ has been found by means of MAGMA and the above mentioned permutation representation of E. Therefore the dihedral group $T_2 = \langle v, n \rangle$ of order 8 is a Sylow 2-subgroup of N_3. In particular, T_2 has a unique irreducible representation of degree 2 over the prime field $GF(3)$. The corresponding irreducible $GF(3)T_2$-module has a basis $\{q_1, q_2\}$ such that v and n are represented by the following matrices:

$$Mv = \begin{pmatrix} 0 & 1 \\ 2 & 0 \end{pmatrix}, \qquad Mn = \begin{pmatrix} 1 & 0 \\ 0 & 2 \end{pmatrix}.$$

In particular, there are two elements $q_1, q_2 \in N_3$ of order 3 such that $q_1^v = q_2, q_2^v = q_1^2, q_1^n = q_1, q_2^n = q_2^2$. Since $|S_3| = 27$ is normal in N_3, the Klein 4-group $W = \langle u, n \rangle$ operates on S_3 by conjugation. From Table 10.7.5 and the Brauer–Wielandt Theorem it follows that t belongs to the center of S_3. Thus $[q_i, t] = 1$ for $i = 1, 2$ and $[q_1, q_2] = t$. Hence (b) holds.

(c) is an immediate consequence of (b).

(d) Using MAGMA and the presentation of N_3 given in (b) we immediately get a faithful permutation representation of N_3 of degree 216. It has been used together with Kratzer's Algorithm 5.3.18 of [92] to get a complete list of representatives of all conjugacy classes of N_3.

(e) Table 10.7.5 also provides information about the power maps of the representatives of three conjugacy classes of elements of order 6. They power onto three elements of order 3 belonging to distinct conjugacy classes of N_3. Furthermore, they are centralized by the three involutions u, n and vn, respectively. These involutions are pairwise non-conjugate in N_3. But in E they are conjugate, as has been checked by means of MAGMA. Now Table 10.7.5 implies that all elements of order 3 of G are conjugate in G, because S_3 is a Sylow 3-subgroup.

(f) The character table of N_3 has been computed by means of the faithful permutation representation of N_3 and MAGMA. This completes the proof. □

Lemma 10.5.2 *Let G be a finite simple group of Tits-type with a 2-central involution z such that $H = C_G(z) = \langle h, s, x \rangle$, where the generators h, s and x are defined in Proposition 10.1.5. Then the following statements hold.*

(a) *x is a representative of the unique conjugacy class of elements of order 5 in H, and $N_H(x) = \langle z, x, h\rangle$ has the following defining relations:*

$$z^2 = h^4 = [z, h] = 1,$$
$$x^5 = 1, \quad x^h = x^2, \quad [z, x] = 1.$$

(b) *$N_5 = N_G(x)$ has a normal Sylow 5-subgroup $S_5 = \langle x, y\rangle$ which is elementary abelian of order 25, and it satisfies the following additional defining relations:*

$$y^5 = 1, \quad [x, y] = 1,$$
$$y^h = y^3, \quad y^z = y^{-1}.$$

(c) *$NS_5 = N_G(S_5) = \langle z, x, h, y, d\rangle$ is generated by N_5 and an element d of order 3. Furthermore, NS_5 has the following additional defining relations:*

$$d^3 = 1, \quad (d^2 z)^2 = h^3 d, \quad [h, d] = 1, \quad (dz)^{12} = 1,$$
$$x^d = x^3 y, \quad y^d = x^2 y.$$

(d) *A system of representatives n_i and the corresponding centralizer orders $|C_{N_5}(n_i)|$ of the 14 conjugacy classes $n_i^{N_5}$ of N_5 is given in Table 10.7.7.*

(e) *The character tables of N_5 and NS_5 are stated in Tables 10.8.6 and 10.8.7, respectively.*

(f) *G has exactly one conjugacy class of elements of order 5.*

Proof (a) By Proposition 10.1.5 $H = C_G(z)$ has a faithful permutation representation of degree 512. MAGMA has been applied to this permutation representation to compute the given systems of generators and defining relations of $N_H(\langle x\rangle)$.

(b) By Proposition 10.4.3 each group G of Tits-type has order $|G| = 2^{11} \cdot 3^3 \cdot 5^2 \cdot 13$. Now (a) and Sylow's Theorem imply that a Sylow 5-subgroup S_5 of G has order 5^2 and that $|NGS_5 = N_G(S_5)| = 2^4 \cdot 3 \cdot 5^2$. Therefore $N_5 = N_G(\langle x\rangle)$ has order $2^3 \cdot 5^2$ by (a) and Tables 10.7.1 and 10.7.5. In particular, N_5 has a normal cyclic subgroup $C_y = \langle y\rangle$ of order 5 such that the following sequence is exact: $C_y \to N_5 \to N_H(\langle x\rangle)$. Using the extension algorithms of MAGMA and its isomorphism test programs it follows that there are six non-isomorphic extensions W_i of $N_H(\langle x\rangle)$ by C_y. Furthermore, two of these groups W_1 and W_2, say, have a center of order 1 and all others have a center of order 2. Now Tables 10.7.1 and 10.7.5 imply that N_5 can only be isomorphic to W_1 or W_2, because there is no involution in G centralizing a subgroup of order 25. In W_1 the following additional relations to (a) hold: $y^5 = 1$, $[x, y] = 1$, $y^z = y^4$ and $[h, y] = 1$. By Proposition 10.4.2 and Table 10.7.5 h^2 is conjugate in G to u. Furthermore, $|C_G(u)| = 2^9 \cdot 3$. Hence $y \notin C_G(u)$, which contradicts $[h, y] = 1$. Therefore $N_5 \cong W_2$, which has the additional relations given in statement (b).

(d) Using MAGMA and the presentation of N_5 given in (b) we immediately get a faithful permutation representation of N_5 of degree 200. It has been used together with Kratzer's Algorithm 5.3.18 of [92] to get a complete list of representatives of all conjugacy classes of N_5 and then the character table of N_5.

(c) Since $S_5 = \langle x, y\rangle$ is normal in N_5 the conjugate action of h and z is described by the following matrices over $F = GF(5)$:

$$Mh = \begin{pmatrix} 2 & 0 \\ 0 & 3 \end{pmatrix}, \quad Mz = \begin{pmatrix} 1 & 0 \\ 0 & 4 \end{pmatrix}.$$

Let Y be the subgroup of $\mathrm{GL}_2(5)$ generated by Mh and Mz. Using MAGMA again we see that there are exactly three subgroups K_i in $\mathrm{GL}_2(5)$ such that $|K_i : Y| = 6$ for $i = 1, 2, 3$. They are generated by Y and the following additional generators Md_i of order 3 and Me_i, where Me_1 has order 4 and Me_j has order 8 and satisfies $Me_j^2 = M_h$ for $j = 2, 3$. In fact we have the following matrices:

$$Md_1 = \begin{pmatrix} 3 & 2 \\ 1 & 1 \end{pmatrix}, \quad Md_2 = \begin{pmatrix} 2 & 4 \\ 2 & 2 \end{pmatrix}, \quad Md_3 = \begin{pmatrix} 2 & 2 \\ 4 & 2 \end{pmatrix},$$

$$Me_1 = \begin{pmatrix} 3 & 0 \\ 0 & 2 \end{pmatrix}, \quad Me_2 = \begin{pmatrix} 1 & 4 \\ 3 & 4 \end{pmatrix}, \quad Me_3 = \begin{pmatrix} 1 & 2 \\ 1 & 4 \end{pmatrix}.$$

Let $K_i = \langle Mh, Mz, Md_i, Me_i\rangle$ for $i = 1, 2, 3$. Then using MAGMA it has been checked that K_1 does not have any elements of order 24. It is easily verified that $Mk_2 = MzMe_2 \in K_2$ and $Mk_3 = MzMe_3 \in K_3$ have order 24 and that $Mk_i^{12} = Mh^2$ for $i = 1, 2$. By Proposition 10.2.2 and Table 10.7.4 h^2 fuses in G to u. But $U = C_G(u)$ has no elements of order 24 by Table 10.7.5.

Therefore $NS_5 = N_G(S_5)$ is a split extension of K_1 by S_5. Since $Me_1 = Mh^3$ we can omit generator e_1 of K_1. Denoting d_1 by just d another application of MAGMA yields that a set of defining relations of $K_1 = \langle z, h, d, e\rangle$ consists of the relations given in the first lines of the relations in (a) and (c). Adding the defining relations corresponding to the conjugate actions of the four generators of K_1 described by the entries of their matrices and the relations of (b) we obtain the presentation of NS_5 given in (c).

(e) The character table of NS_5 has been calculated by means of MAGMA and a faithful regular representation of this group of order 1200.

(f) The character table of NS_5 stated in Table 10.8.7 shows that this group has only one conjugacy class of elements of order 5. Now Burnside's Lemma 1.4.2 of [92] completes the proof. □

Corollary 10.5.3 *Let G be any finite simple group of Tits-type with a 2-central involution z such that $C_G(z) = H$. Then the following assertions hold.*

(a) *The Sylow 13-normalizer is a Frobenius group of order* $13 \cdot 6$.
(b) *G has* 22 *conjugacy classes.*

Proof (a) By Proposition 10.4.3 each group G of Tits-type has order $|G| = 2^{11} \cdot 3^3 \cdot 5^2 \cdot 7 \cdot 11 \cdot 23$. Now the Tables of Section 10.7 and Sylow's Theorem imply that a Sylow 13-subgroup S_{13} of G is cyclic and self-centralizing. Furthermore, $N_G(S_{13})$ has order $13 \cdot 6$.

(b) By the power map information of Tables 10.7.1 and 10.7.5 we see that G has 17 conjugacy classes of elements of even order. Since G has two conjugacy classes of elements of order 13 by (a) and unique conjugacy classes of elements of order 3 and 5 by Lemmas 10.5.1 and 10.5.2, the proof is complete. □

10.6 Uniqueness proof

In this section we show that each finite simple group G of Tits-type has an ordinary irreducible character of degree 26 which is uniquely determined up to complex conjugation. Hence the uniqueness criterion stated in [91] can be applied. It implies that each finite simple group G of Tits-type is isomorphic to the simple group $\mathfrak{G}$ constructed in Theorem 10.3.1.

Proposition 10.6.1 *Let G be a finite simple group of Tits-type. Then the minimal degree of a faithful irreducible complex character χ of G is* 26, *and G has exactly one pair of complex conjugate characters $\chi \in Irr_{\mathbb{C}}(G)$ of this degree.*

Proof Let $H = C_G(z)$ be the centralizer of a 2-central involution of G. Let S be the Sylow 2-subgroup of H and let A be the maximal elementary abelian normal subgroup of S defined in Proposition 10.1.5.

Let $D = N_H(A)$ and $E = N_G(A)$. Then $G = \langle H, E \rangle$ by Proposition 10.4.2. Proposition 10.2.2 states that the amalgam $H \leftarrow D \rightarrow E$ has Goldschmidt index 1. Therefore the free product $P = H *_D E$ of H and E with amalgamated subgroup D is uniquely determined up to isomorphism. By Theorem 10.3.1 the smallest degree of a compatible pair is 26, and there is exactly one such pair, $(\chi, \tau) \in mf\,char_{\mathbb{C}}(H) \times mf\,char_{\mathbb{C}}(E)$ of complex conjugate characters. As G is a simple epimorphic image of P it follows that G has at most one pair of faithful conjugate characters $\chi \in Irr_{\mathbb{C}}(G)$ with minimal degree 26.

Using Brauer's characterization of characters we now prove that any finite simple group G of Tits-type has a pair of complex conjugate irreducible characters $\chi \in Irr_{\mathbb{C}}(G)$ of degree 26. In order to do so we let $x = \sqrt{-2}$ and define a class function $\chi : G \rightarrow \mathbb{C}$ by

i	1_A	2_A	2_B	3_A	4_A	4_B	4_C	5_A	6_A	8_A	8_B	8_C	8_D
$\chi(g_i)$	26	-6	2	-1	-2	-2	2	1	-1	0	0	0	0

i	10_A	12_A	12_B	13_A	13_B	16_A	16_B	16_C	16_D
$\chi(g_i)$	-1	1	1	0	0	x	$-x$	$-x$	x

Let p be any prime divisor of $|G|$. If Y is an over-group of a p-elementary subgroup X, and the restriction $\chi_{|Y}$ of X to Y is a generalized character, then $\chi_{|X}$ is also a generalized character of X. By Proposition 10.2.2 G has a second conjugacy class u^G of involutions. Let $U = C_G(u)$. Let $N_3 = N_G(3_A)$ and $N_5 = N_G(5_A)$ be the normalizers of the cyclic subgroups of order 3 and 5 which are uniquely determined up to conjugation in G by Lemmas 10.5.1 and 10.5.2, respectively.

By Corollary 10.5.3 the Sylow 13-normalizer of G is a Frobenius group of order $13 \cdot 6$. Using its character tables and the fusion of the elements of orders 6 in G it is easy to check that the restrictions of χ to this normalizer N_{13} is equivalent to the character of the sum of two indecomposable projective characters. Therefore by Corollary 2.8.10 of [92] it remains to show that $X_{|Y}$ is a generalized character of Y, whenever Y belongs to the following set of subgroups:

$$\mathfrak{Y} = \{H, U, N_3, N_5\}$$

of G.

In the previous sections the fusion patterns of the conjugacy classes of the groups $Y \in \mathfrak{Y}$ have been determined. Therefore the inner products $(\chi, \xi)_Y$ can be calculated for each $\xi \in Irr_{\mathbb{C}}(Y)$ and each $Y \in \mathfrak{Y}$ by means of the character tables given in Section 10.8. It turns out that

$$\begin{aligned}
\chi_{|H} &= \chi_{12} + \chi_{22}, \text{ where } \chi_{12}, \chi_{22} \in Irr_{\mathbb{C}}(H),\\
\chi_{|U} &= \chi_{6} + \chi_{29} + \chi_{31}, \text{ where } \chi_{3}, \chi_{13}, \chi_{19} \in Irr_{\mathbb{C}}(U),\\
\chi_{|N_3} &= \chi_{6} + \chi_{8} + \chi_{10} + \chi_{11} + \chi_{12}, \text{ where all } \chi_i \in Irr_{\mathbb{C}}(N_3),\\
\chi_{|N_5} &= 2\cdot\chi_{2} + \chi_{10} + \chi_{12} + \chi_{13} + \chi_{14}, \text{ where all } \chi_i \in Irr_{\mathbb{C}}(N_5),
\end{aligned}$$

and that the inner product $(\chi, \chi)_G = 1$. Therefore χ is an irreducible character of G with $\chi(1) = 26$ by Corollary 2.8.10 [92]. Its conjugate complex character is determined similarly. This completes the proof. □

Theorem 10.6.2 (Michler–Wang) *Let H be the finite group of even order defined in Proposition 10.1.5. Then each finite simple group G of Tits-type is isomorphic to the finite simple group $\mathfrak{G} \leq \mathrm{GL}_{26}(73)$ of order $|\mathfrak{G}| = 2^{11}\cdot 3^3\cdot 5^2\cdot 13$ constructed in Theorem 10.3.1.*

Proof By Theorem 10.3.1 there exists a finite simple group $\mathfrak{G}$ of order $|\mathfrak{G}| = 2^{11}\cdot 3^3\cdot 5^2\cdot 13$ which is of Tits-type.

Let G be any finite simple group of Tits-type. By Proposition 10.2.2 G has exactly two conjugacy classes z^G and u^G of involutions. We may assume that $C_G(z) = H$, where H is the group of even order defined in Proposition 10.1.5. It also asserts that the Sylow 2-subgroup S of H contains a uniquely determined maximal elementary abelian normal subgroup A which is not normal in H. Let $D = N_H(A)$. By Theorem 10.3.1 $E = N_G(A)$ is uniquely determined by H up to isomorphism and the amalgam $H \leftarrow D \rightarrow E$ has Goldschmidt index 1. Furthermore, $G = \langle H, E\rangle$ by Proposition 10.4.2, and $\mathfrak{G}$ has a unique pair of dual 73-modular irreducible representations over $GF(73)$ of minimal degree 26 by its character table.

By Proposition 10.4.2 any finite simple group G of Tits-type has order $|G| = |\mathfrak{G}|$. In particular 73 does not divide $|G|$. Therefore Maschke's Theorem and Proposition 2.4.4 imply that each such group G has a unique pair of dual irreducible 73-modular representations of degree 26. Hence $G \cong \mathfrak{G}$ by Theorem 7.5.1 of [92]. This completes the proof. □

Theorem 10.6.3 (Parrott) *Each finite simple group G of Tits-type is isomorphic to the simple Tits group $T = {}^2F_4(2)'$.*

Proof Let G be any simple group of Tits-type. Let z be a 2-central involution of G, and $H = C_G(z)$. Then by the uniqueness theorem we may assume that H is the finitely presented group described in Proposition 10.1.5(a) by generators and relations. Let H_1 be the finitely presented group stated in Proposition 10.1.5(b). As mentioned there, H_1 is Parrott's presentation of the centralizer of a 2-central involution z of Tits' simple group $T = {}^2F_4(2)'$. By Proposition 10.1.5 $H \cong H_1$. Hence $T \cong \mathfrak{G}$ by Theorem 10.6.2. □

10.7 Representatives of conjugacy classes

10.7.1 *Conjugacy classes of* $H = \langle h, s, x\rangle$

Class	*Representative*	|*Class*|	|*Centralizer*|	2P	5P
1	1	1	$2^{11} \cdot 5$	1	1
2_1	$(hs)^8$	1	$2^{11} \cdot 5$	1	2_1
2_2	$(h^2shs)^4$	10	2^{10}	1	2_2
2_3	$(h^2s)^4$	20	2^9	1	2_3
2_4	$(h)^2$	80	2^7	1	2_4
2_5	s	80	2^7	1	2_5
2_6	h^2sx	80	2^7	1	2_6
4_1	$(hs)^4$	80	2^7	2_1	4_1
4_2	$(h^2s)^2$	160	2^6	2_3	4_2
4_3	$(h^2shs)^2$	160	2^6	2_2	4_3
4_4	$h^2xshshsxs$	160	2^6	2_3	4_4
4_5	h	320	2^5	2_4	4_5
4_6	$(h)^3$	320	2^5	2_4	4_6
4_7	$hsxsxs$	320	2^5	2_4	4_7
4_8	$(hsxsxs)^3$	320	2^5	2_4	4_8
4_9	$hsxshsxsx$	320	2^5	2_3	4_9
5	x	1024	$2 \cdot 5$	5	1
8_1	$(hs)^2$	320	2^5	4_1	8_1
8_2	$(hs)^6$	320	2^5	4_1	8_2
8_3	h^2s	640	2^4	4_2	8_3
8_4	h^2shs	640	2^4	4_3	8_4
8_5	$(h^2shs)^3$	640	2^4	4_3	8_5
8_6	h^2sxs	640	2^4	4_2	8_6
10	sx	1024	$2 \cdot 5$	5	2_1
16_1	hs	640	2^4	8_1	16_3
16_2	$(hs)^3$	640	2^4	8_2	16_4
16_3	$(hs)^5$	640	2^4	8_1	16_1
16_4	$(hs)^7$	640	2^4	8_2	16_2

10.7.2 *Conjugacy classes of* $N_G(S_5) = \langle h, z, x, y, d, e\rangle$

Class	*Representative*	$\lvert Class\rvert$	$\lvert Centralizer\rvert$	2P	3P	5P
1	1	1	$2^4 \cdot 3 \cdot 5^2$	1	1	1
2_1	$(h)^2$	25	$2^4 \cdot 3$	1	2_1	2_1
2_2	z	30	$2^3 \cdot 5$	1	2_2	2_2
3_1	d	100	$2^2 \cdot 3$	3_2	1	3_2
3_2	$(d)^2$	100	$2^2 \cdot 3$	3_1	1	3_1
4_1	h	25	$2^4 \cdot 3$	2_1	4_2	4_1
4_2	$(h)^3$	25	$2^4 \cdot 3$	2_1	4_1	4_2
4_3	e	150	2^3	2_1	4_3	4_3
5	x	24	$2 \cdot 5^2$	5	5	1
6_1	$(zd)^2$	100	$2^2 \cdot 3$	3_1	2_1	6_2
6_2	$(zd)^{10}$	100	$2^2 \cdot 3$	3_2	2_1	6_1
10	xz	120	$2 \cdot 5$	5	10	2_2
12_1	zd	100	$2^2 \cdot 3$	6_1	4_2	12_2
12_2	$(zd)^5$	100	$2^2 \cdot 3$	6_2	4_2	12_1
12_3	$(zd)^7$	100	$2^2 \cdot 3$	6_1	4_1	12_4
12_4	$(zd)^{11}$	100	$2^2 \cdot 3$	6_2	4_1	12_3

10.7.3 *Conjugacy classes of* $D = \langle h, s\rangle$

Class	*Representative*	$\lvert Class\rvert$	$\lvert Centralizer\rvert$	2P
1	1	1	2^{11}	1
2_1	$(hs)^8$	1	2^{11}	1
2_2	$(h^2shs)^4$	2	2^{10}	1
2_3	$(h^2s)^4$	4	2^9	1
2_4	$(h^3shs)^2$	8	2^8	1
2_5	$(h^2shshs)^2$	8	2^8	1
2_6	$h^3sh^3sh^2sh^2shshs$	8	2^8	1
2_7	$(h)^2$	16	2^7	1
2_8	$h^3sh^2shsh^2s$	16	2^7	1
2_9	$h^2sh^2sh^2shsh^2shs$	16	2^7	1
2_{10}	s	64	2^5	1
4_1	$(hs)^4$	16	2^7	2_1
4_2	$(h^2s)^2$	32	2^6	2_3
4_3	$(h^2shs)^2$	32	2^6	2_2
4_4	h^3sh^3shshs	32	2^6	2_3
4_5	h	64	2^5	2_7
4_6	$(h)^3$	64	2^5	2_7
4_7	h^3shs	64	2^5	2_4
4_8	h^2shshs	64	2^5	2_5
4_9	$h^3shshshs$	64	2^5	2_3
4_{10}	$h^3shshsh^2s$	64	2^5	2_7
4_{11}	$(h^3shshsh^2s)^3$	64	2^5	2_7
4_{12}	$h^3sh^2sh^2shs$	64	2^5	2_4
4_{13}	$h^3sh^2shshshs$	64	2^5	2_5
4_{14}	h^3shsh^2shshs	64	2^5	2_1
8_1	$(hs)^2$	64	2^5	4_1
8_2	$(hs)^6$	64	2^5	4_1
8_3	h^2s	128	2^4	4_2
8_4	h^2shs	128	2^4	4_3
8_5	$(h^2shs)^3$	128	2^4	4_3
8_6	h^3sh^2shs	128	2^4	4_2
16_1	hs	128	2^4	8_1
16_2	$(hs)^3$	128	2^4	8_2
16_3	$(hs)^5$	128	2^4	8_1
16_4	$(hs)^7$	128	2^4	8_2

10.7.4 *Conjugacy classes of* $E = \langle h, s, r \rangle$

Class	*Representative*	|*Class*|	|*Centralizer*|	2P	3P
1	1	1	$2^{11} \cdot 3$	1	1
2_1	$(hs)^8$	3	2^{11}	1	2_1
2_2	$(hr)^3$	4	$2^9 \cdot 3$	1	2_2
2_3	$(h)^2$	24	2^8	1	2_3
2_4	$(sr)^2$	24	2^8	1	2_4
2_5	h^3rshrs	24	2^8	1	2_5
2_6	s	192	2^5	1	2_6
3	r	512	$2^2 \cdot 3$	3	1
4_1	$(h^2r)^3$	32	$2^6 \cdot 3$	2_2	4_1
4_2	$(hs)^4$	48	2^7	2_1	4_2
4_3	$(h^2s)^2$	96	2^6	2_2	4_3
4_4	h	192	2^5	2_3	4_4
4_5	sr	192	2^5	2_4	4_5
4_6	hsr^2	192	2^5	2_4	4_6
4_7	$hsrsr$	192	2^5	2_3	4_7
4_8	$hrshr^2$	192	2^5	2_1	4_8
6	hr	512	$2^2 \cdot 3$	3	2_2
8_1	$(hs)^2$	192	2^5	4_2	8_2
8_2	$(hs)^6$	192	2^5	4_2	8_1
8_3	h^2s	384	2^4	4_3	8_3
8_4	h^2sr	384	2^4	4_3	8_4
12_1	h^2r	512	$2^2 \cdot 3$	6	4_1
12_2	$(h^2r)^5$	512	$2^2 \cdot 3$	6	4_1
16_1	hs	384	2^4	8_1	16_2
16_2	$(hs)^3$	384	2^4	8_2	16_1
16_3	$(hs)^5$	384	2^4	8_1	16_4
16_4	$(hs)^7$	384	2^4	8_2	16_3

10.7.5 *Conjugacy classes of $U = \langle t, g, s \rangle$*

Class	*Representative*	|*Class*|	|*Centralizer*|	2P	3P
1	1	1	$2^9 \cdot 3$	1	1
2_1	$(tg)^6$	1	$2^9 \cdot 3$	1	2_1
2_2	$(ts)^4$	3	2^9	1	2_2
2_3	$(tgsts)^2$	3	2^9	1	2_3
2_4	g	12	2^7	1	2_4
2_5	$(tgs)^2$	12	2^7	1	2_5
2_6	$tgstgsg$	12	2^7	1	2_6
2_7	$t^2gtgsgs$	12	2^7	1	2_7
2_8	$t^2sgtgts$	12	2^7	1	2_8
2_9	$t^2sgtgtgs$	12	2^7	1	2_9
2_{10}	s	48	2^5	1	2_{10}
2_{11}	t^2stg	48	2^5	1	2_{11}
3	t	128	$2^2 \cdot 3$	3	1
4_1	$(tg)^3$	8	$2^6 \cdot 3$	2_1	4_1
4_2	$(ts)^2$	24	2^6	2_2	4_2
4_3	$(gs)^2$	24	2^6	2_1	4_3
4_4	$tstsg$	24	2^6	2_2	4_4
4_5	tgs	48	2^5	2_5	4_5
4_6	$tgsg$	48	2^5	2_5	4_6
4_7	$tgtgs$	48	2^5	2_5	4_7
4_8	$tgsts$	48	2^5	2_3	4_8
4_9	t^2gtgs	48	2^5	2_3	4_9
4_{10}	$tgstsg$	48	2^5	2_2	4_{10}
4_{11}	$tgtgtgs$	48	2^5	2_3	4_{11}
4_{12}	$tgststs$	48	2^5	2_5	4_{12}
6	$(tg)^2$	128	$2^2 \cdot 3$	3	2_1
8_1	ts	96	2^4	4_2	8_1
8_2	gs	96	2^4	4_3	8_2
8_3	tsg	96	2^4	4_2	8_3
8_4	t^2gstg	96	2^4	4_3	8_4
12_1	tg	128	$2^2 \cdot 3$	6	4_1
12_2	$(tg)^5$	128	$2^2 \cdot 3$	6	4_1

10.7.6 *Conjugacy classes of $N_3 = \langle v, n, t, q_1, q_2 \rangle$*

Class	*Representative*	|*Class*|	|*Centralizer*|	2P	3P
1	1	1	$2^3 \cdot 3^3$	1	1
2_1	$(v)^2$	9	$2^3 \cdot 3$	1	2_1
2_2	n	18	$2^2 \cdot 3$	1	2_2
2_3	vn	18	$2^2 \cdot 3$	1	2_3
3_1	t	2	$2^2 \cdot 3^3$	3_1	1
3_2	q_1	12	$2 \cdot 3^2$	3_2	1
3_3	q_1q_2	12	$2 \cdot 3^2$	3_3	1
4	v	18	$2^2 \cdot 3$	2_1	4
6_1	$(vt)^2$	18	$2^2 \cdot 3$	3_1	2_1
6_2	nq_1	36	$2 \cdot 3$	3_2	2_2
6_3	vnq_1	36	$2 \cdot 3$	3_3	2_3
12_1	vt	18	$2^2 \cdot 3$	6_1	4
12_2	$(vt)^5$	18	$2^2 \cdot 3$	6_1	4

10.7.7 *Conjugacy classes of* $N_5 = \langle x, h, z, y \rangle$

Class	*Representative*	|*Class*|	|*Centralizer*|	2P	5P
1	1	1	$2^3 \cdot 5^2$	1	1
2_1	z	5	$2^3 \cdot 5$	1	2_1
2_2	h^2z	5	$2^3 \cdot 5$	1	2_2
2_3	$(h)^2$	25	2^3	1	2_3
4_1	h	25	2^3	2_3	4_1
4_2	$(h)^3$	25	2^3	2_3	4_2
4_3	hz	25	2^3	2_3	4_3
4_4	$(hz)^3$	25	2^3	2_3	4_4
5_1	x	4	$2 \cdot 5^2$	5_1	1
5_2	y	4	$2 \cdot 5^2$	5_2	1
5_3	xy	8	5^2	5_3	1
5_4	x^2y	8	5^2	5_4	1
10_1	xz	20	$2 \cdot 5$	5_1	2_1
10_2	h^2zy	20	$2 \cdot 5$	5_2	2_2

10.8 Character tables of local subgroups

10.8.1 *Character table of* $H = C_G(z)$

2	11	11	10	9	7	7	7	7	6	6	6	5	5	5	5	5	1	5	5	4	4
5	1	1	.	.	.	.	.	.	.	.	.	.	.	.	.	.	1	.	.	.	.
	1a	2a	2b	2c	2d	2e	2f	4a	4b	4c	4d	4e	4f	4g	4h	4i	5a	8a	8b	8c	8d
2P	1a	1a	1a	1a	1a	1a	1a	2a	2c	2b	2c	2d	2d	2d	2d	2c	5a	4a	4a	4b	4c
3P	1a	2a	2b	2c	2d	2e	2f	4a	4b	4c	4d	4f	4e	4h	4g	4i	5a	8b	8a	8c	8e
5P	1a	2a	2b	2c	2d	2e	2f	4a	4b	4c	4d	4e	4f	4g	4h	4i	1a	8a	8b	8c	8d
7P	1a	2a	2b	2c	2d	2e	2f	4a	4b	4c	4d	4f	4e	4h	4g	4i	5a	8b	8a	8c	8e
11P	1a	2a	2b	2c	2d	2e	2f	4a	4b	4c	4d	4f	4e	4h	4g	4i	5a	8b	8a	8c	8e
13P	1a	2a	2b	2c	2d	2e	2f	4a	4b	4c	4d	4e	4f	4g	4h	4i	5a	8a	8b	8c	8d
X.1	1	1	1	1	1	1	1	1	1	1	1	1	1	1	1	1	1	1	1	1	1
X.2	1	1	1	1	1	1	1	1	1	1	1	−1	−1	−1	−1	1	1	1	1	1	−1
X.3	1	1	1	1	−1	1	−1	1	1	−1	1	A	−A	A	−A	−1	1	−1	−1	−1	−A
X.4	1	1	1	1	−1	1	−1	1	1	−1	1	−A	A	−A	A	−1	1	−1	−1	−1	A
X.5	4	4	4	4	.	4	.	4	4	.	4	.	.	.	.	.	−1	.	.	.	.
X.6	5	5	5	5	1	−3	1	−3	1	1	1	1	1	1	1	1	.	1	1	−1	1
X.7	5	5	5	5	1	−3	1	−3	1	1	1	−1	−1	−1	−1	1	.	1	1	−1	−1
X.8	5	5	5	5	−1	−3	−1	−3	1	−1	1	−A	A	−A	A	−1	.	−1	−1	1	A
X.9	5	5	5	5	−1	−3	−1	−3	1	−1	1	A	−A	A	−A	−1	.	−1	−1	1	−A
X.10	10	10	10	10	−2	2	−2	2	−2	−2	−2	.	.	.	.	−2	.	2	2	.	.
X.11	10	10	10	10	2	2	2	2	−2	2	−2	.	.	.	.	2	.	−2	−2	.	.
X.12	10	10	−6	2	.	2	.	−2	−2	.	2	B	/B	B	/B	.	.	C	−C	.	−/B
X.13	10	10	−6	2	.	2	.	−2	−2	.	2	−B	−/B	−B	−/B	.	.	C	−C	.	/B
X.14	10	10	−6	2	.	2	.	−2	−2	.	2	/B	B	/B	B	.	.	−C	C	.	−B
X.15	10	10	−6	2	.	2	.	−2	−2	.	2	−/B	−B	−/B	−B	.	.	−C	C	.	B
X.16	10	10	−6	2	2	2	2	−2	2	2	−2	.	.	.	.	−2	.	.	.	.	.
X.17	10	10	−6	2	2	2	2	−2	2	2	−2	.	.	.	.	−2	.	.	.	.	.
X.18	10	10	−6	2	−2	2	−2	−2	2	−2	−2	.	.	.	.	2	.	.	.	.	.
X.19	10	10	−6	2	−2	2	−2	−2	2	−2	−2	.	.	.	.	2	.	.	.	.	.
X.20	16	−16	.	.	4	.	−4	.	.	.	.	−2	−2	2	2	.	1	.	.	.	.
X.21	16	−16	.	.	4	.	−4	.	.	.	.	2	2	−2	−2	.	1	.	.	.	.
X.22	16	−16	.	.	−4	.	4	.	.	.	.	C	−C	−C	C	.	1	.	.	.	.
X.23	16	−16	.	.	−4	.	4	.	.	.	.	−C	C	C	−C	.	1	.	.	.	.
X.24	20	20	−12	4	.	−4	.	4	.	.	.	.	.	.	.	.	.	.	.	−2	.
X.25	20	20	−12	4	.	−4	.	4	.	.	.	.	.	.	.	.	.	.	.	2	.
X.26	40	40	8	−8	−4	.	−4	.	.	4	.	.	.	.	.	.	.	.	.	.	.
X.27	40	40	8	−8	4	.	4	.	.	−4	.	.	.	.	.	.	.	.	.	.	.
X.28	64	−64	.	.	.	.	.	.	.	.	.	.	.	.	.	.	−1	.	.	.	.

Character table of $H = C_G(z)$(*continued*)

2	4	4	1	4	4	4	4
5	.	.	1	.	.	.	.
	8e	8f	10a	16a	16b	16c	16d
2P	4c	4b	5a	8a	8b	8a	8b
3P	8d	8f	10a	16b	16a	16d	16c
5P	8e	8f	2a	16c	16d	16a	16b
7P	8d	8f	10a	16d	16c	16b	16a
11P	8d	8f	10a	16b	16a	16d	16c
13P	8e	8f	10a	16c	16d	16a	16b
X.1	1	1	1	1	1	1	1
X.2	−1	1	1	−1	−1	−1	−1
X.3	A	−1	1	A	−A	A	−A
X.4	−A	−1	1	−A	A	−A	A
X.5	.	.	−1	.	.	.	.
X.6	1	−1	.	−1	−1	−1	−1
X.7	−1	−1	.	1	1	1	1
X.8	−A	1	.	A	−A	A	−A
X.9	A	1	.	−A	A	−A	A
X.10	.	.	.	.	.	.	.
X.11	.	.	.	.	.	.	.
X.12	−B	.	.	.	.	.	.
X.13	B	.	.	.	.	.	.
X.14	−/B	.	.	.	.	.	.
X.15	/B	.	.	.	.	.	.
X.16	.	.	.	D	−D	−D	D
X.17	.	.	.	−D	D	D	−D
X.18	.	.	.	E	E	−E	−E
X.19	.	.	.	−E	−E	E	E
X.20	.	.	−1	.	.	.	.
X.21	.	.	−1	.	.	.	.
X.22	.	.	−1	.	.	.	.
X.23	.	.	−1	.	.	.	.
X.24	.	2	.	.	.	.	.
X.25	.	−2	.	.	.	.	.
X.26	.	.	.	.	.	.	.
X.27	.	.	.	.	.	.	.
X.28	.	.	1	.	.	.	.

where $A = -i$, $B = 1 - i$, $C = -2i$, $D = \sqrt{2}$, $E = i\sqrt{2}$.

10.8.2 *Character table of* $D = N_H(A)$

2	11	11	10	9	8	8	8	7	7	7	5	7	6	6	6	5	5	5	5	5	5
	1a	2a	2b	2c	2d	2e	2f	2g	2h	2i	2j	4a	4b	4c	4d	4e	4f	4g	4h	4i	4j
2P	1a	1a	1a	1a	1a	1a	1a	1a	1a	1a	1a	2a	2c	2b	2c	2g	2g	2d	2e	2c	2g
3P	1a	2a	2b	2c	2d	2e	2f	2g	2h	2i	2j	4a	4b	4c	4d	4f	4e	4g	4h	4i	4k
5P	1a	2a	2b	2c	2d	2e	2f	2g	2h	2i	2j	4a	4b	4c	4d	4e	4f	4g	4h	4i	4j
7P	1a	2a	2b	2c	2d	2e	2f	2g	2h	2i	2j	4a	4b	4c	4d	4f	4e	4g	4h	4i	4k
11P	1a	2a	2b	2c	2d	2e	2f	2g	2h	2i	2j	4a	4b	4c	4d	4f	4e	4g	4h	4i	4k
13P	1a	2a	2b	2c	2d	2e	2f	2g	2h	2i	2j	4a	4b	4c	4d	4e	4f	4g	4h	4i	4j
X.1	1	1	1	1	1	1	1	1	1	1	1	1	1	1	1	1	1	1	1	1	1
X.2	1	1	1	1	1	1	1	1	1	1	−1	1	1	1	1	−1	−1	1	−1	1	−1
X.3	1	1	1	1	1	1	1	1	1	1	−1	1	1	1	1	1	1	1	−1	1	1
X.4	1	1	1	1	1	1	1	1	1	1	1	1	1	1	1	−1	−1	1	1	1	−1
X.5	1	1	1	1	1	1	1	−1	1	−1	−1	1	1	−1	1	A	−A	1	−1	−1	−A
X.6	1	1	1	1	1	1	1	−1	1	−1	−1	1	1	−1	1	−A	A	1	−1	−1	A
X.7	1	1	1	1	1	1	1	−1	1	−1	1	1	1	−1	1	A	−A	1	1	−1	−A
X.8	1	1	1	1	1	1	1	−1	1	−1	1	1	1	−1	1	−A	A	1	1	−1	A
X.9	2	2	2	2	2	2	2	−2	2	−2	.	2	2	−2	2	.	.	−2	.	−2	.
X.10	2	2	2	2	2	2	2	2	2	2	.	2	2	2	2	.	.	−2	.	2	.
X.11	2	2	2	2	−2	2	−2	.	2	.	.	−2	−2	.	2	B	/B	.	.	.	/B
X.12	2	2	2	2	−2	2	−2	.	2	.	.	−2	−2	.	2	−B	−/B	.	.	.	−/B
X.13	2	2	2	2	−2	2	−2	.	2	.	.	−2	−2	.	2	/B	B	.	.	.	B
X.14	2	2	2	2	−2	2	−2	.	2	.	.	−2	−2	.	2	−/B	−B	.	.	.	−B
X.15	2	2	2	2	−2	2	−2	−2	2	−2	.	−2	2	−2	−2	.	.	.	.	2	.
X.16	2	2	2	2	−2	2	−2	−2	2	−2	.	−2	2	−2	−2	.	.	.	.	2	.
X.17	2	2	2	2	−2	2	−2	2	2	2	.	−2	2	2	−2	.	.	.	.	−2	.
X.18	2	2	2	2	−2	2	−2	2	2	2	.	−2	2	2	−2	.	.	.	.	−2	.
X.19	4	4	4	4	4	4	4	.	4	.	.	4	−4	.	−4	.	.	.	.	.	.
X.20	4	4	4	4	4	4	4	.	−4	.	−2	−4	.	.	.	.	.	.	2	.	.
X.21	4	4	4	4	4	4	4	.	−4	.	2	−4	.	.	.	.	.	.	−2	.	.
X.22	4	4	4	4	−4	4	−4	.	−4	.	.	4	.	.	.	.	.	.	.	.	.
X.23	4	4	4	4	−4	4	−4	.	−4	.	.	4	.	.	.	.	.	.	.	.	.
X.24	8	8	8	8	.	−8	.	−4	.	−4	.	.	.	4	.	.	.	.	.	.	.
X.25	8	8	−8	.	4	.	−4	.	.	.	2	.	.	.	.	.	.	2	2	.	.
X.26	8	8	−8	.	4	.	−4	.	.	.	−2	.	.	.	.	.	.	2	−2	.	.
X.27	8	8	−8	.	4	.	−4	.	.	.	2	.	.	.	.	.	.	−2	−2	.	.
X.28	8	8	−8	.	4	.	−4	.	.	.	−2	.	.	.	.	.	.	−2	2	.	.
X.29	8	8	8	8	.	−8	.	4	.	4	.	.	.	−4	.	.	.	.	.	.	.
X.30	16	16	16	−16	.	.	.	.	.	.	.	.	.	.	.	.	.	.	.	.	.
X.31	16	−16	.	.	.	.	.	4	.	−4	.	.	.	.	.	−2	−2	.	.	.	2
X.32	16	−16	.	.	.	.	.	4	.	−4	.	.	.	.	.	2	2	.	.	.	−2
X.33	16	16	−16	.	−8	.	8	.	.	.	.	.	.	.	.	.	.	.	.	.	.
X.34	16	−16	.	.	.	.	.	−4	.	4	.	.	.	.	.	C	−C	.	.	.	C
X.35	16	−16	.	.	.	.	.	−4	.	4	.	.	.	.	.	−C	C	.	.	.	−C

Character table of $D = N_H(A)$ *(continued)*

2	5	5	5	5	5	5	4	4	4	4	4	4	4	4
	4k	4l	4m	4n	8a	8b	8c	8d	8e	8f	16a	16b	16c	16d
2P	2g	2d	2e	2a	4a	4a	4b	4c	4c	4b	8a	8b	8a	8b
3P	4j	4l	4m	4n	8b	8a	8c	8e	8d	8f	16b	16a	16d	16c
5P	4k	4l	4m	4n	8a	8b	8c	8d	8e	8f	16c	16d	16a	16b
7P	4j	4l	4m	4n	8b	8a	8c	8e	8d	8f	16d	16c	16b	16a
11P	4j	4l	4m	4n	8b	8a	8c	8e	8d	8f	16b	16a	16d	16c
13P	4k	4l	4m	4n	8a	8b	8c	8d	8e	8f	16c	16d	16a	16b
X.1	1	1	1	1	1	1	1	1	1	1	1	1	1	1
X.2	−1	1	−1	−1	1	1	−1	−1	−1	−1	1	1	1	1
X.3	1	1	−1	−1	1	1	−1	1	1	−1	−1	−1	−1	−1
X.4	−1	1	1	1	1	1	1	−1	−1	1	−1	−1	−1	−1
X.5	A	1	−1	−1	−1	−1	1	−A	A	1	−A	A	−A	A
X.6	−A	1	−1	−1	−1	−1	1	A	−A	1	A	−A	A	−A
X.7	A	1	1	1	−1	−1	−1	−A	A	−1	A	−A	A	−A
X.8	−A	1	1	1	−1	−1	−1	A	−A	−1	−A	A	−A	A
X.9	.	−2	.	.	2	2	.	.	.	.	.	.	.	.
X.10	.	−2	.	.	−2	−2	.	.	.	.	.	.	.	.
X.11	B	.	.	.	−C	C	.	−/B	−B	.	.	.	.	.
X.12	−B	.	.	.	−C	C	.	/B	B	.	.	.	.	.
X.13	/B	.	.	.	C	−C	.	−B	−/B	.	.	.	.	.
X.14	−/B	.	.	.	C	−C	.	B	/B	.	.	.	.	.
X.15	.	.	.	.	.	.	.	.	.	.	D	D	−D	−D
X.16	.	.	.	.	.	.	.	.	.	.	−D	−D	D	D
X.17	.	.	.	.	.	.	.	.	.	.	E	−E	−E	E
X.18	.	.	.	.	.	.	.	.	.	.	−E	E	E	−E
X.19	.	.	.	.	.	.	.	.	.	.	.	.	.	.
X.20	.	.	2	−2	.	.	.	.	.	.	.	.	.	.
X.21	.	.	−2	2	.	.	.	.	.	.	.	.	.	.
X.22	.	.	.	.	.	.	2	.	.	−2	.	.	.	.
X.23	.	.	.	.	.	.	−2	.	.	2	.	.	.	.
X.24	.	.	.	.	.	.	.	.	.	.	.	.	.	.
X.25	.	−2	−2	−2	.	.	.	.	.	.	.	.	.	.
X.26	.	−2	2	2	.	.	.	.	.	.	.	.	.	.
X.27	.	2	2	−2	.	.	.	.	.	.	.	.	.	.
X.28	.	2	−2	2	.	.	.	.	.	.	.	.	.	.
X.29	.	.	.	.	.	.	.	.	.	.	.	.	.	.
X.30	.	.	.	.	.	.	.	.	.	.	.	.	.	.
X.31	2	.	.	.	.	.	.	.	.	.	.	.	.	.
X.32	−2	.	.	.	.	.	.	.	.	.	.	.	.	.
X.33	.	.	.	.	.	.	.	.	.	.	.	.	.	.
X.34	−C	.	.	.	.	.	.	.	.	.	.	.	.	.
X.35	C	.	.	.	.	.	.	.	.	.	.	.	.	.

where $A = i$, $B = 1 - i$, $C = 2i$, $D = i\sqrt{2}$, $E = -\sqrt{2}$.

10.8.3 *Character table of $E = N_G(A)$*

	1a	2a	2b	2c	2d	2e	2f	3a	4a	4b	4c	4d	4e	4f	4g	4h	6a	8a	8b	8c	8d	12a
2	11	11	9	8	8	8	5	2	6	7	6	5	5	5	5	5	2	5	5	4	4	2
3	1	.	1	.	.	.	.	1	1	.	.	.	.	.	.	.	1	.	.	.	.	1
2P	1a	1a	1a	1a	1a	1a	1a	3a	2b	2a	2b	2c	2d	2d	2c	2a	3a	4b	4b	4c	4c	6a
3P	1a	2a	2b	2c	2d	2e	2f	1a	4a	4b	4c	4d	4e	4f	4g	4h	2b	8b	8a	8c	8d	4a
5P	1a	2a	2b	2c	2d	2e	2f	3a	4a	4b	4c	4d	4e	4f	4g	4h	6a	8a	8b	8c	8d	12b
7P	1a	2a	2b	2c	2d	2e	2f	3a	4a	4b	4c	4d	4e	4f	4g	4h	6a	8b	8a	8c	8d	12b
11P	1a	2a	2b	2c	2d	2e	2f	3a	4a	4b	4c	4d	4e	4f	4g	4h	6a	8b	8a	8c	8d	12a
13P	1a	2a	2b	2c	2d	2e	2f	3a	4a	4b	4c	4d	4e	4f	4g	4h	6a	8a	8b	8c	8d	12a
X.1	1	1	1	1	1	1	1	1	1	1	1	1	1	1	1	1	1	1	1	1	1	1
X.2	1	1	1	1	1	1	−1	1	1	1	1	1	−1	−1	1	−1	1	1	1	−1	−1	1
X.3	2	2	2	2	2	2	.	−1	2	2	2	2	.	.	2	.	−1	2	2	.	.	−1
X.4	2	2	2	−2	2	−2	.	−1	−2	−2	2	.	.	.	.	.	−1	.	.	.	.	1
X.5	2	2	2	−2	2	−2	.	−1	−2	−2	2	.	.	.	.	.	−1	.	.	.	.	1
X.6	3	3	3	3	3	3	−1	.	3	3	3	−1	−1	−1	−1	−1	.	−1	−1	−1	−1	.
X.7	3	3	3	3	3	3	1	.	3	3	3	−1	1	1	−1	1	.	−1	−1	1	1	.
X.8	3	3	3	−1	3	−1	−1	.	3	−1	−1	1	−1	−1	1	−1	.	A	/A	1	1	.
X.9	3	3	3	−1	3	−1	1	.	3	−1	−1	1	1	1	1	1	.	A	/A	−1	−1	.
X.10	3	3	3	−1	3	−1	−1	.	3	−1	−1	1	−1	−1	1	−1	.	/A	A	1	1	.
X.11	3	3	3	−1	3	−1	1	.	3	−1	−1	1	1	1	1	1	.	/A	A	−1	−1	.
X.12	4	4	4	−4	4	−4	.	1	−4	−4	4	.	.	.	.	.	1	.	.	.	.	−1
X.13	6	6	6	−2	6	−2	.	.	6	−2	−2	−2	.	.	−2	.	.	2	2	.	.	.
X.14	6	6	6	2	6	2	.	.	−6	2	−2	.	.	.	.	.	.	.	.	.	.	.
X.15	6	6	6	2	6	2	.	.	−6	2	−2	.	.	.	.	.	.	.	.	.	.	.
X.16	12	12	12	−4	−4	−4	.	.	.	4	.	.	.	.	.	.	.	.	.	2	−2	.
X.17	12	12	12	−4	−4	−4	.	.	.	4	.	.	.	.	.	.	.	.	.	−2	2	.
X.18	12	12	12	4	−4	4	2	.	.	−4	.	.	−2	−2	.	2	.	.	.	.	.	.
X.19	12	12	12	4	−4	4	−2	.	.	−4	.	.	2	2	.	−2	.	.	.	.	.	.
X.20	16	16	−16	.	.	.	.	−2	.	.	.	.	.	.	.	.	2	.	.	.	.	.
X.21	16	16	−16	.	.	.	.	1	.	.	.	.	.	.	.	.	−1	.	.	.	.	B
X.22	16	16	−16	.	.	.	.	1	.	.	.	.	.	.	.	.	−1	.	.	.	.	−B
X.23	24	−8	.	4	.	−4	2	.	.	.	.	−2	2	−2	2	−2	.	.	.	.	.	.
X.24	24	−8	.	4	.	−4	−2	.	.	.	.	−2	−2	2	2	2	.	.	.	.	.	.
X.25	24	−8	.	4	.	−4	2	.	.	.	.	2	−2	2	−2	−2	.	.	.	.	.	.
X.26	24	−8	.	4	.	−4	−2	.	.	.	.	2	2	−2	−2	2	.	.	.	.	.	.
X.27	48	−16	.	−8	.	8	.	.	.	.	.	.	.	.	.	.	.	.	.	.	.	.

Character table of $E = N_G(A)$(continued)

	12b	16a	16b	16c	16d
2	2	4	4	4	4
3	1	.	.	.	.
2P	6a	8a	8b	8a	8b
3P	4a	16b	16a	16d	16c
5P	12a	16c	16d	16a	16b
7P	12a	16d	16c	16b	16a
11P	12b	16b	16a	16d	16c
13P	12b	16c	16d	16a	16b
X.1	1	1	1	1	1
X.2	1	−1	−1	−1	−1
X.3	−1	.	.	.	.
X.4	1	C	C	−C	−C
X.5	1	−C	−C	C	C
X.6	.	1	1	1	1
X.7	.	−1	−1	−1	−1
X.8	.	D	−D	D	−D
X.9	.	−D	D	−D	D
X.10	.	−D	D	−D	D
X.11	.	D	−D	D	−D
X.12	−1	.	.	.	.
X.13	.	.	.	.	.
X.14	.	E	−E	−E	E
X.15	.	−E	E	E	−E
X.16	.	.	.	.	.
X.17	.	.	.	.	.
X.18	.	.	.	.	.
X.19	.	.	.	.	.
X.20	.	.	.	.	.
X.21	−B	.	.	.	.
X.22	B	.	.	.	.
X.23	.	.	.	.	.
X.24	.	.	.	.	.
X.25	.	.	.	.	.
X.26	.	.	.	.	.
X.27	.	.	.	.	.

where A $= -1 - 2i$, B $= \sqrt{3}$, C $= -i\sqrt{2}$, D $= -i$, $E = \sqrt{2}$.

10.8.4 *Character table of* $U = C_G(u)$

2	9	9	9	9	7	7	7	7	7	7	5	5	2	6	6	6	6	5	5	5	5	5	5
3	1	1	.	.	.	.	.	.	.	.	.	.	1	1	.	.	.	.	.	.	.	.	.
	1a	2a	2b	2c	2d	2e	2f	2g	2h	2i	2j	2k	3a	4a	4b	4c	4d	4e	4f	4g	4h	4i	4j
2P	1a	1a	1a	1a	1a	1a	1a	1a	1a	1a	1a	1a	3a	2a	2b	2a	2b	2c	2c	2i	2b	2c	2i
3P	1a	2a	2b	2c	2d	2e	2f	2g	2h	2i	2j	2k	1a	4a	4b	4c	4d	4e	4f	4g	4h	4i	4j
5P	1a	2a	2b	2c	2d	2e	2f	2g	2h	2i	2j	2k	3a	4a	4b	4c	4d	4e	4f	4g	4h	4i	4j
7P	1a	2a	2b	2c	2d	2e	2f	2g	2h	2i	2j	2k	3a	4a	4b	4c	4d	4e	4f	4g	4h	4i	4j
11P	1a	2a	2b	2c	2d	2e	2f	2g	2h	2i	2j	2k	3a	4a	4b	4c	4d	4e	4f	4g	4h	4i	4j
X.1	1	1	1	1	1	1	1	1	1	1	1	1	1	1	1	1	1	1	1	1	1	1	1
X.2	1	1	1	1	−1	−1	1	−1	−1	1	−1	1	1	−1	1	1	−1	−1	−1	−1	1	1	1
X.3	1	1	1	1	−1	−1	1	−1	−1	1	1	−1	1	−1	1	1	−1	−1	1	1	1	−1	−1
X.4	1	1	1	1	1	1	1	1	1	1	−1	−1	1	1	1	1	1	1	−1	−1	1	−1	−1
X.5	2	2	2	2	2	2	2	2	2	2	.	.	−1	2	2	2	2	2	.	.	2	.	.
X.6	2	2	2	2	−2	−2	2	−2	−2	2	.	.	−1	−2	2	2	−2	−2	.	.	2	.	.
X.7	3	3	3	3	1	−3	−1	1	−3	−1	1	−1	.	−3	−1	3	1	1	1	−1	−1	−1	1
X.8	3	3	3	3	1	−3	−1	1	−3	−1	−1	1	.	−3	−1	3	1	1	−1	1	−1	1	−1
X.9	3	3	3	3	1	1	3	1	1	3	−1	1	.	−3	−1	−1	−3	1	−1	−1	−1	1	1
X.10	3	3	3	3	1	1	3	1	1	3	1	−1	.	−3	−1	−1	−3	1	1	1	−1	−1	−1
X.11	3	3	3	3	−3	1	−1	−3	1	−1	1	−1	.	−3	3	−1	1	1	1	−1	−1	−1	1
X.12	3	3	3	3	−3	1	−1	−3	1	−1	−1	1	.	−3	3	−1	1	1	−1	1	−1	1	−1
X.13	3	3	3	3	3	−1	−1	3	−1	−1	1	1	.	3	3	−1	−1	−1	1	−1	−1	1	−1
X.14	3	3	3	3	3	−1	−1	3	−1	−1	−1	−1	.	3	3	−1	−1	−1	−1	1	−1	−1	1
X.15	3	3	3	3	−1	3	−1	−1	3	−1	1	1	.	3	−1	3	−1	−1	1	−1	−1	1	−1
X.16	3	3	3	3	−1	3	−1	−1	3	−1	−1	−1	.	3	−1	3	−1	−1	−1	1	−1	−1	1
X.17	3	3	3	3	−1	−1	3	−1	−1	3	−1	−1	.	3	−1	−1	3	−1	−1	−1	−1	−1	−1
X.18	3	3	3	3	−1	−1	3	−1	−1	3	1	1	.	3	−1	−1	3	−1	1	1	−1	1	1
X.19	4	−4	−4	4	.	.	−4	.	.	4	.	.	−2	.	.	.	.	.	.	.	.	.	.
X.20	4	−4	−4	4	.	.	−4	.	.	4	.	.	1	.	.	.	.	.	.	.	.	.	.
X.21	4	−4	−4	4	.	.	−4	.	.	4	.	.	1	.	.	.	.	.	.	.	.	.	.
X.22	6	6	6	6	2	2	−2	2	2	−2	.	.	.	−6	−2	−2	2	−2	.	.	2	.	.
X.23	6	6	6	6	−2	−2	−2	−2	−2	−2	.	.	.	6	−2	−2	−2	2	.	.	2	.	.
X.24	12	12	−4	−4	.	4	.	.	−4	.	−2	−2	.	.	.	.	.	.	2	.	.	2	.
X.25	12	−12	−12	12	.	.	4	.	.	−4	.	.	.	.	.	.	.	.	.	.	.	.	.
X.26	12	−12	4	−4	4	.	.	−4	.	.	.	.	.	.	.	.	.	.	.	−2	.	.	−2
X.27	12	−12	4	−4	−4	.	.	4	.	.	.	.	.	.	.	.	.	.	.	−2	.	.	2
X.28	12	12	−4	−4	.	−4	.	.	4	.	−2	2	.	.	.	.	.	.	2	.	.	−2	.
X.29	12	12	−4	−4	.	4	.	.	−4	.	2	2	.	.	.	.	.	.	−2	.	.	−2	.
X.30	12	−12	4	−4	4	.	.	−4	.	.	.	.	.	.	.	.	.	.	.	2	.	.	2
X.31	12	−12	4	−4	−4	.	.	4	.	.	.	.	.	.	.	.	.	.	.	2	.	.	−2
X.32	12	12	−4	−4	.	−4	.	.	4	.	2	−2	.	.	.	.	.	.	−2	.	.	2	.

Character table of
$U = C_G(u)$ *(continued)*

2	2	4	4	4	4				
3	1	.	.	.	.				
	4k	4l	6a	8a	8b	8c	8d	12a	12b
2P	2i	2i	3a	4b	4c	4b	4c	6a	6a
3P	4k	4l	2a	8a	8b	8c	8d	4a	4a
5P	4k	4l	6a	8a	8b	8c	8d	12b	12a
7P	4k	4l	6a	8a	8b	8c	8d	12b	12a
11P	4k	4l	6a	8a	8b	8c	8d	12a	12b
X.1	1	1	1	1	1	1	1	1	1
X.2	1	−1	1	−1	−1	1	1	−1	−1
X.3	−1	1	1	1	1	−1	−1	−1	−1
X.4	−1	−1	1	−1	−1	−1	−1	1	1
X.5	.	.	−1	.	.	.	.	−1	−1
X.6	.	.	−1	.	.	.	.	1	1
X.7	1	−1	.	−1	1	1	−1	.	.
X.8	−1	1	.	1	−1	−1	1	.	.
X.9	1	−1	.	1	1	−1	−1	.	.
X.10	−1	1	.	−1	−1	1	1	.	.
X.11	1	−1	.	1	−1	−1	1	.	.
X.12	−1	1	.	−1	1	1	−1	.	.
X.13	−1	−1	.	1	−1	1	−1	.	.
X.14	1	1	.	−1	1	−1	1	.	.
X.15	−1	−1	.	−1	1	−1	1	.	.
X.16	1	1	.	1	−1	1	−1	.	.
X.17	−1	−1	.	1	1	1	1	.	.
X.18	1	1	.	−1	−1	−1	−1	.	.
X.19	.	.	2	.	.	.	.	.	.
X.20	.	.	−1	.	.	.	.	A	−A
X.21	.	.	−1	.	.	.	.	−A	A
X.22	.	.	.	.	.	.	.	.	.
X.23	.	.	.	.	.	.	.	.	.
X.24	.	.	.	.	.	.	.	.	.
X.25	.	.	.	.	.	.	.	.	.
X.26	2	2	.	.	.	.	.	.	.
X.27	−2	2	.	.	.	.	.	.	.
X.28	.	.	.	.	.	.	.	.	.
X.29	.	.	.	.	.	.	.	.	.
X.30	−2	−2	.	.	.	.	.	.	.
X.31	2	−2	.	.	.	.	.	.	.
X.32	.	.	.	.	.	.	.	.	.

where $A = -\sqrt{3}$.

10.8.5 *Character table of* N_3

2	3	3	2	2	2	1	1	2	2	1	1	2	2
3	3	1	1	1	3	2	2	1	1	1	1	1	1
	1a	2a	2b	2c	3a	3b	3c	4a	6a	6b	6c	12a	12b
2P	1a	1a	1a	1a	3a	3b	3c	2a	3a	3c	3b	6a	6a
3P	1a	2a	2b	2c	1a	1a	1a	4a	2a	2c	2b	4a	4a
5P	1a	2a	2b	2c	3a	3b	3c	4a	6a	6b	6c	12b	12a
7P	1a	2a	2b	2c	3a	3b	3c	4a	6a	6b	6c	12b	12a
11P	1a	2a	2b	2c	3a	3b	3c	4a	6a	6b	6c	12a	12b
X.1	1	1	1	1	1	1	1	1	1	1	1	1	1
X.2	1	1	1	−1	1	1	1	−1	1	−1	1	−1	−1
X.3	1	1	−1	1	1	1	1	−1	1	1	−1	−1	−1
X.4	1	1	−1	−1	1	1	1	1	1	−1	−1	1	1
X.5	2	−2	.	.	2	2	2	.	−2	.	.	.	.
X.6	4	.	.	2	4	−2	1	.	.	−1	.	.	.
X.7	4	.	.	−2	4	−2	1	.	.	1	.	.	.
X.8	4	.	2	.	4	1	−2	.	.	.	−1	.	.
X.9	4	.	−2	.	4	1	−2	.	.	.	1	.	.
X.10	6	2	.	.	−3	.	.	.	−1	.	.	A	−A
X.11	6	2	.	.	−3	.	.	.	−1	.	.	−A	A
X.12	6	−2	.	.	−3	.	.	−2	1	.	.	1	1
X.13	6	−2	.	.	−3	.	.	2	1	.	.	−1	−1

where $A = \sqrt{3}$.

10.8.6 *Character table of N_5*

2	3	3	3	3	3	3	3	3	1	1	.	.	1	1
5	2	1	1	.	.	.	.	.	2	2	2	2	1	1
	1a	2a	2b	2c	4a	4b	4c	4d	5a	5b	5c	5d	10a	10b
2P	1a	1a	1a	1a	2c	2c	2c	2c	5a	5b	5c	5d	5a	5b
3P	1a	2a	2b	2c	4d	4c	4b	4a	5a	5b	5c	5d	10a	10b
5P	1a	2a	2b	2c	4a	4b	4c	4d	1a	1a	1a	1a	2a	2b
7P	1a	2a	2b	2c	4d	4c	4b	4a	5a	5b	5c	5d	10a	10b
X.1	1	1	1	1	1	1	1	1	1	1	1	1	1	1
X.2	1	−1	−1	1	−1	1	1	−1	1	1	1	1	−1	−1
X.3	1	1	−1	−1	i	i	−i	−i	1	1	1	1	1	−1
X.4	1	−1	1	−1	−i	i	−i	i	1	1	1	1	−1	1
X.5	1	1	1	1	−1	−1	−1	−1	1	1	1	1	1	1
X.6	1	−1	−1	1	1	−1	−1	1	1	1	1	1	−1	−1
X.7	1	1	−1	−1	−i	−i	i	i	1	1	1	1	1	−1
X.8	1	−1	1	−1	i	−i	i	−i	1	1	1	1	−1	1
X.9	4	4	.	.	.	.	.	.	−1	4	−1	−1	−1	.
X.10	4	−4	.	.	.	.	.	.	−1	4	−1	−1	1	.
X.11	4	.	4	.	.	.	.	.	4	−1	−1	−1	.	−1
X.12	4	.	−4	.	.	.	.	.	4	−1	−1	−1	.	1
X.13	8	.	.	.	.	.	.	.	−2	−2	−2	3	.	.
X.14	8	.	.	.	.	.	.	.	−2	−2	3	−2	.	.

10.8.7 *Character table of $NS_5 = N_G(S_5)$*

2	4	4	3	2	2	4	4	3	1	2	2	1	2	2	2	2
3	1	1	.	1	1	1	1	.	.	1	1	.	1	1	1	1
5	2	.	1	.	.	.	.	.	2	.	.	1	.	.	.	.
	1a	2a	2b	3a	3b	4a	4b	4c	5a	6a	6b	10a	12a	12b	12c	12d
2P	1a	1a	1a	3b	3a	2a	2a	2a	5a	3a	3b	5a	6a	6b	6a	6b
3P	1a	2a	2b	1a	1a	4b	4a	4c	5a	2a	2a	10a	4b	4b	4a	4a
5P	1a	2a	2b	3b	3a	4a	4b	4c	1a	6b	6a	2b	12b	12a	12d	12c
7P	1a	2a	2b	3a	3b	4b	4a	4c	5a	6a	6b	10a	12c	12d	12a	12b
11P	1a	2a	2b	3b	3a	4b	4a	4c	5a	6b	6a	10a	12d	12c	12b	12a
X.1	1	1	1	1	1	1	1	1	1	1	1	1	1	1	1	1
X.2	1	1	−1	1	1	−1	−1	1	1	1	1	−1	−1	−1	−1	−1
X.3	1	1	1	A	/A	1	1	1	1	/A	A	1	A	/A	A	/A
X.4	1	1	1	/A	A	1	1	1	1	A	/A	1	/A	A	/A	A
X.5	1	1	−1	/A	A	−1	−1	1	1	A	/A	−1	−/A	−A	−/A	−A
X.6	1	1	−1	A	/A	−1	−1	1	1	/A	A	−1	−A	−/A	−A	−/A
X.7	2	−2	.	−1	−1	B	−B	.	2	1	1	.	C	C	−C	−C
X.8	2	−2	.	−1	−1	−B	B	.	2	1	1	.	−C	−C	C	C
X.9	2	−2	.	−/A	−A	B	−B	.	2	A	/A	.	D	−/D	−D	/D
X.10	2	−2	.	−A	−/A	B	−B	.	2	/A	A	.	−/D	D	/D	−D
X.11	2	−2	.	−A	−/A	−B	B	.	2	/A	A	.	/D	−D	−/D	D
X.12	2	−2	.	−/A	−A	−B	B	.	2	A	/A	.	−D	/D	D	−/D
X.13	3	3	1	.	.	−3	−3	−1	3	.	.	1	.	.	.	.
X.14	3	3	−1	.	.	3	3	−1	3	.	.	−1	.	.	.	.
X.15	24	.	−4	.	.	.	.	.	−1	.	.	1	.	.	.	.
X.16	24	.	4	.	.	.	.	.	−1	.	.	−1	.	.	.	.

where $A = -1/2(3 + i\sqrt{3}), B = -2i, C = i, D = E(12)^7$.

11
McLaughlin's group McL

In 1969 J. McLaughlin constructed his sporadic simple group McL as a primitive rank 3 permutation group; see [89]. According to R. Lyons [88], J. G. Thompson determined the character table of the simple group McL. In particular, he proved that $G = \mathsf{McL}$ has only one conjugacy class of involutions z. Its centralizer $C_G(z)$ is isomorphic to the 2-fold cover $H = 2\mathsf{A}_8$. In [70] Janko and Wong showed that each finite simple group G with a centralizer $C_G(z) \cong H = 2\mathsf{A}_8$ of a 2-central involution z is isomorphic to McLaughlin's simple group McL. Their uniqueness proof establishes the existence of a maximal subgroup $U \cong \mathrm{U}_4(3)$ of G such that the permutation representation $(1_U)^G$ is a rank 3 permutation $\mathbb{C}$-module for G. This argument is involved and does not generalize. Therefore Kratzer, Lempken, Waki and the author gave a more elementary existence and uniqueness proof for McLaughlin's group McL in [80], building on some general ideas of Janko and Wong. It is simplified in this chapter.

In Section 12.2 we apply Algorithm 1.3.8 to the unique conjugacy class of irreducible subgroups T of $\mathrm{GL}_4(2)$ which are isomorphic to the alternating group A_7 and construct a finitely presented group H with center $Z(H) = \langle z \rangle$ of order 2 such that $H/Z(H) \cong \mathsf{A}_8$ and all conditions of Steps 4 and 5 of Algorithm 1.3.8 are satisfied. In Proposition 11.2.1 we show that a Sylow 2-subgroup S of H has a unique normal Klein 4-group V, and that V is the intersection of the two maximal elementary abelian normal subgroups A and B of S. Furthermore, $C = C_G(V) = CH(V) = AB$ for all simple groups G with $C_G(z) = H$.

In Section 12.3 we apply Glauberman's Z^*-theorem and prove that $N = N_G(V)$ is a uniquely determined finitely presented group containing an element $x \notin N_H(V)$ such that $N_G(V) = \langle N_H(V), x \rangle$. Since $C_S(V)$ is a normal subgroup of N, such that A and B are still the only maximal elementary abelian normal subgroups of $C_S(V)$, it follows that

$N_G(A) = \langle N_H(A), x\rangle$ and $N_G(B) = \langle N_H(B), x\rangle$. Both normal subgroups A and B have complements T_1 and T_2 in $N_G(A)$ and $N_G(B)$, respectively. From the presentations of N, $N_H(A)$ and $N_H(B)$ we derive uniquely determined presentations of the complements T_1 and T_2. Thus we are able to get a set $\mathcal{R}(G)$ of defining relations of $G = \langle H, x\rangle$ consisting of $\mathcal{R}(H)$ and some well determined additional relations; see Theorem 11.3.1. Note that G has a faithful permutation representation of degree 22275 with stabilizer H. We also calculated a system of representatives of the conjugacy classes of G. The group G is simple because it has the same character table as McL in the Atlas [19]. Thus Theorem 11.3.1 provides an existence and uniqueness proof for McLaughlin's simple group McL.

In Corollary 11.3.2 we also show that G is a primitive transitive rank 3 extension of the simple unitary group $Q \cong U_4(3)$. It also gives a 22-dimensional 11-modular irreducible representation of G. Its generating matrices are documented on the accompanying DVD.

11.1 Construction of the 2-central involution centralizer

In this section we apply Algorithm 1.3.8 to an irreducible subgroup U of $\mathrm{GL}_4(2)$ which is isomorphic to the alternating group A_7 and construct a presentation of a group H with center $Z(H)$ of order 2 which is isomorphic to the centralizer $C_G(z) \cong 2\mathsf{A}_8$ of a 2-central involution z of the simple McLaughlin group $G = \mathsf{McL}$.

Lemma 11.1.1 *Let U be the subgroup of $\mathrm{GL}_4(2)$ generated by the following matrices:*

$$a_1 = \begin{pmatrix} 1&0&1&0\\0&1&0&0\\0&0&1&0\\0&1&0&1 \end{pmatrix}, \quad a_2 = \begin{pmatrix} 1&0&0&0\\0&1&0&1\\1&0&1&0\\0&0&0&1 \end{pmatrix}, \quad a_3 = \begin{pmatrix} 0&1&1&0\\1&1&0&0\\0&0&1&0\\0&0&0&1 \end{pmatrix}, \quad a_4 = \begin{pmatrix} 0&1&0&1\\1&1&0&1\\0&0&1&0\\0&0&0&1 \end{pmatrix},$$

of orders 2, 2, 3 and 3, respectively. Then the following assertions hold.

(a) *The natural 4-dimensional vector space V over $F = GF(2)$ is an irreducible FU-module.*

(b) *The second cohomological dimension $dim_F H^2(U, V) = 0$.*

(c) *The split extension $E = \langle e_i \mid 1 \le i \le 8\rangle$ of U by V has the following set $\mathcal{R}(E)$ of defining relations:*

$$
\begin{aligned}
&e_1^2 = e_2^2 = e_3^3 = e_4^3 = e_5^2 = e_6^2 = e_7^2 = e_8^2 = 1,\\
&(e_1, e_6) = (e_1, e_7) = (e_2, e_5) = (e_2, e_8) = (e_3, e_7) = 1,\\
&(e_3, e_8) = (e_4, e_7) = (e_4, e_8) = (e_5, e_6) = (e_5, e_7) = 1,\\
&(e_6, e_7) = (e_5, e_8) = (e_6, e_8) = (e_7, e_8) = 1,\\
&e_1^{-1}e_5e_1e_5^{-1}e_7^{-1} = e_1^{-1}e_8e_1e_6^{-1}e_8^{-1} = e_2^{-1}e_6e_2e_6^{-1}e_8^{-1} = 1,\\
&e_2^{-1}e_7e_2e_5^{-1}e_7^{-1} = e_3^{-1}e_5e_3e_6^{-1}e_7^{-1} = e_3^{-1}e_6e_3e_5^{-1}e_6^{-1} = 1,\\
&e_4^{-1}e_5e_4e_6^{-1}e_8^{-1} = e_4^{-1}e_6e_4e_5^{-1}e_6^{-1}e_8^{-1} = 1,\\
&(e_3e_4^{-1})^2 = 1, \quad (e_2e_1)^3 = 1, \quad (e_3^{-1}e_1)^3 = 1,\\
&(e_4^{-1}e_2)^3 = 1, \quad (e_2e_3^{-1})^4 = 1, \quad (e_1e_4e_2e_3^{-1})^2 = 1,\\
&e_3^{-1}e_4^{-1}e_2e_3^{-1}e_2e_4^{-1}e_3^{-1}e_2 = e_4^{-1}e_1e_4^{-1}e_3^{-1}e_1e_4^{-1}e_1e_3^{-1} = 1.
\end{aligned}
$$

Furthermore, E has a faithful permutation representation PE of degree 16 with stabilizer $\langle e_1, e_2, e_3, e_4\rangle$.

(d) *Representatives t_i and the corresponding centralizer orders $|C_E(t_i)|$ of the 15 conjugacy classes t_i^E of E are given in Table 11.4.1.*

(e) *The character table of E is given in Table 11.5.1.*

(f) *$z = e_5$ is a 2-central involution of E and $D = C_E(z) = \langle d_1, d_2, d_3\rangle$ is an indecomposable subgroup of $\mathrm{GL}_4(2)$, where $d_1 = (e_1e_5)^2$, $d_2 = (e_1e_4e_2)^2$ and $d_3 = e_1e_3^2e_4$.*

(g) *$s_1 = (d_1d_2)^2$, $s_2 = d_2d_3^2d_2d_3d_2$ and $s_3 = (d_2d_3^2d_2d_3d_2d_3)^2$ generate a Sylow 2-subgroup S of D, and $V = \langle e_5, e_6, e_7, e_8\rangle$ is a maximal elementary abelian normal subgroup of the Sylow 2-subgroup S of E.*

(h) *The 2-fold cover $2\mathsf{A}_8$ of the alternating group A_8 is isomorphic to the finitely presented group $H = \langle h_1, h_2, h_3\rangle$ with the following set $\mathcal{R}(H)$ of defining relations:*

$$
\begin{aligned}
&h_1^6 = h_2^6 = h_3^2 = 1,\\
&(h_1, h_3) = (h_2, h_3) = h_2^3h_3 = 1,\\
&(h_2^{-1}h_1^{-1})^7 = h_1h_2h_1^{-1}h_2^{-1}h_1h_2h_1^{-1}h_2^{-1}h_3 = 1,\\
&h_1h_2^{-1}h_1^{-2}h_2^{-1}h_1^2h_2^{-1}h_1^{-2}h_2^{-1}h_1h_3 = 1.
\end{aligned}
$$

(i) *$S_1 = \langle t_1, t_2, t_3\rangle$ is a Sylow 2-subgroup of H, where $t_1 = (h_2h_1h_2^2h_1)^2$, $t_2 = h_1h_2h_1h_2^2h_1^2$ and $t_3 = (h_1h_2h_1h_2h_1^2h_2^2)^3$.*

(j) $A = \langle t_3, (t_2t_3)^4, (t_2^2t_3)^2, t_2^3t_3t_2\rangle$ *is a maximal elementary abelian normal subgroup of* S_1 *such that* $D_1 = N_H(A) = \langle n_1, n_2\rangle$ *is isomorphic to* $D = C_E(z)$*, where* $n_1 = h_1h_2^2$ *and* $n_2 = h_1h_2h_1h_2h_1$ *have orders* 7 *and* 6*, respectively.*

(k) H *has a faithful permutation representation* PH *of degree* 240 *with stabilizer* $\langle n_1, (n_2^3n_1n_2)^2\rangle$.

(l) $D_1 \cong D$ *has trivial outer automorphism group.*

In particular, the amalgam $H \leftarrow D \rightarrow E$ *has Goldschmidt index* 1.

Proof (a) Using the presentation of $T = \mathrm{GL}_4(2)$ given in Lemma 3.1.1(a), its permutation representation PT of degree 15 and the MAGMA command `Subgroups(G: OrderEqual :=2520)` the reader can verify by means of a MAGMA calculation that T has a unique conjugacy class of subgroups U of order 2520. Checking the composition factors of such a representative U one finds that $U \cong A_7$. The four generating matrices a_i of U were obtained by means of the command `GetShortGens(T,U)`. Applying then the command `IsIrreducible(U)` MAGMA establishes that the natural vector space $V = 2^4$ is an irreducible FU-module over the prime field $F = GF(2)$.

(b) Let PU be the restricted permutation representation of T to U. The following set $\mathcal{R}(U)$ of defining relations of $U = \langle a_1, a_2, a_3, a_4\rangle$ was found by means of the MAGMA command `U:=FPGroup(PU)`:

$$a_1^2 = a_2^2 = a_3^3 = a_4^3 = 1,$$
$$(a_3a_4^{-1})^2 = (a_2a_1)^3 = (a_3^{-1}a_1)^3 = (a_4^{-1}a_2)^3 = (a_2a_3^{-1})^4 = 1,$$
$$a_3^{-1}a_4^{-1}a_2a_3^{-1}a_2a_4^{-1}a_3^{-1}a_2 = (a_1a_4a_2a_3^{-1})^2 = 1,$$
$$a_4^{-1}a_1a_4^{-1}a_3^{-1}a_1a_4^{-1}a_1a_3^{-1} = 1.$$

The commands `FEalg:=MatrixAlgebra<FiniteField(2),4,a1,a2, a3,a4>` and `nmodQ:=GModule(PU,FEalg)` describe in MAGMA the 4-dimensional FU-module V of the finitely presented group U. By means of the command `CohomologicalDimension(PU,nmodQ,2)` MAGMA calculates the second cohomological dimension of U with coefficients in the FU-module V. It is zero.

(c) Thus U has only the split extension E by V. The presentation of E given in the statement has been calculated by means of the commands `P:=ExtensionProcess(PU,nmodQ,U)` and `E:=Extension(P,[])`. Since the first four generators describe the complement U of V in E it is clear that E has a 16-dimensional faithful permutation representation PE.

(d) Using the faithful permutation PE and Kratzer's Algorithm 5.3.18 of [92] the representatives of the conjugacy classes of E were calculated; see Table 11.4.1.

(e) The character table of E was calculated by means of the permutation representation PE and MAGMA.

(f) By Table 11.4.1 e_5 is the unique 2-central involution of E up to conjugation. The three generators d_i of $D = C_E(e_5)$ were obtained by means of the command `GetShortGens(E,D)`. Let Md_i be the matrix of d_i w.r.t. the basis $\{e_5, e_6, e_7, e_8\}$ of $V = \langle e_j | 5 \leq j \leq 8 \rangle$, and let MD be the subgroup of $\mathrm{GL}_4(2)$ generated by them. MAGMA denied the commands `IsIrreducible(MD)` and `IsDecomposable(MD)`. Thus V is an indecomposable and non-irreducible FD-module.

(g) The Sylow 2-subgroup S was fixed by means of the MAGMA command `SylowSubgroup(D,2)`. Its generators were obtained by means of the program `GetShortGens(D,S)`. Clearly, V is a normal elementary abelian subgroup of S. Using the command

```
Subgroups(S : Al:=Normal, IsElementaryAbelian := true)
```

the reader may verify that S has exactly two maximal elementary abelian normal subgroups, both of order 16. Hence V is one of them.

(h) In order to find an over-group H of D with odd index $|H : D| \neq 1$ the author checked that S is isomorphic to a Sylow 2-subgroup of the covering group $2\mathsf{A}_8$ of the alternating group A_8. MAGMA provides the following set $\mathcal{R}(\mathsf{A}_8)$ of defining relations of A_8:

$$b_1^6 = b_2^3 = 1, \quad (b_1^5, b_2^2)^2 = 1,$$
$$(b_1 b_2^{-1} b_1^{-2} b_2^{-1} b_1)^2 = 1, \quad (b_2^{-1} b_1^5)^7 = 1.$$

Let $P\mathsf{A}_8$ be the permutation representation of A_8 with stabilizer $\langle b_2 \rangle$. Using then the command `pCover(PA8,A8,2)` MAGMA calculates the 2-fold cover $H = 2\mathsf{A}_8$ as a finitely presented group $H = \langle h_1, h_2, h_3 \rangle$ with the set $\mathcal{R}(H)$ of defining relations given in the statement.

(i) The Sylow 2-subgroup S_1 and its given generators were determined using the same methods as in (e) and a permutation representation pH of H with stabilizer $\langle h_1^2 \rangle$.

(j) Running the subgroup command again one sees that S_1 also has two maximal elementary abelian normal subgroups of order 16, one of which is the subgroup A defined by its generators given in the statement. Its normalizer $D_1 = N_H(A)$ has order $2^7 \cdot 3 \cdot 7$. Furthermore, $D_1 = \langle n_1, n_2 \rangle$, where $n_1 = h_1 h_2^2$ and $n_2 = h_1 h_2 h_1 h_2 h_1$.

Using the faithful permutation representations pH and PE a successful application of the MAGMA command `IsIsomorphic(D_1,D)` establishes that there is an isomorphism $h : D_1 \to D$ such that $m_1 = h(n_1) = (e_1e_3e_2e_4e_5e_1e_4e_1)^{10}$ and $m_2 = h(n_2) = (e_3e_2e_3e_5e_3e_4e_3)^5$.

(k) By another application of pH and the command `HasComplement (D_1,A)` MAGMA provides a complement K of A in D_1. It is a simple group of order 168 generated by n_1^2 and $(n_2^3n_1n_2)^2$. In particular, K is a stabilizer of a faithful permutation representation PH of degree 240.

(l) The outer automorphism group of $D_1 = N_H(A)$ was calculated by means of the faithful permutation PH and MAGMA. Its order is 1. Therefore the Goldschmidt index of the amalgam $H \leftarrow D \rightarrow E$ is 1 by Corollary 7.1.9 of [92]. □

Definition 11.1.2 A finite simple group G is said to be of McL-*type* if it has a 2-central involution z such that $C_G(z)$ is isomorphic to the finitely presented group H stated in Lemma 11.1.1(g).

By I. Schur's classic paper [116] the covering groups $2A_n$ of all alternating groups A_n are uniquely determined up to isomorphism provided $n > 7$. Therefore one can replace the presentation of H stated in Lemma 11.1.1(g) by Schur's original presentation of $2A_8$. This is done in [80].

Remark 11.1.3 In view of Lemma 11.1.1 all conditions of Steps 4 and 5 of Algorithm 1.3.8 are satisfied. Thus one could immediately apply all steps of Algorithm 7.4.8 of [92] to provide an existence proof of a simple group of McL-type. This has been done in the author's joint paper [80] with H. Kratzer, W. Lempken, and K. Waki, where McL is realized as a simple subgroup of $GL_{22}(11)$. In the following we instead realize McL as a finitely presented group.

11.2 Structure of the given centralizer $H = 2A_8$

In this section we construct another presentation of the given centralizer $H = C_G(z)$ of any finite simple group G of McL-type which will be extended to a unique presentation of the simple target groups G in Section 11.3.

Proposition 11.2.1 *Let $H = \langle h_i \mid 1 \le i \le 3 \rangle \cong 2A_8$ be the finitely presented group defined in Lemma 11.1.1(g). Then the following assertions hold.*

(a) $H = \langle h, n\rangle$, *where* $h = h_1$ *and* $n = h_1 h_2^2$ *have orders* 6 *and* 7, *respectively.*

(b) *The involution* $z = (hn)^{15}$ *generates the center* $Z(H)$ *of* H.

(c) *The involutions* $t = h^2$ *and* z *generate the unique normal Klein* 4-*group* V *of the Sylow* 2-*subgroup* $S = \langle a_1, a_2, c_1, c_2, r\rangle$ *of* H *generated by the five involutions* $a_1 = (nh^2n^3hn)^3$, $a_2 = (hnhn^2h^3n)^3$, $c_1 = (nh^3nhn^2)^2$, $c_2 = (hnh^3n^2hn)^3$ *and* $r = (h^2nhnh^2n)^3$.

(d) $A = \langle z, t, a_1, a_2\rangle$ *and* $A_2 = \langle z, t, c_1, c_2\rangle$ *are the only elementary abelian subgroups of order* 2^4 *of* S. *They are both maximal elementary abelian normal subgroups of* S. *Furthermore,* $V = A \cap B$.

(e) $C = C_S(V) = AB = \langle z, t, a_1, a_2, c_1, c_2\rangle$ *has the following set* $\mathcal{R}(C)$ *of defining relations:*

$$z^2 = t^2 = a_1^2 = a_2^2 = c_1^2 = c_2^2 = 1,$$
$$(a_1, a_2) = (c_1, c_2) = (z, t) = 1, \quad (z, a_1) = (z, a_2) = (z, c_1) =$$
$$(z, c_2) = 1, (t, a_1) = (t, a_2) = (t, c_1) = (t, c_2) = 1,$$
$$a_1^{c_1} = a_1 t, \quad a_1^{c_2} = z a_1, \quad a_2^{c_1} = z t a_2, \quad a_2^{c_2} = t a_2.$$

In particular, V *is the center* $Z(C)$ *of* C.

(f) C *has a complement* $W = \langle w, r\rangle$ *in* $M = N_H(V)$ *generated by* r *and* $w = (h^2nh^4nh)^2$.

$M = \langle z, t, a_1, a_2, c_1, c_2, w, r\rangle$ *has a set* $\mathcal{R}(M)$ *of defining relations consisting of* $\mathcal{R}(C)$ *and the following relations:*

$$r^2 = w^3 = 1, \quad w^r = w^2, \quad (z, w) = (z, r) = (t, w) = 1, \quad t^r = zt,$$
$$a_1^w = z a_1 a_2, \quad a_1^r = a_1, \quad a_2^w = z a_1, \quad a_2^r = a_1 a_2,$$
$$c_1^w = z c_1 c_2, \quad c_1^r = c_1 c_2, \quad c_2^w = z t c_1, \quad c_2^r = c_2.$$

(g) $N_1 = N_H(A) = A : K_1$, *where* $K_1 = \langle w, r, u_1\rangle \cong \mathrm{GL}_3(2)$, *and* $u_1 = [(nh)^3nh^2n^2]^4$.

(h) $N_2 = N_H(B) = B : K_2$, *where* $K_2 = \langle w, r, u_2\rangle \cong \mathrm{GL}_3(2)$, *and* $u_2 = (nhn^2h^4)^4$.

(i) $H = \langle z, t, a_1, a_2, c_1, c_2, w, r, u_1, u_2\rangle$ *has a set* $\mathcal{R}(H)$ *of defining relations consisting of* $\mathcal{R}(M)$ *and the following relations:*

$$u_1^3 = u_2^3 = 1,$$
$$(u_1 c_1)^4 = (u_1 c_2)^6 = (u_1 r)^3 = 1, \quad (w u_1 w)^3 = u_1^2 w u_1 r,$$
$$(z, u_1) = 1, \quad t^{u_1} = z t a_1, \quad a_1^{u_1} = zt, \quad a_2^{u_1} = z t a_1 a_2,$$

$$(u_2w)^3 = (u_2r)^3 = 1, \quad (u_2^w)^2 = u_2^2wu_2^2,$$
$$(z, u_2) = 1, \quad t^{u_2} = c_2, \quad c_1^{u_2} = c_1, \quad c_2^{u_2} = ztc_2,$$
$$ra_1u_2a_1 = u_2, \quad ru_1u_2 = u_2u_1.$$

(j) *H has a faithful permutation representation of degree 240 with stabilizer $\langle w, r, u_1\rangle$.*

Proof Let PH be the faithful permutation representation of $H = \langle h_1, h_2, h_3\rangle$ of degree 240 defined in Lemma 11.1.1(k). The Sylow 2-subgroup S of H and its elementary abelian normal subgroups V, A and B have been found by means of MAGMA and PH after having checked that $H = \langle h, n\rangle$, where $h = h_1$ and $n = h_1h_2^2$ in H. Using the program `LookupWord(C,x^y)` we established the relations of $C = C_H(V) = AB$. The complement $\langle w, r\rangle$ of C in $M = N_H(V)$ was calculated by means of PH, MAGMA and the command `HasComplement(M,C)`. The new relations of M were established by another application of Algorithm 1.4.3. Using the MAGMA command `CosetAction(M,sub<M|w,r>)` the author verified that $\mathcal{R}(M)$ is a set of defining relations of M.

The complements K_1 and K_2 of the normal subgroups A and B in $N_1 = N_H(A)$ and $N_2 = N_H(B)$ were again found by MAGMA. Their generators were calculated using the program `GetShortGens(H,K_i)`. Employing the MAGMA command `FPGroup(K_i)` we obtained their relations with respect to the given generators. The same method was used for various small subgroups of H in order to get the additional relations of H given in (i).

Employing the Todd–Coxeter Algorithm implemented in MAGMA it has been verified that $H = \langle z, t, a_1, a_2, c_1, c_2, w, r, u_1, u_2\rangle$ is a finite group of order $|H| = 40320$ satisfying $\mathcal{R}(H)$. In particular, it has a faithful permutation representation of degree 240 with stabilizer $\langle w, r, u_1\rangle$. □

Corollary 11.2.2 *Let $H = \langle z, t, a_1, a_2, c_1, c_2, w, r, u_1, u_2\rangle$ be the finitely presented group constructed in Proposition 11.2.1(i). Let E be the finitely presented group constructed in Lemma 11.1.1. Let $d = tu_1$. Then the following assertions hold.*

(a) *d has order 6, and $H = \langle d, u_1, u_2\rangle$.*
(b) *Representatives h_i and corresponding centralizer orders of $|C_H(h_i)|$ of the 23 conjugacy classes h_i^H of H are stated in Table 11.4.2.*
(c) *The character table of H is given in Table 11.5.3.*

(d) *The normalizer D of the maximal elementary abelian normal subgroup $A = \langle z, t, a_1, a_2 \rangle$ of the Sylow 2-subgroup S of H defined in Lemma 11.2.1(c) is generated by d and u_1.*

Representatives d_i and the corresponding centralizer orders of $|C_H(d_i)|$ of the 16 conjugacy classes d_i^D of $D = \langle d, u_1 \rangle$ are stated in Table 11.4.3.

(e) *The character table of D is given in Table 11.5.2.*

(f) *There is exactly one compatible pair $(\chi, \tau) \in mf\,char_{\mathbb{C}}(H) \times mf\,char_{\mathbb{C}}(E)$ of degree 22 of the amalgam $H \leftarrow D \rightarrow E$:*

$$(\chi, \tau) = (\chi_{\mathbf{3}} + \chi_4, \tau_1 + \tau_2 + \tau_{\mathbf{8}})$$

with common restriction

$$\tau_{|D} = \chi_{|D} = \psi_1 + \psi_4 + \psi_7 + \psi_{\mathbf{8}},$$

where irreducible characters with bold face indices denote faithful irreducible characters.

Proof (a), (b) and (d) By Proposition 11.2.1(j) H has a faithful permutation representation PH of degree 240 with stabilizer $\langle w, r, u_1 \rangle$. Using it, the program `GetShortGens(H,D)` and MAGMA the author found the two generators $d = tu_1$ and u_1 of $D = N_H(A)$. The equation $H = \langle D, u_2 \rangle$ is easily verified by means of MAGMA.

The systems of representatives of the conjugacy classes of H and D have been computed by means of PH, MAGMA and Kratzer's Algorithm 5.3.18 of [92].

(c) and (e) The character tables of H and D were calculated by means of PH and MAGMA.

(f) Let E be the finite group defined in Lemma 11.1.1. Then E has a 2-central involution z such that $C_E(z) \cong N_H(A)$ by Lemma 11.1.1(h) and Proposition 11.2.1. Thus the amalgam $H \leftarrow D \rightarrow E$ is well defined.

The authors of [80] found the unique compatible pair by an application of Kratzer's Algorithm 7.3.10 of [92], the character Tables 11.5.1, 11.5.2 and 11.5.3 and the fusion of the classes of $D = N_H(A)$ in H and $D = C_E(z)$ in E. It has been adjusted to the new character tables of the three local subgroups. □

11.3 Existence and uniqueness proof

In this section a new existence and uniqueness proof for McLaughlin's simple group McL is given. It simplifies the one by Kratzer, Lempken,

Waki and the author given in [80]. Both proofs realize a simple group G of McL-type as a uniquely determined finitely presented group having a faithful permutation representation with stabilizer $H \cong C_G(z)$. Whereas the proof in [80] exploits the existence of an automorphism ϕ of $H \cong 2\mathsf{A}_8$ fixing a given Sylow 2-subgroup S of H and permuting the two maximal elementary abelian normal subgroups A and B of S, the new proof is independent of such an assumption, which can be derived from I. Schur's presentation of $2\mathsf{A}_8$; see [116] and [80].

Theorem 11.3.1 (Kratzer–Lempken–Michler–Waki) *Let G be a finite simple group of McL-type. Then the following assertions hold.*

(a) *G has a 2-central involution z such that $C_G(z)$ is isomorphic to the finitely presented group $H = \langle z, t, a_1, a_2, c_1, c_2, w, r, u_1, u_2 \rangle$ with the set $\mathcal{R}(H)$ of defining relations given in Proposition 11.2.1(i). Thus we may assume that $H = C_G(z)$.*

(b) *$S = \langle z, t, a_1, a_2, c_1, c_2, r \rangle$ is a Sylow 2-subgroup of G having a unique normal Klein 4-group $V = \langle z, t \rangle$.*

(c) *The normalizer $N_G(V)$ of V contains an element $x \notin N_H(V)$ of order 3 such that*

$$G = \langle z, t, a_1, a_2, c_1, c_2, w, r, u_1, u_2, x \rangle$$

has a set $\mathcal{R}(G)$ of defining relations consisting of $\mathcal{R}(H)$ and the following relations:

$$x^3 = 1, \quad z^x = t, \quad t^x = zt, \quad a_2^x = zta_2, \quad a_1^x = a_1,$$
$$c_1^x = c_1c_2, \quad c_2^x = c_1, \quad (x, w^2) = 1, \quad (x^{-1}r)^2 = 1,$$
$$xzc_2wx^{-1}zc_2w^2 = 1, \quad (zc_2x^{-1})^3 = 1, \quad (u_1^{-1}x^{-1})^2u_1x = 1,$$
$$(x, za_2) = 1, \quad u_2x^{-1}u_2w^2x^{-1}u_2^{-1}xu_2^{-1}x = 1.$$

(d) *G has order $2^7 \cdot 3^6 \cdot 5^3 \cdot 7 \cdot 11$.*

(e) *G is uniquely determined up to isomorphism.*

(f) *G has a faithful permutation representation PG of degree 22275 with stabilizer H.*

(g) *$G = \langle u_1, d, u_2, x \rangle$ has 24 conjugacy classes g_i^G whose representatives g_i and orders of the corresponding centralizers $|C_G(g_i)|$ are stated in Table 11.4.4.*

(h) *The character table of G coincides with the one of McL in the Atlas [19], p. 101.*

Proof (a) holds by Lemma 11.1.1 and Proposition 11.2.1.

(b) This is a restatement of Proposition 11.2.1(c).

(c) By Table 11.4.2 $H = C_G(z)$ has two conjugacy classes of involutions which can be represented by the generators z and t of the Klein 4-group V. Since G is simple, Glauberman's Z^*-theorem asserts that z and t fuse in G; see Theorem 4.7.3 of [92]. Furthermore, Table 11.4.2 implies that $C_G(V) = C_H(V) = C_H(t)$ has order $2^6 \cdot 3$. Hence $Q = C_S(t)$ has index 2 in the Sylow 2-subgroup S of H given in statement (b). Furthermore, $V = Z(Q)$ by Proposition 11.2.1(e). As t is conjugate to z we also know that Q is properly contained in a Sylow 2-subgroup of $C_G(t)$. Thus Lemma 1.4.4. of [92] implies that $N = N_G(V) > N_H(V)$ and $N/C_G(V) \cong \mathrm{GL}_2(2)$. In particular, there is an element $x \in N$ of order 3 such that $z^x = t$, $t^x = zt$ and $x \notin N_H(V)$. Moreover, $N = \langle N_H(V), x\rangle$.

A presentation of $C = C_S(V) = \langle z, t, a_1, a_2, c_1, c_2, w\rangle$ is given in Proposition 11.2.1(e). Using it and the faithful permutation representation PH of H stated in Proposition 11.2.1(j) MAGMA is able to calculate the automorphism group AC of C and determine generators of the group IC of its inner automorphisms by means of the commands `AutomorphismGroup(C)`, `PermutationRepresentation(AC)` and `InnerGenerators(AC)`. Note that IC is an elementary abelian normal subgroup of order 16 in AC. By Proposition 11.2.1(f) and Theorem 1.4.15 of [92] we know that $NI = N_G(V)/V \geq N_H(V)$ is isomorphic to a semidirect product of IC and a subgroup T of AC of order $|T| = 18$ having a normal Sylow 3-subgroup of order 9 and exponent 3. Using the subgroup command of MAGMA and performing the necessary isomorphism tests the author verified that up to isomorphism there is only one subgroup NI of AC which has all the required properties. Moreover, $NI = \langle c1, c2, c3, c4, k1, k2, u\rangle$ has the following set $\mathcal{R}(NI)$ of defining relations:

$$
\begin{aligned}
&c_1^2 = c_2^2 = c_3^2 = c_4^2 = k_1^3 = k_2^3 = u^2 = 1,\\
&(c_1, c_2) = (c_1, c_3) = (c_1, c_4) = (c_2, c_3) = (c_2, c_4) = (c_3, c_4) =\\
&\quad (k_1, k_2) = 1,\\
&k_1^u = k_1^2, \quad k_2^u = k_2^2, \quad c_1^{k_1} = c_1c_2, \quad c_2^{k_1} = c_1, \quad c_3^{k_1} = c_3c_4,\\
&c_4^{k_1} = c_3, \quad c_1^{k_2} = c_1, \quad c_2^{k_2} = c_2, \quad c_3^{k_2} = c_4, \quad c_4^{k_2} = c_3c_4,\\
&c_1^u = c_1c_2, \quad c_2^u = c_2, \quad c_3^u = c_3, \quad c_4^u = c_3c_4.
\end{aligned}
$$

Furthermore NI has a faithful permutation representation PNI of degree 16 because $T = \langle k_1, k_2, u\rangle$ is a complement in NI. Therefore Holt's

Algorithm can be applied to build the extensions of NI by the $F(NI)$-module V over $F = GF(2)$. Let $W = N_H(V) = \langle C, w, r\rangle$. Then $W/V \cong IC : \langle k_1, u\rangle$. Hence we may identify x with k_2. From Proposition 11.2.1(f) follows that $w = k_1$ operates trivially on V, and $r = u$ operates by $z^r = z$ and $t^r = zt$. Let $M1$ be the identity matrix in $\mathrm{GL}_2(2)$. Let $Mk1$, Mu and $Mk2$ be the matrices in $\mathrm{GL}_2(2)$ corresponding to the actions of k_1, u and $k_2 = x$ on V. The MAGMA commands

```
FEalg :=MatrixAlgebra<FiniteField(2),2|M1,M1,M1,M1,
  M1,Mk2,Mu>
```

and `CM:=GModule(PNI,FEalg)` implement the $F(NI)$-module structure. Applying then the MAGMA command `CohomologicalDimension (PNI, CM, 2)` one sees that this dimension is 1. Since C does not split over V by Proposition 11.2.1(e) it follows that $N_G(V)$ is the uniquely determined non-split extension $E1$ of NI by V. By means of the MAGMA commands `P := ExtensionProcess(PNI, CM, NI)` and `E1 := Extension(P, [1])` we obtained the following set $\mathcal{R}(E_1)$ of defining relations of the extension $E_1 = \langle e_i \mid 1 \le i \le 9\rangle$:

$$\begin{aligned}
&e_1^2 = e_2^2 = e_3^2 = e_4^2 = e_5^3 = e_6^3 = e_7^2 = e_8^2 = e_9^2 = 1,\\
&(e_j, e_8) = (e_j, e_9) = 1 \quad \text{for} \quad 1 \le j \le 5,\\
&(e_1, e_2) = (e_3, e_4) = (e_8, e_9) = (e_7, e_8) = 1,\\
&(e_6, e_1) = (e_6, e_2) = (e_5, e_6) = (e_7, e_2) = (e_7, e_3) = 1,\\
&e_6^{-1}e_8e_6e_9 = 1, \quad e_6^{-1}e_9e_6e_8e_9 = 1, \quad e_7e_9e_7e_8e_9 = 1,\\
&e_1e_3e_1e_3e_9 = 1, \quad e_1e_4e_1e_4e_8e_9 = 1, \quad e_2e_3e_2e_3e_8 = 1,\\
&e_2e_4e_2e_4e_9 = 1, \quad e_7e_5e_7e_5^{-2} = 1, \quad e_7e_6e_7e_6^{-2} = 1,\\
&e_5^{-1}e_1e_5e_2e_1 = 1, \quad e_5^{-1}e_2e_5e_1 = 1, \quad e_5^{-1}e_3e_5e_4e_3 = 1,\\
&e_5^{-1}e_4e_5e_3 = 1, \quad e_6^{-1}e_3e_6e_4 = 1, \quad e_6^{-1}e_4e_6e_4e_3 = 1,\\
&e_7e_1e_7e_2e_1 = 1, \quad e_7e_4e_7e_4e_3 = 1.
\end{aligned}$$

Let $Y_1 = \langle e_1, e_2, e_3, e_4, e_5, e_7, e_8, e_9\rangle$. By means of the MAGMA command `IsIsomorphic(C,Y_1)` we established an explicit isomorphism $\phi : Y_1 \to C$. It has the following images:

$$\begin{aligned}
&\phi(e_1) = za_2, \quad \phi(e_2) = a_1, \quad \phi(e_3) = zc_2, \quad \phi(e_4) = tc_1,\\
&\phi(e_5) = w^2, \quad \phi(e_7) = r, \quad \phi(e_8) = z, \quad \phi(e_9) = t.
\end{aligned}$$

The subgroups $A_1 = \langle e_8, e_9, e_1, e_2\rangle$ and $B_1 = \langle e_8, e_9, e_3, e_4\rangle$ of E_1 are the unique maximal elementary abelian normal subgroups of the Sylow

2-subgroup $S_1 = \langle e_1, e_2, e_3, e_4, e_7 \rangle$ of E_1. Thus $\phi(A_1) = A = \langle z, t, a_1, a_2 \rangle$ and $\phi(B_1) = B = \langle z, t, c_1, c_2 \rangle$ are the unique maximal elementary abelian normal subgroups of the Sylow 2-subgroup S of G by Proposition 11.2.1(d). Using MAGMA and the permutation representation PH it has been checked that A and B are the unique maximal elementary abelian normal subgroups of $C = C_S(V)$. Hence both are normal in $N_G(V)$. In particular, they are invariant under conjugation by $x \in N_G(V)$. The action of $x = e_6$ on the bases $\{e_8, e_9, e_1, e_2\}$ and $\{e_8, e_9, e_3, e_4\}$ of A_1 and B_1 has been determined by application of the program `LookupWord(U, e_j^{e_6})`, where e_j runs through the basis of $U \in \{A_1, B_1\}$. The same method has been applied to determine the action of the remaining generators e_5, e_7, e_1 and e_3.

Similarly, one finds the matrices of $u_1 \in N_H(A)$ and $u_2 \in N_H(B)$ w.r.t. the bases $\{\phi(e_8), \phi(e_9), \phi(e_1), \phi(e_2)\}$ and $\{\phi(e_8), \phi(e_9), \phi(e_3), \phi(e_4)\}$ of A and B, respectively. By Proposition 11.2.1(g) and (h) the subgroups $K_1 = \langle w, r, u_1 \rangle$ and $K_2 = \langle w, r, u_2 \rangle$ are complements of A and B in $N_H(A)$ and $N_H(B)$, respectively. Therefore $L_1 = \langle \phi^{-1}(K_1), e_6 \rangle = \langle e_3, e_5, e_7, \phi^{-1}(u_1), e_6 \rangle$ and $L_2 = \langle \phi^{-1}(K_2), e_6 \rangle = \langle e_1, e_5, e_7, \phi^{-1}(u_2), e_6 \rangle$ are complements of A_1 and B_1 in $Q = N_{E_1}(A_1)$ and $R = N_{E_1}(B_1)$, respectively. The generating matrices of L_1 and L_2 are denoted by Qe_j, Qu_1 and Re_k, Ru_1, respectively. They are:

$$Qe_3 = \begin{pmatrix} 1&0&0&0 \\ 0&1&0&0 \\ 0&1&1&0 \\ 1&0&0&1 \end{pmatrix}, \quad Qe_5 = \begin{pmatrix} 1&0&0&0 \\ 0&1&0&0 \\ 0&0&1&1 \\ 0&0&1&0 \end{pmatrix}, \quad Qe_7 = \begin{pmatrix} 1&0&0&0 \\ 1&1&0&0 \\ 0&0&1&1 \\ 0&0&0&1 \end{pmatrix}, \quad Qe_6 = \begin{pmatrix} 0&1&0&0 \\ 1&1&0&0 \\ 0&0&1&0 \\ 0&0&0&1 \end{pmatrix},$$

$$Qu_1 = \begin{pmatrix} 1&0&0&0 \\ 1&1&0&1 \\ 1&1&1&1 \\ 1&1&0&0 \end{pmatrix}, \quad Re_1 = \begin{pmatrix} 1&0&0&0 \\ 0&1&0&0 \\ 0&1&1&0 \\ 1&1&0&1 \end{pmatrix}, \quad Re_5 = \begin{pmatrix} 1&0&0&0 \\ 0&1&0&0 \\ 0&0&1&1 \\ 0&0&1&0 \end{pmatrix},$$

$$Re_7 = \begin{pmatrix} 1&0&0&0 \\ 1&1&0&0 \\ 0&0&1&0 \\ 0&0&1&1 \end{pmatrix}, \quad Re_6 = \begin{pmatrix} 0&1&0&0 \\ 1&1&0&0 \\ 0&0&0&1 \\ 0&0&1&1 \end{pmatrix}, \quad Ru_2 = \begin{pmatrix} 1&0&0&0 \\ 1&0&1&0 \\ 1&1&1&0 \\ 1&1&1&1 \end{pmatrix}.$$

Using the MAGMA command `FPGroup(X)` for $X \in \{L_1, L_2\}$ one verifies that the matrix groups $ML_1 = \langle Qe_3, Qe_5, Qe_6, Qe_7, Qu_1 \rangle$ and $ML_2 = \langle Re_1, Re_5, Re_6, Re_7, Ru_2 \rangle$ are isomorphic to the finitely presented groups $FL_1 = \langle g_3, g_5, g_6, g_7, g_{10} \rangle$ and $FL_2 = \langle g_1, g_5, g_6, g_7, g_{11} \rangle$, respectively. Here g_{10} and g_{11} correspond to Qu_1 and Ru_2, respectively, and FL_1 has the following set $\mathcal{R}(L_1)$ of defining relations:

$$\begin{aligned}
&g_3^2 = g_5^3 = g_6^3 = g_7^2 = g_{10}^3 = 1, \\
&(g_5^{-1} g_7)^2 = (g_7 g_3)^2 = (g_5^{-1} g_7)^2 = (g_7 g_3)^2 = g_3 g_{10}^{-1} g_7 g_{10} = 1, \\
&g_3 g_{10}^{-1} g_7 g_{10} = g_{10}^{-1} g_3 g_7 g_{10} g_7 = g_5 g_{10}^{-1} g_7 g_5^{-1} g_{10}^{-1} g_5 g_{10} = 1, \\
&(g_6, g_5) = (g_6^{-1} g_7)^2 = (g_3 g_6^{-1})^3 = 1, \\
&g_6 g_3 g_5^{-1} g_6^{-1} g_3 g_5 = 1, \quad (g_{10}^{-1} g_6^{-1})^2 g_{10} g_6 = 1.
\end{aligned}$$

FL_2 has the following set $\mathcal{R}(L_2)$ of defining relations:

$$\begin{aligned}
&g_1^2 = g_5^3 = g_6^3 = g_7^2 = g_{11}^3 = 1,\\
&(g_5^{-1}g_7)^2 = (g_5^{-1}g_1g_7)^2 = (g_1g_5^{-1})^3 = 1,\\
&g_{11}g_1g_7g_{11}g_1 = 1, \quad g_{11}^{-1}g_5^{-1}g_{11}^{-1}g_5g_{11}g_5^{-1} = 1,\\
&(g_6, g_1) = (g_6, g_5) = (g_6^{-1}g_7)^2 = 1, \quad g_{11}g_6^{-1}g_{11}g_5g_6^{-1}(g_{11}^{-1}g_6)^2 = 1.
\end{aligned}$$

Since $V = \langle z, t\rangle$ is a maximal characteristic elementary abelian normal subgroup such that $U = \langle H, N_G(V)\rangle$ has a unique conjugacy class of involutions, Corollary 4.8.10 of [92] asserts that $G = U$. As $N_G(A) \cap N_G(B) = N_G(V)$ it follows that G satisfies all the relations of the respective complements T_1 and T_2 of $N_G(A)$ and $N_G(B)$. The groups L_1 and L_2 are isomorphic to T_1 and T_2 under the isomorphisms $\psi_1 : L_1 \to T_1$ and $\psi_2 : L_1 \to T_1$ defined by $\psi_1(g_k) = \phi(e_k)$ for $k \in \{3, 5, 7\}$, $\psi_1(g_{10}) = u_1$, $\psi_1(g_6) = x$ and $\psi_2(g_j) = \phi(e_j)$ for $j \in \{1, 5, 7\}$, $\psi_1(g_{11}) = u_2$, $\psi_1(g_6) = x$, respectively.

Inserting the values of $\phi(e_s)$ in the basis $\mathcal{A}_1 = \{\phi(e_8), \phi(e_9), \phi(e_1), \phi(e_2)\}$ of the maximal elementary abelian subgroup A of S, and using the action of $x = e_6$ on the corresponding basis of A_1 given by the matrix Qe_6, we obtain the following equations:

$$z^x = t, \quad t^x = zt, \quad a_2^x = zta_2, \quad a_1^x = a_1.$$

Similarly, the coefficients of matrix Re_6 imply the following relations:

$$z^x = t, \quad t^x = zt, \quad c_1^x = c_1c_2, \quad c_2^x = c_1.$$

Furthermore, the relations of $\mathcal{R}(L_1)$ and $\mathcal{R}(L_2)$ involving $g_6 = x$ yield the following additional relations:

$$\begin{aligned}
&(x, w^2) = (x^{-1}r)^2 = (zc_2x^{-1})^3 = 1, \quad xzc_2wx^{-1}zc_2w^2 = 1,\\
&(u_1^{-1}x^{-1})^2u_1x = 1, \quad (x, za_2) = 1, \quad u_2x^{-1}u_2w^2x^{-1}(u_2^{-1}x)^2 = 1.
\end{aligned}$$

Hence $G = \langle H, x\rangle$ satisfies the set $\mathcal{R}(G)$ consisting of $\mathcal{R}(H)$ and these additional relations. Using the Todd–Coxeter Algorithm implemented in MAGMA by means of the command `CosetAction(G,H)` we checked that $\mathcal{R}(G)$ is a defining set of relations for the finitely presented group G.

Thus (c) holds. Furthermore, this application of MAGMA also provides a faithful permutation representation PG of G of degree 22275 with stabilizer H.

In particular, (d) and (f) hold.

(e) Since $N = N_G(V) = \langle N_H(V), x\rangle$ is the uniquely determined extension group of the uniquely determined subgroup NI in the automorphism group of the normal subgroup $C = C_S(V)$ of $N_H(V)$, and since A and B are the uniquely determined maximal elementary abelian normal subgroups of C, the finitely presented group $G = \langle z, t, a_1, a_2, c_1, c_2, w, r, u_1, u_2, x\rangle$ is uniquely determined up to isomorphism by H and the unique action of the Glauberman element x of $N_G(V)$ on V.

(g) Let $d = tu_1$. Then $H = \langle d, u_1, u_2\rangle$ by Corollary 11.2.2. Hence $G = \langle d, u_1, u_2, x\rangle$ by (c). The system of representatives of the conjugacy classes of G has been computed by means of PG, MAGMA and Kratzer's Algorithm 5.3.18 of [92].

(h) The character table of G was computed by means of PG and MAGMA. It agrees with the one of the Atlas [19]. In particular, the finitely presented group G is simple. □

Corollary 11.3.2 *Let $G = \langle d, u_1, u_2, x\rangle$ be the finitely presented group constructed in Theorem 11.3.1. Then the following assertions hold.*

(a) *G has a faithful permutation representation of degree 275 with stabilizer Q generated by the elements $q_1 = u_1x^2$, $q_2 = (u_2xu_1u_2)^{15}$ and $q_3 = (u_2xu_1x)^7$ of orders 3, 2 and 2, respectively. Q is a maximal subgroup of G.*

(b) *The permutation character $\Lambda = 1_Q{}^G$ has three irreducible constituents of degrees 1, 22 and 252, and $Q \cong \mathrm{U}_4(3)$.*

(c) *G has an irreducible 22-dimensional irreducible representation over the prime field $GF(11)$. The four generating matrices $\mathfrak{d}$, $\mathfrak{u}_1$, $\mathfrak{u}_2$ and $\mathfrak{x}$ in $\mathrm{GL}_{22}(11)$ are documented on the accompanying DVD.*

(d) *G has an irreducible 22-dimensional irreducible representation over the prime field $GF(2)$.*

Proof (a) The stabilizer Q of the 275-dimensional permutation representation was found by means of the permutation representation PG of Theorem 11.3.1(d) and the MAGMA command `LowIndexSubgroups(PMcL, 275)`. In particular, Q is a maximal subgroup of G. The three generators q_i of Q were calculated by means of the program `GetShortGens(G,Q)`. The isomorphism type of the simple group $Q \cong \mathrm{U}_4(3)$ was established by applying the command `CompositionFactors(Q)`.

(b) The permutation character $\Lambda = 1_Q{}^G$ has non-trivial inner products with the irreducible characters χ_1, χ_2, χ_4 of the Atlas [19], p. 101.

(c) Let QG be the permutation representation of $G = \langle d, u_1, u_2, x \rangle$ of (a). Using the MAGMA command `PermutationMatrix(FiniteField(11), y)`, $y \in \{d, u_1, u_2, x\}$, we define the subgroup $MQ275 = \langle Md, Mu_1, Mu_2, Mx \rangle$ of $\mathrm{GL}_{275}(11)$. MAGMA's implementation of the Meataxe Algorithm provides four composition factors of degrees 22, 1, 251 and 1 by running the command `MQIrr275 :=IrreducibleRepresentations(MQ275)`. The four generators of the 22-dimensional 11-modular irreducible representation of G are the matrices $\mathfrak{d} = MQIrr275[1].1$, $\mathfrak{u}_1 = MQIrr275[1].2$, $\mathfrak{u}_2 = MQIrr275[1].3$ and $\mathfrak{x} = MQIrr275[1].4$ documented on the accompanying DVD.

(d) This statement is proved as (c). The resulting generating matrices can be calculated by the same commands after replacing the prime 11 by 2. □

11.4 Representatives of conjugacy classes

11.4.1 *Conjugacy classes of* $E = \langle e_1, e_2, e_3, e_4, e_5 \rangle$

Class	*Representative*	*Centralizer*	2P	3P	5P	7P
1	1	$2^7 \cdot 3^2 \cdot 5 \cdot 7$	1	1	1	1
2_1	$e5$	$2^7 \cdot 3 \cdot 7$	1	2_1	2_1	2_1
2_2	$e1$	$2^5 \cdot 3$	1	2_2	2_2	2_2
3_1	$e3$	$2^2 \cdot 3^2$	3_1	1	3_1	3_1
3_2	$e1e2$	$2^2 \cdot 3^2$	3_2	1	3_2	3_2
4_1	$e1e5$	2^5	2_1	4_1	4_1	4_1
4_2	$e1e4$	2^3	2_2	4_2	4_2	4_2
5	$e1e2e3$	5	5	5	1	5
6_1	$e4e5$	$2^2 \cdot 3$	3_1	2_1	6_1	6_1
6_2	$e1e2e3e4$	$2^2 \cdot 3$	3_2	2_2	6_2	6_2
7_1	$e1e3e4$	$2 \cdot 7$	7_1	7_2	7_2	1
7_2	$(e1e3e4)^3$	$2 \cdot 7$	7_2	7_1	7_1	1
8	$e1e4e5$	2^3	4_1	8	8	8
14_1	$e1e4e5e3$	$2 \cdot 7$	7_2	14_2	14_2	2_1
14_2	$(e1e4e5e3)^3$	$2 \cdot 7$	7_1	14_1	14_1	2_1

11.4.2 Conjugacy classes of $H = \langle d, u_1, u_1 \rangle$

Class	*Representative*	*Centralizer*	2P	3P	5P	7P
1	1	$2^7 \cdot 3^2 \cdot 5 \cdot 7$	1	1	1	1
2_1	$(du1)^7$	$2^7 \cdot 3^2 \cdot 5 \cdot 7$	1	2_1	2_1	2_1
2_2	$(d)^3$	$2^6 \cdot 3$	1	2_2	2_2	2_2
3_1	$(u1u2)^4$	$2^3 \cdot 3^2 \cdot 5$	3_1	1	3_1	3_1
3_2	$(d)^2$	$2^2 \cdot 3^2$	3_2	1	3_2	3_2
4_1	$(u1u2)^3$	$2^5 \cdot 3$	2_1	4_1	4_1	4_1
4_2	d^2u1	2^4	2_2	4_2	4_2	4_2
5	$(d^2u1u2)^2$	$2 \cdot 3 \cdot 5$	5	5	1	5
6_1	$(u1u2)^2$	$2^3 \cdot 3^2 \cdot 5$	3_1	2_1	6_1	6_1
6_2	d^3u2	$2^2 \cdot 3^2$	3_2	2_1	6_2	6_2
6_3	d	$2^2 \cdot 3$	3_2	2_2	6_4	6_3
6_4	$(d)^5$	$2^2 \cdot 3$	3_2	2_2	6_3	6_4
7_1	$(du1)^2$	$2 \cdot 7$	7_1	7_2	7_2	1
7_2	$(du1)^6$	$2 \cdot 7$	7_2	7_1	7_1	1
8	$du1^2$	2^3	4_1	8	8	8
10	d^2u1u2	$2 \cdot 3 \cdot 5$	5	10	2_1	10
12	$u1u2$	$2^2 \cdot 3$	6_1	4_1	12	12
14_1	$du1$	$2 \cdot 7$	7_1	14_2	14_2	2_1
14_2	$(du1)^3$	$2 \cdot 7$	7_2	14_1	14_1	2_1
15_1	$(d^2u2u1)^2$	$2 \cdot 3 \cdot 5$	15_1	5	3_1	15_2
15_2	$(d^2u2u1)^{14}$	$2 \cdot 3 \cdot 5$	15_2	5	3_1	15_1
30_1	d^2u2u1	$2 \cdot 3 \cdot 5$	15_1	10	6_1	30_2
30_2	$(d^2u2u1)^7$	$2 \cdot 3 \cdot 5$	15_2	10	6_1	30_1

11.4.3 Conjugacy classes of $D = \langle d, u_1 \rangle$

Class	*Representative*	*Centralizer*	2P	3P	7P
1	1	$2^7 \cdot 3 \cdot 7$	1	1	1
2_1	$(du1)^7$	$2^7 \cdot 3 \cdot 7$	1	2_1	2_1
2_2	$(d)^3$	$2^6 \cdot 3$	1	2_2	2_2
2_3	$(d^2u1)^2$	2^5	1	2_3	2_3
3	$(d)^2$	$2^2 \cdot 3$	3	1	3
4_1	$(du1^2)^2$	2^5	2_1	4_1	4_1
4_2	$d^2u1du1du1$	2^4	2_2	4_2	4_2
4_3	d^2u1	2^3	2_3	4_3	4_3
6_1	d	$2^2 \cdot 3$	3	2_2	6_1
6_2	$(d)^5$	$2^2 \cdot 3$	3	2_2	6_2
6_3	d^3u1du1^2	$2^2 \cdot 3$	3	2_1	6_3
7_1	$(du1)^2$	$2 \cdot 7$	7_1	7_2	1
7_2	$(du1)^6$	$2 \cdot 7$	7_2	7_1	1
8	$du1^2$	2^3	4_1	8	8
14_1	$du1$	$2 \cdot 7$	7_1	14_2	2_1
14_2	$(du1)^3$	$2 \cdot 7$	7_2	14_1	2_1

11.4.4 *Conjugacy classes of* $G = \langle d, u_1, u_2, x\rangle$

Class	*Representative*	*Centralizer*	2P	3P	5P	7P	11P
1	1	$2^7 \cdot 3^6 \cdot 5^3 \cdot 7 \cdot 11$	1	1	1	1	1
2	$(d)^3$	$2^7 \cdot 3^2 \cdot 5 \cdot 7$	1	2	2	2	2
3_1	$(u1u2)^4$	$2^3 \cdot 3^6 \cdot 5$	3_1	1	3_1	3_1	3_1
3_2	$(d)^2$	$2^2 \cdot 3^5$	3_2	1	3_2	3_2	3_2
4	$(u1u2)^3$	$2^5 \cdot 3$	2	4	4	4	4
5_1	$(u1u2x)^6$	$2 \cdot 3 \cdot 5^3$	5_1	5_1	1	5_1	5_1
5_2	$du1x$	5^2	5_2	5_2	1	5_2	5_2
6_1	$(u1u2)^2$	$2^3 \cdot 3^2 \cdot 5$	3_1	2	6_1	6_1	6_1
6_2	d	$2^2 \cdot 3^2$	3_2	2	6_2	6_2	6_2
7_1	$(du1)^2$	$2 \cdot 7$	7_1	7_2	7_2	1	7_1
7_2	$(du1)^6$	$2 \cdot 7$	7_2	7_1	7_1	1	7_2
8	$du1^2$	2^3	4	8	8	8	8
9_1	$du1xu2^2$	3^3	9_2	3_1	9_2	9_1	9_2
9_2	$(du1xu2^2)^2$	3^3	9_1	3_1	9_1	9_2	9_1
10	$(u1u2x)^3$	$2 \cdot 3 \cdot 5$	5_1	10	2	10	10
11_1	$u1xu2$	11	11_2	11_1	11_1	11_2	1
11_2	$(u1xu2)^2$	11	11_1	11_2	11_2	11_1	1
12	$u1u2$	$2^2 \cdot 3$	6_1	4	12	12	12
14_1	$du1$	$2 \cdot 7$	7_1	14_2	14_2	2	14_1
14_2	$(du1)^3$	$2 \cdot 7$	7_2	14_1	14_1	2	14_2
15_1	$(u1u2x)^2$	$2 \cdot 3 \cdot 5$	15_1	5_1	3_1	15_2	15_2
15_2	$(u1u2x)^{14}$	$2 \cdot 3 \cdot 5$	15_2	5_1	3_1	15_1	15_1
30_1	$u1u2x$	$2 \cdot 3 \cdot 5$	15_1	10	6_1	30_2	30_2
30_2	$(u1u2x)^7$	$2 \cdot 3 \cdot 5$	15_2	10	6_1	30_1	30_1

11.5 Character tables of local subgroups

11.5.1 *Character table of* $E = \langle e_1, e_2, e_3, e_4, e_5\rangle$

2	7	7	5	2	2	5	3	.	2	2	1	1	3	1	1
3	2	1	1	2	2	.	.	.	1	1	.	.	.	.	.
5	1	.	.	.	.	.	.	1	.	.	.	.	.	.	.
7	1	1	.	.	.	.	.	.	.	.	1	1	.	1	1
	$1a$	$2a$	$2b$	$3a$	$3b$	$4a$	$4b$	$5a$	$6a$	$6b$	$7a$	$7b$	$8a$	$14a$	$14b$
$2P$	$1a$	$1a$	$1a$	$3a$	$3b$	$2a$	$2b$	$5a$	$3a$	$3b$	$7a$	$7b$	$4a$	$7b$	$7a$
$3P$	$1a$	$2a$	$2b$	$1a$	$1a$	$4a$	$4b$	$5a$	$2a$	$2b$	$7b$	$7a$	$8a$	$14b$	$14a$
$5P$	$1a$	$2a$	$2b$	$3a$	$3b$	$4a$	$4b$	$1a$	$6a$	$6b$	$7b$	$7a$	$8a$	$14b$	$14a$
$7P$	$1a$	$2a$	$2b$	$3a$	$3b$	$4a$	$4b$	$5a$	$6a$	$6b$	$1a$	$1a$	$8a$	$2a$	$2a$
$X.1$	1	1	1	1	1	1	1	1	1	1	1	1	1	1	1
$X.2$	6	6	2	.	3	2	.	1	.	−1	−1	−1	.	−1	−1
$X.3$	10	10	−2	1	1	−2	.	.	1	1	A	$\bar{A}$	.	$\bar{A}$	A
$X.4$	10	10	−2	1	1	−2	.	.	1	1	$\bar{A}$	A	.	A	$\bar{A}$
$X.5$	14	14	2	2	−1	2	.	−1	2	−1	.	.	.	.	.
$X.6$	14	14	2	−1	2	2	.	−1	−1	2	.	.	.	.	.
$X.7$	15	15	−1	.	3	−1	−1	.	.	−1	1	1	−1	1	1
$X.8$	15	−1	3	3	.	−1	1	.	−1	.	1	1	−1	−1	−1
$X.9$	21	21	1	.	−3	1	−1	1	.	1	.	.	−1	.	.
$X.10$	35	35	−1	−1	−1	−1	1	.	−1	−1	.	.	1	.	.
$X.11$	45	−3	−3	.	.	1	1	.	.	.	A	$\bar{A}$	−1	$-\bar{A}$	$-A$
$X.12$	45	−3	−3	.	.	1	1	.	.	.	$\bar{A}$	A	−1	$-A$	$-\bar{A}$
$X.13$	90	−6	6	.	.	−2	.	.	.	.	−1	−1	.	1	1
$X.14$	105	−7	−3	3	.	1	−1	.	−1	.	.	.	1	.	.
$X.15$	120	−8	.	−3	.	.	.	.	1	.	1	1	.	−1	−1

where $A = -\frac{1}{2}(1 - i\sqrt{7})$.

11.5.2 *Character table of* $D = \langle d, u_1 \rangle$

2	7	7	6	5	2	5	4	3	2	2	2	1	1	3	1	1
3	1	1	1	.	1	.	.	.	1	1	1	.	.	.	.	.
7	1	1	.	.	.	.	.	.	.	.	.	1	1	.	1	1
	$1a$	$2a$	$2b$	$2c$	$3a$	$4a$	$4b$	$4c$	$6a$	$6b$	$6c$	$7a$	$7b$	$8a$	$14a$	$14b$
$2P$	$1a$	$1a$	$1a$	$1a$	$3a$	$2a$	$2b$	$2c$	$3a$	$3a$	$3a$	$7a$	$7b$	$4a$	$7a$	$7b$
$3P$	$1a$	$2a$	$2b$	$2c$	$1a$	$4a$	$4b$	$4c$	$2b$	$2b$	$2a$	$7b$	$7a$	$8a$	$14b$	$14a$
$7P$	$1a$	$2a$	$2b$	$2c$	$3a$	$4a$	$4b$	$4c$	$6a$	$6b$	$6c$	$1a$	$1a$	$8a$	$2a$	$2a$
$X.1$	1	1	1	1	1	1	1	1	1	1	1	1	1	1	1	1
$X.2$	3	3	3	−1	.	−1	−1	1	.	.	.	A	$\bar{A}$	1	A	$\bar{A}$
$X.3$	3	3	3	−1	.	−1	−1	1	.	.	.	$\bar{A}$	A	1	$\bar{A}$	A
$X.4$	6	6	6	2	.	2	2	.	.	.	.	−1	−1	.	−1	−1
$X.5$	7	7	7	−1	1	−1	−1	−1	1	1	1	.	.	−1	.	.
$X.6$	7	7	−1	−1	1	3	−1	−1	−1	−1	1	.	.	1	.	.
$X.7$	7	7	−1	3	1	−1	−1	1	−1	−1	1	.	.	−1	.	.
$X.8$	8	−8	.	.	2	.	.	.	.	.	−2	1	1	.	−1	−1
$X.9$	8	8	8	.	−1	.	.	.	−1	−1	−1	1	1	.	1	1
$X.10$	8	−8	.	.	−1	.	.	.	B	$\bar{B}$	1	1	1	.	−1	−1
$X.11$	8	−8	.	.	−1	.	.	.	$\bar{B}$	B	1	1	1	.	−1	−1
$X.12$	14	14	−2	2	−1	2	−2	.	1	1	−1	.	.	.	.	.
$X.13$	21	21	−3	1	.	−3	1	−1	.	.	.	.	.	1	.	.
$X.14$	21	21	−3	−3	.	1	1	1	.	.	.	.	.	−1	.	.
$X.15$	24	−24	.	.	.	.	.	.	.	.	.	A	$\bar{A}$	.	$-A$	$-\bar{A}$
$X.16$	24	−24	.	.	.	.	.	.	.	.	.	$\bar{A}$	A	.	$-\bar{A}$	$-A$

where $A = -\frac{1}{2}(1 - i\sqrt{7})$ *and* $B = -i\sqrt{3}$.

11.5.3 *Character table of* $H = \langle d, u_1, u_2 \rangle$

2	7	7	6	3	2	5	4	1	3	2	2	2	1	1	3	1	2	1	1	1	1	1	1
3	2	2	1	2	2	1	.	1	2	2	1	1	.	.	.	1	1	.	.	1	1	1	1
5	1	1	.	1	.	.	.	1	1	.	.	.	.	.	.	1	.	.	.	1	1	1	1
7	1	1	.	.	.	.	.	.	.	.	.	.	1	1	.	.	.	1	1	.	.	.	.
	$1a$	$2a$	$2b$	$3a$	$3b$	$4a$	$4b$	$5a$	$6a$	$6b$	$6c$	$6d$	$7a$	$7b$	$8a$	$10a$	$12a$	$14a$	$14b$	$15a$	$15b$	$30a$	$30b$
$2P$	$1a$	$1a$	$1a$	$3a$	$3b$	$2a$	$2b$	$5a$	$3a$	$3b$	$3b$	$3b$	$7a$	$7b$	$4a$	$5a$	$6a$	$7a$	$7b$	$15a$	$15b$	$15a$	$15b$
$3P$	$1a$	$2a$	$2b$	$1a$	$1a$	$4a$	$4b$	$5a$	$2a$	$2a$	$2b$	$2b$	$7b$	$7a$	$8a$	$10a$	$4a$	$14b$	$14a$	$5a$	$5a$	$10a$	$10a$
$5P$	$1a$	$2a$	$2b$	$3a$	$3b$	$4a$	$4b$	$1a$	$6a$	$6b$	$6d$	$6c$	$7b$	$7a$	$8a$	$2a$	$12a$	$14b$	$14a$	$3a$	$3a$	$6a$	$6a$
$7P$	$1a$	$2a$	$2b$	$3a$	$3b$	$4a$	$4b$	$5a$	$6a$	$6b$	$6c$	$6d$	$1a$	$1a$	$8a$	$10a$	$12a$	$2a$	$2a$	$15b$	$15a$	$30b$	$30a$
$X.1$	1	1	1	1	1	1	1	1	1	1	1	1	1	1	1	1	1	1	1	1	1	1	1
$X.2$	7	7	−1	4	1	3	−1	2	4	1	−1	−1	.	.	1	2	.	.	.	−1	−1	−1	−1
$X.3$	8	−8	.	−4	2	.	.	−2	4	−2	.	.	1	1	.	2	.	−1	−1	1	1	−1	−1
$X.4$	14	14	6	−1	2	2	2	−1	−1	2	.	.	.	.	.	−1	−1	.	.	−1	−1	−1	−1
$X.5$	20	20	4	5	−1	4	.	.	5	−1	1	1	−1	−1	.	.	1	−1	−1	.	.	.	.
$X.6$	21	21	−3	6	.	1	1	1	6	.	.	.	.	.	−1	1	−2	.	.	1	1	1	1
$X.7$	21	21	−3	−3	.	1	1	1	−3	.	.	.	.	.	−1	1	1	.	.	A	$\bar{A}$	A	$\bar{A}$
$X.8$	21	21	−3	−3	.	1	1	1	−3	.	.	.	.	.	−1	1	1	.	.	$\bar{A}$	A	$\bar{A}$	A
$X.9$	24	−24	.	−6	.	.	.	−1	6	.	.	.	B	$\bar{B}$	.	1	.	$-B$	$-\bar{B}$	−1	−1	1	1
$X.10$	24	−24	.	−6	.	.	.	−1	6	.	.	.	$\bar{B}$	B	.	1	.	$-\bar{B}$	$-B$	−1	−1	1	1
$X.11$	28	28	−4	1	1	4	.	−2	1	1	−1	−1	.	.	.	−2	1	.	.	1	1	1	1
$X.12$	35	35	3	5	2	−5	−1	.	5	2	.	.	.	.	−1	.	1	.	.	.	.	.	.
$X.13$	45	45	−3	.	.	−3	1	.	.	.	.	.	B	$\bar{B}$	1	.	.	B	$\bar{B}$	.	.	.	.
$X.14$	45	45	−3	.	.	−3	1	.	.	.	.	.	$\bar{B}$	B	1	.	.	$\bar{B}$	B	.	.	.	.
$X.15$	48	−48	.	6	.	.	.	−2	−6	.	.	.	−1	−1	.	2	.	1	1	1	1	−1	−1
$X.16$	56	56	8	−4	−1	.	.	1	−4	−1	−1	−1	.	.	.	1	.	.	.	1	1	1	1
$X.17$	56	−56	.	−4	−1	.	.	1	4	1	C	$\bar{C}$	.	.	.	−1	.	.	.	1	1	−1	−1
$X.18$	56	−56	.	−4	−1	.	.	1	4	1	$\bar{C}$	C	.	.	.	−1	.	.	.	1	1	−1	−1
$X.19$	56	−56	.	2	2	.	.	1	−2	−2	.	.	.	.	.	−1	.	.	.	A	$\bar{A}$	$-A$	$-\bar{A}$
$X.20$	56	−56	.	2	2	.	.	1	−2	−2	.	.	.	.	.	−1	.	.	.	$\bar{A}$	A	$-\bar{A}$	$-A$
$X.21$	64	−64	.	4	−2	.	.	−1	−4	2	.	.	1	1	.	1	.	−1	−1	−1	−1	1	1
$X.22$	64	64	.	4	−2	.	.	−1	4	−2	.	.	1	1	.	−1	.	1	1	−1	−1	−1	−1
$X.23$	70	70	−2	−5	1	2	−2	.	−5	1	1	1	.	.	.	.	−1	.	.	.	.	.	.

where $A = -\frac{1}{2}(1 - i\sqrt{15})$, $B = -\frac{1}{2}(1 - i\sqrt{7})$ *and* $C = -i\sqrt{3}$.

12
Rudvalis' group Ru

In this chapter we give a new existence and uniqueness proof for the sporadic simple Rudvalis group Ru using Algorithm 1.3.15. This group was originally discovered by A. Rudvalis [115] as a rank 3 permutation group. The first existence proof for Ru was given by Conway and Wales in [20]. D. Parrott characterized the Rudvalis group by means of the group structure of the centralizer $H = C_{\mathsf{Ru}}(z)$ of a 2-central involution z in [106]. In the course of his uniqueness proof Parrott established a presentation for the Sylow 2-subgroups S of Ru. Using it M. Kratzer [79] determined a presentation of the centralizer $H = C_G(z)$ of a 2-central involution z of $G = \mathsf{Ru}$. Applying then Algorithm 7.4.8 of [92] he gave a self-contained existence proof of a simple group G of Ru-type. He realized G as a simple subgroup of $\mathrm{GL}_{133}(5)$ and calculated a faithful presentation and its concrete character table; see [79]. These results have inspired the following existence and uniqueness proof due to the author.

In Section 12.1 Algorithm 1.3.15 is applied to an irreducible subgroup $T \cong \mathrm{GL}_3(2)$ of $\mathrm{GL}_8(2)$. We construct an iterated extension E of T by its irreducible FT-modules W and V of dimensions 8 and 3 over $F = GF(2)$, respectively. Then we determine the centralizer $DE = C = E(z)$ of a 2-central involution z of E. Now Step 3 of Algorithm 1.3.15 is applied to build a group H with center $Z(H)$ of order 2 having a special subgroup C of order 2^{11} with center $Z(C) \cong V$ such that $DH = N_H(C) \cong DE$.

In Section 12.2 it is shown that the free product $P = H *_D E$ of H and E with amalgamated subgroup D has an irreducible representation $\mathfrak{G}$ in $\mathrm{GL}_{378}(17)$. It is a simple group, having a 2-central involution $\mathfrak{z}$ such that $C_{\mathfrak{G}}(\mathfrak{z}) \cong H$; see Theorem 12.2.1.

In Section 12.3 it is shown that each simple group G of Ru-type has two conjugacy classes of involutions z^G and s^G. The structure of the centralizer $C_G(s)$ of the non-2-central involution s is determined in

Proposition 12.3.4. Furthermore, the group order of any simple group of Ru-type is calculated, and all its conjugacy classes are classified. The group structure of the normalizers of the elements of odd prime orders and the fusion of their conjugacy classes in G are also determined. The uniqueness proof is given in Section 12.4 by means of these results and Theorem 7.5.1 of [92].

12.1 Construction of the 2-central involution centralizer

In this section we apply Algorithm 1.3.15 to an irreducible subgroup $T \cong \mathrm{GL}_3(2)$ of $\mathrm{GL}_8(2)$ and construct a presentation of a group H with center $Z(H)$ of order 2, which finally will be proved to be isomorphic to the centralizer $C_G(z)$ of a 2-central involution z of the simple sporadic group $G \cong$ Ru of Rudvalis [115].

It is well known that $\mathrm{GL}_3(2)$ has exactly three faithful irreducible representations over $F = GF(2)$. Two of them are dual to each other and have dimension 3, and the third has dimension 8. It occurs as the unique non-trivial composition factor of the tensor product of the two non-isomorphic 3-dimensional irreducible $\mathrm{GL}_3(2)$-modules over F. Let V and W be the natural F-vector spaces of dimensions 3 and 8, respectively. By means of Algorithm 1.3.15 we now construct the split extension E_1 of T by W, and then a suitable non-split extension E of E_1 by V.

Lemma 12.1.1 *Let T be the subgroup of $\mathrm{GL}_8(2)$ generated by the following matrices:*

$$Mk1 = \begin{pmatrix} 1&0&0&1&0&0&0&0\\ 1&1&0&1&1&0&0&0\\ 0&0&1&0&0&1&0&0\\ 0&0&0&1&0&0&0&0\\ 0&0&0&1&1&0&0&0\\ 0&0&0&0&0&1&0&0\\ 0&0&0&0&0&0&1&0\\ 0&0&0&0&0&1&1&1 \end{pmatrix}, \quad Mk2 = \begin{pmatrix} 0&0&0&0&1&0&0&0\\ 0&0&0&0&0&0&1&0\\ 0&0&0&0&0&0&0&1\\ 0&0&1&0&0&0&0&0\\ 1&0&0&0&1&0&0&0\\ 0&1&0&0&0&0&0&0\\ 0&0&0&0&0&1&0&0\\ 0&0&0&1&0&0&0&0 \end{pmatrix}.$$

of orders 2 and 3, respectively. Then the following assertions hold.

(a) *The natural 8-dimensional vector space W over F is an irreducible FT-module.*

(b) *T is isomorphic to $L = \mathrm{GL}_3(2)$ generated by the matrices*

$$Lk1 = \begin{pmatrix} 1&1&0\\ 0&1&0\\ 0&0&1 \end{pmatrix}, \quad Lk2 = \begin{pmatrix} 0&0&1\\ 1&0&0\\ 0&1&0 \end{pmatrix}.$$

The isomorphism $\psi : T \to L$ is given by $\psi(Mki) = Lki$ for $i = 1, 2$.

(c) *Both groups T and L are isomorphic to $K = \langle k_1, k_2\rangle$ having the following set $\mathcal{R}(K)$ of defining relations:*

$$k_1^2 = k_2^3 = 1, \quad (k_1k_2^{-1})^7 = 1, \quad (k_2k_1k_2^{-1}k_1)^4 = 1.$$

(d) *The split extension E_1 of T by W is isomorphic to the finitely presented group $E_1 = \langle t_1, t_2, t_3\rangle$ with the following set $\mathcal{R}(E_1)$ of defining relations:*

$$\begin{aligned}
&t_1^2 = t_2^3 = t_3^2 = 1,\\
&(t_2^{-1}t_3t_2t_1)^2 = (t_2t_3t_2^{-1}t_3)^2 = 1, \quad (t_1t_3)^4 = 1,\\
&(t_2^{-1}t_1t_3t_1t_2t_1)^2 = (t_2t_3t_2^{-1}t_1t_3t_1)^2 = 1,\\
&t_2^{-1}t_1t_2t_3t_2^{-1}t_1t_2^{-1}t_1t_3t_1t_3t_2^{-1}t_3 = 1,\\
&(t_1t_2^{-1})^7 = 1, \quad (t_2t_1t_2^{-1}t_1)^4 = 1,\\
&t_2^{-1}t_1t_2t_1t_2t_1t_3t_1t_2^{-1}t_1t_2^{-1}t_1t_2t_1t_2t_1t_3t_1t_3t_2^{-1}t_1 = 1.
\end{aligned}$$

(e) *The cohomological dimension of E_1 with coefficients in the FE_1-module V_1 is 3, and there is a non-split extension E of E_1 by V_1 which is isomorphic to the finitely presented group $E = \langle e_1, e_2, e_3, e_4, e_5, e_6\rangle$ with the following set $\mathcal{R}(E)$ of defining relations:*

$$\begin{aligned}
&e_1^2 = e_2^3 = e_3^2 = e_4^2 = e_5^2 = e_6^2 = 1,\\
&(e_1e_2^{-1})^7 = 1, \quad (e_2^{-1}e_3e_2e_1)^2 = 1, \quad (e_1e_3)^4 = 1,\\
&(e_4, e_5) = (e_4, e_6) = (e_5, e_6) = (e_1, e_5) = (e_1, e_6) = 1,\\
&(e_3, e_4) = (e_3, e_5) = (e_3, e_6) = 1,\\
&e_1^{-1}e_4e_1e_4^{-1}e_5^{-1} = e_2^{-1}e_4e_2e_6^{-1} = 1,\\
&e_2^{-1}e_5e_2e_4^{-1} = e_2^{-1}e_6e_2e_5^{-1} = 1,\\
&e_2e_3e_2^{-1}e_3e_2e_3e_2^{-1}e_3e_4^{-1} = 1,\\
&(e_2^{-1}e_1e_3e_1e_2e_1)^2 = 1, \quad (e_2e_1e_2^{-1}e_1)^4 = 1,\\
&e_2e_3e_2^{-1}e_1e_3e_1e_2e_3e_2^{-1}e_1e_3e_1e_5^{-1}e_6^{-1} = 1,\\
&e_2^{-1}e_1e_2e_3e_2^{-1}e_1e_2^{-1}e_1e_3e_1e_3e_2^{-1}e_3 = 1,\\
&e_2^{-1}e_1e_2e_1e_2e_1e_3e_1e_2^{-1}e_1e_2^{-1}e_1e_2e_1e_2e_1e_3e_1e_3e_2^{-1}e_1 = 1.
\end{aligned}$$

Furthermore, E has a faithful permutation representation of degree 2048 with stabilizer $\langle e_1, e_2\rangle$.

(f) $z = (e_1e_3(e_1e_2e_1e_2e_1e_2^2)^2)^8$ *is a 2-central involution of E and $D = C_E(z) = \langle x, y\rangle$ is a subgroup of E of order $2^{14}\cdot 3$, where*

$$x = e_3e_2^2e_1e_2e_1e_3e_2^2e_1e_2(e_1e_2e_1e_2^2)^2 \text{ and } y = e_3e_2^2e_1e_2.$$

Furthermore, $E = \langle x, y, e\rangle$, where $e = e_2$.

(g) *The Fitting subgroup* C *of* E *has order* 2^{11}. *It is generated by* $c_1 = x^4$, $c_2 = (y)^2$, $c_3 = (x^2y)^4$, $c_4 = (xyx)^4$, $c_5 = (x^3y)^3$, $c_6 = (x^2yx)^3$, $c_7 = (x^2y^2)^2$ *and* $c_8 = (x^3yx)^2$.

(h) C *is the unique maximal special normal subgroup of any Sylow 2-subgroup* S *of* E *with center* $Z(C) \cong V$ *and* $C/Z(C) \cong W$.

(i) *A system of representatives of the* 35 *conjugacy classes of* $E = \langle x, y, e \rangle$ *is given in Table 12.5.3.*

(j) *A system of representatives of the* 52 *conjugacy classes of* $D = \langle x, y \rangle$ *is given in Table 12.5.2.*

(k) *The character table of* E *is given in Table 12.6.2.*

(l) *The character table of* D *is given in Table 12.6.3.*

Proof The first three assertions are easily verified using MAGMA.

(d) Since E_1 is a split extension of the irreducible subgroup T of $\mathrm{GL}_8(2)$ by $W \cong F^8$ one can apply the MAGMA commands `Q:=Getvecs(T)` and `E_1:=Semidir(T,Q)`. They yield the presentation of the split extension E_1 of T by W stated in the assertion.

(e) Using the data of (b) and (d) an application of Holt's algorithm [54] in MAGMA shows that the second cohomological dimension of E_1 in $V = F^3$ over F is 3. Hence there are seven non-split extensions of R by V. For each non-trivial cocycle MAGMA provides a presentation of a non-split extension E of E_1 by V. One of these seven non-split extensions E has the presentation given in the statement. The six other extension groups are not isomorphic to E.

From (c) and the presentation of E it follows immediately that $\langle e_1, e_2 \rangle \cong L$ is a stabilizer of a faithful permutation representation PE of E of degree 2^{11}.

(f) All statements of (f) can be verified by means of PE and MAGMA.

(g) The group E is a split extension of $L \cong \mathrm{GL}_3(2)$ by a special normal subgroup $C = C_E(V)$. The author found its generators c_i, $1 \leq i \leq 8$, by means of PE, MAGMA and the program `GetShortGens(E,C)`.

(h) Since $L \cong \mathrm{GL}_3(2)$ is a complement of C in E, the subgroup C is a maximal special normal subgroup of E. Using PE and the MAGMA command

```
Subgroup(S: Al:=Normal, IsElementaryAbelian := true)
```

one checks that C is the unique maximal special normal subgroup of order 2^{11} in any Sylow 2-subgroup S of E.

(i) and (j) The systems of representatives of the conjugacy classes of E and D have been calculated by means of PE, Kratzer's Algorithm 5.3.18 of [92] and MAGMA.

(k) and (l) The character tables of D and E have been computed by means of the permutation representation of PE described in (e) and MAGMA. □

Proposition 12.1.2 *Let* $E = \langle e_1, e_2, e_3, e_4, e_5, e_6 \rangle = \langle x, y, e \rangle$ *be the finitely presented group constructed in Lemma 12.1.1(e). Then the following assertions hold.*

(a) $D = \langle x, y \rangle$ *has a center* $Z(D)$ *of order* 2 *which is generated by* $z = x^8$.

(b) $D = \langle x, y \rangle$ *has exactly one normal subgroup* N *of order* 2^{11} *with center* $Z(N) = \langle z \rangle$ *such that* $|Aut(N)| = 2^{19} \cdot 3 \cdot 5$.

N is generated by $n_1 = x^4$, $n_2 = (x^2y)^2$, $n_3 = (xyx)^2$, $n_4 = (xy^2)^2$, $n_5 = (yx^2)^2$ *and* $n_6 = (y^2x)^2$.

N has a complement $K = \langle j_1, j_2 \rangle \cong S_4$ *in D, where* $j_1 = (y^3xy^2x^2yx)^4$, $j_2 = x^2yx^2y^2x^2yx^2y$.

(c) *The normal subgroup I of inner automorphisms of N has a complement T in* $Aut(N)$ *such that its Fitting subgroup* $F(T)$ *has a complement R in T isomorphic to* S_5.

The subgroup IR has order $2^{13} \cdot 3 \cdot 5$. *It is isomorphic to the group* $H_1 = \langle y_1, y_2, y_3, y_4, y_5, y_6, y_7, y_8 \rangle$ *having the following set* $\mathcal{R}(H_1)$ *of defining relations:*

$$y_1^2 = y_2^2 = y_3^2 = y_4^4 = y_5^4 = y_6^4 = y_7^2 = y_8^4 = 1,$$
$$(y_1y_2)^2 = 1, \quad (y_1y_3)^2 = 1, \quad (y_2y_5^{-1})^2 = 1,$$
$$y_5^{-1}y_8^{-1}y_6^{-1}y_8 = 1, \quad (y_1y_6^{-1}y_2)^2 = 1, \quad (y_1y_4^{-1}y_3)^2 = 1,$$
$$y_2y_4^{-1}y_3y_2y_3y_4 = y_2y_4y_3y_2y_3y_4^{-1} = 1,$$
$$(y_2y_6^{-1}y_3)^2 = 1, \quad (y_2y_6y_3)^2 = 1,$$
$$y_4^{-1}y_1y_4^{-1}y_1y_6^2 = y_4y_1y_4^{-1}y_1y_5^2 = 1,$$
$$y_7y_1y_4^{-1}y_2y_7y_4^{-1} = y_5^{-1}y_2y_4^{-1}y_2y_5y_4^{-1} = 1,$$
$$(y_1y_5^{-1}y_4^{-1})^2 = 1, \quad (y_2y_6^{-1}y_4^{-1})^2 = 1,$$

$$y_4^{-1}y_7y_4^{-1}y_1y_7y_2 = y_6y_8^{-1}y_4^{-1}y_2y_8y_4 = 1,$$
$$y_7y_1y_5^{-1}y_1y_7y_5^{-1} = y_6^{-1}y_3y_5^{-1}y_3y_6y_5 = 1,$$
$$y_8^{-1}y_2y_6^{-1}y_5y_8y_2 = y_7y_5^{-1}y_6^{-1}y_1y_7y_6 = 1,$$
$$y_5^{-1}y_3y_2y_1y_8y_4^{-1}y_8^{-1} = y_6^{-1}y_4y_3y_1y_8^{-1}y_5^{-1}y_8 = 1,$$
$$y_7y_5^{-1}y_3y_1y_5^{-1}y_7y_3 = y_4^{-1}y_8^{-1}y_4^{-1}y_1y_8y_3y_2 = 1,$$
$$(y_8^{-1}y_7)^5 = 1, \quad (y_7y_8^{-1}y_7y_8)^3 = 1.$$

(d) *There is a central extension H of H_1 with center $Z(H)$ of order 2 having a Sylow 2-subgroup S containing a unique maximal special normal subgroup C of order 2^{11} with elementary abelian center $Z(C)$ of order 8 such that $N_H(C) \cong D$.*

The group $H = \langle h_i \mid 1 \le i \le 9\rangle$ has the following set $\mathcal{R}(H)$ of defining relations:

$$h_1^2 = h_2^2 = h_3^4 = h_4^4 = h_5^4 = h_6^4 = h_7^2 = h_8^4 = h_9^2 = 1,$$
$$(h_1, h_9) = (h_2, h_9) = (h_3, h_9) = (h_4, h_9) = (h_5, h_9) = (h_6, h_9) = 1,$$
$$(h_7, h_9) = (h_8, h_9) = 1, \quad h_3^2h_9^{-1} = 1, \quad (h_1h_2)^2 = (h_1h_3)^2 = 1,$$
$$h_2h_5^{-1}h_2h_5^{-1}h_9^{-1} = h_5^{-1}h_8^{-1}h_6^{-1}h_8h_9^{-1} = 1,$$
$$(h_1h_6^{-1}h_2)^2 = (h_1h_4^{-1}h_3)^2 = (h_2h_6h_3)^2 = 1,$$
$$h_2h_4^{-1}h_3h_2h_3h_4h_9^{-1} = h_2h_4h_3h_2h_3h_4^{-1} = 1,$$
$$h_2h_6^{-1}h_3h_2h_6^{-1}h_3h_9^{-1} = h_4^{-1}h_1h_4^{-1}h_1h_6^2h_9^{-1} = 1,$$
$$h_4h_1h_4^{-1}h_1h_5^2 = h_7h_1h_4^{-1}h_2h_7h_4^{-1}h_9^{-1} = 1,$$
$$h_5^{-1}h_2h_4^{-1}h_2h_5h_4^{-1} = 1, \quad (h_1h_5^{-1}h_4^{-1})^2 = 1,$$
$$h_2h_6^{-1}h_4^{-1}h_2h_6^{-1}h_4^{-1}h_9^{-1} = h_4^{-1}h_7h_4^{-1}h_1h_7h_2h_9^{-1} = 1,$$
$$h_6h_8^{-1}h_4^{-1}h_2h_8h_4 = h_7h_1h_5^{-1}h_1h_7h_5^{-1}h_9^{-1} = 1,$$
$$h_6^{-1}h_3h_5^{-1}h_3h_6h_5 = h_8^{-1}h_2h_6^{-1}h_5h_8h_2h_9^{-1} = 1,$$
$$h_7h_5^{-1}h_6^{-1}h_1h_7h_6 = h_5^{-1}h_3h_2h_1h_8h_4^{-1}h_8^{-1} = 1,$$
$$h_6^{-1}h_4h_3h_1h_8^{-1}h_5^{-1}h_8h_9^{-1} = h_7h_5^{-1}h_3h_1h_5^{-1}h_7h_3 = 1,$$
$$h_4^{-1}h_8^{-1}h_4^{-1}h_1h_8h_3h_2 = 1,$$
$$(h_8^{-1}h_7)^5 = 1, \quad (h_7h_8^{-1}h_7h_8)^3 = 1.$$

Furthermore, H has a faithful permutation representation of degree 2048 with stabilizer $\langle h_7, h_8\rangle$.

(e) *A system of representatives of the 49 conjugacy classes of $H = \langle x, y, h\rangle$ is given in Table 12.5.1.*

(f) *The character table of H is given in Table 12.6.1.*

Proof (a) Using the faithful permutation representation PE of Lemma 12.1.1(e) and MAGMA one verifies that $z = (e_1e_3(e_1e_2e_1e_2e_1e_2^2)^2)^8 = x^4$ generates the center $Z(D)$ of $D = \langle x, y\rangle$.

(b) D has three normal subgroups of order 2^{11}. This can be checked by application of the MAGMA command

```
Subgroups(PD: Al=Normal, OrderEqual:= 2048);
```

C of Lemma 12.1.1(g) is one of them. The others have a cyclic center of order 2 generated by z. Another MAGMA calculation shows that only the normal subgroup N with generators given in the statement has an automorphism whose odd order part is larger than the one of $|D|$. The complement of N in D has also be found by means of MAGMA.

(c) Using the faithful permutation representation PE and the commands `AN := AutomorphisGroup(N)` and `PAN:=PermutationGroup(AN)` MAGMA establishes that PAN is a permutation group acting on a set of cardinality 240 and that $|A| = 2^{19} \cdot 3 \cdot 5$. Furthermore, it gives 11 generators of the subgroup of inner automorphisms I of N in AN and an isomorphism $h : AN \to PAN$. Let $PI = \langle h(I)\rangle$. Then PI is a normal subgroup of PA of order 2^{10}. Using the MAGMA command `PK:=HasComplement(PAN,PI)` one sees that I has a complement K in AN. Furthermore, MAGMA yields that the Fitting subgroup J of K has order 2^6, and K is a split extension of a subgroup $W \cong S_4$ by J. In particular, IW is a subgroup of AN of order $2^{13} \cdot 3 \cdot 5$. Another application of MAGMA yields that the Frattini subgroup of I has order 2^4. Hence it is now easy to find six generators for I and two for W such that IW is generated by eight elements.

Applying then the MAGMA command `FH1:=FPGroup(P[IW])` the author obtained the presentation of the group H_1 given in the statement.

(d) H_1 has a faithful permutation representation PH_1 of degree 1024 with stabilizer $\langle y_7, y_8\rangle \cong S_5$. Let `MGL1_2:=GL(1,2)`, `M1:=MGL1_2![1]` and `FEalg:=MatrixAlgebra<FiniteField(2),1|M1,M1,M1,M1,M1,M1,M1,M1>`. Then `CM:=GModule(PH_1,FEalg)` denotes the cohomology module. Applying the command `CohomologicalDimension(PH_1, CM, 2)` MAGMA establishes that the second cohomological dimension is 4. Thus there are 15 non-split 2-central extensions H of H_1. The command `P :=ExtensionProcess(PH_1,CM,H_1)` provides a presentation for each of them. Only the finitely presented group E stated in the assertion has a special normal subgroup C of order 2^{11} with center $Z(C) \cong V$ such that $N_H(C) \cong D$.

Furthermore, $H = \langle h_i \mid 1 \le i \le 9 \rangle$ has a faithful permutation representation PH of degree 2048 with stabilizer $\langle h_7, h_8 \rangle \cong S_5$.

(e) The system of representatives of the 49 conjugacy classes of H has been calculated by means of PH, Kratzer's Algorithm 5.3.18 of [92] and MAGMA.

(f) The character table of H has been computed by means of the permutation representation of PH and MAGMA. □

Lemma 12.1.3 *Let $H = \langle h_i \mid 1 \le i \le 9 \rangle$ be the group constructed in Proposition 12.1.2(d). Let $d_1 = h_7$, $d_2 = (h_1 h_8)^4$, $d_3 = (h_7 h_8 h_7 h_8^2)^2$, $h = h_8$, $x_1 = (d_1 d_3^2 d_2 d_1 d_3 d_2 d_3 d_2 d_1)^5$ and $y_1 = (d_3 d_1 d_2 d_1 d_3 d_2 d_3 d_2 d_1 d_3 d_2 d_3)^3$.*

Let $s_1 = x_1$, $s_2 = y_1^2$, $s_3 = (x_1^2 y_1)^2$ and $t = (x_1^2 y_1^2)^4$.

Let $c_1 = x_1^4$, $c_2 = y_1^2$, $c_3 = (x_1^2 y_1)^4$, $c_4 = (x_1 y_1 x_1)^4$, $c_5 = (x_1^3 y_1)^3$, $c_6 = (x_1^2 y_1 x_1)^3$, $c_7 = (x_1^2 y_1^2)^2$ and $c_8 = (x_1^3 y_1 x_1)^2$.

Let $E = \langle x, y, e \rangle$ be the finitely presented group constructed in Lemma 12.1.1(e).

Then the following assertions hold.

(a) *$H = \langle x_1, y_1, h \rangle$ has a faithful permutation representation of degree 2^{11} with stabilizer $\langle h, (y_1 x_1^2 y_1^2 x_1)^4 \rangle$.*
(b) *$S = \langle s_1, s_2, s_3 \rangle$ is a Sylow 2-subgroup of H.*
(c) *$C = \langle c_i \mid 1 \le i \le 8 \rangle$ is the unique maximal special normal subgroup of S and $|C| = 2^{11}$.*
(d) *$z = x^8$ is a 2-central involution of E such that $DE = C_E(z) = \langle x, y \rangle$. Furthermore, there is an isomorphism $\psi : DE \to DH = N_H(C)$ such that $\psi(x) = x_1$ and $\psi(y) = y_1$.*
(e) *The center $Z(C)$ of C is the unique elementary abelian normal subgroup $A = \langle (c_1)^2, (c_1 c_2)^2, (c_2 c_3)^2 \rangle$ of order 8 in S. Furthermore, $C_H(A) = C$.*
(f) *$V = \langle z, t \rangle$ is the unique normal Klein 4-subgroup of S, where $t = x_1^8 (x_1^4 y_1^2)^2$. Furthermore, $C_2 = C_H(V)$ has order 2^{13} and its center $Z(C_2) = V$.*
(g) *The automorphism group $Aut(C_2)$ of C_2 has order $2^{19} \cdot 3$.*
(h) *The derived subgroup of C_2 has order 2^9 and its center $Z(C_2') = A$.*
(i) *C has a complement $K = \langle k_1, k_2 \rangle \cong S_4$ in $D = N_H(C)$, where $k_1 = (y_1 x_1^2 y_1^2 x_1)^4$ and $k_2 = y_1 x_1^4 y_1^2 x_1^3 y_1 x_1^2$.*
(j) *The Goldschmidt index of the amalgam $H \leftarrow D \rightarrow E$ is 1.*

Proof All statements (a) through (i) are easily checked using MAGMA and the faithful permutation representation PH of H given in Proposition 12.1.2(d).

(m) The Goldschmidt index has been calculated by means of Kratzer's Algorithm 7.1.10 of [92]. □

Definition 12.1.4 A finite simple group G is called to be of Ru-*type* if it possesses a 2-central involution z such that $C_G(z) \cong H$, where H is the finitely presented group of Proposition 12.1.2.

12.2 Construction of a simple group of Ru-type

In this section the conditions of Steps 1 and 2 of Algorithm 1.3.15 are verified for the amalgam $H \leftarrow D \rightarrow E$ constructed in Section 12.1. Therefore its Step 3 can be applied to give here a self-contained existence proof for a simple group of Ru-type.

Theorem 12.2.1 (Michler) *Keep the notation of Lemma 12.1.1 and Proposition 12.1.2. Using the notation of the three character Tables 12.6.1, 12.6.3 and 12.6.2 of the groups H, D and E, respectively, the following statements hold.*

(a) *There are exactly two dual compatible pairs $(\chi, \tau) \in mf\,char_{\mathbb{C}}(H) \times mf\,char_{\mathbb{C}}(E)$ of degree 378 of the groups $H = \langle D, h\rangle$ and $E = \langle D, e\rangle$:*

$$(\chi, \tau) = (\chi_{\mathbf{48}} + \chi_{44} + \chi_{31} + \chi_{10}, \tau_{24} + \tau_{\mathbf{35}})$$

with common restriction

$$\tau_{|D} = \chi_{|D} = \psi_{19} + \psi_{32} + \psi_{41} + \psi_{46} + \psi_{\mathbf{49}} + \psi_{51} + \psi_{\mathbf{52}},$$

and

$$(\chi, \tau) = (\chi_{\mathbf{48}} + \chi_{44} + \chi_{31} + \chi_{11}, \tau_{23} + \tau_{\mathbf{35}})$$

with common restriction

$$\tau_{|D} = \chi_{|D} = \psi_{20} + \psi_{32} + \psi_{41} + \psi_{46} + \psi_{\mathbf{49}} + \psi_{51} + \psi_{\mathbf{52}},$$

where irreducible characters with bold face indices denote faithful irreducible characters.

(b) *Let* $\mathfrak{V}$ *and* $\mathfrak{W}$ *be the up to isomorphism uniquely determined faithful semi-simple multiplicity-free* 378*-dimensional modules of* H *and* E *over* $F = GF(17)$ *corresponding to the first compatible pair* χ, τ*, respectively.*

Let $\kappa_{\mathfrak{V}} : H \to \mathrm{GL}_{378}(17)$ *and* $\kappa_{\mathfrak{W}} : E \to \mathrm{GL}_{378}(17)$ *be the representations of* H *and* E *afforded by the modules* $\mathfrak{V}$ *and* $\mathfrak{W}$*, respectively.*

Let $\mathfrak{h} = \kappa_{\mathfrak{V}}(h)$, $\mathfrak{x} = \kappa_{\mathfrak{V}}(x)$ *and* $\mathfrak{y} = \kappa_{\mathfrak{V}}(y)$ *in* $\kappa_{\mathfrak{V}}(H) \le \mathrm{GL}_{378}(17)$. *Then the following assertions hold.*

(1) $\mathfrak{V}_{|D} \cong \mathfrak{W}_{|D}$*, and there is a transformation matrix* $\mathcal{T} \in \mathrm{GL}_{378}(17)$ *such that*

$$\mathfrak{x} = \mathcal{T}^{-1}\kappa_{\mathfrak{W}}(x_1)\mathcal{T}, \qquad \mathfrak{y} = \mathcal{T}^{-1}\kappa_{\mathfrak{W}}(y_1)\mathcal{T}.$$

Let $\mathfrak{e} = \mathcal{T}^{-1}\kappa_{\mathfrak{W}}(e)\mathcal{T} \in \mathrm{GL}_{378}(17)$.

(2) $\mathfrak{G} = \langle \mathfrak{h}, \mathfrak{x}, \mathfrak{y}, \mathfrak{e} \rangle$ *has a faithful permutation representation of degree* 424125 *with stabilizer* $\mathfrak{E} = \langle \mathfrak{x}, \mathfrak{y}, \mathfrak{e} \rangle$.

(3) *The four generating matrices of* $\mathfrak{G}$ *are stored in MAGMA format on the accompanying DVD.*

(4) *The character table of* $\mathfrak{G}$ *is stated as Table 12.6.8. It coincides with that of* Ru *in the Atlas [19], p. 127.*

(5) $\mathfrak{G}$ *has* 36 *conjugacy classes. A system of their representatives and their centralizer orders is given in Table 12.5.5.*

(c) $\mathfrak{G}$ *is a finite simple group with* 2*-central involution* $\kappa_{\mathfrak{V}}(z) = (\mathfrak{x})^8$ *such that*

$$C_{\mathfrak{G}}(\kappa_{\mathfrak{V}}(z)) = \kappa_{\mathfrak{V}}(H) \text{ and } |\mathfrak{G}| = 2^{14} \cdot 3^3 \cdot 5^3 \cdot 7 \cdot 13 \cdot 29.$$

Proof (a) The character tables of the groups $H = \langle x_1, y_1, h \rangle$, $DH = \langle x_1, y_1 \rangle$, $DE = \langle x, y \rangle$ and $E = \langle x, y, e \rangle$ are Tables 12.5.1, 12.5.2, and 12.5.1, respectively. In the following we use their notations. Using MAGMA and these character tables and the fusion of the classes of DH in H and DE in E the two compatible pairs stated in assertion (a) have been calculated by means of MAGMA and Kratzer's Algorithm 7.3.10 of [92].

Since they are conjugate complex it suffices to construct the semi-simple representations $\mathfrak{V}$ of $\chi \in Irr_{\mathbb{C}}(H)$ and $\mathfrak{W}$ of $\tau \in Irr_{\mathbb{C}}(H)$ of the first compatible pair(χ, τ) over their common splitting field $GF(17)$.

(b) We first construct the semi-simple faithful representation $\mathfrak{W}$ corresponding to the character $\chi = \tau_{24} + \tau_{48}$. We use the faithful permutation representation PE of E of degree 2048 constructed in Lemma 12.1.1(e). Using PE and the MAGMA command

```
LowIndE2 := LowIndexSubgroups(PE, <100,200>)
```

the author determined all conjugacy classes of subgroups U of E having index $100 \le k \le 200$. For each representative U MAGMA knows generators of each of these permutation subgroups U of E. Furthermore, MAGMA calculates the character table and the fusion of the conjugacy classes of U in E. Thus one can calculate all the inner products $((1_U)^E, \tau_2 4)$. If k is the smallest index for which this inner product is non-zero then we apply the program `GetShortGens(E,U)`. For τ_{24} we found that $k = 168$, and $U = U_{24} = \langle (xe)^7, (ex)^7, (xexe^2)^2, (xexexe^2x)^2 \rangle$. Applying then Kim's implementation of Algorithm 1.5.1 to the permutation representation $(1_{U_2 4})^E$ over the splitting field $GF(17)$ the author obtained the three generating matrices $ME42_x$, $ME42_y$, $ME42_e$ of the corresponding irreducible representation of $E = \langle x, y, e \rangle$ in $\mathrm{GL}_{42}(17)$.

Using PE and the MAGMA command

```
LowIndE3 := LowIndexSubgroups(PE, <600,700>)
```

the author determined all conjugacy classes of subgroups W of E having index $600 \le k \le 700$. We calculated again the inner products $((1_W)^E, \tau_{35})$ for representatives W of conjugacy classes of subgroups of E with index k in the given interval. For $k = 672$ MAGMA established that the permutation character of the subgroup $W_{35} = \langle (ex^2e^2x)^3, e^2x^3e^2x^2, x^2e^2xex^3e^2 \rangle$ has a non-trivial inner product with τ_{35}. Applying now Algorithm 1.5.1 to the permutation representation $(1_{W_{35}})^E$ over the splitting field $GF(17)$ the author obtained the three generating matrices $ME336_x$, $ME336_y$, $ME336_e$ of the corresponding irreducible representation of $E = \langle x, y, e \rangle$ in $\mathrm{GL}_{336}(17)$. These matrices can be reconstructed from the four generating matrices of the simple group $\mathfrak{G}$ stored on the accompanying DVD and the results described in the following.

The semi-simple representation $\mathfrak{W}$ corresponding to the semi-simple character τ of E in $\mathrm{GL}_{378}(17)$ is described by the following three matrices:

$$ME_x := \texttt{DiagonalJoin(ME42\_x, ME336\_x)},$$

$$ME_y := \texttt{DiagonalJoin(ME42\_y, ME336\_y)},$$

$$ME_e := \texttt{DiagonalJoin(ME42\_e, ME336\_e)}.$$

Similarly one shows that the four irreducible characters χ_{10}, χ_{31}, χ_{44}, χ_{48} of $H = \langle x_1, y_1, h\rangle$ with respective degrees 6, 60, 120, 192 are irreducible constituents of the permutation characters $(1_{V_{10}})^H$, $(1_{V_{31}})^H$, $(1_{V_{44}})^H$, $(1_{V_{48}})^H$ of the following subgroups:

$$\begin{aligned}
V_{10} &= \langle x_1^4, x_1 h, h x_1\rangle \text{ of index } 24,\\
V_{31} &= \langle (y_1 h)^5, (x_1^2 y_1)^2, h x_1 y_1 x_1 h, (h x_1 y_1 h y_1)^3\rangle \text{ of index } 120,\\
V_{44} &= \langle (h y_1 x_1)^4, (h^2 y_1)^3, (h^2 y_1 x_1)^5, (x_1 h^2 x_1^2)^2\rangle \text{ of index } 240,\\
V_{48} &= \langle y_1^2 x_1^2 h y_1, (x_1 y_1^3 h x_1^2)^3\rangle \text{ of index } 2048
\end{aligned}$$

in the group H, respectively.

Applying Algorithm 1.5.1 to the permutation representations $(1_{V_{10}})^H$, $(1_{V_{31}})^H$, $(1_{V_{44}})^H$, $(1_{V_{48}})^H$ over the splitting field $GF(17)$ the author obtained the following four of triples-generating matrices:

$MH6_{x1}$, $MH6_{y1}$, $MEH6_h$ in $\mathrm{GL}_6(17)$,
$MH60_{x1}$, $MH60_{y1}$, $MH60_h$ in $\mathrm{GL}_{60}(17)$,
$MH120_{x1}$, $MH120_{y1}$, $MH120_h$ in $\mathrm{GL}_{120}(17)$,
$MH192_{x1}$, $MH192_{y1}$, $MH192_h$ in $\mathrm{GL}_{192}(17)$,

of the irreducible representations of $H = \langle x_1, y_1, h\rangle$ corresponding to the irreducible characters χ_{10}, χ_{31}, χ_{44}, χ_{48}, respectively. These matrices can be reconstructed from the four generating matrices of the simple group $\mathfrak{G}$ stored on the accompanying DVD and the results described in the following.

The semi-simple representation $\mathfrak{V}$ corresponding to the semi-simple character χ of H in $\mathrm{GL}_{378}(17)$ is described by the following three matrices:

MH_{x1} := `DiagonalJoin(MH6_x1, MH60_x1, MH120_x1, MH192_x1)`,
MH_{y1} := `DiagonalJoin(MH6_y1, MH60_y1, MH120_y1, MH192_y1)`,
MH_h := `DiagonalJoin(MH6_h, MH60_h, MH120_h, MH192_h)`.

The matrix groups $\mathfrak{D}H = \langle \mathfrak{V}(x_1), \mathfrak{V}(y_1)\rangle$ and $\mathfrak{D}E = \langle \mathfrak{W}(x), \mathfrak{W}(y)\rangle$ are isomorphic by Proposition 12.1.2(d). So are the semi-simple representations $\mathfrak{V}_{|\mathfrak{D}H}$ and $\mathfrak{W}_{|\mathfrak{D}E}$ by the definition of the compatible pair $(\chi, \tau) \in mfchar_{\mathbb{C}}(H) \times mfchar_{\mathbb{C}}(E)$ and Maschke's Theorem, which is applicable because the orders of H and E are coprime to 17. Let $Y = \mathrm{GL}(278, 17) = \mathrm{GL}_{378}(17)$. By (a) both FD-modules $\mathfrak{V}$ and $\mathfrak{W}$ are a direct sum of pairwise non-isomorphic irreducible FD-modules which can be ordered simultaneously such that the dimensions of the components are 6, 12, 24, 48, 64, 96 and 128. In the following we assume that we have chosen bases in $\mathfrak{V}$ and $\mathfrak{W}$ such that the above matrix generators MH_{x1}, MH_{y1} and ME_x, ME_y are built with respect to these bases.

Applying then Parker's isomorphism test of Proposition 6.1.6 of [92] by means of the MAGMA command

```
IsIsomorphic(GModule(sub<Y|V(x1),V(y1)>),GModule(sub<Y|W(x),W(y)>))
```

we obtain the transformation matrix $\mathcal{T}_1$ satisfying $\mathfrak{V}(x) = (\mathfrak{W}(x_1))^{\mathcal{T}_1}$ and $\mathfrak{V}(y_1) = (\mathfrak{W}(y))^{\mathcal{T}_1}$.

Let $\mathfrak{e}_1 = (\mathfrak{W}(e))^{\mathcal{T}_1}$. Then $\mathfrak{E} = \langle \mathfrak{x}, \mathfrak{y}, \mathfrak{e}_1 \rangle \cong E$. By Corollary 7.2.4 of [92] the transformation matrix $\mathcal{T}$ of the statement is the product matrix $\mathcal{D}\mathcal{T}_1$, where

$$\mathcal{D} = diag(1^6, 1^{12}, 1^{24}, 1^{48}, 16^{64}, 1^{96}, 1^{24}, 2^{128}) \in \mathrm{GL}_{378}(17)$$

and c^n denotes the $n \times n$ diagonal matrix with unique diagonal entry $c \in GF(17)$.

Let $\mathfrak{e} = (\mathfrak{e}_1)^{\mathcal{D}}$. Let $\mathfrak{G} = \langle \mathfrak{h}, \mathfrak{x}, \mathfrak{y}, \mathfrak{e} \rangle$. Using the MAGMA commands `RandomSchreier(G)` and `FPGroupStrong(G)` the author obtained a finitely presented group G with nine generators r_i and a very long list of relations (not restated in this book). The first four generators r_i correspond to $\mathfrak{x}$, $\mathfrak{y}$, $\mathfrak{h}$ and $\mathfrak{e}$. Therefore we applied the program

```
MyCosetAction(G, E:maxsize:=100000000),
```

where $E = \langle r_1, r_2, r_4 \rangle$. Thus we obtained a faithful permutation representation PG of $\mathfrak{G}$ of degree 424125 with stabilizer $\mathfrak{E}$. In particular $\mathfrak{G} = 2^{10} \cdot 3^7 \cdot 5^3 \cdot 7 \cdot 11 \cdot 23$.

The character table of $\mathfrak{G}$ has been calculated by means of the above permutation representation and MAGMA; see Table 12.6.8. It coincides with the one of Ru in [19], p. 127.

The system of representatives of the 36 conjugacy classes of G has been calculated by means of PG, Kratzer's Algorithm 5.3.18 of [92] and MAGMA; see Table 12.5.5.

(c) Let $\mathfrak{z} = (\mathfrak{x})^4$. Then $C_{\mathfrak{G}}(\mathfrak{z})$ contains $\mathfrak{H} = \langle \mathfrak{h}, \mathfrak{x}, \mathfrak{y} \rangle$. Table 12.5.5 implies that $\mathfrak{G}$ is a simple group and that $C_{\mathfrak{G}}(\mathfrak{z}) \cong \mathfrak{H}$. This completes the proof. □

12.3 Fusion

Throughout this section G denotes a finite simple group of Ru-type, and $H = \langle x_1, y_1, h \rangle$ denotes the group constructed in Proposition 12.1.2 by generators and relations. In this section it is proved that G has exactly two conjugacy classes of involutions, and that they are represented by $z = x_1^8$ and $s = x_1^2 h x_1 h^2$. Furthermore, it is shown that $C_G(s) \cong 2^2 \times Sz(8)$ and that $|G| = 2^{14} \cdot 3^3 \cdot 5^3 \cdot 7 \cdot 13 \cdot 29$. By means of these results it is easy to determine representatives for all conjugacy classes of G consisting of elements of even order.

Let X be a group. Then $Z(X)$ denotes its center, and its second center $Z_2(X)$ is defined by $Z_2(X)/Z(X) = Z(X/Z(X))$; $\Omega_1(X)$ is the subgroup of X generated by all elements of prime order in X.

Lemma 12.3.1 *Let z be a 2-central involution of the simple group G of* Ru*-type with $C_G(z) = H = \langle x_1, y_1, h\rangle$ constructed in Lemma 12.1.3. Let $z = x_1^8$, $t = x_1^8(x_1^4 y_1^2)^2$, $v = (y_1^2(x_1^2 y_1)^4)^2$ in H. Let $V = \langle z, t\rangle$ and $A = \langle z, t, v\rangle$.*

Let $E = \langle e_1, e_2, e_3, e_4, e_5, e_6\rangle = \langle x, y, e\rangle$ be the finitely presented group constructed in Lemma 12.1.1(e).

Then the following statements hold.

- (a) $N_G(V)/C_G(V) = N_G(V)/C_H(V) \cong \mathrm{GL}_2(2) \cong S_3$.
- (b) *$t \in H$ is conjugate to z in G.*
- (c) $N_G(A)/C_G(A) = N_G(A)/C \cong \mathrm{GL}_3(2)$.
- (d) *The inverse μ of the isomorphism $\psi : DE = \langle x, y\rangle \to DH = \langle x_1, y_1\rangle$ constructed in Lemma 12.1.3(d) extends to an isomorphism $\nu : N_G(A) \to E$.*

Proof (a) and (b) By Table 12.5.1 t is a representative of a conjugacy class of involutions of H such that $|C_H(t)| = 2^{13}$. By Lemma 12.1.3(f) $V = \langle z, t\rangle$ is the unique normal Klein 4-subgroup of the fixed Sylow 2-subgroup S of H, and $V = Z(C_H(V))$. Furthermore, $|S| = 2^{14}$ and $|S : C_S(t)| = 2$. Since G is a simple group, Glauberman's Theorem 4.7.3 of [92] asserts that that $z^g = t$ for some $g \in G - H$. Thus S^g is a Sylow 2-subgroup of $K = C_G(t)$. Since $X = C_S(t)$ has order 2^{13} Sylow's Theorem implies that $X \le S^{gk}$ for some $k \in K$. Hence $X = C_S(t)$ is properly contained in a Sylow 2-subgroup of $C_G(t)$. Therefore both (a) and (b) hold by Lemma 1.4.4 of [92].

(c) The subgroup $A = \langle z, t, v\rangle = Z(C_2')$ is a characteristic subgroup of $C_2 = C_H(V)$ by Lemma 12.1.3(h). Hence A is normal in $N_G(V)$ because $C_G(V) = C_H(V)$. By (b) there is an element d of order 3 in $N_G(V)$ which is not in $N_H(V)$. In particular, $d \in N_G(A) - N_H(A)$. Lemma 12.1.3(i) asserts that $C = C_G(A) = C_H(A)$ has a complement $K \cong S_4$ in $N_H(A)$. As $|N_G(A) : N_H(A)|$ is odd, Theorem 1.4.15 of [92] implies that $N_G(A)$ splits over C as well. Thus $N_G(A)$ is a semidirect product of C and $K_1 = \langle K, r\rangle \cong \mathrm{GL}_3(2)$ because S_4 is a maximal subgroup of $\mathrm{GL}_3(2)$.

(d) Let $E_1 = N_G(A)$. Then $C_{E_1}(z) = N_H(A) = \langle x_1, y_1\rangle = DH$. Let $DE = \langle x, y\rangle$. Then $\mu(x_1) = x$ and $\mu(y_1)$ define an isomorphism between DH and DE by Lemma 12.1.3(d). Furthermore, $C = C_G(A) = C_H(A) = \psi(C_1)$, where $C_1 = C_E(A) = \langle c_i \mid 1 \le i \le 8\rangle$ and where the c_i are defined in Lemma 12.1.1(g). Using the faithful permutation representation PH of H defined in Proposition 12.1.2(d) it has been checked that $K_1 = \langle k_1, k_2\rangle \cong S_4$ is a complement of C in DH, where $k_1 = (y_1x_1^2y_1^2x_1)^4$ and $k_2 = y_1x^4y_1^2x_1^3y_1x_1^2$. Hence $\mu(K_1) = \langle (yx^2y^2x)^4, yx^4y^2x^3yx^2\rangle$ is a complement of C_1 in DE. It is a subgroup of the complement $L = \langle \mu(K_1), g\rangle \cong \mathrm{GL}_3(2)$ of C_1 in $E = \langle x, y, e\rangle$, where $g = (x^4e^2x^6)^6 \in E$ has order 7, as has been checked by means of the faithful permutation representation PE of E defined in Lemma 12.1.1(e) and MAGMA. In particular, $E_1/C_1 \cong \mathrm{GL}_3(2)$ operates irreducibly on its 8-dimensional module $C_1/A \cong W$ over $GF(2)$ by (c) and Lemma 12.1.1. Now another application of Lemma 12.1.1 yields that $E_1 = N_G(A) \cong E$ because up to isomorphism E is the unique iterated extension of $\mathrm{GL}_3(2)$ by its 8- and 3-dimensional modules such that $DH \cong DE$. This completes the proof. □

Proposition 12.3.2 *Let G be a finite simple group of* Ru*-type with a 2-central involution z such that $H = C_G(z)$. Keep the notation of Table 12.5.1. Then the following statements hold.*

(a) *G has two conjugacy classes of involutions represented by $z = x_1^8$ and $s = x_1^2hx_1h^2$ or 2_A and 2_B, respectively.*

(b) *G has 19 z-special conjugacy classes represented by the conjugacy classes 2_1, 4_1, 4_2, 4_3, 4_6, 6_1, 8_1, 8_2, 8_6, 10, 12_1, 12_2, 16_1, 16_2, 20_1, 20_2, 20_3, 24_1, 24_2 of H.*

Proof (a) By Lemma 12.3.1(d) each simple group G has an elementary abelian subgroup A of order 8 in $H = \langle x_1, y_1, h\rangle$ such that $E = N_G(A) = \langle x, y, e\rangle$ contains a 2-central involution z such that $DE = C_E(z) = N_H(A)$ and $|H : D|$ is odd. By Lemma 12.1.3(d) there is an isomorphism $\psi : DE \to DH = N_H(A)$ such that $\psi(x) = x_1$ and $\psi(y) = y_1$. It has been used to identify the conjugacy classes of DE and DH and their inclusions into E and H, respectively.

Tables 12.5.1, 12.5.2 and 12.5.3 provide systems of representatives of the conjugacy classes of H, D and E, respectively. Let $d = x^6y^2xy^2xy^2 \in D$, $s = x_1^2hx_1h^2 \in H$ and $s_1 = (xe)^7 \in E$. Using the faithful permutation representations PE and PH of E and H defined in Lemma 12.1.1(e) and Proposition 12.1.2(d), respectively, and the MAGMA commands `IsConjugate(PE,d,s_1)` and `IsConjugate(PH, d,s)` the reader can check that d and s_1 are E-conjugate and d and s are H-conjugate. Hence s and s_1 are G-conjugate. By Table 12.5.3 $|C_E(s_1)| = 2^8 \cdot 7$. As $|H = C_G(z)| = 2^{14} \cdot 3 \cdot 7$ the involutions $z = x^8$ and s are not conjugate in G.

The same method can be applied to the representatives of the other seven conjuagcy classes of involutions of D listed in Table 12.5.2. Thus one obtains the following fusion pattern using the notation of Tables 12.5.1 and 12.5.3, respectively. Here the classes of H and E are denoted by the numbers of the representatives of the involutions. The classes of H are stated before the ones of D. The last entries denote the classes of E.

$$2A := \langle [1, 2, 3, 4, 5], [1, 2, 3, 4, 5, 6, 8], [1, 2, 4] \rangle,$$
$$2B := \langle [6], [7], [3] \rangle.$$

In particular, z and s represent all conjugacy classes of involutions of G.

(b) This statement follows immediately from the power map information given in Table 12.5.1. This completes the proof. □

Lemma 12.3.3 *Let G be a finite simple group of* Ru*-type with a 2-central involution z such that $H = C_G(z) = \langle x_1, y_1, h \rangle$. Then the following assertions hold.*

(a) *$n_1 = x_1y_1$ generates a Sylow 3-subgroup T of H.*
(b) *$C_H(T) = \langle n_1, n_2, n_3 \rangle = T \times S$, where $S = \langle n_2, n_3 \rangle$ is semidihedral of order 16 and $n_2 = (h^2y_1x_1)^5$ and $n_3 = (x_1^2hx_1^2y_1^2)^3$ have respective orders 4 and 2.*
(c) *$N_H(T) = \langle C_H(T), n_4 \rangle$, where $n_4 = (x_1hx_1hy_1hx_1)^3$ has order 8. Furthermore, $Q = \langle n_2, n_4^2 \rangle$ is a quaternion subgroup of S.*
(d) *$n_2^2 = n_4^4 = px_1^8 = z$ and n_2 and n_4^2 are not conjugate in G.*
(e) *$C_G(T)/T \cong \mathrm{PGL}_2(9)$.*
(f) *$N_G(T)/T \cong Aut(A_6)$.*
(g) *$|C_G(s)|$ is divisible by 5, where $s = x_1^2hx_1h^2$ is a representative of the unique class of non-2-central involutions of G.*

Proof (a) holds by Table 12.5.1.

(b) Let PH be the faithful permutation representation of H defined in Proposition 12.1.2(d). Using it and MAGMA the reader can check that $C_H(T) = \langle n_1, n_2, n_3\rangle = T \times S$, where $S = \langle n_2, n_3\rangle = \langle v, n_3\rangle$, where $v = n_2n_3$ has order 8. As $v^{n_3} = v^3$ the Sylow 2-subgroup S of $C_H(T)$ is semi-dihedral of order 16.

(c) and (d) These assertions can easily be checked by means of PH and MAGMA. In particular, $Q = \langle n_2, n_4^2\rangle$ is a quaternion subgroup of S with center $z = n_2^2 = n_4^4$. The z-special elements n_2 and n_4^2 of order 4 are not H-conjugate. Thus they are also not G-conjugate by Lemma 2.2.6 of [92].

(e) The semi-dihedral group S is also a Sylow 2-subgroup of $R = C_G(n_1) = C_G(T)$ because z generates the center of S. Clearly, $V_1 = \langle z, n_3\rangle$ is a Klein 4-group of S. Moreover, $C_R(V_1) = C_H(V_1) \cap R = T \times V_1$ and $C_R(Q)Q = T \times Q$ by (b) and Lemma 1.3.9 of [92]. Since R has two conjugacy classes of elements of order 4 by (d), Proposition 1.6.5 of [92] implies that $R = C_G(T)$ has a normal subgroup K of index 2 with a dihedral Sylow 2-subgroup S_1 such that K does not have a normal subgroup of index 2. Furthermore, R has a unique conjugacy class of involutions. Hence Theorem 1.6.4 of [92] due to Gorenstein and Walter implies that $K/T \cong PSL(2, 9) \cong A_6$. Therefore $R/T = C_G(T)/T \cong \mathrm{PGL}_2(9)$.

(f) By (b) and (c) we know that $|N_H(T)/C_H(T)| = 2$. Let $R_1 = N_G(T)$. Then R is normal in R_1 and $|R_1 : R| = 2$. Hence $R_1 = RN_H(T)$. Thus $N_H(T)/T$ is a non-trivial subgroup of $R_1/T \le Aut(C_G(T)/T) \cong Aut(A_6)$. As $|Aut(A_6)| = 1440$ it follows that $R_1/T = N_G(T)/T \cong Aut(A_6)$.

(g) Now (f) implies that $N_G(T)$ contains an element f of order 5 which is centralized by n_1. But f is also centralized by an involution $s_1 \in N_G(T)$ which inverts n_1; see (f) and the Atlas [19], p. 5. By Table 12.5.1 the element $f_1 = (y_1h)^4$ of order 5 has a centralizer $C_H(f_1)$ of order $2^3 \cdot 5$. Using PH and MAGMA the reader can check that a Sylow 2-subgroup S_2 of $C_H(f_1)$ is a quaternion group of order 8. Now Theorem 4.6.5 of [92] asserts that $L = C_G(f_1) = OC_L(z)$, where O is the largest normal subgroup of $L = C_G(f_1)$ of odd order. Operating with the Klein 4-group W generated by z and $w = (x_1h)^2$ on O an application of Theorem 4.2.2 of [92] yields that $|O|$ divides 5^3 because z, w and zw are conjugate in G by Proposition 12.3.2 and Table 12.5.1. Thus $C_G(f_1)$ is a $\{2, 5\}$-group. In particular, f and f_1 are not G-conjugate. Since $HC_G(z)$ has a unique conjugacy class of elements of order 5 the involution $s_1 \in N_G(T)$ is not G-conjugate to z. Thus s_1 is conjugate to $s = x_1^2hx_1h^2$ by Proposition 12.3.2 and 5 divides $|C_G(s)|$. This completes the proof. □

Proposition 12.3.4 *Let G be a finite simple group of Ru-type with a 2-central involution z such that $C_G(z) = H = \langle x_1, y_1, h\rangle$. Keep the notation of Proposition 12.3.2. Then the following assertions hold.*

(a) *$s = x_1^2 h x_1 h^2$ of G represents the only non-2-central conjugacy class of involutions of G. It is G-conjugate to $b = (xe)^7 \in E = \langle x, y, e\rangle$.*

(b) *$P = C_E(b) = \langle p_1, p_2\rangle$, where $p_1 = xe$ and $p_2 = (ex^2ex)^2$. The center $R = Z(P)$ of P is a Klein 4-group generated by b and $c = (p_1p_2)^7$.*

(c) *$P = R \times J$, where J of order $2^6 \cdot 7$ is generated by $j_1 = p_1^2$ and $j_2 = (p_1p_2)^2$.*

The Fitting subgroup FJ of J has order 2^6. Its center $Z(FJ)$ and $FJ/Z(FJ)$ are both elementary abelian of order 8.

(d) *$N = N_E(R) = \langle y_1, y_2\rangle$, where $y_1 = p_1$ and $y_2 = (x^2e^2x^3e)^2$ have orders 14 and 6, respectively. Furthermore, $|N| = 2^6 \cdot 3 \cdot 7$.*

(e) *$U = C_G(s) \cong R \times Y$, where $R = \langle b, c\rangle$ is a Klein 4-group and $Y \cong Sz(8)$, the simple Suzuki group of order $|Y| = 2^6 \cdot 5 \cdot 7 \cdot 13$.*

(f) *$M = N_G(R)$ is isomorphic to the finitely presented group $M = \langle d_1, d_2\rangle$ with the following set $\mathcal{R}(M)$ of defining relations:*

$$
\begin{aligned}
&d_1^2 = d_2^3 = 1, \quad (d_2^2 d_1)^{12} = 1, \quad (d_2 d_1 d_2^2 d_1 d_2^2 d_1)^6 = 1,\\
&d_2 d_1 d_2^2 d_1 d_2 d_1 d_2 d_1 d_2 d_1 d_2^2 d_1 d_2 d_1 d_2 d_1 d_2^2 d_1 d_2^2 d_1 d_2^2 d_1 d_2^2 d_1 d_2 d_1\\
&\quad \cdot d_2^2 d_1 d_2 d_1 d_2 d_1 d_2 d_1 d_2 d_1 d_2^2 d_1 d_2^2 d_1 d_2 d_1 d_2^2 d_1 d_2^2 d_1 d_2^2 d_1 = 1,\\
&d_2^2 d_1 d_2^2 d_1 d_2 d_1 d_2^2 d_1 d_2 d_1 d_2 d_1 d_2^2 d_1 d_2^2 d_1 d_2 d_1 d_2^2 d_1 d_2^2 d_1 d_2^2 d_1 d_2^2 (d_1 d_2)^2\\
&\quad \cdot d_1 d_2^2 d_1 d_2 d_1 d_2^2 d_1 d_2^2 d_1 d_2 d_1 d_2 d_1 d_2 d_1 d_2 d_1 d_2^2 d_1 d_2 d_1 d_2 d_1 = 1,\\
&d_2 d_1 d_2 d_1 d_2^2 d_1 d_2 d_1 d_2^2 d_1 d_2 d_1 d_2^2 d_1 d_2 d_1 d_2^2 d_1 d_2 d_1 d_2^2 d_1 d_2^2 d_1 d_2^2 (d_1 d_2)^2\\
&\quad \cdot d_1 d_2^2 d_1 d_2 d_1 d_2^2 d_1 d_2^2 d_1 d_2 d_1 d_2 d_1 d_2^2 d_1 d_2 d_1 d_2^2 d_1 d_2^2 d_1 d_2 d_1 = 1,\\
&(d_1 d_2^2)^2 d_1 d_2 d_1 d_2^2 d_1 d_2 d_1 d_2^2 d_1 d_2^2 d_1 d_2 d_1 d_2^2 d_1 d_2^2 d_1 d_2^2 d_1 d_2^2 d_1 d_2^2 d_1 d_2 d_1\\
&\quad \cdot d_2^2 d_1 d_2^2 d_1 d_2 d_1 d_2^2 d_1 d_2 d_1 d_2^2 d_1 d_2 d_1 d_2^2 d_1 d_2^2 d_1 d_2 d_1 d_2^2 d_1 d_2 d_1 d_2^2 = 1.
\end{aligned}
$$

Furthermore, M has a faithful permutation representation PM of degree 260 with stabilizer $\langle d_2, (d_1 d_2 d_1 d_2^2 (d_1 d_2)^5)^2\rangle$.

(g) *A system of representatives of the 28 conjugacy classes of $M = \langle d_1, d_2\rangle$ is given in Table 12.5.4.*

(h) *The character table of M is given in Table 12.6.5.*

(i) *G has eight classes of s-special conjugacy classes represented by the following classes of M: 2_b, 10_a, 14_a, 14_b, 14_c, 26_a, 26_b, 26_c.*

Proof (a) holds by Proposition 12.3.2(a). In particular, the involutions $s = x_1^2 h x_1 h^2$ of H and $b = (xe)^7$ of E are conjugate in G.

(b), (c) and (d) The reader can check all three statements by means of MAGMA and the faithful permutation representation PE of $E = \langle x, y, e\rangle$ given in Lemma 12.1.1(e).

(e) Since s and b are G-conjugate $C_G(s)$ is conjugate to $C_G(b)$ in G. Hence we calculate $N = N_E(R)$ containing $C_E(b)$. The stated generators y_1 and y_2 of N have been found by means of the program `GetShortGens(E,N)`.

(g) and (f) By (c) $C_G(b) = R \times Y$, where $Y \geq J$ has a metabelian Sylow 2-subgroup S of order 2^6 with an elementary abelian center of order 2^3. In particular, $|Y : J|$ is odd. Now Lemma 12.3.3 implies that 5 divides $|Y|$. Hence $N_G(S)/R$ is a strongly embedded subgroup of $N_G(R)/R$ by (c) and Theorem 4.8.4 of [92] because all involutions of $W = N_G(R)/R$ are conjugate in W by (a) and Proposition 12.3.2(b). Now Theorem 1.7.3 of [92] asserts that W possesses a normal series

$$W \rhd W_1 \rhd W_2 \unrhd 1 \text{ such that } |W/W_1|,\ |W_2| \text{ are odd, and}$$
$$W_1/W_2 \cong Sz(8).$$

Here $Sz(8)$ denotes the Suzuki group of order $2^6 \cdot 5 \cdot 7 \cdot 13$. Since the outer automorphism group $Out(Sz(8))$ has order 3, and $|N/J| = 3$ by (c) and (d), it follows that $|W/W_1| = 3$. Since $C_H(s)$ has order 2^8 by Table 12.6.1, and all involutions of J are conjugate to z, an application of the Brauer–Wielandt Theorem 4.2.2 of [92] yields that $W_2 = 1$. Therefore $U_1 = C_G(b) = \langle b, c\rangle \times Y$ and $Y \cong Sz(8)$. Thus (e) holds.

Now Theorem 1.4.15 of [92] and (c) imply that the derived subgroup M' of $M = N_G(R) = \langle N, Y\rangle$ is a direct product of R and $Y \cong Sz(8)$ such that there is an element $r \in M - M'$ of order 3 which operates on Y as the generator of order 3 of $Out(Sz(8))$ and $\langle R, r\rangle \cong A_4$. Therefore it is easy to establish a presentation of M. The one of (g) has been found by means of MAGMA.

(d) The system of representatives of the 49 conjugacy classes of M has been calculated by means of PM, Kratzer's Algorithm 5.3.18 of [92] and MAGMA.

(e) The character table of M has been computed by means of the permutation representation of PM and MAGMA. This completes the proof. □

Proposition 12.3.5 *Let G be a finite simple group of* Ru*-type having a 2-central involution z with $C_G(z) = H$. Let $M = \langle d_1, d_2\rangle$ be the finitely* *presented group of Proposition 12.3.4(f). Then the following assertions hold.*

(a) *G has two conjugacy classes of involutions represented by z and $s = x_1^2 h x_1 h^2$, and $C_G(s) \cong R \times Sz$, where $R = \langle r_1, r_2 \rangle$ is a Klein 4-group generated by the involutions $r_1 = (d_1 d_2 d_1 d_2^2)^7$, $r_2 = (d_1 d_2^2 d_1 d_2)^7$ and $Sz = \langle (d_1 d_2)^3, (d_1 (d_1 d_2)^3)^2 \rangle \cong Sz(8)$.*

(b)

$$|G| = 2^{14} \cdot 3^3 \cdot 5^3 \cdot 7 \cdot 13 \cdot 29.$$

Proof. (a) By Propositions 12.3.2 and 12.3.4 each simple group G of Ru-type has two conjugacy classes of involutions z^G and s^G such that

$$|H| = |C_G(z)| = 2^{14} \cdot 3 \cdot 5 \text{ and } \quad |U| = |C_G(s)| = 2^8 \cdot 5 \cdot 7 \cdot 13.$$

Furthermore, $C_G(s) \cong R \times Sz(8)$ by Proposition 12.3.4(e). Using the faithful permutation representation PM of M defined in Proposition 12.3.4(f) and MAGMA the reader can check that the involution $u = r_1 = (d_1 d_2)^3$ of M has a centralizer $C_M(u) = R \times Sz$ generated by the four elements given in the assertion such that $R = \langle r_1, r_2 \rangle$ is a Klein 4-group and $Sz \cong Sz(8)$.

(b) Using the faithful permutation representation PH of H described in Proposition 12.1.2(d) and the fusion pattern of the conjugacy classes of involutions of H in G given in Proposition 12.3.2 an application of MAGMA yields

$$r(z, s, z) = |\{(x, y) \in (z^G \cap H) \times (s^G \cap H) | z \in \langle xy \rangle\}| = 2^6 \cdot 3 \cdot 5 \cdot 856.$$

By (a) we may assume that $U = C_G(s) = C_M(u) = R \times Sz$, $s = u = r_1$, and that z represents the unique conjugacy class of involutions of $Sz \cong Sz(8)$. From Table 12.6.5 and Proposition 12.3.2 it follows that the six conjugacy classes $(r_1)^U$, $(r_2)^U$, $(r_1 r_2)^U$, $(r_1 z)^U$, $(r_2 z)^U$ and $(r_1 r_2 z)^U$ fuse to s^G, and z^U is the only conjugacy class of involutions of U which extends to z^G. Using the faithful permutation representation PM of $M = N_G(R)$ with $R = \langle r_1, r_2 \rangle$ given in Proposition 12.3.4(f) it has been checked that

$$r(z, s, s) = |\{(x, y) \in (z^G \cap U) \times (s^G \cap U) | s \in \langle xy \rangle\}| = 5 \cdot 7 \cdot 13 \cdot 449.$$

Now Thompson's group order formula stated as Theorem 4.2.1 in [92] implies that

$$|G| = r(z,s,z)|U| + r(z,s,s)|H| = 2^{14} \cdot 3^3 \cdot 5^3 \cdot 7 \cdot 13 \cdot 29.$$

This completes the proof.

Proposition 12.3.6 *Let G be a finite simple group of* Ru*-type having a 2-central involution z with $C_G(z) = H\langle x_1, y_1, h\rangle$. Then the following assertions hold.*

(a) *A Sylow 3-subgroup P_3 of G is extra-special of exponent 3 and has order 3^3. Its center $Z(P_3)$ has normalizer $N_G(Z(P_3)) \cong 3Aut(A_6)$. Its character table is given in Table 12.6.4.*
(b) *G has one conjugacy class of elements of order 3. It is represented by $p = x_1y_1 \in H$.*

Proof (a) A Sylow 3-subgroup P_3 of G has order $|P_3| = 3^3$ by Proposition 12.3.5. If it were abelian, then $N_G(P_3) \cong T \times Aut(A_6)$ by Lemma 12.3.3(f), where T is generated by $p = x_1y_1 \in P_3$. By the proof of Lemma 12.3.3(g) $Aut(A_6)$ has two non-conjugate involutions z_1 and z_2 which are G-conjugate to z and s, respectively. Hence $x = ps$ is an element of order 6 which powers onto s. But G does not have such elements by Proposition 12.3.2(b) and Table 12.6.5. Therefore P_3 is extra-special, and $N_G(T)$ is a uniquely determined non-split extension of $Aut(A_6)$ by T. Its character table is given in Table 12.6.4. It asserts that P_3 has exponent 3.

(b) This statement follows now immediately from Tables 12.6.4 and 12.6.1. This completes the proof. □

Proposition 12.3.7 *Let G be a finite simple group of* Ru*-type having a 2-central involution z with $C_G(z) = H = \langle x_1, y_1, h\rangle$. Let $M = \langle d_1, d_2\rangle$ be the finitely presented group of Proposition 12.3.4(f). Let $R = \langle r_1, r_2\rangle$ be the elementary abelian Fitting subgroup of M of order 4 defined in Proposition 12.3.5(a).*

Then the following assertions hold.

(a) *There is a simple subgroup $Sz \cong Sz(8)$ of M such that $U = C_M(r_1) = R \times Sz$ is G-conjugate to $C_G(s)$, where s is a representative of the unique class of non-2-central involutions of G.*

(b) *G has two conjugacy classes 5_A and 5_B of elements of order 5 represented by $f_1 = (y_1h)^4 \in H$ and $f = (d_1d_2)^4 d_1 d_2^2)^2 \in M$, respectively.*

A Sylow 5-subgroup P_5 of G is extra-special of order 125 and exponent 5.

(c) *$N = N_G(\langle f_1 \rangle) \cong P_5 : (\langle y \rangle \times Q_8)$, where y has order 4 and Q_8 denotes the quaternion group of order 8. The character table of N is given in Table 12.6.6.*

(d) *$N_G(\langle f \rangle) \cong (A_5 \times F_{20})$, where F_{20} denotes the Frobenius group of 20. Its character table is given in Table 12.6.7.*

(e) *G has one conjugacy class of elements of order 7. It is represented by $a = (d_1d_2d_1d_2^2)^2 \in M$, and $P_7 = \langle a \rangle$ is a Sylow 7-subgroup of G. Furthermore, there is an element b of order 6 in M such that $N_G(\langle a \rangle) = \left[(\langle a \rangle : \langle b^3 \rangle) \times \langle r_1, r_2 \rangle\right] : \langle b^2 \rangle$.*

(f) *G has one conjugacy class of elements of order 13. It is represented by $k = ((d_1d_2)^2(d_1d_2^2)^2)^2 \in M$, and $P_{13} = \langle k \rangle$ is a Sylow 13-subgroup of G. Furthermore, there is an element c of order 12 in M such that $N_G(\langle k \rangle) = \left[(\langle a \rangle : \langle c^3 \rangle) \times \langle r_1, r_2 \rangle\right] : \langle c^4 \rangle$.*

(g) *A Sylow 29-normalizer $N_G(29_i)$ is a Frobenius group of order $14 \cdot 29$, $i = 1, 2$.*

Proof (a) By Proposition 12.3.4 we know that M has an involution r_1, a Klein 4-subgroup R and a simple subgroup Sz such $C_M(r_1) = R \times Sz$ and $Sz \cong Sz(8)$. Furthermore, Proposition 12.3.4(e) states that $C_M(r_1)$ is G-conjugate to $C_G(s)$, where $s = x_1^2hx_1h^2$ represents the second class of involutions of G.

(b) through (f) Using the faithful permutation representation PM of M defined in Proposition 12.3.4(g) and MAGMA the reader can check that the given elements f, a and k of Sz have respective orders 5, 7 and 13. Furthermore, the automizers $N_M(\langle x \rangle)/C_M(x)$ of the cyclic subgroups generated by them have respective orders 4, 6 and 12. Hence (e) and (f) follow at once from Proposition 12.3.5 and Sylow's Theorem.

By Table 12.5.1 $f_1 = (y_1h)^4$ represents the unique conjugacy class of elements of order 5 in $H = C_G(z)$. Since all involutions of R are G-conjugate to s, Table 12.5.4 implies that f_1 and f are not conjugate in G. By means of PM and MAGMA it has been checked that $N_M(\langle f \rangle)$ is isomorphic to a direct product of an alternating group A_4 and a Frobenius group F_{20} of order 20. Thus (d) holds by Theorem 1.6.2 of [92] and Proposition 12.3.5.

A Sylow 5-subgroup P_5 of G has order 5^3 by Proposition 12.3.5. Assertion (d) implies that a generator f_2 of a Sylow 5-subgroup of A_5 in $N_G(\langle f \rangle)$ is centralized by an involution in $z^G \cap F_{20}$. Hence f_1 and f_2 are G-conjugate. By the proof of Lemma 12.3.3(g) we know that $C_G(f_1)$ is a $\{2,5\}$-group with a normal subgroup O of 5-power order. In particular, Q_8 operates on O. It has an irreducible $GF(5)$-module of dimension 2. As f is not centralized by Q_8 it follows that $O = P_5$. Furthermore, P_5 is extra-special of exponent 5. Thus $N_G(f_1)/P_5$ is a direct product of Q_8 and the cyclic group $F_{20}/\langle f \rangle$ of order 4 because $C_G(f_1) = P_5 : Q_8$ by Table 12.5.1. Therefore (b) holds.

(g) This assertion follows from Propositions 12.3.4, 12.3.5, 12.3.6, Burnside's Theorem 1.4.17 of [92], Sylow's Theorem and the other assertions of this proposition. This completes the proof. □

12.4 Uniqueness proof

In this section we prove the following result due to D. Parrott [106].

Theorem 12.4.1 (D. Parrott) *Let G be a finite simple group which possesses an involution z such that $H = C_G(z)$ satisfies the following properties.*

(a) *$J = O_2(H)$ has order 2^{11} and class at least 3.*
(b) *$H/J \cong S_5$, the symmetric group of degree 5.*
(c) *If Q is a Sylow 3-subgroup of H then $C_J(Q) \cong Q_8$, the quaternion group of order 8.*

Then $G \cong$ Ru, the Rudvalis simple group.

In [106] Parrott does not prove that a finite group H with the above three properties is uniquely determined up to isomorphism. In fact, that need not be true. Therefore we prove here that any simple group G of Ru-type is isomorphic to the finite simple group $\mathfrak{G}$ constructed in Theorem 12.2.1. By means of Lemma 12.1.3 the reader can check that any simple group of Ru-type satisfies the three conditions of Theorem 12.4.1.

In [106] Parrott shows that any simple group G meeting his three conditions is a rank 3 permutation group with stabilizer isomorphic to the finite group of Lie type $F_4(2)$ having index 4060 in G. Thus he is able to identify G with the original Rudvalis group Ru discovered by Rudvalis [115].

This last step can be adapted for $\mathfrak{G} = \langle \mathfrak{x}, \mathfrak{y}, \mathfrak{h}, \mathfrak{e} \rangle$ as follows. Let $P\mathfrak{G}$ be its faithful permutation representation of degree 424125 constructed in the proof of Theorem 12.2.1. In $\mathfrak{H} = \langle \mathfrak{x}, \mathfrak{y}, \mathfrak{h} \rangle$ find a subgroup $\mathfrak{H}_1$ of order $2^{11} \cdot 5$ which is isomorphic to the group H of Proposition 10.1.5. In particular, $\mathfrak{H}_1$ has a uniquely determined elementary abelian normal subgroup $\mathfrak{A}$ which is not normal in $\mathfrak{H}_1$. Let $\mathfrak{D} = N_{\mathfrak{H}_1}(\mathfrak{A})$. Then calculate $\mathfrak{E}_1 = N_{\mathfrak{G}}(\mathfrak{A})$. In $\mathfrak{E}_1$ find a subgroup $\mathfrak{E}$ of order $2^{11} \cdot 3$ which is isomorphic to the finitely presented group E of Lemma 10.1.2. Using MAGMA and $P\mathfrak{G}$ again show that $\mathfrak{T} = \langle \mathfrak{H}_1, \mathfrak{E} \rangle$ is isomorphic to the simple group of order $2^{11} \cdot 3^3 \cdot 5^2 \cdot 13$ constructed in Theorem 10.3.1. Now let $\mathfrak{F} = N_{\mathfrak{G}}(\mathfrak{T})$. Then $|\mathfrak{G} : \mathfrak{F}| = 4060$, and the permutation character $(1_{\mathfrak{T}})^{\mathfrak{G}}$ is a sum of the irreducible characters χ_1, χ_5 and χ_6 of Table 12.6.8. Because of lack of space these calculations are not documented here.

Lemma 12.4.2 *Let G be a finite simple group of* Ru*-type with a 2-central involution z such that $C_G(z) = H$ described in Proposition 12.1.2. Then H is a maximal subgroup of G.*

Proof Lemma 12.4.2 must be true, otherwise there is a proper subgroup X of G containing $H = C_G(z)$ properly such that X is of minimal order with respect to these two properties. Hence $C_X(z) = H$.

Suppose that X has a normal subgroup Y of odd order such that $X = YH$. Then $Y \cap H = 1$ because H does not have any normal subgroups of odd order. Therefore Lemma 1.5.9 of [92] and Proposition 12.3.5 assert that Y is an abelian group of order dividing $q = 3^3 \cdot 5^3 \cdot 7 \cdot 13 \cdot 19$. By Table 12.5.1 and Lemma 12.1.3(f) $V = \langle z, t \rangle$ with $t = x_1^8(x_1^4 y_1^2)^2$ is a Klein 4-subgroup of H such that $|C_H(v)| = 2^{13}$. Lemma 12.3.1(b) asserts that z and t are G-conjugate. Thus $H = C_G(z) \cong C_G(t)$. By Theorem 4.2.2 of [92] we know that $|Y||C_Y(V)|^2 = |C_Y(z)||C_Y(t)|^2$ because t and tz are conjugate in H. But H does not have any normal subgroups of odd order. Hence $|C_Y(z)| = |C_Y(v)| = 1$. Thus $Y = 1$ and $X = H$, a contradiction.

Let S be a Sylow 2-subgroup of H. It is a Sylow 2-subgroup of X also. Now Theorem 4.7.3 of [92] due to G. Glauberman implies that there is an $x \in X$ such that $z^x \in S - \{z\}$. By Lemma 12.1.3(f) we may assume that $V = \langle z, t \rangle$ is the unique normal Klein 4-group of S. Another application of Lemma 12.3.1 asserts that z and t are X-conjugate and $N_X(A)/C_X(A) \cong \mathrm{GL}_3(2)$, where A is the unique elementary abelian normal subgroup of order 8 in S. Now all the arguments of the proof of Proposition 12.3.2 can be applied. Thus $N_X(A) \cong E$, where $E =$

$\langle x, y, e\rangle$ is the group constructed in Lemma 12.1.1. Furthermore, X has two conjugacy classes of involutions represented by z and $s = x_1^2 h x_1 h^2 \in H$. Therefore Propositions 12.3.4 and 12.3.5 yield that $|X| = 2^{14} \cdot 3^3 \cdot 5^3 \cdot 7 \cdot 13 \cdot 29 = |G|$. This contradiction completes the proof. □

Proposition 12.4.3 *Each finite simple group G of* Ru*-type has a unique pair of complex conjugate characters $\chi : G \to \mathbb{C}$ of degree $\chi(1) = 378$.*

Proof By the results of Section 12.3 each simple group G of Ru-type has two conjugacy classes of involutions represented by z and s and order $|G| = 2^{14} \cdot 3^3 \cdot 5^3 \cdot 7 \cdot 13 \cdot 29$. Furthermore, it has 19 z-special and eight s-special classes of even order. Moreover it has one conjugacy class of elements of each of the orders 1, 3, 7, 13, 15 and two conjugacy classes of orders 5 and 29. Thus it has 36 conjugacy classes as $\mathfrak{G}$; see Table 12.5.5. Also the orders of the centralizers $C_G(x)$ of all these representatives x of the conjugacy classes of G coincide with the respective classes of $\mathfrak{G}$. Inspired by Table 12.6.8 we define the class function $\chi : g \to \chi(g)$ for all $g \in G$ as follows:

g	1_A	2_A	2_B	3_A	4_A	4_B	4_C	4_D	5_A	5_B	6_A	7_A	8_A	8_B	8_C
$\chi(g)$	378	−6	14	0	6	−6	2	−2	3	3	0	0	0	−2	0

g	10_A	10_B	12_A	12_B	13_A	14_A	14_B	14_C	15_A	16_A	16_B
$\chi(g)$	−1	−1	0	0	1	0	0	0	0	$2i$	$-2i$

g	20_A	20_B	20_C	24_A	24_B	26_A	26_B	26_C	29_A	29_B
$\chi(g)$	1	−1	−1	0	0	1	1	1	1	1

The complex conjugate class function $\chi^* : G \to \mathbb{C}$ differs from χ only at the classes 16_A and 16_B by a factor of -1.

By Corollary 2.8.10 of [92] it has to be shown that the restriction of χ to any of the normalizers $N_G(\langle p\rangle)$ is a generalized character, where $p \in \{2_A, 2_B, 3_A, 5_A, 5_B, 7_A, 13_A, 29_A, 29_B\}$.

Proposition 12.3.7 states that $C_G(2_B)$, $N_G(7_A)$ and $N_G(13_A)$ are isomorphic to subgroups of $M = \langle d_1, d_2\rangle$. Using the character table of M given in Table 12.6.5, and the fusion between the conjugacy classes of M and the ones of G described in the subsidiary results of Section 12.3, it follows that

$$\chi_{|M} = \chi_{19} + \chi_{28}, \quad \text{where } \chi_{19}, \chi_{28} \in Irr_{\mathbb{C}}(M).$$

Theorem 12.2.1(a) states that

$$\chi_{|H} = \chi_{10} + \chi_{31} + \chi_{44} + \chi_{48}, \quad \text{where } \chi_{10}, \chi_{31}, \chi_{44}, \chi_{48} \in Irr_{\mathbb{C}}(H).$$

By Proposition 12.3.7 a Sylow 29-normalizer N_{29} is a Frobenius group with cyclic Frobenius kernel of order 29 and complement C of order 14. Hence N_{29} has two exceptional irreducible characters τ_1 and τ_2 of degree 14 and 14 linear characters ψ_i, $1 \leq i \leq 14$, with projective indecomposable characters $\Phi_i = \psi_i + (\tau_1 + \tau_2)$ by Proposition 3.17.2(i) of [92].

Let ψ_1 and ψ_2 be the trivial character and the unique non-trivial rational character of C, respectively. Then $(\chi, \psi_i) = 0$ for all ψ_i which do not have the involution of C in their kernel, and $(\chi, \psi_j) = 2$ for all other indices j. Thus by Proposition 3.17.6 of [92] there exists exactly one character ψ_k with $k \in \{j\}$ such that $\chi_{|N(29)} = \psi_k + \Phi_k + \sum_{j \neq k} 2 \cdot \Phi_j$.

Using the character Tables 12.6.4, 12.6.6 and 12.6.7 one checks that also the restrictions of χ to the local subgroups $N_G(3_A)$, $N_G(5_A)$ and $N_G(5_B)$ are positive linear combinations of irreducible characters of the three subgroups. Therefore χ is a generalized character by Theorem 2.8.9 of [92] due to R. Brauer. Since $(\chi, \chi) = 1$ it is an irreducible character. This completes the proof. □

Theorem 12.4.4 *Let H be the finite group of even order defined in Proposition 12.1.2. Then each finite simple group G of* Ru*-type is isomorphic to the finite simple group $\mathfrak{G} \leq \mathrm{GL}_{378}(15)$ of order $|\mathfrak{G}| = 2^{14} \cdot 3^3 \cdot 5^3 \cdot 7 \cdot 13 \cdot 29$ constructed in Theorem 12.2.1.*

Proof Let G be any finite simple group of Ru-type. It suffices to show that G is isomorphic to the finite simple group $\mathfrak{G}$ constructed in Theorem 12.2.1. By Proposition 12.3.5 G has exactly one conjugacy class z^G of 2-central involutions with $C_G(z) = H$ and $|G| = |\mathfrak{G}|$. Proposition 12.1.2 asserts that a Sylow 2-subgroup S of H contains a maximal special normal subgroup C with elementary abelian center $Z(C) = A$ of order 8 such that $D = N_H(C)$ is a proper subgroup of $E = N_G(A)$. Furthermore, E is uniquely determined by H up to isomorphism by Lemma 12.1.1. Lemma 12.4.2 implies that $G = \langle H, E \rangle$. The amalgam $H \leftarrow D \rightarrow E$ has Goldschmidt index 1 by Lemma 12.1.3. Proposition 12.4.3 states that G has a unique pair of conjugate irreducible complex characters of degree 378. As 17 is not a divisor of $|G|$ the group G has an irreducible 17-modular representation of degree 378 by Maschke's Theorem. Hence $G \cong \mathfrak{G}$ by Theorem 7.5.1 of [92]. This completes the proof. □

12.5 Representatives of conjugacy classes

12.5.1 *Conjugacy classes of* $H = \langle x_1, y_1, h \rangle$

Class	*Representative*	$\lvert$*Centralizer*$\rvert$	2P	3P	5P
1	1	$2^{14} \cdot 3 \cdot 5$	1	1	1
2_1	$(x_1)^8$	$2^{14} \cdot 3 \cdot 5$	1	2_1	2_1
2_2	$(x_1 h)^2$	2^{13}	1	2_2	2_2
2_3	$(x_1^2 y_1^3)^2$	2^{10}	1	2_3	2_3
2_4	$(y_1)^2$	2^9	1	2_4	2_4
2_5	$(x_1^2 y_1 x_1 h)^3$	$2^7 \cdot 3$	1	2_5	2_5
2_6	$x_1^2 h x_1 h^2$	2^8	1	2_6	2_6
3	$x_1 y_1$	$2^4 \cdot 3$	3	1	3
4_1	$(y_1 h)^5$	$2^9 \cdot 3 \cdot 5$	2_1	4_1	4_1
4_2	$(x_1^3 y_1)^3$	$2^8 \cdot 3 \cdot 5$	2_1	4_2	4_2
4_3	$(x_1)^4$	2^{10}	2_1	4_3	4_3
4_4	$(x_1^2 y_1)^2$	2^9	2_2	4_4	4_4
4_5	$(x_1^2 y_1^2)^2$	2^9	2_2	4_5	4_5
4_6	$(x_1^4 y_1)^2$	2^9	2_1	4_6	4_6
4_7	$(x_1^3 h y_1)^2$	2^9	2_2	4_7	4_7
4_8	$(x_1^3 h^2)^2$	2^9	2_2	4_8	4_8
4_9	$(x_1^4 h^2)^2$	2^9	2_2	4_9	4_9
4_{10}	$x_1 h$	2^8	2_2	4_{10}	4_{10}
4_{11}	$x_1^4 y_1^2$	2^8	2_2	4_{11}	4_{11}
4_{12}	$x_1^2 y_1 x_1 h^2$	2^8	2_2	4_{12}	4_{12}
4_{13}	$x_1 y_1^2 h x_1 h$	2^8	2_2	4_{13}	4_{13}
4_{14}	$x_1^2 y_1 h y_1 h y_1^2$	2^8	2_2	4_{14}	4_{14}
4_{15}	$x_1^3 y_1 x_1 y_1^2$	2^7	2_2	4_{15}	4_{15}
4_{16}	y_1	2^6	2_4	4_{16}	4_{16}
4_{17}	h	2^6	2_4	4_{17}	4_{17}
4_{18}	$x_1^2 y_1^3$	2^6	2_3	4_{18}	4_{18}
4_{19}	$x_1 y_1 x_1 h y_1^2$	2^6	2_3	4_{19}	4_{19}
4_{20}	$x_1^3 h^2 y_1 h$	2^6	2_3	4_{20}	4_{20}
5	$(y_1 h)^4$	$2^3 \cdot 5$	5	5	1
6_1	$(x_1 y_1 h)^4$	$2^4 \cdot 3$	3	2_1	6_1
6_2	$x_1^2 y_1 x_1 h$	$2^2 \cdot 3$	3	2_5	6_2
8_1	$(x_1 y_1 h)^3$	$2^5 \cdot 3$	4_1	8_1	8_1
8_2	$(x_1)^2$	2^6	4_3	8_2	8_2
8_3	$x_1^2 y_1$	2^5	4_4	8_3	8_3
8_4	$x_1^2 y_1^2$	2^5	4_5	8_4	8_4
8_5	$x_1^4 y_1$	2^5	4_6	8_5	8_5
8_6	$x_1^3 h y_1$	2^5	4_7	8_6	8_6
8_7	$x_1^3 h^2$	2^5	4_8	8_7	8_7
8_8	$x_1^4 h^2$	2^5	4_9	8_8	8_8
10	$(y_1 h)^2$	$2^3 \cdot 5$	5	10	2_1
12_1	$(x_1 y_1 h)^2$	$2^3 \cdot 3$	6_1	4_1	12_1
12_2	$x_1^3 y_1$	$2^2 \cdot 3$	6_1	4_2	12_2
16_1	x_1	2^4	8_2	16_2	16_1
16_2	$x_1 h^2$	2^4	8_2	16_1	16_2
20_1	$y_1 h$	$2^2 \cdot 5$	10	20_1	4_1
20_2	$x_1^3 h$	$2^2 \cdot 5$	10	20_3	4_2
20_3	$x_1 h^2 y_1$	$2^2 \cdot 5$	10	20_2	4_2
24_1	$x_1 y_1 h$	$2^3 \cdot 3$	12_1	8_1	24_1
24_2	$y_1^2 h$	$2^3 \cdot 3$	12_1	8_1	24_2

12.5.2 *Conjugacy classes of* $D = \langle x, y\rangle$

Class	*Representative*	|*Centralizer*|	2P	3P
1	1	$2^{14}\cdot 3$	1	1
2_1	$(x)^8$	$2^{14}\cdot 3$	1	2_1
2_2	$(x^2y^2)^4$	2^{13}	1	2_2
2_3	$(x^2y)^4$	2^{11}	1	2_3
2_4	$(x^2yxy^3)^2$	2^{10}	1	2_4
2_5	$(y)^2$	2^9	1	2_5
2_6	$(x^2y^3)^2$	2^8	1	2_6
2_7	$x^6y^2xy^2xy^2$	2^8	1	2_7
2_8	$x^6yxy^2x^2y$	2^7	1	2_8
3	xy	$2^3\cdot 3$	3	1
4_1	$(x^5y)^3$	$2^9\cdot 3$	2_1	4_1
4_2	$(x)^4$	2^{10}	2_1	4_2
4_3	$(x^3y)^3$	$2^8\cdot 3$	2_1	4_3
4_4	$(x^2y^2)^2$	2^9	2_2	4_4
4_5	$(x^4y)^2$	2^9	2_1	4_5
4_6	$(x^3yx^2y)^2$	2^9	2_2	4_6
4_7	$(x^2y^3xy)^2$	2^9	2_2	4_7
4_8	$(x^2y^3xy^3)^2$	2^9	2_2	4_8
4_9	$(x^3y^2xyx^2y)^2$	2^9	2_2	4_9
4_{10}	x^4y^2	2^8	2_2	4_{10}
4_{11}	$x^2y^2xy^2xy^2$	2^8	2_2	4_{11}
4_{12}	$xyxy^3xy^3$	2^8	2_2	4_{12}
4_{13}	$x^4yx^3y^2xy$	2^8	2_2	4_{13}
4_{14}	$x^3y^2xy^3x^2y$	2^8	2_1	4_{14}
4_{15}	$x^3yx^3yx^2y^2xy$	2^8	2_2	4_{15}
4_{16}	$(x^2y)^2$	2^7	2_3	4_{16}
4_{17}	x^3yxy^2	2^7	2_2	4_{17}
4_{18}	x^6yx^2y	2^7	2_3	4_{18}
4_{19}	x^3yxy^3xy	2^7	2_3	4_{19}
4_{20}	y	2^6	2_5	4_{20}
4_{21}	x^2yxy^3	2^6	2_4	4_{21}
4_{22}	$x^2yx^2y^3$	2^6	2_4	4_{22}
4_{23}	x^8y	2^6	2_5	4_{23}
4_{24}	$x^3yx^2y^3$	2^6	2_3	4_{24}
4_{25}	$x^2yx^2y^2xy$	2^6	2_4	4_{25}
4_{26}	x^2y^3	2^5	2_6	4_{26}
4_{27}	$x^4y^2x^2y^3$	2^5	2_6	4_{27}
6	$(x^3y)^2$	$2^3\cdot 3$	3	2_1
8_1	$(x)^2$	2^6	4_2	8_1
8_2	x^2y^2	2^5	4_4	8_2
8_3	x^4y	2^5	4_5	8_3
8_4	x^3yx^2y	2^5	4_6	8_4
8_5	x^3y^2xy	2^5	4_1	8_5
8_6	x^2y^3xy	2^5	4_7	8_6
8_7	$x^2y^3xy^3$	2^5	4_8	8_7
8_8	$x^3y^2xyx^2y$	2^5	4_9	8_8
8_9	x^2y	2^4	4_{16}	8_9
12_1	x^3y	$2^2\cdot 3$	6	4_3
12_2	x^5y	$2^2\cdot 3$	6	4_1
12_3	x^2y^2xy	$2^2\cdot 3$	6	4_3
16_1	x	2^4	8_1	16_2
16_2	$xyxy^2$	2^4	8_1	16_1

12.5.3 *Conjugacy classes of* $E = \langle x, y, e \rangle$

Class	*Representative*	|*Centralizer*|	2P	3P	7P
1	1	$2^{14} \cdot 3 \cdot 7$	1	1	1
2_1	$(x)^8$	$2^{14} \cdot 3$	1	2_1	2_1
2_2	$(y)^2$	2^{11}	1	2_2	2_2
2_3	$(xe)^7$	$2^8 \cdot 7$	1	2_3	2_3
2_4	$(x^2y^3)^2$	2^8	1	2_4	2_4
3	e	$2^3 \cdot 3$	3	1	3
4_1	$(xey)^3$	$2^9 \cdot 3$	2_1	4_1	4_1
4_2	$(x)^4$	2^{10}	2_1	4_2	4_2
4_3	$(xye)^3$	$2^8 \cdot 3$	2_1	4_3	4_3
4_4	$(x^4y)^2$	2^9	2_1	4_4	4_4
4_5	x^3yxy^2	2^8	2_1	4_5	4_5
4_6	y	2^7	2_2	4_6	4_6
4_7	$(x^2y)^2$	2^7	2_2	4_7	4_7
4_8	y^2e^2ye	2^7	2_2	4_8	4_8
4_9	x^3ye	2^6	2_2	4_9	4_9
4_{10}	x^2y^3	2^5	2_4	4_{10}	4_{10}
4_{11}	x^2yexe	2^5	2_4	4_{11}	4_{11}
6	$(xye)^2$	$2^3 \cdot 3$	3	2_1	6
7_1	$(xe)^2$	$2^2 \cdot 7$	7_1	7_2	1
7_2	$(ye^2)^2$	$2^2 \cdot 7$	7_2	7_1	1
8_1	$(x)^2$	2^6	4_2	8_1	8_1
8_2	x^2y^2	2^5	4_1	8_2	8_2
8_3	x^4y	2^5	4_4	8_3	8_3
8_4	x^2y	2^4	4_7	8_4	8_4
12_1	xye	$2^2 \cdot 3$	6	4_3	12_1
12_2	xey	$2^2 \cdot 3$	6	4_1	12_2
12_3	xe^2	$2^2 \cdot 3$	6	4_3	12_3
14_1	xe	$2^2 \cdot 7$	7_1	14_3	2_3
14_2	ye	$2^2 \cdot 7$	7_1	14_4	2_3
14_3	ye^2	$2^2 \cdot 7$	7_2	14_2	2_3
14_4	x^2ye	$2^2 \cdot 7$	7_2	14_6	2_3
14_5	xye^2	$2^2 \cdot 7$	7_2	14_1	2_3
14_6	y^3e	$2^2 \cdot 7$	7_1	14_5	2_3
16_1	x	2^4	8_1	16_2	16_2
16_2	$xyxy^2$	2^4	8_1	16_1	16_1

12.5.4 *Conjugacy classes of* $M = \langle d_1, d_2 \rangle$

Class	*Representative*	|*Centralizer*|	2P	3P	5P	7P	13P
1	1	$2^8 \cdot 3 \cdot 5 \cdot 7 \cdot 13$	1	1	1	1	1
2_1	$(d_1 d_2 d_1 d_2^2)^7$	$2^8 \cdot 5 \cdot 7 \cdot 13$	1	2_1	2_1	2_1	2_1
2_2	$(d_1 d_2)^6$	$2^8 \cdot 3$	1	2_2	2_2	2_2	2_2
2_3	d_1	2^8	1	2_3	2_3	2_3	2_3
3_1	d_2	$2^2 \cdot 3 \cdot 5$	3_2	1	3_2	3_1	3_1
3_2	$(d_2)^2$	$2^2 \cdot 3 \cdot 5$	3_1	1	3_1	3_2	3_2
4_1	$(d_1 d_2)^3$	$2^6 \cdot 3$	2_2	4_2	4_1	4_2	4_1
4_2	$(d_1 d_2)^9$	$2^6 \cdot 3$	2_2	4_1	4_2	4_1	4_2
4_3	$(d_1 d_2)^4 d_1 d_2^2 d_1 d_2^2 d_1 d_2 d_1 d_2 d_1 d_2^2$	2^6	2_2	4_4	4_3	4_4	4_3
4_4	$((d_1 d_2)^4 d_1 d_2^2 d_1 d_2^2 d_1 d_2 d_1 d_2 d_1 d_2^2)^3$	2^6	2_2	4_3	4_4	4_3	4_4
5	$(d_1 d_2 d_1 d_2 d_1 d_2 d_1 d_2 d_1 d_2^2)^2$	$2^2 \cdot 3 \cdot 5$	5	5	1	5	5
6_1	$(d_1 d_2)^2$	$2^2 \cdot 3$	3_1	2_2	6_2	6_1	6_1
6_2	$(d_1 d_2)^{10}$	$2^2 \cdot 3$	3_2	2_2	6_1	6_2	6_2
7	$(d_1 d_2 d_1 d_2^2)^2$	$2^2 \cdot 7$	7	7	7	1	7
10	$d_1 d_2 d_1 d_2 d_1 d_2 d_1 d_2 d_1 d_2^2$	$2^2 \cdot 5$	5	10	2_1	10	10
12_1	$d_1 d_2$	$2^2 \cdot 3$	6_1	4_1	12_2	12_3	12_1
12_2	$(d_1 d_2)^5$	$2^2 \cdot 3$	6_2	4_1	12_1	12_4	12_2
12_3	$(d_1 d_2)^7$	$2^2 \cdot 3$	6_1	4_2	12_4	12_1	12_3
12_4	$(d_1 d_2)^{11}$	$2^2 \cdot 3$	6_2	4_2	12_3	12_2	12_4
13	$(d_1 d_2 d_1 d_2 d_1 d_2^2 d_1 d_2^2)^2$	$2^2 \cdot 13$	13	13	13	13	1
14_1	$d_1 d_2 d_1 d_2^2$	$2^2 \cdot 7$	7	14_2	14_3	2_1	14_1
14_2	$(d_1 d_2 d_1 d_2^2)^3$	$2^2 \cdot 7$	7	14_3	14_1	2_1	14_2
14_3	$(d_1 d_2 d_1 d_2^2)^5$	$2^2 \cdot 7$	7	14_1	14_2	2_1	14_3
15_1	$d_1 d_2 d_1 d_2 d_1 d_2^2 d_1 d_2 d_1 d_2^2$	$3 \cdot 5$	15_2	5	3_2	15_1	15_1
15_2	$(d_1 d_2 d_1 d_2 d_1 d_2^2 d_1 d_2 d_1 d_2^2)^2$	$3 \cdot 5$	15_1	5	3_1	15_2	15_2
26_1	$d_1 d_2 d_1 d_2 d_1 d_2^2 d_1 d_2^2$	$2^2 \cdot 13$	13	26_2	26_1	26_3	2_1
26_2	$(d_1 d_2 d_1 d_2 d_1 d_2^2 d_1 d_2^2)^3$	$2^2 \cdot 13$	13	26_3	26_2	26_1	2_1
26_3	$(d_1 d_2 d_1 d_2 d_1 d_2^2 d_1 d_2^2)^7$	$2^2 \cdot 13$	13	26_1	26_3	26_2	2_1

12.5.5 *Conjugacy classes of* $G = \langle x, y, h, e \rangle$

Class	*Representative*	$\lvert$*Centralizer*$\rvert$	2P	3P	5P	7P	13P	29P
1	1	$2^{14}\cdot 3^3\cdot 5^3\cdot 7\cdot 13\cdot 29$	1	1	1	1	1	1
2_1	$(x)^8$	$2^{14}\cdot 3\cdot 5$	1	2_1	2_1	2_1	2_1	2_1
2_2	$(xe)^7$	$2^8\cdot 5\cdot 7\cdot 13$	1	2_2	2_2	2_2	2_2	2_2
3	e	$2^4\cdot 3^3\cdot 5$	3	1	3	3	3	3
4_1	y	$2^9\cdot 3\cdot 5$	2_1	4_1	4_1	4_1	4_1	4_1
4_2	$(xye)^3$	$2^8\cdot 3\cdot 5$	2_1	4_2	4_2	4_2	4_2	4_2
4_3	$(x)^4$	2^{10}	2_1	4_3	4_3	4_3	4_3	4_3
4_4	h	2^9	2_1	4_4	4_4	4_4	4_4	4_4
5_1	$(yh)^4$	$2^3\cdot 5^3$	5_1	5_1	1	5_1	5_1	5_1
5_2	$(yhyeh)^2$	$2^2\cdot 3\cdot 5^2$	5_2	5_2	1	5_2	5_2	5_2
6	$(xyh)^4$	$2^4\cdot 3$	3	2_1	6	6	6	6
7	$(xe)^2$	$2^2\cdot 7$	7	7	7	1	7	7
8_1	$(xyh)^3$	$2^5\cdot 3$	4_1	8_1	8_1	8_1	8_1	8_1
8_2	$(x)^2$	2^6	4_3	8_2	8_2	8_2	8_2	8_2
8_3	x^2y	2^5	4_4	8_3	8_3	8_3	8_3	8_3
10_1	$(yh)^2$	$2^3\cdot 5$	5_1	10_1	2_1	10_1	10_1	10_1
10_2	$yhyeh$	$2^2\cdot 5$	5_2	10_2	2_2	10_2	10_2	10_2
12_1	$(xyh)^2$	$2^3\cdot 3$	6	4_1	12_1	12_1	12_1	12_1
12_2	xye	$2^2\cdot 3$	6	4_2	12_2	12_2	12_2	12_2
13	$(xhey)^2$	$2^2\cdot 13$	13	13	13	13	1	13
14_1	xe	$2^2\cdot 7$	7	14_2	14_3	2_2	14_1	14_1
14_2	$(xe)^3$	$2^2\cdot 7$	7	14_3	14_1	2_2	14_2	14_2
14_3	$(xe)^5$	$2^2\cdot 7$	7	14_1	14_2	2_2	14_3	14_3
15	$yhyehe$	$3\cdot 5$	15	5_2	3	15	15	15
16_1	x	2^4	8_2	16_2	16_1	16_2	16_1	16_1
16_2	$(x)^3$	2^4	8_2	16_1	16_2	16_1	16_2	16_2
20_1	yh	$2^2\cdot 5$	10_1	20_1	4_1	20_1	20_1	20_1
20_2	x^3h	$2^2\cdot 5$	10_1	20_3	4_2	20_3	20_3	20_2
20_3	$(x^3h)^3$	$2^2\cdot 5$	10_1	20_2	4_2	20_2	20_2	20_3
24_1	xyh	$2^3\cdot 3$	12_1	8_1	24_1	24_2	24_2	24_1
24_2	$(xyh)^7$	$2^3\cdot 3$	12_1	8_1	24_2	24_1	24_1	24_2
26_1	$xhey$	$2^2\cdot 13$	13	26_2	26_1	26_3	2_2	26_2
26_2	$(xhey)^3$	$2^2\cdot 13$	13	26_3	26_2	26_1	2_2	26_3
26_3	$(xhey)^7$	$2^2\cdot 13$	13	26_1	26_3	26_2	2_2	26_1
29_1	$xehe^2h$	29	29_2	29_2	29_1	29_1	29_1	1
29_2	$(xehe^2h)^2$	29	29_1	29_1	29_2	29_2	29_2	1

12.6 Character tables of local subgroups

12.6.1 *Character table of $H = \langle x, y, h \rangle$*

	1a	2a	2b	2c	2d	2e	2f	3a	4a	4b	4c	4d	4e	4f	4g	4h	4i	4j	4k	4l	4m	4n	4o	4p	4q	4r	4s	4t	5a	6a
2	14	14	13	10	9	7	8	4	9	8	10	9	9	9	9	9	9	8	8	8	8	8	7	6	6	6	6	6	3	4
3	1	1	.	.	.	1	.	1	1	1	.	.	.	.	.	.	.	.	.	.	.	.	.	.	.	.	.	.	.	1
5	1	1	.	.	.	.	.	.	1	1	.	.	.	.	.	.	.	.	.	.	.	.	.	.	.	.	.	.	1	.
2P	1a	1a	1a	1a	1a	1a	1a	3a	2a	2a	2a	2b	2b	2a	2b	2b	2b	2b	2b	2b	2b	2b	2b	2d	2d	2c	2c	2c	5a	3a
3P	1a	2a	2b	2c	2d	2e	2f	1a	4a	4b	4c	4d	4e	4f	4g	4h	4i	4j	4k	4l	4m	4n	4o	4p	4q	4r	4s	4t	5a	2a
5P	1a	2a	2b	2c	2d	2e	2f	3a	4a	4b	4c	4d	4e	4f	4g	4h	4i	4j	4k	4l	4m	4n	4o	4p	4q	4r	4s	4t	1a	6a
X.1	1	1	1	1	1	1	1	1	1	1	1	1	1	1	1	1	1	1	1	1	1	1	1	1	1	1	1	1	1	1
X.2	1	1	1	1	1	−1	1	1	1	1	1	1	1	1	1	1	1	1	1	1	1	1	−1	−1	−1	−1	1	−1	1	1
X.3	4	4	4	4	.	2	.	1	4	4	4	4	4	.	.	.	4	.	.	.	.	.	2	.	.	2	.	2	−1	1
X.4	4	4	4	4	.	−2	.	1	4	4	4	4	4	.	.	.	4	.	.	.	.	.	−2	.	.	−2	.	−2	−1	1
X.5	5	5	5	5	1	1	1	−1	5	5	5	5	5	1	1	1	5	1	1	1	1	1	1	−1	−1	1	1	1	.	−1
X.6	5	5	5	5	1	−1	1	−1	5	5	5	5	5	1	1	1	5	1	1	1	1	1	−1	1	1	−1	1	−1	.	−1
X.7	6	6	6	−2	2	.	−2	.	6	−6	−2	−2	2	2	2	2	2	−2	−2	−2	2	2	.	−2	−2	.	−2	.	1	.
X.8	6	6	6	−2	2	.	−2	.	6	−6	−2	−2	2	2	2	2	2	−2	−2	−2	2	2	.	2	2	.	−2	.	1	.
X.9	6	6	6	6	−2	.	−2	.	6	6	6	6	6	−2	−2	−2	6	−2	−2	−2	−2	−2	.	.	.	.	−2	.	1	.
X.10	6	6	6	−2	−2	.	2	.	6	−6	−2	−2	2	−2	−2	−2	2	2	2	2	−2	−2	.	.	.	.	2	.	1	.
X.11	6	6	6	−2	−2	.	2	.	6	−6	−2	−2	2	−2	−2	−2	2	2	2	2	−2	−2	.	.	.	.	2	.	1	.
X.12	10	10	10	2	−2	2	2	1	10	−10	2	2	−2	−2	−2	−2	−2	2	2	2	−2	−2	2	.	.	−2	−2	−2	.	1
X.13	10	10	10	2	2	−4	−2	1	10	−10	2	2	−2	2	2	2	−2	−2	−2	−2	2	2	−4	.	.	.	2	.	.	1
X.14	10	10	10	2	2	4	−2	1	10	−10	2	2	−2	2	2	2	−2	−2	−2	−2	2	2	4	.	.	.	2	.	.	1
X.15	10	10	10	2	−2	−2	2	1	10	−10	2	2	−2	−2	−2	−2	−2	2	2	2	−2	−2	−2	.	.	2	−2	2	.	1
X.16	12	12	12	4	4	.	.	.	−12	.	4	−4	.	−4	4	−4	.	.	.	.	4	−4	.	.	.	.	.	.	2	.
X.17	12	12	12	4	−4	.	.	.	−12	.	4	−4	.	4	−4	4	.	.	.	.	−4	4	.	.	.	.	.	.	2	.
X.18	15	15	15	−1	−1	−3	−1	.	15	15	−1	−1	−1	−1	−1	−1	−1	−1	−1	−1	−1	−1	−3	1	1	1	3	1	.	.
X.19	15	15	15	−1	3	3	3	.	15	15	−1	−1	−1	3	3	3	−1	3	3	3	3	3	3	1	1	−1	−1	−1	.	.
X.20	15	15	15	−1	−1	3	−1	.	15	15	−1	−1	−1	−1	−1	−1	−1	−1	−1	−1	−1	−1	3	−1	−1	−1	3	−1	.	.
X.21	15	15	15	−1	3	−3	3	.	15	15	−1	−1	−1	3	3	3	−1	3	3	3	3	3	−3	−1	−1	1	−1	1	.	.
X.22	20	20	20	4	.	−2	.	−1	20	−20	4	4	−4	.	.	.	−4	.	.	.	.	.	−2	.	.	−2	.	−2	.	−1
X.23	20	20	20	4	.	2	.	−1	20	−20	4	4	−4	.	.	.	−4	.	.	.	.	.	2	.	.	2	.	2	.	−1
X.24	20	20	20	−4	4	.	.	2	−20	.	−4	4	.	−4	4	−4	.	.	.	.	4	−4	.	.	.	.	.	.	.	2
X.25	20	20	20	−4	−4	.	.	2	−20	.	−4	4	.	4	−4	4	.	.	.	.	−4	4	.	.	.	.	.	.	.	2
X.26	24	24	24	−8	.	.	.	.	24	−24	−8	−8	8	.	.	.	8	.	.	.	.	.	.	.	.	.	.	.	−1	.
X.27	24	24	24	8	.	.	.	.	−24	.	8	−8	.	.	.	.	.	.	.	.	.	.	.	.	.	.	.	.	−1	.
X.28	24	24	24	8	.	.	.	.	−24	.	8	−8	.	.	.	.	.	.	.	.	.	.	.	.	.	.	.	.	−1	.
X.29	30	30	30	−2	−2	.	−2	.	30	30	−2	−2	−2	−2	−2	−2	−2	−2	−2	−2	−2	−2	.	.	.	.	−2	.	.	.
X.30	40	40	40	−8	.	.	.	−2	−40	.	−8	8	.	.	.	.	.	.	.	.	.	.	.	.	.	.	.	.	.	−2
X.31	60	60	−4	4	−4	−6	4	.	.	.	−4	.	4	8	4	.	−4	.	.	−4	.	−4	2	2	2	−2	.	2	.	.
X.32	60	60	−4	4	4	−6	−4	.	.	.	−4	.	4	.	−4	8	−4	.	.	4	.	−4	2	−2	−2	−2	.	2	.	.
X.33	60	60	−4	4	.	−6	.	.	.	.	−4	.	−4	−4	8	4	4	−4	4	.	−4	.	2	.	.	2	.	−2	.	.
X.34	60	60	−4	4	8	−6	.	.	.	.	−4	.	−4	4	.	−4	4	4	−4	.	−4	.	2	.	.	2	.	−2	.	.
X.35	60	60	−4	4	8	6	.	.	.	.	−4	.	−4	4	.	−4	4	4	−4	.	−4	.	−2	.	.	−2	.	2	.	.
X.36	60	60	−4	4	.	6	.	.	.	.	−4	.	−4	−4	8	4	4	−4	4	.	−4	.	−2	.	.	−2	.	2	.	.
X.37	60	60	−4	4	4	6	−4	.	.	.	−4	.	4	.	−4	8	−4	.	.	4	.	−4	−2	2	2	2	.	−2	.	.
X.38	60	60	−4	4	−4	6	4	.	.	.	−4	.	4	8	4	.	−4	.	.	−4	.	−4	−2	−2	−2	2	.	−2	.	.
X.39	120	120	−8	−8	8	.	.	.	.	.	8	.	.	.	−8	.	.	.	8	−8	.	.	.	.	.	.	.	.	.	.
X.40	120	120	−8	8	.	.	.	.	.	.	−8	.	8	−8	.	−8	−8	.	.	.	.	8	.	.	.	.	.	.	.	.
X.41	120	120	−8	8	−8	.	.	.	.	.	−8	.	−8	.	−8	.	8	.	.	.	8	.	.	.	.	.	.	.	.	.
X.42	120	120	−8	−8	−8	.	−8	.	.	.	8	.	.	.	8	.	.	8	.	.	.	.	.	.	.	.	.	.	.	.
X.43	120	120	−8	−8	.	.	.	.	.	.	8	.	.	8	.	−8	.	−8	.	8	.	.	.	.	.	.	.	.	.	.
X.44	120	120	−8	−8	.	.	8	.	.	.	8	.	.	−8	.	8	.	.	−8	.	.	.	.	.	.	.	.	.	.	.
X.45	128	−128	.	.	.	.	.	−4	.	.	.	.	.	.	.	.	.	.	.	.	.	.	.	.	.	.	.	.	−2	4
X.46	128	−128	.	.	.	.	.	2	.	.	.	.	.	.	.	.	.	.	.	.	.	.	.	.	.	.	.	.	−2	−2
X.47	128	−128	.	.	.	.	.	2	.	.	.	.	.	.	.	.	.	.	.	.	.	.	.	.	.	.	.	.	−2	−2
X.48	192	−192	.	.	.	.	.	.	.	.	.	.	.	.	.	.	.	.	.	.	.	.	.	4	−4	.	.	.	2	.
X.49	192	−192	.	.	.	.	.	.	.	.	.	.	.	.	.	.	.	.	.	.	.	.	.	−4	4	.	.	.	2	.

Character table of $H = \langle x, y, h \rangle$ (continued)

	6b	8a	8b	8c	8d	8e	8f	8g	8h	10a	12a	12b	16a	16b	20a	20b	20c	24a	24b
2	2	5	6	5	5	5	5	5	5	3	3	2	4	4	2	2	2	3	3
3	1	1	.	.	.	.	.	.	.	.	1	1	.	.	.	.	.	1	1
5	.	.	.	.	.	.	.	.	.	1	.	.	.	.	1	1	1	.	.
2P	3a	4a	4c	4d	4e	4f	4g	4h	4i	5a	6a	6a	8b	8b	10a	10a	10a	12a	12a
3P	2e	8a	8b	8c	8d	8e	8f	8g	8h	10a	4a	4b	16b	16a	20a	20c	20b	8a	8a
5P	6b	8a	8b	8c	8d	8e	8f	8g	8h	2a	12a	12b	16a	16b	4a	4b	4b	24a	24b
X.1	1	1	1	1	1	1	1	1	1	1	1	1	1	1	1	1	1	1	1
X.2	−1	−1	1	−1	1	−1	−1	−1	1	1	1	1	−1	−1	1	1	1	−1	−1
X.3	−1	2	.	2	.	.	.	.	.	−1	1	1	.	.	−1	−1	−1	−1	−1
X.4	1	−2	.	−2	.	.	.	.	.	−1	1	1	.	.	−1	−1	−1	1	1
X.5	1	1	1	1	1	−1	−1	−1	1	.	−1	−1	−1	−1	.	.	.	1	1
X.6	−1	−1	1	−1	1	1	1	1	1	.	−1	−1	1	1	.	.	.	−1	−1
X.7	.	.	2	.	.	2	−2	2	.	1	.	.	.	.	1	−1	−1	.	.
X.8	.	.	2	.	.	−2	2	−2	.	1	.	.	.	.	1	−1	−1	.	.
X.9	.	.	−2	.	−2	.	.	.	−2	1	.	.	.	.	1	1	1	.	.
X.10	.	.	−2	.	.	.	.	.	.	1	.	.	A	$\bar{A}$	1	−1	−1	.	.
X.11	.	.	−2	.	.	.	.	.	.	1	.	.	$\bar{A}$	A	1	−1	−1	.	.
X.12	−1	−2	2	2	.	.	.	.	.	.	1	−1	.	.	.	.	.	1	1
X.13	−1	4	−2	.	.	.	.	.	.	.	1	−1	.	.	.	.	.	1	1
X.14	1	−4	−2	.	.	.	.	.	.	.	1	−1	.	.	.	.	.	−1	−1
X.15	1	2	2	−2	.	.	.	.	.	.	1	−1	.	.	.	.	.	−1	−1
X.16	.	.	.	.	−2	.	.	.	2	2	.	.	.	.	−2	.	.	.	.
X.17	.	.	.	.	2	.	.	.	−2	2	.	.	.	.	−2	.	.	.	.
X.18	.	−3	3	1	−1	1	1	1	−1	.	.	.	−1	−1	.	.	.	.	.
X.19	.	3	−1	−1	−1	1	1	1	−1	.	.	.	−1	−1	.	.	.	.	.
X.20	.	3	3	−1	−1	−1	−1	−1	−1	.	.	.	1	1	.	.	.	.	.
X.21	.	−3	−1	1	−1	−1	−1	−1	−1	.	.	.	1	1	.	.	.	.	.
X.22	1	2	.	2	.	.	.	.	.	.	−1	1	.	.	.	.	.	−1	−1
X.23	−1	−2	.	−2	.	.	.	.	.	.	−1	1	.	.	.	.	.	1	1
X.24	.	.	.	.	2	.	.	.	−2	.	−2	.	.	.	.	.	.	.	.
X.25	.	.	.	.	−2	.	.	.	2	.	−2	.	.	.	.	.	.	.	.
X.26	.	.	.	.	.	.	.	.	.	−1	.	.	.	.	−1	1	1	.	.
X.27	.	.	.	.	.	.	.	.	.	−1	.	.	.	.	1	B	$-B$	.	.
X.28	.	.	.	.	.	.	.	.	.	−1	.	.	.	.	1	$-B$	B	.	.
X.29	.	.	−2	.	2	.	.	.	2	.	.	.	.	.	.	.	.	.	.
X.30	.	.	.	.	.	.	.	.	.	.	2	.	.	.	.	.	.	.	.
X.31	.	.	.	.	.	.	−2	.	.	.	.	.	.	.	.	.	.	.	.
X.32	.	.	.	.	.	.	2	.	.	.	.	.	.	.	.	.	.	.	.
X.33	.	.	.	.	.	2	.	−2	.	.	.	.	.	.	.	.	.	.	.
X.34	.	.	.	.	.	−2	.	2	.	.	.	.	.	.	.	.	.	.	.
X.35	.	.	.	.	.	2	.	−2	.	.	.	.	.	.	.	.	.	.	.
X.36	.	.	.	.	.	−2	.	2	.	.	.	.	.	.	.	.	.	.	.
X.37	.	.	.	.	.	.	−2	.	.	.	.	.	.	.	.	.	.	.	.
X.38	.	.	.	.	.	.	2	.	.	.	.	.	.	.	.	.	.	.	.
X.39	.	.	.	.	.	.	.	.	.	.	.	.	.	.	.	.	.	.	.
X.40	.	.	.	.	.	.	.	.	.	.	.	.	.	.	.	.	.	.	.
X.41	.	.	.	.	.	.	.	.	.	.	.	.	.	.	.	.	.	.	.
X.42	.	.	.	.	.	.	.	.	.	.	.	.	.	.	.	.	.	.	.
X.43	.	.	.	.	.	.	.	.	.	.	.	.	.	.	.	.	.	.	.
X.44	.	.	.	.	.	.	.	.	.	.	.	.	.	.	.	.	.	.	.
X.45	.	.	.	.	.	.	.	.	.	2	.	.	.	.	.	.	.	.	.
X.46	.	.	.	.	.	.	.	.	.	2	.	.	.	.	.	.	.	C	$-C$
X.47	.	.	.	.	.	.	.	.	.	2	.	.	.	.	.	.	.	$-C$	C
X.48	.	.	.	.	.	.	.	.	.	−2	.	.	.	.	.	.	.	.	.
X.49	.	.	.	.	.	.	.	.	.	−2	.	.	.	.	.	.	.	.	.

where $A = -2\zeta(4)$, $B = -2\zeta(5)^3 - 2\zeta(5)^2 - 1$, $C = 2\zeta(24)_8^3\zeta(24)_3 + \zeta(24)_8^3 + 2\zeta(24)_8\zeta(24)_3 + \zeta(24)_8$.

12.6.2 *Character table of* $E = \langle x, y, e \rangle$

	1a	2a	2b	2c	2d	3a	4a	4b	4c	4d	4e	4f	4g	4h	4i	4j	4k	6a	7a	7b	8a	8b	8c	8d	12a	12b	12c
2	14	14	11	8	8	3	9	10	8	9	8	7	7	7	6	5	5	3	2	2	6	5	5	4	2	2	2
3	1	1	.	.	.	1	1	.	1	.	.	.	.	.	.	.	.	1	.	.	.	.	.	.	1	1	1
7	1	.	.	1	.	.	.	.	.	.	.	.	.	.	.	.	.	.	1	1	.	.	.	.	.	.	.
2P	1a	1a	1a	1a	1a	3a	2a	2a	2a	2a	2a	2b	2b	2b	2b	2d	2d	3a	7a	7b	4b	4a	4d	4g	6a	6a	6a
3P	1a	2a	2b	2c	2d	1a	4a	4b	4c	4d	4e	4f	4g	4h	4i	4j	4k	2a	7b	7a	8a	8b	8c	8d	4c	4a	4c
7P	1a	2a	2b	2c	2d	3a	4a	4b	4c	4d	4e	4f	4g	4h	4i	4j	4k	6a	1a	1a	8a	8b	8c	8d	12a	12b	12c
X.1	1	1	1	1	1	1	1	1	1	1	1	1	1	1	1	1	1	1	1	1	1	1	1	1	1	1	1
X.2	3	3	3	3	−1	.	3	3	3	3	−1	−1	−1	−1	−1	1	1	.	A	$\bar{A}$	−1	−1	−1	1	.	.	.
X.3	3	3	3	3	−1	.	3	3	3	3	−1	−1	−1	−1	−1	1	1	.	$\bar{A}$	A	−1	−1	−1	1	.	.	.
X.4	6	6	6	6	2	.	6	6	6	6	2	2	2	2	2	.	.	.	−1	−1	2	2	2	.	.	.	.
X.5	7	7	7	7	−1	1	7	7	7	7	−1	−1	−1	−1	−1	−1	−1	1	.	.	−1	−1	−1	−1	1	1	1
X.6	8	8	8	8	.	−1	8	8	8	8	.	.	.	.	.	.	.	−1	1	1	.	.	.	.	−1	−1	−1
X.7	21	21	5	−7	1	.	9	−3	−3	1	1	−3	1	−3	5	−1	−1	.	.	.	1	−1	−1	−1	.	.	.
X.8	21	21	5	−7	−3	.	9	−3	−3	1	−3	1	−3	1	1	1	1	.	.	.	5	−1	−1	1	.	.	.
X.9	21	21	5	−7	5	.	9	−3	−3	1	5	1	5	1	1	1	1	.	.	.	−3	−1	−1	1	.	.	.
X.10	21	21	5	−7	1	.	9	−3	−3	1	1	5	1	5	−3	−1	−1	.	.	.	1	−1	−1	−1	.	.	.
X.11	24	24	−8	−8	.	.	.	8	.	.	.	.	.	.	.	.	.	.	3	3	.	.	.	.	.	.	.
X.12	24	24	−8	−8	.	.	.	8	.	.	.	.	.	.	.	.	.	.	$\bar{A}$	A	.	.	.	.	.	.	.
X.13	24	24	−8	−8	.	.	.	8	.	.	.	.	.	.	.	.	.	.	$\bar{A}$	A	.	.	.	.	.	.	.
X.14	24	24	−8	−8	.	.	.	8	.	.	.	.	.	.	.	.	.	.	A	$\bar{A}$	.	.	.	.	.	.	.
X.15	24	24	−8	−8	.	.	.	8	.	.	.	.	.	.	.	.	.	.	A	$\bar{A}$	.	.	.	.	.	.	.
X.16	24	24	−8	−8	.	.	.	8	.	.	.	.	.	.	.	.	.	.	A	$\bar{A}$	.	.	.	.	.	.	.
X.17	24	24	−8	−8	.	.	.	8	.	.	.	.	.	.	.	.	.	.	$\bar{A}$	A	.	.	.	.	.	.	.
X.18	28	28	12	.	4	1	−8	4	−4	.	4	.	−4	.	.	.	.	1	.	.	.	−2	2	.	−1	1	−1
X.19	28	28	12	.	−4	1	−8	4	−4	.	−4	.	4	.	.	.	.	1	.	.	.	2	−2	.	−1	1	−1
X.20	42	42	−6	14	2	.	6	2	−6	−2	2	−2	2	−2	−2	2	2	.	.	.	2	.	.	−2	.	.	.
X.21	42	42	−6	14	2	.	6	2	−6	−2	2	−2	2	−2	−2	−2	−2	.	.	.	2	.	.	2	.	.	.
X.22	42	42	10	−14	−2	.	18	−6	−6	2	−2	−2	−2	−2	−2	.	.	.	.	.	−2	2	2	.	.	.	.
X.23	42	42	−6	14	−2	.	6	2	−6	−2	−2	2	−2	2	2	.	.	.	.	.	−2	.	.	.	.	.	.
X.24	42	42	−6	14	−2	.	6	2	−6	−2	−2	2	−2	2	2	.	.	.	.	.	−2	.	.	.	.	.	.
X.25	56	56	−8	.	.	2	−8	−8	.	8	.	.	.	.	.	.	.	2	.	.	.	.	.	.	.	−2	.
X.26	56	56	24	.	.	−1	−16	8	−8	.	.	.	.	.	.	.	.	−1	.	.	.	.	.	.	1	−1	1
X.27	56	56	−8	.	.	−1	−8	−8	.	8	.	.	.	.	.	.	.	−1	.	.	.	.	.	.	F	1	$\bar{F}$
X.28	56	56	−8	.	.	−1	−8	−8	.	8	.	.	.	.	.	.	.	−1	.	.	.	.	.	.	$\bar{F}$	1	F
X.29	84	84	4	.	4	.	.	−4	12	−8	4	.	−4	.	.	.	.	.	.	.	.	2	−2	.	.	.	.
X.30	84	84	4	.	−4	.	.	−4	12	−8	−4	.	4	.	.	.	.	.	.	.	.	−2	2	.	.	.	.
X.31	112	−16	.	.	4	−2	.	.	.	.	−4	−4	.	4	.	−2	2	2	.	.	.	.	.	.	.	.	.
X.32	112	−16	.	.	4	−2	.	.	.	.	−4	4	.	−4	.	2	−2	2	.	.	.	.	.	.	.	.	.
X.33	224	−32	.	.	8	2	.	.	.	.	−8	.	.	.	.	.	.	−2	.	.	.	.	.	.	.	.	.
X.34	336	−48	.	.	−4	.	.	.	.	.	4	−4	.	4	.	2	−2	.	.	.	.	.	.	.	.	.	.
X.35	336	−48	.	.	−4	.	.	.	.	.	4	4	.	−4	.	−2	2	.	.	.	.	.	.	.	.	.	.

Character table of $E = \langle x, y, e \rangle$ *(continued)*

	14a	14b	14c	14d	14e	14f	16a	16b
2	2	2	2	2	2	2	4	4
3	.	.	.	.	.	.	.	.
7	1	1	1	1	1	1	.	.
2P	7a	7a	7b	7b	7b	7a	8a	8a
3P	14c	14d	14b	14f	14a	14e	16b	16a
7P	2c	2c	2c	2c	2c	2c	16b	16a
X.1	1	1	1	1	1	1	1	1
X.2	A	A	$\bar{A}$	$\bar{A}$	$\bar{A}$	A	1	1
X.3	$\bar{A}$	$\bar{A}$	A	A	A	$\bar{A}$	1	1
X.4	−1	−1	−1	−1	−1	−1	.	.
X.5	.	.	.	.	.	.	−1	−1
X.6	1	1	1	1	1	1	.	.
X.7	.	.	.	.	.	.	1	1
X.8	.	.	.	.	.	.	−1	−1
X.9	.	.	.	.	.	.	−1	−1
X.10	.	.	.	.	.	.	1	1
X.11	−1	−1	−1	−1	−1	−1	.	.
X.12	B	C	D	$\bar{B}$	$\bar{C}$	$\bar{D}$	.	.
X.13	C	$\bar{D}$	$\bar{B}$	$\bar{C}$	D	B	.	.
X.14	$\bar{C}$	D	B	C	$\bar{D}$	$\bar{B}$	.	.
X.15	D	$\bar{B}$	C	$\bar{D}$	B	$\bar{C}$	.	.
X.16	$\bar{B}$	$\bar{C}$	$\bar{D}$	B	C	D	.	.
X.17	$\bar{D}$	B	$\bar{C}$	D	$\bar{B}$	C	.	.
X.18	.	.	.	.	.	.	.	.
X.19	.	.	.	.	.	.	.	.
X.20	.	.	.	.	.	.	.	.
X.21	.	.	.	.	.	.	.	.
X.22	.	.	.	.	.	.	.	.
X.23	.	.	.	.	.	.	E	$\bar{E}$
X.24	.	.	.	.	.	.	$\bar{E}$	E
X.25	.	.	.	.	.	.	.	.
X.26	.	.	.	.	.	.	.	.
X.27	.	.	.	.	.	.	.	.
X.28	.	.	.	.	.	.	.	.
X.29	.	.	.	.	.	.	.	.
X.30	.	.	.	.	.	.	.	.
X.31	.	.	.	.	.	.	.	.
X.32	.	.	.	.	.	.	.	.
X.33	.	.	.	.	.	.	.	.
X.34	.	.	.	.	.	.	.	.
X.35	.	.	.	.	.	.	.	.

where $A = -\zeta(7)^4 - \zeta(7)^2 - \zeta(7) - 1$, $B = -\zeta(7)^4 - \zeta(7)^2 + \zeta(7)$, $C = -\zeta(7)^4 + \zeta(7)^2 - \zeta(7)$, $D = \zeta(7)^4 + 2\zeta(7)^3 + \zeta(7)^2 + \zeta(7) + 1$, $E = 2\zeta(4)$, $F = 2\zeta(3) + 1$.

12.6.3 Character table of $D = \langle x, y \rangle$

	$1a$	$2a$	$2b$	$2c$	$2d$	$2e$	$2f$	$2g$	$2h$	$3a$	4_1	4_2	4_3	4_4	4_5	4_6	4_7	4_8	4_9	4_{10}	4_{11}	4_{12}	4_{13}	4_{14}	4_{15}	4_{16}	4_{17}	4_{18}
2	14	14	13	11	10	9	8	8	7	3	9	10	8	9	9	9	9	9	9	8	8	8	8	8	8	7	7	7
3	1	1	.	.	.	.	.	.	.	1	1	.	1	.	.	.	.	.	.	.	.	.	.	.	.	.	.	.
2P	$1a$	$1a$	$1a$	$1a$	$1a$	$1a$	$1a$	$1a$	$1a$	$3a$	$2a$	$2a$	$2a$	$2b$	$2a$	$2b$	$2b$	$2b$	$2b$	$2b$	$2b$	$2b$	$2b$	$2a$	$2b$	$2c$	$2b$	$2c$
3P	$1a$	$2a$	$2b$	$2c$	$2d$	$2e$	$2f$	$2g$	$2h$	$1a$	4_1	4_2	4_3	4_4	4_5	4_6	4_7	4_8	4_9	4_{10}	4_{11}	4_{12}	4_{13}	4_{14}	4_{15}	4_{16}	4_{17}	4_{18}
$X.1$	1	1	1	1	1	1	1	1	1	1	1	1	1	1	1	1	1	1	1	1	1	1	1	1	1	1	1	1
$X.2$	1	1	1	1	1	1	1	1	−1	1	1	1	1	1	1	1	1	1	1	1	1	1	1	1	1	1	−1	1
$X.3$	2	2	2	2	2	2	2	2	.	−1	2	2	2	2	2	2	2	2	2	2	2	2	2	2	2	2	.	2
$X.4$	3	3	3	3	3	3	−1	3	1	.	3	3	3	3	3	3	3	3	3	3	3	3	3	−1	3	−1	1	−1
$X.5$	3	3	3	3	3	3	−1	3	−1	.	3	3	3	3	3	3	3	3	3	3	3	3	3	−1	3	−1	−1	−1
$X.6$	3	3	3	3	3	−1	−1	−1	−1	.	3	3	3	3	−1	−1	3	−1	3	−1	−1	−1	−1	−1	−1	−1	−1	−1
$X.7$	3	3	3	3	3	−1	−1	−1	1	.	3	3	3	3	−1	−1	3	−1	3	−1	−1	−1	−1	−1	−1	−1	1	−1
$X.8$	3	3	3	3	3	−1	3	−1	1	.	3	3	3	3	−1	−1	3	−1	3	−1	−1	−1	−1	3	−1	3	1	3
$X.9$	3	3	3	3	3	−1	3	−1	−1	.	3	3	3	3	−1	−1	3	−1	3	−1	−1	−1	−1	3	−1	3	−1	3
$X.10$	4	4	4	4	4	.	.	.	−2	1	4	4	−4	−4	.	.	4	.	−4	.	.	.	.	.	.	.	−2	.
$X.11$	4	4	4	4	4	.	.	.	2	1	4	4	−4	−4	.	.	4	.	−4	.	.	.	.	.	.	.	2	.
$X.12$	6	6	6	6	−2	2	2	−2	−2	.	6	−2	−6	2	2	2	−2	2	2	−2	2	−2	2	2	−2	2	−2	−2
$X.13$	6	6	6	6	−2	2	2	−2	2	.	6	−2	−6	2	2	2	−2	2	2	−2	2	−2	2	2	−2	2	2	−2
$X.14$	6	6	6	6	−2	−2	2	2	.	.	6	−2	−6	2	−2	−2	−2	−2	2	2	−2	2	−2	2	2	2	.	−2
$X.15$	6	6	6	6	−2	2	−2	−2	.	.	6	−2	−6	2	2	2	−2	2	2	−2	2	−2	2	−2	−2	−2	.	2
$X.16$	6	6	6	6	−2	−2	2	2	.	.	6	−2	−6	2	−2	−2	−2	−2	2	2	−2	2	−2	2	2	2	.	−2
$X.17$	6	6	6	6	6	−2	−2	−2	.	.	6	6	6	6	−2	−2	6	−2	6	−2	−2	−2	−2	−2	−2	−2	.	−2
$X.18$	6	6	6	6	−2	2	−2	−2	.	.	6	−2	−6	2	2	2	−2	2	2	−2	2	−2	2	−2	−2	−2	.	2
$X.19$	6	6	6	6	−2	−2	−2	2	.	.	6	−2	−6	2	−2	−2	−2	−2	2	2	−2	2	−2	−2	2	−2	.	2
$X.20$	6	6	6	6	−2	−2	−2	2	.	.	6	−2	−6	2	−2	−2	−2	−2	2	2	−2	2	−2	−2	2	−2	.	2
$X.21$	8	8	8	8	8	.	.	.	.	−1	8	8	−8	−8	.	.	8	.	−8	.	.	.	.	.	.	.	.	.
$X.22$	8	8	8	8	−8	.	.	.	.	2	−8	−8	.	.	.	.	8	.	.	.	.	.	.	.	.	.	.	.
$X.23$	8	8	8	8	−8	.	.	.	.	−1	−8	−8	.	.	.	.	8	.	.	.	.	.	.	.	.	.	.	.
$X.24$	8	8	8	8	−8	.	.	.	.	−1	−8	−8	.	.	.	.	8	.	.	.	.	.	.	.	.	.	.	.
$X.25$	12	12	12	−4	4	−4	.	4	2	.	.	−4	.	4	8	4	.	.	−4	.	.	.	−4	.	−4	.	2	.
$X.26$	12	12	12	−4	4	.	.	.	2	.	.	−4	.	−4	−4	8	.	4	4	4	−4	−4	.	.	.	.	2	.
$X.27$	12	12	12	12	4	−4	4	.	.	.	−12	4	.	.	4	−4	−4	4	.	.	−4	.	4	4	.	−4	.	.
$X.28$	12	12	12	−4	4	4	.	−4	2	.	.	−4	.	4	.	−4	.	8	−4	.	.	.	−4	.	4	.	2	.
$X.29$	12	12	12	12	4	4	−4	.	.	.	−12	4	.	.	−4	4	−4	−4	.	.	4	.	−4	−4	.	4	.	.
$X.30$	12	12	12	−4	4	8	.	.	2	.	.	−4	.	−4	4	.	.	−4	4	−4	−4	4	.	.	.	.	2	.
$X.31$	12	12	12	−4	4	8	.	.	−2	.	.	−4	.	−4	4	.	.	−4	4	−4	−4	4	.	.	.	.	−2	.
$X.32$	12	12	12	−4	4	−4	.	4	−2	.	.	−4	.	4	8	4	.	.	−4	.	.	.	−4	.	−4	.	−2	.
$X.33$	12	12	12	12	4	−4	−4	.	.	.	−12	4	.	.	4	−4	−4	4	.	.	−4	.	4	−4	.	4	.	.
$X.34$	12	12	12	12	4	4	4	.	.	.	−12	4	.	.	−4	4	−4	−4	.	.	4	.	−4	4	.	−4	.	.
$X.35$	12	12	12	−4	4	.	.	.	−2	.	.	−4	.	−4	−4	8	.	4	4	4	−4	−4	.	.	.	.	−2	.
$X.36$	12	12	12	12	−4	.	.	.	2	.	12	−4	12	−4	.	.	−4	.	−4	.	.	.	.	.	.	.	2	.
$X.37$	12	12	12	12	−4	.	.	.	−2	.	12	−4	12	−4	.	.	−4	.	−4	.	.	.	.	.	.	.	−2	.
$X.38$	12	12	12	−4	4	4	.	−4	−2	.	.	−4	.	4	.	−4	.	8	−4	.	.	.	−4	.	4	.	−2	.
$X.39$	24	24	24	−8	−8	−8	.	−8	.	.	.	8	.	.	.	8	.	.	.	.	.	8	.	.	.	.	.	.
$X.40$	24	24	24	−8	−8	.	.	.	.	.	.	8	.	.	8	.	.	−8	.	.	.	−8	.	.	8	.	.	.
$X.41$	24	24	24	−8	−8	.	.	8	.	.	.	8	.	.	−8	.	.	8	.	−8	.	.	.	.	.	.	.	.
$X.42$	24	24	24	−8	−8	8	.	.	.	.	.	8	.	.	.	−8	.	.	.	8	.	.	.	.	−8	.	.	.
$X.43$	24	24	24	−8	8	.	.	.	.	.	.	−8	.	8	−8	.	.	−8	−8	.	.	.	8	.	.	.	.	.
$X.44$	24	24	24	−8	8	−8	.	.	.	.	.	−8	.	−8	.	−8	.	.	8	.	8	.	.	.	.	.	.	.
$X.45$	48	48	−16	.	.	.	4	.	4	.	.	.	.	.	.	.	.	.	.	.	.	.	.	−4	.	.	−4	4
$X.46$	48	48	−16	.	.	.	4	.	−4	.	.	.	.	.	.	.	.	.	.	.	.	.	.	−4	.	.	4	−4
$X.47$	48	48	−16	.	.	.	4	.	−4	.	.	.	.	.	.	.	.	.	.	.	.	.	.	−4	.	.	4	4
$X.48$	48	48	−16	.	.	.	4	.	4	.	.	.	.	.	.	.	.	.	.	.	.	.	.	−4	.	.	−4	−4
$X.49$	64	−64	.	.	.	.	.	.	.	−2	.	.	.	.	.	.	.	.	.	.	.	.	.	.	.	.	.	.
$X.50$	64	−64	.	.	.	.	.	.	.	−2	.	.	.	.	.	.	.	.	.	.	.	.	.	.	.	.	.	.
$X.51$	96	96	−32	.	.	.	−8	.	.	.	.	.	.	.	.	.	.	.	.	.	.	.	.	8	.	.	.	.
$X.52$	128	−128	.	.	.	.	.	.	.	2	.	.	.	.	.	.	.	.	.	.	.	.	.	.	.	.	.	.

Character table of $D = \langle x, y \rangle$ *(continued)*

	4_{19}	4_{20}	4_{21}	4_{22}	4_{23}	4_{24}	4_{25}	4_{26}	4_{27}	$6a$	$8a$	$8b$	$8c$	$8d$	$8e$	$8f$	$8g$	$8h$	$8i$	$12a$	$12b$	$12c$	$16a$	$16b$
2	7	6	6	6	6	6	6	5	5	3	6	5	5	5	5	5	5	5	4	2	2	2	4	4
3	.	.	.	.	.	.	.	.	.	1	.	.	.	.	.	.	.	.	.	1	1	1	.	.
2P	$2c$	$2e$	$2d$	$2d$	$2e$	$2c$	$2d$	$2f$	$2f$	$3a$	4_2	4_4	4_5	4_6	4_1	4_7	4_8	4_9	4_{16}	$6a$	$6a$	$6a$	$8a$	$8a$
3P	4_{19}	4_{20}	4_{21}	4_{22}	4_{23}	4_{24}	4_{25}	4_{26}	4_{27}	$2a$	$8a$	$8b$	$8c$	$8d$	$8e$	$8f$	$8g$	$8h$	$8i$	4_3	4_1	4_3	$16b$	$16a$
X.1	1	1	1	1	1	1	1	1	1	1	1	1	1	1	1	1	1	1	1	1	1	1	1	1
X.2	1	−1	−1	1	−1	−1	−1	−1	−1	1	1	1	−1	−1	−1	−1	−1	1	−1	1	1	1	−1	−1
X.3	2	.	.	2	.	.	.	.	.	−1	2	2	.	.	.	.	.	2	.	−1	−1	−1	.	.
X.4	−1	1	1	−1	1	1	1	−1	−1	.	−1	−1	1	1	1	1	1	−1	−1	.	.	.	−1	−1
X.5	−1	−1	−1	−1	−1	−1	−1	1	1	.	−1	−1	−1	−1	−1	−1	−1	−1	1	.	.	.	1	1
X.6	−1	1	−1	3	1	−1	−1	1	1	.	3	−1	1	1	−1	−1	1	−1	1	.	.	.	−1	−1
X.7	−1	−1	1	3	−1	1	1	−1	−1	.	3	−1	−1	−1	1	1	−1	−1	−1	.	.	.	1	1
X.8	3	−1	1	−1	−1	1	1	1	1	.	−1	−1	−1	−1	1	1	−1	−1	1	.	.	.	−1	−1
X.9	3	1	−1	−1	1	−1	−1	−1	−1	.	−1	−1	1	1	−1	−1	1	−1	−1	.	.	.	1	1
X.10	.	.	2	.	.	−2	2	.	.	1	.	.	.	.	2	−2	.	.	.	−1	1	−1	.	.
X.11	.	.	−2	.	.	2	−2	.	.	1	.	.	.	.	−2	2	.	.	.	−1	1	−1	.	.
X.12	−2	.	−2	2	.	−2	−2	.	.	.	−2	.	.	.	2	2	.	.	.	.	.	.	.	.
X.13	−2	.	2	2	.	2	2	.	.	.	−2	.	.	.	−2	−2	.	.	.	.	.	.	.	.
X.14	−2	.	.	−2	.	.	.	2	2	.	2	.	.	.	.	.	.	.	−2	.	.	.	.	.
X.15	2	2	.	−2	2	.	.	.	.	.	2	.	−2	2	.	.	−2	.	.	.	.	.	.	.
X.16	−2	.	.	−2	.	.	.	−2	−2	.	2	.	.	.	.	.	.	.	2	.	.	.	.	.
X.17	−2	.	.	−2	.	.	.	.	.	.	−2	2	.	.	.	.	.	2	.	.	.	.	.	.
X.18	2	−2	.	−2	−2	.	.	.	.	.	2	.	2	−2	.	.	2	.	.	.	.	.	.	.
X.19	2	.	.	2	.	.	.	.	.	.	−2	.	.	.	.	.	.	.	.	.	.	.	A	$\bar{A}$
X.20	2	.	.	2	.	.	.	.	.	.	−2	.	.	.	.	.	.	.	.	.	.	.	$\bar{A}$	A
X.21	.	.	.	.	.	.	.	.	.	−1	.	.	.	.	.	.	.	.	.	1	−1	1	.	.
X.22	.	.	.	.	.	.	.	.	.	2	.	.	.	.	.	.	.	.	.	.	−2	.	.	.
X.23	.	.	.	.	.	.	.	.	.	−1	.	.	.	.	.	.	.	.	.	B	1	$\bar{B}$	.	.
X.24	.	.	.	.	.	.	.	.	.	−1	.	.	.	.	.	.	.	.	.	$\bar{B}$	1	B	.	.
X.25	.	−2	2	.	−2	−2	−2	.	.	.	.	.	.	2	.	.	.	.	.	.	.	.	.	.
X.26	.	.	−2	.	.	−2	2	.	.	.	.	.	−2	.	.	.	2	.	.	.	.	.	.	.
X.27	.	.	.	.	.	.	.	.	.	.	.	2	.	.	.	.	.	−2	.	.	.	.	.	.
X.28	.	2	2	.	2	−2	−2	.	.	.	.	.	.	−2	.	.	.	.	.	.	.	.	.	.
X.29	.	.	.	.	.	.	.	.	.	.	.	2	.	.	.	.	.	−2	.	.	.	.	.	.
X.30	.	.	−2	.	.	−2	2	.	.	.	.	.	2	.	.	.	−2	.	.	.	.	.	.	.
X.31	.	.	2	.	.	2	−2	.	.	.	.	.	−2	.	.	.	2	.	.	.	.	.	.	.
X.32	.	2	−2	.	2	2	2	.	.	.	.	.	.	−2	.	.	.	.	.	.	.	.	.	.
X.33	.	.	.	.	.	.	.	.	.	.	.	−2	.	.	.	.	.	2	.	.	.	.	.	.
X.34	.	.	.	.	.	.	.	.	.	.	.	−2	.	.	.	.	.	2	.	.	.	.	.	.
X.35	.	.	2	.	.	2	−2	.	.	.	.	.	2	.	.	.	−2	.	.	.	.	.	.	.
X.36	.	.	−2	.	.	2	−2	.	.	.	.	.	.	.	2	−2	.	.	.	.	.	.	.	.
X.37	.	.	2	.	.	−2	2	.	.	.	.	.	.	.	−2	2	.	.	.	.	.	.	.	.
X.38	.	−2	−2	.	−2	2	2	.	.	.	.	.	.	2	.	.	.	.	.	.	.	.	.	.
X.39	.	.	.	.	.	.	.	.	.	.	.	.	.	.	.	.	.	.	.	.	.	.	.	.
X.40	.	.	.	.	.	.	.	.	.	.	.	.	.	.	.	.	.	.	.	.	.	.	.	.
X.41	.	.	.	.	.	.	.	.	.	.	.	.	.	.	.	.	.	.	.	.	.	.	.	.
X.42	.	.	.	.	.	.	.	.	.	.	.	.	.	.	.	.	.	.	.	.	.	.	.	.
X.43	.	.	.	.	.	.	.	.	.	.	.	.	.	.	.	.	.	.	.	.	.	.	.	.
X.44	.	.	.	.	.	.	.	.	.	.	.	.	.	.	.	.	.	.	.	.	.	.	.	.
X.45	−4	.	.	.	.	.	.	−2	2	.	.	.	.	.	.	.	.	.	.	.	.	.	.	.
X.46	4	.	.	.	.	.	.	−2	2	.	.	.	.	.	.	.	.	.	.	.	.	.	.	.
X.47	−4	.	.	.	.	.	.	2	−2	.	.	.	.	.	.	.	.	.	.	.	.	.	.	.
X.48	4	.	.	.	.	.	.	2	−2	.	.	.	.	.	.	.	.	.	.	.	.	.	.	.
X.49	.	4	.	.	−4	.	.	.	.	2	.	.	.	.	.	.	.	.	.	.	.	.	.	.
X.50	.	−4	.	.	4	.	.	.	.	2	.	.	.	.	.	.	.	.	.	.	.	.	.	.
X.51	.	.	.	.	.	.	.	.	.	.	.	.	.	.	.	.	.	.	.	.	.	.	.	.
X.52	.	.	.	.	.	.	.	.	.	−2	.	.	.	.	.	.	.	.	.	.	.	.	.	.

where $A = -2\zeta(4)$, $B = -2\zeta(3) - 1$.

12.6.4 *Character table of* $N_G(3_A) \cong 3Aut(A_6)$

	$1a$	$2a$	$2b$	$2c$	$3a$	$3b$	$4a$	$4b$	$4c$	$5a$	$6a$	$6b$	$8a$	$8b$	$10a$	$12a$	$12b$	$15a$	$24a$	$24b$
2	5	5	4	3	4	1	4	3	4	1	4	1	3	3	1	3	2	.	3	3
3	3	1	1	.	3	2	1	1	.	1	1	1	1	.	.	1	1	1	1	1
5	1	.	.	1	1	.	.	.	.	1	.	.	.	.	1	.	.	1	.	.
2P	$1a$	$1a$	$1a$	$1a$	$3a$	$3b$	$2a$	$2a$	$2a$	$5a$	$3a$	$3b$	$4a$	$4a$	$5a$	$6a$	$6a$	$15a$	$12a$	$12a$
3P	$1a$	$2a$	$2b$	$2c$	$1a$	$1a$	$4a$	$4b$	$4c$	$5a$	$2a$	$2b$	$8a$	$8b$	$10a$	$4a$	$4b$	$5a$	$8a$	$8a$
5P	$1a$	$2a$	$2b$	$2c$	$3a$	$3b$	$4a$	$4b$	$4c$	$1a$	$6a$	$6b$	$8a$	$8b$	$2c$	$12a$	$12b$	$3a$	$24a$	$24b$
X.1	1	1	1	1	1	1	1	1	1	1	1	1	1	1	1	1	1	1	1	1
X.2	1	1	−1	1	1	1	1	−1	−1	1	1	−1	−1	1	1	1	−1	1	−1	−1
X.3	1	1	−1	−1	1	1	1	1	−1	1	1	−1	1	−1	−1	1	1	1	1	1
X.4	1	1	1	−1	1	1	1	−1	1	1	1	1	−1	−1	−1	1	−1	1	−1	−1
X.5	9	1	−3	1	9	.	1	1	1	−1	1	.	−1	−1	1	1	1	−1	−1	−1
X.6	9	1	3	1	9	.	1	−1	−1	−1	1	.	1	−1	1	1	−1	−1	1	1
X.7	9	1	−3	−1	9	.	1	−1	1	−1	1	.	1	1	−1	1	−1	−1	1	1
X.8	9	1	3	−1	9	.	1	1	−1	−1	1	.	−1	1	−1	1	1	−1	−1	−1
X.9	10	2	−2	.	10	1	−2	.	−2	.	2	1	.	.	.	−2	.	.	.	.
X.10	10	2	2	.	10	1	−2	.	2	.	2	−1	.	.	.	−2	.	.	.	.
X.11	12	−4	.	.	−6	.	4	.	.	2	2	.	.	.	.	−2	.	−1	.	.
X.12	12	4	.	.	−6	.	.	.	.	2	−2	.	.	.	.	.	.	−1	A	$-A$
X.13	12	4	.	.	−6	.	.	.	.	2	−2	.	.	.	.	.	.	−1	$-A$	A
X.14	16	.	.	4	16	−2	.	.	.	1	.	.	.	.	−1	.	.	1	.	.
X.15	16	.	.	−4	16	−2	.	.	.	1	.	.	.	.	1	.	.	1	.	.
X.16	18	2	.	.	−9	.	2	−2	.	−2	−1	.	2	.	.	−1	1	1	−1	−1
X.17	18	2	.	.	−9	.	2	2	.	−2	−1	.	−2	.	.	−1	−1	1	1	1
X.18	20	−4	.	.	20	2	.	.	.	.	−4	.	.	.	.	.	.	.	.	.
X.19	30	−2	.	.	−15	.	−2	−2	.	.	1	.	−2	.	.	1	1	.	1	1
X.20	30	−2	.	.	−15	.	−2	2	.	.	1	.	2	.	.	1	−1	.	−1	−1

where $A = -\sqrt{6}$.

12.6.5 Character table of $M = N_G(R) = \langle d_1, d_2 \rangle$

2	8	8	8	8	2	2	6	6	6	6	2	2	2	2	2	2	2	2	2	2
3	1	.	1	.	1	1	1	1	.	.	1	1	1	.	.	1	1	1	1	.
5	1	1	.	.	1	1	.	.	.	.	1	.	.	.	1	.	.	.	.	.
7	1	1	.	.	.	.	.	.	.	.	.	.	.	1	.	.	.	.	.	.
13	1	1	.	.	.	.	.	.	.	.	.	.	.	.	.	.	.	.	.	1
	1a	2a	2b	2c	3a	3b	4a	4b	4c	4d	5a	6a	6b	7a	10a	12a	12b	12c	12d	13a
2P	1a	1a	1a	1a	3b	3a	2b	2b	2b	2b	5a	3a	3b	7a	5a	6a	6b	6a	6b	13a
3P	1a	2a	2b	2c	1a	1a	4b	4a	4d	4c	5a	2b	2b	7a	10a	4a	4a	4b	4b	13a
5P	1a	2a	2b	2c	3b	3a	4a	4b	4c	4d	1a	6b	6a	7a	2a	12b	12a	12d	12c	13a
7P	1a	2a	2b	2c	3a	3b	4b	4a	4d	4c	5a	6a	6b	1a	10a	12c	12d	12a	12b	13a
13P	1a	2a	2b	2c	3a	3b	4a	4b	4c	4d	5a	6a	6b	7a	10a	12a	12b	12c	12d	1a
X.1	1	1	1	1	1	1	1	1	1	1	1	1	1	1	1	1	1	1	1	1
X.2	1	1	1	1	A	$\bar{A}$	1	1	1	1	1	$\bar{A}$	A	1	1	A	$\bar{A}$	A	$\bar{A}$	1
X.3	1	1	1	1	$\bar{A}$	A	1	1	1	1	1	A	$\bar{A}$	1	1	$\bar{A}$	A	$\bar{A}$	A	1
X.4	3	−1	3	−1	.	.	3	3	−1	−1	3	.	.	3	−1	.	.	.	.	3
X.5	14	14	−2	−2	−1	−1	B	$\bar{B}$	$\bar{B}$	B	−1	1	1	.	−1	C	C	$\bar{C}$	$\bar{C}$	1
X.6	14	14	−2	−2	−1	−1	$\bar{B}$	B	B	$\bar{B}$	−1	1	1	.	−1	$\bar{C}$	$\bar{C}$	C	C	1
X.7	14	14	−2	−2	$-A$	$-\bar{A}$	B	$\bar{B}$	$\bar{B}$	B	−1	$\bar{A}$	A	.	−1	D	$-\bar{D}$	$-D$	$\bar{D}$	1
X.8	14	14	−2	−2	$-\bar{A}$	$-A$	$\bar{B}$	B	B	$\bar{B}$	−1	A	$\bar{A}$	.	−1	$\bar{D}$	$-D$	$-\bar{D}$	D	1
X.9	14	14	−2	−2	$-\bar{A}$	$-A$	B	$\bar{B}$	$\bar{B}$	B	−1	A	$\bar{A}$	.	−1	$-\bar{D}$	D	$\bar{D}$	$-D$	1
X.10	14	14	−2	−2	$-A$	$-\bar{A}$	$\bar{B}$	B	B	$\bar{B}$	−1	$\bar{A}$	A	.	−1	$-D$	$\bar{D}$	D	$-\bar{D}$	1
X.11	42	−14	−6	2	.	.	E	$\bar{E}$	$\bar{B}$	B	−3	.	.	.	1	.	.	.	.	3
X.12	42	−14	−6	2	.	.	$\bar{E}$	E	B	$\bar{B}$	−3	.	.	.	1	.	.	.	.	3
X.13	64	64	.	.	4	4	.	.	.	.	−1	.	.	1	−1	.	.	.	.	−1
X.14	64	64	.	.	F	$\bar{F}$	.	.	.	.	−1	.	.	1	−1	.	.	.	.	−1
X.15	64	64	.	.	$\bar{F}$	F	.	.	.	.	−1	.	.	1	−1	.	.	.	.	−1
X.16	91	91	−5	−5	1	1	−1	−1	−1	−1	1	1	1	.	1	−1	−1	−1	−1	.
X.17	91	91	−5	−5	A	$\bar{A}$	−1	−1	−1	−1	1	$\bar{A}$	A	.	1	$-A$	$-\bar{A}$	$-A$	$-\bar{A}$	.
X.18	91	91	−5	−5	$\bar{A}$	A	−1	−1	−1	−1	1	A	$\bar{A}$	.	1	$-\bar{A}$	$-A$	$-\bar{A}$	$-A$	.
X.19	105	105	9	9	.	.	−3	−3	−3	−3	.	.	.	.	.	.	.	.	.	1
X.20	105	−35	9	−3	.	.	−3	−3	1	1	.	.	.	.	.	.	.	.	.	1
X.21	105	−35	9	−3	.	.	−3	−3	1	1	.	.	.	.	.	.	.	.	.	1
X.22	105	−35	9	−3	.	.	−3	−3	1	1	.	.	.	.	.	.	.	.	.	1
X.23	192	−64	.	.	.	.	.	.	.	.	−3	.	.	3	1	.	.	.	.	−3
X.24	195	195	3	3	.	.	3	3	3	3	.	.	.	−1	.	.	.	.	.	.
X.25	195	−65	3	−1	.	.	3	3	−1	−1	.	.	.	−1	.	.	.	.	.	.
X.26	195	−65	3	−1	.	.	3	3	−1	−1	.	.	.	−1	.	.	.	.	.	.
X.27	195	−65	3	−1	.	.	3	3	−1	−1	.	.	.	−1	.	.	.	.	.	.
X.28	273	−91	−15	5	.	.	−3	−3	1	1	3	.	.	.	−1	.	.	.	.	.

Character table of $M = \langle d_1, d_2 \rangle$ (continued)

2	2	2	2	.	.	2	2	2
3	.	.	.	1	1	.	.	.
5	.	.	.	1	1	.	.	.
7	1	1	1	.	.	.	.	.
13	.	.	.	.	.	1	1	1
	14a	14b	14c	15a	15b	26a	26b	26c
2P	7a	7a	7a	15b	15a	13a	13a	13a
3P	14b	14c	14a	5a	5a	26b	26c	26a
5P	14c	14a	14b	3b	3a	26a	26b	26c
7P	2a	2a	2a	15a	15b	26c	26a	26b
13P	14a	14b	14c	15a	15b	2a	2a	2a
X.1	1	1	1	1	1	1	1	1
X.2	1	1	1	A	$\bar{A}$	1	1	1
X.3	1	1	1	$\bar{A}$	A	1	1	1
X.4	−1	−1	−1	.	.	−1	−1	−1
X.5	.	.	.	−1	−1	1	1	1
X.6	.	.	.	−1	−1	1	1	1
X.7	.	.	.	$-A$	$-\bar{A}$	1	1	1
X.8	.	.	.	$-\bar{A}$	$-A$	1	1	1
X.9	.	.	.	$-\bar{A}$	$-A$	1	1	1
X.10	.	.	.	$-A$	$-\bar{A}$	1	1	1
X.11	.	.	.	.	.	−1	−1	−1
X.12	.	.	.	.	.	−1	−1	−1
X.13	1	1	1	−1	−1	−1	−1	−1
X.14	1	1	1	$-A$	$-\bar{A}$	−1	−1	−1
X.15	1	1	1	$-\bar{A}$	$-A$	−1	−1	−1
X.16	.	.	.	1	1	.	.	.
X.17	.	.	.	A	$\bar{A}$	.	.	.
X.18	.	.	.	$\bar{A}$	A	.	.	.
X.19	.	.	.	.	.	1	1	1
X.20	.	.	.	.	.	G	H	I
X.21	.	.	.	.	.	H	I	G
X.22	.	.	.	.	.	I	G	H
X.23	−1	−1	−1	.	.	1	1	1
X.24	−1	−1	−1	.	.	.	.	.
X.25	J	K	L	.	.	.	.	.
X.26	L	J	K	.	.	.	.	.
X.27	K	L	J	.	.	.	.	.
X.28	.	.	.	.	.	.	.	.

where $A = -\zeta(3)-1$, $B = 2\zeta(4)$, $C = \zeta(4)$, $D = -\zeta(12)_4\zeta(12)_3-\zeta(12)_4$, $E = -6\zeta(4)$, $F = -4\zeta(3) - 4$, $G = -2\zeta(13)^{11} - 2\zeta(13)^{10} - 2\zeta(13)^3 - 2\zeta(13)^2 - 1$, $H = -2\zeta(13)^9 - 2\zeta(13)^7 - 2\zeta(13)^6 - 2\zeta(13)^4 - 1$, $I = 2\zeta(13)^{11} + 2\zeta(13)^{10} + 2\zeta(13)^9 + 2\zeta(13)^7 + 2\zeta(13)^6 + 2\zeta(13)^4 + 2\zeta(13)^3 + 2\zeta(13)^2 + 1$, $J = 2\zeta(7)^4 + 2\zeta(7)^3 + 1$, $K = 2\zeta(7)^5 + 2\zeta(7)^2 + 1$, $L = -2\zeta(7)^5 - 2\zeta(7)^4 - 2\zeta(7)^3 - 2\zeta(7)^2 - 1$.

12.6.6 *Character table of* $N_G(5_A) \cong 5^{1+2} : (Q_8 \times 4)$

	1a	2a	2b	2c	4a	4b	4c	4d	4e	4f	4g	4h	5a	5b	5c	8a	8b	10a	10b	10c	20a	20b	20c	20d	20e
2	5	5	4	3	4	4	4	3	5	5	4	4	3	2	1	3	3	3	2	1	2	2	2	2	2
5	3	1	1	1	1	1	1	1	.	.	.	.	3	2	2	.	.	1	1	1	1	1	1	1	1
2P	1a	1a	1a	1a	2b	2b	2a	2a	2a	2a	2b	2b	5a	5b	5c	4e	4f	5a	5b	5c	10a	10b	10b	10a	10a
5P	1a	2a	2b	2c	4a	4b	4c	4d	4e	4f	4g	4h	1a	1a	1a	8a	8b	2a	2b	2c	4c	4a	4b	4d	4d
X.1	1	1	1	1	1	1	1	1	1	1	1	1	1	1	1	1	1	1	1	1	1	1	1	1	1
X.2	1	1	−1	1	A	−A	1	−1	−1	−1	A	−A	1	1	1	−A	A	1	−1	1	1	A	−A	−1	−1
X.3	1	1	1	1	−1	−1	1	1	1	1	−1	−1	1	1	1	−1	−1	1	1	1	1	−1	−1	1	1
X.4	1	1	−1	1	−A	A	1	−1	−1	−1	−A	A	1	1	1	A	−A	1	−1	1	1	−A	A	−1	−1
X.5	1	1	1	−1	1	1	1	−1	1	1	1	1	1	1	1	−1	−1	1	1	−1	1	1	1	−1	−1
X.6	1	1	−1	−1	A	−A	1	1	−1	−1	A	−A	1	1	1	A	−A	1	−1	−1	1	A	−A	1	1
X.7	1	1	1	−1	−1	−1	1	−1	1	1	−1	−1	1	1	1	1	1	1	1	−1	1	−1	−1	−1	−1
X.8	1	1	−1	−1	−A	A	1	1	−1	−1	−A	A	1	1	1	−A	A	1	−1	−1	1	−A	A	1	1
X.9	2	2	−2	.	.	.	−2	.	2	2	.	.	2	2	2	.	.	2	−2	.	−2	.	.	.	.
X.10	2	2	2	.	.	.	−2	.	−2	−2	.	.	2	2	2	.	.	2	2	.	−2	.	.	.	.
X.11	2	−2	.	.	B	/B	.	.	D	−D	−B	−/B	2	2	2	.	.	−2	.	.	.	B	/B	.	.
X.12	2	−2	.	.	/B	B	.	.	−D	D	−/B	−B	2	2	2	.	.	−2	.	.	.	/B	B	.	.
X.13	2	−2	.	.	−B	−/B	.	.	D	−D	B	/B	2	2	2	.	.	−2	.	.	.	−B	−/B	.	.
X.14	2	−2	.	.	−/B	−B	.	.	−D	D	/B	B	2	2	2	.	.	−2	.	.	.	−/B	−B	.	.
X.15	8	.	4	.	4	4	.	.	.	.	.	.	8	3	−2	.	.	.	−1	.	.	−1	−1	.	.
X.16	8	.	−4	.	C	−C	.	.	.	.	.	.	8	3	−2	.	.	.	1	.	.	−A	A	.	.
X.17	8	.	4	.	−4	−4	.	.	.	.	.	.	8	3	−2	.	.	.	−1	.	.	1	1	.	.
X.18	8	.	−4	.	−C	C	.	.	.	.	.	.	8	3	−2	.	.	.	1	.	.	A	−A	.	.
X.19	16	.	.	4	.	.	.	.	.	.	.	.	16	−4	1	.	.	.	.	−1	.	.	.	.	.
X.20	16	.	.	−4	.	.	.	.	.	.	.	.	16	−4	1	.	.	.	.	1	.	.	.	.	.
X.21	20	4	.	.	.	.	−4	−4	.	.	.	.	−5	.	.	.	.	−1	.	.	1	.	.	1	1
X.22	20	4	.	.	.	.	−4	4	.	.	.	.	−5	.	.	.	.	−1	.	.	1	.	.	−1	−1
X.23	20	4	.	.	.	.	4	.	.	.	.	.	−5	.	.	.	.	−1	.	.	−1	.	.	E	−E
X.24	20	4	.	.	.	.	4	.	.	.	.	.	−5	.	.	.	.	−1	.	.	−1	.	.	−E	E
X.25	40	−8	.	.	.	.	.	.	.	.	.	.	−10	.	.	.	.	2	.	.	.	.	.	.	.

where $A = -i$, $B = 1 + i$, $C = -4i$, $D = -2i$ *and* $E = \sqrt{5}$.

12.6.7 *Character table of* $N_G(5_B) \cong 5 : 4 \times A_5$

	1a	2a	2b	2c	3a	4a	4b	4c	4d	5a	5b	5c	5d	5e	6a	10a	10b	10c	12a	12b	15a	20a	20b	20c	20d
2	4	4	4	4	2	4	4	4	4	2	2	2	.	.	2	2	2	2	2	2	.	2	2	2	2
3	1	1	.	.	1	1	1	.	.	1	.	.	.	.	1	.	.	.	1	1	1	.	.	.	.
5	2	1	1	.	1	1	1	.	.	2	2	2	2	2	.	1	1	1	.	.	1	1	1	1	1
2P	1a	1a	1a	1a	3a	2a	2a	2a	2a	5a	5c	5b	5e	5d	3a	5a	5c	5b	6a	6a	15a	10c	10b	10c	10b
3P	1a	2a	2b	2c	1a	4b	4a	4d	4c	5a	5c	5b	5e	5d	2a	10a	10c	10b	4b	4a	5a	20d	20c	20b	20a
5P	1a	2a	2b	2c	3a	4a	4b	4c	4d	1a	1a	1a	1a	1a	6a	2b	2a	2a	12a	12b	3a	4a	4a	4b	4b
X.1	1	1	1	1	1	1	1	1	1	1	1	1	1	1	1	1	1	1	1	1	1	1	1	1	1
X.2	1	−1	1	−1	1	A	−A	A	−A	1	1	1	1	1	−1	1	−1	−1	A	−A	1	A	A	−A	−A
X.3	1	1	1	1	1	−1	−1	−1	−1	1	1	1	1	1	1	1	1	1	−1	−1	1	−1	−1	−1	−1
X.4	1	−1	1	−1	1	−A	A	−A	A	1	1	1	1	1	−1	1	−1	−1	−A	A	1	−A	−A	A	A
X.5	3	3	−1	−1	.	3	3	−1	−1	3	E	*E	E	*E	.	−1	E	*E	.	.	.	E	*E	E	*E
X.6	3	3	−1	−1	.	3	3	−1	−1	3	*E	E	*E	E	.	−1	*E	E	.	.	.	*E	E	*E	E
X.7	3	−3	−1	1	.	B	−B	−A	A	3	E	*E	E	*E	.	−1	−E	−*E	.	.	.	G	H	−G	−H
X.8	3	−3	−1	1	.	B	−B	−A	A	3	*E	E	*E	E	.	−1	−*E	−E	.	.	.	H	G	−H	−G
X.9	3	3	−1	−1	.	−3	−3	1	1	3	E	*E	E	*E	.	−1	E	*E	.	.	.	−E	−*E	−E	−*E
X.10	3	3	−1	−1	.	−3	−3	1	1	3	*E	E	*E	E	.	−1	*E	E	.	.	.	−*E	−E	−*E	−E
X.11	3	−3	−1	1	.	−B	B	A	−A	3	E	*E	E	*E	.	−1	−E	−*E	.	.	.	−G	−H	G	H
X.12	3	−3	−1	1	.	−B	B	A	−A	3	*E	E	*E	E	.	−1	−*E	−E	.	.	.	−H	−G	H	G
X.13	4	4	.	.	1	4	4	.	.	4	−1	−1	−1	−1	1	.	−1	−1	1	1	1	−1	−1	−1	−1
X.14	4	−4	.	.	1	C	−C	.	.	4	−1	−1	−1	−1	−1	.	1	1	A	−A	1	−A	−A	A	A
X.15	4	4	.	.	1	−4	−4	.	.	4	−1	−1	−1	−1	1	.	−1	−1	−1	−1	1	1	1	1	1
X.16	4	−4	.	.	1	−C	C	.	.	4	−1	−1	−1	−1	−1	.	1	1	−A	A	1	A	A	−A	−A
X.17	4	.	4	.	4	.	.	.	.	−1	4	4	−1	−1	.	−1	.	.	.	.	−1	.	.	.	.
X.18	5	5	1	1	−1	5	5	1	1	5	.	.	.	.	−1	1	.	.	−1	−1	−1	.	.	.	.
X.19	5	−5	1	−1	−1	D	−D	A	−A	5	.	.	.	.	1	1	.	.	−A	A	−1	.	.	.	.
X.20	5	5	1	1	−1	−5	−5	−1	−1	5	.	.	.	.	−1	1	.	.	1	1	−1	.	.	.	.
X.21	5	−5	1	−1	−1	−D	D	−A	A	5	.	.	.	.	1	1	.	.	A	−A	−1	.	.	.	.
X.22	12	.	−4	.	.	.	.	.	.	−3	F	*F	−E	−*E	.	1	.	.	.	.	.	.	.	.	.
X.23	12	.	−4	.	.	.	.	.	.	−3	*F	F	−*E	−E	.	1	.	.	.	.	.	.	.	.	.
X.24	16	.	.	.	4	.	.	.	.	−4	−4	−4	1	1	.	.	.	.	.	.	−1	.	.	.	.
X.25	20	.	4	.	−4	.	.	.	.	−5	.	.	.	.	.	−1	.	.	.	.	1	.	.	.	.

where $A = i$, $B = 3i$, $C = 4i$, $D = 5i$, $E = \frac{1}{2}(1 + \sqrt{5})$, $F = 1 + \sqrt{5}$, $G = -\theta^{13} - \theta^{17}$, $H = -\theta - \theta^{9}$, *where* θ *is a primitive 20th root of unity.*

12.6.8 Character table of $G = \langle x, y, h, e \rangle$

2	14	14	8	4	9	8	10	9	3	2	4	2	5	6	5	3	2	3	2	2	2	2	2	.
3	3	1	.	3	1	1	.	.	.	1	1	.	1	.	.	.	.	1	1	.	.	.	.	1
5	3	1	1	1	1	1	.	.	3	2	.	.	.	.	.	1	1	.	.	.	.	.	.	1
7	1	.	1	.	.	.	.	.	.	.	.	1	.	.	.	.	.	.	.	.	1	1	1	.
13	1	.	1	.	.	.	.	.	.	.	.	.	.	.	.	.	.	.	.	1	.	.	.	.
29	1	.	.	.	.	.	.	.	.	.	.	.	.	.	.	.	.	.	.	.	.	.	.	.
	1a	2a	2b	3a	4a	4b	4c	4d	5a	5b	6a	7a	8a	8b	8c	10a	10b	12a	12b	13a	14a	14b	14c	15a
2P	1a	1a	1a	3a	2a	2a	2a	2a	5a	5b	3a	7a	4a	4c	4d	5a	5b	6a	6a	13a	7a	7a	7a	15a
3P	1a	2a	2b	1a	4a	4b	4c	4d	5a	5b	2a	7a	8a	8b	8c	10a	10b	4a	4b	13a	14b	14c	14a	5b
5P	1a	2a	2b	3a	4a	4b	4c	4d	1a	1a	6a	7a	8a	8b	8c	2a	2b	12a	12b	13a	14c	14a	14b	3a
7P	1a	2a	2b	3a	4a	4b	4c	4d	5a	5b	6a	1a	8a	8b	8c	10a	10b	12a	12b	13a	2b	2b	2b	15a
13P	1a	2a	2b	3a	4a	4b	4c	4d	5a	5b	6a	7a	8a	8b	8c	10a	10b	12a	12b	1a	14a	14b	14c	15a
29P	1a	2a	2b	3a	4a	4b	4c	4d	5a	5b	6a	7a	8a	8b	8c	10a	10b	12a	12b	13a	14a	14b	14c	15a
X.1	1	1	1	1	1	1	1	1	1	1	1	1	1	1	1	1	1	1	1	1	1	1	1	1
X.2	378	−6	14	.	6	−6	2	−2	3	3	.	.	.	−2	.	−1	−1	.	.	1	.	.	.	.
X.3	378	−6	14	.	6	−6	2	−2	3	3	.	.	.	−2	.	−1	−1	.	.	1	.	.	.	.
X.4	406	22	−14	1	10	−10	−2	2	6	1	1	.	−4	−2	.	2	1	1	−1	3	.	.	.	1
X.5	783	15	−1	.	15	15	−1	−1	8	3	.	−1	3	3	−1	.	−1	.	.	3	−1	−1	−1	.
X.6	3276	76	.	9	16	4	4	8	1	−4	1	.	2	.	2	1	.	1	1	.	.	.	.	−1
X.7	3654	70	14	9	−2	−10	6	−2	4	−1	1	.	−2	2	2	.	−1	1	−1	1	.	.	.	−1
X.8	20475	−5	91	9	−5	−5	−5	−5	.	5	1	.	−1	3	−1	.	1	1	1	.	.	.	.	−1
X.9	21750	−10	−90	15	−10	−10	6	6	.	.	−1	1	2	2	−2	.	.	−1	−1	1	1	1	1	.
X.10	23751	71	91	−9	11	−1	−1	3	1	1	−1	.	1	3	1	1	1	−1	−1	.	.	.	.	1
X.11	27000	120	−40	.	.	.	8	.	.	.	.	1	.	.	.	.	.	.	.	−1	B	C	D	.
X.12	27000	120	−40	.	.	.	8	.	.	.	.	1	.	.	.	.	.	.	.	−1	C	D	B	.
X.13	27000	120	−40	.	.	.	8	.	.	.	.	1	.	.	.	.	.	.	.	−1	D	B	C	.
X.14	27405	141	105	.	9	21	5	1	5	.	.	.	3	−3	−1	1	.	.	.	1	.	.	.	.
X.15	34944	128	.	6	.	.	.	.	−6	4	2	.	.	.	.	−2	.	.	.	.	.	.	.	1
X.16	34944	128	.	6	.	.	.	.	−6	4	2	.	.	.	.	−2	.	.	.	.	.	.	.	1
X.17	43848	72	−56	.	.	.	−8	.	−2	3	.	.	.	.	.	2	−1	.	.	−1	.	.	.	.
X.18	43848	72	−56	.	.	.	−8	.	−2	3	.	.	.	.	.	2	−1	.	.	−1	.	.	.	.
X.19	43848	72	−56	.	.	.	−8	.	−2	3	.	.	.	.	.	2	−1	.	.	−1	.	.	.	.
X.20	45500	60	.	5	40	−20	4	.	.	.	−3	.	−2	.	−2	.	.	1	1	.	.	.	.	.
X.21	52780	−84	.	−5	−16	4	4	8	5	.	3	.	−2	.	−2	1	.	−1	1	.	.	.	.	.
X.22	63336	104	.	−6	16	−24	−8	.	11	−4	2	.	.	.	.	−1	.	−2	.	.	.	.	.	−1
X.23	65975	55	91	14	−5	15	−1	3	.	5	−2	.	−3	−1	1	.	1	−2	.	.	.	.	.	−1
X.24	71253	−171	−91	.	21	21	5	5	3	3	.	.	−3	1	1	−1	−1	.	.	.	.	.	.	.
X.25	75400	−120	104	16	.	.	−8	.	.	−5	.	3	.	.	.	.	−1	.	.	.	−1	−1	−1	1
X.26	76125	−35	105	−15	25	5	5	1	.	.	1	.	−1	1	−1	.	.	1	−1	−3	.	.	.	.
X.27	81432	24	−104	.	24	24	−8	−8	7	−3	.	1	.	.	.	−1	1	.	.	.	1	1	1	.
X.28	91350	−170	−14	9	10	−10	−2	2	.	5	1	.	4	−2	.	.	1	1	−1	−1	.	.	.	−1
X.29	95004	−100	.	−9	8	−20	4	.	4	4	−1	.	2	.	2	.	.	−1	1	.	.	.	.	1
X.30	98280	−24	.	.	−24	.	8	−8	5	.	.	.	.	.	.	1	.	.	.	.	.	.	.	.
X.31	98280	−24	.	.	−24	.	8	−8	5	.	.	.	.	.	.	1	.	.	.	.	.	.	.	.
X.32	102400	.	64	16	.	.	.	.	.	−5	.	−3	.	.	.	.	−1	.	.	−1	1	1	1	1
X.33	105560	88	.	−10	−40	.	−8	8	10	.	−2	.	.	.	.	−2	.	2	.	.	.	.	.	.
X.34	110592	.	−64	.	.	.	.	.	−8	−3	.	−1	.	.	.	.	1	.	.	1	−1	−1	−1	.
X.35	110592	.	−64	.	.	.	.	.	−8	−3	.	−1	.	.	.	.	1	.	.	1	−1	−1	−1	.
X.36	118784	.	64	−16	.	.	.	.	−16	−1	.	1	.	.	.	.	−1	.	.	3	1	1	1	−1

Character table of $G = \langle x, y, h, e \rangle$ (continued)

	16a	16b	20a	20b	20c	24a	24b	26a	26b	26c	29a	29b
2	4	4	2	2	2	3	3	2	2	2	.	.
3	.	.	.	.	.	1	1	.	.	.	.	.
5	.	.	1	1	1	.	.	.	.	.	.	.
7	.	.	.	.	.	.	.	.	.	.	.	.
13	.	.	.	.	.	.	.	1	1	1	.	.
29	.	.	.	.	.	.	.	.	.	.	1	1
2P	8b	8b	10a	10a	10a	12a	12a	13a	13a	13a	29b	29a
3P	16b	16a	20a	20c	20b	8a	8a	26b	26c	26a	29b	29a
5P	16a	16b	4a	4b	4b	24a	24b	26a	26b	26c	29a	29b
7P	16b	16a	20a	20c	20b	24b	24a	26c	26a	26b	29a	29b
13P	16a	16b	20a	20c	20b	24b	24a	2b	2b	2b	29a	29b
29P	16a	16b	20a	20b	20c	24a	24b	26b	26c	26a	1a	1a
X.1	1	1	1	1	1	1	1	1	1	1	1	1
X.2	A	$\bar{A}$	1	−1	−1	.	.	1	1	1	1	1
X.3	$\bar{A}$	A	1	−1	−1	.	.	1	1	1	1	1
X.4	.	.	.	.	.	−1	−1	−1	−1	−1	.	.
X.5	1	1	.	.	.	.	.	−1	−1	−1	.	.
X.6	.	.	1	−1	−1	−1	−1	.	.	.	−1	−1
X.7	.	.	−2	.	.	1	1	1	1	1	.	.
X.8	−1	−1	.	.	.	−1	−1	.	.	.	1	1
X.9	.	.	.	.	.	−1	−1	1	1	1	.	.
X.10	−1	−1	1	−1	−1	1	1	.	.	.	.	.
X.11	.	.	.	.	.	.	.	−1	−1	−1	1	1
X.12	.	.	.	.	.	.	.	−1	−1	−1	1	1
X.13	.	.	.	.	.	.	.	−1	−1	−1	1	1
X.14	−1	−1	−1	1	1	.	.	1	1	1	.	.
X.15	.	.	.	.	.	E	$-E$	.	.	.	−1	−1
X.16	.	.	.	.	.	$-E$	E	.	.	.	−1	−1
X.17	.	.	.	.	.	.	.	F	G	H	.	.
X.18	.	.	.	.	.	.	.	G	H	F	.	.
X.19	.	.	.	.	.	.	.	H	F	G	.	.
X.20	.	.	.	.	.	1	1	.	.	.	−1	−1
X.21	.	.	−1	−1	−1	1	1	.	.	.	.	.
X.22	.	.	1	1	1	.	.	.	.	.	.	.
X.23	1	1	.	.	.	.	.	.	.	.	.	.
X.24	−1	−1	1	1	1	.	.	.	.	.	.	.
X.25	.	.	.	.	.	.	.	.	.	.	.	.
X.26	1	1	.	.	.	−1	−1	1	1	1	.	.
X.27	.	.	−1	−1	−1	.	.	.	.	.	.	.
X.28	.	.	.	.	.	1	1	−1	−1	−1	.	.
X.29	.	.	−2	.	.	−1	−1	.	.	.	.	.
X.30	.	.	1	I	$-I$	.	.	.	.	.	−1	−1
X.31	.	.	1	$-I$	I	.	.	.	.	.	−1	−1
X.32	.	.	.	.	.	.	.	−1	−1	−1	1	1
X.33	.	.	.	.	.	.	.	.	.	.	.	.
X.34	.	.	.	.	.	.	.	1	1	1	J	K
X.35	.	.	.	.	.	.	.	1	1	1	K	J
X.36	.	.	.	.	.	.	.	−1	−1	−1	.	.

where $A = 2\zeta(4)$, $B = 2\zeta(7)^5 + 2\zeta(7)^4 + 2\zeta(7)^3 + 2\zeta(7)^2 + 1$, $C = -2\zeta(7)^4 - 2\zeta(7)^3 - 1$, $D = -(B+C) - 1$, $E = 2\zeta(24)_8^3\zeta(24)_3 + \zeta(24)_8^3 + 2\zeta(24)_8\zeta(24)_3 + \zeta(24)_8$, $F = 2\zeta(13)^9 + 2\zeta(13)^7 + 2\zeta(13)^6 + 2\zeta(13)^4 + 1$, $G = -2\zeta(13)^{11} - 2\zeta(13)^{10} - 2\zeta(13)^9 - 2\zeta(13)^7 - 2\zeta(13)^6 - 2\zeta(13)^4 - 2\zeta(13)^3 - 2\zeta(13)^2 - 1$, $H = 1 - F - G$, $I = 2\zeta(5)^3 + 2\zeta(5)^2 + 1$, $J = \zeta(29)^{27} + \zeta(29)^{26} + \zeta(29)^{21} + \zeta(29)^{19} + \zeta(29)^{18} + \zeta(29)^{17} + \zeta(29)^{15} + \zeta(29)^{14} + \zeta(29)^{12} + \zeta(29)^{11} + \zeta(29)^{10} + \zeta(29)^8 + \zeta(29)^3 + \zeta(29)^2 + 1$, $K = 1 - J$.

13
Lyons' group Ly

The simple group dealt with in this chapter was discovered by Lyons [88] in 1972. He tried to classify all the simple groups G having a 2-central involution z such that $C_G(z) \cong 2A_{11}$, the 2-fold cover of the alternating group A_{11}. He showed that each of these putative groups G has order $2^8 \cdot 3^7 \cdot 5^6 \cdot 7 \cdot 11 \cdot 31 \cdot 37 \cdot 67$. Furthermore, he determined their common character table and many important subgroups, such as the simple Dickson group $\mathsf{G}_2(5)$ and the automorphism group $Aut(3\mathsf{McL})$ of the 3-fold cover of McLaughlin's simple group McL constructed by Griess in [43].

The first construction of such a simple group G is due to Sims [118]. It was constructed by generators and relations. In 1985 Meyer, Neutsch and Parker realized the Lyons group as a simple subgroup of $\mathrm{GL}_{111}(5)$; see [90]. Later several other existence proofs were given by Gollan [35], Havas and Sims [48] and others.

None of these authors used the uniform construction method described in Algorithm 7.4.8 of [92]. Weller, Waki and the author applied it to obtain a short existence proof of Lyons' simple group [95]. In Section 13.4 this existence proof is modified. Instead of realizing the Lyons group Ly as a simple subgroup of $\mathrm{GL}_{111}(5)$ we construct a simple subgroup $\mathfrak{G}$ in $\mathrm{GL}_{2480}(31)$ having a 2-central involution $\mathfrak{z}$ such that $C_{\mathfrak{G}}(\mathfrak{z}) \cong 2A_{11}$; see Theorem 13.5.2 due to Weller, Wang, Kim and the author. Together with Theorem 7.5.1 of [92] it is a main ingredient of the uniqueness proof due to Weller and the author which is presented in Section 13.6. It has not been published before. The uniqueness Theorem 13.5.4 states that each finite simple group G having a 2-central involution z such that $C_G(z) \cong 2A_{11}$ is isomorphic to $\mathfrak{G}$.

13.1 Structure of the given centralizer

By a classical theorem of Schur [117] the covering group $2A_{11}$ of the alternating group A_{11} can be generated by nine elements m_i whose images $\bar{m}_i$ in A_{11} are the following permutations:

$$\bar{m}_1 = (1,2,3), \quad \bar{m}_i = (1,2)(i+1,i+2) \in A_{11} \text{ for } 2 \le i \le 9.$$

Schur's presentation of $2A_{11}$ is restated in the following result.

Lemma 13.1.1 *Let $H = \langle m_i \mid 1 \le i \le 9\rangle$ be the finitely presented group with the following set $\mathcal{R}(H)$ of defining relations:*

$$\begin{aligned} z &= {m_1}^3 = {m_i}^2 \text{ for } 2 \le i \le 9, \\ z^2 &= [z, m_j] = 1 \text{ for } 1 \le j \le 9, \\ (m_i m_{i+1})^3 &= z \text{ for } 1 \le i \le 8, \\ (m_i m_j)^2 &= z \text{ for } 1 \le i+1 < j \le 9. \end{aligned}$$

Then the following assertions hold.

(a) *The center $Z(H) = \langle z\rangle$ and $H/Z(H) \cong A_{11}$.*
(b) *H has a faithful permutation representation of degree 5040 with stabilizer $H_1 = \langle (m_9 m_8 \dots m_2 m_1)^2, (m_4 m_6)^{m_5} m_8^{m_9}\rangle$, and $H_1 \cong M_{11}$.*

Proof (a) This assertion is a special case of Schur's main result [117].

(b) $|H : H_1| = 5040$ has been verified by means of MAGMA. Another application of MAGMA shows that $H_1 \cong M_{11}$. Hence $(1_{H_1})^H$ is faithful. □

The following subsidiary result is closely related to several results of [95] due to Waki, Weller and the author.

Lemma 13.1.2 *Let $H = \langle m_i \mid 1 \le i \le 9\rangle$ be the finitely presented group defined in Lemma 13.1.1. Let $u_j = m_3 m_{j+3}$ for $1 \le j \le 6$. Let $c_0 = m_1^2$, $c_1 = (u_6 u_5 u_4 u_3 u_2)^2$, $v_1 = (u_2 u_1 u_6 u_5 u_4)^2$, $d_1 = (u_4 u_3 u_2)^2$, $w_2 = u_2^{c_1}$, $v_2 = v_1^{c_1}$ and $t = u_2 u_4 u_6$. Let $x = c_0 z t u_2 v_1 v_2 c_1 d_1 m_3$ and $y = z v_1 v_2 c_1 m_3$. Then the following assertions hold.*

(a) *$S = \langle u_2, w_2, v_1, v_2, d_1, m_3\rangle$ is a Sylow 2-subgroup S of H.*
(b) *The Klein 4-group $A = \langle z, t\rangle$ is the unique maximal elementary abelian normal subgroup of S.*

(c) $D = N_H(A) = \langle z, t, u_2, w_2, v_1, v_2, c_1, d_1, m_3, c_0 \rangle$ *has the following set* $\mathcal{R}(D)$ *of defining relations:*

$$\begin{aligned}
&z^2 = t^2 = v_1^2 = v_2^2 = 1, \quad c_0^3 = c_1^3 = 1, \quad u_2^4 = w_2^4 = d_1^4 = m_3^4 = 1,\\
&z^t = d_1^2 = m_3^2 = z, \quad {u_2}^{v_1} = {u_2}^{-1}, \quad w_2^{v_1} = w_2 t,\\
&u_2^{v_2} = {u_2}^{-1} t, \quad w_2^{v_2} = w_2^{-1}, \quad v_1^{v_2} = v_1, \quad u_2^{w_2} = {u_2}^{-1},\\
&v_1^{c_1} = v_2, u_2^{c_1} = w_2, \quad z^{c_1} = z, \quad t^{c_1} = t, \quad w_2^{c_1} = w_2 u_2 t,\\
&v_2^{c_1} = v_1 v_2, \quad c_0^z = c_0^t = c_0^{w_i} = c_0^{v_i} = c_0^{d_1} = c_0^{c_1} = c_0,\\
&c_0^{m_3} = {c_0}^{-1}, \quad z^{d_1} = z, \quad t^{d_1} = tz, \quad c_1^{d_1} = {c_1}^{-1},\\
&u_2^{d_1} = w_2, \quad v_1^{d_1} = v_2, \quad z^{m_3} = z, \quad t^{m_3} = tz,\\
&c_1^{m_3} = c_1, \quad u_2^{m_3} = {u_2}^{-1}, \quad w_2^{m_3} = w_2^{-1},\\
&v_1^{m_3} = u_2 v_1 z, \quad v_2^{m_3} = w_2 v_2 z, \quad d_1^{m_3} = d_1.
\end{aligned}$$

(d) $C_S(A) = \langle z, t, u_2, w_2, v_1, v_2, d_1 m_3, \rangle$ *has center* $Z[C_S(A)] = A$. *Furthermore,* $C_H(A) = \langle C_S(A), c_1, c_0 \rangle$.

(e) *The subgroup* $D = N_H(A)$ *of* H *is generated by the elements* x *and* y *having orders* 8 *and* 6*, respectively.*

(f) $H = \langle x, y, m_2 \rangle$.

(g) *Representatives* h_i *and the corresponding centralizer orders* $|C_H(h_i)|$ *of the conjugacy classes* h_i^H *of* H *are given in Table 13.6.1.*

(h) *The character table of* $H = \langle x, y, m_2 \rangle$ *is stated in Table 13.7.1.*

(i) *Representatives* d_j *and the corresponding centralizer orders* $|C_H(d_j)|$ *of the conjugacy classes* d_j^D *of* $D = \langle x, y \rangle$ *are given in Table 13.6.2.*

(j) *The character table of* $D = \langle x, y \rangle$ *is given in Table 13.7.2.*

(k) *The subgroup* $U = \langle u_j \mid 1 \le j \le 6 \rangle$ *of* H *is isomorphic to the covering group* $2A_8$ *of* A_8.

(l) $c_0 = (xy)^4$, $C_H(c_0) = \langle c_0 \rangle \times U$ *and* $U = \langle x^2, (x m_2 x^2 m_2)^4, (x m_2 (x^2 m_2)^2)^7 \rangle$.

Proof Assertions (a)–(g) and (l) can be verified using the faithful permutation representation of PH of H defined in Lemma 13.1.1(b) and MAGMA.

(g) and (i) The representatives of the conjugacy classes of H and D have been calculated by means of Kratzer's Algorithm 5.3.18 of [92] and PH.

(h) and (j) The character tables of H and D have been calculated by means of PH and MAGMA.

(k) This assertion can be checked by means of PH and the isomorphism test program of MAGMA. □

Lemma 13.1.3 *Keep the notation of Lemma 13.1.2. In* $H = \langle x, y, m_2 \rangle$ *let* $f = (x^2 m_2)^8$, $s = (ym_2)^2$ *and* $p = (xm_2)^2$ *be the elements of orders* 5, 7 *and* 11, *respectively, of Table 13.6.1.*

Then the following assertions hold.

(a) $N_1 = N_H(\langle f \rangle) = \langle x_1, x_2 \rangle$, *where* $x_1 = x^2 m_2$ *and* $x_2 = (xm_2x^2 m_2x^2)^2$ *have respective orders* 40 *and* 4. *Furthermore,* N_1 *has order* $2^6 \cdot 3^2 \cdot 5^2$, *and the subgroup* $\langle x_1^8, (x_1x_2)^3 x_1^2 x_2 x_1 \rangle$ *of* N_1 *is the stabilizer of a faithful permutation representation of* N_1 *of degree* 576.

(b) N_1 *is the semidirect product of its normal subgroup* $Q_1 = \langle x_1^8 \rangle$ *and its subgroup* $K = \langle x_1^5, x_2 \rangle$. *Furthermore,* K *is isomorphic to the central amalgamated product of a cyclic group of order* 4 *and the* 2*-fold cover of the symmetric group* S_6.

(c) $q = (x_1x_2)^3 x_1^2 x_2 x_1$ *represents the second conjugacy class of elements of order* 5 *of* H *and* $J = \langle f, q \rangle$ *is a Sylow* 5*-subgroup of* H *of order* 25. *It has a complement* T_1 *in* N_1, *which is an abelian homocyclic group of order* 16 *generated by the elements* $j_1 = (x_1^2 x_2 x_1^3 x_2^2 x_1^2 x_2^3 x_1 x_2)^3$ *and* $j_2 = x_2^2 x_1^2 x_2 x_1^5 x_2^2$ *of order* 4.

(d) $N_2 = N_H(J)$ *is the semidirect product of* J *and the subgroup* $T = \langle T_1, n_2 \rangle$ *of order* 32, *where* $n_2 = w_2 w_1 w_3$ *has order* 8 *and*

$$w_1 = (m_2x^2m_2x^4m_2x^3)^5, w_2 = (x^4m_2x^2m_2x^2m_2x^3m_2)^4,$$
$$w_3 = (xm_2xm_2x^4m_2xm_2x^3m_2)^3.$$

(e) $H = \langle x_1, x_2, n_2 \rangle$. *Its original generators* x, y *and* m_2 *are the following words in the new generators:*

$$x = (d_3d_1d_4d_1d_2d_4)^7, \quad y = (d_1d_4d_1d_2)^2(d_2d_4d_1d_4)^3,$$
$$m_2 = (x_1n_2^2x_1n_2^3)^{10},$$

where

$$d_1 = x_1^5, \quad d_2 = (n_2^2x_1^2)^6,$$
$$d_3 = (x_1^2n_2x_1n_2^2x_1n_2)^2, \quad d_4 = (n_2^2x_1n_2x_1n_2^2x_1)^7.$$

Proof Table 13.6.1 provides a system of representatives of the conjugacy classes of H. The element f, of order 5, is taken from that list.

(a) Using the faithful permutation representation PH of H defined in Lemma 13.1.1(b) and the program `GetShortGens(H,N1)` MAGMA found the given generators x_1 and x_2 of the subgroup $N_1 = N_H(\langle f \rangle)$ of $H = \langle x, y, m_2 \rangle$. Its defined permutation representation PN_1 of degree 576 can be verified using MAGMA.

(b) The complement $K = \langle x_1^5, x_2 \rangle$ in N_1 of its normal subgroup of Q_1 generated by $f = x_1^8$ has been found by means of PH and the MAGMA command `HasComplement(N1,Q1)`. The structure of K can be verified using the isomorphism test of MAGMA.

(c) This assertion can easily be verified using MAGMA.

(d) Using PH and MAGMA one checks that $N_2 = N_H(J) = \langle J, j_1, j_2, n_2 \rangle$ for the given words of these three generators. In fact, $f^{j_1} = f^2$ and $q^{j_2} = q^2$, and $N_{N_1}(J) = \langle J, j_1, j_2 \rangle$ is isomorphic to a split extension of the homocyclic group $T_1 = \langle j_1, j_2 \rangle$ by J.

Using PH and MAGMA the reader can check that n_2 has order 8 and that it does not belong to N_1.

(e) Similarly, it has been verified that $H = \langle N_1, n_2 \rangle = \langle x_1, x_2, n_2 \rangle$. The words for the original generators x and y, and m_2 in x_1, x_2 and n_2, have been found using PH, MAGMA and the programs `GetShortGens(H1, D1)` and `LookupWord(X,u)`, where $H_1 = \langle x_1, x_2, n_2 \rangle$, $D_1 = \langle x, y \rangle$ in H_1, $X = D_1$ for $u = x$ and $u = y$ and $X = H_1$ for $u = m_2$. It is straightforward to verify all the equations computationally. □

Lemma 13.1.4 *Keep the notation of Lemma 13.1.2. In $H = \langle x, y, m_2 \rangle$ let $s = (ym_2)^2$ and $p = (xm_2)^2$ be the elements of respective orders 7 and 11 of Table 13.6.1. Then the following assertions hold.*

(a) *s^G is the unique conjugacy class of elements of order 7 of H and $N_H(\langle s \rangle)$ is generated by $c_1 = (x^3 m_2 x)^2$, $c_2 = (m_2 x^6 m_2 x^2 m_2 x m_2 x^2)^4$ and $c_3 = (x^3 m_2 x^2 (m_2 x)^4 m_2 x^2 m_2 x)^4$. Furthermore, $C_H(s) \cong \langle s \rangle \times \mathsf{SL}_2(3)$ and $N_G(\langle s \rangle)/C_G(s)$ is cyclic of order 6.*

(b) *p and p^{-1} represent the two conjugacy classes of elements of order 11 of H. Furthermore, $N_H(\langle p \rangle) \cong \mathsf{F}_{55} \times \langle z \rangle$, where F_{55} denotes the Frobenius group of order 55.*

Proof Table 13.6.1 provides a system of representatives of the conjugacy classes. The elements s and e of respective orders 7 and 11 are taken from that list. Tables 13.6.1 and 13.7.1 imply that H has one conjugacy class of elements of order 7 and two of order 11 which are not real.

Using the faithful permutation representation PH of H defined in Lemma 13.1.1(b) and MAGMA the remaining assertions can be verified. □

Definition 13.1.5 A finite simple group G is said to be of Ly-*type* if G possesses a 2-central involution z such that $C_G(z) \cong H$, where H is the group of even order defined by generators and relations in Lemma 13.1.1.

13.2 Conjugacy classes of elements of even order

Throughout this section G denotes a simple group of Ly-type with a 2-central involution z such that $C_G(z) = H$, where H is the group defined in Lemma 13.1.1.

Lemma 13.2.1 *Keep the notation of Lemma 13.1.2. Let G be a finite simple group of Ly-type with a 2-central involution z such that $C_G(z) = H = \langle x, y, m_2 \rangle$. Let $A = \langle z, t \rangle$ be the unique normal Klein 4-subgroup of the Sylow 2-subgroup S of H defined in Lemma 13.1.2(a). Let $D = N_H(A) = \langle z, t, u_2, w_2, v_1, v_2, c_1, d_1, m_3, c_0 \rangle$ be the finitely presented subgroup of H with set $\mathcal{R}(D)$ of defining relations stated in Lemma 13.1.2(c). Let $N = N_G(A)$.*

Then the following assertions hold.

(a) *G has a unique conjugacy class z^G of involutions.*
(b) *$C = C_G(A) = C_H(A)$, and $N/C \cong \mathrm{GL}_2(2)$.*
(c) *There is an element c_2 of order 3 in N such that $N = \langle D, c_2 \rangle$ has order 6912 and a defining set $\mathcal{R}(N)$ of relations consisting of $\mathcal{R}(D)$ and the following relations:*

$$c_0{}^{c_2} = c_0, \quad z^{c_2} = zt, \quad t^{c_2} = z, \quad u_2^{c_2} = zu_2w_2v_2,$$
$$w_2^{c_2} = tu_2v_1v_2, \quad v_1^{c_2} = v_1, \quad v_2^{c_2} = v_2, \quad c_2{}^{d_1} = zc_0{}^2c_2{}^2,$$
$$c_2{}^{m_3} = c_0c_1{}^2c_2{}^2z, \quad c_1{}^{c_2} = c_1c_0.$$

(d) *N has a faithful permutation representation PN of degree 576 with stabilizer $N_0 = \langle v_1, c_1 \rangle$.*
(e) *$N = \langle x, y, c_2 \rangle = \langle D, c_2 \rangle$, where $x = c_0ztu_2v_1v_2c_1d_1m_3$ and $y = zv_1v_2c_1m_3$.*
(f) *The Goldschmidt index of the amalgam $H \leftarrow D \rightarrow N$ is 1.*

(g) *Representatives n_i and the corresponding centralizer orders $|C_N(n_i)|$ of the conjugacy classes n_i^N of $N = \langle x, y, c_2 \rangle$ are stated in Table 13.6.3.*

(h) *The character table of $N = \langle x, y, c_2 \rangle$ is given in Table 13.7.3.*

Proof (a) By Table 13.6.1 H has two conjugacy classes of involutions. They can be represented by the two generators z and t of the normal Klein 4-group A of the Sylow 2-subgroup S of H. As G is simple Theorem 4.7.3 of [92] asserts that $z^g = t$ for some $g \in G - H$.

(b) $C = C_G(A) = C_G(t) \cap C_G(z) = C_G(t) \cap H = C_H(t)$. Thus $|C| = 2^7 \cdot 3^2$ and $X = C_S(A)$ has order 2^7 by Table 13.6.1. Its center $Z(X) = A$ by Lemma 13.1.2(d). Let $K = C_G(t)$. Since $z^g = t$ for some $g \in G - H$ we know that S^g is a Sylow 2-subgroup of K. Clearly, $X \leq K$. Now Sylow's Theorem asserts that there is a $k \in K$ such that $X < S^{gk}$. Hence X is properly contained in a Sylow 2-subgroup of K. Thus assertion (b) holds by Lemma 1.4.4 of [92].

(c) By Lemma 13.1.2(c) $D/C = \Delta = \langle \bar{m}_3 \rangle$, where $\bar{m}_3$ denotes the coset m_3 modulo C. By (b) $D/C < N/C = \Phi \cong Aut(A) \cong \mathrm{GL}_2(2) \cong S_3$. Hence there is an element $c_2 \in \Phi$ of order 3 which is inverted by $\bar{m}_3$ such that $z^{c_2} = tz$, $t^{c_2} = z$.

$P = \langle u_2, w_2, v_1, v_2 \rangle = O_2(C)$ is a characteristic subgroup of C such that P/A is elementary abelian of order 16. By Lemma 13.1.2(c) $Q = \langle c_0, c_1, d_1 m_3 \rangle$ is a normal subgroup of C such that $C = PQ$ and $P \cap Q = \langle z \rangle$.

Since $\Gamma = Aut(P) \cong A_8$ the elements of Γ can be represented as matrices in $\mathrm{GL}_4(2) \cong A_8$ w.r.t. the basis consisting of the cosets of the four generators of P modulo A. From the relations of Lemma 13.1.2(c) it follows that c_1 has an image $\hat{c}_1 \neq 1$ in Γ with $|C_\Gamma(\hat{c}_1)| = 18$. As P is a characteristic subgroup of C it is normal in $N = N_G(A)$. Hence N/C is isomorphic to a subgroup of Γ. Since a Sylow 3-subgroup of Γ has order 9 we may choose $\hat{c}_2$ to be an element of order 3 in the centralizer $C_\Gamma(\hat{c}_1)$ such that $z^{c_2} = tz$, $t^{c_2} = z$ and $c_2^{m_3} c_2 \in AQ$.

The chosen $c_2 \in \Gamma$ acts on the basis elements of P/A such that

$$u_2^{c_2}(u_2 w_2 v_2)^{-1}, \quad w_2^{c_2}(u_2 v_1 v_2)^{-1}, v_1^{c_2} v_1^{-1}, \quad v_2^{c_2} = v_2^{-1} \in A.$$

The values of these four elements in $A = \langle z, t \rangle$ have been established with MAGMA by means of running through all 64 possibilities for a common solution. It follows that $u_2^{c_2} = z u_2 w_2 v_2$, $w_2^{c_2} = t u_2 v_1 v_2$, $v_1^{c_2} = v_1$ and $v_2^{c_2} = v_2$.

The subgroup $\langle c_2, m_3 \rangle$ operates on AQ by conjugation. The relations of Lemma 13.1.2(c) imply that ${c_2}^{m_3} c_2, c_0^{c_2} c_0^2, c_1^{c_2}, c_0^{c_2} c_0^2, c_1^{c_2} \in AQ$. Using the methods described before we found the following common solution: ${c_2}^{d_1} = z{c_0}^2{c_2}^2$, ${c_2}^{m_3} = c_0{c_1}^2{c_2}^2 z$, $c_0^{c_2} = c_0$ and $c_1^{c_2} = c_0 c_1$.

Using MAGMA it has been checked that $N = \langle z, t, u_2, w_2, v_1, v_2, d_1, m_3, c_0, c_1, c_2 \rangle$ has the given set $\mathcal{R}(N)$ of defining relations.

(d) This statement has also been verified computationally.

(e) This assertion can be checked by means of MAGMA using the faithful permutation representation PN of N described in (d).

(f) The Goldschmidt index of the amalgam $H \leftarrow D \rightarrow N$ has been calculated with MAGMA using Kratzer's Algorithm 7.1.10 of [92] and the faithful permutation representations PH and PN of H and N given in Lemma 13.1.1(b) and (d), respectively; it is 1.

(g) The representatives n_i of the conjugacy classes of N have been calculated by means of Kratzer's Algorithm 5.3.18 of [92] and PN.

(h) The character table of N has been calculated by means of the faithful permutation representation PN and MAGMA. □

Lemma 13.2.2 *Let G be a finite simple group of* Ly*-type with a 2-central involution z such that $C_G(z) = H$. Let $A = \langle z, t \rangle$ be the unique normal Klein 4-subgroup of the Sylow 2-subgroup S of H defined in Lemma 13.1.2. Let $N = N_G(A)$ and $D = N_H(A)$. Then the following assertions hold.*

(a) *G has 26 z-special conjugacy classes of elements of even order represented by the following H-classes:*

$$2_a, 4_a, 6_a, 6_b, 6_d, 8_a, 8_b, 10_a, 10_b, 12_a, 12_b, 14_a, 18_a, 20_a,$$

$$22_a, 22_b, 24_a, 24_b, 24_c, 28_a, 30_a, 30_b, 40_a, 40_b, 42_a \text{ and } 42_b.$$

As G-classes they are denoted by $2_A, 4_A, 6_A, 6_B, 6_C$, etc.

(b) *Using the notations of the conjugacy classes of $H = C_G(z)$ and $N = N_G(A)$ as in Tables 13.6.1 and 13.6.3, respectively, their fusion patterns into the z-special conjugacy classes of G are given in the following tables:*

G	2_A					4_A				6_A				6_B		6_C		
H	a	b				a	b			a	c			b	g	d	e	f
N	a	b	c	d	e	a	b	c	d	a	b	c	e	d		i	j	

G	8_A		8_B		12_A				12_B	24_C
H	a	d	b	c	a	c			b	c
N	a	c	b	d	a	b	c	d	e	a

Proof (a) The z-special conjugacy classes of the elements of even order of G are easily determined by means of the power map information given in Table 13.6.1.

(b) By Lemma 13.2.1(e) $D = N_H(A)$ has only one embedding $\mu : D \to N = N_G(A)$. Therefore the fusion patterns of the conjugacy classes of D into the ones of H and N are uniquely determined by H in any finite simple group G of Ly-type. They have been determined by means of the faithful permutation representations of H and N given in Lemmas 13.1.1(b) and 13.2.1(d). Tracing the fusion graph for the fusion of the D-classes into the H- and the N-classes yields the tables given in the statement. □

13.3 Conjugacy classes of p-singular elements, $p \in \{3, 5, 7, 11\}$

In [88] Lyons proved that each finite simple group G of Ly-type has one 3-central conjugacy class c_0^G and that $N_G(\langle c_0 \rangle)$ is isomorphic to the automorphism group $Aut(\mathsf{McL})$ of the 3-fold cover $3\mathsf{McL}$ of McLaughlin's simple group McL dealt with in Chapter 11. The proof of Lyons' result given in this section uses the methods of the author's joint article [95] with Waki and Weller. It follows that G has two conjugacy classes of elements of order 3 and one of order 9.

Using these results we also show in this section that a Sylow 5-subgroup S_5 of a finite simple group G of Ly-type has order 5^6, that its center $Z(S_5)$ is cyclic of order 5 and that $N_G(Z(S_5))$ is a split extension of an extra-special group P of order 5^5 by the amalgamated central product K_1 of the 2-fold cover $2S_6$ of S_6 and a cyclic group of order 4. Furthermore, S_5 has a unique maximal elementary abelian normal subgroup V of order 5^3. Its normalizer $M = N_G(V)$ is isomorphic to a non-split extension of $\mathrm{PSL}_3(5)$ by V constructed in Proposition 13.3.4. These results imply that G has two conjugacy classes of elements of order 5 and one of order 25.

These results and Lemma 13.1.4 imply that G has one conjugacy class of elements of order 7 and two conjugacy classes of order 11. In particular, the Sylow p-subgroups of G are cyclic for $p \in \{7, 11\}$.

Proposition 13.3.1 *Let G be a finite simple group of* Ly*-type with a* 2*-central involution z such that $C_G(z) = H$. Keep the notations of Lemma 13.1.2. In $H = \langle x, y, m_2 \rangle$ let $z = m_2^2$, $c_0 = (xy)^4$ and $v = (m_2y^2)^4$. Then the following assertions hold:*

(a) *$N_H(\langle c_0 \rangle) = \langle x, y, v \rangle$, where v has order* 6 *and commutes with the element c_0 of order* 3.

(b) $C_H(c_0) = \langle c_0 \rangle \times U$, *where* c_0 *has order* 3 *and* $U = \langle x^2, v \rangle \cong 2A_8$.

(c) $C_G(c_0)/\langle c_0 \rangle \cong C$, *where* $C \cong \mathsf{McL}$, *the simple McLaughlin group* McL *of order* $|C| = 2^7 \cdot 3^6 \cdot 5^3 \cdot 7 \cdot 11$.

(d) $E = N_G(c_0) = \langle x, y, c_2, v \rangle \cong 3\mathsf{McL} : 2$.

(e) $E = \langle v, x, y, c_2 \rangle$ *has* 54 *conjugacy classes* $\mathfrak{e}_i^{\mathfrak{E}}$ *with representatives* $\mathfrak{e}_i$ *and corresponding centralizer orders* $|C_{\mathfrak{E}}(\mathfrak{e}_i)|$ *given in Table 13.6.4.*

(f) *The character table of* E *is given in Table 13.7.4.*

Proof (a) The orders of c_0 and v, their relations $v^2 = z$, $vc_0 = c_0 v$ and the equation $N_H(\langle c_0 \rangle) = \langle x, y, v \rangle$ can be checked with MAGMA and the faithful permutation representation PH of H defined in Lemma 13.1.1.

(b) Lemma 13.1.2(k) asserts that $U \cong 2A_8$. Since $c_0 = m_1 = (xy)^4$ it also yields that $C_H(c_0) = \langle c_0 \rangle \times U$. In particular, U is a subgroup of $N_1 = N_H(\langle c_0 \rangle) = \langle x, y, v \rangle$. Using PH and the program `GetShortGens(N_1,U)` a MAGMA calculation yields the assertion.

(c) By Lemma 13.2.2 in each simple group G of Ly-type there is an element c_2 of order 3, $c_2 \notin H$, such that $c_2 \in C_G(c_0)$ and $z^{c_2} = zt$, where $A = \langle z, t \rangle$ is the unique normal Klein 4-subgroup of the Sylow 2-subgroup S of H determined in Lemma 13.1.2. Let $C = C_G(c_0)/\langle c_0 \rangle$. Then by (b) z is a 2-central involution of C such that $H_1 = C_C(z) = C_H(c_0) = \langle c_0 \rangle \times U$ and $U \cong 2A_8$.

Let $N = N_G(\langle c_0 \rangle)$. By Lemma 13.2.1(c) there is an element c_2 such that $z^{c_2} = zt \in S$ and N does not split over $\langle c_0 \rangle$. Hence Theorem 4.7.3 of [92] implies that $C \neq O(C)H_1$. Now Theorem 11.3.1 states that $C \cong \mathsf{McL}$ and $|C| = 2^7 \cdot 3^6 \cdot 5^3 \cdot 7 \cdot 11$. In particular, $C_G(c_0)/\langle c_0 \rangle \cong C$.

(d) Since N does not split over $\langle c_0 \rangle$ C also does not split. Now (a) and (c) imply that $E = N_G(\langle c_0 \rangle) = \langle x, y, v, c_2 \rangle \cong 3\mathsf{McL} : 2$.

(e) Theorem 11.3.1 gives a presentation of McLaughlin's simple group $M = \mathsf{McL}$ and also a faithful permutation representation of degree 22275. Therefore MAGMA is able to calculate a faithful permutation representation of $M_1 = Aut(M)$. It follows that M has a complement of order 2 in M_1. Hence one can extend the presentation of M to one of M_1. Using the faithful permutation representation of M_1 of degree 44550 and Holt's Algorithm of [54] one calculates the extensions of M_1 by the cyclic groups of order 3. It follows that M_1 has only one non-split extension. Therefore it isomorphic to E. Due to lack of space these calculations cannot be documented here. Another construction of E has been described in [95], in which E is realized as a subgroup $\mathfrak{E}$ in $\mathrm{GL}_{111}(5)$ generated by four matrices $\mathfrak{x}$, $\mathfrak{y}$, $\mathfrak{v}$, $\mathfrak{c}_3$ corresponding to the generators of E given in (d). Furthermore, $\mathfrak{E}$ has a faithful permutation representation $P\mathfrak{E}$ of degree 66825 with stabilizer isomorphic to $\langle x^2, y, v\rangle \cong 2S_8$. Using it and Kratzer's Algorithm 5.3.18 of [92] the representatives $\mathfrak{e}_i$ of the conjugacy classes of $\mathfrak{E}$ have been calculated with MAGMA.

(f) The character table of E has been calculated in [95] by means of the faithful permutation representation of $\mathfrak{E}$ and GAP; see [33]. It agrees with the one of the Atlas [19], p. 101. This completes the proof. □

Lemma 13.3.2 *Let G be a finite simple group of Ly-type with a 2-central involution z such that $C_G(z) = H$. Keep the notations of Lemma 13.1.2 and Proposition 13.3.1. In $H = \langle x, y, m_2\rangle$ let $z = m_2^2$, $\mathfrak{c}_0 = (xy)^4$, $v = (m_2 y^2)^4$ and $\mathfrak{c}_3 = (xyv)^5$. Let $E = N_G(\mathfrak{c}_0)$. Then the following assertions hold.*

(a) *$E = \langle x, y, v, \mathfrak{c}_2\rangle$ has two conjugacy classes of highest defect. They are represented by $\mathfrak{c}_0$ and $\mathfrak{c}_3$ in $H \cap E$, whose centralizers $C_E(\mathfrak{c}_0)$ and $C_E(\mathfrak{c}_3)$ have orders $2^7 \cdot 3^7 \cdot 5^3 \cdot 7 \cdot 11$ and $2^3 \cdot 3^7 \cdot 5$.*
(b) *$C_G(\mathfrak{c}_3) = C_E(\mathfrak{c}_3)$ is isomorphic to a split extension of the covering group $2S_5$ of S_5 by a special group Q of order 3^6 with center $Z(Q) = \langle \mathfrak{c}_0, \mathfrak{c}_3\rangle$.*
(c) *$C_G(\mathfrak{c}_3)$ has a complement in $N_G(\langle \mathfrak{c}_3\rangle)$ of order 2.*
(d) *Each Sylow 3-subgroup of G has order 3^7.*
(e) *G has two conjugacy classes 3_A and 3_B of elements of order 3, represented by $\mathfrak{c}_0$ and $\mathfrak{c}_3$.*
(f) *G has seven $\mathfrak{c}_0$-special conjugacy classes of elements of odd order represented by the E-classes 3_a, 15_a, 15_e, 21_a, 21_b, 33_a and 33_b of Table 13.6.4.*
(g) *G has three $\mathfrak{c}_3$-special conjugacy classes of elements of odd order represented by the E-classes 3_c, 9_a and 15_b of Table 13.6.4.*
(h) *Keep the notations of the conjugacy classes of H and E determined in Tables 13.6.1 and 13.6.4, respectively. The fusion of the conjugacy classes of H and E into the conjugacy classes of G is given in the following tables:*

<table>
<tr><td>G</td><td colspan="2">2_A</td><td colspan="2">3_A</td><td colspan="2">3_B</td><td colspan="2">4_A</td><td>5_A</td><td colspan="2">6_A</td><td colspan="2">6_B</td><td colspan="3">6_C</td></tr>
<tr><td>H</td><td>a</td><td>b</td><td>a</td><td></td><td>b</td><td>c</td><td>a</td><td>b</td><td>a</td><td>a</td><td>c</td><td>b</td><td>g</td><td>e</td><td>d</td><td>f</td></tr>
<tr><td>E</td><td>a</td><td>b</td><td>a</td><td>b</td><td>c</td><td>d</td><td>a</td><td>b</td><td>a</td><td>a</td><td>b</td><td>c</td><td>d</td><td>e</td><td>f</td><td></td></tr>
</table>

<table>
<tr><td>G</td><td colspan="2">8_A</td><td colspan="2">8_B</td><td colspan="3">12_A</td><td colspan="2">12_B</td><td>14_A</td><td colspan="2">15_A</td><td colspan="2">15_B</td><td>15_C</td></tr>
<tr><td>H</td><td>a</td><td>d</td><td>b</td><td>c</td><td>a</td><td colspan="2">c</td><td colspan="2">b</td><td>a</td><td colspan="2">a</td><td colspan="2">b</td><td></td></tr>
<tr><td>E</td><td>a</td><td>b</td><td colspan="2">c</td><td>a</td><td>b</td><td>e</td><td>c</td><td>d</td><td>a</td><td>b</td><td>c</td><td>a</td><td>d</td><td>e</td></tr>
</table>

<table>
<tr><td>G</td><td colspan="2">20_A</td><td>21_A</td><td>21_B</td><td>22_A</td><td>22_B</td><td colspan="2">30_A</td><td colspan="2">30_B</td><td>42_a</td><td>42_B</td><td>33_A</td><td>33_B</td></tr>
<tr><td>H</td><td colspan="2">a</td><td>a</td><td>b</td><td>a</td><td>b</td><td colspan="2">a</td><td colspan="2">b</td><td>a</td><td>b</td><td></td><td></td></tr>
<tr><td>E</td><td>a</td><td>b</td><td>a</td><td>b</td><td>a</td><td>b</td><td>a</td><td>b</td><td>c</td><td>d</td><td>b</td><td>a</td><td>a</td><td>b</td></tr>
</table>

Proof (a) By Table 13.6.4 $c_3 = (xyv)^5$ belongs to the conjugacy class 3_c of $E = \langle x, y, v, c_2\rangle$ constructed in [95]; see the proof of Proposition 13.3.1. Hence $|C_E(c_3)| = 2^3 \cdot 3^7 \cdot 5$ by Table 13.6.4. Thus (a) holds by Proposition 13.3.1(c).

(b) Let PH be the faithful permutation representation of H described in Lemma 13.1.1(b). Using it and MAGMA it has been checked that the two elements c_0 and c_3 of H commute and that $N_3 = N_H(\langle c_3\rangle)$ is generated by $n_1 = (m_2x^4m_2x)^6$, $n_2 = (m_2x^2m_2x^2m_2)^2$, $n_3 = (xm_2xm_2xm_2x^5)^2$, $n_4 = (m_2x^3m_2xm_2xm_2x^2)^3$ and $n_5 = (xm_2x^4m_2x^4m_2)^4$. Furthermore, $C_3 = C_H(c_3) = \langle n_1, n_2, n_5, n_3n_4\rangle$ and $W = \langle c_0, c_3\rangle$ has a complement $K_1 = \langle n_1, n_2, n_3n_4\rangle \cong 2S_5$ in C_3. The center of K_1 is generated by z. Therefore (a) and the 3-modular representation theory of K_1 imply that $C_G(c_3)$ is a split extension of K_1 by a special group Q of order 3^6 with center $Z(Q) = W$.

(c) Using PH and the MAGMA command `HasComplement(PN3,PW)` the author found the complement $K = \langle K_1, k\rangle$ of W in N_3 such that the involution $k = (n_1n_5n_4n1)^3$ also generates a complement of K_1 in K. Thus N_3 splits over C_3.

(d) $C_G(W) = C_G(c_0) \cap C_G(c_3) \le C_E(c_3)$ by Proposition 13.3.1(c). By (b) $C_E(c_3) = QK_1$. Thus $Z(Q) = W$ and Q are characteristic subgroups of $C_G(W) = QK_1'$, where $K_1' = \langle n_2, (n_1n_3n_4)^2\rangle$. Since $N_G(W)/C_G(W) \le \mathrm{GL}_2(3)$ it follows from (a) and (b) that each Sylow 3-subgroup S_3 of G has order 3^7 or 3^8. Furthermore, $N_G(W)/C_G(W) \cong N_H(W)/C_H(W)$ because the center of K_1 is generated by the central involution $z = m_2^2$ of H. Using PH and MAGMA again we observed that $N_H(W) = \langle x_i | 1 \le i \le 4\rangle$, where $x_1 = (xm_2x^4m_2)^6$, $x_2 = (m_2x^4m_2x)^6$, $x_3 = m_2x^6m_2x^2m_2$ and $x_4 = (m_2x^3m_2xm_2xm_2x^2)^3$. Furthermore, $|N_H(W)| = 2^6 \cdot 3^3 \cdot 5$. Hence S_3 has order 3^7.

(e) Table 13.6.4 states that E has four conjugacy classes of elements of order 3, two of which are not of highest defect. They are represented by v^2 and y^2. Now both representatives belong to $H \cap E$. Using PH and MAGMA again the reader can check that v^2 and y^2 are H-conjugate to c_0 and c_3, respectively. Therefore (d) implies that G has exactly two conjugacy classes of elements of order 3 because $C_G(c_0)$ and $C_G(c_3)$ have non-isomorphic Sylow 2-subgroups by Proposition 13.3.1(c) and statement (b).

(f) and (g) These assertions are now immediate by the power map information given in Table 13.6.4.

(h) The fusion patterns of the conjugacy classes of $K = N_H(\langle c_0\rangle)$ into the ones of H and E are uniquely determined by H for all simple groups G of Ly-type. Using the faithful permutation representation of H and E the tables of assertion (e) have been calculated by means of MAGMA. □

In [88] Lyons claims that a Sylow 5-subgroup of a simple group G of Ly-type has exponent 5. This is not true, as mentioned already by Harada in [46]. Therefore we now construct a Sylow 5-subgroup and some 5-local subgroups of G which are used to determine the 5-singular conjugacy classes of G.

Proposition 13.3.3 *Keep the notations of Lemmas 13.1.2 and 13.1.3. Let G be a finite simple group of Ly-type with a 2-central involution z such that $C_G(z) = H = \langle x, y, m_2\rangle$. Let $N_1 = N_H(\langle f\rangle) = \langle x_1, x_2\rangle$, where $f = (x^2m_2)^8$, $x_1 = x^2m_2$ and $x_2 = (xm_2x^2m_2x^2)^2$. Let $q = (x_1x_2)^3x_1^2x_2x_1$. Then the following assertions hold.*

(a) *The normal cyclic subgroup $\langle f\rangle$ of N_1 has a complement $K = \langle x_1^5, x_2\rangle$ in N_1 which is isomorphic to the central amalgamated product of the cyclic group of order 4 and the 2-fold cover of the symmetric group S_6.*
(b) *$R = N_G(\langle f\rangle)$ is isomorphic to the split extension of K by an extra-special 5-group Q of order 5^5 with center $Z(Q) = \langle f\rangle$.*
(c) *$P = \langle Q, q\rangle$ is a Sylow 5-subgroup of G of order 5^6 contained in R.*
(d) *$N_G(P)$ is a semidirect product of P and the abelian homocyclic group $T_1 = \langle j_1, j_2\rangle$ of order 16 by P generated by the elements*

$$j_1 = (x_1^2x_2x_1^3x_2^2x_1^2x_2^3x_1x_2)^3 \text{ and } j_2 = x_2^2x_1^2x_2x_1^5x_2^2 \text{ of order } 4.$$

(e) *$K = \langle x_1^5, x_2\rangle$ contains two conjugacy classes of subgroups of order 480 represented by $G_1 = \langle g_i | 1 \le i \le 3\rangle$ and $G_2 = \langle g_i | 4 \le i \le 6\rangle$, where $g_1 = (x_1^5x_2)^3$, $g_2 = (x_2x_1^5)^2$, $g_3 = (x_2^2x_1^5)^2$, $g_4 = (x_1^{10}x_2)^3$, $g_5 = (x_1^5x_2x_1^5)^2$ and $g_6 = (x_2x_1^5x_2)^2$.*
   Furthermore, $G_1 \cong G_2 \cong \mathrm{GL}_2(5)$, $K = \langle G_1, G_2\rangle$, $|G_1 \cap G_2| = 2^4 \cdot 5$ and $N_1 = \langle G_1, j_i\rangle = \langle G_2, j_i\rangle$ for $i = 1, 2$.
(f) *$u = (g_2g_1g_2g_3g_1)^8 \in G_1 \cap G_2$ generates a Sylow 5-subgroup U of K such that $P_1 = QU$ is a Sylow 5-subgroup of G which has an elementary abelian normal subgroup V_1 of order 125 whose normalizer $L = N_R(V_1)$ has a normal subgroup V_2 of order 5^5 with factor group $L/V_2 \cong \mathrm{GL}_2(5)$ and $N_G(V_1)/V_1 \cong \mathrm{PSL}_3(5)$. Furthermore, V_2 does not split over V_1.*

(g) $K = \langle x_1^5, x_2 \rangle$ *is isomorphic to the finitely presented group* $R_1 = \langle r_i | 1 \leq i \leq 5 \rangle$ *with the following set* $\mathcal{R}(R_1)$ *of defining relations:*

$$
\begin{aligned}
&r_1^4 = r_2^4 = r_3^3 = r_4^3 = r_5^5 = 1,\\
&(r_1, r_2) = (r_1, r_3) = (r_1, r_4) = (r_3, r_4) = (r_1, r_5) = 1,\\
&r_5^{-1} r_3 r_2^{-1} r_1^{-1} r_5^{-1} r_3 r_2^{-1} = r_2 r_5^{-1} r_2^{-1} r_4 r_5^{-1} = 1,\\
&r_2^{-1} r_3^{-1} r_2^2 r_3^{-1} r_2^{-1} = r_4^{-1} r_2^{-1} r_1^{-2} r_5 r_4 r_2 = r_2^{-1} r_4^{-1} r_2^2 r_4^{-1} r_2^{-1} = 1,\\
&r_4^{-1} r_5^{-1} r_4^{-1} r_2^{-1} r_5 r_2 = r_5 r_3^{-1} r_5^{-1} r_3^{-1} r_5 r_4 = 1.
\end{aligned}
$$

Proof (a) This is a restatement of Lemma 13.1.3(b).

(b) Let PH be the faithful permutation representation of H defined in Lemma 13.1.1(b). Using it and MAGMA it has been checked that $H_1 = C_H(f)$ has a Sylow 2-subgroup which is a quaternion group of order 16 with center $Z_1 = Z(H_1) = \langle z \rangle$ and that $H_1/Z_1 \cong A_6$. Hence Theorem 4.6.5 of [92] asserts that $C = C_G(f)$ has a normal subgroup Q of odd order such that $C = QC_H(f)$. In particular, f belongs to the center $Z(Q)$ of Q and $R = N_G(\langle f \rangle) = QN_1$.

Now (a) asserts that $N_1 = N_H(\langle f \rangle)$ is a semidirect product of $Q_1 = \langle f \rangle$ and the subgroup $K = \langle x_1^5, x_2 \rangle$. Furthermore, K is an amalgamated product of a cyclic group of order 4 and S_6. Therefore $R = N_G(\langle f \rangle) = QN_1$. By Lemma 13.1.3(c) N_1 contains a Klein 4-group $V = \langle z, z_1 \rangle$, where $z_1 = j_2^2$ as defined there. Now Theorem 4.2.2 of [92] states that $|Q||C_Q(V)|^2 = |C_Q(z)| \cdot |C_Q(z_1)| \cdot |C_Q(zz_1)|$; Q has at least order 5^2 by Table 13.6.4. The central involution z of K_1 operates fixed-point-freely on $Q_2 = Q/\langle f \rangle$. Thus Q_2 is abelian by Lemma 1.5.9 of [92]. By [71], p. 5, the smallest faithful 5-modular representation of K has degree 4. Hence $|Q| \geq 5^5$. By (a) Q_1 has a complement K in N_1 and $|K| = 5 \cdot r$, where $(5, r) = 1$. Therefore Table 13.6.1 implies that 5 divides $|C_Q(z)|$ to the first power only. Thus at least one of the numbers $|C_Q(z_1)|$ or $|C_Q(zz_1)|$ is divisible by 25 because $|C_Q(V)|$ is coprime to 5 by Table 13.6.1. Since all involutions of G are conjugate by Lemma 13.2.1(a), Table 13.6.1 implies that $|Q| = 5^5 \cdot 3^a$, where $1 \leq a \leq 12$. If 3 were a divisor of $|C_Q(z)|$ then $|C_H(f)|$ would be divisible by 3^3, which contradicts Table 13.6.1. By Lemma 13.3.2 the centralizers $C_G(d)$ of elements of order 3 are G-conjugate to centralizers $C_E(k)$ of elements k of order 3 in E. Since $C_Q(t)$ is abelian for each $t \in \{z_1, zz_1\}$ it follows that $C_Q(t)$ is conjugate to a subgroup $C_E(s)$ of some involution $s \in E$. By Table 13.6.4 E has exactly two conjugacy classes of involutions s, and in both cases $|C_E(s)|$ is only divisible by 5 to the first power. Hence $|Q| = 5^5$, and its center

$Z(P)$ has order 5. Furthermore, $K \cap Q = 1$ because K does not have a normal subgroup of odd order. Hence $R = QK$ since $N_1 = Q_1K$ by (a) and $|R/Q| = |N_1/Q_1|$.

(c) By Lemma 13.1.3(c) there is an element q of order 5 in N_1 such that $J = \langle f, q\rangle$ is a Sylow 2-subgroup of N_1. Hence $P = QJ$ is a Sylow 5-subgroup of order 5^6 in $R = QN_1$. Let P_1 be a Sylow 5-subgroup of G of $N_G(P)$. Since $Z(P) = \langle f\rangle$ and $Z(P_1) \cap P \neq 1$ it follows that $f \in Z(P_1)$. Hence $P_1 \leq C_G(f) \leq QN_1$. Therefore $P_1 = P$.

(d) By Lemma 13.1.3(c) $N_2 = N_H(J) = JT_1$, where $T_1 = \langle j_1, j_2\rangle$ is a homocyclic group of order 16 contained in N_1. From $R = QN_1$, $P = QJ$ and $N_H(J) = JT_1 = N_{N_1}(J)$ it follows that $N_R(P) = PT_1$. Thus $N_G(P) = N_R(P)$ because $Z(P) = \langle f\rangle$ is a characteristic subgroup of P.

(e) Using the MAGMA command `Subgroups(PK: OrderEqual:=480)` in the permutation representation PH we observed that PK has exactly two conjugacy classes of subgroups of this order. They are represented by G_1 and G_2. Using the the program `GetShortGens(K,G_i)` we found the given generators. All other statements can now be checked by means of standard MAGMA commands.

(f) Since $G_1 \cong G_2 \cong \mathrm{GL}_2(5)$ we know that both subgroups of K have a cyclic Sylow 5-subgroup of order 5. Furthermore, $C_i = (G_i \cap C_H(f)) \cong \mathsf{SL}_2(5)$ for $i = 1, 2$. The second center Z_2 of the Sylow 5-subgroup $P_1 = QU$ is abelian and its intersection Z_1 with Q is elementary abelian of order at least 25 because $Q/Z(P_1)$ is a proper abelian normal subgroup of $P_1/Z(P_1)$. Hence $Z_2/Z(P_1) \neq 1$. By the modular character tables of $\mathsf{SL}_2(5)$ and $2S_6$ of [71], pp. 2 and 5, the restriction of $Q/Z(P_1)$ to $\mathsf{SL}_2(5)$ is a direct sum of two copies of the unique faithful irreducible $\mathsf{SL}_2(5)$-module of dimension 2 over $GF(2)$. Hence Z_2 is an elementary abelian normal subgroup of order 8 in P_1 which is normal in $N_K(Z_2)$. Thus $L/Q = N_K(Z_2)/Q \cong \mathrm{GL}_2(5)$ and Q/Z_2 is non-split because Q is extra-special.

Since Z_2 is a characteristic subgroup of P_1 it is normal in $N_G(P_1)$. Now (d) implies that $N_G(Z_2)/Z_2 \cong \mathrm{PSL}_3(5)$ because $L/Z_2 \cong 5^2 : \mathrm{GL}_2(5)$ is a maximal subgroup of $Y = \mathrm{PSL}_3(5)$.

(g) Since the finitely presented group $R_1 = \langle r_i\rangle$ has a faithful permutation representation of degree 576 with stabilizer $\langle r_5\rangle$ an isomorphism test with MAGMA verifies the assertion. □

In order to show that the group $R = N_G(\langle f\rangle)$ is uniquely determined (up to isomorphism) in any simple group of Ly-type we first construct a non-split extension E of $\mathrm{PSL}_3(5)$ by its natural vector space V over $GF(5)$. We show that E is uniquely determined. Then we apply an odd analog of Algorithm 1.3.8 to construct the group $R = N_E(\langle f \rangle)$, where $f \in V$ generates the center of a Sylow 5-subgroup of E. Then we show that the amalgam $E \leftarrow L \rightarrow R$ is uniquely determined. It will be shown that R is isomorphic to the split extension $Q : K$ of K by the extra-special group Q constructed in Proposition 13.3.3.

Proposition 13.3.4 *The simple group $T = \mathrm{GL}_3(5)$ has (up to isomorphism) a unique non-split extension M by its natural $GF(5)$-module V of dimension* 3. *The group M has order $2^5 \cdot 3 \cdot 5^6 \cdot 31$ and it is isomorphic to the finitely presented group $M = \langle e_i \mid 1 \leq i \leq 8 \rangle$ having the following set $\mathcal{R}(M)$ of defining relations:*

$$
\begin{aligned}
&e_1^4 = e_2^4 = e_3^{10} = e_4^5 = e_5^{15} = 1, \quad e_6^5 = e_7^5 = e_8^5 = 1,\\
&(e_4, e_6) = (e_4, e_8) = 1, \quad (e_6, e_7) = (e_6, e_8) = (e_7, e_8) = 1,\\
&e_1^{-1} e_6 e_1 e_6 = e_1^{-1} e_7 e_1 e_7^2 = e_1^{-1} e_8 e_1 e_8^2 = 1,\\
&e_2^{-1} e_6 e_2 e_6^2 = e_2^{-1} e_7 e_2 e_7^2 = e_2^{-1} e_8 e_2 e_8 = 1,\\
&e_3^{-1} e_6 e_3 e_6 = e_3^{-1} e_7 e_3 e_8 = e_3^{-1} e_8 e_3 e_7 = e_4^{-1} e_7 e_4 e_7^{-1} e_8^{-2} = 1,\\
&e_5^{-1} e_6 e_5 e_8^{-1} = e_5^{-1} e_7 e_5 e_6^{-1} = e_5^{-1} e_8 e_5 e_7^{-1} = 1,\\
&e_3^2 e_7 e_8^{-1} = e_5^{-3} e_6 e_7 e_8 = e_1^{-1} e_2^{-1} e_1 e_2 e_7^{-2} e_8^2 = 1,\\
&e_1^{-1} e_3 e_1 e_3 e_6^{-2} e_7 e_8^2 = e_1^{-1} e_4^{-1} e_1 e_4 e_6^{-2} e_7^{-2} e_8^{-1} = 1,\\
&e_2^{-1} e_5^{-1} e_1 e_5 e_6^2 e_8^2 = e_3 e_5^{-1} e_3 e_5^{-1} e_6^2 e_7^{-2} = 1,\\
&e_3 e_2^{-1} e_1^{-1} e_3 e_2^{-1} e_6^{-1} e_7^{-1} e_8^{-1} = e_2 e_4^{-1} e_2^{-1} e_4^2 e_8^{-2} = 1,\\
&e_4^{-2} e_2^{-1} e_4^{-1} e_2 e_8^{-1} = e_3 e_4 e_3 e_1^{-1} e_4 e_3 e_4 e_7^{-1} e_8 = 1,\\
&e_5^{-1} e_3 e_4^{-1} e_5^{-1} e_3 e_4^{-1} e_3 e_5 e_4 e_3 e_5 e_4 e_7^{-1} e_8 = 1,\\
&e_5 e_3 e_4^{-1} e_5 e_3 e_4^{-1} e_3 e_5^{-1} e_4 e_3 e_5^{-1} e_4 e_6^{-1} e_7 = 1.
\end{aligned}
$$

Furthermore, the following statements hold.

(a) *M has a faithful permutation representation of degree* 3875 *with stabilizer $M_1 = \langle e_5 e_1 e_3 e_6, (e_2 e_7 e_3 e_5 e_4)^2 \rangle$.*

(b) *$S = \langle (e_2 e_4 e_5 e_1)^4, (e_3 e_4 e_2^2)^2, e_6, e_7, e_8 \rangle$ is a Sylow* 5*-subgroup of M of order 5^6. Its center $Z(S)$ is generated by $f = e_3^2 = e_8 e_7^{-1}$, and $V = \langle e_6, e_7, e_8 \rangle$ is the unique maximal elementary abelian normal subgroup of S of order* 125.

(c) *The subgroup $L = N_M(\langle f \rangle)$ has order $2^5 \cdot 3 \cdot 5^6$; $L = \langle x, y \rangle$ and $M = \langle L, e \rangle$, where $x = e_1 e_3$, $y = e_1 (e_2 e_5 e_4)^4$ and $e = e_2$ have respective orders* 20, 24 *and* 4.

(d) *A system of representatives m_i of the* 39 *conjugacy classes of* $M = \langle x, y, e\rangle$ *and the corresponding centralizer orders* $|C_M(e_i)|$ *are given in Table 13.6.7.*

(e) *A system of representatives n_i of the* 46 *conjugacy classes of* $L = \langle x, y\rangle$ *and the corresponding centralizer orders* $|C_N(n_i)|$ *are given in Table 13.6.6.*

(f) *The character tables of M and L are stated in Tables 13.7.6 and 13.7.7, respectively.*

(g) *L has a unique normal subgroup Q of order* $2^2 \cdot 5^5$. *Its automorphism group $A = Aut(Q)$ has order* $2^9 \cdot 3^2 \cdot 5^9 \cdot 13$. *There is an inner automorphism a of Q of order* 4 *such that $C = C_A(a)$ has order* $2^9 \cdot 3^2 \cdot 5^4 \cdot 13$. *The Fitting subgroup F of the group $IA(Q)$ of inner automorphisms of Q is extra-special of order* 5^5 *and A is the semidirect product of F and C.*

(h) *Up to conjugacy C has a unique subgroup K of order* 2880 *which is isomorphic to the finitely presented group* $R_1 = \langle r_i | 1 \le i \le 5\rangle$ *with the set of defining relations $\mathcal{R}(R_1)$ stated in Proposition 13.3.3(g).*

(i) *The semidirect product $R = FK$ of F and K is isomorphic to the finitely presented group* $R = \langle r_j \mid 1 \le j \le 10\rangle$ *having the set $\mathcal{R}(R)$ of defining relations consisting of $\mathcal{R}(R_1)$ and the following relations:*

$$r_6^5 = 1, \quad (r_6, r_7) = (r_4, r_9) = (r_3, r_{10}) = (r_4, r_{10}) = 1,$$
$$(r_5, r_{10}) = (r_6, r_{10}) = (r_7, r_{10}) = (r_8, r_{10}) = (r_9, r_{10}) = 1,$$
$$r_1^{-1} r_{10}^{-1} r_1 r_{10}^{-1} = r_1 r_8^{-1} r_1^{-1} r_8^2 = r_8^{-2} r_1^{-1} r_8^{-1} r_1 = r_2 r_{10}^{-1} r_2^{-1} r_{10}^{-2} = 1,$$
$$r_{10}^{-2} r_2^{-1} r_{10} r_2 = r_3 r_6^{-1} r_3^{-1} r_2^{-1} r_9^{-1} r_2 = r_1 r_7^{-1} r_5^{-1} r_1^{-1} r_9^{-1} r_5 = 1,$$
$$r_6^{-1} r_1^{-1} r_6^{-1} r_1 r_6^{-1} r_{10}^{-1} = r_7 r_2^{-1} r_6^{-1} r_2 r_6^{-1} r_8^{-1} = 1,$$
$$r_7 r_2 r_6^{-1} r_2^{-1} r_8^{-2} = r_4 r_7^{-1} r_6^{-1} r_4^{-1} r_8 r_6 = 1,$$
$$r_4^{-1} r_8 r_6^{-1} r_4 r_6 r_8^{-1} = r_5^{-1} r_9^{-1} r_6^{-1} r_5 r_6 r_9 = r_6 r_9^{-1} r_6^{-1} r_9 r_{10} = 1,$$
$$r_3^{-1} r_8 r_7^{-1} r_3 r_7 r_8^{-1} = r_8 r_3^{-1} r_8^{-1} r_7 r_3 r_7^{-1} = r_9^{-1} r_3^{-1} r_8^{-1} r_6 r_9 r_3 = 1.$$

R has a faithful permutation representation PR of degree 3125 *with stabilizer R_1.*

(j) *The Sylow* 5*-subgroup* $S_1 = \langle r_s \mid 5 \le s \le 10\rangle$ *of R has a unique elementary abelian normal subgroup of order* 125. *It is*

$$V_1 = \langle (r_5 r_6)^5, r_5^3 r_6 r_5 r_6^3 r_5 r_6, r_5^2 r_6 r_5^2 r_6^2 r_5 r_6^2 \rangle.$$

(k) *$L_1 = N_R(V_1)$ is isomorphic to L, and there is an isomorphism $\sigma : L \to L_1$ such that $\sigma(x_1) = x$ and $\sigma(y_1) = y$, where*

$$x_1 = (r_2^2 r_3 (r_2 r_3 r_4)^3 r_6^2 r_2^2 r_3 r_3 r_4)^{17},$$
$$y_1 = ((r_6 r_3 r_4 r_2^2 r_3)^9 r_6^3 (r_2 r_3 r_4)^3)^{23}$$

have respective orders 20 and 24. Furthermore, $L_1 = \langle x_1, y_1 \rangle$ and $R = \langle L_1, h \rangle$, where $h = r_2$ has order 4.

(l) *A system of representatives r_i of the 55 conjugacy classes of $R = \langle x, y, h \rangle$ and the corresponding centralizer orders $|C_M(e_i)|$ are given in Table 13.6.5.*

(m) *The character table of R is given in Table 13.7.5.*

(n) *The amalgam $R \leftarrow L \rightarrow M$ has Goldschmidt index 1.*

(o) *G has two conjugacy classes 5_A and 5_B of elements of order 5 and one conjugacy class 25_A of elements of order 25.*

Proof The simple group $T = \mathrm{GL}_3(5)$ is generated by the following five matrices:

$$Me_1 = \begin{pmatrix} 4&0&0 \\ 0&3&0 \\ 0&0&3 \end{pmatrix}, \quad Me_2 = \begin{pmatrix} 3&0&0 \\ 0&3&0 \\ 0&0&4 \end{pmatrix}, \quad Me_3 = \begin{pmatrix} 4&0&0 \\ 0&0&4 \\ 0&4&0 \end{pmatrix},$$

$$Me_4 = \begin{pmatrix} 1&0&0 \\ 0&1&2 \\ 0&0&1 \end{pmatrix} \quad \text{and} \quad Me_5 = \begin{pmatrix} 0&0&1 \\ 1&0&0 \\ 0&1&0 \end{pmatrix}$$

of respective orders 4, 4, 2, 5 and 3. It has a faithful permutation representation PL of degree 23250 with stabilizer $\langle Me_1, Me_2 \rangle$. Using the MAGMA command `FPGroup(PL)` we obtain the following set $\mathcal{R}(L)$ of defining relations of the finitely generated group $L = \langle e_i \mid 1 \le i \le 5 \rangle$:

$$e_1^4 = e_2^4 = e_3^2 = e_4^5 = e_5^3 = 1,$$
$$e_1^{-1} e_2^{-1} e_1 e_2 = e_1^{-1} e_3 e_1 e_3 = 1, \quad e_1^{-1} e_4^{-1} e_1 e_4 = 1,$$
$$e_2^{-1} e_5^{-1} e_1 e_5 = e_3 e_5^{-1} e_3 e_5^{-1} = 1, \quad e_3 e_2^{-1} e_1^{-1} e_3 e_2^{-1} = 1,$$
$$e_2 e_4^{-1} e_2^{-1} e_4^2 = e_4^{-2} e_2^{-1} e_4^{-1} e_2 = 1, \quad e_3 e_4 e_3 e_1^{-1} e_4 e_3 e_4 = 1,$$
$$e_5^{-1} e_3 e_4^{-1} e_5^{-1} e_3 e_4^{-1} e_3 e_5 e_4 e_3 e_5 e_4 = 1,$$
$$e_5 e_3 e_4^{-1} e_5 e_3 e_4^{-1} e_3 e_5^{-1} e_4 e_3 e_5^{-1} e_4 = 1,$$

where e_i corresponds to the matrix Me_i for $1 \le i \le 5$.

Thus we can apply Holt's Algorithm to calculate the second cohomological dimension $dim_F[H^2(L, V)]$ of L with coefficients in the 3-dimensional vector space V over the prime field $F = GF(5)$ and all the extensions M_k of L by V. It is 1 and there are four non-split extensions M_j of L by V. Using faithful permutation representations of the four groups M_j and MAGMA's isomorphism test it follows that (up to isomorphism) there is exactly one non-split extension M of T by V. Its presentation is given in the statement.

(a) This can be verified using the MAGMA command `CosetAction (M,M_1)`.

(b) Using the permutation representation PM constructed in (a) and MAGMA one sees that $M = \langle e_j \mid 1 \le j \le 8 \rangle$ has a Sylow 5-subgroup S of order 5^6. Applying the MAGMA command

```
Subgroups(S : Al :=Normal, IsElementaryAbelian:=true)
```

one sees that S has a unique maximal elementary abelian normal subgroup V of order 125. Hence $V = \langle e_6, e_7, e_8 \rangle$ by construction of M.

From the relations of M it follows that $f = e_3^2 = e_8 e_7^{-1}$ generates the center $Z(S)$ of S. Using then the program `GetShortGens(M, S)` we see that $S = \langle (e_2 e_4 e_5 e_1)^4, (e_3 e_4 e_2^2)^2, e_6, e_7, e_8 \rangle$.

(c) $L = N_M(\langle f \rangle)$ has order $2^5 \cdot 3 \cdot 5^6$. Its given generators x and y of N with respective orders 20 and 24 are calculated by MAGMA using the program `GetShortGens(M, L)` and the faithful permutation representation PM of M. Another MAGMA calculation verifies that $M = \langle N, e \rangle$, where $e = e_2$.

(d) and (e) The systems of representatives of the 39 conjugacy classes of $M = \langle x, y, e \rangle$ and the 46 conjugacy classes of $L = \langle x, y \rangle$ have been calculated by means of the faithful permutation representation PM of M, MAGMA and Kratzer's Algorithm 5.3.18 of [92].

(f) The character tables of M and L have been calculated by MAGMA using the faithful permutation representation PM.

(g) Applying the command `Subgroups(pL :Al:=Normal)` MAGMA establishes that L has exactly one normal subgroup Q of order $|Q| = 2^2 \cdot 5^5$ and that Q is a semidirect product of its Sylow 5-subgroup F and a cyclic subgroup $\langle a \rangle$ of order 4 which acts non-trivially on the center $Z(F)$ of F. Hence $Z(Q) = 1$. Using the faithful permutation representation PM and the MAGMA command `AutomorphismGroup(Q)` we see that $A = Aut(Q)$ has order $|A| = 2^9 \cdot 3^2 \cdot 5^9 \cdot 13$. Furthermore, an application of the MAGMA command `PermutationRepresentation(A)` yields a faithful permutation representation pA of A of degree 3125. Let $I(Q)$ be the subgroup of A consisting of all inner automorphisms of Q. Let $p : Q \to pQ$ be the monomorphism from Q into the permutation representation pA of the automorphism group A of Q. Then pa has order 4 in pA. Another application of MAGMA yields that $pC = C_{pA}(pa)$ has order $|pC| = 2^9 \cdot 3^2 \cdot 5^4 \cdot 13$, $|pF| = 5^5$, $pF \cap pC = 1$ and $pA = \langle pF, pC \rangle$.

(h) Applying the command `Subgroups(pC: OrderEqual:=2880)` we find (up to conjugation) exactly one subgroup pK of pC having center $Z(pK) = \langle pa \rangle$ and composition factor A_6. Using the command `FPGroup(pK)` MAGMA also provides a presentation for the permutation group pK. In particular, $pK \cong R_1$, where $R_1 = \langle r_i | 1 \leq i \leq 5 \rangle$ is the finitely presented group R_1 with the set of defining relations stated in Proposition 13.3.3(g).

Since pF is a characteristic subgroup of pQ and pK normalizes pQ, it also normalizes pF. Hence we can form the semidirect product $R = FK$ of order $|R| = 2^6 \cdot 3^2 \cdot 5^6$.

Furthermore, MAGMA produces five new generators r_j of F, $6 \leq j \leq 10$, such that $R = \langle r_i | 1 \leq i \leq 10 \rangle$. Since R is a small permutation group the MAGMA command `FPGroup(R)` provides the set $\mathcal{R}(R)$ of defining relations of R stated in the assertion. The faithful permutation representation PR is well defined because R_1 is a complement of R.

(i) Using the faithful permutation representation PR and MAGMA it is easy to see that $S_1 = \langle r_5, r_6, r_7, r_8, r_9, r_{10} \rangle$ is a Sylow 5-subgroup of R of order 5^6. Applying then the MAGMA command

```
Subgroups(S_1 : Al :=Normal, IsElementaryAbelian:=true)
```

it follows that S_1 has a unique maximal elementary abelian normal subgroup V_1 of order 125. An application of the program `GetShortGens(S_1,V_1)` and MAGMA yields that V_1 is generated by the elements $(r_5 r_6)^5$, $r_5^3 r_6 r_5 r_6^3 r_5 r_6$ and $r_5^2 r_6 r_5^2 r_6^2 r_5 r_6^2$.

(j) Let $L_1 = N_R(V_1)$. Using the faithful permutation representations PM and PR and MAGMA one finds an isomorphism $\varphi : L \to L_1$. Hence $L_1 = \langle \varphi(x), \varphi(y) \rangle$ by (c). The given words for $x_1 = \varphi(x)$ and $y_1 = \varphi(y)$ in the ten generators r_i of R have been calculated by means of the program `LookupWord(R,u)`, where $u \in \{x_1, y_1\}$. Denoting $\sigma = \varphi^{-1}$ we get $\sigma(x_1) = x$ and $\sigma(y_1) = y$. Another application of the permutation representation PR and MAGMA shows that $R = \langle L_1, h \rangle$, where $h = r_2$ has order 4.

(k) The system of representatives of the 55 conjugacy classes of $R = \langle x, y, h \rangle$ has been calculated by means of the faithful permutation representation PR of R, MAGMA and Kratzer's Algorithm 5.3.18 of [92].

(l) The character table of R has been obtained by means of MAGMA using the faithful permutation representation PR.

(m) The Goldschmidt index of the amalgam $R \leftarrow L \rightarrow M$ has been calculated by means of MAGMA using Kratzer's Algorithm 7.1.10 of

[92] and the faithful permutation representations PR and PM of R and M, respectively; it is 1.

(n) By Table 13.6.1 the given centralizer $H = C_G(z) = \langle x, ym_2\rangle$ has two conjugacy classes of elements of order 5 represented by $f_a = (x^2m_2)^8$ and $f_b = xym_2y^2m_2y$. Furthermore, the Sylow 2-subgroups of $C_H(f_a)$ and $C_H(f_b)$ have distinct orders. As all involutions of G are conjugate by Lemma 13.2.1(a) it follows that f_a and f_b do not fuse in G. Now, R, L and M contain a Sylow 5-subgroup of G. Table 13.6.7 states that M has four conjugacy classes of elements of order 5 and that all four representatives are centralized by an involution of M. Hence they are G-conjugate to f_a or to f_b. Another application of Tables 13.6.7 and 13.6.1 implies that the classes 5_1 and 5_2 of M belong to f_a^G, and that 5_3 and 5_4 belong to 5_b^G. Since 5_1^M consists of 5-central conjugacy classes of G, and M has a unique conjugacy class of elements f_c of order 25 which power into f_a^G, the group G has only one conjugacy class of elements of order 25. □

Lemma 13.3.5 (R. Lyons) *Let G be a finite simple group of* Ly*-type with a 2-central involution z such that $C_G(z) = H = \langle x, y, m_2\rangle$. Let $s = (ym_2)^2$ and $p = (xm_2)^2$. Then the following assertions hold.*

(a) *s^G is the unique conjugacy class of elements of order 7 in G.*
(b) *$C_G(s) \cong \langle s\rangle \times \mathsf{SL}_2(3)$, $N_G(\langle s\rangle)/C_G(s)$ is cyclic of order 6 and $N_G(\langle s\rangle) = N_H(\langle s\rangle)$.*
(c) *G has two conjugacy classes of elements of order 11. They are represented by p and p^{-1}.*
(d) *$C_G(p) \cong \langle p\rangle \times S_3$.*
(e) *$N_G(\langle p\rangle) \cong \mathsf{F}_{55} \times S_3$, where F_{55} denotes the Frobenius group of order 55.*

Proof (d) Let $P = C_G(p)$. Let X be the largest normal subgroup of odd order in P. For each prime divisor r of $|X|$ let X_r be a Sylow r-subgroup of X. By Table 13.6.1 P has a cyclic Sylow 2-subgroup $\langle z\rangle$. Hence $P = X\langle z\rangle$ by Corollary 1.4.18 of [92] and z inverts the group $X/\langle p\rangle$. Now Lemma 1.5.9 of [92] implies that $X/\langle p\rangle$ is abelian. Therefore X is nilpotent and all Sylow r-subgroups X_r for $r \neq 11$ are abelian.

By Table 13.6.4 $E = N_G(\langle c_0\rangle)$ has an element p_1 of order 11 which is centralized by c_0 and an involution, z_1 of E, such that p_1 and its inverse represent the conjugacy classes of elements of order 11 in E and c_0 is inverted by z_1. By Lemma 13.2.1(a) G has a unique conjugacy class z^G

of involutions. Hence we may assume that p is G-conjugate to p_1 by Lemma 13.1.4. In particular $X_3 \neq 1$. Hence $X_r = 1$ for $r \notin \{3, 11\}$, $|X_3| = 3$ and $X_{11} = 11$ by Table 13.6.4. Thus (d) holds because c_0 is inverted by z_1.

(e) is an immediate consequence of (d) and Lemma 13.1.4.

(b) By Lemma 13.1.4 $C_H(s)$ has a quaternion Sylow 2-subgroup Q_8 of order 8 with center $\langle z \rangle$. Hence Q_8 is also a Sylow 2-subgroup of $M = C_G(s)$. Thus $M = X\mathsf{SL}_2(3)$ by Theorem 4.6.5 of [92], where X denotes the largest normal subgroup of M with odd order. Again X is nilpotent because z inverts $X_1 = X/\langle s \rangle$.

As $N_H(\langle s \rangle)/C_H(s)$ is cyclic of order 6 by Lemma 13.1.4 there is an element q of order 6 such that $q^3 = t \neq z$. Hence $V = \langle z, t \rangle$ is a Klein 4-subgroup of $N_G(\langle s \rangle)$ which operates on the characteristic subgroup X of $M = C_G(s)$. Now Theorem 4.2.2 of [92] asserts that $|X||C_X(V)|^2 = |C_X(z)||C_X(zt)||C_X(t)|$. Therefore X is a direct product of its Sylow r-subgroups X_r for $r \in \{3, 5, 7, 11\}$ by Table 13.6.1 because G has only one conjugacy class of involutions. Now $X_5 = 1$ and $X_{11} = 1$ by Table 13.6.5 and (d), respectively. Also $X_3 = 1$ by Table 13.6.4 because s is centralized by $q^2 \in H$ which is G-conjugate to c_0. Since $\mathsf{SL}_2(3)$ operates faithfully on $X/\langle s \rangle$ it follows that $|C_X(t)| = |C_X(zt)| = 1$ and $X = X_7 = \langle s \rangle$. In particular, $C_G(s) = C_H(s)$ and $N_G(\langle s \rangle) = N_H(\langle s \rangle)$.

(a) This is an immediate consequence of (b) and Lemma 13.1.4(a). □

13.4 Group order

In this section we show that each simple group G of Ly-type has order $|G| = 2^8 \cdot 3^7 \cdot 5^6 \cdot 7 \cdot 11 \cdot 31 \cdot 37 \cdot 67$. This result is due to Lyons [88]. The new proof given here uses the author's group order formula for groups having a unique conjugacy class of involutions; see Theorem 4.7.3 of [92].

Let G be a finite simple group of Ly-type having a 2-central involution z with $C_G(z) = H = \langle x, y, m_2 \rangle$. Then Lemma 13.2.1(a) states that z^G is the unique conjugacy class of G. By Lemma 4.3.2 of [92] an element r of a subgroup U of G is strongly real if its extended centralizer $C_G^*(r) \leq U$ and the following invariant $d(r)$ is non-zero:

$$d(r) = \left|\{(x, y) \in z^G \times z^G \mid xy = r\}\right| = \frac{1}{|U|} \cdot \sum_{\psi \in Irr_{\mathbb{C}}(U)} \psi(\hat{U})^2 \psi(r^{-1}) \psi(1)^{-1},$$

where $\psi(\hat{U}) = \sum_{i=1}^{k} |U : C_U(z_i)| \psi(z_i)$ and $z_1 = z$, z_2, $\dots$, z_k are representatives of the conjugacy classes of involutions of U.

Proposition 13.4.1 *Let G be a simple group of* Ly*-type with $C_G(z) = H$. Let $\pi = \{2, 3, 5, 7, 11\}$. Then π is the smallest set of primes containing all prime divisors of $|H|$ such that $C_G(g)$ is a π-subgroup for each π-element $1 \neq g \in G$, and the following assertions hold.*

(a) *G has 42 conjugacy classes x^G of π-elements $x \neq 1$ with centralizer orders $|C_G(x)|$ and invariants $d(x)$ given in the following table:*

x	$\vert C_G(x)\vert$	$d(x)$	x	$\vert C_G(x)\vert$	$d(x)$	x	$\vert C_G(x)\vert$	$d(x)$
2_A	$\vert H\vert$	34650	10_B	$2 \cdot 5^2$	50	22_B	$2 \cdot 11$	0
3_A	$\vert E\vert : 2$	340200	11_A	$2 \cdot 3 \cdot 11$	0	24_A	$2^3 \cdot 3$	24
3_B	$2^4 \cdot 3^7 \cdot 5$	2970	11_B	$2 \cdot 3 \cdot 11$	0	24_B	$2^3 \cdot 3$	24
4_A	$2^6 \cdot 3^2 \cdot 5 \cdot 7$	840	12_A	$2^5 \cdot 3^2$	108	24_C	$2^3 \cdot 3$	24
5_A	$2^4 \cdot 3^2 \cdot 5^6$	11250	12_B	$2^3 \cdot 3^2$	48	25_A	5^2	25
5_B	$2 \cdot 3 \cdot 5^4$	500	14_A	$2^3 \cdot 3 \cdot 7$	84	28_A	$2^2 \cdot 7$	28
6_A	$2^7 \cdot 3^3 \cdot 5 \cdot 7$	2520	15_A	$2 \cdot 3^2 \cdot 5^3$	450	30_A	$2 \cdot 3^2 \cdot 5$	90
6_B	$2^4 \cdot 3^3 \cdot 5$	330	15_B	$2 \cdot 3^2 \cdot 5$	90	30_B	$2 \cdot 3^2 \cdot 5$	90
6_C	$2^2 \cdot 3^4$	108	15_C	$3 \cdot 5^2$	75	33_A	$3 \cdot 11$	0
7_A	$2^3 \cdot 3 \cdot 7$	84	18_A	$2 \cdot 3^2$	18	33_B	$3 \cdot 11$	0
8_A	$2^5 \cdot 3 \cdot 5$	120	20_A	$2^3 \cdot 5$	40	40_A	$2^3 \cdot 5$	40
8_B	$2^5 \cdot 3$	48	21_A	$2 \cdot 3 \cdot 7$	42	40_B	$2^3 \cdot 5$	40
9_A	$2 \cdot 3^3$	36	21_B	$2 \cdot 3 \cdot 7$	42	42_A	$2 \cdot 3 \cdot 7$	42
10_A	$2^4 \cdot 3^2 \cdot 5^2$	450	22_A	$2 \cdot 11$	0	42_B	$2 \cdot 3 \cdot 7$	42

(b) *G has 36 strongly real conjugacy classes r_i^G of π-elements represented by the conjugacy classes r_i^G with invariants $d(r_i) \neq 0$, and*

$$c = 1 + \sum_{i=1}^{36} d(r_i) \frac{|H|}{|C_G(r_i)|} = 897658875 = 3^3 \cdot 5^3 \cdot 47 \cdot 5659.$$

(c) *G has ten strongly real conjugacy classes r_j^G of π'-elements.*

(d) $|G| = 2^8 \cdot 3^7 \cdot 5^6 \cdot 7 \cdot 11 \cdot 31 \cdot 37 \cdot 67$.

Proof For any simple group G of Ly-type the character tables of $H = C_G(2_A)$ and the normalizers $E = N_G(3_A)$ and $R = N_G(5_A)$ are given in Tables 13.7.1, 13.7.4 and 13.7.5, respectively. The centralizer $C_G(3_B)$ is contained in $C_E(3_B)$ and it has index 2 in $N_G(3_B)$. The normalizers $N_G(5_B)$, $N_G(7_A)$, $N_G(11_A)$ and $N_G(11_B)$ are contained in R, H, E and E, respectively. From these tables it also follows that $C_G(g)$ is a π-subgroup for each π-element $1 \neq g \in G$. Furthermore, the product g_π of the orders of the Sylow p-subgroups S_p of G with $p \in \pi$ is $g_\pi = 2^8 \cdot 3^7 \cdot 5^6 \cdot 7 \cdot 11$. Let $g_{\pi'} = |G|/g_\pi$.

(a) Let $R = N_G(\langle p \rangle)$ be any of the normalizers of the π-subgroups of prime π-elements p of G. By Lemma 13.2.1 G has only one conjugacy class z^G of involutions. Suppose that $z = z_1, z_2, \ldots, z_k$ are representatives of the conjugacy classes of involutions of R. For each irreducible character ψ of R let $\psi(\hat{R}) = \sum_{i=1}^{k} |R : C_R(2_i)|\psi(z_i)$. Let r be any strongly real element of G with extended centralizer $C_G^*(r) \le R = N_G(p)$ such that $r^{k_1} = p$ for some natural number k_1. Then by Lemma 4.3.2 of [92] the following equation holds:

$$d(r) = \left|\{(x,y) \in z^G \times z^G \mid xy = r\}\right| = \frac{1}{|R|} \cdot \sum_{\psi \in Irr_{\mathbb{C}}(R)} \psi(\hat{R})^2 \psi(r^{-1})\psi(1)^{-1}.$$

This formula has been applied to all normalizers of the representatives of the conjugacy classes p^G of elements of G of prime order p. A MAGMA calculation established the values $d(x)$. They are given in the table in assertion (a). The orders of the centralizers $C_G(r_i)$ can be read off from the above mentioned character tables and their power map information.

(b) Let t be the number of strongly real π'-classes of G. Then Theorem 4.3.7 of [92] asserts that

$$|G| = c|H| + t|H|^2, \quad \text{where} \quad c = 1 + \sum_{i=1}^{36} \frac{|H|}{|C_G(r_i)|} |d(r_i).$$

Inserting the values of the invariants $d(r_i)$ given in the table in assertion (a) and the orders of the centralizers of the 36 strongly real π-classes of G it follows that $c = 3^3 \cdot 5^3 \cdot 47 \cdot 5659$.

(c) By the results of Sections 13.2 and 13.3 we know that $|\Delta| = |C_G(c_0)| = 2^7 \cdot 3^7 \cdot 5^3 \cdot 7 \cdot 11$ is maximal among all the orders of the centralizers of the π-elements of G. Then $t < |\Delta|$ by Theorem 4.3.9 of [92]. Now

$$|G| = g_\pi g_{\pi'} = 2^8 \cdot 3^7 \cdot 5^6 \cdot 7 \cdot 11 \cdot g_{\pi'} = |H|(c + t|H|).$$

Therefore $25g_{\pi'} = 5 \cdot 47 \cdot 5659 + t \cdot 2^8 \cdot 3 \cdot 7 \cdot 11$. Thus 5 is a divisor of t and of $47 \cdot 5659 + t_1 \cdot 2^8 \cdot 3 \cdot 7 \cdot 11$, where $t = 5 \cdot t_1$. Proposition 13.3.4 states that 31 is a divisor of $|G|$. Therefore $5 \cdot g_{\pi'} = 265873 + (2 + s \cdot 155)59136$, where $0 \le s \le 3.476.624$ because $t_1 = (2 + s \cdot 155) < (1/5) \cdot |\Delta|$.

By Theorem 4.5.2 of [92] the factor $g_{\pi'}$ of $|G|$ has to satisfy both Frobenius congruences given by

$$(*) \qquad 1 + g_{\pi'} \left[\sum_{i=1}^{3} \frac{g_\pi}{c(x_i)} \right] \equiv 0 \bmod (3^7),$$

$$(**) \qquad 1 + g_{\pi'} \left[\sum_{j=1}^{3} \frac{g_\pi}{c(y_j)} \right] \equiv 0 \bmod (5^6),$$

where $c(x_i)$ and $c(y_j)$ are the centralizer orders of the three conjugacy classes of 3-power order $3_A, 3_B, 9_A$ and the three conjugacy classes of 5-power order $5_A, 5_B, 25_A$, respectively.

From (a) it follows immediately that

$$c_3 = \sum_{i=1}^{3} \frac{g_\pi}{c(x_i)} = 12477850250,$$

$$c_5 = \sum_{j=1}^{3} \frac{g_\pi}{c(y_j)} = 27123764976.$$

Let $g_s = (1/5) \cdot [265873 + (2 + s \cdot 155)59136]$ for $0 \leq s \leq 3476624$. Applying now simultaneously the MAGMA commands `IsIntegral((g_s * c_3 +1)/3^7)` and `IsIntegral((g_s * c_5 +1)/5^6)` for $0 \leq s \leq 3476624$ MAGMA establishes one common solution for $s = 0$ and confirms that the congruences $(*)$ and $(**)$ do not have a common solution for all other $s > 0$. Hence $t = 2 \cdot 5 = 10$. This completes the proof. □

Corollary 13.4.2 (R. Lyons) *Each finite simple group G of* Ly*-type has self-centralizing Sylow p-subgroups P_p of prime order $|P_p| = p$ for the primes $p \in \{31, 37, 67\}$. Their normalizers $N_p = N_G(P_p)$ have respective orders $|N_{31}| = 31 \cdot 6$, $|N_{37}| = 37 \cdot 18$ and $|N_{67}| = 67 \cdot 22$.*

Proof All assertions are immediate consequences of Proposition 13.4.1, Sylow's Theorem and Burnside's Theorem 1.4.17 of [92]. □

13.5 Existence and uniqueness proofs

In this section we show that each simple group G of Ly-type has a pair of complex conjugate irreducible characters χ_i of degree 2480. Thus G has an irreducible representation $\mathfrak{G}$ in $\mathrm{GL}_{2480}(31)$ because these characters are of defect zero over characteristic 31 by Proposition 13.4.1. The generating matrices of $\mathfrak{G}$ are constructed as well and stored on the accompanying DVD. These results enable us to apply the author's uniqueness criterion stated as Theorem 7.5.1 in [92] to show that each finite simple group G of Ly-type is isomorphic to $\mathfrak{G}$.

Lemma 13.5.1 *Let G be a finite simple group of* Ly*-type with a 2-central involution z such that $C_G(z) = H \cong 2A_{11}$ described in Lemma 13.1.1. Let $L = \langle x, y \rangle$, $M = \langle L, e \rangle$, $R = \langle L, h \rangle$ be the local subgroups of G*

generators defined in Proposition 13.3.4. Then the following statements hold.

(a) *H is a maximal subgroup of G.*
(b) *$G = \langle R, M\rangle$.*

Proof (a) The central involution z of $H \cong 2A_{11}$ generates the only non-trivial proper normal subgroup of H. By Proposition 13.4.1 the order of G is $|G| = 2^8 \cdot 3^7 \cdot 5^6 \cdot 7 \cdot 11 \cdot 31 \cdot 37 \cdot 67$. Therefore the proof of Fendel's Lemma 3.3.11 can be adapted to show that H is a maximal subgroup of G.

(b) By (a) the given centralizer $C_G(z) = H$ of a 2-central involution z in G is a maximal subgroup of G. Proposition 13.3.3 states that H has an element f of order 5 such that $R = N_G(\langle f\rangle)$ contains a Sylow 5-subgroup S_5 of G which has f in its center and that $|S_5| = 5^6$. By Proposition 13.3.4 S_5 has a unique maximal elementary abelian normal subgroup V of order 125 such that $M = N_G(V) \cong V.\mathrm{PSL}_3(5)$ and $L = N_G(\langle f\rangle) = N_R(V)$. The subgroup $G_1 = \langle R, M\rangle$ contains H by Lemma 13.1.3. Hence $G = G_1$ because H is a maximal subgroup of G not containing R. □

Theorem 13.5.2 (Kim–Michler–Wang–Weller) *Let G be a finite simple group of Ly-type. Let $M \leftarrow N \rightarrow R$ be the amalgam constructed in Proposition 13.3.4, where the groups $L = \langle x, y\rangle$, $M = \langle L, e\rangle$ and $R = \langle L, h\rangle$ and their generators are defined in Proposition 13.3.4. Using the notations of the three character tables of the groups R, L and M stated in Tables 13.7.5, 13.7.7 and 13.7.6, respectively, the following statements hold.*

(a) *There is exactly one compatible pair $(\chi, \tau) \in mf\,char_{\mathbb{C}}(R) \times mf\,char_{\mathbb{C}}(M)$ of degree 2480 of the groups $R = \langle L, h\rangle$ and $M = \langle L, e\rangle$ such that*

$$(\chi, \tau) = (\chi_{\mathbf{40}} + \chi_{43} + \chi_{\mathbf{48}} + \chi_{\mathbf{49}}, \quad \tau_{\mathbf{36}})$$

has the common restriction

$$\tau_{\mathbf{36}|\mathbf{L}} = \chi_{|L} = \psi_{\mathbf{36}} + \psi_{38} + \psi_{40} + \psi_{\mathbf{42}} + \psi_{\mathbf{43}} + \psi_{\mathbf{45}} + \psi_{\mathbf{46}},$$

where irreducible characters with bold face indices denote faithful irreducible characters.

(b) *Let $\mathfrak{V}$ and $\mathfrak{W}$ be the up to isomorphism uniquely determined faithful semi-simple multiplicity-free 2480-dimensional modules of R*

and M over $F = GF(31)$ corresponding to the compatible pair χ, τ, respectively.

Let $\kappa_{\mathfrak{V}} : R \to \mathrm{GL}_{2480}(31)$ and $\kappa_{\mathfrak{W}} : M \to \mathrm{GL}_{2480}(31)$ be the representations of R and M afforded by the modules $\mathfrak{V}$ and $\mathfrak{W}$, respectively.

Let $\mathfrak{x} = \kappa_{\mathfrak{V}}(x)$, $\mathfrak{y} = \kappa_{\mathfrak{V}}(y)$ and $\mathfrak{h} = \kappa_{\mathfrak{V}}(h) \in \kappa_{\mathfrak{V}}(R) \le \mathrm{GL}_{2480}(31)$. Then the following assertions hold.

(1) *$\mathfrak{V}_{|L} \cong \mathfrak{W}_{|L}$, and there are two transformation matrices $\mathcal{T}_i \in \mathrm{GL}_{2480}(31)$, $i = 1, 2$, such that*

$$\mathfrak{x} = \mathcal{T}_i^{-1}\kappa_{\mathfrak{W}}(\sigma(x))\mathcal{T}_i \text{ and } \mathfrak{y} = \mathcal{T}_i^{-1}\kappa_{\mathfrak{W}}(\sigma(y))\mathcal{T}_i \quad \text{for} \quad i=1,2.$$

(2) *Let $\mathfrak{e}_i = \mathcal{T}_i^{-1}\kappa_{\mathfrak{W}}(e)\mathcal{T}_i \in \mathrm{GL}_{2480}(31)$ for $i = 1, 2$. Let $\mathfrak{G}_1 = \langle \mathfrak{x}, \mathfrak{y}, \mathfrak{h}, \mathfrak{e}_1 \rangle$ and $\mathfrak{G}_2 = \langle \mathfrak{x}, \mathfrak{y}, \mathfrak{h}, \mathfrak{e}_2 \rangle$.*

The free product $P = R *_N M$ of R and M with amalgamated subgroup N has exactly two irreducible representations $\kappa_i : P \to \mathfrak{G}_i \le \mathrm{GL}_{2480}(F)$ and $G \cong \mathfrak{G}_1 \cong \mathfrak{G}_2$.
The generating matrices $\mathfrak{x}$, $\mathfrak{y}$, $\mathfrak{h}$, $\mathfrak{e} = \mathfrak{e}_1$ of $\mathfrak{G} = \mathfrak{G}_1$ are stored on the accompanying DVD.

Proof (a) Using MAGMA and the faithful permutation representations PR and PM of R and M determined in Proposition 13.3.4, respectively, we determined the fusion of the classes of L in R and in M. Thus we were able to calculate the restrictions of the irreducible characters χ and τ of R and M to their common subgroup L. A MAGMA calculation employing Kratzer's Algorithm 7.3.10 of [92] to the character tables of these three groups then shows that the amalgam $R \leftarrow L \rightarrow M$ has exactly one compatible pair (χ, τ) of degree 2480. Its irreducible constituents are stated in the assertion.

(b) In order to construct the faithful irreducible representation $\mathfrak{W}$ corresponding to the character $\tau = \tau_{\mathbf{36}}$ of M of degree 2480 we applied the MAGMA command

```
LowIndexSubgroups(PE, n)
```

using the faithful permutation presentation PM of $M = \langle x, y, e \rangle$ of degree 3875. For index $n = 15500$ MAGMA and the program `GetShortGens (M, U)` found a subgroup U of M with the following generators:

$$U = \langle (y^2ey^8)^2, (eye^2yey^2eyey^2)^2 \rangle$$

such that τ is a constituent of the permutation character $(1_U)^M$ with multiplicity 1. Using then Algorithm 1.5.1 we calculated the 15 primitive idempotents E_k of the endomorphism ring $\Lambda = End_{FM}[(1_U)^M]$ of the permutation module $(1_U)^M$ corresponding to its 15 distinct irreducible constituents of the permutation character $(1_U)^M$. The projection of the permutation module $(1_U)^M$ corresponding to the idempotent E_{37} of Λ of the irreducible character $\tau = \tau_{36}$ of M constructs the corresponding irreducible representation $M(37)$ of M of dimension 2480 over $F = GF(31)$ as a direct summand of the permutation module $(1_U)^M$. Hence $\mathfrak{W} = M(37)$.

In order to construct the faithful semi-simple representation $\mathfrak{V}$ of $R = \langle x, y, h\rangle$ corresponding to the character $\chi = \chi_{40} + \chi_{43} + \chi_{48} + \chi_{49}$ of degree 2480 we employed the same methods. This time MAGMA found four subgroups:

$$U_{43} = \langle (xh)^8, (h^2x^2h)^3, (xhx^4)^4, (xh^3xhx)^8, (x^4hxhx^2)^2\rangle,$$
$$U_{41} = \langle h, (x^2h^3)^8, (x^4hx^2hxh^2)^8, (x^3h^2x^3h^3)^8\rangle,$$
$$U_{47} = \langle (xh)^{10}, (x^3h)^4, (x^3h^2x^3)^{10}, (h^2xh^3x^2)^5, x^3hx^2hx^2h^3x\rangle,$$
$$U_{48} = \langle (hx^4)^4, (xh^2xhxhxh)^4, hxhx^2hxh^3x, (x^2hxhxh^2x^4)^2\rangle,$$

of R of respective indices $|R : U_{43}| = 1000$, $|R : U_{41}| = 3125$, $|R : U_{47}| = 4256$ and $|R : U_{48}| = 16272$ such that χ_j is an irreducible constituent of the permutation character $(1_{U_j})^R$ of R for $j \in \{43, 40, 49, 48\}$. Four new applications of Algorithm 1.5.1 yield the four irreducible representations $N(43)$, $N(41)$, $N(47)$ and $N(48)$ of respective dimensions 480, 400, 800 and 800 over F belonging to these four irreducible constituents of the four permutation representations $(1_{U_j})^R$, where $j \in \{43, 41, 47, 48\}$. Hence $\mathfrak{V} = N(41) \oplus N(43) \oplus N(47) \oplus N(48)$.

By Proposition 13.3.4(k) the group R has a Sylow 5-subgroup S_1 having a unique elementary abelian normal subgroup V_1 of order 125 such that $L_1 = N_R(V_1) \cong L = N_M(\langle f\rangle) = \langle x, y\rangle$. Furthermore, there is an isomorphism $\sigma : L_1 \to L$ and there are two generators x_1, y_1 of N_1 such that $\sigma(x_1) = x$ and $\sigma(y_1) = y$. Identifying L_1 and L by means of σ it follows from (a) that the FL-modules $\mathfrak{V}_{|L}$ and $\mathfrak{W}_{|L}$ described by the two pairs

$$(\mathfrak{V}(x_1), \mathfrak{V}(y_1)) \quad \text{and} \quad (\mathfrak{W}(\sigma(x_1), \mathfrak{W}(\sigma(y_1)))$$

of matrices in $\mathrm{GL}_{2480}(31)$ are isomorphic.

Let $Y = \mathrm{GL}_{2480}(31)$. Let V and W be the FL-modules described by the first and second pair of matrices of Y, respectively. Applying

Parker's Isomorphism Test stated as Proposition 6.1.6 in [92] by means of the MAGMA command

```
IsIsomorphic(GModule(sub<Y|V>),GModule(sub<Y|W>)),
```

one gets the transformation matrix $\mathcal{T}$ in $\mathrm{GL}_{2480}(31)$ satisfying $\mathfrak{V}(y_1) = (\mathfrak{W}(\sigma(y_1)))^{\mathcal{T}}$ and $\mathfrak{V}(x_1) = (\mathfrak{W}(\sigma(x_1)))^{\mathcal{T}}$.

By Corollary 7.2.4 of [92] and (a) the free product $P = R *_L M$ of the groups R and M with amalgamated subgroup L has two non-isomorphic representations $\mathfrak{V}_i$ of dimension 2480, $i = 1, 2$, with a Sylow 5-subgroup isomorphic to S. In order to construct them we have to determine two double coset representatives D_1 and D_2 of $C_Y(\mathfrak{L})$ in Y, where $\mathfrak{L} = \langle \mathfrak{x}_1, \mathfrak{y}_1 \rangle$. Since $\mathfrak{V}_{|L}$ is multiplicity-free both matrices $\mathcal{D}_1$ and $\mathcal{D}_2$ are blocked diagonal matrices in $\mathrm{GL}_{2480}(31)$ of the form

$$\mathcal{D}_{a,b,c} = diag(1^{400}, 1^{240}, a^{240}, 1^{300}, b^{500}, 1^{200}, c^{600}),$$

for suitable $a, b, c \in F^* = F - 0$, because the irreducible characters χ_j of R have the following restrictions to $L_1 = L$:

$$\chi_{40} = \psi_{43}, \quad \chi_{43} = \psi_{38} + \psi_{40},$$
$$\chi_{48} = \psi_{42} + \psi_{45}, \quad \chi_{49} = \psi_{36} + \psi_{46},$$

and the irreducible characters ψ_k of L have the degrees 400, 240, 240, 300, 500, 200 and 600 for k equal to 43, 37, 38, 42, 45, 35 and 46, respectively. For each triple (a, b, c) let

$$\mathfrak{G}_{a,b,c} = \langle \mathfrak{x}, \mathfrak{y}, (\mathcal{D}_{a,b,c})^{-1}(\mathcal{T})^{-1}\mathfrak{e}\mathcal{T}\mathcal{D}_{a,b,c})\rangle.$$

The Sylow 5-subgroup S of R has order 5^6 and exponent 25 by Table 13.6.5. A Sylow 5-subgroup test running through all 27000 subgroups $\mathfrak{G}_{a,b,c}$ of Y shows that only the groups $\mathfrak{G}_1 = \mathfrak{G}_{17,11,23}$ and $\mathfrak{G}_2 = \mathfrak{G}_{19,15,19}$ satisfy these conditions. Now Corollary 7.2.4 of [92] states that $\kappa_1 : P \to \mathfrak{G}_1$ and $\kappa_2 : P \to \mathfrak{G}_2$ are the two non-isomorphic irreducible representations of the free product $P = R *_L M$ of R and M with amalgamated subgroup L having faithful semi-simple restrictions to R and M which are isomorphic to $\mathfrak{V}$ and $\mathfrak{W}$, respectively.

By Lemma 13.5.1(b) $G = \langle R, M \rangle$. Hence G is an epimorphic image of P. Now the proof of Theorem 7.5.1 of [92] yields that $G \cong \mathfrak{G}_1 \cong \mathfrak{G}_2$ because the amalgam $R \leftarrow L \to M$ has Goldschmidt index 1 by Proposition 13.3.4(m) and G is simple. □

Proposition 13.5.3 *Each finite simple group G of* Ly*-type has 53 conjugacy classes and a unique pair of complex conjugate characters $\chi : G \to \mathbb{C}$ of degree $\chi(1) = 2480$.*

Proof By Proposition 13.4.1 and Corollary 13.4.2 G has 53 conjugacy classes of G. Let $\pi = \{2, 3, 5, 7, 11\}$. Let 1_A be the class of the identity element of G. Keep the notation of the 42 π-classes given in Proposition 13.4.1. Proposition 13.3.1 states that $H = \langle x, y, m_2 \rangle$ has an element c_0 of order 3 such that $E = N_G(\langle c_0 \rangle)$ contains a Sylow 3-subgroup S_3 of G which has c_0 in its center and has order 3^7. The fusion of the even and 3-singular classes of H and $E = N_G(\langle c_0 \rangle)$ is described in Lemmas 13.2.2 and 13.3.2. The remaining ten classes of G are strongly real, and their centralizer orders are given in Corollary 13.4.2. Their representatives are denoted by $31_A, \ldots, 31_E, 37_A, \ldots, 37_C$ and $67_A, \ldots, 67_C$.

By Lemma 13.5.1 the given centralizer $H_G(z) = H$ of a 2-central involution z in G is a maximal subgroup of G. Proposition 13.3.3 states that the given centralizer $H_G(z) = H$ has an element f of order 5 such that $R = N_G(\langle f \rangle)$ contains a Sylow 5-subgroup S_5 of G. By Proposition 13.3.4 and Lemma 13.5.1(b) S_5 has a unique maximal elementary abelian normal subgroup V of order 125 such that $M = N_G(V)$ is isomorphic to a uniquely determined non-split extension M of $\mathrm{PSL}_3(5)$ by V and $G = \langle R, M \rangle$ and $L = N_R(V) = N_M(\langle f \rangle)$. Hence G is an epimorphic image of the free product $P = R *_L M$ of R and M with amalgamated subgroup L. Then Proposition 13.3.4 states also that the amalgam $R \leftarrow L \rightarrow M$ has Goldschmidt index 1. Therefore the fusion of the 5-singular conjugacy classes of R and M in G is uniquely determined. It can easily be deduced from the class representatives and the power map information given in Tables 13.6.5 and 13.6.7.

Keep the notations of the three character tables of R, H and E given in Tables 13.7.5, 13.7.1 and 13.7.4, respectively. By Theorem 13.5.2(a) the semi-simple character $\chi_R = \chi_{\mathbf{40}} + \chi_{43} + \chi_{\mathbf{48}} + \chi_{\mathbf{49}} \in mf\mathrm{char}_{\mathbb{C}}(R)$ is part of a uniquely determined compatible pair (χ, τ) of degree 2480 of the amalgam $R \leftarrow L \rightarrow M$.

By Lemmas 13.2.2, 13.3.2 and 13.3.5 we know that the local subgroups H and E contain the centralizers $C_G(p)$ of all π-elements of prime order. Using the fusion information given in Lemmas 13.2.2, 13.3.2 and the character Tables 13.7.1 and 13.7.4 one deduces that there are two pairs $(\chi, \tau) \in mf\mathrm{char}_{\mathbb{C}}(H) \times mf\mathrm{char}_{\mathbb{C}}(E)$ of degree 2480 with equal restrictions to $H \cap E$. One pair is $(\chi_{\mathbf{3}} + \chi_{39} + \chi_{\mathbf{40}}, \quad \tau_{10} + \tau_{\mathbf{36}})$. The other pair is complex conjugate and well determined by Tables 13.7.1 and 13.7.4.

Let $\chi_H = \chi_{\mathbf{3}} + \chi_{39} + \chi_{\mathbf{40}}$ and $\chi_E = \tau_{10} + \tau_{\mathbf{36}}$. From the fusion information it follows that χ_H and χ_R and χ_E and χ_R agree on $H \cap R$ and $E \cap R$, respectively.

If G has an irreducible character χ of degree 2480 then Theorem 3.17.13 of [92] implies that $\chi(g) = 1$ for all elements of $g \in G$ of orders 37 and 67. Furthermore, $\chi(g) = 0$ for all elements $g \in G$ of orders 31 by Theorem 3.12.4 of [92] because 31 divides 2040.

Let $b_{11} = -\frac{1}{2}(1 + i\sqrt{11})$. In view of all the above information we now define the class function $\chi : g \to \chi(g)$ for all $g \in G$ by the following:

g	1_A	2_A	3_A	3_B	4_A	5_A	5_B	6_A	6_B	6_C	7_A	8_A	8_B	9_A
$\chi(g)$	2480	-16	104	-4	0	-20	5	8	-4	2	2	0	0	-1

g	10_A	10_B	11_A	11_B	12_A	12_B	14_A	15_A	15_B	15_C	18_A	20_A
$\chi(g)$	4	-1	b_{11}	$\overline{b_{11}}$	0	0	-2	4	1	-1	-1	0

g	21_A	21_B	22_A	22_B	24_A	24_B	24_C	25_A	28_A	30_A	30_B
$\chi(g)$	-1	-1	$-b_{11}$	$-\overline{b_{11}}$	0	0	0	0	0	-2	1

g	31_A	31_B	31_C	31_D	31_E	33_A	33_B	37_A	37_B	40_A	40_B
$\chi(g)$	0	0	0	0	0	b_{11}	$\overline{b_{11}}$	1	1	0	0

g	42_A	42_B	67_A	67_B	67_C
$\chi(g)$	1	1	1	1	1

The complex conjugate class function $\chi^* : G \to \mathbb{C}$ differs from χ only at the classes 11_A, 11_B, 22_A, 22_B, 33_A and 33_B, where $\chi^*(x) = \overline{\chi(x)}$. Therefore it suffices to show that χ is an irreducible character of G.

Since all p-elementary subgroups W of G for $p \in \pi$ are G-conjuagate to subgroups of H, E and R the restrictions of χ_W of χ to them is a generalized character because χ_X is a positive sum of irreducible characters of $X \in \{H, E, R\}$.

By Proposition 13.4.1 the primes 31, 37 and 67 are the only π'-prime divisors of $|G|$. Corollary 13.4.2 asserts that $N_p = N_G(\langle p \rangle)$ is a Frobenius group with Frobenius kernel $\langle p \rangle$ and cyclic complement of orders 18 and 22 for the primes 37 and 67, respectively. Hence N_{37} has two exceptional irreducible characters τ_1 and τ_2 of degree 18 and 18 linear characters ψ_i, $1 \leq i \leq 18$, with projective indecomposable characters $\Phi_i = \psi_i + (\tau_1 + \tau_2)$ by Proposition 3.17.2(i) of [92]; N_{67} has 22 linear characters ψ_i and three exceptional characters τ_j. Similarly, the projective characters are $\Phi = \psi_i + (\tau_1 + \tau_2 + \tau_3)$ because 2480 is congruent 1 modulo 67.

If $p = 31$ then the restriction of χ to N_{31} is a positive sum of indecomposable projective characters. If $p \in \{37, 67\}$ then the restriction of χ to N_p is a positive sum of a linear character and some indecomposable projective characters of N_p. This follows from the inner products of the restriction of χ to N_p with the irreducible characters of N_p. Hence Brauer's Theorem 2.8.9 of [92] asserts that χ is a character of G. Since its inner product $(\chi, \chi) = 1$ it is irreducible. This completes the proof. □

Theorem 13.5.4 (Michler–Weller) *Let H be the finite group of even order defined in Lemma 13.1.1. Then each finite simple group G of* Ly*-type is isomorphic to the finite simple group $\mathfrak{G} \leq \mathrm{GL}_{2480}(31)$ of order $|\mathfrak{G}| = 2^8 \cdot 3^7 \cdot 5^6 \cdot 7 \cdot 11 \cdot 31 \cdot 47 \cdot 67$ constructed in Theorem 13.5.2.*

Proof Let G be any finite simple group of Ly-type with $H = C_G(z)$ for an involution z. Proposition 13.4.1 states that $|G| = |\mathfrak{G}|$. By Propositions 13.3.3 and 13.3.4 H has an element f of order 5 such that $R = N_G(\langle f \rangle)$ contains a Sylow 5-subgroup S_5 such that f represents the only 5-central conjugacy class of elements of order 5. Furthermore, S_5 has a unique maximal elementary abelian normal subgroup V of order 125, and $M = N_G(V)$ is the uniquely determined non-split extension of $\mathrm{PSL}_3(5)$ by V and $N_R(V) = N_M(\langle f \rangle)$. Lemma 13.5.1 asserts that $G = \langle R, M \rangle$. Proposition 13.5.3 states that G has a unique pair of conjugate complex characters of degree 2480. Hence G has a faithful 31-modular representation of this degree by Theorem 3.12.4 of [92] because 31 divides 2040. Since the amalgam $R \leftarrow L \rightarrow$ has Goldschmidt index 1 Theorem 7.5.1 of [92] implies that $G \cong \mathfrak{G}$. This completes the proof. □

Remark 13.5.5 In [95] Waki, Weller and the author constructed a simple subgroup $\mathfrak{G}_1 = \langle \mathfrak{x}, \mathfrak{y}, \mathfrak{m}_2, \mathfrak{c}_2 \rangle$ of Ly-type in $\mathrm{GL}_{111}(5)$, where the matrix generators correspond to generators of $H = \langle x, y, m_2 \rangle$ and $N = \langle x, y, c_2 \rangle$ constructed in Lemma 13.2.1. Let $\mathfrak{v} = (\mathfrak{m}_2 \mathfrak{y}^2)^4$. Then Proposition 13.3.1 implies that $\mathfrak{E} = \langle \mathfrak{x}, \mathfrak{y}, \mathfrak{v}, \mathfrak{c}_2 \rangle \cong E$. Hence $|\mathfrak{G}_1 : \mathfrak{E}| = 9606125$ by Proposition 13.4.1 and Table 13.7.4. The corresponding faithful permutation representation $P\mathfrak{G}_1$ has been constructed in [95] by means of the algorithm mentioned in Theorem 6.2.1 of [92]. It has been used to calculate the character table, a system of representatives of the conjugacy classes and a presentation of $\mathfrak{G}$.

13.6 Representatives of conjugacy classes

13.6.1 *Conjugacy classes of* $H = \langle x, y, m_2 \rangle$

Class	*Representative*	*Centralizer*	2P	3P	5P	7P	11P
1	1	$2^8 \cdot 3^4 \cdot 5^2 \cdot 7 \cdot 11$	1	1	1	1	1
2_1	$(m_2)^2$	$2^8 \cdot 3^4 \cdot 5^2 \cdot 7 \cdot 11$	1	2_1	2_1	2_1	2_1
2_2	$(x)^4$	$2^7 \cdot 3^2$	1	2_2	2_2	2_2	2_2
3_1	$(xy)^4$	$2^7 \cdot 3^3 \cdot 5 \cdot 7$	3_1	1	3_1	3_1	3_1
3_2	$(y)^2$	$2^4 \cdot 3^3 \cdot 5$	3_2	1	3_2	3_2	3_2
3_3	$(xyxm_2)^3$	$2^2 \cdot 3^4$	3_3	1	3_3	3_3	3_3
4_1	m_2	$2^6 \cdot 3^2 \cdot 5 \cdot 7$	2_1	4_1	4_1	4_1	4_1
4_2	$(x)^2$	$2^5 \cdot 3$	2_2	4_2	4_2	4_2	4_2
5_1	$(x^2m_2)^8$	$2^4 \cdot 3^2 \cdot 5^2$	5_1	5_1	1	5_1	5_1
5_2	$xym_2y^2m_2y$	$2 \cdot 5^2$	5_2	5_2	1	5_2	5_2
6_1	$(y^2m_2)^4$	$2^7 \cdot 3^3 \cdot 5 \cdot 7$	3_1	2_1	6_1	6_1	6_1
6_2	$(xym_2)^4$	$2^4 \cdot 3^3 \cdot 5$	3_2	2_1	6_2	6_2	6_2
6_3	$(xy)^2$	$2^6 \cdot 3^2$	3_1	2_2	6_3	6_3	6_3
6_4	$(x^2m_2yxm_2)^3$	$2^2 \cdot 3^4$	3_3	2_1	6_4	6_4	6_4
6_5	y	$2^2 \cdot 3^2$	3_2	2_2	6_5	6_5	6_5
6_6	x^3yxm_2	$2^2 \cdot 3^2$	3_3	2_2	6_6	6_6	6_6
6_7	$xyxy^3$	$2^2 \cdot 3^2$	3_3	2_2	6_7	6_7	6_7
7	$(ym_2)^2$	$2^3 \cdot 3 \cdot 7$	7	7	7	1	7
8_1	$(x^2m_2)^5$	$2^5 \cdot 3 \cdot 5$	4_1	8_1	8_1	8_1	8_1
8_2	$(xym_2)^3$	$2^5 \cdot 3$	4_1	8_2	8_2	8_2	8_2
8_3	x	2^4	4_2	8_3	8_3	8_3	8_3
8_4	xm_2^2	2^4	4_2	8_4	8_4	8_4	8_4
9	$xyxm_2$	$2 \cdot 3^2$	9	3_3	9	9	9
10_1	$(x^2m_2)^4$	$2^4 \cdot 3^2 \cdot 5^2$	5_1	10_1	2_1	10_1	10_1
10_2	$x^3m_2xm_2xy$	$2 \cdot 5^2$	5_2	10_2	2_1	10_2	10_2
11_1	$(xm_2)^2$	$2 \cdot 11$	11_2	11_1	11_1	11_2	1
11_2	$(xm_2)^4$	$2 \cdot 11$	11_1	11_2	11_2	11_1	1
12_1	$(y^2m_2)^2$	$2^5 \cdot 3^2$	6_1	4_1	12_1	12_1	12_1
12_2	$(xym_2)^2$	$2^3 \cdot 3^2$	6_2	4_1	12_2	12_2	12_2
12_3	xy	$2^4 \cdot 3$	6_3	4_2	12_3	12_3	12_3
14	ym_2	$2^3 \cdot 3 \cdot 7$	7	14	14	2_1	14
15_1	$(x^3m_2xy)^2$	$2 \cdot 3^2 \cdot 5$	15_1	5_1	3_2	15_1	15_1
15_2	$x^2ym_2ym_2$	$2 \cdot 3^2 \cdot 5$	15_2	5_1	3_1	15_2	15_2
18	$x^2m_2yxm_2$	$2 \cdot 3^2$	9	6_4	18	18	18
20	$(x^2m_2)^2$	$2^3 \cdot 5$	10_1	20	4_1	20	20
21_1	xm_2xm_2y	$2 \cdot 3 \cdot 7$	21_2	7	21_1	3_1	21_2
21_2	$(xm_2xm_2y)^2$	$2 \cdot 3 \cdot 7$	21_1	7	21_2	3_1	21_1
22_1	xm_2	$2 \cdot 11$	11_1	22_1	22_1	22_2	2_1
22_2	$(xm_2)^7$	$2 \cdot 11$	11_2	22_2	22_2	22_1	2_1
24_1	xym_2	$2^3 \cdot 3$	12_2	8_2	24_1	24_2	24_2
24_2	$(xym_2)^7$	$2^3 \cdot 3$	12_2	8_2	24_2	24_1	24_1
24_3	y^2m_2	$2^3 \cdot 3$	12_1	8_1	24_3	24_3	24_3
28	$x^3m_2xm_2$	$2^2 \cdot 7$	14	28	28	4_1	28
30_1	x^3m_2xy	$2 \cdot 3^2 \cdot 5$	15_1	10_1	6_2	30_1	30_1
30_2	$x^2m_2y^2m_2$	$2 \cdot 3^2 \cdot 5$	15_2	10_1	6_1	30_2	30_2
40_1	x^2m_2	$2^3 \cdot 5$	20	40_1	8_1	40_2	40_2
40_2	$(x^2m_2)^7$	$2^3 \cdot 5$	20	40_2	8_1	40_1	40_1
42_1	$xy^2m_2xm_2$	$2 \cdot 3 \cdot 7$	21_2	14	42_1	6_1	42_2
42_2	$(xy^2m_2xm_2)^{11}$	$2 \cdot 3 \cdot 7$	21_1	14	42_2	6_1	42_1

13.6.2 *Conjugacy classes of* $D = N_H(A) = \langle x, y \rangle$

Class	*Representative*	*Centralizer*	2P	3P
1	1	$2^8 \cdot 3^2$	1	1
2_1	$(x^3y)^{12}$	$2^8 \cdot 3^2$	1	2_1
2_2	$(x)^4$	$2^7 \cdot 3^2$	1	2_2
2_3	$(y)^3$	$2^5 \cdot 3$	1	2_3
2_4	$(xy)^6$	$2^5 \cdot 3$	1	2_4
2_5	$(x^3y^2x^2y)^3$	$2^5 \cdot 3$	1	2_5
2_6	xy^2	2^4	1	2_6
3_1	$(xy)^4$	$2^7 \cdot 3^2$	3_1	1
3_2	$(y)^2$	$2^3 \cdot 3^2$	3_2	1
3_3	$(xyxy^3)^2$	$2^2 \cdot 3^2$	3_3	1
4_1	$(x^3y)^6$	$2^6 \cdot 3$	2_1	4_1
4_2	$(x)^2$	$2^5 \cdot 3$	2_2	4_2
4_3	$(x^4y)^3$	$2^5 \cdot 3$	2_1	4_3
4_4	$(x^3yx^2y^2)^3$	$2^5 \cdot 3$	2_1	4_4
4_5	$(xy)^3$	$2^3 \cdot 3$	2_4	4_5
6_1	$(x^3y)^4$	$2^7 \cdot 3^2$	3_1	2_1
6_2	$(x^3yxyxyxy)^2$	$2^6 \cdot 3^2$	3_1	2_2
6_3	$(xy)^2$	$2^5 \cdot 3$	3_1	2_4
6_4	$(xy)^{10}$	$2^5 \cdot 3$	3_1	2_4
6_5	$(x^4y)^2$	$2^3 \cdot 3^2$	3_2	2_1
6_6	$x^3y^2x^2y$	$2^4 \cdot 3$	3_1	2_5
6_7	x^4y^2	$2^2 \cdot 3^2$	3_2	2_2
6_8	$xyxy^3$	$2^2 \cdot 3^2$	3_3	2_2
6_9	x^3yxy^3	$2^2 \cdot 3^2$	3_3	2_1
6_{10}	x^2yxyxy^2	$2^2 \cdot 3^2$	3_3	2_2
6_{11}	y	$2^2 \cdot 3$	3_2	2_3
8_1	$(x^3y)^3$	$2^4 \cdot 3$	4_1	8_1
8_2	x^2yxy^3	2^5	4_1	8_2
8_3	x^2y^3xy	2^5	4_1	8_3
8_4	x	2^4	4_2	8_4
8_5	x^2y^3	2^4	4_1	8_5
8_6	x^4yx^3y	2^4	4_2	8_6
12_1	$(x^3y)^2$	$2^5 \cdot 3$	6_1	4_1
12_2	$x^3yx^2y^2$	$2^4 \cdot 3$	6_1	4_4
12_3	$x^3yxyxyxy$	$2^4 \cdot 3$	6_2	4_2
12_4	xy	$2^3 \cdot 3$	6_3	4_5
12_5	$(xy)^5$	$2^3 \cdot 3$	6_4	4_5
12_6	x^4y	$2^2 \cdot 3$	6_5	4_3
24	x^3y	$2^3 \cdot 3$	12_1	8_1

13.6.3 *Conjugacy classes of $N = N_G(A) = \langle x, y, c_2 \rangle$*

i	n_i	$\|C_N(n_i)\|$	i	n_i	$\|C_N(n_i)\|$
1_a	1	$2^8 \cdot 3^3$	6_e	$xyxc_2xc_2{}^2$	$2^4 \cdot 3$
2_a	$(x)^4$	$2^8 \cdot 3^2$	6_f	x^2c_2	$2^2 \cdot 3^2$
2_b	$(xy)^6$	$2^5 \cdot 3^2$	6_g	$(x^2c_2)^5$	$2^2 \cdot 3^2$
2_c	$(y)^3$	$2^5 \cdot 3$	6_h	$xyxyc_2$	$2^2 \cdot 3^2$
2_d	xy^3c_2	$2^5 \cdot 3$	6_i	x^4y^2	$2^2 \cdot 3^2$
2_e	$(xc_2)^3$	$2^4 \cdot 3$	6_j	y	$2^2 \cdot 3$
3_a	$(xy)^4$	$2^7 \cdot 3^3$	6_k	xc_2	$2 \cdot 3$
3_b	$(y)^2$	$2^3 \cdot 3^2$	8_a	$(xyc_2)^3$	$2^4 \cdot 3$
3_c	c_2	$2^2 \cdot 3^2$	8_b	x	2^5
3_d	$(xc_2)^2$	$2 \cdot 3^2$	8_c	xyc_2yc_2	2^5
4_a	$(x)^2$	$2^6 \cdot 3$	8_d	$yc_2{}^2$	2^4
4_b	$(yc_2)^3$	$2^5 \cdot 3$	12_a	$(xyc_2)^2$	$2^5 \cdot 3$
4_c	$(xc_2{}^2y)^3$	$2^5 \cdot 3$	12_b	$xc_2{}^2y$	$2^4 \cdot 3$
4_d	$(xy)^3$	$2^3 \cdot 3$	12_c	xy	$2^3 \cdot 3$
6_a	$(xyc_2)^4$	$2^7 \cdot 3^2$	12_d	$(xy)^5$	$2^3 \cdot 3$
6_b	$(xy)^2$	$2^5 \cdot 3^2$	12_e	yc_2	$2^2 \cdot 3$
6_c	$(xy)^{10}$	$2^5 \cdot 3^2$	24_a	xyc_2	$2^3 \cdot 3$
6_d	$(yc_2)^2$	$2^3 \cdot 3^2$			

13.6.4 *Conjugacy classes of* $E = N_G(3_A) = \langle x, y, v, c_2 \rangle \cong 3\mathsf{McL}:2$

i	e_i	$\lvert C_E(e_i)\rvert$	2P	3P	5P	7P	11P
1_a	1	$2^8\cdot3^7\cdot5^3\cdot7\cdot11$	1_a	1_a	1_a	1_a	1_a
2_a	$(x)^4$	$2^8\cdot3^3\cdot5\cdot7$	1_a	2_a	2_a	2_a	2_a
2_b	$(y)^3$	$2^5\cdot3^2\cdot5\cdot11$	1_a	2_b	2_b	2_b	2_b
3_a	$(xy)^4$	$2^7\cdot3^7\cdot5^3\cdot7\cdot11$	3_a	1_a	3_a	3_a	3_a
3_b	$(v)^2$	$2^4\cdot3^7\cdot5$	3_b	1_a	3_b	3_b	3_b
3_c	$(vc_2)^4$	$2^3\cdot3^7\cdot5$	3_c	1_a	3_c	3_c	3_c
3_d	$(y)^2$	$2^3\cdot3^5$	3_d	1_a	3_d	3_d	3_d
4_a	$(xv)^5$	$2^5\cdot3^2\cdot5$	2_a	4_a	4_a	4_a	4_a
4_b	$(x)^2$	$2^6\cdot3^2$	2_a	4_b	4_b	4_b	4_b
5_a	$(xv)^4$	$2^2\cdot3^2\cdot5^3$	5_a	5_a	1_a	5_a	5_a
5_b	$(yvc_2)^2$	$2\cdot3\cdot5^2$	5_b	5_b	1_a	5_b	5_b
6_a	$(xy)^2$	$2^7\cdot3^3\cdot5\cdot7$	3_a	2_a	6_a	6_a	6_a
6_b	v	$2^4\cdot3^3\cdot5$	3_b	2_a	6_b	6_b	6_b
6_c	$(vc_2)^2$	$2^3\cdot3^3\cdot5$	3_c	2_a	6_c	6_c	6_c
6_d	$(yc_2)^2$	$2^3\cdot3^3$	3_d	2_a	6_d	6_d	6_d
6_e	x^2c_2	$2^2\cdot3^3$	3_d	2_a	6_e	6_e	6_e
6_f	y	$2^2\cdot3^2$	3_d	2_b	6_f	6_f	6_f
7_a	x^2v	$2\cdot3\cdot7$	7_a	7_a	7_a	1_a	7_a
8_a	xv^2	$2^5\cdot3$	4_b	8_a	8_a	8_a	8_a
8_b	$(xyc_2)^3$	$2^4\cdot3$	4_b	8_b	8_b	8_b	8_b
8_c	x	2^5	4_b	8_c	8_c	8_c	8_c
9_a	$v^2c_2vc_2$	3^3	9_a	3_c	9_a	9_a	9_a
10_a	$(xv)^2$	$2^2\cdot3^2\cdot5$	5_a	10_a	2_a	10_a	10_a
10_b	yvc_2	$2\cdot5$	5_b	10_b	2_b	10_b	10_b
11_a	$(xvc_2)^2$	$2\cdot3\cdot11$	11_b	11_a	11_a	11_b	1
11_b	$(xvc_2)^4$	$2\cdot3\cdot11$	11_a	11_b	11_b	11_a	1
12_a	xy	$2^5\cdot3^2$	6_a	4_b	12_a	12_a	12_a
12_b	$v^2c_2^2$	$2^3\cdot3^2$	6_b	4_b	12_b	12_b	12_b
12_c	yc_2	$2^2\cdot3^2$	6_d	4_a	12_c	12_c	12_c
12_d	vc_2	$2^2\cdot3^2$	6_c	4_b	12_d	12_d	12_d
12_e	x^4yv	$2^2\cdot3^2$	6_b	4_a	12_e	12_e	12_e
14_a	$(xv^2y)^3$	$2\cdot3\cdot7$	7_a	14_a	14_a	2_a	14_a
15_a	x^2yxv^2	$2\cdot3^2\cdot5^3$	15_a	5_a	3_a	15_a	15_a
15_b	xyv	$2\cdot3^2\cdot5$	15_c	5_a	3_c	15_c	15_b
15_c	$(xyv)^2$	$2\cdot3^2\cdot5$	15_b	5_a	3_c	15_b	15_c
15_d	x^2v^2	$2\cdot3^2\cdot5$	15_d	5_a	3_b	15_d	15_d
15_e	$x^2c_2vc_2v$	$3\cdot5^2$	15_e	5_b	3_a	15_e	15_e
20_a	xv	$2^2\cdot5$	10_a	20_a	4_a	20_a	20_b
20_b	$(xv)^{11}$	$2^2\cdot5$	10_a	20_b	4_a	20_b	20_a
21_a	vc_2^2	$2\cdot3\cdot7$	21_b	7_a	21_a	3_a	21_b
21_b	$(vc_2^2)^2$	$2\cdot3\cdot7$	21_a	7_a	21_b	3_a	21_a
22_a	xvc_2	$2\cdot11$	11_a	22_a	22_a	22_b	2_b
22_b	$(xvc_2)^7$	$2\cdot11$	11_b	22_b	22_b	22_a	2_b
24_a	xyc_2	$2^3\cdot3$	12_a	8_b	24_a	24_a	24_a
24_b	x^2yv^2	$2^3\cdot3$	12_b	8_a	24_b	24_b	24_b
24_c	$(x^2yv^2)^{13}$	$2^3\cdot3$	12_b	8_a	24_c	24_c	24_c
30_a	xvy	$2\cdot3^2\cdot5$	15_c	10_a	6_c	30_b	30_a
30_b	$(xvy)^7$	$2\cdot3^2\cdot5$	15_b	10_a	6_c	30_a	30_b
30_c	x^3vxv	$2\cdot3^2\cdot5$	15_d	10_a	6_b	30_c	30_c
30_d	$x^2vc_2vc_2$	$2\cdot3^2\cdot5$	15_a	10_a	6_a	30_d	30_d
33_a	$x^2v^2c_2$	$3\cdot11$	33_b	11_a	33_a	33_b	3_a
33_b	$(x^2v^2c_2)^2$	$3\cdot11$	33_a	11_b	33_b	33_a	3_a
42_a	xv^2y	$2\cdot3\cdot7$	21_a	14_a	42_a	6_a	42_b
42_b	$(xv^2y)^{11}$	$2\cdot3\cdot7$	21_b	14_a	42_b	6_a	42_a

13.6.5 *Conjugacy classes of $R = N_G(f) = \langle x, y, h\rangle$*

Class	*Representative*	*Centralizer*	2P	3P	5P
1	1	$2^6 \cdot 3^2 \cdot 5^6$	1	1	1
2_1	$(y)^{12}$	$2^6 \cdot 3^2 \cdot 5^2$	1	2_1	2_1
2_2	$(x)^{10}$	$2^5 \cdot 5^2$	1	2_2	2_2
3_1	$(x^2h)^8$	$2^3 \cdot 3^2 \cdot 5^3$	3_1	1	3_1
3_2	$(y)^8$	$2^3 \cdot 3^2 \cdot 5$	3_2	1	3_2
4_1	$(y)^6$	$2^6 \cdot 3^2 \cdot 5$	2_1	4_2	4_1
4_2	$(y)^{18}$	$2^6 \cdot 3^2 \cdot 5$	2_1	4_1	4_2
4_3	$(xh)^{10}$	$2^5 \cdot 5$	2_1	4_3	4_3
4_4	$(x)^5$	$2^4 \cdot 5$	2_2	4_5	4_4
4_5	h	$2^4 \cdot 5$	2_2	4_4	4_5
5_1	$(xh)^8$	$2^4 \cdot 3^2 \cdot 5^6$	5_1	5_1	1
5_2	$(x)^4$	$2^2 \cdot 5^5$	5_2	5_2	1
5_3	$(xhyh)^3$	$2 \cdot 3 \cdot 5^4$	5_3	5_3	1
5_4	$(xy)^4$	$2^2 \cdot 5^3$	5_4	5_4	1
5_5	$(x^3y)^2$	$2 \cdot 5^3$	5_5	5_5	1
5_6	$(xh^2y)^2$	$2 \cdot 5^3$	5_6	5_6	1
6_1	$(y)^4$	$2^3 \cdot 3^2 \cdot 5$	3_2	2_1	6_1
6_2	$(x^2h)^4$	$2^3 \cdot 3^2 \cdot 5$	3_1	2_1	6_2
8_1	$(y)^3$	$2^5 \cdot 3$	4_1	8_2	8_1
8_2	$(y)^9$	$2^5 \cdot 3$	4_2	8_1	8_2
8_3	$(x^2h)^3$	$2^5 \cdot 3$	4_2	8_4	8_3
8_4	$(x^2h)^9$	$2^5 \cdot 3$	4_1	8_3	8_4
8_5	$(xh)^5$	$2^4 \cdot 5$	4_3	8_5	8_5
8_6	x^3h	2^4	4_3	8_6	8_6
10_1	$(xh)^4$	$2^4 \cdot 3^2 \cdot 5^2$	5_1	10_1	2_1
10_2	$(x)^2$	$2^2 \cdot 5^2$	5_2	10_2	2_2
10_3	$(xy)^2$	$2^2 \cdot 5^2$	5_4	10_3	2_1
10_4	x^3y	$2 \cdot 5^2$	5_5	10_4	2_1
10_5	xh^2y	$2 \cdot 5^2$	5_6	10_5	2_1
10_6	xy^3h^2	$2 \cdot 5^2$	5_3	10_6	2_2
12_1	$(y)^2$	$2^3 \cdot 3^2$	6_1	4_1	12_1
12_2	$(x^2h)^2$	$2^3 \cdot 3^2$	6_2	4_2	12_2
12_3	$(xy^2)^2$	$2^3 \cdot 3^2$	6_1	4_2	12_3
12_4	$(yh^2)^2$	$2^3 \cdot 3^2$	6_2	4_1	12_4
15_1	$(xyxy^2h)^2$	$2 \cdot 3^2 \cdot 5^3$	15_1	5_1	3_1
15_2	$(x^2y^2hy)^2$	$2 \cdot 3^2 \cdot 5$	15_2	5_1	3_2
15_3	$xhyh$	$3 \cdot 5^2$	15_3	5_3	3_1
20_1	$(xh)^2$	$2^3 \cdot 5$	10_1	20_1	4_3
20_2	x	$2^2 \cdot 5$	10_2	20_5	4_4
20_3	xy	$2^2 \cdot 5$	10_3	20_4	4_2
20_4	x^2hy	$2^2 \cdot 5$	10_3	20_3	4_1
20_5	x^4h	$2^2 \cdot 5$	10_2	20_2	4_5
24_1	y	$2^3 \cdot 3$	12_1	8_1	24_1
24_2	x^2h	$2^3 \cdot 3$	12_2	8_3	24_2
24_3	xy^2	$2^3 \cdot 3$	12_3	8_2	24_3
24_4	xyh	$2^3 \cdot 3$	12_1	8_1	24_4
24_5	yh^2	$2^3 \cdot 3$	12_4	8_4	24_5
24_6	x^3hy	$2^3 \cdot 3$	12_4	8_4	24_6
24_7	x^2hyh	$2^3 \cdot 3$	12_2	8_3	24_7
24_8	xy^2hy	$2^3 \cdot 3$	12_3	8_2	24_8
25	yh	5^2	25	25	5_1
30_1	x^2y^2hy	$2 \cdot 3^2 \cdot 5$	15_2	10_1	6_1
30_2	$xyxy^2h$	$2 \cdot 3^2 \cdot 5$	15_1	10_1	6_2
40_1	xh	$2^3 \cdot 5$	20_1	40_1	8_5
40_2	x^2h^2	$2^3 \cdot 5$	20_1	40_2	8_5

13.6.6 *Conjugacy classes of* $L = N_R(V) = \langle x, y \rangle$

Class	*Representative*	*Centralizer*	2P	3P	5P
1	1	$2^5 \cdot 3 \cdot 5^6$	1	1	1
2_1	$(y)^{12}$	$2^5 \cdot 3 \cdot 5^2$	1	2_1	2_1
2_2	$(x)^{10}$	$2^4 \cdot 5^2$	1	2_2	2_2
3	$(y)^8$	$2^3 \cdot 3 \cdot 5$	3	1	3
4_1	$(y)^6$	$2^5 \cdot 3 \cdot 5$	2_1	4_2	4_1
4_2	$(y)^{18}$	$2^5 \cdot 3 \cdot 5$	2_1	4_1	4_2
4_3	$(x)^5$	$2^4 \cdot 5$	2_2	4_7	4_3
4_4	x^2y^2	$2^4 \cdot 5$	2_1	4_4	4_4
4_5	$(xyxy^2)^5$	$2^4 \cdot 5$	2_2	4_6	4_5
4_6	x^3yx^2y	$2^4 \cdot 5$	2_2	4_5	4_6
4_7	$(x^3yxy^2)^5$	$2^4 \cdot 5$	2_2	4_3	4_7
5_1	$(xy^3)^4$	$2^3 \cdot 3 \cdot 5^6$	5_1	5_1	1
5_2	$(x)^4$	$2^2 \cdot 5^5$	5_2	5_2	1
5_3	$(xyxy^2)^4$	$2^2 \cdot 5^4$	5_3	5_3	1
5_4	$(x^7y^3)^2$	$2 \cdot 5^4$	5_4	5_4	1
5_5	$(x^4y^2x^3y)^2$	$2 \cdot 5^4$	5_5	5_5	1
5_6	$(xy)^4$	$2^2 \cdot 5^3$	5_6	5_6	1
5_7	$(x^3y)^2$	$2 \cdot 5^3$	5_7	5_7	1
5_8	$(x^7y)^2$	$2 \cdot 5^3$	5_8	5_8	1
6	$(y)^4$	$2^3 \cdot 3 \cdot 5$	3	2_1	6
8_1	$(y)^3$	$2^3 \cdot 3$	4_1	8_2	8_1
8_2	$(y)^9$	$2^3 \cdot 3$	4_2	8_1	8_2
10_1	$(xy^3)^2$	$2^3 \cdot 3 \cdot 5^2$	5_1	10_1	2_1
10_2	$(x)^2$	$2^2 \cdot 5^2$	5_2	10_2	2_2
10_3	$(xy)^2$	$2^2 \cdot 5^2$	5_6	10_3	2_1
10_4	$(xyxy^2)^2$	$2^2 \cdot 5^2$	5_3	10_4	2_2
10_5	x^3y	$2 \cdot 5^2$	5_7	10_5	2_1
10_6	x^7y	$2 \cdot 5^2$	5_8	10_6	2_1
10_7	x^7y^3	$2 \cdot 5^2$	5_4	10_7	2_2
10_8	$x^4y^2x^3y$	$2 \cdot 5^2$	5_5	10_8	2_2
12_1	$(y)^2$	$2^3 \cdot 3$	6	4_1	12_1
12_2	$(xy^2)^2$	$2^3 \cdot 3$	6	4_2	12_2
15	$(x^3yxyxy)^2$	$2 \cdot 3 \cdot 5$	15	5_1	3
20_1	x	$2^2 \cdot 5$	10_2	20_5	4_3
20_2	xy	$2^2 \cdot 5$	10_3	20_7	4_2
20_3	xy^3	$2^2 \cdot 5$	10_1	20_3	4_4
20_4	$xyxy^2$	$2^2 \cdot 5$	10_4	20_6	4_5
20_5	x^3yxy^2	$2^2 \cdot 5$	10_2	20_1	4_7
20_6	$x^2y^2xy^2$	$2^2 \cdot 5$	10_4	20_4	4_6
20_7	x^5yxyxy	$2^2 \cdot 5$	10_3	20_2	4_1
24_1	y	$2^3 \cdot 3$	12_1	8_1	24_1
24_2	xy^2	$2^3 \cdot 3$	12_2	8_2	24_2
24_3	x^3y^4	$2^3 \cdot 3$	12_2	8_2	24_3
24_4	$x^3yx^2y^3$	$2^3 \cdot 3$	12_1	8_1	24_4
25	$x^3yx^2y^2$	5^2	25	25	5_1
30	x^3yxyxy	$2 \cdot 3 \cdot 5$	15	10_1	6

13.6.7 *Conjugacy classes of* $M = N_G(V) = \langle x, y, e\rangle$

Class	*Representative*	*Centralizer*	2P	3P	5P	31P
1	1	$2^5 \cdot 3 \cdot 5^6 \cdot 31$	1	1	1	1
2	$(x)^{10}$	$2^5 \cdot 3 \cdot 5^2$	1	2	2	2
3	$(y)^8$	$2^3 \cdot 3 \cdot 5$	3	1	3	3
4_1	$(y)^6$	$2^5 \cdot 3 \cdot 5$	2	4_2	4_1	4_2
4_2	$(y)^{18}$	$2^5 \cdot 3 \cdot 5$	2	4_1	4_2	4_1
4_3	$(x)^5$	$2^4 \cdot 5$	2	4_3	4_3	4_3
5_1	$(x)^4$	$2^3 \cdot 3 \cdot 5^6$	5_1	5_1	1	5_1
5_2	$(xy)^4$	$2^2 \cdot 5^4$	5_2	5_2	1	5_2
5_3	$(x^3y)^2$	$2 \cdot 5^4$	5_3	5_3	1	5_3
5_4	x^2yey	$2 \cdot 5^4$	5_4	5_4	1	5_4
6	$(y)^4$	$2^3 \cdot 3 \cdot 5$	3	2	6	6
8_1	$(y)^3$	$2^3 \cdot 3$	4_1	8_2	8_1	8_2
8_2	$(y)^9$	$2^3 \cdot 3$	4_2	8_1	8_2	8_1
10_1	$(x)^2$	$2^3 \cdot 3 \cdot 5^2$	5_1	10_1	2	10_1
10_2	$(xy)^2$	$2^2 \cdot 5^2$	5_2	10_2	2	10_2
10_3	x^3y	$2 \cdot 5^2$	5_3	10_3	2	10_3
10_4	$yeye^2$	$2 \cdot 5^2$	5_4	10_4	2	10_4
12_1	$(y)^2$	$2^3 \cdot 3$	6	4_1	12_1	12_2
12_2	$(xy^2)^2$	$2^3 \cdot 3$	6	4_2	12_2	12_1
15	x^2yxe	$2 \cdot 3 \cdot 5$	15	5_1	3	15
20_1	x	$2^2 \cdot 5$	10_1	20_1	4_3	20_1
20_2	xy	$2^2 \cdot 5$	10_2	20_3	4_2	20_3
20_3	$xyexe$	$2^2 \cdot 5$	10_2	20_2	4_1	20_2
24_1	y	$2^3 \cdot 3$	12_1	8_1	24_1	24_2
24_2	xy^2	$2^3 \cdot 3$	12_2	8_2	24_2	24_1
24_3	$xeye$	$2^3 \cdot 3$	12_1	8_1	24_3	24_4
24_4	xy^2exe	$2^3 \cdot 3$	12_2	8_2	24_4	24_3
25	xy^2e	5^2	25	25	5_1	25
30	$xyxyey$	$2 \cdot 3 \cdot 5$	15	10_1	6	30
31_1	ye^2	31	31_4	31_{10}	31_1	1
31_2	$xyxe$	31	31_8	31_3	31_2	1
31_3	xey^2	31	31_9	31_1	31_3	1
31_4	$xyxey$	31	31_2	31_6	31_4	1
31_5	xy^2xe	31	31_{10}	31_2	31_5	1
31_6	$xyeye$	31	31_3	31_7	31_6	1
31_7	xye^2y	31	31_1	31_5	31_7	1
31_8	y^3e^2	31	31_7	31_9	31_8	1
31_9	$xyxeye$	31	31_5	31_4	31_9	1
31_{10}	$xyxe^2y$	31	31_6	31_8	31_{10}	1

13.7 Character tables of local subgroups

13.7.1 *Character table of* $H = \langle x, y, m_2 \rangle \cong 2A_{11}$

2	8	8	7	7	4	2	6	5	4	1	7	4	6	2	2	2	2	3	5	5	4	4	1	4	1	1
3	4	4	2	3	3	4	2	1	2	.	3	3	2	4	2	2	2	1	1	1	.	.	2	2	.	.
5	2	2	.	1	1	.	1	.	2	2	1	1	.	.	.	.	.	.	1	.	.	.	.	2	2	.
7	1	1	.	1	.	.	1	.	.	.	1	.	.	.	.	.	.	1	.	.	.	.	.	.	.	.
11	1	1	.	.	.	.	.	.	.	.	.	.	.	.	.	.	.	.	.	.	.	.	.	.	.	1
	1a	2a	2b	3a	3b	3c	4a	4b	5a	5b	6a	6b	6c	6d	6e	6f	6g	7a	8a	8b	8c	8d	9a	10a	10b	11a
2P	1a	1a	1a	3a	3b	3c	2a	2b	5a	5b	3a	3b	3a	3c	3b	3c	3c	7a	4a	4a	4b	4b	9a	5a	5b	11b
3P	1a	2a	2b	1a	1a	1a	4a	4b	5a	5b	2a	2a	2b	2a	2b	2b	2b	7a	8a	8b	8c	8d	3c	10a	10b	11a
5P	1a	2a	2b	3a	3b	3c	4a	4b	1a	1a	6a	6b	6c	6d	6e	6f	6g	7a	8a	8b	8c	8d	9a	2a	2a	11a
7P	1a	2a	2b	3a	3b	3c	4a	4b	5a	5b	6a	6b	6c	6d	6e	6f	6g	1a	8a	8b	8c	8d	9a	10a	10b	11b
11P	1a	2a	2b	3a	3b	3c	4a	4b	5a	5b	6a	6b	6c	6d	6e	6f	6g	7a	8a	8b	8c	8d	9a	10a	10b	1a
X.1	1	1	1	1	1	1	1	1	1	1	1	1	1	1	1	1	1	1	1	1	1	1	1	1	1	1
X.2	10	10	2	7	4	1	6	2	5	.	7	4	−1	1	2	−1	−1	3	4	.	.	.	1	5	.	−1
X.3	16	−16	.	−8	4	−2	.	.	−4	1	8	−4	.	2	.	.	.	2	.	.	.	.	1	4	−1	A
X.4	16	−16	.	−8	4	−2	.	.	−4	1	8	−4	.	2	.	.	.	2	.	.	.	.	1	4	−1	$\bar{A}$
X.5	44	44	4	20	5	−1	16	.	9	−1	20	5	4	−1	1	1	1	2	6	2	.	.	−1	9	−1	.
X.6	45	45	−3	21	6	.	13	1	10	.	21	6	−3	.	.	.	.	3	5	−3	−1	−1	.	10	.	1
X.7	110	110	6	29	2	2	26	−2	5	.	29	2	−3	2	.	.	.	−2	4	.	.	.	−1	5	.	.
X.8	120	120	−8	36	6	3	8	.	10	.	36	6	4	3	−2	1	1	1	.	.	.	.	.	10	.	−1
X.9	126	126	6	21	6	.	−14	−2	1	1	21	6	−3	.	.	.	.	.	−4	.	.	.	.	1	1	A
X.10	126	126	6	21	6	.	−14	−2	1	1	21	6	−3	.	.	.	.	.	−4	.	.	.	.	1	1	$\bar{A}$
X.11	132	132	4	6	6	−3	20	4	−8	2	6	6	−2	−3	−2	1	1	−1	.	.	.	.	.	−8	2	.
X.12	144	−144	.	−48	12	.	.	.	−16	−1	48	−12	.	.	.	.	.	4	.	.	.	.	.	16	1	1
X.13	165	165	5	21	3	3	29	1	−5	.	21	3	5	3	−1	−1	−1	−3	1	1	−1	−1	.	−5	.	.
X.14	210	210	2	42	9	3	−14	−2	5	.	42	9	2	3	−1	−1	−1	.	−6	2	.	.	.	5	.	1
X.15	231	231	−1	63	3	−3	35	−1	16	1	63	3	−1	−3	−1	−1	−1	.	5	1	1	1	.	16	1	.
X.16	330	330	2	−6	9	−3	22	2	−10	.	−6	9	2	−3	−1	−1	−1	1	.	−4	.	.	.	−10	.	.
X.17	385	385	9	49	−8	−2	21	1	5	.	49	−8	9	−2	.	.	.	.	−1	3	−1	−1	1	5	.	.
X.18	462	462	−2	−42	3	3	14	2	7	2	−42	3	−2	3	1	1	1	.	−6	2	.	.	.	7	2	.
X.19	528	−528	.	48	.	6	.	.	−12	−2	−48	.	.	−6	.	.	.	−4	.	.	.	.	.	12	2	.
X.20	550	550	−2	70	−5	1	50	−2	.	.	70	−5	−2	1	1	1	1	−3	.	−4	.	.	1	.	.	.
X.21	560	−560	.	−112	8	2	.	.	−20	.	112	−8	.	−2	.	.	.	.	.	.	.	.	−1	20	.	−1
X.22	594	594	−6	90	.	.	6	2	9	−1	90	.	−6	.	.	.	.	−1	−4	.	.	.	.	9	−1	.
X.23	594	594	−6	−45	.	.	6	2	9	−1	−45	.	3	.	.	.	.	−1	−4	.	.	.	.	9	−1	.
X.24	594	594	−6	−45	.	.	6	2	9	−1	−45	.	3	.	.	.	.	−1	−4	.	.	.	.	9	−1	.
X.25	616	−616	.	−56	−8	4	.	.	−4	1	56	8	.	−4	.	.	.	.	.	.	−2	2	1	4	−1	.
X.26	616	−616	.	−56	−8	4	.	.	−4	1	56	8	.	−4	.	.	.	.	.	.	2	−2	1	4	−1	.
X.27	660	660	−4	−36	−3	3	16	.	5	.	−36	−3	−4	3	−1	−1	−1	2	−6	−2	.	.	.	5	.	.
X.28	672	−672	.	.	12	6	.	.	12	2	.	−12	.	−6	.	.	.	.	.	.	.	.	.	−12	−2	1
X.29	693	693	−3	21	6	.	49	1	−17	−2	21	6	−3	.	.	.	.	.	−1	3	1	1	.	−17	−2	.
X.30	825	825	9	−15	.	6	−15	1	.	.	−15	.	9	6	.	.	.	−1	5	−3	1	1	.	.	.	.
X.31	880	−880	.	64	4	−2	.	.	−10	.	−64	−4	.	2	.	.	.	−2	.	.	.	.	1	10	.	.
X.32	880	−880	.	64	4	−2	.	.	−10	.	−64	−4	.	2	.	.	.	−2	.	.	.	.	1	10	.	.
X.33	924	924	4	84	9	−3	−56	.	−1	−1	84	9	4	−3	1	1	1	.	−6	−2	.	.	.	−1	−1	.
X.34	990	990	6	−42	6	.	34	−2	−5	.	−42	6	6	.	.	.	.	3	−4	.	.	.	.	−5	.	.
X.35	990	990	6	21	−12	.	34	−2	−5	.	21	−12	−3	.	.	.	.	3	−4	.	.	.	.	−5	.	.
X.36	1100	1100	12	50	−10	2	−20	4	.	.	50	−10	−6	2	.	.	.	1	.	.	.	.	−1	.	.	.
X.37	1155	1155	−5	−21	9	3	7	−1	−10	.	−21	9	−5	3	1	1	1	.	5	1	−1	−1	.	−10	.	.
X.38	1200	−1200	.	−120	.	−6	.	.	.	.	120	.	.	6	.	.	.	−4	.	.	.	.	.	.	.	1
X.39	1232	1232	−16	56	−4	−1	.	.	−8	2	56	−4	8	−1	2	−1	−1	.	.	.	.	.	−1	−8	2	.
X.40	1232	−1232	.	56	−4	−1	.	.	−8	2	−56	4	.	1	.	−3	3	.	.	.	.	.	−1	8	−2	.
X.41	1232	−1232	.	56	−4	−1	.	.	−8	2	−56	4	.	1	.	3	−3	.	.	.	.	.	−1	8	−2	.
X.42	1320	1320	8	−84	6	−3	8	.	10	.	−84	6	−4	−3	2	−1	−1	−3	.	.	.	.	.	10	.	.
X.43	1440	−1440	.	−48	12	.	.	.	20	.	48	−12	.	.	.	.	.	−2	.	.	.	.	.	−20	.	−1
X.44	1540	1540	−4	28	1	1	−56	.	−5	.	28	1	−4	1	−1	−1	−1	.	6	2	.	.	1	−5	.	.
X.45	1584	−1584	.	−24	−12	.	.	.	4	−1	24	12	.	.	.	.	.	2	.	.	.	.	.	−4	1	.
X.46	1584	−1584	.	−24	−12	.	.	.	4	−1	24	12	.	.	.	.	.	2	.	.	.	.	.	−4	1	.
X.47	1584	−1584	.	48	6	.	.	.	4	−1	−48	−6	.	.	.	.	.	2	.	.	.	.	.	−4	1	.
X.48	1584	−1584	.	48	6	.	.	.	4	−1	−48	−6	.	.	.	.	.	2	.	.	.	.	.	−4	1	.
X.49	2310	2310	−2	−63	−6	−3	−14	−2	5	.	−63	−6	1	−3	−2	1	1	.	4	.	.	.	.	5	.	.

Character table of $H = \langle x, y, m_2 \rangle$ (continued)

2	1	5	3	4	3	1	1	1	3	1	1	1	1	3	3	3	2	1	1	3	3	1	1
3	.	2	2	1	1	2	2	2	.	1	1	.	.	1	1	1	.	2	2	.	.	1	1
5	.	.	.	.	.	1	1	.	1	.	.	.	.	.	.	.	.	1	1	1	1	.	.
7	.	.	.	.	1	.	.	.	.	1	1	.	.	.	.	.	1	.	.	.	.	1	1
11	1	.	.	.	.	.	.	.	.	.	.	1	1	.	.	.	.	.	.	.	.	.	.
	11b	12a	12b	12c	14a	15a	15b	18a	20a	21a	21b	22a	22b	24a	24b	24c	28a	30a	30b	40a	40b	42a	42b
2P	11a	6a	6b	6c	7a	15a	15b	9a	10a	21b	21a	11a	11b	12b	12b	12a	14a	15a	15b	20a	20a	21b	21a
3P	11b	4a	4a	4b	14a	5a	5a	6d	20a	7a	7a	22a	22b	8b	8b	8a	28a	10a	10a	40a	40b	14a	14a
5P	11b	12a	12b	12c	14a	3b	3a	18a	4a	21a	21b	22a	22b	24a	24b	24c	28a	6b	6a	8a	8a	42a	42b
7P	11a	12a	12b	12c	2a	15a	15b	18a	20a	3a	3a	22b	22a	24b	24a	24c	4a	30a	30b	40b	40a	6a	6a
11P	1a	12a	12b	12c	14a	15a	15b	18a	20a	21b	21a	2a	2a	24b	24a	24c	28a	30a	30b	40b	40a	42b	42a
X.1	1	1	1	1	1	1	1	1	1	1	1	1	1	1	1	1	1	1	1	1	1	1	1
X.2	−1	3	.	−1	3	−1	2	1	1	.	.	−1	−1	.	.	1	−1	−1	2	−1	−1	.	.
X.3	$\bar{A}$	.	.	.	−2	−1	2	−1	.	−1	−1	$-\bar{A}$	$-A$	.	.	.	.	1	−2	.	.	1	1
X.4	A	.	.	.	−2	−1	2	−1	.	−1	−1	$-A$	$-\bar{A}$	.	.	.	.	1	−2	.	.	1	1
X.5	.	4	1	.	2	.	.	−1	1	−1	−1	.	.	−1	−1	.	2	.	.	1	1	−1	−1
X.6	1	1	−2	1	3	1	1	.	−2	.	.	1	1	.	.	−1	−1	1	1	.	.	.	.
X.7	.	5	2	1	−2	2	−1	−1	1	1	1	.	.	.	.	1	−2	2	−1	−1	−1	1	1
X.8	−1	−4	2	.	1	1	1	.	−2	1	1	−1	−1	.	.	.	1	1	1	.	.	1	1
X.9	$\bar{A}$	1	−2	1	.	1	1	.	1	.	.	$\bar{A}$	A	.	.	−1	.	1	1	1	1	.	.
X.10	A	1	−2	1	.	1	1	.	1	.	.	A	$\bar{A}$	.	.	−1	.	1	1	1	1	.	.
X.11	.	2	2	−2	−1	1	1	.	.	−1	−1	.	.	.	.	.	−1	1	1	.	.	−1	−1
X.12	1	.	.	.	−4	2	2	.	.	1	1	−1	−1	.	.	.	.	−2	−2	.	.	−1	−1
X.13	.	5	−1	1	−3	−2	1	.	−1	.	.	.	.	1	1	1	1	−2	1	1	1	.	.
X.14	1	−2	1	−2	.	−1	2	.	1	.	.	1	1	−1	−1	.	.	−1	2	−1	−1	.	.
X.15	.	−1	−1	−1	.	−2	−2	.	.	.	.	.	.	1	1	−1	.	−2	−2	.	.	.	.
X.16	.	−2	1	2	1	−1	−1	.	2	1	1	.	.	−1	−1	.	1	−1	−1	.	.	1	1
X.17	.	−3	.	1	.	2	−1	1	1	.	.	.	.	.	.	−1	.	2	−1	−1	−1	.	.
X.18	.	2	−1	2	.	−2	−2	.	−1	.	.	.	.	−1	−1	.	.	−2	−2	−1	−1	.	.
X.19	.	.	.	.	4	.	3	.	.	−1	−1	.	.	.	.	.	.	.	−3	.	.	1	1
X.20	.	2	−1	−2	−3	.	.	1	.	.	.	.	.	−1	−1	.	1	.	.	.	.	.	.
X.21	−1	.	.	.	.	−2	−2	1	.	.	.	1	1	.	.	.	.	2	2	.	.	.	.
X.22	.	−6	.	2	−1	.	.	.	1	−1	−1	.	.	.	.	2	−1	.	.	1	1	−1	−1
X.23	.	3	.	−1	−1	.	.	.	1	B	C	.	.	.	.	−1	−1	.	.	1	1	B	C
X.24	.	3	.	−1	−1	.	.	.	1	C	B	.	.	.	.	−1	−1	.	.	1	1	C	B
X.25	.	.	.	.	.	2	−1	−1	.	.	.	.	.	.	.	.	.	−2	1	.	.	.	.
X.26	.	.	.	.	.	2	−1	−1	.	.	.	.	.	.	.	.	.	−2	1	.	.	.	.
X.27	.	4	1	.	2	2	−1	.	1	−1	−1	.	.	1	1	.	2	2	−1	−1	−1	−1	−1
X.28	1	.	.	.	.	−3	.	.	.	.	.	−1	−1	.	.	.	.	3	.	.	.	.	.
X.29	.	1	−2	1	.	1	1	.	−1	.	.	.	.	.	.	−1	.	1	1	−1	−1	.	.
X.30	.	−3	.	1	−1	.	.	.	.	−1	−1	.	.	.	.	−1	−1	.	.	.	.	−1	−1
X.31	.	.	.	.	2	−1	−1	−1	.	1	1	.	.	.	.	.	.	1	1	D	$-D$	−1	−1
X.32	.	.	.	.	2	−1	−1	−1	.	1	1	.	.	.	.	.	.	1	1	$-D$	D	−1	−1
X.33	.	4	1	.	.	−1	−1	.	−1	.	.	.	.	1	1	.	.	−1	−1	−1	−1	.	.
X.34	.	−2	−2	−2	3	1	−2	.	−1	.	.	.	.	.	.	2	−1	1	−2	1	1	.	.
X.35	.	1	4	1	3	−2	1	.	−1	.	.	.	.	.	.	−1	−1	−2	1	1	1	.	.
X.36	.	−2	−2	−2	1	.	.	−1	.	1	1	.	.	.	.	.	1	.	.	.	.	1	1
X.37	.	−5	1	−1	.	−1	−1	.	2	.	.	.	.	1	1	−1	.	−1	−1	.	.	.	.
X.38	1	.	.	.	4	.	.	.	.	−1	−1	−1	−1	.	.	.	.	.	.	.	.	1	1
X.39	.	.	.	.	.	1	1	−1	.	.	.	.	.	.	.	.	.	1	1	.	.	.	.
X.40	.	.	.	.	.	1	1	1	.	.	.	.	.	.	.	.	.	−1	−1	.	.	.	.
X.41	.	.	.	.	.	1	1	1	.	.	.	.	.	.	.	.	.	−1	−1	.	.	.	.
X.42	.	−4	2	.	−3	1	1	.	−2	.	.	.	.	.	.	.	1	1	1	.	.	.	.
X.43	−1	.	.	.	2	2	2	.	.	1	1	1	1	.	.	.	.	−2	−2	.	.	−1	−1
X.44	.	4	1	.	.	1	−2	1	−1	.	.	.	.	−1	−1	.	.	1	−2	1	1	.	.
X.45	.	.	.	.	−2	−2	1	.	.	B	C	.	.	.	.	.	.	2	−1	.	.	$-B$	$-C$
X.46	.	.	.	.	−2	−2	1	.	.	C	B	.	.	.	.	.	.	2	−1	.	.	$-C$	$-B$
X.47	.	.	.	.	−2	1	−2	.	.	−1	−1	.	.	E	$-E$	.	.	−1	2	.	.	1	1
X.48	.	.	.	.	−2	1	−2	.	.	−1	−1	.	.	$-E$	E	.	.	−1	2	.	.	1	1
X.49	.	1	−2	1	.	−1	2	.	1	.	.	.	.	.	.	1	.	−1	2	−1	−1	.	.

where $A = -\zeta(11)^9 - \zeta(11)^5 - \zeta(11)^4 - \zeta(11)^3 - \zeta(11) - 1$, $B = 2\zeta(21)_3\zeta(21)_7^4 + 2\zeta(21)_3\zeta(21)_7^2 + 2\zeta(21)_3\zeta(21)_7 + \zeta(21)_3 + \zeta(21)_7^4 + \zeta(21)_7^2 + \zeta(21)_7 + 1$, $C = -B + 1$, $D = 2\zeta(40)_8^3\zeta(40)_5^3 + 2\zeta(40)_8^3\zeta(40)_5^2 + \zeta(40)_8^3 - 2\zeta(40)_8\zeta(40)_5^3 - 2\zeta(40)_8\zeta(40)_5^2 - \zeta(40)_8$, $E = -2\zeta(24)_8^3\zeta(24)_3 - \zeta(24)_8^3 - 2\zeta(24)_8\zeta(24)_3 - \zeta(24)_8$.

13.7.2 *Character table of* $D = \langle x, y \rangle$

	1_a	2_a	2_b	2_c	2_d	2_e	2_f	3_a	3_b	3_c	4_a	4_b	4_c	4_d	4_e	6_a	6_b	6_c	6_d	6_e	6_f	6_g	6_h	6_i	6_j	6_k	8_a	8_b	8_c	8_d	8_e	8_f	12_a	12_b	12_c	12_d	12_e	12_f	24_a
ξ_1	1	1	1	1	1	1	1	1	1	1	1	1	1	1	1	1	1	1	1	1	1	1	1	1	1	1	1	1	1	1	1	1	1	1	1	1	1	1	1
ξ_2	1	1	1	1	1	1	1	1	1	1	1	1	1	1	1	1	1	1	1	1	1	1	1	1	1	1	1	1	1	1	1	1	1	1	1	1			
ξ_3	1	1	1	1	1	1	1	1	1	1	1	1	1	1	1	1	1	1	1	1	1	1	1	1	1	1	1	1	1	1	1	1	1	1	1	1	1	1	1
ξ_4	1	1	1	1	1	1	1	1	1	1	1	1	1	1	1	1	1	1	1	1	1	1	1	1	1	1	1	1	1	1	1	1	1	1	1	1	1	1	1
ξ_5	2	2	2	.	2	2	.	1	2	1	2	2	.	2	2	1	1	1	1	2	1	2	1	1	1	.	2	.	.	.	.	.	1	1	1	1	1	.	1
ξ_6	2	2	2	.	2	2	.	1	2	1	2	2	.	2	2	1	1	1	1	2	1	2	1	1	1	.	2	.	.	.	.	.	1	1	1	1	1	.	1
ξ_7	2	2	2	2	2	.	.	2	1	1	2	2	2	.	.	2	2	2	2	1	.	1	1	1	1	1	.	.	.	.	2	.	2	.	2	.	.	1	.
ξ_8	2	2	2	2	2	.	.	2	1	1	2	2	2	.	.	2	2	2	2	1	.	1	1	1	1	1	.	.	.	.	2	.	2	.	2	.	.	1	.
ξ_9	3	3	3	3	1	1	1	3	.	.	3	1	3	1	1	3	3	1	1	.	1	.	.	.	.	.	1	1	1	1	1	1	3	1	1	1	1	.	1
ξ_{10}	3	3	3	3	1	1	1	3	.	.	3	1	3	1	1	3	3	1	1	.	1	.	.	.	.	.	1	1	1	1	1	1	3	1	1	1	1	.	1
ξ_{11}	3	3	3	3	1	1	1	3	.	.	3	1	3	1	1	3	3	1	1	.	1	.	.	.	.	.	1	1	1	1	1	1	3	1	1	1	1	.	1
ξ_{12}	3	3	3	3	1	1	1	3	.	.	3	1	3	1	1	3	3	1	1	.	1	.	.	.	.	.	1	1	1	1	1	1	3	1	1	1	1	.	1
ξ_{13}	4	4	4	2	.	2	.	4	1	1	.	.	2	2	.	4	4	.	.	1	2	1	1	1	1	1	.	2	2	.	.	.	.	2	.	.	.	1	.
ξ_{14}	4	4	4	2	.	2	.	4	1	1	.	.	2	2	.	4	4	.	.	1	2	1	1	1	1	1	.	2	2	.	.	.	.	2	.	.	.	1	.
ξ_{15}	4	4	4	2	.	2	.	4	1	1	.	.	2	2	.	4	4	.	.	1	2	1	1	1	1	1	.	2	2	.	.	.	.	2	.	.	.	1	.
ξ_{16}	4	4	4	2	.	2	.	4	1	1	.	.	2	2	.	4	4	.	.	1	2	1	1	1	1	1	.	2	2	.	.	.	.	2	.	.	.	1	.
ξ_{17}	4	4	4	.	4	.	.	2	2	1	4	4	.	.	.	2	2	2	2	2	.	2	1	1	1	.	.	.	.	.	.	.	2	.	2	.	.	.	.
ξ_{18}	6	6	6	.	2	2	.	3	.	.	6	2	.	2	2	3	3	1	1	.	1	.	.	.	.	.	2	.	.	.	.	.	3	1	1	1	1	.	1
ξ_{19}	6	6	6	.	2	2	.	3	.	.	6	2	.	2	2	3	3	1	1	.	1	.	.	.	.	.	2	.	.	.	.	.	3	1	1	1	1	.	1
ξ_{20}	6	6	6	.	2	2	.	6	.	.	2	2	.	2	.	6	6	2	2	.	2	.	.	.	.	.	2	.	.	.	.	.	2	2	2	.	.	.	2
ξ_{21}	6	6	6	.	2	2	.	6	.	.	2	2	.	2	.	6	6	2	2	.	2	.	.	.	.	.	2	.	.	.	.	.	2	2	2	.	.	.	2
ξ_{22}	6	6	6	.	2	.	2	6	.	.	2	2	.	.	.	6	6	2	2	.	.	.	.	.	.	.	.	2	2	.	.	.	2	.	2	.	.	.	.
ξ_{23}	6	6	6	.	2	.	2	6	.	.	2	2	.	.	.	6	6	2	2	.	.	.	.	.	.	.	.	2	2	.	.	.	2	.	2	.	.	.	.
ξ_{24}	6	6	6	.	2	2	.	3	.	.	2	2	.	2	.	3	3	α	$\bar{\alpha}$	.	1	.	.	.	.	.	2	.	.	.	.	.	1	1	1	β	β	.	1
ξ_{25}	6	6	6	.	2	2	.	3	.	.	2	2	.	2	.	3	3	$\bar{\alpha}$	α	.	1	.	.	.	.	.	2	.	.	.	.	.	1	1	1	β	β	.	1
ξ_{26}	6	6	6	.	2	2	.	3	.	.	2	2	.	2	.	3	3	α	$\bar{\alpha}$	.	1	.	.	.	.	.	2	.	.	.	.	.	1	1	1	β	β	.	1
ξ_{27}	6	6	6	.	2	2	.	3	.	.	2	2	.	2	.	3	3	$\bar{\alpha}$	α	.	1	.	.	.	.	.	2	.	.	.	.	.	1	1	1	β	β	.	1
ξ_{28}	8	8	.	.	.	.	.	8	2	2	.	.	.	.	.	8	.	.	.	2	.	.	.	2	.	.	.	.	.	2	.	2	.	.	.	.	.	.	.
ξ_{29}	8	8	.	.	.	.	.	8	2	2	.	.	.	.	.	8	.	.	.	2	.	.	.	2	.	.	.	.	.	2	.	2	.	.	.	.	.	.	.
ξ_{30}	8	8	8	.	.	4	.	4	2	1	.	.	.	4	.	4	4	.	.	2	2	2	1	1	1	.	.	.	.	.	.	.	.	2	.	.	.	.	.
ξ_{31}	8	8	8	.	.	4	.	4	2	1	.	.	.	4	.	4	4	.	.	2	2	2	1	1	1	.	.	.	.	.	.	.	.	2	.	.	.	.	.
ξ_{32}	8	8	8	4	.	.	.	8	1	1	.	.	4	.	.	8	8	.	.	1	.	1	1	1	1	1	.	.	.	.	.	.	.	.	.	.	.	1	.
ξ_{33}	8	8	8	4	.	.	.	8	1	1	.	.	4	.	.	8	8	.	.	1	.	1	1	1	1	1	.	.	.	.	.	.	.	.	.	.	.	1	.
ξ_{34}	12	12	12	.	4	.	.	6	.	.	4	4	.	.	.	6	6	2	2	.	.	.	.	.	.	.	.	.	.	.	.	.	2	.	2	.	.	.	.
ξ_{35}	16	16	.	.	.	.	.	16	2	2	.	.	.	.	.	16	.	.	.	2	.	.	.	2	.	.	.	.	.	.	.	.	.	.	.	.	.	.	.
ξ_{36}	16	16	16	.	.	.	.	8	2	1	.	.	.	.	.	8	8	.	.	2	.	2	1	1	1	.	.	.	.	.	.	.	.	.	.	.	.	.	.
ξ_{37}	16	16	.	.	.	.	.	8	4	2	.	.	.	.	.	8	.	.	.	4	.	.	.	2	.	.	.	.	.	.	.	.	.	.	.	.	.	.	.
ξ_{38}	16	16	.	.	.	.	.	8	2	1	.	.	.	.	.	8	.	.	.	2	.	.	3	1	3	.	.	.	.	.	.	.	.	.	.	.	.	.	.
ξ_{39}	16	16	.	.	.	.	.	8	2	1	.	.	.	.	.	8	.	.	.	2	.	.	3	1	3	.	.	.	.	.	.	.	.	.	.	.	.	.	.

where $\alpha = -1 - 2\sqrt{3}i$ *and* $\beta = -\sqrt{3}i$.

13.7.3 *Character table of* $N = \langle x, y, c_2 \rangle$

	1_a	2_a	2_b	2_c	2_d	2_e	3_a	3_b	3_c	3_d	4_a	4_b	4_c	4_d	6_a	6_b	6_c	6_d	6_e	6_f	6_g	6_h	6_i	6_j	6_k	8_a	8_b	8_c	8_d	12_a	12_b	12_c	12_d	12_e	24_a
τ_1	1	1	1	1	1	1	1	1	1	1	1	1	1	1	1	1	1	1	1	1	1	1	1	1	1	1	1	1	1	1	1	1	1	1	1
τ_2	1	1	1	1	1	1	1	1	1	1	1	1	1	1	1	1	1	1	1	1	1	1	1	1	1	1	1	1	1	1	1	1	1	1	1
τ_3	1	1	1	1	1	1	1	1	1	1	1	1	1	1	1	1	1	1	1	1	1	1	1	1	1	1	1	1	1	1	1	1	1	1	1
τ_4	1	1	1	1	1	1	1	1	1	1	1	1	1	1	1	1	1	1	1	1	1	1	1	1	1	1	1	1	1	1	1	1	1	1	1
τ_5	2	2	2	2	.	.	2	1	1	2	2	2	.	.	2	2	2	1	.	1	1	1	1	1	.	.	.	.	2	2	.	.	.	1	.
τ_6	2	2	2	2	.	.	2	1	1	2	2	2	.	.	2	2	2	1	.	1	1	1	1	1	.	.	.	.	2	2	.	.	.	1	.
τ_7	2	2	2	.	.	2	2	2	1	1	2	.	.	.	2	2	2	2	.	1	1	1	2	.	1	.	2	2	.	2	.	.	.	.	.
τ_8	2	2	2	.	.	2	2	2	1	1	2	.	.	.	2	2	2	2	.	1	1	1	2	.	1	.	2	2	.	2	.	.	.	.	.
τ_9	4	4	4	.	.	.	4	2	1	2	4	.	.	.	4	4	4	2	.	1	1	1	2	.	.	.	.	.	.	4	.	.	.	.	.
τ_{10}	6	6	6	.	2	.	3	.	.	.	6	.	2	2	3	3	3	.	1	.	.	.	.	.	.	2	.	.	.	3	1	1	1	.	1
τ_{11}	6	6	6	.	2	.	3	.	.	.	6	.	2	2	3	3	3	.	1	.	.	.	.	.	.	2	.	.	.	3	1	1	1	.	1
τ_{12}	6	6	2	.	2	.	6	.	3	.	2	.	2	.	6	2	2	.	2	1	1	1	.	.	.	2	.	.	.	2	2	.	.	.	2
τ_{13}	6	6	2	.	2	.	6	.	3	.	2	.	2	.	6	2	2	.	2	1	1	1	.	.	.	2	.	.	.	2	2	.	.	.	2
τ_{14}	6	6	2	.	2	.	3	.	.	.	2	.	2	.	3	α	$\bar{\alpha}$	.	1	γ	$\bar{\gamma}$	2	.	.	.	2	.	.	.	1	1	δ	δ	.	1
τ_{15}	6	6	2	.	2	.	3	.	.	.	2	.	2	.	3	α	$\bar{\alpha}$	.	1	γ	$\bar{\gamma}$	2	.	.	.	2	.	.	.	1	1	δ	δ	.	1
τ_{16}	6	6	2	.	2	.	3	.	.	.	2	.	2	.	3	$\bar{\alpha}$	α	.	1	$\bar{\gamma}$	γ	2	.	.	.	2	.	.	.	1	1	δ	δ	.	1
τ_{17}	6	6	2	.	2	.	3	.	.	.	2	.	2	.	3	$\bar{\alpha}$	α	.	1	$\bar{\gamma}$	γ	2	.	.	.	2	.	.	.	1	1	δ	δ	.	1
τ_{18}	9	9	3	3	1	3	9	.	.	.	1	3	1	1	9	3	3	.	1	.	.	.	.	.	.	1	1	1	1	1	1	1	1	.	1
τ_{19}	9	9	3	3	1	3	9	.	.	.	1	3	1	1	9	3	3	.	1	.	.	.	.	.	.	1	1	1	1	1	1	1	1	.	1
τ_{20}	9	9	3	3	1	3	9	.	.	.	1	3	1	1	9	3	3	.	1	.	.	.	.	.	.	1	1	1	1	1	1	1	1	.	1
τ_{21}	9	9	3	3	1	3	9	.	.	.	1	3	1	1	9	3	3	.	1	.	.	.	.	.	.	1	1	1	1	1	1	1	1	.	1
τ_{22}	12	12	4	.	.	.	12	.	3	.	4	.	.	.	12	4	4	.	.	1	1	1	.	.	.	.	.	.	.	4	.	.	.	.	.
τ_{23}	12	4	.	2	2	.	12	3	.	.	.	2	2	.	4	.	.	1	2	.	.	.	1	1	.	.	2	2	.	.	2	.	.	1	.
τ_{24}	12	4	.	2	2	.	12	3	.	.	.	2	2	.	4	.	.	1	2	.	.	.	1	1	.	.	2	2	.	.	2	.	.	1	.
τ_{25}	12	4	.	2	2	.	12	3	.	.	.	2	2	.	4	.	.	1	2	.	.	.	1	1	.	.	2	2	.	.	2	.	.	1	.
τ_{26}	12	4	.	2	2	.	12	3	.	.	.	2	2	.	4	.	.	1	2	.	.	.	1	1	.	.	2	2	.	.	2	.	.	1	.
τ_{27}	12	12	4	.	.	.	6	.	.	.	4	.	.	.	6	β	$\bar{\beta}$	.	.	γ	$\bar{\gamma}$	2	.	.	.	.	.	.	.	2	.	.	.	.	.
τ_{28}	12	12	4	.	.	.	6	.	.	.	4	.	.	.	6	$\bar{\beta}$	β	.	.	$\bar{\gamma}$	γ	2	.	.	.	.	.	.	.	2	.	.	.	.	.
τ_{29}	18	18	6	.	2	.	9	.	.	.	2	.	2	2	9	3	3	.	1	.	.	.	.	.	.	2	.	.	.	1	1	1	1	.	1
τ_{30}	18	18	6	.	2	.	9	.	.	.	2	.	2	2	9	3	3	.	1	.	.	.	.	.	.	2	.	.	.	1	1	1	1	.	1
τ_{31}	24	8	.	.	4	.	12	.	.	.	.	.	4	.	4	.	.	4	2	.	.	.	2	.	.	.	.	.	.	.	2	.	.	.	.
τ_{32}	24	8	.	.	4	.	12	.	.	.	.	.	4	.	4	.	.	4	2	.	.	.	2	.	.	.	.	.	.	.	2	.	.	.	.
τ_{33}	24	8	.	4	.	.	24	3	.	.	.	4	.	.	8	.	.	1	.	.	.	.	1	1	.	.	.	.	.	.	.	.	.	1	.
τ_{34}	24	8	.	4	.	.	24	3	.	.	.	4	.	.	8	.	.	1	.	.	.	.	1	1	.	.	.	.	.	.	.	.	.	1	.
τ_{35}	48	16	.	.	.	.	24	.	.	.	.	.	.	.	8	.	.	4	.	.	.	.	2	.	.	.	.	.	.	.	.	.	.	.	.

where $\alpha = -1 + 2\sqrt{3}i, \beta = -2 + 4\sqrt{3}i, \gamma = -1 + \sqrt{3}i$ *and* $\delta = \sqrt{3}i$.

13.7.4 *Character table of* $E = N_G(3_A) \cong 3\mathsf{McL} : 2$

	1a	2a	2b	3a	3b	3c	3d	4a	4b	5a	5b	6a	6b	6c	6d	6e	6f	7a	8a	8b	8c	9a	10a
2	8	8	5	7	4	3	3	5	6	2	1	7	4	3	3	2	2	1	5	4	5	.	2
3	7	3	2	7	7	7	5	2	2	2	1	3	3	3	3	3	2	1	1	1	.	3	2
5	3	1	1	3	1	1	.	1	.	3	2	1	1	1	.	.	.	.	.	.	.	.	1
7	1	1	.	1	.	.	.	.	.	.	.	1	.	.	.	.	.	1	.	.	.	.	.
11	1	.	1	1	.	.	.	.	.	.	.	.	.	.	.	.	.	.	.	.	.	.	.
$X.1$	1	1	1	1	1	1	1	1	1	1	1	1	1	1	1	1	1	1	1	1	1	1	1
$X.2$	1	1	−1	1	1	1	1	−1	1	1	1	1	1	1	1	1	−1	1	−1	1	−1	1	1
$X.3$	22	6	.	22	−5	−5	4	−4	2	−3	2	6	3	3	.	.	.	1	2	.	−2	1	1
$X.4$	22	6	.	22	−5	−5	4	4	2	−3	2	6	3	3	.	.	.	1	−2	.	2	1	1
$X.5$	231	7	−11	231	15	15	6	5	−1	6	1	7	7	7	−2	−2	−2	.	1	−1	1	.	2
$X.6$	231	7	11	231	15	15	6	−5	−1	6	1	7	7	7	−2	−2	2	.	−1	−1	−1	.	2
$X.7$	252	28	10	252	9	9	9	10	4	2	2	28	1	1	1	1	1	.	2	.	2	.	−2
$X.8$	252	28	−10	252	9	9	9	−10	4	2	2	28	1	1	1	1	−1	.	−2	.	−2	.	−2
$X.9$	896	.	16	896	32	32	−4	.	.	−4	1	.	.	.	.	.	−2	.	.	.	.	−1	.
$X.10$	896	.	−16	896	32	32	−4	.	.	−4	1	.	.	.	.	.	2	.	.	.	.	−1	.
$X.11$	896	.	−16	896	32	32	−4	.	.	−4	1	.	.	.	.	.	2	.	.	.	.	−1	.
$X.12$	896	.	16	896	32	32	−4	.	.	−4	1	.	.	.	.	.	−2	.	.	.	.	−1	.
$X.13$	1540	−28	.	1540	−26	−26	10	.	−4	−10	.	−28	14	14	2	2	.	.	.	.	.	−2	2
$X.14$	1750	70	10	1750	−5	−5	13	−10	2	.	.	70	−5	−5	1	1	1	.	4	.	.	−2	.
$X.15$	1750	70	−10	1750	−5	−5	13	10	2	.	.	70	−5	−5	1	1	−1	.	−4	.	.	−2	.
$X.16$	3520	64	.	3520	−44	−44	10	16	.	−5	.	64	4	4	−2	−2	.	−1	.	.	.	1	−1
$X.17$	3520	64	.	3520	−44	−44	10	−16	.	−5	.	64	4	4	−2	−2	.	−1	.	.	.	1	−1
$X.18$	3520	−64	.	3520	−44	−44	10	.	.	−5	.	−64	−4	−4	2	2	.	−1	.	.	.	1	1
$X.19$	3520	−64	.	3520	−44	−44	10	.	.	−5	.	−64	−4	−4	2	2	.	−1	.	.	.	1	1
$X.20$	4500	20	10	4500	45	45	−9	10	4	.	.	20	5	5	−1	−1	1	−1	−2	.	−2	.	.
$X.21$	4500	20	−10	4500	45	45	−9	−10	4	.	.	20	5	5	−1	−1	−1	−1	2	.	2	.	.
$X.22$	4752	−48	.	4752	54	54	.	.	.	2	2	−48	−6	−6	.	.	.	−1	.	.	.	.	2
$X.23$	4752	−48	.	4752	54	54	.	.	.	2	2	−48	−6	−6	.	.	.	−1	.	.	.	.	2
$X.24$	5103	63	45	5103	.	.	.	9	3	3	−2	63	.	.	.	.	.	.	3	1	−1	.	3
$X.25$	5103	63	−45	5103	.	.	.	−9	3	3	−2	63	.	.	.	.	.	.	−3	1	1	.	3
$X.26$	5544	−56	44	5544	36	36	9	−4	.	19	−1	−56	4	4	1	1	−1	.	.	.	.	.	−1
$X.27$	5544	−56	−44	5544	36	36	9	4	.	19	−1	−56	4	4	1	1	1	.	.	.	.	.	−1
$X.28$	9625	105	55	9625	40	40	−5	−5	−3	.	.	105	.	.	3	3	1	.	−3	−1	1	1	.
$X.29$	9625	105	−55	9625	40	40	−5	5	−3	.	.	105	.	.	3	3	−1	.	3	−1	−1	1	.
$X.30$	16038	−90	.	16038	.	.	.	.	6	−12	−2	−90	.	.	.	.	.	1	.	−2	.	.	.
$X.31$	16500	20	.	16500	30	30	12	.	−4	.	.	20	−10	−10	−4	−4	.	1	.	.	.	.	.
$X.32$	19712	.	.	19712	−160	−160	−16	.	.	12	2	.	.	.	.	.	.	.	.	.	.	−1	.
$X.33$	20790	−42	.	20790	54	54	.	.	−2	−10	.	−42	6	6	.	.	.	.	.	2	.	.	−2
$X.34$	252	28	.	−126	−18	9	.	.	4	2	2	−14	−2	1	4	−2	.	.	.	.	.	.	−2
$X.35$	252	28	.	−126	−18	9	.	.	4	2	2	−14	−2	1	4	−2	.	.	.	.	.	.	−2
$X.36$	1584	−16	.	−792	72	−36	.	.	.	−16	4	8	8	−4	−4	2	.	2	.	.	.	.	4
$X.37$	3960	−72	.	−1980	18	−9	.	.	8	10	.	36	18	−9	.	.	.	−2	.	.	.	.	−2
$X.38$	4752	−48	.	−2376	−108	54	.	.	.	2	2	24	12	−6	.	.	.	−1	.	.	.	.	2
$X.39$	4752	−48	.	−2376	−108	54	.	.	.	2	2	24	12	−6	.	.	.	−1	.	.	.	.	2
$X.40$	5040	112	.	−2520	−36	18	.	.	.	−10	.	−56	4	−2	4	−2	.	.	.	.	.	.	2
$X.41$	5040	112	.	−2520	−36	18	.	.	.	−10	.	−56	4	−2	4	−2	.	.	.	.	.	.	2
$X.42$	5544	168	.	−2772	90	−45	.	.	8	−6	4	−84	−6	3	.	.	.	.	.	.	.	.	−2
$X.43$	9504	−96	.	−4752	108	−54	.	.	.	4	4	48	−12	6	.	.	.	−2	.	.	.	.	4
$X.44$	10206	126	.	−5103	.	.	.	.	6	6	−4	−63	.	.	.	.	.	.	.	2	.	.	6
$X.45$	12672	128	.	−6336	−72	36	.	.	.	22	2	−64	8	−4	−4	2	.	2	.	.	.	.	−2
$X.46$	12672	−128	.	−6336	−72	36	.	.	.	22	2	64	−8	4	4	−2	.	2	.	.	.	.	2
$X.47$	15750	70	.	−7875	90	−45	.	.	−10	.	.	−35	10	−5	4	−2	.	.	.	−2	.	.	.
$X.48$	16038	−90	.	−8019	.	.	.	.	6	−12	−2	45	.	.	.	.	.	1	.	−2	.	.	.
$X.49$	16038	−90	.	−8019	.	.	.	.	6	−12	−2	45	.	.	.	.	.	1	.	−2	.	.	.
$X.50$	16128	.	.	−8064	144	−72	.	.	.	28	−2	.	.	.	.	.	.	.	.	.	.	.	.
$X.51$	20790	−42	.	−10395	−108	54	.	.	−2	−10	.	21	−12	6	.	.	.	.	.	2	.	.	−2
$X.52$	20790	−42	.	−10395	54	−27	.	.	−2	−10	.	21	6	−3	.	.	.	.	.	2	.	.	−2
$X.53$	20790	−42	.	−10395	54	−27	.	.	−2	−10	.	21	6	−3	.	.	.	.	.	2	.	.	−2
$X.54$	24750	110	.	−12375	−90	45	.	.	−2	.	.	−55	−10	5	−4	2	.	−2	.	−2	.	.	.

Character table of $E = N_G(3_A) \cong$ 3McL : 2 *(continued)*

	10b	11a	11b	12a	12b	12c	12d	12e	14a	15a	15b	15c	15d	15e	20a	20b	21a	21b	22a	22b	24a	24b
2	1	1	1	5	3	2	2	2	1	1	1	1	1	.	2	2	1	1	1	1	3	3
3	.	1	1	2	2	2	2	2	1	2	2	2	2	1	.	.	1	1	.	.	1	1
5	1	.	.	.	.	.	.	.	.	3	1	1	1	2	1	1	.	.	.	.	.	.
7	.	.	.	.	.	.	.	.	1	.	.	.	.	.	.	.	1	1	.	.	.	.
11	.	.	1	1	.	.	.	.	.	.	.	.	.	.	.	.	.	.	.	1	1	.
X.1	1	1	1	1	1	1	1	1	1	1	1	1	1	1	1	1	1	1	1	1	1	1
X.2	−1	1	1	1	1	−1	1	−1	1	1	1	1	1	1	−1	−1	1	1	−1	−1	1	−1
X.3	.	.	.	2	−1	2	−1	−1	−1	−3	.	.	.	2	1	1	1	1	.	.	.	−1
X.4	.	.	.	2	−1	−2	−1	1	−1	−3	.	.	.	2	−1	−1	1	1	.	.	.	1
X.5	−1	.	.	−1	−1	2	−1	−1	.	6	.	.	.	1	.	.	.	.	.	.	−1	1
X.6	1	.	.	−1	−1	−2	−1	1	.	6	.	.	.	1	.	.	.	.	.	.	−1	−1
X.7	.	−1	−1	4	1	1	1	1	.	2	−1	−1	−1	2	.	.	.	.	−1	−1	.	−1
X.8	.	−1	−1	4	1	−1	1	−1	.	2	−1	−1	−1	2	.	.	.	.	1	1	.	1
X.9	1	A	/A	.	.	.	.	.	.	−4	2	2	2	1	.	.	.	.	/A	A	.	.
X.10	−1	A	/A	.	.	.	.	.	.	−4	2	2	2	1	.	.	.	.	−/A	−A	.	.
X.11	−1	/A	A	.	.	.	.	.	.	−4	2	2	2	1	.	.	.	.	−A	−/A	.	.
X.12	1	/A	A	.	.	.	.	.	.	−4	2	2	2	1	.	.	.	.	A	/A	.	.
X.13	.	.	.	−4	2	.	2	.	.	−10	−1	−1	−1	.	.	.	.	.	.	.	.	.
X.14	.	1	1	2	−1	−1	−1	−1	.	.	.	.	.	.	.	.	.	.	−1	−1	.	1
X.15	.	1	1	2	−1	1	−1	1	.	.	.	.	.	.	.	.	.	.	1	1	.	−1
X.16	.	.	.	.	.	−2	.	−2	1	−5	1	1	1	.	1	1	−1	−1	.	.	.	.
X.17	.	.	.	.	.	2	.	2	1	−5	1	1	1	.	−1	−1	−1	−1	.	.	.	.
X.18	.	.	.	.	.	.	.	.	−1	−5	1	1	1	.	D	−D	−1	−1	.	.	.	.
X.19	.	.	.	.	.	.	.	.	−1	−5	1	1	1	.	−D	D	−1	−1	.	.	.	.
X.20	.	1	1	4	1	1	1	1	−1	.	.	.	.	.	.	.	−1	−1	−1	−1	.	1
X.21	.	1	1	4	1	−1	1	−1	−1	.	.	.	.	.	.	.	−1	−1	1	1	.	−1
X.22	.	.	.	.	.	.	.	.	1	2	−1	−1	−1	2	.	.	−1	−1	.	.	.	F
X.23	.	.	.	.	.	.	.	.	1	2	−1	−1	−1	2	.	.	−1	−1	.	.	.	−F
X.24	.	−1	−1	3	.	.	.	.	.	3	.	.	.	−2	−1	−1	.	.	1	1	1	.
X.25	.	−1	−1	3	.	.	.	.	.	3	.	.	.	−2	1	1	.	.	−1	−1	1	.
X.26	−1	.	.	.	.	−1	.	2	.	19	1	1	1	−1	1	1	.	.	.	.	.	.
X.27	1	.	.	.	.	1	.	−2	.	19	1	1	1	−1	−1	−1	.	.	.	.	.	.
X.28	.	.	.	−3	.	1	.	−2	.	.	.	.	.	.	.	.	.	.	.	.	−1	.
X.29	.	.	.	−3	.	−1	.	2	.	.	.	.	.	.	.	.	.	.	.	.	−1	.
X.30	.	.	.	6	.	.	.	.	1	−12	.	.	.	−2	.	.	1	1	.	.	−2	.
X.31	.	.	.	−4	2	.	2	.	−1	.	.	.	.	.	.	.	1	1	.	.	.	.
X.32	.	.	.	.	.	.	.	.	.	12	.	.	.	2	.	.	.	.	.	.	.	.
X.33	.	.	.	−2	−2	.	−2	.	.	−10	−1	−1	−1	.	.	.	.	.	.	.	2	.
X.34	.	B	/B	−2	−2	.	1	.	.	−1	−1	−1	2	−1	.	.	.	.	.	.	.	.
X.35	.	/B	B	−2	−2	.	1	.	.	−1	−1	−1	2	−1	.	.	.	.	.	.	.	.
X.36	.	.	.	.	.	.	.	.	−2	8	−1	−1	2	−2	.	.	−1	−1	.	.	.	.
X.37	.	.	.	−4	2	.	−1	.	−2	−5	1	1	−2	.	.	.	1	1	.	.	.	.
X.38	.	.	.	.	.	.	.	.	1	−1	−1	−1	2	−1	.	.	E	*E	.	.	.	.
X.39	.	.	.	.	.	.	.	.	1	−1	−1	−1	2	−1	.	.	*E	E	.	.	.	.
X.40	.	2	2	.	.	.	.	.	.	5	C	*C	−1	.	.	.	.	.	.	.	.	.
X.41	.	2	2	.	.	.	.	.	.	5	*C	C	−1	.	.	.	.	.	.	.	.	.
X.42	.	.	.	−4	2	.	−1	.	.	3	.	.	.	−2	.	.	.	.	.	.	.	.
X.43	.	.	.	.	.	.	.	.	2	−2	1	1	−2	−2	.	.	1	1	.	.	.	.
X.44	.	−2	−2	−3	.	.	.	.	.	−3	.	.	.	2	.	.	.	.	.	.	−1	.
X.45	.	.	.	.	.	.	.	.	2	−11	1	1	−2	−1	.	.	−1	−1	.	.	.	.
X.46	.	.	.	.	.	.	.	.	−2	−11	1	1	−2	−1	.	.	−1	−1	.	.	.	.
X.47	.	−2	−2	5	2	.	−1	.	.	.	.	.	.	.	.	.	.	.	.	.	1	.
X.48	.	.	.	−3	.	.	.	.	1	6	.	.	.	1	.	.	− * E	−E	.	.	1	.
X.49	.	.	.	−3	.	.	.	.	1	6	.	.	.	1	.	.	−E	− * E	.	.	1	.
X.50	.	2	2	.	.	.	.	.	.	−14	−2	−2	4	1	.	.	.	.	.	.	.	.
X.51	.	.	.	1	4	.	−2	.	.	5	−1	−1	2	.	.	.	.	.	.	.	−1	.
X.52	.	.	.	1	−2	.	1	.	.	5	C	*C	−1	.	.	.	.	.	.	.	−1	.
X.53	.	.	.	1	−2	.	1	.	.	5	*C	C	−1	.	.	.	.	.	.	.	−1	.
X.54	.	.	.	1	−2	.	1	.	−2	.	.	.	.	.	.	.	1	1	.	.	1	.

Character table of $E = N_G(3_A) \cong 3\mathsf{McL} : 2$ *(continued)*

2	3	1	1	1	1	.	.	1	1
3	1	2	2	2	2	1	1	1	1
5	.	1	1	1	1	.	.	.	.
7	.	.	.	.	.	.	.	1	1
11	.	.	.	.	.	1	1	.	.
	24c	30a	30b	30c	30d	33a	33b	42a	42b
X.1	1	1	1	1	1	1	1	1	1
X.2	−1	1	1	1	1	1	1	1	1
X.3	−1	−2	−2	−2	1	.	.	−1	−1
X.4	1	−2	−2	−2	1	.	.	−1	−1
X.5	1	2	2	2	2	.	.	.	.
X.6	−1	2	2	2	2	.	.	.	.
X.7	−1	1	1	1	−2	−1	−1	.	.
X.8	1	1	1	1	−2	−1	−1	.	.
X.9	.	.	.	.	.	A	/A	.	.
X.10	.	.	.	.	.	A	/A	.	.
X.11	.	.	.	.	.	/A	A	.	.
X.12	.	.	.	.	.	/A	A	.	.
X.13	.	−1	−1	−1	2	.	.	.	.
X.14	1	.	.	.	.	1	1	.	.
X.15	−1	.	.	.	.	1	1	.	.
X.16	.	−1	−1	−1	−1	.	.	1	1
X.17	.	−1	−1	−1	−1	.	.	1	1
X.18	.	1	1	1	1	.	.	−1	−1
X.19	.	1	1	1	1	.	.	−1	−1
X.20	1	.	.	.	.	1	1	−1	−1
X.21	−1	.	.	.	.	1	1	−1	−1
X.22	−F	−1	−1	−1	2	.	.	1	1
X.23	F	−1	−1	−1	2	.	.	1	1
X.24	.	.	.	.	3	−1	−1	.	.
X.25	.	.	.	.	3	−1	−1	.	.
X.26	.	−1	−1	−1	−1	.	.	.	.
X.27	.	−1	−1	−1	−1	.	.	.	.
X.28	.	.	.	.	.	.	.	.	.
X.29	.	.	.	.	.	.	.	.	.
X.30	.	.	.	.	.	.	.	1	1
X.31	.	.	.	.	.	.	.	−1	−1
X.32	.	.	.	.	.	.	.	.	.
X.33	.	1	1	1	−2	.	.	.	.
X.34	.	1	1	−2	1	−/A	−A	.	.
X.35	.	1	1	−2	1	−A	−/A	.	.
X.36	.	1	1	−2	−2	.	.	1	1
X.37	.	1	1	−2	1	.	.	1	1
X.38	.	−1	−1	2	−1	.	.	− * E	−E
X.39	.	−1	−1	2	−1	.	.	−E	− * E
X.40	.	C	*C	−1	−1	−1	−1	.	.
X.41	.	*C	C	−1	−1	−1	−1	.	.
X.42	.	−2	−2	4	1	.	.	.	.
X.43	.	1	1	−2	−2	.	.	−1	−1
X.44	.	.	.	.	−3	1	1	.	.
X.45	.	1	1	−2	1	.	.	−1	−1
X.46	.	−1	−1	2	−1	.	.	1	1
X.47	.	.	.	.	.	1	1	.	.
X.48	.	.	.	.	.	.	.	−E	− * E
X.49	.	.	.	.	.	.	.	− * E	−E
X.50	.	.	.	.	.	−1	−1	.	.
X.51	.	1	1	−2	1	.	.	.	.
X.52	.	−C	− * C	1	1	.	.	.	.
X.53	.	− * C	−C	1	1	.	.	.	.
X.54	.	.	.	.	.	.	.	1	1

where $A = -\frac{1}{2}(1+i\sqrt{11})$, $B = (-1+i\sqrt{11})$, $C = \frac{1}{2}(1-3\sqrt{5})$, $D = -i\sqrt{5}$, $E = \frac{1}{2}(1+\sqrt{21})$, $F = -i\sqrt{6}$.

13.7.5 *Character table of* $R = N_G(f) \cong 5^{1+4} : 4S_6$

	1a	2a	2b	3a	3b	4a	4b	4c	4d	4e	5a	5b	5c	5d	5e	5f	6a	6b	8a	8b	8c	8d	8e	8f	10a
2	6	6	5	3	3	6	6	5	4	4	4	2	1	2	1	1	3	3	5	5	5	5	4	4	4
3	2	2	.	2	2	2	2	.	.	.	2	.	1	.	.	.	2	2	1	1	1	1	.	.	2
5	6	2	2	3	1	1	1	1	1	1	6	5	4	3	3	3	1	1	.	.	.	.	1	.	2
2P	1a	1a	1a	3a	3b	2a	2a	2a	2b	2b	5a	5b	5c	5d	5e	5f	3b	3a	4b	4b	4a	4a	4c	4c	5a
3P	1a	2a	2b	1a	1a	4b	4a	4c	4e	4d	5a	5b	5c	5d	5e	5f	2a	2a	8d	8c	8b	8a	8e	8f	10a
5P	1a	2a	2b	3a	3b	4a	4b	4c	4d	4e	1a	1a	1a	1a	1a	1a	6a	6b	8a	8b	8c	8d	8e	8f	2a
7P	1a	2a	2b	3a	3b	4b	4a	4c	4e	4d	5a	5b	5c	5d	5e	5f	6a	6b	8d	8c	8b	8a	8e	8f	10a
11P	1a	2a	2b	3a	3b	4b	4a	4c	4e	4d	5a	5b	5c	5d	5e	5f	6a	6b	8d	8c	8b	8a	8e	8f	10a
13P	1a	2a	2b	3a	3b	4a	4b	4c	4d	4e	5a	5b	5c	5d	5e	5f	6a	6b	8a	8b	8c	8d	8e	8f	10a
17P	1a	2a	2b	3a	3b	4a	4b	4c	4d	4e	5a	5b	5c	5d	5e	5f	6a	6b	8a	8b	8c	8d	8e	8f	10a
19P	1a	2a	2b	3a	3b	4b	4a	4c	4e	4d	5a	5b	5c	5d	5e	5f	6a	6b	8d	8c	8b	8a	8e	8f	10a
23P	1a	2a	2b	3a	3b	4b	4a	4c	4e	4d	5a	5b	5c	5d	5e	5f	6a	6b	8d	8c	8b	8a	8e	8f	10a
29P	1a	2a	2b	3a	3b	4a	4b	4c	4d	4e	5a	5b	5c	5d	5e	5f	6a	6b	8a	8b	8c	8d	8e	8f	10a
31P	1a	2a	2b	3a	3b	4b	4a	4c	4e	4d	5a	5b	5c	5d	5e	5f	6a	6b	8d	8c	8b	8a	8e	8f	10a
37P	1a	2a	2b	3a	3b	4a	4b	4c	4d	4e	5a	5b	5c	5d	5e	5f	6a	6b	8a	8b	8c	8d	8e	8f	10a
X.1	1	1	1	1	1	1	1	1	1	1	1	1	1	1	1	1	1	1	1	1	1	1	1	1	1
X.2	1	1	−1	1	1	−1	−1	1	D	−D	1	1	1	1	1	1	1	1	D	−D	D	−D	1	−1	1
X.3	1	1	1	1	1	1	1	1	−1	−1	1	1	1	1	1	1	1	1	−1	−1	−1	−1	1	1	1
X.4	1	1	−1	1	1	−1	−1	1	−D	D	1	1	1	1	1	1	1	1	−D	D	−D	D	1	−1	1
X.5	4	−4	.	−2	1	A	−A	.	.	.	4	4	4	−1	−1	−1	−1	2	.	.	.	.	.	.	−4
X.6	4	−4	.	−2	1	A	−A	.	.	.	4	4	4	−1	−1	−1	−1	2	.	.	.	.	.	.	−4
X.7	4	−4	.	−2	1	−A	A	.	.	.	4	4	4	−1	−1	−1	−1	2	.	.	.	.	.	.	−4
X.8	4	−4	.	−2	1	−A	A	.	.	.	4	4	4	−1	−1	−1	−1	2	.	.	.	.	.	.	−4
X.9	4	−4	.	1	−2	A	−A	.	.	.	4	4	4	−1	−1	−1	2	−1	.	.	.	.	.	.	−4
X.10	4	−4	.	1	−2	A	−A	.	.	.	4	4	4	−1	−1	−1	2	−1	.	.	.	.	.	.	−4
X.11	4	−4	.	1	−2	−A	A	.	.	.	4	4	4	−1	−1	−1	2	−1	.	.	.	.	.	.	−4
X.12	4	−4	.	1	−2	−A	A	.	.	.	4	4	4	−1	−1	−1	2	−1	.	.	.	.	.	.	−4
X.13	5	5	1	−1	2	5	5	1	−1	−1	5	5	5	.	.	.	2	−1	1	−3	−3	1	−1	−1	5
X.14	5	5	1	−1	2	5	5	1	1	1	5	5	5	.	.	.	2	−1	−1	3	3	−1	−1	−1	5
X.15	5	5	−1	−1	2	−5	−5	1	−D	D	5	5	5	.	.	.	2	−1	D	−E	E	−D	−1	1	5
X.16	5	5	−1	−1	2	−5	−5	1	D	−D	5	5	5	.	.	.	2	−1	−D	E	−E	D	−1	1	5
X.17	5	5	1	2	−1	5	5	1	−1	−1	5	5	5	.	.	.	−1	2	−3	1	1	−3	−1	−1	5
X.18	5	5	1	2	−1	5	5	1	1	1	5	5	5	.	.	.	−1	2	3	−1	−1	3	−1	−1	5
X.19	5	5	−1	2	−1	−5	−5	1	−D	D	5	5	5	.	.	.	−1	2	E	−D	D	−E	−1	1	5
X.20	5	5	−1	2	−1	−5	−5	1	D	−D	5	5	5	.	.	.	−1	2	−E	D	−D	E	−1	1	5
X.21	9	9	1	.	.	9	9	1	−1	−1	9	9	9	−1	−1	−1	.	.	3	3	3	3	1	1	9
X.22	9	9	1	.	.	9	9	1	1	1	9	9	9	−1	−1	−1	.	.	−3	−3	−3	−3	1	1	9
X.23	9	9	−1	.	.	−9	−9	1	−D	D	9	9	9	−1	−1	−1	.	.	−E	E	−E	E	1	−1	9
X.24	9	9	−1	.	.	−9	−9	1	D	−D	9	9	9	−1	−1	−1	.	.	E	−E	E	−E	1	−1	9
X.25	10	10	−2	1	1	10	10	−2	.	.	10	10	10	.	.	.	1	1	−2	2	2	−2	.	.	10
X.26	10	10	−2	1	1	10	10	−2	.	.	10	10	10	.	.	.	1	1	2	−2	−2	2	.	.	10
X.27	10	10	2	1	1	−10	−10	−2	.	.	10	10	10	.	.	.	1	1	F	F	−F	−F	.	.	10
X.28	10	10	2	1	1	−10	−10	−2	.	.	10	10	10	.	.	.	1	1	−F	−F	F	F	.	.	10
X.29	16	−16	.	−2	−2	B	−B	.	.	.	16	16	16	1	1	1	2	2	.	.	.	.	.	.	−16
X.30	16	−16	.	−2	−2	−B	B	.	.	.	16	16	16	1	1	1	2	2	.	.	.	.	.	.	−16
X.31	16	16	.	−2	−2	16	16	.	.	.	16	16	16	1	1	1	−2	−2	.	.	.	.	.	.	16
X.32	16	16	.	−2	−2	−16	−16	.	.	.	16	16	16	1	1	1	−2	−2	.	.	.	.	.	.	16
X.33	20	−20	.	2	2	C	−C	.	.	.	20	20	20	.	.	.	−2	−2	.	.	.	.	.	.	−20
X.34	20	−20	.	2	2	−C	C	.	.	.	20	20	20	.	.	.	−2	−2	.	.	.	.	.	.	−20
X.35	100	4	.	−20	4	.	.	4	.	.	−25	.	.	.	5	−5	4	4	.	.	.	.	−4	.	−1
X.36	144	.	8	.	.	.	.	.	−4	−4	144	19	−6	4	4	4	.	.	.	.	.	.	.	.	.
X.37	144	.	8	.	.	.	.	.	4	4	144	19	−6	4	4	4	.	.	.	.	.	.	.	.	.
X.38	144	.	−8	.	.	.	.	.	−A	A	144	19	−6	4	4	4	.	.	.	.	.	.	.	.	.
X.39	144	.	−8	.	.	.	.	.	A	−A	144	19	−6	4	4	4	.	.	.	.	.	.	.	.	.
X.40	400	−16	.	40	4	.	.	.	.	.	−100	.	.	.	−5	5	−4	8	.	.	.	.	.	.	4
X.41	400	−16	.	−20	−8	.	.	.	.	.	−100	.	.	.	−5	5	8	−4	.	.	.	.	.	.	4
X.42	480	.	16	24	.	.	.	.	.	.	480	−20	5	.	.	.	.	.	.	.	.	.	.	.	.
X.43	480	.	−16	24	.	.	.	.	.	.	480	−20	5	.	.	.	.	.	.	.	.	.	.	.	.
X.44	500	20	.	20	8	.	.	4	.	.	−125	.	.	.	.	.	8	−4	.	.	.	.	4	.	−5
X.45	500	20	.	−40	−4	.	.	4	.	.	−125	.	.	.	.	.	−4	8	.	.	.	.	4	.	−5
X.46	576	.	.	.	.	.	.	.	.	.	576	76	−24	−4	−4	−4	.	.	.	.	.	.	.	.	.
X.47	800	32	.	20	−4	.	.	.	.	.	−200	.	.	−10	5	.	−4	−4	.	.	.	.	.	.	−8
X.48	800	32	.	20	−4	.	.	.	.	.	−200	.	.	10	.	−5	−4	−4	.	.	.	.	.	.	−8
X.49	800	−32	.	20	−4	.	.	.	.	.	−200	.	.	−10	5	.	4	4	.	.	.	.	.	.	8
X.50	800	−32	.	20	−4	.	.	.	.	.	−200	.	.	10	.	−5	4	4	.	.	.	.	.	.	8
X.51	900	36	.	.	.	.	.	4	.	.	−225	.	.	.	−5	5	.	.	.	.	.	.	−4	.	−9
X.52	960	.	.	−24	.	.	.	.	.	.	960	−40	10	.	.	.	.	.	.	.	.	.	.	.	.
X.53	1000	40	.	−20	4	.	.	−8	.	.	−250	.	.	.	.	.	4	4	.	.	.	.	.	.	−10
X.54	1000	−40	.	−20	4	.	.	.	.	.	−250	.	.	.	.	.	−4	−4	.	.	.	.	.	.	10
X.55	1000	−40	.	−20	4	.	.	.	.	.	−250	.	.	.	.	.	−4	−4	.	.	.	.	.	.	10

Character table of $R = N_G(f) \cong 5^{1+4} : 4S_6$ (continued)

	10b	10c	10d	10e	10f	12a	12b	12c	12d	15a	15b	15c	20a	20b	20c	20d	20e	24a	24b	24c	24d	24e
2	2	2	1	1	1	3	3	3	3	1	1	.	3	2	2	2	2	3	3	3	3	3
3	.	.	.	.	.	2	2	2	2	2	2	1	.	.	.	.	.	1	1	1	1	1
5	2	2	2	2	2	.	.	.	.	3	1	2	1	1	1	1	1	.	.	.	.	.
2P	5b	5d	5c	5f	5e	6b	6a	6b	6a	15a	15b	15c	10a	10c	10b	10b	10c	12d	12d	12b	12c	12c
3P	10b	10c	10d	10e	10f	4b	4a	4a	4b	5a	5a	5c	20a	20e	20d	20c	20b	8b	8b	8c	8d	8d
5P	2b	2a	2b	2a	2a	12a	12b	12c	12d	3a	3b	3a	4c	4b	4e	4d	4a	24a	24b	24c	24d	24e
7P	10b	10c	10d	10e	10f	12c	12d	12a	12b	15a	15b	15c	20a	20e	20d	20c	20b	24c	24h	24a	24f	24g
11P	10b	10c	10d	10e	10f	12c	12d	12a	12b	15a	15b	15c	20a	20e	20d	20c	20b	24c	24h	24a	24f	24g
13P	10b	10c	10d	10e	10f	12a	12b	12c	12d	15a	15b	15c	20a	20b	20c	20d	20e	24b	24a	24h	24e	24d
17P	10b	10c	10d	10e	10f	12a	12b	12c	12d	15a	15b	15c	20a	20b	20c	20d	20e	24b	24a	24h	24e	24d
19P	10b	10c	10d	10e	10f	12c	12d	12a	12b	15a	15b	15c	20a	20e	20d	20c	20b	24h	24c	24b	24g	24f
23P	10b	10c	10d	10e	10f	12c	12d	12a	12b	15a	15b	15c	20a	20e	20d	20c	20b	24h	24c	24b	24g	24f
29P	10b	10c	10d	10e	10f	12a	12b	12c	12d	15a	15b	15c	20a	20b	20c	20d	20e	24a	24b	24c	24d	24e
31P	10b	10c	10d	10e	10f	12c	12d	12a	12b	15a	15b	15c	20a	20e	20d	20c	20b	24c	24h	24a	24f	24g
37P	10b	10c	10d	10e	10f	12a	12b	12c	12d	15a	15b	15c	20a	20b	20c	20d	20e	24b	24a	24h	24e	24d
X.1	1	1	1	1	1	1	1	1	1	1	1	1	1	1	1	1	1	1	1	1	1	1
X.2	-1	1	-1	1	1	-1	-1	-1	-1	1	1	1	1	-1	-D	D	-1	D	D	-D	D	D
X.3	1	1	1	1	1	1	1	1	1	1	1	1	1	1	-1	-1	1	-1	-1	-1	-1	-1
X.4	-1	1	-1	1	1	-1	-1	-1	-1	1	1	1	1	-1	D	-D	-1	-D	-D	D	-D	-D
X.5	.	1	.	1	1	F	-D	-F	D	-2	1	-2	.	D	.	.	-D	G	-G	-/G	.	.
X.6	.	1	.	1	1	F	-D	-F	D	-2	1	-2	.	D	.	.	-D	-G	G	/G	.	.
X.7	.	1	.	1	1	-F	D	F	-D	-2	1	-2	.	-D	.	.	D	-/G	/G	G	.	.
X.8	.	1	.	1	1	-F	D	F	-D	-2	1	-2	.	-D	.	.	D	/G	-/G	-G	.	.
X.9	.	1	.	1	1	D	-F	-D	F	1	-2	1	.	D	.	.	-D	.	.	.	-/G	/G
X.10	.	1	.	1	1	D	-F	-D	F	1	-2	1	.	D	.	.	-D	.	.	.	/G	-/G
X.11	.	1	.	1	1	-D	F	D	-F	1	-2	1	.	-D	.	.	D	.	.	.	G	-G
X.12	.	1	.	1	1	-D	F	D	-F	1	-2	1	.	-D	.	.	D	.	.	.	-G	G
X.13	1	.	1	.	.	-1	2	-1	2	-1	2	-1	1	.	-1	-1	.	.	.	.	1	1
X.14	1	.	1	.	.	-1	2	-1	2	-1	2	-1	1	.	1	1	.	.	.	.	-1	-1
X.15	-1	.	-1	.	.	1	-2	1	-2	-1	2	-1	1	.	D	-D	.	.	.	.	D	D
X.16	-1	.	-1	.	.	1	-2	1	-2	-1	2	-1	1	.	-D	D	.	.	.	.	-D	-D
X.17	1	.	1	.	.	2	-1	2	-1	2	-1	2	1	.	-1	-1	.	1	1	1	.	.
X.18	1	.	1	.	.	2	-1	2	-1	2	-1	2	1	.	1	1	.	-1	-1	-1	.	.
X.19	-1	.	-1	.	.	-2	1	-2	1	2	-1	2	1	.	D	-D	.	D	D	-D	.	.
X.20	-1	.	-1	.	.	-2	1	-2	1	2	-1	2	1	.	-D	D	.	-D	-D	D	.	.
X.21	1	-1	1	-1	-1	.	.	.	.	.	.	.	1	-1	-1	-1	-1	.	.	.	.	.
X.22	1	-1	1	-1	-1	.	.	.	.	.	.	.	1	-1	1	1	-1	.	.	.	.	.
X.23	-1	-1	-1	-1	-1	.	.	.	.	.	.	.	1	1	D	-D	1	.	.	.	.	.
X.24	-1	-1	-1	-1	-1	.	.	.	.	.	.	.	1	1	-D	D	1	.	.	.	.	.
X.25	-2	.	-2	.	.	1	1	1	1	1	1	1	-2	.	.	.	.	-1	-1	-1	1	1
X.26	-2	.	-2	.	.	1	1	1	1	1	1	1	-2	.	.	.	.	1	1	1	-1	-1
X.27	2	.	2	.	.	-1	-1	-1	-1	1	1	1	-2	.	.	.	.	-D	-D	D	D	D
X.28	2	.	2	.	.	-1	-1	-1	-1	1	1	1	-2	.	.	.	.	D	D	-D	-D	-D
X.29	.	-1	.	-1	-1	F	-F	-F	F	-2	-2	-2	.	-D	.	.	D	.	.	.	.	.
X.30	.	-1	.	-1	-1	-F	F	F	-F	-2	-2	-2	.	D	.	.	-D	.	.	.	.	.
X.31	.	1	.	1	1	-2	-2	-2	-2	-2	-2	-2	.	1	.	.	1	.	.	.	.	.
X.32	.	1	.	1	1	2	2	2	2	-2	-2	-2	.	-1	.	.	-1	.	.	.	.	.
X.33	.	.	.	.	.	-F	F	F	-F	2	2	2	.	.	.	.	.	.	.	.	.	.
X.34	.	.	.	.	.	F	-F	-F	F	2	2	2	.	.	.	.	.	.	.	.	.	.
X.35	.	4	.	-1	-1	.	.	.	.	5	-1	.	-1	.	.	.	.	.	.	.	.	.
X.36	3	.	-2	.	.	.	.	.	.	.	.	.	.	.	1	1	.	.	.	.	.	.
X.37	3	.	-2	.	.	.	.	.	.	.	.	.	.	.	-1	-1	.	.	.	.	.	.
X.38	-3	.	2	.	.	.	.	.	.	.	.	.	.	.	-D	D	.	.	.	.	.	.
X.39	-3	.	2	.	.	.	.	.	.	.	.	.	.	.	D	-D	.	.	.	.	.	.
X.40	.	4	.	-1	-1	.	.	.	.	-10	-1	.	.	.	.	.	.	.	.	.	.	.
X.41	.	4	.	-1	-1	.	.	.	.	5	2	.	.	.	.	.	.	.	.	.	.	.
X.42	-4	.	1	.	.	.	.	.	.	24	.	-1	.	.	.	.	.	.	.	.	.	.
X.43	4	.	-1	.	.	.	.	.	.	24	.	-1	.	.	.	.	.	.	.	.	.	.
X.44	.	.	.	.	.	.	.	.	.	-5	-2	.	-1	.	.	.	.	.	.	.	.	.
X.45	.	.	.	.	.	.	.	.	.	10	1	.	-1	.	.	.	.	.	.	.	.	.
X.46	.	.	.	.	.	.	.	.	.	.	.	.	.	.	.	.	.	.	.	.	.	.
X.47	.	2	.	2	-3	.	.	.	.	-5	1	.	.	.	.	.	.	.	.	.	.	.
X.48	.	2	.	-3	2	.	.	.	.	-5	1	.	.	.	.	.	.	.	.	.	.	.
X.49	.	-2	.	-2	3	.	.	.	.	-5	1	.	.	.	.	.	.	.	.	.	.	.
X.50	.	-2	.	3	-2	.	.	.	.	-5	1	.	.	.	.	.	.	.	.	.	.	.
X.51	.	-4	.	1	1	.	.	.	.	.	.	.	-1	.	.	.	.	.	.	.	.	.
X.52	.	.	.	.	.	.	.	.	.	-24	.	1	.	.	.	.	.	.	.	.	.	.
X.53	.	.	.	.	.	.	.	.	.	5	-1	.	2	.	.	.	.	.	.	.	.	.
X.54	.	.	.	.	.	.	.	.	.	5	-1	.	.	.	.	.	.	.	.	.	.	.
X.55	.	.	.	.	.	.	.	.	.	5	-1	.	.	.	.	.	.	.	.	.	.	.

Character table of $R = N_G(f) \cong 5^{1+4} : 4S_6$ *(continued)*

2	3	3	3	.	1	1	3	3
3	1	1	1	.	2	2	.	.
5	.	.	.	2	1	1	1	1
	24f	24g	24h	25a	30a	30b	40a	40b
2P	12a	12a	12b	25a	15a	15b	20a	20a
3P	8a	8a	8c	25a	10a	10a	40a	40b
5P	24f	24g	24h	5a	6b	6a	8e	8e
7P	24d	24e	24b	25a	30a	30b	40b	40a
11P	24d	24e	24b	25a	30a	30b	40b	40a
13P	24g	24f	24c	25a	30a	30b	40a	40b
17P	24g	24f	24c	25a	30a	30b	40b	40a
19P	24e	24d	24a	25a	30a	30b	40b	40a
23P	24e	24d	24a	25a	30a	30b	40b	40a
29P	24f	24g	24h	25a	30a	30b	40b	40a
31P	24d	24e	24b	25a	30a	30b	40a	40b
37P	24g	24f	24c	25a	30a	30b	40a	40b
X.1	1	1	1	1	1	1	1	1
X.2	−D	−D	−D	1	1	1	1	1
X.3	−1	−1	−1	1	1	1	1	1
X.4	D	D	D	1	1	1	1	1
X.5	.	.	/G	−1	2	−1	.	.
X.6	.	.	−/G	−1	2	−1	.	.
X.7	.	.	−G	−1	2	−1	.	.
X.8	.	.	G	−1	2	−1	.	.
X.9	G	−G	.	−1	−1	2	.	.
X.10	−G	G	.	−1	−1	2	.	.
X.11	−/G	/G	.	−1	−1	2	.	.
X.12	/G	−/G	.	−1	−1	2	.	.
X.13	1	1	.	.	−1	2	−1	−1
X.14	−1	−1	.	.	−1	2	−1	−1
X.15	−D	−D	.	.	−1	2	−1	−1
X.16	D	D	.	.	−1	2	−1	−1
X.17	.	.	1	.	2	−1	−1	−1
X.18	.	.	−1	.	2	−1	−1	−1
X.19	.	.	−D	.	2	−1	−1	−1
X.20	.	.	D	.	2	−1	−1	−1
X.21	.	.	.	−1	.	.	1	1
X.22	.	.	.	−1	.	.	1	1
X.23	.	.	.	−1	.	.	1	1
X.24	.	.	.	−1	.	.	1	1
X.25	1	1	−1	.	1	1	.	.
X.26	−1	−1	1	.	1	1	.	.
X.27	−D	−D	D	.	1	1	.	.
X.28	D	D	−D	.	1	1	.	.
X.29	.	.	.	1	2	2	.	.
X.30	.	.	.	1	2	2	.	.
X.31	.	.	.	1	−2	−2	.	.
X.32	.	.	.	1	−2	−2	.	.
X.33	.	.	.	.	−2	−2	.	.
X.34	.	.	.	.	−2	−2	.	.
X.35	.	.	.	.	−1	−1	1	1
X.36	.	.	.	−1	.	.	.	.
X.37	.	.	.	−1	.	.	.	.
X.38	.	.	.	−1	.	.	.	.
X.39	.	.	.	−1	.	.	.	.
X.40	.	.	.	.	−2	1	.	.
X.41	.	.	.	.	1	−2	.	.
X.42	.	.	.	.	.	.	.	.
X.43	.	.	.	.	.	.	.	.
X.44	.	.	.	.	1	−2	−1	−1
X.45	.	.	.	.	−2	1	−1	−1
X.46	.	.	.	1	.	.	.	.
X.47	.	.	.	.	1	1	.	.
X.48	.	.	.	.	1	1	.	.
X.49	.	.	.	.	−1	−1	.	.
X.50	.	.	.	.	−1	−1	.	.
X.51	.	.	.	.	.	.	1	1
X.52	.	.	.	.	.	.	.	.
X.53	.	.	.	.	−1	−1	.	.
X.54	.	.	.	.	1	1	H	−H
X.55	.	.	.	.	1	1	−H	H

where $A = 4i$, $B = 16i$, $C = 20i$, $D = i$, $E = -3i$, $F = -2i$, $G = -\xi^{11} + \xi^{19}$, $H = \sqrt{10}$ *and* ξ *is a primitive* 24*th root of unity.*

13.7.6 Character table of $M = N_G(V) \cong 5^3.L_3(5)$

	1a	2a	3a	4a	4b	4c	5a	5b	5c	5d	6a	8a	8b	10a	10b	10c	10d	12a	12b	15a	20a	20b	20c	24a
2	5	5	3	5	5	4	3	2	1	1	3	3	3	3	2	1	1	3	3	1	2	2	2	3
3	1	1	1	1	1	.	1	.	.	.	1	1	1	1	.	.	.	1	1	1	.	.	.	1
5	6	2	1	1	1	1	6	4	4	4	1	.	.	2	2	2	2	.	.	1	1	1	1	.
31	1	.	.	.	.	.	.	.	.	.	.	.	.	.	.	.	.	.	.	.	.	.	.	.
2P	1a	1a	3a	2a	2a	2a	5a	5b	5c	5d	3a	4b	4a	5a	5b	5c	5d	6a	6a	15a	10a	10b	10b	12b
3P	1a	2a	1a	4b	4a	4c	5a	5b	5c	5d	2a	8b	8a	10a	10b	10c	10d	4b	4a	5a	20a	20c	20b	8b
5P	1a	2a	3a	4a	4b	4c	1a	1a	1a	1a	6a	8a	8b	2a	2a	2a	2a	12a	12b	3a	4c	4a	4b	24a
7P	1a	2a	3a	4b	4a	4c	5a	5b	5c	5d	6a	8b	8a	10a	10b	10c	10d	12b	12a	15a	20a	20c	20b	24c
11P	1a	2a	3a	4b	4a	4c	5a	5b	5c	5d	6a	8b	8a	10a	10b	10c	10d	12b	12a	15a	20a	20c	20b	24c
13P	1a	2a	3a	4a	4b	4c	5a	5b	5c	5d	6a	8a	8b	10a	10b	10c	10d	12a	12b	15a	20a	20b	20c	24b
17P	1a	2a	3a	4a	4b	4c	5a	5b	5c	5d	6a	8a	8b	10a	10b	10c	10d	12a	12b	15a	20a	20b	20c	24b
19P	1a	2a	3a	4b	4a	4c	5a	5b	5c	5d	6a	8b	8a	10a	10b	10c	10d	12b	12a	15a	20a	20c	20b	24d
23P	1a	2a	3a	4b	4a	4c	5a	5b	5c	5d	6a	8b	8a	10a	10b	10c	10d	12b	12a	15a	20a	20c	20b	24d
29P	1a	2a	3a	4a	4b	4c	5a	5b	5c	5d	6a	8a	8b	10a	10b	10c	10d	12a	12b	15a	20a	20b	20c	24a
31P	1a	2a	3a	4b	4a	4c	5a	5b	5c	5d	6a	8b	8a	10a	10b	10c	10d	12b	12a	15a	20a	20c	20b	24c
X.1	1	1	1	1	1	1	1	1	1	1	1	1	1	1	1	1	1	1	1	1	1	1	1	1
X.2	30	6	.	6	6	2	30	5	5	5	.	.	.	6	1	1	1	.	.	.	2	1	1	.
X.3	31	7	1	−5	−5	−1	31	6	6	6	1	−1	−1	7	2	2	2	1	1	1	−1	.	.	−1
X.4	31	−5	1	A	/A	1	31	6	6	6	1	D	−D	−5	.	.	.	−1	−1	1	1	F	/F	D
X.5	31	−5	1	/A	A	1	31	6	6	6	1	−D	D	−5	.	.	.	−1	−1	1	1	/F	F	−D
X.6	96	.	.	.	.	.	96	−4	−4	−4	.	.	.	.	.	.	.	.	.	.	.	.	.	.
X.7	96	.	.	.	.	.	96	−4	−4	−4	.	.	.	.	.	.	.	.	.	.	.	.	.	.
X.8	96	.	.	.	.	.	96	−4	−4	−4	.	.	.	.	.	.	.	.	.	.	.	.	.	.
X.9	96	.	.	.	.	.	96	−4	−4	−4	.	.	.	.	.	.	.	.	.	.	.	.	.	.
X.10	96	.	.	.	.	.	96	−4	−4	−4	.	.	.	.	.	.	.	.	.	.	.	.	.	.
X.11	96	.	.	.	.	.	96	−4	−4	−4	.	.	.	.	.	.	.	.	.	.	.	.	.	.
X.12	96	.	.	.	.	.	96	−4	−4	−4	.	.	.	.	.	.	.	.	.	.	.	.	.	.
X.13	96	.	.	.	.	.	96	−4	−4	−4	.	.	.	.	.	.	.	.	.	.	.	.	.	.
X.14	96	.	.	.	.	.	96	−4	−4	−4	.	.	.	.	.	.	.	.	.	.	.	.	.	.
X.15	96	.	.	.	.	.	96	−4	−4	−4	.	.	.	.	.	.	.	.	.	.	.	.	.	.
X.16	124	4	1	4	4	.	124	−1	−1	−1	1	−2	−2	4	−1	−1	−1	1	1	1	.	−1	−1	1
X.17	124	4	1	4	4	.	124	−1	−1	−1	1	2	2	4	−1	−1	−1	1	1	1	.	−1	−1	−1
X.18	124	4	1	−4	−4	.	124	−1	−1	−1	1	E	−E	4	−1	−1	−1	−1	−1	1	.	1	1	−D
X.19	124	4	1	−4	−4	.	124	−1	−1	−1	1	−E	E	4	−1	−1	−1	−1	−1	1	.	1	1	D
X.20	124	−4	1	B	−B	.	124	−1	−1	−1	−1	.	.	−4	1	1	1	−D	D	1	.	D	−D	G
X.21	124	−4	1	B	−B	.	124	−1	−1	−1	−1	.	.	−4	1	1	1	−D	D	1	.	D	−D	−G
X.22	124	−4	−2	B	−B	.	124	−1	−1	−1	2	.	.	−4	1	1	1	E	−E	−2	.	D	−D	.
X.23	124	−4	1	−B	B	.	124	−1	−1	−1	−1	.	.	−4	1	1	1	D	−D	1	.	−D	D	/G
X.24	124	−4	1	−B	B	.	124	−1	−1	−1	−1	.	.	−4	1	1	1	D	−D	1	.	−D	D	−/G
X.25	124	−4	−2	−B	B	.	124	−1	−1	−1	2	.	.	−4	1	1	1	−E	E	−2	.	−D	D	.
X.26	125	5	−1	5	5	1	125	.	.	.	−1	−1	−1	5	.	.	.	−1	−1	−1	1	.	.	−1
X.27	155	11	−1	−1	−1	−1	155	5	5	5	−1	1	1	11	1	1	1	−1	−1	−1	−1	−1	−1	1
X.28	155	−1	−1	C	/C	1	155	5	5	5	−1	−D	D	−1	−1	−1	−1	1	1	−1	1	−D	D	−D
X.29	155	−1	−1	/C	C	1	155	5	5	5	−1	D	−D	−1	−1	−1	−1	1	1	−1	1	D	−D	D
X.30	186	−6	.	6	6	−2	186	11	11	11	.	.	.	−6	−1	−1	−1	.	.	.	−2	1	1	.
X.31	620	4	−4	.	.	−4	−5	20	−5	−5	4	.	.	−1	4	−1	−1	.	.	1	1	.	.	.
X.32	1240	−8	4	.	.	.	−10	−10	15	−10	4	.	.	2	2	−3	2	.	.	−1	.	.	.	.
X.33	1240	−8	4	.	.	.	−10	−10	−10	15	4	.	.	2	2	2	−3	.	.	−1	.	.	.	.
X.34	1860	12	.	.	.	4	−15	10	10	−15	.	.	.	−3	2	2	−3	.	.	.	−1	.	.	.
X.35	1860	12	.	.	.	4	−15	10	−15	10	.	.	.	−3	2	−3	2	.	.	.	−1	.	.	.
X.36	2480	−16	−4	.	.	.	−20	−20	5	5	−4	.	.	4	4	−1	−1	.	.	1	.	.	.	.
X.37	2480	16	−4	.	.	.	−20	−20	5	5	4	.	.	−4	−4	1	1	.	.	1	.	.	.	.
X.38	3100	20	4	.	.	−4	−25	.	.	.	−4	.	.	−5	.	.	.	.	.	−1	1	.	.	.
X.39	3720	−24	.	.	.	.	−30	20	−5	−5	.	.	.	6	−4	1	1	.	.	.	.	.	.	.

Character table of $M = N_G(V) \cong 5^3.L_3(5)$ *(continued)*

2	3	3	3	.	1	.	.	.	.	.	.	.	.	.	.
3	1	1	1	.	1	.	.	.	.	.	.	.	.	.	.
5	.	.	.	2	1	.	.	.	.	.	.	.	.	.	.
31	.	.	.	.	.	1	1	1	1	1	1	1	1	1	1
	$24b$	$24c$	$24d$	$25a$	$30a$	$31a$	$31b$	$31c$	$31d$	$31e$	$31f$	$31g$	$31h$	$31i$	$31j$
$2P$	$12b$	$12a$	$12a$	$25a$	$15a$	$31d$	$31i$	$31j$	$31f$	$31h$	$31b$	$31c$	$31g$	$31a$	$31e$
$3P$	$8b$	$8a$	$8a$	$25a$	$10a$	$31e$	$31c$	$31d$	$31h$	$31b$	$31g$	$31a$	$31i$	$31j$	$31f$
$5P$	$24b$	$24c$	$24d$	$5a$	$6a$	$31a$	$31b$	$31c$	$31d$	$31e$	$31f$	$31g$	$31h$	$31i$	$31j$
$7P$	$24d$	$24a$	$24b$	$25a$	$30a$	$31f$	$31a$	$31e$	$31b$	$31g$	$31i$	$31j$	$31c$	$31d$	$31h$
$11P$	$24d$	$24a$	$24b$	$25a$	$30a$	$31c$	$31h$	$31i$	$31j$	$31d$	$31e$	$31b$	$31f$	$31g$	$31a$
$13P$	$24a$	$24d$	$24c$	$25a$	$30a$	$31e$	$31c$	$31d$	$31h$	$31b$	$31g$	$31a$	$31i$	$31j$	$31f$
$17P$	$24a$	$24d$	$24c$	$25a$	$30a$	$31j$	$31g$	$31a$	$31e$	$31f$	$31h$	$31i$	$31b$	$31c$	$31d$
$19P$	$24c$	$24b$	$24a$	$25a$	$30a$	$31d$	$31i$	$31j$	$31f$	$31h$	$31b$	$31c$	$31g$	$31a$	$31e$
$23P$	$24c$	$24b$	$24a$	$25a$	$30a$	$31j$	$31g$	$31a$	$31e$	$31f$	$31h$	$31i$	$31b$	$31c$	$31d$
$29P$	$24b$	$24c$	$24d$	$25a$	$30a$	$31g$	$31e$	$31b$	$31c$	$31a$	$31j$	$31f$	$31d$	$31h$	$31i$
$31P$	$24d$	$24a$	$24b$	$25a$	$30a$	$1a$	$1a$	$1a$	$1a$	$1a$	$1a$	$1a$	$1a$	$1a$	$1a$
$X.1$	1	1	1	1	1	1	1	1	1	1	1	1	1	1	1
$X.2$	.	.	.	.	.	−1	−1	−1	−1	−1	−1	−1	−1	−1	−1
$X.3$	−1	−1	−1	1	1	.	.	.	.	.	.	.	.	.	.
$X.4$	D	$-D$	$-D$	1	1	.	.	.	.	.	.	.	.	.	.
$X.5$	$-D$	D	D	1	1	.	.	.	.	.	.	.	.	.	.
$X.6$	.	.	.	1	.	H	J	$/L$	$/I$	$/K$	L	I	$/H$	K	$/J$
$X.7$	.	.	.	1	.	I	$/K$	J	$/L$	H	$/J$	L	$/I$	$/H$	K
$X.8$	.	.	.	1	.	J	$/I$	$/H$	K	$/L$	H	$/K$	$/J$	L	I
$X.9$	.	.	.	1	.	K	L	I	H	$/J$	$/I$	$/H$	$/K$	J	$/L$
$X.10$	.	.	.	1	.	$/H$	$/J$	L	I	K	$/L$	$/I$	H	$/K$	J
$X.11$	.	.	.	1	.	L	H	$/K$	J	I	K	$/J$	$/L$	$/I$	$/H$
$X.12$	.	.	.	1	.	$/K$	$/L$	$/I$	$/H$	J	I	H	K	$/J$	L
$X.13$	.	.	.	1	.	$/I$	K	$/J$	L	$/H$	J	$/L$	I	H	$/K$
$X.14$	.	.	.	1	.	$/J$	I	H	$/K$	L	$/H$	K	J	$/L$	$/I$
$X.15$	.	.	.	1	.	$/L$	$/H$	K	$/J$	$/I$	$/K$	J	L	I	H
$X.16$	1	1	1	−1	1	.	.	.	.	.	.	.	.	.	.
$X.17$	−1	−1	−1	−1	1	.	.	.	.	.	.	.	.	.	.
$X.18$	$-D$	D	D	−1	1	.	.	.	.	.	.	.	.	.	.
$X.19$	D	$-D$	$-D$	−1	1	.	.	.	.	.	.	.	.	.	.
$X.20$	$-G$	$-/G$	$/G$	−1	−1	.	.	.	.	.	.	.	.	.	.
$X.21$	G	$/G$	$-/G$	−1	−1	.	.	.	.	.	.	.	.	.	.
$X.22$	.	.	.	−1	2	.	.	.	.	.	.	.	.	.	.
$X.23$	$-/G$	$-G$	G	−1	−1	.	.	.	.	.	.	.	.	.	.
$X.24$	$/G$	G	$-G$	−1	−1	.	.	.	.	.	.	.	.	.	.
$X.25$	.	.	.	−1	2	.	.	.	.	.	.	.	.	.	.
$X.26$	−1	−1	−1	.	−1	1	1	1	1	1	1	1	1	1	1
$X.27$	1	1	1	.	−1	.	.	.	.	.	.	.	.	.	.
$X.28$	$-D$	D	D	.	−1	.	.	.	.	.	.	.	.	.	.
$X.29$	D	$-D$	$-D$	.	−1	.	.	.	.	.	.	.	.	.	.
$X.30$	.	.	.	1	.	.	.	.	.	.	.	.	.	.	.
$X.31$	.	.	.	.	−1	.	.	.	.	.	.	.	.	.	.
$X.32$	.	.	.	.	−1	.	.	.	.	.	.	.	.	.	.
$X.33$	.	.	.	.	−1	.	.	.	.	.	.	.	.	.	.
$X.34$	.	.	.	.	.	.	.	.	.	.	.	.	.	.	.
$X.35$	.	.	.	.	.	.	.	.	.	.	.	.	.	.	.
$X.36$	.	.	.	.	1	.	.	.	.	.	.	.	.	.	.
$X.37$	.	.	.	.	−1	.	.	.	.	.	.	.	.	.	.
$X.38$	.	.	.	.	1	.	.	.	.	.	.	.	.	.	.
$X.39$	.	.	.	.	.	.	.	.	.	.	.	.	.	.	.

where $A = -1-6i$, $B = -4i$, $C = -5-6i$, $D = i$, $E = 2i$, $F = -1-i$, $G = \xi^{11} - \xi^{19}$, $H = \rho^{17} + \rho^{22} + \rho^{23}$, $I = \rho^{16} + \rho^{18} + \rho^{28}$, $J = \rho^{12} + \rho^{21} + \rho^{29}$, $K = \rho^{11} + \rho^{24} + \rho^{27}$, $L = \rho^{6} + \rho^{26} + \rho^{30}$, *and* ξ *and* ρ *are primitive* $24th$ *and* $31st$ *roots of unity, respectively.*

13.7.7 *Character table of* $L = N_R(V) \cong 5^3.(5^2 : \mathrm{GL}_2(5))$

2	5	5	4	3	5	5	4	4	4	4	4	3	2	2	1	1	2	1	1	3	3	3	3	2	2
3	1	1	.	1	1	1	.	.	.	.	.	1	.	.	.	.	.	.	.	1	1	1	1	.	.
5	6	2	2	1	1	1	1	1	1	1	1	6	5	4	4	4	3	3	3	1	.	.	2	2	2
	1a	2a	2b	3a	4a	4b	4c	4d	4e	4f	4g	5a	5b	5c	5d	5e	5f	5g	5h	6a	8a	8b	10a	10b	10c
2P	1a	1a	1a	3a	2a	2a	2b	2b	2a	2b	2b	5a	5b	5c	5d	5e	5f	5g	5h	3a	4a	4b	5a	5b	5c
3P	1a	2a	2b	1a	4b	4a	4g	4f	4e	4d	4c	5a	5b	5c	5d	5e	5f	5g	5h	2a	8b	8a	10a	10b	10c
5P	1a	2a	2b	3a	4a	4b	4c	4d	4e	4f	4g	1a	1a	1a	1a	1a	1a	1a	1a	6a	8a	8b	2a	2b	2b
7P	1a	2a	2b	3a	4b	4a	4g	4f	4e	4d	4c	5a	5b	5c	5d	5e	5f	5g	5h	6a	8b	8a	10a	10b	10c
11P	1a	2a	2b	3a	4b	4a	4g	4f	4e	4d	4c	5a	5b	5c	5d	5e	5f	5g	5h	6a	8b	8a	10a	10b	10c
13P	1a	2a	2b	3a	4a	4b	4c	4d	4e	4f	4g	5a	5b	5c	5d	5e	5f	5g	5h	6a	8a	8b	10a	10b	10c
17P	1a	2a	2b	3a	4a	4b	4c	4d	4e	4f	4g	5a	5b	5c	5d	5e	5f	5g	5h	6a	8a	8b	10a	10b	10c
19P	1a	2a	2b	3a	4b	4a	4g	4f	4e	4d	4c	5a	5b	5c	5d	5e	5f	5g	5h	6a	8b	8a	10a	10b	10c
23P	1a	2a	2b	3a	4b	4a	4g	4f	4e	4d	4c	5a	5b	5c	5d	5e	5f	5g	5h	6a	8b	8a	10a	10b	10c
29P	1a	2a	2b	3a	4a	4b	4c	4d	4e	4f	4g	5a	5b	5c	5d	5e	5f	5g	5h	6a	8a	8b	10a	10b	10c
X.1	1	1	1	1	1	1	1	1	1	1	1	1	1	1	1	1	1	1	1	1	1	1	1	1	1
X.2	1	1	−1	1	−1	−1	C	C	1	−C	−C	1	1	1	1	1	1	1	1	1	−C	C	1	−1	−1
X.3	1	1	1	1	1	1	−1	−1	1	−1	−1	1	1	1	1	1	1	1	1	1	−1	−1	1	1	1
X.4	1	1	−1	1	−1	−1	−C	−C	1	C	C	1	1	1	1	1	1	1	1	1	C	−C	1	−1	−1
X.5	4	4	.	1	4	4	.	.	.	.	.	4	4	4	4	4	−1	−1	−1	1	2	2	4	.	.
X.6	4	4	.	1	4	4	.	.	.	.	.	4	4	4	4	4	−1	−1	−1	1	−2	−2	4	.	.
X.7	4	4	.	1	−4	−4	.	.	.	.	.	4	4	4	4	4	−1	−1	−1	1	E	−E	4	.	.
X.8	4	4	.	1	−4	−4	.	.	.	.	.	4	4	4	4	4	−1	−1	−1	1	−E	E	4	.	.
X.9	4	−4	.	1	A	−A	.	.	.	.	.	4	4	4	4	4	−1	−1	−1	−1	.	.	−4	.	.
X.10	4	−4	.	1	A	−A	.	.	.	.	.	4	4	4	4	4	−1	−1	−1	−1	.	.	−4	.	.
X.11	4	−4	.	−2	A	−A	.	.	.	.	.	4	4	4	4	4	−1	−1	−1	2	.	.	−4	.	.
X.12	4	−4	.	1	−A	A	.	.	.	.	.	4	4	4	4	4	−1	−1	−1	−1	.	.	−4	.	.
X.13	4	−4	.	1	−A	A	.	.	.	.	.	4	4	4	4	4	−1	−1	−1	−1	.	.	−4	.	.
X.14	4	−4	.	−2	−A	A	.	.	.	.	.	4	4	4	4	4	−1	−1	−1	2	.	.	−4	.	.
X.15	5	5	1	−1	5	5	−1	−1	1	−1	−1	5	5	5	5	5	.	.	.	−1	1	1	5	1	1
X.16	5	5	1	−1	5	5	1	1	1	1	1	5	5	5	5	5	.	.	.	−1	−1	−1	5	1	1
X.17	5	5	−1	−1	−5	−5	C	C	1	−C	−C	5	5	5	5	5	.	.	.	−1	C	−C	5	−1	−1
X.18	5	5	−1	−1	−5	−5	−C	−C	1	C	C	5	5	5	5	5	.	.	.	−1	−C	C	5	−1	−1
X.19	6	6	2	.	−6	−6	.	.	−2	.	.	6	6	6	6	6	1	1	1	.	.	.	6	2	2
X.20	6	6	−2	.	6	6	.	.	−2	.	.	6	6	6	6	6	1	1	1	.	.	.	6	−2	−2
X.21	6	−6	.	.	B	−B	D	−D	.	−/D	/D	6	6	6	6	6	1	1	1	.	.	.	−6	.	.
X.22	6	−6	.	.	B	−B	−D	D	.	/D	−/D	6	6	6	6	6	1	1	1	.	.	.	−6	.	.
X.23	6	−6	.	.	−B	B	−/D	/D	.	D	−D	6	6	6	6	6	1	1	1	.	.	.	−6	.	.
X.24	6	−6	.	.	−B	B	/D	−/D	.	−D	D	6	6	6	6	6	1	1	1	.	.	.	−6	.	.
X.25	24	.	4	.	.	.	−4	.	.	.	−4	24	24	−1	−1	−1	4	4	4	.	.	.	.	4	−1
X.26	24	.	4	.	.	.	4	.	.	.	4	24	24	−1	−1	−1	4	4	4	.	.	.	.	4	−1
X.27	24	.	−4	.	.	.	−A	.	.	.	A	24	24	−1	−1	−1	4	4	4	.	.	.	.	−4	1
X.28	24	.	−4	.	.	.	A	.	.	.	−A	24	24	−1	−1	−1	4	4	4	.	.	.	.	−4	1
X.29	96	.	.	.	.	.	.	.	.	.	.	96	96	−4	−4	−4	−4	−4	−4	.	.	.	.	.	.
X.30	100	4	.	4	.	.	.	.	4	.	.	−25	.	.	.	.	.	5	−5	4	.	.	−1	.	.
X.31	120	.	4	.	.	.	.	4	.	4	.	120	−5	20	−5	−5	.	.	.	.	.	.	.	−1	4
X.32	120	.	4	.	.	.	.	−4	.	−4	.	120	−5	20	−5	−5	.	.	.	.	.	.	.	−1	4
X.33	120	.	−4	.	.	.	.	A	.	−A	.	120	−5	20	−5	−5	.	.	.	.	.	.	.	1	−4
X.34	120	.	−4	.	.	.	.	−A	.	A	.	120	−5	20	−5	−5	.	.	.	.	.	.	.	1	−4
X.35	200	−8	.	−4	.	.	.	.	.	.	.	−50	.	.	.	.	10	−5	.	4	.	.	2	.	.
X.36	200	−8	.	−4	.	.	.	.	.	.	.	−50	.	.	.	.	−10	.	5	4	.	.	2	.	.
X.37	240	.	8	.	.	.	.	.	.	.	.	240	−10	−10	15	−10	.	.	.	.	.	.	.	−2	−2
X.38	240	.	−8	.	.	.	.	.	.	.	.	240	−10	−10	15	−10	.	.	.	.	.	.	.	2	2
X.39	240	.	8	.	.	.	.	.	.	.	.	240	−10	−10	−10	15	.	.	.	.	.	.	.	−2	−2
X.40	240	.	−8	.	.	.	.	.	.	.	.	240	−10	−10	−10	15	.	.	.	.	.	.	.	2	2
X.41	300	12	.	.	.	.	.	.	−4	.	.	−75	.	.	.	.	10	.	−5	.	.	.	−3	.	.
X.42	300	12	.	.	.	.	.	.	−4	.	.	−75	.	.	.	.	−10	5	.	.	.	.	−3	.	.
X.43	400	−16	.	4	.	.	.	.	.	.	.	−100	.	.	.	.	.	−5	5	−4	.	.	4	.	.
X.44	400	16	.	4	.	.	.	.	.	.	.	−100	.	.	.	.	.	−5	5	4	.	.	−4	.	.
X.45	500	20	.	−4	.	.	.	.	4	.	.	−125	.	.	.	.	.	.	.	−4	.	.	−5	.	.
X.46	600	−24	.	.	.	.	.	.	.	.	.	−150	.	.	.	.	.	5	−5	.	.	.	6	.	.

Character table of $L = N_R(V) \cong 5^3.(5^2 : \mathrm{GL}_2(5))$ *(continued)*

	10d	10e	10f	10g	10h	12a	12b	15a	20a	20b	20c	20d	20e	20f	20g	24a	24b	24c	24d	25a	30a
2	2	1	1	1	1	3	3	1	2	2	2	2	2	2	2	3	3	3	3	.	1
3	.	.	.	.	.	1	1	1	.	.	.	.	.	.	.	1	1	1	1	.	1
5	2	2	2	2	2	.	.	1	1	1	1	1	1	1	1	.	.	.	.	2	1
2P	5f	5g	5d	5e	5h	6a	6a	15a	10a	10b	10d	10c	10c	10b	10d	12a	12b	12b	12a	25a	15a
3P	10d	10e	10f	10g	10h	4b	4a	5a	20a	20f	20g	20e	20d	20b	20c	8b	8a	8a	8b	25a	10a
5P	2a	2a	2b	2b	2a	12a	12b	3a	4e	4f	4b	4g	4c	4d	4a	24a	24b	24c	24d	5a	6a
7P	10d	10e	10f	10g	10h	12b	12a	15a	20a	20f	20g	20e	20d	20b	20c	24c	24d	24a	24b	25a	30a
11P	10d	10e	10f	10g	10h	12b	12a	15a	20a	20f	20g	20e	20d	20b	20c	24c	24d	24a	24b	25a	30a
13P	10d	10e	10f	10g	10h	12a	12b	15a	20a	20b	20c	20d	20e	20f	20g	24d	24c	24b	24a	25a	30a
17P	10d	10e	10f	10g	10h	12a	12b	15a	20a	20b	20c	20d	20e	20f	20g	24d	24c	24b	24a	25a	30a
19P	10d	10e	10f	10g	10h	12b	12a	15a	20a	20f	20g	20e	20d	20b	20c	24b	24a	24d	24c	25a	30a
23P	10d	10e	10f	10g	10h	12b	12a	15a	20a	20f	20g	20e	20d	20b	20c	24b	24a	24d	24c	25a	30a
29P	10d	10e	10f	10g	10h	12a	12b	15a	20a	20b	20c	20d	20e	20f	20g	24a	24b	24c	24d	25a	30a
X.1	1	1	1	1	1	1	1	1	1	1	1	1	1	1	1	1	1	1	1	1	1
X.2	1	1	−1	−1	1	−1	−1	1	1	−C	−1	−C	C	C	−1	−C	C	C	−C	1	1
X.3	1	1	1	1	1	1	1	1	1	−1	1	−1	−1	−1	1	−1	−1	−1	−1	1	1
X.4	1	1	−1	−1	1	−1	−1	1	1	C	−1	C	−C	−C	−1	C	−C	−C	C	1	1
X.5	−1	−1	.	.	−1	1	1	1	.	.	−1	.	.	.	−1	−1	−1	−1	−1	−1	1
X.6	−1	−1	.	.	−1	1	1	1	.	.	−1	.	.	.	−1	1	1	1	1	−1	1
X.7	−1	−1	.	.	−1	−1	−1	1	.	.	1	.	.	.	1	−C	C	C	−C	−1	1
X.8	−1	−1	.	.	−1	−1	−1	1	.	.	1	.	.	.	1	C	−C	−C	C	−1	1
X.9	1	1	.	.	1	−C	C	1	.	.	−C	.	.	.	C	F	/F	−/F	−F	−1	−1
X.10	1	1	.	.	1	−C	C	1	.	.	−C	.	.	.	C	−F	−/F	/F	F	−1	−1
X.11	1	1	.	.	1	E	−E	−2	.	.	−C	.	.	.	C	.	.	.	.	−1	2
X.12	1	1	.	.	1	C	−C	1	.	.	C	.	.	.	−C	−/F	−F	F	/F	−1	−1
X.13	1	1	.	.	1	C	−C	1	.	.	C	.	.	.	−C	/F	F	−F	−/F	−1	−1
X.14	1	1	.	.	1	−E	E	−2	.	.	C	.	.	.	−C	.	.	.	.	−1	2
X.15	.	.	1	1	.	−1	−1	−1	1	−1	.	−1	−1	−1	.	1	1	1	1	.	−1
X.16	.	.	1	1	.	−1	−1	−1	1	1	.	1	1	1	.	−1	−1	−1	−1	.	−1
X.17	.	.	−1	−1	.	1	1	−1	1	−C	.	−C	C	C	.	C	−C	−C	C	.	−1
X.18	.	.	−1	−1	.	1	1	−1	1	C	.	C	−C	−C	.	−C	C	C	−C	.	−1
X.19	1	1	2	2	1	.	.	.	−2	.	−1	.	.	.	−1	.	.	.	.	1	.
X.20	1	1	−2	−2	1	.	.	.	−2	.	1	.	.	.	1	.	.	.	.	1	.
X.21	−1	−1	.	.	−1	.	.	.	.	−/D	−C	/D	D	−D	C	.	.	.	.	1	.
X.22	−1	−1	.	.	−1	.	.	.	.	/D	−C	−/D	−D	D	C	.	.	.	.	1	.
X.23	−1	−1	.	.	−1	.	.	.	.	D	C	−D	−/D	/D	−C	.	.	.	.	1	.
X.24	−1	−1	.	.	−1	.	.	.	.	−D	C	D	/D	−/D	−C	.	.	.	.	1	.
X.25	.	.	−1	−1	.	.	.	.	.	.	.	1	1	.	.	.	.	.	.	−1	.
X.26	.	.	−1	−1	.	.	.	.	.	.	.	−1	−1	.	.	.	.	.	.	−1	.
X.27	.	.	1	1	.	.	.	.	.	.	.	C	−C	.	.	.	.	.	.	−1	.
X.28	.	.	1	1	.	.	.	.	.	.	.	−C	C	.	.	.	.	.	.	−1	.
X.29	.	.	.	.	.	.	.	.	.	.	.	.	.	.	.	.	.	.	.	1	.
X.30	4	−1	.	.	−1	.	.	−1	−1	.	.	.	.	.	.	.	.	.	.	.	−1
X.31	.	.	−1	−1	.	.	.	.	.	−1	.	.	.	−1	.	.	.	.	.	.	.
X.32	.	.	−1	−1	.	.	.	.	.	1	.	.	.	1	.	.	.	.	.	.	.
X.33	.	.	1	1	.	.	.	.	.	−C	.	.	.	C	.	.	.	.	.	.	.
X.34	.	.	1	1	.	.	.	.	.	C	.	.	.	−C	.	.	.	.	.	.	.
X.35	2	−3	.	.	2	.	.	1	.	.	.	.	.	.	.	.	.	.	.	.	−1
X.36	2	2	.	.	−3	.	.	1	.	.	.	.	.	.	.	.	.	.	.	.	−1
X.37	.	.	3	−2	.	.	.	.	.	.	.	.	.	.	.	.	.	.	.	.	.
X.38	.	.	−3	2	.	.	.	.	.	.	.	.	.	.	.	.	.	.	.	.	.
X.39	.	.	−2	3	.	.	.	.	.	.	.	.	.	.	.	.	.	.	.	.	.
X.40	.	.	2	−3	.	.	.	.	.	.	.	.	.	.	.	.	.	.	.	.	.
X.41	2	2	.	.	−3	.	.	.	1	.	.	.	.	.	.	.	.	.	.	.	.
X.42	2	−3	.	.	2	.	.	.	1	.	.	.	.	.	.	.	.	.	.	.	.
X.43	4	−1	.	.	−1	.	.	−1	.	.	.	.	.	.	.	.	.	.	.	.	1
X.44	−4	1	.	.	1	.	.	−1	.	.	.	.	.	.	.	.	.	.	.	.	−1
X.45	.	.	.	.	.	.	.	1	−1	.	.	.	.	.	.	.	.	.	.	.	1
X.46	−4	1	.	.	1	.	.	.	.	.	.	.	.	.	.	.	.	.	.	.	.

where $A = -4i$, $B = 6i$, $C = -i$, $D = -i - 1$, $E = 2i$, $F = 2\xi^{13} + \xi^5$ *and* ξ *is a primitive 24th root of unity.*

14
Suzuki's group Suz

In 1969 M. Suzuki discovered his sporadic simple group Suz. It is a primitive rank 3 permutation group [124]. Soicher's nice presentation of Suz is stated in [110], p. 98. It also provides generators for the stabilizer $T \cong G_2(4)$ of the rank 3 permutation representation of degree 1782 of Suz. In 1974 Wright determined the character table of Suz [139]. In particular, he showed that Suz has two conjugacy classes of involutions. Furthermore, he determined the abstract group structure of the centralizer $H = C_{\mathsf{Suz}}(z)$ of a 2-central involution z and of some other local subgroups of Suz. In [108] Patterson and Wong showed that each finite simple group G with a centralizer $C_G(z) \cong H$ of a 2-central involution $z \in G$ is isomorphic to Suzuki's simple group Suz. Their uniqueness proof is not self-contained.

This chapter provides another existence and uniqueness proof for Suzuki's sporadic group. It is self-contained except for a quotation of Phan's uniqueness theorem of $U_4(3)$ of [109] which can also be proved completely by means of the methods and results of [92].

The sporadic Suzuki group can be constructed by means of Algorithm 1.3.8 from an iterated extension E of an irreducible subgroup $L \cong 3A_6$ of $GL_6(2)$. Let E_1 be the split extension of L by the natural 6-dimensional vector space V_1 over $GF(2)$. Then E_1 has an irreducible 4-dimensional representation V_2 over $GF(2)$ on which the center of L acts trivially. The group E is a non-split extension of E_1 by V_2; see Proposition 14.2.2. It turns out that V_2 is the unique maximal elementary abelian normal subgroup A of any Sylow 2-subgroup of E.

However, the group E is not uniquely determined by the irreducible subgroup L of $GL_6(2)$. In view of the limited space we do not dismiss here all the extensions which do not lead to a simple group. Instead we start our construction of a simple group G of Suz-type from a presentation of a finite group H which is isomorphic to the centralizer $C_G(z)$ of a 2-central involution z of Suzuki's sporadic group Suz. It is stated in Lemma 14.1.1 of Section 14.1. A classification of the conjugacy classes and the character table of H are derived from this presentation also. The presentation of H is due to A. Stein (University of Kiel).

Using Algorithm 7.4.8 of [92] we construct in Section 14.2 the correct iterated extension E which is isomorphic to $N_G(A)$. In order to prove that $N_G(A)$ is uniquely determined by the given presentation of H we also show in Section 14.2 that any simple group G with such a 2-central involution centralizer has two conjugacy classes of involutions represented by z and u. Thus we are able to determine the fusion of the conjugacy classes of H and $E = N_G(A)$ in G; see Proposition 14.2.2. Furthermore, we derive in Proposition 14.2.3 a presentation of $U = C_G(u)$, its character table and the fusion of the conjugacy classes of U in G. Thus Theorem 1.6.4 allows us to show that all simple groups G of Suz-type have order $|G| = 2^{13} \cdot 3^7 \cdot 5^2 \cdot 7 \cdot 11 \cdot 13$; see Proposition 14.2.4.

In Section 14.4 we apply Algorithm 7.4.8 of [92] to H and construct a simple subgroup $\mathfrak{G}$ of $\mathrm{GL}_{143}(13)$ having a 2-central involution $\mathfrak{z}$ such that $C_{\mathfrak{G}}(\mathfrak{z}) \cong H$; see Theorem 14.3.1. Using the algorithm described in Theorem 6.2.1 of [92] we constructed a faithful permutation representation $P\mathfrak{G}$ of $\mathfrak{G}$ of degree 135135. The character table of $\mathfrak{G}$ has been calculated by means of $P\mathfrak{G}$ and MAGMA. It is equivalent to that of Suz given in the Atlas [19], p. 128.

In Section 14.5 it is shown that any finite simple group G having a 2-central involution z such that $C_G(z) \cong H$ is isomorphic to $\mathfrak{G}$; see Theorem 14.4.10. In particular, $\mathfrak{G} \cong$ Suz. Hence Patterson's and Wong's Uniqueness Theorem is a consequence of Theorem 14.4.10.

Sections 14.6 and 14.7 contain the systems of representatives of the conjugacy classes and the character tables of the local subgroups dealt with in this chapter. They are essential tools for the given existence and uniqueness proofs of Suz. The four generating matrices of $\mathfrak{G}$ are documented on the accompanying DVD.

14.1 The centralizer of a 2-central involution

The presentation of the following group H is due to A. Stein (University of Kiel). So is the faithful permutation representation PH of H. Several properties of the abstract group H were observed first by Patterson and Wong [108], Suzuki [124] and Wright [139].

Lemma 14.1.1 *Let $H = \langle h_i \mid 1 \le i \le 16\rangle$ be the finite group with the following set $\mathcal{R}(H)$ of defining relations:*

$$h_1^2 = h_2^2 = h_5^2 = h_6^2 = h_7^2 = h_{10}^2 = h_{11}^2 = h_{12}^2 = h_{13}^3 = h_{14}^2 = h_{15}^3 = 1,$$
$$h_3^2 = h_4^2 = h_{16}, \quad h_8^2 = h_9^2 = h_1,$$
$$h_{16}^2 = 1 \quad \textit{and} \quad (h_{16}, h_i) = 1 \quad \textit{for all} \quad 1 \le i \le 15,$$
$$(h_1,h_2) = (h_1,h_3) = (h_1,h_4) = (h_1,h_5) = (h_1,h_7) = 1, \quad (h_1,h_6) = h_{16},$$
$$(h_2,h_3) = (h_2,h_4) = (h_2,h_6) = (h_2,h_7) = 1, \quad (h_2,h_5) = (h_3,h_4) = h_{16},$$
$$(h_3,h_5) = (h_3,h_6) = (h_3,h_7) = (h_4,h_5) = (h_4,h_6) = (h_4,h_7) = 1,$$
$$(h_5,h_6) = (h_1,h_8) = (h_2,h_8) = (h_3,h_8) = (h_5,h_8) = 1,$$
$$(h_1,h_9) = (h_2,h_9) = (h_4,h_9) = (h_5,h_9) = 1,$$
$$(h_7,h_8) = (h_7,h_9) = (h_8,h_{10}) = (h_9,h_{10}) = (h_{11},h_{12}) = (h_7,h_{11}) = 1,$$
$$(h_7,h_{12}) = (h_7,h_{13}) = (h_{10},h_{13}) = (h_{13},h_{15}) = 1,$$
$$(h_1,h_{10}) = (h_3,h_{10}) = (h_4,h_{10}) = (h_5,h_{10}) = (h_1,h_{11}) = (h_2,h_{11}) = 1,$$
$$(h_6,h_{11} = (h_1,h_{12}) = (h_2,h_{12}) = (h_6,h_{12}) = (h_1,h_{13}) = (h_2,h_{13}) = 1,$$
$$(h_5,h_{13}) = (h_6,h_{13}) = (h_1,h_{14}) = (h_6,h_{14}) = (h_3,h_{15}) = (h_4,h_{15}) = 1,$$
$$(h_7,h_{10}) = (h_8,h_9) = (h_9,h_{11}) = (h_8,h_{12}) = h_1,$$
$$h_{11}^{-1}h_8h_{11}h_7h_8 = h_{11}^{-1}h_{10}h_{11}h_9h_{10}h_7 = h_9^{-1}h_{12}^{-1}h_9h_{12}h_7 = 1,$$
$$h_{10}^{-1}h_{12}^{-1}h_{10}h_{12}h_7h_8 = h_{13}^{-1}h_9h_{13}h_8 = h_{13}^{-1}h_{11}h_{13}h_{12} = 1,$$
$$h_{12}^{-1}h_{13}^{-1}h_{12}h_{13}h_{11} = h_{14}^{-1}h_7h_{14}h_{10} = h_{14}^{-1}h_{10}h_{14}h_7 = 1,$$
$$h_8^{-1}h_{13}^{-1}h_8h_{13}h_9 = h_{14}^{-1}h_8h_{14}h_9 = h_{14}^{-1}h_9h_{14}h_8 = h_1,$$
$$(h_{11}h_{14})^5 = (h_{12}h_{14})^5 = 1, \quad (h_{13}h_{14})^2 = 1, \quad (h_{10}h_{15})^2 = 1,$$
$$h_{14}^{-1}h_{11}^{-1}h_{12}^{-1}h_{11}h_{12}h_{14} = h_{13}^{-1}h_{14}^{-1}h_{11}h_{14}h_{13}h_{14}h_{11}h_{12}h_{14} = 1,$$
$$(h_7,h_{15}) = h_1h_{16}, \quad h_8^{-1}h_{15}^{-1}h_8h_{15}h_7h_{12} = h_4,$$
$$h_9^{-1}h_{15}^{-1}h_9h_{15}h_7h_{11} = h_2h_3h_4, \quad h_{15}^{-1}h_{11}h_{15}h_9 = h_3h_4,$$
$$h_{15}^{-1}h_{12}h_{15}h_8 = h_1h_4, \quad (h_{15},h_{14},h_{15}) = h_1h_{16},$$
$$h_5^{h_7} = h_1h_5h_{16}, \quad h_6^{h_7} = h_2h_6, \quad h_4^{h_8} = h_1h_4h_{16}, \quad h_6^{h_8} = h_1h_3h_6,$$
$$h_3^{h_9} = h_1h_3h_{16}, \quad h_6^{h_9} = h_1h_4h_6, \quad h_2^{h_{10}} = h_1h_2h_{16}, \quad h_6^{h_{10}} = h_5h_6,$$
$$h_3^{h_{11}} = h_2h_3h_{16}, \quad h_4^{h_{11}} = h_4h_{16}, \quad h_5^{h_{11}} = h_2h_4h_5, \quad h_3^{h_{12}} = h_3h_{16},$$
$$h_4^{h_{12}} = h_2h_4, \quad h_5^{h_{12}} = h_2h_3h_5h_{16}, \quad h_3^{h_{13}} = h_3h_4h_{16}, \quad h_4^{h_{13}} = h_3h_{16},$$
$${h_1}^{h_{15}} = h_1h_2, \quad {h_2}^{h_{15}} = h_1h_{16}, \quad {h_5}^{h_{15}} = h_5h_6,$$
$$(h_5,h_{13}) = (h_6,h_{13}) = (h_1,h_{14}) = (h_6,h_{14}) = (h_3,h_{15}) = (h_4,h_{15}) = 1,$$
$$h_2^{h_{14}} = h_5, \quad h_3^{h_{14}} = h_4, \quad h_4^{h_{14}} = h_3, \quad h_5^{h_{14}} = h_2, \quad h_6^{h_{15}} = h_5.$$

Then the following statements hold.

(a) *H has a faithful permutation representation of degree* 1728 *with stabilizer* $H_0 = \langle h_1, h_7, h_8, h_9, h_{11}, h_{12}, h_{13}, h_{14}\rangle$ *of order* $2^7 \cdot 3 \cdot 5$.

(b) *The subgroup* $S = \langle h_i \mid 1 \le i \le 12\rangle$ *is a Sylow* 2*-subgroup of* H *with center* $Z(S) = \langle z\rangle$, *where* $z = h_{16}$. *Furthermore,* $Z(H) = Z(S)$ *is the center of* H.

(c) $V = \langle z, t\rangle$ *is the unique Klein 4-subgroup of* S, *where* $t = h_1$.

(d) $C_H(V)$ *has center* $Z[C_H(V)] = V$ *and it has a complement of order* 2 *in* $N_H(V)$.

(e) $A = \langle z, t, h_2, h_7\rangle$ *is the unique maximal elementary abelian normal subgroup of* S, *and its order is* 16.

(f) $C = C_S(A) = \langle z, h_1, h_2, h_1, h_3, h_4, h_7, h_8, h_{11}, h_{12}\rangle$ *is a special* 2*-group of order* 2^{10} *with center* $Z(C) = A$, *and* V *is the unique normal Klein 4-group of* C.

(g) $D = N_H(A) = \langle C, h_5, h_6, h_{10}, h_{15}\rangle$ *has order* $|D| = 2^{13} \cdot 3^2$, C *is normal in* D *and has a complement* $K = \langle h_5, h_4, h_{13}, h_{15}\rangle$ *in* D; *its center* $Z(K) = \langle h_{13}\rangle$ *has order* 3 *and* $K/Z(K) \cong D/C \cong S_4$.

(h) *H is a perfect non-split extension of* $U_4(2)$ *by its extra-special normal subgroup* $P = O_2(H) = \langle z, h_1, ..., h_6\rangle$ *of order* $|P| = 128$. *Furthermore,* P *and* $Z(H) = Z(P)$ *are the only non-trivial proper normal subgroups of* H.

(i) $D = \langle x, y\rangle$, *where* $x = h_8h_{13}h_5h_{10}h_{15}$ *and* $y = h_6h_{12}h_{11}h_{15}$ *have orders* 24 *and* 12, *respectively.*

(j) $H = \langle x, y, h\rangle$, *where* $h = h_{14}$ *has order 2.*

(k) *A system of representatives* h_i *and the corresponding centralizer orders* $|C_H(h_i)|$ *of the* 57 *conjugacy classes* h_i^H *of* $H = \langle x, y, h\rangle$ *are given in Table 14.5.1.*

(l) *A system of representatives* d_i *and the corresponding centralizer orders* $|C_D(d_i)|$ *of the* 77 *conjugacy classes* d_i^D *of* $D = \langle x, y\rangle$ *are given in Table 14.5.3*

(m) *The character table of H is given in Table 14.6.1.*

(n) *The character table of D is given in Table 14.6.3.*

Proof (a) The permutation representation of H given in (i) is due to A. Stein. Using MAGMA it has been checked that the permutation representation PH of H is faithful.

(c) and (e) The reader can verify these two statements using PH and the MAGMA command

```
Subgroups(X : Al:=Normal,IsElementaryAbelian :=true),
```

where $X \in \{S, C\}$.

(k) and (l) The author used PH and Kratzer's Algorithm 5.3.18 of [92] to calculate the tables of the representatives of the conjugacy classes of H and D.

(m) and (n) Both character tables have been computed by means of MAGMA and the given permutation representations.

All other statements have been established by using standard MAGMA commands and the faithful permutation representation PH of H. □

Definition 14.1.2 A finite simple group G is said to be of Suz-*type* if it has a 2-central involution z such that $C_G(z)$ is isomorphic to the finitely presented group H stated in Lemma 14.1.1.

14.2 Even conjugacy classes and group order

In this section we determine the conjugacy classes of even order and the order of a finite simple group G of Suz-type.

Proposition 14.2.1 *Let G be a finite simple group of* Suz-*type. Let $A = \langle z, h_1, h_2, h_7 \rangle$ be the unique maximal elementary abelian normal subgroup of the Sylow 2-subgroup S of $H = C_G(z)$ defined in Lemma 14.1.1. Then the following statements hold.*

(a) *H has four conjugacy classes of involutions. They are represented by the generators z, $t = h_1$, h_7 and h_{14} of H defined in Lemma 14.1.1.*
(b) *$T = C_H(h_{14})$ is a self-normalizing subgroup of the centralizer $U = C_G(h_{14})$, and $|T| = 2^9$.*
(c) *T is a Sylow 2-subgroup of U.*
(d) *$z, t, h_7 \notin h_{14}^G$ and $z^G = t^G$ or $z^G = h_7^G$.*
(e) *If $z^g = t$ for some $g \in G$ then $g \in N_G(V) - N_H(V)$ and $N_G(V)/C_G(V) = N_G(V)/N_H(V) \cong \mathrm{GL}_2(2)$.*
(f) *If $z^g = t$ for some $g \in G$ then $N_G(V)$ is isomorphic to the finitely presented group $M = \langle m_i | 1 \le i \le 8 \rangle$ with the following set $\mathcal{R}(M)$*

of defining relations:

$$
\begin{aligned}
&m_1^3 = m_2^2 = m_3^3 = m_4^3 = m_5^3 = m_7^2 = m_8^2 = 1, \quad m_6^2 m_7 = 1,\\
&(m_1, m_3) = (m_1, m_4) = (m_1, m_5) = (m_2, m_3) = (m_2, m_4) = 1,\\
&(m_2, m_7) = (m_3, m_7) = (m_4, m_7) = (m_5, m_7) = (m_6, m_7) = 1,\\
&(m_3, m_8) = (m_4, m_)) = (m_5, m_8) = (m_6, m_8) = (m_7, m_8) = 1,\\
&(m_2 m_6)^4 = (m_4^{-1} m_6 m_3^{-1} m_6)^2 = (m_1^{-1} m_2)^2 = 1,\\
&(m_3^{-1} m_4^{-1})^2 = 1, \quad (m_4^{-1} m_5^{-1})^2 = 1, \quad (m_6 m_1^{-1})^3 = 1,\\
&m_1^{-1} m_7 m_1 m_8^{-1} = m_1^{-1} m_8 m_1 m_7^{-1} m_8^{-1} = 1,\\
&m_2^{-1} m_8 m_2 m_7^{-1} m_8^{-1} = 1,\\
&m_2 m_5^{-1} m_2 m_5 = m_3 m_5^{-1} m_3^{-1} m_5 m_4^{-1} = 1,\\
&(m_4 m_6 m_5)^2 = (m_6 m_1^{-1})^3 = m_2 m_6 m_5^{-1} m_4^{-1} m_2 m_6 m_4 m_5 m_7^{-1} = 1,\\
&m_4^{-1} m_6 m_2 m_6 m_4 m_6 m_2 m_6 = m_3 m_6 m_3^{-1} m_6 m_3 m_6 m_3^{-1} m_6 m_8^{-1} = 1,\\
&m_1^{-1} m_6 m_3^{-1} m_6 m_2 m_1^{-1} m_6 m_3 m_6 m_2 m_7^{-1} m_8^{-1} = 1,\\
&m_6 m_2 m_1 m_6 m_3^{-1} m_1^{-1} m_6 m_3^{-1} m_6 m_2 m_3^{-1} m_8^{-1} = 1.
\end{aligned}
$$

(g) *M has a faithful permutation representation PM of degree* 1024 *and stabilizer $\langle m_1, m_2, m_3, m_4, m_5\rangle$.*

(h) *Let $x_1 = (m_3^2 m_5 m_3 m_6)^4$, $x_2 = (m_3 m_6 m_3^2 m_5)^4$, $x_3 = (m_2 m_3^2 m_5 m_6 m_3)^3$, and $r_1 = (m_2 m_6 m_3 m_5 m_3 m_5^2)^2$, $r_2 = m_3 m_6 m_5 m_3 m_5^2 m_6 m_2$, $r_3 = m_3 m_6 m_5 m_6 m_3^2 m_5^2 m_3$. Let $w_1 = x_1^2 x_2 x_3$ and $w_2 = (r_2 r_3^2)^3$. Then $z = w_2^4$ is a* 2*-central involution of M with centralizer $D_M = C_M(z) = \langle w_1, w_2\rangle$ and $M = \langle w_1, w_2, m\rangle$, where $m = m_1$.*

(i) *A system of representatives m_i and the corresponding centralizer orders $|C_M(m_i)|$ of the* 49 *conjugacy classes of m_i^M of $M = \langle w_1, w_2, m\rangle$ are given in Table 14.5.5.*

(j) *If $z^g = t$ for some $g \in G$ then there is an isomorphism $\tau : N_H(V) = \langle v_1, v_2\rangle \to D_M$ such that $w_1 = \tau(v_1)$ and $w_2 = \tau(v_2)$, where $v_1 = d_1 d_2 d_3 d_5$ and $v_2 = d_1^2 d_3 (d_1 d_2 d_3)^2 d_5 d_4 d_5$, and $d_1 = h_8$, $d_2 = h_{14}$, $d_3 = (h_{14} h_{15})^6$, $d_4 = h_{15} h_8 h_{15}^2$, $d_5 = (h_{15} h_8 h_{14} h_{15}^2)^2$.*

(k) *If $z^g = t$ for some $g \in G$ then G has two conjugacy classes of involutions. They are represented by z and $h = h_{14}$.*

(l) *If $z^g = h_7$ for some $g \in G$ then there is a* 2*-element $y \in N_G(A) - N_H(A)$.*

(m) *In any case, $N_G(A) > N_H(A)$.*

Proof (a) Let $w = h_{14}$. Using MAGMA and the faithful permutation representation PH of H given in Lemma 14.1.1(a) one sees that $w \in S$, $|C_H(w)| = 2^9$, $|C_H(h_1)| = 2^{12} \cdot 3 \cdot 5$ and $|C_H(h_1)| = 2^{10} \cdot 3^2$. Furthermore,

$$\{z\} \cup \{h_1^H\} \cup \{h_7^H\} \cup \{w^H\}$$

is the set of all involutions of H.

(b) Let $U = C_G(w)$ and $T = C_G(w)$. Using MAGMA again we checked that $Z(T) \cap T'' = \langle z \rangle$. Therefore each $c \in N_U(T)$ belongs to $C_U(w) \cap C_U(z) = C_H(w) = T$.

(c) This assertion follows immediately from (b).

(d) Let $u \in \{z, h_1, h_7\}$. Suppose that $w \in u^G$. Then $|C_G(w)| = |C_G(u)|$. By Table 14.5.1 a Sylow 2-subgroup of $C_G(u)$ has order at least 2^{10} because $h = h_{14}$. Now (c) provides a contradiction. In particular, G has at least two conjugacy classes of involutions represented by z and h. Now Theorem 4.7.3 of [92] due to Glauberman implies that that $z^g = t$ or $z^t = h_7$ for some $g \in G$.

(e) Suppose that $x^g = t$ and let $Y = C_G(t)$. Then $C_Y(t)$ is properly contained in a Sylow 2-subgroup T of Y. By Lemma 14.1.1 V is the unique normal Klein 4-subgroup of S and it is the center $Z(Y)$ of Y. Hence (e) holds by Lemma 1.4.4 of [92].

(f) By Lemma 14.1.1(d) V is the center of $W = C_G(V) = C_H(V)$. Using the faithful permutation representation PH and MAGMA the author checked that $X = \langle h_2, h_3, h_4, h_5, h_7, h_8, h_9, h_{10} \rangle$ is a unique normal subgroup of order 2^{10} of W. Its derived subgroup X' is the center of V and equals W. Hence X is normal in $N_1 = N_H(V) = \langle h_6, h_7, h_{11}, h_{12}, h_{14} \rangle$. It has a complement $Q = \langle q_1, q_2 \rangle$ with center $Z(Q) = \langle (q_1 q_2)^5 \rangle$ of order 3 and $Q/Z(Q) \cong A_5$, where $q_1 = h_6 h_{14} h_{12} h_{14}$ and $q_2 = (h_6 h_7 h_{11} h_{12} h_{14} h_7)^4$. Now Theorem 1.4.15 of [92] and (e) imply that $N_G(V)$ is isomorphic to the split extension M of the direct product of the symmetric group S_3 of order 6 and the alternating group A_5 by X.

In order to construct M we calculated the automorphism group AW of $W = C_G(V)$ using MAGMA and PH. It has order $2^{12} \cdot 3^2 \cdot 5$. MAGMA provides a faithful permutation representation PAW of degree 320. Using it and MAGMA the author observed that the Frattini subgroup F of the group I of inner automorphisms has a complement K in AW of order $2^4 \cdot 3^2 \cdot 5$. Furthermore, $|K'| = 2^2 \cdot 3^2 \cdot 5$. In particular, K' is isomorphic to a direct product of a cyclic group $\langle c \rangle$ of order 3 and $L \cong A_5$. MAGMA also provided three generators a_j of order 3 for L. Using PAW and the command

```
exists(m){x : x in K| Order(x) eq 2 and c^x eq c^2}
```

we find an involution m of K such that $K = \langle c, m, L\rangle$.

Clearly, $F \cong X/V$ is elementary abelian. Let $\mathcal{B} = \{f_i | 1 \leq i \leq 8\}$ be a basis of F. The conjugation actions of the generators of K_1 induce endomorphisms of M having the following matrices w.r.t. $\mathcal{B}$:

$$Mc = \begin{pmatrix} 0&1&0&0&0&0&0&0 \\ 1&1&0&0&1&0&0&0 \\ 1&0&1&0&1&0&0&0 \\ 0&1&0&0&1&1&1&1 \\ 0&1&1&0&0&0&0&0 \\ 1&1&1&0&1&1&0&1 \\ 0&1&0&1&1&1&1&0 \\ 0&1&0&0&1&1&0&0 \end{pmatrix}, \quad Mm = \begin{pmatrix} 1&0&0&0&0&0&0&0 \\ 1&1&0&0&0&0&0&0 \\ 1&0&1&0&0&0&0&0 \\ 0&0&0&0&1&0&0&0 \\ 0&1&1&0&1&0&0&0 \\ 1&0&0&0&0&1&0&0 \\ 0&1&0&1&0&1&1&0 \\ 0&0&1&0&0&1&0&1 \end{pmatrix}, \quad Ma_1 = \begin{pmatrix} 0&0&1&1&0&1&0&0 \\ 0&0&0&0&1&0&1&0 \\ 1&0&1&1&1&1&1&0 \\ 0&0&0&1&0&0&0&0 \\ 1&1&1&1&1&1&1&0 \\ 0&1&0&0&1&1&1&0 \\ 1&1&0&1&0&1&1&0 \\ 0&0&1&0&0&0&1&1 \end{pmatrix},$$

$$Ma_2 = \begin{pmatrix} 1&1&1&0&0&0&0&0 \\ 0&0&1&0&1&0&0&0 \\ 1&0&1&0&1&0&0&0 \\ 1&1&1&1&0&0&0&0 \\ 1&1&0&0&0&0&0&0 \\ 1&0&0&0&1&1&0&0 \\ 1&1&1&0&0&0&1&0 \\ 1&1&1&0&0&0&0&1 \end{pmatrix}, \quad Ma_3 = \begin{pmatrix} 0&1&0&0&0&1&0&0 \\ 0&0&1&0&1&1&0&1 \\ 1&1&0&0&1&1&0&1 \\ 0&1&0&1&0&1&0&0 \\ 1&1&0&0&1&0&0&0 \\ 1&1&1&0&1&0&0&1 \\ 1&1&1&0&1&1&1&1 \\ 1&0&0&0&0&0&0&1 \end{pmatrix}.$$

The subgroup $MK = \langle Mc, Mm, Ma_1, Ma_2, Ma_3\rangle$ of $\mathrm{GL}_8(2)$ is transformed into a finitely presented group $K = \langle k_i \mid 1 \leq i \leq 5\rangle$ by means of the MAGMA command `FPGroup(MK)`. Its set $\mathcal{R}(K)$ of defining relations consists of the following relations:

$$\begin{aligned}
&k_1^3 = k_2^2 = k_3^3 = k_4^3 = k_5^3 = 1, \\
&(k_1, k_3) = (k_1, k_4) = (k_1, k_5) = (k_2, k_3) = (k_2, k_4) = (k2, k_5) = 1, \\
&(k_1^{-1} k_2)^2 = 1, \quad (k_3^{-1} k_4^{-1})^2 = 1, \quad (k_4^{-1} k_5^{-1})^2 = 1, \\
&k_3 k_5^{-1} k_3^{-1} k_5 k_4^{-1} = 1.
\end{aligned}$$

Then we construct its faithful permutation representation PK of degree 30 with stabilizer $\langle k_3, k_4\rangle$. Thus we can apply Holt's Algorithm to calculate the second cohomological dimension of K with coefficients in F; it is 0. Since K acts irreducibly on F we use the MAGMA commands `Q:=Getvecs(MK)` and `Semidir(MK,Q)` to construct the uniquely (up to isomorphism) determined split extension K_1 of K by F. Its set $\mathcal{R}(K_1)$ of defining relations consists of $\mathcal{R}(K)$ and the following relations:

$$\begin{aligned}
&k_6^2 = 1, \quad (k_4 k_6 k_5)^2 = 1, \quad (k_6 k_1^{-1})^3 = 1, \quad (k_2 k_6)^4 = 1, \\
&k_2 k_6 k_5^{-1} k_4^{-1} k_2 k_6 k_4 k_5 = k_4^{-1} k_6 k_2 k_6 k_4 k_6 k_2 k_6 = 1, \\
&(k_3 k_6 k_3^{-1} k_6)^2 = 1, \quad (k_4^{-1} k_6 k_3^{-1} k_6)^2 = 1, \\
&k_1^{-1} k_6 k_3^{-1} k_6 k_2 k_1^{-1} k_6 k_3 k_6 k_2 k_6 k_2 k_1 k_6 k_3^{-1} k_1^{-1} k_6 k_3^{-1} k_6 k_2 k_3^{-1} = 1.
\end{aligned}$$

Let PK_1 be its natural permutation representation of degree 256 with stabilizer K. In order to construct the extensions of K_1 by V we now use the matrices belonging to the conjugate actions of the generators $m = k_2$ and $c = k_1$ of K on V because all other generators of K_1 act trivially on V:

$$Nk_1 = \begin{pmatrix} 0 & 1 \\ 1 & 1 \end{pmatrix}, \quad Nk_2 = \begin{pmatrix} 1 & 0 \\ 1 & 1 \end{pmatrix}.$$

Let $Nk_i = I$ be the identity matrix in $GL_2(2)$ for $i = 2, 3, ..., 6$ and

```
FEalg :=MatrixAlgebra<FiniteField(2),2|Nk_j, 1 \le j \le 6>,
V1 := GModule(PK_1, FEalg).
```

Application of the MAGMA command `CohomologicalDimension(PK_1, V1, 2)` yields that this dimension is 1. Since the Sylow 2-subgroup S of H does not split over its normal subgroup V by Lemma 14.1.1(c) we know that $N_G(V)$ is the non-split extension $M = \langle m_j | 1 \le j \le 8 \rangle$ of K_1 by V. Its presentation is given in the statement.

(g) This statement can easily be verified by means of MAGMA.

(h) Using the faithful permutation representations PH of H given in Lemma 14.1.1(a) and PM of M all assertions of this statement can be checked computationally by the reader.

(i) The author used PM and Kratzer's Algorithm 5.3.18 of [92] to calculate the table of the representatives of the conjugacy classes of M.

(j) Since $z^g = t$ for some $g \in G$ not in H assertion (f) implies that $N_G(V) \cong M$. Using PH and MAGMA the author verified that $D_1 = N_H(V) = \langle d_k | 1 \le k \le 5 \rangle$ for the elements d_k given in the statement. Furthermore, $D_1 = \langle v_1, v_2 \rangle$. Using PH and PM and MAGMA we found an isomorphism $\tau : D_1 \to D_M$ satisfying $\tau(v_1) = w_1$ and $\tau(v_2) = w_2$.

(k) An application of Kratzer's Algorithm 5.3.18 of [92] provides a natural bijection between the conjugacy classes of D_1 and D_M. In particular, we determined the fusion of the D_1-classes in H and also the fusion of the D_M-classes in M. For the eight conjugacy classes of involutions of $D_1 = D_M$ we obtained the following fusion patterns into the four classes of involutions of H and the six classes of involutions of M using the notation of Tables 14.5.1 and 14.5.5, respectively:

$$\langle [1,2,3], [1,2,3,4,5], [1,2,3] \rangle,$$
$$\langle [4], [6,7,8], [4,5,6] \rangle,$$

where the classes of H and M are denoted by the indices of the involutions of the respective tables, and where the classes of H are given before the ones of D. The final triples denote the classes of M. In particular, (d) and Tables 14.5.1 and 14.5.5 imply that $z \in 2_1$, $t \in 2_2$ and $h_7 \in 2_3$ of H are conjugate in G, because the involutions of M fuse only to z and h in G.

(l) If $z^g = h_7$ for some $g \in G$ then $g \notin H$. By means of another MAGMA calculation with PH the author observed that the derived subgroup S_7 of the centralizer $C_7 = C_H(h_7)$ has order 2^{10}. Therefore it is a Sylow 2-subgroup of C_7 by Table 14.5.1. It also has been verified computationally that its center $Z(S_7) = A$. Hence S_7 is properly contained in a Sylow 2-subgroup T of $C_G(h_7)$ and T contains a subgroup T_1 such that $|T_1 : S_7| = 2$. In particular, the characteristic subgroup A of S_7 is normal in T_1. Hence there is a $y \in T_1$ such that $y \in N_G(A) - N_H(A)$.

(m) By (l) we assume that $z^g = t$ for some $g \in G - H$. Then G has exactly two conjugacy classes of involutions by (k). Hence z and h_7 fuse in G and $N_G(A) > N_H(A)$ by (l). This completes the proof. □

Proposition 14.2.2 *Let G be a finite simple group of* Suz*-type having a 2-central involution z with centralizer $H = C_G(z)$ defined in Lemma 14.1.1. Let $A = \langle z, h_1, h_2, h_1\rangle$ be the unique normal subgroup of order* 16 *of the Sylow 2-subgroup S and let $D = N_H(A) = \langle x, y\rangle$ be its normalizer, where the generators x and y defined in Lemma 14.1.1(i). Then the following statements hold.*

(a) *$N_G(A)$ is uniquely determined up to isomorphism and it is isomorphic to the finitely presented group $E = \langle e_i | 1 \leq i \leq 13\rangle$ with the following set $\mathcal{R}(E)$ of defining relations:*

$$e_1^3 = e_2^4 = e_3^5 = e_6^2 = e_8^2 = e_9^2 = e_{10}^2 = e_{11}^2 = e_{12}^2 = e_{13}^2 = 1,$$
$$e_4^2 e_{11}^{-1} = 1, \quad e_5^2 e_{11}^{-1} = 1, \quad e_7^2 e_{10}^{-1} e_{12}^{-1} = 1,$$
$$(e_8, e_9) = (e_2, e_{11}) = (e_4, e_{10}) = (e_4, e_{11}) = (e_4, e_{12}) = (e_4, e_{13}) = 1,$$
$$(e_5, e_{10}) = (e_5, e_{11}) = (e_5, e_{12}) = (e_5, e_{13}) = (e_6, e_{10}) = (e_6, e_{11}) = 1,$$
$$(e_6, e_{12}) = (e_6, e_{13}) = (e_7, e_{10}) = (e_7, e_{11}) = (e_7, e_{12}) = (e_7, e_{13}) = 1,$$
$$(e_8, e_{10}) = (e_8, e_{11}) = (e_8, e_{12}) = (e_8, e_{13}) = 1,$$

$$
\begin{aligned}
&(e_9, e_{10}) = (e_9, e_{11}) = (e_9, e_{12}) = (e_9, e_{13}) = 1,\\
&(e_{10}, e_{11}) = (e_{10}, e_{12}) = (e_{10}, e_{13}) = (e_{11}, e_{12}) = (e_{11}, e_{13}) = 1,\\
&(e_{12}, e_{13}) = e_1^{-1}e_{10}e_1e_{10}^{-1}e_{11}^{-1} = e_1^{-1}e_{11}e_1e_{10}^{-1} = 1,\\
&e_1^{-1}e_{12}e_1e_{10}^{-1}e_{13}^{-1} = e_1^{-1}e_{13}e_1e_{11}^{-1}e_{12}^{-1}e_{13}^{-1} = e_2^{-1}e_{10}e_2e_{10}^{-1}e_{11}^{-1} = 1,\\
&e_2^{-1}e_{12}e_2e_{10}^{-1}e_{12}^{-1} = e_2^{-1}e_{13}e_2e_{10}^{-1}e_{12}^{-1}e_{13}^{-1} = e_3^{-1}e_{10}e_3e_{12}^{-1}e_{13}^{-1} = 1,\\
&e_3^{-1}e_{11}e_3e_{10}^{-1}e_{11}^{-1} = e_3^{-1}e_{12}e_3e_{11}^{-1} = e_4^{-1}e_6^{-1}e_4e_6e_{10}^{-1}e_{11}^{-1}e_{12}^{-1} = 1,\\
&e_3^{-1}e_{13}e_3e_{10}^{-1}e_{11}^{-1}e_{12}^{-1} = e_4^{-1}e_5^{-1}e_4e_5e_{11}^{-1} = 1,\\
&e_5^{-1}e_6^{-1}e_5e_6e_{11}^{-1} = e_4^{-1}e_7^{-1}e_4e_7e_{10}^{-1}e_{11}^{-1}e_{12}^{-1} = 1,\\
&e_5^{-1}e_7^{-1}e_5e_7e_{10}^{-1}e_{11}^{-1}e_{12}^{-1} = 1,\\
&e_6^{-1}e_7^{-1}e_6e_7e_{11}^{-1} = e_4^{-1}e_8^{-1}e_4e_8e_{11}^{-1} = e_5^{-1}e_8^{-1}e_5e_8e_{10}^{-1}e_{11}^{-1} = 1,\\
&e_6^{-1}e_8^{-1}e_6e_8e_{11}^{-1}e_{13}^{-1} = e_7^{-1}e_8^{-1}e_7e_8e_{10}^{-1}e_{12}^{-1}e_{13}^{-1} = 1,\\
&e_4^{-1}e_9^{-1}e_4e_9e_{10}^{-1} = e_5^{-1}e_9^{-1}e_5e_9e_{11}^{-1} = 1,\\
&e_6^{-1}e_9^{-1}e_6e_9e_{12}^{-1} = e_7^{-1}e_9^{-1}e_7e_9e_{13}^{-1} = 1,\\
&e_1^{-1}e_4e_1e_4^{-1}e_8^{-1}e_9^{-1}e_{12}^{-1}e_{13}^{-1} = e_1^{-1}e_5e_1e_5^{-1}e_8^{-1}e_{11}^{-1} = 1,\\
&e_1^{-1}e_6e_1e_4^{-1}e_5^{-1}e_6^{-1}e_7^{-1}e_{10}^{-1}e_{11}^{-1}e_{12}^{-1} = 1,\\
&e_1^{-1}e_7e_1e_6^{-1}e_8^{-1}e_{13}^{-1} = e_1^{-1}e_8e_1e_9^{-1}e_{11}^{-1}e_{12}^{-1} = 1,\\
&e_1^{-1}e_9e_1e_8^{-1}e_9^{-1}e_{11}^{-1}e_{12}^{-1}e_{13}^{-1} = e_2^{-1}e_4e_2e_4^{-1}e_{10}^{-1} = 1,\\
&e_2^{-1}e_5e_2e_5^{-1}e_{12}^{-1} = e_2^{-1}e_6e_2e_4^{-1}e_5^{-1}e_6^{-1}e_9^{-1}e_{10}^{-1}e_{13}^{-1} = 1,\\
&e_2^{-1}e_7e_2e_4^{-1}e_7^{-1}e_8^{-1}e_9^{-1}e_{11}^{-1}e_{12}^{-1}e_{13}^{-1} = 1,\\
&e_2^{-1}e_8e_2e_4^{-1}e_8^{-1}e_{11}^{-1} = e_2^{-1}e_9e_2e_5^{-1}e_9^{-1}e_{11}^{-1}e_{13}^{-1} = 1,\\
&e_3^{-1}e_4e_3e_4^{-1}e_5^{-1}e_8^{-1} = e_3^{-1}e_5e_3e_4^{-1}e_9^{-1}e_{10}^{-1}e_{11}^{-1}e_{13}^{-1} = 1,\\
&e_3^{-1}e_6e_3e_4^{-1}e_5^{-1}e_6^{-1}e_9^{-1}e_{10}^{-1}e_{11}^{-1} = e_3^{-1}e_7e_3e_5^{-1}e_7^{-1}e_8^{-1}e_{13}^{-1} = 1,\\
&e_3^{-1}e_8e_3e_4^{-1}e_6^{-1}e_7^{-1}e_8^{-1}e_{10}^{-1}e_{11}^{-1}e_{12}^{-1} = 1,\\
&e_3^{-1}e_9e_3e_4^{-1}e_6^{-1}e_9^{-1}e_{10}^{-1}e_{13}^{-1} = e_2^{-1}e_3^{-1}e_2e_1^{-1}e_3^{-1}e_1e_{10}^{-1} = 1,\\
&e_3^{-1}e_1^{-1}e_3^{-1}e_1^{-1}e_3^{-1}e_1^{-1}e_{10}^{-1}e_{13}^{-1} = e_3^{-2}e_1e_2^{-1}e_1e_3^{-2}e_2e_{10}^{-1}e_{12}^{-1}e_{13}^{-1} = 1,\\
&e_2e_3^{-1}e_2^{-1}e_1^{-1}e_3^{-1}e_2e_1e_{10}^{-1}e_{11}^{-1}e_{13}^{-1} = 1.
\end{aligned}
$$

(b) *E has a faithful permutation representation of degree 1536 with stabilizer $\langle e_3, (e_1e_4e_3e_1e_4)^2, (e_1^2e_3e_1e_3^2)^5\rangle$.*

(c) *Let $d_1 = (e_1e_4)^3$, $d_2 = e_3e_1$, $d_3 = (e_1e_3^2)^2$, $d_4 = e_3^2e_1^2e_3^2$. Let*

$$
\begin{aligned}
x_1 &= (d_4d_1d_3)^4(d_1d_3d_2d_1d_2d_4)^{23} \text{ and}\\
y_1 &= (d_1d_3)^2(d_2d_1d_4d_2)^2(d_4d_2d_4c_4c_5)(d_1d_3)(d_4d_2d_4)(d_1d_3).
\end{aligned}
$$

Then $z = x_1^{12}$ *is a* 2*-central involution of* E *with centralizer* $D_E = C_E(z) = \langle x_1, y_1 \rangle$ *and there is an isomorphism* $\sigma : D \to D_E$ *such that* $x_1 = \sigma(x)$ *and* $y_1 = \sigma(y)$*. Furthermore,* $E = \langle x_1, y_1, e_1 \rangle$*.*

(d) *A system of representatives* (e_i) *and the corresponding centralizer orders* $|C_E(e_i)|$ *of the* 47 *conjugacy classes of* $E = \langle x_1, y_1, e_1 \rangle$ *are given in Table 14.5.2.*

(e) *The character table of* E *is given in Table 14.6.2.*

(f) *Each finite simple group* G *of* Suz*-type has* 2 *conjugacy classes of involutions represented by* z *and* $u = h$ *of* $H = \langle x, y, h \rangle$*. The classes* 2_1*,* 2_2 *and* 2_3 *of* H *belong to* z^G *and* 2_4 *belongs to* u^G*.*

(g) *The Goldschmidt index of the amalgam* $H \leftarrow D \rightarrow E$ *is* 1*.*

Proof (a) As in Lemma 14.1.1 we let $H = \langle x, y, h \rangle$. Then $D = N_H(A) = \langle x, y \rangle$, where $A = z, t, h_2, h_7$ is the unique maximal elementary abelian normal subgroup of the fixed Sylow 2-subgroup S of H. Furthermore, $C = C_S(A)$ has a complement K in D. Since D contains S we applied the program `GetShortGens(D,S)` and obtained the four generators of $s_1 = x^6$, $s_2 = (yx)^3$, $s_3 = (x^2y)^3$ and $s_4 = (yx^2)^3$ of S. Similarly, we observed that $z = s_1^2$, $t = s_2^2$, $a_1 = s_3^2$ and $a_2 = (s_1 s_3 s_1)^2$ generate A. The normal subgroup C of D is generated by $b_1 = s_3$, $b_2 = s_4$, $b_3 = (s_1 s_3)^2$, $b_4 = (s_1 s_4)^2$, $b_5 = (s_2 s_3)^2$ and $b_6 = (s_2 s_4)^2$. With respect to these six generators the center $Z(C) = A$ is generated by $z_1 = (b_1)^2$, $z_2 = (b_2)^2$, $z_3 = (b_3)^2$ and $z_4 = (b_1 b_2)^2$. The elements $k_1 = (x^4y^2)^4$ and $k_2 = (x^3yx^2yx^2)^3$ generate the complement K of C in D. Hence $D = \langle b_i, k_j | 1 \le i \le 6, 1 \le j \le 2 \rangle$. Using the permutation representation PH and the command `FPGroup(D)` MAGMA established the following set $\mathcal{R}(D)$ of defining relations of D:

$$
\begin{aligned}
&f_2^4 = f_3^4 = f_5^4 = f_7^3 = f_8^4 = 1,\\
&f_3^2 = f_4^2, \quad (f_4, f_5) = 1, \quad (f_2, f_6) = 1, \quad f_3^{-1} f_4^{-1} f_3 f_4^{-1} = 1,\\
&(f_2^{-1} f_5^{-1})^2 = (f_2 f_5^{-1})^2 = (f_3^{-1} f_5^{-1})^2 = (f_3 f_5^{-1})^2 = 1,\\
&f_6^{-1} f_5^2 f_6^{-1} = (f_3^{-1} f_6^{-1})^2 = (f_4^{-1} f_6^{-1})^2 = f_5^{-1} f_6^{-1} f_5 f_6^{-1} = 1,\\
&f_3^{-1} f_7^{-1} f_1^{-1} f_7 f_1^{-1} = f_1^{-1} f_2^{-1} f_1^2 f_2 f_1^{-1} = 1,\\
&f_4^{-1} f_2^{-1} f_1^2 f_4^{-1} f_2^{-1} = f_4 f_2^{-1} f_1^2 f_4 f_2^{-1} = 1,\\
&f_1^{-1} f_3^{-1} f_1^2 f_3 f_1^{-1} = f_2^{-1} f_7^{-1} f_1^2 f_7 f_2^{-1} = 1,\\
&f_3^{-1} f_1^{-1} f_2^{-1} f_1 f_3 f_2^{-1} = f_1^{-1} f_2^{-} 2 f_1 f_2^2 = 1,
\end{aligned}
$$

$$
\begin{aligned}
&f_1^{-1}f_4^{-1}f_2^{-1}f_1f_4^{-1}f_2^{-1} = f_1^{-1}f_5^{-1}f_2^{-1}f_1^{-1}f_2f_5 = 1,\\
&f_1^{-1}f_6^{-1}f_2^{-1}f_1^{-1}f_2^{-1}f_6^{-1} = f_2^{-1}f_7^{-1}f_2^{-1}f_1f_7f_1 = 1,\\
&f_8^{-1}f_1^{-1}f_3^{-1}f_1f_8f_3^{-1} = f_7f_2^{-1}f_3^{-1}f_2f_7^{-1}f_4 = 1,\\
&f_8f_2^{-1}f_3^{-1}f_2f_8^{-1}f_3^{-1} = f_5^{-1}f_3^{-}2f_1f_5^{-1}f_1^{-1} = 1,\\
&f_8^{-1}f_5f_3^{-1}f_1f_8f_1 = f_8f_6f_3^{-1}f_2^{-1}f_8^{-1}f_2^{-1} = 1,\\
&f_8f_2^{-1}f_4^{-1}f_2f_8^{-1}f_4^{-1} = f_8^{-1}f_6^{-1}f_4^{-1}f_2^{-1}f_8f_2^{-1} = 1,\\
&f_7f_6^{-1}f_5^{-1}f_1f_7^{-1}f_6^{-1} = f_5^{-1}f_8^{-}2f_5^{-1}f_8^2 = 1,\\
&f_8^{-1}f_5^{-1}f_1^{-}2f_8f_4^{-1}f_5 = f_6^{-1}f_4f_2^{-1}f_1^{-1}f_7^{-1}f_5^{-1}f_7 = 1,\\
&f_7^{-1}f_8^{-1}f_7^{-1}f_8^{-1}f_7f_8f_7f_8^{-1} = 1.
\end{aligned}
$$

Therefore a set $\mathcal{R}(D_1)$ of defining relations of $D_1 = D/A$ consists of $\mathcal{R}(D)$ and the relations $f_1^2 = 1$, $f_2^2 = 1$, $f_3^2 = 1$, $(f_1f_2)^2 = 1$.

Let $q_i = f_iA \in D_1$. Then the elementary abelian normal subgroup $B = C/A = \langle q_i | 1 \le i \le 6\rangle$ has a complement $K = \langle k_1, k_2\rangle$ in D_1, where $k_1 = q_7$ and $k_2 = q_8$. Thus D_1 has a faithful permutation representation PD_1 of degree 64. The conjugate actions of k_1 and k_2 induce endomorphisms given by the following matrices w.r.t. the basis $\mathcal{B} := \{q_1, q_2, q_3, q_4, q_5, q_6\}$:

$$
Mk_1 = \begin{pmatrix} 100011\\ 010010\\ 111100\\ 001010\\ 000001\\ 000011 \end{pmatrix}, \quad Mk_2 = \begin{pmatrix} 100000\\ 010000\\ 111001\\ 100111\\ 100010\\ 010001 \end{pmatrix}.
$$

Let MK be the subgroup of $MG = \mathrm{GL}_6(2)$ generated by the matrices Mk_1 and Mk_2. Its center $Z(MK)$ is generated by $M_c = (Mk_1Mk_2)^4$ of order 3. Using MAGMA it has been checked that $MR = C_{MG}(Mc)$ has order $|MR| = 2^6 \cdot 3^4 \cdot 5 \cdot 7$. It has a simple composition factor $\mathrm{PSL}_3(4)$.

By Proposition 14.2.1 $N_G(A) > N_H(A)$ in any simple target group G of Suz-type. Thus $|N_G(A)/A : D_1|$ is a positive odd number. It is not divisible by 7 because z and h_7 in A are G-conjugate by Proposition 14.2.1. Furthermore, A contains eight elements which are H-conjugate to h_7. Therefore all non-trivial $a \in A$ are G-conjugate. Hence an element $y \in N_G(A)$ of order 7 must have a non-trivial centralizer $C_A(y)$. But $|H = C_G(z)| = 2^{13} \cdot 3^4 \cdot 5$.

Using the standard faithful permutation representation $PG63$ of $MG = \mathrm{GL}_6(2)$ of degree $2^6 - 1$ and the MAGMA command

```
Subgroups(MR : OrderEqual := 2^3*3^b*5^c),
```

where $b \in \{2, 3, 4\}$ and $c \in \{0, 1\}$, the author has checked that up to conjugacy there is exactly one subgroup Φ of MR which has a dihedral Sylow 2-subgroup as MK and has an order of the above form. It follows that $b = 3$ and $c = 1$. In particular, $N_G(A)/A$ has a non-trivial Sylow 5-subgroup. Hence it is of order 5. Using the command

```
exists(q){x : x \in MR| Order(x) eq 5 and
  Order(sub<MR|MK,x>) eq 1080}
```

MAGMA established a $q \in PG63$ corresponding to the matrix

$$Mk_3 = \begin{pmatrix} 1&1&0&0&1&0\\ 1&0&0&0&0&1\\ 1&1&1&0&0&1\\ 0&1&0&1&1&0\\ 1&0&1&1&1&0\\ 1&0&1&0&0&1 \end{pmatrix}.$$

The subgroup $MX = \langle Mk_1, Mk_2, Mk_3 \rangle$ has order $2^3 \cdot 3^3 \cdot 5$. As before we transform MX into a finitely presented group FX by means of the command `FPGroup(MX)`. It has the following set $\mathcal{R}(K)$ of defining relations:

$$\begin{aligned} &k_1^3 = k_2^4 = k_3^5 = 1, \\ &k_2^{-1}k_3^{-1}k_2k_1^{-1}k_3^{-1}k_1 = (k_3^{-1}k_1^{-1})^3 = 1, \\ &k_2k_3^{-1}k_2^{-1}k_1^{-1}k_3^{-1}k_2k_1 = k_3^{-2}k_1k_2^{-1}k_1k_3^{-2}k_2 = 1. \end{aligned}$$

Since X is given as a matrix group and as a finitely presented group we can apply Holt's Algorithm stated as Algorithm 7.4 5 in [92] to get the second cohomological dimension of X with coefficients in the $GF(2)X$-module B. It turns out to be zero. Let $E_0 = \langle k_1, k_2, k_3, k_4, k_5, k_6, k_7, k_8, k_9 \rangle$ be the split extension of X by B, where the first three generators correspond to the generators k_1, k_2 and k_3 of X. Then a set $\mathcal{R}(E_0)$ of defining relations of E_0 consists of $\mathcal{R}(K)$ and the following relations:

$$\begin{aligned} &k_4^2 = k_5^2 = k_6^2 = k_7^2 = k_8^2 = k_9^2 = 1, \\ &(k_i, k_j) = 1, \quad \text{for all} \quad 4 \le i, j \le 9, \\ &(k_2, k_4) = (k_2, k_5) = 1, \\ &k_1^{-1}k_4k_1k_4^{-1}k_8^{-1}k_9^{-1} = k_1^{-1}k_5k_1k_5^{-1}k_8^{-1} = 1, \\ &k_1^{-1}k_6k_1k_4^{-1}k_5^{-1}k_6^{-1}k_7^{-1} = k_1^{-1}k_7k_1k_6^{-1}k_8^{-1} = 1, \end{aligned}$$

$$
\begin{aligned}
&k_1^{-1}k_8k_1k_9^{-1} = k_1^{-1}k_9k_1k_8^{-1}k_9^{-1} = 1,\\
&k_2^{-1}k_6k_2k_4^{-1}k_5^{-1}k_6^{-1}k_9^{-1} = k_2^{-1}k_7k_2k_4^{-1}k_7^{-1}k_8^{-1}k_9^{-1} = 1,\\
&k_2^{-1}k_8k_2k_4^{-1}k_8^{-1} = k_2^{-1}k_9k_2k_5^{-1}k_9^{-1} = 1,\\
&k_3^{-1}k_4k_3k_4^{-1}k_5^{-1}k_8^{-1} = k_3^{-1}k_5k_3k_4^{-1}k_9^{-1} = 1,\\
&k_3^{-1}k_6k_3k_4^{-1}k_5^{-1}k_6^{-1}k_9^{-1} = k_3^{-1}k_7k_3k_5^{-1}k_7^{-1}k_8^{-1} = 1,\\
&k_3^{-1}k_8k_3k_4^{-1}k_6^{-1}k_7^{-1}k_8^{-1} = k_3^{-1}k_9k_3k_4^{-1}k_6^{-1}k_9^{-1} = 1.
\end{aligned}
$$

In order to determine all the possible extensions E of E_0 by A we first determine the 4-dimensional irreducible representations of E_0 over $GF(2)$ because $N_G(A)$ acts irreducibly on A. In fact, the exterior square of B contains two irreducible constituents of dimension 4, one of which is described by the following three matrices:

$$
Mv_1 = \begin{pmatrix} 1&1&0&0\\ 1&0&0&0\\ 1&0&0&1\\ 0&1&1&1 \end{pmatrix}, \quad Mv_2 = \begin{pmatrix} 1&1&0&0\\ 0&1&0&0\\ 1&0&1&0\\ 1&0&1&1 \end{pmatrix}, \quad Mv_3 = \begin{pmatrix} 0&0&1&1\\ 1&1&0&0\\ 0&1&0&0\\ 1&1&1&0 \end{pmatrix}.
$$

Let $Mv_i = Id(\mathrm{GL}_4(2))$ for $4 \le i \le 9$ and let N be the $GF(2)E_0$-module described by these nine matrices. The author checked with MAGMA that it is irreducible. Let

```
FEalg:=MatrixAlgebra<FiniteField(2),4|Mvi, i = 1,2...9>
```

and `V := GModule(PE_0, FEalg)`. Then, using the MAGMA command

```
CohomologicalDimension(PE_0, V, 2),
```

MAGMA asserts that this dimension is 1. Since S does not split over its normal subgroup A $N_G(A)$ also does not split over A by Theorem 1.4.15 of [92]. Hence $N_G(A)$ is isomorphic to the uniquely determined non-split extension E of E_0 by N. The MAGMA commands `P:=ExtensionProcess(PE_0,V,E_0)` and `E:=Extension(P, [1])` yield the extension group $E = \langle e_i | 1 \le i \le 13\rangle$ with the set $\mathcal{R}(E)$ of defining relations given in statement (a).

(b) The permutation representation PE has been constructed in two steps. First we took only the cyclic group $\langle e_3 \rangle$ of order 5 as a stabilizer and obtained a permutation representation $PE1$ of degree 221184. Then we used the MAGMA command `DegreeReduction(PE1)` and obtained a faithful permutation representation PE of degree 1536. Its stabilizer is given in the statement.

(c) Using the faithful permutation representations PH of H given in Lemma 14.1.1(a) and PE of E all assertions of this statement can easily be checked by the reader computationally.

(d) The author used PE and Kratzer's Algorithm 5.3.18 of [92] to calculate the table of the representatives of the conjugacy classes of E.

(e) The character table of E was computed with MAGMA and PE.

(f) $D_H = \langle x, y\rangle$ and $D_E = \langle x_1, y_1\rangle$ are subgroups of H and E, respectively. The constructed isomorphism $\sigma : D_H \to D_E$ identifies also the classes of D_H and D_E. We worked out the fusion of the D_H-classes in H and the fusion of the D_E-classes in E. For the eight conjugacy classes of involutions of $D_H = D_E$ we obtained the following fusion patterns into the four classes of involutions of H and the four classes of involutions of E using the notation of Tables 14.5.1 and 14.5.2, respectively:

$$\langle [1, 2, 3], [1, 2, 3, 4, 5], [1, 3]\rangle,$$
$$\langle [4], [6, 7, 8], [2, 4]\rangle,$$

where the classes of H and E are denoted by the indices of the involutions of the respective tables, and where the classes of H are stated before the ones of D. The last entries denote the classes of E. In particular, (d) and Tables 14.5.1 and 14.5.2 imply that $z \in 2_1$, $t \in 2_2$ and $h_7 \in 2_3$ of H are conjugate in G, because the involutions of E fuse only to z and h in G.

(g) The Goldschmidt index has been calculated by means of Kratzer's Algorithm 7.1.10 of [92]. □

Proposition 14.2.3 *Let G be a finite simple group of* Suz*-type having a 2-central involution z with centralizer $H = C_G(z)$ defined in Lemma 14.1.1. Let $A = \langle z, h_1, h_2, h_1\rangle$ be the unique normal subgroup of order 16 of the Sylow 2-subgroup S and let $E = N_G(A) = \langle x, y, e\rangle$ be its normalizer whose generators x, y and e are defined in Lemma 14.1.1(i) and Proposition 14.2.2(c). Then the following assertions hold.*

(a) *$u = (ye)^2$ is a representative of the conjugacy class of non-2-central involutions of G and $U_1 = C_E(u) = \langle s_1, s_2, s_3, s_4\rangle$ has order 2^9, where $s_1 = y_1e_1$, $s_2 = (y_1^4e_1)^2$, $s_3 = (y_1^2e_1y_1e_1^2y_1e_1)^5$ and $s_4 = (y_1^2e_1y_1^2e_1y_1e_1^2)^2$.*

(b) *The involutions $q_1 = s_1^2$, $q_2 = (s_1s_2)^4$, $q_3 = s_1s_2s_3s_1s_2s_4$ and $q_4 = s_4^2$ generate the unique maximal elementary abelian normal subgroup L of U_1.*

(c) *The normal subgroup $J = C_{U_1}(L)$ of U_1 has a complement of order 2 generated by $j = s_4s_3s_1$. Furthermore, J is the direct product of the Klein 4-group $Q = \langle q_1, q_2q_3q_4\rangle$ and the normal* *subgroup* $T = \langle t_1, t_2, t_3, t_4 \rangle$ *of order* 2^6 *of* U_1*, where* $t_1 = s_1 s_2 s_1$, $t_2 = s_1^2 s_2 s_4$, $t_3 = s_1^2 s_2$ *and* $t_4 = s_1 s_2 s_4 s_1$.

(d) T *has two maximal elementary abelian normal subgroups* B_1, B_2 *of order* 16*. Furthermore,* $B_1^j = B_2$*, and* $B_1 \cap B_2$ *is the non-cyclic center of* T *of order* 4.

(e) $C_G(u)$ *is isomorphic to the finitely presented group* $U = \langle r_1, r_2, r_3, r_4, r_5 \rangle$ *with the following set of defining relations:*

$$r_1^2 = r_3^2 = r_2^4 = r_4^2 = r_5^2 = 1, \quad r_5^{r_3} = r_4 r_5,$$
$$(r_4, r_1) = (r_4, r_2) = (r_4, r_3) = (r_4, r_5) = (r_5, r_1) = (r_5, r_2) = 1,$$
$$r_2^{-1} r_1 r_2 r_1 r_2^{-1} r_1 r_2 r_3 r_2 r_1 r_2^{-1} r_3 = 1,$$
$$(r_2^{-1} r_1 r_3 r_1 r_3 r_1)^2 = 1, \quad (r_1 r_2^{-1})^7 = 1,$$
$$r_1 r_2 r_1 r_2 r_1 r_3 r_2^2 r_3 r_1 r_2^{-1} r_3 r_1 r_3 = 1,$$
$$r_2^{-2} r_1 r_2^{-2} r_1 r_2^{-2} r_1 r_2^2 r_1 r_2^2 r_1 = 1,$$
$$r_2^{-2} r_3 r_1 r_2^{-1} r_1 r_2^{-2} r_3 r_1 r_3 r_1 r_2^{-1} r_3 r_1 = 1,$$
$$r_2 r_3 r_2 r_3 r_1 r_2^{-1} r_1 r_2^{-1} r_3 r_2 r_3 r_2 r_3 r_2^{-1} r_3 = 1.$$

(f) U *has a faithful permutation representation* PU *of degree* 480 *with stabilizer* $\langle r_1, r_5, (r_2 r_1 r_2 r_1 r_2^3)^2, (r_1 r_2 r_1 r_2^2 r_1 r_2 r_1 r_2^3)^2 \rangle$.

(g) *A system of representatives* (u_i) *and the corresponding centralizer orders* $|C_U(u_i)|$ *of the* 38 *conjugacy classes of* $U = \langle r_k | 1 \le k \le 5 \rangle$ *are given in Table 14.5.6.*

(h) *The character table of* U *is given in Table 14.6.4.*

(i) U *has six conjugacy classes of involutions. Using the notation of Table 14.5.6 its classes* 2_1, 2_2, 2_4, 2_5, 2_6 *fuse in* G *with the involution* u *in* G*, whereas its class* 2_3 *fuses with the* 2*-central involution* z *of* G.

Proof (a) Table 14.5.2 and the proof of Proposition 14.2.2(f) imply that the involution $u = (ye)^2$ of E is a representative of the second conjugacy class of involutions of G. Let PE be the faithful permutation representation of E with degree 1536 defined in Proposition 14.2.2(b). The four generators s_i of $U_1 = C_E(u)$ have been calculated by means of PE, the program `GetShortGens(E,U_1)` and MAGMA.

(b) By Proposition 14.2.2(f) and Tables 14.6.1 and 14.6.2 the subgroup U_1 is a Sylow 2-subgroup of $U = C_G(u)$ of order 2^9. Using PE and the MAGMA command

```
Subgroups(U_1: Al:=Normal, IsElementaryAbelian := true)
```

it has been verified that U_1 possesses a unique maximal elementary abelian normal subgroup L. It has order 16 and is generated by the four involutions q_j defined in the statement.

(c) All assertions of this statement can easily be checked using *PE* and MAGMA. In particular, $T^j = T$. Hence T is normal in U_1.

(d) Another application of the subgroup command

```
Subgroups(T: Al:=Normal, IsElementaryAbelian := true)
```

yields that T has exactly two-maximal elementary abelian normal subgroups B_1 and B_2 of order 16 and that $B_1^j = B_2$.

(e) The involution $w = (xye)^5$ of E fuses with u in G by Table 14.5.2 and the proof of Proposition 14.2.2(f). Using *PE* and MAGMA the reader can easily verify that $J = C_{U_1}(Q)$ is isomorphic to a Sylow 2-subgroup of $R = C_E(w)$ and that the center Z of R is a Klein 4-group generated by w and w^k, where $k = n_1n_2^3n_1n_2^2n_1n_2^2n_1^2n_2^2$ has order 2. The subgroup W generated by $m_1 = (e_1y_1^3e_1)^2$ and $m_2 = (y_1e_1^2y_1e_1y_1e_1y_1)^3$ is a complement of Z in $N_E(Z)$. The group W has a center $Z(W) = \langle (m_1m_2)^4 \rangle$ of order 3. It has a complement W_1 in W of order $2^6 \cdot 3 \cdot 5$ which is a perfect semidirect product of the alternating group A_5 by an elementary abelian normal subgroup B of order 16.

By Table 14.5.4 W_1 has two conjugacy classes of involutions represented by $z_1 = (m_1m_2)^6$ and $z_2 = (m_1^2m_2m_1m_2^2m_1^2m_2^2)^3$. Using *PE* and MAGMA it has been verified that z_1 and z_2 are E-conjugate to the involutions x_1^{12} and $(x_1e_1^2)^6$ of E, respectively. Therefore they are both G-conjugate to the 2-central involution z of G by Table 14.5.2 and the proof of Proposition 14.2.2(f).

By (d) and Theorem 1.4.15 of [92] the centralizer $C = C_U(Z) = Z \times K$, where $|K : W_1|$ is odd. In particular, K is not solvable and W_1 is a subgroup of the last term M of the derived series of K. By (c) and Theorem 1.4.15 of [92] $N_U(Z)$ is the semidirect product of $C_U(Z)$ and a cyclic group y of order 2 such that $K^y = K$. Let B_1 be the Fitting subgroup B of W_1. Then $B_2 = B^y$ is the Fitting subgroup of $W_2 = W_1^y$. Furthermore, (c) implies that $T = \langle B_1, B_2 \rangle$ is a Sylow 2-subgroup of M such that $T^y = T$. In particular, B_1 and B_2 are unique maximal elementary abelian normal subgroups of T. Thus $D = \langle W_1, W_2 \rangle$, and therefore M have both a unique conjugacy class of involutions. It is represented by z_1 which is G-conjugate to z.

Hence $C_K(x)$ is a 2-group for each involution $x \in M$ by Table 14.5.1. Let X be any normal subgroup of M of odd order. Acting with the Klein 4-group $V = \langle z_1, z_2 \rangle$ on X by conjugation, Theorem 4.2.2 of [92] implies that $X = 1$. Thus any proper normal subgroup Y of M has even order. Thus $z_1^M \leq Y$. Now the Theorem of Feit and Thompson implies that M/Y is solvable. As M is perfect $Y = M$. Hence M is simple.

Since K, M and W_1 have isomorphic Sylow 2 subgroups it follows from (d) that the center $Z(S)$ of S is not cyclic. Suppose that $S \cap S^t = 1$ for all $t \in M - N_M(S)$. Then $N_M(S)$ is a strongly embedded subgroup of T by Theorem 4.8.4 of [92]. Since S has a non-cyclic center Theorem 1.7.3 of [92] due to Bender and Suzuki implies that either $M \cong Sz(q)$ or $M \cong \mathrm{U}_3(q)$, where $q = 2^m$ for some natural number $m > 2$. Hence $M \cong \mathrm{U}_3(4)$ because 3 does not divide $|Sz(q)|$ and $|S| = 2^6$. But $\mathrm{U}_3(4)$ does not have any subgroup isomorphic to W_1. Thus S satisfies the hypothesis of Suzuki's Theorem 1.3.1. Therefore $M \cong \mathrm{PSL}_3(4)$.

Suppose that $M \neq K$. Then K is a subgroup of $Aut(M)$ such that $|K : M| = 3$. In particular, K is a uniquely determined semidirect product of M and a cyclic group $\langle d_1 \rangle$ of order 3. Using a faithful permutation representation of $M = \mathrm{PSL}_3(4)$ and MAGMA we computed a faithful permutation representation of $Aut(M)$ and a presentation of $K = M : \langle d \rangle$. Thus one can verify that K has a unique conjugacy class of involutions z_1^K and that $|C_K(z_1)| = 2^6 \cdot 3$. However, $C_G(z_1) \cap C_G(u)$ has order 2^9 by Table 14.5.1, a contradiction. Hence $K = M$.

Thus $U = C_G(u) \cong (Z \times M) : \langle y \rangle$. MAGMA provides sets of defining relations of $M = \langle r_1, r_2 \rangle$ and $M : \langle y \rangle$, where $y = r_3$ which consist of all relations of the statement which only involve r_1, r_2 and r_3. As $r_4 = u$ and $r_5 = u^y$, and as all elements of $Z = \langle r_4, r_5 \rangle$ commute with the elements of M, the remaining relations are clear.

(f) This statement can easily be checked by the reader using (e) and the MAGMA command `CosetAction(U,T)`, where T is the given stabilizer.

(g) The author used PU and Kratzer's Algorithm 5.3.18 of [92] to calculate the Table 14.5.6 of the representatives of the conjugacy classes of U.

(h) The character table of U was computed with MAGMA and PE.

(i) By Table 14.5.6 the group U has six conjugacy classes of involutions. Using its notation we know that $r_4 = u$ generates the center $Z(U)$ of $U = C_G(u)$, and $Y = \langle r_4, r_5 \rangle$ is a normal Klein 4-group of U commuting with the normal subgroup $X = \langle r_1, r_2 \rangle \cong \mathrm{PSL}_3(4)$. By the construction of X we may assume that the representative r_1 is a 2-central involution of X. Since $C_E(u)$ is a Sylow 2-subgroup S_1 of U containing the representatives $r_1 r_4$ and $r_1 r_5$ one can check that they are E-conjugate to w by means of the faithful permutation representation

PE and MAGMA. Therefore they are both G-conjugate to u. Similarly it has been verified that $N_E(Y)$ contains an element k of order 3 fusing r_4 and r_5. Thus also r_5 is G-conjugate to $u = r_4$. By Table 14.5.6 the representative r_3 is U-conjugate to an involution $x \in S_1$ such that $C_{S_1}(x)$ has order 2^5. Another application of PE and MAGMA yields that any such involution x of S_1 is E-conjugate to $u = (ye)^2$ of E. As we also may assume that the representative r_1 belongs to S_1 the same method can be used to check that r_1 and x_1^{12} are E-conjugate. By Table 14.5.2 and the proof of Proposition 14.2.2(f) we know that x_1^{12} and z are conjugate involutions of G. □

Proposition 14.2.4 *Let G be a finite simple group of* Suz*-type with a 2-central involution z such that $C_G(z) = H = \langle x, y, h\rangle$ described in Lemma 14.1.1. Then $|G| = 2^{13} \cdot 3^7 \cdot 5^2 \cdot 7 \cdot 11 \cdot 13$.*

Proof By Proposition 14.2.2(f) and Tables 14.5.1 and 14.5.6 each simple group G of Suz-type has two conjugacy classes of involutions z^G and u^G such that $H = C_G(z)$ and $U = C_G(u)$ have respective orders $2^{13} \cdot 3^4 \cdot 5$ and $2^9 \cdot 3^2 \cdot 5 \cdot 7$.

Let

$$r(z,u,z) = \left|\{(x,y) \in (z^G \cap H) \times (u^G \cap H) \big| z \in \langle xy\rangle\}\right|$$

and

$$r(z,u,u) = \left|\{(x,y) \in (z^G \cap U) \times (u^G \cap U) \big| u \in \langle xy\rangle\}\right|.$$

Table 14.5.1 and Proposition 14.2.2(f) imply that $z^G \cap H = \{z\} \cup z_2^H \cup z_3^H$, where $z_1 = z$, z_2 and z_3 represent the classes 2_1, 2_2 and 2_3 of H, respectively. Furthermore, $u^G \cap H = u^H$, where u represents the class 2_4 of H. Let $\{t_j \mid 1 \le j \le 20\}$ be representatives of the 20 real z-special conjugacy classes of H. Then Theorem 1.6.4 asserts that

$$r(z,u,z) = \sum_{j=1}^{20} d(z_1,u,t_j) + d(z_2,u,t_j) + d(z_3,u,t_j),$$

where

$$d(z_i,u,t_j) = \frac{|H|^2}{|C_H(z_i)| \cdot |C_H(u)| \cdot |C_H(t_j)|} \times \sum_{\psi \in Irr_{\mathbb{C}}(H)} \psi(z_i)\psi(u)\psi(t_j)\psi(1)^{-1}.$$

Inserting the values of the character Table 14.6.1 of H into these formulas yields $r(z,u,z) = 1529280$.

By Proposition 14.2.3(f) we know that the group U has six conjugacy classes of involutions, five of which fuse in G with u and the remaining one is 2_3; it fuses with z. Let $u_1 = u = 2_1$, $u_2 = 2_2$, $u_3 = 2_4$, $u_4 = 2_5$ and $u_5 = 2_6$ in the notation of Table 14.5.6 of U. Let $\{s_k \mid 1 \leq k \leq 6\}$ be representatives of the six real u-special conjugacy classes of U. Then $u^G \cap U = \{u_1\} \cup u_2^U \cup u_4^U \cup u_5^U \cup u_6^U$ and $z^G \cap U = z^U$. Hence Theorem 1.6.4 asserts that

$$r(z,u,u) = \sum_{k=1}^{6} d(u_2,z,s_k) + d(u_4,z,s_k) + d(u_5,z,s_k) + d(u_6,z,s_k),$$

where

$$d(u_n,z,s_k) = \frac{|U|^2}{|C_U(u_n)| \cdot |C_U(z)| \cdot |C_U(s_k)|} \times \sum_{\psi \in Irr_{\mathbb{C}}(U)} \psi(u_n)\psi(z)\psi(s_k)\psi(1)^{-1}$$

for all $1 \leq n, k \leq 6$. Inserting the values of the character Table 14.6.4 of U into these formulas yields $r(z,u,u) = 60795$. Now Theorem 4.2.1 of [92] implies

$$|G| = r(z,u,z) \cdot |C_G(u)| + r(z,u,u) \cdot |C_G(z)| = 2^{13} \cdot 3^7 \cdot 5^2 \cdot 7 \cdot 11 \cdot 13.$$

□

Corollary 14.2.5 *Let G be a finite simple group of* Suz*-type with a 2-central involution z such that $C_G(z) = H = \langle x,y,h \rangle$ described in Lemma 14.1.1. Let $U = \langle r_1,r_2,r_3,r_4,r_5 \rangle \cong C_G(u)$ be the finitely presented group constructed in Proposition 14.2.3. Then the following statements hold.*

(a) *G has two conjugacy classes 2_A and 2_B of elements of involutions represented by the classes 2_1 of H and 2_1 of U given in Tables 14.5.1 and 14.5.6.*
(b) *G has four conjugacy classes 4_A, 4_B, 4_C and 4_D of elements of order 4 represented by the classes 4_1, 4_2, 4_4 of H and 4_1 of U given in Tables 14.5.1 and 14.5.6, respectively.*
(c) *G has five conjugacy classes 6_A, 6_B, 6_C, 6_D and 6_E of elements of order 6 represented by the classes 6_1, 6_2, 6_3, 6_4 of H and 6_1 of $U = C_G(u)$ given in Tables 14.5.1 and 14.5.6, respectively.*

(d) *G has three conjugacy class* 8_A, 8_B *and* 8_C *of elements of order* 8 *represented by the classes* 8_2, 8_3 *of* H *and* 8_1 *of* $U = C_G(u)$ *given in Tables 14.5.1 and 14.5.6, respectively.*

(e) *G has two conjugacy classes* 10_A *and* 10_B *of elements of order* 10 *represented by the classes* 10_1 *of* H *and* 10_1 *of* U *given in Tables 14.5.1 and 14.5.6, respectively.*

(f) *G has five conjugacy classes* 12_A, 12_B, 12_C, 12_D *and* 12_E *of elements of order* 12 *represented by the classes* 12_1, 12_2, 12_6, 12_7 *of* H *and* 12_1 *of* U *given in Tables 14.5.1 and 14.5.6, respectively.*

(e) *G has one conjugacy class* 14_A *of elements of order* 14 *represented by the class* 14_1 *of* U *given in Table 14.5.6.*

(f) *G has two conjugacy classes* 18_A *and* 18_B *of elements of order* 18 *represented by the classes* 18_1 *and* 18_2 *of* H *given in Table 14.5.1.*

(g) *G has one conjugacy class* 20_A *of elements of order* 20 *represented by the class* 20 *of* H *given in Table 14.5.1.*

(h) *G has one conjugacy class* 24_A *of elements of order* 24 *represented by the conjugacy class* 24 *of* H *given in Table 14.5.1.*

Proof By Proposition 14.2.3 G has two conjugacy classes z^G and u^G of involutions with centralizers $H = C_G(z)$ and $U = C_G(u)$, respectively. Therefore Theorem 2.2.11 of [92] implies that all statements are immediate consequences of Tables 14.5.1 and 14.5.6. □

14.3 Existence proof of Suz inside $\mathrm{GL}_{143}(13)$

In this section we apply Algorithm 7.4.8 of [92] to give an existence proof for a finite simple group G of Suz-type. It is mainly due to Weller, Staszewski and Kim. Throughout, $H = \langle x, y, h\rangle$ is the group of even order defined in Lemma 14.1.1, and $A = \langle z, t, h_2, h_7\rangle$ is the unique maximal elementary abelian normal subgroup of the fixed Sylow 2-subgroup S described there. Let $D_H = N_H(A) = \langle x, y\rangle$ and let $E = \langle x_1, y_1, e_1\rangle$ be the uniquely determined finitely presented group constructed in Proposition 14.2.2. Then $E \cong N_G(A)$ for any simple group of Suz-type.

Using the amalgam $H \leftarrow D \rightarrow E$ we now construct a simple group G of Suz-type. In the course of the construction we will identify D_H and D_E by means of the isomorphism $h : D_H \rightarrow D_E$ satisfying $h(x) = x_1$ and $h(y) = y_1$. Setting $e = e_1$ we will get $E = \langle x, y, e\rangle$.

Theorem 14.3.1 *Keep the notation of Lemma 14.1.1 and Proposition 14.2.2. Using the notation of the three character Tables 14.6.1, 14.6.3 and 14.6.2 of the groups* $H = \langle x, y, h \rangle$, $D = \langle x, y \rangle$ *and* $E = \langle x, y, e \rangle$, *respectively, the following statements hold.*

(a) *There is a compatible pair* $(\chi, \tau) \in mf\,char_{\mathbb{C}}(H) \times mf\,char_{\mathbb{C}}(E)$ *of degree* 143 *of the groups* $H = \langle D, h \rangle$ *and* $E = \langle D, e \rangle$:

$$(\chi, \tau) = (\chi_1 + \chi_7 + \chi_{11} + \chi_{\mathbf{15}} + \chi_{\mathbf{16}} + \chi_{17}, \quad \tau_7 + \tau_{18} + \tau_{\mathbf{32}})$$

with common restriction

$$\tau_{|D} = \chi_{|D} = \psi_1 + \psi_2 + \psi_{11} + \psi_{22} + \psi_{25} + \psi_{34} + \psi_{44} + \psi_{46} + \psi_{\mathbf{62}} + \chi_{\mathbf{64}},$$

where irreducible characters with bold face indices denote faithful irreducible characters.

(b) *Let* $\mathfrak{V}$ *and* $\mathfrak{W}$ *be the up to isomorphism uniquely determined faithful semi-simple multiplicity-free* 143*-dimensional modules of* H *and* E *over* $F = GF(13)$ *corresponding to the compatible pair* χ, τ, *respectively.*

Let $\kappa_{\mathfrak{V}} : H \to \mathrm{GL}_{143}(13)$ *and* $\kappa_{\mathfrak{W}} : E \to \mathrm{GL}_{143}(13)$ *be the representations of* H *and* E *afforded by the modules* $\mathfrak{V}$ *and* $\mathfrak{W}$, *respectively.*

Let $\mathfrak{h} = \kappa_{\mathfrak{V}}(h)$, $\mathfrak{x} = \kappa_{\mathfrak{V}}(x)$, $\mathfrak{y} = \kappa_{\mathfrak{V}}(y)$ *in* $\kappa_{\mathfrak{V}}(H) \le \mathrm{GL}_{143}(13)$. *Then the following assertions hold.*

(1) $\mathfrak{V}_{|D} \cong \mathfrak{W}_{|D}$, *and there is a transformation matrix* $\mathcal{T} \in \mathrm{GL}_{143}(13)$ *such that*

$$\mathfrak{x} = \mathcal{T}^{-1}\kappa_{\mathfrak{W}}(x_1)\mathcal{T}, \qquad \mathfrak{y} = \mathcal{T}^{-1}\kappa_{\mathfrak{W}}(y_1)\mathcal{T}.$$

Let $\mathfrak{e} = \mathcal{T}^{-1}\kappa_{\mathfrak{W}}(e_1)\mathcal{T} \in \mathrm{GL}_{143}(13)$.

(2) $\mathfrak{G} = \langle \mathfrak{h}, \mathfrak{x}, \mathfrak{y}, \mathfrak{e} \rangle$ *has a faithful permutation representation of degree* 135135 *with stabilizer* $\mathfrak{H} = \langle \mathfrak{x}, \mathfrak{y}, \mathfrak{h} \rangle$.

(3) *The generating matrices of* $\mathfrak{G}$ *are presented on the accompanying DVD.*

(4) *The character table of* $\mathfrak{G}$ *is equivalent to that of* Suz *in the Atlas [19], pp. 128–129.*

(c) $\mathfrak{G}$ *is a finite simple group with* 2*-central involution* $\kappa_{\mathfrak{V}}(z) = (\mathfrak{x})^{12}$ *such that*

$$C_{\mathfrak{G}}(\kappa_{\mathfrak{V}}(z)) = \kappa_{\mathfrak{V}}(H) \text{ and } |\mathfrak{G}| = 2^{13} \cdot 3^7 \cdot 5^2 \cdot 7 \cdot 11 \cdot 13.$$

Proof Throughout, we keep the notation of Lemma 14.1.1.

(a) The character tables of the groups $H = \langle x, y, h\rangle$, $D = \langle x, y\rangle$, $D_E = \langle x_1, y_1\rangle$ and E are stated in Section 14.6. In the following we use their notations. Using MAGMA, these character tables and the fusion of the classes of D in H and D_E in E we calculated the compatible pair stated in assertion (a) by means of Kratzer's Algorithm 7.3.10 of [92].

(b) In order to construct the semi-simple faithful representation $\mathfrak{V}$ corresponding to the character χ of H we use the faithful permutation representation PH of H of degree 1728 constructed in Lemma 14.1.1(a). Its stabilizer is the subgroup $U = \langle (x)^8, h, (x^7hxhx)^6\rangle$. The permutation character $((1_U)^H$ contains the irreducible constituents χ_k for $k = 11, 15, 16$.

Using a standard MAGMA command one gets the permutation matrices $P_i(x)$, $P_i(y)$ and $P_i(h)$, for $1 \le i \le 3$, for the three generators of H. Applying Algorithm 5.7.1 of [92] over the field $GF(13)$ to this permutation representation PH of H the 27×27 matrices $M_{11}(x)$, $M_{11}(y)$ and $M_{11}(h)$ corresponding to χ_{11} and the 32×32 matrices $M_k(x)$, $M_k(y)$ and $M_k(h)$ corresponding to χ_k for $k = 15, 16$, respectively.

In order to find the irreducible representations of H corresponding to the irreducible characters χ_7 and χ_{17} of defect zero over $GF(13)$ we applied the MAGMA command `LowIndexSubgroups(PE, m)`. It turned out that χ_7 is a constituent of the permutation character of the subgroup $U_1 = \langle (x^2hxh)^4, (xhxhx)^4, (x^4hx)^3\rangle$ of H of index $m = 40$. Furthermore, χ_{17} is a constituent of the permutation character of $U_2 = \langle h, (x^3hxhx)^6, xhx^5\rangle$ of index $m = 72$. Then we applied Algorithm 5.7.1 of [92] over the field $GF(13)$ to the two permutation representations $(1_{U_i})^H$, $i = 1, 2$. Thus we obtained the 15×15 matrices $M_7(x)$, $M_7(y)$ and $M_7(h)$ over $GF(13)$ corresponding to the irreducible character χ_7 and the three 36×36 matrices $M_{17}(x)$, $M_{17}(y)$ and $M_{17}(h)$ corresponding to χ_{17}.

Hence $\mathfrak{H} = \langle \mathfrak{x}, \mathfrak{y}, \mathfrak{h}\rangle$, where the generators of $\mathfrak{H}$ are the three blocked diagonal matrices

$$\mathfrak{x} = diag(1, M_7(x), M_{11}(x), M_{15}(x), M_{16}(x), M_{17}(x)),$$
$$\mathfrak{y} = diag(1, M_7(y), M_{11}(y), M_{15}(y), M_{16}(y), M_{17}(y)),$$
$$\mathfrak{h} = diag(1, M_7(h), M_{11}(h), M_{15}(h), M_{16}(h), M_{17}(h)),$$

where 1 denotes the 1×1 identity matrix. These three matrices of $\mathrm{GL}_{143}(13)$ describe the semi-simple H-module $\mathfrak{V}$ over $F = GF(13)$.

The faithful semi-simple representation $\mathfrak{W}$ corresponding to the character τ of $E = \langle x_1, y_1, e_1\rangle$ has been obtained in a similar way. Again the MAGMA command `LowIndexSubgroups(PE, 500)` was applied. It turned out that τ_7 and τ_{18} are constituents of the permutation character of the subgroup $W_1 = \langle y_1 e_1 y_1, (e_1^2 y_1 e_1)^2\rangle$ of $E = \langle x_1, y_1, e_1\rangle$. Furthermore, $\tau_{\mathbf{32}}$ is a constituent of $(1_{W_2})^E$, where $W_2 = \langle (e_1 y_1 e_1^2 y_1)^5, (y_1 e_1 y_1^2 e_1^2 y_1 e_1 y_1)^6, (y_1^5 e_1 y_1 e_1 y_1 e_1)^2\rangle$. Another application of Algorithm 5.7.1 of [92] over the field $GF(13)$ to the permutation representation PW_1 yields the three 5×5 matrices $N_7(x)$, $N_7(y)$ and $N_7(h)$ and the three 18×18 matrices $N_{18}(x)$, $N_{18}(y)$ and $N_{18}(h)$ corresponding to τ_7 and τ_{18}, respectively. Similarly we obtained from PW_2 the three 120×120 matrices $N_{32}(x)$, $N_{32}(y)$, $N_{32}(h)$ corresponding to $\tau_{\mathbf{32}}$.

Hence $\mathfrak{E} = \langle \mathfrak{x}_1, \mathfrak{y}_1, \mathfrak{e}_1\rangle$, where the generators of $\mathfrak{E}$ are the three blocked diagonal matrices

$$\mathfrak{x}_1 = diagonal(N_7(x_1), N_{18}(x_1), N_{32}(x_1)),$$
$$\mathfrak{y}_1 = diagonal(N_7(y_1), N_{18}(y_1), N_{32}(y_1)),$$
$$\mathfrak{e}_1 = diagonal(N_7(e_1), N_{18}(e_1), N_{32}(e_1)).$$

These three matrices of $\mathrm{GL}_{143}(13)$ describe the semi-simple E-module $\mathfrak{W}$ over $F = GF(13)$.

The matrix groups $\mathfrak{D}_H = \langle \mathfrak{x}, \mathfrak{y}\rangle$ and $\mathfrak{D} = \langle \mathfrak{W}(x_1), \mathfrak{W}(y_1)\rangle$ are isomorphic, and so are the semi-simple representations $\mathfrak{V}_{|\mathfrak{D}}$ and $\mathfrak{W}_{|\mathfrak{D}_E}$ by the definition of the compatible pair $(\chi, \tau) \in mfchar_{\mathbb{C}}(H) \times mfchar_{\mathbb{C}}(E)$ and Maschke's Theorem, which is applicable because the orders of H and E are not divisible by 13. Let $Y = \mathrm{GL}_{143}(23)$. Applying then Parker's Isomorphism Test stated as Proposition 6.1.6 in [92] by means of the MAGMA command

```
IsIsomorphic(GModule(sub<Y|V(x),V(y)>),
  GModule(sub<Y|W(x1),W(y1)>))
```

we obtained a transformation matrix $\mathcal{T}_1$ in Y such that the first two matrix equations of assertion (1) of (b) hold for $\mathcal{T}_1$ in the place of $\mathcal{T}$.

By Corollary 7.2.4 of [92] the transformation matrix $\mathcal{T}$ of the statement is the product matrix $\mathcal{D}\mathcal{T}_1$, where

$$\mathcal{D} = diag(1, 7, 12^3, 1^6, 10^8, 1^{12}, 11^{24}, 1^{24}, 1^{32}, 1^{32}) \in \mathrm{GL}_{143}(13)$$

and c^n denotes the $n \times n$ diagonal matrix with unique diagonal entry $c \in GF(13)$.

Let $\mathfrak{e} = (\mathfrak{W}(e_1))^{\mathcal{T}}$. Then $\mathfrak{E} = \langle \mathfrak{x}, \mathfrak{y}, \mathfrak{e}\rangle \cong E$.

Let $\mathfrak{G} = \langle \mathfrak{h}, \mathfrak{x}, \mathfrak{y}, \mathfrak{e} \rangle$. Using the algorithm stated in the proof of Theorem 6.2.1 of [92] we obtained a faithful permutation representation $P\mathfrak{G}$ of $\mathfrak{G}$ of degree 135135 with stabilizer $\mathfrak{H}$. In particular, $|\mathfrak{G}| = 2^{13} \cdot 3^7 \cdot 5^2 \cdot 7 \cdot 11 \cdot 13$.

Furthermore, the character table of $\mathfrak{G}$ has been calculated by means of the above permutation representation and MAGMA. It coincides with that of Suz in [19], pp. 128–129.

(c) Let $\mathfrak{z} = (\mathfrak{x})^4$. Then $C_{\mathfrak{G}}(\mathfrak{z})$ contains $\mathfrak{H} = \langle \mathfrak{h}, \mathfrak{x}, \mathfrak{y} \rangle$ which is isomorphic to H. By the character table of $\mathfrak{G}$ we know that $|C_{\mathfrak{G}}(\mathfrak{z})| = |H|$. Hence $C_{\mathfrak{G}}(\mathfrak{z}) \cong \mathfrak{H}$. The character table of $\mathfrak{G}$ implies that $\mathfrak{G}$ is a simple group. □

14.4 Uniqueness proof

The first major step of the uniqueness proof of a finite simple group G of Suz-type is the construction of a Sylow 3-subgroup S of G from the given centralizer $H = C_G(z)$ of a 2-central involution z. It depends on the following theorem due to Phan proved in [109]. For a definition of the 4-dimensional projective special unitary group $\mathrm{U}_4(3)$ over the field $GF(9)$, see [13], pp. 7–8.

Theorem 14.4.1 (K.W. Phan) *Let G be a finite simple group with a 2-central involution z such that $C_G(z)$ is isomorphic to the group $H = \langle a_1, a_2, b_1, b_2, c_1, c_2, v, w \rangle$ satisfying the following set $\mathcal{R}(H)$ of defining relations:*

$$\begin{aligned}
&a_1^2 = a_2^2 = b_1^2 = b_2^2 = v^2 = z, \quad z^2 = c_1^3 = c_2^3 = w^2 = 1,\\
&(a_1, a_2) = (a_1, b_2) = (a_1, c_2) = (b_1, a_2) = (b_1, b_2) = 1,\\
&(b_1, c_2) = (c_1, a_2) = (c_1, b_2) = (c_1, c_2) = 1,\\
&a_i^{b_i} = a_i^3, \quad a_i^{c_i} = b_i, \quad b_i^{c_i} = a_i b_i \quad \mathit{for} \quad 1 \le i \le 2,\\
&a_i^v = a_i^3, \quad b_i^v = b_i a_i, \quad c_i^v = c_i^2 \quad \mathit{for} \quad 1 \le i \le 2,\\
&a_1^w = a_2, \quad b_1^w = b_2, \quad c_1^w = c_2, \quad v^w = v^3.
\end{aligned}$$

Then $G \cong \mathrm{U}_4(3)$.

Remark 14.4.2 Phan's proof of this result is involved. He shows that any simple group of $\mathrm{U}_4(3)$-type has a unique conjugacy class of involutions. His argument could be included into this book. But he also constructs a $(B - N)$-pair from the given centralizer $H = C_G(z)$ and quotes several deep results by J. Tits and J. G. Thompson which cannot be included here. However, using the author's group order formula of

Theorem 4.3.7, Algorithm 7.4.8 and Theorem 7.5.1, all of [92], it is not so difficult to show that each simple group G of $U_4(3)$-type is isomorphic to a well determined irreducible subgroup $\mathfrak{G}$ of $GL_{21}(7)$. This can be proved by means of the adapted arguments used in Chapter 2 for $G_2(3)$.

Proposition 14.4.3 *Let G be a finite simple group of Suz-type with a 2-central involution z such that $C_G(z) = H = \langle x, y, h\rangle$ described in Lemma 14.1.1. For the element $d = x^8$ of order 3 the following statements hold.*

(a) *$C_H(d)$ is a split extension of $K = \langle k_1, k_2, k_3\rangle$ by $\langle d\rangle$, where $k_1 = x^3$, $k_2 = (x^3hxhxh)^6$ and $k_3 = (hx^5hxhx^3h)^6$. Furthermore, K is isomorphic to the centralizer $H_1 = C_{U_1}(z_1)$ of a 2-central involution z_1 of the unitary simple group $U_1 = U_4(3)$.*

(b) *$C_H(d)$ has a complement $\langle u_1\rangle$ of order 2 in $N_H(\langle d\rangle)$ such that $|C_H(u_1)| = 2^9$.*

(c) *$N_1 = N_G(\langle d\rangle)$ is isomorphic to the group $N_1 = \langle g_i \mid 1 \le i \le 4\rangle$ satisfying the following set $\mathcal{R}(N_1)$ of defining relations:*

$$g_1^2 = g_2^6 = g_3^2 = g_4^3 = 1, \quad (g_1, g_4) = (g_2, g_4) = 1,$$
$$g_3g_4g_3g_4 = 1, \quad (g_2g_3g_2^{-1}g_1)^2 = 1, \quad (g_1g_3)^6 = 1,$$
$$g_1g_2g_1g_2g_3g_2^{-2}g_1g_2^{-1}g_1g_3g_1g_2 = g_1g_2^2g_1g_2g_1g_2^2g_3g_1g_2g_3g_1g_2g_4 = 1,$$
$$g_2g_3g_2^{-1}g_3g_2^{-1}g_1g_2^2g_3g_2^{-2}g_1g_2g_3g_4^{-1} = 1,$$
$$g_2^{-1}g_1g_3g_2^{-2}g_3g_1g_2^2g_1g_3g_2^2g_3g_1g_2^{-1} = 1, \quad (g_2g_1g_2^{-1}g_1)^4 = 1,$$
$$g_3g_1g_2^{-1}g_1g_2g_1g_3g_2^{-2}g_1g_2g_1g_2^{-1}g_1g_2g_1g_2 = 1,$$
$$g_3g_2^{-3}g_3g_1g_3g_2^3g_3g_1g_3g_2^3g_3g_1 = 1, \quad (g_2g_3g_2^{-2}g_3g_2^{-2}g_3g_2)^2 = 1,$$
$$g_2g_1g_2^{-1}g_1g_3g_2^{-1}g_3g_1g_2^{-2}g_3g_2g_3g_2g_3g_1g_2^{-2}g_3 = 1.$$

(d) *$N_1 = \langle g_i \mid 1 \le i \le 4\rangle$ has a faithful permutation representation PN_1 of degree 756 with stabilizer $\langle g_2, (g_1g_2g_1g_2^2g_1g_2^2)^6\rangle$.*

(e) *A system of representatives t_i and the corresponding centralizer orders $|C_{N_1}(t_i)|$ of the 38 conjugacy classes of $N_1 = \langle n_1, n_2, n_3\rangle$ are given in Table 14.5.7.*

(f) *The character table of $N_1 = N_G(\langle d\rangle)$ is given in Table 14.6.5.*

(g) *N_1 has two conjugacy classes of involutions represented by the classes 2_1 and 2_2 of Table 14.6.5 fusing in G with z and u, respectively.*

(h) *N_1 has five conjugacy classes of elements of order 3 represented by the classes 3_i, $1 \le i \le 5$, of Table 14.6.5. The classes 3_1, 3_2*

and the classes 3_3, 3_4 *fuse in* G *to the* 3*-central conjugacy classes* 3_A *and* 3_B *of* G*, respectively.*

(i) G *has three conjugacy classes of elements of order* 3*. The third class* 3_C *is represented by the classes* 3_5 *of Table 14.5.7.*

Proof (a) Let PH be the faithful permutation representation of $H = \langle x, y, h\rangle$ of degree 1728 constructed in Lemma 14.1.1(a). By Table 14.5.1 the element $d = x^8$ is a representative of a conjugacy class of elements of order 3 in H such that $C_1 = C_H(d)$ has order $2^7 \cdot 3^3$. Let $D_1 = \langle d\rangle$. Let $k_i \in C_1$ be the elements defined in the statement. Using PH, the MAGMA command `HasComplement(C_1,D_1)` we observed that $K = \langle k_1, k_2, k_3\rangle$ is a complement of D_1 in C_1. Let H_1 be the finitely presented group stated in Theorem 14.4.1. An isomorphism test with MAGMA confirms that $K \cong H_1$. Thus (a) holds.

(b) Since C_1 is a normal subgroup of $M_1 = N_H(D_1)$ the reader can easily verify this assertion by another application of the MAGMA command `HasComplement(M_1,D_1)`.

(c) Let PE be the faithful permutation of degree 1536 of $E = N_G(A)$ stated in Proposition 14.2.2(b), where A is the maximal elementary abelian normal subgroup of the Sylow 2-subgroup S of H defined in Lemma 14.1.1(e). Note that $C_2 = C_E(d)$ has order $2^7 \cdot 3^3 \cdot 5$ by Table 14.5.2. Using PE and MAGMA it is easy to verify that $C_2 = C_E(d)$ does not split over its normal subgroup D_1. Furthermore, C_2/D_1 is isomorphic to a perfect split extension of A_6 by A, and A is an irreducible A_6-module of dimension 4 over $GF(2)$. Let $C = C_G(d)$ and $Q = C/D_1$. Then K is a subgroup of Q containing a Sylow 2-subgroup of Q by Tables 14.5.2 and 14.5.1.

Let P be the last term of the derived series of Q. Since $C_2/D_1 \leq Q$ the index $|Q : P|$ is odd. Now (a) and Table 14.5.1 imply that P has a 2-central involution z_1 such that $C_P(z_1) \cong K$, and that P has a unique conjugacy class of involutions. A standard application of the Brauer–Wielandt Theorem 4.2.2 of [92] shows that P does not have any nontrivial normal subgroups of odd order because $C_P(z_1)$ does not have any normal subgroups by (a) and Theorem 14.4.1. Suppose that X is a proper normal subgroup of P. As all involutions of P are conjugate in P we may assume that $x = z_1$. Hence P/X is a group of odd order and therefore solvable by the Theorem of Feit and Thompson. Thus $X = P$ because P is perfect. Therefore P is simple. Hence $P \cong \mathrm{U}_4(3)$ by Theorem 14.4.1. Since the outer automorphism group of $\mathrm{U}_4(3)$ has order 8 and $|Q : P|$ is odd it follows that $Q = P$. In particular, C is a non-split central extension of Q by D_1.

MAGMA provides two generators t_1 and t_2 for $Q = U_4(3)$ satisfying the following set $\mathcal{R}(Q)$ of defining relations:

$$
\begin{aligned}
&t_1^4 = t_2^9 = 1, \quad (t_1, t_2)^3 = 1,\\
&(t_1^{-1}t_2t_1^{-1}t_2^{-1})^2 = (t_1^{-1}t_2^{-1}t_1t_2^{-1})^3 = 1, \quad (t_1^{-1}t_2^{-1})^8 = (t_1t_2^{-1})^8 = 1,\\
&t_1^{-1}t_2^2t_1t_2^{-1}t_1^{-1}t_2^{-1}t_1t_2^2t_1^{-1}t_2^{-1}t_1t_2^{-1} = t_1^{-1}t_2^{-3}t_1^{-2}t_2^{-3}t_1^2t_2^3t_1^2t_2^3t_1^{-1} = 1,\\
&t_2^{-1}t_1^{-2}t_2^{-2}t_1^{-2}t_2^{-2}t_1^2t_2^{-2}t_1^2t_2^{-2}t_1^2t_2^{-1} = 1,\\
&t_2t_1t_2^{-1}t_1^{-1}t_2^2t_1t_2t_1^{-1}t_2^{-2}t_1^2t_2^{-4}t_1t_2^{-1}t_1t_2^{-1}t_1^{-1}t_2^2 = 1.
\end{aligned}
$$

Q has a faithful permutation representation PQ of degree 280 with stabilizer $\langle t_2, t_1t_2t_1^3t_2t_1\rangle$. Using it and the MAGMA commands

```
AQ:=AutomorphismGroup(Q),

PAQ:=PermutationRepresentation(AQ),
```

and

```
Subgroups(PAQ: OrderEqual := 2^8*3^6*5*7)
```

we determined all conjugacy classes of subgroups PY_i of PAQ which are a split extension of a cyclic subgroup of order 2 by Q. By (b) and Theorem 1.4.15 of [92] we know that N_1/D_1 has a Sylow 2-subgroup S of order 2^8 which is isomorphic to a Sylow 2-subgroup of $M_1 = N_H(D_1) = \langle p_1, p_2, p_3\rangle$, where $p_1 = x$, $p_2 = (x^3hxhxh)^6$ and $p_3 = (x^3hx^4hx^2h)^3$. In particular, S is isomorphic to the subgroup $X = \langle p_1^6, p_2p_3p_2, (p_1p_3p_2p_3)^3\rangle$. For each subgroup PY_i we chose a Sylow 2-subgroup S_i and performed the isomorphism test `IsIsomorphic(X,S_i)`. Thus we found exactly one subgroup PY among the subgroups PY_i satisfying this isomorphism test. Using the MAGMA command `FPGroup(PY)` we obtained the following set $\mathcal{R}(Y)$ of defining relations of $Y = \langle y_1, y_2, y_3\rangle$:

$$
\begin{aligned}
&y_1^2 = y_2^6 = y_3^2 = 1, \quad (y_1y_3)^6 = 1, \quad (y_2y_3y_2^{-1}y_1)^2 = 1,\\
&y_1y_2y_1y_2y_3y_2^{-2}y_1y_2^{-1}y_1y_3y_1y_2 = y_2y_3y_2^{-1}y_3y_2^{-1}y_1y_2^2y_3y_2^{-2}y_1y_2y_3 = 1,\\
&y_1y_2^2y_1y_2y_1y_2^2y_3y_1y_2y_3y_1y_2 = y_2^{-1}y_1y_3y_2^{-2}y_3y_1y_2^2y_1y_3y_2^2y_3y_1y_2^{-1} = 1,\\
&(y_2y_1y_2^{-1}y_1)^4 = 1, \quad (y_2y_3y_2^{-2}y_3y_2^{-2}y_3y_2)^2 = 1,\\
&y_3y_1y_2^{-1}y_1y_2y_1y_3y_2^{-2}y_1y_2y_1y_2^{-1}y_1y_2y_1y_2 = 1,\\
&y_3y_2^{-3}y_3y_1y_3y_2^3y_3y_1y_3y_2^3y_3y_1 = 1,\\
&y_2y_1y_2^{-1}y_1y_3y_2^{-1}y_3y_1y_2^{-2}y_3y_2y_3y_2y_3y_1y_2^{-2}y_3 = 1.
\end{aligned}
$$

It has been checked that $Q \cong \langle y_1, y_2 \rangle$ and that $\langle y_3 \rangle$ is a complement of order 2 of the normal subgroup Q of Y. In order to construct $N_1 = N_G(D_1)$ we built a 1-dimensional Y-module over $F = GF(3)$ by means of the matrices $My_1 = Id(GL(1,F))$, $My_2 = Id(GL(1,F))$ and $My_3 = GL(1,F)![2]$. Using the commands

```
FEalg:=MatrixAlgebra<F,1|My_1,My_2,M_3>,
V:=GModule(PY,FEalg)
```

and

```
CohomologicalDimension(PY,V,2)
```

MAGMA calculated this dimension: it is 1. Applying then the commands

`P:=ExtensionProcess(PY,V,Y)` and `Extension(P,[1])`

we obtained the non-split extension $N_1 = \langle g_1, g_2, g_3, g_4 \rangle$ of Y by D_1 having the set $\mathcal{R}(N_1)$ of defining relations given in the statement. N_1 is uniquely determined by Y up to isomorphism.

(d) This assertion is easily checked by means of a routine MAGMA calculation.

(e) The author used PN_1 and Kratzer's Algorithm 5.3.18 of [92] to calculate Table 14.5.7 of representatives of the conjugacy classes of $N_1 = N_G(D_1)$.

(f) The character table of N_1 was computed with MAGMA and PN_1.

(g) Let $X = C_{N_1}(g_1) \cap C_{N_1}(g_4)$. Using MAGMA and the faithful permutation representation PN_1 of N_1 stated in (d) the reader can verify that $|X| = 2^7 \cdot 3^3$. Furthermore, X is generated by $x_1 = (g_1 g_2^2)^3$, $x_2 = (g_2 g_1 g_2^3 g_1 g_2)^4$ and $x_3 = (g_1 g_2 g_1 g_2^5)^2$.

In $H = \langle x, y, h \rangle$ let $d = x^8$ and $Y = C_H(d)$. By Table 14.5.1 d represents the conjugacy class 3_1 of H and $|Y| = 2^7 \cdot 3^3$. Using MAGMA and the faithful permutation representation PH of H stated in Lemma 14.1.1(a) it can be checked that $Y = \langle j_1, j_2 \rangle$, where $j_1 = x$ and $j_2 = (xhx^3hx^2hxh)^2$. Another application of MAGMA yields the existence of an isomorphism $\sigma : X \to Y$. It identifies the conjugacy classes of X and Y. We worked out the fusion of the X-classes in N_1 and the fusion of the Y-classes in H. Using the notation of Tables 14.5.7 and 14.5.1 we obtained the following fusion pattern:

$$\langle [2_1], [2,3,4,5], [2_1, 2_2, 2_3] \rangle,$$
$$\langle [3_1, 3_2], [6,7,10], [3_1] \rangle,$$
$$\langle [3_3, 3_4], [8,9,11,12,13,14,15,16], [3_2, 3_3, 3_4] \rangle.$$

Here the first entries of the three triples denote the conjugacy classes of involutions and elements of order 3 of $N1$ classified in Table 14.5.7. The last entries of the three triples denote the conjugacy classes of involutions and elements of order 3 of H classified in Table 14.5.1. The middle entries denote the four classes of involutions and 11 classes of 3-elements of the common subgroup $X \cong Y$ of N_1 and H denoted by the class numbers of all conjugacy classes of the common subgroup of N_1 and H.

In particular, the first three classes of involutions of H fuse in G with class 2_1 of N_1. Now Proposition 14.2.2(f) and Table 14.5.7 imply that the class 2_2 of N_1 fuses with the non-2-central involution u of G represented by the conjugacy class 2_4 of Table 14.5.1.

(h) and (i) These statements are immediate consequences of the above fusion pattern and Table 14.5.7 because N_1 contains a Sylow 3-subgroup of G by Proposition 14.2.4. □

Proposition 14.4.4 *Let G be a finite simple group of* Suz*-type with a 2-central involution z such that $C_G(z) = H = \langle x, y, h\rangle$ described in Lemma 14.1.1. Let $N_1 = \langle g_1, g_2, g_3, g_4\rangle$ be the finitely presented group constructed in Proposition 14.4.3(c). Then the following statements hold.*

(a) *The subgroup S of N_1 generated by $s_1 = (g_2 g_1 g_2^3 g_1 g_2^2 g_1 g_2^2)^4$, $s_2 = g_2 g_1 g_2^2 g_1 g_2 g_1 g_2 g_1 g_2^4 g_1 g_2 g_1 g_2^2$ and $s_3 = (g_1 g_2^3 g_1 g_2^3 g_1 g_2 g_1 g_2 g_1 g_2^2 g_1 g_2^3)^2$ is a Sylow 3-subgroup of N_1 and G.*

(b) *The center Z of S is an elementary abelian group of order 9 generated by $z_1 = s_2^3$ and $z_2 = (s_2 s_3)^3$.*

(c) *$Y = N_{N_1}(Z)$ is generated by $n_1 = (g_3 g_2^4 g_3 g_2 g_3)^8$, $n_2 = g_3 g_2^4 g_3 g_2^2 g_3$ and $n_3 = (g_3 g_2 g_3 g_2^2 g_3 g_2^3 g_3 g_2 g_3 g_2)^2$. Its center $Z(Y) = 1$.*

(d) *$N_2 = N_G(Z)$ is isomorphic to the finitely presented group $N_2 = \langle t_1, t_2, t_3, t_4\rangle$ with the following set $\mathcal{R}(N_2)$ of defining relations:*

$$t_1^4 = t_2^6 = t_3^2 = t_4^4 = 1,$$
$$t_2^{-1} t_1^{-2} t_2^{-2} = t_1^{-2} t_3 t_1^2 t_4 t_3 t_4^{-1} = 1,$$
$$t_4^{-1} t_2^{-1} t_3 t_1^2 t_3 t_2 t_4^{-1} = t_4^{-1} t_2 t_4^{-1} t_1^2 t_4 t_2^{-1} t_4^{-1} = 1,$$
$$t_1^{-1} t_2^{-1} t_1^{-1} t_2^{-1} t_1 t_2 t_1 t_2 = t_3 t_4^{-1} t_1^{-1} t_2^{-1} t_1^{-1} t_3 t_2^{-1} t_4^{-1} = 1,$$
$$t_2 t_4^{-1} t_2 t_3 t_1^2 t_3 t_4^{-1} t_2 t_4^{-1} = t_1^{-1} t_4^{-1} t_2^{-1} t_1^{-1} t_2^{-1} t_1 t_4 t_2 t_1 t_2 = 1,$$
$$t_1 t_3 t_1^{-1} t_4 t_2 t_1^{-1} t_3 t_1 t_2^{-1} t_4^{-1} = t_2^{-1} t_4 t_3 t_1^{-1} t_3 t_2^{-1} t_4 t_3 t_1 t_3 = 1,$$
$$(t_1^{-1} t_3 t_4 t_1^{-1} t_3)^2 = 1.$$

(e) *$N_2 = \langle t_i \mid 1 \leq i \leq 4\rangle$ has a faithful permutation representation PN_2 of degree 432 with stabilizer $\langle t_3, t_1 t_2 t_1, (t_1 t_3 t_2)^4, (t_2^2 t_3)^3\rangle$.*

(f) *A system of representatives n_i and the corresponding centralizer orders $|C_{N_2}(n_i)|$ of the* 49 *conjugacy classes of $N_2 = \langle t_1, t_2, t_3, t_4 \rangle$ are given in Table 14.5.8.*

(g) *The character table of $N_2 = N_G(Z)$ is given in Table 14.6.6.*

(h) *N_2 has three conjugacy classes of involutions represented by the classes 2_1, 2_2 and 2_3 of Table 14.6.6. The first two classes fuse in G with z and 2_3 fuses with u.*

(i) *N_2 has eight conjugacy classes of elements of order* 3 *represented by the classes 3_i, $1 \leq i \leq 8$, of Table 14.6.6. The classes 3_1, 3_4, 3_5, 3_6, 3_7 belong to the class 3_A, the classes 3_2 and 3_3 belong to the class 3_B and the class 3_8 belongs to the class 3_C of G defined in Proposition 14.4.3(h) and (i), respectively.*

Proof (a) Using MAGMA and the faithful permutation representation PN_1 of $N_1 = \langle g_i \mid 1 \leq i \leq 4 \rangle$ given in Proposition 14.4.3(d) the reader can verify that $S = \langle s_1, s_2, s_3 \rangle$ is a Sylow 3-subgroup of N_1. Hence S is a Sylow 3-subgroup of G by Proposition 14.2.4.

(b) It is easily checked computationally that $Z = \langle z_1, z_2 \rangle$.

(c) We found the three generators n_i of $Y = N_{N_1}(Z)$ by means of Kim's program `GetShortGens(N_1,Y)`. Another application of MAGMA yielded that $Z(Y) = 1$.

(d) By Proposition 14.4.3 $C = C_G(Z) = C_{N_1}(Z)$. Using PN_1 and MAGMA it has been checked that $|C| = 2^3 \cdot 3^7$, $C_{N_1}(S) = Z$, $|N_{N_1}(S)| = 2^3 \cdot 3^7$ and $C \cap N_{N_1}(S) = S$. Hence $C \cap N_G(S) = S$. Therefore

$$|N_G(S)/S| = |N_G(S)C/C| \leq |N_G(Z)/C| \leq |\,\mathrm{GL}_2(3)| = 2^4 \cdot 3.$$

Now Proposition 14.2.4 and Sylow's Theorem imply that $|N_G(S)| = 2^4 \cdot 3^7$. Hence $N_G(S)$ is a split extension of a semi-dihedral group W of order 16 by S. As $|N_{N_1}(Z)| = 2^5 \cdot 3^7$ it follows that $|N_G(Z)| = 2^6 \cdot 3^7$.

In order to find a presentation of $N_G(Z)$ we calculated the automorphism group $AY = Aut(Y)$ of Y because $Z(Y) = 1$. It has been checked computationally that Z is the unique elementary abelian normal subgroup of Y. Hence it is a characteristic subgroup of Y. Therefore it is normal in AY.

Using PN_1 and the MAGMA commands `AY:=AutomorphismGroup(Y)` and `PAY:=PermutationRepresentation(AY)` we observed that $|AY| = 2^7 \cdot 3^7$. Furthermore, MAGMA provided 12 generators for the normal subgroup PI of PAY corresponding to the group I of inner automorphisms of Y. Applying the MAGMA command `HasComplement(PAY,PI)` we observed that PI has an elementary abelian complement of order 4 in PAY. Thus we got 14 generators of PAY. Using Kim's program `GetShortGens(PAY,PAY)` we saw that AY can be generated by four elements a_i, $1 \leq i \leq 4$. Since the degree of PAY is 5832 we employed the MAGMA command `DegreeReduction(PAY)`. It yielded a faithful permutation representation PA of AY of degree 486. Thus we may assume that $PA = \langle Pa_i \mid 1 \leq i \leq 4 \rangle$.

In order to find the image of Y in PA we determined the set of normal subgroups PU_k of order $2^5 \cdot 3^7$ of PA using the command

```
Subgroups(PA: Al:=Normal, OrderEqual := 2^5*3^7).
```

It turned out that there are exactly seven such normal subgroups PU_k of PA. However, only PU_2 is isomorphic to Y. Let $PU = PU_2$. MAGMA also provides an explicit isomorphism $\eta : Y \to PU$. Hence $\eta(S) = PT$ is a Sylow 3-subgroup of PU. Let $PN = N_{PA}(PT)$. Then $|PN| = 2^4 \cdot 3^7$ and $PN_2 = \langle PU, PN \rangle$ has order $|PN_2| = 2^4 \cdot 3^7$. Furthermore, PN_2 is generated by four elements $Pt_1 = Pa_1$, $Pt_2 = Pa_2$, $Pt_3 = Pa_3$ and $Pt_4 = (Pa_2 Pa_4)^2$ of PA. Applying the MAGMA command `N_2 := FPGroup(PN_2)` we obtained the finitely presented group $N_2 = \langle t_1, t_2, t_3, t_4 \rangle$ with the set $\mathcal{R}(N_2)$ of defining relations given in the statement. In particular, N_2 is uniquely determined up to isomorphism.

(e) This assertion is easily checked by means of the MAGMA command `CosetAction(N_1,sub<N_1|Q>)`, where $Q = \langle t_3, t_1 t_2 t_1, (t_1 t_3 t_2)^4, (t_2^2 t_3)^3 \rangle$.

(f) The author used PN_2 and Kratzer's Algorithm 5.3.18 of [92] to calculate Table 14.5.8 of the representatives of the conjugacy classes of $N_2 = N_G(Z)$.

(g) MAGMA computed the character table of N_2 with PN_2.

(h) and (i) Let $X = C_{N_1}(g_2^2)$. Then X has order $2^3 \cdot 3^7$ by Table 14.5.7. Using MAGMA and the faithful permutation representation PN_2 of N_2 defined in (e) and the MAGMA command

```
Subgroups(PN2: Al:=Normal, OrderEqual := 2^5*3^7)
```

the author verified that N_2 has exactly three normal subgroups T_j, $1 \leq j \leq 3$, of this order. Applying the command `IsIsomorphic(X,T_j)` we observed that there is exactly one positive answer to this question, namely $j = 2$. Using Kim's program `GetShortGens(N_2,T_2)` we saw that $T_2 = \langle t_1, t_2, t_4, (t_1 t_3)^2 \rangle$. Another application of MAGMA yielded the existence of an isomorphism $\sigma : X \to T_2$. It identifies the conjugacy classes of X and T_2. We worked out the fusion of the X-classes in N_1 and the fusion of the T_2-classes in N_2. Using the notation of Tables 14.5.7

and 14.5.8 we obtained the following fusion pattern:

$$\langle [2_1], [2], [2_1] \rangle,$$
$$\langle [2_2], [3], [2_3] \rangle,$$
$$\langle [3_1, 3_2], [4, 5, 7], [3_2, 3_3] \rangle,$$
$$\langle [3_3], [6, 8], [3_1, 3_4] \rangle,$$
$$\langle [3_4], [9, 10, 11, 12], [3_5, 3_6, 3_7] \rangle,$$
$$\langle [3_5], [13], [12] \rangle.$$

Here the first entries of the six triples denote the conjugacy classes of involutions and elements of order 3 of N_1 classified in Table 14.5.7. The last entries of the six triples denote the conjugacy classes of involutions and elements of order 3 of N_2 classified in Table 14.5.8. The middle entries denote the two classes of involutions and ten classes of 3-elements of the common subgroup $X \cong T_2$ of N_1 and N_2 denoted by the class numbers of all conjugacy classes of the common subgroup of N_1 and N_2. Hence (h) holds by Proposition 14.4.3(g) and Proposition 14.2.2(f).

By Proposition 14.4.3(h) the classes 3_1, 3_2 and the classes 3_3, 3_4 of N_1 fuse in G with the classes 3_A and 3_B of G, respectively. Furthermore, the class 3_5 of N_1 represents the non-3-central conjugacy class 3_C of G by Proposition 14.4.3(i). Hence all assertions of (i) follow from the above fusion pattern. □

Proposition 14.4.5 *Let G be a finite simple group of* Suz*-type with a 2-central involution z such that $C_G(z) = H = \langle x, y, h \rangle$ described in Lemma 14.1.1. Let $N_2 = \langle t_1, t_2, t_3, t_4 \rangle$ and $U = \langle r_i \mid 1 \le i \le 5 \rangle$ be the finitely presented groups constructed in Propositions 14.4.4 and 14.2.3, respectively. Then the following statements hold.*

(a) *$d_1 = (t_1t_3t_4)^4$ is an element of order 3 of N_2 whose centralizer $C_{N_2}(d_1)$ is a direct product of an elementary abelian subgroup $Z = \langle z_1, z_2 \rangle$ of order 9 and a group $J = \langle j_1, j_2 \rangle$ of order 36 with an elementary abelian normal Sylow 3-subgroup $\langle j_1, j_1^{j_2} \rangle$ having a cyclic complement $\langle j_2 \rangle$ of order 4, where $z_1 = t_1^2t_3t_1t_3t_1t_3t_1t_3t_1^2t_3t_1^2t_3t_1$, $z_2 = t_1^2t_3t_1^3t_3t_1^2t_3t_1t_3t_1t_3t_1^2t_3t_1$, $j_1 = (t_3t_4)^4$ and $j_2 = (t_1t_3t_4)^3$ are elements of respective orders 3, 3, 3 and 4.*

(b) *The elements z_1, z_2 and d_1 are conjugate in N_2, and j_1 is conjugate to the representative $d_2 = (t_2t_3)^6$ of Table 14.5.8.*

(c) *d_1 is G-conjugate to $d = (r_1r_2^2r_1r_3)^2 \in U$ and $C_U(d)$ is a direct product of an elementary abelian subgroup $\langle d, x \rangle$ of order 9 and a* *dihedral subgroup* $\langle r_5, w \rangle$ *of order* 8, *where* $x = r_2r_3r_2^2r_3r_2^2r_3r_2^3r_3$ *and* w *is an element of order* 4 *which is* U*-conjugate to* r_3r_5.

(d) $C_G(d_1) = Z \times K$, *where* K *is isomorphic to the alternating group* A_6.

Proof (a) By Table 14.5.8 d_1 is a representative of a non-3-central conjugacy class of elements of order 3 such that $|C_{N_2}(d_1)| = 3^4 \cdot 2^2$. The generators and relations of $C = C_{N_2}(d_1)$ have been obtained computationally using the faithful permutation representation PN_2 of N_2 given in Proposition 14.4.4(e), Kim's program `GetShortGens(N_2,C)` and MAGMA.

(b) This statement can easily be checked by means of PN_2, Table 14.5.8 and MAGMA.

(c) The author verified that j_2^2 is N_2-conjugate to the representative $(t_1t_3t_4)^6$ of Table 14.5.8 which is G-conjugate to the involution u of G by Proposition 14.4.4(h). Hence d_1 is G-conjugate to $d = (r_1r_2^2r_1r_3)^2 \in U = C_G(u)$ because U has only one conjugacy class of elements of order 3 by Table 14.5.6. Using the faithful permutation representation PU given in Proposition 14.2.3 and MAGMA the author verified that $Z_1 = \langle d, x \rangle$ is a central subgroup of $C_U(d)$ isomorphic to Z and that Z_1 has a complement W which is a dihedral group of order 8 containing the involutions r_5 and $(r_3r_5)^2$. Thus (c) holds by Table 14.5.6.

(d) By (a) and (c) the centralizer $C_1 = C_G(d_1)$ has a dihedral Sylow 2-subgroup of order 8, a center $Z(C_1)$ of order 9 and $C_2 = C_1/Z(C_1)$ does not have any normal subgroup of odd order. Since C_2 is not a 2-group Theorem 1.6.4 of [92] due to Gorenstein and Walter implies that it is isomorphic to some $\mathrm{PSL}_2(q)$, where $q > 3$ is an odd prime power or $C_2 \cong A_7$. All involutions of C_2 are G-conjugate to u by Table 14.5.6 and (c). Furthermore, $U = C_G(u)$ does not have any elements of order 6 whose centralizer has a Sylow 3-subgroup of order divisible by 27. Hence C_2 cannot have elements of order 6. Therefore $C_2 \neq A_7$. Now Proposition 14.2.4 and statements (a) and (d) complete the proof because J is isomorphic to a maximal subgroup of A_6. □

Proposition 14.4.6 *Let* G *be a finite simple group of* Suz*-type with a* 2*-central involution* z *such that* $C_G(z) = H = \langle x, y, h \rangle$ *described in Lemma 14.1.1. Let* $U = \langle r_1, r_2, r_3, r_4, r_5 \rangle \cong C_G(u)$ *be the finitely presented group constructed in Proposition 14.2.3. Let* $N_1 = \langle g_1, g_2, g_3, g_4 \rangle$ *be the finitely presented group constructed in Proposition 14.4.3(c). Then the following statements hold.*

(a) $e = (x^3hx^2hx^3hx)^4(hx^3hxhx^2hxh)^4 \in H$ *is a representative of the conjugacy class* 5_A *of elements of order* 5 *of* G *and* $C_G(e) \cong \langle e \rangle \times A_6$.

(b) $N_G(\langle e \rangle)/C_G(e)$ *is a cyclic group of order* 4.

(c) *The element* $f = r_1r_2^2 \in U$ *is a representative of the conjugacy class* 5_B *of* G *and* $C_G(f) \cong \langle f \rangle \times A_5$.

(d) $N_G(\langle f \rangle)/C_G(f)$ *is a cyclic group of order* 4.

(e) *The element* $s = r_1r_2$ *of* $U = C_G(u) = \langle r_i | 1 \le i \le 5 \rangle$ *is a representative of the unique conjugacy class of elements of order* 7 *of* G *and* $C_G(s) \cong \langle s \rangle \times A_4$.

(f) *The normalizer* $N_G(\langle s \rangle)$ *is a split extension of a cyclic group of order* 6 *by* $C_G(\langle s \rangle)$.

Proof (a) and (b) Using the faithful permutation representation PH of Lemma 14.1.1(a) and MAGMA the reader can verify that the elements $k_1 = (x^3hx^2hx^3hx)^4$ and $k_2 = (hx^3hxhx^2hxh)^4$ of H both have order 3, $K = \langle k_1, k_2 \rangle \cong A_5$ and that $e = k_1k_2$ has order 5. Furthermore, k_1 is H-conjugate to $d = x^8$ and $C_H(K)$ is a dihedral group D_8 of order 8 generated by the involutions $s_1 = (yxy^2)^6$ and $s_2 = (x^4yx^3yx)^3$. Hence $C_H(e) = \langle e \rangle \times D_8$. Moreover, the element $w = (xhx^2hx^5)^4(hx^5hx^3hx)^3$ has order 4 and satisfies the relation $e^w = e^2$ and normalizes both subgroups K and D_8 of H. Hence $F_A = \langle e, w \rangle$ is a Frobenius group of order 20. In particular, (b) holds.

Another application of MAGMA yields that K has a unique conjugacy class of involutions u_1^K and that u_1 is H-conjugate to the representative $u = h$ of the second conjugacy class u^G of G. Now Proposition 14.2.4 and Table 14.5.6 imply that $C_G(e) = K \times L$, where a Sylow 5-subgroup of L is cyclic of order 5. Since L is not a 2-group Theorem 1.6.4 of [92] due to Gorenstein and Walter implies that L is isomorphic to some $\mathrm{PSL}_2(q)$, where $q > 3$ is an odd prime power or $L \cong A_7$. By Table 14.5.7 $N_G(\langle d \rangle)$ does not have any elements of order 18. Hence L does not have any elements of order 6. Thus $L \cong A_6$.

(c) and (d) By Tables 14.5.1 and 14.5.6 the elements e of H and $f = r_1r_2^2$ of U represent two distinct conjugacy classes 5_A and 5_B of elements of order 5 of G because the involutions z and u are not conjugate in G by Proposition 14.2.2(i). Using the faithful permutation representation PU of U given in Proposition 14.2.3(f) and MAGMA it has been checked that $T = N_U(\langle f \rangle)$ has order $2^4 \cdot 5$ and $C_U(f)$ has a cyclic complement $\langle j \rangle$ of order 4. Hence (d) holds.

As $C_G(e)$ is a subgroup of $C_G(L) \cong A_5 \times L$ and f is G-conjugate to an element f_1 of order 5 of L it follows that $C_G(f) \cong \langle f \rangle \times A_5$.

(e) and (f) By Proposition 14.2.4 the Sylow 7-subgroups of G are cyclic of order 7. By Table 14.5.6 the element $s = r_1 r_2$ of $U = C_G(u) = \langle r_i | 1 \leq i \leq 5 \rangle$ is a representative of a 7-class of G. Using the faithful permutation representation PU of U given in Proposition 14.2.3(f) and MAGMA it can be verified that $w = r_4(r_2 r_3 r_2^2 r_3 r_2 r_3)^4 (r_3 r_2 r_3 r_2^3 r_3 r_2 r_3 r_2 r_3)^3 \in U$ has order 6 and satisfies the relation $s^w = s^3$. Hence G has a unique conjugacy class of elements of order 7 and (f) holds.

Another application of MAGMA shows that the Klein 4-group $V = \langle r_4, r_5 \rangle$ centralizes s. But $(r_5, w) = r_4$. Now Table 14.5.7 implies that s is centralized by an alternating group A_4 and that its Sylow 3-subgroup centralizes w. Furthermore, V is not centralized by w because $(r_5, w) = r_4$. Thus $N_G(\langle s \rangle)$ is a split extension of $\langle w \rangle$ by $C_G(s)$. □

Proposition 14.4.7 *Let G be a finite simple group of* Suz*-type with a 2-central involution z such that $C_G(z) = H = \langle x, y, h \rangle$ described in Lemma 14.1.1. Let $N_1 = \langle g_1, g_2, g_3, g_4 \rangle$ and $N_2 = \langle t_1, t_2, t_3, t_4 \rangle$ be the finitely presented groups constructed in Proposition 14.4.3(c) and Proposition 14.4.4, respectively. Then the following statements hold.*

(a) *G has three conjugacy classes 3_A, 3_B and 3_C of elements of order 3 represented by the classes 3_1, 3_3 and 3_5 of N_1 given in Table 14.5.7.*
(b) *G has two conjugacy classes 9_A and 9_B of elements of order 9 represented by the classes 9_1 and 9_2 of N_2 given in Table 14.5.8.*
(c) *G has two conjugacy classes 5_A and 5_B of elements of order 5 represented by the classes 5 of $H = C_G(z)$ and 5 of $U = C_G(u)$ given in Tables 14.5.1 and 14.5.6, respectively.*
(d) *G has one conjugacy class 7_A of elements of order 7 represented by the class 7 of U given in Table 14.5.6.*
(e) *G has one conjugacy class 11_A of elements of order 11.*
(f) *G has two conjugacy classes 13_A and 13_B of elements of order 13.*
(g) *G has three conjugacy classes 15_A, 15_B and 15_C of elements of order 15.*
(h) *G has two conjugacy classes 21_A and 21_B of elements of order 21 represented by the conjugacy classes 21_1 and 21_2 of N_1 given in Table 14.5.7.*

Proof (a) This is a restatement of Proposition 14.4.3(i).

(b) Applying the same argument to the groups N_1 and N_2 we verified that the classes 3_1, 3_3 and 3_5 of N_1 and the classes 3_1, 3_2 and 3_8 of N_2 fuse in G, respectively. Hence the power map information of Table 14.5.8 implies that G has two conjugacy classes of elements of order 9.

(c) and (d) These statements hold by Proposition 14.4.6.

(e) and (f) These two statements are now obvious by Sylow's Theorem and Proposition 14.2.4.

(g) By Propositions 14.4.3(i), 14.4.4 and 14.4.5 the 3-part x_3 of an element x of order 15 is G-conjugate either to an element of 3_A or to an element of 3_C. By Table 14.5.7 there is a unique conjugacy class 15_A of elements x with $x_3 \in 3_A$. Since A_6 has two conjugacy classes of elements of order 5 Proposition 14.4.5 implies that there are two conjugacy classes 15_B and 15_C of elements x of order 15 with $x_3 \in 3_C$.

(h) This assertion holds by the power map information given in Table 14.5.7.

Finally, Corollary 14.2.5, Proposition 14.2.4, the power map information of the tables in Section 14.5 and the class equation of G yield that G has exactly 43 conjugacy classes. In particular, G has 16 conjugacy classes of elements of odd order with representatives given in the statement. □

Lemma 14.4.8 *Let G be a finite simple group of* Suz*-type with a 2-central involution z such that $C_G(z) = H = \langle x, y, h\rangle$ described in Lemma 14.1.1. Then H is a maximal subgroup of G.*

Proof Lemma 14.4.8 must be true, otherwise there is a proper subgroup X of G containing $H = C_G(z)$ properly such that X is of minimal order with respect to these two properties. Hence $C_X(z) = H$. Therefore Proposition 14.2.4 implies that X has a proper minimal normal subgroup $Y \neq 1$. Thus $H \cap Y < H$. By Lemma 14.1.1 the center $Z(H) = \langle z\rangle$ of order 2 and the Fitting subgroup P of order 2^7 are the only nontrivial proper normal subgroups of H and $H/P \cong U_4(2)$. Therefore $|Y \cap H| \in \{1, 2, 2^7\}$.

Suppose that $|Y|$ is odd. Then Lemma 1.5.9 of [92] and Proposition 14.2.4 assert that Y is an abelian group of order dividing $q = 3^3 \cdot 5 \cdot 7 \cdot 11 \cdot 13$. Operating with the Klein 4-group $V = \langle z, v\rangle$ defined in Proposition 14.2.1 on Y yields by Theorem 4.2.2 of [92] and Table 14.5.1 that $|Y||C_Y(V)|^2 = |C_Y(z)||C_Y(t)|^2$ because v and tz are conjugate in H. Now $C_Y(z) = 1$ as H does not have any normal subgroup of odd order. Furthermore, $|C_Y(V))| \leq 3 \cdot 5$ and $|C_Y(t)| \leq 3^4 \cdot 5$ by Table 14.5.1. Thus $|Y|$ divides 3^2 as $|Y|$ divides q. However, the perfect group H does not have any irreducible 3-modular representations of degree 2.

Suppose that $|Y \cap H| = 2$. Then Corollary 1.4.18 of [92] implies that the cyclic Sylow 2-subgroup of Y has a normal complement Q. As a characteristic subgroup of Y it is normal in X. Hence $Q = 1$ by the previous argument. Furthermore, $Y = Z(H)$ and $X = C_X(z) = H$, a contradiction.

Hence $Y \cap H = P$ and P is a Sylow 2-subgroup of Y. Since Y is a minimal normal subgroup of X its commutator subgroup $Y' = Y$ because P is not abelian. By the previous argument Y does not have any normal subgroups of odd order. Hence any normal subgroup $N \neq 1$ of Y has even order. Thus $P \cap N = N_1 \neq 1$ is a normal subgroup of P. Therefore $z \in N_1$ by Theorem 1.3.4 of [92]. Hence $M = \langle z^Y \rangle$ is a normal subgroup of X contained in N. Thus $M = N = Y$, and Y is a simple group. By Lemma 14.1.1(h) P is extra-special of order 2^7. In particular, P is not a dihedral 2-group of order 8. Therefore Y is not simple by Lemma 4.8.2 of [92]. This contradiction completes the proof. □

Proposition 14.4.9 *Let G be a finite simple group of* Suz*-type. Then G has exactly one complex character $\chi \in Irr_{\mathbb{C}}(G)$ of degree* 143.

Proof Let $H = C_G(z)$ be the centralizer of a 2-central involution of G. Let S be the Sylow 2-subgroup of H defined in Lemma 14.1.1. Let A be its unique maximal elementary abelian normal subgroup. Then $D = N_H(A)$ is a proper subgroup of $E = N_G(A)$ by Proposition 14.2.2. Hence $G = \langle H, E \rangle$ by Lemma 14.4.8. Proposition 14.2.2(g) states that the amalgam $H \leftarrow D \rightarrow E$ has Goldschmidt index 1. Therefore the free product $P = H *_D E$ of H and E with amalgamated subgroup D is uniquely determined up to isomorphism. By Theorem 14.3.1 there is exactly one compatible pair $(\chi, \tau) \in mfchar_{\mathbb{C}}(H) \times mfchar_{\mathbb{C}}(E)$ of degree 143. As G is a simple epimorphic image of P it follows that G has at most one character $\chi \in Irr_{\mathbb{C}}(G)$ with minimal degree 143.

Using Brauer's characterization of characters we now show that any finite simple group G of Suz-type has a complex irreducible character $\chi \in Irr_{\mathbb{C}}(G)$ of degree 143. The simple group of Suz-type $\mathfrak{G}$ constructed in Theorem 14.3.1 has 43 conjugacy classes. Also G has 43 conjugacy classes by Corollary 14.2.5 and Proposition 14.4.7; $\mathfrak{G}$ has an irreducible character of degree 143 by Theorem 14.3.1(b)(4). Inspired by its values we define a class function $\chi : G \to \mathbb{C}$ by

i	1_A	2_A	2_B	3_A	3_B	3_C	4_A	4_B	4_C	4_D	5_A	5_B
$\chi(g_i)$	143	15	-1	35	8	-1	15	-1	-1	-1	8	3

i	6_A	6_B	6_C	6_D	6_E	7_A	8_A	8_B	8_C	9_A	9_B
$\chi(g_i)$	3	6	6	0	-1	3	-1	3	-1	2	2

i	10_A	10_B	11_A	12_A	12_B	12_C	12_D	12_E	13_A	13_B	14_A
$\chi(g_i)$	0	-1	0	3	0	-1	2	-1	0	0	-1

i	15_A	15_B	15_C	18_A	18_B	20_A	21_A	21_B	24_A
$\chi(g_i)$	-1	-1	0	0	0	0	0	0	-1

Let p be any prime divisor of $|G|$. If Y is an over-group of a p-elementary subgroup X, and the restriction $\chi_{|Y}$ of X to Y is a generalized character, then $\chi_{|X}$ is also a generalized character of X.

By Proposition 14.4.7 the Sylow 13-normalizer N_{13} of G and Sylow 11-normalizer $N_G(11)$ of G are Frobenius groups of orders $13 \cdot 6$ and $11 \cdot 12$, respectively. As $\chi(x) = 0$ for all 13-singular and 11-singular elements and $\chi(1) = 143$, its restrictions $\chi_{|N_{13}}$ and $\chi_{|N_{11}}$ are projective characters of N_{13} over $GF(13)$ and N_{11} over $GF(11)$, respectively.

Therefore by Proposition 14.4.7, and Corollary 2.8.10 and Theorem 2.8.9, both of [92], it remains to show that $X_{|Y}$ is a generalized character of Y whenever Y belongs to the following set of subgroups:

$$\mathfrak{Y} = \{H, U, N_1, N_2, N_G(3_C), N_G(5_A), N_G(5_B), N_G(7_A)\}$$

of G.

By means of the results of this section the fusion of the conjugacy classes of the groups $Y \in \mathfrak{Y}$ in G can be determined easily. Therefore the inner products $(\lambda_i, \chi)_Y$ can be calculated for each $\lambda_i \in Irr_{\mathbb{C}}(Y)$ and each $Y \in \{H, U, N_1\}$ by means of the character tables given in Section 14.6. It turns out that

$$\chi_{|H} = \chi_1 + \chi_7 + \chi_{11} + \chi_{\mathbf{15}} + \chi_{\mathbf{16}} + \chi_{17}, \text{ where } \quad \chi_k \in Irr_{\mathbb{C}}(H),$$
$$\chi_{|U} = \chi_3 + \chi_5 + \chi_{15} + \chi_{27}, \text{ where } \quad \chi_k \in Irr_{\mathbb{C}}(U),$$
$$\chi_{|N_1} = \chi_2 + \chi_5 + \chi_6, \text{ where } \quad \chi_k \in Irr_{\mathbb{C}}(N_1).$$

Now $N_1 \cap N_2$ is a normal subgroup of N_2 of index 2 by Proposition 14.4.4. Hence $\chi_{|N_2}$ is a positive sum of irreducible characters of $Irr_{\mathbb{C}}(N_2)$ by Clifford's Theorem.

The character tables of the four remaining local subgroups and the fusion of their conjugacy classes in G can be calculated from the results in Propositions 14.4.5, 14.4.6 and 14.4.7. Therefore the restrictions of χ to these subgroups can be determined similarly. The details are left to the reader.

It has been checked that the inner product $(\chi, \chi)_G = 1$. Therefore χ is an irreducible character of G with $\chi(1) = 143$ by Corollary 2.8.10 of [92]. This completes the proof. □

Theorem 14.4.10 *Each finite simple group G of* Suz*-type is isomorphic to the finite simple group $\mathfrak{G} \le \mathrm{GL}_{23}(23)$ of order $|\mathfrak{G}| = 2^{10} \cdot 3^7 \cdot 5^3 \cdot \cdot 11 \cdot 23$ constructed in Theorem 14.3.1.*

Proof By Theorem 14.3.1 there exists a finite simple group $\mathfrak{G}$ of order $|\mathfrak{G}| = 2^{11} \cdot 3^3 \cdot 5^2 \cdot 7 \cdot 11 \cdot 13$ which is of Suz-type.

Let G be any finite simple group of Suz-type. By Proposition 14.2.4 G has exactly one 2-central conjugacy class z^G of involutions and the same order as $\mathfrak{G}$. Proposition 14.2.2 asserts that the Sylow 2-subgroup S of H contains a maximal elementary abelian normal subgroup A such that $D = N_H(A)$ is a proper subgroup of $E = N_G(A)$. Furthermore, E is uniquely determined by H up to isomorphism. Lemma 14.4.8 implies that $G = \langle H, E \rangle$. The amalgam $H \leftarrow D \rightarrow E$ has Goldschmidt index 1 by Proposition 14.2.2(g). Proposition 14.4.9 states that G has a unique irreducible complex character of degree 143. It is of defect zero for the prime 13. Therefore each finite group G of Suz-type has a unique irreducible 13-modular representation of degree 143 by Theorem 3.12.2 of [92]. Hence $G \cong \mathfrak{G}$ by Theorem 7.5.1 of [92]. This completes the proof. □

14.5 Representatives of conjugacy classes

14.5.1 *Conjugacy classes of* $H = \langle x, y, h \rangle$

Class	*Representative*	*Centralizer*	2P	3P	5P
1	1	$2^{13} \cdot 3^4 \cdot 5$	1	1	1
2_1	$(x)^{12}$	$2^{13} \cdot 3^4 \cdot 5$	1	2_1	2_1
2_2	$(xy)^6$	$2^{12} \cdot 3 \cdot 5$	1	2_2	2_2
2_3	$(x^2y)^6$	$2^{10} \cdot 3^2$	1	2_3	2_3
2_4	h	2^9	1	2_4	2_4
3_1	$(x)^8$	$2^7 \cdot 3^3$	3_1	1	3_1
3_2	$(xh)^3$	$2^4 \cdot 3^4$	3_3	1	3_3
3_3	$(xh)^6$	$2^4 \cdot 3^4$	3_2	1	3_2
3_4	$(y)^4$	$2^4 \cdot 3^3$	3_4	1	3_4
4_1	$(x)^6$	$2^{10} \cdot 3^2 \cdot 5$	2_1	4_1	4_1
4_2	$(y)^3$	$2^{10} \cdot 3$	2_1	4_2	4_2
4_3	$(xy)^3$	$2^9 \cdot 3$	2_2	4_3	4_3
4_4	$(x^3y)^2$	$2^9 \cdot 3$	2_1	4_4	4_4
4_5	$(xy^3)^3$	$2^8 \cdot 3$	2_2	4_5	4_5
4_6	$(y^3h)^2$	2^9	2_2	4_6	4_6
4_7	$(x^2y)^3$	$2^7 \cdot 3$	2_3	4_7	4_7
4_8	$(x^2hxh)^3$	$2^7 \cdot 3$	2_3	4_8	4_8
4_9	$xyxh$	2^5	2_4	4_9	4_9
5	yh	$2^3 \cdot 5$	5	5	1
6_1	$(x)^4$	$2^7 \cdot 3^3$	3_1	2_1	6_1
6_2	$(xhy^2)^3$	$2^4 \cdot 3^4$	3_3	2_1	6_3
6_3	$(xhy^2)^{15}$	$2^4 \cdot 3^4$	3_2	2_1	6_2
6_4	$(y)^2$	$2^4 \cdot 3^3$	3_4	2_1	6_4
6_5	$(xy)^2$	$2^6 \cdot 3$	3_1	2_2	6_5
6_6	$(x^2y)^2$	$2^4 \cdot 3^2$	3_2	2_3	6_7
6_7	$(x^2y)^{10}$	$2^4 \cdot 3^2$	3_3	2_3	6_6
6_8	x^2y^2	$2^4 \cdot 3^2$	3_3	2_3	6_9
6_9	$(x^2y^2)^5$	$2^4 \cdot 3^2$	3_2	2_3	6_8
6_{10}	$xyxhy$	$2^4 \cdot 3^2$	3_1	2_3	6_{11}
6_{11}	$(xyxhy)^5$	$2^4 \cdot 3^2$	3_1	2_3	6_{10}
6_{12}	x^4y^2h	$2^4 \cdot 3^2$	3_4	2_3	6_{13}
6_{13}	$(x^4y^2h)^5$	$2^4 \cdot 3^2$	3_4	2_3	6_{12}
8_1	$(x)^3$	$2^6 \cdot 3$	4_1	8_1	8_1
8_2	x^2y^2hy	2^6	4_2	8_2	8_2
8_3	x^3y	2^5	4_4	8_3	8_3
8_4	y^3h	2^5	4_6	8_4	8_4
8_5	x^3y^2	2^5	4_3	8_5	8_5
9_1	xh	$2 \cdot 3^2$	9_2	3_2	9_2
9_2	$(xh)^2$	$2 \cdot 3^2$	9_1	3_3	9_1
10_1	$(x^2h)^2$	$2^3 \cdot 5$	5	10_1	2_1
10_2	xy^2h	$2^2 \cdot 5$	5	10_3	2_2
10_3	$(xy^2h)^3$	$2^2 \cdot 5$	5	10_2	2_2
12_1	$(x)^2$	$2^5 \cdot 3^2$	6_1	4_1	12_1
12_2	xy^2hxhy	$2^3 \cdot 3^2$	6_4	4_1	12_2
12_3	xy	$2^4 \cdot 3$	6_5	4_3	12_3
12_4	xy^3	$2^4 \cdot 3$	6_5	4_5	12_5
12_5	$(xy^3)^5$	$2^4 \cdot 3$	6_5	4_5	12_4
12_6	x^3yxh	$2^4 \cdot 3$	6_1	4_4	12_6
12_7	y	$2^3 \cdot 3$	6_4	4_2	12_7
12_8	x^2y	$2^3 \cdot 3$	6_6	4_7	12_9
12_9	$(x^2y)^5$	$2^3 \cdot 3$	6_7	4_7	12_8
12_{10}	x^2hxh	$2^3 \cdot 3$	6_6	4_8	12_{11}
12_{11}	$(x^2hxh)^5$	$2^3 \cdot 3$	6_7	4_8	12_{10}
18_1	xhy^2	$2 \cdot 3^2$	9_2	6_2	18_2
18_2	$(xhy^2)^5$	$2 \cdot 3^2$	9_1	6_3	18_1
20	x^2h	$2^2 \cdot 5$	10_1	20	4_1
24	x	$2^3 \cdot 3$	12_1	8_1	24

14.5.2 *Conjugacy classes of* $E = N_G(A) = \langle x_1, y_1, e_1 \rangle$

Class	*Representative*	*Centralizer*	2P	3P	5P
1	1	$2^{13} \cdot 3^3 \cdot 5$	1	1	1
2_1	$(x_1)^{12}$	$2^{13} \cdot 3^2$	1	2_1	2_1
2_2	$(x_1 y_1 e_1)^5$	$2^8 \cdot 3 \cdot 5$	1	2_2	2_2
2_3	$(x_1 e_1^2)^6$	$2^9 \cdot 3$	1	2_3	2_3
2_4	$(y_1 e_1)^2$	2^9	1	2_4	2_4
3_1	$(x_1)^8$	$2^7 \cdot 3^3 \cdot 5$	3_2	1	3_2
3_2	$(x_1)^{16}$	$2^7 \cdot 3^3 \cdot 5$	3_1	1	3_1
3_3	$(y_1)^4$	$2^4 \cdot 3^2$	3_3	1	3_3
3_4	e_1	$2^2 \cdot 3^2$	3_4	1	3_4
4_1	$(y_1)^3$	$2^{10} \cdot 3$	2_1	4_1	4_1
4_2	$(x_1^2 y_1)^3$	$2^{10} \cdot 3$	2_1	4_2	4_2
4_3	$(x_1)^6$	$2^9 \cdot 3$	2_1	4_3	4_3
4_4	$(x_1^2 y_1^3)^3$	$2^8 \cdot 3$	2_1	4_4	4_4
4_5	$(x_1^3 y_1)^2$	2^9	2_1	4_5	4_5
4_6	$x_1 e_1$	2^8	2_1	4_6	4_6
4_7	$(x_1 e_1^2)^3$	$2^5 \cdot 3$	2_3	4_7	4_7
4_8	$y_1 e_1$	2^5	2_4	4_8	4_8
5_1	$(x_1 y_1 e_1)^2$	$2^2 \cdot 3 \cdot 5$	5_2	5_2	1
5_2	$(x_1 y_1 e_1)^4$	$2^2 \cdot 3 \cdot 5$	5_1	5_1	1
6_1	$(x_1)^4$	$2^7 \cdot 3^2$	3_1	2_1	6_2
6_2	$(x_1)^{20}$	$2^7 \cdot 3^2$	3_2	2_1	6_1
6_3	$(y_1)^2$	$2^4 \cdot 3^2$	3_3	2_1	6_3
6_4	$x_1^2 y_1^2$	$2^4 \cdot 3^2$	3_3	2_1	6_5
6_5	$(x_1^2 y_1^2)^5$	$2^4 \cdot 3^2$	3_3	2_1	6_4
6_6	$(x_1 e_1^2)^2$	$2^5 \cdot 3$	3_1	2_3	6_7
6_7	$(x_1 e_1^2)^{10}$	$2^5 \cdot 3$	3_2	2_3	6_6
6_8	$x_1^2 e_1^2$	$2^2 \cdot 3$	3_4	2_2	6_8
8_1	$(x_1)^3$	$2^5 \cdot 3$	4_3	8_1	8_1
8_2	$x_1^3 y_1^2$	2^6	4_2	8_2	8_2
8_3	$x_1 y_1^3 e_1$	2^6	4_1	8_3	8_3
8_4	$x_1^3 y_1$	2^5	4_5	8_4	8_4
10_1	$x_1 y_1 e_1$	$2^2 \cdot 5$	5_1	10_2	2_2
10_2	$(x_1 y_1 e_1)^3$	$2^2 \cdot 5$	5_2	10_1	2_2
12_1	$(x_1)^2$	$2^5 \cdot 3$	6_1	4_3	12_2
12_2	$(x_1)^{10}$	$2^5 \cdot 3$	6_2	4_3	12_1
12_3	$x_1^2 y_1^3$	$2^4 \cdot 3$	6_1	4_4	12_4
12_4	$(x_1^2 y_1^3)^5$	$2^4 \cdot 3$	6_2	4_4	12_3
12_5	y_1	$2^3 \cdot 3$	6_3	4_1	12_5
12_6	$x_1^2 y_1$	$2^3 \cdot 3$	6_3	4_2	12_6
12_7	$x_1 e_1^2$	$2^3 \cdot 3$	6_6	4_7	12_8
12_8	$(x_1 e_1^2)^5$	$2^3 \cdot 3$	6_7	4_7	12_7
15_1	$y_1^2 e_1$	$3 \cdot 5$	15_2	5_1	3_2
15_2	$(y_1^2 e_1)^2$	$3 \cdot 5$	15_1	5_2	3_1
15_3	$(y_1^2 e_1)^7$	$3 \cdot 5$	15_4	5_2	3_2
15_4	$(y_1^2 e_1)^{11}$	$3 \cdot 5$	15_3	5_1	3_1
24_1	x_1	$2^3 \cdot 3$	12_1	8_1	24_2
24_2	$(x_1)^5$	$2^3 \cdot 3$	12_2	8_1	24_1

14.5.3 *Conjugacy classes of* $D = \langle x_1, y_1 \rangle$

Class	*Representative*	*Centralizer*	2P	3P
1	1	$2^{13} \cdot 3^2$	1	1
2_1	$(x_1)^{12}$	$2^{13} \cdot 3^2$	1	2_1
2_2	$(x_1y_1)^6$	$2^{12} \cdot 3$	1	2_2
2_3	$(x_1^2y_1)^6$	$2^{10} \cdot 3^2$	1	2_3
2_4	$(x_1y_1^3)^6$	$2^9 \cdot 3$	1	2_4
2_5	$(x_1y_1^4)^3$	$2^8 \cdot 3$	1	2_5
2_6	$(x_1^2y_1^3x_1y_1)^2$	2^9	1	2_6
2_7	$x_1^5y_1^4x_1y_1$	2^8	1	2_7
2_8	$x_1^5y_1x_1^2y_1^2x_1^2y_1$	2^8	1	2_8
3_1	$(x_1)^8$	$2^7 \cdot 3^2$	3_2	1
3_2	$(x_1)^{16}$	$2^7 \cdot 3^2$	3_1	1
3_3	$(y_1)^4$	$2^4 \cdot 3^2$	3_3	1
3_4	$(x_1^2y_1)^4$	$2^4 \cdot 3^2$	3_5	1
3_5	$(x_1^2y_1)^8$	$2^4 \cdot 3^2$	3_4	1
4_1	$(y_1)^3$	$2^{10} \cdot 3$	2_1	4_1
4_2	$(x_1^6y_1^2)^3$	$2^{10} \cdot 3$	2_1	4_2
4_3	$(x_1)^6$	$2^9 \cdot 3$	2_1	4_3
4_4	$(x_1y_1)^3$	$2^8 \cdot 3$	2_2	4_4
4_5	$(x_1^2y_1^3)^3$	$2^8 \cdot 3$	2_2	4_5
4_6	$(x_1^5y_1)^3$	$2^8 \cdot 3$	2_2	4_6
4_7	$(x_1^3y_1x_1^2y_1^2)^3$	$2^8 \cdot 3$	2_1	4_7
4_8	$(x_1^3y_1)^2$	2^9	2_1	4_8
4_9	$(x_1^3y_1^2)^2$	2^9	2_2	4_9
4_{10}	$x_1^5y_1x_1y_1$	2^9	2_2	4_{10}
4_{11}	$(x_1^2y_1)^3$	$2^7 \cdot 3$	2_3	4_{11}
4_{12}	$(x_1^2y_1^4)^3$	$2^7 \cdot 3$	2_3	4_{12}
4_{13}	$x_1y_1x_1y_1x_1y_1^4$	2^8	2_1	4_{13}
4_{14}	$x_1^3y_1x_1y_1x_1y_1x_1y_1$	2^8	2_2	4_{14}
4_{15}	$x_1^7y_1x_1^2y_1$	2^8	2_2	4_{15}
4_{16}	$x_1y_1x_1y_1^4x_1y_1^4$	2^8	2_2	4_{16}
4_{17}	$(x_1y_1^3)^3$	$2^5 \cdot 3$	2_4	4_{17}
4_{18}	$x_1^2y_1^3x_1y_1$	2^5	2_6	4_{18}
6_1	$(x_1)^4$	$2^7 \cdot 3^2$	3_1	2_1
6_2	$(x_1)^{20}$	$2^7 \cdot 3^2$	3_2	2_1
6_3	$(x_1y_1)^2$	$2^6 \cdot 3$	3_2	2_2
6_4	$(x_1y_1)^{10}$	$2^6 \cdot 3$	3_1	2_2
6_5	$(y_1)^2$	$2^4 \cdot 3^2$	3_3	2_1
6_6	$(x_1^2y_1)^2$	$2^4 \cdot 3^2$	3_4	2_3
6_7	$(x_1^2y_1)^{10}$	$2^4 \cdot 3^2$	3_5	2_3
6_8	$x_1^2y_1^2$	$2^4 \cdot 3^2$	3_5	2_3
6_9	$(x_1^2y_1^2)^5$	$2^4 \cdot 3^2$	3_4	2_3
6_{10}	$x_1^4y_1$	$2^4 \cdot 3^2$	3_4	2_1
6_{11}	$(x_1^4y_1)^5$	$2^4 \cdot 3^2$	3_5	2_1
6_{12}	$x_1^4y_1^3$	$2^4 \cdot 3^2$	3_1	2_3
6_{13}	$(x_1^4y_1^3)^5$	$2^4 \cdot 3^2$	3_2	2_3
6_{14}	$x_1^3y_1x_1^2y_1x_1y_1$	$2^4 \cdot 3^2$	3_3	2_3
6_{15}	$(x_1^3y_1x_1^2y_1x_1y_1)^5$	$2^4 \cdot 3^2$	3_3	2_3
6_{16}	$(x_1y_1^3)^2$	$2^5 \cdot 3$	3_2	2_4
6_{17}	$(x_1y_1^3)^{10}$	$2^5 \cdot 3$	3_1	2_4
6_{18}	$x_1y_1^4$	$2^4 \cdot 3$	3_1	2_5
6_{19}	$(x_1y_1^4)^5$	$2^4 \cdot 3$	3_2	2_5
8_1	$(x_1)^3$	$2^5 \cdot 3$	4_3	8_1
8_2	$x_1^4y_1x_1^2y_1^2$	2^6	4_1	8_2

Conjugacy classes of $D = \langle x_1, y_1 \rangle$ (continued)

Class	*Representative*	*Centralizer*	2P	3P
8_3	$x_1^5 y_1^2 x_1 y_1^2$	2^6	4_2	8_3
8_4	$x_1^3 y_1$	2^5	4_8	8_4
8_5	$x_1^3 y_1^2$	2^5	4_9	8_5
8_6	$x_1^3 y_1^5$	2^5	4_{10}	8_6
12_1	$(x_1)^2$	$2^5 \cdot 3$	6_1	4_3
12_2	$(x_1)^{10}$	$2^5 \cdot 3$	6_2	4_3
12_3	$x_1 y_1$	$2^4 \cdot 3$	6_3	4_4
12_4	$(x_1 y_1)^5$	$2^4 \cdot 3$	6_4	4_4
12_5	$x_1^2 y_1^3$	$2^4 \cdot 3$	6_4	4_5
12_6	$(x_1^2 y_1^3)^5$	$2^4 \cdot 3$	6_3	4_5
12_7	$x_1^5 y_1$	$2^4 \cdot 3$	6_4	4_6
12_8	$(x_1^5 y_1)^5$	$2^4 \cdot 3$	6_3	4_6
12_9	$x_1^3 y_1 x_1^2 y_1^2$	$2^4 \cdot 3$	6_1	4_7
12_{10}	$(x_1^3 y_1 x_1^2 y_1^2)^5$	$2^4 \cdot 3$	6_2	4_7
12_{11}	y_1	$2^3 \cdot 3$	6_5	4_1
12_{12}	$x_1^2 y_1$	$2^3 \cdot 3$	6_6	4_{11}
12_{13}	$(x_1^2 y_1)^5$	$2^3 \cdot 3$	6_7	4_{11}
12_{14}	$x_1 y_1^3$	$2^3 \cdot 3$	6_{16}	4_{17}
12_{15}	$(x_1 y_1^3)^5$	$2^3 \cdot 3$	6_{17}	4_{17}
12_{16}	$x_1^2 y_1^4$	$2^3 \cdot 3$	6_6	4_{12}
12_{17}	$(x_1^2 y_1^4)^5$	$2^3 \cdot 3$	6_7	4_{12}
12_{18}	$x_1^6 y_1^2$	$2^3 \cdot 3$	6_5	4_2
24_1	x_1	$2^3 \cdot 3$	12_1	8_1
24_2	$(x_1)^5$	$2^3 \cdot 3$	12_2	8_1

14.5.4 *Conjugacy classes of $W = \langle m_1, m_2 \rangle$*

Class	*Representative*	*Centralizer*	2P	3P	5P
1	1	$2^6 \cdot 3^2 \cdot 5$	1	1	1
2_1	$(m_1 m_2)^6$	$2^6 \cdot 3$	1	2_1	2_1
2_2	$(m_1^2 m_2 m_1 m_2^2 m_1^2 m_2^2)^3$	$2^4 \cdot 3$	1	2_2	2_2
3_1	$(m_1 m_2)^4$	$2^6 \cdot 3^2 \cdot 5$	3_2	1	3_2
3_2	$(m_1 m_2)^8$	$2^6 \cdot 3^2 \cdot 5$	3_1	1	3_1
3_3	m_1	3^2	3_4	1	3_4
3_4	$(m_1)^2$	3^2	3_3	1	3_3
3_5	$m_1^2 m_2 m_1 m_2$	3^2	3_5	1	3_5
4_1	$(m_1 m_2)^3$	$2^4 \cdot 3$	2_1	4_1	4_1
4_2	$m_1^2 m_2 m_1 m_2^2$	$2^4 \cdot 3$	2_1	4_2	4_2
4_3	$(m_1^2 m_2 m_1^2 m_2 m_1 m_2^3)^3$	$2^4 \cdot 3$	2_1	4_3	4_3
5_1	m_2	$3 \cdot 5$	5_2	5_2	1
5_2	$(m_2)^2$	$3 \cdot 5$	5_1	5_1	1
6_1	$(m_1 m_2)^2$	$2^6 \cdot 3$	3_1	2_1	6_2
6_2	$(m_1 m_2)^{10}$	$2^6 \cdot 3$	3_2	2_1	6_1
6_3	$m_1^2 m_2 m_1 m_2^2 m_1^2 m_2^2$	$2^4 \cdot 3$	3_1	2_2	6_4
6_4	$(m_1^2 m_2 m_1 m_2^2 m_1^2 m_2^2)^5$	$2^4 \cdot 3$	3_2	2_2	6_3
12_1	$m_1 m_2$	$2^4 \cdot 3$	6_1	4_1	12_2
12_2	$(m_1 m_2)^5$	$2^4 \cdot 3$	6_2	4_1	12_1
12_3	$m_1^2 m_2 m_1^2 m_2 m_1 m_2^3$	$2^4 \cdot 3$	6_2	4_3	12_4
12_4	$(m_1^2 m_2 m_1^2 m_2 m_1 m_2^3)^5$	$2^4 \cdot 3$	6_1	4_3	12_3
12_5	$m_1^2 m_2 m_1 m_2 m_1 m_2^2 m_1 m_2$	$2^4 \cdot 3$	6_2	4_2	12_6
12_6	$(m_1^2 m_2 m_1 m_2 m_1 m_2^2 m_1 m_2)^5$	$2^4 \cdot 3$	6_1	4_2	12_5
15_1	$m_1^2 m_2$	$3 \cdot 5$	15_2	5_2	3_1
15_2	$(m_1^2 m_2)^2$	$3 \cdot 5$	15_1	5_1	3_2
15_3	$(m_1^2 m_2)^7$	$3 \cdot 5$	15_4	5_1	3_1
15_4	$(m_1^2 m_2)^{11}$	$3 \cdot 5$	15_3	5_2	3_2

14.5.5 *Conjugacy classes of* $M = N_G(V) = \langle w_1, w_2, m \rangle$

Class	*Representative*	*Centralizer*	2P	3P	5P
1	1	$2^{13} \cdot 3^2 \cdot 5$	1	1	1
2_1	$(w_2)^4$	$2^{13} \cdot 3 \cdot 5$	1	2_1	2_1
2_2	$(w_1)^6$	$2^{11} \cdot 3$	1	2_2	2_2
2_3	$(w_1^2 w_2)^5$	$2^8 \cdot 3 \cdot 5$	1	2_3	2_3
2_4	$(w_1^2 m)^3$	$2^9 \cdot 3$	1	2_4	2_4
2_5	$(w_1^3 w_2 m)^3$	$2^7 \cdot 3$	1	2_5	2_5
2_6	$w_2^3 m w_2^2 m^2$	2^8	1	2_6	2_6
3_1	$(w_1)^4$	$2^7 \cdot 3^2$	3_1	1	3_1
3_2	m	$2^2 \cdot 3^2 \cdot 5$	3_2	1	3_2
3_3	$(w_1^2 m)^2$	$2^2 \cdot 3^2$	3_3	1	3_3
4_1	$(w_1 m^2)^3$	$2^8 \cdot 3 \cdot 5$	2_1	4_1	4_1
4_2	$(w_2 m)^2$	$2^{10} \cdot 3$	2_1	4_2	4_2
4_3	$(w_1^4 w_2^2)^3$	$2^9 \cdot 3$	2_1	4_3	4_3
4_4	$(w_2)^2$	2^{10}	2_1	4_4	4_4
4_5	$(w_1)^3$	$2^7 \cdot 3$	2_2	4_5	4_5
4_6	$w_1^2 w_2 w_1 w_2^2$	2^8	2_2	4_6	4_6
4_7	$w_1^2 w_2 m w_1 m$	2^8	2_2	4_7	4_7
4_8	$w_2^2 m w_2 m^2$	2^8	2_1	4_8	4_8
4_9	$w_1^4 m w_1 w_2 w_1 m$	2^7	2_2	4_9	4_9
4_{10}	$w_1^6 w_2 m$	2^5	2_4	4_{10}	4_{10}
5_1	$w_1 w_2$	$2^3 \cdot 3 \cdot 5$	5_2	5_2	1
5_2	$(w_1^2 w_2)^2$	$2^3 \cdot 3 \cdot 5$	5_1	5_1	1
6_1	$(w_1 m^2)^2$	$2^7 \cdot 3$	3_1	2_1	6_1
6_2	$(w_1)^2$	$2^5 \cdot 3$	3_1	2_2	6_3
6_3	$(w_1 w_2^2)^2$	$2^5 \cdot 3$	3_1	2_2	6_2
6_4	$w_1 w_2^4 m^2$	$2^4 \cdot 3$	3_1	2_3	6_4
6_5	$w_1^2 m$	$2^2 \cdot 3$	3_3	2_4	6_5
6_6	$w_1^3 w_2 m$	$2^2 \cdot 3$	3_2	2_5	6_6
8_1	$w_1^3 m$	$2^6 \cdot 3$	4_2	8_1	8_1
8_2	w_2	2^6	4_4	8_2	8_2
8_3	$w_2 m$	2^6	4_2	8_3	8_3
8_4	$w_2 m^2$	2^6	4_4	8_4	8_4
8_5	$w_1^3 w_2$	2^6	4_2	8_5	8_5
8_6	$w_1^3 w_2^3$	2^6	4_4	8_6	8_6
8_7	$w_1^5 w_2 w_1 m$	2^5	4_3	8_7	8_7
10_1	$(w_1^2 w_2 m)^2$	$2^3 \cdot 5$	5_1	10_2	2_1
10_2	$w_1 w_2^3$	$2^3 \cdot 5$	5_2	10_1	2_1
10_3	$w_1^2 w_2$	$2^2 \cdot 5$	5_2	10_4	2_3
10_4	$w_1^4 w_2 w_1 w_2$	$2^2 \cdot 5$	5_1	10_3	2_3
12_1	$w_1^2 w_2^2$	$2^5 \cdot 3$	6_1	4_2	12_1
12_2	$w_1 m^2$	$2^4 \cdot 3$	6_1	4_1	12_2
12_3	$w_1^4 w_2^2$	$2^4 \cdot 3$	6_1	4_3	12_3
12_4	w_1	$2^3 \cdot 3$	6_2	4_5	12_5
12_5	$w_1 w_2^2$	$2^3 \cdot 3$	6_3	4_5	12_4
15_1	$w_1 w_2 m$	$3 \cdot 5$	15_2	5_2	3_2
15_2	$w_1 w_2 w_1 w_2 m$	$3 \cdot 5$	15_1	5_1	3_2
20_1	$w_1^2 w_2 m$	$2^2 \cdot 5$	10_1	20_2	4_1
20_2	$w_1^4 w_2 m w_1 w_2$	$2^2 \cdot 5$	10_2	20_1	4_1
24	$w_1 w_2^2 m$	$2^3 \cdot 3$	12_1	8_1	24

14.5.6 *Conjugacy classes of $C_G(u) = \langle r_k | 1 \le k \le 5 \rangle$*

Class	*Representative*	*Centralizer*	2P	3P	5P	7P
1	1	$2^9 \cdot 3^2 \cdot 5 \cdot 7$	1	1	1	1
2_1	r_4	$2^9 \cdot 3^2 \cdot 5 \cdot 7$	1	2_1	2_1	2_1
2_2	r_5	$2^8 \cdot 3^2 \cdot 5 \cdot 7$	1	2_2	2_2	2_2
2_3	r_1	2^9	1	2_3	2_3	2_3
2_4	$r_1 r_4$	2^9	1	2_4	2_4	2_4
2_5	r_3	$2^5 \cdot 3^2$	1	2_5	2_5	2_5
2_6	$r_1 r_5$	2^8	1	2_6	2_6	2_6
3	$(r_1 r_2^2 r_1 r_3)^2$	$2^3 \cdot 3^2$	3	1	3	3
4_1	$r_3 r_5$	$2^5 \cdot 3^2$	2_1	4_1	4_1	4_1
4_2	r_2	2^7	2_3	4_2	4_2	4_2
4_3	$(r_1 r_3)^2$	2^7	2_3	4_3	4_3	4_3
4_4	$r_2 r_4$	2^7	2_3	4_4	4_4	4_4
4_5	$(r_1 r_3 r_2)^2$	2^7	2_3	4_5	4_5	4_5
4_6	$(r_1 r_3 r_5)^2$	2^7	2_3	4_6	4_6	4_6
4_7	$(r_1 r_3 r_2 r_5)^2$	2^7	2_3	4_7	4_7	4_7
4_8	$r_2 r_5$	2^6	2_3	4_8	4_8	4_8
4_9	$r_1 r_3 r_1 r_3 r_5$	2^6	2_3	4_9	4_9	4_9
4_{10}	$r_1 r_3 r_2^2 r_3 r_5$	2^6	2_3	4_{10}	4_{10}	4_{10}
4_{11}	$r_1 r_2 r_1 r_2 r_3 r_2$	2^5	2_3	4_{11}	4_{11}	4_{11}
4_{12}	$r_1 r_2 r_1 r_2 r_3 r_2 r_5$	2^5	2_4	4_{12}	4_{12}	4_{12}
5	$r_1 r_2^2$	$2^2 \cdot 5$	5	5	1	5
6_1	$(r_1 r_2^2 r_1 r_3 r_5)^2$	$2^3 \cdot 3^2$	3	2_1	6_1	6_1
6_2	$r_1 r_2^2 r_1 r_3$	$2^2 \cdot 3^2$	3	2_5	6_2	6_2
6_3	$r_1 r_3 r_1 r_3 r_2 r_5$	$2^2 \cdot 3^2$	3	2_2	6_3	6_3
7	$r_1 r_2$	$2^2 \cdot 7$	7	7	7	1
8_1	$r_1 r_3$	2^4	4_3	8_1	8_1	8_1
8_2	$r_1 r_3 r_2$	2^4	4_5	8_2	8_2	8_2
8_3	$r_1 r_3 r_5$	2^4	4_6	8_3	8_3	8_3
8_4	$r_1 r_2 r_1 r_3$	2^4	4_2	8_4	8_4	8_4
8_5	$r_1 r_3 r_2 r_5$	2^4	4_7	8_5	8_5	8_5
8_6	$r_1 r_2 r_1 r_3 r_5$	2^4	4_4	8_6	8_6	8_6
10_1	$r_1 r_2^2 r_4$	$2^2 \cdot 5$	5	10_1	2_1	10_1
10_2	$r_1 r_2^2 r_5$	$2^2 \cdot 5$	5	10_3	2_2	10_3
10_3	$r_1 r_2^2 r_4 r_5$	$2^2 \cdot 5$	5	10_2	2_2	10_2
12	$r_1 r_2^2 r_1 r_3 r_5$	$2^2 \cdot 3^2$	6_1	4_1	12	12
14_1	$r_1 r_2 r_4$	$2^2 \cdot 7$	7	14_1	14_1	2_1
14_2	$r_1 r_2 r_5$	$2^2 \cdot 7$	7	14_3	14_3	2_2
14_3	$r_1 r_2 r_4 r_5$	$2^2 \cdot 7$	7	14_2	14_2	2_2

14.5.7 *Conjugacy classes of* $N_1 = N_G(3_A) = \langle g_1, g_2, g_3, g_4 \rangle$

Class	*Representative*	|*Centralizer*|	2P	3P	5P	7P
1	1	$2^8 \cdot 3^7 \cdot 5 \cdot 7$	1	1	1	1
2_1	g_1	$2^8 \cdot 3^3$	1	2_1	2_1	2_1
2_2	g_3	$2^5 \cdot 3^2 \cdot 5$	1	2_2	2_2	2_2
3_1	g_4	$2^7 \cdot 3^7 \cdot 5 \cdot 7$	3_1	1	3_1	3_1
3_2	$(g_2 g_3)^8$	$2^4 \cdot 3^7$	3_2	1	3_2	3_2
3_3	$(g_2)^2$	$2^3 \cdot 3^7$	3_3	1	3_3	3_3
3_4	$(g_1 g_3 g_2^2 g_3)^2$	$2^2 \cdot 3^5$	3_4	1	3_4	3_4
3_5	$(g_1 g_3)^2$	$2 \cdot 3^4$	3_5	1	3_5	3_5
4_1	$(g_2 g_3)^6$	$2^6 \cdot 3^2$	2_1	4_1	4_1	4_1
4_2	$(g_1 g_3 g_2)^3$	$2^5 \cdot 3$	2_1	4_2	4_2	4_2
4_3	$(g_1 g_2 g_1 g_3)^2$	$2^5 \cdot 3$	2_1	4_3	4_3	4_3
5	$g_1 g_2^3$	$2 \cdot 3 \cdot 5$	5	5	1	5
6_1	$g_1 g_4$	$2^7 \cdot 3^3$	3_1	2_1	6_1	6_1
6_2	$(g_2 g_3)^4$	$2^4 \cdot 3^3$	3_2	2_1	6_2	6_2
6_3	g_2	$2^3 \cdot 3^3$	3_3	2_1	6_3	6_3
6_4	$g_1 g_3 g_2^2 g_3$	$2^2 \cdot 3^3$	3_4	2_1	6_4	6_4
6_5	$g_1 g_2 g_3 g_2 g_1 g_3$	$2^2 \cdot 3^3$	3_4	2_1	6_6	6_5
6_6	$(g_1 g_2 g_3 g_2 g_1 g_3)^5$	$2^2 \cdot 3^3$	3_4	2_1	6_5	6_6
6_7	$g_1 g_3$	$2 \cdot 3^2$	3_5	2_2	6_7	6_7
7	$(g_1 g_2)^3$	$3 \cdot 7$	7	7	7	1
8_1	$(g_2 g_3)^3$	$2^5 \cdot 3$	4_1	8_1	8_1	8_1
8_2	$g_1 g_2 g_3 g_1 g_3 g_2$	$2^4 \cdot 3$	4_1	8_2	8_2	8_2
8_3	$g_2^2 g_3$	2^5	4_1	8_3	8_3	8_3
8_4	$g_1 g_2 g_1 g_3$	2^3	4_3	8_4	8_4	8_4
9_1	$g_1 g_3 g_2 g_3$	3^3	9_2	3_3	9_2	9_1
9_2	$(g_1 g_3 g_2 g_3)^2$	3^3	9_1	3_3	9_1	9_2
10	$g_1 g_3 g_2^2$	$2 \cdot 5$	5	10	2_2	10
12_1	$(g_1 g_2 g_1 g_2^4)^2$	$2^5 \cdot 3^2$	6_1	4_1	12_1	12_1
12_2	$(g_2 g_3)^2$	$2^3 \cdot 3^2$	6_2	4_1	12_2	12_2
12_3	$g_1 g_2 g_1 g_2 g_3 g_2 g_3$	$2^4 \cdot 3$	6_1	4_3	12_3	12_3
12_4	$g_1 g_2^2$	$2^2 \cdot 3^2$	6_3	4_1	12_4	12_4
12_5	$g_1 g_3 g_2$	$2^2 \cdot 3$	6_2	4_2	12_5	12_5
15	$g_1 g_2^3 g_4$	$3 \cdot 5$	15	5	3_1	15
21_1	$g_1 g_2$	$3 \cdot 7$	21_2	7	21_1	3_1
21_2	$(g_1 g_2)^2$	$3 \cdot 7$	21_1	7	21_2	3_1
24_1	$g_2 g_3$	$2^3 \cdot 3$	12_2	8_1	24_1	24_1
24_2	$(g_2 g_3)^{13}$	$2^3 \cdot 3$	12_2	8_1	24_2	24_2
24_3	$g_1 g_2 g_1 g_2^4$	$2^3 \cdot 3$	12_1	8_2	24_3	24_3

14.5.8 *Conjugacy classes of* $N_2 = N_G(Z) = \langle t_1, t_2, t_3, t_4 \rangle$

Class	*Representative*	*Centralizer*	2P	3P
1	1	$2^6 \cdot 3^7$	1	1
2_1	$(t_1)^2$	$2^6 \cdot 3^3$	1	2_1
2_2	t_3	$2^4 \cdot 3^4$	1	2_2
2_3	$(t_1 t_3 t_4)^6$	$2^5 \cdot 3^2$	1	2_3
3_1	$(t_1 t_2)^4$	$2^4 \cdot 3^7$	3_1	1
3_2	$(t_2 t_3)^6$	$2^4 \cdot 3^7$	3_2	1
3_3	$(t_2 t_3 t_4)^2$	$2^3 \cdot 3^5$	3_3	1
3_4	$(t_2^2 t_3)^2$	$2^2 \cdot 3^5$	3_5	1
3_5	$(t_2^2 t_3)^4$	$2^2 \cdot 3^5$	3_4	1
3_6	$(t_2)^2$	$2 \cdot 3^5$	3_6	1
3_7	$(t_1^2 t_3)^2$	$2 \cdot 3^5$	3_7	1
3_8	$(t_1 t_3 t_4)^4$	$2^2 \cdot 3^4$	3_8	1
4_1	t_4	$2^5 \cdot 3^2$	2_1	4_1
4_2	$(t_1 t_3 t_4)^3$	$2^4 \cdot 3^2$	2_3	4_2
4_3	$(t_1 t_3)^2$	$2^5 \cdot 3$	2_1	4_3
4_4	$(t_3 t_4)^3$	$2^4 \cdot 3$	2_1	4_4
4_5	t_1	$2^3 \cdot 3$	2_1	4_5
4_6	$t_1^3 t_3 t_4$	2^4	2_3	4_6
6_1	$(t_2 t_3)^3$	$2^4 \cdot 3^4$	3_2	2_2
6_2	$(t_2 t_3)^{15}$	$2^4 \cdot 3^4$	3_2	2_2
6_3	$(t_1 t_2)^2$	$2^4 \cdot 3^3$	3_1	2_1
6_4	$(t_3 t_4)^2$	$2^4 \cdot 3^3$	3_2	2_1
6_5	$(t_1 t_2 t_3 t_1 t_3)^2$	$2^3 \cdot 3^3$	3_3	2_1
6_6	$t_2^2 t_3$	$2^2 \cdot 3^3$	3_4	2_2
6_7	$(t_2^2 t_3)^5$	$2^2 \cdot 3^3$	3_5	2_2
6_8	$t_2 t_3 t_4$	$2^2 \cdot 3^3$	3_3	2_2
6_9	$(t_2 t_3 t_4)^5$	$2^2 \cdot 3^3$	3_3	2_2
6_{10}	$t_1^2 t_2 t_4$	$2^2 \cdot 3^3$	3_4	2_1
6_{11}	$(t_1^2 t_2 t_4)^5$	$2^2 \cdot 3^3$	3_5	2_1
6_{12}	$t_2 t_3 t_2 t_4$	$2^2 \cdot 3^3$	3_4	2_2
6_{13}	$(t_2 t_3 t_2 t_4)^5$	$2^2 \cdot 3^3$	3_5	2_2
6_{14}	t_2	$2 \cdot 3^3$	3_6	2_1
6_{15}	$t_1^2 t_3$	$2 \cdot 3^3$	3_7	2_2
6_{16}	$(t_1 t_3 t_4)^2$	$2^2 \cdot 3^2$	3_8	2_3
8_1	$t_1 t_4$	$2^3 \cdot 3$	4_1	8_1
8_2	$t_1 t_3$	2^3	4_3	8_2
9_1	$(t_2 t_3)^2$	$2 \cdot 3^3$	9_2	3_2
9_2	$(t_2 t_3)^4$	$2 \cdot 3^3$	9_1	3_2
12_1	$t_1^2 t_4$	$2^3 \cdot 3^2$	6_4	4_1
12_2	$(t_1 t_2 t_4)^2$	$2^3 \cdot 3^2$	6_3	4_1
12_3	$t_1 t_3 t_4$	$2^2 \cdot 3^2$	6_{16}	4_2
12_4	$(t_1 t_3 t_4)^7$	$2^2 \cdot 3^2$	6_{16}	4_2
12_5	$t_3 t_4$	$2^3 \cdot 3$	6_4	4_4
12_6	$t_1 t_2$	$2^2 \cdot 3$	6_3	4_5
12_7	$t_1 t_2 t_3 t_1 t_3$	$2^2 \cdot 3$	6_5	4_3
18_1	$t_2 t_3$	$2 \cdot 3^2$	9_1	6_1
18_2	$(t_2 t_3)^5$	$2 \cdot 3^2$	9_2	6_2
24_1	$t_1 t_2 t_4$	$2^3 \cdot 3$	12_2	8_1
24_2	$(t_1 t_2 t_4)^{13}$	$2^3 \cdot 3$	12_2	8_1

14.6 Character tables of local subgroups

14.6.1 *Character table of $H = \langle x, y, h \rangle$*

	1a	2a	2b	2c	2d	3a	3b	3c	3d	4a	4b	4c	4d	4e	4f	4g	4h	4i	5a	6a	6b	6c	6d	6e
2	13	13	12	10	9	7	4	4	4	10	10	9	9	8	9	7	7	5	3	7	4	4	4	6
3	4	4	1	2	.	3	4	4	3	2	1	1	1	1	.	1	1	.	.	3	4	4	3	1
5	1	1	1	.	.	.	.	.	.	1	.	.	.	.	.	.	.	.	1	.	.	.	.	.
2P	1a	1a	1a	1a	1a	3a	3c	3b	3d	2a	2a	2b	2a	2b	2b	2c	2c	2d	5a	3a	3c	3b	3d	3a
3P	1a	2a	2b	2c	2d	1a	1a	1a	1a	4a	4b	4c	4d	4e	4f	4g	4h	4i	5a	2a	2a	2a	2a	2b
5P	1a	2a	2b	2c	2d	3a	3c	3b	3d	4a	4b	4c	4d	4e	4f	4g	4h	4i	1a	6a	6c	6b	6d	6e
X.1	1	1	1	1	1	1	1	1	1	1	1	1	1	1	1	1	1	1	1	1	1	1	1	1
X.2	5	5	5	−3	1	−1	A	$\bar{A}$	2	5	−3	1	1	−3	1	1	1	−1	.	−1	A	$\bar{A}$	2	−1
X.3	5	5	5	−3	1	−1	$\bar{A}$	A	2	5	−3	1	1	−3	1	1	1	−1	.	−1	$\bar{A}$	A	2	−1
X.4	6	6	6	−2	2	3	−3	−3	.	6	−2	2	2	−2	2	2	2	.	1	3	−3	−3	.	3
X.5	10	10	10	2	−2	1	E	$\bar{E}$	1	10	2	−2	−2	2	−2	2	2	.	.	1	E	$\bar{E}$	1	1
X.6	10	10	10	2	−2	1	$\bar{E}$	E	1	10	2	−2	−2	2	−2	2	2	.	.	1	$\bar{E}$	E	1	1
X.7	15	15	15	−1	−1	3	6	6	.	15	−1	−1	−1	−1	−1	3	3	−1	.	3	6	6	.	3
X.8	15	15	15	7	3	.	−3	−3	3	15	7	3	3	7	3	−1	−1	1	.	.	−3	−3	3	.
X.9	20	20	20	4	4	5	2	2	−1	20	4	4	4	4	4	.	.	.	.	5	2	2	−1	5
X.10	24	24	24	8	.	.	6	6	3	24	8	.	.	8	.	.	.	.	−1	.	6	6	3	.
X.11	27	27	−5	3	3	9	.	.	.	3	3	7	−5	−1	−1	3	3	−1	2	9	.	.	.	1
X.12	30	30	30	−10	2	3	3	3	3	30	−10	2	2	−10	2	−2	−2	.	.	3	3	3	3	3
X.13	30	30	30	6	2	−3	F	$\bar{F}$	.	30	6	2	2	6	2	2	2	.	.	−3	F	$\bar{F}$	.	−3
X.14	30	30	30	6	2	−3	$\bar{F}$	F	.	30	6	2	2	6	2	2	2	.	.	−3	$\bar{F}$	F	.	−3
X.15	32	−32	.	.	.	8	$\bar{A}$	A	2	.	.	.	.	.	.	4	−4	.	2	−8	$-\bar{A}$	$-A$	−2	.
X.16	32	−32	.	.	.	8	A	$\bar{A}$	2	.	.	.	.	.	.	4	−4	.	2	−8	$-A$	$-\bar{A}$	−2	.
X.17	36	36	4	12	−4	6	.	.	3	−4	−4	8	4	.	.	.	.	.	1	6	.	.	3	−2
X.18	36	36	4	−12	.	6	.	.	3	−4	4	4	8	.	−4	.	.	2	1	6	.	.	3	−2
X.19	40	40	40	−8	.	−2	H	$\bar{H}$	1	40	−8	.	.	−8	.	.	.	.	.	−2	H	$\bar{H}$	1	−2
X.20	40	40	40	−8	.	−2	$\bar{H}$	H	1	40	−8	.	.	−8	.	.	.	.	.	−2	$\bar{H}$	H	1	−2
X.21	45	45	45	−3	−3	.	J	$\bar{J}$	.	45	−3	−3	−3	−3	−3	1	1	1	.	.	J	$\bar{J}$	.	.
X.22	45	45	45	−3	−3	.	$\bar{J}$	J	.	45	−3	−3	−3	−3	−3	1	1	1	.	.	$\bar{J}$	J	.	.
X.23	60	60	60	−4	4	−3	6	6	−3	60	−4	4	4	−4	4	.	.	.	.	−3	6	6	−3	−3
X.24	64	64	64	.	.	4	−8	−8	−2	64	.	.	.	.	.	.	.	.	−1	4	−8	−8	−2	4
X.25	81	81	81	9	−3	.	.	.	.	81	9	−3	−3	9	−3	−3	−3	−1	1	.	.	.	.	.
X.26	81	81	−15	9	1	.	.	.	.	9	9	−3	9	−3	5	−3	−3	1	1	.	.	.	.	.
X.27	81	81	−15	9	1	.	.	.	.	9	9	−3	9	−3	5	−3	−3	1	1	.	.	.	.	.
X.28	108	108	−20	12	4	9	.	.	.	12	12	4	4	−4	4	.	.	.	−2	9	.	.	.	1
X.29	135	135	−25	−9	3	18	.	.	.	15	−9	7	−5	3	−1	3	3	1	.	18	.	.	.	2
X.30	135	135	−25	15	7	−9	.	.	.	15	15	11	−1	−5	3	3	3	−1	.	−9	.	.	.	−1
X.31	135	135	−25	−9	3	−9	.	.	.	15	−9	7	−5	3	−1	3	3	1	.	−9	.	.	.	−1
X.32	135	135	−25	−9	3	−9	.	.	.	15	−9	7	−5	3	−1	3	3	1	.	−9	.	.	.	−1
X.33	160	−160	.	.	.	−8	7	7	4	.	.	.	.	.	.	4	−4	.	.	8	−7	−7	−4	.
X.34	160	−160	.	.	.	16	N	$\bar{N}$	−2	.	.	.	.	.	.	4	−4	.	.	−16	$-N$	$-\bar{N}$	2	.
X.35	160	−160	.	.	.	−8	O	$\bar{O}$	4	.	.	.	.	.	.	4	−4	.	.	8	$-O$	$-\bar{O}$	−4	.
X.36	160	−160	.	.	.	16	$\bar{N}$	N	−2	.	.	.	.	.	.	4	−4	.	.	−16	$-\bar{N}$	$-N$	2	.
X.37	160	−160	.	.	.	−8	$\bar{O}$	O	4	.	.	.	.	.	.	4	−4	.	.	8	$-\bar{O}$	$-O$	−4	.
X.38	180	180	20	−12	−8	12	.	.	−3	−20	4	12	.	.	4	.	.	−2	.	12	.	.	−3	−4
X.39	180	180	20	−36	−4	−6	.	.	6	−20	12	8	4	.	.	.	.	.	.	−6	.	.	6	2
X.40	180	180	20	36	.	−6	.	.	6	−20	−12	4	8	.	−4	.	.	−2	.	−6	.	.	6	2
X.41	180	180	20	12	4	12	.	.	−3	−20	−4	.	12	.	−8	.	.	.	.	12	.	.	−3	−4
X.42	270	270	−50	6	2	9	.	.	.	30	6	10	−14	−2	−6	−6	−6	.	.	9	.	.	.	1
X.43	270	270	−50	6	−6	9	.	.	.	30	6	−14	10	−2	2	6	6	.	.	9	.	.	.	1
X.44	288	−288	.	.	.	.	J	$\bar{J}$	.	.	.	.	.	.	.	4	−4	.	−2	.	$-J$	$-\bar{J}$	.	.
X.45	288	−288	.	.	.	.	$\bar{J}$	J	.	.	.	.	.	.	.	4	−4	.	−2	.	$-\bar{J}$	$-J$	.	.
X.46	324	324	36	−36	4	.	.	.	.	−36	12	.	12	.	−8	.	.	.	−1	.	.	.	.	.
X.47	324	324	36	36	−8	.	.	.	.	−36	−12	12	.	.	4	.	.	2	−1	.	.	.	.	.
X.48	360	360	40	−24	.	6	.	.	3	−40	8	−8	−16	.	8	.	.	.	.	6	.	.	3	−2
X.49	360	360	40	24	8	6	.	.	3	−40	−8	−16	−8	.	.	.	.	.	.	6	.	.	3	−2
X.50	405	405	−75	−27	1	.	.	.	.	45	−27	−3	9	9	5	−3	−3	−1	.	.	.	.	.	.
X.51	480	−480	.	.	.	24	3	3	.	.	.	.	.	.	.	−4	4	.	.	−24	−3	−3	.	.
X.52	480	−480	.	.	.	.	F	$\bar{F}$	6	.	.	.	.	.	.	−4	4	.	.	.	$-F$	$-\bar{F}$	−6	.
X.53	480	−480	.	.	.	.	$\bar{F}$	F	6	.	.	.	.	.	.	−4	4	.	.	.	$-\bar{F}$	$-F$	−6	.
X.54	512	−512	.	.	.	−16	8	8	−4	.	.	.	.	.	.	.	.	.	2	16	−8	−8	4	.
X.55	540	540	−100	12	−4	−9	.	.	.	60	12	−4	−4	−4	−4	.	.	.	.	−9	.	.	.	−1
X.56	576	576	64	.	.	−12	.	.	−6	−64	.	.	.	.	.	.	.	.	1	−12	.	.	−6	4
X.57	640	−640	.	.	.	−8	−8	−8	−8	.	.	.	.	.	.	.	.	.	.	8	8	8	8	.

Character table of $H = \langle x, y, h \rangle$ (continued)

	6f	6g	6h	6i	6j	6k	6l	6m	8a	8b	8c	8d	8e	9a	9b	10a	10b	10c	12a	12b	12c	12d	12e	12f	12g
2	4	4	4	4	4	4	4	4	6	6	5	5	5	1	1	3	2	2	5	3	4	4	4	4	3
3	2	2	2	2	2	2	2	2	1	.	.	.	.	2	2	.	.	.	2	2	1	1	1	1	1
5	.	.	.	.	.	.	.	.	.	.	.	.	.	.	.	1	1	1	.	.	.	.	.	.	.
2P	3b	3c	3c	3b	3a	3a	3d	3d	4a	4b	4d	4f	4c	9b	9a	5a	5a	5a	6a	6d	6e	6e	6e	6a	6d
3P	2c	2c	2c	2c	2c	2c	2c	2c	8a	8b	8c	8d	8e	3b	3c	10a	10c	10b	4a	4a	4c	4e	4e	4d	4b
5P	6g	6f	6i	6h	6k	6j	6m	6l	8a	8b	8c	8d	8e	9b	9a	2a	2b	2b	12a	12b	12c	12e	12d	12f	12g
X.1	1	1	1	1	1	1	1	1	1	1	1	1	1	1	1	1	1	1	1	1	1	1	1	1	1
X.2	B	$\bar{B}$	$\bar{B}$	B	C	$\bar{C}$	.	.	1	1	−1	−1	−1	D	$\bar{D}$	.	.	.	−1	2	1	$\bar{C}$	C	1	.
X.3	$\bar{B}$	B	B	$\bar{B}$	$\bar{C}$	C	.	.	1	1	−1	−1	−1	$\bar{D}$	D	.	.	.	−1	2	1	C	$\bar{C}$	1	.
X.4	1	1	1	1	1	1	−2	−2	2	2	.	.	.	.	.	1	1	1	3	.	−1	1	1	−1	−2
X.5	A	$\bar{A}$	$\bar{A}$	A	−1	−1	−1	−1	−2	2	.	.	.	$-D$	$-\bar{D}$	.	.	.	1	1	1	−1	−1	1	−1
X.6	$\bar{A}$	A	A	$\bar{A}$	−1	−1	−1	−1	−2	2	.	.	.	$-\bar{D}$	$-D$	.	.	.	1	1	1	−1	−1	1	−1
X.7	2	2	2	2	−1	−1	2	2	−1	3	−1	−1	−1	.	.	.	.	.	3	.	−1	−1	−1	−1	2
X.8	1	1	1	1	−2	−2	1	1	3	−1	1	1	1	.	.	.	.	.	.	3	.	−2	−2	.	1
X.9	−2	−2	−2	−2	1	1	1	1	4	.	.	.	.	−1	−1	.	.	.	5	−1	1	1	1	1	1
X.10	2	2	2	2	2	2	−1	−1	.	.	.	.	.	.	.	−1	−1	−1	.	3	.	2	2	.	−1
X.11	.	.	.	.	3	3	.	.	−1	−1	−1	1	1	.	.	2	.	.	−3	.	1	−1	−1	1	.
X.12	−1	−1	−1	−1	−1	−1	−1	−1	2	−2	.	.	.	.	.	.	.	.	3	3	−1	−1	−1	−1	−1
X.13	$\bar{B}$	B	B	$\bar{B}$	$\bar{C}$	C	.	.	2	2	.	.	.	.	.	.	.	.	−3	.	−1	C	$\bar{C}$	−1	.
X.14	B	$\bar{B}$	$\bar{B}$	B	C	$\bar{C}$	.	.	2	2	.	.	.	.	.	.	.	.	−3	.	−1	$\bar{C}$	C	−1	.
X.15	$\bar{B}$	B	$-B$	$-\bar{B}$	.	.	G	$\bar{G}$	.	.	.	.	.	$\bar{D}$	D	−2	.	.	.	.	.	.	.	.	.
X.16	B	$\bar{B}$	$-\bar{B}$	$-B$	.	.	$\bar{G}$	G	.	.	.	.	.	D	$\bar{D}$	−2	.	.	.	.	.	.	.	.	.
X.17	.	.	.	.	.	.	3	3	.	.	.	2	−2	.	.	1	−1	−1	2	−1	2	.	.	−2	−1
X.18	.	.	.	.	.	.	−3	−3	.	.	−2	.	.	.	.	1	−1	−1	2	−1	−2	.	.	2	1
X.19	I	$\bar{I}$	$\bar{I}$	I	I	$\bar{I}$	1	1	.	.	.	.	.	$-\bar{D}$	$-D$	.	.	.	−2	1	.	$\bar{I}$	I	.	1
X.20	$\bar{I}$	I	I	$\bar{I}$	$\bar{I}$	I	1	1	.	.	.	.	.	$-D$	$-\bar{D}$	.	.	.	−2	1	.	I	$\bar{I}$	.	1
X.21	K	$\bar{K}$	$\bar{K}$	K	.	.	.	.	−3	1	1	1	1	.	.	.	.	.	.	.	.	.	.	.	.
X.22	$\bar{K}$	K	K	$\bar{K}$	.	.	.	.	−3	1	1	1	1	.	.	.	.	.	.	.	.	.	.	.	.
X.23	2	2	2	2	−1	−1	−1	−1	4	.	.	.	.	.	.	.	.	.	−3	−3	1	−1	−1	1	−1
X.24	.	.	.	.	.	.	.	.	.	.	.	.	.	1	1	−1	−1	−1	4	−2	.	.	.	.	.
X.25	.	.	.	.	.	.	.	.	−3	−3	−1	−1	−1	.	.	1	1	1	.	.	.	.	.	.	.
X.26	.	.	.	.	.	.	.	.	−3	1	1	−1	−1	.	.	1	L	$-L$	.	.	.	.	.	.	.
X.27	.	.	.	.	.	.	.	.	−3	1	1	−1	−1	.	.	1	$-L$	L	.	.	.	.	.	.	.
X.28	.	.	.	.	3	3	.	.	−4	.	.	.	.	.	.	−2	.	.	−3	.	1	−1	−1	1	.
X.29	.	.	.	.	.	.	.	.	−1	−1	1	−1	−1	.	.	.	.	.	−6	.	−2	.	.	−2	.
X.30	.	.	.	.	−3	−3	.	.	−5	−1	−1	1	1	.	.	.	.	.	3	.	−1	1	1	−1	.
X.31	.	.	.	.	M	$\bar{M}$	.	.	−1	−1	1	−1	−1	.	.	.	.	.	3	.	1	C	$\bar{C}$	1	.
X.32	.	.	.	.	$\bar{M}$	M	.	.	−1	−1	1	−1	−1	.	.	.	.	.	3	.	1	$\bar{C}$	C	1	.
X.33	3	3	−3	−3	.	.	.	.	.	.	.	.	.	1	1	.	.	.	.	.	.	.	.	.	.
X.34	$\bar{C}$	C	$\bar{C}$	C	.	.	G	$\bar{G}$	.	.	.	.	.	$-\bar{D}$	$-D$	.	.	.	.	.	.	.	.	.	.
X.35	$-K$	$-\bar{K}$	$\bar{K}$	K	.	.	.	.	.	.	.	.	.	$-\bar{D}$	$-D$	.	.	.	.	.	.	.	.	.	.
X.36	C	$\bar{C}$	C	$\bar{C}$	.	.	$\bar{G}$	G	.	.	.	.	.	$-D$	$-\bar{D}$	.	.	.	.	.	.	.	.	.	.
X.37	$-\bar{K}$	$-K$	K	$\bar{K}$	.	.	.	.	.	.	.	.	.	$-D$	$-\bar{D}$	.	.	.	.	.	.	.	.	.	.
X.38	.	.	.	.	.	.	−3	−3	.	.	2	.	.	.	.	.	.	.	4	1	.	.	.	.	1
X.39	.	.	.	.	.	.	.	.	.	.	.	−2	2	.	.	.	.	.	−2	−2	2	.	.	−2	.
X.40	.	.	.	.	.	.	.	.	.	.	2	.	.	.	.	.	.	.	−2	−2	−2	.	.	2	.
X.41	.	.	.	.	.	.	3	3	.	.	.	−2	2	.	.	.	.	.	4	1	.	.	.	.	−1
X.42	.	.	.	.	−3	−3	.	.	2	2	.	.	.	.	.	.	.	.	−3	.	1	1	1	1	.
X.43	.	.	.	.	−3	−3	.	.	2	−2	.	.	.	.	.	.	.	.	−3	.	1	1	1	1	.
X.44	K	$\bar{K}$	$-\bar{K}$	$-K$	.	.	.	.	.	.	.	.	.	.	.	2	.	.	.	.	.	.	.	.	.
X.45	$\bar{K}$	K	$-K$	$-\bar{K}$	.	.	.	.	.	.	.	.	.	.	.	2	.	.	.	.	.	.	.	.	.
X.46	.	.	.	.	.	.	.	.	.	.	.	2	−2	.	.	−1	1	1	.	.	.	.	.	.	.
X.47	.	.	.	.	.	.	.	.	.	.	−2	.	.	.	.	−1	1	1	.	.	.	.	.	.	.
X.48	.	.	.	.	.	.	3	3	.	.	.	.	.	.	.	.	.	.	2	−1	−2	.	.	2	−1
X.49	.	.	.	.	.	.	−3	−3	.	.	.	.	.	.	.	.	.	.	2	−1	2	.	.	−2	1
X.50	.	.	.	.	.	.	.	.	−3	1	−1	1	1	.	.	.	.	.	.	.	.	.	.	.	.
X.51	3	3	−3	−3	.	.	.	.	.	.	.	.	.	.	.	.	.	.	.	.	.	.	.	.	.
X.52	$\bar{B}$	B	$-B$	$-\bar{B}$	.	.	G	$\bar{G}$	.	.	.	.	.	.	.	.	.	.	.	.	.	.	.	.	.
X.53	B	$\bar{B}$	$-\bar{B}$	$-B$	.	.	$\bar{G}$	G	.	.	.	.	.	.	.	.	.	.	.	.	.	.	.	.	.
X.54	.	.	.	.	.	.	.	.	.	.	.	.	.	−1	−1	−2	.	.	.	.	.	.	.	.	.
X.55	.	.	.	.	3	3	.	.	4	.	.	.	.	.	.	.	.	.	3	.	−1	−1	−1	−1	.
X.56	.	.	.	.	.	.	.	.	.	.	.	.	.	.	.	1	−1	−1	−4	2	.	.	.	.	.
X.57	.	.	.	.	.	.	.	.	.	.	.	.	.	1	1	.	.	.	.	.	.	.	.	.	.

Character table of $H = \langle x, y, h \rangle$ *(continued)*

2	3	3	3	3	1	1	2	3
3	1	1	1	1	2	2	.	1
5	.	.	.	.	.	.	1	.
	12h	12i	12j	12k	18a	18b	20a	24a
2P	6f	6g	6f	6g	9b	9a	10a	12a
3P	4g	4g	4h	4h	6b	6c	20a	8a
5P	12i	12h	12k	12j	18b	18a	4a	24a
X.1	1	1	1	1	1	1	1	1
X.2	$-D$	$-\bar{D}$	$-D$	$-\bar{D}$	D	$\bar{D}$	.	1
X.3	$-\bar{D}$	$-D$	$-\bar{D}$	$-D$	$\bar{D}$	D	.	1
X.4	−1	−1	−1	−1	.	.	1	−1
X.5	$\bar{D}$	D	$\bar{D}$	D	$-D$	$-\bar{D}$	.	1
X.6	D	$\bar{D}$	D	$\bar{D}$	$-\bar{D}$	$-D$	.	1
X.7	.	.	.	.	.	.	.	−1
X.8	−1	−1	−1	−1	.	.	.	.
X.9	.	.	.	.	−1	−1	.	1
X.10	.	.	.	.	.	.	−1	.
X.11	.	.	.	.	.	.	−2	−1
X.12	1	1	1	1	.	.	.	−1
X.13	$\bar{D}$	D	$\bar{D}$	D	.	.	.	−1
X.14	D	$\bar{D}$	D	$\bar{D}$	.	.	.	−1
X.15	$-\bar{D}$	$-D$	$\bar{D}$	D	$-\bar{D}$	$-D$	.	.
X.16	$-D$	$-\bar{D}$	D	$\bar{D}$	$-D$	$-\bar{D}$	.	.
X.17	.	.	.	.	.	.	1	.
X.18	.	.	.	.	.	.	1	.
X.19	.	.	.	.	$-\bar{D}$	$-D$	.	.
X.20	.	.	.	.	$-D$	$-\bar{D}$	.	.
X.21	$-\bar{D}$	$-D$	$-\bar{D}$	$-D$	.	.	.	.
X.22	$-D$	$-\bar{D}$	$-D$	$-\bar{D}$	.	.	.	.
X.23	.	.	.	.	.	.	.	1
X.24	.	.	.	.	1	1	−1	.
X.25	.	.	.	.	.	.	1	.
X.26	.	.	.	.	.	.	−1	.
X.27	.	.	.	.	.	.	−1	.
X.28	.	.	.	.	.	.	2	−1
X.29	.	.	.	.	.	.	.	2
X.30	.	.	.	.	.	.	.	1
X.31	.	.	.	.	.	.	.	−1
X.32	.	.	.	.	.	.	.	−1
X.33	1	1	−1	−1	−1	−1	.	.
X.34	1	1	−1	−1	$\bar{D}$	D	.	.
X.35	$-\bar{D}$	$-D$	$\bar{D}$	D	$\bar{D}$	D	.	.
X.36	1	1	−1	−1	D	$\bar{D}$	.	.
X.37	$-D$	$-\bar{D}$	D	$\bar{D}$	D	$\bar{D}$	.	.
X.38	.	.	.	.	.	.	.	.
X.39	.	.	.	.	.	.	.	.
X.40	.	.	.	.	.	.	.	.
X.41	.	.	.	.	.	.	.	.
X.42	.	.	.	.	.	.	.	−1
X.43	.	.	.	.	.	.	.	−1
X.44	$-\bar{D}$	$-D$	$\bar{D}$	D	.	.	.	.
X.45	$-D$	$-\bar{D}$	D	$\bar{D}$	.	.	.	.
X.46	.	.	.	.	.	.	−1	.
X.47	.	.	.	.	.	.	−1	.
X.48	.	.	.	.	.	.	.	.
X.49	.	.	.	.	.	.	.	.
X.50	.	.	.	.	.	.	.	.
X.51	−1	−1	1	1	.	.	.	.
X.52	$\bar{D}$	D	$-\bar{D}$	$-D$	.	.	.	.
X.53	D	$\bar{D}$	$-D$	$-\bar{D}$	.	.	.	.
X.54	.	.	.	.	1	1	.	.
X.55	.	.	.	.	.	.	.	1
X.56	.	.	.	.	.	.	1	.
X.57	.	.	.	.	−1	−1	.	.

where $A = -3\zeta(3) - 1$, $B = -\zeta(3) - 2$, $C = 2\zeta(3) + 1$, $D = \zeta(3) + 1$, $E = 3\zeta(3) - 2$, $F = -9\zeta(3) - 6$, $G = 4\zeta(3) + 2$, $H = 6\zeta(3) - 2$, $I = -2\zeta(3)$, $J = -9\zeta(3)$, $K = -3\zeta(3) - 3$, $L = 2\zeta(5)^3 + 2\zeta(5)^2 + 1$, $M = 6\zeta(3) + 3$, $N = 6\zeta(3) + 1$, $O = -3\zeta(3) - 8$.

14.6.2 *Character table of* $E = N_G(A) = \langle x_1, y_1, e_1 \rangle$

2	13	13	8	9	9	7	7	4	2	10	10	9	8	9	8	5	5	2	2	7	7	4	4	4	5	5	2	5	6
3	3	2	1	1	.	3	3	2	2	1	1	1	1	.	.	1	.	1	1	2	2	2	2	2	1	1	1	1	.
5	1	.	1	.	.	1	1	.	.	.	.	.	.	.	.	.	.	1	1	.	.	.	.	.	.	.	.	.	.
	1a	2a	2b	2c	2d	3a	3b	3c	3d	4a	4b	4c	4d	4e	4f	4g	4h	5a	5b	6a	6b	6c	6d	6e	6f	6g	6h	8a	8b
2P	1a	1a	1a	1a	1a	3b	3a	3c	3d	2a	2a	2a	2a	2a	2a	2c	2d	5b	5a	3a	3b	3c	3c	3c	3a	3b	3d	4c	4b
3P	1a	2a	2b	2c	2d	1a	1a	1a	1a	4a	4b	4c	4d	4e	4f	4g	4h	5b	5a	2a	2a	2a	2a	2a	2c	2c	2b	8a	8b
5P	1a	2a	2b	2c	2d	3b	3a	3c	3d	4a	4b	4c	4d	4e	4f	4g	4h	1a	1a	6b	6a	6c	6e	6d	6g	6f	6h	8a	8b
X.1	1	1	1	1	1	1	1	1	1	1	1	1	1	1	1	1	1	1	1	1	1	1	1	1	1	1	1	1	1
X.2	3	3	3	-1	-1	A	$\bar{A}$	.	.	3	3	-1	-1	-1	-1	1	1	B	C	$\bar{A}$	A	.	.	.	D	$\bar{D}$	.	1	-1
X.3	3	3	3	-1	-1	$\bar{A}$	A	.	.	3	3	-1	-1	-1	-1	1	1	B	C	A	$\bar{A}$	.	.	.	$\bar{D}$	D	.	1	-1
X.4	3	3	3	-1	-1	$\bar{A}$	A	.	.	3	3	-1	-1	-1	-1	1	1	C	B	A	$\bar{A}$	.	.	.	$\bar{D}$	D	.	1	-1
X.5	3	3	3	-1	-1	A	$\bar{A}$	.	.	3	3	-1	-1	-1	-1	1	1	C	B	$\bar{A}$	A	.	.	.	D	$\bar{D}$	.	1	-1
X.6	5	5	5	1	1	5	5	-1	2	5	5	1	1	1	1	-1	-1	.	.	5	5	-1	-1	-1	1	1	2	-1	1
X.7	5	5	5	1	1	5	5	2	-1	5	5	1	1	1	1	-1	-1	.	.	5	5	2	2	2	1	1	-1	-1	1
X.8	6	6	6	2	2	G	$\bar{G}$	.	.	6	6	2	2	2	2	.	.	1	1	$\bar{G}$	G	.	.	.	H	$\bar{H}$	.	.	2
X.9	6	6	6	2	2	$\bar{G}$	G	.	.	6	6	2	2	2	2	.	.	1	1	G	$\bar{G}$	.	.	.	$\bar{H}$	H	.	.	2
X.10	8	8	8	.	.	8	8	-1	-1	8	8	.	.	.	.	.	.	B	C	8	8	-1	-1	-1	.	.	-1	.	.
X.11	8	8	8	.	.	8	8	-1	-1	8	8	.	.	.	.	.	.	C	B	8	8	-1	-1	-1	.	.	-1	.	.
X.12	9	9	9	1	1	9	9	.	.	9	9	1	1	1	1	1	1	-1	-1	9	9	.	.	.	1	1	.	1	1
X.13	9	9	9	1	1	I	$\bar{I}$	.	.	9	9	1	1	1	1	1	1	-1	-1	$\bar{I}$	I	.	.	.	$-\bar{D}$	$-D$	.	1	1
X.14	9	9	9	1	1	$\bar{I}$	I	.	.	9	9	1	1	1	1	1	1	-1	-1	I	$\bar{I}$	.	.	.	$-D$	$-\bar{D}$	.	1	1
X.15	10	10	10	-2	-2	10	10	1	1	10	10	-2	-2	-2	-2	.	.	.	.	10	10	1	1	1	-2	-2	1	.	-2
X.16	15	15	15	-1	-1	J	$\bar{J}$	.	.	15	15	-1	-1	-1	-1	-1	-1	.	.	$\bar{J}$	J	.	.	.	$\bar{D}$	D	.	-1	-1
X.17	15	15	15	-1	-1	$\bar{J}$	J	.	.	15	15	-1	-1	-1	-1	-1	-1	.	.	J	$\bar{J}$	.	.	.	D	$\bar{D}$	.	-1	-1
X.18	18	18	-6	6	-2	.	.	3	.	2	2	6	6	-2	-2	.	.	3	3	.	.	3	3	3	.	.	.	.	-2
X.19	45	45	5	9	1	.	.	.	3	-3	-3	9	9	1	1	3	-1	.	.	.	.	.	.	.	.	.	-1	3	1
X.20	45	45	5	-3	5	.	.	.	3	-3	-3	-3	-3	5	5	-3	1	.	.	.	.	.	.	.	.	.	-1	-3	-3
X.21	54	54	-18	-6	2	.	.	.	.	6	6	-6	-6	2	2	.	.	K	L	.	.	.	.	.	.	.	.	.	2
X.22	54	54	-18	-6	2	.	.	.	.	6	6	-6	-6	2	2	.	.	L	K	.	.	.	.	.	.	.	.	.	2
X.23	60	-4	.	8	-4	15	15	-3	.	-4	4	.	-4	4	.	2	.	.	.	-1	-1	5	-1	-1	-1	-1	.	-2	.
X.24	60	-4	.	.	4	15	15	-3	.	-4	4	8	-4	-4	.	-2	.	.	.	-1	-1	5	-1	-1	3	3	.	2	.
X.25	60	-4	.	.	4	$\bar{J}$	J	3	.	-4	4	8	-4	-4	.	-2	.	.	.	D	$\bar{D}$	-1	M	$\bar{M}$	$\bar{A}$	A	.	2	.
X.26	60	-4	.	8	-4	J	$\bar{J}$	3	.	-4	4	.	-4	4	.	2	.	.	.	$\bar{D}$	D	-1	$\bar{M}$	M	$\bar{D}$	D	.	-2	.
X.27	60	-4	.	.	4	J	$\bar{J}$	3	.	-4	4	8	-4	-4	.	-2	.	.	.	$\bar{D}$	D	-1	$\bar{M}$	M	A	$\bar{A}$	.	2	.
X.28	60	-4	.	8	-4	$\bar{J}$	J	3	.	-4	4	.	-4	4	.	2	.	.	.	D	$\bar{D}$	-1	M	$\bar{M}$	D	$\bar{D}$	.	-2	.
X.29	72	72	-24	.	.	.	.	3	.	8	8	.	.	.	.	.	.	-3	-3	.	.	3	3	3	.	.	.	.	.
X.30	90	90	10	6	6	.	.	.	-3	-6	-6	6	6	6	6	.	.	.	.	.	.	.	.	.	.	.	1	.	-2
X.31	90	90	-30	6	-2	.	.	-3	.	10	10	6	6	-2	-2	.	.	.	.	.	.	-3	-3	-3	.	.	.	.	-2
X.32	120	-8	.	8	.	30	30	3	.	-8	8	8	-8	.	.	.	.	.	.	-2	-2	-5	1	1	2	2	.	.	.
X.33	120	-8	.	8	.	N	$\bar{N}$	-3	.	-8	8	8	-8	.	.	.	.	.	.	$-H$	$-\bar{H}$	1	$-\bar{M}$	$-M$	H	$\bar{H}$	.	.	.
X.34	120	-8	.	8	.	$\bar{N}$	N	-3	.	-8	8	8	-8	.	.	.	.	.	.	$-\bar{H}$	$-H$	1	$-M$	$-\bar{M}$	$\bar{H}$	H	.	.	.
X.35	135	135	15	-9	-1	.	.	.	.	-9	-9	-9	-9	-1	-1	3	-1	.	.	.	.	.	.	.	.	.	.	3	-1
X.36	135	135	15	3	-5	.	.	.	.	-9	-9	3	3	-5	-5	-3	1	.	.	.	.	.	.	.	.	.	.	-3	3
X.37	180	-12	.	-12	.	.	.	3	.	4	-4	12	.	8	-4	.	2	.	.	.	.	3	-3	-3	.	.	.	.	.
X.38	180	-12	.	12	8	.	.	3	.	4	-4	-12	.	.	-4	.	-2	.	.	.	.	3	-3	-3	.	.	.	.	.
X.39	180	-12	.	.	-4	45	45	.	.	-12	12	-8	4	4	.	-2	.	.	.	-3	-3	.	.	.	-3	-3	.	2	.
X.40	180	-12	.	-8	4	45	45	.	.	-12	12	.	4	-4	.	2	.	.	.	-3	-3	.	.	.	1	1	.	-2	.
X.41	180	-12	.	.	-4	O	$\bar{O}$	.	.	-12	12	-8	4	4	.	-2	.	.	.	$-\bar{A}$	$-A$	.	.	.	$-\bar{A}$	$-A$	.	2	.
X.42	180	-12	.	.	-4	$\bar{O}$	O	.	.	-12	12	-8	4	4	.	-2	.	.	.	$-A$	$-\bar{A}$	.	.	.	$-A$	$-\bar{A}$	.	2	.
X.43	180	-12	.	-8	4	$\bar{O}$	O	.	.	-12	12	.	4	-4	.	2	.	.	.	$-A$	$-\bar{A}$	.	.	.	$-\bar{D}$	$-D$	.	-2	.
X.44	180	-12	.	-8	4	O	$\bar{O}$	.	.	-12	12	.	4	-4	.	2	.	.	.	$-\bar{A}$	$-A$	.	.	.	$-D$	$-\bar{D}$	.	-2	.
X.45	360	-24	.	.	8	.	.	-3	.	8	-8	.	.	8	-8	.	.	.	.	.	.	-3	3	3	.	.	.	.	.
X.46	540	-36	.	12	.	.	.	.	.	12	-12	-12	.	-8	4	.	2	.	.	.	.	.	.	.	.	.	.	.	.
X.47	540	-36	.	-12	-8	.	.	.	.	12	-12	12	.	.	4	.	-2	.	.	.	.	.	.	.	.	.	.	.	.

Character table of $E = N_G(A)$ (continued)

2	6	5	2	2	5	5	4	4	3	3	3	3	.	.	.	.	3	3
3	.	.	.	.	1	1	1	1	1	1	1	1	1	1	1	1	1	1
5	.	.	1	1	.	.	.	.	.	.	.	.	1	1	1	1	.	.
	8c	8d	10a	10b	12a	12b	12c	12d	12e	12f	12g	12h	15a	15b	15c	15d	24a	24b
2P	4a	4e	5a	5b	6a	6b	6a	6b	6c	6c	6f	6g	15b	15a	15d	15c	12a	12b
3P	8c	8d	10b	10a	4c	4c	4d	4d	4a	4b	4g	4g	5a	5b	5b	5a	8a	8a
5P	8c	8d	2b	2b	12b	12a	12d	12c	12e	12f	12h	12g	3b	3a	3b	3a	24b	24a
$X.1$	1	1	1	1	1	1	1	1	1	1	1	1	1	1	1	1	1	1
$X.2$	−1	1	C	B	$\bar{D}$	D	$\bar{D}$	D	.	.	$-\bar{D}$	$-D$	E	F	$\bar{F}$	$\bar{E}$	$-D$	$-\bar{D}$
$X.3$	−1	1	C	B	D	$\bar{D}$	D	$\bar{D}$	.	.	$-D$	$-\bar{D}$	$\bar{E}$	$\bar{F}$	F	E	$-\bar{D}$	$-D$
$X.4$	−1	1	B	C	D	$\bar{D}$	D	$\bar{D}$	.	.	$-D$	$-\bar{D}$	F	E	$\bar{E}$	$\bar{F}$	$-\bar{D}$	$-D$
$X.5$	−1	1	B	C	$\bar{D}$	D	$\bar{D}$	D	.	.	$-\bar{D}$	$-D$	$\bar{F}$	$\bar{E}$	E	F	$-D$	$-\bar{D}$
$X.6$	1	−1	.	.	1	1	1	1	−1	−1	−1	−1	.	.	.	.	−1	−1
$X.7$	1	−1	.	.	1	1	1	1	2	2	−1	−1	.	.	.	.	−1	−1
$X.8$	2	.	1	1	$\bar{H}$	H	$\bar{H}$	H	.	.	.	.	$-D$	$-\bar{D}$	$-D$	$-\bar{D}$	.	.
$X.9$	2	.	1	1	H	$\bar{H}$	H	$\bar{H}$	.	.	.	.	$-\bar{D}$	$-D$	$-\bar{D}$	$-D$	.	.
$X.10$	.	.	C	B	.	.	.	.	−1	−1	.	.	C	B	B	C	.	.
$X.11$	.	.	B	C	.	.	.	.	−1	−1	.	.	B	C	C	B	.	.
$X.12$	1	1	−1	−1	1	1	1	1	.	.	1	1	−1	−1	−1	−1	1	1
$X.13$	1	1	−1	−1	$-D$	$-\bar{D}$	$-D$	$-\bar{D}$	.	.	$-D$	$-\bar{D}$	D	$\bar{D}$	D	$\bar{D}$	$-\bar{D}$	$-D$
$X.14$	1	1	−1	−1	$-\bar{D}$	$-D$	$-\bar{D}$	$-D$	.	.	$-\bar{D}$	$-D$	$\bar{D}$	D	$\bar{D}$	D	$-D$	$-\bar{D}$
$X.15$	−2	.	.	.	−2	−2	−2	−2	1	1	.	.	.	.	.	.	.	.
$X.16$	−1	−1	.	.	D	$\bar{D}$	D	$\bar{D}$	.	.	D	$\bar{D}$	.	.	.	.	$\bar{D}$	D
$X.17$	−1	−1	.	.	$\bar{D}$	D	$\bar{D}$	D	.	.	$\bar{D}$	D	.	.	.	.	D	$\bar{D}$
$X.18$	2	.	−1	−1	.	.	.	.	−1	−1	.	.	.	.	.	.	.	.
$X.19$	−3	−1	.	.	.	.	.	.	.	.	.	.	.	.	.	.	.	.
$X.20$	1	1	.	.	.	.	.	.	.	.	.	.	.	.	.	.	.	.
$X.21$	−2	.	$-B$	$-C$	.	.	.	.	.	.	.	.	.	.	.	.	.	.
$X.22$	−2	.	$-C$	$-B$	.	.	.	.	.	.	.	.	.	.	.	.	.	.
$X.23$	.	.	.	.	3	3	−1	−1	−1	1	−1	−1	.	.	.	.	1	1
$X.24$	.	.	.	.	−1	−1	−1	−1	−1	1	1	1	.	.	.	.	−1	−1
$X.25$	.	.	.	.	$\bar{D}$	D	$\bar{D}$	D	−1	1	$-\bar{D}$	$-D$	.	.	.	.	D	$\bar{D}$
$X.26$	.	.	.	.	$\bar{A}$	A	D	$\bar{D}$	−1	1	D	$\bar{D}$	.	.	.	.	$-\bar{D}$	$-D$
$X.27$	.	.	.	.	D	$\bar{D}$	D	$\bar{D}$	−1	1	$-D$	$-\bar{D}$	.	.	.	.	$\bar{D}$	D
$X.28$	.	.	.	.	A	$\bar{A}$	$\bar{D}$	D	−1	1	$\bar{D}$	D	.	.	.	.	$-D$	$-\bar{D}$
$X.29$	.	.	1	1	.	.	.	.	−1	−1	.	.	.	.	.	.	.	.
$X.30$	−2	.	.	.	.	.	.	.	.	.	.	.	.	.	.	.	.	.
$X.31$	2	.	.	.	.	.	.	.	1	1	.	.	.	.	.	.	.	.
$X.32$	.	.	.	.	2	2	−2	−2	1	−1	.	.	.	.	.	.	.	.
$X.33$	.	.	.	.	$\bar{H}$	H	$-\bar{H}$	$-H$	1	−1	.	.	.	.	.	.	.	.
$X.34$	.	.	.	.	H	$\bar{H}$	$-H$	$-\bar{H}$	1	−1	.	.	.	.	.	.	.	.
$X.35$	3	−1	.	.	.	.	.	.	.	.	.	.	.	.	.	.	.	.
$X.36$	−1	1	.	.	.	.	.	.	.	.	.	.	.	.	.	.	.	.
$X.37$	.	−2	.	.	.	.	.	.	1	−1	.	.	.	.	.	.	.	.
$X.38$	.	2	.	.	.	.	.	.	1	−1	.	.	.	.	.	.	.	.
$X.39$	.	.	.	.	1	1	1	1	.	.	1	1	.	.	.	.	−1	−1
$X.40$	.	.	.	.	−3	−3	1	1	.	.	−1	−1	.	.	.	.	1	1
$X.41$	.	.	.	.	$-\bar{D}$	$-D$	$-\bar{D}$	$-D$	.	.	$-\bar{D}$	$-D$	.	.	.	.	D	$\bar{D}$
$X.42$	.	.	.	.	$-D$	$-\bar{D}$	$-D$	$-\bar{D}$	.	.	$-D$	$-\bar{D}$	.	.	.	.	$\bar{D}$	D
$X.43$	.	.	.	.	$-\bar{A}$	$-A$	$-D$	$-\bar{D}$	.	.	D	$\bar{D}$	.	.	.	.	$-\bar{D}$	$-D$
$X.44$	.	.	.	.	$-A$	$-\bar{A}$	$-\bar{D}$	$-D$	.	.	$\bar{D}$	D	.	.	.	.	$-D$	$-\bar{D}$
$X.45$	.	.	.	.	.	.	.	.	−1	1	.	.	.	.	.	.	.	.
$X.46$	.	−2	.	.	.	.	.	.	.	.	.	.	.	.	.	.	.	.
$X.47$	.	2	.	.	.	.	.	.	.	.	.	.	.	.	.	.	.	.

where $A = -3\zeta(3) - 3$, $B = \zeta(5)^3 + \zeta(5)^2 + 1$, $C = -\zeta(5)^3 - \zeta(5)^2$, $D = \zeta(3) + 1$, $E = \zeta(15)_3\zeta(15)_5^3 + \zeta(15)_3\zeta(15)_5^2 + \zeta(15)_3$, $F = \zeta(15)_3\zeta(15)_5^3 + \zeta(15)_3\zeta(15)_5^2 + \zeta(15)_5^3 + \zeta(15)_5^2$, $G = 6\zeta(3)$, $H = 2\zeta(3)$, $I = -9\zeta(3) - 9$, $J = 15\zeta(3)$, $K = 3\zeta(5)^3 + 3\zeta(5)^2 + 3$, $L = -3\zeta(5)^3 - 3\zeta(5)^2$, $M = 4\zeta(3) + 1$, $N = 30\zeta(3)$, $O = -45\zeta(3) - 45$.

14.6.3 Character table of $D = N_H(A) = \langle x, y \rangle$

	1a	2a	2b	2c	2d	2e	2f	2g	2h	3a	3b	3c	3d	3e	4a	4b	4c	4d	4e	4f	4g	4h	4i	4j	4k	4l	4m	4n	4o
2	13	13	12	10	9	8	9	8	8	7	7	4	4	4	10	10	9	8	8	8	8	9	9	9	7	7	8	8	8
3	2	2	1	2	1	1	.	.	.	2	2	2	2	2	1	1	1	1	1	1	1	.	.	.	1	1	.	.	.
2P	1a	1a	1a	1a	1a	1a	1a	1a	1a	3b	3a	3c	3e	3d	2a	2a	2a	2b	2b	2b	2a	2a	2b	2b	2c	2c	2a	2b	2b
3P	1a	2a	2b	2c	2d	2e	2f	2g	2h	1a	1a	1a	1a	1a	4a	4b	4c	4d	4e	4f	4g	4h	4i	4j	4k	4l	4m	4n	4o
X.1	1	1	1	1	1	1	1	1	1	1	1	1	1	1	1	1	1	1	1	1	1	1	1	1	1	1	1	1	1
X.2	1	1	1	1	1	−1	1	1	−1	1	1	1	1	1	1	1	1	−1	1	−1	−1	1	1	1	1	1	−1	1	−1
X.3	1	1	1	1	1	−1	1	1	−1	A	$\bar{A}$	1	A	$\bar{A}$	1	1	1	−1	1	−1	−1	1	1	1	1	1	−1	1	−1
X.4	1	1	1	1	1	1	1	1	1	$\bar{A}$	A	1	$\bar{A}$	A	1	1	1	1	1	1	1	1	1	1	1	1	1	1	1
X.5	1	1	1	1	1	1	1	1	1	A	$\bar{A}$	1	A	$\bar{A}$	1	1	1	1	1	1	1	1	1	1	1	1	1	1	1
X.6	1	1	1	1	1	−1	1	1	−1	$\bar{A}$	A	1	$\bar{A}$	A	1	1	1	−1	1	−1	−1	1	1	1	1	1	−1	1	−1
X.7	2	2	2	2	2	.	2	2	.	2	2	−1	−1	−1	2	2	2	.	2	.	.	2	2	2	2	2	.	2	.
X.8	2	2	2	2	2	.	2	2	.	B	$\bar{B}$	−1	$-\bar{A}$	$-A$	2	2	2	.	2	.	.	2	2	2	2	2	.	2	.
X.9	2	2	2	2	2	.	2	2	.	$\bar{B}$	B	−1	$-A$	$-\bar{A}$	2	2	2	.	2	.	.	2	2	2	2	2	.	2	.
X.10	3	3	3	3	−1	−1	−1	3	−1	3	3	.	.	.	3	3	−1	−1	−1	−1	−1	−1	3	3	3	3	−1	−1	−1
X.11	3	3	3	3	−1	1	−1	3	1	3	3	.	.	.	3	3	−1	1	−1	1	1	−1	3	3	3	3	1	−1	1
X.12	3	3	3	3	−1	−1	−1	3	−1	C	$\bar{C}$	.	.	.	3	3	−1	−1	−1	−1	−1	−1	3	3	3	3	−1	−1	−1
X.13	3	3	3	3	−1	−1	−1	3	−1	$\bar{C}$	C	.	.	.	3	3	−1	−1	−1	−1	−1	−1	3	3	3	3	−1	−1	−1
X.14	3	3	3	3	−1	1	−1	3	1	C	$\bar{C}$	.	.	.	3	3	−1	1	−1	1	1	−1	3	3	3	3	1	−1	1
X.15	3	3	3	3	−1	1	−1	3	1	$\bar{C}$	C	.	.	.	3	3	−1	1	−1	1	1	−1	3	3	3	3	1	−1	1
X.16	4	4	4	−4	4	2	.	.	−2	1	1	1	−2	−2	−4	4	4	−2	−4	2	−2	.	.	.	.	.	2	.	2
X.17	4	4	4	−4	4	−2	.	.	2	1	1	1	−2	−2	−4	4	4	2	−4	−2	2	.	.	.	.	.	−2	.	−2
X.18	4	4	4	−4	4	2	.	.	−2	$\bar{A}$	A	1	$-B$	$-\bar{B}$	−4	4	4	−2	−4	2	−2	.	.	.	.	.	2	.	2
X.19	4	4	4	−4	4	−2	.	.	2	A	$\bar{A}$	1	$-\bar{B}$	$-B$	−4	4	4	2	−4	−2	2	.	.	.	.	.	−2	.	−2
X.20	4	4	4	−4	4	2	.	.	−2	A	$\bar{A}$	1	$-\bar{B}$	$-B$	−4	4	4	−2	−4	2	−2	.	.	.	.	.	2	.	2
X.21	4	4	4	−4	4	−2	.	.	2	$\bar{A}$	A	1	$-B$	$-\bar{B}$	−4	4	4	2	−4	−2	2	.	.	.	.	.	−2	.	−2
X.22	6	6	6	6	6	.	−2	−2	.	.	.	.	3	3	6	6	6	.	6	.	.	−2	−2	−2	2	2	.	−2	.
X.23	6	6	6	6	6	.	−2	−2	.	.	.	.	$\bar{C}$	C	6	6	6	.	6	.	.	−2	−2	−2	2	2	.	−2	.
X.24	6	6	6	6	6	.	−2	−2	.	.	.	.	C	$\bar{C}$	6	6	6	.	6	.	.	−2	−2	−2	2	2	.	−2	.
X.25	8	8	8	−8	8	.	.	.	.	2	2	−1	2	2	−8	8	8	.	−8	.	.	.	.	.	.	.	.	.	.
X.26	8	8	8	−8	8	.	.	.	.	$\bar{B}$	B	−1	$\bar{B}$	B	−8	8	8	.	−8	.	.	.	.	.	.	.	.	.	.
X.27	8	8	8	−8	8	.	.	.	.	B	$\bar{B}$	−1	B	$\bar{B}$	−8	8	8	.	−8	.	.	.	.	.	.	.	.	.	.
X.28	9	9	9	9	9	−3	1	1	−3	.	.	.	.	.	9	9	9	−3	9	−3	−3	1	1	1	−3	−3	−3	1	−3
X.29	9	9	9	9	−3	3	5	1	3	.	.	.	.	.	9	9	−3	3	−3	3	3	5	1	1	−3	−3	3	5	3
X.30	9	9	9	9	−3	−3	5	1	−3	.	.	.	.	.	9	9	−3	−3	−3	−3	−3	5	1	1	−3	−3	−3	5	−3
X.31	9	9	9	9	9	3	1	1	3	.	.	.	.	.	9	9	9	3	9	3	3	1	1	1	−3	−3	3	1	3
X.32	12	12	12	−12	.	−6	4	.	−2	.	.	3	.	.	4	−4	.	6	.	−6	6	4	.	.	.	.	2	−4	2
X.33	12	12	12	12	.	6	.	−4	−2	.	.	3	.	.	−4	−4	.	6	.	6	6	.	4	4	.	.	−2	.	−2
X.34	12	12	12	12	.	−6	.	−4	2	.	.	3	.	.	−4	−4	.	−6	.	−6	−6	.	4	4	.	.	2	.	2
X.35	12	12	12	−12	.	6	4	.	2	.	.	3	.	.	4	−4	.	−6	.	6	−6	4	.	.	.	.	−2	−4	−2
X.36	12	12	12	−12	−4	−2	.	.	2	3	3	.	.	.	−12	12	−4	2	4	−2	2	.	.	.	.	.	−2	.	−2
X.37	12	12	12	−12	−4	2	.	.	−2	3	3	.	.	.	−12	12	−4	−2	4	2	−2	.	.	.	.	.	2	.	2
X.38	12	12	12	−12	−4	−2	.	.	2	C	$\bar{C}$	.	.	.	−12	12	−4	2	4	−2	2	.	.	.	.	.	−2	.	−2
X.39	12	12	12	−12	−4	−2	.	.	2	$\bar{C}$	C	.	.	.	−12	12	−4	2	4	−2	2	.	.	.	.	.	−2	.	−2
X.40	12	12	12	−12	−4	2	.	.	−2	$\bar{C}$	C	.	.	.	−12	12	−4	−2	4	2	−2	.	.	.	.	.	2	.	2
X.41	12	12	12	−12	−4	2	.	.	−2	C	$\bar{C}$	.	.	.	−12	12	−4	−2	4	2	−2	.	.	.	.	.	2	.	2
X.42	18	18	18	18	−6	.	2	−6	.	.	.	.	.	.	18	18	−6	.	−6	.	.	2	−6	−6	6	6	.	2	.
X.43	18	18	18	18	−6	.	−6	2	.	.	.	.	.	.	18	18	−6	.	−6	.	.	−6	2	2	−6	−6	.	−6	.
X.44	24	24	−8	.	−4	2	4	.	2	6	6	.	.	.	.	.	4	6	.	−2	−6	−4	4	−4	.	.	2	.	−2
X.45	24	24	−8	.	−4	−2	4	.	−2	6	6	.	.	.	.	.	4	−6	.	2	6	−4	4	−4	.	.	−2	.	2
X.46	24	24	−8	.	4	−6	−4	.	2	6	6	.	.	.	.	.	−4	−2	.	6	2	4	4	−4	.	.	2	.	−2
X.47	24	24	24	−24	.	.	8	.	.	.	.	−3	.	.	8	−8	.	.	.	.	.	8	.	.	.	.	.	−8	.
X.48	24	24	−8	.	4	6	−4	.	−2	6	6	.	.	.	.	.	−4	2	.	−6	−2	4	4	−4	.	.	−2	.	2
X.49	24	24	24	24	.	.	.	−8	.	.	.	−3	.	.	−8	−8	.	.	.	.	.	.	8	8	.	.	.	.	.
X.50	24	24	−8	.	4	−6	−4	.	2	D	$\bar{D}$	.	.	.	.	.	−4	−2	.	6	2	4	4	−4	.	.	2	.	−2
X.51	24	24	−8	.	4	6	−4	.	−2	D	$\bar{D}$	.	.	.	.	.	−4	2	.	−6	−2	4	4	−4	.	.	−2	.	2
X.52	24	24	−8	.	−4	2	4	.	2	D	$\bar{D}$	.	.	.	.	.	4	6	.	−2	−6	−4	4	−4	.	.	2	.	−2
X.53	24	24	−8	.	4	−6	−4	.	2	$\bar{D}$	D	.	.	.	.	.	−4	−2	.	6	2	4	4	−4	.	.	2	.	−2
X.54	24	24	−8	.	−4	2	4	.	2	$\bar{D}$	D	.	.	.	.	.	4	6	.	−2	−6	−4	4	−4	.	.	2	.	−2
X.55	24	24	−8	.	−4	−2	4	.	−2	D	$\bar{D}$	.	.	.	.	.	4	−6	.	2	6	−4	4	−4	.	.	−2	.	2
X.56	24	24	−8	.	4	6	−4	.	−2	$\bar{D}$	D	.	.	.	.	.	−4	2	.	−6	−2	4	4	−4	.	.	−2	.	2
X.57	24	24	−8	.	−4	−2	4	.	−2	$\bar{D}$	D	.	.	.	.	.	4	−6	.	2	6	−4	4	−4	.	.	−2	.	2
X.58	32	−32	.	.	.	.	.	.	.	8	8	−4	−1	−1	.	.	.	.	.	.	.	.	.	.	4	−4	.	.	.
X.59	32	−32	.	.	.	.	.	.	.	E	$\bar{E}$	2	F	$\bar{F}$	.	.	.	.	.	.	.	.	.	.	4	−4	.	.	.
X.60	32	−32	.	.	.	.	.	.	.	$\bar{E}$	E	−4	$-A$	$-\bar{A}$	.	.	.	.	.	.	.	.	.	.	4	−4	.	.	.
X.61	32	−32	.	.	.	.	.	.	.	E	$\bar{E}$	2	I	$\bar{I}$	.	.	.	.	.	.	.	.	.	.	4	−4	.	.	.
X.62	32	−32	.	.	.	.	.	.	.	$\bar{E}$	E	2	$\bar{F}$	F	.	.	.	.	.	.	.	.	.	.	4	−4	.	.	.
X.63	32	−32	.	.	.	.	.	.	.	8	8	2	K	$\bar{K}$	.	.	.	.	.	.	.	.	.	.	4	−4	.	.	.
X.64	32	−32	.	.	.	.	.	.	.	$\bar{E}$	E	2	$\bar{I}$	I	.	.	.	.	.	.	.	.	.	.	4	−4	.	.	.
X.65	32	−32	.	.	.	.	.	.	.	E	$\bar{E}$	−4	$-\bar{A}$	$-A$	.	.	.	.	.	.	.	.	.	.	4	−4	.	.	.
X.66	32	−32	.	.	.	.	.	.	.	8	8	2	$\bar{K}$	K	.	.	.	.	.	.	.	.	.	.	4	−4	.	.	.
X.67	36	36	36	−36	.	−6	−4	.	−2	.	.	.	.	.	12	−12	.	6	.	−6	6	−4	.	.	.	.	2	4	2
X.68	36	36	36	−36	.	6	−4	.	2	.	.	.	.	.	12	−12	.	−6	.	6	−6	−4	.	.	.	.	−2	4	−2
X.69	36	36	36	36	.	−6	.	4	2	.	.	.	.	.	−12	−12	.	−6	.	−6	−6	.	−4	−4	.	.	2	.	2
X.70	36	36	36	36	.	6	.	4	−2	.	.	.	.	.	−12	−12	.	6	.	6	6	.	−4	−4	.	.	−2	.	−2
X.71	72	72	−24	.	12	6	4	.	6	.	.	.	.	.	.	.	−12	−6	.	−6	6	−4	−4	4	.	.	−2	.	2
X.72	72	72	−24	.	12	−6	4	.	−6	.	.	.	.	.	.	.	−12	6	.	6	−6	−4	−4	4	.	.	2	.	−2
X.73	72	72	−24	.	−12	6	−4	.	−2	.	.	.	.	.	.	.	12	−6	.	−6	6	4	−4	4	.	.	6	.	−6
X.74	72	72	−24	.	−12	−6	−4	.	2	.	.	.	.	.	.	.	12	6	.	6	−6	4	−4	4	.	.	−6	.	6
X.75	96	−96	.	.	.	.	.	.	.	.	.	.	3	3	.	.	.	.	.	.	.	.	.	.	−4	4	.	.	.
X.76	96	−96	.	.	.	.	.	.	.	.	.	.	C	$\bar{C}$	.	.	.	.	.	.	.	.	.	.	−4	4	.	.	.
X.77	96	−96	.	.	.	.	.	.	.	.	.	.	$\bar{C}$	C	.	.	.	.	.	.	.	.	.	.	−4	4	.	.	.

Character table of $D = N_H(A)$ (continued)

	4p	4q	4r	6a	6b	6c	6d	6e	6f	6g	6h	6i	6j	6k	6l	6m	6n	6o	6p	6q	6r	6s	8a	8b	8c	8d	8e	8f	12a	12b	12c
2	8	5	5	7	7	6	6	4	4	4	4	4	4	4	4	4	4	4	5	5	4	4	5	6	6	5	5	5	5	5	4
3	.	1	.	2	2	1	1	2	2	2	2	2	2	2	2	2	2	2	1	1	1	1	1	.	.	.	.	.	1	1	1
2P	2b	2d	2f	3a	3b	3b	3a	3c	3d	3e	3e	3d	3d	3e	3a	3b	3c	3c	3b	3a	3a	3b	4c	4a	4b	4h	4i	4j	6a	6b	6c
3P	4p	4q	4r	2a	2a	2b	2b	2a	2c	2c	2c	2c	2a	2a	2c	2c	2c	2c	2d	2d	2e	2e	8a	8b	8c	8d	8e	8f	4c	4c	4d
X.1	1	1	1	1	1	1	1	1	1	1	1	1	1	1	1	1	1	1	1	1	1	1	1	1	1	1	1	1	1	1	1
X.2	−1	−1	−1	1	1	1	1	1	1	1	1	1	1	1	1	1	1	1	1	1	−1	−1	−1	1	1	−1	−1	−1	1	1	−1
X.3	−1	−1	−1	$\bar{A}$	A	A	$\bar{A}$	1	$\bar{A}$	A	A	$\bar{A}$	$\bar{A}$	A	$\bar{A}$	A	1	1	A	$\bar{A}$	$-\bar{A}$	$-A$	−1	1	1	−1	−1	−1	A	$\bar{A}$	$-\bar{A}$
X.4	1	1	1	A	$\bar{A}$	$\bar{A}$	A	1	A	$\bar{A}$	$\bar{A}$	A	A	$\bar{A}$	A	$\bar{A}$	1	1	$\bar{A}$	A	A	$\bar{A}$	1	1	1	1	1	1	$\bar{A}$	A	A
X.5	1	1	1	$\bar{A}$	A	A	$\bar{A}$	1	$\bar{A}$	A	A	$\bar{A}$	$\bar{A}$	A	$\bar{A}$	A	1	1	A	$\bar{A}$	$\bar{A}$	A	1	1	1	1	1	1	A	$\bar{A}$	$\bar{A}$
X.6	−1	−1	−1	A	$\bar{A}$	$\bar{A}$	A	1	A	$\bar{A}$	$\bar{A}$	A	A	$\bar{A}$	A	$\bar{A}$	1	1	$\bar{A}$	A	$-A$	$-\bar{A}$	−1	1	1	−1	−1	−1	$\bar{A}$	A	$-A$
X.7	.	.	.	2	2	2	2	−1	−1	−1	−1	−1	−1	−1	2	2	−1	−1	2	2	.	.	.	2	2	.	.	.	2	2	.
X.8	.	.	.	$\bar{B}$	B	B	$\bar{B}$	−1	$-A$	$-\bar{A}$	$-\bar{A}$	$-A$	$-A$	$-\bar{A}$	$\bar{B}$	B	−1	−1	B	$\bar{B}$	.	.	.	2	2	.	.	.	B	$\bar{B}$	.
X.9	.	.	.	B	$\bar{B}$	$\bar{B}$	B	−1	$-\bar{A}$	$-A$	$-A$	$-\bar{A}$	$-\bar{A}$	$-A$	B	$\bar{B}$	−1	−1	$\bar{B}$	B	.	.	.	2	2	.	.	.	$\bar{B}$	B	.
X.10	−1	1	1	3	3	3	3	.	.	.	.	.	.	.	3	3	.	.	−1	−1	−1	−1	1	−1	−1	1	−1	−1	−1	−1	−1
X.11	1	−1	−1	3	3	3	3	.	.	.	.	.	.	.	3	3	.	.	−1	−1	1	1	−1	−1	−1	−1	1	1	−1	−1	1
X.12	−1	1	1	$\bar{C}$	C	C	$\bar{C}$	.	.	.	.	.	.	.	$\bar{C}$	C	.	.	$-\bar{A}$	$-A$	$-A$	$-\bar{A}$	1	−1	−1	1	−1	−1	$-\bar{A}$	$-A$	$-A$
X.13	−1	1	1	C	$\bar{C}$	$\bar{C}$	C	.	.	.	.	.	.	.	C	$\bar{C}$	.	.	$-A$	$-\bar{A}$	$-\bar{A}$	$-A$	1	−1	−1	1	−1	−1	$-A$	$-\bar{A}$	$-\bar{A}$
X.14	1	−1	−1	$\bar{C}$	C	C	$\bar{C}$	.	.	.	.	.	.	.	$\bar{C}$	C	.	.	$-\bar{A}$	$-A$	A	$\bar{A}$	−1	−1	−1	−1	1	1	$-\bar{A}$	$-A$	A
X.15	1	−1	−1	C	$\bar{C}$	$\bar{C}$	C	.	.	.	.	.	.	.	C	$\bar{C}$	.	.	$-A$	$-\bar{A}$	$\bar{A}$	A	−1	−1	−1	−1	1	1	$-A$	$-\bar{A}$	$\bar{A}$
X.16	−2	2	.	1	1	1	1	1	2	2	2	2	−2	−2	−1	−1	−1	−1	1	1	−1	−1	−2	.	.	.	.	.	1	1	1
X.17	2	−2	.	1	1	1	1	1	2	2	2	2	−2	−2	−1	−1	−1	−1	1	1	1	1	2	.	.	.	.	.	1	1	−1
X.18	−2	2	.	A	$\bar{A}$	$\bar{A}$	A	1	$\bar{B}$	B	B	$\bar{B}$	$-\bar{B}$	$-B$	$-A$	$-\bar{A}$	−1	−1	$\bar{A}$	A	$-A$	$-\bar{A}$	−2	.	.	.	.	.	$\bar{A}$	A	A
X.19	2	−2	.	$\bar{A}$	A	A	$\bar{A}$	1	B	$\bar{B}$	$\bar{B}$	B	$-B$	$-\bar{B}$	$-\bar{A}$	$-A$	−1	−1	A	$\bar{A}$	$\bar{A}$	A	2	.	.	.	.	.	A	$\bar{A}$	$-\bar{A}$
X.20	−2	2	.	$\bar{A}$	A	A	$\bar{A}$	1	B	$\bar{B}$	$\bar{B}$	B	$-B$	$-\bar{B}$	$-\bar{A}$	$-A$	−1	−1	A	$\bar{A}$	$-\bar{A}$	$-A$	−2	.	.	.	.	.	A	$\bar{A}$	$\bar{A}$
X.21	2	−2	.	A	$\bar{A}$	$\bar{A}$	A	1	$\bar{B}$	B	B	$\bar{B}$	$-\bar{B}$	$-B$	$-A$	$-\bar{A}$	−1	−1	$\bar{A}$	A	A	$\bar{A}$	2	.	.	.	.	.	$\bar{A}$	A	$-A$
X.22	.	.	.	.	.	.	.	.	3	3	3	3	3	3	.	.	.	.	.	.	.	.	.	2	−2	.	.	.	.	.	.
X.23	.	.	.	.	.	.	.	.	C	$\bar{C}$	$\bar{C}$	C	C	$\bar{C}$	.	.	.	.	.	.	.	.	.	2	−2	.	.	.	.	.	.
X.24	.	.	.	.	.	.	.	.	$\bar{C}$	C	C	$\bar{C}$	$\bar{C}$	C	.	.	.	.	.	.	.	.	.	2	−2	.	.	.	.	.	.
X.25	.	.	.	2	2	2	2	−1	−2	−2	−2	−2	2	2	−2	−2	1	1	2	2	.	.	.	.	.	.	.	.	2	2	.
X.26	.	.	.	B	$\bar{B}$	$\bar{B}$	B	−1	$-B$	$-\bar{B}$	$-\bar{B}$	$-B$	B	$\bar{B}$	$-B$	$-\bar{B}$	1	1	$\bar{B}$	B	.	.	.	.	.	.	.	.	$\bar{B}$	B	.
X.27	.	.	.	$\bar{B}$	B	B	$\bar{B}$	−1	$-\bar{B}$	$-B$	$-B$	$-\bar{B}$	$\bar{B}$	B	$-\bar{B}$	$-B$	1	1	B	$\bar{B}$	.	.	.	.	.	.	.	.	B	$\bar{B}$	.
X.28	−3	−3	1	.	.	.	.	.	.	.	.	.	.	.	.	.	.	.	.	.	.	.	−3	−3	1	1	1	1	.	.	.
X.29	3	−3	1	.	.	.	.	.	.	.	.	.	.	.	.	.	.	.	.	.	.	.	−3	1	−3	1	−1	−1	.	.	.
X.30	−3	3	−1	.	.	.	.	.	.	.	.	.	.	.	.	.	.	.	.	.	.	.	3	1	−3	−1	1	1	.	.	.
X.31	3	3	−1	.	.	.	.	.	.	.	.	.	.	.	.	.	.	.	.	.	.	.	3	−3	1	−1	−1	−1	.	.	.
X.32	−2	.	2	.	.	.	.	3	.	.	.	.	.	.	.	.	−3	−3	.	.	.	.	.	.	.	−2	.	.	.	.	.
X.33	−2	.	.	.	.	.	.	3	.	.	.	.	.	.	.	.	3	3	.	.	.	.	.	.	.	.	−2	2	.	.	.
X.34	2	.	.	.	.	.	.	3	.	.	.	.	.	.	.	.	3	3	.	.	.	.	.	.	.	.	2	−2	.	.	.
X.35	2	.	−2	.	.	.	.	3	.	.	.	.	.	.	.	.	−3	−3	.	.	.	.	.	.	.	2	.	.	.	.	.
X.36	2	2	.	3	3	3	3	.	.	.	.	.	.	.	−3	−3	.	.	−1	−1	1	1	−2	.	.	.	.	.	−1	−1	−1
X.37	−2	−2	.	3	3	3	3	.	.	.	.	.	.	.	−3	−3	.	.	−1	−1	−1	−1	2	.	.	.	.	.	−1	−1	1
X.38	2	2	.	$\bar{C}$	C	C	$\bar{C}$	.	.	.	.	.	.	.	$-\bar{C}$	$-C$	.	.	$-\bar{A}$	$-A$	A	$\bar{A}$	−2	.	.	.	.	.	$-\bar{A}$	$-A$	$-A$
X.39	2	2	.	C	$\bar{C}$	$\bar{C}$	C	.	.	.	.	.	.	.	$-C$	$-\bar{C}$	.	.	$-A$	$-\bar{A}$	$\bar{A}$	A	−2	.	.	.	.	.	$-A$	$-\bar{A}$	$-\bar{A}$
X.40	−2	−2	.	C	$\bar{C}$	$\bar{C}$	C	.	.	.	.	.	.	.	$-C$	$-\bar{C}$	.	.	$-A$	$-\bar{A}$	$-\bar{A}$	$-A$	2	.	.	.	.	.	$-A$	$-\bar{A}$	$\bar{A}$
X.41	−2	−2	.	$\bar{C}$	C	C	$\bar{C}$	.	.	.	.	.	.	.	$-\bar{C}$	$-C$	.	.	$-\bar{A}$	$-A$	$-A$	$-\bar{A}$	2	.	.	.	.	.	$-\bar{A}$	$-A$	A
X.42	.	.	.	.	.	.	.	.	.	.	.	.	.	.	.	.	.	.	.	.	.	.	.	−2	2	.	.	.	.	.	.
X.43	.	.	.	.	.	.	.	.	.	.	.	.	.	.	.	.	.	.	.	.	.	.	.	2	2	.	.	.	.	.	.
X.44	−2	.	.	6	6	−2	−2	.	.	.	.	.	.	.	.	.	.	.	2	2	2	2	.	.	.	.	.	.	−2	−2	.
X.45	2	.	.	6	6	−2	−2	.	.	.	.	.	.	.	.	.	.	.	2	2	−2	−2	.	.	.	.	.	.	−2	−2	.
X.46	−2	.	.	6	6	−2	−2	.	.	.	.	.	.	.	.	.	.	.	−2	−2	.	.	.	.	.	.	.	.	2	2	−2
X.47	.	.	.	.	.	.	.	−3	.	.	.	.	.	.	.	.	3	3	.	.	.	.	.	.	.	.	.	.	.	.	.
X.48	2	.	.	6	6	−2	−2	.	.	.	.	.	.	.	.	.	.	.	−2	−2	.	.	.	.	.	.	.	.	2	2	2
X.49	.	.	.	.	.	.	.	−3	.	.	.	.	.	.	.	.	−3	−3	.	.	.	.	.	.	.	.	.	.	.	.	.
X.50	−2	.	.	$\bar{D}$	D	$-\bar{B}$	$-B$	.	.	.	.	.	.	.	.	.	.	.	$-\bar{B}$	$-B$	.	.	.	.	.	.	.	.	$\bar{B}$	B	$-B$
X.51	2	.	.	$\bar{D}$	D	$-\bar{B}$	$-B$	.	.	.	.	.	.	.	.	.	.	.	$-\bar{B}$	$-B$	.	.	.	.	.	.	.	.	$\bar{B}$	B	B
X.52	−2	.	.	$\bar{D}$	D	$-\bar{B}$	$-B$	.	.	.	.	.	.	.	.	.	.	.	$\bar{B}$	B	B	$\bar{B}$	.	.	.	.	.	.	$-\bar{B}$	$-B$	.
X.53	−2	.	.	D	$\bar{D}$	$-B$	$-\bar{B}$	.	.	.	.	.	.	.	.	.	.	.	$-B$	$-\bar{B}$	.	.	.	.	.	.	.	.	B	$\bar{B}$	$-\bar{B}$
X.54	−2	.	.	D	$\bar{D}$	$-B$	$-\bar{B}$	.	.	.	.	.	.	.	.	.	.	.	B	$\bar{B}$	$\bar{B}$	B	.	.	.	.	.	.	$-B$	$-\bar{B}$	.
X.55	2	.	.	$\bar{D}$	D	$-\bar{B}$	$-B$	.	.	.	.	.	.	.	.	.	.	.	$\bar{B}$	B	$-B$	$-\bar{B}$	.	.	.	.	.	.	$-\bar{B}$	$-B$	.
X.56	2	.	.	D	$\bar{D}$	$-B$	$-\bar{B}$	.	.	.	.	.	.	.	.	.	.	.	$-B$	$-\bar{B}$	.	.	.	.	.	.	.	.	B	$\bar{B}$	$\bar{B}$
X.57	2	.	.	D	$\bar{D}$	$-B$	$-\bar{B}$	.	.	.	.	.	.	.	.	.	.	.	B	$\bar{B}$	$-\bar{B}$	$-B$	.	.	.	.	.	.	$-B$	$-\bar{B}$	.
X.58	.	.	.	−8	−8	.	.	4	3	3	−3	−3	1	1	.	.	.	.	.	.	.	.	.	.	.	.	.	.	.	.	.
X.59	.	.	.	$-\bar{E}$	$-E$	.	.	−2	G	$\bar{G}$	G	$\bar{G}$	$-\bar{F}$	$-F$	.	.	H	$\bar{H}$	.	.	.	.	.	.	.	.	.	.	.	.	.
X.60	.	.	.	$-E$	$-\bar{E}$	.	.	4	C	$\bar{C}$	$-\bar{C}$	$-C$	$\bar{A}$	A	.	.	.	.	.	.	.	.	.	.	.	.	.	.	.	.	.
X.61	.	.	.	$-\bar{E}$	$-E$	.	.	−2	J	$\bar{J}$	$-\bar{J}$	$-J$	$-\bar{I}$	$-I$	.	.	$\bar{H}$	H	.	.	.	.	.	.	.	.	.	.	.	.	.
X.62	.	.	.	$-E$	$-\bar{E}$	.	.	−2	$\bar{G}$	G	$\bar{G}$	G	$-F$	$-\bar{F}$	.	.	$\bar{H}$	H	.	.	.	.	.	.	.	.	.	.	.	.	.
X.63	.	.	.	−8	−8	.	.	−2	$-\bar{J}$	$-J$	J	$\bar{J}$	$-\bar{K}$	$-K$	.	.	$\bar{H}$	H	.	.	.	.	.	.	.	.	.	.	.	.	.
X.64	.	.	.	$-E$	$-\bar{E}$	.	.	−2	$\bar{J}$	J	$-J$	$-\bar{J}$	$-I$	$-\bar{I}$	.	.	H	$\bar{H}$	.	.	.	.	.	.	.	.	.	.	.	.	.
X.65	.	.	.	$-\bar{E}$	$-E$	.	.	4	$\bar{C}$	C	$-C$	$-\bar{C}$	A	$\bar{A}$	.	.	.	.	.	.	.	.	.	.	.	.	.	.	.	.	.
X.66	.	.	.	−8	−8	.	.	−2	$-J$	$-\bar{J}$	$\bar{J}$	J	$-K$	$-\bar{K}$	.	.	H	$\bar{H}$	.	.	.	.	.	.	.	.	.	.	.	.	.
X.67	−2	.	−2	.	.	.	.	.	.	.	.	.	.	.	.	.	.	.	.	.	.	.	.	.	.	2	.	.	.	.	.
X.68	2	.	2	.	.	.	.	.	.	.	.	.	.	.	.	.	.	.	.	.	.	.	.	.	.	−2	.	.	.	.	.
X.69	2	.	.	.	.	.	.	.	.	.	.	.	.	.	.	.	.	.	.	.	.	.	.	.	.	.	−2	2	.	.	.
X.70	−2	.	.	.	.	.	.	.	.	.	.	.	.	.	.	.	.	.	.	.	.	.	.	.	.	.	2	−2	.	.	.
X.71	−6	.	.	.	.	.	.	.	.	.	.	.	.	.	.	.	.	.	.	.	.	.	.	.	.	.	.	.	.	.	.
X.72	6	.	.	.	.	.	.	.	.	.	.	.	.	.	.	.	.	.	.	.	.	.	.	.	.	.	.	.	.	.	.
X.73	2	.	.	.	.	.	.	.	.	.	.	.	.	.	.	.	.	.	.	.	.	.	.	.	.	.	.	.	.	.	.
X.74	−2	.	.	.	.	.	.	.	.	.	.	.	.	.	.	.	.	.	.	.	.	.	.	.	.	.	.	.	.	.	.
X.75	.	.	.	.	.	.	.	.	3	3	−3	−3	−3	−3	.	.	.	.	.	.	.	.	.	.	.	.	.	.	.	.	.
X.76	.	.	.	.	.	.	.	.	$\bar{C}$	C	$-C$	$-\bar{C}$	$-\bar{C}$	$-C$	.	.	.	.	.	.	.	.	.	.	.	.	.	.	.	.	.
X.77	.	.	.	.	.	.	.	.	C	$\bar{C}$	$-\bar{C}$	$-C$	$-C$	$-\bar{C}$	.	.	.	.	.	.	.	.	.	.	.	.	.	.	.	.	.

Character table of $D = N_H(A)$ (continued)

2	4	4	4	4	4	4	4	3	3	3	3	3	3	3	3	3	3
3	1	1	1	1	1	1	1	1	1	1	1	1	1	1	1	1	1
	12d	12e	12f	12g	12h	12i	12j	12k	12l	12m	12n	12o	12p	12q	12r	24a	24b
2P	6d	6d	6c	6d	6c	6a	6b	6e	6f	6g	6p	6q	6f	6g	6e	12a	12b
3P	4d	4e	4e	4f	4f	4g	4g	4a	4k	4k	4q	4q	4l	4l	4b	8a	8a
X.1	1	1	1	1	1	1	1	1	1	1	1	1	1	1	1	1	1
X.2	−1	1	1	−1	−1	−1	−1	1	1	1	−1	−1	1	1	1	−1	−1
X.3	$-A$	A	$\bar{A}$	$-A$	$-\bar{A}$	$-A$	$-\bar{A}$	1	A	$\bar{A}$	$-\bar{A}$	$-A$	A	$\bar{A}$	1	$-\bar{A}$	$-A$
X.4	$\bar{A}$	$\bar{A}$	A	$\bar{A}$	A	$\bar{A}$	A	1	$\bar{A}$	A	A	$\bar{A}$	$\bar{A}$	A	1	A	$\bar{A}$
X.5	A	A	$\bar{A}$	A	$\bar{A}$	A	$\bar{A}$	1	A	$\bar{A}$	$\bar{A}$	A	A	$\bar{A}$	1	$\bar{A}$	A
X.6	$-\bar{A}$	$\bar{A}$	A	$-\bar{A}$	$-A$	$-\bar{A}$	$-A$	1	$\bar{A}$	A	$-A$	$-\bar{A}$	$\bar{A}$	A	1	$-A$	$-\bar{A}$
X.7	.	2	2	.	.	.	.	−1	−1	−1	.	.	−1	−1	−1	.	.
X.8	.	B	$\bar{B}$	.	.	.	.	−1	$-\bar{A}$	$-A$	.	.	$-\bar{A}$	$-A$	−1	.	.
X.9	.	$\bar{B}$	B	.	.	.	.	−1	$-A$	$-\bar{A}$	.	.	$-A$	$-\bar{A}$	−1	.	.
X.10	−1	−1	−1	−1	−1	−1	−1	.	.	.	1	1	.	.	.	1	1
X.11	1	−1	−1	1	1	1	1	.	.	.	−1	−1	.	.	.	−1	−1
X.12	$-\bar{A}$	$-\bar{A}$	$-A$	$-\bar{A}$	$-A$	$-\bar{A}$	$-A$	.	.	.	A	$\bar{A}$	.	.	.	A	$\bar{A}$
X.13	$-A$	$-A$	$-\bar{A}$	$-A$	$-\bar{A}$	$-A$	$-\bar{A}$	.	.	.	$\bar{A}$	A	.	.	.	$\bar{A}$	A
X.14	$\bar{A}$	$-\bar{A}$	$-A$	$\bar{A}$	A	$\bar{A}$	A	.	.	.	$-A$	$-\bar{A}$	.	.	.	$-A$	$-\bar{A}$
X.15	A	$-A$	$-\bar{A}$	A	$\bar{A}$	A	$\bar{A}$	.	.	.	$-\bar{A}$	$-A$	.	.	.	$-\bar{A}$	$-A$
X.16	1	−1	−1	−1	−1	1	1	−1	.	.	−1	−1	.	.	1	1	1
X.17	−1	−1	−1	1	1	−1	−1	−1	.	.	1	1	.	.	1	−1	−1
X.18	$\bar{A}$	$-\bar{A}$	$-A$	$-\bar{A}$	$-A$	$\bar{A}$	A	−1	.	.	$-A$	$-\bar{A}$	.	.	1	A	$\bar{A}$
X.19	$-A$	$-A$	$-\bar{A}$	A	$\bar{A}$	$-A$	$-\bar{A}$	−1	.	.	$\bar{A}$	A	.	.	1	$-\bar{A}$	$-A$
X.20	A	$-A$	$-\bar{A}$	$-A$	$-\bar{A}$	A	$\bar{A}$	−1	.	.	$-\bar{A}$	$-A$	.	.	1	$\bar{A}$	A
X.21	$-\bar{A}$	$-\bar{A}$	$-A$	$\bar{A}$	A	$-\bar{A}$	$-A$	−1	.	.	A	$\bar{A}$	.	.	1	$-A$	$-\bar{A}$
X.22	.	.	.	.	.	.	.	.	−1	−1	.	.	−1	−1	.	.	.
X.23	.	.	.	.	.	.	.	.	$-A$	$-\bar{A}$	.	.	$-A$	$-\bar{A}$	.	.	.
X.24	.	.	.	.	.	.	.	.	$-\bar{A}$	$-A$	.	.	$-\bar{A}$	$-A$	.	.	.
X.25	.	−2	−2	.	.	.	.	1	.	.	.	.	.	.	−1	.	.
X.26	.	$-\bar{B}$	$-B$	.	.	.	.	1	.	.	.	.	.	.	−1	.	.
X.27	.	$-B$	$-\bar{B}$	.	.	.	.	1	.	.	.	.	.	.	−1	.	.
X.28	.	.	.	.	.	.	.	.	.	.	.	.	.	.	.	.	.
X.29	.	.	.	.	.	.	.	.	.	.	.	.	.	.	.	.	.
X.30	.	.	.	.	.	.	.	.	.	.	.	.	.	.	.	.	.
X.31	.	.	.	.	.	.	.	.	.	.	.	.	.	.	.	.	.
X.32	.	.	.	.	.	.	.	1	.	.	.	.	.	.	−1	.	.
X.33	.	.	.	.	.	.	.	−1	.	.	.	.	.	.	−1	.	.
X.34	.	.	.	.	.	.	.	−1	.	.	.	.	.	.	−1	.	.
X.35	.	.	.	.	.	.	.	1	.	.	.	.	.	.	−1	.	.
X.36	−1	1	1	1	1	−1	−1	.	.	.	−1	−1	.	.	.	1	1
X.37	1	1	1	−1	−1	1	1	.	.	.	1	1	.	.	.	−1	−1
X.38	$-\bar{A}$	$\bar{A}$	A	$\bar{A}$	A	$-\bar{A}$	$-A$	.	.	.	$-A$	$-\bar{A}$	.	.	.	A	$\bar{A}$
X.39	$-A$	A	$\bar{A}$	A	$\bar{A}$	$-A$	$-\bar{A}$	.	.	.	$-\bar{A}$	$-A$	.	.	.	$\bar{A}$	A
X.40	A	A	$\bar{A}$	$-A$	$-\bar{A}$	A	$\bar{A}$	.	.	.	$\bar{A}$	A	.	.	.	$-\bar{A}$	$-A$
X.41	$\bar{A}$	$\bar{A}$	A	$-\bar{A}$	$-A$	$\bar{A}$	A	.	.	.	A	$\bar{A}$	.	.	.	$-A$	$-\bar{A}$
X.42	.	.	.	.	.	.	.	.	.	.	.	.	.	.	.	.	.
X.43	.	.	.	.	.	.	.	.	.	.	.	.	.	.	.	.	.
X.44	.	.	.	−2	−2	.	.	.	.	.	.	.	.	.	.	.	.
X.45	.	.	.	2	2	.	.	.	.	.	.	.	.	.	.	.	.
X.46	−2	.	.	.	.	2	2	.	.	.	.	.	.	.	.	.	.
X.47	.	.	.	.	.	.	.	−1	.	.	.	.	.	.	1	.	.
X.48	2	.	.	.	.	−2	−2	.	.	.	.	.	.	.	.	.	.
X.49	.	.	.	.	.	.	.	1	.	.	.	.	.	.	1	.	.
X.50	$-\bar{B}$	.	.	.	.	$\bar{B}$	B	.	.	.	.	.	.	.	.	.	.
X.51	$\bar{B}$	.	.	.	.	$-\bar{B}$	$-B$	.	.	.	.	.	.	.	.	.	.
X.52	.	.	.	$-\bar{B}$	$-B$	.	.	.	.	.	.	.	.	.	.	.	.
X.53	$-B$	.	.	.	.	B	$\bar{B}$	.	.	.	.	.	.	.	.	.	.
X.54	.	.	.	$-B$	$-\bar{B}$	.	.	.	.	.	.	.	.	.	.	.	.
X.55	.	.	.	$\bar{B}$	B	.	.	.	.	.	.	.	.	.	.	.	.
X.56	B	.	.	.	.	$-B$	$-\bar{B}$	.	.	.	.	.	.	.	.	.	.
X.57	.	.	.	B	$\bar{B}$	.	.	.	.	.	.	.	.	.	.	.	.
X.58	.	.	.	.	.	.	.	.	1	1	.	.	−1	−1	.	.	.
X.59	.	.	.	.	.	.	.	.	1	1	.	.	−1	−1	.	.	.
X.60	.	.	.	.	.	.	.	.	A	$\bar{A}$	.	.	$-A$	$-\bar{A}$	.	.	.
X.61	.	.	.	.	.	.	.	.	A	$\bar{A}$	.	.	$-A$	$-\bar{A}$	.	.	.
X.62	.	.	.	.	.	.	.	.	1	1	.	.	−1	−1	.	.	.
X.63	.	.	.	.	.	.	.	.	$\bar{A}$	A	.	.	$-\bar{A}$	$-A$	.	.	.
X.64	.	.	.	.	.	.	.	.	$\bar{A}$	A	.	.	$-\bar{A}$	$-A$	.	.	.
X.65	.	.	.	.	.	.	.	.	$\bar{A}$	A	.	.	$-\bar{A}$	$-A$	.	.	.
X.66	.	.	.	.	.	.	.	.	A	$\bar{A}$	.	.	$-A$	$-\bar{A}$	.	.	.
X.67	.	.	.	.	.	.	.	.	.	.	.	.	.	.	.	.	.
X.68	.	.	.	.	.	.	.	.	.	.	.	.	.	.	.	.	.
X.69	.	.	.	.	.	.	.	.	.	.	.	.	.	.	.	.	.
X.70	.	.	.	.	.	.	.	.	.	.	.	.	.	.	.	.	.
X.71	.	.	.	.	.	.	.	.	.	.	.	.	.	.	.	.	.
X.72	.	.	.	.	.	.	.	.	.	.	.	.	.	.	.	.	.
X.73	.	.	.	.	.	.	.	.	.	.	.	.	.	.	.	.	.
X.74	.	.	.	.	.	.	.	.	.	.	.	.	.	.	.	.	.
X.75	.	.	.	.	.	.	.	.	−1	−1	.	.	1	1	.	.	.
X.76	.	.	.	.	.	.	.	.	$-\bar{A}$	$-A$	.	.	$\bar{A}$	A	.	.	.
X.77	.	.	.	.	.	.	.	.	$-A$	$-\bar{A}$	.	.	A	$\bar{A}$	.	.	.

where $A = -\zeta(3)-1$, $B = 2\zeta(3)$, $C = 3\zeta(3)$, $D = -6\zeta(3)-6$, $E = 8\zeta(3)$, $F = 2\zeta(3)+3$, $G = 2\zeta(3)+1$, $H = 4\zeta(3)+2$, $I = -\zeta(3)-3$, $J = \zeta(3)+2$, $K = 3\zeta(3)+2$.

14.6.4 *Character table of* $U = C_G(u) = \langle r_k | 1 \le k \le 5 \rangle$

	1a	2a	2b	2c	2d	2e	2f	3a	4a	4b	4c	4d	4e	4f	4g	4h	4i	4j	4k	4l	5a	6a	6b	6c	7a	8a	8b	8c	8d	8e
2	9	9	8	9	9	5	8	3	5	7	7	7	7	7	7	6	6	6	5	5	2	3	2	2	2	4	4	4	4	4
3	2	2	2	.	.	2	.	2	2	.	.	.	.	.	.	.	.	.	.	.	.	2	2	2	.	.	.	.	.	.
5	1	1	1	.	.	.	.	.	.	.	.	.	.	.	.	.	.	.	.	.	1	.	.	.	.	.	.	.	.	.
7	1	1	1	.	.	.	.	.	.	.	.	.	.	.	.	.	.	.	.	.	.	.	.	.	1	.	.	.	.	.
2P	1a	1a	1a	1a	1a	1a	1a	3a	2a	2c	2c	2c	2c	2c	2c	2c	2c	2c	2c	2d	5a	3a	3a	3a	7a	4c	4e	4f	4b	4g
3P	1a	2a	2b	2c	2d	2e	2f	1a	4a	4b	4c	4d	4e	4f	4g	4h	4i	4j	4k	4l	5a	2a	2e	2b	7a	8a	8b	8c	8d	8e
5P	1a	2a	2b	2c	2d	2e	2f	3a	4a	4b	4c	4d	4e	4f	4g	4h	4i	4j	4k	4l	1a	6a	6b	6c	7a	8a	8b	8c	8d	8e
7P	1a	2a	2b	2c	2d	2e	2f	3a	4a	4b	4c	4d	4e	4f	4g	4h	4i	4j	4k	4l	5a	6a	6b	6c	1a	8a	8b	8c	8d	8e
X.1	1	1	1	1	1	1	1	1	1	1	1	1	1	1	1	1	1	1	1	1	1	1	1	1	1	1	1	1	1	1
X.2	1	1	1	1	1	−1	1	1	−1	1	1	1	1	1	1	1	1	1	−1	−1	1	1	−1	1	1	−1	−1	−1	−1	−1
X.3	1	1	−1	1	1	−1	−1	1	1	1	1	1	1	1	1	−1	−1	−1	−1	1	1	1	−1	−1	1	−1	−1	1	−1	1
X.4	1	1	−1	1	1	1	−1	1	−1	1	1	1	1	1	1	−1	−1	−1	1	−1	1	1	1	−1	1	1	1	−1	1	−1
X.5	2	−2	.	2	−2	.	.	2	.	2	2	−2	2	−2	−2	.	.	.	.	.	2	−2	.	.	2	.	.	.	.	.
X.6	20	20	−20	4	4	2	−4	2	−2	.	.	.	.	.	.	.	.	.	−2	2	.	2	2	−2	−1	.	.	.	.	.
X.7	20	20	−20	4	4	−2	−4	2	2	.	.	.	.	.	.	.	.	.	2	−2	.	2	−2	−2	−1	.	.	.	.	.
X.8	20	20	20	4	4	−2	4	2	−2	.	.	.	.	.	.	.	.	.	2	2	.	2	−2	2	−1	.	.	.	.	.
X.9	20	20	20	4	4	2	4	2	2	.	.	.	.	.	.	.	.	.	−2	−2	.	2	2	2	−1	.	.	.	.	.
X.10	35	35	−35	3	3	−1	−3	−1	1	3	−1	3	−1	−1	−1	−3	1	1	−1	1	.	−1	−1	1	.	1	1	−1	−1	−1
X.11	35	35	35	3	3	−1	3	−1	−1	−1	−1	−1	3	−1	3	−1	−1	3	−1	−1	.	−1	−1	−1	.	1	−1	1	1	−1
X.12	35	35	−35	3	3	−1	−3	−1	1	−1	3	−1	−1	3	−1	1	−3	1	−1	1	.	−1	−1	1	.	−1	1	1	1	−1
X.13	35	35	−35	3	3	1	−3	−1	−1	−1	3	−1	−1	3	−1	1	−3	1	1	−1	.	−1	1	1	.	1	−1	−1	−1	1
X.14	35	35	35	3	3	1	3	−1	1	−1	−1	−1	3	−1	3	−1	−1	3	1	1	.	−1	1	−1	.	−1	1	−1	−1	1
X.15	35	35	−35	3	3	1	−3	−1	−1	3	−1	3	−1	−1	−1	−3	1	1	1	−1	.	−1	1	1	.	−1	−1	1	1	1
X.16	35	35	35	3	3	1	3	−1	1	3	−1	3	−1	−1	−1	3	−1	−1	1	1	.	−1	1	−1	.	−1	−1	−1	1	−1
X.17	35	35	−35	3	3	1	−3	−1	−1	−1	−1	−1	3	−1	3	1	1	−3	1	−1	.	−1	1	1	.	−1	1	1	−1	−1
X.18	35	35	35	3	3	1	3	−1	1	−1	3	−1	−1	3	−1	−1	3	−1	1	1	.	−1	1	−1	.	1	−1	1	−1	−1
X.19	35	35	35	3	3	−1	3	−1	−1	−1	3	−1	−1	3	−1	−1	3	−1	−1	−1	.	−1	−1	−1	.	−1	1	−1	1	1
X.20	35	35	−35	3	3	−1	−3	−1	1	−1	−1	−1	3	−1	3	1	1	−3	−1	1	.	−1	−1	1	.	1	−1	−1	1	1
X.21	35	35	35	3	3	−1	3	−1	−1	3	−1	3	−1	−1	−1	3	−1	−1	−1	−1	.	−1	−1	−1	.	1	1	1	−1	1
X.22	40	−40	.	8	−8	.	.	4	.	.	.	.	.	.	.	.	.	.	.	.	.	−4	.	.	−2	.	.	.	.	.
X.23	64	64	−64	.	.	8	.	1	−8	.	.	.	.	.	.	.	.	.	.	.	−1	1	−1	−1	1	.	.	.	.	.
X.24	64	64	64	.	.	−8	.	1	−8	.	.	.	.	.	.	.	.	.	.	.	−1	1	1	1	1	.	.	.	.	.
X.25	64	64	−64	.	.	−8	.	1	8	.	.	.	.	.	.	.	.	.	.	.	−1	1	1	−1	1	.	.	.	.	.
X.26	64	64	64	.	.	8	.	1	8	.	.	.	.	.	.	.	.	.	.	.	−1	1	−1	1	1	.	.	.	.	.
X.27	70	−70	.	6	−6	.	.	−2	.	6	−2	−6	−2	2	2	.	.	.	.	.	.	2	.	.	.	.	.	.	.	.
X.28	70	−70	.	6	−6	.	.	−2	.	−2	−2	2	6	2	−6	.	.	.	.	.	.	2	.	.	.	.	.	.	.	.
X.29	70	−70	.	6	−6	.	.	−2	.	−2	6	2	−2	−6	2	.	.	.	.	.	.	2	.	.	.	.	.	.	.	.
X.30	90	90	−90	−6	−6	.	6	.	.	2	2	2	2	2	2	−2	−2	−2	.	.	.	.	.	.	−1	.	.	.	.	.
X.31	90	90	90	−6	−6	.	−6	.	.	2	2	2	2	2	2	2	2	2	.	.	.	.	.	.	−1	.	.	.	.	.
X.32	90	−90	.	−6	6	.	.	.	.	2	2	−2	2	−2	−2	.	.	.	.	.	.	.	.	.	−1	.	.	.	.	.
X.33	90	−90	.	−6	6	.	.	.	.	2	2	−2	2	−2	−2	.	.	.	.	.	.	.	.	.	−1	.	.	.	.	.
X.34	126	126	126	−2	−2	.	−2	.	.	−2	−2	−2	−2	−2	−2	−2	−2	−2	.	.	1	.	.	.	.	.	.	.	.	.
X.35	126	126	−126	−2	−2	.	2	.	.	−2	−2	−2	−2	−2	−2	2	2	2	.	.	1	.	.	.	.	.	.	.	.	.
X.36	126	−126	.	−2	2	.	.	.	.	−2	−2	2	−2	2	2	.	.	.	.	.	1	.	.	.	.	.	.	.	.	.
X.37	126	−126	.	−2	2	.	.	.	.	−2	−2	2	−2	2	2	.	.	.	.	.	1	.	.	.	.	.	.	.	.	.
X.38	128	−128	.	.	.	.	.	2	.	.	.	.	.	.	.	.	.	.	.	.	−2	−2	.	.	2	.	.	.	.	.

Character table of $U = C_G(u)$ *(continued)*

	8f	10a	10b	10c	12a	14a	14b	14c
2	4	2	2	2	2	2	2	2
3	.	.	.	.	2	.	.	.
5	.	1	1	1	.	.	.	.
7	.	.	.	.	.	1	1	1
2P	4d	5a	5a	5a	6a	7a	7a	7a
3P	8f	10a	10c	10b	4a	14a	14c	14b
5P	8f	2a	2b	2b	12a	14a	14c	14b
7P	8f	10a	10c	10b	12a	2a	2b	2b
X.1	1	1	1	1	1	1	1	1
X.2	−1	1	1	1	−1	1	1	1
X.3	1	1	−1	−1	1	1	−1	−1
X.4	−1	1	−1	−1	−1	1	−1	−1
X.5	.	−2	.	.	.	−2	.	.
X.6	.	.	.	.	−2	−1	1	1
X.7	.	.	.	.	2	−1	1	1
X.8	.	.	.	.	−2	−1	−1	−1
X.9	.	.	.	.	2	−1	−1	−1
X.10	1	.	.	.	1	.	.	.
X.11	1	.	.	.	−1	.	.	.
X.12	−1	.	.	.	1	.	.	.
X.13	1	.	.	.	−1	.	.	.
X.14	−1	.	.	.	1	.	.	.
X.15	−1	.	.	.	−1	.	.	.
X.16	1	.	.	.	1	.	.	.
X.17	1	.	.	.	−1	.	.	.
X.18	−1	.	.	.	1	.	.	.
X.19	1	.	.	.	−1	.	.	.
X.20	−1	.	.	.	1	.	.	.
X.21	−1	.	.	.	−1	.	.	.
X.22	.	.	.	.	.	2	.	.
X.23	.	−1	1	1	1	1	−1	−1
X.24	.	−1	−1	−1	1	1	1	1
X.25	.	−1	1	1	−1	1	−1	−1
X.26	.	−1	−1	−1	−1	1	1	1
X.27	.	.	.	.	.	.	.	.
X.28	.	.	.	.	.	.	.	.
X.29	.	.	.	.	.	.	.	.
X.30	.	.	.	.	.	−1	1	1
X.31	.	.	.	.	.	−1	−1	−1
X.32	.	.	.	.	.	1	A	$\bar{A}$
X.33	.	.	.	.	.	1	$\bar{A}$	A
X.34	.	1	1	1	.	.	.	.
X.35	.	1	−1	−1	.	.	.	.
X.36	.	−1	B	$-B$	.	.	.	.
X.37	.	−1	$-B$	B	.	.	.	.
X.38	.	2	.	.	.	−2	.	.

where $A = 2\zeta(7)^4 + 2\zeta(7)^2 + 2\zeta(7) + 1$, $B = 2\zeta(5)^3 + 2\zeta(5)^2 + 1$.

14.6.5 *Character table of* $N_1 = N_G(3_A) = \langle g_1, g_2, g_3, g_4 \rangle$

	1a	2a	2b	3a	3b	3c	3d	3e	4a	4b	4c	5a
2	8	8	5	7	4	3	2	1	6	5	5	1
3	7	3	2	7	7	7	5	4	2	1	1	1
5	1	.	1	1	.	.	.	.	.	.	.	1
7	1	.	.	1	.	.	.	.	.	.	.	.
2P	1a	1a	1a	3a	3b	3c	3d	3e	2a	2a	2a	5a
3P	1a	2a	2b	1a	1a	1a	1a	1a	4a	4b	4c	5a
5P	1a	2a	2b	3a	3b	3c	3d	3e	4a	4b	4c	1a
7P	1a	2a	2b	3a	3b	3c	3d	3e	4a	4b	4c	5a
X.1	1	1	1	1	1	1	1	1	1	1	1	1
X.2	1	1	−1	1	1	1	1	1	1	1	−1	1
X.3	21	5	−1	21	−6	−6	3	3	1	1	3	1
X.4	21	5	1	21	−6	−6	3	3	1	1	−3	1
X.5	70	6	.	70	16	16	7	−2	6	−2	.	.
X.6	72	8	.	−36	18	−9	.	.	8	.	.	2
X.7	90	10	.	90	9	9	9	.	−2	2	4	.
X.8	90	10	.	90	9	9	9	.	−2	2	−4	.
X.9	90	−6	.	−45	−18	9	.	.	2	2	.	.
X.10	90	−6	.	−45	−18	9	.	.	2	2	.	.
X.11	140	12	10	140	5	5	−4	5	4	.	2	.
X.12	140	12	−10	140	5	5	−4	5	4	.	−2	.
X.13	189	−3	−9	189	27	27	.	.	5	1	−1	−1
X.14	189	−3	9	189	27	27	.	.	5	1	1	−1
X.15	210	2	10	210	21	21	3	3	−2	−2	−2	.
X.16	210	2	−10	210	21	21	3	3	−2	−2	2	.
X.17	252	28	.	−126	−18	9	.	.	4	4	.	2
X.18	378	−6	.	−189	54	−27	.	.	10	2	.	−2
X.19	420	4	10	420	−39	−39	6	−3	4	.	2	.
X.20	420	4	−10	420	−39	−39	6	−3	4	.	−2	.
X.21	560	−16	.	560	20	20	11	2	.	.	.	.
X.22	560	−16	.	560	20	20	11	2	.	.	.	.
X.23	560	−16	.	560	−34	−34	2	2	.	.	.	.
X.24	560	−16	.	560	−34	−34	2	2	.	.	.	.
X.25	630	22	.	−315	36	−18	.	.	−2	−2	.	.
X.26	630	22	.	630	−18	−18	9	.	−2	−2	.	.
X.27	630	−10	.	−315	36	−18	.	.	6	−2	.	.
X.28	630	−10	.	−315	36	−18	.	.	6	−2	.	.
X.29	729	9	9	729	.	.	.	.	−3	1	−3	−1
X.30	729	9	−9	729	.	.	.	.	−3	1	3	−1
X.31	896	.	16	896	32	32	−4	−4	.	.	.	1
X.32	896	.	−16	896	32	32	−4	−4	.	.	.	1
X.33	1260	12	.	−630	−90	45	.	.	4	−4	.	.
X.34	1280	.	.	1280	−16	−16	−16	2	.	.	.	.
X.35	1440	32	.	−720	36	−18	.	.	.	.	.	.
X.36	1458	18	.	−729	.	.	.	.	−6	2	.	−2
X.37	1512	−24	.	−756	54	−27	.	.	−8	.	.	2
X.38	1890	−30	.	−945	−54	27	.	.	2	2	.	.

	6a	6b	6c	6d	6e	6f	6g	7a	8a	8b	8c	8d	9a	9b	10a	12a
2	7	4	3	2	2	2	1	.	5	4	5	3	.	.	1	5
3	3	3	3	3	3	3	2	1	1	1	.	.	3	3	.	2
5	.	.	.	.	.	.	.	.	.	.	.	.	.	.	1	.
7	.	.	.	.	.	.	.	1	.	.	.	.	.	.	.	.
2P	3a	3b	3c	3d	3d	3d	3e	7a	4a	4a	4a	4b	9b	9a	5a	6a
3P	2a	2a	2a	2a	2a	2a	2b	7a	8a	8b	8c	8d	3c	3c	10a	4a
5P	6a	6b	6c	6e	6d	6f	6g	7a	8a	8b	8c	8d	9b	9a	2b	12a
7P	6a	6b	6c	6d	6e	6f	6g	1a	8a	8b	8c	8d	9a	9b	10a	12a
X.1	1	1	1	1	1	1	1	1	1	1	1	1	1	1	1	1
X.2	1	1	1	1	1	1	−1	1	−1	1	−1	−1	1	1	−1	1
X.3	5	2	2	−1	−1	−1	−1	.	−3	−1	1	1	.	.	−1	1
X.4	5	2	2	−1	−1	−1	1	.	3	−1	−1	−1	.	.	1	1
X.5	6	.	.	3	3	3	.	.	.	−2	.	.	1	1	.	6
X.6	−4	2	−1	2	2	−4	.	2	.	.	.	.	.	.	.	−4
X.7	10	1	1	1	1	1	.	−1	2	.	−2	.	.	.	.	−2
X.8	10	1	1	1	1	1	.	−1	−2	.	2	.	.	.	.	−2
X.9	3	6	−3	.	.	.	.	−1	.	−2	.	.	.	.	.	−1
X.10	3	6	−3	.	.	.	.	−1	.	−2	.	.	.	.	.	−1
X.11	12	−3	−3	.	.	.	1	.	2	.	2	.	−1	−1	.	4
X.12	12	−3	−3	.	.	.	−1	.	−2	.	−2	.	−1	−1	.	4
X.13	−3	3	3	.	.	.	.	.	−1	1	−1	1	.	.	1	5
X.14	−3	3	3	.	.	.	.	.	1	1	1	−1	.	.	−1	5
X.15	2	5	5	−1	−1	−1	1	.	−4	.	.	.	.	.	.	−2
X.16	2	5	5	−1	−1	−1	−1	.	4	.	.	.	.	.	.	−2
X.17	−14	−2	1	−2	−2	4	.	.	.	.	.	.	.	.	.	−2
X.18	3	6	−3	.	.	.	.	.	.	2	.	.	.	.	.	−5
X.19	4	1	1	−2	−2	−2	1	.	−2	.	−2	.	.	.	.	4
X.20	4	1	1	−2	−2	−2	−1	.	2	.	2	.	.	.	.	4
X.21	−16	−4	−4	−1	−1	−1	.	.	.	.	.	.	C	$\bar{C}$	.	.
X.22	−16	−4	−4	−1	−1	−1	.	.	.	.	.	.	$\bar{C}$	C	.	.
X.23	−16	2	2	2	2	2	.	.	.	.	.	.	−1	−1	.	.
X.24	−16	2	2	2	2	2	.	.	.	.	.	.	−1	−1	.	.
X.25	−11	4	−2	−2	−2	4	.	.	.	2	.	.	.	.	.	1
X.26	22	−2	−2	1	1	1	.	.	.	2	.	.	.	.	.	−2
X.27	5	−4	2	E	$\bar{E}$	2	.	.	.	−2	.	.	.	.	.	−3
X.28	5	−4	2	$\bar{E}$	E	2	.	.	.	−2	.	.	.	.	.	−3
X.29	9	.	.	.	.	.	.	1	3	−1	−1	1	.	.	−1	−3
X.30	9	.	.	.	.	.	.	1	−3	−1	1	−1	.	.	1	−3
X.31	.	.	.	.	.	.	−2	.	.	.	.	.	−1	−1	1	.
X.32	.	.	.	.	.	.	2	.	.	.	.	.	−1	−1	−1	.
X.33	−6	6	−3	.	.	.	.	.	.	.	.	.	.	.	.	−2
X.34	.	.	.	.	.	.	.	−1	.	.	.	.	2	2	.	.
X.35	−16	−4	2	2	2	−4	.	−2	.	.	.	.	.	.	.	.
X.36	−9	.	.	.	.	.	.	2	.	−2	.	.	.	.	.	3
X.37	12	6	−3	.	.	.	.	.	.	.	.	.	.	.	.	4
X.38	15	−6	3	.	.	.	.	.	.	2	.	.	.	.	.	−1

Character table of $N_1 = N_G(3_A)$ *(continued)*

	12b	12c	12d	12e	15a	21a	21b	24a	24b	24c
2	3	4	2	2	.	.	.	3	3	3
3	2	1	2	1	1	1	1	1	1	1
5	.	.	.	.	1	.	.	.	.	.
7	.	.	.	.	.	1	1	.	.	.
2P	6b	6a	6c	6b	15a	21b	21a	12b	12b	12a
3P	4a	4b	4a	4c	5a	7a	7a	8a	8a	8b
5P	12b	12c	12d	12e	3a	21a	21b	24a	24b	24c
7P	12b	12c	12d	12e	15a	3a	3a	24a	24b	24c
X.1	1	1	1	1	1	1	1	1	1	1
X.2	1	1	1	−1	1	1	1	−1	−1	1
X.3	−2	1	−2	.	1	.	.	.	.	−1
X.4	−2	1	−2	.	1	.	.	.	.	−1
X.5	.	−2	.	.	.	.	.	.	.	−2
X.6	2	.	−1	.	−1	−1	−1	.	.	.
X.7	1	2	1	1	.	−1	−1	−1	−1	.
X.8	1	2	1	−1	.	−1	−1	1	1	.
X.9	2	−1	−1	.	.	A	B	.	.	1
X.10	2	−1	−1	.	.	B	A	.	.	1
X.11	1	.	1	−1	.	.	.	−1	−1	.
X.12	1	.	1	1	.	.	.	1	1	.
X.13	−1	1	−1	−1	−1	.	.	−1	−1	1
X.14	−1	1	−1	1	−1	.	.	1	1	1
X.15	1	−2	1	1	.	.	.	−1	−1	.
X.16	1	−2	1	−1	.	.	.	1	1	.
X.17	−2	−2	1	.	−1	.	.	.	.	.
X.18	−2	−1	1	.	1	.	.	.	.	−1
X.19	1	.	1	−1	.	.	.	1	1	.
X.20	1	.	1	1	.	.	.	−1	−1	.
X.21	.	.	.	.	.	.	.	.	.	.
X.22	.	.	.	.	.	.	.	.	.	.
X.23	.	.	.	.	.	.	.	D	$\bar{D}$	.
X.24	.	.	.	.	.	.	.	$\bar{D}$	D	.
X.25	4	1	−2	.	.	.	.	.	.	−1
X.26	−2	−2	−2	.	.	.	.	.	.	2
X.27	.	1	.	.	.	.	.	.	.	1
X.28	.	1	.	.	.	.	.	.	.	1
X.29	.	1	.	.	−1	1	1	.	.	−1
X.30	.	1	.	.	−1	1	1	.	.	−1
X.31	.	.	.	.	1	.	.	.	.	.
X.32	.	.	.	.	1	.	.	.	.	.
X.33	−2	2	1	.	.	.	.	.	.	.
X.34	.	.	.	.	.	−1	−1	.	.	.
X.35	.	.	.	.	.	1	1	.	.	.
X.36	.	−1	.	.	1	−1	−1	.	.	1
X.37	−2	.	1	.	−1	.	.	.	.	.
X.38	2	−1	−1	.	.	.	.	.	.	−1

where $A = 2\zeta(21)_3\zeta(21)_7^4 + 2\zeta(21)_3\zeta(21)_7^2 + 2\zeta(21)_3\zeta(21)_7 + \zeta(21)_3 + \zeta(21)_7^4 + \zeta(21)_7^2 + \zeta(21)_7 + 1$, $B = -A + 1$, $C = 3\zeta(3) + 2$, $D = 2\zeta(24)_8^3\zeta(24)_3 + \zeta(24)_8^3 - 2\zeta(24)_8\zeta(24)_3 - \zeta(24)_8$, $E = -2C$.

14.6.6 *Character table of* $N_2 = N_G(Z) = \langle t_1, t_2, t_3, t_4 \rangle$

	1a	2a	2b	2c	3a	3b	3c	3d	3e	3f	3g	3h	4a	4b	4c	4d	4e	4f	6a	6b	6c	6d	6e	6f	6g	6h	6i	6j	6k
2	6	6	4	5	4	4	3	2	2	1	1	2	5	4	5	4	3	4	4	4	4	4	3	2	2	2	2	2	2
3	7	3	4	2	7	7	5	5	5	5	5	4	2	2	1	1	1	.	4	4	3	3	3	3	3	3	3	3	3
2P	1a	1a	1a	1a	3a	3b	3c	3e	3d	3f	3g	3h	2a	2c	2a	2a	2a	2c	3b	3b	3a	3b	3c	3d	3e	3c	3c	3d	3e
3P	1a	2a	2b	2c	1a	1a	1a	1a	1a	1a	1a	1a	4a	4b	4c	4d	4e	4f	2b	2b	2a	2a	2a	2b	2b	2b	2b	2a	2a
X.1	1	1	1	1	1	1	1	1	1	1	1	1	1	1	1	1	1	1	1	1	1	1	1	1	1	1	1	1	1
X.2	1	1	−1	1	1	1	1	1	1	1	1	1	1	−1	1	−1	1	−1	−1	−1	1	1	1	−1	−1	−1	−1	1	1
X.3	1	1	−1	1	1	1	1	1	1	1	1	1	1	1	1	−1	−1	1	−1	−1	1	1	1	−1	−1	−1	−1	1	1
X.4	1	1	1	1	1	1	1	1	1	1	1	1	1	−1	1	1	−1	−1	1	1	1	1	1	1	1	1	1	1	1
X.5	2	2	.	−2	2	2	2	2	2	2	2	2	2	.	−2	.	.	.	.	.	2	2	2	.	.	.	.	2	2
X.6	2	2	−2	2	2	2	−1	−1	−1	−1	2	2	2	.	2	−2	.	.	−2	−2	2	2	−1	1	1	1	1	−1	−1
X.7	2	2	2	2	2	2	−1	−1	−1	−1	2	2	2	.	2	2	.	.	2	2	2	2	−1	−1	−1	−1	−1	−1	−1
X.8	2	2	.	−2	2	2	−1	−1	−1	−1	2	2	2	.	−2	.	.	.	.	.	2	2	−1	A	$\bar{A}$	$\bar{A}$	A	−1	−1
X.9	2	2	.	−2	2	2	−1	−1	−1	−1	2	2	2	.	−2	.	.	.	.	.	2	2	−1	$\bar{A}$	A	A	$\bar{A}$	−1	−1
X.10	3	3	3	−1	3	3	.	.	.	.	3	3	−1	−1	3	−1	1	−1	3	3	3	3	.	.	.	.	.	.	.
X.11	3	3	−3	−1	3	3	.	.	.	.	3	3	−1	−1	3	1	−1	−1	−3	−3	3	3	.	.	.	.	.	.	.
X.12	3	3	−3	−1	3	3	.	.	.	.	3	3	−1	1	3	1	1	1	−3	−3	3	3	.	.	.	.	.	.	.
X.13	3	3	3	−1	3	3	.	.	.	.	3	3	−1	1	3	−1	−1	1	3	3	3	3	.	.	.	.	.	.	.
X.14	4	−4	.	.	4	4	−2	−2	−2	−2	4	4	.	−2	.	.	.	2	.	.	−4	−4	2	.	.	.	.	2	2
X.15	4	−4	.	.	4	4	−2	−2	−2	−2	4	4	.	2	.	.	.	−2	.	.	−4	−4	2	.	.	.	.	2	2
X.16	6	6	.	2	6	6	.	.	.	.	6	6	−2	.	−6	.	.	.	.	.	6	6	.	.	.	.	.	.	.
X.17	8	−8	.	.	8	8	2	2	2	2	8	8	.	.	.	.	.	.	.	.	−8	−8	−2	.	.	.	.	−2	−2
X.18	32	.	8	.	32	32	8	8	8	8	5	−4	.	.	.	.	.	.	8	8	.	.	.	2	2	2	2	.	.
X.19	32	.	−8	.	32	32	8	8	8	8	5	−4	.	.	.	.	.	.	−8	−8	.	.	.	−2	−2	−2	−2	.	.
X.20	32	.	−8	.	32	32	−4	−4	−4	−4	5	−4	.	.	.	.	.	.	−8	−8	.	.	.	B	$\bar{B}$	$\bar{B}$	B	.	.
X.21	32	.	8	.	32	32	−4	−4	−4	−4	5	−4	.	.	.	.	.	.	8	8	.	.	.	$-B$	$-\bar{B}$	$-\bar{B}$	$-B$	.	.
X.22	32	.	−8	.	32	32	−4	−4	−4	−4	5	−4	.	.	.	.	.	.	−8	−8	.	.	.	$\bar{B}$	B	B	$\bar{B}$	.	.
X.23	32	.	8	.	32	32	−4	−4	−4	−4	5	−4	.	.	.	.	.	.	8	8	.	.	.	$-\bar{B}$	$-B$	$-B$	$-\bar{B}$	.	.
X.24	36	4	.	.	9	−18	−12	6	6	−3	.	.	4	.	.	.	2	.	.	.	1	−2	4	.	.	.	.	−2	−2
X.25	36	4	−6	.	−18	9	12	3	3	−6	.	.	4	.	.	2	.	.	3	3	−2	1	4	3	3	.	.	1	1
X.26	36	4	6	.	−18	9	12	3	3	−6	.	.	4	.	.	−2	.	.	−3	−3	−2	1	4	−3	−3	.	.	1	1
X.27	36	4	.	.	9	−18	−12	6	6	−3	.	.	4	.	.	.	−2	.	.	.	1	−2	4	.	.	.	.	−2	−2
X.28	36	4	6	.	−18	9	−6	E	$\bar{E}$	3	.	.	4	.	.	−2	.	.	−3	−3	−2	1	−2	F	$\bar{F}$	.	.	$-C$	$-\bar{C}$
X.29	36	4	−6	.	−18	9	−6	$\bar{E}$	E	3	.	.	4	.	.	2	.	.	3	3	−2	1	−2	$-\bar{F}$	$-F$	.	.	$-\bar{C}$	$-C$
X.30	36	4	6	.	−18	9	−6	$\bar{E}$	E	3	.	.	4	.	.	−2	.	.	−3	−3	−2	1	−2	$\bar{F}$	F	.	.	$-\bar{C}$	$-C$
X.31	36	4	−6	.	−18	9	−6	E	$\bar{E}$	3	.	.	4	.	.	2	.	.	3	3	−2	1	−2	$-F$	$-\bar{F}$	.	.	$-C$	$-\bar{C}$
X.32	48	.	.	8	48	48	.	.	.	.	−6	3	.	8	.	.	.	.	.	.	.	.	.	.	.	.	.	.	.
X.33	48	.	.	8	48	48	.	.	.	.	−6	3	.	−8	.	.	.	.	.	.	.	.	.	.	.	.	.	.	.
X.34	48	.	.	−8	48	48	.	.	.	.	−6	3	.	.	.	.	.	.	.	.	.	.	.	.	.	.	.	.	.
X.35	48	.	.	−8	48	48	.	.	.	.	−6	3	.	.	.	.	.	.	.	.	.	.	.	.	.	.	.	.	.
X.36	72	8	.	.	18	−36	12	−6	−6	3	.	.	8	.	.	.	.	.	.	.	2	−4	−4	.	.	.	.	2	2
X.37	72	−8	.	.	−36	18	6	$-\bar{E}$	$-E$	−3	.	.	.	.	.	.	.	.	H	$\bar{H}$	4	−2	−2	I	$\bar{I}$	J	$\bar{J}$	$-\bar{C}$	$-C$
X.38	72	−8	.	.	−36	18	6	$-\bar{E}$	$-E$	−3	.	.	.	.	.	.	.	.	$\bar{H}$	H	4	−2	−2	$-I$	$-\bar{I}$	$-J$	$-\bar{J}$	$-\bar{C}$	$-C$
X.39	72	−8	.	.	−36	18	−12	−3	−3	6	.	.	.	.	.	.	.	.	$\bar{H}$	H	4	−2	4	A	$\bar{A}$	K	$\bar{K}$	1	1
X.40	72	−8	.	.	−36	18	−12	−3	−3	6	.	.	.	.	.	.	.	.	H	$\bar{H}$	4	−2	4	$\bar{A}$	A	$\bar{K}$	K	1	1
X.41	72	−8	.	.	−36	18	6	$-E$	$-\bar{E}$	−3	.	.	.	.	.	.	.	.	H	$\bar{H}$	4	−2	−2	$-\bar{I}$	$-I$	$-\bar{J}$	$-J$	$-C$	$-\bar{C}$
X.42	72	−8	.	.	−36	18	6	$-E$	$-\bar{E}$	−3	.	.	.	.	.	.	.	.	$\bar{H}$	H	4	−2	−2	$\bar{I}$	I	$\bar{J}$	J	$-C$	$-\bar{C}$
X.43	72	−8	.	.	18	−36	12	−6	−6	3	.	.	.	.	.	.	.	.	.	.	−2	4	4	.	.	.	.	−2	−2
X.44	72	−8	.	.	18	−36	12	−6	−6	3	.	.	.	.	.	.	.	.	.	.	−2	4	4	.	.	.	.	−2	−2
X.45	108	12	.	.	27	−54	.	.	.	.	.	.	−4	.	.	.	2	.	.	.	3	−6	.	.	.	.	.	.	.
X.46	108	12	18	.	−54	27	.	.	.	.	.	.	−4	.	.	2	.	.	−9	−9	−6	3	.	.	.	.	.	.	.
X.47	108	12	−18	.	−54	27	.	.	.	.	.	.	−4	.	.	−2	.	.	9	9	−6	3	.	.	.	.	.	.	.
X.48	108	12	.	.	27	−54	.	.	.	.	.	.	−4	.	.	.	−2	.	.	.	3	−6	.	.	.	.	.	.	.
X.49	144	−16	.	.	36	−72	−12	6	6	−3	.	.	.	.	.	.	.	.	.	.	−4	8	−4	.	.	.	.	2	2

Character table of $N_2 = N_G(Z)$ *(continued)*

2	2	2	1	1	2	3	3	1	1	3	3	2	2	3	2	2	1	1	3	3
3	3	3	3	3	2	1	.	3	3	2	2	2	2	1	1	1	2	2	1	1
	6l	6m	6n	6o	6p	8a	8b	9a	9b	12a	12b	12c	12d	12e	12f	12g	18a	18b	24a	24b
2P	3d	3e	3f	3g	3h	4a	4c	9b	9a	6d	6c	6p	6p	6d	6c	6e	9a	9b	12b	12b
3P	2b	2b	2a	2b	2c	8a	8b	3b	3b	4a	4a	4b	4b	4d	4e	4c	6a	6b	8a	8a
X.1	1	1	1	1	1	1	1	1	1	1	1	1	1	1	1	1	1	1	1	1
X.2	−1	−1	1	−1	1	1	−1	1	1	1	1	−1	−1	−1	1	1	−1	−1	1	1
X.3	−1	−1	1	−1	1	−1	1	1	1	1	1	1	1	−1	−1	1	−1	−1	−1	−1
X.4	1	1	1	1	1	−1	−1	1	1	1	1	−1	−1	1	−1	1	1	1	−1	−1
X.5	.	.	2	.	−2	.	.	2	2	2	2	.	.	.	.	−2	.	.	.	.
X.6	1	1	−1	−2	2	.	.	−1	−1	2	2	.	.	−2	.	−1	1	1	.	.
X.7	−1	−1	−1	2	2	.	.	−1	−1	2	2	.	.	2	.	−1	−1	−1	.	.
X.8	A	$\bar{A}$	−1	.	−2	.	.	−1	−1	2	2	.	.	.	.	1	$\bar{A}$	A	.	.
X.9	$\bar{A}$	A	−1	.	−2	.	.	−1	−1	2	2	.	.	.	.	1	A	$\bar{A}$	.	.
X.10	.	.	.	3	−1	−1	1	.	.	−1	−1	−1	−1	−1	1	.	.	.	−1	−1
X.11	.	.	.	−3	−1	1	1	.	.	−1	−1	−1	−1	1	−1	.	.	.	1	1
X.12	.	.	.	−3	−1	−1	−1	.	.	−1	−1	1	1	1	1	.	.	.	−1	−1
X.13	.	.	.	3	−1	1	−1	.	.	−1	−1	1	1	−1	−1	.	.	.	1	1
X.14	.	.	2	.	.	.	.	−2	−2	.	.	−2	−2	.	.	.	.	.	.	.
X.15	.	.	2	.	.	.	.	−2	−2	.	.	2	2	.	.	.	.	.	.	.
X.16	.	.	.	.	2	.	.	.	.	−2	−2	.	.	.	.	.	.	.	.	.
X.17	.	.	−2	.	.	.	.	2	2	.	.	.	.	.	.	.	.	.	.	.
X.18	2	2	.	−1	.	.	.	−1	−1	.	.	.	.	.	.	.	−1	−1	.	.
X.19	−2	−2	.	1	.	.	.	−1	−1	.	.	.	.	.	.	.	1	1	.	.
X.20	B	$\bar{B}$	.	1	.	.	.	C	$\bar{C}$	.	.	.	.	.	.	.	D	$\bar{D}$	.	.
X.21	$-B$	$-\bar{B}$	.	−1	.	.	.	C	$\bar{C}$	.	.	.	.	.	.	.	$-D$	$-\bar{D}$	.	.
X.22	$\bar{B}$	B	.	1	.	.	.	$\bar{C}$	C	.	.	.	.	.	.	.	$\bar{D}$	D	.	.
X.23	$-\bar{B}$	$-B$	.	−1	.	.	.	$\bar{C}$	C	.	.	.	.	.	.	.	$-\bar{D}$	$-D$	.	.
X.24	.	.	1	.	.	−2	.	.	.	−2	1	.	.	.	−1	.	.	.	1	1
X.25	−3	−3	−2	.	.	.	.	.	.	1	−2	.	.	−1	.	.	.	.	.	.
X.26	3	3	−2	.	.	.	.	.	.	1	−2	.	.	1	.	.	.	.	.	.
X.27	.	.	1	.	.	2	.	.	.	−2	1	.	.	.	1	.	.	.	−1	−1
X.28	$-F$	$-\bar{F}$	1	.	.	.	.	.	.	1	−2	.	.	1	.	.	.	.	.	.
X.29	$\bar{F}$	F	1	.	.	.	.	.	.	1	−2	.	.	−1	.	.	.	.	.	.
X.30	$-\bar{F}$	$-F$	1	.	.	.	.	.	.	1	−2	.	.	1	.	.	.	.	.	.
X.31	F	$\bar{F}$	1	.	.	.	.	.	.	1	−2	.	.	−1	.	.	.	.	.	.
X.32	.	.	.	.	−1	.	.	.	.	.	.	−1	−1	.	.	.	.	.	.	.
X.33	.	.	.	.	−1	.	.	.	.	.	.	1	1	.	.	.	.	.	.	.
X.34	.	.	.	.	1	.	.	.	.	.	.	G	$\bar{G}$	.	.	.	.	.	.	.
X.35	.	.	.	.	1	.	.	.	.	.	.	$\bar{G}$	G	.	.	.	.	.	.	.
X.36	.	.	−1	.	.	.	.	.	.	−4	2	.	.	.	.	.	.	.	.	.
X.37	I	$\bar{I}$	1	.	.	.	.	.	.	.	.	.	.	.	.	.	.	.	.	.
X.38	$-I$	$-\bar{I}$	1	.	.	.	.	.	.	.	.	.	.	.	.	.	.	.	.	.
X.39	A	$\bar{A}$	−2	.	.	.	.	.	.	.	.	.	.	.	.	.	.	.	.	.
X.40	$\bar{A}$	A	−2	.	.	.	.	.	.	.	.	.	.	.	.	.	.	.	.	.
X.41	$-\bar{I}$	$-I$	1	.	.	.	.	.	.	.	.	.	.	.	.	.	.	.	.	.
X.42	$\bar{I}$	I	1	.	.	.	.	.	.	.	.	.	.	.	.	.	.	.	.	.
X.43	.	.	1	.	.	.	.	.	.	.	.	.	.	.	.	.	.	.	L	$\bar{L}$
X.44	.	.	1	.	.	.	.	.	.	.	.	.	.	.	.	.	.	.	$\bar{L}$	L
X.45	.	.	.	.	.	2	.	.	.	2	−1	.	.	.	−1	.	.	.	−1	−1
X.46	.	.	.	.	.	.	.	.	.	−1	2	.	.	−1	.	.	.	.	.	.
X.47	.	.	.	.	.	.	.	.	.	−1	2	.	.	1	.	.	.	.	.	.
X.48	.	.	.	.	.	−2	.	.	.	2	−1	.	.	.	1	.	.	.	1	1
X.49	.	.	−1	.	.	.	.	.	.	.	.	.	.	.	.	.	.	.	.	.

where $A = 2\zeta(3) + 1$, $B = 2\zeta(3) + 2$, $C = -3\zeta(3) - 1$, $D = \zeta(3)$, $E = -9\zeta(3) - 6$, $F = 3\zeta(3) + 3$, $G = 3\zeta(4)$, $H = -12\zeta(3) - 6$, $I = \zeta(3) + 2$, $J = 2\zeta(3) - 2$, $K = 4\zeta(3) + 2$, $L = -2\zeta(24)_8^3\zeta(24)_3 - \zeta(24)_8^3 + 2\zeta(24)_8\zeta(24)_3 + \zeta(24)_8$

15
O'Nan's group ON

The simple group ON was discovered by M. O'Nan in [104]. He determined its order $|\mathsf{ON}| = 2^9 \cdot 3^4 \cdot 5 \cdot 7^3 \cdot 11 \cdot 19 \cdot 31$, character table and the group structure of the normalizers $N_G(\langle p \rangle)$ of elements p of prime order. O'Nan also described a stabilizer of a faithful permutation representation of degree 122760 of ON. He derived all his results from a presentation of a given Sylow 2-subgroup of ON due to Alperin [1]. However, he neither proved the existence nor the uniqueness of ON. The first published existence proof is due to L. H. Soicher [121]. He also showed that a simple group G is isomorphic to ON if and only if a Sylow 2-subgroup of G is isomorphic to the one of ON. Soicher's proofs are not self-contained. They depend on O'Nan's work.

It is the purpose of this chapter to present Previtali's and the author's existence and uniqueness proof for the sporadic simple group ON; see [94]. A finite simple group G is called a simple group of ON-type if it possesses a 2-central involution z such that $C_G(z) \cong H$, where H is the finitely presented group given in Lemma 15.1.1 of Section 15.2. This lemma also describes the group structure of H. In particular, a Sylow 2-subgroup S of H contains a unique maximal elementary abelian normal subgroup $A = \langle z, t, u \rangle$ of order 8 such that $C = C_H(A)$ is homocyclic of order 64.

In Section 15.2 it is shown that in any simple group G of ON-type there exists an element g of order 3 which is not contained in H such that $z^g = t$, $t^g = zt$ and $u^g = u$ and $E = N_G(A) = \langle r, s, g \rangle$ is uniquely determined by H up to isomorphism. A set $\mathcal{R}(E)$ of defining relations for E with respect to the generating set $\{r, s, g\}$ is also given in Proposition 15.2.2. In particular, E is an iterated extension E of $T = \mathrm{GL}_3(2)$ by the natural F-vector space V of dimension 3 over $F = GF(2)$ such that $C = C_E(V)$ is a homocyclic group of order 64 with invariants $(4, 4, 4)$.

Furthermore, all conditions of the first three steps Algorithm 1.3.15 are satisfied; see Proposition 15.2.1. Therefore the centralizer H can also be constructed from the non-split extension E of $T = \mathrm{PSL}_3(2)$ by C by application of Algorithm 1.3.15. This construction is not presented here.

The subgroup $E_1 = \langle z, t, u, r, g\rangle$ of E is a split extension of A by a Frobenius group F_{21} of order 21; see Lemma 15.4.1. Applying Algorithm 1.3.8 to E_1 it follows that there is an element f of order 5 in H such that $H_1 = \langle z\rangle \times \langle r, uz, tz, f\rangle$ is the unique subgroup of H which is isomorphic to $\langle z\rangle \times A_5$ in H and contains the subgroup $D_1 = \langle A, r\rangle \cong S_4$ such that the amalgam $H_1 \leftarrow D_1 \rightarrow E_1$ satisfies all conditions of Step 5 of Algorithm 1.3.8.

Proposition 15.4.2 due to Previtali and the author [94] provides a set $\mathcal{R}(J)$ of defining relations for the subgroup $J = \langle H_1, E_1\rangle = \langle r, f, g\rangle$ of any finite simple group G of ON-type. By Janko's uniqueness Theorem 9.4.2 of [92] it follows that J is isomorphic to the smallest Janko group J_1; see also [66].

In Section 15.5 it is shown that $H = \langle r, s, c\rangle = \langle r, s, f\rangle$. This fact enables us to derive a set $\mathcal{R}_1(H)$ of defining relations of $H = \langle r, s, f\rangle$. Theorem 15.5.2 due to Previtali and the author [94] states that any finite simple group G of ON-type is generated by r, s, f and g and has the set of defining relations

$$\mathcal{R}(G) = \mathcal{R}_1(H) \cup \mathcal{R}(E) \cup \mathcal{R}(J).$$

Furthermore, the finitely presented group G has a faithful permutation representation of degree 2624832 with stabilizer $J \cong \mathsf{J}_1$. It is simple and $C_G(z) = \{r, s, f\} \cong H$. Hence, this theorem provides an existence and uniqueness proof for the O'Nan group ON.

In Corollary 15.6.1 a permutation representation of G of degree 122760 is derived from this theorem. It is used to calculate the character table of G, determine the conjugacy classes and local subgroups of G.

Soicher's uniqueness theorem is stronger than Previtali's and the author's uniqueness theorem. But it is an immediate consequence of Theorem 15.5.2 and O'Nan's Lemma 4.8 of [104], because Previtali has shown that there are exactly 36 non-isomorphic groups H_i of shape $4L_3(4):2$, and that the isomorphism type of H_i is uniquely determined by the isomorphism type of a Sylow 2-subgroup S_i of H_i. Furthermore, these examples show that the given centralizer $H = C_G(z)$ of a 2-central involution of a simple group G of ON-type has to be defined explicitly by generators and relations.

15.1 The centralizer of a 2-central involution

In this section we give a presentation of the finite group H which is assumed to be the centralizer of a 2-central involution in a group of ON-type. We determine a Sylow 2-subgroup S of H and an elementary abelian normal subgroup A of S such that $D = N_H(A)$ is of maximal order among all normalizers $N_H(B)$ of the elementary abelian maximal subgroups B of S.

Lemma 15.1.1 *Let $H = \langle r, s, c\rangle$ be the finitely presented group with the following set $\mathcal{R}(H)$ of defining relations:*

$$r^3 = s^{16} = c^5 = (sc^{-1})^4 = (rs^2r^{-1}s^{-1})^2 = (r^{-1}s^2rs^{-1})^2 = 1,$$
$$s^{-1}r^{-1}c^{-2}s^{-1}r^{-1}crc^2 = (r^{-1}c^{-1}s^{-1}cs)^2 = rc^2rs^{-1}r^{-1}s^{-1}c^{-2}rc^{-1} = 1,$$
$$cr^{-1}s^{-3}r^{-1}c^{-1}r^{-1}scr^{-1} = sc^{-1}sr^{-1}sr^{-1}c^{-1}sc^{-1}r^{-1}c = 1,$$
$$s^5r^{-1}cr^{-1}c^{-1}sc^{-1} = sr^{-1}s^{-1}c^{-1}sr^{-1}s^{-1}rcr^{-1}c = 1,$$
$$cs^2c^{-1}sr^{-1}cr^{-1}scr = (rsr^{-1}s^{-1}r^{-1}s^{-1})^2 = (r^{-1}s^{-3}r^{-1}s^{-1})^2 = 1,$$
$$s^2r^{-1}s^{-1}r^{-1}s^{-1}r^{-1}s^3r^{-1}s = r^{-1}c^{-2}sr^{-1}s^{-1}r^{-1}sc^{-1}r^{-1}s^{-1}c = 1.$$

Then $z = (rsrs^{-1})^2$, $t = (rs^2)^3s^{-4}$ and $u = (sr)^4$ are involutions and $v_1 = (cr)^{-3}$, $v_2 = r^{-1}s^{-1}rs^{-1}r^{-1}s^{-2}$ and $v_3 = (sr)^{-2}$ have order 4. Furthermore, the following assertions hold.

(a) *$Aut(H) = Inn(H) : K$, where K is a Klein 4-group and there is $\gamma \in Aut(H) - Inn(H)$ fixing r.*
(b) *$S = \langle s, s^{-1}rsr, s^2r^{-1}s^{-2}rs^{-1}\rangle$ is a Sylow 2-subgroup of H with center $Z(S) = \langle z\rangle$ such that $\gamma(S) = S$. Moreover, $s^8 = v_1^2 = z$, $v_2^2 = t$ and $v_3^2 = u$, z, t, u are involutions.*
(c) *$P = O_2(H) = \langle v_1\rangle$ is cyclic of order 4.*
(d) *$A = \langle z, t, u\rangle$ is the unique elementary abelian normal subgroup of rank 3 of S.*
(e) *$V = \langle z, t\rangle$ is the unique elementary abelian normal 4-group of S.*
(f) *$C_H(V) = C_S(V)$ is a maximal subgroup of S, and $V = Z(C_S(V))$.*
(g) *$S = N_H(V)$.*
(h) *$C = C_H(A) = C_S(A) = \langle v_1, v_2, v_3\rangle$ is a homocyclic subgroup of rank 3 and order 64 of S.*
(i) *C is the unique abelian subgroup of its order in S.*
(j) *$D = N_H(A) = \langle r, s\rangle$ is a non-split extension of C by S_4.*
(k) *H has a faithful permutation representation of degree 448 with stabilizer $\langle sr^{-1}c^2sr^{-1}c^{-1}rc, cs^{-1}r^{-1}cs^{-3}c\rangle \cong A_6$.*

(l) *A system of representatives* h_i *and the corresponding centralizer orders* $|C_H(h_i)|$ *of the* 31 *conjugacy classes* h_i^H *of* H *is given in Table 15.7.1.*

(m) *The character table of* H *is given in Table 15.8.1.*

(n) $Inn(D)$ *has an elementary abelian complement of order* 8 *in* $Aut(D)$.

(o) *A system of representatives* d_i *and the corresponding centralizer orders* $|C_D(d_i)|$ *of the classes of* $D = \langle r, s\rangle$ *is given in Table 15.7.2.*

(p) *The character table of* D *is given in Table 15.8.2.*

Proof All these statements have been checked by means of MAGMA and the given presentation of H. The system of representatives of the conjugacy classes of H and D have been calculated by means of Kratzer's Algorithm 5.3.18 of [92]. □

Definition 15.1.2 A finite simple group G is said to be of ON-*type*, if it has a 2-central involution z such that $C_G(z)$ is isomorphic to the finitely presented group H stated in Lemma 15.1.1.

15.2 Fusion

In this section it is shown that any finite simple group G of ON-type has one conjugacy class of involutions and the conjugacy classes of elements of even order are classified. Most of the results are originally due to O'Nan; see [104].

Proposition 15.2.1 *Let* G *be a finite simple group having a* 2*-central involution* z *with centralizer* $C_G(z) = H$ *given in Lemma 15.1.1. The following statements hold.*

(a) *There exists an element* g *of order* 3 *in* $N_G(A)$, $g \notin H$, *such that* $z^g = t$, $t^g = zt$ *and* $u^g = u$. *Such an element is unique up to conjugation in* $C = C_H(A)$.

(b) $E = N_G(A)$ *is the unique non-split extension of* $C = C_G(A) = C_H(A)$ *by* $\mathrm{GL}_3(2)$ *which has a Sylow* 2*-subgroup isomorphic to the ones of* H.

(c) *The Goldschmidt index of the amalgam* $H \leftarrow D \rightarrow E$ *is* 1.

(d) *$E = \langle r, s, g\rangle$ has the following set $\mathcal{R}(E)$ of defining relations:*

$$\begin{aligned} &r^3 = g^3 = s^{16} = 1,\\ &sr^{-1}sr^{-1}g^{-1}s^{-1}g^{-1} = (r^{-1}g^{-1}r^{-1}s^{-1})^2 = (sgs^2)^2 = 1,\\ &(sg^{-1}r^{-1}g^{-1})^2 = (gr^{-1}s^{-1})^3 = (rs^2r^{-1}s^{-1})^2 = 1,\\ &s^{-2}g^{-1}r^{-1}s^{-1}rs^{-2}rg = s^{-1}r^{-1}gr^{-1}s^{-1}rsgr^{-1}g^{-1} = 1. \end{aligned}$$

(e) *$z = s^8$ is a 2-central involution of E such that $C_E(z) \cong D = N_H(A)$. In particular, all conditions of Step 5 of Algorithm 1.3.8 are satisfied by the amalgam $H \leftarrow D \rightarrow E$.*

(f) *E has a faithful permutation representation PE of degree 112 with stabilizer $\langle r^{-1}sgs^2r, s^3g^{-1}s^{-1}rg^{-1}\rangle$ whose generators have order 8.*

(g) *A system of representatives e_k and the corresponding centralizer orders $|C_E(e_k)|$ of the 18 conjugacy classes of e_k^E in E are given in Table 15.7.3.*

(h) *The character table of E is stated in Table 15.8.3.*

Proof (a) Let S be the Sylow 2-subgroup of H given in Lemma 15.1.1(b). Using the faithful permutation representation of H described in Lemma 15.1.1(k) and MAGMA it has been checked that $w = rs^{-2}r^2s$ is an involution of S that does not belong to H'. Therefore $z \in H'$, $t \in H'$ and $w \in H - H'$ are representatives of all conjugacy classes of involutions of H and $|C_H(w)| = 144$ by Lemma 15.1.1(k).

Suppose that z and t are not conjugate in G. Then Glauberman's Z^*-Theorem implies that $z^q = w$ for some $q \in G$. Hence $C_H(w) = H \cap H^q$. Now $\frac{|H\cap H^q|}{|H'\cap H'^q|}$ divides 4 because $|H : H'| = 2$. As 16 divides $|C_H(w)|$ it follows that $|H' \cap H'^q|$ is even. Hence there is an involution $b_1 \in H' \cap H'^q$ and $v = wb_1$ is an involution in $H - H'$. Thus v and w are H-conjugate. Furthermore, $v \in H'^q$, because $w = z^q$ and b_1 belong to H'^q. Therefore $v^h = t^q$ for some $h \in H^q$. In particular, z and t are G-conjugate. This contradiction to our assumption proves that there is an x in G such that $z^x = t$ and $x \notin H$.

Let $K = C_G(t)$, then S^x is a Sylow 2-subgroup of $K = H^x$. By Lemma 15.1.1 $X = C_H(V) = C_H(t)$ has index 2 in S. Let Y be a Sylow 2-subgroup of K containing X. Then X is normal in Y. By Sylow's Theorem there is a $y \in K$ such that $Y = S^{xy}$. Now $V = Z(X)$ *char* $X \lhd Y$ implies that $V \lhd Y$. Also $V^{xy} \lhd Y = S^{xy}$. By Lemma 15.1.1 Y has only one normal elementary abelian subgroup of order 4. Thus $V = V^{xy}$ and

$xy \in N_G(V)$. Furthermore, Lemma 15.1.1 asserts that

$$C = C_H(A) = C_S(A) \le C_G(V) \trianglelefteq N_G(V),$$

and that C is the unique homocyclic subgroup of S of order 64. Hence $C, C^{xy} \le C_G(V) \le S$ implies that $C = C^{xy}$. Since $A = \Omega_1(C)$ it follows that $xy \in N_G(A)$. Clearly $z^{xy} = t^y = t$. In particular, $xy \notin H$. Now $C_G(A) = C_H(A) = C$. Since $D/C \cong S_4$ is a maximal subgroup of $Aut(A) \cong \mathrm{GL}_3(2)$ and $xy \notin D$, it follows that $N_G(A)/C \cong \mathrm{GL}_3(2)$. Thus $E = N_G(A)$ is an extension of C by $\mathrm{GL}_3(2)$. By Lemma 15.1.1(j) and the Theorem of Gaschütz E is a non-split extension and $C_E(C) \le C_E(A) = C$ is 2-constrained.

Using the MAGMA command `ExtensionOfSolubleGroup(C,T)` Previtali and the author obtained in [94] four extensions of $T = \mathrm{GL}_3(2)$ by C. But only the finitely presented group E stated in assertion (d) has a Sylow 2-subgroup isomorphic to the one of H.

Its permutation representation PE of degree 112 given in (f) has been found by means of MAGMA.

(e) Using it and the MAGMA command `IsIsomorphic(C_E(z),D)` one verifies that $C_E(z) \cong D$. This isomorphism has been used in [94] to identify the elements r and s of $C_E(z)$ and the corresponding ones in $D = N_H(A)$. Thus $C_E(z) \cong D = N_H(A)$.

The element g of assertion (a) has been obtained by means of PE and the MAGMA command

```
exists(g){x: x in E|Order(x) eq 3, z^x eq t, t^g eq zt and
  u^g= u}.
```

Let g, q be a pair of distinct elements of order 3 in E having the same action on A. Then $x = gq^{-1} \in C_G(A) = C$. By Lemma 15.1.1(h) C is abelian, so $a^{g^c} = a^g$ for any $c \in C$, $a \in A$. So all C-conjugates of g act as g on A. Since $|C_C(g)| = 4$ and $|C| = 64$ there are 16 such elements. MAGMA shows there are no others. As $r \in D$ acts on A by $z^r = z$, $t^r = ztu$ and $u^r = t$ it follows that their matrices $(g), (r) \in \mathrm{GL}_3(2)$ generate a Frobenius subgroup F_{21} inside $\mathrm{GL}_3(2)$. Let (s) be the matrix of s with respect to its action on A. Then $\mathrm{GL}_3(2) = \langle (r), (s), (g) \rangle$. Thus (a), (b), (d), (e) and (f) hold.

(c) An application of Kratzer's Algorithm 7.1.10 of [92] proves that the number of (H^*, E^*)-double cosets in $Aut(D)$ is 1. Thus the Goldschmidt index the amalgam (H, D, E) is 1.

(g) and (h) Representatives of conjugacy classes and character tables were calculated in [94] by means of MAGMA using Kratzer's, Weller's and Previtali's programs.

□

Proposition 15.2.2 *Let G be any finite simple group of* ON*-type having a 2-central involution z with centralizer $H = C_G(z)$. Then the following statements hold.*

(a) *G has a unique class of involutions represented by z, denoted $2A$.*
(b) *G has 17 z-special H-conjugacy classes represented by*

$$2_a, 4_a, 4_b, 6_a, 8_a, 8_b, 10_a, 12_a, 14_a, 16_a, 16_b, 16_c, 16_d, 20_a, 20_b, 28_a, 28_b.$$

(c) *Using the notation of the conjugacy classes of H and $E = N_G(A)$ as in Tables 15.7.1 and 15.7.3, respectively, the fusion of their elements of order 2, 4 and 8 into the special conjugacy classes of G is given by the following table:*

G	$2A$			$4A$		$4B$				$8A$		$8B$	
H	2_a	2_b	2_c	4_a	4_d	4_b	4_c	4_e	4_f	8_a	8_d	8_b	8_c
E	2_a	2_b		4_a		4_b	4_c			8_a		8_b	

Proof (a) The fusion of involutions can be read off from the following diagram:

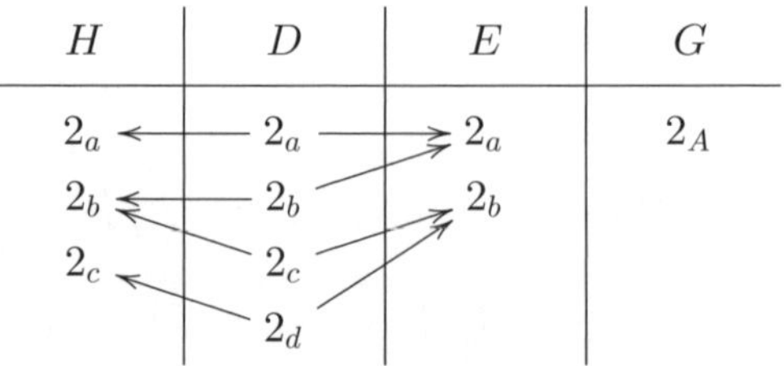

Each fusion relation has been checked using MAGMA and Tables 15.7.1, 15.7.2 and 15.7.3 classifying the conjugacy classes of H, D and E, respectively.

(b) z-special classes of H are easily read off from Table 15.7.1.

(c) The fusion graph for the other 2-elements is obtained in the same way as in (a). By Theorem 7.1.8 of [92] and Proposition 15.2.2 the fusion of the conjugacy classes of elements of even order of H in G is uniquely determined by H. □

15.3 3-singular classes

In this section a proof is given for O'Nan's result [104], asserting that any simple group G of ON-type has a unique conjugacy class of elements of order 3. It follows then that G has exactly three 3-singular classes.

Lemma 15.3.1 *Let G be any finite simple group of ON-type having a 2-central involution z with centralizer $C_G(z) = H = \langle r, s, c\rangle$ given in Lemma 15.1.1. Then the following assertions hold.*

(a) *$C_H(r) = \langle r\rangle \times \langle r_1\rangle \times \langle w, \ell\rangle$, where $r_1 = (s^{-1}r^{-1})^2c^{-1}r^{-1}c$, $w = rsrc^{-2}sc^{-1}r^{-1}s^{-1}$ and $\ell = (rc)^3$ have respective orders 3, 2 and 4. Furthermore, $D_1 = \langle \ell, w\rangle$ is a dihedral group of order 8 and $\ell^w = \ell^{-1}$.*

(b) *$N_1 = N_H(r) = \langle C_H(r), n_1\rangle$, where $n_1 = (s^{-1}c[c^2, s])^2$ is an involution satisfying the following relations: $r^{n_1} = r^2$, $r_1^{n_1} = r_1^2$, $[x, n_1] = 1$ for all $x \in D_1$.*

(c) *$S_2 = D_1 \times \langle n_1\rangle$ is a Sylow 2-subgroup of N_1.*

(d) *$R = \langle r\rangle \times \langle r_1\rangle$ is a Sylow 3-subgroup of H.*

(e) *Let $q_1 = s^3r^{-1}c^{-1}rc^{-1}s^{-1}$ and $q_2 = c^{-1}s^{-1}cr^{-1}c^{-1}r^{-1}c^{-1}$ in H. Then Q_8 is a quaternion group of order 8 and $N_H(R) = (R\times D_1) : Q_8$ has the following set $\mathcal{R}(N_H(R)$ of defining relations:*

$$
\begin{aligned}
&r^3 = r_1^3 = 1, \quad \ell^4 = w^2 = q_1^4 = q_2^4 = 1,\\
&(\ell^{-1}w)^2 = 1, \quad [\ell, q_1] = 1, \quad q_2^{-1}q_1^2q_2^{-1} = 1,\\
&\ell^{-1}q_2^{-1}\ell^{-1}q_2 = 1, \quad q_1^{-1}q_2^{-1}q_1q_2^{-1} = 1,\\
&wq_2^{-1}\ell^{-1}wq_2 = 1, \quad q_1^{-1}w\ell^2q_1w = 1,\\
&r^{q_1} = r_1, r_1^{q_1} = r^2, \quad r^{q_2} = rr_1^2, \quad r_1^{q_2} = (rr_1)^2.
\end{aligned}
$$

Proof Using the faithful permutation representation of H given in Lemma 15.1.1 and MAGMA all these statements have been deduced from Lemma 15.1.1 computationally. □

The following result is due to O'Nan; see [104], Lemma 5.13. Its proof is taken from [94].

Proposition 15.3.2 *Let G be a simple ON-type group having a 2-central involution z such that $C_G(z) = H = \langle r, s, c\rangle$. Let $N_1 = N_H(r) = \langle r, r_1, w, \ell, n_1\rangle$ be as in Lemma 15.3.1 and let $R = \langle r, r_1\rangle$. Then the following assertions hold.*

(a) *$C_G(r) \cong R \times A_6 \cong C_G(R)$.*

(b) *There is an element* f_1 *of order* 5 *in* G *such that* $N_G(r) = \langle r, r_1, l, n_1, w, f_1 \rangle$ *is isomorphic to* $(R \times A_6).2$ *and has the following presentation:*

$$\begin{aligned}
&r^3 = r_1^3 = n_1^2 = 1, \quad \ell^4 = w^2 = f_1^5 = 1,\\
&[r, r_1] = 1, \quad (r^{-1}n_1)^2 = 1, \quad (r_1^{-1}n_1)^2 = 1,\\
&[r, \ell] = [r_1, \ell] = 1, \quad n_1\ell^{-1}n_1\ell = 1,\\
&r^{-1}wrw = 1, \quad r_1^{-1}wr_1w = 1,\\
&(n_1w)^2 = 1, \quad (\ell^{-1}w)^2 = 1, \quad [r, f_1] = 1, \quad [r_1, f_1] = 1,\\
&n_1f_1^{-1}n_1f_1 = 1, \quad (wf_1^{-1})^2 = 1, \quad (f_1^{-1}\ell^{-1})^3 = 1,\\
&\ell^{-2}f_1^{-1}\ell^2f_1^{-1}\ell^2f_1^{-1} = 1,\\
&\ell^{-1}f_1\ell^{-1}f_1^{-1}w\ell^{-1}f_1^{-2}\ell f_1 = 1.
\end{aligned}$$

(c) G *has three* r*-special conjugacy classes represented by the classes* 3_a, 15_a *and* 15_b *of the character table of* $N = N_G(r)$ *given in Table 15.8.4.*

(d) *A system of representatives* n_i *and the corresponding centralizer orders* $|C_N(n_i)|$ *of the* 42 *conjugacy classes* n_i^N *of* $N = N_G(r)$ *are given in Table 15.7.4.*

(e) $n_1 \in N - C_G(r)$ *inverts* r *and centralizes the subgroup* $U_1 = \langle \ell, w, f_1 \rangle \cong A_6$ *of* N.

Proof (a) By Lemma 15.3.1 $C_H(r) = R \times D_1$, where R is elementary abelian of order 9 and $D_1 = \langle \ell, w \rangle$ is a dihedral Sylow 2-subgroup of $C = C_G(r)$.

Let j be any involution of $C_G(r)$. By Proposition 15.2.2(a) there is an $x \in G$ such that $z = j^x$. Hence

$$r^x \in C_G(j^x) = C_G(z) = H.$$

By Lemma 15.1.1(l) H has a unique conjugacy class of elements of order 3 represented by r. Hence there is an $h \in H$ such that $r^{xh} = r$ and $z = z^h = j^{xh}$. Therefore $xh \in C_G(r)$. Thus all involutions of $C = C_G(r)$ are conjugate in C.

Let $Q = O(C)$. Then $C/Q \neq D_1$, because D_1 has three conjugacy classes of involutions. Now Theorem 1.6.4 of [92] asserts that

$$C/Q \in \{A_6, A_7, L_2(7)\}.$$

In the case of A_7 a 2-central involution would have a centralizer in A_7 of order divisible by 3. Hence $|H| = |C_G(z)|$ would be divisible by 3^3, a contradiction to the order of H. Thus $C/Q \in \{A_6, \mathrm{GL}_2(3)\}$.

Suppose that $C/Q \cong \mathrm{GL}_3(2)$. By Lemma 15.3.1 $|N_H(r)/C_H(r)| = 2$, and there is an involution $n_1 \in N_H(r)$ satisfying $r^{n_1} = r^2$. If the residue class of n_1 in C/Q were operating like an outer automorphism of $L_2(7)$, then the Sylow 2-subgroup of $N_G(r)$ would be a dihedral group of order 16. Since $N = N_G(r)$ and $N_H(r)$ have conjugate Sylow 2-subgroups, this is impossible by Lemma 15.3.1(c).

Hence we may assume that n_1 centralizes the normal subgroup $\mathrm{GL}_3(2)$ of N/Q. If the residue class of n_1 were not generating a direct factor of $\mathrm{GL}_3(2)$ in N/Q, then $N/Q = \mathsf{SL}_2(7)$. In particular, N would have a generalized quaternion Sylow 2-subgroup of order 16, which is a contradiction to Lemma 15.3.1(c). Therefore $N/Q = \langle n_1 \rangle \times \mathrm{GL}_3(2)$. As G has a unique conjugacy class of involutions, n_1 is conjugate to z and $\mathrm{GL}_3(2)$ is a subgroup of H up to conjugation.

Another application of MAGMA yields that H has a unique conjugacy class of subgroups $U \cong \mathrm{GL}_3(2)$ and that their dihedral Sylow 2-subgroups D_2 of order 8 contain a unique class of elements v of order 4 which are H-conjugate to $(rs)^2 \in 4_c$. By Lemma 15.1.1(l) $\ell = (rc)^3 \in D_1$ belongs to 4_a. Proposition 15.2.1(c) implies that ℓ and $(rs)^2$ are not conjugate in G. Hence the Sylow 2-subgroup D_1 of $C = C_G(r)$ and the Sylow 2-subgroup D_2 of $L_2(7) \le C$ cannot be conjugate in C, a contradiction to Sylow's Theorem. Therefore $C/Q = A_6$.

Applying Theorem 4.2.2 of [92] to the Klein 4-subgroup $V = \langle z, w \rangle$ of D_1 it follows that

$$(*) \qquad |Q||C_Q(V)|^2 = |C_Q(z)|^3$$

because C has a unique class of involutions. Furthermore, $r \in C_Q(z)$, and $C_Q(z) \le H \cap C_G(r) = C_H(r)$. Since $|C_H(r)| = 2^3 3^2$ by Lemma 15.1.1, we obtain $|C_Q(z)| \in \{3, 9\}$.

Suppose that $|C_Q(z)| = 3$. As $r \in C_Q(z) \cap C_Q(w)$ it follows from $(*)$ that $|C_Q(V)| = |C_Q(z)| = |Q| = 3$. Hence $Q = \langle r \rangle$, and $C/Q \cong A_6$. Therefore either $C \cong 3A_6$ or $C \cong \langle r \rangle \times A_6$.

Let $C_1 = C_G(R)$. As $C_H(r) = R \times D_1 = C_H(R)\, C_1$ also has a dihedral Sylow 2-subgroup D_1. Let $W = O(C_1)$. Then Theorem 1.6.4 of [92] asserts that

$$C_1/W \in \{\mathrm{GL}_3(2), A_6, A_7, D_1\}.$$

As $C_1 \le C$, and 7 does not divide $|C| = 3|A_6|$, the simple groups $\mathrm{GL}_3(2)$ and A_7 are not involved in C_1. Clearly, $R \le W$. If $C_1/W \cong A_6$, then

$$3|A_6| = |C| \ge |C_1| = |W||A_6| \ge 9|A_6|,$$

a contradiction. Thus $C_1/W \cong D_8$, the dihedral group of order 8.

Applying Theorem 4.2.2 of [92] again to the Klein 4-subgroup $V = \langle z, w\rangle$ of the Sylow 2-subgroup D_1 of C_1 it follows that

$$|W||C_W(V)|^2 = |C_W(z)||C_W(w)|^2$$

because the involutions w and wz are conjugate in D_1. As $|C_W(z)| = |R| \le |C_W(w)|$, we obtain $|C_W(V)| = |R|$. Therefore $|W| = |R|^{-1}| C_W(w)|^2$. Since $C_1 \le C$ and $|C|$ has a Sylow 3-subgroup of order 27 we see that $W = R$. Hence

$$C_1 = C_G(R) \cong W \times D_8 \cong R \times D_1.$$

In particular, $|R| = 9$ divides the order of $C_G(D_1)$. However, if $C \in \{3A_6, 3 \times A_6\}$, then $|C_C(D_1)| = 6$. This contradiction proves that $Q = R$ and $C/R \cong A_6$. Hence $C_G(R)/R \cong A_6$. By Lemma 15.3.1(e) $N_H(R) = (R \times D_8) : Q_8$. Furthermore, Q_8 acts irreducibly on R, because n_1 inverts both generators r and r_1 of R. Since the Schur multiplicator of A_6 is cyclic of order 6 this implies that $C_G(R) \cong R \times A_6$, and so

$$C = C_G(r) = C_G(R) \cong R \times A_6,$$

which completes the proof of (a). The remaining assertions are now easy to verify computationally by means of MAGMA. In particular, statement (e) follows from the table of representatives of the conjugacy classes given in Table 15.7.4. □

15.4 Embedding Janko's group J_1 into ON-type groups

In this section it is shown that each finite simple group G of ON-type with centralizer $C_G(z) = H = \langle c, r, s\rangle$ has a subgroup $J = \langle r, f, g\rangle$ isomorphic to the small Janko group J_1 of order $|\mathrm{J}_1| = 175560$, where g is the element of G defined in Proposition 15.2.2, and f is a suitable element of order 5 in H.

Lemma 15.4.1 *Let G be a finite simple group of ON-type with a 2-central involution z having centralizer $C_G(z) = H = \langle r, s, c\rangle$. Let $A = \langle z, t, u\rangle$ be the unique elementary abelian normal subgroup of the finite Sylow 2-subgroup S of H defined in Lemma 15.1.1. Let $g \in G$ be the* *element of order 3 defined in Proposition 15.2.2. Then the following assertions hold.*

(a) *The element $f = r^{-1}c^{-1}r^{-1}c^{-2}rc^{-1}r^{-1} \in H$ has order 5.*
(b) *$H_1 = \langle z, t, u, r, f\rangle$ is the only subgroup of H of shape $\langle z\rangle \times A_5$ containing the subgroup $\langle A, r\rangle$.*
(c) *H_1 has the following set $\mathcal{R}(H_1)$ of defining relations:*

$$t^2 = u^2 = r^3 = f^5 = (zt)^2 = (zu)^2 = (tu)^2 = 1,$$
$$zr^{-1}zr = tr^{-1}ur = zf^{-1}zf = (tf^{-1})^2 = 1,$$
$$r^{-1}utzru = f^{-1}r^{-1}utfr^{-1} = f^{-2}utf^{-1}r^{-1} = 1.$$

(d) *$E_1 = \langle r, g\rangle$ is the split extension of the elementary abelian group $A = \langle z, t, u\rangle$ with a Frobenius group $\mathsf{F}_{21} = \langle [r, r^g]\rangle : \langle r\rangle \cong 7 : 3$. Moreover, $C_A(r) = \langle z\rangle$ and $C_A(g) = \langle u\rangle$.*
(e) *E_1 has the following set $\mathcal{R}(E_1)$ of defining relations:*

$$t^2 = u^2 = r^3 = g^3 = (zt)^2 = (zu)^2 = (tu)^2 = 1,$$
$$zr^{-1}zr = tr^{-1}ur = tg^{-1}zg = ug^{-1}ug = 1,$$
$$g^{-1}tzgz = r^{-1}utzru = r^{-1}gr^{-1}zgrg^{-1} = 1.$$

Proof (a) and (b) have been checked by means of MAGMA and the faithful permutation representation of H described in Lemma 15.1.1(k).

(c) The presentation of H_1 has been calculated by means of MAGMA.

(d) $E_1 = \langle z, t, u, r, g\rangle \leq E = N_G(A) = \langle r, s, g\rangle$ by Proposition 15.2.2. Using the faithful permutation representation of E given in Proposition 15.2.2(f) and MAGMA it follows that E_1 is generated by r and g, and that $A = \langle z, t, u\rangle$ is a normal Sylow 2-subgroup of E_1 having a Frobenius group $\mathsf{F}_{21} = \langle r, r^g\rangle$ as complement, where

$$z = rg^{-1}rgr^{-1}g^{-1}, \qquad u = (rg^{-1})^2rgr, \qquad t = z^g.$$

The remaining statements are now trivial.

(e) The presentation of E_1 has been determined by means of MAGMA.

□

Proposition 15.4.2 *Each finite simple group G of ON-type has a simple subgroup J isomorphic to Janko's sporadic group J_1 of order $|\mathsf{J}_1| = 175560$. Furthermore, $J = \langle r, f, g\rangle$ has the following set $\mathcal{R}(J)$ of defining relations:*

$$f^5 = r^3 = g^3 = (rf)^3 = 1, \quad (f^2r^{-1})^2 = rgfr^{-1}f^{-1}grg = 1,$$
$$rg^{-1}r^{-1}frf^{-1}gr^{-1}g^{-1} = rgf^{-2}g^{-1}r^{-1}fgrf^{-1}g^{-1}f^{-1} = 1.$$

Proof Let z be a 2-central involution with centralizer $C_G(z) = H = \langle r, s, c\rangle$ defined in Lemma 15.1.1. Let S be the fixed Sylow 2-subgroup of H chosen in Lemma 15.1.1. Let $A = \langle z, t, u\rangle$ be the unique maximal elementary abelian normal subgroup of S. Proposition 15.2.2 asserts that there is an element $g \in N_G(A) - N_H(A)$ of order 3 such that $z^g = t$, $t^g = zt$ and $u^g = u$.

By Lemma 15.4.1(a) and (b) the element $f = r^{-1}c^{-1}r^{-1}c^{-2}rc^{-1}r^{-1} \in H$ has order 5 and the subgroup $H_1 = \langle z, t, u, r, f\rangle \cong \langle z\rangle \times A_5$. Let $\mathcal{R}(H_1)$ be the set of defining relations of H given in Lemma 15.4.1(c). Let $\mathcal{R}(E_1)$ be the set of defining relations of the subgroup $E_1 = \langle z, t, u, r, g\rangle \cong A : \mathsf{F}_{21}$ of G given in Lemma 15.4.1(e).

Keeping the notations of Proposition 15.3.2 we know that, in any group G of ON-type,

$$N := N_G(r) = [\langle r, r_1\rangle : \langle n_1\rangle] \times \langle l, w, f_1\rangle,$$

where $X = \langle l, w, f_1\rangle \cong A_6$ and f_1 is of order 5.

Using MAGMA and the faithful permutation representation of H given in Lemma 15.1.1(k) one can see that $h = (frs^2)^2 s^{12} \in H$ has order 8 and satisfies the equations $n_1^h = zt$ and $r^h = [r^2, f]$. Hence $n_1^{hg} = z$ by Proposition 15.2.2. Therefore

$$X^{hg} \leq C_G(n_1^{hg}) = C_G(z) = H,$$

$$N^{hg} = N_G(r^{hg}) = [\langle r^{hg}, r_1^{hg}\rangle : \langle z\rangle] \times X^{hg}.$$

Using MAGMA again we see that $w^h \in D = N_H(A)$. As $l^h = l$ it follows that the dihedral Sylow 2-subgroup $Y = \langle l, w\rangle$ of X is mapped under conjugation by hg to

$$Y^{hg} = \langle l, w^h\rangle^g \leq H \cap E = D$$

because $g \in E = N_G(A)$ by Proposition 15.2.2.

Another application of MAGMA then yields that there are exactly two subgroups $L_i \cong \langle z\rangle \times A_6$ of H containing Y^{hg}. Hence $X^{hg} \in \{L_1, L_2\}$. Furthermore it has been checked that $f \in L_1$ and $f^u \in L_2$ up to choosing the indices of the L_i. Hence exactly one of the following relations holds:

$$(*) \qquad [r^{hg}, f] = [[r^2, f]^g, f] = 1,$$

$$(**) \qquad [r^{hg}, f^u] = [[r^2, f]^g, f^u] = 1.$$

Using the Todd–Coxeter Algorithm implemented in MAGMA it has been checked in [94] that the subgroup $J = \langle H_1, E_1\rangle = \langle z, u, t, r, f, g\rangle$ of G having the set

$$\mathcal{R}(J) = \mathcal{R}(H_1) \cup \mathcal{R}(E_1) \cup (**)$$

as the defining relations with respect to its generating set $\{z, u, t, r, f, g\}$ is the identity group. This contradiction shows that $f \in L_1 = X^{hg}$. In particular, relation $(*)$ holds in J. Another application of the Todd–Coxeter Algorithm now shows that the subgroup $J = \langle H_1, E_1\rangle = \langle z, u, t, r, f, g\rangle$ of G having the set

$$\mathcal{R}(J) = \mathcal{R}(H_1) \cup \mathcal{R}(E_1) \cup (*)$$

as defining relations has a faithful permutation representation PJ of degree $|J : H_1| = 1463$ with stabilizer H_1. Using PJ and MAGMA one can verify that J is simple and $|J| = 175560$. Therefore $J \cong \mathsf{J}_1$ by Theorem 9.4.2 of [92]. The presentation of J given in the statement has been obtained in [94] by means of PJ and MAGMA. □

15.5 Existence and uniqueness proof

In this section the theorem of Previtali and the author [94] is proved. It provides a presentation for all finite simple groups of ON-type. Therefore this presentation yields an existence and uniqueness proof for O'Nan's simple group $G \cong \mathsf{ON}$. It follows that G has a faithful permutation representation of degree 2624832 with stabilizer J isomorphic to Janko's smallest sporadic simple group J_1.

Lemma 15.5.1 *Let G be a finite simple group of ON-type having a 2-central involution z with centralizer $C_G(z) = H$. Let S be the fixed Sylow 2-subgroup of H and let $A = \langle z, t, u\rangle$ be the unique maximal elementary abelian normal subgroup of S given in Lemma 15.1.1. Then $G = \langle H, N_G(A)\rangle$.*

Proof By Lemma 15.1.1 the Sylow 2-subgroup S of H has derived length 3. Thus a simple group G of ON-type cannot be isomorphic to $PSL_2(q)$, $Sz(q)$ or $PSU_3(q)$, where q is a power of 2. Therefore the Bender–Suzuki theorem stated as Theorem 1.7.3 in [92] asserts that G does not have a strongly embedded subgroup. By Proposition 15.2.1(a) G has a unique conjugacy class of involutions. Hence Corollary 4.8.10 of [92] implies that $G = \langle H, E\rangle$, which completes the proof. □

Theorem 15.5.2 *Let G be a finite simple group of* ON*-type with a 2-central involution z having centralizer $C_G(z) = H = \langle r, s, c\rangle$ and fixed Sylow 2-subgroup $S \le H$ defined in Lemma 15.1.1. Let $A = \langle z, t, u\rangle$ be the unique maximal elementary abelian normal subgroup of S. Then the following assertions hold.*

(a) *There exists an element $g \in N_G(A) - N_H(A)$ of order 3 such that $z^g = t$, $t^g = zt$ and $u^g = u$, and $E = N_G(A) = \langle r, s, g\rangle$.*

(b) *The elements $f = r^{-1}c^{-1}r^{-1}c^{-2}rc^{-1}r^{-1}$ and $c = s^2fr^2frs^2r^2$ of H have order 5, and*

$$H = \langle r, s, c\rangle = \langle r, s, f\rangle.$$

(c) *$G = \langle r, s, f, g\rangle$ has the set*

$$\mathcal{R}(G) = \mathcal{R}_1(H) \cup \mathcal{R}(J) \cup \mathcal{R}(E)$$

of defining relations, where $H = \langle r, s, f \mid \mathcal{R}_1(H)\rangle$. Leaving out repetitions, $\mathcal{R}(G)$ consists of the following relations:

$$\begin{aligned}
&r^3 = s^{16} = f^5 = g^3 = 1,\\
&r^{-1}g^{-1}f^{-1}rgf^2g^{-1}r^{-1}fgf = s^{-1}r^{-1}gr^{-1}s^{-1}rsgr^{-1}g^{-1} = 1,\\
&frf^{-1}s^{-2}r^{-1}s^2r^{-1}s^{-2} = (r^{-1}g^{-1}r^{-1}s^{-1})^2 = 1,\\
&s^{-1}f^{-1}s^4f^{-1}s^{-3} = 1,\\
&(rf^{-1}sf^{-1}s^{-1})^2 = (r^{-1}s^2rs^{-1})^2,\\
&f^{-2}gf^{-1}g^{-1}f^2r^{-1}gfg^{-1}r = 1,\\
&(gr^{-1}s^{-1})^3 = (sg^{-1}r^{-1}g^{-1})^2 = (sgs^2)^2 = (s^{-1}f^{-1})^4 = 1,\\
&r^{-1}sfrs^{-1}r^{-1}sf^2s^{-1}f^{-1} = (f^{-1}r^{-1})^3,\\
&(rf^{-2})^2 = (sf^{-1})^4 = sr^{-1}sr^{-1}g^{-1}s^{-1}g^{-1} = 1\\
&(fr^{-1}f)^2 = (rs^2r^{-1}s^{-1})^2 = 1,\\
&s^{-2}g^{-1}r^{-1}s^{-1}rs^{-2}rg = fr^{-1}f^{-1}r^{-1}s^{-1}r^{-1}fr^{-1}f^{-1}s = 1,\\
&rgfr^{-1}f^{-1}grg = rg^{-1}fr^{-1}f^{-1}rgr^{-1}g = 1.
\end{aligned}$$

(d) *The simple subgroup $J = \langle r, f, g\rangle$ of order 175560 has index $|G : J| = 2624832$.*

(e) *G is uniquely determined by H up to isomorphism and has order*

$$|G| = 2^9 \cdot 3^4 \cdot 5 \cdot 7^3 \cdot 11 \cdot 19 \cdot 31.$$

(f) *The finitely presented group* $G = \langle r, s, f, g \rangle$ *with set* $\mathcal{R}(G)$ *of defining relations is a simple group with 2-central involution* $z = s^8$ *such that* $C_G(z) = \langle r, s, f \rangle \cong H$.

Proof Assertion (a) holds by Proposition 15.2.2(a) and (d).

(b) Lemma 15.4.1(a) states that $f = r^{-1}c^{-1}r^{-1}c^{-2}rc^{-1}r^{-1} \in H$ has order 5. Using MAGMA and the faithful permutation representation of H given in Lemma 15.1.1(k) it has been checked that

$$H = \langle r, s, c \rangle = \langle r, s, f \rangle$$

and that $c = s^2fr^2frs^2r^2$.

(c) Furthermore, Previtali and the author verified in [94] that $H = \langle r, s, f \rangle$ has the following set $\mathcal{R}_1(H)$ of defining relations:

$$\begin{aligned}
&r^3 = s^{16} = f^5 = 1,\\
&(f^{-1}r^{-1})^3 = (rf^{-2})^2 = (s^{-1}f^{-1})^4 = (sf^{-1})^4 = 1,\\
&(rs^2r^{-1}s^{-1})^2 = fr^{-1}f^{-1}r^{-1}s^{-1}r^{-1}fr^{-1}f^{-1}s = 1,\\
&(r^{-1}s^2rs^{-1})^2 = (rf^{-1}sf^{-1}s^{-1})^2 = 1,\\
&s^{-1}f^{-1}s^4f^{-1}s^{-3} = r^{-1}sfrs^{-1}r^{-1}sf^2s^{-1}f^{-1} = 1,\\
&rf^{-1}s^{-2}r^{-1}s^2r^{-1}s^{-2} = 1.
\end{aligned}$$

By Proposition 15.4.2 the subgroup $J = \langle r, f, g \rangle$ has the following set $\mathcal{R}(J)$ of defining relations:

$$\begin{aligned}
&f^5 = r^3 = g^3 = (fr^{-1}f)^2 = rgfr^{-1}f^{-1}grg = rg^{-1}fr^{-1}f^{-1}rgr^{-1}g = 1,\\
&f^{-2}gf^{-1}g^{-1}f^2r^{-1}gfg^{-1}r = r^{-1}g^{-1}f^{-1}rgf^2g^{-1}r^{-1}fgf = 1.
\end{aligned}$$

Proposition 15.2.2 asserts that $E = N_G(A) = \langle r, s, g \rangle$ has the following set $\mathcal{R}(E)$ of defining relations:

$$\begin{aligned}
&r^3 = g^3 = s^{16} = 1,\\
&sr^{-1}sr^{-1}g^{-1}s^{-1}g^{-1} = (r^{-1}g^{-1}r^{-1}s^{-1})^2 = (sgs^2)^2 = 1,\\
&(sg^{-1}r^{-1}g^{-1})^2 = (gr^{-1}s^{-1})^3 = (rs^2r^{-1}s^{-1})^2 = 1,\\
&s^{-2}g^{-1}r^{-1}s^{-1}rs^{-2}rg = s^{-1}r^{-1}gr^{-1}s^{-1}rsgr^{-1}g^{-1} = 1.
\end{aligned}$$

Hence $G_1 = \langle H, J, E \rangle = \langle r, s, c, f, g \rangle = \langle r, s, f, g \rangle$ is a subgroup of G with the set

$$\mathcal{R}(G_1) = \mathcal{R}_1(H) \cup \mathcal{R}(J) \cup \mathcal{R}(E)$$

as a set of defining relations. As J is a subgroup of G_1, Previtali and the author applied in [94] the Todd–Coxeter algorithm implemented in MAGMA. It returned that

$$|G_1 : J| = 2624832.$$

Hence $\mathcal{R}(G_1)$ is a set of defining relation of G_1. Now, Lemma 15.5.1 asserts that

$$G = \langle H, E\rangle \le \langle H, J, E\rangle = G_1,$$

which completes the proof of (c).

Assertions (d) and (e) are immediate consequences of Proposition 15.4.2 and statement (c).

(f) Let $G = \langle r, s, f, g\rangle$ be the finitely presented group with set $\mathcal{R}(G)$ of defining relations. By (b), (e) and Lemma 15.1.1 $z = s^8$ is a 2-central involution with centralizer $C_G(z) \ge \langle r, s, f\rangle \cong H$. Using the faithful permutation representation of G with stabilizer $J = \langle r, f, g\rangle$ and MAGMA we can check that z has 1344 fixed points on the permutation module $(1_J)^G$. By Propositions 15.4.2 and 15.2.1 both groups J and G have a unique conjugacy class of involutions represented by z, respectively. Hence Proposition 2.6.6 of [92] asserts that $1344 = C_G(z)/C_J(z)$. Thus $|C_G(z)| = 161280$ and $C_G(z) = \langle r, s, f\rangle \cong H$.

Let $X \ne 1$ be a proper normal subgroup of G. If $|X|$ is even, then $z^G \subseteq X$ by Proposition 15.2.1. As $H = C_G(z) = \langle r, s, f\rangle$ we can apply all statements of Lemma 15.1.1. Therefore $y = (sc^2)^3 \in H - H'$ by Lemma 15.1.1(l). Using the faithful permutation representation of H of degree 448 and MAGMA one can verify that $H = \langle y^H\rangle$. Hence $H \le X$. In particular, $r \in X$. By Proposition 15.2.2 we know that $E = \langle r, s, g\rangle$ has a unique conjugacy class r^E of elements of order 3. Hence $g \in r^E \le X$ and $G = \langle H, g\rangle = X$. This contradiction shows that $|X|$ is odd. By Lemma 15.1.1(l) $W = \langle z, t\rangle$, with $t = (rs)^4$, is a Klein 4-group and $C_X(z) = X \cap C_G(z) = 1$. As W acts on X by conjugation Theorem 4.2.2 of [92] asserts that

$$|X||C_X(W)|^2 = |C_X(z)||C_X(t)||C_X(zt)| = |C_X(z)|^3$$

because all involutions are conjugate in G. As $C_X(W) \le X \cap C_H(t) = 1$ it follows that $|X| = |H \cap X|^3 \le O(H) = 1$. This completes the proof. □

Lemma 15.5.3 *The normalizers $N_G(p)$ of the Sylow p-subgroups of G for the primes* 11, 19 *and* 31 *are Frobenius groups of orders* $11 \cdot 10$, $19 \cdot 6$ *and* $31 \cdot 15$, *respectively.*

Proof Using Theorem 15.5.2 and Sylow's Theorem it is easy to get the orders of $N_G(p)$. □

15.6 Local subgroups, fusion and character table

Using the faithful permutation representation PG of the finitely presented simple group G constructed in Theorem 15.5.2 of degree 2624832 Previtali and the author constructed in [94] another permutation representation PLG of G of degree 122760 with stabilizer $L \cong \mathrm{PSL}_3(7) : 2$. Applying MAGMA to it one gets, fairly easily, a system of representatives of all the conjugacy classes and the character table of G. Moreover, all results of O'Nan [104], and of Previtali and the author [94], on the normalizers of elements of odd prime order of G, and the fusion of the classes of their classes, and the ones of H into the conjugacy classes of G, are deduced from this small permutation representation PLG of G in this section.

Corollary 15.6.1 *Let $G = \langle r, s, f, g\rangle = \langle c, r, s, f, g\rangle$ be the finitely presented group with the set $\mathcal{R}(G)$ of defining relations given in Theorem 15.5.2. Then the following statements hold.*

(a) *G has a faithful permutation representation PLG of degree 122760 with stabilizer $L = \langle p_1, p_2, n_1, x_1\rangle$, where $p_1 = rs^3c^{-1}r^{-1}s^{-1}c^{-1}$, $p_2 = s^2c^{-1}rc^2s^{-1}$, $n_1 = (s^{-1}c[c^2, s])^2$ and $x_1 = [r^{-1}, f]^{g^2(rfr)^{-1}}$.*

(b) *Table 15.7.8 contains a complete set of representatives of the 30 conjugacy classes of G which are given in terms of short words in the generators of G. Furthermore, their power maps and the orders of their centralizers are also stated.*

(c) *The character table of G coincides with the one for* ON *stated in the Atlas [19], p. 133.*

(d) $L \cong \mathrm{PSL}_3(7) : 2$.

(e) *G has a unique conjugacy class of elements r of order 7 which is of highest 7-defect. $R = N_G(7_A) = N_G(\langle r\rangle) = \langle a, d\rangle$, where $a = p_1p_2p_1x_1^{-1}p_1x_1^{-1}n_1p_1p_2^{-1}x_1$ and $d = p_2^{-1}x_1^{-1}p_1p_2^{-1}p_1x_1^{-1}p_1n_1p_2^{-1}x_1$.*

(f) *A system of representatives r_i and corresponding centralizer orders $|C_R(r_i)|$ of the 24 conjugacy classes r_i^R of R is given in Table 15.7.6.*

(g) *G has two conjugacy classes of order 7 represented by r and $s = dada^5$, where $|C_G(s)| = 49$.*

(h) *The character table of R is given in Table 15.8.6.*

Proof (a) By Theorem 15.5.2 $G = \langle r, s, f, g\rangle$ has a faithful permutation representation PG of degree 2624832 with stabilizer $J = \langle r, f, g\rangle$.

Using PG and MAGMA the reader can verify that the subgroup $L = \langle p_1, p_2, x_1, n_1 \rangle$ of G has index 122760. Since G is simple the cosets of L in G provide a faithful permutation representation PLG of G.

(b) The representatives x of the conjugacy classes of G and their centralizer orders $|C_G(x)|$ given in Table 15.7.8 have been calculated by means of PLG, MAGMA and Kratzer's Algorithm 5.3.18 of [92].

(c) The character table of G can now be calculated by means of PLG and MAGMA. It has been checked that it coincides with the one of ON given in the Atlas; see also [94] for another proof.

(d) By (a) and Theorem 15.5.2 $|L| = 2^6 \cdot 3^2 \cdot 7^3 \cdot 19$. The statement has been verified using PLG and the MAGMA command `CompositionFactors(L)`.

(e) Using PLG and MAGMA it follows that $N_L(d^4) = \langle a, d \rangle = R \cong N_L(7_A) \cong 7^{1+2} : (3 \times D_8)$. By Theorem 15.5.2(e) and Sylow's Theorem we have $N_G(7_A) = N_L(7_A)$.

The tables in (f) and (h) have been calculated by means of the permutation representation of PLG, MAGMA and Kratzer's Algorithm mentioned above.

(g) This assertion holds by Tables 15.7.8 and 15.7.6. □

Proposition 15.6.2 *Let $G = \langle r, s, f, g \rangle = \langle c, r, s, f, g \rangle$ be the finitely presented group with the set $\mathcal{R}(G)$ of defining relations given in Theorem 15.5.2. Then the following assertions hold.*

(a) *The normalizer $Y = N_G(\langle f \rangle) = \langle a_1, d_1, q_1 \rangle$ of the cyclic subgroup $\langle f \rangle$ of order 5 is generated by the elements $a_1 = rfr^{-1}s^{-4}r$, $d_1 = g^{-1}rf^{-1}r^{-1}fg$ and $q_1 = s^{-1}grs^2r^{-1}g$. It has the following set $\mathcal{R}(Y)$ of defining relations:*

$$d_1^3 = q_1^4 = 1,$$
$$q_1 a_1^3 q_1^{-1} a_1^{-1} = a_1 d_1 a_1^{-2} d_1 a_1 = 1,$$
$$a_1 q_1^{-2} a_1^2 q_1^2 a_1 = d_1 q_1^{-2} d_1 q_1^2 = 1.$$

(b) *G has a unique conjugacy class of elements of order 5.*

(c) *A system of representatives y_i and corresponding centralizer orders $|C_Y(y_i)|$ of the 18 conjugacy classes y_i^Y of Y is given in Table 15.7.5.*

Proof (b) holds by Table 15.7.8.

(a) and (c) These two assertions have been proved by means of the faithful permutation representation PLG constructed in Corollary 15.6.1, MAGMA and Kratzer's Algorithm 5.3.18 of [92]. □

Corollary 15.6.3 *Let $G = \langle r, s, f, g \rangle$ be the finitely presented simple group with set $\mathcal{R}(G)$ of defining relations given in Theorem 15.5.2. Representatives of the conjugacy classes of the local subgroups $H = C_G(z)$, $N = N_G(3_A)$, $Y = N_G(5_A)$, $R = N_G(7_A)$ and the subgroup $J = \langle r, s, g \rangle$ of G are given in Tables 15.7.1, 15.7.4, 15.7.5, 15.7.6 and 15.7.7, respectively. Using their abbreviations for the conjugacy classes of these subgroups the fusion of the conjugacy classes of these six subgroups of G into the 30 conjugacy classes of G is given in the following table:*

x	$\|C_G(x)\|$	$C_G(2_A)$	$N_G(3_A)$	$N_G(5_A)$	$N_G(7_A)$	J
1_A	$\|G\|$	a	a	a	a	a
2_A	$2^9 \cdot 3^2 \cdot 5 \cdot 7$	a, b, c	a, b, c	a, b, c	a, b, c	a
3_A	$2^3 \cdot 3^4 \cdot 5$	a	$a, \ldots, n$	a	a, b	a
4_A	$2^8 \cdot 3^2 \cdot 5 \cdot 7$	a, d	a	a, b	a	
4_B	2^8	b, c, e, f	b	c, d, e, f		
5_A	$2^2 \cdot 3^2 \cdot 5$	a	a, b	a		a, b
6_A	$2^3 \cdot 3^2$	a, b, c	$a, \ldots, f$	a	$a, \ldots, f$	
7_A	$2^2 \cdot 7^3$	a			a, b, c	a
7_B	7^2				d	
8_A	2^5	a, d				
8_B	2^5	b, c				
10_A	$2^5 \cdot 5$	a	a, b	a		a, b
11_A	11					a
12_A	$2^2 \cdot 3^2$	a	a, b, c, d		a, b	
14_A	$2^2 \cdot 7$	a			a, b, c	
15_A	$3^2 \cdot 5$		a, c, e, g	a		a
15_B	$3^2 \cdot 5$		b, d, f, h	b		b
16_A	2^4	a				
16_B	2^4	b				
16_C	2^4	c				
16_D	2^4	d				
19_A	19	a				a
19_B	19	b				b
19_C	19	c				c
20_A	$2^2 \cdot 5$	a		a		
20_B	$2^2 \cdot 5$	b		b		
28_A	$2^2 \cdot 7$	a			a, b	
28_B	$2^2 \cdot 7$	b				
31_A	31					
31_B	31					

Proof The fusion of the conjugacy classes of H and N is given in the previous sections. The fusion of the classes of the other subgroups has been determined computationally by means of MAGMA and the faithful permutation representation of G constructed in Proposition 15.6.2. □

15.7 Representatives of conjugacy classes

15.7.1 *Conjugacy classes of* $H = C_G(z) = \langle c, r, s\rangle$

i	h_i	$\lvert C_H(h_i)\rvert$	2P	3P	5P	7P
1_a	1	$2^9 \cdot 3^2 \cdot 5 \cdot 7$	1_a	1_a	1_a	1_a
2_a	$(s)^8$	$2^9 \cdot 3^2 \cdot 5 \cdot 7$	1_a	2_a	2_a	2_a
2_b	$(rs)^4$	2^8	1_a	2_b	2_b	2_b
2_c	$(sc^2)^3$	$2^4 \cdot 3^2$	1_a	2_c	2_c	2_c
3_a	r	$2^3 \cdot 3^2$	3_a	1_a	3_a	3_a
4_a	$(rc)^3$	$2^8 \cdot 3^2 \cdot 5 \cdot 7$	2_a	4_a	4_a	4_a
4_b	$(s)^4$	2^8	2_a	4_b	4_b	4_b
4_c	$(rs)^2$	2^7	2_b	4_c	4_c	4_c
4_d	$rcscs$	2^7	2_b	4_d	4_d	4_d
4_e	$rsrsrc$	2^6	2_b	4_e	4_e	4_e
4_f	rcs	2^4	2_b	4_f	4_f	4_f
5_a	c	$2^2 \cdot 5$	5	5	1	5
6_a	$(rc)^2$	$2^3 \cdot 3^2$	3	2_a	6_a	6_a
6_b	sc^2	$2^2 \cdot 3^2$	3	2_c	6_b	6_b
6_c	rsc^2	$2^2 \cdot 3^2$	3	2_c	6_c	6_c
7_a	$(r^2c)^2$	$2^2 \cdot 7$	7_a	7_a	7_a	1_a
8_a	$(s)^2$	2^5	4_b	8_a	8_a	8_a
8_b	$(sc)^2$	2^5	4_b	8_b	8_b	8_b
8_c	rs	2^4	4_c	8_c	8_c	8_c
8_d	rsc	2^4	4_c	8_d	8_d	8_d
10_a	$(s^2c^2)^2$	$2^2 \cdot 5$	5_a	10_a	2_a	10_a
12_a	rc	$2^2 \cdot 3^2$	6_a	4_a	12_a	12_a
14_a	r^2c	$2^2 \cdot 7$	7_a	14_a	14_a	2_a
16_a	s	2^4	8_a	16_b	16_b	16_a
16_b	$(s)^3$	2^4	8_a	16_a	16_a	16_b
16_c	sc	2^4	8_b	16_d	16_d	16_c
16_d	$(sc)^3$	2^4	8_b	16_c	16_c	16_d
20_a	s^2c^2	$2^2 \cdot 5$	10_a	20_a	4_a	20_a
20_b	$(s^2c^2)^{11}$	$2^2 \cdot 5$	10_a	20_b	4_a	20_b
28_a	rc^2	$2^2 \cdot 7$	14_a	28_a	28_b	4_a
28_b	$(rc^2)^5$	$2^2 \cdot 7$	14_a	28_b	28_a	4_a

15.7.2 *Conjugacy classes of* $D = N_H(A) = \langle r, s\rangle$

i	d_i	$\lvert C_D(d_i)\rvert$
1_a	1	$2^9 \cdot 3$
2_a	$(s)^8$	$2^9 \cdot 3$
2_b	$(rs)^4$	2^8
2_c	$rsrs^3$	2^5
2_d	$r^2s^2rs^3$	2^4
3_a	r	$2^2 \cdot 3$
4_a	$(rs^2)^3$	$2^8 \cdot 3$
4_b	$(s)^4$	2^8
4_c	$(rs)^2$	2^7
4_d	$r^2sr^2srs^3rs$	2^7
4_e	r^2srs^2rs	2^6
4_f	$r^2sr^2s^3$	2^5

i	d_i	$\lvert C_D(d_i)\rvert$
4_g	r^2srs^2	2^4
6_a	$(rs^2)^2$	$2^2 \cdot 3$
8_a	$(s)^2$	2^5
8_b	$(r^2s)^2$	2^5
8_c	rs	2^4
8_d	r^2s^3	2^4
12_a	rs^2	$2^2 \cdot 3$
12_b	$(rs^2)^5$	$2^2 \cdot 3$
16_a	s	2^4
16_b	$(s)^3$	2^4
16_c	r^2s	2^4
16_d	$(r^2s)^3$	2^4

15.7.3 *Conjugacy classes of* $E = N_G(A) = \langle r, s, g \rangle$

i	e_i	$\vert C_E(e_i)\vert$
1_a	1	$2^9 \cdot 3 \cdot 7$
2_a	$(s)^8$	$2^9 \cdot 3$
2_b	s^2gsg	2^5
3_a	r	$2^2 \cdot 3$
4_a	$(rg)^3$	$2^8 \cdot 3$
4_b	$(s)^4$	2^8
4_c	$rsrg$	2^5
6_a	$(rg)^2$	$2^2 \cdot 3$
7_a	r^2g	7

i	e_i	$\vert C_E(e_i)\vert$
7_b	$(r^2g)^3$	7
8_a	$(s)^2$	2^5
8_b	$(r^2s)^2$	2^5
12_a	rg	$2^2 \cdot 3$
12_b	$(rg)^5$	$2^2 \cdot 3$
16_a	s	2^4
16_b	$(s)^3$	2^4
16_c	r^2s	2^4
16_d	$(r^2s)^3$	2^4

15.7.4 *Conjugacy classes of* $N = N_G(3_A) = N_G(r) = \langle r, r_1, n_1, l, w, f_1 \rangle$

i	n_i	$\vert C_{N_1}(n_i)\vert$	2P	3P	5P
1_a	1	$2^4 \cdot 3^4 \cdot 5$	1_a	1_a	1_a
2_a	n_1	$2^4 \cdot 3^2 \cdot 5$	1_a	2_a	2_a
2_b	$(l)^2$	$2^4 \cdot 3^2$	1	2_b	2_b
2_c	n_1w	2^4	1	2_c	2_c
3_a	r	$2^3 \cdot 3^4 \cdot 5$	3_a	1_a	3_a
3_b	r_1	$2^3 \cdot 3^4 \cdot 5$	3_b	1_a	3_b
3_c	rr_1	$2^3 \cdot 3^4 \cdot 5$	3_c	1_a	3_c
3_d	r^2r_1	$2^3 \cdot 3^4 \cdot 5$	3_d	1_a	3_d
3_e	lf_1	$2 \cdot 3^4$	3_e	1_a	3_e
3_f	l^2f_1	$2 \cdot 3^4$	3_{f_1}	1	3_{f_1}
3_g	rlf_1	3^4	3_g	1_a	3_g
3_h	r_1lf_1	3^4	3_h	1_a	3_h
3_i	rr_1lf_1	3^4	3_i	1_a	3_i
3_j	rl^2f_1	3^4	3_j	1_a	3_j
3_k	$r_1l^2f_1$	3^4	3_k	1_a	3_k
3_l	$r^2r_1lf_1$	3^4	3_l	1_a	3_l
3_m	$rr_1l^2f_1$	3^4	3_m	1_a	3_m
3_n	$r^2r_1l^2f_1$	3^4	3_n	1_a	3_n
4_a	l	$2^3 \cdot 3^2$	2_b	4_a	4_a
4_b	n_1l	2^3	2_b	4_b	4_b
5_a	f_1	$2 \cdot 3^2 \cdot 5$	5_b	5_b	1
5_b	$(f_1)^2$	$2 \cdot 3^2 \cdot 5$	5_a	5_a	1
6_a	$(rl)^2$	$2^3 \cdot 3^2$	3_a	2_b	6_a
6_b	$(r_1l)^2$	$2^3 \cdot 3^2$	3_b	2_b	6_b
6_c	$(rr_1l)^2$	$2^3 \cdot 3^2$	3_c	2_b	6_c
6_d	$(r^2r_1l)^2$	$2^3 \cdot 3^2$	3_d	2_b	6_d
6_e	n_1lf_1	$2 \cdot 3^2$	3_e	2_a	6_e
6_{f_1}	$n_1l^2f_1$	$2 \cdot 3^2$	3_{f_1}	2_a	6_{f_1}
10_a	n_1f_1	$2 \cdot 5$	5_b	10_b	2_a
10_b	$(n_1f_1)^3$	$2 \cdot 5$	5_a	10_a	2_a
12_a	rl	$2^2 \cdot 3^2$	6_a	4_a	12_a
12_b	r_1l	$2^2 \cdot 3^2$	6_b	4_a	12_b
12_c	rr_1l	$2^2 \cdot 3^2$	6_c	4_a	12_c
12_d	r^2r_1l	$2^2 \cdot 3^2$	6_d	4_a	12_d
15_a	rf_1	$3^2 \cdot 5$	15_b	5_b	3_a
15_b	$(rf_1)^2$	$3^2 \cdot 5$	15_a	5_a	3_a
15_c	r_1f_1	$3^2 \cdot 5$	15_d	5_b	3_b
15_d	$(r_1f_1)^2$	$3^2 \cdot 5$	15_c	5_a	3_b
15_e	rr_1f_1	$3^2 \cdot 5$	15_{f_1}	5_b	3_c
15_f	$(rr_1f_1)^2$	$3^2 \cdot 5$	15_e	5_a	3_c
15_g	$r^2r_1f_1$	$3^2 \cdot 5$	15_h	5_b	3_d
15_h	$(r^2r_1f_1)^2$	$3^2 \cdot 5$	15_g	5_a	3_d

15.7.5 *Conjugacy classes of* $Y = N_G(5_A) = \langle a_1, d_1, q_1 \rangle$

i	y_i	$\lvert C_Y(y_i)\rvert$	2P	3P	5P
1	1	$2^4 \cdot 3^2 \cdot 5$	1	1	1
2_a	$a_1^2 q_1^2$	$2^4 \cdot 3^2$	1	2_a	2_a
2_b	$(a_1)^{10}$	$2^4 \cdot 5$	1	2_b	2_b
2_c	$(q_1)^2$	2^4	1	2_c	2_c
3	d_1	$2 \cdot 3^2 \cdot 5$	3	1	3
4_a	$(a_1)^5$	$2^3 \cdot 5$	2_b	4_a	4_a
4_b	q_1	2^3	2_c	4_c	4_b
4_c	$(q_1)^3$	2^3	2_c	4_b	4_c
4_d	$a_1 q_1$	2^3	2_c	4_e	4_d
4_e	$(a_1 q_1)^3$	2^3	2_c	4_d	4_e
4_f	$a_1 q_1^2$	2^3	2_b	4_f	4_f
5	$(a_1)^4$	$2^2 \cdot 3^2 \cdot 5$	5	5	1
6	$a_1^2 d_1 q_1^2$	$2 \cdot 3^2$	3	2_a	6
10	$(a_1)^2$	$2^2 \cdot 5$	5	10	2_b
15_a	$a_1^4 d_1$	$3^2 \cdot 5$	15_b	5	3
15_b	$(a_1^4 d_1)^2$	$3^2 \cdot 5$	15_a	5	3
20_a	a_1	$2^2 \cdot 5$	10	20_a	4_a
20_b	$(a_1)^{11}$	$2^2 \cdot 5$	10	20_b	4_a

where the generators a_1, d_1 *and* q_1 *have orders* 20, 3 *and* 4, *respectively.*

15.7.6 *Conjugacy classes of* $N_G(7_A) \simeq R = \langle d, a \rangle$

i	r_i	$\lvert C_R(r_i)\rvert$	2P	3P	7P
1_a	1	$2^3 \cdot 3 \cdot 7^3$	1_a	1_a	1_a
2_a	$(d)^{14}$	$2^3 \cdot 3 \cdot 7$	1_a	2_a	2_a
2_b	$(a)^3$	$2^2 \cdot 3 \cdot 7$	1_a	2_b	2_b
2_c	$(da)^3$	$2^2 \cdot 3 \cdot 7$	1_a	2_c	2_c
3_a	$(a)^2$	$2^3 \cdot 3$	3_b	1_a	3_a
3_b	$(a)^4$	$2^3 \cdot 3$	3_a	1_a	3_b
4_a	$(d)^7$	$2^2 \cdot 3 \cdot 7$	2_a	4_a	4_a
6_a	$(da^2)^2$	$2^3 \cdot 3$	3_a	2_a	6_a
6_b	$(da^2)^{10}$	$2^3 \cdot 3$	3_b	2_a	6_b
6_c	a	$2^2 \cdot 3$	3_a	2_b	6_c
6_d	$(a)^5$	$2^2 \cdot 3$	3_b	2_b	6_d
6_e	da	$2^2 \cdot 3$	3_a	2_c	6_e
6_f	$(da)^5$	$2^2 \cdot 3$	3_b	2_c	6_f
7_a	$(d)^4$	$2^2 \cdot 7^3$	7_a	7_a	1_a
7_b	$(da^3)^2$	$2 \cdot 7^2$	7_b	7_b	1_a
7_c	$(d^2a^3)^2$	$2 \cdot 7^2$	7_c	7_c	1_a
7_d	$dada^5$	7^2	7_d	7_d	1_a
12_a	da^2	$2^2 \cdot 3$	6_a	4_a	12_a
12_b	$(da^2)^5$	$2^2 \cdot 3$	6_b	4_a	12_b
14_a	$(d)^2$	$2^2 \cdot 7$	7_a	14_a	2_a
14_b	da^3	$2 \cdot 7$	7_b	14_b	2_c
14_c	d^2a^3	$2 \cdot 7$	7_c	14_c	2_b
28_a	d	$2^2 \cdot 7$	14_a	28_a	4_a
28_b	$(d)^5$	$2^2 \cdot 7$	14_a	28_b	4_a

15.7.7 *Conjugacy classes of Janko subgroup* $J = \langle z, t, u, f, r, g \rangle$

i	j_i	$\lvert C_J(j_i)\rvert$	2P	3P	5P	7P	11P	19P
1_a	1	$2^3 \cdot 3 \cdot 5 \cdot 7 \cdot 11 \cdot 19$	1_a	1_a	1_a	1_a	1_a	1_a
2_a	z	$2^3 \cdot 3 \cdot 5$	1_a	2_a	2_a	2_a	2_a	2_a
3_a	r	$2 \cdot 3 \cdot 5$	3_a	1_a	3_a	3_a	3_a	3_a
5_a	f	$2 \cdot 3 \cdot 5$	5_b	5_b	1_a	5_b	5_a	5_a
5_b	$(f)^2$	$2 \cdot 3 \cdot 5$	5_a	5_a	1_a	5_a	5_b	5_b
6_a	zr	$2 \cdot 3$	3_a	2_a	6_a	6_a	6_a	6_a
7_a	r^2g	7	7_a	7_a	7_a	1_a	7_a	7_a
10_a	zf	$2 \cdot 5$	5_b	10_b	2	10_b	10_a	10_a
10_b	$(zf)^3$	$2 \cdot 5$	5_a	10_a	2_a	10_a	10_b	10_b
11_a	frg	11_a	11_a	11_a	11_a	11_a	1_a	11_a
15_a	zfg	$3 \cdot 5$	15_b	5_b	3_a	15_b	15_a	15_a
15_b	$(zfg)^2$	$3 \cdot 5$	15_a	5_a	3_a	15_a	15_b	15_b
19_a	$trgf$	19	19_b	19_b	19_b	19_a	19_a	1_a
19_b	$(trgf)^2$	19	19_c	19_c	19_c	19_b	19_b	1_a
19_c	$(trgf)^4$	19	19_a	19_a	19_a	19_c	19_c	1_a

15.7.8 *Conjugacy classes of* $\mathsf{ON} \simeq G = \langle r, s, f, g \rangle$

i	g_i	$\lvert g_i^G \rvert$	$\lvert C_G(g_i) \rvert$	2P	3P	5P	7P	11P	19P	31P
1	1	1	$2^9 \cdot 3^4 \cdot 5 \cdot 7^3 \cdot 11 \cdot 19 \cdot 31$	1	1	1	1	1	1	1
2	s^8	2857239	$2^9 \cdot 3^2 \cdot 5 \cdot 7$	1	2	2	2	2	2	2
3	r	142227008	$2^3 \cdot 3^4 \cdot 5$	3	1	3	3	3	3	3
4_a	$(rs^2)^3$	5714478	$2^8 \cdot 3^2 \cdot 5 \cdot 7$	2	4_a	4_a	4_a	4_a	4_a	4_a
4_b	s^4	1800060570	2^8	2	4_b	4_b	4_b	4_b	4_b	4_b
5	f	2560086144	$2^2 \cdot 3^2 \cdot 5$	5	5	1	5	5	5	5
6	rg	6400215360	$2^3 \cdot 3^2$	3	2	6	6	6	6	6
7_a	$(rs^2f)^2$	335871360	$2^2 \cdot 7^3$	7_a	7_a	7_a	1	7_a	7_a	7_a
7_b	r^2g	9404398080	7^2	7_b	7_b	7_b	1	7_b	7_b	7_b
8_a	s^2	14400484560	2^5	4_b	8_a	8_a	8_a	8_a	8_a	8_a
8_b	rs	14400484560	2^5	4_b	8_b	8_b	8_b	8_b	8_b	8_b
10	fg	23040775296	$2^2 \cdot 5$	5	10	2	10	10	10	10
11	rgf	41892318720	11	11	11	11	11	1	11	11
12	rs^2	12800430720	$2^2 \cdot 3^2$	6	4_a	12	12	12	12	12
14	rs^2f	16457696640	$2^2 \cdot 7$	7_a	14	14	2	14	14	14
15_a	fg^2	10240344576	$3^2 \cdot 5$	15_b	5	3	15_b	15_a	15_a	15_a
15_b	$(fg^2)^2$	10240344576	$3^2 \cdot 5$	15_a	5	3	15_a	15_b	15_b	15_b
16_a	s	28800969120	2^4	8_a	16_b	16_b	16_a	16_b	16_b	16_a
16_b	s^3	28800969120	2^4	8_a	16_a	16_a	16_b	16_a	16_a	16_b
16_c	r^2s	28800969120	2^4	8_b	16_d	16_d	16_c	16_d	16_d	16_c
16_d	$(r^2s)^3$	28800969120	2^4	8_b	16_c	16_c	16_d	16_c	16_c	16_d
19_a	r^2fg	24253447680	19	19_b	19_b	19_b	19_a	19_a	1	19_a
19_b	$(r^2fg)^2$	24253447680	19	19_c	19_c	19_c	19_b	19_b	1	19_b
19_c	$(r^2fg)^4$	24253447680	19	19_a	19_a	19_a	19_c	19_c	1	19_c
20_a	s^2f	23040775296	$2^2 \cdot 5$	10	20_a	4_a	20_a	20_b	20_b	20_b
20_b	$(s^2f)^{11}$	23040775296	$2^2 \cdot 5$	10	20_b	4_a	20_b	20_a	20_a	20_a
28_a	$rsfg$	16457696640	$2^2 \cdot 7$	14	28_a	28_b	4_a	28_b	28_a	28_a
28_b	$(rsfg)^5$	16457696640	$2^2 \cdot 7$	14	28_b	28_a	4_a	28_a	28_b	28_b
31_a	sfg	14865016320	31	31_a	31_b	31_a	31_a	31_b	31_a	1
31_b	$(sfg)^3$	14865016320	31	31_b	31_a	31_b	31_b	31_a	31_b	1

15.8 Character tables of local subgroups

15.8.1 *Character table of $H = C_G(z)$*

2	9	9	8	4	3	8	8	7	7	6	4	2	3	2	2	2	5	5	4	4	2	2	2	4	4	4	4	2	2	2	2
3	2	2	.	2	2	2	.	.	.	.	.	.	2	2	2	.	.	.	.	.	.	2	.	.	.	.	.	.	.	.	.
5	1	1	.	.	.	1	.	.	.	.	.	1	.	.	.	.	.	.	.	.	1	.	.	.	.	.	.	1	1	.	.
7	1	1	.	.	.	1	.	.	.	.	.	.	.	.	.	1	.	.	.	.	.	.	1	.	.	.	.	.	.	1	1
	1a	2a	2b	2c	3a	4a	4b	4c	4d	4e	4f	5a	6a	6b	6c	7a	8a	8b	8c	8d	10a	12a	14a	16a	16b	16c	16d	20a	20b	28a	28b
2P	1a	1a	1a	1a	3a	2a	2a	2b	2b	2b	2b	5a	3a	3a	3a	7a	4b	4b	4d	4d	5a	6a	7a	8a	8b	8a	8b	10a	10a	14a	14a
3P	1a	2a	2b	2c	1a	4a	4b	4c	4d	4e	4f	5a	2a	2c	2c	7a	8a	8b	8c	8d	10a	4a	14a	16c	16d	16a	16b	20a	20b	28a	28b
5P	1a	2a	2b	2c	3a	4a	4b	4c	4d	4e	4f	1a	6a	6b	6c	7a	8a	8b	8c	8d	2a	12a	14a	16c	16d	16a	16b	4a	4a	28b	28a
7P	1a	2a	2b	2c	3a	4a	4b	4c	4d	4e	4f	5a	6a	6b	6c	1a	8a	8b	8c	8d	10a	12a	2a	16a	16b	16c	16d	20a	20b	4a	4a
11P	1a	2a	2b	2c	3a	4a	4b	4c	4d	4e	4f	5a	6a	6b	6c	7a	8a	8b	8c	8d	10a	12a	14a	16c	16d	16a	16b	20b	20a	28b	28a
13P	1a	2a	2b	2c	3a	4a	4b	4c	4d	4e	4f	5a	6a	6b	6c	7a	8a	8b	8c	8d	10a	12a	14a	16c	16d	16a	16b	20b	20a	28b	28a
17P	1a	2a	2b	2c	3a	4a	4b	4c	4d	4e	4f	5a	6a	6b	6c	7a	8a	8b	8c	8d	10a	12a	14a	16a	16b	16c	16d	20b	20a	28b	28a
19P	1a	2a	2b	2c	3a	4a	4b	4c	4d	4e	4f	5a	6a	6b	6c	7a	8a	8b	8c	8d	10a	12a	14a	16c	16d	16a	16b	20b	20a	28a	28b
23P	1a	2a	2b	2c	3a	4a	4b	4c	4d	4e	4f	5a	6a	6b	6c	7a	8a	8b	8c	8d	10a	12a	14a	16a	16b	16c	16d	20a	20b	28b	28a
X.1	1	1	1	1	1	1	1	1	1	1	1	1	1	1	1	1	1	1	1	1	1	1	1	1	1	1	1	1	1	1	1
X.2	1	1	1	-1	1	1	1	1	1	1	-1	1	1	-1	-1	1	1	1	-1	-1	1	1	1	-1	-1	-1	-1	1	1	1	1
X.3	20	20	4	.	2	-20	-4	4	4	-4	.	.	2	.	.	-1	.	.	.	.	.	-2	-1	.	.	.	.	.	.	1	1
X.4	20	20	4	-2	2	20	4	.	.	.	2	.	2	-2	-2	-1	.	.	.	.	.	2	-1	.	.	.	.	.	.	-1	-1
X.5	20	20	4	2	2	20	4	.	.	.	-2	.	2	2	2	-1	.	.	.	.	.	2	-1	.	.	.	.	.	.	-1	-1
X.6	35	35	3	-1	-1	35	3	3	3	3	-1	.	-1	-1	-1	.	-1	-1	-1	-1	.	-1	.	1	1	1	1	.	.	.	.
X.7	35	35	3	-1	-1	35	3	-1	-1	-1	-1	.	-1	-1	-1	.	3	-1	1	1	.	-1	.	-1	1	-1	1	.	.	.	.
X.8	35	35	3	-1	-1	35	3	-1	-1	-1	-1	.	-1	-1	-1	.	-1	3	1	1	.	-1	.	1	-1	1	-1	.	.	.	.
X.9	35	35	3	1	-1	35	3	3	3	3	1	.	-1	1	1	.	-1	-1	1	1	.	-1	.	-1	-1	-1	-1	.	.	.	.
X.10	35	35	3	1	-1	35	3	-1	-1	-1	1	.	-1	1	1	.	3	-1	-1	-1	.	-1	.	1	-1	1	-1	.	.	.	.
X.11	35	35	3	1	-1	35	3	-1	-1	-1	1	.	-1	1	1	.	-1	3	-1	-1	.	-1	.	-1	1	-1	1	.	.	.	.
X.12	36	36	4	.	.	-36	-4	.	.	.	.	1	.	.	.	1	.	.	2	-2	1	.	1	.	.	.	.	-1	-1	-1	-1
X.13	36	36	4	.	.	-36	-4	.	.	.	.	1	.	.	.	1	.	.	-2	2	1	.	1	.	.	.	.	-1	-1	-1	-1
X.14	40	-40	.	.	4	.	.	-4	4	.	.	.	-4	.	.	-2	.	.	.	.	.	.	2	.	.	.	.	.	.	.	.
X.15	56	56	-8	.	2	-56	8	.	.	.	.	1	2	.	.	.	.	.	.	.	1	-2	.	.	.	.	.	-1	-1	.	.
X.16	56	-56	.	.	2	.	.	4	-4	.	.	1	-2	.	.	.	.	.	.	.	-1	.	.	.	.	.	.	B	-B	.	.
X.17	56	-56	.	.	2	.	.	4	-4	.	.	1	-2	.	.	.	.	.	.	.	-1	.	.	.	.	.	.	-B	B	.	.
X.18	64	64	.	.	1	-64	.	.	.	.	.	-1	1	3	-3	1	.	.	.	.	-1	-1	1	.	.	.	.	1	1	-1	-1
X.19	64	64	.	.	1	-64	.	.	.	.	.	-1	1	-3	3	1	.	.	.	.	-1	-1	1	.	.	.	.	1	1	-1	-1
X.20	64	64	.	8	1	64	.	.	.	.	.	-1	1	-1	-1	1	.	.	.	.	-1	1	1	.	.	.	.	-1	-1	1	1
X.21	64	64	.	-8	1	64	.	.	.	.	.	-1	1	1	1	1	.	.	.	.	-1	1	1	.	.	.	.	-1	-1	1	1
X.22	70	70	-2	.	-2	-70	2	2	2	-2	.	.	-2	.	.	.	.	.	.	.	.	2	.	A	-A	-A	A	.	.	.	.
X.23	70	70	-2	.	-2	-70	2	2	2	-2	.	.	-2	.	.	.	.	.	.	.	.	2	.	-A	A	A	-A	.	.	.	.
X.24	72	-72	.	.	.	.	.	-4	4	.	.	2	.	.	.	2	.	.	.	.	-2	.	-2	.	.	.	.	.	.	.	.
X.25	90	90	2	.	.	-90	-2	-2	-2	2	.	.	.	.	.	-1	.	.	.	.	.	.	-1	A	A	-A	-A	.	.	1	1
X.26	90	90	2	.	.	-90	-2	-2	-2	2	.	.	.	.	.	-1	.	.	.	.	.	.	-1	-A	-A	A	A	.	.	1	1
X.27	90	90	-6	.	.	90	-6	2	2	2	.	.	.	.	.	-1	2	2	.	.	.	.	-1	.	.	.	.	.	.	-1	-1
X.28	126	126	-2	.	.	126	-2	-2	-2	-2	.	1	.	.	.	.	-2	-2	.	.	1	.	.	.	.	.	.	1	1	.	.
X.29	128	-128	.	.	2	.	.	.	.	.	.	-2	-2	.	.	2	.	.	.	.	2	.	-2	.	.	.	.	.	.	.	.
X.30	160	-160	.	.	-2	.	.	.	.	.	.	.	2	.	.	-1	.	.	.	.	.	.	1	.	.	.	.	.	.	C	-C
X.31	160	-160	.	.	-2	.	.	.	.	.	.	.	2	.	.	-1	.	.	.	.	.	.	1	.	.	.	.	.	.	-C	C

where $A = \sqrt{2}$, $B = i\sqrt{5}$ *and* $C = \sqrt{7}$.

15.8.2 *Character table of $D = N_H(A)$*

2	9	9	8	5	4	2	8	8	7	7	6	5	4	2	5	5	4	4	2	2	4	4	4	4
3	1	1	.	.	.	1	1	.	.	.	.	.	.	1	.	.	.	.	1	1	.	.	.	.
	1a	2a	2b	2c	2d	3a	4a	4b	4c	4d	4e	4f	4g	6a	8a	8b	8c	8d	12a	12b	16a	16b	16c	16d
2P	1a	1a	1a	1a	1a	3a	2a	2a	2b	2b	2b	2a	2b	3a	4b	4b	4d	4d	6a	6a	8a	8a	8b	8b
3P	1a	2a	2b	2c	2d	1a	4a	4b	4c	4d	4e	4f	4g	2a	8a	8b	8c	8d	4a	4a	16b	16a	16d	16c
X.1	1	1	1	1	1	1	1	1	1	1	1	1	1	1	1	1	1	1	1	1	1	1	1	1
X.2	1	1	1	1	-1	1	1	1	1	1	1	1	-1	1	1	1	-1	-1	1	1	-1	-1	-1	-1
X.3	2	2	2	2	.	-1	2	2	2	2	2	2	.	-1	2	2	.	.	-1	-1	.	.	.	.
X.4	3	3	3	-1	-1	.	3	3	3	3	3	-1	-1	.	-1	-1	-1	-1	.	.	1	1	1	1
X.5	3	3	3	-1	1	.	3	3	3	3	3	-1	1	.	-1	-1	1	1	.	.	-1	-1	-1	-1
X.6	3	3	3	-1	-1	.	3	3	-1	-1	-1	-1	-1	.	-1	3	1	1	.	.	1	1	-1	-1
X.7	3	3	3	-1	1	.	3	3	-1	-1	-1	-1	1	.	-1	3	-1	-1	.	.	-1	-1	1	1
X.8	3	3	3	-1	-1	.	3	3	-1	-1	-1	-1	-1	.	3	-1	1	1	.	.	-1	-1	1	1
X.9	3	3	3	-1	1	.	3	3	-1	-1	-1	-1	1	.	3	-1	-1	-1	.	.	1	1	-1	-1
X.10	4	4	4	.	.	1	-4	-4	.	.	.	.	.	1	.	.	-2	2	-1	-1	.	.	.	.
X.11	4	4	4	.	.	1	-4	-4	.	.	.	.	.	1	.	.	2	-2	-1	-1	.	.	.	.
X.12	6	6	6	2	.	.	6	6	-2	-2	-2	2	.	.	-2	-2	.	.	.	.	.	.	.	.
X.13	6	6	-2	-2	.	.	-6	2	2	2	-2	2	.	.	.	.	.	.	.	.	B	-B	-B	B
X.14	6	6	-2	-2	.	.	-6	2	2	2	-2	2	.	.	.	.	.	.	.	.	-B	B	B	-B
X.15	6	6	-2	2	.	.	-6	2	2	2	-2	-2	.	.	.	.	.	.	.	.	B	-B	B	-B
X.16	6	6	-2	2	.	.	-6	2	2	2	-2	-2	.	.	.	.	.	.	.	.	-B	B	-B	B
X.17	8	-8	.	.	.	2	.	.	-4	4	.	.	.	-2	.	.	.	.	.	.	.	.	.	.
X.18	8	-8	.	.	.	-1	.	.	-4	4	.	.	.	1	.	.	.	.	A	-A	.	.	.	.
X.19	8	-8	.	.	.	-1	.	.	-4	4	.	.	.	1	.	.	.	.	-A	A	.	.	.	.
X.20	8	8	8	.	.	-1	-8	-8	.	.	.	.	.	-1	.	.	.	.	1	1	.	.	.	.
X.21	12	12	-4	.	.	.	-12	4	-4	-4	4	.	.	.	.	.	.	.	.	.	.	.	.	.
X.22	12	12	-4	.	2	.	12	-4	.	.	.	.	-2	.	.	.	.	.	.	.	.	.	.	.
X.23	12	12	-4	.	-2	.	12	-4	.	.	.	.	2	.	.	.	.	.	.	.	.	.	.	.
X.24	24	-24	.	.	.	.	.	.	4	-4	.	.	.	.	.	.	.	.	.	.	.	.	.	.

where $A = -\sqrt{3}$ *and* $B = -\sqrt{2}$.

15.8.3 *Character table of $E = N_G(A)$*

2	9	9	5	2	8	8	5	2	.	.	5	5	2	2	4	4	4	4
3	1	1	.	1	1	.	.	1	.	.	.	.	1	1	.	.	.	.
7	1	.	.	.	.	.	.	.	1	1	.	.	.	.	.	.	.	.
	1a	2a	2b	3a	4a	4b	4c	6a	7a	7b	8a	8b	12a	12b	16a	16b	16c	16d
2P	1a	1a	1a	3a	2a	2a	2a	3a	7a	7b	4b	4b	6a	6a	8a	8a	8b	8b
3P	1a	2a	2b	1a	4a	4b	4c	2a	7b	7a	8a	8b	4a	4a	16b	16a	16d	16c
5P	1a	2a	2b	3a	4a	4b	4c	6a	7b	7a	8a	8b	12b	12a	16b	16a	16d	16c
7P	1a	2a	2b	3a	4a	4b	4c	6a	1a	1a	8a	8b	12b	12a	16a	16b	16c	16d
11P	1a	2a	2b	3a	4a	4b	4c	6a	7a	7b	8a	8b	12a	12b	16b	16a	16d	16c
13P	1a	2a	2b	3a	4a	4b	4c	6a	7b	7a	8a	8b	12a	12b	16b	16a	16d	16c
X.1	1	1	1	1	1	1	1	1	1	1	1	1	1	1	1	1	1	1
X.2	3	3	−1	.	3	3	−1	.	A	/A	−1	−1	.	.	1	1	1	1
X.3	3	3	−1	.	3	3	−1	.	/A	A	−1	−1	.	.	1	1	1	1
X.4	6	6	2	.	6	6	2	.	−1	−1	2	2	.	.	.	.	.	.
X.5	7	7	−1	1	7	7	−1	1	.	.	−1	−1	1	1	−1	−1	−1	−1
X.6	7	7	−1	1	−1	−1	−1	1	.	.	−1	3	−1	−1	1	1	−1	−1
X.7	7	7	−1	1	−1	−1	−1	1	.	.	3	−1	−1	−1	−1	−1	1	1
X.8	8	8	.	−1	8	8	.	−1	1	1	.	.	−1	−1	.	.	.	.
X.9	14	14	−2	−1	−2	−2	−2	−1	.	.	2	2	1	1	.	.	.	.
X.10	21	21	1	.	−3	−3	1	.	.	.	−3	1	.	.	−1	−1	1	1
X.11	21	21	1	.	−3	−3	1	.	.	.	1	−3	.	.	1	1	−1	−1
X.12	28	−4	.	−2	−12	4	.	2	.	.	.	.	.	.	.	.	.	.
X.13	28	−4	.	1	−12	4	.	−1	.	.	.	.	B	−B	.	.	.	.
X.14	28	−4	.	1	−12	4	.	−1	.	.	.	.	−B	B	.	.	.	.
X.15	42	−6	−2	.	6	−2	2	.	.	.	.	.	.	.	C	−C	C	−C
X.16	42	−6	−2	.	6	−2	2	.	.	.	.	.	.	.	−C	C	−C	C
X.17	42	−6	2	.	6	−2	−2	.	.	.	.	.	.	.	C	−C	−C	C
X.18	42	−6	2	.	6	−2	−2	.	.	.	.	.	.	.	−C	C	C	−C

where $A = \frac{1}{2}(-1 + i\sqrt{7})$, $B = -\sqrt{3}$ *and* $C = \sqrt{2}$.

15.8.4 *Character table of $N_1 = N_G(r)$*

2	4	4	4	4	3	3	3	3	1	1	.	.	.	.	.	.	.	.	3	3	.	3
3	4	2	1	1	4	4	4	4	4	4	4	4	4	4	4	4	4	4	2	.	2	2
5	1	.	.	.	1	1	1	1	.	.	.	.	.	.	.	.	.	.	.	.	1	.
	1a	2a	2b	2c	3a	3b	3c	3d	3e	3f	3g	3h	3i	3j	3k	3l	3m	3n	4a	4b	5a	6a
2P	1a	1a	1a	1a	3a	3b	3c	3d	3e	3f	3g	3h	3i	3j	3k	3l	3m	3n	2a	2a	5a	3b
3P	1a	2a	2b	2c	1a	1a	1a	1a	1a	1a	1a	1a	1a	1a	1a	1a	1a	1a	4a	4b	5a	2a
5P	1a	2a	2b	2c	3a	3b	3c	3d	3e	3f	3g	3h	3i	3j	3k	3l	3m	3n	4a	4b	1a	6a
X.1	1	1	1	1	1	1	1	1	1	1	1	1	1	1	1	1	1	1	1	1	1	1
X.2	1	1	−1	−1	1	1	1	1	1	1	1	1	1	1	1	1	1	1	1	−1	1	1
X.3	2	2	.	.	−1	−1	−1	2	2	2	2	−1	−1	2	−1	−1	−1	−1	2	.	2	−1
X.4	2	2	.	.	−1	−1	2	−1	2	2	−1	2	−1	−1	−1	2	−1	−1	2	.	2	−1
X.5	2	2	.	.	2	−1	−1	−1	2	2	−1	−1	−1	−1	2	−1	−1	2	2	.	2	−1
X.6	2	2	.	.	−1	2	−1	−1	2	2	−1	−1	2	−1	−1	−1	2	−1	2	.	2	2
X.7	5	1	3	−1	5	5	5	5	−1	2	2	2	−1	−1	−1	−1	2	2	−1	1	.	1
X.8	5	1	−3	1	5	5	5	5	−1	2	2	2	−1	−1	−1	−1	2	2	−1	−1	.	1
X.9	5	1	1	−3	5	5	5	5	2	−1	−1	−1	2	2	2	2	−1	−1	−1	−1	.	1
X.10	5	1	−1	3	5	5	5	5	2	−1	−1	−1	2	2	2	2	−1	−1	−1	1	.	1
X.11	9	1	3	3	9	9	9	9	.	.	.	.	.	.	.	.	.	.	1	−1	−1	1
X.12	9	1	−3	−3	9	9	9	9	.	.	.	.	.	.	.	.	.	.	1	1	−1	1
X.13	10	2	.	.	−5	−5	−5	10	−2	4	4	−2	1	−2	1	1	−2	−2	−2	.	.	−1
X.14	10	2	.	.	−5	−5	−5	10	4	−2	−2	1	−2	4	−2	−2	1	1	−2	.	.	−1
X.15	10	−2	2	−2	10	10	10	10	1	1	1	1	1	1	1	1	1	1	.	.	.	−2
X.16	10	−2	−2	2	10	10	10	10	1	1	1	1	1	1	1	1	1	1	.	.	.	−2
X.17	10	2	.	.	−5	−5	10	−5	−2	4	−2	4	1	1	1	−2	−2	−2	−2	.	.	−1
X.18	10	2	.	.	−5	−5	10	−5	4	−2	1	−2	−2	−2	−2	4	1	1	−2	.	.	−1
X.19	10	2	.	.	10	−5	−5	−5	−2	4	−2	−2	1	1	−2	1	−2	4	−2	.	.	−1
X.20	10	2	.	.	10	−5	−5	−5	4	−2	1	1	−2	−2	4	−2	1	−2	−2	.	.	−1
X.21	10	2	.	.	−5	10	−5	−5	−2	4	−2	−2	−2	1	1	1	4	−2	−2	.	.	2
X.22	10	2	.	.	−5	10	−5	−5	4	−2	1	1	4	−2	−2	−2	−2	1	−2	.	.	2
X.23	16	.	.	.	−8	−8	−8	16	−2	−2	−2	1	1	−2	1	1	1	1	.	.	1	.
X.24	16	.	.	.	−8	−8	−8	16	−2	−2	−2	1	1	−2	1	1	1	1	.	.	1	.
X.25	16	.	.	.	16	16	16	16	−2	−2	−2	−2	−2	−2	−2	−2	−2	−2	.	.	1	.
X.26	16	.	.	.	−8	−8	16	−8	−2	−2	1	−2	1	1	1	−2	1	1	.	.	1	.
X.27	16	.	.	.	−8	−8	16	−8	−2	−2	1	−2	1	1	1	−2	1	1	.	.	1	.
X.28	16	.	.	.	16	−8	−8	−8	−2	−2	1	1	1	1	−2	1	1	−2	.	.	1	.
X.29	16	.	.	.	16	−8	−8	−8	−2	−2	1	1	1	1	−2	1	1	−2	.	.	1	.
X.30	16	.	.	.	−8	16	−8	−8	−2	−2	1	1	−2	1	1	1	−2	1	.	.	1	.
X.31	16	.	.	.	−8	16	−8	−8	−2	−2	1	1	−2	1	1	1	−2	1	.	.	1	.
X.32	18	2	.	.	−9	−9	−9	18	.	.	.	.	.	.	.	.	.	.	2	.	−2	−1
X.33	18	2	.	.	−9	−9	18	−9	.	.	.	.	.	.	.	.	.	.	2	.	−2	−1
X.34	18	2	.	.	18	−9	−9	−9	.	.	.	.	.	.	.	.	.	.	2	.	−2	−1
X.35	18	2	.	.	−9	18	−9	−9	.	.	.	.	.	.	.	.	.	.	2	.	−2	2
X.36	20	−4	.	.	−10	−10	−10	20	2	2	2	−1	−1	2	−1	−1	−1	−1	.	.	.	2
X.37	20	−4	.	.	−10	−10	20	−10	2	2	−1	2	−1	−1	−1	2	−1	−1	.	.	.	2
X.38	20	−4	.	.	20	−10	−10	−10	2	2	−1	−1	−1	−1	2	−1	−1	2	.	.	.	2
X.39	20	−4	.	.	−10	20	−10	−10	2	2	−1	−1	2	−1	−1	−1	2	−1	.	.	.	−4

Character table of $N_1 = N_G(r)$ *(continued)*

	6b	6c	6d	6e	6f	12a	12b	12c	12d	15a	15b	15c	15d	15e	15f	15g	15h
2	3	3	3	1	1	2	2	2	2	.	.	.	.	.	.	.	.
3	2	2	2	1	1	2	2	2	2	2	2	2	2	2	2	2	2
5	.	.	.	.	.	.	.	.	.	1	1	1	1	1	1	1	1
2P	3a	3d	3c	3f	3e	6d	6c	6b	6a	15a	15b	15c	15d	15e	15f	15g	15h
3P	2a	2a	2a	2b	2c	4a	4a	4a	4a	5a	5a	5a	5a	5a	5a	5a	5a
5P	6b	6c	6d	6e	6f	12a	12b	12c	12d	3d	3a	3c	3b	3c	3a	3b	3d
X.1	1	1	1	1	1	1	1	1	1	1	1	1	1	1	1	1	1
X.2	1	1	1	−1	−1	1	1	1	1	1	1	1	1	1	1	1	1
X.3	−1	2	−1	.	.	−1	2	−1	−1	2	−1	−1	−1	−1	−1	−1	2
X.4	−1	−1	2	.	.	2	−1	−1	−1	−1	−1	2	−1	2	−1	−1	−1
X.5	2	−1	−1	.	.	−1	−1	2	−1	−1	2	−1	−1	−1	2	−1	−1
X.6	−1	−1	−1	.	.	−1	−1	−1	2	−1	−1	−1	2	−1	−1	2	−1
X.7	1	1	1	.	−1	−1	−1	−1	−1	.	.	.	.	.	.	.	.
X.8	1	1	1	.	1	−1	−1	−1	−1	.	.	.	.	.	.	.	.
X.9	1	1	1	1	.	−1	−1	−1	−1	.	.	.	.	.	.	.	.
X.10	1	1	1	−1	.	−1	−1	−1	−1	.	.	.	.	.	.	.	.
X.11	1	1	1	.	.	1	1	1	1	−1	−1	−1	−1	−1	−1	−1	−1
X.12	1	1	1	.	.	1	1	1	1	−1	−1	−1	−1	−1	−1	−1	−1
X.13	−1	2	−1	.	.	1	−2	1	1	.	.	.	.	.	.	.	.
X.14	−1	2	−1	.	.	1	−2	1	1	.	.	.	.	.	.	.	.
X.15	−2	−2	−2	−1	1	.	.	.	.	.	.	.	.	.	.	.	.
X.16	−2	−2	−2	1	−1	.	.	.	.	.	.	.	.	.	.	.	.
X.17	−1	−1	2	.	.	−2	1	1	1	.	.	.	.	.	.	.	.
X.18	−1	−1	2	.	.	−2	1	1	1	.	.	.	.	.	.	.	.
X.19	2	−1	−1	.	.	1	1	−2	1	.	.	.	.	.	.	.	.
X.20	2	−1	−1	.	.	1	1	−2	1	.	.	.	.	.	.	.	.
X.21	−1	−1	−1	.	.	1	1	1	−2	.	.	.	.	.	.	.	.
X.22	−1	−1	−1	.	.	1	1	1	−2	.	.	.	.	.	.	.	.
X.23	.	.	.	.	.	.	.	.	.	1	A	/A	A	A	/A	/A	1
X.24	.	.	.	.	.	.	.	.	.	1	/A	A	/A	/A	A	A	1
X.25	.	.	.	.	.	.	.	.	.	1	1	1	1	1	1	1	1
X.26	.	.	.	.	.	.	.	.	.	A	A	1	/A	1	/A	A	/A
X.27	.	.	.	.	.	.	.	.	.	/A	/A	1	A	1	A	/A	A
X.28	.	.	.	.	.	.	.	.	.	A	1	A	A	/A	1	/A	/A
X.29	.	.	.	.	.	.	.	.	.	/A	1	/A	/A	A	1	A	A
X.30	.	.	.	.	.	.	.	.	.	A	/A	/A	1	A	A	1	/A
X.31	.	.	.	.	.	.	.	.	.	/A	A	A	1	/A	/A	1	A
X.32	−1	2	−1	.	.	−1	2	−1	−1	−2	1	1	1	1	1	1	−2
X.33	−1	−1	2	.	.	2	−1	−1	−1	1	1	−2	1	−2	1	1	1
X.34	2	−1	−1	.	.	−1	−1	2	−1	1	−2	1	1	1	−2	1	1
X.35	−1	−1	−1	.	.	−1	−1	−1	2	1	1	1	−2	1	1	−2	1
X.36	2	−4	2	.	.	.	.	.	.	.	.	.	.	.	.	.	.
X.37	2	2	−4	.	.	.	.	.	.	.	.	.	.	.	.	.	.
X.38	−4	2	2	.	.	.	.	.	.	.	.	.	.	.	.	.	.
X.39	2	2	2	.	.	.	.	.	.	.	.	.	.	.	.	.	.

where $A = -\frac{1}{2}(1 + i\sqrt{15})$.

15.8.5 *Character table of* $Y = N_G(f) = N_G(5A)$

	1_a	2_a	2_b	2_c	3_a	4_a	4_b	4_c	4_d	4_e	4_f	5_a	6_a	10_a	15_a	15_b	20_a	20_b
2	4	4	4	4	1	3	3	3	3	3	3	2	1	2	.	.	2	2
3	2	2	.	.	2	.	.	.	.	.	.	2	2	.	2	2	.	.
5	1	.	1	.	1	1	.	.	.	.	.	1	.	1	1	1	1	1
2P	1_a	1_a	1_a	1_a	3_a	2_b	2_c	2_b	2_c	2_c	2_c	5_a	3_a	5_a	15_b	15_a	10_a	10_a
3P	1_a	2_a	2_b	2_c	1_a	4_a	4_d	4_c	4_b	4_f	4_e	5_a	2_a	10_a	5_a	5_a	20_a	20_b
5P	1_a	2_a	2_b	2_c	3_a	4_a	4_b	4_c	4_d	4_e	4_f	1_a	6_a	2_b	3_a	3_a	4_a	4_a
X.1	1	1	1	1	1	1	1	1	1	1	1	1	1	1	1	1	1	1
X.2	1	1	1	1	1	−1	1	−1	1	−1	−1	1	1	1	1	1	−1	−1
X.3	1	−1	1	−1	1	1	−A	−1	A	A	−A	1	−1	1	1	1	1	1
X.4	1	−1	1	−1	1	−1	−A	1	A	−A	A	1	−1	1	1	1	−1	−1
X.5	1	1	1	1	1	1	−1	1	−1	−1	−1	1	1	1	1	1	1	1
X.6	1	1	1	1	1	−1	−1	−1	−1	1	1	1	1	1	1	1	−1	−1
X.7	1	−1	1	−1	1	1	A	−1	−A	−A	A	1	−1	1	1	1	1	1
X.8	1	−1	1	−1	1	−1	A	1	−A	A	−A	1	−1	1	1	1	−1	−1
X.9	2	2	−2	−2	2	.	.	.	.	.	.	2	2	−2	2	2	.	.
X.10	2	−2	−2	2	2	.	.	.	.	.	.	2	−2	−2	2	2	.	.
X.11	4	.	4	.	4	4	.	.	.	.	.	−1	.	−1	−1	−1	−1	−1
X.12	4	.	4	.	4	−4	.	.	.	.	.	−1	.	−1	−1	−1	1	1
X.13	4	.	−4	.	4	.	.	.	.	.	.	−1	.	1	−1	−1	−B	B
X.14	4	.	−4	.	4	.	.	.	.	.	.	−1	.	1	−1	−1	B	−B
X.15	8	8	.	.	−1	.	.	.	.	.	.	8	−1	.	−1	−1	.	.
X.16	8	−8	.	.	−1	.	.	.	.	.	.	8	1	.	−1	−1	.	.
X.17	16	.	.	.	−2	.	.	.	.	.	.	−4	.	.	C	/C	.	.
X.18	16	.	.	.	−2	.	.	.	.	.	.	−4	.	.	/C	C	.	.

where $A = i$, $B = i\sqrt{5}$ *and* $C = \frac{1}{2}(1 + 3\sqrt{5})$.

15.8.6 *Character table of $R = N_G(7A)$*

	1_a	2_a	2_b	2_c	3_a	3_b	4_a	6_a	6_b	6_c	6_d	6_e	6_f	7_a	7_b	7_c	7_d	12_a	12_b	14_a	14_b	14_c	28_a	28_b
2	3	3	2	2	3	3	2	3	3	2	2	2	2	2	1	1	.	2	2	2	1	1	2	2
3	1	1	1	1	1	1	1	1	1	1	1	1	1	.	.	.	.	1	1	.	.	.	.	.
7	3	1	1	1	.	.	1	.	.	.	.	.	.	3	2	2	2	.	.	1	1	1	1	1
2P	1_a	1_a	1_a	1_a	3_b	3_a	2_a	3_b	3_a	3_b	3_a	3_b	3_a	7_a	7_b	7_c	7_d	6_b	6_a	7_a	7_c	7_b	14_a	14_a
3P	1_a	2_a	2_b	2_c	1_a	1_a	4_a	2_a	2_a	2_b	2_b	2_c	2_c	7_a	7_b	7_c	7_d	4_a	4_a	14_a	14_b	14_c	28_a	28_b
7P	1_a	2_a	2_b	2_c	3_a	3_b	4_a	6_a	6_b	6_c	6_d	6_e	6_f	1_a	1_a	1_a	1_a	12_a	12_b	2_a	2_b	2_c	4_a	4_a
X.1	1	1	1	1	1	1	1	1	1	1	1	1	1	1	1	1	1	1	1	1	1	1	1	1
X.2	1	1	−1	1	1	1	−1	1	1	−1	−1	1	1	1	1	1	1	−1	−1	1	−1	1	−1	−1
X.3	1	1	1	−1	1	1	−1	1	1	1	1	−1	−1	1	1	1	1	−1	−1	1	1	−1	−1	−1
X.4	1	1	−1	−1	1	1	1	1	1	−1	−1	−1	−1	1	1	1	1	1	1	1	−1	−1	1	1
X.5	1	1	1	1	A	/A	1	A	/A	A	/A	A	/A	1	1	1	1	A	/A	1	1	1	1	1
X.6	1	1	−1	1	A	/A	−1	A	/A	−A	−/A	A	/A	1	1	1	1	−A	−/A	1	−1	1	−1	−1
X.7	1	1	1	−1	A	/A	−1	A	/A	A	/A	−A	−/A	1	1	1	1	−A	−/A	1	1	−1	−1	−1
X.8	1	1	−1	−1	A	/A	1	A	/A	−A	−/A	−A	−/A	1	1	1	1	A	/A	1	−1	−1	1	1
X.9	1	1	1	1	/A	A	1	/A	A	/A	A	/A	A	1	1	1	1	/A	A	1	1	1	1	1
X.10	1	1	−1	1	/A	A	−1	/A	A	−/A	−A	/A	A	1	1	1	1	−/A	−A	1	−1	1	−1	−1
X.11	1	1	1	−1	/A	A	−1	/A	A	/A	A	−/A	−A	1	1	1	1	−/A	−A	1	1	−1	−1	−1
X.12	1	1	−1	−1	/A	A	1	/A	A	−/A	−A	−/A	−A	1	1	1	1	/A	A	1	−1	−1	1	1
X.13	2	−2	.	.	2/A	2A	.	−2/A	−2A	.	.	.	.	2	2	2	2	.	.	−2	.	.	.	.
X.14	2	−2	.	.	2A	2/A	.	−2A	−2/A	.	.	.	.	2	2	2	2	.	.	−2	.	.	.	.
X.15	2	−2	.	.	2	2	.	−2	−2	.	.	.	.	2	2	2	2	.	.	−2	.	.	.	.
X.16	12	.	6	.	.	.	.	.	.	.	.	.	.	12	−2	5	−2	.	.	.	−1	.	.	.
X.17	12	.	−6	.	.	.	.	.	.	.	.	.	.	12	−2	5	−2	.	.	.	1	.	.	.
X.18	12	.	.	6	.	.	.	.	.	.	.	.	.	12	5	−2	−2	.	.	.	.	−1	.	.
X.19	12	.	.	−6	.	.	.	.	.	.	.	.	.	12	5	−2	−2	.	.	.	.	1	.	.
X.20	24	.	.	.	.	.	.	.	.	.	.	.	.	24	−4	−4	3	.	.	.	.	.	.	.
X.21	42	6	.	.	.	.	.	.	.	.	.	.	.	−7	.	.	.	.	.	−1	.	.	−B	B
X.22	42	6	.	.	.	.	.	.	.	.	.	.	.	−7	.	.	.	.	.	−1	.	.	B	−B
X.23	42	−6	.	.	.	.	6	.	.	.	.	.	.	−7	.	.	.	.	.	1	.	.	−1	−1
X.24	42	−6	.	.	.	.	−6	.	.	.	.	.	.	−7	.	.	.	.	.	1	.	.	1	1

where $A = \frac{1}{2}(-1 + i\sqrt{3})$ and $B = \sqrt{7}$.

16

Concluding remarks and open problems

In [92] and in this book it is shown that 23 known sporadic simple groups G different from M_{11}, the baby monster B and the monster M can be constructed uniformly by means of Algorithms 1.3.8 or 1.3.15 from well defined indecomposable subgroups T of fairly small general linear groups $\mathrm{GL}_n(2)$ with $3 \leq n \leq 11$. Each of these groups G is realized either as an irreducible subgroup of some general linear group $\mathrm{GL}_n(p)$, where p is a suitable odd prime number, or as a uniquely determined finitely presented group with a documented presentation. Furthermore, the author and his collaborators calculated the character tables of all of them. They all agree with their counterparts in the Atlas [19].

For technical but no theoretical reasons the monster M and the baby monster B cannot be constructed as such a matrix group at this time. In Section 16.1 we give more reasons for this statement. Furthermore, we mention some relevant literature and open problems about the monster M.

Section 16.2 summarizes the uniqueness problems of some finite simple groups that we either left open in the previous chapters or have not been dealt with in the mathematical literature.

In Section 16.3 we quote an open problem mentioned in Thompson's article [127] published in 1976. It asks for the existence of a 27th sporadic simple group and provides some specific information about a certain local subgroup. We describe also Griess' related results [41]. Thus we are able to show that an application of Algorithm 1.3.15 to Thompson's local subgroup would either prove the existence of a new sporadic group or provide a negative answer to Thompson's question.

In Section 16.4 we address Brauer's problems summarized in Remark 1.1.4(c) asking for a general classification scheme of the finite simple groups. We illustrate his problems by considering a special classification problem that asks for an abstract classification of all finite simple groups G that can be constructed by an application of Algorithm 1.3.8 to all the indecomposable subgroups of the infinite sequence of general linear groups $\mathrm{GL}_n(2)$. For a fixed integer n we also describe an exhaustive search method for finding the corresponding simple groups. A similar classification problem is mentioned in which Algorithm 1.3.8 is replaced by Algorithm 1.3.15. By Higman's Theorem 1.2.3 there are infinitely many non-isomorphic iterated extensions of a given indecomposable subgroup T of a fixed $\mathrm{GL}_n(2)$. Therefore there is no finite search method for the possible simple groups of this classification problem.

16.1 On the monster and the baby monster

There are two sporadic simple groups which have not been discussed so far. B. Fischer's unpublished work on finite groups generated by a non-degenerate conjugacy class of $\{3,4\}$-transpositions laid the foundation of the discovery of two simple groups B and M of respective orders

$$|\mathsf{B}| = 21^{41} \cdot 3^{13} \cdot 5^6 \cdot 7^2 \cdot 11 \cdot 13 \cdot 17 \cdot 19 \cdot 23 \cdot 47,$$
$$|\mathsf{M}| = 21^{46} \cdot 3^{20} \cdot 5^9 \cdot 7^6 \cdot 11^2 \cdot 13^3 \cdot 17 \cdot 19 \cdot 23 \cdot 29 \cdot 31 \cdot 41 \cdot 47 \cdot 59 \cdot 71.$$

He called his presumed groups *baby monster* and *monster*, respectively.

Definition 16.1.1 Let K be a conjugacy class of involutions of a finite group G. Then K is a *non-degenerate conjugacy class of* $\{3,4\}$-*transpositions* if the product xy of any pair $x, y \in K$ has order 1, 2, 3 or 4 and at least some product xy has order 4.

Fischer's early (unpublished) results on a presumed simple group B generated by a non-degenerate conjugacy class $K = x^B$ of $\{3,4\}$-transpositions and minimal order convinced him that $C_B(x)/\langle x\rangle \cong Aut({}^2\mathrm{E}_6(2))$. He saw that B would also have a 2-central involution z whose centralizer $H_B = C_B(z)$ has an extra-special Fitting subgroup O_B of order 2^{23} such that $H_B/O_B \cong \mathsf{Co}_2$. According to Gorenstein [38], p. 120, Leon and Sims used Fischer's almost complete information on the internal structure of his presumed group B to prove its existence and uniqueness. They announced their result in [87]. But their proof for the assertion that B has a faithful permutation representation of degree 13571955000 with stabilizer H_B appears to be unpublished. Recently, Robinson and Cooperman [114] constructed a permutation presentation for B of that degree using two matrix generators of B in $\mathrm{GL}_{4370}(2)$ constructed by R. A. Wilson [136]; see also [137]. Furthermore, Leon's work [86] on the irreducible characters of B was used in the calculation of the complete character table of B; see [19], pp. 208–216.

Definition 16.1.2 A finite simple group G is said to be of B-*type* if it has a 2-central involution z such that its centralizer $H = C_G(z)$ has an extra-special Fitting subgroup O of order 2^{23} and $H/O \cong \mathsf{Co}_2$.

In [60] A. Ivanov published a presentation for the baby monster B. But at the present time it does not seem possible to construct a faithful permutation representation of degree 13571955000 of this finitely presented group B and a system of generators of its stabilizer. The author is also unaware of a presentation of the centralizer H_{B} of a 2-central involution z of B.

By Conway's work [16] on the large Conway group Co_1 it is known that the Leech lattice Λ of rank 24 reduces to an irreducible 24-dimensional $GF(2)$-module V of Co_1. Thus the obvious question arises as to whether there is also a finite simple group M of M-type in the sense of the following definition.

Definition 16.1.3 A finite simple group G is said to be of M-*type* if it has a 2-central involution z such that its centralizer $H = C_G(z)$ has an extra-special Fitting subgroup O of order 2^{25} and $H/O \cong \mathsf{Co}_1$.

R. L. Griess and B. Fischer studied the structure of a simple group G of M-type independently. They also assumed that G has a second involution u with centralizer $U = C_G(u)$ isomorphic to a 2-fold cover $2\mathsf{B}$ of the baby monster B. In collaboration with J. G. Thompson, J. Conway, S. Norton and several other mathematicians they established a long list of conjectural properties of such a group G; see Thompson's article [129]. There he proves the uniqueness of G up to isomorphism provided that all 15 properties stated in [129] are satisfied. The most important one asserts that G has a faithful irreducible character of degree 196883. Assuming the existence of such a character along with several other conditions of Thompson's article, B. Fischer, D. Livingstone and M. Thorne proved that G has 194 conjugacy classes. Furthermore, they calculated the rational values of all irreducible characters of the group G; see Thorne's thesis [130]. A completed version of this character table is contained in the Atlas [19], pp. 222–227.

In his seminal paper [42], R. L. Griess established the existence of a simple group M of M-type. Inspired by previous work of S. Norton he constructed a non-associative algebra $\mathcal{B}$ of dimension 196884 over the rational field $\mathbb{Q}$ with a non-degenerate, symmetric, associative form $<,>$. He proved that his constructed centralizer H satisfies the properties of Definition 16.1.3 and that it acts faithfully on $\mathcal{B}$ as a group of automorphisms preserving the form $<,>$. Then he builds an automorphism σ of $\mathcal{B}$ of order 2 which also preserves the form $<,>$. It allows him to define the group $M = \langle H, \sigma \rangle$ which he proves to be a simple group of M-type having the order $|\mathsf{M}|$ stated above. Griess' proof for the important isomorphism $C_M(z) \cong H$ quotes several results from the literature on the classification of finite simple groups.

The following theorem is due to Griess, Meierfrankenfeld and Segev [44]. In its proof the authors apply Thompson's group order formula to calculate the order of M.

Theorem 16.1.4 *Let G be a finite simple group of M-type containing an involution u such that $C_G(u)$ is isomorphic to the 2-fold cover $2\mathsf{B}$ of the baby monster B. Then G is uniquely determined up to isomorphism and* $|G| = 2^{46} \cdot 3^{20} \cdot 5^9 \cdot 7^6 \cdot 11^2 \cdot 13^3 \cdot 17 \cdot 19 \cdot 23 \cdot 29 \cdot 31 \cdot 41 \cdot 47 \cdot 59 \cdot 71$.

The Griess algebra $\mathcal{B}$ and his construction of the group M have received great attention in the literature.

Remark 16.1.5 (a) Proposition 10.3.6 of the book by Frenkel, Lepowsky and Meurman [32] gives another construction of a similar algebra $\mathcal{B}_1$. It also uses the Leech lattice Λ. Proposition 10.4.1 of that book describes a construction of a group H_M with center $Z(H_M) = \langle z \rangle$ of order 2 and extra-special Fitting subgroup O_M of order 2^{25} such that H_M/O_M is isomorphic to the sporadic Conway group Co_1 and $O_M/Z(H_M) \cong \Lambda \otimes GF(2)$. This construction of H_M uses only the results of Conway's classical articles [16] and [17] and elementary homological algebra.

(b) Frenkel, Lepowsky and Meurman define an action of H_M on their constructed Griess-type algebra $\mathcal{B}_1$. They give another existence proof of a monster group M_1 in the context of automorphism groups of vertex operator algebras; see Theorems 12.3.1(d) and Theorem 12.3.4 of [32]. However, they calculate neither the order nor the character table of their group M_1. Instead they quote Tits' article [133] and Griess' proof [42] for the isomorphism $C_{M_1}(z) \cong H_M$; see [32], p. 407.

(c) Frenkel, Lepowsky and Meurman say in [32], p. 407, that their Griess-type algebra $\mathcal{B}_1$ is a *deformation* of the original Griess algebra $\mathcal{B}$. The authors of the Atlas [19] write on p. 228: *"The multiplication of the algebra described below differs slightly from that of Griess, so care is necessary in comparing results."* Therefore we denote the monster groups of [42] and [32] by M and M_1, respectively.

Remark 16.1.6 Neither [32] nor [42] provide a presentation of the extension group H_M of Remark 16.1.5. They also do not give a faithful permutation representation for this group. Furthermore, an arbitrary extension group H of Definition 16.1.3 is not necessarily isomorphic to H_M; see statement (2.6) on p. 575 of [44]. Therefore an independent construction of the character table of the well defined group H_M should be given from scratch. See Remark 16.1.11.

Remark 16.1.7 (a) The mathematical literature does not contain self-contained proofs of the character tables of the monster groups M and M_1. *Why should they be equivalent to the table in the Atlas [19], pp. 222–227?*

The monster group M_1 constructed in [32], p. 407, acts on a well defined semi-simple representation $\mathcal{B}_1$ which is the direct sum of the trivial representation and a 196883-dimensional representation of M_1. So all values of the corresponding faithful character χ of degree 196883 of M_1 are uniquely determined. If it were possible to find representatives m of the conjugacy classes of M_1 then one could calculate their values $\chi(m)$ and construct the whole character table of M_1 by the Burnside–Brauer Theorem 2.5.4 of [92], at least in theory. Of course, a *proven* character table of $H_M \cong H_{M_1}$ would simplify that computation; see Remark 16.1.6.

If one constructs the character table of M_1 then one can read off the orders $|M_1|$ and $|C_{M_1}(z)|$. If $|C_{M_1}(z)| = |H_M|$ then $C_{M_1}(z) \cong H_M$. In particular, this existence proof for the monster group M_1 would be self-contained and free from quotations of [42]. If true, the simplicity of M_1 would follow from its character table.

Of course, this character table method can also be applied to Griess' constructed monster group M of [42].

Remark 16.1.8 For other coherent accounts describing our present knowledge about these two huge sporadic simple groups B and M the reader is referred to A. Ivanov's books [62] and [63], which study all sporadic simple groups from a geometric point of view.

Recent matrix constructions of the monster group are described by P. A. Holmes and R. A. Wilson in [51] and the references therein.

The main question in this section is: *Can the baby monster* B *and the monster* M *be constructed by means of Algorithms 1.3.8* or *1.3.15?*

The author conjectures that the answer is *yes*, at least in theory. The following two remarks support this statement.

Remark 16.1.9 Inspired by the Atlas [19], p. 217, we conjecture that a Sylow 2-subgroup of B has an elementary abelian normal subgroup A of order 2^9 with special centralizer $C = C_{\mathsf{B}}(A)$ of order 2^{25} such that C/A and A are irreducible and indecomposable $GF(2)\mathsf{Sp}_8(2)$-modules of dimensions 16 and 9 over $GF(2)$, respectively, and $N_{\mathsf{B}}(A)/C \cong \mathsf{Sp}_8(2)$. In particular, $E = N_{\mathsf{B}}(A)$ is an iterated extension of the indecomposable subgroup $T = \mathsf{Sp}_8(2)$ of $\mathrm{GL}_9(2)$.

This conjecture would follow from a construction of a simple group G of B-type by means of an application of Algorithm 1.3.15 to an indecomposable subgroup T of $\mathrm{GL}_9(2)$. In theory, such groups T and G can be constructed as follows.

(a) Determine the conjugacy classes of indecomposable subgroups T of $\mathrm{GL}_9(2)$ of order $|T| = 2^{16} \cdot 3^5 \cdot 5^2 \cdot 7 \cdot 17$. At least one subgroup T should be isomorphic to $\mathsf{Sp}_8(2)$. Let $A = GF(2)^9$ be the $GF(2)T$-module defined by the generating matrices of $T \cong \mathsf{Sp}_8(2)$.
(b) Construct the non-isomorphic simple $GF(2)T$-modules V_i of dimension 16 over $GF(2)$. Compute all $GF(2)T$-module extensions X of T by V_i up to isomorphism by means of Holt's Algorithm [54] and the Cannon–Holt Algorithm [12].
(c) For each group X compute all $GF(2)T$-module extensions E of X by A by means of Holt's Algorithm [54] such that the Fitting subgroup V of X acts trivially on A. Determine a complete set $\mathfrak{S}$ of non-isomorphic extension groups E by means of the Cannon–Holt Algorithm [12].
(d) Let $E \in \mathfrak{S}$. Apply Step 2 of Algorithm 1.3.15 until you have found an extension E having a 2-central involution z not contained in the center $Z(E)$ of E such that A is an elementary abelian normal subgroup in a Sylow 2-subgroup S of $D = C_E(z)$ and $O = C_E(A)$ is a maximal special normal subgroup of E with center $Z(O) = A$, respectively.

(e) Check that D has an extra-special normal subgroup P of order $|P| = 2^{23}$ with center $Z(P) = \langle z \rangle = Z(D)$ such that the elementary abelian normal subgroup $P_1 = P/Z(D)$ of $D_1 = D/Z(D)$ has a quotient group $D_2 = D_1/P_1 \cong H(\mathsf{Co}_2)$, where $H(\mathsf{Co}_2)$ is the finitely presented group H with center $Z(H)$ of order 2 constructed in Proposition 4.2.1(n). By Theorem 4.3.2 $H(\mathsf{Co}_2) \cong C_{\mathsf{Co}_2}(t)$, where t is a 2-central involution of Co_2.

(f) The conjugate action of D_1 on its elementary abelian normal subgroup P_1 induces a monomorphism $\psi : D_2 \to \mathrm{GL}_{22}(2)$ such that $D_3 = \psi(D_2) \cong H(\mathsf{Co}_2)$. Hence we can apply Algorithm 7.4.8 of [92] as in the proof of Proposition 4.2.1 to construct a subgroup C of $\mathrm{GL}_{22}(2)$ such that $C \cong \mathsf{Co}_2$ and P_1 is an irreducible $GF(2)C$-module of dimension 22.

(g) Using the permutation representation and the presentation of the group $C \cong \mathsf{Co}_2$ given in Corollary 4.3.3 compute all $GF(2)C$-module extensions Y of C by P_1 up to isomorphism by means of Holt's Algorithm [54] and the Cannon–Holt Algorithm [12]. For each extension group Y construct a faithful permutation representation PY. Check that at least one extension group Y has a Sylow 2-subgroup S_Y such that S_Y is isomorphic to a Sylow 2-subgroup of D_1.

(h) For such an extension group Y construct a faithful permutation representation PY. Using the presentation of Y constructed by MAGMA (or other tools) in (g) compute all central extensions W of Y by $Z(D)$ by means of Holt's Algorithm [54] up to isomorphism. For each group W construct a faithful permutation representation PW and a Sylow 2-subgroup S_W. Check that there is at least one extension group W such that S_W has an elementary abelian normal subgroup B_W of order 2^9 satisfying $N_W(B_W) \cong D$. If so, let $H = W$ and let $B_W = A_H$.

(i) This group H satisfies the conditions of Definition 16.1.2. Moreover, by (h) all conditions of Step 3 of Algorithm 1.3.15 are satisfied. Hence we can apply all further steps of Algorithm 1.3.15 and assume that $D = H \cap E$ and so $A_H = A$.

Now we calculate the character tables of H, D and E. A successful application of Step 7 of Algorithm 1.3.8 will deliver an irreducible subgroup $\mathfrak{B}$ of $\mathrm{GL}_{4371}(p)$, where $p \in \{31, 47\}$. The final choice of the prime p depends on the values of the irreducible constituents of the constructed compatible pair (χ, τ) of degree 4371 of semi-simple characters of H and E.

(j) At this time it is technically not possible to perform Step 8 of Algorithm 1.3.8 to construct a faithful permutation representation of degree 13571955000 of the matrix group $\mathfrak{B}$ and show that its order equals $|\mathsf{B}|$ given above. But this may change in the future.

Remark 16.1.10 Let H be the finitely presented group constructed in Remark 16.1.9(i). One can try to show that $G \cong \mathfrak{B}$ for each simple group G having a 2-central involution z such that $C_G(z) = H$ by the following steps.

Using the faithful permutation representation PH of the given finitely presented group H constructed in Remark 16.1.9(h) one tries to settle the fusion problem of all the conjugaccy classes of involutions of H in such a putative simple group G. Then one constructs the centralizers $H_i = C_G(z_i)$ of all non-conjugate classes z_i of G, where $z_1 = z$ and $H_1 = H$. Check that they are uniquely determined by H also. After calculating the fusion of the conjugacy classes of involutions of each group H_i in G, and also the character tables of all H_i, we can apply Theorem 1.6.4 to obtain the group order $|G|$ of each simple group G of B-type. Determining then the structure and the character tables of all normalizers $N_G(\langle q\rangle)$ of the non-conjugate cyclic subgroups of prime order q in G we should be able to apply Brauer's characterization of characters stated as Theorem 2.8.9 in [92]. If so, then it will assert that each simple group of B-type has an irreducible character of degree 4371. Hence Theorem 7.5.1 of [92] will imply that $G \cong \mathfrak{B}$.

Remark 16.1.11 Inspired by the Atlas [19], p. 234, we conjecture that a Sylow 2-subgroup of M has an elementary abelian normal subgroup A of order 2^{10} with special centralizer $C = C_{\mathsf{M}}(A)$ of order 2^{26} such that C/A and A are irreducible $GF(2)\mathsf{O}^+_{10}(2)$-modules of dimensions 16 and 10 over $GF(2)$, respectively, and that $N_{\mathsf{M}}(A)/C \cong \mathsf{O}^+_{10}(2)$. In particular, $E = N_{\mathsf{M}}(A)$ is an iterated extension of the irreducible subgroup $T = \mathsf{O}^+_{10}(2)$ of $\mathrm{GL}_{10}(2)$.

This conjecture would follow from a construction of a simple group G of M-type by means of an application of Algorithm 1.3.15 to an iterated extension E of the irreducible subgroup T of $\mathrm{GL}_{10}(2)$. In theory, such groups T, E and G can be constructed as follows.

(a) Determine the conjugacy classes of irreducible subgroups T of $\mathrm{GL}_{10}(2)$ of order $|T| = 2^{20} \cdot 3^5 \cdot 5^2 \cdot 7 \cdot 17 \cdot 31$. At least one subgroup T should be isomorphic to $\mathsf{O}^+_{10}(2)$. Let $A = GF(2)^{10}$ be the $GF(2)T$-module defined by the generating matrices of $T \cong \mathsf{O}^+_{10}(2)$.

(b) Now apply steps (b), (c) and (d) of Remark 16.1.9 until you obtain an extension group E having a 2-central involution z such that A is an elementary abelian normal subgroup in a Sylow 2-subgroup S of $D = C_E(z)$ and $O = C_E(A)$ is the special Fitting subgroup of E with center $Z(O) = A$.

(c) Check that D has an extra-special normal subgroup P of order $|P| = 2^{25}$ with center $Z(P) = \langle z \rangle = Z(D)$ such that the elementary abelian normal subgroup $P_1 = P/Z(D)$ of $D_1 = D/Z(D)$ has a quotient group $D_2 = D_1/P_1 \cong H(\mathsf{Co}_1)$, where $H(\mathsf{Co}_1)$ is the finitely presented group H with center $Z(H)$ of order 2 constructed in Proposition 7.2.2(i). By Theorem 7.3.2, $H \cong C_{\mathsf{Co}_1}(t)$, where t is a 2-central involution of Co_1.

(d) The conjugate action of D_1 on its elementary abelian normal subgroup P_1 induces a monomorphism $\psi : D_2 \to \mathrm{GL}_{24}(2)$ such that $D_3 = \psi(D_2) \cong H(\mathsf{Co}_1)$. Hence we can apply Algorithm 7.4.8 of [92] as in Proposition 7.2.2 to construct a subgroup C of $\mathrm{GL}_{24}(2)$ such that $C \cong \mathsf{Co}_1$ and P_1 is an irreducible $GF(2)C$-module of dimension 24.

(e) Using the permutation representation of Theorem 7.3.2 and the presentation of Theorem 7.3.3 compute all $GF(2)C$-module extensions Y of C by P_1 up to isomorphism by means of Holt's Algorithm [54] and the Cannon–Holt Algorithm [12]. For each extension group Y construct a faithful permutation representation PY. Check that at least one extension group Y has a Sylow 2-subgroup S_Y which is isomorphic to a Sylow 2-subgroup of D_1.

(f) For such an extension group Y compute all central extensions W of Y by $Z(D)$ by means of Holt's Algorithm [54] up to isomorphism. For each group W construct a faithful permutation representation PW and a Sylow 2-subgroup S_W. Check that there is at least one extension group W such that S_W has an elementary abelian normal subgroup B_W of order 2^{10} such that $N_W(B_W) \cong D$. If so, let $H = W$ and let $B_W = A_H$.

(g) This group H satisfies the conditions of Definition 16.1.3. Moreover, by (f) all conditions of Step 5 of Algorithm 1.3.8 are satisfied. Hence we can apply all further steps of Algorithm 1.3.8 and assume that $D = H \cap E$ and so $A_H = A$.

In particular, we have to calculate the character tables of H, D and E. Most likely, a successful application of Step 7 of Algorithm 1.3.8 will deliver an irreducible subgroup $\mathfrak{M}$ of $\mathrm{GL}_{196883}(p)$, where $p \in \{47, 59, 71\}$. The final choice of the prime p depends on the values of the irreducible constituents of the constructed compatible pair (χ, τ) of degree 196883 of well determined semi-simple characters of H and E. The characters χ and τ will be the restrictions of an irreducible character ψ of $\mathfrak{M}$ of p-defect zero to the isomorphic images of H and E in $\mathfrak{M}$, respectively. So one should try to calculate the values of ψ. This problem is related to that of Remark 16.1.7.

Remark 16.1.12 For technical reasons it is not possible to perform Step 8 of Algorithm 1.3.8 and construct a faithful permutation representation of degree $|\mathsf{M}| : |\mathsf{B}|$ of the matrix group $\mathfrak{M}$ and show that its order equals $|\mathsf{M}|$ given above.

One might be able to show, however, that any simple group G having a 2-central involution z such that $C_G(z) \cong H$ described in Remark 16.1.11 has an irreducible character of degree 196883 as follows.

Using the faithful permutation representation PH of the given finitely presented group H constructed in Remark 16.1.11 one should try to settle the fusion problem of all the conjugacy classes of involutions of H in the presumed simple group G with $C_G(z) \cong H$. In view of Theorem 16.1.4 it should follow that G has exactly two conjugacy classes of involutions represented by z and u. Then one tries to construct $U = C_G(u)$ from $H = C_G(z)$ as a finitely presented group. The structure and the presentation of U should be uniquely determined by H up to isomorphism. Now calculate the fusion of the conjugacy classes of involutions of the group U in G. Compute the character tables of H and U. Then apply Theorem 1.6.4 to obtain the group order of G. It will be of great help to determine the structure and the character tables of all normalizers $N_G(\langle q \rangle)$ of the non-conjugate cyclic subgroups of prime order q in G. Then one should be able to apply Brauer's characterization of characters stated as Theorem 2.8.9 in [92]. If so, then it will assert that each simple group G with $C_G(z) \cong H$ has an irreducible character of degree 196883.

16.2 Uniqueness problems

Let H be a fixed finite group with a center $Z(H)$ of even order. In order to avoid ambiguities we assume throughout this section that H is given as a finitely presented group. The literature contains many uniqueness theorems. But, in general, their definitions of *uniqueness* do not agree. By the Brauer–Fowler Theorem 2.5.12 of [92] the following definition is justified.

Definition 16.2.1 A simple group G having a 2-central involution z with centralizer $C_G(Z)$ is said to be *uniquely determined* if each simple group Y having a 2-central involution y such that $C_Y(y) \cong H$ is isomorphic to G.

A not necessarily simple group Y having a 2-central involution y with $C_Y(y) \cong H$ is said to be a *satellite of* G if Y is not isomorphic to G.

Example 16.2.2 By Theorem 8.1.9 of [92] due to W. J. Wong [138] we know that the Mathieu group M_{12} has up to isomorphism exactly one non-simple satellite. Therefore M_{12} is uniquely determined in the sense of Definition 16.2.1.

M. Suzuki's survey article [125] shows that many simple groups are uniquely determined. However, that paper also mentions several examples of simple groups which have simple satellites. Suzuki's list is restated in the introduction of [92] (see p. 5). It contains the five sporadic simple groups M_{11}, M_{24}, J_2, J_3 and He, several classical groups and an infinite sequence of pairs of alternating groups. But Suzuki's article does *not* claim that his list of simple groups with simple satellites is complete.

In [92] and this book we constructed a presentation of the centralizer $H = C_G(z)$ of at least one 2-central involution for each sporadic simple group G different from the baby monster B and the monster M. Furthermore, we proved in [92] that the sporadic simple groups J_1, HS, Ha and Th are uniquely determined. This is also true for the Mathieu groups M_{22} and M_{23} by Theorems 8.2.5 and 8.2.6 of [92] due to Z. Janko. In this book we provide such uniqueness proofs for the sporadic groups Co_3, J_4, Ru, McL, Suz and ON. We give fairly detailed outlines of uniqueness proofs for Fi_{22} and Fi'_{24} in Chapters 5 and 9, respectively.

Remark 16.2.3 In [72] Kim and the author showed that Co_2 does not satisfy condition (4) of the author's sufficient uniqueness criterion stated as Theorem 7.5.1 in [92] because the amalgam $H \leftarrow D \rightarrow E$ constructed in Lemma 4.3.1 has Goldschmidt index 2, *not* 1. Since the uniqueness criterion is not *necessary* the author mentions in Chapter 4 that Smith's article [119] contains a uniqueness proof for Conway's sporadic simple group Co_2. But it quotes a result on the structure of the centralizer H of a 2-central involution of Co_2 of an article which has not been published. By the proof of Proposition 4.2.1 the group H is not uniquely determined by its composition factors. Hence in [119] the centralizer $H = C_G(z)$ of a 2-central involution is not well defined. Therefore we pose the question:

Is each simple group G of Co_2-type isomorphic to the simple subgroup $\mathfrak{G}_3$ of $\mathrm{GL}_{23}(13)$ constructed in Theorem 4.3.2?

Remark 16.2.4 We mention in Chapter 6 that the literature does not contain a complete proof for the uniqueness of Fischer's simple group $G = \mathsf{Fi}_{23}$. Since it has three conjugacy classes of 2-central involutions z_i Theorem 7.5.1 of [92] cannot be applied because we have to prove uniqueness of G with respect to each of the three non-isomorphic centralizers $H_i = C_G(z_i)$, $1 \leq i \leq 3$. At least the uniqueness problem with respect to the centralizer H_2 of G has not been considered in the literature; see Remark 6.4.4.

This is a special problem of a more difficult *uniqueness problem of simple groups G having several conjugacy classes of 2-central involutions z_i, $1 \leq i \leq k$*. In fact, in general the number k of 2-central involutions of a simple group G may be large; see Kondo's articles [76] and [77] on the infinite series of alternating groups A_{4k+r}, with $k > 3$ and $0 \leq r \leq 3$.

Remark 16.2.5 Whenever a finite simple group G has several conjugacy classes of 2-central involutions z_i, $1 \leq i \leq k$, it has k centralizers $H_i = C_G(z_i)$ of 2-central involutions, each of which can be taken as the input of Algorithm 7.4.8 of [92], at least in theory.

For any fixed group H_j one has to construct all possible isomorphism types of simple groups X having a 2-central involution x such that $C_X(x) \cong H_j$. In order to construct these groups X one has to settle the fusion problem of the conjugacy classes of involutions of the given centralizer H_j in a putative group X. For each conjugacy class of involutions y of X one has to construct its centralizer $C_X(y)$ in X as an *abstract* group. If one does not obtain the same set $\{H_i \mid 1 \leq i \leq k\}$ of isomorphism types of centralizers of 2-central involutions of X in each of the k cases $j = 1, 2, \ldots k$, then one cannot prove that G is uniquely determined up to isomorphism.

T. Kondo and H. Yamaki studied this problem for the alternating groups in their articles [76], [77] and [140], respectively. Kondo's works show that the alternating groups $G \in \{A_{4k+2}, A_{4k+3}\}$ with $k \geq 3$ are uniquely determined by any given centralizer H of a 2-central involution of G, and that A_{4k} and A_{4k+1} are the only pairs of simple satellites provided $k \geq 4$. In [76] Kondo gives a presentation for the centralizer of each 2-central involution of the alternating groups.

Example 16.2.6 In [140] Yamaki proved that the simple group A_{12} has two satellites in the sense of Definition 16.2.1. Moreover, A_{12} has precisely three conjugacy classes of involutions z_i with centralizers H_i of orders $|H_1| = 2^9 \cdot 3^2 \cdot 5 \cdot 7$, $|H_2| = 2^9 \cdot 3^2 \cdot 5$ and $|H_3| = 2^9 \cdot 3^2$. Hence all involutions of A_{12} are 2-central. They also represent all the conjugacy classes of involutions of $G_1 = A_{13}$. Hence they are 2-central in G_1 as well. Let $K_i = C_{G_1}(z_i)$. Then $|K_1| = 2^9 \cdot 3^4 \cdot 5 \cdot 7$, $K_2 = H_2$ and $|K_3| = 2^9 \cdot 3^2 \cdot 5$ by [19], p. 102. Yamaki showed that $G_2 = \mathsf{Sp}_6(2)$ has a 2-central involution t whose centralizer $H = C_{G_2}(t) \cong H_2$. He proved that A_{12}, G_1 and G_2 are up to isomorphism all the simple groups G having a 2-central involution z with centralizer $C_G(z) \cong H_2$. His proof is based on a highly original classification of all possible fusion patterns of the distinct conjugacy classes of involutions of any simple group with given centralizer H_2. He showed that G_2 is the only satellite of A_{12} which has four conjugacy class of involutions; see also Proposition 8.6.5 and Theorem 8.6.6, both of [92].

Furthermore, Example 16.2.6 shows that any of the three satellites of A_{12} is uniquely determined by any given pair of its centralizers of 2-central involutions. Therefore theorems such as Theorem 16.1.4 *do not* solve the general uniqueness problem of the M-type simple groups G completely, although they are important contributions.

Remark 16.2.7 Proving the uniqueness in the sense of Definition 16.2.1 for all simple groups of Lie type requires more methods than those developed in [92] and this book; see P. Fong's article [31] for an example. There is a huge amount of literature written about the classification of their conjugacy classes and irreducible characters; see [14] and its bibliography.

At this time, however, the mathematical literature may not contain an accessible proof for the assertion that any simple group G of Lie type is *uniquely determined* in the sense of Definition 16.2.1 if G is not an item in Suzuki's list of simple groups with the proper simple satellites mentioned before. Held's Theorem 8.3.3 of [92] shows that a finite simple group G of Lie type may have satellites which are sporadic simple groups. In this case $G = \mathrm{GL}_5(2)$ and the sporadic simple group He was unknown before Held's discovery in [49]. Therefore *the above question on the uniqueness of all simple groups of Lie type* must find an answer which is independent of the announced classification theorem.

This is a very demanding problem. It certainly would require new general uniqueness criteria. Furthermore, one would have to find presentations of the centralizers of all 2-central involutions of the finite simple groups of Lie type in terms of their classical generators and Steinberg relations described in [13].

16.3 Is there a 27th sporadic simple group?

According to the announced classification theorem the answer to this critical question is *no*. Therefore the reader may be surprised to find it in R. Solomon's recent survey article [122] entitled "A brief history of the classification of the finite simple groups" published in 2002. On its p. 347 Solomon writes: *"Is there a 27th sporadic simple group? I seriously doubt it, but it would be chutzpahdich to assert that a* 5000-*page 40-year human endeavor is beyond the possibility of human error."* Then he gives several references and important reasons supporting his opinion. One of them is: *"For almost two decades in the* 60*s and* 70*s, keen-eyed seekers scoured the terrain searching for new sporadic treasures."*

He does not mention, however, that the mathematical literature contains a suggestion for such an unfinished experiment. In [127] Thompson announced the existence and uniqueness of his sporadic simple group Th and sketched an existence proof realizing Th as a subgroup of the simple group of Lie type $E_8(3)$. Another existence proof and a uniqueness proof for Th are given in chapter 12 of [92]. In [127] Thompson also constructs a finite subgroup K of $E_8(\mathbb{C})$ with the following properties: K has a normal subgroup L of order 2^{15} such that $K/L \cong \mathrm{GL}_5(2)$; see [127], p. 115, and also [78], p. 442. Then he writes: "*Since* $|L| = 2^{15}$ *(and not* 2^5*), I wonder if there might be a new simple finite group in* $E_8(\mathbb{C})$*. I do not recommend hunting for such a thing, but if one happens to come your way, you might want to examine it."*

Remark 16.3.1 Theorem 12.3.1 of [92] realizes Thompson's sporadic simple group Th inside $\mathrm{GL}_{248}(11)$ and not as a subgroup of $E_8(\mathbb{C})$. Therefore it is not necessary to search for such a new simple group only inside $E_8(\mathbb{C})$.

In [41] Griess describes the above iterated extension K of $\mathrm{GL}_5(2)$ in more detail as an abstract group and proves the following result.

Theorem 16.3.2 *Let D be the unique non-split extension of* $\mathrm{GL}_5(2)$ *by its natural $GF(2)$-vector space V of dimension* 5. *Let R be the split extension of* $\mathrm{GL}_5(2)$ *by the* 10*-dimensional exterior power $\Lambda^2 V$ of V. Then Thompson's group K is isomorphic to a non-split extension of R by V with the following properties.*

(a) *The Fitting subgroup L of K is a special group of order 2^{15} such that $L' = \Phi(L) = Z(L)$ is elementary abelian of order 2^5.*

(b) *K contains a subgroup $U \cong D$ such that $K = UL$, $L \cap U = Z(L)$, $K/L \cong \mathrm{GL}_5(2)$ and $C_K(Z(L)) = L$.*

Proof See [44], pp. 275–277. □

By Remark 1.3.12 we know that Thompson's sporadic simple group Th can be constructed by an application of Algorithm 1.3.8 to the uniquely determined non-split extension $E \cong D$ of $T = \mathrm{GL}_5(2)$ by V. For a presentation and a faithful permutation representation PD of D of degree 34560 see Lemma 12.1.2 of [92].

The existence and uniqueness of the extension group D is due to U. Dempwolff [22]. Nowadays his result can easily be checked by means of MAGMA and Holt's Algorithm [54].

Remark 16.3.3 Similarly, it can be shown computationally that $M = \Lambda^2 V$ is an irreducible $GF(2)T$-module and that all extensions R of T by M split. Hence R is uniquely determined. MAGMA provides a fairly short presentation and a faithful permutation representation PR for R. Given a large enough computer with an efficient implementation of Holt's algorithm [54] it should be possible to calculate $\dim_{GF(2)}[H^2(R, M)]$. If so, then one searches for a non-split extension E of R by M such that $E \cong K$. Furthermore, MAGMA will automatically provide a presentation of E, but *not* a faithful permutation representation. If a suitable permutation representation of E can be constructed then it is theoretically possible to apply Steps 2 and 3 of Algorithm 1.3.15 to this iterated extension group E. Since a Sylow 2-subgroup of E has order 2^{25} this is a technically demanding experiment.

If no further technical difficulties occur, then this application of Algorithm 1.3.15 will either prove or disprove Thompson's question stated above.

16.4 Is there a general classification scheme?

This question is closely related to Brauer's statements quoted in Remark 1.1.4. One of them asks whether there is an infinite sequence of unknown non-isomorphic sporadic simple groups G_n with orders $|G_n|$ strictly increasing. In such a situation he finds it necessary to have a precise definition for "what we mean by a classification of the simple groups." By Brauer's Remark 1.1.4(c) such a classification scheme may not exist at all. The following *special* classification problem illustrates some evidence for this alternative to the announced classification Theorem 1.1.2.

In this section we introduce an exhaustive search method for all finite simple groups G which can be constructed by an application of Algorithm 1.3.8 to some indecomposable subgroup of the strictly increasing sequence of general linear groups $\mathrm{GL}_n(2)$, $n \geq 3$. By Remarks 1.3.11, 1.3.12 and Chapters 3–12 we know that the sporadic simple groups M_{12}, M_{22}, M_{23}, M_{24}, J_1, He, Ha, Th, Co_3, Co_2, Co_1, Fi_{22}, Fi_{23}, Fi_{24}', J_4, McL and the Tits simple group ${}^2F_4(2)'$ can be constructed by means of Algorithm 1.3.8 from well determined indecomposable subgroups T of some general linear groups $\mathrm{GL}_n(2)$, where $3 \leq n \leq 11$ is an invariant of the particular simple group. However, Algorithm 1.3.8 can be applied (in theory) to all indecomposable subgroups of *any* general linear group $\mathrm{GL}_n(2)$.

There are several classical *special* classification theorems dealing with the simple groups G having a strongly embedded subgroup or a dihedral or semi-dihedral Sylow 2-subgroup; see chapter 1 of [92]. They were used in the proof of Theorem 4.8.5 of [92]. This theorem and Lemma 1.3.2 are the theoretical foundations of Algorithms 1.3.8 and 1.3.15.

All the preceding examples satisfy the conditions of the following special classification problem.

Question 16.4.1 What are the finite simple groups G having a Sylow 2-subgroup S with the following properties?

(a) S has a maximal elementary abelian normal subgroup A of order 2^n which contains a non-cyclic characteristic subgroup of S.

(b) $C_G(A) = A$, and $T = N_G(A)/A$ is isomorphic to an indecomposable subgroup of $\mathrm{GL}_n(2)$.

(c) S has a central involution z such that A is not normal in $H = C_G(z)$ and $G = \langle H, N_G(A)\rangle$.

In particular, Question 16.4.1 asks for all the finite simple groups G with property (a) which can be constructed by means of an application of Algorithm 1.3.8 to some indecomposable subgroup T of $\mathrm{GL}_n(2)$. By Remark 1.3.14 we know that the condition $C_G(A) = A$ implies that $n \geq 3$.

Remark 16.4.2 Fong's Theorem 2.1.2 stated in Chapter 2 deals with an *infinite* series of finite groups of Lie type $\mathsf{G}_2(q)$, where $q \equiv 3, 5 \bmod (8)$. All of them have a Sylow 2-subgroup S of order 2^6 such that $N_{\mathsf{G}_2(q)}(A)$ is isomorphic to a unique non-split extension E of $T = \mathrm{GL}_3(2)$ by the maximal elementary abelian normal subgroup A of S of order 8. Theorem 2.1.2 asserts that each of these simple groups $\mathsf{G}_2(q)$ is uniquely determined by the centralizer $H(q) = C_{\mathsf{G}_2(q)}(z)$ of a 2-central involution $z \in \mathsf{G}_2(q)$. Furthermore, each $H(q)$ has a presentation which is generic in q. Thus the whole infinite series can be detected whenever one of the centralizers $H(q)$ has been constructed by a successful application of Algorithm 1.3.8 to T.

Remark 16.4.3 Let $F = GF(2)$ and let n be a fixed integer n. There are only finitely many conjugacy classes of indecomposable subgroups T_k in $\mathrm{GL}_n(2)$. Let $\mathcal{T}_n$ be the set of all representatives T_k of indecomposable subgroups which are *not* 2-groups.

Let $\mathcal{S}_n$ be the set of representatives of the Sylow 2-subgroups S_k of all extensions E of all the groups $T_k \in \mathcal{T}_n$ by $V = F^n$. Hence each $S \in \mathcal{S}_n$ has an order divisible by 2^n and dividing 2^{n+r}, where $r = \frac{1}{2}n \cdot (n-1)$.

This order restriction allows us to determine all *known* finite simple groups K having a Sylow 2-group of order divisible by 2^n and being a divisor of 2^{n+r}. Let $\mathcal{K}_n$ be the set of all known simple groups whose Sylow 2-subgroups have such an order.

In general $\mathcal{K}_n$ is *not* a finite set because there are some infinite series of finite groups of Lie-type such that all members G of the series have isomorphic Sylow 2-subgroups; see D. Gorenstein's book [39], p. 75.

Whenever there is a group $G \in \mathcal{K}_n$ that also belongs to such an infinite sequence $\mathcal{L}_i$ of groups of Lie type $G_m(q)$ defined over finite fields $GF(q)$ of odd characteristic, we choose a group $G(\mathcal{L}_i)$ of minimal order in $\mathcal{K}_n \cap \mathcal{L}_i$. Since the number ℓ of such infinite sequences $\mathcal{L}_i$ is finite, the set $\mathcal{N}_n = \{G(\mathcal{L}_i) \mid 1 \leq i \leq \ell\}$ is finite as well. Let

$$I_n = \{G \in \mathcal{K}_n | G \notin \mathcal{L}_i \quad \text{for all} \quad 1 \leq i \leq \ell\}.$$

Then for each n the set $\mathcal{M}_n = \mathcal{N}_n \cup I_n$ is finite.

Remark 16.4.4 Keep the notations of Remark 16.4.3 and Question 16.4.1. Let $n \geq 3$ be a fixed integer. Let $\mathcal{M}_n$ be the finite set of known finite simple groups defined in Remark 16.4.3.

For each known simple group $K \in \mathcal{M}_n$ determine a Sylow 2-subgroup S, its maximal elementary abelian normal subgroups A containing a non-cyclic characteristic subgroup of S, $N_K(A)$ and $C_K(A)$. Let

$$\mathcal{R}_n = \{K \in \mathcal{M}_n \| A| = 2^n,\, C_K(A) = A \quad \text{and} \quad N_K(A)/A \cong T_k \in \mathcal{T}_n\}.$$

For each simple group K in the finite set $\mathcal{R}_n$ determine a Sylow 2-subgroup $S(K)$. Let $\mathcal{S}(R_n)$ be a set of representatives of the isomorphism types of all the groups $S(K)$.

Let $\mathcal{G}_n$ be the set of all finite simple groups G having a Sylow 2-subgroup S satisfying conditions (a)–(c) of Question 16.4.1.

Then all groups G in $\mathcal{G}_n$ are known finite simple groups if and only if either $G \in \mathcal{R}_n$ or G belongs to an infinite series of finite simple groups of Lie-type $\mathsf{G}(q)$ defined over finite fields $GF(q)$ of odd characteristic such that all its groups $\mathsf{G}(q)$ have a Sylow 2-subgroup which is isomorphic to some fixed group $S(K) \in \mathcal{S}(R_n)$.

In view of Remark 16.4.4 the problem of Question 16.4.1 can be answered for each integer n by the following *exhaustive search method.*

Remark 16.4.5 Keep the notation of Remark 16.4.4. Let n be the smallest integer for which Question 16.4.1 has yet not been answered. Then perform the following steps.

(a) For each $K \in \mathcal{R}_n$ determine $T(K) = N_K(A)/A$ as a subgroup of $\mathrm{GL}_n(2)$.

(b) Using a faithful permutation representation P_n of $\mathrm{GL}_n(2)$ and MAGMA determine all conjugacy classes of indecomposable subgroups T_k of $\mathrm{GL}_n(2)$ which are not 2-groups.

(c) Using P_n and the Cannon–Holt Algorithm [12] determine a complete set $\mathcal{C}_n$ of non-isomorphic indecomposable subgroups T_k of $\mathrm{GL}_n(2)$ which are not isomorphic to any indecomposable subgroup $T(K)$ with $K \in \mathcal{R}_n$.

(d) Apply Algorithm 1.3.8 to each indecomposable subgroup $T_k \in \mathcal{C}_n$.

If the application of Algorithm 1.3.8 to all $T_k \in \mathcal{C}_n$ fails to construct a simple group G which is not a finite simple group of Lie-type $\mathsf{G}(q)$ belonging to an infinite series of finite simple groups with isomorphic

Sylow 2-subgroups, then $\mathcal{G}_n$ consists of known simple groups described in Remark 16.4.4.

Otherwise, there is a new sporadic simple group G such that $N_G(A)/A = T_k$ for some indecomposable subgroup T_k of $\mathrm{GL}_n(2)$.

Remark 16.4.6 Step (c) of the exhaustive search method described in Remark 16.4.5 is demanding. Already for $n = 4$ there are 75 conjugacy classes of indecomposable subgroups in $\mathrm{GL}_4(2)$ which are not 2-groups. The author does not know the complete answer to Question 16.4.1, even for $n = 4$.

In Chapter 12 we constructed the Rudvalis simple group Ru by application of Algorithm 1.3.15 to the irreducible subgroup $T = \mathrm{GL}_3(2)$ of $\mathrm{GL}_8(2)$. Thus one may try to classify all finite simple groups G which can be constructed by an application of Algorithm 1.3.15 to some indecomposable subgroup T of $\mathrm{GL}_m(2)$. This classification problem can be restated as follows.

Question 16.4.7 What are the finite simple groups G having a Sylow 2-subgroup S with the following properties?

(a) S has an elementary abelian normal subgroup in A of order 2^n which contains a non-cyclic characteristic subgroup of S.
(b) $C = C_G(A) = C_S(A)$ is a maximal special or homocyclic normal subgroup of S with center $Z(C) = A$ or $\Omega(C) = \{c \in C | c^2 = 1\} = A$, respectively.
(c) $C/A = V$ is a self-centralizing elementary abelian normal subgroup of order 2^m in $T_1 = N_G(A)/A$ such that $T = N_G(A)/C$ is isomorphic to an indecomposable subgroup of $\mathrm{GL}_m(2)$.
(d) $T = N_G(A)/C$ is isomorphic to an indecomposable subgroup of $\mathrm{GL}_n(2)$.
(e) S has a central involution z such that A is not normal in $H = C_G(z)$ and $G = \langle H, N_G(A)\rangle$.

Remark 16.4.8 Keep the notation and conditions of Question 16.4.7. Let $m > 1$ be a fixed number. Then the modified classification problem of Question 16.4.7 *cannot* be solved by a finite exhaustion method. This is true because one cannot construct all integers n for which a given indecomposable subgroup T_k of $\mathrm{GL}_m(2)$ also has an indecomposable image in $\mathrm{GL}_n(2)$ since there are infinitely many such integers k by Higman's Theorem 1.2.3.

The special classification problems described above illustrate R. Brauer's Remark 1.1.4(c). In particular, Remarks 16.4.4, 16.4.8 and all the examples of known sporadic simple groups dealt with in [92] and in this book provide some evidence that it might be possible that there are some infinite sequences of non-isomorphic *unknown* simple groups G_n with strictly increasing orders $|G_n|$.

Appendix: Table of contents of the accompanying DVD

The accompanying DVD consists of two folders, DVD.1 and DVD.2, containing the pdf files of quoted tables and the MAGMA files of the generating matrices and permutations, respectively. Their contents are described below.

A.1 Folder DVD.1: Pdf files of quoted tables

- *Conway's group* Co_3; see Theorem 3.2.2.
  (1) Table `DVD.1.1.1` is a table of representatives of the conjugacy classes of $\mathfrak{G} \cong \mathsf{Co}_3$.
  (2) Table `DVD.1.1.2` is the character table of $\mathfrak{G} \cong \mathsf{Co}_3$.
- *Conway's group* Co_2; see Theorem 4.3.2.
  (1) Table `DVD.1.2.1` is a table of representatives of the conjugacy classes of $\mathfrak{G}_3 \cong \mathsf{Co}_2$.
  (2) Table `DVD.1.2.2` is the character table of $D(\mathsf{Co}_2)$; see Lemma 4.3.1.
- *Fischer's group* Fi_{22}; see Theorem 5.2.1.
  (1) Table `DVD.1.3.1` is a table of representatives of the conjugacy classes of $\mathfrak{G}_2 \cong \mathsf{Fi}_{22}$.
  (2) Table `DVD.1.3.2` is the character table of $D(\mathsf{Fi}_{22})$; see Lemma 5.1.2.
- *Fischer's group* $\mathfrak{G} \cong \mathsf{Fi}_{23}$; see Theorem 6.3.1.
  (1) Table `DVD.1.4.1` is a table of representatives of the conjugacy classes of $H_2 \cong H(2\mathsf{Fi}_{22})$; see Proposition 6.2.1.
  (2) Table `DVD.1.4.2` is a table of representatives of the conjugacy classes of $D_2 \cong D(2\mathsf{Fi}_{22})$; see Proposition 6.2.1.

(3) Table DVD.1.4.3 is the character table of $D_2 = D(2\mathsf{Fi}_{22})$; see Proposition 6.2.1.
(4) Table DVD.1.4.4 is the character table of $H_2 \cong H(2\mathsf{Fi}_{22})$; see Proposition 6.2.1.

- *Conway's group* $\mathfrak{G} \cong \mathsf{Co}_1$; see Theorem 7.3.2.

  (1) Table DVD.1.5.1 is a table of representatives of the conjugacy classes of $D(\mathsf{Co}_1) = \langle x, y \rangle$; see Proposition 7.2.2.
  (2) Table DVD.1.5.2 is a table of representatives of the conjugacy classes of $H_1(\mathsf{Co}_1)$; see Proposition 7.2.2.
  (3) Table DVD.1.5.3 is a table of representatives of the conjugacy classes of $U(\mathsf{Co}_1)$; see Proposition 7.2.2.
  (4) Table DVD.1.5.4 is a table of representatives of the conjugacy classes of $C_{\mathsf{Co}_1}(2b)$; see Corollary 7.4.4.
  (5) Table DVD.1.5.5 is a table of representatives of the conjugacy classes of $C_{\mathsf{Co}_1}(2c)$; see Corollary 7.4.4.
  (6) Table DVD.1.5.6 is the character table of $D = D(\mathsf{Co}_1)$; see Proposition 7.2.2.
  (7) Table DVD.1.5.7 is the character table of $H_1(\mathsf{Co}_1)$; see Proposition 7.2.2.
  (8) Table DVD.1.5.8 is the character table of $U(\mathsf{Co}_1)$; see Proposition 7.2.2.
  (9) Table DVD.1.5.9 is the character table of $C_{\mathsf{Co}_1}(2b)$; see Corollary 7.4.4.
  (10) Table DVD.1.5.10 is the character table of $C_{\mathsf{Co}_1}(2c)$; see Corollary 7.4.4.

- *Janko's group* $\mathfrak{G} \cong \mathsf{J}_4$; see Theorem 8.3.2.

  (1) Table DVD.1.6.1 is a table of representatives of the conjugacy classes of $D = D(\mathsf{J}_4) = \langle x, y \rangle$; see Proposition 8.1.1(1).
  (2) Table DVD.1.6.2 is a table of representatives of the conjugacy classes of $C_{\mathsf{J}_4}(2b)$; see Lemma 8.2.2.
  (3) Table DVD.1.6.3 is the character table of $D = D(\mathsf{J}_4) = \langle x, y \rangle$; see Proposition 8.1.1(1).
  (4) Table DVD.1.6.4 is the character table of $C_{\mathsf{J}_4}(2b)$; see Lemma 8.2.2.

- *Fischer's group* $\mathfrak{G} \cong \mathsf{Fi}'_{24}$; see Theorem 9.4.3.

  (1) Table DVD.1.7.1 is a table of representatives of the conjugacy classes of $D = D(\mathsf{Fi}'_{24})$; see Lemma 7.1.1(1).

(2) Table DVD.1.7.2 is the character table of mH; see Lemma 9.2.1.
(3) Table DVD.1.7.3 is the character table of mE; see Lemma 9.2.1.
(4) Table DVD.1.7.4 is the character table of $D = D(\mathsf{Fi}_{24}')$; see Lemma 7.1.1(1).

A.2 Folder DVD.2: MAGMA files of generating matrices and permutations

All files mentioned below are in MAGMA format. For unusually large matrices we give the storage space of their files. In MAGMA, load each of the files by means of "file-name.mag."

- *Conway's group* $G \cong \mathsf{Co}_3$; see Theorem. 3.2.2.
  (a) File 040907_MCo3.mag contains the four generating matrices $MCo3x$, $MCo3y$, $MCo3h$, $MCo3e$ of $G \cong \mathsf{Co}_3$ in $\mathrm{GL}_{23}(23)$.
- *Conway's group* $G_3 \cong \mathsf{Co}_2$; see Theorem 4.3.2.
  (a) File 011307_Co2.mag contains the four generating matrices $MCo2x$, $MCo2y$, $MCo2h$, $MCo2e$ of $G \cong \mathsf{Co}_2$ in $\mathrm{GL}_{23}(13)$.
- *Fischer's group* $G_2 \cong \mathsf{Fi}_{22}$; see Theorem 5.2.1.
  (a) File 012807_M78Fi22.mag contains the four generating matrices $MFi22x$, $MFi22y$, $MFi22h$, $MFi22e$ of $G_2 \cong \mathsf{Fi}_{22}$ in $\mathrm{GL}_{78}(13)$.
- *Fischer's group* Fi_{23}; see Theorem 6.3.1.
  (a) File 052908_M782Fi23.mag contains the four generating matrices $MFi23x$, $MFi23y$, $MFi23h$, $MFi23e$ of $G \cong \mathsf{Fi}_{23}$ in $\mathrm{GL}_{782}(17)$.
- *Conway's group* $G \cong \mathsf{Co}_1$; see Theorem 7.3.2.
  (a) File 070407_MCo1.mag contains the four generating matrices $MCo1x$, $MCo1y$, $MCo1h$, $MCo1e$ of $G \cong \mathsf{Co}_1$ in $\mathrm{GL}_{276}(23)$.
  (b) File 071607_Co1.perm.mag contains the four permutations $PCo1x$, $PCo1y$, $PCo1h$, $PCo1e$ of degree 98280 corresponding to the above matrix generators of $G \cong \mathsf{Co1}$.
- *Janko's group* J_4; see Theorem 8.3.2.
  (a) File 052908_MJ4.GL(1333,43).mag contains the four generating matrices $MJ4x$, $MJ4y$, $MJ4h$, $MJ4e$ in $\mathrm{GL}_{1333}(43)$, size 23 MB.

(b) The files

`J4GF11_1333matrizen.mag`, size 10 MB,

`J4GF2_112matrizen.mag`, size 74 KB

each contain two matrix generators of $\mathsf{J}_4 = \langle x, y \rangle \in \mathrm{GL}(1333, 11)$ and of $\mathsf{J}_4 = \langle x, y \rangle \in \mathrm{GL}(112, 2)$, respectively.

(c) The files

`XX.perm.0`, size 660 MB,

`YY.perm.0`, size 660 MB

contain M. Weller's generating permutations of degree 173067389 corresponding to the above matrix generators of $G \cong \mathsf{J}_4$.

(d) File `readme.txt` contains M. Weller's instructions for using the files of (c) in a new MAGMA session on a suitable workstation.

- *Fischer's group* $\mathfrak{G} \cong \mathsf{Fi}_{24}'$; see Theorem 9.4.3.

(a) File `052908_MFi24t.mag` contains the generating matrix $\mathfrak{t}$ in $\mathrm{GL}_{8671}(13)$ in MAGMA format, size 215 MB.

(b) File `052908_MFi24y.mag` contains the generating matrix $\mathfrak{y}$ in $\mathrm{GL}_{8671}(13)$ in MAGMA format, size 218 MB.

(c) File `052908_MFi24q.mag` contains the generating matrix $\mathfrak{q}$ in $\mathrm{GL}_{8671}(13)$ in MAGMA format, size 218 MB.

(d) File `052908_MFi24w.mag` contains the generating matrix $\mathfrak{w}$ in $\mathrm{GL}_{8671}(13)$ in MAGMA format, size 224 MB.

(e) File `052908_MFi24s.mag` gives a word in y and t for s.

(f) File `052908_MFi24x.mag` gives a word in y and q for x.

(g) The files `Fi24load.mag` and `readme.txt` contain H. Kim's instructions for using these files in a new MAGMA session on a suitable workstation.

- *McLaughlin's group* McL; see Theorem 11.3.1.

(a) File `052908_MMcL.mag` contains the four generating matrices $McLd$, $McLu1$, $McLu2$, $McLx$ in $\mathrm{GL}_{22}(11)$; see Corollary 11.3.2(c).

- *Rudvalis' group* Ru; see Theorem 12.2.1.

(a) File `102308_MRu.mag` contains the four generating matrices $MRux$, $MRuy$, $MRuh$, $MRue$ of $G \cong \mathsf{Ru}$ in $\mathrm{GL}_{378}(17)$.

- *Lyons' group* Ly; see Theorem 13.5.2.

(a) File `072207_MLyD.mag` contains the four generating matrices $MLyx$, $MLyy$, $MLyh$, $MLye$ of $G \cong \mathsf{Ly}$ in $\mathrm{GL}_{2480}(31)$.

- *Suzuki's group* Suz; see Theorem 14.3.1.

  (a) File `052208_MSuz.mag` contains the four generating matrices $MSuzx$, $MSuzy$, $MSuzh$, $MSuze$ in $\mathrm{GL}_{143}(13)$.

References

[1] Alperin, J. L., Sylow 2-subgroups of rank 3, in T. Gagen, M. P. Hale and E. Shult (eds.) *Finite Groups '72*. North Holland Mathematics Studies **7** (Amsterdam: North Holland, 1973), pp. 3–5.

[2] Alperin, J. L., R. Brauer and D. Gorenstein, Finite groups with quasidihedral and wreathed Sylow 2-subgroups, *Trans. Amer. Math. Soc.* **151** (1970), 1–262.

[3] Aschbacher, M., *3-Transposition Groups*, Cambridge Tracts in Mathematics **124** (Cambridge: Cambridge University Press, 1997).

[4] Aschbacher, M., *Sporadic Groups*, Cambridge Tracts in Mathematics **104** (Cambridge: Cambridge University Press, 1994).

[5] Aschbacher, M., Finite groups of $G_2(3)$-type, *J. Algebra* **257** (2002), 197–214.

[6] Assem, I., D. Simson and A. Skowronski, *Elements of Representation Theory of Associative Algebras I*, London Math. Soc. Student Texts **65** (Cambridge: Cambridge University Press, 2006).

[7] Bender, H., Transitive Gruppen gerader Ordnung, in denen jede Involution genau einen Punkt festlässt, *J. Algebra* **17** (1971), 527–554.

[8] Benson, D. J., The simple group J_4, Ph. D. Thesis, Trinity College, Cambridge, 1980.

[9] Bosma, W. and J. Cannon, *MAGMA V2.11 Handbook* (Sydney: University of Sydney), 2004. http://magma.maths.usyd.edu.au/magma/

[10] Brauer, R., Blocks of characters and structure of finite groups, *Bull. Amer. Math. Soc.* **1** (1979), 21–38.

[11] Cannon, J. J., B. Cox and D. F. Holt, Computing the subgroups of a permutation group, *J. Symbolic Computation* **31** (2001), 149–161.

[12] Cannon, J. J. and D. F. Holt, Automorphism group computation and isomorphism testing in finite groups, *J. Symbolic Computation* **35** (2003), 241–267.

[13] Carter, R. W., *Simple Groups of Lie Type* (London: J. Wiley and Sons, 1972).

[14] Carter, R. W., *Finite Groups of Lie Type: Conjugacy Classes and Complex Characters* (New York: J. Wiley and Sons, 1985).

[15] Crawley-Boevey, W. W., On tame algebras and bocses, *Proc. London Math. Soc.* **56** (1988), 451–483.

[16] Conway, J. H., A group of order 8,315,553,613,086,720,000, *Bull. London Math. Soc.* **1** (1969), 79–88.

[17] Conway, J. H., Three lectures on exceptional groups, in M. B. Powel and G. Higman (eds.) *Finite Simple Groups* (London: Academic Press, 1971), pp. 215–247.

[18] Conway, J. H., A characterization of the Leech lattice, *Inventiones Math.* **7** (1969), 137–142.

[19] Conway, J. H., R. T. Curtis, S. P. Norton, R. A. Parker and R. A. Wilson, *Atlas of Finite Groups* (Oxford: Clarendon, 1985).

[20] Conway, J. H. and D. B. Wales, Construction of the Rudvalis group of order $2^{14} \cdot 3^3 \cdot 5^3 \cdot 7 \cdot 13 \cdot 29$, *J. Algebra* **27** (1973), 538–548.

[21] Cooperman, G., W. Lempken, G. O. Michler and M. Weller, A new existence proof of Janko's simple group J_4, *Progress in Math.* **173**, (1999), 161–175.

[22] Dempwolff, U., On extensions of an elementary abelian group of order 2^5 by $\mathrm{GL}_5(2)$, *Rend. Sem. Mat. Univ. Padua* **48** (1973), 359–364.

[23] Dennis, R. K. and M. R. Stein, Injective stability for K_2 of local rings, *Bull. Amer. Math. Soc.* **80** (1974), 1010–1013.

[24] Dickson, L. E., A new system of simple groups, *Math. Annalen*, **60**, (1905), 137–150.

[25] Erdmann, K., *Blocks of Tame Representation Type and Related Algebras*, Lecture Notes in Mathematics **1248** (Heidelberg: Springer Verlag, 1990).

[26] Feit, W., *The Representation Theory of Finite Groups* (Amsterdam: North-Holland, 1982).

[27] Feit, W., On integral representations of finite groups, *Proc. London Math. Soc.* **29** (1974), 633–683.

[28] Fendel D., A characterization of Conway's group .3, *J. Algebra* **24** (1973), 159–196.

[29] Fischer, B., Finite groups generated by 3-transpositions, *Inventiones Math.* **13** (1971), 232–246.

[30] Fischer, B., Finite groups generated by 3-transpositions. Lecture Notes, University of Warwick, Coventry, England (1979).

[31] Fong, P., A characterization of the finite simple groups $PS_p(4,q)$, $G_2(q)$, $D_4^2(q)$, II, *Nagoya J. Math.* **39** (1970), 39–79.

[32] Frenkel, I., J. Lepowsky and A. Meurman, Vertex operator algebras and the monster (San Diego: Academic Press, 1988).

[33] GAP group, GAP: Groups, Algorithms and Programming. Version 4.4 (http://www.gap-system.org).

[34] Gebhardt, V., *An Algorithm for the Construction of a Defining Set of Relations for a Finite Group*, Vorlesungen aus dem Fachbereich Mathematik der Universität Essen **27** (Essen: University of Essen, 1999).

[35] Gollan, H. W., A new existence proof for *Ly*, the sporadic simple group of R. Lyons, *J. Symbolic Computation* **31** (2001), 203–209.

[36] Gorenstein, D., *Finite Groups* (New York: Harper & Row, 1968).

[37] Gorenstein, D., The classification of finite simple groups I. Simple groups and local analysis, *Bull. Amer. Math. Soc.* **1**, (1979), 43–199.

[38] Gorenstein, D., *The Classification of Finite Simple Groups: An Introduction to their Classification* (New York: Plenum Press, 1982)

[39] Gorenstein, D., *The Classification of Finite Simple Groups I: Groups of Noncharacteristic 2 type.* (New York: Plenum Press, 1983).

[40] Gorenstein, D., R. Lyons and R. Solomon, *The Classification of Finite Simple Groups*, Mathematical Surveys and Monographs **40**, pt. 1 (Providence, Rhode Island: American Mathematical Society, 1994).

[41] Griess, R. L., On a subgroup of order $2^{15}|GL(5,2)|$, the Dempwolff group and Aut($D_8 * D_8 * D_8$), *J. Algebra* **40** (1976), 271–279.

[42] Griess, R. L., The friendly giant, *Inventiones Math.* **69** (1982), 1–102.

[43] Griess, R. L., The Schur multiplier of McLaughlin's simple group, *Arch. Math* **48** (1987), 31.

[44] Griess, R. L., U. Meierfrankenfeld and Y. Segev, A uniqueness proof for the Monster, *Annals of Math.* **130** (1989), 567–602.

[45] Hall, J. and L. H. Soicher, Presenations of some 3-transposition groups, *Comm. Algebra,* **23** (1995), 232–246.

[46] Harada, K., On the simple group F of order $2^{14} \cdot 3^6 \cdot 5^6 \cdot 7 \cdot 11 \cdot 19$, in W. R. Scott, and F. Gross (eds.) *Proceedings of Conference on Finite Groups, Utah, 1975* (New York: Academic Press, 1976), pp. 119–276.

[47] Harada, K. and H. Yamaki, Irreducible subgroups of $\mathrm{GL}_n(2), n \leq 6$, *Proceedings of the 18th Kusatsu Symposium on Group Theory,* 2006.

[48] Havas, G. and C. Sims, A presentation for the Lyons simple group, *Progress in Math.* **173** (1999), 241–249.

[49] Held, D., The simple groups related to M_{24}, *J. Algebra* **13** (1969), 253–296.

[50] Higman, D. G., Indecomposable representations at characteristic p, *Duke Math. J.* **21** (1954), 377–381.

[51] Holmes, P. E. and R. A. Wilson, A new computer construction of the monster using 2-local subgroups, *J. London Math. Soc.* **67** (2003), 349–364.

[52] Holt, D. F., The mechanical computation of first and second cohomology groups, *J. Symbolic Computation* **1** (1985), 351–361.

[53] Holt, D. F., B. Eick and E. A. O'Brien, *Handbook of Computational Group Theory* (Boca Raton, Florida: Chapman and Hall/CRC, 2005).

[54] Holt, D. F., Cohomology and group extensions in Magma, in W. Bosma and J. Cannon (eds.), *Discovering Mathematics with Magma* (Berlin: Springer, 2006), pp. 221–241.

[55] Hunt, D., A characterization of the finite simple group $M(22)$, *J. Algebra* **21** (1972), 103–112.

[56] Hunt, D., A characterization of the finite simple group $M(23)$, *J. Algebra* **26** (1973), 431–439.

[57] Hunt, D., Character tables of certain finite simple groups, *Bull. Aust. Math. Soc.* **5** (1971), 1–42.

[58] Isaacs, I. M., *Character Theory of Finite Group* (New York: Academic Press, 1976).

[59] Ivanov, A. A., A presentation of J_4, *Proc. London Math. Soc.* **64** (1992), 369–396.

[60] Ivanov, A. A., Presenting the baby monster, *J. Algebra* **163** (1994), 88–108.

[61] Ivanov, A. A. and U. Meierfrankenfeld, A computer free construction of J_4, *J. Algebra* **219** (1999), 113–172.

[62] Ivanov, A. A., *Geometry of Sporadic Groups I* (Cambridge: Cambridge University Press, 1999).

[63] Ivanov, A. A., *Geometry of Sporadic Groups II* (Cambridge: Cambridge University Press, 2002).

[64] Ivanov, A. A., *The Fourth Janko Group* (Oxford: Clarendon Press, 2004).

[65] James, G., The modular characters of the Mathieu groups, *J. Algebra* **27** (1973), 57–111.

[66] Janko, Z., A new finite simple group with abelian Sylow 2-subgroups and its characterization, *J. Algebra* **3** (1966), 147–186.

[67] Janko, Z., A characterization of the simple Mathieu groups, I, *J. Algebra* **9** (1968), 1–19.

[68] Janko, Z., A characterization of the simple group $G_2(3)$, *J. Algebra* **12** (1969), 260–396.

[69] Janko, Z., A new finite simple group of order 86.775.571.046.077.562.880 which possesses M_{24} and the full covering group of M_{22} as subgroups, *J. Algebra* **42** (1972), 564–596.

[70] Janko, Z. and S. K. Wong, A characterization of the McLaughlin's simple group, *J. Algebra* **20** (1972), 203–225.

[71] Jansen, C., K. Lux, R. Parker and R. Wilson, *An Atlas of Brauer Characters* (Oxford: Clarendon Press, 1995).

[72] Kim, H. and G. Michler, Simultaneous constructions of the sporadic groups Co_2 and Fi_{22}, in L. -C. Kappe, A. Magidin and R. F. Morse (eds.) *Computational Group Theory and the Theory of Groups*, Contemporary Mathematics **470** (Providence, Rhode Island: American Mathematical Society, 2008).

[73] Kim, H., Representation theoretic existence proof for Fischer group Fi_{23}. Senior Thesis, Mathematics Department Cornell University, Ithaca, New York, 2008. See also http://arxiv.org/abs/0904.0639v1

[74] Kim, H. and G. Michler, Construction of Fischer's sporadic group Fi_{24}' inside $\mathrm{GL}_{8671}(13)$, http://arxiv.org/abs/0906.1064v1

[75] Kim, H. and G. Michler, Construction of Co_1 from an irreducible subgroup M_{24} of $\mathrm{GL}_{11}(2)$, http://arxiv.org/abs/0908.1393v1

[76] Kondo, T., On the alternating groups II, *J. Math. Soc. Japan* **21** (1969), 116–139.

[77] Kondo, T., On the alternating groups III, *J. Algebra* **14** (1970), 35–69.

[78] Kostrikin A. I. and P. H. Tiep, *Orthogonal Decompositions and Integral Lattices* (Berlin: de Gruyter Verlag, 1994).

[79] Kratzer, M., *Konkrete Charaktertafeln und kompatible Charaktere*, Vorlesungen aus dem Fachbereich Mathematik der Universität Essen **30** (Essen: University of Essen, 2001).

[80] Kratzer M., W. Lempken, G. O. Michler and K. Waki, Another existence and uniqueness proof for McLaughlin's simple group, *J. Group Theory* **6** (2003), 443–459.

[81] Kurzweil, H. and B. Stellmacher, *Theorie der endlichen Gruppen* (Berlin: Springer-Verlag, 1998).

[82] Leech, J., Some sphere packings in higher space, *Canad. J. Math.* **16** (1964), 657–682.

[83] Lempken, W., A 2-local characterization of Janko's simple group J_4, *J. Algebra* **55** (1978), 403–445.

[84] Lempken, W., Constructing J_4 in GL(1333, 11), *Commun. Algebra* **21** (1993), 4311–4351.

[85] Lenstra, A. K., H. W. Lenstra and L. Lovász, Factoring polynomials with rational coefficients, *Math. Ann.* **261** (1982), 515–534.

[86] Leon, J. S., On the irreducible characters of a simple group of order $21^{41} \cdot 3^{13} \cdot 5^6 \cdot 7^2 \cdot 11 \cdot 13 \cdot 17 \cdot 19 \cdot 23 \cdot 47$, in W. R. Scott and F. Gross (eds.), *Proceedings of the Conference on Finite Groups* (New York: Academic Press, 1976), pp. 285–299.

[87] Leon, J. S. and C. C. Sims, The existence and uniqueness of a simple group generated by 3, 4-transpositions, *Bull. Amer. Math. Soc.* **83** (1977), 1039–1049.

[88] Lyons, R., Evidence for a new finite simple group, *J. Algebra* **20** (1972), 540–569.

[89] McLaughlin, J., A Simple Group of Order 898128000, *Symposium on "Theory of Finite Groups"* (New York: Benjamin, 1969), pp. 109–111.

[90] Meyer, W, W. Neutsch, and R. Parker, The minimal 5-representation of Lyons' sporadic group, *Math. Annalen* **272** (1985), 29–39.

[91] Michler, G., On the uniqueness of the finite simple groups with a given centralizer of a 2-central involution, *Illinois J. Math.* **47** (2003), 419–444.

[92] Michler, G., *Theory of Finite Simple Groups* (Cambridge: Cambridge University Press, 2006).

[93] Michler, G., Constructing finite simple groups from irreducible subgroups of $\mathrm{GL}_n(2)$, in L. -C. Kappe, A. Magidin and R. F. Morse (eds.), *Computational Group Theory and the Theory of Groups*, Contemporary Mathematics **470** (Providence, Rhode Island: American Mathematical Society), pp. 235–262.

[94] Michler, G. and A. Previtali, O'Nan group uniquely determined by centralizer of a 2-central involution, *J. Algebra Appl.*, **6** (2007), 135–171.

[95] Michler, G., K. Waki and M. Weller, Natural existence proof for Lyons simple group, *J. Algebra Appl.* **2** (2003), 277–315.

[96] Michler, G. and L. Wang, Another existence and uniqueness proof of the Tits group, *Algebra Colloquium* **15** (2008), 241–278.

[97] Michler, G. and L. Wang, Another existence and uniqueness proof of the Dickson simple group $\mathsf{G}_2(3)$, *Algebra Colloquium* (to appear in 2010).

[98] Michler, G. and M. Weller, A new computer construction of the irreducible 112-dimensional representation of Janko's group J_4, *Commun. Algebra* **29** (2001), 1773–1806.

[99] Milnor, J. W. *Introduction to Algebraic K-theory* (Princeton, New Jersey: Princeton University Press, 1971).

[100] Nagao, H. and Y. Tsushima, Representations of Finite Groups (San Diego, California: Academic Press, 1988).

[101] Niemeyer, H. V., Definite quadratische Formen der Dimension 24 und Diskriminante 1. Dissertation Universitaet Goettingen, 1968.

[102] Norton, S., The construction of J_4, in B. Cooperstein and G. Mason (eds.), *The Santa Cruz Conference on Finite Groups*, Proceedings of Symposia in Pure Mathematics **37** (Providence, Rhode Island: American Mathematical Society, 1980), pp. 271–282.

[103] O'Meara, O. T., *Introduction to Quadratic Forms* (Heidelberg: Springer-Verlag, 1963).

[104] O'Nan, M. E., Some evidence for the existence of a new simple group, *Proc. London Math. Soc.* **32** (1976), 421–479.

[105] Parrott, D., A characterization of the Tits simple group, *Canad. J. Math.* **24** (1972), 672–685.

[106] Parrott, D., A characterization of the Rudvalis simple group, *Proc. London Math. Soc.* **32** (1976), 25–51.

[107] Parrott, D., Characterization of the Fischer groups I, II, III, *Trans. Amer. Math. Soc.* **265** (1981), 303–347.

[108] Patterson, N. J. and S. K. Wong, A characterization of the Suzuki sporadic simple group of order 448345497600, *J. Algebra* **39** (1976), 277–286.

[109] Phan, K. W., A characterization of the finite simple group $U_4(3)$, *J. Australian Math. Soc.* **10**, (1969), 60–81.

[110] Praeger C. E. and L. H. Soicher, *Low Rank Representations and Graphs for Sporadic Groups*, Australian Mathematical Society Lecture Series **8** (Cambridge: Cambridge University Press, 1997).

[111] Ree, R., A family of simple groups associated with the simple Lie algebra of type G_2, *Amer. J. Math.* **83** (1961), 432–462.

[112] Reifart, A., A remark on Conway's group .1, *Archiv d. Math.* **29** (1977), 389–391.

[113] Reifart, A., A 2-local characterization of the simple groups $M(24)'$, .1 and J_4, *J. Algebra* **50** (1978), 213–227.

[114] Robinson, E. and G. Cooperman, A parallel architecture for disk-based computing over the Baby Monster and other large finite simple groups, *ISSAC 06*, July 9–12, 2006, Genova, Italy.

[115] Rudvalis, A., A new simple group of order $2^{14} \cdot 3^3 \cdot 5^3 \cdot 7 \cdot 13 \cdot 29$, *Notices Amer. Math. Soc.* **20** (1973), 1–95.

[116] Schur, I., Über die Darstellung der endlichen Gruppen durch gebrochene lineare Substitutionen, *J. reine angew. Math.* **127** (1904), 20–50.

[117] Schur, I., Über die Darstellung der symmetrischen und der alternierenden Gruppen durch gebrochene lineare Substitutionen, *J. reine angew. Math.* **139** (1911), 155–250.

[118] Sims, C., The existence and uniqueness of Lyons' group, in T. Gagen, M. P. Hale and E. Shult (eds.), *Finite Groups* '72, North Holland Mathematics Studies **7** (Amsterdam: North Holland, 1973), pp. 138–141.

[119] Smith, F., A characterization of the .2 Conway simple group, *J. Algebra* **31** (1974), 91–116.

[120] Soicher L., Presentations for Conway's group Co_1, *Math. Proc. Cambridge Phil. Soc.* **102** (1987), 1–3.

[121] Soicher L., A new existence and uniqueness proof for O'Nan's simple sporadic group, *Bull. London Math. Soc.* **22** (1990), 148–152.

[122] Solomon R., A brief history of the classification of the finite simple groups, *Bull. Amer. Math. Soc.* **38** (2001), 315–352.

[123] Suzuki, M., Finite groups in which the centralizer of any element of order 2 is 2-closed, *Ann. Math.* **82** (1965), 191–212.

[124] Suzuki, M., A simple group of order 448345497600, in R. Brauer and C. H. Sar (eds.), *Theory of Finite Groups* (New York: Benjamin, 1969).

[125] Suzuki, M., Characterizations of linear groups, *Bull. Amer. Math. Soc.* **75** (1969), 1043–1091.

[126] Suzuki, M., *Group Theory II*, Grundlehren der mathematischen Wissenschaften **248** (Heidelberg: Springer Verlag, 1986).

[127] Thompson, J. G., A simple subgroup of $E_8(3)$, in N. Iwahori (ed.), *Finite Groups* (Tokyo: Japan Society for the Promotion of Science, 1976).

[128] Thompson, J. G., Nonsolvable finite groups all whose local subgroups are solvable II, *Pac. J. Math.* **33** (1970), 451–536.

[129] Thompson, J. G., The uniqueness of the Fischer-Griess monster, *Bull. London Math. Soc.* **11** (1979), 340–346.

[130] Thorne, M. P., On Fischer's "Monster", Ph. D. thesis, University of Birmingham, England, 1979.

[131] Tits, J., Théorème de Bruhat et sous-groupes paraboliques, *C. R. Acad. Sci. Paris* **254** (1962), 2910–2912.

[132] Tits, J., Algebraic and abstract simple groups, *Inventiones Math.* **80** (1964), 313–329.

[133] Tits, J., On R. Griess' "Friendly Giant", *Ann. of Math.* **78** (1984), 491–499.

[134] Weller, M., G. Michler and A. Previtali, Thompson's sporadic group uniquely determined by centralizer of a 2-central involution, *J. Algebra* **298** (2006), 371–459.

[135] Weller, M., Construction of large permutation representations for matrix groups II, *Applicable Algebra in Eng., Commun. Computation* **11** (2001), 463–488.

[136] Wilson R. A., A new construction of the baby Monster and its applications, *Bull. London Math. Soc.* **25** (1993), 431–437.

[137] Wilson R. A., P. Walsh, J. Tripp, I. Suleiman, S. Rogers, R. Parker, S. Norton, S. Nickerson, S. Linton, J. Bray and R. Abbott, ATLAS of Finite Group Representations, http://web.mat.bham.ac.uk/atlas/v2.0/

[138] Wong, W. J., A characterization of the Mathieu group M_{12}, *Math. Z.* **84** (1964), 378–388.

[139] Wright, D., Irreducible characters of the Suzuki group. *J. Algebra* **29** (1974), 303–323.

[140] Yamaki, H., A characterization of the finite simple group $Sp(6,2)$, *J. Math. Soc. Japan* **21** (1969), 334–356.

[141] Yoshida, T., A characterization of the .2 Conway simple group. *J. Algebra* **46** (1977), 405–414.

Index